MOLECULAR CELL BIOLOGY

ABOUT THE AUTHORS

 **HARVEY LODISH** is Professor of Biology at the Massachusetts Institute of Technology and a member of the Whitehead Institute for Biomedical Research. Dr. Lodish is also a member of the National Academy of Sciences and the American Academy of Arts and Sciences and President (2004) of the American Society for Cell Biology. He is well known for his work on cell membrane physiology, particularly the biosynthesis of many cell-surface proteins, and on the cloning and functional analysis of several cell-surface receptor proteins, such as the erythropoietin and TGFβ receptors, and transport proteins, including those for glucose and fatty acids. Dr. Lodish teaches undergraduate and graduate courses in cell biology.

 **ARNOLD BERK** is Professor of Microbiology, Immunology and Molecular Genetics and a member of the Molecular Biology Institute at the University of California, Los Angeles. Dr. Berk is also a fellow of the American Academy of Arts and Sciences. He is one of the original discoverers of RNA splicing and of mechanisms for gene control in viruses. His laboratory studies the molecular interactions that regulate transcription initiation in mammalian cells, focusing particular attention on transcription factors encoded by oncogenes and tumor suppressors. He teaches introductory courses in molecular biology and virology and an advanced course in cell biology of the nucleus.

 **PAUL MATSUDAIRA** is a member of the Whitehead Institute for Biomedical Research, Professor of Biology and Bioengineering at the Massachusetts Institute of Technology, and Director of the WI/MIT BioImaging Center. His laboratory studies the mechanics and biochemistry of cell motility and adhesion and has developed high-speed, high-through-put DNA analysis methods based on microfabricated chips. He organized the first biology course required of all MIT undergraduates and teaches courses in undergraduate biology and graduate bioengineering at MIT.

 **CHRIS A. KAISER** is a cell biologist and geneticist who has made fundamental contributions to understanding the basic processes of intracellular protein folding and membrane protein trafficking. His laboratory at the Massachusetts Institute of Technology, where he is Professor of Biology, studies how newly synthesized membrane and secretory proteins are folded and sorted in the compartments of the secretory pathway. Dr. Kaiser teaches genetics to undergraduates and graduate students at MIT.

 **MONTY KRIEGER** is Thomas D. & Virginia W. Cabot Professor in the Department of Biology at the Massachusetts Institute of Technology. For his innovative teaching of undergraduate biology and human physiology as well as graduate cell biology courses, he has received numerous awards. His laboratory has made contributions to our understanding of membrane trafficking through the Golgi apparatus and has cloned and characterized receptor proteins important for the movement of cholesterol into and out of cells.

 **MATTHEW P. SCOTT** is Professor of Developmental Biology and Genetics at Stanford University School of Medicine and Investigator at the Howard Hughes Medical Institute. He is a member of the National Academy of Sciences and the American Academy of Arts and Sciences and a past president of the Society for Developmental Biology. He is known for his work in developmental biology and genetics, particularly in areas of cell-cell signaling and homeobox genes and for discovering the roles of developmental regulators in cancer. Dr. Scott teaches development and disease mechanisms to medical students and developmental biology to graduate students at Stanford University.

MOLECULAR CELL BIOLOGY
FIFTH EDITION

Harvey Lodish

Arnold Berk

Paul Matsudaira

Chris A. Kaiser

Monty Krieger

Matthew P. Scott

S. Lawrence Zipursky

James Darnell

W. H. Freeman and Company
New York

PUBLISHER: SARA TENNEY

ACQUISITIONS EDITOR: KATHERINE AHR

DEVELOPMENT EDITORS: RUTH STEYN, SONIA DiVITTORIO

EDITORIAL ASSISTANTS: MELANIE MAYS, JENNIFER ZARR

DIRECTOR OF MARKETING: JOHN BRITCH

PROJECT EDITOR: MARY LOUISE BYRD

TEXT DESIGNER: VICTORIA TOMASELLI

PAGE MAKEUP: MARSHA COHEN

COVER ILLUSTRATION: SONIA DiVITTORIO

COVER DESIGN: VICTORIA TOMASELLI, STEVEN MOSKOWITZ

ILLUSTRATION COORDINATOR: CECILIA VARAS

ILLUSTRATIONS: NETWORK GRAPHICS, ERICA BEADE, SONIA DiVITTORIO

PHOTO EDITOR: PATRICIA MARX

PHOTO RESEARCHERS: TOBI ZANSNER, BRIAN DONNELLY

PRODUCTION COORDINATOR: SUSAN WEIN

MEDIA AND SUPPLEMENTS EDITORS: TANYA AWABDY, JOY OHM, JENNIFER VAN HOVE

MEDIA DEVELOPERS: BIOSTUDIO, INC., SUMANAS, INC.

COMPOSITION: THE GTS COMPANIES/YORK, PA CAMPUS

MANUFACTURING: RR DONNELLEY & SONS COMPANY

About the cover: The illustration depicts a diverse array of integral and peripheral membrane proteins. The phospholipid bilayer is derived from a molecular dynamics model. The protein structures (determined by x-ray crystallography), left to right from back cover to front, are aquaporin (bovine; PDB ID 1j4n), rhodopsin (bovine; 1f88), G protein (chimeric bovine and rat, $\alpha\beta$ subunits; 1got), light-harvesting complex (bacterial; 1kzu), potassium channel (bacterial; 1bl8), photosynthetic reaction center (bacterial; 1prc), photosystem 1 (bacterial; 1jb0), aquaporin (1j4n), ATP synthase (composite bovine [1e79] and bacterial [1c17]). [Molecular dynamics model after H. Heller, M. Schaefer, and K. Schulten, 1993, Molecular dynamics simulation of 200 lipids in the gel and in the liquid-crystal phases, *Phys. Chem.* **97**:8343; H. Heller, 1993, Simulation einer Lipidmembran auf einem Parallelrechner, Ph.D. diss., University of Munich, Germany.]

Library of Congress Cataloging-in-Publication Data

Molecular cell biology/Harvey Lodish ... [et al.].—5th ed.
    p. cm.
    Includes bibliographical references and index.
    ISBN 0-7167-4366-3
    1. Cytology. 2. Molecular biology. I. Lodish, Harvey F.

QH581.2.M655 2003
571.6—dc21   2003049089

Printed in the United States of America

First printing 2003

W. H. Freeman and Company
41 Madison Avenue, New York, New York 10010
Houndsmills, Basingstoke RG21 6XS, England

www.whfreeman.com

To our students and to our teachers,

from whom we continue to learn,

and to our families, for their support,

encouragement, and love

PREFACE

Spectacular advances have been made in many areas of molecular cell biology since the publication of the fourth edition that contribute to our growing understanding of the beauty and wonder of this field. The complete sequences of the worm, human, mouse, rice, fly, and many other genomes have led to novel insights into the evolution of life forms, the regulation of gene expression, and the functions of individual members of multiprotein families. The simultaneous expression of thousands of genes can now be analyzed with the use of newly developed DNA "chip" microarray technology, enhancing our understanding of gene control during development and disease. Applications abound, both realized and potential, to medical science as well as to animal and plant science.

All these applications hinge on an understanding of the internal and external workings of cells, especially on the functions of the many different types of proteins and lipids synthesized by each eukaryotic cell, and how cells interact with their environment. Understanding such complexity has driven the discipline to the point where molecular cell biologists are bringing together previously unconnected pools of knowledge as integrated systems. Given this integration, we can now understand biology at new levels of complexity, such as large multiprotein signaling complexes in cells and the mechanisms by which cells interact with other cells.

▲ Mitotic spindle in dividing cell

New Author Team

▲ From left to right: Matt Scott, Chris Kaiser, Paul Matsudaira, Harvey Lodish, Arnie Berk, Monty Krieger

Three new authors for the fifth edition have been instrumental in focusing this book toward these exciting new developments:

Chris A. Kaiser is a cell biologist and geneticist who has made fundamental contributions to our understanding of the basic processes of intracellular protein folding and membrane protein trafficking.

Monty Krieger has made major contributions to our understanding of the organization and function of the Golgi apparatus and has cloned and characterized scavenger and HDL receptor proteins important for the movement of cholesterol into and out of cells and for their influence on heart disease.

Matthew P. Scott is known for his work in developmental biology and genetics, particularly in areas of cell-cell signaling and homeobox genes, and for discovering the roles of developmental regulators in cancer.

We are grateful to Jim Darnell, Larry Zipursky, and David Baltimore for their exceptional contributions to the preceding editions of *Molecular Cell Biology*. Much of their vision and insight is apparent at many places in this book.

New Discoveries, New Methodologies

Molecular Cell Biology provides a clear introduction to the techniques and experiments of scientists past and present, showing how important discoveries led to the formation of the field's key concepts. A number of experimental organisms, from yeasts to worms to mice, are used throughout so that the student can see how discoveries made with them can lead immediately to insights even about human biology and disease. The following list includes just a few of the new experimental methodologies and concepts introduced in this edition:

• Explanation of how mass spectrometry is used to identify the components of large multiprotein complexes (Chapter 3)

• New material on the structure of ion-channel proteins, explaining their ion selectivity and gating properties and their roles in the conduction of action potentials by nerve cells (Chapter 7)

• The experimental use of RNA interference (RNAi) to block the expression of specific genes in many types of eukaryotic cells and organisms (Chapter 9)

• The use of transcriptional profiling by DNA microarrays (DNA chips) to uncover patterns of gene control in several systems and to differentiate otherwise similar types of human cancers (Chapters 9, 11, and 23)

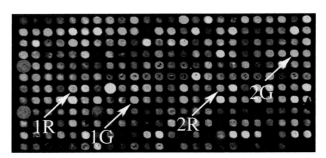

▲ **DNA microarray analysis**

• New discoveries concerning the role of chromatin structure in controlling transcription, and the technique of chromatin immunoprecipitation (ChIP) to analyze proteins associated with specific genes in vivo (Chapter 11)

• New discoveries concerning the role of naturally occurring 21–23 base RNAs in controlling gene expression (Chapter 12)

• The use of fluorescence energy transfer to monitor protein-protein interactions in living cells (Chapter 13)

• New information about regulated intermembrane proteolytic cleavage in several signaling pathways (Chapters 14 and 18)

• Discussion of a newly identified pathway for vesicle budding away from the cytosol that is also used by lipid-enveloped viruses such as HIV (Chapter 17)

• Treatment of forces produced by polymerization and motor proteins during cell movement (Chapter 19)

• Description of how speckle microscopy reveals the flow of tubulin subunits toward the spindle poles at metaphase (Chapter 20)

• New insights into the mechanisms of chromosome condensation and segregation in mitosis, the regulation of origin firing in the S phase, and modifications of the cell cycle in meiosis (Chapter 21)

• New material on the role of asymmetric cell division in early development and the participating protein complexes (Chapter 22)

• The use of developmental genetics to uncover new proteins essential for stem cell function and cell lineage determination (Chapter 22)

Focus on Fundamental Principles

Since the publication of the fourth edition, fundamental principles have emerged from our understanding of molecular cell biology. This fifth edition strives to present these principles clearly while providing essential experimental information. These changes have resulted in a reorganized table of contents, including several new chapters and the reorganization of topics within chapters. The content changes in the fifth edition are as follows:

• New Chapter 1, "Life Begins with Cells," provides a conceptual overview of the text.

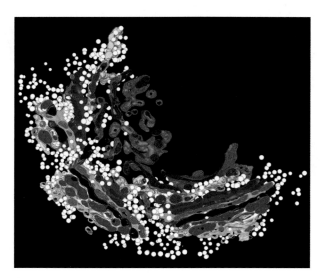

▲ **Model of the Golgi complex and transport vesicles**

• Coverage of basic chemical concepts in Chapter 2, "Chemical Foundations," focuses on those most relevant to molecular cell biology, and a new section introducing the cellular building blocks (amino acids, nucleotides, carbohydrates, fatty acids, and phospholipids) has been added.

• Earlier coverage of protein machines and motor proteins in Chapter 3 and the cytoskeleton in Chapter 5.

• Chapter 4, "Basic Molecular Genetic Mechanisms," has been restructured to describe the basic mechanisms of transcription, translation, and DNA replication and the molecular machines that carry out these processes. The concept of transcriptional control is introduced, and a brief discussion of gene control in bacteria is presented.

• Earlier expanded and reorganized treatment of adhesive cell-cell and cell-matrix interactions in Chapter 6, "Integrating Cells into Tissues," which focuses on adhesion as well as outside-in and inside-out signaling, prepares students to think about how cells relate to one another and to their immediate surroundings.

• Earlier treatment of transport across cell membranes and of cellular energetics, now in Chapters 7 and 8.

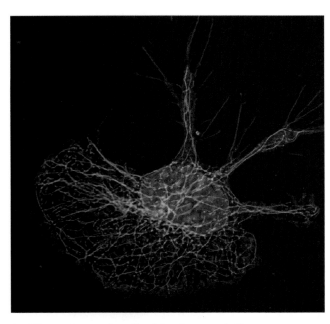

▲ **Macrophage visualized by deconvolution fluorescence microscopy**

• Coverage of genetic and recombinant DNA techniques has been reorganized and streamlined in new Chapter 9, "Molecular Genetic Techniques and Genomics." Several examples of DNA microarray analysis to determine genome-wide expression patterns are presented here and throughout the book.

• Chapters 11 and 12, "Transcriptional Control of Gene Expression" and "Post-transcriptional Controls and Nuclear Transport," are focused entirely on eukaryotic cells. Coverage of prokaryotic gene control has been shifted to Chapter 4.

• Expanded treatment of signal-transduction pathways and their integration within the whole cell and organism in Chapters 13 through 15, "Signaling at the Cell Surface," "Signaling Pathways That Control Gene Activity," and "Integration of Signals and Gene Controls." These chapters contain much information obtained through genetic and molecular analysis of the development of many organisms.

• Treatment of protein sorting has been entirely rewritten and separated into two chapters, 16 and 17, "Moving Proteins into Membranes and Organelles" and "Vesicular Trafficking, Secretion, and Endocytosis."

• A new chapter, 18, "Metabolism and Movement of Lipids," presents expanded coverage of this sometimes overlooked class of molecules and an elegant case study of the two-way interplay between basic molecular cell biology and medicine.

• A new chapter, 22, "Cell Birth, Lineage, and Death," includes a discussion of stem cells and cell lineage, new coverage of *C. elegans* cell lineage, a section on the importance and regulation of apoptosis (regulated cell death) in development, and descriptions of cell-type specification in yeast and muscle.

• Coverage of the mechanisms of DNA replication and recombination has been reduced, and material on DNA damage and repair is now covered in Chapter 23, "Cancer."

• In the illustration program, we have enhanced the legends for experimental figures by emphasizing the experimental result in the title.

• We have added end-of-chapter questions titled "Analyze the Data," which ask students to answer a research problem by looking at real experimental data. The data included in the question are taken from an experiment covered in the chapter.

• Each chapter has an up-to-date list of references of landmark studies and comprehensive review articles that direct students and instructors to much additional information.

We hope that these changes in organization, the emphasis on the experimental basis of current understanding, and coverage of exciting new discoveries will make this fifth edition a useful tool for conveying the complexity of biological systems and the astonishing beauty and wonder of molecular cell biology.

New Ways of Seeing Molecular Cell Biology

Every illustration in the book has been revisited, scrutinized, and, if necessary, changed to ensure clarity and visual consistency throughout the book. The shapes and, as much as possible, the colors of structures in illustrations are consistent throughout. In addition, many illustrations have been simplified. Less detail makes it easier for students to grasp the central points being made. New figures open each chapter providing the reader with an overview of details to come. Wherever possible, we pair illustrations with corresponding experimental images, such as micrographs, to bridge the gap between models and reality.

► New overview figures
open each chapter

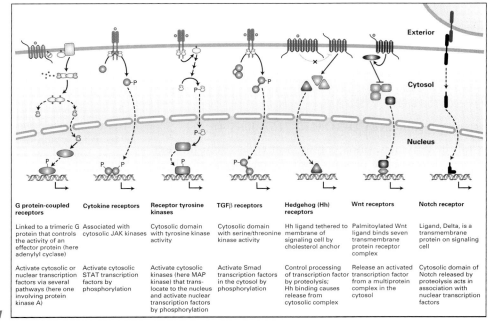

Figure 13-1

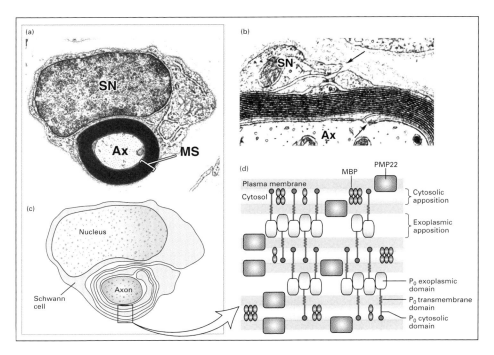

◄ Pairing illustrations
with experimental
images

Figure 7-39

SUPPLEMENTS

For the Instructors

Access

▶ **Instructor's Resource Web Site (www.whfreeman.com/lodish)**

• The instructor's resource Web site contains all the resources from the Instructor's Resource CD-ROM, including all the illustrations from the text, all the Flash animations, all the videos, the test bank files, and the solutions manual files. Contact your W. H. Freeman and Company sales representative for the instructor access code necessary to register for instructor's resources.

Presentation Visuals

▶ **Instructor's Resource CD-ROM (0-7167-0065-4)**

• All the drawings, photographs, and tables from the text optimized for lecture presentations. Images are available as high-resolution chapter-by-chapter PowerPoint™ presentations or as individual jpeg files. For optimized classroom projection, labels have been enlarged and are in boldface type; complicated multistep illustrations have been separated into parts, and colors are enhanced.

• More than 50 Flash animations of key cell processes that can be presented step by step or through continuous play. The animations are called out in the text next to the illustration to which they correspond. They fall into three categories: overview animations, focus animations, and technique animations. The animations were storyboarded by the textbook authors in conjunction with BioStudio, Inc., and programmed by Sumanas, Inc.

• **NEW** Sixty research videos are available to show your students what cells and processes really look like. These videos come from cutting-edge laboratories throughout the world. Additional videos are provided to instructors on an annual basis throughout the life of this edition. Please contact your sales representative for more information.

▶ **Overhead Transparency Set (0-7167-0069-7)**

• Two hundred seventy images from the text. For optimized classroom projection, labels have been enlarged and are in boldface type; complicated multistep illustrations have been separated into parts, and colors are enhanced.

Assessment

▶ **Test Bank, available on the Instructor's Resource CD-ROM (0-7167-0065-4) and Web site (www.whfreeman.com/lodish)**

• The test bank, written by Brian Storrie, Eric A. Wong, Richard Walker, Glenda Gillaspy, and Jill Sible of Virginia Polytechnic Institute and State University, is available as chapter-by-chapter Microsoft Word files. Each file can be opened and modified in Word just like any other Word document.

Further Research

Instructor's Online Update Service, powered by the npg Nature Publishing Group.

• Use this resource to access the latest research from NATURE Reviews Molecular Cell Biology.

For the Student

Visualization

▶ **Student Resource Web Site (www.whfreeman.com/lodish)**

• More than 50 Flash animations of key cell processes that can be presented step by step or through continuous play. The animations are called out in the text next to the illustration to which they correspond. They fall into three categories: overview animations, focus animations, and technique animations. The animations were storyboarded by the textbook authors in conjunction with BioStudio, Inc., and programmed by Sumanas, Inc.

• More than 50 research videos will show you what cells and biological processes really look like. These videos come from cutting-edge laboratories throughout the world.

Test Yourself

▶ *Working with Molecular Cell Biology Student Companion and Solutions Manual* **(0-7167-5993-4)**

• The Student Companion, written by Brian Storrie, Eric A. Wong, Richard Walker, Glenda Gillaspy, and Jill Sible of Virginia Polytechnic Institute and State University, contains three parts: Chapter Summary, Reviewing Concepts, and Analyzing Experiments. It also includes complete worked-out solutions to all the end-of-chapter problems in the textbook, including the Analyze the Data problems. The organization of the *Companion* follows that of the textbook, and you are encouraged to draw together the text, the media resources, and their lecture notes to answer the questions.

▶ *MCAT® Practice Test II* **(0-7167-5907-1)**

A full-length authentic MCAT® practice test in booklet form, made up of items previously used in "live" administrations of the MCAT. Revised to reflect the content and format changes to the MCAT for 2003, and including updated scoring tables. Authentic MCAT practice tests provide the best estimate of likely MCAT scores, given your level of preparation at the time you take the practice test.

▶ *MCAT/GRE-style prep exam* **(www.whfreeman.com/lodish)**

A brief MCAT/GRE-style prep exam, specifically designed for use with *Molecular Cell Biology,* is available on the Web site.

Further Research

► **Classic Experiment Essays** (www.whfreeman.com/lodish)
  • Twenty essays written by Lisa Rezende, Harvard Medical School, treat the investigative side of classic groundbreaking experiments by exploring the process of asking questions and devising tests.

ACKNOWLEDGMENTS

We could not have prepared this edition without the input of many colleagues. We thank

Susan M. Abmayr, *Pennsylvania State University*
Chris Akey, *Boston University*
Lizabeth Allison, *The College of William and Mary*
Marsha Altschuler, *Williams College*
Richard Anderson, *University of Texas Southwestern Medical Center*
Rosalie Anderson, *Tulane University*
Nigel S. Atkinson, *University of Texas at Austin*
Jnanankur Bag, *University of Guelph*
Robert Baker, *University of Southern California*
Lisa Banner, *California State University at Northridge*
Margarida Barroso, *University of Virginia*
Greg J. Beitel, *Northwestern University*
John D. Bell, *Brigham Young University*
Bill Bement, *University of Wisconsin at Madison*
Sanford Bernstein, *San Diego State University*
Stephen Blacklow, *Harvard Medical School*
Larry Blanton, *Texas Tech University*
Subbarao Bondada, *University of Kentucky*
Roger Bradley, *Montana State University*
William S. Bradshaw, *Brigham Young University*
Gail Breen, *University of Texas at Austin*
Tony Bretscher, *Cornell University*
Robert J. Brooker, *University of Minnesota*
Michael S. Brown, *University of Texas Southwestern Medical Center*
Chris Burge, *Massachusetts Institute of Technology*
Peter Byers, *University of Washington*
Francisco Carrapico, *University of Lisbon*
Lynne Cassimeris, *Lehigh University*
T. Y. Chang, *Dartmouth University*
Randy W. Cohen, *California State University at Northridge*
John Colicelli, *University of California at Los Angeles, School of Medicine*
Nathan L. Collie, *Texas Tech University*
Kathleen Collins, *University of California at Berkeley*
Duane Compton, *Dartmouth University*
Reid Compton, *University of Maryland*
Andrew Conery, *University of California at Berkeley*

Scott Cooper, *University of Wisconsin at La Crosse*
Anne L. Cordon, *University of Toronto*
Gerald Crabtree, *Stanford University*
Donald B. DeFranco, *University of Pittsburgh*
Virginia Ann Dell, *Oklahoma School of Science and Mathematics*
Dave Denhardt, *Rutgers University*
Claude Deplan, *New York University*
Joyce Diwan, *Rensselaer Polytechnic Institute*
Robert S. Dotson, *Tulane University*
William Dowhan, *University of Texas at Houston*
Michael Edidin, *Johns Hopkins University*
Matt Elrod-Erickson, *Middle Tennessee State University*
Bevin P. Engelward, *Massachusetts Institute of Technology*
R. Paul Evans, *Brigham Young University*
Wayne Fagerberg, *University of New Hampshire*
Guy E. Farish, *Adams State College*
Richard Fehon, *Duke University*
Andrew Fire, *Carnegie Institute of Washington*
Michelle French, *University of Toronto*
Terrence G. Frey, *San Diego State University*
Marilyn Gist Farquhar, *University of California at San Diego*
David S. Goldfarb, *University of Rochester*
Elliot S. Goldstein, *Arizona State University*
Joseph Goldstein, *University of Texas Southwestern Medical Center*
Lawrence Goldstein, *University of San Diego*
Stephen Gould, *Johns Hopkins University*
Carla B. Green, *University of Virginia*
Bruce Greenberg, *University of Waterloo*
Paul Greenwood, *Colby College*
Barry M. Gumbiner, *University of Virginia*
Dipak Haldar, *St. John's University*
Vincent Hascall, *The Cleveland Clinic*
Jesse C. Hay, *University of Michigan*
Michele Heath, *University of Toronto*
Merrill Hille, *University of Washington*
Richard Holdeman, *Indiana University*
J. E. Honts, *Drake University*
H. Robert Horvitz, *Massachusetts Institute of Technology*
Richard Hynes, *Massachusetts Institute of Technology*
Carlos Jaramillo, *Los Andes University*
Margaret Johnson, *University of Arizona*
Margaret Dean Johnson, *University of Alabama*
Ross G. Johnson, *University of Minnesota*
Herbert M. Kagan, *Boston University*
Teh-hui Kao, *Pennsylvania State University*
Thomas C. S. Keller III, *Florida State University*
Greg M. Kelly, *University of Western Ontario*
Mike Klymkowsky, *University of Colorado at Boulder*
Jurgen Knoblich, *Institute for Molecular Pathology, Vienna*

Donna J. Koslowsky, *Michigan State University*

Balagurunathan Kuberan, *Massachusetts Institute of Technology*

Arthur D. Lander, *University of California at Irvine*

Torvard Laurent, *University of Uppsala*

David Leaf, *Western Washington University*

Jackie Lee, *University of Colorado*

Robert L. Levine, *McGill University*

Haifan Lin, *Duke University*

Troy Littleton, *Massachusetts Institute of Technology*

Xuan Liu, *University of California at Riverside*

Elizabeth Lord, *University of California at Riverside*

Ponzy Lu, *University of Pennsylvania*

Paula M. Lutz, *University of Missouri at Rolla*

Thomas H. MacRae, *Dalhousie University*

Tom Maniatis, *Harvard University*

Ruthann Mararacchia, *University of North Texas*

Joan Massagué, *Memorial Sloan-Kettering Cancer Center*

Andreas Matouschek, *Northwestern University*

Maryanne McClellan, *Reed College*

Sara McCowen, *Virginia Commonwealth University*

Jose Mejia, *California State University at Northridge*

Stephanie Mel, *University of California at San Diego*

Hsiao-Ping Moore, *University of California at Berkeley*

James Moroney, *Louisiana State University*

Donald O. Natvig, *University of New Mexico at Albuquerque*

Martin Nemeroff, *Rutgers University*

Jeffrey Newman, *Lycoming College*

Alan Nighorn, *University of Arizona*

Laura J. Olsen, *University of Michigan*

Charlotte Omoto, *Washington State University*

Rekha Patel, *University of South Carolina*

Mark Peifer, *University of North Carolina at Chapel Hill*

Jacques Perrault, *San Diego State University*

Dorothy Pocock, *McGill University*

Tom Rapoport, *Harvard University*

Tal Raveh, *Stanford University*

Terrie Rife, *James Madison University*

Austen F. Riggs II, *University of Texas at Austin*

Peter J. Rizzo, *Texas A & M*

Daniel M. Roberts, *University of Tennessee*

Jane Rossant, *University of Toronto*

Robert D. Rosenberg, *Massachusetts Institute of Technology*

Gary Ruvkun, *Harvard University*

James Sellers, *National Institutes of Health*

Florence Schmieg, *University of Delaware*

Diane Shakes, *The College of William and Mary*

Ellen Shibuya, *University of Alberta at Edmonton*

Ke Shuai, *University of California at Los Angeles*

Roger Sloboda, *Dartmouth College*

Douglas Smith, *University of California at San Diego*

Walter D. Sotero-Esteva, *University of Central Florida*

Domenico Spadafora, *San Diego State College*

Philip Stanford, *Dartmouth College*

Jackie Stephens, *Louisiana State University*

Paul Sternberg, *California Institute of Technology*

Brian Storrie, *Virginia Polytechnic Institute and State University*

Jerome Strauss, *University of Pennsylvania*

Robert M. Stroud, *University of California, San Francisco*

Wes Sundquist, *University of Utah*

Markku Tammi, *University of Kuopio*

Robert M. Tombes, *Virginia Commonwealth University*

John L. Tymoczko, *Carleton College*

Elizabeth Vallen, *Swarthmore College*

Volker M. Vogt, *Cornell University*

Dennis R. Voelker, *National Jewish Medical Research Center*

Charles Walker, *University of New Hampshire*

Kenneth Walsh, *University of Washington*

Christopher Watters, *Middlebury College*

Steve Weiner, *Weizmann Institute of Science*

Patrick Weir, *Felician College*

Matt Welch, *University of California at Berkeley*

Beverly Wendland, *Johns Hopkins University*

Bruce Wightman, *Muhlenberg College*

David Worcester, *University of Missouri*

Michael Wormington, *University of Virginia*

Lijuan Zhang, *Washington University in St. Louis*

R. Andrew Zoeller, *Boston University*

Nor would this edition have been published without the careful and committed collaboration of our publishing partners at W. H. Freeman and Company. We thank Kate Ahr, Tanya Awabdy, John Britch, Mary Louise Byrd, Marsha Cohen, Brian Donnelly, Patricia Marx, Melanie Mays, Joy Ohm, Bill O'Neal, Sara Tenney, Vicki Tomaselli, Jennifer Van Hove, Cecilia Varas, Susan Wein, Jennifer Zarr, Tobi Zausner, and Patty Zimmerman for their labor and for their willingness to work overtime to produce a book that excels in every way.

Ruth Steyn

Sonia DiVittorio

In particular we would like to acknowledge the talent and commitment of our text and art editors, Ruth Steyn and Sonia DiVittorio. They are remarkable editors. Thank you both for all you've done in this edition.

Thanks to our own staff: Erica Beade of MBC Graphics (www.MBCGraphics.com) for her work in developing the art program, Sally Bittancourt, Mary Anne Donovan, Carol Eng, James Evans, George Kokkinogenis, Kathy Sweeney, Ketsada Syhakhom, Guicky Waller, Nicki Watson, and Rob Welsh.

Finally, special thanks to our families for inspiring us and for granting us the time it takes to work on such a book.

CONTENTS IN BRIEF

CONTENTS

II   Cell Organization and Biochemistry

8 | Cellular Energetics   301

III Genetics and Molecular Biology

9 Molecular Genetic Techniques and Genomics   351

10 | Molecular Structure of Genes and Chromosomes   405

11 | Transcriptional Control of Gene Expression   447

IV Cell Signaling

13 | Signaling at the Cell Surface   533

V Membrane Trafficking

16 Moving Proteins into Membranes and Organelles   657

17 Vesicular Traffic, Secretion, and Endocytosis   701

18 | Metabolism and Movement of Lipids   743

VI Cytoskeleton

19 | Microfilaments and Intermediate Filaments   779

1

LIFE BEGINS WITH CELLS

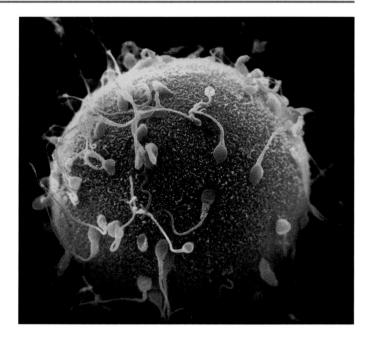

A single ~200 micrometer (μm) cell, the human egg, with sperm, which are also single cells. From the union of an egg and sperm will arise the 10 trillion cells of a human body.
[Photo Researchers, Inc.]

Like ourselves, the individual cells that form our bodies can grow, reproduce, process information, respond to stimuli, and carry out an amazing array of chemical reactions. These abilities define life. We and other multicellular organisms contain billions or trillions of cells organized into complex structures, but many organisms consist of a single cell. Even simple unicellular organisms exhibit all the hallmark properties of life, indicating that the cell is the fundamental unit of life. As the twenty-first century opens, we face an explosion of new data about the components of cells, what structures they contain, how they touch and influence each other. Still, an immense amount remains to be learned, particularly about how information flows through cells and how they decide on the most appropriate ways to respond.

Molecular cell biology is a rich, integrative science that brings together biochemistry, biophysics, molecular biology, microscopy, genetics, physiology, computer science, and developmental biology. Each of these fields has its own emphasis and style of experimentation. In the following chapters, we will describe insights and experimental approaches drawn from all of these fields, gradually weaving the multifaceted story of the birth, life, and death of cells. We start in this prologue chapter by introducing the diversity of cells, their basic constituents and critical functions, and what we can learn from the various ways of studying cells.

1.1 The Diversity and Commonality of Cells

Cells come in an amazing variety of sizes and shapes (Figure 1-1). Some move rapidly and have fast-changing structures, as we can see in movies of amoebae and rotifers. Others are largely stationary and structurally stable. Oxygen kills some cells but is an absolute requirement for others. Most cells in multicellular organisms are intimately involved with other cells. Although some unicellular organisms live in isolation, others form colonies or live in close association with other types of organisms, such as the bacteria that help plants to extract nitrogen from the air or the bacteria that live in our intestines and help us digest food. Despite these and numerous

1

▲ **FIGURE 1-1 Cells come in an astounding assortment of shapes and sizes.** Some of the morphological variety of cells is illustrated in these photographs. In addition to morphology, cells differ in their ability to move, internal organization (prokaryotic versus eukaryotic cells), and metabolic activities. (a) Eubacteria; note dividing cells. These are *Lactococcus lactis,* which are used to produce cheese such as Roquefort, Brie, and Camembert. (b) A mass of archaebacteria (*Methanosarcina*) that produce their energy by converting carbon dioxide and hydrogen gas to methane. Some species that live in the rumen of cattle give rise to >150 liters of methane gas/day. (c) Blood cells, shown in false color. The red blood cells are oxygen-bearing erythrocytes, the white blood cells (leukocytes) are part of the immune system and fight infection, and the green cells are platelets that provide substances to make blood clot at a wound. (d) Large single cells: fossilized dinosaur eggs. (e) A colonial single-celled green alga, *Volvox aureus.* The large spheres are made up of many individual cells, visible as blue or green dots. The yellow masses inside are daughter colonies, each made up of many cells. (f) A single

Purkinje neuron of the cerebellum, which can form more than a hundred thousand connections with other cells through the branched network of dendrites. The cell was made visible by introduction of a fluorescent protein; the cell body is the bulb at the bottom. (g) Cells can form an epithelial sheet, as in the slice through intestine shown here. Each finger-like tower of cells, a villus, contains many cells in a continuous sheet. Nutrients are transferred from digested food through the epithelial sheet to the blood for transport to other parts of the body. New cells form continuously near the bases of the villi, and old cells are shed from the top. (h) Plant cells are fixed firmly in place in vascular plants, supported by a rigid cellulose skeleton. Spaces between the cells are joined into tubes for transport of water and food. [Part (a) Gary Gaugler/ Photo Researchers, Inc. Part (b) Ralph Robinson/ Visuals Inlimited, Inc. Part (c) NIH/Photo Researchers, Inc. Part (d) John D. Cunningham/Visuals Unlimited, Inc. Part (e) Carolina Biological/Visuals Unlimited, Inc. Part (f) Helen M. Blau, Stanford University. Part (g) Jeff Gordon, Washington University School of Medicine. Part (h) Richard Kessel and C. Shih/Visuals Unlimited, Inc.]

other differences, all cells share certain structural features and carry out many complicated processes in basically the same way. As the story of cells unfolds throughout this book, we will focus on the molecular basis of both the differences and similarities in the structure and function of various cells.

All Cells Are Prokaryotic or Eukaryotic

The biological universe consists of two types of cells—prokaryotic and eukaryotic. Prokaryotic cells consist of a sin-

gle closed compartment that is surrounded by the **plasma membrane,** lacks a defined **nucleus,** and has a relatively simple internal organization (Figure 1-2a). All **prokaryotes** have cells of this type. Bacteria, the most numerous prokaryotes, are single-celled organisms; the cyanobacteria, or blue-green algae, can be unicellular or filamentous chains of cells. Although bacterial cells do not have membrane-bounded compartments, many proteins are precisely localized in their aqueous interior, or **cytosol,** indicating the presence of internal organization. A single *Escherichia coli* bacterium has a dry weight of about

(a) Prokaryotic cell

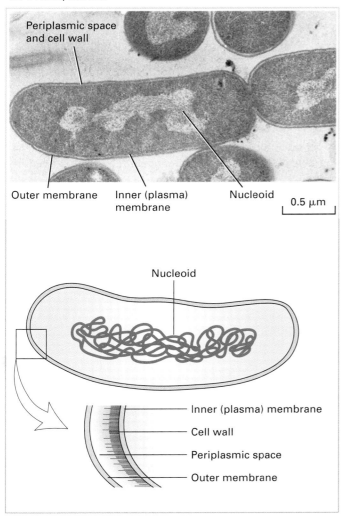

(b) Eukaryotic cell

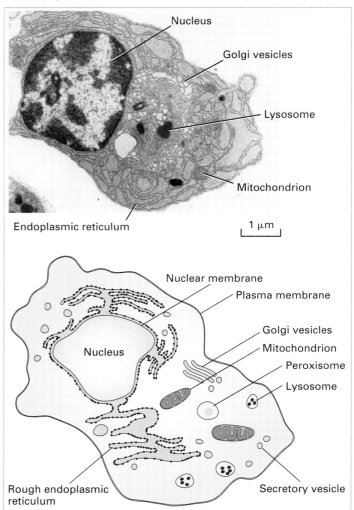

▲ **FIGURE 1-2 Prokaryotic cells have a simpler internal organization than eukaryotic cells.** (a) Electron micrograph of a thin section of *Escherichia coli,* a common intestinal bacterium. The nucleoid, consisting of the bacterial DNA, is not enclosed within a membrane. *E. coli* and some other bacteria are surrounded by two membranes separated by the periplasmic space. The thin cell wall is adjacent to the inner membrane. (b) Electron micrograph of a plasma cell, a type of white blood cell that secretes antibodies. Only a single membrane (the plasma membrane) surrounds the cell, but the interior contains many membrane-limited compartments, or organelles. The defining characteristic of eukaryotic cells is segregation of the cellular DNA within a defined nucleus, which is bounded by a double membrane. The outer nuclear membrane is continuous with the rough endoplasmic reticulum, a factory for assembling proteins. Golgi vesicles process and modify proteins, mitochondria generate energy, lysosomes digest cell materials to recycle them, peroxisomes process molecules using oxygen, and secretory vesicles carry cell materials to the surface to release them. [Part (a) courtesy of I. D. J. Burdett and R. G. E. Murray. Part (b) from P. C. Cross and K. L. Mercer, 1993, *Cell and Tissue Ultrastructure: A Functional Perspective,* W. H. Freeman and Company.]

25×10^{-14} g. Bacteria account for an estimated 1–1.5 kg of the average human's weight. The estimated number of bacteria on earth is 5×10^{30}, weighing a total of about 10^{12} kg. Prokaryotic cells have been found 7 miles deep in the ocean and 40 miles up in the atmosphere; they are quite adaptable! The carbon stored in bacteria is nearly as much as the carbon stored in plants.

Eukaryotic cells, unlike prokaryotic cells, contain a defined membrane-bound nucleus and extensive internal membranes that enclose other compartments, the **organelles** (Figure 1-2b). The region of the cell lying between the plasma membrane and the nucleus is the **cytoplasm,** comprising the cytosol (aqueous phase) and the organelles. **Eukaryotes** comprise all members of the plant and animal kingdoms, including the fungi, which exist in both multicellular forms (molds) and unicellular forms (yeasts), and the protozoans (*proto,* primitive; *zoan,* animal), which are exclusively unicellular. Eukaryotic cells are commonly about 10–100 μm across,

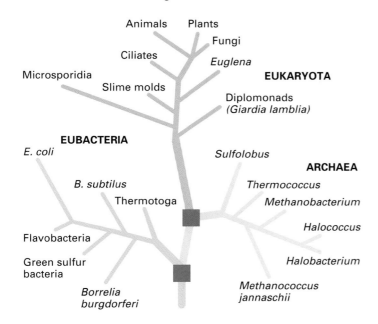

Presumed common progenitor
of all extant organisms

Presumed common progenitor
of archaebacteria and eukaryotes

▲ **FIGURE 1-3 All organisms from simple bacteria to complex mammals probably evolved from a common, single-celled progenitor.** This family tree depicts the evolutionary relations among the three major lineages of organisms. The structure of the tree was initially ascertained from morphological criteria: Creatures that look alike were put close together. More recently the sequences of DNA and proteins have been examined as a more information-rich criterion for assigning relationships. The greater the similarities in these macromolecular sequences, the more closely related organisms are thought to be. The trees based on morphological comparisons and the fossil record generally agree well with those those based on molecular data. Although all organisms in the eubacterial and archaean lineages are prokaryotes, archaea are more similar to eukaryotes than to eubacteria ("true" bacteria) in some respects. For instance, archaean and eukaryotic genomes encode homologous histone proteins, which associate with DNA; in contrast, bacteria lack histones. Likewise, the RNA and protein components of archaean ribosomes are more like those in eukaryotes than those in bacteria.

generally much larger than bacteria. A typical human fibroblast, a connective tissue cell, might be about 15 μm across with a volume and dry weight some thousands of times those of an *E. coli* bacterial cell. An amoeba, a single-celled protozoan, can be more than 0.5 mm long. An ostrich egg begins as a single cell that is even larger and easily visible to the naked eye.

All cells are thought to have evolved from a common progenitor because the structures and molecules in all cells have

so many similarities. In recent years, detailed analysis of the DNA sequences from a variety of prokaryotic organisms has revealed two distinct types: the so-called "true" bacteria, or **eubacteria,** and **archaea** (also called archaebacteria or archaeans). Working on the assumption that organisms with more similar genes evolved from a common progenitor more recently than those with more dissimilar genes, researchers have developed the evolutionary lineage tree shown in Figure 1-3. According to this tree, the archaea and the eukaryotes diverged from the true bacteria before they diverged from each other.

Many archaeans grow in unusual, often extreme, environments that may resemble ancient conditions when life first appeared on earth. For instance, halophiles ("salt loving") require high concentrations of salt to survive, and thermoacidophiles ("heat and acid loving") grow in hot (80° C) sulfur springs, where a pH of less than 2 is common. Still other archaeans live in oxygen-free milieus and generate methane (CH_4) by combining water with carbon dioxide.

Unicellular Organisms Help and Hurt Us

Bacteria and archaebacteria, the most abundant single-celled organisms, are commonly 1–2 μm in size. Despite their small size and simple architecture, they are remarkable biochemical factories, converting simple chemicals into complex biological molecules. Bacteria are critical to the earth's ecology, but some cause major diseases: bubonic plague (Black Death) from *Yersinia pestis,* strep throat from *Streptomyces,* tuberculosis from *Mycobacterium tuberculosis,* anthrax from *Bacillus anthracis,* cholera from *Vibrio cholerae,* food poisoning from certain types of *E. coli* and *Salmonella.*

Humans are walking repositories of bacteria, as are all plants and animals. We provide food and shelter for a staggering number of "bugs," with the greatest concentration in our intestines. Bacteria help us digest our food and in turn are able to reproduce. A common gut bacterium, *E. coli* is also a favorite experimental organism. In response to signals from bacteria such as *E. coli,* the intestinal cells form appropriate shapes to provide a niche where bacteria can live, thus facilitating proper digestion by the combined efforts of the bacterial and the intestinal cells. Conversely, exposure to intestinal cells changes the properties of the bacteria so that they participate more effectively in digestion. Such communication and response is a common feature of cells.

The normal, peaceful mutualism of humans and bacteria is sometimes violated by one or both parties. When bacteria begin to grow where they are dangerous to us (e.g., in the bloodstream or in a wound), the cells of our immune system fight back, neutralizing or devouring the intruders. Powerful antibiotic medicines, which selectively poison prokaryotic cells, provide rapid assistance to our relatively slow-developing immune response. Understanding the molecular biology of bacterial cells leads to an understanding of how bacteria are normally poisoned by antibiotics, how they become resistant to antibiotics, and what processes or structures present in bacterial but not human cells might be usefully targeted by new drugs.

(a)

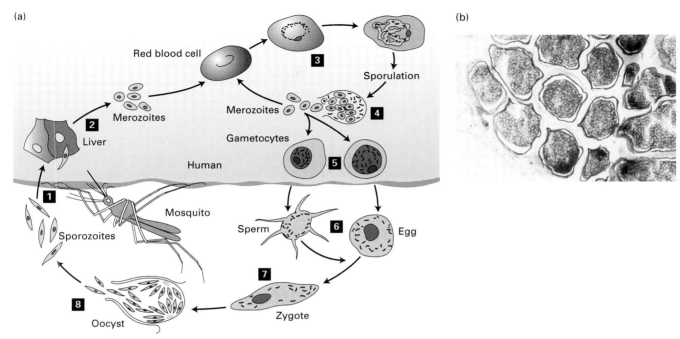

(b)

▲ **FIGURE 1-4 *Plasmodium* organisms, the parasites that cause malaria, are single-celled protozoans with a remarkable life cycle.** Many *Plasmodium* species are known, and they can infect a variety of animals, cycling between insect and vertebrate hosts. The four species that cause malaria in humans undergo several dramatic transformations within their human and mosquito hosts. (a) Diagram of the life cycle. Sporozoites enter a human host when an infected *Anopheles* mosquito bites a person **1**. They migrate to the liver where they develop into merozoites, which are released into the blood **2**. Merozoites differ substantially from sporozoites, so this transformation is a metamorphosis (Greek, "to transform" or "many shapes"). Circulating merozoites invade red blood cells (RBCs) and reproduce within them **3**. Proteins produced by some *Plasmodium* species move to the surface of infected RBCs, causing the cells to adhere to the walls of blood vessels. This prevents infected RBCs cells from circulating to the spleen where cells of the immune system would destroy the RBCs and the *Plasmodium* organisms they harbor. After growing and reproducing in RBCs for a period of time characteristic of each *Plasmodium* species, the merozoites suddenly burst forth in synchrony from large numbers of infected cells **4**. It is this event that brings on the fevers and shaking chills that are the well-known symptoms of malaria. Some of the released merozoites infect additional RBCs, creating a cycle of production and infection. Eventually, some merozoites develop into male and female gametocytes **5**, another metamorphosis. These cells, which contain half the usual number of chromosomes, cannot survive for long unless they are transferred in blood to an *Anopheles* mosquito. In the mosquito's stomach, the gametocytes are transformed into sperm or eggs (gametes), yet another metamorphosis marked by development of long hairlike flagella on the sperm **6**. Fusion of sperm and eggs generates zygotes **7**, which implant into the cells of the stomach wall and grow into oocysts, essentially factories for producing sporozoites. Rupture of an oocyst releases thousands of sporozoites **8**; these migrate to the salivary glands, setting the stage for infection of another human host. (b) Scanning electron micrograph of mature oocysts and emerging sporozoites. Oocysts abut the external surface of stomach wall cells and are encased within a membrane that protects them from the host immune system. [Part (b) courtesy of R. E. Sinden.]

Like bacteria, protozoa are usually beneficial members of the food chain. They play key roles in the fertility of soil, controlling bacterial populations and excreting nitrogenous and phosphate compounds, and are key players in waste treatment systems—both natural and man-made. These unicellular eukaryotes are also critical parts of marine ecosystems, consuming large quantities of phytoplankton and harboring photosynthetic algae, which use sunlight to produce biologically useful energy forms and small fuel molecules.

However, some protozoa do give us grief: *Entamoeba histolytica* causes dysentery; *Trichomonas vaginalis,* vagini-tis; and *Trypanosoma brucei,* sleeping sickness. Each year the worst of the protozoa, *Plasmodium falciparum* and related species, is the cause of more than 300 million new cases of malaria, a disease that kills 1.5 to 3 million people annually. These protozoans inhabit mammals and mosquitoes alternately, changing their morphology and behavior in response to signals in each of these environments. They also recognize receptors on the surfaces of the cells they infect. The complex life cycle of *Plasmodium* dramatically illustrates how a single cell can adapt to each new challenge it encounters (Figure 1-4). All of the transformations in cell type that occur during the *Plasmodium* life cycle are governed by instructions encoded in the genetic material of this parasite and triggered by environmental inputs.

The other group of single-celled eukaryotes, the yeasts, also have their good and bad points, as do their multicellular cousins, the molds. Yeasts and molds, which collectively constitute the fungi, have an important ecological role in breaking down plant and animal remains for reuse. They also

make numerous antibiotics and are used in the manufacture of bread, beer, wine, and cheese. Not so pleasant are fungal diseases, which range from relatively innocuous skin infections, such as jock itch and athlete's foot, to life-threatening *Pneumocystis carinii* pneumonia, a common cause of death among AIDS patients.

Even Single Cells Can Have Sex

The common yeast used to make bread and beer, *Saccharomyces cerevisiae*, appears fairly frequently in this book because it has proven to be a great experimental organism. Like many other unicellular organisms, yeasts have two mating types that are conceptually like the male and female gametes (eggs and sperm) of higher organisms. Two yeast cells of opposite mating type can fuse, or mate, to produce a third cell type containing the genetic material from each cell (Figure 1-5). Such sexual life cycles allow more rapid changes in genetic inheritance than would be possible without sex, resulting in valuable adaptations while quickly eliminating detrimental mutations. That, and not just Hollywood, is probably why sex is so ubiquitous.

Viruses Are the Ultimate Parasites

Virus-caused diseases are numerous and all too familiar: chicken pox, influenza, some types of pneumonia, polio, measles, rabies, hepatitis, the common cold, and many others. Smallpox, once a worldwide scourge, was eradicated by a decade-long global immunization effort beginning in the mid-1960s. Viral infections in plants (e.g., dwarf mosaic virus in corn) have a major economic impact on crop production. Planting of virus-resistant varieties, developed by traditional breeding methods and more recently by genetic engineering techniques, can reduce crop losses significantly. Most viruses have a rather limited host range, infecting certain bacteria, plants, or animals (Figure 1-6).

Because viruses cannot grow or reproduce on their own, they are not considered to be alive. To survive, a virus must infect a host cell and take over its internal machinery to synthesize viral proteins and in some cases to replicate the viral genetic material. When newly made viruses are released, the cycle starts anew. Viruses are much smaller than cells, on the order of 100 nanometer (nm) in diameter; in comparison, bacterial cells are usually >1000 nm (1 nm=10^{-9} meters). A virus is typically composed of a protein coat that encloses a core containing the genetic material, which carries the information for producing more viruses (Chapter 4). The coat protects a virus from the environment and allows it to stick to, or enter, specific host cells. In some viruses, the protein coat is surrounded by an outer membrane-like envelope.

The ability of viruses to transport genetic material into cells and tissues represents a medical menace and a medical opportunity. Viral infections can be devastatingly destructive, causing cells to break open and tissues to fall apart. However, many methods for manipulating cells depend upon using

(a)

1 *Mating between haploid cells of opposite mating type*

Diploid cells (a/α)

2 *Vegetative growth of diploid cells*

Bud

5 *Vegetative growth of haploid cells*

Four haploid ascospores within ascus

4 *Ascus ruptures, spores germinate*

3 *Starvation causes ascus formation, meiosis*

(b)

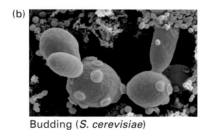

Budding (*S. cerevisiae*)

▲ **FIGURE 1-5 The yeast *Saccharomyces cerevisiae* reproduces sexually and asexually.** (a) Two cells that differ in mating type, called **a** and α, can mate to form an **a**/α cell **1**. The **a** and α cells are haploid, meaning they contain a single copy of each yeast chromosome, half the usual number. Mating yields a diploid **a**/α cell containing two copies of each chromosome. During vegetative growth, diploid cells multiply by mitotic budding, an asexual process **2**. Under starvation conditions, diploid cells undergo meiosis, a special type of cell division, to form haploid ascospores **3**. Rupture of an ascus releases four haploid spores, which can germinate into haploid cells **4**. These also can multiply asexually **5**. (b) Scanning electron micrograph of budding yeast cells. After each bud breaks free, a scar is left at the budding site so the number of previous buds can be counted. The orange cells are bacteria. [Part (b) M. Abbey/Visuals Unlimited, Inc.]

(a) T4 bacteriophage

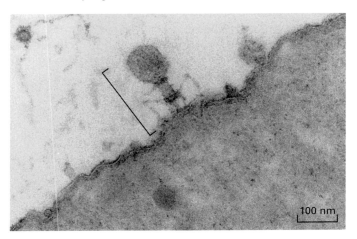

100 nm

(b) Tobacco mosaic virus

50 nm

(c) Adenovirus

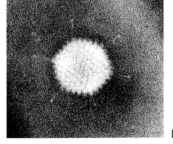

50 nm

▲ **FIGURE 1-6 Viruses must infect a host cell to grow and reproduce.** These electron micrographs illustrate some of the structural variety exhibited by viruses. (a) T4 bacteriophage (bracket) attaches to a bacterial cell via a tail structure. Viruses that infect bacteria are called bacteriophages, or simply phages. (b) Tobacco mosaic virus causes a mottling of the leaves of infected tobacco plants and stunts their growth. (c) Adenovirus causes eye and respiratory tract infections in humans. This virus has an outer membranous envelope from which long glycoprotein spikes protrude. [Part (a) from A. Levine, 1991, *Viruses*, Scientific American Library, p. 20. Part (b) courtesy of R. C. Valentine. Part (c) courtesy of Robley C. Williams, University of California.]

viruses to convey genetic material into cells. To do this, the portion of the viral genetic material that is potentially harmful is replaced with other genetic material, including human genes. The altered viruses, or **vectors**, still can enter cells toting the introduced genes with them (Chapter 9). One day, diseases caused by defective genes may be treated by using viral vectors to introduce a normal copy of a defective gene into patients. Current research is dedicated to overcoming the considerable obstacles to this approach, such as getting the introduced genes to work at the right places and times.

We Develop from a Single Cell

In 1827, German physician Karl von Baer discovered that mammals grow from eggs that come from the mother's ovary. Fertilization of an egg by a sperm cell yields a **zygote,** a visually unimpressive cell 200 μm in diameter. Every human being begins as a zygote, which houses all the necessary instructions for building the human body containing about 100 trillion (10^{14}) cells, an amazing feat. **Development** begins with the fertilized egg cell dividing into two, four, then eight cells, forming the very early embryo (Figure 1-7). Continued cell proliferation and then **differentiation** into distinct cell types gives rise to every tissue in the body. One initial cell, the fertilized egg (zygote), generates hundreds of different kinds of cells that differ in contents, shape, size, color, mobility, and surface composition. We will see how genes and signals control cell diversification in Chapters 15 and 22.

Making different kinds of cells—muscle, skin, bone, neuron, blood cells—is not enough to produce the human body. The cells must be properly arranged and organized into tissues, organs, and appendages. Our two hands have the same kinds of cells, yet their different arrangements—in a mirror image—are critical for function. In addition, many cells exhibit distinct functional and/or structural asymmetries, a property often called **polarity.** From such polarized cells arise

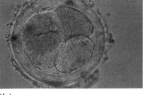

(a)

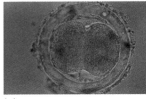

(b)

(c)

◄ **FIGURE 1-7 The first few cell divisions of a fertilized egg set the stage for all subsequent development.** A developing mouse embryo is shown at (a) the two-cell, (b) four-cell, and (c) eight-cell stages. The embryo is surrounded by supporting membranes. The corresponding steps in human development occur during the first few days after fertilization. [Claude Edelmann/Photo Researchers, Inc.]

MEDIA CONNECTIONS

Video: Early Embryonic Development

asymmetric, polarized tissues such as the lining of the intestines and structures like hands and hearts. The features that make some cells polarized, and how they arise, also are covered in later chapters.

Stem Cells, Cloning, and Related Techniques Offer Exciting Possibilities but Raise Some Concerns

Identical twins occur naturally when the mass of cells composing an early embryo divides into two parts, each of which develops and grows into an individual animal. Each cell in an eight-cell-stage mouse embryo has the potential to give rise to any part of the entire animal. Cells with this capability are referred to as *embryonic stem (ES) cells.* As we learn in Chapter 22, ES cells can be grown in the laboratory (cultured) and will develop into various types of differentiated cells under appropriate conditions.

The ability to make and manipulate mammalian embryos in the laboratory has led to new medical opportunities as well as various social and ethical concerns. In vitro fertilization, for instance, has allowed many otherwise infertile couples to have children. A new technique involves extraction of nuclei from defective sperm incapable of normally fertilizing an egg, injection of the nuclei into eggs, and implantation of the resulting fertilized eggs into the mother.

In recent years, nuclei taken from cells of adult animals have been used to produce new animals. In this procedure, the nucleus is removed from a body cell (e.g., skin or blood cell) of a donor animal and introduced into an unfertilized mammalian egg that has been deprived of its own nucleus. This manipulated egg, which is equivalent to a fertilized egg, is then implanted into a foster mother. The ability of such a donor nucleus to direct the development of an entire animal suggests that all the information required for life is retained in the nuclei of some adult cells. Since all the cells in an animal produced in this way have the genes of the single original donor cell, the new animal is a **clone** of the donor (Figure 1-8). Repeating the process can give rise to many clones. So far, however, the majority of embryos produced by this technique of *nuclear-transfer cloning* do not survive due to birth defects. Even those animals that are born live have shown abnormalities, including accelerated aging. The "rooting" of plants, in contrast, is a type of cloning that is readily accomplished by gardeners, farmers, and laboratory technicians.

The technical difficulties and possible hazards of nuclear-transfer cloning have not deterred some individuals from pursuing the goal of human cloning. However, cloning of humans per se has very limited scientific interest and is opposed by most scientists because of its high risk. Of greater scientific and medical interest is the ability to generate specific cell types starting from embryonic or adult stem cells. The scientific interest comes from learning the signals that can unleash the potential of the genes to form a certain cell type. The medical interest comes from the possibility of treating the nu-

▲ **FIGURE 1-8 Five genetically identical cloned sheep.** An early sheep embryo was divided into five groups of cells and each was separately implanted into a surrogate mother, much like the natural process of twinning. At an early stage the cells are able to adjust and form an entire animal; later in development the cells become progressively restricted and can no longer do so. An alternative way to clone animals is to replace the nuclei of multiple single-celled embryos with donor nuclei from cells of an adult sheep. Each embryo will be genetically identical to the adult from which the nucleus was obtained. Low percentages of embryos survive these procedures to give healthy animals, and the full impact of the techniques on the animals is not yet known. [Geoff Tompkinson/Science Photo Library/Photo Researchers, Inc.]

merous diseases in which particular cell types are damaged or missing, and of repairing wounds more completely.

1.2 The Molecules of a Cell

Molecular cell biologists explore how all the remarkable properties of the cell arise from underlying molecular events: the assembly of large molecules, binding of large molecules to each other, catalytic effects that promote particular chemical reactions, and the deployment of information carried by giant molecules. Here we review the most important kinds of molecules that form the chemical foundations of cell structure and function.

Small Molecules Carry Energy, Transmit Signals, and Are Linked into Macromolecules

Much of the cell's contents is a watery soup flavored with small molecules (e.g., simple sugars, amino acids, vitamins) and ions (e.g., sodium, chloride, calcium ions). The locations and concentrations of small molecules and ions within the cell are controlled by numerous proteins inserted in cellular membranes. These pumps, transporters, and ion channels move nearly all small molecules and ions into or out of the cell and its organelles (Chapter 7).

One of the best-known small molecules is **adenosine triphosphate (ATP),** which stores readily available chemical energy in two of its chemical bonds (see Figure 2-24). When cells split apart these energy-rich bonds in ATP, the released energy can be harnessed to power an energy-requiring process like muscle contraction or protein biosynthesis. To obtain energy for making ATP, cells break down food molecules. For instance, when sugar is degraded to carbon dioxide and water, the energy stored in the original chemical bonds is released and much of it can be "captured" in ATP (Chapter 8). Bacterial, plant, and animal cells can all make ATP by this process. In addition, plants and a few other organisms can harvest energy from sunlight to form ATP in **photosynthesis.**

Other small molecules act as signals both within and between cells; such signals direct numerous cellular activities (Chapters 13–15). The powerful effect on our bodies of a frightening event comes from the instantaneous flooding of the body with epinephrine, a small-molecule **hormone** that mobilizes the "fight or flight" response. The movements needed to fight or flee are triggered by nerve impulses that flow from the brain to our muscles with the aid of **neurotransmitters,** another type of small-molecule signal that we discuss in Chapter 7.

Certain small molecules (**monomers**) in the cellular soup can be joined to form **polymers** through repetition of a single type of chemical-linkage reaction (see Figure 2-11). Cells produce three types of large polymers, commonly called **macromolecules:** polysaccharides, proteins, and nucleic acids. Sugars, for example, are the monomers used to form

polysaccharides. These macromolecules are critical structural components of plant cell walls and insect skeletons. A typical polysaccharide is a linear or branched chain of repeating identical sugar units. Such a chain carries information: the number of units. However if the units are *not* identical, then the order and type of units carry additional information. As we see in Chapter 6, some polysaccharides exhibit the greater informational complexity associated with a linear code made up of different units assembled in a particular order. This property, however, is most typical of the two other types of biological macromolecules—**proteins** and **nucleic acids.**

Proteins Give Cells Structure and Perform Most Cellular Tasks

The varied, intricate structures of proteins enable them to carry out numerous functions. Cells string together 20 different **amino acids** in a linear chain to form a protein (see Figure 2-13). Proteins commonly range in length from 100 to 1000 amino acids, but some are much shorter and others longer. We obtain amino acids either by synthesizing them from other molecules or by breaking down proteins that we eat. The "essential" amino acids, from a dietary standpoint, are the eight that we cannot synthesize and must obtain from food. Beans and corn together have all eight, making their combination particularly nutritious. Once a chain of amino acids is formed, it folds into a complex shape, conferring a distinctive three-dimensional structure and function on each protein (Figure 1-9).

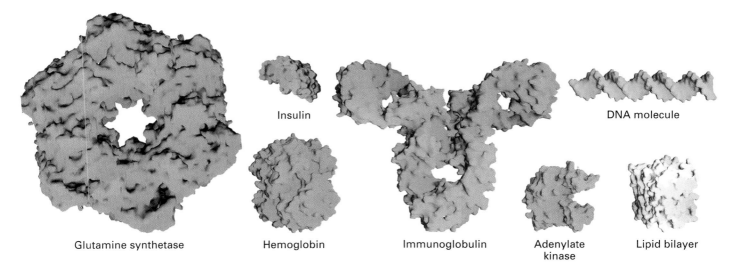

Insulin

DNA molecule

Glutamine synthetase   Hemoglobin   Immunoglobulin   Adenylate kinase   Lipid bilayer

▲ **FIGURE 1-9 Proteins vary greatly in size, shape, and function.** These models of the water-accessible surface of some representative proteins are drawn to a common scale and reveal the numerous projections and crevices on the surface. Each protein has a defined three-dimensional shape (conformation) that is stabilized by numerous chemical interactions discussed in Chapters 2 and 3. The illustrated proteins include enzymes (glutamine synthetase and adenylate kinase), an antibody (immunoglobulin), a hormone (insulin), and the blood's oxygen carrier (hemoglobin). Models of a segment of the nucleic acid DNA and a small region of the lipid bilayer that forms cellular membranes (see Section 1.3) demonstrate the relative width of these structures compared with typical proteins. [Courtesy of Gareth White.]

Some proteins are similar to one another and therefore can be considered members of a **protein family.** A few hundred such families have been identified. Most proteins are designed to work in particular places within a cell or to be released into the extracellular (*extra,* "outside") space. Elaborate cellular pathways ensure that proteins are transported to their proper intracellular (*intra,* within) locations or secreted (Chapters 16 and 17).

Proteins can serve as structural components of a cell, for example, by forming an internal skeleton (Chapters 5, 19, and 20). They can be sensors that change shape as temperature, ion concentrations, or other properties of the cell change. They can import and export substances across the plasma membrane (Chapter 7). They can be **enzymes,** causing chemical reactions to occur much more rapidly than they would without the aid of these protein **catalysts** (Chapter 3). They can bind to a specific gene, turning it on or off (Chapter 11). They can be extracellular signals, released from one cell to communicate with other cells, or intracellular signals, carrying information within the cell (Chapters 13–15). They can be motors that move other molecules around, burning chemical energy (ATP) to do so (Chapters 19 and 20).

How can 20 amino acids form all the different proteins needed to perform these varied tasks? Seems impossible at first glance. But if a "typical" protein is about 400 amino acids long, there are 20^{400} possible different protein sequences. Even assuming that many of these would be functionally equivalent, unstable, or otherwise discountable, the number of possible proteins is well along toward infinity.

Next we might ask how many protein molecules a cell needs to operate and maintain itself. To estimate this number, let's take a typical eukaryotic cell, such as a hepatocyte (liver cell). This cell, roughly a cube 15 μm (0.0015 cm) on a side, has a volume of 3.4×10^{-9} cm^3 (or milliliters). Assuming a cell density of 1.03 g/ml, the cell would weigh 3.5×10^{-9} g. Since protein accounts for approximately 20 percent of a cell's weight, the total weight of cellular protein is 7×10^{-10} g. The average yeast protein has a molecular weight of 52,700 (g/mol). Assuming this value is typical of eukaryotic proteins, we can calculate the total number of protein molecules per liver cell as about 7.9×10^9 from the total protein weight and Avogadro's number, the number of molecules per mole of any chemical compound (6.02×10^{23}). To carry this calculation one step further, consider that a liver cell contains about 10,000 different proteins; thus, a cell contains close to a million molecules of each type of protein on average. In actuality the abundance of different proteins varies widely, from the quite rare insulin-binding receptor protein (20,000 molecules) to the abundant structural protein actin (5×10^8 molecules).

Nucleic Acids Carry Coded Information for Making Proteins at the Right Time and Place

The information about how, when, and where to produce each kind of protein is carried in the genetic material, a polymer called **deoxyribonucleic acid (DNA).** The three-dimensional structure of DNA consists of two long helical strands that are coiled around a common axis, forming a **double helix.** DNA strands are composed of monomers called **nucleotides;** these often are referred to as *bases* because their structures contain cyclic organic bases (Chapter 4).

Four different nucleotides, abbreviated A, T, C, and G, are joined end to end in a DNA strand, with the base parts projecting out from the helical backbone of the strand. Each DNA double helix has a simple construction: wherever there is an A in one strand there is a T in the other, and each C is matched with a G (Figure 1-10). This **complementary** matching of the two strands is so strong that if complementary strands are separated, they will spontaneously zip back together in the right salt and temperature conditions. Such **hybridization** is extremely useful for detecting one strand using the other. For example, if one strand is purified and attached to a piece of paper, soaking the paper in a solution containing the other complementary strand will lead to zipping,

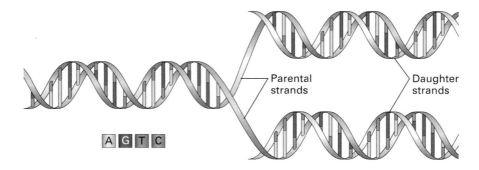

▲ **FIGURE 1-10 DNA consists of two complementary strands wound around each other to form a double helix.** *(Left)* The double helix is stabilized by weak hydrogen bonds between the A and T bases and between the C and G bases. *(Right)* During replication, the two strands are unwound and used as templates to produce complementary strands. The outcome is two copies of the original double helix, each containing one of the original strands and one new daughter (complementary) strand.

even if the solution also contains many other DNA strands that do not match.

The genetic information carried by DNA resides in its sequence, the linear order of nucleotides along a strand. The information-bearing portion of DNA is divided into discrete functional units, the **genes,** which typically are 5000 to 100,000 nucleotides long. Most bacteria have a few thousand genes; humans, about 40,000. The genes that carry instructions for making proteins commonly contain two parts: a *coding region* that specifies the amino acid sequence of a protein and a *regulatory region* that controls when and in which cells the protein is made.

Cells use two processes in series to convert the coded information in DNA into proteins (Figure 1-11). In the first, called **transcription,** the coding region of a gene is copied into a single-stranded **ribonucleic acid (RNA)** version of the double-stranded DNA. A large enzyme, **RNA polymerase,** catalyzes the linkage of nucleotides into a RNA chain using DNA as a template. In eukaryotic cells, the initial RNA product is processed into a smaller **messenger RNA (mRNA)** molecule, which moves to the cytoplasm. Here the **ribosome,** an enormously complex molecular machine composed of both RNA and protein, carries out the second process, called **translation.** During translation, the ribosome assembles and links together amino acids in the precise order dictated by the mRNA sequence according to the nearly universal **genetic code.** We examine the cell components that carry out transcription and translation in detail in Chapter 4.

All organisms have ways to control when and where their genes can be transcribed. For instance, nearly all the cells in our bodies contain the full set of human genes, but in each cell type only some of these genes are active, or turned on, and used to make proteins. That's why liver cells produce some proteins that are not produced by kidney cells, and vice versa. Moreover, many cells can respond to external signals or changes in external conditions by turning specific genes on or off, thereby adapting their repertoire of proteins to meet current needs. Such control of gene activity depends on DNA-binding proteins called **transcription factors,** which bind to DNA and act as switches, either activating or repressing transcription of particular genes (Chapter 11).

Transcription factors are shaped so precisely that they are able to bind preferentially to the regulatory regions of just a few genes out of the thousands present in a cell's DNA. Typically a DNA-binding protein will recognize short DNA sequences about 6–12 base pairs long. A segment of DNA containing 10 base pairs can have 4^{10} possible sequences (1,048,576) since each position can be any of four nucleotides. Only a few copies of each such sequence will occur in the DNA of a cell, assuring the specificity of gene activation and repression. Multiple copies of one type of transcription factor can coordinately regulate a set of genes if binding sites for that factor exist near each gene in the set. Transcription factors often work as multiprotein complexes, with more than one protein contributing its own DNA-binding specificity to selecting the regulated genes. In complex organisms,

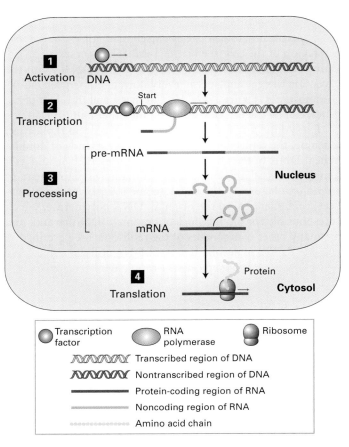

▲ **FIGURE 1-11 The coded information in DNA is converted into the amino acid sequences of proteins by a multistep process.** Step **1**: Transcription factors bind to the regulatory regions of the specific genes they control and activate them. Step **2**: Following assembly of a multiprotein initiation complex bound to the DNA, RNA polymerase begins transcription of an activated gene at a specific location, the start site. The polymerase moves along the DNA linking nucleotides into a single-stranded pre-mRNA transcript using one of the DNA strands as a template. Step **3**: The transcript is processed to remove noncoding sequences. Step **4**: In a eukaryotic cell, the mature messenger RNA (mRNA) moves to the cytoplasm, where it is bound by ribosomes that read its sequence and assemble a protein by chemically linking amino acids into a linear chain.

hundreds of different transcription factors are employed to form an exquisite control system that activates the right genes in the right cells at the right times.

The Genome Is Packaged into Chromosomes and Replicated During Cell Division

Most of the DNA in eukaryotic cells is located in the nucleus, extensively folded into the familiar structures we know as **chromosomes** (Chapter 10). Each chromosome contains a single linear DNA molecule associated with certain proteins. In prokaryotic cells, most or all of the genetic information resides in a single circular DNA molecule about a millimeter in length; this molecule lies, folded back on itself many times, in the central region of the cell (see Figure 1-2a). The **genome** of an organism comprises its entire complement of DNA. With the exception of eggs and sperm, every normal human cell has 46 chromosomes (Figure 1-12). Half of these, and thus half of the genes, can be traced back to Mom; the other half, to Dad.

Every time a cell divides, a large multiprotein replication machine, the replisome, separates the two strands of double-helical DNA in the chromosomes and uses each strand as a template to assemble nucleotides into a new complementary strand (see Figure 1-10). The outcome is a pair of double helices, each identical to the original. **DNA polymerase,** which is responsible for linking nucleotides into a DNA strand, and the many other components of the replisome are described in Chapter 4. The molecular design of DNA and the remarkable properties of the replisome assure rapid, highly accurate copying. Many DNA polymerase molecules work in concert, each one copying part of a chromosome. The entire genome of fruit flies, about 1.2×10^8 nucleotides long, can be copied in three minutes! Because of the accuracy of DNA replication, nearly all the cells in our bodies carry the same genetic instructions, and we can inherit Mom's brown hair and Dad's blue eyes.

A rather dramatic example of gene control involves inactivation of an entire chromosome in human females. Women have two X chromosomes, whereas men have one X chromosome and one Y chromosome, which has different genes than the X chromosome. Yet the genes on the X chromosome must, for the most part, be equally active in female cells (XX) and male cells (XY). To achieve this balance, one of the X chromosomes in female cells is chemically modified and condensed into a very small mass called a Barr body, which is inactive and never transcribed.

Surprisingly, we inherit a small amount of genetic material entirely and uniquely from our mothers. This is the circular DNA present in mitochondria, the organelles in eukaryotic cells that synthesize ATP using the energy released by the breakdown of nutrients. Mitochondria contain multiple copies of their own DNA genomes, which code for some of the mitochondrial proteins (Chapter 10). Because each human inherits mitochondrial DNA only from his or her mother (it comes with the egg but not the sperm), the distinctive features of a particular mitochondrial DNA can be used to trace maternal history. Chloroplasts, the organelles that carry out photosynthesis in plants, also have their own circular genomes.

Mutations May Be Good, Bad, or Indifferent

Mistakes occasionally do occur spontaneously during DNA replication, causing changes in the sequence of nucleotides. Such changes, or **mutations,** also can arise from radiation

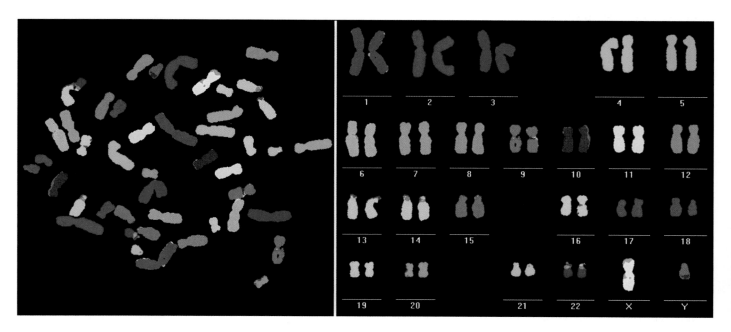

▲ **FIGURE 1-12 Chromosomes can be "painted" for easy identification.** A normal human has 23 pairs of morphologically distinct chromosomes; one member of each pair is inherited from the mother and the other member from the father. *(Left)* A chromosome spread from a human body cell midway through mitosis, when the chromosomes are fully condensed. This preparation was treated with fluorescent-labeled staining reagents that allow each of the 22 pairs and the X and Y chromosomes to appear in a different color when viewed in a fluorescence microscope. This technique of multiplex fluorescence in situ hybridization (M-FISH) sometimes is called chromosome painting (Chapter 10). *(Right)* Chromosomes from the preparation on the left arranged in pairs in descending order of size, an array called a karyotype. The presence of X and Y chromosomes identifies the sex of the individual as male. [Courtesy of M. R. Speicher.]

that causes damage to the nucleotide chain or from chemical poisons, such as those in cigarette smoke, that lead to errors during the DNA-copying process (Chapter 23). Mutations come in various forms: a simple swap of one nucleotide for another; the deletion, insertion, or inversion of one to millions of nucleotides in the DNA of one chromosome; and translocation of a stretch of DNA from one chromosome to another.

In sexually reproducing animals like ourselves, mutations can be inherited only if they are present in cells that potentially contribute to the formation of offspring. Such **germ-line cells** include eggs, sperm, and their precursor cells. Body cells that do not contribute to offspring are called **somatic cells.** Mutations that occur in these cells never are inherited, although they may contribute to the onset of cancer. Plants have a less distinct division between somatic and germ-line cells, since many plant cells can function in both capacities.

Mutated genes that encode altered proteins or that cannot be controlled properly cause numerous inherited diseases. For example, sickle cell disease is attributable to a single nucleotide substitution in the hemoglobin gene, which encodes the protein that carries oxygen in red blood cells. The single amino acid change caused by the sickle cell mutation reduces the ability of red blood cells to carry oxygen from the lungs to the tissues. Recent advances in detecting disease-causing mutations and in understanding how they affect cell functions offer exciting possibilities for reducing their often devastating effects.

Sequencing of the human genome has shown that a very large proportion of our DNA does not code for any RNA or have any discernible regulatory function, a quite unexpected finding. Mutations in these regions usually produce no immediate effects—good or bad. However, such "indifferent" mutations in nonfunctional DNA may have been a major player in evolution, leading to creation of new genes or new regulatory sequences for controlling already existing genes. For instance, since binding sites for transcription factors typically are only 10–12 nucleotides long, a few single-nucleotide mutations might convert a nonfunctional bit of DNA into a functional protein-binding regulatory site.

Much of the nonessential DNA in both eukaryotes and prokaryotes consists of highly repeated sequences that can move from one place in the genome to another. These **mobile DNA elements** can jump (transpose) into genes, most commonly damaging but sometimes activating them. Jumping generally occurs rarely enough to avoid endangering the host organism. Mobile elements, which were discovered first in plants, are responsible for leaf color variegation and the diverse beautiful color patterns of Indian corn kernels. By jumping in and out of genes that control pigmentation as plant development progresses, the mobile elements give rise to elaborate colored patterns. Mobile elements were later found in bacteria in which they often carry and, unfortunately, disseminate genes for antibiotic resistance.

Now we understand that mobile elements have multiplied and slowly accumulated in genomes over evolutionary time, becoming a universal property of genomes in present-day organisms. They account for an astounding 45 percent of the human genome. Some of our own mobile DNA elements are copies—often highly mutated and damaged—of genomes from viruses that spend part of their life cycle as DNA segments inserted into host-cell DNA. Thus we carry in our chromosomes the genetic residues of infections acquired by our ancestors. Once viewed only as molecular parasites, mobile DNA elements are now thought to have contributed significantly to the evolution of higher organisms (Chapter 10).

1.3 The Work of Cells

In essence, any cell is simply a compartment with a watery interior that is separated from the external environment by a surface membrane (the plasma membrane) that prevents the free flow of molecules in and out of cells. In addition, as we've noted, eukaryotic cells have extensive internal membranes that further subdivide the cell into various compartments, the organelles. The plasma membrane and other cellular membranes are composed primarily of two layers of **phospholipid** molecules. These bipartite molecules have a "water-loving" (**hydrophilic**) end and a "water-hating" (**hydrophobic**) end. The two phospholipid layers of a membrane are oriented with all the hydrophilic ends directed toward the inner and outer surfaces and the hydrophobic ends buried within the interior (Figure 1-13). Smaller amounts of

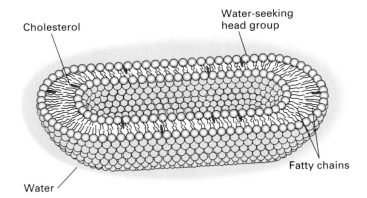

▲ **FIGURE 1-13 The watery interior of cells is surrounded by the plasma membrane, a two-layered shell of phospholipids.** The phospholipid molecules are oriented with their fatty acyl chains (black squiggly lines) facing inward and their water-seeking head groups (white spheres) facing outward. Thus both sides of the membrane are lined by head groups, mainly charged phosphates, adjacent to the watery spaces inside and outside the cell. All biological membranes have the same basic phospholipid bilayer structure. Cholesterol (red) and various proteins (not shown) are embedded in the bilayer. In actuality, the interior space is much larger relative to the volume of the plasma membrane depicted here.

other lipids, such as cholesterol, and many kinds of proteins are inserted into the phospholipid framework. The lipid molecules and some proteins can float sidewise in the plane of the membrane, giving membranes a fluid character. This fluidity allows cells to change shape and even move. However, the attachment of some membrane proteins to other molecules inside or outside the cell restricts their lateral movement. We learn more about membranes and how molecules cross them in Chapters 5 and 7.

The cytosol and the internal spaces of organelles differ from each other and from the cell exterior in terms of acidity, ionic composition, and protein contents. For example, the composition of salts inside the cell is often drastically different from what is outside. Because of these different "microclimates," each cell compartment has its own assigned tasks in the overall work of the cell (Chapter 5). The unique functions and micro-climates of the various cell compartments are due largely to the proteins that reside in their membranes or interior.

We can think of the entire cell compartment as a factory dedicated to sustaining the well-being of the cell. Much cellular work is performed by molecular machines, some housed in the cytosol and some in various organelles. Here we quickly review the major tasks that cells carry out in their pursuit of the good life.

Cells Build and Degrade Numerous Molecules and Structures

As chemical factories, cells produce an enormous number of complex molecules from simple chemical building blocks. All of this synthetic work is powered by chemical energy extracted primarily from sugars and fats or sunlight, in the case of plant cells, and stored primarily in ATP, the universal "currency" of chemical energy (Figure 1-14). In animal and plant cells, most ATP is produced by large molecular machines located in two organelles, **mitochondria** and **chloroplasts.** Similar machines for generating ATP are located in the plasma membrane of bacterial cells. Both mitochondria and chloroplasts are thought to have originated as bacteria that took up residence inside eukaryotic cells and then became welcome collaborators (Chapter 8). Directly or indirectly, all of our food is created by plant cells using sunlight to build complex macromolecules during photosynthesis. Even underground oil supplies are derived from the decay of plant material.

Cells need to break down worn-out or obsolete parts into small molecules that can be discarded or recycled. This housekeeping task is assigned largely to **lysosomes,** organelles crammed with degradative enzymes. The interior of lysosomes has a pH of about 5.0, roughly 100 times more acidic than that of the surrounding cytosol. This aids in the breakdown of materials by lysosomal enzymes, which are specially designed to function at such a low pH. To create the low pH environment, proteins located in the lysosomal membrane pump hydrogen ions into the lysosome using energy supplied from ATP (Chapter 7). Lysosomes are assisted in the cell's cleanup work by **peroxisomes.** These small organelles are specialized for breaking down the lipid components of membranes and rendering various toxins harmless.

Most of the structural and functional properties of cells depend on proteins. Thus for cells to work properly, the nu-

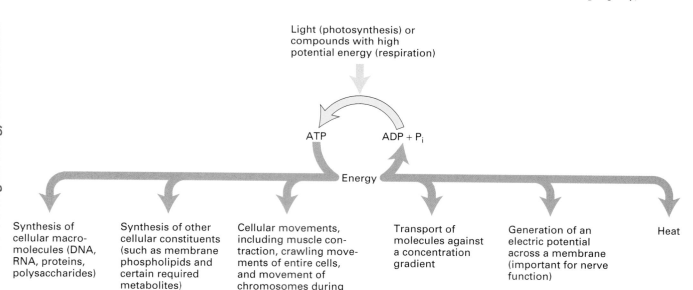

MEDIA CONNECTIONS

Overview Animation: Biological Energy Interconversions

▲ **FIGURE 1-14 ATP is the most common molecule used by cells to capture and transfer energy.** ATP is formed from ADP and inorganic phosphate (P_i) by photosynthesis in plants and by the breakdown of sugars and fats in most cells. The energy released by the splitting (hydrolysis) of P_i from ATP drives many cellular processes.

merous proteins composing the various working compartments must be transported from where they are made to their proper locations (Chapters 16 and 17). Some proteins are made on ribosomes that are free in the cytosol. Proteins secreted from the cell and most membrane proteins, however, are made on ribosomes associated with the **endoplasmic reticulum (ER)**. This organelle produces, processes, and ships out both proteins and lipids. Protein chains produced on the ER move to the **Golgi apparatus,** where they are further modified before being forwarded to their final destinations. Proteins that travel in this way contain short sequences of amino acids or attached sugar chains (oligosaccharides) that serve as addresses for directing them to their correct destinations. These addresses work because they are recognized and bound by other proteins that do the sorting and shipping in various cell compartments.

Animal Cells Produce Their Own External Environment and Glues

The simplest multicellular animals are single cells embedded in a jelly of proteins and polysaccharides called the **extracellular matrix**. Cells themselves produce and secrete these materials, thus creating their own immediate environment (Chapter 6). **Collagen,** the single most abundant protein in the animal kingdom, is a major component of the extracellular matrix in most tissues. In animals, the extracellular matrix cushions and lubricates cells. A specialized, especially tough matrix, the **basal lamina,** forms a supporting layer underlying sheetlike cell layers and helps prevent the cells from ripping apart.

The cells in animal tissues are "glued" together by **cell-adhesion molecules (CAMs)** embedded in their surface membranes. Some CAMs bind cells to one another; other types bind cells to the extracellular matrix, forming a cohesive unit. The cells of higher plants contain relatively few such molecules; instead, plants cells are rigidly tied together by extensive interlocking of the cell walls of neighboring

cells. The cytosols of adjacent animal or plant cells often are connected by functionally similar but structurally different "bridges" called **gap junctions** in animals and **plasmodesmata** in plants. These structures allow cells to exchange small molecules including nutrients and signals, facilitating coordinated functioning of the cells in a tissue.

Cells Change Shape and Move

Although cells sometimes are spherical, they more commonly have more elaborate shapes due to their internal skeletons and external attachments. Three types of protein filaments, organized into networks and bundles, form the **cytoskeleton** within animal cells (Figure 1-15). The cytoskeleton prevents the plasma membrane of animal cells from relaxing into a sphere (Chapter 5); it also functions in cell locomotion and the intracellular transport of vesicles, chromosomes, and macromolecules (Chapters 19 and 20). The cytoskeleton can be linked through the cell surface to the extracellular matrix or to the cytoskeleton of other cells, thus helping to form tissues (Chapter 6).

All cytoskeletal filaments are long polymers of protein subunits. Elaborate systems regulate the assembly and disassembly of the cytoskeleton, thereby controlling cell shape. In some cells the cytoskeleton is relatively stable, but in others it changes shape continuously. Shrinkage of the cytoskeleton in some parts of the cell and its growth in other parts can produce coordinated changes in shape that result in cell locomotion. For instance, a cell can send out an extension that attaches to a surface or to other cells and then retract the cell body from the other end. As this process continues due to coordinated changes in the cytoskeleton, the cell moves forward. Cells can move at rates on the order of 20 μm/second. Cell locomotion is used during embryonic development of multicellular animals to shape tissues and during adulthood to defend against infection, to transport nutrients, and to heal wounds. This process does not play a role in the growth and development of multicellular plants because new plant cells

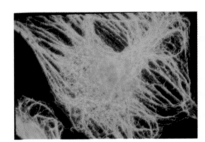
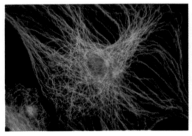

| **Intermediate filaments** | **Microtubules** | **Microfilaments** |

▲ **FIGURE 1-15 The three types of cytoskeletal filaments have characteristic distributions within cells.** Three views of the same cell. A cultured fibroblast was treated with three different antibody preparations. Each antibody binds specifically to the protein monomers forming one type of filament and is chemically linked to a differently colored fluorescent dye (green,

blue, or red). Visualization of the stained cell in a fluorescence microscope reveals the location of filaments bound to a particular dye-antibody preparation. In this case, intermediate filaments are stained green; microtubules, blue; and microfilaments, red. All three fiber systems contribute to the shape and movements of cells. [Courtesy of V. Small.]

are generated by the division of existing cells that share cell walls. As a result, plant development involves cell enlargement but not movement of cells from one position to another.

Cells Sense and Send Information

A living cell continuously monitors its surroundings and adjusts its own activities and composition accordingly. Cells also communicate by deliberately sending signals that can be received and interpreted by other cells. Such signals are common not only within an individual organism, but also between organisms. For instance, the odor of a pear detected by us and other animals signals a food source; consumption of the pear by an animal aids in distributing the pear's seeds. Everyone benefits! The signals employed by cells include simple small chemicals, gases, proteins, light, and mechanical movements. Cells possess numerous receptor proteins for detecting signals and elaborate pathways for transmitting them within the cell to evoke a response. At any time, a cell may be able to sense only some of the signals around it, and how a cell responds to a signal may change with time. In some cases, receiving one signal primes a cell to respond to a subsequent different signal in a particular way.

Both changes in the environment (e.g., an increase or decrease in a particular nutrient or the light level) and signals received from other cells represent external information that cells must process. The most rapid responses to such signals generally involve changes in the location or activity of preexisting proteins. For instance, soon after you eat a carbohydrate-rich meal, glucose pours into your bloodstream. The rise in blood glucose is sensed by β cells in the pancreas, which respond by releasing their stored supply of the protein hormone **insulin**. The circulating insulin signal causes glucose transporters in the cytoplasm of fat and muscle cells to move to the cell surface, where they begin importing glucose. Meanwhile, liver cells also are furiously taking in glucose via a different glucose transporter. In both liver and muscle cells, an intracellular signaling pathway triggered by binding of insulin to cell-surface receptors activates a key enzyme needed to make **glycogen**, a large glucose polymer (Figure 1-16a). The net result of these cell responses is that your blood glucose level falls and extra glucose is stored as glycogen, which your cells can use as a glucose source when you skip a meal to cram for a test.

The ability of cells to send and respond to signals is crucial to development. Many developmentally important signals are secreted proteins produced by specific cells at specific times and places in a developing organism. Often a receiving cell integrates multiple signals in deciding how to behave, for example, to differentiate into a particular tissue type, to extend a process, to die, to send back a confirming signal (yes, I'm here!), or to migrate.

The functions of about half the proteins in humans, roundworms, yeast, and several other eukaryotic organisms have been predicted based on analyses of genomic sequences (Chapter 9). Such analyses have revealed that at least 10–15 percent of the proteins in eukaryotes function as secreted ex-

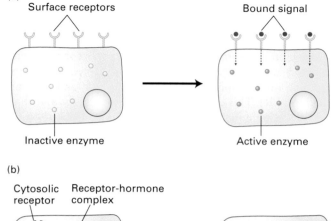

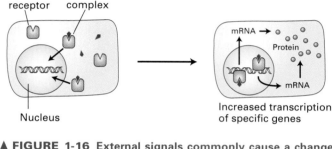

▲ **FIGURE 1-16 External signals commonly cause a change in the activity of preexisting proteins or in the amounts and types of proteins that cells produce.** (a) Binding of a hormone or other signaling molecule to its specific receptors can trigger an intracellular pathway that increases or decreases the activity of a preexisting protein. For example, binding of insulin to receptors in the plasma membrane of liver and muscle cells leads to activation of glycogen synthase, a key enzyme in the synthesis of glycogen from glucose. (b) The receptors for steroid hormones are located within cells, not on the cell surface. The hormone-receptor complexes activate transcription of specific target genes, leading to increased production of the encoded proteins. Many signals that bind to receptors on the cell surface also act, by more complex pathways, to modulate gene expression.

tracellular signals, signal receptors, or intracellular **signal-transduction** proteins, which pass along a signal through a series of steps culminating in a particular cellular response (e.g., increased glycogen synthesis). Clearly, signaling and signal transduction are major activities of cells.

Cells Regulate Their Gene Expression to Meet Changing Needs

In addition to modulating the activities of existing proteins, cells often respond to changing circumstances and to signals from other cells by altering the amount or types of proteins they contain. **Gene expression,** the overall process of selectively reading and using genetic information, is commonly controlled at the level of transcription, the first step in the production of proteins. In this way cells can produce a particular mRNA only when the encoded protein is needed, thus minimizing wasted energy. Producing a mRNA is, however, only the first in a chain of regulated events that together determine whether an active protein product is produced from a particular gene.

Transcriptional control of gene expression was first decisively demonstrated in the response of the gut bacterium *E. coli* to different sugar sources. *E. coli* cells prefer glucose as a sugar source, but they can survive on lactose in a pinch. These bacteria use both a DNA-binding *repressor* protein and a DNA-binding *activator* protein to change the rate of transcription of three genes needed to metabolize lactose depending on the relative amounts of glucose and lactose present (Chapter 4). Such dual positive/negative control of gene expression fine tunes the bacterial cell's enzymatic equipment for the job at hand.

Like bacterial cells, unicellular eukaryotes may be subjected to widely varying environmental conditions that require extensive changes in cellular structures and function. For instance, in starvation conditions yeast cells stop growing and form dormant spores (see Figure 1-4). In multicellular organisms, however, the environment around most cells is relatively constant. The major purpose of gene control in us and in other complex organisms is to tailor the properties of various cell types to the benefit of the entire animal or plant.

Control of gene activity in eukaryotic cells usually involves a balance between the actions of transcriptional activators and repressors. Binding of activators to specific DNA regulatory sequences called **enhancers** turns on transcription, and binding of repressors to other regulatory sequences called **silencers** turns off transcription. In Chapters 11 and 12, we take a close look at transcriptional activators and repressors and how they operate, as well as other mechanisms for controlling gene expression. In an extreme case, expression of a particular gene could occur only in part of the brain, only during evening hours, only during a certain stage of development, only after a large meal, and so forth.

Many external signals modify the activity of transcriptional activators and repressors that control specific genes. For example, lipid-soluble steroid hormones, such as estrogen and testosterone, can diffuse across the plasma membrane and bind to their specific receptors located in the cytoplasm or nucleus (Figure 1-16b). Hormone binding changes the shape of the receptor so that it can bind to specific enhancer sequences in the DNA, thus turning the receptor into a transcriptional activator. By this rather simple signal-transduction pathway, steroid hormones cause cells to change which genes they transcribe (Chapter 11). Since steroid hormones can circulate in the bloodstream, they can affect the properties of many or all cells in a temporally coordinated manner. Binding of many other hormones and of growth factors to receptors on the cell surface triggers different signal-transduction pathways that also lead to changes in the transcription of specific genes (Chapters 13–15). Although these pathways involve multiple components and are more complicated than those transducing steroid hormone signals, the general idea is the same.

Cells Grow and Divide

The most remarkable feature of cells and entire organisms is their ability to reproduce. Biological reproduction, combined with continuing evolutionary selection for a highly functional body plan, is why today's horseshoe crabs look much as they did 300 million years ago, a time span during which entire mountain ranges have risen or fallen. The Teton Mountains in Wyoming, now about 14,000 feet high and still growing, did not exist a mere 10 million years ago. Yet horseshoe crabs, with a life span of about 19 years, have faithfully reproduced their ancient selves more than half a million times during that period. The common impression that biological structure is transient and geological structure is stable is the exact opposite of the truth. Despite the limited duration of our individual lives, reproduction gives us a potential for immortality that a mountain or a rock does not have.

The simplest type of reproduction entails the division of a "parent" cell into two "daughter" cells. This occurs as part of the **cell cycle,** a series of events that prepares a cell to divide followed by the actual division process, called **mitosis.** The eukaryotic cell cycle commonly is represented as four stages (Figure 1-17). The chromosomes and the DNA they carry are copied during the **S (synthesis) phase.** The replicated chromosomes separate during the **M (mitotic) phase,** with each daughter cell getting a copy of each chromosome during cell division. The M and S phases are separated by two gap stages, the G_1 **phase** and G_2 **phase,** during which mRNAs and proteins are made. In single-celled organisms, both daughter cells

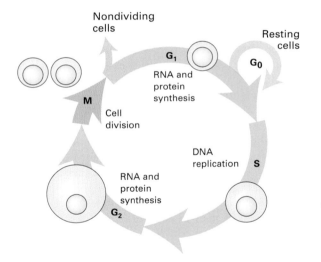

▲ **FIGURE 1-17 During growth, eukaryotic cells continually progress through the four stages of the cell cycle, generating new daughter cells.** In most proliferating cells, the four phases of the cell cycle proceed successively, taking from 10–20 hours depending on cell type and developmental state. During interphase, which consists of the G_1, S, and G_2 phases, the cell roughly doubles its mass. Replication of DNA during S leaves the cell with four copies of each type of chromosome. In the mitotic (M) phase, the chromosomes are evenly partitioned to two daughter cells, and the cytoplasm divides roughly in half in most cases. Under certain conditions such as starvation or when a tissue has reached its final size, cells will stop cycling and remain in a waiting state called G_0. Most cells in G_0 can reenter the cycle if conditions change.

often (though not always) resemble the parent cell. In multicellular organisms, **stem cells** can give rise to two different cells, one that resembles the parent cell and one that does not. Such asymmetric cell division is critical to the generation of different cell types in the body (Chapter 22).

During growth the cell cycle operates continuously, with newly formed daughter cells immediately embarking on their own path to mitosis. Under optimal conditions bacteria can divide to form two daughter cells once every 30 minutes. At this rate, in an hour one cell becomes four; in a day one cell becomes more than 10^{14}, which if dried would weigh about 25 grams. Under normal circumstances, however, growth cannot continue at this rate because the food supply becomes limiting.

Most eukaryotic cells take considerably longer than bacterial cells to grow and divide. Moreover, the cell cycle in adult plants and animals normally is highly regulated (Chapter 21). This tight control prevents imbalanced, excessive growth of tissues while assuring that worn-out or damaged cells are replaced and that additional cells are formed in response to new circumstances or developmental needs. For instance, the proliferation of red blood cells increases substantially when a person ascends to a higher altitude and needs more capacity to capture oxygen. Some highly specialized cells in adult animals, such as nerve cells and striated muscle cells, rarely divide, if at all. The fundamental defect in cancer is loss of the ability to control the growth and division of cells. In Chapter 23, we examine the molecular and cellular events that lead to inappropriate, uncontrolled proliferation of cells.

Mitosis is an *asexual* process since the daughter cells carry the exact same genetic information as the parental cell. In *sexual* reproduction, fusion of two cells produces a third cell that contains genetic information from each parental cell. Since such fusions would cause an ever-increasing number of chromosomes, sexual reproductive cycles employ a special type of cell division, called **meiosis**, that reduces the number of chromosomes in preparation for fusion (see Figure 9-3). Cells with a full set of chromosomes are called **diploid** cells. During meiosis, a diploid cell replicates its chromosomes as usual for mitosis but then divides twice without copying the chromosomes in-between. Each of the resulting four daughter cells, which has only half the full number of chromosomes, is said to be **haploid**.

Sexual reproduction occurs in animals and plants, and even in unicellular organisms such as yeasts (see Figure 1-5). Animals spend considerable time and energy generating eggs and sperm, the haploid cells, called **gametes**, that are used for sexual reproduction. A human female will produce about half a million eggs in a lifetime, all these cells form before she is born; a young human male, about 100 million sperm each day. Gametes are formed from diploid precursor germ-line cells, which in humans contain 46 chromosomes. In humans the X and Y chromosomes are called sex chromosomes because they determine whether an individual is male or female. In human diploid cells, the 44 remaining chromosomes, called **autosomes**, occur as pairs of 22 different kinds. Through meiosis, a man produces sperm that have 22 chromosomes plus either an X or a Y, and a woman produces ova (unfertilized eggs) with

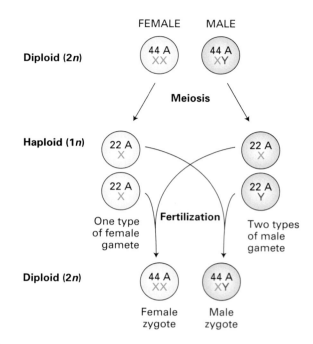

▲ **FIGURE 1-18 Dad made you a boy or girl.** In animals, meiosis of diploid precursor cells forms eggs and sperm (gametes). The male parent produces two types of sperm and determines the sex of the zygote. In humans, as shown here, X and Y are the sex chromosomes; the zygote must receive a Y chromosome from the male parent to develop into a male. A=autosomes (non-sex chromosomes).

22 chromosomes plus an X. Fusion of an egg and sperm (fertilization) yields a fertilized egg, the zygote, with 46 chromosomes, one pair of each of the 22 kinds and a pair of X's in females or an X and a Y in males (Figure 1-18). Errors during meiosis can lead to disorders resulting from an abnormal number of chromosomes. These include Down's syndrome, caused by an extra chromosome 21, and Klinefelter's syndrome, caused by an extra X chromosome.

Cells Die from Aggravated Assault or an Internal Program

When cells in multicellular organisms are badly damaged or infected with a virus, they die. Cell death resulting from such a traumatic event is messy and often releases potentially toxic cell constituents that can damage surrounding cells. Cells also may die when they fail to receive a life-maintaining signal or when they receive a death signal. In this type of programmed cell death, called **apoptosis**, a dying cell actually produces proteins necessary for self-destruction. Death by apoptosis avoids the release of potentially toxic cell constituents (Figure 1-19).

Programmed cell death is critical to the proper development and functioning of our bodies (Chapter 22). During fetal life, for instance, our hands initially develop with "webbing" between the fingers; the cells in the webbing subsequently die in an orderly and precise pattern that leaves the

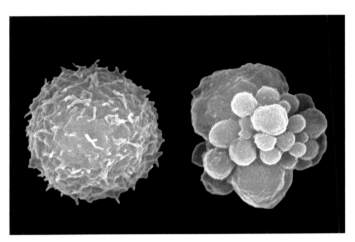

▲ **FIGURE 1-19 Apoptotic cells break apart without spewing forth cell constituents that might harm neighboring cells.** White blood cells normally look like the cell on the left. Cells undergoing programmed cell death (apoptosis), like the cell on the right, form numerous surface blebs that eventually are released. The cell is dying because it lacks certain growth signals. Apoptosis is important to eliminate virus-infected cells, remove cells where they are not needed (like the webbing that disappears as fingers develop), and to destroy immune system cells that would react with our own bodies. [Gopal Murti/Visuals Unlimited, Inc.]

fingers and thumb free to play the piano. Nerve cells in the brain soon die if they do not make proper or useful electrical connections with other cells. Some developing **lymphocytes,** the immune-system cells intended to recognize foreign proteins and polysaccharides, have the ability to react against our own tissues. Such self-reactive lymphocytes become programmed to die before they fully mature. If these cells are not weeded out before reaching maturity, they can cause autoimmune diseases in which our immune system destroys the very tissues it is meant to protect.

1.4 Investigating Cells and Their Parts

To build an integrated understanding of how the various molecular components that underlie cellular functions work together in a living cell, we must draw on various perspectives. Here, we look at how five disciplines—cell biology, biochemistry, genetics, genomics, and developmental biology—can contribute to our knowledge of cell structure and function. The experimental approaches of each field probe the cell's inner workings in different ways, allowing us to ask different types of questions about cells and what they do. Cell division provides a good example to illustrate the role of different perspectives in analyzing a complex cellular process.

The realm of biology ranges in scale more than a billion-fold (Figure 1-20). Beyond that, it's ecology and earth science

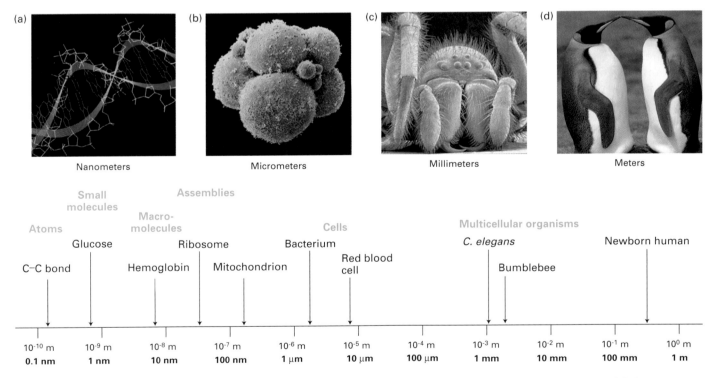

▲ **FIGURE 1-20 Biologists are interested in objects ranging in size from small molecules to the tallest trees.** A sampling of biological objects aligned on a logarithmic scale. (a) The DNA double helix has a diameter of about 2 nm. (b) Eight-cell-stage human embryo three days after fertilization, about 200 μm across. (c) A wolf spider, about 15 mm across. (d) Emperor penguins are about 1 m tall. [Part (a) Will and Deni McIntyre. Part (b) Yorgas Nikas/Photo Researchers, Inc. Part (c) Gary Gaugler/Visuals Unlimited, Inc. Part (d) Hugh S. Rose/Visuals Unlimited, Inc.]

at the "macro" end, chemistry and physics at the "micro" end. The visible plants and animals that surround us are measured in meters (10^0–10^2 m). By looking closely, we can see a biological world of millimeters (1 mm = 10^{-3} m) and even tenths of millimeters (10^{-4} m). Setting aside oddities like chicken eggs, most cells are 1–100 micrometers (1 μm = 10^{-6} m) long and thus clearly visible only when magnified. To see the structures within cells, we must go farther down the size scale to 10–100 nanometers (1 nm = 10^{-9} m).

Cell Biology Reveals the Size, Shape, and Location of Cell Components

Actual observation of cells awaited development of the first, crude microscopes in the early 1600s. A compound microscope, the most useful type of light microscope, has two lenses. The total magnifying power is the product of the magnification by each lens. As better lenses were invented, the magnifying power and the ability to distinguish closely spaced objects, the **resolution**, increased greatly. Modern compound microscopes magnify the view about a thousandfold, so that a bacterium 1 micrometer (1 μm) long looks like it's a millimeter long. Objects about 0.2 μm apart can be discerned in these instruments.

Microscopy is most powerful when particular components of the cell are stained or labeled specifically, enabling them to be easily seen and located within the cell. A simple example is staining with dyes that bind specifically to DNA to visualize the chromosomes. Specific proteins can be detected by harnessing the binding specificity of **antibodies,** the proteins whose normal task is to help defend animals against infection and foreign substances. In general, each type of antibody binds to one protein or large polysaccharide and no other (Chapter 3). Purified antibodies can be chemically linked to a fluorescent molecule, which permits their detection in a special fluorescence microscope (Chapter 5). If a cell or tissue is treated with a detergent that partially dissolves cell membranes, fluorescent antibodies can drift in and bind to the specific protein they recognize. When the sample is viewed in the microscope, the bound fluorescent antibodies identify the location of the target protein (see Figure 1-15).

Better still is pinpointing proteins in living cells with intact membranes. One way of doing this is to introduce an engineered gene that codes for a hybrid protein: part of the hybrid protein is the cellular protein of interest; the other part is a protein that fluoresces when struck by ultraviolet light. A common fluorescent protein used for this purpose is *green fluorescent protein (GFP)*, a natural protein that makes some jellyfish colorful and fluorescent. GFP "tagging" could reveal, for instance, that a particular protein is first made on the endoplasmic reticulum and then is moved by the cell into the lysosomes. In this case, first the endoplasmic reticulum and later the lysosomes would glow in the dark.

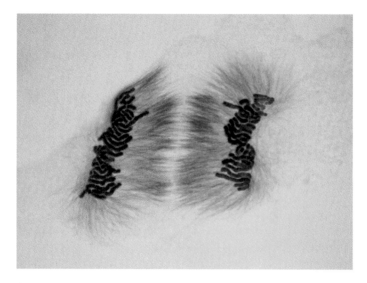

▲ **FIGURE 1-21 During the later stages of mitosis, microtubules (red) pull the replicated chromosomes (black) toward the ends of a dividing cell.** This plant cell is stained with a DNA-binding dye (ethidium) to reveal chromosomes and with fluorescent-tagged antibodies specific for tubulin to reveal microtubules. At this stage in mitosis, the two copies of each replicated chromosome (called chromatids) have separated and are moving away from each other. [Courtesy of Andrew Bajer.]

Chromosomes are visible in the light microscope only during mitosis, when they become highly condensed. The extraordinary behavior of chromosomes during mitosis first was discovered using the improved compound microscopes of the late 1800s. About halfway through mitosis, the replicated chromosomes begin to move apart. Microtubules, one of the three types of cytoskeletal filaments, participate in this movement of chromosomes during mitosis. Fluorescent tagging of tubulin, the protein subunit that polymerizes to form microtubules, reveals structural details of cell division that otherwise could not be seen and allows observation of chromosome movement (Figure 1-21).

Electron microscopes use a focused beam of electrons instead of a beam of light. In transmission electron microscopy, specimens are cut into very thin sections and placed under a high vacuum, precluding examination of living cells. The resolution of transmission electron microscopes, about 0.1 nm, permits fine structural details to be distinguished, and their powerful magnification would make a 1-μm-long bacterial cell look like a soccer ball. Most of the organelles in eukaryotic cells and the double-layered structure of the plasma membrane were first observed with electron microscopes (Chapter 5). With new specialized electron microscopy techniques, three-dimensional models of organelles and large protein complexes can be constructed from multiple images. But to obtain a more detailed look at the individual macromolecules within cells, we must turn to techniques within the purview of biochemistry.

Biochemistry Reveals the Molecular Structure and Chemistry of Purified Cell Constituents

Biochemists extract the contents of cells and separate the constituents based on differences in their chemical or physical properties, a process called *fractionation*. Of particular interest are proteins, the workhorses of many cellular processes. A typical fractionation scheme involves use of various separation techniques in a sequential fashion. These separation techniques commonly are based on differences in the size of molecules or the electrical charge on their surface (Chapter 3). To purify a particular protein of interest, a purification scheme is designed so that each step yields a preparation with fewer and fewer contaminating proteins, until finally only the protein of interest remains (Figure 1-22).

The initial purification of a protein of interest from a cell extract often is a tedious, time-consuming task. Once a small amount of purified protein is obtained, antibodies to it can be produced by methods discussed in Chapter 6. For a biochemist, antibodies are near-perfect tools for isolating larger amounts of a protein of interest for further analysis. In effect, antibodies can "pluck out" the protein they specifically recognize and bind from a semipure sample containing numerous different proteins. An increasingly common alternative is to engineer a gene that encodes a protein of interest with a small attached protein "tag," which can be used to pull out the protein from whole cell extracts.

Purification of a protein is a necessary prelude to studies on how it catalyzes a chemical reaction or carries out other functions and how its activity is regulated. Some enzymes are made of multiple protein chains (subunits) with one chain catalyzing a chemical reaction and other chains regulating when and where that reaction occurs. The molecular machines that perform many critical cell processes constitute even larger assemblies of proteins. By separating the individual proteins composing such assemblies, their individual catalytic or other activities can be assessed. For example, purification and study of the activity of the individual proteins composing the DNA replication machine provided clues about how they work together to replicate DNA during cell division (Chapter 4).

The folded, three-dimensional structure, or **conformation,** of a protein is vital to its function. To understand the relation between the function of a protein and its form, we need to know both what it does and its detailed structure. The most widely used method for determining the complex structures of proteins, DNA, and RNA is **x-ray crystallography.** Computer-assisted analysis of the data often permits the location of every atom in a large, complex molecule to be determined. The double-helix structure of DNA, which is key to its role in heredity, was first proposed based on x-ray crystallographic studies. Throughout this book you will encounter numerous examples of protein structures as we zero in on how proteins work.

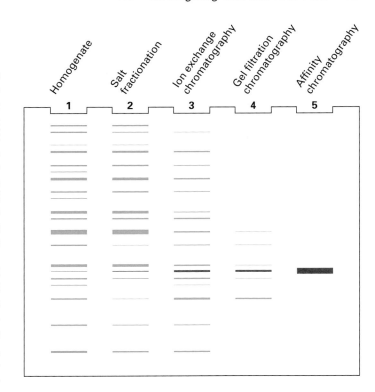

▲ **FIGURE 1-22 Biochemical purification of a protein from a cell extract often requires several separation techniques.** The purification can be followed by gel electrophoresis of the starting protein mixture and the fractions obtained from each purification step. In this procedure, a sample is applied to wells in the top of a gelatin-like slab and an electric field is applied. In the presence of appropriate salt and detergent concentrations, the proteins move through the fibers of the gel toward the anode, with larger proteins moving more slowly through the gel than smaller ones (see Figure 3-32). When the gel is stained, separated proteins are visible as distinct bands whose intensities are roughly proportional to the protein concentration. Shown here are schematic depictions of gels for the starting mixture of proteins (lane 1) and samples taken after each of several purification steps. In the first step, salt fractionation, proteins that precipitated with a certain amount of salt were re-dissolved; electrophoresis of this sample (lane 2) shows that it contains fewer proteins than the original mixture. The sample then was subjected in succession to three types of column chromatography that separate proteins by electrical charge, size, or binding affinity for a particular small molecule (see Figure 3-34). The final preparation is quite pure, as can be seen from the appearance of just one protein band in lane 5. [After J. Berg et al., 2002, *Biochemistry,* W. H. Freeman and Company, p. 87.]

Genetics Reveals the Consequences of Damaged Genes

Biochemical and crystallographic studies can tell us much about an individual protein, but they cannot prove that it is required for cell division or any other cell process. The importance of a protein is demonstrated most firmly if a mutation that prevents its synthesis or makes it nonfunctional adversely affects the process under study.

We define the **genotype** of an organism as its composition of genes; the term also is commonly used in reference to different versions of a single gene or a small number of genes of interest in an individual organism. A diploid organism generally carries two versions (**alleles**) of each gene, one derived from each parent. There are important exceptions, such as the genes on the X and Y chromosomes in males of some species including our own. The **phenotype** is the visible outcome of a gene's action, like blue eyes versus brown eyes or the shapes of peas. In the early days of genetics, the location and chemical identity of genes were unknown; all that could be followed were the observable characteristics, the phenotypes. The concept that genes are like "beads" on a long "string," the chromosome, was proposed early in the 1900s based on genetic work with the fruit fly *Drosophila*.

In the classical genetics approach, mutants are isolated that lack the ability to do something a normal organism can do. Often large genetic "screens" are done, looking for many different mutant individuals (e.g., fruit flies, yeast cells) that are unable to complete a certain process, such as cell division or muscle formation. In experimental organisms or cultured cells, mutations usually are produced by treatment with a **mutagen,** a chemical or physical agent that promotes mutations in a largely random fashion. But how can we isolate and maintain mutant organisms or cells that are defective in some process, such as cell division, that is necessary for survival? One way is to look for **temperature-sensitive mutants.** These mutants are able to grow at one temperature, the *permissive* temperature, but not at another, usually higher temperature, the *nonpermissive* temperature. Normal cells can grow at either temperature. In most cases, a temperature-sensitive mutant produces an altered protein that works at the permissive temperature but unfolds and is nonfunctional at the nonpermissive temperature. Temperature-sensitive screens are readily done with viruses, bacteria, yeast, roundworms, and fruit flies.

By analyzing the effects of numerous different temperature-sensitive mutations that altered cell division, geneticists discovered all the genes necessary for cell division without knowing anything, initially, about which proteins they encode or how these proteins participate in the process. The great power of genetics is to reveal the existence and relevance of proteins without prior knowledge of their biochemical identity or molecular function. Eventually these "mutation-defined" genes were isolated and replicated (cloned) with **recombinant DNA** techniques discussed in Chapter 9. With the isolated genes in hand, the encoded proteins could be produced in the test tube or in engineered bacteria or cultured cells. Then the biochemists could investigate whether the proteins associate with other proteins or DNA or catalyze particular chemical reactions during cell division (Chapter 21).

The analysis of genome sequences from various organisms during the past decade has identified many previously unknown DNA regions that are likely to encode proteins

(i.e., protein-coding genes). The general function of the protein encoded by a sequence-identified gene may be deduced by analogy with known proteins of similar sequence. Rather than randomly isolating mutations in novel genes, several techniques are now available for inactivating specific genes by engineering mutations into them (Chapter 9). The effects of such deliberate gene-specific mutations provide information about the role of the encoded proteins in living organisms. This application of genetic techniques starts with a gene/protein sequence and ends up with a mutant phenotype; traditional genetics starts with a mutant phenotype and ends up with a gene/protein sequence.

Genomics Reveals Differences in the Structure and Expression of Entire Genomes

Biochemistry and genetics generally focus on one gene and its encoded protein at a time. While powerful, these traditional approaches do not give a comprehensive view of the structure and activity of an organism's genome, its entire set of genes. The field of **genomics** does just that, encompassing the molecular characterization of whole genomes and the determination of global patterns of gene expression. The recent completion of the genome sequences for more than 80 species of bacteria and several eukaryotes now permits comparisons of entire genomes from different species. The results provide overwhelming evidence of the molecular unity of life and the evolutionary processes that made us what we are (see Section 1.5). Genomics-based methods for comparing thousands of pieces of DNA from different individuals all at the same time are proving useful in tracing the history and migrations of plants and animals and in following the inheritance of diseases in human families.

New methods using **DNA microarrays** can simultaneously detect all the mRNAs present in a cell, thereby indicating which genes are being transcribed. Such global patterns of gene expression clearly show that liver cells transcribe a quite different set of genes than do white blood cells or skin cells. Changes in gene expression also can be monitored during a disease process, in response to drugs or other external signals, and during development. For instance, the recent identification of all the mRNAs present in cultured fibroblasts before, during, and after they divide has given us an overall view of transcriptional changes that occur during cell division (Figure 1-23). Cancer diagnosis is being transformed because previously indistinguishable cancer cells have distinct gene expression patterns and prognoses (Chapter 23). Similar studies with different organisms and cell types are revealing what is universal about the genes involved in cell division and what is specific to particular organisms.

The entire complement of proteins in a cell, its **proteome,** is controlled in part by changes in gene transcription. The regulated synthesis, processing, localization, and degradation of specific proteins also play roles in determining the proteome of a particular cell, and the association of certain proteins with one another is critical to the functional abilities

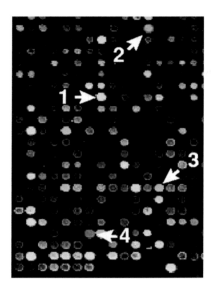

▲ **FIGURE 1-23 DNA microarray analysis gives a global view of changes in transcription following addition of serum to cultured human cells.** Serum contains growth factors that stimulate nondividing cells to begin growing and dividing. DNA microarray analysis can detect the relative transcription of genes in two different cell populations (see Figure 9-35). The microarray consists of tiny spots of DNA attached to a microscope slide. Each spot contains many copies of a DNA sequence from a single human gene. One preparation of RNA, containing all the different types of RNA being made in nongrowing cells cultured without serum, is labeled with green fluorescent molecules. Another RNA population from growing, serum-treated, cells is labeled with red. The two are mixed and hybridized to the slide, where they "zipper up" with their corresponding genes. Green spots (e.g., spot 3) therefore indicate genes that are transcribed in nondividing (serum-deprived) cells; red spots (e.g., spot 4) indicate genes that are transcribed in dividing cells, and yellow spots (e.g., spots 1 and 2) indicate genes that are transcribed equally in dividing and nondividing cells. [From V. R. Iyer et al., 1999, *Science* **283**:83.]

bodies require an enormous amount of communication and division of labor. During the development of multicellular organisms, differentiation processes form hundreds of cell types, each specialized for a particular task: transmission of electrical signals by neurons, transport of oxygen by red blood cells, destruction of infecting bacteria by macrophages, contraction by muscle cells, chemical processing by liver cells.

Many of the differences among differentiated cells are due to production of specific sets of proteins needed to carry out the unique functions of each cell type. That is, only a subset of an organism's genes is transcribed at any given time or in any given cell. Such **differential gene expression** at different times or in different cell types occurs in bacteria, fungi, plants, animals, and even viruses. Differential gene expression is readily apparent in an early fly embryo in which all the cells look alike until they are stained to detect the proteins encoded by particular genes (Figure 1-24). Transcription can change within one cell type in response to an external signal or in accordance with a biological clock; some genes, for instance, undergo a daily cycle between low and high transcription rates.

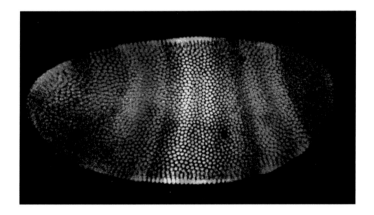

▲ **FIGURE 1-24 Differential gene expression can be detected in early fly embryos before cells are morphologically different.** An early *Drosophila* embryo has about 6000 cells covering its surface, most of which are indistinguishable by simple light microscopy. If the embryo is made permeable to antibodies with a detergent that partially dissolves membranes, the antibodies can find and bind to the proteins they recognize. In this embryo we see antibodies tagged with a fluorescent label bound to proteins that are in the nuclei; each small sphere corresponds to one nucleus. Three different antibodies were used, each specific for a different protein and each giving a distinct color (yellow, green, or blue) in a fluorescence microscope. The red color is added to highlight overlaps between the yellow and blue stains. The locations of the different proteins show that the cells are in fact different at this early stage, with particular genes turned on in specific stripes of cells. These genes control the subdivision of the body into repeating segments, like the black and yellow stripes of a hornet. [Courtesy of Sean Carroll, University of Wisconsin.]

of cells. New techniques for monitoring the presence and interactions of numerous proteins simultaneously, called **proteomics,** are one way of assembling a comprehensive view of the proteins and molecular machines important for cell functioning. The field of proteomics will advance dramatically once high-throughput x-ray crystallography, currently under development, permits researchers to rapidly determine the structures of hundreds or thousands of proteins.

Developmental Biology Reveals Changes in the Properties of Cells as They Specialize

Another approach to viewing cells comes from studying how they change during development of a complex organism. Bacteria, algae, and unicellular eukaryotes (protozoans, yeasts) often, but by no means always, can work solo. The concerted actions of the trillions of cells that compose our

Producing different kinds of cells is not enough to make an organism, any more than collecting all the parts of a truck in one pile gives you a truck. The various cell types must be organized and assembled into all the tissues and organs. Even more remarkable, these body parts must work almost immediately after their formation and continue working during the growth process. For instance, the human heart begins to beat when it is less than 3 mm long, when we are mere 23-day-old embryos, and continues beating as it grows into a fist-size muscle. From a few hundred cells to billions, and still ticking.

In the developing organism, cells grow and divide at some times and not others, they assemble and communicate, they prevent or repair errors in the developmental process, and they coordinate each tissue with others. In the adult organism, cell division largely stops in most organs. If part of an organ such as the liver is damaged or removed, cell division resumes until the organ is regenerated. The legend goes that Zeus punished Prometheus for giving humans fire by chaining him to a rock and having an eagle eat his liver. The punishment was eternal because, as the Greeks evidently knew, the liver regenerates.

Developmental studies involve watching where, when, and how different kinds of cells form, discovering which signals trigger and coordinate developmental events, and understanding the differential gene action that underlies differentiation (Chapters 15 and 22). During development we can see cells change in their normal context of other cells. Cell biology, biochemistry, cell biology, genetics, and genomics approaches are all employed in studying cells during development.

Choosing the Right Experimental Organism for the Job

Our current understanding of the molecular functioning of cells rests on studies with viruses, bacteria, yeast, protozoa, slime molds, plants, frogs, sea urchins, worms, insects, fish, chickens, mice, and humans. For various reasons, some organisms are more appropriate than others for answering particular questions. Because of the evolutionary conservation of genes, proteins, organelles, cell types, and so forth, discoveries about biological structures and functions obtained with one experimental organism often apply to others. Thus researchers generally conduct studies with the organism that is most suitable for rapidly and completely answering the question being posed, knowing that the results obtained in one organism are likely to be broadly applicable. Figure 1-25 summarizes the typical experimental uses of various organisms whose genomes have been sequenced completely or nearly so. The availability of the genome sequences for these organisms makes them particularly useful for genetics and genomics studies.

Bacteria have several advantages as experimental organisms: They grow rapidly, possess elegant mechanisms for controlling gene activity, and have powerful genetics. This

▶ **FIGURE 1-25 Each experimental organism used in cell biology has advantages for certain types of studies.** Viruses and bacteria have small genomes amenable to genetic dissection. Many insights into gene control initially came from studies with these organisms. The yeast *Saccharomyces cerevisiae* has the cellular organization of a eukaryote but is a relatively simple single-celled organism that is easy to grow and to manipulate genetically. In the nematode worm *Caenorhabditis elegans*, which has a small number of cells arranged in a nearly identical way in every worm, the formation of each individual cell can be traced. The fruit fly *Drosophila melanogaster*, first used to discover the properties of chromosomes, has been especially valuable in identifying genes that control embryonic development. Many of these genes are evolutionarily conserved in humans. The zebrafish *Danio rerio* is used for rapid genetic screens to identify genes that control development and organogenesis. Of the experimental animal systems, mice (*Mus musculus*) are evolutionarily the closest to humans and have provided models for studying numerous human genetic and infectious diseases. The mustard-family weed *Arabidopsis thaliana*, sometimes described as the *Drosophila* of the plant kingdom, has been used for genetic screens to identify genes involved in nearly every aspect of plant life. Genome sequencing is completed for many viruses and bacterial species, the yeast *Saccharomyces cerevisiae*, the roundworm *C. elegans*, the fruit fly *D. melanogaster*, humans, and the plant *Arabidopsis thaliana*. It is mostly completed for mice and in progress for zebrafish. Other organisms, particularly frogs, sea urchins, chickens, and slime molds, continue to be immensely valuable for cell biology research. Increasingly, a wide variety of other species are used, especially for studies of evolution of cells and mechanisms. [Part (a) Visuals Unlimited, Inc. Part (b) Kari Lountmaa/Science Photo Library/Photo Researchers, Inc. Part (c) Scimat/Photo Researchers, Inc. Part (d) Photo Researchers, Inc. Part (e) Darwin Dale/Photo Researchers, Inc. Part (f) Inge Spence/Visuals Unlimited, Inc. Part (g) J. M. Labat/Jancana/Visuals Unlimited, Inc. Part (h) Darwin Dale/Photo Researchers, Inc.]

latter property relates to the small size of bacterial genomes, the ease of obtaining mutants, the availability of techniques for transferring genes into bacteria, an enormous wealth of knowledge about bacterial gene control and protein functions, and the relative simplicity of mapping genes relative to one another in the genome. Single-celled yeasts not only have some of the same advantages as bacteria, but also possess the cell organization, marked by the presence of a nucleus and organelles, that is characteristic of all eukaryotes.

Studies of cells in specialized tissues make use of animal and plant "models," that is, experimental organisms with attributes typical of many others. Nerve cells and muscle cells, for instance, traditionally were studied in mammals or in creatures with especially large or accessible cells, such as the giant neural cells of the squid and sea hare or the flight muscles of birds. More recently, muscle and nerve development have been extensively studied in fruit flies *(Drosophila melanogaster)*, roundworms *(Caenorhabditis elegans)*, and zebrafish in which mutants can be readily isolated. Organisms with large-celled embryos that develop outside the

(a)

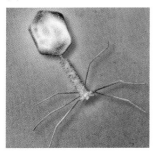

Viruses

Proteins involved in DNA, RNA,
    protein synthesis
Gene regulation
Cancer and control of cell
    proliferation
Transport of proteins and
    organelles inside cells
Infection and immunity
Possible gene therapy approaches

(b)

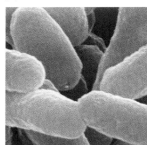

Bacteria

Proteins involved in DNA, RNA,
    protein synthesis,
    metabolism
Gene regulation
Targets for new antibiotics
Cell cycle
Signaling

(c)

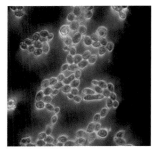

Yeast (*Saccharomyces cerevisiae*)

Control of cell cycle and cell division
Protein secretion and membrane
    biogenesis
Function of the cytoskeleton
Cell differentiation
Aging
Gene regulation and chromosome
    structure

(d)

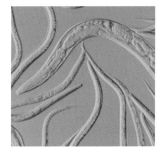

Roundworm (*Caenorhabditis
elegans*)

Development of the body plan
Cell lineage
Formation and function of the
    nervous system
Control of programmed cell death
Cell proliferation and cancer genes
Aging
Behavior
Gene regulation and chromosome
    structure

(e)

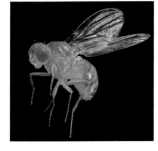

Fruit fly (*Drosophila melanogaster*)

Development of the body plan
Generation of differentiated cell
    lineages
Formation of the nervous system,
    heart, and musculature
Programmed cell death
Genetic control of behavior
Cancer genes and control of cell
    proliferation
Control of cell polarization
Effects of drugs, alcohol, pesticides

(f)

Zebrafish

Development of vertebrate body
    tissues
Formation and function of brain and
    nervous system
Birth defects
Cancer

(g)

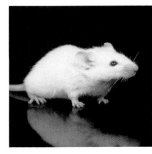

Mice, including cultured cells

Development of body tissues
Function of mammalian immune
    system
Formation and function of brain
    and nervous system
Models of cancers and other
    human diseases
Gene regulation and inheritance
Infectious disease

(h)

Plant (*Arabidopsis thaliana*)

Development and patterning of
    tissues
Genetics of cell biology
Agricultural applications
Physiology
Gene regulation
Immunity
Infectious disease

mother (e.g., frogs, sea urchins, fish, and chickens) are extremely useful for tracing the fates of cells as they form different tissues and for making extracts for biochemical studies. For instance, a key protein in regulating mitosis was first identified in studies with frog and sea urchin embryos and subsequently purified from extracts (Chapter 21).

Using recombinant DNA techniques researchers can engineer specific genes to contain mutations that inactivate or increase production of their encoded proteins. Such genes can be introduced into the embryos of worms, flies, frogs, sea urchins, chickens, mice, a variety of plants, and other organisms, permitting the effects of activating a gene abnormally or inhibiting a normal gene function to be assessed. This approach is being used extensively to produce mouse versions of human genetic diseases. New techniques specifically for inactivating particular genes by injecting short pieces of RNA are making quick tests of gene functions possible in many organisms.

Mice have one enormous advantage over other experimental organisms: they are the closest to humans of any animal for which powerful genetic approaches are feasible. Engineered mouse genes carrying mutations similar to those associated with a particular inherited disease in humans can be introduced into mouse embryonic stem (ES) cells. These cells can be injected into an early embryo, which is then implanted into a pseudopregnant female mouse (Chapter 9). If the mice that develop from the injected ES cells exhibit diseases similar to the human disease, then the link between the disease and mutations in a particular gene or genes is supported. Once mouse models of a human disease are available, further studies on the molecular defects causing the disease can be done and new treatments can be tested, thereby minimizing human exposure to untested treatments.

A continuous unplanned genetic screen has been performed on human populations for millennia. Thousands of inherited traits have been identified and, more recently, mapped to locations on the chromosomes. Some of these traits are inherited propensities to get a disease; others are eye color or other minor characteristics. Genetic variations in virtually every aspect of cell biology can be found in human populations, allowing studies of normal and disease states and of variant cells in culture.

Less-common experimental organisms offer possibilities for exploring unique or exotic properties of cells and for studying standard properties of cells that are exaggerated in a useful fashion in a particular animal. For example, the ends of chromosomes, the **telomeres,** are extremely dilute in most cells. Human cells typically contain 92 telomeres (46 chromosomes × 2 ends per chromosome). In contrast, some protozoa with unusual "fragmented" chromosomes contain millions of telomeres per cell. Recent discoveries about telomere structure have benefited greatly from using this natural variation for experimental advantage.

1.5 A Genome Perspective on Evolution

Comprehensive studies of genes and proteins from many organisms are giving us an extraordinary documentation of the history of life. We share with other eukaryotes thousands of individual proteins, hundreds of macromolecular machines, and most of our organelles, all as a result of our shared evolutionary history. New insights into molecular cell biology arising from genomics are leading to a fuller appreciation of the elegant molecular machines that arose during billions of years of genetic tinkering and evolutionary selection for the most efficient, precise designs. Despite all that we currently know about cells, many new proteins, new macromolecular assemblies, and new activities of known ones remain to be discovered. Once a more complete description of cells is in hand, we will be ready to fully investigate the rippling, flowing dynamics of living systems.

Metabolic Proteins, the Genetic Code, and Organelle Structures Are Nearly Universal

Even organisms that look incredibly different share many biochemical properties. For instance, the enzymes that catalyze degradation of sugars and many other simple chemical reactions in cells have similar structures and mechanisms in most living things. The genetic code whereby the nucleotide sequences of mRNA specifies the amino acid sequences of proteins can be read equally well by a bacterial cell and a human cell. Because of the universal nature of the genetic code, bacterial "factories" can be designed to manufacture growth factors, insulin, clotting factors, and other human proteins with therapeutic uses. The biochemical similarities among organisms also extend to the organelles found in eukaryotic cells. The basic structures and functions of these subcellular components are largely conserved in all eukaryotes.

Computer analysis of DNA sequence data, now available for numerous bacterial species and several eukaryotes, can locate protein-coding genes within genomes. With the aid of the genetic code, the amino acid sequences of proteins can be deduced from the corresponding gene sequences. Although simple conceptually, "finding" genes and deducing the amino acid sequences of their encoded proteins is complicated in practice because of the many noncoding regions in eukaryotic DNA (Chapter 9). Despite the difficulties and occasional ambiguities in analyzing DNA sequences, comparisons of the genomes from a wide range of organisms provide stunning, compelling evidence for the conservation of the molecular mechanisms that build and change organisms and for the common evolutionary history of all species.

Many Genes Controlling Development Are Remarkably Similar in Humans and Other Animals

As humans, we probably have a biased and somewhat exaggerated view of our status in the animal kingdom. Pride in our swollen forebrain and its associated mental capabilities may blind us to the remarkably sophisticated abilities of other species: navigation by birds, the sonar system of bats, homing by salmon, or the flight of a fly.

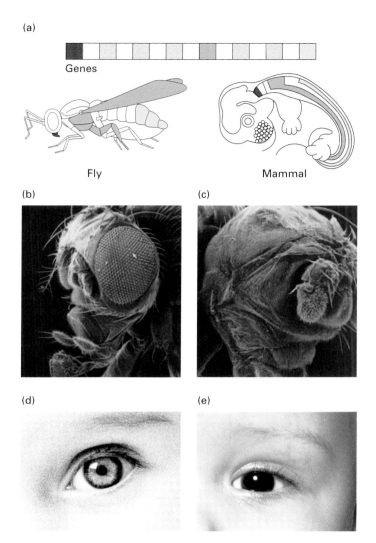

(a)

Genes

Fly   Mammal

(b)   (c)

(d)   (e)

◀ **FIGURE 1-26 Similar genes, conserved during evolution, regulate many developmental processes in diverse animals.** Insects and mammals are estimated to have had a common ancestor about half a billion years ago. They share genes that control similar processes, such as growth of heart and eyes and organization of the body plan, indicating conservation of function from ancient times. (a) *Hox* genes are found in clusters on the chromosomes of most or all animals. *Hox* genes encode related proteins that control the activities of other genes. *Hox* genes direct the development of different segments along the head-to-tail axis of many animals as indicated by corresponding colors. Each gene is activated (transcriptually) in a specific region along the head-to-toe axis and controls the growth of tissues there. For example, in mice the *Hox* genes are responsible for the distinctive shapes of vertebrae. Mutations affecting *Hox* genes in flies cause body parts to form in the wrong locations, such as legs in lieu of antennae on the head. These genes provide a head-to-tail address and serve to direct formation of the right structures in the right places. (b) Development of the large compound eyes in fruit flies requires a gene called *eyeless* (named for the mutant phenotype). (c) Flies with inactivated *eyeless* genes lack eyes. (d) Normal human eyes require the human gene, called *Pax6*, that corresponds to *eyeless*. (e) People lacking adequate *Pax6* function have the genetic disease *aniridia*, a lack of irises in the eyes. *Pax6* and *eyeless* encode highly related proteins that regulate the activities of other genes, and are descended from the same ancestral gene. [Parts (a) and (b) Andreas Hefti, Interdepartmental Electron Microscopy (IEM) Biocenter, University of Basel. Part (d) © Simon Fraser/Photo Researchers, Inc.]

This is not to say that all genes or proteins are evolutionarily conserved. Many striking examples exist of proteins that, as far as we can tell, are utterly absent from certain lineages of animals. Plants, not surprisingly, exhibit many such differences from animals after a billion-year separation in their evolution. Yet certain DNA-binding proteins differ between peas and cows at only two amino acids out of 102!

Darwin's Ideas About the Evolution of Whole Animals Are Relevant to Genes

Darwin did not know that genes exist or how they change, but we do: the DNA replication machine makes an error, or a mutagen causes replacement of one nucleotide with another or breakage of a chromosome. Some changes in the genome are innocuous, some mildly harmful, some deadly; a very few are beneficial. Mutations can change the sequence of a gene in a way that modifies the activity of the encoded protein or alters when, where, and in what amounts the protein is produced in the body.

Gene-sequence changes that are harmful will be lost from a population of organisms because the affected individuals cannot survive as well as their relatives. This selection process is exactly what Darwin described without knowing the underlying mechanisms that cause organisms to vary. Thus the selection of whole organisms for survival is really a selection of genes, or more accurately sets of genes. A population of organisms often contains many variants that are

Despite all the evidence for evolutionary unity at the cellular and physiological levels, everyone expected that genes regulating animal development would differ greatly from one phylum to the next. After all, insects and sea urchins and mammals look *so* different. We must have many unique proteins to create a brain like ours . . . or must we? The fruits of research in developmental genetics during the past two decades reveal that insects and mammals, which have a common ancestor about half a billion years ago, possess many similar development-regulating genes (Figure 1-26). Indeed, a large number of these genes appear to be conserved in many and perhaps all animals. Remarkably, the developmental functions of the proteins encoded by these genes are also often preserved. For instance, certain proteins involved in eye development in insects are related to protein regulators of eye development in mammals. Same for development of the heart, gut, lungs, and capillaries and for placement of body parts along the head-to-tail and back-to-front body axes (Chapter 15).

all roughly equally well-suited to the prevailing conditions. When conditions change—a fire, a flood, loss of preferred food supply, climate shift—variants that are better able to adapt will survive, and those less suited to the new conditions will begin to die out. In this way, the genetic composition of a population of organisms can change over time.

Human Medicine Is Informed by Research on Other Organisms

Mutations that occur in certain genes during the course of our lives contribute to formation of various human cancers. The normal, wild-type forms of such "cancer-causing" genes generally encode proteins that help regulate cell proliferation or death (Chapter 23). We also can inherit from our parents mutant copies of genes that cause all manner of genetic diseases, such as cystic fibrosis, muscular dystrophy, sickle cell anemia, and Huntington's disease. Happily we can also inherit genes that make us robustly resist disease. A remarkable number of genes associated with cancer and other human diseases are present in evolutionarily distant animals. For example, a recent study shows that more than three-quarters of the known human disease genes are related to genes found in the fruit fly *Drosophila*.

With the identification of human disease genes in other organisms, experimental studies in experimentally tractable organisms should lead to rapid progress in understanding the normal functions of the disease-related genes and what occurs when things go awry. Conversely, the disease states themselves constitute a genetic analysis with well-studied phenotypes. All the genes that can be altered to cause a certain disease may encode a group of functionally related proteins. Thus clues about the normal cellular functions of proteins come from human diseases and can be used to guide initial research into mechanism. For instance, genes initially identified because of their link to cancer in humans can be studied in the context of normal development in various model organisms, providing further insight about the functions of their protein products.

2

CHEMICAL FOUNDATIONS

Polysaccharide chains on the surface of cellulose visualized by atomic force microscopy. [Courtesy of M. Miles from A. A. Baker et al., 2000, *Biophys J.* **79**:1139–1145.]

The life of a cell depends on thousands of chemical interactions and reactions exquisitely coordinated with one another in time and space and under the influence of the cell's genetic instructions and its environment. How does a cell extract critical nutrients and information from its environment? How does a cell convert the energy stored in nutrients into work (movement, synthesis of critical components)? How does a cell transform nutrients into the fundamental structures required for its survival (cell wall, nucleus, nucleic acids, proteins, cytoskeleton)? How does a cell link itself to other cells to form a tissue? How do cells communicate with one another so that the organism as a whole can function? One of the goals of molecular cell biology is to answer such questions about the structure and function of cells and organisms in terms of the properties of individual molecules and ions.

Life first arose in a watery environment, and the properties of this ubiquitous substance have a profound influence on the chemistry of life. Constituting 70–80 percent by weight of most cells, water is the most abundant molecule in biological systems. About 7 percent of the weight of living matter is composed of inorganic ions and small molecules such as amino acids (the building blocks of proteins), nucleotides (the building blocks of DNA and RNA), lipids (the building blocks of biomembranes), and sugars (the building blocks of starches and cellulose), the remainder being the macromolecules and macromolecular aggregates composed of these building blocks.

Many biomolecules (e.g., sugars) readily dissolve in water; these water-liking molecules are described as **hydrophilic.** Other biomolecules (e.g., fats like triacylglycerols) shun water; these are said to be **hydrophobic** (water-fearing). Still other biomolecules (e.g., phospholipids), referred to as **amphipathic,** are a bit schizophrenic, containing both hydrophilic and hydrophobic regions. These are used to build the membranes that surround cells and their internal organelles (Chapter 5). The smooth functioning of cells, tissues, and organisms depends on all these molecules, from the smallest to the largest. Indeed, the chemistry of the simple proton (H^+) with a mass of 1 dalton (Da) can be as important to the survival of a human cell as that of each gigantic DNA molecule with a mass as large as 8.6×10^{10} Da (single strand of DNA from human chromosome 1).

A relatively small number of principles and facts of chemistry are essential for understanding cellular processes at the molecular level (Figure 2-1). In this chapter we review some of these key principles and facts, beginning with the covalent bonds that connect atoms into a molecule and the noncovalent forces that stabilize groups of atoms within and between molecules. We then consider the key properties of the basic building blocks of cellular structures. After reviewing those aspects of chemical equilibrium that are most relevant to biological systems, we end the chapter with basic

OUTLINE

2.1 **Atomic Bonds and Molecular Interactions**

2.2 **Chemical Building Blocks of Cells**

2.3 **Chemical Equilibrium**

2.4 **Biochemical Energetics**

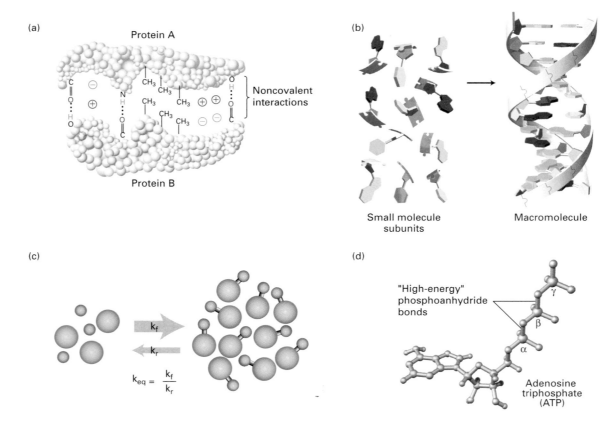

▲ **FIGURE 2-1 Chemistry of life: key concepts.** (a) Covalent and noncovalent interactions lie at the heart of all biomolecules, as when two proteins with complementary shapes and chemical properties come together to form a tightly bound complex. In addition to the covalent bonds that hold the atoms of an amino acid together and link amino acids together, noncovlent interactions help define the structure of each individual protein and serve to help hold the complementary structures together. (b) Small molecules serve as building blocks for larger structures. For example, to generate the information-carrying macromolecule DNA, the four small nucleotide building blocks deoxyadenylate (A), deoxythymidylate (T), deoxyguanylate (G), and deoxycytidylate (C) are covalently linked together into long strings (polymers), which then dimerize into the double helix. (c) Chemical reactions are reversible, and the distribution of the chemicals between starting compounds (*left*) and the products of the reactions (*right*) depends on the rate constants of the forward (k_f, upper arrow) and reverse (k_r, lower arrow) reactions. In the reaction shown, the forward reaction rate constant is faster than the reverse reaction, indicated by the thickness of the arrows. The ratio of these K_{eq}, provides an informative measure of the relative amounts of products and reactants that will be present at equilibrium. (d) In many cases, the source of energy for chemical reactions in cells is the hydrolysis of the molecule ATP. This energy is released when a high-energy phosphoanhydride bond linking the α and β or the β and γ phosphates in the ATP molecule (yellow) is broken by the addition of a water molecule. Proteins can efficiently transfer the energy of ATP hydrolysis to other chemicals, thus fueling other chemical reactions, or to other biomolecules for physical work.

principles of biochemical energetics, including the central role of ATP (adenosine triphosphate) in capturing and transferring energy in cellular metabolism.

2.1   Atomic Bonds and Molecular Interactions

Strong and weak attractive forces between atoms are the glue that holds them together in individual molecules and permits interactions between different biological molecules. Strong forces form a **covalent bond** when two atoms share one pair of electrons ("single" bond) or multiple pairs of electrons ("double" bond, "triple" bond, etc.). The weak attractive forces of **noncovalent interactions** are equally important in

determining the properties and functions of biomolecules such as proteins, nucleic acids, carbohydrates, and lipids. There are four major types of noncovalent interactions: ionic interactions, hydrogen bonds, van der Waals interactions, and the hydrophobic effect.

Each Atom Has a Defined Number and Geometry of Covalent Bonds

Hydrogen, oxygen, carbon, nitrogen, phosphorus, and sulfur are the most abundant elements found in biological molecules. These atoms, which rarely exist as isolated entities, readily form covalent bonds with other atoms, using electrons that reside in the outermost electron orbitals surrounding their nuclei. As a rule, each type of atom forms a characteristic number of covalent bonds with other atoms, with a well-defined geometry determined by the atom's size and by both the distribution of electrons around the nucleus and the number of electrons that it can share. In some cases (e.g., carbon), the number of stable covalent bonds formed is fixed; in other cases (e.g., sulfur), different numbers of stable covalent bonds are possible.

All the biological building blocks are organized around the carbon atom, which normally forms four covalent bonds with two to four other atoms. As illustrated by the methane (CH_4) molecule, when carbon is bonded to four other atoms, the angle between any two bonds is 109.5° and the positions of bonded atoms define the four points of a tetrahedron (Figure 2-2a). This geometry helps define the structures of many biomolecules. A carbon (or any other) atom bonded to four dissimilar atoms or groups in a nonplanar configuration is said to be asymmetric. The tetrahedral orientation of bonds formed by an **asymmetric carbon atom** can be arranged in three-dimensional space in two different ways, producing molecules that are mirror images of each other, a property called *chirality.* Such molecules are called *optical isomers,* or

TABLE 2-1	Bonding Properties of Atoms Most Abundant in Biomolecules		
Atom and Outer Electrons	Usual Number of Covalent Bonds		Bond Geometry
Ḣ	1		H
·Ö·	2		Ö
·S̈·	2, 4, or 6		S̈
·N̈·	3 or 4		N̈
·P̈·	5		P
·C̈·	4		C

stereoisomers. Many molecules in cells contain at least one asymmetric carbon atom, often called a **chiral carbon** atom. The different stereoisomers of a molecule usually have completely different biological activities because the arrangement of atoms within their structures differs, yielding their unique abilities to interact and chemically react with other molecules.

Carbon can also bond to three other atoms in which all atoms are in a common plane. In this case, the carbon atom forms two typical single bonds with two atoms and a double bond (two shared electron pairs) with the third atom (Figure 2-2b). In the absence of other constraints, atoms joined by a single bond generally can rotate freely about the bond axis, while those connected by a double bond cannot. The rigid planarity imposed by double bonds has enormous significance for the shapes and flexibility of large biological molecules such as proteins and nucleic acids.

The number of covalent bonds formed by other common atoms is shown in Table 2-1. A hydrogen atom forms only one bond. An atom of oxygen usually forms only two covalent bonds, but has two additional pairs of electrons that can participate in noncovalent interactions. Sulfur forms two covalent bonds in hydrogen sulfide (H_2S), but also can accommodate six covalent bonds, as in sulfuric acid (H_2SO_4) and its sulfate derivatives. Nitrogen and phosphorus each have five electrons to share. In ammonia (NH_3), the nitrogen atom forms three covalent bonds; the pair of electrons around the atom not involved in a covalent bond can take part in noncovalent interactions. In the ammonium ion (NH_4^+), nitrogen forms four covalent bonds, which have a tetrahedral geometry. Phosphorus commonly forms five covalent bonds, as in phosphoric acid (H_3PO_4) and its phosphate derivatives, which form the backbone of nucleic acids. Phosphate groups attached to proteins play a key role in regulating the activity of many proteins (Chapter 3), and the central molecule in cellular energetics, ATP, contains three phosphate groups (see Section 2.4).

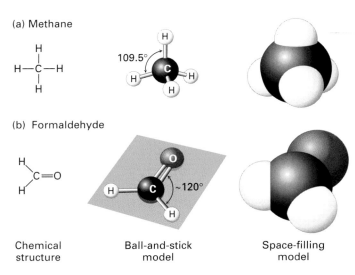

(a) Methane

(b) Formaldehyde

Chemical structure   Ball-and-stick model   Space-filling model

▲ **FIGURE 2-2 Geometry of bonds when carbon is covalently linked to four or three other atoms.** (a) If a carbon atom forms four single bonds, as in methane (CH_4), the bonded atoms (all H in this case) are oriented in space in the form of a tetrahedron. The letter representation on the left clearly indicates the atomic composition of the molecule and the bonding pattern. The ball-and-stick model in the center illustrates the geometric arrangement of the atoms and bonds, but the diameters of the balls representing the atoms and their nonbonding electrons are unrealistically small compared with the bond lengths. The sizes of the electron clouds in the space-filling model on the right more accurately represent the structure in three dimensions. (b) A carbon atom also can be bonded to three, rather than four, other atoms, as in formaldehyde (CH_2O). In this case, the carbon bonding electrons participate in two single bonds and one double bond, which all lie in the same plane. Unlike atoms connected by a single bond, which usually can rotate freely about the bond axis, those connected by a double bond cannot.

Electrons Are Shared Unequally in Polar Covalent Bonds

In many molecules, the bonded atoms exert different attractions for the electrons of the covalent bond, resulting in unequal sharing of the electrons. The extent of an atom's ability to attract an electron is called its **electronegativity**. A bond between atoms with identical or similar electronegativities is said to be **nonpolar**. In a nonpolar bond, the bonding electrons are essentially shared equally between the two atoms, as is the case for most C—C and C—H bonds. However, if two atoms differ in their electronegativities, the bond between them is said to be **polar.**

One end of a polar bond has a partial negative charge (δ^-), and the other end has a partial positive charge (δ^+). In an O—H bond, for example, the greater electronegativity of the oxygen atom relative to hydrogen results in the electrons spending more time around the oxygen atom than the hydrogen. Thus the O—H bond possesses an **electric dipole**, a positive charge separated from an equal but opposite negative charge. We can think of the oxygen atom of the O—H bond as having, on average, a charge of 25 percent of an electron, with the H atom having an equivalent positive charge. Because of its two O—H bonds, water molecules (H_2O) are dipoles that form electrostatic, noncovalent interactions with one another and with other molecules (Figure 2-3). These interactions play a critical role in almost every biochemical interaction and are thus fundamental to cell biology.

The polarity of the O═P double bond in H_3PO_4 results in a "resonance hybrid," a structure between the two forms shown below in which nonbonding electrons are shown as pairs of dots:

In the resonance hybrid on the right, one of the electrons from the P═O double bond has accumulated around the O atom, giving it a negative charge and leaving the P atom with a positive charge. These charges are important in noncovalent interactions.

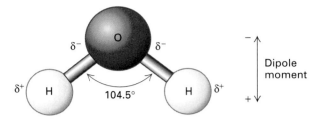

▲ **FIGURE 2-3 The dipole nature of a water molecule.** The symbol δ represents a partial charge (a weaker charge than the one on an electron or a proton). Because of the difference in the electronegativities of H and O, each of the polar H—O bonds in water has a dipole moment. The sizes and directions of the dipole moments of each of the bonds determine the net dipole moment of the molecule.

Covalent Bonds Are Much Stronger and More Stable Than Noncovalent Interactions

Covalent bonds are very stable because the energies required to break them are much greater than the thermal energy available at room temperature (25 °C) or body temperature (37 °C). For example, the thermal energy at 25 °C is approximately 0.6 kilocalorie per mole (kcal/mol), whereas the energy required to break the carbon-carbon single bond (C—C) in ethane is about 140 times larger (Figure 2-4). Consequently at room temperature (25 °C), fewer than 1 in 10^{12} ethane molecules is broken into a pair of ·CH_3 radicals, each containing an unpaired, nonbonding electron.

Covalent single bonds in biological molecules have energies similar to that of the C—C bond in ethane. Because more electrons are shared between atoms in double bonds, they require more energy to break than single bonds. For instance, it takes 84 kcal/mol to break a single C—O bond, but 170 kcal/mol to break a C═O double bond. The most common double bonds in biological molecules are C═O, C═N, C═C, and P═O.

The energy required to break noncovalent interactions is only 1–5 kcal/mol, much less than the bond energies of covalent bonds (see Figure 2-4). Indeed, noncovalent interactions are weak enough that they are constantly being

▶ **FIGURE 2-4 Relative energies of covalent bonds and noncovalent interactions.** Bond energies are determined as the energy required to break a particular type of linkage. Covalent bonds are one to two powers of 10 stronger than noncovalent interactions. The latter are somewhat greater than the thermal energy of the environment at normal room temperature (25 °C). Many biological processes are coupled to the energy released during hydrolysis of a phosphoanhydride bond in ATP.

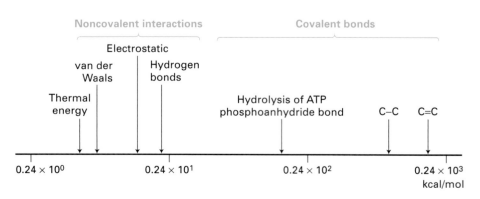

formed and broken at room temperature. Although these interactions are weak and have a transient existence at physiological temperatures (25–37 °C), multiple noncovalent interactions can act together to produce highly stable and specific associations between different parts of a large molecule or between different macromolecules. We first review the four main types of noncovalent interactions and then consider their role in the binding of biomolecules to one another and to other molecules.

Ionic Interactions Are Attractions Between Oppositely Charged Ions

Ionic interactions result from the attraction of a positively charged ion—a **cation**—for a negatively charged ion—an **anion**. In sodium chloride (NaCl), for example, the bonding electron contributed by the sodium atom is completely transferred to the chlorine atom. Unlike covalent bonds, ionic interactions do not have fixed or specific geometric orientations, because the electrostatic field around an ion—its attraction for an opposite charge—is uniform in all directions.

In aqueous solutions, simple ions of biological significance, such as Na^+, K^+, Ca^{2+}, Mg^{2+}, and Cl^-, do not exist as free, isolated entities. Instead, each is hydrated, surrounded by a stable shell of water molecules, which are held in place by ionic interactions between the central ion and the oppositely charged end of the water dipole (Figure 2-5). Most ionic compounds dissolve readily in water because the energy of hydration, the energy released when ions tightly bind water molecules, is greater than the lattice energy that stabilizes the crystal structure. Parts or all of the aqueous hydration shell must be removed from ions when they directly interact with proteins. For example, water of hydration is lost when ions pass through protein pores in the cell membrane during nerve conduction (Chapter 7).

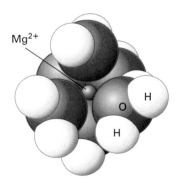

▲ **FIGURE 2-5 Electrostatic interaction between water and a magnesium ion (Mg^{2+}).** Water molecules are held in place by electrostatic interactions between the two positive charges on the ion and the partial negative charge on the oxygen of each water molecule. In aqueous solutions, all ions are surrounded by a similar hydration shell.

The relative strength of the interaction between two ions, A^- and C^+, depends on the concentration of other ions in a solution. The higher the concentration of other ions (e.g., Na^+ and Cl^-), the more opportunities A^- and C^+ have to interact ionically with these other ions, and thus the lower the energy required to break the interaction between A^- and C^+. As a result, increasing the concentrations of salts such as NaCl in a solution of biological molecules can weaken and even disrupt the ionic interactions holding the biomolecules together.

Hydrogen Bonds Determine Water Solubility of Uncharged Molecules

A **hydrogen bond** is the interaction of a partially positively charged hydrogen atom in a molecular dipole (e.g., water) with unpaired electrons from another atom, either in the same (intramolecular) or in a different (intermolecular) molecule. Normally, a hydrogen atom forms a covalent bond with only one other atom. However, a hydrogen atom covalently bonded to an electronegative donor atom D may form an additional weak association, the hydrogen bond, with an acceptor atom A, which must have a nonbonding pair of electrons available for the interaction:

$$D^{\delta-}\!-\!H^{\delta+} + :A^{\delta-} \rightleftharpoons D^{\delta-}\!-\!H^{\delta+}\cdots\cdots:A^{\delta-}$$
Hydrogen bond

The length of the covalent D—H bond is a bit longer than it would be if there were no hydrogen bond, because the acceptor "pulls" the hydrogen away from the donor. An important feature of all hydrogen bonds is directionality. In the strongest hydrogen bonds, the donor atom, the hydrogen atom, and the acceptor atom all lie in a straight line. Nonlinear hydrogen bonds are weaker than linear ones; still, multiple nonlinear hydrogen bonds help to stabilize the three-dimensional structures of many proteins.

Hydrogen bonds are both longer and weaker than covalent bonds between the same atoms. In water, for example, the distance between the nuclei of the hydrogen and oxygen atoms of adjacent, hydrogen-bonded molecules is about 0.27 nm, about twice the length of the covalent O—H bonds within a single water molecule (Figure 2-6a). The strength of a hydrogen bond between water molecules (approximately 5 kcal/mol) is much weaker than a covalent O—H bond (roughly 110 kcal/mol), although it is greater than that for many other hydrogen bonds in biological molecules (1–2 kcal/mol). The extensive hydrogen bonding between water molecules accounts for many of the key properties of this compound, including its unusually high melting and boiling points and its ability to interact with many other molecules.

The solubility of uncharged substances in an aqueous environment depends largely on their ability to form hydrogen bonds with water. For instance, the hydroxyl group (—OH) in methanol (CH_3OH) and the amino group (—NH_2) in methylamine (CH_3NH_2) can form several hydrogen bonds with water, enabling these molecules to dissolve in water to

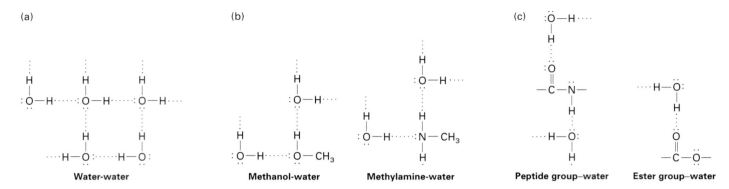

Water-water **Methanol-water** **Methylamine-water** **Peptide group–water** **Ester group–water**

▲ **FIGURE 2-6 Hydrogen bonding of water with itself and with other compounds.** Each pair of nonbonding outer electrons in an oxygen or nitrogen atom can accept a hydrogen atom in a hydrogen bond. The hydroxyl and the amino groups can also form hydrogen bonds with water. (a) In liquid water, each water molecule apparently forms transient hydrogen bonds with several others, creating a dynamic network of hydrogen-bonded molecules. (b) Water also can form hydrogen bonds with methanol and methylamine, accounting for the high solubility of these compounds. (c) The peptide group and ester group, which are present in many biomolecules, commonly participate in hydrogen bonds with water or polar groups in other molecules.

high concentrations (Figure 2-6b). In general, molecules with polar bonds that easily form hydrogen bonds with water can readily dissolve in water; that is, they are hydrophilic. Many biological molecules contain, in addition to hydroxyl and amino groups, peptide and ester groups, which form hydrogen bonds with water (Figure 2-6c). X-ray crystallography combined with computational analysis permits an accurate depiction of the distribution of electrons in covalent bonds and the outermost unbonded electrons of atoms, as illustrated in Figure 2-7. These unbonded electrons can form hydrogen bonds with donor hydrogens.

Van der Waals Interactions Are Caused by Transient Dipoles

When any two atoms approach each other closely, they create a weak, nonspecific attractive force called a **van der Waals interaction**. These nonspecific interactions result from the momentary random fluctuations in the distribution of the electrons of any atom, which give rise to a transient unequal distribution of electrons. If two noncovalently bonded atoms are close enough together, electrons of one atom will perturb the electrons of the other. This perturbation generates a transient dipole in the second atom, and the two dipoles will attract each other weakly (Figure 2-8). Similarly, a polar covalent bond in one molecule will attract an oppositely oriented dipole in another.

Van der Waals interactions, involving either transiently induced or permanent electric dipoles, occur in all types of molecules, both polar and nonpolar. In particular, van der Waals interactions are responsible for the cohesion between molecules of nonpolar liquids and solids, such as heptane, CH_3—$(CH_2)_5$—CH_3, that cannot form hydrogen bonds or ionic interactions with other molecules. The strength of van der Waals interactions decreases rapidly with increasing distance; thus these noncovalent bonds can form only when

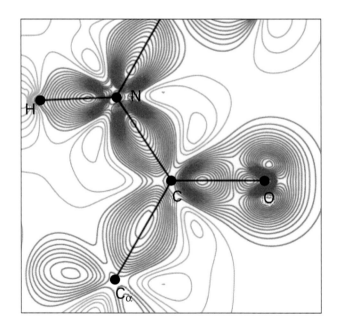

▲ **FIGURE 2-7 Distribution of bonding and outer nonbonding electrons in the peptide group.** Shown here is one amino acid within a protein called crambin. The black lines diagrammatically represent the covalent bonds between atoms. The red (negative) and blue (positive) lines represent contours of charge. The greater the number of contour lines, the higher the charge. The high density of red contour lines between atoms represents the covalent bonds (shared electron pairs). The two sets of red contour lines emanating from the oxygen (O) and not falling on a covalent bond (black line) represent the two pairs of nonbonded electrons on the oxygen that are available to participate in hydrogen bonding. The high density of blue contour lines near the hydrogen (H) bonded to nitrogen (N) represents a partial positive charge, indicating that this H can act as a donor in hydrogen bonding. [From C. Jelsch et al., 2000, *Proc. Nat'l. Acad. Sci. USA* **97**:3171. Courtesy of M. M. Teeter.]

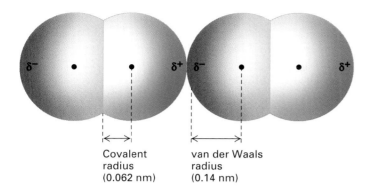

▲ FIGURE 2-8 Two oxygen molecules in van der Waals contact. In this space-filling model, red indicates negative charge and blue indicates positive charge. Transient dipoles in the electron clouds of all atoms give rise to weak attractive forces, called *van der Waals interactions*. Each type of atom has a characteristic van der Waals radius at which van der Waals interactions with other atoms are optimal. Because atoms repel one another if they are close enough together for their outer electrons to overlap, the van der Waals radius is a measure of the size of the electron cloud surrounding an atom. The covalent radius indicated here is for the double bond of O=O; the single-bond covalent radius of oxygen is slightly longer.

atoms are quite close to one another. However, if atoms get too close together, they become repelled by the negative charges of their electrons. When the van der Waals attraction between two atoms exactly balances the repulsion between their two electron clouds, the atoms are said to be in van der Waals contact. The strength of the van der Waals interaction is about 1 kcal/mol, weaker than typical hydrogen bonds and only slightly higher than the average thermal energy of molecules at 25 °C. Thus multiple van der Waals interactions, a van der Waals interaction in conjunction with other noncovalent interactions, or both are required to significantly influence intermolecular contacts.

The Hydrophobic Effect Causes Nonpolar Molecules to Adhere to One Another

Because nonpolar molecules do not contain charged groups, possess a dipole moment, or become hydrated, they are insoluble or almost insoluble in water; that is, they are hydrophobic. The covalent bonds between two carbon atoms and between carbon and hydrogen atoms are the most common nonpolar bonds in biological systems. **Hydrocarbons**—molecules made up only of carbon and hydrogen—are virtually insoluble in water. Large triacylglycerols (or triglycerides), which comprise animal fats and vegetable oils, also are insoluble in water. As we see later, the major portion of these molecules consists of long hydrocarbon chains. After being shaken in water, triacylglycerols form a separate phase. A familiar example is the separation of oil from the water-based vinegar in an oil-and-vinegar salad dressing.

Nonpolar molecules or nonpolar portions of molecules tend to aggregate in water owing to a phenomenon called the **hydrophobic effect.** Because water molecules cannot form hydrogen bonds with nonpolar substances, they tend to form "cages" of *relatively* rigid hydrogen-bonded pentagons and hexagons around nonpolar molecules (Figure 2-9, *left*). This state is energetically unfavorable because it decreases the randomness (entropy) of the population of water molecules. (The role of entropy in chemical systems is discussed in a later section.) If nonpolar molecules in an aqueous environment aggregate with their hydrophobic surfaces facing each other, there is a reduction in the hydrophobic surface area exposed to water (Figure 2-9, *right*). As a consequence, less water is needed to form the cages surrounding the nonpolar molecules, and entropy increases (an energetically more favorable state) relative to the unaggregated state. In a sense, then, water squeezes the nonpolar molecules into spontaneously forming aggregates. Rather than constituting an attractive force such as in hydrogen bonds, the hydrophobic effect results from an avoidance of an unstable state (extensive water cages around individual nonpolar molecules).

Nonpolar molecules can also associate, albeit weakly, through van der Waals interactions. The net result of the hydrophobic and van der Waals interactions is a very powerful tendency for hydrophobic molecules to interact with one another, not with water. Simply put, *like dissolves like*. Polar molecules dissolve in polar solvents such as water; nonpolar molecules dissolve in nonpolar solvents such as hexane.

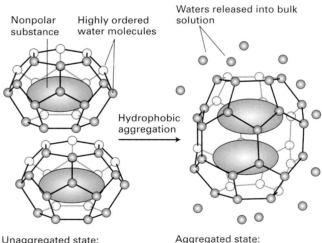

▲ FIGURE 2-9 Schematic depiction of the hydrophobic effect. Cages of water molecules that form around nonpolar molecules in solution are more ordered than water molecules in the surrounding bulk liquid. Aggregation of nonpolar molecules reduces the number of water molecules involved in highly ordered cages, resulting in a higher-entropy, more energetically favorable state (*right*) compared with the unaggregated state (*left*).

Molecular Complementarity Permits Tight, Highly Specific Binding of Biomolecules

Both inside and outside of cells, ions and molecules are constantly bumping into one another. The greater the number of copies of any two types of molecules per unit volume (i.e., the higher their concentration), the more likely they are to encounter one another. When two molecules encounter each other, they most likely will simply bounce apart because the noncovalent interactions that would bind them together are weak and have a transient existence at physiological temperatures. However, molecules that exhibit **molecular complementarity**, a lock-and-key kind of fit between their shapes, charges, or other physical properties, can form multiple noncovalent interactions at close range. When two such structurally **complementary** molecules bump into each other, they can bind (stick) together.

Figure 2-10 illustrates how multiple, different weak bonds can bind two proteins together. Almost any other arrangement of the same groups on the two surfaces would not allow the molecules to bind so tightly. Such multiple, specific interactions between complementary regions within a protein molecule allow it to fold into a unique three-dimensional shape (Chapter 3) and hold the two chains of DNA together in a double helix (Chapter 4). Similar interactions underlie the association of groups of more than two molecules into multimolecular complexes, leading to formation of muscle fibers, to the gluelike associations between cells in solid tissues, and to numerous other cellular structures.

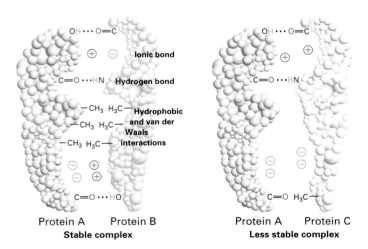

Protein A   Protein B
Stable complex

Protein A   Protein C
Less stable complex

▲ **FIGURE 2-10 Molecular complementarity and the binding of proteins via multiple noncovalent interactions.** The complementary shapes, charges, polarity, and hydrophobicity of two protein surfaces permit multiple weak interactions, which in combination produce a strong interaction and tight binding. Because deviations from molecular complementarity substantially weaken binding, any given biomolecule usually can bind tightly to only one or a very limited number of other molecules. The complementarity of the two protein molecules on the left permits them to bind much more tightly than the two noncomplementary proteins on the right.

Depending on the number and strength of the noncovalent interactions between the two molecules and on their environment, their binding may be tight (strong) or loose (weak) and, as a consequence, either long-lasting or transient. The higher the **affinity** of two molecules for each other, the better the molecular "fit" between them, the more noncovalent interactions can form, and the tighter they can bind together. An important quantitative measure of affinity is the binding dissociation constant K_d, described later.

As we discuss in Chapter 3, nearly all the chemical reactions that occur in cells also depend on the binding properties of enzymes. These proteins not only speed up reactions but also do so with a high degree of **specificity**, a reflection of their ability to bind tightly to only one or a few related molecules. Indeed, the binding specificity of large biological molecules, particularly proteins and nucleic acids, is one of the distinctive features that distinguish biochemistry from typical solution chemistry. Clearly, molecular complementarity and noncovalent interactions underlie the structures of biomolecules and many processes critical to life.

KEY CONCEPTS OF SECTION 2.1

Atomic Bonds and Molecular Interactions

■ Covalent bonds, which bind the atoms composing a molecule in a fixed orientation, consist of pairs of electrons shared by two atoms. Relatively high energies are required to break them (50–200 kcal/mol).

■ In polar bonds, which link atoms that differ in electronegativity, the bonding electrons are distributed unequally. One end of a polar bond has a partial positive charge and the other end has a partial negative charge (see Figure 2-3).

■ Noncovalent interactions between atoms are considerably weaker than covalent bonds, with bond energies ranging from about 1–5 kcal/mol (see Figure 2-4).

■ Four main types of noncovalent interactions occur in biological systems: ionic bonds, hydrogen bonds, van der Waals interactions, and interactions due to the hydrophobic effect.

■ Ionic bonds result from the electrostatic attraction between the positive and negative charges of ions. In aqueous solutions, all cations and anions are surrounded by a shell of bound water molecules (see Figure 2-5). Increasing the salt (e.g., NaCl) concentration of a solution can weaken the relative strength of and even break the ionic bonds between biomolecules.

■ In a hydrogen bond, a hydrogen atom covalently bonded to an electronegative atom associates with an acceptor atom whose nonbonding electrons attract the hydrogen (see Figure 2-6).

■ Weak and relatively nonspecific van der Waals interactions are created whenever any two atoms approach each other closely. They result from the attraction between transient dipoles associated with all molecules (see Figure 2-8).

- In an aqueous environment, nonpolar molecules or nonpolar portions of larger molecules are driven together by the hydrophobic effect, thereby reducing the extent of their direct contact with water molecules (see Figure 2-9).

- Molecular complementarity is the lock-and-key fit between molecules whose shapes, charges, and other physical properties are complementary. Multiple noncovalent interactions can form between complementary molecules, causing them to bind tightly (see Figure 2-10), but not between molecules that are not complementary.

- The high degree of binding specificity that results from molecular complementarity is one of the features that distinguish biochemistry from typical solution chemistry.

2.2 Chemical Building Blocks of Cells

The three most abundant biological macromolecules—**proteins**, **nucleic acids**, and **polysaccharides**—are all **polymers** composed of multiple covalently linked identical or nearly identical small molecules, or **monomers** (Figure 2-11). The covalent bonds between monomer molecules usually are formed by **dehydration reactions** in which a water molecule is lost:

$$H—X_1—OH + H—X_2—OH \rightarrow H—X_1—X_2—OH + H_2O$$

Proteins are linear polymers containing ten to several thousand amino acids linked by **peptide bonds.** Nucleic acids

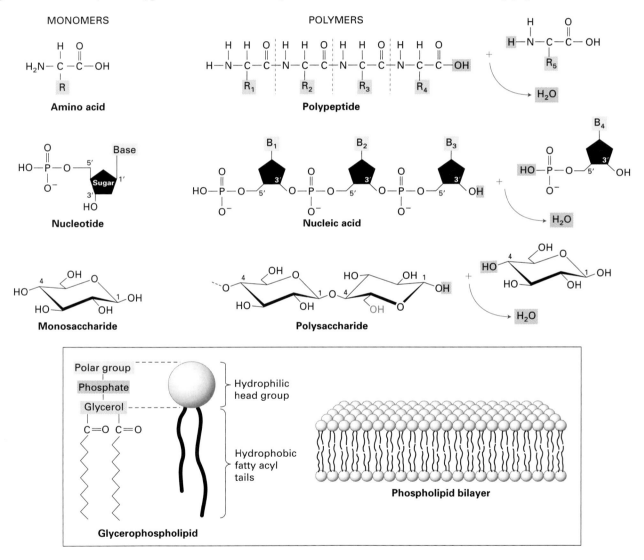

▲ **FIGURE 2-11 Covalent and noncovalent linkage of monomers to form biopolymers and membranes.** Overview of the cell's chemical building blocks and the macrostructures formed from them. (*Top*) The three major types of biological macromolecules are each assembled by the polymerization of multiple small molecules (monomers) of a particular type: proteins from amino acids (Chapter 3), nucleic acids from nucleotides (Chapter 4), and polysaccharides from monosaccharides (sugars). The monomers are covalently linked into polymers by coupled reactions whose net result is condensation through the dehydration reaction shown. (*Bottom*) In contrast, phospholipid monomers noncovalently assemble into bilayer structure, which forms the basis of all cellular membranes (Chapter 5).

are linear polymers containing hundreds to millions of nucleotides linked by **phosphodiester bonds.** Polysaccharides are linear or branched polymers of monosaccharides (sugars) such as glucose linked by **glycosidic bonds.**

A similar approach is used to form various large structures in which the repeating components associate by noncovalent interactions. For instance, the fibers of the cytoskeleton are composed of many repeating protein molecules. And, as we discuss below, phospholipids assemble noncovalently to form a two-layered (bilayer) structure that is the basis of all cellular membranes (see Figure 2-11). Thus a repeating theme in biology is the construction of large molecules and structures by the covalent or noncovalent association of many similar or identical smaller molecules.

Amino Acids Differing Only in Their Side Chains Compose Proteins

The monomeric building blocks of proteins are 20 **amino acids,** all of which have a characteristic structure consisting of a central **α carbon atom (C_α)** bonded to four different chemical groups: an amino (NH_2) group, a carboxyl (COOH) group, a hydrogen (H) atom, and one variable group, called a **side chain,** or R group. Because the α carbon in all amino acids except glycine is asymmetric, these molecules can exist in two mirror-image forms called by convention the D (dextro) and the L (levo) isomers (Figure 2-12). The two isomers cannot be interconverted (one made identical with the other) without breaking and then re-forming a chemical bond in one of them. With rare exceptions, only the L forms of amino acids are found in proteins. We discuss the properties of the covalent peptide bond that links amino acids into long chains in Chapter 3.

To understand the structures and functions of proteins, you must be familiar with some of the distinctive properties of the amino acids, which are determined by their side chains. The side chains of different amino acids vary in size, shape, charge, hydrophobicity, and reactivity. Amino acids can be classified into several broad categories based primarily on their solubility in water, which is influenced by the polarity of their side chains (Figure 2-13). Amino acids with polar side chains are hydrophilic and tend to be on the surfaces of proteins; by interacting with water, they make proteins soluble in aqueous solutions and can form noncovalent interactions with other water-soluble molecules. In contrast, amino acids with nonpolar side chains are hydrophobic; they avoid water and often aggregate to help form the water-insoluble cores of many proteins. The polarity of amino acid side chains thus is responsible for shaping the final three-dimensional structure of proteins.

A subset of the hydrophilic amino acids are charged (ionized) at the pH (≈ 7) typical of physiological conditions (see Section 2.3). **Arginine** and **lysine** are positively charged; **aspartic acid** and **glutamic acid** are negatively charged (their charged forms are called *aspartate* and *glutamate*). These four amino acids are the prime contributors to the overall charge of a protein. A fifth amino acid, **histidine,** has an im-

idazole side chain, which can shift from being positively charged to uncharged with small changes in the acidity of its environment:

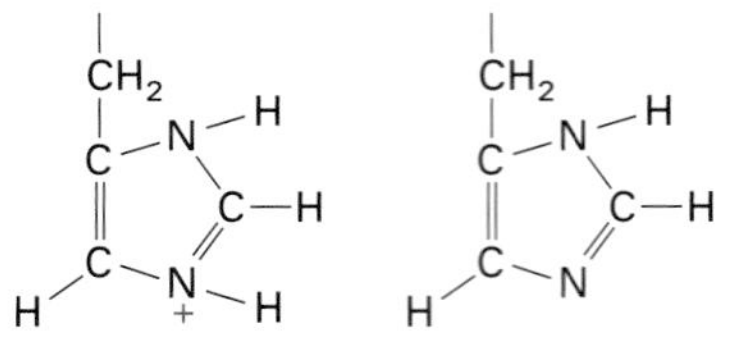

The activities of many proteins are modulated by shifts in environmental acidity through protonation of histidine side chains. **Asparagine** and **glutamine** are uncharged but have polar side chains containing amide groups with extensive hydrogen-bonding capacities. Similarly, **serine** and **threonine** are uncharged but have polar hydroxyl groups, which also participate in hydrogen bonds with other polar molecules.

The side chains of hydrophobic amino acids are insoluble or only slightly soluble in water. The noncyclic side chains of **alanine, valine, leucine, isoleucine,** and **methionine** consist entirely of hydrocarbons, except for the one sulfur atom in methionine, and all are nonpolar. **Phenylalanine, tyrosine,** and **tryptophan** have large bulky aromatic side chains. In later chapters, we will see in detail how hydrophobic residues line the *surface* of proteins that are embedded within biomembranes.

Lastly, cysteine, glycine, and proline exhibit special roles in proteins because of the unique properties of their side chains. The side chain of **cysteine** contains a reactive **sulfhydryl group** (—SH), which can oxidize to form a covalent **disulfide bond** (—S—S—) to a second cysteine:

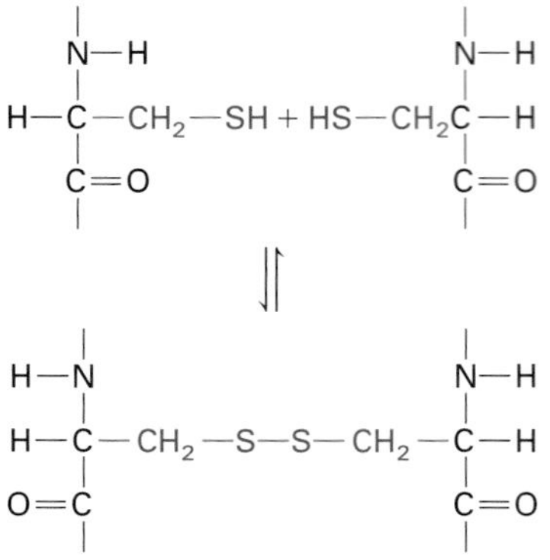

Regions within a protein chain or in separate chains sometimes are cross-linked through disulfide bonds. Disulfide bonds are commonly found in extracellular proteins, where they help stabilize the folded structure. The smallest amino acid, **glycine,** has a single hydrogen atom as its R group. Its small size allows it to fit into tight spaces. Unlike the other common amino acids, the side chain of **proline** bends around to form a ring by covalently bonding to the nitrogen atom (amino group) attached to the C_α. As a result, proline is very rigid and creates a fixed kink in a protein chain, limiting how a protein can fold in the region of proline residues.

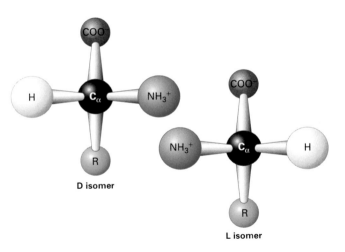

D isomer

L isomer

◀ **FIGURE 2-12 Common structure of amino acids.** The α carbon atom ($C_α$) of each amino acid is bonded to four chemical groups. The side chain, or R group, is unique to each type of amino acid (see Figure 2-13). Because the $C_α$ in all amino acids, except glycine, is asymmetric, these molecules have two mirror-image forms, designated L and D. Although the chemical properties of such optical isomers are identical, their biological activities are distinct. Only L amino acids are found in proteins.

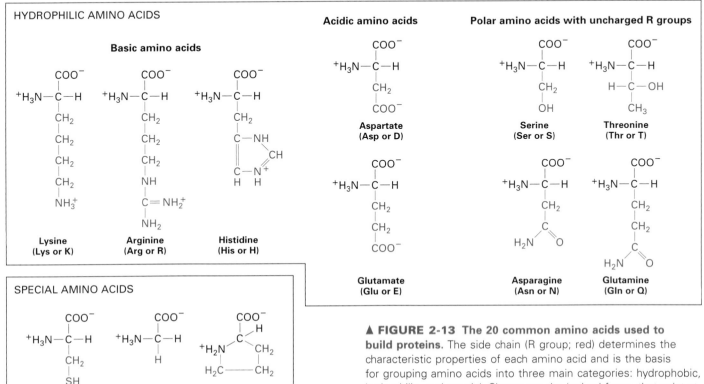

HYDROPHOBIC AMINO ACIDS

Alanine (Ala or A) Valine (Val or V) Isoleucine (Ile or I) Leucine (Leu or L) Methionine (Met or M) Phenylalanine (Phe or F) Tyrosine (Tyr or Y) Tryptophan (Trp or W)

HYDROPHILIC AMINO ACIDS

Basic amino acids

Lysine (Lys or K) Arginine (Arg or R) Histidine (His or H)

Acidic amino acids

Aspartate (Asp or D) Glutamate (Glu or E)

Polar amino acids with uncharged R groups

Serine (Ser or S) Threonine (Thr or T) Asparagine (Asn or N) Glutamine (Gln or Q)

SPECIAL AMINO ACIDS

Cysteine (Cys or C) Glycine (Gly or G) Proline (Pro or P)

▲ **FIGURE 2-13 The 20 common amino acids used to build proteins.** The side chain (R group; red) determines the characteristic properties of each amino acid and is the basis for grouping amino acids into three main categories: hydrophobic, hydrophilic, and special. Shown are the ionized forms that exist at the pH (≈7) of the cytosol. In parentheses are the three-letter and one-letter abbreviations for each amino acid.

Some amino acids are more abundant in proteins than other amino acids. Cysteine, tryptophan, and methionine are rare amino acids; together they constitute approximately 5 percent of the amino acids in a protein. Four amino acids—leucine, serine, lysine, and glutamic acid—are the most abundant amino acids, totaling 32 percent of all the amino acid residues in a typical protein. However, the amino acid composition of proteins can vary widely from these values.

Five Different Nucleotides Are Used to Build Nucleic Acids

Two types of chemically similar nucleic acids, **DNA** (deoxyribonucleic acid) and **RNA** (ribonucleic acid), are the principal information-carrying molecules of the cell. The monomers from which DNA and RNA are built, called **nucleotides**, all have a common structure: a phosphate group linked by a phosphoester bond to a pentose (a five-carbon sugar molecule) that in turn is linked to a nitrogen- and carbon-containing ring structure commonly referred to as a "base" (Figure 2-14a). In RNA, the pentose is ribose; in DNA, it is deoxyribose (Figure 2-14b). The bases **adenine, guanine,** and **cytosine** are found in both DNA and RNA; **thymine** is found only in DNA, and **uracil** is found only in RNA.

Adenine and guanine are **purines**, which contain a pair of fused rings; cytosine, thymine, and uracil are **pyrimidines**, which contain a single ring (Figure 2-15). The bases are often abbreviated A, G, C, T, and U, respectively; these same single-letter abbreviations are also commonly used to denote the entire nucleotides in nucleic acid polymers. In nucleotides,

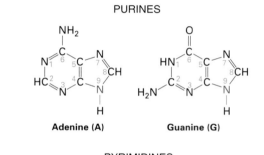

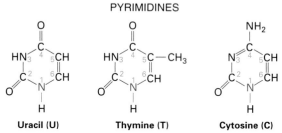

▲ **FIGURE 2-15 Chemical structures of the principal bases in nucleic acids.** In nucleic acids and nucleotides, nitrogen 9 of purines and nitrogen 1 of pyrimidines (red) are bonded to the 1' carbon of ribose or deoxyribose. U is only in RNA, and T is only in DNA. Both RNA and DNA contain A, G, and C.

the 1' carbon atom of the sugar (ribose or deoxyribose) is attached to the nitrogen at position 9 of a purine (N_9) or at position 1 of a pyrimidine (N_1). The acidic character of nucleotides is due to the phosphate group, which under normal intracellular conditions releases a hydrogen ion (H^+), leaving the phosphate negatively charged (see Figure 2-14a). Most nucleic acids in cells are associated with proteins, which form ionic interactions with the negatively charged phosphates.

Cells and extracellular fluids in organisms contain small concentrations of **nucleosides,** combinations of a base and a sugar without a phosphate. Nucleotides are nucleosides that have one, two, or three phosphate groups esterified at the 5' hydroxyl. Nucleoside monophosphates have a single esterified phosphate (see Figure 2-14a); diphosphates contain a pyrophosphate group:

and triphosphates have a third phosphate. Table 2-2 lists the names of the nucleosides and nucleotides in nucleic acids and the various forms of nucleoside phosphates. The nucleoside triphosphates are used in the synthesis of nucleic acids, which we cover in Chapter 4. Among their other functions in the cell, GTP participates in intracellular signaling and acts as an energy reservoir, particularly in protein synthesis, and ATP, discussed later in this chapter, is the most widely used biological energy carrier.

(a)

(b)

Adenosine 5'-monophosphate (AMP)

Ribose

2-Deoxyribose

▲ **FIGURE 2-14 Common structure of nucleotides.** (a) Adenosine 5'-monophosphate (AMP), a nucleotide present in RNA. By convention, the carbon atoms of the pentose sugar in nucleotides are numbered with primes. In natural nucleotides, the 1' carbon is joined by a β linkage to the base (in this case adenine); both the base (blue) and the phosphate on the 5' hydroxyl (red) extend above the plane of the furanose ring. (b) Ribose and deoxyribose, the pentoses in RNA and DNA, respectively.

TABLE 2-2	Terminology of Nucleosides and Nucleotides			
		Bases		
		Purines		Pyrimidines	
		Adenine (A)	**Guanine (G)**	**Cytosine (C)**	**Uracil (U) Thymine [T]**
Nucleosides	in RNA	Adenosine	Guanosine	Cytidine	Uridine
	in DNA	Deoxyadenosine	Deoxyguanosine	Deoxycytidine	Deoxythymidine
Nucleotides	in RNA	Adenylate	Guanylate	Cytidylate	Uridylate
	in DNA	Deoxyadenylate	Deoxyguanylate	Deoxycytidylate	Deoxythymidylate
Nucleoside monophosphates		AMP	GMP	CMP	UMP
Nucleoside diphosphates		ADP	GDP	CDP	UDP
Nucleoside triphosphates		ATP	GTP	CTP	UTP
Deoxynucleoside mono-, di-, and triphosphates		dAMP, etc.			

Monosaccharides Joined by Glycosidic Bonds Form Linear and Branched Polysaccharides

The building blocks of the polysaccharides are the simple sugars, or **monosaccharides**. Monosaccharides are **carbohydrates**, which are literally covalently bonded combinations of carbon and water in a one-to-one ratio $(CH_2O)_n$, where n equals 3, 4, 5, 6, or 7. **Hexoses** ($n = 6$) and **pentoses** ($n = 5$) are the most common monosaccharides. All monosaccharides contain hydroxyl (—OH) groups and either an aldehyde or a keto group:

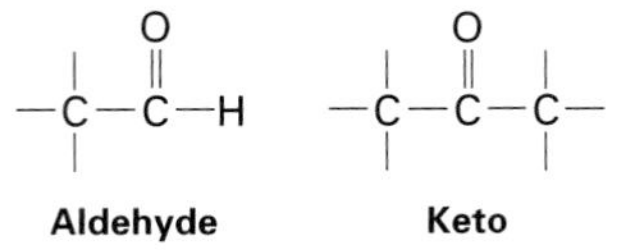

Aldehyde   **Keto**

D-Glucose $(C_6H_{12}O_6)$ is the principal external source of energy for most cells in higher organisms and can exist in three different forms: a linear structure and two different hemiacetal ring structures (Figure 2-16a). If the aldehyde group on carbon 1 reacts with the hydroxyl group on carbon 5, the resulting hemiacetal, D-glucopyranose, contains a six-member ring. In the α anomer of D-glucopyranose, the hydroxyl group attached to carbon 1 points "downward" from the ring as shown in Figure 2-16a; in the β anomer, this hydroxyl points "upward." In aqueous solution the α and β anomers readily interconvert spontaneously; at equilibrium there is about one-third α anomer and two-thirds β, with very little of the open-chain form. Because enzymes can distinguish between the α and β anomers of D-glucose, these forms have distinct biological roles. Condensation of the hydroxyl group on carbon 4 of the linear glucose with its alde-

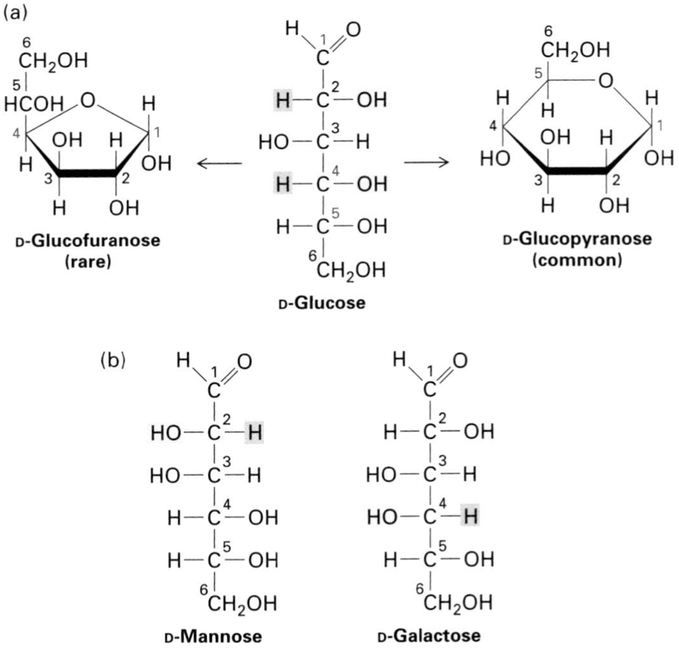

▲ **FIGURE 2-16 Chemical structures of hexoses.** All hexoses have the same chemical formula $(C_6H_{12}O_6)$ and contain an aldehyde or keto group. (a) The ring forms of D-glucose are generated from the linear molecule by reaction of the aldehyde at carbon 1 with the hydroxyl on carbon 5 or carbon 4. The three forms are readily interconvertible, although the pyranose form (*right*) predominates in biological systems. (b) In D-mannose and D-galactose, the configuration of the H (green) and OH (blue) bound to one carbon atom differs from that in glucose. These sugars, like glucose, exist primarily as pyranoses.

hyde group results in the formation of D-glucofuranose, a hemiacetal containing a five-member ring. Although all three forms of D-glucose exist in biological systems, the pyranose form is by far the most abundant.

Many biologically important sugars are six-carbon sugars that are structurally related to D-glucose (Figure 2-16b). Mannose is identical with glucose except that the orientation of the groups bonded to carbon 2 is reversed. Similarly, galactose, another hexose, differs from glucose only in the orientation of the groups attached to carbon 4. Interconversion of glucose and mannose or galactose requires the breaking and making of covalent bonds; such reactions are carried out by enzymes called **epimerases.**

The pyranose ring in Figure 2-16a is depicted as planar. In fact, because of the tetrahedral geometry around carbon atoms, the most stable conformation of a pyranose ring has a nonplanar, chairlike shape. In this conformation, each bond from a ring carbon to a nonring atom (e.g., H or O) is either nearly perpendicular to the ring, referred to as axial (a), or nearly in the plane of the ring, referred to as equatorial (e):

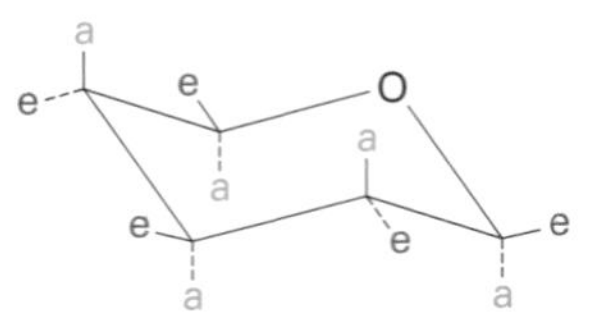

Pyranoses 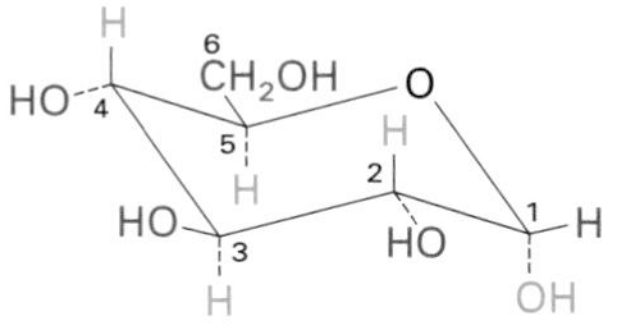   **α-D-Glucopyranose**

The enzymes that make the glycosidic bonds linking monosaccharides into polysaccharides are specific for the α or β anomer of one sugar and a particular hydroxyl group on the other. In principle, any two sugar molecules can be linked in a variety of ways because each monosaccharide has multiple hydroxyl groups that can participate in the formation of glycosidic bonds. Furthermore, any one monosaccharide has the potential of being linked to more than two other monosaccharides, thus generating a branch point and nonlinear polymers. Glycosidic bonds are usually formed between a covalently modified sugar and the growing polymer chain. Such modifications include a phosphate (e.g., glucose 6-phosphate) or a nucleotide (e.g., UDP-galactose):

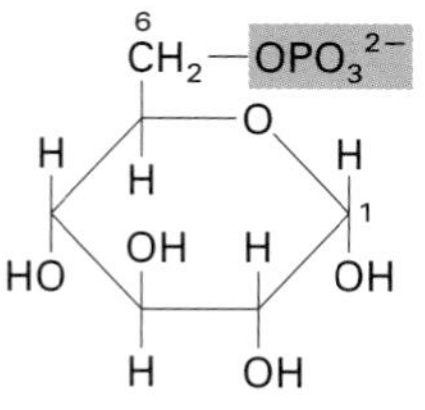

Glucose 6-phosphate 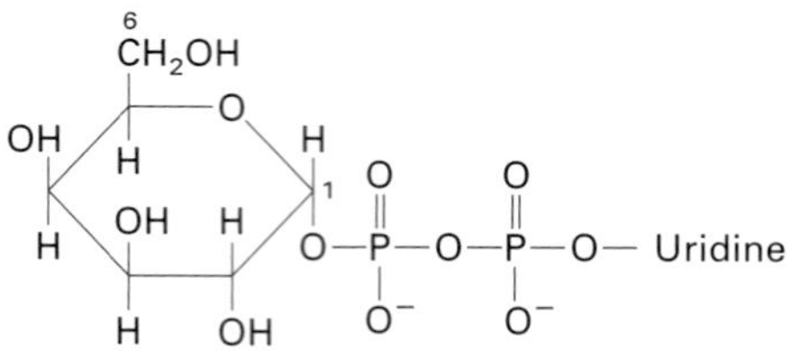   **UDP-galactose**

The epimerase enzymes that interconvert different monosaccharides often do so using the nucleotide sugars rather than the unsubstituted sugars.

Disaccharides, formed from two monosaccharides, are the simplest polysaccharides. The disaccharide lactose, composed of galactose and glucose, is the major sugar in milk; the disaccharide sucrose, composed of glucose and fructose, is a principal product of plant photosynthesis and is refined into common table sugar (Figure 2-17).

Larger polysaccharides, containing dozens to hundreds of monosaccharide units, can function as reservoirs for glucose, as structural components, or as adhesives that help hold cells together in tissues. The most common storage carbohydrate in animal cells is **glycogen,** a very long, highly branched polymer of glucose. As much as 10 percent by weight of the liver can be glycogen. The primary storage carbohydrate in plant cells, **starch,** also is a glucose polymer. It occurs in an unbranched form (amylose) and lightly branched form (amylopectin). Both glycogen and starch are composed of the α anomer of glucose. In contrast, **cellulose,** the major constituent of plant cell walls, is an unbranched polymer of the β anomer of glucose. Human digestive enzymes can hydrolyze the α glycosidic bonds in starch, but not the β glycosidic bonds in cellulose. Many species of plants, bacteria, and molds produce cellulose-degrading enzymes.

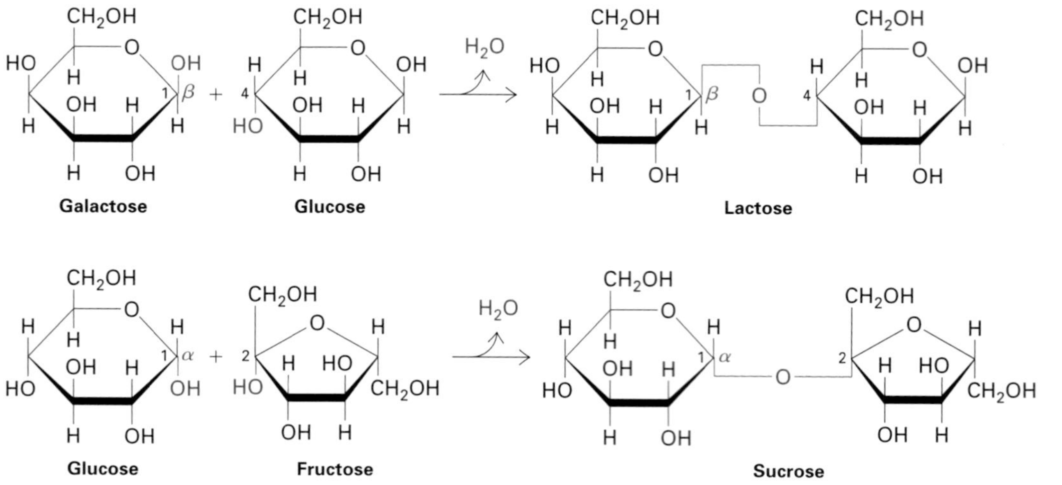

▶ **FIGURE 2-17 Formation of the disaccharides lactose and sucrose.** In any glycosidic linkage, the anomeric carbon of one sugar molecule (in either the α or β conformation) is linked to a hydroxyl oxygen on another sugar molecule. The linkages are named accordingly: thus lactose contains a β(1 → 4) bond, and sucrose contains an α(1 → 2) bond.

Cows and termites can break down cellulose because they harbor cellulose-degrading bacteria in their gut.

Many complex polysaccharides contain modified sugars that are covalently linked to various small groups, particularly amino, sulfate, and acetyl groups. Such modifications are abundant in **glycosaminoglycans,** major polysaccharide components of the extracellular matrix that we describe in Chapter 6.

Fatty Acids Are Precursors for Many Cellular Lipids

Before considering phospholipids and their role in the structure of biomembranes, we briefly review the properties of **fatty acids.** Like glucose, fatty acids are an important energy source for many cells and are stored in the form of triacylglycerols within adipose tissue (Chapter 8). Fatty acids also are precursors for phospholipids and many other lipids with a variety of functions (Chapter 18).

Fatty acids consist of a hydrocarbon chain attached to a carboxyl group (—COOH). They differ in length, although the predominant fatty acids in cells have an even number of carbon atoms, usually 14, 16, 18, or 20. The major fatty acids in phospholipids are listed in Table 2-3. Fatty acids often are designated by the abbreviation $Cx:y$, where x is the number of carbons in the chain and y is the number of double bonds. Fatty acids containing 12 or more carbon atoms are nearly insoluble in aqueous solutions because of their long hydrophobic hydrocarbon chains.

Fatty acids with no carbon-carbon double bonds are said to be **saturated;** those with at least one double bond are **unsaturated.** Unsaturated fatty acids with more than one carbon-carbon double bond are referred to as **polyunsaturated.** Two "essential" polyunsaturated fatty acids, linoleic acid (C18:2) and linolenic acid (C18:3), cannot be synthesized by mammals and must be supplied in their diet. Mammals can synthesize other common fatty acids. Two stereoisomeric configurations, cis and trans, are possible around each carbon-carbon double bond:

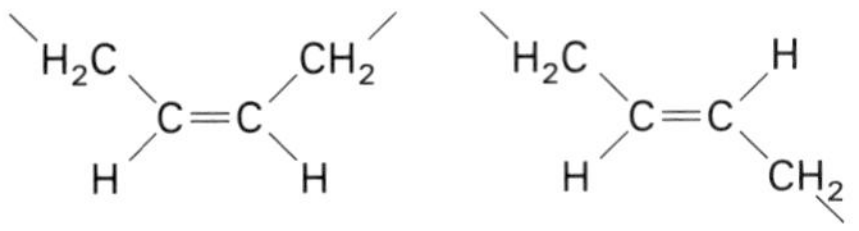

Cis   **Trans**

A cis double bond introduces a rigid kink in the otherwise flexible straight chain of a fatty acid (Figure 2-18). In general, the fatty acids in biological systems contain only cis double bonds.

Fatty acids can be covalently attached to another molecule by a type of dehydration reaction called **esterification,** in which the OH from the carboxyl group of the fatty acid and a H from a hydroxyl group on the other molecule are lost. In the combined molecule formed by this reaction, the portion derived from the fatty acid is called an **acyl** *group,* or *fatty acyl group.* This is illustrated by **triacylglycerols,** which contain three acyl groups esterfied to glycerol:

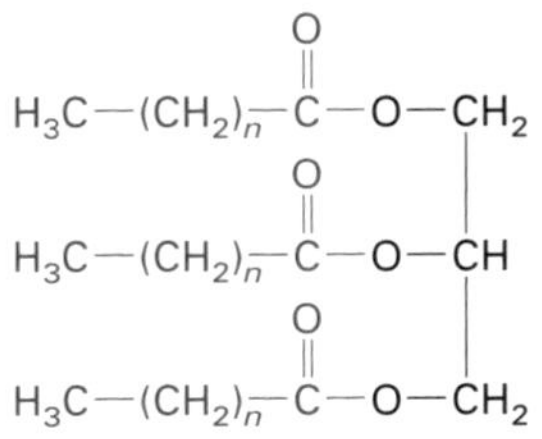

Triacylglycerol

TABLE 2-3	Fatty Acids That Predominate in Phospholipids	
Common Name of Acid (Ionized Form in Parentheses)	Abbreviation	Chemical Formula
SATURATED FATTY ACIDS		
Myristic (myristate)	C14:0	$CH_3(CH_2)_{12}COOH$
Palmitic (palmitate)	C16:0	$CH_3(CH_2)_{14}COOH$
Stearic (stearate)	C18:0	$CH_3(CH_2)_{16}COOH$
UNSATURATED FATTY ACIDS		
Oleic (oleate)	C18:1	$CH_3(CH_2)_7CH{=}CH(CH_2)_7COOH$
Linoleic (linoleate)	C18:2	$CH_3(CH_2)_4CH{=}CHCH_2CH{=}CH(CH_2)_7COOH$
Arachidonic (arachidonate)	C20:4	$CH_3(CH_2)_4(CH{=}CHCH_2)_3CH{=}CH(CH_2)_3COOH$

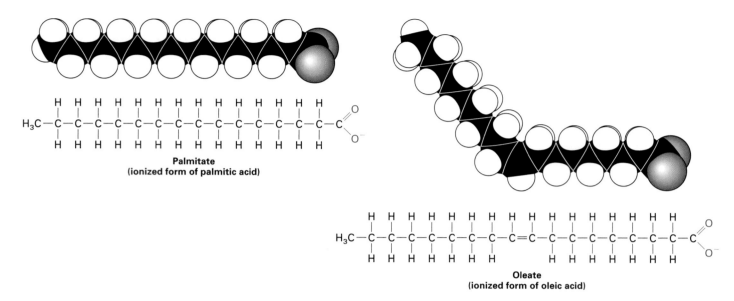

Palmitate
(ionized form of palmitic acid)

Oleate
(ionized form of oleic acid)

▲ **FIGURE 2-18 The effect of a double bond on the shape of fatty acids.** Shown are space-filling models and chemical structures of the ionized form of palmitic acid, a saturated fatty acid with 16 C atoms, and oleic acid, an unsaturated one with 18 C atoms. In saturated fatty acids, the hydrocarbon chain is often linear; the cis double bond in oleate creates a rigid kink in the hydrocarbon chain. [After L. Stryer, 1994, *Biochemistry*, 4th ed., W. H. Freeman and Company, p. 265.]

If the acyl groups are long enough, these molecules are insoluble in water even though they contain three polar ester bonds. Fatty acyl groups also form the hydrophobic portion of phospholipids, which we discuss next.

Phospholipids Associate Noncovalently to Form the Basic Bilayer Structure of Biomembranes

Biomembranes are large flexible sheets that serve as the boundaries of cells and their intracellular organelles and form the outer surfaces of some viruses. Membranes literally define what is a cell (the outer membrane and the contents within the membrane) and what is not (the extracellular space outside the membrane). Unlike the proteins, nucleic acids, and polysaccharides, membranes are assembled by the *noncovalent* association of their component building blocks.

The primary building blocks of all biomembranes are **phospholipids,** whose physical properties are responsible for the formation of the sheetlike structure of membranes.

Phospholipids consist of two long-chain, nonpolar fatty acyl groups linked (usually by an ester bond) to small, highly polar groups, including a phosphate. In **phosphoglycerides,** the major class of phospholipids, fatty acyl side chains are esterified to two of the three hydroxyl groups in glycerol. The third hydroxyl group is esterified to phosphate. The simplest phospholipid, phosphatidic acid, contains only these components. In most phospholipids found in membranes, the phosphate group is esterified to a hydroxyl group on another hydrophilic compound. In phosphatidylcholine, for example, choline is attached to the phosphate (Figure 2-19). The negative charge on the phosphate as well as the charged or polar groups esterified to it can interact strongly with water. The

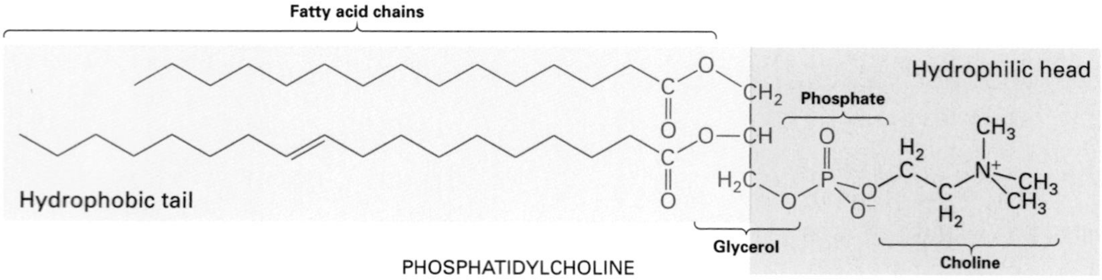

Fatty acid chains

Hydrophilic head

Phosphate

Hydrophobic tail

CH3

Glycerol   **Choline**

PHOSPHATIDYLCHOLINE

▲ **FIGURE 2-19 Phosphatidylcholine, a typical phosphoglyceride.** All phosphoglycerides are amphipathic, having a hydrophobic tail (yellow) and a hydrophilic head (blue) in which glycerol is linked via a phosphate group to an alcohol. Either of or both the fatty acyl side chains in a phosphoglyceride may be saturated or unsaturated. In phosphatidic acid (red), the simplest phospholipid, the phosphate is not linked to an alcohol.

phosphate and its associated esterified group, the "head" group of a phospholipid, is hydrophilic, whereas the fatty acyl chains, the "tails," are hydrophobic.

The amphipathic nature of phospholipids, which governs their interactions, is critical to the structure of biomembranes. When a suspension of phospholipids is mechanically dispersed in aqueous solution, the phospholipids aggregate into one of three forms: spherical **micelles** and **liposomes** and sheetlike, two-molecule-thick **phospholipid bilayers** (Figure 2-20). The type of structure formed by a pure phospholipid or a mixture of phospholipids depends on several factors, including the length of the fatty acyl chains, their degree of saturation, and temperature. In all three structures, the hydrophobic effect causes the fatty acyl chains to aggregate and exclude water molecules from the "core." Micelles are rarely formed from natural phosphoglycerides, whose fatty acyl chains generally are too bulky to fit into the interior of a micelle. If one of the two fatty acyl chains is removed by hydrolysis, forming a lysophospholipid, the predominant type of aggregate that forms is the micelle. Common detergents and soaps form micelles in aqueous solution that behave as tiny ball bearings, thus giving soap solutions their slippery feel and lubricating properties.

Under suitable conditions, phospholipids of the composition present in cells spontaneously form symmetric phospholipid bilayers. Each phospholipid layer in this lamellar

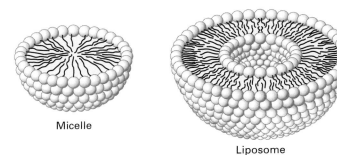

Micelle

Liposome

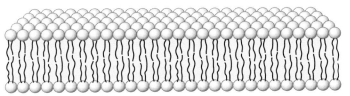

Phospholipid bilayer

▲ **FIGURE 2-20 Cross-sectional views of the three structures formed by phospholipids in aqueous solutions.** The white spheres depict the hydrophilic heads of the phospholipids, and the squiggly black lines (in the yellow regions) represent the hydrophobic tails. Shown are a spherical micelle with a hydrophobic interior composed entirely of fatty acyl chains; a spherical liposome, which has two phospholipid layers and an aqueous center; and a two-molecule-thick sheet of phospholipids, or bilayer, the basic structural unit of biomembranes.

structure is called a **leaflet.** The fatty acyl chains in each leaflet minimize contact with water by aligning themselves tightly together in the center of the bilayer, forming a hydrophobic core that is about 3 nm thick (see Figure 2-20). The close packing of these nonpolar tails is stabilized by the hydrophobic effect and van der Waals interactions between them. Ionic and hydrogen bonds stabilize the interaction of the phospholipid polar head groups with one another and with water.

A phospholipid bilayer can be of almost unlimited size— from micrometers (μm) to millimeters (mm) in length or width—and can contain tens of millions of phospholipid molecules. Because of their hydrophobic core, bilayers are virtually impermeable to salts, sugars, and most other small hydrophilic molecules. The phospholipid bilayer is the basic structural unit of nearly all biological membranes; thus, although they contain other molecules (e.g., cholesterol, glycolipids, proteins), biomembranes have a hydrophobic core that separates two aqueous solutions and acts as a permeability barrier. The structural organization of biomembranes and the general properties of membrane proteins are described in Chapter 5.

KEY CONCEPTS OF SECTION 2.2

Chemical Building Blocks of Cells

■ Three major biopolymers are present in cells: proteins, composed of amino acids linked by peptide bonds; nucleic acids, composed of nucleotides linked by phosphodiester bonds; and polysaccharides, composed of monosaccharides (sugars) linked by glycosidic bonds (see Figure 2-11).

■ Many molecules in cells contain at least one asymmetric carbon atom, which is bonded to four dissimilar atoms. Such molecules can exist as optical isomers (mirror images), designated D and L, which have different biological activities. In biological systems, nearly all sugars are D isomers, while nearly all amino acids are L isomers.

■ Differences in the size, shape, charge, hydrophobicity, and reactivity of the side chains of amino acids determine the chemical and structural properties of proteins (see Figure 2-13).

■ Amino acids with hydrophobic side chains tend to cluster in the interior of proteins away from the surrounding aqueous environment; those with hydrophilic side chains usually are toward the surface.

■ The bases in the nucleotides composing DNA and RNA are heterocyclic rings attached to a pentose sugar. They form two groups: the purines—adenine (A) and guanine (G)—and the pyrimidines—cytosine (C), thymine (T), and uracil (U) (see Figure 2-15). A, G, T, and C are in DNA, and A, G, U, and C are in RNA.

■ Glucose and other hexoses can exist in three forms: an open-chain linear structure, a six-member (pyranose) ring, and a five-member (furanose) ring (see Figure 2-16). In biological systems, the pyranose form of D-glucose predominates.

■ Glycosidic bonds are formed between either the α or β anomer of one sugar and a hydroxyl group on another sugar, leading to formation of disaccharides and other poly-saccharides (see Figure 2-17).

■ The long hydrocarbon chain of a fatty acid may contain no carbon-carbon double bond (saturated) or one or more double bonds (unsaturated), which bends the chain.

■ Phospholipids are amphipathic molecules with a hydrophobic tail (often two fatty acyl chains) and a hydrophilic head (see Figure 2-19).

■ In aqueous solution, the hydrophobic effect and van der Waals interactions organize and stabilize phospholipids into one of three structures: a micelle, liposome, or sheetlike bilayer (see Figure 2-20).

■ In a phospholipid bilayer, which constitutes the basic structure of all biomembranes, fatty acyl chains in each leaflet are oriented toward one another, forming a hydrophobic core, and the polar head groups line both surfaces and directly interact with the aqueous solution.

2.3 Chemical Equilibrium

We now shift our discussion to chemical reactions in which bonds, primarily covalent bonds in **reactant** chemicals, are broken and new bonds are formed to generate reaction **products**. At any one time several hundred different kinds of chemical reactions are occurring simultaneously in every cell, and many chemicals can, in principle, undergo multiple chemical reactions. Both the *extent* to which reactions can proceed and the *rate* at which they take place determine the chemical composition of cells.

When reactants first mix together—before any products have been formed—their rate of reaction is determined in part by their initial concentrations. As the reaction products accumulate, the concentration of each reactant decreases and so does the reaction rate. Meanwhile, some of the product molecules begin to participate in the reverse reaction, which re-forms the reactants (microscopic reversibility). This reverse reaction is slow at first but speeds up as the concentration of product increases. Eventually, the rates of the forward and reverse reactions become equal, so that the concentrations of reactants and products stop changing. The system is then said to be in **chemical equilibrium.**

At equilibrium the ratio of products to reactants, called the **equilibrium constant,** is a fixed value that is independent of the rate at which the reaction occurs. The rate of a chemical reaction can be increased by a **catalyst,** which brings reactants together and accelerates their interactions, but is not permanently changed during a reaction. In this section, we discuss several aspects of chemical equilibria; in the

next section, we examine energy changes during reactions and their relationship to equilibria.

Equilibrium Constants Reflect the Extent of a Chemical Reaction

The equilibrium constant K_{eq} depends on the nature of the reactants and products, the temperature, and the pressure (particularly in reactions involving gases). Under standard physical conditions (25 °C and 1 atm pressure, for biological systems), the K_{eq} is always the same for a given reaction, whether or not a catalyst is present.

For the general reaction

$$aA + bB + cC + \cdots \rightleftharpoons zZ + yY + xX + \cdots \quad (2\text{-}1)$$

where capital letters represent particular molecules or atoms and lowercase letters represent the number of each in the reaction formula, the equilibrium constant is given by

$$K_{eq} = \frac{[X]^x [Y]^y [Z]^z}{[A]^a [B]^b [C]^c} \quad (2\text{-}2)$$

where brackets denote the concentrations of the molecules. The rate of the forward reaction (left to right in Equation 2-1) is

$$\text{Rate}_{forward} = k_f[A]^a[B]^b[C]^c$$

where k_f is the rate constant for the forward reaction. Similarly, the rate of the reverse reaction (right to left in Equation 2-1) is

$$\text{Rate}_{reverse} = k_r[X]^x[Y]^y[Z]^z$$

where k_r is the rate constant for the reverse reaction. At equilibrium the forward and reverse rates are equal, so $\text{Rate}_{forward}/\text{Rate}_{reverse} = 1$. By rearranging these equations, we can express the equilibrium constant as the ratio of the rate constants

$$K_{eq} = \frac{k_f}{k_r} \quad (2\text{-}3)$$

Chemical Reactions in Cells Are at Steady State

Under appropriate conditions and given sufficient time, individual biochemical reactions carried out in a test tube eventually will reach equilibrium. Within cells, however, many reactions are linked in pathways in which a product of one reaction serves as a reactant in another or is pumped out of the cell. In this more complex situation, when the rate of formation of a substance is equal to the rate of its consumption, the concentration of the substance remains constant, and the system of linked reactions for producing and consuming that substance is said to be in a **steady state** (Figure 2-21). One consequence of such linked reactions is that they prevent the accumulation of excess intermediates, protecting cells from the harmful effects of intermediates that have the potential of being toxic at high concentrations.

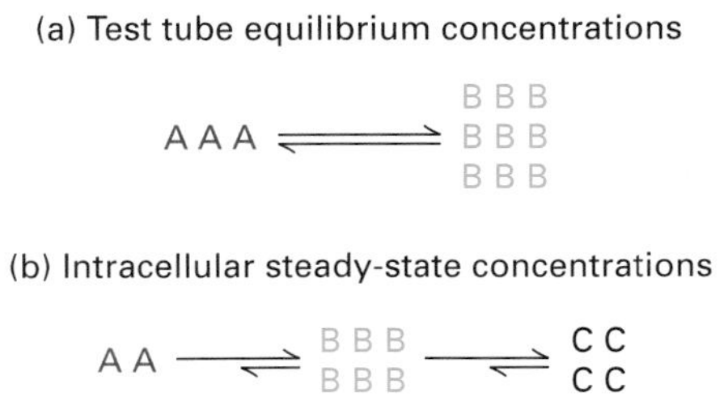

(a) Test tube equilibrium concentrations

$$A\ A\ A \rightleftharpoons \begin{array}{c} B\ B\ B \\ B\ B\ B \\ B\ B\ B \end{array}$$

(b) Intracellular steady-state concentrations

$$A\ A \rightleftharpoons \begin{array}{c} B\ B\ B \\ B\ B\ B \end{array} \rightleftharpoons \begin{array}{c} C\ C \\ C\ C \end{array}$$

▲ **FIGURE 2-21 Comparison of reactions at equilibrium and steady state.** (a) In the test tube, a biochemical reaction (A → B) eventually will reach equilibrium in which the rates of the forward and reverse reactions are equal (as indicated by the reaction arrows of equal length). (b) In metabolic pathways within cells, the product B commonly would be consumed, in this example by conversion to C. A pathway of linked reactions is at steady state when the rate of formation of the intermediates (e.g., B) equals their rate of consumption. As indicated by the unequal length of the arrows, the individual reversible reactions constituting a metabolic pathway do not reach equilibrium. Moreover, the concentrations of the intermediates at steady state differ from what they would be at equilibrium.

Dissociation Constants for Binding Reactions Reflect the Affinity of Interacting Molecules

The concept of chemical equilibrium also applies to the binding of one molecule to another. Many important cellular processes depend on such binding "reactions," which involve the making and breaking of various noncovalent interactions rather than covalent bonds, as discussed above. A common example is the binding of a **ligand** (e.g., the hormone insulin or adrenaline) to its **receptor** on the surface of a cell, triggering a biological response. Another example is the binding of a protein to a specific sequence of base pairs in a molecule of DNA, which frequently causes the expression of a nearby gene to increase or decrease (Chapter 11). If the equilibrium constant for a binding reaction is known, the intracellular stability of the resulting complex can be predicted. To illustrate the general approach for determining the concentration of noncovalently associated complexes, we will calculate the extent to which a protein is bound to DNA in a cell.

Most commonly, binding reactions are described in terms of the **dissociation constant** K_d, which is the reciprocal of the equilibrium constant. For the binding reaction P + D $\rightleftharpoons$ PD, where PD is the specific complex of a protein (P) and DNA (D), the dissociation constant is given by

$$K_d = \frac{[P][D]}{[PD]} \qquad (2\text{-}4)$$

Typical reactions in which a protein binds to a specific DNA sequence have a K_d of 10^{-10} M, where M symbolizes molarity, or moles per liter (mol/L). To relate the magnitude of this dissociation constant to the intracellular ratio of bound to unbound DNA, let's consider the simple example of a bacterial cell having a volume of 1.5×10^{-15} L and containing

1 molecule of DNA and *10 molecules* of the DNA-binding protein P. In this case, given a K_d of 10^{-10} M, 99 percent of the time this specific DNA sequence will have a molecule of protein bound to it, and 1 percent of the time it will not, even though the cell contains only 10 molecules of the protein! Clearly, P and D bind very tightly (have a high affinity), as reflected by the low value of the dissociation constant for their binding reaction.

The large size of biological macromolecules, such as proteins, can result in the availability of multiple surfaces for complementary intermolecular interactions. As a consequence, many macromolecules have the capacity to bind multiple other molecules simultaneously. In some cases, these binding reactions are independent, with their own distinct K_d values that are constant. In other cases, binding of a molecule at one site on a macromolecule can change the three-dimensional shape of a distant site, thus altering the binding interactions at that distant site. This is an important mechanism by which one molecule can alter (regulate) the activity of a second molecule (e.g., a protein) by changing its capacity to interact with a third molecule. We examine this regulatory mechanism in more detail in Chapter 3.

Biological Fluids Have Characteristic pH Values

The solvent inside cells and in all extracellular fluids is water. An important characteristic of any aqueous solution is the concentration of positively charged hydrogen ions (H^+) and negatively charged hydroxyl ions (OH^-). Because these ions are the dissociation products of H_2O, they are constituents of all living systems, and they are liberated by many reactions that take place between organic molecules within cells.

When a water molecule dissociates, one of its polar H—O bonds breaks. The resulting hydrogen ion, often referred to as a **proton**, has a short lifetime as a free particle and quickly combines with a water molecule to form a hydronium ion (H_3O^+). For convenience, however, we refer to the concentration of hydrogen ions in a solution, $[H^+]$, even though this really represents the concentration of hydronium ions, $[H_3O^+]$. Dissociation of H_2O generates one OH^- ion along with each H^+. The dissociation of water is a reversible reaction,

$$H_2O \rightleftharpoons H^+ + OH^-$$

At 25 °C, $[H^+][OH^-] = 10^{-14}$ M^2, so that in pure water, $[H^+] = [OH^-] = 10^{-7}$ M.

The concentration of hydrogen ions in a solution is expressed conventionally as its **pH,** defined as the negative log of the hydrogen ion concentration. The pH of pure water at 25 °C is 7:

$$pH = -\log[H^+] = \log\frac{1}{[H^+]} = \log\frac{1}{10^{-7}} = 7$$

It is important to keep in mind that a 1 unit difference in pH represents a tenfold difference in the concentration of protons. On the pH scale, 7.0 is considered neutral: pH values below 7.0 indicate acidic solutions (higher [H$^+$]), and values above 7.0 indicate basic (alkaline) solutions. For instance, gastric juice, which is rich in hydrochloric acid (HCl), has a pH of about 1. Its [H$^+$] is roughly a millionfold greater than that of cytoplasm with a pH of about 7.

Although the cytosol of cells normally has a pH of about 7.2, the pH is much lower (about 4.5) in the interior of lysosomes, one type of organelle in eukaryotic cells. The many degradative enzymes within lysosomes function optimally in an acidic environment, whereas their action is inhibited in the near neutral environment of the cytoplasm. This illustrates that maintenance of a specific pH is imperative for proper functioning of some cellular structures. On the other hand, dramatic shifts in cellular pH may play an important role in controlling cellular activity. For example, the pH of the cytoplasm of an unfertilized sea urchin egg is 6.6. Within 1 minute of fertilization, however, the pH rises to 7.2; that is, the H$^+$ concentration decreases to about one-fourth its original value, a change that is necessary for subsequent growth and division of the egg.

Hydrogen Ions Are Released by Acids and Taken Up by Bases

In general, an **acid** is any molecule, ion, or chemical group that tends to release a hydrogen ion (H$^+$), such as hydrochloric acid (HCl) and the carboxyl group (—COOH), which tends to dissociate to form the negatively charged carboxylate ion (—COO$^-$). Likewise, a **base** is any molecule, ion, or chemical group that readily combines with a H$^+$, such as the hydroxyl ion (OH$^-$), ammonia (NH$_3$), which forms an ammonium ion (NH$_4{}^+$), and the amino group (—NH$_2$).

When acid is added to an aqueous solution, the [H$^+$] increases (the pH goes down). Conversely, when a base is added to a solution, the [H$^+$] decreases (the pH goes up). Because [H$^+$][OH$^-$] = 10^{-14}M^2, any increase in [H$^+$] is coupled with a decrease in [OH$^-$], and vice versa.

Many biological molecules contain both acidic and basic groups. For example, in neutral solutions (pH = 7.0), amino acids exist predominantly in the doubly ionized form in which the carboxyl group has lost a proton and the amino group has accepted one:

$$H—\underset{\underset{R}{|}}{\overset{\overset{NH_3{}^+}{|}}{C}}—COO^-$$

where R represents the side chain. Such a molecule, containing an equal number of positive and negative ions, is called a **zwitterion.** Zwitterions, having no net charge, are neutral. At extreme pH values, only one of these two ionizable groups of an amino acid will be charged.

The dissociation reaction for an acid (or acid group in a larger molecule) HA can be written as HA $\rightleftharpoons$ H$^+$ + A$^-$.

The equilibrium constant for this reaction, denoted K_a (subscript a for "acid"), is defined as $K_a = $ [H$^+$][A$^-$]/[HA]. Taking the logarithm of both sides and rearranging the result yields a very useful relation between the equilibrium constant and pH:

$$pH = pK_a + \log \frac{[A^-]}{[HA]} \qquad (2\text{-}5)$$

where pK_a equals –log K_a.

From this expression, commonly known as the **Henderson-Hasselbalch equation,** it can be seen that the pK_a of any acid is equal to the pH at which half the molecules are dissociated and half are neutral (undissociated). This is because when pK_a = pH, then log ([A$^-$]/[HA]) = 0, and therefore [A$^-$] = [HA]. The Henderson-Hasselbalch equation allows us to calculate the degree of dissociation of an acid if both the pH of the solution and the pK_a of the acid are known. Experimentally, by measuring the [A$^-$] and [HA] as a function of the solution's pH, one can calculate the pK_a of the acid and thus the equilibrium constant K_a for the dissociation reaction.

Buffers Maintain the pH of Intracellular and Extracellular Fluids

A growing cell must maintain a constant pH in the cytoplasm of about 7.2–7.4 despite the metabolic production of many acids, such as lactic acid and carbon dioxide; the latter reacts with water to form carbonic acid (H$_2$CO$_3$). Cells have a reservoir of weak bases and weak acids, called **buffers,** which ensure that the cell's pH remains relatively constant despite small fluctuations in the amounts of H$^+$ or OH$^-$ being generated by metabolism or by the uptake or secretion of molecules and ions by the cell. Buffers do this by "soaking up" excess H$^+$ or OH$^-$ when these ions are added to the cell or are produced by metabolism.

If additional acid (or base) is added to a solution that contains a buffer at its pK_a value (a 1:1 mixture of HA and A$^-$), the pH of the solution changes, but it changes less than it would if the buffer had not been present. This is because protons released by the added acid are taken up by the ionized form of the buffer (A$^-$); likewise, hydroxyl ions generated by the addition of base are neutralized by protons released by the undissociated buffer (HA). The capacity of a substance to release hydrogen ions or take them up depends partly on the extent to which the substance has already taken up or released protons, which in turn depends on the pH of the solution. The ability of a buffer to minimize changes in pH, its **buffering capacity,** depends on the relationship between its pK_a value and the pH, which is expressed by the Henderson-Hasselbalch equation.

The titration curve for acetic acid shown in Figure 2-22 illustrates the effect of pH on the fraction of molecules in the un-ionized (HA) and ionized forms (A$^-$). At one pH unit below the pK_a of an acid, 91 percent of the molecules are in the HA form; at one pH unit above the pK_a, 91 percent are

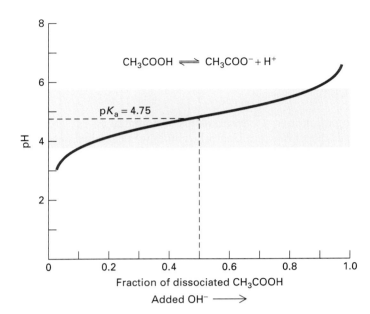

▲ **FIGURE 2-22 The titration curve of acetic acid (CH₃COOH).** The pK_a for the dissociation of acetic acid to hydrogen and acetate ions is 4.75. At this pH, half the acid molecules are dissociated. Because pH is measured on a logarithmic scale, the solution changes from 91 percent CH_3COOH at pH 3.75 to 9 percent CH_3COOH at pH 5.75. The acid has maximum buffering capacity in this pH range.

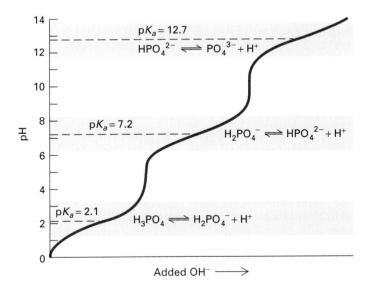

▲ **FIGURE 2-23 The titration curve of phosphoric acid (H₃PO₄).** This biologically ubiquitous molecule has three hydrogen atoms that dissociate at different pH values; thus, phosphoric acid has three pK_a values, as noted on the graph. The shaded areas denote the pH ranges—within one pH unit of the three pK_a values—where the buffering capacity of phosphoric acid is high. In these regions the addition of acid (or base) will cause relatively small changes in the pH.

in the A^- form. At pH values more than one unit above or below the pK_a, the buffering capacity of weak acids and bases declines rapidly. In other words, the addition of the same number of moles of acid to a solution containing a mixture of HA and A^- that is at a pH near the pK_a will cause less of a pH change than it would if the HA and A^- were not present or if the pH were far from the pK_a value.

All biological systems contain one or more buffers. Phosphate ions, the ionized forms of phosphoric acid, are present in considerable quantities in cells and are an important factor in maintaining, or buffering, the pH of the cytoplasm. Phosphoric acid (H_3PO_4) has three protons that are capable of dissociating, but they do not dissociate simultaneously. Loss of each proton can be described by a discrete dissociation reaction and pK_a as shown in Figure 2-23. The titration curve for phosphoric acid shows that the pK_a for the dissociation of the second proton is 7.2. Thus at pH 7.2, about 50 percent of cellular phosphate is $H_2PO_4^-$ and about 50 percent is HPO_4^{2-} according to the Henderson-Hasselbalch equation. For this reason, phosphate is an excellent buffer at pH values around 7.2, the approximate pH of the cytoplasm of cells, and at pH 7.4, the pH of human blood.

KEY CONCEPTS OF SECTION 2.3

Chemical Equilibrium

■ A chemical reaction is at equilibrium when the rate of the forward reaction is equal to the rate of the reverse reaction (no net change in the concentration of the reactants or products).

■ The equilibrium constant K_{eq} of a reaction reflects the ratio of products to reactants at equilibrium and thus is a measure of the extent of the reaction and the relative stabilities of the reactants and products.

■ The K_{eq} depends on the temperature, pressure, and chemical properties of the reactants and products, but is independent of the reaction rate and of the initial concentrations of reactants and products.

■ For any reaction, the equilibrium constant K_{eq} equals the ratio of the forward rate constant to the reverse rate constant (k_f/k_r).

■ Within cells, the linked reactions in metabolic pathways generally are at steady state, not equilibrium, at which rate of formation of the intermediates equals their rate of consumption (see Figure 2-21).

■ The dissociation constant K_d for a reaction involving the noncovalent binding of two molecules is a measure of the stability of the complex formed between the molecules (e.g., ligand-receptor or protein-DNA complexes).

■ The pH is the negative logarithm of the concentration of hydrogen ions ($-\log [H+]$). The pH of the cytoplasm is normally about 7.2–7.4, whereas the interior of lysosomes has a pH of about 4.5.

■ Acids release protons (H^+) and bases bind them. In biological molecules, the carboxyl and phosphate groups are the most common acidic groups; the amino group is the most common basic group.

■ Buffers are mixtures of a weak acid (HA) and its corresponding base form (A^-), which minimize the change in pH of a solution when acid or alkali is added. Biological systems use various buffers to maintain their pH within a very narrow range.

2.4 Biochemical Energetics

The production of energy, its storage, and its use are central to the economy of the cell. Energy may be defined as the ability to do work, a concept applicable to automobile engines and electric power plants in our physical world and to cellular engines in the biological world. The energy associated with chemical bonds can be harnessed to support chemical work and the physical movements of cells.

Several Forms of Energy Are Important in Biological Systems

There are two principal forms of energy: kinetic and potential. **Kinetic energy** is the energy of movement—the motion of molecules, for example. The second form of energy, **potential energy,** or stored energy, is particularly important in the study of biological or chemical systems.

Kinetic Energy Heat, or **thermal energy,** is a form of kinetic energy—the energy of the motion of molecules. For heat to do work, it must flow from a region of higher temperature—where the average speed of molecular motion is greater—to one of lower temperature. Although differences in temperature can exist between the internal and external environments of cells, these thermal gradients do not usually serve as the source of energy for cellular activities. The thermal energy in warm-blooded animals, which have evolved a mechanism for thermoregulation, is used chiefly to maintain constant organismic temperatures. This is an important function, since the rates of many cellular activities are temperature-dependent. For example, cooling mammalian cells from their normal body temperature of 37 °C to 4 °C can virtually "freeze" or stop many cellular processes (e.g., intracellular membrane movements).

Radiant energy is the kinetic energy of photons, or waves of light, and is critical to biology. Radiant energy can be converted to thermal energy, for instance when light is absorbed by molecules and the energy is converted to molecular motion. During photosynthesis, light energy absorbed by specialized molecules (e.g., chlorophyll) is subsequently converted into the energy of chemical bonds (Chapter 8).

Mechanical energy, a major form of kinetic energy in biology, usually results from the conversion of stored chemical energy. For example, changes in the lengths of cytoskeletal filaments generates forces that push or pull on membranes and organelles (Chapter 19).

Electric energy—the energy of moving electrons or other charged particles—is yet another major form of kinetic energy.

Potential Energy Several forms of potential energy are biologically significant. Central to biology is **chemical potential energy,** the energy stored in the bonds connecting atoms in molecules. Indeed, most of the biochemical reactions described in this book involve the making or breaking of at least one covalent chemical bond. We recognize this energy when chemicals undergo energy-releasing reactions. For example, the high potential energy in the covalent bonds of glucose can be released by controlled enzymatic combustion in cells (see later discussion). This energy is harnessed by the cell to do many kinds of work.

A second biologically important form of potential energy is the energy in a **concentration gradient.** When the concentration of a substance on one side of a barrier, such as a membrane, is different from that on the other side, a concentration gradient exists. All cells form concentration gradients between their interior and the external fluids by selectively exchanging nutrients, waste products, and ions with their surroundings. Also, organelles within cells (e.g., mitochondria, lysosomes) frequently contain different concentrations of ions and other molecules; the concentration of protons within a lysosome, as we saw in the last section, is about 500 times that of the cytoplasm.

A third form of potential energy in cells is an **electric potential**—the energy of charge separation. For instance, there is a gradient of electric charge of ≈200,000 volts per cm across the plasma membrane of virtually all cells. We discuss how concentration gradients and the potential difference across cell membranes are generated and maintained in Chapter 7.

Cells Can Transform One Type of Energy into Another

According to the first law of thermodynamics, energy is neither created nor destroyed, but can be converted from one form to another. (In nuclear reactions mass is converted to energy, but this is irrelevant to biological systems.) In photosynthesis, for example, the radiant energy of light is transformed into the chemical potential energy of the covalent bonds between the atoms in a sucrose or starch molecule. In muscles and nerves, chemical potential energy stored in covalent bonds is transformed, respectively, into the kinetic energy of muscle contraction and the electric energy of nerve transmission. In all cells, potential energy, released by breaking certain chemical bonds, is used to generate potential energy in the form of concentration and electric potential gradients. Similarly, energy stored in chemical concentration gradients or electric potential gradients is used to synthesize chemical bonds or to transport molecules from one side of a membrane to another to generate a concentration gradient. This latter process occurs during the transport of nutrients such as glucose into certain cells and transport of many waste products out of cells.

Because all forms of energy are interconvertible, they can be expressed in the same units of measurement. Although the standard unit of energy is the joule, biochemists have traditionally used an alternative unit, the **calorie** (1 joule = 0.239 calories). Throughout this book, we use the kilocalorie to measure energy changes (1 kcal = 1000 cal).

The Change in Free Energy Determines the Direction of a Chemical Reaction

Because biological systems are generally held at constant temperature and pressure, it is possible to predict the direction of a chemical reaction from the change in the **free energy G**, named after J. W. Gibbs, who showed that "all systems change in such a way that free energy [G] is minimized." In the case of a chemical reaction, reactants $\rightleftharpoons$ products, the change in free energy ΔG is given by

$$\Delta G = G_{products} - G_{reactants}$$

The relation of ΔG to the direction of any chemical reaction can be summarized in three statements:

■ If ΔG is negative, the forward reaction (from left to right as written) will tend to occur spontaneously.

■ If ΔG is positive, the reverse reaction (from right to left as written) will tend to occur.

■ If ΔG is zero, both forward and reverse reactions occur at equal rates; the reaction is at equilibrium.

The standard free-energy change of a reaction $\Delta G^{\circ\prime}$ is the value of the change in free energy under the conditions of 298 K (25 °C), 1 atm pressure, pH 7.0 (as in pure water), and initial concentrations of 1 M for all reactants and products except protons, which are kept at 10^{-7} M (pH 7.0). Most biological reactions differ from standard conditions, particularly in the concentrations of reactants, which are normally less than 1 M.

The free energy of a chemical system can be defined as $G = H - TS$, where H is the bond energy, or **enthalpy**, of the system; T is its temperature in degrees Kelvin (K); and S is the **entropy**, a measure of its randomness or disorder. If temperature remains constant, a reaction proceeds spontaneously only if the free-energy change ΔG in the following equation is negative:

$$\Delta G = \Delta H - T \Delta S \qquad (2\text{-}6)$$

In an **exothermic** reaction, the products contain less bond energy than the reactants, the liberated energy is usually converted to heat (the energy of molecular motion), and ΔH is negative. In an **endothermic** reaction, the products contain

more bond energy than the reactants, heat is absorbed, and ΔH is positive. The combined effects of the changes in the enthalpy and entropy determine if the ΔG for a reaction is positive or negative. An exothermic reaction ($\Delta H < 0$) in which entropy increases ($\Delta S > 0$) occurs spontaneously ($\Delta G < 0$). An endothermic reaction ($\Delta H > 0$) will occur spontaneously if ΔS increases enough so that the $T \Delta S$ term can overcome the positive ΔH.

Many biological reactions lead to an increase in order, and thus a decrease in entropy ($\Delta S < 0$). An obvious example is the reaction that links amino acids together to form a protein. A solution of protein molecules has a lower entropy than does a solution of the same amino acids unlinked, because the free movement of any amino acid in a protein is restricted when it is bound into a long chain. Often cells compensate for decreases in entropy by "coupling" such synthetic reactions with independent reactions that have a very highly negative ΔG (see below). In this fashion cells can convert sources of energy in their environment into the building of highly organized structures and metabolic pathways that are essential for life.

The actual change in free energy ΔG during a reaction is influenced by temperature, pressure, and the initial concentrations of reactants and products and usually differs from $\Delta G^{\circ\prime}$. Most biological reactions—like others that take place in aqueous solutions—also are affected by the pH of the solution. We can estimate free-energy changes for different temperatures and initial concentrations, using the equation

$$\Delta G = \Delta G^{\circ\prime} + RT \ln Q = \Delta G^{\circ\prime} + RT \ln \frac{[products]}{[reactants]} \qquad (2\text{-}7)$$

where R is the gas constant of 1.987 cal/(degree·mol), T is the temperature (in degrees Kelvin), and Q is the *initial* ratio of products to reactants. For a reaction A + B $\rightleftharpoons$ C, in which two molecules combine to form a third, Q in Equation 2-7 equals [C]/[A][B]. In this case, an increase in the initial concentration of either [A] or [B] will result in a large negative value for ΔG and thus drive the reaction toward more formation of C.

Regardless of the $\Delta G^{\circ\prime}$ for a particular biochemical reaction, it will proceed spontaneously within cells only if ΔG is negative, given the usual intracellular concentrations of reactants and products. For example, the conversion of glyceraldehyde 3-phosphate (G3P) to dihydroxyacetone phosphate (DHAP), two intermediates in the breakdown of glucose,

$$\text{G3P} \rightleftharpoons \text{DHAP}$$

has a $\Delta G^{\circ\prime}$ of -1840 cal/mol. If the initial concentrations of G3P and DHAP are equal, then $\Delta G = \Delta G^{\circ\prime}$, because $RT \ln 1 = 0$; in this situation, the reversible reaction G3P $\rightleftharpoons$ DHAP will proceed in the direction of DHAP formation until equilibrium is reached. However, if the initial [DHAP] is 0.1 M and the initial [G3P] is 0.001 M, with other conditions being standard, then Q in Equation 2-7 equals 0.1/0.001 = 100, giving a ΔG of $+887$ cal/mole. Under these conditions, the reaction will proceed in the direction of formation of G3P.

The ΔG for a reaction is independent of the reaction rate. Indeed, under usual physiological conditions, few, if any, of the biochemical reactions needed to sustain life would occur without some mechanism for increasing reaction rates. As we describe in Chapter 3, the rates of reactions in biological systems are usually determined by the activity of **enzymes**, the protein catalysts that accelerate the formation of products from reactants without altering the value of ΔG.

The $\Delta G^{\circ\prime}$ of a Reaction Can Be Calculated from Its K_{eq}

A chemical mixture at equilibrium is already in a state of minimal free energy; that is, no free energy is being generated or released. Thus, for a system at equilibrium ($\Delta G = 0$, $Q = K_{eq}$), we can write

$$\Delta G^{\circ\prime} = -2.3RT \log K_{eq} = -1362 \log K_{eq} \qquad (2\text{-}8)$$

under standard conditions (note the change to base 10 logarithms). Thus, if the concentrations of reactants and products at equilibrium (i.e., the K_{eq}) are determined, the value of $\Delta G^{\circ\prime}$ can be calculated. For example, K_{eq} for the interconversion of glyceraldehyde 3-phosphate to dihydroxyacetone phosphate (G3P $\rightleftharpoons$ DHAP) is 22.2 under standard conditions. Substituting this value into Equation 2-8, we can easily calculate the $\Delta G^{\circ\prime}$ for this reaction as -1840 cal/mol.

By rearranging Equation 2-8 and taking the antilogarithm, we obtain

$$K_{eq} = 10^{-(\Delta G^{\circ\prime}/2.3RT)} \qquad (2\text{-}9)$$

From this expression, it is clear that if $\Delta G^{\circ\prime}$ is negative, the exponent will be positive and hence K_{eq} will be greater than 1. Therefore at equilibrium there will be more products than reactants; in other words, the formation of products from reactants is favored. Conversely, if $\Delta G^{\circ\prime}$ is positive, the exponent will be negative and K_{eq} will be less than 1.

An Unfavorable Chemical Reaction Can Proceed If It Is Coupled with an Energetically Favorable Reaction

Many processes in cells are energetically unfavorable ($\Delta G > 0$) and will not proceed spontaneously. Examples include the synthesis of DNA from nucleotides and transport of a substance across the plasma membrane from a lower to a higher concentration. Cells can carry out an energy-requiring reaction ($\Delta G_1 > 0$) by coupling it to an energy-releasing reaction ($\Delta G_2 < 0$) if the sum of the two reactions has a net negative ΔG.

Suppose, for example, that the reaction A $\rightleftharpoons$ B + X has a ΔG of $+5$ kcal/mol and that the reaction X $\rightleftharpoons$ Y + Z has a ΔG of -10 kcal/mol.

(1) $\quad$ A $\rightleftharpoons$ B + X $\qquad\qquad \Delta G = +5$ kcal/mol

(2) $\quad$ X $\rightleftharpoons$ Y + Z $\qquad\qquad \Delta G = -10$ kcal/mol

Sum: $\quad$ A $\rightleftharpoons$ B + Y + Z $\qquad \Delta G^{\circ\prime} = -5$ kcal/mol

In the absence of the second reaction, there would be much more A than B at equilibrium. However, because the conversion of X to Y + Z is such a favorable reaction, it will pull the first process toward the formation of B and the consumption of A. Energetically unfavorable reactions in cells often are coupled to the hydrolysis of ATP, as we discuss next.

Hydrolysis of ATP Releases Substantial Free Energy and Drives Many Cellular Processes

In almost all organisms, **adenosine triphosphate**, or **ATP**, is the most important molecule for capturing, transiently storing, and subsequently transferring energy to perform work (e.g., biosynthesis, mechanical motion). The useful energy in an ATP molecule is contained in **phosphoanhydride bonds**, which are covalent bonds formed from the condensation of two molecules of phosphate by the loss of water:

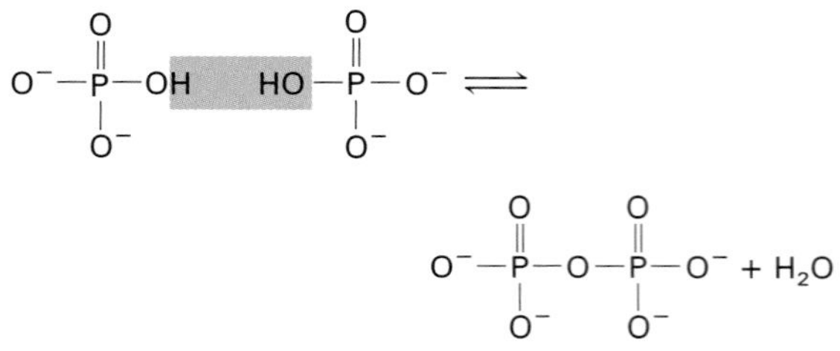

An ATP molecule has two key phosphoanhydride bonds (Figure 2-24). Hydrolysis of a phosphoanhydride bond ($\sim$) in each of the following reactions has a highly negative $\Delta G^{\circ\prime}$ of about -7.3 kcal/mol:

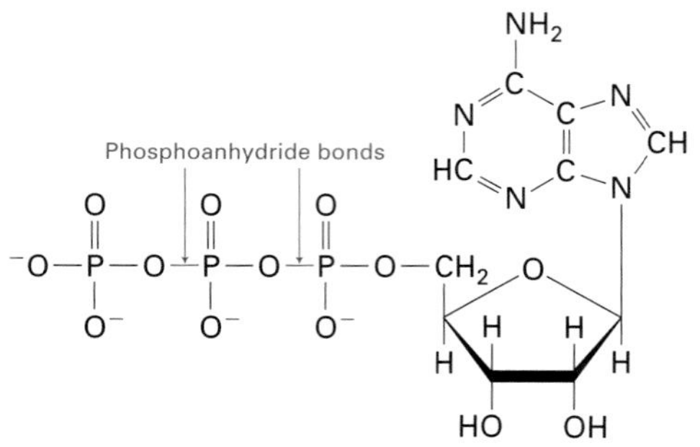

▲ **FIGURE 2-24 Adenosine triphosphate (ATP).** The two phosphoanhydride bonds (red) in ATP, which link the three phosphate groups, each has a $\Delta G^{\circ\prime}$ of -7.3 kcal/mol for hydrolysis. Hydrolysis of these bonds, especially the terminal one, drives many energy-requiring reactions in biological systems.

$$Ap\sim p\sim p + H_2O \longrightarrow Ap\sim p + P_i + H^+$$
$$\textbf{(ATP)} \qquad\qquad\qquad \textbf{(ADP)}$$

$$Ap\sim p\sim p + H_2O \longrightarrow Ap + PP_i + H^+$$
$$\textbf{(ATP)} \qquad\qquad\qquad \textbf{(AMP)}$$

$$Ap\sim p + H_2O \longrightarrow Ap + P_i + H^+$$
$$\textbf{(ADP)} \qquad\qquad\qquad \textbf{(AMP)}$$

In these reactions, P_i stands for inorganic phosphate (PO_4^{3-}) and PP_i for inorganic pyrophosphate, two phosphate groups linked by a phosphodiester bond. As the top two reactions show, the removal of a phosphate or a pyrophosphate group from ATP leaves adenosine diphosphate (ADP) or adenosine monophosphate (AMP), respectively.

A phosphoanhydride bond or other **high-energy bond** (commonly denoted by ~) is not intrinsically different from other covalent bonds. High-energy bonds simply release especially large amounts of energy when broken by addition of water (hydrolyzed). For instance, the $\Delta G^{o'}$ for hydrolysis of a phosphoanhydride bond in ATP (-7.3 kcal/mol) is more than three times the $\Delta G^{o'}$ for hydrolysis of the phosphoester bond (red) in glycerol 3-phosphate (-2.2 kcal/mol):

$$HO-\overset{\overset{\displaystyle O}{\|}}{\underset{\underset{\displaystyle O^-}{|}}{P}}-O-CH_2-\overset{\overset{\displaystyle OH}{|}}{CH}-CH_2OH$$

Glycerol 3-phosphate

A principal reason for this difference is that ATP and its hydrolysis products ADP and P_i are highly charged at neutral pH. During synthesis of ATP, a large input of energy is required to force the negative charges in ADP and P_i together. Conversely, much energy is released when ATP is hydrolyzed to ADP and P_i. In comparison, formation of the phosphoester bond between an uncharged hydroxyl in glycerol and P_i requires less energy, and less energy is released when this bond is hydrolyzed.

Cells have evolved protein-mediated mechanisms for transferring the free energy released by hydrolysis of phosphoanhydride bonds to other molecules, thereby driving reactions that would otherwise be energetically unfavorable. For example, if the ΔG for the reaction B + C $\longrightarrow$ D is positive but less than the ΔG for hydrolysis of ATP, the reaction can be driven to the right by coupling it to hydrolysis of the terminal phosphoanhydride bond in ATP. In one common mechanism of such **energy coupling,** some of the energy stored in this phosphoanhydride bond is transferred to the one of the reactants by breaking the bond in ATP and forming a covalent bond between the released phosphate group and one of the reactants. The phosphorylated intermediate generated in this fashion can then react with C to form D + P_i in a reaction that has a negative ΔG :

$$B + Ap\sim p\sim p \longrightarrow B\sim p + Ap\sim p$$
$$B\sim p + C \longrightarrow D + P_i$$

The overall reaction

$$B + C + ATP \rightleftharpoons D + ADP + P_i$$

is energetically favorable ($\Delta G < 0$).

An alternative mechanism of energy coupling is to use the energy released by ATP hydrolysis to change the conformation of the molecule to an "energy-rich" stressed state. In turn, the energy stored as conformational stress can be released as the molecule "relaxes" back into its unstressed conformation. If this relaxation process can be mechanistically coupled to another reaction, the released energy can be harnessed to drive important cellular processes.

As with many biosynthetic reactions, transport of molecules into or out of the cell often has a positive ΔG and thus requires an input of energy to proceed. Such simple transport reactions do not *directly* involve the making or breaking of covalent bonds; thus the $\Delta G^{o'}$ is 0. In the case of a substance moving into a cell, Equation 2-7 becomes

$$\Delta G = RT \ln\frac{[C_{in}]}{[C_{out}]} \qquad (2\text{-}10)$$

where $[C_{in}]$ is the initial concentration of the substance inside the cell and $[C_{out}]$ is its concentration outside the cell. We can see from Equation 2-10 that ΔG is positive for transport of a substance into a cell against its concentration gradient (when $[C_{in}] > [C_{out}]$); the energy to drive such "uphill" transport often is supplied by the hydrolysis of ATP. Conversely, when a substance moves down its concentration gradient ($[C_{out}] > [C_{in}]$), ΔG is negative. Such "downhill" transport releases energy that can be coupled to an energy-requiring reaction, say, the movement of another substance uphill across a membrane or the synthesis of ATP itself (see Chapter 7).

ATP Is Generated During Photosynthesis and Respiration

Clearly, to continue functioning cells must constantly replenish their ATP supply. The initial energy source whose energy is ultimately transformed into the phosphoanhydride bonds of ATP and bonds in other compounds in nearly all cells is sunlight. In **photosynthesis,** plants and certain microorganisms can trap the energy in light and use it to synthesize ATP from ADP and P_i. Much of the ATP produced in photosynthesis is hydrolyzed to provide energy for the conversion of carbon dioxide to six-carbon sugars, a process called **carbon fixation:**

$$6\,CO_2 + 6\,H_2O \overset{\displaystyle ATP \quad ADP + P_i}{\underset{\displaystyle }{\searrow\;\nearrow}} C_6H_{12}O_6 + 6\,O_2$$

In animals, the free energy in sugars and other molecules derived from food is released in the process of **respiration.** All synthesis of ATP in animal cells and in nonphotosynthetic microorganisms results from the chemical transformation of energy-rich compounds in the diet (e.g., glucose, starch). We discuss the mechanisms of photosynthesis and cellular respiration in Chapter 8.

The complete oxidation of glucose to yield carbon dioxide

$$C_6H_{12}O_6 + 6\ O_2 \longrightarrow 6\ CO_2 + 6\ H_2O$$

has a $\Delta G^{o\prime}$ of -686 kcal/mol and is the reverse of photosynthetic carbon fixation. Cells employ an elaborate set of enzyme-catalyzed reactions to couple the metabolism of 1 molecule of glucose to the synthesis of as many as 30 molecules of ATP from 30 molecules of ADP. This oxygen-dependent (**aerobic**) degradation (**catabolism**) of glucose is the major pathway for generating ATP in all animal cells, nonphotosynthetic plant cells, and many bacterial cells.

Light energy captured in photosynthesis is not the only source of chemical energy for all cells. Certain microorganisms that live in deep ocean vents, where sunlight is completely absent, derive the energy for converting ADP and P_i into ATP from the oxidation of reduced inorganic compounds. These reduced compounds originate in the center of the earth and are released at the vents.

NAD⁺ and FAD Couple Many Biological Oxidation and Reduction Reactions

In many chemical reactions, electrons are transferred from one atom or molecule to another; this transfer may or may not accompany the formation of new chemical bonds. The loss of electrons from an atom or a molecule is called **oxidation,** and the gain of electrons by an atom or a molecule is called **reduction**. Because electrons are neither created nor destroyed in a chemical reaction, if one atom or molecule is oxidized, another must be reduced. For example, oxygen draws electrons from Fe^{2+} (ferrous) ions to form Fe^{3+} (fer-

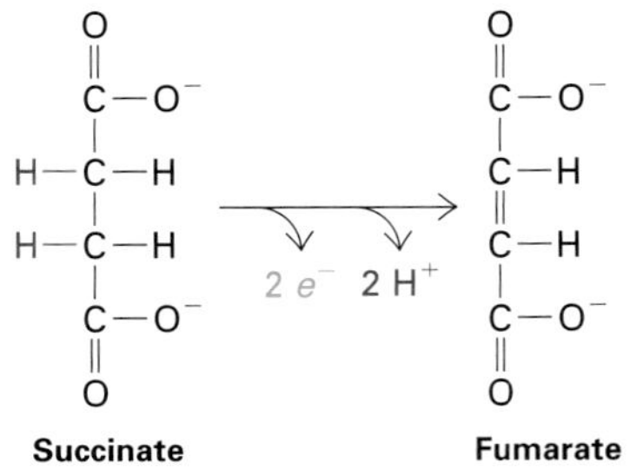

▲ **FIGURE 2-25 Conversion of succinate to fumarate.** In this oxidation reaction, which occurs in mitochondria as part of the citric acid cycle, succinate loses two electrons and two protons. These are transferred to FAD, reducing it to $FADH_2$.

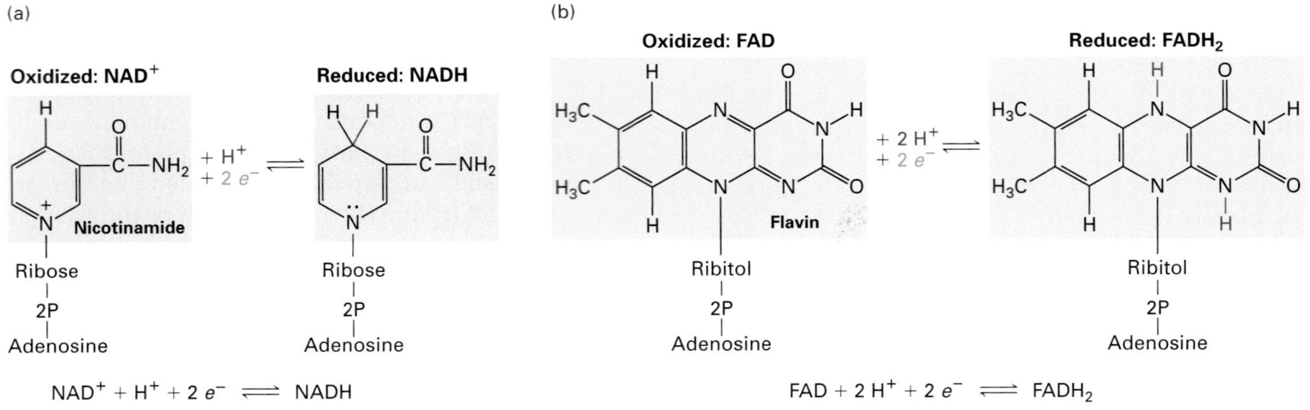

ric) ions, a reaction that occurs as part of the process by which carbohydrates are degraded in mitochondria. Each oxygen atom receives two electrons, one from each of two Fe^{2+} ions:

$$2\ Fe^{2+} + \tfrac{1}{2}\ O_2 \longrightarrow 2\ Fe^{3+} + O^{2-}$$

Thus Fe^{2+} is oxidized, and O_2 is reduced. Such reactions in which one molecule is reduced and another oxidized often are referred to as **redox reactions**. Oxygen is an electron acceptor in many redox reactions in aerobic cells.

Many biologically important oxidation and reduction reactions involve the removal or the addition of hydrogen atoms (protons plus electrons) rather than the transfer of isolated electrons on their own. The oxidation of succinate to fumarate, which also occurs in mitochondria, is an example (Figure 2-25). Protons are soluble in aqueous solutions (as H_3O^+), but electrons are not and must be transferred di-

▲ **FIGURE 2-26 The electron-carrying coenzymes NAD⁺ and FAD.** (a) NAD⁺ (nicotinamide adenine dinucleotide) is reduced to NADH by addition of two electrons and one proton simultaneously. In many biological redox reactions (e.g., succinate → fumarate), a pair of hydrogen atoms (two protons and two electrons) are removed from a molecule. One of the protons and both electrons are transferred to NAD⁺; the other proton is released into solution. (b) FAD (flavin adenine dinucleotide) is reduced to FADH₂ by addition of two electrons and two protons. In this two-step reaction, addition of one electron together with one proton first generates a short-lived semiquinone intermediate (not shown), which then accepts a second electron and proton.

rectly from one atom or molecule to another without a water-dissolved intermediate. In this type of oxidation reaction, electrons often are transferred to small electron-carrying molecules, sometimes referred to as coenzymes. The most common of these electron carriers are NAD^+ (nicotinamide adenine dinucleotide), which is reduced to NADH, and **FAD** (flavin adenine dinucleotide), which is reduced to $FADH_2$ (Figure 2-26). The reduced forms of these coenzymes can transfer protons and electrons to other molecules, thereby reducing them.

To describe redox reactions, such as the reaction of ferrous ion (Fe^{2+}) and oxygen (O_2), it is easiest to divide them into two **half-reactions:**

$$Oxidation \ of \ Fe^{2+}: \quad 2 \ Fe^{2+} \longrightarrow 2 \ Fe^{3+} + 2 \ e^-$$
$$Reduction \ of \ O_2: \quad 2 \ e^- + \frac{1}{2} \ O_2 \longrightarrow O^{2-}$$

In this case, the reduced oxygen (O^{2-}) readily reacts with two protons to form one water molecule (H_2O). The readiness with which an atom or a molecule *gains* an electron is its **reduction potential** E. The tendency to *lose* electrons, the **oxidation potential,** has the same magnitude but opposite sign as the reduction potential for the reverse reaction.

Reduction potentials are measured in volts (V) from an arbitrary zero point set at the reduction potential of the following half-reaction under standard conditions (25 °C, 1 atm, and reactants at 1 M):

$$H^+ + e^- \underset{oxidation}{\overset{reduction}{\rightleftharpoons}} \frac{1}{2} H_2$$

The value of E for a molecule or an atom under standard conditions is its standard reduction potential E'_0. A molecule or ion with a positive E'_0 has a higher affinity for electrons than the H^+ ion does under standard conditions. Conversely, a molecule or ion with a negative E'_0 has a lower affinity for electrons than the H^+ ion does under standard conditions. Like the values of $\Delta G^{\circ\prime}$, standard reduction potentials may differ somewhat from those found under the conditions in a cell because the concentrations of reactants in a cell are not 1 M.

In a redox reaction, electrons move spontaneously toward atoms or molecules having *more positive* reduction potentials. In other words, a compound having a more negative reduction potential can transfer electrons to (i.e., reduce) a compound with a more positive reduction potential. In this type of reaction, the change in electric potential ΔE is the sum of the reduction and oxidation potentials for the two half-reactions. The ΔE for a redox reaction is related to the change in free energy ΔG by the following expression:

$$\Delta G \ (cal/mol) = -n \ (23,064) \ \Delta E \ (volts) \quad (2\text{-}11)$$

where n is the number of electrons transferred. Note that a redox reaction with a positive ΔE value will have a negative ΔG and thus will tend to proceed from left to right.

KEY CONCEPTS OF SECTION 2.4

Biochemical Energetics

■ The change in free energy ΔG is the most useful measure for predicting the direction of chemical reactions in biological systems. Chemical reactions tend to proceed in the direction for which ΔG is negative.

■ Directly or indirectly, light energy captured by photosynthesis in plants and photosynthetic bacteria is the ultimate source of chemical energy for almost all cells.

■ The chemical free-energy change $\Delta G^{\circ\prime}$ equals $-2.3 \ RT$ log K_{eq}. Thus the value of $\Delta G^{\circ\prime}$ can be calculated from the experimentally determined concentrations of reactants and products at equilibrium.

■ A chemical reaction having a positive ΔG can proceed if it is coupled with a reaction having a negative ΔG of larger magnitude.

■ Many otherwise energetically unfavorable cellular processes are driven by hydrolysis of phosphoanhydride bonds in ATP (see Figure 2-24).

■ An oxidation reaction (loss of electrons) is always coupled with a reduction reaction (gain of electrons).

■ Biological oxidation and reduction reactions often are coupled by electron-carrying coenzymes such as NAD^+ and FAD (see Figure 2-26).

■ Oxidation-reduction reactions with a positive ΔE have a negative ΔG and thus tend to proceed spontaneously.

KEY TERMS

REVIEW THE CONCEPTS

1. The gecko is a reptile with an amazing ability to climb smooth surfaces, including glass. Recent discoveries indicate that geckos stick to smooth surfaces via van der Waals interactions between septae on their feet and the smooth surface. How is this method of stickiness advantageous over covalent interactions? Given that van der Waals forces are among the weakest molecular interactions, how can the gecko's feet stick so effectively?

2. The K^+ channel is an example of a transmembrane protein (a protein that spans the phospholipid bilayer of the plasma membrane). What types of amino acids are likely to be found (a) lining the channel through which K^+ passes; (b) in contact with the phospholipid bilayer containing fatty acid; (c) in the cytosolic domain of the protein; and (d) in the extracellular domain of the protein?

3. V-M-Y-Y-E-N: This is the single-letter amino acid abbreviation for a peptide. Draw the structure of this peptide. What is the net charge of this peptide at pH 7.0? An enzyme called a protein tyrosine kinase can attach phosphates to the hydroxyl groups of tyrosine. What is the net charge of the peptide at pH 7.0 after it has been phosphorylated by a tyrosine kinase? What is the likely source of phosphate utilized by the kinase for this reaction?

4. Disulfide bonds help to stabilize the three-dimensional structure of proteins. What amino acids are involved in the formation of disulfide bonds? Does the formation of a disulfide bond increase or decrease entropy (ΔS)?

5. In the 1960s, the drug thalidomide was prescribed to pregnant women to treat morning sickness. However, thalidomide caused severe limb defects in the children of some women who took the drug, and its use for morning sickness was discontinued. It is now known that thalidomide was administered as a mixture of two stereoisomeric compounds, one of which relieved morning sickness and the other of which was responsible for the birth defects. What are stereoisomers? Why might two such closely related compounds have such different physiologic effects?

6. Name the compound shown below. Is this nucleotide a component of DNA, RNA, or both? Name one other function of this compound.

7. The chemical basis of blood-group specificity resides in the carbohydrates displayed on the surface of red blood cells. Carbohydrates have the potential for great structural diversity. Indeed, the structural complexity of the oligosaccharides that can be formed from four sugars is greater than that for oligopeptides from four amino acids. What properties of carbohydrates make this great structural diversity possible?

8. Ammonia (NH_3) is a weak base that under acidic conditions becomes protonated to the ammonium ion in the following reaction:

$$NH_3 + H^+ \rightarrow NH_4^+$$

NH_3 freely permeates biological membranes, including those of lysosomes. The lysosome is a subcellular organelle with a pH of about 5.0; the pH of cytoplasm is 7.0. What is the effect on the pH of the fluid content of lysosomes when cells are exposed to ammonia? *Note:* Protonated ammonia does not diffuse freely across membranes.

9. Consider the binding reaction L + R → LR, where L is a ligand and R is its receptor. When 1×10^{-3} M L is added to a solution containing 5×10^{-2} M R, 90% of the L binds to form LR. What is the K_{eq} of this reaction? How will the K_{eq} be affected by the addition of a protein that catalyzes this binding reaction? What is the K_d?

10. What is the ionization state of phosphoric acid in the cytoplasm? Why is phosphoric acid such a physiologically important compound?

11. The $\Delta G^{\circ\prime}$ for the reaction X + Y → XY is −1000 cal/mol. What is the ΔG at 25°C (298 Kelvin) starting with 0.01 M each X, Y, and XY? Suggest two ways one could make this reaction energetically favorable.

REFERENCES

Alberty, R. A., and R. J. Silbey. 2000. *Physical Chemistry*, 3d ed. Wiley.

Atkins, P. W. 2000. *The Elements of Physical Chemistry*, 3d ed. W. H. Freeman and Company.

Berg, J. M., J. L. Tymoczko, and L. Stryer. 2002. *Biochemistry*, 5th ed. W. H. Freeman and Company.

Cantor, P. R., and C. R. Schimmel. 1980. *Biophysical Chemistry*. W. H. Freeman and Company.

Davenport, H. W. 1974. *ABC of Acid-Base Chemistry*, 6th ed. University of Chicago Press.

Edsall, J. T., and J. Wyman. 1958. *Biophysical Chemistry*, vol. 1. Academic Press.

Eisenberg, D., and D. Crothers. 1979. *Physical Chemistry with Applications to the Life Sciences*. Benjamin-Cummings.

Gennis, R. B. 1989. *Biomembranes: Molecular Structure and Function*. Springer-Verlag, New York.

Guyton, A. C., and J. E. Hall. 2000. *Textbook of Medical Physiology*, 10th ed. Saunders.

Hill, T. J. 1977. *Free Energy Transduction in Biology*. Academic Press.

Klotz, I. M. 1978. *Energy Changes in Biochemical Reactions*. Academic Press.

Lehninger, A. L., D. L. Nelson, and M. M. Cox. 2000. *Principles of Biochemistry*, 3d ed. Worth.

Murray, R. K., et al. 1999. *Harper's Biochemistry*, 25th ed. Lange.

Nicholls, D. G., and S. J. Ferguson. 1992. *Bioenergetics 2*. Academic Press.

Oxtoby, D., H. Gillis, and N. Nachtrieb. 2003. *Principles of Modern Chemistry*, 5th ed. Saunders.

Sharon, N. 1980. Carbohydrates. *Sci. Am.* **243**(5):90–116.

Tanford, C. 1980. *The Hydrophobic Effect: Formation of Micelles and Biological Membranes*, 2d ed. Wiley.

Tinoco, I., K. Sauer, and J. Wang. 2001. *Physical Chemistry—Principles and Applications in Biological Sciences*, 4th ed. Prentice Hall.

Van Holde, K., W. Johnson, and P. Ho. 1998. *Principles of Physical Biochemistry*. Prentice Hall.

Voet, D., and J. Voet. 1995. *Biochemistry*, 2d ed. Wiley.

Watson, J. D., et al. 2003. *Molecular Biology of the Gene*, 5th ed. Benjamin-Cummings.

Wood, W. B., et al. 1981. *Biochemistry: A Problems Approach*, 2d ed. Benjamin-Cummings.

3

PROTEIN STRUCTURE AND FUNCTION

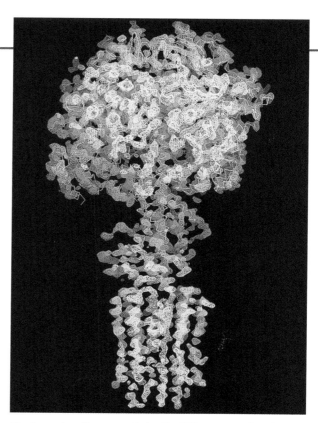

Electron density map of the F₁-ATPase associated with a ring of 10 c-subunits from the F₀ domain of ATP synthase, a molecular machine that carries out the synthesis of ATP in eubacteria, chloroplasts, and mitochondria. [Courtesy of Andrew Leslie, MRC Laboratory of Molecular Biology, Cambridge, UK.]

Proteins, the working molecules of a cell, carry out the program of activities encoded by genes. This program requires the coordinated effort of many different types of proteins, which first evolved as rudimentary molecules that facilitated a limited number of chemical reactions. Gradually, many of these primitive proteins evolved into a wide array of **enzymes** capable of catalyzing an incredible range of intracellular and extracellular chemical reactions, with a speed and specificity that is nearly impossible to attain in a test tube. With the passage of time, other proteins acquired specialized abilities and can be grouped into several broad functional classes: *structural proteins,* which provide structural rigidity to the cell; *transport proteins,* which control the flow of materials across cellular membranes; *regulatory proteins,* which act as sensors and switches to control protein activity and gene function; *signaling proteins,* including cell-surface receptors and other proteins that transmit external signals to the cell interior; and *motor proteins,* which cause motion.

A key to understanding the functional design of proteins is the realization that many have "moving" parts and are capable of transmitting various forces and energy in an orderly fashion. However, several critical and complex cell processes—synthesis of nucleic acids and proteins, signal transduction, and photosynthesis—are carried out by huge macromolecular assemblies sometimes referred to as *molecular machines.*

A fundamental goal of molecular cell biologists is to understand how cells carry out various processes essential for life. A major contribution toward achieving this goal is the identification of all of an organism's proteins—that is, a list of the parts that compose the cellular machinery. The compilation of such lists has become feasible in recent years with the sequencing of entire **genomes**—complete sets of genes—of more and more organisms. From a computer analysis of

OUTLINE

genome sequences, researchers can deduce the number and primary structure of the encoded proteins (Chapter 9). The term **proteome** was coined to refer to the entire protein complement of an organism. For example, the proteome of the yeast *Saccharomyces cerevisiae* consists of about 6000 different proteins; the human proteome is only about five times as large, comprising about 32,000 different proteins. By comparing protein sequences and structures, scientists can classify many proteins in an organism's proteome and deduce their functions by homology with proteins of known function. Although the three-dimensional structures of relatively few proteins are known, the function of a protein whose structure has not been determined can often be inferred from its interactions with other proteins, from the effects result-

ing from genetically mutating it, from the biochemistry of the complex to which it belongs, or from all three.

In this chapter, we begin our study of how the structure of a protein gives rise to its function, a theme that recurs throughout this book (Figure 3-1). The first section examines how chains of amino acid building blocks are arranged and the various higher-order folded forms that the chains assume. The next section deals with special proteins that aid in the folding of proteins, modifications that take place after the protein chain has been synthesized, and mechanisms that degrade proteins. The third section focuses on proteins as catalysts and reviews the basic properties exhibited by all enzymes. We then introduce molecular motors, which convert chemical energy into motion. The structure and function of these functional classes of proteins and others are detailed in numerous later chapters. Various mechanisms that cells use to control the activity of proteins are covered next. The chapter concludes with a section on commonly used techniques in the biologist's tool kit for isolating proteins and characterizing their properties.

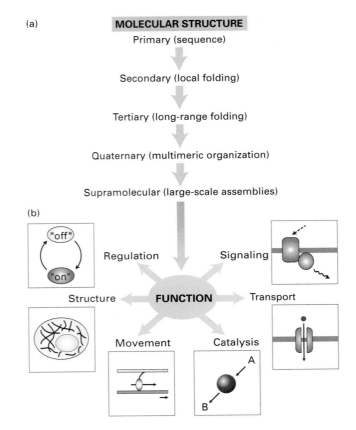

▲ **FIGURE 3-1 Overview of protein structure and function.** (a) The linear sequence of amino acids (primary structure) folds into helices or sheets (secondary structure) which pack into a globular or fibrous domain (tertiary structure). Some individual proteins self-associate into complexes (quaternary structure) that can consist of tens to hundreds of subunits (supramolecular assemblies). (b) Proteins display functions that include catalysis of chemical reactions (enzymes), flow of small molecules and ions (transport), sensing and reaction to the environment (signaling), control of protein activity (regulation), organization of the genome, lipid bilayer membrane, and cytoplasm (structure), and generation of force for movement (motor proteins). These functions and others arise from specific binding interactions and conformational changes in the structure of a properly folded protein.

3.1 Hierarchical Structure of Proteins

Although constructed by the polymerization of only 20 different **amino acids** into linear chains, proteins carry out an incredible array of diverse tasks. A protein chain folds into a unique shape that is stabilized by noncovalent interactions between regions in the linear sequence of amino acids. This *spatial* organization of a protein—its shape in three dimensions—is a key to understanding its function. Only when a protein is in its correct three-dimensional structure, or **conformation,** is it able to function efficiently. A key concept in understanding how proteins work is that *function is derived from three-dimensional structure, and three-dimensional structure is specified by amino acid sequence.* Here, we consider the structure of proteins at four levels of organization, starting with their monomeric building blocks, the amino acids.

The Primary Structure of a Protein Is Its Linear Arrangement of Amino Acids

We reviewed the properties of the amino acids used in synthesizing proteins and their linkage by **peptide bonds** into linear chains in Chapter 2. The repeated amide N, α carbon (C_α), and carbonyl C atoms of each amino acid residue form the *backbone* of a protein molecule from which the various side-chain groups project (Figure 3-2). As a consequence of the peptide linkage, the backbone exhibits directionality because all the amino groups are located on the same side of the C_α atoms. Thus one end of a protein has a free (unlinked) amino group (the *N-terminus*) and the other end has a free carboxyl group (the *C-terminus*). The sequence of a protein chain is conventionally written with its N-terminal amino acid on the left and its C-terminal amino acid on the right.

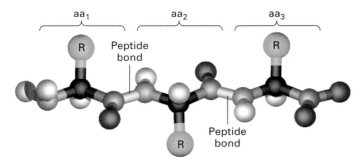

▲ **FIGURE 3-2 Structure of a tripeptide.** Peptide bonds (yellow) link the amide nitrogen atom (blue) of one amino acid (aa) with the carbonyl carbon atom (gray) of an adjacent one in the linear polymers known as peptides or polypeptides, depending on their length. Proteins are polypeptides that have folded into a defined three-dimensional structure (conformation). The side chains, or R groups (green), extending from the α carbon atoms (black) of the amino acids composing a protein largely determine its properties. At physiological pH values, the terminal amino and carboxyl groups are ionized.

The **primary structure** of a protein is simply the linear arrangement, or *sequence*, of the amino acid residues that compose it. Many terms are used to denote the chains formed by the polymerization of amino acids. A short chain of amino acids linked by peptide bonds and having a defined sequence is called a **peptide;** longer chains are referred to as **polypeptides.** Peptides generally contain fewer than 20–30 amino acid residues, whereas polypeptides contain as many as 4000 residues. We generally reserve the term **protein** for a polypeptide (or for a complex of polypeptides) that has a well-defined three-dimensional structure. It is implied that proteins and peptides are the natural products of a cell.

The size of a protein or a polypeptide is reported as its mass in **daltons** (a dalton is 1 atomic mass unit) or as its molecular weight (MW), which is a dimensionless number. For example, a 10,000-MW protein has a mass of 10,000 daltons (Da), or 10 kilodaltons (kDa). In the last section of this chapter, we will consider different methods for measuring the sizes and other physical characteristics of proteins. The known and predicted proteins encoded by the yeast genome have an average molecular weight of 52,728 and contain, on average, 466 amino acid residues. The average molecular weight of amino acids in proteins is 113, taking into account their average relative abundance. This value can be used to estimate the number of residues in a protein from its molecular weight or, conversely, its molecular weight from the number of residues.

Secondary Structures Are the Core Elements of Protein Architecture

The second level in the hierarchy of protein structure consists of the various spatial arrangements resulting from the folding of localized parts of a polypeptide chain; these arrangements are referred to as **secondary structures.** A single polypeptide may exhibit multiple types of secondary structure depending on its sequence. In the absence of stabilizing noncovalent interactions, a polypeptide assumes a *random-coil* structure. However, when stabilizing hydrogen bonds form between certain residues, parts of the backbone fold into one or more well-defined periodic structures: the **alpha (α) helix,** the **beta (β) sheet,** or a short U-shaped *turn.* In an average protein, 60 percent of the polypeptide chain exist as α helices and β sheets; the remainder of the molecule is in random coils and turns. Thus, α helices and β sheets are the major internal supportive elements in proteins. In this section, we explore forces that favor the formation of secondary structures. In later sections, we examine how these structures can pack into larger arrays.

The α Helix In a polypeptide segment folded into an α helix, the carbonyl oxygen atom of each peptide bond is hydrogen-bonded to the amide hydrogen atom of the amino acid four residues toward the C-terminus. This periodic arrangement of bonds confers a directionality on the helix because all the hydrogen-bond donors have the same orientation (Figure 3-3).

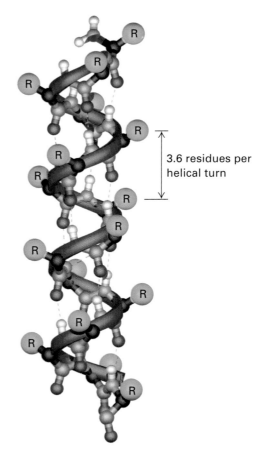

3.6 residues per helical turn

▲ **FIGURE 3-3 The α helix, a common secondary structure in proteins.** The polypeptide backbone (red) is folded into a spiral that is held in place by hydrogen bonds between backbone oxygen and hydrogen atoms. The outer surface of the helix is covered by the side-chain R groups (green).

▶ **FIGURE 3-4 The β sheet, another common secondary structure in proteins.** (a) Top view of a simple two-stranded β sheet with antiparallel β strands. The stabilizing hydrogen bonds between the β strands are indicated by green dashed lines. The short turn between the β strands also is stabilized by a hydrogen bond. (b) Side view of a β sheet. The projection of the R groups (green) above and below the plane of the sheet is obvious in this view. The fixed angle of the peptide bond produces a pleated contour.

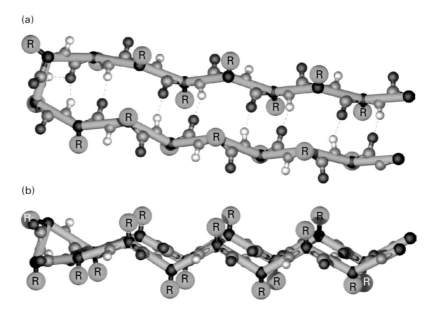

(a)

(b)

The stable arrangement of amino acids in the α helix holds the backbone in a rodlike cylinder from which the side chains point outward. The hydrophobic or hydrophilic quality of the helix is determined entirely by the side chains because the polar groups of the peptide backbone are already engaged in hydrogen bonding in the helix.

The β Sheet Another type of secondary structure, the β sheet, consists of laterally packed β strands. Each β strand is a short (5- to 8-residue), nearly fully extended polypeptide segment. Hydrogen bonding between backbone atoms in adjacent β strands, within either the same polypeptide chain or between different polypeptide chains, forms a β sheet (Figure 3-4a). The planarity of the peptide bond forces a β sheet to be pleated; hence this structure is also called a *β pleated sheet*, or simply a *pleated sheet*. Like α helices, β strands have a directionality defined by the orientation of the peptide bond. Therefore, in a pleated sheet, adjacent β strands can be oriented in the same (parallel) or opposite (antiparallel) directions with respect to each other. In both arrangements, the side chains project from both faces of the sheet (Figure 3-4b). In some proteins, β sheets form the floor of a binding pocket; the hydrophobic core of other proteins contains multiple β sheets.

Turns Composed of three or four residues, turns are located on the surface of a protein, forming sharp bends that redirect the polypeptide backbone back toward the interior. These short, U-shaped secondary structures are stabilized by a hydrogen bond between their end residues (see Figure 3-4a). Glycine and proline are commonly present in turns. The lack of a large side chain in glycine and the presence of a built-in bend in proline allow the polypeptide backbone to fold into a tight U shape. Turns allow large proteins to fold into highly compact structures. A polypeptide backbone also may contain longer bends, or *loops*. In contrast with turns, which ex-

hibit just a few well-defined structures, loops can be formed in many different ways.

Overall Folding of a Polypeptide Chain Yields Its Tertiary Structure

Tertiary structure refers to the overall conformation of a polypeptide chain—that is, the three-dimensional arrangement of all its amino acid residues. In contrast with secondary structures, which are stabilized by hydrogen bonds, tertiary structure is primarily stabilized by hydrophobic interactions between the nonpolar side chains, hydrogen bonds between polar side chains, and peptide bonds. These stabilizing forces hold elements of secondary structure—α helices, β strands, turns, and random coils—compactly together. Because the stabilizing interactions are weak, however, the tertiary structure of a protein is not rigidly fixed but undergoes continual and minute fluctuation. This variation in structure has important consequences in the function and regulation of proteins.

Different ways of depicting the conformation of proteins convey different types of information. The simplest way to represent three-dimensional structure is to trace the course of the backbone atoms with a solid line (Figure 3-5a); the most complex model shows every atom (Figure 3-5b). The former, a C_α trace, shows the overall organization of the polypeptide chain without consideration of the amino acid side chains; the latter, a ball-and-stick model, details the interactions between side-chain atoms, which stabilize the protein's conformation, as well as the atoms of the backbone. Even though both views are useful, the elements of secondary structure are not easily discerned in them. Another type of representation uses common shorthand symbols for depicting secondary structure—for example, coiled ribbons or solid cylinders for α helices, flat ribbons or arrows for β strands, and flexible

(a) C$_\alpha$ backbone trace

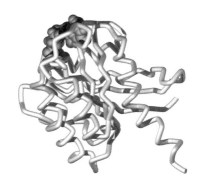

(b) Ball and stick

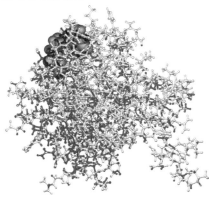

(c) Ribbons

(d) Solvent-accessible surface

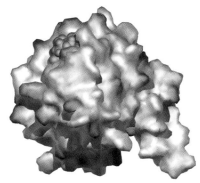

◀ **FIGURE 3-5 Various graphic representations of the structure of Ras, a monomeric guanine nucleotide-binding protein.** The inactive, guanosine diphosphate (GDP)–bound form is shown in all four panels, with GDP always depicted in blue spacefill. (a) The C$_\alpha$ backbone trace demonstrates how the polypeptide is packed into the smallest possible volume. (b) A ball-and-stick representation reveals the location of all atoms. (c) A ribbon representation emphasizes how β strands (blue) and α helices (red) are organized in the protein. Note the turns and loops connecting pairs of helices and strands. (d) A model of the water-accessible surface reveals the numerous lumps, bumps, and crevices on the protein surface. Regions of positive charge are shaded blue; regions of negative charge are shaded red.

thin strands for turns and loops (Figure 3-5c). This type of representation makes the secondary structures of a protein easy to see.

However, none of these three ways of representing protein structure convey much information about the protein surface, which is of interest because it is where other molecules bind to a protein. Computer analysis can identify the surface atoms that are in contact with the watery environment. On this water-accessible surface, regions having a common chemical character (hydrophobicity or hydrophilicity) and electrical character (basic or acidic) can be mapped. Such models reveal the topography of the protein surface and the distribution of charge, both important features of binding sites, as well as clefts in the surface where small molecules often bind (Figure 3-5d). This view represents a protein as it is "seen" by another molecule.

Motifs Are Regular Combinations of Secondary Structures

Particular combinations of secondary structures, called **motifs** or folds, build up the tertiary structure of a protein. In some cases, motifs are signatures for a specific function. For example, the **helix-loop-helix** is a Ca^{2+}-binding motif marked by the presence of certain hydrophilic residues at invariant positions in the loop (Figure 3-6a). Oxygen atoms in

the invariant residues bind a Ca^{2+} ion through ionic bonds. This motif, also called the *EF hand*, has been found in more than 100 calcium-binding proteins. In another common motif, the **zinc finger**, three secondary structures—an α helix and two β strands with an antiparallel orientation—form a fingerlike bundle held together by a zinc ion (Figure 3-6b). This motif is most commonly found in proteins that bind RNA or DNA.

Many proteins, especially fibrous proteins, self-associate into oligomers by using a third motif, the **coiled coil.** In these proteins, each polypeptide chain contains α-helical segments in which the hydrophobic residues, although apparently randomly arranged, are in a regular pattern—a repeated heptad sequence. In the heptad, a hydrophobic residue—sometimes valine, alanine, or methionine—is at position 1 and a leucine residue is at position 4. Because hydrophilic side chains extend from one side of the helix and hydrophobic side chains extend from the opposite side, the overall helical structure is **amphipathic.** The amphipathic character of these α helices permits two, three, or four helices to wind around each other, forming a coiled coil; hence the name of this motif (Figure 3-6c).

We will encounter numerous additional motifs in later discussions of other proteins in this chapter and other chapters. The presence of the same motif in different proteins with similar functions clearly indicates that these useful

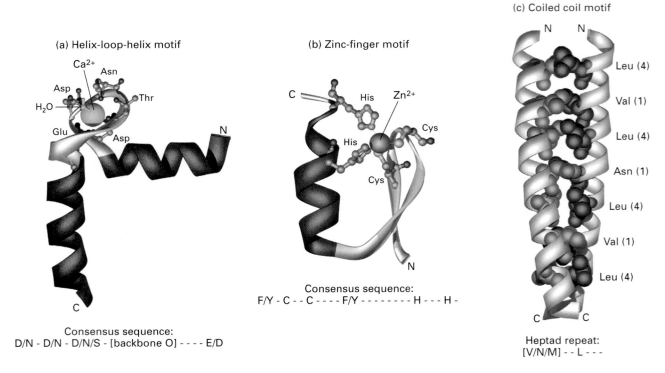

(a) Helix-loop-helix motif

Consensus sequence:
D/N - D/N - D/N/S - [backbone O] - - - - E/D

(b) Zinc-finger motif

Consensus sequence:
F/Y - C - - C - - - - F/Y - - - - - - - - H - - - H -

(c) Coiled coil motif

Heptad repeat:
[V/N/M] - - L - - -

▲ **FIGURE 3-6 Motifs of protein secondary structure.**
(a) Two helices connected by a short loop in a specific conformation constitute a helix-loop-helix motif. This motif exists in many calcium-binding and DNA-binding regulatory proteins. In calcium-binding proteins such as calmodulin, oxygen atoms from five loop residues and one water molecule form ionic bonds with a Ca^{2+} ion. (b) The zinc-finger motif is present in many DNA-binding proteins that help regulate transcription. A Zn^{2+} ion is held between a pair of β strands (blue) and a single α helix (red) by a pair of cysteine residues and a pair of histidine residues. The two invariant cysteine residues are usually at positions 3 and 6 and the two invariant histidine residues are at positions 20 and 24 in this 25-residue motif. (c) The parallel two-stranded coiled-coil motif found in the transcription factor Gcn4 is characterized by two α helices wound around one another. Helix packing is stabilized by interactions between hydrophobic side chains (red and blue) present at regular intervals along the surfaces of the intertwined helices. Each α helix exhibits a characteristic heptad repeat sequence with a hydrophobic residue at positions 1 and 4. [See A. Lewit-Bentley and S. Rety, 2000, EF-hand calcium-binding proteins, *Curr. Opin. Struct. Biol.* **10**:637–643; S. A. Wolfe, L. Nekludova, and C. O. Pabo, 2000, DNA recognition by Cys2His2 zinc finger proteins, *Ann. Rev. Biophys. Biomol. Struct.* **29**:183–212.]

combinations of secondary structures have been conserved in evolution. To date, hundreds of motifs have been cataloged and proteins are now classified according to their motifs.

Structural and Functional Domains Are Modules of Tertiary Structure

The tertiary structure of proteins larger than 15,000 MW is typically subdivided into distinct regions called **domains.** Structurally, a domain is a compactly folded region of polypeptide. For large proteins, domains can be recognized in structures determined by x-ray crystallography or in images captured by electron microscopy. Although these discrete regions are well distinguished or physically separated from one another, they are connected by intervening segments of the polypeptide chain. Each of the subunits in hemagglutinin, for example, contains a globular domain and a fibrous domain (Figure 3-7a).

A structural domain consists of 100–150 residues in various combinations of motifs. Often a domain is characterized by some interesting structural feature: an unusual abundance of a particular amino acid (e.g., a proline-rich domain, an acidic domain), sequences common to (conserved in) many proteins (e.g., SH3, or Src homology region 3), or a particular secondary-structure motif (e.g., zinc-finger motif in the kringle domain).

Domains are sometimes defined in functional terms on the basis of observations that an activity of a protein is localized to a small region along its length. For instance, a particular region or regions of a protein may be responsible for its catalytic activity (e.g., a kinase domain) or binding ability (e.g., a DNA-binding domain, a membrane-binding domain). Functional domains are often identified experimentally by whittling down a protein to its smallest active fragment with the aid of proteases, enzymes that cleave the polypeptide backbone. Alternatively, the DNA encoding a protein can be

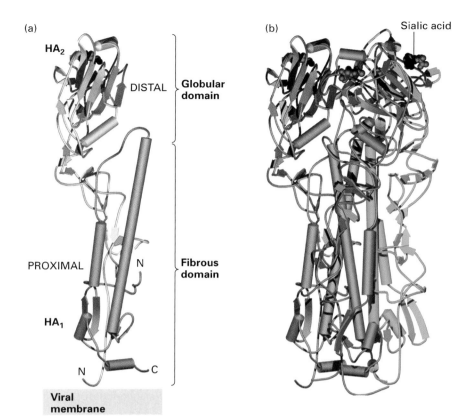

(a)

HA$_2$

DISTAL — Globular domain

PROXIMAL   N — Fibrous domain

HA$_1$

N   C

Viral membrane

(b)

Sialic acid

◄ **FIGURE 3-7 Tertiary and quaternary levels of structure in hemagglutinin (HA), a surface protein on influenza virus.** This long multimeric molecule has three identical subunits, each composed of two polypeptide chains, HA$_1$ and HA$_2$. (a) Tertiary structure of each HA subunit constitutes the folding of its helices and strands into a compact structure that is 13.5 nm long and divided into two domains. The membrane-distal domain is folded into a globular conformation. The membrane-proximal domain has a fibrous, stemlike conformation owing to the alignment of two long α helices (cylinders) of HA$_2$ with β strands in HA$_1$. Short turns and longer loops, which usually lie at the surface of the molecule, connect the helices and strands in a given chain. (b) Quaternary structure of HA is stabilized by lateral interactions between the long helices (cylinders) in the fibrous domains of the three subunits (yellow, blue, and green), forming a triple-stranded coiled-coil stalk. Each of the distal globular domains in HA binds sialic acid (red) on the surface of target cells. Like many membrane proteins, HA contains several covalently linked carbohydrate chains (not shown).

---

subjected to mutagenesis so that segments of the protein's backbone are removed or changed. The activity of the truncated or altered protein product synthesized from the mutated gene is then monitored and serves as a source of insight about which part of a protein is critical to its function.

The organization of large proteins into multiple domains illustrates the principle that complex molecules are built from simpler components. Like motifs of secondary structure, domains of tertiary structure are incorporated as modules into different proteins. In Chapter 10 we consider the mechanism by which the gene segments that correspond to domains became shuffled in the course of evolution, resulting in their appearance in many proteins. The modular approach to protein architecture is particularly easy to recognize in large proteins, which tend to be mosaics of different domains and thus can perform different functions simultaneously.

The epidermal growth factor (EGF) domain is one example of a module that is present in several proteins (Figure 3-8). EGF is a small, soluble peptide hormone that binds to cells in the embryo and in skin and connective tissue in adults, causing them to divide. It is generated by proteolytic cleavage between repeated EGF domains in the EGF precursor protein, which is anchored in the cell membrane by a membrane-spanning domain. EGF modules are also present in other proteins and are liberated by proteolysis; these proteins include tissue plasminogen activator (TPA), a protease that is used to dissolve blood clots in heart attack victims;

Neu protein, which takes part in embryonic differentiation; and Notch protein, a receptor protein in the plasma membrane that functions in developmentally important signaling (Chapter 14). Besides the EGF domain, these proteins contain domains found in other proteins. For example, TPA possesses a trypsin domain, a common feature in enzymes that degrade proteins.

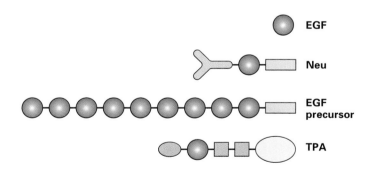

EGF

Neu

EGF precursor

TPA

▲ **FIGURE 3-8 Schematic diagrams of various proteins illustrating their modular nature.** Epidermal growth factor (EGF) is generated by proteolytic cleavage of a precursor protein containing multiple EGF domains (green) and a membrane-spanning domain (blue). The EGF domain is also present in Neu protein and in tissue plasminogen activator (TPA). These proteins also contain other widely distributed domains indicated by shape and color. [Adapted from I. D. Campbell and P. Bork, 1993, *Curr. Opin. Struct. Biol.* **3**:385.]

Proteins Associate into Multimeric Structures and Macromolecular Assemblies

Multimeric proteins consist of two or more polypeptides or subunits. A fourth level of structural organization, **quaternary structure,** describes the number (stoichiometry) and relative positions of the subunits in multimeric proteins. Hemagglutinin, for example, is a trimer of three identical subunits held together by noncovalent bonds (Figure 3-7b). Other multimeric proteins can be composed of any number of identical or different subunits. The multimeric nature of many proteins is critical to mechanisms for regulating their function. In addition, enzymes in the same pathway may be associated as subunits of a large multimeric protein within the cell, thereby increasing the efficiency of pathway operation.

The highest level of protein structure is the association of proteins into macromolecular assemblies. Typically, such structures are very large, exceeding 1 mDa in mass, approaching 30–300 nm in size, and containing tens to hundreds of polypeptide chains, as well as nucleic acids in some cases. Macromolecular assemblies with a structural function include the **capsid** that encases the viral genome and bundles of cytoskeletal filaments that support and give shape to the plasma membrane. Other macromolecular assemblies act as molecular machines, carrying out the most complex cellular processes by integrating individual functions into one coordinated process. For example, the transcriptional machine that initiates the synthesis of messenger RNA (mRNA) consists of RNA polymerase, itself a multimeric protein, and at least 50 additional components including general transcription factors, promoter-binding proteins, helicase, and other protein complexes (Figure 3-9). The transcription factors and promoter-binding proteins correctly position a polymerase molecule at a **promoter,** the DNA site that determines where transcription of a specific gene begins. After helicase unwinds the double-stranded DNA molecule, polymerase simultaneously translocates along the DNA template strand and synthesizes an mRNA strand. The operational details of this complex machine and of others listed in Table 3-1 are discussed elsewhere.

TABLE 3-1	Selected Molecular Machines		
Machine[*]	**Main Components**	**Cellular Location**	**Function**
Replisome (4)	Helicase, primase, DNA polymerase	Nucleus	DNA replication
Transcription initiation complex (11)	Promoter-binding protein, helicase, general transcription factors (TFs), RNA polymerase, large multisubunit mediator complex	Nucleus	RNA synthesis
Spliceosome (12)	Pre-mRNA, small nuclear RNAs (snRNAs), protein factors	Nucleus	mRNA splicing
Nuclear pore complex (12)	Nucleoporins (50–100)	Nuclear membrane	Nuclear import and export
Ribosome (4)	Ribosomal proteins (>50) and four rRNA molecules (eukaryotes) organized into large and small subunits; associated mRNA and protein factors (IFs, EFs)	Cytoplasm/ER membrane	Protein synthesis
Chaperonin (3)	GroEL, GroES (bacteria)	Cytoplasm, mitochondria, endoplasmic reticulum	Protein folding
Proteasome (3)	Core proteins, regulatory (cap) proteins	Cytoplasm	Protein degradation
Photosystem (8)	Light-harvesting complex (multiple proteins and pigments), reaction center (multisubunit protein with associated pigments and electron carriers)	Thylakoid membrane in plant chloroplasts, plasma membrane of photosynthetic bacteria	Photosynthesis (initial stage)
MAP kinase cascades (14)	Scaffold protein, multiple different protein kinases	Cytoplasm	Signal transduction
Sarcomere (19)	Thick (myosin) filaments, thin (actin) filaments, Z lines, titin/nebulin	Cytoplasm of muscle cells	Contraction

[*]Numbers in parentheses indicate chapters in which various machines are discussed.

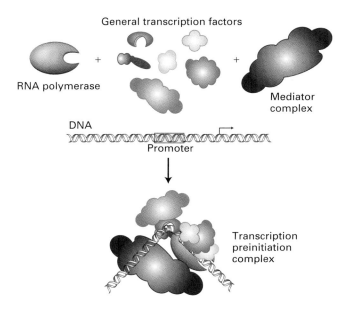

▲ **FIGURE 3-9 The mRNA transcription-initiation machinery.** The core RNA polymerase, general transcription factors, a mediator complex containing about 20 subunits, and other protein complexes not depicted here assemble at a promoter in DNA. The polymerase carries out transcription of DNA; the associated proteins are required for initial binding of polymerase to a specific promoter, thereby initiating transcription.

Members of Protein Families Have a Common Evolutionary Ancestor

Studies on myoglobin and hemoglobin, the oxygen-carrying proteins in muscle and blood, respectively, provided early evidence that function derives from three-dimensional structure, which in turn is specified by amino acid sequence. X-ray crystallographic analysis showed that the three-dimensional structures of myoglobin and the α and β subunits of hemoglobin are remarkably similar. Subsequent sequencing of myoglobin and the hemoglobin subunits revealed that many identical or chemically similar residues are found in identical positions throughout the primary structures of both proteins.

Similar comparisons between other proteins conclusively confirmed the relation between the amino acid sequence, three-dimensional structure, and function of proteins. This principle is now commonly employed to predict, on the basis of sequence comparisons with proteins of known structure and function, the structure and function of proteins that have not been isolated (Chapter 9). This use of sequence comparisons has expanded substantially in recent years as the genomes of more and more organisms have been sequenced.

The molecular revolution in biology during the last decades of the twentieth century also created a new scheme

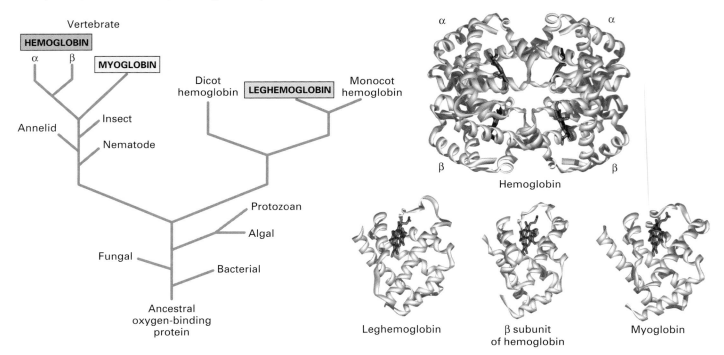

▲ **FIGURE 3-10 Evolution of the globin protein family.** (*Left*) A primitive monomeric oxygen-binding globin is thought to be the ancestor of modern-day blood hemoglobins, muscle myoglobins, and plant leghemoglobins. Sequence comparisons have revealed that evolution of the globin proteins parallels the evolution of animals and plants. Major junctions occurred with the divergence of plant globins from animal globins and of myoglobin from hemoglobin. Later gene duplication gave rise to the α and β subunits of hemoglobin. (*Right*) Hemoglobin is a tetramer of two α and two β subunits. The structural similarity of these subunits with leghemoglobin and myoglobin, both of which are monomers, is evident. A heme molecule (red) noncovalently associated with each globin polypeptide is the actual oxygen-binding moiety in these proteins. [(*Left*) Adapted from R. C. Hardison, 1996, *Proc. Natl. Acad. Sci. USA* **93**:5675.]

of biological classification based on similarities and differences in the amino acid sequences of proteins. Proteins that have a common ancestor are referred to as *homologs*. The main evidence for **homology** among proteins, and hence their common ancestry, is similarity in their sequences or structures. We can therefore describe homologous proteins as belonging to a "family" and can trace their lineage from comparisons of their sequences. The folded three-dimensional structures of homologous proteins are similar even if parts of their primary structure show little evidence of homology.

The kinship among homologous proteins is most easily visualized by a tree diagram based on sequence analyses. For example, the amino acid sequences of globins from bacteria, plants, and animals suggest that they evolved from an ancestral monomeric, oxygen-binding protein (Figure 3-10). With the passage of time, the gene for this ancestral protein slowly changed, initially diverging into lineages leading to animal and plant globins. Subsequent changes gave rise to myoglobin, a monomeric oxygen-storing protein in muscle, and to the α and β subunits of the tetrameric hemoglobin molecule ($\alpha_2\beta_2$) of the circulatory system.

KEY CONCEPTS OF SECTION 3.1

Hierarchical Structure of Proteins

■ A protein is a linear polymer of amino acids linked together by peptide bonds. Various, mostly noncovalent, interactions between amino acids in the linear sequence stabilize a specific folded three-dimensional structure (conformation) for each protein.

■ The α helix, β strand and sheet, and turn are the most prevalent elements of protein secondary structure, which is stabilized by hydrogen bonds between atoms of the peptide backbone.

■ Certain combinations of secondary structures give rise to different motifs, which are found in a variety of proteins and are often associated with specific functions (see Figure 3-6).

■ Protein tertiary structure results from hydrophobic interactions between nonpolar side groups and hydrogen bonds between polar side groups that stabilize folding of the secondary structure into a compact overall arrangement, or conformation.

■ Large proteins often contain distinct domains, independently folded regions of tertiary structure with characteristic structural or functional properties or both (see Figure 3-7).

■ The incorporation of domains as modules in different proteins in the course of evolution has generated diversity in protein structure and function.

■ Quaternary structure encompasses the number and organization of subunits in multimeric proteins.

■ Cells contain large macromolecular assemblies in which all the necessary participants in complex cellular processes (e.g., DNA, RNA, and protein synthesis; photosynthesis; signal transduction) are integrated to form molecular machines (see Table 3-1).

■ The sequence of a protein determines its three-dimensional structure, which determines its function. In short, function derives from structure; structure derives from sequence.

■ Homologous proteins, which have similar sequences, structures, and functions, evolved from a common ancestor.

3.2 Folding, Modification, and Degradation of Proteins

A polypeptide chain is synthesized by a complex process called **translation** in which the assembly of amino acids in a particular sequence is dictated by **messenger RNA (mRNA)**. The intricacies of translation are considered in Chapter 4. Here, we describe how the cell promotes the proper folding of a nascent polypeptide chain and, in many cases, modifies residues or cleaves the polypeptide backbone to generate the final protein. In addition, the cell has error-checking processes that eliminate incorrectly synthesized or folded proteins. Incorrectly folded proteins usually lack biological activity and, in some cases, may actually be associated with disease. Protein misfolding is suppressed by two distinct mechanisms. First, cells have systems that reduce the chances for misfolded proteins to form. Second, any misfolded proteins that do form, as well as cytosolic proteins no longer needed by a cell, are degraded by a specialized cellular garbage-disposal system.

The Information for Protein Folding Is Encoded in the Sequence

Any polypeptide chain containing *n* residues could, in principle, fold into 8^n conformations. This value is based on the fact that only eight bond angles are stereochemically allowed in the polypeptide backbone. In general, however, all molecules of any protein species adopt a single conformation, called the *native state;* for the vast majority of proteins, the native state is the most stably folded form of the molecule.

What guides proteins to their native folded state? The answer to this question initially came from in vitro studies on protein refolding. Thermal energy from heat, extremes of pH that alter the charges on amino acid side chains, and chemicals such as urea or guanidine hydrochloride at concentrations of 6–8 M can disrupt the weak noncovalent interactions that stabilize the native conformation of a protein. The **denaturation** resulting from such treatment causes a protein to lose both its native conformation and its biological activity.

Many proteins that are completely unfolded in 8 M urea and β-mercaptoethanol (which reduces disulfide bonds) spontaneously *renature* (refold) into their native states when the denaturing reagents are removed by dialysis. Because no cofactors or other proteins are required, in vitro protein folding is a self-directed process. In other words, sufficient information must be contained in the protein's primary sequence to direct correct refolding. The observed similarity in the folded, three-dimensional structures of proteins with similar amino acid sequences, noted in Section 3.1, provided other evidence that the primary sequence also determines protein folding in vivo.

Folding of Proteins in Vivo Is Promoted by Chaperones

Although protein folding occurs in vitro, only a minority of unfolded molecules undergo complete folding into the native conformation within a few minutes. Clearly, cells require a faster, more efficient mechanism for folding proteins into their correct shapes; otherwise, cells would waste much energy in the synthesis of nonfunctional proteins and in the degradation of misfolded or unfolded proteins. Indeed, more than 95 percent of the proteins present within cells have been shown to be in their native conformation, despite high protein concentrations (200–300 mg/ml), which favor the precipitation of proteins in vitro.

The explanation for the cell's remarkable efficiency in promoting protein folding probably lies in **chaperones,** a class of proteins found in all organisms from bacteria to humans. Chaperones are located in every cellular compartment, bind a wide range of proteins, and function in the general protein-folding mechanism of cells. Two general families of chaperones are reconized:

- *Molecular chaperones,* which bind and stabilize unfolded or partly folded proteins, thereby preventing these proteins from aggregating and being degraded

- *Chaperonins,* which directly facilitate the folding of proteins

Molecular chaperones consist of Hsp70 and its homologs: Hsp70 in the cytosol and mitochondrial matrix, BiP in the endoplasmic reticulum, and DnaK in bacteria. First identified by their rapid appearance after a cell has been stressed by heat shock, Hsp70 and its homologs are the major chaperones in all organisms. (Hsc70 is a constitutively expressed homolog of Hsp70.) When bound to ATP, Hsp70-like proteins assume an open form in which an exposed hydrophobic pocket transiently binds to exposed hydrophobic regions of the unfolded target protein. Hydrolysis of the bound ATP causes molecular chaperones to assume a closed form in which a target protein can undergo folding. The exchange of ATP for ADP releases the target protein (Figure 3-11a, *top*). This cycle is

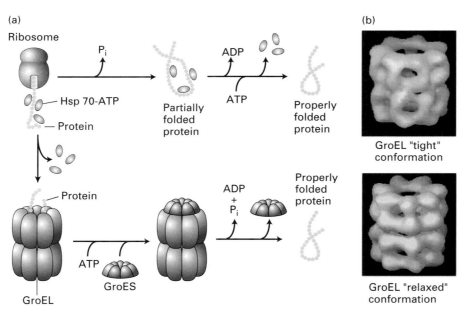

(a)

(b)

GroEL "tight" conformation

GroEL "relaxed" conformation

▲ **FIGURE 3-11 Chaperone- and chaperonin-mediated protein folding.** (a) Many proteins fold into their proper three-dimensional structures with the assistance of Hsp70-like proteins (*top*). These molecular chaperones transiently bind to a nascent polypeptide as it emerges from a ribosome. Proper folding of other proteins (*bottom*) depends on chaperonins such as the prokaryotic GroEL, a hollow, barrel-shaped complex of 14 identical 60,000-MW subunits arranged in two stacked rings.

One end of GroEL is transiently blocked by the co-chaperonin GroES, an assembly of 10,000-MW subunits. (b) In the absence of ATP or presence of ADP, GroEL exists in a "tight" conformational state that binds partly folded or misfolded proteins. Binding of ATP shifts GroEL to a more open, "relaxed" state, which releases the folded protein. See text for details. [Part (b) from A. Roseman et al., 1996, *Cell* **87**:241; courtesy of H. Saibil.]

speeded by the co-chaperone Hsp40 in eukaryotes. In bacteria, an additional protein called GrpE also interacts with DnaK, promoting the exchange of ATP for the bacterial co-chaperone DnaJ and possibly its dissociation. Molecular chaperones are thought to bind all nascent polypeptide chains as they are being synthesized on ribosomes. In bacteria, 85 percent of the proteins are released from their chaperones and proceed to fold normally; an even higher percentage of proteins in eukaryotes follow this pathway.

The proper folding of a large variety of newly synthesized or translocated proteins also requires the assistance of chaperonins. These huge cylindrical macromolecular assemblies are formed from two rings of oligomers. The eukaryotic chaperonin *TriC* consists of eight subunits per ring. In the bacterial, mitochondrial, and chloroplast chaperonin, known as *GroEL,* each ring contains seven identical subunits (Figure 3-11b). The GroEL folding mechanism, which is better understood than TriC-mediated folding, serves as a general model (Figure 3-11a, *bottom*). In bacteria, a partly folded or misfolded polypeptide is inserted into the cavity of GroEL, where it binds to the inner wall and folds into its native conformation. In an ATP-dependent step, GroEL undergoes a conformational change and releases the folded protein, a process assisted by a co-chaperonin, GroES, which caps the ends of GroEL.

Many Proteins Undergo Chemical Modification of Amino Acid Residues

Nearly every protein in a cell is chemically modified after its synthesis on a ribosome. Such modifications, which may alter the activity, life span, or cellular location of proteins, entail the linkage of a chemical group to the free –NH$_2$ or –COOH group at either end of a protein or to a reactive side-chain group in an internal residue. Although cells use the 20 amino acids shown in Figure 2-13 to synthesize proteins, analysis of cellular proteins reveals that they contain upward of 100 different amino acids. Chemical modifications after synthesis account for this difference.

Acetylation, the addition of an acetyl group (CH$_3$CO) to the amino group of the N-terminal residue, is the most common form of chemical modification, affecting an estimated 80 percent of all proteins:

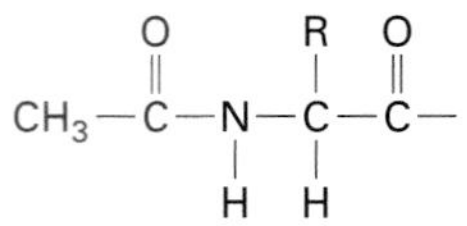

Acetylated N-terminus

This modification may play an important role in controlling the life span of proteins within cells because nonacetylated proteins are rapidly degraded by intracellular proteases. Residues at or near the termini of some membrane proteins are chemically modified by the addition of long lipidlike groups. The attachment of these hydrophobic "tails," which function to anchor proteins to the lipid bilayer, constitutes one way that cells localize certain proteins to membranes (Chapter 5).

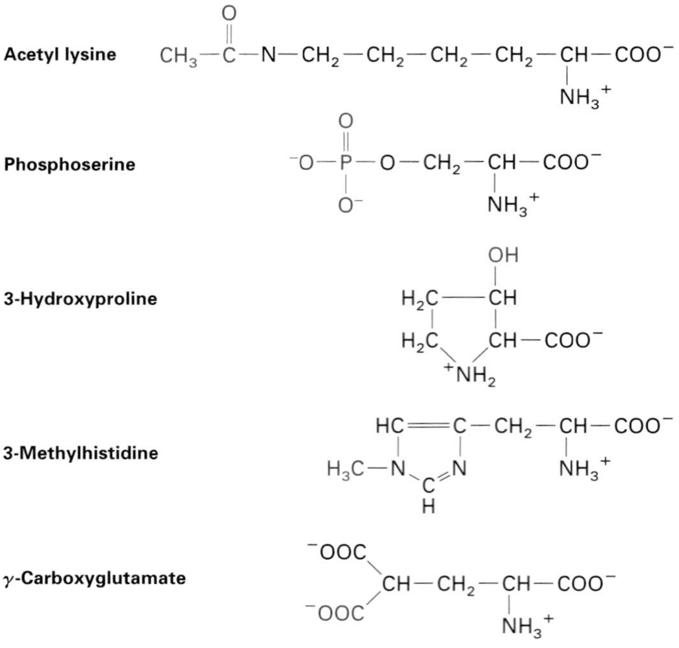

▲ **FIGURE 3-12 Common modifications of internal amino acid residues found in proteins.** These modified residues and numerous others are formed by addition of various chemical groups (red) to the amino acid side chains after synthesis of a polypeptide chain.

Acetyl groups and a variety of other chemical groups can also be added to specific internal residues in proteins (Figure 3-12). An important modification is the *phosphorylation* of serine, threonine, tyrosine, and histidine residues. We will encounter numerous examples of proteins whose activity is regulated by reversible phosphorylation and dephosphorylation. The side chains of asparagine, serine, and threonine are sites for *glycosylation,* the attachment of linear and branched carbohydrate chains. Many secreted proteins and membrane proteins contain glycosylated residues; the synthesis of such proteins is described in Chapters 16 and 17. Other post-translational modifications found in selected proteins include the *hydroxylation* of proline and lysine residues in collagen, the *methylation* of histidine residues in membrane receptors, and the *γ-carboxylation* of glutamate in prothrombin, an essential blood-clotting factor. A special modification, discussed shortly, marks cytosolic proteins for degradation.

Peptide Segments of Some Proteins Are Removed After Synthesis

After their synthesis, some proteins undergo irreversible changes that do not entail changes in individual amino acid residues. This type of post-translational alteration is sometimes called *processing*. The most common form is enzymatic cleavage of a backbone peptide bond by proteases, resulting in the removal of residues from the C- or N-terminus of a polypeptide chain. Proteolytic cleavage is a common mechanism for activating enzymes that function in blood coagulation, digestion, and programmed cell death (Chapter 22). Proteolysis also generates active peptide hormones, such as EGF and insulin, from larger precursor polypeptides.

An unusual and rare type of processing, termed *protein self-splicing*, takes place in bacteria and some eukaryotes. This process is analogous to editing film: an internal segment of a polypeptide is removed and the ends of the polypeptide are rejoined. Unlike proteolytic processing, protein self-splicing is an autocatalytic process, which proceeds by itself without the participation of enzymes. The excised peptide appears to eliminate itself from the protein by a mechanism similar to that used in the processing of some RNA molecules (Chapter 12). In vertebrate cells, the processing of some proteins includes self-cleavage, but the subsequent ligation step is absent. One such protein is Hedgehog, a membrane-bound signaling molecule that is critical to a number of developmental processes (Chapter 15).

Ubiquitin Marks Cytosolic Proteins for Degradation in Proteasomes

In addition to chemical modifications and processing, the activity of a cellular protein depends on the amount present, which reflects the balance between its rate of synthesis and rate of degradation in the cell. The numerous ways that cells regulate protein synthesis are discussed in later chapters. In this section, we examine protein degradation, focusing on the major pathways for degrading cytosolic proteins.

The life span of intracellular proteins varies from as short as a few minutes for mitotic cyclins, which help regulate passage through mitosis, to as long as the age of an organism for proteins in the lens of the eye. Eukaryotic cells have several intracellular proteolytic pathways for degrading misfolded or denatured proteins, normal proteins whose concentration must be decreased, and extracellular proteins taken up by the cell. One major intracellular pathway is degradation by enzymes within **lysosomes,** membrane-limited organelles whose acidic interior is filled with hydrolytic enzymes. Lysosomal degradation is directed primarily toward extracellular proteins taken up by the cell and aged or defective organelles of the cell (see Figure 5-20).

Distinct from the lysosomal pathway are cytosolic mechanisms for degrading proteins. Chief among these mechanisms is a pathway that includes the chemical modification of a lysine side chain by the addition of **ubiquitin,** a 76-residue polypeptide, followed by degradation of the ubiquitin-tagged protein by a specialized proteolytic machine. Ubiquitination is a three-step process (Figure 3-13a):

■ Activation of *ubiquitin-activating enzyme* (E1) by the addition of a ubitiquin molecule, a reaction that requires ATP

■ Transfer of this ubiquitin molecule to a cysteine residue in *ubiquitin-conjugating enzyme* (E2)

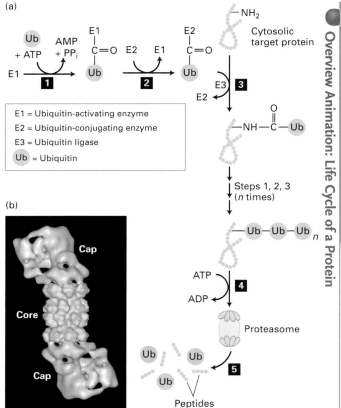

▲ **FIGURE 3-13 Ubiquitin-mediated proteolytic pathway.** (a) Enzyme E1 is activated by attachment of a ubiquitin (Ub) molecule (step **1**) and then transfers this Ub molecule to E2 (step **2**). Ubiquitin ligase (E3) transfers the bound Ub molecule on E2 to the side-chain —NH₂ of a lysine residue in a target protein (step **3**). Additional Ub molecules are added to the target protein by repeating steps **1**–**3**, forming a polyubiquitin chain that directs the tagged protein to a proteasome (step **4**). Within this large complex, the protein is cleaved into numerous small peptide fragments (step **5**). (b) Computer-generated image reveals that a proteasome has a cylindrical structure with a cap at each end of a core region. Proteolysis of ubiquitin-tagged proteins occurs along the inner wall of the core. [Part (b) from W. Baumeister et al., 1998, *Cell* **92**:357; courtesy of W. Baumeister.]

■ Formation of a peptide bond between the ubiquitin molecule bound to E2 and a lysine residue in the target protein, a reaction catalyzed by *ubiquitin ligase* (E3)

This process is repeated many times, with each subsequent ubiquitin molecule being added to the preceding one. The resulting polyubiquitin chain is recognized by a **proteasome,** another of the cell's molecular machines (Figure 3-13b). The numerous proteasomes dispersed throughout the cell cytosol proteolytically cleave ubiquitin-tagged proteins in an ATP-dependent process that yields short (7- to 8-residue) peptides and intact ubiquitin molecules.

Cellular proteins degraded by the ubiquitin-mediated pathway fall into one of two general categories: (1) native cytosolic proteins whose life spans are tightly controlled and (2) proteins that become misfolded in the course of their synthesis in the endoplasmic reticulum (ER). Both contain sequences recognized by the ubiquitinating enzyme complex. The cyclins, for example, are cytosolic proteins whose amounts are tightly controlled throughout the cell cycle. These proteins contain the internal sequence Arg-X-X-Leu-Gly-X-Ile-Gly-Asp/Asn (X can be any amino acid), which is recognized by specific ubiquitinating enzyme complexes. At a specific time in the cell cycle, each cyclin is phosphorylated by a cyclin kinase. This phosphorylation is thought to cause a conformational change that exposes the recognition sequence to the ubiquitinating enzymes, leading to degradation of the tagged cyclin (Chapter 21). Similarly, the misfolding of proteins in the endoplasmic reticulum exposes hydrophobic sequences normally buried within the folded protein. Such proteins are transported to the cytosol, where ubiquitinating enzymes recognize the exposed hydrophobic sequences.

The immune system also makes use of the ubiquitin-mediated pathway in the response to altered self-cells, particularly virus-infected cells. Viral proteins within the cytosol of infected cells are ubiquitinated and then degraded in proteasomes specially designed for this role. The resulting antigenic peptides are transported to the endoplasmic reticulum, where they bind to class I major histocompatibility complex (MHC) molecules within the ER membrane. Subsequently, the peptide-MHC complexes move to the cell membrane where the antigenic peptides can be recognized by cytotoxic T lymphocytes, which mediate the destruction of the infected cells.

Digestive Proteases Degrade Dietary Proteins

The major extracellular pathway for protein degradation is the system of digestive proteases that breaks down ingested proteins into peptides and amino acids in the intestinal tract. Three classes of proteases function in digestion. *Endoproteases* attack selected peptide bonds within a polypeptide chain. The principal endoproteases are pepsin, which preferentially cleaves the backbone adjacent to phenylalanine and leucine residues, and trypsin and chymotrypsin, which cleave the backbone adjacent to basic and aromatic residues. *Exopeptidases* sequentially remove residues from the N-terminus (aminopeptidases) or C-terminus (carboxypeptidases) of a protein. *Peptidases* split oligopeptides containing as many as about 20 amino acids into di- and tripeptides and individual amino acids. These small molecules are then transported across the intestinal lining into the bloodstream.

To protect a cell from degrading itself, endoproteases and carboxypeptidases are synthesized and secreted as inactive forms (zymogens): pepsin by chief cells in the lining of the stomach; the others by pancreatic cells. Proteolytic cleavage of the zymogens within the gastic or intestinal lumen yields the active enzymes. Intestinal epithelial cells produce aminopeptidases and the di- and tripeptidases.

Alternatively Folded Proteins Are Implicated in Slowly Developing Diseases

 As noted earlier, each protein species normally folds into a single, energetically favorable conformation that is specified by its amino acid sequence. Recent evidence suggests, however, that a protein may fold into an alternative three-dimensional structure as the result of mutations, inappropriate post-translational modification, or other as-yet-unidentified reasons. Such "misfolding" not only leads to a loss of the normal function of the protein but also marks it for proteolytic degradation. The subsequent accumulation of proteolytic fragments contributes to certain degenerative diseases characterized by the presence of insoluble protein plaques in various organs, including the liver and brain. ∎

Some neurodegenerative diseases, including Alzheimer's disease and Parkinson's disease in humans and transmissible spongiform encephalopathy ("mad cow" disease) in cows

(a)

(b)

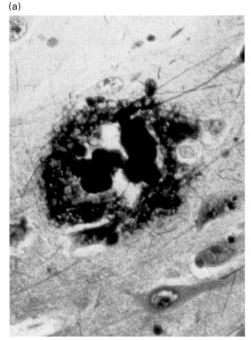

20 μm   100 nm

▲ **EXPERIMENTAL FIGURE 3-14 Alzheimer's disease is characterized by the formation of insoluble plaques composed of amyloid protein.** (a) At low resolution, an amyloid plaque in the brain of an Alzheimer's patient appears as a tangle of filaments. (b) The regular structure of filaments from plaques is revealed in the atomic force microscope. Proteolysis of the naturally occurring amyloid precursor protein yields a short fragment, called β-amyloid protein, that for unknown reasons changes from an α-helical to a β-sheet conformation. This alternative structure aggregates into the highly stable filaments (amyloid) found in plaques. Similar pathologic changes in other proteins cause other degenerative diseases. [Courtesy of K. Kosik.]

and sheep, are marked by the formation of tangled filamentous plaques in a deteriorating brain (Figure 3-14). The *amyloid filaments* composing these structures derive from abundant natural proteins such as amyloid precursor protein, which is embedded in the plasma membrane, Tau, a microtubule-binding protein, and prion protein, an "infectious" protein whose inheritance follows Mendelian genetics. Influenced by unknown causes, these α helix–containing proteins or their proteolytic fragments fold into alternative β sheet–containing structures that polymerize into very stable filaments. Whether the extracellular deposits of these filaments or the soluble alternatively folded proteins are toxic to the cell is unclear.

KEY CONCEPTS OF SECTION 3.2

Folding, Modification, and Degradation of Proteins

■ The amino acid sequence of a protein dictates its folding into a specific three-dimensional conformation, the native state.

■ Protein folding in vivo occurs with assistance from molecular chaperones (Hsp70 proteins), which bind to nascent polypeptides emerging from ribosomes and prevent their misfolding (see Figure 3-11). Chaperonins, large complexes of Hsp60-like proteins, shelter some partly folded or misfolded proteins in a barrel-like cavity, providing additional time for proper folding.

■ Subsequent to their synthesis, most proteins are modified by the addition of various chemical groups to amino acid residues. These modifications, which alter protein structure and function, include acetylation, hydroxylation, glycosylation, and phosphorylation.

■ The life span of intracellular proteins is largely determined by their susceptibility to proteolytic degradation by various pathways.

■ Viral proteins produced within infected cells, normal cytosolic proteins, and misfolded proteins are marked for destruction by the covalent addition of a polyubiquitin chain and then degraded within proteasomes, large cylindrical complexes with multiple proteases in their interiors (see Figure 3-13).

■ Some neurodegenerative diseases are caused by aggregates of proteins that are stably folded in an alternative conformation.

3.3 Enzymes and the Chemical Work of Cells

Proteins are designed to bind every conceivable molecule—from simple ions and small metabolites (sugars, fatty acids) to large complex molecules such as other proteins and nucleic acids. Indeed, the function of nearly all proteins depends on their ability to bind other molecules, or **ligands**, with a high degree of specificity. For instance, an enzyme must first bind specifically to its target molecule, which may be a small molecule (e.g., glucose) or a macromolecule, before it can execute its specific task. Likewise, the many different types of hormone receptors on the surface of cells display a high degree of sensitivity and discrimination for their ligands. And, as we will examine in Chapter 11, the binding of certain regulatory proteins to specific sequences in DNA is a major mechanism for controlling genes. Ligand binding often causes a change in the shape of a protein. Ligand-driven conformational changes are integral to the mechanism of action of many proteins and are important in regulating protein activity. After considering the general properties of protein–ligand binding, we take a closer look at how enzymes are designed to function as the cell's chemists.

Specificity and Affinity of Protein–Ligand Binding Depend on Molecular Complementarity

Two properties of a protein characterize its interaction with ligands. *Specificity* refers to the ability of a protein to bind one molecule in preference to other molecules. *Affinity* refers to the strength of binding. The K_d for a protein–ligand complex, which is the inverse of the equilibrium constant K_{eq} for the binding reaction, is the most common quantitative measure of affinity (Chapter 2). The stronger the interaction between a protein and ligand, the lower the value of K_d. Both the specificity and the affinity of a protein for a ligand depend on the structure of the ligand-binding site, which is designed to fit its partner like a mold. For high-affinity and highly specific interactions to take place, the shape and chemical surface of the binding site must be complementary to the ligand molecule, a property termed **molecular complementarity**.

The ability of proteins to distinguish different molecules is perhaps most highly developed in the blood proteins called **antibodies**, which animals produce in response to **antigens**, such as infectious agents (e.g., a bacterium or a virus), and certain foreign substances (e.g., proteins or polysaccharides in pollens). The presence of an antigen causes an organism to make a large quantity of different antibody proteins, each of which may bind to a slightly different region, or **epitope**, of the antigen. Antibodies act as specific sensors for antigens, forming antibody–antigen complexes that initiate a cascade of protective reactions in cells of the immune system.

All antibodies are Y-shaped molecules formed from two identical heavy chains and two identical light chains (Figure 3-15a). Each arm of an antibody molecule contains a single light chain linked to a heavy chain by a disulfide bond. Near the end of each arm are six highly variable loops, called *complementarity-determining regions (CDRs),* which form the antigen-binding sites. The sequences of the six loops are highly variable among antibodies, making them specific for different antigens. The interaction between an antibody and an epitope in an antigen is complementary in all cases; that is, the surface of the antibody's antigen-binding site physically matches the corresponding epitope like a glove

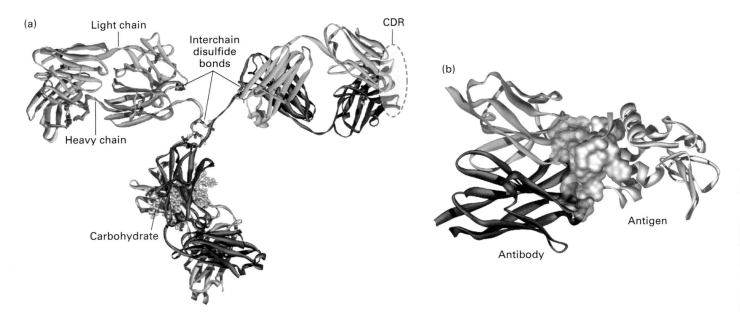

▲ **FIGURE 3-15 Antibody structure and antibody-antigen interaction.** (a) Ribbon model of an antibody. Every antibody molecule consists of two identical heavy chains (red) and two identical light chains (blue) covalently linked by disulfide bonds. (b) The hand-in-glove fit between an antibody and an epitope on its antigen—in this case, chicken egg-white lysozyme. Regions where the two molecules make contact are shown as surfaces. The antibody contacts the antigen with residues from all its complementarity-determining regions (CDRs). In this view, the complementarity of the antigen and antibody is especially apparent where "fingers" extending from the antigen surface are opposed to "clefts" in the antibody surface.

(Figure 3-15b). The intimate contact between these two surfaces, stabilized by numerous noncovalent bonds, is responsible for the exquisite binding specificity exhibited by an antibody.

The specificity of antibodies is so precise that they can distinguish between the cells of individual members of a species and in some cases can distinguish between proteins that differ by only a single amino acid. Because of their specificity and the ease with which they can be produced, antibodies are highly useful reagents in many of the experiments discussed in subsequent chapters.

Enzymes Are Highly Efficient and Specific Catalysts

In contrast with antibodies, which bind and simply present their ligands to other components of the immune system, enzymes promote the chemical alteration of their ligands, called **substrates.** Almost every chemical reaction in the cell is catalyzed by a specific enzyme. Like all catalysts, enzymes do not affect the extent of a reaction, which is determined by the change in free energy ΔG between reactants and products (Chapter 2). For reactions that are energetically favorable ($-\Delta G$), enzymes increase the reaction rate by lowering the **activation energy** (Figure 3-16). In the test tube, catalysts such as charcoal and platinum facilitate reactions but usually only at high temperatures or pressures, at extremes of high

or low pH, or in organic solvents. As the cell's protein catalysts, however, enzymes must function effectively in aqueous environment at 37°C, 1 atmosphere pressure, and pH 6.5–7.5.

Two striking properties of enzymes enable them to function as catalysts under the mild conditions present in cells: their enormous *catalytic power* and their high degree of *specificity.* The immense catalytic power of enzymes causes the rates of enzymatically catalyzed reactions to be 10^6–10^{12} times that of the corresponding uncatalyzed reactions under otherwise similar conditions. The exquisite specificity of enzymes—their ability to act selectively on one substrate or a small number of chemically similar substrates—is exemplified by the enzymes that act on amino acids. As noted in Chapter 2, amino acids can exist as two stereoisomers, designated L and D, although only L isomers are normally found in biological systems. Not surprisingly, enzyme-catalyzed reactions of L-amino acids take place much more rapidly than do those of D-amino acids, even though both stereoisomers of a given amino acid are the same size and possess the same R groups (see Figure 2-12).

Approximately 3700 different types of enzymes, each of which catalyzes a single chemical reaction or set of closely related reactions, have been classified in the enzyme database. Certain enzymes are found in the majority of cells because they catalyze the synthesis of common cellular products (e.g., proteins, nucleic acids, and phospholipids) or take part in the

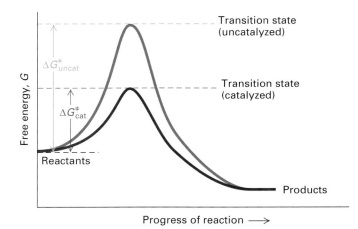

▲ FIGURE 3-16 Effect of a catalyst on the activation energy of a chemical reaction. This hypothetical reaction pathway depicts the changes in free energy G as a reaction proceeds. A reaction will take place spontaneously only if the total G of the products is less than that of the reactants ($-\Delta G$). However, all chemical reactions proceed through one or more high-energy transition states, and the rate of a reaction is inversely proportional to the activation energy ($\Delta G^{\ddagger}$), which is the difference in free energy between the reactants and the highest point along the pathway. Enzymes and other catalysts accelerate the rate of a reaction by reducing the free energy of the transition state and thus $\Delta G^{\ddagger}$.

production of energy by the conversion of glucose and oxygen into carbon dioxide and water. Other enzymes are present only in a particular type of cell because they catalyze chemical reactions unique to that cell type (e.g., the enzymes that convert tyrosine into dopamine, a neurotransmitter, in nerve cells). Although most enzymes are located within cells, some are secreted and function in extracellular sites such as the blood, the lumen of the digestive tract, or even outside the organism.

The catalytic activity of some enzymes is critical to cellular processes other than the synthesis or degradation of molecules. For instance, many regulatory proteins and intracellular signaling proteins catalyze the phosphorylation of proteins, and some transport proteins catalyze the hydrolysis of ATP coupled to the movement of molecules across membranes.

An Enzyme's Active Site Binds Substrates and Carries Out Catalysis

Certain amino acid side chains of an enzyme are important in determining its specificity and catalytic power. In the native conformation of an enzyme, these side chains are brought into proximity, forming the **active site.** Active sites thus consist of two functionally important regions: one that recognizes and binds the substrate (or substrates) and another that catalyzes the reaction after the substrate has been

bound. In some enzymes, the catalytic region is part of the substrate-binding region; in others, the two regions are structurally as well as functionally distinct.

To illustrate how the active site binds a specific substrate and then promotes a chemical change in the bound substrate, we examine the action of cyclic AMP–dependent protein kinase, now generally referred to as **protein kinase A (PKA).** This enzyme and other protein kinases, which add a phosphate group to serine, threonine, or tyrosine residues in proteins, are critical for regulating the activity of many cellular proteins, often in response to external signals. Because the eukaryotic protein kinases belong to a common superfamily, the structure of the active site and mechanism of phosphorylation are very similar in all of them. Thus protein kinase A can serve as a general model for this important class of enzymes.

The active site of protein kinase A is located in the 240-residue "kinase core" of the catalytic subunit. The kinase core, which is largely conserved in all protein kinases, is responsible for the binding of substrates (ATP and a target peptide sequence) and the subsequent transfer of a phosphate group from ATP to a serine, threonine, or tyrosine residue in the target sequence. The kinase core consists of a large domain and small one, with an intervening deep cleft; the active site comprises residues located in both domains.

Substrate Binding by Protein Kinases The structure of the ATP-binding site in the catalytic kinase core complements the structure of the nucleotide substrate. The adenine ring of ATP sits snugly at the base of the cleft between the large and the small domains. A highly conserved sequence, Gly-X-Gly-X-X-Gly-X-Val (X can be any amino acid), dubbed the "glycine lid," closes over the adenine ring and holds it in position (Figure 3-17a). Other conserved residues in the binding pocket stabilize the highly charged phosphate groups.

Although ATP is a common substrate for all protein kinases, the sequence of the target peptide varies among different kinases. The peptide sequence recognized by protein kinase A is Arg-Arg-X-Ser-Y, where X is any amino acid and Y is a hydrophobic amino acid. The part of the polypeptide chain containing the target serine or threonine residue is bound to a shallow groove in the large domain of the kinase core. The peptide specificity of protein kinase A is conferred by several glutamic acid residues in the large domain, which form salt bridges with the two arginine residues in the target peptide. Different residues determine the specificity of other protein kinases.

The catalytic core of protein kinase A exists in an "open" and "closed" conformation (Figure 3-17b). In the open conformation, the large and small domains of the core region are separated enough that substrate molecules can enter and bind. When the active site is occupied by substrate, the domains move together into the closed position. This change in tertiary structure, an example of *induced fit,* brings the target peptide sequence sufficiently close to accept a phosphate

(a)

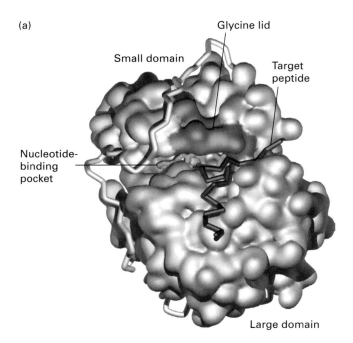

(b)

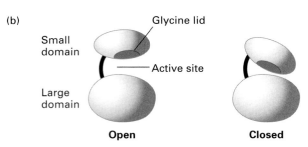

▲ **FIGURE 3-17 Protein kinase A and conformational change induced by substrate binding.** (a) Model of the catalytic subunit of protein kinase A with bound substrates; the conserved kinase core is indicated as a molecular surface. An overhanging glycine-rich sequence (blue) traps ATP (green) in a deep cleft between the large and small domains of the core. Residues in the large domain bind the target peptide (red). The structure of the kinase core is largely conserved in other eukaryotic protein kinases. (b) Schematic diagrams of open and closed conformations of the kinase core. In the absence of substrate, the kinase core is in the open conformation. Substrate binding causes a rotation of the large and small domains that brings the ATP- and peptide-binding sites closer together and causes the glycine lid to move over the adenine residue of ATP, thereby trapping the nucleotide in the binding cleft. The model in part (a) is in the closed conformation.

the active site. In the open position, ATP can enter and bind the active site cleft; in the closed position, the glycine lid prevents ATP from leaving the cleft. Subsequent to phosphoryl transfer from the bound ATP to the bound peptide sequence, the glycine lid must rotate back to the open position before ADP can be released. Kinetic measurements show that the rate of ADP release is 20-fold slower than that of phosphoryl transfer, indicating the influence of the glycine lid on the rate of kinase reactions. Mutations in the glycine lid that inhibit its flexibility slow catalysis by protein kinase A even further.

Phosphoryl Transfer by Protein Kinases After substrates have bound and the catalytic core of protein kinase A has assumed the closed conformation, the phosphorylation of a serine or threonine residue on the target peptide can take place (Figure 3-18). As with all chemical reactions, phosphoryl transfer catalyzed by protein kinase A proceeds through a transition state in which the phosphate group to be transferred and the acceptor hydroxyl group are brought into close proximity. Binding and stabilization of the intermediates by protein kinase A reduce the activation energy of the phosphoryl transfer reaction, permitting it to take place at measurable rates under the mild conditions present within cells (see Figure 3-16). Formation of the products induces the enzyme to revert to its open conformational state, allowing ADP and the phosphorylated target peptide to diffuse from the active site.

V_{max} and K_m Characterize an Enzymatic Reaction

The catalytic action of an enzyme on a given substrate can be described by two parameters: V_{max}, the maximal velocity of the reaction at saturating substrate concentrations, and K_m (the **Michaelis constant**), a measure of the affinity of an enzyme for its substrate (Figure 3-19). The K_m is defined as the substrate concentration that yields a half-maximal reaction rate (i.e., $\frac{1}{2} V_{max}$). The smaller the value of K_m, the more avidly an enzyme can bind substrate from a dilute solution and the smaller the substrate concentration needed to reach half-maximal velocity.

The concentrations of the various small molecules in a cell vary widely, as do the K_m values for the different enzymes that act on them. Generally, the intracellular concentration of a substrate is approximately the same as or greater than the K_m value of the enzyme to which it binds.

Enzymes in a Common Pathway Are Often Physically Associated with One Another

Enzymes taking part in a common metabolic process (e.g., the degradation of glucose to pyruvate) are generally located in the same cellular compartment (e.g., in the cytosol, at a membrane, within a particular organelle). Within a compartment, products from one reaction can move by diffusion to the next enzyme in the pathway. However, diffusion entails random movement and is a slow, inefficient process for

group from the bound ATP. After the phosphorylation reaction has been completed, the presence of the products causes the domains to rotate to the open position, from which the products are released.

The rotation from the open to the closed position also causes movement of the glycine lid over the ATP-binding cleft. The glycine lid controls the entry of ATP and release of ADP at

Initial state

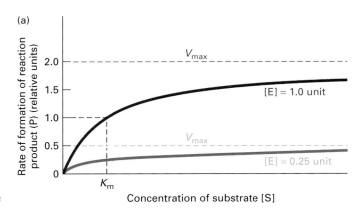

Formation of transition state

↓

Intermediate state

Phosphate transfer

↓

End state

Phosphorylated peptide

▲ **FIGURE 3-18 Mechanism of phosphorylation by protein kinase A.** (*Top*) Initially, ATP and the target peptide bind to the active site (see Figure 3-17a). Electrons of the phosphate group are delocalized by interactions with lysine side chains and Mg^{2+}. Colored circles represent the residues in the kinase core critical to substrate binding and phosphoryl transfer. Note that these residues are not adjacent to one another in the amino acid sequence. (*Middle*) A new bond then forms between the serine or threonine side-chain oxygen atom and γ phosphate, yielding a pentavalent intermediate. (*Bottom*) The phosphoester bond between the β and γ phosphates is broken, yielding the products ADP and a peptide with a phosphorylated serine or threonine side chain. The catalytic mechanism of other protein kinases is similar.

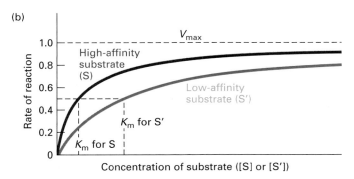

▲ **EXPERIMENTAL FIGURE 3-19 The K_m and V_{max} for an enzyme-catalyzed reaction are determined from plots of the initial velocity versus substrate concentration.** The shape of these hypothetical kinetic curves is characteristic of a simple enzyme-catalyzed reaction in which one substrate (S) is converted into product (P). The initial velocity is measured immediately after addition of enzyme to substrate before the substrate concentration changes appreciably. (a) Plots of the initial velocity at two different concentrations of enzyme [E] as a function of substrate concentration [S]. The [S] that yields a half-maximal reaction rate is the Michaelis constant K_m, a measure of the affinity of E for S. Doubling the enzyme concentration causes a proportional increase in the reaction rate, and so the maximal velocity V_{max} is doubled; the K_m, however, is unaltered. (b) Plots of the initial velocity versus substrate concentration with a substrate S for which the enzyme has a high affinity and with a substrate S' for which the enzyme has a low affinity. Note that the V_{max} is the same with both substrates but that K_m is higher for S', the low-affinity substrate.

moving molecules between widely dispersed enzymes (Figure 3-20a). To overcome this impediment, cells have evolved mechanisms for bringing enzymes in a common pathway into close proximity.

In the simplest such mechanism, polypeptides with different catalytic activities cluster closely together as subunits of a multimeric enzyme or assemble on a common "scaffold" (Figure 3-20b). This arrangement allows the products of one reaction to be channeled directly to the next enzyme in the pathway. The first approach is illustrated by pyruvate

(a)

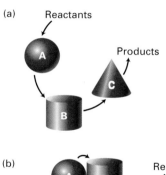

(b)

OR

(c)

▲ FIGURE 3-20 Evolution of multifunctional enzyme.
In the hypothetical reaction pathways illustrated here the initial
reactants are converted into final products by the sequential
action of three enzymes: A, B, and C. (a) When the enzymes are
free in solution or even constrained within the same cellular
compartment, the intermediates in the reaction sequence must
diffuse from one enzyme to the next, an inherently slow process.
(b) Diffusion is greatly reduced or eliminated when the enzymes
associate into multisubunit complexes. (c) The closest integration
of different catalytic activities occurs when the enzymes are
fused at the genetic level, becoming domains in a single protein.

(a)

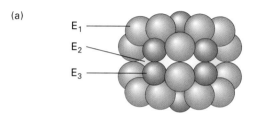

(b)

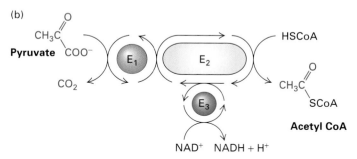

Net reaction:
$$\text{Pyruvate} + \text{NAD}^+ + \text{CoA} \longrightarrow \text{CO}_2 + \text{NADH} + \text{acetyl CoA}$$

**▲ FIGURE 3-21 Structure and function of pyruvate
dehydrogenase, a large multimeric enzyme complex that
converts pyruvate into acetyl CoA.** (a) The complex consists of
24 copies of pyruvate decarboxylase (E_1), 24 copies of lipoamide
transacetylase (E_2), and 12 copies of dihydrolipoyl dehydrogenase
(E_3). The E_1 and E_3 subunits are bound to the outside of the core
formed by the E_2 subunits. (b) The reactions catalyzed by the
complex include several enzyme-bound intermediates (not
shown). The tight structural integration of the three enzymes
increases the rate of the overall reaction and minimizes possible
side reactions.

dehydrogenase, a complex of three distinct enzymes that converts pyruvate into acetyl CoA in mitochondria (Figure 3-21).
The scaffold approach is employed by MAP kinase signal-
transduction pathways, discussed in Chapter 14. In yeast,
three protein kinases assembled on the Ste5 scaffold protein
form a kinase cascade that transduces the signal triggered by
the binding of mating factor to the cell surface.

In some cases, separate proteins have been fused together
at the genetic level to create a single multidomain, multifunctional enzyme (Figure 3-20c). For instance, the isomerization of citrate to isocitrate in the citric acid cycle is
catalyzed by aconitase, a single polypeptide that carries out
two separate reactions: (1) the dehydration of citrate to form
cis-aconitate and then (2) the hydration of *cis*-aconitate to
yield isocitrate (see Figure 8-9).

KEY CONCEPTS OF SECTION 3.3

Enzymes and the Chemical Work of Cells

■ The function of nearly all proteins depends on their ability to bind other molecules (ligands). Ligand-binding sites

on proteins and the corresponding ligands are chemically
and topologically complementary.

■ The affinity of a protein for a particular ligand refers to
the strength of binding; its specificity refers to the preferential binding of one or a few closely related ligands.

■ Enzymes are catalytic proteins that accelerate the rate
of cellular reactions by lowering the activation energy
and stabilizing transition-state intermediates (see Figure
3-16).

■ An enzyme active site comprises two functional parts: a
substrate-binding region and a catalytic region. The amino
acids composing the active site are not necessarily adjacent
in the amino acid sequence but are brought into proximity in the native conformation.

■ From plots of reaction rate versus substrate concentration, two characteristic parameters of an enzyme can
be determined: the Michaelis constant K_m, a measure of
the enzyme's affinity for substrate, and the maximal velocity V_{max}, a measure of its catalytic power (see Figure
3-19).

■ Enzymes in a common pathway are located within specific cell compartments and may be further associated as domains of a monomeric protein, subunits of a multimeric protein, or components of a protein complex assembled on a common scaffold (see Figure 3-20).

3.4 Molecular Motors and the Mechanical Work of Cells

A common property of all cells is motility, the ability to move in a specified direction. Many cell processes exhibit some type of movement at either the molecular or the cellular level; all movements result from the application of a force. In Brownian motion, for instance, thermal energy constantly buffets molecules and organelles in *random* directions and for very short distances. On the other hand, materials within a cell are transported in specific directions and for longer distances. This type of movement results from the mechanical work carried out by proteins that function as motors. We first briefly describe the types and general properties of molecular motors and then look at how one type of motor protein generates force for movement.

Molecular Motors Convert Energy into Motion

At the nanoscale of cells and molecules, movement is effected by much different forces from those in the macroscopic world. For example, the high protein concentration (200–300 mg/ml) of the cytoplasm prevents organelles and vesicles from diffusing faster than 100 μm/3 hours. Even a micrometer-sized bacterium experiences a drag force from water that stops its forward movement within a fraction of a nanometer when it stops actively swimming. To generate the forces necessary for many cellular movements, cells depend on specialized enzymes commonly called **motor proteins.** These *mechanochemical enzymes* convert energy released by the hydrolysis of ATP or from ion gradients into a mechanical force.

Motor proteins generate either linear or rotary motion (Table 3-2). Some motor proteins are components of macro-molecular assemblies, but those that move along cytoskeletal fibers are not. This latter group comprises the **myosins, kinesins,** and **dyneins**—linear motor proteins that carry attached "cargo" with them as they proceed along either **microfilaments** or **microtubules** (Figure 3-22a). DNA and RNA polymerases also are linear motor proteins because they translocate along DNA during replication and transcription. In contrast, rotary motors revolve to cause the beat of bacterial flagella, to pack DNA into the capsid of a virus, and to synthesize ATP. The propulsive force for bacterial swimming, for instance, is generated by a rotary motor protein complex in the bacterial membrane. Ions flow down an electrochemical gradient through an immobile ring of proteins, the *stator*, which is located in the membrane. Torque generated by the stator rotates an inner ring of proteins and the attached flagellum (Figure 3-22b). Similarly, in the mitochondrial ATP synthase, or F_0F_1 complex, a flux of ions across the inner mitochondrial membrane is transduced by the F_0 part into rotation of the γ subunit, which projects into a surrounding ring of α and β subunits in the F_1 part. Interactions between the γ subunit and the β subunits directs the synthesis of ATP (Chapter 8).

From the observed activities of motor proteins, we can infer three general properties that they possess:

■ The ability to transduce a source of energy, either ATP or an ion gradient, into linear or rotary movement

■ The ability to bind and translocate along a cytoskeletal filament, nucleic acid strand, or protein complex

■ Net movement in a given direction

The motor proteins that attach to cytoskeletal fibers also bind to and carry along cargo as they translocate. The cargo in muscle cells and eukaryotic flagella consists of thick filaments and B tubules, respectively (see Figure 3-22a). These motor proteins can also transport cargo chromosomes and membrane-limited vesicles as they move along microtubules or microfilaments (Figure 3-23).

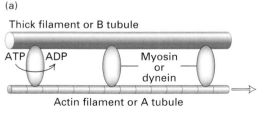

(a)

Thick filament or B tubule

ATP ADP   Myosin or dynein

Actin filament or A tubule

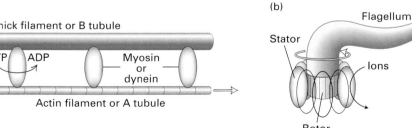

(b)

Flagellum

Stator

Ions

Rotor

▲ **FIGURE 3-22 Comparison of linear and rotary molecular motors.** (a) In muscle and eukaryotic flagella, the head domains of motor proteins (blue) bind to an actin thin filament (muscle) or the A tubule of a doublet microtubule (flagella). ATP hydrolysis in the head causes linear movement of the cytoskeletal fiber (orange) relative to the attached thick filament or B tubule of an

adjacent doublet microtubule. (b) In the rotary motor in the bacterial membrane, the stator (blue) is immobile in the membrane. Ion flow through the stator generates a torque that powers rotation of the rotor (orange) and the flagellum attached to it.

TABLE 3-2	Selected Molecular Motors			
Motor*	**Energy Source**	**Structure/Components**	**Cellular Location**	**Movement Generated**
LINEAR MOTORS				
DNA polymerase (4)	ATP	Multisubunit polymerase δ within replisome	Nucleus	Translocation along DNA during replication
RNA polymerase (4)	ATP	Multisubunit polymerase within transcription elongation complex	Nucleus	Translocation along DNA during transcription
Ribosome (4)	GTP	Elongation factor 2 (EF2) bound to ribosome	Cytoplasm/ER membrane	Translocation along mRNA during translation
Myosins (3, 19)	ATP	Heavy and light chains; head domains with ATPase activity and microfilament-binding site	Cytoplasm	Transport of cargo vesicles; contraction
Kinesins (20)	ATP	Heavy and light chains; head domains with ATPase activity and microtubule-binding site	Cytoplasm	Transport of cargo vesicles and chromosomes during mitosis
Dyneins (20)	ATP	Multiple heavy, intermediate, and light chains; head domains with ATPase activity and microtubule-binding site	Cytoplasm	Transport of cargo vesicles; beating of cilia and eukaryotic flagella
ROTARY MOTORS				
Bacterial flagellar motor	H^+/Na^+ gradient	Stator and rotor proteins, flagellum	Plasma membrane	Rotation of flagellum attached to rotor
ATP synthase, F_0F_1(8)	H^+ gradient	Multiple subunits forming F_0 and F_1 particles	Inner mitochondrial membrane, thylakoid membrane, bacterial plasma membrane	Rotation of γ subunit leading to ATP synthesis
Viral capsid motor	ATP	Connector, prohead RNA, ATPase	Capsid	Rotation of connector leading to DNA packaging

*Numbers in parentheses indicate chapters in which various motors are discussed.

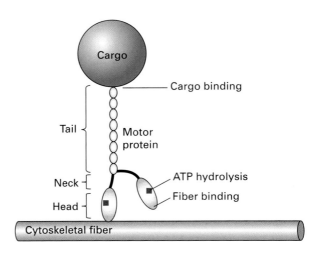

▶ **FIGURE 3-23 Motor protein-dependent movement of cargo.** The head domains of myosin, dynein, and kinesin motor proteins bind to a cytoskeletal fiber (microfilaments or microtubules), and the tail domain attaches to one of various types of cargo—in this case, a membrane-limited vesicle. Hydrolysis of ATP in the head domain causes the head domain to "walk" along the track in one direction by a repeating cycle of conformational changes.

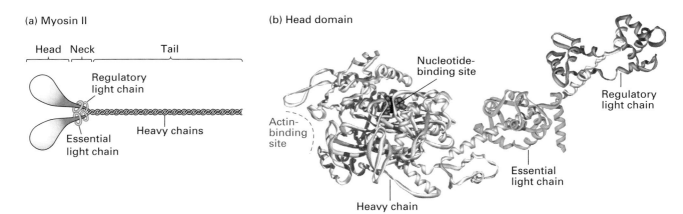

▲ FIGURE 3-24 Structure of myosin II. (a) Myosin II is a dimeric protein composed of two identical heavy chains (white) and four light chains (blue and green). Each of the head domains transduces the energy from ATP hydrolysis into movement. Two light chains are associated with the neck domain of each heavy chain. The coiled-coil sequence of the tail domain organizes myosin II into a dimer. (b) Three-dimensional model of a single head domain shows that it has a curved, elongated shape and is bisected by a large cleft. The nucleotide-binding pocket lies on one side of this cleft, and the actin-binding site lies on the other side near the tip of the head. Wrapped around the shaft of the α-helical neck are the two light chains. These chains stiffen the neck so that it can act as a lever arm for the head. Shown here is the ADP-bound conformation.

All Myosins Have Head, Neck, and Tail Domains with Distinct Functions

To further illustrate the properties of motor proteins, we consider myosin II, which moves along actin filaments in muscle cells during contraction. Other types of myosin can transport vesicles along actin filaments in the cytoskeleton. Myosin II and other members of the myosin superfamily are composed of one or two heavy chains and several light chains. The heavy chains are organized into three structurally and functionally different types of domains (Figure 3-24a).

The two globular *head domains* are specialized ATPases that couple the hydrolysis of ATP with motion. A critical feature of the myosin ATPase activity is that it is *actin activated*. In the absence of actin, solutions of myosin slowly convert ATP into ADP and phosphate. However, when myosin is complexed with actin, the rate of myosin ATPase activity is four to five times as fast as it is in the absence of actin. The actin-activation step ensures that the myosin ATPase operates at its maximal rate only when the myosin head domain is bound to actin. Adjacent to the head domain lies the α-helical *neck region*, which is associated with the light chains. These light chains are crucial for converting small conformational changes in the head into large movements of the molecule and for regulating the activity of the head domain. The rodlike *tail domain* contains the binding sites that determine the specific activities of a particular myosin.

The results of studies of myosin fragments produced by proteolysis helped elucidate the functions of the domains. X-ray crystallographic analysis of the S1 fragment of myosin II, which consists of the head and neck domains, revealed its shape, the positions of the light chains, and the locations of the ATP-binding and actin-binding sites. The elongated myosin head is attached at one end to the α-helical neck (Figure 3-24b). Two light-chain molecules lie at the base of the head, wrapped around the neck like C-clamps. In this position, the light chains stiffen the neck region and are therefore able to regulate the activity of the head domain.

Conformational Changes in the Myosin Head Couple ATP Hydrolysis to Movement

The results of studies of muscle contraction provided the first evidence that myosin heads slide or walk along actin filaments. Unraveling the mechanism of muscle contraction was greatly aided by the development of in vitro motility assays and single-molecule force measurements. On the basis of information obtained with these techniques and the three-dimensional structure of the myosin head, researchers developed a general model for how myosin harnesses the energy released by ATP hydrolysis to move along an actin filament. Because all myosins are thought to use the same mechanism to generate movement, we will ignore whether the myosin tail is bound to a vesicle or is part of a thick filament as it is in muscle. One assumption in this model is that the hydrolysis of a single ATP molecule is coupled to each step taken by a myosin molecule along an actin filament. Evidence supporting this assumption is discussed in Chapter 19.

As shown in Figure 3-25, myosin undergoes a series of events during each step of movement. In the course of one cycle, myosin must exist in at least three conformational states: an ATP state unbound to actin, an ADP-P_i state bound to actin, and a state after the power-generating stroke has been completed. The major question is how the nucleotide-binding pocket and the distant actin-binding site are mutually influenced and how changes at these sites are converted into force. The results of structural studies of myosin in the presence of nucleotides and nucleotide analogs that mimic the various steps in the cycle indicate that the binding and hydrolysis of a nucleotide cause a

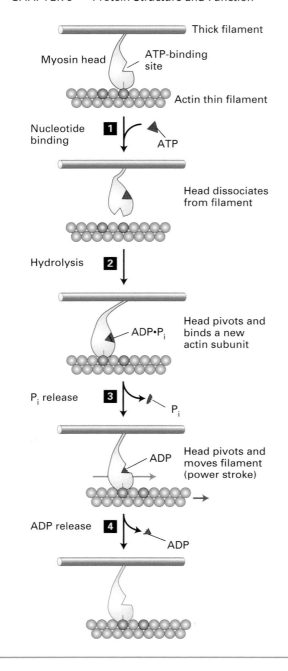

Thick filament

Myosin head

ATP-binding site

Actin thin filament

Nucleotide binding   **1**   ATP

Head dissociates from filament

Hydrolysis   **2**

ADP·P$_i$ Head pivots and binds a new actin subunit

P$_i$ release   **3**   P$_i$

ADP   Head pivots and moves filament (power stroke)

ADP release   **4**   ADP

MEDIA CONNECTIONS

Focus Animation: Myosin Crossbridge Cycle

◄ **FIGURE 3-25 Operational model for the coupling of ATP hydrolysis to movement of myosin along an actin filament.** Shown here is the cycle for a myosin II head that is part of a thick filament in muscle, but other myosins that attach to other cargo (e.g., the membrane of a vesicle) are thought to operate according to the same cyclical mechanism. In the absence of bound nucleotide, a myosin head binds actin tightly in a "rigor" state. Step **1**: Binding of ATP opens the cleft in the myosin head, disrupting the actin-binding site and weakening the interaction with actin. Step **2**: Freed of actin, the myosin head hydrolyzes ATP, causing a conformational change in the head that moves it to a new position, closer to the (+) end of the actin filament, where it rebinds to the filament. Step **3**: As phosphate (P$_i$) dissociates from the ATP-binding pocket, the myosin head undergoes a second conformational change—the power stroke—which restores myosin to its rigor conformation. Because myosin is bound to actin, this conformational change exerts a force that causes myosin to move the actin filament. Step **4**: Release of ADP completes the cycle. [Adapted from R. D. Vale and R. A. Milligan, 2002, *Science* **288**:88.]

KEY CONCEPTS OF SECTION 3.4

Molecular Motors and the Mechanical Work of Cells

■ Motor proteins are mechanochemical enzymes that convert energy released by ATP hydrolysis into either linear or rotary movement (see Figure 3-22).

■ Linear motor proteins (myosins, kinesins, and dyneins) move along cytoskeletal fibers carrying bound cargo, which includes vesicles, chromosomes, thick filaments in muscle, and microtubules in eukaryotic flagella.

■ Myosin II consists of two heavy chains and several light chains. Each heavy chain has a head (motor) domain, which is an actin-activated ATPase; a neck domain, which is associated with light chains; and a long rodlike tail domain that organizes the dimeric molecule and binds to thick filaments in muscle cells (see Figure 3-24).

■ Movement of myosin relative to an actin filament results from the attachment of the myosin head to an actin filament, rotation of the neck region, and detachment in a cyclical ATP-dependent process (see Figure 3-25). The same general mechanism is thought to account for all myosin- and kinesin-mediated movement.

small conformational change in the head domain that is amplified into a large movement of the neck region. The small conformational change in the head domain is localized to a "switch" region consisting of the nucleotide- and actin-binding sites. A "converter" region at the base of the head acts like a fulcrum that causes the leverlike neck to bend and rotate.

Homologous switch, converter, and lever arm structures in kinesin are responsible for the movement of kinesin motor proteins along microtubules. The structural basis for dynein movement is unknown because the three-dimensional structure of dynein has not been determined.

3.5 Common Mechanisms for Regulating Protein Function

Most processes in cells do not take place independently of one another or at a constant rate. Instead, the catalytic activity of enzymes or the assembly of a macromolecular complex is so regulated that the amount of reaction product or the appearance of the complex is just sufficient to meet the needs of the cell. As a result, the steady-state concentrations of substrates and products will vary, depending on cellular conditions. The flow of material in an enzymatic pathway is controlled by several mechanisms, some of which also regulate the functions of nonenzymatic proteins.

One of the most important mechanisms for regulating protein function entails **allostery.** Broadly speaking, allostery refers to any change in a protein's tertiary or quaternary structure or both induced by the binding of a ligand, which may be an activator, inhibitor, substrate, or all three. Allosteric regulation is particularly prevalent in multimeric enzymes and other proteins. We first explore several ways in which allostery influences protein function and then consider other mechanisms for regulating proteins.

Cooperative Binding Increases a Protein's Response to Small Changes in Ligand Concentration

In many cases, especially when a protein binds several molecules of one ligand, the binding is graded; that is, the binding of one ligand molecule affects the binding of subsequent ligand molecules. This type of allostery, often called **cooper-**

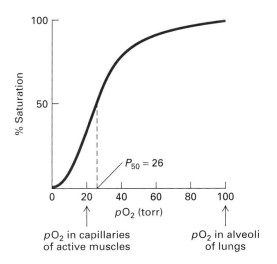

▲ **EXPERIMENTAL FIGURE 3-26 Sequential binding of oxygen to hemoglobin exhibits positive cooperativity.** Each hemoglobin molecule has four oxygen-binding sites; at saturation all the sites are loaded with oxygen. The oxygen concentration is commonly measured as the partial pressure (pO_2). P_{50} is the pO_2 at which half the oxygen-binding sites at a given hemoglobin concentration are occupied; it is equivalent to the K_m for an enzymatic reaction. The large change in the amount of oxygen bound over a small range of pO_2 values permits efficient unloading of oxygen in peripheral tissues such as muscle. The sigmoidal shape of a plot of percent saturation versus ligand concentration is indicative of cooperative binding. In the absence of cooperative binding, a binding curve is a hyperbola, similar to the simple kinetic curves in Figure 3-19. [Adapted from L. Stryer, *Biochemistry,* 4th ed., 1995, W. H. Freeman and Company.]

ativity, permits many multisubunit proteins to respond more efficiently to small changes in ligand concentration than would otherwise be possible. In positive cooperativity, sequential binding is enhanced; in negative cooperativity, sequential binding is inhibited.

Hemoglobin presents a classic example of positive cooperative binding. Each of the four subunits in hemoglobin contains one heme molecule, which consists of an iron atom held within a porphyrin ring (see Figure 8-16a). The heme groups are the oxygen-binding components of hemoglobin (see Figure 3-10). The binding of oxygen to the heme molecule in one of the four hemoglobin subunits induces a local conformational change whose effect spreads to the other subunits, lowering the K_m for the binding of additional oxygen molecules and yielding a sigmoidal oxygen-binding curve (Figure 3-26). Consequently, the sequential binding of oxygen is facilitated, permitting hemoglobin to load more oxygen in peripheral tissues than it otherwise could at normal oxygen concentrations.

Ligand Binding Can Induce Allosteric Release of Catalytic Subunits or Transition to a State with Different Activity

Previously, we looked at protein kinase A to illustrate binding and catalysis by the active site of an enzyme. This enzyme can exist as an inactive tetrameric protein composed of two catalytic subunits and two regulatory subunits. Each regulatory subunit contains a *pseudosubstrate* sequence that binds to the active site in a catalytic subunit. By blocking substrate binding, the regulatory subunit inhibits the activity of the catalytic subunit.

Inactive protein kinase A is turned on by **cyclic AMP (cAMP),** a small second-messenger molecule. The binding of cAMP to the regulatory subunits induces a conformational change in the pseudosubstrate sequence so that it can no longer bind the catalytic subunit. Thus, in the presence of cAMP, the inactive tetramer dissociates into two monomeric active catalytic subunits and a dimeric regulatory subunit (Figure 3-27). As discussed in Chapter 13, the binding of various hormones to cell-surface receptors induces a rise in the intracellular concentration of cAMP, leading to the activation of protein kinase A. When the signaling ceases and the cAMP level decreases, the activity of protein kinase A is turned off by reassembly of the inactive tetramer. The binding of cAMP to the regulatory subunits exhibits positive cooperativity; thus small changes in the concentration of this allosteric molecule produce a large change in the activity of protein kinase A.

Many multimeric enzymes undergo allosteric transitions that alter the relation of the subunits to one another but do not cause dissociation as in protein kinase A. In this type of allostery, the activity of a protein in the ligand-bound state differs from that in the unbound state. An example is the GroEL chaperonin discussed earlier. This barrel-shaped

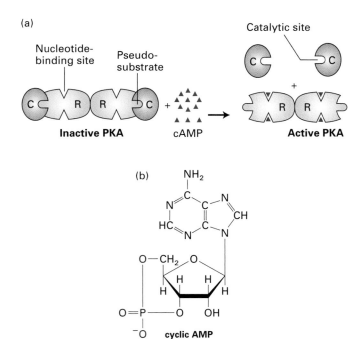

▲ **FIGURE 3-27 Ligand-induced activation of protein kinase A (PKA).** At low concentrations of cyclic AMP (cAMP), the PKA is an inactive tetramer. Binding of cAMP to the regulatory (R) subunits causes a conformational change in these subunits that permits release of the active, monomeric catalytic (C) subunits. (b) Cyclic AMP is a derivative of adenosine monophosphate. This intracellular signaling molecule, whose concentration rises in response to various extracellular signals, can modulate the activity of many proteins.

protein-folding machine comprises two back-to-back multisubunit rings, which can exist in a "tight" peptide-binding state and a "relaxed" peptide-releasing state (see Figure 3-11). The binding of ATP and the co-chaperonin GroES to one of the rings in the tight state causes a twofold expansion of the GroEL cavity, shifting the equilibrium toward the relaxed peptide-folding state.

Calcium and GTP Are Widely Used to Modulate Protein Activity

In the preceding examples, oxygen, cAMP, and ATP cause allosteric changes in the activity of their target proteins (hemoglobin, protein kinase A, and GroEL, respectively). Two additional allosteric ligands, Ca^{2+} and GTP, act through two types of ubiquitous proteins to regulate many cellular processes.

Calmodulin-Mediated Switching The concentration of Ca^{2+} free in the cytosol is kept very low ($\approx 10^{-7}$ M) by membrane transport proteins that continually pump Ca^{2+} out of the cell or into the endoplasmic reticulum. As we learn in Chapter 7, the cytosolic Ca^{2+} level can increase from 10- to

100-fold by the release of Ca^{2+} from ER stores or by its import from the extracellular environment. This rise in cytosolic Ca^{2+} is sensed by Ca^{2+}-binding proteins, particularly those of the *EF hand family*, all of which contain the helix-loop-helix motif discussed earlier (see Figure 3-6a).

The prototype EF hand protein, **calmodulin**, is found in all eukaryotic cells and may exist as an individual monomeric protein or as a subunit of a multimeric protein. A dumbbell-shaped molecule, calmodulin contains four Ca^{2+}-binding sites with a K_D of $\approx 10^{-6}$ M. The binding of Ca^{2+} to calmodulin causes a conformational change that permits Ca^{2+}/calmodulin to bind various target proteins, thereby switching their activity on or off (Figure 3-28). Calmodulin and similar EF hand proteins thus function as *switch proteins*, acting in concert with Ca^{2+} to modulate the activity of other proteins.

Switching Mediated by Guanine Nucleotide–Binding Proteins Another group of intracellular switch proteins constitutes the **GTPase superfamily**. These proteins include monomeric **Ras protein** (see Figure 3-5) and the G_α subunit of the trimeric **G proteins**. Both Ras and G_α are bound to the plasma membrane, function in cell signaling, and play a key role in cell proliferation and differentiation. Other members

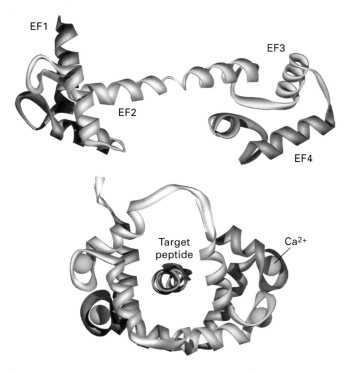

▲ **FIGURE 3-28 Switching mediated by Ca^{2+}/calmodulin.** Calmodulin is a widely distributed cytosolic protein that contains four Ca^{2+}-binding sites, one in each of its EF hands. Each EF hand has a helix-loop-helix motif. At cytosolic Ca^{2+} concentrations above about 5×10^{-7} M, binding of Ca^{2+} to calmodulin changes the protein's conformation. The resulting Ca^{2+}/calmodulin wraps around exposed helices of various target proteins, thereby altering their activity.

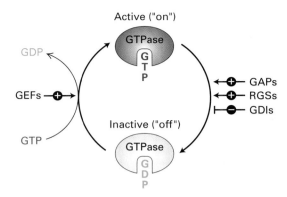

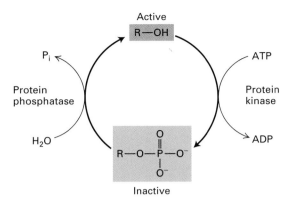

▲ **FIGURE 3-29 Cycling of GTPase switch proteins between the active and inactive forms.** Conversion of the active into the inactive form by hydrolysis of the bound GTP is accelerated by GAPs (GTPase-accelerating proteins) and RGSs (regulators of G protein–signaling) and inhibited by GDIs (guanine nucleotide dissociation inhibitors). Reactivation is promoted by GEFs (guanine nucleotide–exchange factors).

▲ **FIGURE 3-30 Regulation of protein activity by kinase/phosphatase switch.** The cyclic phosphorylation and dephosphorylation of a protein is a common cellular mechanism for regulating protein activity. In this example, the target protein R is inactive (light orange) when phosphorylated and active (dark orange) when dephosphorylated; some proteins have the opposite pattern.

of the GTPase superfamily function in protein synthesis, the transport of proteins between the nucleus and the cytoplasm, the formation of coated vesicles and their fusion with target membranes, and rearrangements of the actin cytoskeleton.

All the GTPase switch proteins exist in two forms (Figure 3-29): (1) an active ("on") form with bound GTP (guanosine triphosphate) that modulates the activity of specific target proteins and (2) an inactive ("off") form with bound GDP (guanosine diphosphate). The GTPase activity of these switch proteins hydrolyzes bound GTP to GDP slowly, yielding the inactive form. The subsequent exchange of GDP with GTP to regenerate the active form occurs even more slowly. Activation is temporary and is enhanced or depressed by other proteins acting as allosteric regulators of the switch protein. We examine the role of various GTPase switch proteins in regulating intracellular signaling and other processes in several later chapters.

Cyclic Protein Phosphorylation and Dephosphorylation Regulate Many Cellular Functions

As noted earlier, one of the most common mechanisms for regulating protein activity is phosphorylation, the addition and removal of phosphate groups from serine, threonine, or tyrosine residues. Protein **kinases** catalyze phosphorylation, and **phosphatases** catalyze dephosphorylation. Although both reactions are essentially irreversible, the counteracting activities of kinases and phosphatases provide cells with a "switch" that can turn on or turn off the function of various proteins (Figure 3-30). Phosphorylation changes a protein's charge and generally leads to a conformational change; these effects can significantly alter ligand binding by a protein, leading to an increase or decrease in its activity.

Nearly 3 percent of all yeast proteins are protein kinases or phosphatases, indicating the importance of phosphorylation and dephosphorylation reactions even in simple cells. All classes of proteins—including structural proteins, enzymes, membrane channels, and signaling molecules—are regulated by kinase/phosphatase switches. Different protein kinases and phosphatases are specific for different target proteins and can thus regulate a variety of cellular pathways, as discussed in later chapters. Some of these enzymes act on one or a few target proteins, whereas others have multiple targets. The latter are useful in integrating the activities of proteins that are coordinately controlled by a single kinase/phosphatase switch. Frequently, another kinase or phosphatase is a target, thus creating a web of interdependent controls.

Proteolytic Cleavage Irreversibly Activates or Inactivates Some Proteins

The regulatory mechanisms discussed so far act as switches, reversibly turning proteins on and off. The regulation of some proteins is by a distinctly different mechanism: the irreversible activation or inactivation of protein function by proteolytic cleavage. This mechanism is most common in regard to some hormones (e.g., insulin) and digestive proteases. Good examples of such enzymes are trypsin and chymotrypsin, which are synthesized in the pancreas and secreted into the small intestine as the inactive zymogens *trypsinogen* and *chymotrypsinogen,* respectively. Enterokinase, an aminopeptidase secreted from cells lining the small intestine, converts trypsinogen into trypsin, which in turn cleaves chymotrypsinogen to form chymotrypsin. The delay in the activation of these proteases until they reach the intestine prevents them from digesting the pancreatic tissue in which they are made.

Higher-Order Regulation Includes Control of Protein Location and Concentration

The activities of proteins are extensively regulated in order that the numerous proteins in a cell can work together harmoniously. For example, all metabolic pathways are closely controlled at all times. Synthetic reactions take place when the products of these reactions are needed; degradative reactions take place when molecules must be broken down. All the regulatory mechanisms heretofore described affect a protein locally at its site of action, turning its activity on or off.

Normal functioning of a cell, however, also requires the segregation of proteins to particular compartments such as the mitochondria, nucleus, and lysosomes. In regard to enzymes, compartmentation not only provides an opportunity for controlling the delivery of substrate or the exit of product but also permits competing reactions to take place simultaneously in different parts of a cell. We describe the mechanisms that cells use to direct various proteins to different compartments in Chapters 16 and 17.

In addition to compartmentation, cellular processes are regulated by protein synthesis and degradation. For example, proteins are often synthesized at low rates when a cell has little or no need for their activities. When the cell faces increased demand (e.g., appearance of substrate in the case of enzymes, stimulation of B lymphocytes by antigen), the cell responds by synthesizing new protein molecules. Later, the protein pool is lowered when levels of substrate decrease or the cell becomes inactive. Extracellular signals are often instrumental in inducing changes in the rates of protein synthesis and degradation (Chapters 13–15). Such regulated changes play a key role in the cell cycle (Chapter 21) and in cell differentiation (Chapter 22).

KEY CONCEPTS OF SECTION 3.5

Common Mechanisms for Regulating Protein Function

■ In allostery, the binding of one ligand molecule (a substrate, activator, or inhibitor) induces a conformational change, or allosteric transition, that alters a protein's activity or affinity for other ligands.

■ In multimeric proteins, such as hemoglobin, that bind multiple ligand molecules, the binding of one ligand molecule may modulate the binding affinity for subsequent ligand molecules. Enzymes that cooperatively bind substrates exhibit sigmoidal kinetics similar to the oxygen-binding curve of hemoglobin (see Figure 3-26).

■ Several allosteric mechanisms act as switches, turning protein activity on and off in a reversible fashion.

■ The binding of allosteric ligand molecules may lead to the conversion of a protein from one conformational/ activity state into another or to the release of active subunits (see Figure 3-27).

■ Two classes of intracellular switch proteins regulate a variety of cellular processes: (1) calmodulin and related Ca^{2+}-binding proteins in the EF hand family and (2) members of the GTPase superfamily (e.g., Ras and G_α), which cycle between active GTP-bound and inactive GDP-bound forms (see Figure 3-29).

■ The phosphorylation and dephosphorylation of amino acid side chains by protein kinases and phosphatases provide reversible on/off regulation of numerous proteins.

■ Nonallosteric mechanisms for regulating protein activity include proteolytic cleavage, which irreversibly converts inactive zymogens into active enzymes, compartmentation of proteins, and signal-induced modulation of protein synthesis and degradation.

3.6 Purifying, Detecting, and Characterizing Proteins

A protein must be purified before its structure and the mechanism of its action can be studied. However, because proteins vary in size, charge, and water solubility, no single method can be used to isolate all proteins. To isolate one particular protein from the estimated 10,000 different proteins in a cell is a daunting task that requires methods both for separating proteins and for detecting the presence of specific proteins.

Any molecule, whether protein, carbohydrate, or nucleic acid, can be separated, or *resolved*, from other molecules on the basis of their differences in one or more physical or chemical characteristics. The larger and more numerous the differences between two proteins, the easier and more efficient their separation. The two most widely used characteristics for separating proteins are *size*, defined as either length or mass, and *binding affinity* for specific ligands. In this section, we briefly outline several important techniques for separating proteins; these techniques are also useful for the separation of nucleic acids and other biomolecules. (Specialized methods for removing membrane proteins from membranes are described in the next chapter after the unique properties of these proteins are discussed.) We then consider general methods for detecting, or *assaying*, specific proteins, including the use of radioactive compounds for tracking biological activity. Finally, we consider several techniques for characterizing a protein's mass, sequence, and three-dimensional structure.

Centrifugation Can Separate Particles and Molecules That Differ in Mass or Density

The first step in a typical protein purification scheme is centrifugation. The principle behind centrifugation is that two particles in suspension (cells, organelles, or molecules) with different masses or densities will settle to the bottom of a tube at different rates. Remember, *mass* is the weight of a sample (measured in grams), whereas *density* is the ratio of its weight to volume (grams/liter). Proteins vary greatly in mass but not in density. Unless a protein has an attached lipid or carbohydrate, its density will not vary by more than 15 percent from 1.37 g/cm^3, the average protein density. Heavier or more dense molecules settle, or sediment, more quickly than lighter or less dense molecules.

A centrifuge speeds sedimentation by subjecting particles in suspension to centrifugal forces as great as 1,000,000 times the force of gravity *g*, which can sediment particles as small as 10 kDa. Modern ultracentrifuges achieve these forces by reaching speeds of 150,000 revolutions per minute (rpm) or greater. However, small particles with masses of 5 kDa or less will not sediment uniformly even at such high rotor speeds.

Centrifugation is used for two basic purposes: (1) as a preparative technique to separate one type of material from others and (2) as an analytical technique to measure physical properties (e.g., molecular weight, density, shape, and equilibrium binding constants) of macromolecules. The *sedimentation constant, s,* of a protein is a measure of its sedimentation rate. The sedimentation constant is commonly expressed in svedbergs (S): 1 S = 10^{-13} seconds.

Differential Centrifugation The most common initial step in protein purification is the separation of soluble proteins from insoluble cellular material by *differential centrifugation*. A starting mixture, commonly a cell homogenate, is poured into a tube and spun at a rotor speed and for a period of time that forces cell organelles such as nuclei to collect as a pellet at the bottom; the soluble proteins remain in the supernatant (Figure 3-31a). The supernatant fraction then is poured off and can be subjected to other purification methods to separate the many different proteins that it contains.

Rate-Zonal Centrifugation On the basis of differences in their masses, proteins can be separated by centrifugation through a solution of increasing density called a *density gradient*. A concentrated sucrose solution is commonly used to form density gradients. When a protein mixture is layered on top of a sucrose gradient in a tube and subjected to centrifugation, each protein in the mixture migrates down the tube at a rate controlled by the factors that affect the sedimentation constant. All the proteins start from a thin zone at the top of the tube and separate into bands, or zones (actually disks), of proteins of different masses. In this separation technique, called *rate-zonal centrifugation,* samples are centrifuged just long enough to separate the molecules of interest into discrete zones (Figure 3-31b). If a sample is centrifuged for too short a time, the different protein molecules will not separate sufficiently. If a sample is centrifuged much

longer than necessary, all the proteins will end up in a pellet at the bottom of the tube.

Although the sedimentation rate is strongly influenced by particle mass, rate-zonal centrifugation is seldom effective in determining *precise* molecular weights because variations in shape also affect sedimentation rate. The exact effects of shape are hard to assess, especially for proteins and single-stranded nucleic acid molecules that can assume many complex shapes. Nevertheless, rate-zonal centrifugation has proved to be the most practical method for separating many different types of polymers and particles. A second density-gradient technique, called *equilibrium density-gradient centrifugation,* is used mainly to separate DNA or organelles (see Figure 5-37).

Electrophoresis Separates Molecules on the Basis of Their Charge : Mass Ratio

Electrophoresis is a technique for separating molecules in a mixture under the influence of an applied electric field. Dissolved molecules in an electric field move, or migrate, at a speed determined by their charge:mass ratio. For example, if two molecules have the same mass and shape, the one with the greater net charge will move faster toward an electrode.

SDS-Polyacrylamide Gel Electrophoresis Because many proteins or nucleic acids that differ in size and shape have nearly identical charge:mass ratios, electrophoresis of these macromolecules in solution results in little or no separation of molecules of different lengths. However, successful separation of proteins and nucleic acids can be accomplished by electrophoresis in various gels (semisolid suspensions in water) rather than in a liquid solution. Electrophoretic separation of proteins is most commonly performed in *polyacrylamide gels*. When a mixture of proteins is applied to a gel and an electric current is applied, smaller proteins migrate faster through the gel than do larger proteins.

Gels are cast between a pair of glass plates by polymerizing a solution of acrylamide monomers into polyacrylamide chains and simultaneously cross-linking the chains into a semisolid matrix. The *pore size* of a gel can be varied by adjusting the concentrations of polyacrylamide and the cross-linking reagent. The rate at which a protein moves through a gel is influenced by the gel's pore size and the strength of the electric field. By suitable adjustment of these parameters, proteins of widely varying sizes can be separated.

In the most powerful technique for resolving protein mixtures, proteins are exposed to the ionic detergent SDS (sodium dodecylsulfate) before and during gel electrophoresis (Figure 3-32). SDS denatures proteins, causing multimeric proteins to dissociate into their subunits, and all polypeptide chains are forced into extended conformations with similar charge:mass ratios. SDS treatment thus

(a) Differential centrifugation

1 **Sample is poured into tube**

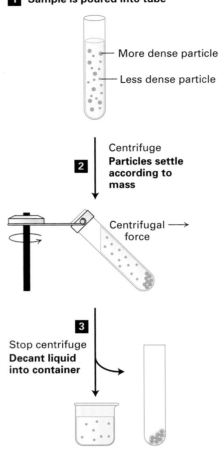

(b) Rate-zonal centrifugation

1 **Sample is layered on top of gradient**

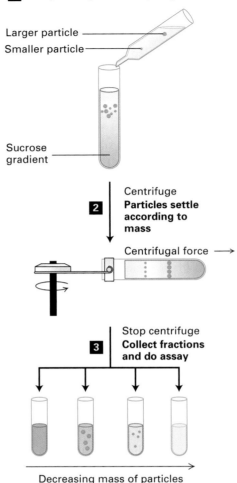

Decreasing mass of particles

▲ **EXPERIMENTAL FIGURE 3-31 Centrifugation techniques separate particles that differ in mass or density.** (a) In differential centrifugation, a cell homogenate or other mixture is spun long enough to sediment the denser particles (e.g., cell organelles, cells), which collect as a pellet at the bottom of the tube (step **2**). The less dense particles (e.g., soluble proteins, nucleic acids) remain in the liquid supernatant, which can be transferred to another tube (step **3**). (b) In rate-zonal centrifugation, a mixture is spun just long enough to separate molecules that differ in mass but may be similar in shape and density (e.g., globular proteins, RNA molecules) into discrete zones within a density gradient commonly formed by a concentrated sucrose solution (step **2**). Fractions are removed from the bottom of the tube and assayed (step **5**).

eliminates the effect of differences in shape, and so chain length, which corresponds to mass, is the sole determinant of the migration rate of proteins in SDS-polyacrylamide electrophoresis. Even chains that differ in molecular weight by less than 10 percent can be separated by this technique. Moreover, the molecular weight of a protein can be estimated by comparing the distance that it migrates through a gel with the distances that proteins of known molecular weight migrate.

Two-Dimensional Gel Electrophoresis Electrophoresis of all cellular proteins through an SDS gel can separate proteins having relatively large differences in mass but cannot resolve proteins having similar masses (e.g., a 41-kDa protein from a 42-kDa protein). To separate proteins of similar masses, another physical characteristic must be exploited. Most commonly, this characteristic is electric charge, which is determined by the number of acidic and basic residues in a protein. Two unrelated proteins having similar masses are

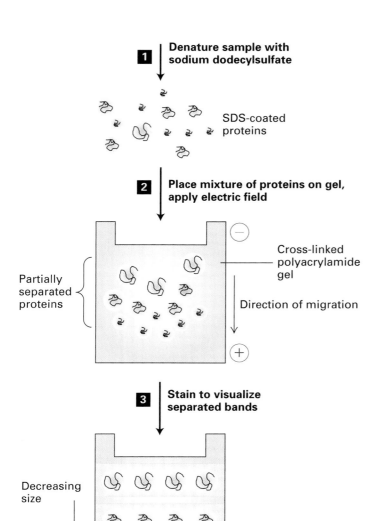

1 | Denature sample with sodium dodecylsulfate

SDS-coated proteins

2 | Place mixture of proteins on gel, apply electric field

Partially separated proteins

Cross-linked polyacrylamide gel

Direction of migration

3 | Stain to visualize separated bands

Decreasing size

◄ **EXPERIMENTAL FIGURE 3-32 SDS-polyacrylamide gel electrophoresis separates proteins solely on the basis of their masses.** Initial treatment with SDS, a negatively charged detergent, dissociates multimeric proteins and denatures all the polypeptide chains (step **1**). During electrophoresis, the SDS-protein complexes migrate through the polyacrylamide gel (step **2**). Small proteins are able to move through the pores more easily, and faster, than larger proteins. Thus the proteins separate into bands according to their sizes as they migrate through the gel. The separated protein bands are visualized by staining with a dye (step **3**).

unlikely to have identical net charges because their sequences, and thus the number of acidic and basic residues, are different.

In two-dimensional electrophoresis, proteins are separated sequentially, first by their charges and then by their masses (Figure 3-33a). In the first step, a cell extract is fully denatured by high concentrations (8 M) of urea and then layered on a gel strip that contains an continuous pH gradient. The gradient is formed by *ampholytes,* a mixture of polyanionic and polycationic molecules, that are cast into the gel, with the most acidic ampholyte at one end and the most basic ampholyte at the opposite end. A charged protein will migrate through the gradient until it reaches its **isoelectric point (pI),** the pH at which the net charge of the protein is zero. This technique, called iso-electric focusing (IEF), can resolve proteins that differ by only one charge unit. Proteins that have been separated on an IEF gel can then be separated in a second dimension on the basis of their molecular weights. To accomplish this separation, the IEF gel strip is placed lengthwise on a polyacrylamide slab gel, this time saturated with SDS. When an electric field is imposed, the proteins will migrate from the IEF gel into the SDS slab gel and then separate according to their masses.

The sequential resolution of proteins by charge and mass can achieve excellent separation of cellular proteins (Figure 3-33b). For example, two-dimensional gels have been very useful in comparing the proteomes in undifferentiated and differentiated cells or in normal and cancer cells because as many as 1000 proteins can be resolved simultaneously.

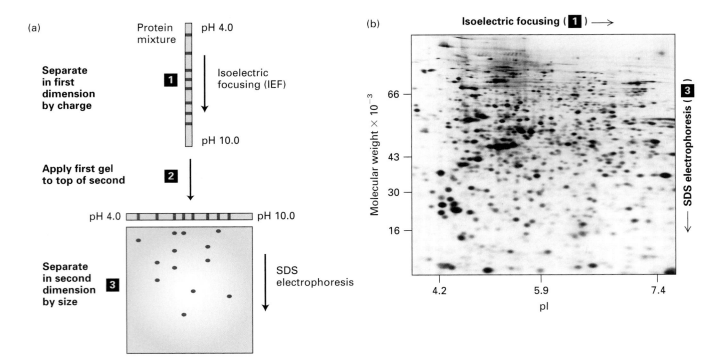

▲ EXPERIMENTAL FIGURE 3-33 Two-dimensional gel electrophoresis can separate proteins of similar mass. (a) In this technique, proteins are first separated on the basis of their charges by isoelectric focusing (step **1**). The resulting gel strip is applied to an SDS-polyacrylamide gel and the proteins are separated into bands by mass (step **3**). (b) In this two-dimensional gel of a protein extract from cultured cells, each spot represents a single polypeptide. Polypeptides can be detected by dyes, as here, or by other techniques such as autoradiography. Each polypeptide is characterized by its isoelectric point (pI) and molecular weight. [Part (b) courtesy of J. Celis.]

Liquid Chromatography Resolves Proteins by Mass, Charge, or Binding Affinity

A third common technique for separating mixtures of proteins, as well as other molecules, is based on the principle that molecules dissolved in a solution will interact (bind and dissociate) with a solid surface. If the solution is allowed to flow across the surface, then molecules that interact frequently with the surface will spend more time bound to the surface and thus move more slowly than molecules that interact infrequently with the surface. In this technique, called **liquid chromatography,** the sample is placed on top of a tightly packed column of spherical beads held within a glass cylinder. The nature of these beads determines whether the separation of proteins depends on differences in mass, charge, or binding affinity.

Gel Filtration Chromatography Proteins that differ in mass can be separated on a column composed of porous beads made from polyacrylamide, dextran (a bacterial polysaccharide), or agarose (a seaweed derivative), a technique called *gel filtration chromatography.* Although proteins *flow around* the spherical beads in gel filtration chromatography, they spend some time within the large depressions that cover a bead's surface. Because smaller proteins can penetrate into these depressions more easily than can larger proteins, they travel through a gel filtration column more slowly than do larger proteins (Figure 3-34a). (In contrast, proteins migrate *through* the pores in an electrophoretic gel; thus smaller proteins move faster than larger ones.) The total volume of liquid required to elute a protein from a gel filtration column depends on its mass: the smaller the mass, the greater the elution volume. By use of proteins of known mass, the elution volume can be used to estimate the mass of a protein in a mixture.

Ion-Exchange Chromatography In a second type of liquid chromatography, called *ion-exchange chromatography,* proteins are separated on the basis of differences in their charges. This technique makes use of specially modified beads whose surfaces are covered by amino groups or carboxyl groups and thus carry either a positive charge (NH_3^+) or a negative charge (COO^-) at neutral pH.

The proteins in a mixture carry various net charges at any given pH. When a solution of a protein mixture flows through a column of positively charged beads, only proteins with a net negative charge (acidic proteins) adhere to the beads; neutral and positively charged (basic) proteins flow unimpeded through the column (Figure 3-34b). The acidic proteins are then eluted selectively by passing a gradient of increasing concentrations of salt through the column. At low

(a) Gel filtration chromatography

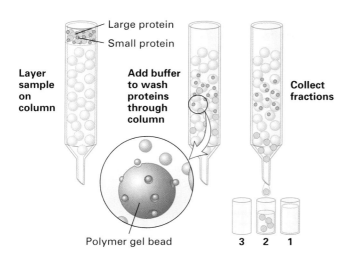

Layer
sample
on
column

Add buffer
to wash
proteins
through
column

Collect
fractions

Large protein
Small protein

Polymer gel bead

3　2　1

(c) Antibody-affinity chromatography

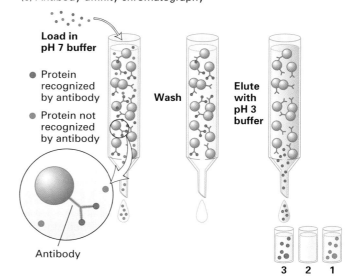

Load in
pH 7 buffer

● Protein
recognized
by antibody
● Protein not
recognized
by antibody

Wash

Elute
with
pH 3
buffer

Antibody

3　2　1

(b) Ion-exchange chromatography

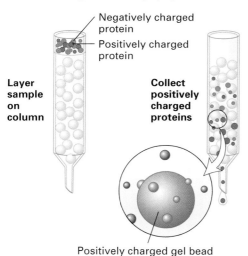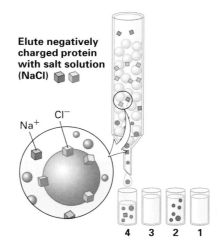

Layer
sample
on
column

Negatively charged
protein
Positively charged
protein

Collect
positively
charged
proteins

Positively charged gel bead

Elute negatively
charged protein
with salt solution
(NaCl)

Na$^+$　Cl$^-$

4　3　2　1

▲ **EXPERIMENTAL FIGURE 3-34 Three commonly used
liquid chromatographic techniques separate proteins on the
basis of mass, charge, or affinity for a specific ligand.** (a) Gel
filtration chromatography separates proteins that differ in size.
A mixture of proteins is carefully layered on the top of a glass
cylinder packed with porous beads. Smaller proteins travel
through the column more slowly than larger proteins. Thus
different proteins have different elution volumes and can be
collected in separate liquid fractions from the bottom. (b) Ion-
exchange chromatography separates proteins that differ in net
charge in columns packed with special beads that carry either a
positive charge (shown here) or a negative charge. Proteins
having the same net charge as the beads are repelled and flow
through the column, whereas proteins having the opposite
charge bind to the beads. Bound proteins—in this case,
negatively charged—are eluted by passing a salt gradient (usually
of NaCl or KCl) through the column. As the ions bind to the
beads, they desorb the protein. (c) In antibody-affinity
chromatography, a specific antibody is covalently attached to
beads packed in a column. Only protein with high affinity for the
antibody is retained by the column; all the nonbinding proteins
flow through. The bound protein is eluted with an acidic solution,
which disrupts the antigen–antibody complexes.

salt concentrations, protein molecules and beads are attracted by their opposite charges. At higher salt concentrations, negative salt ions bind to the positively charged beads,
displacing the negatively charged proteins. In a gradient of
increasing salt concentration, weakly charged proteins are
eluted first and highly charged proteins are eluted last. Similarly, a negatively charged column can be used to retain and
fractionate basic proteins.

Affinity Chromatography The ability of proteins to bind specifically to other molecules is the basis of *affinity chromatography*. In this technique, ligand molecules that bind to the protein of interest are covalently attached to the beads used to form the column. Ligands can be enzyme substrates or other small molecules that bind to specific proteins. In a widely used form of this technique, *antibody-affinity chromatography,* the attached ligand is an antibody specific for the desired protein (Figure 3-34c).

An affinity column will retain only those proteins that bind the ligand attached to the beads; the remaining proteins, regardless of their charges or masses, will pass through the column without binding to it. However, if a retained protein interacts with other molecules, forming a complex, then the entire complex is retained on the column. The proteins bound to the affinity column are then eluted by adding an excess of ligand or by changing the salt concentration or pH. The ability of this technique to separate particular proteins depends on the selection of appropriate ligands.

Highly Specific Enzyme and Antibody Assays Can Detect Individual Proteins

The purification of a protein, or any other molecule, requires a specific assay that can detect the molecule of interest in column fractions or gel bands. An assay capitalizes on some highly distinctive characteristic of a protein: the ability to bind a particular ligand, to catalyze a particular reaction, or to be recognized by a specific antibody. An assay must also

be simple and fast to minimize errors and the possibility that the protein of interest becomes denatured or degraded while the assay is performed. The goal of any purification scheme is to isolate sufficient amounts of a given protein for study; thus a useful assay must also be sensitive enough that only a small proportion of the available material is consumed. Many common protein assays require just from 10^{-9} to 10^{-12} g of material.

Chromogenic and Light-Emitting Enzyme Reactions Many assays are tailored to detect some functional aspect of a protein. For example, enzyme assays are based on the ability to detect the loss of substrate or the formation of product. Some enzyme assays utilize *chromogenic* substrates, which change color in the course of the reaction. (Some substrates are naturally chromogenic; if they are not, they can be linked to a chromogenic molecule.) Because of the specificity of an enzyme for its substrate, only samples that contain the enzyme will change color in the presence of a chromogenic substrate and other required reaction components; the rate of the reaction provides a measure of the quantity of enzyme present.

Such chromogenic enzymes can also be fused or chemically linked to an antibody and used to "report" the presence or location of the antigen. Alternatively, *luciferase,* an enzyme present in fireflies and some bacteria, can be linked to an antibody. In the presence of ATP and luciferin, luciferase catalyzes a light-emitting reaction. In either case, after the antibody binds to the protein of interest, substrates of the linked enzyme are added and the appearance of color or

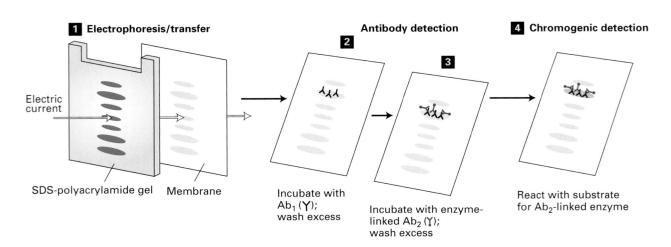

▲ **EXPERIMENTAL FIGURE 3-35 Western blotting (immunoblotting) combines several techniques to resolve and detect a specific protein.** Step **1**: After a protein mixture has been electrophoresed through an SDS gel, the separated bands are transferred (blotted) from the gel onto a porous membrane. Step **2**: The membrane is flooded with a solution of antibody (Ab$_1$) specific for the desired protein. Only the band containing this protein binds the antibody, forming a layer of antibody molecules (although their position

cannot be seen at this point). After sufficient time for binding, the membrane is washed to remove unbound Ab$_1$. Step **3**: The membrane is incubated with a second antibody (Ab$_2$) that binds to the bound Ab$_1$. This second antibody is covalently linked to alkaline phosphatase, which catalyzes a chromogenic reaction. Step **4**: Finally, the substrate is added and a deep purple precipitate forms, marking the band containing the desired protein.

emitted light is monitored. A variation of this technique, particularly useful in detecting specific proteins within living cells, makes use of green fluorescent protein (GFP), a naturally fluorescent protein found in jellyfish (see Figure 5-46).

Western Blotting A powerful method for detecting a particular protein in a complex mixture combines the superior resolving power of gel electrophoresis, the specificity of antibodies, and the sensitivity of enzyme assays. Called **Western blotting**, or immunoblotting, this multistep procedure is commonly used to separate proteins and then identify a specific protein of interest. As shown in Figure 3-35, two different antibodies are used in this method, one specific for the desired protein and the other linked to a reporter enzyme.

Radioisotopes Are Indispensable Tools for Detecting Biological Molecules

A sensitive method for tracking a protein or other biological molecule is by detecting the radioactivity emitted from radioisotopes introduced into the molecule. At least one atom in a radiolabeled molecule is present in a radioactive form, called a **radioisotope.**

Radioisotopes Useful in Biological Research Hundreds of biological compounds (e.g., amino acids, nucleosides, and numerous metabolic intermediates) labeled with various radioisotopes are commercially available. These preparations vary considerably in their *specific activity,* which is the amount of radioactivity per unit of material, measured in disintegrations per minute (dpm) per millimole. The specific activity of a labeled compound depends on the probability of decay of the radioisotope, indicated by its *half-life,* which is the time required for half the atoms to undergo radioactive decay. In general, the shorter the half-life of a radioisotope, the higher its specific activity (Table 3-3).

The specific activity of a labeled compound must be high enough that sufficient radioactivity is incorporated into cellular molecules to be accurately detected. For example, methionine and cysteine labeled with sulfur-35 (^{35}S) are widely used to label cellular proteins because preparations of these

amino acids with high specific activities (>10^{15} dpm/mmol) are available. Likewise, commercial preparations of ^{3}H-labeled nucleic acid precursors have much higher specific activities than those of the corresponding ^{14}C-labeled preparations. In most experiments, the former are preferable because they allow RNA or DNA to be adequately labeled after a shorter time of incorporation or require a smaller cell sample. Various phosphate-containing compounds in which every phosphorus atom is the radioisotope phosphorus-32 are readily available. Because of their high specific activity, ^{32}P-labeled nucleotides are routinely used to label nucleic acids in cell-free systems.

Labeled compounds in which a radioisotope replaces atoms normally present in the molecule have the same chemical properties as the corresponding nonlabeled compounds. Enzymes, for instance, cannot distinguish between substrates labeled in this way and their nonlabeled substrates. In contrast, labeling with the radioisotope iodine-125 (^{125}I) requires the covalent addition of ^{125}I to a protein or nucleic acid. Because this labeling procedure modifies the chemical structure of a protein or nucleic acid, the biological activity of the labeled molecule may differ somewhat from that of the nonlabeled form.

Labeling Experiments and Detection of Radiolabeled Molecules Whether labeled compounds are detected by **autoradiography,** a semiquantitative visual assay, or their radioactivity is measured in an appropriate "counter," a highly quantitative assay that can determine the concentration of a radiolabeled compound in a sample, depends on the nature of the experiment. In some experiments, both types of detection are used.

In one use of autoradiography, a cell or cell constituent is labeled with a radioactive compound and then overlaid with a photographic emulsion sensitive to radiation. Development of the emulsion yields small silver grains whose distribution corresponds to that of the radioactive material. Autoradiographic studies of whole cells were crucial in determining the intracellular sites where various macromolecules are synthesized and the subsequent movements of these macromolecules within cells. Various techniques employing fluorescent microscopy, which we describe in the next chapter, have largely supplanted autoradiography for studies of this type. However, autoradiography is commonly used in various assays for detecting specific isolated DNA or RNA sequences (Chapter 9).

Quantitative measurements of the amount of radioactivity in a labeled material are performed with several different instruments. A *Geiger counter* measures ions produced in a gas by the β particles or γ rays emitted from a radioisotope. In a *scintillation counter,* a radiolabeled sample is mixed with a liquid containing a fluorescent compound that emits a flash of light when it absorbs the energy of the β particles or γ rays released in the decay of the radioisotope; a phototube in the instrument detects and counts these light flashes. *Phosphorimagers* are used to detect radiolabeled compounds on a surface, storing digital data on the number of decays in

TABLE 3-3	Radioisotopes Commonly Used in Biological Research
Isotope	**Half-Life**
Phosphorus-32	14.3 days
Iodine-125	60.4 days
Sulfur-35	87.5 days
Tritium (hydrogen-3)	12.4 years
Carbon-14	5730.4 years

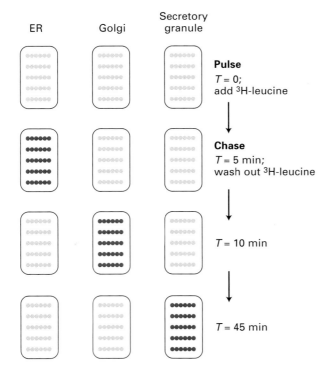

ER　　Golgi　　Secretory granule

Pulse
$T = 0$;
add ^{3}H-leucine

Chase
$T = 5$ min;
wash out ^{3}H-leucine

$T = 10$ min

$T = 45$ min

▲ **EXPERIMENTAL FIGURE 3-36 Pulse-chase experiments can track the pathway of protein movement within cells.** To determine the pathway traversed by secreted proteins subsequent to their synthesis on the rough endoplasmic reticulum (ER), cells are briefly incubated in a medium containing a radiolabeled amino acid (e.g., [^{3}H]leucine), the pulse, which will label any protein synthesized during this period. The cells are then washed with buffer to remove the pulse and transferred to medium lacking a radioactive precursor, the chase. Samples taken periodically are analyzed by autoradiography to determine the cellular location of labeled protein. At the beginning of the experiment ($t = 0$), no protein is labeled, as indicated by the green dotted lines. At the end of the pulse ($t = 5$ minutes), all the labeled protein (red lines) appears in the ER. At subsequent times, this newly synthesized labeled protein is visualized first in the Golgi complex and then in secretory vesicles. Because any protein synthesized during the chase period is not labeled, the movement of the labeled protein can be defined quite precisely.

disintegrations per minute per small pixel of surface area. These instruments, which can be thought of as a kind of reusable electronic film, are commonly used to quantitate radioactive molecules separated by gel electrophoresis and are replacing photographic emulsions for this purpose.

A combination of labeling and biochemical techniques and of visual and quantitative detection methods is often employed in labeling experiments. For instance, to identify the major proteins synthesized by a particular cell type, a sample of the cells is incubated with a radioactive amino acid (e.g., [^{35}S]methionine) for a few minutes. The mixture of cellular proteins is then resolved by gel electrophoresis, and the gel

is subjected to autoradiography or phosphorimager analysis. The radioactive bands correspond to newly synthesized proteins, which have incorporated the radiolabeled amino acid. Alternatively, the proteins can be resolved by liquid chromatography, and the radioactivity in the eluted fractions can be determined quantitatively with a counter.

Pulse-chase experiments are particularly useful for tracing changes in the intracellular location of proteins or the transformation of a metabolite into others over time. In this experimental protocol, a cell sample is exposed to a radiolabeled compound—the "pulse"—for a brief period of time, then washed with buffer to remove the labeled pulse, and finally incubated with a nonlabeled form of the compound—the "chase" (Figure 3-36). Samples taken periodically are assayed to determine the location or chemical form of the radiolabel. A classic use of the pulse-chase technique was in studies to elucidate the pathway traversed by secreted proteins from their site of synthesis in the endoplasmic reticulum to the cell surface (Chapter 17).

Mass Spectrometry Measures the Mass of Proteins and Peptides

A powerful technique for measuring the mass of molecules such as proteins and peptides is *mass spectrometry*. This

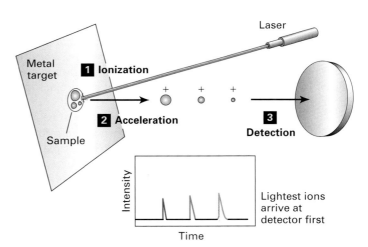

Laser

Metal target

1 Ionization

2 Acceleration

3 Detection

Sample

Intensity

Time

Lightest ions arrive at detector first

▲ **EXPERIMENTAL FIGURE 3-37 The molecular weight of proteins and peptides can be determined by time-of-flight mass spectrometry.** In a laser-desorption mass spectrometer, pulses of light from a laser ionize a protein or peptide mixture that is absorbed on a metal target (**1**). An electric field accelerates the molecules in the sample toward the detector (**2** and **3**). The time to the detector is inversely proportional to the mass of a molecule. For molecules having the same charge, the time to the detector is inversely proportional to the mass. The molecular weight is calculated using the time of flight of a standard.

technique requires a method for ionizing the sample, usually a mixture of peptides or proteins, accelerating the molecular ions, and then detecting the ions. In a laser desorption mass spectrometer, the protein sample is mixed with an organic acid and then dried on a metal target. Energy from a laser ionizes the proteins, and an electric field accelerates the ions down a tube to a detector (Figure 3-37). Alternatively, in an electrospray mass spectrometer, a fine mist containing the sample is ionized and then introduced into a separation chamber where the positively charged molecules are accelerated by an electric field. In both instruments, the time of flight is inversely proportional to a protein's mass and directly proportional to its charge. As little as 1×10^{-15} mol (1 femtomole) of a protein as large as 200,000 MW can be measured with an error of 0.1 percent.

Protein Primary Structure Can Be Determined by Chemical Methods and from Gene Sequences

The classic method for determining the amino acid sequence of a protein is *Edman degradation*. In this procedure, the free amino group of the N-terminal amino acid of a polypeptide is labeled, and the labeled amino acid is then cleaved from the polypeptide and identified by high-pressure liquid chromatography. The polypeptide is left one residue shorter, with a new amino acid at the N-terminus. The cycle is repeated on the ever shortening polypeptide until all the residues have been identified.

Before about 1985, biologists commonly used the Edman chemical procedure for determining protein sequences. Now, however, protein sequences are determined primarily by analysis of genome sequences. The complete genomes of several organisms have already been sequenced, and the database of genome sequences from humans and numerous model organisms is expanding rapidly. As discussed in Chapter 9, the sequences of proteins can be deduced from DNA sequences that are predicted to encode proteins.

A powerful approach for determining the primary structure of an isolated protein combines mass spectroscopy and the use of sequence databases. First, mass spectrometry is used to determine the *peptide mass fingerprint* of the protein. A peptide mass fingerprint is a compilation of the molecular weights of peptides that are generated by a specific protease. The molecular weights of the parent protein and its proteolytic fragments are then used to search genome databases for any similarly sized protein with identical or similar peptide mass maps.

Peptides with a Defined Sequence Can Be Synthesized Chemically

Synthetic peptides that are identical with peptides synthesized in vivo are useful experimental tools in studies of proteins and cells. For example, short synthetic peptides of 10–15 residues can function as antigens to trigger the production of antibodies in animals. A synthetic peptide, when coupled to a large protein carrier, can trick an animal into producing antibodies that bind the full-sized, natural protein antigen. As we'll see throughout this book, antibodies are extremely versatile reagents for isolating proteins from mixtures by affinity chromatography (see Figure 3-34c), for separating and detecting proteins by Western blotting (see Figure 3-35), and for localizing proteins in cells by microscopic techniques described in Chapter 5.

Peptides are routinely synthesized in a test tube from monomeric amino acids by condensation reactions that form peptide bonds. Peptides are constructed sequentially by coupling the C-terminus of a monomeric amino acid with the N-terminus of the growing peptide. To prevent unwanted reactions entailing the amino groups and carboxyl groups of the side chains during the coupling steps, a protecting (blocking) group is attached to the side chains. Without these protecting groups, branched peptides would be generated. In the last steps of synthesis, the side chain–protecting groups are removed and the peptide is cleaved from the resin on which synthesis takes place.

Protein Conformation Is Determined by Sophisticated Physical Methods

In this chapter, we have emphasized that protein function is dependent on protein structure. Thus, to figure out how a protein works, its three-dimensional structure must be known. Determining a protein's conformation requires sophisticated physical methods and complex analyses of the experimental data. We briefly describe three methods used to generate three-dimensional models of proteins.

X-Ray Crystallography The use of **x-ray crystallography** to determine the three-dimensional structures of proteins was pioneered by Max Perutz and John Kendrew in the 1950s. In this technique, beams of x-rays are passed through a protein crystal in which millions of protein molecules are precisely aligned with one another in a rigid array characteristic of the protein. The wavelengths of x-rays are about 0.1–0.2 nm, short enough to resolve the atoms in the protein crystal. Atoms in the crystal scatter the x-rays, which produce a diffraction pattern of discrete spots when they are intercepted by photographic film (Figure 3-38). Such patterns are extremely complex—composed of as many as 25,000 diffraction spots for a small protein. Elaborate calculations and modifications of the protein (such as the binding of heavy metals) must be made to interpret the diffraction pattern and to solve the structure of the protein. The process is analogous to reconstructing the precise shape of a rock from the ripples that it creates in a pond. To date, the detailed three-dimensional structures of more than 10,000 proteins have been established by x-ray crystallography.

Cryoelectron Microscopy Although some proteins readily crystallize, obtaining crystals of others—particularly large multisubunit proteins—requires a time-consuming trial-and-

(a)

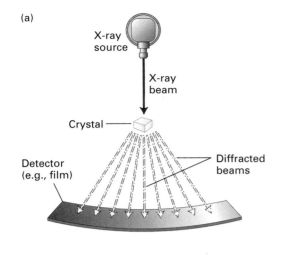

(b)

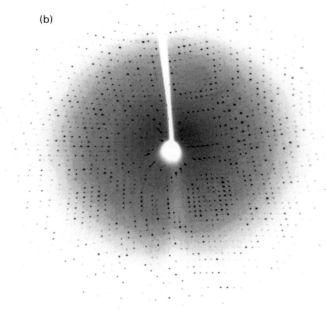

▲ **EXPERIMENTAL FIGURE 3-38 X-ray crystallography provides diffraction data from which the three-dimensional structure of a protein can be determined.** (a) Basic components of an x-ray crystallographic determination. When a narrow beam of x-rays strikes a crystal, part of it passes straight through and the rest is scattered (diffracted) in various directions. The intensity of the diffracted waves is recorded on an x-ray film or with a solid-state electronic detector. (b) X-ray diffraction pattern for a topoisomerase crystal collected on a solid-state detector. From complex analyses of patterns like this one, the location of every atom in a protein can be determined. [Part (a) adapted from L. Stryer, 1995, *Biochemistry,* 4th ed., W. H. Freeman and Company, p. 64; part (b) courtesy of J. Berger.]

error effort to find just the right conditions. The structures of such difficult-to-crystallize proteins can be obtained by *cryoelectron microscopy.* In this technique, a protein sample is rapidly frozen in liquid helium to preserve its structure and then examined in the frozen, hydrated state in a cryoelectron microscope. Pictures are recorded on film by using a low dose of electrons to prevent radiation-induced damage to the structure. Sophisticated computer programs analyze the images and reconstruct the protein's structure in three dimensions. Recent advances in cryoelectron microscopy permit researchers to generate molecular models that compare with those derived from x-ray crystallography. The use of cryoelectron microscopy and other types of electron microscopy for visualizing cell structures are discussed in Chapter 5.

NMR Spectroscopy The three-dimensional structures of small proteins containing about as many as 200 amino acids can be studied with nuclear magnetic resonance (NMR) spectroscopy. In this technique, a concentrated protein solution is placed in a magnetic field and the effects of different radio frequencies on the spin of different atoms are measured. The behavior of any atom is influenced by neighboring atoms in adjacent residues, with closely spaced residues being more perturbed than distant residues. From the magnitude of the effect, the distances between residues can be calculated; these distances are then used to generate a model of the three-dimensional structure of the protein.

Although NMR does not require the crystallization of a protein, a definite advantage, this technique is limited to proteins smaller than about 20 kDa. However, NMR analysis can also be applied to protein domains, which tend to be small enough for this technique and can often be obtained as stable structures.

KEY CONCEPTS OF SECTION 3.6

Purifying, Detecting, and Characterizing Proteins

■ Proteins can be separated from other cell components and from one another on the basis of differences in their physical and chemical properties.

■ Centrifugation separates proteins on the basis of their rates of sedimentation, which are influenced by their masses and shapes.

■ Gel electrophoresis separates proteins on the basis of their rates of movement in an applied electric field. SDS-polyacrylamide gel electrophoresis can resolve polypeptide chains differing in molecular weight by 10 percent or less (see Figure 3-32).

■ Liquid chromatography separates proteins on the basis of their rates of movement through a column packed with spherical beads. Proteins differing in mass are resolved on gel filtration columns; those differing in charge, on ion-exchange columns; and those differing in ligand-binding properties, on affinity columns (see Figure 3-34).

■ Various assays are used to detect and quantify proteins. Some assays use a light-producing reaction or radioactivity to generate a signal. Other assays produce an amplified colored signal with enzymes and chromogenic substrates.

■ Antibodies are powerful reagents used to detect, quantify, and isolate proteins. They are used in affinity chromatography and combined with gel electrophoresis in

Western blotting, a powerful method for separating and detecting a protein in a mixture (see Figure 3-35).

■ Autoradiography is a semiquantitative technique for detecting radioactively labeled molecules in cells, tissues, or electrophoretic gels.

■ Pulse-chase labeling can determine the intracellular fate of proteins and other metabolites (see Figure 3-36).

■ Three-dimensional structures of proteins are obtained by x-ray crystallography, cryoelectron microscopy, and NMR spectroscopy. X-ray crystallography provides the most detailed structures but requires protein crystallization. Cryoelectron microscopy is most useful for large protein complexes, which are difficult to crystallize. Only relatively small proteins are amenable to NMR analysis.

teins, especially in the early stages in the formation of insoluble filaments.

Understanding the operation of protein machines will require the measurement of many new characteristics of proteins. For example, because many machines do nonchemical work of some type, biologists will have to identify the energy sources (mechanical, electrical, or thermal) and measure the amounts of energy to determine the limits of a particular machine. Because most activities of machines include movement of one type or another, the force powering the movement and its relation to biological activity can be a source of insight into how force generation is coupled to chemistry. Improved tools such as optical traps and atomic force microscopes will enable detailed studies of the forces and chemistry pertinent to the operation of individual protein machines.

PERSPECTIVES FOR THE FUTURE

Impressive expansion of the computational power of computers is at the core of advances in determining the three-dimensional structures of proteins. For example, vacuum tube computers running on programs punched on cards were used to solve the first protein structures on the basis of x-ray crystallography. In the future, researchers aim to predict the structures of proteins only on the basis of amino acid sequences deduced from gene sequences. This computationally challenging problem requires supercomputers or large clusters of computers working in synchrony. Currently, only the structures of very small domains containing 100 residues or fewer can be predicted at a low resolution. However, continued developments in computing and models of protein folding, combined with large-scale efforts to solve the structures of all protein motifs by x-ray crystallography, will allow the prediction of the structures of larger proteins. With an exponentially expanding database of motifs, domains, and proteins, scientists will be able to identify the motifs in an unknown protein, match the motif to the sequence, and use this head start in predicting the three-dimensional structure of the entire protein.

New combined approaches will also help in in determining high-resolution structures of molecular machines such as those listed in Table 3-1. Although these very large macromolecular assemblies usually are difficult to crystallize and thus to solve by x-ray crystallography, they can be imaged in a cryoelectron microscope at liquid helium temperatures and high electron energies. From millions of individual "particles," each representing a random view of the protein complex, the three-dimensional structure can be built. Because subunits of the complex may already be solved by crystallography, a composite structure consisting of the x-ray-derived subunit structures fitted to the EM-derived model will be generated. An interesting application of this type of study would be the solution of the structures of amyloid and prion pro-

KEY TERMS

α helix *61*	molecular machine *59*
activation energy *74*	motif *63*
active site *75*	motor protein *79*
allostery *83*	peptide bond *60*
amyloid filament *73*	polypeptide *61*
autoradiography *93*	primary structure *61*
β sheet *61*	proteasome *71*
chaperone *69*	protein *61*
conformation *60*	proteome *60*
cooperativity *83*	quaternary structure *66*
domain *63*	rate-zonal centrifugation *87*
electrophoresis *87*	secondary structure *61*
homology *68*	tertiary structure *62*
K_m *76*	ubiquitin *71*
ligand *73*	V_{max} *76*
liquid chromatography *90*	x-ray crystallography *95*

REVIEW THE CONCEPTS

1. The three-dimensional structure of a protein is determined by its primary, secondary, and tertiary structures. Define the primary, secondary, and tertiary structures. What are some of the common secondary structures? What are the forces that hold together the secondary and tertiary structures? What is the quaternary structure?

2. Proper folding of proteins is essential for biological activity. Describe the roles of molecular chaperones and chaperonins in the folding of proteins.

3. Proteins are degraded in cells. What is ubiquitin, and what role does it play in tagging proteins for degradation? What is the role of proteasomes in protein degradation?

4. Enzymes can catalyze chemical reactions. How do enzymes increase the rate of a reaction? What constitutes the active site of an enzyme? For an enzyme-catalyzed reaction, what are K_m and V_{max}? For enzyme X, the K_m for substrate A is 0.4 mM and for substrate B is 0.01 mM. Which substrate has a higher affinity for enzyme X?

5. Motor proteins, such as myosin, convert energy into a mechanical force. Describe the three general properties characteristic of motor proteins. Describe the biochemical events that occur during one cycle of movement of myosin relative to an actin filament.

6. The function of proteins can be regulated in a number of ways. What is cooperativity, and how does it influence protein function? Describe how protein phosphorylation and proteolytic cleavage can modulate protein function.

7. A number of techniques can separate proteins on the basis of their differences in mass. Describe the use of two of these techniques, centrifugation and gel electrophoresis. The blood proteins transferrin (MW 76 kDa) and lysozyme (MW 15 kDa) can be separated by rate zonal centrifugation or SDS polyacrylamide gel electrophoresis. Which of the two proteins will sediment faster during centrifugation? Which will migrate faster during electrophoresis?

8. Chromatography is an analytical method used to separate proteins. Describe the principles for separating proteins by gel filtration, ion-exchange, and affinity chromatography.

9. Various methods have been developed for detecting proteins. Describe how radioisotopes and autoradiography can be used for labeling and detecting proteins. How does Western blotting detect proteins?

10. Physical methods are often used to determine protein conformation. Describe how x-ray crystallography, cryoelectron microscopy, and NMR spectroscopy can be used to determine the shape of proteins.

ANALYZE THE DATA

Proteomics involves the global analysis of protein expression. In one approach, all the proteins in control cells and treated cells are extracted and subsequently separated using two-dimensional gel electrophoresis. Typically, hundreds or thousands of protein spots are resolved and the steady-state levels of each protein are compared between control and treated cells. In the following example, only a few protein spots are shown for simplicity. Proteins are separated in the first dimension on the basis of charge by isoelectric focusing (pH 4–10) and then separated by size by SDS polyacrylamide gel electrophoresis. Proteins are detected with a stain such as Coomassie blue and assigned numbers for identification.

a. Cells are treated with a drug ("+ Drug") or left untreated ("Control") and then proteins are extracted and separated by two-dimensional gel electrophoresis. The stained gels are shown below. What do you conclude about the effect of the drug on the steady-state levels of proteins 1–7?

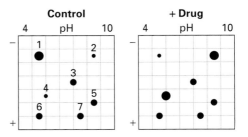

b. You suspect that the drug may be inducing a protein kinase and so repeat the experiment in part a in the presence of ^{32}P-labeled inorganic phosphate. In this experiment the two-dimensional gels are exposed to x-ray film to detect the presence of ^{32}P-labeled proteins. The x-ray films are shown below. What do you conclude from this experiment about the effect of the drug on proteins 1–7?

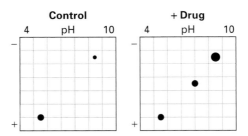

c. To determine the cellular localization of proteins 1–7, the cells from part a were separated into nuclear and cytoplasmic fractions by differential centrifugation. Two-dimensional gels were run and the stained gels are shown below. What do you conclude about the cellular localization of proteins 1–7?

d. Summarize the overall properties of proteins 1–7, combining the data from parts a, b, and c. Describe how you could determine the identity of any one of the proteins.

REFERENCES

General References

Berg, J. M., J. L. Tymoczko, and L. Stryer. 2002. *Biochemistry,* 5th ed. W. H. Freeman and Company, chaps. 2–4, 7–10.

Nelson, D. L., and M. M. Cox. 2000. *Lehninger Principles of Biochemistry,* 3d ed. Worth Publishers, chaps. 5–8.

Web Sites

Entry site into the proteins, structures, genomes, and taxonomy: http://www.ncbi.nlm.nih.gov/Entrez/

The protein 3D structure database: http://www.rcsb.org/

Structural classifications of proteins: http://scop.mrclmb.cam.ac. uk/scop/

Sites containing general information about proteins: http://www. expasy.ch/; http://www.proweb.org/

Sites for specific protein families: http://www.pkr.sdsc. edu/html/ index.shtml The protein kinase resource; http://www.mrc-lmb.cam. ac.uk/myosin/myosin.html The myosin home page; http://www. proweb.org/kinesin// The kinesin home page

Hierarchical Structure of Proteins

Branden, C., and J. Tooze. 1999. *Introduction to Protein Structure.* Garland.

Creighton, T. E. 1993. *Proteins: Structures and Molecular Properties,* 2d ed. W. H. Freeman and Company.

Hardison, R. 1998. Hemoglobins from bacteria to man: Evolution of different patterns of gene expression. *J. Exp. Biol.* **201:** 1099.

Lesk, A. M. 2001. *Introduction to Protein Architecture.* Oxford.

Macromolecular Machines. 1998. *Cell* **92:**291–423. A special review issue on protein machines.

Patthy, L. 1999. *Protein Evolution.* Blackwell Science.

Folding, Modification, and Degradation of Proteins

Cohen, F. E. 1999. Protein misfolding and prion diseases. *J. Mol. Biol.* **293:**313–320.

Dobson, C. M. 1999. Protein misfolding, evolution, and disease. *Trends Biochem. Sci.* **24:**329–332.

Hartl, F. U., and M. Hayer-Hartl. 2002. Molecular chaperones in the cytosol: From nascent chain to folded protein. *Science* **295:**1852–1858.

Kirschner, M. 1999. Intracellular proteolysis. *Trends Cell Biol.* **9:**M42–M45.

Kornitzer, D., and A. Ciechanover. 2000. Modes of regulation of ubiqutin-mediated protein degradation. *J. Cell Physiol.* **182:**1–11.

Laney, J. D., and M. Hochstrasser. 1999. Substrate targeting in the ubiquitin system. *Cell* **97:**427–430.

Rochet, J.-C., and P. T. Landsbury. 2000. Amyloid fibrillogenesis: Themes and variations. *Curr. Opin. Struct. Biol.* **10:**60–68.

Weissman, A. M. 2001. Themes and variations on ubiquitylation. *Nature Cell Biol.* **2:**169–177.

Zhang, X., F. Beuron, and P. S. Freemont. 2002. Machinery of protein folding and unfolding. *Curr. Opin. Struct. Biol.* **12:**231–238.

Zwickil, P., W. Baumeister, and A. Steven. 2000. Dis-assembly lines: The proteasome and related ATPase-assisted proteases. *Curr. Opin. Struct. Biol.* **10:**242–250.

Enzymes and the Chemical Work of Cells

Dressler, D. H., and H. Potter. 1991. *Discovering Enzymes.* Scientific American Library.

Fersht, A. 1999. *Enzyme Structure and Mechanism,* 3d ed. W. H. Freeman and Company.

Smith, C. M., et al. 1997. The protein kinase resource. *Trends Biochem. Sci.* **22:**444–446.

Taylor, S. S., and E. Radzio-Andzelm. 1994. Three protein kinase structures define a common motif. *Structure* **2:**345–355.

Molecular Motors and the Mechanical Work of Cells

Cooke, R. 2001. Motor proteins. *Encyclopedia Life Sciences.* Nature Publishing Group.

Spudich, J. A. 2001. The myosin swinging cross-bridge model. *Nature Rev. Mol. Cell Biol.* **2:**387–392.

Vale, R. D., and R. A. Milligan. 2000. The way things move: Looking under the hood of molecular motor proteins. *Science* **288:**88–95.

Common Mechanisms for Regulating Protein Function

Ackers, G. K. 1998. Deciphering the molecular code of hemoglobin allostery. *Adv. Protein Chem.* **51:**185–253.

Austin, D. J., G. R. Crabtree, and S. L. Schreiber. 1994. Proximity versus allostery: The role of regulated protein dimerization in biology. *Chem. Biol.* **1:**131–136.

Burack, W. R., and A. S. Shaw. 2000. Signal transduction: Hanging on a scaffold. *Curr. Opin. Cell Biol.* **12:**211–216.

Cox, S., E. Radzio-Andzelm, and S. S. Taylor. 1994. Domain movements in protein kinases. *Curr. Opin. Struct. Biol.* **4:**893–901.

Horovitz, A., Y. Fridmann, G. Kafri, and O. Yifrach. 2001. Review: Allostery in chaperonins. *J. Struct. Biol.* **135:**104–114.

Kawasaki, H., S. Nakayama, and R. H. Kretsinger. 1998. Classification and evolution of EF-hand proteins. *Biometals* **11:**277–295.

Lim, W. A. 2002. The modular logic of signaling proteins: Building allosteric switches from simple binding domains. *Curr. Opin. Struct. Biol.* **12:**61–68.

Ptashne, M., and A. Gann. 1998. Imposing specificity by localization: Mechanism and evolvability. *Curr. Biol.* **8:**R812–R822.

Saibil, H. R., A. L. Horwich, and W. A. Fenton. 2001. Allostery and protein substrate conformational change during GroEL/GroES-mediated protein folding. *Adv. Protein Chem.* **59:**45–72.

Yap, K. L., J. A. B. Ames, M. B. Sindells, and M. Ikura. 1999. Diversity of conformational states and changes within the EF-hand protein superfamily. *Proteins* **37:**499–507.

Purifying, Detecting, and Characterizing Proteins

Hames, B. D. *A Practical Approach.* Oxford University Press. A methods series that describes protein purification methods and assays.

4

BASIC MOLECULAR GENETIC MECHANISMS

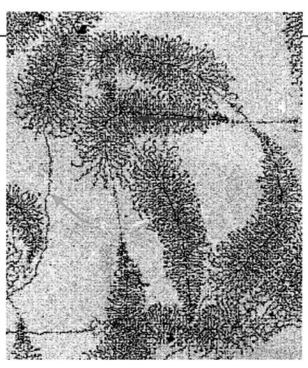

Electron micrograph of DNA (green arrow) being transcribed into RNA (red arrow). [O. L. Miller, Jr., and Barbara R. Beatty, Oak Ridge National Laboratory.]

The extraordinary versatility of proteins as molecular machines and switches, cellular catalysts, and components of cellular structures was described in Chapter 3. In this chapter we consider the **nucleic acids.** These macromolecules (1) contain the information for determining the amino acid sequence and hence the structure and function of all the proteins of a cell, (2) are part of the cellular structures that select and align amino acids in the correct order as a polypeptide chain is being synthesized, and (3) catalyze a number of fundamental chemical reactions in cells, including formation of peptide bonds between amino acids during protein synthesis.

Deoxyribonucleic acid (DNA) contains all the information required to build the cells and tissues of an organism. The exact replication of this information in any species assures its genetic continuity from generation to generation and is critical to the normal development of an individual. The information stored in DNA is arranged in hereditary units, now known as **genes,** that control identifiable traits of an organism. In the process of **transcription,** the information stored in DNA is copied into **ribonucleic acid (RNA),** which has three distinct roles in protein synthesis.

Messenger RNA (mRNA) carries the instructions from DNA that specify the correct order of amino acids during protein synthesis. The remarkably accurate, stepwise assembly of amino acids into proteins occurs by **translation** of mRNA. In this process, the information in mRNA is interpreted by a second type of RNA called transfer RNA (tRNA) with the aid of a third type of RNA, ribosomal RNA (rRNA), and its associated proteins. As the correct amino acids are brought into sequence by tRNAs, they are linked by peptide bonds to make proteins.

Discovery of the structure of DNA in 1953 and subsequent elucidation of how DNA directs synthesis of RNA, which then directs assembly of proteins—the so-called *central dogma*—were monumental achievements marking the early days of molecular biology. However, the simplified representation of the central dogma as DNA→RNA→protein does not reflect the role of proteins in the synthesis of nucleic acids. Moreover, as discussed in later chapters, proteins are largely responsible for *regulating* **gene expression,** the entire process whereby the information encoded in DNA is decoded into the proteins that characterize various cell types.

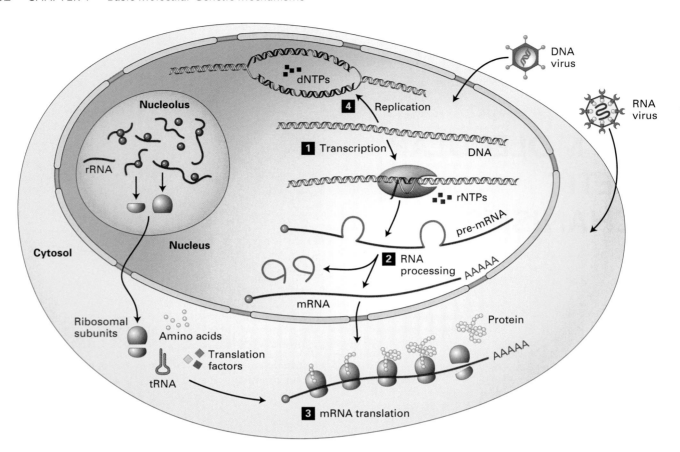

▲ **FIGURE 4-1 Overview of four basic molecular genetic processes.** In this chapter we cover the three processes that lead to production of proteins (**1**–**3**) and the process for replicating DNA (**4**). Because viruses utilize host-cell machinery, they have been important models for studying these processes. During transcription of a protein-coding gene by RNA polymerase (**1**), the four-base DNA code specifying the amino acid sequence of a protein is copied into a precursor messenger RNA (pre-mRNA) by the polymerization of ribonucleoside triphosphate monomers (rNTPs). Removal of extraneous sequences and other modifications to the pre-mRNA (**2**), collectively known as *RNA processing*, produce a functional mRNA, which is transported to the cytoplasm. During translation (**3**), the four-base code of the mRNA is decoded into the 20–amino acid "language" of proteins. Ribosomes, the macromolecular machines that translate the mRNA code, are composed of two subunits assembled in the nucleolus from ribosomal RNAs (rRNAs) and multiple proteins (*left*). After transport to the cytoplasm, ribosomal subunits associate with an mRNA and carry out protein synthesis with the help of transfer RNAs (tRNAs) and various translation factors. During DNA replication (**4**), which occurs only in cells preparing to divide, deoxyribonucleoside triphosphate monomers (dNTPs) are polymerized to yield two identical copies of each chromosomal DNA molecule. Each daughter cell receives one of the identical copies.

In this chapter, we first review the basic structures and properties of DNA and RNA. In the next several sections we discuss the basic processes summarized in Figure 4-1: transcription of DNA into RNA precursors, processing of these precursors to make functional RNA molecules, translation of mRNAs into proteins, and the replication of DNA. Along the way we compare gene structure in prokaryotes and eukaryotes and describe how bacteria control transcription, setting the stage for the more complex eukaryotic transcription-control mechanisms discussed in Chapter 11. After outlining the individual roles of mRNA, tRNA, and rRNA in protein synthesis, we present a detailed description of the components and biochemical steps in translation. We also consider the molecular problems involved in DNA repli-cation and the complex cellular machinery for ensuring ac-curate copying of the genetic material. The final section of the chapter presents basic information about viruses, which are important model organisms for studying macromolecular synthesis and other cellular processes.

4.1 Structure of Nucleic Acids

DNA and RNA are chemically very similar. The primary structures of both are linear **polymers** composed of **monomers** called **nucleotides.** Cellular RNAs range in length from less than one hundred to many thousands of nu-cleotides. Cellular DNA molecules can be as long as several hundred million nucleotides. These large DNA units in association with proteins can be stained with dyes and visualized in the light microscope as chromosomes, so named because of their stainability.

A Nucleic Acid Strand Is a Linear Polymer with End-to-End Directionality

DNA and RNA each consist of only four different nucleotides. Recall from Chapter 2 that all nucleotides consist of an organic base linked to a five-carbon sugar that has a phosphate group attached to carbon 5. In RNA, the sugar is ribose; in DNA, deoxyribose (see Figure 2-14). The nucleotides used in synthesis of DNA and RNA contain five different bases. The bases *adenine* (A) and *guanine* (G) are **purines,** which con-

tain a pair of fused rings; the bases *cytosine* (C), *thymine* (T), and *uracil* (U) are **pyrimidines,** which contain a single ring (see Figure 2-15). Both DNA and RNA contain three of these bases—A, G, and C; however, T is found only in DNA, and U only in RNA. (Note that the single-letter abbreviations for these bases are also commonly used to denote the entire nucleotides in nucleic acid polymers.)

A single nucleic acid strand has a *backbone* composed of repeating pentose-phosphate units from which the purine and pyrimidine bases extend as side groups. Like a polypeptide, a nucleic acid strand has an end-to-end chemical orientation: the *5′ end* has a hydroxyl or phosphate group on the 5′ carbon of its terminal sugar; the *3′ end* usually has a hydroxyl group on the 3′ carbon of its terminal sugar (Figure 4-2). This directionality, plus the fact that synthesis proceeds 5′ to 3′, has given rise to the convention that polynucleotide sequences are written and read in the 5′→3′ direction (from left to right); for example, the sequence AUG is assumed to be (5′)AUG(3′). As we will see, the 5′→3′ directionality of a nucleic acid strand is an important property of the molecule. The chemical linkage between adjacent nucleotides, commonly called a **phosphodiester bond,** actually consists of two phosphoester bonds, one on the 5′ side of the phosphate and another on the 3′ side.

The linear sequence of nucleotides linked by phosphodiester bonds constitutes the primary structure of nucleic acids. Like polypeptides, polynucleotides can twist and fold into three-dimensional conformations stabilized by noncovalent bonds. Although the primary structures of DNA and RNA are generally similar, their three-dimensional conformations are quite different. These structural differences are critical to the different functions of the two types of nucleic acids.

Native DNA Is a Double Helix of Complementary Antiparallel Strands

The modern era of molecular biology began in 1953 when James D. Watson and Francis H. C. Crick proposed that DNA has a double-helical structure. Their proposal, based on analysis of x-ray diffraction patterns coupled with careful model building, proved correct and paved the way for our modern understanding of how DNA functions as the genetic material.

DNA consists of two associated polynucleotide strands that wind together to form a **double helix.** The two sugar-phosphate backbones are on the outside of the double helix, and the bases project into the interior. The adjoining bases in each strand stack on top of one another in parallel planes (Figure 4-3a). The orientation of the two strands is *antiparallel*; that is, their 5′→3′ directions are opposite. The strands are held in precise register by formation of **base pairs** between the two strands: A is paired with T through two hydrogen bonds; G is paired with C through three hydrogen bonds (Figure 4-3b). This base-pair complementarity is a consequence of the size, shape, and chemical composition of the bases. The presence of thousands of such hydrogen bonds in a DNA molecule contributes greatly to the stability

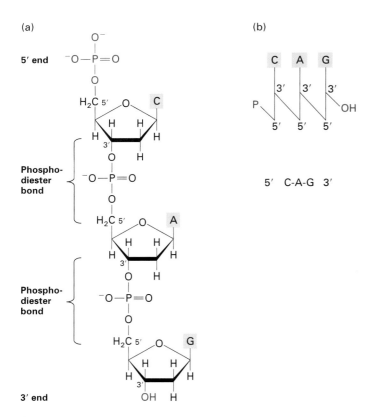

▲ **FIGURE 4-2 Alternative representations of a nucleic acid strand illustrating its chemical directionality.** Shown here is a single strand of DNA containing only three bases: cytosine (C), adenine (A), and guanine (G). (a) The chemical structure shows a hydroxyl group at the 3′ end and a phosphate group at the 5′ end. Note also that two phosphoester bonds link adjacent nucleotides; this two-bond linkage commonly is referred to as a *phosphodiester bond.* (b) In the "stick" diagram (*top*), the sugars are indicated as vertical lines and the phosphodiester bonds as slanting lines; the bases are denoted by their single-letter abbreviations. In the simplest representation (*bottom*), only the bases are indicated. By convention, a polynucleotide sequence is always written in the 5′→3′ direction (left to right) unless otherwise indicated.

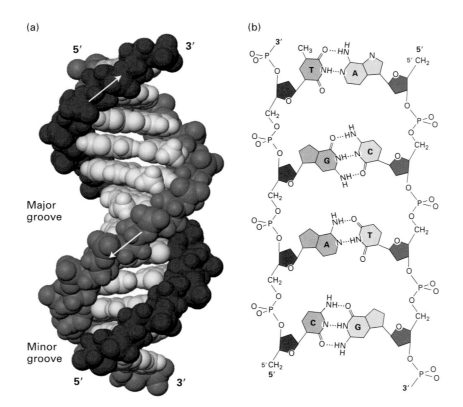

▲ **FIGURE 4-3 The DNA double helix.** (a) Space-filling model of B DNA, the most common form of DNA in cells. The bases (light shades) project inward from the sugar-phosphate backbones (dark red and blue) of each strand, but their edges are accessible through major and minor grooves. Arrows indicate the 5'→3' direction of each strand. Hydrogen bonds between the bases are in the center of the structure. The major and minor grooves are lined by potential hydrogen bond donors and acceptors (highlighted in yellow). (b) Chemical structure of DNA double helix. This extended schematic shows the two sugar-phosphate backbones and hydrogen bonding between the Watson-Crick base pairs, A·T and G·C. [Part (a) from R. Wing et al., 1980, *Nature* **287**:755; part (b) from R. E. Dickerson, 1983, *Sci. Am.* **249**:94.]

of the double helix. Hydrophobic and van der Waals interactions between the stacked adjacent base pairs further stabilize the double-helical structure.

In natural DNA, A always hydrogen bonds with T and G with C, forming A·T and G·C base pairs as shown in Figure 4-3b. These associations between a larger purine and smaller pyrimidine are often called *Watson-Crick base pairs*. Two polynucleotide strands, or regions thereof, in which all the nucleotides form such base pairs are said to be **complementary**. However, in theory and in synthetic DNAs other base pairs can form. For example, a guanine (a purine) could theoretically form hydrogen bonds with a thymine (a pyrimidine), causing only a minor distortion in the helix. The space available in the helix also would allow pairing between the two pyrimidines cytosine and thymine. Although the nonstandard G·T and C·T base pairs are normally not found in DNA, G·U base pairs are quite common in double-helical regions that form within otherwise single-stranded RNA.

Most DNA in cells is a *right-handed* helix. The x-ray diffraction pattern of DNA indicates that the stacked bases are regularly spaced 0.36 nm apart along the helix axis. The helix makes a complete turn every 3.6 nm; thus there are about 10.5 pairs per turn. This is referred to as the *B form* of DNA, the normal form present in most DNA stretches in cells. On the outside of B-form DNA, the spaces between the intertwined strands form two helical grooves of different widths described as the *major* groove and the *minor* groove (see Figure 4-3a). As a consequence, the atoms on the edges of each base within these grooves are accessible from outside the helix, forming two types of binding surfaces. DNA-binding proteins can "read" the sequence of bases in duplex DNA by contacting atoms in either the major or the minor grooves.

In addition to the major B form, three additional DNA structures have been described. Two of these are compared to B DNA in Figure 4-4. In very low humidity, the crystallographic structure of B DNA changes to the *A form*; RNA-DNA and RNA-RNA helices exist in this form in cells and in vitro. Short DNA molecules composed of alternating purine-pyrimidine nucleotides (especially Gs and Cs) adopt an alternative left-handed configuration instead of the normal right-handed helix. This structure is called *Z DNA* because the bases seem to zigzag when viewed from the side. Some evidence suggests that Z DNA may occur in cells, although its function is unknown. Finally, a triple-stranded DNA structure is formed when synthetic polymers of poly(A) and

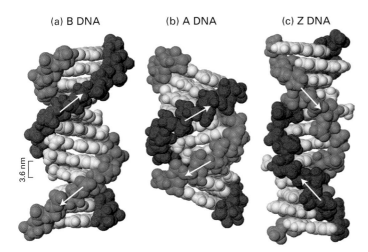

(a) B DNA   (b) A DNA   (c) Z DNA

3.6 nm

▲ **FIGURE 4-4 Models of various known DNA structures.** The sugar-phosphate backbones of the two strands, which are on the outside in all structures, are shown in red and blue; the bases (lighter shades) are oriented inward. (a) The B form of DNA has ≈10.5 base pairs per helical turn. Adjacent stacked base pairs are 0.36 nm apart. (b) The more compact A form of DNA has 11 base pairs per turn and exhibits a large tilt of the base pairs with respect to the helix axis. (c) Z DNA is a left-handed double helix.

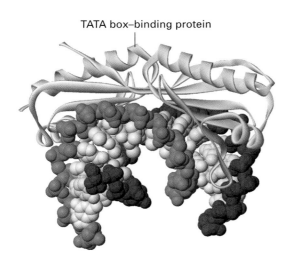

TATA box–binding protein

▲ **FIGURE 4-5 Bending of DNA resulting from protein binding.** The conserved C-terminal domain of the TATA box–binding protein (TBP) binds to the minor groove of specific DNA sequences rich in A and T, untwisting and sharply bending the double helix. Transcription of most eukaryotic genes requires participation of TBP. [Adapted from D. B. Nikolov and S. K. Burley, 1997, *Proc. Nat'l. Acad. Sci. USA* **94**:15.]

polydeoxy(U) are mixed in the test tube. In addition, homopolymeric stretches of DNA composed of C and T residues in one strand and A and G residues in the other can form a triple-stranded structure by binding matching lengths of synthetic poly(C+T). Such structures probably do not occur naturally in cells but may prove useful as therapeutic agents.

By far the most important modifications in the structure of standard B-form DNA come about as a result of protein binding to specific DNA sequences. Although the multitude of hydrogen and hydrophobic bonds between the bases provide stability to DNA, the double helix is flexible about its long axis. Unlike the α helix in proteins (see Figure 3-3), there are no hydrogen bonds parallel to the axis of the DNA helix. This property allows DNA to bend when complexed with a DNA-binding protein (Figure 4-5). Bending of DNA is critical to the dense packing of DNA in chromatin, the protein-DNA complex in which nuclear DNA occurs in eukaryotic cells (Chapter 10).

DNA Can Undergo Reversible Strand Separation

During replication and transcription of DNA, the strands of the double helix must separate to allow the internal edges of the bases to pair with the bases of the nucleotides to be polymerized into new polynucleotide chains. In later sections, we describe the cellular mechanisms that separate and subsequently reassociate DNA strands during replication and transcription. Here we discuss factors influencing the in vitro separation and reassociation of DNA strands.

The unwinding and separation of DNA strands, referred to as **denaturation,** or "melting," can be induced experimentally by increasing the temperature of a solution of DNA. As the thermal energy increases, the resulting increase in molecular motion eventually breaks the hydrogen bonds and other forces that stabilize the double helix; the strands then separate, driven apart by the electrostatic repulsion of the negatively charged deoxyribose-phosphate backbone of each strand. Near the denaturation temperature, a small increase in temperature causes a rapid, near simultaneous loss of the multiple weak interactions holding the strands together along the entire length of the DNA molecules, leading to an abrupt change in the absorption of ultraviolet (UV) light (Figure 4-6a).

The *melting temperature* T_m at which DNA strands will separate depends on several factors. Molecules that contain a greater proportion of G·C pairs require higher temperatures to denature because the three hydrogen bonds in G·C pairs make these base pairs more stable than A·T pairs, which have only two hydrogen bonds. Indeed, the percentage of G·C base pairs in a DNA sample can be estimated from its T_m (Figure 4-6b). The ion concentration also influences the T_m because the negatively charged phosphate groups in the

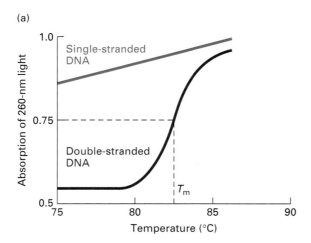

▲ **EXPERIMENTAL FIGURE 4-6 The temperature at which DNA denatures increases with the proportion of G·C pairs.** (a) Melting of doubled-stranded DNA can be monitored by the absorption of ultraviolet light at 260 nm. As regions of double-stranded DNA unpair, the absorption of light by those regions increases almost twofold. The temperature at which half the bases in a double-stranded DNA sample have denatured is denoted T_m (for temperature of melting). Light absorption by single-stranded DNA changes much less as the temperature is increased. (b) The T_m is a function of the G·C content of the DNA; the higher the G+C percentage, the greater the T_m.

two strands are shielded by positively charged ions. When the ion concentration is low, this shielding is decreased, thus increasing the repulsive forces between the strands and reducing the T_m. Agents that destabilize hydrogen bonds, such as formamide or urea, also lower the T_m. Finally, extremes of pH denature DNA at low temperature. At low (acid) pH, the bases become protonated and thus positively charged, repelling each other. At high (alkaline) pH, the bases lose protons and become negatively charged, again repelling each other because of the similar charge.

The single-stranded DNA molecules that result from denaturation form random coils without an organized structure. Lowering the temperature, increasing the ion concentration, or neutralizing the pH causes the two complementary strands to reassociate into a perfect double helix. The extent of such *renaturation* is dependent on time, the DNA concentration, and the ionic concentration. Two DNA strands not related in sequence will remain as random coils and will not renature; most importantly, they will not inhibit complementary DNA partner strands from finding each other and renaturing. Denaturation and renaturation of DNA are the basis of nucleic acid **hybridization,** a powerful technique used to study the relatedness of two DNA samples and to detect and isolate specific DNA molecules in a mixture containing numerous different DNA sequences (see Figure 9-16).

Many DNA Molecules Are Circular

Many prokaryotic genomic DNAs and many viral DNAs are circular molecules. Circular DNA molecules also occur in mitochondria, which are present in almost all eukaryotic cells, and in chloroplasts, which are present in plants and some unicellular eukaryotes.

Each of the two strands in a circular DNA molecule forms a closed structure without free ends. Localized unwinding of a circular DNA molecule, which occurs during DNA replication, induces torsional stress into the remaining portion of the molecule because the ends of the strands are not free to rotate. As a result, the DNA molecule twists back on itself, like a twisted rubber band, forming *supercoils* (Figure 4-7b). In other words, when part of the DNA helix is underwound, the remainder of the molecule becomes overwound. Bacterial and eukaryotic cells, however, contain *topoisomerase I,* which can relieve any torsional stress that develops in cellular DNA molecules during replication or other processes. This enzyme binds to DNA at random sites and breaks a phosphodiester bond in one strand. Such a one-strand break in DNA is called a *nick.* The broken end then winds around the uncut strand, leading to loss of supercoils (Figure 4-7a). Finally, the same enzyme joins (ligates) the two ends of the broken strand. Another type of enzyme, *topoisomerase II,* makes breaks in both strands of a double-stranded DNA and then religates them. As a result, topoisomerase II can both relieve torsional stress and link together two circular DNA molecules as in the links of a chain.

Although eukaryotic nuclear DNA is linear, long loops of DNA are fixed in place within chromosomes (Chapter 10). Thus torsional stress and the consequent formation of supercoils also could occur during replication of nuclear DNA. As in bacterial cells, abundant topoisomerase I in eukaryotic nuclei relieves any torsional stress in nuclear DNA that would develop in the absence of this enzyme.

(a) Supercoiled

(b) Relaxed circle

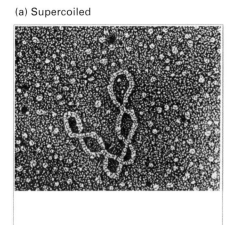

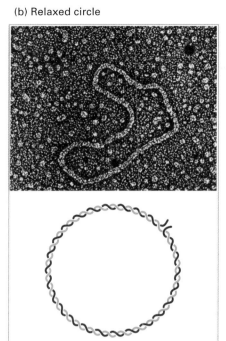

◄ **EXPERIMENTAL FIGURE 4-7** DNA **supercoils can be removed by cleavage of one strand.** (a) Electron micrograph of SV40 viral DNA. When the circular DNA of the SV40 virus is isolated and separated from its associated protein, the DNA duplex is underwound and assumes the supercoiled configuration. (b) If a supercoiled DNA is nicked (i.e., one strand cleaved), the strands can rewind, leading to loss of a supercoil. Topoisomerase I catalyzes this reaction and also reseals the broken ends. All the supercoils in isolated SV40 DNA can be removed by the sequential action of this enzyme, producing the relaxed-circle conformation. For clarity, the shapes of the molecules at the bottom have been simplified.

Different Types of RNA Exhibit Various Conformations Related to Their Functions

As noted earlier, the primary structure of RNA is generally similar to that of DNA with two exceptions: the sugar component of RNA, ribose, has a hydroxyl group at the 2′ position (see Figure 2-14b), and thymine in DNA is replaced by uracil in RNA. The hydroxyl group on C_2 of ribose makes RNA more chemically labile than DNA and provides a chemically reactive group that takes part in RNA-mediated catalysis. As a result of this lability, RNA is cleaved into mononucleotides by alkaline solution, whereas DNA is not. Like DNA, RNA is a long polynucleotide that can be double-stranded or single-stranded, linear or circular. It can also participate in a hybrid helix composed of one RNA strand and one DNA strand. As noted above, RNA-RNA and RNA-DNA double helices have a compact conformation like the A form of DNA (see Figure 4-4b).

Unlike DNA, which exists primarily as a very long double helix, most cellular RNAs are single-stranded and exhibit a variety of conformations (Figure 4-8). Differences in the sizes and conformations of the various types of RNA permit them to carry out specific functions in a cell. The simplest secondary structures in single-stranded RNAs are formed by pairing of complementary bases. "Hairpins" are formed by pairing of bases within ≈5–10 nucleotides of each other, and "stem-loops" by pairing of bases that are separated by >10 to

several hundred nucleotides. These simple folds can cooperate to form more complicated tertiary structures, one of which is termed a "pseudoknot."

As discussed in detail later, tRNA molecules adopt a well-defined three-dimensional architecture in solution that is crucial in protein synthesis. Larger rRNA molecules also have locally well-defined three-dimensional structures, with more flexible links in between. Secondary and tertiary structures also have been recognized in mRNA, particularly near the ends of molecules. Clearly, then, RNA molecules are like proteins in that they have structured domains connected by less structured, flexible stretches.

The folded domains of RNA molecules not only are structurally analogous to the α helices and β strands found in proteins, but in some cases also have catalytic capacities. Such catalytic RNAs are called **ribozymes**. Although ribozymes usually are associated with proteins that stabilize the ribozyme structure, it is the RNA that acts as a catalyst. Some ribozymes can catalyze splicing, a remarkable process in which an internal RNA sequence is cut and removed, and the two resulting chains then ligated. This process occurs during formation of the majority of functional mRNA molecules in eukaryotic cells, and also occurs in bacteria and archaea. Remarkably, some RNAs carry out *self-splicing*, with the catalytic activity residing in the sequence that is removed. The mechanisms of splicing and self-splicing are discussed in detail in Chapter 12. As noted later in this chapter, rRNA

▶ **FIGURE 4-8 RNA secondary and tertiary structures.** (a) Stem-loops, hairpins, and other secondary structures can form by base pairing between distant complementary segments of an RNA molecule. In stem-loops, the single-stranded loop between the base-paired helical stem may be hundreds or even thousands of nucleotides long, whereas in hairpins, the short turn may contain as few as four nucleotides. (b) Pseudoknots, one type of RNA tertiary structure, are formed by interaction of secondary loops through base pairing between complementary bases (green and blue). Only base-paired bases are shown. A secondary structure diagram is shown at right. [Part (b) adapted from P. J. A. Michiels et al., 2001, *J. Mol. Biol.* **310**:1109.]

(a) Secondary structure

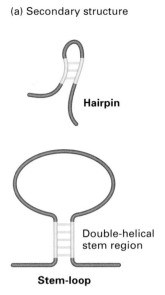

Hairpin

Double-helical stem region

Stem-loop

(b) Tertiary structure

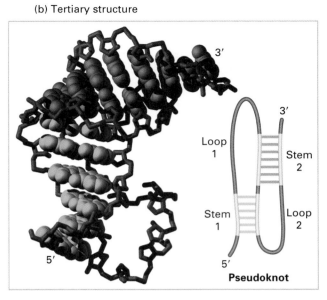

Pseudoknot

plays a catalytic role in the formation of peptide bonds during protein synthesis.

In this chapter, we focus on the functions of mRNA, tRNA, and rRNA in gene expression. In later chapters we will encounter other RNAs, often associated with proteins, that participate in other cell functions.

KEY CONCEPTS OF SECTION 4.1

Structure of Nucleic Acids

■ Deoxyribonucleic acid (DNA), the genetic material, carries information to specify the amino acid sequences of proteins. It is transcribed into several types of ribonucleic acid (RNA), including messenger RNA (mRNA), transfer RNA (tRNA), and ribosomal RNA (rRNA), which function in protein synthesis (see Figure 4-1).

■ Both DNA and RNA are long, unbranched polymers of nucleotides, which consist of a phosphorylated pentose linked to an organic base, either a purine or pyrimidine.

■ The purines adenine (A) and guanine (G) and the pyrimidine cytosine (C) are present in both DNA and RNA. The pyrimidine thymine (T) present in DNA is replaced by the pyrimidine uracil (U) in RNA.

■ Adjacent nucleotides in a polynucleotide are linked by phosphodiester bonds. The entire strand has a chemical directionality: the 5′ end with a free hydroxyl or phosphate group on the 5′ carbon of the sugar, and the 3′ end with a free hydroxyl group on the 3′ carbon of the sugar (see Figure 4-2).

■ Natural DNA (B DNA) contains two complementary antiparallel polynucleotide strands wound together into a regular right-handed double helix with the bases on the in-

side and the two sugar-phosphate backbones on the outside (see Figure 4-3). Base pairing between the strands and hydrophobic interactions between adjacent bases in the same strand stabilize this native structure.

■ The bases in nucleic acids can interact via hydrogen bonds. The standard Watson-Crick base pairs are G·C, A·T (in DNA), and A·U (in RNA). Base pairing stabilizes the native three-dimensional structures of DNA and RNA.

■ Binding of protein to DNA can deform its helical structure, causing local bending or unwinding of the DNA molecule.

■ Heat causes the DNA strands to separate (denature). The melting temperature T_m of DNA increases with the percentage of G·C base pairs. Under suitable conditions, separated complementary nucleic acid strands will renature.

■ Circular DNA molecules can be twisted on themselves, forming supercoils (see Figure 4-7). Enzymes called *topoisomerases* can relieve torsional stress and remove supercoils from circular DNA molecules.

■ Cellular RNAs are single-stranded polynucleotides, some of which form well-defined secondary and tertiary structures (see Figure 4-8). Some RNAs, called *ribozymes*, have catalytic activity.

4.2 Transcription of Protein-Coding Genes and Formation of Functional mRNA

The simplest definition of a gene is a "unit of DNA that contains the information to specify synthesis of a single polypeptide chain or functional RNA (such as a tRNA)." The vast majority of genes carry information to build protein molecules, and it is the RNA copies of such *protein-coding genes* that constitute the mRNA molecules of cells. The DNA molecules of small viruses contain only a few genes, whereas the single DNA molecule in each of the chromosomes of higher animals and plants may contain several thousand genes.

During synthesis of RNA, the four-base language of DNA containing A, G, C, and T is simply copied, or *transcribed*, into the four-base language of RNA, which is identical except that U replaces T. In contrast, during protein synthesis the four-base language of DNA and RNA is *translated* into the 20–amino acid language of proteins. In this section we focus on formation of functional mRNAs from protein-coding genes (see Figure 4-1, step 1). A similar process yields the precursors of rRNAs and tRNAs encoded by rRNA and tRNA genes; these precursors are then further modified to yield functional rRNAs and tRNAs (Chapter 12).

A Template DNA Strand Is Transcribed into a Complementary RNA Chain by RNA Polymerase

During transcription of DNA, one DNA strand acts as a *template*, determining the order in which ribonucleoside triphosphate (rNTP) monomers are polymerized to form a complementary RNA chain. Bases in the template DNA strand base-pair with complementary incoming rNTPs, which then are joined in a polymerization reaction catalyzed by **RNA polymerase.** Polymerization involves a nucleophilic attack by the 3′ oxygen in the growing RNA chain on the α phosphate of the next nucleotide precursor to be added, resulting in formation of a phosphodiester bond and release of pyrophosphate (PP$_i$). As a consequence of this mechanism, RNA molecules are always synthesized in the 5′→3′ direction (Figure 4-9).

The energetics of the polymerization reaction strongly favors addition of ribonucleotides to the growing RNA chain because the high-energy bond between the α and β phosphate of rNTP monomers is replaced by the lower-energy phosphodiester bond between nucleotides. The equilibrium for the reaction is driven further toward chain elongation by pyrophosphatase, an enzyme that catalyzes cleavage of the released PP$_i$ into two molecules of inorganic phosphate. Like the two strands in DNA, the template DNA strand and the growing RNA strand that is base-paired to it have opposite 5′→3′ directionality.

By convention, the site at which RNA polymerase begins transcription is numbered +1. **Downstream** denotes the direction in which a template DNA strand is transcribed (or mRNA translated); thus a downstream sequence is toward the 3′ end relative to the start site, considering the DNA strand with the same polarity as the transcribed RNA. **Upstream** denotes the opposite direction. Nucleotide positions in the DNA sequence downstream from a start site are indicated by a positive (+) sign; those upstream, by a negative (−) sign.

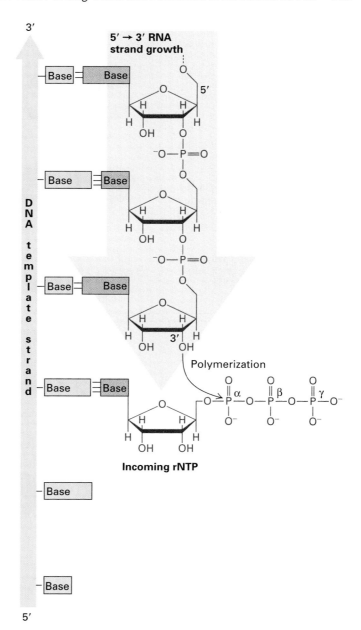

▲ **FIGURE 4-9 Polymerization of ribonucleotides by RNA polymerase during transcription.** The ribonucleotide to be added at the 3′ end of a growing RNA strand is specified by base pairing between the next base in the template DNA strand and the complementary incoming ribonucleoside triphosphate (rNTP). A phosphodiester bond is formed when RNA polymerase catalyzes a reaction between the 3′ O of the growing strand and the α phosphate of a correctly base-paired rNTP. RNA strands always are synthesized in the 5′→3′ direction and are opposite in polarity to their template DNA strands.

Stages in Transcription To carry out transcription, RNA polymerase performs several distinct functions, as depicted in Figure 4-10. During transcription *initiation*, RNA polymerase recognizes and binds to a specific site, called a **promoter,** in double-stranded DNA (step 1). Nuclear RNA

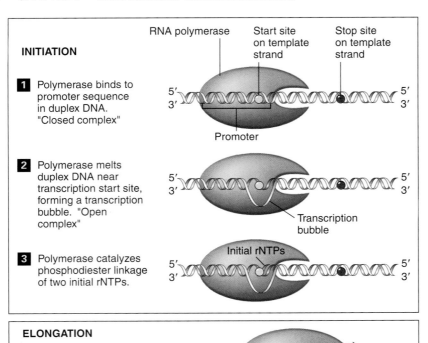

INITIATION

1 Polymerase binds to promoter sequence in duplex DNA. "Closed complex"

RNA polymerase | Start site on template strand | Stop site on template strand

Promoter

2 Polymerase melts duplex DNA near transcription start site, forming a transcription bubble. "Open complex"

Transcription bubble

3 Polymerase catalyzes phosphodiester linkage of two initial rNTPs.

Initial rNTPs

ELONGATION

4 Polymerase advances 3′ → 5′ down template strand, melting duplex DNA and adding rNTPs to growing RNA.

Nascent RNA   5′   DNA-RNA hybrid region

TERMINATION

5 At transcription stop site, polymerase releases completed RNA and dissociates from DNA.

Completed RNA strand

◄ **FIGURE 4-10 Three stages in transcription.** During initiation of transcription, RNA polymerase forms a transcription bubble and begins polymerization of ribonucleotides (rNTPs) at the start site, which is located within the promoter region. Once a DNA region has been transcribed, the separated strands reassociate into a double helix, displacing the nascent RNA except at its 3′ end. The 5′ end of the RNA strand exits the RNA polymerase through a channel in the enzyme. Termination occurs when the polymerase encounters a specific termination sequence (stop site). See the text for details.

polymerases require various protein factors, called general **transcription factors,** to help them locate promoters and initiate transcription. After binding to a promoter, RNA polymerase melts the DNA strands in order to make the bases in the template strand available for base pairing with the bases of the ribonucleoside triphosphates that it will polymerize together. Cellular RNA polymerases melt approximately 14 base pairs of DNA around the transcription start site, which is located on the template strand within the promoter region (step **2**). Transcription initiation is considered complete when the first two ribonucleotides of an RNA chain are linked by a phosphodiester bond (step **3**).

After several ribonucleotides have been polymerized, RNA polymerase dissociates from the promoter DNA and general transcription factors. During the stage of *strand elongation,* RNA polymerase moves along the template DNA one base at a time, opening the double-stranded DNA in front of its direction of movement and hybridizing the strands behind

it (Figure 4-10, step **4**). One ribonucleotide at a time is added to the 3′ end of the growing (*nascent*) RNA chain during strand elongation by the polymerase. The enzyme maintains a melted region of approximately 14 base pairs, called the *transcription bubble.* Approximately eight nucleotides at the 3′ end of the growing RNA strand remain base-paired to the template DNA strand in the transcription bubble. The elongation complex, comprising RNA polymerase, template DNA, and the growing (nascent) RNA strand, is extraordinarily stable. For example, RNA polymerase transcribes the longest known mammalian genes, containing $\approx 2 \times 10^6$ base pairs, without dissociating from the DNA template or releasing the nascent RNA. Since RNA synthesis occurs at a rate of about 1000 nucleotides per minute at 37 °C, the elongation complex must remain intact for more than 24 hours to assure continuous RNA synthesis.

During transcription *termination,* the final stage in RNA synthesis, the completed RNA molecule, or **primary transcript,** is released from the RNA polymerase and the polymerase dissociates from the template DNA (Figure 4-10, step ⑤). Specific sequences in the template DNA signal the bound RNA polymerase to terminate transcription. Once released, an RNA polymerase is free to transcribe the same gene again or another gene.

Structure of RNA Polymerases The RNA polymerases of bacteria, archaea, and eukaryotic cells are fundamentally similar in structure and function. Bacterial RNA polymerases are composed of two related large subunits (β′ and β), two copies of a smaller subunit (α), and one copy of a fifth subunit (ω) that is not essential for transcription or cell viability but stabilizes the enzyme and assists in the assembly of its subunits. Archaeal and eukaryotic RNA polymerases have several additional small subunits associated with this core complex, which we describe in Chapter 11. Schematic dia-

grams of the transcription process generally show RNA polymerase bound to an unbent DNA molecule, as in Figure 4-10. However, according to a current model of the interaction between bacterial RNA polymerase and promoter DNA, the DNA bends sharply following its entry into the enzyme (Figure 4-11).

Organization of Genes Differs in Prokaryotic and Eukaryotic DNA

Having outlined the process of transcription, we now briefly consider the large-scale arrangement of information in DNA and how this arrangement dictates the requirements for RNA synthesis so that information transfer goes smoothly. In recent years, sequencing of the entire **genomes** from several organisms has revealed not only large variations in the number of protein-coding genes but also differences in their organization in prokaryotes and eukaryotes.

The most common arrangement of protein-coding genes in all prokaryotes has a powerful and appealing logic: genes devoted to a single metabolic goal, say, the synthesis of the amino acid tryptophan, are most often found in a contiguous array in the DNA. Such an arrangement of genes in a functional group is called an **operon**, because it operates as a unit from a single promoter. Transcription of an operon produces a continuous strand of mRNA that carries the message for a related series of proteins (Figure 4-12a). Each section of the mRNA represents the unit (or gene) that encodes one of the proteins in the series. In prokaryotic DNA the genes are closely packed with very few noncoding gaps, and the DNA is transcribed directly into colinear mRNA, which then is translated into protein.

This economic clustering of genes devoted to a single metabolic function does not occur in eukaryotes, even simple ones like yeasts, which can be metabolically similar to bacteria. Rather, eukaryotic genes devoted to a single pathway are most often physically separated in the DNA; indeed such genes usually are located on different chromosomes. Each gene is transcribed from its own promoter, producing one mRNA, which generally is translated to yield a single polypeptide (Figure 4-12b).

When researchers first compared the nucleotide sequences of eukaryotic mRNAs from multicellular organisms with the DNA sequences encoding them, they were surprised to find that the uninterrupted protein-coding sequence of a given mRNA was broken up (discontinuous) in its corresponding section of DNA. They concluded that the eukaryotic gene existed in pieces of coding sequence, the **exons,** separated by non-protein-coding segments, the **introns.** This astonishing finding implied that the long initial primary transcript—the RNA copy of the entire transcribed DNA sequence—had to be clipped apart to remove the introns and then carefully stitched back together to produce many eukaryotic mRNAs.

Although introns are common in multicellular eukaryotes, they are extremely rare in bacteria and archaea and

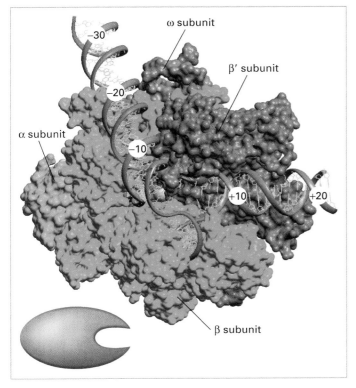

▲ **FIGURE 4-11 Current model of bacterial RNA polymerase bound to a promoter.** This structure corresponds to the polymerase molecule as schematically shown in step ② of Figure 4-10. The β′ subunit is in orange; β is in green. Part of one of the two α subunits can be seen in light blue; the ω subunit is in gray. The DNA template and nontemplate strands are shown, respectively, as gray and pink ribbons. A Mg^{2+} ion at the active center is shown as a gray sphere. Numbers indicate positions in the DNA sequence relative to the transcription start site, with positive (+) numbers in the direction of transcription and negative (−) numbers in the opposite direction. [Courtesy of R. H. Ebright, Waksman Institute.]

(a) Prokaryotes

E. coli genome

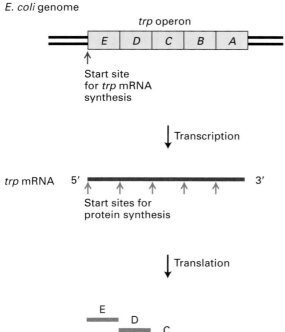

(b) Eukaryotes

Yeast chromosomes

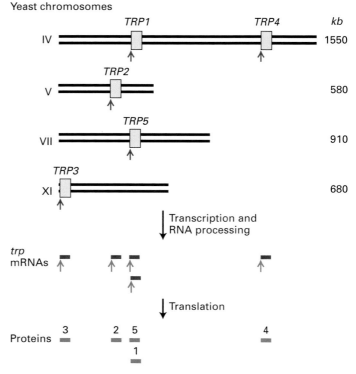

▲ **FIGURE 4-12 Comparison of gene organization, transcription, and translation in prokaryotes and eukaryotes.** (a) The tryptophan (*trp*) operon is a continuous segment of the *E. coli* chromosome, containing five genes (blue) that encode the enzymes necessary for the stepwise synthesis of tryptophan. The entire operon is transcribed from one promoter into one long continuous *trp* mRNA (red). Translation of this mRNA begins at five different start sites, yielding five proteins (green). The order of the genes in the bacterial genome parallels the sequential function of the encoded proteins in the tryptophan pathway. (b) The five genes encoding the enzymes required for tryptophan synthesis in yeast (*Saccharomyces cerevisiae*) are carried on four different chromosomes. Each gene is transcribed from its own promoter to yield a primary transcript that is processed into a functional mRNA encoding a single protein. The lengths of the yeast chromosomes are given in kilobases (10^3 bases).

uncommon in many unicellular eukaryotes such as baker's yeast. However, introns are present in the DNA of viruses that infect eukaryotic cells. Indeed, the presence of introns was first discovered in such viruses, whose DNA is transcribed by host-cell enzymes.

Eukaryotic Precursor mRNAs Are Processed to Form Functional mRNAs

In prokaryotic cells, which have no nuclei, translation of an mRNA into protein can begin from the 5′ end of the mRNA even while the 3′ end is still being synthesized by RNA polymerase. In other words, transcription and translation can occur concurrently in prokaryotes. In eukaryotic cells, however, not only is the nucleus separated from the cytoplasm where translation occurs, but also the primary transcripts of protein-coding genes are precursor mRNAs (**pre-mRNAs**) that must undergo several modifications, collectively termed *RNA processing,* to yield a functional mRNA (see Figure 4-1, step ②). This mRNA then must be exported to the cytoplasm before it can be translated into protein. Thus transcription and translation cannot occur concurrently in eukaryotic cells.

All eukaryotic pre-mRNAs initially are modified at the two ends, and these modifications are retained in mRNAs. As the 5′ end of a nascent RNA chain emerges from the surface of RNA polymerase II, it is immediately acted on by several enzymes that together synthesize the *5′ cap*, a 7-methylguanylate that is connected to the terminal nucleotide of the RNA by an unusual 5′,5′ triphosphate linkage (Figure 4-13). The cap protects an mRNA from enzymatic degradation and assists in its export to the cytoplasm. The cap also is bound by a protein factor required to begin translation in the cytoplasm.

Processing at the 3′ end of a pre-mRNA involves cleavage by an endonuclease to yield a free 3′-hydroxyl group to which a string of adenylic acid residues is added one at a time by an enzyme called *poly(A) polymerase.* The resulting *poly(A) tail* contains 100–250 bases, being shorter in yeasts and invertebrates than in vertebrates. Poly(A) polymerase is part of a complex of proteins that can locate and cleave a transcript at a specific site and then add the correct number of A residues, in a process that does not require a template.

The final step in the processing of many different eukaryotic mRNA molecules is **RNA splicing**: the internal cleavage of a transcript to excise the introns, followed by ligation of the coding exons. Figure 4-14 summarizes the basic steps in eukaryotic mRNA processing, using the β-globin gene as an example. We examine the cellular machinery for carrying out processing of mRNA, as well as tRNA and rRNA, in Chapter 12.

The functional eukaryotic mRNAs produced by RNA processing retain noncoding regions, referred to as *5′* and *3′*

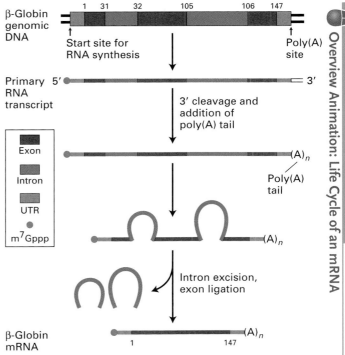

▲ **FIGURE 4-14 Overview of RNA processing to produce functional mRNA in eukaryotes.** The β-globin gene contains three protein-coding exons (coding region, red) and two intervening noncoding introns (blue). The introns interrupt the protein-coding sequence between the codons for amino acids 31 and 32 and 105 and 106. Transcription of eukaryotic protein-coding genes starts before the sequence that encodes the first amino acid and extends beyond the sequence encoding the last amino acid, resulting in noncoding regions (gray) at the ends of the primary transcript. These untranslated regions (UTRs) are retained during processing. The 5′ cap (m^7Gppp) is added during formation of the primary RNA transcript, which extends beyond the poly(A) site. After cleavage at the poly(A) site and addition of multiple A residues to the 3′ end, splicing removes the introns and joins the exons. The small numbers refer to positions in the 147–amino acid sequence of β-globin.

7-Methylguanylate

$5′ → 5′$ linkage

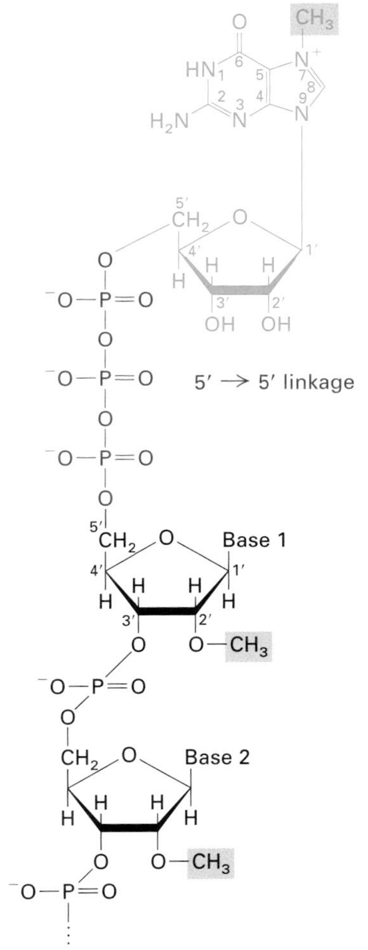

▲ **FIGURE 4-13 Structure of the 5′ methylated cap of eukaryotic mRNA.** The distinguishing chemical features are the 5′→5′ linkage of 7-methylguanylate to the initial nucleotide of the mRNA molecule and the methyl group on the 2′ hydroxyl of the ribose of the first nucleotide (base 1). Both these features occur in all animal cells and in cells of higher plants; yeasts lack the methyl group on nucleotide 1. The ribose of the second nucleotide (base 2) also is methylated in vertebrates. [See A. J. Shatkin, 1976, *Cell* **9**:645.]

untranslated regions (UTRs), at each end. In mammalian mRNAs, the 5′ UTR may be a hundred or more nucleotides long, and the 3′ UTR may be several kilobases in length. Prokaryotic mRNAs also usually have 5′ and 3′ UTRs, but these are much shorter than those in eukaryotic mRNAs, generally containing fewer than 10 nucleotides.

Alternative RNA Splicing Increases the Number of Proteins Expressed from a Single Eukaryotic Gene

In contrast to bacterial and archaeal genes, the vast majority of genes in higher, multicellular eukaryotes contain multiple introns. As noted in Chapter 3, many proteins from

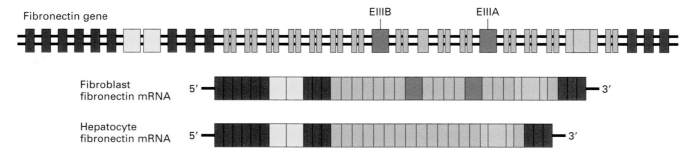

▲ **FIGURE 4-15 Cell type–specific splicing of fibronectin pre-mRNA in fibroblasts and hepatocytes.** The ≈75-kb fibronectin gene *(top)* contains multiple exons. The EIIIB and EIIIA exons (green) encode binding domains for specific proteins on the surface of fibroblasts. The fibronectin mRNA produced in fibroblasts includes the EIIIA and EIIIB exons, whereas these exons are spliced out of fibronectin mRNA in hepatocytes. In this diagram, introns (black lines) are not drawn to scale; most of them are much longer than any of the exons.

higher eukaryotes have a multidomain tertiary structure (see Figure 3-8). Individual repeated protein domains often are encoded by one exon or a small number of exons that code for identical or nearly identical amino acid sequences. Such repeated exons are thought to have evolved by the accidental multiple duplication of a length of DNA lying between two sites in adjacent introns, resulting in insertion of a string of repeated exons, separated by introns, between the original two introns. The presence of multiple introns in many eukaryotic genes permits expression of multiple, related proteins from a single gene by means of **alternative splicing.** In higher eukaryotes, alternative splicing is an important mechanism for production of different forms of a protein, called **isoforms,** by different types of cells.

Fibronectin, a multidomain extracellular adhesive protein found in mammals, provides a good example of alternative splicing (Figure 4-15). The fibronectin gene contains numerous exons, grouped into several regions corresponding to specific domains of the protein. Fibroblasts produce fibronectin mRNAs that contain exons EIIIA and EIIIB; these exons encode amino acid sequences that bind tightly to proteins in the fibroblast plasma membrane. Consequently, this fibronectin isoform adheres fibroblasts to the extracellular matrix. Alternative splicing of the fibronectin primary transcript in hepatocytes, the major type of cell in the liver, yields mRNAs that lack the EIIIA and EIIIB exons. As a result, the fibronectin secreted by hepatocytes into the blood does not adhere tightly to fibroblasts or most other cell types, allowing it to circulate. During formation of blood clots, however, the fibrin-binding domains of hepatocyte fibronectin binds to fibrin, one of the principal constituents of clots. The bound fibronectin then interacts with integrins on the membranes of passing, activated platelets, thereby expanding the clot by addition of platelets.

More than 20 different isoforms of fibronectin have been identified, each encoded by a different, alternatively spliced mRNA composed of a unique combination of fibronectin gene exons. Recent sequencing of large numbers of mRNAs isolated from various tissues and comparison of their sequences with genomic DNA has revealed that nearly 60 percent of all human genes are expressed as alternatively spliced mRNAs. Clearly, alternative RNA splicing greatly expands the number of proteins encoded by the genomes of higher, multicellular organisms.

KEY CONCEPTS OF SECTION 4.2

Transcription of Protein-Coding Genes and Formation of Functional mRNA

■ Transcription of DNA is carried out by RNA polymerase, which adds one ribonucleotide at a time to the 3′ end of a growing RNA chain (see Figure 4-10). The sequence of the template DNA strand determines the order in which ribonucleotides are polymerized to form an RNA chain.

■ During transcription initiation, RNA polymerase binds to a specific site in DNA (the promoter), locally melts the double-stranded DNA to reveal the unpaired template strand, and polymerizes the first two nucleotides.

■ During strand elongation, RNA polymerase moves along the DNA, melting sequential segments of the DNA and adding nucleotides to the growing RNA strand.

■ When RNA polymerase reaches a termination sequence in the DNA, the enzyme stops transcription, leading to release of the completed RNA and dissociation of the enzyme from the template DNA.

■ In prokaryotic DNA, several protein-coding genes commonly are clustered into a functional region, an operon, which is transcribed from a single promoter into one mRNA encoding multiple proteins with related functions (see Figure 4-12a). Translation of a bacterial mRNA can begin before synthesis of the mRNA is complete.

■ In eukaryotic DNA, each protein-coding gene is transcribed from its own promoter. The initial primary transcript very often contains noncoding regions (introns) interspersed among coding regions (exons).

■ Eukaryotic primary transcripts must undergo RNA processing to yield functional RNAs. During processing, the ends of nearly all primary transcripts from protein-coding genes are modified by addition of a 5′ cap and 3′ poly(A) tail. Transcripts from genes containing introns undergo splicing, the removal of the introns and joining of the exons (see Figure 4-14).

■ The individual domains of multidomain proteins found in higher eukaryotes are often encoded by individual exons or a small number of exons. Distinct isoforms of such proteins often are expressed in specific cell types as the result of alternative splicing of exons.

4.3 Control of Gene Expression in Prokaryotes

Since the structure and function of a cell are determined by the proteins it contains, the control of gene expression is a fundamental aspect of molecular cell biology. Most commonly, the "decision" to initiate transcription of the gene encoding a particular protein is the major mechanism for controlling production of the encoded protein in a cell. By controlling transcription initiation, a cell can regulate which proteins it produces and how rapidly. When transcription of a gene is *repressed,* the corresponding mRNA and encoded protein or proteins are synthesized at low rates. Conversely, when transcription of a gene is *activated,* both the mRNA and encoded protein or proteins are produced at much higher rates.

In most bacteria and other single-celled organisms, gene expression is highly regulated in order to adjust the cell's enzymatic machinery and structural components to changes in the nutritional and physical environment. Thus, at any given time, a bacterial cell normally synthesizes only those proteins of its entire proteome required for survival under the particular conditions. In multicellular organisms, control of gene expression is largely directed toward assuring that the right gene is expressed in the right cell at the right time during embryological development and tissue differentiation. Here we describe the basic features of transcription control in bacteria, using the *lac* operon in *E. coli* as our primary example. Many of the same processes, as well as others, are involved in eukaryotic transcription control, which is discussed in Chapter 11.

In *E. coli,* about half the genes are clustered into operons each of which encodes enzymes involved in a particular metabolic pathway or proteins that interact to form one multisubunit protein. For instance, the *trp* operon mentioned earlier encodes five enzymes needed in the biosynthesis of tryptophan (see Figure 4-12). Similarly, the *lac* operon encodes three enzymes required for the metabolism of lactose, a sugar present in milk. Since a bacterial operon is transcribed from one start site into a single mRNA, all the genes within an operon are coordinately regulated; that is, they are all activated or repressed to the same extent.

Transcription of operons, as well as of isolated genes, is controlled by an interplay between RNA polymerase and specific *repressor* and *activator* proteins. In order to initiate transcription, however, *E. coli* RNA polymerase must be associated with one of a small number of σ *(sigma) factors,* which function as initiation factors. The most common one in bacterial cells is σ^{70}.

Initiation of *lac* Operon Transcription Can Be Repressed and Activated

When *E. coli* is in an environment that lacks lactose, synthesis of *lac* mRNA is repressed, so that cellular energy is not wasted synthesizing enzymes the cells cannot use. In an environment containing both lactose and glucose, *E. coli* cells preferentially metabolize glucose, the central molecule of carbohydrate metabolism. Lactose is metabolized at a high rate only when lactose is present and glucose is largely depleted from the medium. This metabolic adjustment is achieved by repressing transcription of the *lac* operon until lactose is present, and synthesis of only low levels of *lac* mRNA until the cytosolic concentration of glucose falls to low levels. Transcription of the *lac* operon under different conditions is controlled by *lac* repressor and catabolite activator protein (CAP), each of which binds to a specific DNA sequence in the *lac* transcription-control region (Figure 4-16, *top*).

For transcription of the *lac* operon to begin, the σ^{70} subunit of the RNA polymerase must bind to the *lac* promoter, which lies just upstream of the start site. When no lactose is present, binding of the *lac* repressor to a sequence called the *lac* **operator,** which overlaps the transcription start site, blocks transcription initiation by the polymerase (Figure 4-16a). When lactose is present, it binds to specific binding sites in each subunit of the tetrameric *lac* repressor, causing a conformational change in the protein that makes it dissociate from the *lac* operator. As a result, the polymerase can initiate transcription of the *lac* operon. However, when glucose also is present, the rate of transcription initiation (i.e., the number of times per minute different polymerase molecules initiate transcription) is very low, resulting in synthesis of only low levels of *lac* mRNA and the proteins encoded in the *lac* operon (Figure 4-16b).

Once glucose is depleted from the media and the intracellular glucose concentration falls, *E. coli* cells respond by synthesizing cyclic AMP, cAMP (see Figure 3-27b). As the concentration of cAMP increases, it binds to a site in each subunit of the dimeric CAP protein, causing a conformational change that allows the protein to bind to the CAP site in the *lac* transcription-control region. The bound CAP-cAMP complex interacts with the polymerase bound to the promoter, greatly stimulating the rate of transcription initiation. This activation leads to synthesis of high levels of *lac*

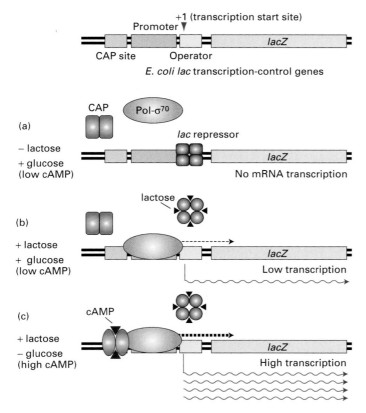

▲ FIGURE 4-16 **Regulation of transcription from the**
***lac* operon of *E. coli.** (*Top*) The transcription-control region,
composed of ≈100 base pairs, includes three protein-binding
regions: the CAP site, which binds catabolite activator protein;
the *lac* promoter, which binds the RNA polymerase–σ^{70} complex;
and the *lac* operator, which binds *lac* repressor. The *lacZ* gene,
the first of three genes in the operon, is shown to the right.
(a) In the absence of lactose, very little *lac* mRNA is produced
because the *lac* repressor binds to the operator, inhibiting
transcription initiation by RNA polymerase–σ^{70}. (b) In the
presence of glucose and lactose, *lac* repressor binds lactose
and dissociates from the operator, allowing RNA polymerase–σ^{70}
to initiate transcription at a low rate. (c) Maximal transcription of
the *lac* operon occurs in the presence of lactose and absence of
glucose. In this situation, cAMP increases in response to the low
glucose concentration and forms the CAP-cAMP complex, which
binds to the CAP site, where it interacts with RNA polymerase
to stimulate the rate of transcription initiation.

mRNA and subsequently of the enzymes encoded by the *lac*
operon (Figure 4-16c).

Although the promoters for different *E. coli* genes exhibit
considerable homology, their exact sequences differ. The promoter sequence determines the intrinsic rate at which an
RNA polymerase–σ complex initiates transcription of a gene
in the absence of a repressor or activator protein. Promoters
that support a high rate of transcription initiation are called
strong promoters. Those that support a low rate of transcription initiation are called *weak promoters*. The *lac*
operon, for instance, has a weak promoter; its low intrinsic

rate of initiation is further reduced by the *lac* repressor and
substantially increased by the cAMP-CAP activator.

Small Molecules Regulate Expression of Many Bacterial Genes via DNA-Binding Repressors

Transcription of most *E. coli* genes is regulated by processes
similar to those described for the *lac* operon. The general
mechanism involves a specific repressor that binds to the operator region of a gene or operon, thereby blocking transcription initiation. A small molecule (or molecules), called an
inducer, binds to the repressor, controlling its DNA-binding
activity and consequently the rate of transcription as appropriate for the needs of the cell.

For example, when the tryptophan concentration in the
medium and cytosol is high, the cell does not synthesize the
several enzymes encoded in the *trp* operon. Binding of tryptophan to the *trp* repressor causes a conformational change
that allows the protein to bind to the *trp* operator, thereby
repressing expression of the enzymes that synthesize tryptophan. Conversely, when the tryptophan concentration in the
medium and cytosol is low, tryptophan dissociates from the
trp repressor, causing a conformational change in the protein
that causes it to dissociate from the *trp* operator, allowing
transcription of the *trp* operon. In the case of the *lac* operon,
binding of the inducer lactose to the *lac* repressor reduces
binding of the repressor to the operator, thereby promoting
transcription.

Specific activator proteins, such as CAP in the lac operon,
also control transcription of some but not all bacterial genes.
These activators bind to DNA together with the RNA polymerase, stimulating transcription from a specific promoter.
The DNA-binding activity of an activator is modulated in response to cellular needs by the binding of specific small molecules (e.g., cAMP) that alter the conformation of the
activator.

Transcription by σ^{54}-RNA Polymerase Is Controlled by Activators That Bind Far from the Promoter

Most *E. coli* promoters interact with σ^{70}-RNA polymerase,
the major form of the bacterial enzyme. Transcription of certain groups of genes, however, is carried out by *E. coli* RNA
polymerases containing one of several alternative sigma factors that recognize different consensus promoter sequences
than σ^{70} does. All but one of these are related to σ^{70} in sequence. Transcription initiation by RNA polymerases containing these σ^{70}-like factors is regulated by repressors and
activators that bind to DNA near the region where the polymerase binds, similar to initiation by σ^{70}-RNA polymerase
itself.

The sequence of one *E. coli* sigma factor, σ^{54}, is distinctly
different from that of all the σ^{70}-like factors. Transcription
of genes by RNA polymerases containing σ^{54} is regulated

solely by activators whose binding sites in DNA, referred to as **enhancers,** generally are located 80–160 base pairs upstream from the start site. Even when enhancers are moved more than a kilobase away from a start site, σ^{54}-activators can activate transcription.

The best-characterized σ^{54}-activator—the NtrC protein (nitrogen regulatory protein C)—stimulates transcription from the promoter of the *glnA* gene. This gene encodes the enzyme glutamine synthetase, which synthesizes the amino acid glutamine from glutamic acid and ammonia. The σ^{54}-RNA polymerase binds to the *glnA* promoter but does not melt the DNA strands and initiate transcription until it is activated by NtrC, a dimeric protein. NtrC, in turn, is regulated by a protein kinase called NtrB. In response to low levels of glutamine, NtrB phosphorylates dimeric NtrC, which then binds to an enhancer upstream of the *glnA* pro-

(a)   NtrC σ^{54} polymerase

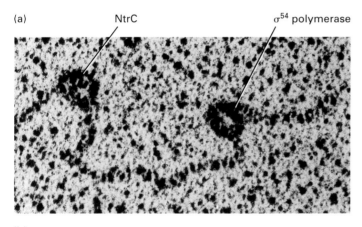

(b)

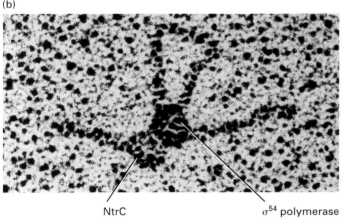

NtrC σ^{54} polymerase

▲ **EXPERIMENTAL FIGURE 4-17 DNA looping permits interaction of bound NtrC and σ^{54}-polymerase.** (a) Electron micrograph of DNA restriction fragment with phosphorylated NtrC dimer binding to the enhancer region near one end and σ^{54}–RNA polymerase bound to the *glnA* promoter near the other end. (b) Electron micrograph of the same fragment preparation showing NtrC dimers and σ^{54}-polymerase binding to each other with the intervening DNA forming a loop between them. [From W. Su et al., 1990, *Proc. Nat'l. Acad. Sci. USA* **87**:5505; courtesy of S. Kustu.]

moter. Enhancer-bound phosphorylated NtrC then stimulates the σ^{54}-polymerase bound at the promoter to separate the DNA strands and initiate transcription. Electron microscopy studies have shown that phosphorylated NtrC bound at enhancers and σ^{54}-polymerase bound at the promoter directly interact, forming a loop in the DNA between the binding sites (Figure 4-17). As discussed in Chapter 11, this activation mechanism is somewhat similar to the predominant mechanism of transcriptional activation in eukaryotes.

NtrC has ATPase activity, and ATP hydrolysis is required for activation of bound σ^{54}-polymerase by phosphorylated NtrC. Evidence for this is that mutants with an NtrC defective in ATP hydrolysis are invariably defective in stimulating the σ^{54}-polymerase to melt the DNA strands at the transcription start site. It is postulated that ATP hydrolysis supplies the energy required for melting the DNA strands. In contrast, the σ^{70}-polymerase does not require ATP hydrolysis to separate the strands at a start site.

Many Bacterial Responses Are Controlled by Two-Component Regulatory Systems

As we've just seen, control of the *E. coli glnA* gene depends on two proteins, NtrC and NtrB. Such two-component regulatory systems control many responses of bacteria to changes in their environment. Another example involves the *E. coli* proteins PhoR and PhoB, which regulate transcription in response to the concentration of free phosphate. PhoR is a transmembrane protein, located in the inner (plasma) membrane, whose periplasmic domain binds phosphate with moderate affinity and whose cytosolic domain has protein kinase activity; PhoB is a cytosolic protein.

Large protein pores in the *E. coli* outer membrane allow ions to diffuse freely between the external environment and the periplasmic space. Consequently, when the phosphate concentration in the environment falls, it also falls in the periplasmic space, causing phosphate to dissociate from the PhoR periplasmic domain, as depicted in Figure 4-18. This causes a conformational change in the PhoR cytoplasmic domain that activates its protein kinase activity. The activated PhoR initially transfers a γ-phosphate from ATP to a histidine side chain in the PhoR kinase domain itself. The same phosphate is then transferred to a specific aspartic acid side chain in PhoB, converting PhoB from an inactive to an active transcriptional activator. Phosphorylated, active PhoB then induces transcription from several genes that help the cell cope with low phosphate conditions.

Many other bacterial responses are regulated by two proteins with homology to PhoR and PhoB. In each of these regulatory systems, one protein, called a *sensor,* contains a transmitter domain homologous to the PhoR protein kinase domain. The transmitter domain of the sensor protein is regulated by a second unique protein domain (e.g., the periplasmic domain of PhoR) that senses environmental changes. The second protein, called a *response regulator,* contains a

▶ **FIGURE 4-18 The PhoR/PhoB two-component regulatory system in *E. coli*.** In response to low phosphate concentrations in the environment and periplasmic space, a phosphate ion dissociates from the periplasmic domain of the inactive sensor protein PhoR. This causes a conformational change that activates a protein kinase transmitter domain in the cytosolic region of PhoR. The activated transmitter domain transfers an ATP γ phosphate to a conserved histidine in the transmitter domain. This phosphate is then transferred to an aspartic acid in the receiver domain of the response regulator PhoB. Several PhoB proteins can be phosphorylated by one activated PhoR. Phosphorylated PhoB proteins then activate transcription from genes encoding proteins that help the cell to respond to low phosphate, including *phoA, phoS, phoE,* and *ugpB*.

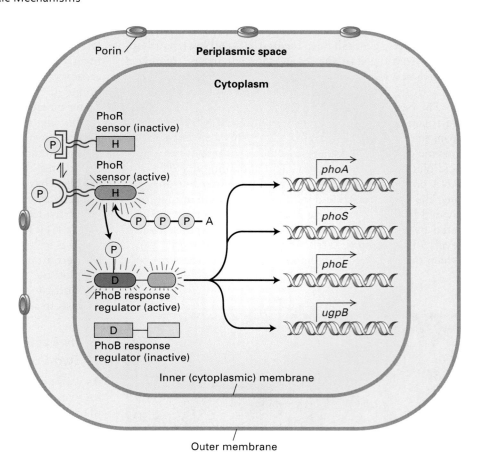

receiver domain homologous to the region of PhoB that is phosphorylated by activated PhoR. The receiver domain of the response regulator is associated with a second domain that determines the protein's function. The activity of this second functional domain is regulated by phosphorylation of the receiver domain. Although all transmitter domains are homologous (as are receiver domains), the transmitter domain of a specific sensor protein will phosphorylate only specific receiver domains of specific response regulators, allowing specific responses to different environmental changes. Note that NtrB and NtrC, discussed above, function as sensor and response regulator proteins, respectively, in the two-component regulatory system that controls transcription of *glnA*. Similar two-component histidyl-aspartyl phosphorelay regulatory systems are also found in plants.

KEY CONCEPTS OF SECTION 4.3

Control of Gene Expression in Prokaryotes

■ Gene expression in both prokaryotes and eukaryotes is regulated primarily by mechanisms that control the initiation of transcription.

■ Binding of the σ subunit in an RNA polymerase to a promoter region is the first step in the initiation of transcription in *E. coli*.

■ The nucleotide sequence of a promoter determines its strength, that is, how frequently different RNA polymerase molecules can bind and initiate transcription per minute.

■ Repressors are proteins that bind to operator sequences, which overlap or lie adjacent to promoters. Binding of a repressor to an operator inhibits transcription initiation.

■ The DNA-binding activity of most bacterial repressors is modulated by small effector molecules (inducers). This allows bacterial cells to regulate transcription of specific genes in response to changes in the concentration of various nutrients in the environment.

■ The *lac* operon and some other bacterial genes also are regulated by activator proteins that bind next to promoters and increase the rate of transcription initiation by RNA polymerase.

■ The major sigma factor in *E. coli* is σ^{70}, but several other less abundant sigma factors are also found, each recognizing different consensus promoter sequences.

■ Transcription initiation by all *E. coli* RNA polymerases, except those containing σ^{54}, can be regulated by repressors and activators that bind near the transcription start site (see Figure 4-16).

■ Genes transcribed by σ^{54}–RNA polymerase are regulated by activators that bind to enhancers located ≈100 base pairs upstream from the start site. When the activator and σ^{54}–RNA polymerase interact, the DNA between their binding sites forms a loop (see Figure 4-17).

■ In two-component regulatory systems, one protein acts as a sensor, monitoring the level of nutrients or other components in the environment. Under appropriate conditions, the γ-phosphate of an ATP is transferred first to a histidine in the sensor protein and then to an aspartic acid in a second protein, the response regulator. The phosphorylated response regulator then binds to DNA regulatory sequences, thereby stimulating or repressing transcription of specific genes (see Figure 4-18).

4.4 The Three Roles of RNA in Translation

Although DNA stores the information for protein synthesis and mRNA conveys the instructions encoded in DNA, most biological activities are carried out by proteins. As we saw in Chapter 3, the linear order of amino acids in each protein determines its three-dimensional structure and activity. For this reason, assembly of amino acids in their correct order, as encoded in DNA, is critical to production of functional proteins and hence the proper functioning of cells and organisms.

Translation is the whole process by which the nucleotide sequence of an mRNA is used to order and to join the amino acids in a polypeptide chain (see Figure 4-1, step 3). In eukaryotic cells, protein synthesis occurs in the cytoplasm, where three types of RNA molecules come together to perform different but cooperative functions (Figure 4-19):

1. **Messenger RNA (mRNA)** carries the genetic information transcribed from DNA in the form of a series of three-nucleotide sequences, called **codons,** each of which specifies a particular amino acid.

2. **Transfer RNA (tRNA)** is the key to deciphering the codons in mRNA. Each type of amino acid has its own subset of tRNAs, which bind the amino acid and carry it to the growing end of a polypeptide chain if the next codon in the mRNA calls for it. The correct tRNA with its attached amino acid is selected at each step because each specific tRNA molecule contains a three-nucleotide sequence, an **anticodon,** that can base-pair with its complementary codon in the mRNA.

3. **Ribosomal RNA (rRNA)** associates with a set of proteins to form **ribosomes.** These complex structures, which physically move along an mRNA molecule, catalyze the assembly of amino acids into polypeptide chains. They also bind tRNAs and various accessory proteins necessary for protein synthesis. Ribosomes are composed of a large and a small subunit, each of which contains its own rRNA molecule or molecules.

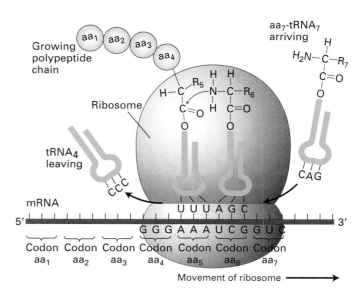

▲ **FIGURE 4-19 The three roles of RNA in protein synthesis.** Messenger RNA (mRNA) is translated into protein by the joint action of transfer RNA (tRNA) and the ribosome, which is composed of numerous proteins and two major ribosomal RNA (rRNA) molecules (not shown). Note the base pairing between tRNA anticodons and complementary codons in the mRNA. Formation of a peptide bond between the amino group N on the incoming aa-tRNA and the carboxyl-terminal C on the growing protein chain (purple) is catalyzed by one of the rRNAs. aa = amino acid; R = side group. [Adapted from A. J. F. Griffiths et al., 1999, *Modern Genetic Analysis,* W. H. Freeman and Company.]

These three types of RNA participate in translation in all cells. Indeed, development of three functionally distinct RNAs was probably the molecular key to the origin of life. How the structure of each RNA relates to its specific task is described in this section; how the three types work together, along with required protein factors, to synthesize proteins is detailed in the following section. Since translation is essential for protein synthesis, the two processes commonly are referred to interchangeably. However, the polypeptide chains resulting from translation undergo post-translational folding and often other changes (e.g., chemical modifications, association with other chains) that are required for production of mature, functional proteins (Chapter 3).

Messenger RNA Carries Information from DNA in a Three-Letter Genetic Code

As noted above, the **genetic code** used by cells is a *triplet* code, with every three-nucleotide sequence, or codon, being "read" from a specified starting point in the mRNA. Of the 64 possible codons in the genetic code, 61 specify individual amino acids and three are stop codons. Table 4-1 shows that most amino acids are encoded by more than one codon. Only two—methionine and tryptophan—have a single

TABLE 4-1   The Genetic Code (RNA to Amino Acids)*

First Position (5' end)	Second Position				Third Position (3' end)
	U	**C**	**A**	**G**	
U	Phe	Ser	Tyr	Cys	U
	Phe	Ser	Tyr	Cys	C
	Leu	Ser	Stop	Stop	A
	Leu	Ser	Stop	Trp	G
C	Leu	Pro	His	Arg	U
	Leu	Pro	His	Arg	C
	Leu	Pro	Gln	Arg	A
	Leu (Met)*	Pro	Gln	Arg	G
A	Ile	Thr	Asn	Ser	U
	Ile	Thr	Asn	Ser	C
	Ile	Thr	Lys	Arg	A
	Met (start)	Thr	Lys	Arg	G
G	Val	Ala	Asp	Gly	U
	Val	Ala	Asp	Gly	C
	Val	Ala	Glu	Gly	A
	Val (Met)*	Ala	Glu	Gly	G

*AUG is the most common initiator codon; GUG usually codes for valine, and CUG for leucine, but, rarely, these codons can also code for methionine to initiate a protein chain.

codon; at the other extreme, leucine, serine, and arginine are each specified by six different codons. The different codons for a given amino acid are said to be synonymous. The code itself is termed *degenerate,* meaning that more than one codon can specify the same amino acid.

Synthesis of all polypeptide chains in prokaryotic and eukaryotic cells begins with the amino acid methionine. In most mRNAs, the *start (initiator) codon* specifying this amino-terminal methionine is AUG. In a few bacterial mRNAs, GUG is used as the initiator codon, and CUG occasionally is used as an initiator codon for methionine in eukaryotes. The three codons UAA, UGA, and UAG do not specify amino acids but constitute *stop (termination) codons* that mark the carboxyl terminus of polypeptide chains in almost all cells. The sequence of codons that runs from a specific start codon to a stop codon is called a **reading frame.** This precise linear array of ribonucleotides in groups of three in mRNA specifies the precise linear sequence of amino acids in a polypeptide chain and also signals where synthesis of the chain starts and stops.

Because the genetic code is a comma-less, non-overlapping triplet code, a particular mRNA theoretically could be translated in three different reading frames. Indeed some mRNAs have been shown to contain overlapping information that can be translated in different reading frames, yielding different polypeptides (Figure 4-20). The vast majority of mRNAs, however, can be read in only one frame because stop codons encountered in the other two possible reading frames terminate translation before a functional protein is produced. Another unusual coding arrangement occurs because of *frame-*

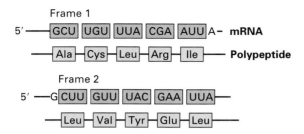

▲ **FIGURE 4-20 Example of how the genetic code—a non-overlapping, comma-less triplet code—can be read in different frames.** If translation of the mRNA sequence shown begins at two different upstream start sites (not shown), then two overlapping reading frames are possible. In this example, the codons are shifted one base to the right in the lower frame. As a result, the same nucleotide sequence specifies different amino acids during translation. Although they are rare, many instances of such overlaps have been discovered in viral and cellular genes of prokaryotes and eukaryotes. It is theoretically possible for the mRNA to have a third reading frame.

shifting. In this case the protein-synthesizing machinery may read four nucleotides as one amino acid and then continue reading triplets, or it may back up one base and read all succeeding triplets in the new frame until termination of the chain occurs. These frameshifts are not common events, but a few dozen such instances are known.

The meaning of each codon is the same in most known organisms—a strong argument that life on earth evolved only once. However, the genetic code has been found to differ for a few codons in many mitochondria, in ciliated protozoans, and in *Acetabularia,* a single-celled plant. As shown in Table 4-2, most of these changes involve reading of normal stop codons as amino acids, not an exchange of one amino acid for another. These exceptions to the general code probably were later evolutionary developments; that is, at no single time was the code immutably fixed, although massive changes were not tolerated once a general code began to function early in evolution.

The Folded Structure of tRNA Promotes Its Decoding Functions

Translation, or decoding, of the four-nucleotide language of DNA and mRNA into the 20–amino acid language of proteins requires tRNAs and enzymes called *aminoacyl-tRNA synthetases.* To participate in protein synthesis, a tRNA molecule must become chemically linked to a particular amino acid via a high-energy bond, forming an **aminoacyl-tRNA;** the anticodon in the tRNA then base-pairs with a codon in mRNA so that the activated amino acid can be added to the growing polypeptide chain (Figure 4-21).

Some 30–40 different tRNAs have been identified in bacterial cells and as many as 50–100 in animal and plant cells. Thus the number of tRNAs in most cells is more than the number of amino acids used in protein synthesis (20) and also differs from the number of amino acid codons in the genetic code (61). Consequently, many amino acids have more than one tRNA to which they can attach (explaining how there can be more tRNAs than amino acids); in addition, many tRNAs can pair with more than one codon (explaining how there can be more codons than tRNAs).

The function of tRNA molecules, which are 70–80 nucleotides long, depends on their precise three-dimensional structures. In solution, all tRNA molecules fold into a similar stem-loop arrangement that resembles a cloverleaf when drawn in two dimensions (Figure 4-22a). The four stems are short double helices stabilized by Watson-Crick base pairing; three of the four stems have loops containing seven or eight bases at their ends, while the remaining, unlooped stem contains the free 3′ and 5′ ends of the chain. The three nucleotides composing the anticodon are located at the center of the middle loop, in an accessible position that facilitates codon-anticodon base pairing. In all tRNAs, the 3′ end of the unlooped amino acid *acceptor stem* has the sequence CCA, which in most cases is added after synthesis and processing of the tRNA are complete. Several bases in most tRNAs also are modified after synthesis. Viewed in three

Codon	Universal Code	Unusual Code*	Occurrence
UGA	Stop	Trp	*Mycoplasma, Spiroplasma,* mitochondria of many species
CUG	Leu	Thr	Mitochondria in yeasts
UAA, UAG	Stop	Gln	*Acetabularia, Tetrahymena,* Paramecium, etc.
UGA	Stop	Cys	*Euplotes*

TABLE 4-2 Known Deviations from the Universal Genetic Code

*"Unusual code" is used in nuclear genes of the listed organisms and in mitochondrial genes as indicated.
SOURCE: S. Osawa et al., 1992, *Microbiol. Rev.* **56**:229.

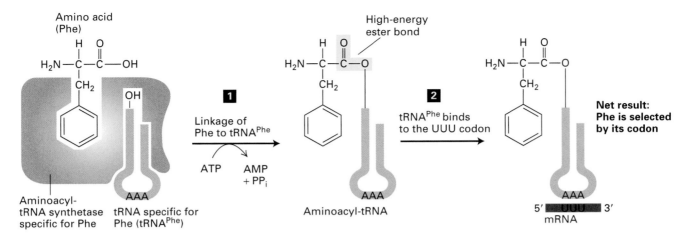

▲ **FIGURE 4-21 Two-step decoding process for translating nucleic acid sequences in mRNA into amino acid sequences in proteins.** Step **1**: An aminoacyl-tRNA synthetase first couples a specific amino acid, via a high-energy ester bond (yellow), to either the 2′ or 3′ hydroxyl of the terminal adenosine in the corresponding tRNA. Step **2**: A three-base sequence in the tRNA (the anticodon) then base-pairs with a codon in the mRNA specifying the attached amino acid. If an error occurs in either step, the wrong amino acid may be incorporated into a polypeptide chain. Phe = phenylalanine.

dimensions, the folded tRNA molecule has an L shape with the anticodon loop and acceptor stem forming the ends of the two arms (Figure 4-22b).

Nonstandard Base Pairing Often Occurs Between Codons and Anticodons

If perfect Watson-Crick base pairing were demanded between codons and anticodons, cells would have to contain exactly 61 different tRNA species, one for each codon that specifies an amino acid. As noted above, however, many cells contain fewer than 61 tRNAs. The explanation for the smaller number lies in the capability of a single tRNA anticodon to recognize more than one, but not necessarily every, codon corresponding to a given amino acid. This broader recognition can occur because of nonstandard pairing between bases in the so-called *wobble* position: that is, the third (3′) base in an mRNA codon and the corresponding first (5′) base in its tRNA anticodon.

The first and second bases of a codon almost always form standard Watson-Crick base pairs with the third and

▶ **FIGURE 4-22 Structure of tRNAs.**
(a) Although the exact nucleotide sequence varies among tRNAs, they all fold into four base-paired stems and three loops. The CCA sequence at the 3′ end also is found in all tRNAs. Attachment of an amino acid to the 3′ A yields an aminoacyl-tRNA. Some of the A, C, G, and U residues are modified in most tRNAs (see key). Dihydrouridine (D) is nearly always present in the D loop; likewise, ribothymidine (T) and pseudouridine (Ψ) are almost always present in the TΨCG loop. Yeast alanine tRNA, represented here, also contains other modified bases. The triplet at the tip of the anticodon loop base-pairs with the corresponding codon in mRNA. (b) Three-dimensional model of the generalized backbone of all tRNAs. Note the L shape of the molecule. [Part (a) see R. W. Holly et al., 1965, *Science* **147**:1462; part (b) from J. G. Arnez and D. Moras, 1997, *Trends Biochem. Sci.* **22**:211.]

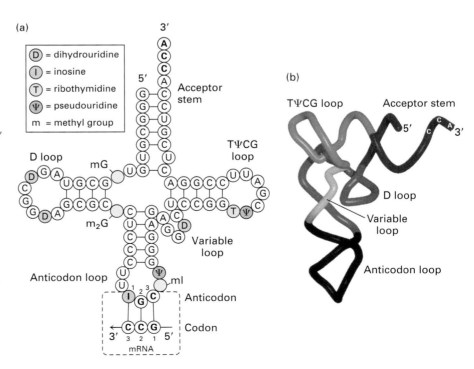

second bases, respectively, of the corresponding anticodon, but four nonstandard interactions can occur between bases in the wobble position. Particularly important is the G·U base pair, which structurally fits almost as well as the standard G·C pair. Thus, a given anticodon in tRNA with G in the first (wobble) position can base-pair with the two corresponding codons that have either pyrimidine (C or U) in the third position (Figure 4-23). For example, the phenylalanine codons UUU and UUC (5′→3′) are both recognized by the tRNA that has GAA (5′→3′) as the anticodon. In fact, any two codons of the type NNPyr (N = any base; Pyr = pyrimidine) encode a single amino acid and are decoded by a single tRNA with G in the first (wobble) position of the anticodon.

Although adenine rarely is found in the anticodon wobble position, many tRNAs in plants and animals contain inosine

(I), a deaminated product of adenine, at this position. Inosine can form nonstandard base pairs with A, C, and U. A tRNA with inosine in the wobble position thus can recognize the corresponding mRNA codons with A, C, or U in the third (wobble) position (see Figure 4-23). For this reason, inosine-containing tRNAs are heavily employed in translation of the synonymous codons that specify a single amino acid. For example, four of the six codons for leucine (CUA, CUC, CUU, and UUA) are all recognized by the same tRNA with the anticodon 3′-GAI-5′; the inosine in the wobble position forms nonstandard base pairs with the third base in the four codons. In the case of the UUA codon, a nonstandard G·U pair also forms between position 3 of the anticodon and position 1 of the codon.

Aminoacyl-tRNA Synthetases Activate Amino Acids by Covalently Linking Them to tRNAs

Recognition of the codon or codons specifying a given amino acid by a particular tRNA is actually the second step in decoding the genetic message. The first step, attachment of the appropriate amino acid to a tRNA, is catalyzed by a specific aminoacyl-tRNA synthetase. Each of the 20 different synthetases recognizes *one* amino acid and *all* its compatible, or *cognate*, tRNAs. These coupling enzymes link an amino acid to the free 2′ or 3′ hydroxyl of the adenosine at the 3′ terminus of tRNA molecules by an ATP-requiring reaction. In this reaction, the amino acid is linked to the tRNA by a high-energy bond and thus is said to be *activated*. The energy of this bond subsequently drives formation of the peptide bonds linking adjacent amino acids in a growing polypeptide chain. The equilibrium of the aminoacylation reaction is driven further toward activation of the amino acid by hydrolysis of the high-energy phosphoanhydride bond in the released pyrophosphate (see Figure 4-21).

Because some amino acids are so similar structurally, aminoacyl-tRNA synthetases sometimes make mistakes. These are corrected, however, by the enzymes themselves, which have a *proofreading* activity that checks the fit in their amino acid–binding pocket. If the wrong amino acid becomes attached to a tRNA, the bound synthetase catalyzes removal of the amino acid from the tRNA. This crucial function helps guarantee that a tRNA delivers the correct amino acid to the protein-synthesizing machinery. The overall error rate for translation in *E. coli* is very low, approximately 1 per 50,000 codons, evidence of the importance of proofreading by aminoacyl-tRNA synthetases.

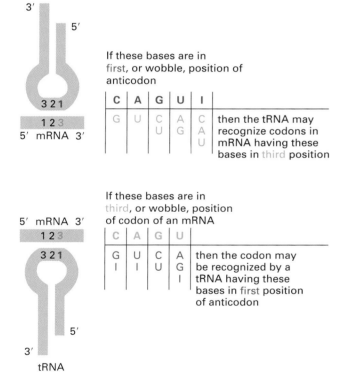

▲ **FIGURE 4-23 Nonstandard codon-anticodon base pairing at the wobble position.** The base in the third (or wobble) position of an mRNA codon often forms a nonstandard base pair with the base in the first (or wobble) position of a tRNA anticodon. Wobble pairing allows a tRNA to recognize more than one mRNA codon *(top)*; conversely, it allows a codon to be recognized by more than one kind of tRNA *(bottom)*, although each tRNA will bear the same amino acid. Note that a tRNA with I (inosine) in the wobble position can "read" (become paired with) three different codons, and a tRNA with G or U in the wobble position can read two codons. Although A is theoretically possible in the wobble position of the anticodon, it is almost never found in nature.

Ribosomes Are Protein-Synthesizing Machines

If the many components that participate in translating mRNA had to interact in free solution, the likelihood of simultaneous collisions occurring would be so low that the rate of amino acid polymerization would be very slow. The efficiency of translation is greatly increased by the binding of the mRNA and the individual aminoacyl-tRNAs to the most abundant RNA-protein complex in the cell, the ribosome, which directs elongation of a polypeptide at a rate of three to five amino acids added per second. Small proteins of 100–200 amino acids are therefore made in a minute or less. On the other hand, it takes 2–3 hours to make the largest known protein, titin, which is found in muscle and contains about 30,000 amino acid residues. The cellular machine that accomplishes this task must be precise and persistent.

With the aid of the electron microscope, ribosomes were first discovered as small, discrete, RNA-rich particles in cells that secrete large amounts of protein. However, their role in protein synthesis was not recognized until reasonably pure ribosome preparations were obtained. In vitro radiolabeling experiments with such preparations showed that radioactive amino acids first were incorporated into growing polypeptide chains that were associated with ribosomes before appearing in finished chains.

A ribosome is composed of three (in bacteria) or four (in eukaryotes) different rRNA molecules and as many as 83 proteins, organized into a large subunit and a small subunit (Figure 4-24). The ribosomal subunits and the rRNA molecules are commonly designated in Svedberg units (S), a measure of the sedimentation rate of suspended particles centrifuged under standard conditions. The small ribosomal subunit contains a single rRNA molecule, referred to as *small rRNA*. The large subunit contains a molecule of *large rRNA* and one molecule of 5S rRNA, plus an additional molecule of 5.8S rRNA in vertebrates. The lengths of the rRNA molecules, the quantity of proteins in each subunit, and consequently the sizes of the subunits differ in bacterial and eukaryotic cells. The assembled ribosome is 70S in bacteria and 80S in vertebrates. But more interesting than these differences are the great structural and functional similarities between ribosomes from all species. This consistency is another reflection of the common evolutionary origin of the most basic constituents of living cells.

The sequences of the small and large rRNAs from several thousand organisms are now known. Although the primary nucleotide sequences of these rRNAs vary considerably, the same parts of each type of rRNA theoretically can form base-paired stem-loops, which would generate a similar three-dimensional structure for each rRNA in all organisms. The actual three-dimensional structures of bacterial rRNAs from *Thermus thermopolis* recently have been determined by x-ray crystallography of the 70S ribosome. The multiple, much smaller ribosomal proteins for the most part are associated

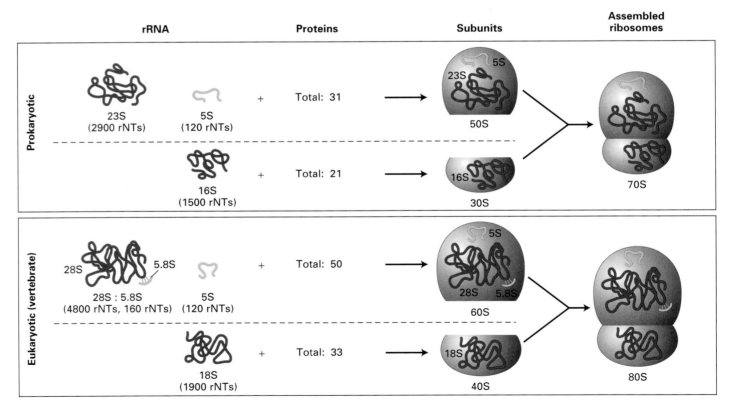

▲ **FIGURE 4-24 The general structure of ribosomes in prokaryotes and eukaryotes.** In all cells, each ribosome consists of a large and a small subunit. The two subunits contain rRNAs (red) of different lengths, as well as a different set of proteins. All ribosomes contain two major rRNA molecules (23S and 16S rRNA in bacteria; 28S and 18S rRNA in vertebrates) and a 5S rRNA. The large subunit of vertebrate ribosomes also contains a 5.8S rRNA base-paired to the 28S rRNA. The number of ribonucleotides (rNTs) in each rRNA type is indicated.

with the surface of the rRNAs. Although the number of protein molecules in ribosomes greatly exceeds the number of RNA molecules, RNA constitutes about 60 percent of the mass of a ribosome.

During translation, a ribosome moves along an mRNA chain, interacting with various protein factors and tRNAs and very likely undergoing large conformational changes. Despite the complexity of the ribosome, great progress has been made in determining the overall structure of bacterial ribosomes and in identifying various reactive sites. X-ray crystallographic studies on the *T. thermophilus* 70S ribosome, for instance, not only have revealed the dimensions and overall shape of the ribosomal subunits but also have localized the positions of tRNAs bound to the ribosome during elongation of a growing protein chain. In addition, powerful chemical techniques such as **footprinting**, which is described in Chapter 11, have been used to identify specific nucleotide sequences in rRNAs that bind to protein or another RNA. Some 40 years after the initial discovery of ribosomes, their overall structure and functioning during protein synthesis are finally becoming clear, as we describe in the next section.

KEY CONCEPTS OF SECTION 4.4

The Three Roles of RNA in Translation

■ Genetic information is transcribed from DNA into mRNA in the form of a comma-less, overlapping, degenerate triplet code.

■ Each amino acid is encoded by one or more three-nucleotide sequences (codons) in mRNA. Each codon specifies one amino acid, but most amino acids are encoded by multiple codons (see Table 4-1).

■ The AUG codon for methionine is the most common start codon, specifying the amino acid at the NH_2-terminus of a protein chain. Three codons (UAA, UAG, UGA) function as stop codons and specify no amino acids.

■ A reading frame, the uninterrupted sequence of codons in mRNA from a specific start codon to a stop codon, is translated into the linear sequence of amino acids in a polypeptide chain.

■ Decoding of the nucleotide sequence in mRNA into the amino acid sequence of proteins depends on tRNAs and aminoacyl-tRNA synthetases.

■ All tRNAs have a similar three-dimensional structure that includes an acceptor arm for attachment of a specific amino acid and a stem-loop with a three-base anticodon sequence at its ends (see Figure 4-22). The anticodon can base-pair with its corresponding codon in mRNA.

■ Because of nonstandard interactions, a tRNA may base-pair with more than one mRNA codon; conversely, a particular codon may base-pair with multiple tRNAs. In each case, however, only the proper amino acid is inserted into a growing polypeptide chain.

■ Each of the 20 aminoacyl-tRNA synthetases recognizes a single amino acid and covalently links it to a cognate tRNA, forming an aminoacyl-tRNA (see Figure 4-21). This reaction activates the amino acid, so it can participate in peptide bond formation.

■ Both prokaryotic and eukaryotic ribosomes—the large ribonucleoprotein complexes on which translation occurs—consist of a small and a large subunit (see Figure 4-24). Each subunit contains numerous different proteins and one major rRNA molecule (small or large). The large subunit also contains one accessory 5S rRNA in bacteria and two accessory rRNAs in eukaryotes (5S and 5.8S in vertebrates).

■ Analogous rRNAs from many different species fold into quite similar three-dimensional structures containing numerous stem-loops and binding sites for proteins, mRNA, and tRNAs. Much smaller ribosomal proteins are associated with the periphery of the rRNAs.

4.5 Stepwise Synthesis of Proteins on Ribosomes

The previous sections have introduced the major participants in protein synthesis—mRNA, aminoacylated tRNAs, and ribosomes containing large and small rRNAs. We now take a detailed look at how these components are brought together to carry out the biochemical events leading to formation of polypeptide chains on ribosomes. Similar to transcription, the complex process of translation can be divided into three stages—initiation, elongation, and termination—which we consider in order. We focus our description on translation in eukaryotic cells, but the mechanism of translation is fundamentally the same in all cells.

Methionyl-tRNA$_i^{Met}$ Recognizes the AUG Start Codon

As noted earlier, the AUG codon for methionine functions as the start codon in the vast majority of mRNAs. A critical aspect of translation initiation is to begin protein synthesis at the start codon, thereby establishing the correct reading frame for the entire mRNA. Both prokaryotes and eukaryotes contain two different methionine tRNAs: tRNA$_i^{Met}$ can initiate protein synthesis, and tRNA^{Met} can incorporate methionine only into a growing protein chain. The same aminoacyl-tRNA synthetase (MetRS) charges both tRNAs with methionine. But *only* Met-tRNA$_i^{Met}$ (i.e., activated methionine attached to tRNA$_i^{Met}$) can bind at the appropriate site on the small ribosomal subunit, the *P site*, to begin synthesis of a polypeptide chain. The regular Met-tRNA^{Met} and all other charged tRNAs bind only to another ribosomal site, the *A site*, as described later.

Translation Initiation Usually Occurs Near the First AUG Closest to the 5′ End of an mRNA

During the first stage of translation, a ribosome assembles, complexed with an mRNA and an activated initiator tRNA, which is correctly positioned at the start codon. Large and small ribosomal subunits not actively engaged in translation are kept apart by binding of two **initiation factors,** designated eIF3 and eIF6 in eukaryotes. A translation *preinitiation complex* is formed when the 40S subunit–eIF3 complex is bound by eIF1A and a ternary complex of the Met-tRNA$_i^{Met}$, eIF2, and GTP (Figure 4-25, step ⊡). Cells can regulate protein synthesis by phosphorylating a serine residue on the eIF2 bound to GDP; the phosphorylated complex is unable to exchange the bound GDP for GTP and cannot bind Met-tRNA$_i^{Met}$, thus inhibiting protein synthesis.

During translation initiation, the 5′ cap of an mRNA to be translated is bound by the eIF4E subunit of the eIF4 cap-binding complex. The mRNA-eIF4 complex then associates with the preinitiation complex through an interaction of the eIF4G subunit and eIF3, forming the *initiation complex* (Figure 4-25, step ⊡). The initiation complex then probably slides along, or *scans,* the associated mRNA as the **helicase** activity of eIF4A uses energy from ATP hydrolysis to unwind the RNA secondary structure. Scanning stops when the tRNA$_i^{Met}$ anticodon recognizes the start codon, which is the first AUG downstream from the 5′ end in most eukaryotic mRNAs (step ⊡). Recognition of the start codon leads to hydrolysis of the GTP associated with eIF2, an irreversible step that prevents further scanning. Selection of the initiating AUG is facilitated by specific surrounding nucleotides called the *Kozak sequence,* for Marilyn Kozak, who defined it: (5′) ACC**AUG**G (3′). The A preceding the AUG (underlined) and the G immediately following it are the most important nucleotides affecting translation initiation efficiency. Once the small ribosomal subunit with its bound Met-tRNA$_i^{Met}$ is correctly positioned at the start codon, union with the large (60S) ribosomal subunit completes formation of an 80S ribosome. This requires the action of another factor (eIF5) and hydrolysis

▶ **FIGURE 4-25 Initiation of translation in eukaryotes.** *(Inset)* When a ribosome dissociates at the termination of translation, the 40S and 60S subunits associate with initiation factors eIF3 and eIF6, forming complexes that can initiate another round of translation. Steps ⊡ and ⊡: Sequential addition of the indicated components to the 40S subunit–eIF3 complex forms the initiation complex. Step ⊡: Scanning of the mRNA by the associated initiation complex leads to positioning of the small subunit and bound Met-tRNA$_i^{Met}$ at the start codon. Step ⊡: Association of the large subunit (60S) forms an 80S ribosome ready to translate the mRNA. Two initiation factors, eIF2 (step ⊡) and eIF5 (step ⊡) are GTP-binding proteins, whose bound GTP is hydrolyzed during translation initiation. The precise time at which particular initiation factors are released is not yet well characterized. See the text for details. [Adapted from R. Mendez and J. D. Richter, 2001, *Nature Rev. Mol. Cell Biol.* **2**:521.]

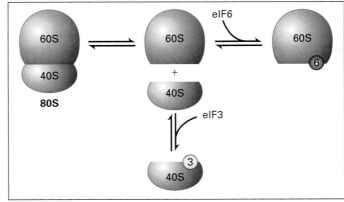

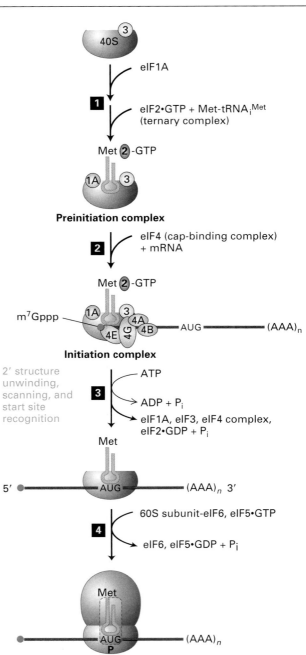

of a GTP associated with it (step 4). Coupling the joining reaction to GTP hydrolysis makes this an irreversible step, so that the ribosomal subunits do not dissociate until the entire mRNA is translated and protein synthesis is terminated. As discussed later, during chain elongation, the growing polypeptide remains attached to the tRNA at this P site in the ribosome.

The eukaryotic protein-synthesizing machinery begins translation of most cellular mRNAs within about 100 nucleotides of the 5' capped end as just described. However, some cellular mRNAs contain an internal ribosome entry site (IRES) located far downstream of the 5' end. In addition, translation of some viral mRNAs, which lack a 5' cap, is initiated at IRESs by the host-cell machinery of infected eukaryotic cells. Some of the same translation initiation factors that assist in ribosome scanning from a 5' cap are required for locating an internal AUG start codon, but exactly how an IRES is recognized is less clear. Recent results indicate that some IRESs fold into an RNA structure that binds to a third site on the ribosome, the *E site*, thereby positioning a nearby internal AUG start codon in the P site.

During Chain Elongation Each Incoming Aminoacyl-tRNA Moves Through Three Ribosomal Sites

The correctly positioned eukaryotic 80S ribosome–Met-tRNA$_i^{Met}$ complex is now ready to begin the task of stepwise addition of amino acids by the in-frame translation of the mRNA. As is the case with initiation, a set of special proteins, termed **elongation factors (EFs)**, are required to carry out this process of chain elongation. The key steps in elongation are entry of each succeeding aminoacyl-tRNA, formation of a peptide bond, and the movement, or *translocation*, of the ribosome one codon at a time along the mRNA.

▶ **FIGURE 4-26 Cycle of peptidyl chain elongation during translation in eukaryotes.** Once the 80S ribosome with Met-tRNA$_i^{Met}$ in the ribosome P site is assembled *(top)*, a ternary complex bearing the second amino acid (aa$_2$) coded by the mRNA binds to the A site (step **1**). Following a conformational change in the ribosome induced by hydrolysis of GTP in EF1α·GTP (step **2**), the large rRNA catalyzes peptide bond formation between Met$_i$ and aa$_2$ (step **3**). Hydrolysis of GTP in EF2·GTP causes another conformational change in the ribosome that results in its translocation one codon along the mRNA and shifts the unacylated tRNA$_i^{Met}$ to the E site and the tRNA with the bound peptide to the P site (step **4**). The cycle can begin again with binding of a ternary complex bearing aa$_3$ to the now-open A site. In the second and subsequent elongation cycles, the tRNA at the E site is ejected during step **2** as a result of the conformational change induced by hydrolysis of GTP in EF1α·GTP. See the text for details. [Adapted from K. H. Nierhaus et al., 2000, in R. A. Garrett et al., eds., *The Ribosome: Structure, Function, Antibiotics, and Cellular Interactions,* ASM Press, p. 319.]

At the completion of translation initiation, as noted already, Met-tRNA$_i^{Met}$ is bound to the P site on the assembled 80S ribosome (Figure 4-26, *top*). This region of the ribosome is called the *P* site because the tRNA chemically linked to the growing polypeptide chain is located here. The second

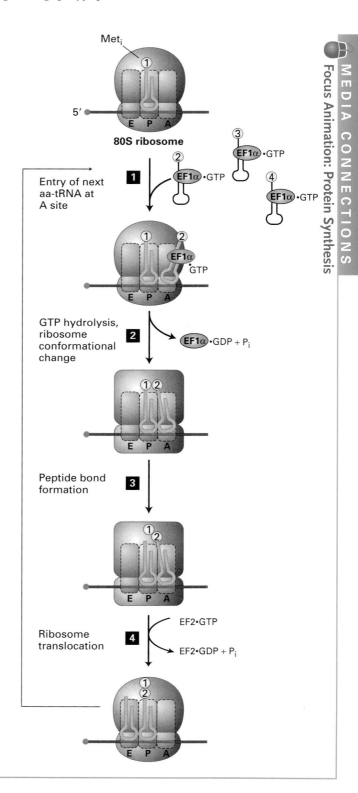

aminoacyl-tRNA is brought into the ribosome as a ternary complex in association with EF1α·GTP and becomes bound to the A site, so named because it is where *a*minoacylated tRNAs bind (step $\boxed{1}$). If the anticodon of the incoming (second) aminoacyl-tRNA correctly base-pairs with the second codon of the mRNA, the GTP in the associated EF1α·GTP is hydrolyzed. The hydrolysis of GTP promotes a conformational change in the ribosome that leads to tight binding of the aminoacyl-tRNA in the A site and release of the resulting EF1α·GDP complex (step $\boxed{2}$). This conformational change also positions the aminoacylated 3′ end of the tRNA in the A site in close proximity to the 3′ end of the Met-tRNA$_i^{Met}$ in the P site. GTP hydrolysis, and hence tight binding, does not occur if the anticodon of the incoming aminoacyl-tRNA cannot base-pair with the codon at the A site. In this case, the ternary complex diffuses away, leaving an empty A site that can associate with other aminoacyltRNA–EF1α·GTP complexes until a correctly base-paired tRNA is bound. This phenomenon contributes to the fidelity with which the correct aminoacyl-tRNA is loaded into the A site.

With the initiating Met-tRNA$_i^{Met}$ at the P site and the second aminoacyl-tRNA tightly bound at the A site, the α amino group of the second amino acid reacts with the "activated" (ester-linked) methionine on the initiator tRNA, forming a peptide bond (Figure 4-26, step $\boxed{3}$; see Figures 4-19 and 4-21). This *peptidyltransferase reaction* is catalyzed by the large rRNA, which precisely orients the interacting atoms, permitting the reaction to proceed. The catalytic ability of the large rRNA in bacteria has been demonstrated by carefully removing the vast majority of the protein from large ribosomal subunits. The nearly pure bacterial 23S rRNA can catalyze a peptidyltransferase reaction between analogs of aminoacylated-tRNA and peptidyl-tRNA. Further support for the catalytic role of large rRNA in protein synthesis comes from crystallographic studies showing that no proteins lie near the site of peptide bond synthesis in the crystal structure of the bacterial large subunit.

Following peptide bond synthesis, the ribosome is translocated along the mRNA a distance equal to one codon. This translocation step is promoted by hydrolysis of the GTP in eukaryotic EF2·GTP. As a result of translocation, tRNA$_i^{Met}$, now without its activated methionine, is moved to the *E* (*e*xit) site on the ribosome; concurrently, the second tRNA, now covalently bound to a dipeptide (a peptidyl-tRNA), is moved to the P site (Figure 4-26, step $\boxed{4}$). Translocation thus returns the ribosome conformation to a state in which the A site is open and able to accept another aminoacylated tRNA complexed with EF1α·GTP, beginning another cycle of chain elongation.

Repetition of the elongation cycle depicted in Figure 4-26 adds amino acids one at a time to the C-terminus of the growing polypeptide as directed by the mRNA sequence until a stop codon is encountered. In subsequent cycles, the conformational change that occurs in step $\boxed{2}$ ejects the

unacylated tRNA from the E site. As the nascent polypeptide chain becomes longer, it threads through a channel in the large ribosomal subunit, exiting at a position opposite the side that interacts with the small subunit (Figure 4-27).

The locations of tRNAs bound at the A, P, and E sites are visible in the recently determined crystal structure of the bacterial ribosome (Figure 4-28). Base pairing is also apparent between the tRNAs in the A and P sites with their respective codons in mRNA (see Figure 4-28, *inset*). An RNA-RNA hybrid of only three base pairs is not stable under physio-

(a)

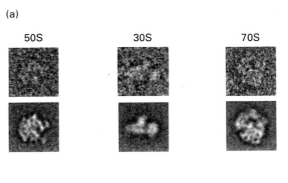

(b)

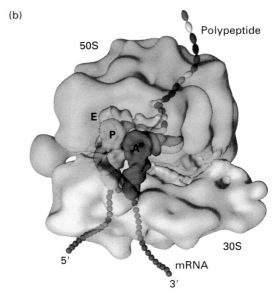

▲ **FIGURE 4-27 Low-resolution model of *E. coli* 70S ribosome.** (a) Top panels show cryoelectron microscopic images of *E. coli* 70S ribosomes and 50S and 30S subunits. Bottom panels show computer-derived averages of many dozens of images in the same orientation. (b) Model of a 70S ribosome based on the computer-derived images and on chemical cross-linking studies. Three tRNAs are superimposed on the A (pink), P (green), and E (yellow) sites. The nascent polypeptide chain is buried in a tunnel in the large ribosomal subunit that begins close to the acceptor stem of the tRNA in the P site. [See I. S. Gabashvili et al., 2000, *Cell* **100**:537; courtesy of J. Frank.]

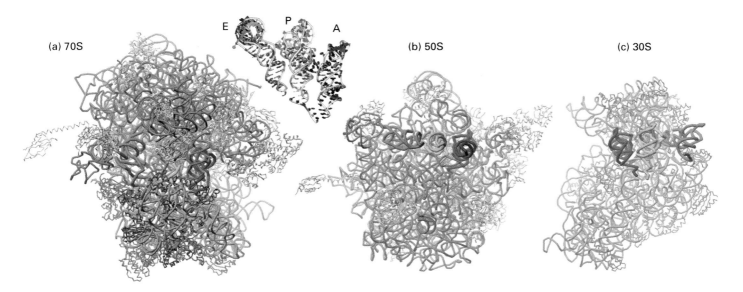

▲ FIGURE 4-28 Structure of *T. thermophilus* 70S ribosome as determined by x-ray crystallography. (a) Model of the entire ribosome viewed from the side diagrammed in Figure 4-26 with large subunit on top and small subunit below. The tRNAs positioned at the A (blue), P (yellow), and E (green) sites are visible in the interface between the subunits with their anticodon loops pointing down into the small subunit. 16S rRNA is cyan; 23S rRNA, purple; 5S rRNA, pink; mRNA, red; small ribosomal proteins, dark gray; and large ribosomal proteins, light gray. Note that the ribosomal proteins are located primarily on the surface of the ribosome and the rRNAs on the inside. (b) View of the large subunit rotated 90° about the horizontal from the view in (a) showing the face that interacts with the small subunit. The tRNA anticodon loops point out of the page. In the intact ribosome, these extend into the small subunit where the anticodons of the tRNAs in the A and P sites base-pair with codons in the mRNA. (c) View of the face of the small subunit that interacts with the large subunit in (b). Here the tRNA anticodon loops point into the page. The TΨCG loops and acceptor stems extend out of the page and the 3' CCA ends of the tRNAs in the A and P sites point downward. Note the close opposition of the acceptor stems of tRNAs in the A and P sites, which allows the amino group of the acylated tRNA in the A site to react with the carboxyl-terminal C of the peptidyl-tRNA in the P site (see Figure 4-19). In the intact ribosome, these are located at the peptidyltransferase active site of the large subunit. [Adapted from M. M. Yusupov et al., 2001, *Science* **292**:883.]

logical conditions. However, multiple interactions between the large and small rRNAs and general domains of tRNAs (e.g., the D and TΨCG loops) stabilize the tRNAs in the A and P sites, while other RNA-RNA interactions sense correct codon-anticodon base pairing, assuring that the genetic code is read properly.

Translation Is Terminated by Release Factors When a Stop Codon Is Reached

The final stage of translation, like initiation and elongation, requires highly specific molecular signals that decide the fate of the mRNA–ribosome–tRNA-peptidyl complex. Two types of specific protein **release factors** (RFs) have been discovered. Eukaryotic eRF1, whose shape is similar to that of tRNAs, apparently acts by binding to the ribosomal A site and recognizing stop codons directly. Like some of the initiation and elongation factors discussed previously, the second eukaryotic release factor, eRF3, is a GTP-binding protein. The eRF3·GTP acts in concert with eRF1 to promote cleavage of the peptidyl-tRNA, thus releasing the completed protein

chain (Figure 4-29). Bacteria have two release factors (RF1 and RF2) that are functionally analogous to eRF1 and a GTP-binding factor (RF3) that is analogous to eRF3.

After its release from the ribosome, a newly synthesized protein folds into its native three-dimensional conformation, a process facilitated by other proteins called **chaperones** (Chapter 3). Additional release factors then promote dissociation of the ribosome, freeing the subunits, mRNA, and terminal tRNA for another round of translation.

We can now see that one or more GTP-binding proteins participate in each stage of translation. These proteins belong to the **GTPase superfamily** of switch proteins that cycle between a GTP-bound active form and GDP-bound inactive form (see Figure 3-29). Hydrolysis of the bound GTP is thought to cause conformational changes in the GTPase itself or other associated proteins that are critical to various complex molecular processes. In translation initiation, for instance, hydrolysis of eIF2·GTP to eIF2·GDP prevents further scanning of the mRNA once the start site is encountered and allows binding of the large ribosomal subunit to the small subunit (see Figure 4-25, step [3]). Similarly, hydrolysis of

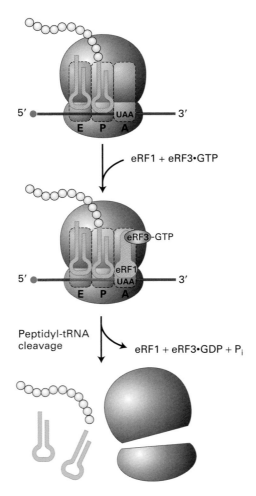

▲ **FIGURE 4-29 Termination of translation in eukaryotes.**
When a ribosome bearing a nascent protein chain reaches a
stop codon (UAA, UGA, UAG), release factor eRF1 enters the
ribosomal complex, probably at or near the A site together with
eRF3·GTP. Hydrolysis of the bound GTP is accompanied by
cleavage of the peptide chain from the tRNA in the P site and
release of the tRNAs and the two ribosomal subunits.

disengage from the 3′ end of an mRNA. Simultaneous translation of an mRNA by multiple ribosomes is readily observable in electron micrographs and by sedimentation analysis,
revealing mRNA attached to multiple ribosomes bearing
nascent growing polypeptide chains. These structures, referred to as **polyribosomes** or *polysomes,* were seen to be circular in electron micrographs of some tissues. Subsequent
studies with yeast cells explained the circular shape of polyribosomes and suggested the mode by which ribosomes recycle efficiently.

These studies revealed that multiple copies of a cytosolic
protein found in all eukaryotic cells, *poly(A)-binding protein*
(PABPI), can interact with both an mRNA poly(A) tail and
the 4G subunit of yeast eIF4. Moreover, the 4E subunit of
yeast eIF4 binds to the 5′ end of an mRNA. As a result of
these interactions, the two ends of an mRNA molecule can
be bridged by the intervening proteins, forming a "circular"
mRNA (Figure 4-30). Because the two ends of a polysome
are relatively close together, ribosomal subunits that disengage from the 3′ end are positioned near the 5′ end, facilitating re-initiation by the interaction of the 40S subunit with
eIF4 bound to the 5′ cap. The circular pathway depicted in
Figure 4-31, which may operate in many eukaryotic cells,
would enhance ribosome recycling and thus increase the efficiency of protein synthesis.

EF2·GTP to EF2·GDP during chain elongation leads to
translocation of the ribosome along the mRNA (see Figure
4-26, step 4).

Polysomes and Rapid Ribosome Recycling Increase the Efficiency of Translation

As noted earlier, translation of a single eukaryotic mRNA
molecule to yield a typical-sized protein takes 30–60 seconds. Two phenomena significantly increase the overall rate
at which cells can synthesize a protein: the simultaneous
translation of a single mRNA molecule by multiple ribosomes and rapid recycling of ribosomal subunits after they

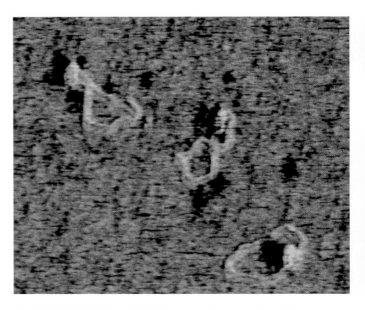

▲ **EXPERIMENTAL FIGURE 4-30 Eukaryotic mRNA forms
a circular structure owing to interactions of three proteins.**
In the presence of purified poly(A)-binding protein I (PABPI),
eIF4E, and eIF4G, eukaryotic mRNAs form circular structures,
visible in this force-field electron micrograph. In these structures,
protein-protein and protein-mRNA interactions form a bridge
between the 5′ and 3′ ends of the mRNA as diagrammed in
Figure 4-31. [Courtesy of A. Sachs.]

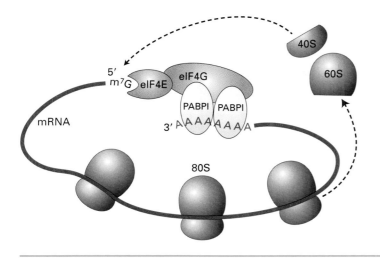

◀ **FIGURE 4-31 Model of protein synthesis on circular polysomes and recycling of ribosomal subunits.** Multiple individual ribosomes can simultaneously translate a eukaryotic mRNA, shown here in circular form stabilized by interactions between proteins bound at the 3′ and 5′ ends. When a ribosome completes translation and dissociates from the 3′ end, the separated subunits can rapidly find the nearby 5′ cap (m⁷G) and initiate another round of synthesis.

KEY CONCEPTS OF SECTION 4.5

Stepwise Synthesis of Proteins on Ribosomes

■ Of the two methionine tRNAs found in all cells, only one (tRNA$_i^{Met}$) functions in initiation of translation.

■ Each stage of translation—initiation, chain elongation, and termination—requires specific protein factors including GTP-binding proteins that hydrolyze their bound GTP to GDP when a step has been completed successfully.

■ During initiation, the ribosomal subunits assemble near the translation start site in an mRNA molecule with the tRNA carrying the amino-terminal methionine (Met-tRNA$_i^{Met}$) base-paired with the start codon (Figure 4-25).

■ Chain elongation entails a repetitive four-step cycle: loose binding of an incoming aminoacyl-tRNA to the A site on the ribosome; tight binding of the correct aminoacyl-tRNA to the A site accompanied by release of the previously used tRNA from the E site; transfer of the growing peptidyl chain to the incoming amino acid catalyzed by large rRNA; and translocation of the ribosome to the next codon, thereby moving the peptidyl-tRNA in the A site to the P site and the now unacylated tRNA in the P site to the E site (see Figure 4-26).

■ In each cycle of chain elongation, the ribosome undergoes two conformational changes monitored by GTP-binding proteins. The first permits tight binding of the incoming aminoacyl-tRNA to the A site and ejection of a tRNA from the E site, and the second leads to translocation.

■ Termination of translation is carried out by two types of termination factors: those that recognize stop codons and those that promote hydrolysis of peptidyl-tRNA (see Figure 4-29).

■ The efficiency of protein synthesis is increased by the simultaneous translation of a single mRNA by multiple ribosomes. In eukaryotic cells, protein-mediated interactions bring the two ends of a polyribosome close together, thereby promoting the rapid recycling of ribosomal subunits, which further increases the efficiency of protein synthesis (see Figure 4-31).

4.6 DNA Replication

Now that we have seen how genetic information encoded in the nucleotide sequences of DNA is translated into the structures of proteins that perform most cell functions, we can appreciate the necessity of the precise copying of DNA sequences during DNA replication (see Figure 4-1, step 4). The regular pairing of bases in the double-helical DNA structure suggested to Watson and Crick that new DNA strands are synthesized by using the existing (*parental*) strands as **templates** in the formation of new, *daughter* strands complementary to the parental strands.

This base-pairing template model theoretically could proceed either by a *conservative* or a *semiconservative* mechanism. In a conservative mechanism, the two daughter strands would form a new double-stranded (*duplex*) DNA molecule and the parental duplex would remain intact. In a semiconservative mechanism, the parental strands are permanently separated and each forms a duplex molecule with the daughter strand base-paired to it. Definitive evidence that duplex DNA is replicated by a semiconservative mechanism came from a now classic experiment conducted by M. Meselson and W. F. Stahl, outlined in Figure 4-32.

Copying of a DNA template strand into a complementary strand thus is a common feature of DNA replication and transcription of DNA into RNA. In both cases, the information in the template is preserved. In some viruses, single-stranded RNA molecules function as templates for synthesis of complementary RNA or DNA strands. However, the vast preponderance of RNA and DNA in cells is synthesized from preexisting duplex DNA.

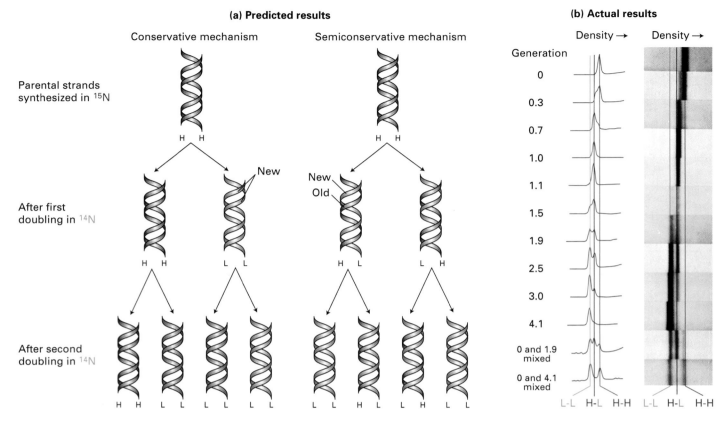

(a) Predicted results

Conservative mechanism

Semiconservative mechanism

(b) Actual results

Parental strands synthesized in ¹⁵N

After first doubling in ¹⁴N

After second doubling in ¹⁴N

Density →   Density →

Generation

0
0.3
0.7
1.0
1.1
1.5
1.9
2.5
3.0
4.1
0 and 1.9 mixed
0 and 4.1 mixed

L-L   H-L   H-H   L-L   H-L   H-H

▲ **EXPERIMENTAL FIGURE 4-32 The Meselson-Stahl experiment showed that DNA replicates by a semiconservative mechanism.** In this experiment, *E. coli* cells initially were grown in a medium containing ammonium salts prepared with "heavy" nitrogen (¹⁵N) until all the cellular DNA was labeled. After the cells were transferred to a medium containing the normal "light" isotope (¹⁴N), samples were removed periodically from the cultures and the DNA in each sample was analyzed by equilibrium density-gradient centrifugation (see Figure 5-37). This technique can separate heavy-heavy (H-H), light-light (L-L), and heavy-light (H-L) duplexes into distinct bands. (a) Expected composition of daughter duplex molecules synthesized from ¹⁵N-labeled DNA after *E. coli* cells are shifted to ¹⁴N-containing medium if DNA replication occurs by a conservative or semiconservative mechanism. Parental heavy (H) strands are in red; light (L) strands synthesized after shift to ¹⁴N-containing medium are in blue. Note that the conservative mechanism never generates H-L DNA and that the semiconservative mechanism never generates H-H DNA but does generate H-L DNA during the second and subsequent doublings. With additional replication cycles, the ¹⁵N-labeled (H) strands from the original DNA are diluted, so that the vast bulk of the DNA would consist of L-L duplexes with either

mechanism. (b) Actual banding patterns of DNA subjected to equilibrium density-gradient centrifugation before and after shifting ¹⁵N-labeled *E. coli* cells to ¹⁴N-containing medium. DNA bands were visualized under UV light and photographed. The traces on the left are a measure of the density of the photographic signal, and hence the DNA concentration, along the length of the centrifuge cells from left to right. The number of generations (far left) following the shift to ¹⁴N-containing medium was determined by counting the concentration of *E. coli* cells in the culture. This value corresponds to the number of DNA replication cycles that had occurred at the time each sample was taken. After one generation of growth, all the extracted DNA had the density of H-L DNA. After 1.9 generations, approximately half the DNA had the density of H-L DNA; the other half had the density of L-L DNA. With additional generations, a larger and larger fraction of the extracted DNA consisted of L-L duplexes; H-H duplexes never appeared. These results match the predicted pattern for the semiconservative replication mechanism depicted in (a). The bottom two centrifuge cells contained mixtures of H-H DNA and DNA isolated at 1.9 and 4.1 generations in order to clearly show the positions of H-H, H-L, and L-L DNA in the density gradient. [Part (b) from M. Meselson and F. W. Stahl, 1958, *Proc. Nat'l. Acad. Sci. USA* **44**:671.]

DNA Polymerases Require a Primer to Initiate Replication

Analogous to RNA, DNA is synthesized from deoxynucleoside 5′-triphosphate precursors (dNTPs). Also like RNA synthesis, DNA synthesis always proceeds in the 5′→3′

direction because chain growth results from formation of a phosphoester bond between the 3′ oxygen of a growing strand and the α phosphate of a dNTP (see Figure 4-9). As discussed earlier, an RNA polymerase can find an appropriate transcription start site on duplex DNA and initiate the synthesis of an RNA complementary to the template DNA strand (see Figure 4-10). In contrast, **DNA polymerases** cannot initiate chain synthesis de novo; instead, they require a short, preexisting RNA or DNA strand, called a **primer**, to begin chain growth. With a primer base-paired to the template strand, a DNA polymerase adds deoxynucleotides to the free hydroxyl group at the 3′ end of the primer as directed by the sequence of the template strand:

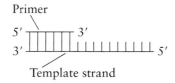

When RNA is the primer, the daughter strand that is formed is RNA at the 5′ end and DNA at the 3′ end.

Duplex DNA Is Unwound, and Daughter Strands Are Formed at the DNA Replication Fork

In order for duplex DNA to function as a template during replication, the two intertwined strands must be unwound, or melted, to make the bases available for base pairing with the bases of the dNTPs that are polymerized into the newly synthesized daughter strands. This unwinding of the parental DNA strands is by specific **helicases**, beginning at unique segments in a DNA molecule called *replication origins*, or simply *origins*. The nucleotide sequences of origins from different organisms vary greatly, although they usually contain A·T-rich sequences. Once helicases have unwound the parental DNA at an origin, a specialized RNA polymerase called **primase** forms a short RNA primer complementary to the unwound template strands. The primer, still base-paired to its complementary DNA strand, is then elongated by a DNA polymerase, thereby forming a new daughter strand.

The DNA region at which all these proteins come together to carry out synthesis of daughter strands is called the **replication fork,** or growing fork. As replication proceeds, the growing fork and associated proteins move away from the origin. As noted earlier, local unwinding of duplex DNA produces torsional stress, which is relieved by topoisomerase I. In order for DNA polymerases to move along and copy a duplex DNA, helicase must sequentially unwind the duplex and topoisomerase must remove the supercoils that form.

A major complication in the operation of a DNA replication fork arises from two properties: the two strands of the parental DNA duplex are antiparallel, and DNA polymerases (like RNA polymerases) can add nucleotides to the growing new strands only in the 5′→3′ direction. Synthesis of one daughter strand, called the **leading strand,** can proceed continuously from a single RNA primer in the 5′→3′ direction, *the same direction as movement of the replication fork* (Figure 4-33). The problem comes in synthesis of the other daughter strand, called the **lagging strand.**

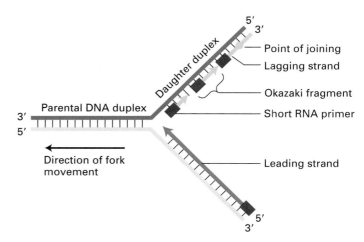

▲ **FIGURE 4-33 Schematic diagram of leading-strand and lagging-strand DNA synthesis at a replication fork.** Nucleotides are added by a DNA polymerase to each growing daughter strand in the 5′→3′ direction (indicated by arrowheads). The leading strand is synthesized continuously from a single RNA primer (red) at its 5′ end. The lagging strand is synthesized discontinuously from multiple RNA primers that are formed periodically as each new region of the parental duplex is unwound. Elongation of these primers initially produces Okazaki fragments. As each growing fragment approaches the previous primer, the primer is removed and the fragments are ligated. Repetition of this process eventually results in synthesis of the entire lagging strand.

Because growth of the lagging strand must occur in the 5′→3′ direction, copying of its template strand must somehow occur in the *opposite* direction from the movement of the replication fork. A cell accomplishes this feat by synthesizing a new primer every few hundred bases or so on the second parental strand, as more of the strand is exposed by unwinding. Each of these primers, base-paired to their template strand, is elongated in the 5′→3′ direction, forming discontinuous segments called **Okazaki fragments** after their discoverer Reiji Okazaki (see Figure 4-33). The RNA primer of each Okazaki fragment is removed and replaced by DNA chain growth from the neighboring Okazaki fragment; finally an enzyme called *DNA ligase* joins the adjacent fragments.

Helicase, Primase, DNA Polymerases, and Other Proteins Participate in DNA Replication

Detailed understanding of the eukaryotic proteins that participate in DNA replication has come largely from studies with small viral DNAs, particularly SV40 DNA, the circular genome of a small virus that infects monkeys. Figure 4-34 depicts the multiple proteins that coordinate copying of SV40 DNA at a replication fork. The assembled proteins at a replication fork further illustrate the concept of molecular machines introduced in Chapter 3. These multicomponent

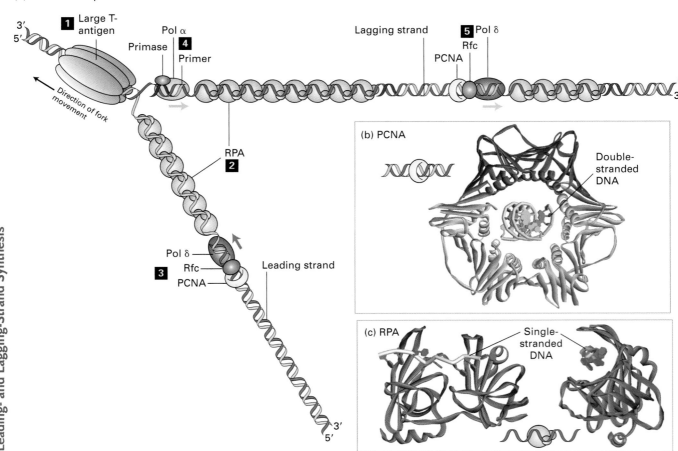

▲ FIGURE 4-34 Model of an SV40 DNA replication fork and assembled proteins. (a) A hexamer of large T-antigen (**1**), a viral protein, functions as a helicase to unwind the parental DNA strands. Single-strand regions of the parental template unwound by large T-antigen are bound by multiple copies of the heterotrimeric protein RPA (**2**). The leading strand is synthesized by a complex of DNA polymerase δ (Pol δ), PCNA, and Rfc (**3**). Primers for lagging-strand synthesis (red, RNA; light blue, DNA) are synthesized by a complex of DNA polymerase α (Pol α) and primase (**4**). The 3′ end of each primer synthesized by Pol α–primase is then bound by a PCNA-Rfc–Pol δ complex, which proceeds to extend the primer and synthesize most of each Okazaki fragment (**5**). See the text for details. (b) The three subunits of PCNA, shown in different colors, form a circular structure with a central hole through which double-stranded DNA passes. A diagram of DNA is shown in the center of a ribbon model of the PCNA trimer. (c) The large subunit of RPA contains two domains that bind single-stranded DNA. On the left, the two DNA-binding domains of RPA are shown perpendicular to the DNA backbone (white backbone with blue bases). Note that the single DNA strand is extended with the bases exposed, an optimal conformation for replication by a DNA polymerase. On the right, the view is down the length of the single DNA strand, revealing how RPA β strands wrap around the DNA. [Part (a) adapted from S. J. Flint et al., 2000, *Virology: Molecular Biology, Pathogenesis, and Control,* ASM Press; part (b) after J. M. Gulbis et al., 1996, *Cell* **87**:297; and part (c) after A. Bochkarev et al., 1997, *Nature* **385**:176.]

complexes permit the cell to carry out an ordered sequence of events that accomplish essential cell functions.

In the molecular machine that replicates SV40 DNA, a hexamer of a viral protein called *large T-antigen* unwinds the parental strands at a replication fork. All other proteins involved in SV40 DNA replication are provided by the host cell. Primers for leading and lagging daughter-strand DNA are synthesized by a complex of primase, which synthesizes a short RNA primer, and *DNA polymerase α* (Pol α), which extends the RNA primer with deoxynucleotides, forming a mixed RNA-DNA primer.

The primer is extended into daughter-strand DNA by *DNA polymerase δ* (Pol δ), which is less likely to make errors during copying of the template strand than is Pol α. Pol δ forms a complex with *Rfc* (replication factor C) and *PCNA* (proliferating cell nuclear antigen), which displaces the primase–Pol α complex following primer synthesis. As illustrated in Figure 4-34b, PCNA is a homotrimeric protein that has a central hole through which the daughter duplex DNA passes, thereby preventing the PCNA-Rfc–Pol δ complex from dissociating from the template.

After parental DNA is separated into single-stranded templates at the replication fork, it is bound by multiple copies of RPA (*replication protein A*), a heterotrimeric protein (Figure 4-34c). Binding of RPA maintains the template in a uniform conformation optimal for copying by DNA polymerases. Bound RPA proteins are dislodged from the parental strands by Pol α and Pol δ as they synthesize the complementary strands base-paired with the parental strands.

Several eukaryotic proteins that function in DNA replication are not depicted in Figure 4-34. A topoisomerase associates with the parental DNA ahead of the helicase to remove torsional stress introduced by the unwinding of the parental strands. Ribonuclease H and FEN I remove the ribonucleotides at the 5′ ends of Okazaki fragments; these are replaced by deoxynucleotides added by DNA polymerase δ as it extends the upstream Okazaki fragment. Successive Okazaki fragments are coupled by DNA ligase through standard 5′→3′ phosphoester bonds.

DNA Replication Generally Occurs Bidirectionally from Each Origin

As indicated in Figures 4-33 and 4-34, both parental DNA strands that are exposed by local unwinding at a replication fork are copied into a daughter strand. In theory, DNA replication from a single origin could involve one replication fork that moves in one direction. Alternatively, two replication forks might assemble at a single origin and then move in opposite directions, leading to *bidirectional growth* of both daughter strands. Several types of experiments, including the one shown in Figure 4-35, provided early evidence in support of bidirectional strand growth.

The general consensus is that all prokaryotic and eukaryotic cells employ a bidirectional mechanism of DNA replication. In the case of SV40 DNA, replication is initiated by binding of two large T-antigen hexameric helicases to the single SV40 origin and assembly of other proteins to form two replication forks. These then move away from the SV40 origin in opposite directions with leading- and lagging-strand synthesis occurring at both forks. As shown in Figure 4-36, the left replication fork extends DNA synthesis in the leftward direction; similarly, the right replication fork extends DNA synthesis in the rightward direction.

Unlike SV40 DNA, eukaryotic chromosomal DNA molecules contain multiple replication origins separated by tens to hundreds of kilobases. A six-subunit protein called *ORC*, for *origin recognition complex*, binds to each origin and associates with other proteins required to load cellular hexameric helicases composed of six homologous *MCM* proteins.

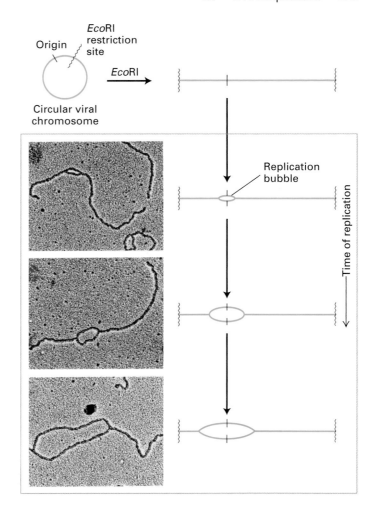

▲ **EXPERIMENTAL FIGURE 4-35 Electron microscopy of replicating SV40 DNA indicates bidirectional growth of DNA strands from an origin.** The replicating viral DNA from SV40-infected cells was cut by the restriction enzyme *Eco*RI, which recognizes one site in the circular DNA. Electron micrographs of treated samples showed a collection of cut molecules with increasingly longer replication "bubbles," whose centers are a constant distance from each end of the cut molecules. This finding is consistent with chain growth in two directions from a common origin located at the center of a bubble, as illustrated in the corresponding diagrams. [See G. C. Fareed et al., 1972, *J. Virol.* **10**:484; photographs courtesy of N. P. Salzman.]

Two opposed MCM helicases separate the parental strands at an origin, with RPA proteins binding to the resulting single-stranded DNA. Synthesis of primers and subsequent steps in replication of cellular DNA are thought to be analogous to those in SV40 DNA replication (see Figures 4-34 and 4-36).

Replication of cellular DNA and other events leading to proliferation of cells are tightly regulated, so that the appropriate numbers of cells constituting each tissue are produced during development and throughout the life of an organism. As in transcription of most genes, control of the initiation

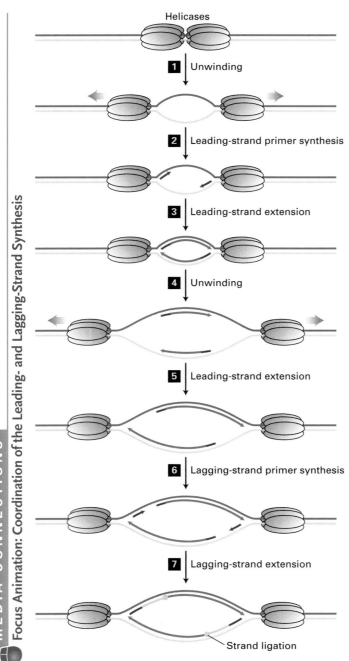

MEDIA CONNECTIONS

Focus Animation: Coordination of the Leading- and Lagging-Strand Synthesis

◀ **FIGURE 4-36 Bidirectional mechanism of DNA replication.** The left replication fork here is comparable to the replication fork diagrammed in Figure 4-34, which also shows proteins other than large T-antigen. (*Top*) Two large T-antigen hexameric helicases first bind at the replication origin in opposite orientations. Step **1**: Using energy provided from ATP hydrolysis, the helicases move in opposite directions, unwinding the parental DNA and generating single-strand templates that are bound by RPA proteins. Step **2**: Primase–Pol α complexes synthesize short primers base-paired to each of the separated parental strands. Step **3**: PCNA-Rfc–Pol δ complexes replace the primase–Pol α complexes and extend the short primers, generating the leading strands (dark green) at each replication fork. Step **4**: The helicases further unwind the parental strands, and RPA proteins bind to the newly exposed single-strand regions. Step **5**: PCNA-Rfc–Pol δ complexes extend the leading strands further. Step **6**: Primase–Pol α complexes synthesize primers for lagging-strand synthesis at each replication fork. Step **7**: PCNA-Rfc–Pol δ complexes displace the primase–Pol α complexes and extend the lagging-strand Okazaki fragments (light green), which eventually are ligated to the 5′ ends of the leading strands. The position where ligation occurs is represented by a circle. Replication continues by further unwinding of the parental strands and synthesis of leading and lagging strands as in steps **4**–**7**. Although depicted as individual steps for clarity, unwinding and synthesis of leading and lagging strands occur concurrently.

KEY CONCEPTS OF SECTION 4.6

DNA Replication

■ Each strand in a parental duplex DNA acts as a template for synthesis of a daughter strand and remains base-paired to the new strand, forming a daughter duplex (semiconservative mechanism). New strands are formed in the 5′→3′ direction.

■ Replication begins at a sequence called an *origin*. Each eukaryotic chromosomal DNA molecule contains multiple replication origins.

■ DNA polymerases, unlike RNA polymerases, cannot unwind the strands of duplex DNA and cannot initiate synthesis of new strands complementary to the template strands.

■ At a replication fork, one daughter strand (the leading strand) is elongated continuously. The other daughter strand (the lagging strand) is formed as a series of discontinuous Okazaki fragments from primers synthesized every few hundred nucleotides (Figure 4-33).

■ The ribonucleotides at the 5′ end of each Okazaki fragment are removed and replaced by elongation of the 3′ end of the next Okazaki fragment. Finally, adjacent Okazaki fragments are joined by DNA ligase.

■ Helicases use energy from ATP hydrolysis to separate the parental (template) DNA strands. Primase synthesizes

step is the primary mechanism for regulating cellular DNA replication. Activation of MCM helicase activity, which is required to initiate cellular DNA replication, is regulated by specific protein kinases called S-phase **cyclin-dependent kinases.** Other cyclin-dependent kinases regulate additional aspects of cell proliferation, including the complex process of mitosis by which a eukaryotic cell divides into two daughter cells. We discuss the various regulatory mechanisms that determine the rate of cell division in Chapter 21.

a short RNA primer, which remains base-paired to the template DNA. This initially is extended at the 3′ end by DNA polymerase α (Pol α), resulting in a short (5′)RNA-(3′)DNA daughter strand.

■ Most of the DNA in eukaryotic cells is synthesized by Pol δ, which takes over from Pol α and continues elongation of the daughter strand in the 5′→3′ direction. Pol δ remains stably associated with the template by binding to Rfc protein, which in turn binds to PCNA, a trimeric protein that encircles the daughter duplex DNA (see Figure 4-34).

■ DNA replication generally occurs by a bidirectional mechanism in which two replication forks form at an origin and move in opposite directions, with both template strands being copied at each fork (see Figure 4-36).

■ Synthesis of eukaryotic DNA in vivo is regulated by controlling the activity of the MCM helicases that initiate DNA replication at multiple origins spaced along chromosomal DNA.

4.7 Viruses: Parasites of the Cellular Genetic System

Viruses cannot reproduce by themselves and must commandeer a host cell's machinery to synthesize viral proteins and in some cases to replicate the viral genome. RNA viruses, which usually replicate in the host-cell cytoplasm, have an RNA genome, and DNA viruses, which commonly replicate in the host-cell nucleus, have a DNA genome (see Figure 4-1). Viral genomes may be single- or double-stranded, depending on the specific type of virus. The entire infectious virus particle, called a **virion,** consists of the nucleic acid and an outer shell of protein. The simplest viruses contain only enough RNA or DNA to encode four proteins; the most complex can encode 100–200 proteins. In addition to their obvious importance as causes of disease, viruses are extremely useful as research tools in the study of basic biological processes.

Most Viral Host Ranges Are Narrow

The surface of a virion contains many copies of one type of protein that binds specifically to multiple copies of a receptor protein on a host cell. This interaction determines the *host range*—the group of cell types that a virus can infect—and begins the infection process. Most viruses have a rather limited host range.

A virus that infects only bacteria is called a **bacteriophage,** or simply a *phage.* Viruses that infect animal or plant cells are referred to generally as animal viruses or plant viruses. A few viruses can grow in both plants and the insects that feed on them. The highly mobile insects serve as vectors for transferring such viruses between susceptible plant hosts. Wide host ranges are also characteristic of some strictly animal viruses, such as vesicular stomatitis virus, which grows in insect vectors and in many different types of mammals. Most animal viruses, however, do not cross phyla, and some (e.g., poliovirus) infect only closely related species such as primates. The host-cell range of some animal viruses is further restricted to a limited number of cell types because only these cells have appropriate surface receptors to which the virions can attach.

Viral Capsids Are Regular Arrays of One or a Few Types of Protein

The nucleic acid of a virion is enclosed within a protein coat, or **capsid,** composed of multiple copies of one protein or a few different proteins, each of which is encoded by a single viral gene. Because of this structure, a virus is able to encode all the information for making a relatively large capsid in a small number of genes. This efficient use of genetic information is important, since only a limited amount of RNA or DNA, and therefore a limited number of genes, can fit into a virion capsid. A capsid plus the enclosed nucleic acid is called a **nucleocapsid.**

Nature has found two basic ways of arranging the multiple capsid protein subunits and the viral genome into a nucleocapsid. In some viruses, multiple copies of a single coat protein form a *helical* structure that encloses and protects the viral RNA or DNA, which runs in a helical groove within the protein tube. Viruses with such a helical nucleocapsid, such as tobacco mosaic virus, have a rodlike shape. The other major structural type is based on the *icosahedron,* a solid, approximately spherical object built of 20 identical faces, each of which is an equilateral triangle.

The number and arrangement of coat proteins in icosahedral, or quasi-spherical, viruses differ somewhat depending on their size. In small viruses of this type, each of the 20 triangular faces is constructed of three identical capsid protein subunits, making a total of 60 subunits per capsid. All the protein subunits are in *equivalent* contact with one another (Figure 4-37a). In large quasi-spherical viruses, each face of the icosahedron is composed of more than three subunits. As a result, the contacts between subunits not at the vertices are *quasi-equivalent* (Figure 4-37b). Models of several quasi-spherical viruses, based on cryoelectron microscopy, are shown in Figure 4-37. In the smaller viruses (e.g., poliovirus), clefts that encircle each of the vertices of the icosahedral structure interact with receptors on the surface of host cells during infection. In the larger viruses (e.g., adenovirus), long fiberlike proteins extending from the nucleocapsid interact with cell-surface receptors on host cells.

In many DNA bacteriophages, the viral DNA is located within an icosahedral "head" that is attached to a rodlike "tail." During infection, viral proteins at the tip of the tail bind to host-cell receptors, and then the viral DNA passes down the tail into the cytoplasm of the host cell.

In some viruses, the symmetrically arranged nucleocapsid is covered by an external membrane, or **envelope,** which

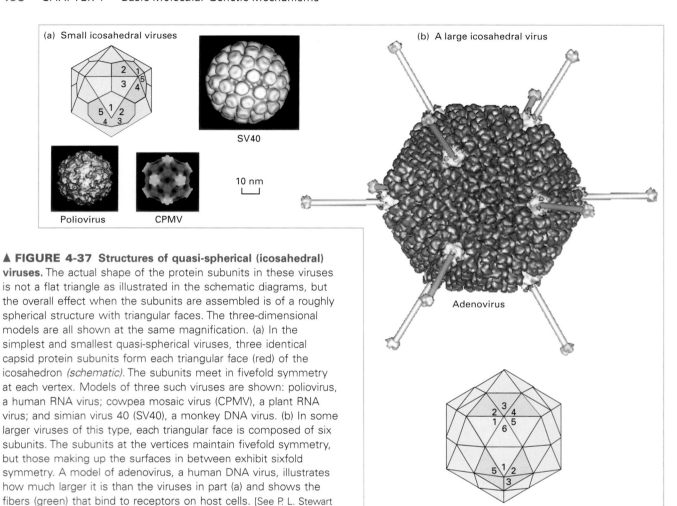

(a) Small icosahedral viruses

SV40

10 nm

Poliovirus   CPMV

(b) A large icosahedral virus

Adenovirus

▲ **FIGURE 4-37 Structures of quasi-spherical (icosahedral) viruses.** The actual shape of the protein subunits in these viruses is not a flat triangle as illustrated in the schematic diagrams, but the overall effect when the subunits are assembled is of a roughly spherical structure with triangular faces. The three-dimensional models are all shown at the same magnification. (a) In the simplest and smallest quasi-spherical viruses, three identical capsid protein subunits form each triangular face (red) of the icosahedron *(schematic)*. The subunits meet in fivefold symmetry at each vertex. Models of three such viruses are shown: poliovirus, a human RNA virus; cowpea mosaic virus (CPMV), a plant RNA virus; and simian virus 40 (SV40), a monkey DNA virus. (b) In some larger viruses of this type, each triangular face is composed of six subunits. The subunits at the vertices maintain fivefold symmetry, but those making up the surfaces in between exhibit sixfold symmetry. A model of adenovirus, a human DNA virus, illustrates how much larger it is than the viruses in part (a) and shows the fibers (green) that bind to receptors on host cells. [See P. L. Stewart et al., 1997, *EMBO J.* **16**:1189. Models of CPMV, poliovirus, and SV40 courtesy of T. S. Baker; model of adenovirus courtesy of P. L. Stewart.]

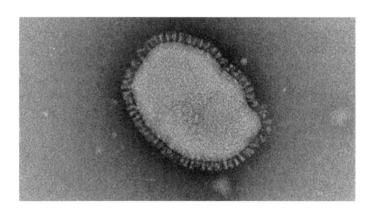

▲ **EXPERIMENTAL FIGURE 4-38 Viral protein spikes protrude from the surface of an influenza virus virion.** Influenza viruses are surrounded by an envelope consisting of a phospholipid bilayer and embedded viral proteins. The large spikes seen in this electron micrograph of a negatively stained influenza virion are composed of neuraminidase, a tetrameric protein, or hemagglutinin, a trimeric protein (see Figure 3-7). Inside is the helical nucleocapsid. [Courtesy of A. Helenius and J. White.]

consists mainly of a phospholipid bilayer but also contains one or two types of virus-encoded glycoproteins (Figure 4-38). The phospholipids in the viral envelope are similar to those in the plasma membrane of an infected host cell. The viral envelope is, in fact, derived by budding from that membrane, but contains mainly viral glycoproteins, as we discuss shortly.

Viruses Can Be Cloned and Counted in Plaque Assays

The number of infectious viral particles in a sample can be quantified by a **plaque assay.** This assay is performed by culturing a dilute sample of viral particles on a plate covered with host cells and then counting the number of local lesions, called *plaques*, that develop (Figure 4-39). A plaque develops on the plate wherever a single virion initially infects a single cell. The virus replicates in this initial host cell and then lyses (ruptures) the cell, releasing many progeny virions that infect the neighboring cells on the plate. After a few such cycles of infection, enough cells are lysed to pro-

(a)

Confluent layer of susceptible host cells growing on surface of a plate

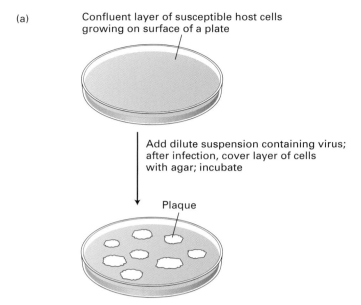

Add dilute suspension containing virus; after infection, cover layer of cells with agar; incubate

Plaque

Each plaque represents cell lysis initiated by one viral particle (agar restricts movement so that virus can infect only contiguous cells)

◀ **EXPERIMENTAL FIGURE 4-39** **Plaque assay determines the number of infectious particles in a viral suspension.** (a) Each lesion, or plaque, which develops where a single virion initially infected a single cell, constitutes a pure viral clone. (b) Plate illuminated from behind shows plaques formed by bacteriophage λ plated on *E. coli.* (c) Plate showing plaques produced by poliovirus plated on HeLa cells. [Part (b) courtesy of Barbara Morris; part (c) from S. E. Luria et al., 1978, *General Virology*, 3d ed., Wiley, p. 26.]

(b) Plaque   (c) Plaque

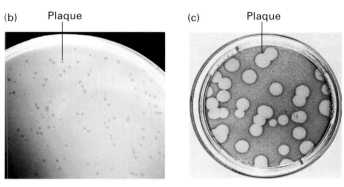

duce a visible clear area, or plaque, in the layer of remaining uninfected cells.

Since all the progeny virions in a plaque are derived from a single parental virus, they constitute a virus **clone**. This type of plaque assay is in standard use for bacterial and animal viruses. Plant viruses can be assayed similarly by counting local lesions on plant leaves inoculated with viruses. Analysis of viral mutants, which are commonly isolated by plaque assays, has contributed extensively to current understanding of molecular cellular processes. The plaque assay also is critical in isolating bacteriophage λ clones carrying segments of cellular DNA, as discussed in Chapter 9.

Lytic Viral Growth Cycles Lead to Death of Host Cells

Although details vary among different types of viruses, those that exhibit a *lytic cycle* of growth proceed through the following general stages:

1. *Adsorption*—Virion interacts with a host cell by binding of multiple copies of capsid protein to specific receptors on the cell surface.

2. *Penetration*—Viral genome crosses the plasma membrane. For animal and plant viruses, viral proteins also enter the host cell.

3. *Replication*—Viral mRNAs are produced with the aid of the host-cell transcription machinery (DNA viruses) or by viral enzymes (RNA viruses). For both types of viruses, viral mRNAs are translated by the host-cell translation machinery. Production of multiple copies of the viral

genome is carried out either by viral proteins alone or with the help of host-cell proteins.

4. *Assembly*—Viral proteins and replicated genomes associate to form progeny virions.

5. *Release*—Infected cell either ruptures suddenly (**lysis**), releasing all the newly formed virions at once, or disintegrates gradually, with slow release of virions.

Figure 4-40 illustrates the lytic cycle for T4 bacteriophage, a nonenveloped DNA virus that infects *E. coli.* Viral capsid proteins generally are made in large amounts because many copies of them are required for the assembly of each progeny virion. In each infected cell, about 100–200 T4 progeny virions are produced and released by lysis.

The lytic cycle is somewhat more complicated for DNA viruses that infect eukaryotic cells. In most such viruses, the DNA genome is transported (with some associated proteins) into the cell nucleus. Once inside the nucleus, the viral DNA is transcribed into RNA by the host's transcription machinery. Processing of the viral RNA primary transcript by host-cell enzymes yields viral mRNA, which is transported to the cytoplasm and translated into viral proteins by host-cell ribosomes, tRNA, and translation factors. The viral proteins are then transported back into the nucleus, where some of them either replicate the viral DNA directly or direct cellular proteins to replicate the viral DNA, as in the case of SV40 discussed in the last section. Assembly of the capsid proteins with the newly replicated viral DNA occurs in the nucleus, yielding hundreds to thousands of progeny virions.

Most plant and animal viruses with an RNA genome do not require nuclear functions for lytic replication. In some

▶ **FIGURE 4-40 Lytic replication cycle of *E. coli* bacteriophage T4, a nonenveloped virus with a double-stranded DNA genome.** After viral coat proteins at the tip of the tail in T4 interact with specific receptor proteins on the exterior of the host cell, the viral genome is injected into the host (step **1**). Host-cell enzymes then transcribe viral "early" genes into mRNAs and subsequently translate these into viral "early" proteins (step **2**). The early proteins replicate the viral DNA and induce expression of viral "late" proteins by host-cell enzymes (step **3**). The viral late proteins include capsid and assembly proteins and enzymes that degrade the host-cell DNA, supplying nucleotides for synthesis of more viral DNA. Progeny virions are assembled in the cell (step **4**) and released (step **5**) when viral proteins lyse the cell. Newly liberated viruses initiate another cycle of infection in other host cells.

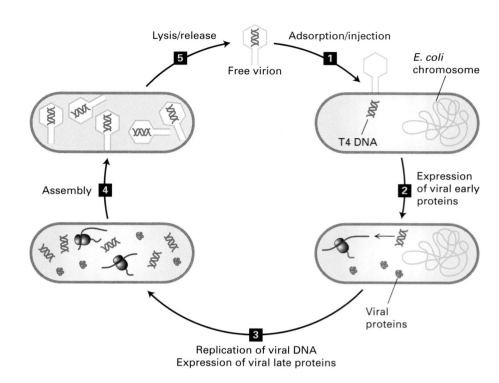

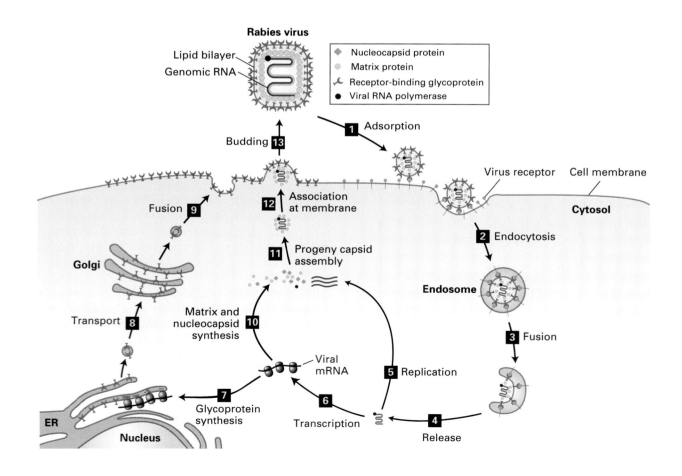

of these viruses, a virus-encoded enzyme that enters the host during penetration transcribes the genomic RNA into mRNAs in the cell cytoplasm. The mRNA is directly translated into viral proteins by the host-cell translation machinery. One or more of these proteins then produces additional copies of the viral RNA genome. Finally, progeny genomes are assembled with newly synthesized capsid proteins into progeny virions in the cytoplasm.

After the synthesis of hundreds to thousands of new virions has been completed, most infected bacterial cells and some infected plant and animal cells are lysed, releasing all the virions at once. In many plant and animal viral infections, however, no discrete lytic event occurs; rather, the dead host cell releases the virions as it gradually disintegrates.

As noted previously, enveloped animal viruses are surrounded by an outer phospholipid layer derived from the plasma membrane of host cells and containing abundant viral glycoproteins. The processes of adsorption and release of enveloped viruses differ substantially from these processes in nonenveloped viruses. To illustrate lytic replication of enveloped viruses, we consider the rabies virus, whose nucleocapsid consists of a single-stranded RNA genome surrounded by multiple copies of nucleocapsid protein. Like

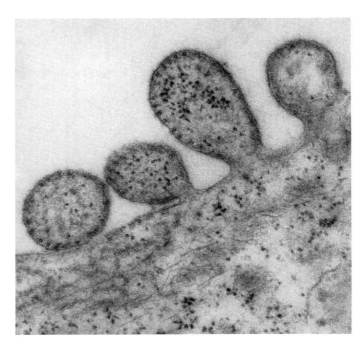

▲ **EXPERIMENTAL FIGURE 4-42** **Progeny virions of enveloped viruses are released by budding from infected cells.** In this transmission electron micrograph of a cell infected with measles virus, virion buds are clearly visible protruding from the cell surface. Measles virus is an enveloped RNA virus with a helical nucleocapsid, like rabies virus, and replicates as illustrated in Figure 4-41. [From A. Levine, 1991, *Viruses*, Scientific American Library, p. 22.]

◄ **FIGURE 4-41** **Lytic replication cycle of rabies virus, an enveloped virus with a single-stranded RNA genome.** The structural components of this virus are depicted at the top. Note that the nucleocapsid is helical rather than icosahedral. After a virion adsorbs to multiple copies of a specific host membrane protein (step **1**), the cell engulfs it in an endosome (step **2**). A cellular protein in the endosome membrane pumps H$^+$ ions from the cytosol into the endosome interior. The resulting decrease in endosomal pH induces a conformational change in the viral glycoprotein, leading to fusion of the viral envelope with the endosomal lipid bilayer membrane and release of the nucleocapsid into the cytosol (steps **3** and **4**). Viral RNA polymerase uses ribonucleoside triphosphates in the cytosol to replicate the viral RNA genome (step **5**) and to synthesize viral mRNAs (step **6**). One of the viral mRNAs encodes the viral transmembrane glycoprotein, which is inserted into the membrane of the endoplasmic reticulum (ER) as it is synthesized on ER-bound ribosomes (step **7**). Carbohydrate is added to the large folded domain inside the ER lumen and is modified as the membrane and the associated glycoprotein pass through the Golgi apparatus (step **8**). Vesicles with mature glycoprotein fuse with the host plasma membrane, depositing viral glycoprotein on the cell surface with the large receptor-binding domain outside the cell (step **9**). Meanwhile, other viral mRNAs are translated on host-cell ribosomes into nucleocapsid protein, matrix protein, and viral RNA polymerase (step **10**). These proteins are assembled with replicated viral genomic RNA (bright red) into progeny nucleocapsids (step **11**), which then associate with the cytosolic domain of viral transmembrane glycoproteins in the plasma membrane (step **12**). The plasma membrane is folded around the nucleocapsid, forming a "bud" that eventually is released (step **13**).

other lytic RNA viruses, rabies virions are replicated in the cytoplasm and do not require host-cell nuclear enzymes. As shown in Figure 4-41, a rabies virion is adsorbed by endocytosis, and release of progeny virions occurs by *budding* from the host-cell plasma membrane. Budding virions are clearly visible in electron micrographs of infected cells, as illustrated in Figure 4-42. Many tens of thousands of progeny virions bud from an infected host cell before it dies.

Viral DNA Is Integrated into the Host-Cell Genome in Some Nonlytic Viral Growth Cycles

Some bacterial viruses, called *temperate phages*, can establish a nonlytic association with their host cells that does not kill the cell. For example, when λ bacteriophage infects *E. coli*, the viral DNA may be integrated into the host-cell chromosome rather than being replicated. The integrated viral DNA, called a *prophage*, is replicated as part of the cell's DNA from one host-cell generation to the next. This phenomenon is referred to as **lysogeny.** Under certain conditions, the prophage DNA is activated, leading to its excision from the host-cell chromosome, entrance into the lytic cycle, and subsequent production and release of progeny virions.

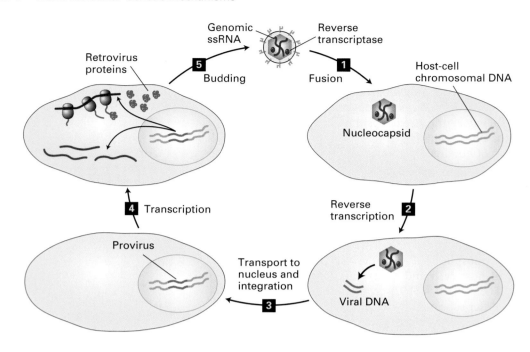

▲ **FIGURE 4-43 Retroviral life cycle.** Retroviruses have a genome of two identical copies of single-stranded RNA and an outer envelope. Step **1**: After viral glycoproteins in the envelope interact with a specific host-cell membrane protein, the retroviral envelope fuses directly with the plasma membrane, allowing entry of the nucleocapsid into the cytoplasm of the cell. Step **2**: Viral reverse transcriptase and other proteins copy the viral ssRNA genome into a double-stranded DNA. Step **3**: The viral dsDNA is transported into the nucleus and integrated into one of many possible sites in the host-cell chromosomal DNA. For simplicity, only one host-cell chromosome is depicted. Step **4**: The integrated viral DNA (provirus) is transcribed by the host-cell RNA polymerase, generating mRNAs (dark red) and genomic RNA molecules (bright red). The host-cell machinery translates the viral mRNAs into glycoproteins and nucleocapsid proteins. Step **5**: Progeny virions then assemble and are released by budding as illustrated in Figure 4-41.

The genomes of a number of animal viruses also can integrate into the host-cell genome. Probably the most important are the **retroviruses,** which are enveloped viruses with a genome consisting of two identical strands of RNA. These viruses are known as *retroviruses* because their RNA genome acts as a template for formation of a DNA molecule—the opposite flow of genetic information compared with the more common transcription of DNA into RNA. In the retroviral life cycle (Figure 4-43), a viral enzyme called **reverse transcriptase** initially copies the viral RNA genome into single-stranded DNA complementary to the virion RNA; the same enzyme then catalyzes synthesis of a complementary DNA strand. (This complex reaction is detailed in Chapter 10 when we consider closely related intracellular parasites called retrotransposons.) The resulting double-stranded DNA is integrated into the chromosomal DNA of the infected cell. Finally, the integrated DNA, called a *provirus,* is transcribed by the cell's own machinery into RNA, which either is translated into viral proteins or is packaged within virion coat proteins to form progeny virions that are released by budding from the host-cell membrane. Because most retroviruses do not kill their host cells, infected cells can replicate, producing daughter cells with integrated proviral DNA. These daughter cells continue to transcribe the proviral DNA and bud progeny virions.

 Some retroviruses contain cancer-causing genes (**oncogenes**), and cells infected by such retroviruses are oncogenically transformed into tumor cells. Studies of oncogenic retroviruses (mostly viruses of birds and mice) have revealed a great deal about the processes that lead to transformation of a normal cell into a cancer cell (Chapter 23).

Among the known human retroviruses are human T-cell lymphotrophic virus (HTLV), which causes a form of leukemia, and human immunodeficiency virus (HIV), which causes acquired immune deficiency syndrome (AIDS). Both of these viruses can infect only specific cell types, primarily certain cells of the immune system and, in the case of HIV, some central nervous system neurons and glial cells. Only these cells have cell-surface receptors that interact with viral envelope proteins, accounting for the host-cell specificity of these viruses. Unlike most other retroviruses, HIV eventually kills its host cells. The eventual death of large numbers of immune-system cells results in the defective immune response characteristic of AIDS.

Some DNA viruses also can integrate into a host-cell chromosome. One example is the human papillomaviruses (HPVs), which most commonly cause warts and other benign skin lesions. The genomes of certain HPV serotypes, however, occasionally integrate into the chromosomal DNA of infected cervical epithelial cells, initiating development of cervical cancer. Routine Pap smears can detect cells in the early stages of the transformation process initiated by HPV integration, permitting effective treatment. ∎

KEY CONCEPTS OF SECTION 4.7

Viruses: Parasites of the Cellular Genetic System

■ Viruses are small parasites that can replicate only in host cells. Viral genomes may be either DNA (DNA viruses) or RNA (RNA viruses) and either single- or double-stranded.

■ The capsid, which surrounds the viral genome, is composed of multiple copies of one or a small number of virus-encoded proteins. Some viruses also have an outer envelope, which is similar to the plasma membrane but contains viral transmembrane proteins.

■ Most animal and plant DNA viruses require host-cell nuclear enzymes to carry out transcription of the viral genome into mRNA and production of progeny genomes. In contrast, most RNA viruses encode enzymes that can transcribe the RNA genome into viral mRNA and produce new copies of the RNA genome.

■ Host-cell ribosomes, tRNAs, and translation factors are used in the synthesis of all viral proteins in infected cells.

■ Lytic viral infection entails adsorption, penetration, synthesis of viral proteins and progeny genomes (replication), assembly of progeny virions, and release of hundreds to thousands of virions, leading to death of the host cell (see Figure 4-40). Release of enveloped viruses occurs by budding through the host-cell plasma membrane (see Figure 4-41).

■ Nonlytic infection occurs when the viral genome is integrated into the host-cell DNA and generally does not lead to cell death.

■ Retroviruses are enveloped animal viruses containing a single-stranded RNA genome. After a host cell is penetrated, reverse transcriptase, a viral enzyme carried in the virion, converts the viral RNA genome into double-stranded DNA, which integrates into chromosomal DNA (see Figure 4-43).

■ Unlike infection by other retroviruses, HIV infection eventually kills host cells, causing the defects in the immune response characteristic of AIDS.

■ Tumor viruses, which contain oncogenes, may have an RNA genome (e.g., human T-cell lymphotrophic virus) or a DNA genome (e.g., human papillomaviruses). In the case of these viruses, integration of the viral genome into a host-cell chromosome can cause transformation of the cell into a tumor cell.

PERSPECTIVES FOR THE FUTURE

In this chapter we first reviewed the basic structure of DNA and RNA and then described fundamental aspects of the transcription of DNA by RNA polymerases. Eukaryotic RNA polymerases are discussed in greater detail in Chapter 11, along with additional factors required for transcription initiation in eukaryotic cells and interactions with regulatory transcription factors that control transcription initiation. Next, we discussed the genetic code and the participation of tRNA and the protein-synthesizing machine, the ribosome, in decoding the information in mRNA to allow accurate assembly of protein chains. Mechanisms that regulate protein synthesis are considered further in Chapter 12. Finally, we considered the molecular details underlying the accurate replication of DNA required for cell division. Chapter 21 covers the mechanisms that regulate when a cell replicates its DNA and that coordinate DNA replication with the complex process of mitosis that distributes the daughter DNA molecules equally to each daughter cell.

These basic cellular processes form the foundation of molecular cell biology. Our current understanding of these processes is grounded in a wealth of experimental results and is not likely to change. However, the depth of our understanding will continue to increase as additional details of the structures and interactions of the macromolecular machines involved are uncovered. The determination in recent years of the three-dimensional structures of RNA polymerases, ribosomal subunits, and DNA replication proteins has allowed researchers to design ever more penetrating experimental approaches for revealing how these macromolecules operate at the molecular level. The detailed level of understanding that results may allow the design of new and more effective drugs for treating human illnesses. For example, the recent high-resolution structures of ribosomes are providing insights into the mechanism by which antibiotics inhibit bacterial protein synthesis without affecting the function of mammalian ribosomes. This new knowledge may allow the design of even more effective antibiotics. Similarly, detailed understanding of the mechanisms regulating transcription of specific human genes may lead to therapeutic strategies that can reduce or prevent inappropriate immune responses that lead to multiple sclerosis and arthritis, the inappropriate cell division that is the hallmark of cancer, and other pathological processes.

Much of current biological research is focused on discovering how molecular interactions endow cells with decision-making capacity and their special properties. For this reason several of the following chapters describe current knowledge about how such interactions regulate transcription and protein synthesis in multicellular organisms and how such regulation endows cells with the capacity to become specialized and grow into complicated organs. Other chapters deal with how protein-protein interactions underlie the construction of specialized organelles in cells, and how they determine cell shape and movement. The rapid advances in molecular cell biology in recent years hold promise that in the not too distant future we will understand how the regulation of specialized cell function, shape, and mobility coupled with regulated cell replication and cell death (apoptosis) lead to the growth of complex organisms like trees and human beings.

KEY TERMS

anticodon *119*	plaque assay *138*
codons *119*	polyribosomes *130*
complementary *104*	primary transcript *110*
DNA polymerases *133*	primer *133*
double helix *103*	promoter *109*
envelope (viral) *137*	reading frame *120*
exons *111*	replication fork *133*
genetic code *119*	reverse transcriptase *142*
introns *111*	ribosomal RNA (rRNA) *119*
lagging strand *133*	ribosomes *119*
leading strand *133*	RNA polymerase *109*
messenger RNA (mRNA) *119*	transcription *101*
Okazaki fragments *133*	transfer RNA (tRNA) *119*
operon *111*	translation *101*
phosphodiester bond *103*	Watson-Crick base pairs *103*

REVIEW THE CONCEPTS

1. What are Watson-Crick base pairs? Why are they important?

2. TATA box–binding protein binds to the minor groove of DNA, resulting in the bending of the DNA helix (see Figure 4-5). What property of DNA allows the TATA box–binding protein to recognize the DNA helix?

3. Preparing plasmid (double-stranded, circular) DNA for sequencing involves annealing a complementary, short, single-stranded oligonucleotide DNA primer to one strand of the plasmid template. This is routinely accomplished by heating the plasmid DNA and primer to 90 °C and then slowly bringing the temperature down to 25 °C. Why does this protocol work?

4. What difference between RNA and DNA helps to explain the greater stability of DNA? What implications does this have for the function of DNA?

5. What are the major differences in the synthesis and structure of prokaryotic and eukaryotic mRNAs?

6. While investigating the function of a specific growth factor receptor gene from humans, it was found that two types of proteins are synthesized from this gene. A larger protein containing a membrane-spanning domain functions to recognize growth factors at the cell surface, stimulating a specific downstream signaling pathway. In contrast, a related, smaller protein is secreted from the cell and functions to bind available growth factor circulating in the blood, thus inhibiting the downstream signaling pathway. Speculate on how the cell synthesizes these disparate proteins.

7. Describe the molecular events that occur at the *lac* operon when *E. coli* cells are shifted from a glucose-containing medium to a lactose-containing medium.

8. The concentration of free phosphate affects transcription of some *E. coli* genes. Describe the mechanism for this.

9. Contrast how selection of the translational start site occurs on bacterial, eukaryotic, and poliovirus mRNAs.

10. What is the evidence that the 23S rRNA in the large rRNA subunit has a peptidyl transferase activity?

11. How would a mutation in the poly(A)-binding protein I gene affect translation? How would an electron micrograph of polyribosomes from such a mutant differ from the normal pattern?

12. What characteristic of DNA results in the requirement that some DNA synthesis is discontinuous? How are Okazaki fragments and DNA ligase utilized by the cell?

13. What gene is unique to retroviruses? Why is the protein encoded by this gene absolutely necessary for maintaining the retroviral life cycle, but not that of other viruses?

ANALYZE THE DATA

NASA has identified a new microbe present on Mars and requests that you determine the genetic code of this organism. To accomplish this goal, you isolate an extract from this microbe that contains all the components necessary for protein synthesis except mRNA. Synthetic mRNAs are added to this extract and the resulting polypeptides are analyzed:

Synthetic mRNA	*Resulting Polypeptides*
AAAAAAAAAAAAAAAAA	Lysine-Lysine-Lysine etc.
CACACACACACACACA	Threonine-Histidine-Threonine-Histidine etc.
AACAACAACAACAACA	Threonine-Threonine-Threonine etc.
	Glutamine-Glutamine-Glutamine etc.
	Asparagine-Asparagine-Asparagine etc.

From these data, what specifics can you conclude about the microbe's genetic code? What is the sequence of the anticodon loop of a tRNA carrying a threonine? If you found that this microbe contained 61 different tRNAs, what could you speculate about the fidelity of translation in this organism?

REFERENCES

Structure of Nucleic Acids

Dickerson, R. E. 1983. The DNA helix and how it is read. *Sci. Am.* **249**:94–111.

Doudna, J. A., and T. R. Cech. 2002. The chemical repertoire of natural ribozymes. *Nature* **418**:222–228.

Kornberg, A., and T. A. Baker. 1992. *DNA Replication,* 2d ed. W. H. Freeman and Company, chap. 1. A good summary of the principles of DNA structure.

Wang, J. C. 1980. Superhelical DNA. *Trends Biochem. Sci.* **5**:219–221.

Transcription of Protein-Coding Genes and Formation of Functional mRNA

Brenner, S., F. Jacob, and M. Meselson. 1961. An unstable intermediate carrying information from genes to ribosomes for protein synthesis. *Nature* **190**:576–581.

Young, B. A., T. M. Gruber, and C. A. Gross. 2002. Views of transcription initiation. *Cell* **109**:417–420.

Control of Gene Expression in Prokaryotes

Bell, C. E., and M. Lewis. 2001. The Lac repressor: a second generation of structural and functional studies. *Curr. Opin. Struc. Biol.* **11**:19–25.

Busby, S., and R. H. Ebright. 1999. Transcription activation by catabolite activator protein (CAP). *J. Mol. Biol.* **293**:199–213.

Darst, S. A. 2001. Bacterial RNA polymerase. *Curr. Opin. Struc. Biol.* **11**:155–162.

Muller-Hill, B. 1998. Some repressors of bacterial transcription. *Curr. Opin. Microbiol.* **1**:145–151.

The Three Roles of RNA in Translation

Alexander, R. W., and P. Schimmel. 2001. Domain-domain communication in aminoacyl-tRNA synthetases. *Prog. Nucleic Acid Res. Mol. Biol.* **69**:317–349.

Bjork, G. R., et al. 1987. Transfer RNA modification. *Ann. Rev. Biochem.* **56**:263–287.

Garrett, R. A., et al., eds. 2000. *The Ribosome: Structure, Function, Antibiotics, and Cellular Interactions.* ASM Press.

Hatfield, D. L., and V. N. Gladyshev. 2002. How selenium has altered our understanding of the genetic code. *Mol. Cell Biol.* **22**:3565–3576.

Hoagland, M. B., et al. 1958. A soluble ribonucleic acid intermediate in protein synthesis. *J. Biol. Chem.* **231**:241–257.

Holley, R. W., et al. 1965. Structure of a ribonucleic acid. *Science* **147**:1462–1465.

Ibba, M., and D. Soll. 2001. The renaissance of aminoacyl-tRNA synthesis. *EMBO Rep.* **2**:382–387.

Khorana, G. H., et al. 1966. Polynucleotide synthesis and the genetic code. *Cold Spring Harbor Symp. Quant. Biol.* **31**:39–49.

Maguire, B. A., and R. A. Zimmermann. 2001. The ribosome in focus. *Cell* **104**:813–816.

Nirenberg, M., et al. 1966. The RNA code in protein synthesis. *Cold Spring Harbor Symp. Quant. Biol.* **31**:11–24.

Ramakrishnan, V. 2002. Ribosome structure and the mechanism of translation. *Cell* **108**:557–572.

Rich, A., and S.-H. Kim. 1978. The three-dimensional structure of transfer RNA. *Sci. Am.* **240**(1):52–62 (offprint 1377).

Stepwise Synthesis of Proteins on Ribosomes

Gingras, A. C., R. Raught, and N. Sonenberg. 1999. eIF4 initiation factors: effectors of mRNA recruitment to ribosomes and regulators of translation. *Ann. Rev. Biochem.* **68**:913–963.

Green, R. 2000. Ribosomal translocation: EF-G turns the crank. *Curr. Biol.* **10**:R369–R373.

Hellen, C. U., and P. Sarnow. 2001. Internal ribosome entry sites in eukaryotic mRNA molecules. *Genet. Devel.* **15**:1593–1612.

Kisselev, L. L., and R. H. Buckingham. 2000. Translational termination comes of age. *Trends Biochem. Sci.* **25**:561–566.

Kozak, M. 1999. Initiation of translation in prokaryotes and eukaryotes. *Gene* **234**:187–208.

Noller, H. F., et al. 2002. Translocation of tRNA during protein synthesis. *FEBS Lett.* **514**:11–16.

Pestova, T. V., et al. 2001. Molecular mechanisms of translation initiation in eukaryotes. *Proc. Nat'l. Acad. Sci. USA* **98**:7029–7036.

Poole, E., and W. Tate. 2000. Release factors and their role as decoding proteins: specificity and fidelity for termination of protein synthesis. *Biochim. Biophys. Acta* **1493**:1–11.

Ramakrishnan, V. 2002. Ribosome structure and the mechanism of translation. *Cell* **108**:557–572.

Sonenberg, N., J. W. B. Hershey, and M. B. Mathews, eds. 2000. *Translational Control of Gene Expression.* Cold Spring Harbor Laboratory Press.

DNA Replication

Bullock, P. A. 1997. The initiation of simian virus 40 DNA replication in vitro. *Crit. Rev. Biochem. Mol. Biol.* **32**:503–568.

Kornberg, A., and T. A. Baker. 1992. *DNA Replication,* 2d ed. W. H. Freeman and Company

Waga, S., and B. Stillman. 1998. The DNA replication fork in eukaryotic cells. *Ann. Rev. Biochem.* **67**:721–751.

Viruses: Parasites of the Cellular Genetic System

Flint, S. J., et al. 2000. *Principles of Virology: Molecular Biology, Pathogenesis, and Control.* ASM Press.

Hull, R. 2002. *Mathews' Plant Virology.* Academic Press.

Knipe, D. M., and P. M. Howley, eds. 2001. *Fields Virology.* Lippincott Williams & Wilkins.

Kornberg, A., and T. A. Baker. 1992. *DNA Replication,* 2d ed. W. H. Freeman and Company. Good summary of bacteriophage molecular biology.

5

BIOMEMBRANES AND CELL ARCHITECTURE

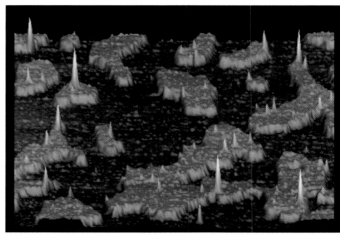

Atomic force microscopy reveals sphyingomyelin rafts (orange) protruding from a dioleoylphosphatidylcholine background (black) in a mica-supported lipid bilayer. Placental alkaline phosphatase (yellow peaks), a glycosylphosphatidylinositol-anchored protein, is shown to be almost exclusively raft associated. [From D. E. Saslowsky et al., 2002, *J. Biol. Chem.* **277**:26966–26970.]

Prokaryotes, which represent the simplest and smallest cells, about 1–2 μm in length, are surrounded by a **plasma membrane** but contain no internal membrane-limited subcompartments (see Figure 1-2a). Although DNA is concentrated in the center of these unicellular organisms, most enzymes and metabolites are thought to diffuse freely within the single internal aqueous compartment. Certain metabolic reactions, including protein synthesis and anaerobic glycolysis, take place there; others, such as the replication of DNA and the production of ATP, take place at the plasma membrane.

In the larger cells of eukaryotes, however, the rates of chemical reactions would be limited by the diffusion of small molecules if a cell were not partitioned into smaller subcompartments termed **organelles.** Each organelle is surrounded by one or more **biomembranes,** and each type of organelle contains a unique complement of proteins—some embedded in its membrane(s), others in its aqueous interior space, or **lumen.** These proteins enable each organelle to carry out its characteristic cellular functions. The **cytoplasm** is the part of the cell outside the largest organelle, the nucleus. The **cytosol,** the aqueous part of the cytoplasm outside all of the organelles, also contains its own distinctive proteins.

All biomembranes form closed structures, separating the lumen on the inside from the outside, and are based on a similar bilayer structure. They control the movement of molecules between the inside and the outside of a cell and into and out of the organelles of eukaryotic cells. In accord with the importance of internal membranes to cell function, the total surface area of these membranes is roughly tenfold as great as that of the plasma membrane (Figure 5-1).

Although the basic architecture of all eukaryotic cells is constructed from membranes, organelles, and the cytosol, each type of cell exhibits a distinctive design defined by the shape of the cell and the location of its organelles. The structural basis of the unique design of each cell type lies in the **cytoskeleton,** a dense network of three classes of protein filaments that permeate the cytosol and mechanically support cellular membranes. Cytoskeletal proteins are among the most abundant proteins in a cell, and the enormous surface area of the cytoskeleton (see Figure 5-1) constitutes a scaffold to which particular sets of proteins and membranes are bound.

We begin our examination of cell architecture by considering the basic structure of biomembranes. The lipid components of membranes not only affect their shape and

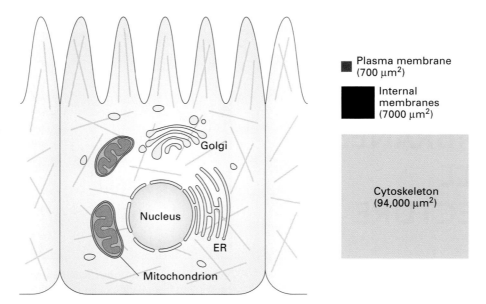

Plasma membrane
(700 μm²)

Internal
membranes
(7000 μm²)

Cytoskeleton
(94,000 μm²)

▲ **FIGURE 5-1 Schematic overview of the major
components of eukaryotic cell architecture.** The plasma
membrane (red) defines the exterior of the cell and controls the
movement of molecules between the cytosol and the
extracellular medium. Different types of organelles and smaller
vesicles enclosed within their own distinctive membranes (black)
carry out special functions such as gene expression, energy
production, membrane synthesis, and intracellular transport.

Fibers of the cytoskeleton (green) provide structural support for
the cell and its internal compartments. The internal membranes
of organelles and vesicles possess more surface area than that
of the plasma membrane but less area than that of the
cytoskeleton, as schematically represented by the red, black, and
green boxes. The enormous surface area of the cytoskeleton
allows it to function as a scaffold on which cellular reactions can
take place.

function but also play important roles in anchoring proteins
to the membrane, modifying membrane protein activities,
and transducing signals to the cytoplasm. We then consider
the general structure of membrane proteins and how they
can relate to different membranes. The unique function of
each membrane is determined largely by the complement of
proteins within and adjacent to it. The theme of membrane-
limited compartments is continued with a review of the functions of various organelles. We then introduce the structure
and function of the cytoskeleton, which is intimately associated with all biomembranes; changes in the organization of
this filamentous network affect the structure and function
of the attached membranes. In the remainder of the chapter,
we describe common methods for isolating particular types
of cells and subcellular structures and various microscopic
techniques for studying cell structure and function.

▶ **FIGURE 5-2 The bilayer structure of biomembranes.**
(a) Electron micrograph of a thin section through an erythrocyte
membrane stained with osmium tetroxide. The characteristic
"railroad track" appearance of the membrane indicates the
presence of two polar layers, consistent with the bilayer
structure for phospholipid membranes. (b) Schematic
interpretation of the phospholipid bilayer in which polar groups
face outward to shield the hydrophobic fatty acyl tails from
water. The hydrophobic effect and van der Waals interactions
between the fatty acyl tails drive the assembly of the bilayer
(Chapter 2). [Part (a) courtesy of J. D. Robertson.]

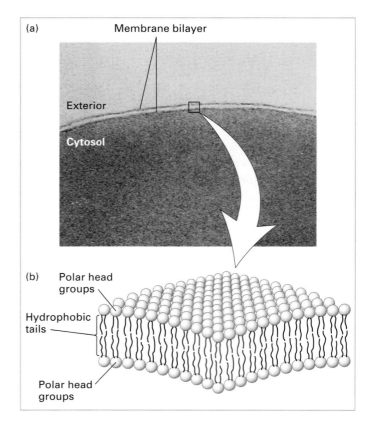

(a)   Membrane bilayer

Exterior

Cytosol

(b)   Polar head
       groups

Hydrophobic
tails

Polar head
groups

5.1 Biomembranes: Lipid Composition and Structural Organization

Phospholipids of the composition present in cells spontaneously form sheetlike **phospholipid bilayers,** which are two molecules thick. The hydrocarbon chains of the phospholipids in each layer, or *leaflet,* form a **hydrophobic** core that is 3–4 nm thick in most biomembranes. Electron microscopy of thin membrane sections stained with osmium tetroxide, which binds strongly to the polar head groups of phospholipids, reveals the bilayer structure (Figure 5-2). A cross section of all single membranes stained with osmium tetroxide looks like a railroad track: two thin dark lines (the stain–head group complexes) with a uniform light space of about 2 nm (the hydrophobic tails) between them.

The lipid bilayer has two important properties. First, the hydrophobic core is an impermeable barrier that prevents the diffusion of water-soluble (**hydrophilic**) solutes across the membrane. Importantly, this simple barrier function is modulated by the presence of membrane proteins that mediate the transport of specific molecules across this otherwise impermeable bilayer. The second property of the bilayer is its stability. The bilayer structure is maintained by hydrophobic and van der Waals interactions between the lipid chains. Even though the exterior aqueous environment can vary widely in ionic strength and pH, the bilayer has the strength to retain its characteristic architecture.

Natural membranes from different cell types exhibit a variety of shapes, which complement a cell's function (Figure 5-3). The smooth flexible surface of the erythrocyte plasma membrane allows the cell to squeeze through narrow blood capillaries. Some cells have a long, slender extension of the plasma membrane, called a cilium or flagellum, which beats in a whiplike manner. This motion causes fluid to flow across the surface of an epithelium or a sperm cell to swim through the medium. The axons of many neurons are encased by multiple layers of modified plasma membrane called the **myelin sheath.** This membranous structure is elaborated by

(a)

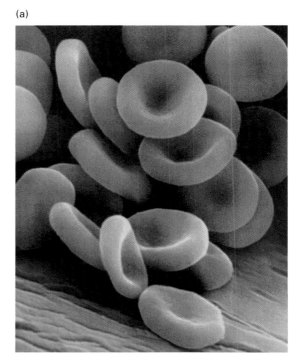

10 μm

(b)

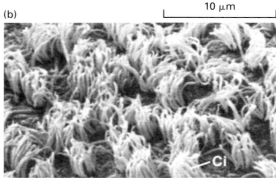

Ci

(c)

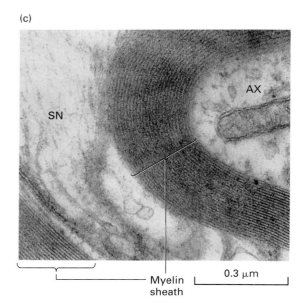

AX

SN

Myelin sheath

0.3 μm

▶ **FIGURE 5-3 Variation in biomembranes in different cell types.** (a) A smooth, flexible membrane covers the surface of the discoid erythrocyte cell. (b) Tufts of cilia (Ci) project from the ependymal cells that line the brain ventricles. (c) Many nerve axons are enveloped in a myelin sheath composed of multiple layers of modified plasma membrane. The individual myelin layers can be seen in this electron micrograph of a cross section of an axon (AX). The myelin sheath is formed by an adjacent supportive (glial) cell (SC). [Parts (a) and (b) from R. G. Kessel and R. H. Kardon, 1979, *Tissues and Organs: A Text-Atlas of Scanning Electron Microscopy,* W. H. Freeman and Company. Part (c) from P. C. Cross and K. L. Mercer, 1993, *Cell and Tissue Ultrastructure: A Functional Perspective,* W. H. Freeman and Company, p. 137.]

▶ **FIGURE 5-4 The faces of cellular membranes.** The plasma membrane, a single bilayer membrane, encloses the cell. In this highly schematic representation, internal cytosol (green stipple) and external environment (purple) define the cytosolic (red) and exoplasmic (black) faces of the bilayer. Vesicles and some organelles have a single membrane and their internal aqueous space (purple) is topologically equivalent to the outside of the cell. Three organelles—the nucleus, mitochondrion, and chloroplast (which is not shown)—are enclosed by two membranes separated by a small intermembrane space. The exoplasmic faces of the inner and outer membranes around these organelles border the intermembrane space between them. For simplicity, the hydrophobic membrane interior is not indicated in this diagram.

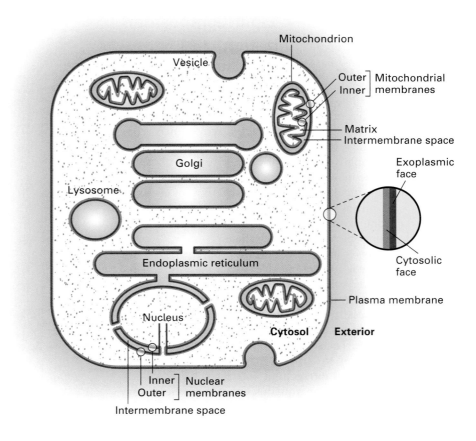

an adjacent supportive cell and facilitates the conduction of nerve impulses over long distances (Chapter 7). Despite their diverse shapes and functions, these biomembranes and all other biomembranes have a common bilayer structure.

Because all cellular membranes enclose an entire cell or an internal compartment, they have an *internal face* (the surface oriented toward the interior of the compartment) and an *external face* (the surface presented to the environment). More commonly, the surfaces of a cellular membrane are designated as the **cytosolic face** and the **exoplasmic face**. This nomenclature is useful in highlighting the topological equivalence of the faces in different membranes, as diagrammed in Figure 5-4. For example, the exoplasmic face of the plasma membrane is directed away from the cytosol, toward the extracellular space or external environment, and defines the outer limit of the cell. For organelles and vesicles surrounded by a single membrane, however, the face directed away from the cytosol—the exoplasmic face—is on the inside in contact with an internal aqueous space equivalent to the extracellular space. This equivalence is most easily understood for vesicles that arise by invagination of the plasma membrane; this process results in the external face of the plasma membrane becoming the internal face of the vesicle membrane. Three organelles—the nucleus, mitochondrion, and chloroplast—are surrounded by two membranes; the exoplasmic

surface of each membrane faces the space between the two membranes.

Three Classes of Lipids Are Found in Biomembranes

A typical biomembrane is assembled from phosphoglycerides, sphingolipids, and steroids. All three classes of lipids are **amphipathic** molecules having a polar (hydrophilic) head group and hydrophobic tail. The hydrophobic effect and van der Waals interactions, discussed in Chapter 2, cause the tail groups to self-associate into a bilayer with the polar head groups oriented toward water (see Figure 5-2). Although the common membrane lipids have this amphipathic character in common, they differ in their chemical structures, abundance, and functions in the membrane.

Phosphoglycerides, the most abundant class of lipids in most membranes, are derivatives of glycerol 3-phosphate (Figure 5-5a). A typical phosphoglyceride molecule consists of a hydrophobic tail composed of two fatty acyl chains esterified to the two hydroxyl groups in glycerol phosphate and a polar head group attached to the phosphate group. The two fatty acyl chains may differ in the number of carbons that they contain (commonly 16 or 18) and their degree of saturation (0, 1, or 2 double bonds). A phosphogyceride is classified according to the nature of its head group. In phosphatidylcholines, the most abundant phospholipids in the plasma membrane, the head group consists of choline, a positively charged alcohol, esterified to the negatively charged phosphate. In other phosphoglycerides, an OH-containing molecule such as ethanolamine, serine, and the sugar derivative inositol is linked to the phosphate group. The negatively charged phosphate group and the positively charged groups or the hydroxyl groups on the head group interact strongly with water.

The *plasmalogens* are a group of phosphoglycerides that contain one fatty acyl chain, attached to glycerol by an ester linkage, and one long hydrocarbon chain, attached to glycerol by an ether linkage (C—O—C). These molecules constitute about 20 percent of the total phosphoglyceride content in humans. Their abundance varies among tissues and species but is especially high in human brain and heart tissue. The additional chemical stability of the ether linkage

in plasmalogens or the subtle differences in their three-dimensional structure compared with that of other phosphoglycerides may have as-yet unrecognized physiologic significance.

A second class of membrane lipid is the **sphingolipids.** All of these compounds are derived from sphingosine, an amino alcohol with a long hydrocarbon chain, and contain a long-chain fatty acid attached to the sphingosine amino group. In sphingomyelin, the most abundant sphingolipid, phosphocholine is attached to the terminal hydroxyl group of sphingosine (Figure 5-5b). Thus sphingomyelin is a phospholipid, and its overall structure is quite similar to that of phosphatidylcholine. Other sphingolipids are amphipathic **glycolipids** whose polar head groups are sugars. Glucosylcerebroside, the simplest glycosphingolipid, contains a single glucose unit attached to sphingosine. In the complex glycosphingolipids called *gangliosides,* one or two branched sugar chains containing sialic acid groups are attached to

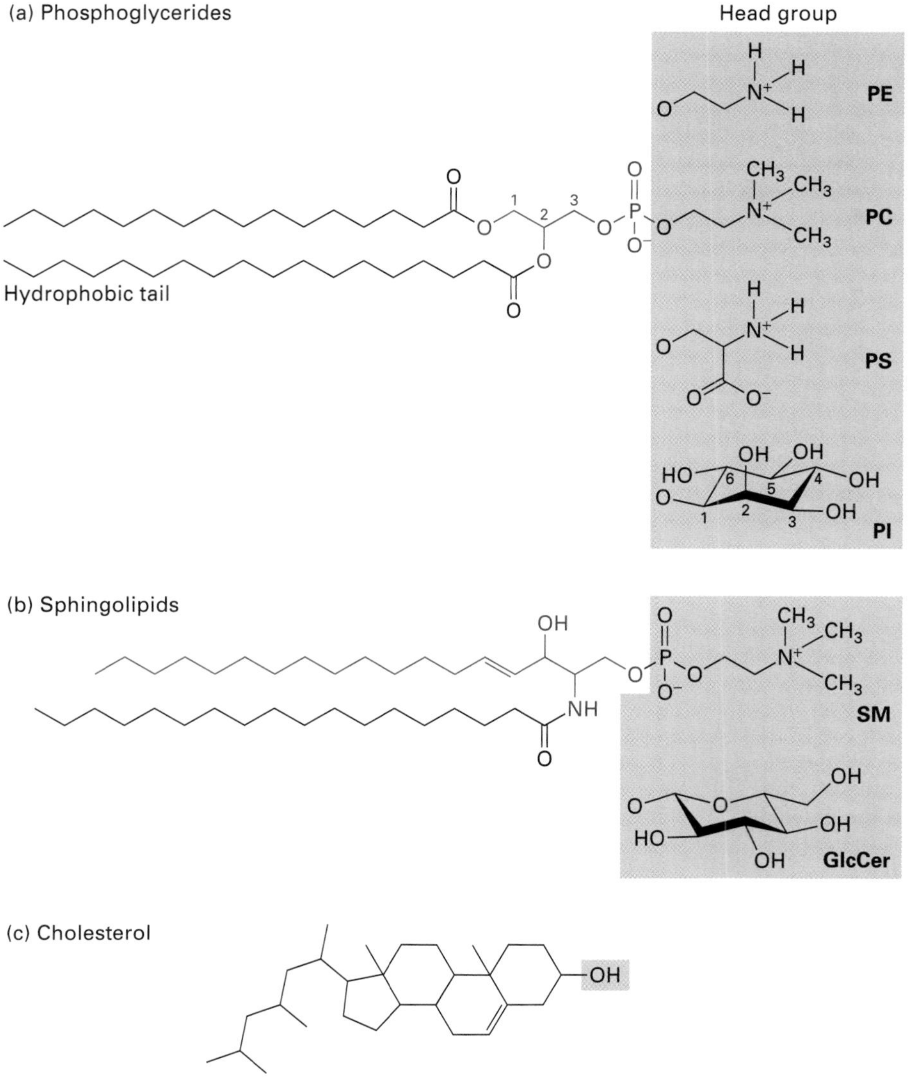

(a) Phosphoglycerides

Head group

Hydrophobic tail

PE
PC
PS
PI

(b) Sphingolipids

SM
GlcCer

(c) Cholesterol

◄ **FIGURE 5-5 Three classes of membrane lipids.** (a) Most phosphoglycerides are derivatives of glycerol 3-phosphate (red) containing two esterified fatty acyl chains, constituting the hydrophobic "tail" and a polar "head group" esterified to the phosphate. The fatty acids can vary in length and be saturated (no double bonds) or unsaturated (one, two, or three double bonds). In phosphatidylcholine (PC), the head group is choline. Also shown are the molecules attached to the phosphate group in three other common phosphoglycerides: phosphatidylethanolamine (PE), phosphatidylserine (PS), and phosphatidylinositol (PI). (b) Sphingolipids are derivatives of sphingosine (red), an amino alcohol with a long hydrocarbon chain. Various fatty acyl chains are connected to sphingosine by an amide bond. The sphingomyelins (SM), which contain a phosphocholine head group, are phospholipids. Other sphingolipids are glycolipids in which a single sugar residue or branched oligosaccharide is attached to the sphingosine backbone. For instance, the simple glycolipid glucosylcerebroside (GlcCer) has a glucose head group. (c) Like other membrane lipids, the steroid cholesterol is amphipathic. Its single hydroxyl group is equivalent to the polar head group in other lipids; the conjugated ring and short hydrocarbon chain form the hydrophobic tail. [See H. Sprong et al., 2001, *Nature Rev. Mol. Cell Biol.* **2**:504.]

sphingosine. Glycolipids constitute 2–10 percent of the total lipid in plasma membranes; they are most abundant in nervous tissue.

Cholesterol and its derivatives constitute the third important class of membrane lipids, the **steroids.** The basic structure of steroids is a four-ring hydrocarbon. Cholesterol, the major steroidal constituent of animal tissues, has a hydroxyl substituent on one ring (Figure 5-5c). Although cholesterol is almost entirely hydrocarbon in composition, it is amphipathic because its hydroxyl group can interact with water. Cholesterol is especially abundant in the plasma membranes of mammalian cells but is absent from most prokaryotic cells. As much as 30–50 percent of the lipids in plant plasma membranes consist of certain steroids unique to plants.

At neutral pH, some phosphoglycerides (e.g., phosphatidylcholine and phosphatidylethanolamine) carry no net electric charge, whereas others (e.g., phosphatidylinositol and phosphatidylserine) carry a single net negative charge. Nonetheless, the polar head groups in all phospholipids can pack together into the characteristic bilayer structure. Sphingomyelins are similar in shape to phosphoglycerides and can form mixed bilayers with them. Cholesterol and other steroids are too hydrophobic to form a bilayer structure unless they are mixed with phospholipids.

Most Lipids and Many Proteins Are Laterally Mobile in Biomembranes

In the two-dimensional plane of a bilayer, thermal motion permits lipid molecules to rotate freely around their long axes and to diffuse laterally within each leaflet. Because such movements are lateral or rotational, the fatty acyl chains remain in the hydrophobic interior of the bilayer. In both natural and ar-

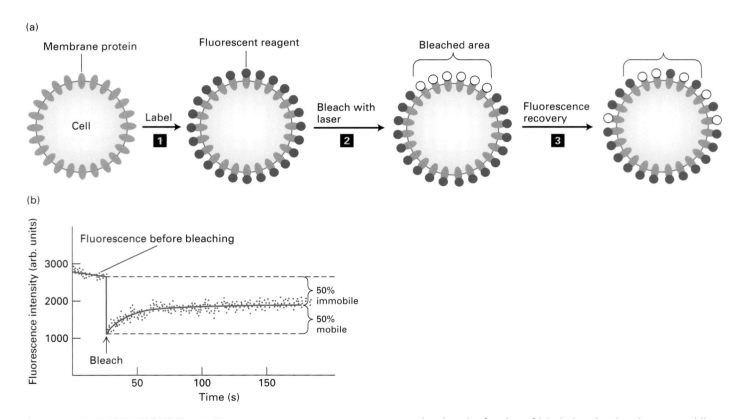

(a)

(b)

▲ **EXPERIMENTAL FIGURE 5-6 Fluorescence recovery after photobleaching (FRAP) experiments can quantify the lateral movement of proteins and lipids within the plasma membrane.** (a) Experimental protocol. Step **1**: Cells are first labeled with a fluorescent reagent that binds uniformly to a specific membrane lipid or protein. Step **2**: A laser light is then focused on a small area of the surface, irreversibly bleaching the bound reagent and thus reducing the fluorescence in the illuminated area. Step **3**: In time, the fluorescence of the bleached patch increases as unbleached fluorescent surface molecules diffuse into it and bleached ones diffuse outward. The extent of recovery of fluorescence in the bleached patch is proportional to the fraction of labeled molecules that are mobile in the membrane. (b) Results of FRAP experiment with human hepatoma cells treated with a fluorescent antibody specific for the asialoglycoprotein receptor protein. The finding that 50 percent of the fluorescence returned to the bleached area indicates that 50 percent of the receptor molecules in the illuminated membrane patch were mobile and 50 percent were immobile. Because the rate of fluorescence recovery is proportional to the rate at which labeled molecules move into the bleached region, the diffusion coefficient of a protein or lipid in the membrane can be calculated from such data. [See Y. I. Henis et al., 1990, *J. Cell Biol.* **111**:1409.]

tificial membranes, a typical lipid molecule exchanges places with its neighbors in a leaflet about 10^7 times per second and diffuses several micrometers per second at 37 °C. These diffusion rates indicate that the viscosity of the bilayer is 100 times as great as that of water—about the same as the viscosity of olive oil. Even though lipids diffuse more slowly in the bilayer than in an aqueous solvent, a membrane lipid could diffuse the length of a typical bacterial cell (1 μm) in only 1 second and the length of an animal cell in about 20 seconds.

The lateral movements of specific plasma-membrane proteins and lipids can be quantified by a technique called *fluorescence recovery after photobleaching (FRAP)*. With this method, described in Figure 5-6, the rate at which membrane lipid or protein molecules move—the diffusion coefficient—can be determined, as well as the proportion of the molecules that are laterally mobile.

The results of FRAP studies with fluorescence-labeled phospholipids have shown that, in fibroblast plasma membranes, all the phospholipids are freely mobile over distances of about 0.5 μm, but most cannot diffuse over much longer distances. These findings suggest that protein-rich regions of the plasma membrane, about 1 μm in diameter, separate lipid-rich regions containing the bulk of the membrane phospholipid. Phospholipids are free to diffuse within such a region but not from one lipid-rich region to an adjacent one. Furthermore, the rate of lateral diffusion of lipids in the plasma membrane is nearly an order of magnitude slower than in pure phospholipid bilayers: diffusion constants of 10^{-8} cm²/s and 10^{-7} cm²/s are characteristic of the plasma membrane and a lipid bilayer, respectively. This difference suggests that lipids may be tightly but not irreversibly bound to certain integral proteins in some membranes.

Lipid Composition Influences the Physical Properties of Membranes

A typical cell contains myriad types of membranes, each with unique properties bestowed by its particular mix of lipids and proteins. The data in Table 5-1 illustrate the variation in lipid composition among different biomembranes. Several phenomena contribute to these differences. For instance, differences between membranes in the endoplasmic reticulum (ER) and the Golgi are largely explained by the fact that phospholipids are synthesized in the ER, whereas sphingolipids are synthesized in the Golgi. As a result, the proportion of sphingomyelin as a percentage of total membrane lipid phosphorus is about six times as high in Golgi membranes as it is in ER membranes. In other cases, the translocation of membranes from one cellular compartment to another can selectively enrich membranes in certain lipids.

Differences in lipid composition may also correspond to specialization of membrane function. For example, the plasma membrane of absorptive epithelial cells lining the intestine exhibits two distinct regions: the **apical** surface faces the lumen of the gut and is exposed to widely varying external conditions; the **basolateral** surface interacts with other epithelial cells and with underlying extracellular structures (see Figure 6-5). In these **polarized cells,** the ratio of sphingolipid to phosphoglyceride to cholesterol in the basolateral membrane is 0.5:1.5:1, roughly equivalent to that in the plasma membrane of a typical unpolarized cell subjected to mild stress. In contrast, the apical membrane of intestinal cells, which is subjected to considerable stress, exhibits a 1:1:1 ratio of these lipids. The relatively high concentration of sphingolipid in this membrane may increase its stability

---

TABLE 5-1   Major Lipid Components of Selected Biomembranes

Source/Location	Composition (mol %)			
	PC	PE + PS	SM	Cholesterol
Plasma membrane (human erythrocytes)	21	29	21	26
Myelin membrane (human neurons)	16	37	13	34
Plasma membrane (*E. coli*)	0	85	0	0
Endoplasmic reticulum membrane (rat)	54	26	5	7
Golgi membrane (rat)	45	20	13	13
Inner mitochondrial membrane (rat)	45	45	2	7
Outer mitochondrial membrane (rat)	34	46	2	11
Primary leaflet location	Exoplasmic	Cytosolic	Exoplasmic	Both

PC = phosphatidylcholine; PE = phosphatidylethanolamine; PS = phosphatidylserine; SM = sphingomyelin.
SOURCE: W. Dowhan and M. Bogdanov, 2002, in D. E. Vance and J. E. Vance, eds., *Biochemistry of Lipids, Lipoproteins, and Membranes,* Elsevier.

because of extensive hydrogen bonding by the free —OH group in the sphingosine moiety (see Figure 5-5).

The ability of lipids to diffuse laterally in a bilayer indicates that it can act as a fluid. The degree of bilayer fluidity depends on the lipid composition, structure of the phospholipid hydrophobic tails, and temperature. As already noted, van der Waals interactions and the hydrophobic effect cause the nonpolar tails of phospholipids to aggregate. Long, saturated fatty acyl chains have the greatest tendency to aggregate, packing tightly together into a gel-like state. Phospholipids with short fatty acyl chains, which have less surface area for interaction, form more fluid bilayers. Likewise, the kinks in unsaturated fatty acyl chains result in their forming less stable van der Waals interactions with other lipids than do saturated chains and hence more fluid bilayers. When a highly ordered, gel-like bilayer is heated, the increased molecular motions of the fatty acyl tails cause it to undergo a transition to a more fluid, disordered state (Figure 5-7).

At usual physiologic temperatures, the hydrophobic interior of natural membranes generally has a low viscosity and a fluidlike, rather than gel-like, consistency. Cholesterol is important in maintaining the fluidity of natural membranes, which appears to be essential for normal cell growth and reproduction. As noted previously, cholesterol cannot form a sheetlike bilayer on its own. At concentrations found in natural membranes, cholesterol is interca-

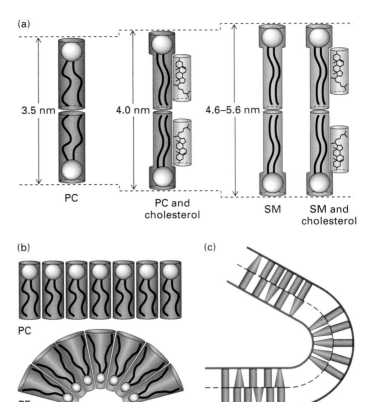

▲ **FIGURE 5-8 Effect of lipid composition on bilayer thickness and curvature.** (a) A pure sphingomyelin (SM) bilayer is thicker than one formed from a phosphoglyceride such as phosphatidylcholine (PC). Cholesterol has a lipid-ordering effect on phosphoglyceride bilayers that increases their thickness but does not affect the thickness of the more ordered SM bilayer. (b) Phospholipids such as PC have a cylindrical shape and form more or less flat monolayers, whereas those with smaller head groups such as phosphatidylethanolamine (PE) have a conical shape. (c) A bilayer enriched with PC in the exoplasmic leaflet and with PE in the cytosolic face, as in many plasma membranes, would have a natural curvature. [Adapted from H. Sprong et al., 2001, *Nature Rev. Mol. Cell Biol.* **2**:504.]

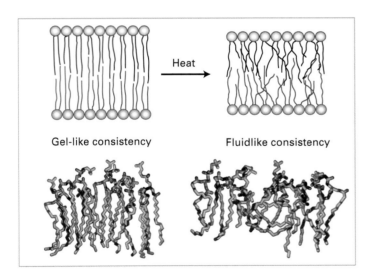

▲ **FIGURE 5-7 Gel and fluid forms of the phospholipid bilayer.** (*Top*) Depiction of gel-to-fluid transition. Phospholipids with long saturated fatty acyl chains tend to assemble into a highly ordered, gel-like bilayer in which there is little overlap of the nonpolar tails in the two leaflets. Heat disorders the nonpolar tails and induces a transition from a gel to a fluid within a temperature range of only a few degrees. As the chains become disordered, the bilayer also decreases in thickness. (*Bottom*) Molecular models of phospholipid monolayers in gel and fluid states, as determined by molecular dynamics calculations. [Bottom based on H. Heller et al., 1993, *J. Phys. Chem.* **97**:8343.]

lated (inserted) among phospholipids. Cholesterol restricts the random movement of phospholipid head groups at the outer surfaces of the leaflets, but its effect on the movement of long phospholipid tails depends on concentration. At the usual cholesterol concentrations, the interaction of the steroid ring with the long hydrophobic tails of phospholipids tends to immobilize these lipids and thus decrease biomembrane fluidity. At lower cholesterol concentrations, however, the steroid ring separates and disperses phospholipid tails, causing the inner regions of the membrane to become slightly more fluid.

The lipid composition of a bilayer also influences its thickness, which in turn may play a role in localizing proteins to a particular membrane. The results of studies on artificial membranes demonstrate that sphingomyelin associates into a more gel-like and thicker bilayer than phospholipids do (Figure 5-8a). Similarly, cholesterol and other molecules that decrease membrane fluidity increase membrane thickness. Because sphingomyelin tails are already optimally stabilized, the addition of cholesterol has no effect on the thickness of a sphingomyelin bilayer.

Another property dependent on the lipid composition of a bilayer is its local curvature, which depends on the relative sizes of the polar head groups and nonpolar tails of its constituent phospholipids. Lipids with long tails and large head groups are cylindrical in shape; those with small head groups are cone shaped (Figure 5-8b). As a result, bilayers composed of cylindrical lipids are relatively flat, whereas those containing large amounts of cone-shaped lipids form curved bilayers (Figure 5-8c). This effect of lipid composition on bilayer curvature may play a role in the formation of highly curved membrane pits and blebs, internal membrane vesicles, and specialized membrane structures such as microvilli.

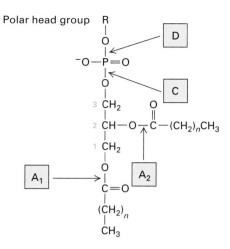

▲ **FIGURE 5-9 Specificity of phospholipases.** Each type of phospholipase cleaves one of the susceptible bonds shown in red. The glycerol carbon atoms are indicated by small numbers. In intact cells, only phospholipids in the exoplasmic leaflet of the plasma membrane are cleaved by phospholipases in the surrounding medium. Phospholipase C, a cytosolic enzyme, cleaves certain phospholipids in the cytosolic leaflet of the plasma membrane.

Membrane Lipids Are Usually Distributed Unequally in the Exoplasmic and Cytosolic Leaflets

A characteristic of all membranes is an asymmetry in lipid composition across the bilayer. Although most phospholipids are present in both membrane leaflets, they are commonly more abundant in one or the other leaflet. For instance, in plasma membranes from human erythrocytes and certain canine kidney cells grown in culture, almost all the sphingomyelin and phosphatidylcholine, both of which form less fluid bilayers, are found in the exoplasmic leaflet. In contrast, phosphatidylethanolamine, phosphatidylserine, and phosphatidylinositol, which form more fluid bilayers, are preferentially located in the cytosolic leaflet. This segregation of lipids across the bilayer may influence membrane curvature (see Figure 5-8c). Unlike phospholipids, cholesterol is relatively evenly distributed in both leaflets of cellular membranes.

The relative abundance of a particular phospholipid in the two leaflets of a plasma membrane can be determined on the basis of its susceptibility to hydrolysis by **phospholipases,** enzymes that cleave various bonds in the hydrophilic ends of phospholipids (Figure 5-9). Phospholipids in the cytosolic leaflet are resistant to hydrolysis by phospholipases added to the external medium because the enzymes cannot penetrate to the cytosolic face of the plasma membrane.

How the asymmetric distribution of phospholipids in membrane leaflets arises is still unclear. In pure bilayers, phospholipids do not spontaneously migrate, or flip-flop, from one leaflet to the other. Energetically, such flip-flopping is extremely unfavorable because it entails movement of the polar phospholipid head group through the hydrophobic interior of the membrane. To a first approximation, the asymmetry in phospholipid distribution results from the vectorial synthesis of lipids in the endoplasmic reticulum and Golgi. Sphingomyelin is synthesized on the luminal (exoplasmic) face of the Golgi, which becomes the exoplasmic face of the plasma membrane. In contrast, phosphoglycerides are synthesized on the cytosolic face of the ER membrane, which is topologically identical with the cytosolic face of the plasma membrane (see Figure 5-4). Clearly, this explanation does not account for the preferential location of phosphatidylcholine in the exoplasmic leaflet. Movement of this phosphoglyceride and perhaps others from one leaflet to the other in some natural membranes is catalyzed by certain ATP-powered transport proteins called **flippases** discussed in Chapters 7 and 18.

The preferential location of lipids to one face of the bilayer is necessary for a variety of membrane-based functions. For example, the head groups of all phosphorylated forms of phosphatidylinositol face the cytosol. Certain of them are cleaved by phospholipase C located in the cytosol; this enzyme in turn is activated as a result of cell stimulation by many hormones. These cleavages generate cytosol-soluble phosphoinositols and membrane-soluble diacylglycerol. As we see in later chapters, these molecules participate in intracellular signaling pathways that affect many aspects of cellular metabolism. Phosphatidylserine also is normally most abundant in the cytosolic leaflet of the plasma membrane. In the initial stages of platelet stimulation by serum, phosphatidylserine is briefly translocated to the exoplasmic face, presumably by a flippase enzyme, where it activates enzymes participating in blood clotting.

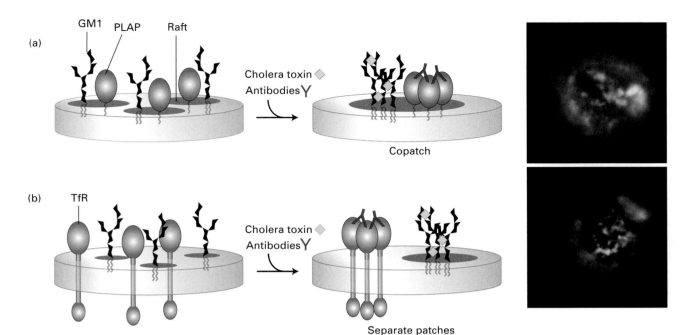

▲ **EXPERIMENTAL FIGURE 5-10 Some membrane lipids and proteins colocalize in lipid rafts.** The results of biochemical studies suggested that GM1, a glycosphingolipid, and placental alkaline phosphatase (PLAP), a lipid-anchored membrane protein, aggregate together into lipid rafts, whereas the transferrin receptor (TfR), which traverses the entire membrane, does not. To locate these components in the intact plasma membrane, cells were treated with fluorescence-labeled cholera toxin (green), which cross-links closely spaced GM1 molecules, and with fluorescence-labeled antibodies (red) specific for either PLAP or TfR. Each antibody can cross-link closely spaced molecules of the protein that it recognizes. Cross-linking causes the proteins or lipids to form larger patches that can be detected by fluorescence microscopy (see Figure 5-42). (a) Micrograph of a cell treated with toxin and with anti-PLAP antibody shows GM1 and PLAP colocalized in the same patches (yellow). This copatching suggests that both GM1 and PLAP are present in lipid rafts that coalesce in the presence of the cross-linking reagents. (b) Micrograph of a cell treated with toxin and with anti-TfR antibody shows that GM1 and TfR reside in separate patches (i.e., red and green), indicating that TfR is not a raft-resident protein. [Micrographs from T. Harder et al., 1998, *J. Cell Biol.* **141**:929.]

Cholesterol and Sphingolipids Cluster with Specific Proteins in Membrane Microdomains

The results of recent studies have challenged the long-held belief that lipids are randomly mixed in each leaflet of a bilayer. The first hint that lipids may be organized within the leaflets was the discovery that the residues remaining after the extraction of plasma membranes with detergents contain two lipids: cholesterol and sphingomyelin. Because these two lipids are found in more ordered, less fluid bilayers, researchers hypothesized that they form microdomains, termed **lipid rafts,** surrounded by other more fluid phospholipids that are easily extracted by detergents.

Biochemical and microscopic evidence supports the existence of lipid rafts in natural membranes. For instance, fluorescence microscopy reveals aggregates of lipids and raft-specific proteins in the membrane (Figure 5-10). The rafts are heterogeneous in size but are typically 50 nm in diameter. Rafts can be disrupted by methyl-β-cyclodextrin, which depletes the membrane of cholesterol, or by antibiotics, such as filipin, that sequester cholesterol; such findings indicate the importance of cholesterol in maintaining the integrity of these rafts. Besides their enrichment by cholesterol and sphingolipids, lipid rafts are enriched for many types of cell-surface receptor proteins, as well as many signaling proteins that bind to the receptors and are activated by them. These lipid–protein complexes can form only in the two-dimensional environment of a hydrophobic bilayer and, as discussed in later chapters, they are thought to facilitate the detection of chemical signals from the external environment and the subsequent activation of cytosolic events.

KEY CONCEPTS OF SECTION 5.1

Biomembranes: Lipid Composition and Structural Organization

■ The eukaryotic cell is demarcated from the external environment by the plasma membrane and organized into membrane-limited internal compartments (organelles and vesicles).

■ The total surface area of internal membranes far exceeds that of the plasma membrane.

■ The phospholipid bilayer, the basic structural unit of all biomembranes, is a two-dimensional lipid sheet with hydrophilic faces and a hydrophobic core, which is impermeable to water-soluble molecules and ions (see Figure 5-2).

■ Certain proteins present in biomembranes make them selectively permeable to water-soluble molecules and ions.

■ The primary lipid components of biomembranes are phosphoglycerides, sphingolipids, and steroids (see Figure 5-5).

■ Most lipids and many proteins are laterally mobile in biomembranes.

■ Different cellular membranes vary in lipid composition (see Table 5-1). Phospholipids and sphingolipids are asymmetrically distributed in the two leaflets of the bilayer, whereas cholesterol is fairly evenly distributed in both leaflets.

■ Natural biomembranes generally have a fluidlike consistency. In general, membrane fluidity is decreased by sphingolipids and cholesterol and increased by phosphoglycerides. The lipid composition of a membrane also influences its thickness and curvature (see Figure 5-8).

■ Lipid rafts are microdomains containing cholesterol, sphingolipids, and certain membrane proteins that form in the plane of the bilayer. These aggregates are sites for signaling across the plasma membrane.

5.2 Biomembranes: Protein Components and Basic Functions

Membrane proteins are defined by their location within or at the surface of a phospholipid bilayer. Although every biological membrane has the same basic bilayer structure, the proteins associated with a particular membrane are responsible for its distinctive activities. The density and complement of proteins associated with biomembranes vary, depending on cell type and subcellular location. For example, the inner mitochondrial membrane is 76 percent protein; the myelin membrane, only 18 percent. The high phospholipid content of myelin allows it to electrically insulate a nerve cell from its environment. The importance of membrane proteins is suggested from the finding that approximately a third of all yeast genes encode a membrane protein. The relative abundance of genes for membrane proteins is even greater in multicellular organisms in which membrane proteins have additional functions in cell adhesion.

The lipid bilayer presents a unique two-dimensional hydrophobic environment for membrane proteins. Some proteins are buried within the lipid-rich bilayer; other proteins are associated with the exoplasmic or cytosolic leaflet of the bilayer. Protein domains on the extracellular surface of the plasma membrane generally bind to other molecules, including external signaling proteins, ions, and small metabolites (e.g., glucose, fatty acids), and to adhesion molecules on other cells or in the external environment. Domains within the plasma membrane, particularly those that form channels and pores, move molecules in and out of cells. Domains lying along the cytosolic face of the plasma membrane have a wide range of functions, from anchoring cytoskeletal proteins to the membrane to triggering intracellular signaling pathways.

In many cases, the function of a membrane protein and the topology of its polypeptide chain in the membrane can be predicted on the basis of its homology with another, well-characterized protein. In this section, we examine the characteristic structural features of membrane proteins and some of their basic functions. More complete characterization of the structure and function of various types of membrane proteins is presented in several later chapters; the synthesis and processing of this large, diverse group of proteins are discussed in Chapters 16 and 17.

Proteins Interact with Membranes in Three Different Ways

Membrane proteins can be classified into three categories—integral, lipid-anchored, and peripheral—on the basis of the nature of the membrane–protein interactions (Figure 5-11).

Integral membrane proteins, also called *transmembrane proteins*, span a phospholipid bilayer and are built of three segments. The cytosolic and exoplasmic domains have hydrophilic exterior surfaces that interact with the aqueous solutions on the cytosolic and exoplasmic faces of the membrane. These domains resemble other water-soluble proteins in their amino acid composition and structure. In contrast, the 3-nm-thick membrane-spanning domain contains many hydrophobic amino acids whose side chains protrude outward and interact with the hydrocarbon core of the phospholipid bilayer. In all transmembrane proteins examined to date, the membrane-spanning domains consist of one or more α helices or of multiple β strands. In addition, most transmembrane proteins are glycosylated with a complex branched sugar group attached to one or several amino acid side chains. Invariably these sugar chains are localized to the exoplasmic domains.

Lipid-anchored membrane proteins are bound covalently to one or more lipid molecules. The hydrophobic carbon chain of the attached lipid is embedded in one leaflet of the membrane and anchors the protein to the membrane. The polypeptide chain itself does not enter the phospholipid bilayer.

Peripheral membrane proteins do not interact with the hydrophobic core of the phospholipid bilayer. Instead they are usually bound to the membrane indirectly by interactions with integral membrane proteins or directly by interactions with lipid head groups. Peripheral proteins are localized to either the cytosolic or the exoplasmic face of the plasma membrane.

In addition to these proteins, which are closely associated with the bilayer, cytoskeletal filaments are more loosely associated with the cytosolic face, usually through one or more

▶ **FIGURE 5-11 Diagram of how various classes of proteins associate with the lipid bilayer.** Integral (transmembrane) proteins span the bilayer. Lipid-anchored proteins are tethered to one leaflet by a long covalently attached hydrocarbon chain. Peripheral proteins associate with the membrane primarily by specific noncovalent interactions with integral proteins or membrane lipids. Farther from the membrane are membrane-associated proteins including the cytoskeleton, extracellular matrix in animal cells, and cell wall in plant and bacterial cells (not depicted). Carbohydrate chains are attached to many extracellular proteins and to the exoplasmic domains of many transmembrane proteins.

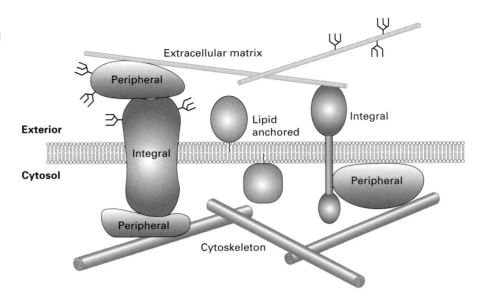

peripheral (adapter) proteins (see Figure 5-11). Such associations with the cytoskeleton provide support for various cellular membranes (see Section 5.4); they also play a role in the two-way communication between the cell interior and the cell exterior, as we learn in Chapter 6. Finally, peripheral proteins on the outer surface of the plasma membrane and the exoplasmic domains of integral membrane proteins are often attached to components of the **extracellular matrix** or to the **cell wall** surrounding bacterial and plant cells.

Membrane-Embedded α Helices Are the Primary Secondary Structures in Most Transmembrane Proteins

Soluble proteins exhibit hundreds of distinct localized folded structures, or motifs (see Figure 3-6). In comparison, the repertoire of folded structures in integral membrane proteins is quite limited, with the hydrophobic α helix predominating. Integral proteins containing membrane-spanning α-helical domains are embedded in membranes by hydrophobic interactions with specific lipids and probably also by ionic interactions with the polar head groups of the phospholipids.

Glycophorin A, the major protein in the erythrocyte plasma membrane, is a representative *single-pass* transmembrane protein, which contains only one membrane-spanning α helix (Figure 5-12). Typically, a membrane-embedded α helix is composed of 20–25 hydrophobic (uncharged) amino acids (see Figure 2-13). The predicted length of such a helix (3.75 nm) is just sufficient to span the hydrocarbon core of a phospholipid bilayer. The hydrophobic side chains protrude outward from the helix and form van der Waals interactions with the fatty acyl chains in the bilayer. In contrast, the carbonyl (C═O) and imino (NH) groups taking part in the formation of backbone peptide bonds through

hydrogen bonding are in the interior of the α helix (see Figure 3-3); thus these polar groups are shielded from the hydrophobic interior of the membrane. The transmembrane helix of one glycophorin A molecule associates with the helix in another to form a coiled-coil dimer (see Figure 5-12b). Such interaction of membrane-spanning helices is a common mechanism for creating dimeric membrane proteins. Many cell-surface receptors, for instance, are activated by dimerization.

A large and important family of integral proteins is defined by the presence of seven membrane-spanning α helices. Among the more than 150 such "seven spanning" *multipass* proteins that have been identified are the G protein–coupled receptors described in Chapter 13. The structure of bacteriorhodopsin, a protein found in the membrane of certain photosynthetic bacteria, illustrates the general structure of all these proteins (Figure 5-13). Absorption of light by the retinal group covalently attached to bacteriorhodopsin causes a conformational change in the protein that results in the pumping of protons from the cytosol across the bacterial membrane to the extracellular space. The proton concentration gradient thus generated across the membrane is used to synthesize ATP (Chapter 8). In the high-resolution structure of bacteriorhodopsin now available, the positions of all the individual amino acids, retinal, and the surrounding lipids are determined. As might be expected, virtually all of the amino acids on the exterior of the membrane-spanning segments of bacteriorhodopsin are hydrophobic and interact with the hydrocarbon core of the surrounding lipid bilayer.

Ion channels compose a second large and important family of multipass transmembrane proteins. As revealed by the crystal structure of a resting K^+ channel, ion channels are typically tetrameric proteins. Each of the four subunits has a pair of membrane-spanning helices that bundle with helices

(a)

Extracellular
domain

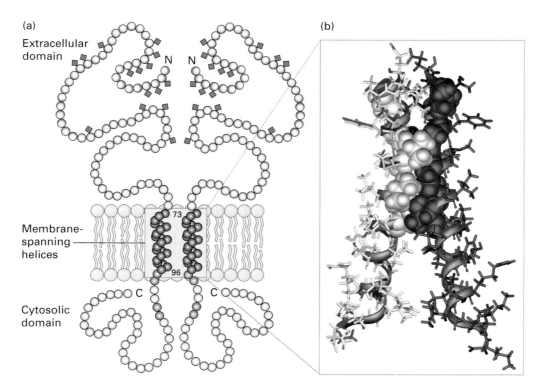

(b)

Membrane-
spanning
helices

Cytosolic
domain

▲ **FIGURE 5-12 Structure of glycophorin A, a typical single-pass transmembrane protein.** (a) Diagram of dimeric glycophorin showing major sequence features and its relation to the membrane. The single 23-residue membrane-spanning α helix in each monomer is composed of amino acids with hydrophobic (uncharged) side chains (red spheres). By binding negatively charged phospholipid head groups, the positively charged arginine and lysine residues (blue spheres) near the cytosolic side of the helix help anchor glycophorin in the membrane. Both the extracellular and the cytosolic domains are rich in charged residues and polar uncharged residues; the extracellular domain is heavily glycosylated, with the carbohydrate side chains (green diamonds) attached to specific serine, threonine, and asparagine residues. (b) Molecular model of the transmembrane domain of dimeric glycophorin corresponding to residues 73–96. The side chains of the α helix in one monomer are shown in red; those in the other monomer, in gray. Residues depicted as space-filling structures participate in intermonomer van der Waals interactions that stabilize the coiled-coil dimer. [Part (b) adapted from K. R. MacKenzie et al., 1997, *Science* **276**:131.]

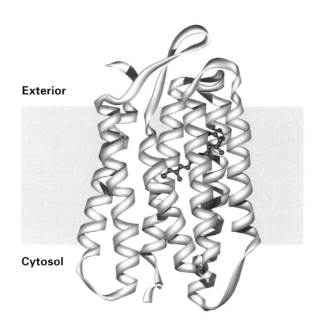

Exterior

Cytosol

of other subunits, forming a central channel (see Figure 7-15). Polar and hydrophobic residues lining the center of the bundle form a channel in the membrane, but as with bacteriorhodopsin virtually all of the amino acids on the exterior of the membrane-spanning domain are hydrophobic. In many ion channels, external factors (e.g., a ligand, voltage, or mechanical strain) regulate ion flow across the bilayer by reorienting the helices. Details of ion channels and their structures are discussed in Chapter 7.

◀ **FIGURE 5-13 Structural model of bacteriorhodopsin, a multipass transmembrane protein that functions as a photoreceptor in certain bacteria.** The seven hydrophobic α helices in bacteriorhodopsin traverse the lipid bilayer. A retinal molecule (red) covalently attached to one helix absorbs light. The large class of G protein–coupled receptors in eukaryotic cells also has seven membrane-spanning α helices; their three-dimensional structure is similar to that of bacteriorhodopsin. [After H. Luecke et al., 1999, *J. Mol. Biol.* **291**:899.]

Multiple β Strands in Porins Form Membrane-Spanning "Barrels"

The **porins** are a class of transmembrane proteins whose structure differs radically from that of other integral proteins. Several types of porin are found in the outer membrane of gram-negative bacteria such as *E. coli* and in the outer membranes of mitochondria and chloroplasts. The outer membrane protects an intestinal bacterium from harmful agents (e.g., antibiotics, bile salts, and proteases) but permits the uptake and disposal of small hydrophilic molecules including nutrients and waste products. The porins in the outer membrane of an *E. coli* cell provide channels for the passage of disaccharides and other small molecules as well as phosphate.

The amino acid sequences of porins are predominantly polar and contain no long hydrophobic segments typical of integral proteins with α-helical membrane-spanning domains. X-ray crystallography has revealed that porins are trimers of identical subunits. In each subunit, 16 β strands form a barrel-shaped structure with a pore in the center (Fig-

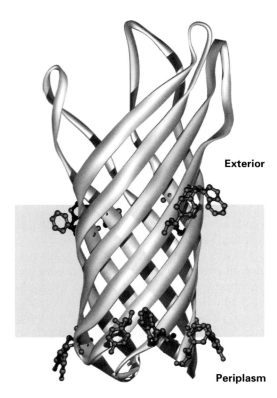

▲ **FIGURE 5-14 Structural model of one subunit of OmpX, a porin found in the *E. coli* outer membrane.** All porins are trimeric transmembrane proteins. Each subunit is barrel shaped, with β strands forming the wall and a transmembrane pore in the center. A band of aliphatic (noncyclic) side chains (yellow) and a border of aromatic (ring-containing) side chains (red) position the protein in the bilayer. [After G. E. Schulz, 2000, *Curr. Opin. Struc. Biol.* **10**:443.]

ure 5-14). Unlike a typical water-soluble globular protein, a porin has a hydrophilic inside and a hydrophobic exterior; in this sense, porins are inside-out. In a porin monomer, the outward-facing side groups on each of the β strands are hydrophobic and form a nonpolar ribbonlike band that encircles the outside of the barrel. This hydrophobic band interacts with the fatty acyl groups of the membrane lipids or with other porin monomers. The side groups facing the inside of a porin monomer are predominantly hydrophilic; they line the pore through which small water-soluble molecules cross the membrane.

As discussed in Chapter 7, the plasma membranes of animal cells contain a water channel called *aquaporin*. Like most other integral proteins, aquaporin contains multiple transmembrane α helices. Thus, despite its name, aquaporin differs structurally from the porins as well as functionally in that it mediates transport of a single molecule—namely, water.

Covalently Attached Hydrocarbon Chains Anchor Some Proteins to Membranes

In eukaryotic cells, several types of covalently attached lipids anchor some proteins to one or the other leaflet of the plasma membrane and certain other cellular membranes. In these lipid-anchored proteins, the lipid hydrocarbon chains are embedded in the bilayer, but the protein itself does not enter the bilayer.

A group of cytosolic proteins are anchored to the cytosolic face of a membrane by a fatty acyl group (e.g., myristate or palmitate) attached to the N-terminal glycine residue (Figure 5-15a). Retention of such proteins at the membrane by the N-terminal acyl anchor may play an important role in a membrane-associated function. For example, v-Src, a mutant form of a cellular tyrosine kinase, is oncogenic and can transform cells only when it has a myristylated N-terminus.

A second group of cytosolic proteins are anchored to membranes by an unsaturated fatty acyl group attached to a cysteine residue at or near the C-terminus (Figure 5-15b). In these proteins, a farnesyl or geranylgeranyl group is bound through a thioether bond to the —SH group of a C-terminal cysteine residue. These *prenyl anchors* are built from isoprene units (C_5), which are also used in the synthesis of cholesterol (Chapter 18). In some cases, a second geranylgeranyl group or a palmitate group is linked to a nearby cysteine residue. The additional anchor is thought to reinforce the attachment of the protein to the membrane. Ras, a GTPase superfamily protein that functions in intracellular signaling, is localized to the cytosolic face of the plasma membrane by such a double anchor. Rab proteins, which also belong to the GTPase superfamily, are similarly bound to the cytosolic surface of intracellular vesicles by prenyl-type anchors; these proteins are required for the fusion of vesicles with their target membranes (Chapter 17).

Some cell-surface proteins and heavily glycosylated proteoglycans of the extracellular matrix are bound to the exo-

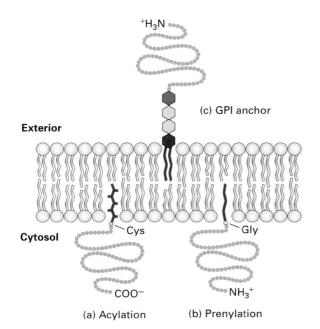

▲ **FIGURE 5-15 Anchoring of plasma-membrane proteins to the bilayer by covalently linked hydrocarbon groups.**
(a) Cytosolic proteins such as v-Src are associated with the plasma membrane through a single fatty acyl chain attached to the N-terminal glycine (Gly) residue of the polypeptide. Myristate (C_{14}) and palmitate (C_{16}) are common acyl anchors. (b) Other cytosolic proteins (e.g., Ras and Rab proteins) are anchored to the membrane by prenylation of one or two cysteine (Cys) residues, at or near the C-terminus. The anchors are farnesyl (C_{15}) and geranylgeranyl (C_{20}) groups, both of which are unsaturated. (c) The lipid anchor on the exoplasmic surface of the plasma membrane is glycosylphosphatidylinositol (GPI). The phosphatidylinositol part (red) of this anchor contains two fatty acyl chains that extend into the bilayer. The phosphoethanolamine unit (purple) in the anchor links it to the protein. The two green hexagons represent sugar units, which vary in number and arrangement in different GPI anchors. The complete structure of a yeast GPI anchor is shown in Figure 16-14. [Adapted from H. Sprong et al., 2001, *Nature Rev. Mol. Cell Biol.* **2**:504.]

plasmic face of the plasma membrane by a third type of anchor group, glycosylphosphatidylinositol (GPI). The exact structures of *GPI anchors* vary greatly in different cell types, but they always contain phosphatidylinositol (PI), whose two fatty acyl chains extend into the lipid bilayer; phosphoethanolamine, which covalently links the anchor to the C-terminus of a protein; and several sugar residues (Figure 5-15c). Various experiments have shown that the GPI anchor is both necessary and sufficient for binding proteins to the membrane. For instance, the enzyme phospholipase C cleaves the phosphate–glycerol bond in phospholipids and in GPI anchors (see Figure 5-9). Treatment of cells with phospholipase C releases GPI-anchored proteins such as Thy-1 and placental alkaline phosphatase (PLAP) from the cell surface.

As already discussed, PLAP is concentrated in lipid rafts, the more ordered bilayer microdomains that are enriched in sphingolipids and cholesterol (see Figure 5-10). Although PLAP and other GPI-anchored proteins lie in the opposite membrane leaflet from acyl-anchored proteins, both types of membrane proteins are concentrated in lipid rafts. In contrast, prenylated proteins are not found in lipid rafts.

All Transmembrane Proteins and Glycolipids Are Asymmetrically Oriented in the Bilayer

Lipid-anchored proteins are just one example of membrane proteins that are asymmetrically located with respect to the faces of cellular membranes. Each type of transmembrane protein also has a specific orientation with respect to the membrane faces. In particular, the same part(s) of a particular protein always faces the cytosol, whereas other parts face the exoplasmic space. This asymmetry in protein orientation confers different properties on the two membrane faces. (We describe how the orientation of different types of transmembrane proteins is established during their synthesis in Chapter 16.) Membrane proteins have never been observed to flip-flop across a membrane; such movement, requiring a transient movement of hydrophilic amino acid residues through the hydrophobic interior of the membrane, would be energetically unfavorable. Accordingly, the asymmetry of a transmembrane protein, which is established during its biosynthesis and insertion into a membrane, is maintained throughout the protein's lifetime.

Many transmembrane proteins contain carbohydrate chains covalently linked to serine, threonine, or asparagine side chains of the polypeptide. Such transmembrane glycoproteins are always oriented so that the carbohydrate chains are in the exoplasmic domain (see Figures 5-11 and 5-12). Likewise, glycolipids, in which a carbohydrate chain is attached to the glycerol or sphingosine backbone, are always located in the exoplasmic leaflet with the carbohydrate chain protruding from the membrane surface. Both glycoproteins and glycolipids are especially abundant in the plasma membranes of eukaryotic cells; they are absent from the inner mitochondrial membrane, chloroplast lamellae, and several other intracellular membranes. Because the carbohydrate chains of glycoproteins and glycolipids in the plasma membrane extend into the extracellular space, they are available to interact with components of the extracellular matrix as well as lectins, growth factors, and antibodies.

 One important consequence of such interactions is illustrated by the A, B, and O blood-group antigens. These three structurally related oligosaccharide components of certain glycoproteins and glycolipids are expressed on the surfaces of human erythrocytes and many other cell types (Figure 5-16). All humans have the enzymes for synthesizing O antigen. Persons with type A blood also have a glycosyltransferase that adds an extra

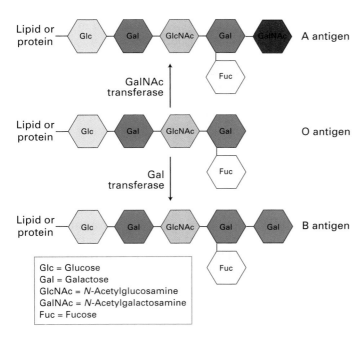

▲ **FIGURE 5-16 Human ABO blood-group antigens.** These antigens are oligosaccharide chains covalently attached to glycolipids or glycoproteins in the plasma membrane. The terminal oligosaccharide sugars distinguish the three antigens. The presence or absence of the glycosyltransferases that add galactose (Gal) or N-acetylgalactosamine (GalNAc) to O antigen determine a person's blood type.

N-acetylgalactosamine to O antigen to form A antigen. Those with type B blood have a different transferase that adds an extra galactose to O antigen to form B antigen. People with both transferases produce both A and B antigen (AB blood type); those who lack these transferases produce O antigen only (O blood type).

Persons whose erythrocytes lack the A antigen, B antigen, or both on their surface normally have antibodies against the missing antigen(s) in their serum. Thus if a type A or O person receives a transfusion of type B blood, antibodies against the B epitope will bind to the introduced red cells and trigger their destruction. To prevent such harmful reactions, blood-group typing and appropriate matching of blood donors and recipients are required in all transfusions (Table 5-2). ∎

Interactions with the Cytoskeleton Impede the Mobility of Integral Membrane Proteins

The results of experiments like the one depicted in Figure 5-6 and other types of studies have shown that many transmembrane proteins and lipid-anchored proteins, like phospholipids, float quite freely within the plane of a natural membrane. From 30 to 90 percent of all integral proteins in the plasma membrane are freely mobile, depending on the cell type. The lateral diffusion rate of a mobile protein in a pure phospholipid bilayer or isolated plasma membrane is similar to that of lipids. However, the diffusion rate of a protein in the plasma membrane of intact cells is generally 10–30 times lower than that of the same protein embedded in synthetic spherical bilayer structures (liposomes). These findings suggest that the mobility of integral proteins in the plasma membrane of living cells is restricted by interactions with the rigid submembrane cytoskeleton. Some integral proteins are permanently linked to the underlying cytoskeleton; these proteins are completely immobile in the membrane. In regard to mobile proteins, such interactions are broken and remade as the proteins diffuse laterally in the plasma membrane, slowing down their rate of diffusion. We consider the nature and functional consequences of linkages between integral membrane proteins and the cytoskeleton in Chapter 6.

Lipid-Binding Motifs Help Target Peripheral Proteins to the Membrane

Until the past decade or so, the interaction of peripheral proteins with integral proteins was thought to be the major

TABLE 5-2	ABO Blood Groups		
Blood-Group Type	**Antigens on RBCs***	**Serum Antibodies**	**Can Receive Blood Types**
A	A	Anti-A	A and O
B	B	Anti-B	B and O
AB	A and B	None	All
O	O	Anti-A and anti-B	O

*See Figure 5-16 for antigen structures.

TABLE 5-3	Selected Lipid-Binding Motifs	
Motif	**Ligand**[*]	**Selected Proteins with Motif**
PH	PIP_2, PIP_3	Phospholipase $C_\gamma 1$, protein kinase B, pleckstrin
C2	Acidic phospholipids	Protein kinase C, PI-3 kinase, phospholipase, PTEN phosphatase
Ankyrin repeat	PS	Ankyrin[†]
FERM	PIP_2	Band 4.1 protein; ezrin, radixin, moesin (ERM)[†]

[*] PIP_2, PIP_3, and PI-3P = phosphatidylinositol derivatives with additional phosphate groups on the inositol ring (see Figure 14-26); PH = pleckstrin homology; PS = phosphatidylserine;.
[†] These proteins have roles in linking the actin cytoskeleton to the plasma membrane.

mechanism by which peripheral proteins were bound to membranes. The results of more recent research indicate that protein–lipid interactions are equally important in localizing peripheral proteins to cellular membranes (see Figure 5-11).

Analyses of genome sequences have revealed several widely distributed lipid-binding motifs in proteins (Table 5-3). For instance, the *pleckstrin homology (PH) domain,* which binds two types of phosphorylated phosphatidylinositols, is the eleventh most common protein domain encoded in the human genome. This domain was initially recognized in pleckstrin, a protein found in platelets. The high frequency of the PH domain indicates that proteins localized to membrane surfaces carry out many important functions. Other common lipid-binding motifs include the C2 domain, the ankyrin-repeat domain, and the FERM domain. Originally discovered in protein kinase C, the C2 domain is a membrane-targeting domain for various kinases, phosphatases, and phospholipases.

The phospholipases are representative of those water-soluble enzymes that associate with the polar head groups of membrane phospholipids to carry out their catalytic functions. As noted earlier, phospholipases hydrolyze various bonds in the head groups of phospholipids (see Figure 5-9). These enzymes have an important role in the degradation of damaged or aged cell membranes and are active molecules in many snake venoms. The mechanism of action of phospholipase A_2 illustrates how such water-soluble enzymes can reversibly interact with membranes and catalyze reactions at the interface of an aqueous solution and lipid surface. When this enzyme is in aqueous solution, its Ca^{2+}-containing active site is buried in a channel lined with hydrophobic amino acids. The enzyme binds with greatest affinity to bilayers composed of negatively charged phospholipids (e.g., phosphotidylethanolamine). This finding suggests that a rim of positively charged lysine and arginine residues around the entrance catalytic channel is particularly important in interfacial binding (Figure 5-17a). Binding

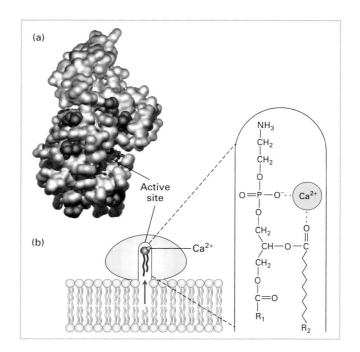

▲ **FIGURE 5-17 Interfacial binding surface and mechanism of action of phospholipase A_2.** (a) A structural model of the enzyme showing the surface that interacts with a membrane. This interfacial binding surface contains a rim of positively charged arginine and lysine residues shown in blue surrounding the cavity of the catalytic active site in which a substrate lipid (red stick structure) is bound. (b) Diagram of catalysis by phospholipase A_2. When docked on a model lipid membrane, positively charged residues of the interfacial binding site bind to negatively charged polar groups at the membrane surface. This binding triggers a small conformational change, opening a channel lined with hydrophobic amino acids that leads from the bilayer to the catalytic site. As a phospholipid moves into the channel, an enzyme-bound Ca^{2+} ion (green) binds to the head group, positioning the ester bond to be cleaved next to the catalytic site. [Part (a) adapted from M. H. Gelb et al., 1999, *Curr. Opin. Struc. Biol.* **9**:428. Part (b), see D. Blow, 1991, *Nature* **351**:444.]

induces a small conformational change in phospholipase A_2 that fixes the protein to the phospholipid heads and opens the hydrophobic channel. As a phospholipid molecule diffuses from the bilayer into the channel, the enzyme-bound Ca^{2+} binds to the phosphate in the head group, thereby positioning the ester bond to be cleaved next to the catalytic site (Figure 5-17b).

The Plasma Membrane Has Many Common Functions in All Cells

Although the lipid composition of a membrane largely determines its physical characteristics, its complement of proteins is primarily responsible for a membrane's functional properties. We have alluded to many functions of the plasma membrane in the preceding discussion and briefly consider its major functions here.

In all cells, the plasma membrane acts as a permeability barrier that prevents the entry of unwanted materials from the extracellular milieu and the exit of needed metabolites. Specific **membrane transport proteins** in the plasma membrane permit the passage of nutrients into the cell and metabolic wastes out of it; others function to maintain the proper ionic composition and pH (≈ 7.2) of the cytosol. The structure and function of proteins that make the plasma membrane selectively permeable to different molecules are discussed in Chapter 7.

The plasma membrane is highly permeable to water but poorly permeable to salts and small molecules such as sugars and amino acids. Owing to **osmosis,** water moves across such a semipermeable membrane from a solution of low solute (high water) concentration to one of high solute (low water) concentration until the total solute concentrations and thus the water concentrations on both sides are equal. Figure 5-18 illustrates the effect on animal cells of different external ion concentrations. When most animal cells are placed in an **isotonic** solution (i.e., one with total concentration of solutes equal to that of the cell interior), there is no net movement of water into or out of cells. However, when cells are placed in a **hypotonic** solution (i.e., one with a lower solute concentration than that of the cell interior), water flows into the cells, causing them to swell. Conversely, in a **hypertonic** solution (i.e., one with a higher solute concentration than that of the cell interior), water flows out of cells, causing them to shrink. Under normal in vivo conditions, ion channels in the plasma membrane control the movement of ions into and out of cells so that there is no net movement of water and the usual cell volume is maintained.

Unlike animal cells, bacterial, fungal, and plant cells are surrounded by a rigid cell wall and lack the extracellular matrix found in animal tissues. The plasma membrane is intimately engaged in the assembly of cell walls, which in plants are built primarily of **cellulose.** The cell wall prevents the swelling or shrinking of a cell that would otherwise occur when it is placed in a hypotonic or hyper-

(a) Isotonic medium

(b) Hypotonic medium

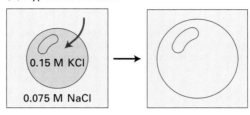

(c) Hypertonic medium

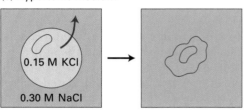

▲ **FIGURE 5-18 Effect of external ion concentration on water flow across the plasma membrane of an animal cell.** Sodium, potassium, and chloride ions do not move freely across the plasma membrane, but water channels (aquaporins) in the membrane permit the flow of water in the direction dictated by the ion concentration of the surrounding medium. (a) When the medium is isotonic, there is no net flux of water into or out of the cell. (b) When the medium is hypotonic, water flows into the cell (red arrow) until the ion concentration inside and outside the cell is the same. Because of the influx of water, the cell volume increases. (c) When the medium is hypertonic, water flows out of the cell until the ion concentration inside and outside the cell is the same. Because water is lost, the cell volume decreases.

tonic medium, respectively. For this reason, cells surrounded by a wall can grow in media having an osmotic strength much less than that of the cytosol. The properties, function, and formation of the plant cell wall are covered in Chapter 6.

In addition to these universal functions, the plasma membrane has other crucial roles in multicellular organisms. Few of the cells in multicellular plants and animals exist as isolated entities; rather, groups of cells with related specializations combine to form tissues. In animal cells, specialized areas of the plasma membrane contain proteins and glycolipids that form specific junctions between cells to strengthen tissues and to allow the exchange of metabolites between cells. Certain plasma-membrane proteins anchor cells to components of the extracellular matrix, the mixture of fibrous proteins and polysaccharides that provides a bedding on which most sheets of epithelial cells or small glands lie. We examine both of these membrane functions in Chapter 6. Still other proteins in the plasma membrane act as anchoring points for many of the cytoskeletal fibers that permeate the cytosol, imparting shape and strength to cells (see Section 5.4).

The plasma membranes of many types of eukaryotic cells also contain receptor proteins that bind specific signaling molecules (e.g., hormones, growth factors, neurotransmitters), leading to various cellular responses. These proteins, which are critical for cell development and functioning, are described in several later chapters. Finally, peripheral cytosolic proteins that are recruited to the membrane surface function as enzymes, intracellular signal transducers, and structural proteins for stabilizing the membrane.

Like the plasma membrane, the membrane surrounding each organelle in eukaryotic cells contains a unique set of proteins essential for its proper functioning. In the next section, we provide a brief overview of the main eukaryotic organelles.

KEY CONCEPTS OF SECTION 5.2

Biomembranes: Protein Components and Basic Functions

■ Biological membranes usually contain both integral (transmembrane) and peripheral membrane proteins, which do not enter the hydrophobic core of the bilayer (see Figure 5-11).

■ Most integral membrane proteins contain one or more membrane-spanning hydrophobic α helices and hydrophilic domains that extend from the cytosolic and exoplasmic faces of the membrane (see Figure 5-12).

■ The porins, unlike other integral proteins, contain membrane-spanning β sheets that form a barrel-like channel through the bilayer.

■ Long-chain lipids attached to certain amino acids anchor some proteins to one or the other membrane leaflet (see Figure 5-15).

■ Some peripheral proteins associate with the membrane by interactions with integral proteins. Lipid-binding motifs in other peripheral proteins interact with the polar head groups of membrane phospholipids (see Table 5-3).

■ The binding of a water-soluble enzyme (e.g., a phospholipase, kinase, or phosphatase) to a membrane surface brings the enzyme close to its substrate and in some cases activates it. Such interfacial binding is due to the attraction between positive charges on basic residues in the protein and negative charges on phospholipid head groups in the bilayer.

5.3 Organelles of the Eukaryotic Cell

The cell is in a dynamic flux. In the light microscope, a live cell exhibits myriad movements ranging from the translocation of chromosomes and vesicles to the changes in shape associated with cell crawling and swimming. Investigation of intracellular structures begins with micrographs of fixed, sectioned cells in which all cell movements are frozen. Such static pictures of the cell reveal the organization of the cytoplasm into compartments and the stereotypic location of each type of organelle within the cell. In this section, we describe the basic structures and functions of the major organelles in animal and plant cells (Figure 5-19). Plant and fungal cells contain most of the organelles found in an animal cell but lack lysosomes. Instead, they contain a large central vacuole that subserves many of the functions of a lysosome. A plant cell also contains chloroplasts, and its membrane is strengthened by a rigid cell wall. Unique proteins in the interior and membranes of each type of organelle largely determine its specific functional characteristics, which are examined in more detail in later chapters. Those organelles bounded by a single membrane are covered first, followed by the three types that have a double membrane—the nucleus, mitochondrion, and chloroplast.

Endosomes Take Up Soluble Macromolecules from the Cell Exterior

Although transport proteins in the plasma membrane mediate the movement of ions and small molecules across the lipid bilayer, proteins and some other soluble macromolecules in the extracellular milieu are internalized by **endocytosis**. In this process, a segment of the plasma membrane invaginates into a "coated pit," whose cytosolic face is lined by a specific set of proteins including **clathrin**. The pit pinches from the membrane into a small membrane-bounded vesicle that contains extracellular material and is delivered to an early **endosome**, a sorting station of membrane-limited tubules and vesicles (Figure 5-20a, b). From this compartment, some membrane proteins are recycled back to the plasma membrane; other membrane proteins are transported to a late endosome where further sorting takes place. The endocytic pathway ends when a late endosome delivers its membrane and internal contents to lysosomes for degradation. The entire endocytic pathway is described in some detail in Chapter 17.

Lysosomes Are Acidic Organelles That Contain a Battery of Degradative Enzymes

Lysosomes provide an excellent example of the ability of intracellular membranes to form closed compartments in which the composition of the lumen (the aqueous interior of the compartment) differs substantially from that of the surrounding cytosol. Found exclusively in animal cells,

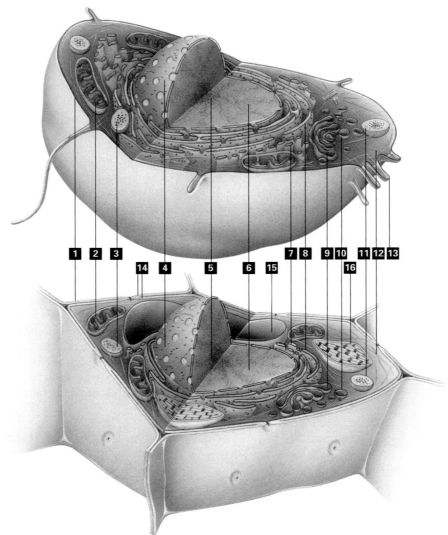

1 Plasma membrane controls movement of molecules in and out of the cell and functions in cell-cell signaling and cell adhesion.

2 Mitochondria, which are surrounded by a double membrane, generate ATP by oxidation of glucose and fatty acids.

3 Lysosomes, which have an acidic lumen, degrade material internalized by the cell and worn-out cellular membranes and organelles.

4 Nuclear envelope, a double membrane, encloses the contents of the nucleus; the outer nuclear membrane is continuous with the rough ER.

5 Nucleolus is a nuclear subcompartment where most of the cell's rRNA is synthesized.

6 Nucleus is filled with chromatin composed of DNA and proteins; in dividing cells is site of mRNA and tRNA synthesis.

7 Smooth endoplasmic reticulum (ER) synthesizes lipids and detoxifies certain hydrophobic compounds.

8 Rough endoplasmic reticulum (ER) functions in the synthesis, processing, and sorting of secreted proteins, lysosomal proteins, and certain membrane.

9 Golgi complex processes and sorts secreted proteins, lysosomal proteins, and membrane proteins synthesized on the rough ER.

10 Secretory vesicles store secreted proteins and fuse with the plasma membrane to release their contents.

11 Peroxisomes detoxify various molecules and also break down fatty acids to produce acetyl groups for biosynthesis.

12 Cytoskeletal fibers form networks and bundles that support cellular membranes, help organize organelles, and participate in cell movement.

13 Microvilli increase surface area for absorption of nutrients from surrounding medium.

14 Cell wall, composed largely of cellulose, helps maintain the cell's shape and provides protection against mechanical stress.

15 Vacuole stores water, ions, and nutrients, degrades macromolecules, and functions in cell elongation during growth.

16 Chloroplasts, which carry out photosynthesis, are surrounded by a double membrane and contain a network of internal membrane-bounded sacs.

▲ **FIGURE 5-19 Schematic overview of a "typical" animal cell and plant cell and their major substructures.** Not every cell will contain all the organelles, granules, and fibrous structures shown here, and other substructures can be present in some. Cells also differ considerably in shape and in the prominence of various organelles and substructures.

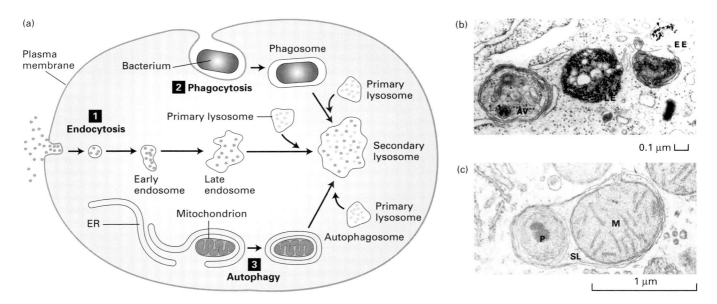

▲ **FIGURE 5-20 Cellular structures that participate in delivering materials to lysosomes.** (a) Schematic overview of three pathways by which materials are moved to lysosomes. Soluble macromolecules are taken into the cell by invagination of coated pits in the plasma membrane and delivered to lysosomes through the endocytic pathway (**1**). Whole cells and other large, insoluble particles move from the cell surface to lysosomes through the phagocytic pathway (**2**). Worn-out organelles and bulk cytoplasm are delivered to lysosomes through the autophagic pathway (**3**). Within the acidic lumen of lysosomes, hydrolytic enzymes degrade proteins, nucleic acids, and other large molecules. (b) An electron micrograph of a section of a cultured mammalian cell that had taken up small gold particles coated with the egg protein ovalbumin. Gold-labeled ovalbumin (black spots) is found in early endosomes (EE) and late endosomes (LE), but very little is present in autophagosomes (AV). (c) Electron micrograph of a section of a rat liver cell showing a secondary lysosome containing fragments of a mitochondrion (M) and a peroxisome (P). [Part (b) from T. E. Tjelle et al., 1996, *J. Cell Sci.* **109**:2905. Part (c) courtesy of D. Friend.]

lysosomes are responsible for degrading certain components that have become obsolete for the cell or organism. The process by which an aged organelle is degraded in a lysosome is called *autophagy* ("eating oneself"). Materials taken into a cell by endocytosis or **phagocytosis** also may be degraded in lysosomes (see Figure 5-20a). In phagocytosis, large, insoluble particles (e.g., bacteria) are enveloped by the plasma membrane and internalized.

Lysosomes contain a group of enzymes that degrade polymers into their monomeric subunits. For example, nucleases degrade RNA and DNA into their mononucleotide building blocks; proteases degrade a variety of proteins and peptides; phosphatases remove phosphate groups from mononucleotides, phospholipids, and other compounds; still other enzymes degrade complex polysaccharides and glycolipids into smaller units. All the lysosomal enzymes work most efficiently at acid pH values and collectively are termed *acid hydrolases.* Two types of transport proteins in the lysosomal membrane work together to pump H^+ and Cl^- ions (HCl) from the cytosol across the membrane, thereby acidifying the lumen (see Figure 7-10b). The acid pH helps to denature proteins, making them accessible to the action of the lysosomal hydrolases, which themselves are resistant to acid denaturation. Lysosomal enzymes are poorly active at the neutral pH of cells and most extracellular fluids. Thus, if a lysosome releases its enzymes into the cytosol, where the pH is between 7.0 and 7.3, they cause little degradation of cytosolic components. Cytosolic and nuclear proteins generally are not degraded in lysosomes but rather in proteasomes, large multiprotein complexes in the cytosol (see Figure 3-13).

Lysosomes vary in size and shape, and several hundred may be present in a typical animal cell. In effect, they function as sites where various materials to be degraded collect. *Primary lysosomes* are roughly spherical and do not contain obvious particulate or membrane debris. *Secondary lysosomes,* which are larger and irregularly shaped, appear to result from the fusion of primary lysosomes with other membrane-bounded organelles and vesicles. They contain particles or membranes in the process of being digested (Figure 5-20c).

 Tay-Sachs disease is caused by a defect in one enzyme catalyzing a step in the lysosomal breakdown of gangliosides. The resulting accumulation of these glycolipids, especially in nerve cells, has devastating consequences. The symptoms of this inherited disease are usually evident before the age of 1. Affected children commonly become demented and blind by age 2 and die before their third birthday. Nerve cells from such children are greatly enlarged with swollen lipid-filled lysosomes. ■

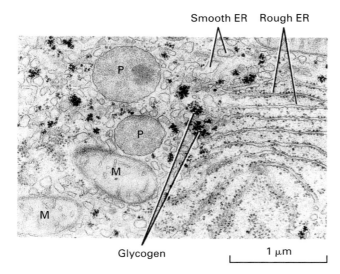

Smooth ER   Rough ER

Glycogen   1 μm

▲ **FIGURE 5-21 Electron micrograph showing various organelles in a rat liver cell.** Two peroxisomes (P) lie in close proximity to mitochondria (M) and the rough and smooth endoplasmic reticulum (ER). Also visible are accumulations of glycogen, a polysaccharide that is the primary glucose-storage molecule in animals. [Courtesy of P. Lazarow.]

Peroxisomes Degrade Fatty Acids and Toxic Compounds

All animal cells (except erythrocytes) and many plant cells contain **peroxisomes,** a class of roughly spherical organelles, 0.2–1.0 μm in diameter (Figure 5-21). Peroxisomes contain several *oxidases*—enzymes that use molecular oxygen to oxidize organic substances, in the process forming hydrogen peroxide (H_2O_2), a corrosive substance. Peroxisomes also contain copious amounts of the enzyme *catalase,* which degrades hydrogen peroxide to yield water and oxygen:

$$2\,H_2O_2 \xrightarrow{\text{Catalase}} 2\,H_2O + O_2$$

In contrast with the oxidation of fatty acids in mitochondria, which produces CO_2 and is coupled to the generation of ATP, peroxisomal oxidation of fatty acids yields acetyl groups and is not linked to ATP formation (see Figure 8-11). The energy released during peroxisomal oxidation is converted into heat, and the acetyl groups are transported into the cytosol, where they are used in the synthesis of cholesterol and other metabolites. In most eukaryotic cells, the peroxisome is the principal organelle in which fatty acids are oxidized, thereby generating precursors for important biosynthetic pathways. Particularly in liver and kidney cells, various toxic molecules that enter the bloodstream also are degraded in peroxisomes, producing harmless products.

 In the human genetic disease *X-linked adrenoleukodystrophy (ADL),* peroxisomal oxidation of very long chain fatty acids is defective. The *ADL* gene encodes the peroxisomal membrane protein that transports into peroxisomes an enzyme required for the oxidation of these fatty acids. Persons with the severe form of ADL are unaffected until midchildhood, when severe neurological disorders appear, followed by death within a few years. ∎

 Plant seeds contain *glyoxisomes,* small organelles that oxidize stored lipids as a source of carbon and energy for growth. They are similar to peroxisomes and contain many of the same types of enzymes as well as additional ones used to convert fatty acids into glucose precursors. ∎

The Endoplasmic Reticulum Is a Network of Interconnected Internal Membranes

Generally, the largest membrane in a eukaryotic cell encloses the **endoplasmic reticulum (ER)**—an extensive network of closed, flattened membrane-bounded sacs called **cisternae** (see Figure 5-19). The endoplasmic reticulum has a number of functions in the cell but is particularly important in the synthesis of lipids, membrane proteins, and secreted proteins. The *smooth endoplasmic reticulum* is smooth because it lacks ribosomes. In contrast, the cytosolic face of the *rough endoplasmic reticulum* is studded with ribosomes.

The Smooth Endoplasmic Reticulum   The synthesis of fatty acids and phospholipids takes place in the smooth ER. Although many cells have very little smooth ER, this organelle is abundant in hepatocytes. Enzymes in the smooth ER of the liver also modify or detoxify hydrophobic chemicals such as pesticides and carcinogens by chemically converting them into more water-soluble, conjugated products that can be excreted from the body. High doses of such compounds result in a large proliferation of the smooth ER in liver cells.

The Rough Endoplasmic Reticulum   Ribosomes bound to the rough ER synthesize certain membrane and organelle proteins and virtually all proteins to be secreted from the cell (Chapter 16). A ribosome that fabricates such a protein is bound to the rough ER by the nascent polypeptide chain of the protein. As the growing polypeptide emerges from the ribosome, it passes through the rough ER membrane, with the help of specific proteins in the membrane. Newly made membrane proteins remain associated with the rough ER membrane, and proteins to be secreted accumulate in the lumen of the organelle.

All eukaryotic cells contain a discernible amount of rough ER because it is needed for the synthesis of plasma-membrane proteins and proteins of the extracellular matrix. Rough ER is particularly abundant in specialized cells that produce an abundance of specific proteins to be secreted. For example, plasma cells produce antibodies, pancreatic acinar cells synthesize digestive enzymes, and cells in the pancreatic islets of Langerhans produce the polypeptide hormones insulin and glucagon. In these secretory cells and others, a large part of the cytosol is filled with rough ER and secretory vesicles (Figure 5-22).

(a)

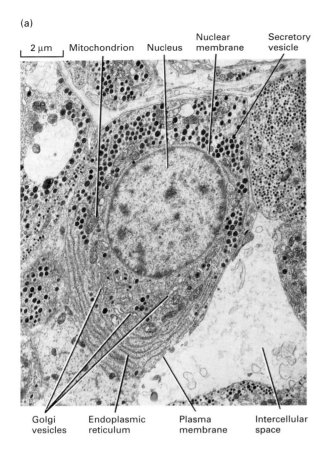

2 μm   Mitochondrion   Nucleus   Nuclear membrane   Secretory vesicle

Golgi vesicles   Endoplasmic reticulum   Plasma membrane   Intercellular space

(b)

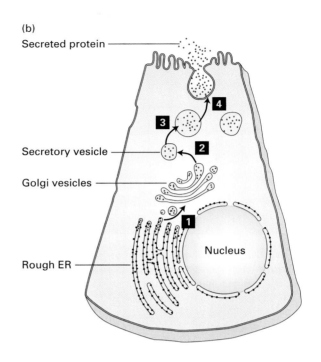

Secreted protein

Secretory vesicle

Golgi vesicles

Rough ER

Nucleus

▲ FIGURE 5-22 **Charateristic features of cells specialized to secrete large amounts of particular proteins (e.g., hormones, antibodies).** (a) Electron micrograph of a thin section of a hormone-secreting cell from the rat pituitary. One end of the cell (*top*) is filled with abundant rough ER and Golgi sacs, where polypeptide hormones are synthesized and packaged. At the opposite end of the cell (*bottom*) are numerous secretory vesicles, which contain recently made hormones eventually to be secreted. (b) Diagram of a typical secretory cell tracing the pathway followed by a protein (small red dots) to be secreted. Immediately after their synthesis on ribosomes (blue dots) of the rough ER, secreted proteins are found in the lumen of the rough ER. Transport vesicles bud off and carry these proteins to the Golgi complex (**1**), where the proteins are concentrated and packaged into immature secretory vesicles (**2**). These vesicles then coalesce to form larger mature secretory vesicles that lose water to the cytosol, leaving an almost crystalline mixture of secreted proteins in the lumen (**3**). After these vesicles accumulate under the apical surface, they fuse with the plasma membrane and release their contents (exocytosis) in response to appropriate hormonal or nerve stimulation (**4**). [Part (a) courtesy of Biophoto Associates.]

The Golgi Complex Processes and Sorts Secreted and Membrane Proteins

Several minutes after proteins are synthesized in the rough ER, most of them leave the organelle within small membrane-bounded transport vesicles. These vesicles, which bud from regions of the rough ER not coated with ribosomes, carry the proteins to another membrane-limited organelle, the **Golgi complex** (see Figure 5-22).

Three-dimensional reconstructions from serial sections of a Golgi complex reveal this organelle to be a series of flattened membrane vesicles or sacs (cisternae), surrounded by a number of more or less spherical membrane-limited vesicles (Figure 5-23). The stack of Golgi cisternae has three de-

fined regions—the *cis*, the *medial*, and the *trans*. Transport vesicles from the rough ER fuse with the *cis* region of the Golgi complex, where they deposit their protein contents. As detailed in Chapter 17, these proteins then progress from the *cis* to the *medial* to the *trans* region. Within each region are different enzymes that modify proteins to be secreted and membrane proteins differently, depending on their structures and their final destinations.

After proteins to be secreted and membrane proteins are modified in the Golgi complex, they are transported out of the complex by a second set of vesicles, which seem to bud from the trans side of the Golgi complex. Some vesicles carry membrane proteins destined for the plasma membrane or soluble proteins to be released from the cell surface; others

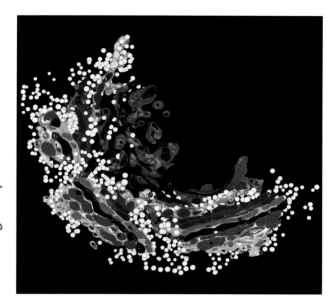

▲ **FIGURE 5-23 Model of the Golgi complex based on three-dimensional reconstruction of electron microscopy images.** Transport vesicles (white spheres) that have budded off the rough ER fuse with the *cis* membranes (light blue) of the Golgi complex. By mechanisms described in Chapter 17, proteins move from the *cis* region to the *medial* region and finally to the *trans* region of the Golgi complex. Eventually, vesicles bud off the *trans*-Golgi membranes (orange and red); some move to the cell surface and others move to lysosomes. The Golgi complex, like the rough endoplasmic reticulum, is especially prominent in secretory cells. [From B. J. Marsh et al., 2001, *Proc Nat'l. Acad. Sci USA* **98**:2399.]

carry soluble or membrane proteins to lysosomes or other organelles. How intracellular transport vesicles "know" with which membranes to fuse and where to deliver their contents is also discussed in Chapter 17.

Plant Vacuoles Store Small Molecules and Enable a Cell to Elongate Rapidly

Most plant cells contain at least one membrane-limited internal vacuole. The number and size of vacuoles depend on both the type of cell and its stage of development; a single vacuole may occupy as much as 80 percent of a mature plant cell (Figure 5-24). A variety of transport proteins in the vacuolar membrane allow plant cells to accumulate and store water, ions, and nutrients (e.g., sucrose, amino acids) within vacuoles (Chapter 7). Like a lysosome, the lumen of a vacuole contains a battery of degradative enzymes and has an acidic pH, which is maintained by similar transport proteins in the vacuolar membrane. Thus plant vacuoles may also have a degradative

function similar to that of lysosomes in animal cells. Similar storage vacuoles are found in green algae and many microorganisms such as fungi.

Like most cellular membranes, the vacuolar membrane is permeable to water but is poorly permeable to the small molecules stored within it. Because the solute concentration is much higher in the vacuole lumen than in the cytosol or extracellular fluids, water tends to move by osmotic flow into vacuoles, just as it moves into cells placed in a hypotonic medium (see Figure 5-18). This influx of water causes both the vacuole to expand and water to move into the cell, creating hydrostatic pressure, or *turgor*, inside the cell. This pressure is balanced by the mechanical resistance of the cellulose-containing cell walls that surround plant cells. Most plant cells have a turgor of 5–20 atmospheres (atm); their cell walls must be strong enough to react to this pressure in a controlled way. Unlike animal cells, plant cells can elongate extremely rapidly, at rates of 20–75 μm/h. This elongation,

(a)

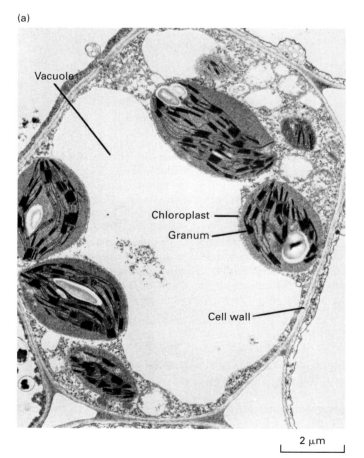

2 μm

▲ **FIGURE 5-24 Electron micrograph of a thin section of a leaf cell.** In this cell, a single large vacuole occupies much of the cell volume. Parts of five chloroplasts and the cell wall also are visible. Note the internal subcompartments in the chloroplasts. [Courtesy of Biophoto Associates/Myron C. Ledbetter/Brookhaven National Laboratory.]

which usually accompanies plant growth, occurs when a segment of the somewhat elastic cell wall stretches under the pressure created by water taken into the vacuole. ∎

The Nucleus Contains the DNA Genome, RNA Synthetic Apparatus, and a Fibrous Matrix

The **nucleus,** the largest organelle in animal cells, is surrounded by two membranes, each one a phospholipid bilayer containing many different types of proteins. The inner nuclear membrane defines the nucleus itself. In most cells, the outer nuclear membrane is continuous with the rough endoplasmic reticulum, and the space between the inner and outer nuclear membranes is continuous with the lumen of the rough endoplasmic reticulum (see Figure 5-19). The two nuclear membranes appear to fuse at **nuclear pores,** the ringlike complexes composed of specific membrane proteins through which material moves between the nucleus and the cytosol. The structure of nuclear pores and the regulated transport of material through them are detailed in Chapter 12.

In a growing or differentiating cell, the nucleus is metabolically active, replicating DNA and synthesizing rRNA,

tRNA, and mRNA. Within the nucleus mRNA binds to specific proteins, forming ribonucleoprotein particles. Most of the cell's ribosomal RNA is synthesized in the **nucleolus,** a subcompartment of the nucleus that is not bounded by a phospholipid membrane (Figure 5-25). Some ribosomal proteins are added to ribosomal RNAs within the nucleolus as well. The finished or partly finished ribosomal subunits, as well as tRNAs and mRNA-containing particles, pass through a nuclear pore into the cytosol for use in protein synthesis (Chapter 4). In mature erythrocytes from nonmammalian vertebrates and other types of "resting" cells, the nucleus is inactive or dormant and minimal synthesis of DNA and RNA takes place.

How nuclear DNA is packaged into chromosomes is described in Chapter 10. In a nucleus that is not dividing, the chromosomes are dispersed and not dense enough to be observed in the light microscope. Only during cell division are individual chromosomes visible by light microscopy. In the electron microscope, the nonnucleolar regions of the nucleus, called the *nucleoplasm,* can be seen to have dark- and light-staining areas. The dark areas, which are often closely associated with the nuclear membrane, contain condensed concentrated DNA, called **heterochromatin** (see Figure 5-25). Fibrous proteins called **lamins** form a two-dimensional network along the inner surface of the inner membrane, giving it shape and apparently binding DNA to it. The breakdown of this network occurs early in cell division, as we detail in Chapter 21.

Mitochondria Are the Principal Sites of ATP Production in Aerobic Cells

Most eukaryotic cells contain many **mitochondria,** which occupy up to 25 percent of the volume of the cytoplasm. These complex organelles, the main sites of ATP production during aerobic metabolism, are generally exceeded in size only by the nucleus, vacuoles, and chloroplasts.

The two membranes that bound a mitochondrion differ in composition and function. The outer membrane, composed of about half lipid and half protein, contains porins (see Figure 5-14) that render the membrane permeable to molecules having molecular weights as high as 10,000. In this respect, the outer membrane is similar to the outer membrane of gram-negative bacteria. The inner membrane, which is much less permeable, is about 20 percent lipid and 80 percent protein—a higher proportion of protein than exists in other cellular membranes. The surface area of the inner membrane is greatly increased by a large number of infoldings, or *cristae,* that protrude into the *matrix,* or central space (Figure 5-26).

In nonphotosynthetic cells, the principal fuels for ATP synthesis are fatty acids and glucose. The complete aerobic degradation of glucose to CO_2 and H_2O is coupled to the synthesis of as many as 30 molecules of ATP. In eukaryotic cells, the initial stages of glucose degradation take place in

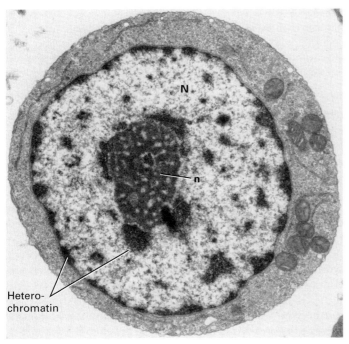

▲ **FIGURE 5-25 Electron micrograph of a thin section of a bone marrow stem cell.** The nucleolus (n) is a subcompartment of the nucleus (N) and is not surrounded by a membrane. Most ribosomal RNA is produced in the nucleolus. Darkly staining areas in the nucleus outside the nucleolus are regions of heterochromatin. [From P. C. Cross and K. L. Mercer, 1993, *Cell and Tissue Ultrastructure,* W. H. Freeman and Company, p. 165.]

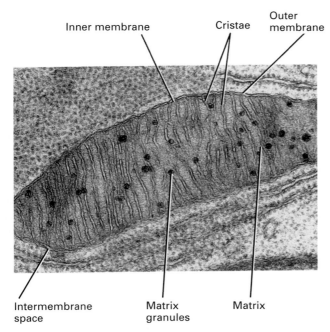

Inner membrane   Cristae   Outer membrane

Intermembrane space   Matrix granules   Matrix

▲ **FIGURE 5-26 Electron micrograph of a mitochondrion.** Most ATP production in nonphotosynthetic cells takes place in mitochondria. The inner membrane, which surrounds the matrix space, has many infoldings, called cristae. Small calcium-containing matrix granules also are evident. [From D. W. Fawcett, 1981, *The Cell*, 2d ed., Saunders, p. 421.]

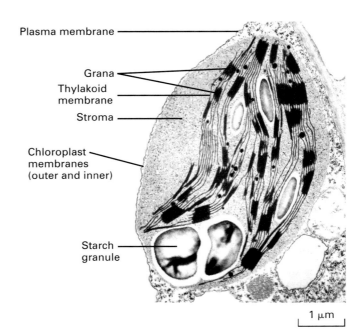

Plasma membrane

Grana

Thylakoid membrane

Stroma

Chloroplast membranes (outer and inner)

Starch granule

1 µm

▲ **FIGURE 5-27 Electron micrograph of a plant chloroplast.** The internal membrane vesicles (thylakoids) are fused into stacks (grana), which reside in a matrix (the stroma). All the chlorophyll in the cell is contained in the thylakoid membranes, where the light-induced production of ATP takes place during photosynthesis. [Courtesy of Biophoto Associates/M. C. Ledbetter/Brookhaven National Laboratory.]

the cytosol, where 2 ATP molecules per glucose molecule are generated. The terminal stages of oxidation and the coupled synthesis of ATP are carried out by enzymes in the mitochondrial matrix and inner membrane (Chapter 8). As many as 28 ATP molecules per glucose molecule are generated in mitochondria. Similarly, virtually all the ATP formed in the oxidation of fatty acids to CO_2 is generated in mitochondria. Thus mitochondria can be regarded as the "power plants" of the cell.

Chloroplasts Contain Internal Compartments in Which Photosynthesis Takes Place

Except for vacuoles, **chloroplasts** are the largest and the most characteristic organelles in the cells of plants and green algae. They can be as long as 10 µm and are typically 0.5–2 µm thick, but they vary in size and shape in different cells, especially among the algae. In addition to the double membrane that bounds a chloroplast, this organelle also contains an extensive internal system of interconnected membrane-limited sacs called **thylakoids,** which are flattened to form disks (Figure 5-27). Thylakoids often form stacks called *grana* and are embedded in a matrix, the *stroma*. The thylakoid membranes contain green pigments

(chlorophylls) and other pigments that absorb light, as well as enzymes that generate ATP during photosynthesis. Some of the ATP is used to convert CO_2 into three-carbon intermediates by enzymes located in the stroma; the intermediates are then exported to the cytosol and converted into sugars. ∎

The molecular mechanisms by which ATP is formed in mitochondria and chloroplasts are very similar, as explained in Chapter 8. Chloroplasts and mitochondria have other features in common: both often migrate from place to place within cells, and they contain their own DNA, which encodes some of the key organellar proteins (Chapter 10). The proteins encoded by mitochondrial or chloroplast DNA are synthesized on ribosomes within the organelles. However, most of the proteins in each organelle are encoded in nuclear DNA and are synthesized in the cytosol; these proteins are then incorporated into the organelles by processes described in Chapter 16.

KEY CONCEPTS OF SECTION 5.3

Organelles of the Eukaryotic Cell

■ All eukaryotic cells contain a nucleus and numerous other organelles in their cytosols (see Figure 5-19).

- The nucleus, mitochondrion, and chloroplast are bounded by two bilayer membranes separated by an intermembrane space. All other organelles are surrounded by a single membrane.

- Endosomes internalize plasma-membrane proteins and soluble materials from the extracellular medium, and they sort them back to the membranes or to lysosomes for degradation.

- Lysosomes have an acidic interior and contain various hydrolases that degrade worn-out or unneeded cellular components and some ingested materials (see Figure 5-20).

- Peroxisomes are small organelles containing enzymes that oxidize various organic compounds without the production of ATP. By-products of oxidation are used in biosynthetic reactions.

- Secreted proteins and membrane proteins are synthesized on the rough endoplasmic reticulum, a network of flattened membrane-bounded sacs studded with ribosomes.

- Proteins synthesized on the rough ER first move to the Golgi complex, where they are processed and sorted for transport to the cell surface or other destination (see Figure 5-22).

- Plant cells contain one or more large vacuoles, which are storage sites for ions and nutrients. Osmotic flow of water into vacuoles generates turgor pressure that pushes the plasma membrane against the cell wall.

- The nucleus houses the genome of a cell. The inner and outer nuclear membranes are fused at numerous nuclear pores, through which materials pass between the nucleus and the cytosol. The outer nuclear membrane is continuous with that of the rough endoplasmic reticulum.

- Mitochondria have a highly permeable outer membrane and a protein-enriched inner membrane that is extensively folded. Enzymes in the inner mitochondrial membrane and central matrix carry out the terminal stages of sugar and lipid oxidation coupled to ATP synthesis.

- Chloroplasts contain a complex system of thylakoid membranes in their interiors. These membranes contain the pigments and enzymes that absorb light and produce ATP during photosynthesis.

5.4 The Cytoskeleton: Components and Structural Functions

The cytosol is a major site of cellular metabolism and contains a large number of different enzymes. Proteins constitute about 20–30 percent of the cytosol by weight, and from a quarter to half of the total protein within cells is in the cytosol. Estimates of the protein concentration in the cytosol range from 200 to 400 mg/ml. Because of the high concentration of cytosolic proteins, complexes of proteins can form even if the energy that stabilizes them is weak. Many inves-

tigators believe that the cytosol is highly organized, with most soluble proteins either bound to filaments or otherwise localized in specific regions. In an electron micrograph of a typical animal cell, soluble proteins packing the cell interior conceal much of the internal structure. If a cell is pretreated with a nonionic detergent (e.g., Triton X-100), which permeabilizes the membrane, soluble cytosolic proteins diffuse away. In micrographs of detergent-extracted animal cells, two types of structures stand out—membrane-limited organelles and the filaments of the cytoskeleton, which fill the cytosol (Figure 5-28).

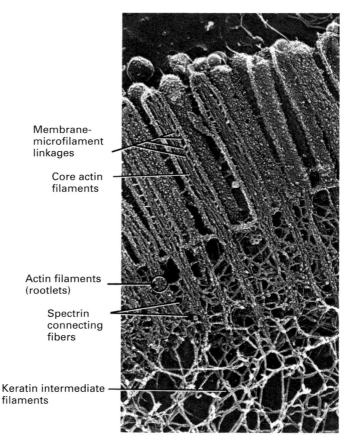

Membrane-microfilament linkages

Core actin filaments

Actin filaments (rootlets)

Spectrin connecting fibers

Keratin intermediate filaments

▲ **FIGURE 5-28 Electron micrograph of the apical part of a detergent-extracted intestinal epithelial cell.** Microvilli, fingerlike projections of the plasma membrane, cover the apical surface of an intestinal epithelial cell. A bundle of microfilaments in the core of each microvillus stabilizes the structure. The plasma membrane surrounding a microvillus is attached to the sides of the bundle by evenly spaced membrane–microfilament linkages (yellow). The bundle continues into the cell as a short rootlet. The rootlets of multiple microvilli are cross-braced by connecting fibers (red) composed of an intestinal isoform of spectrin. This fibrous actin-binding protein is found in a narrow band just below the plasma membrane in many animal cells. The bases of the rootlets are attached to keratin intermediate filaments. These numerous connections anchor the rootlets in a meshwork of filaments and thereby support the upright orientation of the microvilli. [Courtesy of N. Hirokawa.]

In this section, we introduce the protein filaments that compose the cytoskeleton and then describe how they support the plasma and nuclear membranes and organize the contents of the cell. Later chapters will deal with the dynamic properties of the cytoskeleton—its assembly and disassembly and its role in cellular movements.

Three Types of Filaments Compose the Cytoskeleton

The cytosol of a eukaryotic cell contains three types of filaments that can be distinguished on the bases of their diameter, type of subunit, and subunit arrangment (Figure 5-29). **Actin filaments,** also called **microfilaments,** are 8–9 nm in diameter and have a twisted two-stranded structure. **Microtubules** are hollow tubelike structures, 24 nm in diameter, whose walls are formed by adjacent protofilaments. **Intermediate filaments (IFs)** have the structure of a 10-nm-diameter rope.

Each type of cytoskeletal filament is a polymer of protein subunits (Table 5-4). Monomeric **actin** subunits assemble into microfilaments; dimeric subunits composed of **α-** and **β-tubulin** polymerize into microtubules. Unlike microfilaments and microtubules, which are assembled from one or two proteins, intermediate filaments are assembled from a large diverse family of proteins. The most common intermediate filaments, found in the nucleus, are composed of *lamins*. Intermediate filaments constructed from other proteins are expressed preferentially in certain tissues: for example, *keratin*-containing filaments in epithelial cells, *desmin*-containing filaments in muscle cells, and *vimentin*-containing filaments in mesenchymal cells.

▶ **FIGURE 5-29 Comparison of the three types of filaments that form the cytoskeleton.** (a) Diagram of the basic structures of an actin filament (AF), intermediate filament (IF), and microtubule (MT). The beadlike structure of an actin filament shows the packing of actin subunits. Intermediate filament subunits pack to form ropes in which the individual subunits are difficult to distinguish. The walls of microtubules are formed from protofilaments of tubulin subunits. (b) Micrograph of a mixture of actin filaments, microtubules, and vimentin intermediate filaments showing the differences in their shape, size, and flexibility. Purified preparations of actin, tubulin, and vimentin subunits were separately polymerized in a test tube to form the corresponding filaments. A mixture of the filaments was applied to a carbon film on a microscope grid and then rinsed with a dilute solution of uranyl acetate (UC), which surrounds but does not penetrate the protein (c). Because uranyl acetate is a heavy metal that easily scatters electrons, areas of the microscope grid occupied by protein produce a "negative" image in metal film when projected onto a photographic plate, as seen in part (b). [Part (b) courtesy of G. Waller and P. Matsudaira.]

Most eukaryotic cells contain all three types of cytoskeletal filaments, often concentrated in distinct locations. For example, in the absorptive epithelial cells that line the lumen of the intestine, actin microfilaments are abundant in the apical region, where they are associated with cell–cell junctions and support a dense carpet of microvilli (Figure 5-30a). Actin filaments are also present in a narrow zone adjacent to the plasma membrane in the lateral regions of these cells. Keratin intermediate filaments,

TABLE 5-4	Protein Subunits in Cytoskeletal Filaments			
Protein Subunits	**MW**	**Expression**	**Function**	
MICROFILAMENTS				
Actin	42,000	Fungi, plant, animal	Structural support, motility	
MreB	36,000	Rod-shaped bacteria	Width control	
MICROTUBULES				
Tubulin (α and β)	58,000	Fungi, plant, animal	Structural support, motility, cell polarity	
FtsZ	58,000	Bacteria	Cell division	
INTERMEDIATE FILAMENTS				
Lamins	Various	Plant, animal	Support for nuclear membrane	
Desmin, keratin, vimentin, others	Various	Animal	Cell adhesion	
OTHER				
MSP	50,000	Nematode sperm	Motility	

(a)

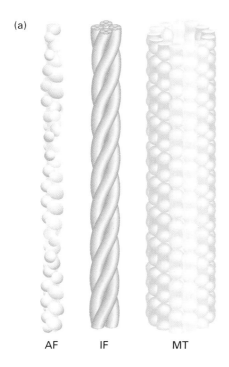

AF   IF   MT

(b)

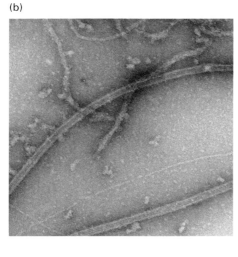

(c)

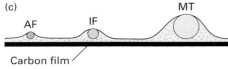

Carbon film

conservation is explained by the variety of critical functions that depend on the cytoskeleton. A mutation in a cytoskeleton protein subunit could disrupt the assembly of filaments and their binding to other proteins. Analyses of gene sequences and protein structures have identified bacterial homologs of actin and tubulin. The absence of IF-like proteins in bacteria and unicellular eukaryotes is evidence that intermediate filaments appeared later in the evolution of the cytoskeletal system. The first IF protein to arise was most likely a nuclear lamin from which cytosolic IF proteins later evolved.

The simple bacterial cytoskeleton controls cell length, width, and the site of cell division. The FtsZ protein, a bacterial homolog of tubulin, is localized around the neck of dividing bacterial cells, suggesting that FtsZ participates in cell division (Figure 5-30b). The results of biochemical experiments with purified FtsZ demonstrate that it can polymerize into protofilaments, but these protofilaments do not assemble into intact microtubules. Another bacterial protein, MreB, has been found to be similar to actin in atomic structure and filament structure—strong evidence that actin evolved from MreB. Clues to the function of MreB include its localization in a filament that girdles rod-shaped bacterial cells, its absence from spherical bacteria, and the finding that mutant cells lacking MreB become wider but not longer. These observations suggest MreB controls the width of rod-shaped bacteria.

(a)   Actin   MTs   IFs

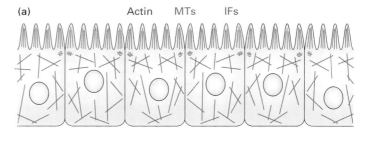

(b)

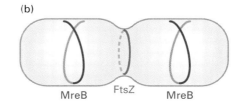

MreB   FtsZ   MreB

▲ **FIGURE 5-30 Schematic depiction of the distribution of cytoskeletal filaments in eukaryotic cells and bacterial cells.** (a) In absorptive epithelial cells, actin filaments (red) are concentrated in the apical region and in a narrow band in the basolateral region. Microtubules (blue) are oriented with the long axis of the cell, and intermediate filaments (green) are concentrated along the cell periphery especially at specialized junctions with neighboring cells and lining the nuclear membrane. (b) In a rod-shaped bacterial cell, filaments of MreB, the bacterial actin homolog, ring the cell and constrict its width. The bacterial tubulin homolog, FtsZ, forms filaments at the site of cell division.

forming a meshwork, connect microvilli and are tethered to junctions between cells. Lamin intermediate filaments support the inner nuclear membrane. Finally, microtubules, aligned with the long axis of the cell, are in close proximity to major cell organelles such as the endoplasmic reticulum, Golgi complex, and vesicles.

The cytoskeleton has been highly conserved in evolution. A comparison of gene sequences shows only a small percentage of differences in sequence between yeast actin and tubulin and human actin and tubulin. This structural

Cytoskeletal Filaments Are Organized into Bundles and Networks

On first looking at micrographs of a cell, one is struck by the dense, seemingly disorganized mat of filaments present in the cytosol. However, a keen eye will start to pick out areas—generally where the membrane protrudes from the cell surface or where a cell adheres to the surface or another cell—in which the filaments are concentrated into bundles. From these bundles, the filaments continue into the cell interior, where they fan out and become part of a network of filaments. These two structures, *bundles* and *networks,* are the most common arrangements of cytoskeletal filaments in a cell.

Structurally, bundles differ from networks mainly in the organization of the filaments. In bundles, the filaments are closely packed in parallel arrays. In a network, the filaments crisscross, often at right angles, and are loosely packed. Networks can be further subdivided. One type, associated with the nuclear and plasma membranes, is planar (two-dimensional), like a net or a web; the other type, present within the cell, is three-dimensional, giving the cytosol gel-like properties. In all bundles and networks, the filaments are held together by various cross-linking proteins.

We will consider various cytoskeletal cross-linking proteins and their functions in Chapters 19 and 20.

Microfilaments and Membrane-Binding Proteins Form a Skeleton Underlying the Plasma Membrane

The distinctive shape of a cell depends on the organization of actin filaments and proteins that connect microfilaments to the membrane. These proteins, called *membrane–microfilament binding proteins,* act as spot welds that tack the actin cytoskeleton framework to the overlying membrane. When attached to a bundle of filaments, the membrane acquires the fingerlike shape of a microvillus or similar projection (see Figure 5-28). When attached to a planar network of filaments, the membrane is held flat like the red blood cell membrane. The simplest membrane–cytoskeleton connections entail the binding of integral membrane proteins directly to actin filaments. More common are complex linkages that connect actin filaments to integral membrane proteins through peripheral membrane proteins that function as adapter proteins. Such linkages between the cytoskeleton and certain plasma-membrane proteins are considered in Chapter 6.

(a)

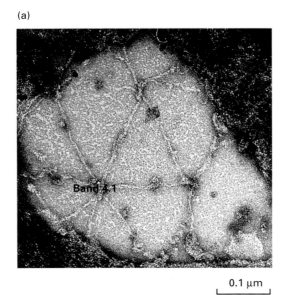

0.1 µm

(b)

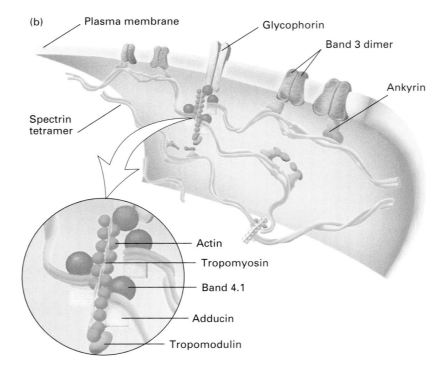

▲ **FIGURE 5-31 Cortical cytoskeleton supporting the plasma membrane in human erythrocytes.** (a) Electron micrograph of the erythrocyte membrane showing the spoke-and-hub organization of the cytoskeleton. The long spokes are composed mainly of spectrin and can be seen to intersect at the hubs, or membrane-attachment sites. The darker spots along the spokes are ankyrin molecules, which cross-link spectrin to integral membrane proteins. (b) Diagram of the erythrocyte cytoskeleton showing the various components. See text for discussion. [Part (a) from T. J. Byers and D. Branton, 1985, *Proc. Nat'l. Acad. Sci. USA* **82**:6153. Courtesy of D. Branton. Part (b) adapted from S. E. Lux, 1979, *Nature* **281**:426, and E. J. Luna and A. L. Hitt, 1992, *Science* **258**:955.]

The richest area of actin filaments in many cells lies in the *cortex*, a narrow zone just beneath the plasma membrane. In this region, most actin filaments are arranged in a network that excludes most organelles from the cortical cytoplasm. Perhaps the simplest cytoskeleton is the two-dimensional network of actin filaments adjacent to the erythrocyte plasma membrane. In more complicated cortical cytoskeletons, such as those in platelets, epithelial cells, and muscle, actin filaments are part of a three-dimensional network that fills the cytosol and anchors the cell to the substratum.

A red blood cell must squeeze through narrow blood capillaries without rupturing its membrane. The strength and flexibility of the erythrocyte plasma membrane depend on a dense cytoskeletal network that underlies the entire membrane and is attached to it at many points. The primary component of the erythrocyte cytoskeleton is *spectrin*, a 200-nm-long fibrous protein. The entire cytoskeleton is arranged in a spoke-and-hub network (Figure 5-31a). Each spoke is composed of a single spectrin molecule, which extends from two hubs and cross-links them. Each hub comprises a short (14-subunit) actin filament plus adducin, tropomyosin, and tropomodulin (Figure 5-31b, *inset*). The last two proteins strengthen the network by preventing the actin filament from depolymerizing. Six or seven spokes radiate from each hub, suggesting that six or seven spectrin molecules are bound to the same actin filament.

To ensure that the erythrocyte retains its characteristic shape, the spectrin-actin cytoskeleton is firmly attached to the overlying erythrocyte plasma membrane by two peripheral membrane proteins, each of which binds to a specific integral membrane protein and to membrane phospholipids. *Ankyrin* connects the center of spectrin to band 3 protein, an anion-transport protein in the membrane. *Band 4.1 protein*, a component of the hub, binds to the integral membrane protein glycophorin, whose structure was discussed previously (see Figure 5-12). Both ankyrin and band 4.1 protein also contain lipid-binding motifs, which help bind them to the membrane (see Table 5-3). The dual binding by ankyrin and band 4.1 ensures that the membrane is connected to both the spokes and the hubs of the spectrin-actin cytoskeleton (see Figure 5-31b).

Intermediate Filaments Support the Nuclear Membrane and Help Connect Cells into Tissues

Intermediate filaments typically crisscross the cytosol, forming an internal framework that stretches from the nuclear envelope to the plasma membrane (Figure 5-32). A network of intermediate filaments is located adjacent to some cellular membranes, where it provides mechanical support. For example, lamin A and lamin C filaments form an orthogonal lattice that is associated with lamin B. The entire supporting structure, called the **nuclear lamina,** is anchored to the inner nuclear membrane by prenyl anchors on lamin B.

At the plasma membrane, intermediate filaments are attached by adapter proteins to specialized cell junctions called

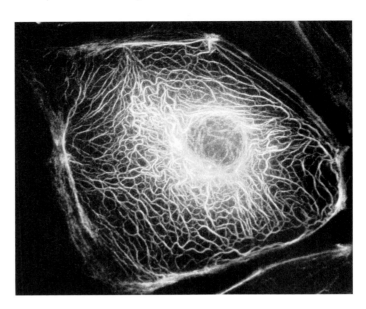

▲ **FIGURE 5-32 Fluorescence micrograph of a PtK2 fibroblast cell stained to reveal keratin intermediate filaments.** A network of filaments crisscrosses the cell from the nucleus to the plasma membrane. At the plasma membrane, the filaments are linked by adapter proteins to two types of anchoring junctions: desmosomes between adjacent cells and hemidesmosomes between the cell and the matrix. [Courtesy of R. D. Goldman.]

desmosomes and *hemidesmosomes,* which mediate cell–cell adhesion and cell–matrix adhesion, respectively, particularly in epithelial tissues. In this way, intermediate filaments in one cell are indirectly connected to intermediate filaments in a neighboring cell or to the extracellular matrix. Because of the important role of cell junctions in cell adhesion and the stability of tissues, we consider their structure and relation to cytoskeletal filaments in detail in Chapter 6.

Microtubules Radiate from Centrosomes and Organize Certain Subcellular Structures

Like microfilaments and intermediate filaments, microtubules are not randomly distributed in cells. Rather, microtubules radiate from the **centrosome,** which is the primary **microtubule-organizing center (MTOC)** in animal cells (Figure 5-33). As detailed in Chapter 20, the two ends of a microtubule differ in their dynamic properties and are commonly designated as the (+) and (−) ends. For this reason, microtubules can have two distinct orientations relative to one another and to other cell structures. In many nondividing animal cells, the MTOC is located at the center of the cell near the nucleus, and the radiating microtubules are all oriented with their (+) ends directed toward the cell periphery. Although most interphase animal cells contain a single perinuclear MTOC, epithelial cells and plant cells contain hundreds of MTOCs. Both of these cell types exhibit distinct

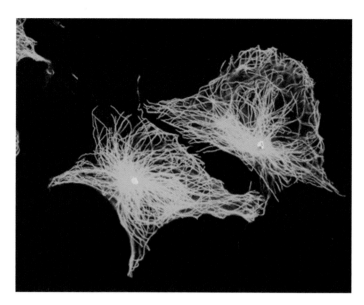

▲ **FIGURE 5-33 Fluorescence micrograph of a Chinese hamster ovary cell stained to reveal microtubles and the MTOC.** The microtubules (green), detected with an antibody to tubulin, are seen to radiate from a central point, the microtubule-organizing center (MTOC), near the nucleus. The MTOC (yellow) is detected with an antibody to a protein localized to the centrosome. [Courtesy of R. Kuriyame.]

functional or structural properties or both in different regions of the cell. The functional and structural **polarity** of these cells is linked to the orientation of microtubules within them.

Findings from studies discussed in Chapter 20 show that the association of microtubules with the endoplasmic reticulum and other membrane-bounded organelles may be critical to the location and organization of these organelles within the cell. For instance, if microtubules are destroyed by drugs such as nocodazole or colcemid, the ER loses its networklike organization. Microtubules are also critical to the formation of the mitotic apparatus—the elaborate, transient structure that captures and subsequently separates replicated chromosomes in cell division.

KEY CONCEPTS OF SECTION 5.4

The Cytoskeleton: Components and Structural Functions

- The cytosol is the internal aqueous medium of a cell exclusive of all organelles and the cytoskeleton. It contains numerous soluble enzymes responsible for much of the cell's metabolic activity.

- Three major types of protein filaments—actin filaments, microtubules, and intermediate filaments—make up the cytoskeleton (see Figure 5-29).

- Microfilaments are assembled from monomeric actin subunits; microtubules, from α,β-tubulin subunits; and intermediate filaments, from lamin subunits and other tissue-specific proteins.

- In all animal and plant cells, the cytoskeleton provides structural stability for the cell and contributes to cell movement. Some bacteria have a primitive cytoskeleton.

- Actin bundles form the core of microvilli and other fingerlike projections of the plasma membrane.

- Cortical spectrin-actin networks are attached to the cell membrane by bivalent membrane–microfilament binding proteins such as ankyrin and band 4.1 (see Figure 5-31).

- Intermediate filaments are assembled into networks and bundles by various intermediate filament–binding proteins, which also cross-link intermediate filaments to the plasma and nuclear membranes, microtubules, and microfilaments.

- In some animal cells, microtubules radiate out from a single microtubule-organizing center lying at the cell center (see Figure 5-33). Intact microtubules appear to be necessary for endoplasmic reticulum and Golgi membranes to form into organized structures.

5.5 Purification of Cells and Their Parts

Many studies on cell structure and function require samples of a particular type of cell or subcellular organelle. Most animal and plant tissues, however, contain a mixture of cell types; likewise, most cells are filled with a variety of organelles. In this section, we describe several commonly used techniques for separating different cell types and organelles. The purification of membrane proteins presents some unique problems also considered here.

Flow Cytometry Separates Different Cell Types

Some cell types differ sufficiently in density that they can be separated on the basis of this physical property. White blood cells (leukocytes) and red blood cells (erythrocytes), for instance, have very different densities because erythrocytes have no nucleus; thus these cells can be separated by equilibrium density centrifugation (described shortly). Because most cell types cannot be differentiated so easily, other techniques such as flow cytometry must be used to separate them.

A flow cytometer identifies different cells by measuring the light that they scatter and the fluorescence that they emit as they flow through a laser beam; thus it can sort out cells of a particular type from a mixture. Indeed, a *fluorescence-activated cell sorter (FACS)*, an instrument based on flow cytometry, can select one cell from thousands of other cells (Figure 5-34). For example, if an antibody specific to a certain cell-surface molecule is linked to a fluorescent dye, any

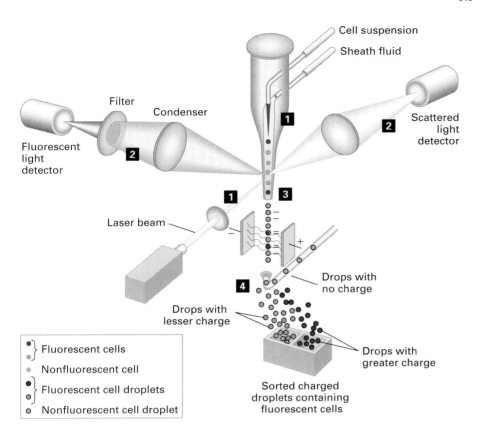

◀ **EXPERIMENTAL FIGURE 5-34**
Fluorescence-activated cell sorter (FACS)
separates cells that are labeled
differentially with a fluorescent reagent.
Step **1**: A concentrated suspension of
labeled cells is mixed with a buffer (the
sheath fluid) so that the cells pass single-file
through a laser light beam. Step **2**: Both the
fluorescent light emitted and the light
scattered by each cell are measured; from
measurements of the scattered light, the size
and shape of the cell can be determined.
Step **3**: The suspension is then forced
through a nozzle, which forms tiny droplets
containing at most a single cell. At the time
of formation, each droplet is given a negative
electric charge proportional to the amount of
fluorescence of its cell. Step **4**: Droplets
with no charge and those with different
electric charges are separated by an electric
field and collected. It takes only milliseconds
to sort each droplet, and so as many as 10
million cells per hour can pass through the
machine. In this way, cells that have desired
properties can be separated and then grown.
[Adapted from D. R. Parks and L. A. Herzenberg,
1982, *Meth. Cell Biol.* **26**:283.]

cell bearing this molecule will bind the antibody and will then be separated from other cells when it fluoresces in the FACS. Having been sorted from other cells, the selected cells can be grown in culture.

The FACS procedure is commonly used to purify the different types of white blood cells, each of which bears on its surface one or more distinctive proteins and will thus bind monoclonal antibodies specific for that protein. Only the T cells of the immune system, for instance, have both CD3 and Thy1.2 proteins on their surfaces. The presence of these surface proteins allows T cells to be separated easily from other types of blood cells or spleen cells (Figure 5-35). In a variation of the use of monoclonal antibodies for separating cells, small magnetic beads are coated with a monoclonal antibody specific for a surface protein such as CD3 or Thy1.2. Only cells with these proteins will stick to the beads and can be

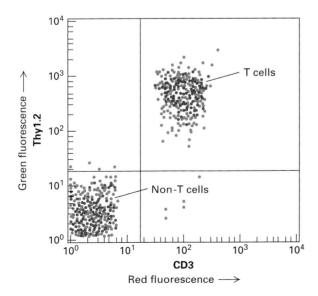

◀ **EXPERIMENTAL FIGURE 5-35** **T cells bound to**
fluorescence-tagged antibodies to two cell-surface proteins
are separated from other white blood cells by FACS. Spleen
cells from a mouse reacted with a fluorescence-tagged
monoclonal antibody (green) specific for the CD3 cell-surface
protein and with a fluorescence-tagged monoclonal antibody (red)
specific for a second cell-surface protein, Thy1.2. As cells were
passed through a FACS machine, the intensity of the green and
red fluorescence emitted by each cell was recorded. This plot of
the red fluorescence (vertical axis) versus green fluorescence
(horizontal axis) for thousands of cells shows that about half of
the cells—the T cells—express both CD3 and Thy1.2 proteins on
their surfaces (upper-right quadrant). The remaining cells, which
exhibit low fluorescence (lower-left quadrant), express only
background levels of these proteins and are other types of white
blood cells. Note the logarithmic scale on both axes. [Courtesy of
Chengcheng Zhang.]

recovered from the preparation by adhesion to a small magnet on the side of the test tube.

Other uses of flow cytometry include the measurement of a cell's DNA and RNA content and the determination of its general shape and size. The FACS can make simultaneous measurements of the size of a cell (from the amount of scattered light) and the amount of DNA that it contains (from the amount of fluorescence emitted from a DNA-binding dye).

Disruption of Cells Releases Their Organelles and Other Contents

The initial step in purifying subcellular structures is to rupture the plasma membrane and the cell wall, if present. First, the cells are suspended in a solution of appropriate pH and salt content, usually isotonic sucrose (0.25 M) or a combination of salts similar in composition to those in the cell's interior. Many cells can then be broken by stirring the cell suspension in a high-speed blender or by exposing it to ultrahigh-frequency sound (*sonication*). Plasma membranes can also be sheared by special pressurized tissue homogeniz-

ers in which the cells are forced through a very narrow space between the plunger and the vessel wall. As noted earlier, water flows into cells when they are placed in a hypotonic solution (see Figure 5-18). This osmotic flow causes cells to swell, weakening the plasma membrane and facilitating its rupture. Generally, the cell solution is kept at 0 °C to best preserve enzymes and other constituents after their release from the stabilizing forces of the cell.

Disrupting the cell produces a mix of suspended cellular components, the homogenate, from which the desired organelles can be retrieved. Homogenization of the cell and dilution of the cytosol cause the depolymerization of actin microfilaments and microtubules, releasing their monomeric subunits, and shear intermediate filaments into short fragments. Thus other procedures, described in Chapters 19 and 20, are used to study these important constituents. Because rat liver contains an abundance of a single cell type, this tissue has been used in many classic studies of cell organelles. However, the same isolation principles apply to virtually all cells and tissues, and modifications of these cell-fractionation techniques can be used to separate and purify any desired components.

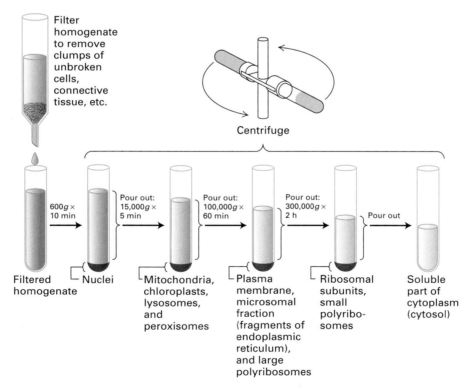

▲ **EXPERIMENTAL FIGURE 5-36 Differential centrifugation is a common first step in fractionating a cell homogenate.** The homogenate resulting from disrupting cells is usually filtered to remove unbroken cells and then centrifuged at a fairly low speed to selectively pellet the nucleus—the largest organelle. The undeposited material (the supernatant) is next centrifuged at a higher speed to sediment the mitochondria, chloroplasts, lysosomes, and peroxisomes. Subsequent centrifugation in the ultracentrifuge at 100,000g for 60 minutes results in deposition of the plasma membrane, fragments of the endoplasmic reticulum, and large polyribosomes. The recovery of ribosomal subunits, small polyribosomes, and particles such as complexes of enzymes requires additional centrifugation at still higher speeds. Only the cytosol—the soluble aqueous part of the cytoplasm—remains in the supernatant after centrifugation at 300,000g for 2 hours.

Centrifugation Can Separate Many Types of Organelles

In Chapter 3, we considered the principles of centrifugation and the uses of centrifugation techniques for separating proteins and nucleic acids. Similar approaches are used for separating and purifying the various organelles, which differ in both size and density and thus undergo sedimentation at different rates.

Most cell-fractionation procedures begin with *differential centrifugation* of a filtered cell homogenate at increasingly higher speeds (Figure 5-36). After centrifugation at each speed for an appropriate time, the supernatant is poured off and centrifuged at higher speed. The pelleted fractions obtained by differential centrifugation generally contain a mixture of organelles, although nuclei and viral particles can sometimes be purified completely by this procedure. An impure organelle fraction obtained by differential centrifugation can be further purified by *equilibrium density-gradient centrifugation*, which separates cellular components according to their density. After the fraction is resuspended, it is layered on top of a solution that contains a gradient of a dense nonionic substance (e.g., sucrose or glycerol). The tube is centrifuged at a high speed (about 40,000 rpm) for several hours, allowing each particle to migrate to an equilibrium position where the density of the surrounding liquid is equal to the density of the particle (Figure 5-37).

Because each organelle has unique morphological features, the purity of organelle preparations can be assessed by examination in an electron microscope. Alternatively, organelle-specific marker molecules can be quantified. For example, the protein cytochrome *c* is present only in mitochondria; so the presence of this protein in a fraction of lysosomes would indicate its contamination by mitochondria. Similarly, catalase is present only in peroxisomes; acid phosphatase, only in lysosomes; and ribosomes, only in the rough endoplasmic reticulum or the cytosol.

Organelle-Specific Antibodies Are Useful in Preparing Highly Purified Organelles

Cell fractions remaining after differential and equilibrium density-gradient centrifugation may still contain more than one type of organelle. Monoclonal antibodies for various organelle-specific membrane proteins are a powerful tool for further purifying such fractions. One example is the purification of coated vesicles whose outer surface is covered with clathrin (Figure 5-38). An antibody to clathrin, bound to a bacterial carrier, can selectively bind these vesicles in a crude preparation of membranes, and the whole antibody complex can then be isolated by low-speed centrifugation. A related technique uses tiny metallic beads coated with specific antibodies. Organelles that bind to the antibodies, and are thus linked to the metallic beads, are recovered from the preparation by adhesion to a small magnet on the side of the test tube.

All cells contain a dozen or more different types of small membrane-limited vesicles of about the same size (50–100 nm in diameter) and density. Because of their similar size and density, these vesicles are difficult to separate from one another by centrifugation techniques. Immunological techniques are particularly useful for purifying specific classes of such vesicles. Fat and muscle cells, for instance, contain a particular glucose transporter (GLUT4) that is localized to the membrane of a specific kind of vesicle. When insulin is added to the cells, these vesicles fuse with the cell-surface membrane and increase the number of glucose transporters able to take up glucose from the blood. As will be seen in Chapter 15, this process is critical to maintaining the appropriate concentration of sugar in the blood. The GLUT4-containing vesicles can be purified by using an antibody that binds to a segment of the GLUT4 protein that faces the cytosol. Likewise, the various transport vesicles discussed in Chapter 17 are characterized by unique surface proteins that permit their separation with the aid of specific antibodies.

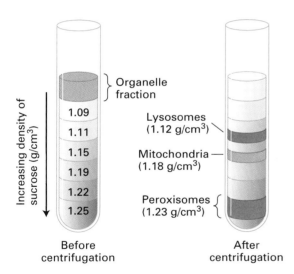

▲ **EXPERIMENTAL FIGURE 5-37 A mixed organelle fraction can be further separated by equilibrium density-gradient centrifugation.** In this example, material in the pellet from centrifugation at 15,000*g* (see Figure 5-36) is resuspended and layered on a gradient of increasingly more dense sucrose solutions in a centrifuge tube. During centrifugation for several hours, each organelle migrates to its appropriate equilibrium density and remains there. To obtain a good separation of lysosomes from mitochondria, the liver is perfused with a solution containing a small amount of detergent before the tissue is disrupted. During this perfusion period, detergent is taken into the cells by endocytosis and transferred to the lysosomes, making them less dense than they would normally be and permitting a "clean" separation of lysosomes from mitochondria.

(a)

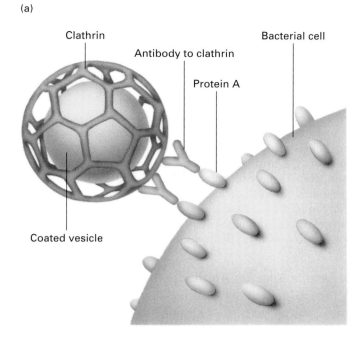

(b)

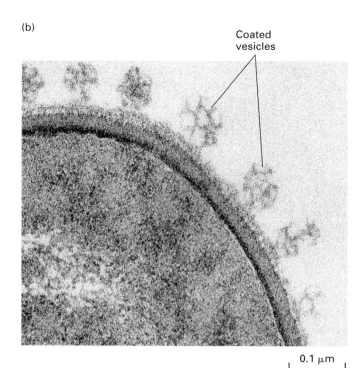

0.1 μm

▲ **EXPERIMENTAL FIGURE 5-38 Small vesicles can be purified by binding of antibody specific for a vesicle surface protein and linkage to bacterial cells.** In this example, a suspension of membranes from rat liver is incubated with an antibody specific for clathrin, a protein that coats the outer surface of certain cytosolic vesicles. To this mixture is added a suspension of *Staphylococcus aureus* bacteria whose surface membrane contains protein A, which binds to the Fc constant region of antibodies. (a) Interaction of protein A with antibodies bound to clathrin-coated vesicles links the vesicles to the bacterial cells. The vesicle–bacteria complexes can then be recovered by low-speed centrifugation. (b) A thin-section electron micrograph reveals clathrin-coated vesicles bound to an *S. aureus* cell. [See E. Merisko et al., 1982, *J. Cell Biol.* **93**:846. Micrograph courtesy of G. Palade.]

Proteins Can Be Removed from Membranes by Detergents or High-Salt Solutions

Detergents are amphipathic molecules that disrupt membranes by intercalating into phospholipid bilayers and solubilizing lipids and proteins. The hydrophobic part of a detergent molecule is attracted to hydrocarbons and mingles with them readily; the hydrophilic part is strongly attracted to water. Some detergents are natural products, but most are synthetic molecules developed for cleaning and for dispersing mixtures of oil and water (Figure 5-39). Ionic detergents, such as sodium deoxycholate and sodium dodecylsulfate (SDS), contain a charged group; nonionic detergents, such as Triton X-100 and octylglucoside, lack a charged group. At very low concentrations, detergents dissolve in pure water as isolated molecules. As the concentration increases, the molecules begin to form **micelles**—small, spherical aggregates in which hydrophilic parts of the molecules face outward and the hydrophobic parts cluster in the center (see Figure 2-20). The *critical micelle concentration (CMC)* at which micelles form is characteristic of each detergent and is a function of the structures of its hydrophobic and hydrophilic parts.

Ionic detergents bind to the exposed hydrophobic regions of membrane proteins as well as to the hydrophobic cores of water-soluble proteins. Because of their charge, these detergents also disrupt ionic and hydrogen bonds. At high concentrations, for example, sodium dodecylsulfate completely denatures proteins by binding to every side chain, a property that is exploited in SDS gel electrophoresis (see Figure 3-32). *Nonionic detergents* do not denature proteins and are thus useful in extracting proteins from membranes before purifying them. These detergents act in different ways at different concentrations. At high concentrations (above the CMC), they solubilize biological membranes by forming mixed micelles of detergent, phospholipid, and integral membrane proteins (Figure 5-40). At low concentrations (below the CMC), these detergents bind to the hydrophobic regions of most integral membrane proteins, making them soluble in aqueous solution.

Treatment of cultured cells with a buffered salt solution containing a nonionic detergent such as Triton X-100 extracts water-soluble proteins as well as integral membrane proteins. As noted earlier, the exoplasmic and cytosolic domains of integral membrane proteins are generally hydrophilic and sol-

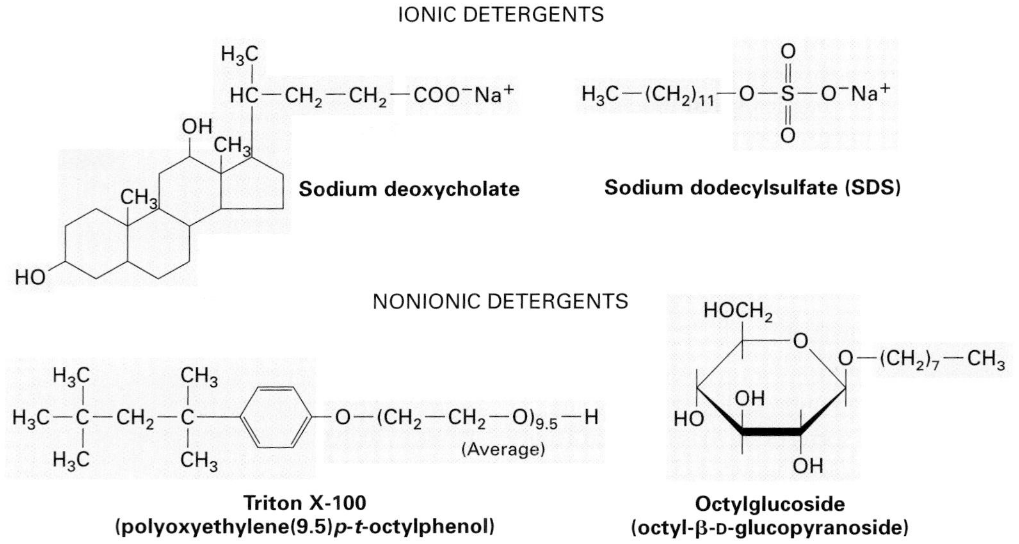

IONIC DETERGENTS

Sodium deoxycholate

Sodium dodecylsulfate (SDS)

NONIONIC DETERGENTS

**Triton X-100
(polyoxyethylene(9.5)*p-t*-octylphenol)**

**Octylglucoside
(octyl-β-D-glucopyranoside)**

▲ **FIGURE 5-39 Structures of four common detergents.** The hydrophobic part of each molecule is shown in yellow; the hydrophilic part, in blue. The bile salt sodium deoxycholate is a natural product; the others are synthetic. Although ionic detergents commonly cause denaturation of proteins, nonionic detergents do not and are thus useful in solubilizing integral membrane proteins.

uble in water. The membrane-spanning domains, however, are rich in hydrophobic and uncharged residues (see Figure 5-12). When separated from membranes, these exposed hydrophobic segments tend to interact with one another, causing the protein molecules to aggregate and precipitate from aqueous solutions. The hydrophobic parts of nonionic detergent molecules preferentially bind to the hydrophobic seg-

ments of transmembrane proteins, preventing protein aggregation and allowing the proteins to remain in the aqueous solution. Detergent-solubilized transmembrane proteins can then be purified by affinity chromatography and other techniques used in purifying water-soluble proteins (Chapter 3).

As discussed previously, most peripheral proteins are bound to specific transmembrane proteins or membrane

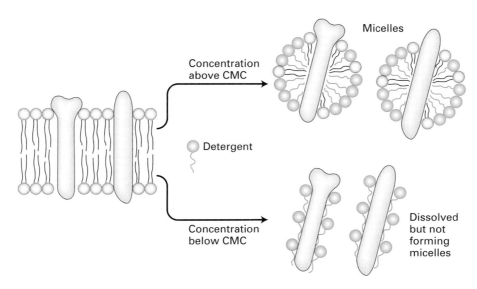

▲ **FIGURE 5-40 Solubilization of integral membrane proteins by nonionic detergents.** At a concentration higher than its critical micelle concentration (CMC), a detergent solubilizes lipids and integral membrane proteins, forming mixed micelles containing detergent, protein, and lipid molecules. At concentrations below the CMC, nonionic detergents (e.g., octylglucoside, Triton X-100) can dissolve membrane proteins without forming micelles by coating the membrane-spanning regions.

phospholipids by ionic or other weak interactions. Generally, peripheral proteins can be removed from the membrane by solutions of high ionic strength (high salt concentrations), which disrupt ionic bonds, or by chemicals that bind divalent cations such as Mg^{2+}. Unlike integral proteins, most peripheral proteins are soluble in aqueous solution and need not be solubilized by nonionic detergents.

KEY CONCEPTS OF SECTION 5.5

Purification of Cells and Their Parts

■ Flow cytometry can identify different cells on the basis of the light that they scatter and the fluorescence that they emit. The fluorescence-activated cell sorter (FACS) is useful in separating different types of cells (see Figures 5-34 and 5-35).

■ Disruption of cells by vigorous homogenization, sonication, or other techniques releases their organelles. Swelling of cells in a hypotonic solution weakens the plasma membrane, making it easier to rupture.

■ Sequential differential centrifugation of a cell homogenate yields fractions of partly purified organelles that differ in mass and density (see Figure 5-36).

■ Equilibrium density-gradient centrifugation, which separates cellular components according to their densities, can further purify cell fractions obtained by differential centrifugation.

■ Immunological techniques, using antibodies against organelle-specific membrane proteins, are particularly useful in purifying organelles and vesicles of similar sizes and densities.

■ Transmembrane proteins are selectively solubilized and purified with the use of nonionic detergents.

5.6 Visualizing Cell Architecture

In the 1830s, Matthias Schleiden and Theodore Schwann proposed that individual cells constitute the fundamental unit of life. This first formulation of the cell theory was based on observations made with rather primitive light microscopes. Modern cell biologists have many more-powerful tools for revealing cell architecture. For example, variations of standard light microscopy permit scientists to view objects that were undetectable several decades ago. Electron microscopy, which can reveal extremely small objects, has yielded much information about subcellular particles and the organization of plant and animal tissues. Each technique is most suitable for detecting and imaging particular structural features of the cell (Figure 5-41). Digital recording systems and appropriate computer algorithms represent another advance in visualizing cell architecture that has spread widely in the past decade. Digital systems not only

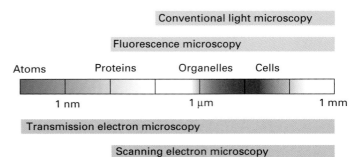

▲ **FIGURE 5-41 The range in sizes of objects imaged by different microscopy techniques.** The smallest object that can be imaged by a particular technique is limited by the resolving power of the equipment and other factors.

can provide microscopic images of improved quality but also permit three-dimensional reconstructions of cell components from two-dimensional images.

A Microscope Detects, Magnifies, and Resolves Small Objects

All microscopes produce a magnified image of a small object, but the nature of the images depends on the type of microscope employed and on the way in which the specimen is prepared. The compound microscope, used in conventional *bright-field light microscopy*, contains several lenses that magnify the image of a specimen under study (Figure 5-42a, b). The total magnification is a product of the magnification of the individual lenses: if the objective lens magnifies 100-fold (a 100× lens, the maximum usually employed) and the projection lens, or eyepiece, magnifies 10-fold, the final magnification recorded by the human eye or on film will be 1000-fold.

However, the most important property of any microscope is not its magnification but its resolving power, or **resolution**—the ability to distinguish between two very closely positioned objects. Merely enlarging the image of a specimen accomplishes nothing if the image is blurred. The resolution of a microscope lens is numerically equivalent to D, the minimum distance between two distinguishable objects. The smaller the value of D, the better the resolution. The value of D is given by the equation

$$D = \frac{0.61\lambda}{N \sin\alpha} \qquad (5-1)$$

where α is the angular aperture, or half-angle, of the cone of light entering the objective lens from the specimen; N is the refractive index of the medium between the specimen and the objective lens (i.e., the relative velocity of light in the medium compared with the velocity in air); and λ is the wavelength of the incident light. Resolution is improved by using shorter wavelengths of light (decreasing the value of

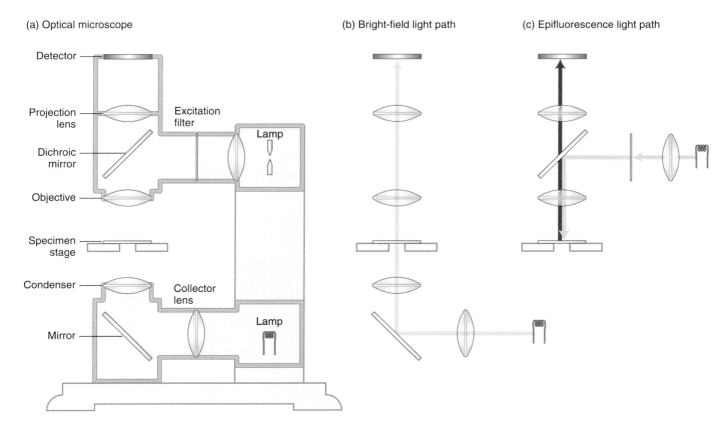

(a) Optical microscope

Detector

Projection lens

Excitation filter

Lamp

Dichroic mirror

Objective

Specimen stage

Condenser

Collector lens

Mirror

Lamp

(b) Bright-field light path

(c) Epifluorescence light path

▲ **EXPERIMENTAL FIGURE 5-42 Optical microscopes are commonly configured for both bright-field (transmitted) and epifluorescence microscopy.** (a) In a typical light microscope, the specimen is usually mounted on a transparent glass slide and positioned on the movable specimen stage. The two imaging methods require separate illumination systems but use the same light gathering and detection systems. (b) In bright-field light microscopy, light from a tungsten lamp is focused on the specimen by a condenser lens below the stage; the light travels the pathway shown. (c) In epifluorescence microscopy, ultraviolet light from a mercury lamp positioned above the stage is focused on the specimen by the objective lens. Filters in the light path select a particular wavelength of ultraviolet light for illumination and are matched to capture the wavelength of the emitted light by the specimen.

λ) or gathering more light (increasing either N or α). Note that the magnification is not part of this equation.

Owing to limitations on the values of α, λ, and N, the *limit of resolution* of a light microscope using visible light is about 0.2 μm (200 nm). No matter how many times the image is magnified, the microscope can never resolve objects that are less than ≈0.2 μm apart or reveal details smaller than ≈0.2 μm in size. Despite this limit on resolution, the light microscope can be used to track the location of a small bead of known size to a precision of only a few nanometers. If we know the precise size and shape of an object—say, a 5-nm sphere of gold—and if we use a video camera to record the microscopic image as a digital image, then a computer can calculate the position of the center of the object to within a few nanometers. This technique has been used to measure nanometer-size steps as molecules and vesicles move along cytoskeletal filaments (see Figures 19-17, 19-18, and 20-18).

Samples for Microscopy Must Be Fixed, Sectioned, and Stained to Image Subcellular Details

Live cells and tissues lack compounds that absorb light and are thus nearly invisible in a light microscope. Although such specimens can be visualized by special techniques to be discussed shortly, these methods do not reveal the fine details of structure and require cells to be housed in special glass-faced chambers, called culture chambers, that can be mounted on a microscope stage. For these reasons, cells are often fixed, sectioned, and stained to reveal subcellular structures.

Specimens for light and electron microscopy are commonly fixed with a solution containing chemicals that cross-link most proteins and nucleic acids. Formaldehyde, a common fixative, cross-links amino groups on adjacent molecules; these covalent bonds stabilize protein–protein and protein–nucleic acid interactions and render the molecules

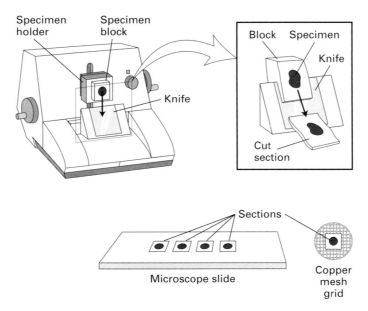

▲ **EXPERIMENTAL FIGURE 5-43 Tissues for microscopy are commonly fixed, embedded in a solid medium, and cut into thin sections.** A fixed tissue is dehydrated by soaking in a series of alcohol-water solutions, ending with an organic solvent compatible with the embedding medium. To embed the tissue for sectioning, the tissue is placed in liquid paraffin for light microscopy or in liquid plastic for electron microscopy; after the block containing the specimen has hardened, it is mounted on the arm of a microtome and slices are cut with a knive. Typical sections cut for electron microscopy 50–100 nm thick; sections cut for light microscopy are 0.5–50 μm thick. The sections are collected either on microscope slides (light microscopy) or copper mesh grids (electron microscopy) and stained with an appropriate agent.

insoluble and stable for subsequent procedures. After fixation, a sample is usually embedded in paraffin or plastic and cut into sections 0.5–50 μm thick (Figure 5-43). Alternatively, the sample can be frozen without prior fixation and then sectioned; such treatment preserves the activity of enzymes for later detection by cytochemical reagents.

A final step in preparing a specimen for light microscopy is to stain it so as to visualize the main structural features of the cell or tissue. Many chemical stains bind to molecules that have specific features. For example, *hematoxylin* binds to basic amino acids (lysine and arginine) on many different kinds of proteins, whereas *eosin* binds to acidic molecules (such as DNA and side chains of aspartate and glutamate). Because of their different binding properties, these dyes stain various cell types sufficiently differently that they are distinguishable visually. If an enzyme catalyzes a reaction that produces a colored or otherwise visible precipitate from a colorless precursor, the enzyme may be detected in cell sections by their colored reaction products. Such staining techniques, although once quite common, have been largely replaced by other techniques for visualizing particular proteins or structures as discussed next.

Phase-Contrast and Differential Interference Contrast Microscopy Visualize Unstained Living Cells

Two common methods for imaging live cells and unstained tissues generate contrast by taking advantage of differences in the refractive index and thickness of cellular materials. These methods, called *phase-contrast microscopy* and *differential interference contrast (DIC) microscopy* (or Nomarski interference microscopy), produce images that differ in appearance and reveal different features of cell architecture. Figure 5-44 compares images of live, cultured cells obtained with these two methods and standard bright-field microscopy.

In phase-contrast images, the entire object and subcellular structures are highlighted by interference rings—concentric halos of dark and light bands. This artifact is inherent in the method, which generates contrast by interference between diffracted and undiffracted light by the specimen. Because the interference rings around an object obscure many details, this technique is suitable for observing only single cells or thin cell layers but not thick tissues. It is particularly useful for examining the location and movement of larger organelles in live cells.

DIC microscopy is based on interference between polarized light and is the method of choice for visualizing extremely small details and thick objects. Contrast is generated by differences in the index of refraction of the object and its surrounding medium. In DIC images, objects appear to cast a shadow to one side. The "shadow" primarily represents a difference in the refractive index of a specimen rather than its topography. DIC microscopy easily defines the outlines of large organelles, such as the nucleus and vacuole. In addition to having a "relief"-like appearance, a DIC image is a thin *optical section*, or slice, through the object. Thus details of the nucleus

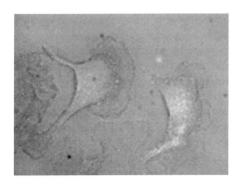 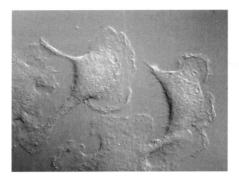 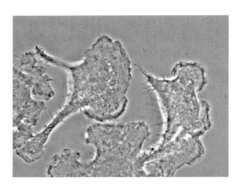

▲ **EXPERIMENTAL FIGURE 5-44 Live cells can be visualized by microscopy techniques that generate contrast by interference.** These micrographs show live, cultured macrophage cells viewed by bright-field microscopy (*left*), phase-contrast microscopy (*middle*), and differential interference contrast (DIC) microscopy (*right*). In a phase-contrast image, cells are surrounded by alternating dark and light bands; in-focus and out-of-focus details are simultaneously imaged in a phase-contrast microscope. In a DIC image, cells appear in pseudorelief. Because only a narrow in-focus region is imaged, a DIC image is an optical slice through the object. [Courtesy of N. Watson and J. Evans.]

in thick specimens (e.g., an intact *Caenorhabditis elegans* roundworm) can be observed in a series of such optical sections, and the three-dimensional structure of the object can be reconstructed by combining the individual DIC images.

Fluorescence Microscopy Can Localize and Quantify Specific Molecules in Fixed and Live Cells

Perhaps the most versatile and powerful technique for localizing proteins within a cell by light microscopy is **fluorescent staining** of cells and observation by *fluorescence microscopy.* A chemical is said to be fluorescent if it absorbs light at one wavelength (the excitation wavelength) and emits light (fluoresces) at a specific and longer wavelength. Most fluorescent dyes, or **flurochromes,** emit visible light, but some (such as Cy5 and Cy7) emit infrared light. In modern fluorescence microscopes, only fluorescent light emitted by the sample is used to form an image; light of the exciting wavelength induces the fluorescence but is then not allowed to pass the filters placed between the objective lens and the eye or camera (see Figure 5-42a, c).

Immunological Detection of Specific Proteins in Fixed Cells
The common chemical dyes just mentioned stain nucleic acids or broad classes of proteins. However, investigators often want to detect the presence and location of specific proteins. A widely used method for this purpose employs specific antibodies covalently linked to flurochromes. Commonly used flurochromes include rhodamine and Texas red, which emit red light; Cy3, which emits orange light; and fluorescein, which emits green light. These flurochromes can be chemically coupled to purified antibodies specific for almost any desired macromolecule. When a flurochrome–antibody complex is added to a permeabilized cell or tissue section, the complex will bind to the corresponding antigens, which then light up when illuminated by the exciting wavelength, a technique called *immunfluorescence microscopy* (Figure 5-45). Staining a specimen with two or three dyes that fluoresce at different wavelengths allows multiple proteins to be localized within a cell (see Figure 5-33).

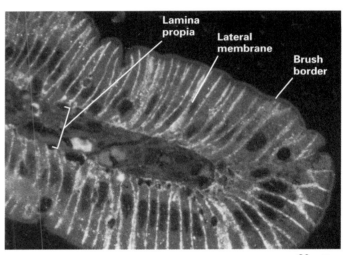

▲ **EXPERIMENTAL FIGURE 5-45 One or more specific proteins can be localized in fixed tissue sections by immunofluorescence microscopy.** A section of the rat intestinal wall was stained with Evans blue, which generates a nonspecific red fluorescence, and with a yellow green–fluorescing antibody specific for GLUT2, a glucose transport protein. As evident from this fluorescence micrograph, GLUT2 is present in the basal and lateral sides of the intestinal cells but is absent from the brush border, composed of closely packed microvilli on the apical surface facing the intestinal lumen. Capillaries run through the lamina propria, a loose connective tissue beneath the epithelial layer. [See B. Thorens et al., 1990, *Am. J. Physio.* **259**:C279; courtesy of B. Thorens.]

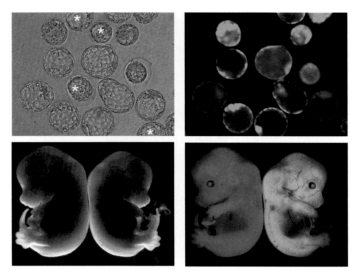

▲ **EXPERIMENTAL FIGURE 5-46 Expression of fluorescent proteins in early and late mouse embryos is detected by emitted blue and yellow light.** The genes encoding blue fluorescent protein (ECFP) and yellow fluorescent protein (EYFP) were introduced into mouse embryonic stem cells, which then were grown into early-stage embryos (*top*) and late-stage embryos (*bottom*). These bright-field (*left*) and fluorescence (*right*) micrographs reveal that all but four of the early-stage embryos display a blue or yellow fluorescence, indicating expression of the introduced ECFP and EYFP genes. Of the two late-stage embryos shown, one expressed the ECFP gene (*left*) and one expressed the EYFP gene (*right*). [From A.-K. Hadjantonakis et al., 2002, *BMC Biotechnol.* **2**:11.]

Expression of Fluorescent Proteins in Live Cells A naturally fluorescent protein found in the jellyfish *Aequorea victoria* can be exploited to visualize live cells and specific proteins within them. This 238-residue protein, called green fluorescent protein (GFP), contains a serine, tyrosine, and glycine sequence whose side chains have spontaneously cyclized to form a green-fluorescing chromophore. With the use of recombinant DNA techniques discussed in Chapter 9, the GFP gene can be introduced into living cultured cells or into specific cells of an entire animal. Cells containing the introduced gene will express GFP and thus emit a green fluorescence when irradiated; this GFP fluorescence can be used to localize the cells within a tissue. Figure 5-46 illustrates the results of this approach, in which a variant of GFP that emits blue fluorescence was used.

In a particularly useful application of GFP, a cellular protein of interest is "tagged" with GFP to localize it. In this technique, the gene for GFP is fused to the gene for a particular cellular protein, producing a recombinant DNA encoding one long chimeric protein that contains the entirety of both proteins. Cells in which this recombinant DNA has been introduced will synthesize the chimeric protein whose green fluorescence reveals the subcellular location of the protein of interest. This GFP-tagging technique, for example, has been

used to visualize the expression and distribution of specific proteins that mediate cell–cell adhesion (see Figure 6-8).

In some cases, a purified protein chemically linked to a fluorescent dye can be microinjected into cells and followed by fluorescence microscopy. For example, findings from careful biochemical studies have established that purified actin "tagged" with a flurochrome is indistinguishable in function from its normal counterpart. When the tagged protein is microinjected into a cultured cell, the endogenous cellular and injected tagged actin monomers copolymerize into normal long actin fibers. This technique can also be used to study individual microtubules within a cell.

Determination of Intracellular Ca^{2+} and H^+ Levels with Ion-Sensitive Fluoresent Dyes Flurochromes whose fluorescence depends on the concentration of Ca^{2+} or H^+ have proved useful in measuring the concentration of these ions within live cells. As discussed in later chapters, intracellular Ca^{2+} and H^+ concentrations have pronounced effects on many cellular processes. For instance, many hormones or other stimuli cause a rise in cytosolic Ca^{2+} from the resting level of about 10^{-7} M to 10^{-6} M, which induces various cellular responses including the contraction of muscle.

The fluorescent dye *fura-2*, which is sensitive to Ca^{2+}, contains five carboxylate groups that form ester linkages with ethanol. The resulting fura-2 ester is lipophilic and can

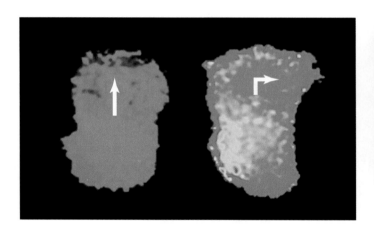

▲ **EXPERIMENTAL FIGURE 5-47 Fura-2, a Ca^{2+}-sensitive flurochrome, can be used to monitor the relative cytosolic Ca^{2+} concentrations in different regions of live cells.** (*Left*) In a moving leukocyte, a Ca^{2+} gradient is established. The highest levels (green) are at the rear of the cell, where cortical contractions take place, and the lowest levels (blue) are at the cell front, where actin undergoes polymerization. (*Right*) When a pipette filled with chemotactic molecules placed to the side of the cell induces the cell to turn, the Ca^{2+} concentration momentarily increases throughout the cytoplasm and a new gradient is established. The gradient is oriented such that the region of lowest Ca^{2+} (blue) lies in the direction that the cell will turn, whereas a region of high Ca^{2+} (yellow) always forms at the site that will become the rear of the cell. [From R. A. Brundage et al., 1991, *Science* **254**:703; courtesy of F. Fay.]

diffuse from the medium across the plasma membrane into cells. Within the cytosol, esterases hydrolyze fura-2 ester, yielding fura-2, whose free carboxylate groups render the molecule nonlipophilic, and so it cannot cross cellular membranes and remains in the cytosol. Inside cells, each fura-2 molecule can bind a single Ca^{2+} ion but no other cellular cation. This binding, which is proportional to the cytosolic Ca^{2+} concentration over a certain range, increases the fluorescence of fura-2 at one particular wavelength. At a second wavelength, the fluorescence of fura-2 is the same whether or not Ca^{2+} is bound and provides a measure of the total amount of fura-2 in a region of the cell. By examining cells continuously in the fluorescence microscope and measuring rapid changes in the ratio of fura-2 fluorescence at these two wavelengths, one can quantify rapid changes in the fraction of fura-2 that has a bound Ca^{2+} ion and thus in the concentration of cytosolic Ca^{2+} (Figure 5-47).

Similarly to fura-2, fluorescent dyes (e.g., SNARF-1) that are sensitive to the H^+ concentration can be used to monitor the cytosolic pH of living cells.

Confocal Scanning and Deconvolution Microscopy Provide Sharp Images of Three-Dimensional Objects

Conventional fluorescence microscopy has two major limitations. First, the physical process of cutting a section destroys material, and so in consecutive (serial) sectioning a small part of a cell's structure is lost. Second, the fluorescent light emitted by a sample comes from molecules above and below the plane of focus; thus the observer sees a blurred image caused by the superposition of fluorescent images from molecules at many depths in the cell. The blurring effect makes it difficult to determine the actual three-dimensional molecular arrangement (Figure 5-48a). Two powerful refinements of fluorescence microscopy produce much sharper images by reducing the image-degrading effects of out-of-focus light.

In *confocal scanning microscopy*, exciting light from a focused laser beam illuminates only a single small part of a sample for an instant and then rapidly moves to different spots in the sample focal plane. The emitted fluorescent light passes through a pinhole that rejects out-of-focus light, thereby producing a sharp image. Because light in focus with the image is collected by the pinhole, the scanned area is an optical section through the specimen. The intensity of light from these in-focus areas is recorded by a photomultiplier tube, and the image is stored in a computer (Figure 5-48b).

Deconvolution microscopy achieves the same image-sharpening effect as confocal scanning microscopy but through a different process. In this method, images from consecutive focal planes of the specimen are collected. A separate focal series of images from a test slide of subresolution size (i.e., 0.2 μm diameter) bead are also collected. Each bead represents a pinpoint of light that becomes an object blurred by the imperfect optics of the microscope. Deconvolution

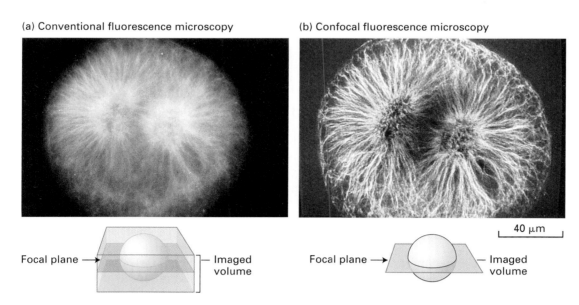

(a) Conventional fluorescence microscopy

(b) Confocal fluorescence microscopy

Focal plane → — Imaged volume

Focal plane → — Imaged volume

40 μm

▲ **EXPERIMENTAL FIGURE 5-48 Confocal microscopy produces an in-focus optical section through thick cells.** A mitotic fertilized egg from a sea urchin (*Psammechinus*) was lysed with a detergent, exposed to an anti-tubulin antibody, and then exposed to a fluorescein-tagged antibody that binds to the first antibody. (a) When viewed by conventional fluorescence microscopy, the mitotic spindle is blurred. This blurring occurs because background fluorescence is detected from tubulin above and below the focal plane as depicted in the sketch. (b) The confocal microscopic image is sharp, particularly in the center of the mitotic spindle. In this case, fluorescence is detected only from molecules in the focal plane, generating a very thin optical section. [Micrographs from J. G. White et al., 1987, *J. Cell Biol.* **104**:41.]

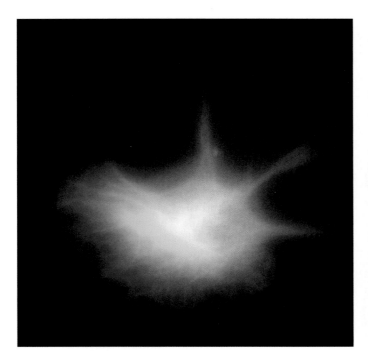
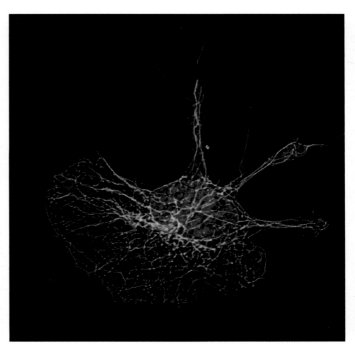

▲ **EXPERIMENTAL FIGURE 5-49 Deconvolution fluorescence microscopy yields high-resolution optical sections that can be reconstructed into one three-dimensional image.** A macrophage cell was stained with fluorochrome-labeled reagents specific for DNA (blue), microtubules (green), and actin microfilaments (red). The series of fluorescent images obtained at consecutive focal planes (optical sections) through the cell were recombined in three dimensions. (a) In this three-dimensional reconstruction of the raw images, the DNA, microtubules, and actin appear as diffuse zones in the cell. (b) After application of the deconvolution algorithm to the images, the fibrillar organization of microtubules and the localization of actin to adhesions become readily visible in the reconstruction. [Courtesy of J. Evans.]

reverses the degradation of the image by using the blurred beads as a reference object. The out-of-focus light is mathematically reassigned with the aid of deconvolution algorithms. Images restored by deconvolution display impressive detail without any blurring (Figure 5-49). Astronomers use deconvolution algorithms to sharpen images of distant stars.

Resolution of Transmission Electron Microscopy Is Vastly Greater Than That of Light Microscopy

The fundamental principles of electron microscopy are similar to those of light microscopy; the major difference is that electromagnetic lenses, rather than optical lenses, focus a high-velocity electron beam instead of visible light. In the *transmission electron microscope (TEM)*, electrons are emitted from a filament and accelerated in an electric field. A condenser lens focuses the electron beam onto the sample; objective and projector lenses focus the electrons that pass through the specimen and project them onto a viewing screen or other detector (Figure 5-50, *left*). Because electrons are absorbed by atoms in air, the entire tube between the electron source and the detector is maintained under an ultrahigh vacuum.

The short wavelength of electrons means that the limit of resolution for the transmission electron microscope is the-oretically 0.005 nm (less than the diameter of a single atom), or 40,000 times better than the resolution of the light microscope and 2 million times better than that of the unaided human eye. However, the effective resolution of the transmission electron microscope in the study of biological systems is considerably less than this ideal. Under optimal conditions, a resolution of 0.10 nm can be obtained with transmission electron microscopes, about 2000 times better than the best resolution of light microscopes. Several examples of cells and subcellular structures imaged by TEM are included in Section 5.3.

Because TEM requires very thin, fixed sections (about 50 nm), only a small part of a cell can be observed in any one section. Sectioned specimens are prepared in a manner similar to that for light microscopy, by using a knife capable of producing sections 50–100 nm in thickness (see Figure 5-43). The generation of the image depends on differential scattering of the incident electrons by molecules in the preparation. Without staining, the beam of electrons passes through a specimen uniformly, and so the entire sample appears uniformly bright with little differentiation of components. To obtain useful images by TEM, sections are commonly stained with heavy metals such as gold or osmium. Metal-stained areas appear dark on a micrograph because the metals scatter (diffract) most of the incident

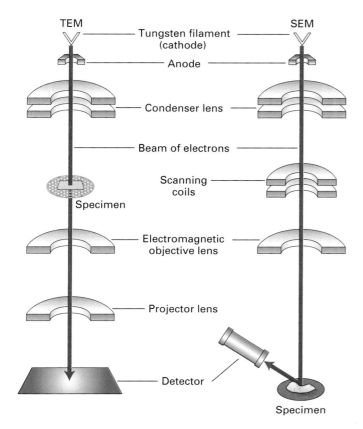

▲ **EXPERIMENTAL FIGURE 5-50 In electron microscopy, images are formed from electrons that pass through a specimen or are released from a metal-coated specimen.** In a transmission electron microscope (TEM), electrons are extracted from a heated filament, accelerated by an electric field, and focused on the specimen by a magnetic condenser lens. Electrons that pass through the specimen are focused by a series of magnetic objective and projector lenses to form a magnified image of the specimen on a detector, which may be a fluorescent viewing screen, photographic film, or a charged-couple-device (CCD) camera. In a scanning electron microscope (SEM), electrons are focused by condensor and objective lenses on a metal-coated specimen. Scanning coils move the beam across the specimen, and electrons from the metal are collected by a photomultiplier tube detector. In both types of microscopes, because electrons are easily scattered by air molecules, the entire column is maintained at a very high vacuum.

electrons; scattered electrons are not focused by the electromagnetic lenses and do not contribute to the image. Areas that take up less stain appear lighter. Osmium tetroxide preferentially stains certain cellular components, such as membranes (see Figure 5-2a). Specific proteins can be detected in thin sections by the use of electron-dense gold particles coated with protein A, a bacterial protein that binds antibody molecules nonspecifically (Figure 5-51).

Standard electron microscopy cannot be used to study live cells because they are generally too vulnerable to the required conditions and preparatory techniques. In particular, the absence of water causes macromolecules to become de-

natured and nonfunctional. However, the technique of *cryo-electron microscopy* allows examination of hydrated, unfixed, and unstained biological specimens directly in a transmission electron microscope. In this technique, an aqueous suspension of a sample is applied in an extremely thin film to a grid. After it has been frozen in liquid nitrogen and maintained in this state by means of a special mount, the sample is observed in the electron microscope. The very low temperature (−196 °C) keeps water from evaporating, even in a vacuum, and the sample can be observed in detail in its native, hydrated state without fixing or heavy metal

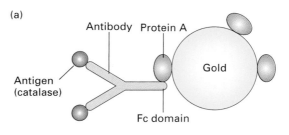

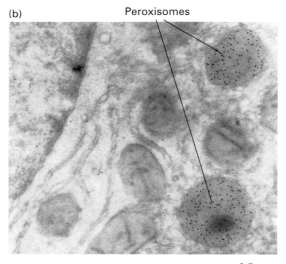

▲ **EXPERIMENTAL FIGURE 5-51 Gold particles coated with protein A are used to detect an antibody-bound protein by transmission electron microscopy.** (a) First antibodies are allowed to interact with their specific antigen (e.g., catalase) in a section of fixed tissue. Then the section is treated with a complex of protein A from the bacterium *S. aureus* and electron-dense gold particles. Binding of this complex to the Fc domains of the antibody molecules makes the location of the target protein, catalase in this case, visible in the electron microscope. (b) A slice of liver tissue was fixed with glutaraldehyde, sectioned, and then treated as described in part (a) to localize catalase. The gold particles (black dots) indicating the presence of catalase are located exclusively in peroxisomes. [From H. J. Geuze et al., 1981, *J. Cell Biol.* **89**:653. Reproduced from the *Journal of Cell Biology* by copyright permission of The Rockefeller University Press.]

staining. By computer-based averaging of hundreds of images, a three-dimensional model almost to atomic resolution can be generated. For example, this method has been used to generate models of ribosomes (see Figure 4-27), the muscle calcium pump discussed in Chapter 7, and other large proteins that are difficult to crystallize.

Electron Microscopy of Metal-Coated Specimens Can Reveal Surface Features of Cells and Their Components

Transmission electron microscopy is also used to obtain information about the shapes of purified viruses, fibers, enzymes, and other subcellular particles by using a technique, called *metal shadowing*, in which a thin layer of metal, such

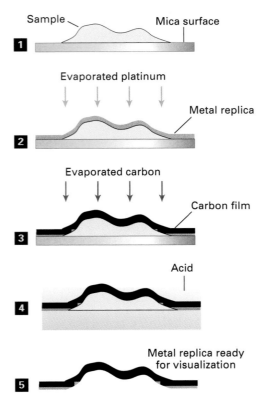

▲ **EXPERIMENTAL FIGURE 5-52** Metal shadowing makes surface details on very small particles visible by transmission electron microscopy. The sample is spread on a mica surface and then dried in a vacuum evaporator (**1**). A filament of a heavy metal, such as platinum or gold, is heated electrically so that the metal evaporates and some of it falls over the sample grid in a very thin film (**2**). To stabilize the replica, the specimen is then coated with a carbon film evaporated from an overhead electrode (**3**). The biological material is then dissolved by acid (**4**), leaving a metal replica of the sample (**5**), which is viewed in a TEM. In electron micrographs of such preparations, the carbon-coated areas appear light—the reverse of micrographs of simple metal-stained preparations in which the areas of heaviest metal staining appear the darkest.

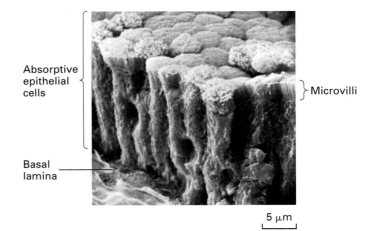

5 μm

▲ **EXPERIMENTAL FIGURE 5-53** Scanning electron microscopy (SEM) produces a three-dimensional image of the surface of an unsectioned specimen. Shown here is an SEM image of the epithelium lining the lumen of the intestine. Abundant fingerlike microvilli extend from the lumen-facing surface of each cell. The basal lamina beneath the epithelium helps support and anchor it to the underlying connective tissue (Chapter 6). Compare this image of intestinal cells with those in Figure 5-28, a transmission electron micrograph, and in Figure 5-45, a fluorescence micrograph. [From R. Kessel and R. Kardon, 1979, *Tissues and Organs, A Text-Atlas of Scanning Electron Microscopy*, W. H. Freeman and Company, p. 176.]

as platinum, is evaporated on a fixed and sectioned or rapidly frozen biological sample (Figure 5-52). Acid treatment dissolves away the cell, leaving a metal replica that is viewed in a transmission electron microscope.

Alternatively, the *scanning electron microscope* allows investigators to view the surfaces of unsectioned metal-coated specimens. An intense electron beam inside the microscope scans rapidly over the sample. Molecules in the coating are excited and release secondary electrons that are focused onto a scintillation detector; the resulting signal is displayed on a cathode-ray tube (see Figure 5-50, *right*). Because the number of secondary electrons produced by any one point on the sample depends on the angle of the electron beam in relation to the surface, the scanning electron micrograph has a three-dimensional appearance (Figure 5-53). The resolving power of scanning electron microscopes, which is limited by the thickness of the metal coating, is only about 10 nm, much less than that of transmission instruments.

Three-Dimensional Models Can Be Constructed from Microscopy Images

In the past decade, digital cameras have largely replaced optical cameras to record microscopy images. Digital images can be stored in a computer and manipulated by conventional photographic software as well as specialized algorithms. As mentioned earlier, the deconvolution algorithm can sharpen an image by restoring out-of-focus photons to their origin—an example of a computational method that improves the quality of the image. The details in stored digital images also can be quantified, and objects in images can be reconstructed in three dimensions. For example, the three-dimensional model of an object can be calculated by tomographic methods from a collection of images that cover different views of the object. In light microscopy, a stack of optical sections collected with either a confocal or a deconvolution microscope can be recombined into one three-dimensional image (see Figure 5-49). If a TEM specimen is tilted through various degrees, the resulting images also can be recombined to generate a three-dimensional view of the object (see Figure 5-23).

KEY CONCEPTS OF SECTION 5.6

Visualizing Cell Architecture

■ The limit of resolution of a light microscope is about 200 nm; of a scanning electron microscope, about 10 nm; and of a transmission electron microscope, about 0.1 nm.

■ Because cells and tissues are almost transparent, various types of stains and optical techniques are used to generate sufficient contrast for imaging.

■ Phase-contrast and differential interference contrast (DIC) microscopy are used to view the details of live, unstained cells and to monitor cell movement.

■ In immunofluorescence microscopy, specific proteins and organelles in fixed cells are stained with fluorescence-labeled monoclonal antibodies. Multiple proteins can be localized in the same sample by staining with antibodies labeled with different fluorochromes.

■ When proteins tagged with naturally occurring green fluorescent protein (GFP) or its variants are expressed in live cells, they can be visualized in a fluorescence microscope.

■ With the use of dyes whose fluorescence is proportional to the concentration of Ca^{2+} or H^+ ions, fluorescence microscopy can measure the local concentration of Ca^{2+} ions and intracellular pH in living cells.

■ Confocal microscopy and deconvolution microscopy use different methods to optically section a specimen, thereby reducing the blurring due to out-of-focus fluorescence light. Both methods provide much sharper images, particularly of thick specimens, than does standard fluorescence microscopy.

■ Specimens for electron microscopy generally must be fixed, sectioned, dehydrated, and then stained with electron-dense heavy metals.

■ Surface details of objects can be revealed by transmission electron microscopy of metal-coated specimens. Scanning electron microscopy of metal-coated unsectioned cells or tissues produces images that appear to be three-dimensional.

PERSPECTIVES FOR THE FUTURE

Advances in bioengineering will make major contributions not only to our understanding of cell and tissue function but also to the quality of human health. In a glass slide consisting of microfabricated wells and channels, for example, reagents can be introduced and exposed to selected parts of individual cells; the responses of the cells can then be detected by light microscopy and analyzed by powerful image-processing software. These types of studies will lead to discovery of new drugs, detection of subtle phenotypes of mutant cells (e.g., tumor cells), and development of comprehensive models of cellular processes. Bioengineers also are fabricating artificial tissues based on a synthetic three-dimensional architecture incorporating layers of different cells. Eventually such artificial tissues will provide replacements for defective tissues in sick, injured, or aging individuals.

Microscopy will continue to be a major tool in cell biology, providing images that relate to both the chemistry (i.e., interactions among proteins) and the mechanics (i.e., movements) involved in various cell processes. The forces causing molecular and cellular movements will be directly detected by fluorescent sensors in cells and the extracellular matrix. Improvements to high-resolution imaging methods will permit studies of single molecules in live cells, something that is currently possible only in vitro. Finally, cells will be studied in more natural contexts, not on glass coverslips but in 3D gels of extracellular matrix molecules. To aid in the imaging, the use of more fluorescent labels and tags will allow visualization of five or six different types of molecules simultaneously. With more labeled proteins, the complex interactions among proteins and organelles will become better understood.

Finally, the electron microscope will become the dominant instrument for studying protein machines in vitro and in situ. Tomographic methods applied to single cells and molecules combined with automated reconstruction methods will generate models of protein-based structures that cannot be determined by x-ray crystallography. High resolution three-dimensional models of molecules in cells will help explain the intricate biochemical interactions among proteins.

KEY TERMS

REVIEW THE CONCEPTS

1. When viewed by electron microscopy, the lipid bilayer is often described as looking like a railroad track. Explain how the structure of the bilayer creates this image.

2. Biomembranes contain many different types of lipid molecules. What are the three main types of lipid molecules found in biomembranes? How are the three types similar, and how are they different?

3. Lipid bilayers are considered to be two-dimensional fluids; what does this mean? What drives the movement of lipid molecules and proteins within the bilayer? How can such movement be measured? What factors affect the degree of membrane fluidity?

4. Explain the following statement: The structure of all biomembranes depends on the chemical properties of phospholipids, whereas the function of each specific biomembrane depends on the specific proteins associated with that membrane.

5. Name the three groups into which membrane-associated proteins may be classified. Explain the mechanism by which each group associates with a biomembrane.

6. Although both faces of a biomembrane are composed of the same general types of macromolecules, principally lipids and proteins, the two faces of the bilayer are not identical. What accounts for the asymmetry between the two faces?

7. One of the defining features of eukaryotic cells is the presence of organelles. What are the major organelles of eukaryotic cells, and what is the function of each? What is the cytosol? What cellular processes occur within the cytosol?

8. Cell organelles such as mitochondria, chloroplasts, and the Golgi apparatus each have unique structures. How is the structure of each organelle related to its function?

9. Much of what we know about cellular function depends on experiments utilizing specific cells and specific parts (e.g., organelles) of cells. What techniques do scientists commonly use to isolate cells and organelles from complex mixtures, and how do these techniques work?

10. Isolation of some membrane proteins requires the use of detergents; isolation of others can be accomplished with the use of high-salt solutions. What types of membrane proteins require detergents as part of the isolation procedure? What types of membrane proteins may be isolated with high-salt solutions? Describe how the chemical properties of detergents and high salt facilitate the isolation process of each type of membrane protein.

11. Three systems of cytoskeletal filaments exist in most eukaryotic cells. Compare them in terms of composition, function, and structure.

12. Individual cytoskeletal filaments are typically organized into more complex structures within the cytosol. What two general types of structures do individual filaments combine to form in the cytosol? How are these structures created and maintained?

13. Both light and electron microscopy are commonly used to visualize cells, cell structures, and the location of specific molecules. Explain why a scientist may choose one or the other microscopy technique for use in research.

14. Why are chemical stains required for visualizing cells and tissues with the basic light microscope? What advantage does fluorescent microscopy provide in comparison to the chemical dyes used to stain specimens for light microscopy? What advantages do confocal scanning microscopy and deconvolution microscopy provide in comparison to conventional fluorescence microscopy?

15. In certain electron microscopy methods, the specimen is not directly imaged. How do these methods provide information about cellular structure, and what types of structures do they visualize?

ANALYZE THE DATA

Mouse liver cells were homogenized and the homogenate subjected to equilibrium density-gradient centrifugation with sucrose gradients. Fractions obtained from these gradients were assayed for *marker molecules* (i.e., molecules that are limited to specific organelles). The results of these assays are shown in the figure. The marker molecules have the following functions: Cytochrome oxidase is an enzyme involved in the process by which ATP is formed in the complete aerobic degradation of glucose or fatty acids; ribosomal RNA forms part of the protein-synthesizing ribosomes; catalase catalyzes decomposition of hydrogen peroxide; acid phosphatase hydrolyzes monophosphoric esters at acid pH; cytidylyl transferase is involved in phospholipid biosynthesis; and amino acid permease aids in transport of amino acids across membranes.

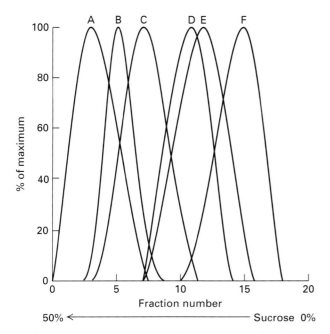

Curve A = cytochrome oxidase

Curve B = ribosomal RNA

Curve C = catalase

Curve D = acid phosphatase

Curve E = cytidylyl transferase

Curve F = amino acid permease

a. Name the marker molecule and give the number of the fraction that is *most* enriched for each of the following: lysosomes; peroxisomes; mitochondria; plasma membrane; rough endoplasmic reticulum; smooth endoplasmic reticulum.

b. Is the rough endoplasmic reticulum more or less dense than the smooth endoplasmic reticulum? Why?

c. Describe an alternative approach by which you could identify which fraction was enriched for which organelle.

d. How would addition of a detergent to the homogenate affect the equilibrium density-gradient results?

REFERENCES

General Histology Texts and Atlases

Cross, P. A., and K. L. Mercer. 1993. *Cell and Tissue Ultrastructure: A Functional Perspective.* W. H. Freeman and Company.

Fawcett, D. W. 1993. *Bloom and Fawcett: A Textbook of Histology,* 12th ed. Chapman & Hall.

Kessel, R., and R. Kardon. 1979. *Tissues and Organs: A Text-Atlas of Scanning Electron Microscopy.* W. H. Freeeman and Company.

Biomembranes: Lipid Composition and Structural Organization

Simons, K., and D. Toomre. 2000. Lipid rafts and signal transduction. *Nature Rev. Mol. Cell Biol.* **1**:31–41.

Sprong, H., P. van der Sluijs, and G. van Meer. 2001. How proteins move lipids and lipids move proteins. *Nature Rev. Mol. Cell Biol.* **2**:504–513.

Tamm, L. K., V. K. Kiessling, and M. L. Wagner. 2001. Membrane dynamics. *Encyclopedia of Life Sciences.* Nature Publishing Group.

Vance, D. E., and J. E. Vance. 2002. *Biochemistry of Lipids, Lipoproteins, and Membranes,* 4th ed. Elsevier.

Yeager, P. L. 2001. Lipids. *Encyclopedia of Life Sciences.* Nature Publishing Group.

Biomembranes: Protein Components and Basic Functions

Cullen, P. J., G. E. Cozier, G. Banting, and H. Mellor. 2001. Modular phosphoinositide-binding domains: their role in signalling and membrane trafficking. *Curr. Biol.* **11**:R882–R893.

Lanyi, J. K., and H. Luecke. 2001. Bacteriorhodopsin. *Curr. Opin. Struc. Biol.* **11**:415–519.

MacKenzie, K. R., J. H. Prestegard, and D. M. Engelman. 1997. A transmembrane helix dimer: structure and implications. *Science* **276**:131–133.

Minor, D. L. 2001. Potassium channels: life in the post-structural world. *Curr. Opin. Struc. Biol.* **11**:408–414.

Schulz, G. E. 2000. β-Barrel membrane proteins. *Curr. Opin. Struc. Biol.* **10**:443–447.

Organelles of the Eukaryotic Cell

Bainton, D. 1981. The discovery of lysosomes. *J. Cell Biol.* **91**:66s–76s.

Cuervo, A. M., and J. F. Dice. 1998. Lysosomes: a meeting point of proteins, chaperones, and proteases. *J. Mol. Med.* **76**:6–12.

de Duve, C. 1996. The peroxisome in retrospect. *Ann. NY Acad. Sci.* **804**:1–10.

Holtzman, E. 1989. *Lysosomes.* Plenum Press.

Lamond, A., and W. Earnshaw. 1998. Structure and function in the nucleus. *Science* **280**:547–553.

Masters, C., and D. Crane. 1996. Recent developments in peroxisome biology. *Endeavour* **20**:68–73.

Palade, G. 1975. Intracellular aspects of the process of protein synthesis. *Science* **189**:347–358. The Nobel Prize lecture of a pioneer in the study of cellular organelles. (See also de Duve, 1996.)

Subramani, S. 1998. Components involved in peroxisome import, biogenesis, proliferation, turnover, and movement. *Physiol. Rev.* **78**:171–188.

The Cytoskeleton: Components and Structural Functions

Bray, D. 2001. *Cell Movements: From Molecules to Motility.* Garland. Excellent overview of the cytoskeleton and motility.

Various authors. *Curr. Topics Cell Biol.* February issue is always devoted to the cytoskeleton.

Purification of Cells and Their Parts

Battye, F. L., and K. Shortman. 1991. Flow cytometry and cell-separation procedures. *Curr. Opin. Immunol.* **3**:238–241.

de Duve, C. 1975. Exploring cells with a centrifuge. *Science* **189**:186–194. The Nobel Prize lecture of a pioneer in the study of cellular organelles.

de Duve, C., and H. Beaufay. 1981. A short history of tissue fractionation. *J. Cell Biol.* **91**:293s–299s.

Howell, K. E., E. Devaney, and J. Gruenberg. 1989. Subcellular fractionation of tissue culture cells. *Trends Biochem. Sci.* **14**:44–48.

Ormerod, M. G., ed. 1990. *Flow Cytometry: A Practical Approach*. IRL Press.

Rickwood, D. 1992. *Preparative Centrifugation: A Practical Approach*. IRL Press.

Visualizing Cell Architecture

Bastiaens, P. I. H., and R. Pepperkok. 2000. Observing proteins in their natural habitat: the living cell. *Trends Biochem. Sci.* 25:631–637.

Baumeister, W., and A. C. Steven. 2000. Macromolecular electron microscopy in the era of structural genomics. *Trends Biochem. Sci.* 25:624–630.

Bozzola, J. J., and L. D. Russell. 1992. *Electron Microscopy*. Jones and Bartlett.

Dykstra, M. J. 1992. *Biological Electron Microscopy: Theory, Techniques, and Troubleshooting*. Plenum Press.

Gilroy, S. 1997. Fluorescence microscopy of living plant cells. *Ann. Rev. Plant Physiol. Plant Mol. Biol.* 48:165–190.

Inoué, S., and K. Spring. 1997. *Video Microscopy*, 2d ed. Plenum Press.

Lippincott-Schwartz, J., and C. L Smith. 1997. Insights into secretory and endocytic membrane traffic using green fluorescent protein chimeras. *Curr. Opin. Neurobiol.* 7:631–639.

Mason, W. T. 1999. *Fluorescent and Luminescent Probes for Biological Activity*, 2d ed. Academic Press.

Matsumoto, B., ed. 2002. *Methods in Cell Biology*, Vol. 70: *Cell Biological Applications of Confocal Microscopy*. Academic Press.

Misteli, T., and D. L. Spector. 1997. Applications of the green fluorescent protein in cell biology and biotechnology. *Nature Biotech.* 15:961–964.

Sluder, G., and D. Wolf, eds. 1998. *Methods in Cell Biology*, Vol. 56: *Video Microscopy*. Academic Press.

6

INTEGRATING CELLS INTO TISSUES

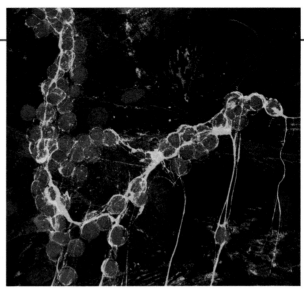

Model of inflammatory bowel disease in which cultured flat colonic smooth muscle cells were induced to secrete cables of hyaluronan (green) that bind to spheroidal mononuclear leukocytes via their CD44 receptors (red). Nuclei are stained blue. [Courtesy of C. de la Motte et al., Lerner Research Institute.]

In the development of complex multicellular organisms such as plants and animals, progenitor cells differentiate into distinct "types" that have characteristic compositions, structures, and functions. Cells of a given type often aggregate into a *tissue* to cooperatively perform a common function: muscle contracts; nervous tissues conduct electrical impulses; xylem tissue in plants transports water. Different tissues can be organized into an *organ,* again to perform one or more specific functions. For instance, the muscles, valves, and blood vessels of a heart work together to pump blood through the body. The coordinated functioning of many types of cells within tissues, as well as of multiple specialized tissues, permits the organism as a whole to move, metabolize, reproduce, and carry out other essential activities.

The adult form of the roundworm *Caenorhabditis elegans* contains a mere 959 cells, yet these cells fall into 12 different general cell types and many distinct subtypes. Vertebrates have hundreds of different cell types, including leukocytes (white blood cells), erythrocytes, and macrophages in the blood; photoreceptors in the retina; adipocytes that store fat; secretory α and β cells in the pancreas; fibroblasts in connective tissue; and hundreds of different subtypes of neurons in the human brain. Despite their diverse forms and functions, all animal cells can be classified as being components of just five main classes of tissue: *epithelial tissue, connective tissue, muscular tissue, nervous tissue,* and *blood.* Various cell types are arranged in precise patterns of staggering complexity to generate the different tissues and organs. The costs of such complexity include increased requirements for information,

material, energy, and time during the development of an individual organism. Although the physiological costs of complex tissues and organs are high, they provide organisms with the ability to thrive in varied and variable environments, a major evolutionary advantage.

The complex and diverse morphologies of plants and animals are examples of the whole being greater than the sum of the individual parts, more technically described as the emergent properties of a complex system. For example, the root-stem-leaf organization of plants permits them to simultaneously obtain energy (sunlight) and carbon (CO_2) from the atmosphere and water and nutrients (e.g., minerals) from the soil. The distinct mechanical properties of rigid bones, flexible joints, and contracting muscles permit vertebrates to move efficiently and achieve substantial size. Sheets of tightly attached epithelial cells can act as regulatable, selective permeability barriers, which permit the generation of chemically and functionally distinct compartments in an organism (e.g., stomach, bloodstream). As a result, distinct and sometimes opposite functions (e.g., digestion and synthesis) can efficiently proceed simultaneously within an organism. Such compartmentalization also permits more sophisticated regulation of diverse biological functions. In many ways, the roles of complex tissues and organs in an organism are analogous to those of organelles and membranes in individual cells.

The assembly of distinct tissues and their organization into organs are determined by molecular interactions at the cellular level and would not be possible without the temporally and spatially regulated expression of a wide array of adhesive molecules. Cells in tissues can adhere directly to one another (*cell–cell adhesion*) through specialized integral membrane proteins called **cell-adhesion molecules (CAMs)** that often cluster into specialized cell junctions (Figure 6-1). Cells in animal tissues also adhere indirectly (*cell–matrix*

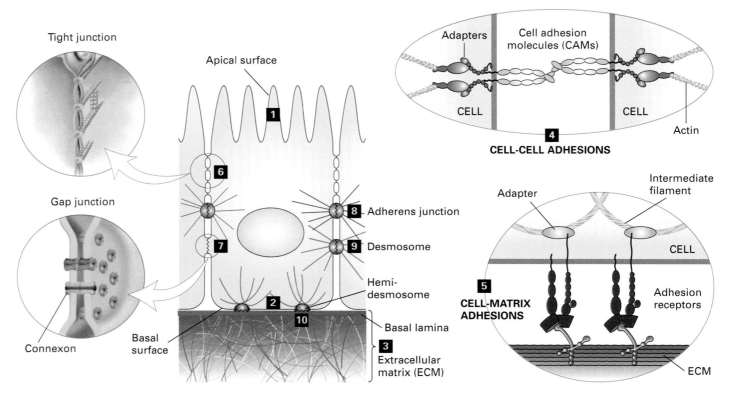

▲ **FIGURE 6-1 Schematic overview of major adhesive interactions that bind cells to each other and to the extracellular matrix.** Schematic cutaway drawing of a typical epithelial tissue, such as the intestines. The apical (upper) surface of these cells is packed with fingerlike microvilli **1** that project into the intestinal lumen, and the basal (bottom) surface **2** rests on extracellular matrix (ECM). The ECM associated with epithelial cells is usually organized into various interconnected layers (e.g., the basal lamina, connecting fibers, connective tissue), in which large, interdigitating ECM macromolecules bind to one another and to the cells **3**. Cell-adhesion molecules (CAMs) bind to CAMs on other cells, mediating cell–cell adhesions **4**, and adhesion receptors bind to various components of the ECM, mediating cell–matrix adhesions **5**. Both types of cell-surface adhesion molecules are usually integral membrane proteins whose cytosolic domains often bind to multiple intracellular adapter proteins. These adapters, directly or indirectly, link the CAM to the cytoskeleton (actin or intermediate filaments) and to

intracellular signaling pathways. As a consequence, information can be transferred by CAMs and the macromolecules to which they bind from the cell exterior into the intracellular environment, and vice versa. In some cases, a complex aggregate of CAMs, adapters, and associated proteins is assembled. Specific localized aggregates of CAMs or adhesion receptors form various types of cell junctions that play important roles in holding tissues together and facilitating communication between cells and their environment. Tight junctions **6**, lying just under the microvilli, prevent the diffusion of many substances through the extracellular spaces between the cells. Gap junctions **7** allow the movement through connexon channels of small molecules and ions between the cytosols of adjacent cells. The remaining three types of junctions, adherens junctions **8**, spot desmosomes **9**, and hemidesmosomes **10**, link the cytoskeleton of a cell to other cells or the ECM. [See V. Vasioukhin and E. Fuchs, 2001, *Curr. Opin. Cell Biol.* **13**:76.]

adhesion) through the binding of **adhesion receptors** in the plasma membrane to components of the surrounding **extracellular matrix (ECM),** a complex interdigitating meshwork of proteins and polysaccharides secreted by cells into the spaces between them. These two basic types of interactions not only allow cells to aggregate into distinct tissues but also provide a means for the bidirectional transfer of information between the exterior and the interior of cells.

In this chapter, we examine the various types of adhesive molecules and how they interact. The evolution of plants and animals is thought to have diverged before multicellular organisms arose. Thus multicellularity and the molecular means for assembling tissues and organs must have arisen independently in animal and plant lineages. Not surprisingly, then, animals and plants exhibit many differences in the organization and development of tissues. For this reason, we first consider the organization of epithelial and nonepithelial tissues in animals and then deal separately with plant tissues. Although most cells in living organisms exist within tissues, our understanding about cells depends greatly on the study of isolated cells. Hence, we present some general features of working with populations of cells removed from tissues and organisms in the last section of this chapter.

6.1 Cell–Cell and Cell–Matrix Adhesion: An Overview

We begin with a brief orientation to the various types of adhesive molecules, their major functions in organisms, and their evolutionary origin. In subsequent sections, we examine in detail the unique structures and properties of the various participants in cell–cell and cell–matrix interactions in animals.

Cell-Adhesion Molecules Bind to One Another and to Intracellular Proteins

A large number of CAMs fall into four major families: the **cadherins, immunoglobulin (Ig) superfamily, integrins,** and **selectins.** As the schematic structures in Figure 6-2 illustrate, many CAMs are mosaics of multiple distinct domains, many

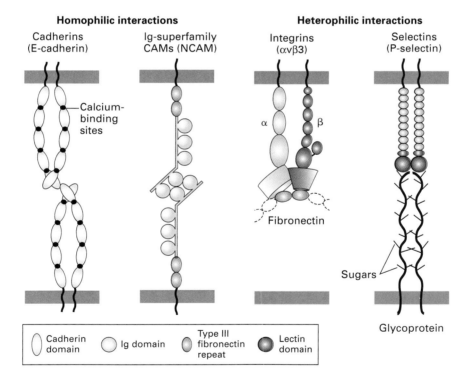

Homophilic interactions

Cadherins (E-cadherin)   Ig-superfamily CAMs (NCAM)

Calcium-binding sites

Heterophilic interactions

Integrins (αvβ3)   Selectins (P-selectin)

α   β

Fibronectin

Sugars

Glycoprotein

| ⬭ Cadherin domain | ◯ Ig domain | ◯ Type III fibronectin repeat | ● Lectin domain |

▲ **FIGURE 6-2 Major families of cell-adhesion molecules (CAMs) and adhesion receptors.** Dimeric E-cadherins most commonly form homophilic (self) cross-bridges with E-cadherins on adjacent cells. Members of the immunoglobulin (Ig) superfamily of CAMs can form both homophilic linkages (shown here) and heterophilic (nonself) linkages. Selectins, shown as dimers, contain a carbohydrate-binding lectin domain that recognizes specialized sugar structures on glycoproteins (shown here) and glycolipids on adjacent cells. Heterodimeric integrins (for example, αv and β3 chains) function as CAMs or as adhesion receptors (shown here) that bind to very large, multiadhesive matrix proteins such as fibronectin, only a small part of which is shown here (see also Figure 6-25). Note that CAMs often form higher-order oligomers within the plane of the plasma membrane. Many adhesive molecules contain multiple distinct domains, some of which are found in more than one kind of CAM. The cytoplasmic domains of these proteins are often associated with adapter proteins that link them to the cytoskeleton or to signaling pathways. [See R. O. Hynes, 1999, *Trends Cell Biol.* **9**(12):M33, and R. O. Hynes, 2002, *Cell* **110**:673–687.]

of which can be found in more than one kind of CAM. They are called "repeats" when they exist multiple times in the same molecule. Some of these domains confer the binding specificity that characterizes a particular protein. Some other membrane proteins, whose structures do not belong to any of the major classes of CAMs, also participate in cell–cell adhesion in various tissues.

CAMs mediate, through their extracellular domains, adhesive interactions between cells of the same type (*homotypic* adhesion) or between cells of different types (*heterotypic* adhesion). A CAM on one cell can directly bind to the same kind of CAM on an adjacent cell (*homophilic* binding) or to a different class of CAM (*heterophilic* binding). CAMs can be broadly distributed along the regions of plasma membranes that contact other cells or clustered in discrete patches or spots called **cell junctions**. Cell–cell adhesions can be tight and long lasting or relatively weak and transient. The associations between nerve cells in the spinal cord or the metabolic cells in the liver exhibit tight adhesion. In contrast, immune-system cells in the blood can exhibit only weak, short-lasting interactions, allowing them to roll along and pass through a blood vessel wall on their way to fight an infection within a tissue.

The cytosol-facing domains of CAMs recruit sets of multifunctional adapter proteins (see Figure 6-1). These adapters act as linkers that directly or indirectly connect CAMs to elements of the cytoskeleton (Chapter 5); they can also recruit intracellular molecules that function in signaling pathways to control protein activity and gene expression (Chapters 13 and 14). In some cases, a complex aggregate of CAMs, adapter proteins, and other associated proteins is assembled at the inner surface of the plasma membrane. Because cell–cell adhesions are intrinsically associated with the cytoskeleton and signaling pathways, a cell's surroundings influence its shape and functional properties ("outside-in"

effects); likewise, cellular shape and function influence a cell's surroundings ("inside-out" effects). Thus *connectivity* and *communication* are intimately related properties of cells in tissues.

The formation of many cell–cell adhesions entails two types of molecular interactions (Figure 6-3). First, CAMs on one cell associate laterally through their extracellular domains or cytosolic domains or both into homodimers or higher-order oligomers in the plane of the cell's plasma membrane; these interactions are called intracellular, lateral, or cis interactions. Second, CAM oligomers on one cell bind to the same or different CAMs on an adjacent cell; these interactions are called intercellular or trans interactions. Trans interactions sometimes induce additional cis interactions and, as a consequence, yet even more trans interactions.

Adhesive interactions between cells vary considerably, depending on the particular CAMs participating and the tissue. Just like Velcro, very tight adhesion can be generated when many weak interactions are combined together in a small, well-defined area. Furthermore, the association of intracellular molecules with the cytosolic domains of CAMs can dramatically influence the intermolecular interactions of CAMs by promoting their cis association (clustering) or by altering their conformation. Among the many variables that determine the nature of adhesion between two cells are the binding affinity of the interacting molecules (thermodynamic properties); the overall "on" and "off" rates of association and dissociation for each interacting molecule (kinetic properties); the spatial distribution (clustering, high or low density) of adhesion molecules (geometric properties); the active versus inactive states of CAMs with respect to adhesion (biochemical properties); and external forces such as the laminar and turbulent flow of cells in the circulatory system (mechanical properties).

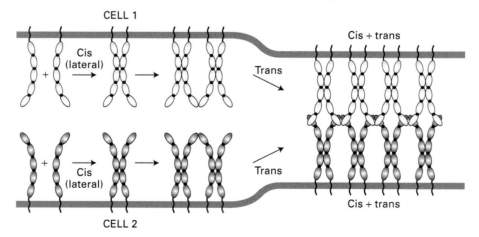

▲ **FIGURE 6-3 Schematic model for the generation of cell–cell adhesions**. Lateral interactions between cell-adhesion molecules (CAMs) within the plasma membrane of a cell form dimers and larger oligomers. The parts of the molecules that participate in these cis interactions vary among the different CAMs. Subsequent trans interactions between distal domains of CAMs on adjacent cells generate a zipperlike strong adhesion between the cells. [Adapted from M. S. Steinberg and P. M. McNutt, 1999, *Curr. Opin. Cell Biol.* **11**:554.]

The Extracellular Matrix Participates in Adhesion and Other Functions

Certain cell-surface receptors, including some integrins, can bind components of the extracellular matrix (ECM), thereby indirectly adhering cells to each other through their interactions with the matrix. Three abundant ECM components are proteoglycans, a unique type of glycoprotein; collagens, proteins that often form fibers; and soluble multiadhesive matrix proteins (e.g., fibronectin). The relative volumes of cells versus matrix vary greatly among different animal tissues and organs. Some connective tissue, for instance, is mostly matrix, whereas many organs are composed of very densely packed cells with relatively little matrix.

Although the extracellular matrix generally provides mechanical support to tissues, it serves several other functions as well. Different combinations of ECM components tailor the extracellular matrix for specific purposes: strength in a tendon, tooth, or bone; cushioning in cartilage; and adhesion in most tissues. In addition, the composition of the matrix, which can vary, depending on the anatomical site and physiological status of a tissue, can let a cell know where it is and what it should do (environmental cues). Changes in ECM components, which are constantly being remodeled, degraded, and resynthesized locally, can modulate the interactions of a cell with its environment. The matrix also serves as a reservoir for many extracellular signaling molecules that control cell growth and differentiation. In addition, the matrix provides a lattice through or on which cells can move, particularly in the early stages of tissue assembly. Morphogenesis—the later stage of embryonic development in which tissues, organs, and body parts are formed by cell movements and rearrangements—also is critically dependent on cell–matrix adhesion as well as cell–cell adhesion.

Diversity of Animal Tissues Depends on Evolution of Adhesion Molecules with Various Properties

Cell–cell adhesions and cell–matrix adhesions are responsible for the formation, composition, architecture, and function of animal tissues. Not surprisingly, adhesion molecules of animals are evolutionarily ancient and are some of the most highly conserved proteins among multicellular (metazoan) organisms. Sponges, the most primitive metazoans, express certain CAMs and multiadhesive ECM molecules whose structures are strikingly similar to those of the corresponding human proteins. The evolution of organisms with complex tissues and organs has depended on the evolution of diverse CAMs, adhesion receptors, and ECM molecules with novel properties and functions, whose levels of expression differ in different types of cells.

The diversity of adhesive molecules arises in large part from two phenomena that can generate numerous closely related proteins, called **isoforms,** that constitute a protein family. In some cases, the different members of a protein family are encoded by multiple genes that arose from a common ancestor by gene duplication and divergent evolution (Chapter 9). Analyses of gene and cDNA sequences can provide evidence for the existence of such a set of related genes, or gene family. In other cases, a single gene produces an RNA transcript that can undergo alternative splicing to yield multiple mRNAs, each encoding a distinct isoform (Chapter 4). Alternative splicing thus increases the number of proteins that can be expressed from one gene. Both of these phenomena contribute to the diversity of some protein families such as the cadherins. Particular isoforms of an adhesive protein are often expressed in some cell types but not others, accounting for their differential distribution in various tissues.

KEY CONCEPTS OF SECTION 6.1

Cell–Cell and Cell–Matrix Adhesion: An Overview

■ Cell-adhesion molecules (CAMs) mediate direct cell–cell adhesions (homotypic and heterotypic), and cell-surface adhesion receptors mediate cell–matrix adhesions (see Figure 6-1). These interactions bind cells into tissues and facilitate communication between cells and their environments.

■ The cytosolic domains of CAMs and adhesion receptors bind multifunctional adapter proteins that mediate interaction with cytoskeletal fibers and intracellular signaling proteins.

■ The major families of cell-surface adhesion molecules are the cadherins, selectins, Ig-superfamily CAMs, and integrins (see Figure 6-2).

■ Tight cell–cell adhesions entail both cis (lateral or intracellular) oligomerization of CAMs and trans (intercellular) interaction of like (homophilic) or different (heterophilic) CAMs (see Figure 6-3).

■ The extracellular matrix (ECM) is a complex meshwork of proteins and polysaccharides that contributes to the structure and function of a tissue.

■ The evolution of CAMs, adhesion receptors, and ECM molecules with specialized structures and functions permits cells to assemble into diverse classes of tissues with varying functions.

6.2 Sheetlike Epithelial Tissues: Junctions and Adhesion Molecules

In general, the external and internal surfaces of organs are covered by a sheetlike layer of epithelial tissue called an **epithelium.** Cells that form epithelial tissues are said to be *polarized* because their plasma membranes are organized into at least two discrete regions. Typically, the distinct surfaces of a polarized epithelial cell are called the **apical** (top), basal

(a) Simple columnar

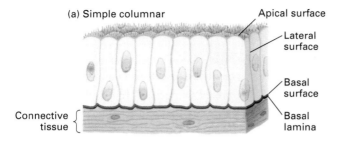

Apical surface

Lateral surface

Basal surface

Basal lamina

Connective tissue

(b) Simple squamous

(c) Transitional

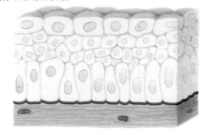

(d) Stratified squamous (nonkeratinized)

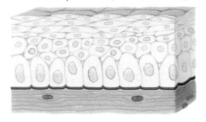

▲ **FIGURE 6-4 Principal types of epithelium.** The apical and basolateral surfaces of epithelial cells exhibit distinctive characteristics. (a) Simple columnar epithelia consist of elongated cells, including mucus-secreting cells (in the lining of the stomach and cervical tract) and absorptive cells (in the lining of the small intestine). (b) Simple squamous epithelia, composed of thin cells, line the blood vessels (endothelial cells/endothelium) and many body cavities. (c) Transitional epithelia, composed of several layers of cells with different shapes, line certain cavities subject to expansion and contraction (e.g., the urinary bladder). (d) Stratified squamous (nonkeratinized) epithelia line surfaces such as the mouth and vagina; these linings resist abrasion and generally do not participate in the absorption or secretion of materials into or out of the cavity. The basal lamina, a thin fibrous network of collagen and other ECM components, supports all epithelia and connects them to the underlying connective tissue.

(base or bottom), and lateral (side) surfaces (Figure 6-4). The basal surface usually contacts an underlying extracellular matrix called the **basal lamina,** whose composition and function are discussed in Section 6.3. Often the basal and lateral surfaces are similar in composition and together are called

the **basolateral** surface. The basolateral surfaces of most epithelia are usually on the side of the cell closest to the blood vessels. In animals with closed circulatory systems, blood flows through vessels whose inner lining is composed of flattened epithelial cells called endothelial cells. The apical side of endothelial cells, which faces the blood, is usually called the *luminal* surface, and the opposite basal side, the *abluminal* surface.

Epithelia in different body locations have characteristic morphologies and functions (see Figure 6-4). Stratified (multilayered) epithelia commonly serve as barriers and protective surfaces (e.g., the skin), whereas simple (single-layer) epithelia often selectively move ions and small molecules from one side of the layer to the other. For instance, the simple columnar epithelium lining the stomach secretes hydrochloric acid into the stomach lumen; a similar epithelium lining the small intestine transports products of digestion (e.g., glucose and amino acids) from the lumen of the intestine across the basolateral surface into the blood (Chapter 7). The simple columnar epithelium lining the small intestine has numerous fingerlike projections (100 nm in diameter) called **microvilli** (singular, microvillus) that extend from the luminal (apical) surface (see Figure 5-45). The upright orientation of a microvillus is maintained by numerous connections between the surrounding plasma membrane and a central bundle of actin microfilaments, which extend into the cell and interact with keratin intermediate filaments (see Figure 5-28). Microvilli greatly increase the area of the apical surface and thus the number of proteins that it can contain, enhancing the absorptive capacity of the intestinal epithelium.

Here we describe the various cell junctions and CAMs that play key roles in the assembly and functioning of epithelial sheets. In Section 6.3, we consider the components of the extracellular matrix intimately associated with epithelia.

Specialized Junctions Help Define the Structure and Function of Epithelial Cells

All epithelial cells in a sheet are connected to one another and the extracellular matrix by specialized cell junctions consisting of dense clusters of CAMs. Although hundreds of individual CAM-mediated interactions are sufficient to cause cells to adhere, junctions play special roles in imparting strength and rigidity to a tissue, transmitting information between the extracellular and the intracellular space, controlling the passage of ions and molecules across cell layers, and serving as conduits for the movement of ions and molecules from the cytoplasm of one cell to that of its immediate neighbor.

Three major classes of animal cell junctions are prominent features of the intestinal epithelium (Figure 6-5; see also Figure 6-1). **Anchoring junctions** and **tight junctions** perform the key task of holding cells together into tissues. These junctions are organized into three parts: adhesive proteins in the plasma membrane that connect one cell to another cell (CAMs) or to the extracellular matrix (adhesion receptors); adapter proteins, which connect the CAMs or adhesion re-

(a)   (b)

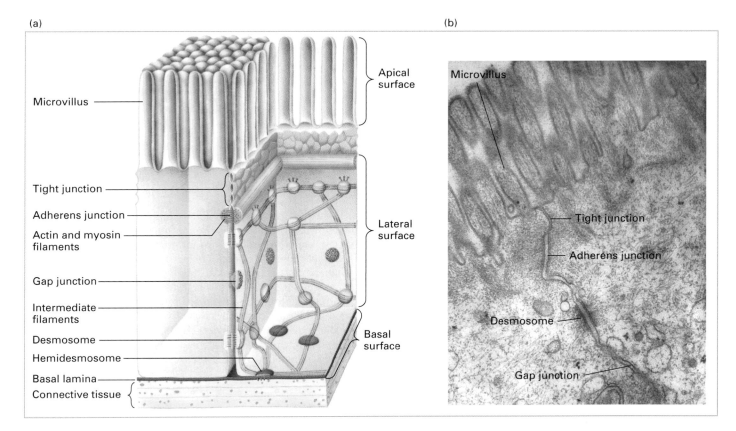

▲ **FIGURE 6-5 The principal types of cell junctions that connect the columnar epithelial cells lining the small intestine.** (a) Schematic cutaway drawing of intestinal epithelial cells. The basal surface of the cells rests on a basal lamina, and the apical surface is packed with fingerlike microvilli that project into the intestinal lumen. Tight junctions, lying just under the microvilli, prevent the diffusion of many substances between the intestinal lumen and the blood through the extracellular space between cells. Gap junctions allow the movement of small molecules and ions between the cytosols of adjacent cells. The remaining three types of junctions—adherens junctions, spot desmosomes, and hemidesmosomes—are critical to cell–cell and cell–matrix adhesion and signaling. (b) Electron micrograph of a thin section of intestinal epithelial cells, showing relative locations of the different junctions. [Part (b) C. Jacobson et al., 2001, *Journal Cell Biol.* **152**:435–450.]

ceptors to cytoskeletal filaments and signaling molecules; and the cytoskeletal filaments themselves. Tight junctions also control the flow of solutes between the cells forming an epithelial sheet. **Gap junctions** permit the rapid diffusion of small, water-soluble molecules between the cytoplasm of adjacent cells. Although present in epithelia, gap junctions are also abundant in nonepithelial tissues and structurally are very different from anchoring junctions and tight junctions; they also bear some resemblance to an important cell–cell junction in plants. For these reasons, we wait to consider gap junctions at the end of Section 6.5.

Of the three types of anchoring junctions present in epithelial cells, two participate in cell–cell adhesion, whereas the third participates in cell–matrix adhesion. *Adherens junctions*, which connect the lateral membranes of adjacent epithelial cells, are usually located near the apical surface, just below the tight junctions (see Figures 6-1 and 6-5). A circumferential belt of actin and myosin filaments in a complex with the adherens junction functions as a tension cable that can internally brace the cell and thereby control its shape.

Epithelial and some other types of cells, such as smooth muscle, are also bound tightly together by *desmosomes*, button-like points of contact sometimes called spot desmosomes. *Hemidesmosomes*, found mainly on the basal surface of epithelial cells, anchor an epithelium to components of the underlying extracellular matrix, much like nails holding down a carpet. Bundles of intermediate filaments, running parallel to the cell surface or through the cell, rather than actin filaments, interconnect spot desmosomes and hemidesmosomes, imparting shape and rigidity to the cell.

Desmosomes and hemidesmosomes also transmit shear forces from one region of a cell layer to the epithelium as a whole, providing strength and rigidity to the entire epithelial cell layer. These junctions are especially important in maintaining the integrity of skin epithelia. For instance, mutations that interfere with hemidesmosomal anchoring in the skin can lead to blistering in which the epithelium becomes detached from its matrix foundation and extracellular fluid accumulates at the basolateral surface, forcing the skin to balloon outward.

Ca²⁺-Dependent Homophilic Cell–Cell Adhesion in Adherens Junctions and Desmosomes Is Mediated by Cadherins

The primary CAMs in adherens junctions and desmosomes belong to the **cadherin** family. In vertebrates and invertebrates, this protein family of more than 100 members can be grouped into at least six subfamilies. The diversity of cadherins arises from the presence of multiple cadherin genes and alternative RNA splicing, which generates multiple mRNAs from one gene.

Cadherins are key molecules in cell–cell adhesion and cell signaling, and they play a critical role during tissue differentiation. The "classical" E-, P-, and N-cadherins are the most widely expressed, particularly during early differentiation. Sheets of polarized epithelial cells, such as those that line the small intestine or kidney tubules, contain abundant E-cadherin along their lateral surfaces. Although E-cadherin is concentrated in adherens junctions, it is present throughout the lateral surfaces where it is thought to link adjacent cell membranes. The brain expresses the largest number of different cadherins, presumably owing to the necessity of forming many very specific cell–cell contacts to help establish its complex wiring diagram.

Classical Cadherins The results of experiments with L cells, a line of cultured mouse fibroblasts grown in the laboratory, demonstrated that E-cadherin and P-cadherin preferentially mediate homophilic interactions. L cells express no cadherins and adhere poorly to themselves or to other types of cultured cells. When genes encoding either E-cadherin or P-cadherin were introduced into L cells with the use of techniques described in Chapter 9, the resulting engineered L cells expressed the encoded cadherin. These cadherin-expressing L cells were found to adhere preferentially to cells expressing the same type

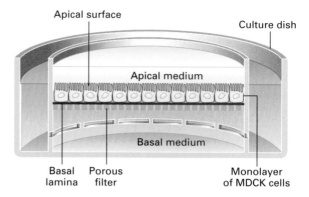

▲ **EXPERIMENTAL FIGURE 6-6 Madin-Darby canine kidney (MDCK) cells grown in specialized containers provide a useful experimental system for studying epithelial cells.** MDCK cells form a polarized epithelium when grown on a porous membrane filter coated on one side with collagen and other components of the basal lamina. With the use of the special culture dish shown here, the medium on each side of the filter (apical and basal sides of the monolayer) can be experimentally manipulated and the movement of molecules across the layer monitored. Anchoring junctions and tight junctions form only if the growth medium contains sufficient Ca²⁺.

of cadherin molecules; that is, they mediate homophilic interactions. The L cells expressing E-cadherin also exhibited the polarized distribution of a membrane protein similar to that in epithelial cells, and they formed epithelial-like aggregates with one another and with epithelial cells isolated from lungs.

The adhesiveness of cadherins depends on the presence of extracellular Ca²⁺, the property that gave rise to their name (calcium *adhering*). For example, the adhesion of engineered L cells expressing E-cadherin is prevented when the cells are bathed in a solution (growth medium) that is low in Ca²⁺. The role of E-cadherin in adhesion can also be demonstrated

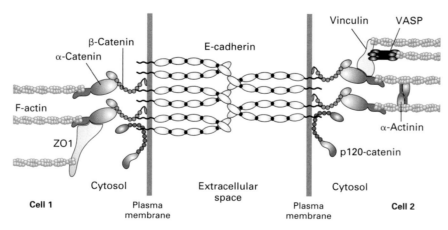

▲ **FIGURE 6-7 Protein constituents of typical adherens junctions.** The exoplasmic domains of E-cadherin dimers clustered at adherens junctions on adjacent cells (1 and 2) form Ca⁺²-dependent homophilic interactions. The cytosolic domains of the E-cadherins bind directly or indirectly to multiple adapter proteins that connect the junctions to actin filaments (F-actin) of the cytoskeleton and participate in intracellular signaling pathways (e.g., β-catenin). Somewhat different sets of adapter proteins are illustrated in the two cells shown to emphasize that a variety of adapters can interact with adherens junctions, which can thereby participate in diverse activities. [Adapted from V. Vasioukhin and E. Fuchs, 2001, *Curr. Opin. Cell Biol.* **13**:76.]

in experiments with cultured cells called *Madin-Darby canine kidney (MDCK) cells*. When grown in specialized containers, these cells form a continuous one-cell-thick sheet (monolayer) of polarized kidneylike epithelial cells (Figure 6-6). In this experimental system, the addition of an antibody that binds to E-cadherin, preventing its homophilic interactions, blocks the Ca^{2+}-dependent attachment of suspended MDCK cells to a substrate and the subsequent formation of intercellular adherens junctions.

Each classical cadherin contains a single transmembrane domain, a relatively short C-terminal cytosolic domain, and five extracellular "cadherin" domains (see Figure 6-2). The extracellular domains are necessary for Ca^{2+} binding and cadherin-mediated cell–cell adhesion. Cadherin-mediated adhesion entails both lateral (intracellular) and trans (intercellular) molecular interactions as described previously (see Figure 6-3). The Ca^{2+}-binding sites, located between the cadherin repeats, serve to rigidify the cadherin oligomers. The cadherin oligomers subsequently form intercellular complexes to generate cell–cell adhesion and then additional lateral contacts, resulting in a "zipping up" of cadherins into clusters. In this way, multiple low-affinity interactions sum to produce a very tight intercellular adhesion.

The results of domain swap experiments, in which an extracellular domain of one kind of cadherin is replaced with the corresponding domain of a different cadherin, have indicated that the specificity of binding resides, at least in part, in the most distal extracellular domain, the N-terminal domain. In the past, cadherin-mediated adhesion was commonly thought to require only head-to-head interactions between the N-terminal domains of cadherin oligomers on adjacent cells, as depicted in Figure 6-3. However, the results of some experiments suggest that under some experimental conditions at least three cadherin domains from each molecule, not just the N-terminal domains, participate by interdigitation in trans associations.

The C-terminal cytosolic domain of classical cadherins is linked to the actin cytoskeleton by a number of cytosolic adapter proteins (Figure 6-7). These linkages are essential for strong adhesion, apparently owing primarily to their contributing to increased lateral associations. For example, disruption of the interactions between classical cadherins and α- or β-catenin—two common adapter proteins that link these cadherins to actin filaments—dramatically reduces cadherin-mediated cell–cell adhesion. This disruption occurs spontaneously in tumor cells, which sometimes fail to express α-catenin, and can be induced experimentally by depleting the cytosolic pool of accessible β-catenin. The cytosolic domains of cadherins also interact with intracellular signaling molecules such as β-catenin and p120-catenin. Interestingly, β-catenin not only mediates cytoskeletal attachment but can also translocate to the nucleus and alter gene transcription (see Figure 15-32).

Although E-cadherins exhibit primarily homophilic binding, some cadherins mediate heterophilic interactions. Importantly, each classical cadherin has a characteristic tissue distribution. In the course of differentiation, the amount or nature of the cell-surface cadherins changes, affecting many

aspects of cell–cell adhesion and cell migration. For instance, the reorganization of tissues during morphogenesis is often accompanied by the conversion of nonmotile epithelial cells into motile precursor cells for other tissues (mesenchymal cells). Such epithelial-to-mesenchymal transitions are associated with a reduction in the expression of E-cadherin. The conversion of epithelial cells into cancerous melanoma cells also is marked by a loss of E-cadherin activity. The resulting decrease in cell–cell adhesion permits melanoma cells to invade the underlying tissue and spread throughout the body.

Desmosomal Cadherins Desmosomes (Figure 6-8) contain two specialized cadherin proteins, *desmoglein* and

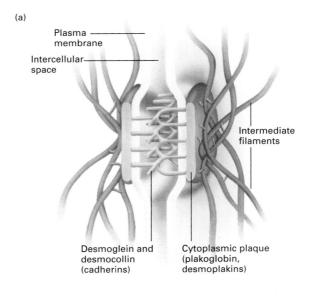

(a)

Plasma membrane

Intercellular space

Intermediate filaments

Desmoglein and desmocollin (cadherins)

Cytoplasmic plaque (plakoglobin, desmoplakins)

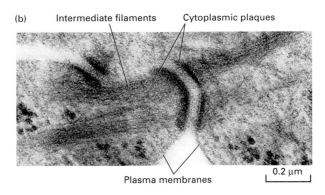

(b)   Intermediate filaments   Cytoplasmic plaques

Plasma membranes   0.2 μm

▲ **FIGURE 6-8 Desmosomes.** (a) Schematic model showing components of a desmosome between epithelial cells and attachments to the sides of keratin intermediate filaments, which crisscross the interior of cells. The transmembrane CAMs, desmoglein and desmocollin, belong to the cadherin family. (b) Electron micrograph of a thin section of a desmosome connecting two cultured differentiated human keratinocytes. Bundles of intermediate filaments radiate from the two darkly staining cytoplasmic plaques that line the inner surface of the adjacent plasma membranes. [Part (a) see B. M. Gumbiner, 1993, *Neuron* **11**:551, and D. R. Garrod, 1993, *Curr. Opin. Cell Biol.* **5**:30. Part (b) courtesy of R. van Buskirk.]

desmocollin, whose cytosolic domains are distinct from those in the classical cadherins. The cytosolic domains of desmosomal cadherins interact with plakoglobin (similar in structure to β-catenin) and the plakophilins. These adapter proteins, which form the thick cytoplasmic plaques characteristic of desmosomes, in turn interact with intermediate filaments. Thus desmosomes and adherens junctions are linked to different cytoskeletal fibers.

 The cadherin desmoglein was first identified by an unusual, but revealing, skin disease called *pemphigus vulgaris,* an autoimmune disease. Patients with autoimmune disorders synthesize antibodies that bind to a normal body protein. In this case, the autoantibodies disrupt adhesion between epithelial cells, causing blisters of the skin and mucous membranes. The predominant autoantibody was shown to be specific for desmoglein; indeed, the addition of such antibodies to normal skin induces the formation of blisters and disruption of cell adhesion.∎

Tight Junctions Seal Off Body Cavities and Restrict Diffusion of Membrane Components

For polarized epithelial cells to carry out their functions as barriers and mediators of selective transport, extracellular fluids surrounding their apical and basolateral membranes must be kept separate. The tight junctions between adjacent epithelial cells are usually located just below the apical surface and help establish and maintain cell polarity (see Figures 6-1 and 6-5). These specialized regions of the plasma membrane form a barrier that seals off body cavities such as the intestine, the stomach lumen, the blood (e.g., the blood–brain barrier), and the bile duct in the liver.

▶ **FIGURE 6-9 Tight junctions.** (a) Freeze-fracture preparation of tight junction zone between two intestinal epithelial cells. The fracture plane passes through the plasma membrane of one of the two adjacent cells. A honeycomb-like network of ridges and grooves below the microvilli constitutes the tight junction zone. (b) Schematic drawing shows how a tight junction might be formed by the linkage of rows of protein particles in adjacent cells. In the inset micrograph of an ultrathin sectional view of a tight junction, the adjacent cells can be seen in close contact where the rows of proteins interact. (c) As shown in these schematic drawings of the major proteins in tight junctions, both occludin and claudin-1 contain four transmembrane helices, whereas the junction adhesion molecule (JAM) has a single transmembrane domain and a large extracellular region. See text for discussion. [Part (a) courtesy of L. A. Staehelin. Drawing in part (b) adapted from L. A. Staehelin and B. E. Hull, 1978, *Sci. Am.* **238**(5):140, and D. Goodenough, 1999, *Proc. Nat'l. Acad. Sci. USA* **96**:319. Photograph in part (b) courtesy of S. Tsukita et al., 2001, *Nature Rev. Mol. Cell Biol.* **2**:285. Drawing in part (c) adapted from S. Tsukita et al., 2001, *Nature Rev. Mol. Cell Biol.* **2**:285.]

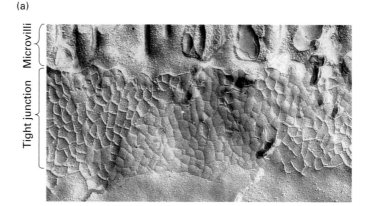

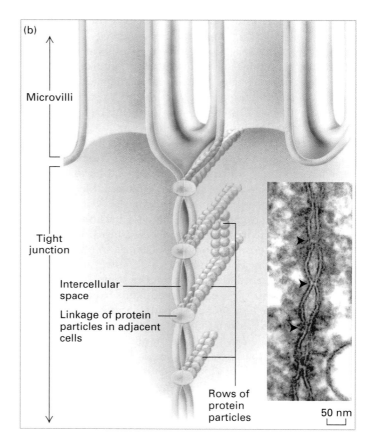

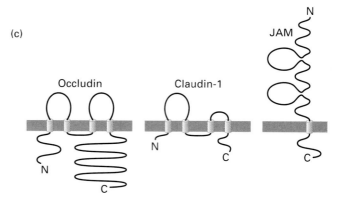

Tight junctions prevent the diffusion of macromolecules and to varying degrees impede the diffusion of small water-soluble molecules and ions across an epithelial sheet in the spaces between cells. They also maintain the polarity of epithelial cells by preventing the diffusion of membrane proteins and glycolipids (lipids with covalently attached sugars) between the apical and the basolateral regions of the plasma membrane, ensuring that these regions contain different membrane components. As a consequence, movement of many nutrients across the intestinal epithelium is in large part through the *transcellular pathway*. In this pathway, specific transport proteins in the apical membrane import small molecules from the intestinal lumen into cells; other transport proteins located in the basolateral membrane then export these molecules into the extracellular space. Such transcellular transport is covered in detail in Chapter 7.

Tight junctions are composed of thin bands of plasma-membrane proteins that completely encircle a polarized cell and are in contact with similar thin bands on adjacent cells. When thin sections of cells are viewed in an electron microscope, the lateral surfaces of adjacent cells appear to touch each other at intervals and even to fuse in the zone just below the apical surface (see Figure 6-5b). In freeze-fracture preparations, tight junctions appear as an interlocking network of ridges in the plasma membrane (Figure 6-9a). More specifically, there appear to be ridges on the cytosolic face of the plasma membrane of each of the two contacting cells. Corresponding grooves are found on the exoplasmic face.

Very high magnification reveals that rows of protein particles 3–4 nm in diameter form the ridges seen in freeze-fracture micrographs of tight junctions. In the model shown in Figure 6-9b, the tight junction is formed by a double row of these particles, one row donated by each cell. The two principal integral-membrane proteins found in tight junctions are *occludin* and *claudin*. Initially, investigators thought that occludin was the only essential protein component of tight junctions. However, when investigators engineered mice with mutations inactivating the occludin gene, the mice still had morphologically distinct tight junctions. (This technique, called **gene knockout,** is described in Chapter 9.) Further analysis led to the discovery of claudin. Each of these proteins has four membrane-spanning α helices (Figure 6-9c). The claudin multigene family encodes numerous homologous proteins (isoforms) that exhibit distinct tissue-specific patterns of expression. Recently, a group of *junction adhesion molecules (JAMs)* have been found to contribute to homophilic adhesion and other functions of tight junctions. These molecules, which contain a single transmembrane α helix, belong to the Ig superfamily of CAMs. The extracellular domains of rows of occludin, claudin, and JAM proteins in the plasma membrane of one cell apparently form extremely tight links with similar rows of the same proteins in an adjacent cell, creating a tight seal. Treatment of an epithelium with the protease trypsin destroys the tight junctions, supporting the proposal that proteins are essential structural components of these junctions.

The long C-terminal cytosolic segment of occludin binds to PDZ domains in certain large cytosolic adapter proteins. These domains are found in various cytosolic proteins and mediate binding to the C-termini of particular plasma-membrane proteins. PDZ-containing adapter proteins associated with occludin are bound, in turn, to other cytoskeletal and signaling proteins and to actin fibers. These interactions appear to stabilize the linkage between occludin and claudin molecules that is essential for maintaining the integrity of tight junctions.

A simple experiment demonstrates the impermeability of certain tight junctions to many water-soluble substances. In this experiment, lanthanum hydroxide (an electron-dense colloid of high molecular weight) is injected into the pancreatic blood vessel of an experimental animal; a few minutes later, the pancreatic acinar cells, which are specialized epithelial cells, are fixed and prepared for microscopy. As shown in Figure 6-10, the lanthanum hydroxide diffuses from the blood into the space that separates the lateral surfaces of adjacent acinar cells, but cannot penetrate past the tight junction.

The importance of Ca^{2+} to the formation and integrity of tight junctions has been demonstrated in studies with MDCK cells in the experimental system described previously (see Figure 6-7). If the growth medium in the chamber contains very low concentrations of Ca^{2+}, MDCK cells form a monolayer in which the cells are not connected by tight junctions. As a result, fluids and salts flow freely across the cell layer. When sufficient Ca^{2+} is added to the medium, tight junctions form within an hour, and the cell layer becomes impermeable

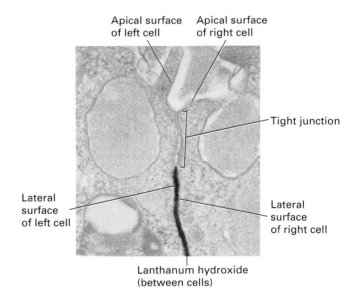

▲ **EXPERIMENTAL FIGURE 6-10 Tight junctions prevent passage of large molecules through extracellular space between epithelial cells.** This experiment, described in the text, demonstrates the impermeability of tight junctions in the pancreas to the large water-soluble colloid lanthanum hydroxide. [Courtesy of D. Friend.]

to fluids and salts. Thus Ca^{2+} is required for the formation of tight junctions as well as for cell–cell adhesion mediated by cadherins.

Plasma-membrane proteins cannot diffuse in the plane of the membrane past tight junctions. These junctions also restrict the lateral movement of lipids in the exoplasmic leaflet of the plasma membrane in the apical and basolateral regions of epithelial cells. Indeed, the lipid compositions of the exoplasmic leaflet in these two regions are distinct. Essentially all glycolipids are present in the exoplasmic face of the apical membrane, as are all proteins linked to the membrane by a glycosylphosphatidylinositol (GPI) anchor (see Figure 5-15). In contrast, lipids in the cytosolic leaflet in the apical and basolateral regions of epithelial cells have the same composition and can apparently diffuse laterally from one region of the membrane to the other.

Differences in Permeability of Tight Junctions Can Control Passage of Small Molecules Across Epithelia

The barrier to diffusion provided by tight junctions is not absolute. Owing at least in part to the varying properties of the different isoforms of claudin located in different tight junctions, their permeability to ions, small molecules, and water varies enormously among different epithelial tissues. In epithelia with "leaky" tight junctions, small molecules can move from one side of the cell layer to the other through the *paracellular pathway* in addition to the transcellular pathway (Figure 6-11).

The leakiness of tight junctions can be altered by intracellular signaling pathways, especially G protein–coupled pathways entailing cyclic AMP and protein kinase C (Chapter 13). The regulation of tight junction permeability is often

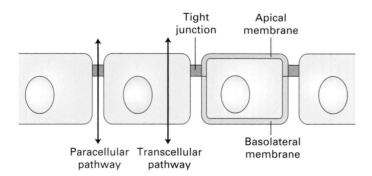

▲ FIGURE 6-11 Transcellular and paracellular pathways of transepithelial transport. Transcellular transport requires the cellular uptake of molecules on one side and subsequent release on the opposite side by mechanisms discussed in Chapters 7 and 17. In paracellular transport, molecules move extracellularly through parts of tight junctions, whose permeability to small molecules and ions depends on the composition of the junctional components and the physiologic state of the epithelial cells. [Adapted from S. Tsukita et al., 2001, *Nature Rev. Mol. Cell Biol.* **2**:285.]

studied by measuring ion flux (electrical resistance) or the movement of radioactive or fluorescent molecules across monolayers of MDCK cells.

 The importance of paracellular transport is illustrated in several human diseases. In hereditary hypomagnesemia, defects in the *claudin16* gene prevent the normal paracellular flow of magnesium through tight junctions in the kidney. This results in an abnormally low blood level of magnesium, which can lead to convulsions. Furthermore, a mutation in the *claudin14* gene causes hereditary deafness, apparently by altering transport around hair cells in the cochlea of the inner ear.

Toxins produced by *Vibrio cholerae*, which causes cholera, and several other enteric (gastrointestinal tract) bacteria alter the permeability barrier of the intestinal epithelium by altering the composition or activity of tight junctions. Other bacterial toxins can affect the ion-pumping activity of membrane transport proteins in intestinal epithelial cells. Toxin-induced changes in tight junction permeability (increased paracellular transport) and in protein-mediated ion-pumping proteins (increased transcellular transport) can result in massive loss of internal body ions and water into the gastrointestinal tract, which in turn leads to diarrhea and potentially lethal dehydration. ∎

Many Cell–Matrix and Some Cell–Cell Interactions Are Mediated by Integrins

The **integrin** family comprises heterodimeric integral membrane proteins that function as adhesion receptors, mediating many cell–matrix interactions (see Figure 6-2). In vertebrates, at least 24 integrin heterodimers, composed of 18 types of α subunits and 8 types of β subunits in various combinations, are known. A single β chain can interact with any one of multiple α chains, forming integrins that bind different ligands. This phenomenon of *combinatorial diversity,* which is found throughout the biological world, allows a relatively small number of components to serve a large number of distinct functions.

In epithelial cells, integrin α6β4 is concentrated in hemidesmosomes and plays a major role in adhering cells to matrix in the underlying basal lamina, as discussed in detail in Section 6.3. Some integrins, particularly those expressed by certain blood cells, participate in heterophilic cell–cell interactions. The members of this large family play important roles in adhesion and signaling in both epithelial and nonepithelial tissues.

Integrins typically exhibit low affinities for their ligands with dissociation constants K_D between 10^{-6} and 10^{-8} mol/L. However, the multiple weak interactions generated by the binding of hundreds or thousands of integrin molecules to their ligands on cells or in the extracellular matrix allow a cell to remain firmly anchored to its ligand-expressing target. Moreover, the weakness of individual integrin-mediated interactions facilitates cell migration.

Parts of both the α and the β subunits of an integrin molecule contribute to the primary extracellular ligand-binding site (see Figure 6-2). Ligand binding to integrins also requires the simultaneous binding of divalent cations (positively charged ions). Like other cell-surface adhesive molecules, the cytosolic region of integrins interacts with adapter proteins that in turn bind to the cytoskeleton and intracellular signaling molecules. Although most integrins are linked to the actin cytoskeleton, the cytosolic domain of the β4 chain in the α6β4 integrin in hemidesmosomes, which is much longer than those of other β integrins, binds to specialized adapter proteins (e.g., plectin) that in turn interact with keratin-based intermediate filaments.

In addition to their adhesion function, integrins can mediate outside-in and inside-out transfer of information (signaling). In outside-in signaling, the engagement of integrins with their extracellular ligands can, through adapter proteins bound to the integrin cytosolic region, influence the cytoskeleton and intracellular signaling pathways. Conversely, in inside-out signaling, intracellular signaling pathways can alter, from the cytoplasm, the structure of integrins and consequently their abilities to adhere to their extracellular ligands and mediate cell–cell and cell–matrix interactions. Integrin-mediated signaling pathways influence processes as diverse as cell survival, cell proliferation, and programmed cell death (Chapter 22). Many cells express several different integrins that bind the same ligand. By selectively regulating the activity of each type of integrin, these cells can fine-tune their cell–cell and cell–matrix interactions and the associated signaling processes.

We will consider various integrins and the regulation of their activity in detail in Section 6.5.

KEY CONCEPTS OF SECTION 6.2

Sheetlike Epithelial Tissues: Junctions and Adhesion Molecules

■ Polarized epithelial cells have distinct apical, basal, and lateral surfaces. Microvilli projecting from the apical surfaces of many epithelial cells considerably expand their surface areas.

■ Three major classes of cell junctions—anchoring junctions, tight junctions, and gap junctions—assemble epithelial cells into sheets and mediate communication between them (see Figures 6-1 and 6-5).

■ Adherens junctions and desmosomes are cadherin-containing anchoring junctions that bind the membranes of adjacent cells, giving strength and rigidity to the entire tissue. Hemidesmosomes are integrin-containing anchoring junctions that attach cells to elements of the underlying extracellular matrix.

■ Cadherins are cell-adhesion molecules (CAMs) responsible for Ca^{2+}-dependent interactions between cells in epithelial and other tissues. They promote strong cell–cell adhesion by mediating both lateral and intercellular interactions.

■ Adapter proteins that bind to the cytosolic domain of cadherins and other CAMs mediate the association of cytoskeletal and signaling molecules with the plasma membrane (see Figure 6-9). Strong cell–cell adhesion depends on the linkage of the interacting CAMs to the cytoskeleton.

■ Tight junctions block the diffusion of proteins and some lipids in the plane of the plasma membrane, contributing to the polarity of epithelial cells. They also limit and regulate the extracellular (paracellular) flow of water and solutes from one side of the epithelium to the other (see Figure 6-11).

■ Integrins are a large family of αβ heterodimeric cell-surface proteins that mediate both cell–cell and cell–matrix adhesions and inside-out and outside-in signaling in numerous tissues.

6.3 The Extracellular Matrix of Epithelial Sheets

In animals, the extracellular matrix helps organize cells into tissues and coordinates their cellular functions by activating intracellular signaling pathways that control cell growth, proliferation, and gene expression. Many functions of the matrix require transmembrane adhesion receptors that bind directly to ECM components and that also interact, through adapter proteins, with the cytoskeleton. The principal class of adhesion receptors that mediate cell–matrix adhesion are integrins, which were introduced in Section 6.2. However, other types of molecules also function as important adhesion receptors in some nonepithelial tissues.

Three types of molecules are abundant in the extracellular matrix of all tissues.

■ Highly viscous **proteoglycans,** a group of glycoproteins that cushion cells and bind a wide variety of extracellular molecules

■ **Collagen** fibers, which provide mechanical strength and resilience

■ Soluble **multiadhesive matrix proteins,** which bind to and cross-link cell-surface adhesion receptors and other ECM components

We begin our description of the structures and functions of these major ECM components in this section, focusing on the molecular components and organization of the basal lamina—the specialized extracellular matrix that helps determine the overall architecture of an epithelial tissue. In Section 6.4, we extend our discussion to specific ECM molecules that are commonly present in nonepithelial tissues.

(a)

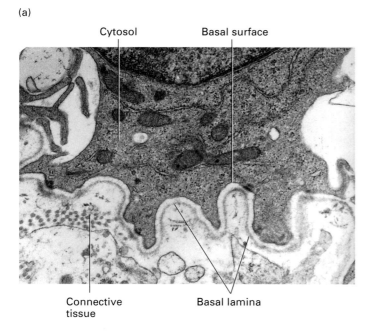

Cytosol   Basal surface

Connective   Basal lamina
tissue

(b)

Plasma membrane

Basal lamina

Cell-surface
receptor proteins   Collagen fibers

▲ **EXPERIMENTAL FIGURE 6-12 The basal lamina
separates epithelial cells and some other cells from
connective tissue.** (a) Transmission electron micrograph of a thin
section of cells (*top*) and underlying connective tissue (*bottom*).
The electron-dense layer of the basal lamina can be seen to
follow the undulation of the basal surface of the cells.
(b) Electron micrograph of a quick-freeze deep-etch preparation of
skeletal muscle showing the relation of the plasma membrane,
basal lamina, and surrounding connective tissue. In this
preparation, the basal lamina is revealed as a meshwork of
filamentous proteins that associate with the plasma membrane
and the thicker collagen fibers of the connective tissue. [Part (a)
courtesy of P. FitzGerald. Part (b) from D. W. Fawcett, 1981, *The Cell*,
2d ed., Saunders/Photo Researchers; courtesy of John Heuser.]

The Basal Lamina Provides a Foundation
for Epithelial Sheets

In animals, epithelia and most organized groups of cells are
underlain or surrounded by the basal lamina, a sheetlike
meshwork of ECM components usually no more than
60–120 nm thick (Figure 6-12; see also Figures 6-1 and
6-4). The basal lamina is structured differently in different
tissues. In columnar and other epithelia (e.g., intestinal
lining, skin), it is a foundation on which only one surface
of the cells rests. In other tissues, such as muscle or fat,
the basal lamina surrounds each cell. Basal laminae play important roles in regeneration after tissue damage and in embryonic development. For instance, the basal lamina helps

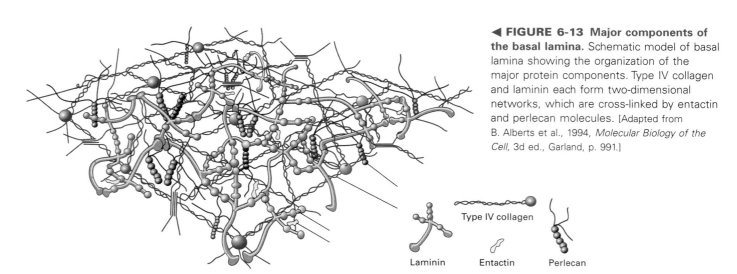

◄ **FIGURE 6-13 Major components of
the basal lamina.** Schematic model of basal
lamina showing the organization of the
major protein components. Type IV collagen
and laminin each form two-dimensional
networks, which are cross-linked by entactin
and perlecan molecules. [Adapted from
B. Alberts et al., 1994, *Molecular Biology of the
Cell*, 3d ed., Garland, p. 991.]

Type IV collagen

Laminin   Entactin   Perlecan

four- and eight-celled embryos adhere together in a ball. In the development of the nervous system, neurons migrate along ECM pathways that contain basal lamina components. Thus the basal lamina is important not only for organizing cells into tissues but also for tissue repair and for guiding migrating cells during tissue formation.

Most of the ECM components in the basal lamina are synthesized by the cells that rest on it. Four ubiquitous protein components are found in basal laminae (Figure 6-13):

■ *Type IV collagen,* trimeric molecules with both rodlike and globular domains that form a two-dimensional network

■ *Laminins,* a family of multiadhesive proteins that form a fibrous two-dimensional network with type IV collagen and that also bind to integrins

■ *Entactin* (also called nidogen), a rodlike molecule that cross-links type IV collagen and laminin and helps incorporate other components into the ECM

■ *Perlecan,* a large multidomain proteoglycan that binds to and cross-links many ECM components and cell-surface molecules

As depicted in Figure 6-1, one side of the basal lamina is linked to cells by adhesion receptors, including α6β4 integrin that binds to laminin in the basal lamina. The other side of the basal lamina is anchored to the adjacent connective tissue by a layer of fibers of collagen embedded in a proteoglycan-rich matrix. In stratified squamous epithelia (e.g., skin), this linkage is mediated by anchoring fibrils of type VII collagen. Together, the basal lamina and this collagen-containing layer (see the micrograph on page 197) form the structure called the *basement membrane.*

Sheet-Forming Type IV Collagen Is a Major Structural Component in Basal Laminae

Type IV collagen, the principal component of all basal lamina, is one of more than 20 types of collagen that participate in the formation of the extracellular matrix in various tissues. Although they differ in certain structural features and tissue distribution, all collagens are trimeric proteins made from three polypeptides called collagen α chains. All three α chains can be identical (homotrimeric) or different (heterotrimeric). A trimeric collagen molecule contains one or more three-stranded segments, each with a similar triple-helical structure (Figure 6-14a). Each strand contributed by one of the α chains is twisted into a left-handed helix, and three such strands from the three α chains wrap around each other to form a right-handed triple helix.

The collagen triple helix can form because of an unusual abundance of three amino acids: glycine, proline, and a modified form of proline called hydroxyproline (see Figure 3-12). They make up the characteristic repeating motif Gly-X-Y, where X and Y can be any amino

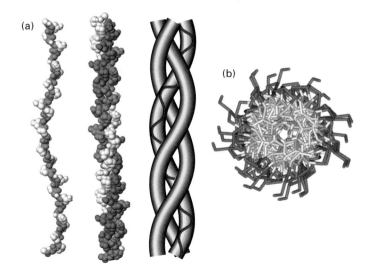

▲ **FIGURE 6-14 The collagen triple helix.** (a) (*Left*) Side view of the crystal structure of a polypeptide fragment whose sequence is based on repeating sets of three amino acids, Gly-X-Y, characteristic of collagen α chains. (*Center*) Each chain is twisted into a left-handed helix, and three chains wrap around each other to form a right-handed triple helix. The schematic model (*right*) clearly illustrates the triple helical nature of the structure. (b) View down the axis of the triple helix. The proton side chains of the glycine residues (orange) point into the very narrow space between the polypeptide chains in the center of the triple helix. In mutations in collagen in which other amino acids replace glycine, the proton in glycine is replaced by larger groups that disrupt the packing of the chains and destablize the triple-helical structure. [Adapted from R. Z. Kramer et al., 2001, *J. Mol. Biol.* **311**(1):131.]

acid but are often proline and hydroxyproline and less often lysine and hydroxylysine. Glycine is essential because its small side chain, a hydrogen atom, is the only one that can fit into the crowded center of the three-stranded helix (Figure 6-14b). Hydrogen bonds help hold the three chains together. Although the rigid peptidyl-proline and peptidyl-hydroxyproline linkages are not compatible with formation of a classic single-stranded α helix, they stabilize the distinctive three-stranded collagen helix. The hydroxyl group in hydroxyproline helps hold its ring in a conformation that stabilizes the three-stranded helix.

The unique properties of each type of collagen are due mainly to differences in (1) the number and lengths of the collagenous, triple-helical segments; (2) the segments that flank or interrupt the triple-helical segments and that fold into other kinds of three-dimensional structures; and (3) the covalent modification of the α chains (e.g., hydroxylation, glycosylation, oxidation, cross-linking). For example, the chains in type IV collagen, which is unique to basal laminae, are designated IVα chains. Mammals express six homologous IVα chains, which assemble into a series of type IV

(a)

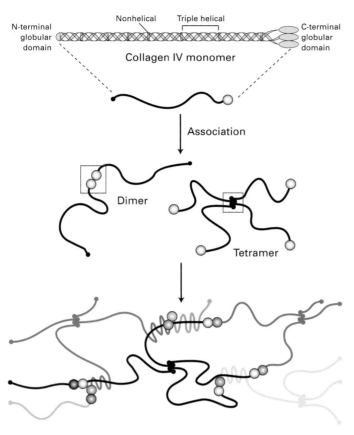

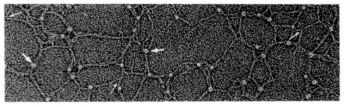

(b) Type IV network

250 nm

▲ FIGURE 6-15 Structure and assembly of type IV collagen.
(a) Schematic representation of type IV collagen. This 400-nm-
long molecule has a small noncollagenous globular domain at the
N-terminus and a large globular domain at the C-terminus. The
triple helix is interrupted by nonhelical segments that introduce
flexible kinks in the molecule. Lateral interactions between triple
helical segments, as well as head-to-head and tail-to-tail
interactions between the globular domains, form dimers,
tetramers, and higher-order complexes, yielding a sheetlike
network. (b) Electron micrograph of type IV collagen network
formed in vitro. The lacy appearance results from the flexibility of
the molecule, the side-to-side binding between triple-helical
segments (thin arrows), and the interactions between C-terminal
globular domains (thick arrows). [Part (a) adapted from A. Boutaud,
2000, *J. Biol. Chem.* **275**:30716. Part (b) courtesy of P. Yurchenco; see
P. Yurchenco and G. C. Ruben, 1987, *J. Cell Biol.* **105**:2559.]

collagens with distinct properties. All subtypes of type IV
collagen, however, form a 400-nm-long triple helix that is interrupted about 24 times with nonhelical segments and
flanked by large globular domains at the C-termini of the
chains and smaller globular domains at the N-termini. The
nonhelical regions introduce flexibility into the molecule.
Through both lateral associations and interactions entailing
the globular N- and C-termini, type IV collagen molecules
assemble into a branching, irregular two-dimensional fibrous
network that forms the lattice on which the basal lamina is
built (Figure 6-15).

 In the kidney, a double basal lamina, the glomerular basement membrane, separates the epithelium
MEDICINE that lines the urinary space from the endothelium
that lines the surrounding blood-filled capillaries. Defects in
this structure, which is responsible for ultrafiltration of the
blood and initial urine formation, can lead to renal failure.
For instance, mutations that alter the C-terminal globular
domain of certain IVα chains are associated with progressive renal failure as well as sensorineural hearing loss and
ocular abnormalities, a condition known as *Alport's
syndrome*. In *Goodpasture's syndrome*, a relatively rare
autoimmune disease, self-attacking, or "auto," antibodies
bind to the α3 chains of type IV collagen found in the
glomerular basement membrane and lungs. This binding
sets off an immune response that causes cellular damage
resulting in progressive renal failure and pulmonary
hemorrhage.∎

Laminin, a Multiadhesive Matrix Protein, Helps Cross-link Components of the Basal Lamina

Multiadhesive matrix proteins are long, flexible molecules
that contain multiple domains responsible for binding various types of collagen, other matrix proteins, polysaccharides,
cell-surface adhesion receptors, and extracellular signaling
molecules (e.g., growth factors and hormones). These proteins are important for organizing the other components of
the extracellular matrix and for regulating cell–matrix adhesion, cell migration, and cell shape in both epithelial and
nonepithelial tissues.

Laminin, the principal multiadhesive matrix protein in
basal laminae, is a heterotrimeric, cross-shaped protein with
a total molecular weight of 820,000 (Figure 6-16). Many
laminin isoforms, containing slightly different polypeptide
chains, have been identified. Globular *LG domains* at the C-
terminus of the laminin α subunit mediate Ca^{2+}-dependent
binding to specific carbohydrates on certain cell-surface
molecules such as syndecan and dystroglycan. LG domains
are found in a wide variety of proteins and can mediate
binding to steroids and proteins as well as carbohydrates.
For example, LG domains in the α chain of laminin can mediate binding to certain integrins, including α6β4 integrin
on epithelial cells.

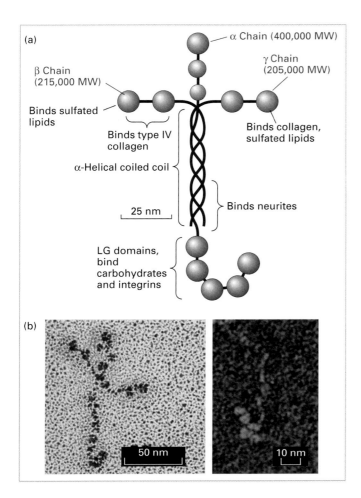

▲ **FIGURE 6-16 Laminin, a heterotrimeric multiadhesive matrix protein found in all basal laminae.** (a) Schematic model showing the general shape, location of globular domains, and coiled-coil region in which laminin's three chains are covalently linked by several disulfide bonds. Different regions of laminin bind to cell-surface receptors and various matrix components. (b) Electron micrographs of intact laminin molecule, showing its characteristic cross appearance (*left*) and the carbohydrate-binding LG domains near the C-terminus (*right*). [Part (a) adapted from G. R. Martin and R. Timpl, 1987, *Ann. Rev. Cell Biol.* **3**:57, and K. Yamada, 1991, *J. Biol. Chem.* **266**:12809. Part (b) from R. Timpl et al., 2000, *Matrix Biol.* **19**:309; photograph at right courtesy of Jürgen Engel.]

Secreted and Cell-Surface Proteoglycans Are Expressed by Many Cell Types

Proteoglycans are a subset of glycoproteins containing co-valently linked specialized polysaccharide chains called **glycosaminoglycans (GAGs),** which are long linear polymers of specific repeating disaccharides. Usually one sugar is ei-ther a uronic acid (D-glucuronic acid or L-iduronic acid) or D-galactose; the other sugar is N-acetylglucosamine or N-acetylgalactosamine (Figure 6-17). One or both of the sug-ars contain at least one anionic group (carboxylate or sul-fate). Thus each GAG chain bears many negative charges.

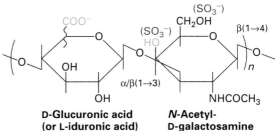

(a) Hyaluronan ($n \leq 25,000$)

D-Glucuronic acid **N-Acetyl-D-glucosamine**

(b) Chondroitin (or dermatan) sulfate ($n \leq 250$)

D-Glucuronic acid (or L-iduronic acid) **N-Acetyl-D-galactosamine**

(c) Heparin/Heparan sulfate ($n = 200$)

D-Glucuronic or L-iduronic acid **N-Acetyl- or N-sulfo-D-glucosamine**

(d) Keratan sulfate ($n = 20$–40)

D-Galactose **N-Acetyl-D-glucosamine**

▲ **FIGURE 6-17 The repeating disaccharides of glycosaminoglycans (GAGs), the polysaccharide components of proteoglycans.** Each of the four classes of GAGs is formed by polymerization of monomer units into repeats of a particular disaccharide and subsequent modifications, including addition of sulfate groups and inversion (epimerization) of the carboxyl group on carbon 5 of D-glucuronic acid to yield L-iduronic acid. Heparin is generated by hypersulfation of heparan sulfate, whereas hyaluronan is unsulfated. The number (*n*) of disaccharides typically found in each glycosaminoglycan chain is given. The squiggly lines represent covalent bonds that are oriented either above (D-glucuronic acid) or below (L-iduronic acid) the ring.

GAGs are classified into several major types based on the nature of the repeating disaccharide unit: heparan sulfate, chondroitin sulfate, dermatan sulfate, keratan sulfate, and hyaluronan. A hypersulfated form of heparan sulfate called heparin, produced mostly by mast cells, plays a key role in allergic reactions. It is also used medically as an anticlotting drug because of its ability to activate a natural clotting inhibitor called antithrombin III.

As we will see in later chapters, complex signaling pathways direct the emergence of various cell types in the proper position and at the proper time in normal embryonic development. Laboratory generation and analysis of mutants with defects in proteoglycan production in *Drosophila melanogaster* (fruit fly), *C. elegans* (roundworm), and mice have clearly shown that proteoglycans play critical roles in development, most likely as modulators of various signaling pathways.

Biosynthesis of Proteoglycans With the exception of hyaluronan, which is discussed in the next section, all the major GAGs occur naturally as components of proteoglycans. Like other secreted and transmembrane glycoproteins, proteoglycan core proteins are synthesized on the endoplasmic reticulum (Chapter 16). The GAG chains are assembled on these cores in the Golgi complex. To generate heparan or chondroitin sulfate chains, a three-sugar "linker" is first attached to the hydroxyl side chains of certain serine residues in a core protein (Figure 6-18). In contrast, the linkers for the addition of keratan sulfate chains are oligosaccharide chains attached to asparagine residues; such **N-linked oligosaccharides** are present in most glycoproteins, although only a subset carry GAG chains. All GAG chains are elongated by the alternating addition of sugar monomers to form the disaccharide repeats characteristic of a particular GAG; the chains are often modified subsequently by the covalent linkage of small molecules such as sulfate. The mechanisms responsible for determining which proteins are modified with GAGs, the sequence of disaccharides to be added, the sites to be sulfated, and the lengths of the GAG chains are unknown. The ratio of polysaccharide to protein in all proteoglycans is much higher than that in most other glycoproteins.

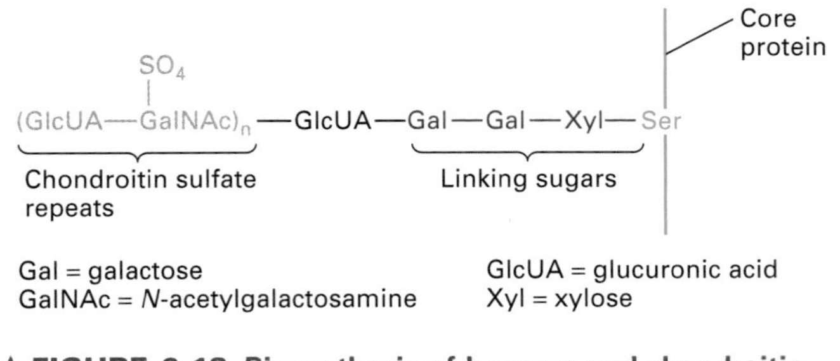

Diversity of Proteoglycans The proteoglycans constitute a remarkably diverse group of molecules that are abundant in the extracellular matrix of all animal tissues and are also expressed on the cell surface. For example, of the five major classes of heparan sulfate proteoglycans, three are located in the extracellular matrix (perlecan, agrin, and type XVIII collagen) and two are cell-surface proteins. The latter include integral membrane proteins (syndecans) and GPI-anchored proteins (glypicans); the GAG chains in both types of cell-surface proteoglycans extend into the extracellular space. The sequences and lengths of proteoglycan core proteins vary considerably, and the number of attached GAG chains ranges from just a few to more than 100. Moreover, a core protein is often linked to two different types of GAG chains (e.g., heparan sulfate and chondroitin sulfate), generating a "hybrid" proteoglycan. Thus, the molecular weight and charge density of a population of proteoglycans can be expressed only as an average; the composition and sequence of individual molecules can differ considerably.

Perlecan, the major secreted proteoglycan in basal laminae, consists of a large multidomain core protein ($\approx$400 kDa) with three or four specialized GAG chains. Both the protein and the GAG components of perlecan contribute to its ability to incorporate into and define the structure and function of basal laminae. Because of its multiple domains with distinctive binding properties, perlecan can cross-link not only ECM components to one another but also certain cell-surface molecules to ECM components.

Syndecans are expressed by epithelial cells and many other cell types. These cell-surface proteoglycans bind to collagens and multiadhesive matrix proteins such as the fibronectins, which are discussed in Section 6.4. In this way, cell-surface proteoglycans can anchor cells to the extracellular matrix. Like that of many integral membrane proteins, the cytosolic domain of syndecan interacts with the actin cytoskeleton and in some cases with intracellular regulatory molecules. In addition, cell-surface proteoglycans bind many protein growth factors and other external signaling molecules, thereby helping to regulate cellular metabolism and function. For instance, syndecans in the hypothalamic region of the brain modulate feeding behavior in response to food deprivation (fasted state). They do so by participating in the binding of antisatiety peptides to cell-surface receptors that help control feeding behavior. In the fed state, the syndecan extracellular domain decorated with heparan sulfate chains is released from the surface by proteolysis, thus suppressing the activity of the antisatiety peptides and feeding behavior. In mice engineered to overexpress the syndecan-1 gene in the hypothalamic region of the brain and other tissues, normal control of feeding by antisatiety peptides is disrupted and the animals overeat and become obese. Other examples of proteoglycans interacting with external signaling molecules are described in Chapter 14.

▲ **FIGURE 6-18 Biosynthesis of heparan and chondroitin sulfate chains in proteoglycans.** Synthesis of a chondroitin sulfate chain (shown here) is initiated by transfer of a xylose residue to a serine residue in the core protein, most likely in the Golgi complex, followed by sequential addition of two galactose residues. Glucuronic acid and *N*-acetylgalactosamine residues are then added sequentially to these linking sugars, forming the chondroitin sulfate chain. Heparan sulfate chains are connected to core proteins by the same three-sugar linker.

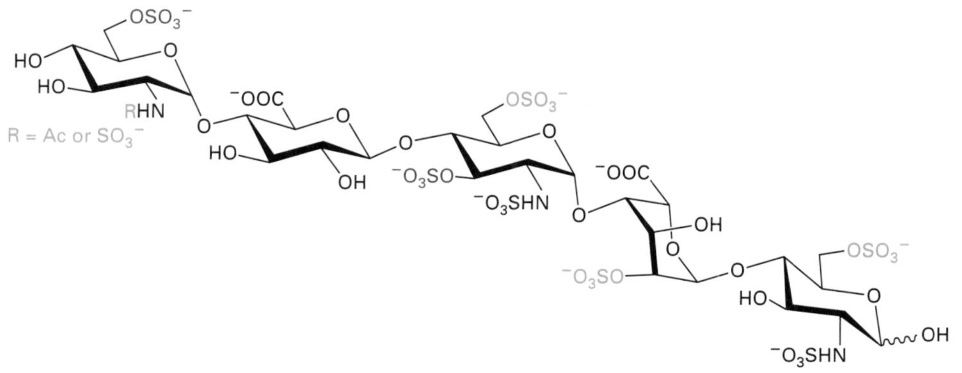

◀ **FIGURE 6-19 Pentasaccharide GAG sequence that regulates the activity of antithrombin III (ATIII).** Sets of modified five-residue sequences in heparin with the composition shown here bind to ATIII and activate it, thereby inhibiting blood clotting. The sulfate groups in red type are essential for this heparin function; the modifications in blue type may be present but are not essential. Other sets of modified GAG sequences are thought to regulate the activity of other target proteins. [Courtesy of Robert Rosenberg and Balagurunathan Kuberan.]

Modifications in Glycosaminoglycan (GAG) Chains Can Determine Proteoglycan Functions

As is the case with the sequence of amino acids in proteins, the arrangement of the sugar residues in GAG chains and the modification of specific sugars (e.g., addition of sulfate) in the chains can determine their function and that of the proteoglycans containing them. For example, groupings of certain modified sugars in the GAG chains of heparin sulfate proteoglycans can control the binding of growth factors to certain receptors, the activities of proteins in the blood-clotting cascade, and the activity of lipoprotein lipase, a membrane-associated enzyme that hydrolyzes triglycerides to fatty acids (Chapter 18).

For years, the chemical and structural complexity of proteoglycans posed a daunting barrier to an analysis of their structures and an understanding of their many diverse functions. In recent years, investigators employing classical and new state-of-the-art biochemical techniques (e.g., capillary high-pressure liquid chromatography), mass spectrometry, and genetics have begun to elucidate the detailed structures and functions of these ubiquitous ECM molecules. The results of ongoing studies suggest that sets of sugar-residue sequences containing some modifications in common, rather than single unique sequences, are responsible for specifying distinct GAG functions. A case in point is a set of five-residue (pentasaccharide) sequences found in a subset of heparin GAGs that control the activity of antithrombin III (ATIII), an inhibitor of the key blood-clotting protease thrombin. When these pentasaccharide sequences in heparin are sulfated at two specific positions, heparin can activate ATIII, thereby inhibiting clot formation (Figure 6-19). Several other sulfates can be present in the active pentasaccharide in various combinations, but they are not essential for the anticlotting activity of heparin. The rationale for generating sets of similar active sequences rather than a single unique sequence and the mechanisms that control GAG biosynthetic pathways, permitting the generation of such active sequences, are not well understood.

KEY CONCEPTS OF SECTION 6.3

The Extracellular Matrix of Epithelial Sheets

■ The basal lamina, a thin meshwork of extracellular matrix (ECM) molecules, separates most epithelia and other organized groups of cells from adjacent connective tissue. Together, the basal lamina and collagenous reticular lamina form a structure called the basement membrane.

■ Four ECM proteins are found in all basal laminae (see Figure 6-13): type IV collagen, laminin (a multiadhesive matrix protein), entactin (nidogen), and perlecan (a proteoglycan).

■ Cell-surface adhesion receptors (e.g., $\alpha 6 \beta 4$ integrin in hemidesmosomes) anchor cells to the basal lamina, which in turn is connected to other ECM components (see Figure 6-1).

■ Repeating sequences of Gly-X-Y give rise to the collagen triple-helical structure (see Figure 6-14). Different collagens are distinguished by the length and chemical modifications of their α chains and by the segments that interrupt or flank their triple-helical regions.

■ The large, flexible molecules of type IV collagen interact end to end and laterally to form a meshlike scaffold to which other ECM components and adhesion receptors can bind (see Figure 6-15).

■ Laminin and other multiadhesive matrix proteins are multidomain molecules that bind multiple adhesion receptors and ECM components.

■ Proteoglycans consist of membrane-associated or secreted core proteins covalently linked to one or more glycosaminoglycan (GAG) chains, which are linear polymers of sulfated disaccharides.

■ Perlecan, a large secreted proteoglycan present primarily in the basal lamina, binds many ECM components and adhesion receptors.

■ Cell-surface proteoglycans such as the syndecans facilitate cell–matrix interactions and help present certain external signaling molecules to their cell-surface receptors.

6.4 The Extracellular Matrix of Nonepithelial Tissues

We have seen how diverse CAMs and adhesion receptors participate in the assembly of animal cells into epithelial sheets that rest on and adhere to a well-defined ECM structure, the basal lamina. The same or similar molecules mediate and control cell–cell and cell–matrix interactions in connective, muscle, and neural tissues and between blood cells and the surrounding vessels. In this section, we consider some of the ECM molecules characteristic of these nonepithelial tissues. We also describe the synthesis of fibrillar collagens, which are the most abundant proteins in animals. The interactions entailing CAMs and adhesion receptors expressed by various nonepithelial cells, which serve a wide variety of distinctive functions, are covered in Section 6.5.

TABLE 6-1	Selected Collagens		
Type	Molecule Composition	Structural Features	Representative Tissues
FIBRILLAR COLLAGENS			
I	$[\alpha 1(I)]_2[\alpha 2(I)]$	300-nm-long fibrils	Skin, tendon, bone, ligaments, dentin, interstitial tissues
II	$[\alpha 1(II)]_3$	300-nm-long fibrils	Cartilage, vitreous humor
III	$[\alpha 1(III)]_3$	300-nm-long fibrils; often with type I	Skin, muscle, blood vessels
V	$[\alpha 1(V)_2 \, \alpha 2(V)]$, $[\alpha 1(V)_3]$	390-nm-long fibrils with globular N-terminal extension; often with type I	Cornea, teeth, bone, placenta, skin, smooth muscle
FIBRIL-ASSOCIATED COLLAGENS			
VI	$[\alpha 1(VI)][\alpha 2(VI)]$	Lateral association with type I; periodic globular domains	Most interstitial tissues
IX	$[\alpha 1(IX)][\alpha 2(IX)][\alpha 3(IX)]$	Lateral association with type II; N-terminal globular domain; bound GAG	Cartilage, vitreous humor
SHEET-FORMING AND ANCHORING COLLAGENS			
IV	$[\alpha 1(IV)]_2[\alpha 2(IV)]$	Two-dimensional network	All basal laminae
VII	$[\alpha 1(VII)]_3$	Long fibrils	Below basal lamina of the skin
XV	$[\alpha 1(XV)]_3$	Core protein of chondroitin sulfate proteoglycan	Widespread; near basal lamina in muscle
TRANSMEMBRANE COLLAGENS			
XIII	$[\alpha 1(XIII)]_3$	Integral membrane protein	Hemidesmosomes in skin
XVII	$[\alpha 1(XVII)]_3$	Integral membrane protein	Hemidesmosomes in skin
HOST DEFENSE COLLAGENS			
Collectins		Oligomers of triple helix; lectin domains	Blood, alveolar space
C1q		Oligomers of triple helix	Blood (complement)
Class A scavenger receptors		Homotrimeric membrane proteins	Macrophages

SOURCES: K. Kuhn, 1987, in R. Mayne and R. Burgeson, eds., *Structure and Function of Collagen Types*, Academic Press, p. 2; and M. van der Rest and R. Garrone, 1991, *FASEB J.* 5:2814.

Fibrillar Collagens Are the Major Fibrous Proteins in the Extracellular Matrix of Connective Tissues

Connective tissue, such as tendon and cartilage, differs from other solid tissues in that most of its volume is made up of extracellular matrix rather than cells. This matrix is packed with insoluble protein fibers and contains proteoglycans, various multiadhesive proteins, and **hyaluronan**, a very large, nonsulfated GAG. The most abundant fibrous protein in connective tissue is collagen. Rubberlike elastin fibers, which can be stretched and relaxed, also are present in deformable sites (e.g., skin, tendons, heart). As discussed later, the fibronectins, a family of multiadhesive matrix proteins, form their own distinct fibrils in the matrix of some connective tissues. Although several types of cells are found in connective tissues, the various ECM components are produced largely by cells called **fibroblasts.**

About 80–90 percent of the collagen in the body consists of types I, II, and III collagens, located primarily in connective tissues. Because of its abundance in tendon-rich tissue such as rat tail, type I collagen is easy to isolate and was the first collagen to be characterized. Its fundamental structural unit is a long (300-nm), thin (1.5-nm-diameter) triple helix consisting of two $\alpha 1$(I) chains and one $\alpha 2$(I) chain, each precisely 1050 amino acids in length (see Figure 6-14). The triple-stranded molecules associate into higher-order poly-

mers called collagen *fibrils*, which in turn often aggregate into larger bundles called collagen *fibers*.

The minor classes of collagen include *fibril-associated collagens*, which link the fibrillar collagens to one another or to other ECM components; *sheet-forming and anchoring collagens*, which form two-dimensional networks in basal laminae (type IV) and connect the basal lamina in skin to the underlying connective tissue (type VII); *transmembrane collagens*, which function as adhesion receptors; and *host defense collagens*, which help the body recognize and eliminate pathogens. Table 6-1 lists specific examples in the various classes of collagens. Interestingly, several collagens (e.g., types XVIII and XV) function as core proteins in proteoglycans.

Formation of Collagen Fibrils Begins in the Endoplasmic Reticulum and Is Completed Outside the Cell

Collagen biosynthesis and secretion follow the normal pathway for a secreted protein, which is described in detail in Chapters 16 and 17. The collagen α chains are synthesized as longer precursors, called pro-α chains, by ribosomes attached to the endoplasmic reticulum (ER). The pro-α chains undergo a series of covalent modifications and fold into triple-helical *procollagen* molecules before their release from cells (Figure 6-20).

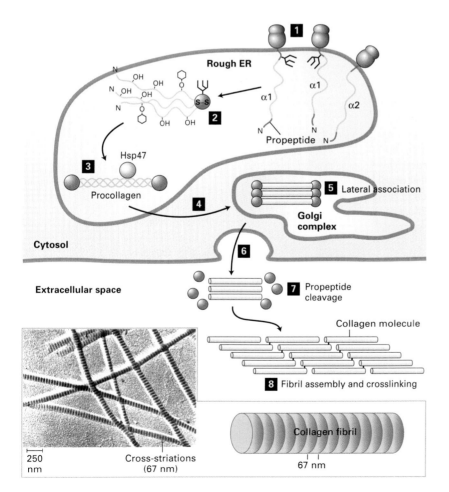

◀ **FIGURE 6-20 Major events in biosynthesis of fibrillar collagens.** Step **1**: Procollagen α chains are synthesized on ribosomes associated with the endoplasmic reticulum (ER) membrane, and asparagine-linked oligosaccharides are added to the C-terminal propeptide. Step **2**: Propeptides associate to form trimers and are covalently linked by disulfide bonds, and selected residues in the Gly-X-Y triplet repeats are covalently modified [certain prolines and lysines are hydroxylated, galactose (Gal) or galactose-glucose (hexagons) is attached to some hydroxylysines, prolines are cis → trans isomerized]. Step **3**: The modifications facilitate zipperlike formation, stabilization of triple helices, and binding by the chaperone protein Hsp47 (Chapter 16), which may stabilize the helices or prevent premature aggregation of the trimers or both. Steps **4** and **5**: The folded procollagens are transported to and through the Golgi apparatus, where some lateral association into small bundles takes place. The chains are then secreted (step **6**), the N- and C-terminal propeptides are removed (step **7**), and the trimers assemble into fibrils and are covalently cross-linked (step **8**). The 67-nm staggering of the trimers gives the fibrils a striated appearance in electron micrographs (*inset*). [Adapted from A. V. Persikov and B. Brodsky, 2002, *Proc. Nat'l. Acad. Sci. USA* **99**(3):1101–1103.]

After the secretion of procollagen from the cell, extracellular peptidases (e.g., bone morphogenetic protein-1) remove the N-terminal and C-terminal propeptides. In regard to fibrillar collagens, the resulting molecules, which consist almost entirely of a triple-stranded helix, associate laterally to generate fibrils with a diameter of 50–200 nm. In fibrils, adjacent collagen molecules are displaced from one another by 67 nm, about one-quarter of their length. This staggered array produces a striated effect that can be seen in electron micrographs of collagen fibrils (see Figure 6-20, *inset*). The unique properties of the fibrous collagens (e.g., types I, II, III) are mainly due to the formation of fibrils.

Short non-triple-helical segments at either end of the collagen α chains are of particular importance in the formation of collagen fibrils. Lysine and hydroxylysine side chains in these segments are covalently modified by extracellular lysyl oxidases to form aldehydes in place of the amine group at the end of the side chain. These reactive aldehyde groups form covalent cross-links with lysine, hydroxylysine, and histidine residues in adjacent molecules. These cross-links stabilize the side-by-side packing of collagen molecules and generate a strong fibril. The removal of the propeptides and covalent cross-linking take place in the extracellular space to prevent the potentially catastrophic assembly of fibrils within the cell.

 The post-translational modifications of pro-α chains are crucial for the formation of mature collagen molecules and their assembly into fibrils. Defects in these modifications have serious consequences, as ancient mariners frequently experienced. For example, ascorbic acid (vitamin C) is an essential cofactor for the hydroxylases responsible for adding hydroxyl groups to proline and lysine residues in pro-α chains. In cells deprived of ascorbate, as in the disease *scurvy,* the pro-α chains are not hydroxylated sufficiently to form stable triple-helical procollagen at normal body temperature, and the procollagen that forms cannot assemble into normal fibrils. Without the structural support of collagen, blood vessels, tendons, and skin become fragile. Because fresh fruit in the diet can supply sufficient vitamin C to support the formation of normal collagen, early British sailors were provided with limes to prevent scurvy, leading to their being called "limeys."

Rare mutations in lysyl hydroxylase genes cause Bruck syndrome and one form of Ehlers-Danlos syndrome. Both disorders are marked by connective-tissue defects, although their clinical symptoms differ. ▌

Type I and II Collagens Form Diverse Structures and Associate with Different Nonfibrillar Collagens

Collagens differ in their ability to form fibers and to organize the fibers into networks. In tendons, for instance, long type I collagen fibrils are packed side by side in parallel bundles, forming thick collagen fibers. Tendons connect muscles to bones and must withstand enormous forces. Because type

I collagen fibers have great tensile strength, tendons can be stretched without being broken. Indeed, gram for gram, type I collagen is stronger than steel. Two quantitatively minor fibrillar collagens, type V and type XI, coassemble into fibers with type I collagen, thereby regulating the structures and properties of the fibers. Incorporation of type V collagen, for example, results in smaller-diameter fibers.

Type I collagen fibrils are also used as the reinforcing rods in the construction of bone. Bones and teeth are hard and strong because they contain large amounts of dahllite, a crystalline calcium- and phosphate-containing mineral. Most bones are about 70 percent mineral and 30 percent protein, the vast majority of which is type I collagen. Bones form when certain cells (chondrocytes and osteoblasts) secrete collagen fibrils that are then mineralized by deposition of small dahllite crystals.

In many connective tissues, type VI collagen and proteoglycans are noncovalently bound to the sides of type I fibrils and may bind the fibrils together to form thicker collagen fibers (Figure 6-21a). Type VI collagen is unusual in that the molecule consists of a relatively short triple helix with glob-

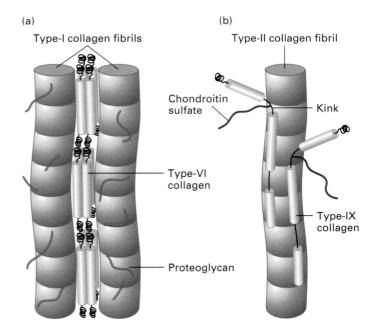

▲ **FIGURE 6-21 Interactions of fibrous collagens with nonfibrous fibril-associated collagens.** (a) In tendons, type I fibrils are all oriented in the direction of the stress applied to the tendon. Proteoglycans and type VI collagen bind noncovalently to fibrils, coating the surface. The microfibrils of type VI collagen, which contain globular and triple-helical segments, bind to type I fibrils and link them together into thicker fibers. (b) In cartilage, type IX collagen molecules are covalently bound at regular intervals along type II fibrils. A chondroitin sulfate chain, covalently linked to the α2(IX) chain at the flexible kink, projects outward from the fibril, as does the globular N-terminal region. [Part (a), see R. R. Bruns et al., 1986, *J. Cell Biol.* **103**:393. Part (b), see L. M. Shaw and B. Olson, 1991, *Trends Biochem. Sci.* **18**:191.]

ular domains at both ends. The lateral association of two type VI monomers generates an "antiparallel" dimer. The end-to-end association of these dimers through their globular domains forms type VI "microfibrils." These microfibrils have a beads-on-a-string appearance, with about 60-nm-long triple-helical regions separated by 40-nm-long globular domains.

The fibrils of type II collagen, the major collagen in cartilage, are smaller in diameter than type I fibrils and are oriented randomly in a viscous proteoglycan matrix. The rigid collagen fibrils impart a strength and compressibility to the matrix and allow it to resist large deformations in shape. This property allows joints to absorb shocks. Type II fibrils are cross-linked to matrix proteoglycans by type IX collagen, another fibril-associated collagen. Type IX collagen and several related types have two or three triple-helical segments connected by flexible kinks and an N-terminal globular segment (Figure 6-22b). The globular N-terminal segment of type IX collagen extends from the fibrils at the end of one of its helical segments, as does a GAG chain that is sometimes linked to one of the type IX chains. These protruding nonhelical structures are thought to anchor the type II fibril to proteoglycans and other components of the matrix. The interrupted triple-helical structure of type IX and related collagens prevents them from assembling into fibrils, although they can associate with fibrils formed from other collagen types and form covalent cross-links to them.

 Certain mutations in the genes encoding collagen α1(I) or α2(I) chains, which form type I collagen, lead to *osteogenesis imperfecta*, or brittle-bone disease. Because every third position in a collagen α chain must be a glycine for the triple helix to form (see Figure 6-14), mutations of glycine to almost any other amino acid are deleterious, resulting in poorly formed and unstable helices. Only one defective α chain of the three in a collagen molecule can disrupt the whole molecule's triple-helical structure and function. A mutation in a single copy (allele) of either the α1(I) gene or the α2(I) gene, which are located on nonsex chromosomes (autosomes), can cause this disorder. Thus it normally shows autosomal dominant inheritance (Chapter 9). ∎

Hyaluronan Resists Compression and Facilitates Cell Migration

Hyaluronan, also called hyaluronic acid (HA) or hyaluronate, is a nonsulfated GAG formed as a disaccharide repeat composed of glucuronic acid and N-acetylglucosamine (see Figure 6-17a) by a plasma-membrane-bound enzyme (HA synthase) and is directly secreted into the extracellular space. It is a major component of the extracellular matrix that surrounds migrating and proliferating cells, particularly in embryonic tissues. In addition, as will be described shortly, hyaluronan forms the backbone of complex proteoglycan aggregates found in many extracellular matrices, particularly cartilage. Because of its remarkable physical properties, hyaluronan imparts stiffness and resilience as well as a lu-

bricating quality to many types of connective tissue such as joints.

Hyaluronan molecules range in length from a few disaccharide repeats to ≈25,000. The typical hyaluronan in joints such as the elbow has 10,000 repeats for a total mass of 4×10^6 Da and length of 10 μm (about the diameter of a small cell). Individual segments of a hyaluronan molecule fold into a rodlike conformation because of the β glycosidic linkages between the sugars and extensive intrachain hydrogen bonding. Mutual repulsion between negatively charged carboxylate groups that protrude outward at regular intervals also contributes to these local rigid structures. Overall, however, hyaluronan is not a long, rigid rod as is fibrillar collagen; rather, in solution it is very flexible, bending and twisting into many conformations, forming a random coil.

Because of the large number of anionic residues on its surface, the typical hyaluronan molecule binds a large amount of water and behaves as if it were a large hydrated sphere with a diameter of ≈500 nm. As the concentration of hyaluronan increases, the long chains begin to entangle, forming a viscous gel. Even at low concentrations, hyaluronan forms a hydrated gel; when placed in a confining space, such as in a matrix between two cells, the long hyaluronan molecules will tend to push outward. This outward pushing creates a swelling, or *turgor pressure*, within the extracellular space. In addition, the binding of cations by COO⁻ groups on the surface of hyaluronan increases the concentration of ions and thus the osmotic pressure in the gel. As a result, large amounts of water are taken up into the matrix, contributing to the turgor pressure. These swelling forces give connective tissues their ability to resist compression forces, in contrast with collagen fibers, which are able to resist stretching forces.

Hyaluronan is bound to the surface of many migrating cells by a number of adhesion receptors (e.g., one called CD44) containing HA-binding domains, each with a similar three-dimensional conformation. Because of its loose, hydrated, porous nature, the hyaluronan "coat" bound to cells appears to keep cells apart from one another, giving them the freedom to move about and proliferate. The cessation of cell movement and the initiation of cell–cell attachments are frequently correlated with a decrease in hyaluronan, a decrease in HA-binding cell-surface molecules, and an increase in the extracellular enzyme hyaluronidase, which degrades hyaluronan in the matrix. These functions of hyaluronan are particularly important during the many cell migrations that facilitate differentiation and in the release of a mammalian egg cell (oocyte) from its surrounding cells after ovulation.

Association of Hyaluronan and Proteoglycans Forms Large, Complex Aggregates

The predominant proteoglycan in cartilage, called *aggrecan*, assembles with hyaluronan into very large aggregates, illustrative of the complex structures that proteoglycans sometimes form. The backbone of the cartilage proteoglycan aggregate is a long molecule of hyaluronan to which multiple aggrecan molecules are bound tightly but noncovalently (Figure 6-22a). A single aggrecan aggregate, one of the largest macromolecular complexes known, can be more than 4 mm long and have a volume larger than that of a bacterial cell.

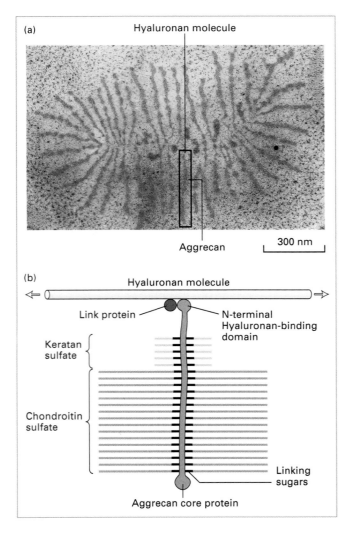

▲ **FIGURE 6-22 Structure of proteoglycan aggregate from cartilage.** (a) Electron micrograph of an aggrecan aggregate from fetal bovine epiphyseal cartilage. Aggrecan core proteins are bound at ≈40-nm intervals to a molecule of hyaluronan. (b) Schematic representation of an aggrecan monomer bound to hyaluronan. In aggrecan, both keratan sulfate and chondroitin sulfate chains are attached to the core protein. The N-terminal domain of the core protein binds noncovalently to a hyaluronan molecule. Binding is facilitated by a link protein, which binds to both the hyaluronan molecule and the aggrecan core protein. Each aggrecan core protein has 127 Ser-Gly sequences at which GAG chains can be added. The molecular weight of an aggrecan monomer averages 2×10^6. The entire aggregate, which may contain upward of 100 aggrecan monomers, has a molecular weight in excess of 2×10^8. [Part (a) from J. A. Buckwalter and L. Rosenberg, 1983, *Coll. Rel. Res.* **3**:489; courtesy of L. Rosenberg.]

These aggregates give cartilage its unique gel-like properties and its resistance to deformation, essential for distributing the load in weight-bearing joints.

The aggrecan core protein (≈250,000 MW) has one N-terminal globular domain that binds with high affinity to a specific decasaccharide sequence within hyaluronan. This specific sequence is generated by covalent modification of some of the repeating disaccharides in the hyaluronan chain. The interaction between aggrecan and hyaluronan is facilitated by a link protein that binds to both the aggrecan core protein and hyaluronan (Figure 6-22b). Aggrecan and the link protein have in common a "link" domain, ≈100 amino acids long, that is found in numerous matrix and cell-surface hyaluronan-binding proteins in both cartilaginous and noncartilaginous tissues. Almost certainly these proteins arose in the course of evolution from a single ancestral gene that encoded just this domain.

The importance of the GAG chains that are part of various matrix proteoglycans is illustrated by the rare humans who have a genetic defect in one of the enzymes required for synthesis of the GAG dermatan sulfate. These persons have many defects in their bones, joints, and muscles; do not grow to normal height; and have wrinkled skin, giving them a prematurely aged appearance. ∎

Fibronectins Connect Many Cells to Fibrous Collagens and Other Matrix Components

Many different cell types synthesize **fibronectin,** an abundant multiadhesive matrix protein found in all vertebrates. The discovery that fibronectin functions as an adhesive molecule stemmed from observations that it is present on the surfaces of normal fibroblastic cells, which adhere tightly to petri dishes in laboratory experiments, but is absent from the surfaces of tumorigenic cells, which adhere weakly. The 20 or so isoforms of fibronectin are generated by alternative splicing of the RNA transcript produced from a single gene (see Figure 4-15). Fibronectins are essential for the migration and differentiation of many cell types in embryogenesis. These proteins are also important for wound healing because they promote blood clotting and facilitate the migration of macrophages and other immune cells into the affected area.

Fibronectins help attach cells to the extracellular matrix by binding to other ECM components, particularly fibrous collagens and heparan sulfate proteoglycans, and to cell-surface adhesion receptors such as integrins (see Figure 6-2). Through their interactions with adhesion receptors (e.g., $\alpha5\beta1$ integrin), fibronectins influence the shape and movement of cells and the organization of the cytoskeleton. Conversely, by regulating their receptor-mediated attachments to fibronectin and other ECM components, cells can sculpt the immediate ECM environment to suit their needs.

Fibronectins are dimers of two similar polypeptides linked at their C-termini by two disulfide bonds; each chain is about 60–70 nm long and 2–3 nm thick. Partial digestion

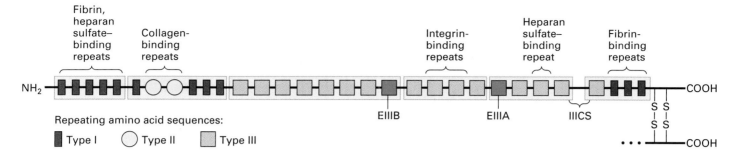

▲ **FIGURE 6-23 Organization of fibronectin chains.** Only one of the two chains present in the dimeric fibronectin molecule is shown; both chains have very similar sequences. Each chain contains about 2446 amino acids and is composed of three types of repeating amino acid sequences. Circulating fibronectin lacks one or both of the type III repeats designated EIIIA and EIIIB owing to alternative mRNA splicing (see Figure 4-15). At least five different sequences may be present in the IIICS region as a result of alternative splicing. Each chain contains six domains (tan boxes), some of which contain specific binding sites for heparan sulfate, fibrin (a major constituent of blood clots), collagen, and cell-surface integrins. The integrin-binding domain is also known as the cell-binding domain. [Adapted from G. Paolella, M. Barone, and F. Baralle, 1993, in M. Zern and L. Reid, eds., *Extracellular Matrix*, Marcel Dekker, pp. 3–24.]

of fibronectin with low amounts of proteases and analysis of the fragments showed that each chain comprises six functional regions with different ligand-binding specificities (Figure 6-23). Each region, in turn, contains multiple copies of certain sequences that can be classified into one of three types. These classifications are designated fibronectin type I, II, and III repeats, on the basis of similarities in amino acid sequence, although the sequences of any two repeats of a given type are not always identical. These linked repeats give the molecule the appearance of beads on a string. The combination of different repeats composing the regions, another example of combinatorial diversity, confers on fibronectin its ability to bind multiple ligands.

One of the type III repeats in the cell-binding region of fibronectin mediates binding to certain integrins. The results of studies with synthetic peptides corresponding to parts of this repeat identified the tripeptide sequence Arg-Gly-Asp, usually called the *RGD sequence*, as the minimal sequence within this repeat required for recognition by those integrins. In one study, heptapeptides containing the RGD sequence or a variation of this sequence were tested for their ability to mediate the adhesion of rat kidney cells to a culture dish. The results showed that heptapeptides containing the RGD sequence mimicked intact fibronectin's ability to stimulate integrin-mediated adhesion, whereas variant heptapeptides lacking this sequence were ineffective (Figure 6-24).

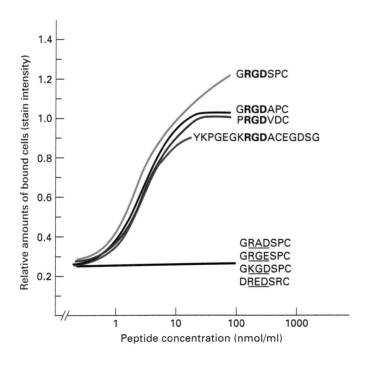

◄ **EXPERIMENTAL FIGURE 6-24 A specific tripeptide sequence (RGD) in the cell-binding region of fibronectin is required for adhesion of cells.** The cell-binding region of fibronectin contains an integrin-binding heptapeptide sequence, GRDSPC in the single-letter amino acid code (see Figure 2-13). This heptapeptide and several variants were synthesized chemically. Different concentrations of each synthetic peptide were added to polystyrene dishes that had the protein immunoglobulin G (IgG) firmly attached to their surfaces; the peptides were then chemically cross-linked to the IgG. Subsequently, cultured normal rat kidney cells were added to the dishes and incubated for 30 minutes to allow adhesion. After the nonbound cells were washed away, the relative amounts of cells that had adhered firmly were determined by staining the bound cells with a dye and measuring the intensity of the staining with a spectrophotometer. The plots shown here indicate that cell adhesion increased above the background level with increasing peptide concentration for those peptides containing the RGD sequence but not for the variants lacking this sequence (modification underlined). [From M. D. Pierschbacher and E. Ruoslahti, 1984, *Proc. Nat'l. Acad. Sci. USA* **81**:5985.]

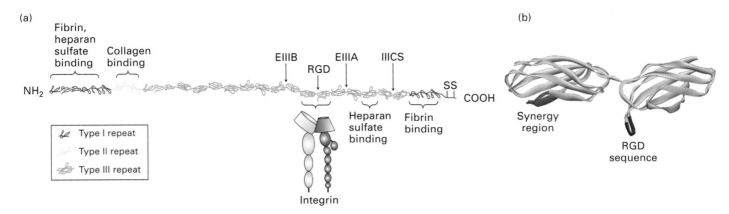

▲ FIGURE 6-25 Model of fibronectin binding to integrin through its RGD-containing type III repeat. (a) Scale model of fibronectin is shown docked by two type III repeats to the extracellular domains of integrin. Structures of fibronectin's domains were determined from fragments of the molecule. The EIIIA, EIIIB, and IIICS domains (not shown; see Figure 6-23) are variably spliced into the structure at locations indicated by arrows. (b) A high-resolution structure shows that the RGD binding sequence (red) extends outward in a loop from its compact type III domain on the same side of fibronectin as the synergy region (blue), which also contributes to high-affinity binding to integrins. [Adapted from D. J. Leahy et al., 1996, *Cell* **84**:161.]

A three-dimensional model of fibronectin binding to integrin based on structures of parts of both fibronectin and integrin has been assembled (Figure 6-25a). In a high-resolution structure of the integrin-binding fibronectin type III repeat and its neighboring type III domain, the RGD sequence is at the apex of a loop that protrudes outward from the molecule, in a position facilitating binding to integrins (Figure 6-25a, b). Although the RGD sequence is required for binding to several integrins, its affinity for integrins is substantially less than that of intact fibronectin or of the entire cell-binding region in fibronectin. Thus structural features near to the RGD sequence in fibronectins (e.g., parts of adjacent repeats, such as the synergy region; see Figure 6-25b) and in other RGD-containing proteins enhance their binding to certain integrins. Moreover, the simple soluble dimeric forms of fibronectin produced by the liver or fibroblasts are initially in a nonfunctional closed conformation that binds poorly to integrins because the RGD sequence is not readily accessible. The adsorption of fibronectin to a col-

▶ EXPERIMENTAL FIGURE 6-26 Integrins mediate linkage between fibronectin in the extracellular matrix and the cytoskeleton. (a) Immunofluorescent micrograph of a fixed cultured fibroblast showing colocalization of the α5β1 integrin and actin-containing stress fibers. The cell was incubated with two types of monoclonal antibody: an integrin-specific antibody linked to a green fluorescing dye and an actin-specific antibody linked to a red fluorescing dye. Stress fibers are long bundles of actin microfilaments that radiate inward from points where the cell contacts a substratum. At the distal end of these fibers, near the plasma membrane, the coincidence of actin (red) and fibronectin-binding integrin (green) produces a yellow fluorescence. (b) Electron micrograph of the junction of fibronectin and actin fibers in a cultured fibroblast. Individual actin-containing 7-nm microfilaments, components of a stress fiber, end at the obliquely sectioned cell membrane. The microfilaments appear in close proximity to the thicker, densely stained fibronectin fibrils on the outside of the cell. [Part (a) from J. Duband et al., 1988, *J. Cell Biol.* **107**:1385. Part (b) from I. J. Singer, 1979, *Cell* **16**:675; courtesy of I. J. Singer; copyright 1979, MIT.]

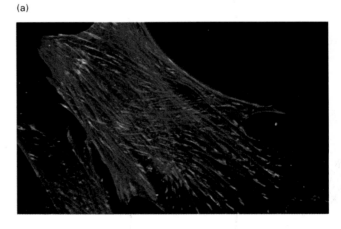

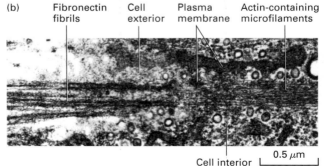

lagen matrix or the basal lamina or, experimentally, to a plastic tissue-culture dish results in a conformational change that enhances its ability to bind to cells. Most likely, this conformational change increases the accessibility of the RGD sequence for integrin binding.

Microscopy and other experimental approaches (e.g., biochemical binding experiments) have demonstrated the role of integrins in cross-linking fibronectin and other ECM components to the cytoskeleton. For example, the colocalization of cytoskeletal actin filaments and integrins within cells can be visualized by fluorescence microscopy (Figure 6-26a). The binding of cell-surface integrins to fibronectin in the matrix induces the actin cytoskeleton–dependent movement of some integrin molecules in the plane of the membrane. The ensuing mechanical tension due to the relative movement of different integrins bound to a single fibronectin dimer stretches the fibronectin. This stretching promotes self-association of the fibronectin into multimeric fibrils.

The force needed to unfold and expose functional self-association sites in fibronectin is much less than that needed to disrupt fibronectin–integrin binding. Thus fibronectin molecules remain bound to integrin while cell-generated mechanical forces induce fibril formation. In effect, the integrins through adapter proteins transmit the intracellular forces generated by the actin cytoskeleton to extracellular fibronectin. Gradually, the initially formed fibronectin fibrils mature into highly stable matrix components by covalent cross-linking. In some electron micrographic images, exterior fibronectin fibrils appear to be aligned in a seemingly continuous line with bundles of actin fibers within the cell (Figure 6-26b). These observations and the results from other studies provided the first example of a molecularly well defined adhesion receptor (i.e., an integrin) forming a bridge between the intracellular cytoskeleton and the extracellular matrix components—a phenomenon now known to be widespread.

KEY CONCEPTS OF SECTION 6.4

The Extracellular Matrix of Nonepithelial Tissues

■ Connective tissue, such as tendon and cartilage, differs from other solid tissues in that most of its volume is made up of extracellular matrix (ECM) rather than cells.

■ The synthesis of fibrillar collagen (e.g., types I, II, and III) begins inside the cell with the chemical modification of newly made α chains and their assembly into triple-helical procollagen within the endoplasmic reticulum. After secretion, procollagen molecules are cleaved, associate laterally, and are covalently cross-linked into bundles called fibrils, which can form larger assemblies called fibers (see Figure 6-20).

■ The various collagens are distinguished by the ability of their helical and nonhelical regions to associate into fibrils, to form sheets, or to cross-link other collagen types (see Table 6-1).

■ Hyaluronan, a highly hydrated GAG, is a major component of the ECM of migrating and proliferating cells. Certain cell-surface adhesion receptors bind hyaluronan to cells.

■ Large proteoglycan aggregates containing a central hyaluronan molecule noncovalently bound to the core protein of multiple proteoglycan molecules (e.g., aggrecan) contribute to the distinctive mechanical properties of the matrix (see Figure 6-22).

■ Fibronectins are abundant multiadhesive matrix proteins that play a key role in migration and cellular differentiation. They contain binding sites for integrins and ECM components (collagens, proteoglycans) and can thus attach cells to the matrix (see Figure 6-23).

■ The tripeptide RGD sequence (Arg-Gly-Asp), found in fibronectins and some other matrix proteins, is recognized by several integrins.

6.5 Adhesive Interactions and Nonepithelial Cells

After adhesive interactions in epithelia form during differentiation, they often are very stable and can last throughout the life span of epithelial cells or until the cells undergo differentiation into loosely associated nonpolarized mesenchymal cells, the epithelial–mesenchymal transition. Although such long-lasting (nonmotile) adhesion also exists in nonepithelial tissues, some nonepithelial cells must be able to crawl across or through a layer of extracellular matrix or other cells. In this section, we describe various cell-surface structures in nonepithelial cells that mediate long-lasting adhesion and transient adhesive interactions that are especially adapted for the movement of cells. The detailed intracellular mechanisms used to generate the mechanical forces that propel cells and modify their shapes are covered in Chapter 19.

Integrin-Containing Adhesive Structures Physically and Functionally Connect the ECM and Cytoskeleton in Nonepithelial Cells

As already discussed in regard to epithelia, integrin-containing hemidesmosomes connect epithelial cells to the basal lamina and, through adapter proteins, to intermediate filaments of the cytoskeleton (see Figure 6-1). In nonepithelial cells, integrins in the plasma membrane also are clustered with other molecules in various adhesive structures called focal adhesions, focal contacts, focal complexes, 3D adhesions, and fibrillar adhesions and in circular adhesions called podosomes (Chapter 14). These structures are readily observed by fluorescence microscopy with the use of antibodies that recognize integrins or other coclustered molecules (Figure 6-27). Like cell–matrix anchoring junctions in epithelial cells, the various adhesive structures attach nonepithelial cells to the extracellular matrix;

▶ **EXPERIMENTAL FIGURE 6-27**

Integrins cluster into adhesive structures with various morphologies in nonepithelial cells. Immunofluorescence methods were used to detect adhesive structures (green) on cultured cells. Shown here are focal adhesions (a) and 3D adhesions (b) on the surfaces of human fibroblasts. Cells were grown directly on the flat surface of a culture dish (a) or on a three-dimensional matrix of ECM components (b). The shape, distribution, and composition of the integrin-based adhesions formed by cells vary, depending on culture conditions. [Part (a) from B. Geiger et al., 2001, *Nature Rev. Mol. Cell Biol.* **2**:793. Part (b) courtesy of K. Yamada and E. Cukierman; see E. Cukierman et al., 2001, *Science* **294**:1708–12.]

(a) Focal adhesion

(b) 3D adhesion

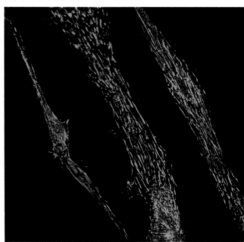

they also contain dozens of intracellular adapter and associated proteins that mediate attachment to cytoskeletal actin filaments and activate adhesion-dependent signals for cell growth and cell motility.

Although found in many nonepithelial cells, integrin-containing adhesive structures have been studied most frequently in fibroblasts grown in cell culture on flat glass or plastic surfaces (substrata). These conditions only poorly approximate the three-dimensional ECM environment that normally surrounds such cells in vivo. When fibroblasts are cultured in three-dimensional ECM matrices derived from cells or tissues, they form adhesions to the three-dimensional ECM substratum, called 3D adhesions. These structures differ somewhat in composition, shape, distribution, and activity from the focal or fibrillar adhesions seen in cells growing on the flat substratum typically used in cell-culture experiments (see Figure 6-27). Cultured fibroblasts with these "more natural" anchoring junctions display greater adhesion and mobility, increased rates of cell proliferation, and spindle-shaped morphologies more like those of fibroblasts in tissues than do cells cultured on flat surfaces. These observations indicate that the topological, compositional, and mechanical (e.g., flexibility) properties of the extracellular matrix all play a role in controlling the shape and activity of a cell. Tissue-specific differences in these matrix characteristics probably contribute to the tissue-specific properties of cells.

TABLE 6-2 Selected Vertebrate Integrins*		
Subunit Composition	**Primary Cellular Distribution**	**Ligands**
α1β1	Many types	Mainly collagens; also laminins
α2β1	Many types	Mainly collagens; also laminins
α4β1	Hematopoietic cells	Fibronectin; VCAM-1
α5β1	Fibroblasts	Fibronectin
αLβ2	T lymphocytes	ICAM-1, ICAM-2
αMβ2	Monocytes	Serum proteins (e.g., C3b, fibrinogen, factor X); ICAM-1
αIIbβ3	Platelets	Serum proteins (e.g., fibrinogen, von Willebrand factor, vitronectin); fibronectin
α6β4	Epithelial cells	Laminin

*The integrins are grouped into subfamilies having a common β subunit. Ligands shown in red are CAMs; all others are ECM or serum proteins. Some subunits can have multiply spliced isoforms with different cytosolic domains.

SOURCE: R. O. Hynes, 1992, *Cell* **69**:11.

Diversity of Ligand–Integrin Interactions Contributes to Numerous Biological Processes

Although most cells express several distinct integrins that bind the same ligand or different ligands, many integrins are expressed predominantly in certain types of cells. Table 6-2 lists a few of the numerous integrin-mediated interactions with ECM components or CAMs or both. Not only do many integrins bind more than one ligand, but several of their ligands bind to multiple integrins.

All integrins appear to have evolved from two ancient general subgroups: those that bind RGD-containing molecules (e.g., fibronectin) and those that bind laminin. For example, $\alpha5\beta1$ integrin binds fibronectin, whereas the widely expressed $\alpha1\beta1$ and $\alpha2\beta1$ integrins, as well as the $\alpha6\beta4$ integrin expressed by epithelial cells, bind laminin. The $\alpha1$, $\alpha2$, and several other integrin α subunits contain a distinctive inserted domain, the *I-domain*. The I-domain in some integrins (e.g., $\alpha1\beta1$ and $\alpha2\beta1$) mediates binding to various collagens. Other integrins containing α subunits with I-domains are expressed exclusively on leukocytes and hematopoietic cells; these integrins recognize cell-adhesion molecules on other cells, including members of the Ig superfamily (e.g., ICAMs, VCAMs), and thus participate in cell–cell adhesion.

The diversity of integrins and their ECM ligands enables integrins to participate in a wide array of key biological processes, including the migration of cells to their correct locations in the formation the body plan of an embryo (morphogenesis) and in the inflammatory response. The importance of integrins in diverse processes is highlighted by the defects exhibited by knockout mice engineered to have mutations in each of almost all of the integrin subunit genes. These defects include major abnormalities in development,

blood vessel formation, leukocyte function, the response to infection (inflammation), bone remodeling, and hemostasis.

Cell–Matrix Adhesion Is Modulated by Changes in the Binding Activity and Numbers of Integrins

Cells can exquisitely control the strength of integrin-mediated cell–matrix interactions by regulating the ligand-binding activity of integrins or their expression or both. Such regulation is critical to the role of these interactions in cell migration and other functions.

Many, if not all, integrins can exist in two conformations: a low-affinity (inactive) form and a high-affinity (active) form (Figure 6-28). The results of structural studies and experiments investigating the binding of ligands by integrins have provided a model of the changes that take place when integrins are activated. In the inactive state, the $\alpha\beta$ heterodimer is bent, the conformation of the ligand-binding site at the tip of the molecule allows only low-affinity ligand binding, and the cytoplasmic C-terminal tails of the two subunits are closely bound together. In the "straight," active state, alterations in the conformation of the domains that form the binding site permit tighter (high-affinity) ligand binding, and the cytoplasmic tails separate.

These structural models also provide an attractive explanation for the ability of integrins to mediate outside-in and inside-out signaling. The binding of certain ECM molecules or CAMs on other cells to the bent, low-affinity structure would force the molecule to straighten and consequently separate the cytoplasmic tails. Intracellular adapters could "sense" the separation of the tails and, as a result, either bind or dissociate from the tails. The changes in these adapters

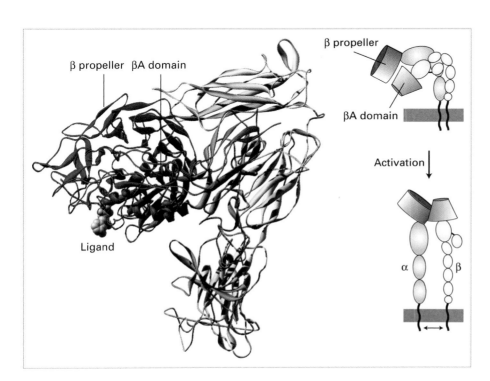

◀ **FIGURE 6-28 Model for integrin activation.** (*Left*) The molecular model is based on the x-ray crystal structure of the extracellular region of $\alpha v\beta3$ integrin in its inactive, low-affinity ("bent") form, with the α subunit in shades of blue and the β subunit in shades of red. The major ligand-binding sites are at the tip of the molecule where the β propeller domain (dark blue) and βA domain (dark red) interact. An RGD peptide ligand is shown in yellow.
(*Right*) Activation of integrins is thought to be due to conformational changes that include straightening of the molecule, key movements near the β propeller and βA domains, which increases the affinity for ligands, and separation of the cytoplasmic domains, resulting in altered interactions with adapter proteins. See text for further discussion. [Adapted from M. Arnaout et al., 2002, *Curr. Opin. Cell Biol.* **14**:641, and R. O. Hynes, 2002, *Cell* **110**:673.]

could then alter the cytoskeleton and activate or inhibit intracellular signaling pathways. Conversely, changes in the metabolic state of the cells (e.g., changes in the platelet cytoskeleton that accompany platelet activation; see Figure 19-5) could cause intracellular adapters to bind to the tails or to dissociate from them and thus force the tails to either separate or associate. As a consequence, the integrin would either bend (inactivate) or straighten (activate), thereby altering its interaction with the ECM or other cells.

Platelet function provides a good example of how cell–matrix interactions are modulated by controlling integrin binding activity. In its basal state, the $\alpha IIb\beta 3$ integrin present on the plasma membranes of platelets normally cannot bind tightly to its protein ligands (e.g., fibrinogen, fibronectin), all of which participate in the formation of a blood clot, because it is in the inactive (bent) conformation. The binding of a platelet to collagen or thrombin in a forming clot induces from the cytoplasm an activating conformational change in $\alpha IIb\beta 3$ integrin that permits it to tightly bind clotting proteins and participate in clot formation. Persons with genetic defects in the $\beta 3$ integrin subunit are prone to excessive bleeding, attesting to the role of this integrin in the formation of blood clots.

The attachment of cells to ECM components can also be modulated by altering the number of integrin molecules exposed on the cell surface. The $\alpha 4\beta 1$ integrin, which is found on many hematopoietic cells (precursors of red and white blood cells), offers an example of this regulatory mechanism. For these hematopoietic cells to proliferate and differentiate, they must be attached to fibronectin synthesized by supportive ("stromal") cells in the bone marrow. The $\alpha 4\beta 1$ integrin on hematopoietic cells binds to a Glu-Ile-Leu-Asp-Val (EILDV) sequence in fibronectin, thereby anchoring the cells to the matrix. This integrin also binds to a sequence in a CAM called vascular CAM-1 (VCAM-1), which is present on stromal cells of the bone marrow. Thus hematopoietic cells directly contact the stromal cells, as well as attach to the matrix. Late in their differentiation, hematopoietic cells decrease their expression of $\alpha 4\beta 1$ integrin; the resulting reduction in the number of $\alpha 4\beta 1$ integrin molecules on the cell surface is thought to allow mature blood cells to detach from the matrix and stromal cells in the bone marrow and subsequently enter the circulation.

Molecular Connections Between the ECM and Cytoskeleton Are Defective in Muscular Dystrophy

The importance of the adhesion receptor–mediated linkage between ECM components and the cytoskeleton is highlighted by a set of hereditary muscle-wasting diseases, collectively called muscular dystrophies. Duchenne muscular dystrophy (DMD), the most common type, is a sex-linked disorder, affecting 1 in 3300 boys, that results in cardiac or respiratory failure in the late teens or early twenties. The first clue to understanding the molecular basis of this disease came from the discovery that persons with DMD carry mutations in the gene encoding a protein named *dystrophin*. This very large protein was found to be a cytosolic adapter protein, binding to actin filaments and to an adhesion receptor called *dystroglycan*. ▪

Dystroglycan is synthesized as a large glycoprotein precursor that is proteolytically cleaved into two subunits. The α subunit is a peripheral membrane protein, and the β subunit is a transmembrane protein whose extracellular domain associates with the α subunit (Figure 6-29). Multiple **O-linked oligosaccharides** are attached covalently to side-chain hydroxyl groups of serine and threonine residues in the α subunit. These O-linked oligosaccharides bind to various basal lamina components, including the multiadhesive matrix protein laminin and the proteoglycans perlecan and

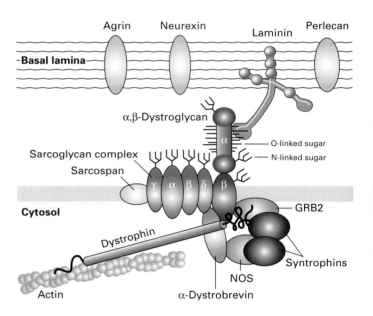

▲ **FIGURE 6-29 Schematic model of the dystrophin glycoprotein complex (DGC) in skeletal muscle cells.** The DGC comprises three subcomplexes: the α,β dystroglycan subcomplex; the sarcoglycan/sarcospan subcomplex of integral membrane proteins; and the cytosolic adapter subcomplex comprising dystrophin, other adapter proteins, and signaling molecules. Through its O-linked sugars, β-dystroglycan binds to components of the basal lamina, such as laminin. Dystrophin—the protein defective in Duchenne muscular dystrophy—links β-dystroglycan to the actin cytoskeleton, and α-dystrobrevin links dystrophin to the sarcoglycan/sarcospan subcomplex. Nitric oxide synthase (NOS) produces nitric oxide, a gaseous signaling molecule, and GRB2 is a component of signaling pathways activated by certain cell-surface receptors (Chapter 14). See text for further discussion. [Adapted from S. J. Winder, 2001, *Trends Biochem. Sci.* **26**:118, and D. E. Michele and K. P. Campbell, 2003, *J. Biol. Chem.*]

agrin. The neurexins, a family of adhesion molecules expressed by neurons, also are bound by the α subunit.

The transmembrane segment of the dystroglycan β subunit associates with a complex of integral membrane proteins; its cytosolic domain binds dystrophin and other adapter proteins, as well as various intracellular signaling proteins. The resulting large, heterogeneous assemblage, the *dystrophin glycoprotein complex (DGC)*, links the extracellular matrix to the cytoskeleton and signaling pathways within muscle cells (see Figure 6-29). For instance, the signaling enzyme nitric oxide synthase (NOS) is associated through syntrophin with the cytosolic dystrophin subcomplex in skeletal muscle. The rise in intracellular Ca^{2+} during muscle contraction activates NOS to produce nitric oxide (NO), which diffuses into smooth muscle cells surrounding nearby blood vessels. By a signaling pathway described in Chapter 13, NO promotes smooth muscle relaxation, leading to a local rise in the flow of blood supplying nutrients and oxygen to the skeletal muscle.

Mutations in dystrophin, other DGC components, laminin, or enzymes that add the O-linked sugars to dystroglycan disrupt the DGC-mediated link between the exterior and the interior of muscle cells and cause muscular dystrophies. In addition, dystroglycan mutations have been shown to greatly reduce the clustering of acetylcholine receptors on muscle cells at the neuromuscular junctions (Chapter 7), which also is dependent on the basal lamina proteins laminin and agrin. These and possibly other effects of DGC defects apparently lead to a cumulative weakening of the mechanical stability of muscle cells as they undergo contraction and relaxation, resulting in deterioration of the cells and muscular dystrophy.

Ca^{2+}-Independent Cell–Cell Adhesion in Neuronal and Other Tissues Is Mediated by CAMs in the Immunoglobulin Superfamily

Numerous transmembrane proteins characterized by the presence of multiple immunoglobulin domains (repeats) in their extracellular regions constitute the Ig superfamily of CAMs, or **IgCAMs**. The Ig domain is a common protein motif, containing 70–110 residues, that was first identified in antibodies, the antigen-binding immunoglobulins. The human, *D. melanogaster,* and *C. elegans* genomes include about 765, 150, and 64 genes, respectively, that encode proteins containing Ig domains. Immunoglobin domains are found in a wide variety of cell-surface proteins including T-cell receptors produced by lymphocytes and many proteins that take part in adhesive interactions. Among the IgCAMs are neural CAMs; intercellular CAMs (ICAMs), which function in the movement of leukocytes into tissues; and junction adhesion molecules (JAMs), which are present in tight junctions.

As their name implies, neural CAMs are of particular importance in neural tissues. One type, the NCAMs, primarily mediate homophilic interactions. First expressed during morphogenesis, NCAMs play an important role in the differentiation of muscle, glial, and nerve cells. Their role in cell adhesion has been directly demonstrated by the inhibition of adhesion with anti-NCAM antibodies. Numerous NCAM isoforms, encoded by a single gene, are generated by alternative mRNA splicing and by differences in glycosylation. Other neural CAMs (e.g., L1-CAM) are encoded by different genes. In humans, mutations in different parts of the L1-CAM gene cause various neuropathologies (e.g., mental retardation, congenital hydrocephalus, and spasticity).

An NCAM comprises an extracellular region with five Ig repeats and two fibronectin type III repeats, a single membrane-spanning segment, and a cytosolic segment that interacts with the cytoskeleton (see Figure 6-2). In contrast, the extracellular region of L1-CAM has six Ig repeats and four fibronectin type III repeats. As with cadherins, cis (intracellular) interactions and trans (intercellular) interactions probably play key roles in IgCAM-mediated adhesion (see Figure 6-3).

The covalent attachment of multiple chains of sialic acid, a negatively charged sugar derivative, to NCAMs alters their adhesive properties. In embryonic tissues such as brain, polysialic acid constitutes as much as 25 percent of the mass of NCAMs. Possibly because of repulsion between the many negatively charged sugars in these NCAMs, cell–cell contacts are fairly transient, being made and then broken, a property necessary for the development of the nervous system. In contrast, NCAMs from adult tissues contain only one-third as much sialic acid, permitting more stable adhesions.

Movement of Leukocytes into Tissues Depends on a Precise Sequence of Combinatorially Diverse Sets of Adhesive Interactions

In adult organisms, several types of white blood cells (leukocytes) participate in the defense against infection caused by foreign invaders (e.g., bacteria and viruses) and tissue damage due to trauma or inflammation. To fight infection and clear away damaged tissue, these cells must move rapidly from the blood, where they circulate as unattached, relatively quiescent cells, into the underlying tissue at sites of infection, inflammation, or damage. We know a great deal about the movement into tissue, termed *extravasation,* of four types of leukocytes: neutrophils, which release several antibacterial proteins; monocytes, the precursors of macrophages, which can engulf and destroy foreign particles; and T and B lymphocytes, the antigen-recognizing cells of the immune system.

Extravasation requires the successive formation and breakage of cell–cell contacts between leukocytes in the blood and endothelial cells lining the vessels. Some of these contacts are mediated by **selectins**, a family of CAMs that mediate leukocyte–vascular cell interactions. A key player in these interactions is *P-selectin*, which is localized to the blood-facing surface of endothelial cells. All selectins contain a Ca^{2+}-dependent *lectin domain,* which is located at the distal end of the extracellular region of the molecule and recognizes oligosaccharides in glycoproteins or glycolipids (see Figure 6-2). For example, the primary ligand for P- and E-selectins is an oligosaccharide called the *sialyl Lewis-x antigen,* a part of longer oligosaccharides present in abundance on leukocyte glycoproteins and glycolipids.

Figure 6-30 illustrates the basic sequence of cell–cell interactions leading to the extravasation of leukocytes. Various inflammatory signals released in areas of infection or inflammation first cause activation of the endothelium. P-selectin exposed on the surface of activated endothelial cells mediates the weak adhesion of passing leukocytes. Because of the force of the blood flow and the rapid "on" and "off" rates of P-selectin binding to its ligands, these "trapped" leukocytes are slowed but not stopped and literally roll along the surface of the endothelium. Among the

signals that promote activation of the endothelium are *chemokines,* a group of small secreted proteins (8–12 kDa) produced by a wide variety of cells, including endothelial cells and leukocytes.

For tight adhesion to occur between activated endothelial cells and leukocytes, β2-containing integrins on the surfaces of leukocytes also must be activated by chemokines or other local activation signals such as platelet-activating factor (PAF). Platelet-activating factor is unusual in that it is a phospholipid, rather than a protein; it is exposed on the surface of activated endothelial cells at the same time that P-selectin is exposed. The binding of PAF or other activators to their receptors on leukocytes leads to activation of the leukocyte integrins to their high-affinity form (see Figure 6-28). (Most of the receptors for chemokines and PAF are members of the G protein–coupled receptor superfamily discussed in Chapter 13.) Activated integrins on leukocytes then bind to each of two distinct IgCAMs on the surface of en-

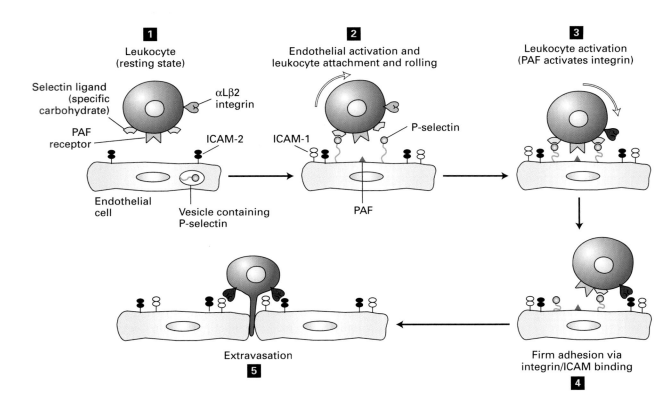

▲ **FIGURE 6-30 Sequence of cell–cell interactions leading to tight binding of leukocytes to activated endothelial cells and subsequent extravasation.** Step **1**: In the absence of inflammation or infection, leukocytes and endothelial cells lining blood vessels are in a resting state. Step **2**: Inflammatory signals released only in areas of inflammation or infection or both activate resting endothelial cells to move vesicle-sequestered selectins to the cell surface. The exposed selectins mediate loose binding of leukocytes by interacting with carbohydrate ligands on leukocytes. Activation of the endothelium also causes

synthesis of platelet-activating factor (PAF) and ICAM-1, both expressed on the cell surface. PAF and other usually secreted activators, including chemokines, then induce changes in the shapes of the leukocytes and activation of leukocyte integrins such as αLβ2, which is expressed by T lymphocytes (**3**). The subsequent tight binding between activated integrins on leukocytes and CAMs on the endothelium (e.g., ICAM-2 and ICAM-1) results in firm adhesion (**4**) and subsequent movement (extravasation) into the underlying tissue (**5**). See text for further discussion. [Adapted from R. O. Hynes and A. Lander, 1992, *Cell* **68**:303.]

dothelial cells: ICAM-2, which is expressed constitutively, and ICAM-1. ICAM-1, whose synthesis along with that of E-selectin and P-selectin is induced by activation, does not usually contribute substantially to leukocyte endothelial cell adhesion immediately after activation, but rather participates at later times in cases of chronic inflammation. The resulting tight adhesion mediated by the Ca^{2+}-independent integrin–ICAM interactions leads to the cessation of rolling and to the spreading of leukocytes on the surface of the endothelium; soon the adhered cells move between adjacent endothelial cells and into the underlying tissue.

The selective adhesion of leukocytes to the endothelium near sites of infection or inflammation thus depends on the sequential appearance and activation of several different CAMs on the surfaces of the interacting cells. Different types of leukocytes express specific integrins containing the β2 subunit: for example, αLβ2 by T lymphocytes and αMβ2 by monocytes, the circulating precursors of tissue macrophages. Nonetheless, all leukocytes move into tissues by the same general mechanism depicted in Figure 6-30.

Many of the CAMs used to direct leukocyte adhesion are shared among different types of leukocytes and target tissues. Yet often only a particular type of leukocyte is directed to a particular tissue. A three-step model has been proposed to account for the cell-type specificity of such leukocyte–endothelial cell interactions. First, endothelium activation promotes initial relatively weak, transient, and reversible binding (e.g., the interaction of selectins and their carbohydrate ligands). Without additional local activation signals, the leukocyte will quickly move on. Second, cells in the immediate vicinity of the site of infection or inflammation release or express on their surfaces chemical signals (e.g., chemokines, PAF) that activate only special subsets of the transiently attached leukocytes. Third, additional activation-dependent CAMs (e.g., integrins) engage their binding partners, leading to strong sustained adhesion. Only if the proper combination of CAMs, binding partners, and activation signals are engaged in the right order at a specific site will a given leukocyte adhere strongly. This additional example of combinatorial diversity and cross talk allows parsimonious exploitation of a small set of CAMs for diverse functions throughout the body.

 Leukocyte-adhesion deficiency is caused by a genetic defect in the synthesis of the integrin β2 subunit. Persons with this disorder are susceptible to repeated bacterial infections because their leukocytes cannot extravasate properly and thus fight the infection within the tissue.

Some pathogenic viruses have evolved mechanisms to exploit for their own purposes cell-surface proteins that participate in the normal response to inflammation. For example, many of the RNA viruses that cause the common cold (rhinoviruses) bind to and enter cells through ICAM-1, and chemokine receptors can be important entry sites for human immunodeficiency virus (HIV), the cause of AIDS. ∎

Gap Junctions Composed of Connexins Allow Small Molecules to Pass Between Adjacent Cells

Early electron micrographs of virtually all animal cells that were in contact revealed sites of cell–cell contact with a characteristic intercellular gap (Figure 6-31a). This feature prompted early morphologists to call these regions gap junctions. In retrospect, the most important feature of these junctions is not the gap itself but a well-defined set of cylindrical particles that cross the gap and compose pores connecting the cytoplasms of adjacent cells—hence their alternate name of *intercytoplasmic* junctions. In epithelia, gap junctions are distributed along the lateral surfaces of adjacent cells (see Figures 6-1 and 6-5).

In many tissues (e.g., the liver), large numbers of individual cylindrical particles cluster together in patches. This property has enabled researchers to separate gap junctions from other components of the plasma membrane. When the plasma membrane is purified and then sheared into small fragments, some pieces mainly containing patches of gap junctions are generated. Owing to their relatively high protein content, these fragments have a higher density than that of the bulk of the plasma membrane and can be purified on an equilibrium density gradient (see Figure 5-37). When these

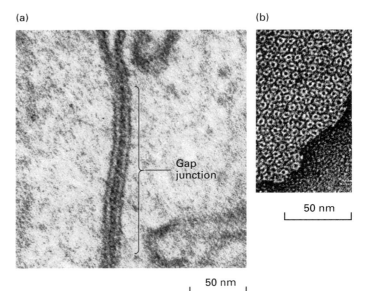

(a)   (b)

Gap junction

50 nm

50 nm

▲ **EXPERIMENTAL FIGURE 6-31 Gap junctions have a characteristic appearance in electron micrographs.** (a) In this thin section through a gap junction connecting two mouse liver cells, the two plasma membranes are closely associated for a distance of several hundred nanometers, separated by a "gap" of 2–3 nm. (b) Numerous roughly hexagonal particles are visible in this perpendicular view of the cytosolic face of a region of plasma membrane enriched in gap junctions. Each particle aligns with a similar particle on an adjacent cell, forming a channel connecting the two cells. [Part (a) courtesy of D. Goodenough. Part (b) courtesy of N. Gilula.]

preparations are viewed in cross section, the gap junctions appear as arrays of hexagonal particles that enclose water-filled channels (Figure 6-31b). Such pure preparations of gap junctions have permitted the detailed biophysical and functional analysis of these structures.

The effective pore size of gap junctions can be measured by injecting a cell with a fluorescent dye covalently linked to molecules of various sizes and observing with a fluorescence microscope whether the dye passes into neighboring cells. Gap junctions between mammalian cells permit the passage of molecules as large as 1.2 nm in diameter. In insects, these junctions are permeable to molecules as large as 2 nm in diameter. Generally speaking, molecules smaller than 1200 Da pass freely, and those larger than 2000 Da do not pass; the passage of intermediate-sized molecules is variable and limited. Thus ions, many low-molecular-weight precursors of cellular macromolecules, products of intermediary metabolism, and small intracellular signaling molecules can pass from cell to cell through gap junctions.

In nervous tissue, some neurons are connected by gap junctions through which ions pass rapidly, thereby allowing very rapid transmission of electrical signals. Impulse transmission through these connections, called electrical synapses, is almost a thousandfold as rapid as at chemical synapses (Chapter 7). Gap junctions are also present in many non-neuronal tissues where they help to integrate the electrical and metabolic activities of many cells. In the heart, for instance, gap junctions rapidly pass ionic signals among muscle cells and thus contribute to the electrically stimulated coordinate contraction of cardiac muscle cells during a beat. As discussed in Chapter 13, some extracellular hormonal signals induce the production or release of small intracellular signaling molecules called **second messengers** (e.g., cyclic AMP and Ca^{2+}) that regulate cellular metabolism. Because second messengers can be transferred between cells through gap junctions, hormonal stimulation of one cell can trigger a coordinated response by that same cell and many of its neighbors. Such gap junction–mediated signaling plays an important role, for example, in the secretion of digestive enzymes by the pancreas and in the coordinated muscular contractile waves (peristalsis) in the intestine. Another vivid example of gap junction–mediated transport is the phenomenon of *metabolic coupling*, or *metabolic cooperation*, in which a cell transfers nutrients or intermediary metabolites to a neighboring cell that is itself unable to synthesize them. Gap junctions play critical roles in the development of egg cells in the ovary by mediating the movement of both metabolites and signaling molecules between an oocyte and its surrounding granulosa cells as well as between neighboring granulosa cells.

A current model of the structure of the gap junction is shown in Figure 6-32. Vertebrate gap junctions are composed

▶ **FIGURE 6-32 Molecular structure of gap junctions.** (a) Schematic model of a gap junction, which comprises a cluster of channels between two plasma membranes separated by a gap of about 2–3 nm. Both membranes contain connexon hemichannels, cylinders of six dumbbell-shaped connexin molecules. Two connexons join in the gap between the cells to form a gap-junction channel, 1.5–2.0 nm in diameter, that connects the cytosols of the two cells. (b) Electron density of a recombinant gap-junction channel determined by electron crystallography. Shown here are side views of the complete structure (*top*) and the same structure with several chains removed to show the channel's interior (*center*); on the bottom are perpendicular cross sections through the gap junction within and between the membrane bilayers. There appear to be 24 transmembrane α helices per connexon hemichannel, consistent with each of the six connexin subunits having four α helices. The narrowest part of the channel is ≈1.5 nm in diameter. M = membrane bilayer; E = extracellular gap; C = cytosol. [Part (b) from V. M. Unger et al., 1999, *Science* **283**:1176.]

(a)

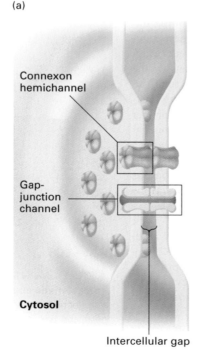

Connexon hemichannel

Gap-junction channel

Cytosol

Intercellular gap

(b)

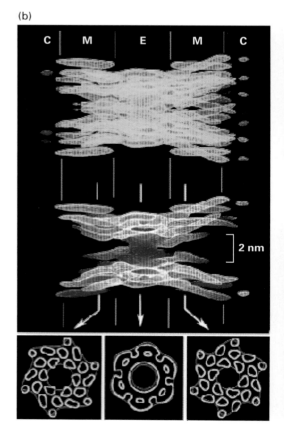

of **connexins,** a family of structurally related transmembrane proteins with molecular weights between 26,000 and 60,000. A completely different family of proteins, the innexins, forms the gap junctions in invertebrates. Each vertebrate hexagonal particle consists of 12 connexin molecules: 6 of the molecules are arranged in a connexon hemichannel—a hexagonal cylinder in one plasma membrane—and joined to a connexon hemichannel in the adjacent cell membrane, forming the continuous aqueous channel between the cells. Each connexin molecule spans the plasma membrane four times; one conserved transmembrane α helix from each subunit apparently lines the aqueous channel.

There are probably more than 20 different connexin genes in vertebrates, and different sets of connexins are expressed in different cell types. Some cells express a single connexin; consequently their gap-junction channels are homotypic, consisting of identical connexons. Most cells, however, express at least two connexins; these different proteins assemble into hetero-oligomeric connexons, which in turn form heterotypic gap-junction channels. This diversity in channel composition leads to differences in the permeability of channels to various molecules. For example, channels made from a 43-kDa connexin isoform, Cx43, are more than a hundredfold as permeable to ADP and ATP as those made from Cx32 (32 kDa). Moreover, the permeability of gap junctions can be altered by changes in the intracellular pH and Ca^{2+} concentration, as well as by the phosphorylation of connexin, providing numerous mechanisms for regulating transport through them.

The generation of mutant mice with inactivating mutations in connexin genes has highlighted the importance of connexins in a wide variety of cellular systems. For instance, Cx43-defective mice exhibit numerous defects including defective oocyte maturation due to decreased gap-junctional communication between granulosa cells in the ovary.

 Mutations in several connexin genes are related to human diseases, including neurosensory deafness (Cx26 and Cx31), cataract or heart malformations (Cx43, Cx46, and Cx50), and the X-linked form of Charcot-Marie-Tooth disease (Cx32), which is marked by progressive degeneration of peripheral nerves. ∎

KEY CONCEPTS OF SECTION 6.5

Adhesive Interactions and Nonepithelial Cells

■ Many nonepithelial cells have integrin-containing aggregates (e.g., focal adhesions, 3D adhesions, podosomes) that physically and functionally connect cells to the extracellular matrix and facilitate inside-out and outside-in signaling.

■ Integrins exist in two conformations that differ in the affinity for ligands and interactions with cytosolic adapter proteins (see Figure 6-28).

■ Dystroglycan, an adhesion receptor expressed by muscle cells, forms a large complex with dystrophin, other adapter proteins, and signaling molecules (see Figure 6-29). This complex links the actin cytoskeleton to the surrounding matrix, providing mechanical stability to muscle. Mutations in various components of this complex cause different types of muscular dystrophy.

■ Neural cell-adhesion molecules (CAMs), which belong to the immunoglobulin (Ig) family of CAMs, mediate Ca^{2+}-independent cell–cell adhesion, predominantly in neural tissue and muscle.

■ The combinatorial and sequential interaction of several types of CAMs (e.g., selectins, integrins, and ICAMs) is critical for the specific and tight adhesion of different types of leukocytes to endothelial cells in response to local signals induced by infection or inflammation (see Figure 6-30).

■ Gap junctions are constructed of multiple copies of connexin proteins, assembled into a transmembrane channel that interconnects the cytoplasm of two adjacent cells (see Figure 6-32). Small molecules and ions can pass through gap junctions, permitting metabolic and electrical coupling of adjacent cells.

6.6 Plant Tissues

 We turn now to the assembly of plant cells into tissues. The overall structural organization of plants is generally simpler than that of animals. For instance, plants have only four broad types of cells, which in mature plants form four basic classes of tissue: *dermal tissue* interacts with the environment; *vascular tissue* transports water and dissolved substances (e.g., sugars, ions); space-filling *ground tissue* constitutes the major sites of metabolism; and *sporogenous tissue* forms the reproductive organs. Plant tissues are organized into just four main organ systems: *stems* have support and transport functions; *roots* provide anchorage and absorb and store nutrients; *leaves* are the sites of photosynthesis; and *flowers* enclose the reproductive structures. Thus at the cell, tissue, and organ levels, plants are generally less complex than most animals.

Moreover, unlike animals, plants do not replace or repair old or damaged cells or tissues; they simply grow new organs. Most importantly for this chapter and in contrast with animals, few cells in plants directly contact one another through molecules incorporated into their plasma membranes. Instead, plant cells are typically surrounded by a rigid **cell wall** that contacts the cell walls of adjacent cells (Figure 6-33). Also in contrast with animal cells, a plant cell rarely changes its position in the organism relative to other cells. These features of plants and their organization have determined the distinctive molecular mechanisms by which their cells are incorporated into tissues. ∎

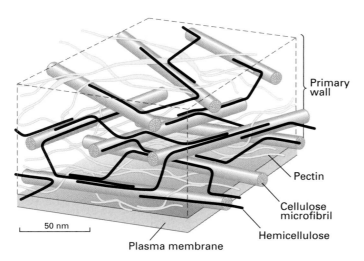

Primary wall

Pectin

Cellulose microfibril

Hemicellulose

Plasma membrane

50 nm

▲ **FIGURE 6-33 Schematic representation of the cell wall of an onion.** Cellulose and hemicellulose are arranged into at least three layers in a matrix of pectin polymers. The size of the polymers and their separations are drawn to scale. To simplify the diagram, most of the hemicellulose cross-links and other matrix constituents (e.g., extensin, lignin) are not shown. [Adapted from M. McCann and K. R. Roberts, 1991, in C. Lloyd, ed., *The Cytoskeletal Basis of Plant Growth and Form*, Academic Press, p. 126.]

The Plant Cell Wall Is a Laminate of Cellulose Fibrils in a Matrix of Glycoproteins

The plant cell wall is ≈0.2 μm thick and completely coats the outside of the plant cell's plasma membrane. This structure serves some of the same functions as those of the extracellular matrix produced by animal cells, even though the two structures are composed of entirely different macromolecules and have a different organization. Like the extracellular matrix, the plant cell wall connects cells into tissues, signals a plant cell to grow and divide, and controls the shape of plant organs. Just as the extracellular matrix helps define the shapes of animal cells, the cell wall defines the shapes of plant cells. When the cell wall is digested away from plant cells by hydrolytic enzymes, spherical cells enclosed by a plasma membrane are left. In the past, the plant cell wall was viewed as an inanimate rigid box, but it is now recognized as a dynamic structure that plays important roles in controlling the differentiation of plant cells during embryogenesis and growth.

Because a major function of a plant cell wall is to withstand the osmotic turgor pressure of the cell, the cell wall is built for lateral strength. It is arranged into layers of **cellulose** microfibrils—bundles of long, linear, extensively hydrogen-bonded polymers of glucose in β glycosidic linkages. The cellulose microfibrils are embedded in a matrix composed of *pectin,* a polymer of D-galacturonic acid and other monosaccharides, and *hemicellulose,* a short, highly branched

polymer of several five- and six-carbon monosaccharides. The mechanical strength of the cell wall depends on cross-linking of the microfibrils by hemicellulose chains (see Figure 6-33). The layers of microfibrils prevent the cell wall from stretching laterally. Cellulose microfibrils are synthesized on the exoplasmic face of the plasma membrane from UDP-glucose and ADP-glucose formed in the cytosol. The polymerizing enzyme, called *cellulose synthase,* moves within the plane of the plasma membrane as cellulose is formed, in directions determined by the underlying microtubule cytoskeleton.

Unlike cellulose, pectin and hemicellulose are synthesized in the Golgi apparatus and transported to the cell surface where they form an interlinked network that helps bind the walls of adjacent cells to one another and cushions them. When purified, pectin binds water and forms a gel in the presence of Ca^{2+} and borate ions—hence the use of pectins in many processed foods. As much as 15 percent of the cell wall may be composed of *extensin,* a glycoprotein that contains abundant hydroxyproline and serine. Most of the hydroxyproline residues are linked to short chains of arabinose (a five-carbon monosaccharide), and the serine residues are linked to galactose. Carbohydrate accounts for about 65 percent of extensin by weight, and its protein backbone forms an extended rodlike helix with the hydroxyl or O-linked carbohydrates protruding outward. *Lignin*—a complex, insoluble polymer of phenolic residues—associates with cellulose and is a strengthening material. Like cartilage proteoglycans, lignin resists compression forces on the matrix.

The cell wall is a selective filter whose permeability is controlled largely by pectins in the wall matrix. Whereas water and ions diffuse freely across cell walls, the diffusion of large molecules, including proteins larger than 20 kDa, is limited. This limitation may account for why many plant hormones are small, water-soluble molecules, which can diffuse across the cell wall and interact with receptors in the plasma membrane of plant cells.

Loosening of the Cell Wall Permits Elongation of Plant Cells

Because the cell wall surrounding a plant cell prevents the cell from expanding, its structure must be loosened when the cell grows. The amount, type, and direction of plant cell growth are regulated by small-molecule hormones (e.g., indoleacetic acid) called *auxins.* The auxin-induced weakening of the cell wall permits the expansion of the intracellular vacuole by uptake of water, leading to elongation of the cell. We can grasp the magnitude of this phenomenon by considering that, if all cells in a redwood tree were reduced to the size of a typical liver cell, the tree would have a maximum height of only 1 meter.

The cell wall undergoes its greatest changes at the **meristem** of a root or shoot tip. These sites are where cells divide and expand. Young meristematic cells are connected by thin primary cell walls, which can be loosened and stretched to allow subsequent cell elongation. After cell elongation ceases, the cell wall is generally thickened, either by the secretion of additional macromolecules into the primary wall or, more usually, by the formation of a secondary cell wall composed of several layers. Most of the cell eventually degenerates, leaving only the cell wall in mature tissues such as the xylem—the tubes that conduct salts and water from the roots through the stems to the leaves (see Figure 8-45). The unique properties of wood and of plant fibers such as cotton are due to the molecular properties of the cell walls in the tissues of origin.

Plasmodesmata Directly Connect the Cytosols of Adjacent Cells in Higher Plants

Plant cells can communicate directly through specialized cell–cell junctions called **plasmodesmata,** which extend through the cell wall. Like gap junctions, plasmodesmata are open channels that connect the cytosol of a cell with that of an adjacent cell. The diameter of the cytosol-filled channel is about 30–60 nm, and plasmodesmata can traverse cell walls as much as 90 nm thick. The density of plasmodesmata varies depending on the plant and cell type, and even the smallest meristematic cells have more than 1000 interconnections with their neighbors.

Molecules smaller than about 1000 Da, including a variety of metabolic and signaling compounds, generally can diffuse through plasmodesmata. However, the size of the channel through which molecules pass is highly regulated. In some circumstances, the channel is clamped shut; in others, it is dilated sufficiently to permit the passage of molecules larger than 10,000 Da. Among the factors that affect the permeability of plasmodesmata is the cytosolic Ca^{2+} concentration, with an increase in cytosolic Ca^{2+} reversibly inhibiting movement of molecules through these structures.

Although plasmodesmata and gap junctions resemble each other functionally, their structures differ in two significant ways (Figure 6-34). The plasma membranes of the adjacent plant cells merge to form a continuous channel, the *annulus,* at each plasmodesma, whereas the membranes of cells at a gap junction are not continuous with each other. In addition, an extension of the endoplasmic reticulum called a *desmotubule* passes through the annulus, which connects the cytosols of adjacent plant cells. Many types of molecules spread from cell to cell through plasmodesmata, including proteins, nucleic acids, metabolic products, and plant viruses. Soluble molecules pass through the cytosolic annulus, whereas membrane-bound molecules can pass from cell to cell through the desmotubule.

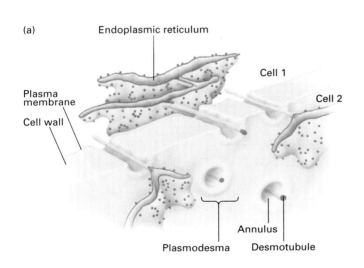

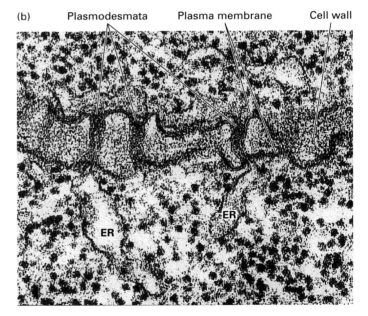

▲ **FIGURE 6-34 Structure of a plasmodesma.**
(a) Schematic model of a plasmodesma showing the desmotubule, an extension of the endoplasmic reticulum, and the annulus, a plasma membrane–lined channel filled with cytosol that interconnects the cytosols of adjacent cells. Not shown is a gating complex that fills the channel and controls the transport of materials through the plasmodesma.
(b) Electron micrograph of thin section of plant cell and cell wall containing multiple plasmodesmata. [E. H. Newcomb and W. P. Wergin/Biological Photo Service.]

Only a Few Adhesive Molecules Have Been Identified in Plants

Systematic analysis of the *Arabidopsis* genome and biochemical analysis of other plant species provide no evidence for the existence of plant homologs of most animal CAMs, adhesion receptors, and ECM components. This finding is not surprising, given the dramatically different nature of cell–cell and cell–matrix/wall interactions in animals and plants.

Among the adhesive-type proteins apparently unique to plants are five wall-associated kinases (WAKs) and WAK-like proteins expressed in the plasma membrane of *Arabidopsis* cells. The extracellular regions in all these proteins contain multiple epidermal growth factor (EGF) repeats, which may

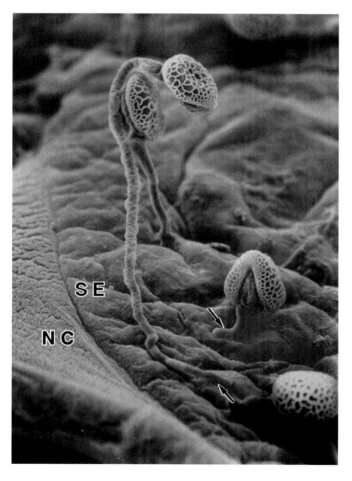

▲ **EXPERIMENTAL FIGURE 6-35 An in vitro assay used to identify molecules required for adherence of pollen tubes to the stylar matrix.** In this assay, extracellular stylar matrix collected from lily styles (SE) or an artificial matrix is dried onto nitrocellulose membranes (NC). Pollen tubes containing sperm are then added and their binding to the dried matrix is assessed. In this scanning electron micrograph, the tips of pollen tubes (arrows) can be seen binding to dried stylar matrix. This type of assay has shown that pollen adherence depends on stigma/stylar cysteine-rich adhesin (SCA) and a pectin that binds to SCA. [From G. Y. Jauh et al., 1997, *Sex Plant Reprod.* **10**:173.]

directly participate in binding to other molecules. Some WAKs have been shown to bind to glycine-rich proteins in the cell wall, thereby mediating membrane–wall contacts. These *Arabidopsis* proteins have a single transmembrane domain and an intracellular cytosolic tyrosine kinase domain, which may participate in signaling pathways somewhat like the receptor tyrosine kinases discussed in Chapter 14.

The results of in vitro binding assays combined with in vivo studies and analyses of plant mutants have identified several macromolecules in the ECM that are important for adhesion. For example, normal adhesion of pollen, which contains sperm cells, to the stigma or style in the female reproductive organ of the Easter lily requires a cysteine-rich protein called stigma/stylar cysteine-rich adhesin (SCA) and a specialized pectin that can bind to SCA (Figure 6-35).

Disruption of the gene encoding glucuronyltransferase 1, a key enzyme in pectin biosynthesis, has provided a striking illustration of the importance of pectins in intercellular adhesion in plant meristems. Normally, specialized pectin molecules help hold the cells in meristems tightly together. When grown in culture as a cluster of relatively undifferentiated cells, called a callus, normal meristematic cells adhere tightly and can differentiate into chlorophyll-producing cells, giving the callus a green color. Eventually the callus will generate shoots. In contrast, mutant cells with an inactivated glucuronyltransferase 1 gene are large, associate loosely with each other, and do not differentiate normally, forming a yellow callus. The introduction of a normal glucuronyltransferase 1 gene into the mutant cells by methods discussed in Chapter 9 restores their ability to adhere and differentiate normally.

The paucity of plant adhesive molecules identified to date, in contrast with the many well-defined animal adhesive molecules, may be due to the technical difficulties in working with the ECM/cell wall of plants. Adhesive interactions are often likely to play different roles in plant and animal biology, at least in part because of their differences in development and physiology.

KEY CONCEPTS OF SECTION 6.6

Plant Tissues

■ The integration of cells into tissues in plants is fundamentally different from the assembly of animal tissues, primarily because each plant cell is surrounded by a relatively rigid cell wall.

■ The plant cell wall comprises layers of cellulose microfibrils embedded within a matrix of hemicellulose, pectin, extensin, and other less abundant molecules.

■ Cellulose, a large, linear glucose polymer, assembles spontaneously into microfibrils stabilized by hydrogen bonding.

■ The cell wall defines the shapes of plant cells and restricts their elongation. Auxin-induced loosening of the cell wall permits elongation.

■ Adjacent plant cells can communicate through plasmodesmata, which allow small molecules to pass between the cells (see Figure 6-34).

■ Plants do not produce homologs of the common adhesive molecules found in animals. Only a few adhesive molecules unique to plants have been well documented to date.

6.7 Growth and Use of Cultured Cells

Many technical constraints hamper studies on specific cells or subsets of cells in intact animals and plants. One alternative is the use of intact organs that are removed from animals and perfused with an appropriately buffered solution to maintain their physiologic integrity and function. Such organ perfusion systems have been widely used by physiologists. However, the organization of organs, even isolated ones, is sufficiently complex to pose numerous problems for research on many fundamental aspects of cell biology. Thus molecular cell biologists often conduct experimental studies on cells isolated from an organism and maintained in conditions that permit their survival and growth, a procedure known as *culturing*.

Cultured cells have several advantages over intact organisms for cell biology research. First, most animal and plant tissues consist of a variety of different types of cells, whereas cells of a single specific type with homogeneous properties can be grown in culture. Second, experimental conditions (e.g., composition of the extracellular environment) can be controlled far better in culture than in an intact organism. Third, in many cases a single cell can be readily grown into a colony of many identical cells, a process called cell cloning, or simply cloning (Figure 6-36). The resulting strain of cells, which is genetically homogeneous, is called a **clone**. This simple technique, which is commonly used with many bacteria, yeasts, and mammalian cell types, makes it easy to isolate genetically distinct clones of cells.

A major disadvantage of cultured cells is that they are not in their normal environment and hence their activities are not regulated by the other cells and tissues as they are in an intact organism. For example, insulin produced by the pancreas has an enormous effect on liver glucose metabolism; however, this normal regulatory mechanism does not operate in a purified population of liver cells (called hepatocytes) grown in culture. In addition, as already described, the three-dimensional distribution of cells and extracellular matrix around a cell influences its shape and behavior. Because the immediate environment of cultured cells differs radically from this "normal" environment, their properties may be affected in various ways. Thus care must always be exercised in drawing conclusions about the normal properties of cells in complex tissues and organisms only on the basis of experiments with isolated, cultured cells.

Culture of Animal Cells Requires Nutrient-Rich Media and Special Solid Surfaces

In contrast with most bacterial cells, which can be cultured quite easily, animal cells require many specialized nutrients and often specially coated dishes for successful culturing. To permit the survival and normal function of cultured tissues or cells, the temperature (37 °C for mammalian cells), pH, ionic strength, and access to essential nutrients must simulate as closely as possible the conditions within an intact organism. Isolated animal cells are typically placed in a nutrient-rich liquid, the culture medium, within specially treated plastic dishes or flasks. The cultures are kept in incubators in which the temperature, atmosphere, and humidity can be controlled. To reduce the chances of bacterial or fungal contamination, antibiotics are often added to the culture medium. To further guard against contamination, investigators usually transfer cells between dishes, add reagents to the culture medium, and otherwise manipulate the specimens within special cabinets

(a)

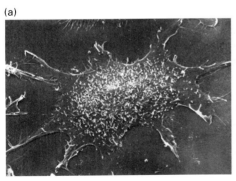

(b)

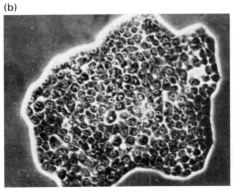

(c)

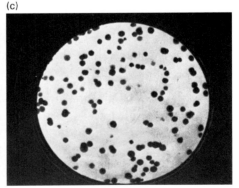

▲ **FIGURE 6-36 Cultured mammalian cells viewed at three magnifications.** (a) A single mouse cell attached to a plastic petri dish, viewed through a scanning electron microscope. (b) A single colony of human HeLa cells about 1 mm in diameter, produced from a single cell after growth for 2 weeks. (c) After cells initially introduced into a 6-cm-diameter petri dish have grown for several days and then been stained, individual colonies can easily be seen and counted. All the cells in a colony are progeny of a single precursor cell and thus genetically identical. [Part (a) courtesy of N. K. Weller. Parts (b) and (c) courtesy of T. T. Puck.]

containing circulating air that is filtered to remove microorganisms and other airborne contaminants.

Media for culturing animal cells must supply histidine, isoleucine, leucine, lysine, methionine, phenylalanine, threonine, tryptophan, and valine; no cells in adult vertebrate animals can synthesize these nine essential amino acids. In addition, most cultured cells require three other amino acids (cysteine, glutamine, and tyrosine) that are synthesized only by specialized cells in intact animals. The other necessary components of a medium for culturing animal cells are vitamins, various salts, fatty acids, glucose, and serum—the fluid remaining after the noncellular part of blood (plasma) has been allowed to clot. Serum contains various protein factors that are needed for the proliferation of mammalian cells in culture. These factors include the polypeptide hormone insulin; transferrin, which supplies iron in a bioaccessible form; and numerous growth factors. In addition, certain cell types require specialized protein growth factors not present in serum. For instance, hematopoietic cells require erythropoietin, and T lymphocytes require interleukin 2 (Chapter 14). A few mammalian cell types can be grown in a chemically defined, serum-free medium containing amino acids, glucose, vitamins, and salts plus certain trace minerals, specific protein growth factors, and other components.

Unlike bacterial and yeast cells, which can be grown in suspension, most animal cells will grow only on a solid surface. This highlights the importance of cell adhesion molecules. Many types of cells can grow on glass or on specially treated plastics with negatively charged groups on the surface (e.g., SO_3^{2-}). The cells secrete ECM components, which adhere to these surfaces, and then attach and grow on the secreted matrix. A single cell cultured on a glass or a plastic dish proliferates to form a visible mass, or *colony,* containing thousands of genetically identical cells in 4–14 days, depending on the growth rate (see Figure 6-36c). Some specialized blood cells and tumor cells can be maintained or grown in suspension as single cells.

Primary Cell Cultures and Cell Strains Have a Finite Life Span

Normal animal tissues (e.g., skin, kidney, liver) or whole embryos are commonly used to establish *primary cell cultures.* To prepare tissue cells for a primary culture, the cell–cell and cell–matrix interactions must be broken. To do so, tissue fragments are treated with a combination of a protease (e.g., trypsin or the collagen-hydrolyzing enzyme collagenase or both) and a divalent cation chelator (e.g., EDTA) that depletes the medium of usable Ca^{2+} or Mg^{2+}. The released cells are then placed in dishes in a nutrient-rich, serum-supplemented medium, where they can adhere to the surface and one another. The same protease/chelator solution is used to remove adherent cells from a culture dish for biochemical studies or subculturing (transfer to another dish).

Often connective tissue fibroblasts divide in culture more rapidly than other cells in a tissue, eventually becoming the predominant type of cells in the primary culture, unless special precautions are taken to remove them when isolating other types of cells. Certain cells from blood, spleen, or bone marrow adhere poorly, if at all, to a culture dish but nonetheless grow well. In the body, such *nonadherent* cells are held in suspension (in the blood) or they are loosely adherent (in the bone marrow and spleen). Because these cells often come from immature stages in the development of differentiated blood cells, they are very useful for studying normal blood cell differentiation and the abnormal development of leukemias.

When cells removed from an embryo or an adult animal are cultured, most of the adherent ones will divide a finite number of times and then cease growing (cell senescence, Figure 6-38a). For instance, human fetal fibroblasts divide about 50 times before they cease growth. Starting with 10^6 cells, 50 doublings can produce $10^6 \times 2^{50}$, or more than 10^{20} cells, which is equivalent to the weight of about 10^5 people. Normally, only a very small fraction of these cells are used in any one experiment. Thus, even though its lifetime is limited, a single culture, if carefully maintained, can be studied through many generations. Such a lineage of cells originating from one initial primary culture is called a **cell strain.**

Cell strains can be frozen in a state of suspended animation and stored for extended periods at liquid nitrogen temperature, provided that a preservative that prevents the formation of damaging ice crystals is used. Although some cells do not survive thawing, many do survive and resume growth. Research with cell strains is simplified by the ability to freeze and successfully thaw them at a later time for experimental analysis.

Transformed Cells Can Grow Indefinitely in Culture

To be able to clone individual cells, modify cell behavior, or select mutants, biologists often want to maintain cell cultures for many more than 100 doublings. Such prolonged growth is exhibited by cells derived from some tumors. In addition, rare cells in a population of primary cells that undergo certain spontaneous genetic changes, called oncogenic **transformation,** are able to grow indefinitely. These cells are said to be oncogenically transformed or simply *transformed.* A culture of cells with an indefinite life span is considered immortal and is called a **cell line.**

The HeLa cell line, the first human cell line, was originally obtained in 1952 from a malignant tumor (carcinoma) of the uterine cervix. Although primary cell cultures of normal human cells rarely undergo transformation into a cell line, rodent cells commonly do. After rodent cells are grown in culture for several generations, the culture goes into senescence (Figure 6-37b). During this period, most of the cells stop growing, but often a rapidly dividing transformed cell arises spontaneously and takes over, or overgrows, the culture. A cell line derived from such a transformed variant will grow indefinitely if provided with the necessary nutrients.

Regardless of the source, cells in immortalized lines often have chromosomes with abnormal structures. In addition, the number of chromosomes in such cells is usually greater than that in the normal cell from which they arose, and the chromosome number expands and contracts as the cells continue to divide in culture. A noteworthy exception is the Chinese hamster ovary (CHO) line and its derivatives, which have fewer chromosomes than their hamster progenitors. Cells with an abnormal number of chromosomes are said to be *aneuploid.*

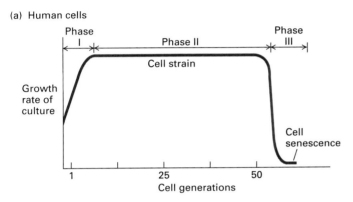

(a) Human cells

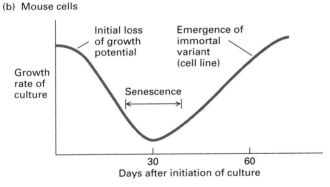

(b) Mouse cells

▲ **FIGURE 6-37 Stages in the establishment of a cell culture.** (a) When cells isolated from human tissue are initially cultured, some cells die and others (mainly fibroblasts) start to grow; overall, the growth rate increases (phase I). If the remaining cells are harvested, diluted, and replated into dishes again and again, the cell strain continues to divide at a constant rate for about 50 cell generations (phase II), after which the growth rate falls rapidly. In the ensuing period (phase III), all the cells in the culture stop growing (senescence). (b) In a culture prepared from mouse or other rodent cells, initial cell death (not shown) is coupled with the emergence of healthy growing cells. As these dividing cells are diluted and allowed to continue growth, they soon begin to lose growth potential, and most stop growing (i.e., the culture goes into senescence). Very rare cells survive and continue dividing until their progeny overgrow the culture. These cells constitute a cell line, which will grow indefinitely if it is appropriately diluted and fed with nutrients: such cells are said to be immortal.

Most cell lines have lost some or many of the functions characteristic of the differentiated cells from which they were derived. Such relatively undifferentiated cells are poor models for investigating the normal functions of specific cell types. Better in this regard are several more-differentiated cell lines that exhibit many properties of normal nontransformed cells. These lines include the liver tumor (hepatoma) HepG2 line, which synthesizes most of the serum proteins made by normal liver cells (hepatocytes). Another example consists of cells from a certain cultured fibroblast line, which under certain experimental conditions behave as muscle precursor cells, or myoblasts. These cells can be induced to fuse to form myotubes, which resemble differentiated multinucleated muscle cells and synthesize many of the specialized proteins associated with contraction. The results of studies with this cell line have provided valuable information about the differentiation of muscle (Chapter 22). Finally, as discussed previously, the MDCK cell line retains many properties of highly differentiated epithelial cells and forms well-defined epithelial sheets in culture (see Figure 6-6).

Hybrid Cells Called Hybridomas Produce Abundant Monoclonal Antibodies

In addition to serving as research models for studies on cell function, cultured cells can be converted into "factories" for producing specific proteins. In Chapter 9, we describe how it is done by introducing genes encoding insulin, growth factors, and other therapeutically useful proteins into bacterial or eukaryotic cells. Here we consider the use of special cultured cells to generate **monoclonal antibodies,** which are widely used experimental tools and increasingly are being used for diagnostic and therapeutic purposes in medicine.

To understand the challenge of generating monoclonal antibodies, we need to briefly review how antibodies are produced by mammals. Each normal B lymphocyte in a mammal is capable of producing a single type of antibody directed against (can bind to) a specific chemical structure (called a determinant or epitope) on an antigen molecule. If an animal is injected with an antigen, B lymphocytes that make antibodies recognizing the antigen are stimulated to grow and secrete the antibodies. Each antigen-activated B lymphocyte forms a clone of cells in the spleen or lymph nodes, with each cell of the clone producing the identical antibody—that is, a monoclonal antibody. Because most natural antigens contain multiple epitopes, exposure of an animal to an antigen usually stimulates the formation of multiple different B-lymphocyte clones, each producing a different antibody. The resulting mixture of antibodies that recognize different epitopes on the same antigen is said to be *polyclonal.* Such polyclonal antibodies circulate in the blood and can be isolated as a group and used for a variety of experiments. However, monoclonal antibodies are required for many types of experiments or medical applications. Unfortunately, the biochemical purification of any

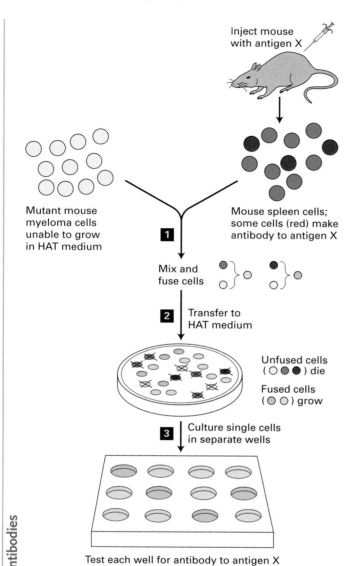

Inject mouse with antigen X

Mutant mouse myeloma cells unable to grow in HAT medium

Mouse spleen cells; some cells (red) make antibody to antigen X

1 Mix and fuse cells

2 Transfer to HAT medium

Unfused cells (○ ○ ● ●) die

Fused cells (○ ○) grow

3 Culture single cells in separate wells

Test each well for antibody to antigen X

▲ **EXPERIMENTAL FIGURE 6-38 Use of cell fusion and selection to obtain hybridomas producing monoclonal antibody to a specific protein.** Step **1**: Immortal myeloma cells that lack HGPRT, an enzyme required for growth on HAT selection medium, are fused with normal antibody-producing spleen cells from an animal that was immunized with antigen X. The spleen cells can make HGPRT. Step **2**: When plated in HAT medium, the unfused cells do not grow; neither do the mutant myeloma cells, because they cannot make purines through an HGPRT-dependent metabolic "salvage" pathway (see Figure 6-41), and the spleen cells, because they have a limited life span in culture. Thus only fused cells formed from a myeloma cell and a spleen cell survive on HAT medium, proliferating into clones called hybridomas. Each hybridoma produces a single antibody. Step **3**: Testing of individual clones identifies those that recognize antigen X. After a hybridoma that produces a desired antibody has been identified, the clone can be cultured to yield large amounts of that antibody.

one type of monoclonal antibody from blood is not feasible, in part because the concentration of any given antibody is quite low.

Because of their limited life span, primary cultures of normal B lymphocytes are of limited usefulness for the production of monoclonal antibody. Thus the first step in producing a monoclonal antibody is to generate immortal, antibody-producing cells. This immortality is achieved by fusing normal B lymphocytes from an immunized animal with transformed, immortal lymphocytes called *myeloma cells.* During cell fusion, the plasma membranes of two cells fuse together, allowing their cytosols and organelles to intermingle. Treatment with certain viral glycoproteins or the chemical polyethylene glycol promotes cell fusion. Some of the fused cells can undergo division and their nuclei eventually coalesce, producing viable *hybrid cells* with a single nucleus that contains chromosomes from both "parents." The fusion of two cells that are genetically different can yield a hybrid cell with novel characteristics. For instance, the fusion of a myeloma cell with a normal antibody-producing cell from a rat or mouse spleen yields a hybrid that proliferates into a clone called a **hybridoma.** Like myeloma cells, hybridoma cells grow rapidly and are immortal. Each hybridoma produces the monoclonal antibody encoded by its B-lymphocyte parent.

The second step in this procedure for producing monoclonal antibody is to separate, or select, the hybridoma cells from the unfused parental cells and the self-fused cells generated by the fusion reaction. This selection is usually performed by incubating the mixture of cells in a special culture medium, called *selection medium,* that permits the growth of only the hybridoma cells because of their novel characteristics. Such a selection is readily performed if the myeloma cells used for the fusion carry a mutation that blocks a metabolic pathway and renders them, but not their lymphocyte fusion partners that do not have the mutation, sensitive to killing by the selection medium. In the immortal hybrid cells, the functional gene from the lymphocyte can supply the gene product missing because of the mutation in the myeloma cell, and thus the hybridoma cells but not the myeloma cells, will be able to grow in the selection medium. Because the lymphocytes used in the fusion are not immortalized and do not divide rapidly, only the hybridoma cells will proliferate rapidly in the selection medium and can thus be readily isolated from the initial mixture of cells.

Figure 6-38 depicts the general procedure for generating and selecting hybridomas. In this case, normal B lymphocytes are fused with myeloma cells that cannot grow in *HAT medium,* the most common selection medium used in the production of hybridomas. Only the myeloma-lymphocyte hybrids can survive and grow for an extended period in HAT medium for reasons described shortly. Thus, this selection medium permits the separation of hybridoma cells from both types of parental cells and any self-fused cells. Finally, each selected hybridoma is then tested for the production of the desired antibody; any clone producing that antibody is then grown in large cultures, from which a substantial quantity of pure monoclonal antibody can be obtained.

 Monoclonal antibodies are commonly employed in affinity chromatography to isolate and purify proteins from complex mixtures (see Figure 3-34c). They can also be used to label and thus locate a particular protein in specific cells of an organ and within cultured cells with the use of immunofluorescence microscopy techniques (see Figures 6-26a and 6-27) or in specific cell fractions with the use of immunoblotting (see Figure 3-35). Monoclonal antibodies also have become important diagnostic and therapeutic tools in medicine. For example, monoclonal antibodies that bind to and inactivate toxic proteins (toxins) secreted by bacterial pathogens are used to treat diseases caused by these pathogens. Other monoclonal antibodies are specific for cell-surface proteins expressed by certain types of tumor cells; chemical complexes of such monoclonal antibodies with toxic drugs or simply the antibodies themselves have been developed for cancer chemotherapy. ∎

HAT Medium Is Commonly Used to Isolate Hybrid Cells

The principles underlying HAT selection are important not only for understanding how hybridoma cells are isolated but also for understanding several other frequently used selection methods, including selection of the ES cells used in generating knockout mice (Chapter 9). HAT medium contains *hy*poxanthine (a purine), *a*minopterin, and *t*hymidine. Most animal cells can synthesize the purine and pyrimidine nucleotides from simpler carbon and nitrogen compounds (Figure 6-39, *top*). The folic acid antagonists amethopterin and aminopterin interfere with the donation of methyl and formyl groups by tetrahydrofolic acid in the early stages of the synthesis of glycine, purine nucleoside monophosphates, and thymidine monophosphate. These drugs are called *antifolates* because they block reactions of tetrahydrofolate, an active form of folic acid.

Many cells, however, are resistant to antifolates because they contain enzymes that can synthesize the necessary nucleotides from purine bases and thymidine (Figure 6-39, *bottom*). Two key enzymes in these *nucleotide salvage pathways* are thymidine kinase (TK) and hypoxanthine-guanine phosphoribosyl transferase (HGPRT). Cells that produce these enzymes can grow on HAT medium, which supplies a

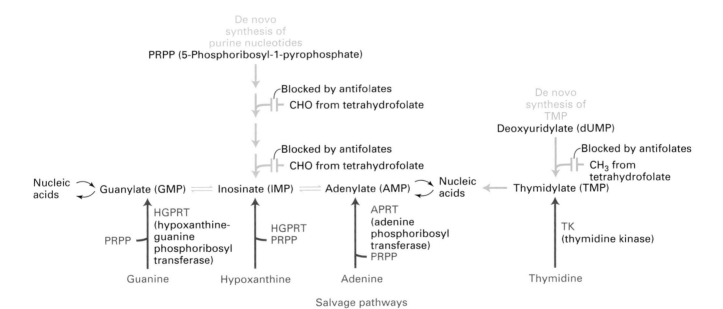

▲ **FIGURE 6-39 De novo and salvage pathways for nucleotide synthesis.** Animal cells can synthesize purine nucleotides (AMP, GMP, IMP) and thymidylate (TMP) from simpler compounds by de novo pathways (blue). They require the transfer of a methyl or formyl ("CHO") group from an activated form of tetrahydrofolate (e.g., N^5,N^{10}-methylenetetrahydrofolate), as shown in the upper part of the diagram. Antifolates, such as aminopterin and amethopterin, block the reactivation of

tetrahydrofolate, preventing purine and thymidylate synthesis. Many animal cells can also use salvage pathways (red) to incorporate purine bases or nucleosides and thymidine. If these precursors are present in the medium, normal cells will grow even in the presence of antifolates. Cultured cells lacking one of the enzymes—HGPRT, APRT, or TK—of the salvage pathways will not survive in media containing antifolates.

salvageable purine and thymidine, whereas those lacking one of them cannot.

Cells with a TK mutation that prevents the production of the functional TK enzyme can be isolated because such cells are resistant to the otherwise toxic thymidine analog 5-bromodeoxyuridine. Cells containing TK convert this compound into 5-bromodeoxyuridine monophosphate, which is then converted into a nucleoside triphosphate by other enzymes. The triphosphate analog is incorporated by DNA polymerase into DNA, where it exerts its toxic effects. This pathway is blocked in TK⁻ mutants, and thus they are resistant to the toxic effects of 5-bromodeoxyuridine. Similarly, cells lacking the HGPRT enzyme, such as the HGPRT⁻ myeloma cell lines used in producing hybridomas, can be isolated because they are resistant to the otherwise toxic guanine analog 6-thioguanine.

Normal cells can grow in HAT medium because even though the aminopterin in the medium blocks de novo synthesis of purines and TMP, the thymidine in the medium is transported into the cell and converted into TMP by TK and the hypoxanthine is transported and converted into usable purines by HGPRT. On the other hand, neither TK⁻ nor HGPRT⁻ cells can grow in HAT medium because each lacks an enzyme of the salvage pathway. However, hybrids formed by the fusion of these two mutants will carry a normal *TK* gene from the HGPRT⁻ parent and a normal *HGPRT* gene from the TK⁻ parent. The hybrids will thus produce both functional salvage-pathway enzymes and will grow on HAT medium.

KEY CONCEPTS OF SECTION 6.7

Growth and Use of Cultured Cells

■ Growth of vertebrate cells in culture requires rich media containing essential amino acids, vitamins, fatty acids, and peptide or protein growth factors; the last are frequently provided by serum.

■ Most cultured vertebrate cells will grow only when attached to a negatively charged substratum that is coated with components of the extracellular matrix.

■ Primary cells, which are derived directly from animal tissue, have limited growth potential in culture and may give rise to a cell strain. Transformed cells, which are derived from animal tumors or arise spontaneously from primary cells, grow indefinitely in culture, forming cell lines (see Figure 6-37).

■ The fusion of an immortal myeloma cell and a single B lymphocyte yields a hybrid cell that can proliferate indefinitely, forming a clone called a hybridoma (see Figure 6-38). Because each individual B lymphocyte produces antibodies specific for one antigenic determinant (epitope), a hybridoma produces only the mono-

clonal antibody synthesized by its original B-lymphocyte parental cell.

■ HAT medium is commonly used to isolated hybridoma cells and other types of hybrid cells.

PERSPECTIVES FOR THE FUTURE

A deeper understanding of the integration of cells into tissues in complex organisms will draw on insights and techniques from virtually all subdisciplines of molecular cell biology: biochemistry, biophysics, microscopy, genetics, genomics, proteomics, and developmental biology. An important set of questions for the future deals with the mechanisms by which cells detect mechanical forces on them and the extracellular matrix, as well as the influence of their three-dimensional arrangements and interactions. A related question is how this information is used to control cell and tissue structure and function. Shear stresses can induce distinct patterns of gene expression and cell growth and can greatly alter cell metabolism and responses to extracellular stimuli. Future research should give us a far more sophisticated understanding of the roles of the three-dimensional organization of cells and ECM components in controlling the structures and activities of tissues.

Numerous questions relate to intracellular signaling from CAMs and adhesion receptors. Such signaling must be integrated with other cellular signaling pathways that are activated by various external signals (e.g., growth factors) so that the cell responds appropriately and in a single coordinated fashion to many different simultaneous internal and external stimuli. How are the logic circuits constructed that allow cross-talk between diverse signaling pathways? How do these circuits integrate the information from these pathways? How is the combination of outside-in and inside-out signaling mediated by CAMs and adhesion receptors merged into such circuits?

The importance of specialized GAG sequences in controlling cellular activities, especially interactions between some growth factors and their receptors, is now clear. With the identification of the biosynthetic mechanisms by which these complex structures are generated and the development of tools to manipulate GAG structures and test their functions in cultured systems and intact animals, we can expect a dramatic increase in our understanding of the cell biology of GAGs in the next several years.

A structural hallmark of CAMs, adhesion receptors, and ECM proteins is the presence of multiple domains that impart diverse functions to a single polypeptide chain. It is generally agreed that such multidomain proteins arose evolutionarily by the assembly of distinct DNA sequences encoding the distinct domains. Genes encoding multiple domains provide opportunities to generate enormous sequence and functional diversity by alternative splicing and the use of alternate promoters within a gene. Thus, even though the number of independent genes in the human genome seems surprisingly small in comparison with other organisms, far more distinct protein molecules can be produced than predicted from the number of genes. Such diversity seems especially well suited to the generation of proteins that take part in specifying adhesive connections in the nervous system, especially the brain. Indeed, several groups of proteins expressed by neurons appear to have just such combinatorial diversity of structure. They include the protocadherins, a family of cadherins with as many as 70 proteins encoded per gene; the neurexins, which comprise more than 1000 proteins encoded by three genes; and the Dscams, members of the IgCAM superfamily encoded by a *Drosophila* gene that has the potential to express 38,016 distinct proteins owing to alternative splicing. A continuing goal for future work will be to describe and understand the molecular basis of functional cell–cell and cell–matrix attachments—the "wiring"—in the nervous system and how that wiring ultimately permits complex neuronal control and, indeed, the intellect required to understand molecular cell biology.

KEY TERMS

adhesion receptor *199*
anchoring junction *202*
basal lamina *202*
cadherin *199*
cell-adhesion
    molecule (CAM) *198*
cell line *236*
cell strain *236*
cell wall *231*
connexin *231*
dystrophin glycoprotein
    complex (DGC) *227*
epithelium *201*
extracellular matrix
    (ECM) *199*
fibril-associated
    collagen *217*
fibrillar collagen *217*
fibronectin *220*
glycosaminoglycan
    (GAG) *213*

HAT medium *238*
hyaluronan *217*
hybridoma *238*
immunoglobulin
    cell-adhesion molecule
    (IgCAM) *227*
integrin *199*
laminin *211*
monoclonal antibody *237*
multiadhesive matrix
    protein *209*
paracellular pathway *208*
plasmodesma *233*
proteoglycan *209*
RGD sequence *221*
selectin *199*
syndecan *214*
tight junction *202*

REVIEW THE CONCEPTS

1. Using specific examples, describe the two phenomena that give rise to the diversity of adhesive molecules.

2. Cadherins are known to mediate homophilic interactions between cells. What is a homophilic interaction, and how can it be demonstrated experimentally for E-cadherins?

3. What is the normal function of tight junctions? What can happen to tissues when tight junctions do not function properly?

4. What is collagen, and how is it synthesized? How do we know that collagen is required for tissue integrity?

5. You have synthesized an oligopeptide containing an RGD sequence surrounded by other amino acids. What is the effect of this peptide when added to a fibroblast cell culture grown on a layer of fibronectin absorbed to the tissue culture dish? Why does this happen?

6. Blood clotting is a crucial function for mammalian survival. How do the multiadhesive properties of fibronectin lead to the recruitment of platelets to blood clots?

7. Using structural models, explain how integrins mediate outside-in and inside-out signaling.

8. How do changes in molecular connections between the extracellular matrix (ECM) and cytoskeleton give rise to Duchenne muscular dystrophy?

9. What is the difference between a cell strain, a cell line, and a clone?

10. Explain why the process of cell fusion is necessary to produce monoclonal antibodies used for research.

ANALYZE THE DATA

Researchers have isolated two E-cadherin mutant isoforms that are hypothesized to function differently from that of the wild-type E-cadherin. An E-cadherin negative mammary carcinoma cell line was transfected with the mutant E-cadherin genes A (part a in the figure) and B (part b) (diamonds) and the wild-type E-cadherin gene (black circles) and compared to untransfected cells (open circles) in an aggregation assay. In this assay, cells are first dissociated by trypsin treatment and then allowed to aggregate in solution over a period of minutes. Aggregating cells from mutants A and B are presented in panels a and b respectively. To demonstrate that the observed adhesion was cadherin-mediated, the cells were pretreated with a nonspecific antibody (left panel) or a function-blocking anti-E-cadherin monoclonal antibody (right panel).

a. Why do cells transfected with the wild-type E-cadherin gene have greater aggregation than control, nontransfected cells?

b. From these data, what can be said about the function of mutants A and B?

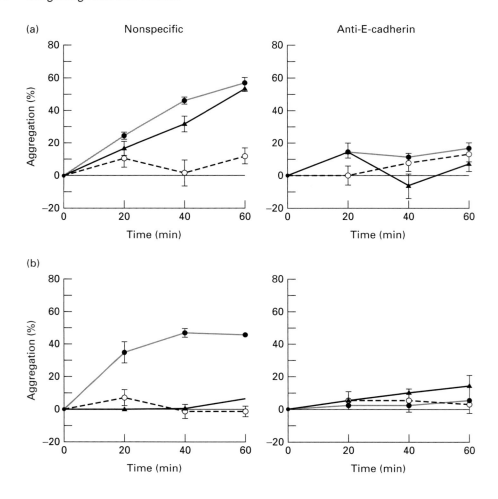

c. Why does the addition of the anti-E-cadherin monoclonal antibody, but not the nonspecific antibody, block aggregation?

d. What would happen to the aggregation ability of the cells transfected with the wild-type E-cadherin gene if the assay were performed in media low in Ca^{2+}?

REFERENCES

Cell–Cell and Cell–Matrix Adhesion: An Overview

Gumbiner, B. M. 1996. Cell adhesion: the molecular basis of tissue architecture and morphogenesis. *Cell* 84:345–357.

Hynes, R. O. 1999. Cell adhesion: old and new questions. *Trends Cell Biol.* 9(12):M33–M337. Millennium issue.

Hynes, R. O. 2002. Integrins: bidirectional, allosteric signaling machines. *Cell* 110:673–687.

Jamora, C., and E. Fuchs. 2002. Intercellular adhesion, signalling and the cytoskeleton. *Nature Cell Biol.* 4(4):E101–E108.

Juliano, R. L. 2002. Signal transduction by cell adhesion receptors and the cytoskeleton: functions of integrins, cadherins, selectins, and immunoglobulin-superfamily members. *Ann. Rev. Pharmacol. Toxicol.* 42:283–323.

Leahy, D. J. 1997. Implications of atomic resolution structures for cell adhesion. *Ann. Rev. Cell Devel. Biol.* 13:363–393.

Sheetlike Epithelial Tissues: Junctions and Cell-Adhesion Molecules

Boggon, T. J., et al. 2002. C-cadherin ectodomain structure and implications for cell adhesion mechanisms. *Science* 296:1308–1313.

Clandinin, T. R., and S. L. Zipursky. 2002. Making connections in the fly visual system. *Neuron* 35:827–841.

Conacci-Sorrell, M., J. Zhurinsky, and A. Ben-Ze'ev. 2002. The cadherin-catenin adhesion system in signaling and cancer. *J. Clin. Invest.* 109:987–991.

Fuchs, E., and S. Raghavan. 2002. Getting under the skin of epidermal morphogenesis. *Nature Rev. Genet.* 3(3):199–209.

Leckband, D. 2002. The structure of the C-cadherin ectodomain resolved. *Structure (Camb).* 10:739–740.

Pierschbacher, M. D., and E. Ruoslahti. 1984. Cell attachment activity of fibronectin can be duplicated by small synthetic fragments of the molecule. *Nature* 309(5963):30–33.

Schöck, F., and N. Perrimon. 2002. Molecular mechanisms of epithelial morphogenesis. *Ann. Rev. Cell Devel. Biol.* 18:463–493.

Shimaoka, M., J. Takagi, and T. A. Springer. 2002. Conformational regulation of integrin structure and function. *Ann. Rev. Biophys. Biomol. Struc.* 31:485–516.

Tsukita, S., M. Furuse, and M. Itoh. 2001. Multifunctional strands in tight junctions. *Nature Rev. Mol. Cell Biol.* 2:285–293.

Xiong, J. P., et al. 2001. Crystal structure of the extracellular segment of integrin αVβ3. *Science* **294**:339–345.

The Extracellular Matrix of Epithelial Sheets

Boutaud, A., et al. 2000. Type IV collagen of the glomerular basement membrane: evidence that the chain specificity of network assembly is encoded by the noncollagenous NC1 domains. *J. Biol. Chem.* **275**:30716–30724.

Esko, J. D., and U. Lindahl. 2001. Molecular diversity of heparan sulfate. *J. Clin. Invest.* **108**:169–173.

Hohenester, E., and J. Engel. 2002. Domain structure and organisation in extracellular matrix proteins. *Matrix Biol.* **21**(2): 115–128.

Nakato, H., and K. Kimata. 2002. Heparan sulfate fine structure and specificity of proteoglycan functions. *Biochim. Biophys. Acta* **1573**:312–318.

Perrimon, N., and M. Bernfield. 2001. Cellular functions of proteoglycans: an overview. *Semin. Cell Devel. Biol.* **12**(2):65–67.

Rosenberg, R. D., et al. 1997. Heparan sulfate proteoglycans of the cardiovascular system: specific structures emerge but how is synthesis regulated? *J. Clin. Invest.* **99**:2062–2070.

The Extracellular Matrix of Nonepithelial Tissues

Kramer, R. Z., J. Bella, B. Brodsky, and H. M. Berman. 2001. The crystal and molecular structure of a collagen-like peptide with a biologically relevant sequence. *J. Mol. Biol.* **311**:131–147.

Mao, J. R., and J. Bristow. 2001. The Ehlers-Danlos syndrome: on beyond collagens. *J. Clin. Invest.* **107**:1063–1069.

Shaw, L. M., and B. R. Olsen. 1991. FACIT collagens: diverse molecular bridges in extracellular matrices. *Trends Biochem. Sci.* **16**(5):191–194.

Weiner, S., W. Traub, and H. D. Wagner. 1999. Lamellar bone: structure–function relations. *J. Struc. Biol.* **126**:241–255.

Adhesive Interactions Involving Nonepithelial Cells

Bartsch, U. 2003. Neural CAMs and their role in the development and organization of myelin sheaths. *Front. Biosci.* **8**:D477–D490.

Brummendorf, T., and V. Lemmon. 2001. Immunoglobulin superfamily receptors: cis-interactions, intracellular adapters and alternative splicing regulate adhesion. *Curr. Opin. Cell Biol.* **13**:611–618.

Cukierman, E., R. Pankov, and K. M. Yamada. 2002. Cell interactions with three-dimensional matrices. *Curr. Opin. Cell Biol.* **14**:633–639.

Durbeej, M., and K. P. Campbell. 2002. Muscular dystrophies involving the dystrophin-glycoprotein complex: an overview of current mouse models. *Curr. Opin. Genet. Devel.* **12**:349–361.

Geiger, B., A. Bershadsky, R. Pankov, and K. M. Yamada. 2001. Transmembrane crosstalk between the extracellular matrix and the cytoskeleton. *Nature Rev. Mol. Cell Biol.* **2**:793–805.

Hobbie, L., et al. 1987. Restoration of LDL receptor activity in mutant cells by intercellular junctional communication. *Science* **235**:69–73.

Lawrence, M. B., and T. A. Springer. 1991. Leukocytes roll on a selectin at physiologic flow rates: distinction from and prerequisite for adhesion through integrins. *Cell* **65**:859–873.

Lo, C. Gap junctions in development and disease. *Ann. Rev. Cell. Devel. Biol.* 10.1146/annurev.cellbio.19.111301.144309. (Expected to be published in 2003.)

Reizes, O., et al. 2001. Transgenic expression of syndecan-1 uncovers a physiological control of feeding behavior by syndecan-3. *Cell* **106**:105–116.

Somers, W. S., J. Tang, G. D. Shaw, and R. T. Camphausen. 2000. Insights into the molecular basis of leukocyte tethering and rolling revealed by structures of P- and E-selectin bound to SLe(X) and PSGL-1. *Cell* **103**:467–479.

Stein, E., and M. Tessier-Lavigne. 2001. Hierarchical organization of guidance receptors: silencing of netrin attraction by Slit through a Robo/DCC receptor complex. *Science* **291**:1928–1938.

Plant Tissues

Delmer, D. P., and C. H. Haigler. 2002. The regulation of metabolic flux to cellulose, a major sink for carbon in plants. *Metab. Eng.* **4**:22–28.

Iwai, H., N. Masaoka, T. Ishii, and S. Satoh. 2000. A pectin glucuronyltransferase gene is essential for intercellular attachment in the plant meristem. *Proc. Nat'l. Acad. Sci. USA* **99**:16319–16324.

Lord, E. M. 2003. Adhesion and guidance in compatible pollination. *J. Exp. Bot.* **54**(380):47–54.

Lord, E. M., and J. C. Mollet. 2002. Plant cell adhesion: a bioassay facilitates discovery of the first pectin biosynthetic gene. *Proc. Nat'l. Acad. Sci. USA* **99**:15843–15845.

Lord, E. M., and S. D. Russell. 2002. The mechanisms of pollination and fertilization in plants. *Ann. Rev. Cell Devel. Biol.* **18**:81–105.

Pennell, R. 1998. Cell walls: structures and signals. *Curr. Opin. Plant Biol.* **1**:504–510.

Ross, W. Whetten, J. J. MacKay, and R. R. Sederoff. 1998. Recent advances in understanding lignin biosynthesis. *Ann. Rev. Plant Physiol. Plant Mol. Biol.* **49**:585–609.

Zambryski, P., and K. Crawford. 2000. Plasmodesmata: gatekeepers for cell-to-cell transport of developmental signals in plants. *Ann. Rev. Cell Devel. Biol.* **16**:393–421.

Growth and Use of Cultured Cells

Davis, J. M., ed. 1994. *Basic Cell Culture: A Practical Approach.* IRL Press.

Goding, J. W. 1996. *Monoclonal Antibodies: Principles and Practice. Production and Application of Monoclonal Antibodies in Cell Biology, Biochemistry, and Immunology,* 3d ed. Academic Press.

Kohler, G., and C. Milstein. 1975. Continuous cultures of fused cells secreting antibody of predefined specificity. *Nature* **256**:495–497.

Shaw, A. J., ed. 1996. *Epithelial Cell Culture.* IRL Press.

Tyson, C. A., and J. A. Frazier, eds. 1993. *Methods in Toxicology.* Vol. I (Part A): *In Vitro Biological Systems.* Academic Press. Describes methods for growing many types of primary cells in culture.

7

TRANSPORT OF IONS AND SMALL MOLECULES ACROSS CELL MEMBRANES

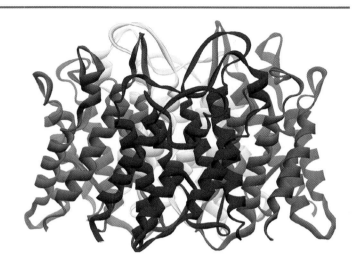

Aquaporin, the water channel, consists of four identical transmembrane polypeptides.

The plasma membrane is a selectively permeable barrier between the cell and the extracellular environment. Its permeability properties ensure that essential molecules such as ions, glucose, amino acids, and lipids readily enter the cell, metabolic intermediates remain in the cell, and waste compounds leave the cell. In short, the selective permeability of the plasma membrane allows the cell to maintain a constant internal environment. In Chapter 5, we learned about the components and structural organization of cell membranes. Movement of virtually all molecules and ions across cellular membranes is mediated by selective **membrane transport proteins** embedded in the phospholipid bilayer. Because different cell types require different mixtures of low-molecular-weight compounds, the plasma membrane of each cell type contains a specific set of transport proteins that allow only certain ions and molecules to cross. Similarly, organelles within the cell often have a different internal environment from that of the surrounding cytosol, and organelle membranes contain specific transport proteins that maintain this difference.

We begin our discussion by reviewing some general principles of transport across membranes and distinguishing three major classes of transport proteins. In subsequent sections, we describe the structure and operation of specific examples of each class and show how members of families of homologous transport proteins have different properties that enable different cell types to function appropriately. We also explain how specific combinations of transport proteins in different subcellular membranes enable cells to carry out essential physiological processes, including the maintenance of cytosolic pH, the accumulation of sucrose and salts in plant cell vacuoles, and the directed flow of water in both plants and animals. Epithelial cells, such as those lining the small intestine, transport ions, sugars and other small molecules, and water from one side to the other. We shall see how, in order to do this, their plasma membranes are organized into at least two discrete regions, each with its own set of transport proteins. The last two sections of the chapter focus on the panoply of transport proteins that allow nerve cells to generate and conduct the type of electric signal called an **action potential** along their entire length and to transmit these signals to other cells, inducing a change in the electrical properties of the receiving cells.

7.1 Overview of Membrane Transport

The phospholipid bilayer, the basic structural unit of biomembranes, is essentially impermeable to most water-soluble molecules, ions, and water itself. After describing the factors that influence the permeability of lipid membranes, we briefly compare the three major classes of membrane proteins that increase the permeability of biomembranes. We then examine operation of the simplest type of transport protein to illustrate basic features of protein-mediated transport. Finally, two common experimental systems used in studying the functional properties of transport proteins are described.

Few Molecules Cross Membranes by Passive Diffusion

Gases, such as O_2 and CO_2, and small, uncharged polar molecules, such as urea and ethanol, can readily move by **passive (simple) diffusion** across an artificial membrane composed of pure phospholipid or of phospholipid and cholesterol (Figure 7-1). Such molecules also can diffuse across cellular membranes without the aid of transport proteins. No metabolic energy is expended because movement is from a high to a low concentration of the molecule, down its chemical con-

centration gradient. As noted in Chapter 2, such transport reactions are spontaneous because they have a positive ΔS value (increase in entropy) and thus a negative ΔG (decrease in free energy).

The relative diffusion rate of any substance across a pure phospholipid bilayer is proportional to its concentration gradient across the layer and to its hydrophobicity and size; charged molecules are also affected by any electric potential across the membrane (see below). When a phospholipid bilayer separates two aqueous compartments, membrane permeability can be easily determined by adding a small amount of radioactive material to one compartment and measuring its rate of appearance in the other compartment. The greater the concentration gradient of the substance, the faster its rate of diffusion across a bilayer.

The hydrophobicity of a substance is measured by its partition coefficient K, the equilibrium constant for its partition between oil and water. The higher a substance's partition coefficient, the more lipid-soluble it is. The first and rate-limiting step in transport by passive diffusion is movement of a molecule from the aqueous solution into the hydrophobic interior of the phospholipid bilayer, which resembles oil in its chemical properties. This is the reason that the more hydrophobic a molecule is, the faster it diffuses across a pure phospholipid bilayer. For example, diethylurea, with an ethyl group (CH_3CH_2—) attached to each nitrogen atom of urea, has a K of 0.01, whereas urea has a K of 0.0002 (see Figure 7-1). Diethylurea, which is 50 times (0.01/0.0002) more hydrophobic than urea, will diffuse through phospholipid bilayer membranes about 50 times faster than urea. Diethylurea also enters cells about 50 times faster than urea. Similarly, fatty acids with longer hydrocarbon chains are more hydrophobic than those with shorter chains and will diffuse more rapidly across a pure phospholipid bilayer at all concentrations.

If a transported substance carries a net charge, its movement is influenced by both its concentration gradient and the **membrane potential,** the electric potential (voltage) across the membrane. The combination of these two forces, called the **electrochemical gradient,** determines the energetically favorable direction of transport of a charged molecule across a membrane. The electric potential that exists across most cellular membranes results from a small imbalance in the concentration of positively and negatively charged ions on the two sides of the membrane. We discuss how this ionic imbalance, and resulting potential, arise and are maintained in Sections 7.2 and 7.3.

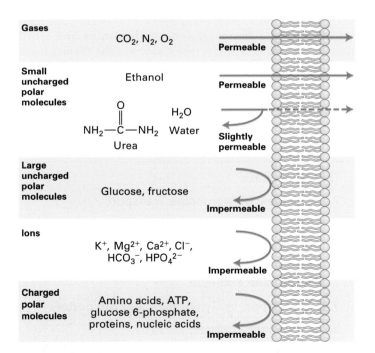

▲ **FIGURE 7-1 Relative permeability of a pure phospholipid bilayer to various molecules.** A bilayer is permeable to small hydrophobic molecules and small uncharged polar molecules, slightly permeable to water and urea, and essentially impermeable to ions and to large polar molecules.

Membrane Proteins Mediate Transport of Most Molecules and All Ions Across Biomembranes

As is evident from Figure 7-1, very few molecules and no ions can cross a pure phospholipid bilayer at appreciable rates by passive diffusion. Thus transport of most molecules

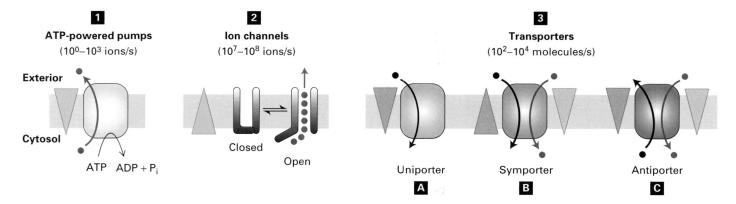

▲ **FIGURE 7-2 Overview of membrane transport proteins.** Gradients are indicated by triangles with the tip pointing toward lower concentration, electrical potential, or both. **1** Pumps utilize the energy released by ATP hydrolysis to power movement of specific ions (red circles) or small molecules against their electrochemical gradient. **2** Channels permit movement of specific ions (or water) down their electrochemical gradient. Transporters, which fall into three groups, facilitate movement of specific small molecules or ions. Uniporters transport a single type of molecule down its concentration gradient **3A**. Cotransport proteins (symporters, **3B**, and antiporters, **3C**) catalyze the movement of one molecule *against* its concentration gradient (black circles), driven by movement of one or more ions down an electrochemical gradient (red circles). Differences in the mechanisms of transport by these three major classes of proteins account for their varying rates of solute movement.

into and out of cells requires the assistance of specialized membrane proteins. Even transport of molecules with a relatively large partition coefficient (e.g., water and urea) is frequently accelerated by specific proteins because their transport by passive diffusion usually is not sufficiently rapid to meet cellular needs.

All transport proteins are transmembrane proteins containing multiple membrane-spanning segments that generally are α helices. By forming a protein-lined pathway across the membrane, transport proteins are thought to allow movement of hydrophilic substances without their coming into contact with the hydrophobic interior of the membrane. Here we introduce the various types of transport proteins covered in this chapter (Figure 7-2).

ATP-powered pumps (or simply **pumps**) are **ATPases** that use the energy of ATP hydrolysis to move ions or small molecules across a membrane *against* a chemical concentration gradient or electric potential or both. This process, referred to as **active transport**, is an example of a coupled chemical reaction (Chapter 2). In this case, transport of ions or small molecules "uphill" against an electrochemical gradient, which requires energy, is coupled to the hydrolysis of ATP, which releases energy. The overall reaction—ATP hydrolysis and the "uphill" movement of ions or small molecules—is energetically favorable.

Channel proteins transport water or specific types of ions and hydrophilic small molecules *down* their concentration or electric potential gradients. Such protein-assisted transport sometimes is referred to as **facilitated diffusion**. Channel proteins form a hydrophilic passageway across the membrane through which multiple water molecules or ions move simultaneously, single file at a very rapid rate. Some ion chan-

nels are open much of the time; these are referred to as *non-gated* channels. Most ion channels, however, open only in response to specific chemical or electrical signals; these are referred to as *gated* channels.

Transporters (also called *carriers*) move a wide variety of ions and molecules across cell membranes. Three types of transporters have been identified. *Uniporters* transport a single type of molecule *down* its concentration gradient via facilitated diffusion. Glucose and amino acids cross the plasma membrane into most mammalian cells with the aid of uniporters. In contrast, *antiporters* and *symporters* couple the movement of one type of ion or molecule *against* its concentration gradient with the movement of one or more different ions *down* its concentration gradient. These proteins often are called *cotransporters,* referring to their ability to transport two different solutes simultaneously.

Like ATP pumps, cotransporters mediate coupled reactions in which an energetically unfavorable reaction (i.e., uphill movement of molecules) is coupled to an energetically favorable reaction. Note, however, that the nature of the energy-supplying reaction driving active transport by these two classes of proteins differs. ATP pumps use energy from hydrolysis of ATP, whereas cotransporters use the energy stored in an electrochemical gradient. This latter process sometimes is referred to as *secondary* active transport.

Table 7-1 summarizes the four mechanisms by which small molecules and ions are transported across cellular membranes. In this chapter, we focus on the properties and operation of the membrane proteins that mediate the three protein-dependent transport mechanisms. Conformational changes are essential to the function of all transport proteins.

TABLE 7-1   Mechanisms for Transporting Ions and Small Molecules Across Cell Membranes

Property	Transport Mechanism			
	Passive Diffusion	Facilitated Diffusion	Active Transport	Cotransport*
Requires specific protein	−	+	+	+
Solute transported against its gradient	−	−	+	+
Coupled to ATP hydrolysis	−	−	+	−
Driven by movement of a cotransported ion down its gradient	−	−	−	+
Examples of molecules transported	O_2, CO_2, steroid hormones, many drugs	Glucose and amino acids (uniporters); ions and water (channels)	Ions, small hydrophilic molecules, lipids (ATP-powered pumps)	Glucose and amino acids (symporters); various ions and sucrose (antiporters)

*Also called *secondary active transport*.

ATP-powered pumps and transporters undergo a cycle of conformational change exposing a binding site (or sites) to one side of the membrane in one conformation and to the other side in a second conformation. Because each such cycle results in movement of only one (or a few) substrate molecules, these proteins are characterized by relatively slow rates of transport ranging from 10^0 to 10^4 ions or molecules per second (see Figure 7-2). Ion channels shuttle between a closed state and an open state, but many ions can pass through an open channel without any further conformational change. For this reason, channels are characterized by very fast rates of transport, up to 10^8 ions per second.

Several Features Distinguish Uniport Transport from Passive Diffusion

The protein-mediated movement of glucose and other small hydrophilic molecules across a membrane, known as **uniport** transport, exhibits the following distinguishing properties:

1. The rate of facilitated diffusion by uniporters is far higher than passive diffusion through a pure phospholipid bilayer.

2. Because the transported molecules never enter the hydrophobic core of the phospholipid bilayer, the partition coefficient K is irrelevant.

3. Transport occurs via a limited number of uniporter molecules, *rather than throughout the phospholipid bilayer.* Consequently, there is a maximum transport rate V_{max} that is achieved when the concentration gradient across the membrane is very large and each uniporter is working at its maximal rate.

4. Transport is specific. Each uniporter transports only a single species of molecule or a single group of closely related molecules. A measure of the affinity of a transporter for its substrate is K_m, which is the concentration of substrate at which transport is half-maximal.

These properties also apply to transport mediated by the other classes of proteins depicted in Figure 7-2.

One of the best-understood uniporters is the glucose transporter *GLUT1* found in the plasma membrane of erythrocytes. The properties of GLUT1 and many other transport proteins from mature erythrocytes have been extensively studied. These cells, which have no nucleus or other internal organelles, are essentially "bags" of hemoglobin containing relatively few other intracellular proteins and a single membrane, the plasma membrane (see Figure 5-3a). Because the erythrocyte plasma membrane can be isolated in high purity, isolating and purifying a transport protein from mature erythrocytes is a straightforward procedure.

Figure 7-3 shows that glucose uptake by erythrocytes and liver cells exhibits kinetics characteristic of a simple enzyme-catalyzed reaction involving a single substrate. The kinetics of transport reactions mediated by other types of proteins are more complicated than for uniporters. Nonetheless, all protein-assisted transport reactions occur faster than allowed by passive diffusion, are substrate-specific as reflected in lower K_m values for some substrates than others, and exhibit a maximal rate (V_{max}).

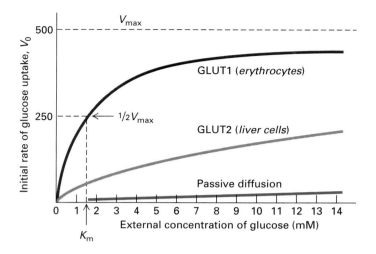

◀ **EXPERIMENTAL FIGURE 7-3 Cellular uptake of glucose mediated by GLUT proteins exhibits simple enzyme kinetics and greatly exceeds the calculated rate of glucose entry solely by passive diffusion.** The initial rate of glucose uptake (measured as micromoles per milliliter of cells per hour) in the first few seconds is plotted against increasing glucose concentration in the extracellular medium. In this experiment, the initial concentration of glucose in the cells is always zero. Both GLUT1, expressed by erythrocytes, and GLUT2, expressed by liver cells, greatly increase the rate of glucose uptake (red and orange curves) compared with that associated with passive diffusion (blue curve) at all external concentrations. Like enzyme-catalyzed reactions, GLUT-facilitated uptake of glucose exhibits a maximum rate (V_{max}). The K_m is the concentration at which the rate of glucose uptake is half maximal. GLUT2, with a K_m of about 20 mM, has a much lower affinity for glucose than GLUT1, with a K_m of about 1.5 mM.

GLUT1 Uniporter Transports Glucose into Most Mammalian Cells

Most mammalian cells use blood glucose as the major source of cellular energy and express GLUT1. Since the glucose concentration usually is higher in the extracellular medium (blood in the case of erythrocytes) than in the cell, GLUT1 generally catalyzes the net import of glucose from the extracellular medium into the cell. Under this condition, V_{max} is achieved at high external glucose concentrations.

Like other uniporters, GLUT1 alternates between two conformational states: in one, a glucose-binding site faces the outside of the membrane; in the other, a glucose-binding site faces the inside. Figure 7-4 depicts the sequence of events occurring during the unidirectional transport of glucose from the cell exterior inward to the cytosol. GLUT1 also can catalyze the net export of glucose from the cytosol to the extra-

cellular medium exterior when the glucose concentration is higher inside the cell than outside.

The kinetics of the unidirectional transport of glucose from the outside of a cell inward via GLUT1 can be described by the same type of equation used to describe a simple enzyme-catalyzed chemical reaction. For simplicity, let's assume that the substrate glucose, S, is present initially only on the outside of the membrane. In this case, we can write:

$$S_{out} + GLUT1 \underset{}{\overset{K_m}{\rightleftharpoons}} S_{out} - GLUT1 \underset{}{\overset{V_{max}}{\rightleftharpoons}} S_{in} + GLUT1$$

where $S_{out} - GLUT1$ represents GLUT1 in the outward-facing conformation with a bound glucose. By a similar derivation used to arrive at the Michaelis-Menten equation

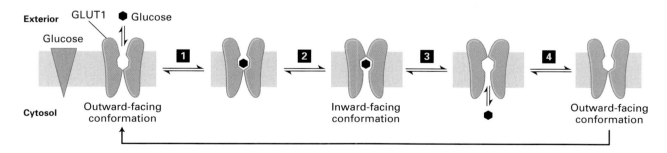

▲ **FIGURE 7-4 Model of uniport transport by GLUT1.** In one conformation, the glucose-binding site faces outward; in the other, the binding site faces inward. Binding of glucose to the outward-facing site (step **1**) triggers a conformational change in the transporter that results in the binding site's facing inward toward the cytosol (step **2**). Glucose then is released to the inside of the cell (step **3**). Finally, the transporter undergoes the

reverse conformational change, regenerating the outward-facing binding site (step **4**). If the concentration of glucose is higher inside the cell than outside, the cycle will work in reverse (step **4** → step **1**), resulting in net movement of glucose from inside to out. The actual conformational changes are probably smaller than those depicted here.

in Chapter 3, we can derive the following expression for ν, the initial transport rate for S into the cell catalyzed by GLUT1:

$$\nu = \frac{V_{max}}{1 + \dfrac{K_m}{C}} \qquad (7\text{-}1)$$

where C is the concentration of S_{out} (initially, the concentration of $S_{in} = 0$). V_{max}, the rate of transport when all molecules of GLUT1 contain a bound S, occurs at an infinitely high S_{out} concentration. The lower the value of K_m, the more tightly the substrate binds to the transporter, and the greater the transport rate at a fixed concentration of substrate. Equation 7-1 describes the curve for glucose uptake by erythrocytes shown in Figure 7-3 as well as similar curves for other uniporters.

For GLUT1 in the erythrocyte membrane, the K_m for glucose transport is 1.5 millimolar (mM); at this concentration roughly half the transporters with outward-facing binding sites would have a bound glucose and transport would occur at 50 percent of the maximal rate. Since blood glucose is normally 5 mM, the erythrocyte glucose transporter usually is functioning at 77 percent of the maximal rate, as can be seen from Figure 7-3. GLUT1 and the very similar GLUT3 are expressed by erythrocytes and other cells that need to take up glucose from the blood continuously at high rates; the rate of glucose uptake by such cells will remain high regardless of small changes in the concentration of blood glucose.

In addition to glucose, the isomeric sugars D-mannose and D-galactose, which differ from D-glucose in the configuration at only one carbon atom, are transported by GLUT1 at measurable rates. However, the K_m for glucose (1.5 mM) is much lower than the K_m for D-mannose (20 mM) or D-galactose (30 mM). Thus GLUT1 is quite specific, having a much higher affinity (indicated by a lower K_m) for the normal substrate D-glucose than for other substrates.

GLUT1 accounts for 2 percent of the protein in the plasma membrane of erythrocytes. After glucose is transported into the erythrocyte, it is rapidly phosphorylated, forming glucose 6-phosphate, which cannot leave the cell. Because this reaction, the first step in the metabolism of glucose (see Figure 8-4), is rapid, the intracellular concentration of glucose does not increase as glucose is taken up by the cell. Consequently, the glucose concentration gradient across the membrane is maintained, as is the rate of glucose entry into the cell.

The Human Genome Encodes a Family of Sugar-Transporting GLUT Proteins

The human genome encodes 12 proteins, GLUT1–GLUT12, that are highly homologous in sequence, and all are thought to contain 12 membrane-spanning α helices. Detailed studies on GLUT1 have shown that the amino acid residues in the transmembrane α helices are predominantly hydrophobic; several helices, however, bear amino acid residues (e.g., serine, threonine, asparagine, and glutamine) whose side chains can form hydrogen bonds with the hydroxyl groups on glucose. These residues are thought to form the inward-facing and outward-facing glucose-binding sites in the interior of the protein (see Figure 7-4).

The structures of all GLUT isoforms are quite similar, and all transport sugars. Nonetheless, their differential expression in various cell types and isoform-specific functional properties enable different body cells to regulate glucose metabolism independently and at the same time maintain a constant concentration of glucose in the blood. For instance, GLUT2, expressed in liver and the insulin-secreting β cells of the pancreas, has a K_m of ≈20 mM, about 13 times higher than the K_m of GLUT1. As a result, when blood glucose rises from its basal level of 5 mM to 10 mM or so after a meal, the rate of glucose influx will almost double in GLUT2-expressing cells, whereas it will increase only slightly in GLUT1-expressing cells (see Figure 7-3). In liver, the "excess" glucose brought into the cell is stored as the polymer **glycogen**. In islet β cells, the rise in glucose triggers secretion of the hormone **insulin,** which in turn lowers blood glucose by increasing glucose uptake and metabolism in muscle and by inhibiting glucose production in liver.

Another GLUT isoform, GLUT4, is expressed only in fat and muscle cells, the cells that respond to insulin by increasing their uptake of glucose, thereby removing glucose from the blood. In the absence of insulin, GLUT4 is found in intracellular membranes, not on the plasma membrane, and obviously is unable to facilitate glucose uptake. By a process detailed in Chapter 15, insulin causes these GLUT4-rich internal membranes to fuse with the plasma membrane, increasing the number of GLUT4 molecules on the cell surface and thus the rate of glucose uptake. Defects in this process, one principal mechanism by which insulin lowers blood glucose, lead to diabetes, a disease marked by continuously high blood glucose.

In contrast to GLUT1–GLUT4, which all transport glucose at physiological concentrations, GLUT5 transports fructose. The properties of other members of the GLUT family have not yet been studied in detail.

Transport Proteins Can Be Enriched Within Artificial Membranes and Cells

Although transport proteins can be isolated from membranes and purified, the functional properties of these proteins can be studied only when they are associated with a membrane. Most cellular membranes contain many different types of transport proteins but a relatively low concentration of any particular one, making functional studies of a single protein difficult. To facilitate such studies, researchers use two ap-

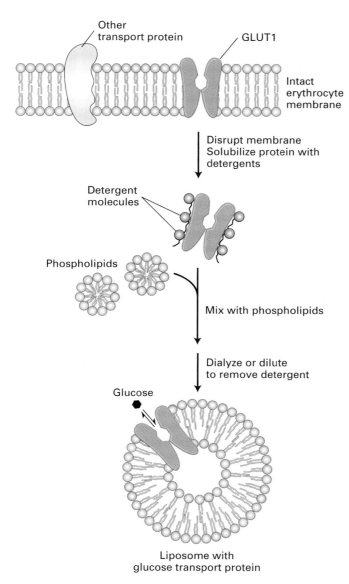

▲ **EXPERIMENTAL FIGURE 7-5 Liposomes containing a single type of transport protein are very useful in studying functional properties of transport proteins.** Here, all the integral proteins of the erythrocyte membrane are solubilized by a nonionic detergent, such as octylglucoside. The glucose uniporter GLUT1 can be purified by chromatography on a column containing a specific monoclonal antibody and then incorporated into liposomes made of pure phospholipids.

proaches for enriching a transport protein of interest so that it predominates in the membrane.

In one common approach, a specific transport protein is extracted and purified; the purified protein then is reincorporated into pure phospholipid bilayer membranes, such as **liposomes** (Figure 7-5). Alternatively, the gene encoding a specific transport protein can be expressed at high levels in

a cell type that normally does not express it. The difference in transport of a substance by the transfected and by control nontransfected cells will be due to the expressed transport protein. In these systems, the functional properties of the various membrane proteins can be examined without ambiguity.

KEY CONCEPTS OF SECTION 7.1

Overview of Membrane Transport

■ The plasma membrane regulates the traffic of molecules into and out of the cell.

■ With the exception of gases (e.g., O_2 and CO_2) and small hydrophobic molecules, most molecules cannot diffuse across the phospholipid bilayer at rates sufficient to meet cellular needs.

■ Three classes of transmembrane proteins mediate transport of ions, sugars, amino acids, and other metabolites across cell membranes: ATP-powered pumps, channels, and transporters (see Figure 7-2).

■ In active transport, a transport protein couples movement of a substrate against its concentration gradient to ATP hydrolysis.

■ In facilitated diffusion, a transport protein assists in the movement of a specific substrate (molecule or ion) down its concentration gradient.

■ In secondary active transport, or cotransport, a transport protein couples movement of a substrate against its concentration gradient to the movement of a second substrate down its concentration gradient (see Table 7-1).

■ Protein-catalyzed transport of a solute across a membrane occurs much faster than passive diffusion, exhibits a V_{max} when the limited number of transporter molecules are saturated with substrate, and is highly specific for substrate (see Figure 7-3).

■ Uniport proteins, such as the glucose transporters (GLUTs), are thought to shuttle between two conformational states, one in which the substrate-binding site faces outward and one in which the binding site faces inward (see Figure 7-4).

■ All members of the GLUT protein family transport sugars and have similar structures. Differences in their K_m values, expression in different cell types, and substrate specificities are important for proper sugar metabolism in the body.

■ Two common experimental systems for studying the functions of transport proteins are liposomes containing a purified transport protein (see Figure 7-5) and cells transfected with the gene encoding a particular transport protein.

7.2 ATP-Powered Pumps and the Intracellular Ionic Environment

We turn now to the ATP-powered pumps, which transport ions and various small molecules against their concentration gradients. All ATP-powered pumps are transmembrane proteins with one or more binding sites for ATP located on the cytosolic face of the membrane. Although these proteins commonly are called *ATPases*, they normally do not hydrolyze ATP into ADP and P_i unless ions or other molecules are simultaneously transported. Because of this tight coupling between ATP hydrolysis and transport, the energy stored in the phosphoanhydride bond is not dissipated but rather used to move ions or other molecules uphill against an electrochemical gradient.

Different Classes of Pumps Exhibit Characteristic Structural and Functional Properties

The general structures of the four classes of ATP-powered pumps are depicted in Figure 7-6, with specific examples in each class listed below. Note that the members of three classes (P, F, and V) transport ions only, whereas members of the ABC superfamily primarily transport small molecules.

All *P-class ion pumps* possess two identical catalytic α subunits that contain an ATP-binding site. Most also have two smaller β subunits that usually have regulatory functions. During the transport process, at least one of the α subunits is phosphorylated (hence the name "P" class), and the transported ions are thought to move through the phosphorylated subunit. The sequence around the phosphorylated residue is homologous in different pumps. This class includes the Na^+/K^+ ATPase in the plasma membrane, which main-

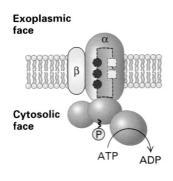
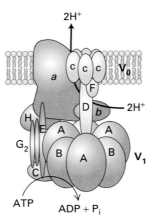
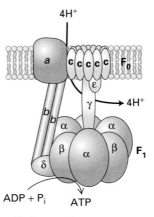
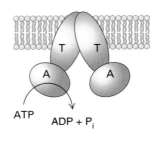

P-class pumps

Plasma membrane of plants, fungi, bacteria (H^+ pump)

Plasma membrane of higher eukaryotes (Na^+/K^+ pump)

Apical plasma membrane of mammalian stomach (H^+/K^+ pump)

Plasma membrane of all eukaryotic cells (Ca^{2+} pump)

Sarcoplasmic reticulum membrane in muscle cells (Ca^{2+} pump)

V-class proton pumps

Vacuolar membranes in plants, yeast, other fungi

Endosomal and lysosomal membranes in animal cells

Plasma membrane of osteoclasts and some kidney tubule cells

F-class proton pumps

Bacterial plasma membrane

Inner mitochondrial membrane

Thylakoid membrane of chloroplast

ABC superfamily

Bacterial plasma membranes (amino acid, sugar, and peptide transporters)

Mammalian plasma membranes (transporters of phospholipids, small lipophilic drugs, cholesterol, other small molecules)

▲ **FIGURE 7-6 The four classes of ATP-powered transport proteins.** The location of specific pumps are indicated below each class. P-class pumps are composed of a catalytic α subunit, which becomes phosphorylated as part of the transport cycle. A β subunit, present in some of these pumps, may regulate transport. F-class and V-class pumps do not form phosphoprotein intermediates and transport only protons. Their structures are similar and contain similar proteins, but none of their subunits are related to those of P-class pumps. V-class pumps couple ATP hydrolysis to transport of protons against a concentration gradient, whereas F-class pumps normally operate in the reverse direction to utilize energy in a proton concentration or electrochemical gradient to synthesize ATP. All members of the large ABC superfamily of proteins contain two transmembrane (T) domains and two cytosolic ATP-binding (A) domains, which couple ATP hydrolysis to solute movement. These core domains are present as separate subunits in some ABC proteins (depicted here), but are fused into a single polypeptide in other ABC proteins. [See T. Nishi and M. Forgac, 2002, *Nature Rev. Mol. Cell Biol.* **3**:94; G. Chang and C. Roth, 2001, *Science* **293**:1793; C. Toyoshima et al., 2000, *Nature* **405**:647; D. McIntosh, 2000, *Nature Struc. Biol.* **7**:532; and T. Elston, H. Wang, and G. Oster, 1998, *Nature* **391**:510.]

tains the low cytosolic Na^+ and high cytosolic K^+ concentrations typical of animal cells. Certain Ca^{2+} ATPases pump Ca^{2+} ions out of the cytosol into the external medium; others pump Ca^{2+} from the cytosol into the endoplasmic reticulum or into the specialized ER called the *sarcoplasmic reticulum*, which is found in muscle cells. Another member of the P class, found in acid-secreting cells of the mammalian stomach, transports protons (H^+ ions) out of and K^+ ions into the cell. The H^+ pump that generates and maintains the membrane electric potential in plant, fungal, and bacterial cells also belongs to this class.

The structures of *F-class* and *V-class ion pumps* are similar to one another but unrelated to and more complicated than P-class pumps. F- and V-class pumps contain several different transmembrane and cytosolic subunits. All known V and F pumps transport only protons, in a process that does not involve a phosphoprotein intermediate. V-class pumps generally function to maintain the low pH of plant vacuoles and of lysosomes and other acidic vesicles in animal cells by pumping protons from the cytosolic to the exoplasmic face of the membrane against a proton electrochemical gradient. F-class pumps are found in bacterial plasma membranes and in mitochondria and chloroplasts. In contrast to V pumps, they generally function to power the synthesis of ATP from ADP and P_i by movement of protons from the exoplasmic to the cytosolic face of the membrane down the proton electrochemical gradient. Because of their importance in ATP synthesis in chloroplasts and mitochondria, F-class proton pumps, commonly called ATP synthases, are treated separately in Chapter 8.

The final class of ATP-powered pumps contains more members and is more diverse than the other classes. Referred to as the **ABC** (*ATP-binding cassette*) **superfamily,** this class includes several hundred different transport proteins found in organisms ranging from bacteria to humans. Each ABC protein is specific for a single substrate or group of related substrates, which may be ions, sugars, amino acids, phospholipids, peptides, polysaccharides, or even proteins. All ABC transport proteins share a structural organization consisting of four "core" domains: two transmembrane (T) domains, forming the passageway through which transported molecules cross the membrane, and two cytosolic ATP-binding (A) domains. In some ABC proteins, mostly in bacteria, the core domains are present in four separate polypeptides; in others, the core domains are fused into one or two multidomain polypeptides.

ATP-Powered Ion Pumps Generate and Maintain Ionic Gradients Across Cellular Membranes

The specific ionic composition of the cytosol usually differs greatly from that of the surrounding extracellular fluid. In virtually all cells—including microbial, plant, and animal cells—the cytosolic pH is kept near 7.2 regardless of the extracellular pH. Also, the cytosolic concentration of K^+ is

much higher than that of Na^+. In addition, in both invertebrates and vertebrates, the concentration of K^+ is 20–40 times higher in cells than in the blood, while the concentration of Na^+ is 8–12 times lower in cells than in the blood (Table 7-2). Some Ca^{2+} in the cytosol is bound to the negatively charged groups in ATP and other molecules, but it is the concentration of free, unbound Ca^{2+} that is critical to its functions in signaling pathways and muscle contraction. The concentration of free Ca^{2+} in the cytosol is generally less than 0.2 micromolar (2×10^{-7} M), a thousand or more times lower than that in the blood. Plant cells and many microorganisms maintain similarly high cytosolic concentrations of K^+ and low concentrations of Ca^{2+} and Na^+ even if the cells are cultured in very dilute salt solutions.

The ion pumps discussed in this section are largely responsible for establishing and maintaining the usual ionic gradients across the plasma and intracellular membranes. In carrying out this task, cells expend considerable energy. For

TABLE 7-2	**Typical Intracellular and Extracellular Ion Concentrations**	
Ion	Cell (mM)	Blood (mM)
SQUID AXON (INVERTEBRATE)*		
K^+	400	20
Na^+	50	440
Cl^-	40–150	560
Ca^{2+}	0.0003	10
$X^{-\dagger}$	300–400	5–10
MAMMALIAN CELL (VERTEBRATE)		
K^+	139	4
Na^+	12	145
Cl^-	4	116
HCO_3^-	12	29
X^-	138	9
Mg^{2+}	0.8	1.5
Ca^{2+}	<0.0002	1.8

*The large nerve axon of the squid has been widely used in studies of the mechanism of conduction of electric impulses.

†X^- represents proteins, which have a net negative charge at the neutral pH of blood and cells.

example, up to 25 percent of the ATP produced by nerve and kidney cells is used for ion transport, and human erythrocytes consume up to 50 percent of their available ATP for this purpose; in both cases, most of this ATP is used to power the Na^+/K^+ pump.

In cells treated with poisons that inhibit the aerobic production of ATP (e.g., 2,4-dinitrophenol in aerobic cells), the ion concentrations inside the cell gradually approach those of the exterior environment as ions move through channels in the plasma membrane down their electrochemical gradients. Eventually treated cells die: partly because protein synthesis requires a high concentration of K^+ ions and partly because in the absence of a Na^+ gradient across the cell membrane, a cell cannot import certain nutrients such as amino acids. Studies on the effects of such poisons provided early evidence for the existence of ion pumps.

Muscle Ca^{2+} ATPase Pumps Ca^{2+} Ions from the Cytosol into the Sarcoplasmic Reticulum

In skeletal muscle cells, Ca^{2+} ions are concentrated and stored in the sarcoplasmic reticulum (SR); release of stored Ca^{2+} ions from the SR lumen into the cytosol causes contraction, as discussed in Chapter 19. A P-class Ca^{2+} ATPase located in the SR membrane of skeletal muscle pumps Ca^{2+} from the cytosol into the lumen of the SR, thereby inducing muscle relaxation. Because this *muscle calcium pump* constitutes more than 80 percent of the integral protein in SR membranes, it is easily purified and has been studied extensively.

In the cytosol of muscle cells, the free Ca^{2+} concentration ranges from 10^{-7} M (resting cells) to more than 10^{-6} M (contracting cells), whereas the *total* Ca^{2+} concentration in the SR lumen can be as high as 10^{-2} M. However, two soluble proteins in the lumen of SR vesicles bind Ca^{2+} and serve as a reservoir for intracellular Ca^{2+}, thereby reducing the concentration of free Ca^{2+} ions in the SR vesicles and consequently the energy needed to pump Ca^{2+} ions into them from the cytosol. The activity of the muscle Ca^{2+} ATPase increases as the free Ca^{2+} concentration in the cytosol rises. Thus in skeletal muscle cells, the calcium pump in the SR membrane can supplement the activity of a similar Ca^{2+} pump located in the plasma membrane to assure that the cytosolic concentration of free Ca^{2+} in resting muscle remains below 1 μM.

The current model for the mechanism of action of the Ca^{2+} ATPase in the SR membrane involves two conformational states of the protein termed E1 and E2. Coupling of

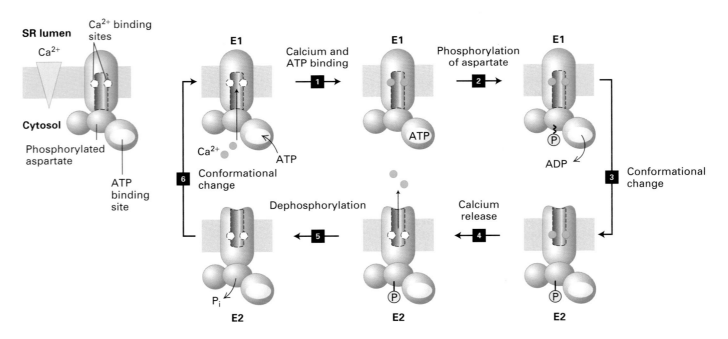

▲ **FIGURE 7-7 Operational model of the Ca^{2+} ATPase in the SR membrane of skeletal muscle cells.** Only one of the two catalytic α subunits of this P-class pump is depicted. E1 and E2 are alternative conformations of the protein in which the Ca^{2+}-binding sites are accessible to the cytosolic and exoplasmic faces, respectively. An ordered sequence of steps (**1**–**6**, as diagrammed here, is essential for coupling ATP hydrolysis and the transport of Ca^{2+} ions across the membrane. In the figure,

~P indicates a high-energy acyl phosphate bond; –P indicates a low-energy phosphoester bond. Because the affinity of Ca^{2+} for the cytosolic-facing binding sites in E1 is a thousandfold greater than the affinity of Ca^{2+} for the exoplasmic-facing sites in E2, this pump transports Ca^{2+} unidirectionally from the cytosol to the SR lumen. See the text and Figure 7-8 for more details. [See C. Toyoshima et al., 2000, *Nature* **405**:647; P. Zhang et al., 1998, *Nature* **392**:835; and W. P. Jencks, 1989, *J. Biol. Chem.* **264**:18855.]

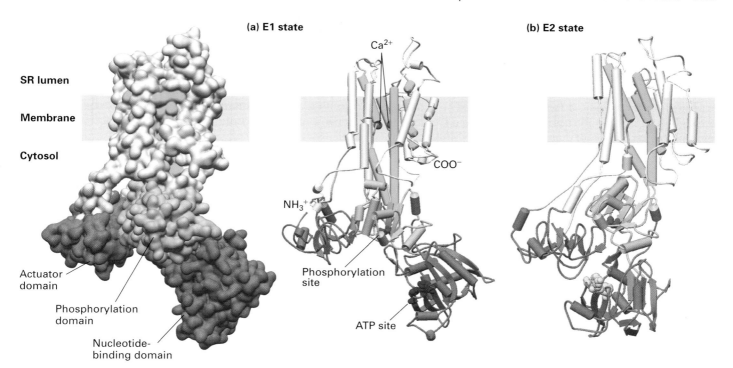

(a) E1 state

SR lumen

Membrane

Cytosol

Actuator domain

Phosphorylation domain

Nucleotide-binding domain

Ca²⁺

COO⁻

NH₃⁺

Phosphorylation site

ATP site

(b) E2 state

▲ **FIGURE 7-8 Structure of the catalytic α subunit of the muscle Ca²⁺ ATPase.** (a) Three-dimensional models of the protein in the E1 state based on the structure determined by x-ray crystallography. There are 10 transmembrane α helices, four of which (green) contain residues that site-specific mutagenesis studies have identified as participating in Ca²⁺ binding. The cytosolic segment forms three domains: the nucleotide-binding domain (orange), the phosphorylation domain (yellow), and the actuator domain (pink) that connects two of the membrane-spanning helices. (b) Hypothetical model of the pump in the E2 state, based on a lower-resolution structure determined by electron microscopy of frozen crystals of the pure protein. Note the differences between the E1 and E2 states in the conformations of the nucleotide-binding and actuator domains; these changes probably power the conformational changes of the membrane-spanning α helices (green) that constitute the Ca²⁺-binding sites, converting them from one in which the Ca²⁺-binding sites are accessible to the cytosolic face (E1 state) to one in which they are accessible to the exoplasmic face (E2 state). [Adapted from C. Xu, 2002, *J. Mol. Biol.* **316**:201, and D. McIntosh, 2000, *Nature Struc. Biol.* **7**:532.]

ATP hydrolysis with ion pumping involves several steps that must occur in a defined order, as shown in Figure 7-7. When the protein is in the E1 conformation, two Ca²⁺ ions bind to two high-affinity binding sites accessible from the cytosolic side and an ATP binds to a site on the cytosolic surface (step 1). The bound ATP is hydrolyzed to ADP in a reaction that requires Mg²⁺, and the liberated phosphate is transferred to a specific aspartate residue in the protein, forming the high-energy acyl phosphate bond denoted by E1 ~ P (step 2). The protein then undergoes a conformational change that generates E2, which has two low-affinity Ca²⁺-binding sites accessible to the SR lumen (step 3). The free energy of hydrolysis of the aspartyl-phosphate bond in E1 ~ P is greater than that in E2−P, and this reduction in free energy of the aspartyl-phosphate bond can be said to power the E1 → E2 conformational change. The Ca²⁺ ions spontaneously dissociate from the low-affinity sites to enter the SR lumen (step 4), following which the aspartyl-phosphate bond is hydrolyzed (step 5). Dephosphorylation powers the E2 → E1 conformational change (step 6), and E1 is ready to transport two more Ca²⁺ ions.

Much evidence supports the model depicted in Figure 7-7. For instance, the muscle calcium pump has been isolated with phosphate linked to an aspartate residue, and spectroscopic studies have detected slight alterations in protein conformation during the E1 → E2 conversion. The 10 membrane-spanning α helices in the catalytic subunit are thought to form the passageway through which Ca²⁺ ions move, and mutagenesis studies have identified amino acids in four of these helices that are thought to form the two Ca²⁺-binding sites (Figure 7-8). Cryoelectron microscopy and x-ray crystallography of the protein in different conformational states also revealed that the bulk of the catalytic subunit consists of cytosolic globular domains that are involved in ATP binding, phosphorylation of aspartate, and transduction of the energy released by hydrolysis of the aspartyl phosphate into conformational changes in the protein. These domains are connected by a "stalk" to the membrane-embedded domain.

All P-class ion pumps, regardless of which ion they transport, are phosphorylated on a highly conserved aspartate residue during the transport process. Thus the operational model in Figure 7-7 is generally applicable to all these ATP-powered ion pumps. In addition, the catalytic α subunits of all the P pumps examined to date have a similar molecular weight and, as deduced from their amino acid sequences derived from cDNA clones, have a similar arrangement of transmembrane α helices (see Figure 7-8). These findings strongly suggest that all these proteins evolved from a common precursor, although they now transport different ions.

Calmodulin-Mediated Activation of Plasma-Membrane Ca²⁺ ATPase Leads to Rapid Ca²⁺ Export

As we explain in Chapter 13, small increases in the concentration of free Ca^{2+} ions in the cytosol trigger a variety of cellular responses. In order for Ca^{2+} to function in intracellular signaling, the concentration of Ca^{2+} ions free in the cytosol usually must be kept below $0.1 - 0.2 \mu M$. Animal, yeast, and probably plant cells express plasma-membrane Ca^{2+} ATPases that transport Ca^{2+} out of the cell against its electrochemical gradient. The catalytic α subunit of these P-class pumps is similar in structure and sequence to the α subunit of the muscle SR Ca^{2+} pump.

The activity of plasma-membrane Ca^{2+} ATPases is regulated by **calmodulin**, a cytosolic Ca^{2+}-binding protein (see Figure 3-28). A rise in cytosolic Ca^{2+} induces the binding of Ca^{2+} ions to calmodulin, which triggers allosteric activation of the Ca^{2+} ATPase. As a result, the export of Ca^{2+} ions from the cell accelerates, quickly restoring the low concentration of free cytosolic Ca^{2+} characteristic of the resting cell.

Na⁺/K⁺ ATPase Maintains the Intracellular Na⁺ and K⁺ Concentrations in Animal Cells

A second important P-class ion pump present in the plasma membrane of all animal cells is the **Na⁺/K⁺ ATPase.** This ion pump is a tetramer of subunit composition $\alpha_2\beta_2$. (Classic Experiment 7.1 describes the discovery of this enzyme.) The small, glycosylated β polypeptide helps newly synthesized α subunits to fold properly in the endoplasmic reticulum but apparently is not involved directly in ion pumping. The amino acid sequence and predicted secondary structure of the catalytic α subunit are very similar to those of the muscle SR Ca^{2+} ATPase (see Figure 7-8). In particular, the Na⁺/K⁺ ATPase has a stalk on the cytosolic face that links

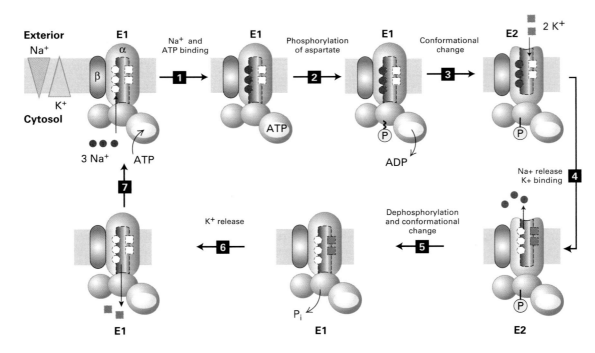

▲ **FIGURE 7-9 Operational model of the Na⁺/K⁺ ATPase in the plasma membrane.** Only one of the two catalytic α subunits of this P-class pump is depicted. It is not known whether just one or both subunits in a single ATPase molecule transport ions. Ion pumping by the Na⁺/K⁺ ATPase involves phosphorylation, dephosphorylation, and conformational changes similar to those in the muscle

Ca^{2+} ATPase (see Figure 7-7). In this case, hydrolysis of the E2–P intermediate powers the E2 → E1 conformational change and concomitant transport of two ions (K⁺) inward. Na⁺ ions are indicated by red circles; K⁺ ions, by purple squares; high-energy acyl phosphate bond, by ~P; low-energy phosphoester bond, by –P. [See K. Sweadner and C. Donnet, 2001, *Biochem. J.* **356**:6875, for details of the structure of the α subunit.]

MEDIA CONNECTIONS   Overview Animation: Biological Energy Interconversions domains containing the ATP-binding site and the phosphorylated aspartate to the membrane-embedded domain. The overall transport process moves three Na^+ ions out of and two K^+ ions into the cell per ATP molecule hydrolyzed.

The mechanism of action of the Na^+/K^+ ATPase, outlined in Figure 7-9, is similar to that of the muscle calcium pump, except that ions are pumped in both directions across the membrane. In its E1 conformation, the Na^+/K^+ ATPase has three high-affinity Na^+-binding sites and two low-affinity K^+-binding sites accessible to the cytosolic surface of the protein. The K_m for binding of Na^+ to these cytosolic sites is 0.6 mM, a value considerably lower than the intracellular Na^+ concentration of $\approx$12 mM; as a result, Na^+ ions normally will fully occupy these sites. Conversely, the affinity of the cytosolic K^+-binding sites is low enough that K^+ ions, transported inward through the protein, dissociate from E1 into the cytosol despite the high intracellular K^+ concentration. During the E1 → E2 transition, the three bound Na^+ ions become accessible to the exoplasmic face, and simultaneously the affinity of the three Na^+-binding sites becomes reduced. The three Na^+ ions, transported outward through the protein and now bound to the low-affinity Na^+ sites exposed to the exoplasmic face, dissociate one at a time into the extracellular medium despite the high extracellular Na^+ concentration. Transition to the E2 conformation also generates two high-affinity K^+ sites accessible to the exoplasmic face. Because the K_m for K^+ binding to these sites (0.2 mM) is lower than the extracellular K^+ concentration (4 mM), these sites will fill with K^+ ions. Similarly, during the E2 → E1 transition, the two bound K^+ ions are transported inward and then released into the cytosol.

Certain drugs (e.g., ouabain and digoxin) bind to the exoplasmic domain of the plasma-membrane Na^+/K^+ ATPase and specifically inhibit its ATPase activity. The resulting disruption in the Na^+/K^+ balance of cells is strong evidence for the critical role of this ion pump in maintaining the normal K^+ and Na^+ ion concentration gradients.

V-Class H^+ ATPases Pump Protons Across Lysosomal and Vacuolar Membranes

All V-class ATPases transport only H^+ ions. These proton pumps, present in the membranes of lysosomes, endosomes, and plant vacuoles, function to acidify the lumen of these organelles. The pH of the lysosomal lumen can be measured precisely in living cells by use of particles labeled with a pH-sensitive fluorescent dye. After these particles are phagocytosed by cells and transferred to lysosomes, the lysosomal pH can be calculated from the spectrum of the fluorescence emitted. Maintenance of the 100-fold or more proton gradient between the lysosomal lumen (pH $\approx$4.5–5.0) and the cytosol (pH $\approx$7.0) depends on ATP production by the cell.

The ATP-powered proton pumps in lysosomal and vacuolar membranes have been isolated, purified, and incorporated into liposomes. As illustrated in Figure 7-6 (center), these V-class proton pumps contain two discrete domains: a cytosolic hydrophilic domain (V_1) and a transmembrane domain (V_0) with multiple subunits in each domain. Binding and hydrolysis of ATP by the B subunits in V_1 provide the energy for pumping of H^+ ions through the proton-conducting channel formed by the c and a subunits in V_0. Unlike P-class ion pumps, V-class proton pumps are not phosphorylated and dephosphorylated during proton transport. The structurally similar F-class proton pumps, which we describe in the next chapter, normally operate in the "reverse" direction to generate ATP rather than pump protons and their mechanism of action is understood in great detail.

Pumping of relatively few protons is required to acidify an intracellular vesicle. To understand why, recall that a solution of pH 4 has a H^+ ion concentration of 10^{-4} moles per liter, or 10^{-7} moles of H^+ ions per milliliter. Since there are 6.02×10^{23} molecules per mole (Avogadro's number), then a milliliter of a pH 4 solution contains 6.02×10^{16} H^+ ions. Thus at pH 4, a primary spherical lysosome with a volume of 4.18×10^{-15} ml (diameter of 0.2 μm) will contain just 252 protons.

By themselves ATP-powered proton pumps cannot acidify the lumen of an organelle (or the extracellular space) because these pumps are *electrogenic*; that is, a net movement of electric charge occurs during transport. Pumping of just a few protons causes a buildup of positively charged H^+ ions on the exoplasmic (inside) face of the organelle membrane. For each H^+ pumped across, a negative ion (e.g., OH^- or Cl^-) will be "left behind" on the cytosolic face, causing a buildup of negatively charged ions there. These oppositely charged ions attract each other on opposite faces of the membrane, generating a charge separation, or electric potential, across the membrane. As more and more protons are pumped, the excess of positive charges on the exoplasmic face repels other H^+ ions, soon preventing pumping of additional protons long before a significant transmembrane H^+ concentration gradient had been established (Figure 7-10a). In fact, this is the way that P-class H^+ pumps generate a cytosol-negative potential across plant and yeast plasma membranes.

In order for an organelle lumen or an extracellular space (e.g., the lumen of the stomach) to become acidic, movement of protons must be accompanied either by (1) movement of an equal number of anions (e.g., Cl^-) in the same direction or by (2) movement of equal numbers of a different cation in the opposite direction. The first process occurs in lysosomes and plant vacuoles whose membranes contain V-class H^+ ATPases and anion channels through which accompanying Cl^- ions move (Figure 7-10b). The second process occurs in the lining of the stomach, which contains a P-class H^+/K^+ ATPase that is not electrogenic and pumps one H^+ outward and one K^+ inward. Operation of this pump is discussed later in the chapter.

(a)

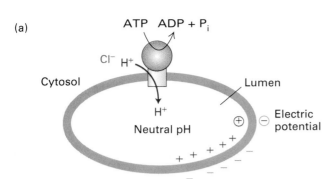

(b)

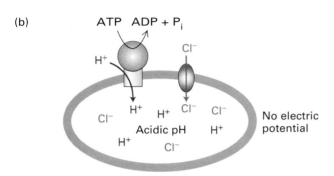

▲ **FIGURE 7-10 Effect of proton pumping by V-class ion pumps on H⁺ concentration gradients and electric potential gradients across cellular membranes.** (a) If an intracellular organelle contains only V-class pumps, proton pumping generates an electric potential across the membrane, luminal-side positive, but no significant change in the intraluminal pH. (b) If the organelle membrane also contains Cl⁻ channels, anions passively follow the pumped protons, resulting in an accumulation of H⁺ ions (low luminal pH) but no electric potential across the membrane.

Bacterial Permeases Are ABC Proteins That Import a Variety of Nutrients from the Environment

As noted earlier, all members of the very large and diverse ABC superfamily of transport proteins contain two transmembrane (T) domains and two cytosolic ATP-binding (A) domains (see Figure 7-6). The T domains, each built of six membrane-spanning α helices, form the pathway through which the transported substance (substrate) crosses the membrane and determine the substrate specificity of each ABC protein. The sequences of the A domains are ≈30–40 percent homologous in all members of this superfamily, indicating a common evolutionary origin. Some ABC proteins also contain an additional exoplasmic substrate-binding subunit or regulatory subunit.

The plasma membrane of many bacteria contains numerous *permeases* that belong to the ABC superfamily. These proteins use the energy released by hydrolysis of ATP to transport specific amino acids, sugars, vitamins, or even peptides into the cell. Since bacteria frequently grow in soil or pond water where the concentration of nutrients is low, these

ABC transport proteins enable the cells to import nutrients against substantial concentration gradients. Bacterial permeases generally are inducible; that is, the quantity of a transport protein in the cell membrane is regulated by both the concentration of the nutrient in the medium and the metabolic needs of the cell.

In *E. coli* histidine permease, a typical bacterial ABC protein, the two transmembrane domains and two cytosolic ATP-binding domains are formed by four separate subunits. In gram-negative bacteria such as *E. coli*, the outer membrane contains **porins** that render them highly permeable to most small molecules (see Figure 5-14). A soluble histidine-binding protein is located in the periplasmic space between the outer membrane and plasma membrane. This soluble protein binds histidine tightly and directs it to the T subunits of the permease, through which histidine crosses the plasma membrane powered by ATP hydrolysis. Mutant *E. coli* cells that are defective in any of the histidine permease subunits or the soluble binding protein are unable to transport histidine into the cell, but are able to transport other amino acids whose uptake is facilitated by other transport proteins. Such genetic analyses provide strong evidence that histidine permease and similar ABC proteins function to transport various solutes into bacterial cells.

About 50 ABC Small-Molecule Pumps Are Known in Mammals

Discovery of the first eukaryotic ABC protein to be recognized came from studies on tumor cells and cultured cells that exhibited resistance to several drugs with unrelated chemical structures. Such cells eventually were shown to express elevated levels of a *multidrug-resistance (MDR) transport protein* known as *MDR1*. This protein uses the energy derived from ATP hydrolysis to *export* a large variety of drugs from the cytosol to the extracellular medium. The *Mdr1* gene is frequently amplified in multidrug-resistant cells, resulting in a large overproduction of the MDR1 protein.

Most drugs transported by MDR1 are small hydrophobic molecules that diffuse from the medium across the plasma membrane, unaided by transport proteins, into the cell cytosol, where they block various cellular functions. Two such drugs are colchicine and vinblastine, which block assembly of microtubules. ATP-powered export of such drugs by MDR1 reduces their concentration in the cytosol. As a result, a much higher extracellular drug concentration is required to kill cells that express MDR1 than those that do not. That MDR1 is an ATP-powered small-molecule pump has been demonstrated with liposomes containing the purified protein (see Figure 7-5). The ATPase activity of these liposomes is enhanced by different drugs in a dose-dependent manner corresponding to their ability to be transported by MDR1.

About 50 different mammalian ABC transport proteins are now recognized (see Table 18-2). These are expressed in abundance in the liver, intestines, and kidney—sites where natural toxic and waste products are removed from the body. Substrates for these ABC proteins include sugars, amino acids, cholesterol, peptides, proteins, toxins, and xenobiotics. Thus the normal function of MDR1 most likely is to transport various natural and metabolic toxins into the bile, intestinal lumen, or forming urine. During the course of its evolution, MDR1 appears to have acquired the ability to transport drugs whose structures are similar to those of these endogenous toxins. Tumors derived from MDR-expressing cell types, such as hepatomas (liver cancers), frequently are resistant to virtually all chemotherapeutic agents and thus difficult to treat, presumably because the tumors exhibit increased expression of the MDR1 or the related MDR2.

 Several human genetic diseases are associated with defective ABC proteins. X-linked adrenoleukodystrophy (ALD), for instance, is characterized by a defective ABC transport protein (ABCD1) that is localized to peroxisomal membranes. This protein normally regulates import of very long chain fatty acids into peroxisomes, where they undergo oxidation; in its absence these fatty acids accumulate in the cytosol and cause cellular damage. Tangiers disease is marked by a deficiency of the plasma-membrane ABC protein (ABCA1) that transports phospholipids and possibly cholesterol (Chapter 18).

A final example is cystic fibrosis (CF), which is caused by a mutation in the gene encoding the *cystic fibrosis transmembrane regulator (CFTR)*. This Cl^- transport protein is expressed in the apical plasma membranes of epithelial cells in the lung, sweat glands, pancreas, and other tissues. For instance, CFTR protein is important for resorption of Cl^- into cells of sweat glands, and babies with cystic fibrosis, if licked, often taste "salty." An increase in cyclic AMP (cAMP), a small intracellular signaling molecule, causes phosphorylation of CFTR and stimulates Cl^- transport by such cells from normal individuals, but not from CF individuals who have a defective CFTR protein. (The role of cAMP in numerous signaling pathways is covered in Chapter 13.) The sequence and predicted structure of the CFTR protein, based on analysis of the cloned gene, are very similar to those of MDR1 protein except for the presence of an additional domain, the regulatory (R) domain, on the cytosolic face. Moreover, the Cl^--transport activity of CFTR protein is enhanced by the binding of ATP. Given its similarity to other ABC proteins, CFTR may also function as an ATP-powered pump of some still unidentified molecule. ∎

ABC Proteins That Transport Lipid-Soluble Substrates May Operate by a Flippase Mechanism

The substrates of mammalian MDR1 are primarily planar, lipid-soluble molecules with one or more positive charges;

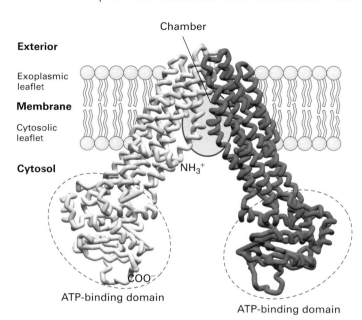

▲ **FIGURE 7-11 Structural model of *E. coli* lipid flippase, an ABC protein homologous to mammalian MDR1.** The V-shaped protein encloses a "chamber" within the bilayer where it is hypothesized that bound substrates are flipped across the membrane, as depicted in Figure 7-12. Each identical subunit in this homodimeric protein has one transmembrane domain, comprising six α helices, and one cytosolic domain where ATP binding occurs. [Adapted from G. Chang and C. Roth, 2001, *Science* **293**:1793.]

they all compete with one another for transport by MDR1, suggesting that they bind to the same site or sites on the protein. In contrast to bacterial ABC proteins, all four domains of MDR1 are fused into a single 170,000-MW protein. The recently determined three-dimensional structure of a homologous *E. coli* lipid-transport protein reveals that the molecule is V shaped, with the apex in the membrane and the arms containing the ATP-binding sites protruding into the cytosol (Figure 7-11).

Although the mechanism of transport by MDR1 and similar ABC proteins has not been definitively demonstrated, a likely candidate is the *flippase model* depicted in Figure 7-12. According to this model, MDR1 "flips" a charged substrate molecule from the cytosolic to the exoplasmic leaflet, an energetically unfavorable reaction powered by the coupled ATPase activity of the protein. Support for the flippase model of transport by MDR1 comes from MDR2, a homologous protein present in the region of the liver cell plasma membrane that faces the bile duct. As detailed in Chapter 18, MDR2 has been shown to flip phospholipids from the cytosolic-facing leaflet of the plasma membrane to the exoplasmic leaflet, thereby generating an excess of phospholipids in the exoplasmic leaflet; these phospholipids then peel off into the bile duct and form an essential part of the bile.

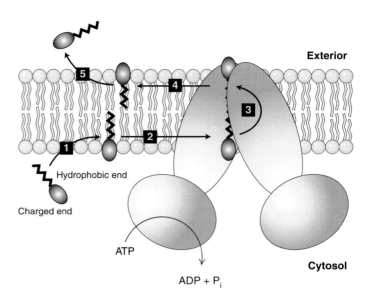

▶ **FIGURE 7-12 Flippase model of transport by MDR1 and similar ABC proteins.** Step **1**: The hydrophobic portion (black) of a substrate molecule moves spontaneously from the cytosol into the cytosolic-facing leaflet of the lipid bilayer, while the charged end (red) remains in the cytosol. Step **2**: The substrate diffuses laterally until encountering and binding to a site on the MDR1 protein within the bilayer. Step **3**: The protein then "flips" the charged substrate molecule into the exoplasmic leaflet, an energetically unfavorable reaction powered by the coupled hydrolysis of ATP by the cytosolic domain. Steps **4** and **5**: Once in the exoplasmic face, the substrate again can diffuse laterally in the membrane and ultimately moves into the aqueous phase on the outside of the cell. [Adapted from P. Borst, N. Zelcer, and A. van Helvoort, 2000, *Biochim. Biophys. Acta* **1486**:128.]

KEY CONCEPTS OF SECTION 7.2

ATP-Powered Pumps and the Intracellular Ionic Environment

■ Four classes of transmembrane proteins couple the energy-releasing hydrolysis of ATP with the energy-requiring transport of substances against their concentration gradient: P-, V-, and F-class pumps and ABC proteins (see Figure 7-6).

■ The combined action of P-class Na^+/K^+ ATPases in the plasma membrane and homologous Ca^{2+} ATPases in the plasma membrane or sarcoplasmic reticulum creates the usual ion milieu of animal cells: high K^+, low Ca^{2+}, and low Na^+ in the cytosol; low K^+, high Ca^{2+}, and high Na^+ in the extracellular fluid.

■ In P-class pumps, phosphorylation of the α (catalytic) subunit and a change in conformational states are essential for coupling ATP hydrolysis to transport of H^+, Na^+, K^+, or Ca^{2+} ions (see Figures 7-7 and 7-9).

■ V- and F-class ATPases, which transport protons exclusively, are large, multisubunit complexes with a proton-conducting channel in the transmembrane domain and ATP-binding sites in the cytosolic domain.

■ V-class H^+ pumps in animal lysosomal and endosomal membranes and plant vacuole membranes are responsible for maintaining a lower pH inside the organelles than in the surrounding cytosol (see Figure 7-10b).

■ All members of the large and diverse ABC superfamily of transport proteins contain four core domains: two transmembrane domains, which form a pathway for solute movement and determine substrate specificity, and two cytosolic ATP-binding domains (see Figure 7-11).

■ The ABC superfamily includes bacterial amino acid and sugar permeases and about 50 mammalian proteins (e.g.,

MDR1, ABCA1) that transport a wide array of substrates including toxins, drugs, phospholipids, peptides, and proteins.

■ According to the flippase model of MDR activity, a substrate molecule diffuses into the cytosolic leaflet of the plasma membrane, then is flipped to the exoplasmic leaflet in an ATP-powered process, and finally diffuses from the membrane into the extracellular space (see Figure 7-12).

7.3 Nongated Ion Channels and the Resting Membrane Potential

In addition to ATP-powered ion pumps, which transport ions against their concentration gradients, the plasma membrane contains channel proteins that allow the principal cellular ions (Na^+, K^+, Ca^{2+}, and Cl^-) to move through them at different rates down their concentration gradients. Ion concentration gradients generated by pumps and selective movements of ions through channels constitute the principal mechanism by which a difference in voltage, or electric potential, is generated across the plasma membrane. The magnitude of this electric potential generally is ≈70 millivolts (mV) with the inside of the cell always negative with respect to the outside. This value does not seem like much until we consider the thickness of the plasma membrane (3.5 nm). Thus the voltage gradient across the plasma membrane is 0.07 V per 3.5×10^{-7} cm, or 200,000 volts per centimeter! (To appreciate what this means, consider that high-voltage transmission lines for electricity utilize gradients of about 200,000 volts per kilometer.)

The ionic gradients and electric potential across the plasma membrane play a role in many biological processes. As noted previously, a rise in the cytosolic Ca^{2+} concentration is an important regulatory signal, initiating contraction in muscle cells and triggering secretion of digestive enzymes in the exocrine pancreatic cells. In many animal cells, the combined force of the Na^+ concentration gradient and membrane electric potential drives the uptake of amino acids and other molecules against their concentration gradient by ion-linked symport and antiport proteins (see Section 7.4). And the conduction of action potentials by nerve cells depends on the opening and closing of ion channels in response to changes in the membrane potential (see Section 7.7).

Here we discuss the origin of the membrane electric potential in resting cells, how ion channels mediate the selective movement of ions across a membrane, and useful experimental techniques for characterizing the functional properties of channel proteins.

Selective Movement of Ions Creates a Transmembrane Electric Potential Difference

To help explain how an electric potential across the plasma membrane can arise, we first consider a set of simplified experimental systems in which a membrane separates a 150 mM NaCl/15 mM KCl solution on the right from a 15 mM NaCl/150 mM KCl solution on the left. A potentiometer (voltmeter) is connected to both solutions to measure any difference in electric potential across the membrane. If the membrane is impermeable to all ions, no ions will flow across it and no electric potential will be generated, as shown in Figure 7-13a.

Now suppose that the membrane contains Na^+-channel proteins that accommodate Na^+ ions but exclude K^+ and Cl^- ions (Figure 7-13b). Na^+ ions then tend to move down their concentration gradient from the right side to the left, leaving an excess of negative Cl^- ions compared with Na^+ ions on the right side and generating an excess of positive Na^+ ions compared with Cl^- ions on the left side. The excess Na^+ on the left and Cl^- on the right remain near the respective surfaces of the membrane because the excess positive charges on one side of the membrane are attracted to the excess negative charges on the other side. The resulting separation of charge across the membrane constitutes an electric

▶ **EXPERIMENTAL FIGURE 7-13 Generation of a transmembrane electric potential (voltage) depends on the selective movement of ions across a semipermeable membrane.** In this experimental system, a membrane separates a 15 mM NaCl/150 mM KCl solution (*left*) from a 150 mM NaCl/15 mM KCl solution (*right*); these ion concentrations are similar to those in cytosol and blood, respectively. If the membrane separating the two solutions is impermeable to all ions (a), no ions can move across the membrane and no difference in electric potential is registered on the potentiometer connecting the two solutions. If the membrane is selectively permeable only to Na^+ (b) or to K^+ (c), then diffusion of ions through their respective channels leads to a separation of charge across the membrane. At equilibrium, the membrane potential caused by the charge separation becomes equal to the Nernst potential E_{Na} or E_K registered on the potentiometer. See the text for further explanation.

(a) Membrane impermeable to Na^+, K^+, and Cl^-

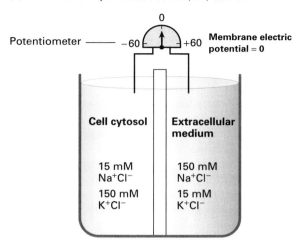

(b) Membrane permeable only to Na^+

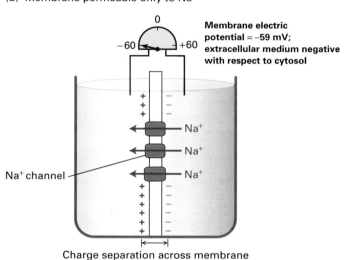

Charge separation across membrane

(c) Membrane permeable only to K^+

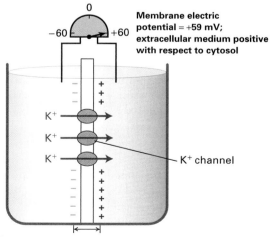

Charge separation across membrane

potential, or voltage, with the right side of the membrane having excess negative charge with respect to the left.

As more and more Na^+ ions move through channels across the membrane, the magnitude of this charge difference (i.e., voltage) increases. However, continued right-to-left movement of the Na^+ ions eventually is inhibited by the mutual repulsion between the excess positive (Na^+) charges accumulated on the left side of the membrane and by the attraction of Na^+ ions to the excess negative charges built up on the right side. The system soon reaches an equilibrium point at which the two opposing factors that determine the movement of Na^+ ions—the membrane electric potential and the ion concentration gradient—balance each other out. At equilibrium, no net movement of Na^+ ions occurs across the membrane. Thus this *semipermeable* membrane, like all biological membranes, acts like a *capacitor*—a device consisting of a thin sheet of nonconducting material (the hydrophobic interior) surrounded on both sides by electrically conducting material (the polar phospholipid head groups and the ions in the surrounding aqueous solution)—that can store positive charges on one side and negative charges on the other.

If a membrane is permeable only to Na^+ ions, then at equilibrium the measured electric potential across the membrane equals the sodium equilibrium potential in volts, E_{Na}. The magnitude of E_{Na} is given by the *Nernst equation*, which is derived from basic principles of physical chemistry:

$$E_{Na} = \frac{RT}{ZF} \ln \frac{[Na_1]}{[Na_r]} \qquad (7\text{-}2)$$

where R (the gas constant) = 1.987 cal/(degree · mol), or 8.28 joules/(degree · mol); T (the absolute temperature in degrees Kelvin) = 293 K at 20 °C; Z (the charge, also called the valency) here equal to $+1$; F (the Faraday constant) 23,062 cal/(mol · V), or 96,000 coulombs/(mol · V); and $[Na_1]$ and $[Na_r]$ are the Na^+ concentrations on the left and right sides, respectively, at equilibrium. At 20 °C, Equation 7-2 reduces to

$$E_{Na} = 0.059 \log_{10} \frac{[Na_1]}{[Na_r]} \qquad (7\text{-}3)$$

If $[Na_1]/[Na_r] = 0.1$, a tenfold ratio of concentrations as in Figure 7-13b, then $E_{Na} = -0.059$ V (or -59 mV), with the right side negative with respect to the left.

If the membrane is permeable only to K^+ ions and not to Na^+ or Cl^- ions, then a similar equation describes the potassium equilibrium potential E_K:

$$E_K = \frac{RT}{ZF} \ln \frac{[K_1]}{[K_r]} \qquad (7\text{-}4)$$

The *magnitude* of the membrane electric potential is the same (59 mV for a tenfold difference in ion concentrations), except that the right side is now *positive* with respect to the left (Figure 7-13c), opposite to the polarity obtained across a membrane selectively permeable to Na^+ ions.

The Membrane Potential in Animal Cells Depends Largely on Resting K^+ Channels

The plasma membranes of animal cells contain many open K^+ channels but few open Na^+, Cl^-, or Ca^{2+} channels. As a result, the major ionic movement across the plasma membrane is that of K^+ from the inside outward, powered by the K^+ concentration gradient, leaving an excess of negative charge on the inside and creating an excess of positive charge on the outside, similar to the experimental system shown in Figure 7-13c. Thus the outward flow of K^+ ions through these channels, called **resting K^+ channels**, is the major determinant of the inside-negative membrane potential. Like all channels, these alternate between an open and a closed state, but since their opening and closing are not affected by the membrane potential or by small signaling molecules, these channels are called *nongated*. The various gated channels discussed in later sections open only in response to specific ligands or to changes in membrane potential.

Quantitatively, the usual resting membrane potential of -70 mV is close to but lower in magnitude than that of the potassium equilibrium potential calculated from the Nernst equation because of the presence of a few open Na^+ channels. These open Na^+ channels allow the net inward flow of Na^+ ions, making the cytosolic face of the plasma membrane more positive, that is, less negative, than predicted by the

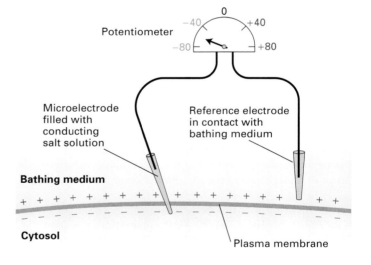

▲ **EXPERIMENTAL FIGURE 7-14 The electric potential across the plasma membrane of living cells can be measured.** A microelectrode, constructed by filling a glass tube of extremely small diameter with a conducting fluid such as a KCl solution, is inserted into a cell in such a way that the surface membrane seals itself around the tip of the electrode. A reference electrode is placed in the bathing medium. A potentiometer connecting the two electrodes registers the potential, in this case -60 mV. A potential difference is registered only when the microelectrode is inserted into the cell; no potential is registered if the microelectrode is in the bathing fluid.

Nernst equation for K$^+$. The K$^+$ concentration gradient that drives the flow of ions through resting K$^+$ channels is generated by the Na$^+$/K$^+$ ATPase described previously (see Figure 7-9). In the absence of this pump, or when it is inhibited, the K$^+$ concentration gradient cannot be maintained and eventually the magnitude of the membrane potential falls to zero.

Although resting K$^+$ channels play the dominant role in generating the electric potential across the plasma membrane of animal cells, this is not the case in plant and fungal cells. The inside-negative membrane potential in these cells is generated by transport of H$^+$ ions out of the cell by P-class proton pumps (see Figure 7-10a).

The potential across the plasma membrane of large cells can be measured with a microelectrode inserted inside the cell and a reference electrode placed in the extracellular fluid. The two are connected to a potentiometer capable of measuring small potential differences (Figure 7-14). In virtually all cells the inside (cytosolic face) of the cell membrane is negative relative to the outside; typical membrane potentials range between -30 and -70 mV. The potential across the surface membrane of most animal cells generally does not vary with time. In contrast, neurons and muscle cells—the principal types of electrically active cells—undergo controlled changes in their membrane potential that we discuss later.

Ion Channels Contain a Selectivity Filter Formed from Conserved Transmembrane α Helices and P Segments

All ion channels exhibit specificity for particular ions: K$^+$ channels allow K$^+$ but not closely related Na$^+$ ions to enter, whereas Na$^+$ channels admit Na$^+$ but not K$^+$. Determination of the three-dimensional structure of a bacterial K$^+$ channel first revealed how this exquisite ion selectivity is achieved. As the sequences of other K$^+$, Na$^+$, and Ca^{2+} channels subsequently were determined, it became apparent that all such proteins share a common structure and probably evolved from a single type of channel protein.

Like all other K$^+$ channels, bacterial K$^+$ channels are built of four identical subunits symmetrically arranged around a central pore (Figure 7-15). Each subunit contains two membrane-spanning α helices (S5 and S6) and a short P (pore domain) segment that partly penetrates the membrane bilayer. In the tetrameric K$^+$ channel, the eight transmembrane α helices (two from each subunit) form an "inverted teepee," generating a water-filled cavity called the vestibule in the central portion of the channel. Four extended loops that are part of the P segments form the actual *ion-selectivity filter* in the narrow part of the pore near the exoplasmic surface above the vestibule.

(a) Single subunit

(b) Tetrameric channel

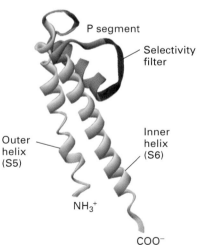
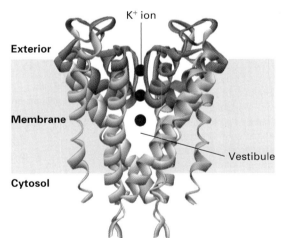

▲ **Figure 7-15 Structure of resting K$^+$ channel from the bacterium *Streptomyces lividans*.** All K$^+$ channel proteins are tetramers comprising four identical subunits each containing two conserved membrane-spanning α helices, called by convention S5 and S6 (yellow), and a shorter P, or pore segment (pink). (a) One of the subunits, viewed from the side, with key structural features indicated. (b) The complete tetrameric channel viewed from the side (*left*) and the top, or extracellular, end (*right*). The P segments are located near the exoplasmic surface and connect the S5 and S6 α helices; they consist of a nonhelical "turret," which lines the upper part of the pore; a short α helix; and an extended loop that protrudes into the narrowest part of the pore and forms the ion-selectivity filter. This filter allows K$^+$ (purple spheres) but not other ions to pass. Below the filter is the central cavity or vestibule lined by the inner, or S6 α, helixes. The subunits in gated K$^+$ channels, which open and close in response to specific stimuli, contain additional transmembrane helices not shown here. [See Y. Zhou et al., 2001, *Nature* **414**:43.]

Several types of evidence support the role of P segments in ion selection. First, the amino acid sequence of the P segment is highly homologous in all known K^+ channels and different from that in other ion channels. Second, mutation of amino acids in this segment alters the ability of a K^+ channel to distinguish Na^+ from K^+. Finally, replacing the P segment of a bacterial K^+ channel with the homologous segment from a mammalian K^+ channel yields a chimeric protein that exhibits normal selectivity for K^+ over other ions. Thus all K^+ channels are thought to use the same mechanism to distinguish K^+ over other ions.

The ability of the ion-selectivity filter in K^+ channels to select K^+ over Na^+ is due mainly to backbone carbonyl oxygens on glycine residues located in a Gly-Tyr-Gly sequence that is found in an analogous position in the P segment in every known K^+ channel. As a K^+ ion enters the narrow selectivity filter, it loses its water of hydration but becomes bound to eight backbone carbonyl oxygens, two from the extended loop in each P segment lining the channel (Figure 7-16a, *left*). As a result, a relatively low activation energy is required for passage of K^+ ions through the channel. Because a dehydrated Na^+ ion is too small to bind to all eight carbonyl oxygens that line the selectivity filter, the activation energy for passage of Na^+ ions is relatively high (Figure 7-16a, *right*). This difference in activation energies favors

▶ **Figure 7-16 Mechanism of ion selectivity and transport in resting K^+ channels.** (a) Schematic diagrams of K^+ and Na^+ ions hydrated in solution and in the pore of a K^+ channel. As K^+ ions pass through the selectivity filter, they lose their bound water molecules and become coordinated instead to eight backbone carbonyl oxygens, four of which are shown, that are part of the conserved amino acids in the channel-lining loop of each P segment. The smaller Na^+ ions cannot perfectly coordinate with these oxygens and therefore pass through the channel only rarely. (b) High-resolution electron-density map obtained from x-ray crystallography showing K^+ ions (purple spheres) passing through the selectivity filter. Only two of the diagonally opposed channel subunits are shown. Within the selectivity filter each unhydrated K^+ ion interacts with eight carbonyl oxygen atoms (red sticks) lining the channel, two from each of the four subunits, as if to mimic the eight waters of hydration. (c) Interpretation of the electron-density map showing the two alternating states by which K^+ ions move through the channel. In State 1, moving from the exoplasmic side of the channel inward one sees a hydrated K^+ ion with its eight bound water molecules, K^+ ions at positions 1 and 3 within the selectivity filter, and a fully hydrated K^+ ion within the vestibule. During K^+ movement each ion in State 1 moves one step inward, forming State 2. Thus in State 2 the K^+ ion on the exoplasmic side of the channel has lost four of its eight waters, the ion at position 1 in State 1 has moved to position 2, and the ion at position 3 in State 1 has moved to position 4. In going from State 2 to State 1 the K^+ at position 4 moves into the vestibule and picks up eight water molecules, while another hydrated K^+ ion moves into the channel opening and the other K^+ ions move down one step. [Part (a) adapted from C. Armstrong, 1998, Science **280**:56. Parts (b) and (c) adapted from Y. Zhou et al., 2001, Nature **414**:43.]

(a) K^+ and Na^+ ions in the pore of a K^+ channel (top view)

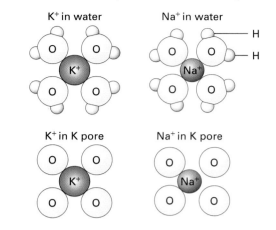

(b) K^+ ions in the pore of a K^+ channel (side view)

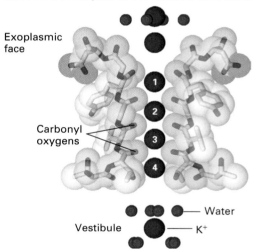

(c) Ion movement through selectivity filter

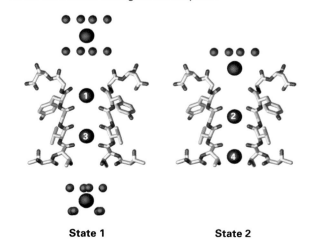

passage of K^+ ions over Na^+ by a factor of thousand. Calcium ions are too large to pass through a K^+ channel with or without their bound water.

Recent x-ray crystallographic studies reveal that the channel contains K^+ ions within the selectivity filter even when it is closed; without these ions the channel probably would collapse. These ions are thought to be present either at positions 1 and 3 or at 2 and 4, each boxed by eight carbonyl oxygen atoms (Figure 7-16b and c). K^+ ions move simultaneously through the channel such that when the ion on the exoplasmic face that has been partially stripped of its water of hydration moves into position 1, the ion at position 2 jumps to position 3 and the one at position 4 exits the channel (Figure 7-16c).

Although the amino acid sequences of the P segment in Na^+ and K^+ channels differ somewhat, they are similar enough to suggest that the general structure of the ion-selectivity filters are comparable in both types of channels. Presumably the diameter of the filter in Na^+ channels is small enough that it permits dehydrated Na^+ ions to bind to the backbone carbonyl oxygens but excludes the larger K^+ ions from entering.

Patch Clamps Permit Measurement of Ion Movements Through Single Channels

The technique of **patch clamping** enables workers to investigate the opening, closing, regulation, and ion conductance of a *single* ion channel. In this technique, the inward or outward movement of ions across a patch of membrane is quantified from the amount of electric current needed to maintain the membrane potential at a particular "clamped" value (Figure 7-17a, b). To preserve electroneutrality and to keep the membrane potential constant, the entry of each positive ion (e.g., a Na^+ ion) into the cell through a channel in the patch of membrane is balanced by the addition of an electron

into the cytosol through a microelectrode inserted into the cytosol; an electronic device measures the numbers of electrons (current) required to counterbalance the inflow of ions through the membrane channels. Conversely, the exit of each positive ion from the cell (e.g., a K^+ ion) is balanced by the withdrawal of an electron from the cytosol. The patch-clamping technique can be employed on whole cells or isolated membrane patches to measure the effects of different substances and ion concentrations on ion flow (Figure 7-17c).

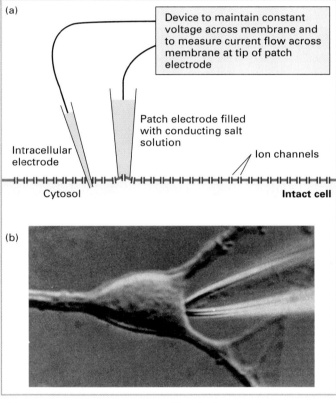

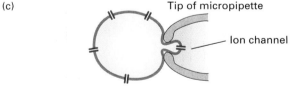

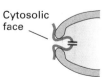
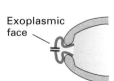

▶ **EXPERIMENTAL FIGURE 7-17 Current flow through individual ion channels can be measured by patch-clamping technique.** (a) Basic experimental arrangement for measuring current flow through individual ion channels in the plasma membrane of a living cell. The patch electrode, filled with a current-conducting saline solution, is applied, with a slight suction, to the plasma membrane. The 0.5-μm-diameter tip covers a region that contains only one or a few ion channels. The second electrode is inserted through the membrane into the cytosol. A recording device measures current flow only through the channels in the patch of plasma membrane. (b) Photomicrograph of the cell body of a cultured neuron and the tip of a patch pipette touching the cell membrane. (c) Different patch-clamping configurations. Isolated, detached patches are the best configurations for studying the effects on channels of different ion concentrations and solutes such as extracellular hormones and intracellular second messengers (e.g., cAMP). [Part (b) from B. Sakmann, 1992, *Neuron* **8**:613 (Nobel lecture); also published in E. Neher and B. Sakmann, 1992, *Sci. Am.* **266**(3):44. Part c adapted from B. Hille, 1992, *Ion Channels of Excitable Membranes*, 2d ed., Sinauer Associates, p. 89.]

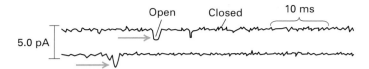

▲ **EXPERIMENTAL FIGURE 7-18 Ion flux through individual Na⁺ channels can be calculated from patch-clamp tracings.** Two inside-out patches of muscle plasma membrane were clamped at a potential of slightly less than that of the resting membrane potential. The patch electrode contained NaCl. The transient pulses of electric current in picoamperes (pA), recorded as large downward deviations (blue arrows), indicate the opening of a Na⁺ channel and movement of Na⁺ ions inward across the membrane. The smaller deviations in current represent background noise. The average current through an open channel is 1.6 pA, or 1.6×10^{-12} amperes. Since 1 ampere = 1 coulomb (C) of charge per second, this current is equivalent to the movement of about 9900 Na⁺ ions per channel per millisecond: $(1.6 \times 10^{-12}\ C/s)(10^{-3}\ s/ms)(6 \times 10^{23}\ molecules/mol) \div 96,500\ C/mol.$ [See F. J. Sigworth and E. Neher, 1980, *Nature* **287**:447.]

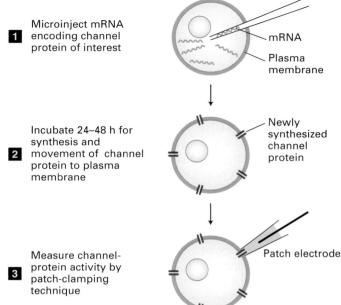

1 Microinject mRNA encoding channel protein of interest — mRNA, Plasma membrane

2 Incubate 24–48 h for synthesis and movement of channel protein to plasma membrane — Newly synthesized channel protein

3 Measure channel-protein activity by patch-clamping technique — Patch electrode

▲ **EXPERIMENTAL FIGURE 7-19 Oocyte expression assay is useful in comparing the function of normal and mutant forms of a channel protein.** A follicular frog oocyte is first treated with collagenase to remove the surrounding follicle cells, leaving a denuded oocyte, which is microinjected with mRNA encoding the channel protein under study. [Adapted from T. P. Smith, 1988, *Trends Neurosci.* **11**:250.]

The patch-clamp tracings in Figure 7-18 illustrate the use of this technique to study the properties of voltage-gated Na⁺ channels in the plasma membrane of muscle cells. As we discuss later, these channels normally are closed in resting muscle cells and open following nervous stimulation. Patches of muscle membrane, each containing one Na⁺ channel, were clamped at a voltage slightly less than the resting membrane potential. Under these circumstances, transient pulses of current cross the membrane as individual Na⁺ channels open and then close. Each channel is either fully open or completely closed. From such tracings, it is possible to determine the time that a channel is open and the ion flux through it. For the channels measured in Figure 7-18, the flux is about 10 million Na⁺ ions per channel per second, a typical value for ion channels. Replacement of the NaCl within the patch pipette (corresponding to the outside of the cell) with KCl or choline chloride abolishes current through the channels, confirming that they conduct only Na⁺ ions.

Novel Ion Channels Can Be Characterized by a Combination of Oocyte Expression and Patch Clamping

Cloning of human disease-causing genes and sequencing of the human genome have identified many genes encoding putative channel proteins, including 67 putative K⁺ channel proteins. One way of characterizing the function of these proteins is to transcribe a cloned cDNA in a cell-free system to produce the corresponding mRNA. Injection of this mRNA into frog oocytes and patch-clamp measurements on the newly synthesized channel protein can often reveal its function (Figure 7-19). This experimental approach is especially useful because frog oocytes normally do not express any channel proteins, so only the channel under study is

present in the membrane. In addition, because of the large size of frog oocytes, patch-clamping studies are technically easier to perform on them than on smaller cells.

 This approach has provided insight into the underlying defect in polycystic kidney disease, the most common single-gene disorder leading to kidney failure. Mutations in either of two proteins, PKD1 or PKD2, produce the clinical symptoms of polycystic kidney disease in which fluid-filled cysts accumulate throughout the organ. The amino acid sequence of PDK2 is consistent with its being an ion-channel protein, and it contains a conserved P segment. When expressed in oocytes, PDK2 mediates transport of Na⁺, K⁺, and Ca²⁺ ions. In contrast, the sequence of PKD1 differs substantially from that of channel proteins, and it has a long extracellular domain that probably binds to a component of the extracellular matrix. Coexpression of PKD1 with PKD2 in frog oocyte eggs modifies the cation-transporting activity of PDK2. These findings provided the first, albeit partial, molecular understanding of cyst formation characteristic of polycystic kidney disease and also suggest that some channel proteins may be regulated in complex ways. Indeed, most Na⁺ and K⁺ channel proteins are associated with other transmembrane or cytosolic proteins that are thought to regulate their opening, closing, or ion conductivity. ∎

Na$^+$ Entry into Mammalian Cells Has a Negative Change in Free Energy (ΔG)

As mentioned earlier, two forces govern the movement of ions across selectively permeable membranes: the voltage and the ion concentration gradient across the membrane. The sum of these forces, which may act in the same direction or in opposite directions, constitutes the electrochemical gradient. To calculate the **free-energy change ΔG** corresponding to the transport of any ion across a membrane, we need to consider the independent contributions from each of the forces to the electrochemical gradient.

For example, when Na$^+$ moves from outside to inside the cell, the free-energy change generated from the Na$^+$ concentration gradient is given by

$$\Delta G_c = RT \ln \frac{[\mathrm{Na_{in}}]}{[\mathrm{Na_{out}}]} \qquad (7\text{-}5)$$

At the concentrations of Na$_{in}$ and Na$_{out}$ shown in Figure 7-20, which are typical for many mammalian cells, ΔG_c, the change in free energy due to the concentration gradient, is -1.45 kcal for transport of 1 mol of Na$^+$ ions from outside to inside the cell, assuming there is no membrane electric potential. The free-energy change generated from the membrane electric potential is given by

$$\Delta G_m = FE \qquad (7\text{-}6)$$

where F is the Faraday constant and E is the membrane electric potential. If $E = -70$ mV, then ΔG_m, the free-energy change due to the membrane potential, is -1.61 kcal for transport of 1 mol of Na$^+$ ions from outside to inside the cell, assuming there is no Na$^+$ concentration gradient. Since both forces do in fact act on Na$^+$ ions, the total ΔG is the sum of the two partial values:

$$\Delta G = \Delta G_c + \Delta G_m = (-1.45) + (-1.61) = -3.06 \text{ kcal/mol}$$

In this example, the Na$^+$ concentration gradient and the membrane electric potential contribute almost equally to the total ΔG for transport of Na$^+$ ions. Since ΔG is <0, the inward movement of Na$^+$ ions is thermodynamically favored. As discussed in the next section, certain cotransport proteins use the inward movement of Na$^+$ to power the uphill movement of other ions and several types of small molecules into or out of animal cells. The rapid, energetically favorable movement of Na$^+$ ions through gated Na$^+$ channels also is critical in generating action potentials in nerve and muscle cells.

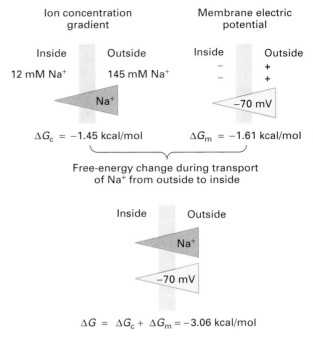

▲ **FIGURE 7-20 Transmembrane forces acting on Na$^+$ ions.** As with all ions, the movement of Na$^+$ ions across the plasma membrane is governed by the sum of two separate forces—the ion concentration gradient and the membrane electric potential. At the internal and external Na$^+$ concentrations typical of mammalian cells, these forces usually act in the same direction, making the inward movement of Na$^+$ ions energetically favorable.

KEY CONCEPTS OF SECTION 7.3

Nongated Ion Channels and the Resting Membrane Potential

■ An inside-negative electric potential (voltage) of 50–70 mV exists across the plasma membrane of all cells.

■ In animal cells, the membrane potential is generated primarily by movement of cytosolic K$^+$ ions through resting K$^+$ channels to the external medium. Unlike the more common gated ion channels, which open only in response to various signals, these nongated K$^+$ channels are usually open.

■ In plants and fungi, the membrane potential is maintained by the ATP-driven pumping of protons from the cytosol to the exterior of the cell.

■ K$^+$ channels are assembled from four identical subunits, each of which has at least two conserved membrane-spanning α helices and a nonhelical P segment that lines the ion pore (see Figure 7-15).

■ The ion specificity of K$^+$ channel proteins is due mainly to coordination of the selected ion with the carbonyl oxygen atoms of specific amino acids in the P segments, thus lowering the activation energy for passage of the selected K$^+$ compared with other ions (see Figure 7-16).

■ Patch-clamping techniques, which permit measurement of ion movements through single channels, are used to determine the ion conductivity of a channel and the effect of various signals on its activity (see Figure 7-17).

■ Recombinant DNA techniques and patch clamping allow the expression and functional characterization of channel proteins in frog oocytes (see Figure 7-19).

■ The electrochemical gradient across a semipermeable membrane determines the direction of ion movement through channel proteins. The two forces constituting the electrochemical gradient, the membrane electric potential and the ion concentration gradient, may act in the same or opposite directions (see Figure 7-20).

7.4 Cotransport by Symporters and Antiporters

Besides ATP-powered pumps, cells have a second, discrete class of proteins that transport ions and small molecules, such as glucose and amino acids, against a concentration gradient. As noted previously, cotransporters use the energy stored in the electrochemical gradient of Na^+ or H^+ ions to power the uphill movement of another substance, which may be a small organic molecule or a different ion. For instance, the energetically favored movement of a Na^+ ion (the cotransported ion) into a cell across the plasma membrane, driven both by its concentration gradient and by the transmembrane voltage gradient, can be coupled to movement of the transported molecule (e.g., glucose) against its concentration gradient. An important feature of such **cotransport** is that neither molecule can move alone; movement of both molecules together is obligatory, or *coupled*.

Cotransporters share some features with uniporters such as the GLUT proteins. The two types of transporters exhibit certain structural similarities, operate at equivalent rates, and undergo cyclical conformational changes during transport of their substrates. They differ in that uniporters can only accelerate thermodynamically favorable transport down a concentration gradient, whereas cotransporters can harness the energy of a coupled favorable reaction to actively transport molecules against a concentration gradient.

When the transported molecule and cotransported ion move in the same direction, the process is called **symport**; when they move in opposite directions, the process is called **antiport** (see Figure 7-2). Some cotransporters transport only positive ions (cations), while others transport only negative ions (anions). An important example of a cation cotransporter is the *Na^+/H^+ antiporter*, which exports H^+ from cells coupled to the energetically favorable import of Na^+. An example of an anion cotransporter is the *AE1 anion antiporter protein*, which catalyzes the one-for-one exchange of Cl^- and HCO_3^- across the plasma membrane. Yet other cotransporters mediate movement of both cations and anions together. In this section, we describe the operation and physiological role of several widely distributed symporters and antiporters.

Na⁺-Linked Symporters Import Amino Acids and Glucose into Animal Cells Against High Concentration Gradients

Most body cells import glucose from the blood down its concentration gradient, utilizing one or another GLUT protein to facilitate this transport. However, certain cells, such as those lining the small intestine and the kidney tubules, need to import glucose from the intestinal lumen or forming urine against a very large concentration gradient. Such cells utilize a *two-Na^+/one-glucose symporter*, a protein that couples import of one glucose molecule to the import of two Na^+ ions:

$$2\,Na^+_{\text{out}} + \text{glucose}_{\text{out}} \rightleftharpoons 2\,Na^+_{\text{in}} + \text{glucose}_{\text{in}}$$

Quantitatively, the free-energy change for the symport transport of two Na^+ ions and one glucose molecule can be written

$$\Delta G = RT \ln \frac{[\text{glucose}_{\text{in}}]}{[\text{glucose}_{\text{out}}]} + 2RT \ln \frac{[Na^+_{\text{in}}]}{[Na^+_{\text{out}}]} + 2FE \quad (7\text{-}7)$$

Thus the ΔG for the overall reaction is the sum of the free-energy changes generated by the glucose concentration gradient (1 molecule transported), the Na^+ concentration gradient (2 Na^+ ions transported), and the membrane potential (2 Na^+ ions transported). At equilibrium $\Delta G = 0$. As illustrated in Figure 7-20, the free energy released by movement of Na^+ into mammalian cells down its electrochemical gradient has a free-energy change ΔG of about -3 kcal per mole of Na^+ transported. Thus the ΔG for transport of two moles of Na^+ inward is about -6 kcal. By substituting this value into Equation 7-7 and setting $\Delta G = 0$, we see that

$$0 = RT \ln \frac{[\text{glucose}_{\text{in}}]}{[\text{glucose}_{\text{out}}]} - 6 \text{ kcal}$$

and we can calculate that at equilibrium the ratio $\text{glucose}_{\text{in}}/\text{glucose}_{\text{out}} = \approx 30{,}000$. Thus the inward flow of two moles of Na^+ can generate an intracellular glucose concentration that is $\approx 30{,}000$ times greater than the exterior concentration. If only one Na^+ ion were imported (ΔG of -3 kcal/mol) per glucose molecule, then the available energy could generate a glucose concentration gradient (inside > outside) of only about 170-fold. Thus by coupling the transport of two Na^+ ions to the transport of one glucose, the two-Na^+/one-glucose symporter permits cells to accumulate a very high concentration of glucose relative to the external concentration.

The two-Na^+/glucose symporter is thought to contain 14 transmembrane α helices with both its N- and C-termini extending into the cytosol. A truncated recombinant protein consisting of only the five C-terminal transmembrane α helices can transport glucose independently of Na^+ across the plasma membrane, *down* its concentration gradient. This portion of the molecule thus functions as a glucose uniporter. The N-terminal portion of the protein, including helices 1–9, is required to couple Na^+ binding and influx to the transport of glucose against a concentration gradient.

Figure 7-21 depicts the current model of transport by Na^+/glucose symporters. This model entails conformational changes in the protein analogous to those that occur in uniport transporters, such as GLUT1, which do not require a cotransported ion (see Figure 7-4). Binding of all substrates to their sites on the extracellular domain is required before

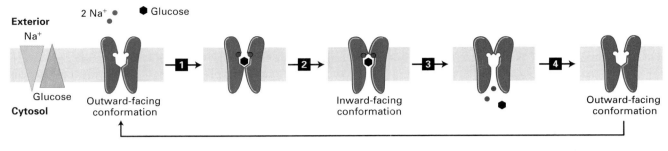

▲ FIGURE 7-21 Operational model for the two-Na⁺/one-glucose symporter. Simultaneous binding of Na⁺ and glucose to the conformation with outward-facing binding sites (step **1**) generates a second conformation with inward-facing sites (step **2**). Dissociation of the bound Na⁺ and glucose into the cytosol (step **3**) allows the protein to revert to its original outward-facing conformation (step **4**), ready to transport additional substrate. [See M. Panayotova-Heiermann et al., 1997, *J. Biol. Chem.* **272**:20324, for details on the structure and function of this and related transporters.]

the protein undergoes the conformational change that changes the substrate-binding sites from outward- to inward-facing; this ensures that inward transport of glucose and Na⁺ ions are coupled.

Na⁺-Linked Antiporter Exports Ca²⁺ from Cardiac Muscle Cells

In cardiac muscle cells a *three-Na⁺/one-Ca²⁺ antiporter,* rather than the plasma membrane Ca²⁺ ATPase discussed earlier, plays the principal role in maintaining a low concentration of Ca²⁺ in the cytosol. The transport reaction mediated by this *cation antiporter* can be written

$$3\ Na^+{}_{out} + Ca^{2+}{}_{in} \rightleftharpoons 3Na^+{}_{in} + Ca^{2+}{}_{out}$$

Note that the movement of three Na⁺ ions is required to power the export of one Ca²⁺ ion from the cytosol with a [Ca²⁺] of $\approx 2 \times 10^{-7}$ M to the extracellular medium with a [Ca²⁺] of 2×10^{-3} M, a gradient of some 10,000-fold. As in other muscle cells, a rise in the cytosolic Ca²⁺ concentration in cardiac muscle triggers contraction. By lowering cytosolic Ca²⁺, operation of the Na⁺/Ca²⁺ antiporter reduces the strength of heart muscle contraction.

MEDICINE The Na⁺/K⁺ ATPase in the plasma membrane of cardiac cells, as in other body cells, creates the Na⁺ concentration gradient necessary for export of Ca²⁺ by the Na⁺-linked Ca²⁺ antiporter. As mentioned earlier, inhibition of the Na⁺/K⁺ ATPase by the drugs ouabain and digoxin lowers the cytosolic K⁺ concentration and, more important, increases cytosolic Na⁺. The resulting reduced Na⁺ electrochemical gradient across the membrane causes the Na⁺-linked Ca²⁺ antiporter to function less efficiently. As a result, fewer Ca²⁺ ions are exported and the cytosolic Ca²⁺ concentration increases, causing the muscle to contract more strongly. Because of their ability to increase the force of heart muscle contractions, inhibitors of the Na⁺/K⁺ ATPase are widely used in the treatment of congestive heart failure. ∎

Several Cotransporters Regulate Cytosolic pH

The anaerobic metabolism of glucose yields lactic acid, and aerobic metabolism yields CO_2, which adds water to form carbonic acid (H_2CO_3). These weak acids dissociate, yielding H⁺ ions (protons); if these excess protons were not removed from cells, the cytosolic pH would drop precipitously, endangering cellular functions. Two types of cotransport proteins help remove some of the "excess" protons generated during metabolism in animal cells. One is a *Na⁺HCO₃⁻/Cl⁻ antiporter,* which imports one Na⁺ ion down its concentration gradient, together with one HCO₃⁻, in exchange for export of one Cl⁻ ion against its concentration gradient. The cytosolic enzyme *carbonic anhydrase* catalyzes dissociation of the imported HCO₃⁻ ions into CO_2 and an OH⁻ (hydroxyl) ion:

$$HCO_3{}^- \rightleftharpoons CO_2 + OH^-$$

The CO_2 diffuses out of the cell, and the OH⁻ ions combine with intracellular protons, forming water. Thus the overall action of this transporter is to consume cytosolic H⁺ ions, thereby raising the cytosolic pH. Also important in raising cytosolic pH is a *Na⁺/H⁺ antiporter,* which couples entry of one Na⁺ ion into the cell down its concentration gradient to the export of one H⁺ ion.

Under certain circumstances the cytosolic pH can rise beyond the normal range of 7.2–7.5. To cope with the excess OH⁻ ions associated with elevated pH, many animal cells utilize an *anion antiporter* that catalyzes the one-for-one exchange of HCO₃⁻ and Cl⁻ across the plasma membrane. At high pH, this *Cl⁻/HCO₃⁻ antiporter* exports HCO₃⁻ (which can be viewed as a "complex" of OH⁻ and CO_2) in exchange for Cl⁻, thus lowering the cytosolic pH. The import of Cl⁻ down its concentration gradient (Cl⁻ₘₑₐᵢᵤₘ > Cl⁻꜀ᵧₜₒₛₒₗ) powers the reaction.

The activity of all three of these antiport proteins depends on pH, providing cells with a fine-tuned mechanism

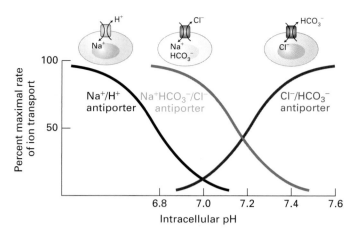

▲ EXPERIMENTAL FIGURE 7-22 The activity of membrane transport proteins that regulate the cytosolic pH of mammalian cells changes with pH. Direction of ion transport is indicated above the curve for each protein. See the text for discussion. [See S. L. Alper et al., 2001, *J. Pancreas* **2**:171, and S. L. Alper, 1991, *Ann. Rev. Physiol.* **53**:549.]

for controlling the cytosolic pH (Figure 7-22). The two antiporters that operate to increase cytosolic pH are activated when the pH of the cytosol falls. Similarly, a rise in pH above 7.2 stimulates the Cl^-/HCO_3^- antiporter, leading to a more rapid export of HCO_3^- and decrease in the cytosolic pH. In this manner the cytosolic pH of growing cells is maintained very close to pH 7.4.

Numerous Transport Proteins Enable Plant Vacuoles to Accumulate Metabolites and Ions

The lumen of plant vacuoles is much more acidic (pH 3 to 6) than is the cytosol (pH 7.5). The acidity of vacuoles is maintained by a V-class ATP-powered proton pump (see Figure 7-6) and by a PP_i-powered pump that is unique to plants. Both of these pumps, located in the vacuolar membrane, import H^+ ions into the vacuolar lumen against a concentration gradient. The vacuolar membrane also contains Cl^- and NO_3^- channels that transport these anions from the cytosol into the vacuole. Entry of these anions against their concentration gradients is driven by the inside-positive potential generated by the H^+ pumps. The combined operation of these proton pumps and anion channels produces an inside-positive electric potential of about 20 mV across the vacuolar membrane and also a substantial pH gradient (Figure 7-23).

The proton electrochemical gradient across the plant vacuole membrane is used in much the same way as the Na^+ electrochemical gradient across the animal-cell plasma membrane: to power the selective uptake or extrusion of ions and small molecules by various antiporters. In the leaf, for example, excess sucrose generated during photosynthesis in the day is stored in the vacuole; during the night the stored sucrose moves into the cytoplasm and is metabolized to CO_2 and H_2O with concomitant generation of ATP from ADP and P_i. A *proton/sucrose antiporter* in the vacuolar membrane operates to accumulate sucrose in plant vacuoles. The inward movement of sucrose is powered by the outward movement of H^+, which is favored by its concentration gradient (lumen > cytosol) and by the cytosolic-negative potential across the vacuolar membrane (see Figure 7-23). Uptake of Ca^{2+} and Na^+ into the vacuole from the cytosol against their concentration gradients is similarly mediated by proton antiporters. ∎

Understanding of the transport proteins in plant vacuolar membranes has the potential for increasing agricultural production in high-salt (NaCl) soils, which are found throughout the world. Because most agriculturally useful crops cannot grow in such saline soils, agricultural scientists have long sought to develop salt-tolerant plants by traditional breeding methods. With the availability of the cloned gene encoding the vacuolar Na^+/H^+ antiporter, researchers can now produce transgenic plants that overexpress this transport protein, leading to in-

▲ FIGURE 7-23 Concentration of ions and sucrose by the plant vacuole. The vacuolar membrane contains two types of proton pumps (orange): a V-class H^+ ATPase (*left*) and a pyrophosphate-hydrolyzing proton pump (*right*) that differs from all other ion-transport proteins and probably is unique to plants. These pumps generate a low luminal pH as well as an inside-positive electric potential across the vacuolar membrane owing to the inward pumping of H^+ ions. The inside-positive potential powers the movement of Cl^- and NO_3^- from the cytosol through separate channel proteins (purple). Proton antiporters (green), powered by the H^+ gradient, accumulate Na^+, Ca^{2+}, and sucrose inside the vacuole. [After P. Rea and D. Sanders, 1987, *Physiol. Plant* **71**:131; J. M. Maathuis and D. Sanders, 1992, *Curr. Opin. Cell Biol.* **4**:661; P. A. Rea et al., 1992, *Trends Biochem. Sci.* **17**:348.]

creased sequestration of Na^+ in the vacuole. For instance, transgenic tomato plants that overexpress the vacuolar Na^+/H^+ antiporter have been shown to grow, flower, and produce fruit in the presence of soil NaCl concentrations that kill wild-type plants. Interestingly, although the leaves of these transgenic tomato plants accumulate large amounts of salt, the fruit has a very low salt content. ∎

KEY CONCEPTS OF SECTION 7.4

Cotransport by Symporters and Antiporters

■ Cotransporters use the energy released by movement of an ion (usually H^+ or Na^+) down its electrochemical gradient to power the import or export of a small molecule or different ion against its concentration gradient.

■ The cells lining the small intestine and kidney tubules express symport proteins that couple the energetically favorable entry of Na^+ to the import of glucose and amino acids against their concentration gradients (see Figure 7-21).

■ In cardiac muscle cells, the export of Ca^{2+} is coupled to and powered by the import of Na^+ by a cation antiporter, which transports 3 Na^+ ions inward for each Ca^{2+} ion exported.

■ Two cotransporters that are activated at low pH help maintain the cytosolic pH in animal cells very close to 7.4 despite metabolic production of carbonic and lactic acids. One, a Na^+/H^+ antiporter, exports excess protons. The other, a $Na^+HCO_3^-/Cl^-$ cotransporter, imports HCO_3^-, which dissociates in the cytosol to yield pH-raising OH^- ions.

■ A Cl^-/HCO_3^- antiporter that is activated at high pH functions to export HCO_3^- when the cytosolic pH rises above normal, and causes a decrease in pH.

■ Uptake of sucrose, Na^+, Ca^{2+}, and other substances into plant vacuoles is carried out by proton antiporters in the vacuolar membrane. Ion channels and proton pumps in the membrane are critical in generating a large enough proton concentration gradient to power accumulation of ions and metabolites in vacuoles by these proton antiporters (see Figure 7-23).

7.5 Movement of Water

In this section, we describe the factors that influence the movement of water in and out of cells, an important feature of the life of both plants and animals. The following section discusses other transport phenomena that are critical to essential physiological processes, focusing on the asymmetrical distribution of certain transport proteins in epithelial cells. We will see how this permits absorption of nutrients from the intestinal lumen and acidification of the stomach lumen.

Osmotic Pressure Causes Water to Move Across Membranes

Water tends to move across a semipermeable membrane from a solution of low solute concentration to one of high concentration, a process termed **osmosis,** or osmotic flow. In other words, since solutions with a high concentration of dissolved solute have a lower concentration of water, water will spontaneously move from a solution of high water concentration to one of lower. In effect, osmosis is equivalent to "diffusion" of water. **Osmotic pressure** is defined as the hydrostatic pressure required to stop the net flow of water across a membrane separating solutions of different compositions (Figure 7-24). In this context, the "membrane" may be a layer of cells or a plasma membrane that is permeable to water but not to the solutes. The osmotic pressure is directly proportional to the difference in the concentration of the total number of solute molecules on each side of the membrane. For example, a 0.5 M NaCl solution is actually 0.5 M Na^+ ions and 0.5 M Cl^- ions and has the same osmotic pressure as a 1 M solution of glucose or sucrose.

Pure phospholipid bilayers are essentially impermeable to water, but most cellular membranes contain water-channel proteins that facilitate the rapid movement of water in and out of cells. Such movement of water across the epithelial layer lining the kidney tubules of vertebrates is responsible for concentrating the urine. If this did not happen, one would excrete several liters of urine a day! In higher plants, water and minerals are absorbed from the soil by the roots and move up the plant through conducting tubes (the xylem); water loss from the plant, mainly by evaporation

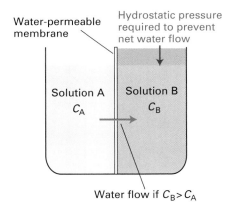

▲ **FIGURE 7-24 Osmotic pressure.** Solutions A and B are separated by a membrane that is permeable to water but impermeable to all solutes. If C_B (the total concentration of solutes in solution B) is greater than C_A, water will tend to flow across the membrane from solution A to solution B. The osmotic pressure π between the solutions is the hydrostatic pressure that would have to be applied to solution B to prevent this water flow. From the van't Hoff equation, osmotic pressure is given by $\pi = RT(C_B - C_A)$, where R is the gas constant and T is the absolute temperature.

from the leaves, drives these movements of water. The movement of water across the plasma membrane also determines the volume of individual cells, which must be regulated to avoid damage to the cell. In all cases, osmotic pressure is the force powering the movement of water in biological systems.

Different Cells Have Various Mechanisms for Controlling Cell Volume

When placed in a **hypotonic** solution (i.e., one in which the concentration of solutes is *lower* than in the cytosol), animal cells swell owing to the osmotic flow of water inward. Conversely, when placed in a **hypertonic** solution (i.e., one in which the concentration of solutes is *higher* than in the cytosol), animal cells shrink as cytosolic water leaves the cell by osmotic flow. Consequently, cultured animal cells must be maintained in an **isotonic** medium, which has a solute concentration identical with that of the cell cytosol (see Figure 5-18).

Even in an isotonic environment, however, animal cells face a problem in maintaining their cell volume within a limited range, thereby avoiding lysis. Not only do cells contain a large number of charged macromolecules and small metabolites, which attract oppositely charge ions from the exterior, but there also is a slow inward leakage of extracellular ions, particularly Na^+ and Cl^-, down their concentration gradients. In the absence of some countervailing mechanism, the osmolarity of the cytosol would increase beyond that of the surrounding fluid, causing an osmotic influx of water and eventual cell lysis. To prevent this, animal cells actively export inorganic ions. The net export of cations by the ATP-powered Na^+/K^+ pump (3 Na^+ out for 2 K^+ in) plays the major role

in this mechanism for preventing cell swelling. If cultured cells are treated with an inhibitor that prevents production of ATP, they swell and eventually burst, demonstrating the importance of active transport in maintaining cell volume.

 Unlike animal cells, plant, algal, fungal, and bacterial cells are surrounded by a rigid cell wall. Because of the cell wall, the osmotic influx of water that occurs when such cells are placed in a hypotonic solution (even pure water) leads to an increase in intracellular pressure but not in cell volume. In plant cells, the concentration of solutes (e.g., sugars and salts) usually is higher in the vacuole than in the cytosol, which in turn has a higher solute concentration than the extracellular space. The osmotic pressure, called *turgor pressure*, generated from the entry of water into the cytosol and then into the vacuole pushes the cytosol and the plasma membrane against the resistant cell wall. Cell elongation during growth occurs by a hormone-induced localized loosening of a region of the cell wall, followed by influx of water into the vacuole, increasing its size. ∎

Although most protozoans (like animal cells) do not have a rigid cell wall, many contain a contractile vacuole that permits them to avoid osmotic lysis. A contractile vacuole takes up water from the cytosol and, unlike a plant vacuole, periodically discharges its contents through fusion with the plasma membrane. Thus, even though water continuously enters the protozoan cell by osmotic flow, the contractile vacuole prevents too much water from accumulating in the cell and swelling it to the bursting point.

| 0.5 min | 1.5 min | 2.5 min | 3.5 min |

▲ **EXPERIMENTAL FIGURE 7-25 Expression of aquaporin by frog oocytes increases their permeability to water.** Frog oocytes, which normally do not express aquaporin, were microinjected with mRNA encoding aquaporin. These photographs show control oocytes (bottom cell in each panel) and microinjected oocytes (top cell in each panel) at the indicated times after transfer from an isotonic salt solution (0.1 mM) to a hypotonic salt solution (0.035 M). The volume of the control oocytes remained unchanged because they are poorly permeable to water. In contrast, the microinjected oocytes expressing aquaporin swelled because of an osmotic influx of water, indicating that aquaporin is a water-channel protein. [Courtesy of Gregory M. Preston and Peter Agre, Johns Hopkins University School of Medicine.]

Aquaporins Increase the Water Permeability of Cell Membranes

As just noted, small changes in extracellular osmotic strength cause most animal cells to swell or shrink rapidly. In contrast, frog oocytes and eggs do not swell when placed in pond water of very low osmotic strength even though their internal salt (mainly KCl) concentration is comparable to that of other cells ($\approx$150 mM KCl). These observations first led investigators to suspect that the plasma membranes of erythrocytes and other cell types, but not frog oocytes, contain water-channel proteins that accelerate the osmotic flow of water. The experimental results shown in Figure 7-25 demonstrate that the erythrocyte cell-surface protein known as *aquaporin* functions as a water channel.

In its functional form, aquaporin is a tetramer of identical 28-kDa subunits (Figure 7-26a). Each subunit contains six membrane-spanning α helices that form a central pore through which water moves (Figure 7-26b, c). At its center the $\approx$2-nm-long water-selective gate, or pore, is only 0.28 nm in diameter, which is only slightly larger than the diameter of a water molecule. The molecular sieving properties of the constriction are determined by several conserved hydrophilic amino acid residues whose side-chain and carbonyl groups extend into the middle of the channel. Several water molecules move simultaneously through the channel, each of which sequentially forms specific hydrogen bonds and displaces another water molecule downstream. The formation of hydrogen bonds between the oxygen atom of water and the amino groups of the side chains ensures that only water passes through the channel; even protons cannot pass through.

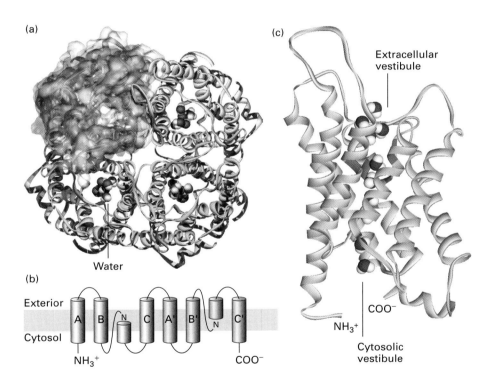

▲ **FIGURE 7-26 Structure of the water-channel protein aquaporin.** (a) Structural model of the tetrameric protein comprising four identical subunits. Each subunit forms a water channel, as seen in this end-on view from the exoplasmic surface. One of the monomers is shown with a molecular surface in which the pore entrance can be seen. (b) Schematic diagram of the topology of a single aquaporin subunit in relation to the membrane. Three pairs of homologous transmembrane α helices (A and A', B and B', and C and C') are oriented in the opposite direction with respect to the membrane and are connected by two hydrophilic loops containing short non-membrane-spanning helices and conserved asparagine (N) residues. The loops bend into the cavity formed by the six transmembrane helices, meeting in the middle to form part of the water-selective gate. (c) Side view of the pore in a single aquaporin subunit in which several water molecules (red oxygens and white hydrogens) are seen within the 2-nm-long water-selective gate that separates the water filled cytosolic and extracellular vestibules. The gate contains highly conserved arginine and histidine residues, as well as the two asparagine residues (blue) whose side chains form hydrogen bonds with transported waters. (Key gate residues are highlighted in blue.) Transported waters also form hydrogen bonds to the main-chain carbonyl group of a cysteine residue. The arrangement of these hydrogen bonds and the narrow pore diameter of 0.28 nm prevent passage of protons (i.e., H_3O^+) or other ions. [After H. Sui et al., 2001, *Nature* **414**:872. See also T. Zeuthen, 2001, *Trends Biochem. Sci.* **26**:77, and K. Murata et al., 2000, *Nature* **407**:599.]

As is the case for glucose transporters, mammals express a family of aquaporins. Aquaporin 1 is expressed in abundance in erythrocytes; the homologous aquaporin 2 is found in the kidney epithelial cells that resorb water from the urine. Inactivating mutations in both alleles of the aquaporin 2 gene cause diabetes insipidus, a disease marked by excretion of large volumes of dilute urine. This finding establishes the etiology of the disease and demonstrates that the level of aquaporin 2 is rate-limiting for water transport by the kidney. Other members of the aquaporin family transport hydroxyl-containing molecules such as glycerol rather than water. ∎

KEY CONCEPTS OF SECTION 7.5

Movement of Water

■ Most biological membranes are semipermeable, more permeable to water than to ions or most other solutes. Water moves by osmosis across membranes from a solution of lower solute concentration to one of higher solute concentration.

■ Animal cells swell or shrink when placed in hypotonic or hypertonic solutions, respectively. By maintaining the normal osmotic balance inside and outside cells, the Na^+/K^+ ATPase and other ion-transporting proteins in the plasma membrane control cell volume.

■ The rigid cell wall surrounding plant cells prevents their swelling and leads to generation of turgor pressure in response to the osmotic influx of water.

■ In response to the entry of water, protozoans maintain their normal cell volume by extruding water from contractile vacuoles.

■ Aquaporins are water-channel proteins that specifically increase the permeability of biomembranes for water (see Figure 7-26). Aquaporin 2 in the plasma membrane of certain kidney cells is essential for resorption of water from the forming urine; its absence leads to a form of diabetes.

7.6 Transepithelial Transport

We saw in Chapter 6 that several types of specialized regions of the plasma membrane, called *cell junctions*, connect epithelial cells forming the sheetlike lining of the intestines and other body surfaces (see Figure 6-5). Of concern to us here are **tight junctions,** which prevent water-soluble materials on one side of an epithelium from moving across to the other side through the extracellular space between cells. For this reason, absorption of nutrients from the intestinal lumen occurs by the import of molecules on the luminal side of intestinal epithelial cells and their export on the blood-

facing (serosal) side, a two-stage process called *transcellular transport*.

An intestinal epithelial cell, like all epithelial cells, is said to be **polarized** because the apical and basolateral domains of the plasma membrane contain different sets of proteins. These two plasma-membrane domains are separated by the tight junctions between cells. The apical portion of the plasma membrane, which faces the intestinal lumen, is specialized for absorption of sugars, amino acids, and other molecules that are produced from food by various digestive enzymes. Numerous fingerlike projections (100 nm in diameter) called **microvilli** greatly increase the area of the apical surface and thus the number of transport proteins it can contain, enhancing the cell's absorptive capacity.

Multiple Transport Proteins Are Needed to Move Glucose and Amino Acids Across Epithelia

Figure 7-27 depicts the proteins that mediate absorption of glucose from the intestinal lumen into the blood. In the first stage of this process, a two-Na^+/one-glucose symporter located in microvillar membranes imports glucose, against its concentration gradient, from the intestinal lumen across the apical surface of the epithelial cells. As noted above, this symporter couples the energetically unfavorable inward

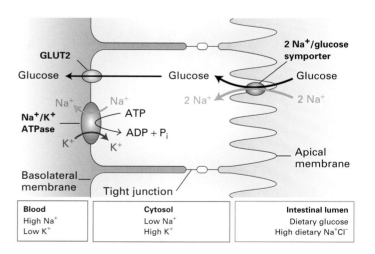

▲ **FIGURE 7-27 Transcellular transport of glucose from the intestinal lumen into the blood.** The Na^+/K^+ ATPase in the basolateral surface membrane generates Na^+ and K^+ concentration gradients. The outward movement of K^+ ions through nongated K^+ channels (not shown) generates an inside-negative membrane potential. Both the Na^+ concentration gradient and the membrane potential are used to drive the uptake of glucose from the intestinal lumen by the two-Na^+/one-glucose symporter located in the apical surface membrane. Glucose leaves the cell via facilitated diffusion catalyzed by GLUT2, a glucose uniporter located in the basolateral membrane.

movement of one glucose molecule to the energetically favorable inward transport of two Na^+ ions (see Figure 7-21). In the steady state, all the Na^+ ions transported from the intestinal lumen into the cell during Na^+/glucose symport, or the similar process of Na^+/amino acid symport, are pumped out across the basolateral membrane, which faces the underlying tissue. Thus the low intracellular Na^+ concentration is maintained. The Na^+/K^+ ATPase that accomplishes this is found exclusively in the basolateral membrane of intestinal epithelial cells. The coordinated operation of these two transport proteins allows uphill movement of glucose and amino acids from the intestine into the cell. This first stage in transcellular transport ultimately is powered by ATP hydrolysis by the Na^+/K^+ ATPase.

In the second stage, glucose and amino acids concentrated inside intestinal cells by symporters are exported down their concentration gradients into the blood via uniport proteins in the basolateral membrane. In the case of glucose, this movement is mediated by GLUT2 (see Figure 7-27). As noted earlier, this GLUT isoform has a relatively low affinity for glucose but increases its rate of transport substantially when the glucose gradient across the membrane rises (see Figure 7-3).

The net result of this two-stage process is movement of Na^+ ions, glucose, and amino acids from the intestinal lumen across the intestinal epithelium into the extracellular medium that surrounds the basolateral surface of intestinal epithelial cells. Tight junctions between the epithelial cells prevent these molecules from diffusing back into the intestinal lumen, and eventually they move into the blood. The increased osmotic pressure created by transcellular transport of salt, glucose, and amino acids across the intestinal epithelium draws water from the intestinal lumen into the extracellular medium that surrounds the basolateral surface. In a sense, salts, glucose, and amino acids "carry" the water along with them.

Simple Rehydration Therapy Depends on the Osmotic Gradient Created by Absorption of Glucose and Na^+

 An understanding of osmosis and the intestinal absorption of salt and glucose forms the basis for a simple therapy that saves millions of lives each year, particularly in less-developed countries. In these countries, cholera and other intestinal pathogens are major causes of death of young children. A toxin released by the bacteria activates chloride secretion by the intestinal epithelial cells into the lumen; water follows osmotically, and the resultant massive loss of water causes diarrhea, dehydration, and ultimately death. A cure demands not only killing the bacteria with antibiotics, but also *rehydration*—replacement of the water that is lost from the blood and other tissues.

Simply drinking water does not help, because it is excreted from the gastrointestinal tract almost as soon as it enters. However, as we have just learned, the coordinated transport of glucose and Na^+ across the intestinal epithelium creates a transepithelial osmotic gradient, forcing movement of water from the intestinal lumen across the cell layer. Thus, giving affected children a solution of sugar and salt to drink (but not sugar or salt alone) causes the osmotic flow of water into the blood from the intestinal lumen and leads to rehydration. Similar sugar/salt solutions are the basis of popular drinks used by athletes to get sugar as well as water into the body quickly and efficiently. ∎

Parietal Cells Acidify the Stomach Contents While Maintaining a Neutral Cytosolic pH

The mammalian stomach contains a 0.1 M solution of hydrochloric acid (HCl). This strongly acidic medium kills many ingested pathogens and denatures many ingested proteins before they are degraded by proteolytic enzymes (e.g., pepsin) that function at acidic pH. Hydrochloric acid is secreted into the stomach by specialized epithelial cells called *parietal cells* (also known as *oxyntic cells*) in the gastric lining. These cells contain a H^+/K^+ *ATPase* in their apical membrane, which faces the stomach lumen and generates a millionfold H^+ concentration gradient: pH = 1.0 in the stomach lumen versus pH = 7.0 in the cell cytosol. This transport protein is a P-class ATP-powered ion pump similar in structure and function to the plasma-membrane Na^+/K^+ ATPase discussed earlier. The numerous mitochondria in parietal cells produce abundant ATP for use by the H^+/K^+ ATPase.

If parietal cells simply exported H^+ ions in exchange for K^+ ions, the loss of protons would lead to a rise in the concentration of OH^- ions in the cytosol and thus a marked increase in cytosolic pH. (Recall that $[H^+] \times [OH^-]$ always is a constant, $10^{-14} M^2$.) Parietal cells avoid this rise in cytosolic pH in conjunction with acidification of the stomach lumen by using Cl^-/HCO_3^- antiporters in the basolateral membrane to export the "excess" OH^- ions in the cytosol into the blood. As noted earlier, this anion antiporter is activated at high cytosolic pH (see Figure 7-22).

The overall process by which parietal cells acidify the stomach lumen is illustrated in Figure 7-28. In a reaction catalyzed by carbonic anhydrase the "excess" cytosolic OH^- combines with CO_2 that diffuses in from the blood, forming HCO_3^-. Catalyzed by the basolateral anion antiporter, this bicarbonate ion is exported across the basolateral membrane (and ultimately into the blood) in exchange for a Cl^- ion. The Cl^- ions then exit through Cl^- channels in the apical membrane, entering the stomach lumen. To preserve electroneutrality, each Cl^- ion that moves into the stomach lumen across the apical membrane is accompanied by a K^+ ion that moves outward through a separate K^+ channel. In this way, the excess K^+ ions pumped inward by the H^+/K^+

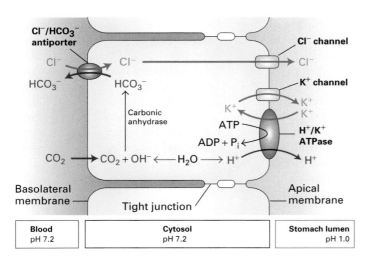

▲ **FIGURE 7-28 Acidification of the stomach lumen by parietal cells in the gastric lining.** The apical membrane of parietal cells contains an H⁺/K⁺ ATPase (a P-class pump) as well as Cl⁻ and K⁺ channel proteins. Note the cyclic K⁺ transport across the apical membrane: K⁺ ions are pumped inward by the H⁺/K⁺ ATPase and exit via a K⁺ channel. The basolateral membrane contains an anion antiporter that exchanges HCO_3^- and Cl⁻ ions. The combined operation of these four different transport proteins and carbonic anhydrase acidifies the stomach lumen while maintaining the neutral pH and electroneutrality of the cytosol. See the text for more details.

ATPase are returned to the stomach lumen, thus maintaining the normal intracellular K⁺ concentration. The net result is secretion of equal amounts of H⁺ and Cl⁻ ions (i.e., HCl) into the stomach lumen, while the pH of the cytosol remains neutral and the excess OH⁻ ions, as HCO_3^-, are transported into the blood.

KEY CONCEPTS OF SECTION 7.6

Transepithelial Transport

■ The apical and basolateral plasma membrane domains of epithelial cells contain different transport proteins and carry out quite different transport processes.

■ In the intestinal epithelial cell, the coordinated operation of Na⁺-linked symporters in the apical membrane with Na⁺/K⁺ ATPases and uniporters in the basolateral membrane mediates transcellular transport of amino acids and glucose from the intestinal lumen to the blood (see Figure 7-27).

■ The combined action of carbonic anhydrase and four different transport proteins permits parietal cells in the stomach lining to secrete HCl into the lumen while maintaining their cytosolic pH near neutrality (see Figure 7-28).

7.7 Voltage-Gated Ion Channels and the Propagation of Action Potentials in Nerve Cells

In the previous section, we examined how different transport proteins work together to absorb nutrients across the intestinal epithelium and to acidify the stomach. The nervous system, however, provides the most striking example of the interplay of various ion channels, transporters, and ion pumps in carrying out physiological functions. **Neurons** (nerve cells) and certain muscle cells are specialized to generate and conduct a particular type of electric impulse, the **action potential**. This alteration of the electric potential across the cell membrane is caused by the opening and closing of certain voltage-gated ion channels.

In this section, we first introduce some of the key properties of neurons and action potentials, which move down the axon very rapidly. We then describe how the voltage-gated channels responsible for propagating action potentials in neurons operate. In the following section, we examine how arrival of an action potential at the axon terminus causes secretion of chemicals called **neurotransmitters**. These chemicals, in turn, bind to receptors on adjacent cells and cause changes in the membrane potential of these cells. Thus electric signals carry information within a nerve cell, while chemical signals transmit information from one neuron to another or from a neuron to a muscle or other target cell.

Specialized Regions of Neurons Carry Out Different Functions

Although the morphology of various types of neurons differs in some respects, they all contain four distinct regions with differing functions: the cell body, the axon, the axon terminals, and the dendrites (Figure 7-29).

The *cell body* contains the nucleus and is the site of synthesis of virtually all neuronal proteins and membranes. Some proteins are synthesized in dendrites, but none are made in axons or axon terminals. Special transport processes involving microtubules move proteins and membranes from their sites of synthesis in the cell body down the length of the axon to the terminals (Chapter 20).

Axons, whose diameter varies from a micrometer in certain nerves of the human brain to a millimeter in the giant fiber of the squid, are specialized for conduction of action potentials. An action potential is a series of sudden changes in the voltage, or equivalently the electric potential, across the plasma membrane. When a neuron is in the resting (nonstimulated) state, the electric potential across the axonal membrane is approximately −60 mV (the inside negative relative to the outside); this *resting potential* is similar to that of the membrane potential in most non-neuronal cells. At the peak of an action potential, the membrane potential can be as much as +50 mV (inside positive), a net change of ≈110 mV.

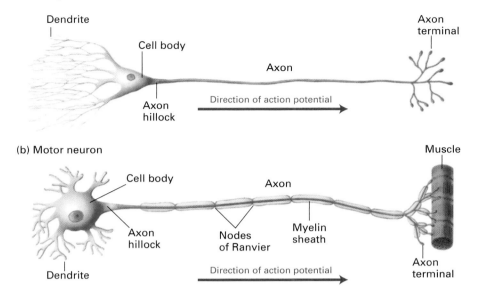

(a) Multipolar interneuron

Dendrite

Cell body

Axon hillock

Axon

Direction of action potential

Axon terminal

(b) Motor neuron

Cell body

Axon hillock

Axon

Nodes of Ranvier

Myelin sheath

Dendrite

Direction of action potential

Muscle

Axon terminal

◀ **FIGURE 7-29 Typical morphology of two types of mammalian neurons.** Action potentials arise in the axon hillock and are conducted toward the axon terminus. (a) A multipolar interneuron has profusely branched dendrites, which *receive* signals at synapses with several hundred other neurons. A single long axon that branches laterally at its terminus *transmits* signals to other neurons. (b) A motor neuron innervating a muscle cell typically has a single long axon extending from the cell body to the effector cell. In mammalian motor neurons, an insulating sheath of myelin usually covers all parts of the axon except at the nodes of Ranvier and the axon terminals.

This **depolarization** of the membrane is followed by a rapid repolarization, returning the membrane potential to the resting value (Figure 7-30).

An action potential originates at the *axon hillock,* the junction of the axon and cell body, and is actively conducted down the axon to the *axon terminals,* small branches of the axon that form the **synapses,** or connections, with other cells.

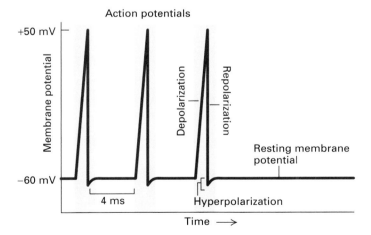

Action potentials

+50 mV

Membrane potential

Depolarization

Repolarization

Resting membrane potential

−60 mV

4 ms

Hyperpolarization

Time ⟶

▲ **EXPERIMENTAL FIGURE 7-30 Recording of an axonal membrane potential over time reveals the amplitude and frequency of action potentials.** An action potential is a sudden, transient depolarization of the membrane, followed by repolarization to the resting potential of about −60 mV. The axonal membrane potential can be measured with a small electrode placed into it (see Figure 7-14). This recording of the axonal membrane potential in this neuron shows that it is generating one action potential about every 4 milliseconds.

Action potentials move at speeds up to 100 meters per second. In humans, for instance, axons may be more than a meter long, yet it takes only a few milliseconds for an action potential to move along their length. Arrival of an action potential at an axon terminal leads to opening of voltage-sensitive Ca^{2+} channels and an influx of Ca^{2+}, causing a localized rise in the cytosolic Ca^{2+} concentration in the axon terminus. The rise in Ca^{2+} in turn triggers fusion of small vesicles containing neurotransmitters with the plasma membrane, releasing neurotransmitters from this *presynaptic cell* into the synaptic cleft, the narrow space separating it from *postsynaptic cells* (Figure 7-31).

It takes about 0.5 millisecond (ms) for neurotransmitters to diffuse across the synaptic cleft and bind to a receptor on the postsynaptic cells. Binding of neurotransmitter triggers opening or closing of specific ion channels in the plasma membrane of postsynaptic cells, leading to changes in the membrane potential at this point. A single axon in the central nervous system can synapse with many neurons and induce responses in all of them simultaneously.

Most neurons have multiple **dendrites,** which extend outward from the cell body and are specialized to receive chemical signals from the axon termini of other neurons. Dendrites convert these signals into small electric impulses and conduct them toward the cell body. Neuronal cell bodies can also form synapses and thus receive signals. Particularly in the central nervous system, neurons have extremely long dendrites with complex branches. This allows them to form synapses with and receive signals from a large number of other neurons, perhaps up to a thousand (see Figure 7-29a). Membrane depolarizations or hyperpolarizations generated in the dendrites or cell body spread to the axon hillock. If the membrane depolarization at that

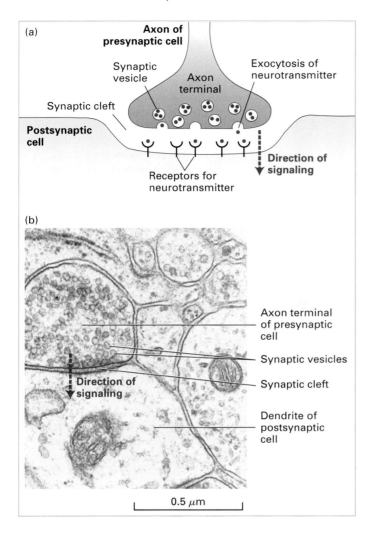

(a)

Axon of
presynaptic cell

Synaptic
vesicle

Axon
terminal

Exocytosis of
neurotransmitter

Synaptic cleft

**Postsynaptic
cell**

Receptors for
neurotransmitter

Direction of
signaling

(b)

Axon terminal
of presynaptic
cell

Synaptic vesicles

Synaptic cleft

Direction of
signaling

Dendrite of
postsynaptic
cell

0.5 μm

▲ **FIGURE 7-31 A chemical synapse.** (a) A narrow region—the
synaptic cleft—separates the plasma membranes of the
presynaptic and postsynaptic cells. Arrival of action potentials at a
synapse causes release of neurotransmitters (red circles) by the
presynaptic cell, their diffusion across the synaptic cleft, and their
binding by specific receptors on the plasma membrane of the
postsynaptic cell. Generally these signals depolarize the postsynaptic
membrane (making the potential inside less negative), tending to
induce an action potential in it. (b) Electron micrograph shows a
dendrite synapsing with an axon terminal filled with synaptic
vesicles. In the synaptic region, the plasma membrane of the
presynaptic cell is specialized for vesicle exocytosis; synaptic
vesicles containing a neurotransmitter are clustered in these regions.
The opposing membrane of the postsynaptic cell (in this case, a
neuron) contains receptors for the neurotransmitter. [Part (b) from
C. Raine et al., eds., 1981, *Basic Neurochemistry,* 3d ed., Little, Brown, p. 32.]

point is great enough, an action potential will originate and
will be actively conducted down the axon.

Thus neurons use changes in the membrane potential, the
action potentials, to conduct signals along their length, and
small molecules, the neurotransmitters, to send signals from
cell to cell.

Magnitude of the Action Potential Is Close to E_{Na}

Earlier in this chapter we saw how operation of the Na^+/K^+
pump generates a high concentration of K^+ and a low concentration of Na^+ in the cytosol, relative to those in the extracellular medium. The subsequent outward movement of
K^+ ions through nongated K^+ channels is driven by the K^+
concentration gradient (cytosol > medium), generating the
resting membrane potential. The entry of Na^+ ions into the
cytosol from the medium also is thermodynamically favored,
driven by the Na^+ concentration gradient (medium >
cytosol) and the inside-negative membrane potential (see
Figure 7-20). However, most Na^+ channels in the plasma
membrane are closed in resting cells, so little inward movement of Na^+ ions can occur (Figure 7-32a).

If enough Na^+ channels open, the resulting influx of Na^+
ions will overwhelm the efflux of K^+ ions though open resting K^+ channels. The result would be a *net* inward movement of cations, generating an excess of positive charges on
the cytosolic face and a corresponding excess of negative
charges (due to the Cl^- ions "left behind" in the extracellular medium after influx of Na^+ ions) on the extracellular face
(Figure 7-32b). In other words, the plasma membrane is
depolarized to such an extent that the inside face becomes
positive.

The magnitude of the membrane potential at the peak of
depolarization in an action potential is very close to the Na^+

(a) Resting state (cytosolic face negative)

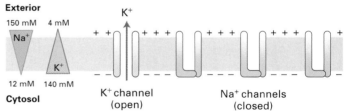

(b) Depolarized state (cytosolic face positive)

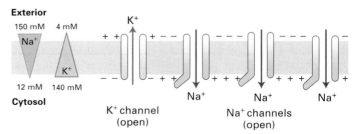

▲ **FIGURE 7-32 Depolarization of the plasma membrane
due to opening of gated Na^+ channels.** (a) In resting neurons,
nongated K^+ channels are open, but the more numerous gated
Na^+ channels are closed. The movement of K^+ ions outward
establishes the inside-negative membrane potential characteristic
of most cells. (b) Opening of gated Na^+ channels permits an
influx of sufficient Na^+ ions to cause a reversal of the membrane
potential. See text for details.

equilibrium potential E_{Na} given by the Nernst equation (Equation 7-2), as would be expected if opening of voltage-gated Na^+ channels is responsible for generating action potentials. For example, the measured peak value of the action potential for the squid giant axon is 35 mV, which is close to the calculated value of E_{Na} (55 mV) based on Na^+ concentrations of 440 mM outside and 50 mM inside. The relationship between the magnitude of the action potential and the concentration of Na^+ ions inside and outside the cell has been confirmed experimentally. For instance, if the concentration of Na^+ ions in the solution bathing the squid axon is reduced to one-third of normal, the magnitude of the depolarization is reduced by 40 mV, nearly as predicted.

Sequential Opening and Closing of Voltage-Gated Na^+ and K^+ Channels Generate Action Potentials

The cycle of membrane depolarization, hyperpolarization, and return to the resting value that constitutes an action potential lasts 1–2 milliseconds and can occur hundreds of times a second in a typical neuron (see Figure 7-30). These cyclical changes in the membrane potential result from the sequential opening and closing first of *voltage-gated Na^+ channels* and then of *voltage-gated K^+ channels*. The role of these channels in the generation of action potentials was elucidated in classic studies done on the giant axon of the squid, in which multiple microelectrodes can be inserted without causing damage to the integrity of the plasma membrane. However, the same basic mechanism is used by all neurons.

Voltage-Gated Na^+ Channels As just discussed, voltage-gated Na^+ channels are closed in resting neurons. A small depolarization of the membrane causes a conformational change in these channel proteins that opens a gate on the cytosolic surface of the pore, permitting Na^+ ions to pass through the pore into the cell. The greater the initial membrane depolarization, the more voltage-gated Na^+ channels that open and the more Na^+ ions enter.

As Na^+ ions flow inward through opened channels, the excess positive charges on the cytosolic face and negative charges on the exoplasmic face diffuse a short distance away from the initial site of depolarization. This *passive spread* of positive and negative charges depolarizes (makes the inside less negative) adjacent segments of the plasma membrane, causing opening of additional voltage-gated Na^+ channels in these segments and an increase in Na^+ influx. As more Na^+ ions enter the cell, the inside of the cell membrane becomes more depolarized, causing the opening of yet more voltage-gated Na^+ channels and even more membrane depolarization, setting into motion an explosive entry of Na^+ ions. For a fraction of a millisecond, the permeability of this region of the membrane to Na^+ becomes vastly greater than that for K^+, and the membrane potential approaches E_{Na}, the equilibrium potential for a membrane permeable only to Na^+ ions. As the membrane potential approaches E_{Na}, however, further net inward movement of Na^+ ions ceases, since the concentration gradient of Na^+ ions (outside > inside) is now offset by the inside-positive membrane potential E_{Na}. The action potential is at its peak, close to the value of E_{Na}.

Figure 7-33 schematically depicts the critical structural features of voltage-gated Na^+ channels and the conformational changes that cause their opening and closing. In the resting state, a segment of the protein on the cytosolic face—the *"gate"*—obstructs the central pore, preventing passage of ions. A small depolarization of the membrane triggers movement of positively charged *voltage-sensing* α helices toward the exoplasmic surface, causing a conformational change in the gate that opens the channel and allows ion flow. After about 1 millisecond, further Na^+ influx is prevented by movement of the cytosol-facing *channel-inactivating segment* into the open channel. As long as the membrane remains depolarized, the channel-inactivating segment remains in the channel opening; during this *refractory period*, the channel is inactivated and cannot be reopened. A few milliseconds after the inside-negative resting potential is reestablished, the channel-inactivating segment swings away from the pore and the channel returns to the closed resting state, once again able to be opened by depolarization.

Voltage-Gated K^+ Channels The repolarization of the membrane that occurs during the refractory period is due largely to opening of voltage-gated K^+ channels. The subsequent increased efflux of K^+ from the cytosol removes the excess positive charges from the cytosolic face of the plasma membrane (i.e., makes it more negative), thereby restoring the inside-negative resting potential. Actually, for a brief instant the membrane becomes hyperpolarized, with the potential approaching E_K, which is more negative than the resting potential (see Figure 7-30).

Opening of the voltage-gated K^+ channels is induced by the large depolarization of the action potential. Unlike voltage-gated Na^+ channels, most types of voltage-gated K^+ channels remain open as long as the membrane is depolarized, and close only when the membrane potential has returned to an inside-negative value. Because the voltage-gated K^+ channels open slightly after the initial depolarization, at the height of the action potential, they sometimes are called *delayed K^+ channels*. Eventually all the voltage-gated K^+ and Na^+ channels return to their closed resting state. The only open channels in this baseline condition are the nongated K^+ channels that generate the resting membrane potential, which soon returns to its usual value (see Figure 7-32a).

The patch-clamp tracings in Figure 7-34 reveal the essential properties of voltage-gated K^+ channels. In this experiment, small segments of a neuronal plasma membrane were held "clamped" at different potentials, and the flux of electric charges through the patch due to flow of K^+ ions through open K^+ channels was measured. At the depolarizing voltage of −10 mV, the channels in the membrane patch open infrequently and remain open for only a few milliseconds, as judged, respectively, by the number and width of the "upward blips" on the tracings. Further, the ion flux through

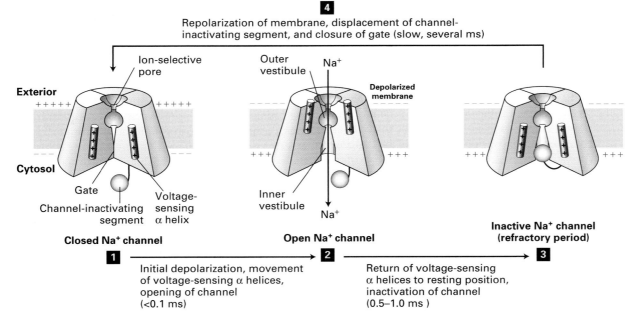

4 Repolarization of membrane, displacement of channel-inactivating segment, and closure of gate (slow, several ms)

Ion-selective pore

Outer vestibule

Na⁺

Exterior

+ + + + +   + + + + +

Depolarized membrane

Cytosol

+ + +   + + +

Gate

Channel-inactivating segment

Voltage-sensing α helix

Inner vestibule

Na⁺

Closed Na⁺ channel

Open Na⁺ channel

Inactive Na⁺ channel (refractory period)

1 Initial depolarization, movement of voltage-sensing α helices, opening of channel (<0.1 ms)

2 Return of voltage-sensing α helices to resting position, inactivation of channel (0.5–1.0 ms)

3

▲ **FIGURE 7-33 Operational model of the voltage-gated Na⁺ channel.** Four transmembrane domains in the protein contribute to the central pore through which ions move. The critical components that control movement of Na⁺ ions are shown here in the cutaway views depicting three of the four transmembrane domains. **1** In the closed, resting state, the voltage-sensing α helices, which have positively charged side chains every third residue, are attracted to the negative charges on the cytosolic side of the resting membrane. This keeps the gate segment in a position that blocks the channel, preventing entry of Na⁺ ions. **2** In response to a small depolarization, the voltage-sensing helices rotate in a screwlike manner toward the outer membrane surface, causing an immediate conformational change in the gate segment that opens the channel. **3** The voltage-sensing helices rapidly return to the resting position and the channel-inactivating segment moves into the open channel, preventing passage of further ions. **4** Once the membrane is repolarized, the channel-inactivating segment is displaced from the channel opening and the gate closes; the protein reverts to the closed, resting state and can be opened again by depolarization. [See W. A. Catterall, 2001, *Nature* **409**:988; M. Zhou et al., 2001, *Nature* **411**:657; and B. A. Yi and L. Y. Jan, 2000, *Neuron* **27**:423.]

them is rather small, as measured by the electric current passing through each open channel (the height of the blips). Depolarizing the membrane further to +20 mV causes these channels to open about twice as frequently. Also, more K⁺ ions move through each open channel (the height of the blips is greater) because the force driving cytosolic K⁺ ions outward is greater at a membrane potential of +20 mV than at −10 mV. Depolarizing the membrane further to +50 mV, the value at the peak of an action potential, causes opening of more K⁺ channels and also increases the flux of K⁺ through them. Thus, by opening during the peak of the action potential, these K⁺ channels permit the outward movement of K⁺

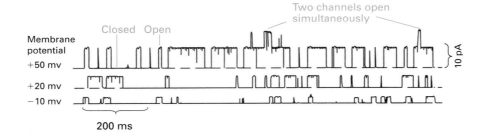

Two channels open simultaneously

Membrane potential

Closed   Open

+50 mv

+20 mv

−10 mv

10 pA

200 ms

▲ **EXPERIMENTAL FIGURE 7-34 Probability of channel opening and current flux through individual voltage-gated K⁺ channels increases with the extent of membrane depolarization.** These patch-clamp tracings were obtained from patches of neuronal plasma membrane clamped at three different potentials, +50, +20, and −10 mV. The upward deviations in the current indicate the opening of K⁺ channels and movement of K⁺ ions outward (cytosolic to exoplasmic face) across the membrane. Increasing the membrane depolarization (i.e., the clamping voltage) from −10 mV to +50 mV increases the probability a channel will open, the time it stays open, and the amount of electric current (numbers of ions) that pass through it. [From B. Pallota et al., 1981, *Nature* **293**:471, as modified by B. Hille, 1992, *Ion Channels of Excitable Membranes,* 2d ed., Sinauer Associates, p. 122.]

ions and repolarization of the membrane potential while the voltage-gated Na$^+$ channels are closed and inactivated.

More than 100 voltage-gated K$^+$ channel proteins have been isolated from humans and other vertebrates. As we discuss later, all these channel proteins have a similar overall structure, but they exhibit different voltage dependencies, conductivities, channel kinetics, and other functional properties. However, many open only at strongly depolarizing voltages, a property required for generation of the maximal depolarization characteristic of the action potential before repolarization of the membrane begins.

Action Potentials Are Propagated Unidirectionally Without Diminution

The generation of an action potential just described relates to the changes that occur in a small patch of the neuronal plasma membrane. At the peak of the action potential, passive spread of the membrane depolarization is sufficient to depolarize a neighboring segment of membrane. This causes a few voltage-gated Na$^+$ channels in this region to open, thereby increasing the extent of depolarization in this region and causing an explosive opening of more Na$^+$ channels and generation of an action potential. This depolarization soon triggers opening of voltage-gated K$^+$ channels and restoration of the resting potential. The action potential thus spreads as a traveling wave away from its initial site without diminution.

As noted earlier, during the refractory period voltage-gated Na$^+$ channels are inactivated for several milliseconds. Such previously opened channels cannot open during this period even if the membrane is depolarized due to passive spread. As illustrated in Figure 7-35, the inability of Na$^+$ channels to reopen during the refractory period ensures that

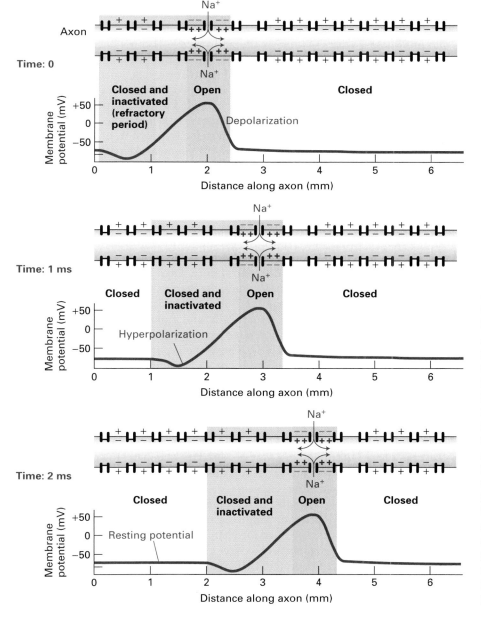

◀ **FIGURE 7-35 Unidirectional conduction of an action potential due to transient inactivation of voltage-gated Na$^+$ channels.** At time 0, an action potential (red) is at the 2-mm position on the axon; the Na$^+$ channels at this position are open and Na$^+$ ions are flowing inward. The excess Na$^+$ ions diffuse in both directions along the inside of the membrane, passively spreading the depolarization. Because the Na$^+$ channels at the 1-mm position are still inactivated (green), they cannot yet be reopened by the small depolarization caused by passive spread; the Na$^+$ channels at the 3-mm position, in contrast, begin to open. Each region of the membrane is refractory (inactive) for a few milliseconds after an action potential has passed. Thus, the depolarization at the 2-mm site at time 0 triggers action potentials downstream only; at 1 ms an action potential is passing the 3-mm position, and at 2 ms, an action potential is passing the 4-mm position.

action potentials are propagated only in one direction, from the axon hillock where they originate to the axon terminus. This property of the Na^+ channels also limits the number of action potentials per second that a neuron can conduct. Reopening of Na^+ channels "upstream" of an action potential (i.e., closer to the cell body) also is delayed by the membrane hyperpolarization that results from opening of voltage-gated K^+ channels.

Nerve Cells Can Conduct Many Action Potentials in the Absence of ATP

It is important to note that the depolarization of the membrane characteristic of an action potential results from movement of just a small number of Na^+ ions into a neuron and does not significantly affect the intracellular Na^+ concentration gradient. A typical nerve cell has about 10 voltage-gated Na^+ channels per square micrometer (μm^2) of plasma membrane. Since each channel passes $\approx 5000–10,000$ ions during the millisecond it is open (see Figure 7-18), a maximum of 10^5 ions per μm^2 of plasma membrane will move inward during each action potential.

To assess the effect of this ion flux on the cytosolic Na^+ concentration of 10 mM (0.01 mol/L), typical of a resting axon, we focus on a segment of axon 1 micrometer (μm) long and 10 μm in diameter. The volume of this segment is 78 μm^3 or 7.8×10^{-14} liters, and it contains 4.7×10^8 Na^+ ions: $(10^{-2} \text{ mol/L}) (7.8 \times 10^{-14} \text{ L}) (6 \times 10^{23} \text{ } Na^+/\text{mol})$. The surface area of this segment of the axon is 31 μm^2, and during passage of one action potential, 10^5 Na^+ ions will enter per μm^2 of membrane. Thus this Na^+ influx increases the number of Na^+ ions in this segment by only one part in about 150: $(4.7 \times 10^8) \div (3.1 \times 10^6)$. Likewise, the repolarization of the membrane due to the efflux of K^+ ions through voltage-gated K^+ channels does not significantly change the intracellular K^+ concentration.

Because so few Na^+ and K^+ ions move across the plasma membrane during each action potential, the Na^+/K^+ pump that maintains the usual ion gradients plays no direct role in impulse conduction. When this pump is experimentally inhibited by dinitrophenol or another inhibitor of ATP production, the concentrations of Na^+ and K^+ gradually become the same inside and outside the cell, and the membrane potential falls to zero. This elimination of the usual Na^+ and K^+ electrochemical gradients occurs extremely slowly in large squid neurons but takes about 5 minutes in smaller mammalian neurons. In either case, the electrochemical gradients are essentially independent of the supply of ATP over the short time spans required for nerve cells to generate and conduct action potentials. Since the ion movements during each action potential involve only a minute fraction of the cell's K^+ and Na^+ ions, a nerve cell can fire hundreds or even thousands of times in the absence of ATP.

All Voltage-Gated Ion Channels Have Similar Structures

Having explained how the action potential is dependent on regulated opening and closing of voltage-gated channels, we turn to a molecular dissection of these remarkable proteins. After describing the basic structure of these channels, we focus on three questions:

■ How do these proteins sense changes in membrane potential?

■ How is this change transduced into opening of the channel?

■ What causes these channels to become inactivated shortly after opening?

The initial breakthrough in understanding voltage-gated ion channels came from analysis of fruit flies (*Drosophila melanogaster*) carrying the *shaker* mutation. These flies shake vigorously under ether anesthesia, reflecting a loss of motor control and a defect in certain motor neurons that have an abnormally prolonged action potential. This phenotype suggested that the *shaker* mutation causes a defect in voltage-gated K^+ channels that prevents them from opening normally immediately upon depolarization. To show that the wild-type *shaker* gene encoded a K^+ channel, cloned wild-type *shaker* cDNA was used as a template to produce *shaker* mRNA in a cell-free system. Expression of this mRNA in frog oocytes and patch-clamp measurements on the newly synthesized channel protein showed that its functional properties were identical to those of the voltage-gated K^+ channel in the neuronal membrane, demonstrating conclusively that the *shaker* gene encodes this K^+-channel protein.

The Shaker K^+ channel and most other voltage-gated K^+ channels that have been identified are tetrameric proteins composed of four identical subunits arranged in the membrane around a central pore. Each subunit is constructed of six membrane-spanning α helices, designated S1–S6, and a P segment (Figure 7-36a). The S5 and S6 helices and the P segment are structurally and functionally homologous to those in the nongated resting K^+ channel discussed earlier (see Figure 7-15). The S4 helix, which contains numerous positively charged lysine and arginine residues, acts as a voltage sensor; the N-terminal "ball" extending into the cytosol from S1 is the channel-inactivating segment.

Voltage-gated Na^+ channels and Ca^{2+} channels are monomeric proteins organized into four homologous domains, I–IV (Figure 7-36b). Each of these domains is similar to a subunit of a voltage-gated K^+ channel. However, in contrast to voltage-gated K^+ channels, which have four channel-inactivating segments, the monomeric voltage-gated channels have a single channel-inactivating segment. Except for this minor structural difference and their varying ion permeabilities, all voltage-gated ion channels are thought to function in a similar manner and to have evolved from a monomeric ancestral channel protein that contained six transmembrane α helices.

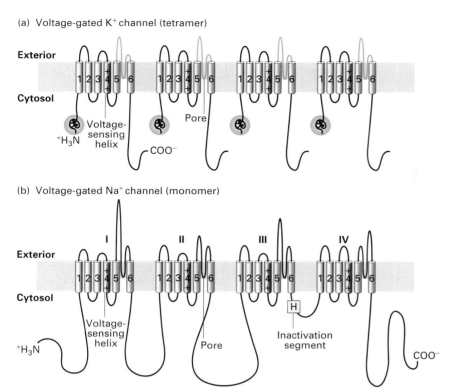

(a) Voltage-gated K⁺ channel (tetramer)

Exterior

Cytosol

^+H_3N Voltage-
sensing
helix

Pore

COO⁻

(b) Voltage-gated Na⁺ channel (monomer)

I   II   III   IV

Exterior

Cytosol

^+H_3N Voltage-
sensing
helix

Pore

Inactivation
segment

COO⁻

◀ FIGURE 7-36 Schematic depictions of the secondary structures of voltage-gated K⁺ and Na⁺ channels. (a) Voltage-gated K⁺ channels are composed of four identical subunits, each containing 600–700 amino acids, and six membrane-spanning α helices, S1–S6. The N-terminus of each subunit, located in the cytosol, forms a globular domain (orange ball) essential for inactivation of the open channel. The S5 and S6 helices (green) and the P segment (blue) are homologous to those in nongated resting K⁺ channels, but each subunit contains four additional transmembrane α helices. One of these, S4 (red), is the voltage-sensing α helix. (b) Voltage-gated Na⁺ channels are monomers containing 1800–2000 amino acids organized into four trans-membrane domains (I–IV) that are similar to the subunits in voltage-gated K⁺ channels. The single channel-inactivating segment, located in the cytosol between domains III and IV, contains a conserved hydrophobic motif (H). Voltage-gated Ca²⁺ channels have a similar overall structure. Most voltage-gated ion channels also contain regulatory (β) subunits that are not depicted here. [Part (a) adapted from C. Miller, 1992, *Curr. Biol.* **2**:573, and H. Larsson et al., 1996, *Neuron* **16**:387. Part (b) adapted from W. A. Catterall, 2001, *Nature* **409**:988.]

Voltage-Sensing S4 α Helices Move in Response to Membrane Depolarization

Sensitive electric measurements suggest that the opening of a voltage-gated Na⁺ or K⁺ channel is accompanied by the movement of 10 to 12 protein-bound positive charges from the cytosolic to the exoplasmic surface of the membrane; alternatively, a larger number of charges may move a shorter distance across the membrane. The movement of these gating charges (or voltage sensors) under the force of the electric field triggers a conformational change in the protein that opens the channel. The four voltage-sensing S4 transmembrane α helices, often called *gating helices*, are present in all voltage-gated channels and generally have a positively charged lysine or arginine every third or fourth residue. In the closed resting state, the C-terminal half of each S4 helix is exposed to the cytosol; when the membrane is depolarized, these amino acids move outward, probably rotate 180°, and become exposed to the exoplasmic surface of the channel (see Figure 7-33).

Studies with mutant Shaker K⁺ channels support this model for operation of the S4 helix in voltage sensing. When one or more arginine or lysine residues in the S4 helix of the Shaker K⁺ channel were replaced with neutral or acidic residues, fewer positive charges than normal moved across the membrane in response to a membrane depolarization, indicating that arginine and lysine residues in the S4 helix do indeed move across the membrane. In other studies, mutant Shaker proteins in which various S4 residues were converted to cysteine were tested for their reactivity with a water-soluble cysteine-modifying chemical agent that cannot cross the membrane. On the basis of whether the cysteines reacted with the agent added to one side or other of the membrane, the results indicated that in the resting state amino acids near the C-terminus of the S4 helix face the cytosol; after the membrane is depolarized, some of these same amino acids become exposed to the exoplasmic surface of the channel. These experiments directly demonstrate movement of the S4 helix across the membrane, as schematically depicted in Figure 7-33 for voltage-gated Na⁺ channels.

The gate itself is thought to comprise the cytosolic facing N-termini of the four S5 helixes and the C-termini of the four S6 helixes (see Figure 7-36). By analogy to the bacterial K⁺ channel, the ends of these helixes most likely come together just below the actual ion-selectivity filter, thereby blocking the pore in the closed state (see Figure 7-15). X-ray crystallographic studies of bacterial channels indicate that in the open state, the ends of these helixes have moved outward, leaving a hole in the middle sufficient for ions to move through. Thus the gates probably are composed of the cytosol-facing ends of the S5 and S6 helixes. However, since the molecular structure of no voltage-gated channel has been determined to date, we do not yet know how this putative conformational change is induced by movements of the voltage-sensing S4 helices.

Movement of the Channel-Inactivating Segment into the Open Pore Blocks Ion Flow

An important characteristic of most voltage-gated channels is inactivation; that is, soon after opening they close spontaneously, forming an inactive channel that will not reopen until the membrane is repolarized. In the resting state, the positively charged globular balls at the N-termini of the four subunits in a voltage-gated K^+ channel are free in the cytosol. Several milliseconds after the channel is opened by depolarization, one ball moves through an opening ("lateral window") between two of the subunits and binds in a hydrophobic pocket in the pore's central cavity, blocking the flow of K^+ ions (Figure 7-37). After a few milliseconds, the ball is displaced from the pore, and the protein reverts to the closed, resting state. The ball-and-chain domains in K^+ channels are functionally equivalent to the channel-inactivating segment in Na^+ channels.

The experimental results shown in Figure 7-38 demonstrate that inactivation of K^+ channels depends on the ball domains, occurs after channel opening, and does not require the ball domains to be covalently linked to the channel protein. In other experiments, mutant K^+ channels lacking portions of the ≈40-residue chain connecting the ball to the S1 helix were expressed in frog oocytes. Patch-clamp measurements of channel activity showed that the shorter the chain,

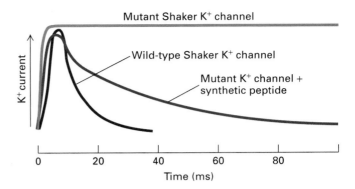

▲ **EXPERIMENTAL FIGURE 7-38 Experiments with a mutant K^+ channel lacking the N-terminal globular domains support the ball-and-chain inactivation model.** The wild-type Shaker K^+ channel and a mutant form lacking the amino acids composing the N-terminal ball were expressed in *Xenopus* oocytes. The activity of the channels then was monitored by the patch-clamp technique. When patches were depolarized from −0 to +30 mV, the wild-type channel opened for ≈5 ms and then closed (red curve), whereas the mutant channel opened normally, but could not close (green curve). When a chemically synthesized ball peptide was added to the cytosolic face of the patch, the mutant channel opened normally and then closed (blue curve). This demonstrated that the added peptide inactivated the channel after it opened and that the ball does not have to be tethered to the protein in order to function. [From W. N. Zagotta et al., 1990, *Science* **250**:568.]

the more rapid the inactivation, as if a ball attached to a shorter chain can move into the open channel more readily. Conversely, addition of random amino acids to lengthen the normal chain slows channel inactivation.

The single channel-inactivating segment in voltage-gated Na^+ channels contains a conserved hydrophobic motif composed of isoleucine, phenylalanine, methionine, and threonine (see Figure 7-36b). Like the longer ball-and-chain domain in K^+ channels, this segment folds into and blocks the Na^+-conducting pore until the membrane is repolarized (see Figure 7-33).

Myelination Increases the Velocity of Impulse Conduction

As we have seen, action potentials can move down an axon without diminution at speeds up to 1 meter per second. But even such fast speeds are insufficient to permit the complex movements typical of animals. In humans, for instance, the cell bodies of motor neurons innervating leg muscles are located in the spinal cord, and the axons are about a meter in length. The coordinated muscle contractions required for walking, running, and similar movements would be impossible if it took one second for an action potential to move from the spinal cord down the axon of a motor neuron to a leg muscle. The presence of a **myelin sheath** around an axon increases the velocity of impulse conduction to 10–100 meters

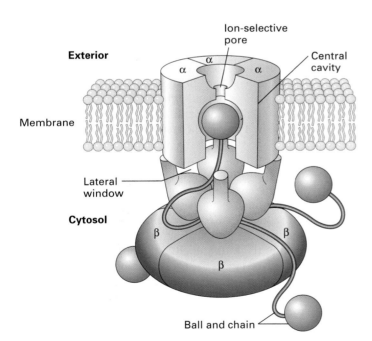

▲ **FIGURE 7-37 Ball-and-chain model for inactivation of voltage-gated K^+ channels.** Three-dimensional cutaway view of the channel in the inactive state. The globular domain (green ball) at the terminus of one subunit has moved through a lateral window to block the central pore. In addition to the four α subunits (orange) that form the channel, these channel proteins also have four regulatory β subunits (blue). See text for discussion. [Adapted from R. Aldrich, 2001, *Nature* **411**:643, and M. Zhou et al., 2001, *Nature* **411**:657.]

per second. As a result, in a typical human motor neuron, an action potential can travel the length of a 1-meter-long axon and stimulate a muscle to contract within 0.01 seconds.

In nonmyelinated neurons, the conduction velocity of an action potential is roughly proportional to the diameter of the axon, because a thicker axon will have a greater number of ions that can diffuse. The human brain is packed with relatively small, myelinated neurons. If the neurons in the human brain were not myelinated, their axonal diameters would have to increase about 10,000-fold to achieve the same conduction velocities as myelinated neurons. Thus vertebrate brains, with their densely packed neurons, never could have evolved without myelin.

The myelin sheath is a stack of specialized plasma membrane sheets produced by a glial cell that wraps itself around the axon (Figure 7-39). In the peripheral nervous system,

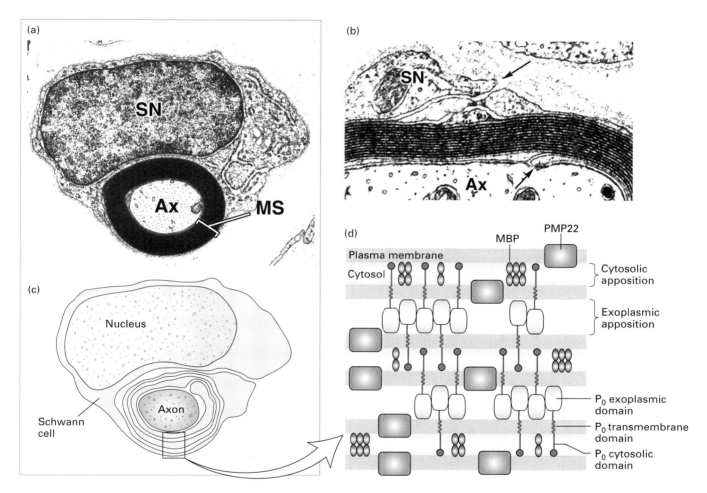

▲ **FIGURE 7-39 Formation and structure of a myelin sheath in the peripheral nervous system.** Myelinated axons are surrounded by an insulating layer of compressed membranes. (a) Electron micrograph of a cross section through an axon (Ax) surrounded by a myelin sheath (MS) and Schwann cell (SN). (b) At higher magnification this specialized spiral membrane appears as a series of layers, or lamellae, of phospholipid bilayers. In this image, the termini of both the outer and innermost wraps are evident (arrows). (c) As a Schwann cell repeatedly wraps around an axon, all the spaces between its plasma membranes, both cytosolic and exoplasmic, are reduced. Eventually all the cytosol is forced out and a structure of compact stacked plasma membranes, the myelin sheath, is formed. (d) The two most abundant membrane proteins, P_0 and PMP22, in peripheral myelin are expressed only by Schwann cells. The exoplasmic domain of a P_0 protein, which has an immunoglobulin fold, associates with similar domains emanating from P_0 proteins in the opposite membrane surface, thereby "zippering" together the exoplasmic membrane surfaces in close apposition. These interactions are stabilized by binding of a tryptophan residue on the tip of the exoplasmic domain to lipids in the opposite membrane. Close apposition of the cytosolic faces of the membrane may result from binding of the cytosolic tail of each P_0 protein to phospholipids in the opposite membrane. PMP22 may also contribute to membrane compaction. Myelin basic protein (MBP), a cytosolic protein, remains between the closely apposed membranes as the cytosol is squeezed out. [Parts (a) and (b) courtesy Grahame Kidd, Lerner Research Institute. Part (d) adapted from L. Shapiro et al., 1996, *Neuron* **17**:435, and E. J. Arroyo and S. S. Scherer, 2000, *Histochem. Cell Biol.* **113**:1.]

these glial cells are called *Schwann cells*. In the central nervous system, they are called *oligodendrocytes*. In both vertebrates and some invertebrates, glial cells accompany axons along their length, but specialization of these glial cells to form myelin occurs predominantly in vertebrates. Vertebrate glial cells that will later form myelin have on their surface a myelin-associated glycoprotein and other proteins that bind to adjacent axons and trigger the formation of myelin.

Figure 7-39d illustrates the formation and basic structure of a myelin sheath, which contains both membrane and cytosolic components. A myelin membrane, like all biomembranes, has a basic phospholipid bilayer structure, but it contains far fewer types of proteins than found in most other membranes. Two proteins predominate in the myelin membrane around peripheral axons: P_0, which causes adjacent plasma membranes to stack tightly together, and *PMP22*. Gene knockout studies in mice have recently identified PMP22 as essential for myelination. In the central nervous system, a different membrane protein and a proteolipid together function similarly to P_0. The major cytosolic protein in all myelin sheaths is *myelin basic protein* (MBP). Mice that contain the *shiverer* mutation in the MBP gene exhibit severe neurological problems, evidence for the importance of myelination in the normal functioning of the nervous system.

Action Potentials "Jump" from Node to Node in Myelinated Axons

The myelin sheath surrounding an axon is formed from many glial cells. Each region of myelin formed by an individual glial cell is separated from the next region by an unmyelinated area of axonal membrane about 1 μm in length called the *node of Ranvier* (or simply, node). The axonal membrane is in direct contact with the extracellular fluid only at the nodes. Moreover, all the voltage-gated Na$^+$ channels and all the Na$^+$/K$^+$ pumps, which maintain the ionic gradients in the axon, are located in the nodes.

As a consequence of this localization, the inward movement of Na$^+$ ions that generates the action potential can occur only at the myelin-free nodes (Figure 7-40). The excess cytosolic positive ions generated at a node during the membrane depolarization associated with an action potential spread passively through the axonal cytosol to the next node with very little loss or attenuation, since they cannot cross the myelinated axonal membrane. This causes a depolarization at one node to spread rapidly to the next node, permitting, in effect, the action potential to "jump" from node to node. This phenomenon explains why the conduction velocity of myelinated neurons is about the same as that of much larger diameter unmyelinated neurons. For instance, a 12-μm-diameter myelinated vertebrate axon and a 600-μm-diameter unmyelinated squid axon both conduct impulses at 12 m/s.

Several factors contribute to the clustering of voltage-gated Na$^+$ channels and Na$^+$/K$^+$ pumps at the nodes of

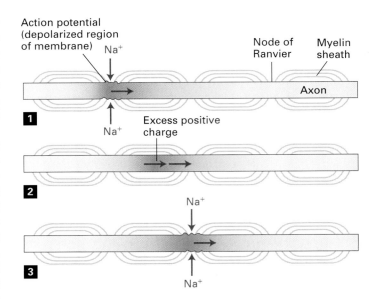

▲ **FIGURE 7-40 Conduction of action potentials in myelinated axons.** Because voltage-gated Na$^+$ channels are localized to the axonal membrane at the nodes of Ranvier, the influx of Na$^+$ ions associated with an action potential can occur only at nodes. When an action potential is generated at one node (step **1**), the excess positive ions in the cytosol, which cannot move outward across the sheath, diffuse rapidly down the axon, causing sufficient depolarization at the next node (step **2**) to induce an action potential at that node (step **3**). By this mechanism the action potential jumps from node to node along the axon.

Ranvier. Both of these transport proteins interact with two cytoskeletal proteins, ankyrin and spectrin, similar to those in the erythrocyte membrane (see Figure 5-31). The extracellular domain of the β1 subunit of the Na$^+$ channel also binds to the extracellular domain of Nr-CAM, a type of adhesive protein that is localized to the node. As a result of these multiple protein-protein interactions, the concentration of Na$^+$ channels is roughly a hundredfold higher in the nodal membrane of myelinated axons than in the axonal membrane of nonmyelinated neurons. In addition, glial cells secrete protein hormones that somehow trigger the clustering of these nerve membrane proteins at the nodes. Finally, tight junctions between the axon and the glial cell plasma membrane in the paranodal junctions immediately adjacent to the nodes may prevent diffusion of Na$^+$ channels and Na$^+$/K$^+$ pumps away from the nodes.

 One prevalent neurological disease among human adults is multiple sclerosis (MS), usually characterized by spasms and weakness in one or more limbs, bladder dysfunction, local sensory losses, and visual disturbances. This disorder—the prototype *demyelinating disease*—is caused by patchy loss of myelin in areas of the brain and spinal cord. In MS patients, conduction of action potentials by the demyelinated neurons is slowed, and the

Na$^+$ channels spread outward from the nodes, lowering their nodal concentration. The cause of the disease is not known but appears to involve either the body's production of auto-antibodies (antibodies that bind to normal body proteins) that react with myelin basic protein or the secretion of proteases that destroy myelin proteins. ∎

KEY CONCEPTS OF SECTION 7.7

Voltage-Gated Ion Channels and the Propagation of Action Potentials in Nerve Cells

■ Action potentials are sudden membrane depolarizations followed by a rapid repolarization. They originate at the axon hillock and move down the axon toward the axon terminals, where the electric impulse is transmitted to other cells via a synapse (see Figures 7-29 and 7-31).

■ An action potential results from the sequential opening and closing of voltage-gated Na$^+$ and K$^+$ channels in the plasma membrane of neurons and muscle cells.

■ Opening of voltage-gated Na$^+$ channels permits influx of Na$^+$ ions for about 1 ms, causing a sudden large depolarization of a segment of the membrane. The channels then close and become unable to open (refractory) for several milliseconds, preventing further Na$^+$ flow (see Figure 7-33).

■ As the action potential reaches its peak, opening of voltage-gated K$^+$ channels permits efflux of K$^+$ ions, which repolarizes and then hyperpolarizes the membrane. As these channels close, the membrane returns to its resting potential (see Figure 7-30).

■ The excess cytosolic cations associated with an action potential generated at one point on an axon spread passively to the adjacent segment, triggering opening of voltage-gated Na$^+$ channels and movement of the action potential along the axon.

■ Because of the absolute refractory period of the voltage-gated Na$^+$ channels and the brief hyperpolarization resulting from K$^+$ efflux, the action potential is propagated in one direction only, toward the axon terminus.

■ Voltage-gated Na$^+$ and Ca^{2+} channels are monomeric proteins containing four domains that are structurally and functionally similar to each of the subunits in the tetrameric voltage-gated K$^+$ channels.

■ Each domain or subunit in voltage-gated cation channels contains six transmembrane α helices and a nonhelical P segment that forms the ion-selectivity pore (see Figure 7-36).

■ Opening of voltage-gated channels results from outward movement of the positively charged S4 α helices in response to a depolarization of sufficient magnitude.

■ Closing and inactivation of voltage-gated cation chan-nels result from movement of a cytosolic segment into the open pore (see Figure 7-37).

■ Myelination, which increases the rate of impulse conduction up to a hundredfold, permits the close packing of neurons characteristic of vertebrate brains.

■ In myelinated neurons, voltage-gated Na$^+$ channels are concentrated at the nodes of Ranvier. Depolarization at one node spreads rapidly with little attenuation to the next node, so that the action potential "jumps" from node to node (see Figure 7-40).

7.8 Neurotransmitters and Receptor and Transport Proteins in Signal Transmission at Synapses

As noted earlier, synapses are the junctions where neurons release a chemical neurotransmitter that acts on a postsynaptic target cell, which can be another neuron or a muscle or gland cell (see Figure 7-31). In this section, we focus on several key issues related to impulse transmission:

■ How neurotransmitters are packaged in membrane-bounded *synaptic vesicles* in the axon terminus

■ How arrival of an action potential at axon termini in presynaptic cells triggers secretion of neurotransmitters

■ How binding of neurotransmitters by receptors on postsynaptic cells leads to changes in their membrane potential

■ How neurotransmitters are removed from the synaptic cleft after stimulating postsynaptic cells

Neurotransmitter receptors fall into two broad classes: ligand-gated ion channels, which open immediately upon neurotransmitter binding, and G protein–coupled receptors. Neurotransmitter binding to a G protein–coupled receptor induces the opening or closing of a *separate* ion channel protein over a period of seconds to minutes. These "slow" neurotransmitter receptors are discussed in Chapter 13 along with G protein–coupled receptors that bind different types of ligands and modulate the activity of cytosolic proteins other than ion channels. Here we examine the structure and operation of the *nicotinic acetylcholine receptor* found at many nerve-muscle synapses. The first ligand-gated ion channel to be purified, cloned, and characterized at the molecular level, this receptor provides a paradigm for other neurotransmitter-gated ion channels.

Neurotransmitters Are Transported into Synaptic Vesicles by H$^+$-Linked Antiport Proteins

Numerous small molecules function as neurotransmitters at various synapses. With the exception of **acetylcholine,** the

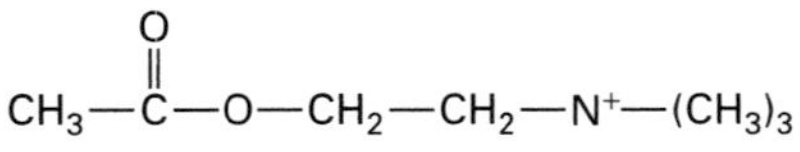

$$CH_3—\overset{\overset{O}{\|}}{C}—O—CH_2—CH_2—N^+—(CH_3)_3$$

Acetylcholine

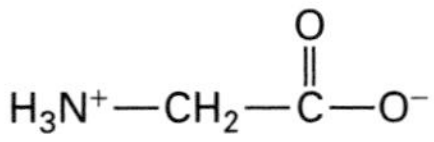

$$H_3N^+—CH_2—\overset{\overset{O}{\|}}{C}—O^-$$

Glycine

$$H_3N^+—\underset{\underset{|}{\overset{|}{C=O}}}{CH}—CH_2—CH_2—\overset{\overset{O}{\|}}{C}—O^-$$

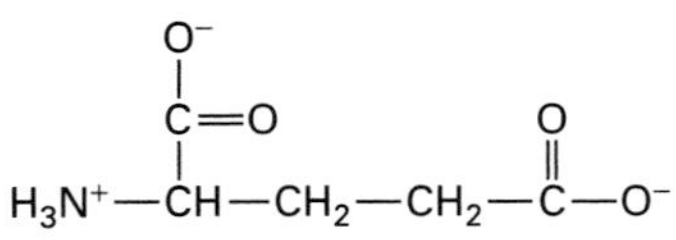

Glutamate

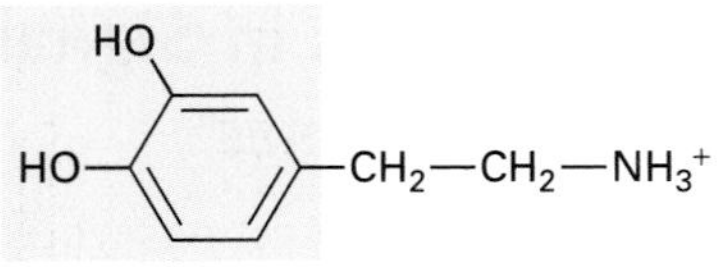

Dopamine
(derived from tyrosine)

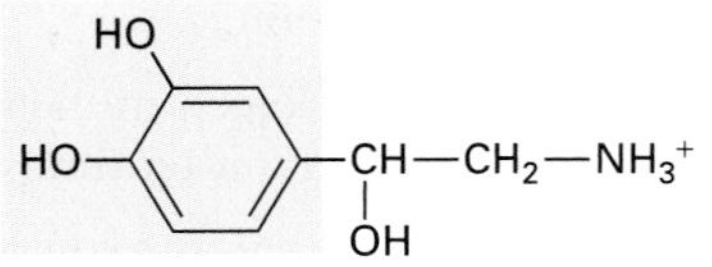

Norepinephrine
(derived from tyrosine)

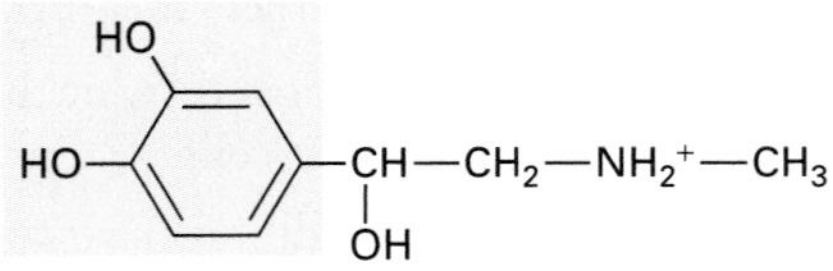

Epinephrine
(derived from tyrosine)

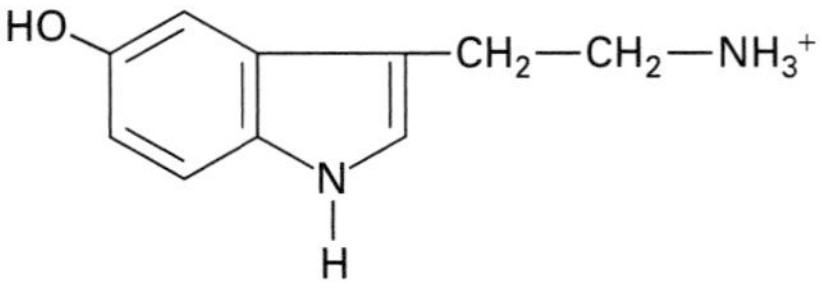

Serotonin, or **5-hydroxytryptamine**
(derived from tryptophan)

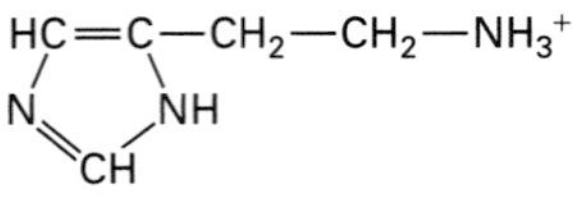

Histamine
(derived from histidine)

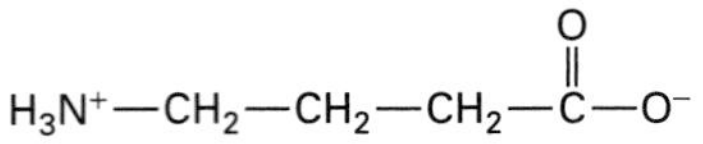

$$H_3N^+—CH_2—CH_2—CH_2—\overset{\overset{O}{\|}}{C}—O^-$$

γ-Aminobutyric acid, or **GABA**
(derived from glutamate)

neurotransmitters shown in Figure 7-41 are amino acids or derivatives of amino acids. Nucleotides such as ATP and the corresponding nucleosides, which lack phosphate groups, also function as neurotransmitters. Each neuron generally produces just one type of neurotransmitter.

All the "classic" neurotransmitters are synthesized in the cytosol and imported into membrane-bound synaptic vesicles within axon terminals, where they are stored. These vesicles are 40–50 nm in diameter, and their lumen has a low pH, generated by operation of a V-class proton pump in the vesicle membrane. Similar to the accumulation of metabolites in plant vacuoles (see Figure 7-23), this proton concentration gradient (vesicle lumen > cytosol) powers neurotransmitter import by ligand-specific H^+-linked antiporters in the vesicle membrane.

For example, acetylcholine is synthesized from acetyl coenzyme A (acetyl CoA), an intermediate in the degradation of glucose and fatty acids, and choline in a reaction catalyzed by choline acetyltransferase:

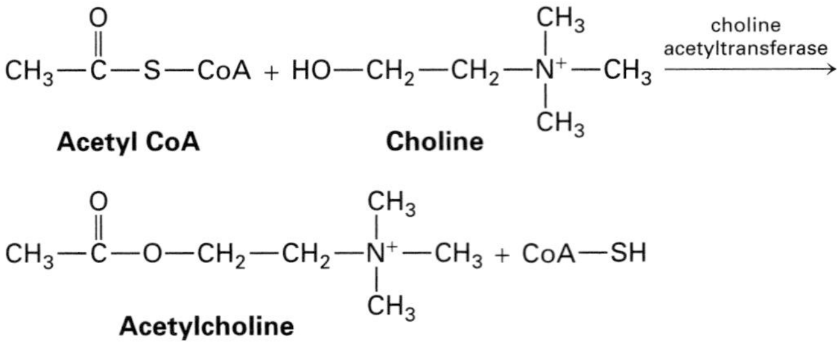

Synaptic vesicles take up and concentrate acetylcholine from the cytosol against a steep concentration gradient, using an H^+/acetylcholine antiporter in the vesicle membrane. Curiously, the gene encoding this antiporter is contained entirely within the first intron of the gene encoding choline acetyltransferase, a mechanism conserved throughout evolution for ensuring coordinate expression of these two proteins. Different H^+/neurotransmitter antiport proteins are used for import of other neurotransmitters into synaptic vesicles.

Influx of Ca^{2+} Through Voltage-Gated Ca^{2+} Channels Triggers Release of Neurotransmitters

Neurotransmitters are released by **exocytosis,** a process in which neurotransmitter-filled synaptic vesicles fuse with the axonal membrane, releasing their contents into the synaptic cleft. The exocytosis of neurotransmitters from synaptic vesicles involves vesicle-targeting and fusion events similar to those that occur during the intracellular transport of secreted and plasma-membrane proteins (Chapter 17). Two features

◀ **FIGURE 7-41 Structures of several small molecules that function as neurotransmitters.** Except for acetylcholine, all these are amino acids (glycine and glutamate) or derived from the indicated amino acids. The three transmitters synthesized from tyrosine, which contain the catechol moiety (blue highlight), are referred to as catecholamines.

critical to synapse function differ from other secretory pathways: (a) secretion is tightly coupled to arrival of an action potential at the axon terminus, and (b) synaptic vesicles are recycled locally to the axon terminus after fusion with the plasma membrane. Figure 7-42 shows the entire cycle whereby synaptic vesicles are filled with neurotransmitter, release their contents, and are recycled.

Depolarization of the plasma membrane cannot, by itself, cause synaptic vesicles to fuse with the plasma membrane. In order to trigger vesicle fusion, an action potential must be converted into a chemical signal—namely, a localized rise in the cytosolic Ca^{2+} concentration. The transducers of the electric signals are *voltage-gated Ca^{2+} channels* localized to the region of the plasma membrane adjacent to the synaptic vesicles. The membrane depolarization due to arrival of an action potential opens these channels, permitting an influx of Ca^{2+} ions from the extracellular medium into the axon terminal. This ion flux raises the local cytosolic Ca^{2+} concentration near the synaptic vesicles from <0.1 μM, characteristic of the resting state, to $1–100$ μM. Binding of Ca^{2+} ions to proteins that connect the synaptic vesicle with the plasma membrane induces membrane fusion and thus exocytosis of the neurotransmitter. The subsequent export of extra Ca^{2+} ions by ATP-powered Ca^{2+} pumps in the plasma membrane rapidly lowers the cytosolic Ca^{2+} level to that of the resting state, enabling the axon terminus to respond to the arrival of another action potential.

A simple experiment demonstrates the importance of voltage-gated Ca^{2+} channels in release of neurotransmitters. A preparation of neurons in a Ca^{2+}-containing medium is treated with tetrodotoxin, a drug that blocks voltage-gated Na^+ channels and thus prevents conduction of action

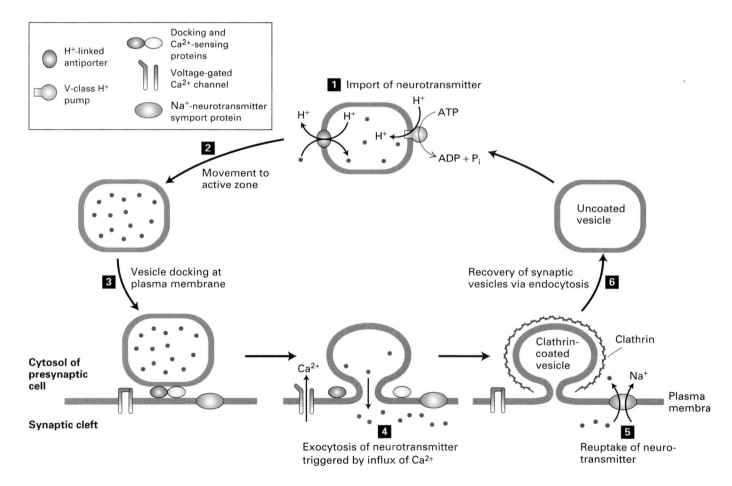

▲ **FIGURE 7-42 Cycling of neurotransmitters and of synaptic vesicles in axon terminals.** The entire cycle depicted here typically takes about 60 seconds. Note that several transport proteins participate in the filling of synaptic vesicles with neurotransmitter (red circles), its release by exocytosis, and subsequent reuptake from the synaptic cleft. Once synaptic-vesicle membrane proteins (e.g., pumps, antiporters, and fusion proteins needed for exocytosis) are specifically recovered by endocytosis in clathrin-coated vesicles, the clathrin coat is depolymerized, yielding vesicles that can be filled with neurotransmitter. Unlike most neurotransmitters, acetylcholine is not recycled. See text for details. [See T. Südhof and R. Jahn, 1991, *Neuron* **6**:665; K. Takei et al., 1996, *J. Cell. Biol.* **133**:1237; and V. Murthy and C. Stevens, 1998, *Nature* **392**:497.]

potentials. As expected, no neurotransmitters are secreted into the culture medium. If the axonal membrane then is artificially depolarized by making the medium ≈100 mM KCl, neurotransmitters are released from the cells because of the influx of Ca^{2+} through open voltage-gated Ca^{2+} channels. Indeed, patch-clamping experiments show that voltage-gated Ca^{2+} channels, like voltage-gated Na^+ channels, open transiently upon depolarization of the membrane.

Two pools of neurotransmitter-filled synaptic vesicles are present in axon terminals: those "docked" at the plasma membrane, which can be readily exocytosed, and those in reserve in the *active zone* near the plasma membrane. Each rise in Ca^{2+} triggers exocytosis of about 10 percent of the docked vesicles. Membrane proteins unique to synaptic vesicles then are specifically internalized by **endocytosis,** usually via the same types of clathrin-coated vesicles used to recover other plasma-membrane proteins by other types of cells. After the endocytosed vesicles lose their clathrin coat, they are rapidly refilled with neurotransmitter. The ability of many neurons to fire 50 times a second is clear evidence that the recycling of vesicle membrane proteins occurs quite rapidly.

Signaling at Synapses Usually Is Terminated by Degradation or Reuptake of Neurotransmitters

Following their release from a presynaptic cell, neurotransmitters must be removed or destroyed to prevent continued stimulation of the postsynaptic cell. Signaling can be terminated by diffusion of a transmitter away from the synaptic cleft, but this is a slow process. Instead, one of two more rapid mechanisms terminates the action of neurotransmitters at most synapses.

Signaling by acetylcholine is terminated when it is hydrolyzed to acetate and choline by *acetylcholinesterase,* an enzyme localized to the synaptic cleft. Choline released in this reaction is transported back into the presynaptic axon terminal by a Na^+/choline symporter and used in synthesis of more acetylcholine. The operation of this transporter is similar to that of the Na^+-linked symporters used to transport glucose into cells against a concentration gradient (see Figure 7-21).

With the exception of acetylcholine, all the neurotransmitters shown in Figure 7-41 are removed from the synaptic cleft by transport into the axon terminals that released them. Thus these transmitters are recycled intact, as depicted in Figure 7-42 (step ⑤). Transporters for GABA, norepinephrine, dopamine, and serotonin were the first to be cloned and studied. These four transport proteins are all Na^+-linked symporters. They are 60–70 percent identical in their amino acid sequences, and each is thought to contain 12 transmembrane α helices. As with other Na^+ symporters, the movement of Na^+ into the cell down its electrochemical gradient provides the energy for uptake of the neurotransmitter. To maintain electroneutrality, Cl^- often is transported via an ion channel along with the Na^+ and neurotransmitter.

 Cocaine inhibits the transporters for norepinephrine, serotonin, and dopamine. Binding of cocaine to the dopamine transporter inhibits reuptake of dopamine, thus prolonging signaling at key brain synapses; indeed, the dopamine transporter is the principal brain "cocaine receptor." Therapeutic agents such as the antidepressant drugs fluoxetine (Prozac) and imipramine block serotonin uptake, and the tricyclic antidepressant desipramine blocks norepinephrine uptake. ∎

Opening of Acetylcholine-Gated Cation Channels Leads to Muscle Contraction

Acetylcholine is the neurotransmitter at synapses between motor neurons and muscle cells, often called *neuromuscular junctions.* A single axon terminus of a frog motor neuron may contain a million or more synaptic vesicles, each containing 1000–10,000 molecules of acetylcholine; these vesicles often accumulate in rows in the active zone (Figure 7-43). Such a neuron can form synapses with a single skeletal muscle cell at several hundred points.

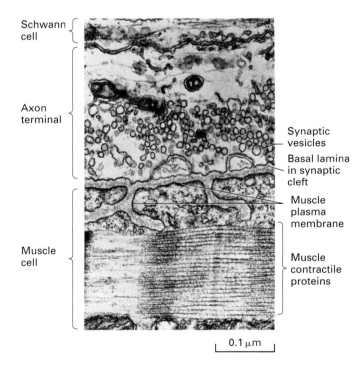

▲ **FIGURE 7-43 Synaptic vesicles in the axon terminal near the region where neurotransmitter is released.** In this longitudinal section through a neuromuscular junction, the basal lamina lies in the synaptic cleft separating the neuron from the muscle membrane, which is extensively folded. Acetylcholine receptors are concentrated in the postsynaptic muscle membrane at the top and part way down the sides of the folds in the membrane. A Schwann cell surrounds the axon terminal. [From J. E. Heuser and T. Reese, 1977, in E. R. Kandel, ed., *The Nervous System,* vol. 1, *Handbook of Physiology,* Williams & Wilkins, p. 266.]

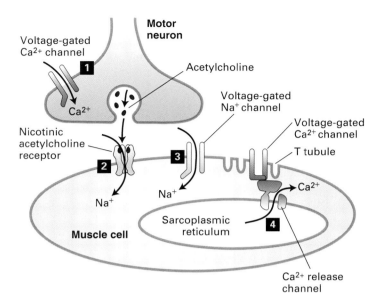

▲ FIGURE 7-44 Sequential activation of gated ion channels at a neuromuscular junction. Arrival of an action potential at the terminus of a presynaptic motor neuron induces opening of voltage-gated Ca^{2+} channels (step **1**) and subsequent release of acetylcholine, which triggers opening of the ligand-gated acetylcholine receptors in the muscle plasma membrane (step **2**). The resulting influx of Na^+ produces a localized depolarization of the membrane, leading to opening of voltage-gated Na^+ channels and generation of an action potential (step **3**). When the spreading depolarization reaches T tubules, it is sensed by voltage-gated Ca^{2+} channels in the plasma membrane. This leads to opening of Ca^{2+}-release channels in the sarcoplasmic reticulum membrane, releasing stored Ca^{2+} into the cytosol (step **4**). The resulting rise in cytosolic Ca^{2+} causes muscle contraction by mechanisms discussed in Chapter 19.

The nicotinic acetylcholine receptor, which is expressed in muscle cells, is a ligand-gated channel that admits both K^+ and Na^+. The effect of acetylcholine on this receptor can be determined by patch-clamping studies on isolated outside-out patches of muscle plasma membranes (see Figure 7-17c). Such measurements have shown that acetylcholine causes opening of a cation channel in the receptor capable of transmitting 15,000–30,000 Na^+ or K^+ ions per millisecond. However, since the resting potential of the muscle plasma membrane is near E_K, the potassium equilibrium potential, opening of acetylcholine receptor channels causes little increase in the efflux of K^+ ions; Na^+ ions, on the other hand, flow into the muscle cell driven by the Na^+ electrochemical gradient.

The simultaneous increase in permeability to Na^+ and K^+ ions following binding of acetylcholine produces a net depolarization to about −15 mV from the muscle resting potential of −85 to −90 mV. As shown in Figure 7-44, this localized depolarization of the muscle plasma membrane triggers opening of voltage-gated Na^+ channels, leading to generation and conduction of an action potential in the muscle cell surface membrane by the same mechanisms described

previously for neurons. When the membrane depolarization reaches T tubules, specialized invaginations of the plasma membrane, it affects Ca^{2+} channels in the plasma membrane apparently without causing them to open. Somehow this causes opening of adjacent Ca^{2+}-release channels in the sarcoplasmic reticulum membrane. The subsequent flow of stored Ca^{2+} ions from the sarcoplasmic reticulum into the cytosol raises the cytosolic Ca^{2+} concentration sufficiently to induce muscle contraction.

Careful monitoring of the membrane potential of the muscle membrane at a synapse with a cholinergic motor neuron has demonstrated spontaneous, intermittent, and random ≈2-ms depolarizations of about 0.5–1.0 mV in the absence of stimulation of the motor neuron. Each of these depolarizations is caused by the spontaneous release of acetylcholine from a single synaptic vesicle. Indeed, demonstration of such spontaneous small depolarizations led to the notion of the quantal release of acetylcholine (later applied to other neurotransmitters) and thereby led to the hypothesis of vesicle exocytosis at synapses. The release of one acetylcholine-containing synaptic vesicle results in the opening of about 3000 ion channels in the postsynaptic membrane, far short of the number needed to reach the threshold depolarization that induces an action potential. Clearly, stimulation of muscle contraction by a motor neuron requires the nearly simultaneous release of acetylcholine from numerous synaptic vesicles.

All Five Subunits in the Nicotinic Acetylcholine Receptor Contribute to the Ion Channel

The acetylcholine receptor from skeletal muscle is a pentameric protein with a subunit composition of $\alpha_2\beta\gamma\delta$. The α, β, γ, and δ subunits have considerable sequence homology; on average, about 35–40 percent of the residues in any two subunits are similar. The complete receptor has fivefold symmetry, and the actual cation channel is a tapered central pore lined by homologous segments from each of the five subunits (Figure 7-45).

The channel opens when the receptor cooperatively binds two acetylcholine molecules to sites located at the interfaces of the $\alpha\delta$ and $\alpha\gamma$ subunits. Once acetylcholine is bound to a receptor, the channel is opened within a few microseconds. Studies measuring the permeability of different small cations suggest that the open ion channel is, at its narrowest, about 0.65–0.80 nm in diameter, in agreement with estimates from electron micrographs. This would be sufficient to allow passage of both Na^+ and K^+ ions with their shell of bound water molecules. Thus the acetylcholine receptor probably transports hydrated ions, unlike Na^+ and K^+ channels, both of which allow passage only of nonhydrated ions (see Figure 7-16).

Although the structure of the central ion channel is not known in molecular detail, much evidence indicates that it is lined by five homologous transmembrane M2 α helices, one from each of the five subunits. The M2 helices are

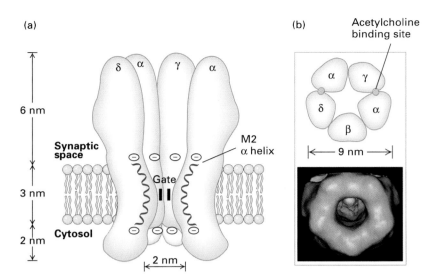

▲ **FIGURE 7-45 Three-dimensional structure of the nicotinic acetylcholine receptor.** (a) Schematic cutaway model of the pentameric receptor in the membrane; for clarity, the β subunit is not shown. Each subunit contains an M2 α helix (red) that faces the central pore. Aspartate and glutamate side chains at both ends of the M2 helices form two rings of negative charges that help exclude anions from and attract cations to the channel. The gate, which is opened by binding of acetylcholine, lies within the pore. (b) *Top:* Cross section of the exoplasmic face of the receptor showing the arrangement of subunits around the central pore. The two acetylcholine binding sites are located about 3 nm from the membrane surface. *Bottom:* Top-down view looking into the synaptic entrance of the channel. The tunnel-like entrance narrows abruptly after a distance of about 6 nm. These models are based on amino acid sequence data, computer-generated averaging of high-resolution electron micrographs, and information from site-specific mutations. [Part (b) from N. Unwin, 1993, *Cell* **72**, and *Neuron,* **10** (suppl.), p. 31.]

composed largely of hydrophobic or uncharged polar amino acids, but negatively charged aspartate or glutamate residues are located at each end, near the membrane faces, and several serine or threonine residues are near the middle. Mutant acetylcholine receptors in which a single negatively charged glutamate or aspartate in one M2 helix is replaced by a positively charged lysine have been expressed in frog oocytes. Patch-clamping measurements indicate that such altered proteins can function as channels, but the number of ions that pass through during the open state is reduced. The greater the number of glutamate or aspartate residues mutated (in one or multiple M2 helices), the greater the reduction in ion conductivity. These findings suggest that aspartate and glutamate residues form a ring of negative charges on the external surface of the pore that help to screen out anions and attract Na$^+$ or K$^+$ ions as they enter the channel. A similar ring of negative charges lining the cytosolic pore surface also helps select cations for passage (see Figure 7-45).

The two acetylcholine binding sites in the extracellular domain of the receptor lie ≈4 to 5 nm from the center of the pore. Binding of acetylcholine thus must trigger conformational changes in the receptor subunits that can cause channel opening at some distance from the binding sites. Receptors in isolated postsynaptic membranes can be trapped in the open or closed state by rapid freezing in liquid nitrogen. Images of such preparations suggest that the five M2 helices rotate relative to the vertical axis of the channel during opening and closing (Figure 7-46).

▶ **FIGURE 7-46 Schematic models of the pore-lining M2 helices in the closed and opened states.** In the closed state, the kink in the center of each M2 helix points inward, constricting the passageway, whose perimeter is indicated by the blue spheres. In the open state, the kinks rotate to one side, so that the helices are farther apart. The green spheres denote the hydroxyl groups of serine (S) and threonine (T) residues in the center of the M2 helices; in the open state these are parallel to the channel axis and allow ions to flow. [Adapted from N. Unwin, 1995, *Nature* **373**:37.]

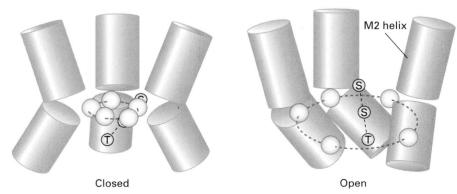

Closed   Open

Nerve Cells Make an All-or-None Decision to Generate an Action Potential

At the neuromuscular junction, virtually every action potential in the presynaptic motor neuron triggers an action potential in the postsynaptic muscle cell. The situation at synapses between neurons, especially those in the brain, is much more complex because the postsynaptic neuron commonly receives signals from many presynaptic neurons (Figure 7-47). The neurotransmitters released from presynaptic neurons may bind to an *excitatory receptor* on the postsynaptic neuron, thereby opening a channel that admits Na^+ ions or both Na^+ and K^+ ions. The acetylcholine receptor just discussed is one of many excitatory receptors, and opening of such ion channels leads to depolarization of the postsynaptic plasma membrane, promoting generation of an action potential. In contrast, binding of a neurotransmitter to an *inhibitory receptor* on the postsynaptic cell causes opening of K^+ or Cl^- channels, leading to an efflux of additional K^+ ions from the cytosol or an influx of Cl^- ions. In either case, the ion flow tends to hyperpolarize the plasma membrane, which inhibits generation of an action potential in the postsynaptic cell.

A single neuron can be affected simultaneously by signals received at multiple excitatory and inhibitory synapses. The neuron continuously integrates these signals and determines whether or not to generate an action potential. In this process, the various small depolarizations and hyperpolarizations generated at synapses move along the plasma membrane from the dendrites to the cell body and then to the axon hillock, where they are summed together. An action potential is generated whenever the membrane at the axon hillock becomes depolarized to a certain voltage called the *threshold potential* (Figure 7-48). Thus an action potential

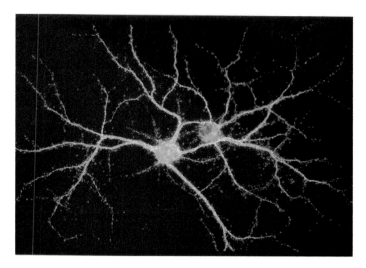

▲ **EXPERIMENTAL FIGURE 7-47 A fluorescent micrograph of two interneurons reveals that many other neurons synapse with them.** These cells, from the hippocampal region of the brain, were stained with two fluorescent antibodies: one specific for the microtubule-associated protein MAP2 (green), which is found only in dendrites and cell bodies, and the other specific for synaptotagmin (orange-red), a protein found in presynaptic axon terminals. The numerous orange-red dots, which represent presynaptic axon terminals from neurons that are not visible in this field, indicate that these interneurons receive signals from many other cells. [Courtesy of O. Mundigl and P. deCamilli.]

is generated in an all-or-nothing fashion: Depolarization to the threshold always leads to an action potential, whereas any depolarization that does not reach the threshold potential never induces it.

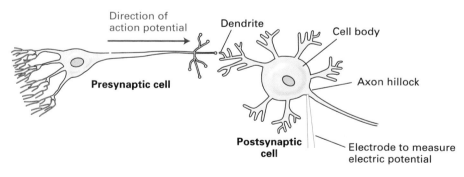

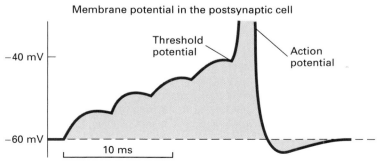

◄ **EXPERIMENTAL FIGURE 7-48 Incoming signals must reach the threshold potential to trigger an action potential in a postsynaptic cell.** In this example, the presynaptic neuron is generating about one action potential every 4 milliseconds. Arrival of each action potential at the synapse causes a small change in the membrane potential at the axon hillock of the postsynaptic cell, in this example a depolarization of ≈5 mV. When multiple stimuli cause the membrane of this postsynaptic cell to become depolarized to the threshold potential, here approximately −40 mV, an action potential is induced in it.

Whether a neuron generates an action potential in the axon hillock depends on the balance of the timing, amplitude, and localization of all the various inputs it receives; this signal computation differs for each type of neuron. In a sense, each neuron is a tiny computer that averages all the receptor activations and electric disturbances on its membrane and makes a decision whether to trigger an action potential and conduct it down the axon. An action potential will always have the same *magnitude* in any particular neuron. The *frequency* with which action potentials are generated in a particular neuron is the important parameter in its ability to signal other cells.

The Nervous System Uses Signaling Circuits Composed of Multiple Neurons

In complex multicellular animals, such as insects and mammals, various types of neurons form signaling circuits. In the simple type of circuit, called a *reflex arc*, interneurons connect multiple sensory and motor neurons, allowing one sensory neuron to affect multiple motor neurons and one motor neuron to be affected by multiple sensory neurons; in this way interneurons integrate and enhance reflexes. For example, the knee-jerk reflex in humans involves a complex reflex arc in which one muscle is stimulated to contract while another is inhibited from contracting (Figure 7-49). Such circuits allow an organism to respond to a sensory input by the coordinated action of sets of muscles that together achieve a single purpose.

These simple signaling circuits, however, do not directly explain higher-order brain functions such as reasoning, computation, and memory development. Typical neurons in the brain receive signals from up to a thousand other neurons and, in turn, can direct chemical signals to many other neurons. The output of the nervous system depends on its circuit properties, that is, the wiring, or interconnections, between neurons and the strength of these interconnections. Complex aspects of the nervous system, such as vision and consciousness, cannot be understood at the single-cell level, but only at the level of networks of nerve cells that can be studied by techniques of systems analysis. The nervous system is constantly changing; alterations in the number and nature of the interconnections between individual neurons occur, for example, in the development of new memories.

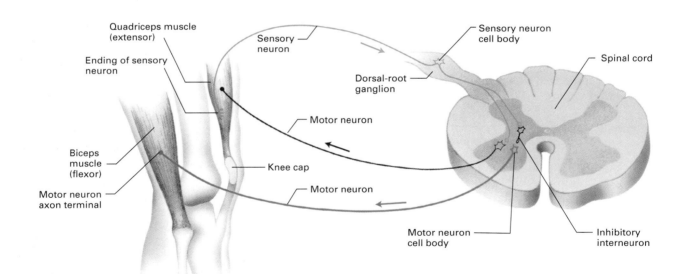

▲ **FIGURE 7-49 The knee-jerk reflex arc in the human.** Positioning and movement of the knee joint are accomplished by two muscles that have opposite actions: Contraction of the quadriceps muscle straightens the leg, whereas contraction of the biceps muscle bends the leg. The knee-jerk response, a sudden extension of the leg, is stimulated by a blow just below the kneecap. The blow directly stimulates sensory neurons (blue) located in the tendon of the quadriceps muscle. The axon of each sensory neuron extends from the tendon to its cell body in a dorsal root ganglion. The sensory axon then continues to the spinal cord, where it branches and synapses with two neurons: (1) a motor neuron (red) that innervates the quadriceps muscle and (2) an inhibitory interneuron (black) that synapses with a motor neuron (green) innervating the biceps muscle. Stimulation of the sensory neuron causes a contraction of the quadriceps and, via the inhibitory neuron, a simultaneous inhibition of contraction of the biceps muscle. The net result is an extension of the leg at the knee joint. Each cell illustrated here actually represents a nerve, that is, a population of neurons.

KEY CONCEPTS OF SECTION 7.8

Neurotransmitters and Receptor and Transport Proteins in Signal Transmission at Synapses

■ Neurotransmitter receptors fall into two classes: ligand-gated ion channels, which permit ion passage when open, and G protein–coupled receptors, which are linked to a separate ion channel.

■ At synapses impulses are transmitted by neurotransmitters released from the axon terminal of the presynaptic cell and subsequently bound to specific receptors on the postsynaptic cell (see Figure 7-31).

■ Low-molecular-weight neurotransmitters (e.g., acetylcholine, dopamine, epinephrine) are imported from the cytosol into synaptic vesicles by H^+-linked antiporters. V-class proton pumps maintain the low intravesicular pH that drives neurotransmitter import against a concentration gradient.

■ Arrival of an action potential at a presynaptic axon terminal opens voltage-gated Ca^{2+} channels, leading to a localized rise in the cytosolic Ca^{2+} level that triggers exocytosis of synaptic vesicles. Following neurotransmitter release, vesicles are formed by endocytosis and recycled (see Figure 7-42).

■ Coordinated operation of four gated ion channels at the synapse of a motor neuron and striated muscle cell leads to release of acetylcholine from the axon terminal, depolarization of the muscle membrane, generation of an action potential, and then contraction (see Figure 7-44).

■ The nicotinic acetylcholine receptor, a ligand-gated cation channel, contains five subunits, each of which has a transmembrane α helix (M2) that lines the channel (see Figure 7-45).

■ A postsynaptic neuron generates an action potential only when the plasma membrane at the axon hillock is depolarized to the threshold potential by the summation of small depolarizations and hyperpolarizations caused by activation of multiple neuronal receptors (see Figure 7-48).

PERSPECTIVES FOR THE FUTURE

In this chapter, we have explained certain aspects of human physiology in terms of the action of specific membrane transport proteins. Such a molecular physiology approach has many medical applications. Even today, specific inhibitors or activators of channels, pumps, and transporters constitute the largest single class of drugs. For instance, an inhibitor of the gastric H^+/K^+ ATPase that acidifies the stomach is the most widely used drug for treating stomach ulcers. Inhibitors of channel proteins in the kidney are widely used to control

hypertension (high blood pressure). By blocking resorption of water from the forming urine into the blood, these drugs reduce blood volume and thus blood pressure. Calcium-channel blockers are widely employed to control the intensity of contraction of the heart. And the Na^+-linked symport proteins that are used by nerve cells for reuptake of neurotransmitters are specifically inhibited by many drugs of abuse (e.g., cocaine) and antidepression medications (e.g., Prozac).

With the completion of the human genome project, we are positioned to learn the sequences of all human membrane transport proteins. Already we know that mutations in many of them cause disease. For example, mutation in a voltage-gated Na^+ channel that is expressed in the heart causes ventricular fibrillation and heart attacks. Mutations in other Na^+ channels, expressed mainly in the brain, cause epilepsy and febrile seizures. In some cases the molecular mechanisms are known. One type of missense mutation in a Na^+ channel that causes epilepsy affects the voltage dependence of channel opening and closing; another slows inactivation of the channel at depolarizing potentials, prolonging the influx of Na^+ ions. Studies in mice expressing mutant forms of these and many other membrane transport proteins are continuing to provide clues to their role in human physiology and disease.

This exploding basic knowledge will enable researchers to identify new types of compounds that inhibit or activate just one of these membrane transport proteins and not its homologs. An important challenge, however is to understand the role of an individual transport protein in each of the several tissues in which it is expressed. As an example, a drug that inhibits a particular ion channel in sensory neurons might be useful in treatment of chronic pain, but if this channel also is expressed in certain areas of the brain, its inhibition may have serious undesired actions ("side effects").

A real understanding of the function of nerve cells requires knowledge of the three-dimensional structures of many different channels, neurotransmitter receptors, and other membrane proteins. Determination of the structure of the first voltage-gated K^+ channel should illuminate the mechanisms of channel gating, opening, and inactivation that may also apply to other voltage-gated channels. Similarly, the nicotinic acetylcholine, glutamate, GABA, and glycine receptors are all ligand-gated ion channels, but it is disputed whether they all have the same overall structures in the membrane. Resolving this issue will also require knowledge of their three-dimensional structures, which, in addition, should tell us in detail how neurotransmitter binding leads to channel opening.

How does a neuron achieve its very long, branching structure? Why does one part of a neuron become a dendrite and another an axon? Why are certain key membrane proteins clustered at particular points—neurotransmitter receptors in postsynaptic densities in dendrites, Ca^{2+} channels in axon termini, and Na^+ channels in myelinated neurons at the nodes of Ranvier? Such questions of cell shape and protein targeting also apply to other types of cells, but the morphological diversity of different types of neurons makes these particularly intriguing questions in the nervous system. Development of pure cultures of specific types of neurons that maintain their normal properties would enable many of these problems to be studied by techniques of molecular cell biology. Perhaps the most difficult questions concern the formation of specific synapses within the nervous system; that is, how does a neuron "know" to synapse with one type of cell and not another? Ongoing research on the development of the nervous system, which we briefly discuss in Chapter 15, is beginning to provide more complete answers about the complex wiring among neurons and how this relates to brain function.

KEY TERMS

ABC superfamily *253*

action potential *245*

active transport *247*

antiport *268*

cotransport *268*

depolarization *274*

electrochemical
    gradient *246*

facilitated diffusion *247*

gated channel *247*

hypertonic *272*

hypotonic *272*

isotonic *272*

membrane potential *246*

myelin sheath *284*

Na^+/K^+ ATPase *256*

neurotransmitter *276*

passive diffusion *246*

patch clamping *265*

resting K^+ channels *262*

symport *268*

synapses *277*

tight junctions *274*

transcellular transport
    274

uniport *248*

REVIEW THE CONCEPTS

1. The basic structural unit of biomembranes is the phospholipid bilayer. Acetic acid and ethanol are composed each of two carbons, hydrogen and oxygen, and both enter cells by passive diffusion. At pH 7, one is much more membrane permeant than the other. Which is the more permeable, and why? Predict how the permeability of each is altered when pH is reduced to 1.0, a value typical of the stomach.

2. Uniporters and ion channels support facilitated diffusion across biomembranes. Although both are examples of facilitated diffusion, the rates of ion movement via a channel are roughly 10^4- to 10^5-fold faster than that of molecules via a uniporter. What key mechanistic difference results in this large difference in transport rate?

3. Name the three classes of transporters. Explain which of these classes is able to move glucose or bicarbonate (HCO_3^-), for example, against an electrochemical gradient. In the case of bicarbonate, but not glucose, the ΔG of the transport process has two terms. What are these two terms, and why does the second not apply to glucose? Why are cotransporters often referred to as examples of secondary active transport?

4. GLUT1, found in the plasma membrane of erythrocytes, is a classic example of a uniporter. Design a set of experiments to prove that GLUT1 is indeed a glucose-specific uniporter rather than a galactose- or mannose-specific uniporter. Glucose is a 6-carbon sugar while ribose is a 5-carbon sugar. Despite this smaller size, ribose is not efficiently transported by GLUT1. How can this be explained?

5. Name the four classes of ATP-powered pumps that produce active transport of ions and molecules. Indicate which of these classes transport ions only and which transport primarily small molecules. In the case of one class of these ATP-powered pumps, the initial discovery of the class came from studying not the transport of a natural substrate but rather artificial substrates used as cancer chemotherapy drugs. What do investigators now think are common examples of the natural substrates of this particular class of ATP-powered pumps?

6. Genome sequencing projects continue, and for an increasing number of organisms the complete genome sequence of that organism is known. How does this information allow us to state the total number of transporters or pumps of a given type in either mice or humans? Many of the sequence-identified transporters or pumps are "orphan" proteins, in the sense that their natural substrate or physiological role is not known. How can this be, and how might one establish the physiological role of an orphan protein?

7. As cited in the section Perspectives for the Future, specific inhibitors or activators of channels, pumps, and transporters constitute the largest single class of drugs produced by the pharmaceutical industry. Skeletal muscle contraction is caused by elevation of Ca^{2+} concentration in the cytosol. What is the expected effect on muscle contraction of selective drug inhibition of sarcoplasmic reticulum (SR) P-class Ca^{2+} ATPase?

8. The membrane potential in animal cells, but not in plants, depends largely on resting K^+ channels. How do these channels contribute to the resting potential? Why are these channels considered to be nongated channels? How do these channels achieve selectivity for K^+ versus Na^+?

9. Patch clamping can be used to measure the conductance properties of individual ion channels. Describe how patch clamping can be used to determine whether or not the gene coding for a putative K^+ channel actually codes for a K^+ or Na^+ channel.

10. Plants use the proton electrochemical gradient across the vacuole membrane to power the accumulation of salts and sugars in the organelle. This creates a hypertonic situation. Why does this not result in the plant cell bursting? How does the plasma membrane Na$^+$/K$^+$ ATPase allow animal cells to avoid osmotic lysis even under isotonic conditions?

11. Movement of glucose from one side to the other side of the intestinal epithelium is a major example of transcellular transport. How does the Na$^+$/K$^+$ ATPase power the process? Why are tight junctions essential for the process? Rehydration supplements such as sport drinks include a sugar and a salt. Why are both important to rehydration?

12. Name the three phases of an action potential. Describe for each the underlying molecular basis and the ion involved. Why is the term *voltage-gated channel* applied to Na$^+$ channels involved in the generation of an action potential?

13. Myelination increases the velocity of action potential propagation along an axon. What is myelination? Myelination causes clustering of voltage-gated Na$^+$ channels and Na$^+$/K$^+$ pumps at nodes of Ranvier along the axon. Predict the consequences to action potential propagation of increasing the spacing between nodes of Ranvier by a factor of 10.

14. Compare the role of H$^+$-linked antiporters in the accumulation of neurotransmitters in synaptic vesicles and sucrose in the plant vacuole. Acetylcholine is a common neurotransmitter released at the synapse. Predict the consequences for muscle activation of decreased acetylcholine esterase activity at nerve-muscle synapses.

15. Neurons, particularly those in the brain, receive multiple excitatory and inhibitory signals. What is the name of the extension of the neuron at which such signals are received? How does the neuron integrate these signals to determine whether or not to generate an action potential?

ANALYZE THE DATA

Imagine that you are evaluating candidates for the glutamate transporter resident in the membrane of synaptic vesicles of the brain. Glutamate is a major neurotransmitter. You conduct a thorough search of protein databases and literature and find that the *Caenorhabditis elegans* protein EAT-4 has many of the properties expected of a brain glutamate transporter. The protein is presynaptic in *C. elegans* and mutations in it are associated with glutamatergic defects. By sequence comparison, the mammalian homolog is BNP1. Related type I transporters are known to transport organic anions.

a. To determine whether BNP1 mediates the transport of glutamate into synaptic vesicles, you transfect BNP1 cDNA into PC12 cells, which lack detectable endogenous BNP1 protein. You then prepare a synaptic vesicle–like microvesicle population from transfected and untransfected cells. The glutamate transport properties of the isolated vesicles are shown in the figure at the right. How do these results indi-

cate transporter-dependent uptake of glutamate in transfected versus nontransfected, wild-type PC12 cell vesicles? What is the apparent K_m of the glutamate transporter? The literature reports K_m values of ~10–100 μM for plasma membrane excitatory amino acid transporters. Is the K_m observed consistent with BNP1 being a plasma membrane amino acid transporter?

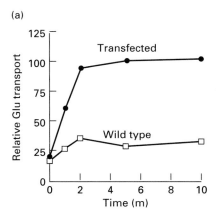

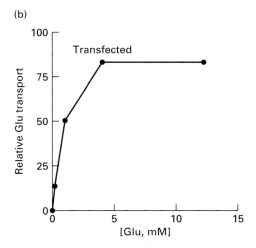

Effect of BNP1 transfection on glutamate uptake by PC12 cell vesicles.

b. You next decide to test the specificity of BNP1 for individual amino acids and whether or not glutamate uptake is dependent on the electrical gradient as predicted. For these experiments, you set up a series of incubation mixes with various amino acids as competitors and separately a mix with valinomycin (V) and nigericin (N) to dissipate selectively the electrical gradient. Based on the data shown in the following figure, does BNP1-dependent transport display the expected properties? Why are the chosen amino acids the logical choices for potential competitors?

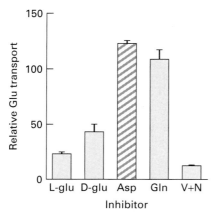

Effect of amino acid competitors and valinomycin plus nigericin (V + N) on BNP1 transport.

c. As a final test, you determine the dependence of BNP1 transport on chloride concentration. Vesicular glutamate transport is known to exhibit a biphasic dependence on chloride concentration. The results of your experiment are shown in the figure below. Do these data indicate a biphasic chloride dependence? How may such a chloride dependence be explained?

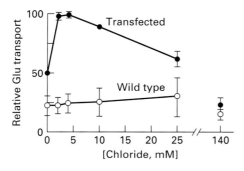

Chloride dependence of transport by BNP1.

d. You conclude on the basis of the data that BNP1 is the synaptic vesicle glutamate transporter. How could you test in PC12 cells whether mutations modeled on the *C. elegans* EAT-4 protein were important in the mammalian protein?

REFERENCES

Overview of Membrane Transport

Hruz, P. W., and M. M. Mueckler. 2001. Structural analysis of the GLUT1 facilitative glucose transporter (review). *Mol. Memb. Biol.* **18**:183–193.

Malandro, M., and M. Kilberg. 1996. Molecular biology of mammalian amino acid transporters. *Ann. Rev. Biochem.* **66**:305–336.

Mueckler, M. 1994. Facilitative glucose transporters. *Eur. J. Biochem.* **219**:713–725.

ATP-Powered Pumps and the Intracellular Ionic Environment

Borst, P., N. Zelcer, and A. van Helvoort. 2000. ABC transporters in lipid transport. *Biochim. Biophys. Acta* **1486**:128–144.

Carafoli, E., and M. Brini. 2000. Calcium pumps: structural basis for and mechanism of calcium transmembrane transport. *Curr. Opin. Chem. Biol.* **4**:152–161.

Davies, J., F. Chen, and Y. Ioannou. 2000. Transmembrane molecular pump activity of Niemann-Pick C1 protein. *Science* **290**:2295–2298.

Doige, C. A., and G. F. Ames. 1993. ATP-dependent transport systems in bacteria and humans: relevance to cystic fibrosis and multidrug resistance. *Ann. Rev. Microbiol.* **47**:291–319.

Gottesman, M. M. 2002. Mechanisms of cancer drug resistance. *Ann. Rev. Med.* **53**:615–627.

Gottesman, M. M., and S. V. Ambudkar. 2001. Overview: ABC transporters and human disease. *J. Bioenerg. Biomemb.* **33**:453–458.

Higgins, C. F., and K. J. Linton. 2001. Structural biology. The xyz of ABC transporters. *Science* **293**:1782–1784.

Holmgren, M., et al. 2000. Three distinct and sequential steps in the release of sodium ions by the Na^+/K^+-ATPase. *Nature* **403**:898–901.

Jencks, W. P. 1995. The mechanism of coupling chemical and physical reactions by the calcium ATPase of sarcoplasmic reticulum and other coupled vectorial systems. *Biosci. Rept.* **15**:283–287.

Nishi, T., and M. Forgac. 2002. The vacuolar (H^+)-ATPases—nature's most versatile proton pumps. *Nature. Rev. Mol. Cell Biol.* **3**:94–103.

Ostedgaard, L. S., O. Baldursson, and M. J. Welsh. 2001. Regulation of the cystic fibrosis transmembrane conductance regulator Cl^- channel by its R domain. *J. Biol. Chem.* **276**:7689–7692.

Raggers, R. J., et al. 2000. Lipid traffic: the ABC of transbilayer movement. *Traffic* **1**:226–234.

Rea, P. A., et al. 1992. Vacuolar H^+-translocating pyrophosphatases: a new category of ion translocase. *Trends Biochem. Sci.* **17**:348–353.

Sweadner, K. J., and C. Donnet. 2001. Structural similarities of Na,K-ATPase and SERCA, the Ca(2+)-ATPase of the sarcoplasmic reticulum. *Biochem. J.* **356**:685–704.

Toyoshima, C., M. Nakasako, H. Nomura, and H. Ogawa. 2000. Crystal structure of the calcium pump of sarcoplasmic reticulum at 2.6 Å resolution. *Nature* **405**:647–655.

Xu, C., W. J. Rice, W. He, and D. L. Stokes. 2002. A structural model for the catalytic cycle of Ca(2+)-ATPase. *J. Mol. Biol.* **316**:201–211.

Nongated Ion Channels and the Resting Membrane Potential

Clapham, D. 1999. Unlocking family secrets: K^+ channel transmembrane domains. *Cell* **97**:547–550.

Cooper, E. C., and L. Y. Jan. 1999. Ion channel genes and human neurological disease: recent progress, prospects, and challenges. *Proc. Nat'l. Acad. Sci. USA* **96**:4759–4766.

Dutzler, R., et al. 2002. X-ray structure of a ClC chloride channel at 3.0 Å reveals the molecular basis of anion selectivity. *Nature* **415**:287–294.

Gulbis, J. M., M. Zhou, S. Mann, and R. MacKinnon. 2000. Structure of the cytoplasmic beta subunit-T1 assembly of voltage-dependent K+ channels. *Science* **289**:123–127.

Hanaoka, K., et al. 2000. Co-assembly of polycystin-1 and -2 produces unique cation-permeable currents. *Nature* **408**:990–994.

Hille, B. 2001. *Ion Channels of Excitable Membranes*, 3d ed. Sinauer Associates.

Montell, C., L. Birnbaumer, and V. Flockerzi. 2002. The TRP channels, a remarkably functional family. *Cell* 108:595–598.

Neher, E. 1992. Ion channels for communication between and within cells. Nobel Lecture reprinted in *Neuron* 8:605–612 and *Science* 256:498–502.

Neher, E., and B. Sakmann. 1992. The patch clamp technique. *Sci. Am.* 266(3):28–35.

Nichols, C., and A. Lopatin. 1997. Inward rectifier potassium channels. *Ann. Rev. Physiol.* 59:171–192.

Yi, B. A., et al. 2001. Controlling potassium channel activities: interplay between the membrane and intracellular factors. *Proc. Nat'l. Acad. Sci. USA* 98:11016–11023.

Zhou, M., J. H. Morais-Cabral, S. Mann, and R. MacKinnon. 2001. Potassium channel receptor site for the inactivation gate and quaternary amine inhibitors. *Nature* 411:657–661.

Zhou, Y., J. Morais-Cabral, A. Kaufman, and R. MacKinnon. 2001. Chemistry of ion coordination and hydration revealed by a K^+ channel–Fab complex at 2 Å resolution. *Nature* 414:43–48.

Cotransport by Symporters and Antiporters

Alper, S. L., M. N. Chernova, and A. K. Stewart. 2001. Regulation of Na^+-independent Cl^-/HCO_3^- exchangers by pH. *J. Pancreas* 2:171–175.

Barkla, B., and O. Pantoja. 1996. Physiology of ion transport across the tonoplast of higher plants. *Ann. Rev. Plant Physiol. Plant Mol. Biol.* 47:159–184.

Orlowski, J., and S. Grinstein. 1997. Na^+/H^+ exchangers of mammalian cells. *J. Biol. Chem.* 272:22373–22376.

Shrode, L. D., H. Tapper, and S. Grinstein. 1997. Role of intracellular pH in proliferation, transformation, and apoptosis. *J. Bioenerg. Biomemb.* 29:393–399.

Wakabayashi, S., M. Shigekawa, and J. Pouyssegur. 1997. Molecular physiology of vertebrate Na^+/H^+ exchangers. *Physiol. Rev.* 77:51–74.

Wright, E. M. 2001. Renal $Na(+)$-glucose cotransporters. *Am. J. Physiol. Renal Physiol.* 280:F10–F18.

Wright, E. M., and D. D. Loo. 2000. Coupling between Na^+, sugar, and water transport across the intestine. *Ann. NY Acad. Sci.* 915:54–66.

Zhang, H-X., and E. Blumwald. 2001. Transgenic salt-tolerant tomato plants accumulate salt in foliage but not in fruit. *Nature Biotech.* 19:765–769.

Movement of Water

Agre, P., M. Bonhivers, and M. Borgina. 1998. The aquaporins, blueprints for cellular plumbing systems. *J. Biol. Chem.* 273:14659.

Engel, A., Y. Fujiyoshi, and P. Agre. 2000. The importance of aquaporin water channel protein structures. *EMBO J.* 19:800–806.

Maurel, C. 1997. Aquaporins and water permeability of plant membranes. *Ann. Rev. Plant Physiol. Plant Mol. Biol.* 48:399–430.

Nielsen, S., et al. 2002. Aquaporins in the kidney: from molecules to medicine. *Physiol. Rev.* 82:205–244.

Schultz, S. G. 2001. Epithelial water absorption: osmosis or cotransport? *Proc. Nat'l. Acad. Sci. USA* 98:3628–3630.

Sui, H., et al. 2001. Structural basis of water-specific transport through the AQP1 water channel. *Nature* 414:872–878.

Verkman, A. S. 2000. Physiological importance of aquaporins: lessons from knockout mice. *Curr. Opin. Nephrol. Hypertens.* 9:517–522.

Transepithelial Transport

Cereijido, M., J. Valdés, L. Shoshani, and R. Conteras. 1998. Role of tight junctions in establishing and maintaining cell polarity. *Ann. Rev. Physiol.* 60:161–177.

Goodenough, D. A. 1999. Plugging the leaks. *Proc. Nat'l. Acad. Sci. USA* 96:319.

Mitic, L., and J. Anderson. 1998. Molecular architecture of tight junctions. *Ann. Rev. Physiol.* 60:121–142.

Schultz, S., et al., eds. 1997. *Molecular Biology of Membrane Transport Disorders*. Plenum Press.

Propagation of Electric Impulses by Gated Ion Channels

Aldrich, R. W. 2001. Fifty years of inactivation. *Nature* 411:643–644.

Armstrong, C., and B. Hille. 1998. Voltage-gated ion channels and electrical excitability. *Neuron* 20:371–380.

Catterall, W. A. 2000. From ionic currents to molecular mechanisms: the structure and function of voltage-gated sodium channels. *Neuron* 26:13–25.

Catterall, W. A. 2000. Structure and regulation of voltage-gated Ca^{2+} channels. *Ann. Rev. Cell Dev. Biol.* 16:521–555.

Catterall, W. A. 2001. A 3D view of sodium channels. *Nature* 409:988–989.

del Camino, D., and G. Yellen. 2001. Tight steric closure at the intracellular activation gate of a voltage-gated $K(+)$ channel. *Neuron* 32:649–656.

Doyle, D. A., et al. 1998. The structure of the potassium channel: molecular basis of K^+ conduction and selectivity. *Science* 280:69–77.

Hille, B. 2001. *Ion Channels of Excitable Membranes*, 3d ed. Sinauer Associates.

Jan, Y. N., and L. Y. Jan. 2001. Dendrites. *Genes Dev.* 15:2627–2641.

Miller, C. 2000. Ion channel surprises: prokaryotes do it again! *Neuron* 25:7–9.

Sato, C., et al. 2001. The voltage-sensitive sodium channel is a bell-shaped molecule with several cavities. *Nature* 409:1047–1051.

Yi, B. A., and L. Y. Jan. 2000. Taking apart the gating of voltage-gated K^+ channels. *Neuron* 27:423–425.

Myelin

Arroyo, E. J., and S. S. Scherer. 2000. On the molecular architecture of myelinated fibers. *Histochem. Cell Biol.* 113:1–18.

Pedraza, L., J. K. Huang, and D. R. Colman. 2001. Organizing principles of the axoglial apparatus. *Neuron* 30:335–344.

Salzer, J. L. 1997. Clustering sodium channels at the node of Ranvier: close encounters of the axon-glia kind. *Neuron* 18:843–846.

Trapp, B. D., and G. J. Kidd. 2000. Axo-glial septate junctions. The maestro of nodal formation and myelination? *J. Cell Biol.* 150:F97–F100.

Neurotransmitters and Transport Proteins in the Transmission of Electric Impulses

Amara, S. G., and M. J. Kuhar. 1993. Neurotransmitter transporters: recent progress. *Ann. Rev. Neurosci.* 16:73–93.

Bajjalieh, S. M., and R. H. Scheller. 1995. The biochemistry of neurotransmitter secretion. *J. Biol. Chem.* 270:1971–1974.

Betz, W., and J. Angleson. 1998. The synaptic vesicle cycle. *Ann. Rev. Physiol.* 60:347–364.

Brejc, K., et al. 2001. Crystal structure of an ACh-binding protein reveals the ligand-binding domain of nicotinic receptors. *Nature* 411:269–276.

Fernandez, J. M. 1997. Cellular and molecular mechanics by atomic force microscopy: capturing the exocytotic fusion pore in vivo? *Proc. Nat'l. Acad. Sci. USA* 94:9–10.

Ikeda, S. R. 2001. Signal transduction. Calcium channels—link locally, act globally. *Science* 294:318–319.

Jan, L. Y., and C. F. Stevens. 2000. Signaling mechanisms: a decade of signaling. *Curr. Opin. Neurobiol.* **10**:625–630.

Karlin, A. 2002. Emerging structure of the nicotinic acetylcholine receptors. *Nature Rev. Neurosci.* **3**:102–114.

Kavanaugh, M. P. 1998. Neurotransmitter transport: models in flux. *Proc. Nat'l. Acad. Sci. USA* **95**:12737–12738.

Lin, R. C., and R. H. Scheller. 2000. Mechanisms of synaptic vesicle exocytosis. *Ann. Rev. Cell Dev. Biol.* **16**:19–49.

Neher, E. 1998. Vesicle pools and Ca^{2+} microdo-mains: new tools for understanding their roles in neurotransmitter release. *Neuron* **20**:389–399.

Reith, M., ed. 1997. *Neurotransmitter Transporters: Structure, Function, and Regulation.* Humana Press.

Sakmann, B. 1992. Elementary steps in synaptic transmission revealed by currents through single ion channels. Nobel Lecture reprinted in *EMBO J.* **11**:2002–2016 and *Science* **256**:503–512.

Sudhof, T. C. 1995. The synaptic vesicle cycle: a cascade of protein-protein interactions. *Nature* **375**:645–653.

Usdin, T. B., L. E. Eiden, T. I. Bonner, and J. D. Erickson. 1995. Molecular biology of the vesicular ACh transporter. *Trends Neurosci.* **18**:218–224.

8

CELLULAR ENERGETICS

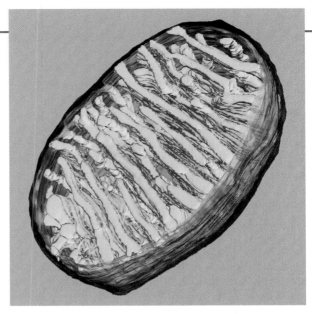

Computer-generated model of a section of a mitochondrion from chicken brain, based on a three-dimensional electron tomogram. [T. Frey and C. Mannella, 2000, *Trends Biochem. Sci.* **25**:319.]

The most important molecule for capturing and transferring free energy in biological systems is **adenosine triphosphate,** or **ATP** (see Figure 2-24). Cells use the energy released during hydrolysis of the terminal "high-energy" phosphoanhydride bond in ATP to power many energetically unfavorable processes. Examples include the synthesis of proteins from amino acids and of nucleic acids from nucleotides (Chapter 4), transport of molecules against a concentration gradient by ATP-powered pumps (Chapter 7), contraction of muscle (Chapter 19), and movement (beating) of cilia (Chapter 20). Although other high-energy molecules occur in cells, ATP is the universal "currency" of chemical energy; it is found in all types of organisms and must have occurred in the earliest life-forms.

This chapter focuses on how cells generate the high-energy phosphoanhydride bond of ATP from ADP and inorganic phosphate (HPO_4^{2-}). This endergonic reaction, which is the reverse of ATP hydrolysis and requires an input of 7.3 kcal/mol to proceed, can be written as

$$P_i^{2-} + H^+ + ADP^{3-} \longrightarrow ATP^{4-} + H_2O$$

where P_i^{2-} represents inorganic phosphate (HPO_4^{2-}). The energy to drive this reaction is produced primarily by two main processes: **aerobic oxidation,** which occurs in nearly all cells, and **photosynthesis,** which occurs only in leaf cells of plants and certain single-celled organisms.

In aerobic oxidation, fatty acids and sugars, principally glucose, are metabolized to carbon dioxide (CO_2) and water (H_2O), and the released energy is converted to the chemical energy of phosphoanhydride bonds in ATP. In animal cells and most other nonphotosynthetic cells, ATP is generated mainly by this process. The initial steps in the oxidation of glucose, called **glycolysis,** occur in the cytosol in both eukaryotes and prokaryotes and do not require oxygen (O_2). The final steps, which require oxygen, generate most of the ATP. In eukaryotes, these later stages of aerobic oxidation occur in **mitochondria;** in prokaryotes, which contain only a plasma membrane and lack internal organelles, many of the final steps occur on the plasma membrane. The final stages of fatty acid metabolism sometimes occur in mitochondria and generate ATP; in most eukaryotic cells, however, fatty acids are metabolized to CO_2 and H_2O in **peroxisomes** without production of ATP.

In photosynthesis, light energy is converted to the chemical energy of phosphoanhydride bonds in ATP and stored in the chemical bonds of carbohydrates (primarily sucrose and starch). Oxygen also is formed during photosynthesis. In plants and eukaryotic single-celled algae, photosynthesis occurs in **chloroplasts.** Several prokaryotes also carry out photosynthesis on their plasma membrane or its invaginations by a mechanism similar to that in chloroplasts. The oxygen generated during photosynthesis is the source of virtually all the oxygen in the air, and the carbohydrates produced are the ultimate source of energy for virtually all nonphotosynthetic organisms. Bacteria living in deep ocean vents, where there is no sunlight, disprove the popular view that sunlight is the ultimate source of energy for all organisms on earth. These bacteria obtain energy for converting carbon dioxide into carbohydrates and other cellular constituents by oxidation of reduced inorganic compounds in dissolved vent gas.

At first glance, photosynthesis and aerobic oxidation appear to have little in common. However, a revolutionary discovery in cell biology is that bacteria, mitochondria, and chloroplasts all use the same basic mechanism, called **chemiosmosis** (or chemiosmotic coupling), to generate ATP from ADP and P_i. In chemiosmosis, a proton (H^+) concentration gradient and an electric potential (voltage gradient) across the membrane, collectively termed the **proton-motive force,** drive an energy-requiring process such as ATP synthesis (Figure 8-1, *bottom*).

Chemiosmosis can occur only in sealed, membrane-limited compartments that are impermeable to H^+. The proton-motive force is generated by the stepwise movement of electrons from higher to lower energy states via membrane-bound **electron carriers.** In mitochondria and nonphotosynthetic bacterial cells, electrons from NADH (produced during the metabolism of sugars, fatty acids, and other substances) are transferred to O_2, the ultimate electron acceptor. In the thylakoid membrane of chloroplasts, energy absorbed from light strips electrons from water (forming O_2) and pow-

ers their movement to other electron carriers, particularly $NADP^+$; eventually these electrons are donated to CO_2 to synthesize carbohydrates. All these systems, however, contain some similar carriers that couple **electron transport** to the pumping of protons across the membrane—always from the **cytosolic face** to the **exoplasmic face** of the membrane—thereby generating the proton-motive force (Figure 8-1, *top*). Invariably, the cytosolic face has a negative electric potential relative to the exoplasmic face.

Moreover, mitochondria, chloroplasts, and bacteria utilize essentially the same kind of membrane protein, the **F_0F_1 complex,** to synthesize ATP. The F_0F_1 complex, now commonly called **ATP synthase,** is a member of the F class of ATP-powered proton pumps (see Figure 7-6). In all cases, ATP synthase is positioned with the globular F_1 domain, which catalyzes ATP synthesis, on the cytosolic face of the membrane, so ATP is always formed on the cytosolic face of the membrane (Figure 8-2). Protons always flow through ATP synthase from the exoplasmic to the cytosolic face of the membrane, driven by a combination of the proton concentration gradient ($[H^+]_{exoplasmic} > [H^+]_{cytosolic}$) and the membrane electric potential (exoplasmic face positive with respect to the cytosolic face).

These commonalities between mitochondria, chloroplasts, and bacteria undoubtedly have an evolutionary origin. In bacteria both photosynthesis and oxidative phosphorylation occur on the plasma membrane. Analysis of the sequences and transcription of mitochondrial and chloroplast DNAs (Chapters 10 and 11) has given rise to the popular hypothesis that these organelles arose early in the evolution of eukaryotic cells by endocytosis of bacteria capable of oxidative phosphorylation or photosynthesis, respectively (Figure 8-3). According to this *endosymbiont*

▶ **FIGURE 8-1 Overview of the generation and utilization of a proton-motive force.** A transmembrane proton concentration gradient and a voltage gradient, collectively called the *proton-motive force,* are generated during photosynthesis and the aerobic oxidation of carbon compounds in mitochondria and aerobic bacteria. In chemiosmotic coupling, a proton-motive force powers an energy-requiring process such as ATP synthesis (A), transport of metabolites across the membrane against their concentration gradient (B), or rotation of bacterial flagella (C).

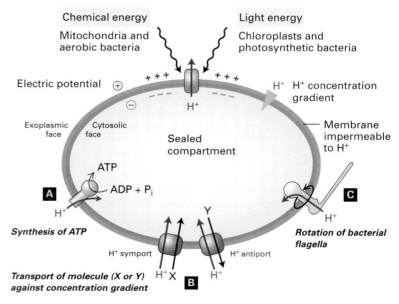

GENERATION OF PROTON-MOTIVE FORCE

CHEMIOSMOTIC COUPLING

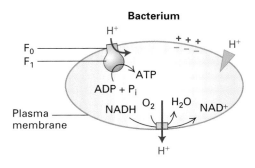

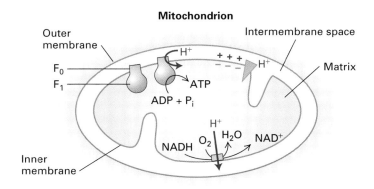

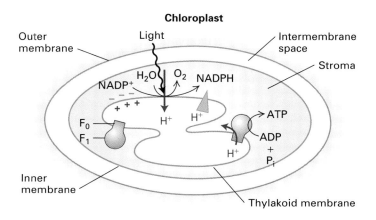

▲ FIGURE 8-2 Membrane orientation and the direction of proton movement during chemiosmotically coupled ATP synthesis in bacteria, mitochondria, and chloroplasts. The membrane surface facing a shaded area is a cytosolic face; the surface facing an unshaded area is an exoplasmic face. Note that the cytosolic face of the bacterial plasma membrane, the matrix face of the inner mitochondrial membrane, and the stromal face of the thylakoid membrane are all equivalent. During electron transport, protons are always pumped from the cytosolic face to the exoplasmic face, creating a proton concentration gradient (exoplasmic face > cytosolic face) and an electric potential (negative cytosolic face and positive exoplasmic face) across the membrane. During the coupled synthesis of ATP, protons flow in the reverse direction (down their electrochemical gradient) through ATP synthase (F$_0$F$_1$ complex), which protrudes from the cytosolic face in all cases.

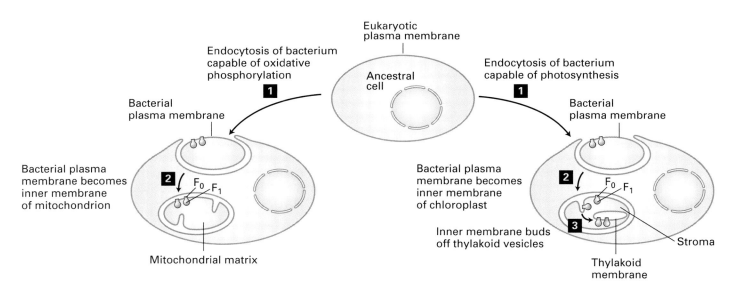

▲ FIGURE 8-3 Evolutionary origin of mitochondria and chloroplasts according to endosymbiont hypothesis. Membrane surfaces facing a shaded area are cytosolic faces; surfaces facing an unshaded area are exoplasmic faces. Endocytosis of a bacterium by an ancestral eukaryotic cell would generate an organelle with two membranes, the outer membrane derived from the eukaryotic plasma membrane and the inner one from the bacterial membrane. The F$_1$ subunit of

ATP synthase, localized to the cytosolic face of the bacterial membrane, would then face the matrix of the evolving mitochondrion (*left*) or chloroplast (*right*). Budding of vesicles from the inner chloroplast membrane, such as occurs during development of chloroplasts in contemporary plants, would generate the thylakoid vesicles with the F$_1$ subunit remaining on the cytosolic face, facing the chloroplast stroma.

hypothesis, the inner mitochondrial membrane would be derived from the bacterial plasma membrane with the globular F_1 domain still on its cytosolic face pointing toward the matrix space of the mitochondrion. Similarly, the globular F_1 domain would be on the cytosolic face of the thylakoid membrane facing the stromal space of the chloroplast.

In addition to powering ATP synthesis, the proton-motive force can supply energy for the transport of small molecules across a membrane against a concentration gradient (see Figure 8-1). For example, a H^+/sugar symport protein catalyzes the uptake of lactose by certain bacteria, and proton-driven antiporters catalyze the accumulation of ions and sucrose by plant vacuoles (Chapter 7). The proton-motive force also powers the rotation of bacterial flagella. (The beating of eukaryotic cilia, however, is powered by ATP hydrolysis.) Conversely, hydrolysis of ATP by V-class ATP-powered proton pumps, which are similar in structure to F-class pumps (see Figure 7-6), provides the energy for transporting protons against a concentration gradient. Chemiosmotic coupling thus illustrates an important principle introduced in our discussion of active transport in Chapter 7: *the membrane potential, the concentration gradients of protons (and other ions) across a membrane, and the phosphoanhydride bonds in ATP are equivalent and interconvertible forms of chemical potential energy.*

In this brief overview, we've seen that oxygen and carbohydrates are produced during photosynthesis, whereas they are consumed during aerobic oxidation. In both processes, the flow of electrons creates a H^+ electrochemical gradient, or proton-motive force, that can power ATP synthesis. As we examine these two processes at the molecular level, focusing first on aerobic oxidation and then on photosynthesis, the striking parallels between them will become evident.

8.1 Oxidation of Glucose and Fatty Acids to CO_2

The complete aerobic oxidation of each molecule of glucose yields 6 molecules of CO_2 and is coupled to the synthesis of as many as 30 molecules of ATP:

$$C_6H_{12}O_6 + 6\ O_2 + 30\ P_i^{2-} + 30\ ADP^{3-} + 30\ H^+$$
$$\longrightarrow 6\ CO_2 + 30\ ATP^{4-} + 36\ H_2O$$

Glycolysis, the initial stage of glucose metabolism, takes place in the cytosol and does not involve molecular O_2. It produces a small amount of ATP and the three-carbon compound *pyruvate.* In aerobic cells, pyruvate formed in glycolysis is transported into the mitochondria, where it is oxidized by O_2 to CO_2. Via chemiosmotic coupling, the oxidation of pyruvate in the mitochondria generates the

bulk of the ATP produced during the conversion of glucose to CO_2. In this section, we discuss the biochemical pathways that oxidize glucose and fatty acids to CO_2 and H_2O; the fate of the released electrons is described in the next section.

Cytosolic Enzymes Convert Glucose to Pyruvate During Glycolysis

A set of 10 water-soluble cytosolic enzymes catalyze the reactions constituting the *glycolytic pathway,* in which one molecule of glucose is converted to two molecules of pyruvate (Figure 8-4). All the metabolic intermediates between glucose and pyruvate are water-soluble phosphorylated compounds.

Four molecules of ATP are formed from ADP during glycolysis via **substrate-level phosphorylation,** which is catalyzed by enzymes in the cytosol (reactions 7 and 10). Unlike ATP formation in mitochondria and chloroplasts, a proton-motive force is not involved in substrate-level phosphorylation. Early in the glycolytic pathway, two ATP molecules are consumed: one by the addition of a phosphate residue to glucose in the reaction catalyzed by *hexokinase* (reaction 1), and another by the addition of a second phosphate to fructose 6-phosphate in the reaction catalyzed by *phosphofructokinase-1* (reaction 3). Thus glycolysis yields a net of only two ATP molecules per glucose molecule.

The balanced chemical equation for the conversion of glucose to pyruvate shows that four hydrogen atoms (four protons and four electrons) are also formed:

$$C_6H_{12}O_6 \longrightarrow 2\ CH_3\!-\!\overset{\overset{\displaystyle O}{\|}}{C}\!-\!\overset{\overset{\displaystyle O}{\|}}{C}\!-\!OH + 4\ H^+ + 4\ e^-$$
$$\text{Glucose} \qquad\qquad \text{Pyruvate}$$

(For convenience, we show pyruvate here in its un-ionized form, pyruvic acid, although at physiological pH it would be largely dissociated.) All four electrons and two of the four protons are transferred to two molecules of the oxidized form of the electron carrier **nicotinamide adenine dinucleotide (NAD^+)** to produce the reduced form, **NADH** (see Figure 2-26):

$$2\ H^+ + 4\ e^- + 2\ NAD^+ \longrightarrow 2\ NADH$$

The reaction that generates these hydrogen atoms and transfers them to NAD^+ is catalyzed by *glyceraldehyde 3-phosphate dehydrogenase* (see Figure 8-4, reaction 6). The overall chemical equation for this first stage of glucose metabolism is

$$C_6H_{12}O_6 + 2\ NAD^+ + 2\ ADP^{3-} + 2\ P_i^{2-}$$
$$\longrightarrow 2\ C_3H_4O_3 + 2\ NADH + 2\ ATP^{4-}$$

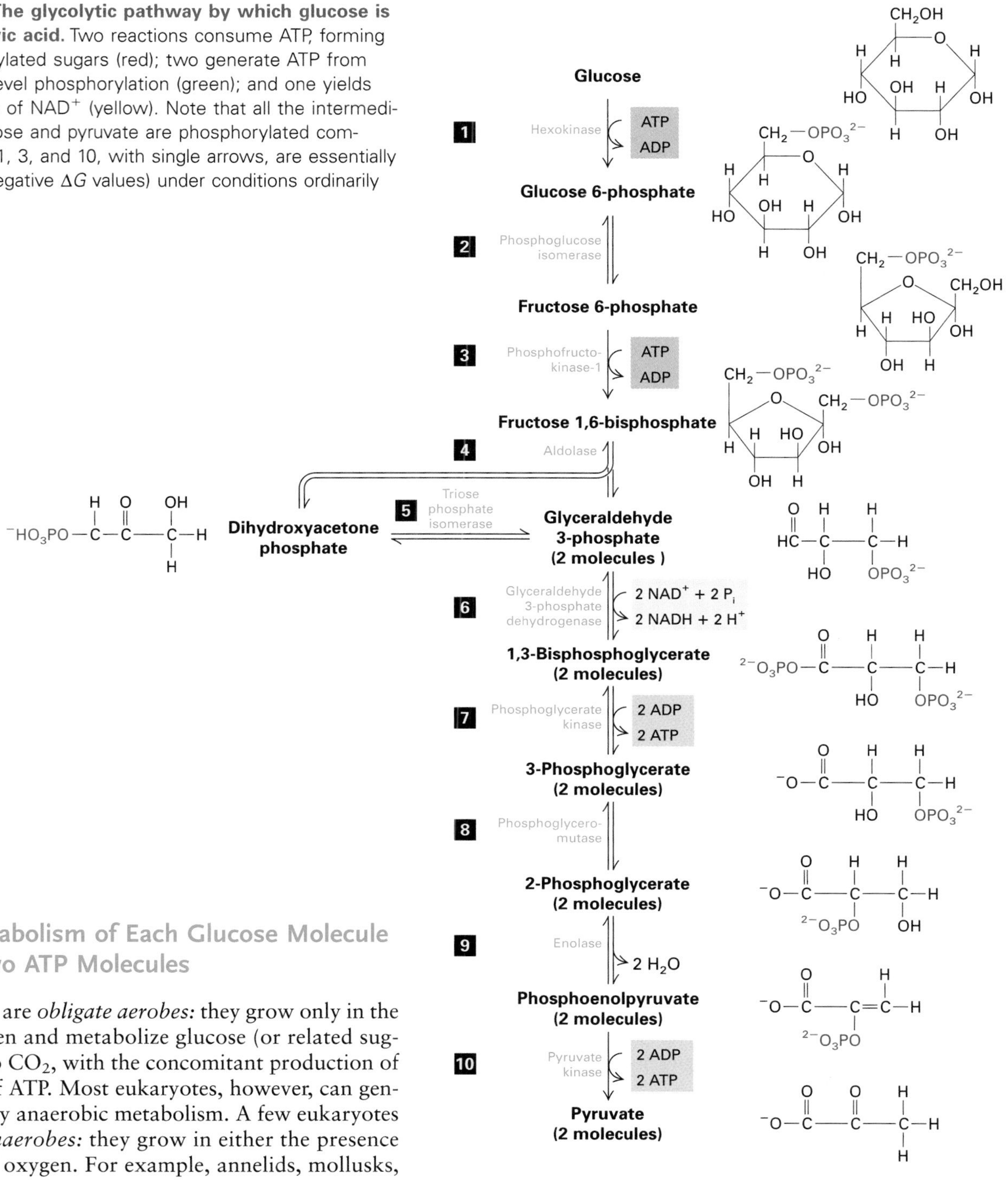

▶ **FIGURE 8-4 The glycolytic pathway by which glucose is degraded to pyruvic acid.** Two reactions consume ATP, forming ADP and phosphorylated sugars (red); two generate ATP from ADP by substrate-level phosphorylation (green); and one yields NADH by reduction of NAD^+ (yellow). Note that all the intermediates between glucose and pyruvate are phosphorylated compounds. Reactions 1, 3, and 10, with single arrows, are essentially irreversible (large negative ΔG values) under conditions ordinarily obtaining in cells.

Anaerobic Metabolism of Each Glucose Molecule Yields Only Two ATP Molecules

Many eukaryotes are *obligate aerobes:* they grow only in the presence of oxygen and metabolize glucose (or related sugars) completely to CO_2, with the concomitant production of a large amount of ATP. Most eukaryotes, however, can generate some ATP by anaerobic metabolism. A few eukaryotes are *facultative anaerobes:* they grow in either the presence or the absence of oxygen. For example, annelids, mollusks, and some yeasts can live and grow for days without oxygen.

In the absence of oxygen, facultative anaerobes convert glucose to one or more two- or three-carbon compounds, which are generally released into the surrounding medium. For instance, yeasts degrade glucose to two pyruvate molecules via glycolysis, generating a net of two ATP. In this process two NADH molecules are formed from NAD^+ per glucose molecule. In the absence of oxygen, yeasts convert pyruvate to one molecule each of ethanol and CO_2; in these

reactions two NADH molecules are oxidized to NAD^+ for each two pyruvates converted to ethanol, thereby regenerating the supply of NAD^+ (Figure 8-5a, *left*). This anaerobic degradation of glucose, called *fermentation,* is the basis of beer and wine production.

During the prolonged contraction of mammalian skeletal muscle cells, when oxygen becomes limited, muscle cells

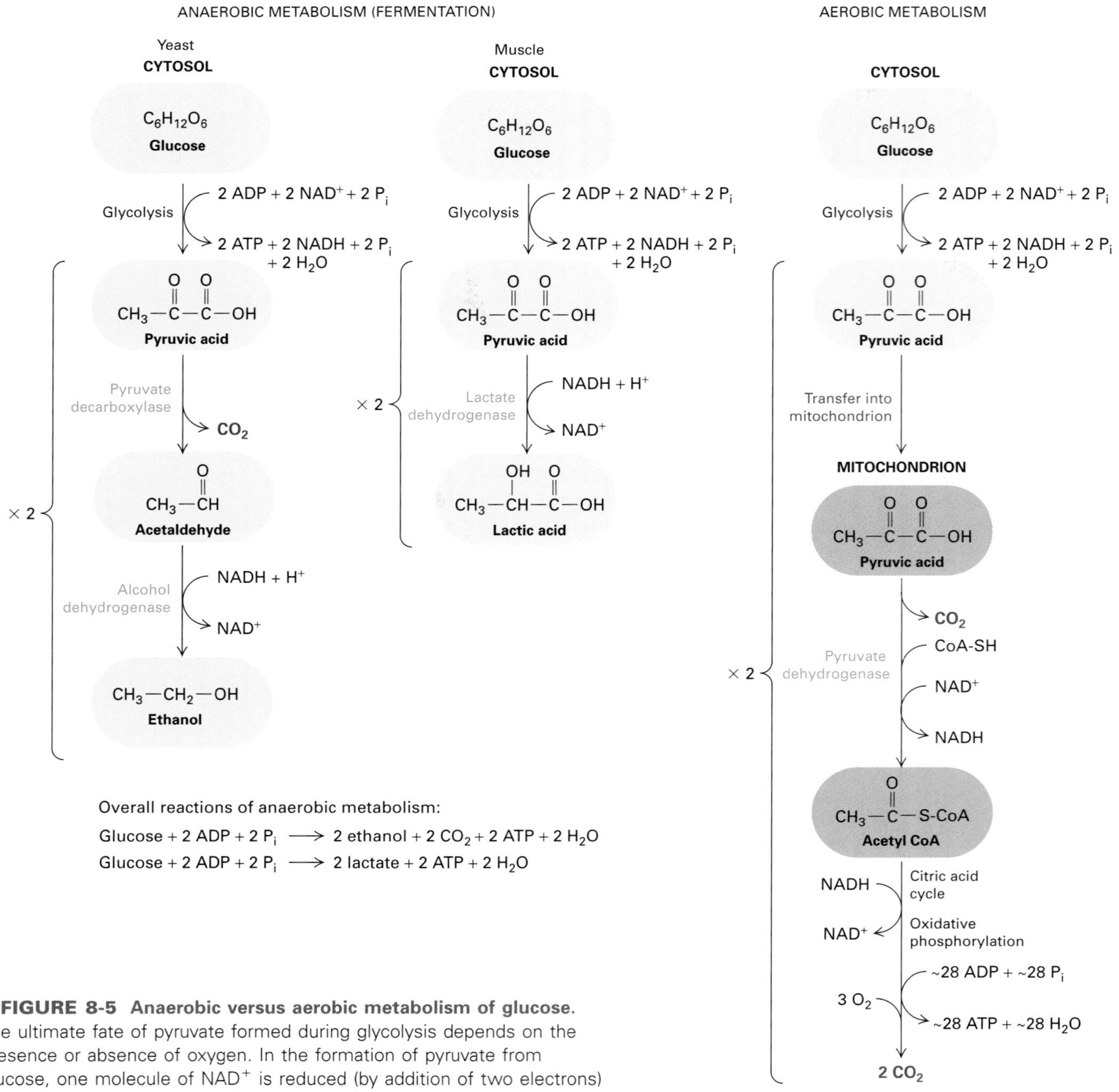

ANAEROBIC METABOLISM (FERMENTATION)

AEROBIC METABOLISM

▲ **FIGURE 8-5 Anaerobic versus aerobic metabolism of glucose.**
The ultimate fate of pyruvate formed during glycolysis depends on the
presence or absence of oxygen. In the formation of pyruvate from
glucose, one molecule of NAD^+ is reduced (by addition of two electrons)
to NADH for each molecule of pyruvate formed (see Figure 8-4, reaction
6). (*Left*) In the absence of oxygen, two electrons are transferred from
each NADH molecule to an acceptor molecule to regenerate NAD^+, which
is required for continued glycolysis. In yeasts, acetaldehyde is the acceptor
and ethanol is the product. This process is called *alcoholic fermentation.*
When oxygen is limiting in muscle cells, NADH reduces pyruvate to form
lactic acid, regenerating NAD^+. (*Right*) In the presence of oxygen, pyruvate
is transported into mitochondria. First it is converted by pyruvate
dehydrogenase into one molecule of CO_2 and one of acetic acid, the
latter linked to coenzyme A (CoA-SH) to form acetyl CoA, concomitant
with reduction of one molecule of NAD^+ to NADH. Further metabolism
of acetyl CoA and NADH generates approximately an additional 28
molecules of ATP per glucose molecule oxidized.

ferment glucose to two molecules of lactic acid—again, with the net production of only two molecules of ATP per glucose molecule (Figure 8-5a, *right*). The lactic acid causes muscle and joint aches. It is largely secreted into the blood; some passes into the liver, where it is reoxidized to pyruvate and either further metabolized to CO$_2$ aerobically or converted to glucose. Much lactate is metabolized to CO$_2$ by the heart, which is highly perfused by blood and can continue aerobic metabolism at times when exercising skeletal muscles secrete lactate. Lactic acid bacteria (the organisms that "spoil" milk) and other prokaryotes also generate ATP by the fermentation of glucose to lactate.

In the presence of oxygen, however, pyruvate formed in glycolysis is transported into mitochondria, where it is oxidized by O$_2$ to CO$_2$ in a series of oxidation reactions collectively termed **cellular respiration** (Figure 8-5b). These reactions generate an estimated 28 additional ATP molecules per glucose molecule, far outstripping the ATP yield from anaerobic glucose metabolism. To understand how mitochondria operate as ATP-generating factories, we first describe their structure and then the reactions they employ to degrade pyruvate.

Mitochondria Possess Two Structurally and Functionally Distinct Membranes

Mitochondria are among the larger organelles in the cell, each one being about the size of an *E. coli* bacterium. Most eukaryotic cells contain many mitochondria, which collectively can occupy as much as 25 percent of the volume of the cytoplasm. They are large enough to be seen under a light microscope, but the details of their structure can be viewed only with the electron microscope (see Figure 5-26). The outer membrane defines the smooth outer perimeter of the mitochondrion. In contrast, the inner membrane has numerous invaginations called *cristae*. These membranes define two submitochondrial compartments: the *intermembrane space* between the outer membrane and the inner membrane with its cristae, and the *matrix*, or central compartment (Figure 8-6). The fractionation and purification of these membranes and compartments have made it possible to determine their protein and phospholipid compositions and to localize each enzyme-catalyzed reaction to a specific membrane or space.

The outer membrane contains mitochondrial *porin*, a transmembrane channel protein similar in structure to

(a)

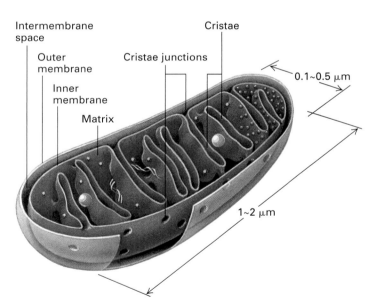

(b)

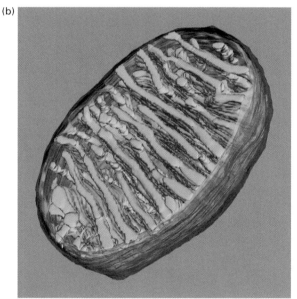

▲ FIGURE 8-6 Internal structure of a mitochondrion.
(a) Schematic diagram showing the principal membranes and compartments. The cristae form sheets and tubes by invagination of the inner membrane and connect to the inner membrane through relatively small uniform tubular structures called *crista junctions*. The intermembrane space appears continuous with the lumen of each crista. The F$_0$F$_1$ complexes (small red spheres), which synthesize ATP, are intramembrane particles that protrude from the cristae and inner membrane into the matrix. The matrix contains the mitochondrial DNA (blue strand), ribosomes (small blue spheres), and granules (large yellow spheres).

(b) Computer-generated model of a section of a mitochondrion from chicken brain. This model is based on a three-dimensional electron tomogram calculated from a series of two-dimensional electron micrographs recorded at regular angular intervals. This technique is analogous to a three-dimensional X-ray tomogram or CAT scan. Note the tightly packed cristae (yellow-green), the inner membrane (light blue), and the outer membrane (dark blue). [Part (a) courtesy of T. Frey; part (b) from T. Frey and C. Mannella, 2000, *Trends Biochem. Sci.* **25**:319.]

bacterial porins (see Figure 5-14). Ions and most small molecules (up to about 5000 Da) can readily pass through these channel proteins. Although the flow of metabolites across the outer membrane may limit their rate of mitochondrial oxidation, the inner membrane and cristae are the major permeability barriers between the cytosol and the mitochondrial matrix.

Freeze-fracture studies indicate that mitochondrial cristae contain many protein-rich intramembrane particles. Some are the F_0F_1 complexes that synthesize ATP; others function in transporting electrons to O_2 from NADH or other electron carriers. Various transport proteins located in the inner membrane and cristae allow otherwise impermeable molecules, such as ADP and P_i, to pass from the cytosol to the matrix, and other molecules, such as ATP, to move from the matrix into the cytosol. Protein constitutes 76 percent of the total weight of the inner membrane—a higher fraction than in any other cellular membrane. Cardiolipin (diphosphatidyl glycerol), a lipid concentrated in the inner membrane, sufficiently reduces the membrane's permeability to protons that a proton-motive force can be established across it.

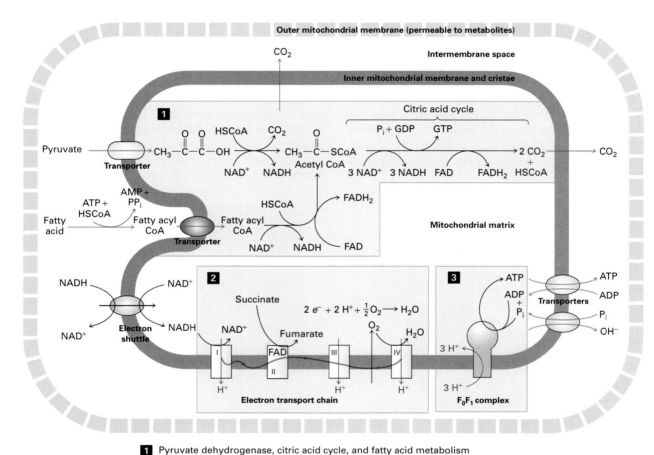

1 Pyruvate dehydrogenase, citric acid cycle, and fatty acid metabolism

2 Electron transport from NADH and $FADH_2$ to oxygen; generation of proton-motive force

3 ATP synthesis by F_0F_1 using proton-motive force

▲ **FIGURE 8-7 Summary of the aerobic oxidation of pyruvate and fatty acids in mitochondria.** The outer membrane is freely permeable to all metabolites, but specific transport proteins (colored ovals) in the inner membrane are required to import pyruvate (yellow), ADP (green), and P_i (purple) into the matrix and to export ATP (green). NADH generated in the cytosol is not transported directly to the matrix because the inner membrane is impermeable to NAD^+ and NADH; instead, a shuttle system (red) transports electrons from cytosolic NADH to NAD^+ in the matrix. O_2 diffuses into the matrix and CO_2 diffuses out. Stage 1: Fatty acyl groups are transferred from fatty acyl CoA and transported across the inner membrane via a special carrier (blue oval) and then reattached to CoA on the matrix side. Pyruvate is converted to acetyl CoA with the formation of NADH, and fatty acids (attached to CoA) are also converted to acetyl CoA with formation of NADH and FADH. Oxidation of acetyl CoA in the citric acid cycle generates NADH and $FADH_2$. Stage 2: Electrons from these reduced coenzymes are transferred via electron-transport complexes (blue boxes) to O_2 concomitant with transport of H^+ ions from the matrix to the intermembrane space, generating the proton-motive force. Electrons from NADH flow directly from complex I to complex III, bypassing complex II. Stage 3: ATP synthase, the F_0F_1 complex (orange), harnesses the proton-motive force to synthesize ATP. Blue arrows indicate electron flow; red arrows transmembrane movement of protons; and green arrows transport of metabolites.

The mitochondrial inner membrane, cristae, and matrix are the sites of most reactions involving the oxidation of pyruvate and fatty acids to CO_2 and H_2O and the coupled synthesis of ATP from ADP and P_i. These processes involve many steps but can be subdivided into three groups of reactions, each of which occurs in a discrete membrane or space in the mitochondrion (Figure 8-7):

1. Oxidation of pyruvate and fatty acids to CO_2 coupled to reduction of NAD^+ to NADH and of **flavin adenine dinucleotide (FAD)**, another oxidized electron carrier, to its reduced form, $FADH_2$ (see Figure 2-26). These electron carriers are often referred to as *coenzymes*. NAD^+, NADH, FAD, and $FADH_2$ are diffusible and not permanently bound to proteins. Most of the reactions occur in the matrix; two are catalyzed by inner-membrane enzymes that face the matrix.

2. Electron transfer from NADH and $FADH_2$ to O_2, regenerating the oxidized electron carriers NAD^+ and FAD. These reactions occur in the inner membrane and are coupled to the generation of a proton-motive force across it.

3. Harnessing of the energy stored in the electrochemical proton gradient for ATP synthesis by the F_0F_1 complex in the inner membrane.

The cristae greatly expand the surface area of the inner mitochondrial membrane, enhancing its ability to generate ATP (see Figure 8-6). In typical liver mitochondria, for example, the area of the inner membrane including cristae is about five times that of the outer membrane. In fact, the total area of all inner mitochondrial membranes in liver cells is about 17 times that of the plasma membrane. The mitochondria in heart and skeletal muscles contain three times as many cristae as are found in typical liver mitochondria—presumably reflecting the greater demand for ATP by muscle cells.

 In plants, stored carbohydrates, mostly in the form of starch, are hydrolyzed to glucose. Glycolysis then produces pyruvate, which is transported into mitochondria, as in animal cells. Mitochondrial oxidation of pyruvate and concomitant formation of ATP occur in photosynthetic cells during dark periods when photosynthesis is not possible, and in roots and other nonphotosynthetic tissues all the time. ∎

Acetyl CoA Derived from Pyruvate Is Oxidized to Yield CO_2 and Reduced Coenzymes in Mitochondria

Immediately after pyruvate is transported from the cytosol across the mitochondrial membranes to the matrix, it reacts with coenzyme A, forming CO_2 and the intermediate **acetyl CoA** (Figure 8-8). This reaction, catalyzed by *pyruvate dehydrogenase,* is highly exergonic ($\Delta G^{\circ\prime} = -8.0$ kcal/mol) and essentially irreversible. Note that during the pyruvate dehydrogenase reaction, NAD^+ is reduced, forming NADH; in contrast, during the reactions catalyzed by lactate dehydrogenase and alcohol dehydrogenase, NADH is oxidized, forming NAD^+ (see Figure 8-5).

As discussed later, acetyl CoA plays a central role in the oxidation of fatty acids and many amino acids. In addition, it is an intermediate in numerous biosynthetic reactions, such as the transfer of an acetyl group to lysine residues in histone proteins and to the N-termini of many mammalian proteins. Acetyl CoA also is a biosynthetic precursor of cholesterol and other steroids and of the farnesyl and related groups that form the lipid anchors used to attach some proteins (e.g., Ras) to membranes (see Figure 5-15). In respiring mitochondria, however, the acetyl group of acetyl CoA is almost always oxidized to CO_2.

The final stage in the oxidation of glucose entails a set of nine reactions in which the acetyl group of acetyl CoA is oxidized to CO_2. These reactions operate in a cycle that is referred to by several names: the **citric acid cycle,** the **tricarboxylic acid cycle,** and the **Krebs cycle.** The net result is that for each acetyl group entering the cycle as acetyl CoA, two molecules of CO_2 are produced.

As shown in Figure 8-9, the cycle begins with condensation of the two-carbon acetyl group from acetyl CoA with the four-carbon molecule *oxaloacetate* to yield the six-carbon *citric acid.* The two-step conversion of citrate to iso-citrate (reactions 2 and 3) is carried out by a single multifunctional enzyme. In both reactions 4 and 5, a CO_2 molecule is released. Reaction 5, catalyzed by the enzyme α-ketoglutarate dehydrogenase, also results in reduction of NAD^+ to NADH. This reaction is chemically similar to that catalyzed by pyruvate dehydrogenase, and indeed these two large enzyme complexes are similar in structure and mechanism. Reduction of NAD^+ to NADH also occurs during reactions 4 and 9; thus three molecules of NADH are generated per turn of the cycle. In reaction 7, two electrons and two protons are transferred

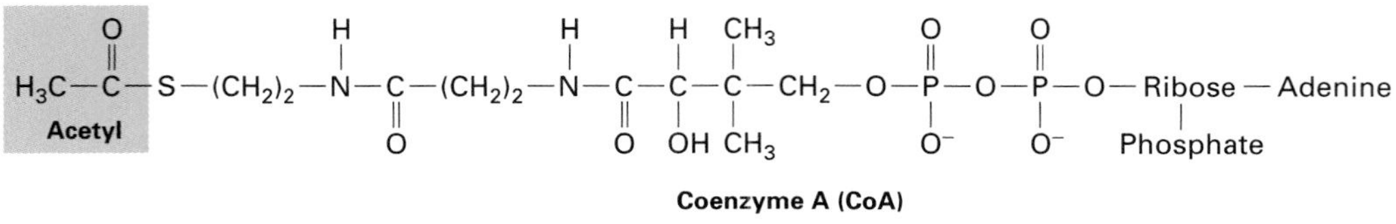

▲ **FIGURE 8-8 The structure of acetyl CoA.** This compound is an important intermediate in the aerobic oxidation of pyruvate, fatty acids, and many amino acids. It also contributes acetyl groups in many biosynthetic pathways.

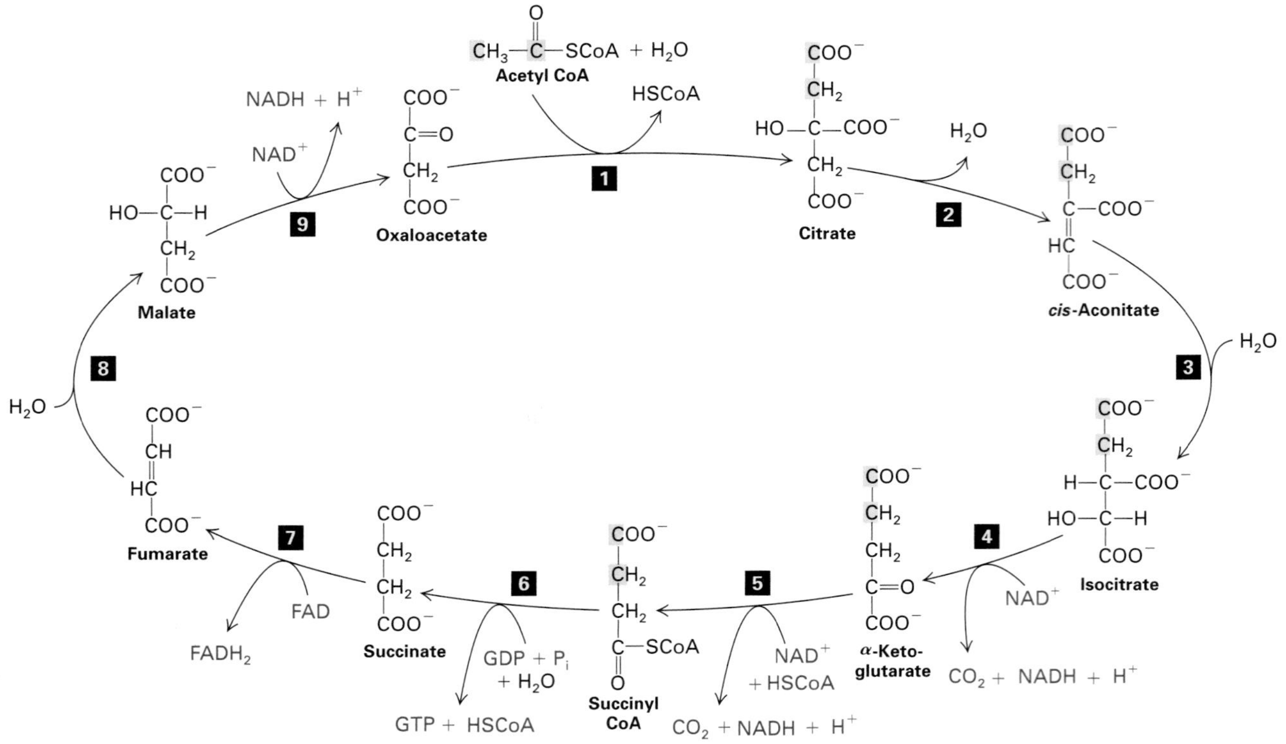

▲ **FIGURE 8-9 The citric acid cycle, in which acetyl groups transferred from acetyl CoA are oxidized to CO₂.** In reaction 1, a two-carbon acetyl residue from acetyl CoA condenses with the four-carbon molecule oxaloacetate to form the six-carbon molecule citrate. In the remaining reactions (2–9) each molecule of citrate is eventually converted back to oxaloacetate, losing two CO₂ molecules in the process. In each turn of the cycle, four pairs of electrons are removed from carbon atoms, forming three molecules of NADH and one molecule of FADH₂. The two carbon atoms that enter the cycle with acetyl CoA are highlighted in blue through succinyl CoA. In succinate and fumarate, which are symmetric molecules, they can no longer be specifically denoted. Isotope labeling studies have shown that these carbon atoms are *not* lost in the turn of the cycle in which they enter; on average one will be lost as CO₂ during the next turn of the cycle and the other in subsequent turns.

to FAD, yielding the reduced form of this coenzyme, FADH₂. In reaction 6, hydrolysis of the high-energy thioester bond in succinyl CoA is coupled to synthesis of one GTP by substrate-level phosphorylation (GTP and ATP are interconvertible). Reaction 9, the final one, also regenerates oxaloacetate, so the cycle can begin again. Note that molecular O₂ does not participate in the citric acid cycle.

Most enzymes and small molecules involved in the citric acid cycle are soluble in aqueous solution and are localized to the mitochondrial matrix. These include CoA, acetyl CoA, succinyl CoA, NAD⁺, and NADH, as well as six of the eight cycle enzymes. *Succinate dehydrogenase,* (reaction 7) and *α-ketoglutarate dehydrogenase* (reaction 5) are integral proteins in the inner membrane, with their active sites facing the matrix. When mitochondria are disrupted by gentle ultrasonic vibration or osmotic lysis, the six non-membrane-bound enzymes in the citric acid cycle are released as a very large multiprotein complex. The reaction product of one enzyme is thought to pass directly to the next enzyme without diffusing through the solution. However, much work is needed to determine the structure of this enzyme complex as it exists in the cell.

Since glycolysis of one glucose molecule generates two acetyl CoA molecules, the reactions in the glycolytic pathway and citric acid cycle produce six CO₂ molecules, ten NADH molecules, and two FADH₂ molecules per glucose molecule (Table 8-1). Although these reactions also generate four high-energy phosphoanhydride bonds in the form of two ATP and two GTP molecules, this represents only a small fraction of the available energy released in the complete aerobic oxidation of glucose. The remaining energy is stored in the reduced coenzymes NADH and FADH₂.

Synthesis of most of the ATP generated in aerobic oxidation is coupled to the reoxidation of NADH and FADH₂ by O₂ in a stepwise process involving the **respiratory chain,** also called the *electron transport chain.* Even though molecular O₂ is not involved in any reaction of the citric acid cycle, in the absence of O₂ the cycle soon stops operating as the supply of NAD⁺ and FAD dwindles. Before considering electron transport and the coupled formation of ATP in detail, we discuss first how the supply of NAD⁺ in the cytosol is regenerated and then the oxidation of fatty acids to CO₂.

TABLE 8-1 Net Result of the Glycolytic Pathway and the Citric Acid Cycle

Reaction	CO_2 Molecules Produced	NAD^+ Molecules Reduced to NADH	FAD Molecules Reduced to $FADH_2$
1 glucose molecule to 2 pyruvate molecules	0	2	0
2 pyruvates to 2 acetyl CoA molecules	2	2	0
2 acetyl CoA to 4 CO_2 molecules	4	6	2
Total	6	10	2

Transporters in the Inner Mitochondrial Membrane Allow the Uptake of Electrons from Cytosolic NADH

For aerobic oxidation to continue, the NADH produced during glycolysis in the cytosol must be oxidized to NAD^+. As with NADH generated in the mitochondrial matrix, electrons from cytosolic NADH are ultimately transferred to O_2 via the respiratory chain, concomitant with the generation of a proton-motive force. Although the inner mitochondrial membrane is impermeable to NADH itself, several *electron shuttles* can transfer electrons from cytosolic NADH to the matrix.

Operation of the most widespread shuttle—the *malate-aspartate shuttle*—is depicted in Figure 8-10. Critical to the shuttle are two antiport proteins in the inner mitochondrial membrane, a malate/α-ketoglutarate antiporter and a glutamate/aspartate antiporter, that permit transport of their

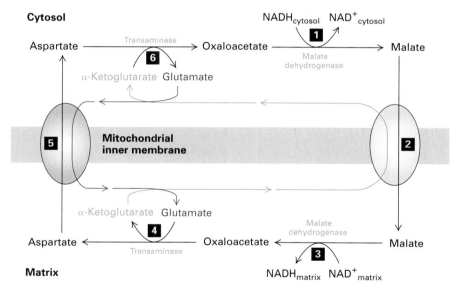

▲ **FIGURE 8-10 The malate shuttle.** This cyclical series of reactions transfers electrons from NADH in the cytosol (intermembrane space) across the inner mitochondrial membrane, which is impermeable to NADH itself. Step **1**: Cytosolic malate dehydrogenase transfers electrons from cytosolic NADH to oxaloacetate, forming malate. Step **2**: An antiporter (blue oval) in the inner mitochondrial membrane transports malate into the matrix in exchange for α-ketoglutarate. Step **3**: Mitochondrial malate dehydrogenase converts malate back to oxaloacetate, reducing NAD^+ in the matrix to NADH in the process. Step **4**: Oxaloacetate, which cannot directly cross the inner membrane, is converted to aspartate by addition of an amino group from glutamate. In this transaminase-catalyzed reaction in the matrix, glutamate is converted to α-ketoglutarate. Step **5**: A second antiporter (red oval) exports aspartate to the cytosol in exchange for glutamate. Step **6**: A cytosolic transaminase converts aspartate to oxaloacetate, completing the cycle. The blue and red arrows reflect the movement of the α-ketoglutarate and glutamate, respectively. In step **4** glutamate is deaminated to α-ketoglutarate, which is transported to the cytosol by an antiporter (step **2**; in step **6**, the α-ketoglutarate is aminated, converting it back to glutamate, which is transported to the matrix by the antiporter in step **5**.

substrates into and out of the matrix. Because oxaloacetate, one component of the shuttle, cannot directly cross the inner membrane, it is converted to the amino acid aspartate in the matrix and to malate in the cytosol. The net effect of the reactions constituting the malate-aspartate shuttle is oxidation of cytosolic NADH to NAD^+ and reduction of matrix NAD^+ to NADH:

$$NADH_{cytosol} + NAD^+_{matrix}$$
$$\longrightarrow NAD^+_{cytosol} + NADH_{matrix}$$

Mitochondrial Oxidation of Fatty Acids Is Coupled to ATP Formation

Fatty acids are stored as **triacylglycerols,** primarily as droplets in adipose (fat-storing) cells. In response to hormones such as adrenaline, triacylglycerols are hydrolyzed in the cytosol to free fatty acids and glycerol:

Fatty acids released into the blood are taken up and oxidized by most other cells, constituting the major energy source for many tissues, particularly heart muscle. In humans, the oxidation of fats is quantitatively more important than the oxidation of glucose as a source of ATP. The oxidation of

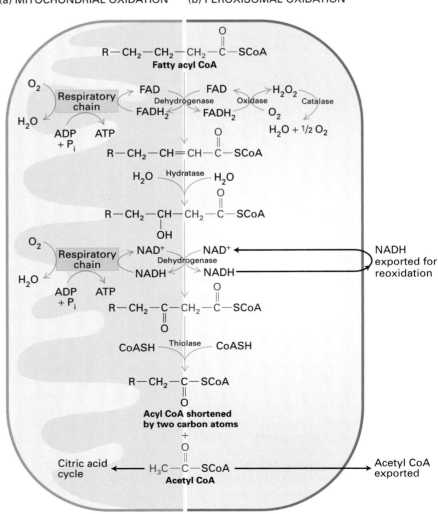

▶ **FIGURE 8-11 Oxidation of fatty acids in mitochondria and peroxisomes.** In both mitochondrial oxidation (a) and peroxisomal oxidation (b), four identical enzyme-catalyzed reactions (shown down the center of the figure) convert a fatty acyl CoA molecule to acetyl CoA and a fatty acyl CoA shortened by two carbon atoms. Concomitantly (in reactions moving to the left of center for mitochondria and to the right of center for peroxisomes), one FAD molecule is reduced to $FADH_2$, and one NAD^+ molecule is reduced to NADH. The cycle is repeated on the shortened acyl CoA until fatty acids with an even number of carbon atoms are completely converted to acetyl CoA. In mitochondria, electrons from $FADH_2$ and NADH enter the respiratory chain and ultimately are used to generate ATP; the acetyl CoA generated is oxidized in the citric acid cycle, resulting in synthesis of additional ATP. Because peroxisomes lack the electron-transport complexes composing the respiratory chain and the enzymes of the citric acid cycle, oxidation of fatty acids in these organelles yields no ATP. [Adapted from D. L. Nelson and M. M. Cox, *Lehninger Principles of Biochemistry,* 3d ed., 2000, Worth Publishers.]

1 g of triacylglycerol to CO_2 generates about six times as much ATP as does the oxidation of 1 g of hydrated glycogen, the polymeric storage form of glucose in muscle and liver. Triglycerides are more efficient for storage of energy because they are stored in anhydrous form and are much more reduced (have more hydrogens) than carbohydrates and therefore yield more energy when oxidized.

In the cytosol, free fatty acids are esterified to coenzyme A to form a fatty acyl CoA in an exergonic reaction coupled to the hydrolysis of ATP to AMP and PP_i (inorganic pyrophosphate):

$$R-\overset{\overset{O}{\|}}{C}-O^- + HSCoA + ATP \longrightarrow$$

Fatty acid

$$R-\overset{\overset{O}{\|}}{C}-SCoA + AMP + PP_i$$

Fatty acyl CoA

Subsequent hydrolysis of PP_i to two molecules of phosphate (P_i) drives this reaction to completion. Then the fatty acyl group is transferred to carnitine and moved across the inner mitochondrial membrane by an acylcarnitine transporter protein (see Figure 8-7, blue oval); on the matrix side, the fatty acyl group is released from carnitine and reattached to another CoA molecule.

Each molecule of a fatty acyl CoA in the mitochondrion is oxidized in a cyclical sequence of four reactions in which all the carbon atoms are converted to acetyl CoA with generation of NADH and $FADH_2$ (Figure 8-11a). For example, mitochondrial oxidation of each molecule of the 18-carbon stearic acid, $CH_3(CH_2)_{16}COOH$, yields nine molecules of acetyl CoA and eight molecules each of NADH and $FADH_2$. As with acetyl CoA generated from pyruvate, these acetyl groups enter the citric acid cycle and are oxidized to CO_2. Electrons from the reduced coenzymes produced in the oxidation of fatty acyl CoA to acetyl CoA and in the subsequent oxidation of acetyl CoA in the citric acid cycle move via the respiratory chain to O_2. This electron movement is coupled to generation of a proton-motive force that is used to power ATP synthesis as described previously for the oxidation of pyruvate (see Figure 8-7).

Peroxisomal Oxidation of Fatty Acids Generates No ATP

Mitochondrial oxidation of fatty acids is the major source of ATP in mammalian liver cells, and biochemists at one time believed this was true in all cell types. However, rats treated with clofibrate, a drug used to reduce the level of blood lipoproteins, were found to exhibit an increased rate of fatty acid oxidation and a large increase in the number of peroxisomes in their liver cells. This finding suggested that peroxisomes, as well as mitochondria, can oxidize fatty acids. These small organelles, ≈0.2–1 μm in diameter, are lined by a single membrane (see Figure 5-21). They are present in all mammalian cells except erythrocytes and are also found in plant cells, yeasts, and probably most other eukaryotic cells.

The peroxisome is now recognized as the principal organelle in which fatty acids are oxidized in most cell types. Indeed, very long chain fatty acids containing more than about 20 CH_2 groups are degraded only in peroxisomes; in mammalian cells, mid-length fatty acids containing 10–20 CH_2 groups can be degraded in both peroxisomes and mitochondria. In contrast to mitochondrial oxidation of fatty acids, which is coupled to generation of ATP, peroxisomal oxidation of fatty acids is not linked to ATP formation, and the released energy is converted to heat.

The reaction pathway by which fatty acids are degraded to acetyl CoA in peroxisomes is similar to that used in liver mitochondria (Figure 8-11b). However, peroxisomes lack a respiratory chain, and electrons from the $FADH_2$ produced during the oxidation of fatty acids are immediately transferred to O_2 by *oxidases*, regenerating FAD and forming hydrogen peroxide (H_2O_2). In addition to oxidases, peroxisomes contain abundant *catalase*, which quickly decomposes the H_2O_2, a highly cytotoxic metabolite. NADH produced during oxidation of fatty acids is exported and reoxidized in the cytosol. Peroxisomes also lack the citric acid cycle, so acetyl CoA generated during peroxisomal degradation of fatty acids cannot be oxidized further; instead it is transported into the cytosol for use in the synthesis of cholesterol and other metabolites.

 Before fatty acids can be degraded in the peroxisome, they must first be transported into the organelle from the cytosol. Mid-length fatty acids are esterified to coenzyme A in the cytosol; the resulting fatty acyl CoAs are then transported into the peroxisome by a specific transporter. However, very long chain fatty acids enter the peroxisome by another transporter and then are esterified to CoA once inside. In the human genetic disease X-linked adrenoleukodystrophy (ALD), peroxisomal oxidation of very long chain fatty acids is specifically defective, while the oxidation of mid-length fatty acids is normal. The most common peroxisomal disorder, ALD is marked by elevated levels of very long chain fatty acids in the plasma and tissues. Patients with the most severe form of ALD are unaffected until mid-childhood, when severe neurological disorders appear, followed by death within a few years. In recent years, the gene that is defective in ALD patients has been identified and cloned by techniques described in Chapter 9. Sequence analysis shows that the gene encodes an ABC transport protein (ABCD1) that is localized to peroxisomal membranes and is thought to mediate the import of very long chain fatty acids into the organelle. ∎

The Rate of Glucose Oxidation Is Adjusted to Meet the Cell's Need for ATP

All enzyme-catalyzed reactions and metabolic pathways are regulated by cells so as to produce the needed amounts of metabolites but not an excess. The primary function of the oxidation of glucose to CO_2 via the glycolytic pathway, the pyruvate dehydrogenase reaction, and the citric acid cycle is to produce NADH and $FADH_2$, whose oxidation in mitochondria generates ATP. Operation of the glycolytic pathway and citric acid cycle is continuously regulated, primarily by **allosteric** mechanisms, to meet the cell's need for ATP (see Chapter 3 for general principles of allosteric control).

Three allosterically controlled glycolytic enzymes play a key role in regulating the entire glycolytic pathway (see Figure 8-4). *Hexokinase* (step 1) is inhibited by its reaction product, glucose 6-phosphate. *Pyruvate kinase* (step 10) is inhibited by ATP, so glycolysis slows down if too much ATP is present. The third enzyme, *phosphofructokinase-1*, which converts fructose 6-phosphate to fructose 1,6-bisphosphate (step 3), is the principal rate-limiting enzyme of the glycolytic pathway. Emblematic of its critical role in regulating the rate of glycolysis, this enzyme is allosterically controlled by several molecules (Figure 8-12).

Phosphofructokinase-1 is allosterically *inhibited* by ATP and allosterically *activated* by AMP. As a result, the rate of glycolysis is very sensitive to the cell's energy charge, reflected in the ATP:AMP ratio. The allosteric inhibition of phosphofructokinase-1 by ATP may seem unusual, since ATP is also a substrate of this enzyme. But the affinity of the substrate-binding site for ATP is much higher (has a lower K_m) than that of the allosteric site. Thus at low concentrations, ATP binds to the catalytic but not to the inhibitory allosteric site, and enzymatic catalysis proceeds at near maximal rates. At high concentrations, ATP also binds to the allosteric site, inducing a conformational change that reduces the affinity of the enzyme for the other substrate, fructose 6-phosphate, and thus inhibits the rate of this reaction and the overall rate of glycolysis.

Another important allosteric activator of phosphofructokinase-1 is *fructose 2,6-bisphosphate*. This metabolite is formed from fructose 6-phosphate by *phosphofructokinase-2*, an enzyme different from phosphofructokinase-1. Fructose 6-phosphate accelerates the formation of fructose 2,6-bisphosphate, which, in turn, activates phosphofructokinase-1. This type of control, by analogy with feedback control, is known as *feed-forward activation,* in which the abundance of a metabolite (here, fructose 6-phosphate) induces an acceleration in its metabolism. Fructose 2,6-bisphosphate allosterically activates phosphofructokinase-1 in liver cells by decreasing the inhibitory effect of high ATP and by increasing the affinity of phosphofructokinase-1 for one of its substrates, fructose 6-phosphate.

The three glycolytic enzymes that are regulated by allosteric molecules catalyze reactions with large negative $\Delta G^{\circ\prime}$ values—reactions that are essentially irreversible under ordinary conditions. These enzymes thus are particularly suitable for regulating the entire glycolytic pathway. Additional control is exerted by glyceraldehyde 3-phosphate dehydrogenase, which catalyzes the reduction of NAD^+ to NADH (see Figure 8-4, step 6). If cytosolic NADH builds up owing to a slowdown in mitochondrial oxidation, this reaction will be slowed by mass action. As we discuss later, mitochondrial oxidation of NADH and $FADH_2$, produced in the glycolytic pathway and citric acid cycle, also is tightly controlled to produce the appropriate amount of ATP required by the cell.

Glucose metabolism is controlled differently in various mammalian tissues to meet the metabolic needs of the organism as a whole. During periods of carbohydrate starvation, for instance, glycogen in the liver is converted directly to glucose 6-phosphate (without involvement of hexokinase). Under these conditions, there is a reduction in fructose 2,6-bisphos-

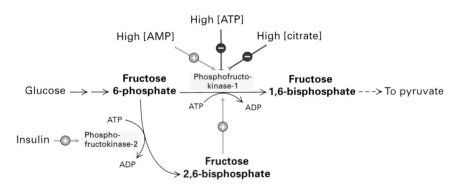

▲ **FIGURE 8-12 Allosteric control of glucose metabolism in the cytosol at the level of fructose 6-phosphate.** The key regulatory enzyme in glycolysis, phosphofructokinase-1, is allosterically activated by AMP and fructose 2,6-bisphosphate, which are elevated when the cell's energy stores are low. The enzyme is inhibited by ATP and citrate, which are elevated when the cell is actively oxidizing glucose to CO_2. Phosphofructokinase-2 (PFK2) is a bifunctional enzyme: its kinase activity forms fructose 2,6-bisphosphate from fructose 6-phosphate, and its phosphatase activity catalyzes the reverse reaction. Insulin, which is released by the pancreas when blood glucose levels are high, promotes PFK2 kinase activity and thus stimulates glycolysis. At low blood glucose, glucagon is released by the pancreas and promotes PFK2 phosphatase activity in the liver, indirectly slowing down glycolysis. We describe the role of insulin and glucagon in the integrated control of blood glucose levels in Chapter 15.

phate levels and decreased phosphofructokinase-1 activity (see Figure 8-12). As a result, glucose 6-phosphate derived from glycogen is not metabolized to pyruvate; rather, it is converted to glucose by a phosphatase and released into the blood to nourish the brain and red blood cells, which depend primarily on glucose as an energy fuel. (Chapter 13 contains a more detailed discussion of hormonal control of glucose metabolism in liver and muscle.) In all cases, the activity of these regulated enzymes is controlled by the level of small-molecule metabolites, generally by allosteric interactions or by hormone-mediated phosphorylation and dephosphorylation.

KEY CONCEPTS OF SECTION 8.1

Oxidation of Glucose and Fatty Acids to CO_2

■ In the cytosol of eukaryotic cells, glucose is converted to pyruvate via the glycolytic pathway, with the net formation of two ATPs and the net reduction of two NAD^+ molecules to NADH (see Figure 8-4). ATP is formed by two substrate-level phosphorylation reactions in the conversion of glyceraldehyde 3-phosphate to pyruvate.

■ In anaerobic conditions, cells can metabolize pyruvate to lactate or to ethanol plus CO_2 (in the case of yeast), with the reoxidation of NADH. In aerobic conditions, pyruvate is transported into the mitochondrion, where pyruvate dehydrogenase converts it into acetyl CoA and CO_2 (see Figure 8-5).

■ Mitochondria have a permeable outer membrane and an inner membrane, which is the site of electron transport and ATP synthesis.

■ In each turn of the citric acid cycle, acetyl CoA condenses with the four-carbon molecule oxaloacetate to form the six-carbon citrate, which is converted back to oxaloacetate by a series of reactions that release two molecules of CO_2 and generate three NADH molecules, one $FADH_2$ molecule and one GTP (see Figure 8-9).

■ Although cytosolic NADH generated during glycolysis cannot enter mitochondria directly, the malate-aspartate shuttle indirectly transfers electrons from the cytosol to the mitochondrial matrix, thereby regenerating cytosolic NAD^+ for continued glycolysis.

■ The flow of electrons from NADH and $FADH_2$ to O_2, via a series of electron carriers in the inner mitochondrial membrane, is coupled to pumping of protons across the inner membrane (see Figure 8-7). The resulting proton-motive force powers ATP synthesis and generates most of the ATP resulting from aerobic oxidation of glucose.

■ Oxidation of fatty acids in mitochondria yields acetyl CoA, which enters the citric acid cycle, and the reduced coenzymes NADH and $FADH_2$. Subsequent oxidation of these metabolites is coupled to formation of ATP.

■ In most eukaryotic cells, oxidation of fatty acids, especially very long chain fatty acids, occurs primarily in per-oxisomes and is not linked to ATP production; the released energy is converted to heat.

■ The rate of glucose oxidation via glycolysis and the citric acid cycle is controlled by the inhibition or stimulation of several enzymes, depending on the cell's need for ATP. This complex regulation coordinates the activities of the glycolytic pathway and the citric acid cycle and results in the storage of glucose (as glycogen) or fat when ATP is abundant.

8.2 Electron Transport and Generation of the Proton-Motive Force

As noted in the previous section, most of the free energy released during the oxidation of glucose to CO_2 is retained in the reduced coenzymes NADH and $FADH_2$ generated during glycolysis and the citric acid cycle. During respiration, electrons are released from NADH and $FADH_2$ and eventually are transferred to O_2, forming H_2O according to the following overall reactions:

$$NADH + H^+ + \tfrac{1}{2}\, O_2 \longrightarrow NAD^+ + H_2O$$

$$FADH_2 + \tfrac{1}{2}\, O_2 \longrightarrow FAD + H_2O$$

The $\Delta G^{\circ\prime}$ values for these strongly exergonic reactions are -52.6 kcal/mol (NADH) and -43.4 kcal/mol ($FADH_2$). Recall that the conversion of 1 glucose molecule to CO_2 via the glycolytic pathway and citric acid cycle yields 10 NADH and 2 $FADH_2$ molecules (see Table 8-1). Oxidation of these reduced coenzymes has a total $\Delta G^{\circ\prime}$ of -613 kcal/mol [10 (-52.6) + 2 (-43.4)]. Thus, of the potential free energy present in the chemical bonds of glucose (-680 kcal/mol), about 90 percent is conserved in the reduced coenzymes.

The free energy released during oxidation of a single NADH or $FADH_2$ molecule by O_2 is sufficient to drive the synthesis of several molecules of ATP from ADP and P_i, a reaction with a $\Delta G^{\circ\prime}$ of $+7.3$ kcal/mol. The mitochondrion maximizes the production of ATP by transferring electrons from NADH and $FADH_2$ through a series of electron carriers, all but one of which are integral components of the inner membrane. This step-by-step transfer of electrons via the respiratory (electron-transport) chain allows the free energy in NADH and $FADH_2$ to be released in small increments and stored as the proton-motive force.

At several sites during electron transport from NADH to O_2, protons from the mitochondrial matrix are pumped across the inner mitochondrial membrane; this "uphill" transport generates a proton concentration gradient across the inner membrane (Figure 8-13). Because the outer membrane is freely permeable to protons, whereas the inner membrane is not, this pumping causes the pH of the mitochondrial matrix to become higher (i.e., the H^+ concentration is lower) than that of the cytosol and intermembrane space. An electric potential across the inner membrane also

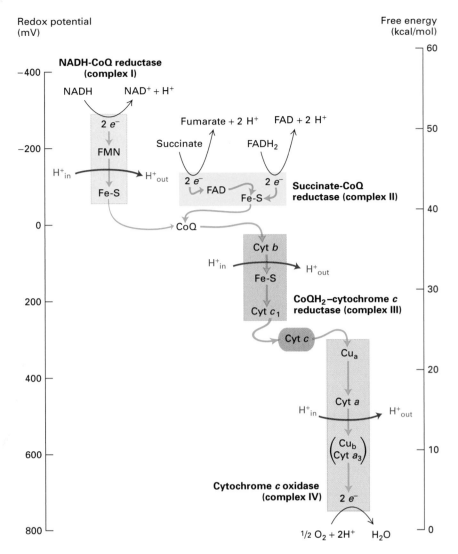

▲ **FIGURE 8-13 Changes in redox potential and free energy during stepwise flow of electrons through the respiratory chain.** Blue arrows indicate electron flow; red arrows, translocation of protons across the inner mitochondrial membrane. Four large multiprotein complexes located in the inner membrane contain several electron-carrying prosthetic groups. Coenzyme Q (CoQ) and cytochrome c transport electrons between the complexes. Electrons pass through the multiprotein complexes from those at a lower reduction potential to those with a higher (more positive) potential (left scale), with a corresponding reduction in free energy (right scale). The energy released as electrons flow through three of the complexes is sufficient to power the pumping of H^+ ions across the membrane, establishing a proton-motive force.

results from the uphill pumping of H^+ outward from the matrix, which becomes negative with respect to the intermembrane space. Thus free energy released during the oxidation of NADH or $FADH_2$ is stored both as an electric potential and a proton concentration gradient—collectively, the proton-motive force—across the inner membrane. The movement of protons back across the inner membrane, driven by this force, is coupled to the synthesis of ATP from ADP and P_i by ATP synthase (see Figure 8-7).

The synthesis of ATP from ADP and P_i, driven by the transfer of electrons from NADH or $FADH_2$ to O_2, is the major source of ATP in aerobic nonphotosynthetic cells. Much evidence shows that in mitochondria and bacteria this process, called **oxidative phosphorylation**, depends on generation of a proton-motive force across the inner membrane, with electron transport, proton pumping, and ATP formation occurring simultaneously. In the laboratory, for instance, addition of O_2 and an oxidizable substrate such as pyruvate or succinate to isolated intact mitochondria results in a net synthesis of ATP if the inner mitochondrial membrane is intact. In the presence of minute amounts of detergents that make the membrane leaky, electron transport and the oxidation of these metabolites by O_2 still occurs, but no ATP is made. Under these conditions, no transmembrane proton concentration gradient or membrane electric potential can be maintained.

In this section we first discuss the magnitude of the proton-motive force, then the components of the respiratory chain and the pumping of protons across the inner membrane. In the following section we describe the structure of the ATP synthase and how it uses the proton-motive force to synthesize ATP. We also consider how mitochondrial oxidation of NADH and $FADH_2$ is controlled to meet the cell's need for ATP.

The Proton-Motive Force in Mitochondria Is Due Largely to a Voltage Gradient Across the Inner Membrane

As we've seen, the proton-motive force (pmf) is the sum of a transmembrane proton concentration (pH) gradient and electric potential, or voltage gradient. The relative contribution of the two components to the total pmf depends on the permeability of the membrane to ions other than H^+. A significant voltage gradient can develop only if the membrane is poorly permeable to other cations and to anions, as is the inner mitochondrial membrane. In this case, the developing voltage gradient (i.e., excess H^+ ions on the intermembrane face and excess anions on the matrix face) soon prevents further proton movement, so only a small pH gradient is generated. In contrast, a significant pH gradient can develop only if the membrane is also permeable to a major anion (e.g., Cl^-) or if the H^+ ions are exchanged for another cation (e.g., K^+). In either case, proton movement does not lead to a voltage gradient across the membrane because there is always an equal concentration of positive and negative ions on each side of the membrane. This is the situation in the chloroplast thylakoid membrane during photosynthesis, as we discuss later. Compared with chloroplasts, then, a greater portion of the pmf in mitochondria is due to the membrane electric potential, and the actual pH gradient is smaller.

Since a difference of one pH unit represents a tenfold difference in H^+ concentration, a pH gradient of one unit across a membrane is equivalent to an electric potential of 59 mV at 20 °C according to the Nernst equation (Chapter 7). Thus we can define the proton-motive force, pmf, as

$$\text{pmf} = \Psi - \left(\frac{RT}{F} \times \Delta\text{pH} \right) = \Psi - 59\,\Delta\text{pH}$$

where R is the gas constant of 1.987 cal/(degree·mol), T is the temperature (in degrees Kelvin), F is the Faraday constant [23,062 cal/(V·mol)], and Ψ is the transmembrane electric potential; Ψ and pmf are measured in millivolts. Measurements on respiring mitochondria have shown that the electric potential Ψ across the inner membrane is −160 mV (negative inside matrix) and that ΔpH is ≈1.0 (equivalent to ≈60 mV). Thus the total pmf is −220 mV, with the transmembrane electric potential responsible for about 73 percent.

Because mitochondria are much too small to be impaled with electrodes, the electric potential and pH gradient across the inner mitochondrial membrane cannot be determined by direct measurement. However, researchers can measure the inside pH by trapping fluorescent pH-sensitive dyes inside vesicles formed from the inner mitochondrial membrane. They also can determine the electric potential by adding radioactive $^{42}K^+$ ions and a trace amount of valinomycin to a suspension of respiring mitochondria. Although the inner membrane is normally impermeable to K^+, valinomycin is an **ionophore**, a small lipid-soluble molecule that selectively binds a specific ion (in this case, K^+) in its hydrophilic interior and carries it across otherwise impermeable membranes. In the presence of valinomycin, $^{42}K^+$ equilibrates across the inner membrane of isolated mitochondria in accordance with the electric potential; the more negative the matrix side of the membrane, the more $^{42}K^+$ will accumulate in the matrix.

Addition of small amounts of valinomycin and radioactive K^+ has little effect on oxidative phosphorylation by a suspension of respiring mitochondria. At equilibrium, the measured concentration of radioactive K^+ ions in the matrix, $[K_{in}]$, is about 500 times greater than that in the surrounding medium, $[K_{out}]$. Substitution of this value into the Nernst equation shows that the electric potential E (in mV) across the inner membrane in respiring mitochondria is −160 mV, with the inside negative:

$$E = -59 \log \frac{[K_{in}]}{[K_{out}]} = -59 \log 500 = -160 \text{ mV}$$

Electron Transport in Mitochondria Is Coupled to Proton Translocation

The coupling between electron transport from NADH (or $FADH_2$) to O_2 and proton transport across the inner mitochondrial membrane, which generates the proton-motive force, also can be demonstrated experimentally with isolated mitochondria (Figure 8-14). As soon as O_2 is added to a suspension of mitochondria, the medium outside the mitochondria becomes acidic. During electron transport from NADH to O_2, protons translocate from the matrix to the intermembrane space; since the outer membrane is freely permeable to protons, the pH of the outside medium is lowered briefly. The measured change in pH indicates that about 10 protons are transported out of the matrix for every electron pair transferred from NADH to O_2.

When this experiment is repeated with succinate rather than NADH as the reduced substrate, the medium outside the mitochondria again becomes acidic, but less so. Recall that oxidation of succinate to fumarate in the citric acid cycle generates $FADH_2$ (see Figure 8-9). Because electrons in $FADH_2$ have less potential energy (43.4 kcal/mol) than electrons in NADH (52.6 kcal/mole), $FADH_2$ transfers electrons to the respiratory chain at a later point than NADH does. As a result, electron transport from $FADH_2$ (or succinate) results in translocation of fewer protons from the matrix, and thus a smaller change in pH (see Figure 8-13).

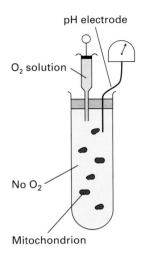
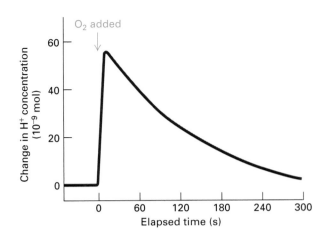

▲ **EXPERIMENTAL FIGURE 8-14 Electron transfer from NADH or FADH₂ to O₂ is coupled to proton transport across the mitochondrial membrane.** If NADH is added to a suspension of mitochondria depleted of O_2, no NADH is oxidized. When a small amount of O_2 is added to the system (arrow), the pH of the surrounding medium drops sharply—a change that corresponds to an *increase* in protons outside the mitochondria. (The presence of a large amount of valinomycin and K^+ in the reaction dissipates the voltage gradient generated by H^+ translocation, so that all pumped H^+ ions contribute to the pH change.) Thus the oxidation of NADH by O_2 is coupled to the movement of protons out of the matrix. Once the O_2 is depleted, the excess protons slowly move back into the mitochondria (powering the synthesis of ATP) and the pH of the extracellular medium returns to its initial value.

Electrons Flow from FADH₂ and NADH to O₂ Through a Series of Four Multiprotein Complexes

We now examine more closely the energetically favored movement of electrons from NADH and $FADH_2$ to the final electron acceptor, O_2. In respiring mitochondria, each NADH molecule releases two electrons to the respiratory chain; these electrons ultimately reduce one oxygen atom (half of an O_2 molecule), forming one molecule of water:

$$NADH \longrightarrow NAD^+ + H^+ + 2\ e^-$$

$$2\ e^- + 2\ H^+ + \tfrac{1}{2}\ O_2 \longrightarrow H_2O$$

As electrons move from NADH to O_2, their potential declines by 1.14 V, which corresponds to 26.2 kcal/mol of electrons transferred, or $\approx$ 53 kcal/mol for a pair of electrons. As noted earlier, much of this energy is conserved in the proton-motive force generated across the inner mitochondrial membrane.

Each of the four large multiprotein complexes in the respiratory chain spans the inner mitochondrial membrane and contains several *prosthetic groups* that participate in moving electrons. These small nonpeptide organic molecules or metal ions are tightly and specifically associated with the multiprotein complexes (Table 8-2). Before considering the function of each complex, we examine several of these electron carriers.

Several types of *heme*, an iron-containing prosthetic group similar to that in hemoglobin and myoglobin, are tightly bound or covalently linked to mitochondrial proteins, forming the **cytochromes** (Figure 8-15a). Electron flow through the cytochromes occurs by oxidation and reduction of the Fe atom in the center of the heme molecule:

$$Fe^{3+} + e^- \rightleftharpoons Fe^{2+}$$

TABLE 8-2	Electron-Carrying Prosthetic Groups in the Respiratory Chain
Protein Component	**Prosthetic Groups***
NADH-CoQ reductase (complex I)	FMN Fe-S
Succinate-CoQ reductase (complex II)	FAD Fe-S
CoQH₂–cytochrome c reductase (complex III)	Heme b_L Heme b_H Fe-S Heme c_1
Cytochrome c	Heme c
Cytochrome c oxidase (complex IV)	Cu_a^{2+} Heme a Cu_b^{2+} Heme a_3

*Not included is coenzyme Q, an electron carrier that is not permanently bound to a protein complex.
SOURCE: J. W. De Pierre and L. Ernster, 1977, *Ann. Rev. Biochem.* **46**:201.

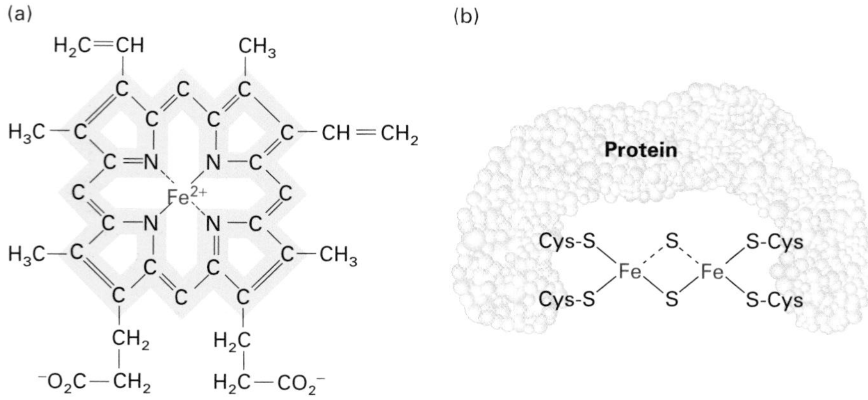

(a)   (b)

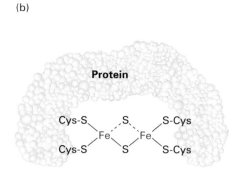

▲ **FIGURE 8-15 Heme and iron-sulfur prosthetic groups in the respiratory (electron-transport) chain.** (a) Heme portion of cytochromes b_L and b_H, which are components of the CoQH$_2$–cytochrome c reductase complex. The same porphyrin ring (yellow) is present in all hemes. The chemical substituents attached to the porphyrin ring differ in the other cytochromes in the respiratory chain. All hemes accept and release one electron at a time. (b) Dimeric iron-sulfur cluster (2Fe-2S). Each Fe atom is bonded to four S atoms: two are inorganic sulfur and two are in cysteine side chains of the associated protein. (Note that only the two inorganic S atoms are counted in the chemical formula.) All Fe-S clusters accept and release one electron at a time.

In the respiratory chain, electrons move through the cytochromes in the following order: b, c_1, c, a, and a_3 (see Figure 8-13). The various cytochromes have slightly different heme groups and axial ligands, which generate different environments for the Fe ion. Therefore, each cytochrome has a different reduction potential, or tendency to accept an electron—an important property dictating the unidirectional electron flow along the chain. Because the heme ring in cytochromes consists of alternating double- and single-bonded atoms, a large number of resonance forms exist, and the extra electron is delocalized to the heme carbon and nitrogen atoms as well as to the Fe ion. All the cytochromes, except cytochrome c, are components of multiprotein complexes in the inner mitochondrial membrane. Although cytochrome c comprises a heme-protein complex, it moves freely by diffusion in the intermembrane space.

Iron-sulfur clusters are nonheme, iron-containing prosthetic groups consisting of Fe atoms bonded both to inorganic S atoms and to S atoms on cysteine residues on a protein (Figure 8-15b). Some Fe atoms in the cluster bear a +2 charge; others have a +3 charge. However, the net charge of each Fe atom is actually between +2 and +3 because electrons in the outermost orbits are dispersed among the Fe atoms and move rapidly from one atom to another. Iron-sulfur clusters accept and release electrons one at a time; the additional electron is also dispersed over all the Fe atoms in the cluster.

Coenzyme Q (CoQ), also called *ubiquinone*, is the only electron carrier in the respiratory chain that is not a protein-bound prosthetic group. It is a carrier of hydrogen atoms, that is, protons plus electrons. The oxidized quinone form of CoQ can accept a single electron to form a semiquinone, a charged free radical denoted by CoQ⁻·. Addition of a second electron

and two protons to CoQ⁻· forms dihydroubiquinone (CoQH$_2$), the fully reduced form (Figure 8-16). Both CoQ and CoQH$_2$ are soluble in phospholipids and diffuse freely in the inner mitochondrial membrane.

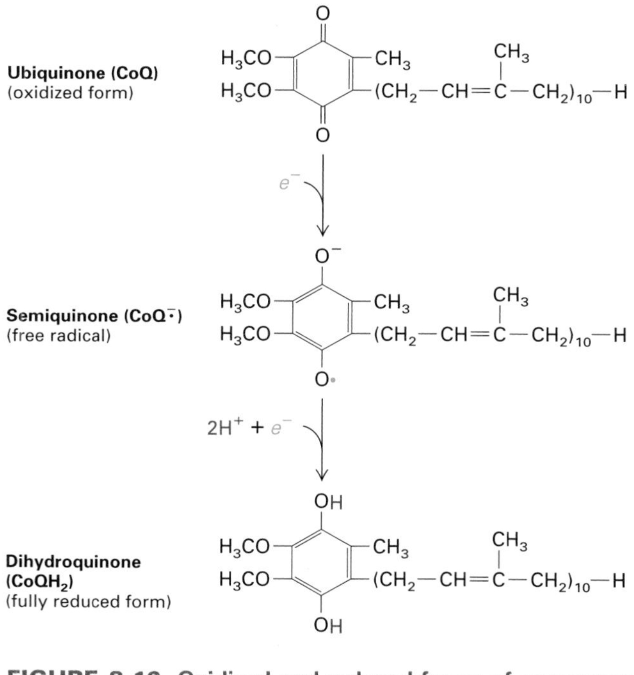

▲ **FIGURE 8-16 Oxidized and reduced forms of coenzyme Q (CoQ), which carries two protons and two electrons.** Because of its long hydrocarbon "tail" of isoprene units, CoQ is soluble in the hydrophobic core of phospholipid bilayers and is very mobile. Reduction of CoQ to the fully reduced form, QH$_2$, occurs in two steps with a half-reduced free-radical intermediate, called *semiquinone*.

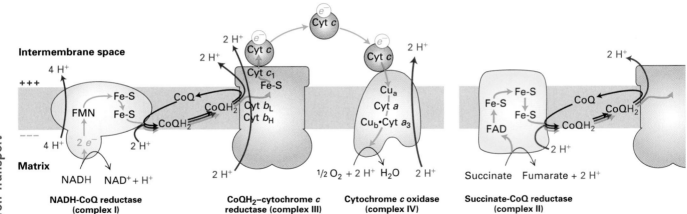

▲ FIGURE 8-17 Overview of multiprotein complexes, bound prosthetic groups, and associated mobile carriers in the respiratory chain. Blue arrows indicate electron flow; red arrows indicate proton translocation. (*Left*) Pathway from NADH. A total of 10 protons are translocated per pair of electrons that flow from NADH to O_2. The protons released into the matrix space during oxidation of NADH by NADH-CoQ reductase are consumed in the formation of water from O_2 by cytochrome c oxidase, resulting in no net proton translocation from these reactions. (*Right*) Pathway from succinate. During oxidation of succinate to fumarate and reduction of CoQ by the succinate-CoQ reductase complex, no protons are translocated across the membrane. The remainder of electron transport from $CoQH_2$ proceeds by the same pathway as in the left diagram. Thus for every pair of electrons transported from succinate to O_2, six protons are translocated from the matrix into the intermembrane space. Coenzyme Q and cytochrome c function as mobile carriers in electron transport from both NADH and succinate. See the text for details.

As shown in Figure 8-17, CoQ accepts electrons released from the NADH-CoQ reductase complex (I) and the succinate-CoQ reductase complex (II) and donates them to the $CoQH_2$–cytochrome c reductase complex (III). Importantly, reduction and oxidation of CoQ are coupled to pumping of protons. Whenever CoQ accepts electrons, it does so at a binding site on the cytosolic (matrix) face of the protein complex, always picking up protons from the medium facing the cytosolic face. Whenever $CoQH_2$ releases its electrons, it does so at a binding site on the exoplasmic face of the protein complex, releasing protons into the exoplasmic medium (intermembrane space). Thus transport of each pair of electrons by CoQ is obligatorily coupled to movement of two protons from the cytosolic to the exoplasmic medium.

NADH-CoQ Reductase (Complex I) Electrons are carried from NADH to CoQ by the NADH-CoQ reductase complex. NAD^+ is exclusively a two-electron carrier: it accepts or releases a pair of electrons at a time. In the NADH-CoQ reductase complex, electrons first flow from NADH to FMN (flavin mononucleotide), a cofactor related to FAD, then to an iron-sulfur cluster, and finally to CoQ (see Figure 8-17). FMN, like FAD, can accept two electrons, but does so one electron at a time.

The overall reaction catalyzed by this complex is

$$NADH + CoQ + 2\ H^+ \longrightarrow NAD^+ + H^+ + CoQH_2$$
$$\text{(Reduced)}\quad\text{(Oxidized)}\qquad\qquad\text{(Oxidized)}\qquad\quad\text{(Reduced)}$$

Each transported electron undergoes a drop in potential of ≈360 mV, equivalent to a $\Delta G^{\circ\prime}$ of -16.6 kcal/mol for the two electrons transported (see Figure 8-13). Much of this released energy is used to transport four protons across the inner membrane per molecule of NADH oxidized by the NADH-CoQ reductase complex.

Succinate-CoQ Reductase (Complex II) Succinate dehydrogenase, the enzyme that oxidizes a molecule of succinate to fumarate in the citric acid cycle, is an integral component of the succinate-CoQ reductase complex. The two electrons released in conversion of succinate to fumarate are transferred first to FAD, then to an iron-sulfur cluster, and finally to CoQ (see Figure 8-17). The overall reaction catalyzed by this complex is

$$\text{Succinate} + CoQ \longrightarrow \text{fumarate} + CoQH_2$$
$$\text{(Reduced)}\qquad\text{(Oxidized)}\qquad\text{(Oxidized)}\qquad\text{(Reduced)}$$

Although the $\Delta G^{\circ\prime}$ for this reaction is negative, the released energy is insufficient for proton pumping. Thus no protons are translocated across the membrane by the succinate-CoQ reductase complex, and no proton-motive force is generated in this part of the respiratory chain.

$CoQH_2$–Cytochrome c Reductase (Complex III) A $CoQH_2$ generated either by complex I or complex II donates two electrons to the $CoQH_2$–cytochrome c reductase complex, regenerating oxidized CoQ. Concomitantly it releases two protons picked up on the cytosolic face into the intermembrane space, generating part of the proton-motive force. Within complex III, the released electrons first are transferred to an iron-sulfur cluster within complex III and then to two *b*-type cytochromes (b_L and b_H) or cytochrome c_1. Finally, the two electrons are transferred to two molecules of the oxidized form of cytochrome *c*, a water-soluble peripheral protein that diffuses in the intermembrane space (see Figure 8-17). For each pair of electrons transferred, the overall reaction catalyzed by the CoQH$_2$–cytochrome *c* reductase complex is

$$\text{CoQH}_2 + 2\text{ Cyt }c^{3+} \longrightarrow \text{CoQ} + 2\text{ H}^+ + 2\text{ Cyt }c^{2+}$$
(Reduced)   (Oxidized)   (Oxidized)   (Reduced)

The $\Delta G^{\circ\prime}$ for this reaction is sufficiently negative that two additional protons are translocated from the mitochondrial matrix across the inner membrane for each pair of electrons transferred; this involves the proton-motive Q cycle discussed later.

Cytochrome *c* Oxidase (Complex IV) Cytochrome *c*, after being reduced by the CoQH$_2$–cytochrome *c* reductase complex, transports electrons, one at a time, to the cytochrome *c* oxidase complex (Figure 8-18). Within this complex, electrons are transferred, again one at a time, first to a pair of copper ions called Cu_a^{2+}, then to cytochrome *a*, next to a complex of another copper ion (Cu_b^{2+}) and cytochrome a_3, and finally to O$_2$, the ultimate electron acceptor, yielding H$_2$O. For each pair of electrons transferred, the overall reaction catalyzed by the cytochrome *c* oxidase complex is

$$2\text{ Cyt }c^{2+} + 2\text{ H}^+ + \tfrac{1}{2}\text{ O}_2 \longrightarrow 2\text{ Cyt }c^{3+} + \text{H}_2\text{O}$$
(Reduced)   (Oxidized)

During transport of each pair of electrons through the cytochrome *c* oxidase complex, two protons are translocated across the membrane.

CoQ and Cytochrome *c* as Mobile Electron Shuttles The four electron-transport complexes just described are laterally mobile in the inner mitochondrial membrane; moreover, they are present in unequal amounts and do not form stable contacts with one another. These properties preclude the direct transfer of electrons from one complex to the next. Instead, electrons are transported from one complex to another by diffusion of CoQ in the membrane and by cytochrome *c* in the intermembrane space, as depicted in Figure 8-17.

Reduction Potentials of Electron Carriers Favor Electron Flow from NADH to O$_2$

As we saw in Chapter 2, the **reduction potential *E*** for a partial reduction reaction

$$\text{Oxidized molecule} + \text{e}^- \rightleftharpoons \text{reduced molecule}$$

is a measure of the equilibrium constant of that partial reaction. With the exception of the *b* cytochromes in the CoQH$_2$–cytochrome *c* reductase complex, the standard reduction potential $E^{\circ\prime}$ of the carriers in the mitochondrial respiratory chain increases steadily from NADH to O$_2$. For instance, for the partial reaction

$$\text{NAD}^+ + \text{H}^+ + 2\text{ e}^- \rightleftharpoons \text{NADH}$$

the value of the standard reduction potential is -320 mV, which is equivalent to a $\Delta G^{\circ\prime}$ of $+14.8$ kcal/mol for transfer of two electrons. Thus this partial reaction tends to proceed toward the left, that is, toward the oxidation of NADH to NAD$^+$.

By contrast, the standard reduction potential for the partial reaction

$$\text{Cytochrome }c_{\text{ox}}\text{ (Fe}^{3+}) + \text{e}^- \rightleftharpoons \text{cytochrome }c_{\text{red}}\text{ (Fe}^{2+})$$

is $+220$ mV ($\Delta G^{\circ\prime} = -5.1$ kcal/mol) for transfer of one electron. Thus this partial reaction tends to proceed toward the right, that is, toward the reduction of cytochrome *c* (Fe^{3+}) to cytochrome *c* (Fe^{2+}).

The final reaction in the respiratory chain, the reduction of O$_2$ to H$_2$O

$$2\text{ H}^+ + \tfrac{1}{2}\text{ O}_2 + 2\text{ e}^- \longrightarrow \text{H}_2\text{O}$$

has a standard reduction potential of $+816$ mV ($\Delta G^{\circ\prime} = -37.8$ kcal/mol for transfer of two electrons), the most

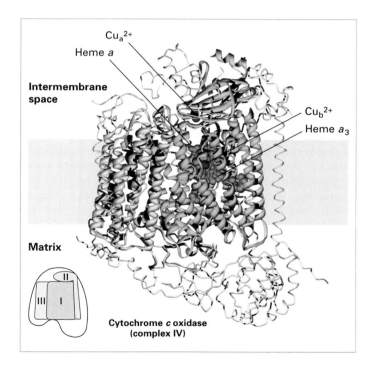

▲ **FIGURE 8-18 Molecular structure of the core of the cytochrome *c* oxidase complex in the inner mitochondrial membrane.** Mitochondrial cytochrome *c* oxidases contain 13 different subunits, but the catalytic core of the enzyme consists of only three subunits: I (green), II (blue), and III (yellow). The function of the remaining subunits (white) is not known. Bacterial cytochrome *c* oxidases contain only the three catalytic subunits. Hemes *a* and a_3 are shown as purple and orange space-filling models, respectively; the three copper atoms are dark blue spheres. [Adapted from T. Tsukihara et al., 1996, *Science* **272**:1136.]

positive in the whole series; thus this reaction also tends to proceed toward the right.

As illustrated in Figure 8-13, the steady increase in $E^{\circ\prime}$ values, and the corresponding decrease in $\Delta G^{\circ\prime}$ values, of the carriers in the respiratory chain favors the flow of electrons from NADH and succinate to oxygen.

CoQ and Three Electron-Transport Complexes Pump Protons Out of the Mitochondrial Matrix

The multiprotein complexes responsible for proton pumping coupled to electron transport have been identified by selectively extracting mitochondrial membranes with detergents, isolating each of the complexes in near purity, and then preparing artificial phospholipid vesicles (liposomes) containing each complex (see Figure 7-5). When an appropriate electron donor and electron acceptor are added to such liposomes, a change in pH of the medium will occur if the embedded complex transports protons (Figure 8-19). Studies of this type indicate that the NADH-CoQ reductase complex translocates four protons per pair of electrons transported, whereas the cytochrome c oxidase complex translocates two protons per electron pair transported (or, equivalently, for every two molecules of cytochrome c oxidized).

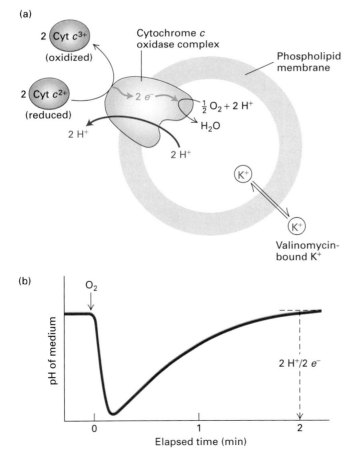

(a)

(b)

Current evidence suggests that a total of 10 protons are transported from the matrix space across the inner mitochondrial membrane for every electron pair that is transferred from NADH to O_2 (see Figure 8-17). Since the succinate-CoQ reductase complex does not transport protons, only six protons are transported across the membrane for every electron pair that is transferred from succinate (or $FADH_2$) to O_2. Relatively little is known about the coupling of electron flow and proton translocation by the NADH-CoQ reductase complex. More is known about operation of the cytochrome c oxidase complex, which we discuss here. The coupled electron and proton movements mediated by the $CoQH_2$–cytochrome c reductase complex, which involves a unique mechanism, are described separately.

After cytochrome c is reduced by the QH_2–cytochrome c reductase complex, it is reoxidized by the cytochrome c oxidase complex, which transfers electrons to oxygen. As noted earlier, cytochrome c oxidase contains three copper ions and two heme groups (see Figure 8-18). The flow of electrons through these carriers is depicted in Figure 8-20. Four molecules of reduced cytochrome c bind, one at a time, to a site on subunit II of the oxidase. An electron is transferred from the heme of each cytochrome c, first to Cu_a^{2+} bound to subunit II, then to the heme a bound to subunit I, and finally to the Cu_b^{2+} and heme a_3 that make up the oxygen reduction center.

The cyclic oxidation and reduction of the iron and copper in the oxygen reduction center of cytochrome c oxidase, together with the uptake of four protons from the matrix space, are coupled to the transfer of the four electrons to oxygen and the formation of water. Proposed intermediates in oxygen reduction include the peroxide anion (O_2^{2-}) and probably the hydroxyl radical (OH·) as well as unusual complexes of iron and oxygen atoms. These intermediates would be harmful to the cell if they escaped from the reaction center, but they do so only rarely.

◄ **EXPERIMENTAL FIGURE 8-19 Electron transfer from reduced cytochrome c (Cyt c^{2+}) to O_2 via the cytochrome c oxidase complex is coupled to proton transport.** The oxidase complex is incorporated into liposomes with the binding site for cytochrome c positioned on the outer surface. (a) When O_2 and reduced cytochrome c are added, electrons are transferred to O_2 to form H_2O and protons are transported from the inside to the outside of the vesicles. Valinomycin and K^+ are added to the medium to dissipate the voltage gradient generated by the translocation of H^+, which would otherwise reduce the number of protons moved across the membrane. (b) Monitoring of the medium pH reveals a sharp drop in pH following addition of O_2. As the reduced cytochrome c becomes fully oxidized, protons leak back into the vesicles, and the pH of the medium returns to its initial value. Measurements show that two protons are transported per O atom reduced. Two electrons are needed to reduce one O atom, but cytochrome c transfers only one electron; thus two molecules of Cyt c^{2+} are oxidized for each O reduced. [Adapted from B. Reynafarje et al., 1986, *J. Biol. Chem.* **261**:8254.]

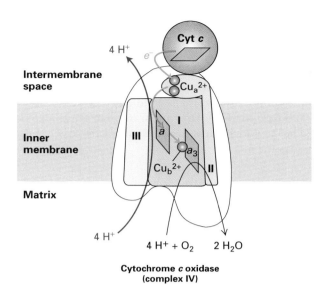

▲ **FIGURE 8-20 Schematic depiction of the cytochrome *c* oxidase complex showing the pathway of electron flow from reduced cytochrome *c* to O$_2$.** Heme groups are denoted by red diamonds. Blue arrows indicate electron flow. Four electrons, sequentially released from four molecules of reduced cytochrome *c*, together with four protons from the matrix, combine with one O$_2$ molecule to form two water molecules. Additionally, for each electron transferred from cytochrome *c* to oxygen, one proton is transported from the matrix to the intermembrane space, or a total of four for each O$_2$ molecule reduced to two H$_2$O molecules.

For every four electrons transferred from reduced cytochrome *c* through cytochrome *c* oxidase (i.e., for every molecule of O$_2$ reduced to two H$_2$O molecules), four protons are translocated from the matrix space to the intermembrane space (two protons per electron pair). However, the mechanism by which these protons are translocated is not known.

The Q Cycle Increases the Number of Protons Translocated as Electrons Flow Through the CoQH$_2$–Cytochrome *c* Reductase Complex

Experiments like the one depicted in Figure 8-19 have shown that four protons are translocated across the membrane per electron pair transported from CoQH$_2$ through the CoQH$_2$–cytochrome *c* reductase complex. Thus this complex transports two protons per electron transferred, whereas the cytochrome *c* oxidase complex transports only one proton per electron transferred. An evolutionarily conserved mechanism, called the *Q cycle*, accounts for the two-for-one transport of protons and electrons by the CoQH$_2$–cytochrome *c* reductase complex.

CoQH$_2$ is generated both by the NADH-CoQ reductase and succinate-CoQ reductase complexes and, as we shall see, by the CoQH$_2$–cytochrome *c* reductase complex itself. In all cases, a molecule from the pool of reduced CoQH$_2$ in the membrane binds to the Q$_o$ site on the *intermembrane*

space (outer) side of the CoQH$_2$–cytochrome *c* reductase complex (Figure 8-21, step ①). Once bound to this site, CoQH$_2$ releases two protons into the intermembrane space (step ②a); these represent two of the four protons pumped per pair of electrons transferred. Operation of the Q cycle results in pumping of the other two protons. To understand this cycle, we first need to focus on the fates of the two electrons released by CoQH$_2$ at the Q$_o$ site. One of the two

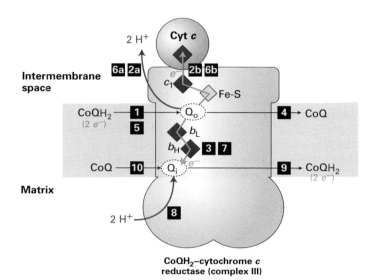

At Q$_o$ site: 2 CoQH$_2$ + 2 Cyt c^{3+} $\longrightarrow$
(4 H$^+$, 4 e^-)   2 CoQ + 2 Cyt c^{2+} + 2 e^- + 4 H$^+$(outside)
                                            (2 e^-)

At Q$_i$ site: CoQ + 2 e^- + 2 H$^+$(matrix side) $\longrightarrow$ CoQH$_2$
                                                (2 H$^+$, 2 e^-)

Net Q cycle (sum of reactions at Q$_o$ and Q$_i$):
   CoQH$_2$ + 2 Cyt c^{3+} + 2 H$^+$(matrix side) $\longrightarrow$
(2 H$^+$, 2 e^-)   CoQ + 2 Cyt c^{2+} + 4 H$^+$(outside)
                                                (2 e^-)

Per 2 e^- transferred through complex III to cytochrome *c*, 4 H$^+$ released to the intermembrane space

▲ **FIGURE 8-21 The Q cycle.** CoQH$_2$ binds to the Q$_o$ site on the intermembrane space (outer) side of CoQ–cytochrome *c* reductase complex and CoQ binds to the Q$_i$ site on the matrix (inner) side. One electron from the CoQH$_2$ bound to Q$_o$ travels directly to cytochrome *c* via an Fe-S cluster and cytochrome c_1. The other electron moves through the *b* cytochromes to CoQ at the Q$_i$ site, forming the partially reduced semiquinone (Q$^-$·). Simultaneously, CoQH$_2$ releases its two protons into the intermembrane space. The CoQ now at the Q$_o$ site dissociates and a second CoQH$_2$ binds there. As before, one electron moves directly to cytochrome c_1 and the other to the Q$^-$· at the Q$_i$ site forming, together with two protons picked up from the matrix space, CoQH$_2$, which then dissociates. The net result is that four protons are translocated from the matrix to the intermembrane space for each pair of electrons transported through the CoQH$_2$–cytochrome *c* reductase complex *(bottom)*. See the text for details. [Adapted from B. Trumpower, 1990, *J. Biol. Chem.* **265**:11409, and E. Darrouzet et al., 2001, *Trends Biochem. Sci.* **26**:445.]

electrons from $CoQH_2$ is transported, via an iron-sulfur protein and cytochrome c_1, directly to cytochrome c (step 2b). The other electron released from the $CoQH_2$ moves through cytochromes b_L and b_H and partially reduces an oxidized CoQ molecule bound to the Q_i site on the *matrix (inner) side* of the complex, forming a CoQ semiquinone anion $Q^{-} \cdot$ (step 3).

After the loss of two protons and two electrons, the now oxidized CoQ at the Q_o site dissociates (step 4), and a second $CoQH_2$ binds to the site (step 5). As before, the bound $CoQH_2$ releases two protons into the intermembrane space (step 6a), while simultaneously one electron from $CoQH_2$ moves directly to cytochrome c (step 6b), and the other electron moves through the b cytochromes to the $Q^{-}\cdot$ bound at the Q_i site (step 7). There the addition of two protons from the matrix yields a fully reduced $CoQH_2$ molecule at the Q_i site (step 8). This $CoQH_2$ molecule then dissociates from the $CoQH_2$–cytochrome c reductase complex (step 9), freeing the Q_i to bind a new molecule of CoQ (step 10) and begin the Q cycle over again.

In the Q cycle, two molecules of $CoQH_2$ are oxidized to CoQ at the Q_o site and release a total of four protons into the intermembrane space, but one molecule of $CoQH_2$ is regenerated from CoQ at the Q_i site (see Figure 8-21, *bottom*). Thus the *net* result of the Q cycle is that four protons are translocated to the intermembrane space for every two electrons transported through the $CoQH_2$–cytochrome c reductase complex and accepted by two molecules of cytochrome c. The translocated protons are all derived from the matrix, taken up either by the NADH-CoQ reductase complex or by the $CoQH_2$–cytochrome c reductase complex during reduction of CoQ. While seemingly cumbersome, the Q cycle increases the numbers of protons pumped per pair of electrons moving through the $CoQH_2$–cytochrome c reductase complex. The Q cycle is found in all plants and animals as well as in bacteria. Its formation at a very early stage of cellular evolution was likely essential for the success of all life-forms as a way of converting the potential energy in reduced coenzyme Q into the maximum proton-motive force across a membrane.

Although the model presented in Figure 8-21 is consistent with a great deal of mutagenesis and spectroscopic studies on the $CoQH_2$–cytochrome c reductase complex, it raises a number of questions. For instance, how are the two electrons released from $CoQH_2$ at the Q_o site directed to different acceptors (cytochromes c_1 and b_L)? Previous studies implicated an iron-sulfur (2Fe-2S) cluster in the transfer of electrons, one at a time, from $CoQH_2$ at the Q_o site to cytochrome c_1. Yet the recently determined three-dimensional structure of the $CoQH_2$–cytochrome c reductase complex, which is a dimeric protein, initially suggested that the 2Fe-2S cluster is positioned too far away from the Q_o site for an electron to "jump" to it. Subsequently, researchers discovered that the subunit containing the 2Fe-2S cluster has a flexible hinge that permits it to exist in two conformational states

(Figure 8-22). In one conformation, the 2Fe-2S cluster is close enough to the Q_o site to pick up an electron from $CoQH_2$ bound there. Movement of the hinge then positions the 2Fe-2S cluster near enough to the heme on cytochrome c_1 for electron transfer to occur. With the Fe-S subunit in this alternative conformation, the second electron released from $CoQH_2$ bound to the Q_o site cannot move to the 2Fe-2S cluster and has to take the less thermodynamically favored route to cytochrome b_L.

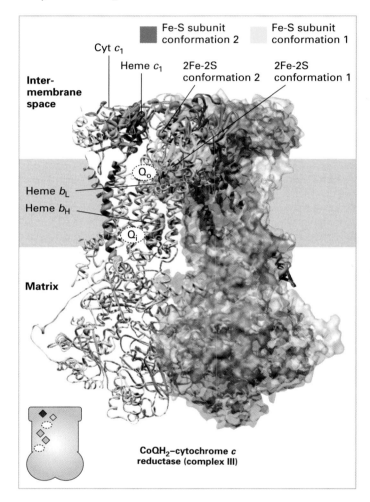

▲ **FIGURE 8-22 Alternative three-dimensional conformations of the Fe-S subunit of the CoQ–cytochrome c reductase complex.** In the dimeric complex, cytochromes b_L and b_H are associated with one subunit and the 2Fe-2S cluster with the other subunit. The subunit containing the 2Fe-2S cluster is shown in its two alternative conformational states, which differ primarily in the portion of the protein toward the intermembrane space. In one conformation (yellow), the 2Fe-2S cluster (green) is positioned near the Q_o site on the intermembrane side of the protein, able to pick up an electron from $CoQH_2$. In the alternative conformation (blue), the 2Fe-2S cluster is located adjacent to the c_1 heme on cytochrome c_1 and able to transfer an electron to it. [Adapted from Z. Zhang et al., 1998, *Nature* **392**:678; see also E. Darrouzet et al., 2001, *Trends Biochem. Sci.* **26**:445.]

Electron Transport and Generation of the Proton-Motive Force

■ The proton-motive force is a combination of a proton concentration (pH) gradient (exoplasmic face > cytosolic face) and an electric potential (negative cytosolic face) across a membrane.

■ In the mitochondrion, the proton-motive force is generated by coupling electron flow from NADH and $FADH_2$ to O_2 to the uphill transport of protons from the matrix across the inner membrane to the intermembrane space.

■ The major components of the mitochondrial respiratory chain are four inner membrane multiprotein complexes: NADH-CoQ reductase (I), succinate-CoQ reductase (II), $CoQH_2$–cytochrome c reductase (III), and cytochrome c oxidase (IV). The last complex transfers electrons to O_2 to form H_2O.

■ Each complex contains one or more electron-carrying prosthetic groups: iron-sulfur clusters, flavins, heme groups, and copper ions (see Table 8-2). Cytochrome c, which contains heme, and coenzyme Q (CoQ) are mobile electron carriers.

■ Each electron carrier accepts an electron or electron pair from a carrier with a less positive reduction potential and transfers the electron to a carrier with a more positive reduction potential. Thus the reduction potentials of electron carriers favor unidirectional electron flow from NADH and $FADH_2$ to O_2 (see Figure 8-13).

■ A total of 10 H^+ ions are translocated from the matrix across the inner membrane per electron pair flowing from NADH to O_2 (see Figure 8-17).

■ The Q cycle allows four protons (rather than two) to be translocated per pair of electrons moving through the $CoQH_2$–cytochrome c reductase complex (see Figure 8-21).

8.3 Harnessing the Proton-Motive Force for Energy-Requiring Processes

The hypothesis that a proton-motive force across the inner mitochondrial membrane is the immediate source of energy for ATP synthesis was proposed in 1961 by Peter Mitchell. Virtually all researchers working in oxidative phosphorylation and photosynthesis initially opposed this chemiosmotic mechanism. They favored a mechanism similar to the well-elucidated substrate-level phosphorylation in glycolysis, in which oxidation of a substrate molecule is directly coupled to ATP synthesis. By analogy, electron transport through the membranes of chloroplasts or mitochondria was believed to generate an intermediate containing a high-energy chemical bond (e.g., a phosphate linked to an enzyme by an ester bond), which was then used to convert P_i and ADP to ATP. Despite intense efforts by a large number of investigators, however, no such intermediate was ever identified.

Definitive evidence supporting the role of the proton-motive force in ATP synthesis awaited development of techniques to purify and reconstitute organelle membranes and membrane proteins. The experiment with chloroplast thylakoid vesicles containing F_0F_1 particles, outlined in Figure 8-23, was one of several demonstrating that the F_0F_1 complex is an ATP-generating enzyme and that ATP generation is

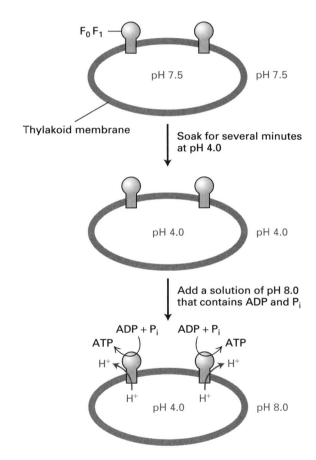

▲ **EXPERIMENTAL FIGURE 8-23 Synthesis of ATP by F_0F_1 depends on a pH gradient across the membrane.** Isolated chloroplast thylakoid vesicles containing F_0F_1 particles were equilibrated in the dark with a buffered solution at pH 4.0. When the pH in the thylakoid lumen became 4.0, the vesicles were rapidly mixed with a solution at pH 8.0 containing ADP and P_i. A burst of ATP synthesis accompanied the transmembrane movement of protons driven by the 10,000-fold H^+ concentration gradient (10^{-4} M versus 10^{-8} M). In similar experiments using "inside-out" preparations of submitochondrial vesicles, an artificially generated membrane electric potential also resulted in ATP synthesis.

dependent on proton movement down an electrochemical gradient. With general acceptance of Mitchell's chemiosmotic mechanism, researchers turned their attention to the structure and operation of the F_0F_1 complex.

Bacterial Plasma-Membrane Proteins Catalyze Electron Transport and Coupled ATP Synthesis

Although bacteria lack any internal membranes, aerobic bacteria nonetheless carry out oxidative phosphorylation by the same processes that occur in eukaryotic mitochondria. Enzymes that catalyze the reactions of both the glycolytic pathway and the citric acid cycle are present in the cytosol of bacteria; enzymes that oxidize NADH to NAD^+ and transfer the electrons to the ultimate acceptor O_2 are localized to the bacterial plasma membrane.

The movement of electrons through these membrane carriers is coupled to the pumping of protons out of the cell (see Figure 8-2). The movement of protons back into the cell, down their concentration gradient, is coupled to the synthesis of ATP. Bacterial F_0F_1 complexes are essentially identical in structure and function with the mitochondrial F_0F_1 complex, but are simpler to purify and study. The proton-motive force across the bacterial plasma membrane is also used to power the uptake of nutrients such as sugars, using proton/sugar symporters, and the rotation of bacterial flagella (see Figure 8-1). As we noted earlier, a primitive aerobic bacterium was probably the progenitor of mitochondria in eukaryotic cells (see Figure 8-3).

ATP Synthase Comprises Two Multiprotein Complexes Termed F_0 and F_1

The F_0F_1 complex, or ATP synthase, has two principal components, F_0 and F_1, both of which are multimeric proteins (Figure 8-24). The F_0 component contains three types of integral membrane proteins, designated **a**, **b**, and **c**. In bacteria and in yeast mitochondria the most common subunit composition is $a_1b_2c_{10}$, but F_0 complexes in animal mitochondria have 12 **c** subunits and those in chloroplasts have 14. In all cases the **c** subunits form a donut-shaped ring in the plane of the membrane. The **a** and two **b** subunits are rigidly linked to one another but not to the ring of **c** subunits.

The F_1 portion is a water-soluble complex of five distinct polypeptides with the composition $\alpha_3\beta_3\gamma\delta\varepsilon$. The lower part of the F_1 γ subunit is a coiled coil that fits into the center of the **c**-subunit ring of F_0 and appears rigidly attached to it. The F_1 ε subunit is rigidly attached to γ and also forms rigid contacts with several of the **c** subunits of F_0. The F_1 α and β subunits associate in alternating order to form a hexamer, $\alpha\beta\alpha\beta\alpha\beta$, or $(\alpha\beta)_3$, which rests atop the single long γ subunit. The F_1 δ subunit is permanently linked to one of the F_1 α subunits and also to the **b** subunit of F_0. Thus the F_0 **a** and **b** subunits and the δ subunit and $(\alpha\beta)_3$ hexamer of the F_1 complex form a rigid structure anchored in the membrane. The rodlike **b** subunits form a "stator" that prevents the $(\alpha\beta)_3$ hexamer from moving while it rests on the γ subunit (see Figure 8-24).

When ATP synthase is embedded in a membrane, the F_1 component forms a knob that protrudes from the cytosolic

▶ **FIGURE 8-24 Model of the structure and function of ATP synthase (the F_0F_1 complex) in the bacterial plasma membrane.** The F_0 portion is built of three integral membrane proteins: one copy of **a**, two copies of **b**, and on average 10 copies of **c** arranged in a ring in the plane of the membrane. Two proton half-channels lie at the interface between the **a** subunit and the **c** ring. Half-channel I allows protons to move one at a time from the exoplasmic medium and bind to aspartate-61 in the center of a **c** subunit near the middle of the membrane. Half-channel II (after rotation of the **c** ring) permits protons to dissociate from the aspartate and move into the cytosolic medium. The F_1 portion contains three copies each of subunits α and β that form a hexamer resting atop the single rod-shaped γ subunit, which is inserted into the **c** ring of F_0. The ε subunit is rigidly attached to the γ subunit and also to several of the **c** subunits. The δ subunit permanently links one of the α subunits in the F_1 complex to the **b** subunit of F_0. Thus the F_0 **a** and **b** subunits and the F_1 δ subunit and $(\alpha\beta)_3$ hexamer form a rigid structure anchored in the membrane (orange). During proton flow, the **c** ring and the attached F_1 ε and γ subunits rotate as a unit (green), causing conformation changes in the F_1 β subunits leading to ATP synthesis. [Adapted from M. J. Schnitzer, 2001, *Nature* **410**:878, and P. D. Boyer, 1999, *Nature* **402**:247.]

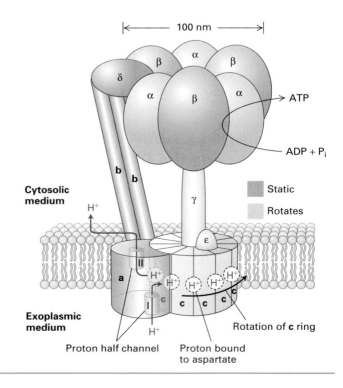

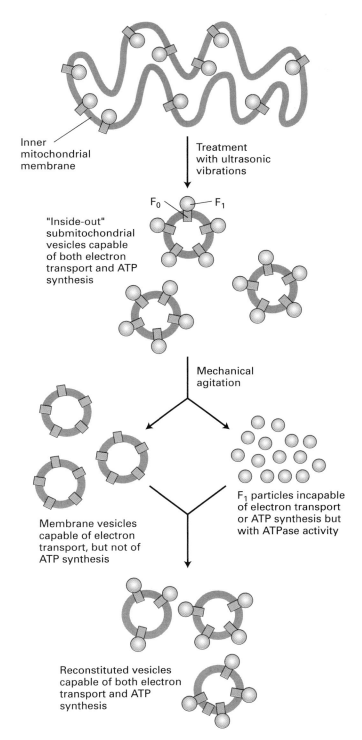

Inner
mitochondrial
membrane

Treatment
with ultrasonic
vibrations

F_0 F_1

"Inside-out"
submitochondrial
vesicles capable
of both electron
transport and ATP
synthesis

Mechanical
agitation

Membrane vesicles
capable of electron
transport, but not of
ATP synthesis

F_1 particles incapable
of electron transport
or ATP synthesis but
with ATPase activity

Reconstituted vesicles
capable of both electron
transport and ATP
synthesis

▲ **EXPERIMENTAL FIGURE 8-25 Mitochondrial F_1
particles are required for ATP synthesis, but not for electron
transport.** "Inside-out" membrane vesicles that lack F_1 and
retain the electron transport complexes are prepared as
indicated. Although these can transfer electrons from NADH to
O_2, they cannot synthesize ATP. The subsequent addition of F_1
particles reconstitutes the native membrane structure, restoring
the capacity for ATP synthesis. When detached from the
membrane, F_1 particles exhibit ATPase activity.

face. As demonstrated in the experiment depicted in Figure
8-25, submitochondrial vesicles from which F_1 is removed by
mechanical agitation cannot catalyze ATP synthesis; when F_1
particles reassociate with these vesicles, they once again become
fully active in ATP synthesis. Because F_1 separated from membranes is capable of catalyzing ATP hydrolysis, it has been
called the F_1 ATPase; however, its function in the cells is to synthesize ATP. We examine how it does so in the next section.

Rotation of the F_1 γ Subunit, Driven by Proton Movement Through F_0, Powers ATP Synthesis

Each of the three β subunits in the complete F_0F_1 complex
can bind ADP and P_i and catalyze ATP synthesis. However,
the coupling between proton flow and ATP synthesis must be
indirect, since the nucleotide-binding sites on the β subunits
of F_1, where ATP synthesis occurs, are 9–10 nm from the surface of the mitochondrial membrane. The most widely accepted model for ATP synthesis by the F_0F_1 complex—the
binding-change mechanism—posits just such an indirect coupling (Figure 8-26).

According to this mechanism, energy released by the
"downhill" movement of protons through F_0 directly powers
rotation of the c-subunit ring together with its attached γ
and ε subunits (see Figure 8-24). The γ subunit acts as a cam,
or rotating shaft, whose movement within F_1 causes cyclical
changes in the conformations of the β subunits. As schematically depicted in Figure 8-26, rotation of the γ subunit relative to the fixed (αβ)$_3$ hexamer causes the nucleotide-binding
site of each β subunit to cycle through three conformational
states in the following order:

1. An O state that binds ATP very poorly and ADP and P_i
weakly

2. An L state that binds ADP and P_i more strongly

3. A T state that binds ADP and P_i so tightly that they
spontaneously form ATP and that binds ATP very strongly

A final rotation of γ returns the β subunit to the O state,
thereby releasing ATP and beginning the cycle again. ATP or
ADP also binds to regulatory or allosteric sites on the three α
subunits; this binding modifies the rate of ATP synthesis according to the level of ATP and ADP in the matrix, but is
not directly involved in synthesis of ATP from ADP and P_i.

Several types of evidence support the binding-change
mechanism, which is now generally accepted. First, biochemical studies showed that one of the three β subunits on isolated
F_1 particles can tightly bind ADP and P_i and then form ATP,
which remains tightly bound. The measured ΔG for this reaction is near zero, indicating that once ADP and P_i are bound to
what is now called the T state of a β subunit, they spontaneously form ATP. Importantly, dissociation of the bound ATP
from the β subunit on isolated F_1 particles occurs extremely
slowly. This finding suggested that dissociation of ATP would
have to be powered by a conformational change in the β subunit, which, in turn, would be caused by proton movement.

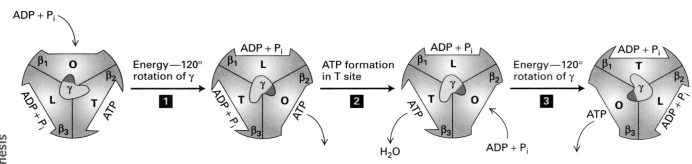

▲ **FIGURE 8-26 The binding-change mechanism of ATP synthesis from ADP and P$_i$ by the F$_0$F$_1$ complex.** This view is looking up at F$_1$ from the membrane surface (see Figure 8-24). Each of the F$_1$ β subunits alternate between three conformational states that differ in their binding affinities for ATP, ADP, and P$_i$. Step **1**: After ADP and P$_i$ bind to one of the three β subunits (here, arbitrarily designated β$_1$) whose nucleotide-binding site is in the O (open) conformation, proton flux powers a 120° rotation of the γ subunit (relative to the fixed β subunits). This causes an increase in the binding affinity of the β$_1$ subunit for ADP and P$_i$ to L (low), an increase in the binding affinity of the β$_3$ subunit for ADP

and P$_i$ from L to T (tight), and a decrease in the binding affinity of the β$_2$ subunit for ATP from T to O, causing release of the bound ATP. Step **2**: The ADP and P$_i$ in the T site (here the β$_3$ subunit) form ATP, a reaction that does not require an input of energy, and ADP and P$_i$ bind to the β$_2$ subunit, which is in the O state. This generates an F$_1$ complex identical with that which started the process (*left*) except that it is rotated 120°. Step **3**: Another 120° rotation of γ again causes the O → L → T → O conformational changes in the β subunits described above. Repetition of steps **1** and **2** leads to formation of three ATP molecules for every 360° rotation of γ. [Adapted from P. Boyer, 1989, *FASEB J.* **3**:2164, and Y. Zhou et al., 1997, *Proc. Nat'l. Acad. Sci. USA* **94**:10583.]

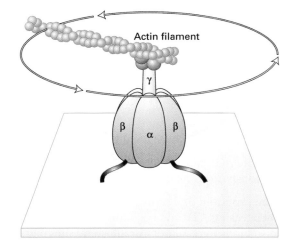

Actin filament

▲ **EXPERIMENTAL FIGURE 8-27 Rotation of the γ subunit of the F$_1$ complex relative to the (αβ)$_3$ hexamer can be observed microscopically.** F$_1$ complexes were engineered that contained β subunits with an additional His$_6$ sequence, which causes them to adhere to a glass plate coated with a metal reagent that binds histidine. The γ subunit in the engineered F$_1$ complexes was linked covalently to a fluorescently labeled actin filament. When viewed in a fluorescence microscope, the actin filaments were seen to rotate counterclockwise in discrete 120° steps in the presence of ATP, powered by ATP hydrolysis by the β subunits. [Adapted from H. Noji et al., 1997, *Nature* **386**:299, and R. Yasuda et al., 1998, *Cell* **93**:1117.] See also K. Nishio et al., 2002, *Proc. Nat'l. Acad. Sci.* **97**:13448, for another way to demonstrate α and β subunit rotation relative to the **c** subunit ring.]

Later x-ray crystallographic analysis of the (αβ)$_3$ hexamer yielded a striking conclusion: although the three β subunits are identical in sequence and overall structure, the ADP/ATP-binding sites have different conformations in each subunit. The most reasonable conclusion was that the three β subunits cycle between three conformational states, with different nucleotide-binding sites, in an energy-dependent reaction.

In other studies, intact F$_0$F$_1$ complexes were treated with chemical cross-linking agents that covalently linked the γ and ε subunits and the c-subunit ring. The observation that such treated complexes could synthesize ATP or use ATP to power proton pumping indicates that the cross-linked proteins normally rotate together.

Finally, rotation of the γ subunit relative to the fixed (αβ)$_3$ hexamer, as proposed in the binding-change mechanism, was observed directly in the clever experiment depicted in Figure 8-27. In one modification of this experiment in which tiny gold particles were attached to the γ subunit, rotation rates of 134 revolutions per second were observed. Recalling that hydrolysis of 3 ATPs is thought to power one revolution (see Figure 8-26), this result is close to the experimentally determined rate of ATP hydrolysis by F$_0$F$_1$ complexes: about 400 ATPs per second. In a related experiment, a γ subunit linked to an ε subunit and a ring of c subunits was seen to rotate relative to the fixed (αβ)$_3$ hexamer. Rotation of the γ subunit in these experiments was powered by ATP hydrolysis, the reverse of the normal process in which proton movement through the F$_0$ complex drives rotation of the γ subunit. Nonetheless, these observations established that the γ subunit, along with the attached c ring and ε subunit, does indeed rotate, thereby driving the conformational changes in the β subunits that are required for binding of ADP and P_i, followed by synthesis and subsequent release of ATP.

Number of Translocated Protons Required for ATP Synthesis

A simple calculation indicates that the passage of more than one proton is required to synthesize one molecule of ATP from ADP and P_i. Although the ΔG for this reaction under standard conditions is +7.3 kcal/mol, at the concentrations of reactants in the mitochondrion, ΔG is probably higher (+10 to +12 kcal/mol). We can calculate the amount of free energy released by the passage of 1 mol of protons down an electrochemical gradient of 220 mV (0.22 V) from the Nernst equation, setting $n = 1$ and measuring ΔE in volts:

$$\Delta G \text{ (cal/mol)} = -nF\Delta E = -(23,062 \text{ cal·V}^{-1}\text{·mol}^{-1}) \Delta E$$
$$= (23,062 \text{ cal·V}^{-1}\text{·mol}^{-1})(0.22 \text{ V})$$
$$= -5074 \text{ cal/mol, or } -5.1 \text{ kcal/mol}$$

Since the downhill movement of 1 mol of protons releases just over 5 kcal of free energy, the passage of at least two protons is required for synthesis of each molecule of ATP from ADP and P_i.

Proton Movement Through F_0 and Rotation of the c Ring

Each copy of subunit c contains two membrane-spanning α helices that form a hairpin. An aspartate residue, Asp61, in the center of one of these helices is thought to participate in proton movement. Chemical modification of this aspartate by the poison dicyclohexylcarbodiimide or its mutation to alanine specifically blocks proton movement through F_0. According to one current model, two proton half-channels lie at the interface between the a subunit and c ring (see Figure 8-24). Protons are thought to move one at a time through half-channel I from the exoplasmic medium and bind to the carboxylate side chain on Asp61 of one c subunit. Binding of a proton to this aspartate would result in a conformational change in the c subunit, causing it to move relative to the fixed a subunit, or equivalently to rotate in the membrane plane. This rotation would bring the adjacent c subunit, with its ionized aspartyl side chain, into channel I, thereby allowing it to receive a proton and subsequently move relative to the a subunit. Continued rotation of the c ring, due to binding of protons to additional c subunits, eventually would align the first c subunit containing a protonated Asp61 with the second half-channel (II), which is connected to the cytosol. Once this occurs, the proton on the aspartyl residue could dissociate (forming ionized aspartate) and move into the cytosolic medium.

Since the γ subunit of F_1 is tightly attached to the c ring of F_0, rotation of the c ring associated with proton movement causes rotation of the γ subunit. According to the binding-change mechanism, a 120° rotation of γ powers synthesis of one ATP (see Figure 8-26). Thus complete rotation of the c ring by 360° would generate three ATPs. In *E. coli*, where the F_0 composition is $a_1b_2c_{10}$, movement of 10 protons drives one complete rotation and thus synthesis of three

ATPs. This value is consistent with experimental data on proton flux during ATP synthesis, providing indirect support for the model coupling proton movement to c-ring rotation depicted in Figure 8-24. The F_0 from chloroplasts contains 14 c subunits per ring, and movement of 14 protons would be needed for synthesis of three ATPs. Why these otherwise similar F_0F_1 complexes have evolved to have different H^+:ATP ratios in not clear.

ATP-ADP Exchange Across the Inner Mitochondrial Membrane Is Powered by the Proton-Motive Force

In addition to powering ATP synthesis, the proton-motive force across the inner mitochondrial membrane also powers the exchange of ATP formed by oxidative phosphorylation inside the mitochondrion for ADP and P_i in the cytosol. This exchange, which is required for oxidative phosphorylation to continue, is mediated by two proteins in the inner membrane: a *phosphate transporter* (HPO_4^{2-}/OH^- antiporter) and an *ATP/ADP antiporter* (Figure 8-28).

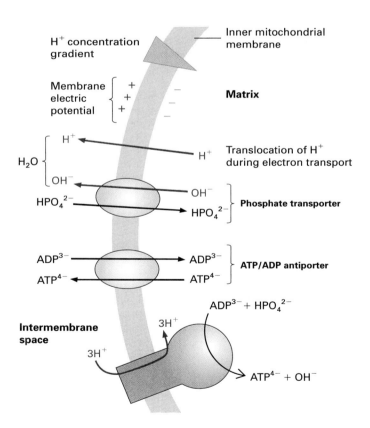

▲ **FIGURE 8-28 The phosphate and ATP/ADP transport system in the inner mitochondrial membrane.** The coordinated action of two antiporters (purple and green) results in the uptake of one ADP^{3-} and one HPO_4^{2-} in exchange for one ATP^{4-}, powered by the outward translocation of one proton during electron transport. The outer membrane is not shown here because it is permeable to molecules smaller than 5000 Da.

The phosphate transporter catalyzes the import of one HPO_4^{2-} coupled to the export of one OH^-. Likewise, the ATP/ADP antiporter allows one molecule of ADP to enter only if one molecule of ATP exits simultaneously. The ATP/ADP antiporter, a dimer of two 30,000-Da subunits, makes up 10–15 percent of the protein in the inner membrane, so it is one of the more abundant mitochondrial proteins. Functioning of the two antiporters together produces an influx of one ADP and one P_i and efflux of one ATP together with one OH^-. Each OH^- transported outward combines with a proton, translocated during electron transport to the intermembrane space, to form H_2O. This drives the overall reaction in the direction of ATP export and ADP and P_i import.

Because some of the protons translocated out of the mitochondrion during electron transport provide the power (by combining with the exported OH^-) for the ATP-ADP exchange, fewer protons are available for ATP synthesis. It is estimated that for every four protons translocated out, three are used to synthesize one ATP molecule and one is used to power the export of ATP from the mitochondrion in exchange for ADP and P_i. This expenditure of energy from the proton concentration gradient to export ATP from the mitochondrion in exchange for ADP and P_i ensures a high ratio of ATP to ADP in the cytosol, where hydrolysis of the high-energy phosphoanhydride bond of ATP is utilized to power many energy-requiring reactions.

Rate of Mitochondrial Oxidation Normally Depends on ADP Levels

If intact isolated mitochondria are provided with NADH (or $FADH_2$), O_2, and P_i, but not with ADP, the oxidation of NADH and the reduction of O_2 rapidly cease, as the amount of endogenous ADP is depleted by ATP formation. If ADP then is added, the oxidation of NADH is rapidly restored. Thus mitochondria can oxidize $FADH_2$ and NADH only as long as there is a source of ADP and P_i to generate ATP. This phenomenon, termed *respiratory control*, occurs because oxidation of NADH, succinate, or $FADH_2$ is obligatorily coupled to proton transport across the inner mitochondrial membrane. If the resulting proton-motive force is not dissipated in the synthesis of ATP from ADP and P_i (or for some other purpose), both the transmembrane proton concentration gradient and the membrane electric potential will increase to very high levels. At this point, pumping of additional protons across the inner membrane requires so much energy that it eventually ceases, thus blocking the coupled oxidation of NADH and other substrates.

Certain poisons, called **uncouplers,** render the inner mitochondrial membrane permeable to protons. One example is the lipid-soluble chemical 2,4-dinitrophenol (DNP), which can reversibly bind and release protons and shuttle protons across the inner membrane from the intermembrane space into the matrix. As a result, DNP dissipates the proton-motive force by short-circuiting both the transmembrane proton concentration gradient and the membrane electric potential. Uncouplers such as DNP abolish ATP synthesis and overcome respiratory control, allowing NADH oxidation to occur regardless of the ADP level. The energy released by the oxidation of NADH in the presence of DNP is converted to heat.

Brown-Fat Mitochondria Contain an Uncoupler of Oxidative Phosphorylation

Brown-fat tissue, whose color is due to the presence of abundant mitochondria, is specialized for the generation of heat. In contrast, *white-fat tissue* is specialized for the storage of fat and contains relatively few mitochondria.

The inner membrane of brown-fat mitochondria contains *thermogenin*, a protein that functions as a natural uncoupler of oxidative phosphorylation. Like synthetic uncouplers, thermogenin dissipates the proton-motive force across the inner mitochondrial membrane, converting energy released by NADH oxidation to heat. Thermogenin is a proton transporter, not a proton channel, and shuttles protons across the membrane at a rate that is a millionfold slower than that of typical ion channels. Its amino acid sequence is similar to that of the mitochondrial ATP/ADP antiporter, and it functions at a rate that is characteristic of other transporters (see Figure 7-2).

Environmental conditions regulate the amount of thermogenin in brown-fat mitochondria. For instance, during the adaptation of rats to cold, the ability of their tissues to generate heat is increased by the induction of thermogenin synthesis. In cold-adapted animals, thermogenin may constitute up to 15 percent of the total protein in the inner mitochondrial membrane.

Adult humans have little brown fat, but human infants have a great deal. In the newborn, thermogenesis by brown-fat mitochondria is vital to survival, as it also is in hibernating mammals. In fur seals and other animals naturally acclimated to the cold, muscle-cell mitochondria contain thermogenin; as a result, much of the proton-motive force is used for generating heat, thereby maintaining body temperature.

KEY CONCEPTS OF SECTION 8.3

Harnessing the Proton-Motive Force for Energy-Requiring Processes

■ The multiprotein F_0F_1 complex catalyzes ATP synthesis as protons flow back through the inner mitochondrial membrane (plasma membrane in bacteria) down their electrochemical proton gradient.

■ F_0 contains a ring of 10–14 c subunits that is rigidly linked to the rod-shaped γ subunit and ε subunit of F_1. Resting atop the γ subunit is the hexameric knob of F_1 [$(\alpha\beta)_3$], which protrudes into the mitochondrial matrix (cytosol in bacteria). The three β subunits are the sites of ATP synthesis (see Figure 8-24).

■ Movement of protons across the membrane via two half-channels at the interface of the F_0 **a** subunit and the **c** ring powers rotation of the **c** ring with its attached F_1 ε and γ subunits.

■ Rotation of the F_1 γ subunit leads to changes in the conformation of the nucleotide-binding sites in the F_1 β subunits (see Figure 8-26). By means of this binding-change mechanism, the β subunits bind ADP and P_i, condense them to form ATP, and then release the ATP.

■ The proton-motive force also powers the uptake of P_i and ADP from the cytosol in exchange for mitochondrial ATP and OH^-, thus reducing some of the energy available for ATP synthesis.

■ Continued mitochondrial oxidation of NADH and the reduction of O_2 are dependent on sufficient ADP being present. This phenomenon, termed *respiratory control*, is an important mechanism for coordinating oxidation and ATP synthesis in mitochondria.

■ In brown fat, the inner mitochondrial membrane contains thermogenin, a proton transporter that converts the proton-motive force into heat. Certain chemicals (e.g., DNP) have the same effect, uncoupling oxidative phosphorylation from electron transport.

8.4 Photosynthetic Stages and Light-Absorbing Pigments

We now shift our attention to photosynthesis, the second main process for synthesizing ATP. Photosynthesis in plants occurs in chloroplasts, large organelles found mainly in leaf cells. The principal end products are two carbohydrates that are polymers of hexose (six-carbon) sugars: sucrose, a glucose-fructose disaccharide (see Figure 2-17), and leaf **starch**, a large insoluble glucose polymer that is the primary storage carbohydrate in higher plants (Figure 8-29). Leaf starch is synthesized and stored in the chloroplast. Sucrose is synthesized in the leaf cytosol from three-carbon precursors generated in the chloroplast; it is transported to nonphotosynthetic (nongreen) plant tissues (e.g., roots and seeds), which metabolize sucrose for energy by the pathways described in the previous sections. Photosynthesis in plants, as well as in eukaryotic single-celled algae and in several photosynthetic bacteria (e.g., the cyanobacteria and prochlorophytes), also generates oxygen. The overall reaction of oxygen-generating photosynthesis,

$$6\ CO_2 + 6\ H_2O \longrightarrow 6\ O_2 + C_6H_{12}O_6$$

is the reverse of the overall reaction by which carbohydrates are oxidized to CO_2 and H_2O.

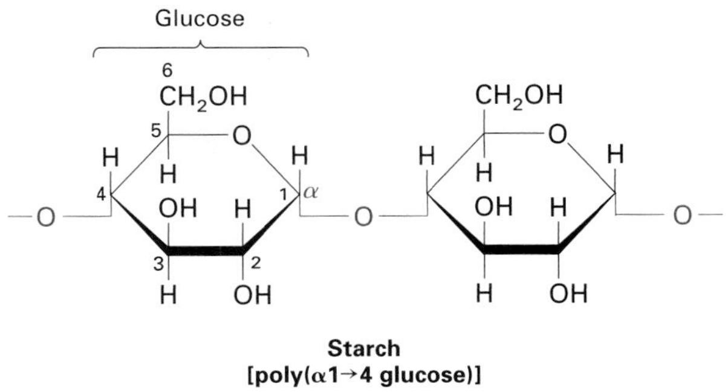

▲ FIGURE 8-29 Structure of starch. This large glucose polymer and the disaccharide sucrose (see Figure 2-17) are the principal end products of photosynthesis. Both are built of six-carbon sugars.

Although green and purple bacteria also carry out photosynthesis, they use a process that does not generate oxygen. As discussed in Section 8.5, detailed analysis of the photosynthetic system in these bacteria has provided insights about the first stages in the more common process of oxygen-generating photosynthesis. In this section, we provide an overview of the stages in oxygen-generating photosynthesis and introduce the main components, including the **chlorophylls**, the principal light-absorbing pigments.

Thylakoid Membranes Are the Sites of Photosynthesis in Plants

Chloroplasts are bounded by two membranes, which do not contain chlorophyll and do not participate directly in photosynthesis (Figure 8-30). As in mitochondria, the outer membrane of chloroplasts contains porins and thus is permeable to metabolites of small molecular weight. The inner membrane forms a permeability barrier that contains transport proteins for regulating the movement of metabolites into and out of the organelle.

Unlike mitochondria, chloroplasts contain a third membrane—the *thylakoid membrane*—on which photosynthesis occurs. The chloroplast thylakoid membrane is believed to constitute a single sheet that forms numerous small, interconnected flattened vesicles, the **thylakoids,** which commonly are arranged in stacks termed *grana* (see Figure 8-30). The spaces within all the thylakoids constitute a single continuous compartment, the *thylakoid lumen.* The thylakoid membrane contains a number of integral membrane proteins to which are bound several important prosthetic groups and light-absorbing pigments, most notably chlorophyll. Carbohydrate synthesis occurs in the *stroma,* the soluble phase between the thylakoid membrane and the inner membrane. In photosynthetic bacteria extensive invaginations of the plasma membrane form a set of internal membranes, also termed *thylakoid membranes,* where photosynthesis occurs.

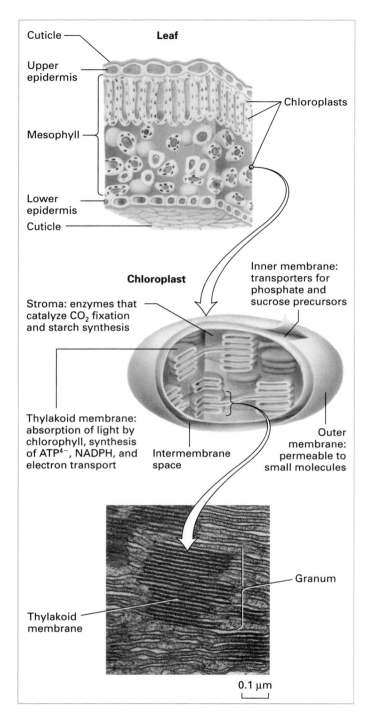

▲ **FIGURE 8-30 Cellular structure of a leaf and chloroplast.** Like mitochondria, plant chloroplasts are bounded by a double membrane separated by an intermembrane space. Photosynthesis occurs on the thylakoid membrane, which forms a series of flattened vesicles (thylakoids) that enclose a single interconnected luminal space. The green color of plants is due to the green color of chlorophyll, all of which is localized to the thylakoid membrane. A granum is a stack of adjacent thylakoids. The stroma is the space enclosed by the inner membrane and surrounding the thylakoids. [Photomicrograph courtesy of Katherine Esau, University of California, Davis.]

Three of the Four Stages in Photosynthesis Occur Only During Illumination

The photosynthetic process in plants can be divided into four stages, each localized to a defined area of the chloroplast: (1) absorption of light, (2) electron transport leading to formation of O_2 from H_2O, reduction of $NADP^+$ to NADPH, and generation of a proton-motive force, (3) synthesis of ATP, and (4) conversion of CO_2 into carbohydrates, commonly referred to as **carbon fixation**. All four stages of photosynthesis are tightly coupled and controlled so as to produce the amount of carbohydrate required by the plant. All the reactions in stages 1–3 are catalyzed by proteins in the thylakoid membrane. The enzymes that incorporate CO_2 into chemical intermediates and then convert them to starch are soluble constituents of the chloroplast stroma. The enzymes that form sucrose from three-carbon intermediates are in the cytosol.

Stage 1: Absorption of Light The initial step in photosynthesis is the absorption of light by chlorophylls attached to proteins in the thylakoid membranes. Like the heme component of cytochromes, chlorophylls consist of a porphyrin ring attached to a long hydrocarbon side chain (Figure 8-31). In contrast to the hemes, chlorophylls contain a central

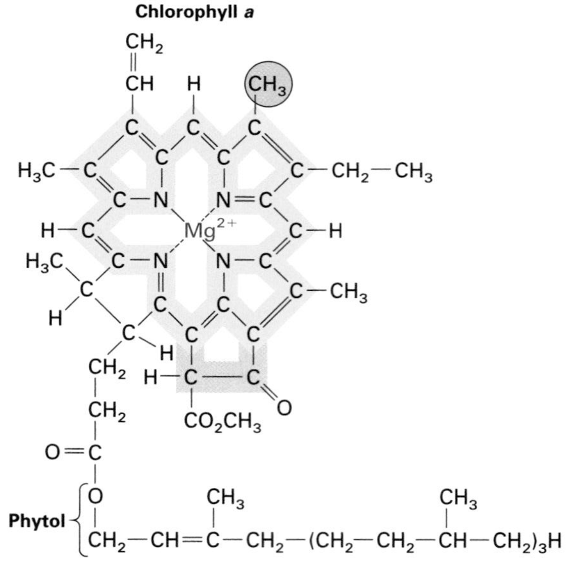

▲ **FIGURE 8-31 Structure of chlorophyll *a*, the principal pigment that traps light energy.** The CH_3 group (green) is replaced by a CHO group in chlorophyll *b*. In the porphyrin ring (yellow), electrons are delocalized among three of the four central rings and the atoms that interconnect them in the molecule. In chlorophyll, an Mg^{2+} ion, rather than an Fe^{3+} ion, is in the center of the porphyrin ring and an additional five-membered ring (blue) is present; otherwise, its structure is similar to that of heme found in molecules such as hemoglobin and cytochromes (see Figure 8-15a). The hydrocarbon phytol "tail" facilitates binding of chlorophyll to hydrophobic regions of chlorophyll-binding proteins.

Mg^{2+} ion (rather than Fe atom) and have an additional five-membered ring. The energy of the absorbed light is used to remove electrons from an unwilling donor (water, in green plants), forming oxygen,

$$2\ H_2O \xrightarrow{\text{light}} O_2 + 4\ H^+ + 4\ e^-$$

and then to transfer the electrons to a *primary electron acceptor,* a quinone designated Q, which is similar to CoQ.

Stage 2: Electron Transport and Generation of a Proton-Motive Force Electrons move from the quinone primary electron acceptor through a series of electron carriers until they reach the ultimate electron acceptor, usually the oxidized form of **nicotinamide adenine dinucleotide phosphate** (**NADP⁺**), reducing it to NADPH. (NADP is identical in structure with NAD except for the presence of an additional phosphate group. Both molecules gain and lose electrons in the same way; see Figure 2-26a.) The transport of electrons in the thylakoid membrane is coupled to the movement of protons from the stroma to the thylakoid lumen, forming a pH gradient across the membrane ($pH_{lumen} < pH_{stroma}$). This process is analogous to generation of a proton-motive force across the inner mitochondrial membrane during electron transport (see Figure 8-2).

Thus the overall reaction of stages 1 and 2 can be summarized as

$$2\ H_2O + 2\ NADP^+ \xrightarrow{\text{light}} 2\ H^+ + 2\ NADPH + O_2$$

Stage 3: Synthesis of ATP Protons move down their concentration gradient from the thylakoid lumen to the stroma through the F_0F_1 complex (ATP synthase), which couples proton movement to the synthesis of ATP from ADP and P_i. The mechanism whereby chloroplast F_0F_1 harnesses the proton-motive force to synthesize ATP is identical with that used by ATP synthase in the inner mitochondrial membrane and bacterial plasma membrane (see Figures 8-24 and 8-26).

Stage 4: Carbon Fixation The ATP and NADPH generated by the second and third stages of photosynthesis provide the energy and the electrons to drive the synthesis of polymers of six-carbon sugars from CO_2 and H_2O. The overall balanced chemical equation is written as

$$6\ CO_2 + 18\ ATP^{4-} + 12\ NADPH + 12\ H_2O \longrightarrow$$
$$C_6H_{12}O_6 + 18\ ADP^{3-} + 18\ P_i^{2-} + 12\ NADP^+ + 6\ H^+$$

The reactions that generate the ATP and NADPH used in carbon fixation are directly dependent on light energy; thus stages 1–3 are called the *light reactions* of photosynthesis. The reactions in stage 4 are indirectly dependent on light energy; they are sometimes called the *dark reactions* of photosynthesis because they can occur in the dark, utilizing the supplies of ATP and NADPH generated by light energy. However, the reactions in stage 4 are not confined to the dark; in fact, they occur primarily during illumination.

Each Photon of Light Has a Defined Amount of Energy

Quantum mechanics established that light, a form of electromagnetic radiation, has properties of both waves and particles. When light interacts with matter, it behaves as discrete packets of energy (quanta) called *photons*. The energy of a photon, ε, is proportional to the frequency of the light wave: $\varepsilon = h\gamma$, where h is Planck's constant (1.58×10^{-34} cal·s, or 6.63×10^{-34} J·s) and γ is the frequency of the light wave. It is customary in biology to refer to the wavelength of the light wave, λ, rather than to its frequency γ. The two are related by the simple equation $\gamma = c \div \lambda$, where c is the velocity of light (3×10^{10} cm/s in a vacuum). Note that photons of *shorter* wavelength have *higher* energies.

Also, the energy in 1 mol of photons can be denoted by $E = N\varepsilon$, where N is Avogadro's number (6.02×10^{23} molecules or photons/mol). Thus

$$E = Nh\gamma = \frac{Nhc}{\lambda}$$

The energy of light is considerable, as we can calculate for light with a wavelength of 550 nm (550×10^{-7} cm), typical of sunlight:

$$E = \frac{(6.02 \times 10^{23}\ \text{photons/mol})(1.58 \times 10^{-34}\text{cal·s})(3 \times 10^{10}\ \text{cm/s})}{550 \times 10^{-7}\text{cm}}$$
$$= 51,881\ \text{cal/mol}$$

or about 52 kcal/mol. This is enough energy to synthesize several moles of ATP from ADP and P_i if all the energy were used for this purpose.

Photosystems Comprise a Reaction Center and Associated Light-Harvesting Complexes

The absorption of light energy and its conversion into chemical energy occurs in multiprotein complexes called **photosystems.** Found in all photosynthetic organisms, both eukaryotic and prokaryotic, photosystems consist of two closely linked components: a *reaction center,* where the primary events of photosynthesis occur, and an antenna complex consisting of numerous protein complexes, termed *light-harvesting complexes (LHCs),* which capture light energy and transmit it to the reaction center.

Both reaction centers and antennas contain tightly bound light-absorbing pigment molecules. Chlorophyll *a* is the principal pigment involved in photosynthesis, being present in both reaction centers and antennas. In addition to chlorophyll *a*, antennas contain other light-absorbing pigments: *chlorophyll b* in vascular plants and *carotenoids* in both plants and photosynthetic bacteria. Carotenoids consist of long hydrocarbon chains with alternating single and double bonds; they are similar in structure to the visual pigment retinal, which absorbs light in the eye (Chapter 7). The presence of various antenna pigments, which absorb light at

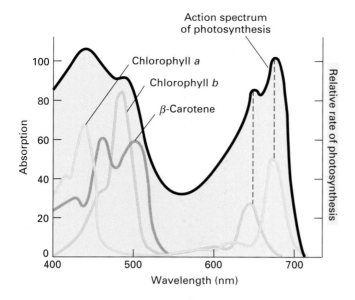

▲ **EXPERIMENTAL FIGURE 8-32** The rate of photosynthesis is greatest at wavelengths of light absorbed by three pigments. The action spectrum of photosynthesis in plants, that is, the ability of light of different wavelengths to support photosynthesis, is shown in black. Absorption spectra for three photosynthetic pigments present in the antennas of plant photosystems are shown in color. Each absorption spectrum shows how well light of different wavelengths is absorbed by one of the pigments. A comparison of the action spectrum with the individual absorption spectra suggests that photosynthesis at 680 nm is primarily due to light absorbed by chlorophyll a; at 650 nm, to light absorbed by chlorophyll b; and at shorter wavelengths, to light absorbed by chlorophylls a and b and by carotenoid pigments, including β-carotene.

different wavelengths, greatly extends the range of light that can be absorbed and used for photosynthesis.

One of the strongest pieces of evidence for the involvement of chlorophylls and carotenoids in photosynthesis is that the absorption spectrum of these pigments is similar to the action spectrum of photosynthesis (Figure 8-32). The latter is a measure of the relative ability of light of different wavelengths to support photosynthesis.

When chlorophyll a (or any other molecule) absorbs visible light, the absorbed light energy raises the chlorophyll a to a higher energy (excited) state. This differs from the ground (unexcited) state largely in the distribution of electrons around the C and N atoms of the porphyrin ring. Excited states are unstable and return to the ground state by one of several competing processes. For chlorophyll a molecules dissolved in organic solvents such as ethanol, the principal reactions that dissipate the excited-state energy are the emission of light (fluorescence and phosphorescence) and thermal emission (heat). When the same chlorophyll a is bound to the unique protein environment of the reaction center, dissipation of excited-state energy occurs by a quite different process that is the key to photosynthesis.

Photoelectron Transport from Energized Reaction-Center Chlorophyll a Produces a Charge Separation

The absorption of a photon of light of wavelength ≈680 nm by chlorophyll a increases its energy by 42 kcal/mol (the first excited state). Such an energized chlorophyll a molecule in a plant reaction center rapidly donates an electron to an intermediate acceptor, and the electron is rapidly passed on to the primary electron acceptor, quinone Q, on the stromal surface of the thylakoid membrane. This light-driven electron transfer, called **photoelectron transport,** depends on the unique environment of both the chlorophylls and the acceptor within the reaction center. Photoelectron transport, which occurs nearly every time a photon is absorbed, leaves a positive charge on the chlorophyll a close to the luminal surface and generates a reduced, negatively charged acceptor (Q^-) near the stromal surface (Figure 8-33).

The Q^- produced by photoelectron transport is a powerful reducing agent with a strong tendency to transfer an electron to another molecule, ultimately to $NADP^+$. The positively charged chlorophyll a^+, a strong oxidizing agent, attracts an electron from an electron donor on the luminal surface to regenerate the original chlorophyll a. In plants, the oxidizing power of four chlorophyll a^+ molecules is used, by way of intermediates, to remove four electrons from $2 H_2O$ molecules bound to a site on the luminal surface to form O_2:

$$2 H_2O + 4 \text{ chlorophyll } a^+ \longrightarrow 4 H^+ + O_2 + 4 \text{ chlorophyll } a$$

These potent biological reductants and oxidants provide all the energy needed to drive all subsequent reactions of photosynthesis: electron transport, ATP synthesis, and CO_2 fixation.

Chlorophyll a also absorbs light at discrete wavelengths shorter than 680 nm (see Figure 8-32). Such absorption raises

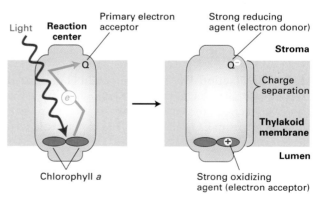

▲ **FIGURE 8-33 Photoelectron transport, the primary event in photosynthesis.** After absorption of a photon of light, one of the excited special pair of chlorophyll a molecules in the reaction center (*left*) donates an electron to a loosely bound acceptor molecule, the quinone Q, on the stromal surface of the thylakoid membrane, creating an essentially irreversible charge separation across the membrane (*right*). The electron cannot easily return through the reaction center to neutralize the positively charged chlorophyll a.

the molecule into one of several higher excited states, which decay within 10^{-12} seconds (1 picosecond, ps) to the first excited state with loss of the extra energy as heat. Because photoelectron transport and the resulting charge separation occur only from the first excited state of the reaction-center chlorophyll *a*, the quantum yield—the amount of photosynthesis per absorbed photon—is the same for all wavelengths of visible light shorter (and, therefore, of higher energy) than 680 nm.

Light-Harvesting Complexes Increase the Efficiency of Photosynthesis

Although chlorophyll *a* molecules within a reaction center are capable of directly absorbing light and initiating photosynthesis, they most commonly are energized indirectly by energy transferred from light-harvesting complexes (LHCs) in an associated antenna. Even at the maximum light intensity encountered by photosynthetic organisms (tropical noontime sunlight), each reaction-center chlorophyll *a* molecule absorbs only about one photon per second, which is not enough to support photosynthesis sufficient for the needs of the plant. The involvement of LHCs greatly increases the efficiency of photosynthesis, especially at more typical light intensities, by increasing absorption of 680-nm light and by extending the range of wavelengths of light that can be absorbed by other antenna pigments.

Photons can be absorbed by any of the pigment molecules in an LHC. The absorbed energy is then rapidly transferred (in $< 10^{-9}$ seconds) to one of the two "special-pair" chlorophyll *a* molecules in the associated reaction center, where it promotes the primary photosynthetic charge sepa-

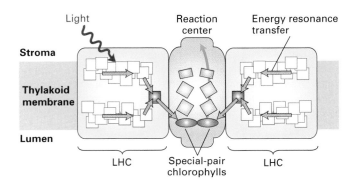

▲ **FIGURE 8-34 Energy transfer from light-harvesting complexes to associated reaction center in photosystem I of cyanobacteria.** The multiprotein light-harvesting complex binds 90 chlorophyll molecules (white and blue) and 31 other small molecules, all held in a specific geometric arrangement for optimal light absorption. Of the six chlorophyll molecules (green) in the reaction center, two constitute the special-pair chlorophylls (ovals) that can initiate photoelectron transport when excited (blue arrows). Resonance transfer of energy (red arrows) rapidly funnels energy from absorbed light to one of two "bridging" chlorophylls (blue) and thence to chlorophylls in the reaction center. [Adapted from W. Kühlbrandt, 2001, *Nature* **411**:896, and P. Jordan et al., 2001, *Nature* **411**:909.]

ration (see Figure 8-33). LHC proteins maintain the pigment molecules in the precise orientation and position optimal for light absorption and energy transfer, thereby maximizing the very rapid and efficient *resonance transfer* of energy from antenna pigments to reaction-center chlorophylls. Recent studies on one of the two photosystems in cyanobacteria, which are similar to those in higher plants, suggest that energy from absorbed light is funneled first to a "bridging" chlorophyll in each LHC and then to the special pair of reaction-center chlorophylls (Figure 8-34). Surprisingly, however, the molecular structures of LHCs from plants and cyanobacteria are completely different from those in green and purple bacteria, even though both types contain carotenoids and chlorophylls in a clustered geometric arrangement within the membrane.

Although LHC antenna chlorophylls can transfer light energy absorbed from a photon, they cannot release an electron. As we've seen already, this function resides in the two reaction-center chlorophylls. To understand their electron-releasing ability, we examine the structure and function of the reaction center in bacterial and plant photosystems in the next section.

KEY CONCEPTS OF SECTION 8.4

Photosynthetic Stages and Light-Absorbing Pigments

■ The principal end products of photosynthesis in plants are oxygen and polymers of six-carbon sugars (starch and sucrose).

■ The light-capturing and ATP-generating reactions of photosynthesis occur in the thylakoid membrane located within chloroplasts. The permeable outer membrane and inner membrane surrounding chloroplasts do not participate in photosynthesis (see Figure 8-30).

■ In stage 1 of photosynthesis, light is absorbed by chlorophyll *a* molecules bound to reaction-center proteins in the thylakoid membrane. The energized chlorophylls donate an electron to a quinone on the opposite side of the membrane, creating a charge separation (see Figure 8-33). In green plants, the positively charged chlorophylls then remove electrons from water, forming oxygen.

■ In stage 2, electrons are transported from the reduced quinone via carriers in the thylakoid membrane until they reach the ultimate electron acceptor, usually $NADP^+$, reducing it to NADPH. Electron transport is coupled to movement of protons across the membrane from the stroma to the thylakoid lumen, forming a pH gradient (proton-motive force) across the thylakoid membrane.

■ In stage 3, movement of protons down their electrochemical gradient through F_0F_1 complexes powers the synthesis of ATP from ADP and P_i.

■ In stage 4, the ATP and NADPH generated in stages 2 and 3 provide the energy and the electrons to drive the fixation of CO_2 and synthesis of carbohydrates. These reactions occur in the thylakoid stroma and cytosol.

■ Associated with each reaction center are multiple light-harvesting complexes (LHCs), which contain chlorophylls *a* and *b*, carotenoids, and other pigments that absorb light at multiple wavelengths. Energy is transferred from the LHC chlorophyll molecules to reaction-center chlorophylls by resonance energy transfer (see Figure 8-34).

8.5 Molecular Analysis of Photosystems

As noted in the previous section, photosynthesis in the green and purple bacteria does not generate oxygen, whereas photosynthesis in cyanobacteria, algae, and higher plants does.* This difference is attributable to the presence of two types of photosystem (PS) in the latter organisms: PSI reduces $NADP^+$ to NADPH, and PSII forms O_2 from H_2O. In contrast, the green and purple bacteria have only one type of photosystem, which cannot form O_2. We first discuss the simpler photosystem of purple bacteria and then consider the more complicated photosynthetic machinery in chloroplasts.

The Single Photosystem of Purple Bacteria Generates a Proton-Motive Force but No O_2

The three-dimensional structures of the photosynthetic reaction centers from two purple bacteria have been determined, permitting scientists to trace the detailed paths of electrons during and after the absorption of light. Similar proteins and pigments compose photosystem II of plants as well, and the conclusions drawn from studies on this simple photosystem have proven applicable to plant systems.

The reaction center of purple bacteria contains three protein subunits (L, M, and H) located in the plasma membrane (Figure 8-35). Bound to these proteins are the prosthetic groups that absorb light and transport electrons during photosynthesis. The prosthetic groups include a "special pair" of bacteriochlorophyll *a* molecules equivalent to the reaction-center chlorophyll *a* molecules in plants, as well as several other pigments and two quinones, termed Q_A and Q_B, that are structurally similar to mitochondrial ubiquinone.

Initial Charge Separation The mechanism of charge separation in the photosystem of purple bacteria is identical with that in plants outlined earlier; that is, energy from absorbed light is used to strip an electron from a reaction-center bacteriochlorophyll *a* molecule and transfer it, via several different pigments, to the primary electron acceptor Q_B, which is loosely bound to a site on the cytosolic membrane face. The chlorophyll thereby acquires a positive charge, and Q_B acquires a negative charge. To determine the pathway traversed by electrons through the bacterial reaction center, re-

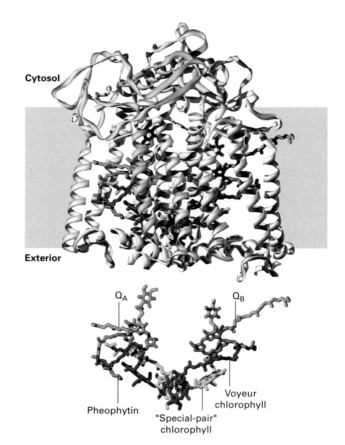

▲ FIGURE 8-35 **Three-dimensional structure of the photosynthetic reaction center from the purple bacterium** *Rhodobacter spheroides.* (*Top*) The L subunit (yellow) and M subunit (white) each form five transmembrane α helices and have a very similar structure overall; the H subunit (light blue) is anchored to the membrane by a single transmembrane α helix. A fourth subunit (not shown) is a peripheral protein that binds to the exoplasmic segments of the other subunits. (*Bottom*) Within each reaction center is a special pair of bacteriochlorophyll *a* molecules (green), capable of initiating photoelectron transport; two voyeur chlorophylls (purple); two pheophytins (dark blue), and two quinones, Q_A and Q_B (orange). Q_B is the primary electron acceptor during photosynthesis. [After M. H. Stowell et al., 1997, *Science* **276**:812.]

searchers exploited the fact that each pigment absorbs light of only certain wavelengths, and its absorption spectrum changes when it possesses an extra electron. Because these electron movements are completed in less than 1 millisecond (ms), a special technique called *picosecond absorption spectroscopy* is required to monitor the changes in the absorption spectra of the various pigments as a function of time shortly after the absorption of a light photon.

When a preparation of bacterial membrane vesicles is exposed to an intense pulse of laser light lasting less than 1 ps, each

---

*A very different type of bacterial photosynthesis, which occurs only in certain archaebacteria, is not discussed here because it is very different from photosynthesis in higher plants. In this type of photosynthesis, the plasma-membrane protein bacteriorhodopsin pumps one proton from the

cytosol to the extracellular space for every quantum of light absorbed. This small protein has seven membrane-spanning segments and a covalently attached retinal pigment (see Figure 5-13).

reaction center absorbs one photon. Light absorbed by the chlorophyll *a* molecules in each reaction center converts them to the excited state, and the subsequent electron transfer processes are synchronized in all reaction centers. Within 4×10^{-12} seconds (4 ps), an electron moves to one of the pheophytin molecules (Ph), leaving a positive charge on the chlorophyll *a*. It takes 200 ps for the electron to move to Q_A, and then, in the slowest step, 200μs for it to move to Q_B. This pathway of electron flow is traced in the left portion of Figure 8-36.

Subsequent Electron Flow and Coupled Proton Movement

After the primary electron acceptor, Q_B, in the bacterial reaction center accepts one electron, forming Q_B^-·, it accepts a second electron from the same reaction-center chlorophyll following its absorption of a second photon. The quinone then binds two protons from the cytosol, forming the reduced quinone (QH_2), which is released from the reaction center (see Figure 8-36). QH_2 diffuses within the bacterial membrane to the Q_o site on the exoplasmic face of a cytochrome bc_1 complex, where it releases its two protons into the periplasmic space (the space between the plasma membrane and the bacterial cell wall). This process moves protons from the cytosol to the outside of the cell, generating a proton-motive force across the plasma membrane. Simultaneously, QH_2 releases its two electrons, which move through the cytochrome bc_1 complex exactly as depicted for the mitochondrial $CoQH_2$–cytochrome *c* reductase complex in Figure 8-21. The Q cycle in the bacterial reaction center, like the Q cycle in mitochondria, pumps additional protons from the cytosol to the intermembrane space, thereby increasing the proton-motive force.

The acceptor for electrons transferred through the cytochrome bc_1 complex is a soluble cytochrome, a one-electron carrier, in the periplasmic space, which is reduced from the Fe^{3+} to the Fe^{2+} state. The reduced cytochrome (analogous to cytochrome *c* in mitochondria) then diffuses to a reaction center, where it releases its electron to a positively charged chlorophyll a^+, returning the chlorophyll to the ground state and the cytochrome to the Fe^{3+} state. This *cyclic* electron flow generates no oxygen and no reduced coenzymes.

Electrons also can flow through the single photosystem of purple bacteria via a *linear* (noncyclic) pathway. In this case, electrons removed from reaction-center chlorophylls ultimately are transferred to NAD^+ (rather than $NADP^+$ as in plants), forming NADH. To reduce the oxidized reaction-center chlorophyll *a* back to its ground state, an electron is transferred from a reduced cytochrome *c*; the oxidized cytochrome *c* that is formed is reduced by electrons removed from hydrogen sulfide (H_2S), forming elemental sulfur (S), or from hydrogen gas (H_2). Since H_2O is not the electron donor, no O_2 is formed.

Both the cyclic and linear pathways of electron flow in the bacterial photosystem generate a proton-motive force. As

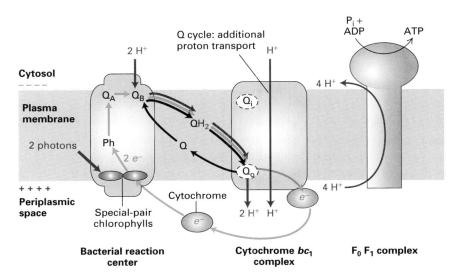

▲ **FIGURE 8-36 Cyclic electron flow in the single photosystem of purple bacteria.** Blue arrows indicate flow of electrons; red arrows indicate proton movement. *(Left)* Energy funneled from an associated LHC (not illustrated here) energizes one of the special-pair chlorophylls in the reaction center. Photoelectron transport from the energized chlorophyll, via pheophytin (Ph) and quinone A (Q_A), to quinone B (Q_B) forms the semiquinone Q⁻· and leaves a positive charge on the chlorophyll. Following absorption of a second photon and transfer of a second electron to the semiquinone, it rapidly picks up two protons from the cytosol to form QH_2. *(Center)* After diffusing through the membrane and binding to the Q_o site on the

exoplasmic face of the cytochrome bc_1 complex, QH_2 donates two electrons and simultaneously gives up two protons to the external medium, generating a proton-motive force ($H^+_{exoplasmic} > H^+_{cytosolic}$). Electrons are transported back to the reaction-center chlorophyll via a soluble cytochrome, which diffuses in the periplasmic space. Operation of a Q cycle in the cytochrome bc_1 complex pumps additional protons across the membrane to the external medium, as in mitochondria. The proton-motive force is used by the F_0F_1 complex to synthesize ATP and, as in other bacteria, to transport molecules in and out of the cell. [Adapted from J. Deisenhofer and H. Michael, 1991, *Ann. Rev. Cell Biol.* **7**:1.]

in other systems, this proton-motive force is used by the F_0F_1 complex located in the plasma membrane to synthesize ATP and also to transport molecules across the membrane against a concentration gradient.

Chloroplasts Contain Two Functionally and Spatially Distinct Photosystems

In the 1940s, biophysicist R. Emerson discovered that the rate of plant photosynthesis generated by light of wavelength 700 nm can be greatly enhanced by adding light of shorter wavelength. He found that a combination of light at, say, 600 and 700 nm supports a greater rate of photosynthesis than the sum of the rates for the two separate wavelengths. This so-called *Emerson effect* led researchers to conclude that photosynthesis in plants involves the interaction of two separate photosystems, referred to as *PSI* and *PSII*. PSI is driven by light of wavelength 700 nm or less; PSII, only by light of shorter wavelength <680 nm.

As in the reaction center of green and purple bacteria, each chloroplast photosystem contains a pair of specialized reaction-center chlorophyll *a* molecules, which are capable of initiating photoelectron transport. The reaction-center chlorophylls in PSI and PSII differ in their light-absorption maxima because of differences in their protein environment. For this reason, these chlorophylls are often denoted P_{680}

(PSII) and P_{700} (PSI). Like a bacterial reaction center, each chloroplast reaction center is associated with multiple light-harvesting complexes (LHCs); the LHCs associated with PSII and PSI contain different proteins.

The two photosystems also are distributed differently in thylakoid membranes: PSII primarily in stacked regions (grana) and PSI primarily in unstacked regions. The stacking of the thylakoid membranes may be due to the binding properties of the proteins in PSII. Evidence for this distribution came from studies in which thylakoid membranes were gently fragmented into vesicles by ultrasound. Stacked and unstacked thylakoid vesicles were then fractionated by density-gradient centrifugation. The stacked fractions contained primarily PSII protein and the unstacked fraction, PSI.

Finally, and most importantly, the two chloroplast photosystems differ significantly in their functions: only PSII splits water to form oxygen, whereas only PSI transfers electrons to the final electron acceptor, $NADP^+$. Photosynthesis in chloroplasts can follow a linear or cyclic pathway, again like green and purple bacteria. The linear pathway, which we discuss first, can support carbon fixation as well as ATP synthesis. In contrast, the cyclic pathway supports only ATP synthesis and generates no reduced NADPH for use in carbon fixation. Photosynthetic algae and cyanobacteria contain two photosystems analogous to those in chloroplasts.

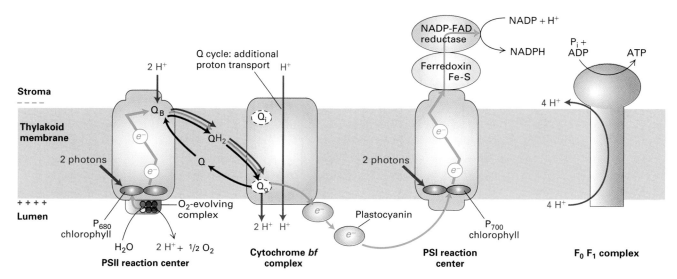

▲ **FIGURE 8-37 Linear electron flow in plants, which requires both chloroplast photosystems, PSI and PSII.** Blue arrows indicate flow of electrons; red arrows indicate proton movement. LHCs are not shown. *(Left)* In the PSII reaction center, two sequential light-induced excitations of the same P_{680} chlorophylls result in reduction of the primary electron acceptor Q_B to QH_2. On the luminal side of PSII, electrons removed from H_2O in the thylakoid lumen are transferred to P_{680}^+, restoring the reaction-center chlorophylls to the ground state and generating O_2. *(Center)* The cytochrome *bf* complex then accepts electrons from QH_2, coupled to the release of two protons into the lumen. Operation of a Q cycle in the cytochrome *bf* complex

translocates additional protons across the membrane to the thylakoid lumen, increasing the proton-motive force generated. *(Right)* In the PSI reaction center, each electron released from light-excited P_{700} chlorophylls moves via a series of carriers in the reaction center to the stromal surface, where soluble ferredoxin (an Fe-S protein) transfers the electron to FAD and finally to $NADP^+$, forming NADPH. P_{700}^+ is restored to its ground state by addition of an electron carried from PSII via the cytochrome *bf* complex and plastocyanin, a soluble electron carrier. The proton-motive force generated by linear electron flow from PSII to NADP-FAD reductase powers ATP synthesis by the F_0F_1 complex.

Linear Electron Flow Through Both Plant Photosystems, PSII and PSI, Generates a Proton-Motive Force, O_2, and NADPH

Linear electron flow in chloroplasts involves PSII and PSI in an obligate series in which electrons are transferred from H_2O to $NADP^+$. The process begins with absorption of a photon by PSII, causing an electron to move from a P_{680} chlorophyll *a* to an acceptor plastoquinone (Q_B) on the stromal surface (Figure 8-37). The resulting oxidized P_{680}^+ strips one electron from the highly unwilling donor H_2O, forming an intermediate in O_2 formation and a proton, which remains in the thylakoid lumen and contributes to the proton-motive force. After P_{680} absorbs a second photon, the semiquinone $Q^-\cdot$ accepts a second electron and picks up two protons from the stromal space, generating QH_2. After diffusing in the membrane, QH_2 binds to the Q_o site on a cytochrome *bf* complex, which is analogous in structure and function to the cytochrome bc_1 complex in purple bacteria and the $CoQH_2$–cytochrome *c* reductase complex in mitochondria. As in these systems a Q cycle operates in the cytochrome *bf* complex in association with the PSII reaction center, thereby increasing the proton-motive force generated by electron transport.

Absorption of a photon by PSI leads to removal of an electron from the reaction-center chlorophyll *a*, P_{700} (see Figure 8-37). The resulting oxidized P_{700}^+ is reduced by an electron passed from the PSII reaction center via the cytochrome *bf* complex and *plastocyanin*, a soluble electron carrier that contains a single copper (Cu) atom. After the cytochrome *bf* complex accepts electrons from QH_2, it transfers them, one at a time, to the Cu^{2+} form of plastocyanin, reducing it to the Cu^+ form. Reduced plastocyanin then diffuses in the thylakoid lumen, carrying the electron to P_{700}^+ in PSI. The electron taken up at the luminal surface by P_{700} moves via several carriers to the stromal surface of the thylakoid membrane, where it is accepted by ferredoxin, an iron-sulfur (Fe-S) protein. Electrons excited in PSI can be transferred from ferredoxin via the electron carrier FAD to $NADP^+$, forming, together with one proton picked up from the stroma, the reduced molecule NADPH.

F_0F_1 complexes in the thylakoid membrane use the proton-motive force generated during linear electron flow to synthesize ATP on the stromal side of membrane. Thus this pathway yields both NADPH and ATP in the stroma of the chloroplast, where they are utilized for CO_2 fixation.

An Oxygen-Evolving Complex Is Located on the Luminal Surface of the PSII Reaction Center

Somewhat surprisingly, the structure of the PSII reaction center, which removes electrons from H_2O to form O_2, resembles that of the reaction center of photosynthetic purple bacteria, which does not form O_2. Like the bacterial reaction center, the PSII reaction center contains two molecules of chlorophyll *a* (P_{680}), as well as two other chlorophylls, two pheophytins, two quinones (Q_A and Q_B), and

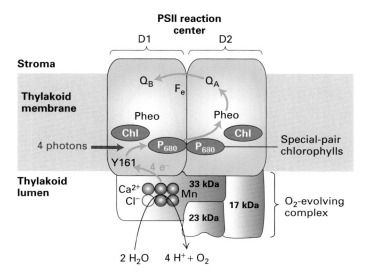

▲ **FIGURE 8-38 Electron flow and O_2 evolution in chloroplast PSII.** The PSII reaction center, comprising two integral proteins, D1 and D2, special-pair chlorophylls (P_{680}), and other electron carriers, is associated with an oxygen-evolving complex on the luminal surface. Bound to the three extrinsic proteins (33, 23, and 17 kDa) of the oxygen-evolving complex are four manganese ions (red), a Ca^{2+} ion (blue), and a Cl^- ion (yellow). These bound ions function in the splitting of H_2O and maintain the environment essential for high rates of O_2 evolution. Tyrosine-161 (Y161) of the D1 polypeptide conducts electrons from the Mn ions to the oxidized reaction-center chlorophyll (P_{680}^+), reducing it to the ground state P_{680}. See the text for details. [Adapted from C. Hoganson and G. Babcock, 1997, *Science* **277**:1953.]

one nonheme iron atom. These small molecules are bound to two PSII proteins, called D1 and D2, whose sequences are remarkably similar to the sequences of the L and M peptides of the bacterial reaction center, attesting to their common evolutionary origins (see Figure 8-35). When PSII absorbs a photon with a wavelength of <680 nm, it triggers the loss of an electron from a P_{680} molecule, generating P_{680}^+. As in photosynthetic purple bacteria, the electron is transported via a pheophytin and a quinone (Q_A) to the primary electron acceptor, Q_B, on the outer (stromal) surface of the thylakoid membrane (Figures 8-37 and 8-38).

The photochemically oxidized reaction-center chlorophyll of PSII, P_{680}^+, is the strongest biological oxidant known. The reduction potential of P_{680}^+ is more positive than that of water, and thus it can oxidize water to generate O_2 and H^+ ions. Photosynthetic bacteria cannot oxidize water because the excited chlorophyll a^+ in the bacterial reaction center is not a sufficiently strong oxidant. (As noted earlier, purple bacteria use H_2S and H_2 as electron donors to reduce chlorophyll a^+ in linear electron flow.)

The splitting of H_2O, which provides the electrons for reduction of P_{680}^+ in PSII, is catalyzed by a three-protein complex, the *oxygen-evolving complex*, located on the luminal surface of the thylakoid membrane. The oxygen-evolving complex contains four manganese (Mn) ions as well as

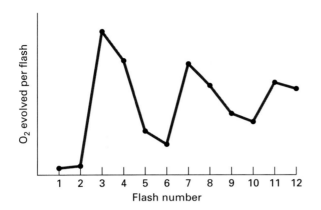

▲ **EXPERIMENTAL FIGURE 8-39 A single PSII absorbs a photon and transfers an electron four times to generate one O_2.** Dark-adapted chloroplasts were exposed to a series of closely spaced, short (5-μs) pulses of light that activated virtually all the PSIIs in the preparation. The peaks in O_2 evolution occurred after every fourth pulse, indicating that absorption of four photons by one PSII is required to generate each O_2 molecule. Because the dark-adapted chloroplasts were initially in a partially reduced state, the peaks in O_2 evolution occurred after flashes 3, 7, and 11. [From J. Berg et al., 2002, *Biochemistry*, 5th ed., W. H. Freeman and Company.]

bound Cl^- and Ca^{2+} ions (see Figure 8-38); this is one of the very few cases in which Mn plays a role in a biological system. These Mn ions together with the three extrinsic proteins can be removed from the reaction center by treatment with solutions of concentrated salts; this abolishes O_2 formation but does not affect light absorption or the initial stages of electron transport.

The oxidation of two molecules of H_2O to form O_2 requires the removal of four electrons, but absorption of each photon by PSII results in the transfer of just one electron. A simple experiment, described in Figure 8-39, resolved whether the formation of O_2 depends on a single PSII or multiple ones acting in concert. The results indicated that a single PSII must lose an electron and then oxidize the oxygen-evolving complex four times in a row for an O_2 molecule to be formed.

Manganese is known to exist in multiple oxidation states with from two to five positive charges. Subsequent spectroscopic studies indeed showed that the bound Mn ions in the oxygen-evolving complex cycle through five different oxidation states, S_0–S_4. In this S cycle, a total of two H_2O molecules are split into four protons, four electrons, and one O_2 molecule. The electrons released from H_2O are transferred, one at a time, via the Mn ions and a nearby tyrosine side chain on the D1 subunit, to the reaction-center P_{680}^+, where they regenerate the reduced chlorophyll, P_{680}. The protons released from H_2O remain in the thylakoid lumen. Despite a great deal of experimentation we still do not know the details of how O_2 is formed by this Mn-containing complex.

Herbicides that inhibit photosynthesis not only are very important in agriculture but also have proved useful in dissecting the pathway of photoelectron transport in plants. One such class of herbicides, the *s*-triazines (e.g., atrazine), binds specifically to D1 in the PSII reaction center, thus inhibiting binding of oxidized Q_B to its site on the stromal surface of the thylakoid membrane. When added to illuminated chloroplasts, *s*-triazines cause all downstream electron carriers to accumulate in the oxidized form, since no electrons can be released from PSII. In atrazine-resistant mutants, a single amino acid change in D1 renders it unable to bind the herbicide, so photosynthesis proceeds at normal rates. Such resistant weeds are prevalent and present a major agricultural problem. ∎

Cyclic Electron Flow Through PSI Generates a Proton-Motive Force but No NADPH or O_2

As we've seen, electrons from reduced ferredoxin in PSI are transferred to $NADP^+$ during linear electron flow (see Figure 8-37). Alternatively, reduced ferredoxin can donate two electrons to a quinone (Q) bound to a site on the stromal surface of PSI; the quinone then picks up two protons from the stroma, forming QH_2. This QH_2 then diffuses through the thylakoid membrane to the Q_o binding site on the luminal surface of the cytochrome *bf* complex. There it releases two electrons to the cytochrome *bf* complex and two protons to the thylakoid lumen, generating a proton-motive force. As in linear electron flow, these electrons return to PSI via plastocyanin. This cyclic electron flow, which does not involve PSII, is similar to the cyclic process that occurs in the single photosystem of purple bacteria (see Figure 8-36). A Q cycle operates in the cytochrome *bf* complex during cyclic electron flow, leading to transport of two additional protons into the lumen for each pair of electrons transported and a greater proton-motive force.

The proton-motive force generated during cyclic electron flow in chloroplasts powers ATP synthesis by F_0F_1 complexes in the thylakoid membrane. This process, however, generates no NADPH, and no O_2 evolves.

Relative Activity of Photosystems I and II Is Regulated

In order for PSII and PSI to act in sequence during linear electron flow, the amount of light energy delivered to the two reaction centers must be controlled so that each center activates the same number of electrons. If the two photosystems are not equally excited, then cyclic electron flow occurs in PSI, and PSII becomes less active. One mechanism for regulating the relative contribution of linear and cyclic electron flow in chloroplasts entails the reversible phosphorylation and dephosphorylation of the proteins associated with the PSII light-harvesting complex (LHCII).

According to this mechanism, a membrane-bound protein kinase senses the relative activities of the two photosys-

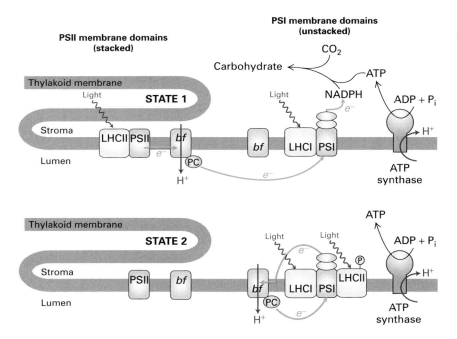

PSII membrane domains (stacked)

PSI membrane domains (unstacked)

◀ **FIGURE 8-40 Distribution of multiprotein complexes in the thylakoid membrane and the regulation of linear versus cyclic electron flow.** *(Top)* In sunlight, PSI and PSII are equally activated, and the photosystems are organized in state I. In this arrangement, light-harvesting complex II (LHCII) is not phosphorylated and is tightly associated with the PSII reaction center in the grana. As a result, PSII and PSI can function in parallel in linear electron flow. *(Bottom)* When light excitation of the two photosystems is unbalanced, LHCII becomes phosphorylated, dissociates from PSII, and diffuses into the unstacked membranes, where it associates with PSI and its permanently associated LHCI. In this alternative supramolecular organization (state II), most of the absorbed light energy is transferred to PSI, supporting cyclic electron flow and ATP production but no formation of NADPH and thus no CO_2 fixation. PC = plastocyanin. [Adapted from F. A. Wollman, 2001, *EMBO J.* **20**:3623.]

tems. This monitoring may involve recognition of the oxidized-reduced state of the plastoquinone pool that transfers electrons from PSII to the cytochrome *bf* complex en route to PSI. Balanced excitation of the two photosystems is associated with linear electron flow (Figure 8-40, state I). However, if too much plastoquinone is reduced (indicating excessive activation of PSII relative to PSI), the kinase becomes activated and some LHCIIs are phosphorylated. These phosphorylated LHCIIs dissociate from PSII, which is preferentially located in the grana, and diffuse in the membrane to the unstacked thylakoid membranes, where they can activate PSI. This redistribution of LHCIIs, which decreases the antenna size of PSII and increases that of PSI, and the concomitant redistribution of cytochrome *bf* complexes from PSII-rich to PSI-rich membrane domains promote cyclic electron flow (Figure 8-40, state II).

The supramolecular organization of the photosystems in plants thus has the effect of directing them toward ATP production (state II) or toward generation of reducing equivalents (NADPH) and ATP (state I). Both NADPH and ATP are required to convert CO_2 to sucrose or starch, the fourth stage in photosynthesis, which we cover in the last section of this chapter.

KEY CONCEPTS OF SECTION 8.5

Molecular Analysis of Photosystems

■ In the single photosystem of purple bacteria, cyclic electron flow from light-excited chlorophyll *a* molecules in the reaction center generates a proton-motive force, which is used mainly to power ATP synthesis by the F_0F_1 complex in the plasma membrane (Figure 8-36).

■ Plants contain two photosystems, PSI and PSII, which have different functions and are physically separated in the thylakoid membrane. PSII splits H_2O into O_2. PSI reduces $NADP^+$ to NADPH. Cyanobacteria have two analogous photosystems.

■ In chloroplasts, light energy absorbed by light-harvesting complexes (LHCs) is transferred to chlorophyll *a* molecules in the reaction centers (P_{680} in PSII and P_{700} in PSI).

■ Electrons flow through PSII via the same carriers that are present in the bacterial photosystem. In contrast to the bacterial system, photochemically oxidized P_{680}^+ in PSII is regenerated to P_{680} by electrons derived from the splitting of H_2O with evolution of O_2 (see Figure 8-37, *left*).

■ In linear electron flow, photochemically oxidized P_{700}^+ in PSI is reduced, regenerating P_{700}, by electrons transferred from PSII via the cytochrome *bf* complex and soluble plastocyanin. Electrons lost from P_{700} following excitation of PSI are transported via several carriers ultimately to $NADP^+$, generating NADPH (see Figure 8-37, *right*).

■ In contrast to linear electron flow, which requires both PSII and PSI, cyclic electron flow in plants involves only PSI. In this pathway, neither NADPH nor O_2 is formed, although a proton-motive force is generated.

■ The proton-motive force generated by photoelectron transport in plant and bacterial photosystems is augmented by operation of the Q cycle in cytochrome *bf* complexes associated with each of the photosystems.

■ Reversible phosphorylation and dephosphorylation of the PSII light-harvesting complex control the functional organization of the photosynthetic apparatus in thylakoid membranes. State I favors linear electron flow, whereas state II favors cyclic electron flow (see Figure 8-40).

8.6 CO_2 Metabolism During Photosynthesis

PLANTS Chloroplasts perform many metabolic reactions in green leaves. In addition to CO_2 fixation, the synthesis of almost all amino acids, all fatty acids and carotenes, all pyrimidines, and probably all purines occurs in chloroplasts. However, the synthesis of sugars from CO_2 is the most extensively studied biosynthetic pathway in plant cells. We first consider the unique pathway, known as the **Calvin cycle** (after discoverer Melvin Calvin), that fixes CO_2 into three-carbon compounds, powered by energy released during ATP hydrolysis and oxidation of NADPH.

CO_2 Fixation Occurs in the Chloroplast Stroma

The reaction that actually fixes CO_2 into carbohydrates is catalyzed by *ribulose 1,5-bisphosphate carboxylase* (often called *rubisco*), which is located in the stromal space of the chloroplast. This enzyme adds CO_2 to the five-carbon sugar ribulose 1,5-bisphosphate to form two molecules of 3-phosphoglycerate (Figure 8-41). Rubisco is a large enzyme (≈500 kDa) composed of eight identical large subunits and eight identical small subunits. One subunit is encoded in chloroplast DNA; the other, in nuclear DNA. Because the catalytic rate of rubisco is quite slow, many copies of the enzyme are needed to fix sufficient CO_2. Indeed, this enzyme makes up almost 50 percent of the chloroplast protein and is believed to be the most abundant protein on earth.

When photosynthetic algae are exposed to a brief pulse of ^{14}C-labeled CO_2 and the cells are then quickly disrupted, 3-phosphoglycerate is radiolabeled most rapidly, and all the radioactivity is found in the carboxyl group. This finding establishes that ribulose 1,5-bisphosphate carboxylase fixes the CO_2. Because CO_2 is initially incorporated into a three-carbon compound, the Calvin cycle is also called the *C_3 pathway* of carbon fixation.

The fate of 3-phosphoglycerate formed by rubisco is complex: some is converted to starch or sucrose, but some is used to regenerate ribulose 1,5-bisphosphate. At least nine enzymes are required to regenerate ribulose 1,5-bisphosphate from 3-phosphoglycerate. Quantitatively, for every 12 molecules of 3-phosphoglycerate generated by rubisco (a total of 36 C atoms), 2 molecules (6 C atoms) are converted to 2 molecules of glyceraldehyde 3-phosphate (and later to 1 hexose), while 10 molecules (30 C atoms) are converted to 6 molecules of ribulose 1,5-bisphosphate (Figure 8-42, *top*). The fixation of six CO_2 molecules and the net formation of two glyceraldehyde 3-phosphate molecules require the consumption of 18 ATPs and 12 NADPHs, generated by the light-requiring processes of photosynthesis.

▶ **FIGURE 8-42 The pathway of carbon during photosynthesis.** *(Top)* Six molecules of CO_2 are converted into two molecules of glyceraldehyde 3-phosphate. These reactions, which constitute the Calvin cycle, occur in the stroma of the chloroplast. Via the phosphate/triosephosphate antiporter, some glyceraldehyde 3-phosphate is transported to the cytosol in exchange for phosphate. *(Bottom)* In the cytosol, an exergonic series of reactions converts glyceraldehyde 3-phosphate to fructose 1,6-bisphosphate and, ultimately, to the disaccharide sucrose. Some glyceraldehyde 3-phosphate (not shown here) is also converted to amino acids and fats, compounds essential to plant growth.

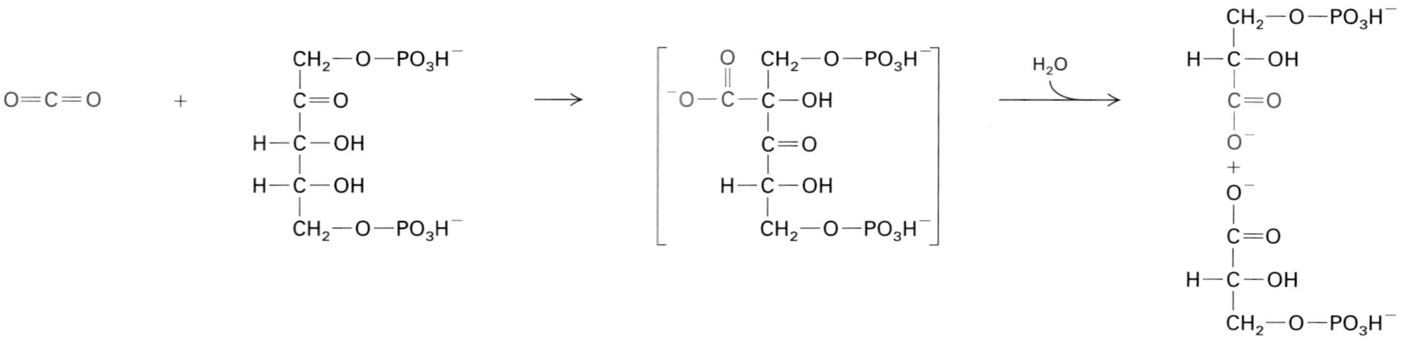

CO_2 Ribulose 1,5-bisphosphate Enzyme-bound intermediate 3-Phosphoglycerate (two molecules)

▲ **FIGURE 8-41 The initial reaction that fixes CO_2 into organic compounds.** In this reaction, catalyzed by ribulose 1,5-bisphosphate carboxylase, CO_2 condenses with the five-carbon sugar ribulose 1,5-bisphosphate. The products are two molecules of 3-phosphoglycerate.

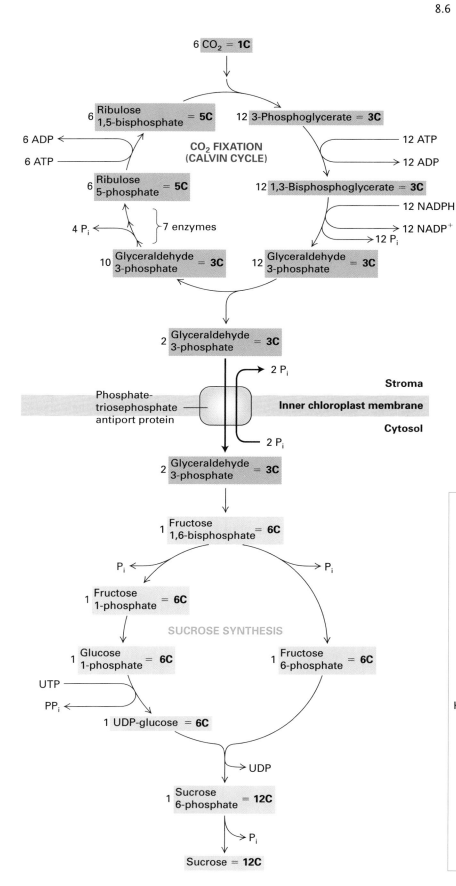

Synthesis of Sucrose Incorporating Fixed CO_2 Is Completed in the Cytosol

After its formation in the chloroplast stroma, glyceraldehyde 3-phosphate is transported to the cytosol in exchange for phosphate. The final steps of sucrose synthesis occur in the cytosol of leaf cells. In these reactions, one molecule of glyceraldehyde 3-phosphate is isomerized to dihydroxyacetone phosphate. This compound condenses with a second molecule of glyceraldehyde 3-phosphate to form fructose 1,6-bisphosphate, which is the reverse of the aldolase reaction in glycolysis (see Figure 8-4, step 4). Fructose 1,6-bisphosphate is converted primarily to sucrose by the reactions shown in the bottom portion of Figure 8-42.

The transport protein in the chloroplast membrane that brings fixed CO_2 (as glyceraldehyde 3-phosphate) into the cytosol when the cell is exporting sucrose vigorously is a strict antiporter: No fixed CO_2 leaves the chloroplast unless phosphate is fed into it. The phosphate is generated in the cytosol, primarily during the formation of sucrose, from phosphorylated three-carbon intermediates. Thus the synthesis of sucrose and its export from leaf cells to other cells encourages the transport of additional glyceraldehyde 3-phosphate from the chloroplast to the cytosol.

Light and Rubisco Activase Stimulate CO_2 Fixation

The Calvin cycle enzymes that catalyze CO_2 fixation are rapidly inactivated in the dark, thereby conserving ATP that is generated in the dark for other synthetic reactions, such as lipid and amino acid biosynthesis. One mechanism that contributes to this control is the pH dependence of several Calvin cycle enzymes. Because protons are transported from the stroma into the thylakoid lumen during photoelectron transport (see Figure 8-37), the pH of the stroma increases from ≈ 7 in the dark to ≈ 8 in the light. The increased activity of several Calvin cycle enzymes at the higher pH promotes CO_2 fixation in the light.

A stromal protein called *thioredoxin (Tx)* also plays a role in controlling some Calvin cycle enzymes. In the dark, thioredoxin contains a disulfide bond; in the light, electrons are transferred from PSI, via ferredoxin, to thioredoxin, reducing its disulfide bond:

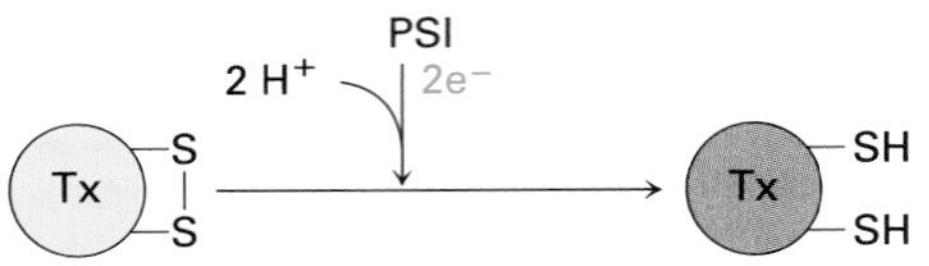

Reduced thioredoxin then activates several Calvin cycle enzymes by reducing disulfide bonds in them. In the dark, when thioredoxin becomes reoxidized, these enzymes are reoxidized and thus inactivated.

Rubisco is spontaneously activated in the presence of high CO_2 and Mg^{2+} concentrations. The activating reaction entails covalent addition of CO_2 to the side-chain amino group of lysine-191, forming a carbamate group that then binds a Mg^{2+} ion. Under normal conditions, however, with ambient levels of CO_2, the reaction requires catalysis by rubisco activase, an enzyme that simultaneously hydrolyzes ATP and uses the energy to attach a CO_2 to the lysine. This enzyme was discovered during the study of a mutant strain of *Arabidopsis thaliana* that required high CO_2 levels to grow and that did not exhibit light activation of ribulose 1,5-bisphosphate carboxylase; the mutant had a defective rubisco activase.

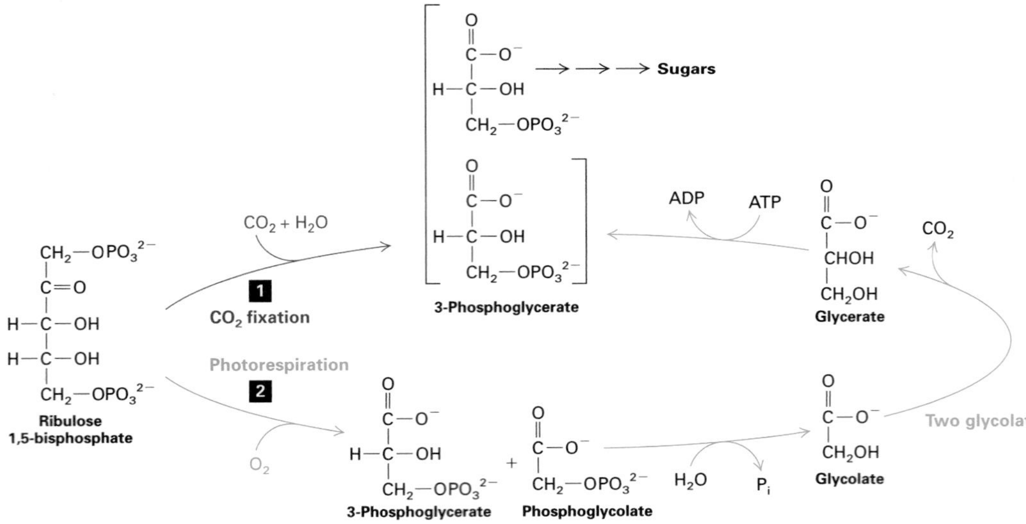

▲ **FIGURE 8-43 CO_2 fixation and photorespiration.** These competing pathways are both initiated by ribulose 1,5-bisphosphate carboxylase (rubisco), and both utilize ribulose 1,5-bisphosphate. CO_2 fixation, pathway (1), is favored by high CO_2 and low O_2 pressures; photorespiration, pathway (2), occurs at low CO_2 and high O_2 pressures (that is, under normal atmospheric conditions). Phosphoglycolate is recycled via a complex set of reactions that take place in peroxisomes and mitochondria, as well as chloroplasts. The net result: for every two molecules of phosphoglycolate formed by photorespiration (four C atoms), one molecule of 3-phosphoglycerate is ultimately formed and recycled, and one molecule of CO_2 is lost.

Photorespiration, Which Competes with Photosynthesis, Is Reduced in Plants That Fix CO₂ by the C₄ Pathway

Photosynthesis is always accompanied by **photorespiration**—a process that takes place in light, consumes O_2, and converts ribulose 1,5-bisphosphate in part to CO_2. As Figure 8-43 shows, rubisco catalyzes two competing reactions: the addition of CO_2 to ribulose 1,5-bisphosphate to form two molecules of 3-phosphoglycerate and the addition of O_2 to form one molecule of 3-phosphoglycerate and one molecule of the two-carbon compound phosphoglycolate. Photorespiration is wasteful to the energy economy of the plant: it consumes ATP and O_2, and it generates CO_2. It is surprising, therefore, that all known rubiscos catalyze photorespiration. Probably the necessary structure of the active site of rubisco precluded evolution of an enzyme that does not catalyze photorespiration.

In a hot, dry environment, plants must keep the gas-exchange pores (stomata) in their leaves closed much of the time to prevent excessive loss of moisture. This causes the CO_2 level inside the leaf to fall below the K_m of rubisco for CO_2. Under these conditions, the rate of photosynthesis is slowed and photorespiration is greatly favored. Corn, sugar cane, crabgrass, and other plants that can grow in hot, dry environments have evolved a way to avoid this problem by utilizing a two-step pathway of CO_2 fixation in which a CO_2-hoarding step precedes the Calvin cycle. The pathway has been named the *C₄ pathway* because [¹⁴C] CO_2 labeling showed that the first radioactive molecules formed during photosynthesis in this pathway are four-carbon compounds, such as oxaloacetate and malate, rather than the three-carbon molecules that begin the Calvin cycle (C₃ pathway).

The C₄ pathway involves two types of cells: *mesophyll cells,* which are adjacent to the air spaces in the leaf interior, and *bundle sheath cells,* which surround the vascular tissue (Figure 8-44a). In the mesophyll cells of C₄ plants, phosphoenolpyruvate, a three-carbon molecule derived from pyruvate, reacts with CO_2 to generate oxaloacetate, a four-

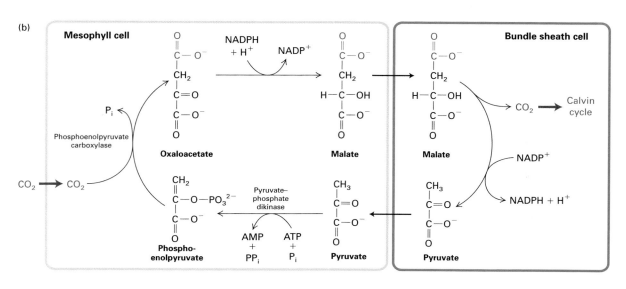

(a)

Vascular bundle (xylem, phloem)

Mesophyll cells

Bundle sheath cells

Epidermis

Air space

CO₂ O₂ Stoma

Chloroplast

◀ **FIGURE 8-44 Leaf anatomy of C₄ plants and the C₄ pathway.** (a) In C₄ plants, bundle sheath cells line the vascular bundles containing the xylem and phloem. Mesophyll cells, which are adjacent to the substomal air spaces, can assimilate CO₂ into four-carbon molecules at low ambient CO₂ and deliver it to the interior bundle sheath cells. Bundle sheath cells contain abundant chloroplasts and are the sites of photosynthesis and sucrose synthesis. Sucrose is carried to the rest of the plant via the phloem. In C₃ plants, which lack bundle sheath cells, the Calvin cycle operates in the mesophyll cells to fix CO₂. (b) The key enzyme in the C₄ pathway is phosphoenolpyruvate carboxylase, which assimilates CO₂ to form oxaloacetate in mesophyll cells. Decarboxylation of malate or other C₄ intermediates in bundle sheath cells releases CO₂, which enters the standard Calvin cycle (see Figure 8-42, *top*).

carbon compound. The enzyme that catalyzes this reaction, *phosphoenolpyruvate carboxylase*, is found almost exclusively in C_4 plants and unlike rubisco is insensitive to O_2. The overall reaction from pyruvate to oxaloacetate involves the hydrolysis of one phosphoanhydride bond in ATP and has a negative ΔG. Therefore, CO_2 fixation will proceed even when the CO_2 concentration is low. The oxaloacetate formed in mesophyll cells is reduced to malate, which is transferred, by a special transporter, to the bundle sheath cells, where the CO_2 is released by decarboxylation and enters the Calvin cycle (Figure 8-44b).

Because of the transport of CO_2 from mesophyll cells, the CO_2 concentration in the bundle sheath cells of C_4 plants is much higher than it is in the normal atmosphere. Bundle sheath cells are also unusual in that they lack PSII and carry out only cyclic electron flow catalyzed by PSI, so no O_2 is evolved. The high CO_2 and reduced O_2 concentrations in the bundle sheath cells favor the fixation of CO_2 by rubisco to

form 3-phosphoglycerate and inhibit the utilization of ribulose 1,5-bisphosphate in photorespiration.

In contrast, the high O_2 concentration in the atmosphere favors photorespiration in the mesophyll cells of C_3 plants (pathway 2 in Figure 8-43); as a result, as much as 50 percent of the carbon fixed by rubisco may be reoxidized to CO_2 in C_3 plants. C_4 plants are superior to C_3 plants in utilizing the available CO_2, since the C_4 enzyme phosphoenolpyruvate carboxylase has a higher affinity for CO_2 than does rubisco in the Calvin cycle. However, one phosphodiester bond of ATP is consumed in the cyclic C_4 process (to generate phosphoenolpyruvate from pyruvate); thus the overall efficiency of the photosynthetic production of sugars from NADPH and ATP is lower than it is in C_3 plants, which use only the Calvin cycle for CO_2 fixation. Nonetheless, the net rates of photosynthesis for C_4 grasses, such as corn or sugar cane, can be two to three times the rates for otherwise similar C_3 grasses, such as wheat, rice, or oats, owing to the elimination of losses from photorespiration.

(a)

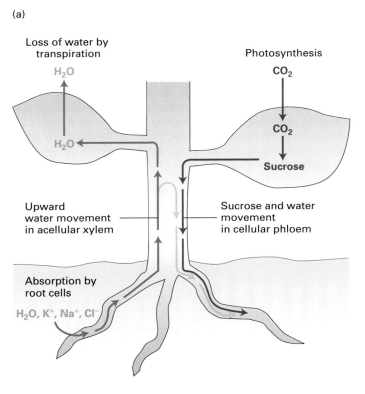

(b)

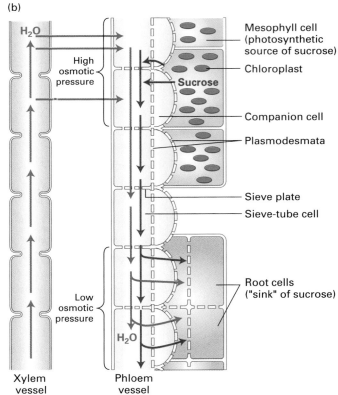

▲ **FIGURE 8-45 Schematic diagrams of the two vascular systems—xylem and phloem—in higher plants, showing the transport of water (blue) and sucrose (red).** (a) Water and salts enter the xylem through the roots. Water is lost by evaporation, mainly through the leaves, creating a suction pressure that draws the water and dissolved salts upward through the xylem. The phloem is used to conduct dissolved sucrose, produced in the leaves, to other parts of the plant. (b) Enlarged view illustrates the mechanism of sucrose flow in a higher plant. Sucrose is actively transported from mesophyll cells into companion cells, and then moves through plasmodesmata into the sieve-tube cells that constitute the phloem vessels. The resulting increase in osmotic pressure within the phloem causes water carried in xylem vessels to enter the phloem by osmotic flow. Root cells and other nonphotosynthetic cells remove sucrose from the phloem by active transport and metabolize it. This lowers the osmotic pressure in the phloem, causing water to exit the phloem. These differences in osmotic pressure in the phloem between the source and the sink of sucrose provide the force that drives sucrose through the phloem.

Sucrose Is Transported from Leaves Through the Phloem to All Plant Tissues

Of the two carbohydrate products of photosynthesis, starch remains in the mesophyll cells of C_3 plants and the bundle sheaf cells in C_4 plants. In these cells, starch is subjected to glycolysis, mainly in the dark, forming ATP, NADH, and small molecules that are used as building blocks for the synthesis of amino acids, lipids, and other cellular constituents. Sucrose, in contrast, is exported from the photosynthetic cells and transported throughout the plant. The vascular system used by higher plants to transport water, ions, sucrose, and other water-soluble substances has two components: the *xylem* and the *phloem*, which generally are grouped together in the *vascular bundle* (see Figure 8-44).

As illustrated in Figure 8-45a, the xylem conducts salts and water from the roots through the stems to the leaves. Water transported upward through the xylem is lost from the plant by evaporation, primarily from the leaves. In young plants the xylem is built of cells interconnected by plasmodesmata, but in mature tissues the cell body degenerates, leaving only the cell walls. The phloem, in contrast, transports dissolved sucrose and organic molecules such as amino acids from their sites of origin in leaves to tissues throughout the plant; water also is transported downward in the phloem.

A phloem vessel consists of long, narrow cells, called *sieve-tube cells*, interconnected by *sieve plates*, a type of cell wall that contains many **plasmodesmata** and is highly perforated (Figure 8-45b). Numerous plasmodesmata also connect the sieve-tube cells to *companion cells*, which line the phloem vessels, and mesophyll cells to companion cells. Sieve-tube cells have lost their nuclei and most other organelles but retain a water-permeable plasma membrane and cytoplasm, through which sucrose and water move. In effect, the sieve-tube cells form one continuous tube of cytosol that extends throughout the plant. Differences in osmotic strength cause the movement of sucrose from the photosynthetic mesophyll cells in the leaves, through the phloem, to the roots and other nonphotosynthetic tissues that catabolize sucrose.

KEY CONCEPTS OF SECTION 8.6

CO₂ Metabolism During Photosynthesis

■ In the Calvin cycle, CO_2 is fixed into organic molecules in a series of reactions that occur in the chloroplast stroma. The initial reaction, catalyzed by rubisco, forms a three-carbon intermediate. Some of the glyceraldehyde 3-phosphate generated in the cycle is transported to the cytosol and converted to sucrose (see Figure 8-42).

■ The light-dependent activation of several Calvin cycle enzymes and other mechanisms increases fixation of CO_2 in the light.

■ In C_3 plants, much of the CO_2 fixed by the Calvin cycle is lost as the result of photorespiration, a wasteful reaction catalyzed by rubisco that is favored at low CO_2 and high O_2 pressures (see Figure 8-43).

■ In C_4 plants, CO_2 is fixed initially in the outer mesophyll cells by reaction with phosphoenolpyruvate. The four-carbon molecules so generated are shuttled to the interior bundle sheath cells, where the CO_2 is released and then used in the Calvin cycle. The rate of photorespiration in C_4 plants is much lower than in C_3 plants.

■ Sucrose from photosynthetic cells is transported through the phloem to nonphotosynthetic parts of the plant. Osmotic pressure differences provide the force that drives sucrose transport (see Figure 8-45).

PERSPECTIVES FOR THE FUTURE

Although the overall processes of photosynthesis and mitochondrial oxidation are well understood, many important details remain to be uncovered by a new generation of scientists. For example, little is known about how complexes I and IV in mitochondria couple proton and electron movements to create a proton-motive force. Similarly, although the binding-change mechanism for ATP synthesis by the F_0F_1 complex is now generally accepted, we do not understand how conformational changes in each β subunit are coupled to the cyclical binding of ADP and P_i, formation of ATP, and then release of ATP. Many questions also remain about the structure and function of transport proteins in the inner mitochondrial and chloroplast membranes that play key roles in oxidative phosphorylation and photosynthesis. Molecular analysis of such membrane proteins is difficult, and new types of techniques will be needed to elucidate the details of their structure and operation.

We now know that release of cytochrome c and other proteins from the intermembrane space of mitochondria into the cytosol plays a major role in triggering apoptosis (Chapter 22). Certain members of the Bcl-2 family of apoptotic proteins and ion channels localized in part to the outer mitochondrial membrane participate in this process. Yet much remains to be learned about the structure of these proteins in the mitochondrial membrane, their normal functions in cell metabolism, and the alterations that lead to apoptosis.

KEY TERMS

aerobic oxidation *301*	citric acid cycle *309*
ATP synthase *302*	cytochromes *318*
C_4 pathway *345*	electron carriers *302*
Calvin cycle *342*	endosymbiont
cellular respiration *307*	hypothesis *302*
chemiosmosis *302*	F_0F_1 complex *302*
chlorophylls *331*	glycolysis *301*
chloroplasts *302*	mitochondria *301*

REVIEW THE CONCEPTS

1. The proton motive force (pmf) is essential for both mitochondrial and chloroplast function. What produces the pmf, and what is its relationship to ATP?

2. The mitochondrial inner membrane exhibits all of the fundamental characteristics of a typical cell membrane, but it also has several unique characteristics that are closely associated with its role in oxidative phosphorylation. What are these unique characteristics? How does each contribute to the function of the inner membrane?

3. Maximal production of ATP from glucose involves the reactions of glycolysis, the citric acid cycle, and the electron transport chain. Which of these reactions requires O_2, and why? Which, in certain organisms or physiological conditions, can proceed in the absence of O_2?

4. Describe how the electrons produced by glycolysis are delivered to the electron transport chain. What role do amino acids play in this process? What would be the consequence for overall ATP yield per glucose molecule if a mutation inactivated this delivery system? What would be the longer-term consequence for the activity of the glycolytic pathway?

5. Each of the cytochromes in the mitochondria contains prosthetic groups. What is a prosthetic group? Which type of prosthetic group is associated with the cytochromes? What property of the various cytochromes ensures unidirectional electron flow along the electron transport chain?

6. It is estimated that each electron pair donated by NADH leads to the synthesis of approximately three ATP molecules, while each electron pair donated by $FADH_2$ leads to the synthesis of approximately two ATP molecules. What is the underlying reason for the difference in yield for electrons donated by $FADH_2$ versus NADH?

7. Much of our understanding of ATP synthase is derived from research on aerobic bacteria. What makes these organisms useful for this research? Where do the reactions of glycolysis, the citric acid cycle, and the electron transport chain occur in these organisms? Where is the pmf generated in aerobic bacteria? What other cellular processes depend on the pmf in these organisms?

8. An important function of the mitochondrial inner membrane is to provide a selectively permeable barrier to the movement of water-soluble molecules and thus generate different chemical environments on either side of the membrane. However, many of the substrates and products of oxidative phosphorylation are water-soluble and must cross the inner membrane. How does this transport occur?

9. The Q cycle plays a major role in the electron transport chain of mitochondria, chloroplasts, and bacteria. What is the function of the Q cycle, and how does it carry out this function? What electron transport train components participate in the Q cycle in mitochondria, in purple bacteria, and in chloroplasts?

10. Write the overall reaction of oxygen-generating photosynthesis. Explain the following statement: the O_2 generated by photosynthesis is simply a by-product of the reaction's generation of carbohydrates and ATP.

11. Photosynthesis can be divided into multiple stages. What are the stages of photosynthesis, and where does each occur within the chloroplast? Where is the sucrose produced by photosynthesis generated?

12. The photosystems responsible for absorption of light energy are composed of two linked components, the reaction center and an antenna complex. What is the pigment composition and role of each in the process of light absorption? What evidence exists that the pigments found in these components are involved in photosynthesis?

13. Photosynthesis in green and purple bacteria does not produce O_2. Why? How can these organisms still use photosynthesis to produce ATP? What molecules serve as electron donors in these organisms?

14. Chloroplasts contain two photosystems. What is the function of each? For linear electron flow, diagram the flow of electrons from photon absorption to NADPH formation. What does the energy stored in the form of NADPH synthesize?

15. The Calvin cycle reactions that fix CO_2 do not function in the dark. What are the likely reasons for this? How are these reactions regulated by light?

ANALYZE THE DATA

A proton gradient can be analyzed with fluorescent dyes whose emission intensity profiles depend on pH. One of the most useful dyes for measuring the pH gradient across mitochondrial membranes is the membrane-impermeant, water-soluble fluorophore 2′,7′-bis-(2-carboxyethyl)-5(6)-carboxyfluorescein (BCECF). The effect of pH on the emission intensity of BCECF, excited at 505 nm, is shown in the accompanying figure. In one study, sealed vesicles containing this compound were prepared by mixing unsealed, isolated inner mitochondrial membranes with BCECF; after resealing of the membranes, the vesicles were collected by centrifugation and then resuspended in nonfluorescent medium.

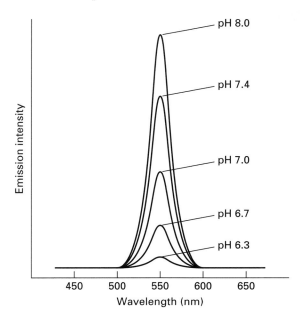

a. When these vesicles were incubated in a physiological buffer containing NADH, ADP, P_i, and O_2, the fluorescence of BCECF trapped inside gradually decreased in intensity. What does this decrease in fluorescent intensity suggest about this vesicular preparation?

b. How would you expect the concentrations of ADP, P_i, and O_2 to change during the course of the experiment described in part a? Why?

c. After the vesicles were incubated in buffer containing ADP, P_i, and O_2 for a period of time, addition of dinitrophenol caused an increase in BCECF fluorescence. In contrast, addition of valinomycin produced only a small transient effect. Explain these findings.

d. Predict the outcome of an experiment performed as described in part a if brown-fat tissue was used as a source of unsealed, isolated inner mitochondrial membranes. Explain your answer.

REFERENCES

Oxidation of Glucose and Fatty Acids to CO_2

Berg, J., J. Tymoczko, and L. Stryer. 2002. *Biochemistry,* 5th ed. W. H. Freeman and Company, chaps. 16 and 17.

Depre, C., M. Rider, and L. Hue. 1998. Mechanisms of control of heart glycolysis. *Eur. J. Biochem.* **258**:277–290.

Fell, D. 1997. *Understanding the Control of Metabolism.* Portland Press.

Fersht, A. 1999. *Structure and Mechanism in Protein Science: A Guide to Enzyme Catalysis and Protein Folding.* W. H. Freeman and Company. Contains an excellent discussion of the reaction mechanisms of key enzymes.

Fothergill-Gilmore, L. A., and P. A. Michels. 1993. Evolution of glycolysis. *Prog. Biophys. Mol. Biol.* **59**:105–135.

Guest, J. R., and G. C. Russell. 1992. Complexes and complexities of the citric acid cycle in *Escherichia coli. Curr. Top. Cell Reg.* **33**:231–247.

Krebs, H. A. 1970. The history of the tricarboxylic acid cycle. *Perspect. Biol. Med.* **14**:154–170.

Mannaerts, G. P., and P. P. Van Veldhoven. 1993. Metabolic pathways in mammalian peroxisomes. *Biochimie* **75**:147–158.

Nelson, D. L., and M. M. Cox. 2000. *Lehninger Principles of Biochemistry.* Worth, chaps. 14–17, 19.

Pilkis, S. J., T. H. Claus, I. J. Kurland, and A. J. Lange. 1995. 6-Phosphofructo-2-kinase/fructose-2,6-bisphosphatase: a metabolic signaling enzyme. *Ann. Rev. Biochem.* **64**:799–835.

Rasmussen, B., and R. Wolfe. 1999. Regulation of fatty acid oxidation in skeletal muscle. *Ann. Rev. Nutrition* **19**:463–484.

Velot, C., M. Mixon, M. Teige, and P. Srere. 1997. Model of a quinary structure between Krebs TCA cycle enzymes: a model for the metabolon. *Biochemistry* **36**:14271–14276.

Electron Transport and Generation of the Proton-Motive Force

Babcock, G. 1999. How oxygen is activated and reduced in respiration. *Proc. Nat'l. Acad. Sci. USA* **96**:12971–12973.

Beinert, H., R. Holm, and E. Münck. 1997. Iron-sulfur clusters: nature's modular, multipurpose structures. *Science* **277**:653–659.

Brandt, U., and B. Trumpower. 1994. The protonmotive Q cycle in mitochondria and bacteria. *Crit. Rev. Biochem. Mol. Biol.* **29**:165–197.

Darrouzet, E., C. Moser, P. L. Dutton, and F. Daldal. 2001. Large scale domain movement in cytochrome bc1: a new device for electron transfer in proteins. *Trends Biochem. Sci.* **26**:445–451.

Grigorieff, N. 1999. Structure of the respiratory NADH:ubiquinone oxidoreductase (complex I). *Curr. Opin. Struc. Biol.* **9**:476–483.

Michel, H., J. Behr, A. Harrenga, and A. Kannt. 1998. Cytochrome *c* oxidase. *Ann. Rev. Biophys. Biomol. Struc.* **27**:329–356.

Mitchell, P. 1979. Keilin's respiratory chain concept and its chemiosmotic consequences. *Science* **206**:1148–1159. (Nobel Prize Lecture.)

Ramirez, B. E., B. Malmström, J. R. Winkler, and H. B. Gray. 1995. The currents of life: the terminal electron-transfer complex of respiration. *Proc. Nat'l. Acad. Sci. USA* **92**:11949–11951.

Ruitenberg, M., et al. 2002. Reduction of cytochrome *c* oxidase by a second electron leads to proton translocation. *Nature* **417**:99–102.

Saraste, M. 1999. Oxidative phosphorylation at the fin de siecle. *Science* **283**:1488–1492.

Scheffler, I. 1999. *Mitochondria.* Wiley.

Schultz, B., and S. Chan. 2001. Structures and proton-pumping strategies of mitochondrial respiratory enzymes. *Ann. Rev. Biophys. Biomol. Struc.* **30**:23–65.

Tsukihara, T., et al. 1996. The whole structure of the 13-subunit oxidized cytochrome *c* oxidase at 2.8 Å. *Science* **272**:1136–1144.

Walker, J. E. 1995. Determination of the structures of respiratory enzyme complexes from mammalian mitochondria. *Biochim. Biophys. Acta* **1271**:221–227.

Xia, D., et al. 1997. Crystal structure of the cytochrome bc1 complex from bovine heart mitochondria. *Science* **277**:60–66.

Zaslavsky, D., and R. Gennis. 2000. Proton pumping by cytochrome oxidase: progress and postulates. *Biochim. Biophys. Acta* **1458**:164–179.

Zhang, Z., et al. 1998. Electron transfer by domain movement in cytochrome bc_1. *Nature* **392**:677–684.

Harnessing the Proton-Motive Force for Energy-Requiring Processes

Bianchet, M. A., J. Hullihen, P. Pedersen, and M. Amzel. The 2.8-Å structure of rat liver F1-ATPase: configuration of a critical intermediate in ATP synthesis/hydrolysis. *Proc. Nat'l. Acad. Sci. USA* **95**:11065–11070.

Boyer, P. D. 1989. A perspective of the binding change mechanism for ATP synthesis. *FASEB J.* **3**:2164–2178.

Boyer, P. D. 1997. The ATP synthase—a splendid molecular machine. *Ann. Rev. Biochem.* **66**:717–749.

Capaldi, R., and R. Aggeler. 2002. Mechanism of the F_0F_1-type ATP synthase—a biological rotary motor. *Trends Biochem. Sci.* **27**:154–160.

Elston, T., H. Wang, and G. Oster. 1998. Energy transduction in ATP synthase. *Nature* **391**:510–512.

Kinosita, K., et al. 1998. F_1-ATPase: a rotary motor made of a single molecule. *Cell* **93**:21–24.

Klingenberg, M., and S. Huang. 1999. Structure and function of the uncoupling protein from brown adipose tissue. *Biochim. Biophys. Acta* **1415**:271–296.

Tsunoda, S., R. Aggeler, M. Yoshida, and R. Capaldi. 2001. Rotation of the **c** subunit oligomer in fully functional F_0F_1 ATP synthase. *Proc. Nat'l. Acad. Sci. USA* **98**:898–902.

Yasuda, R., et al. 2001. Resolution of distinct rotational substeps by submillisecond kinetic analysis of F1- ATPase. *Nature* **410**:898–904.

Photosynthetic Stages and Light-Absorbing Pigments

Blankenship, R. E. 2002. *Molecular Mechanisms of Photosynthesis.* Blackwell.

Deisenhofer, J., and J. R. Norris, eds. 1993. *The Photosynthetic Reaction Center,* vols. 1 and 2. Academic Press.

Govindjee, and W. J. Coleman. 1990. How plants make oxygen. *Sci. Am.* **262**(2):50–58.

Harold, F. M. 1986. *The Vital Force: A Study of Bioenergetics.* W. H. Freeman and Company, chap. 8.

McDermott, G., et al. 1995. Crystal structure of an integral membrane light-harvesting complex from photosynthetic bacteria. *Nature* **364**:517.

Prince, R. 1996. Photosynthesis: the Z-scheme revisited. *Trends Biochem. Sci.* **21**:121–122.

Wollman, F. A. 2001. State transitions reveal the dynamics and flexibility of the photosynthetic apparatus. *EMBO J.* **20**:3623–3630.

Molecular Analysis of Photosystems

Allen, J. F. 2002. Photosynthesis of ATP—electrons, proton pumps, rotors, and poise. *Cell* **110**:273–276.

Aro, E. M., I. Virgin, and B. Andersson. 1993. Photoinhibition of photosystem II. Inactivation, protein damage, and turnover. *Biochim. Biophys. Acta* **1143**:113–134.

Deisenhofer, J., and H. Michel. 1989. The photosynthetic reaction center from the purple bacterium *Rhodopseudomonas viridis. Science* **245**:1463–1473. (Nobel Prize Lecture.)

Deisenhofer, J., and H. Michel. 1991. Structures of bacterial photosynthetic reaction centers. *Ann. Rev. Cell Biol.* **7**:1–23.

Golbeck, J. H. 1993. Shared thematic elements in photochemical reaction centers. *Proc. Nat'l. Acad. Sci. USA* **90**:1642–1646.

Haldrup, A., P. Jensen, C. Lunde, and H. Scheller. 2001. Balance of power: a view of the mechanism of photosynthetic state transitions. *Trends Plant Sci.* **6**:301–305.

Hankamer, B., J. Barber, and E. Boekema. 1997. Structure and membrane organization of photosystem II from green plants. *Ann. Rev. Plant Physiol. Plant Mol. Biol.* **48**:641–672.

Heathcote, P., P. Fyfe, and M. Jones. 2002. Reaction centres: the structure and evolution of biological solar power. *Trends Biochem. Sci.* **27**:79–87.

Horton, P., A. Ruban, and R. Walters. 1996. Regulation of light harvesting in green plants. *Ann. Rev. Plant Physiol. Plant Mol. Biol.* **47**:655–684.

Jordan, P., et al. 2001 Three-dimensional structure of cyanobacterial photosystem I at 2.5 Å resolution. *Nature* **411**:909–917.

Kühlbrandt, W. 2001. Chlorophylls galore. *Nature* **411**:896–898.

Martin, J. L., and M. H. Vos. 1992. Femtosecond biology. *Ann. Rev. Biophys. Biomol. Struc.* **21**:199–222.

Penner-Hahn, J. 1998. Structural characterization of the Mn site in the photosynthetic oxygen-evolving complex. *Struc. and Bonding* **90**:1–36.

Tommos, C., and G. Babcock. 1998. Oxygen production in nature: a light-driven metalloradical enzyme process. *Accounts Chem. Res.* **31**:18–25.

CO_2 Metabolism During Photosynthesis

Bassham, J. A. 1962. The path of carbon in photosynthesis. *Sci. Am.* **206**(6):88–100.

Buchanan, B. B. 1991. Regulation of CO_2 assimilation in oxygenic photosynthesis: the ferredoxin/thioredoxin system. Perspective on its discovery, present status, and future development. *Arch. Biochem. Biophys.* **288**:1–9.

Portis, A. 1992. Regulation of ribulose 1,5-bisphosphate carboxylase/oxygenase activity. *Ann. Rev. Plant Physiol. Plant Mol. Biol.* **43**:415–437.

Rawsthorne, S. 1992. Towards an understanding of C_3-C_4 photosynthesis. *Essays Biochem.* **27**:135–146.

Rokka, A., I. Zhang, and E.-M. Aro. 2001. Rubisco activase: an enzyme with a temperature-dependent dual function? *Plant J.* **25**:463–472.

Sage, R., and J. Colemana. 2001. Effects of low atmospheric CO_2 on plants: more than a thing of the past. *Trends Plant Sci.* **6**:18–24.

Schneider, G., Y. Lindqvist, and C. I. Branden. 1992. Rubisco: structure and mechanism. *Ann. Rev. Biophys. Biomol. Struc.* **21**:119–153.

Wolosiuk, R. A., M. A. Ballicora, and K. Hagelin. 1993. The reductive pentose phosphate cycle for photosynthetic CO_2 assimilation: enzyme modulation. *FASEB J.* **7**:622–637.

9

MOLECULAR GENETIC TECHNIQUES AND GENOMICS

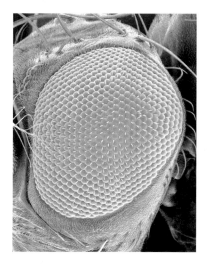 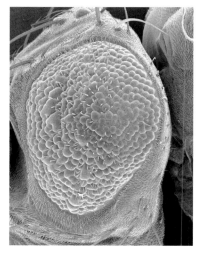 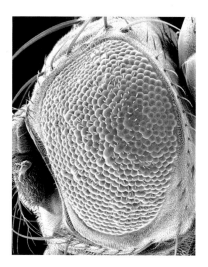

The effect of mutations on *Drosophila* development. Scanning electron micrographs of the eye from (*left*) a wild-type fly, (*middle*) a fly carrying a dominant developmental mutation produced by recombinant DNA methods, and (*right*) a fly carrying a suppresor mutation that partially reverses the effect of the dominant mutation. [Courtesy of Ilaria Rebay, Whitehead Institute, MIT.]

In previous chapters, we were introduced to the variety of tasks that proteins perform in biological systems. How some proteins carry out their specific tasks is described in detail in later chapters. In studying a newly discovered protein, cell biologists usually begin by asking what is its function, where is it located, and what is its structure? To answer these questions, investigators employ three tools: the gene that encodes the protein, a mutant cell line or organism that lacks the function of the protein, and a source of the purified protein for biochemical studies. In this chapter we consider various aspects of two basic experimental strategies for obtaining all three tools (Figure 9-1).

The first strategy, often referred to as *classical genetics*, begins with isolation of a mutant that appears to be defective in some process of interest. Genetic methods then are used to

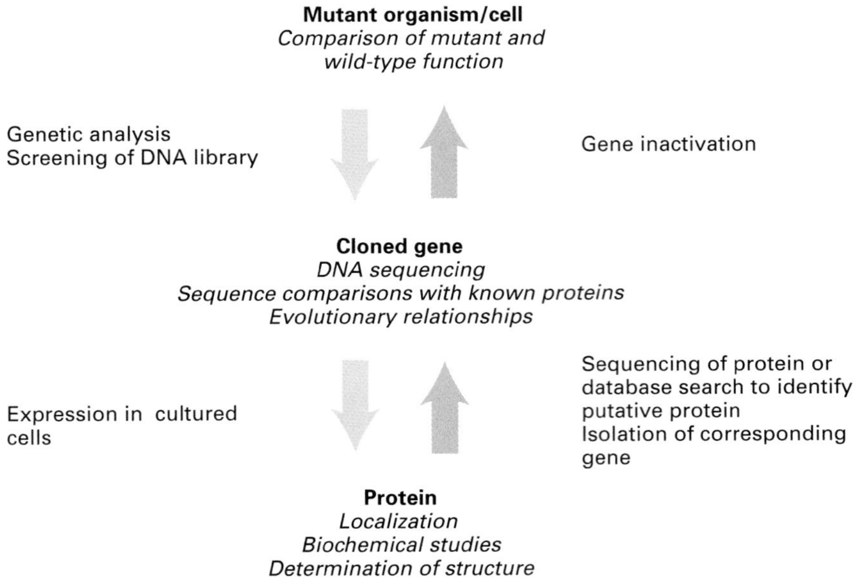

▶ **FIGURE 9-1 Overview of two strategies for determining the function, location, and primary structure of proteins.** A mutant organism is the starting point for the classical genetic strategy (green arrows). The reverse strategy (orange arrows) begins with biochemical isolation of a protein or identification of a putative protein based on analysis of stored gene and protein sequences. In both strategies, the actual gene is isolated from a DNA library, a large collection of cloned DNA sequences representing an organism's genome. Once a cloned gene is isolated, it can be used to produce the encoded protein in bacterial or eukaryotic expression systems. Alternatively, a cloned gene can be inactivated by one of various techniques and used to generate mutant cells or organisms.

Mutant organism/cell
Comparison of mutant and wild-type function

Genetic analysis
Screening of DNA library

Gene inactivation

Cloned gene
DNA sequencing
Sequence comparisons with known proteins
Evolutionary relationships

Expression in cultured cells

Sequencing of protein or database search to identify putative protein
Isolation of corresponding gene

Protein
Localization
Biochemical studies
Determination of structure

identify the affected gene, which subsequently is isolated from an appropriate **DNA library,** a large collection of individual DNA sequences representing all or part of an organism's genome. The isolated gene can be manipulated to produce large quantities of the protein for biochemical experiments and to design probes for studies of where and when the encoded protein is expressed in an organism. The second strategy follows essentially the same steps as the classical approach but in reverse order, beginning with isolation of an interesting protein or its identification based on analysis of an organism's genomic sequence. Once the corresponding gene has been isolated from a DNA library, the gene can be altered and then reinserted into an organism. By observing the effects of the altered gene on the organism, researchers often can infer the function of the normal protein.

An important component in both strategies for studying a protein and its biological function is isolation of the corresponding gene. Thus we discuss various techniques by which researchers can isolate, sequence, and manipulate specific regions of an organism's DNA. The extensive collections of DNA sequences that have been amassed in recent years has given birth to a new field of study called **genomics,** the molecular characterization of whole genomes and overall patterns of gene expression. Several examples of the types of information available from such genome-wide analysis also are presented.

9.1 Genetic Analysis of Mutations to Identify and Study Genes

As described in Chapter 4, the information encoded in the DNA sequence of genes specifies the sequence and therefore

the structure and function of every protein molecule in a cell. The power of genetics as a tool for studying cells and organisms lies in the ability of researchers to selectively alter every copy of just one type of protein in a cell by making a change in the gene for that protein. Genetic analyses of mutants defective in a particular process can reveal (a) new genes required for the process to occur; (b) the order in which gene products act in the process; and (c) whether the proteins encoded by different genes interact with one another. Before seeing how genetic studies of this type can provide insights into the mechanism of complicated cellular or developmental process, we first explain some basic genetic terms used throughout our discussion.

The different forms, or variants, of a gene are referred to as **alleles.** Geneticists commonly refer to the numerous naturally occurring genetic variants that exist in populations, particularly human populations, as alleles. The term **mutation** usually is reserved for instances in which an allele is known to have been newly formed, such as after treatment of an experimental organism with a **mutagen,** an agent that causes a heritable change in the DNA sequence.

Strictly speaking, the particular set of alleles for all the genes carried by an individual is its **genotype.** However, this term also is used in a more restricted sense to denote just the alleles of the particular gene or genes under examination. For experimental organisms, the term **wild type** often is used to designate a standard genotype for use as a reference in breeding experiments. Thus the normal, nonmutant allele will usually be designated as the wild type. Because of the enormous naturally occurring allelic variation that exists in human populations, the term *wild type* usually denotes an allele that is present at a much higher frequency than any of the other possible alternatives.

Geneticists draw an important distinction between the *genotype* and the **phenotype** of an organism. The phenotype refers to all the physical attributes or traits of an individual that are the consequence of a given genotype. In practice, however, the term *phenotype* often is used to denote the physical consequences that result from just the alleles that are under experimental study. Readily observable phenotypic characteristics are critical in the genetic analysis of mutations.

Recessive and Dominant Mutant Alleles Generally Have Opposite Effects on Gene Function

A fundamental genetic difference between experimental organisms is whether their cells carry a single set of chromosomes or two copies of each chromosome. The former are referred to as **haploid;** the latter, as **diploid.** Complex multicellular organisms (e.g., fruit flies, mice, humans) are diploid, whereas many simple unicellular organisms are haploid. Some organisms, notably the yeast *Saccharomyces,* can exist in either haploid or diploid states. Many cancer cells and the normal cells of some organisms, both plants and animals, carry more than two copies of each chromosome. However, our discussion of genetic techniques and analysis relates to diploid organisms, including diploid yeasts.

Since diploid organisms carry two copies of each gene, they may carry identical alleles, that is, be **homozygous** for a gene, or carry different alleles, that is, be **heterozygous** for a gene. A **recessive** mutant allele is defined as one in which both alleles must be mutant in order for the mutant phenotype to be observed; that is, the individual must be homozygous for the mutant allele to show the mutant phenotype. In contrast, the phenotypic consequences of a **dominant** mutant allele are observed in a heterozygous individual carrying one mutant and one wild-type allele (Figure 9-2).

Whether a mutant allele is recessive or dominant provides valuable information about the function of the affected gene and the nature of the causative mutation. Recessive alleles usually result from a mutation that inactivates the affected gene, leading to a partial or complete *loss of function.* Such recessive mutations may remove part of or the entire gene from the chromosome, disrupt expression of the gene, or alter the structure of the encoded protein, thereby altering its function. Conversely, dominant alleles are often the consequence of a mutation that causes some kind of *gain of function.* Such dominant mutations may increase the ac-

tivity of the encoded protein, confer a new activity on it, or lead to its inappropriate spatial or temporal pattern of expression.

Dominant mutations in certain genes, however, are associated with a loss of function. For instance, some genes are *haplo-insufficient,* meaning that both alleles are required for normal function. Removing or inactivating a single allele in such a gene leads to a mutant phenotype. In other rare instances a dominant mutation in one allele may lead to a structural change in the protein that interferes with the function of the wild-type protein encoded by the other allele. This type of mutation, referred to as a *dominant negative,* produces a phenotype similar to that obtained from a loss-of-function mutation.

 Some alleles can exhibit both recessive and dominant properties. In such cases, statements about whether an allele is dominant or recessive must specify the phenotype. For example, the allele of the hemoglobin gene in humans designated Hb^s has more than one phenotypic consequence. Individuals who are homozygous for this allele (Hb^s/Hb^s) have the debilitating disease sickle-cell anemia, but heterozygous individuals (Hb^s/Hb^a) do not have the disease. Therefore, Hb^s is *recessive* for the trait of sickle-cell disease. On the other hand, heterozygous (Hb^s/Hb^a) individuals are more resistant to malaria than homozygous (Hb^a/Hb^a) individuals, revealing that Hb^s is *dominant* for the trait of malaria resistance. ∎

A commonly used agent for inducing mutations (mutagenesis) in experimental organisms is ethylmethane sulfonate (EMS). Although this mutagen can alter DNA sequences in several ways, one of its most common effects is to chemically modify guanine bases in DNA, ultimately leading to the conversion of a $G \cdot C$ base pair into an $A \cdot T$ base pair. Such an alteration in the sequence of a gene, which involves only a single base pair, is known as a **point mutation.** A *silent* point mutation causes no change in the amino acid sequence or activity of a gene's encoded protein. However, observable phenotypic consequences due to changes in a protein's activity can arise from point mutations that result in substitution of one amino acid for another (*missense* mutation), introduction of a premature stop codon (*nonsense* mutation), or a change in the reading

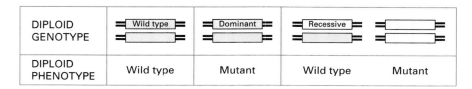

▲ **FIGURE 9-2 Effects of recessive and dominant mutant alleles on phenotype in diploid organisms.** Only one copy of a dominant allele is sufficient to produce a mutant phenotype, whereas both copies of a recessive allele must be present to cause a mutant phenotype. Recessive mutations usually cause a loss of function; dominant mutations usually cause a gain of function or an altered function.

frame of a gene (*frameshift* mutation). Because alterations in the DNA sequence leading to a decrease in protein activity are much more likely than alterations leading to an increase or qualitative change in protein activity, mutagenesis usually produces many more recessive mutations than dominant mutations.

Segregation of Mutations in Breeding Experiments Reveals Their Dominance or Recessivity

Geneticists exploit the normal life cycle of an organism to test for the dominance or recessivity of alleles. To see how

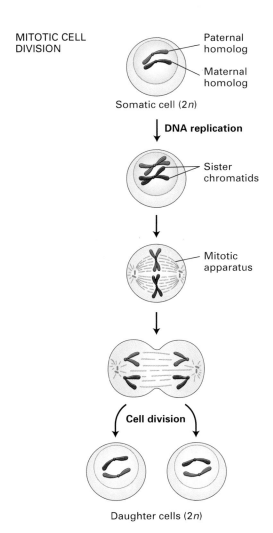

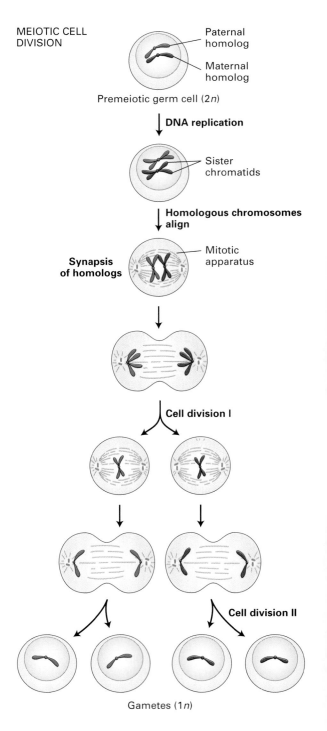

▲ **FIGURE 9-3 Comparison of mitosis and meiosis.** Both somatic cells and premeiotic germ cells have two copies of each chromosome (2*n*), one maternal and one paternal. In mitosis, the replicated chromosomes, each composed of two sister chromatids, align at the cell center in such a way that both daughter cells receive a maternal and paternal homolog of each morphologic type of chromosome. During the first *meiotic* division, however, each replicated chromosome pairs with its homologous partner at the cell center; this pairing off is referred to as *synapsis*. One replicated chromosome of each morphologic type then goes into one daughter cell, and the other goes into the other cell in a random fashion. The resulting cells undergo a second division without intervening DNA replication, with the sister chromatids of each morphologic type being apportioned to the daughter cells. Each diploid cell that undergoes meiosis produces four haploid (1*n*) cells.

this is done, we need first to review the type of cell division that gives rise to **gametes** (sperm and egg cells in higher plants and animals). Whereas the body (somatic) cells of most multicellular organisms divide by **mitosis,** the **germ cells** that give rise to gametes undergo **meiosis.** Like somatic cells, premeiotic germ cells are diploid, containing two **homologs** of each morphologic type of chromosome. The two homologs constituting each pair of **homologous chromosomes** are descended from different parents, and thus their genes may exist in different allelic forms. Figure 9-3 depicts the major events in mitotic and meiotic cell division. In mitosis DNA replication is always followed by cell division, yielding two diploid daughter cells. In meiosis *one* round of DNA replication is followed by *two* separate cell divisions, yielding four haploid ($1n$) cells that contain only one chromosome of each homologous pair. The apportionment, or **segregation,** of the replicated homologous chromosomes to daughter cells during the first meiotic division is random; that is, maternally and paternally derived homologs segregate independently, yielding daughter cells with different mixes of paternal and maternal chromosomes.

As a way to avoid unwanted complexity, geneticists usually strive to begin breeding experiments with strains that are homozygous for the genes under examination. In such *true-breeding* strains, every individual will receive the same allele from each parent and therefore the composition of alleles will not change from one generation to the next. When a true-breeding mutant strain is mated to a true-breeding wild-type strain, all the first filial (F_1) progeny will be heterozygous (Figure 9-4). If the F_1 progeny exhibit the mutant trait, then the mutant allele is dominant; if the F_1 progeny exhibit the wild-type trait, then the mutant is recessive. Further crossing between F_1 individuals will also reveal different patterns of inheritance according to whether the mutation is dominant or recessive. When F_1 individuals that are heterozygous for a dominant allele are crossed among themselves, three-fourths of the resulting F_2 progeny will exhibit the mutant trait. In contrast, when F_1 individuals that are heterozygous for a recessive allele are crossed among themselves, only one-fourth of the resulting F_2 progeny will exhibit the mutant trait.

As noted earlier, the yeast *Saccharomyces,* an important experimental organism, can exist in either a haploid or a diploid state. In these unicellular eukaryotes, crosses between haploid cells can determine whether a mutant allele is dominant or recessive. Haploid yeast cells, which carry one copy of each chromosome, can be of two different mating types known as **a** and **α**. Haploid cells of opposite mating type can mate to produce **a**/**α** diploids, which carry two copies of each chromosome. If a new mutation with an observable phenotype is isolated in a haploid strain, the mutant strain can be mated to a wild-type strain of the opposite mating type to produce **a**/**α** diploids that are heterozygous for the mutant allele. If these diploids exhibit the mutant trait, then the mutant allele is dominant, but if the diploids appear as wild-type, then the mutant allele is recessive. When **a**/**α** diploids are placed under starvation conditions, the cells

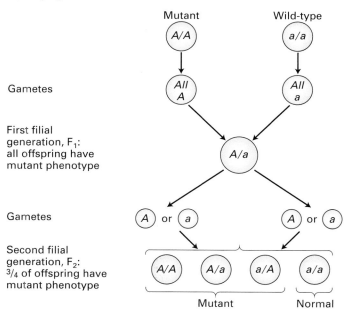

(a) Segregation of **dominant** mutation

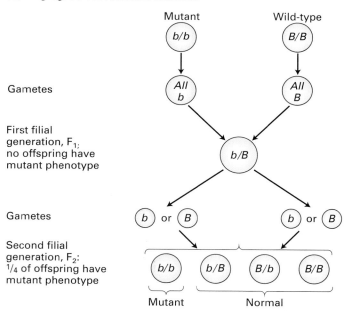

(b) Segregation of **recessive** mutation

▲ **FIGURE 9-4 Segregation patterns of dominant and recessive mutations in crosses between true-breeding strains of diploid organisms.** All the offspring in the first (F_1) generation are heterozygous. If the mutant allele is dominant, the F_1 offspring will exhibit the mutant phenotype, as in part (a). If the mutant allele is recessive, the F_1 offspring will exhibit the wild-type phenotype, as in part (b). Crossing of the F_1 heterozygotes among themselves also produces different segregation ratios for dominant and recessive mutant alleles in the F_2 generation.

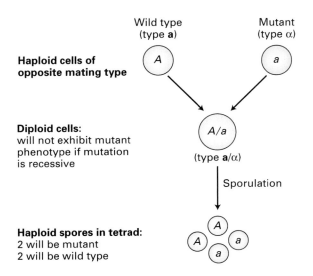

Wild type
(type **a**)

Mutant
(type α)

**Haploid cells of
opposite mating type**

Diploid cells:
will not exhibit mutant
phenotype if mutation
is recessive

(type **a**/α)

Sporulation

Haploid spores in tetrad:
2 will be mutant
2 will be wild type

▲ **FIGURE 9-5 Segregation of alleles in yeast.** Haploid *Saccharomyces* cells of opposite mating type (i.e., one of mating type α and one of mating type **a**) can mate to produce an **a**/α diploid. If one haploid carries a dominant mutant allele and the other carries a recessive wild-type allele of the same gene, the resulting heterozygous diploid will express the dominant trait. Under certain conditions, a diploid cell will form a tetrad of four haploid spores. Two of the spores in the tetrad will express the recessive trait and two will express the dominant trait.

(a)

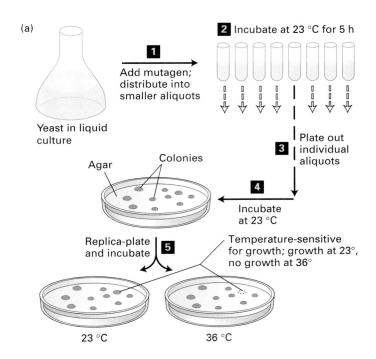

1 Add mutagen; distribute into smaller aliquots

2 Incubate at 23 °C for 5 h

Yeast in liquid culture

3 Plate out individual aliquots

Agar   Colonies

4 Incubate at 23 °C

Replica-plate and incubate **5**

Temperature-sensitive for growth; growth at 23°, no growth at 36°

23 °C   36 °C

(b)
Wild type

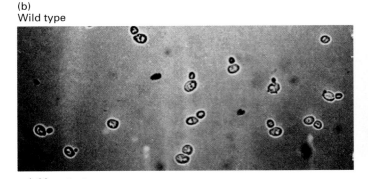

cdc28 mutants

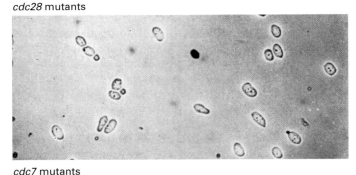

cdc7 mutants

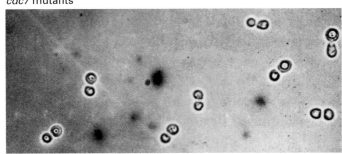

undergo meiosis, giving rise to a tetrad of four haploid spores, two of type **a** and two of type α. Sporulation of a heterozygous diploid cell yields two spores carrying the mutant allele and two carrying the wild-type allele (Figure 9-5). Under appropriate conditions, yeast spores will germinate, producing vegetative haploid strains of both mating types.

Conditional Mutations Can Be Used to Study Essential Genes in Yeast

The procedures used to identify and isolate mutants, referred to as *genetic screens*, depend on whether the experimental organism is haploid or diploid and, if the latter, whether the mutation is recessive or dominant. Genes that encode proteins essential for life are among the most interesting and important ones to study. Since phenotypic expression of mutations in essential genes leads to death of the individual, ingenious genetic screens are needed to isolate and maintain organisms with a lethal mutation.

In haploid yeast cells, essential genes can be studied through the use of *conditional mutations*. Among the most common conditional mutations are **temperature-sensitive mutations,** which can be isolated in bacteria and lower eukaryotes but not in warm-blooded eukaryotes. For instance, a mutant protein may be fully functional at one temperature (e.g., 23 °C) but completely inactive at another temperature (e.g., 36 °C), whereas the normal protein would be fully functional at both temperatures. A temperature at which the

◄ **EXPERIMENTAL FIGURE 9-6 Haploid yeasts carrying temperature-sensitive lethal mutations are maintained at permissive temperature and analyzed at nonpermissive temperature.** (a) Genetic screen for temperature-sensitive cell-division cycle (*cdc*) mutants in yeast. Yeasts that grow and form colonies at 23 °C (permissive temperature) but not at 36 °C (nonpermissive temperature) may carry a lethal mutation that blocks cell division. (b) Assay of temperature-sensitive colonies for blocks at specific stages in the cell cycle. Shown here are micrographs of wild-type yeast and two different temperature-sensitive mutants after incubation at the nonpermissive temperature for 6 h. Wild-type cells, which continue to grow, can be seen with all different sizes of buds, reflecting different stages of the cell cycle. In contrast, cells in the lower two micrographs exhibit a block at a specific stage in the cell cycle. The *cdc28* mutants arrest at a point before emergence of a new bud and therefore appear as unbudded cells. The *cdc7* mutants, which arrest just before separation of the mother cell and bud (emerging daughter cell), appear as cells with large buds. [Part (a) see L. H. Hartwell, 1967, *J. Bacteriol.* **93**:1662; part (b) from L. M. Hereford and L. H. Hartwell, 1974, *J. Mol. Biol.* **84**:445.]

mutant phenotype is observed is called *nonpermissive;* a *permissive* temperature is one at which the mutant phenotype is not observed even though the mutant allele is present. Thus mutant strains can be maintained at a permissive temperature and then subcultured at a nonpermissive temperature for analysis of the mutant phenotype.

An example of a particularly important screen for temperature-sensitive mutants in the yeast *Saccharomyces cerevisiae* comes from the studies of L. H. Hartwell and colleagues in the late 1960s and early 1970s. They set out to identify genes important in regulation of the **cell cycle** during which a cell synthesizes proteins, replicates its DNA, and then undergoes mitotic cell division, with each daughter cell receiving a copy of each chromosome. Exponential growth of a single yeast cell for 20–30 cell divisions forms a visible yeast colony on solid agar medium. Since mutants with a complete block in the cell cycle would not be able to form a colony, conditional mutants were required to study mutations that affect this basic cell process. To screen for such mutants, the researchers first identified mutagenized yeast cells that could grow normally at 23 °C but that could not form a colony when placed at 36 °C (Figure 9-6a).

Once temperature-sensitive mutants were isolated, further analysis revealed that they indeed were defective in cell division. In *S. cerevisiae,* cell division occurs through a budding process, and the size of the bud, which is easily visualized by light microscopy, indicates a cell's position in the cell cycle. Each of the mutants that could not grow at 36 °C was examined by microscopy after several hours at the nonpermissive temperature. Examination of many different temperature-sensitive mutants revealed that about 1 percent exhibited a distinct block in the cell cycle. These mutants were therefore designated *cdc* (*cell-division cycle*) mutants. Importantly, these yeast mutants did not simply fail to grow, as they might

if they carried a mutation affecting general cellular metabolism. Rather, at the nonpermissive temperature, the mutants of interest grew normally for part of the cell cycle but then arrested at a particular stage of the cell cycle, so that many cells at this stage were seen (Figure 9-6b). Most *cdc* mutations in yeast are recessive; that is, when haploid *cdc* strains are mated to wild-type haploids, the resulting heterozygous diploids are neither temperature-sensitive nor defective in cell division.

Recessive Lethal Mutations in Diploids Can Be Identified by Inbreeding and Maintained in Heterozygotes

In diploid organisms, phenotypes resulting from recessive mutations can be observed only in individuals homozygous for the mutant alleles. Since mutagenesis in a diploid organism typically changes only one allele of a gene, yielding heterozygous mutants, genetic screens must include inbreeding steps to generate progeny that are homozygous for the mutant alleles. The geneticist H. Muller developed a general and efficient procedure for carrying out such inbreeding experiments in the fruit fly *Drosophila*. Recessive lethal mutations in *Drosophila* and other diploid organisms can be maintained in heterozygous individuals and their phenotypic consequences analyzed in homozygotes.

The Muller approach was used to great effect by C. Nüsslein-Volhard and E. Wieschaus, who systematically screened for recessive lethal mutations affecting embryogenesis in *Drosophila*. Dead homozygous embryos carrying recessive lethal mutations identified by this screen were examined under the microscope for specific morphological defects in the embryos. Current understanding of the molecular mechanisms underlying development of multicellular organisms is based, in large part, on the detailed picture of embryonic development revealed by characterization of these *Drosophila* mutants. We will discuss some of the fundamental discoveries based on these genetic studies in Chapter 15.

Complementation Tests Determine Whether Different Recessive Mutations Are in the Same Gene

In the genetic approach to studying a particular cellular process, researchers often isolate multiple recessive mutations that produce the same phenotype. A common test for determining whether these mutations are in the same gene or in different genes exploits the phenomenon of genetic **complementation,** that is, the restoration of the wild-type phenotype by mating of two different mutants. If two recessive mutations, *a* and *b*, are in the *same* gene, then a diploid organism heterozygous for both mutations (i.e., carrying one *a* allele and one *b* allele) will exhibit the mutant phenotype because neither allele provides a functional copy of the gene. In contrast, if mutation *a* and *b* are in *separate* genes, then heterozygotes carrying a single copy of each mutant allele

► **EXPERIMENTAL FIGURE 9-7**
Complementation analysis determines whether recessive mutations are in the same or different genes. Complementation tests in yeast are performed by mating haploid **a** and α cells carrying different recessive mutations to produce diploid cells. In the analysis of *cdc* mutations, pairs of different haploid temperature-sensitive *cdc* strains were systematically mated and the resulting diploids tested for growth at the permissive and nonpermissive temperatures. In this hypothetical example, the *cdcX* and *cdcY* mutants complement each other and thus have mutations in different genes, whereas the *cdcX* and *cdcZ* mutants have mutations in the same gene.

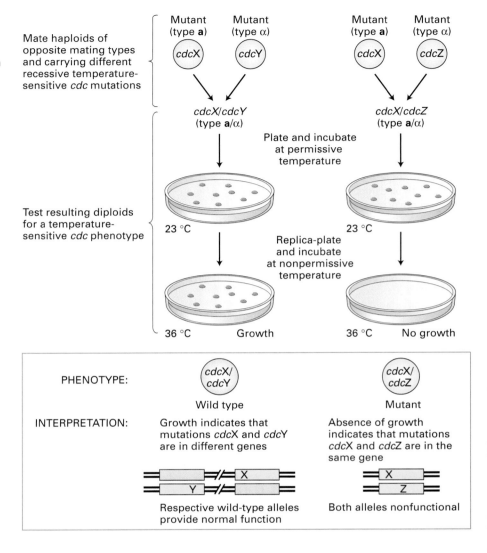

will not exhibit the mutant phenotype because a wild-type allele of each gene will also be present. In this case, the mutations are said to *complement* each other.

Complementation analysis of a set of mutants exhibiting the same phenotype can distinguish the individual genes in a set of functionally related genes, all of which must function to produce a given phenotypic trait. For example, the screen for *cdc* mutations in *Saccharomyces* described above yielded many recessive temperature-sensitive mutants that appeared arrested at the same cell-cycle stage. To determine how many genes were affected by these mutations, Hartwell and his colleagues performed complementation tests on all of the pair-wise combinations of *cdc* mutants following the general protocol outlined in Figure 9-7. These tests identified more than 20 different *CDC* genes. The subsequent molecular characterization of the *CDC* genes and their encoded proteins, as described in detail in Chapter 21, has provided a framework for understanding how cell division is regulated in organisms ranging from yeast to humans.

Double Mutants Are Useful in Assessing the Order in Which Proteins Function

Based on careful analysis of mutant phenotypes associated with a particular cellular process, researchers often can deduce the order in which a set of genes and their protein products function. Two general types of processes are amenable to such analysis: (a) biosynthetic pathways in which a precursor material is converted via one or more intermediates to a final product and (b) signaling pathways that regulate other processes and involve the flow of information rather than chemical intermediates.

Ordering of Biosynthetic Pathways A simple example of the first type of process is the biosynthesis of a metabolite such as the amino acid tryptophan in bacteria. In this case, each of the enzymes required for synthesis of tryptophan catalyzes the conversion of one of the intermediates in the pathway to the next. In *E. coli*, the genes encoding these enzymes lie adjacent to one another in the genome, constituting the

(a) Analysis of a biosynthetic pathway

A mutation in **A** accumulates intermediate 1.

A mutation in **B** accumulates intermediate 2.

PHENOTYPE OF DOUBLE MUTANT:	A double mutation in A and B accumulates intermediate 1.
INTERPRETATION:	*The reaction catalyzed by A precedes the reaction catalyzed by B.*

(b) Analysis of a signaling pathway

A mutation in **A** gives repressed reporter expression.

A mutation in **B** gives constitutive reporter expression.

PHENOTYPE OF DOUBLE MUTANT:	A double mutation in A and B gives repressed reporter expression.
INTERPRETATION:	*A positively regulates reporter expression and is negatively regulated by B.*

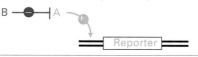

PHENOTYPE OF DOUBLE MUTANT:	A double mutation in A and B gives constitutive reporter expression.
INTERPRETATION:	*B negatively regulates reporter expression and is negatively regulated by A.*

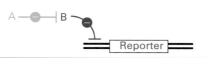

▶ **EXPERIMENTAL FIGURE 9-8 Analysis of double mutants often can order the steps in biosynthetic or signaling pathways.** When mutations in two different genes affect the same cellular process but have distinctly different phenotypes, the phenotype of the double mutant can often reveal the order in which the two genes must function. (a) In the case of mutations that affect the same biosynthetic pathway, a double mutant will accumulate the intermediate immediately preceding the step catalyzed by the protein that acts earlier in the wild-type organism. (b) Double-mutant analysis of a signaling pathway is possible if two mutations have opposite effects on expression of a reporter gene. In this case, the observed phenotype of the double mutant provides information about the order in which the proteins act and whether they are positive or negative regulators.

trp operon (see Figure 4-12a). The order of action of the different genes for these enzymes, hence the order of the biochemical reactions in the pathway, initially was deduced from the types of intermediate compounds that accumulated in each mutant. In the case of complex synthetic pathways, however, phenotypic analysis of mutants defective in a single step may give ambiguous results that do not permit con-

clusive ordering of the steps. Double mutants defective in two steps in the pathway are particularly useful in ordering such pathways (Figure 9-8a).

In Chapter 17 we discuss the classic use of the double-mutant strategy to help elucidate the secretory pathway. In this pathway proteins to be secreted from the cell move from their site of synthesis on the rough endoplasmic reticulum (ER) to the Golgi complex, then to secretory vesicles, and finally to the cell surface.

Ordering of Signaling Pathways As we learn in later chapters, expression of many eukaryotic genes is regulated by signaling pathways that are initiated by extracellular hormones, growth factors, or other signals. Such signaling pathways may include numerous components, and double-mutant analysis often can provide insight into the functions and interactions of these components. The only prerequisite for obtaining useful information from this type of analysis is that the two mutations must have opposite effects on the output of the same regulated pathway. Most commonly, one mutation represses expression of a particular reporter gene even when the signal is present, while another mutation results in reporter gene expression even when the signal is absent (i.e., constitutive expression). As illustrated in Figure 9-8b, two simple regulatory mechanisms are consistent with such single mutants, but the double-mutant phenotype can distinguish between them. This general approach has enabled geneticists to delineate many of the key steps in a variety of different regulatory pathways, setting the stage for more specific biochemical assays.

Genetic Suppression and Synthetic Lethality Can Reveal Interacting or Redundant Proteins

Two other types of genetic analysis can provide additional clues about how proteins that function in the same cellular process may interact with one another in the living cell. Both of these methods, which are applicable in many experimental organisms, involve the use of double mutants in which the phenotypic effects of one mutation are changed by the presence of a second mutation.

Suppressor Mutations The first type of analysis is based on *genetic suppression*. To understand this phenomenon, suppose that point mutations lead to structural changes in one protein (A) that disrupt its ability to associate with another protein (B) involved in the same cellular process. Similarly, mutations in protein B lead to small structural changes that inhibit its ability to interact with protein A. Assume, furthermore, that the normal functioning of proteins A and B depends on their interacting. In theory, a specific structural change in protein A might be suppressed by compensatory changes in protein B, allowing the mutant proteins to interact. In the rare cases in which such **suppressor mutations** occur, strains carrying both mutant alleles would be normal, whereas strains carrying only one or the other mutant allele would have a mutant phenotype (Figure 9-9a).

The observation of genetic suppression in yeast strains carrying a mutant actin allele (*act1-1*) and a second mutation (*sac6*) in another gene provided early evidence for a direct interaction in vivo between the proteins encoded by the two genes. Later biochemical studies showed that these two proteins—Act1 and Sac6—do indeed interact in the construction of functional actin structures within the cell.

(a) Suppression

Genotype	*AB*	*aB*	*Ab*	*ab*
Phenotype	Wild type	Mutant	Mutant	Suppressed mutant

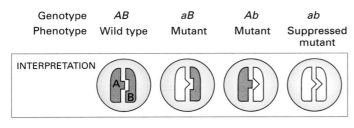

(b) Synthetic lethality 1

Genotype	*AB*	*aB*	*Ab*	*ab*
Phenotype	Wild type	Partial defect	Partial defect	Severe defect

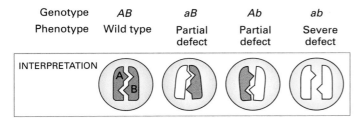

(c) Synthetic lethality 2

Genotype	*AB*	*aB*	*Ab*	*ab*
Phenotype	Wild type	Wild type	Wild type	Mutant

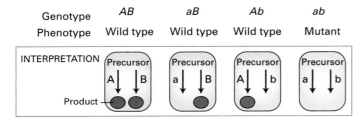

▲ **EXPERIMENTAL FIGURE 9-9 Mutations that result in genetic suppression or synthetic lethality reveal interacting or redundant proteins.** (a) Observation that double mutants with two defective proteins (A and B) have a wild-type phenotype but that single mutants give a mutant phenotype indicates that the function of each protein depends on interaction with the other. (b) Observation that double mutants have a more severe phenotypic defect than single mutants also is evidence that two proteins (e.g., subunits of a heterodimer) must interact to function normally. (c) Observation that a double mutant is nonviable but that the corresponding single mutants have the wild-type phenotype indicates that two proteins function in redundant pathways to produce an essential product.

Synthetic Lethal Mutations Another phenomenon, called *synthetic lethality,* produces a phenotypic effect opposite to that of suppression. In this case, the deleterious effect of one mutation is greatly exacerbated (rather than suppressed) by a second mutation in the same or a related gene. One situation in which such **synthetic lethal mutations** can occur is illustrated in Figure 9-9b. In this example, a heterodimeric protein is partially, but not completely, inactivated by mutations in either one of the nonidentical subunits. However, in double mutants carrying specific mutations in the genes encoding both subunits, little interaction between subunits occurs, resulting in severe phenotypic effects.

Synthetic lethal mutations also can reveal nonessential genes whose encoded proteins function in redundant pathways for producing an essential cell component. As depicted in Figure 9-9c, if either pathway alone is inactivated by a mutation, the other pathway will be able to supply the needed product. However, if both pathways are inactivated at the same time, the essential product cannot be synthesized, and the double mutants will be nonviable.

KEY CONCEPTS OF SECTION 9.1

Genetic Analysis of Mutations to Identify and Study Genes

■ Diploid organisms carry two copies (alleles) of each gene, whereas haploid organisms carry only one copy.

■ Recessive mutations lead to a loss of function, which is masked if a normal allele of the gene is present. For the mutant phenotype to occur, both alleles must carry the mutation.

■ Dominant mutations lead to a mutant phenotype in the presence of a normal allele of the gene. The phenotypes associated with dominant mutations often represent a gain of function but in the case of some genes result from a loss of function.

■ In meiosis, a diploid cell undergoes one DNA replication and two cell divisions, yielding four haploid cells in which maternal and paternal alleles are randomly assorted (see Figure 9-3).

■ Dominant and recessive mutations exhibit characteristic segregation patterns in genetic crosses (see Figure 9-4).

■ In haploid yeast, temperature-sensitive mutations are particularly useful for identifying and studying genes essential to survival.

■ The number of functionally related genes involved in a process can be defined by complementation analysis (see Figure 9-7).

■ The order in which genes function in either a biosynthetic or a signaling pathway can be deduced from the phenotype of double mutants defective in two steps in the affected process.

■ Functionally significant interactions between proteins can be deduced from the phenotypic effects of allele-specific suppressor mutations or synthetic lethal mutations.

9.2 DNA Cloning by Recombinant DNA Methods

Detailed studies of the structure and function of a gene at the molecular level require large quantities of the individual gene in pure form. A variety of techniques, often referred to as *recombinant DNA technology,* are used in **DNA cloning,** which permits researchers to prepare large numbers of identical DNA molecules. **Recombinant DNA** is simply any DNA molecule composed of sequences derived from different sources.

The key to cloning a DNA fragment of interest is to link it to a **vector** DNA molecule, which can replicate within a host cell. After a single recombinant DNA molecule, composed of a vector plus an inserted DNA fragment, is introduced into a host cell, the inserted DNA is replicated along with the vector, generating a large number of identical DNA molecules. The basic scheme can be summarized as follows:

Vector + DNA fragment

↓

Recombinant DNA

↓

Replication of recombinant DNA within host cells

↓

Isolation, sequencing, and manipulation
of purified DNA fragment

Although investigators have devised numerous experimental variations, this flow diagram indicates the essential steps in DNA cloning. In this section, we cover the steps in this basic scheme, focusing on the two types of vectors most commonly used in *E. coli* host cells: plasmid vectors, which replicate along with their host cells, and bacteriophage λ vectors, which replicate as lytic viruses, killing the host cell and packaging their DNA into virions. We discuss the characterization and various uses of cloned DNA fragments in subsequent sections.

Restriction Enzymes and DNA Ligases Allow Insertion of DNA Fragments into Cloning Vectors

A major objective of DNA cloning is to obtain discrete, small regions of an organism's DNA that constitute specific genes. In addition, only relatively small DNA molecules can be cloned in any of the available vectors. For these reasons, the very long DNA molecules that compose an organism's genome must be cleaved into fragments that can be inserted into the vector DNA. Two types of enzymes—**restriction enzymes** and **DNA ligases**—facilitate production of such recombinant DNA molecules.

Cutting DNA Molecules into Small Fragments Restriction enzymes are endonucleases produced by bacteria that typically recognize specific 4- to 8-bp sequences, called *restriction sites,* and then cleave both DNA strands at this site. Restriction sites commonly are short *palindromic* sequences; that is, the restriction-site sequence is the same on each DNA strand when read in the $5' \rightarrow 3'$ direction (Figure 9-10).

For each restriction enzyme, bacteria also produce a *modification enzyme,* which protects a bacterium's own DNA from cleavage by modifying it at or near each potential cleavage site. The modification enzyme adds a methyl group to one or two bases, usually within the restriction site. When a methyl group is present there, the restriction endonuclease is prevented from cutting the DNA. Together with the restriction endonuclease, the methylating enzyme forms a restriction-modification system that protects the host DNA while it destroys incoming foreign DNA (e.g., bacteriophage DNA or DNA taken up during transformation) by cleaving it at all the restriction sites in the DNA.

Many restriction enzymes make staggered cuts in the two DNA strands at their recognition site, generating fragments that have a single-stranded "tail" at both ends (see Figure 9-10). The tails on the fragments generated at a given restriction site are complementary to those on all other fragments generated by the same restriction enzyme. At room temperature, these single-stranded regions, often called "sticky ends," can transiently base-pair with those on other DNA fragments generated with the same restriction enzyme. A few restriction enzymes, such as *Alu*I and *Sma*I, cleave both DNA strands at the same point within the restriction site, generating fragments with "blunt" (flush) ends in which all the nucleotides at the fragment ends are base-paired to nucleotides in the complementary strand.

The DNA isolated from an individual organism has a specific sequence, which purely by chance will contain a specific

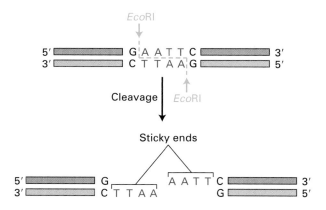

▲ **FIGURE 9-10 Cleavage of DNA by the restriction enzyme *Eco*RI.** This restriction enzyme from *E. coli* makes staggered cuts at the specific 6-bp inverted repeat (palindromic) sequence shown, yielding fragments with single-stranded, complementary "sticky" ends. Many other restriction enzymes also produce fragments with sticky ends.

TABLE 9-1   Selected Restriction Enzymes and Their Recognition Sequences

Enzyme	Source Microorganism	Recognition Site*	Ends Produced
BamHI	*Bacillus amyloliquefaciens*	↓ -G-G-A-T-C-C- -C-C-T-A-G-G- ↑	Sticky
EcoRI	*Escherichia coli*	↓ -G-A-A-T-T-C- -C-T-T-A-A-G- ↑	Sticky
HindIII	*Haemophilus influenzae*	↓ -A-A-G-C-T-T- -T-T-C-G-A-A- ↑	Sticky
KpnI	*Klebsiella pneumonia*	↓ -G-G-T-A-C-C- -C-C-A-T-G-G- ↑	Sticky
PstI	*Providencia stuartii*	↓ -C-T-G-C-A-G- -G-A-C-G-T-C- ↑	Sticky
SacI	*Streptomyces achromogenes*	↓ -G-A-G-C-T-C- -C-T-C-G-A-G- ↑	Sticky
SalI	*Streptomyces albue*	↓ -G-T-C-G-A-C- -C-A-G-C-T-G- ↑	Sticky
SmaI	*Serratia marcescens*	↓ -C-C-C-G-G-G- -G-G-G-C-C-C- ↑	Blunt
SphI	*Streptomyces phaeochromogenes*	↓ -G-C-A-T-G-C- -C-G-T-A-C-G- ↑	Sticky
XbaI	*Xanthomonas badrii*	↓ -T-C-T-A-G-A- -A-G-A-T-C-T- ↑	Sticky

*These recognition sequences are included in a common polylinker sequence (see Figure 9-12).

set of restriction sites. Thus a given restriction enzyme will cut the DNA from a particular source into a reproducible set of fragments called **restriction fragments**. Restriction enzymes have been purified from several hundred different species of bacteria, allowing DNA molecules to be cut at a large number of different sequences corresponding to the recognition sites of these enzymes (Table 9-1).

Inserting DNA Fragments into Vectors DNA fragments with either sticky ends or blunt ends can be inserted into vec-

tor DNA with the aid of DNA ligases. During normal DNA replication, DNA ligase catalyzes the end-to-end joining (ligation) of short fragments of DNA, called *Okazaki fragments*. For purposes of DNA cloning, purified DNA ligase is used to covalently join the ends of a restriction fragment and vector DNA that have complementary ends (Figure 9-11). The vector DNA and restriction fragment are covalently ligated together through the standard 3′ → 5′ phosphodiester bonds of DNA. In addition to ligating complementary sticky ends, the DNA ligase from bacteriophage T4 can ligate any two

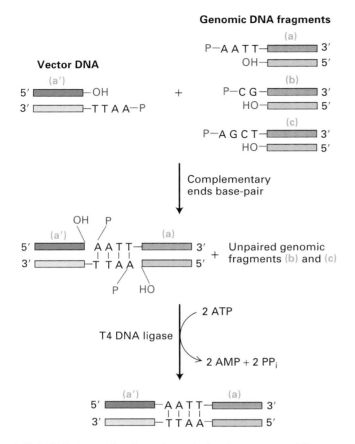

▲ FIGURE 9-11 Ligation of restriction fragments with complementary sticky ends. In this example, vector DNA cut with *Eco*RI is mixed with a sample containing restriction fragments produced by cleaving genomic DNA with several different restriction enzymes. The short base sequences composing the sticky ends of each fragment type are shown. The sticky end on the cut vector DNA (a′) base-pairs only with the complementary sticky ends on the *Eco*RI fragment (a) in the genomic sample. The adjacent 3′-hydroxyl and 5′-phosphate groups (red) on the base-paired fragments then are covalently joined (ligated) by T4 DNA ligase.

blunt DNA ends. However, blunt-end ligation is inherently inefficient and requires a higher concentration of both DNA and DNA ligase than for ligation of sticky ends.

E. coli Plasmid Vectors Are Suitable for Cloning Isolated DNA Fragments

Plasmids are circular, double-stranded DNA (dsDNA) molecules that are separate from a cell's chromosomal DNA. These extrachromosomal DNAs, which occur naturally in bacteria and in lower eukaryotic cells (e.g., yeast), exist in a parasitic or symbiotic relationship with their host cell. Like the host-cell chromosomal DNA, plasmid DNA is duplicated before every cell division. During cell division, copies of the plasmid DNA segregate to each daughter cell, assuring con-

tinued propagation of the plasmid through successive generations of the host cell.

The plasmids most commonly used in recombinant DNA technology are those that replicate in *E. coli*. Investigators have engineered these plasmids to optimize their use as vectors in DNA cloning. For instance, removal of unneeded portions from naturally occurring *E. coli* plasmids yields plasmid vectors, ≈1.2–3 kb in circumferential length, that contain three regions essential for DNA cloning: a replication origin; a marker that permits selection, usually a drug-resistance gene; and a region in which exogenous DNA fragments can be inserted (Figure 9-12). Host-cell enzymes replicate a plasmid beginning at the replication origin (ORI), a specific DNA sequence of 50–100 base pairs. Once DNA replication is initiated at the ORI, it continues around the circular plasmid regardless of its nucleotide sequence. Thus any DNA sequence inserted into such a plasmid is replicated along with the rest of the plasmid DNA.

Figure 9-13 outlines the general procedure for cloning a DNA fragment using *E. coli* plasmid vectors. When *E. coli* cells are mixed with recombinant vector DNA under certain conditions, a small fraction of the cells will take up the plasmid DNA, a process known as **transformation.** Typically, 1 cell in about 10,000 incorporates a *single* plasmid DNA molecule and thus becomes transformed. After plasmid vectors are incubated with *E. coli*, those cells that take up the plasmid can be easily selected from the much larger number of cells. For instance, if the plasmid carries a gene that confers resistance to the antibiotic ampicillin, transformed cells

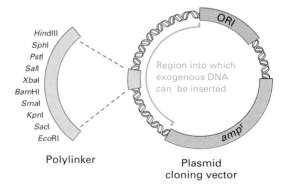

Polylinker — Plasmid cloning vector

▲ FIGURE 9-12 Basic components of a plasmid cloning vector that can replicate within an *E. coli* cell. Plasmid vectors contain a selectable gene such as *amp*ʳ, which encodes the enzyme β-lactamase and confers resistance to ampicillin. Exogenous DNA can be inserted into the bracketed region without disturbing the ability of the plasmid to replicate or express the *amp*ʳ gene. Plasmid vectors also contain a replication origin (ORI) sequence where DNA replication is initiated by host-cell enzymes. Inclusion of a synthetic polylinker containing the recognition sequences for several different restriction enzymes increases the versatility of a plasmid vector. The vector is designed so that each site in the polylinker is unique on the plasmid.

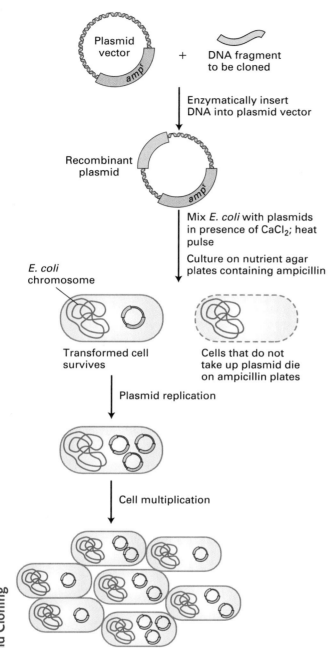

Plasmid vector

+ DNA fragment to be cloned

↓ Enzymatically insert DNA into plasmid vector

Recombinant plasmid

↓ Mix *E. coli* with plasmids in presence of CaCl₂; heat pulse

Culture on nutrient agar plates containing ampicillin

E. coli chromosome

Transformed cell survives

Cells that do not take up plasmid die on ampicillin plates

↓ Plasmid replication

↓ Cell multiplication

Colony of cells, each containing copies of the same recombinant plasmid

▲ **EXPERIMENTAL FIGURE 9-13 DNA cloning in a plasmid vector permits amplification of a DNA fragment.** A fragment of DNA to be cloned is first inserted into a plasmid vector containing an ampicillin-resistance gene (*amp*ʳ), such as that shown in Figure 9-12. Only the few cells transformed by incorporation of a plasmid molecule will survive on ampicillin-containing medium. In transformed cells, the plasmid DNA replicates and segregates into daughter cells, resulting in formation of an ampicillin-resistant colony.

can be selected by growing them in an ampicillin-containing medium.

DNA fragments from a few base pairs up to ≈20 kb commonly are inserted into plasmid vectors. If special precautions are taken to avoid manipulations that might mechanically break DNA, even longer DNA fragments can be inserted into a plasmid vector. When a recombinant plasmid with an inserted DNA fragment transforms an *E. coli* cell, all the antibiotic-resistant progeny cells that arise from the initial transformed cell will contain plasmids with the same inserted DNA. The inserted DNA is replicated along with the rest of the plasmid DNA and segregates to daughter cells as the colony grows. In this way, the initial fragment of DNA is replicated in the colony of cells into a large number of identical copies. Since all the cells in a colony arise from a single transformed parental cell, they constitute a **clone** of cells, and the initial fragment of DNA inserted into the parental plasmid is referred to as *cloned DNA* or a *DNA clone*.

The versatility of an *E. coli* plasmid vector is increased by incorporating into it a *polylinker*, a synthetically generated sequence containing one copy of several different restriction sites that are not present elsewhere in the plasmid sequence (see Figure 9-12). When such a vector is treated with a restriction enzyme that recognizes a restriction site in the polylinker, the vector is cut only once within the polylinker. Subsequently any DNA fragment of appropriate length produced with the same restriction enzyme can be inserted into the cut plasmid with DNA ligase. Plasmids containing a polylinker permit a researcher to clone DNA fragments generated with different restriction enzymes using the same plasmid vector, which simplifies experimental procedures.

Bacteriophage λ Vectors Permit Efficient Construction of Large DNA Libraries

Vectors constructed from bacteriophage λ are about a thousand times more efficient than plasmid vectors in cloning large numbers of DNA fragments. For this reason, phage λ vectors have been widely used to generate DNA libraries, comprehensive collections of DNA fragments representing the genome or expressed mRNAs of an organism. Two factors account for the greater efficiency of phage λ as a cloning vector: infection of *E. coli* host cells by λ virions occurs at about a thousandfold greater frequency than transformation by plasmids, and many more λ clones than transformed colonies can be grown and detected on a single culture plate.

When a λ virion infects an *E. coli* cell, it can undergo a cycle of lytic growth during which the phage DNA is replicated and assembled into more than 100 complete progeny phage, which are released when the infected cell lyses (see Figure 4-40). If a sample of λ phage is placed on a lawn of *E. coli* growing on a petri plate, each virion will infect a single cell. The ensuing rounds of phage growth will give rise to a visible cleared region, called a *plaque*, where the cells have been lysed and phage particles released (see Figure 4-39).

(a) λ Phage genome

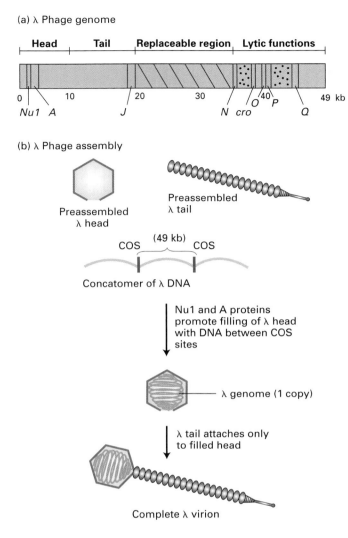

(b) λ Phage assembly

Preassembled
λ head

Preassembled
λ tail

COS   (49 kb)   COS

Concatomer of λ DNA

Nu1 and A proteins
promote filling of λ head
with DNA between COS
sites

λ genome (1 copy)

λ tail attaches only
to filled head

Complete λ virion

▲ FIGURE 9-14 The bacteriophage λ genome and packaging of bacteriophage λ DNA. (a) Simplified map of the λ phage genome. There are about 60 genes in the λ genome, only a few of which are shown in this diagram. Genes encoding proteins required for assembly of the head and tail are located at the left end; those encoding additional proteins required for the lytic cycle, at the right end. Some regions of the genome can be replaced by exogenous DNA (diagonal lines) or deleted (dotted) without affecting the ability of λ phage to infect host cells and assemble new virions. Up to ≈25 kb of exogenous DNA can be stably inserted between the *J* and *N* genes. (b) In vivo assembly of λ virions. Heads and tails are formed from multiple copies of several different λ proteins. During the late stage of λ infection, long DNA molecules called *concatomers* are formed; these multimeric molecules consist of multiple copies of the 49-kb λ genome linked end to end and separated by COS sites (red), protein-binding nucleotide sequences that occur once in each copy of the λ genome. Binding of λ head proteins Nu1 and A to COS sites promotes insertion of the DNA segment between two adjacent COS sites into an empty head. After the heads are filled with DNA, assembled λ tails are attached, producing complete λ virions capable of infecting *E. coli* cells.

A λ virion consists of a head, which contains the phage DNA genome, and a tail, which functions in infecting *E. coli* host cells. The λ genes encoding the head and tail proteins, as well as various proteins involved in phage DNA replication and cell lysis, are grouped in discrete regions of the ≈50-kb viral genome (Figure 9-14a). The central region of the λ genome, however, contains genes that are not essential for the lytic pathway. Removing this region and replacing it with a foreign DNA fragment up to ≈25 kb long yields a recombinant DNA that can be packaged in vitro to form phage capable of replicating and forming plaques on a lawn of *E. coli* host cells. In vitro packaging of recombinant λ DNA, which mimics the in vivo assembly process, requires preassembled heads and tails as well as two viral proteins (Figure 9-14b).

It is technically feasible to use λ phage cloning vectors to generate a *genomic library*, that is, a collection of λ clones that collectively represent all the DNA sequences in the genome of a particular organism. However, such genomic libraries for higher eukaryotes present certain experimental difficulties. First, the genes from such organisms usually contain extensive intron sequences and therefore are too large to be inserted intact into λ phage vectors. As a result, the sequences of individual genes are broken apart and carried in more than one λ clone (this is also true for plasmid clones). Moreover, the presence of introns and long intergenic regions in genomic DNA often makes it difficult to identify the important parts of a gene that actually encode protein sequences. Thus for many studies, cellular mRNAs, which lack the noncoding regions present in genomic DNA, are a more useful starting material for generating a DNA library. In this approach, DNA copies of mRNAs, called **complementary DNAs (cDNAs)**, are synthesized and cloned in phage vectors. A large collection of the resulting cDNA clones, representing all the mRNAs expressed in a cell type, is called a *cDNA library*.

cDNAs Prepared by Reverse Transcription of Cellular mRNAs Can Be Cloned to Generate cDNA Libraries

The first step in preparing a cDNA library is to isolate the total mRNA from the cell type or tissue of interest. Because of their poly(A) tails, mRNAs are easily separated from the much more prevalent rRNAs and tRNAs present in a cell extract by use of a column to which short strings of thymidylate (oligo-dTs) are linked to the matrix.

The general procedure for preparing a λ phage cDNA library from a mixture of cellular mRNAs is outlined in Figure 9-15. The enzyme **reverse transcriptase,** which is found in retroviruses, is used to synthesize a strand of DNA complementary to each mRNA molecule, starting from an oligo-dT primer (steps 1 and 2). The resulting cDNA-mRNA hybrid molecules are converted in several steps to double-stranded cDNA molecules corresponding to all the mRNA molecules in the original preparation (steps 3–5). Each double-stranded

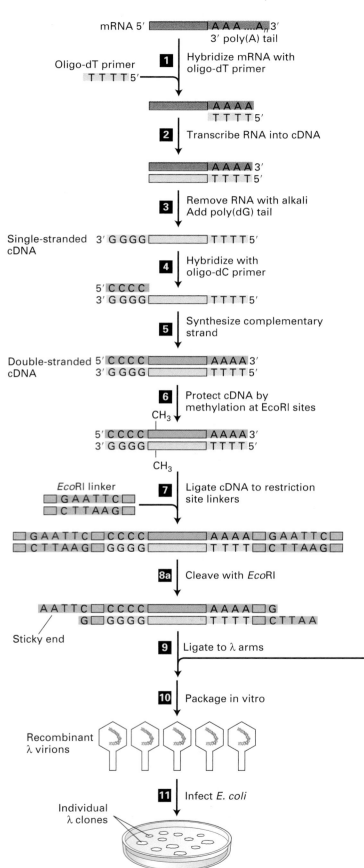

cDNA contains an oligo-dC · oligo-dG double-stranded region at one end and an oligo-dT·oligo-dA double-stranded region at the other end. Methylation of the cDNA protects it from subsequent restriction enzyme cleavage (step 6).

To prepare double-stranded cDNAs for cloning, short double-stranded DNA molecules containing the recognition site for a particular restriction enzyme are ligated to both ends of the cDNAs using DNA ligase from bacteriophage T4 (Figure 9-15, step 7). As noted earlier, this ligase can join "blunt-ended" double-stranded DNA molecules lacking sticky ends. The resulting molecules are then treated with the restriction enzyme specific for the attached linker, generating cDNA molecules with sticky ends at each end (step 8a). In a separate procedure, λ DNA first is treated with the same restriction enzyme to produce fragments called λ *vector arms*, which have sticky ends and together contain all the genes necessary for lytic growth (step 8b).

The λ arms and the collection of cDNAs, all containing complementary sticky ends, then are mixed and joined covalently by DNA ligase (Figure 9-15, step 9). Each of the resulting recombinant DNA molecules contains a cDNA located between the two arms of the λ vector DNA. Virions containing the ligated recombinant DNAs then are assembled in vitro as described above (step 10). Only DNA molecules of the correct size can be packaged to produce fully infectious recombinant λ phage. Finally, the recombinant λ phages are plated on a lawn of *E. coli* cells to generate a large number of individual plaques (step 11).

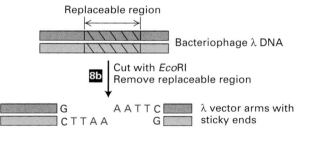

▲ **EXPERIMENTAL FIGURE 9-15 A cDNA library can be constructed using a bacteriophage λ vector.** A mixture of mRNAs is the starting point for preparing recombinant λ virions each containing a cDNA. To maximize the size of the exogenous DNA that can be inserted into the λ genome, the nonessential regions of the λ genome (diagonal lines in Figure 9-14) usually are deleted. Plating of the recombinant phage on a lawn of *E. coli* generates a set of cDNA clones representing all the cellular mRNAs. See the text for a step-by-step discussion.

Since each plaque arises from a single recombinant phage, all the progeny λ phages that develop are genetically identical and constitute a clone carrying a cDNA derived from a single mRNA; collectively they constitute a λ cDNA library. One feature of cDNA libraries arises because different genes are transcribed at very different rates. As a result, cDNA clones corresponding to rapidly transcribed genes will be represented many times in a cDNA library, whereas cDNAs corresponding to slowly transcribed genes will be extremely rare or not present at all. This property is advantageous if an investigator is interested in a gene that is transcribed at a high rate in a particular cell type. In this case, a cDNA library prepared from mRNAs expressed in that cell type will be enriched in the cDNA of interest, facilitating screening of the library for λ clones carrying that cDNA. However, to have a reasonable chance of including clones corresponding to slowly transcribed genes, mammalian cDNA libraries must contain 10^6–10^7 individual recombinant λ phage clones.

DNA Libraries Can Be Screened by Hybridization to an Oligonucleotide Probe

Both genomic and cDNA libraries of various organisms contain hundreds of thousands to upwards of a million individual clones in the case of higher eukaryotes. Two general approaches are available for screening libraries to identify clones carrying a gene or other DNA region of interest: (1) detection with oligonucleotide **probes** that bind to the clone of interest and (2) detection based on expression of the encoded protein. Here we describe the first method; an example of the second method is presented in the next section.

The basis for screening with oligonucleotide probes is **hybridization,** the ability of complementary single-stranded DNA or RNA molecules to associate (hybridize) specifically with each other via base pairing. As discussed in Chapter 4, double-stranded (duplex) DNA can be denatured (melted) into single strands by heating in a dilute salt solution. If the temperature then is lowered and the ion concentration raised, complementary single strands will reassociate (hybridize) into duplexes. In a mixture of nucleic acids, only complementary single strands (or strands containing complementary regions) will reassociate; moreover, the extent of their reassociation is virtually unaffected by the presence of noncomplementary strands.

In the *membrane-hybridization assay* outlined in Figure 9-16, a single-stranded nucleic acid probe is used to detect those DNA fragments in a mixture that are complementary to the probe. The DNA sample first is denatured and the single strands attached to a solid support, commonly a nitrocellulose filter or treated nylon membrane. The membrane is then incubated in a solution containing a radioactively labeled probe. Under hybridization conditions (near neutral pH, 40–65 °C, 0.3–0.6 M NaCl), this labeled probe hybridizes to any complementary nucleic acid strands bound to

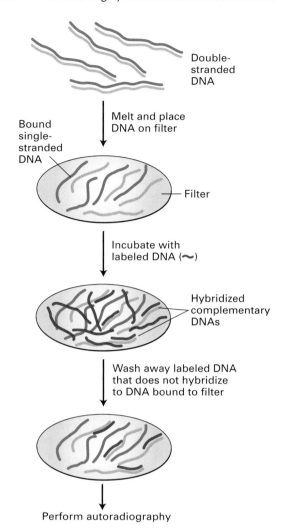

▲ **EXPERIMENTAL FIGURE 9-16 Membrane-hybridization assay detects nucleic acids complementary to an oligonucleotide probe.** This assay can be used to detect both DNA and RNA, and the radiolabeled complementary probe can be either DNA or RNA.

the membrane. Any excess probe that does not hybridize is washed away, and the labeled hybrids are detected by autoradiography of the filter.

Application of this procedure for screening a λ cDNA library is depicted in Figure 9-17. In this case, a *replica* of the petri dish containing a large number of individual λ clones initially is reproduced on the surface of a nitrocellulose membrane. The membrane is then assayed using a radiolabeled probe specific for the recombinant DNA containing the fragment of interest. Membrane hybridization with radiolabeled oligonucleotides is most commonly used to screen λ cDNA libraries. Once a cDNA clone encoding a particular protein is obtained, the full-length cDNA can be radiolabeled and used to probe a genomic library for clones containing fragments of the corresponding gene.

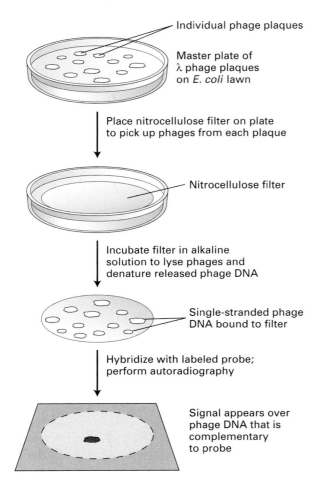

▲ **EXPERIMENTAL FIGURE 9-17 Phage cDNA libraries can be screened with a radiolabeled probe to identify a clone of interest.** In the initial plating of a library, the λ phage plaques are not allowed to develop to a visible size so that up to 50,000 recombinants can be analyzed on a single plate. The appearance of a spot on the autoradiogram indicates the presence of a recombinant λ clone containing DNA complementary to the probe. The position of the spot on the autoradiogram is the mirror image of the position on the original petri dish of that particular clone. Aligning the autoradiogram with the original petri dish will locate the corresponding clone from which infectious phage particles can be recovered and replated at low density, resulting in well-separated plaques. Pure isolates eventually are obtained by repeating the hybridization assay.

Oligonucleotide Probes Are Designed Based on Partial Protein Sequences

Clearly, identification of specific clones by the membrane-hybridization technique depends on the availability of complementary radiolabeled probes. For an oligonucleotide to be useful as a probe, it must be long enough for its sequence to occur uniquely in the clone of interest and not in any other clones. For most purposes, this condition is satisfied by oligonucleotides containing about 20 nucleotides. This is be-

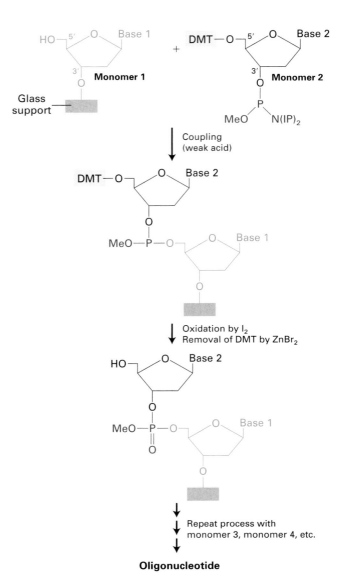

▲ **FIGURE 9-18 Chemical synthesis of oligonucleotides by sequential addition of reactive nucleotide derivatives.** The first (3′) nucleotide in the sequence (monomer 1) is bound to a glass support by its 3′ hydroxyl; its 5′ hydroxyl is available for addition of the second nucleotide. The second nucleotide in the sequence (monomer 2) is derivatized by addition of 4′,4′-dimethoxytrityl (DMT) to its 5′ hydroxyl, thus blocking this hydroxyl from reacting; in addition, a highly reactive group (red letters) is attached to the 3′ hydroxyl. When the two monomers are mixed in the presence of a weak acid, they form a 5′ → 3′ phosphodiester bond with the phosphorus in the trivalent state. Oxidation of this intermediate increases the phosphorus valency to 5, and subsequent removal of the DMT group with zinc bromide ($ZnBr_2$) frees the 5′ hydroxyl. Monomer 3 then is added, and the reactions are repeated. Repetition of this process eventually yields the entire oligonucleotide. Finally, all the methyl groups on the phosphates are removed at the same time at alkaline pH, and the bond linking monomer 1 to the glass support is cleaved. [See S. L. Beaucage and M. H. Caruthers, 1981, *Tetrahedron Lett.* **22**:1859.]

cause a specific 20-nucleotide sequence occurs once in every 4^{20} ($\approx 10^{12}$) nucleotides. Since all genomes are much smaller ($\approx 3 \times 10^9$ nucleotides for humans), a specific 20-nucleotide sequence in a genome usually occurs only once. Oligonucleotides of this length with a specific sequence can be synthesized chemically and then radiolabeled by using polynucleotide kinase to transfer a ^{32}P-labeled phosphate group from ATP to the 5' end of each oligonucleotide.

How might an investigator design an oligonucleotide probe to identify a cDNA clone encoding a particular protein? If all or a portion of the amino acid sequence of the protein is known, then a DNA probe corresponding to a small region of the gene can be designed based on the genetic code. However, because the genetic code is degenerate (i.e., many amino acids are encoded by more than one codon), a probe based on an amino acid sequence must include *all* the possible oligonucleotides that could theoretically encode that peptide sequence. Within this mixture of oligonucleotides will be one that hybridizes perfectly to the clone of interest.

In recent years, this approach has been simplified by the availability of the complete genomic sequences for humans and some important model organisms such as the mouse, *Drosophila*, and the roundworm *Caenorhabditis elegans*. Using an appropriate computer program, a researcher can search the genomic sequence database for the coding sequence that corresponds to a specific portion of the amino acid sequence of the protein under study. If a match is found, then a single, unique DNA probe based on this known genomic sequence will hybridize perfectly with the clone encoding the protein under study.

Chemical synthesis of single-stranded DNA probes of defined sequence can be accomplished by the series of reactions shown in Figure 9-18. With automated instruments now available, researchers can program the synthesis of oligonucleotides of specific sequence up to about 100 nucleotides long. Alternatively, these probes can be prepared by the polymerase chain reaction (PCR), a widely used technique for amplifying specific DNA sequences that is described later.

▶ **EXPERIMENTAL FIGURE 9-19 Yeast genomic library can be constructed in a plasmid shuttle vector that can replicate in yeast and E. coli.** (a) Components of a typical plasmid shuttle vector for cloning *Saccharomyces* genes. The presence of a yeast origin of DNA replication (ARS) and a yeast centromere (CEN) allows, stable replication and segregation in yeast. Also included is a yeast selectable marker such as *URA3*, which allows a *ura3⁻* mutant to grow on medium lacking uracil. Finally, the vector contains sequences for replication and selection in *E. coli* (ORI and *amp*ʳ) and a polylinker for easy insertion of yeast DNA fragments. (b) Typical protocol for constructing a yeast genomic library. Partial digestion of total yeast genomic DNA with *Sau*3A is adjusted to generate fragments with an average size of about 10 kb. The vector is prepared to accept the genomic fragments by digestion with *Bam*HI, which produces the same sticky ends as *Sau*3A. Each transformed clone of *E. coli* that grows after selection for ampicillin resistance contains a single type of yeast DNA fragment.

Yeast Genomic Libraries Can Be Constructed with Shuttle Vectors and Screened by Functional Complementation

In some cases a DNA library can be screened for the ability to express a functional protein that complements a recessive mutation. Such a screening strategy would be an efficient way to isolate a cloned gene that corresponds to an interesting recessive mutation identified in an experimental organism. To illustrate this method, referred to as **functional complementation,** we describe how yeast genes cloned in special *E. coli*

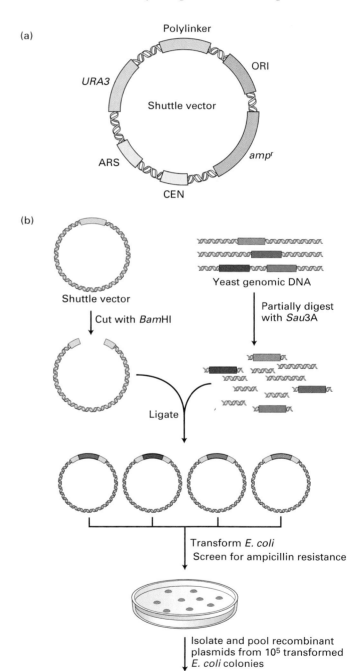

plasmids can be introduced into mutant yeast cells to identify the wild-type gene that is defective in the mutant strain.

Libraries constructed for the purpose of screening among yeast gene sequences usually are constructed from genomic DNA rather than cDNA. Because *Saccharomyces* genes do not contain multiple introns, they are sufficiently compact so that the entire sequence of a gene can be included in a genomic DNA fragment inserted into a plasmid vector. To construct a plasmid genomic library that is to be screened by functional complementation in yeast cells, the plasmid vector must be capable of replication in both *E. coli* cells and yeast cells. This type of vector, capable of propagation in two different hosts, is called a **shuttle vector**. The structure of a typical yeast shuttle vector is shown in Figure 9-19a (see page 369). This vector contains the basic elements that permit cloning of DNA fragments in *E. coli*. In addition, the shuttle vector contains an autonomously replicating sequence (ARS), which functions as an origin for DNA replication in yeast; a yeast centromere (called CEN), which allows faithful segregation of the plasmid during yeast cell division; and a yeast gene encoding an enzyme for uracil synthesis (*URA3*), which serves as a selectable marker in an appropriate yeast mutant.

To increase the probability that all regions of the yeast genome are successfully cloned and represented in the plasmid library, the genomic DNA usually is only partially digested to yield overlapping restriction fragments of ≈10 kb. These fragments are then ligated into the shuttle vector in which the polylinker has been cleaved with a restriction enzyme that produces sticky ends complementary to those on the yeast DNA fragments (Figure 9-19b). Because the 10-kb restriction fragments of yeast DNA are incorporated into the shuttle vectors randomly, at least 10^5 *E. coli* colonies, each containing a particular recombinant shuttle vector, are necessary to assure that each region of yeast DNA has a high probability of being represented in the library at least once.

Figure 9-20 outlines how such a yeast genomic library can be screened to isolate the wild-type gene corresponding to one of the temperature-sensitive *cdc* mutations mentioned earlier in this chapter. The starting yeast strain is a double mutant that requires uracil for growth due to a *ura3* mutation and is temperature-sensitive due to a *cdc28* mutation identified by its phenotype (see Figure 9-6). Recombinant plasmids isolated from the yeast genomic library are mixed with yeast cells under conditions that promote transformation of the cells with foreign DNA. Since transformed yeast cells carry a plasmid-borne copy of the wild-type *URA3* gene, they can be selected by their ability to grow in the absence of uracil. Typically, about 20 petri dishes, each containing about 500 yeast transformants, are sufficient to represent the entire yeast genome. This collection of yeast transformants can be maintained at 23 °C, a temperature permissive for growth of the *cdc28* mutant. The entire collection on 20 plates is then transferred to replica plates, which are placed at 36 °C, a nonpermissive temperature for

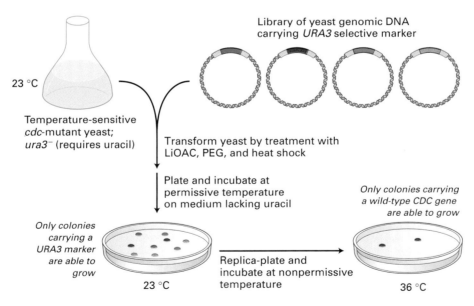

▲ **EXPERIMENTAL FIGURE 9-20 Screening of a yeast genomic library by functional complementation can identify clones carrying the normal form of mutant yeast gene.** In this example, a wild-type *CDC* gene is isolated by complementation of a *cdc* yeast mutant. The *Saccharomyces* strain used for screening the yeast library carries *ura3⁻* and a temperature-sensitive *cdc* mutation. This mutant strain is grown and maintained at a permissive temperature (23 °C). Pooled recombinant plasmids prepared as shown in Figure 9-19 are incubated with the mutant yeast cells under conditions that promote transformation. The relatively few transformed yeast cells, which contain recombinant plasmid DNA, can grow in the absence of uracil at 23 °C. When transformed yeast colonies are replica-plated and placed at 36 °C (a nonpermissive temperature), only clones carrying a library plasmid that contains the wild-type copy of the *CDC* gene will survive. LiOAC = lithium acetate; PEG = polyethylene glycol.

cdc mutants. Yeast colonies that carry recombinant plasmids expressing a wild-type copy of the *CDC28* gene will be able to grow at 36 °C. Once temperature-resistant yeast colonies have been identified, plasmid DNA can be extracted from the cultured yeast cells and analyzed by subcloning and DNA sequencing, topics we take up in the next section.

KEY CONCEPTS OF SECTION 9.2

DNA Cloning by Recombinant DNA Methods

■ In DNA cloning, recombinant DNA molecules are formed in vitro by inserting DNA fragments into vector DNA molecules. The recombinant DNA molecules are then introduced into host cells, where they replicate, producing large numbers of recombinant DNA molecules.

■ Restriction enzymes (endonucleases) typically cut DNA at specific 4- to 8-bp palindromic sequences, producing defined fragments that often have self-complementary single-stranded tails (sticky ends).

■ Two restriction fragments with complementary ends can be joined with DNA ligase to form a recombinant DNA (see Figure 9-11).

■ *E. coli* cloning vectors are small circular DNA molecules (plasmids) that include three functional regions: an origin of replication, a drug-resistance gene, and a site where a DNA fragment can be inserted. Transformed cells carrying a vector grow into colonies on the selection medium (see Figure 9-13).

■ Phage cloning vectors are formed by replacing nonessential parts of the λ genome with DNA fragments up to ≈25 kb in length and packaging the resulting recombinant DNAs with preassembled heads and tails in vitro.

■ In cDNA cloning, expressed mRNAs are reverse-transcribed into complementary DNAs, or cDNAs. By a series of reactions, single-stranded cDNAs are converted into double-stranded DNAs, which can then be ligated into a λ phage vector (see Figure 9-15).

■ A cDNA library is a set of cDNA clones prepared from the mRNAs isolated from a particular type of tissue. A genomic library is a set of clones carrying restriction fragments produced by cleavage of the entire genome.

■ The number of clones in a cDNA or genomic library must be large enough so that all or nearly all of the original nucleotide sequences are present in at least one clone.

■ A particular cloned DNA fragment within a library can be detected by hybridization to a radiolabeled oligonucleotide whose sequence is complementary to a portion of the fragment (see Figures 9-16 and 9-17).

■ Shuttle vectors that replicate in both yeast and *E. coli* can be used to construct a yeast genomic library. Specific genes can be isolated by their ability to complement the corresponding mutant genes in yeast cells (see Figure 9-20).

9.3 Characterizing and Using Cloned DNA Fragments

Now that we have described the basic techniques for using recombinant DNA technology to isolate specific DNA clones, we consider how cloned DNAs are further characterized and various ways in which they can be used. We begin here with several widely used general techniques and examine some more specific applications in the following sections.

Gel Electrophoresis Allows Separation of Vector DNA from Cloned Fragments

In order to manipulate or sequence a cloned DNA fragment, it first must be separated from the vector DNA. This can be accomplished by cutting the recombinant DNA clone with the same restriction enzyme used to produce the recombinant vectors originally. The cloned DNA and vector DNA then are subjected to gel **electrophoresis**, a powerful method for separating DNA molecules of different size.

Near neutral pH, DNA molecules carry a large negative charge and therefore move toward the positive electrode during gel electrophoresis. Because the gel matrix restricts random diffusion of the molecules, molecules of the same length migrate together as a band whose width equals that of the well into which the original DNA mixture was placed at the start of the electrophoretic run. Smaller molecules move through the gel matrix more readily than larger molecules, so that molecules of different length migrate as distinct bands (Figure 9-21). DNA molecules composed of up to ≈2000 nucleotides usually are separated electrophoretically on *polyacrylamide gels*, and molecules from about 200 nucleotides to more than 20 kb on *agarose gels*.

A common method for visualizing separated DNA bands on a gel is to incubate the gel in a solution containing the fluorescent dye ethidium bromide. This planar molecule binds to DNA by intercalating between the base pairs. Binding concentrates ethidium in the DNA and also increases its intrinsic fluorescence. As a result, when the gel is illuminated with ultraviolet light, the regions of the gel containing DNA fluoresce much more brightly than the regions of the gel without DNA.

Once a cloned DNA fragment, especially a long one, has been separated from vector DNA, it often is treated with various restriction enzymes to yield smaller fragments. After separation by gel electrophoresis, all or some of these smaller fragments can be ligated individually into a plasmid vector and cloned in *E. coli* by the usual procedure. This process, known as *subcloning*, is an important step in rearranging parts of genes into useful new configurations. For instance, an investigator who wants to change the conditions under which a gene is expressed might use subcloning to replace the normal promoter associated with a cloned gene with a DNA segment containing a different promoter. Subcloning also can be used to obtain cloned DNA fragments that are of an appropriate length for determining the nucleotide sequence.

DNA restriction fragments

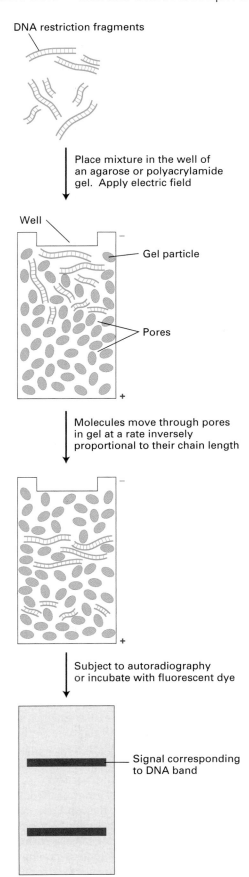

Place mixture in the well of an agarose or polyacrylamide gel. Apply electric field

Well

Gel particle

Pores

Molecules move through pores in gel at a rate inversely proportional to their chain length

Subject to autoradiography or incubate with fluorescent dye

Signal corresponding to DNA band

◄ **EXPERIMENTAL FIGURE 9-21 Gel electrophoresis separates DNA molecules of different lengths.** A gel is prepared by pouring a liquid containing either melted agarose or unpolymerized acrylamide between two glass plates a few millimeters apart. As the agarose solidifies or the acrylamide polymerizes into polyacrylamide, a gel matrix (orange ovals) forms consisting of long, tangled chains of polymers. The dimensions of the interconnecting channels, or pores, depend on the concentration of the agarose or acrylamide used to form the gel. The separated bands can be visualized by autoradiography (if the fragments are radiolabeled) or by addition of a fluorescent dye (e.g., ethidium bromide) that binds to DNA.

Cloned DNA Molecules Are Sequenced Rapidly by the Dideoxy Chain-Termination Method

The complete characterization of any cloned DNA fragment requires determination of its nucleotide sequence. F. Sanger and his colleagues developed the method now most commonly used to determine the exact nucleotide sequence of DNA fragments up to ≈500 nucleotides long. The basic idea behind this method is to synthesize from the DNA fragment to be sequenced a set of daughter strands that are labeled at one end and differ in length by one nucleotide. Separation of the truncated daughter strands by gel electrophoresis can then establish the nucleotide sequence of the original DNA fragment.

Synthesis of truncated daughter stands is accomplished by use of 2′,3′-dideoxyribonucleoside triphosphates (ddNTPs). These molecules, in contrast to normal deoxyribonucleotides (dNTPs), lack a 3′ hydroxyl group (Figure 9-22). Although ddNTPs can be incorporated into a growing DNA chain by

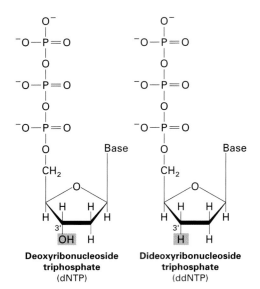

Deoxyribonucleoside triphosphate (dNTP)   **Dideoxyribonucleoside triphosphate** (ddNTP)

▲ **FIGURE 9-22 Structures of deoxyribonucleoside triphosphate (dNTP) and dideoxyribonucleoside triphosphate (ddNTP).** Incorporation of a ddNTP residue into a growing DNA strand terminates elongation at that point.

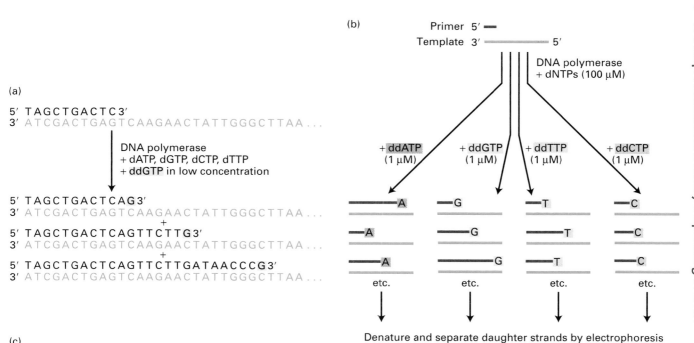

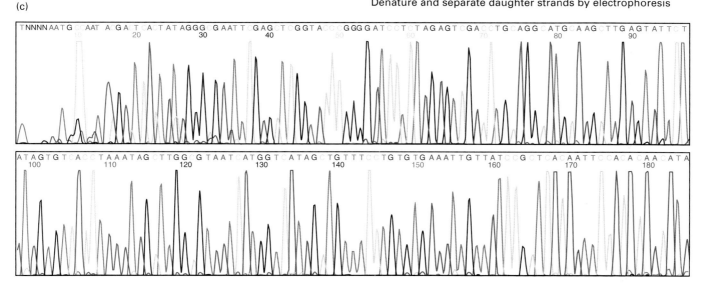

▲ **EXPERIMENTAL FIGURE 9-23 Cloned DNAs can be sequenced by the Sanger method, using fluorescent-tagged dideoxyribonucleoside triphosphates (ddNTPs).** (a) A single (template) strand of the DNA to be sequenced (blue letters) is hybridized to a synthetic deoxyribonucleotide primer (black letters). The primer is elongated in a reaction mixture containing the four normal deoxyribonucleoside triphosphates plus a relatively small amount of one of the four dideoxyribonucleoside triphosphates. In this example, ddGTP (yellow) is present. Because of the relatively low concentration of ddGTP, incorporation of a ddGTP, and thus chain termination, occurs at a given position in the sequence only about 1 percent of the time. Eventually the reaction mixture will contain a mixture of prematurely terminated (truncated) daughter fragments ending at every occurrence of ddGTP. (b) To obtain the complete sequence of a template DNA, four separate reactions are performed, each with a different dideoxyribonucleoside triphosphate (ddNTP). The ddNTP that terminates each truncated fragment can be identified by use of ddNTPs tagged with four different fluorescent dyes (indicated by colored highlights). (c) In an automated sequencing machine, the four reaction mixtures are subjected to gel electrophoresis and the order of appearance of each of the four different fluorescent dyes at the end of the gel is recorded. Shown here is a sample printout from an automated sequencer from which the sequence of the original template DNA can be read directly. N = nucleotide that cannot be assigned. [Part (c) from Griffiths et al., Figure 14-27.]

DNA polymerase, once incorporated they cannot form a phosphodiester bond with the next incoming nucleotide triphosphate. Thus incorporation of a ddNTP terminates chain synthesis, resulting in a truncated daughter strand.

Sequencing using the Sanger *dideoxy chain-termination method* begins by denaturing a double-stranded DNA fragment to generate template strands for in vitro DNA synthesis. A synthetic oligodeoxynucleotide is used as the primer for *four* separate polymerization reactions, each with a low concen-

tration of one of the four ddNTPs in addition to higher concentrations of the normal dNTPs. In each reaction, the ddNTP is randomly incorporated at the positions of the corresponding dNTP, causing termination of polymerization at those positions in the sequence (Figure 9-23a). Inclusion of fluorescent tags of different colors on each of the ddNTPs allows each set of truncated daughter fragments to be distinguished by their corresponding fluorescent label (Figure 9-23b). For example, all truncated fragments that end with a G would fluoresce one color (e.g., yellow), and those ending with an A would fluoresce another color (e.g., red), regardless of their lengths. The mixtures of truncated daughter fragments from each of the four reactions are subjected to electrophoresis on special polyacrylamide gels that can separate single-stranded DNA molecules differing in length by only 1 nucleotide. In automated DNA sequencing machines, a fluorescence detector that can distinguish the four fluorescent tags is located at the end of the gel. The sequence of the original DNA template strand can be determined from the order in which different labeled fragments migrate past the fluorescence detector (Figure 9-23c).

In order to sequence a long continuous region of genomic DNA, researchers often start with a collection of cloned DNA fragments whose sequences overlap. Once the sequence of one of these fragments is determined, oligonucleotides based on that sequence can be chemically synthesized for use as primers in sequencing the adjacent overlapping fragments. In this way, the sequence of a long stretch of DNA is determined incrementally by sequencing of the overlapping cloned DNA fragments that compose it.

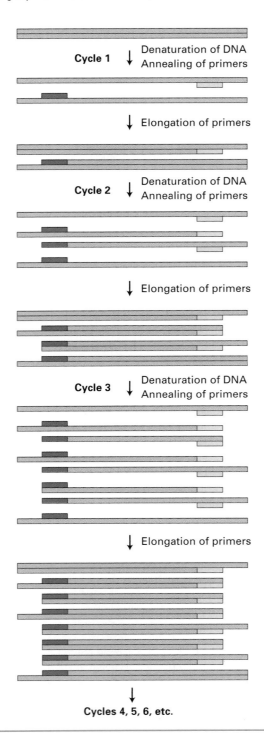

◀ **EXPERIMENTAL FIGURE 9-24 The polymerase chain reaction (PCR) is widely used to amplify DNA regions of known sequences.** To amplify a specific region of DNA, an investigator will chemically synthesize two different oligonucleotide primers complementary to sequences of approximately 18 bases flanking the region of interest (designated as light blue and dark blue bars). The complete reaction is composed of a complex mixture of double-stranded DNA (usually genomic DNA containing the target sequence of interest), a stoichiometric excess of both primers, the four deoxynucleoside triphosphates, and a heat-stable DNA polymerase known as Taq *polymerase.* During each PCR cycle, the reaction mixture is first heated to separate the strands and then cooled to allow the primers to bind to complementary sequences flanking the region to be amplified. *Taq* polymerase then extends each primer from its 3′ end, generating newly synthesized strands that extend in the 3′ direction to the 5′ end of the template strand. During the third cycle, two double-stranded DNA molecules are generated equal in length to the sequence of the region to be amplified. In each successive cycle the target segment, which will anneal to the primers, is duplicated, and will eventually vastly outnumber all other DNA segments in the reaction mixture. Successive PCR cycles can be automated by cycling the reaction for timed intervals at high temperature for DNA melting and at a defined lower temperature for the annealing and elongation portions of the cycle. A reaction that cycles 20 times will amplify the specific target sequence 1-million-fold.

The Polymerase Chain Reaction Amplifies a Specific DNA Sequence from a Complex Mixture

If the nucleotide sequences at the ends of a particular DNA region are known, the intervening fragment can be amplified directly by the **polymerase chain reaction (PCR)**. Here we describe the basic PCR technique and three situations in which it is used.

The PCR depends on the ability to alternately denature (melt) double-stranded DNA molecules and renature (anneal) complementary single strands in a controlled fashion. As in the membrane-hybridization assay described earlier, the presence of noncomplementary strands in a mixture has little effect on the base pairing of complementary single DNA strands or complementary regions of strands. The second requirement for PCR is the ability to synthesize oligonucleotides at least 18–20 nucleotides long with a defined sequence. Such synthetic nucleotides can be readily produced with automated instruments based on the standard reaction scheme shown in Figure 9-18.

As outlined in Figure 9-24, a typical PCR procedure begins by heat-denaturation of a DNA sample into single strands. Next, two synthetic oligonucleotides complementary to the 3′ ends of the target DNA segment of interest are added in great excess to the denatured DNA, and the temperature is lowered to 50–60 °C. These specific oligonucleotides, which are at a very high concentration, will hybridize with their complementary sequences in the DNA sample, whereas the long strands of the sample DNA remain apart because of their low concentration. The hybridized oligonucleotides then serve as primers for DNA chain synthesis in the presence of deoxynucleotides (dNTPs) and a temperature-resistant DNA polymerase such as that from *Thermus aquaticus* (a bacterium that lives in hot springs). This enzyme, called *Taq polymerase*, can remain active even after being heated to 95 °C and can extend the primers at temperatures up to 72 °C. When synthesis is complete, the whole mixture is then heated to 95 °C to melt the newly formed DNA duplexes. After the temperature is lowered again, another cycle of synthesis takes place because excess primer is still present. Repeated cycles of melting (heating) and synthesis (cooling) quickly amplify the sequence of interest. At each cycle, the number of copies of the sequence between the primer sites is doubled; therefore, the desired sequence increases exponentially—about a million-fold after 20 cycles—whereas all other sequences in the original DNA sample remain unamplified.

Direct Isolation of a Specific Segment of Genomic DNA

For organisms in which all or most of the genome has been sequenced, PCR amplification starting with the total genomic DNA often is the easiest way to obtain a specific DNA region of interest for cloning. In this application, the two oligonucleotide primers are designed to hybridize to sequences flanking the genomic region of interest and to include sequences that are recognized by specific restriction enzymes (Figure 9-25). After amplification of the desired

target sequence for about 20 PCR cycles, cleavage with the appropriate restriction enzymes produces sticky ends that allow efficient ligation of the fragment into a plasmid vector cleaved by the same restriction enzymes in the polylinker. The resulting recombinant plasmids, all carrying the identical genomic DNA segment, can then be cloned in

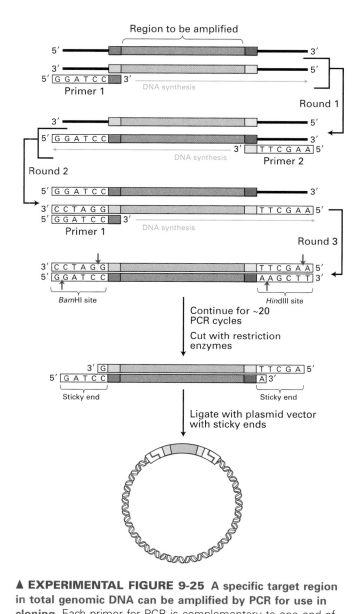

▲ **EXPERIMENTAL FIGURE 9-25 A specific target region in total genomic DNA can be amplified by PCR for use in cloning.** Each primer for PCR is complementary to one end of the target sequence and includes the recognition sequence for a restriction enzyme that does not have a site within the target region. In this example, primer 1 contains a *Bam*HI sequence, whereas primer 2 contains a *Hind*III sequence. (Note that for clarity, in any round, amplification of only one of the two strands is shown, the one in brackets.) After amplification, the target segments are treated with appropriate restriction enzymes, generating fragments with sticky ends. These can be incorporated into complementary plasmid vectors and cloned in *E. coli* by the usual procedure (see Figure 9-13).

E. coli cells. With certain refinements of the PCR, DNA segments >10 kb in length can be amplified and cloned in this way.

Note that this method does not involve cloning of large numbers of restriction fragments derived from genomic DNA and their subsequent screening to identify the specific fragment of interest. In effect, the PCR method inverts this traditional approach and thus avoids its most tedious aspects. The PCR method is useful for isolating gene sequences to be manipulated in a variety of useful ways described later. In addition the PCR method can be used to isolate gene sequences from mutant organisms to determine how they differ from the wild-type.

Preparation of Probes Earlier we discussed how oligonucleotide probes for hybridization assays can be chemically synthesized. Preparation of such probes by PCR amplification requires chemical synthesis of only two relatively short primers corresponding to the two ends of the target sequence. The starting sample for PCR amplification of the target sequence can be a preparation of genomic DNA. Alternatively, if the target sequence corresponds to a mature mRNA sequence, a complete set of cellular cDNAs synthesized from the total cellular mRNA using reverse transcriptase or obtained by pooling cDNA from all the clones in a λ cDNA library can be used as a source of template DNA. To generate a radiolabeled product from PCR, ^{32}P-labeled dNTPs are included during the last several amplification cycles. Because probes prepared by PCR are relatively long and have many radioactive ^{32}P atoms incorporated into them, these probes usually give a stronger and more specific signal than chemically synthesized probes.

Tagging of Genes by Insertion Mutations Another useful application of the PCR is to amplify a "tagged" gene from the genomic DNA of a mutant strain. This approach is a simpler method for identifying genes associated with a particular mutant phenotype than screening of a library by functional complementation (see Figure 9-20).

The key to this use of PCR is the ability to produce mutations by insertion of a known DNA sequence into the genome of an experimental organism. Such insertion mutations can be generated by use of **mobile DNA elements,** which can move (or transpose) from one chromosomal site to another. As discussed in more detail in Chapter 10, these DNA sequences occur naturally in the genomes of most organisms and may give rise to loss-of-function mutations if they transpose into a protein-coding region.

For example, researchers have modified a *Drosophila* mobile DNA element, known as the **P element,** to optimize its use in the experimental generation of insertion mutations. Once it has been demonstrated that insertion of a P element causes a mutation with an interesting phenotype, the genomic sequences adjacent to the insertion site can be amplified by a variation of the standard PCR protocol that uses synthetic primers complementary to the known P-element sequence but that allows unknown neighboring sequences to be amplified. Again, this approach avoids the cloning of large numbers of DNA fragments and their screening to detect a cloned DNA corresponding to a mutated gene of interest.

Similar methods have been applied to other organisms for which insertion mutations can be generated using either mobile DNA elements or viruses with sequenced genomes that can insert randomly into the genome.

Blotting Techniques Permit Detection of Specific DNA Fragments and mRNAs with DNA Probes

Two very sensitive methods for detecting a particular DNA or RNA sequence within a complex mixture combine separation by gel electrophoresis and hybridization with a complementary radiolabeled DNA probe. We will encounter

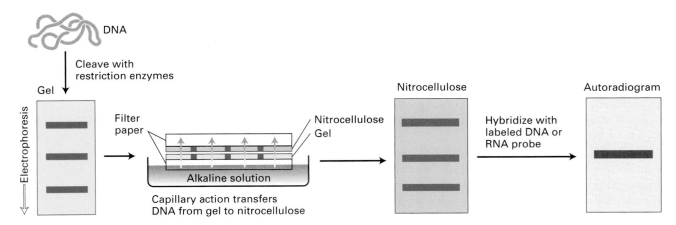

▲ **EXPERIMENTAL FIGURE 9-26 Southern blot technique can detect a specific DNA fragment in a complex mixture of restriction fragments.** The diagram depicts three different restriction fragments in the gel, but the procedure can be applied to a mixture of millions of DNA fragments. Only fragments that hybridize to a labeled probe will give a signal on an autoradiogram. A similar technique called *Northern blotting* detects specific mRNAs within a mixture. [See E. M. Southern, 1975, *J. Mol. Biol.* **98**:508.]

references to both these techniques, which have numerous applications, in other chapters.

Southern Blotting   The first blotting technique to be devised is known as **Southern blotting** after its originator E. M. Southern. This technique is capable of detecting a single specific restriction fragment in the highly complex mixture of fragments produced by cleavage of the entire human genome with a restriction enzyme. In such a complex mixture, many fragments will have the same or nearly the same length and thus migrate together during electrophoresis. Even though all the fragments are not separated completely by gel electrophoresis, an individual fragment within one of the bands can be identified by hybridization to a specific DNA probe. To accomplish this, the restriction fragments present in the gel are denatured with alkali and transferred onto a nitrocellulose filter or nylon membrane by blotting (Figure 9-26). This procedure preserves the distribution of the fragments in the gel, creating a replica of the gel on the filter, much like the replica filter produced from clones in a λ library. (The blot is used because probes do not readily diffuse into the original gel.) The filter then is incubated under hybridization conditions with a specific radiolabeled DNA probe, which usually is generated from a cloned restriction frag-

ment. The DNA restriction fragment that is complementary to the probe hybridizes, and its location on the filter can be revealed by autoradiography.

Northern Blotting   One of the most basic ways to characterize a cloned gene is to determine when and where in an organism the gene is expressed. Expression of a particular gene can be followed by assaying for the corresponding mRNA by **Northern blotting,** named, in a play on words, after the related method of Southern blotting. An RNA sample, often the total cellular RNA, is denatured by treatment with an agent such as formaldehyde that disrupts the hydrogen bonds between base pairs, ensuring that all the RNA molecules have an unfolded, linear conformation. The individual RNAs are separated according to size by gel electrophoresis and transferred to a nitrocellulose filter to which the extended denatured RNAs adhere. As in Southern blotting, the filter then is exposed to a labeled DNA probe that is complementary to the gene of interest; finally, the labeled filter is subjected to autoradiography. Because the amount of a specific RNA in a sample can be estimated from a Northern blot, the procedure is widely used to compare the amounts of a particular mRNA in cells under different conditions (Figure 9-27).

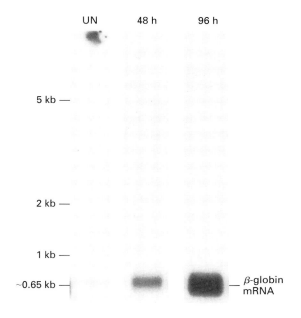

▲ **EXPERIMENTAL FIGURE 9-27 Northern blot analysis reveals increased expression of β-globin mRNA in differentiated erythroleukemia cells.** The total mRNA in extracts of erythroleukemia cells that were growing but uninduced and in cells induced to stop growing and allowed to differentiate for 48 hours or 96 hours was analyzed by Northern blotting for β-globin mRNA. The density of a band is proportional to the amount of mRNA present. The β-globin mRNA is barely detectable in uninduced cells (UN lane) but increases more than 1000-fold by 96 hours after differentiation is induced. [Courtesy of L. Kole.]

E. coli Expression Systems Can Produce Large Quantities of Proteins from Cloned Genes

MEDICINE   Many protein hormones and other signaling or regulatory proteins are normally expressed at very low concentrations, precluding their isolation and purification in large quantities by standard biochemical techniques. Widespread therapeutic use of such proteins, as well as basic research on their structure and functions, depends on efficient procedures for producing them in large amounts at reasonable cost. Recombinant DNA techniques that turn *E. coli* cells into factories for synthesizing low-abundance proteins now are used to commercially produce factor VIII (a blood-clotting factor), granulocyte colony-stimulating factor (G-CSF), insulin, growth hormone, and other human proteins with therapeutic uses. For example, G-CSF stimulates the production of granulocytes, the phagocytic white blood cells critical to defense against bacterial infections. Administration of G-CSF to cancer patients helps offset the reduction in granulocyte production caused by chemotherapeutic agents, thereby protecting patients against serious infection while they are receiving chemotherapy. ∎

The first step in producing large amounts of a low-abundance protein is to obtain a cDNA clone encoding the full-length protein by methods discussed previously. The second step is to engineer plasmid vectors that will express large amounts of the encoded protein when it is inserted into *E. coli* cells. The key to designing such **expression vectors** is

(a)

(b)

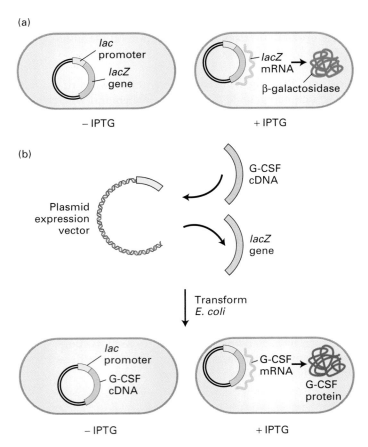

▲ **EXPERIMENTAL FIGURE 9-28 Some eukaryotic proteins can be produced in *E. coli* cells from plasmid vectors containing the *lac* promoter.** (a) The plasmid expression vector contains a fragment of the *E. coli* chromosome containing the *lac* promoter and the neighboring *lacZ* gene. In the presence of the lactose analog IPTG, RNA polymerase normally transcribes the *lacZ* gene, producing *lacZ* mRNA, which is translated into the encoded protein, β-galactosidase. (b) The *lacZ* gene can be cut out of the expression vector with restriction enzymes and replaced by a cloned cDNA, in this case one encoding granulocyte colony-stimulating factor (G-CSF). When the resulting plasmid is transformed into *E. coli* cells, addition of IPTG and subsequent transcription from the *lac* promoter produce G-CSF mRNA, which is translated into G-CSF protein.

inclusion of a **promoter,** a DNA sequence from which transcription of the cDNA can begin. Consider, for example, the relatively simple system for expressing G-CSF shown in Figure 9-28. In this case, G-CSF is expressed in *E. coli* transformed with plasmid vectors that contain the *lac* promoter adjacent to the cloned cDNA encoding G-CSF. Transcription from the *lac* promoter occurs at high rates only when lactose, or a lactose analog such as isopropylthiogalactoside (IPTG), is added to the culture medium. Even larger quantities of a desired protein can be produced in more complicated *E. coli* expression systems.

To aid in purification of a eukaryotic protein produced in an *E. coli* expression system, researchers often modify the cDNA encoding the recombinant protein to facilitate its separation from endogenous *E. coli* proteins. A commonly used modification of this type is to add a short nucleotide sequence to the end of the cDNA, so that the expressed protein will have six histidine residues at the C-terminus. Proteins modified in this way bind tightly to an affinity matrix that contains chelated nickel atoms, whereas most *E. coli* proteins will not bind to such a matrix. The bound proteins can be released from the nickel atoms by decreasing the pH of the surrounding medium. In most cases, this procedure yields a pure recombinant protein that is functional, since addition of short amino acid sequences to either the C-terminus or the N-terminus of a protein usually does not interfere with the protein's biochemical activity.

Plasmid Expression Vectors Can Be Designed for Use in Animal Cells

One disadvantage of bacterial expression systems is that many eukaryotic proteins undergo various modifications (e.g., glycosylation, hydroxylation) after their synthesis on ribosomes (Chapter 3). These post-translational modifications generally are required for a protein's normal cellular function, but they cannot be introduced by *E. coli* cells, which lack the necessary enzymes. To get around this limitation, cloned genes are introduced into cultured animal cells, a process called **transfection.** Two common methods for transfecting animal cells differ in whether the recombinant vector DNA is or is not integrated into the host-cell genomic DNA.

In both methods, cultured animal cells must be treated to facilitate their initial uptake of a recombinant plasmid vector. This can be done by exposing cells to a preparation of lipids that penetrate the plasma membrane, increasing its permeability to DNA. Alternatively, subjecting cells to a brief electric shock of several thousand volts, a technique known as *electroporation,* makes them transiently permeable to DNA. Usually the plasmid DNA is added in sufficient concentration to ensure that a large proportion of the cultured cells will receive at least one copy of the plasmid DNA.

Transient Transfection The simplest of the two expression methods, called *transient transfection,* employs a vector similar to the yeast shuttle vectors described previously. For use in mammalian cells, plasmid vectors are engineered also to carry an origin of replication derived from a virus that infects mammalian cells, a strong promoter recognized by mammalian RNA polymerase, and the cloned cDNA encoding the protein to be expressed adjacent to the promoter (Figure 9-29a). Once such a plasmid vector enters a mammalian cell, the viral origin of replication allows it to replicate efficiently, generating numerous plasmids from which the protein is ex-

(a) Transient transfection

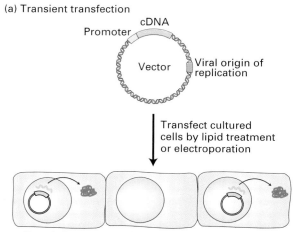

Transfect cultured cells by lipid treatment or electroporation

Protein is expressed from cDNA in plasmid DNA

(b) Stable transfection (transformation)

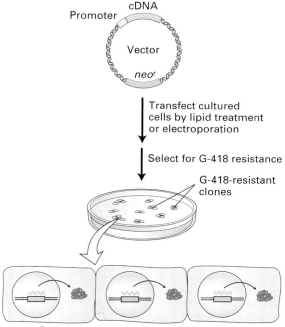

Transfect cultured cells by lipid treatment or electroporation

Select for G-418 resistance

G-418-resistant clones

Protein is expressed from cDNA integrated into host chromosome

◀ **EXPERIMENTAL FIGURE 9-29 Transient and stable transfection with specially designed plasmid vectors permit expression of cloned genes in cultured animal cells.** Both methods employ plasmid vectors that contain the usual elements—ORI, selectable marker (e.g., *amp*r), and polylinker—that permit propagation in *E. coli* and insertion of a cloned cDNA with an adjacent animal promoter. For simplicity, these elements are not depicted. (a) In transient transfection, the plasmid vector contains an origin of replication for a virus that can replicate in the cultured animal cells. Since the vector is not incorporated into the genome of the cultured cells, production of the cDNA-encoded protein continues only for a limited time. (b) In stable transfection, the vector carries a selectable marker such as *neo*r, which confers resistance to G-418. The relatively few transfected animal cells that integrate the exogenous DNA into their genomes are selected on medium containing G-418. These stably transfected, or transformed, cells will continue to produce the cDNA-encoded protein as long as the culture is maintained. See the text for discussion.

lectable marker in order to identify the small fraction of cells that integrate the plasmid DNA. A commonly used selectable marker is the gene for neomycin phosphotransferase (designated *neo*r), which confers resistance to a toxic compound chemically related to neomycin known as G-418. The basic procedure for expressing a cloned cDNA by *stable transfection* is outlined in Figure 9-29b. Only those cells that have integrated the expression vector into the host chromosome will survive and give rise to a clone in the presence of a high concentration of G-418. Because integration occurs at random sites in the genome, individual transformed clones resistant to G-418 will differ in their rates of transcribing the inserted cDNA. Therefore, the stable transfectants usually are screened to identify those that produce the protein of interest at the highest levels.

Epitope Tagging In addition to their use in producing proteins that are modified after translation, eukaryotic expression vectors provide an easy way to study the intracellular localization of eukaryotic proteins. In this method, a cloned cDNA is modified by fusing it to a short DNA sequence encoding an amino acid sequence recognized by a known monoclonal antibody. Such a short peptide that is bound by an antibody is called an **epitope**; hence this method is known as *epitope tagging*. After transfection with a plasmid expression vector containing the fused cDNA, the expressed epitope-tagged form of the protein can be detected by immunofluorescence labeling of the cells with the monoclonal antibody specific for the epitope. Figure 9-30 illustrates the use of this method to localize AP1 adapter proteins, which participate in formation of clathrin-coated vesicles involved in intracellular protein trafficking (Chapter 17). Epitope tagging of a protein so it is detectable with an available monoclonal antibody obviates the time-consuming task of producing a new monoclonal antibody specific for the natural protein.

pressed. However, during cell division such plasmids are not faithfully segregated into both daughter cells and in time a substantial fraction of the cells in a culture will not contain a plasmid, hence the name *transient transfection*.

Stable Transfection (Transformation) If an introduced vector integrates into the genome of the host cell, the genome is permanently altered and the cell is said to be *transformed*. Integration most likely is accomplished by mammalian enzymes that function normally in DNA repair and recombination. Because integration is a rare event, plasmid expression vectors designed to transform animal cells must carry a se-

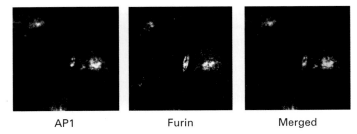

AP1   Furin   Merged

▲ **EXPERIMENTAL FIGURE 9-30 Epitope tagging facilitates cellular localization of proteins expressed from cloned genes.** In this experiment, the cloned cDNA encoding one subunit of the AP1 adapter protein was modified by addition of a sequence encoding an epitope for a known monoclonal antibody. Plasmid expression vectors, similar to those shown in Figure 9-29, were constructed to contain the epitope-tagged AP1 cDNA. After cells were transfected and allowed to express the epitope-tagged version of the AP1 protein, they were fixed and labeled with monoclonal antibody to the epitope and with antibody to furin, a marker protein for the late Golgi and endosomal membranes. Addition of a green fluorescently labeled secondary antibody specific for the anti-epitope antibody visualized the AP1 protein (*left*). Another secondary antibody with a different (red) fluorescent signal was used to visualize furin (*center*). The colocalization of epitope-tagged AP1 and furin to the same intracellular compartment is evident when the two fluorescent signals are merged (*right*). [Courtesy of Ira Mellman, Yale University School of Medicine.]

KEY CONCEPTS OF SECTION 9.3

Characterizing and Using Cloned DNA Fragments

■ Long cloned DNA fragments often are cleaved with restriction enzymes, producing smaller fragments that then are separated by gel electrophoresis and subcloned in plasmid vectors prior to sequencing or experimental manipulation.

■ DNA fragments up to about 500 nucleotides long are most commonly sequenced in automated instruments based on the Sanger (dideoxy chain termination) method (see Figure 9-23).

■ The polymerase chain reaction (PCR) permits exponential amplification of a specific segment of DNA from just a single initial template DNA molecule if the sequence flanking the DNA region to be amplified is known (see Figure 9-24).

■ Southern blotting can detect a single, specific DNA fragment within a complex mixture by combining gel electrophoresis, transfer (blotting) of the separated bands to a filter, and hybridization with a complementary radio-labeled DNA probe (see Figure 9-26). The similar technique of Northern blotting detects a specific RNA within a mixture.

■ Expression vectors derived from plasmids allow the production of abundant amounts of a protein of interest once a cDNA encoding it has been cloned. The unique feature of these vectors is the presence of a promoter fused to the cDNA that allows high-level transcription in host cells.

■ Eukaryotic expression vectors can be used to express cloned genes in yeast or mammalian cells (see Figure 9-29). An important application of these methods is the tagging of proteins with an epitope for antibody detection.

9.4 Genomics: Genome-wide Analysis of Gene Structure and Expression

Using specialized recombinant DNA techniques, researchers have determined vast amounts of DNA sequence including the entire genomic sequence of humans and many key experimental organisms. This enormous volume of data, which is growing at a rapid pace, has been stored and organized in two primary data banks: the GenBank at the National Institutes of Health, Bethesda, Maryland, and the EMBL Sequence Data Base at the European Molecular Biology Laboratory in Heidelberg, Germany. These databases continuously exchange newly reported sequences and make them available to scientists throughout the world on the Internet. In this section, we examine some of the ways researchers use this treasure trove of data to provide insights about gene function and evolutionary relationships, to identify new genes whose encoded proteins have never been isolated, and to determine when and where genes are expressed.

Stored Sequences Suggest Functions of Newly Identified Genes and Proteins

As discussed in Chapter 3, proteins with similar functions often contain similar amino acid sequences that correspond to important functional domains in the three-dimensional structure of the proteins. By comparing the amino acid sequence of the protein encoded by a newly cloned gene with the sequences of proteins of known function, an investigator can look for sequence similarities that provide clues to the function of the encoded protein. Because of the degeneracy in the genetic code, related proteins invariably exhibit more sequence similarity than the genes encoding them. For this reason, protein sequences rather than the corresponding DNA sequences are usually compared.

The computer program used for this purpose is known as BLAST (*b*asic *l*ocal *a*lignment *s*earch *t*ool). The BLAST algorithm divides the new protein sequence (known as the *query sequence*) into shorter segments and then searches the database for significant matches to any of the stored sequences. The matching program assigns a high score to identically matched amino acids and a lower score to matches between amino acids that are related (e.g., hydrophobic, polar, positively charged, negatively charged). When a significant match is found for a segment, the BLAST algorithm will search locally to extend the region of similarity. After searching is completed, the program ranks the matches between the query protein and various known proteins according to their *p-values*. This parameter is a measure of the probability of finding such a degree of similarity between two protein sequences by chance. The lower the *p*-value, the greater the sequence similarity between two sequences. A *p*-value less than about 10^{-3} usually is considered as significant evidence that two proteins share a common ancestor.

 To illustrate the power of this approach, we consider *NF1*, a human gene identified and cloned by methods described later in this chapter. Mutations in *NF1* are associated with the inherited disease neurofibromatosis 1, in which multiple tumors develop in the peripheral nervous system, causing large protuberances in the skin (the "elephant-man" syndrome). After a cDNA clone of *NF1* was isolated and sequenced, the deduced sequence of the NF1 protein was checked against all other protein sequences in GenBank. A region of NF1 protein was discovered to have considerable homology to a portion of the yeast protein called Ira (Figure 9-31). Previous studies had shown that Ira is a GTPase-accelerating protein (GAP) that modulates the GTPase activity of the monomeric G protein called Ras (see Figure 3-E). As we examine in detail in Chapters 14 and 15, GAP and Ras proteins normally function to control cell replication and differentiation in response to signals from neighboring cells. Functional studies on the normal NF1 protein, obtained by expression of the cloned wild-type gene, showed that it did, indeed, regulate Ras activity, as suggested by its homology with Ira. These findings suggest that individuals with neurofibromatosis express a mutant NF1 protein in cells of the peripheral nervous system, leading to inappropriate cell division and formation of the tumors characteristic of the disease. ∎

Even when a protein shows no significant similarity to other proteins with the BLAST algorithm, it may nevertheless share a short sequence with other proteins that is functionally important Such short segments recurring in many different proteins, referred to as **motifs**, generally have similar functions. Several such motifs are described in Chapter 3 (see Figure 3-6). To search for these and other motifs in a new protein, researchers compare the query protein sequence with a database of known motif sequences. Table 9-2 summarizes several of the more commonly occurring motifs.

```
NF1   841  T R A T F M E V L T K I L Q Q G T E F D T L A E T V L A D R F E R L V E L V T M M G D Q G E L P I A  890
Ira  1500  I R I A F L R V F I D I V . . . T N Y P V N P E K H E M D K M L A I D D F L K Y I I K N P I L A F F  1546

      891  M A L A N V V P C S Q W D E L A R V L V T L F D S R H L L Y Q L L W N M F S K E V E L A D S M Q T L  940
     1547  G S L A . . C S P A D V D L Y A G G F L N A F D T R N A S H I L V T E L L K Q E I K R A A R S D D I  1594

      941  F R G N S L A S K I M T F C F K V Y G A T Y L Q K L L D P L L R I V I T S S D W Q H V S F E V D P T  990
     1595  L R R N S C A T R A L S L Y T R S R G N K Y L I K T L R P V L Q G I V D N K E . . . . S F E I D . .  1638

      991  R L E P S E S L E E N Q R N L L Q M T E K F . . . . F H A I I S S S S E F P P Q L R S V C H C L Y Q  1036
     1639  K M K P G . . . S E N S E K M L D L F E K Y M T R L I D A I T S S I D D F P I E L V D I C K T I Y N  1685

     1037  V V S Q R F P Q N S I G A V G S A M F L R F I N P A I V S P Y E A G I L D K K P P P R I E R G L K L  1086
     1686  A A S V N F P E Y A Y I A V G S F V F L R F I G P A L V S P D S E N I I . I V T H A H D R K P F I T  1734

     1087  M S K I L Q S I A N . . . . . . . H V L F T K E E H M R P F N D . . . . F V K S N F D A A R R F F  1124
     1735  L A K V I Q S L A N G R E N I F K K D I L V S K E E F L K T C S D K I F N F L S E L C K I P T N N F  1784

     1125  L D I A S D C P T S D A V N H S L . . . . . . . . . . . . . . S F I S D G N V L A L H R L L W N N .  1159
     1785  T V N V R E D P T P I S F D Y S F L H K F F Y L N E F T I R K E I I N E S K L P G E F S F L K N T V  1834

     1160  . . Q E K I G Q Y L S S N R D H K A V G R R P F . . . . D K M A T L L A Y L G P P E H K P V A  1200
     1835  M L N D K I L G V L G Q P S M E I K N E I P P F V V E N R E K Y P S L Y E F M S R Y A F K K V D  1882
```

▲ **FIGURE 9-31 Comparison of the regions of human NF1 protein and *S. cerevisiae* Ira protein that show significant sequence similarity.** The NF1 and the Ira sequences are shown on the top and bottom lines of each row, respectively, in the one-letter amino acid code (see Figure 2-13). Amino acids that are identical in the two proteins are highlighted in yellow. Amino acids with chemically similar but nonidentical side chains are connected by a blue dot. Amino acid numbers in the protein sequences are shown at the left and right ends of each row. Dots indicate "gaps" in the protein sequence inserted in order to maximize the alignment of homologous amino acids. The BLAST *p*-value for these two sequences is 10^{-28}, indicating a high degree of similarity. [From Xu et al., 1990, *Cell* **62**:599.]

TABLE 9-2	Protein Sequence Motifs	
Name	**Sequence***	**Function**
ATP/GTP binding	[A,G]-X$_4$-G-K-[S,T]	Residues within a nucleotide-binding domain that contact the nucleotide
Prenyl-group binding site	C-Ø-Ø-X (C-terminus)	C-terminal sequence covalently attached to isoprenoid lipids in some lipid-anchored proteins (e.g., Ras)
Zinc finger (C$_2$H$_2$ type)	C-X$_{2-4}$-C-X$_3$-Ø-X$_8$-H-X$_{3-5}$-H	Zn^{2+}-binding sequence within DNA- or RNA-binding domain of some proteins
DEAD box	Ø$_2$-D-E-A-D-[R,K,E,N]-Ø	Sequence present in many ATP-dependent RNA helicases
Heptad repeat	(Ø-X$_2$-Ø-X$_3$)$_n$	Repeated sequence in proteins that form coiled-coil structures

*Single-letter amino acid abbreviations used for sequences (see Figure 2-13). X = any residue; Ø = hydrophobic residue. Brackets enclose alternative permissible residues.

Comparison of Related Sequences from Different Species Can Give Clues to Evolutionary Relationships Among Proteins

BLAST searches for related protein sequences may reveal that proteins belong to a **protein family**. (The corresponding genes constitute a **gene family**.) Protein families are thought to arise by two different evolutionary processes, *gene duplication* and *speciation*, discussed in Chapter 10. Consider, for example, the tubulin family of proteins, which constitute the basic subunits of microtubules. According to the simplified scheme in Figure 9-32a, the earliest eukaryotic cells are thought to have contained a single tubulin gene that was duplicated early in evolution; subsequent divergence of the different copies of the

(a)

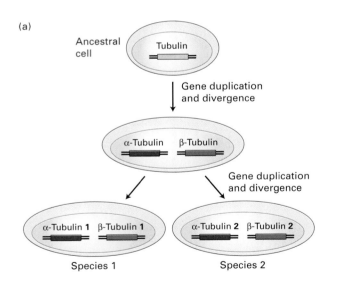

(b)

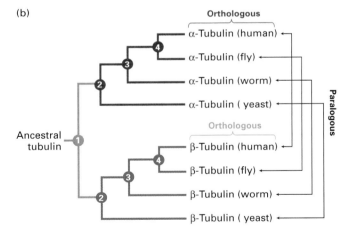

▲ **FIGURE 9-32 The generation of diverse tubulin sequences during the evolution of eukaryotes.** (a) Probable mechanism giving rise to the tubulin genes found in existing species. It is possible to deduce that a gene duplication event occurred before speciation because the α-tubulin sequences from different species (e.g., humans and yeast) are more alike than are the α-tubulin and β-tubulin sequences within a species. (b) A phylogenetic tree representing the relationship between the tubulin sequences. The branch points (nodes), indicated by small numbers, represent common ancestral genes at the time that two sequences diverged. For example, node 1 represents the duplication event that gave rise to the α-tubulin and β-tubulin families, and node 2 represents the divergence of yeast from multicellular species. Braces and arrows indicate, respectively, the orthologous tubulin genes, which differ as a result of speciation, and the paralogous genes, which differ as a result of gene duplication. This diagram is simplified somewhat because each of the species represented actually contains multiple α-tubulin and β-tubulin genes that arose from later gene duplication events.

original tubulin gene formed the ancestral versions of the α- and β-tubulin genes. As different species diverged from these early eukaryotic cells, each of these gene sequences further diverged, giving rise to the slightly different forms of α-tubulin and β-tubulin now found in each species.

All the different members of the tubulin family are sufficiently similar in sequence to suggest a common ancestral sequence. Thus all these sequences are considered to be *homologous*. More specifically, sequences that presumably diverged as a result of gene duplication (e.g., the α- and β-tubulin sequences) are described as *paralogous*. Sequences that arose because of speciation (e.g., the α-tubulin genes in different species) are described as *orthologous*. From the degree of sequence relatedness of the tubulins present in different organisms today, evolutionary relationships can be deduced, as illustrated in Figure 9-32b. Of the three types of sequence relationships, orthologous sequences are the most likely to share the same function.

Genes Can Be Identified Within Genomic DNA Sequences

The complete genomic sequence of an organism contains within it the information needed to deduce the sequence of every protein made by the cells of that organism. For organisms such as bacteria and yeast, whose genomes have few introns and short intergenic regions, most protein-coding

sequences can be found simply by scanning the genomic sequence for **open reading frames** (**ORFs**) of significant length. An ORF usually is defined as a stretch of DNA containing at least 100 codons that begins with a start codon and ends with a stop codon. Because the probability that a random DNA sequence will contain no stop codons for 100 codons in a row is very small, most ORFs encode a protein.

ORF analysis correctly identifies more than 90 percent of the genes in yeast and bacteria. Some of the very shortest genes are missed by this method, and occasionally long open reading frames that are not actually genes arise by chance. Both types of miss assignments can be corrected by more sophisticated analysis of the sequence and by genetic tests for gene function. Of the *Saccharomyces* genes identified in this manner, about half were already known by some functional criterion such as mutant phenotype. The functions of some of the proteins encoded by the remaining putative genes identified by ORF analysis have been assigned based on their sequence similarity to known proteins in other organisms.

Identification of genes in organisms with a more complex genome structure requires more sophisticated algorithms than searching for open reading frames. Figure 9-33 shows a comparison of the genes identified in a representative 50-kb segment from the genomes of yeast, *Drosophila*, and humans. Because most genes in higher eukaryotes, including humans and *Drosophila*, are composed of multiple, relatively short coding regions (**exons**) separated by noncoding

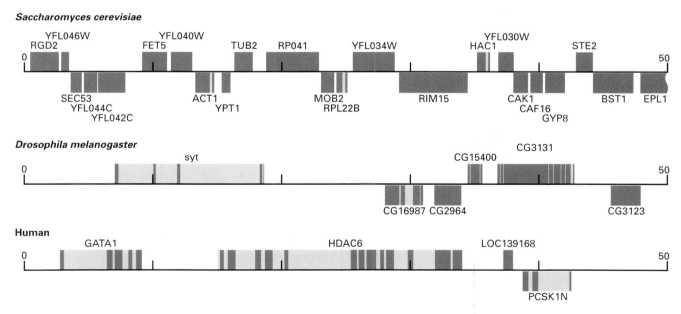

▲ **FIGURE 9-33 Arrangement of gene sequences in representative 50-kb segments of yeast, fruit fly, and human genomes.** Genes above the line are transcribed to the right; genes below the line are transcribed to the left. Blue blocks represent exons (coding sequences); green blocks represent introns (noncoding sequences). Because yeast genes contain few if any introns, scanning genomic sequences for open reading frames (ORFs) correctly identifies most gene sequences. In

contrast, the genes of higher eukaryotes typically comprise multiple exons separated by introns. ORF analysis is not effective in identifying genes in these organisms. Likely gene sequences for which no functional data are available are designated by numerical names: in yeast, these begin with Y; in *Drosophila*, with CG; and in humans, with LOC. The other genes shown here encode proteins with known functions.

regions (**introns**), scanning for ORFs is a poor method for finding genes. The best gene-finding algorithms combine all the available data that might suggest the presence of a gene at a particular genomic site. Relevant data include alignment or hybridization to a full-length cDNA; alignment to a partial cDNA sequence, generally 200–400 bp in length, known as an *expressed sequence tag (EST)*; fitting to models for exon, intron, and splice site sequences; and sequence similarity to other organisms. Using these methods computational biologists have identified approximately 35,000 genes in the human genome, although for as many as 10,000 of these putative genes there is not yet conclusive evidence that they actually encode proteins or RNAs.

A particularly powerful method for identifying human genes is to compare the human genomic sequence with that of the mouse. Humans and mice are sufficiently related to have most genes in common; however, largely nonfunctional DNA sequences, such as intergenic regions and introns, will tend to be very different because they are not under strong selective pressure. Thus corresponding segments of the human and mouse genome that exhibit high sequence similarity are likely to be functional coding regions (i.e., exons).

The Size of an Organism's Genome Is Not Directly Related to Its Biological Complexity

The combination of genomic sequencing and gene-finding computer algorithms has yielded the complete inventory of protein-coding genes for a variety of organisms. Figure 9-34 shows the total number of protein-coding genes in several eukaryotic genomes that have been completely sequenced. The functions of about half the proteins encoded in these genomes are known or have been predicted on the basis of sequence comparisons. One of the surprising features of this comparison is that the number of protein-coding genes within different organisms does not seem proportional to our intuitive sense of their biological complexity. For example, the roundworm *C. elegans* apparently has more genes than the fruit fly *Drosophila*, which has a much more complex body plan and more complex behavior. And humans have

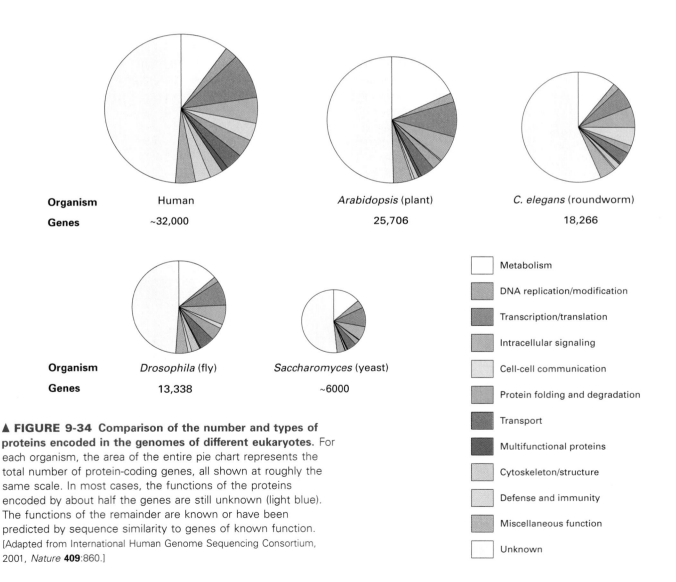

Organism	Human	*Arabidopsis* (plant)	*C. elegans* (roundworm)
Genes	~32,000	25,706	18,266

Organism	*Drosophila* (fly)	*Saccharomyces* (yeast)
Genes	13,338	~6000

- Metabolism
- DNA replication/modification
- Transcription/translation
- Intracellular signaling
- Cell-cell communication
- Protein folding and degradation
- Transport
- Multifunctional proteins
- Cytoskeleton/structure
- Defense and immunity
- Miscellaneous function
- Unknown

▲ **FIGURE 9-34 Comparison of the number and types of proteins encoded in the genomes of different eukaryotes.** For each organism, the area of the entire pie chart represents the total number of protein-coding genes, all shown at roughly the same scale. In most cases, the functions of the proteins encoded by about half the genes are still unknown (light blue). The functions of the remainder are known or have been predicted by sequence similarity to genes of known function. [Adapted from International Human Genome Sequencing Consortium, 2001, *Nature* **409**:860.]

fewer than twice the number of genes as *C. elegans*, which seems completely inexplicable given the enormous differences between these organisms.

Clearly, simple quantitative differences in the genomes of different organisms are inadequate for explaining differences in biological complexity. However, several phenomena can generate more complexity in the expressed proteins of higher eukaryotes than is predicted from their genomes. First, alternative splicing of a pre-mRNA can yield multiple functional mRNAs corresponding to a particular gene (Chapter 12). Second, variations in the post-translational modification of some proteins may produce functional differences. Finally, qualitative differences in the interactions between proteins and their integration into pathways may contribute significantly to the differences in biological complexity among organisms. The specific functions of many genes and proteins identified by analysis of genomic sequences still have not been determined. As researchers unravel the functions of individual proteins in different organisms and further detail their interactions, a more sophisticated understanding of the genetic basis of complex biological systems will emerge.

DNA Microarrays Can Be Used to Evaluate the Expression of Many Genes at One Time

Monitoring the expression of thousands of genes simultaneously is possible with **DNA microarray** analysis. A DNA microarray consists of thousands of individual, closely packed gene-specific sequences attached to the surface of a glass microscopic slide. By coupling microarray analysis with the results from genome sequencing projects, researchers can analyze the global patterns of gene expression of an organism during specific physiological responses or developmental processes.

Preparation of DNA Microarrays In one method for preparing microarrays, a ≈1-kb portion of the coding region of each gene analyzed is individually amplified by PCR. A robotic device is used to apply each amplified DNA sample to the surface of a glass microscope slide, which then is chemically processed to permanently attach the DNA sequences to the glass surface and to denature them. A typical array might contain ≈6000 spots of DNA in a 2 × 2 cm grid.

In an alternative method, multiple DNA oligonucleotides, usually at least 20 nucleotides in length, are synthesized from an initial nucleotide that is covalently bound to the surface of a glass slide. The synthesis of an oligonucleotide of specific sequence can be programmed in a small region on the surface of the slide. Several oligonucleotide sequences from a single gene are thus synthesized in neighboring regions of the slide to analyze expression of that gene. With this method, oligonucleotides representing thousands of genes can be produced on a single glass slide. Because the methods for constructing these arrays of synthetic oligonucleotides were adapted from methods for manufacturing microscopic integrated circuits used in computers, these types of oligonucleotide microarrays are often called *DNA chips.*

Effect of Carbon Source on Gene Expression in Yeast The initial step in a microarray expression study is to prepare fluorescently labeled cDNAs corresponding to the mRNAs expressed by the cells under study. When the cDNA preparation is applied to a microarray, spots representing genes

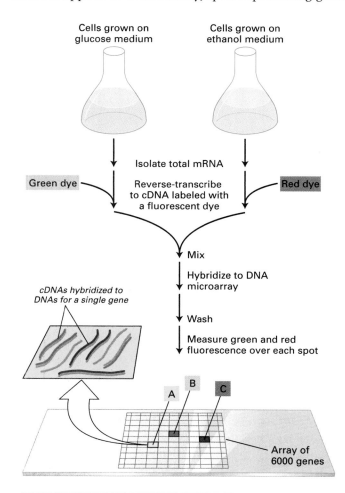

A If a spot is yellow, expression of that gene is the same in cells grown either on glucose or ethanol

B If a spot is green, expression of that gene is greater in cells grown in glucose

C If a spot is red, expression of that gene is greater in cells grown in ethanol

▲ **EXPERIMENTAL FIGURE 9-35 DNA microarray analysis can reveal differences in gene expression in yeast cells under different experimental conditions.** In this example, cDNA prepared from mRNA isolated from wild-type *Saccharomyces* cells grown on glucose or ethanol is labeled with different fluorescent dyes. A microarray composed of DNA spots representing each yeast gene is exposed to an equal mixture of the two cDNA preparations under hybridization conditions. The ratio of the intensities of red and green fluorescence over each spot, detected with a scanning confocal laser microscope, indicates the relative expression of each gene in cells grown on each of the carbon sources. Microarray analysis also is useful for detecting differences in gene expression between wild-type and mutant strains.

that are expressed will hybridize under appropriate conditions to their complementary cDNAs and can subsequently be detected in a scanning laser microscope.

Figure 9-35 depicts how this method can be applied to compare gene expression in yeast cells growing on glucose versus ethanol as the source of carbon and energy. In this type of experiment, the separate cDNA preparations from glucose-grown and ethanol-grown cells are labeled with differently colored fluorescent dyes. A DNA array comprising all 6000 genes then is incubated with a mixture containing equal amounts of the two cDNA preparations under hybridization conditions. After unhybridized cDNA is washed away, the intensity of green and red fluorescence at each DNA spot is measured using a fluorescence microscope and stored in computer files under the name of each gene according to its known position on the slide. The relative intensities of red and green fluorescence signals at each spot are a measure of the relative level of expression of that gene in cells grown in glucose or ethanol. Genes that are not transcribed under these growth conditions give no detectable signal.

Hybridization of fluorescently labeled cDNA preparations to DNA microarrays provides a means for analyzing gene expression patterns on a genomic scale. This type of analysis has shown that as yeast cells shift from growth on glucose to growth on ethanol, expression of 710 genes increases by a factor of two or more, while expression of 1030 genes decreases by a factor of two or more. Although about 400 of the differentially expressed genes have no known function, these results provide the first clue as to their possible function in yeast biology.

Cluster Analysis of Multiple Expression Experiments Identifies Co-regulated Genes

Firm conclusions rarely can be drawn from a single microarray experiment about whether genes that exhibit similar changes in expression are co-regulated and hence likely to be closely related functionally. For example, many of the observed differences in gene expression just described in yeast growing on glucose or ethanol could be indirect consequences of the many different changes in cell physiology that occur when cells are transferred from one medium to another. In other words, genes that appear to be co-regulated in a single microarray expression experiment may undergo changes in expression for very different reasons and may actually have very different biological functions. A solution to this problem is to combine the information from a set of expression array experiments to find genes that are similarly regulated under a variety of conditions or over a period of time.

This more informative use of multiple expression array experiments is illustrated by the changes in gene expression observed after starved human fibroblasts are transferred to a rich, serum-containing, growth medium. In one study, the relative expression of 8600 genes was determined at different

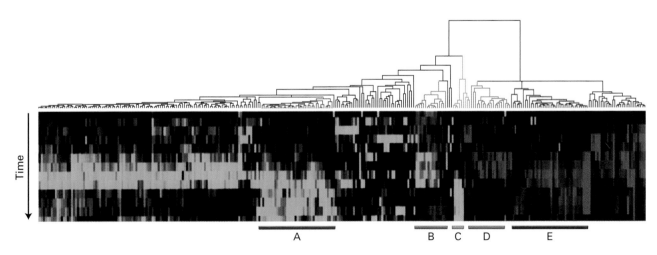

▲ **EXPERIMENTAL FIGURE 9-36 Cluster analysis of data from multiple microarray expression experiments can identify co-regulated genes.** In this experiment, the expression of 8600 mammalian genes was detected by microarray analysis at time intervals over a 24-hour period after starved fibroblasts were provided with serum. The cluster diagram shown here is based on a computer algorithm that groups genes showing similar changes in expression compared with a starved control sample over time. Each column of colored boxes represents a single gene, and each row represents a time point. A red box indicates an increase in expression relative to the control; a green box, a decrease in expression; and a black box, no significant change in expression. The "tree" diagram at the top shows how the expression patterns for individual genes can be organized in a hierarchical fashion to group together the genes with the greatest similarity in their patterns of expression over time. Five clusters of coordinately regulated genes were identified in this experiment, as indicated by the bars at the bottom. Each cluster contains multiple genes whose encoded proteins function in a particular cellular process: cholesterol biosynthesis (A), the cell cycle (B), the immediate-early response (C), signaling and angiogenesis (D), and wound healing and tissue remodeling (E). [Courtesy of Michael B. Eisen, Lawrence Berkeley National Laboratory.]

times after serum addition, generating more than 10^4 individual pieces of data. A computer program, related to the one used to determine the relatedness of different protein sequences, can organize these data and cluster genes that show similar expression over the time course after serum addition. Remarkably, such *cluster analysis* groups sets of genes whose encoded proteins participate in a common cellular process, such as cholesterol biosynthesis or the cell cycle (Figure 9-36).

Since genes with identical or similar patterns of regulation generally encode functionally related proteins, cluster analysis of multiple microarray expression experiments is another tool for deducing the functions of newly identified genes. This approach allows any number of different experiments to be combined. Each new experiment will refine the analysis, with smaller and smaller cohorts of genes being identified as belonging to different clusters.

KEY CONCEPTS OF SECTION 9.4

Genomics: Genome-wide Analysis of Gene Structure and Expression

■ The function of a protein that has not been isolated often can be predicted on the basis of similarity of its amino acid sequence to proteins of known function.

■ A computer algorithm known as BLAST rapidly searches databases of known protein sequences to find those with significant similarity to a new (query) protein.

■ Proteins with common functional motifs may not be identified in a typical BLAST search. These short sequences may be located by searches of motif databases.

■ A protein family comprises multiple proteins all derived from the same ancestral protein. The genes encoding these proteins, which constitute the corresponding gene family, arose by an initial gene duplication event and subsequent divergence during speciation (see Figure 9-32).

■ Related genes and their encoded proteins that derive from a gene duplication event are paralogous; those that derive from speciation are orthologous. Proteins that are orthologous usually have a similar function.

■ Open reading frames (ORFs) are regions of genomic DNA containing at least 100 codons located between a start codon and stop codon.

■ Computer search of the entire bacterial and yeast genomic sequences for open reading frames (ORFs) correctly identifies most protein-coding genes. Several types of additional data must be used to identify probable genes in the genomic sequences of humans and other higher eukaryotes because of the more complex gene structure in these organisms.

■ Analysis of the complete genome sequences for several different organisms indicates that biological complexity is not directly related to the number of protein-coding genes (see Figure 9-34).

■ DNA microarray analysis simultaneously detects the relative level of expression of thousands of genes in different types of cells or in the same cells under different conditions (see Figure 9-35).

■ Cluster analysis of the data from multiple microarray expression experiments can identify genes that are similarly regulated under various conditions. Such co-regulated genes commonly encode proteins that have biologically related functions.

9.5 Inactivating the Function of Specific Genes in Eukaryotes

The elucidation of DNA and protein sequences in recent years has led to identification of many genes, using sequence patterns in genomic DNA and the sequence similarity of the encoded proteins with proteins of known function. As discussed in the previous section, the general functions of proteins identified by sequence searches may be predicted by analogy with known proteins. However, the precise in vivo roles of such "new" proteins may be unclear in the absence of mutant forms of the corresponding genes. In this section, we describe several ways for disrupting the normal function of a specific gene in the genome of an organism. Analysis of the resulting mutant phenotype often helps reveal the in vivo function of the normal gene and its encoded protein.

Three basic approaches underlie these gene-inactivation techniques: (1) replacing a normal gene with other sequences; (2) introducing an allele whose encoded protein inhibits functioning of the expressed normal protein; and (3) promoting destruction of the mRNA expressed from a gene. The normal endogenous gene is modified in techniques based on the first approach but is not modified in the other approaches.

Normal Yeast Genes Can Be Replaced with Mutant Alleles by Homologous Recombination

Modifying the genome of the yeast *Saccharomyces* is particularly easy for two reasons: yeast cells readily take up exogenous DNA under certain conditions, and the introduced DNA is efficiently exchanged for the homologous chromosomal site in the recipient cell. This specific, targeted **recombination** of identical stretches of DNA allows any gene in yeast chromosomes to be replaced with a mutant allele. (As we discuss in Section 9.6, recombination between homologous chromosomes also occurs naturally during meiosis.)

In one popular method for disrupting yeast genes in this fashion, PCR is used to generate a *disruption construct* containing a selectable marker that subsequently is transfected into yeast cells. As shown in Figure 9-37a, primers for PCR amplification of the selectable marker are designed to include about 20 nucleotides identical with sequences flanking the yeast gene to be replaced. The resulting amplified construct comprises the selectable marker (e.g., the *kanMX* gene,

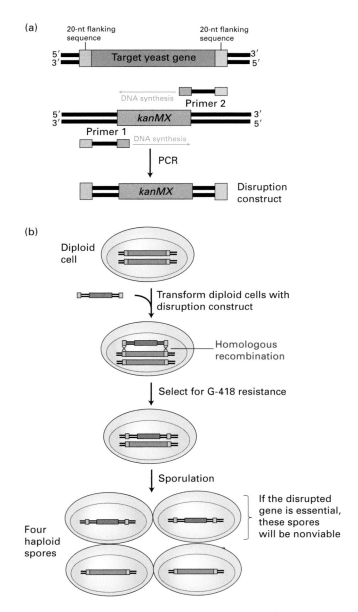

▲ **EXPERIMENTAL FIGURE 9-37** Homologous recombination with transfected disruption constructs can inactivate specific target genes in yeast. (a) A suitable construct for disrupting a target gene can be prepared by the PCR. The two primers designed for this purpose each contain a sequence of about 20 nucleotides (nt) that is homologous to one end of the target yeast gene as well as sequences needed to amplify a segment of DNA carrying a selectable marker gene such as *kanMX*, which confers resistance to G-418. (b) When recipient diploid *Saccharomyces* cells are transformed with the gene disruption construct, homologous recombination between the ends of the construct and the corresponding chromosomal sequences will integrate the *kanMX* gene into the chromosome, replacing the target gene sequence. The recombinant diploid cells will grow on a medium containing G-418, whereas nontransformed cells will not. If the target gene is essential for viability, half the haploid spores that form after sporulation of recombinant diploid cells will be nonviable.

which like *neo*[r] confers resistance to G-418) flanked by about 20 base pairs that match the ends of the target yeast gene. Transformed diploid yeast cells in which one of the two copies of the target endogenous gene has been replaced by the disruption construct are identified by their resistance to G-418 or other selectable phenotype. These heterozygous diploid yeast cells generally grow normally regardless of the function of the target gene, but half the haploid spores derived from these cells will carry only the disrupted allele (Figure 9-37b). If a gene is essential for viability, then spores carrying a disrupted allele will not survive.

Disruption of yeast genes by this method is proving particularly useful in assessing the role of proteins identified by ORF analysis of the entire genomic DNA sequence. A large consortium of scientists has replaced each of the approximately 6000 genes identified by ORF analysis with the *kanMX* disruption construct and determined which gene disruptions lead to nonviable haploid spores. These analyses have shown that about 4500 of the 6000 yeast genes are not required for viability, an unexpectedly large number of apparently nonessential genes. In some cases, disruption of a particular gene may give rise to subtle defects that do not compromise the viability of yeast cells growing under laboratory conditions. Alternatively, cells carrying a disrupted gene may be viable because of operation of backup or compensatory pathways. To investigate this possibility, yeast geneticists currently are searching for synthetic lethal mutations that might reveal nonessential genes with redundant functions (see Figure 9-9c).

Transcription of Genes Ligated to a Regulated Promoter Can Be Controlled Experimentally

Although disruption of an essential gene required for cell growth will yield nonviable spores, this method provides little information about what the encoded protein actually does in cells. To learn more about how a specific gene contributes to cell growth and viability, investigators must be able to selectively inactivate the gene in a population of growing cells. One method for doing this employs a regulated promoter to selectively shut off transcription of an essential gene.

A useful promoter for this purpose is the yeast *GAL1* promoter, which is active in cells grown on galactose but completely inactive in cells grown on glucose. In this approach, the coding sequence of an essential gene (*X*) ligated to the *GAL1* promoter is inserted into a yeast shuttle vector (see Figure 9-19a). The recombinant vector then is introduced into haploid yeast cells in which gene *X* has been disrupted. Haploid cells that are transformed will grow on galactose medium, since the normal copy of gene *X* on the vector is expressed in the presence of galactose. When the cells are transferred to a glucose-containing medium, gene *X* no longer is transcribed; as the cells divide, the amount of the encoded protein X gradually declines, eventually reaching a state of depletion that mimics a complete loss-of-function mutation. The observed changes in the phenotype of these cells after the shift to glucose medium may suggest which cell processes depend on the protein encoded by the essential gene *X*.

In an early application of this method, researchers explored the function of cytosolic *Hsc70* genes in yeast. Haploid cells with a disruption in all four redundant *Hsc70* genes were nonviable, unless the cells carried a vector containing a copy of the *Hsc70* gene that could be expressed from the *GAL1* promoter on galactose medium. On transfer to glucose, the vector-carrying cells eventually stopped growing because of insufficient Hsc70 activity. Careful examination of these dying cells revealed that their secretory proteins could no longer enter the endoplasmic reticulum (ER). This study provided the first evidence for the unexpected role of Hsc70 protein in translocation of secretory proteins into the ER, a process examined in detail in Chapter 16.

Specific Genes Can Be Permanently Inactivated in the Germ Line of Mice

Many of the methods for disrupting genes in yeast can be applied to genes of higher eukaryotes. These genes can be introduced into the **germ line** via homologous recombination to produce animals with a **gene knockout,** or simply "knockout." Knockout mice in which a specific gene is disrupted are a powerful experimental system for studying mammalian development, behavior, and physiology. They also are useful in studying the molecular basis of certain human genetic diseases.

Gene-targeted knockout mice are generated by a two-stage procedure. In the first stage, a DNA construct contain-

ing a disrupted allele of a particular target gene is introduced into *embryonic stem (ES) cells.* These cells, which are derived from the blastocyst, can be grown in culture through many generations (see Figure 22-3). In a small fraction of transfected cells, the introduced DNA undergoes homologous recombination with the target gene, although recombination at nonhomologous chromosomal sites occurs much more frequently. To select for cells in which homologous gene-targeted insertion occurs, the recombinant DNA construct introduced into ES cells needs to include two selectable marker genes (Figure 9-38). One of these genes (*neo*^{r}), which

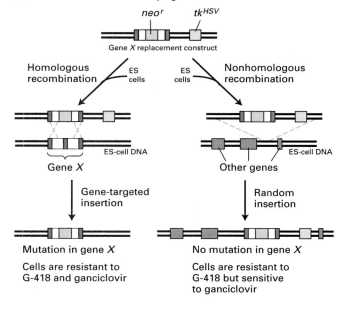

(a) Formation of ES cells carrying a knockout mutation

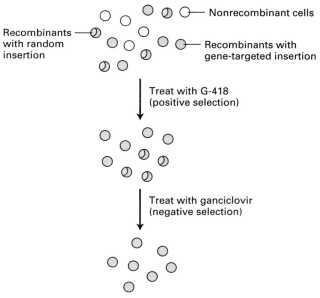

(b) Positive and negative selection of recombinant ES cells

▶ **EXPERIMENTAL FIGURE 9-38 Isolation of mouse ES cells with a gene-targeted disruption is the first stage in production of knockout mice.** (a) When exogenous DNA is introduced into embryonic stem (ES) cells, random insertion via nonhomologous recombination occurs much more frequently than gene-targeted insertion via homologous recombination. Recombinant cells in which one allele of gene *X* (orange and white) is disrupted can be obtained by using a recombinant vector that carries gene *X* disrupted with *neo*^{r} (green), which confers resistance to G-418, and, outside the region of homology, *tk*^{HSV} (yellow), the thymidine kinase gene from herpes simplex virus. The viral thymidine kinase, unlike the endogenous mouse enzyme, can convert the nucleotide analog ganciclovir into the monophosphate form; this is then modified to the triphosphate form, which inhibits cellular DNA replication in ES cells. Thus ganciclovir is cytotoxic for recombinant ES cells carrying the *tk*^{HSV} gene. Nonhomologous insertion includes the *tk*^{HSV} gene, whereas homologous insertion does not; therefore, only cells with nonhomologous insertion are sensitive to ganciclovir. (b) Recombinant cells are selected by treatment with G-418, since cells that fail to pick up DNA or integrate it into their genome are sensitive to this cytotoxic compound. The surviving recombinant cells are treated with ganciclovir. Only cells with a targeted disruption in gene *X*, and therefore lacking the *tk*^{HSV} gene, will survive. [See S. L. Mansour et al., 1988, *Nature* **336**:348.]

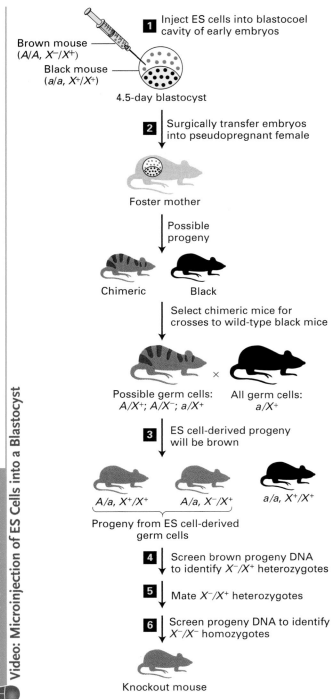

◄ **EXPERIMENTAL FIGURE 9-39 ES cells heterozygous for a disrupted gene are used to produce gene-targeted knockout mice.** Step **1**: Embryonic stem (ES) cells heterozygous for a knockout mutation in a gene of interest (X) and homozygous for a dominant allele of a marker gene (here, brown coat color, A) are transplanted into the blastocoel cavity of 4.5-day embryos that are homozygous for a recessive allele of the marker (here, black coat color, a). Step **2**: The early embryos then are implanted into a pseudopregnant female. Those progeny containing ES-derived cells are chimeras, indicated by their mixed black and brown coats. Step **3**: Chimeric mice then are backcrossed to black mice; brown progeny from this mating have ES-derived cells in their germ line. Steps **4**–**6**: Analysis of DNA isolated from a small amount of tail tissue can identify brown mice heterozygous for the knockout allele. Intercrossing of these mice produces some individuals homozygous for the disrupted allele, that is, knockout mice. [Adapted from M. R. Capecchi, 1989, *Trends Genet.* **5**:70.]

In the second stage in production of knockout mice, ES cells heterozygous for a knockout mutation in gene X are injected into a recipient wild-type mouse blastocyst, which subsequently is transferred into a surrogate pseudopregnant female mouse (Figure 9-39). The resulting progeny will be **chimeras,** containing tissues derived from both the transplanted ES cells and the host cells. If the ES cells also are homozygous for a visible marker trait (e.g., coat color), then chimeric progeny in which the ES cells survived and proliferated can be identified easily. Chimeric mice are then mated with mice homozygous for another allele of the marker trait to determine if the knockout mutation is incorporated into the germ line. Finally, mating of mice, each heterozygous for the knockout allele, will produce progeny homozygous for the knockout mutation.

Development of knockout mice that mimic certain human diseases can be illustrated by cystic fibrosis. By methods discussed in Section 9.6, the recessive mutation that causes this disease eventually was shown to be located in a gene known as *CFTR*, which encodes a chloride channel. Using the cloned wild-type human *CFTR* gene, researchers isolated the homologous mouse gene and subsequently introduced mutations in it. The gene-knockout technique was then used to produce homozygous mutant mice, which showed symptoms (i.e., a phenotype), including disturbances to the functioning of epithelial cells, similar to those of humans with cystic fibrosis. These knockout mice are currently being used as a model system for studying this genetic disease and developing effective therapies. ∎

Somatic Cell Recombination Can Inactivate Genes in Specific Tissues

Investigators often are interested in examining the effects of knockout mutations in a particular tissue of the mouse, at a specific stage in development, or both. However, mice car-

confers G-418 resistance, is inserted within the target gene (X), thereby disrupting it. The other selectable gene, the thymidine kinase gene from herpes simplex virus (tk^{HSV}), confers sensitivity to ganciclovir, a cytotoxic nucleotide analog; it is inserted into the construct outside the target-gene sequence. Only ES cells that undergo homologous recombination can survive in the presence of both G-418 and ganciclovir. In these cells one allele of gene X will be disrupted.

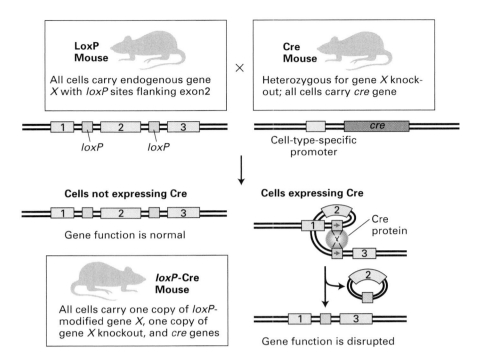

▲ **EXPERIMENTAL FIGURE 9-40** The *loxP-Cre* recombination system can knock out genes in specific cell types. Two *loxP* sites are inserted on each side of an essential exon (2) of the target gene *X* (blue) by homologous recombination, producing a *loxP* mouse. Since the *loxP* sites are in introns, they do not disrupt the function of *X*. The Cre mouse carries one gene *X* knockout allele and an introduced *cre* gene (orange) from bacteriophage P1 linked to a cell-type-specific promoter (yellow). The *cre* gene is incorporated into the mouse genome by nonhomologous recombination and does not affect the function of other genes. In the *loxP*-Cre mice that result from crossing, Cre protein is produced only in those cells in which the promoter is active. Thus these are the only cells in which recombination between the *loxP* sites catalyzed by Cre occurs, leading to deletion of exon 2. Since the other allele is a constitutive gene *X* knockout, deletion between the *loxP* sites results in complete loss of function of gene *X* in all cells expressing Cre. By using different promoters, researchers can study the effects of knocking out gene *X* in various types of cells.

rying a germ-line knockout may have defects in numerous tissues or die before the developmental stage of interest. To address this problem, mouse geneticists have devised a clever technique to inactivate target genes in specific types of somatic cells or at particular times during development.

This technique employs site-specific DNA recombination sites (called *loxP* sites) and the enzyme Cre that catalyzes recombination between them. The *loxP-Cre recombination system* is derived from bacteriophage P1, but this site-specific recombination system also functions when placed in mouse cells. An essential feature of this technique is that expression of Cre is controlled by a cell-type-specific promoter. In *loxP*-Cre mice generated by the procedure depicted in Figure 9-40, inactivation of the gene of interest (*X*) occurs only in cells in which the promoter controlling the *cre* gene is active.

An early application of this technique provided strong evidence that a particular neurotransmitter receptor is important for learning and memory. Previous pharmacological and physiological studies had indicated that normal learning requires the NMDA class of glutamate receptors in the hippocampus, a region of the brain. But mice in which the gene encoding an NMDA receptor subunit was knocked out died neonatally, precluding analysis of the receptor's role in learning. Following the protocol in Figure 9-40, researchers generated mice in which the receptor subunit gene was inactivated in the hippocampus but expressed in other tissues. These mice survived to adulthood and showed learning and memory defects, confirming a role for these receptors in the ability of mice to encode their experiences into memory.

Dominant-Negative Alleles Can Functionally Inhibit Some Genes

In diploid organisms, as noted in Section 9.1, the phenotypic effect of a recessive allele is expressed only in homozygous individuals, whereas dominant alleles are expressed in heterozygotes. That is, an individual must carry two copies of a recessive allele but only one copy of a dominant allele to exhibit the corresponding phenotypes. We have seen how strains of mice that are homozygous for a given recessive knockout mutation can be produced by crossing individuals that are heterozygous for the same knockout mutation (see Figure 9-39). For experiments with cultured animal cells, however, it is usually difficult to disrupt both copies of a gene in order to produce a mutant phenotype. Moreover, the difficulty in producing strains with both copies of a gene mutated is often compounded by the presence of related genes of similar function that must also be inactivated in order to reveal an observable phenotype.

For certain genes, the difficulties in producing homozygous knockout mutants can be avoided by use of an allele carrying a **dominant-negative mutation.** These alleles are genetically dominant; that is, they produce a mutant phenotype even in cells carrying a wild-type copy of the gene. But unlike other

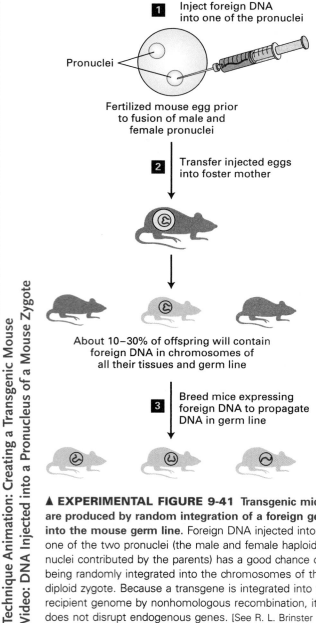

▲ **EXPERIMENTAL FIGURE 9-41 Transgenic mice are produced by random integration of a foreign gene into the mouse germ line.** Foreign DNA injected into one of the two pronuclei (the male and female haploid nuclei contributed by the parents) has a good chance of being randomly integrated into the chromosomes of the diploid zygote. Because a transgene is integrated into the recipient genome by nonhomologous recombination, it does not disrupt endogenous genes. [See R. L. Brinster et al., 1981, *Cell* **27**:223.]

▲ **FIGURE 9-42 Inactivation of the function of a wild-type GTPase by the action of a dominant-negative mutant allele.** (a) Small (monomeric) GTPases (purple) are activated by their interaction with a guanine-nucleotide exchange factor (GEF), which catalyzes the exchange of GDP for GTP. (b) Introduction of a dominant-negative allele of a small GTPase gene into cultured cells or transgenic animals leads to expression of a mutant GTPase that binds to and inactivates the GEF. As a result, endogenous wild-type copies of the same small GTPase are trapped in the inactive GDP-bound state. A single dominant-negative allele thus causes a loss-of-function phenotype in heterozygotes similar to that seen in homozygotes carrying two recessive loss-of-function alleles.

types of dominant alleles, dominant-negative alleles produce a phenotype equivalent to that of a loss-of-function mutation.

Useful dominant-negative alleles have been identified for a variety of genes and can be introduced into cultured cells by transfection or into the germ line of mice or other organisms. In both cases, the introduced gene is integrated into the genome by nonhomologous recombination. Such randomly inserted genes are called **transgenes;** the cells or organisms carrying them are referred to as **transgenic.** Transgenes carrying a dominant-negative allele usually are engineered so that the allele is controlled by a regulated promoter, allowing expression of the mutant protein in different tissues at different times. As noted above, the random integration of exogenous DNA via nonhomologous recombination occurs at a much higher frequency than insertion via homologous recombination. Because of this phenomenon, the production of transgenic mice is an efficient and straightforward process (Figure 9-41).

Among the genes that can be functionally inactivated by introduction of a dominant-negative allele are those encoding small (monomeric) GTP-binding proteins belonging to the GTPase superfamily. As we will examine in several later chapters, these proteins (e.g., Ras, Rac, and Rab) act as intracellular switches. Conversion of the small GTPases from an inactive GDP-bound state to an active GTP-bound state depends on their interacting with a corresponding guanine nucleotide exchange factor (GEF). A mutant small GTPase that permanently binds to the GEF protein will block conversion of endogenous wild-type small GTPases to the active GTP-bound state, thereby inhibiting them from performing their switching function (Figure 9-42).

Double-Stranded RNA Molecules Can Interfere with Gene Function by Targeting mRNA for Destruction

Researchers are exploiting a recently discovered phenomenon known as **RNA interference (RNAi)** to inhibit the function of specific genes. This approach is technically simpler than the methods described above for disrupting genes. First observed in the roundworm *C. elegans*, RNAi refers to the ability of a double-stranded (ds) RNA to block expression of its corresponding single-stranded mRNA but not that of mRNAs with a different sequence.

To use RNAi for intentional silencing of a gene of interest, investigators first produce dsRNA based on the sequence of the gene to be inactivated (Figure 9-43a). This dsRNA is injected into the gonad of an adult worm, where it has access to the developing embryos. As the embryos develop, the mRNA molecules corresponding to the injected dsRNA are rapidly destroyed. The resulting worms display a phenotype similar to the one that would result from disruption of the corresponding gene itself. In some cases, entry of just a few molecules of a particular dsRNA into a cell is sufficient to inactivate many copies of the corresponding mRNA. Figure 9-43b illustrates the ability of an injected dsRNA to interfere with production of the corresponding endogenous mRNA in *C. elegans* embryos. In this experiment, the mRNA levels in embryos were determined by incubating the embryos with a fluorescently labeled probe specific for the mRNA of interest. This technique, **in situ hybridization,** is useful in assaying expression of a particular mRNA in cells and tissue sections.

Initially, the phenomenon of RNAi was quite mysterious to geneticists. Recent studies have shown that specialized RNA-processing enzymes cleave dsRNA into short segments, which base-pair with endogenous mRNA. The resulting hybrid molecules are recognized and cleaved by specific nucleases at these hybridization sites. This model accounts for the specificity of RNAi, since it depends on base pairing, and for its potency in silencing gene function, since the complementary mRNA is permanently destroyed by nucleolytic degradation. Although the normal cellular function of RNAi is not understood, it may provide a defense against viruses with dsRNA genomes or help regulate certain endogenous genes. (For a more detailed discussion of the mechanism of RNA interference, see Section 12.4.)

Other organisms in which RNAi-mediated gene inactivation has been successful include *Drosophila*, many kinds of plants, zebrafish, spiders, the frog *Xenopus*, and mice. Although most other organisms do not appear to be as sensitive to the effects of RNAi as *C. elegans,* the method does have general use when the dsRNA is injected directly into embryonic tissues.

(a) In vitro production of double-stranded RNA

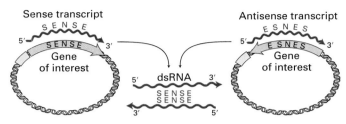

(b)

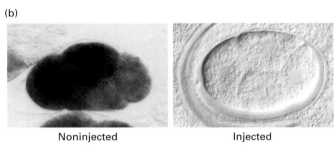

Noninjected   Injected

▲ **EXPERIMENTAL FIGURE 9-43 RNA interference (RNAi) can functionally inactivate genes in *C. elegans* and some other organisms.** (a) Production of double-stranded RNA (dsRNA) for RNAi of a specific target gene. The coding sequence of the gene, derived from either a cDNA clone or a segment of genomic DNA, is placed in two orientations in a plasmid vector adjacent to a strong promoter. Transcription of both constructs in vitro using RNA polymerase and ribonucleotide triphosphates yields many RNA copies in the sense orientation (identical with the mRNA sequence) or complementary antisense orientation. Under suitable conditions, these complementary RNA molecules will hybridize to form dsRNA. (b) Inhibition of *mex3* RNA expression in worm embryos by RNAi (see the text for the mechanism). (*Left*) Expression of *mex3* RNA in embryos was assayed by in situ hybridization with a fluorescently labeled probe (purple) specific for this mRNA. (*Right*) The embryo derived from a worm injected with double-stranded *mex3* mRNA produces little or no endogenous *mex3* mRNA, as indicated by the absence of color. Each four-cell stage embryo is ≈50 μm in length. [Part (b) from A. Fire et al., 1998, *Nature* **391**:806.]

KEY CONCEPTS OF SECTION 9.5

Inactivating the Function of Specific Genes in Eukaryotes

■ Once a gene has been cloned, important clues about its normal function in vivo can be deduced from the observed phenotypic effects of mutating the gene.

■ Genes can be disrupted in yeast by inserting a selectable marker gene into one allele of a wild-type gene via homologous recombination, producing a heterozygous mutant. When such a heterozygote is sporulated, disruption of an essential gene will produce two nonviable haploid spores (Figure 9-37).

■ A yeast gene can be inactivated in a controlled manner by using the *GAL1* promoter to shut off transcription of a gene when cells are transferred to glucose medium.

- In mice, modified genes can be incorporated into the germ line at their original genomic location by homologous recombination, producing knockouts (see Figures 9-38 and 9-39). Mouse knockouts can provide models for human genetic diseases such as cystic fibrosis.

- The *loxP*-Cre recombination system permits production of mice in which a gene is knocked out in a specific tissue.

- In the production of transgenic cells or organisms, exogenous DNA is integrated into the host genome by nonhomologous recombination (see Figure 9-41). Introduction of a dominant-negative allele in this way can functionally inactivate a gene without altering its sequence.

- In some organisms, including the roundworm *C. elegans,* double-stranded RNA triggers destruction of the all the mRNA molecules with the same sequence (see Figure 9-43). This phenomenon, known as *RNAi* (RNA interference), provides a specific and potent means of functionally inactivating genes without altering their structure.

9.6 Identifying and Locating Human Disease Genes

Inherited human diseases are the phenotypic consequence of defective human genes. Table 9-3 lists several of the most commonly occurring inherited diseases. Although a "disease" gene may result from a new mutation that arose in the preceding generation, most cases of inherited diseases are caused by preexisting mutant alleles that have been passed from one generation to the next for many generations.

Nowadays, the typical first step in deciphering the underlying cause for any inherited human disease is to identify the affected gene and its encoded protein. Comparison of the sequences of a disease gene and its product with those of genes and proteins whose sequence and function are known can provide clues to the molecular and cellular cause of the disease. Historically, researchers have used whatever pheno-

TABLE 9-3   Common Inherited Human Diseases

Disease	Molecular and Cellular Defect	Incidence
AUTOSOMAL RECESSIVE		
Sickle-cell anemia	Abnormal hemoglobin causes deformation of red blood cells, which can become lodged in capillaries; also confers resistance to malaria.	1/625 of sub-Saharan African origin
Cystic fibrosis	Defective chloride channel (CFTR) in epithelial cells leads to excessive mucus in lungs.	1/2500 of European origin
Phenylketonuria (PKU)	Defective enzyme in phenylalanine metabolism (tyrosine hydroxylase) results in excess phenylalanine, leading to mental retardation, unless restricted by diet.	1/10,000 of European origin
Tay-Sachs disease	Defective hexosaminidase enzyme leads to accumulation of excess sphingolipids in the lysosomes of neurons, impairing neural development.	1/1000 Eastern European Jews
AUTOSOMAL DOMINANT		
Huntington's disease	Defective neural protein (huntingtin) may assemble into aggregates causing damage to neural tissue.	1/10,000 of European origin
Hypercholesterolemia	Defective LDL receptor leads to excessive cholesterol in blood and early heart attacks.	1/122 French Canadians
X-LINKED RECESSIVE		
Duchenne muscular dystrophy (DMD)	Defective cytoskeletal protein dystrophin leads to impaired muscle function.	1/3500 males
Hemophilia A	Defective blood clotting factor VIII leads to uncontrolled bleeding.	1–2/10,000 males

typic clues might be relevant to make guesses about the molecular basis of inherited diseases. An early example of successful guesswork was the hypothesis that sickle-cell anemia, known to be a disease of blood cells, might be caused by a defective hemoglobin. This idea led to identification of a specific amino acid substitution in hemoglobin that causes polymerization of the defective hemoglobin molecules, causing the sickle-like deformation of red blood cells in individuals who have inherited two copies of the Hb^s allele for sickle-cell hemoglobin.

Most often, however, the genes responsible for inherited diseases must be found without any prior knowledge or reasonable hypotheses about the nature of the affected gene or its encoded protein. In this section, we will see how human geneticists can find the gene responsible for an inherited disease by following the segregation of the disease in families. The segregation of the disease can be correlated with the segregation of many other **genetic markers,** eventually leading to identification of the chromosomal position of the affected gene. This information, along with knowledge of the sequence of the human genome, can ultimately allow the affected gene and the disease-causing mutations to be pinpointed. ∎

Many Inherited Diseases Show One of Three Major Patterns of Inheritance

Human genetic diseases that result from mutation in one specific gene exhibit several inheritance patterns depending on the nature and chromosomal location of the alleles that cause them. One characteristic pattern is that exhibited by a dominant allele in an **autosome** (that is, one of the 22 human chromosomes that is not a sex chromosome). Because an *autosomal dominant* allele is expressed in the heterozygote, usually at least one of the parents of an affected individual will also have the disease. It is often the case that the diseases caused by dominant alleles appear later in life after the reproductive age. If this were not the case, natural selection would have eliminated the allele during human evolution. An example of an autosomal dominant disease is Huntington's disease, a neural degenerative disease that generally strikes in mid- to late life. If either parent carries a mutant *HD* allele, each of his or her children (regardless of sex) has a 50 percent chance of inheriting the mutant allele and being affected (Figure 9-44a).

A recessive allele in an autosome exhibits a quite different segregation pattern. For an *autosomal recessive* allele, both parents must be heterozygous **carriers** of the allele in order for their children to be at risk of being affected with the disease. Each child of heterozygous parents has a 25 percent chance of receiving both recessive alleles and thus being affected, a 50 percent chance of receiving one normal and one mutant allele and thus being a carrier, and a 25 percent chance of receiving two normal alleles. A clear example of an autosomal recessive disease is cystic fibrosis, which results from a defective chloride channel gene known as *CFTR* (Fig-

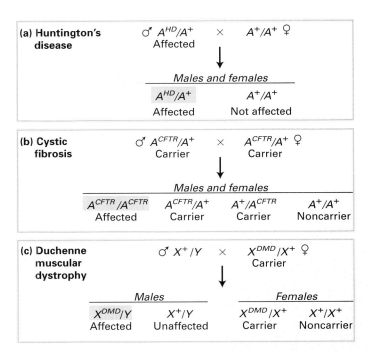

▲ **FIGURE 9-44 Three common inheritance patterns for human genetic diseases.** Wild-type autosomal (A) and sex chromosomes (X and Y) are indicated by superscript plus signs. (a) In an autosomal dominant disorder such as Huntington's disease, only one mutant allele is needed to confer the disease. If either parent is heterozygous for the mutant *HD* allele, his or her children have a 50 percent chance of inheriting the mutant allele and getting the disease. (b) In an autosomal recessive disorder such as cystic fibrosis, two mutant alleles must be present to confer the disease. Both parents must be heterozygous carriers of the mutant *CFTR* gene for their children to be at risk of being affected or being carriers. (c) An X-linked recessive disease such as Duchenne muscular dystrophy is caused by a recessive mutation on the X chromosome and exhibits the typical sex-linked segregation pattern. Males born to mothers heterozygous for a mutant *DMD* allele have a 50 percent chance of inheriting the mutant allele and being affected. Females born to heterozygous mothers have a 50 percent chance of being carriers.

ure 9-44b). Related individuals (e.g., first or second cousins) have a relatively high probability of being carriers for the same recessive alleles. Thus children born to related parents are much more likely than those born to unrelated parents to be homozygous for, and therefore affected by, an autosomal recessive disorder.

The third common pattern of inheritance is that of an *X-linked recessive* allele. A recessive allele on the X-chromosome will most often be expressed in males, who receive only one X chromosome from their mother, but not in females who receive an X chromosome from both their mother and father. This leads to a distinctive sex-linked segregation pattern where the disease is exhibited much more frequently in males than in females. For example, Duchenne muscular dystrophy (DMD), a muscle degenerative disease that specifically affects males, is caused by a recessive allele on the X chromosome. DMD exhibits the typical sex-linked segregation pattern in which mothers who are heterozygous and therefore phenotypically normal can act as carriers, transmitting the DMD allele, and therefore the disease, to 50 percent of their male progeny (Figure 9-44c).

Recombinational Analysis Can Position Genes on a Chromosome

The independent segregation of chromosomes during meiosis provides the basis for determining whether genes are on the same or different chromosomes. Genetic traits that segregate together during meiosis more frequently than expected from random segregation are controlled by genes located on the same chromosome. (The tendency of genes on the same chromosome to be inherited together is referred to as genetic **linkage**.) However, the occurrence of recombination during meiosis can separate linked genes; this phenomenon provides a means for locating (mapping) a particular gene relative to other genes on the *same* chromosome.

Recombination takes place before the first meiotic cell division in germ cells when the replicated chromosomes of each homologous pair align with each other, an act called *synapsis* (see Figure 9-3). At this time, homologous DNA sequences on maternally and paternally derived chromatids

can exchange with each other, a process known as **crossing over**. The sites of recombination occur more or less at random along the length of chromosomes; thus the closer together two genes are, the less likely that recombination will occur between them during meiosis (Figure 9-45). In other words, *the less frequently recombination occurs between two genes on the same chromosome, the more tightly they are linked and the closer together they are.* The frequency of recombination between two genes can be determined from the proportion of recombinant progeny, whose phenotypes differ from the parental phenotypes, produced in crosses of parents carrying different alleles of the genes.

The presence of many different already mapped genetic traits, or markers, distributed along the length of a chromosome facilitates the mapping of a new mutation by assessing its possible linkage to these marker genes in appropriate crosses. The more markers that are available, the more precisely a mutation can be mapped. As more and more mutations are mapped, the linear order of genes along the length of a chromosome can be constructed. This ordering of genes along a chromosome is called a *genetic map*, or *linkage map*. By convention, one genetic map unit is defined as the distance between two positions along a chromosome that results in one recombinant individual in 100 progeny. The distance corresponding to this 1 percent recombination frequency is called a *centimorgan (cM)*. Comparison of the actual physical distances between known genes, determined by molecular analysis, with their recombination frequency indicates that in humans 1 centimorgan on average represents a distance of about 7.5×10^5 base pairs.

DNA Polymorphisms Are Used in Linkage-Mapping Human Mutations

Many different genetic markers are needed to construct a high-resolution genetic map. In the experimental organisms commonly used in genetic studies, numerous markers with easily detectable phenotypes are readily available for genetic mapping of mutations. This is not the case for mapping genes whose mutant alleles are associated with inherited diseases in humans. However, recombinant DNA technology has made available a wealth of useful DNA-based *molecular markers*. Because most of the human genome does not code for protein, a large amount of sequence variation exists between individuals. Indeed, it has been estimated that nucleotide differences between unrelated individuals can be detected on an average of every 10^3 nucleotides. If these variations in DNA sequence, referred to as *DNA polymorphisms*, can be followed from one generation to the next, they can serve as genetic markers for linkage studies. Currently, a panel of as many as 10^4 different known polymorphisms whose locations have been mapped in the human genome is used for genetic linkage studies in humans.

Restriction fragment length polymorphisms (RFLPs) were the first type of molecular markers used in linkage studies. RFLPs arise because mutations can create or destroy the

(a) Homologous chromosomes undergoing crossing over

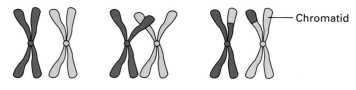

Chromatid

(b)

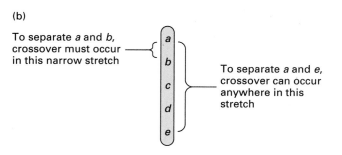

To separate *a* and *b*, crossover must occur in this narrow stretch

To separate *a* and *e*, crossover can occur anywhere in this stretch

▲ **FIGURE 9-45 Recombination during meiosis.** (a) Crossing over can occur between chromatids of homologous chromosomes before the first meiotic division (see Figure 9-3). (b) The longer the distance between two genes on a chromatid, the more likely they are to be separated by recombination.

sites recognized by specific restriction enzymes, leading to variations between individuals in the length of restriction fragments produced from identical regions of the genome. Differences in the sizes of restriction fragments between individuals can be detected by Southern blotting with a probe specific for a region of DNA known to contain an RFLP (Figure 9-46a). The segregation and meiotic recombination of such DNA polymorphisms can be followed like typical genetic markers. Figure 9-46b illustrates how RFLP analysis of a family can detect the segregation of an RFLP that can be used to test for statistically significant linkage to the allele for an inherited disease or some other human trait of interest.

The amassing of vast amounts of genomic sequence information from different humans in recent years has led to identification of other useful DNA polymorphisms. *Single nucleotide polymorphisms (SNPs)* constitute the most abundant type and are therefore useful for constructing high-resolution genetic maps. Another useful type of DNA polymorphism consists of a variable number of repetitions of a one- two-, or three-base sequence. Such polymorphisms, known as simple sequence repeats (SSRs), or *microsatellites*, presumably are formed by recombination or a slippage mechanism of either the template or newly synthesized

strands during DNA replication. A useful property of SSRs is that different individuals will often have different numbers of repeats. The existence of multiple versions of an SSR makes it more likely to produce an informative segregation pattern in a given pedigree and therefore be of more general use in mapping the positions of disease genes. If an SNP or SSR alters a restriction site, it can be detected by RFLP analysis. More commonly, however, these polymorphisms do not alter restriction fragments and must be detected by PCR amplification and DNA sequencing.

Linkage Studies Can Map Disease Genes with a Resolution of About 1 Centimorgan

Without going into all the technical considerations, let's see how the allele conferring a particular dominant trait (e.g., familial hypercholesterolemia) might be mapped. The first step is to obtain DNA samples from all the members of a family containing individuals that exhibit the disease. The DNA from each affected and unaffected individual then is analyzed to determine the identity of a large number of known DNA polymorphisms (either SSR or SNP markers can be used). The segregation pattern of each DNA polymorphism within the family is then compared with the segregation of the

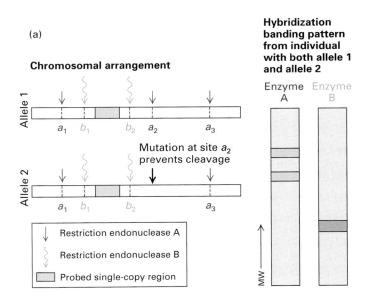

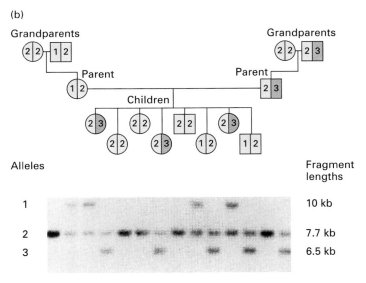

▲ **EXPERIMENTAL FIGURE 9-46 Restriction fragment length polymorphisms (RFLPs) can be followed like genetic markers.** (a) In the example shown, DNA from an individual is treated with two different restriction enzymes (A and B), which cut DNA at different sequences *(a and b)*. The resulting fragments are subjected to Southern blot analysis (see Figure 9-26) with a radioactive probe that binds to the indicated DNA region (green) to detect the fragments. Since no differences between the two homologous chromosomes occur in the sequences recognized by the B enzyme, only one fragment is recognized by the probe, as indicated by a single hybridization band. However, treatment with enzyme A produces fragments of

two different lengths (two bands are seen), indicating that a mutation has caused the loss of one of the *a* sites in one of the two chromosomes. (b) Pedigree based on RFLP analysis of the DNA from a region known to be present on chromosome 5. The DNA samples were cut with the restriction enzyme *Taq*I and analyzed by Southern blotting. In this family, this region of the genome exists in three allelic forms characterized by *Taq*I sites spaced 10, 7.7, or 6.5 kb apart. Each individual has two alleles; some contain allele 2 (7.7 kb) on both chromosomes, and others are heterozygous at this site. Circles indicate females; squares indicate males. The gel lanes are aligned below the corresponding subjects. [After H. Donis-Keller et al., 1987, *Cell* **51**:319.]

disease under study to find those polymorphisms that tend to segregate along with the disease. Finally, computer analysis of the segregation data is used to calculate the likelihood of linkage between each DNA polymorphism and the disease-causing allele.

In practice, segregation data are collected from different families exhibiting the same disease and pooled. The more families exhibiting a particular disease that can be examined, the greater the statistical significance of evidence for linkage that can be obtained and the greater the precision with which the distance can be measured between a linked DNA polymorphism and a disease allele. Most family studies have a maximum of about 100 individuals in which linkage between a disease gene and a panel of DNA polymorphisms can be tested. This number of individuals sets the practical upper limit on the resolution of such a mapping study to about 1 centimorgan, or a physical distance of about 7.5×10^5 base pairs.

A phenomenon called *linkage disequilibrium* is the basis for an alternative strategy, which in some cases can afford a higher degree of resolution in mapping studies. This approach depends on the particular circumstance in which a genetic disease commonly found in a particular population results from a single mutation that occurred many generations in the past. This ancestral chromosome will carry closely linked DNA polymorphisms that will have been conserved through many generations. Polymorphisms that are farthest away on the chromosome will tend to become separated from the disease gene by recombination, whereas those closest to the disease gene will remain associated with it. By assessing the distribution of specific markers in all the affected individuals in a population, geneticists can identify DNA markers tightly associated with the disease, thus localizing the disease-associated gene to a relatively small region. The resolving power of this method comes from the ability to determine whether a polymorphism and the disease allele were ever separated by a meiotic recombination event at any time since the disease allele first appeared on the ancestral chromosome. Under ideal circumstances linkage disequilibrium studies can improve the resolution of mapping studies to less than 0.1 centimorgan.

Further Analysis Is Needed to Locate a Disease Gene in Cloned DNA

Although linkage mapping can usually locate a human disease gene to a region containing about 7.5×10^5 base pairs, as many as 50 different genes may be located in a region of this size. The ultimate objective of a mapping study is to locate the gene within a cloned segment of DNA and then to determine the nucleotide sequence of this fragment.

One strategy for further localizing a disease gene within the genome is to identify mRNA encoded by DNA in the region of the gene under study. Comparison of gene expression in tissues from normal and affected individuals may suggest tissues in which a particular disease gene normally is expressed. For instance, a mutation that phenotypically affects muscle, but no other tissue, might be in a gene that is expressed only in muscle tissue. The expression of mRNA in both normal and affected individuals generally is determined by Northern blotting or in situ hybridization of labeled DNA or RNA to tissue sections. Northern blots permit comparison of both the level of expression and the size of mRNAs in mutant and wild-type tissues (see Figure 9-27). Although the sensitivity of in situ hybridization is lower than that of Northern blot analysis, it can be very helpful in identifying an mRNA that is expressed at low levels in a given tissue but at very high levels in a subclass of cells within that tissue. An mRNA that is altered or missing in various individuals affected with a disease compared with wild-type individuals would be an excellent candidate for encoding the protein whose disrupted function causes that disease.

In many cases, point mutations that give rise to disease-causing alleles may result in no detectable change in the level of expression or electrophoretic mobility of mRNAs. Thus if comparison of the mRNAs expressed in normal and affected individuals reveals no detectable differences in the candidate mRNAs, a search for point mutations in the DNA regions encoding the mRNAs is undertaken. Now that highly efficient methods for sequencing DNA are available, researchers frequently determine the sequence of candidate regions of DNA isolated from affected individuals to identify point mutations. The overall strategy is to search for a coding sequence that consistently shows possibly deleterious alterations in DNA from individuals that exhibit the disease. A limitation of this approach is that the region near the affected gene may carry naturally occurring polymorphisms unrelated to the gene of interest. Such polymorphisms, not functionally related to the disease, can lead to misidentification of the DNA fragment carrying the gene of interest. For this reason, the more mutant alleles available for analysis, the more likely that a gene will be correctly identified.

Many Inherited Diseases Result from Multiple Genetic Defects

Most of the inherited human diseases that are now understood at the molecular level are *monogenetic traits*. That is, a clearly discernible disease state is produced by the presence of a defect in a single gene. Monogenic diseases caused by mutation in one specific gene exhibit one of the characteristic inheritance patterns shown in Figure 9-44. The genes associated with most of the common monogenic diseases have already been mapped using DNA-based markers as described previously.

However, many other inherited diseases show more complicated patterns of inheritance, making the identification of the underlying genetic cause much more difficult. One type of added complexity that is frequently encountered is *genetic heterogeneity*. In such cases, mutations in any one of multiple different genes can cause the same disease. For example, retinitis pigmentosa, which is characterized by degeneration of the retina usually leading to blindness, can be caused by mutations in any one of more than 60 different genes. In human linkage studies, data from multiple families usually must be combined to determine whether a statistically significant linkage exists between a disease gene and known molecular markers. Genetic heterogeneity such as that exhibited by retinitis pigmentosa can confound such an approach because any statistical trend in the mapping data from one family tends to be canceled out by the data obtained from another family with an unrelated causative gene.

Human geneticists used two different approaches to identify the many genes associated with retinitis pigmentosa. The first approach relied on mapping studies in exceptionally large single families that contained a sufficient number of affected individuals to provide statistically significant evidence for linkage between known DNA polymorphisms and a single causative gene. The genes identified in such studies showed that several of the mutations that cause retinitis pigmentosa lie within genes that encode abundant proteins of the retina. Following up on this clue, geneticists concentrated their attention on those genes that are highly expressed in the retina when screening other individuals with retinitis pigmentosa. This approach of using additional information to direct screening efforts to a subset of candidate genes led to identification of additional rare causative mutations in many different genes encoding retinal proteins.

A further complication in the genetic dissection of human diseases is posed by diabetes, heart disease, obesity, predisposition to cancer, and a variety of mental disorders that have at least some heritable properties. These and many other diseases can be considered to be *polygenic traits* in the sense that alleles of multiple genes, acting together within an individual, contribute to both the occurrence and the severity of disease. A systematic solution to the problem of mapping complex polygenic traits in humans does not yet exist. Future progress may come from development of refined diagnostic methods that can distinguish the different forms of diseases resulting from multiple causes.

Models of human disease in experimental organisms may also contribute to unraveling the genetics of complex traits such as obesity or diabetes. For instance, large-scale controlled breeding experiments in mice can identify mouse genes associated with diseases analogous to those in humans. The human orthologs of the mouse genes identified in such studies would be likely candidates for involvement in the corresponding human disease. DNA from human populations then could be examined to determine if particular alleles of the candidate genes show a tendency to be present in individuals affected with the disease but absent from unaffected individuals. This "candidate gene" approach is currently being used intensively to search for genes that may contribute to the major polygenic diseases in humans.

KEY CONCEPTS OF SECTION 9.6

Identifying and Locating Human Disease Genes

■ Inherited diseases and other traits in humans show three major patterns of inheritance: autosomal dominant, autosomal recessive, and X-linked recessive (see Figure 9-44).

■ Genes located on the same chromosome can be separated by crossing over during meiosis, thus producing new recombinant genotypes in the next generation (see Figure 9-45).

■ Genes for human diseases and other traits can be mapped by determining their cosegregation with markers whose locations in the genome are known. The closer a gene is to a particular marker, the more likely they are to cosegregate.

■ Mapping of human genes with great precision requires thousands of molecular markers distributed along the chromosomes. The most useful markers are differences in the DNA sequence (polymorphisms) among individuals in noncoding regions of the genome.

■ DNA polymorphisms useful in mapping human genes include restriction fragment length polymorphisms (RFLPs), single-nucleotide polymorphisms (SNPs), and simple sequence repeats (SSRs).

■ Linkage mapping often can locate a human disease gene to a chromosomal region that includes as many as 50 genes. To identify the gene of interest within this candidate region typically requires expression analysis and comparison of DNA sequences between wild-type and disease-affected individuals.

■ Some inherited diseases can result from mutations in different genes in different individuals (genetic heterogeneity). The occurrence and severity of other diseases depend on the presence of mutant alleles of multiple genes in the same individuals (polygenic traits). Mapping of the genes associated with such diseases is particularly difficult because the occurrence of the disease cannot readily be correlated to a single chromosomal locus.

PERSPECTIVES FOR THE FUTURE

As the examples in this chapter and throughout the book illustrate, genetic analysis is the foundation of our understanding of many fundamental processes in cell biology. By examining the phenotypic consequences of mutations that inactivate a particular gene, geneticists are able to connect knowledge about the sequence, structure, and biochemical activity of the encoded protein to its function in the context of a living cell or multicellular organism. The classical approach to making these connections in both humans and simpler, experimentally accessible organisms has been to identify new mutations of interest based on their phenotypes and then to isolate the affected gene and its protein product.

Although scientists continue to use this classical genetic approach to dissect fundamental cellular processes and biochemical pathways, the availability of complete genomic sequence information for most of the common experimental organisms has fundamentally changed the way genetic experiments are conducted. Using various computational methods, scientists have identified most of the protein-coding gene sequences in *E. coli*, yeast, *Drosophila*, *Arabidopsis*, mouse, and humans. The gene sequences, in turn, reveal the primary amino acid sequence of the encoded protein products, providing us with a nearly complete list of the proteins found in each of the major experimental organisms.

The approach taken by most researchers has thus shifted from discovering new genes and proteins to discovering the functions of genes and proteins whose sequences are already known. Once an interesting gene has been identified, genomic sequence information greatly speeds subsequent genetic manipulations of the gene, including its designed inactivation, to learn more about its function. Already all the ≈6000 possible gene knockouts in yeast have been produced; this relatively small but complete collection of mutants has become the preferred starting point for many genetic screens in yeast. Similarly, sets of vectors for RNAi inactivation of a large number of defined genes in the nematode *C. elegans* now allow efficient genetic screens to be performed in this multicellular organism. Following the trajectory of recent advances, it seems quite likely that in the foreseeable future either RNAi or knockout methods will have been used to inactivate every gene in the principal model organisms, including the mouse.

In the past, a scientist might spend many years studying only a single gene, but nowadays scientists commonly study whole sets of genes at once. For example, with DNA microarrays the level of expression of all genes in an organism can be measured almost as easily as the expression of a single gene. One of the great challenges facing geneticists in the twenty-first century will be to exploit the vast amount of available data on the function and regulation of individual genes to gain fundamental insights into the organization of complex biochemical pathways and regulatory networks.

KEY TERMS

REVIEW THE CONCEPTS

1. Genetic mutations can provide insights into the mechanisms of complex cellular or developmental processes. What is the difference between recessive and dominant mutations? What is a temperature-sensitive mutation, and how is this type of mutation useful?

2. A number of experimental approaches can be used to analyze mutations. Describe how complementation analysis can be used to reveal whether two mutations are in the same or in different genes. What are suppressor mutations and synthetic lethal mutations?

3. Restriction enzymes and DNA ligase play essential roles in DNA cloning. How is it that a bacterium that produces a restriction enzyme does not cut its own DNA? Describe some general features of restriction enzyme sites. What are the three types of DNA ends that can be generated after cutting DNA with restriction enzymes? What reaction is catalyzed by DNA ligase?

4. Bacterial plasmids and λ phage serve as cloning vectors. Describe the essential features of a plasmid and a λ phage vector. What are the advantages and applications of plasmids and λ phage as cloning vectors?

5. A DNA library is a collection of clones, each containing a different fragment of DNA, inserted into a cloning vector. What is the difference between a cDNA and a genomic DNA library? How can you use hybridization or expression to screen a library for a specific gene? What oligonucleotide primers could be synthesized as probes to screen a library for the gene encoding the peptide Met-Pro-Glu-Phe-Tyr?

6. In 1993, Kerry Mullis won the Nobel Prize in Chemistry for his invention of the PCR process. Describe the three steps in each cycle of a PCR reaction. Why was the discovery of a thermostable DNA polymerase (e.g., Taq polymerase) so important for the development of PCR?

7. Southern and Northern blotting are powerful tools in molecular biology; describe the technique of each. What are the applications of these two blotting techniques?

8. A number of foreign proteins have been expressed in bacterial and mammalian cells. Describe the essential features of a recombinant plasmid that are required for expression of a foreign gene. How can you modify the foreign protein to facilitate its purification? What is the advantage of expressing a protein in mammalian cells versus bacteria?

9. Why is the screening for genes based on the presence of ORFs (open reading frames) more useful for bacterial genomes than for eukaryotic genomes? What are paralogous and orthologous genes? What are some of the explanations for the finding that humans are a much more complex organism than the roundworm C. *elegans,* yet have only less than twice the number of genes (35,000 versus 19,000)?

10. A global analysis of gene expression can be accomplished by using a DNA microarray. What is a DNA microarray? How are DNA microarrays used for studying gene expression? How do experiments with microarrays differ from Northern botting experiments described in question 7?

11. The ability to selectively modify the genome in the mouse has revolutionized mouse genetics. Outline the procedure for generating a knockout mouse at a specific genetic locus. How can the *loxP*-Cre system be used to conditionally knock out a gene? What is an important medical application of knockout mice?

12. Two methods for functionally inactivating a gene without altering the gene sequence are by dominant negative mutations and RNA interference (RNAi). Describe how each method can inhibit expression of a gene.

13. DNA polymorphisms can be used as DNA markers. Describe the differences among RFLP, SNP, and SSR polymorphisms. How can these markers be used for DNA mapping studies?

14. Genetic linkage studies can roughly locate the chromosomal position of a "disease" gene. Describe how expression analysis and DNA sequence analysis can be used to identify a "disease" gene.

ANALYZE THE DATA

RNA interference (RNAi) is a process of post-transcriptional gene silencing mediated by short double-stranded RNA molecules called siRNA (small interfering RNAs). In mammalian cells, transfection of 21–22 nucleotide siRNAs leads to degradation of mRNA molecules that contain the same sequence as the siRNA. In the following experiment, siRNA and knockout mice are used to investigate two related cell surface proteins designated p24 and p25 that are suspected to be cellular receptors for the uptake of a newly isolated virus.

a. To test the efficacy of RNAi in cells, siRNAs specific to cell surface proteins p24 (siRNA–p24) and p25 (siRNA–p25) are transfected individually into cultured mouse cells. RNA is extracted from these transfected cells and the mRNA for proteins p24 and p25 are detected on Northern blots using labeled p24 cDNA or p25 cDNA as probes. The control for this experiment is a mock transfection with no siRNA. What do you conclude from this Northern blot about the specificity of the siRNAs for their target mRNAs?

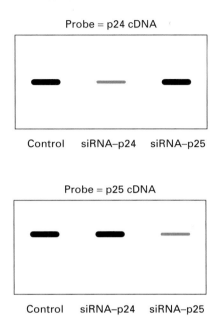

b. Next, the ability of siRNAs to inhibit viral replication is investigated. Cells are transfected with siRNA–p24 or siRNA–p25 or with siRNA to an essential viral protein. Twenty hours later, transfected cells are infected with the virus. After a further incubation period, the cells are collected and lysed. The number of viruses produced by each culture is shown below. The control is a mock transfection with no siRNA. What do you conclude about the role of p24 and p25 in the uptake of the virus? Why might the siRNA to the viral protein be more effective than siRNA to the receptors in reducing the number of viruses?

Cell Treatment	Number of Viruses/ml
Control	1×10^7
siRNA–p24	3×10^6
siRNA–p25	2×10^6
siRNA–p24 and siRNA–p25	1×10^4
siRNA to viral protein	1×10^2

c. To investigate the role of proteins p24 and p25 for viral replication in live mice, transgenic mice that lack genes for p24 or p25 are generated. The *loxP*-Cre conditional knockout system is used to selectively delete the genes in cells of either the liver or the lung. Wild type and knockout mice are infected with virus. After a 24-hour incubation period, mice are killed and lung and liver tissues are removed and examined for the presence (infected) or absence (normal) of virus by immunohistochemistry. What do these data indicate about the cellular requirements for viral infection in different tissues?

Mouse	Tissue Examined	
	Liver	Lung
Wild type	infected	infected
Knockout of p24 in liver	normal	infected
Knockout of p24 in lung	infected	infected
Knockout of p25 in liver	infected	infected
Knockout of p25 in lung	infected	normal

d. By performing Northern blots on different tissues from wild-type mice, you find that p24 is expressed in the liver but not in the lung, whereas p25 is expressed in the lung but not the liver. Based on all the data you have collected, propose a model to explain which protein(s) are involved in the virus entry into liver and lung cells? Would you predict that the cultered mouse cells used in parts (a) and (b) express p24, p25, or both proteins?

REFERENCES

Genetic Analysis of Mutations to Identify and Study Genes

Adams, A. E. M., D. Botstein, and D. B. Drubin. 1989. A yeast actin-binding protein is encoded by *sac6*, a gene found by suppression of an actin mutation. *Science* **243**:231.

Griffiths, A. G. F., et al. 2000. *An Introduction to Genetic Analysis*, 7th ed. W. H. Freeman and Company.

Guarente, L. 1993. Synthetic enhancement in gene interaction: a genetic tool comes of age. *Trends Genet.* **9**:362–366.

Hartwell, L. H. 1967. Macromolecular synthesis of temperature-sensitive mutants of yeast. *J. Bacteriol.* **93**:1662.

Hartwell, L. H. 1974. Genetic control of the cell division cycle in yeast. *Science* **183**:46.

Nüsslein-Volhard, C., and E. Wieschaus. 1980. Mutations affecting segment number and polarity in *Drosophila*. *Nature* **287**:795–801.

Simon, M. A., et al. 1991. Ras1 and a putative guanine nucleotide exchange factor perform crucial steps in signaling by the sevenless protein tyrosine kinase. *Cell* **67**:701–716.

Tong, A. H., et al. 2001. Systematic genetic analysis with ordered arrays of yeast deletion mutants. *Science* **294**:2364–2368.

DNA Cloning by Recombinant DNA Methods

Ausubel, F. M., et al. 2002. *Current Protocols in Molecular Biology*. Wiley.

Gubler, U., and B. J. Hoffman. 1983. A simple and very efficient method for generating cDNA libraries. *Gene* **25**:263–289.

Han, J. H., C. Stratowa, and W. J. Rutter. 1987. Isolation of full-length putative rat lysophospholipase cDNA using improved methods for mRNA isolation and cDNA cloning. *Biochem.* **26**:1617–1632.

Itakura, K., J. J. Rossi, and R. B. Wallace. 1984. Synthesis and use of synthetic oligonucleotides. *Ann. Rev. Biochem.* **53**:323–356.

Maniatis, T., et al. 1978. The isolation of structural genes from libraries of eucaryotic DNA. *Cell* **15**:687–701.

Nasmyth, K. A., and S. I. Reed. 1980. Isolation of genes by complementation in yeast: molecular cloning of a cell-cycle gene. *Proc. Nat'l. Acad. Sci. USA* **77**:2119–2123.

Nathans, D., and H. O. Smith. 1975. Restriction endonucleases in the analysis and restructuring of DNA molecules. *Ann. Rev. Biochem.* **44**:273–293.

Roberts, R. J., and D. Macelis. 1997. REBASE—restriction enzymes and methylases. *Nucl. Acids Res.* **25**:248–262. Information on accessing a continuously updated database on restriction and modification enzymes at http://www.neb.com/rebase.

Thomas, M., J. R. Cameron, and R. W. Davis. 1974. Viable molecular hybrids of bacteriophage lambda and eukaryotic DNA. *Proc. Nat'l. Acad. Sci. USA* **71**:4579–4583.

Sambrook, J., and D. Russell. 2001. *Molecular Cloning: A Laboratory Manual*. Cold Spring Harbor Laboratory.

Characterizing and Using Cloned DNA Fragments

Andrews, A. T. 1986. *Electrophoresis*, 2d ed. Oxford University Press.

Erlich, H., ed. 1992. *PCR Technology: Principles and Applications for DNA Amplification*. W. H. Freeman and Company.

Pellicer, A., M. Wigler, R. Axel, and S. Silverstein. 1978. The transfer and stable integration of the HSV thymidine kinase gene into mouse cells. *Cell* **41**:133–141.

Saiki, R. K., et al. 1988. Primer-directed enzymatic amplification of DNA with a thermostable DNA polymerase. *Science* **239**:487–491.

Sanger, F. 1981. Determination of nucleotide sequences in DNA. *Science* **214**:1205–1210.

Souza, L. M., et al. 1986. Recombinant human granulocyte-colony stimulating factor: effects on normal and leukemic myeloid cells. *Science* **232**:61–65.

Wahl, G. M., J. L. Meinkoth, and A. R. Kimmel. 1987. Northern and Southern blots. *Meth. Enzymol.* **152**:572–581.

Wallace, R. B., et al. 1981. The use of synthetic oligonucleotides as hybridization probes. II: Hybridization of oligonucleotides of mixed sequence to rabbit β-globin DNA. *Nucl. Acids Res.* **9**:879–887.

Genomics: Genome-wide Analysis of Gene Structure and Expression

BLAST Information can be found at: http://www.ncbi.nlm.nih.gov/Education/BLASTinfo/information3.htm

Ballester, R., et al. 1990. The NF1 locus encodes a protein functionally related to mammalian GAP and yeast IRA proteins. *Cell* **63**:851–859.

Chervitz, S. A., et al. 1998. Comparison of the complete protein sets of worm and yeast: orthology and divergence. *Science* **282**:2022–2028.

Gene Ontology Consortium. 2000. Gene ontology: tool for the unification of biology. *Nature Gen.* **25**:25–29.

Lander, E. S., et al. 2001 Initial sequencing and analysis of the human genome. *Nature* **409**:860–921.

Rubin, G. M., et al. 2000. Comparative genomics of the eukaryotes. *Science* **287**:2204–2215.

Waterston, R. H., et al. 2002. Initial sequencing and comparative analysis of the mouse genome. *Nature* **420**:520–562.

Inactivating the Function of Specific Genes in Eukaryotes

Capecchi, M. R. 1989. Altering the genome by homologous recombination. *Science* **244**:1288–1292.

Deshaies, R. J., et al. 1988. A subfamily of stress proteins facilitates translocation of secretory and mitochondrial precursor polypeptides. *Nature* **332**:800–805.

Fire, A., et al. 1998. Potent and specific genetic interference by double-stranded RNA in *Caenorhabditis elegans*. *Nature* **391**:806–811.

Gu, H., et al. 1994. Deletion of a DNA polymerase beta gene segment in T cells using cell type-specific gene targeting. *Science* **265**:103–106.

Zamore, P. D., T. Tuschl, P. A. Sharp, and D. P. Bartel. 2000. RNAi: double-stranded RNA directs the ATP-dependent cleavage of mRNA at 21 to 23 nucleotide intervals. *Cell* **101**:25–33.

Zimmer, A. 1992. Manipulating the genome by homologous recombination in embryonic stem cells. *Ann. Rev. Neurosci.* **15**:115.

Identifying and Locating Human Disease Genes

Botstein, D., et al. 1980. Construction of a genetic linkage map in man using restriction fragment length polymorphisms. *Am. J. Genet.* **32**:314–331.

Donis-Keller, H., et al. 1987. A genetic linkage map of the human genome. *Cell* **51**:319–337.

Hartwell, et al. 2000. *Genetics: From Genes to Genomes.* McGraw-Hill.

Hastbacka, T., et al. 1994. The diastrophic dysplasia gene encodes a novel sulfate transporter: positional cloning by fine-structure linkage disequilibrium mapping. *Cell* **78**:1073.

Orita, M., et al. 1989. Rapid and sensitive detection of point mutations and DNA polymorphisms using the polymerase chain reaction. *Genomics* **5**:874.

Tabor, H. K., N. J. Risch, and R. M. Myers. 2002. Opinion: candidate-gene approaches for studying complex genetic traits: practical considerations. *Nat. Rev. Genet.* **3**:391–397.

10

MOLECULAR STRUCTURE OF GENES AND CHROMOSOMES

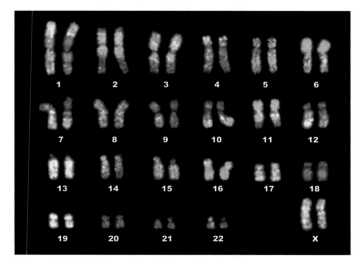

These brightly colored RxFISH-painted chromosomes are both beautiful and useful in revealing chromosome anomalies and in comparing karyotypes of different species. [© Department of Clinical Cytogenetics, Addenbrookes Hospital/Photo Researchers, Inc.]

By the beginning of the twenty-first century, molecular biologists had completed sequencing the entire genomes of hundreds of viruses, scores of bacteria, and the budding yeast *S. cerevisiae,* a unicellular eukaryote. In addition, the vast majority of the genome sequence is now known for several multicellular eukaryotes including the roundworm *C. elegans,* the fruit fly *D. melanogaster,* and humans (see Figure 9-34). Detailed analysis of these sequencing data has revealed that a large portion of the genomes of higher eukaryotes does not encode mRNAs or any other RNAs required by the organism. Remarkably, such noncoding DNA constitutes more than 95 percent of human chromosomal DNA.

The noncoding DNA in multicellular organisms contains many regions that are similar but not identical. Variations within some stretches of this *repetitious DNA* are so great that each single person can be distinguished by a DNA "fingerprint" based on these sequence variations. Moreover, some repetitious DNA sequences are not found in constant positions in the DNA of individuals of the same species. Such "mobile" DNA elements, which are present in both prokaryotic and eukaryotic organisms, can cause mutations when they move to new sites in the genome. Even though they generally have no function in the life cycle of an individual organism, mobile elements probably have played an important role in evolution.

In higher eukaryotes, DNA regions encoding proteins—that is, **genes**—lie amidst this expanse of apparently nonfunctional DNA. In addition to the nonfunctional DNA *between* genes, noncoding **introns** are common *within* genes of multicellular plants and animals. Introns are less common, but sometimes present, in single-celled eukaryotes and very

rare in bacteria. Sequencing of the same protein-coding gene in a variety of eukaryotic species has shown that evolutionary pressure selects for maintenance of relatively similar sequences in the coding regions, or **exons**. In contrast, wide sequence variation, even including total loss, occurs among introns, suggesting that most intron sequences have little functional significance.

The sheer length of cellular DNA is a significant problem with which cells must contend. The DNA in a single human cell, which measures about 2 meters in total length, must be contained within cells with diameters of less than 10 μm, a compaction ratio of greater than 10^5. Specialized eukaryotic proteins associated with nuclear DNA fold and organize it into the structures of DNA and protein visualized as individual **chromosomes** during mitosis. Mitochondria and

405

▶ **FIGURE 10-1 Overview of the structure of genes and chromosomes.** DNA of higher eukaryotes consists of unique and repeated sequences. Only ~5% of human DNA encodes proteins and functional RNAs and the regulatory sequences that control their expression; the remainder is merely spacer DNA between genes and introns within genes. Much of this DNA, ~50% in humans, is derived from mobile DNA elements, genetic symbiots that have contributed to the evolution of contemporary genomes. Each chromosome consists of a single, long molecule of DNA up to ~280 Mb in humans, organized into increasing levels of condensation by the histone and nonhistone proteins with which it is intricately complexed. Much smaller DNA molecules are localized in mitochondria and chloroplasts.

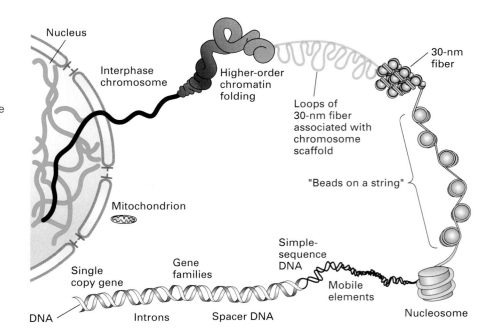

chloroplasts also contain DNA, probably evolutionary remnants of their origins, that encode essential components of these vital organelles.

In this chapter we first present a molecular definition of genes and the complexities that arise in higher organisms from the processing of mRNA precursors into alternatively spliced mRNAs. Next we discuss the main classes of eukaryotic DNA and the special properties of mobile DNA. We also consider the packaging of DNA and proteins into compact complexes, the large-scale structure of chromosomes, and the functional elements required for chromosome duplication and segregation. In the final section, we consider organelle DNA and how it differs from nuclear DNA. Figure 10-1 provides an overview of these interrelated subjects.

10.1 Molecular Definition of a Gene

In molecular terms, a gene commonly is defined as *the entire nucleic acid sequence that is necessary for the synthesis of a functional gene product (polypeptide or RNA).* According to this definition, a gene includes more than the nucleotides encoding the amino acid sequence of a protein, referred to as the *coding region.* A gene also includes all the DNA sequences required for synthesis of a particular RNA transcript. In eukaryotic genes, transcription-control regions known as **enhancers** can lie 50 kb or more from the coding region. Other critical noncoding regions in eukaryotic genes are the sequences that specify 3′ cleavage and polyadenylation, known as *poly(A) sites,* and splicing of primary RNA transcripts, known as *splice sites* (see Figure 4-14). Mutations in these RNA-processing signals prevent expression of a functional mRNA and thus of the encoded polypeptide.

Although most genes are transcribed into mRNAs, which encode proteins, clearly some DNA sequences are transcribed into RNAs that do not encode proteins (e.g., tRNAs and rRNAs). However, because the DNA that encodes tRNAs and rRNAs can cause specific phenotypes when it is mutated, these DNA regions generally are referred to as tRNA and rRNA *genes,* even though the final products of these genes are RNA molecules and not proteins. Many other RNA molecules described in later chapters also are transcribed from non-protein-coding genes.

Most Eukaryotic Genes Produce Monocistronic mRNAs and Contain Lengthy Introns

As discussed in Chapter 4, many bacterial mRNAs are *polycistronic;* that is, a single mRNA molecule (e.g., the mRNA encoded by the *trp* operon) includes the coding region for several proteins that function together in a biological process. In contrast, most eukaryotic mRNAs are *monocistronic;* that is, each mRNA molecule encodes a single protein. This difference between polycistronic and monocistronic mRNAs correlates with a fundamental difference in their translation.

Within a bacterial polycistronic mRNA a ribosome-binding site is located near the start site for each of the protein-coding regions, or cistrons, in the mRNA. Translation initiation can begin at any of these multiple internal sites, producing multiple proteins (see Figure 4-12a). In most eukaryotic mRNAs, however, the 5′-cap structure directs ribosome binding, and translation begins at the closest AUG start codon (see Figure 4-12b). As a result, translation begins only at this site. In many cases, the primary transcripts of eukaryotic protein-coding genes are processed into a single type of mRNA, which is translated to give a single type of polypeptide (see Figure 4-14).

Unlike bacterial and yeast genes, which generally lack introns, most genes in multicellular animals and plants contain introns, which are removed during RNA processing. In many cases, the introns in a gene are considerably longer than the exons. For instance, of the ≈50,000 base pairs composing many human genes encoding average-size proteins, more than 95 percent are present in introns and noncoding 5′ and 3′ regions. Many large proteins in higher organisms have repeated domains and are encoded by genes consisting of repeats of similar exons separated by introns of variable length. An example of this is fibronectin, a component of the extracellular matrix that is encoded by a gene containing multiple copies of three types of exons (see Figure 4-15).

Simple and Complex Transcription Units Are Found in Eukaryotic Genomes

The cluster of genes that form a bacterial operon comprises a single **transcription unit,** which is transcribed from a particular promoter into a single primary transcript. In other words, genes and transcription units often are distinguishable in prokaryotes. In contrast, most eukaryotic genes and transcription units generally are identical, and the two terms commonly are used interchangeably. Eukaryotic transcription units, however, are classified into two types, depending on the fate of the primary transcript.

The primary transcript produced from a *simple* transcription unit is processed to yield a single type of mRNA, encoding a single protein. Mutations in exons, introns, and transcription-control regions all may influence expression of the protein encoded by a simple transcription unit (Figure 10-2a).

(a) Simple transcription unit

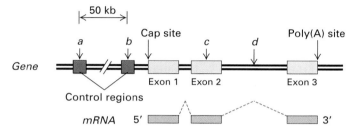

(b) Complex transcription units

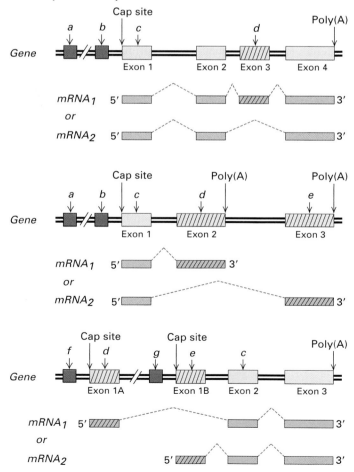

▶ **FIGURE 10-2 Comparison of simple and complex eukaryotic transcription units.** (a) A simple transcription unit includes a region that encodes one protein, extending from the 5′ cap site to the 3′ poly(A) site, and associated control regions. Introns lie between exons (blue rectangles) and are removed during processing of the primary transcripts (dashed red lines); they thus do not occur in the functional monocistronic mRNA. Mutations in a transcription-control region (*a, b*) may reduce or prevent transcription, thus reducing or eliminating synthesis of the encoded protein. A mutation within an exon (*c*) may result in an abnormal protein with diminished activity. A mutation within an intron (*d*) that introduces a new splice site results in an abnormally spliced mRNA encoding a nonfunctional protein. (b) Complex transcription units produce primary transcripts that can be processed in alternative ways. (*Top*) If a primary transcript contains alternative splice sites, it can be processed into mRNAs with the same 5′ and 3′ exons but different internal exons. (*Middle*) If a primary transcript has two poly(A) sites, it can be processed into mRNAs with alternative 3′ exons. (*Bottom*) If alternative promoters (*f* or *g*) are active in different cell types, mRNA₁, produced in a cell type in which *f* is activated, has a different exon (1A) than mRNA₂ has, which is produced in a cell type in which *g* is activated (and where exon 1B is used). Mutations in control regions (*a* and *b*) and those designated *c* within exons shared by the alternative mRNAs affect the proteins encoded by both alternatively processed mRNAs. In contrast, mutations (designated *d* and *e*) within exons unique to one of the alternatively processed mRNAs affect only the protein translated from that mRNA. For genes that are transcribed from different promoters in different cell types (*bottom*), mutations in different control regions (*f* and *g*) affect expression only in the cell type in which that control region is active.

In the case of *complex* transcription units, which are quite common in multicellular organisms, the primary RNA transcript can be processed in more than one way, leading to formation of mRNAs containing different exons. Each mRNA, however, is monocistronic, being translated into a single polypeptide, with translation usually initiating at the first AUG in the mRNA. Multiple mRNAs can arise from a primary transcript in three ways (Figure 10-2b):

1. Use of different splice sites, producing mRNAs with the same 5′ and 3′ exons but different internal exons. Figure 10-2b (*top*) shows one example of this type of alternative RNA processing, *exon skipping.*

2. Use of alternative poly(A) sites, producing mRNAs that share the same 5′ exons but have different 3′ exons (Figure 10-2b [*middle*]).

3. Use of alternative promoters, producing mRNAs that have different 5′ exons and common 3′ exons. A gene expressed selectively in two or more types of cells is often transcribed from distinct cell-type-specific promoters (Figure 10-2b [*bottom*]).

Examples of all three types of alternative RNA processing occur during sexual differentiation in *Drosophila* (see Figure 12-14). Commonly, one mRNA is produced from a complex transcription unit in some cell types, and an alternative mRNA is made in other cell types. For example, differences in RNA splicing of the primary fibronectin transcript in fibroblasts and hepatocytes determines whether or not the secreted protein includes domains that adhere to cell surfaces (see Figure 4-15).

The relationship between a mutation and a gene is not always straightforward when it comes to complex transcription units. A mutation in the control region or in an exon shared by alternative mRNAs will affect all the alternative proteins encoded by a given complex transcription unit. On the other hand, mutations in an exon present in only one of the alternative mRNAs will affect only the protein encoded by that mRNA. As explained in Chapter 9, **genetic complementation** tests commonly are used to determine if two mutations are in the same or different genes (see Figure 9-7). However, in the complex transcription unit shown in Figure 10-2b (*middle*), mutations *d* and *e* would complement each other in a genetic complementation test, even though they occur in the same gene. This is because a chromosome with mutation *d* can express a normal protein encoded by mRNA$_2$ and a chromosome with mutation *e* can express a normal protein encoded by mRNA$_1$. However, a chromosome with mutation *c* in an exon common to both mRNAs would not complement either mutation *d* or *e*. In other words, mutation *c* would be in the same complementation groups as mutations *d* and *e*, even though *d* and *e* themselves would not be in the same complementation group!

KEY CONCEPTS OF SECTION 10.1

Molecular Definition of a Gene

■ In molecular terms, a gene is the entire DNA sequence required for synthesis of a functional protein or RNA molecule. In addition to the coding regions (exons), a gene includes control regions and sometimes introns.

■ Most bacterial and yeast genes lack introns, whereas most genes in multicellular organisms contain introns. The total length of intron sequences often is much longer than that of exon sequences.

■ A simple eukaryotic transcription unit produces a single monocistronic mRNA, which is translated into a single protein.

■ A complex eukaryotic transcription unit is transcribed into a primary transcript that can be processed into two or more different monocistronic mRNAs depending on the choice of splice sites or polyadenylation sites (see Figure 10-2b).

■ Many complex transcription units (e.g., the fibronectin gene) express one mRNA in one cell type and an alternative mRNA in a different cell type.

10.2 Chromosomal Organization of Genes and Noncoding DNA

Having reviewed the relation between transcription units and genes, we now consider the organization of genes on chromosomes and the relationship of noncoding DNA sequences to coding sequences.

Genomes of Many Organisms Contain Much Nonfunctional DNA

Comparisons of the total chromosomal DNA per cell in various species first suggested that much of the DNA in certain organisms does not encode RNA or have any apparent regulatory or structural function. For example, yeasts, fruit flies, chickens, and humans have successively more DNA in their haploid chromosome sets (12; 180; 1300; and 3300 Mb, respectively), in keeping with what we perceive to be the increasing complexity of these organisms. Yet the vertebrates with the greatest amount of DNA per cell are amphibians, which are surely less complex than humans in their structure and behavior. Even more surprising, the unicellular protozoal species *Amoeba dubia* has 200 times more DNA per cell than humans. Many plant species also have considerably more DNA per cell than humans have. For example, tulips have 10 times as much DNA per cell as humans. The DNA content per cell also varies considerably between closely related species. All insects or all amphibians would appear to be similarly complex, but the amount of haploid DNA in

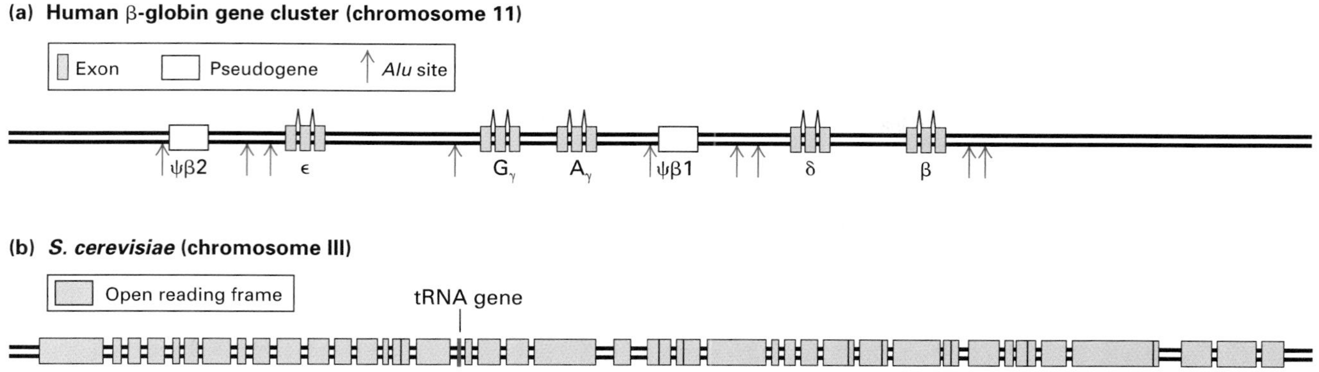

(a) Human β-globin gene cluster (chromosome 11)

| | Exon | | Pseudogene | ↑ *Alu* site |

ψβ2   ε   G$_\gamma$ A$_\gamma$ ψβ1   δ   β

(b) *S. cerevisiae* (chromosome III)

| | Open reading frame |

tRNA gene

▲ **FIGURE 10-3 Comparative density of genes in ≈80-kb regions of genomic DNA from humans and the yeast *S. cerevisiae*.** (a) In the diagram of the β-globin gene cluster on human chromosome 11, the green boxes represent exons of β-globin–related genes. Exons spliced together to form one mRNA are connected by caret-like spikes. The human β-globin gene cluster contains two pseudogenes (white); these regions are related to the functional globin-type genes but are not transcribed. Each red arrow indicates the location of an *Alu* sequence, an ≈300-bp noncoding repeated sequence that is abundant in the human genome. (b) In the diagram of yeast DNA from chromosome III, the green boxes indicate open reading frames. Most of these potential protein-coding sequences probably are functional genes without introns. Note the much higher proportion of noncoding-to-coding sequences in the human DNA than in the yeast DNA. [Part (a), see F. S. Collins and S. M. Weissman, 1984, *Prog. Nucl. Acid Res. Mol. Biol.* **31**:315; part (b), see S. G. Oliver et al., 1992, *Nature* **357**:28.]

species within each of these phylogenetic classes varies by a factor of 100.

Detailed sequencing and identification of exons in chromosomal DNA have provided direct evidence that the genomes of higher eukaryotes contain large amounts of noncoding DNA. For instance, only a small portion of the β-globin gene cluster of humans, about 80 kb long, encodes protein (Figure 10-3a). Moreover, compared with other regions of vertebrate DNA, the β-globin gene cluster is unusually rich in protein-coding sequences, and the introns in globin genes are considerably shorter than those in many human genes. In contrast, a typical 80-kb stretch of DNA from the yeast *S. cerevisiae*, a single-celled eukaryote (Figure 10-3b) contains many closely spaced protein-coding sequences without introns and relatively much less noncoding DNA.

The density of genes varies greatly in different regions of human chromosomal DNA, from "gene-rich" regions, such as the β-globin cluster, to large gene-poor "deserts." Of the 94 percent of human genomic DNA that has been sequenced, only ≈1.5 percent corresponds to protein-coding sequences (exons). Most human exons contain 50–200 base pairs, although the 3′ exon in many transcription units is much longer. Human introns vary in length considerably. Although many are ≈90 bp long, some are much longer; their median length is 3.3 kb. Approximately one-third of human genomic DNA is thought to be transcribed into pre-mRNA precursors, but some 95 percent of these sequences are in introns, which are removed by RNA splicing.

Different selective pressures during evolution may account, at least in part, for the remarkable difference in the amount of nonfunctional DNA in unicellular and multicellu-lar organisms. For example, microorganisms must compete for limited amounts of nutrients in their environment, and metabolic economy thus is a critical characteristic. Since synthesis of nonfunctional (i.e., noncoding) DNA requires time and energy, presumably there was selective pressure to lose nonfunctional DNA during the evolution of microorganisms. On the other hand, natural selection in vertebrates depends largely on their behavior. The energy invested in DNA synthesis is trivial compared with the metabolic energy required for the movement of muscles; thus there was little selective pressure to eliminate nonfunctional DNA in vertebrates.

Protein-Coding Genes May Be Solitary or Belong to a Gene Family

The nucleotide sequences within chromosomal DNA can be classified on the basis of structure and function, as shown in Table 10-1. We will examine the properties of each class, beginning with protein-coding genes, which comprise two groups.

In multicellular organisms, roughly 25–50 percent of the protein-coding genes are represented only once in the haploid genome and thus are termed *solitary* genes. A well-studied example of a solitary protein-coding gene is the chicken lysozyme gene. The 15-kb DNA sequence encoding chicken lysozyme constitutes a simple transcription unit containing four exons and three introns. The flanking regions, extending for about 20 kb upstream and downstream from the transcription unit, do not encode any detectable mRNAs. Lysozyme, an enzyme that cleaves the polysaccharides in bacterial cell walls, is an abundant

TABLE 10-1	Major Classes of Eukaryotic DNA and Their Representation in the Human Genome		
Class	Length	Copy Number in Human Genome	Fraction of Human Genome, %
Protein-coding genes			
Solitary genes	Variable	1	≈15* (0.8)†
Duplicated or diverged genes in gene families	Variable	2–≈1000	≈15* (0.8)†
Tandemly repeated genes encoding rRNAs, tRNAs, snRNAs, and histones	Variable	20–300	0.3
Repetitious DNA			
Simple-sequence DNA	1–500 bp	Variable	3
Interspersed repeats			
DNA transposons	2–3 kb	300,000	3
LTR retrotransposons	6–11 kb	440,000	8
Non-LTR retrotransposons			
LINEs	6–8 kb	860,000	21
SINEs	100–300 bp	1,600,000	13
Processed pseudogenes	Variable	1–≈100	≈0.4
Unclassified spacer DNA	Variable	n.a.‡	≈25

*Complete transcription units, including introns.
†Protein-coding exons. The total number of human protein-coding genes is estimated to be 30,000–35,000, but this number is based on current methods for identifying genes in the human genome sequence and may be an underestimate.
‡Not applicable.
SOURCE: E. S. Lander et al., 2001, *Nature* **409**:860.

component of chicken egg-white protein and also is found in human tears. Its activity helps to keep the surface of the eye and the chicken egg sterile.

Duplicated genes constitute the second group of protein-coding genes. These are genes with close but nonidentical sequences that generally are located within 5–50 kb of one another. In vertebrate genomes, duplicated genes probably constitute half the protein-coding DNA sequences. A set of duplicated genes that encode proteins with similar but nonidentical amino acid sequences is called a **gene family**; the encoded, closely related, homologous proteins constitute a **protein family**. A few protein families, such as protein kinases, transcription factors, and vertebrate immunoglobulins, include hundreds of members. Most protein families, however, include from just a few to 30 or so members; common examples are cytoskeletal proteins, 70-kDa heat-shock proteins, the myosin heavy chain, chicken ovalbumin, and the α- and β-globins in vertebrates.

The genes encoding the β-like globins are a good example of a gene family. As shown in Figure 10-3a, the β-like globin gene family contains five functional genes designated β, δ, A_γ, G_γ, and ε; the encoded polypeptides are similarly designated. Two identical β-like globin polypeptides combine with two identical α-globin polypeptides (encoded by another gene family) and four small heme groups to form a hemoglobin molecule (see Figure 3-10). All the hemoglobins formed from the different β-like globins carry oxygen in the blood, but they exhibit somewhat different properties that are suited to specific roles in human physiology. For example, hemoglobins containing either the A_γ or G_γ polypeptides are expressed only during fetal life. Because these fetal hemoglobins have a higher affinity for oxygen than adult hemoglobins, they can effectively extract oxygen from the maternal circulation in the placenta. The lower oxygen affinity of adult hemoglobins, which are expressed after birth, permits better release of oxygen to the tissues, espe-

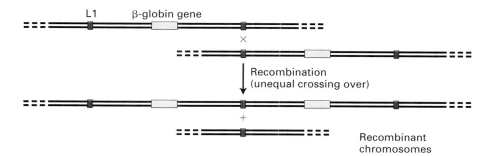

▲ **FIGURE 10-4 Gene duplication resulting from unequal crossing over.** Each parental chromosome (*top*) contains one ancestral β-globin gene containing three exons and two introns. Homologous noncoding L1 repeated sequences lie 5′ and 3′ of the β-globin gene. The parental chromosomes are shown displaced relative to each other, so that the L1 sequences are aligned. Homologous recombination between L1 sequences as shown would generate one recombinant chromosome with two copies of the β-globin gene and one chromosome with a deletion of the β-globin gene. Subsequent independent mutations in the duplicated genes could lead to slight changes in sequence that might result in slightly different functional properties of the encoded proteins. Unequal crossing over also can result from rare recombinations between unrelated sequences. [See D. H. A. Fitch et al., 1991, *Proc. Nat'l. Acad. Sci. USA* **88**:7396.]

cially muscles, which have a high demand for oxygen during exercise.

The different β-globin genes probably arose by duplication of an ancestral gene, most likely as the result of an "unequal crossover" during meiotic recombination in a developing germ cell (egg or sperm) (Figure 10-4). Over evolutionary time the two copies of the gene that resulted accumulated random mutations; beneficial mutations that conferred some refinement in the basic oxygen-carrying function of hemoglobin were retained by natural selection, resulting in *sequence drift*. Repeated gene duplications and subsequent sequence drift are thought to have generated the contemporary globin-like genes observed in humans and other complex species today.

Two regions in the human β-like globin gene cluster contain nonfunctional sequences, called **pseudogenes,** similar to those of the functional β-like globin genes (see Figure 10-3a). Sequence analysis shows that these pseudogenes have the same apparent exon-intron structure as the functional β-like globin genes, suggesting that they also arose by duplication of the same ancestral gene. However, sequence drift during evolution generated sequences that either terminate translation or block mRNA processing, rendering such regions nonfunctional even if they were transcribed into RNA. Because such pseudogenes are not deleterious, they remain in the genome and mark the location of a gene duplication that occurred in one of our ancestors. As discussed in a later section, other nonfunctional gene copies can arise by reverse transcription of mRNA into cDNA and integration of this intron-less DNA into a chromosome.

Several different gene families encode the various proteins that make up the cytoskeleton. These proteins are present in varying amounts in almost all cells. In vertebrates, the major cytoskeletal proteins are the actins, tubulins, and intermediate filament proteins like the keratins. We examined the origin of one such family, the tubulin

family, in the last chapter (see Figure 9-32). Although the physiological rationale for the cytoskeletal protein families is not as obvious as it is for the globins, the different members of a family probably have similar but subtly different functions suited to the particular type of cell in which they are expressed.

Tandemly Repeated Genes Encode rRNAs, tRNAs, and Histones

In vertebrates and invertebrates, the genes encoding rRNAs and some other noncoding RNAs such as some of the snRNAs involved in RNA splicing occur as *tandemly repeated arrays*. These are distinguished from the duplicated genes of gene families in that the multiple tandemly repeated genes encode identical or nearly identical proteins or functional RNAs. Most often copies of a sequence appear one after the other, in a head-to-tail fashion, over a long stretch of DNA. Within a tandem array of rRNA genes, each copy is exactly, or almost exactly, like all the others. Although the transcribed portions of rRNA genes are the same in a given individual, the non-transcribed spacer regions between the transcribed regions can vary.

The tandemly repeated rRNA, tRNA, and histone genes are needed to meet the great cellular demand for their transcripts. To understand why, consider that a fixed maximal number of RNA copies can be produced from a single gene during one cell generation when the gene is fully loaded with RNA polymerase molecules. If more RNA is required than can be transcribed from one gene, multiple copies of the gene are necessary. For example, during early embryonic development in humans, many embryonic cells have a doubling time of ≈24 hours and contain 5–10 million ribosomes. To produce enough rRNA to form this many ribosomes, an embryonic human cell needs at least 100 copies of the large and small subunit rRNA genes, and most of these must be close

TABLE 10-2	Effect of Gene Copy Number and RNA Polymerase Loading on Rate of Pre-rRNA Synthesis in Humans	
Copies of Pre-rRNA Gene	RNA Polymerase Molecules per Gene	Molecules of Pre-rRNA Produced in 24 Hours
1	1	288
1	≈250	≈70,000
100	≈250	≈7,000,000

to maximally active for the cell to divide every 24 hours (Table 10-2). That is, multiple RNA polymerases must be loaded onto and transcribing each rRNA gene at the same time (see Figure 12-32).

All eukaryotes, including yeasts, contain 100 or more copies of the genes encoding 5S rRNA and the large and small subunit rRNAs. The importance of repeated rRNA genes is illustrated by *Drosophila* mutants called *bobbed* (because they have stubby wings), which lack a full complement of the tandemly repeated **pre-rRNA** genes. A *bobbed* mutation that reduces the number of pre-rRNA genes to less than ≈50 is a recessive lethal mutation.

Multiple copies of tRNA and histone genes also occur, often in clusters, but generally not in tandem arrays.

Most Simple-Sequence DNAs Are Concentrated in Specific Chromosomal Locations

Besides duplicated protein-coding genes and tandemly repeated genes, eukaryotic cells contain multiple copies of other DNA sequences in the genome, generally referred to as *repetitious DNA* (see Table 10-1). Of the two main types of repetitious DNA, the less prevalent is **simple-sequence DNA**, which constitutes about 3 percent of the human genome and is composed of perfect or nearly perfect repeats of relatively short sequences. The more common type of repetitious DNA, composed of much longer sequences, is discussed in Section 10.3.

Simple-sequence DNA is commonly called *satellite* DNA because in early studies of DNAs from higher organisms using equilibrium buoyant-density ultracentrifugation some simple-sequence DNAs banded at a different position from the bulk of cellular DNA. These were called *satellite bands* to distinguish them from the main band of DNA in the buoyant-density gradient. Simple-sequence DNAs in which the repeats contain 1–13 base pairs are often called *microsatellites*. Most have repeat lengths of 1–4 base pairs and usually occur in tandem repeats of 150 base pairs or fewer. Microsatellites are thought to have originated by "backward slippage" of a daughter strand on its template strand during DNA replication so that the same short sequence is copied twice.

 Microsatellites occasionally occur within transcription units. Some individuals are born with a larger number of repeats in specific genes than observed in the general population, presumably because of daughter-strand slippage during DNA replication in a germ cell from which they developed. Such expanded microsatellites have been found to cause at least 14 different types of neuromuscular diseases, depending on the gene in which they occur. In some cases expanded microsatellites behave like a recessive mutation because they interfere with the function or expression of the encoded gene. But in the more common types of diseases associated with expanded microsatellite repeats, *myotonic dystrophy* and *spinocerebellar ataxia,* the expanded repeats behave like dominant mutations because they interfere with RNA processing in general in the neurons where the affected genes are expressed. ∎

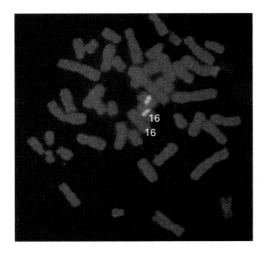

▲ **EXPERIMENTAL FIGURE 10-5 Simple-sequence DNAs are useful chromosomal markers.** Human metaphase chromosomes stained with a fluorescent dye were hybridized in situ with a particular simple-sequence DNA labeled with a fluorescent biotin derivative. When viewed under the appropriate wavelength of light, the DNA appears red and the hybridized simple-sequence DNA appears as a yellow band on chromosome 16, thus locating this particular simple sequence to one site in the genome. [See R. K. Moyzis et al., 1987, *Chromosoma* **95**:378; courtesy of R. K. Moyzis.]

Most satellite DNA is composed of repeats of 14–500 base pairs in tandem repeats of 20–100 kb. In situ hybridization studies with metaphase chromosomes have localized these satellite DNAs to specific chromosomal regions. In most mammals, much of this satellite DNA lies near **centromeres,** the discrete chromosomal regions that attach to spindle microtubules during mitosis and meiosis. Satellite DNA is also located at **telomeres,** the ends of chromosomes, and at specific locations within chromosome arms in some organisms. These latter sequences can be useful for identifying particular chromosomes by **fluorescence in situ hybridization (FISH),** as illustrated in Figure 10-5.

Simple-sequence DNA located at centromeres may assist in attaching chromosomes to spindle microtubules during mitosis. As yet, however, there is little clear-cut experimental evidence demonstrating any function for most simple-sequence DNA, with the exception of the short repeats at the very ends of chromosomes discussed in a later section.

DNA Fingerprinting Depends on Differences in Length of Simple-Sequence DNAs

Within a species, the nucleotide sequences of the repeat units composing simple-sequence DNA tandem arrays are highly conserved among individuals. In contrast, differences in the *number* of repeats, and thus in the length of simple-sequence tandem arrays containing the same repeat unit, are quite common among individuals. These differences in length are

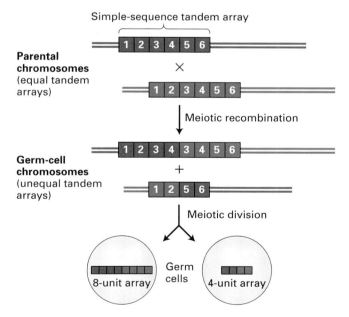

▲ **FIGURE 10-6 Generation of differences in lengths of a simple-sequence DNA by unequal crossing over during meiosis.** In this example, unequal crossing over within a stretch of DNA containing six copies (1–6) of a particular simple-sequence repeat unit yields germ cells containing either an eight-unit or a four-unit tandem array.

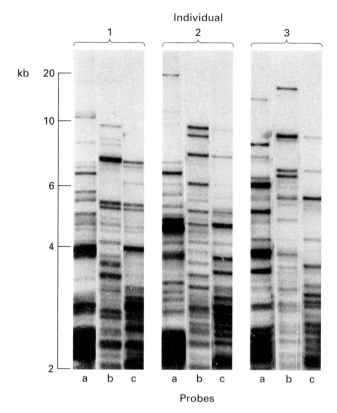

▲ **EXPERIMENTAL FIGURE 10-7 Probes for minisatellite DNA can reveal unique restriction fragments (DNA fingerprints) that distinguish individuals.** DNA samples from three individuals (1, 2, and 3) were subjected to Southern blot analysis using the restriction enzyme *Hin*f1 and three different labeled minisatellites as probes (lanes a, b, and c). DNA from each individual produced a unique band pattern with each probe. Conditions of electrophoresis can be adjusted so that for each person at least 50 bands can be resolved with this restriction enzyme. The nonidentity of these three samples is easily distinguished. [From A. J. Jeffreys et al., 1985, *Nature* **316**:76; courtesy of A. J. Jeffreys.]

thought to result from unequal crossing over within regions of simple-sequence DNA during meiosis (Figure 10-6). As a consequence of this unequal crossing over, the lengths of some tandem arrays are unique in each individual.

In humans and other mammals, some of the satellite DNA exists in relatively short 1- to 5-kb regions made up of 20–50 repeat units, each containing 15 to about 100 base pairs. These regions are called *minisatellites* to distinguish them from the more common regions of tandemly repeated satellite DNA, which are ≈20–100 kb in length. They differ from microsatellites mentioned earlier, which have very short repeat units. Even slight differences in the total lengths of various minisatellites from different individuals can be detected by **Southern blotting** of cellular DNA treated with a restriction enzyme that cuts outside the repeat sequence (Figure 10-7). The polymerase chain reaction (PCR), using primers that hybridize to the unique sequences flanking each minisatellite, also can detect differences in minisatellite lengths between individuals. These DNA polymorphisms form the basis of *DNA fingerprinting,* which is superior to conventional fingerprinting for identifying individuals.

KEY CONCEPTS OF SECTION 10.2

Chromosomal Organization of Genes and Noncoding DNA

■ In the genomes of prokaryotes and most lower eukaryotes, which contain few nonfunctional sequences, coding regions are densely arrayed along the genomic DNA.

■ In contrast, vertebrate genomes contain many sequences that do not code for RNAs or have any structural or regulatory function. Much of this nonfunctional DNA is composed of repeated sequences. In humans, only about 1.5 percent of total DNA (the exons) actually encodes proteins or functional RNAs.

■ Variation in the amount of nonfunctional DNA in the genomes of various species is largely responsible for the lack of a consistent relationship between the amount of DNA in the haploid chromosomes of an animal or plant and its phylogenetic complexity.

■ Eukaryotic genomic DNA consists of three major classes of sequences: genes encoding proteins and functional RNAs, including gene families and tandemly repeated genes; repetitious DNA; and spacer DNA (see Table 10-1).

■ About half the protein-coding genes in vertebrate genomic DNA are solitary genes, each occurring only once in the haploid genome. The remainder are duplicated genes, which arose by duplication of an ancestral gene and subsequent independent mutations (see Figure 10-4).

■ Duplicated genes encode closely related proteins and generally appear as a cluster in a particular region of DNA. The proteins encoded by a gene family have homologous but nonidentical amino acid sequences and exhibit similar but slightly different properties.

■ In invertebrates and vertebrates, rRNAs are encoded by multiple copies of genes located in tandem arrays in genomic DNA. Multiple copies of tRNA and histone genes also occur, often in clusters, but not generally in tandem arrays.

■ Simple-sequence DNA, which consists largely of quite short sequences repeated in long tandem arrays, is preferentially located in centromeres, telomeres, and specific locations within the arms of particular chromosomes.

■ The length of a particular simple-sequence tandem array is quite variable between individuals in a species, probably because of unequal crossing over during meiosis (see Figure 10-6). Differences in the lengths of some simple-sequence tandem arrays form the basis for DNA fingerprinting.

10.3 Mobile DNA

The second type of repetitious DNA in eukaryotic genomes, termed *interspersed repeats* (also known as *moderately repeated DNA,* or *intermediate-repeat DNA*) is composed of a very large number of copies of relatively few sequence families (see Table 10-1). These sequences, which are interspersed throughout mammalian genomes, make up ≈25–50 percent of mammalian DNA (≈45 percent of human DNA).

Because moderately repeated DNA sequences have the unique ability to "move" in the genome, they are called **mobile DNA elements** (or *transposable elements*). Although mobile DNA elements, ranging from hundreds to a few thousand base pairs in length, originally were discovered in eukaryotes, they also are found in prokaryotes. The process by which these sequences are copied and inserted into a new site in the genome is called **transposition.** Mobile DNA elements (or simply *mobile elements*) are essentially molecular symbiots that in most cases appear to have no specific function in the biology of their host organisms, but exist only to maintain themselves. For this reason, Francis Crick referred to them as "selfish DNA."

When transposition of eukaryotic mobile elements occurs in germ cells, the transposed sequences at their new sites can be passed on to succeeding generations. In this way, mobile elements have multiplied and slowly accumulated in eukaryotic genomes over evolutionary time. Since mobile elements are eliminated from eukaryotic genomes very slowly, they now constitute a significant portion of the genomes of many eukaryotes. Transposition also may occur within a somatic cell; in this case the transposed sequence is transmitted only to the daughter cells derived from that cell. In rare cases, this may lead to a somatic-cell mutation with detrimental phenotypic effects, for example, the inactivation of a tumor-suppressor gene (Chapter 23).

Movement of Mobile Elements Involves a DNA or an RNA Intermediate

Barbara McClintock discovered the first mobile elements while doing classical genetic experiments in maize (corn) during the 1940s. She characterized genetic entities that could move into and back out of genes, changing the phenotype of corn kernels. Her theories were very controversial until similar mobile elements were discovered in bacteria, where they were characterized as specific DNA sequences, and the molecular basis of their transposition was deciphered.

As research on mobile elements progressed, they were found to fall into two categories: (1) those that transpose directly as DNA and (2) those that transpose via an RNA intermediate transcribed from the mobile element by an RNA polymerase and then converted back into double-stranded DNA by a **reverse transcriptase** (Figure 10-8). Mobile elements that transpose through a DNA intermediate are generally referred to as **DNA transposons.** Mobile elements that transpose to new sites in the genome via an RNA intermedi-

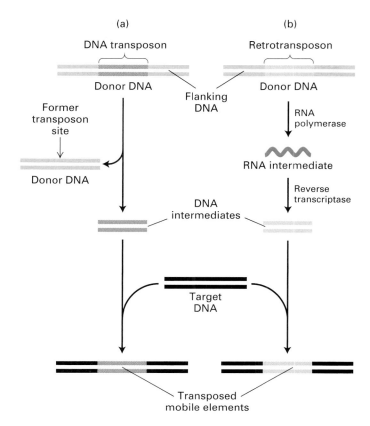

▲ FIGURE 10-8 Classification of mobile elements into two major classes. (a) Eukaryotic DNA transposons (orange) move via a DNA intermediate, which is excised from the donor site. (b) Retrotransposons (green) are first transcribed into an RNA molecule, which then is reverse-transcribed into double-stranded DNA. In both cases, the double-stranded DNA intermediate is integrated into the target-site DNA to complete movement. Thus DNA transposons move by a cut-and-paste mechanism, whereas retrotransposons move by a copy-and-paste mechanism.

ate are called **retrotransposons** because their movement is analogous to the infectious process of retroviruses. Indeed, retroviruses can be thought of as retrotransposons that evolved genes encoding viral coats, thus allowing them to transpose between cells. Retrotransposons can be further classified on the basis of their specific mechanism of transposition. We describe the structure and movement of the major types of mobile elements and then consider their likely role in evolution.

Mobile Elements That Move as DNA Are Present in Prokaryotes and Eukaryotes

Most mobile elements in bacteria transpose directly as DNA. In contrast, most mobile elements in eukaryotes are retrotransposons, but eukaryotic DNA transposons also occur. Indeed, the original mobile elements discovered by Barbara McClintock are DNA transposons.

Bacterial Insertion Sequences The first molecular understanding of mobile elements came from the study of certain *E. coli* mutations caused by the spontaneous insertion of a DNA sequence, ≈1–2 kb long, into the middle of a gene. These inserted stretches of DNA are called *insertion sequences*, or *IS elements*. So far, more than 20 different IS elements have been found in *E. coli* and other bacteria.

Transposition of an IS element is a very rare event, occurring in only one in 10^5–10^7 cells per generation, depending on the IS element. Many transpositions inactivate essential genes, killing the host cell and the IS elements it carries. Therefore, higher rates of transposition would probably result in too great a mutation rate for the host cell to survive. However, since IS elements transpose more or less randomly, some transposed sequences enter nonessential regions of the genome (e.g., regions between genes), allowing the cell to survive. At a very low rate of transposition, most host cells survive and therefore propagate the symbiotic IS element. IS elements also can insert into plasmids or lysogenic viruses, and thus be transferred to other cells. When this happens, IS elements can transpose into the chromosomes of virgin cells.

The general structure of IS elements is diagrammed in Figure 10-9. An *inverted repeat*, usually containing ≈50 base pairs, invariably is present at each end of an insertion sequence. In an inverted repeat the 5′ → 3′ sequence on one strand is repeated on the other strand, as:

$$5'\ \overrightarrow{\text{GAGC}}\text{———GCTC}\ 3'$$
$$3'\ \text{CTCG———}\overleftarrow{\text{CGAG}}\ 5'$$

Between the inverted repeats is a region that encodes *transposase*, an enzyme required for transposition of the IS ele-

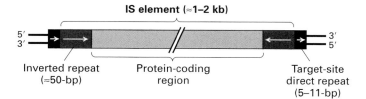

▲ FIGURE 10-9 General structure of bacterial IS elements. The relatively large central region of an IS element, which encodes one or two enzymes required for transposition, is flanked by an inverted repeat at each end. The sequences of the inverted repeats are nearly identical, but they are oriented in opposite directions. The sequence is characteristic of a particular IS element. The 5′ and 3′ short *direct* (as opposed to *inverted*) repeats are not transposed with the insertion element; rather, they are insertion-site sequences that become duplicated, with one copy at each end, during insertion of a mobile element. The length of the direct repeats is constant for a given IS element, but their sequence depends on the site of insertion and therefore varies with each transposition of the IS element. Arrows indicate sequence orientation. The regions in this diagram are not to scale; the coding region makes up most of the length of an IS element.

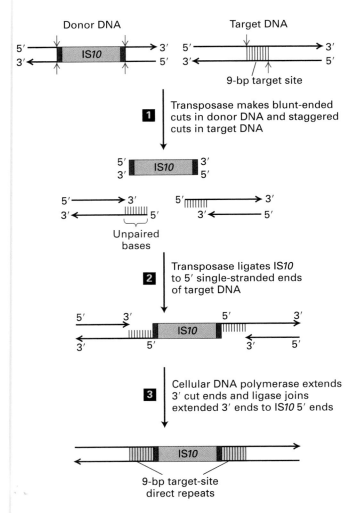

▲ FIGURE 10-10 Model for transposition of bacterial insertion sequences. Step **1**: Transposase, which is encoded by the IS element (IS10 in this example), cleaves both strands of the donor DNA next to the inverted repeats (dark red), excising the IS10 element. At a largely random target site, transposase makes staggered cuts in the target DNA. In the case of IS10, the two cuts are 9 bp apart. Step **2**: Ligation of the 3′ ends of the excised IS element to the staggered sites in the target DNA also is catalyzed by transposase. Step **3**: The 9-bp gaps of single-stranded DNA left in the resulting intermediate are filled in by a cellular DNA polymerase; finally cellular DNA ligase forms the 3′→5′ phosphodiester bonds between the 3′ ends of the extended target DNA strands and the 5′ ends of the IS10 strands. This process results in duplication of the target-site sequence on each side of the inserted IS element. Note that the length of the target site and IS10 are not to scale. [See H. W. Benjamin and N. Kleckner, 1989, *Cell* **59**:373, and 1992, *Proc. Nat'l. Acad. Sci. USA* **89**:4648.]

ment to a new site. The transposase is expressed at a very low rate, accounting for the very low frequency of transposition. An important hallmark of IS elements is the presence of a short *direct-repeat sequence*, containing 5–11 base pairs, immediately adjacent to both ends of the inserted element.

The *length* of the direct repeat is characteristic of each type of IS element, but its *sequence* depends on the target site where a particular copy of the IS element is inserted. When the sequence of a mutated gene containing an IS element is compared with the sequence of the wild-type gene before insertion, only one copy of the short direct-repeat sequence is found in the wild-type gene. Duplication of this target-site sequence to create the second direct repeat adjacent to an IS element occurs during the insertion process.

As depicted in Figure 10-10, transposition of an IS element is similar to a "cut-and-paste" operation in a word-processing program. Transposase performs three functions in this process: it (1) precisely excises the IS element in the donor DNA, (2) makes staggered cuts in a short sequence in the target DNA, and (3) ligates the 3′ termini of the IS element to the 5′ ends of the cut donor DNA. Finally, a host-cell DNA polymerase fills in the single-stranded gaps, generating the short direct repeats that flank IS elements, and DNA ligase joins the free ends.

Eukaryotic DNA Transposons McClintock's original discovery of mobile elements came from observation of certain spontaneous mutations in maize that affect production of any of the several enzymes required to make anthocyanin, a purple pigment in maize kernels. Mutant kernels are white, and wild-type kernels are purple. One class of these mutations is revertible at high frequency, whereas a second class of mutations does not revert unless they occur in the presence of the first class of mutations. McClintock called the agent responsible for the first class of mutations the *activator (Ac) element* and those responsible for the second class *dissociation (Ds) elements* because they also tended to be associated with chromosome breaks.

Many years after McClintock's pioneering discoveries, cloning and sequencing revealed that Ac elements are equivalent to bacterial IS elements. Like IS elements, they contain inverted terminal repeat sequences that flank the coding region for a transposase, which recognizes the terminal repeats and catalyzes transposition to a new site in DNA. Ds elements are deleted forms of the Ac element in which a portion of the sequence encoding transposase is missing. Because it does not encode a functional transposase, a Ds element cannot move by itself. However, in plants that carry the Ac element, and thus express a functional transposase, Ds elements can move.

Since McClintock's early work on mobile elements in corn, transposons have been identified in other eukaryotes. For instance, approximately half of all the spontaneous mutations observed in *Drosophila* are due to the insertion of mobile elements. Although most of the mobile elements in *Drosophila* function as retrotransposons, at least one—the **P element**—functions as a DNA transposon, moving by a cut-and-paste mechanism similar to that used by bacterial insertion sequences. Current methods for constructing transgenic *Drosophila* depend on engineered, high-level expression of the P-element transposase and use of the P-element inverted terminal repeats as targets for transposition.

DNA transposition by the cut-and-paste mechanism can result in an increase in the copy number of a transposon when it occurs during S phase, the period of the cell cycle when DNA synthesis occurs. This happens when the donor DNA is from one of the two daughter DNA molecules in a region of a chromosome that has replicated and the target DNA is in the region that has not yet replicated. When DNA replication is complete at the end of the S phase, the target DNA in its new location is also replicated. This results in a net increase by one in the total number of these transposons in the cell. When this occurs during the S phase preceding meiosis, two of the four germ cells produced have the extra copy. Repetition of this process over evolutionary time has resulted in the accumulation of large numbers of DNA transposons in the genomes of some organisms. Human DNA contains about 300,000 copies of full-length and deleted DNA tranposons, amounting to ≈3 percent of human DNA.

Some Retrotransposons Contain LTRs and Behave Like Intracellular Retroviruses

The genomes of all eukaryotes studied from yeast to humans contain retrotransposons, mobile DNA elements that transpose through an RNA intermediate utilizing a reverse transcriptase (see Figure 10-8b). These mobile elements are divided into two major categories, those containing and those lacking **long terminal repeats (LTRs)**. LTR retrotransposons, which we discuss in this section, are common in yeast (e.g., Ty elements) and in *Drosophila* (e.g., *copia* elements). Although less abundant in mammals than non-LTR retrotransposons, LTR retrotransposons nonetheless constitute ≈8 percent of human genomic DNA. Because they exhibit some similarities with retroviruses, these mobile elements sometimes are called *viral retrotransposons*. In mammals, retrotransposons lacking LTRs are the most common type of mobile element; these are described in the next section.

The general structure of LTR retrotransposons found in eukaryotes is depicted in Figure 10-11. In addition to short 5′ and 3′ direct repeats typical of all mobile elements, these retrotransposons are marked by the presence of LTRs flanking the central protein-coding region. These *l*ong direct *t*erminal *r*epeats, containing ≈250–600 base pairs, are characteristic of integrated retroviral DNA and are critical to the life cycle of retroviruses. In addition to sharing LTRs with retroviruses, LTR retrotransposons encode all the proteins of the most common type of retroviruses, except for the envelope proteins. Lacking these envelope proteins, LTR retrotransposons cannot bud from their host cell and infect other cells; however, they can transpose to new sites in the DNA of their host cell.

A key step in the retroviral life cycle is formation of retroviral genomic RNA from integrated retroviral DNA (see Figure 4-43). This process serves as a model for generation of the RNA intermediate during transposition of LTR retrotransposons. As depicted in Figure 10-12, the leftward retroviral LTR functions as a promoter that directs host-cell RNA polymerase II to initiate transcription at the 5′ nucleotide of the R sequence. After the entire downstream retroviral DNA has been transcribed, the RNA sequence corresponding to the rightward LTR directs host-cell RNA-processing enzymes to cleave the primary transcript and add a poly(A) tail at the 3′ end of the R sequence.

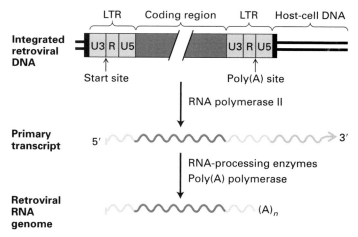

▲ **FIGURE 10-12 Generation of retroviral genomic RNA from integrated retroviral DNA.** The left LTR directs cellular RNA polymerase II to initiate transcription at the first nucleotide of the left R region. The resulting primary transcript extends beyond the right LTR. The right LTR, now present in the RNA primary transcript, directs cellular enzymes to cleave the primary transcript at the last nucleotide of the right R region and to add a poly(A) tail, yielding a retroviral RNA genome with the structure shown at the top of Figure 10-13. A similar mechanism is thought to generate the RNA intermediate during transposition of retrotransposons. The short direct-repeat sequences (black) of target-site DNA are generated during integration of the retroviral DNA into the host-cell genome.

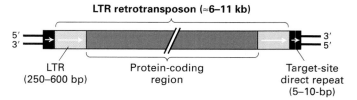

▲ **FIGURE 10-11 General structure of eukaryotic LTR retrotransposons.** The central protein-coding region is flanked by two long terminal repeats (LTRs), which are element-specific direct repeats. Like other mobile elements, integrated retrotransposons have short target-site direct repeats at each end. Note that the different regions are not drawn to scale. The protein-coding region constitutes 80 percent or more of a retrotransposon and encodes reverse transcriptase, integrase, and other retroviral proteins.

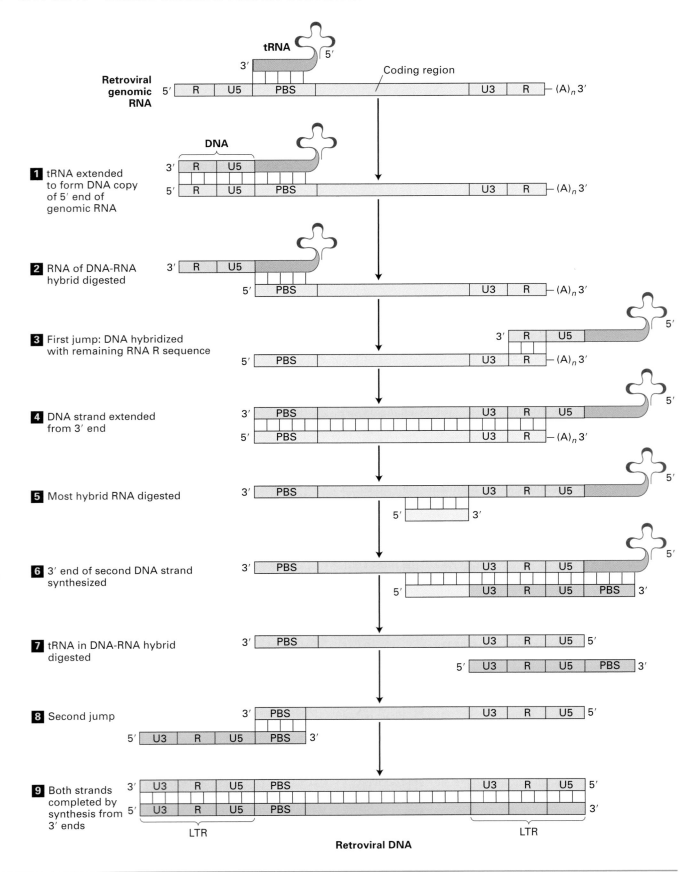

Retroviral DNA

◀ **FIGURE 10-13 Model for reverse transcription of retroviral genomic RNA into DNA.** In this model, a complicated series of nine events generates a double-stranded DNA copy of the single-stranded RNA genome of a retrovirus (*top*). The genomic RNA is packaged in the virion with a retrovirus-specific cellular tRNA hybridized to a complementary sequence near its 5′ end called the *primer-binding site* (PBS). The retroviral RNA has a short direct-repeat terminal sequence (R) at each end. The overall reaction is carried out by reverse transcriptase, which catalyzes polymerization of deoxyribonucleotides. RNaseH digests the RNA strand in a DNA-RNA hybrid. The entire process yields a double-stranded DNA molecule that is longer than the template RNA and has a long terminal repeat (LTR) at each end. The different regions are not shown to scale. The PBS and R regions are actually much shorter than the U5 and U3 regions, and the central coding region is very much longer than the other regions. [See E. Gilboa et al., 1979, *Cell* **18**:93.]

The resulting retroviral RNA genome, which lacks a complete LTR, is packaged into a virion that buds from the host cell.

After a retrovirus infects a cell, reverse transcription of its RNA genome by the retrovirus-encoded reverse transcriptase yields a double-stranded DNA containing complete LTRs (Figure 10-13). Integrase, another enzyme encoded by retroviruses that is closely related to the transposase of some DNA transposons, uses a similar mechanism to insert the double-stranded retroviral DNA into the host-cell genome. In this process, short direct repeats of the target-site sequence are generated at either end of the inserted viral DNA sequence.

As noted above, LTR retrotransposons encode reverse transcriptase and integrase. By analogy with retroviruses, these mobile elements are thought to move by a "copy-and-paste" mechanism whereby reverse transcriptase converts an RNA copy of a donor-site element into DNA, which is inserted into a target site by integrase. The experiments depicted in Figure 10-14 provided strong evidence for the role of an RNA intermediate in transposition of Ty elements.

Sequencing of the human genome has revealed that the most common LTR retrotransposon–related sequences in humans are derived from endogenous retroviruses (ERVs). Most of the 443,000 ERV-related DNA sequences in the human genome consist only of isolated LTRs. These are derived from full-length proviral DNA by homologous recombination between the two LTRs, resulting in deletion of the internal retroviral sequences.

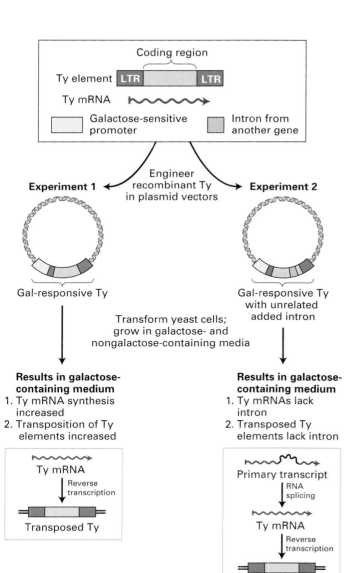

◀ **EXPERIMENTAL FIGURE 10-14 Recombinant plasmids demonstrate that the yeast Ty element transposes through an RNA intermediate.** When yeast cells are transformed with a Ty-containing plasmid, the Ty element can transpose to new sites, although normally this occurs at a low rate. Using the elements diagrammed at the top, researchers engineered two different plasmid vectors containing recombinant Ty elements adjacent to a galactose-sensitive promoter. These plasmids were transformed into yeast cells, which were grown in a galactose-containing and a nongalactose medium. In experiment 1, growth of cells in galactose-containing medium resulted in many more transpositions than in nongalactose medium, indicating that transcription into an mRNA intermediate is required for Ty transposition. In experiment 2, an intron from an unrelated yeast gene was inserted into the putative protein-coding region of the recombinant galactose-responsive Ty element. The observed absence of the intron in transposed Ty elements is strong evidence that transposition involves an mRNA intermediate from which the intron was removed by RNA splicing, as depicted in the box on the right. In contrast, eukaryotic DNA transposons, like the Ac element of maize, contain introns within the transposase gene, indicating that they do not transpose via an RNA intermediate. [See J. Boeke et al., 1985, *Cell* **40**:491.]

Retrotransposons That Lack LTRs Move by a Distinct Mechanism

The most abundant mobile elements in mammals are retrotransposons that lack LTRs, sometimes called *nonviral retrotransposons*. These moderately repeated DNA sequences form two classes in mammalian genomes: *long interspersed elements (LINEs)* and *short interspersed elements (SINEs)*. In humans, full-length LINEs are ≈6 kb long, and SINEs are ≈300 bp long. Repeated sequences with the characteristics of LINEs have been observed in protozoans, insects, and plants, but for unknown reasons they are particularly abundant in the genomes of mammals. SINEs also are found primarily in mammalian DNA. Large numbers of LINEs and SINEs in higher eukaryotes have accumulated over evolutionary time by repeated copying of sequences at a few positions in the genome and insertion of the copies into new positions. Although these mobile elements do not contain LTRs, the available evidence indicates that they transpose through an RNA intermediate.

LINEs Human DNA contains three major families of LINE sequences that are similar in their mechanism of transposition, but differ in their sequences: L1, L2, and L3. Only members of the L1 family transpose in the contemporary human genome. LINE sequences are present at ≈900,000 sites in the human genome, accounting for a staggering 21 percent of total human DNA. The general structure of a complete LINE is diagrammed in Figure 10-15. LINEs usually are flanked by short direct repeats, the hallmark of mobile elements, and contain two long open reading frames (ORFs). ORF1, ≈1 kb long, encodes an RNA-binding protein. ORF2, ≈4 kb long, encodes a protein that has a long region of homology with the reverse transcriptases of retroviruses and viral retrotransposons, but also exhibits DNA endonuclease activity.

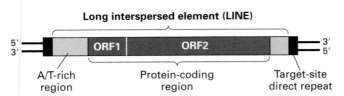

Long interspersed element (LINE)

A/T-rich region | Protein-coding region | Target-site direct repeat

▲ **FIGURE 10-15 General structure of a LINE, one of the two classes of non-LTR retrotransposons in mammalian DNA.** The length of the target-site direct repeats varies among copies of the element at different sites in the genome. Although the full-length L1 sequence is ≈6 kb long, variable amounts of the left end are absent at over 90 percent of the sites where this mobile element is found. The shorter open reading frame (ORF1), ≈1 kb in length, encodes an RNA-binding protein. The longer ORF2, ≈4 kb in length, encodes a bifunctional protein with reverse transcriptase and DNA endonuclease activity.

  Evidence for the mobility of L1 elements first came from analysis of DNA cloned from humans with certain genetic diseases. DNA from these patients was found to carry mutations resulting from insertion of an L1 element into a gene, whereas no such element occurred within this gene in either parent. About 1 in 600 mutations that cause significant disease in humans are due to L1 transpositions or SINE transpositions that are catalyzed by L1-encoded proteins. Later experiments similar to those just described with yeast Ty elements (see Figure 10-14) confirmed that L1 elements transpose through an RNA intermediate. In these experiments, an intron was introduced into a cloned mouse L1 element, and the recombinant L1 element was stably transformed into cultured hamster cells. After several cell doublings, a PCR-amplified fragment corresponding to the L1 element but lacking the inserted intron was detected in the cells. This finding strongly suggests that over time the recombinant L1 element containing the inserted intron had transposed to new sites in the hamster genome through an RNA intermediate that underwent RNA splicing to remove the intron. ∎

Since LINEs do not contain LTRs, their mechanism of transposition through an RNA intermediate differs from that of LTR retrotransposons. ORF1 and ORF2 proteins are translated from a LINE RNA. In vitro studies indicate that transcription by RNA polymerase II is directed by promoter sequences at the left end of integrated LINE DNA. LINE RNA is polyadenylated by the same post-transcriptional mechanism that polyadenylates other mRNAs. The LINE RNA then is transported into the cytoplasm, where it is translated into ORF1 and ORF2 proteins. Multiple copies of ORF1 protein then bind to the LINE RNA, and ORF2 protein binds to the poly(A) tail.

The LINE RNA is then transported back into the nucleus as a complex with ORF1 and ORF2. ORF2 then makes staggered nicks in chromosomal DNA on either side of any A/T-rich sequence in the genome (Figure 10-16, step 1). Reverse transcription of LINE RNA by ORF2 is primed by the single-stranded T-rich sequence generated by the nick in the bottom strand, which hybridizes to the LINE poly(A) tail (step 2). ORF2 then reverse-transcribes the LINE RNA (step 3) and then continues this new DNA strand, switching to the single-stranded region of the upper chromosomal strand as a template (steps 4 and 5). Cellular enzymes then hydrolyze the RNA and extend the 3' end of the chromosomal DNA top strand, replacing the LINE RNA strand with DNA (step 6).

▶ **FIGURE 10-16 Proposed mechanism of LINE reverse transcription and integration.** Only ORF2 protein is represented. Newly synthesized LINE DNA is shown in black. See the text for explanation. [Adapted from D. D. Luan et al., 1993, *Cell* **72**:595.]

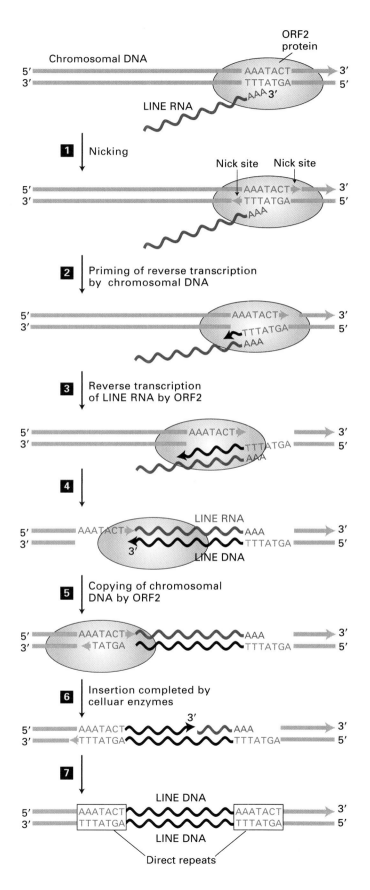

ORF2 protein

Chromosomal DNA

5′ ——— AAATACT ——— 3′
3′ ——— TTTATGA ——— 5′
AAA 3′

LINE RNA

1 Nicking

Nick site   Nick site

5′ ——— AAATACT ——— 3′
3′ ——— TTTATGA ——— 5′
AAA

2 Priming of reverse transcription by chromosomal DNA

5′ ——— AAATACT ——— 3′
3′ ——— TTTATGA ——— 5′
AAA

3 Reverse transcription of LINE RNA by ORF2

5′ ——— AAATACT ——— 3′
3′ ——— TTTATGA ——— 5′
AAA

4

LINE RNA

5′ ——— AAATACT ~~~~~~ AAA ——— 3′
3′ ——— TTTATGA ——— 5′
3′
LINE DNA

5 Copying of chromosomal DNA by ORF2

5′ ——— AAATACT ~~~~~~ AAA ——— 3′
3′ ——— TATGA ~~~~~~ TTTATGA ——— 5′

6 Insertion completed by cellular enzymes

3′
5′ ——— AAATACT ~~~~~~ AAA ——— 3′
3′ ——— TTTATGA ~~~~~~ TTTATGA ——— 5′

7

LINE DNA

5′ ——— AAATACT ~~~~~~ AAATACT ——— 3′
3′ ——— TTTATGA ~~~~~~ TTTATGA ——— 5′
LINE DNA

Direct repeats

Finally, 5′ and 3′ ends of DNA strands are ligated, completing the insertion (step ⑦). These last steps (⑥ and ⑦) probably are catalyzed by the same cellular enzymes that remove RNA primers and ligate Okazaki fragments during DNA replication (see Figure 4-33). The complete process results in insertion of a copy of the original LINE retrotransposon into a new site in chromosomal DNA. A short direct repeat is generated at the insertion site because of the initial staggered cleavage of the two chromosomal DNA strands (step ①).

The vast majority of LINEs in the human genome are truncated at their 5′ end, suggesting that reverse transcription terminated before completion and the resulting fragments extending variable distances from the poly(A) tail were inserted. Because of this shortening, the average size of LINE elements is only about 900 base pairs, whereas the full-length sequence is ≈6 kb long. In addition, nearly all the full-length elements contain stop codons and frameshift mutations in ORF1 and ORF2; these mutations probably have accumulated in most LINE sequences over evolutionary time. As a result of truncation and mutation, only ≈0.01 percent of the LINE sequences in the human genome are full-length with intact open reading frames for ORF1 and ORF2, ≈60–100 in total.

SINEs The second most abundant class of mobile elements in the human genome, SINEs constitute ≈13 percent of total human DNA. Varying in length from about 100 to 400 base pairs, these retrotransposons do not encode protein, but most contain a 3′ A/T-rich sequence similar to that in LINEs. SINEs are transcribed by RNA polymerase III, the same nuclear RNA polymerase that transcribes genes encoding tRNAs, 5S rRNAs, and other small stable RNAs (Chapter 11). Most likely, the ORF1 and ORF2 proteins expressed from full-length LINEs mediate transposition of SINEs by the retrotransposition mechanism depicted in Figure 10-16.

SINEs occur at about 1.6 million sites in the human genome. Of these, ≈1.1 million are *Alu elements,* so named because most of them contain a single recognition site for the restriction enzyme *Alu*I. *Alu* elements exhibit considerable sequence homology with and may have evolved from 7SL RNA, a component of the signal-recognition particle. This abundant cytosolic ribonucleoprotein particle aids in targeting certain polypeptides, as they are being synthesized, to the membranes of the endoplasmic reticulum (Chapter 16). *Alu* elements are scattered throughout the human genome at sites where their insertion has not disrupted gene expression: between genes, within introns, and in the 3′ untranslated regions of some mRNAs. For instance, nine *Alu* elements are located within the human β-globin gene cluster (see Figure 10-3b). The overall frequency of L1 and SINE retrotranspositions in humans is estimated to be about one new retrotransposition in very eight individuals, with ≈40 percent being L1 and 60 percent SINEs, of which ≈90 percent are *Alu* elements.

Similar to other mobile elements, most SINEs have accumulated mutations from the time of their insertion in the germ line of an ancient ancestor of modern humans. Like LINEs, many SINEs also are truncated at their 5′ end. Table 10-1 summarizes the major types of interspersed repeats derived from mobile elements in the human genome.

In addition to the mobile elements listed in Table 10-1, DNA copies of a wide variety of mRNAs appear to have integrated into chromosomal DNA. Since these sequences lack introns and do not have flanking sequences similar to those of the functional gene copies, they clearly are not simply duplicated genes that have drifted into nonfunctionality and become pseudogenes, as discussed earlier (Figure 10-3a). Instead, these DNA segments appear to be retrotransposed copies of spliced and polyadenylated (processed) mRNA. Compared with normal genes encoding mRNAs, these inserted segments generally contain multiple mutations, which are thought to have accumulated since their mRNAs were first reverse-transcribed and randomly integrated into the genome of a germ cell in an ancient ancestor. These nonfunctional genomic copies of mRNAs are referred to as *processed pseudogenes*. Most processed pseudogenes are flanked by short direct repeats, supporting the hypothesis that they were generated by rare retrotransposition events involving cellular mRNAs.

Other moderately repetitive sequences representing partial or mutant copies of genes encoding small nuclear RNAs (snRNAs) and tRNAs are found in mammalian genomes. Like processed pseudogenes derived from mRNAs, these nonfunctional copies of small RNA genes are flanked by short direct repeats and most likely result from rare retrotransposition events that have accumulated through the course of evolution. Enzymes expressed from a LINE are thought to have carried out all these retrotransposition events involving mRNAs, snRNAs, and tRNAs.

Mobile DNA Elements Probably Had a Significant Influence on Evolution

Although mobile DNA elements appear to have no direct function other than to maintain their own existence, their presence probably had a profound impact on the evolution of modern-day organisms. As mentioned earlier, about half the spontaneous mutations in *Drosophila* result from insertion of a mobile DNA element into or near a transcription unit. In mammals, however, mobile elements cause a much smaller proportion of spontaneous mutations: ≈10 percent in mice and only 0.1–0.2 percent in humans. Still, mobile elements have been found in mutant alleles associated with several human genetic diseases.

In lineages leading to higher eukaryotes, homologous recombination between mobile DNA elements dispersed throughout ancestral genomes may have generated gene duplications and other DNA rearrangements during evolution (see Figure 10-4). For instance, cloning and sequencing of the β-globin gene cluster from various primate species has provided strong evidence that the human G_γ and A_γ genes arose from an unequal homologous crossover between two L1 sequences flanking an ancestral globin gene. Subsequent divergence of such duplicated genes could lead to acquisition of distinct, beneficial functions associated with each member of a gene family. Unequal crossing over between mobile elements located within introns of a particular gene could lead to the duplication of exons within that gene. This process most likely influenced the evolution of genes that contain multiple copies of similar exons encoding similar protein domains, such as the fibronectin gene (see Figure 4-15).

Some evidence suggests that during the evolution of higher eukaryotes, recombination between interspersed repeats in introns of *two separate* genes also occurred, generating new genes made from novel combinations of preexisting exons (Figure 10-17). This evolutionary process, termed **exon shuffling,** may have occurred during evolution of the genes encoding tissue plasminogen activator, the Neu receptor, and epidermal growth factor, which all contain an EGF domain (see Figure 3-8). In this case, exon shuffling presumably resulted in insertion of an EGF domain–encoding exon into an intron of the ancestral form of each of these genes.

Both DNA transposons and LINE retrotransposons have been shown to occasionally carry unrelated flanking sequences when they insert into new sites by the mechanisms diagrammed in Figure 10-18. These mechanisms likely also contributed to exon shuffling during the evolution of contemporary genes.

▶ **FIGURE 10-17 Exon shuffling via recombination between homologous interspersed repeats.** Recombination between interspersed repeats in the introns of separate genes produces transcription units with a new combination of exons. In the example shown here, a double crossover between two sets of *Alu* repeats results in an exchange of exons between the two genes.

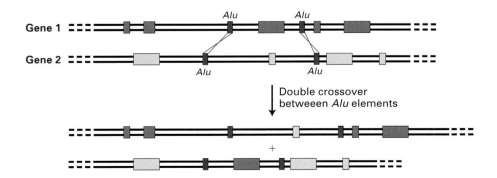

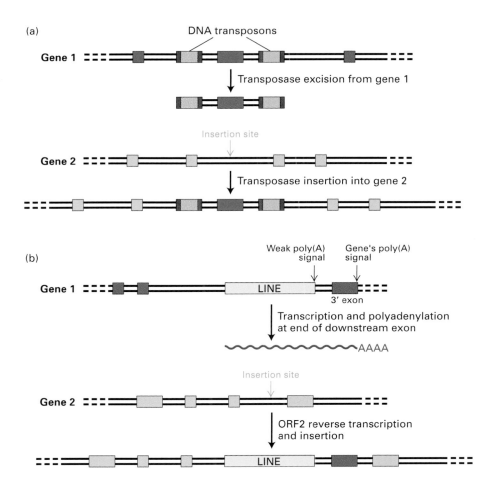

(a)

DNA transposons

Gene 1

Transposase excision from gene 1

Insertion site

Gene 2

Transposase insertion into gene 2

(b)

Weak poly(A) signal   Gene's poly(A) signal

Gene 1

LINE

3' exon

Transcription and polyadenylation at end of downstream exon

AAAA

Insertion site

Gene 2

ORF2 reverse transcription and insertion

LINE

◀ **FIGURE 10-18 Exon shuffling by transposition.** (a) Transposition of an exon flanked by homologous DNA transposons into an intron on a second gene. As we saw in Figure 10-10, step ❶, transposase can recognize and cleave the DNA at the ends of the transposon inverted repeats. In gene 1, if the transposase cleaves at the left end of the transposon on the left and at the right end of the transposon on the right, it can transpose all the intervening DNA, including the exon from gene 1, to a new site in an intron of gene 2. The net result is an insertion of the exon from gene 1 into gene 2. (b) Integration of an exon into another gene via LINE transposition. Some LINEs have weak poly(A) signals. If such a LINE is in the 3'-most intron of gene 1, during transposition its transcription may continue beyond its own poly(A) signals and extend into the 3' exon, transcribing the cleavage and polyadenylation signals of gene 1 itself. This RNA can then be reverse-transcribed and integrated by the LINE ORF2 protein (Figure 10-16) into an intron on gene 2, introducing a new 3' exon (from gene 1) into gene 2.

In addition to causing changes in coding sequences in the genome, recombination between mobile elements and transposition of DNA adjacent to DNA transposons and retrotransposons likely played a significant role in the evolution of regulatory sequences that control gene expression. As noted earlier, eukaryotic genes have transcription-control regions, called enhancers, that can operate over distances of tens of thousands of base pairs. Transcription of many genes is controlled through the combined effects of several enhancer elements. Insertion of mobile elements near such transcription-control regions probably contributed to the evolution of new combinations of enhancer sequences. These in turn control which specific genes are expressed in particular cell types and the amount of the encoded protein produced in modern organisms, as we discuss in the next chapter.

These considerations suggest that the early view of mobile DNA elements as completely selfish molecular parasites misses the mark. Rather, they have likely contributed profoundly to the evolution of higher organisms by promoting (1) the generation of gene families via gene duplication, (2) the creation of new genes via shuffling of preexisting exons, and (3) formation of more complex regulatory regions that provide multifaceted control of gene expression.

KEY CONCEPTS OF SECTION 10.3

Mobile DNA

■ Mobile DNA elements are moderately repeated DNA sequences interspersed at multiple sites throughout the genomes of higher eukaryotes. They are present less frequently in prokaryotic genomes.

■ Mobile DNA elements that transpose to new sites directly as DNA are called *DNA transposons;* those that first are transcribed into an RNA copy of the element, which then is reverse-transcribed into DNA, are called *retrotransposons* (see Figure 10-8).

■ A common feature of all mobile elements is the presence of short direct repeats flanking the sequence.

■ Enzymes encoded by mobile elements themselves catalyze insertion of these sequences at new sites in genomic DNA.

■ Although DNA transposons, similar in structure to bacterial IS elements, occur in eukaryotes (e.g., the *Drosophila* P element), retrotransposons generally are much more abundant, especially in vertebrates.

■ LTR retrotransposons are flanked by long terminal repeats (LTRs), similar to those in retroviral DNA; like retroviruses, they encode reverse transcriptase and integrase. They move in the genome by being transcribed into RNA, which then undergoes reverse transcription and integration into the host-cell chromosome (see Figure 10-13).

- Long interspersed elements (LINEs) and short interspersed elements (SINEs) are the most abundant mobile elements in the human genome. LINEs account for ≈21 percent of human DNA; SINEs, for ≈13 percent.

- Both LINEs and SINEs lack LTRs and have an A/T-rich stretch at one end. They are thought to move by a nonviral retrotransposition mechanism mediated by LINE-encoded proteins involving priming by chromosomal DNA (see Figure 10-16).

- SINE sequences exhibit extensive homology with small cellular RNAs transcribed by RNA polymerase III. *Alu* elements, the most common SINEs in humans, are ≈300-bp sequences found scattered throughout the human genome.

- Some moderately repeated DNA sequences are derived from cellular RNAs that were reverse-transcribed and inserted into genomic DNA at some time in evolutionary history. Those derived from mRNAs, called *processed pseudogenes*, lack introns, a feature that distinguishes them from pseudogenes, which arose by sequence drift of duplicated genes.

- Mobile DNA elements most likely influenced evolution significantly by serving as recombination sites and by mobilizing adjacent DNA sequences.

10.4 Structural Organization of Eukaryotic Chromosomes

We turn now to the question of how DNA molecules are organized within eukaryotic cells. Because the total length of cellular DNA is up to a hundred thousand times a cell's length, the packing of DNA is crucial to cell architecture. During interphase, when cells are not dividing, the genetic material exists as a nucleoprotein complex called **chromatin,** which is dispersed through much of the nucleus. Further folding and compaction of chromatin during mitosis produces the visible **metaphase chromosomes,** whose morphology and staining characteristics were detailed by early cytogeneticists. In this section, we consider the properties of chromatin and its organization into chromosomes. Important features of chromosomes in their entirety are covered in the next section.

Eukaryotic Nuclear DNA Associates with Histone Proteins to Form Chromatin

When the DNA from eukaryotic nuclei is isolated in isotonic buffers (i.e., buffers with the same salt concentration found in cells, ≈0.15 M KCl), it is associated with an equal mass of protein as chromatin. The general structure of chromatin has been found to be remarkably similar in the cells of all eukaryotes, including fungi, plants, and animals.

The most abundant proteins associated with eukaryotic DNA are **histones,** a family of small, basic proteins present in all eukaryotic nuclei. The five major types of histone proteins—termed *H1, H2A, H2B, H3,* and *H4*—are rich in positively charged basic amino acids, which interact with the negatively charged phosphate groups in DNA.

The amino acid sequences of four histones (H2A, H2B, H3, and H4) are remarkably similar among distantly related species. For example, the sequences of histone H3 from sea urchin tissue and calf thymus differ by only a single amino acid, and H3 from the garden pea and calf thymus differ only in four amino acids. Minor histone variants encoded by genes that differ from the highly conserved major types also exist, particularly in vertebrates.

The amino acid sequence of H1 varies more from organism to organism than do the sequences of the other major histones. In certain tissues, H1 is replaced by special histones. For example, in the nucleated red blood cells of birds, a histone termed *H5* is present in place of H1. The similarity in sequence among histones from all eukaryotes suggests that they fold into very similar three-dimensional conformations, which were optimized for histone function early in evolution in a common ancestor of all modern eukaryotes.

Chromatin Exists in Extended and Condensed Forms

When chromatin is extracted from nuclei and examined in the electron microscope, its appearance depends on the salt concentration to which it is exposed. At low salt concentration in the absence of divalent cations such as Mg^{+2}, isolated chromatin resembles "beads on a string" (Figure 10-19a). In this extended form, the string is composed of free DNA called "linker" DNA connecting the beadlike structures termed **nucleosomes.** Composed of DNA and histones, nucleosomes are about 10 nm in diameter and are the primary structural units of chromatin. If chromatin is isolated at physiological salt concentration (≈0.15 M KCl, 0.004 M Mg^{+2}), it assumes a more condensed fiberlike form that is 30 nm in diameter (Figure 10-19b).

Structure of Nucleosomes The DNA component of nucleosomes is much less susceptible to nuclease digestion than is the linker DNA between them. If nuclease treatment is carefully controlled, all the linker DNA can be digested, releasing individual nucleosomes with their DNA component. A nucleosome consists of a protein core with DNA wound around its surface like thread around a spool. The core is an octamer containing two copies each of histones H2A, H2B, H3, and H4. X-ray crystallography has shown that the octameric histone core is a roughly disk-shaped molecule made of interlocking histone subunits

(a)

(b)

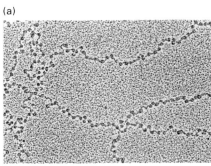

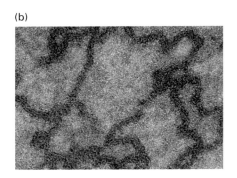

▲ **EXPERIMENTAL FIGURE 10-19** The extended and condensed forms of extracted chromatin have very different appearances in electron micrographs. (a) Chromatin isolated in low-ionic-strength buffer has an extended "beads-on-a-string" appearance. The "beads" are nucleosomes (10-nm diameter) and the "string" is connecting (linker) DNA. (b) Chromatin isolated in buffer with a physiological ionic strength (0.15 M KCl) appears as a condensed fiber 30 nm in diameter. [Part (a) courtesy of S. McKnight and O. Miller, Jr.; part (b) courtesy of B. Hamkalo and J. B. Rattner.]

(Figure 10-20). Nucleosomes from all eukaryotes contain 147 base pairs of DNA wrapped slightly less than two turns around the protein core. The length of the linker DNA is more variable among species, ranging from about 15 to 55 base pairs.

In cells, newly replicated DNA is assembled into nucleosomes shortly after the replication fork passes, but when isolated histones are added to DNA in vitro at physiological salt concentration, nucleosomes do not spontaneously form. However, nuclear proteins that bind histones and assemble them with DNA into nucleosomes in vitro have been characterized. Proteins of this type are thought to assemble histones and newly replicated DNA into nucleosomes in vivo as well.

Structure of Condensed Chromatin When extracted from cells in isotonic buffers, most chromatin appears as fibers ≈30 nm in diameter (see Figure 10-19b). In these condensed

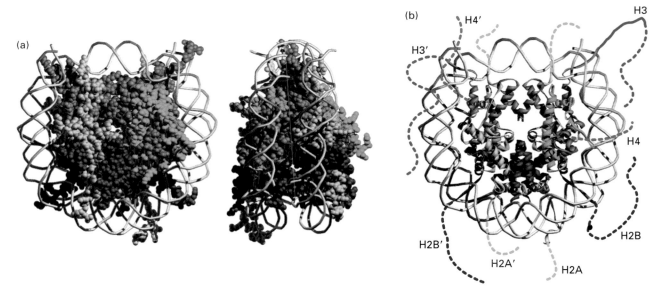

▲ **FIGURE 10-20 Structure of the nucleosome based on x-ray crystallography.** (a) Space-filling model shown from the front (*left*) and from the side (*right*, rotated clockwise 90°). H2A is yellow; H2B is red; H3 is blue; H4 is green. The sugar-phosphate backbone of the DNA strands is shown as white tubes. The N-terminal tails of the eight histones and the two H2A C-terminal tails, involved in condensation of the chromatin, are not visible because they are disordered in the crystal. (b) Ribbon diagram of the histones showing the lengths of the histone tails (dotted lines) not visible in the crystal structure. H2A N-terminal tails at the bottom, C-terminal tails at the top. [Part (a) after K. Luger et al., 1997, *Nature* **389**:251; part (b) K. Luger and T. J. Richmond, 1998, *Curr. Opin. Gen. Dev.* **8**:140.]

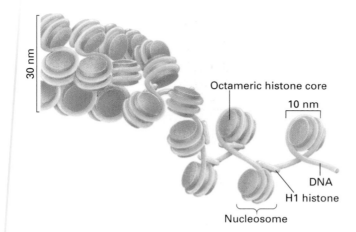

30 nm

Octameric histone core

10 nm

DNA

H1 histone

Nucleosome

▲ **FIGURE 10-21 Solenoid model of the 30-nm condensed chromatin fiber in a side view.** The octameric histone core (see Figure 10-20) is shown as an orange disk. Each nucleosome associates with one H1 molecule, and the fiber coils into a solenoid structure with a diameter of 30 nm. [Adapted from M. Grunstein, 1992, *Sci. Am.* **267**:68.]

fibers, nucleosomes are thought to be packed into an irregular spiral or solenoid arrangement, with approximately six nucleosomes per turn (Figure 10-21). H1, the fifth major histone, is bound to the DNA on the inside of the solenoid, with one H1 molecule associated with each nucleosome. Recent electron microscopic studies suggest that the 30-nm fiber is less uniform than a perfect solenoid. Condensed chromatin may in fact be quite dynamic, with regions occasionally partially unfolding and then refolding into a solenoid structure.

The chromatin in chromosomal regions that are not being transcribed exists predominantly in the condensed, 30-nm fiber form and in higher-order folded structures whose detailed conformation is not currently understood. The regions of chromatin actively being transcribed are thought to assume the extended beads-on-a-string form.

Modification of Histone Tails Controls Chromatin Condensation

Each of the histone proteins making up the nucleosome core contains a flexible amino terminus of 11–37 residues extending from the fixed structure of the nucleosome; these termini are called *histone tails*. Each H2A also contains a flexible C-terminal tail (see Figure 10-20b). The histone tails are required for chromatin to condense from the beads-on-a-string conformation into the 30-nm fiber. Several positively charged lysine side chains in the histone tails may interact with linker DNA, and the tails of one nucleosome likely interact with neighboring nucleosomes. The histone tail lysines, especially those in H3 and H4, undergo reversible acetylation and deacetylation by enzymes that act on specific lysines in the N-termini. In the acetylated form, the positive

charge of the lysine ε-amino group is neutralized, thereby eliminating its interaction with a DNA phosphate group. Thus the greater the acetylation of histone N-termini, the less likely chromatin is to form condensed 30-nm fibers and possibly higher-order folded structures.

The histone tails can also bind to other proteins associated with chromatin that influence chromatin structure and processes such as transcription and DNA replication. The interaction of histone tails with these proteins can be regulated by a variety of covalent modifications of histone tail amino acid side chains. These include acetylation of lysine ε-amino groups, as mentioned earlier, as well as methylation of these groups, a process that prevents acetylation, thus maintaining their positive charge. Arginine side chains can also be methylated. Serine and threonine side chains can be phosphorylated, introducing a negative charge. Finally, a single 76-amino-acid ubiquitin molecule can be added to some lysines. Recall that addition of multiple linked ubiquitin molecules to a protein can mark it for degradation by the proteasome (Chapter 3). In this case, the addition of a single ubiquitin does not affect the stability of a histone, but influences chromatin structure. In summary, multiple types of covalent modifications of histone tails can influence chromatin structure by altering histone-DNA interactions and interactions between nucleosomes and by controlling interactions with additional proteins that participate in the regulation of transcription, as discussed in the next chapter.

The extent of histone acetylation is correlated with the relative resistance of chromatin DNA to digestion by nucleases. This phenomenon can be demonstrated by digesting isolated nuclei with DNase I. Following digestion, the DNA is completely separated from chromatin protein, digested to completion with a restriction enzyme, and analyzed by Southern blotting (see Figure 9-26). An intact gene treated with a restriction enzyme yields characteristic fragments. When a gene is exposed first to DNase, it is cleaved at random sites within the boundaries of the restriction enzyme cut sites. Consequently, any Southern blot bands normally seen with that gene will be lost. This method has been used to show that the transcriptionally inactive β-globin gene in nonerythroid cells, where it is associated with relatively unacetylated histones, is much more resistant to DNase I than is the active, transcribed β-globin gene in erythroid precursor cells, where it is associated with acetylated histones (Figure 10-22). These results indicate that the chromatin structure of nontranscribed DNA is more condensed, and therefore more protected from DNase digestion, than that of transcribed DNA. In condensed chromatin, the DNA is largely inaccessible to DNase I because of its close association with histones and other less abundant chromatin proteins. In contrast, actively transcribed DNA is much more accessible to DNase I digestion because it is present in the extended, beads-on-a-string form of chromatin.

Genetic studies in yeast indicate that specific histone acetylases are required for the full activation of transcription of a number of genes. Consequently, as discussed in

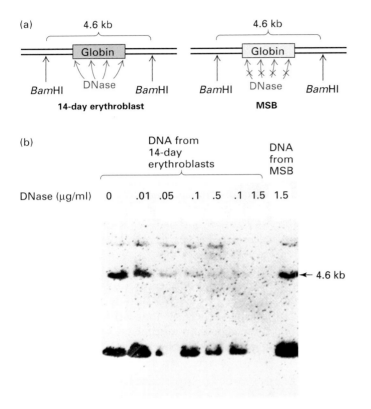

(a)

4.6 kb — 14-day erythroblast: *Bam*HI ... Globin ... DNase ... *Bam*HI

4.6 kb — MSB: *Bam*HI ... Globin ... DNase ... *Bam*HI

(b)

DNA from 14-day erythroblasts | DNA from MSB

DNase (µg/ml) 0 .01 .05 .1 .5 .1 1.5 1.5

← 4.6 kb

▲ **EXPERIMENTAL FIGURE 10-22 Nontranscribed genes are less susceptible to DNase I digestion than active genes.** Chick embryo erythroblasts at 14 days actively synthesize globin, whereas cultured undifferentiated MSB cells do not. (a) Nuclei from each type of cell were isolated and exposed to increasing concentrations of DNase I. The nuclear DNA was then extracted and treated with the restriction enzyme *Bam*HI, which cleaves the DNA around the globin sequence and normally releases a 4.6-kb globin fragment. (b) The DNase I- and *Bam*HI-digested DNA was subjected to Southern blot analysis with a probe of labeled cloned adult globin DNA, which hybridizes to the 4.6-kb *Bam*HI fragment. If the globin gene is susceptible to the initial DNase digestion, it would be cleaved repeatedly and would not be expected to show this fragment. As seen in the Southern blot, the transcriptionally active DNA from the 14-day globin-synthesizing cells was sensitive to DNase I digestion, indicated by the absence of the 4.6-kb band at higher nuclease concentrations. In contrast, the inactive DNA from MSB cells was resistant to digestion. These results suggest that the inactive DNA is in a more condensed form of chromatin in which the globin gene is shielded from DNase digestion. [See J. Stalder et al., 1980, *Cell* **19**:973; photograph courtesy of H. Weintraub.]

Chapter 11, the control of acetylation of histone N-termini in specific chromosomal regions is thought to contribute to gene control by regulating the strength of the interaction of histones with DNA and the folding of chromatin into condensed structures. Genes in condensed, folded regions of chromatin are inaccessible to RNA polymerase and other proteins required for transcription.

Nonhistone Proteins Provide a Structural Scaffold for Long Chromatin Loops

Although histones are the predominant proteins in chromosomes, nonhistone proteins are also involved in organizing chromosome structure. Electron micrographs of histone-depleted metaphase chromosomes from HeLa cells reveal long loops of DNA anchored to a *chromosome scaffold* composed of nonhistone proteins (Figure 10-23). This scaffold has the shape of the metaphase chromosome and persists even when the DNA is digested by nucleases. As depicted schematically in Figure 10-24, loops of the 30-nm chromatin fiber a few megabases in length have been proposed to associate with a flexible chromosome scaffold, yielding an extended form characteristic of chromosomes during interphase. Folding of the scaffold has been proposed to produce the highly condensed structure characteristic of metaphase chromosomes. But the geometry of scaffold folding in metaphase chromosomes has not yet been determined.

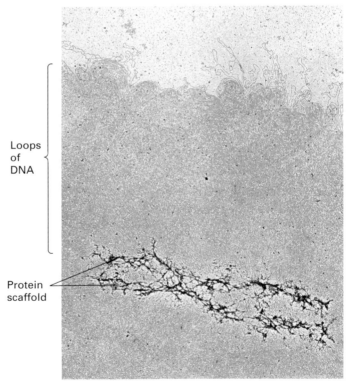

Loops of DNA

Protein scaffold

▲ **EXPERIMENTAL FIGURE 10-23 An electron micrograph of a histone-depleted metaphase chromosome reveals the scaffold around which the DNA is organized.** The long loops of DNA are visible extending from the nonhistone protein scaffold (the dark structure). The scaffold shape reflects that of the metaphase chromosome itself. The chromosome was prepared from HeLa cells by treatment with a mild detergent. [From J. R. Paulson and U. K. Laemmli, 1977, *Cell* **12**:817. Copyright 1977 MIT.]

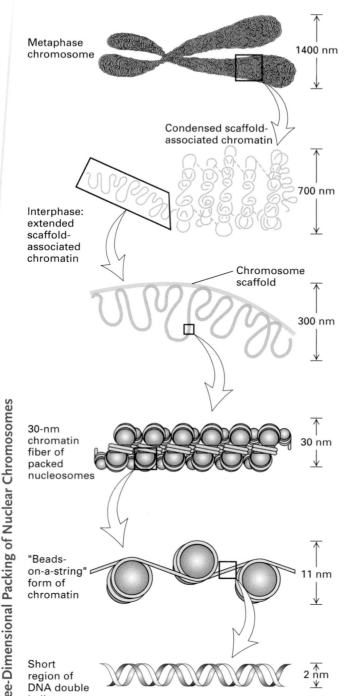

◀ FIGURE 10-24 Model for the packing of chromatin and the chromosome scaffold in metaphase chromosomes. In interphase chromosomes, long stretches of 30-nm chromatin loop out from extended scaffolds. In metaphase chromosomes, the scaffold is folded further into a highly compacted structure, whose precise geometry has not been determined.

In situ hybridization experiments with several different fluorescent-labeled probes to DNA in human interphase cells support the loop model shown in Figure 10-24. In these experiments, some probe sequences separated by millions of base pairs in linear DNA appeared reproducibly very close to one another in interphase nuclei from different cells (Figure 10-25). These closely spaced probe sites are postulated to lie close to specific sequences in the DNA, called *scaffold-associated regions (SARs)* or *matrix-attachment regions (MARs)*, that are bound to the chromosome scaffold. SARs have been mapped by digesting histone-depleted chromosomes with restriction enzymes and then recovering the fragments that are bound to scaffold proteins.

In general, SARs are found between transcription units. In other words, genes are located primarily within chromatin loops, which are attached at their bases to a chromosome scaffold. Experiments with transgenic mice indicate that in some cases SARs are required for transcription of neighboring genes. In *Drosophila*, some SARs can insulate transcription units from each other, so that proteins regulating transcription of one gene do not influence the transcription of a neighboring gene separated by a SAR.

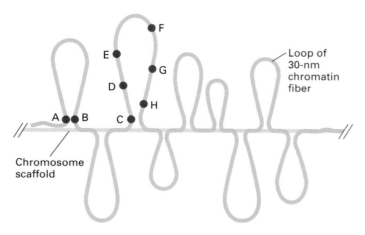

▲ EXPERIMENTAL FIGURE 10-25 Fluorescent-labeled probes hybridized to interphase chromosomes demonstrate chromatin loops and permit their measurement. In situ hybridization of interphase cells was carried out with several different probes specific for sequences separated by known distances in linear, cloned DNA. Lettered circles represent probes. Measurement of the distances between different hybridized probes, which could be distinguished by their color, showed that some sequences (e.g., A, B, and C), separated from one another by millions of base pairs, appear located near one another within nuclei. For some sets of sequences, the measured distances in nuclei between one probe (e.g., C) and sequences successively farther away initially appear to increase (e.g., D, E, and F) and then appear to decrease (e.g., G and H). The measured distances between probes are consistent with loops ranging in size from 1 million to 4 million base pairs. [Adapted from H. Yokota et al., 1995, *J. Cell Biol.* **130**:1239.]

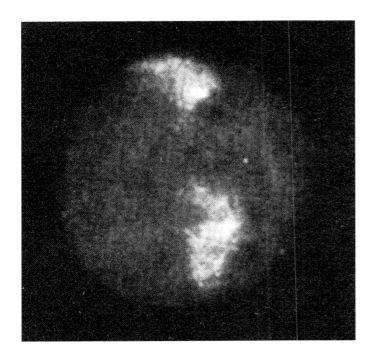

▲ **EXPERIMENTAL FIGURE 10-26 During interphase human chromosomes remain in specific domains in the nucleus.** Fixed interphase human lymphocytes were hybridized in situ to biotin-labeled probes specific for sequences along the full length of human chromosome 7 and visualized with fluorescently labeled avidin. In the diploid cell shown here, each of the two chromosome 7s is restricted to a territory or domain within the nucleus, rather than stretching throughout the entire nucleus. [From P. Lichter et al., 1988, *Hum. Genet.* **80**:224.]

Individual interphase chromosomes, which are less condensed than metaphase chromosomes, cannot be resolved by standard microscopy or electron microscopy. Nonetheless, the chromatin of interphase cells is associated with extended scaffolds and is further organized into specific domains. This can be demonstrated by the in situ hybridization of interphase nuclei with a large mixture of fluorescent-labeled probes specific for sequences along the length of a particular chromosome. As illustrated in Figure 10-26, the bound probes are visualized within restricted regions or domains of the nucleus rather than appearing throughout the nucleus. Use of probes specific for different chromosomes shows that there is little overlap between chromosomes in interphase nuclei. However, the precise positions of chromosomes are not reproducible between cells.

Chromatin Contains Small Amounts of Other Proteins in Addition to Histones and Scaffold Proteins

The total mass of the histones associated with DNA in chromatin is about equal to that of the DNA. Interphase chromatin and metaphase chromosomes also contain small amounts of a complex set of other proteins. For instance, a growing list of DNA-binding **transcription factors** have been identified associated with interphase chromatin. The structure and function of these critical nonhistone proteins, which help regulate transcription, are examined in Chapter 11. Other low-abundance nonhistone proteins associated with chromatin regulate DNA replication during the eukaryotic cell cycle (Chapter 21).

A few other nonhistone DNA-binding proteins are present in much larger amounts than the transcription or replication factors. Some of these exhibit high mobility during electrophoretic separation and thus have been designated *HMG (high-mobility group) proteins*. When genes encoding the most abundant HMG proteins are deleted from yeast cells, normal transcription is disturbed in most other genes examined. Some HMG proteins have been found to bind to DNA cooperatively with transcription factors that bind to specific DNA sequences, stabilizing multiprotein complexes that regulate transcription of a neighboring gene.

Eukaryotic Chromosomes Contain One Linear DNA Molecule

In lower eukaryotes, the sizes of the largest DNA molecules that can be extracted indicate that each chromosome contains a single DNA molecule. For example, the DNA from each of the quite small chromosomes of *S. cerevisiae* (2.3×10^5 to 1.5×10^6 base pairs) can be separated and individually identified by pulsed-field gel electrophoresis. Physical analysis of the largest DNA molecules extracted from several genetically different *Drosophila* species and strains shows that they are from 6×10^7 to 1×10^8 base pairs long. These sizes match the DNA content of single stained metaphase chromosomes of these *Drosophila* species, as measured by the amount of DNA-specific stain absorbed. The longest DNA molecules in human chromosomes are too large (2.8×10^8 base pairs, or almost 10 cm long) to extract without breaking. Nonetheless, the observations with lower eukaryotes support the conclusion that each chromosome visualized during mitosis contains a single DNA molecule.

To summarize our discussion so far, we have seen that the eukaryotic chromosome is a linear structure composed of an immensely long, single DNA molecule that is wound around histone octamers about every 200 bp, forming strings of closely packed nucleosomes. Nucleosomes fold to form a 30-nm chromatin fiber, which is attached to a flexible protein scaffold at intervals of millions of base pairs, resulting in long loops of chromatin extending from the scaffold (see Figure 10-24). In addition to this general chromosomal structure, a complex set of thousands of low-abundance regulatory proteins are associated with specific sequences in chromosomal DNA.

KEY CONCEPTS OF SECTION 10.4

Structural Organization of Eukaryotic Chromosomes

■ In eukaryotic cells, DNA is associated with about an equal mass of histone proteins in a highly condensed nucleoprotein complex called *chromatin*. The building block of chromatin is the nucleosome, consisting of a histone octamer around which is wrapped 147 bp of DNA (see Figure 10-21).

■ The chromatin in transcriptionally inactive regions of DNA within cells is thought to exist in a condensed, 30-nm fiber form and higher-order structures built from it (see Figure 10-19).

■ The chromatin in transcriptionally active regions of DNA within cells is thought to exist in an open, extended form.

■ The reversible acetylation and deacetylation of lysine residues in the N-termini of histones H2A, H2B, H3, and H4 controls how tightly DNA is bound by the histone octamer and affects the assembly of nucleosomes into the condensed forms of chromatin (see Figure 10-20b).

■ Histone tails can also be modified by methylation, phosphorylation, and monoubiquitination. These modifications influence chromatin structure by regulating the binding of histone tails to other less abundant chromatin-associated proteins.

■ Hypoacetylated, transcriptionally inactive chromatin assumes a more condensed structure and is more resistant to DNase I than hyperacetylated, transcriptionally active chromatin.

■ Each eukaryotic chromosome contains a single DNA molecule packaged into nucleosomes and folded into a 30-nm chromatin fiber, which is attached to a protein scaffold at specific sites (see Figure 10-24). Additional folding of the scaffold further compacts the structure into the highly condensed form of metaphase chromosomes.

10.5　Morphology and Functional Elements of Eukaryotic Chromosomes

Having examined the detailed structural organization of chromosomes in the previous section, we now view them from a more global perspective. Early microscopic observations on the number and size of chromosomes and their staining patterns led to the discovery of many important general characteristics of chromosome structure. Researchers subsequently identified specific chromosomal regions critical to their replication and segregation to daughter cells during cell division.

Chromosome Number, Size, and Shape at Metaphase Are Species-Specific

As noted previously, in nondividing cells individual chromosomes are not visible, even with the aid of histologic stains for DNA (e.g., Feulgen or Giemsa stains) or electron microscopy. During mitosis and meiosis, however, the chromosomes condense and become visible in the light microscope. Therefore, almost all cytogenetic work (i.e., studies of chromosome morphology) has been done with condensed metaphase chromosomes obtained from dividing cells—either somatic cells in mitosis or dividing gametes during meiosis.

The condensation of metaphase chromosomes probably results from several orders of folding and coiling of 30-nm chromatin fibers (see Figure 10-24). At the time of mitosis, cells have already progressed through the S phase of the cell cycle and have replicated their DNA. Consequently, the chromosomes that become visible during metaphase are duplicated structures. Each metaphase chromosome consists of two sister **chromatids,** which are attached at the centromere (Figure 10-27). The number, sizes, and shapes of the metaphase chromosomes constitute the **karyotype,** which is distinctive for each species. In most organisms, all cells have the same karyotype. However, species that appear quite similar can have very different karyotypes, indicating that similar genetic potential can be organized on chromosomes in very different ways. For example, two species of small deer—the Indian muntjac and Reeves muntjac—contain about the same total amount of genomic DNA. In one species, this

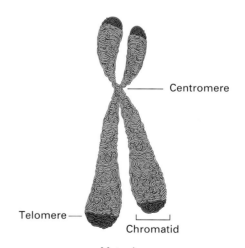

▲ **FIGURE 10-27 Microscopic appearance of typical metaphase chromosome.** Each chromosome has replicated and comprises two chromatids, each containing one of two identical DNA molecules. The centromere, where the chromatids are attached, is required for their separation late in mitosis. Special telomere sequences at the ends function in preventing chromosome shortening.

DNA is organized into 22 pairs of homologous **autosomes** and two physically separate sex chromosomes. In contrast, the other species contains only three pairs of autosomes; one sex chromosome is physically separate, but the other is joined to the end of one autosome.

During Metaphase, Chromosomes Can Be Distinguished by Banding Patterns and Chromosome Painting

Certain dyes selectively stain some regions of metaphase chromosomes more intensely than other regions, producing characteristic banding patterns that are specific for individual chromosomes. Although the molecular basis for the regularity of chromosomal bands remains unknown, they serve as useful visible landmarks along the length of each chromosome and can help to distinguish chromosomes of similar size and shape.

G bands are produced when metaphase chromosomes are subjected briefly to mild heat or proteolysis and then stained with Giemsa reagent, a permanent DNA dye (Figure 10-28). G bands correspond to large regions of the human genome that have an unusually low G+C content. Treatment of chromosomes with a hot alkaline solution before staining with Giemsa reagent produces *R bands* in a pattern that is approximately the reverse of the G-band pattern. The dis-

tinctiveness of these banding patterns permits cytologists to identify specific parts of a chromosome and to locate the sites of chromosomal breaks and translocations (Figure 10-29a). In addition, cloned DNA probes that have hybridized to specific sequences in the chromosomes can be located in particular bands.

A recently developed method for visualizing each of the human chromosomes in distinct, bright colors, called *chromosome painting*, greatly simplifies differentiating chromosomes of similar size and shape. This technique, a variation of fluorescence in situ hybridization, makes use of probes specific for sites scattered along the length of each chromosome. The probes are labeled with one of two dyes that fluoresce at different wavelengths. Probes specific for each chromosome are labeled with a predetermined fraction of each of the two dyes. After the probes are hybridized to chromosomes and the excess removed, the sample is observed with a fluorescent microscope in which a detector determines the fraction of each dye present at each fluorescing position in the microscopic field. This information is conveyed to a computer, and a special program assigns a false color image to each type of chromosome. A related technique called *multicolor FISH* can detect chromosomal translocations (Figure 10-29b). The much more detailed analysis possible with this technique permits detection of chromosomal translocations that banding analysis does not reveal.

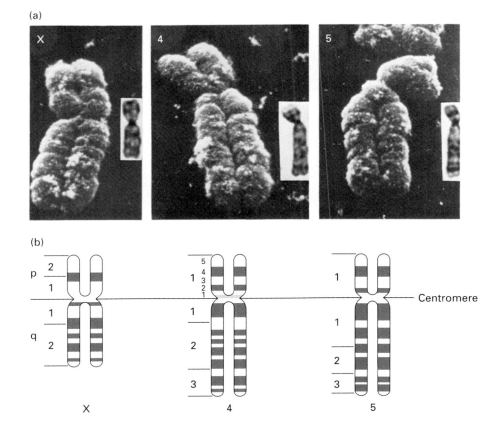

(a)

(b)

Centromere

X   4   5

◀ **EXPERIMENTAL FIGURE 10-28**
G bands produced with Giemsa stains are useful markers for identifying specific chromosomes. The chromosomes are subjected to brief proteolytic treatment and then stained with Giemsa reagent, producing distinctive bands at characteristic places. (a) Scanning electron micrographs of chromosomes X, 4, and 5 show constrictions at the sites where bands appear in light micrographs (*insets*). (b) Standard diagrams of G bands (purple). Regions of variable length (green) are most common in the centromere region (e.g., on chromosome 4). By convention, p denotes the short arm; q, the long arm. Each arm is divided into major sections (1, 2, etc.) and subsections. The short arm of chromosome 4, for example, has five subsections. DNA located in the fourth subsection would be said to be in p14. [Part (a) from C. J. Harrison et al., 1981, *Exp. Cell Res.* **134**:141; courtesy of C. J. Harrison.]

(a)

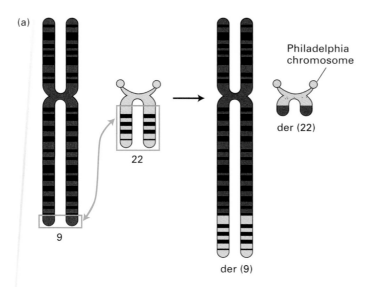

(b)

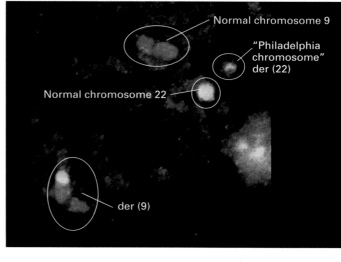

▲ **EXPERIMENTAL FIGURE 10-29 Chromosomal translocations can be analyzed using banding patterns and multicolor FISH.** Characteristic chromosomal translocations are associated with certain genetic disorders and specific types of cancers. For example, in nearly all patients with chronic myelogenous leukemia, the leukemic cells contain the Philadelphia chromosome, a shortened chromosome 22 [der

(22)], and an abnormally long chromosome 9 [der (9)]. These result from a translocation between normal chromosomes 9 and 22. This translocation can be detected by classical banding analysis (a) and by multicolor FISH (b). [Part (a) from J. Kuby, 1997, *Immunology*, 3d ed., W. H. Freeman and Company, p. 578; part (b) courtesy of J. Rowley and R. Espinosa.]

(a)
Chromocenter

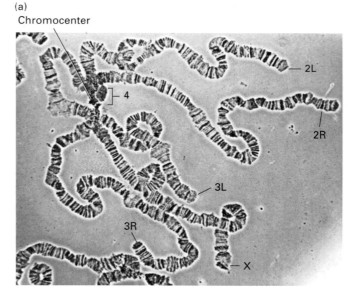

▶ **EXPERIMENTAL FIGURE 10-30 Banding on *Drosophila* polytene salivary gland chromosomes and in situ hybridization are used together to localize gene sequences.** (a) In this light micrograph of *Drosophila melanogaster* larval salivary gland chromosomes, four chromosomes can be observed (X, 2, 3, and 4), with a total of approximately 5000 distinguishable bands. The centromeres of all four chromosomes often appear fused at the chromocenter. The tips of chromosomes 2 and 3 are labeled (L = left arm; R = right arm), as is the tip of the X chromosome. (b) A particular DNA sequence can be mapped on *Drosophila* salivary gland chromosomes by in situ hybridization. This photomicrograph shows a portion of a chromosome that was hybridized with a cloned DNA sequence labeled with biotin-derivatized nucleotides. Hybridization is detected with the biotin-binding protein avidin that is covalently bound to the enzyme alkaline phosphatase. On addition of a soluble substrate, the enzyme catalyzes a reaction that results in formation of an insoluble colored precipitate at the site of hybridization (asterisk). Since the very reproducible banding patterns are characteristic of each *Drosophila* polytene chromosome, the hybridized sequence can be located on a particular chromosome. The numbers indicate major bands. Bands between those indicated are designated with numbers and letters (not shown). [Part (a) courtesy of J. Gall; part (b) courtesy of F. Pignoni.]

(b)

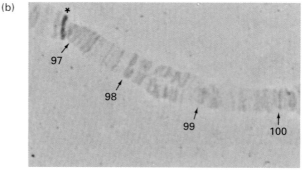

Interphase Polytene Chromosomes Arise by DNA Amplification

The larval salivary glands of *Drosophila* species and other dipteran insects contain enlarged interphase chromosomes that are visible in the light microscope. When fixed and stained, these **polytene chromosomes** are characterized by a large number of reproducible, well-demarcated bands that have been assigned standardized numbers (Figure 10-30a). The highly reproducible banding pattern seen in *Drosophila* salivary gland chromosomes provides an extremely powerful method for locating specific DNA sequences along the lengths of the chromosomes in this species. For example, the chromosomal location of a cloned DNA sequence can be accurately determined by hybridizing a labeled sample of the cloned DNA to polytene chromosomes prepared from larval salivary glands (Figure 10-30b).

A generalized amplification of DNA gives rise to the polytene chromosomes found in the salivary glands of *Drosophila.* This process, termed *polytenization,* occurs when the DNA repeatedly replicates, but the daughter chromosomes do not separate. The result is an enlarged chromosome composed of many parallel copies of itself (Figure 10-31). The amplification of chromosomal DNA greatly increases gene copy number, presumably to supply sufficient mRNA for protein synthesis in the massive salivary gland cells. Although the bands seen in human metaphase chromosomes probably represent very long folded or compacted stretches of DNA containing about 10^7 base pairs, the bands in *Drosophila* polytene chromosomes represent much shorter stretches of only 50,000–100,000 base pairs.

Heterochromatin Consists of Chromosome Regions That Do Not Uncoil

As cells exit from mitosis and the condensed chromosomes uncoil, certain sections of the chromosomes remain dark-staining. The dark-staining areas, termed **heterochromatin,** are regions of condensed chromatin. The light-staining, less condensed portions of chromatin are called **euchromatin.** Heterochromatin appears most frequently—but not exclusively—at the centromere and telomeres of chromosomes and is mostly simple-sequence DNA.

In mammalian cells, heterochromatin appears as darkly staining regions of the nucleus, often associated with the nuclear envelope. Pulse labeling with ^{3}H-uridine and autoradiography have shown that most transcription occurs in regions of euchromatin and the nucleolus. Because of this and because heterochromatic regions apparently remain condensed throughout the life cycle of the cell, they have been regarded as sites of inactive genes. However, some transcribed genes have been located in regions of heterochromatin. Also, not all inactive genes and nontranscribed regions of DNA are visible as heterochromatin.

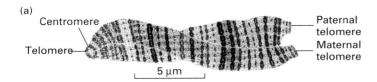

▲ **FIGURE 10-31 Amplification of DNA along the polytene fourth chromosome of *Drosophila melanogaster.*** (a) Morphology of the stained polytene fourth chromosome as it appears in the light microscope at high magnification. The fourth chromosome is by far the smallest (see Figure 10-30). The homologous chromatids (paternal and maternal) are paired. The banding pattern results from reproducible packing of DNA and protein within each amplified site along the chromosome. Dark bands are regions of more highly compacted chromatin. (b) The pattern of amplification of one of the two homologous chromatids during five replications. Double-stranded DNA is represented by a single line. Telomere and centromere DNA are not amplified. In salivary gland polytene chromosomes, each parental chromosome undergoes ≈10 replications (2^{10} = 1024 strands). [Part (a) from C. Bridges, 1935, *J. Hered.* **26**:60; part (b) adapted from C. D. Laird et al., 1973, *Cold Spring Harbor Symp. Quant. Biol.* **38**:311.]

Three Functional Elements Are Required for Replication and Stable Inheritance of Chromosomes

Although chromosomes differ in length and number between species, cytogenetic studies have shown that they all behave similarly at the time of cell division. Moreover, any eukaryotic chromosome must contain three functional elements in order to replicate and segregate correctly: (1) **replication origins** at which DNA polymerases and other proteins initiate synthesis of DNA (see Figures 4-34 and 4-36), (2) the centromere, and (3) the two ends, or telomeres. The yeast transformation studies depicted in Figure 10-32 demonstrated the functions of these three chromosomal elements and established their importance for chromosome function.

As discussed in Chapter 4, replication of DNA begins from sites that are scattered throughout eukaryotic chromosomes. The yeast genome contains many ≈100-bp sequences, called *autonomously replicating sequences (ARSs),* that act as replication origins. The observation that insertion of an ARS into a circular plasmid allows the plasmid to replicate in yeast cells provided the first functional identification of origin sequences in eukaryotic DNA (see Figure 10-32a).

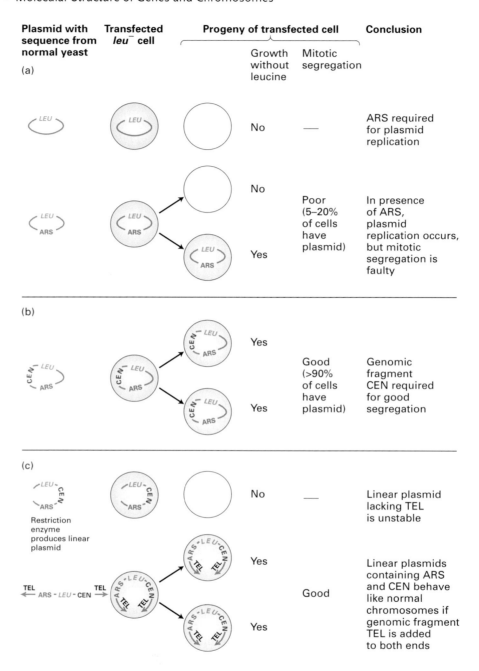

▲ **EXPERIMENTAL FIGURE 10-32 Yeast transfection experiments identify the functional chromosomal elements necessary for normal chromosome replication and segregation.** In these experiments, plasmids containing the *LEU* gene from normal yeast cells are constructed and introduced into *leu⁻* cells by transfection. If the plasmid is maintained in the *leu⁻* cells, they are transformed to *LEU⁺* by the *LEU* gene on the plasmid and can form colonies on medium lacking leucine. (a) Sequences that allow autonomous replication (ARS) of a plasmid were identified because their insertion into a plasmid vector containing a cloned *LEU* gene resulted in a high frequency of transformation to *LEU⁺*. However, even plasmids with ARS exhibit poor segregation during mitosis, and therefore do not appear in each of the daughter cells. (b) When randomly broken

pieces of genomic yeast DNA are inserted into plasmids containing ARS and *LEU*, some of the subsequently transfected cells produce large colonies, indicating that a high rate of mitotic segregation among their plasmids is facilitating the continuous growth of daughter cells. The DNA recovered from plasmids in these large colonies contains yeast centromere (CEN) sequences. (c) When *leu⁻* yeast cells are transfected with linearized plasmids containing *LEU*, ARS, and CEN, no colonies grow. Addition of telomere (TEL) sequences to the ends of the linear DNA gives the linearized plasmids the ability to replicate as new chromosomes that behave very much like a normal chromosome in both mitosis and meiosis. [See A. W. Murray and J. W. Szostak, 1983, *Nature* **305**:89, and L. Clarke and J. Carbon, 1985, *Ann. Rev. Genet.* **19**:29.]

Even though circular ARS-containing plasmids can replicate in yeast cells, only about 5–20 percent of progeny cells contain the plasmid because mitotic segregation of the plasmids is faulty. However, plasmids that also carry a CEN sequence, derived from the centromeres of yeast chromosomes, segregate equally or nearly so to both mother and daughter cells during mitosis (see Figure 10-32b).

If circular plasmids containing an ARS and CEN sequence are cut once with a restriction enzyme, the resulting linear plasmids do not produce *LEU⁺* colonies unless they contain special telomeric (TEL) sequences ligated to their ends (see Figure 10-32c). The first successful experiments involving transfection of yeast cells with linear plasmids were achieved by using the ends of a DNA molecule that was known to replicate as a linear molecule in the ciliated protozoan *Tetrahymena*. During part of the life cycle of *Tetrahymena*, much of the nuclear DNA is repeatedly copied in short pieces to form a so-called *macronucleus*. One of these repeated fragments was identified as a dimer of ribosomal DNA, the ends of which contained a repeated sequence $(G_4T_2)_n$. When a section of this repeated TEL sequence was ligated to the ends of linear yeast plasmids containing ARS and CEN, replication and good segregation of the linear plasmids occurred.

Centromere Sequences Vary Greatly in Length

Once the yeast centromere regions that confer mitotic segregation were cloned, their sequences could be determined and compared, revealing three regions (I, II, and III) conserved between them (Figure 10-33a). Short, fairly well conserved nucleotide sequences are present in regions I and III. Although region II seems to have a fairly constant length, it contains no definite consensus sequence; however, it is rich in A and T residues. Regions I and III are bound by proteins that interact with a set of more than 30 proteins that bind the short *S. cerevisiae* chromosome to one microtubule of the spindle apparatus during mitosis. Region II is bound to a nucleosome that has a variant form of histone H3 replacing the usual H3. Centromeres from all eukaryotes similarly are bound by nucleosomes with a specialized, centromere-specific form of histone H3 called *CENP-A* in humans.

S. cerevisiae has by far the simplest centromere sequence known in nature.

In the fission yeast *S. pombe*, centromeres are ≈40 kb in length and are composed of repeated copies of sequences similar those in *S. cerevisiae* centromeres. Multiple copies of proteins homologous to those that interact with the *S. cerevisiae* centromeres bind to these complex *S. pombe* centromeres and in turn bind the much longer *S. pombe* chromosomes to several microtubules of the mitotic spindle apparatus. In plants and animals, centromeres are megabases in length and are composed of multiple repeats of simple-sequence DNA. One *Drosophila* simple-sequence DNA, which comes from a centromeric region, has a repeat unit that bears some similarity to yeast CEN regions I and III (see Figure 10-33b). In humans, centromeres contain 2- to 4-megabase arrays of a 171-bp simple-sequence DNA called *alphoid* DNA that is bound by nucleosomes with the CENP-A histone H3 variant, as well as other repeated simple-sequence DNA.

In higher eukaryotes, a complex protein structure called the **kinetochore** assembles at centromeres and associates with multiple mitotic spindle fibers during mitosis. Homologs of most of the centromeric proteins found in the yeasts occur in humans and other higher eukaryotes and are thought to be components of kinetochores. The role of the centromere and proteins that bind to it in the segregation of sister chromatids during mitosis is described in Chapters 20 and 21.

Addition of Telomeric Sequences by Telomerase Prevents Shortening of Chromosomes

Sequencing of telomeres from a dozen or so organisms, including humans, has shown that most are repetitive oligomers with a high G content in the strand with its 3′ end at the end of the chromosome. The telomere repeat sequence in humans and other vertebrates is TTAGGG. These simple sequences are repeated at the very termini of chromosomes for a total of a few hundred base pairs in yeasts and protozoans and a few thousand base pairs in vertebrates. The 3′ end of the G-rich strand extends 12–16 nucleotides beyond the 5′ end of the complementary C-rich strand. This region is bound by specific proteins that both protect the ends of

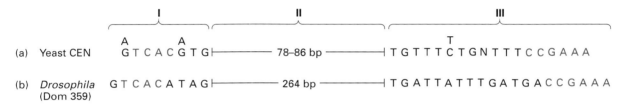

▲ **FIGURE 10-33 Comparison of yeast CEN sequence and *Drosophila* simple-sequence DNA.** (a) Consensus yeast CEN sequence, based on analysis of 10 yeast centromeres, includes three conserved regions. Region II, although variable in sequence, is fairly constant in length and is rich in A and T residues. (b) One *Drosophila* simple-sequence DNA located near the centromere has a repeat unit with some homology to the yeast consensus CEN, including two identical 4-bp and 6-bp stretches (red). [See L. Clarke and J. Carbon, 1985, *Ann. Rev. Genet.* **19**:29.]

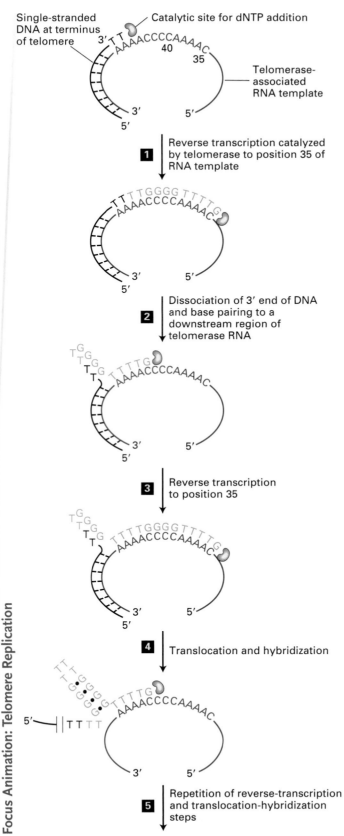

Single-stranded DNA at terminus of telomere

Catalytic site for dNTP addition

Telomerase-associated RNA template

1 Reverse transcription catalyzed by telomerase to position 35 of RNA template

2 Dissociation of 3′ end of DNA and base pairing to a downstream region of telomerase RNA

3 Reverse transcription to position 35

4 Translocation and hybridization

5 Repetition of reverse-transcription and translocation-hybridization steps

linear chromosomes from attack by exonucleases and associate telomeres in specific domains within the nucleus.

The need for a specialized region at the ends of eukaryotic chromosomes is apparent when we consider that all known DNA polymerases elongate DNA chains at the 3′ end, and all require an RNA or DNA primer. As the growing fork approaches the end of a linear chromosome, synthesis of the leading strand continues to the end of the DNA template strand, completing one daughter DNA double helix. However, because the lagging-strand template is copied in a discontinuous fashion, it cannot be replicated in its entirety (see Figure 4-33). When the final RNA primer is removed, there is no upstream strand onto which DNA polymerase can build to fill the resulting gap. Without some special mechanism, the daughter DNA strand resulting from lagging-strand synthesis would be shortened at each cell division.

The problem of telomere shortening is solved by an enzyme that adds telomeric sequences to the ends of each chromosome. The enzyme is a protein and RNA complex called *telomere terminal transferase*, or *telomerase*. Because the sequence of the telomerase-associated RNA, as we will see, serves as the template for addition of deoxyribonucleotides to the ends of telomeres, the source of the enzyme and not the source of the telomeric DNA primer determines the sequence added. This was proved by transforming *Tetrahymena* with a mutated form of the gene encoding the

◄ **FIGURE 10-34 Mechanism of action of telomerase.** The single-stranded 3′ terminus of a telomere is extended by telomerase, counteracting the inability of the DNA replication mechanism to synthesize the extreme terminus of linear DNA. Telomerase elongates this single-stranded end by a reiterative reverse-transcription mechanism. The action of the telomerase from the protozoan *Oxytricha*, which adds a T_4G_4 repeat unit, is depicted; other telomerases add slightly different sequences. The telomerase contains an RNA template (red) that base-pairs to the 3′ end of the lagging-strand template. The telomerase catalytic site (green) then adds deoxyribonucleotides (blue) using the RNA molecule as a template; this reverse transcription proceeds to position 35 of the RNA template (step **1**). The strands of the resulting DNA-RNA duplex are then thought to slip relative to each other, leading to displacement of a single-stranded region of the telomeric DNA strand and to uncovering of part of the RNA template sequence (step **2**). The lagging-strand telomeric sequence is again extended to position 35 by telomerase, and the DNA-RNA duplex undergoes translocation and hybridization as before (steps **3** and **4**). The slippage mechanism is thought to be facilitated by the unusual base pairing (black dots) between the displaced G residues, which is less stable than Watson-Crick base pairing. Telomerases can add multiple repeats by repetition of steps **3** and **4**. DNA polymerase α-primase can prime synthesis of new Okazaki fragments on this extended template strand. The net result prevents shortening of the lagging strand at each cycle of DNA replication [Adapted from D. Shippen-Lentz and E. H. Blackburn, 1990, *Nature* **247**:550.]

telomerase-associated RNA. The resulting telomerase added a DNA sequence complementary to the mutated RNA sequence to the ends of telomeric primers. Thus telomerase is a specialized form of a reverse transcriptase that carries its own internal RNA template to direct DNA synthesis.

Figure 10-34 depicts how telomerase, by reverse transcription of its associated RNA, elongates the 3′ end of the single-stranded DNA at the end of the G-rich strand mentioned above. Cells from knockout mice that cannot produce the telomerase-associated RNA exhibit no telomerase activity, and their telomeres shorten successively with each cell generation. Such mice can breed and reproduce normally for three generations before the long telomere repeats become substantially eroded. Then, the absence of telomere DNA results in adverse effects, including fusion of chromosome termini and chromosomal loss. By the fourth generation, the reproductive potential of these knockout mice declines, and they cannot produce offspring after the sixth generation.

 The human genes expressing the telomerase protein and the telomerase-associated RNA are active in germ cells but are turned off in most adult tissues, in which cells replicate only rarely. However, these genes are reactivated in most human cancer cells, where telomerase is required for the multiple cell divisions necessary to form a tumor. This phenomenon has stimulated a search for inhibitors of human telomerase as potential therapeutic agents for treating cancer. ∎

While telomerase prevents telomere shortening in most eukaryotes, some organisms use alternative strategies. *Drosophila* species maintain telomere lengths by the regulated insertion of non-LTR retrotransposons into telomeres. This is one of the few instances in which a mobile element has a specific function in its host organism.

Yeast Artificial Chromosomes Can Be Used to Clone Megabase DNA Fragments

The research on circular and linear plasmids in yeast identified all the basic components of a *yeast artificial chromosome (YAC)*. To construct YACs, TEL sequences from yeast cells or from the protozoan *Tetrahymena* are combined with yeast CEN and ARS sequences; to these are added DNA with selectable yeast genes and enough DNA from any source to make a total of more than 50 kb. (Smaller DNA segments do not work as well.) Such artificial chromosomes replicate in yeast cells and segregate almost perfectly, with only 1 daughter cell in 1000 to 10,000 failing to receive an artificial chromosome. During meiosis, the two sister chromatids of the artificial chromosome separate correctly to produce haploid spores. The successful propagation of YACs and studies such as those depicted in Figure 10-32 strongly support the conclusion that yeast chromosomes, and probably all eukaryotic chromosomes, are linear, double-stranded DNA molecules containing special regions that ensure replication and proper segregation.

 YACs are valuable experimental tools that can be used to clone very long chromosomal pieces from other species. For instance, YACs have been used extensively for cloning fragments of human DNA up to 1000 kb (1 megabase) in length, permitting the isolation of overlapping YAC clones that collectively encompass nearly the entire length of individual human chromosomes. Currently, research is in progress to construct human artificial chromosomes using the much longer centromeric DNA of human chromosomes and the longer human telomeres. Such human artificial chromosomes might be useful in gene therapy to treat human genetic diseases such as sickle-cell anemia, cystic fibrosis, and inborn errors of metabolism. ∎

KEY CONCEPTS OF SECTION 10.5

Morphology and Functional Elements of Eukaryotic Chromosomes

■ During metaphase, eukaryotic chromosomes become sufficiently condensed that they can be visualized individually in the light microscope.

■ The karyotype, the set of metaphase chromosomes, is characteristic of each species. Closely related species can have dramatically different karyotypes, indicating that similar genetic information can be organized on chromosomes in different ways.

■ Banding analysis and chromosome painting, a more precise method, are used to identify the different human metaphase chromosomes and to detect translocations and deletions (see Figure 10-29).

■ In certain cells of the fruit fly *Drosophila melanogaster* and related insects, interphase chromosomes are reduplicated 10 times, generating polytene chromosomes that are visible in the light microscope (see Figure 10-31).

■ The highly reproducible banding patterns of polytene chromosomes make it possible to localize cloned *Drosophila* DNA on a *Drosophila* chromosome by in situ hybridization (Figure 10-30) and to visualize chromosomal deletions and rearrangements as changes in the normal pattern of bands.

■ When metaphase chromosomes decondense during interphase, certain regions, termed *heterochromatin*, remain much more condensed than the bulk of chromatin, called *euchromatin*.

■ Three types of DNA sequences are required for a long linear DNA molecule to function as a chromosome: a replication origin, called *ARS* in yeast; a centromere (CEN) sequence; and two telomere (TEL) sequences at the ends of the DNA (see Figure 10-32).

■ DNA fragments up to 10^6 base pairs long can be cloned in yeast artificial chromosome (YAC) vectors.

10.6 Organelle DNAs

Although the vast majority of DNA in most eukaryotes is found in the nucleus, some DNA is present within the mitochondria of animals, plants, and fungi and within the chloroplasts of plants. These organelles are the main cellular sites for ATP formation, during oxidative phosphorylation in mitochondria and photosynthesis in chloroplasts (Chapter 8). Many lines of evidence indicate that mitochondria and chloroplasts evolved from bacteria that were endocytosed into ancestral cells containing a eukaryotic nucleus, forming *endosymbionts*. Over evolutionary time, most of the bacterial genes encoding components of the present-day organelles were transferred to the nucleus. However, mitochondria and chloroplasts in today's eukaryotes retain DNAs encoding proteins essential for organellar function as well as the ribosomal and transfer RNAs required for their translation. Thus eukaryotic cells have multiple genetic systems: a predominant nuclear system and secondary systems with their own DNA in the mitochondria and chloroplasts.

Mitochondria Contain Multiple mtDNA Molecules

Individual mitochondria are large enough to be seen under the light microscope, and even the mitochondrial DNA

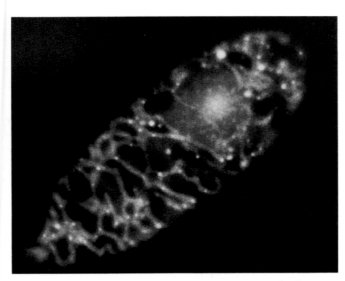

10 μm

▲ **EXPERIMENTAL FIGURE 10-35 Dual staining reveals the multiple mitochondrial DNA molecules in a growing *Euglena gracilis* cell.** Cells were treated with a mixture of two dyes: ethidium bromide, which binds to DNA and emits a red fluorescence, and DiOC6, which is incorporated specifically into mitochondria and emits a green fluorescence. Thus the nucleus emits a red fluorescence, and areas rich in mitochondrial DNA fluoresce yellow—a combination of red DNA and green mitochondrial fluorescence. [From Y. Hayashi and K. Ueda, 1989, *J. Cell Sci.* **93**:565.]

(mtDNA) can be detected by fluorescence microscopy. The mtDNA is located in the interior of the mitochondrion, the region known as the matrix (see Figure 5-26). As judged by the number of yellow fluorescent "dots" of mtDNA, a *Euglena gracilis* cell contains at least 30 mtDNA molecules (Figure 10-35).

Since the dyes used to visualize nuclear and mitochondrial DNA do not affect cell growth or division, replication of mtDNA and division of the mitochondrial network can be followed in living cells using time-lapse microscopy. Such studies show that in most organisms mtDNA replicates throughout interphase. At mitosis each daughter cell receives approximately the same number of mitochondria, but since there is no mechanism for apportioning exactly equal numbers of mitochondria to the daughter cells, some cells contain more mtDNA than others. By isolating mitochondria from cells and analyzing the DNA extracted from them, it can be seen that each mitochondrion contains multiple mtDNA molecules. Thus the total amount of mtDNA in a cell depends on the number of mitochondria, the size of the mtDNA, and the number of mtDNA molecules per mitochondrion. Each of these parameters varies greatly between different cell types.

mtDNA Is Inherited Cytoplasmically and Encodes rRNAs, tRNAs, and Some Mitochondrial Proteins

Studies of mutants in yeasts and other single-celled organisms first indicated that mitochondria exhibit *cytoplasmic inheritance* and thus must contain their own genetic system (Figure 10-36). For instance, *petite* yeast mutants exhibit structurally abnormal mitochondria and are incapable of oxidative phosphorylation. As a result, petite cells grow more slowly than wild-type yeasts and form smaller colonies (hence the name "petite"). Genetic crosses between different (haploid) yeast strains showed that the *petite* mutation does not segregate with any known nuclear gene or chromosome. In later studies, most petite mutants were found to contain deletions of mtDNA.

In the mating by fusion of haploid yeast cells, both parents contribute equally to the cytoplasm of the resulting diploid; thus inheritance of mitochondria is biparental (see Figure 10-36a). In mammals and most other multicellular organisms, however, the sperm contributes little (if any) cytoplasm to the zygote, and virtually all the mitochondria in the embryo are derived from those in the egg, not the sperm. Studies in mice have shown that 99.99 percent of mtDNA is maternally inherited, but a small part (0.01 percent) is inherited from the male parent. In higher plants, mtDNA is inherited exclusively in a uniparental fashion through the female parent (egg), not the male (pollen).

The entire mitochondrial genome from a number of different organisms has now been cloned and sequenced, and

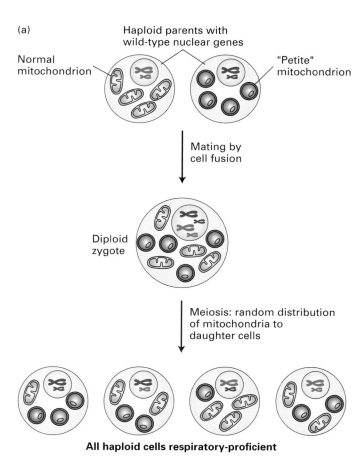

mtDNAs from all these sources have been found to encode rRNAs, tRNAs, and essential mitochondrial proteins. All proteins encoded by mtDNA are synthesized on mitochondrial ribosomes. All mitochondrially synthesized polypeptides identified thus far (with one possible exception) are not complete enzymes but subunits of multimeric complexes used in electron transport or ATP synthesis. Most proteins localized in mitochondria, such as the mitochondrial RNA and DNA polymerases, are synthesized on cytosolic ribosomes and are imported into the organelle by processes discussed in Chapter 16.

The Size and Coding Capacity of mtDNA Vary Considerably in Different Organisms

Surprisingly, the size of the mtDNA, the number and nature of the proteins it encodes, and even the mitochondrial genetic code itself vary greatly between different organisms. Human mtDNA, a circular molecule that has been completely sequenced, is among the smallest known mtDNAs, containing 16,569 base pairs (Figure 10-37). It encodes the two rRNAs found in mitochondrial ribosomes and the 22 tRNAs used to translate mitochondrial mRNAs. Human mtDNA has 13 sequences that begin with an ATG (methionine) codon, end with a stop codon, and are long enough to encode a polypeptide of more than 50 amino acids; all the possible proteins encoded by these open reading frames have been identified. Mammalian mtDNA, in contrast to nuclear DNA, lacks introns and contains no long noncoding sequences.

The mtDNA from most multicellular animals (metazoans) is about the same size as human mtDNA and encodes similar gene products. In contrast, yeast mtDNA is almost five times as large (≈78,000 bp). The mtDNAs from yeast and other fungi encode many of the same gene products as mammalian mtDNA, as well as others whose genes are found in the nuclei of metazoan cells.

◄ **FIGURE 10-36 Cytoplasmic inheritance of the *petite* mutation in yeast.** Petite-strain mitochondria are defective in oxidative phosphorylation owing to a deletion in mtDNA. (a) Haploid cells fuse to produce a diploid cell that undergoes meiosis, during which random segregation of parental chromosomes and mitochondria containing mtDNA occurs. Since yeast normally contain ≈50 mtDNA molecules per cell, all products of meiosis usually contain both normal and petite mtDNAs and are capable of respiration. (b) As these haploid cells grow and divide mitotically, the cytoplasm (including the mitochondria) is randomly distributed to the daughter cells. Occasionally, a cell is generated that contains only defective petite mtDNA and yields a petite colony. Thus formation of such petite cells is independent of any nuclear genetic marker.

▶ **FIGURE 10-37 The coding capacity of human mitochondrial DNA (mtDNA).** Proteins and RNAs encoded by each of the two strands are shown separately. Transcription of the outer (H) strand occurs in the clockwise direction and of the inner (L) strand in the counterclockwise direction. The abbreviations for amino acids denote the corresponding tRNA genes. *ND1*, *ND2*, etc., denote genes encoding subunits of the NADH-CoQ reductase complex. The 207-bp gene encoding F_0 ATPase subunit 8 overlaps, out of frame, with the N-terminal portion of the segment encoding F_0 ATPase subunit 6. Mammalian mtDNA genes do not contain introns, although intervening DNA lies between some genes. [See D. A. Clayton, 1991, *Ann. Rev. Cell Biol.* **7**:453.]

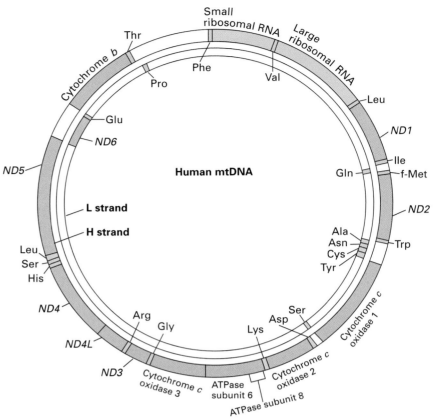

 In contrast to other eukaryotes, which contain a single type of mtDNA, plants contain several types of mtDNA that appear to recombine with one another. Plant mtDNAs are much larger and more variable in size than the mtDNAs of other organisms. Even in a single family of plants, mtDNAs can vary as much as eightfold in size (watermelon = 330 kb; muskmelon = 2500 kb). Unlike animal, yeast, and fungal mtDNAs, plant mtDNAs contain genes encoding a 5S mitochondrial rRNA, which is present only in the mitochondrial ribosomes of plants, and the α subunit of the F_1 ATPase. The mitochondrial rRNAs of plants are also considerably larger than those of other eukaryotes. Long, noncoding regions and duplicated sequences are largely responsible for the greater length of plant mtDNAs. ∎

Differences in the size and coding capacity of mtDNA from various organisms most likely reflect the movement of DNA between mitochondria and the nucleus during evolution. Direct evidence for this movement comes from the observation that several proteins encoded by mtDNA in some species are encoded by nuclear DNA in others. It thus appears that entire genes moved from the mitochondrion to the nucleus, or vice versa, during evolution.

The most striking example of this phenomenon involves the gene *cox II*, which encodes subunit 2 of cytochrome *c* oxidase. This gene is found in mtDNA in all organisms studied except for one species of legume, the mung bean: in this organism only, the *cox II* gene is nuclear. Many RNA transcripts of plant mitochondrial genes are edited, mainly by the enzyme-catalyzed conversion of selected C residues to U, and occasionally U to C. (RNA editing is discussed in Chapter 12.) The nuclear *cox II* gene of mung bean corresponds more closely to the edited cox II RNA transcripts than to the mitochondrial *cox II* genes found in other legumes. These observations are strong evidence that the *cox II* gene moved from the mitochondrion to the nucleus during mung bean evolution by a process that involved an RNA intermediate. Presumably this movement involved a reverse-transcription mechanism similar to that by which processed pseudogenes are generated in the nuclear genome from nuclear-encoded mRNAs.

Products of Mitochondrial Genes Are Not Exported

As far as is known, all RNA transcripts of mtDNA and their translation products remain in the mitochondrion, and all mtDNA-encoded proteins are synthesized on mitochondrial ribosomes. Mitochondria encode the rRNAs that form mitochondrial ribosomes, although all but one or two of the ribosomal proteins (depending on the species) are imported from the cytosol. In most eukaryotes, all the tRNAs used for protein synthesis in mitochondria are encoded by mtDNAs. However, in wheat, in the parasitic protozoan *Trypanosoma brucei* (the cause of African sleeping sickness), and in ciliated protozoans, most mitochondrial tRNAs are encoded by the nuclear DNA and imported into the mitochondrion.

 Reflecting the bacterial ancestry of mitochondria, mitochondrial ribosomes resemble bacterial ribosomes and differ from eukaryotic cytosolic ribosomes in their RNA and protein compositions, their size, and their sensitivity to certain antibiotics (see Figure 4-24). For instance, chloramphenicol blocks protein synthesis by bacterial and mitochondrial ribosomes from most organisms, but cycloheximide does not. This sensitivity of mitochondrial ribosomes to the important aminoglycoside class of antibiotics is the main cause of the toxicity that these antibiotics can cause. Conversely, cytosolic ribosomes are sensitive to cycloheximide and resistant to chloramphenicol. In cultured mammalian cells the only proteins synthesized in the presence of cycloheximide are encoded by mtDNA and produced by mitochondrial ribosomes. ∎

Mitochondrial Genetic Codes Differ from the Standard Nuclear Code

The genetic code used in animal and fungal mitochondria is different from the standard code used in all prokaryotic and eukaryotic nuclear genes; remarkably, the code even differs in mitochondria from different species (Table 10-3). Why and how these differences arose during evolution is mysterious. UGA, for example, is normally a stop codon, but is read as tryptophan by human and fungal mitochondrial translation systems; however, in plant mitochondria, UGA is still recognized as a stop codon. AGA and AGG, the standard nuclear codons for arginine, also code for arginine in fungal and plant mtDNA, but they are stop codons in mammalian mtDNA and serine codons in *Drosophila* mtDNA.

 As shown in Table 10-3, plant mitochondria appear to utilize the standard genetic code. However, comparisons of the amino acid sequences of plant mitochondrial proteins with the nucleotide sequences of plant mtDNAs suggested that CGG could code for *either* arginine (the "standard" amino acid) or tryptophan. This apparent nonspecificity of the plant mitochondrial code is explained by editing of mitochondrial RNA transcripts, which can convert cytosine residues to uracil residues. If a CGG sequence is edited to UGG, the codon specifies tryptophan, the standard amino acid for UGG, whereas unedited CGG codons encode the standard arginine. Thus the translation system in plant mitochondria does utilize the standard genetic code. ∎

Mutations in Mitochondrial DNA Cause Several Genetic Diseases in Humans

The severity of disease caused by a mutation in mtDNA depends on the nature of the mutation and on the proportion of mutant and wild-type mtDNAs present in a particular cell type. Generally, when mutations in mtDNA are found, cells contain mixtures of wild-type and mutant mtDNAs—a condition known as *heteroplasmy*. Each time a mammalian somatic or germ-line cell divides, the mutant and wild-type mtDNAs will segregate randomly into the daughter cells, as occurs in yeast cells (see Figure 10-36). Thus, the mtDNA genotype, which fluctuates from one generation and from one cell division to the next, can drift toward predominantly wild-type or predominantly mutant mtDNAs. Since all enzymes for the replication and growth of mitochondria, such as DNA and RNA polymerases, are imported from the cytosol, a mutant mtDNA should not be at a "replication disadvantage"; mutants that involve large

TABLE 10-3	Alterations in the Standard Genetic Code in Mitochondria

Codon	Standard Code: Nuclear-Encoded Proteins	Mammals	*Drosophila*	*Neurospora*	Yeasts	Plants
UGA	Stop	Trp	Trp	Trp	Trp	Stop
AGA, AGG	Arg	Stop	Ser	Arg	Arg	Arg
AUA	Ile	Met	Met	Ile	Met	Ile
AUU	Ile	Met	Met	Met	Met	Ile
CUU, CUC, CUA, CUG	Leu	Leu	Leu	Leu	Thr	Leu

SOURCES: S. Anderson et al., 1981, *Nature* **290**:457; P. Borst, in *International Cell Biology 1980–1981*, H. G. Schweiger, ed., Springer-Verlag, p. 239; C. Breitenberger and U. L. Raj Bhandary, 1985, *Trends Biochem. Sci.* **10**:478–483; V. K. Eckenrode and C. S. Levings, 1986, *In Vitro Cell Dev. Biol.* **22**:169–176; J. M. Gualber et al., 1989, *Nature* **341**:660–662; and P. S. Covello and M. W. Gray, 1989, *Nature* **341**:662–666.

deletions of mtDNA might even be at a selective advantage in replication because they can replicate faster.

All cells have mitochondria, yet mutations in mtDNA affect only some tissues. Those most usually affected are tissues that have a high requirement for ATP produced by oxidative phosphorylation and tissues that require most of or all the mtDNA in the cell to synthesize sufficient amounts of functional mitochondrial proteins. *Leber's hereditary optic neuropathy* (degeneration of the optic nerve, accompanied by increasing blindness), for instance, is caused by a missense mutation in the mtDNA gene encoding subunit 4 of the NADH-CoQ reductase. Any of several large deletions in mtDNA causes another set of diseases including *chronic progressive external ophthalmoplegia* and *Kearns-Sayre syndrome,* which are characterized by eye defects and, in Kearns-Sayre syndrome, also by abnormal heartbeat and central nervous system degeneration. A third condition, causing "ragged" muscle fibers (with improperly assembled mitochondria) and associated uncontrolled jerky movements, is due to a single mutation in the TψCG loop of the mitochondrial lysine tRNA. As a result of this mutation, the translation of several mitochondrial proteins apparently is inhibited ∎

Chloroplasts Contain Large Circular DNAs Encoding More Than a Hundred Proteins

As we discuss in Chapter 8, the structure of chloroplasts is similar in many respects to that of mitochondria. Like mitochondria, chloroplasts contain multiple copies of the organellar DNA and ribosomes, which synthesize some chloroplast-encoded proteins using the "standard" genetic code. Other chloroplast proteins are fabricated on cytosolic ribosomes and are incorporated into the organelle after translation (Chapter 16).

Chloroplast DNAs are circular molecules of 120,000–160,000 bp, depending on the species. The complete sequences of several chloroplast DNAs have been determined. Of the ≈120 genes in chloroplast DNA, about 60 are involved in RNA transcription and translation, including genes for rRNAs, tRNAs, RNA polymerase subunits, and ribosomal proteins. About 20 genes encode subunits of the chloroplast photosynthetic electron-transport complexes and the F_0F_1 ATPase complex. Also encoded in the chloroplast genome is the larger of the two subunits of ribulose 1,5-bisphosphate carboxylase, which is involved in the fixation of carbon dioxide during photosynthesis.

Reflecting the endosymbiotic origin of chloroplasts, some regions of chloroplast DNA are strikingly similar to the DNA of present-day bacteria. For instance, chloroplast DNA encodes four subunits of RNA polymerase that are highly homologous to the subunits of *E. coli* RNA polymerase. One segment of chloroplast DNA encodes eight

proteins that are homologous to eight *E. coli* ribosomal proteins; moreover, the order of these genes is the same in the two DNAs.

Although the overall organization of chloroplast DNAs from different species is similar, some differences in gene composition occur. For instance, liverwort chloroplast DNA has some genes that are not detected in the larger tobacco chloroplast DNA, and vice versa. Since chloroplasts in both species contain virtually the same set of proteins, these data suggest that some genes are present in the chloroplast DNA of one species and in the nuclear DNA of the other, indicating that some exchange of genes between chloroplast and nucleus occurred during evolution. ∎

Methods similar to those used for the transformation of yeast cells (Chapter 9) have been developed for stably introducing foreign DNA into the chloroplasts of higher plants. The large number of chloroplast DNA molecules per cell permits the introduction of thousands of copies of an engineered gene into each cell, resulting in extraordinarily high levels of foreign protein production. Chloroplast transformation has recently led to the engineering of plants that are resistant to bacterial and fungal infections, drought, and herbicides. The level of production of foreign proteins is comparable with that achieved with engineered bacteria, making it likely that chloroplast transformation will be used for the production of human pharmaceuticals and possibly for the engineering of food crops containing high levels of all the amino acids essential to humans. ∎

KEY CONCEPTS OF SECTION 10.6

Organelle DNAs

∎ Mitochondria and chloroplasts most likely evolved from bacteria that formed a symbiotic relationship with ancestral cells containing a eukaryotic nucleus. Most of the genes originally within these organelles moved to the nuclear genome over evolutionary time, leaving different gene sets in the mtDNAs of different organisms.

∎ Most mitochondrial and chloroplast DNAs are circular molecules, reflecting their probable bacterial origin. Plant mtDNAs and chloroplast DNAs generally are longer than mtDNAs from other eukaryotes, largely because they contain more noncoding regions and repetitive sequences.

∎ All mtDNAs and chloroplast DNAs appear to encode rRNAs, tRNAs, and some of the proteins involved in mitochondrial or photosynthetic electron transport and ATP synthesis.

∎ Because most mtDNA is inherited from egg cells rather than sperm, mutations in mtDNA exhibit a maternal cytoplasmic pattern of inheritance.

- Mitochondrial ribosomes resemble bacterial ribosomes in their structure, sensitivity to chloramphenicol, and resistance to cycloheximide.

- The genetic code of animal and fungal mtDNAs differs slightly from that of bacteria and the nuclear genome and varies between different animals and plants (see Table 10-3). In contrast, plant mtDNAs and chloroplast DNAs appear to conform to the standard genetic code.

- Several human neuromuscular disorders result from mutations in mtDNA. Patients generally have a mixture of wild-type and mutant mtDNA in their cells (heteroplasmy): the higher the fraction of mutant mtDNA, the more severe the mutant phenotype.

PERSPECTIVES FOR THE FUTURE

The human genome sequence is a goldmine for new discoveries about molecular cell biology, new proteins that may be the basis of effective therapies of human diseases, and even for discoveries concerning early human history and evolution. However, because only ≈1.5 percent of the sequence encodes proteins, finding new genes is like finding a needle in a haystack. Identification of genes in bacterial genome sequences is relatively simple because of the scarcity of introns. Simply searching for long open reading frames free of stop codons identifies most genes. In contrast, the search for human genes is vastly complicated by the structure of human genes, most of which are composed of multiple, relatively short exons separated by much longer, noncoding introns. Some have estimated that only approximately half of all human genes have been identified. And the identification of complex transcription units by examining DNA sequence alone is even more challenging. Future improvements in bioinformatic methods for gene identification, and further characterization of cDNA copies of mRNAs isolated from the hundreds of human cell types, will likely lead to the discovery of new proteins and, through their study, to a new understanding of biological processes and applications to medicine and agriculture.

We have seen that although most transposons do not function directly in cellular processes, they have helped to shape modern genomes by promoting gene duplications, exon shuffling, the generation of new combinations of transcription control sequences, and other aspects of contemporary genomes. They also have the potential today to teach us about our own history and origins. This is because L1 and *Alu* retrotransposons have inserted into new sites in individuals throughout our history. Large numbers of these interspersed repeats are polymorphic in the population—they occur at a particular site in some individuals and not others. Individuals sharing an insertion at a particular site descended from the same common ancestor that developed from an egg or sperm in which that insertion occurred. The

time elapsed from the initial insertion can be estimated by the differences in sequences of the element among individuals that carry them because these differences arose from the accumulation of random mutations. Further studies of these retrotransposon polymorphisms will undoubtedly add immensely to our understanding both of human migrations since *Homo sapiens* first evolved and of the history of contemporary populations.

KEY TERMS

centromere *413*
chromatid *430*
chromatin *424*
chromosome scaffold *427*
cytoplasmic inheritance *438*
euchromatin *433*
exon shuffling *422*
fluorescence in situ hybridization (FISH) *413*
G bands *431*
gene family *410*
heterochromatin *433*
histones *424*
karyotype *430*
LINEs *420*

long terminal repeats (LTRs) *417*
mobile DNA elements *414*
nucleosome *424*
P element *416*
polytene chromosomes *433*
pseudogene *411*
retrotransposons *415*
scaffold-associated regions (SARs) *428*
simple-sequence DNA *412*
SINEs *420*
telomere *413*
transcription unit *407*
transposition *414*
transposons *414*

REVIEW THE CONCEPTS

1. Genes can be transcribed into mRNA for protein-coding genes or RNA for genes such as ribosomal or transfer RNAs. Define a gene. Describe how a complex transcription unit can be alternatively processed to generate a variety of mRNAs and ultimately proteins.

2. Sequencing of the human genome has revealed much about the organization of genes. Describe the differences between single genes, gene families, pseudogenes, and tandemly repeated genes.

3. Much of the human genome consists of repetitive DNA. Describe the difference between microsatellite and minisatellite DNA. How is this repetitive DNA useful for identifying individuals by the technique of DNA fingerprinting?

4. Mobile DNA elements that can move or transpose to a new site directly as DNA are called DNA transposons. Describe the mechanism by which a bacterial insertion sequence can transpose.

5. Retrotransposons are a class of mobile elements that use a RNA intermediate. Contrast the mechanism of transposition between retrotransposons that contain and those that lack long terminal repeats (LTRs).

6. Describe the role that mobile DNA elements may have played in the evolution of modern organisms. What is the process known as exon shuffling, and what role do mobile DNA elements play in this process?

7. DNA in a cell associates with proteins to form chromatin. What is a nucleosome? What role do histones play in nucleosomes? How are nucleosomes arranged in condensed 30-nm fibers?

8. Describe the general organization of a eukaryotic chromosome. What structural role do scaffold associated regions (SARs) or matrix attachment regions (MARs) play? Where are genes primarily located relative to chromosome structure?

9. Metaphase chromosomes can be identified by characteristic banding patterns. What are G bands and R bands? What is chromosome painting, and how is this technique useful?

10. Replication and segregation of eukaryotic chromosomes require three functional elements: replication origins, a centromere, and telomeres. Describe how these three elements function. What is the role of telomerase in maintaining chromosome structure? What is a yeast artificial chromosome (YAC)?

11. Mitochondria contain their own DNA molecules. Describe the types of genes encoded in the mitochondrial genome. How do the mitochondrial genomes of plants, fungi, and animals differ?

12. Mitochondria and chloroplasts are thought to have evolved from symbiotic bacteria present in nucleated cells. Review the experimental evidence that supports this hypothesis.

ANALYZE THE DATA

In culture, normal human cells undergo a finite number of cell divisions until they no longer proliferate and enter a state known as replicative senescence. The inability to maintain normal telomere length is thought to play an important role in this process. Telomerase is a ribonucleoprotein complex that regenerates the ends of telomeres lost during each round of DNA replication. Human telomerase consists of a template containing RNA subunit and a catalytic protein subunit known as human telomerase reverse transcriptase (hTERT). Most normal cells do not express telomerase; most cancer cells do express telomerase. Thus, telomerase is proposed to play a key role in the transformation of cells from a normal to a malignant state.

a. In the following experiments, the role of telomerase in the growth of human cancer cells was investigated (see W. C. Hahn et al., 1999, *Nature Medicine* 5:1164–1170). Immortal, telomerase-positive cells (cell A) and immortal, telomerase-negative cells (cell B) were transfected with a plasmid ex-

pressing a wild-type or mutated hTERT. Telomerase activity in cell extracts was measured using the telomeric repeat amplification protocol (TRAP) assay, a PCR-based assay that measures the addition of telomere repeat units onto a DNA fragment. A six-base-pair ladder pattern is typically seen. Control indicates transfection of cells with just the plasmid. Wild type and mutant indicate transfection with a plasmid expressing a wild-type hTERT or the mutated hTERT, respectively. What do you conclude about the effect of the mutant hTERT on telomerase activity in the transfected cells? What type of mutation would this represent?

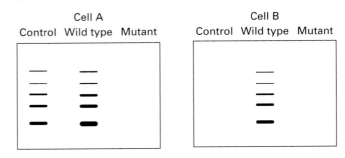

b. Telomere length in these transfected cells was examined by Southern blot analysis of total genomic DNA digested with a restriction enzyme and probed with a telomere-specific DNA sequence. What do you conclude about the lengths of the telomeres in cell A and cell B after transfection with the wild-type or mutant hTERT? By what mechanism do cells A and B maintain their telomere lengths?

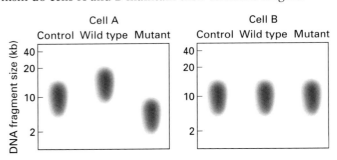

c. The proliferation of transfected cells A and B was assayed by measuring the number of cells versus time in culture. What do you conclude about the effect of the mutant hTERT on cell proliferation in transfected cells A and B?

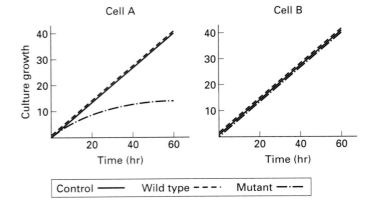

d. Do the results of the three assays support or refute the proposed role of telomerase in long-term survival of cells in culture? What is the value of comparing transfection of cells A with cells B?

e. Tumorigenic cells, which can replicate continuously in culture, can cause tumors when injected into mice that lack a functional immune system (nude mice). From the above assays, what would you predict would happen if wild-type or mutant hTERT transfected cells A were injected into nude mice?

REFERENCES

Molecular Definition of a Gene

Ayoubi, T. A., and W. J. Van De Ven. 1996. Regulation of gene expression by alternative promoters. *FASEB J.* **10**:453–460.

Maniatis, T., and B. Tasic. 2002. Alternative pre-mRNA splicing and proteome expansion in metazoans. *Nature* **418**:236–243.

Mathe, C., et al. 2002. Current methods of gene prediction, their strengths and weaknesses. *Nucl. Acids Res.* **30**:4103–4117.

Chromosomal Organization of Genes and Noncoding DNA

Cummings, C. J., and H. Y. Zoghbi. 2000. Trinucleotide repeats: mechanisms and pathophysiology. *Ann. Rev. Genomics Hum. Genet.* **1**:281–328.

Goff, S. A., et al. 2002. A draft sequence of the rice genome (*Oryza sativa* L. ssp. *japonica*). *Science* **296**:92–100.

Lander, E. S., et al. 2001. Initial sequencing and analysis of the human genome. *Nature* **409**:860–921.

Ranum, L. P., and J. W. Day. 2002. Dominantly inherited, noncoding microsatellite expansion disorders. *Curr. Opin. Genet. Dev.* **12**:266–271.

Mobile DNA

Feschotte, C., N. Jiang, and S. R. Wessler. 2002. Plant transposable elements: where genetics meets genomics. *Nat. Rev. Genet.* **3**:329–341.

Gray, Y. H. 2000. It takes two transposons to tango: transposable-element-mediated chromosomal rearrangements. *Trends Genet.* **16**:461–468.

Kazazian, H. H., Jr. 1999. An estimated frequency of endogenous insertional mutations in humans. *Nature Genet.* **22**:130.

Ostertag, E. M., and H. H. Kazazian, Jr. 2001. Biology of mammalian L1 retrotransposons. *Ann. Rev. Genet.* **35**:501–538.

Structural Organization of Eukaryotic Chromosomes

Horn, P. J., and C. L. Peterson. 2002. Molecular biology. Chromatin higher order folding—wrapping up transcription. *Science* **297**:1824–1827.

Luger, K., and T. J. Richmond. 1998. The histone tails of the nucleosome. *Curr. Opin. Genet. Dev.* **8**:140–146.

Morphology and Functional Elements of Eukaryotic Chromosomes

Cheeseman, I. M., D. G. Drubin, and G. Barnes. 2002. Simple centromere, complex kinetochore: linking spindle microtubules and centromeric DNA in budding yeast. *J. Cell Biol.* **157**:199–203.

Kelleher, C., et al. 2002. Telomerase: biochemical considerations for enzyme and substrate. *Trends Biochem, Sci.* **27**:572–579.

Kitagawa, K., and P. Hieter. 2001. Evolutionary conservation between budding yeast and human kinetochores. *Nature Rev. Mol. Cell Biol.* **2**:678–687.

Larin, Z., and J. E. Mejia. 2002. Advances in human artificial chromosome technology. *Trends Genet.* **18**:313–319.

McEachern, M. J., A. Krauskopf, and E. H. Blackburn. 2000. Telomeres and their control. *Ann. Rev. Genet.* **34**:331–358.

Organelle DNA

Clayton, D. A. 2000. Transcription and replication of mitochondrial DNA. *Hum. Reprod.* **2**(Suppl.):11–17.

Dahl, H.-H. M., and D. R. Thorburn. 2001. Mitochondrial diseases: beyond the magic circle. *Am. J. Med. Genet.* **106**:1–3.

Daniell, H., M. S. Khan, and L. Allison. 2002. Milestones in chloroplast genetic engineering: an environmentally friendly era in biotechnology. *Trends Plant Sci.* **7**:84–91.

Gray, M. W. 1999. Evolution of organellar genomes. *Curr. Opin. Genet. Dev.* **9**:678–687.

Palmer, J. D., et al. 2000. Dynamic evolution of plant mitochondrial genomes: mobile genes and introns and highly variable mutation rates. *Proc. Nat'l. Acad. Sci. USA* **97**:6960–6966.

Sugiura, M., T. Hirose, and M. Sugita. 1998. Evolution and mechanism of translation in chloroplasts. *Ann. Rev. Genet.* **32**:437–459.

11

TRANSCRIPTIONAL CONTROL OF GENE EXPRESSION

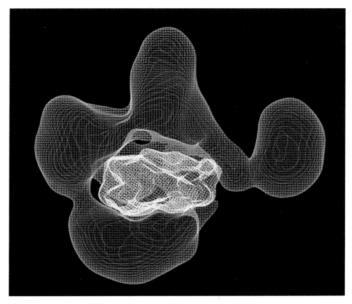

Yeast RSC chromatin remodeling complex structure determined by electron microscopy. The complex (red) is shown bound to a nucleosome (yellow). [Courtesy of Francisco J. Asturias, 2002, *PNAS* **99**:13477.]

In previous chapters we've seen that the actions and properties of each cell type are determined by the proteins it contains. In this and the next chapter, we consider how the kinds and amounts of the various proteins produced by a particular cell type in a multicellular organism are regulated. The basic steps in **gene expression,** the entire process whereby the information encoded in a particular gene is decoded into a particular protein, are reviewed in Chapter 4. Synthesis of mRNA requires that an **RNA polymerase** initiate transcription, polymerize ribonucleoside triphosphates complementary to the DNA coding strand, and then terminate transcription (see Figure 4-10). In prokaryotes, ribosomes and translation-initiation factors have immediate access to newly formed RNA transcripts, which function as mRNA without further modification. In eukaryotes, however, the initial RNA transcript is subjected to processing that yields a functional mRNA (see Figure 4-14). The mRNA then is transported from its site of synthesis in the nucleus to the cytoplasm, where it is translated into protein with the aid of ribosomes, tRNAs, and translation factors (see Figure 4-26).

Theoretically, regulation at any one of the various steps in gene expression could lead to *differential* production of proteins in different cell types or developmental stages or in response to external conditions. Although examples of regulation at each step in gene expression have been found, control of transcription initiation—the first step—is the most important mechanism for determining whether most genes are expressed and how much of the encoded mRNAs and, consequently, proteins are produced. The molecular mecha-

nisms that regulate transcription initiation are critical to numerous biological phenomena, including the development of a multicellular organism from a single fertilized egg cell, the immune responses that protect us from pathogenic microorganisms, and neurological processes such as learning and memory. When these regulatory mechanisms function improperly, pathological processes such as cancer and aging may occur.

In this chapter, we focus on the molecular events that determine when transcription of eukaryotic genes is initiated and also briefly consider control of premature termination of transcription. RNA processing and various posttranscriptional mechanisms for controlling eukaryotic gene expression are covered in the next chapter. Subsequent chapters, particularly Chapters 13, 14, and 15, provide examples of how transcription is regulated by interactions between cells and how the resulting **gene control** contributes to the development and function of specific types of cells in multicellular organisms.

11.1 Overview of Eukaryotic Gene Control and RNA Polymerases

In bacteria, gene control serves mainly to allow a single cell to adjust to changes in its environment so that its growth and division can be optimized. In multicellular organisms, environmental changes also induce changes in gene expression. An example is the response to low oxygen (hypoxia) that is described in Chapter 15. However, the most characteristic and biologically far-reaching purpose of gene control in multicellular organisms is execution of the genetic program that underlies embryological development. Generation of the many different cell types that collectively form a multicellular organism depends on the right genes being activated in the right cells at the right time during the developmental period.

In most cases, once a developmental step has been taken by a cell, it is not reversed. Thus these decisions are fundamentally different from the reversible activation and repression of bacterial genes in response to environmental conditions. In executing their genetic programs, many differentiated cells (e.g., skin cells, red blood cells, and antibody-producing cells) march down a pathway to final cell death, leaving no progeny behind. The fixed patterns of gene con-

trol leading to differentiation serve the needs of the whole organism and not the survival of an individual cell. Despite the differences in the purposes of gene control in bacteria and eukaryotes, two key features of transcription control first discovered in bacteria and described in Chapter 4 also apply to eukaryotic cells. First, protein-binding regulatory DNA sequences, or control elements, are associated with genes. Second, specific proteins that bind to a gene's regulatory sequences determine where transcription will start, and either activate or repress its transcription. As represented in Figure 11-1, in multicellular eukaryotes, inactive genes are assembled into condensed chromatin, which inhibits the binding of RNA polymerases and general transcription factors required for transcription initiation. Activator proteins bind to control elements near the transcription start site of a gene as well as kilobases away and promote chromatin decondensation and binding of RNA polymerase to the promoter. Repressor proteins bind to alternative control elements, causing condensation of chromatin and inhibition of polymerase binding. In this chapter we consider how activators and repressors control chromatin structure and stimulate or inhibit transcription initiation by RNA polymerase.

Most Genes in Higher Eukaryotes Are Regulated by Controlling Their Transcription

Direct measurements of the transcription rates of multiple genes in different cell types have shown that regulation of transcription initiation is the most widespread form of gene control in eukaryotes, as it is in bacteria. *Nascent-chain analysis* is a common method for determining the relative rates of transcription of different genes in cultured cells. In this method, also called *run-on transcription analysis*, isolated nuclei are incubated with ^{32}P-labeled ribonucleoside triphosphates for a brief time (e.g., 5 minutes or less). During

► **FIGURE 11-1 Overview of transcription control in multicellular eukaryotes.** Activator proteins bind to specific DNA control elements in chromatin and interact with multiprotein co-activator machines, such as mediator, to decondense chromatin and assemble RNA polymerase and general transcription factors on promoters. Inactive genes are assembled into regions of condensed chromatin that inhibit RNA polymerases and their associated general transcription factors (GTFs) from interacting with promoters. Alternatively, repressor proteins bind to other control elements to inhibit initiation by RNA polymerase and interact with multiprotein co-repressor complexes to condense chromatin.

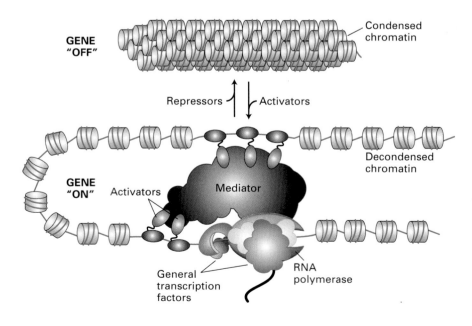

this incubation period, RNA polymerase molecules that were actively transcribing genes when the nuclei were isolated incorporate several hundred labeled nucleotides into each nascent (growing) RNA chain, but little initiation of new chains occurs. The total radioactive label incorporated into RNA is a measure of the overall transcription rate. The fraction of the total labeled RNA produced by transcription of a particular gene—that is, its relative transcription rate—is determined by hybridizing the labeled RNA to the cloned DNA of that gene attached to a membrane.

The results of run-on transcription analyses illustrated in Figure 11-2 show that transcription of genes encoding proteins expressed specifically in hepatocytes (the major cell type in mammalian liver) is readily detected in nuclei pre-

pared from liver, but not in nuclei from brain or kidney. Since the run-on transcription assay measures RNA synthesis, these results indicate that differential synthesis of liver-specific proteins is regulated by controlling transcription of the corresponding genes in different tissues. Similar results have been obtained in run-on transcription experiments with other cell types and a wide variety of tissue-specific proteins, indicating that transcriptional control is the primary mechanism of gene control in complex organisms.

Regulatory Elements in Eukaryotic DNA Often Are Many Kilobases from Start Sites

In eukaryotes, as in bacteria, a DNA sequence that specifies where RNA polymerase binds and initiates transcription of a gene is called a **promoter.** Transcription from a particular promoter is controlled by DNA-binding proteins, termed **transcription factors,** that are equivalent to bacterial *repressors* and *activators.* However, the DNA control elements in eukaryotic genomes that bind transcription factors often are located much farther from the promoter they regulate than is the case in prokaryotic genomes. In some cases, transcription factors that regulate expression of protein-coding genes in higher eukaryotes bind at regulatory sites tens of thousands of base pairs either **upstream** (opposite to the direction of transcription) or **downstream** (in the same direction as transcription) from the promoter. As a result of this arrangement, transcription from a single promoter may be regulated by binding of multiple transcription factors to alternative control elements, permitting complex control of gene expression.

For example, alternative transcription-control elements regulate expression of the mammalian gene that encodes transthyretin (*TTR*), which transports thyroid hormone in blood and the cerebrospinal fluid that surrounds the brain

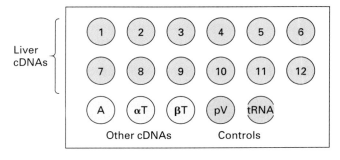
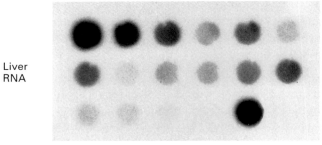
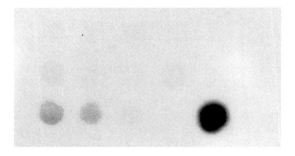
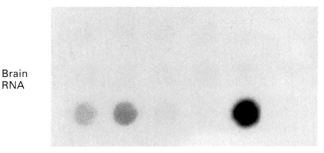

◄ **EXPERIMENTAL FIGURE 11-2 Measurement of RNA production in various tissues demonstrates that transcription of a given gene generally occurs only in the cell types in which it is expressed.** Nuclei from mouse liver, kidney, and brain cells were exposed to ^{32}P-UTP. The three resulting labeled RNA samples were hybridized to separate nitrocellulose membranes; on each membrane was fixed an identical pattern of various cDNAs (*top*). The cDNAs labeled 1–12 (green) encode proteins synthesized actively in liver (e.g., albumin, transferrin) but not in most other tissues. The other cDNAs tested were actin (A) and α- and β-tubulin (αT, βT) (yellow), which are proteins found in almost all cell types. A gene encoding methionine tRNA and the plasmid vector DNA (pV) in which the cDNAs were cloned were included as controls (purple). After removal of unhybridized RNAs, the labeled RNAs complementary to the cDNAs were revealed by autoradiography. RNA from the liver sample hybridized extensively with the cDNAs. Little hybridization is seen for the kidney and brain samples, indicating that genes encoding proteins found in hepatocytes are transcribed in liver cells but not in the cells of the other tissues. [See E. Derman et al., 1981, *Cell* **23**:731, and D. J. Powell et al., 1984, *J. Mol. Biol.* **197**:21.]

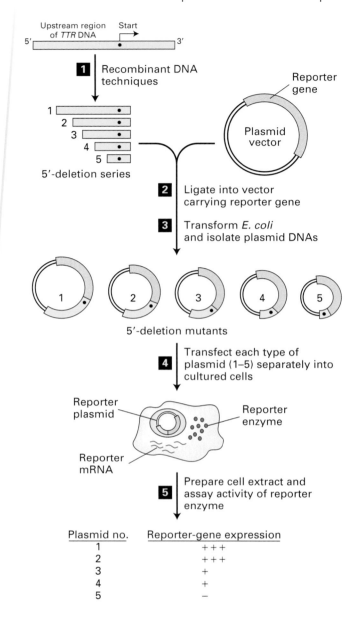

▲ **EXPERIMENTAL FIGURE 11-3** **5′-Deletion analysis can identify transcription-control sequences in DNA upstream of a eukaryotic gene.** Step **1**: Recombinant DNA techniques are used to prepare a series of DNA fragments that extend from the 5′-untranslated region of a gene various distances upstream. Step **2**: The DNA fragments are ligated into a reporter plasmid upstream of an easily assayed reporter gene. Step **3**: The DNA is transformed into *E. coli* to isolate plasmids with deletions of various lengths 5′ to the transcription start site. Step **4**: Each plasmid is then transfected into cultured cells (or used to prepare transgenic organisms) and expression of the reporter gene is assayed (step **5**). The results of this hypothetical example (*bottom*) indicate that the test fragment contains two control elements. The 5′ end of one lies between deletions 2 and 3; the 5′ end of the other lies between deletions 4 and 5.

and spinal cord. Transthyretin is expressed in hepatocytes, which synthesize and secrete most of the blood serum proteins, and in choroid plexus cells in the brain, which secrete cerebrospinal fluid and its constituent proteins. The control elements required for transcription of the *TTR* gene were identified by the procedure outlined in Figure 11-3. In this experimental approach, DNA fragments with varying extents of sequence upstream of a start site are cloned in front of a **reporter gene** in a bacterial plasmid using recombinant DNA techniques. Reporter genes express enzymes that are easily assayed in cell extracts. Commonly used reporter genes include the *E. coli lacZ* gene encoding β-galactosidase; the firefly gene encoding luciferase, which converts energy from ATP hydrolysis into light; and the jellyfish gene encoding green fluorescent protein (GFP).

By constructing and analyzing a *5′-deletion series* upstream of the *TTR* gene, researchers identified two control elements that stimulate reporter-gene expression in hepatocytes, but not in other cell types. One region mapped between ≈2.01 and 1.85 kb upstream of the *TTR* gene start site; the other mapped between ≈200 base pairs upstream and the start site. Further studies demonstrated that alternative DNA sequences control *TTR* transcription in choroid plexus cells. Thus, alternative control elements regulate transcription of the *TTR* gene in two different cell types. We examine the basic molecular events underlying this type of eukaryotic transcriptional control in later sections.

Three Eukaryotic Polymerases Catalyze Formation of Different RNAs

The nuclei of all eukaryotic cells examined so far (e.g., vertebrate, *Drosophila*, yeast, and plant cells) contain three different RNA polymerases, designated I, II, and III. These enzymes are eluted at different salt concentrations during ion-exchange chromatography and also differ in their sensitivity to α-amanitin, a poisonous cyclic octapeptide produced by some mushrooms (Figure 11-4). Polymerase I is very insensitive to α-amanitin; polymerase II is very sensitive; and polymerase III has intermediate sensitivity.

Each eukaryotic RNA polymerase catalyzes transcription of genes encoding different classes of RNA. *RNA polymerase I*, located in the nucleolus, transcribes genes encoding precursor rRNA (**pre-rRNA**), which is processed into 28S, 5.8S, and 18S rRNAs. *RNA polymerase III* transcribes genes encoding tRNAs, 5S rRNA, and an array of small, stable RNAs, including one involved in RNA splicing (U6) and the RNA component of the signal-recognition particle (SRP) involved in directing nascent proteins to the endoplasmic reticulum (Chapter 16). *RNA polymerase II* transcribes all protein-coding genes; that is, it functions in production of mRNAs. RNA polymerase II also produces four of the five small nuclear RNAs that take part in RNA splicing.

Each of the three eukaryotic RNA polymerases is more complex than *E. coli* RNA polymerase, although their structures are similar (Figure 11-5). All three contain two large

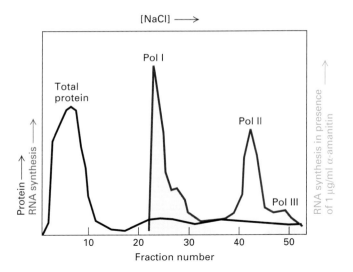

◀ **EXPERIMENTAL FIGURE 11-4** Column chromatography separates and identifies the three eukaryotic RNA polymerases, each with its own sensitivity to α-amanitin. A protein extract from the nuclei of cultured eukaryotic cells is passed through a DEAE Sephadex column and adsorbed protein eluted (black curve) with a solution of constantly increasing NaCl concentration. Three fractions from the eluate subsequently showed RNA polymerase activity (red curve). At a concentration of 1 μg/ml, α-amanitin inhibits polymerase II activity but has no effect on polymerases I and III (green shading). Polymerase III is inhibited by 10 μg/ml of α-amanitin, whereas polymerase I is unaffected even at this higher concentration. [See R. G. Roeder, 1974, *J. Biol. Chem.* **249**:241.]

subunits and 10–14 smaller subunits, some of which are present in two or all three of the polymerases. The best-characterized eukaryotic RNA polymerases are from the yeast *S. cerevisiae*. Each of the yeast genes encoding the polymerase subunits has been cloned and sequenced and the effects of gene-knockout mutations have been characterized. In addition, the three-dimensional structure of yeast RNA

polymerase II missing two nonessential subunits has been determined (see Figure 11-5). The three nuclear RNA polymerases from all eukaryotes so far examined are very similar to those of yeast.

The two large subunits (RPB1 and RPB2) of all three eukaryotic RNA polymerases are related to each other and are similar to the *E. coli* β′ and β subunits, respectively

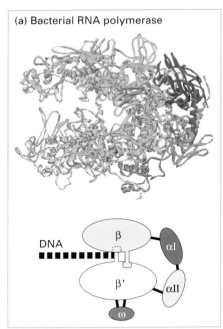

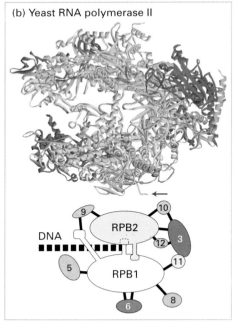

▲ **FIGURE 11-5 Comparison of three-dimensional structures of bacterial and eukaryotic RNA polymerases.** These Cα trace models are based on x-ray crystallographic analysis of RNA polymerase from the bacterium *T. aquaticus* and RNA polymerase II from *S. cerevisiae*. (a) The five subunits of the bacterial enzyme are distinguished by color. Only the N-terminal domains of the α subunits are included in this model. (b) Ten of the twelve subunits constituting yeast RNA polymerase II are shown in this model.

Subunits that are similar in conformation to those in the bacterial enzyme are shown in the same colors. The C-terminal domain of the large subunit RPB1 was not observed in the crystal structure, but it is known to extend from the position marked with a red arrow. (RPB is the abbreviation for "*R*NA polymerase *B*," which is an alternative way of referring to RNA polymerase II.) [Part (a) based on crystal structures from G. Zhang et al.,1999, *Cell* **98**:811. Part (b) from P. Cramer et al., 2001, *Science* **292**:1863.]

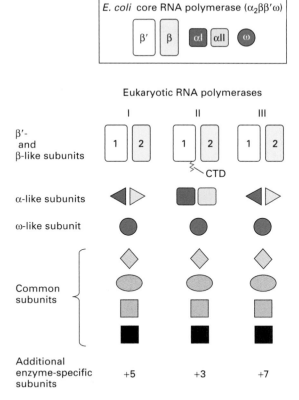

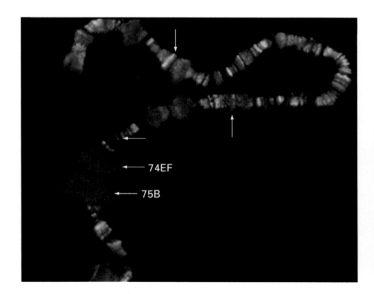

▲ **FIGURE 11-6 Schematic representation of the subunit structure of the *E. coli* RNA core polymerase and yeast nuclear RNA polymerases.** All three yeast polymerases have five core subunits homologous to the β, β′, two α, and ω subunits of *E. coli* RNA polymerase. The largest subunit (RPB1) of RNA polymerase II also contains an essential C-terminal domain (CTD). RNA polymerases I and III contain the same two nonidentical α-like subunits, whereas RNA polymerase II contains two other nonidentical α-like subunits. All three polymerases share the same ω-like subunit and four other common subunits. In addition, each yeast polymerase contains three to seven unique smaller subunits.

(Figure 11-6). Each of the eukaryotic polymerases also contains an ω-like and two nonidentical α-like subunits. The extensive similarity in the structures of these core subunits in RNA polymerases from various sources indicates that this enzyme arose early in evolution and was largely conserved. This seems logical for an enzyme catalyzing a process so basic as copying RNA from DNA.

In addition to their core subunits related to the *E. coli* RNA polymerase subunits, all three yeast RNA polymerases contain four additional small subunits, common to them but not to the bacterial RNA polymerase. Finally, each eukaryotic polymerase has several enzyme-specific subunits that are not present in the other two polymerases. Gene-knockout experiments in yeast indicate that most of these subunits are essential for cell viability. Disruption of the few polymerase subunit genes that are not absolutely essential for viability nevertheless results in very poorly growing cells. Thus it

seems likely that all the subunits are necessary for eukaryotic RNA polymerases to function normally.

The Largest Subunit in RNA Polymerase II Has an Essential Carboxyl-Terminal Repeat

The carboxyl end of the largest subunit of RNA polymerase II (RPB1) contains a stretch of seven amino acids that is nearly precisely repeated multiple times. Neither RNA polymerase I nor III contains these repeating units. This heptapeptide repeat, with a consensus sequence of Tyr-Ser-Pro-Thr-Ser-Pro-Ser, is known as the *carboxyl-terminal domain (CTD)*. Yeast RNA polymerase II contains 26 or more repeats, the mammalian enzyme has 52 repeats, and an intermediate number of repeats occur in RNA polymerase II from nearly all other eukaryotes. The CTD is critical for viability, and at least 10 copies of the repeat must be present for yeast to survive.

▲ **EXPERIMENTAL FIGURE 11-7 Antibody staining demonstrates that the carboxyl-terminal domain (CTD) of RNA polymerase II is phosphorylated during in vivo transcription.** Salivary gland polytene chromosomes were prepared from *Drosophila* larvae just before molting. The preparation was treated with a rabbit antibody specific for phosphorylated CTD and with a goat antibody specific for unphosphorylated CTD. The preparation then was stained with fluorescein-labeled anti-goat antibody (green) and rhodamine-labeled anti-rabbit antibody (red). Thus polymerase molecules with an unphosphorylated CTD stain green, and those with a phosphorylated CTD stain red. The molting hormone ecdysone induces very high rates of transcription in the puffed regions labeled 74EF and 75B; note that only phosphorylated CTD is present in these regions. Smaller puffed regions transcribed at high rates also are visible. Nonpuffed sites that stain red (up arrow) or green (horizontal arrow) also are indicated, as is a site staining both red and green, producing a yellow color (down arrow). [From J. R. Weeks et al., 1993, *Genes & Dev.* **7**:2329; courtesy of J. R. Weeks and A. L. Greenleaf.]

In vitro experiments with model promoters first showed that RNA polymerase II molecules that initiate transcription have an unphosphorylated CTD. Once the polymerase initiates transcription and begins to move away from the promoter, many of the serine and some tyrosine residues in the CTD are phosphorylated. Analysis of polytene chromosomes from *Drosophila* salivary glands prepared just before molting of the larva indicate that the CTD also is phosphorylated during in vivo transcription. The large chromosomal "puffs" induced at this time in development are regions where the genome is very actively transcribed. Staining with antibodies specific for the phosphorylated or unphosphorylated CTD demonstrated that RNA polymerase II associated with the highly transcribed puffed regions contains a phosphorylated CTD (Figure 11-7).

RNA Polymerase II Initiates Transcription at DNA Sequences Corresponding to the 5′ Cap of mRNAs

Several experimental approaches have been used to identify DNA sequences at which RNA polymerase II initiates transcription. Approximate mapping of the transcription start site is possible by exposing cultured cells or isolated nuclei to ^{32}P-labeled ribonucleotides for very brief times, as described earlier. After the resulting labeled nascent transcripts are separated on the basis of chain length, each size fraction is incubated under hybridization conditions with overlapping restriction fragments that encompass the DNA region of interest. The restriction fragment that hybridizes to the shortest labeled nascent chain, as well as all the longer ones, contains the transcription start site. One of the first transcription units to be analyzed in this way was the major late transcription unit of adenovirus. This analysis indicated that transcription was initiated in a region ≈6 kb from the left end of the viral genome.

The precise base pair where RNA polymerase II initiates transcription in the adenovirus late transcription unit was determined by analyzing the RNAs synthesized during in vitro transcription of adenovirus DNA restriction fragments that extended somewhat upstream and downstream of the approximate initiation region determined by nascent-transcript analysis. The rationale of this experiment and typical results are illustrated in Figure 11-8. The RNA transcripts synthesized in vitro by RNA polymerase II from the

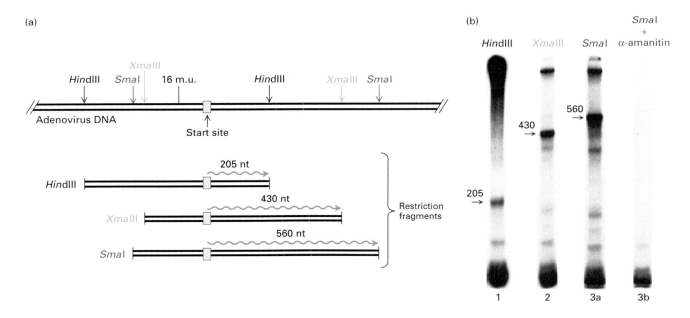

▲ **EXPERIMENTAL FIGURE 11-8 In vitro transcription of restriction fragments and measurement of the RNA lengths localize the initiation site of the adenovirus major late transcription unit.** (a) The top line shows restriction sites for *Hind*III (black), *Xma*III (blue), and *Sma*I (red) in the region of the adenovirus genome where the transcription-initiation site was located by nascent-transcript analysis (near 16 map units). The *Hind*III, *Xma*III, and *Sma*I restriction fragments that encompass the initiation site were individually incubated with a nuclear extract prepared from cultured cells and ^{32}P-labeled ribonucleoside triphosphates. Transcription of each fragment begins at the start site and ends when an RNA polymerase II molecule "runs off" the cut end of the fragment template, producing a run-off transcript (wavy red lines). (b) The run-off transcripts synthesized from each fragment were then subjected to gel electrophoresis and autoradiography to determine their exact lengths. Since the positions of the restriction sites in the adenovirus DNA were known, the lengths of the run-off transcripts in nucleotides (nt) produced from the restriction fragments precisely map the initiation site on the adenovirus genome, as diagrammed in part (a). In the gels shown here, the bands at the top and bottom represent high- and low-molecular-weight RNA transcripts that are formed under the conditions of the experiment. The sample in lane 3b is the same as that in lane 3a, except that α-amanitin, an inhibitor of RNA polymerase II, was included in the transcription mixture. See the text for further discussion. [See R. M. Evans and E. Ziff, 1978, *Cell* **15**:1463, and P. A. Weil et al., 1979, *Cell* **18**:469. Autoradiogram courtesy of R. G. Roeder.]

start site determined in this experiment contained an RNA cap structure identical with that present at the 5′ end of nearly all eukaryotic mRNAs (see Figure 4-13). This 5′ cap was added by enzymes in the nuclear extract, which can add a cap only to an RNA that has a 5′ tri- or diphosphate. Because a 5′ end generated by cleavage of a longer RNA would have a 5′ monophosphate, it could not be capped. Consequently, researchers concluded that the capped nucleotide generated in the in vitro transcription reaction must have been the nucleotide with which transcription was initiated. The finding that the sequence at the 5′ end of the RNA transcripts produced in vitro is the same as that at the 5′ end of late adenovirus mRNAs isolated from cells confirmed that the capped nucleotide of adenovirus late mRNAs coincides with the transcription-initiation site.

Similar in vitro transcription assays with other cloned eukaryotic genes have produced similar results. In each case, the start site was found to be equivalent to the capped 5′ sequence of the corresponding mRNA. Thus synthesis of eukaryotic precursors of mRNAs by RNA polymerase II begins at the DNA sequence encoding the capped 5′ end of the mRNA. Today, the transcription start site for a newly characterized mRNA generally is determined simply by identifying the DNA sequence encoding the 5′ end of the mRNA.

KEY CONCEPTS OF SECTION 11.1

Overview of Eukaryotic Gene Control and RNA Polymerases

■ The primary purpose of gene control in multicellular organisms is the execution of precise developmental decisions so that the proper genes are expressed in the proper cells during development and cellular differentiation.

■ Transcriptional control is the primary means of regulating gene expression in eukaryotes, as it is in bacteria.

■ In eukaryotic genomes, DNA transcription control elements may be located many kilobases away from the promoter they regulate. Different control regions can control transcription of the same gene in different cell types.

■ Eukaryotes contain three types of nuclear RNA polymerases. All three contain two large and three smaller core subunits with homology to the β′, β, α, and ω subunits of *E. coli* RNA polymerase, as well several additional small subunits (see Figure 11-6).

■ RNA polymerase I synthesizes only pre-rRNA. RNA polymerase II synthesizes mRNAs and some of the small nuclear RNAs that participate in mRNA splicing. RNA polymerase III synthesizes tRNAs, 5S rRNA, and several other relatively short, stable RNAs.

■ The carboxyl-terminal domain (CTD) in the largest subunit of RNA polymerase II becomes phosphorylated dur-

ing transcription initiation and remains phosphorylated as the enzyme transcribes the template.

■ RNA polymerase II initiates transcription of genes at the nucleotide in the DNA template that corresponds to the 5′ nucleotide that is capped in the encoded mRNA.

11.2 Regulatory Sequences in Protein-Coding Genes

As noted in the previous section, expression of eukaryotic protein-coding genes is regulated by multiple protein-binding DNA sequences, generically referred to as **transcription-control regions**. These include promoters and other types of control elements located near transcription start sites, as well as sequences located far from the genes they regulate. In this section, we take a closer look at the properties of various control elements found in eukaryotic protein-coding genes and some techniques used to identify them.

The TATA Box, Initiators, and CpG Islands Function as Promoters in Eukaryotic DNA

The first genes to be sequenced and studied in in vitro transcription systems were viral genes and cellular protein-coding genes that are very actively transcribed either at particular times of the cell cycle or in specific differentiated cell types. In all these rapidly transcribed genes, a conserved sequence called the **TATA box** was found ≈25–35 base pairs upstream of the start site (Figure 11-9). Mutagenesis studies have shown that a single-base change in this nucleotide sequence drastically decreases in vitro transcription by RNA polymerase II of genes adjacent to a TATA box. In most cases, sequence changes between the TATA box and start site do not significantly affect the transcription rate. If the base pairs between the TATA box and the normal start site are deleted, transcription of the altered, shortened template begins at a new site ≈25 base pairs downstream from the TATA box. Consequently, the TATA box acts similarly to an *E. coli* promoter to position RNA polymerase II for transcription initiation (see Figure 4-11).

Instead of a TATA box, some eukaryotic genes contain an alternative promoter element called an *initiator*. Most naturally occurring initiator elements have a cytosine (C) at the −1 position and an adenine (A) residue at the transcription start site (+1). Directed mutagenesis of mammalian genes with an initiator-containing promoter has revealed that the nucleotide sequence immediately surrounding the start site determines the strength of such promoters. Unlike the conserved TATA box sequence, however, only an extremely degenerate initiator consensus sequence has been defined:

$$(5') \text{ Y-Y-A}^{+1}\text{-N-T/A-Y-Y-Y } (3')$$

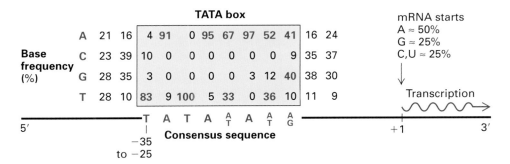

▲ FIGURE 11-9 Determination of consensus TATA box sequence. The nucleotide sequences upstream of the start site in 900 different eukaryotic protein-coding genes were aligned to maximize homology in the region from −35 to −26. The tabulated numbers are the percentage frequency of each base at each position. Maximum homology occurs over an eight-base region, referred to as the *TATA box*, whose consensus sequence is shown at the bottom. The initial base in mRNAs encoded by genes containing a TATA box most frequently is an A. [See P. Bucher, 1990, *J. Mol. Biol.* **212**:563, and http://www.epd.isb-sib.ck/promoter_elements.]

where A^{+1} is the base at which transcription starts, Y is a pyrimidine (C or T), N is any of the four bases, and T/A is T or A at position +3.

Transcription of genes with promoters containing a TATA box or initiator element begins at a well-defined initiation site. However, transcription of many protein-coding genes has been shown to begin at any one of multiple possible sites over an extended region, often 20–200 base pairs in length. As a result, such genes give rise to mRNAs with multiple alternative 5′ ends. These genes, which generally are transcribed at low rates (e.g., genes encoding the enzymes of intermediary metabolism, often called "housekeeping genes"), do not contain a TATA box or an initiator. Most genes of this type contain a CG-rich stretch of 20–50 nucleotides within ≈100 base pairs upstream of the start-site region. The dinucleotide CG is statistically underrepresented in vertebrate DNAs, and the presence of a CG-rich region, or *CpG island,* just upstream from a start site is a distinctly nonrandom distribution. For this reason, the presence of a CpG island in genomic DNA suggests that it may contain a transcription-initiation region.

Promoter-Proximal Elements Help Regulate Eukaryotic Genes

Recombinant DNA techniques have been used to systematically mutate the nucleotide sequences upstream of the start sites of various eukaryotic genes in order to identify transcription-control regions. By now, hundreds of eukaryotic genes have been analyzed, and scores of transcription-control regions have been identified. These control elements, together with the TATA-box or initiator, often are referred to as the *promoter* of the gene they regulate. However, we prefer to reserve the term *promoter* for the TATA-box or initiator sequences that determine the initiation site in the template. We use the term **promoter-proximal elements** for

control regions lying within 100–200 base pairs upstream of the start site. In some cases, promoter-proximal elements are cell-type-specific; that is, they function only in specific differentiated cell types.

One approach frequently taken to determine the upstream border of a transcription-control region for a mammalian gene involves constructing a set of 5′ deletions as discussed earlier (see Figure 11-3). Once the 5′ border of a transcription-control region is determined, analysis of *linker scanning mutations* can pinpoint the sequences with regulatory functions that lie between the border and the transcription start site. In this approach, a set of constructs with contiguous overlapping mutations are assayed for their effect on expression of a reporter gene or production of a specific mRNA (Figure 11-10a). One of the first uses of this type of analysis identified promoter-proximal elements of the thymidine kinase (*tk*) gene from herpes simplex virus (HSV). The results demonstrated that the DNA region upstream of the HSV *tk* gene contains three separate transcription-control sequences: a TATA box in the interval from −32 to −16, and two other control elements farther upstream (Figure 11-10b).

To test the spacing constraints on control elements in the HSV *tk* promoter region identified by analysis of linker scanning mutations, researchers prepared and assayed constructs containing small deletions and insertions between the elements. Changes in spacing between the promoter and promoter-proximal control elements of 20 nucleotides or fewer had little effect. However, insertions of 30 to 50 base pairs between a promoter-proximal element and the TATA box was equivalent to deleting the element. Similar analyses of other eukaryotic promoters have also indicated that considerable flexibility in the spacing between promoter-proximal elements is generally tolerated, but separations of several tens of base pairs may decrease transcription.

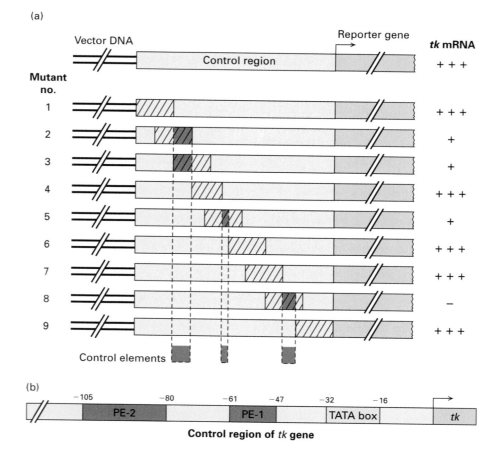

▲ **EXPERIMENTAL FIGURE 11-10** Linker scanning mutations identify transcription-control elements. (a) A region of eukaryotic DNA (orange) that supports high-level expression of a reporter gene (light blue) is cloned in a plasmid vector as diagrammed at the top. Overlapping linker scanning (LS) mutations (crosshatch) are introduced from one end of the region being analyzed to the other. These mutations result from scrambling the nucleotide sequence in a short stretch of the DNA. After the mutant plasmids are transfected separately into cultured cells, the activity of the reporter-gene product is assayed. In the hypothetical example shown here, LS mutations 1, 4, 6, 7, and 9 have little or no effect on expression of the reporter gene, indicating that the regions altered in these mutants contain no control elements. Reporter-gene expression is significantly reduced in mutants 2, 3, 5, and 8, indicating that control elements (brown) lie in the intervals shown at the bottom. (b) Analysis of LS mutations in the transcription-control region of the thymidine kinase (*tk*) gene from herpes simplex virus (HSV) identified a TATA box and two promoter-proximal elements (PE-1 and PE-2). [Part (b) see S. L. McKnight and R. Kingsbury, 1982, *Science* **217**:316.]

Distant Enhancers Often Stimulate Transcription by RNA Polymerase II

As noted earlier, transcription from many eukaryotic promoters can be stimulated by control elements located thousands of base pairs away from the start site. Such long-distance transcription-control elements, referred to as **enhancers,** are common in eukaryotic genomes but fairly rare in bacterial genomes. The first enhancer to be discovered that stimulates transcription of eukaryotic genes was in a 366-bp fragment of the simian virus 40 (SV40) genome (Figure 11-11). Further analysis of this region of SV40 DNA revealed that an ≈100-bp sequence lying ≈100 base pairs upstream of the SV40 early transcription start site was responsible for its ability to enhance transcription. In SV40, this enhancer sequence functions to stimulate transcription from viral promoters. The SV40 enhancer, however, stimulates transcription from all mammalian promoters that have been tested when it is inserted in either orientation anywhere on a plasmid carrying the test promoter, even when it is thousands of base pairs from the start site. An extensive linker scanning mutational analysis of the SV40 enhancer indicated that it is composed of multiple individual elements, each of which contributes to the total activity of the enhancer. As discussed later, each of these regulatory elements is a protein-binding site.

Soon after discovery of the SV40 enhancer, enhancers were identified in other viral genomes and in eukaryotic

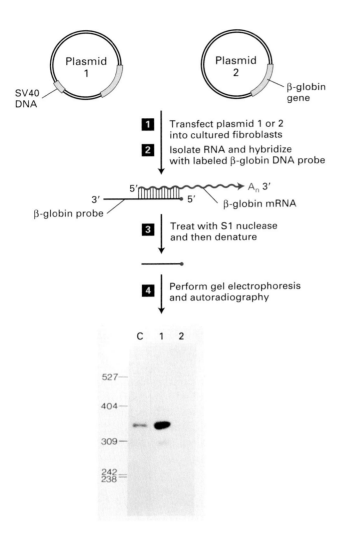

▲ **EXPERIMENTAL FIGURE 11-11 Plasmids containing a particular SV40 DNA fragment showed marked increase in mRNA production compared with plasmids lacking this enhancer.** Plasmids containing the β-globin gene with or without a 366-bp fragment of SV40 DNA were constructed. These plasmids were transfected into cultured cells, and any resulting RNA was hybridized to a β-globin DNA probe (steps **1** and **2**). The amount of β-globin mRNA synthesized by cells transfected with one or the other plasmid was assayed by the S1 nuclease–protection method (step **3**). The restriction-fragment probe, generated from a β-globin cDNA clone, was complementary to the 5′ end of β-globin mRNA. The 5′ end of the probe was labeled with ^{32}P (red dot). Hybridization of β-globin mRNA to the probe protected an ≈340-nucleotide fragment of the probe from digestion by S1 nuclease, which digests single-stranded DNA but not DNA in an RNA-DNA hybrid. Autoradiography of electrophoresed S1-protected fragments (step **4**) revealed that cells transfected with plasmid 1 (lane 1) produced much more β-globin mRNA than those transfected with plasmid 2 (lane 2). Lane C is a control assay of β-globin mRNA isolated from reticulocytes, which actively synthesize β-globin. These results show that the SV40 DNA fragment in plasmid 1 contains an element, the enhancer, that greatly stimulates synthesis of β-globin mRNA. [Adapted from J. Banerji et al., 1981, *Cell* **27**:299.]

cellular DNA. Some of these control elements are located 50 or more kilobases from the promoter they control. Analyses of many different eukaryotic cellular enhancers have shown that they can occur upstream from a promoter, downstream from a promoter within an intron, or even downstream from the final exon of a gene. Like promoter-proximal elements, many enhancers are cell-type-specific. For example, the genes encoding antibodies (immunoglobulins) contain an enhancer within the second intron that can stimulate transcription from all promoters tested, but only in B lymphocytes, the type of cells that normally express antibodies. Analyses of the effects of deletions and linker scanning mutations in cellular enhancers have shown that, like the SV40 enhancer, they generally are composed of multiple elements that contribute to the overall activity of the enhancer.

Most Eukaryotic Genes Are Regulated by Multiple Transcription-Control Elements

Initially, enhancers and promoter-proximal elements were thought to be distinct types of transcription-control elements. However, as more enhancers and promoter-proximal elements were analyzed, the distinctions between them became less clear. For example, both types of element generally can stimulate transcription even when inverted, and both types often are cell-type-specific. The general consensus now is that a spectrum of control elements regulates transcription by RNA polymerase II. At one extreme are enhancers, which can stimulate transcription from a promoter tens of thousands of base pairs away (e.g., the SV40 enhancer). At the other extreme are promoter-proximal elements, such as the upstream elements controlling the HSV *tk* gene, which lose their influence when moved an additional 30–50 base pairs farther from the promoter. Researchers have identified a large number of transcription-control elements that can stimulate transcription from distances between these two extremes.

Figure 11-12a summarizes the locations of transcription-control sequences for a hypothetical mammalian gene. The start site at which transcription initiates encodes the first (5′) nucleotide of the first exon of an mRNA, the nucleotide that is capped. For many genes, especially those encoding abundantly expressed proteins, a TATA box located approximately 25–35 base pairs upstream from the start site directs RNA polymerase II to begin transcription at the proper nucleotide. Promoter-proximal elements, which are relatively short (≈10–20 base pairs), are located within the first ≈200 base pairs upstream of the start site. Enhancers, in contrast, usually are ≈100 base pairs long and are composed of multiple elements of ≈10–20 base pairs. Enhancers may be located up to 50 kilobases upstream or downstream from the start site or within an intron. Many mammalian genes are controlled by more than one enhancer region.

The *S. cerevisiae* genome contains regulatory elements called **upstream activating sequences (UASs)**, which function

(a) **Mammalian gene**

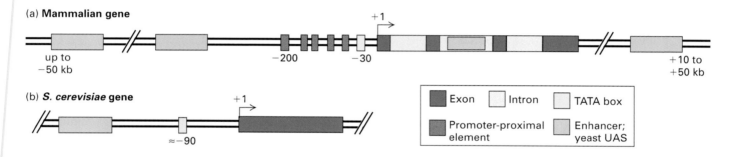

(b) *S. cerevisiae* gene

▲ **FIGURE 11-12 General pattern of control elements that regulate gene expression in multicellular eukaryotes and yeast.** (a) Genes of multicellular organisms contain both promoter-proximal elements and enhancers, as well as a TATA box or other promoter element. The promoter elements position RNA polymerase II to initiate transcription at the start site and influence the rate of transcription. Enhancers may be either upstream or downstream and as far away as 50 kb from the transcription start site. In some cases, enhancers lie within introns. For some genes, promoter-proximal elements occur downstream from the start site as well as upstream. (b) Most *S. cerevisiae* genes contain only one regulatory region, called an *upstream activating sequence (UAS)*, and a TATA box, which is ≈90 base pairs upstream from the start site.

similarly to enhancers and promoter-proximal elements in higher eukaryotes. Most yeast genes contain only one UAS, which generally lies within a few hundred base pairs of the start site. In addition, *S. cerevisiae* genes contain a TATA box ≈90 base pairs upstream from the transcription start site (Figure 11-12b).

KEY CONCEPTS OF SECTION 11.2

Regulatory Sequences in Protein-Coding Genes

■ Expression of eukaryotic protein-coding genes generally is regulated through multiple protein-binding control regions that are located close to or distant from the start site (Figure 11-12).

■ Promoters direct binding of RNA polymerase II to DNA, determine the site of transcription initiation, and influence transcription rate.

■ Three principal types of promoter sequences have been identified in eukaryotic DNA. The TATA box, the most common, is prevalent in rapidly transcribed genes. Initiator promoters are found in some genes, and CpG islands are characteristic of genes transcribed at a low rate.

■ Promoter-proximal elements occur within ≈200 base pairs upstream of a start site. Several such elements, containing ≈10–20 base pairs, may help regulate a particular gene.

■ Enhancers, which contain multiple short control elements, may be located from 200 base pairs to tens of kilobases upstream or downstream from a promoter, within an intron, or downstream from the final exon of a gene.

■ Promoter-proximal elements and enhancers often are cell-type-specific, functioning only in specific differentiated cell types.

11.3 Activators and Repressors of Transcription

The various transcription-control elements found in eukaryotic DNA are binding sites for regulatory proteins. In this section, we discuss the identification, purification, and structures of these transcription factors, which function to activate or repress expression of eukaryotic protein-coding genes.

Footprinting and Gel-Shift Assays Detect Protein-DNA Interactions

In yeast, *Drosophila*, and other genetically tractable eukaryotes, numerous genes encoding transcriptional activators and repressors have been identified by classical genetic analyses like those described in Chapter 9. However, in mammals and other vertebrates, which are less amenable to such genetic analysis, most transcription factors have been detected initially and subsequently purified by biochemical techniques. In this approach, a DNA regulatory element that has been identified by the kinds of mutational analyses described in the previous section is used to identify *cognate* proteins that bind specifically to it. Two common techniques for detecting such cognate proteins are **DNase I footprinting** and the **electrophoretic mobility shift assay.**

DNase I footprinting takes advantage of the fact that when a protein is bound to a region of DNA, it protects that DNA sequence from digestion by nucleases. As illustrated in Figure 11-13, when samples of a DNA fragment that is labeled at one end are digested in the presence and absence of a DNA-binding protein and then denatured, electrophoresed, and the resulting gel subjected to autoradiography, the region protected by the bound protein appears as a gap, or "footprint," in the array of bands resulting from digestion in the absence of protein. When footprinting is performed

▶ **EXPERIMENTAL FIGURE 11-13** DNase I footprinting reveals control-element sequences and can be used as an assay in transcription factor purification. (a) DNase I footprinting can identify control element sequences. A DNA fragment known to contain the control-element is labeled at one end with ³²P (red dot). Portions of the labeled DNA sample then are digested with DNase I in the presence and absence of protein samples thought to contain a cognate protein. DNase I randomly hydrolyzes the phosphodiester bonds of DNA between the 3′ oxygen on the deoxyribose of one nucleotide and the 5′ phosphate of the next nucleotide. A low concentration of DNase I is used so that on average each DNA molecule is cleaved just once (vertical arrows). If the protein sample does not contain a cognate DNA-binding protein, the DNA fragment is cleaved at multiple positions between the labeled and unlabeled ends of the original fragment, as in sample A on the left. If the protein sample contains a cognate protein, as in sample B on the right, the protein binds to the DNA, thereby protecting a portion of the fragment from digestion. Following DNase treatment, the DNA is separated from protein, denatured to separate the strands, and electrophoresed. Autoradiography of the resulting gel detects only labeled strands and reveals fragments extending from the labeled end to the site of cleavage by DNase I. Cleavage fragments containing the control sequence show up on the gel for sample A, but are missing in sample B because the bound cognate protein blocked cleavages within that sequence and thus production of the corresponding fragments. The missing bands on the gel constitute the footprint. (b) A protein fraction containing a sequence-specific DNA-binding protein can be purified by column chromatography. DNase I footprinting can then identify which of the eluted fractions contain the cognate protein. In the absence of added protein (NE, *no* extract), DNase I cleaves the DNA fragment at multiple sites, producing multiple bands on the gel shown here. A cognate protein present in the nuclear extract applied to the column (O, *o*nput) generated a footprint. This protein was bound to the column, since footprinting activity was not detected in the flow-through protein fraction (FT). After applying a salt gradient to the column, most of the cognate protein eluted in fractions 9–12, as evidenced by the missing bands (footprints). The sequence of the protein-binding region can be determined by comparison with marker DNA fragments of known length analyzed on the same gel (M). [Part (b) from S. Yoshinaga et al., 1989, *J. Biol. Chem.* **264**:10529.]

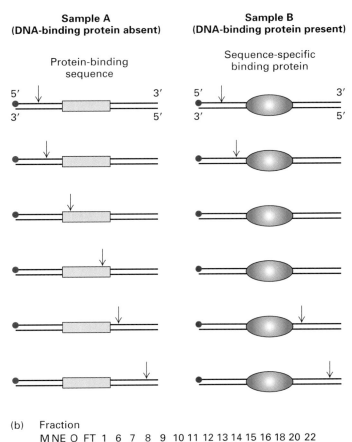

with a DNA fragment containing a known DNA control element, the appearance of a footprint indicates the presence of a transcription factor that binds that control element in the protein sample being assayed. Footprinting also identifies the specific DNA sequence to which the transcription factor binds.

The electrophoretic mobility shift assay (EMSA), also called the *gel-shift* or *band-shift* assay, is more useful than the footprinting assay for quantitative analysis of DNA-binding proteins. In general, the electrophoretic mobility of a DNA fragment is reduced when it is complexed to protein, causing a shift in the location of the fragment band. This assay can be used to detect a transcription factor in protein

fractions incubated with a radiolabeled DNA fragment containing a known control element (Figure 11-14).

In the biochemical isolation of a transcription factor, an extract of cell nuclei commonly is subjected sequentially to several types of column chromatography (Chapter 3). Fractions eluted from the columns are assayed by DNase I footprinting or EMSA using DNA fragments containing an identified regulatory element (see Figures 11-13 and 11-14). Fractions containing protein that binds to the regulatory element in these assays probably contain a putative transcription factor. A powerful technique commonly used for the final step in purifying transcription factors is *sequence-specific DNA affinity chromatography*, a particular type of

▶ **EXPERIMENTAL FIGURE 11-14 Electrophoretic mobility shift assay can be used to detect transcription factors during purification.** In this example, protein fractions separated by column chromatography were assayed for their ability to bind to a radiolabeled DNA-fragment probe containing a known regulatory element. After an aliquot of the protein sample loaded onto the column (ON) and successive column fractions (numbers) were incubated with the labeled probe, the samples were electrophoresed under conditions that do not denature proteins. The free probe not bound to protein migrated to the bottom of the gel. A protein in the preparation applied to the column and in fractions 7 and 8 bound to the probe, forming a DNA-protein complex that migrated more slowly than the free probe. These fractions therefore likely contain the regulatory protein being sought. [From S. Yoshinaga et al., 1989, *J. Biol. Chem.* **264**:10529.]

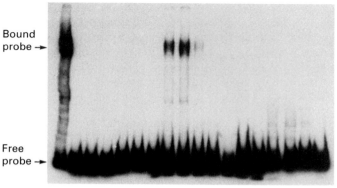

affinity chromatography in which long DNA strands containing multiple copies of the transcription factor–binding site are coupled to a column matrix. As a final test that an isolated protein is in fact a transcription factor, its ability to modulate transcription of a template containing the corresponding protein-binding sites is assayed in an in vitro transcription reaction. Figure 11-15 shows the results of such an assay for SP1, a transcription factor that binds to GC-rich sequences, thereby activating transcription from nearby promoters.

Once a transcription factor is isolated and purified, its partial amino acid sequence can be determined and used to clone the gene or cDNA encoding it, as outlined in Chapter 9. The isolated gene can then be used to test the ability of the encoded protein to activate or repress transcription in an in vivo transfection assay (Figure 11-16).

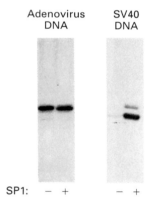

▲ **EXPERIMENTAL FIGURE 11-15 Transcription factors can be identified by in vitro assay for transcription activity.** SP1 was identified based on its ability to bind to a region of the SV40 genome that contains six copies of a GC-rich promoter-proximal element and was purified by column chromatography. To test the transcription-activating ability of purified SP1, it was incubated in vitro with template DNA, a protein fraction containing RNA polymerase II and associated general transcription factors, and labeled ribonucleoside triphosphates. The labeled RNA products were subjected to electrophoresis and autoradiography. Shown here are autoradiograms from assays with adenovirus and SV40 DNA in the absence (−) and presence (+) of SP1. SP1 had no significant effect on transcription from the adenovirus promoter, which contains no SP1-binding sites. In contrast, SP1 stimulated transcription from the SV40 promoter about tenfold. [Adapted from M. R. Briggs et al., 1986, *Science* **234**:47.]

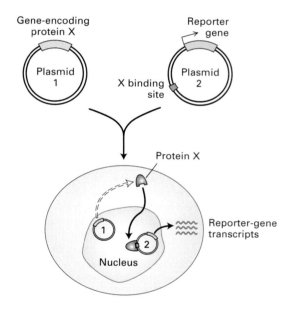

▲ **EXPERIMENTAL FIGURE 11-16 In vivo transfection assay measures transcription activity to evaluate proteins believed to be transcription factors.** The assay system requires two plasmids. One plasmid contains the gene encoding the putative transcription factor (protein X). The second plasmid contains a reporter gene (e.g., *lacZ*) and one or more binding sites for protein X. Both plasmids are simultaneously introduced into cells that lack the gene encoding protein X. The production of reporter-gene RNA transcripts is measured; alternatively, the activity of the encoded protein can be assayed. If reporter-gene transcription is greater in the presence of the X-encoding plasmid, then the protein is an activator; if transcription is less, then it is a repressor. By use of plasmids encoding a mutated or rearranged transcription factor, important domains of the protein can be identified.

Activators Are Modular Proteins Composed of Distinct Functional Domains

Studies with a yeast transcription activator called GAL4 provided early insight into the domain structure of transcription factors. The gene encoding the GAL4 protein, which promotes expression of enzymes needed to metabolize galactose, was identified by complementation analysis of *gal4* mutants (Chapter 9). Directed mutagenesis studies like those described previously identified UASs for the genes activated by GAL4. Each of these UASs was found to contain one or more copies of a related 17-bp sequence called UAS$_{GAL}$. DNase I footprinting assays with recombinant GAL4 protein produced in *E. coli* from the yeast GAL4 gene showed that

GAL4 protein binds to UAS$_{GAL}$ sequences. When a copy of UAS$_{GAL}$ was cloned upstream of a TATA box followed by a *lacZ* reporter gene, expression of *lacZ* was activated in galactose media in wild-type cells, but not in *gal4* mutants. These results showed that UAS$_{GAL}$ is a transcription-control element activated by the GAL4 protein in galactose media.

A remarkable set of experiments with *gal4* deletion mutants demonstrated that the GAL4 transcription factor is composed of separable functional domains: an N-terminal **DNA-binding domain**, which binds to specific DNA sequences, and a C-terminal **activation domain**, which interacts with other proteins to stimulate transcription from a nearby promoter (Figure 11-17). When the N-terminal DNA-binding domain of GAL4 was fused directly to various

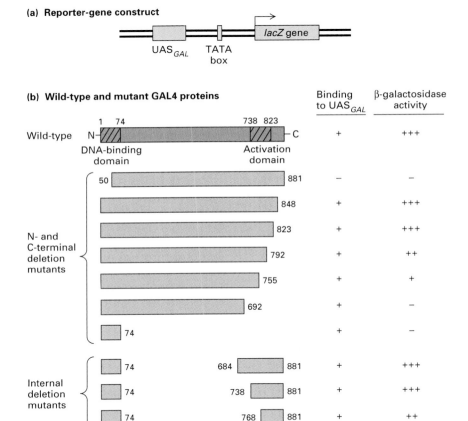

▲ **EXPERIMENTAL FIGURE 11-17 Deletion mutants of the GAL4 gene in yeast with a UAS$_{GAL}$ reporter-gene construct demonstrate the separate functional domains in an activator.** (a) Diagram of DNA construct containing a *lacZ* reporter gene and TATA box ligated to UAS$_{GAL}$, a regulatory element that contains several GAL4-binding sites. The reporter-gene construct and DNA encoding wild-type or mutant (deleted) GAL4 were simultaneously introduced into mutant (*gal4*) yeast cells, and the activity of β-galactosidase expressed from *lacZ* was assayed. Activity will be high if the introduced GAL4 DNA encodes a functional protein. (b) Schematic diagrams of wild-type GAL4 and various mutant forms. Small numbers refer to positions in the wild-type sequence. Deletion of 50 amino acids from the N-terminal end destroyed the

ability of GAL4 to bind to UAS$_{GAL}$ and to stimulate expression of β-galactosidase from the reporter gene. Proteins with extensive deletions from the C-terminal end still bound to UAS$_{GAL}$. These results localize the DNA-binding domain to the N-terminal end of GAL4. The ability to activate β-galactosidase expression was not entirely eliminated unless somewhere between 126–189 or more amino acids were deleted from the C-terminal end. Thus the activation domain lies in the C-terminal region of GAL4. Proteins with internal deletions (*bottom*) also were able to stimulate expression of β-galactosidase, indicating that the central region of GAL4 is not crucial for its function in this assay. [See J. Ma and M. Ptashne, 1987, *Cell* **48**:847; I. A. Hope and K. Struhl, 1986, *Cell* **46**:885; and R. Brent and M. Ptashne, 1985, *Cell* **43**:729.]

of its C-terminal fragments, the resulting truncated proteins retained the ability to stimulate expression of a reporter gene in an in vivo assay like that depicted in Figure 11-16. Thus the internal portion of the protein is not required for functioning of GAL4 as a transcription factor. Similar experiments with another yeast transcription factor, GCN4, which regulates genes required for synthesis of many amino acids, indicated that it contains an ≈60-aa DNA-binding domain at its C-terminus and an ≈20-aa activation domain near the middle of its sequence.

Further evidence for the existence of distinct activation domains in GAL4 and GCN4 came from experiments in which their activation domains were fused to a DNA-binding domain from an entirely unrelated E. coli DNA-binding protein. When these fusion proteins were assayed in vivo, they activated transcription of a reporter gene containing the cognate site for the E. coli protein. Thus functional transcription factors can be constructed from entirely novel combinations of prokaryotic and eukaryotic elements.

Studies such as these have now been carried out with many eukaryotic activators. The structural model of eukaryotic activators that has emerged from these studies is a modular one in which one or more activation domains are connected to a sequence-specific DNA-binding domain through flexible protein domains (Figure 11-18). In some cases, amino acids included in the DNA-binding domain also contribute to transcriptional activation. As discussed in a later section, activation domains are thought to function by

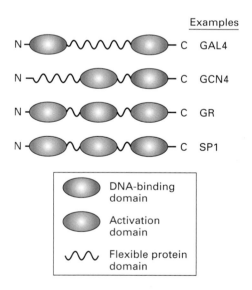

Examples

GAL4

GCN4

GR

SP1

DNA-binding domain

Activation domain

Flexible protein domain

▲ **FIGURE 11-18 Schematic diagrams illustrating the modular structure of eukaryotic transcription activators.** These transcription factors may contain more than one activation domain but rarely contain more than one DNA-binding domain. GAL4 and GCN4 are yeast transcription activators. The glucocorticoid receptor (GR) promotes transcription of target genes when certain hormones are bound to the C-terminal activation domain. SP1 binds to GC-rich promoter elements in a large number of mammalian genes.

binding other proteins involved in transcription. The presence of flexible domains connecting the DNA-binding domains to activation domains may explain why alterations in the spacing between control elements are so well tolerated in eukaryotic control regions. Thus even when the positions of transcription factors bound to DNA are shifted relative to each other, their activation domains may still be able to interact because they are attached to their DNA-binding domains through flexible protein regions.

Repressors Are the Functional Converse of Activators

Eukaryotic transcription is regulated by repressors as well as activators. For example, geneticists have identified mutations in yeast that result in continuously high expression of certain genes. This type of unregulated, abnormally high expression is called **constitutive** expression and results from the inactivation of a repressor that normally inhibits the transcription of these genes. Similarly, mutants of *Drosophila* and *C. elegans* have been isolated that are defective in embryonic development because they express genes in embryonic cells where they are normally repressed. The mutations in these mutants inactivate repressors, leading to abnormal development.

Repressor-binding sites in DNA have been identified by systematic linker scanning mutation analysis similar to that depicted in Figure 11-10. In this type of analysis, mutation of an activator-binding site leads to decreased expression of the linked reporter gene, whereas mutation of a repressor-binding site leads to increased expression of a reporter gene. Repressor proteins that bind such sites can be purified and assayed using the same biochemical techniques described earlier for activator proteins.

Eukaryotic transcription repressors are the functional converse of activators. They can inhibit transcription from a gene they do not normally regulate when their cognate binding sites are placed within a few hundred base pairs of the gene's start site. Like activators, most eukaryotic repressors are modular proteins that have two functional domains: a DNA-binding domain and a **repression domain**. Similar to activation domains, repression domains continue to function when fused to another type of DNA-binding domain. If binding sites for this second DNA-binding domain are inserted within a few hundred base pairs of a promoter, expression of the fusion protein inhibits transcription from the promoter. Also like activation domains, repression domains function by interacting with other proteins as discussed later.

 The absence of appropriate repressor activity can have devastating consequences. For instance, the protein encoded by the *Wilms' tumor (WT1)* gene is a repressor that is expressed preferentially in the developing kidney. Children who inherit mutations in both the maternal and paternal *WT1* genes, so that they produce no functional WT1 protein, invariably develop kidney tumors

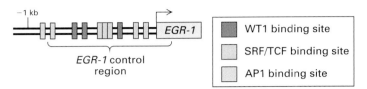

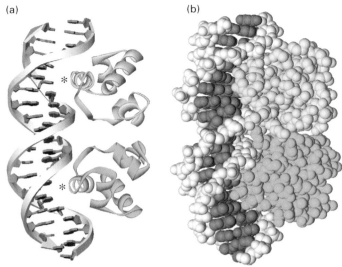

▲ FIGURE 11-19 Diagram of the control region of the gene encoding EGR-1, a transcription activator. The binding sites for WT1, a eukaryotic repressor protein, do not overlap the binding sites for the activator AP1 or the composite binding site for the activators SRF and TCF. Thus repression by WT1 does not involve direct interference with binding of other proteins as in the case of bacterial repressors.

early in life. The WT1 protein binds to the control region of the gene encoding a transcription activator called EGR-1 (Figure 11-19). This gene, like many other eukaryotic genes, is subject to both repression and activation. Binding by WT1 represses transcription of the *EGR-1* gene without inhibiting binding of the activators that normally stimulate expression of this gene. ∎

DNA-Binding Domains Can Be Classified into Numerous Structural Types

The DNA-binding domains of eukaryotic activators and repressors contain a variety of structural **motifs** that bind specific DNA sequences. The ability of DNA-binding proteins to bind to specific DNA sequences commonly results from noncovalent interactions between atoms in an α helix in the DNA-binding domain and atoms on the edges of the bases within a major groove in the DNA. Interactions with sugar-phosphate backbone atoms and, in some cases, with atoms in a DNA minor groove also contribute to binding.

The principles of specific protein-DNA interactions were first discovered during the study of bacterial repressors. Many bacterial repressors are dimeric proteins in which an α helix from each monomer inserts into a major groove in the DNA helix (Figure 11-20). This α helix is referred to as the *recognition helix* or *sequence-reading helix* because most of the amino acid side chains that contact DNA extend from this helix. The recognition helix that protrudes from the surface of bacterial repressors to enter the DNA major groove and make multiple, specific interactions with atoms in the DNA is usually supported in the protein structure in part by hydrophobic interactions with a second α helix just N-terminal to it. This structural element, which is present in many bacterial repressors, is called a *helix-turn-helix* motif.

Many additional motifs that can present an α helix to the major groove of DNA are found in eukaryotic transcription factors, which often are classified according to the type of DNA-binding domain they contain. Because most of these motifs have characteristic consensus amino acid sequences, newly characterized transcription factors frequently can be classified once the corresponding genes or cDNAs are cloned

▲ FIGURE 11-20 Interaction of bacteriophage 434 repressor with DNA. (a) Ribbon diagram of 434 repressor bound to its specific operator DNA. Repressor monomers are in yellow and green. The recognition helices are indicated by asterisks. A space filling model of the repressor-operator complex (b) shows how the protein interacts intimately with one side of the DNA molecule over a length of 1.5 turns. [Adapted from A. K. Aggarwal et al., 1988, *Science* **242**:899.]

and sequenced. The genomes of higher eukaryotes encode dozens of classes of DNA-binding domains and hundreds to thousands of transcription factors. The human genome, for instance, encodes ≈2000 transcription factors.

Here we introduce several common classes of DNA-binding proteins whose three-dimensional structures have been determined. In all these examples and many other transcription factors, at least one α helix is inserted into a major groove of DNA. However, some transcription factors contain alternative structural motifs (e.g., β strands and loops) that interact with DNA.

Homeodomain Proteins Many eukaryotic transcription factors that function during development contain a conserved 60-residue DNA-binding motif that is similar to the helix-turn-helix motif of bacterial repressors. Called **homeodomain** proteins, these transcription factors were first identified in *Drosophila* mutants in which one body part was transformed into another during development (Chapter 15). The conserved homeodomain sequence has also been found in vertebrate transcription factors, including those that have similar master control functions in human development.

Zinc-Finger Proteins A number of different eukaryotic proteins have regions that fold around a central Zn^{2+} ion, producing a compact domain from a relatively short length of the polypeptide chain. Termed a **zinc finger**, this structural motif was first recognized in DNA-binding domains but now is known to occur also in proteins that do not bind to DNA. Here we describe two of the several classes of zinc-finger motifs that have been identified in eukaryotic transcription factors.

The C_2H_2 *zinc finger* is the most common DNA-binding motif encoded in the human genome and the genomes of most other multicellular animals. It is also common in multicellular plants, but is not the dominant type of DNA-binding domain in plants as it is in animals. This motif has a 23- to 26-residue consensus sequence containing two conserved cysteine (C) and two conserved histidine (H) residues, whose side chains bind one Zn^{2+} ion (see Figure 3-6b). The name "zinc finger" was coined because a two-dimensional diagram of the structure resembles a finger. When the three-dimensional structure was solved, it became clear that the binding of the Zn^{2+} ion by the two cysteine and two histidine residues folds the relatively short polypeptide sequence into a compact domain, which can insert its α helix into the major groove of DNA. Many transcription factors contain multiple C_2H_2 zinc fingers, which interact with successive groups of base pairs, within the major groove, as the protein wraps around the DNA double helix (Figure 11-21a).

A second type of zinc-finger structure, designated the C_4 *zinc finger* (because it has four conserved cysteines in contact with the Zn^{2+}), is found in ≈50 human transcription factors. The first members of this class were identified as specific intracellular high-affinity binding proteins, or "receptors," for steroid hormones, leading to the name *steroid receptor superfamily*. Because similar intracellular receptors for nonsteroid hormones subsequently were found, these transcrip-

tion factors are now commonly called **nuclear receptors.** The characteristic feature of C_4 zinc fingers is the presence of two groups of four critical cysteines, one toward each end of the 55- or 56-residue domain. Although the C_4 zinc finger initially was named by analogy with the C_2H_2 zinc finger, the three-dimensional structures of proteins containing these DNA-binding motifs later were found to be quite distinct. A particularly important difference between the two is that C_2H_2 zinc-finger proteins generally contain three or more repeating finger units and bind as monomers, whereas C_4 zinc-finger proteins generally contain only two finger units and generally bind to DNA as homodimers or heterodimers. Homodimers of C_4 zinc-finger DNA-binding domains have twofold rotational symmetry (Figure 11-21b). Consequently, homodimeric nuclear receptors bind to consensus DNA sequences that are inverted repeats.

Leucine-Zipper Proteins Another structural motif present in the DNA-binding domains of a large class of transcription factors contains the hydrophobic amino acid leucine at every seventh position in the sequence. These proteins bind to DNA as dimers, and mutagenesis of the leucines showed that they were required for dimerization. Consequently, the name **leucine zipper** was coined to denote this structural motif.

The DNA-binding domain of the yeast GCN4 transcription factor mentioned earlier is a leucine-zipper domain. X-ray crystallographic analysis of complexes between DNA and the GCN4 DNA-binding domain has shown that the dimeric protein contains two extended α helices that "grip" the DNA molecule, much like a pair of scissors, at two ad-

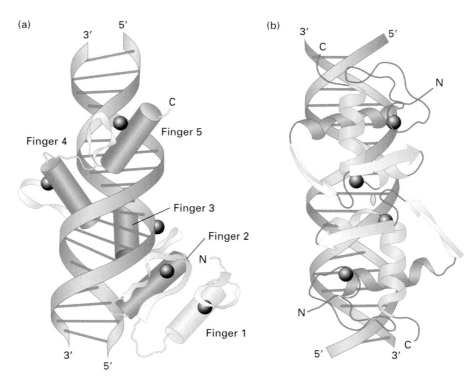

▶ **FIGURE 11-21 Interaction between DNA and proteins containing zinc fingers.** (a) GL1 is a monomeric protein that contains five C_2H_2 zinc fingers. α-Helices are shown as cylinders, Zn^{+2} ions as spheres. Finger 1 does not interact with DNA, whereas the other four fingers do. (b) The glucocorticoid receptor is a homodimeric C_4 zinc-finger protein. α-Helices are shown as purple ribbons, β-strands as green arrows, Zn^{+2} ions as spheres. Two α helices (darker shade), one in each monomer, interact with the DNA. Like all C_4 zinc-finger homodimers, this transcription factor has twofold rotational symmetry; the center of symmetry is shown by the yellow ellipse. In contrast, heterodimeric nuclear receptors do not exhibit rotational symmetry. [See N. P. Pavletich and C. O. Pabo, 1993, *Science* **261**:1701, and B. F. Luisi et al., 1991, *Nature* **352**:497.]

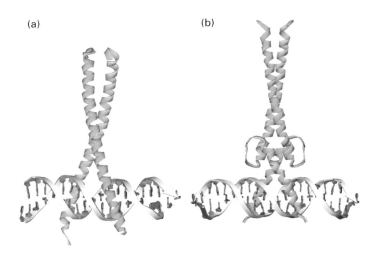

▲ **FIGURE 11-22 Interaction of homodimeric leucine-zipper and basic helix-loop-helix (bHLH) proteins with DNA.** (a) In leucine-zipper proteins, basic residues in the extended α-helical regions of the monomers interact with the DNA backbone at adjacent major grooves. The coiled-coil dimerization domain is stabilized by hydrophobic interactions between the monomers. (b) In bHLH proteins, the DNA-binding helices at the bottom (N-termini of the monomers) are separated by nonhelical loops from a leucine-zipper-like region containing a coiled-coil dimerization domain. [Part (a) see T. E. Ellenberger et al., 1992, *Cell* **71**:1223; part (b) see A. R. Ferre-D'Amare et al., 1993, *Nature* **363**:38.]

jacent major grooves separated by about half a turn of the double helix (Figure 11-22a). The portions of the α helices contacting the DNA include positively charged (basic) residues that interact with phosphates in the DNA backbone and additional residues that interact with specific bases in the major groove.

GCN4 forms dimers via hydrophobic interactions between the C-terminal regions of the α helices, forming a **coiled-coil** structure. This structure is common in proteins containing amphipathic α helices in which hydrophobic amino acid residues are regularly spaced alternately three or four positions apart in the sequence, forming a stripe down one side of the α helix. These hydrophobic stripes make up the interacting surfaces between the α-helical monomers in a coiled-coil dimer (see Figure 3-6c).

Although the first leucine-zipper transcription factors to be analyzed contained leucine residues at every seventh position in the dimerization region, additional DNA-binding proteins containing other hydrophobic amino acids in these positions subsequently were identified. Like leucine-zipper proteins, they form dimers containing a C-terminal coiled-coil dimerization region and an N-terminal DNA-binding domain. The term *basic zipper (bZip)* now is frequently used to refer to all proteins with these common structural features. Many basic-zipper transcription factors are heterodimers of two different polypeptide chains, each containing one basic-zipper domain.

Basic Helix-Loop-Helix (bHLH) Proteins The DNA-binding domain of another class of dimeric transcription factors contains a structural motif very similar to the basic-zipper motif except that a nonhelical loop of the polypeptide chain separates two α-helical regions in each monomer (Figure 11-22b). Termed a **basic helix-loop-helix (bHLH),** this motif was predicted from the amino acid sequences of these proteins, which contain an N-terminal α helix with basic residues that interact with DNA, a middle loop region, and a C-terminal region with hydrophobic amino acids spaced at intervals characteristic of an amphipathic α helix. As with basic-zipper proteins, different bHLH proteins can form heterodimers.

Transcription-Factor Interactions Increase Gene-Control Options

Two types of DNA-binding proteins discussed in the previous section—basic-zipper proteins and bHLH proteins—often exist in alternative heterodimeric combinations of monomers. Other classes of transcription factors not discussed here also form heterodimeric proteins. In some heterodimeric transcription factors, each monomer has a DNA-binding domain with equivalent sequence specificity. In these proteins, the formation of alternative heterodimers does not influence DNA-binding specificity, but rather allows the activation domains associated with each monomer to be brought together in alternative combinations in a single transcription factor. As we shall see later, and in subsequent chapters, the activities of individual transcription factors can be regulated by multiple mechanisms. Consequently, a single bZIP or bHLH DNA regulatory element in the control region of a gene may elicit different transcriptional responses depending on which bZIP or bHLH monomers that bind to that site are expressed in a particular cell at a particular time and how their activities are regulated.

In some heterodimeric transcription factors, however, each monomer has a different DNA-binding specificity. The resulting combinatorial possibilities increase the number of potential DNA sequences that a family of transcription factors can bind. Three different factor monomers theoretically could combine to form six homo- and heterodimeric factors, as illustrated in Figure 11-23a. Four different factor monomers could form a total of 10 dimeric factors; five monomers, 16 dimeric factors; and so forth. In addition, inhibitory factors are known that bind to some basic-zipper and bHLH monomers, thereby blocking their binding to DNA. When these inhibitory factors are expressed, they repress transcriptional activation by the factors with which they interact (Figure 11-23b). The rules governing the interactions of members of a heterodimeric transcription-factor class are complex. This combinatorial complexity expands both the number of DNA sites from which these factors can activate transcription and the ways in which they can be regulated.

Similar combinatorial transcriptional regulation is achieved through the interaction of structurally unrelated

▶ **FIGURE 11-23 Combinatorial possibilities due to formation of heterodimeric transcription factors.** (a) In the hypothetical example shown, transcription factors A, B, and C can all interact with one another, permitting the three factors to bind to six different DNA sequences (sites 1–6) and creating six combinations of activation domains. Each composite binding site is divided into two half-sites, and each heterodimeric factor contains the activation domains of its two constituent monomers. (b) Expression of an inhibitory factor (green) that interacts only with factor A inhibits binding; hence, transcriptional activation at sites 1, 4, and 5 is inhibited, but activation at sites 2, 3, and 6 is unaffected.

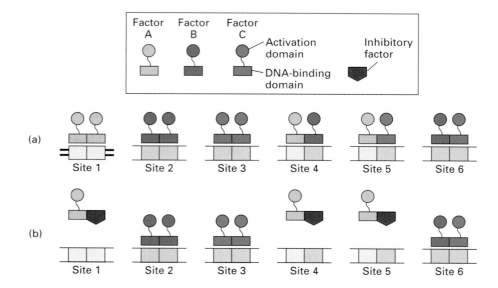

transcription factors bound to closely spaced binding sites in DNA. An example is the interaction of two transcription factors, NFAT and AP1, which bind to neighboring sites in a composite promoter-proximal element regulating the gene encoding interleukon-2 (IL-2). Expression of the *IL-2* gene is critical to the immune response, but abnormal expression of IL-2 can lead to autoimmune diseases such as rheumatoid arthritis. Neither NFAT nor AP1 binds to its site in the *IL-2* control region in the absence of the other. The affinities of the factors for these particular DNA sequences are too low for the individual factors to form a stable complex with DNA. However, when both NFAT and AP1 are present, protein-protein interactions between them stabilize the DNA ternary complex composed of NFAT, AP1, and DNA (Figure 11-24). Such *cooperative DNA binding* of various transcription factors results in considerable combinatorial complexity of transcription control. As a result, the approx-

imately 2000 transcription factors encoded in the human genome can bind to DNA through a much larger number of cooperative interactions, resulting in unique transcriptional control for each of the several tens of thousands of human genes. In the case of *IL-2*, transcription occurs only when both NFAT is activated, resulting in its transport from cytoplasm to the nucleus, and the two subunits of AP1 are synthesized. These events are controlled by distinct signal transduction pathways (Chapters 13 and 14), allowing stringent control of IL-2 expression.

Cooperative binding by NFAT and AP1 occurs only when their weak binding sites are located at a precise distance, quite close to each other in DNA. Recent studies have shown that the requirements for cooperative binding are not so stringent in the case of some other transcription factors and control regions. For example, the *EGR-1* control region contains a composite binding site to which the SRF and TCF

▶ **FIGURE 11-24 Cooperative binding of two unrelated transcription factors to neighboring sites in a composite control element.** By themselves, both monomeric NFAT and heterodimeric AP1 transcription factors have low affinity for their respective binding sites in the *IL-2* promoter-proximal region. Protein-protein interactions between NFAT and AP1 add to the overall stability of the NFAT-AP1-DNA complex, so that the two proteins bind to the composite site cooperatively. [See L. Chen et al., 1998, *Nature* **392**:42.]

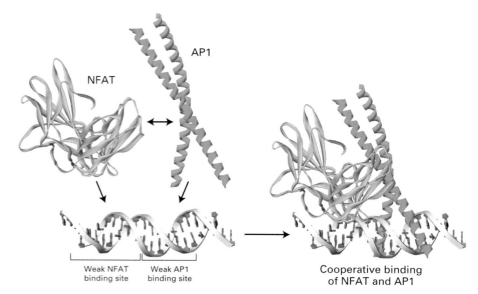

transcription factors bind cooperatively (see Figure 11-19). Because a TCF has a long, flexible domain that interacts with SRF, the two proteins can bind cooperatively when their individual sites in DNA are separated by any distance up to 10 base pairs or are inverted relative to each other.

Structurally Diverse Activation and Repression Domains Regulate Transcription

Experiments with fusion proteins composed of the GAL4 DNA-binding domain and random segments of *E. coli* proteins demonstrated that a diverse group of amino acid sequences can function as activation domains, ~1% of all *E. coli* sequences, even though they evolved to perform other functions. Many transcription factors contain activation domains marked by an unusually high percentage of particular amino acids. GAL4, GCN4, and most other yeast transcription factors, for instance, have activation domains that are rich in acidic amino acids (aspartic and glutamic acids). These so-called *acidic activation domains* generally are capable of stimulating transcription in nearly all types of eukaryotic cells—fungal, animal, and plant cells. Activation domains from some *Drosophila* and mammalian transcription factors are glutamine-rich, and some are proline-rich; still others are rich in the closely related amino acids serine and threonine, both of which have hydroxyl groups. However, some strong activation domains are not particularly rich in any specific amino acid.

Biophysical studies indicate that acidic activation domains have an unstructured, random-coil conformation. These domains stimulate transcription when they are bound to a protein *co-activator*. The interaction with a co-activator causes the activation domain to assume a more structured α-helical conformation in the activation domain–co-activator complex. A well-studied example of a transcription factor with an acidic activation domain is the mammalian CREB protein, which is phosphorylated in response to increased levels of cAMP. This regulated phosphorylation is required for CREB to bind to its co-activator CBP (CREB *b*inding *p*rotein), resulting in the transcription of genes whose control regions contain a CREB-binding site (see Figure 13-32). When the phosphorylated random coil activation domain of CREB interacts with CBP, it undergoes a conformational change to form two α helices that wrap around the interacting domain of CBP.

Some activation domains are larger and more highly structured than acidic activation domains. For example, the ligand-binding domains of nuclear receptors function as activation domains when they bind their specific ligand (Figure 11-25). Binding of ligand induces a large conformational change that allows the ligand-binding domain with bound hormone to interact with a short α helix in nuclear-receptor co-activators; the resulting complex then can activate transcription of genes whose control regions bind the nuclear receptor.

Thus the acidic activation domain in CREB and the ligand-binding activation domains in nuclear receptors represent

two structural extremes. The CREB acidic activation domain is a random coil that folds into two α-helices when it binds to the surface of a globular domain in a co-activator. In contrast, the nuclear-receptor ligand-binding activation domain is a structured globular domain that interacts with a short α helix in a co-activator, which probably is a random coil before it is bound. In both cases, however, specific protein-protein interactions between co-activators and the activation domains permit the transcription factors to stimulate gene expression.

Currently, less is known about the structure of repression domains. The globular ligand-binding domains of some nuclear receptors function as repression domains in the absence of their specific hormone ligand (see Figure 11-25b). Like activation domains, repression domains may be relatively short, comprising 15 or fewer amino acids. Biochemical and

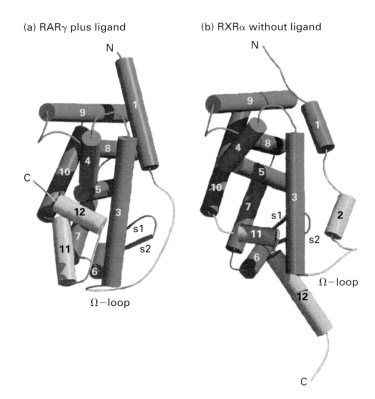

▲ **FIGURE 11-25 Effect of ligand binding on conformation of the ligand-binding activation domains in two related human nuclear receptors.** Cylinders represent α helices. Regions that do not change significantly in conformation after ligand binding are shown in green; regions that do, in yellow. (a) Ligand-binding domain of human RARγ monomer when bound to its ligand, all-*trans* retinoic acid. In this fairly compact conformation, RARγ can stimulate transcription. (b) Ligand-binding domain of human RXRα monomer in the absence of its ligand, 9-*cis* retinoic acid. As discussed later, RXR associates with several different nuclear-receptor monomers (e.g., RAR), forming heterodimeric transcription factors. [From J. M. Wurtz et al., 1996, *Nature Struc.* **3**:87; courtesy of Hinrich Gronemeyer.]

genetic studies indicate that repression domains also mediate protein-protein interactions and bind to *co-repressor* proteins, forming a complex that inhibits transcription initiation by mechanisms that are discussed later in the chapter.

Multiprotein Complexes Form on Enhancers

As noted previously, enhancers generally range in length from about 50 to 200 base pairs and include binding sites for several transcription factors. The multiple transcription factors that bind to a single enhancer are thought to interact. Analysis of the ≈70-bp enhancer that regulates expression of β-interferon, an important protein in defense against viral infections in humans, provides a good example of such transcription-factor interactions. The β-interferon enhancer contains four control elements that bind four different transcription factors simultaneously. In the presence of a small, abundant protein associated with chromatin called *HMGI*, binding of the transcription factors is highly cooperative, similar to the binding of NFAT and AP1 to the composite promoter-proximal site in the *IL-2* control region (see Figure 11-24). This cooperative binding produces a multiprotein complex on the β-interferon enhancer DNA (Figure 11-26). The term **enhancesome** has been coined to describe such large nucleoprotein complexes that assemble from transcription factors as they bind cooperatively to their multiple binding sites in an enhancer.

HMGI binds to the minor groove of DNA regardless of the sequence and, as a result, bends the DNA molecule sharply. This bending of the enhancer DNA permits bound transcription factors to interact properly. The inherently

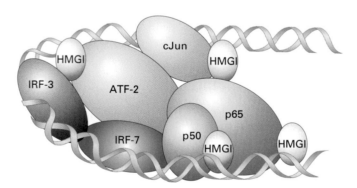

▲ **FIGURE 11-26 Model of the enhancesome that forms on the β-interferon enhancer.** Two monomeric transcription factors, IRF-3 and IRF-7, and two heterodimeric factors, Jun/ATF-2 and p50/ p65 (NF-κB) bind to the four control elements in this enhancer. Cooperative binding of these transcription factors is facilitated by HMGI, which binds to the minor groove of DNA and also interacts directly with the dimeric factors. Bending of the enhancer sequence resulting from HMGI binding is critical to formation of an enhancesome. Different DNA-bending proteins act similarly at other enhancers. [Adapted from D. Thanos and T. Maniatis, 1995, *Cell* **83**:1091, and M. A. Wathel et al., 1998, *Mol. Cell* **1**:507.]

weak, noncovalent protein-protein interactions between transcription factors are strengthened by their binding to neighboring DNA sites, which keeps the proteins at very high relative concentrations.

Because of the presence of flexible regions connecting the DNA-binding domains and activation or repression domains in transcription factors (see Figure 11-18) and the ability of bound proteins to bend DNA, considerable leeway in the spacing between regulatory elements in transcription-control regions is permissible. This property probably contributed to rapid evolution of gene control in eukaryotes. Transposition of DNA sequences and recombination between repeated sequences over evolutionary time likely created new combinations of control elements that were subjected to natural selection and retained if they proved beneficial. The latitude in spacing between regulatory elements probably allowed many more functional combinations to be subjected to this evolutionary experimentation than would be the case if constraints on the spacing between regulatory elements were strict.

KEY CONCEPTS OF SECTION 11.3

Activators and Repressors of Transcription

■ Transcription factors, which stimulate or repress transcription, bind to promoter-proximal regulatory elements and enhancers in eukaryotic DNA.

■ Transcription activators and repressors are generally modular proteins containing a single DNA-binding domain and one or a few activation domains (for activators) or repression domains (for repressors). The different domains frequently are linked through flexible polypeptide regions (see Figure 11-18).

■ Among the most common structural motifs found in the DNA-binding domains of eukaryotic transcription factors are the C_2H_2 zinc finger, homeodomain, basic helix-loop-helix (bHLH), and basic zipper (leucine zipper). All these and many other DNA-binding motifs contain one or more α helices that interact with major grooves in their cognate site in DNA.

■ The transcription-control regions of most genes contain binding sites for multiple transcription factors. Transcription of such genes varies depending on the particular repertoire of transcription factors that are expressed and activated in a particular cell at a particular time.

■ Combinatorial complexity in transcription control results from alternative combinations of monomers that form heterodimeric transcription factors (see Figure 11-23) and from cooperative binding of transcription factors to composite control sites (see Figure 11-24).

■ Activation and repression domains in transcription factors exhibit a variety of amino acid sequences and three-dimensional structures. In general, these functional domains interact with co-activators or co-repressors, which are critical to the ability of transcription factors to modulate gene expression.

■ Cooperative binding of multiple activators to nearby sites in an enhancer forms a multiprotein complex called an *enhancesome* (see Figure 11-26). Assembly of enhancesomes often requires small proteins that bind to the DNA minor groove and bend the DNA sharply, allowing bound proteins on either side of the bend to interact more readily.

11.4 Transcription Initiation by RNA Polymerase II

In previous sections many of the eukaryotic proteins and DNA sequences that participate in transcription and its control have been introduced. In this section, we focus on assembly of *transcription preinitiation complexes* involving RNA polymerase II (Pol II). Recall that this eukaryotic RNA polymerase catalyzes synthesis of mRNAs and a few small nuclear RNAs (snRNAs). Mechanisms that control the assembly of Pol II transcription preinitiation complexes, and hence the rate of transcription of protein-coding genes, are considered in the next section.

General Transcription Factors Position RNA Polymerases II at Start Sites and Assist in Initiation

In vitro transcription by purified RNA polymerase II requires the addition of several initiation factors that are separated from the polymerase during purification. These initiation factors, which position polymerase molecules at transcription start sites and help to melt the DNA strands so that the template strand can enter the active site of the enzyme, are called *general transcription factors*. In contrast to the transcription factors discussed in the previous section, which bind to specific sites in a limited number of genes, general transcription factors are required for synthesis of RNA from most genes.

The general transcription factors that assist Pol II in initiation of transcription from most TATA-box promoters in vitro have been isolated and characterized. These proteins are designated *TFIIA, TFIIB,* etc., and most are multimeric proteins. The largest is TFIID, which consists of a single 38-kDa *TATA box–binding protein (TBP)* and thirteen TBP-associated factors (TAFs). General transcription factors with similar activities have been isolated from cultured human cells, rat liver, *Drosophila* embryos, and yeast. The genes encoding these proteins in yeast have been sequenced as part of the complete yeast genome sequence, and many of the cDNAs encoding human and *Drosophila* Pol II general transcription factors have been cloned and sequenced. In all cases, equivalent general transcription factors from different eukaryotes are highly conserved.

Sequential Assembly of Proteins Forms the Pol II Transcription Preinitiation Complex in Vitro

Detailed biochemical studies revealed how the Pol II preinitiation complex, comprising a Pol II molecule and general transcription factors bound to a promoter region of DNA, is assembled. In these studies DNase I footprinting and electrophoretic mobility shift assays were used to determine the order in which Pol II and general transcription factors bound to TATA-box promoters. Because the complete, multisubunit TFIID is difficult to purify, researchers used only the isolated TBP component of this general transcription factor in these experiments. Pol II can initiate transcription in vitro in the absence of the other TFIID subunits.

Figure 11-27 summarizes our current understanding of the stepwise assembly of the Pol II transcription preinitiation complex in vitro. TBP is the first protein to bind to a TATA-box promoter. All eukaryotic TBPs analyzed to date have very similar C-terminal domains of 180 residues. The sequence of this region is 80 percent identical in the yeast and human proteins, and most differences are conservative substitutions. This conserved C-terminal domain functions as well as the full-length protein does in in vitro transcription. (The N-terminal domain of TBP, which varies greatly in sequence and length among different eukaryotes, functions in the Pol II–catalyzed transcription of genes encoding snRNAs.) TBP is a monomer that folds into a saddle-shape structure; the two halves of the molecule exhibit an overall dyad symmetry but are not identical. Like the HMGI and other DNA-bending proteins that participate in formation of enhancesomes, TBP interacts with the minor groove in DNA, bending the helix considerably (see Figure 4-5). The DNA-binding surface of TBP is conserved in all eukaryotes, explaining the high conservation of the TATA-box promoter element (see Figure 11-9).

Once TBP has bound to the TATA box, TFIIB can bind. TFIIB is a monomeric protein, slightly smaller than TBP. The C-terminal domain of TFIIB makes contact with both TBP and DNA on either side of the TATA-box, while its N-terminal domain extends toward the transcription start site. Following TFIIB binding, a preformed complex of tetrameric TFIIF and Pol II binds, positioning the polymerase over the start site. At most promoters, two more general transcription factors must bind before the DNA duplex can be separated to expose the template strand. First to bind is tetrameric TFIIE, creating a docking site for TFIIH, another multimeric factor containing nine subunits. Binding of TFIIH completes assembly of the transcription preinitiation complex in vitro (see Figure 11-27).

The **helicase** activity of one of the TFIIH subunits uses energy from ATP hydrolysis to unwind the DNA duplex at the start site, allowing Pol II to form an *open* complex in which the DNA duplex surrounding the start site is melted and the template strand is bound at the polymerase active site. If the remaining ribonucleoside triphosphates are present, Pol II begins transcribing the template strand. As the polymerase transcribes away from the promoter region, another subunit of TFIIH phosphorylates the Pol II CTD at multiple sites (see Figure 11-27). In the minimal in vitro transcription assay containing only these general transcription factors and purified RNA polymerase II, TBP remains bound

to the TATA box as the polymerase transcribes away from the promoter region, but the other general transcription factors dissociate.

 The first subunits of TFIIH to be cloned from humans were identified because mutations in the genes encoding them cause defects in the repair of damaged DNA. In normal individuals, when a transcribing RNA polymerase becomes stalled at a region of damaged template DNA, a subcomplex of TFIIH is thought to recognize the stalled polymerase and then recruit other proteins that function with TFIIH in repairing the damaged DNA region. In the absence of functional TFIIH, such repair of damaged DNA in transcriptionally active genes is impaired. As a result, affected individuals have extreme skin sensitivity to sunlight (a common cause of DNA damage) and exhibit a high incidence of cancer. Depending on the severity of the defect in TFIIH function, these individuals may suffer from diseases such as xeroderma pigmentosum and Cockayne's syndrome (Chapter 23). ∎

In Vivo Transcription Initiation by Pol II Requires Additional Proteins

Although the general transcription factors discussed above allow Pol II to initiate transcription in vitro, another general transcription factor, TFIIA, is required for initiation by Pol II in vivo. Purified TFIIA forms a complex with TBP and TATA-box DNA. X-ray crystallography of this complex shows that TFIIA interacts with the side of TBP that is upstream from the direction of transcription. Biochemical experiments suggest that in cells of higher eukaryotes TFIIA and TFIID, with its multiple TAF subunits, bind first to TATA-box DNA and then the other general transcription factors subsequently bind as indicated in Figure 11-27.

The TAF subunits of TFIID appear to play a role in initiating transcription from promoters that lack a TATA box. For instance, some TAF subunits contact the initiator element in promoters where it occurs, probably explaining how such sequences can replace a TATA box. Additional TFIID TAF subunits can bind to a consensus sequence A/G-G-A/T-C-G-T-G centered ≈30 base pairs downstream from the tran-

◄ **FIGURE 11-27 In vitro assembly of RNA polymerase II preinitiation complex.** The indicated general transcription factors and purified RNA polymerase II (Pol II) bind sequentially to TATA-box DNA to form a preinitiation complex. ATP hydrolysis then provides the energy for unwinding of DNA at the start site by a TFIIH subunit. As Pol II initiates transcription in the resulting open complex, the polymerase moves away from the promoter and its CTD becomes phosphorylated. In vitro, the general transcription factors (except for TBP) dissociate from the TBP-promoter complex, but it is not yet known which factors remain associated with promoter regions following each round of transcription initiation in vivo.

scription start site in many genes that lack a TATA-box promoter. Because of its position, this regulatory sequence is called the *downstream promoter element (DPE)*. The DPE facilitates transcription of TATA-less genes that contain it by increasing TFIID binding.

In addition to general transcription factors, specific activators and repressors regulate transcription of genes by Pol II. In the next section we will examine how these regulatory proteins influence Pol II transcription initiation.

KEY CONCEPTS OF SECTION 11.4

Transcription Initiation by RNA Polymerase II

■ Transcription of protein-coding genes by Pol II can be initiated in vitro by sequential binding of the following in the indicated order: TBP, which binds to TATA-box DNA; TFIIB; a complex of Pol II and TFIIF; TFIIE; and finally TFIIH (see Figure 11-27).

■ The helicase activity of a TFIIH subunit separates the template strands at the start site in most promoters, a process that requires hydrolysis of ATP. As Pol II begins transcribing away from the start site, its CTD is phosphorylated by another TFIIH subunit.

■ In vivo transcription initiation by Pol II also requires TFIIA and, in metazoans, a complete TFIID protein, including its multiple TAF subunits as well as the TBP subunit.

11.5 Molecular Mechanisms of Transcription Activation and Repression

The activators and repressors that bind to specific sites in DNA and regulate expression of the associated protein-coding genes do so by two general mechanisms. First, these regulatory proteins act in concert with other proteins to modulate chromatin structure, thereby influencing the ability of general transcription factors to bind to promoters. Recall from Chapter 10 that the DNA in eukaryotic cells is not free, but is associated with a roughly equal mass of protein in the form of **chromatin**. The basic structural unit of chromatin is the **nucleosome**, which is composed of ≈147 base pairs of DNA wrapped tightly around a disk-shaped core of **histone** proteins. Residues within the N-terminal region of each histone, and the C-terminal region of histone H2A, called *histone tails*, extend from the surface of the nucleosome, and can be reversibly modified (see Figure 10-20). Such modifications, especially the acetylation of histone H3 and H4 tails, influence the relative condensation of chromatin and thus its accessibility to proteins required for transcription initiation. In addition to their role in such chromatin-mediated transcriptional control, activators and repressors interact with a large multiprotein complex called the *mediator of transcription complex*, or simply **mediator**.

This complex in turn binds to Pol II and directly regulates assembly of transcription preinitiation complexes.

In this section, we review current understanding of how activators and repressors control chromatin structure and preinitiation complex assembly. In the next section of the chapter, we discuss how the concentrations and activities of activators and repressors themselves are controlled, so that gene expression is precisely attuned to the needs of the cell and organism.

Formation of Heterochromatin Silences Gene Expression at Telomeres, near Centromeres, and in Other Regions

For many years it has been clear that inactive genes in eukaryotic cells are often associated with **heterochromatin**, regions of chromatin that are more highly condensed and stain more darkly with DNA dyes than **euchromatin**, where most transcribed genes are located (see Figure 5-25). Regions of chromosomes near the centromeres and telomeres and additional specific regions that vary in different cell types are organized into heterochromatin. The DNA in heterochromatin is less accessible to externally added proteins than DNA in euchromatin, and consequently often referred to as "closed" chromatin. For instance, in an experiment described in the last chapter, the DNA of inactive genes was found to be far more resistant to digestion by DNase I than the DNA of transcribed genes (see Figure 10-22).

Study of DNA regions in *S. cerevisiae* that behave like the heterochromatin of higher eukaryotes provided early insight into the *chromatin-mediated repression* of transcription. This yeast can grow either as haploid or diploid cells. Haploid cells exhibit one of two possible mating types, called **a** and α. Cells of different mating type can "mate," or fuse, to generate a diploid cell (see Figure 1-5). When a haploid cell divides by budding, the larger "mother" cell switches its mating type (see Figure 22-21). Genetic and molecular analyses have revealed that three genetic loci on yeast chromosome III control the mating type of yeast cells (Figure 11-28). Only the central mating-type locus, termed *MAT*, is actively transcribed. How the proteins encoded at the *MAT* locus determine whether a cell has the **a** or α phenotype is explained in Chapter 22. The two additional loci, termed *HML* and *HMR*, near the left and right telomere, respectively, contain "silent" (nontranscribed) copies of the **a** or α genes. These sequences are transferred alternately from *HMLα* or *HMRa* into the *MAT* locus by a type of nonreciprocal recombination between sister chromatids during cell division. When the *MAT* locus contains the DNA sequence from *HMLα*, the cells behave as α cells. When the *MAT* locus contains the DNA sequence from *HMRa*, the cells behave like **a** cells.

Our interest here is how transcription of the silent mating-type loci at *HML* and *HMR* are repressed. If the genes at these loci are expressed, as they are in yeast mutants with defects in the repressing mechanism, both **a** and α proteins are expressed, causing the cells to behave like diploid cells,

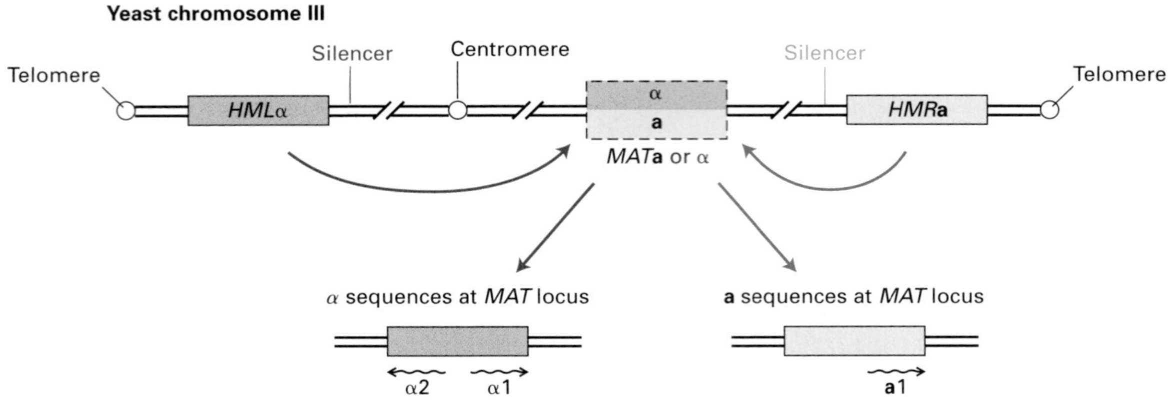

Yeast chromosome III

▲ **FIGURE 11-28 Arrangement of mating-type loci on chromosome III in the yeast *S. cerevisiae*.** Silent (unexpressed) mating-type genes (either **a** or α, depending on the strain) are located at the *HML* locus. The opposite mating-type genes are present at the silent *HMR* locus. When the α or **a** sequences are present at the *MAT* locus, they can be transcribed into mRNAs whose encoded proteins specify the mating-type phenotype of the cell. The silencer sequences near *HML* and *HMR* bind proteins that are critical for repression of these silent loci. Haploid cells can switch mating types in a process that transfers the DNA sequence from *HML* or *HMR* to the transcriptionally active *MAT* locus.

which cannot mate. The promoters and UASs controlling transcription of the **a** and α genes lie near the center of the DNA sequence that is transferred and are identical whether the sequences are at the *MAT* locus or at one of the silent loci. Consequently, the function of the transcription factors that interact with these sequences is somehow blocked at *HML* and *HMR*. This repression of the silent loci depends on **silencer sequences** located next to the region of transferred DNA at *HML* and *HMR* (see Figure 11-28). If the silencer is deleted, the adjacent silent locus is transcribed. Remarkably, any gene placed near the yeast mating-type silencer sequence by recombinant DNA techniques is repressed, or "silenced," even a tRNA gene transcribed by RNA polymerase III, which uses a different set of general transcription factors than RNA polymerase II uses.

Several lines of evidence indicate that repression of the *HML* and *HMR* loci results from a condensed chromatin structure that sterically blocks transcription factors from interacting with the DNA. In one telling experiment, the gene encoding an *E. coli* enzyme that methylates adenine residues in GATC sequences was introduced into yeast cells under the control of a yeast promoter so that the enzyme was expressed. Researchers found that GATC sequences within the *MAT* locus and most other regions of the genome in these cells were methylated, but not those within the *HML* and *HMR* loci. These results indicate that the DNA of the silent loci is inaccessible to the *E. coli* methylase and presumably to proteins in general, including transcription factors and RNA polymerase. Similar experiments conducted with various yeast histone mutants indicated that specific interactions involving the histone tails of H3 and H4 are required for formation of a fully repressing chromatin structure. Other studies have shown that the telomeres of every yeast chromosome also behave like silencer sequences. For instance, when a gene is placed within a few kilobases of any yeast telomere, its expression is repressed. In addition, this repression is relieved by the same mutations in the H3 and H4 histone tails that interfere with repression at the silent mating-type loci.

Genetic studies led to identification of several proteins, RAP1 and three SIR proteins, that are required for repression of the silent mating-type loci and the telomeres in yeast. RAP1 was found to bind within the DNA silencer sequences associated with *HML* and *HMR* and to a sequence that is repeated multiple times at each yeast chromosome telomere. Further biochemical studies showed that the SIR proteins bind to one another and that two bind to the N-terminal tails of histones H3 and H4 that are maintained in a largely unacetylated state by the deacetylase activity of SIR2. Several experiments using fluorescence confocal microscopy of yeast cells either stained with fluorescent-labeled antibody to any one of the SIR proteins or RAP1 or hybridized to a labeled telomere-specific DNA probe revealed that these proteins form large, condensed telomeric nucleoprotein structures resembling the heterochromatin found in higher eukaryotes (Figure 11-29). Figure 11-30 depicts a model for the chromatin-mediated silencing at yeast telomeres based on these and other studies. Formation of heterochromatin at telomeres is nucleated by multiple RAP1 proteins bound to repeated sequences in a nucleosome-free region at the extreme end of a telomere. A network of protein-protein interactions involving telomere-bound RAP1, three SIR proteins (2, 3, and 4), and hypo-acetylated histones H3 and H4 creates a stable, higher-order nucleoprotein complex that includes several telomeres and in which the DNA is largely inaccessible to external proteins. One additional protein, SIR1, is also required for silencing of the silent mating-type loci. Although the function of SIR1 is not yet well understood, it is known

(a) Nuclei and telomeres   (b) Telomeres   (c) SIR3 protein

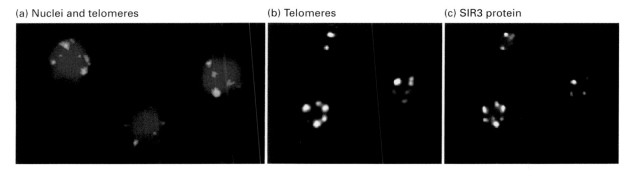

▲ **EXPERIMENTAL FIGURE 11-29 Antibody and DNA probes colocalize SIR3 protein with telomeric heterochromatin in yeast nuclei.** (a) Confocal micrograph 0.3 μm thick through three diploid yeast cells, each containing 68 telomeres. Telomeres were labeled by hybridization to a fluorescent telomere-specific probe (yellow). DNA was stained red to reveal the nuclei. The 68 telomeres coalesce into a much smaller number of regions near the nuclear periphery. (b, c) Confocal micrographs of yeast cells labeled with a telomere-specific hybridization probe (b) and a fluorescent-labeled antibody specific for SIR3 (c). Note that SIR3 is localized in the repressed telomeric heterochromatin. Similar experiments with RAP1, SIR2, and SIR4 have shown that these proteins also colocalize with the repressed telomeric heterochromatin. [From M. Gotta et al., 1996, *J. Cell Biol.* **134**:1349; courtesy of M. Gotta, T. Laroche, and S. M. Gasser.]

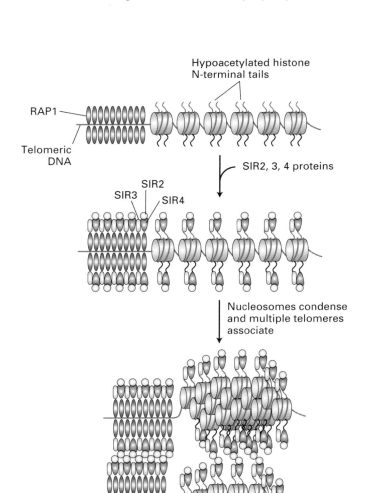

◄ **FIGURE 11-30 Schematic model of silencing mechanism at yeast telomeres.** Multiple copies of RAP1 bind to a simple repeated sequence at each telomere region, which lacks nucleosomes (*top*). This nucleates the assembly of a multiprotein complex (*bottom*) through protein-protein interactions between RAP1, SIR2, SIR3, SIR4, and the hypoacetylated N-terminal tails of histones H3 and H4 of nearby nucleosomes. SIR2 deacetylates the histone tails. The heterochromatin structure at each telomere encompasses ≈4 kb of DNA neighboring the RAP1-binding sites, irrespective of its sequence. Association of several condensed telomeres forms higher-order heterochromatin complexes, such as those shown in Figure 11-29, that sterically block other proteins from interacting with the DNA. See the text for more details. [Adapted from M. Grunstein, 1997, *Curr. Opin. Cell Biol.* **9**:383.]

to associate with the silencer region, where it is thought to cause further assembly of the multiprotein telomeric silencing complex so that it spreads farther from the end of the chromosome, encompassing *HML* and *HMR*.

An important feature of this model is the dependence of silencing on hypoacetylation of the histone tails. This was shown in experiments with yeast mutants expressing histones in which lysines in histone N-termini were substituted with arginines or glutamines. Arginine is positively charged like lysine, but cannot be acetylated. It is thought to function in histone N-terminal tails like an unacetylated lysine. Glutamine on the other hand simulates an acetylated lysine. Repression at telomeres and at the silent mating-type loci was defective in the mutants with glutamine substitutions, but not in mutants with arginine substitutions. Hyperacetylation of the H3 and H4 tails subsequently was found to interfere with binding by SIR3 and SIR4.

Although chromatin-mediated repression of transcription is also important in multicellular eukaryotes, the mechanism of this repression is still being worked out. Genetic and biochemical analyses in *Drosophila* have revealed that multiple proteins associated in large multiprotein complexes participate in the process. These *Polycomb complexes* are discussed further in Chapter 15. As in the case of SIR proteins binding to yeast telomeres (see Figure 11-29), these *Drosophila* Polycomb proteins can be visualized binding to the genes they repress at multiple, specific locations in the genome by in situ binding of specific-labeled antibodies to salivary gland polytene chromosomes.

Repressors Can Direct Histone Deacetylation at Specific Genes

The importance of *histone deacetylation* in chromatin-mediated gene repression has been further supported by studies of eukaryotic repressors that regulate genes at internal chromosomal positions. These proteins are now known to act in part by causing deacetylation of histone tails in nucleosomes that bind to the TATA box and promoter-proximal region of the genes they repress. In vitro studies have shown that when promoter DNA is assembled onto a nucleosome with unacetylated histones, the general transcription factors cannot bind to the TATA box and initiation region. In un-

acetylated histones, the N-terminal lysines are positively charged and interact strongly with DNA phosphates. The unacetylated histone tails also interact with neighboring histone octamers, favoring the folding of chromatin into condensed, higher-order structures whose precise conformation is not well understood. The net effect is that general transcription factors cannot assemble into a preinitiation complex on a promoter associated with hypoacetylated histones. In contrast, binding of general transcription factors is repressed much less by histones with hyperacetylated tails in which the positively charged lysines are neutralized and electrostatic interactions with DNA phosphates are eliminated.

The connection between histone deacetylation and repression of transcription at nearby yeast promoters became clearer when the cDNA encoding a human *histone deacetylase* was found to have high homology to the yeast *RPD3* gene, known to be required for the normal repression of a number of yeast genes. Further work showed that RPD3 protein has histone deacetylase activity. The ability of RPD3 to deacetylate histones at a number of promoters depends on two other proteins: UME6, a repressor that binds to a specific upstream regulatory sequence (URS1), and SIN3, which is part of a large, multiprotein complex that also contains

▶ **EXPERIMENTAL FIGURE 11-31**

The chromatin immunoprecipitation method can reveal the acetylation state of histones in chromatin. Histones are lightly cross-linked to DNA in vivo using a cell-permeable, reversible, chemical cross-linking agent. Nucleosomes with acetylated histone tails are shown in green. Step **1**: Cross-linked chromatin is then isolated and sheared to an average length of two to three nucleosomes. Step **2**: An antibody against a particular acetylated histone tail sequence is added, and (step **3**) bound nucleosomes are immunoprecipitated. Step **4**: DNA in the immunoprecipitated chromatin fragments is released by reversing the cross-link and then is quantitated using a sensitive PCR method. The method can be used to analyze the in vivo association of any protein with a specific sequence of DNA by using an antibody against the protein of interest in step **2**. [See S. E. Rundlett et al., 1998, *Nature* **392**:831.]

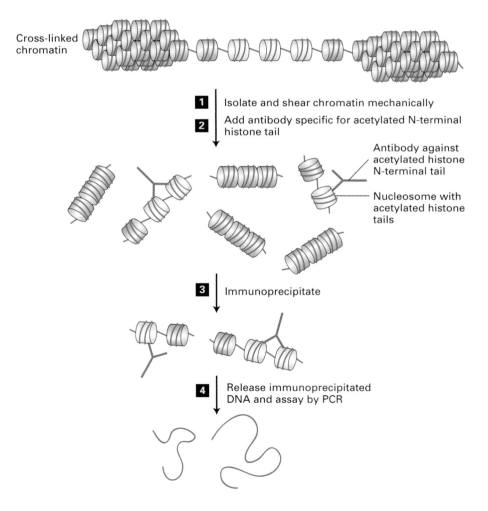

Cross-linked chromatin

1 Isolate and shear chromatin mechanically

2 Add antibody specific for acetylated N-terminal histone tail

Antibody against acetylated histone N-terminal tail

Nucleosome with acetylated histone tails

3 Immunoprecipitate

4 Release immunoprecipitated DNA and assay by PCR

(a) Repressor-directed histone deacetylation

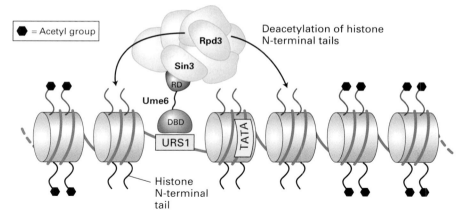

(b) Activator-directed histone hyperacetylation

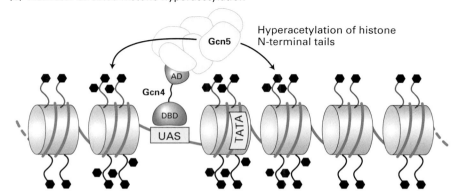

◀ **FIGURE 11-32 Proposed mechanism of histone deacetylation and hyperacetylation in yeast transcription control.** (a) Repressor-directed deacetylation of histone N-terminal tails. The DNA-binding domain (DBD) of the repressor UME6 interacts with a specific upstream control element (URS1) of the genes it regulates. The UME6 repression domain (RD) binds SIN3, a subunit of a multiprotein complex that includes RPD3, a histone deacetylase. Deacetylation of histone N-terminal tails on nucleosomes in the region of the UME6-binding site inhibits binding of general transcription factors at the TATA box, thereby repressing gene expression. (b) Activator-directed hyperacetylation of histone N-terminal tails. The DNA-binding domain of the activator GCN4 interacts with specific upstream activating sequences (UAS) of the genes it regulates. The GCN4 activation domain (AD) then interacts with a multiprotein histone acetylase complex that includes the GCN5 catalytic subunit. Subsequent hyperacetylation of histone N-terminal tails on nucleosomes in the vicinity of the GCN4-binding site facilitates access of the general transcription factors required for initiation. Repression and activation of many genes in higher eukaryotes occurs by similar mechanisms.

RPD3. SIN3 also binds to the repression domain of UME6, thus positioning the RPD3 histone deacetylase in the complex so it can interact with nearby promoter-associated nucleosomes and remove acetyl groups from histone N-terminal lysines. Additional experiments, using the *chromatin immunoprecipitation* technique outlined in Figure 11-31, demonstrated that in wild-type yeast, one or two nucleosomes in the immediate vicinity of UME6-binding sites are hypoacetylated. These DNA regions include the promoters of genes repressed by UME6. In *sin3* and *rpd3* deletion mutants, not only were these promoters derepressed, but the nucleosomes near the UME6-binding sites were hyperacetylated.

All these findings provide considerable support for the model of repressor-directed deacetylation shown in Figure 11-32a. In this model, the SIN3-RPD3 complex functions as a co-repressor. Co-repressor complexes containing histone deacetylases also have been found associated with many repressors from mammalian cells. Some of these complexes contain the mammalian homolog of SIN3 (mSin3), which interacts with the repressor protein. Other histone deacetylase complexes identified in mammalian cells appear to contain additional or different repressor-binding proteins. These various repressor and co-repressor combinations are thought to mediate histone deacetylation at specific promoters by a mechanism similar to the yeast mechanism (see

Figure 11-32a). However, the observation that a number of eukaryotic repressor proteins inhibit in vitro transcription in the absence of histones indicates that more direct repression mechanisms, not involving histone deacetylation, also operate.

The discovery of mSin3-containing histone deacetylase complexes provides an explanation for earlier observations that in vertebrates transcriptionally inactive DNA regions often contain the modified cytidine residue *5-methylcytidine* (mC) followed immediately by a G, whereas transcriptionally active DNA regions lack mC residues. DNA containing 5-methylcytidine has been found to bind a specific protein that in turn interacts specifically with mSin3. This finding suggests that association of mSin3-containing co-repressors with methylated sites in DNA leads to deacetylation of histones in neighboring nucleosomes, making these regions inaccessible to general transcription factors and Pol II, and hence transcriptionally inactive.

Activators Can Direct Histone Acetylation at Specific Genes

Genetic and biochemical studies in yeast led to discovery of a large multiprotein complex containing the protein GCN5, which has histone acetylase activity. Another subunit of this histone acetylase complex binds to acidic activation domains in yeast activator proteins such as GCN4. Maximal transcription activation by GCN4 depends on these histone acetylase complexes, which thus function as co-activators. The model shown in Figure 11-32b is consistent with the observation that nucleosomes near the promoter region of a gene regulated by the GCN4 activator are specifically hyperacetylated, as determined by the chromatin immunoprecipitation method. The activator-directed hyperacetylation of nucleosomes near a promoter region changes ("opens") the chromatin structure so as to facilitate the binding of other proteins required for transcription initiation.

A similar activation mechanism operates in higher eukaryotes. For example, mammals express two related ≈400-kD, multidomain proteins called *CBP* and *P300*, which are thought to function similarly. As noted earlier, one domain of CBP binds the phosphorylated acidic activation domain in the CREB transcription factor. Other domains of CBP interact with different activation domains in other transcription factors. Yet another domain of CBP has histone acetylase activity, and another CBP domain associates with a multiprotein histone acetylase complex that is homologous to the yeast GCN5-containing complex. CREB and many other mammalian activators are thought to function in part by directing CBP and the associated histone acetylase complex to specific nucleosomes, where they acetylate histone tails, facilitating the interaction of general transcription factors with promoter DNA. In addition, the largest TFIID subunit also has histone acetylase activity and may function as a co-activator by acetylating histone N-terminal tails in the vicinity of the TATA box.

Modifications of Specific Residues in Histone Tails Control Chromatin Condensation

In addition to reversible acetylation, histone tails in chromatin can undergo reversible phosphorylation of serine and threonine residues, reversible monoubiquitination of a lysine residue in the H2A C-terminal tail, and irreversible methylation of lysine residues. Evidence is accumulating that it is not simply the overall level of histone acetylation that controls the condensation of chromatin and hence the accessibility of DNA. Rather the precise amino acids in the tails that are acetylated or otherwise modified may constitute a "*histone code*" that helps control the condensation of chromatin (Figure 11-33). For instance, the lysine at position 9 in histone H3 often is methylated in heterochromatin.

The histone code is "read" by proteins that bind to these specific modifications and in turn promote condensation or decondensation of chromatin, forming "closed" or "open" chromatin structures. For example, higher eukaryotes express a number of heterochromatin-associated proteins containing a so-called *chromodomain*, which binds to the histone H3 tail when it is methylated at lysine 9. These proteins are postulated to contribute to the higher-order folding characteristic of heterochromatin, somewhat like the SIR proteins at yeast telomeres (see Figure 11-30). Alternatively, the *bromodomain* found in a number of euchromatin-associated proteins binds to acetylated histone tails. The largest subunit of TFIID, for example, contains two closely spaced bromodomains, which may help it to associate with chromatin containing an active code, while the histone acetylase activity of this same subunit maintains the chromatin in a hyperacetylated state.

Chromatin-Remodeling Factors Help Activate or Repress Some Genes

In addition to histone acetylase complexes, another type of multiprotein complex, called the *SWI/SNF chromatin-remodeling complex*, is required for activation at some yeast promoters. Several of the SWI/SNF subunits have homology to DNA helicases, enzymes that use energy from ATP hydrolysis to disrupt interactions between base-paired nucleic acids or between nucleic acids and proteins. The SWI/SNF complex is thought to transiently dissociate DNA from the surface of nucleosomes, permitting nucleosomes to "slide" along the DNA and promoting the unfolding of condensed,

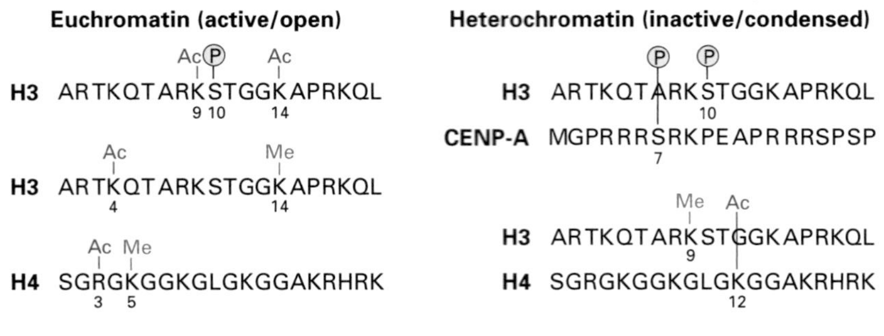

Euchromatin (active/open)

H3 ARTKQTARKSTGGKAPRKQL
      Ac P   Ac
      9 10   14

H3 ARTKQTARKSTGGKAPRKQL
      Ac   Me
      4   14

H4 SGRGKGGKGLGKGGAKRHRK
      Ac Me
      3 5

Heterochromatin (inactive/condensed)

H3 ARTKQTARKSTGGKAPRKQL
          P P
          10

CENP-A MGPRRRSRKPEAPRRRSPSP
               7

H3 ARTKQTARKSTGGKAPRKQL
              Me Ac
              9

H4 SGRGKGGKGLGKGGAKRHRK
                   12

▲ **FIGURE 11-33 Examples of the histone code.** Specific post-translational modifications of the N-terminal tails in histones H3 and H4 are found in euchromatin, which is accessible to proteins and transcriptionally active. Different modifications are found in heterochromatin, which is condensed and thus largely inaccessible to proteins and transcriptionally inactive. Histone tail sequences are shown in the one-letter amino acid code. CENP-A is a variant form for H3 found in nucleosomes associated with the centromeres of mammalian chromosomes. [Adapted from T. Jenuwein and C. D. Allis, 2001, *Science* **293**:1074.]

(a)   (b)

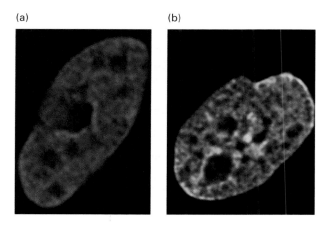

▲ **EXPERIMENTAL FIGURE 11-34 Expression of fusion proteins demonstrates chromatin decondensation in response to an activation domain.** A cultured hamster cell line was engineered to contain multiple copies of a tandem array of *E. coli lac operator* sequences integrated into a chromosome in a region of heterochromatin. (a) When an expression vector for the *lac* repressor was transfected into these cells, *lac* repressors bound to the *lac* operator sites could be visualized in a region of condensed chromatin using an antibody against the *lac* repressor (red). DNA was visualized by staining with DAPI (blue), revealing the nucleus. (b) When an expression vector for the *lac* repressor fused to an activation domain was transfected into these cells, staining as in (a) revealed that the activation domain causes this region of chromatin to decondense into a thinner chromatin fiber that fills a much larger volume of the nucleus. Bar = 1 μm. [Courtesy of Andrew S. Belmont, 1999, *J. Cell Biol.* **145**:1341.]

higher-order chromatin structures. The net result of such chromatin remodeling is to facilitate the binding of transcription factors to DNA in chromatin. Some activation domains have been shown to bind to the SWI/SNF complex, and this binding stimulates in vitro transcription from chromatin templates (DNA bound to nucleosomes). Thus the SWI/SNF complex represents another type of co-activator complex. Other multi-protein complexes with similar chromatin-remodeling activities have been identified in yeast, raising the possibility that different chromatin-remodeling complexes may be required by distinct families of activators (see chapter opening figure).

Higher eukaryotes also contain multiprotein complexes with homology to the yeast SWI/SNF complex. These complexes isolated from nuclear extracts of mammalian and *Drosophila* cells have been found to assist binding of transcription factors to their cognate sites in nucleosomal DNA in an ATP-requiring process. The experiment shown in Figure 11-34 dramatically demonstrates how an activation domain can cause decondensation of a region of chromatin. This is thought to result from the interaction of the activation domain with chromatin remodeling and histone acetylase complexes.

Surprisingly, SWI/SNF complexes are also required for the repression of some genes, perhaps because they help expose histone tails to deacetylases or because they assist in the

folding of chromatin into condensed, higher-order structures. Much remains to be learned about how this important class of co-activators and co-repressors alters chromatin structure to influence gene expression.

The Mediator Complex Forms a Molecular Bridge Between Activation Domains and Pol II

Still another type of co-activator, the multiprotein mediator complex, assists more directly in assembly of Pol II preinitiation complexes (Figure 11-35). Some of the ≈20 mediator subunits binds to RNA polymerase II, and other mediator subunits bind to activation domains in various activator proteins. Thus mediator can form a molecular bridge between

(a) **Yeast mediator–Pol II complex**

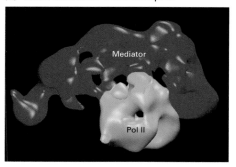

(b) **Human mediator**

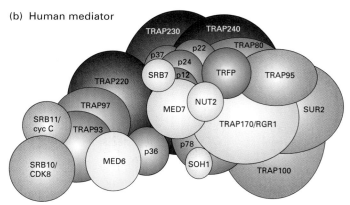

▲ **FIGURE 11-35 Structure of yeast and human mediator complexes.** (a) Reconstructed image of mediator from *S. cerevisiae* bound to Pol II. Multiple electron microscopy images were aligned and computer-processed to produce this average image in which the three-dimensional Pol II structure (light blue) is shown associated with the yeast mediator complex (dark blue). (b) Diagrammatic representation of mediator subunits from human cells. Subunits shown in the same color are thought to form a module. Subunits in orange, yellow, and green are homologous with subunits in the yeast mediator complex. Genetic studies in yeast show that mutations in one of the subunits in a module inhibit the association of other subunits in the same module with the rest of the complex. [Part (a) courtesy of Francisco J. Asturias, 2002, *Mol. Cell* **10**:409. Part (b) adapted from S. Malik and R. G. Roeder, 2000, *Trends Biochem. Sci.* **25**:277.]

an activator bound to its cognate site in DNA and Pol II at a promoter. In addition, one of the mediator subunits has histone acetylase activity and may function to maintain a promoter region in a hyperacetylated state.

Experiments with temperature-sensitive yeast mutants indicate that some mediator subunits are required for transcription of virtually all yeast genes. These subunits most likely help maintain the overall structure of the mediator complex or bind to Pol II and therefore are required for activation by all activators. In contrast, other mediator subunits are required for activation of specific subsets of genes. DNA microarray analysis of gene expression in mutants with defects in these mediator subunits indicates that each such subunit influences transcription of ≈3–10 percent of all genes (see Figure 9-35 for DNA microarray technique). These mediator subunits are thought to interact with specific activation domains; thus when one subunit is defective, transcription of genes regulated by activators that bind to that subunit is severely depressed but transcription of other genes is unaffected. Consistent with this explanation are binding studies showing that some activation domains do indeed interact with specific mediator subunits.

Large mediator complexes, isolated from cultured mammalian cells, are required for mammalian activators to stimulate transcription by Pol II in vitro. Since genes encoding homologs of the mammalian mediator subunits have been identified in the genome sequences of *C. elegans* and *Drosophila,* it appears that most multicellular animals (metazoans) have homologous mediator complexes. About one third of the metazoan mediator subunits are clearly homologous to yeast mediator subunits (see Figure 11-35b). But the remaining subunits, which appear to be distinct from any yeast proteins, may interact with activation domains that

are not found in yeast. As with yeast mediator, some of the mammalian mediator subunits have been shown to interact with specific activation domains. For example, the Sur2 subunit of mammalian mediator binds to the activation domain of a TCF transcription factor that controls expression of the *EGR-1* gene (see Figure 11-19). The function of this TCF activator in vivo normally is regulated in response to specific protein hormones present in serum. Mouse embryonic stem cells with a knockout of the *sur2* gene fail to induce expression of EGR-1 protein in response to serum, whereas multiple other activators function normally in the mutant cells. This finding implicates the mediator Sur2 subunit in the activating function of TCF.

The various experimental results indicating that individual mediator subunits bind to specific activation domains suggest that multiple activators influence transcription from a single promoter by interacting with a mediator complex simultaneously (Figure 11-36). Activators bound at enhancers or promoter-proximal elements can interact with mediator associated with a promoter because DNA is flexible and can form a loop bringing the regulatory regions and the promoter close together. Such loops have been observed in experiments with the *E. coli* NtrC activator and σ^{54}-RNA polymerase (see Figure 4-17). The multiprotein nucleoprotein complexes that form on eukaryotic promoters may comprise as many as 100 polypeptides with a total mass of ≈3 megadaltons (MDa), as large as a ribosome.

Transcription of Many Genes Requires Ordered Binding of Activators and Action of Co-Activators

We can now extend the model of Pol II transcription initiation in Figure 11-27 to take into account the role of activators and co-activators. These accessory proteins function not only to make genes within nucleosomal DNA accessible to general transcription factors and Pol II but also directly recruit Pol II to promoter regions.

Recent studies have analyzed the order in which activators bind to a transcription-control region and interact with co-activators as a gene is induced. Such studies show that assembly of preinitiation complexes depends on multiple protein-DNA and protein-protein interactions, as illustrated in Figure 11-37 depicting activation of the yeast *HO* gene. This gene encodes a sequence-specific nuclease that initiates mating-type switching in haploid yeast cells (see Figure 11-28). Activation of the *HO* gene begins with binding of the SWI5 activator to an upstream enhancer. Bound SWI5 then interacts with the SWI/SNF chromatin-remodeling complex and GCN5-containing histone acetylase complex. Once the chromatin in the *HO* control region is decondensed and hyperacetylated, a second activator, SBF, can bind to several sites in the promoter-proximal region. Subsequent binding of the mediator complex by SBF then leads to assembly of the transcription preinitiation complex containing Pol II and the general transcription factors shown in Figure 11-36.

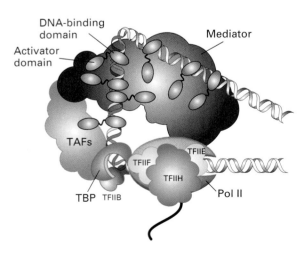

▲ **FIGURE 11-36 Model of several DNA-bound activators interacting with a single mediator complex.** The ability of different mediator subunits to interact with specific activation domains may contribute to the integration of signals from several activators at a single promoter. See the text for discussion.

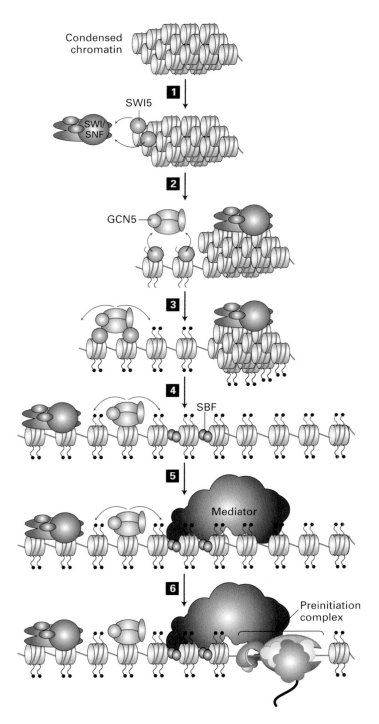

◄ **FIGURE 11-37** Ordered binding and interaction of activators and co-activators leading to transcription of the yeast *HO* gene. Step **1**: Initially, the HO gene is packaged into condensed chromatin. Activation begins when the SWI5 activator binds to enhancer sites 1200–1400 base pairs upstream of the start site and interacts with the SWI/SNF chromatin-remodeling complex. Step **2**: The SWI/SNF complex acts to decondense the chromatin, thereby exposing histone tails. Step **3**: A GCN5-containing histone acetylase complex associates with bound SWI5 and acetylates histone tails in the *HO* locus as SWI/SNF continues to decondense adjacent chromatin. Step **4**: SWI5 is released from the DNA, but the SWI/SNF and GCN5 complexes remain associated with the *HO* control region (in the case of GCN5, by poorly understood interactions). Their action allows the SBF activator to bind several sites in the promoter-proximal region. Step **5**: SBF then binds the mediator complex. Step **6**: Subsequent binding of Pol II and general transcription factors results in assembly of a transcription preinitiation complex whose components are detailed in Figure 11-37. [Adapted from C. J. Fry and C. L. Peterson, 2001, *Curr. Biol.* **11**:R185. See also M. P. Cosma et al., 1999, *Cell* **97**:299, and M. P. Cosma et al., 2001, *Mol. Cell* **7**:1213.]

ated by several activators. This allows genes to be regulated in a cell-type-specific manner by specific combinations of transcription factors. The *TTR* gene, which encodes transthyretin in mammals, is a good example of this. As noted earlier, transthyretin is expressed in hepatocytes and in choroid plexus cells. Transcription of the *TTR* gene in hepatocytes is controlled by at least five different transcriptional activators (Figure 11-38). Even though three of these activators—HNF4, C/EBP, and AP1— are also expressed in cells of the intestine and kidney, *TTR* transcription does not occur in these cells, because all five activators are required but HNF1 and HNF3 are missing. Other hepatocyte-

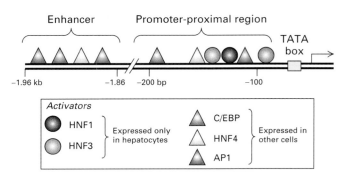

▲ **FIGURE 11-38 Transcription-control region of the mouse transthyretin (*TTR*) gene.** Binding sites for the five activators required for transcription of *TTR* in hepatocytes are indicated. The complete set of activators is expressed at the required concentrations to stimulate transcription only in hepatocytes. A different set of activators stimulates transcription in choroid plexus cells. [See R. Costa et al., 1989, *Mol. Cell Biol.* **9**:1415, and K. Xanthopoulus et al., 1989, *Proc. Nat'l. Acad. Sci. USA* **86**:4117.]

We can now see that the assembly of a preinitiation complex and stimulation of transcription at a promoter results from the interaction of several activators with various multiprotein co-activator complexes. These include chromatin-remodeling complexes, histone acetylase complexes, and a mediator complex. While much remains to be learned about these processes, it is clear that the net result of these multiple molecular events is that activation of transcription at a promoter depends on highly cooperative interactions initi-

specific enhancers and promoter-proximal regions that regulate additional genes expressed only in hepatocytes contain
binding sites for other specific combinations of transcription
factors found only in these cells, together with those expressed ubiquitously.

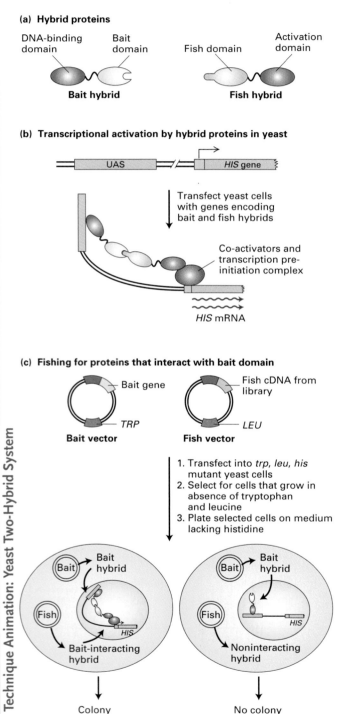

(a) Hybrid proteins

DNA-binding   Bait   Activation
domain   domain   Fish domain domain

Bait hybrid   **Fish hybrid**

(b) Transcriptional activation by hybrid proteins in yeast

UAS   HIS gene

Transfect yeast cells
with genes encoding
bait and fish hybrids

Co-activators and
transcription pre-
initiation complex

HIS mRNA

(c) Fishing for proteins that interact with bait domain

Bait gene   Fish cDNA from
                       library

TRP   LEU

Bait vector   **Fish vector**

1. Transfect into *trp, leu, his*
   mutant yeast cells
2. Select for cells that grow in
   absence of tryptophan
   and leucine
3. Plate selected cells on medium
   lacking histidine

Bait   Bait
Bait   hybrid   Bait   hybrid

Fish   Fish

HIS   HIS

Bait-interacting   Noninteracting
hybrid   hybrid

Colony   No colony
formation   formation

The Yeast Two-Hybrid System Exploits Activator Flexibility to Detect cDNAs That Encode Interacting Proteins

A powerful molecular genetic method called the **yeast two-hybrid system** exploits the flexibility in activator structures to identify genes whose products bind to a specific protein of interest. Because of the importance of protein-protein inter-actions in virtually every biological process, the yeast two-hybrid system is used widely in biological research.

This method employs a yeast vector for expressing a DNA-binding domain and flexible linker region without the associated activation domain, such as the deleted GAL4 con-taining amino acids 1–692 (see Figure 11-17). A cDNA se-quence encoding a protein or protein domain of interest, called the *bait domain*, is fused in frame to the flexible linker region so that the vector will express a hybrid protein com-posed of the DNA-binding domain, linker region, and bait domain (Figure 11-39a, *left*). A cDNA library is cloned into multiple copies of a second yeast vector that encodes a strong activation domain and flexible linker, to produce a vector li-brary expressing multiple hybrid proteins, each containing a different *fish domain* (Figure 11-39a, *right*).

The bait vector and library of fish vectors are then trans-fected into engineered yeast cells in which the only copy of

◀ **EXPERIMENTAL FIGURE 11-39 The yeast two-hybrid system provides a way of screening a cDNA library for clones encoding proteins that interact with a specific protein of interest.** This is a common technique for screening a cDNA library for clones encoding proteins that interact with a specific protein of interest. (a) Two vectors are constructed containing genes that encode hybrid (chimeric) proteins. In one vector (*left*), coding sequence for the DNA-binding domain of a transcription factor is fused to the sequences for a known protein, referred to as the "bait" domain (light blue). The second vector (*right*) expresses an activation domain fused to a "fish" domain (green) that interacts with the bait domain. (b) If yeast cells are transformed with vectors expressing both hybrids, the bait and fish portions of the chimeric proteins interact to produce a functional transcriptional activator. In this example, the activator promotes transcription of a *HIS* gene. One end of this protein complex binds to the upstream activating sequence (UAS) of the *HIS3* gene; the other end, consisting of the activation domain, stimulates assembly of the transcription preinitiation complex (orange) at the promoter (yellow). (c) To screen a cDNA library for clones encoding proteins that interact with a particular bait protein of interest, the library is cloned into the vector encoding the activation domain so that hybrid proteins are expressed. The bait vector and fish vectors contain wild-type selectable genes (e.g., a *TRP* or *LEU* gene). The only transformed cells that survive the indicated selection scheme are those that express the bait hybrid and a fish hybrid that interacts with it. See the text for discussion. [See S. Fields and O. Song, 1989, *Nature* **340**:245.]

a gene required for histidine synthesis (*HIS*) is under control of a UAS with binding sites for the DNA-binding domain of the hybrid bait protein. Transcription of the *HIS* gene requires activation by proteins bound to the UAS. Transformed cells that express the bait hybrid and an *interacting* fish hybrid will be able to activate transcription of the *HIS* gene (Figure 11-39b). This system works because of the flexibility in the spacing between the DNA-binding and activation domains of eukaryotic activators.

A two-step selection process is used (Figure 11-39c). The bait vector also expresses a wild-type *TRP* gene, and the hybrid vector expresses a wild-type *LEU* gene. Transfected cells are first grown in a medium that lacks tryptophan and leucine but contains histidine. Only cells that have taken up the bait vector and one of the fish plasmids will survive in this medium. The cells that survive then are plated on a medium that lacks histidine. Those cells expressing a fish hybrid that does not bind to the bait hybrid cannot transcribe the *HIS* gene and consequently will not form a colony on medium lacking histidine. The few cells that express a bait-binding fish hybrid will grow and form colonies in the absence of histidine. Recovery of the fish vectors from these colonies yields cDNAs encoding protein domains that interact with the bait domain.

KEY CONCEPTS OF SECTION 11.5

Molecular Mechanisms of Transcription Activation and Repression

■ Eukaryotic transcription activators and repressors exert their effects largely by binding to multisubunit co-activators or co-repressors that influence assembly of Pol II transcription preinitiation complexes either by modulating chromatin structure (indirect effect) or by interacting with Pol II and general transcription factors (direct effect).

■ The DNA in condensed regions of chromatin (heterochromatin) is relatively inaccessible to transcription factors and other proteins, so that gene expression is repressed.

■ The interactions of several proteins with each other and with the hypoacetylated N-terminal tails of histones H3 and H4 are responsible for the chromatin-mediated repression of transcription that occurs in the telomeres and the silent mating-type loci in *S. cerevisiae*. (see Figure 11-30).

■ Some repression domains function by interacting with co-repressors that are histone deacetylase complexes. The subsequent deacetylation of histone N-terminal tails in nucleosomes near the repressor-binding site inhibits interaction between the promoter DNA and general transcription factors, thereby repressing transcription initiation (see Figure 11-32a).

■ Some activation domains function by binding multiprotein co-activator complexes such as histone acetylase complexes. The subsequent hyperacetylation of histone N-terminal tails in nucleosomes near the activator-binding site facilitates interactions between the promoter DNA and general transcription factors, thereby stimulating transcription initiation (see Figure 11-32b).

■ SWI/SNF chromatin-remodeling factors constitute another type of co-activator. These multisubunit complexes can transiently dissociate DNA from histone cores in an ATP-dependent reaction and may also decondense regions of chromatin, thereby promoting the binding of DNA-binding proteins needed for initiation to occur at some promoters.

■ Mediator, another type of co-activator, is an ≈20-subunit complex that forms a molecular bridge between activation domains and RNA polymerase II by binding directly to the polymerase and activation domains. By binding to several different activators simultaneously, mediator probably helps integrate the effects of multiple activators on a single promoter (see Figure 11-36).

■ Activators bound to a distant enhancer can interact with transcription factors bound to a promoter because DNA is flexible and the intervening DNA can form a large loop.

■ The highly cooperative assembly of preinitiation complexes in vivo generally requires several activators. A cell must produce the specific set of activators required for transcription of a particular gene in order to express that gene.

■ The yeast two-hybrid system is widely used to detect cDNAs encoding protein domains that bind to a specific protein of interest (see Figure 11-39).

11.6 Regulation of Transcription-Factor Activity

We have seen in the preceding discussion how combinations of activators and repressors that bind to specific DNA regulatory sequences control transcription of eukaryotic genes. Whether or not a specific gene in a multicellular organism is expressed in a particular cell at a particular time is largely a consequence of the concentrations and activities of the transcription factors that interact with the regulatory sequences of that gene. Which transcription factors are expressed in a particular cell type, and the amounts produced, is determined by multiple regulatory interactions between transcription-factor genes that occur during the development and differentiation of a particular cell type. In Chapters 15 and 22, we present examples of such regulatory interactions during development and discuss the principles of development and differentiation that have emerged from these examples.

Not only is the expression of transcription factors by a cell regulated, but the activities of those factors expressed in a particular cell type commonly are further controlled

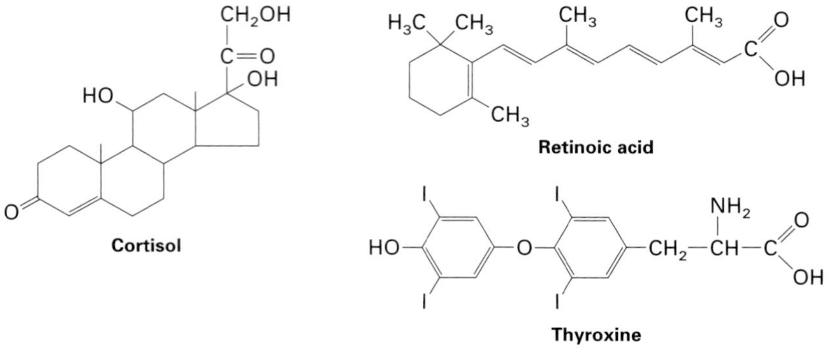

▶ **FIGURE 11-40 Examples of hormones that bind to nuclear receptors.** These and related lipid-soluble hormones bind to receptors located in the cytosol or nucleus. The ligand-receptor complex functions as a transcription activator.

Cortisol

Retinoic acid

Thyroxine

indirectly as the result of interactions between proteins on the surfaces of neighboring cells and by extracellular hormones and growth factors. In multicellular organisms, these latter signaling molecules are secreted from one cell type and affect the function of cells that may be nearby or at a different location in the organism. One major group of extracellular signals comprises peptides and proteins, which bind to receptors in the plasma membrane. Ligand binding to these receptors triggers intracellular **signal-transduction** pathways. In Chapters 14 and 15, we describe the major types of cell-surface receptors and intracellular signaling pathways that regulate transcription-factor activity.

In this section, we discuss the second major group of extracellular signals, the small, lipid-soluble hormones—including many different steroid hormones, retinoids, and thyroid hormones—that can diffuse through plasma and nuclear membranes and interact directly with the transcription factors they control (Figure 11-40). As noted earlier, the intracellular receptors for most of these lipid-soluble hormones,

which constitute the *nuclear-receptor superfamily*, function as transcription activators when bound to their ligands.

All Nuclear Receptors Share a Common Domain Structure

Cloning and sequencing of the genes encoding various nuclear receptors revealed a remarkable conservation in their amino acid sequences and three functional regions (Figure 11-41). All the nuclear receptors have a unique N-terminal region of variable length (100–500 amino acids). Portions of this variable region function as activation domains in some nuclear receptors. The DNA-binding domain maps near the center of the primary sequence and has a repeat of the C_4 zinc-finger motif. The hormone-binding domain, located near the C-terminal end, contains a hormone-dependent activation domain. In some nuclear receptors, the hormone-binding domain functions as a repression domain in the absence of ligand.

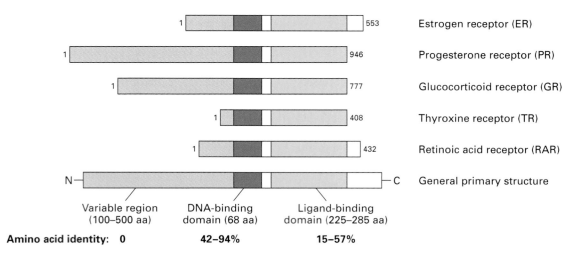

▲ **FIGURE 11-41 General design of transcription factors in nuclear-receptor superfamily.** The centrally located DNA-binding domain exhibits considerable sequence homology among different receptors and contains two copies of the C_4 zinc-finger motif. The C-terminal hormone-binding domain exhibits somewhat less homology. The N-terminal regions in various receptors vary in length, have unique sequences, and may contain one or more activation domains. [See R. M. Evans, 1988, *Science* **240**:889.]

Nuclear-Receptor Response Elements Contain Inverted or Direct Repeats

The characteristic nucleotide sequences of the DNA sites, called *response elements*, that bind several nuclear receptors have been determined. The sequences of the consensus response elements for the glucocorticoid and estrogen receptors are 6-bp inverted repeats separated by any three base pairs (Figure 11-42a, b). This finding suggested that the cognate steroid hormone receptors would bind to DNA as symmetrical dimers, as was later shown from the x-ray crystallographic analysis of the homodimeric glucocorticoid receptor's C_4 zinc-finger DNA-binding domain (see Figure 11-21b).

Some nuclear-receptor response elements, such as those for the receptors that bind vitamin D_3, thyroid hormone, and retinoic acid, are direct repeats of the same sequence recognized by the estrogen receptor, separated by three to five base pairs (Figure 11-42c–e). The specificity for responding to these different hormones by binding distinct receptors is determined by the spacing between the repeats. The receptors that bind to such direct-repeat response elements do so as heterodimers with a common nuclear-receptor monomer called RXR. The vitamin D_3 response element, for example, is bound by the RXR-VDR heterodimer, and the retinoic acid response element is bound by RXR-RAR. The monomers composing these heterodimers interact with each other in such a way that the two DNA-binding domains lie in the same rather than inverted orientation, allowing the RXR heterodimers to bind to direct repeats of the binding site for each monomer. In contrast, the monomers in homodimeric nuclear receptors (e.g., GRE and ERE) have an inverted orientation.

Hormone Binding to a Nuclear Receptor Regulates Its Activity as a Transcription Factor

The mechanism whereby hormone binding controls the activity of nuclear receptors differs for heterodimeric and homodimeric receptors. Heterodimeric nuclear receptors (e.g., RXR-VDR, RXR-TR, and RXR-RAR) are located exclusively in the nucleus. In the absence of their hormone ligand, they repress transcription when bound to their cognate sites in DNA. They do so by directing histone deacetylation at nearby nucleosomes by the mechanism described earlier (see Figure 11-32a). As we saw earlier, in the presence of hormone, the ligand-binding domain of the RAR monomer undergoes a dramatic conformational change compared with the ligand-binding domain of a nuclear receptor in the absence of hormone (see Figure 11-25). In the ligand-bound conformation, heterodimeric nuclear receptors containing RXR can direct hyperacetylation of histones in nearby nucleosomes, thereby reversing the repressing effects of the free ligand-binding domain. In the presence of ligand, ligand-binding domains of nuclear receptors also bind mediator, stimulating preinitiation complex assembly.

In contrast to heterodimeric nuclear receptors, homodimeric receptors are found in the cytoplasm in the absence of their ligands. Hormone binding to these receptors leads to their translocation to the nucleus. The hormone-dependent translocation of the homodimeric glucocorticoid receptor (GR) was demonstrated in the transfection experiments shown in Figure 11-43. The GR hormone-binding domain alone mediates this transport. Subsequent studies showed that, in the absence of hormone, GR is anchored in the cytoplasm as a large protein aggregate complexed with inhibitor proteins, including Hsp90, a protein related to Hsp70, the major heat-shock chaperone in eukaryotic cells. As long as the receptor is confined to the cytoplasm, it cannot interact with target genes and hence cannot activate transcription. Hormone binding to a homodimeric nuclear receptor releases the inhibitor proteins, allowing the receptor to enter the nucleus, where it can bind to response elements associated with target genes (Figure 11-44). Once the receptor with bound hormone binds to a response element, it activates transcription by interacting with chromatin-remodeling and histone acetylase complexes and mediator.

(a) **GRE**   5' AGAACA(N)₃ TGTTCT 3'
             3' TCTTGT(N)₃ ACAAGA 5'

(b) **ERE**   5' AGGTCA(N)₃ TGACCT 3'
             3' TCCAGT(N)₃ ACTGGA 5'

(c) **VDRE** 5' AGGTCA(N)₃ AGGTCA 3'
             3' TCCAGT(N)₃ TCCAGT 5'

(d) **TRE**   5' AGGTCA(N)₄ AGGTCA 3'
             3' TCCAGT(N)₄ TCCAGT 5'

(e) **RARE** 5' AGGTCA(N)₅ AGGTCA 3'
             3' TCCAGT(N)₅ TCCAGT 5'

▲ **FIGURE 11-42 Consensus sequences of DNA response elements that bind three nuclear receptors.** The response elements for the glucocorticoid receptor (GRE) and estrogen receptor (ERE) contain inverted repeats that bind these homodimeric proteins. The response elements for heterodimeric receptors contain a common direct repeat separated by three to five base pairs, for the vitamin D_3 receptor (VDRE), thyroid hormone receptor (TRE), and retinoic acid receptor (RARE). The repeat sequences are indicated by red arrows. [See K. Umesono et al., 1991, *Cell* **65**:1255, and A. M. Naar et al., 1991, *Cell* **65**:1267.]

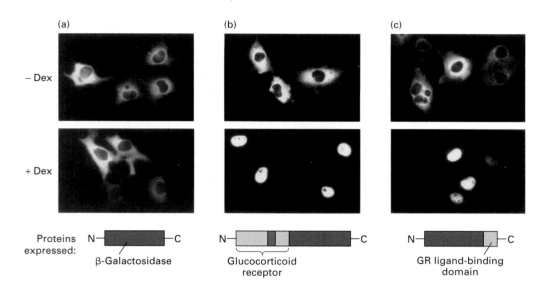

(a)

(b)

(c)

– Dex

+ Dex

Proteins expressed:

N—[⬛]—C
β-Galactosidase

N—[⬛⬛⬛⬛]—C
Glucocorticoid receptor

N—[⬛⬛]—C
GR ligand-binding domain

▲ **EXPERIMENTAL FIGURE 11-43 Fusion proteins from expression vectors demonstrate that the hormone-binding domain of the glucocorticoid receptor (GR) mediates translocation to the nucleus in the presence of hormone.** Cultured animal cells were transfected with expression vectors encoding the proteins diagrammed at the bottom. Immunofluorescence with a labeled antibody specific for β-galactosidase was used to detect the expressed proteins in transfected cells. (a) In cells that expressed β-galactosidase alone, the enzyme was localized to the cytoplasm in the presence and absence of the glucocorticoid hormone dexamethasone (Dex). (b) In cells that expressed a fusion protein consisting of β-galactosidase and the entire glucocorticoid receptor (GR), the fusion protein was present in the cytoplasm in the absence of hormone but was transported to the nucleus in the presence of hormone. (c) Cells that expressed a fusion protein composed of β-galactosidase and just the GR ligand-binding domain (light purple) also exhibited hormone-dependent transport of the fusion protein to the nucleus. [From D. Picard and K. R. Yamamoto, 1987, *EMBO J.* **6**:3333; courtesy of the authors.]

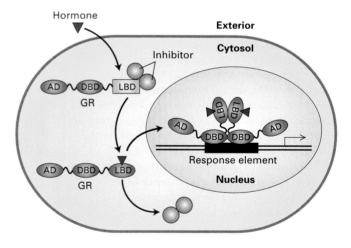

▲ **FIGURE 11-44 Model of hormone-dependent gene activation by a homodimeric nuclear receptor.** In the absence of hormone, the receptor is kept in the cytoplasm by interaction between its ligand-binding domain (LBD) and inhibitor proteins. When hormone is present, it diffuses through the plasma membrane and binds to the ligand-binding domain, causing a conformational change that releases the receptor from the inhibitor proteins. The receptor with bound ligand is then translocated into the nucleus, where its DNA-binding domain (DBD) binds to response elements, allowing the ligand-binding domain and an additional activation domain (AD) at the N-terminus to stimulate transcription of target genes.

KEY CONCEPTS OF SECTION 11.6

Regulation of Transcription-Factor Activity

■ The activities of many transcription factors are indirectly regulated by binding of extracellular proteins and peptides to cell-surface receptors. These receptors activate intracellular signal transduction pathways that regulate specific transcription factors through a variety of mechanisms discussed in Chapters 13 and 14.

■ Nuclear receptors constitute a superfamily of dimeric C_4 zinc-finger transcription factors that bind lipid-soluble hormones and interact with specific response elements in DNA (see Figure 11-41).

■ Hormone binding to nuclear receptors induces conformational changes that modify their interactions with other proteins.

■ Heterodimeric nuclear receptors (e.g., those for retinoids, vitamin D, and thyroid hormone) are found only in the nucleus. In the absence of hormone, they repress transcription of target genes with the corresponding response element. When bound to their ligands, they activate transcription.

■ Steroid hormone receptors are homodimeric nuclear receptors. In the absence of hormone, they are trapped in the cytoplasm by inhibitor proteins. When bound to their ligands, they can translocate to the nucleus and activate transcription of target genes (see Figure 11-44).

11.7 Regulated Elongation and Termination of Transcription

In eukaryotes, the mechanisms for terminating transcription differ for each of the three RNA polymerases. Transcription of pre-rRNA genes by RNA polymerase I is terminated by a mechanism that requires a polymerase-specific termination factor. This DNA-binding protein binds to a specific DNA sequence downstream of the transcription unit. Efficient termination requires that the termination factor bind to the template DNA in the correct orientation. Purified RNA polymerase III terminates after polymerizing a series of U residues. The deoxy$(A)_n$-ribo$(U)_n$ DNA-RNA hybrid that results when a stretch of U's are synthesized is particularly unstable compared with all other base-paired sequences. The ease with which this hybrid can be melted probably contributes to the mechanism of termination by RNA polymerase III.

In most mammalian protein-coding genes transcribed by RNA polymerase II, once the polymerase has transcribed beyond about fifty bases, further elongation is highly processive and does not terminate until after a sequence is transcribed that directs cleavage and polyadenylation of the RNA at the sequence that forms the 3′ end of the encoded mRNA. RNA polymerase II then can terminate at multiple sites located over a distance of 0.5–2 kb beyond this poly(A) addition site. Experiments with mutant genes show that termination is coupled to the process that cleaves and polyadenylates the 3′ end of a transcript, which is discussed in the next chapter. Biochemical and chromatin immunoprecipitation experiments suggest that the protein complex that cleaves and polyadenylates the nascent mRNA transcript at specific sequences associates with the phosphorylated carboxyl-terminal domain (CTD) of RNA polymerase II following initiation (see Figure 11-27). This cleavage/polyadenylation complex may suppress termination by RNA polymerase II until the sequence signaling cleavage and polyadenylation is transcribed by the polymerase.

While transcription termination is unregulated for most genes, for some specific genes, a choice is made between elongation and termination or pausing within a few tens of bases from the transcription start site. This choice between elongation and termination or pausing can be regulated; thus expression of the encoded protein is controlled not only by transcription initiation, but also by control of transcription elongation early in the transcription unit. We discuss two examples of such regulation next.

Transcription of the HIV Genome Is Regulated by an Antitermination Mechanism

Currently, transcription of the human immunodeficiency virus (HIV) genome by RNA polymerase II provides the best-understood example of regulated transcription termination in eukaryotes. Efficient expression of HIV genes requires a small viral protein encoded at the *tat* locus. Cells infected with *tat⁻* mutants produce short viral transcripts that hybridize to restriction fragments containing promoter-proximal regions of the HIV DNA but not to restriction fragments farther downstream from the promoter. In contrast, cells infected with wild-type HIV synthesize long viral transcripts that hybridize to restriction fragments throughout the single HIV transcription unit. Thus Tat protein functions as an *antitermination factor*, permitting RNA polymerase II to read through a transcriptional block. Since antitermination by Tat protein is required for HIV replication, further understanding of this gene-control mechanism may offer possibilities for designing effective therapies for acquired immunodeficiency syndrome (AIDS).

Tat is a sequence-specific RNA-binding protein. It binds to the RNA copy of a sequence called TAR, which is located near the 5′ end of the HIV transcript. The TAR sequence folds into an RNA hairpin with a bulge in the middle of the stem (Figure 11-45). TAR contains two binding sites: one that interacts with Tat and one that interacts with a cellular protein called *cyclin T*. As depicted in Figure 11-45, the HIV Tat protein and cellular cyclin T each bind to TAR RNA and also interact directly with each other so that they bind cooperatively, much like the cooperative binding of DNA-binding transcription factors (see Figure 11-24). Interaction of cyclin T with a protein kinase called CDK9 activates the kinase, whose substrate is the CTD of RNA polymerase II. In vitro transcription studies using a specific inhibitor of CDK9 suggest that RNA polymerase II molecules that initiate transcription on the HIV promoter terminate after transcribing ≈50 bases unless the CTD is hyperphosphorylated by CDK9. Cooperative binding of cyclin T and Tat to the TAR sequence at the 5′ end of the HIV transcript positions CDK9 so that it can phosphorylate the CTD, thereby preventing termination and permitting the polymerase to continue chain elongation.

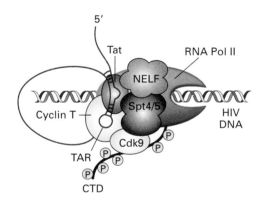

▲ **FIGURE 11-45 Model of antitermination complex composed of HIV Tat protein and several cellular proteins.** The *TAR* element in the HIV transcript contains sequences recognized by Tat and the cellular protein cyclin T. Cyclin T activates and helps position the protein kinase CDK9 near its substrate, the CTD of RNA polymerase II. See the text for a more detailed discussion. [See P. Wei et al., 1998, *Cell* **92**:451; T. Wada et al., 1998, *Genes & Devel.* **12**:357; and Y. Yamaguchi et al., 1999, *Cell* **97**:41.]

Several additional cellular proteins, including Spt4 and Spt5 and the NELF complex, participate in the process by which HIV Tat controls elongation versus termination (see Figure 11-45). Experiments with the specific inhibitor of CDK9 mentioned above and with *spt4* and *spt5* yeast mutants indicate that these cellular proteins are required for transcription elongation beyond ≈50 bases for most cellular genes. But for most genes, these proteins appear to function constitutively, that is, without being regulated. As discussed in Chapter 12, RNA polymerase II pausing instigated by Spt4/5 and NELF is thought to delay elongation until mRNA processing factors associate with the phosphorylated CTD. Further phosphorylation of the CTD by cyclin T–CDK9 (also known as pTEFb) appears to reverse this pause and allow elongation to continue. Currently, it is not clear why this process is not constitutive for the HIV promoter, where cooperative binding of HIV Tat and cyclin T to the TAR RNA sequence is required for efficient elongation.

Promoter-Proximal Pausing of RNA Polymerase II Occurs in Some Rapidly Induced Genes

The *heat-shock genes* (e.g., *hsp70*) illustrate another mechanism for regulating RNA chain elongation in eukaryotes. During transcription of these genes, RNA polymerase II pauses after transcribing ≈25 nucleotides but does not terminate transcription (as it does when transcribing the HIV genome in the absence of Tat protein). The paused polymerase remains associated with the nascent RNA and template DNA, until conditions occur that lead to activation of HSTF (heat-shock transcription factor). Subsequent binding of activated HSTF to specific sites in the promoter-proximal region of heat-shock genes stimulates the paused polymerase to continue chain elongation and promotes rapid re-initiation by additional RNA polymerase II molecules.

The pausing during transcription of heat-shock genes initially was discovered in *Drosophila,* but a similar mechanism most likely occurs in other eukaryotes. Heat-shock genes are induced by intracellular conditions that denature proteins (such as elevated temperature, "heat shock"). Some encode proteins that are relatively resistant to denaturing conditions and act to protect other proteins from denaturation; others are chaperonins that refold denatured proteins (see Chapter 3). The mechanism of transcriptional control that evolved to regulate expression of these genes permits a rapid response: these genes are already paused in a state of suspended transcription and therefore, when an emergency arises, require no time to remodel and acetylate chromatin over the promoter and assemble a transcription preinitiation complex.

KEY CONCEPTS FOR SECTION 11.7

Regulated Elongation and Termination of Transcription

■ Different mechanisms of transcription termination are employed by each of the eukaryotic nuclear RNA polymerases. Transcription of most protein-coding genes is not terminated until an RNA sequence is synthesized that specifies a site of cleavage and polyadenylation.

■ Transcription of the HIV genome by RNA polymerase II is regulated by an antitermination mechanism that requires cooperative binding by the virus-encoded Tat protein and cyclin T to the *TAR* sequence near the 5′ end of the HIV RNA.

■ During transcription of *Drosophila* heat-shock genes, RNA polymerase II pauses within the downstream promoter-proximal region; this interruption in transcription is released when the HSTF transcription factor is activated, resulting in very rapid transcription of the heat-shock genes in response to the accumulation of denatured proteins.

11.8 Other Eukaryotic Transcription Systems

We conclude this chapter with a brief discussion of transcription initiation by the other two eukaryotic nuclear RNA polymerases, Pol I and Pol III, and by the distinct polymerases that transcribe mitochondrial and chloroplast DNA. Although these systems, particularly their regulation, are less thoroughly understood than transcription by RNA polymerase II, they are also fundamental to the life of eukaryotic cells.

Transcription Initiation by Pol I and Pol III Is Analogous to That by Pol II

The formation of transcription-initiation complexes involving Pol I and Pol III is similar in some respects to assembly of Pol II initiation complexes (see Figure 11-27). However, each of the three eukaryotic nuclear RNA polymerases requires its own polymerase-specific general transcription factors and recognizes different DNA control elements. Moreover, neither Pol I nor Pol III requires ATP hydrolysis to initiate transcription, whereas Pol II does.

Transcription initiation by Pol I, which synthesizes pre-rRNA, and by Pol III, which synthesizes tRNAs, 5S rRNA, and other short, stable RNAs, has been characterized most extensively in *S. cerevisiae* using both biochemical and genetic approaches. It is clear that synthesis of tRNAs and of rRNAs, which are incorporated into ribosomes, is tightly coupled to the rate of cell growth and proliferation. However, much remains to be learned about how transcription initiation by Pol I and Pol III is regulated so that synthesis of pre-rRNA, 5S rRNA, and tRNAs is coordinated with the growth and replication of cells.

Initiation by Pol I The regulatory elements directing Pol I initiation are similarly located relative to the transcription start site in both yeast and mammals. A *core element* spanning the transcription start site from −40 to +5 is essential for Pol I transcription. An additional *upstream element* extending from roughly −155 to −60 stimulates in vitro Pol I transcription tenfold.

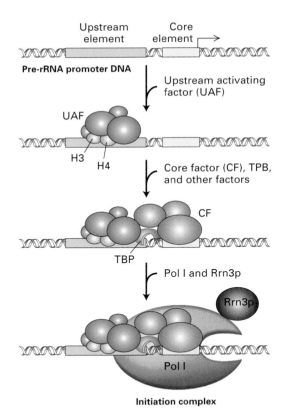

▲ FIGURE 11-46 In vitro assembly of the yeast Pol I transcription initiation complex. UAF and CF, both multimeric general transcription factors, bind to the upstream element (UE) and core element, respectively, in the promoter DNA. TBP and a monomeric factor (Rrn3p) associated with RNA polymerase I (Pol I) also participate in forming the initiation complex. [Adapted from N. Nomura, 1998, in R. M. Paule, *Transcription of Ribosomal RNA Genes by Eukaryotic RNA Polymerase I*, Landes Bioscience, pp. 157–172.]

Assembly of a fully active Pol I initiation complex begins with binding of a multimeric upstream activating factor (UAF) to the upstream element (Figure 11-46). Two of the six subunits composing UAF are histones, which probably participate in DNA binding. Next, a trimeric core factor binds to the core element together with TBP, which makes contact with both the bound UAF and the core factor. Finally, a preformed complex of Pol I and Rrn3p associates with the bound proteins, positioning Pol I near the start site. In human cells, TBP is stably bound to three other polypeptides, forming an initiation factor called *SL1* that binds to the core promoter element and is functionally equivalent to yeast core factor plus TBP.

Initiation by Pol III Unlike protein-coding genes and pre-rRNA genes, the promoter regions of tRNA and 5S-rRNA genes lie entirely within the transcribed sequence (Figure 11-47). Two such *internal* promoter elements, termed the *A box* and *B box*, are present in all tRNA genes. These highly conserved sequences not only function as promoters but also encode two invariant portions of eukaryotic tRNAs that are required for protein synthesis. In 5S-rRNA genes, a single internal control region, the *C box*, acts as a promoter.

Three general transcription factors are required for Pol III to initiate transcription of tRNA and 5S-rRNA genes in vitro. Two multimeric factors, TFIIIC and TFIIIB, participate in initiation at both tRNA and 5S-rRNA promoters; a third factor, TFIIIA, is required for initiation at 5S-rRNA promoters. As with assembly of Pol I and Pol II initiation complexes, the Pol III general transcription factors bind to promoter DNA in a defined sequence.

The N-terminal half of one TFIIIB subunit, called *BRF* (for TFIIB-*related factor*), is similar in sequence to TFIIB (a Pol II factor). This similarity suggests that BRF and TFIIB perform a similar function in initiation, namely, to direct the polymerase to the correct start site. Once TFIIIB has bound to either a tRNA or 5S-rRNA gene, Pol III can bind and initiate transcription in the presence of ribonucleoside triphosphates. The BRF subunit of TFIIIB interacts specifically with one of the polymerase subunits unique to Pol III, accounting for initiation by this specific nuclear RNA polymerase.

Another of the three subunits composing TFIIIB is TBP, which we can now see is a component of a general transcription factor for all three eukaryotic nuclear RNA polymerases. The finding that TBP participates in transcription initiation by Pol I and Pol III was surprising, since the promoters recognized by these enzymes often do not contain TATA boxes. Nonetheless, recent studies indicate that the TBP subunit of TFIIIB interacts with DNA similarly to the way it interacts with TATA boxes.

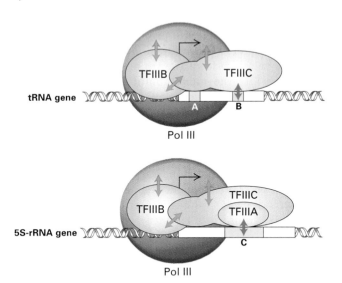

▲ FIGURE 11-47 Transcription-control elements in genes transcribed by RNA polymerase III. Both tRNA and 5S-rRNA genes contain internal promoter elements (yellow) located downstream from the start site and named A-, B-, and C-boxes, as indicated. Assembly of transcription initiation complexes on these genes begins with the binding of Pol III-specific general transcription factors TFIIIA, TFIIIB, and TFIIIC to these control elements. Green arrows indicate strong, sequence-specific protein-DNA interactions. Blue arrows indicate interactions between general transcription factors. Purple arrows indicate interactions between general transcription factors and Pol III. [From L. Schramm and N. Hernandez, 2002, *Genes Dev.* **16**:2593.]

Mitochondrial and Chloroplast DNAs Are Transcribed by Organelle-Specific RNA Polymerases

As discussed in Chapter 10, mitochondria and chloroplasts probably evolved from bacteria that were endocytosed into ancestral cells containing a eukaryotic nucleus. In modern-day eukaryotes, both organelles contain distinct circular DNAs that encode some of the proteins essential to their specific functions. The RNA polymerases that transcribe mitochondrial (mt) DNA and chloroplast DNA are similar to polymerases from bacteria and bacteriophages.

Mitochondrial RNA Polymerase The RNA polymerase that transcribes mtDNA is encoded in nuclear DNA. After synthesis of the enzyme in the cytosol, it is imported into the mitochondrial matrix by mechanisms described in Chapter 16. The mitochondrial RNA polymerases from *S. cerevisiae* and the frog *Xenopus laevis* both consist of a large subunit with ribonucleotide-polymerizing activity and a small subunit, or specificity factor, essential for initiating transcription at the start sites in mtDNA used in the cell. The large subunit of yeast mitochondrial RNA polymerase clearly is related to the monomeric RNA polymerases of bacteriophage T7 and similar bacteriophages. However, the mitochondrial enzyme is functionally distinct from the bacteriophage enzyme in its dependence on the small subunit for transcription from the proper start sites. This small subunit is related to the σ factors in bacterial RNA polymerases, which interact with promoter DNA and function as initiation factors. Thus mitochondrial RNA polymerase appears to be a hybrid of the simple bacteriophage RNA polymerases and the multisubunit bacterial RNA polymerases of intermediate complexity.

The promoter sequences recognized by mitochondrial RNA polymerases include the transcription start site. These promoter sequences, which are rich in A residues, have been characterized in the mtDNA from yeast, plants, and animals. The circular, human mitochondrial genome contains two related 15-bp promoter sequences, one for the transcription of each strand. Each strand is transcribed in its entirety; the long primary transcripts are then processed to yield mitochondrial mRNAs, rRNAs, and tRNAs. A small basic protein called *mtTF1*, which binds immediately upstream from the two mitochondrial promoters, greatly stimulates transcription. A homologous protein found in yeast mitochondria is required for maintenance of mtDNA and probably performs a similar function.

Chloroplast RNA Polymerase In contrast to mitochondrial RNA polymerase, the enzyme that transcribes chloroplast DNA is encoded in the chloroplast genome itself. This RNA polymerase has subunits with considerable homology to the *E. coli* RNA polymerase α, β, and β′ subunits, but apparently lacks a subunit equivalent to the *E. coli* σ factor.

Some chloroplast promoters are quite reminiscent of the *E. coli* σ^{70}-promoter, with similar sequences in the −10 and −35 regions. Transcription from one chloroplast promoter, however, depends on sequences from about −20 to +60, quite different from most *E. coli* promoters. This promoter may be recognized by a second RNA polymerase, which most likely is encoded in the nuclear genome and imported into the organelle. Analysis of chloroplast transcription is still in its infancy, but at this point it is clear that at least one transcription system is highly homologous to transcription in *E. coli* and other bacteria from which chloroplasts evolved.

KEY CONCEPTS FOR SECTION 11.8

Other Eukaryotic Transcription Systems

■ The process of transcription initiation by Pol I and Pol III is similar to that by Pol II but requires different general transcription factors, is directed by different promoter elements, and does not require ATP hydrolysis.

■ Mitochondrial DNA is transcribed by a nuclear-encoded RNA polymerase composed of two subunits. One subunit is homologous to the monomeric RNA polymerase from bacteriophage T7; the other resembles bacterial σ factors.

■ Chloroplast DNA is transcribed by a chloroplast-encoded RNA polymerase homologous to bacterial RNA polymerases, except that it lacks a σ factor.

PERSPECTIVES FOR THE FUTURE

A great deal has been learned in recent years about transcription control in eukaryotes. Genes encoding about 2000 activators and repressors can be recognized in the human genome. We now have a glimpse of how the astronomical number of possible combinations of these transcription factors can generate the complexity of gene control required to produce organisms as remarkable as those we see around us. But very much remains to be understood. While we now have some understanding of what processes turn a gene on and off, we have very little understanding of how the frequency of transcription is controlled in order to provide a cell with the appropriate amounts of its various proteins. In a red blood cell precursor, for example, the globin genes are transcribed at a far greater rate than the genes encoding the enzymes of intermediary metabolism. How are the vast differences in the frequency of transcription initiation at various genes achieved? What happens to the multiple interactions between activation domains, co-activator complexes, general transcription factors, and RNA polymerase II when the polymerase initiates transcription and transcribes away from the promoter region? Do these completely dissociate at promoters that are transcribed infrequently, so that the combination of multiple factors required for transcription must be reassembled anew for each round of transcription? Do complexes of activators with their multiple interacting co-activators remain assembled at promoters from which re-initiation takes place at a high rate, so that the entire assembly does not have to be reconstructed each time a polymerase initiates?

Much remains to be learned about the structure of chromatin and how that structure influences transcription. What directs certain regions of chromatin to form heterochromatin where transcription is repressed? Precisely how is the structure of chromatin changed by activators and repressors and how does this promote or inhibit transcription? Once chromatin-remodeling complexes and histone acetylase complexes become associated with a promoter region, how do they remain associated? Do certain subunits of these complexes associate with modified histone tails so that additional complexes associate along the length of a chromatin fiber as additional histones are modified?

Single activation domains have been discovered to interact with several co-activator complexes. Are these interactions transient, so that the same activation domain can interact with several co-activators sequentially? Is a specific order of co-activator interaction required? How does the interaction of activation domains with mediator stimulate transcription? Do these interactions simply stimulate the assembly of a preinitiation complex, or do they also influence the rate at which RNA polymerase II initiates transcription from an assembled preinitiation complex?

Transcriptional activation is a highly cooperative process so that genes expressed in a specific type of cell are expressed only when the complete set of activators that control that gene are expressed and activated. As mentioned earlier, some of the transcription factors that control expression of the *TTR* gene in the liver are also expressed in intestinal and kidney cells. Yet the *TTR* gene is not expressed in these other tissues, since its transcription requires two additional transcription factors expressed only in the liver. What mechanisms account for this highly cooperative action of transcription factors that is critical to cell-type-specific gene expression?

A thorough understanding of normal development and of abnormal processes associated with disease will require answers to these and many other questions. As further understanding of the principles of transcription control are discovered, applications of the knowledge will likely be made. This understanding may allow fine control of the expression of therapeutic genes introduced by gene therapy vectors as they are developed. Detailed understanding of the molecular interactions that regulate transcription may provide new targets for the development of therapeutic drugs that inhibit or stimulate the expression of specific genes. A more complete understanding of the mechanisms of transcriptional control may allow improved engineering of crops with desirable characteristics. Certainly, further advances in the area of transcription control will help to satisfy our desire to understand how complex organisms such as ourselves develop and function.

KEY TERMS

activation domain *461*

activators *449*

antitermination factor *485*

carboxyl-terminal domain (CTD) *452*

chromatin-mediated repression *471*

co-activator *467*

DNase I footprinting *458*

enhancers *456*

enhancesome *468*

general transcription factors *469*

heat-shock genes *486*

histone code *476*

histone deacetylation *474*

leucine zipper *464*

MAT locus (in yeast) *471*

mediator *471*

nuclear receptors *464*

promoter *449*

promoter-proximal elements *455*

repression domain *462*

repressors *449*

RNA polymerase II *450*

silencer sequences *472*

TATA box *454*

upstream activating sequences (UASs) *457*

yeast two-hybrid system *480*

zinc finger *463*

REVIEW THE CONCEPTS

1. What is the evidence that transcriptional initiation is the primary mechanism of gene control in complex organisms?

2. What types of genes are transcribed by RNA polymerases I, II, and III? Design an experiment to determine whether a specific gene is transcribed by RNA polymerase II.

3. The CTD of the largest subunit of RNA polymerase II can be phosphorylated and hyperphosphorylated at various serine and tyrosine residues. What are the conditions that lead to phosphorylation versus hyperphosphorylation?

4. What do TATA boxes, initiators, and CpG islands have in common? Which was the first of these to be identified? Why?

5. Describe the methods used to identify the location of DNA control elements in regulatory regions of genes.

6. What is the difference between a promoter-proximal element and a distal enhancer?

7. Describe the methods used to identify the location of DNA-binding proteins in the regulatory regions of genes.

8. Describe the structural features of transcriptional activator and repressor proteins.

9. What happens to transcription of the *EGR-1* gene in patients with Wilm's tumor? Why?

10. Using CREB and nuclear receptors as examples, compare and contrast the structural changes that take place when these transcription factors bind to their co-activators.

11. What structural change takes place on polymerase II promoters during preinitiation complex formation?

12. Expression of recombinant proteins in yeast is an important tool for biotechnology companies that produce new drugs for human use. In an attempt to get a new gene *X* expressed in yeast, a researcher has integrated gene *X* into the yeast genome near a telomere. Will this strategy result in good expression of gene *X*? Why or why not? Would the outcome of this experiment differ if the experiment had been performed in a yeast line containing mutations in the H3 or H4 histone tails?

13. You have isolated a new protein called STICKY. You can predict from comparisons with other known proteins that STICKY contains a bHLH domain and a Sin3-interacting domain. Predict the function of STICKY and rationale for the importance of these domains in STICKY function.

14. Describe at least one gene you would expect to be able to clone using the following genes as bait in a yeast two-hybrid experiment: alpha-globin; the catalytic subunit of protein kinase A; and the catalytic subunit of aspartate transcarbamylase.

ANALYZE THE DATA

An electrophoretic mobility shift assay (EMSA) was performed using a radiolabeled DNA fragment from the sequence upstream of gene *X*. This DNA probe was incubated with (+) or without (−) nuclear extract isolated from tissues A (bone); B (lung); C (brain); and D (skin). The DNA:protein complexes were then fractionated on nondenaturing polyacrylamide gels. The gels were exposed to autoradiographic film; the results are presented in the figure.

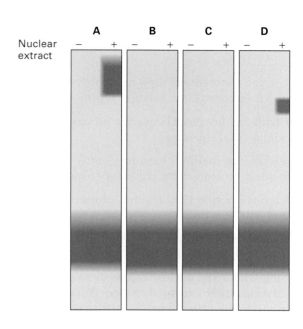

a. Which tissues contain a binding activity that recognizes the sequence upstream of gene *X*? Is the transcription factor the same in each tissue?

b. If the binding activity was purified, what test could be done to verify that this factor is in fact a transcription factor?

c. What type of assay would be performed to determine the specific DNA sequence(s) to which the transcription factor binds?

d. If gene *X* is transcribed in lung and brain tissue but not in bone and skin tissue, what type of transcription factor is the binding activity? Speculate as to the identity of other factors that might be complexed at the gene *X* promoter in bone and skin tissue.

REFERENCES

Overview of Eukaryotic Gene Control and RNA Polymerases

Cramer, P. 2002. Multisubunit RNA polymerases. *Curr. Opin. Struc. Biol.* **12**:89–97.

Regulatory Sequences in Protein-Coding Genes

Blackwood, E. M., and J. T. Kadonaga. 1998. Going the distance: a current view of enhancer action. *Science* **281**:60–63.

Butler, J. E., and J. T. Kadonaga. 2002. The RNA polymerase II core promoter: a key component in the regulation of gene expression. *Genes Dev.* **16**:2583–2592.

http://www.epd.isb-sib.ch/promoter_elements

Activators and Repressors of Transcription

Brivanlou, A. H., and J. E. Darnell, Jr. 2002. Signal transduction and the control of gene expression. *Science* **295**:813–818.

Luscombe, N. M., et al. 2000. An overview of the structures of protein-DNA complexes. *Genome Biol.* **1**:1–37.

Riechmann, J. L., et al. 2000. *Arabidopsis* transcription factors: genome-wide comparative analysis among eukaryotes. *Science* **290**:2105–2110.

Tupler, R., G. Perini, and M. R. Green. 2001. Expressing the human genome. *Nature* **409**:832–833.

Transcription Initiation by RNA Polymerase II

Woychik, N. A., and M. Hampsey. 2002. The RNA polymerase II machinery: structure illuminates function. *Cell* **108**:453–463.

Molecular Mechanisms of Transcription Activation and Repression

Berger, S. L. 2002. Histone modifications in transcriptional regulation. *Curr. Opin. Genet. Dev.* **12**:142–148.

Boube, M., et al. 2002. Evidence for a mediator of RNA polymerase II transcriptional regulation conserved from yeast to man. *Cell* **110**:143–151.

Courey, A. J., and S. Jia. 2001. Transcriptional repression: the long and the short of it. *Genes Dev.* **15**:2786–2796.

Horn, P. J., and C. L. Peterson. 2002. Chromatin Higher Order Folding—Wrapping up Transcription. *Science* **297**:1824–1827.

Richards, E. J., and S. C. Elgin. 2002. Epigenetic codes for heterochromatin formation and silencing: rounding up the usual suspects. *Cell* **108**:489–500.

Control of Transcription-Factor Activity

Hermanson, O., C. K. Glass, and M. G. Rosenfeld. 2002. Nuclear receptor coregulators: multiple modes of modification. *Trends Endocrinol. Metab.* **13**:55–60.

Lin, R. J., et al. 1998. The transcriptional basis of steroid physiology. *Cold Spring Harbor Symp. Quant. Biol.* **63**:577–585.

McKenna, N. J., and B. W. O'Malley. 2002. Minireview: nuclear receptor coactivators—an update. *Endocrinol.* **143**:2461–2465.

Yamamoto, K. R. 1995. Multilayered control of intracellular receptor function. *Harvey Lect.* **91**:1–19.

Regulated Elongation and Termination of Transcription

Garber, M. E., and K. A. Jones. 1999. HIV-1 Tat: coping with negative elongation factors. *Curr. Opin. Immunol.* **11**:460–465.

Kim, D. K., et al. 2001. The regulation of elongation by eukaryotic RNA polymerase II: a recent view. *Mol. Cells* **11**:267–274.

Lis, J. 1998. Promoter-associated pausing in promoter architecture and postinitiation transcriptional regulation. *Cold Spring Harbor Symp. Quant. Biol.* **63**:347–356.

Price, D. H. 2000. P-TEFb, a cyclin-dependent kinase controlling elongation by RNA polymerase II. *Mol. Cell Biol.* **20**:2629–2634.

Other Eukaryotic Transcription Systems

Paule, M. R., and R. J. White. 2000. Survey and summary: transcription by RNA polymerases I and III. *Nucl. Acids Res.* **28**:1283–1298.

Schramm, L., and N. Hernandez. 2002. Recruitment of RNA polymerase III to its target promoters. *Genes Dev.* **16**:2593–2620.

12

POST-TRANSCRIPTIONAL GENE CONTROL AND NUCLEAR TRANSPORT

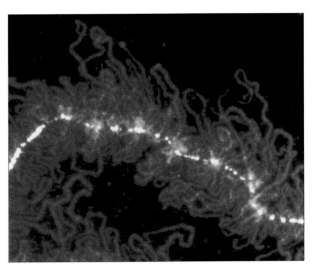

Portion of a "lampbrush chromosome" from an oocyte of the newt *Nophthalmus viridescens*; hnRNP protein associated with nascent RNA transcripts fluoresces red after staining with a monoclonal antibody. [Courtesy of M. Roth and J. Gall.]

n the previous chapter, we saw that most genes are regulated at the first step in gene expression, namely, the initiation of transcription. However, once transcription has been initiated, synthesis of the encoded RNA requires that RNA polymerase transcribe the entire gene and not terminate prematurely. Moreover, the initial **primary transcripts** produced from eukaryotic genes must undergo various processing reactions to yield the corresponding functional RNAs. Once formed in the nucleus, mature, functional RNAs are transported to the cytoplasm as components of ribonucleoproteins. Both processing of RNAs and their export from the nucleus offer opportunities for further regulating gene expression at stages subsequent to transcription initiation. Additional control of gene expression can occur in the cytoplasm. In the case of protein-coding genes, for instance, the amount of protein produced depends on the stability of the corresponding mRNAs in the cytoplasm and the rate of their translation. In addition, the cellular locations of some mRNAs are regulated, so that newly synthesized protein is concentrated where it is needed.

We refer to all the mechanisms that regulate gene expression following transcription as *post-transcriptional gene control* (Figure 12-1). Although these mechanisms generally play a smaller overall role in **gene control** than activation or repression of transcription initiation, they are critical in regulating expression of some genes. In this chapter, we consider the events that occur in the processing of mRNA following transcription initiation and the various mechanisms that are known to regulate these events. In the last section, we briefly discuss the processing of primary transcripts produced from genes encoding rRNAs and tRNAs.

12.1 Processing of Eukaryotic Pre-mRNA

In this section, we take a closer look at how eukaryotic cells convert the initial primary transcript synthesized by RNA polymerase II into a functional mRNA. Three major events occur during the process: *5′ capping, 3′ cleavage/polyadenylation*, and *RNA splicing* (Figure 12-2). Processing occurs in the nucleus as the nascent mRNA precursor is being transcribed, and the functional mRNA produced is transported to the cytoplasm by mechanisms discussed later.

The 5′ Cap Is Added to Nascent RNAs Shortly After Initiation by RNA Polymerase II

After nascent RNA molecules produced by RNA polymerase II reach a length of 25–30 nucleotides, 7-methylguanosine and the other components of the *5′ cap* found on eukaryotic

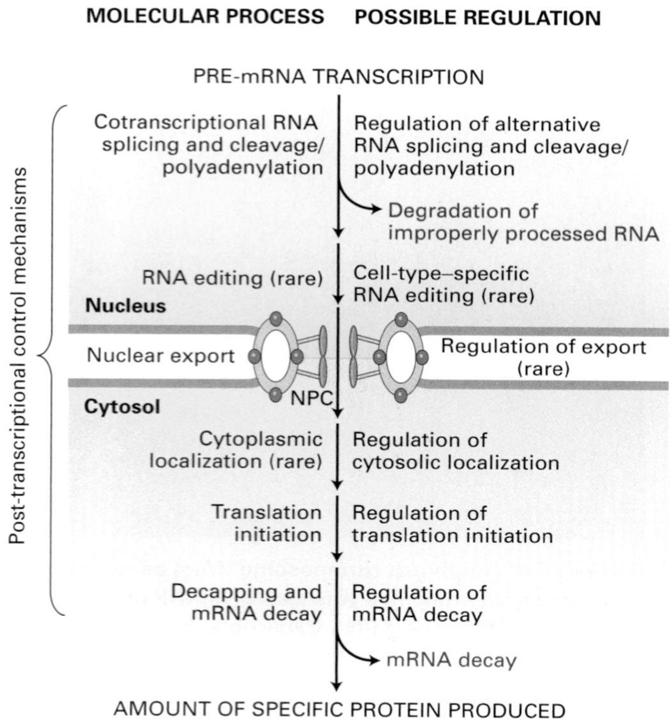

MOLECULAR PROCESS **POSSIBLE REGULATION**

PRE-mRNA TRANSCRIPTION

Post-transcriptional control mechanisms

Cotranscriptional RNA splicing and cleavage/polyadenylation

Regulation of alternative RNA splicing and cleavage/polyadenylation

→ Degradation of improperly processed RNA

Nucleus
RNA editing (rare)

Cell-type–specific RNA editing (rare)

Nuclear export

Regulation of export (rare)

NPC

Cytosol
Cytoplasmic localization (rare)

Regulation of cytosolic localization

Translation initiation

Regulation of translation initiation

Decapping and mRNA decay

Regulation of mRNA decay

→ mRNA decay

AMOUNT OF SPECIFIC PROTEIN PRODUCED

▲ **FIGURE 12-1 Overview of post-transcriptional control of protein-coding genes.** Control may be exerted as a primary transcript is processed in the nucleus, during export of an mRNA to the cytoplasm, or in the cytoplasm. Any one gene would likely be regulated by only one or a few of the possible control mechanisms.

mRNAs are added to their 5′ end (see Figure 4-13). This initial step in RNA processing is catalyzed by a dimeric capping enzyme, which associates with the phosphorylated *carboxyl-terminal domain (CTD)* of RNA polymerase II. Recall that the CTD becomes phosphorylated during transcription initiation (see Figure 11-27). Because the capping enzyme does not associate with polymerase I or III, which do not contain a CTD, capping is specific for transcripts produced by RNA polymerase II.

One subunit of the capping enzyme removes the γ phosphate from the 5′ end of the nascent RNA emerging from the surface of an RNA polymerase II. Another domain of this subunit transfers the GMP moiety from GTP to the 5′-diphosphate of the nascent transcript, creating the unusual guanosine 5′-5′-triphosphate structure. In the final steps, separate enzymes transfer methyl groups from *S*-adenosyl-methionine to the N_7 position of the guanine and the 2′ oxygens of riboses at the 5′ end of the nascent RNA.

Pre-mRNAs Are Associated with hnRNP Proteins Containing Conserved RNA-Binding Domains

Nascent RNA transcripts from protein-coding genes and mRNA processing intermediates, collectively referred to as **pre-mRNA,** do not exist as free RNA molecules in the nuclei of eukaryotic cells. From the time nascent transcripts first emerge from RNA polymerase II until mature mRNAs are transported into the cytoplasm, the RNA molecules are associated with an abundant set of nuclear proteins. These

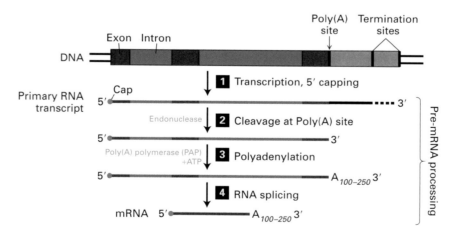

▲ **FIGURE 12-2 Overview of mRNA processing in eukaryotes.** Shortly after RNA polymerase II initiates transcription at the first nucleotide of the first exon of a gene, the 5′ end of the nascent RNA is capped with 7-methylguanylate (step **1**). Transcription by RNA polymerase II terminates at any one of multiple termination sites downstream from the poly(A) site, which is located at the 3′ end of the final exon. After the primary transcript is cleaved at the poly(A) site (step **2**),

a string of adenosine (A) residues is added (step **3**). The poly(A) tail contains ≈250 A residues in mammals, ≈150 in insects, and ≈100 in yeasts. For short primary transcripts with few introns, splicing (step **4**) usually follows cleavage and polyadenylation, as shown. For large genes with multiple introns, introns often are spliced out of the nascent RNA during its transcription, i.e., before transcription of the gene is complete. Note that the 5′ cap and sequence adjacent to the poly(A) tail are retained in mature mRNAs.

proteins are the major protein components of *heterogeneous ribonucleoprotein particles (hnRNPs)*, which contain *heterogeneous nuclear RNA (hnRNA)*, a collective term referring to pre-mRNA and other nuclear RNAs of various sizes. The proteins in these ribonucleoprotein particles can be dramatically visualized with fluorescent-labeled monoclonal antibodies (see the chapter opening figure).

Researchers identified hnRNP proteins by first exposing cultured cells to high-dose UV irradiation, which causes covalent cross-links to form between RNA bases and closely associated proteins. Chromatography of nuclear extracts from treated cells on an oligo-dT cellulose column, which binds RNAs with a poly(A) tail, was used to recover the proteins that had become cross-linked to nuclear polyadenylated RNA. Subsequent treatment of cell extracts from unirradiated cells with monoclonal antibodies specific for the major proteins identified by this cross-linking technique revealed a complex set of abundant hnRNP proteins ranging in size from 34 to 120 kDa.

Like transcription factors, most hnRNP proteins have a modular structure. They contain one or more RNA-binding domains and at least one other domain that is thought to interact with other proteins. Several different RNA-binding motifs have been identified by constructing deletions of hnRNP proteins and testing their ability to bind RNA.

Conserved RNA-Binding Motifs The *RNA recognition motif (RRM)*, also called the RNP motif and the RNA-binding domain (RBD), is the most common RNA-binding domain in hnRNP proteins. This ≈80-residue domain, which occurs in many other RNA-binding proteins, contains two highly conserved sequences (RNP1 and RNP2) that allow the motif to be recognized in newly sequenced genes. X-ray crystallographic analysis has shown that the RRM domain consists of a four-stranded β sheet flanked on one side by two α helices. The conserved RNP1 and RNP2 sequences lie side by side on the two central β strands, and their side chains make multiple contacts with a single-stranded region of RNA (Figure 12-3). The single-stranded RNA lies across the surface of the β sheet, with the central RNP1 and RNP2 strands forming a positively charged surface that interacts with the negatively charged RNA phosphates.

The *RGG box*, another RNA-binding motif found in hnRNP proteins, contains five Arg-Gly-Gly (RGG) repeats with several interspersed aromatic amino acids. Although the structure of this motif has not yet been determined, its arginine-rich nature is similar to the RNA-binding domains of the HIV Tat protein. The 45-residue *KH motif* is found in the hnRNP K protein and several other RNA-binding proteins; commonly two or more copies of the KH motif are interspersed with RGG repeats. The three-dimensional structure of representative KH domains is similar to that of the RRM domain but smaller, consisting of a three-stranded β sheet supported from one side by two α helices. RNA binds to the KH motif by interacting with a hydrophobic surface formed by the α helices and one

(a) RNA recognition motif (RRM)

(b) Sex-lethal RRM domains

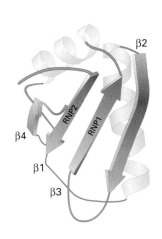

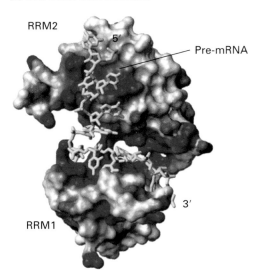

▲ **FIGURE 12-3 Structure of the RRM domain and its interaction with RNA.** (a) Diagram of the RRM domain showing the two α helices and four β strands that characterize this motif. The conserved RNP1 and RNP2 regions are located in the two central β strands. (b) Surface representation of the two RRM domains in *Drosophila* Sex-lethal (Sxl) protein, which binds a nine-base sequence in *transformer* pre-mRNA (green). The two RRMs are oriented like the two parts of an open pair of castanets, with the β sheet of RRM1 facing upward and the β sheet of RRM2 facing downward. Positively charged regions in Sxl protein are shown in shades of blue; negatively charged regions, in shades of red. The pre-mRNA is bound to the surfaces of the positively charged β sheets, making most of its contacts with the RNP1 and RNP2 regions of each RRM. [Part (a) adapted from K. Nagai et al., 1995, *Trends Biochem. Sci.* **20**: 235; part (b) after N. Harada et al., 1999, *Nature* **398**:579.]

β-strand. Thus, despite the similarity in their structures, RRM and KH domains interact differently with RNA.

Functions of hnRNP Proteins The association of pre-mRNAs with hnRNP proteins prevents formation of short secondary structures dependent on base-pairing of complementary regions, thereby making the pre-mRNAs accessible for interaction with other RNA molecules or proteins. Pre-mRNAs associated with hnRNP proteins present a more uniform substrate for further processing steps than would free, unbound pre-mRNAs, each type of which forms a unique secondary structure dependent on its specific sequence.

Binding studies with purified hnRNP proteins suggest that different hnRNP proteins associate with different regions of a newly made pre-mRNA molecule. For example, the hnRNP proteins A1, C, and D bind preferentially to the pyrimidine-rich sequences at the 3′ ends of introns (see later discussion). This observation suggests that some hnRNP proteins may interact with the RNA sequences that specify RNA splicing or cleavage/polyadenylation and contribute to the structure recognized by RNA-processing factors. Finally, cell-fusion experiments have shown that some hnRNP proteins remain localized in the nucleus, whereas others cycle in and out of the cytoplasm, suggesting that they function in the transport of mRNA. We discuss the role of these proteins in nuclear transport in Section 12.3.

3′ Cleavage and Polyadenylation of Pre-mRNAs Are Tightly Coupled

In eukaryotic cells, all mRNAs, except histone mRNAs, have a 3′ *poly(A) tail*. Early studies of pulse-labeled adenovirus and SV40 RNA demonstrated that the viral primary transcripts extend beyond the site in the viral mRNAs from which the poly(A) tail extends. These results suggested that A residues are added to a 3′ hydroxyl generated by endonucleolytic cleavage of a longer transcript, but the predicted downstream RNA fragments never were detected in vivo, presumably because of their rapid degradation. That cleavage of a primary transcript precedes its polyadenylation was firmly established by detection of both predicted cleavage

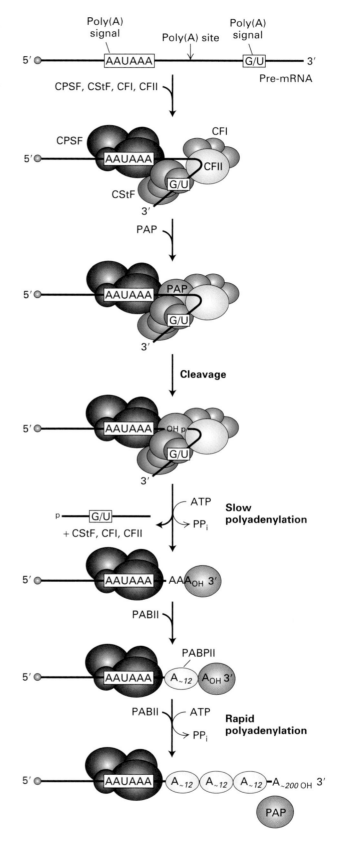

▶ **FIGURE 12-4 Model for cleavage and polyadenylation of pre-mRNAs in mammalian cells.** Cleavage and polyadenylation specificity factor (CPSF) binds to the upstream AAUAAA poly(A) signal. CStF interacts with a downstream GU- or U-rich sequence and with bound CPSF, forming a loop in the RNA; binding of CFI and CFII help stabilize the complex. Binding of poly(A) polymerase (PAP) then stimulates cleavage at a poly(A) site, which usually is 10–35 nucleotides 3′ of the upstream poly(A) signal. The cleavage factors are released, as is the downstream RNA cleavage product, which is rapidly degraded. Bound PAP then adds ≈12 A residues at a slow rate to the 3′-hydroxyl group generated by the cleavage reaction. Binding of poly(A)-binding protein II (PABPII) to the initial short poly(A) tail accelerates the rate of addition by PAP. After 200–250 A residues have been added, PABPII signals PAP to stop polymerization.

products in in vitro processing reactions performed with nuclear extracts of HeLa cells.

Early sequencing of cDNA clones from animal cells showed that nearly all mRNAs contain the sequence AAUAAA 10–35 nucleotides *upstream* from the poly(A) tail. Polyadenylation of RNA transcripts is virtually eliminated when the corresponding sequence in the template DNA is mutated to any other sequence except one encoding a closely related sequence (AUUAAA). The unprocessed RNA transcripts produced from such mutant templates do not accumulate in nuclei, but are rapidly degraded. Further mutagenesis studies revealed that a second signal *downstream* from the cleavage site is required for efficient cleavage and polyadenylation of most pre-mRNAs in animal cells. This downstream signal is not a specific sequence but rather a GU-rich or simply a U-rich region within ≈50 nucleotides of the cleavage site.

Identification and purification of the proteins required for cleavage and polyadenylation of pre-mRNA have led to the model shown in Figure 12-4. According to this model, a 360-kDa *cleavage and polyadenylation specificity factor (CPSF)*, composed of four different polypeptides, first forms an unstable complex with the upstream AAUAAA poly(A) signal. Then at least three additional proteins bind to the CPSF-RNA complex: a 200-kDa heterotrimer called *cleavage stimulatory factor (CStF)*, which interacts with the G/U-rich sequence; a 150-kDa heterotrimer called *cleavage factor I (CFI)*; and a second, poorly characterized cleavage factor (CFII). Finally, a *poly(A) polymerase (PAP)* binds to the complex *before* cleavage can occur. This requirement for PAP binding links cleavage and polyadenylation, so that the free 3′ end generated is rapidly polyadenylated.

Assembly of the large, multiprotein **cleavage/polyadenylation complex** around the AU-rich poly(A) signal in a pre-mRNA is analogous in many ways to formation of the transcription-preinitiation complex at the AT-rich TATA box of a template DNA molecule (see Figure 11-27). In both cases, multiprotein complexes assemble cooperatively through a network of specific protein–nucleic acid and protein-protein interactions.

Following cleavage at the poly(A) site, polyadenylation proceeds in two phases. Addition of the first 12 or so A residues occurs slowly, followed by rapid addition of up to 200–250 more A residues. The rapid phase requires the binding of multiple copies of a *poly(A)-binding protein* containing the RRM motif. This protein is designated **PABPII** to distinguish it from the poly(A)-binding protein present in the cytoplasm. PABPII binds to the short A tail initially added by PAP, stimulating polymerization of additional A residues by PAP (see Figure 12-4). PABPII is also responsible for signaling poly(A) polymerase to terminate polymerization when the poly(A) tail reaches a length of 200–250 residues, although the mechanism for controlling the length of the tail is not yet understood.

Splicing Occurs at Short, Conserved Sequences in Pre-mRNAs via Two Transesterification Reactions

During formation of a mature, functional mRNA, the **introns** are removed and **exons** are spliced together. For short transcription units, **RNA splicing** usually follows cleavage and polyadenylation of the 3′ end of the primary transcript, as depicted in Figure 12-2. However, for long transcription units

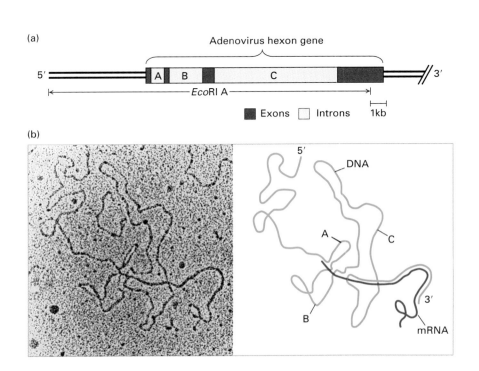

(a)
Adenovirus hexon gene

■ Exons   □ Introns   1kb

(b)

◄ **EXPERIMENTAL FIGURE 12-5** RNA-DNA hybridization studies show that introns are spliced out during pre-mRNA processing. Electron microscopy of an RNA-DNA hybrid between adenovirus DNA and the mRNA encoding hexon, a major viral protein, reveals DNA segments (introns) that are absent from the hexon mRNA. (a) Diagram of the EcoRI A fragment of adenovirus DNA, which extends from the left end of the genome to just before the end of the final exon of the hexon gene. The gene consists of three short exons and one long (≈3.5 kb) exon separated by three introns of ≈1, 2.5, and 9 kb. (b) Electron micrograph (*left*) and schematic drawing (*right*) of hybrid between an EcoRI A fragment and hexon mRNA. The loops marked A, B, and C correspond to the introns indicated in (a). Since these intron sequences in the viral genomic DNA are not present in mature hexon mRNA, they loop out between the exon sequences that hybridize to their complementary sequences in the mRNA. [Micrograph from S. M. Berget et al., 1977, *Proc. Nat'l. Acad. Sci. USA* **74**:3171; courtesy of P. A. Sharp.]

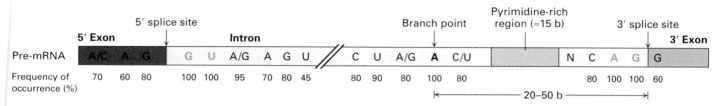

		5' splice site				Intron							Branch point			Pyrimidine-rich region (≈15 b)			3' splice site		
	5' Exon																			**3' Exon**	
Pre-mRNA	A/C	A	G		G	U	A/G	A	G	U	//	C	U	A/G	A	C/U	▨▨▨	N	C	A	G	G
Frequency of occurrence (%)	70	60	80		100	100	95	70	80	45		80	90	80	100	80		80	100	100	60

← 20–50 b →

▲ **FIGURE 12-6 Consensus sequences around 5′ and 3′ splice sites in vertebrate pre-mRNAs.** The only nearly invariant bases are the 5′ GU and the 3′ AG of the intron (blue), although the flanking bases indicated are found at frequencies higher than expected based on a random distribution. A pyrimidine-rich region (hatch marked) near the 3′ end of the intron is found in most cases. The branch-point adenosine, also invariant, usually is 20–50 bases from the 3′ splice site. The central region of the intron, which may range from 40 bases to 50 kilobases in length, generally is unnecessary for splicing to occur. [See R. A. Padgett et al., 1986, *Ann. Rev. Biochem.* **55**:1119, and E. B. Keller and W. A. Noon, 1984, *Proc. Nat'l. Acad. Sci. USA* **81**:7417.]

containing multiple exons, splicing of exons in the nascent RNA usually begins before transcription of the gene is complete.

Early evidence that introns are removed during splicing came from electron microscopy of RNA-DNA hybrids between adenovirus DNA and the mRNA encoding hexon, a major virion capsid protein (Figure 12-5). Other studies revealed nuclear viral RNAs that were colinear with the viral DNA (primary transcripts) and RNAs with one or two of the introns removed (processing intermediates). These results, together with the findings that the 5′ cap and 3′ poly(A) tail at each end of long mRNA precursors are retained in shorter mature cytoplasmic mRNAs, led to the realization that introns are removed from primary transcripts as exons are spliced together.

The location of *splice sites*—that is, exon-intron junctions—in a pre-mRNA can be determined by comparing the sequence of genomic DNA with that of the cDNA prepared from the corresponding mRNA. Sequences that are present in the genomic DNA but absent from the cDNA represent introns and indicate the positions of splice sites. Such analysis of a large number of different mRNAs revealed moderately conserved, short consensus sequences at the splice sites flanking introns in eukaryotic pre-mRNAs; in higher organisms, a pyrimidine-rich region just upstream of the 3′ splice site also is common (Figure 12-6). Studies with deletion mutants have shown that much of the center portion of introns can be removed without affecting splicing; generally only 30–40 nucleotides at each end of an intron are necessary for splicing to occur at normal rates.

Analysis of the intermediates formed during splicing of pre-mRNAs in vitro led to the discovery that splicing of exons proceeds via two sequential *transesterification reactions* (Figure 12-7). Introns are removed as a lariat-like structure in which the 5′ G of the intron is joined in an unusual 2′,5′-phosphodiester bond to an adenosine near the 3′ end of the intron. This A residue is called the *branch point* because it forms an RNA branch in the lariat structure. In each transesterification reaction, one phosphoester bond is exchanged for another. Since the number of phosphoester bonds in the molecule is not changed in either reaction, no energy is consumed. The net result of these two reactions is that two exons are ligated and the intervening intron is released as a branched lariat structure.

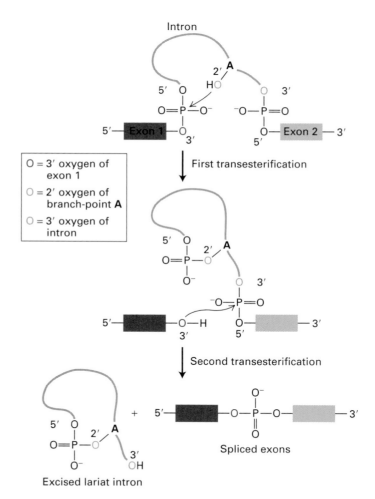

O = 3′ oxygen of exon 1
O = 2′ oxygen of branch-point **A**
O = 3′ oxygen of intron

▲ **FIGURE 12-7 Two transesterification reactions that result in splicing of exons in pre-mRNA.** In the first reaction, the ester bond between the 5′ phosphorus of the intron and the 3′ oxygen (dark red) of exon 1 is exchanged for an ester bond with the 2′ oxygen (blue) of the branch-site **A** residue. In the second reaction, the ester bond between the 5′ phosphorus of exon 2 and the 3′ oxygen (light red) of the intron is exchanged for an ester bond with the 3′ oxygen of exon 1, releasing the intron as a lariat structure and joining the two exons. Arrows show where the activated hydroxyl oxygens react with phosphorus atoms.

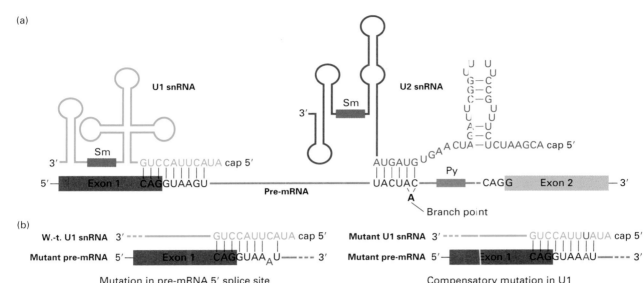

▲ **FIGURE 12-8 Base pairing between pre-mRNA, U1 snRNA, and U2 snRNA early in the splicing process.** (a) In this diagram, secondary structures in the snRNAs that are not altered during splicing are depicted schematically. The yeast branch-point sequence is shown here. Note that U2 snRNA base-pairs with a sequence that includes the branch-point A, although this residue is not base-paired. The purple rectangles represent sequences that bind snRNP proteins recognized by anti-Sm antibodies. For unknown reasons, antisera from patients with the autoimmune disease systemic lupus erythematosus (SLE) contain these antibodies. Such antisera have been useful in characterizing components of the splicing reaction. (b) Only the 5' ends of U1 snRNAs and 5' splice sites in pre-mRNAs are shown. (*Left*) A mutation (A) in a pre-mRNA splice site that interferes with base-pairing to the 5' end of U1 snRNA blocks splicing. Expression of a U1 snRNA with a compensating mutation (U) that restores base-pairing also restores splicing of the mutant pre-mRNA. [Part (a) adapted from M. J. Moore et al., 1993, in R. Gesteland and J. Atkins, eds., *The RNA World*, Cold Spring Harbor Press, pp. 303–357; part (b) see Y. Zhuang and A. M. Weiner, 1986, *Cell* **46**:827.]

snRNAs Base-Pair with Pre-mRNA and with One Another During Splicing

Five U-rich **small nuclear RNAs (snRNAs),** designated U1, U2, U4, U5, and U6, participate in pre-mRNA splicing. Ranging in length from 107 to 210 nucleotides, these snRNAs are associated with 6 to 10 proteins in *small nuclear ribonucleoprotein particles (snRNPs)* in the nucleus of eukaryotic cells.

Definitive evidence for the role of U1 snRNA in splicing came from experiments which indicated that base pairing between the 5' splice site of a pre-mRNA and the 5' region of U1 snRNA is required for RNA splicing (Figure 12-8a). In vitro experiments showed that a synthetic oligonucleotide that hybridizes with the 5'-end region of U1 snRNA blocks RNA splicing. In vivo experiments showed that mutations in the pre-mRNA 5' splice site that disrupt the base pairing block RNA splicing, but that splicing can be restored by expression of a mutant U1 snRNA with a compensating mutation that restores base pairing to the mutant pre-mRNA 5' splice site (Figure 12-8b). Involvement of U2 snRNA in splicing initially was suspected when it was found to have an internal sequence that is largely complementary to the consensus sequence flanking the branch point in pre-mRNAs (see Figure 12-6). Compensating mutation experiments, similar to those conducted with U1 snRNA and 5' splice sites, demonstrated that base pairing between U2 snRNA and the branch-point sequence in pre-mRNA also is critical to splicing.

Figure 12-8a illustrates the general structures of the U1 and U2 snRNAs and how they base-pair with pre-mRNA during splicing. Significantly, the branch-point A itself, which is not base-paired to U2 snRNA, "bulges out," allowing its 2' hydroxyl to participate in the first transesterification reaction of RNA splicing (see Figure 12-7).

Similar studies with other snRNAs demonstrated that base pairing between them also occurs during splicing. Moreover, rearrangements in these RNA-RNA interactions are critical in the splicing pathway, as we describe next.

Spliceosomes, Assembled from snRNPs and a Pre-mRNA, Carry Out Splicing

According to the current model of pre-mRNA splicing, the five splicing snRNPs are thought to assemble on the pre-mRNA, forming a large ribonucleoprotein complex called a **spliceosome** (Figure 12-9). Assembly of a spliceosome begins with the base pairing of the snRNAs of the U1 and U2 snRNPs to the pre-mRNA (see Figure 12-8). Extensive base pairing between the snRNAs in the U4 and U6 snRNPs forms a complex that associates with U5 snRNP. The U4/U6/U5 complex then associates with the previously formed U1/U2/pre-mRNA complex to yield a spliceosome.

After formation of the spliceosome, extensive rearrangements in the pairing of snRNAs and the pre-mRNA lead to release of the U1 and U4 snRNPs. The catalytically active rearranged spliceosome then mediates the first transesterification reaction that forms the 2′,5′-phosphodiester bond between the 2′ hydroxyl on the branch point A and the phos-

phate at the 5′ end of the intron. Following another rearrangement of the snRNPs, the second transesterification reaction ligates the two exons in a standard 3′,5′-phosphodiester bond, releasing the intron as a lariat structure associated with the snRNPs. This final intron-snRNP complex rapidly dissociates, and the individual snRNPs released can participate in a new cycle of splicing. The excised intron is then rapidly degraded by a debranching enzyme and other nuclear RNases discussed later.

A spliceosome is roughly the size of a small ribosomal subunit, and at least 70 proteins are thought to participate in RNA splicing, making this process comparable in complexity to initiation of transcription and protein synthesis. Some of these splicing factors are associated with snRNPs, but others are not. For instance, the 65-kD subunit of the U2-associated factor (U2AF) binds to the pyrimidine-rich region near the 3′ end of introns and to the U2 snRNP. The 35-kD subunit of U2AF binds to the AG dinucleotide at the 3′ end of the intron and also interacts with the larger U2AF subunit bound nearby. These two U2AF subunits act together to help specify the 3′ splice site by promoting interaction of U2 snRNP with the branch point (see Figure 12-8). Some splicing factors also exhibit sequence homologies to known RNA helicases; these are probably necessary for the base-pairing rearrangements that occur in snRNAs during the spliceosomal splicing cycle.

Some pre-mRNAs contain introns whose splice sites do not conform to the standard consensus sequence. This class of introns begins with AU and ends with AC rather than following the usual "GU–AG rule" (see Figure 12-6). Splicing of this special class of introns appears to occur via a splicing cycle analogous to that shown in Figure 12-9, except that four novel, low-abundance snRNPs, together with the standard U5 snRNP, are involved.

Nearly all functional mRNAs in vertebrate, insect, and plant cells are derived from a single molecule of the corresponding pre-mRNA by removal of internal introns and splicing of exons. However, in two types of protozoans—trypanosomes and euglenoids—mRNAs are constructed by

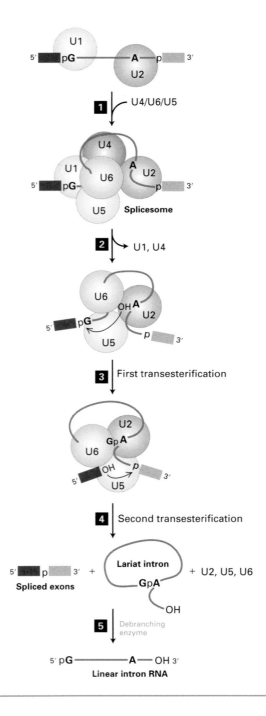

◀ **FIGURE 12-9 Model of spliceosome-mediated splicing of pre-mRNA.** Step **1**: After U1 and U2 snRNPs associate with the pre-mRNA, via base-pairing interactions shown in Figure 12-8, a trimeric snRNP complex of U4, U5, and U6 joins the initial complex to form the spliceosome. Step **2**: Rearrangements of base-pairing interactions between snRNAs converts the spliceosome into a catalytically active conformation and destabilizes the U1 and U4 snRNPs, which are released. Step **3**: The catalytic core, thought to be formed by U6 and U2, then catalyzes the first transesterification reaction, forming the intermediate containing a 2′,5′-phosphodiester bond shown in Figure 12-7. Step **4**: Following further rearrangements between the snRNPs, the second transesterification reaction joins the two exons by a standard 3′,5′-phosphodiester bond and releases the intron as a lariat structure and the remaining snRNPs. Step **5**: The excised lariat intron is converted into a linear RNA by a debranching enzyme. [Adapted from T. Villa et al., 2002, *Cell* **109**:149.]

RNA polymerase II

Linker

CTD

|← 28 nm →|← 65 nm →|

▲ **FIGURE 12-10 Schematic diagram of RNA polymerase II with the CTD extended.** The length of the fully extended yeast RNA polymerase II carboxyl-terminal domain (CTD) and the linker region that connects it to the polymerase is shown relative to the globular domain of the polymerase. The CTD of mammalian RNA polymerase II is twice as long. In its extended form, the CTD can associate with multiple RNA-processing factors simultaneously. [From P. Cramer, D. A. Bushnell, and R. D. Kornberg, 2001, *Science* **292**:1863.]

splicing together separate RNA molecules. This process, referred to as *trans-splicing,* is also used in the synthesis of 10–15 percent of the mRNAs in the nematode (round worm) *Caenorhabditis elegans,* an important model organism for studying embryonic development. Trans-splicing is carried out by snRNPs by a process similar to the splicing of exons in a single pre-mRNA.

Chain Elongation by RNA Polymerase II Is Coupled to the Presence of RNA-Processing Factors

The carboxyl-terminal domain (CTD) of RNA polymerase II is composed of multiple repeats of a seven-residue (heptapeptide) sequence. When fully extended, the CTD domain in the yeast enzyme is about 65 nm long (Figure 12-10); the CTD in human polymerase is about twice as long. The remarkable length of the CTD apparently allows multiple proteins to associate simultaneously with a single RNA polymerase II molecule. For instance, as mentioned earlier, the enzyme that adds the 5′ cap to nascent transcripts associates with the phosphorylated CTD shortly after transcription initiation. In addition, RNA splicing and polyadenylation factors have been found to associate with the phosphorylated CTD. As a consequence, these processing factors are present at high local concentrations when splice sites and poly(A) signals are transcribed by the polymerase, enhancing the rate and specificity of RNA processing.

Recent results indicate that the association of RNA splicing factors with the phosphorylated CTD also stimulates transcription elongation. Thus chain elongation is coupled to the binding of RNA-processing factors to the phosphorylated CTD. This mechanism may ensure that a pre-mRNA is not synthesized unless the machinery for processing it is properly positioned.

SR Proteins Contribute to Exon Definition in Long Pre-mRNAs

The average length of an exon in the human genome is ≈150 bases, whereas the average length of an intron is much longer (≈3500 bases). The longest introns contain upwards of 500 kb! Because the sequences of 5′ and 3′ splice sites and branch points are so degenerate, multiple copies are likely to occur randomly in long introns. Consequently, additional sequence information is required to define the exons that should be spliced together in higher organisms with long introns.

The more sophisticated mechanism of exon recognition required in higher organisms entails a family of RNA-binding proteins, the *SR proteins,* that interact with sequences within exons called *exonic splicing enhancers.* SR proteins are characterized by having one or more RRM RNA-binding domains and several protein-protein interaction domains rich in serine and arginine residues. When bound to exonic splicing enhancers, SR proteins mediate the cooperative binding of U1 snRNP to a true 5′ splice site and U2 snRNP to a branch point through a network of protein-protein interactions that span across an exon (Figure 12-11). The complex of SR proteins, snRNPs, and other splicing factors (e.g., U2AF) that assemble across an exon, which has been called a **cross-exon recognition complex,** permits precise specification of exons in long pre-mRNAs.

 In the transcription units of higher organisms with long introns, exons not only encode the amino acid sequences of different portions of a protein but also contain binding sites for SR proteins. Mutations that interfere with the binding of an SR protein to an exonic splicing enhancer, even if they do not change the encoded amino acid sequence, would prevent formation of the cross-exon recognition complex. As a result, the affected exon is "skipped" during splicing and not included in the final processed mRNA. The truncated mRNA produced in this case is either degraded or translated into a mutant, abnormally functioning protein. Recent studies have implicated this type of mutation in human genetic diseases. For example, *spinal muscle atrophy* is one of the most common genetic causes of childhood mortality. This disease results from mutations in a region of the genome containing two closely related genes, *SMN1* and *SMN2,* that arose by gene duplication. *SMN2* encodes a protein identical with *SMN1,* but it is expressed at much lower level because a silent mutation in one exon interferes with the binding of an SR protein, leading to exon skipping in most of the *SMN2* mRNAs. The homologous *SMN* gene in the mouse, where there is only a single copy, is essential for cell viability. Spinal muscle atrophy in humans results from homozygous mutations that inactivate *SMN1.* The low level of protein translated from the small

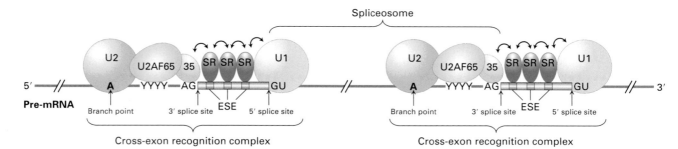

▲ **FIGURE 12-11 Exon recognition through cooperative binding of SR proteins and splicing factors to pre-mRNA.** The correct 5' GU and 3' AG splice sites are recognized by splicing factors on the basis of their proximity to exons. The exons contain exonic splicing enhancers (ESEs) that are binding sites for SR proteins. When bound to ESEs, the SR proteins interact with one another and promote the cooperative binding of the U1 snRNP to the 5' splice site of the downstream intron, the U2 snRNP to the branch point of the upstream intron, the 65- and 35-kD subunits of U2AF to the pyrimidine-rich region and AG 3' splice site of the upstream intron, and other splicing factors (not shown). The resulting RNA-protein cross-exon recognition complex spans an exon and activates the correct splice sites for RNA splicing. Note that the U1 and U2 snRNPs in this unit are not part of the same spliceosome. The U2 snRNP on the right forms a spliceosome with the U1 snRNP bound to the 5'-end of the same intron. The U1 snRNP shown on the right forms a spliceosome with the U2 snRNP bound to the branch point of the downstream intron (not shown), and the U2 snRNP on the left forms a spliceosome with a U1 snRNP bound to the 5' splice site of the upstream intron (not shown). Double-headed arrows indicate protein-protein interactions. [Adapted from T. Maniatis, 2002, *Nature* **418**:236; see also S. M. Berget, 1995, *J. Biol. Chem.* **270**:2411.]

fraction of *SMN2* mRNAs that are correctly spliced is sufficient to maintain cell viability during embryogenesis and fetal development, but it is not sufficient to maintain viability of spinal cord motor neurons in childhood, resulting in their death and the associated disease.

Approximately 15 percent of the single-base-pair mutations that cause human genetic diseases interfere with proper exon definition. Some of these mutations occur in 5' or 3' splice sites, often resulting in the use of nearby alternative "cryptic" splice sites present in the normal gene sequence. In the absence of the normal splice site, the cross-exon recognition complex recognizes these alternative sites. Other mutations that cause abnormal splicing result in a new consensus splice site sequence that becomes recognized in place of the normal splice site. Finally, some mutations can interfere with the binding of specific hnRNP proteins to pre-mRNAs that enhance or repress splicing at normal splice sites, as in the case of the *SMN2* gene. ∎

Self-Splicing Group II Introns Provide Clues to the Evolution of snRNAs

Under certain nonphysiological in vitro conditions, pure preparations of some RNA transcripts slowly splice out introns in the absence of any protein. This observation led to recognition that some introns are *self-splicing*. Two types of self-splicing introns have been discovered: *group I introns*, present in nuclear rRNA genes of protozoans, and *group II introns*, present in protein-coding genes and some rRNA and tRNA genes in mitochondria and chloroplasts of plants and fungi. Discovery of the catalytic activity of self-splicing introns revolutionized concepts about the functions of RNA. As discussed in Chapter 4, RNA is now thought to catalyze

peptide-bond formation during protein synthesis in ribosomes. Here we discuss the probable role of group II introns, now found only in mitochondrial and chloroplast DNA, in the evolution of snRNAs; the functioning of group I introns is considered in the later section on rRNA processing.

Even though their precise sequences are not highly conserved, all group II introns fold into a conserved, complex secondary structure containing numerous stem-loops (Figure 12-12a). Self-splicing by a group II intron occurs via two transesterification reactions, involving intermediates and products analogous to those found in nuclear pre-mRNA splicing. The mechanistic similarities between group II intron self-splicing and spliceosomal splicing led to the hypothesis that snRNAs function analogously to the stem-loops in the secondary structure of group II introns. According to this hypothesis, snRNAs interact with 5' and 3' splice sites of pre-mRNAs and with each other to produce a three-dimensional RNA structure functionally analogous to that of group II self-splicing introns (Figure 12-12b).

An extension of this hypothesis is that introns in ancient pre-mRNAs evolved from group II self-splicing introns through the progressive loss of internal RNA structures, which concurrently evolved into trans-acting snRNAs that perform the same functions. Support for this type of evolutionary model comes from experiments with group II intron mutants in which domain V and part of domain I are deleted. RNA transcripts containing such mutant introns are defective in self-splicing, but when RNA molecules equivalent to the deleted regions are added to the in vitro reaction, self-splicing occurs. This finding demonstrates that these domains in group II introns can be trans-acting, like snRNAs.

The similarity in the mechanisms of group II intron self-splicing and spliceosomal splicing of pre-mRNAs also

(a) Group II intron (b) U snRNAs in spliceosome

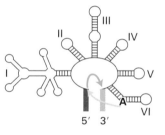

▲ FIGURE 12-12 Schematic diagrams comparing the secondary structures of group II self-splicing introns (a) and U snRNAs present in the spliceosome (b). The first transesterification reaction is indicated by green arrows; the second reaction, by blue arrows. The branch-point A is boldfaced. The similarity in these structures suggests that the spliceosomal snRNAs evolved from group II introns, with the trans-acting snRNAs being functionally analogous to the corresponding domains in group II introns. The colored bars flanking the introns in (a) and (b) represent exons [Adapted from P. A. Sharp, 1991, *Science* **254**:663.]

thought to stabilize the precise geometry of snRNAs and intron nucleotides required to catalyze pre-mRNA splicing.

The evolution of snRNAs may have been an important step in the rapid evolution of higher eukaryotes. As internal intron sequences were lost and their functions in RNA splicing supplanted by trans-acting snRNAs, the remaining intron sequences would be free to diverge. This in turn likely facilitated the evolution of new genes through **exon shuffling** since there are few constraints on the sequence of new introns generated in the process (see Figure 10-17). It also permitted the increase in protein diversity that results from alternative RNA splicing and an additional level of gene control resulting from regulated RNA splicing.

Most Transcription and RNA Processing Occur in a Limited Number of Domains in Mammalian Cell Nuclei

Several kinds of studies suggest that transcription and RNA processing occur largely within discrete foci within the nucleus of eukaryotic cells. For instance, digital imaging microscopy of human fibroblasts reveals that most of the nuclear polyadenylated RNA (i.e., unspliced and partially spliced pre-mRNA and nuclear mRNA) is localized in about 100 foci (Figure 12-13). An SR protein involved in exon definition (SC-35) is localized to the center of these same loci.

These data imply that the nucleus is not simply an unorganized container for chromatin and the proteins that carry out chromosome replication, transcription, and RNA processing. Rather, there is a highly organized underlying nuclear substructure. A poorly understood fibrillar network comprising multiple proteins called the *nuclear matrix* can be

suggests that the splicing reaction is catalyzed by the snRNA, not the protein, components of spliceosomes. Although group II introns can self-splice in vitro at elevated temperatures and Mg^{2+} concentrations, under in vivo conditions proteins called *maturases*, which bind to group II intron RNA, are required for rapid splicing. Maturases are thought to stabilize the precise three-dimensional interactions of the intron RNA required to catalyze the two splicing transesterification reactions. By analogy, snRNP proteins in spliceosomes are

(a)

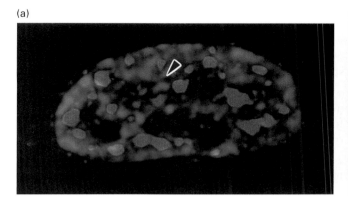

(b)

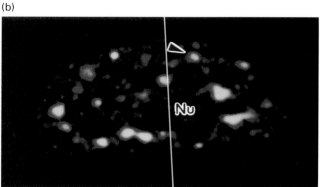

▲ EXPERIMENTAL FIGURE 12-13 Staining shows that polyadenylated RNA and RNA splicing factors are sequestered in discrete areas of the mammalian fibroblast nucleus. Digital imaging microscopy was used to reconstruct a 1-μm-thick section of a stained human fibroblast nucleus. (a) Section stained with red rhodamine-labeled poly(dT) to detect polyadenylated RNA (red) and with DAPI to detect DNA (blue). Polyadenylated RNA is localized to a limited number of discrete foci (speckles) between regions of chromatin, although not all regions containing low levels of DNA contained detectable polyadenylated RNA (arrow). (b) The same section shown in (a) stained to detect polyadenylated RNA (red) and SR protein SC-35, which was visualized with a green fluorescein-labeled monoclonal antibody. Regions where the stains overlap appear yellow. SC-35 is present in the center of many foci (arrow). Nu = nucleolus. [From K. C. Carter et al., 1993, *Science* **259**:1330.]

observed in the nucleus after digestion of most DNA and extraction with high salt. This network of fibrous proteins probably contributes to nuclear organization just as cytoskeletal fibers contribute to the organization of the cytoplasm. Much remains to be learned about the molecular mechanisms underlying this organization and its consequences for nuclear processes.

Nuclear Exonucleases Degrade RNA That Is Processed out of Pre-mRNAs

Because the human genome contains long introns, only ≈5 percent of the nucleotides that are polymerized by RNA polymerase II during transcription are retained in the mature, processed mRNA. The introns that are spliced out and the region downstream from the cleavage and polyadenylation site are degraded by nuclear exonucleases that hydrolyze one base at a time from either the 5′ or 3′ end of an RNA strand.

As mentioned earlier, the 2′,5′-phosphodiester bond in excised introns is hydrolyzed by a debranching enzyme, yielding a linear molecule with unprotected ends that can be attacked by exonucleases (see Figure 12-9). The predominant nuclear decay pathway is 3′→5′ hydrolysis by 11 exonucleases that associate with one another in a large protein complex called the **exosome.** Other proteins in the complex include RNA helicases that disrupt base pairing and RNA-protein interactions that would otherwise impede the exonucleases. Exosomes also function in the cytoplasm as discussed later. In addition the exosome appears to degrade pre-mRNAs that have not been properly spliced or polyadenylated. It is not yet clear how the exosome recognizes improperly processed pre-mRNAs.

To avoid being degraded by nuclear exonucleases, nascent transcripts, pre-mRNA processing intermediates, and mature mRNAs in the nucleus must have their ends protected. The 5′ end of a nascent transcript is protected by addition of the 5′ cap structure as soon as the 5′ end emerges from the polymerase. Further protection is provided by interaction of the cap with a *nuclear cap-binding complex,* which also functions in export of mRNA to the cytoplasm. The 3′ end of a nascent transcript lies within the RNA polymerase and thus is inaccessible to exonucleases (see Figure 4-10). As discussed previously, the free 3′ end generated by cleavage of a pre-mRNA downstream from the poly(A) signal is rapidly polyadenylated by the poly(A) polymerase associated with the other 3′ processing factors, and the resulting poly(A) tail is bound by PABPII (see Figure 12-4). This tight coupling of cleavage and polyadenylation protects the 3′ end from exonuclease attack.

KEY CONCEPTS OF SECTION 12.1

Processing of Eukaryotic Pre-mRNA

■ In the nucleus of eukaryotic cells, pre-mRNAs are associated with hnRNP proteins and processed by 5′ capping, 3′ cleavage and polyadenylation, and splicing before being transported to the cytoplasm (see Figure 12-2).

■ Shortly after transcription initiation, a capping enzyme associated with the phosphorylated CTD of RNA polymerase II adds the 5′ cap to the nascent transcript.

■ In most protein-coding genes, a conserved AAUAAA poly(A) signal lies slightly upstream from a poly(A) site where cleavage and polyadenylation occur. A GU- or U-rich sequence downstream from the poly(A) site contributes to the efficiency of cleavage and polyadenylation.

■ A multiprotein complex that includes poly(A) polymerase (PAP) carries out the cleavage and polyadenylation of a pre-mRNA. A nuclear poly(A)-binding protein, PABPII, stimulates addition of A residues by PAP and stops addition once the poly(A) tail reaches 200–250 residues (see Figure 12-4).

■ Five different snRNPs interact via base pairing with one another and with pre-mRNA to form the spliceosome (see Figure 12-9). This very large ribonucleoprotein complex catalyzes two transesterification reactions that join two exons and remove the intron as a lariat structure, which is subsequently degraded (see Figure 12-7).

■ The association of pre-mRNA processing factors with the CTD of RNA polymerase II stimulates chain elongation. This coupling ensures that a pre-mRNA is not synthesized until the factors required for its processing are in place to interact with splicing and cleavage and polyadenylation signals in the pre-mRNA as they emerge from the polymerase.

■ SR proteins that bind to exonic splicing enhancer sequences in exons are critical in defining exons in the large pre-mRNAs of higher organisms. A network of interactions between SR proteins, snRNPs, and splicing factors forms a cross-exon recognition complex that specifies correct splice sites (see Figure 12-11).

■ The snRNAs in the spliceosome are thought to have an overall tertiary structure similar to that of group II self-splicing introns.

■ For long transcription units in higher organisms, splicing of exons usually begins as the pre-mRNA is still being formed. Cleavage and polyadenylation to form the 3′ end of the mRNA occur after the poly(A) site is transcribed.

■ Excised introns are degraded primarily by exosomes, multiprotein complexes that contain eleven 3′→5′ exonucleases as well as RNA helicases. Exosomes also degrade improperly processed pre-mRNAs.

12.2 Regulation of Pre-mRNA Processing

Now that we've seen how pre-mRNAs are processed into mature, functional mRNAs, we consider how regulation of this process can contribute to gene control. Recall from Chapter 10 that higher eukaryotes contain both simple and complex transcription units. The primary transcripts produced from the former contain one poly(A) site and exhibit only one pattern of RNA splicing, even if multiple introns are present; thus simple transcription units encode a single mRNA. In contrast, the primary transcripts produced from complex transcription units can be processed in alternative ways to yield different mRNAs that encode distinct proteins (see Figure 10-2).

Alternative Splicing Is the Primary Mechanism for Regulating mRNA Processing

The results from recent genomic studies suggest that the number of complex transcription units may be much greater than previously realized. Researchers estimate that up to ≈60 percent of all transcription units in the human genome are complex. This estimate comes from comparison of the genomic DNA sequence with expressed sequence tags (ESTs), which are randomly sequenced segments of cDNA (Chapter 9). Comparison of these cDNA sequences with genomic DNA sequence shows the locations of exons and introns in genomic DNA. The finding of alternative combinations of exons expressed from a region of genomic DNA reveals the existence of a complex transcription unit. Even though this estimate may be elevated somewhat by inclusion in the EST database of rare RNAs that result from mistakes in RNA processing, it is clear that complex transcription units account for a large portion of the protein-coding sequences in humans. This finding and the discovery of multiple examples of regulated RNA processing indicate that alternative RNA processing of complex pre-mRNAs produced from such transcription units is a significant gene-control mechanism in higher eukaryotes.

Although examples of cleavage at alternative poly(A) sites in pre-mRNAs are known, *alternative splicing* of different exons is the more common mechanism for expressing different proteins from one complex transcription unit. Such alternative processing pathways usually are regulated, often in a cell-type-specific manner; that is, one possible mRNA is expressed in one type of cell or tissue, while another is expressed in different cells or tissues. In Chapter 4, for example, we mentioned that fibroblasts produce one type of the extracellular protein fibronectin, whereas hepatocytes produce another type. Both fibronectin **isoforms** are encoded by the same transcription unit, which is spliced differently in the two cell types to yield two different mRNAs (see Figure 4-15). In other cases, alternative processing may occur in the same cell type in response to different developmental or environmental signals. First we discuss one of the best-understood examples of regulated RNA processing and then consider its consequences in development of the nervous system.

A Cascade of Regulated RNA Splicing Controls *Drosophila* Sexual Differentiation

One of the earliest examples of regulated alternative splicing of pre-mRNA came from studies of sexual differentiation

in *Drosophila*. Genes required for normal *Drosophila* sexual differentiation were first characterized by isolating *Drosophila* mutants defective in the process. When the proteins encoded by the wild-type genes were characterized biochemically, two of them were found to regulate a cascade of alternative RNA splicing in *Drosophila* embryos. More recent research has provided insight into how these proteins regulate RNA processing.

The Sxl protein, encoded by the *sex-lethal* gene, is the first protein to act in the cascade. Early in development, this gene is transcribed from a promoter that functions only in female embryos. Later in development, this female-specific promoter is shut off and another promoter for *sex-lethal* becomes active in both male and female embryos. However, in the absence of early Sxl protein, the *sex-lethal* pre-mRNA in male embryos is spliced to produce an mRNA that contains a stop codon early in the sequence. The net result is that male embryos produce no functional Sxl protein either early or later in development.

In contrast, the Sxl protein expressed in early female embryos directs splicing of the *sex-lethal* pre-mRNA so that a functional *sex-lethal* mRNA is produced (Figure 12-14a). Sxl accomplishes this by binding to a sequence in the pre-mRNA near the 3' end of the intron between exon 2 and exon 3, thereby blocking the proper association of U2AF and U2 snRNP. As a consequence, the U1 snRNP bound to the 3' end of exon 2 assembles into a spliceosome with U2 snRNP bound to the branch point at the 3' end of the intron between exons 3 and 4, leading to splicing of exon 2 to 4 and skipping of exon 3. The resulting female-specific *sex-lethal* mRNA is translated into functional Sxl protein, which reinforces its own expression in female embryos by continuing to cause skipping of exon 3. The absence of Sxl protein in male embryos allows the inclusion of exon 3 and, consequently, of the stop codon that prevents translation of functional Sxl protein.

Sxl protein also regulates alternative RNA splicing of the *transformer* gene pre-mRNA (Figure 12-14b). In male embryos, where no Sxl is expressed, exon 1 is spliced to exon 2, which contains a stop codon that prevents synthesis of a functional protein. In female embryos, however, binding of Sxl protein to the 3' end of the intron between exons 1 and 2 blocks binding of U2AF at this site. The interaction of Sxl with *transformer* pre-mRNA is mediated by two RRM domains in the protein (see Figure 12-3). When Sxl is bound, U2AF binds to a lower-affinity site farther 3' in the pre-mRNA; as a result exon 1 is spliced to this alternative 3' splice site, eliminating exon 2 with its stop codon. The resulting female-specific *transformer* mRNA, which contains additional constitutively spliced exons, is translated into functional Transformer (Tra) protein.

Finally, Tra protein regulates the alternative processing of pre-mRNA transcribed from the *double-sex* gene (Figure 12-14c). In female embryos, a complex of Tra and two constitutively expressed proteins, Rbp1 and Tra2, directs splicing of exon 3 to exon 4 and also promotes cleavage/

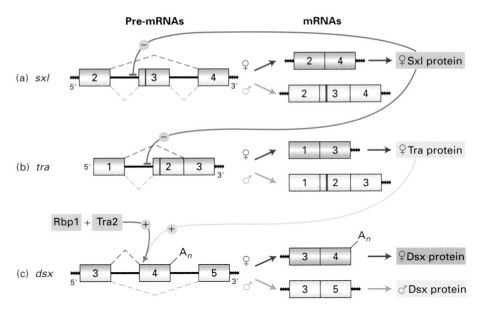

▲ FIGURE 12-14 Cascade of regulated splicing that controls sex determination via expression of *sex-lethal (sxl)*, *transformer (tra)*, and *double-sex (dsx)* genes in *Drosophila* embryos. For clarity, only the exons (boxes) and introns (black lines) where regulated splicing occurs are shown. Splicing is indicated by red dashed lines above (female) and blue dashed lines below (male) the pre-mRNAs. Vertical red lines in exons indicate in-frame stop codons, which prevent synthesis of functional protein. Only female embryos produce functional Sxl protein, which *represses* splicing between exons 2 and 3 in *sxl* pre-mRNA (a) and between exons 1 and 2 in *tra* pre-mRNA (b).

(c) In contrast, the cooperative binding of Tra protein and two SR proteins, Rbp1 and Tra2, *activates* splicing between exons 3 and 4 and cleavage/polyadenylation A_n at the 3′ end of exon 4 in *dsx* pre-mRNA in female embryos. In male embryos, which lack functional Tra, the SR proteins do not bind to exon 4, and consequently exon 3 is spliced to exon 5. The distinct Dsx proteins produced in female and male embryos as the result of this cascade of regulated splicing repress transcription of genes required for sexual differentiation of the opposite sex. [Adapted from M. J. Moore et al., 1993, in R. Gesteland and J. Atkins, eds., *The RNA World*, Cold Spring Harbor Press, pp. 303–357.]

polyadenylation at the alternative poly(A) site at the 3′ end of exon 4. In male embryos, which produce no Tra protein, exon 4 is skipped, so that exon 3 is spliced to exon 5. Exon 5 is constitutively spliced to exon 6, which is polyadenylated at its 3′ end. As a result of the cascade of regulated RNA processing depicted in Figure 12-14, different Dsx proteins are expressed in male and female embryos. The male Dsx protein is a transcriptional repressor that inhibits the expression of genes required for female development. Conversely, the female Dsx protein represses transcription of genes required for male development.

Figure 12-15 illustrates how the Tra/Tra2/Rbp1 complex is thought to interact with *double-sex* pre-mRNA. Recent studies have shown that Rbp1 and Tra2 are SR proteins, similar to those discussed previously. They mediate the cooperative binding of the Tra/Tra2/Rbp1 complex to six exonic splicing enhancers in exon 4. The bound Tra2 and Rbp1 proteins then promote the binding of U2AF and U2 snRNP to the 3′ end of the intron between exons 3 and 4, just as other SR proteins do for constitutively spliced exons (see Figure 12-11). The Tra/Tra2/Rbp1 complexes may also enhance binding of the cleavage/polyadnylation complex to the 3′ end of exon 4.

► FIGURE 12-15 Model of splicing activation by Tra protein and the SR proteins Rbp1 and Tra2. In female *Drosophila* embryos, splicing of exons 3 and 4 in *dsx* pre-mRNA is activated by binding of Tra/Tra2/Rbp1 complexes to six sites in exon 4. Because Rbp1 and Tra2 cannot bind to the pre-mRNA in the absence of Tra, exon 4 is skipped in male embryos. See the text for discussion. A_n = polyadenylation. [Adapted from T. Maniatis and B. Tasic, 2002, *Nature* **418**:236.]

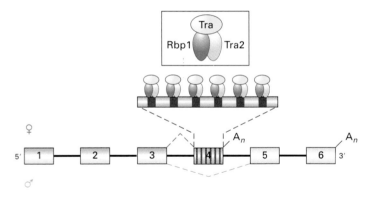

Splicing Repressors and Activators Control Splicing at Alternative Sites

As is evident from Figure 12-14, the *Drosophila* Sxl protein and Tra protein have opposite effects: Sxl prevents splicing, causing exons to be skipped, whereas Tra promotes splicing. The action of similar proteins may explain the cell-type-specific expression of fibronectin isoforms in humans. For instance, an Sxl-like splicing repressor expressed in hepatocytes might bind to splice sites for the EIIIA and EIIIB exons in the fibronectin pre-mRNA, causing them to be skipped during RNA splicing (see Figure 4-15). Alternatively, a Tra-like splicing activator expressed in fibroblasts might activate the splice sites associated with the fibronectin EIIIA and EIIIB exons, leading to inclusion of these exons in the mature mRNA. Experimental examination in some systems has revealed that inclusion of an exon in some cell types versus skipping of the same exon in other cell types results from the combined influence of several splicing repressors and enhancers.

Alternative splicing of exons is especially common in the nervous system, generating multiple isoforms of many proteins required for neuronal development and function in both vertebrates and invertebrates. The primary transcripts from these genes often show quite complex splicing patterns that can generate several different mRNAs, with different spliced forms expressed in different anatomical locations within the central nervous system. We consider two remarkable examples that illustrate the critical role of this process in neural function.

Expression of *Slo* Isoforms in Vertebrate Hair Cells In the inner ear of vertebrates, individual "hair cells," which are ciliated neurons, respond most strongly to a specific frequency of sound. Cells tuned to low frequency ($\approx$50 Hz) are found at one end of the tubular cochlea that makes up the inner ear; cells responding to high frequency ($\approx$5000 Hz) are found at the other end (Figure 12-16a). Cells in between respond to a gradient of frequencies between these extremes. One component in the tuning of hair cells in reptiles and birds is the opening of K^+ ion channels in response to increased intracellular Ca^{2+} concentrations. The Ca^{2+} concentration at which the channel opens determines the frequency with which the membrane potential oscillates, and hence the frequency to which the cell is tuned.

The gene encoding this channel (called *slo*, after the homologous *Drosophila* gene) is expressed as multiple, alternatively spliced mRNAs. The various Slo proteins encoded by these alternative mRNAs open at different Ca^{2+} concentrations. Hair cells with different response frequencies express different isoforms of the Slo channel protein depending on their position along the length of the cochlea. The sequence variation in the protein is very complex: there are at least eight regions in the mRNA where alternative exons are utilized, permitting the expression of 576 possible isoforms (Figure 12-16b). PCR analysis of *slo* mRNAs from individual hair cells has shown that each hair cell expresses a mixture of

different alternative *slo* mRNAs, with different forms predominating in different cells according to their position along the cochlea. This remarkable arrangement suggests that splicing of the *slo* pre-mRNA is regulated in response to extracellular signals that inform the cell of its position along the cochlea.

Recent results have shown that splicing at one of the alternative splice sites in the *slo* pre-mRNA in the rat is suppressed when a specific protein kinase is activated by depolarization of the neuron in response to synaptic activity from interacting neurons. This observation raises the possibility that a splicing repressor specific for this site may be activated when it is phosphorylated by this protein kinase,

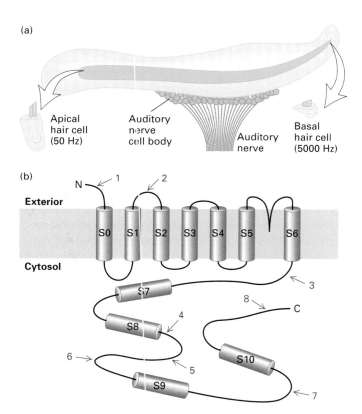

▲ **FIGURE 12-16 Role of alternative splicing of *slo* mRNA in the perception of sounds of different frequency.** (a) The chicken cochlea, a 5-mm-long tube, contains an epithelium of auditory hair cells that are tuned to a gradient of vibrational frequencies from 50 Hz at the apical end (*left*) to 5000 Hz at the basal end (*right*). (b) The Slo protein contains seven transmembrane α helices (S0–S6), which associate to form the K^+ channel. The cytosolic domain, which includes four hydrophobic regions (S7–S10), regulates opening of the channel in response to Ca^{2+}. Isoforms of the Slo channel, encoded by alternatively spliced mRNAs produced from the same primary transcript, open at different Ca^{2+} concentrations and thus respond to different frequencies. Red numbers refer to regions where alternative splicing produces different amino acid sequences in the various Slo isoforms. [Adapted from K. P. Rosenblatt et al., 1997, *Neuron* **19**:1061.]

whose activity in turn is regulated by synaptic activity. Since hnRNP and SR proteins are extensively modified by phosphorylation and other post-translational modifications, it seems likely that complex regulation of alternative RNA splicing through post-translational modifications of splicing factors plays a significant role in modulating neuron function.

Expression of *Dscam* Isoforms in *Drosophila* Retinal Neurons The most extreme example of regulated alternative RNA processing yet uncovered occurs in expression of the *Dscam* gene in *Drosophila*. Mutations in this gene interfere with the normal connections made by the axons of retinal neurons with neurons in a specific region of the brain during fly development. Analysis of the *Dscam* gene shows that it contains 95 alternatively spliced exons that could be spliced to generate over 38,000 possible isoforms! These results raise the possibility that the expression of different *Dscam* isoforms through regulated RNA splicing helps to specify the tens of thousands of different specific synaptic connections made between retinal and brain neurons. In other words, the correct wiring of neurons in the brain may depend on regulated RNA splicing.

RNA Editing Alters the Sequences of Pre-mRNAs

Sequencing of numerous cDNA clones and of the corresponding genomic DNAs from multiple organisms led in the mid-1980s to the unexpected discovery of a previously unrecognized type of pre-mRNA processing. In this type of processing, called **RNA editing**, the sequence of a pre-mRNA is altered; as a result, the sequence of the corresponding mature mRNA differs from the exons encoding it in genomic DNA.

RNA editing is widespread in the mitochondria of protozoans and plants and also in chloroplasts; in these organelles, more than half the sequence of some mRNAs is altered from the sequence of the corresponding primary transcripts. In higher eukaryotes, RNA editing is much rarer, and thus far only single-base changes have been observed. Such minor editing, however, turns out to have important functional consequences in some cases.

An important example of RNA editing in mammals involves the *apoB* gene, which encodes two alternative forms of the serum protein apolipoprotein B (apoB): apoB-100 expressed in hepatocytes and apoB-48 expressed in intestinal epithelial cells. The ≈240-kDa apoB-48 corresponds to the N-terminal region of the ≈500-kDa apoB-100. As we detail in Chapter 18, both apoB proteins are components of large lipoprotein complexes that transport lipids in the serum. However, only low-density lipoprotein (LDL) complexes, which contain apoB-100 on their surface, deliver cholesterol to body tissues by binding to the LDL receptor present on all cells.

The cell-type-specific expression of the two forms of apoB results from editing of *apoB* pre-mRNA so as to change the nucleotide at position 6666 in the sequence from a C to a U. This alteration, which occurs only in intestinal cells, converts a CAA codon for glutamine to a UAA stop codon, leading to synthesis of the shorter apoB-48 (Figure 12-17). Studies with the partially purified enzyme that performs the post-transcriptional deamination of C_{6666} to U shows that it can recognize and edit an RNA as short as 26 nucleotides with the sequence surrounding C_{6666} in the *apoB* primary transcript.

KEY CONCEPTS OF SECTION 12.2

Regulation of Pre-mRNA Processing

■ Because of alternative splicing of primary transcripts and cleavage at different poly(A) sites, different mRNAs may be expressed from the same gene in different cell types or at different developmental stages (see Figure 12-14).

■ Alternative splicing can be regulated by RNA-binding proteins that bind to specific sequences near regulated

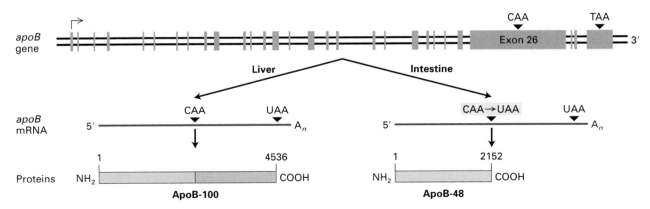

▲ **FIGURE 12-17 RNA editing of *apo-B* pre-mRNA.** The *apoB* mRNA produced in the liver has the same sequence as the exons in the primary transcript. This mRNA is translated into apoB-100, which has two functional domains: an N-terminal domain (green) that associates with lipids and a C-terminal domain (orange) that binds to LDL receptors on cell membranes.

In the *apo-B* mRNA produced in the intestine, the CAA codon in exon 26 is edited to a UAA stop codon. As a result, intestinal cells produce apoB-48, which corresponds to the N-terminal domain of apoB-100. [Adapted from P. Hodges and J. Scott, 1992, *Trends Biochem. Sci.* **17**:77.]

splice sites. Splicing repressors may sterically block the binding of splicing factors to specific sites in pre-mRNAs or inhibit their function. Splicing activators enhance splicing by interacting with splicing factors, thus promoting their association with a regulated splice site.

■ In RNA editing the nucleotide sequence of a pre-mRNA is altered in the nucleus. In vertebrates, this process is fairly rare and entails deamination of a single base in the mRNA sequence, resulting in a change in the amino acid specified by the corresponding codon and production of a functionally different protein (see Figure 12-17).

12.3 Macromolecular Transport Across the Nuclear Envelope

Once the processing of an mRNA is completed in the nucleus, it remains associated with specific hnRNP proteins in a *messenger ribonuclear protein complex*, or *mRNP*. Before it can be translated into the encoded protein, it must be exported out of the nucleus into the cytoplasm. The nucleus is separated from the cytoplasm by two membranes, which form the **nuclear envelope** (see Figure 5-19). Like the plasma membrane surrounding cells, each nuclear membrane consists of a water-impermeable phospholipid bilayer and various associated proteins. Transport of macromolecules including mRNPs, tRNAs, and ribosomal subunits out of the nucleus and transport of all nuclear proteins translated in the cytoplasm into the nucleus occur through *nuclear pores* in a process that differs fundamentally from the transport of small molecules and ions across other cellular membranes (Chapter 7). Insight into the mechanisms of transport through nuclear pores first came from studies of the nuclear pores and of import and export of individual proteins, which we consider first, before returning to the export of mRNPs.

Large and Small Molecules Enter and Leave the Nucleus via Nuclear Pore Complexes

Numerous pores perforate the nuclear envelope in all eukaryotic cells. Each nuclear pore is formed from an elaborate structure termed the **nuclear pore complex (NPC)**, which is immense by molecular standards, ≈125 million daltons in vertebrates, or about 30 times larger than a ribosome. An NPC is made up of multiple copies of some 50 (in yeast) to 100 (in vertebrates) different proteins called **nucleoporins**. Electron micrographs of nuclear pore complexes reveal a roughly octagonal, membrane-embedded structure from which eight ≈100-nm-long filaments extend into the nucleoplasm (Figure 12-18). The distal ends of these filaments are joined by the terminal ring, forming a structure called the

(a)

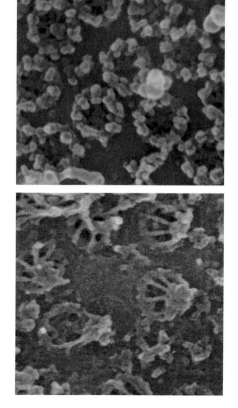

(b)

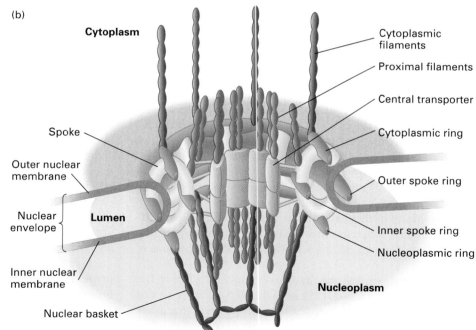

▲ **FIGURE 12-18 Nuclear pore complex.** (a) Nuclear envelopes microdissected from the large nuclei of *Xenopus* oocytes visualized by field emission in-lens scanning electron microscopy. *Top:* View of the cytoplasmic face reveals octagonal shape of membrane-embedded portion of nuclear pore complexes. *Bottom:* View of the nucleoplasmic face shows the nuclear basket that extends from the membrane portion. (b) Cut-away model of the pore complex. [Part (a) from V. Doye and E. Hurt, 1997, *Curr. Opin. Cell Biol.* **9**:401; courtesy of M. W. Goldberg and T. D. Allen. Part (b) adapted from M. P. Rout and J. D. Atchison, 2001, *J. Biol. Chem.* **276**:16593.]

nuclear basket. The membrane-embedded portion also is attached directly to the **nuclear lamina,** a network of lamin intermediate filaments that extends over the inner surface of the nuclear envelope (see Figure 21-16). Cytoplasmic filaments extend from the cytoplasmic side of the NPC into the cytosol.

Ions, small metabolites, and globular proteins up to ≈60 kDa can diffuse through a water-filled channel in the nuclear pore complex; these channels behave as if they are ≈0.9 nm in diameter. However, large proteins and ribonucleoprotein complexes cannot diffuse in and out of the nucleus. Rather, these macromolecules are selectively transported in and out of the nucleus with the assistance of soluble transporter proteins that bind macromolecules and also interact with certain nucleoporins. The principles underlying macromolecular transport through nuclear pore complexes were first determined for the import of individual proteins into the nucleus, which we discuss first before turning to the question of how fully processed mRNAs are transported into the cytoplasm.

Importins Transport Proteins Containing Nuclear-Localization Signals into the Nucleus

All proteins found in the nucleus are synthesized in the cytoplasm and imported into the nucleus through nuclear pore complexes. Such proteins contain a *nuclear-localization signal (NLS)* that directs their selective transport into the nucleus. NLSs were first discovered through the analysis of mutants of simian virus 40 (SV40) that produced an abnormal form of the viral protein called large T-antigen. The wild-type form of this protein is localized to the nucleus in virus-infected cells, whereas some mutated forms of large T-antigen accumulate in the cytoplasm. The mutations responsible for this altered cellular localization all occur within a specific seven-residue sequence rich in basic amino acids near the C-terminus of the protein: Pro-Lys-Lys-Lys-Arg-Lys-Val. Experiments with engineered hybrid proteins in which this sequence was fused to a cytosolic protein demonstrated that it directs transport into the nucleus, and consequently functions as an NLS (Figure 12-19). NLS sequences subsequently were identified in numerous other proteins imported into the nucleus. Many of these are similar to the basic NLS in SV40 large T-antigen, whereas other NLSs are chemically quite different. For instance, an NLS in the hnRNP A1 protein is hydrophobic.

Early work on the mechanism of nuclear import focused on proteins containing a basic NLS, similar to the one in SV40 large T-antigen. A digitonin-permeabilized cell system provided an in vitro assay for analyzing soluble cytosolic components required for nuclear import (Figure 12-20). Using this assay system, four required proteins were purified: Ran, nuclear transport factor 2 (NTF2), importin α, and importin β. *Ran* is a monomeric **G protein** that exists in two conformations, one when complexed with GTP and an alternative one when the GTP is hydrolyzed to GDP (see Figure 3-29). The two **importins**

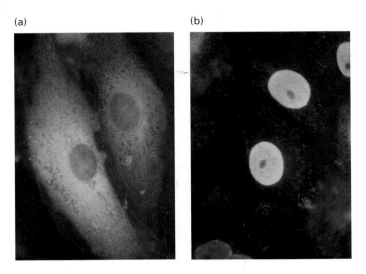

(a)　　　　　　　(b)

▲ **EXPERIMENTAL FIGURE 12-19** Fusion of a nuclear-localization signal (NLS) to a cytoplasmic protein causes the protein to enter the cell nucleus. (a) Normal pyruvate kinase, visualized by immunofluorescence after treating cultured cells with a specific antibody (yellow), is localized to the cytoplasm. This very large cytosolic protein functions in carbohydrate metabolism. (b) When a chimeric pyruvate kinase protein containing the SV40 NLS at its N-terminus was expressed in cells, it was localized to the nucleus. The chimeric protein was expressed from a transfected engineered gene produced by fusing a viral gene fragment encoding the SV40 NLS to the pyruvate kinase gene. [From D. Kalderon et al., 1984, *Cell* **39**:499; courtesy of Dr. Alan Smith.]

form a heterodimeric *nuclear-import receptor:* the α subunit binds to a basic NLS in a "cargo" protein to be transported into the nucleus, and the β subunit interacts with a class of nucleoporins called *FG-nucleoporins.* These nucleoporins, which line the channel of the nuclear pore complex and also are found in the nuclear basket and the cytoplasmic filaments, contain multiple repeats of short hydrophobic sequences rich in phenylalanine (F) and glycine (G) residues (FG-repeats). More recently, several importin β homologs have been found that function in the nuclear import of proteins with other classes of NLSs. These importins interact directly with their cognate NLSs without the need for an adapter protein like importin α.

A current model for the import of cytoplasmic cargo proteins mediated by a monomeric importin is shown in Figure 12-21. Free importin in the cytoplasm binds to its cognate NLS in a cargo protein, forming a *bimolecular cargo complex.* The cargo complex then translocates through the NPC channel as the importin binds transiently to successive individual FG-repeats in the FG-nucleoporins that line the channel. The FG-repeats are thought to act like "stepping stones" as the cargo complex diffuses from one FG-nucleoporin to another on its way through the channel by a process that does not require a direct input of energy from ATP hydrolysis or other mechanisms. The

(a) Effect of digitonin

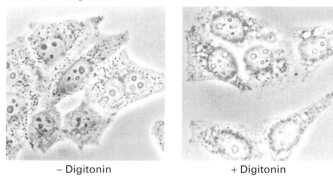

− Digitonin   + Digitonin

(b) Nuclear import by permeabilized cells

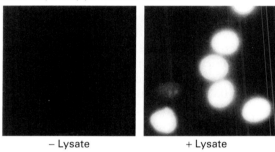

− Lysate   + Lysate

▲ **EXPERIMENTAL FIGURE 12-20 The failure of nuclear transport to occur in permeabilized cultured cells in the absence of lysate demonstrates the involvement of soluble cytosolic components in the process.** (a) Phase-contrast micrographs of untreated and digitonin-permeabilized HeLa cells. Treatment of a monolayer of cultured cells with the mild, nonionic detergent digitonin permeabilizes the plasma membrane so that cytosolic constituents leak out, but leaves the nuclear envelope and NPCs intact. (b) Fluorescence micrographs of digitonin-permeabilized HeLa cells incubated with a fluorescent protein chemically coupled to a synthetic SV40 T-antigen NLS peptide in the presence and absence of cytosol (lysate). Accumulation of this transport substrate in the nucleus occurred only when cytosol was included in the incubation (*lower right*). [From S. Adam et al., 1990, *J. Cell. Biol.* **111**:807; courtesy of Dr. Larry Gerace.]

hydrophobic FG-repeats are thought to occur in regions of extended, otherwise hydrophilic polypeptide chains that fill the central transporter channel. The extended chains are proposed to form a meshwork of hydrophilic strands associated through hydrophobic interactions between the FG-repeats. This molecular meshwork is thought to restrict the diffusion of proteins larger than ≈60 kD. However, importins carrying cargo proteins are thought to penetrate the molecular meshwork by making successive interactions with the FG-repeats.

When the cargo complex reaches the nucleoplasm, the importin interacts with Ran·GTP, causing a conformational change in the importin that decreases its affinity for the NLS, releasing the cargo protein into the nucleoplasm.

The importin-Ran·GTP complex then diffuses back through the NPC, again, through transient interactions of the importin with FG repeats. Once the importin-Ran·GTP complex reaches the cytoplasmic side of the NPC, Ran interacts with a specific *GTPase-accelerating protein (Ran-GAP)* that is a component of the NPC cytoplasmic filaments. This stimulates Ran to hydrolyze its bound GTP to GDP, causing it to convert to a conformation that has low affinity for the importin, so that the free importin is released into the cytoplasm, where it can participate in another cycle of import.

Ran is returned to the nucleus by NTF2. The NTF2 dimer binds specifically to Ran·GDP and also interacts with the FG repeats of FG-nucleoporins. Consequently, the NTF2-Ran·GDP complex can diffuse through the pore via transient

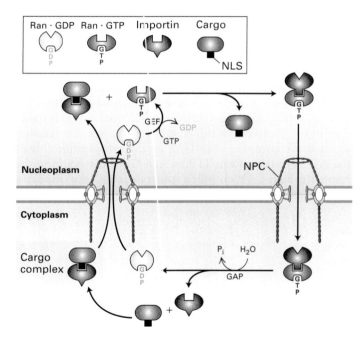

▲ **FIGURE 12-21 Mechanism for nuclear import of "cargo" proteins.** In the cytoplasm (*bottom*), a free importin binds to the NLS of a cargo protein, forming a bimolecular cargo complex. In the case of a basic NLS, the adapter protein importin α bridges the NLS and importin β, forming a trimolecular cargo complex (not shown). The cargo complex diffuses through the NPC by interacting with successive FG-nucleoporins. In the nucleoplasm, interaction of Ran·GTP with the importin causes a conformational change that decreases its affinity for the NLS, releasing the cargo. To support another cycle of import, the importin-Ran·GTP complex is transported back to the cytoplasm. A GTPase-accelerating protein (GAP) associated with the cytoplasmic filaments of the NPC stimulates Ran to hydrolyze the bound GTP. This generates a conformational change causing dissociation from the importin, which can then initiate another round of import. Ran·GDP is bound by NTF2 (not shown) and returned to the nucleoplasm, where a guanine nucleotide–exchange factor (GEF) causes release of GDP and rebinding of GTP.

interactions between NTF2 and the FG repeats. When this complex reaches the nucleoplasm, it encounters a specific *guanine nucleotide–exchange factor (Ran-GEF)* that causes Ran to release its bound GDP and rebind GTP that is present at much higher concentration. The resulting conformational change in Ran decreases its affinity for NTF2 so that free Ran·GTP is released into the nucleoplasm and NTF2 is free to diffuse back through the pore.

The import complex travels through the pore by diffusion, a random process. Yet transport is unidirectional. The direction of transport is a consequence of the rapid dissociation of the import complex when it reaches the nucleoplasm. As a result, there is a concentration gradient of the importin-cargo complex across the NPC: high in the cytoplasm where the complex assembles and low in the nucleoplasm where it dissociates. This concentration gradient is responsible for the unidirectional nature of nuclear import. A similar concentration gradient is responsible for driving the importin in the nucleus back into the cytoplasm. The concentration of the importin-Ran·GTP complex is higher in the nucleoplasm, where it assembles, than on the cytoplasmic side of the NPC, where it dissociates. Ultimately, the direction of the transport processes is dependent on the asymmetric distribution of the Ran-GEF and the Ran-GAP. Ran-GEF in the nucleoplasm maintains Ran in the Ran·GTP state, where it promotes dissociation of the cargo complex. Ran-GAP on the cytoplasmic side of the NPC converts Ran·GTP to Ran·GDP, dissociating the importin-Ran·GTP complex and releasing free importin into the cytosol.

Exportins Transport Proteins Containing Nuclear-Export Signals out of the Nucleus

A very similar mechanism is used to export proteins, tRNAs, and ribosomal subunits from the nucleus to the cytoplasm. This mechanism initially was elucidated from studies of certain hnRNP proteins that "shuttle" between the nucleus and cytoplasm. For instance, the cell-fusion experiments described in Figure 12-22 first showed that some hnRNP proteins cycle in and out of the cytoplasm, whereas others remain localized in the nucleus. Such "shuttling" proteins contain a *nuclear-export signal (NES)* that stimulates their export from the nucleus to the cytoplasm through nuclear pores, in addition to an NLS that results in their reuptake into the nucleus. Experiments with engineered hybrid genes encoding a nucleus-restricted protein fused to various segments of a protein that shuttles in and out of the nucleus have identified at least three different classes of NESs: a leucine-rich sequence found in PKI (an inhibitor of protein kinase A) and in the Rev protein of human immunodeficiency virus (HIV), a 38-residue sequence in hnRNP A1, and a sequence in hnRNP K. To date, no functionally significant structural features (e.g., repeated amino acids) have been identified in the latter two classes.

The mechanism whereby shuttling proteins are exported from the nucleus is best understood for those containing a leucine-rich NES. According to the current model shown in Figure 12-23, a specific *nuclear-export receptor* in the nucleus, **exportin 1**, first forms a complex with Ran·GTP and then binds the NES in a cargo protein. Binding of exportin 1

(a)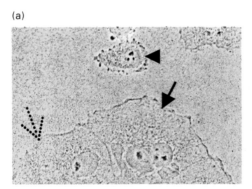

(b)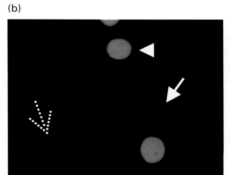

(c)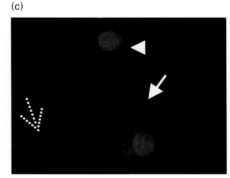

▲ **EXPERIMENTAL FIGURE 12-22 The movement of human hnRNP A1 protein between nuclei in a heterokaryon shows that it can cycle in and out of the cytoplasm, but human hnRNP C protein, which showed no such movement, cannot.** Cultured HeLa cells and *Xenopus* cells were fused by treatment with polyethylene glycol, producing heterokaryons containing nuclei from each cell type. The hybrid cells were treated with cycloheximide immediately after fusion to prevent protein synthesis. After 2 hours, the cells were fixed and stained with fluorescent-labeled antibodies specific for human hnRNP C and A1 proteins. These antibodies do not bind to the homologous *Xenopus* proteins. (a) A fixed preparation viewed by phase-contrast microscopy includes unfused HeLa cells (arrowhead) and *Xenopus* cells (dotted arrow), as well as fused heterokaryons (solid arrow). In the heterokaryon in this micrograph, the round HeLa-cell nucleus is to the right of the oval-shaped *Xenopus* nucleus. (b, c) When the same preparation was viewed by fluorescence microscopy, the stained hnRNP C protein appeared green and the stained hnRNP A1 protein appeared red. Note that the unfused *Xenopus* cell on the left is unstained, confirming that the antibodies are specific for the human proteins. In the heterokaryon, hnRNP C protein appears only in HeLa-cell nuclei (b), whereas the A1 protein appears in both nuclei (c). Since protein synthesis was blocked after cell fusion, some of the human hnRNP A1 protein must have left the HeLa-cell nucleus, moved through the cytoplasm, and entered the *Xenopus* nucleus in the heterokaryon. [See S. Pinol-Roma and G. Dreyfuss, 1992, *Nature* **355**:730; courtesy of G. Dreyfuss.]

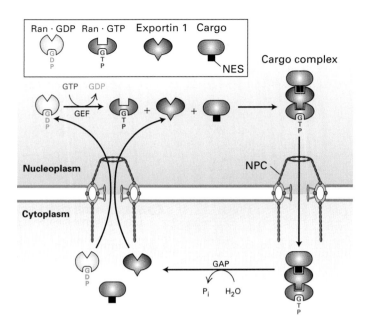

▲ FIGURE 12-23 Mechanism for nuclear export of cargo proteins containing a leucine-rich nuclear-export signal (NES). In the nucleoplasm, the protein exportin 1 binds cooperatively to the NES of the cargo protein to be transported and to Ran·GTP. After the resulting cargo complex diffuses through an NPC via transient interactions with FG repeats in FG-nucleoporins, the Ran GAP associated with the NPC cytoplasmic filaments stimulates conversion of Ran·GTP to Ran·GDP. The accompanying conformational change in Ran leads to dissociation of the complex. The NES-containing cargo protein is released into the cytosol, while exportin 1 and Ran·GDP are transported back into the nucleus through NPCs. Ran·GDP is transported through its interaction with NTF2. Ran-GEF in the nucleoplasm then stimulates conversion of Ran·GDP to Ran·GTP.

to Ran·GTP causes a conformational change in exportin 1 that increases its affinity for the NES so that a *trimolecular cargo complex* is formed. Like importins, exportin 1 interacts transiently with FG repeats in FG-nucleoporins and diffuses through the NPC. The cargo complex dissociates when it encounters the Ran-GAP in the NPC cytoplasmic filaments that stimulates Ran to hydrolyze the bound GTP, shifting it into a conformation that has low affinity for exportin 1. The released exportin 1 changes conformation to a structure that has low affinity for the NES, releasing the cargo into the cytosol. The direction of the export process is driven by this dissociation of the cargo from exportin 1 in the cytoplasm that causes a concentration gradient of the cargo complex across the NPC so that it is high in the nucleoplasm and low in the cytoplasm. Exportin 1 and the Ran·GDP are then transported back into the nucleus through an NPC. Ran·GDP is transported by NTF2, as discussed earlier in regard to nuclear import. It is then converted into Ran·GTP by the Ran-GEF in the nucleoplasm.

By comparing this model for nuclear export with that in Figure 12-21 for nuclear import, we can see one obvious dif-

ference: Ran·GTP is part of the cargo complex during export but not during import. Apart from this difference, the two transport processes are remarkably similar. In both processes, association of a transport signal receptor with Ran·GTP in the nucleoplasm causes a conformational change that affects its affinity for the transport signal. During import, the interaction causes release of the cargo, whereas, during export, the interaction promotes association with the cargo. In both export and import, stimulation of Ran·GTP hydrolysis in the cytoplasm by the Ran-GAP associated with NPC cytoplasmic filaments produces a conformational change in Ran that releases the transport signal receptor. During nuclear export, the cargo is also released. Importins and exportins both are thought to diffuse through the NPC channel by successive interactions with FG-repeats in FG-nucleoporins. Localization of the Ran-GAP and -GEF to the cytoplasm and nucleus, respectively, is the basis for the unidirectional transport of cargo proteins across the NPC.

In keeping with the similarity in function of importins and exportins, the two types of transport proteins are highly homologous in sequence and structure. The entire family is called the importin β family, or **karyopherins.** There are 14 karyopherins in yeast and more than 20 in mammalian cells. The NESs or NLSs to which they bind have been determined for only a fraction of them. Remarkably, some individual karyopherins function as both an importin and an exportin.

A similar shuttling mechanism has been shown to export other cargoes from the nucleus. For example, *exportin-t* functions to export tRNAs. Exportin-t binds fully processed tRNAs in a complex with Ran·GTP that diffuses through NPCs and dissociates when it interacts with Ran-GAP in the NPC cytoplasmic filaments, releasing the tRNA into the cytosol. A Ran-dependent process is also required for the nuclear export of ribosomal subunits through NPCs. Likewise, certain specific mRNAs that associate with particular hnRNP proteins (e.g., HIV Rev discussed later) can be exported by a Ran-dependent mechanism. However, most mRNAs are exported from the nucleus by another type of transporter that does not interact with Ran. This will be discussed after we examine gene regulation through control of transcription-factor import and export.

Control of Some Genes Is Achieved by Regulating Transport of Transcription Factors

Regulation of some genes is achieved by regulating the nuclear transport of specific transcription factors that control their transcription. For example, in Chapter 11 we learned that homodimeric **nuclear receptors** such as the glucocorticoid receptor are sequestered in the cytoplasm in the absence of ligand. Binding of ligand stimulates their translocation into the nucleus, where they can bind to hormone response elements in target genes and activate transcription (see Figures 11-43 and 11-44). The ligand-binding domain of these transcription factors contains an NLS, but in the absence of ligand, the NLS is masked by interactions with a cytoplasmic tethering complex. The conformational change that results from binding hormone releases the domain from the tethering complex, exposing the NLS so that it is free to interact with its cognate importin and undergo nuclear import, as discussed above. Other examples of regulated nuclear import of transcription factors through controlled masking and exposure of their associated NLS are discussed in subsequent chapters on signal transduction.

In some cases, regulated nuclear transport of transcription factors is achieved by expressing them as fusions to domains that localize them to cytoplasmic membranes, such as the plasma membrane or ER. In response to specific signals, specific proteases are activated that cleave them, releasing the transcription-factor domains so that they can be transported into the nucleus by an importin. In other cases, transcription-factor nuclear export is regulated by post-translational modifications such as phosphorylation that increase their affinity for a specific exportin. Regulation of nuclear import or export is a powerful mechanism for controlling transcription-factor activity, since a transcription factor can bind to its cognate sites in the genes it regulates only when it is in the nucleus.

Most mRNPs Are Exported from the Nucleus with the Aid of an mRNA-Exporter

Recent studies of temperature-sensitive yeast mutants defective in nuclear transport identified a heterodimeric **mRNA-exporter** protein that appears to direct most mRNPs through nuclear pores. Yeast cells with mutations in the subunits of this protein accumulate polyadenylated RNA in the nucleus at the nonpermissive temperature, indicating that the protein is required for the transport of most mRNPs. Highly conserved homologs of the large and small subunits of the yeast mRNA-exporter are found in all eukaryotes, and several lines of evidence indicate that the mRNA-exporter is required for the export of most mRNPs in vertebrate cells as well as in yeast cells. Surprisingly, as mentioned earlier, experiments in both yeast and vertebrate cells indicate that Ran is not required for export of the vast majority of mRNPs.

The small subunit of the mRNA-exporter is homologous to NTF2 and interacts with a region of the larger subunit that also shares homology to NTF2. Together they form a domain that interacts with FG repeats in FG-nucleoporins similarly to NTF2, which is a dimer. A C-terminal domain of the large subunit also interacts with FG repeats. Although several other proteins are also required for mRNP transport, researchers consider this mRNP-exporter to be the primary transport protein because of the importance of FG-repeat binding for the mechanism of cargo transport by karyopherins. The large mRNA-exporter subunit also contains RNA-binding domains that appear to bind to RNA cooperatively with specific mRNP proteins. The finding that the mRNA-exporter is associated with mRNPs and binds directly to FG repeats led to the model of mRNP transport depicted in Figure 12-24. This model proposes that the mRNA-exporter translocates mRNPs through the nuclear pore channel similarly to the karyopherins, that is, by binding transiently to successive individual FG-repeats as it diffuses through the channel.

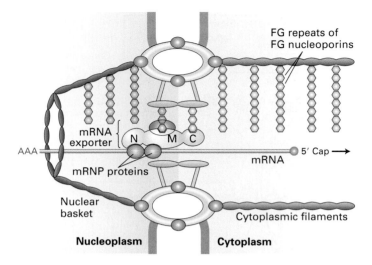

▲ **FIGURE 12-24 Proposed mechanism of mRNP transport from the nucleus via the mRNA-exporter.** The large subunit (light purple) of the mRNA-exporter contains three key domains: a middle domain (M) and a carboxyl domain (C), both of which bind to hydrophobic phenylalanine-glycine (FG) repeats in FG-nucleoporins, and an N-terminal region that has weak RNA-binding activity for most sequences. Binding to most mRNPs requires cooperative binding with other specific hnRNP proteins. The small subunit (dark purple) binds to the middle domain of the large subunit and contributes to the binding of FG repeats. For simplicity, FG repeats are not shown on the lower half of the nuclear pore complex. Note that the mRNA-exporter is not drawn to scale; it is far smaller than the nuclear pore complex. The mRNA-exporter is proposed to diffuse through the pore by making transient interactions with adjacent FG-nucleoporins as it progresses. [Adapted from R. Reed and E. Hurt, 2002, *Cell* **108**:523.]

Currently, it is not clear what provides directionality to this mechanism of mRNP transport through nuclear pores. By analogy with karyopherin transport, one hypothesis is that the mRNA-exporter-mRNP complex dissociates as it reaches the cytoplasm. This would result in a high concentration of the mRNA-exporter-mRNP complex at the nuclear basket, where it associates with FG-nucleoporins, and a low concentration of the complex in the cytoplasm, where it dissociates. As for karyopherin transport, such a concentration gradient could drive vectorial mRNP translocation. However, much remains to be learned about the hnRNP proteins associated with mRNPs during transport through the pore and the mechanism by which they dissociate from the mRNP on the cytoplasmic side of the pore.

Other Proteins That Assist in mRNP Export In addition to the mRNA-exporter and FG-nucleoporins, several other types of proteins are involved in the transport of mRNPs by this mechanism. As mentioned earlier, the mRNA-exporter is thought to bind to mRNAs cooperatively with specific mRNP proteins. For example, SR proteins associated with exons appear to stimulate the binding of the mRNA-exporter to processed mRNAs in mRNPs. Thus SR proteins not only function to define exons during RNA splicing (see Figure 12-11) but also participate in the export process that translocates most mRNPs into the cytoplasm.

Other nuclear RNA-binding proteins that probably function in transport of mRNPs recently have been identified. These proteins, which initially were found associated with mRNAs spliced in vitro, bind to a region about 20 bases 5′ of the exon sequences joined by splicing. Analysis of the proteins present in these *exon-junction complexes* revealed that several are homologous to yeast proteins that are altered in mutants defective in mRNP export. Moreover, one of these proteins has been shown to bind directly to the large subunit of the mRNA-exporter. These results suggest that exon-junction complexes deposited on mRNAs during RNA splicing stimulate mRNP transport through nuclear pores. Some of the proteins in exon-junction complexes also function as splicing factors. As we discuss in the next section, on mechanisms of cytoplasmic post-transcriptional control, the exon-junction complex also participates in a type of post-transcriptional control that prevents translation of improperly spliced mRNAs.

Another participant in mRNP transport to the cytoplasm is the nuclear cap-binding complex, mentioned earlier as protection against exonuclease attack on the 5′ end of nascent transcripts and pre-mRNAs. Electron microscopy experiments discussed below have demonstrated that the 5′ end of mRNAs lead the way through the nuclear pore complex. Recent experiments in yeast indicate that the 3′ poly(A) tail plays an important role in mRNP transport, suggesting that a poly(A)-binding protein participates. Nucleoporins associated with the NPC cytoplasmic filaments in addition to FG-nucleoporins are required for mRNA export and may function to dissociate the mRNA-exporter and other mRNP proteins that accompany the mRNP through the pore.

Once the mRNP reaches the cytoplasm, most of the mRNP proteins that associated with the mRNA in the nucleus, the nuclear cap-binding complex, and the nuclear poly(A)-binding protein (PABPII) dissociate and are shuttled back to the nucleus. In the cytoplasm, the 5′ cap of an exported mRNA is bound by the eIF4E translation initiation factor, the poly(A) tail is bound by multiple copies of the cytoplasmic poly(A)-binding protein (PABPI), and other RNA-binding proteins associate with the body of the mRNA, forming a cytoplasmic mRNP that has a lower ratio of protein to RNA than nuclear mRNPs.

Nuclear Export of Balbiani Ring mRNPs The salivary glands of larvae of the insect *Chironomous tentans* have provided a good model system for EM studies of the formation of hnRNPs and the export of mRNPs. In these larvae, genes in large chromosomal puffs called *Balbiani rings* are abundantly transcribed into nascent pre-mRNAs that associate with hnRNP proteins and are processed into coiled mRNPs with an mRNA of ≈75 kb (Figure 12-25a, b). These giant mRNAs encode large glue proteins that adhere the developing larvae to a leaf. After processing of the pre-mRNA in Balbiani ring hnRNPs, the resulting mRNPs move through nuclear pores to

the cytoplasm. Electron micrographs of sections of these cells show mRNPs that appear to uncoil during their passage through nuclear pores and then bind to ribosomes as they enter the cytoplasm. The observation that mRNPs become associated with ribosomes during transport indicates that the 5′ end leads the way through the nuclear pore complex, as mentioned earlier. Detailed electron microscopic studies of the transport of Balbiani ring mRNPs through nuclear pore complexes led to the model depicted in Figure 12-25c.

Pre-mRNAs in Spliceosomes Are Not Exported from the Nucleus

It is critical that only fully processed mature mRNAs be exported from the nucleus because translation of incompletely processed pre-mRNAs containing introns would produce defective proteins that might interfere with the functioning of the cell. By mechanisms that are not fully understood, pre-mRNAs associated with snRNPs in spliceosomes usually are prevented from being transported to the cytoplasm.

In one type of experiment demonstrating this restriction, a gene encoding a pre-mRNA with a single intron that normally is spliced out was mutated to introduce deviations from the consensus splice-site sequences. Mutation of *either* the 5′ or the 3′ invariant splice-site bases at the ends of the intron resulted in pre-mRNAs that were bound by snRNPs to form spliceosomes; however, RNA splicing was blocked, and the pre-mRNA was retained in the nucleus. In contrast, mutation of *both* the 5′ and 3′ splice sites in the same pre-mRNA resulted in export of the unspliced pre-mRNA, although less efficiently than for the spliced mRNA. When both splice sites were mutated, the pre-mRNAs were not efficiently bound by snRNPs, and, consequently, their export was not blocked.

 Many cases of thalassemia, an inherited disease that results in abnormally low levels of globin proteins, are due to mutations in globin-gene splice sites that decrease the efficiency of splicing but do not prevent association of the pre-mRNA with snRNPs. The resulting unspliced globin pre-mRNAs are retained in reticulocyte nuclei and are rapidly degraded. ∎

HIV Rev Protein Regulates the Transport of Unspliced Viral mRNAs

As discussed earlier, transport of mRNPs containing mature, functional mRNAs from the nucleus to the cytoplasm entails a complex mechanism that is crucial to gene expression (see Figures 12-24 and 12-25). Regulation of this transport theoretically could provide another means of gene control, although it appears to be relatively rare. Indeed, the only examples of regulated mRNA export discovered to date occur during the cellular response to conditions (e.g., heat shock) that cause protein denaturation and during viral infection when virus-induced alterations in nuclear transport maximize viral replication. Here we describe the regulation

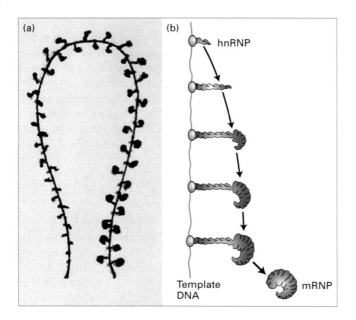

(a)

(b)

hnRNP

Template DNA

mRNP

◀ **FIGURE 12-25 Formation of heterogeneous ribonucleoprotein particles (hnRNPs) and export of mRNPs from the nucleus.** (a) Model of a single chromatin transcription loop and assembly of Balbiani ring (BR) mRNP in *Chironomous tentans*. Nascent RNA transcripts produced from the template DNA rapidly associate with proteins, forming hnRNPs. The gradual increase in size of the hnRNPs reflects the increasing length of RNA transcripts at greater distances from the transcription start site. The model was reconstructed from electron micrographs of serial thin sections of salivary gland cells. (b) Schematic diagram of the biogenesis of hnRNPs. Following processing of the pre-mRNA, the resulting ribonucleoprotein particle is referred to as an mRNP. (c) Model for the transport of BR mRNPs through the nuclear pore complex (NPC) based on electron microscopic studies. Note that the curved mRNPs appear to uncoil as they pass through nuclear pores. As the mRNA enters the cytoplasm, it rapidly associates with ribosomes, indicating that the 5′ end passes through the NPC first. [Part (a) from C. Erricson et al., 1989, *Cell* **56**:631; courtesy of B. Daneholt. Parts (b) and (c) adapted from B. Daneholt, 1997, *Cell* **88**:585. See also B. Daneholt, 2001, *Proc. Nat'l. Acad. Sci. USA* **98**:7012.]

(c)   Nuclear envelope

Nucleoplasm   **Cytoplasm**

mRNP

mRNA

NPC

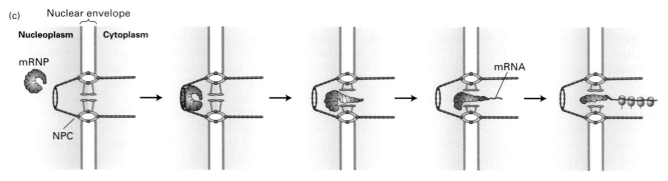

of mRNP export mediated by a protein encoded by human immunodeficiency virus (HIV).

A retrovirus, HIV integrates a DNA copy of its RNA genome into the host-cell DNA (see Figure 4-43). The integrated viral DNA, or provirus, contains a single transcription unit, which is transcribed into a single primary transcript by cellular RNA polymerase II. The HIV transcript can be spliced in alternative ways to yield three classes of mRNAs: a 9-kb unspliced mRNA; ≈4-kb mRNAs formed by removal of one intron; and ≈2-kb mRNAs formed by removal of two or more introns (Figure 12-26). After their synthesis in the host-cell nucleus, all three classes of HIV mRNAs are transported to the cytoplasm and translated into viral proteins; some of the 9-kb unspliced RNA is used as the viral genome in progeny virions that bud from the cell surface.

Since the 9-kb and 4-kb HIV mRNAs contain splice sites, they can be viewed as incompletely spliced mRNAs. However, as discussed earlier, association of such incompletely spliced mRNAs with snRNPs in spliceosomes normally blocks their export from the nucleus. Thus HIV, as well as other retroviruses, must have some mechanism for overcoming this block, permitting export of the longer viral mRNAs. Some retroviruses have evolved a sequence called the *constitutive transport element (CTE)* that binds to the mRNA-exporter with

high affinity, thereby permitting export of unspliced retroviral RNA into the cytoplasm. HIV solved the problem differently.

Studies with HIV mutants showed that transport of unspliced 9-kb and singly spliced 4-kb viral mRNAs from the nucleus to the cytoplasm requires the virus-encoded *Rev protein*. Subsequent biochemical experiments demonstrated that Rev binds to a specific Rev-response element (RRE) present in HIV RNA. In cells infected with HIV mutants lacking the RRE, unspliced and singly spliced viral mRNAs remain in the nucleus, demonstrating that the RRE is required for Rev-mediated stimulation of transport. Rev contains a leucine-rich NES that interacts with exportin 1 complexed with Ran·GTP. As a result, Rev exports unspliced and singly spliced HIV mRNAs through interactions with exportin 1 and the nuclear pore complex.

KEY CONCEPTS OF SECTION 12.3

Macromolecular Transport Across the Nuclear Envelope

■ The nuclear envelope contains numerous nuclear pore complexes (NPCs), large, complicated structures composed of multiple copies of ≈50–100 proteins called *nucleoporins* (see Figure 12-18). FG-nucleoporins, which contain multiple

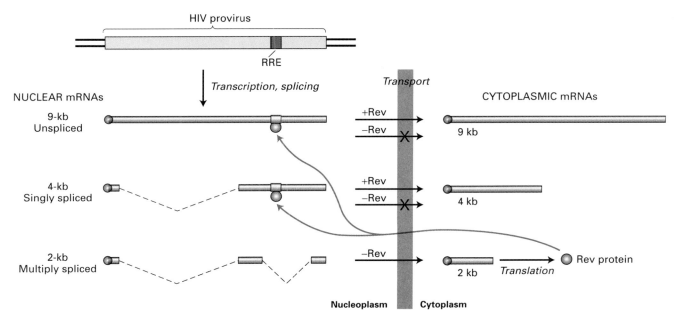

▲ FIGURE 12-26 Role of Rev protein in transport of HIV mRNAs from the nucleus to the cytoplasm. The HIV genome, which contains several coding regions, is transcribed into a single 9-kb primary transcript. Several ≈4-kb mRNAs result from alternative splicing out of any one of several introns (dashed lines), and several ≈2-kb mRNAs from splicing out of two or more alternative introns. After transport to the cytoplasm, the various RNA species are translated into different viral proteins. Rev protein, encoded by a 2-kb mRNA, interacts with the Rev-response element (RRE) in the unspliced and singly spliced mRNAs, stimulating their transport to the cytoplasm. [Adapted from B. R. Cullen and M. H. Malim, 1991, *Trends Biochem. Sci.* **16**:346.]

repeats of a short hydrophobic sequence (FG-repeats), line the central transporter channel and play a role in transport of all macromolecules through nuclear pores.

■ Transport of macromolecules larger than ≈60 kDa through nuclear pores requires the assistance of proteins that interact with both the transported molecule and with FG-repeats of FG-nucleoporins.

■ Proteins imported to or exported from the nucleus contain a specific amino acid sequence that functions as a nuclear-localization signal (NLS) or a nuclear-export signal (NES). Nucleus-restricted proteins contain an NLS but not an NES, whereas proteins that shuttle between the nucleus and cytoplasm contain both signals.

■ Several different types of NES and NLS have been identified. Each type of nuclear-transport signal is thought to interact with a specific receptor protein (exportin or importin) belonging to a family of homologous proteins termed *karyopharins*.

■ A "cargo" protein bearing an NES or NLS translocates through nuclear pores bound to its cognate receptor protein (karyopharin), which also interacts with FG-nucleoporins. Importins and exportins are thought to diffuse through the channel by binding transiently to different FG-repeats that act like "stepping stones" through the pore. Both transport processes also require participation of Ran, a monomeric G protein that exists in different conformations when bound to GTP or GDP.

■ After a cargo complex reaches its destination (the cytoplasm during export and the nucleus during import), it dissociates, freeing the cargo protein and other components. The latter then are transported through nuclear pores in the reverse direction to participate in transporting additional molecules of cargo protein (see Figures 12-21 and 12-23).

■ The unidirectional nature of protein export and import through nuclear pores results from localization of the Ran guanine nucleotide–exchange factor (GEF) in the nucleus and of Ran GTPase-accelerating protein (GAP) in the cytoplasm. The interaction of import cargo complexes with the Ran-GEF in the nucleoplasm causes dissociation of the complex, releasing the cargo into the nucleoplasm (see Figure 12-21). Export cargo complexes dissociate in the cytoplasm when they interact with Ran GAP localized to the NPC cytoplasmic filaments (see Figure 12-23).

■ Most mRNPs are exported from the nucleus by a heterodimeric mRNA-exporter that interacts with FG-repeats (see Figure 12-24). The direction of transport (nucleus → cytoplasm) may result from dissociation of the exporter-mRNP complex in the cytoplasm by an as yet uncharacterized mechanism that does not depend on Ran.

■ The mRNA-exporter binds to most mRNAs cooperatively with SR proteins bound to exons, to exon-junction complexes that associate with mRNAs following RNA splicing, and to additional mRNP proteins that remain to be characterized.

■ Pre-mRNAs bound by a spliceosome normally are not exported from the nucleus, assuring that only fully processed, functional mRNAs reach the cytoplasm for translation.

12.4 Cytoplasmic Mechanisms of Post-transcriptional Control

Before proceeding, let's quickly review the steps in gene expression at which control is exerted. We saw in the previous chapter that regulation of transcription initiation is the principal mechanism for controlling expression of genes. In preceding sections of this chapter we learned that transcription of some genes is regulated by controlling the transport into and out of the nucleus of transcription factors that regulate them. We also learned that expression of protein isoforms is controlled by regulating alternative RNA splicing. Although transport of fully processed mRNAs to the cytoplasm rarely is regulated, the transport of unspliced retroviral RNAs is specifically controlled.

In this section we consider other mechanisms of post-transcriptional control that contribute to regulating the expression of some genes. Most of these mechanisms operate in the cytoplasm, controlling the stability or localization of mRNA or its translation into protein. We begin by discussing two recently discovered and related mechanisms of gene control.

Micro RNAs Repress Translation of Specific mRNAs

Micro RNAs (miRNAs) were first discovered during analysis of mutations in the *lin-4* and *let-7* genes of the nematode *C. elegans*. Cloning and analysis of wild-type *lin-4* and *let-7* revealed that they encode no protein products, but rather RNAs only 21 and 22 nucleotides long, respectively, that hybridize to the 3′ untranslated regions of specific target mRNAs. For example, the *lin-4* miRNA, which is expressed early in embryogenesis, hybridizes to the 3′ untranslated regions of both the *lin-14* and *lin-28* mRNAs, thereby repressing translation of these mRNAs by an as yet unknown mechanism. Expression of *lin-4* miRNA ceases later in development, allowing translation of newly synthesized *lin-14* and *lin-28* mRNAs at that time. Expression of *let-7* miRNA occurs at comparable times during embryogenesis of all bilaterally symmetric animals. The role of *lin-4* and *let-7* miRNAs in coordinating the timing of early developmental events in *C. elegans* is discussed in Chapter 22. Here we focus on what is currently understood about how miRNAs repress translation.

This form of translational regulation is probably not limited to the *lin-4* and *let-7* miRNAs in *C. elegans*; about 100 different miRNAs have been found in *C. elegans*, and at least as many in humans. All miRNAs appear to be formed by processing of ≈70-nucleotide precursor RNAs that form hairpin structures with a few base-pair mismatches in the stem of the hairpin. A ribonuclease called *Dicer*, which cleaves double-stranded RNA, is required for production of miRNAs from these precursors. The base pairing between a miRNA and the 3′ untranslated region of its target mRNAs is not precisely complementary, so that some base-pair mismatches occur in the hybridized region. This mismatching distinguishes miRNA-mediated translational repression from the related phenomenon of RNA interference, which we describe next.

RNA Interference Induces Degradation of mRNAs with Sequences Complementary to Double-Stranded RNAs

RNA interference (RNAi) was discovered unexpectedly during attempts to experimentally manipulate the expression of specific genes. Researchers tried to inhibit the expression of a gene in *C. elegans* by microinjecting a single-stranded, complementary RNA that would hybridize to the encoded mRNA and prevent its translation, a method called *antisense inhibition*. But in control experiments, perfectly base-paired double-stranded RNA a few hundred base pairs long was much more effective at inhibiting expression of the gene than the antisense strand alone. Similar inhibition of gene expression by an introduced double-stranded RNA soon was observed in plants. In each case, the double-stranded RNA induced degradation of all cellular RNAs containing a sequence that was exactly the same as one strand of the double-stranded RNA. Because of the specificity of RNA interference in targeting mRNAs for destruction, it has become a powerful experimental tool for studying gene function (see Figure 9-43).

Subsequent biochemical studies with extracts of *Drosophila* embryos showed that a long double-stranded RNA that mediates interference is initially processed into a double-stranded intermediate referred to as *short interfering RNA (siRNA)*. The strands in siRNA contain 21–23 nucleotides hybridized to each other so that the two bases at the 3′ end of each strand are single-stranded. The finding that Dicer ribonuclease is required for formation of siRNAs suggested that RNA interference and miRNA-mediated translational repression are related processes.

Recent studies indicate that double-stranded siRNAs and miRNAs are further processed into a multiprotein complex containing only one of the RNA strands (Figure 12-27). This **RNA-induced silencing complex (RISC)** then cleaves target RNAs that are *precisely* complementary to their corresponding single-stranded siRNAs. These complexes also appear to function in the inhibition of translation by miRNAs. The human *let-7* miRNA, for instance, is found in an RNA-induced silencing complex that can cleave a synthetic target RNA that is precisely complementary to *let-7* miRNA. However, the same complex does not cleave an RNA whose

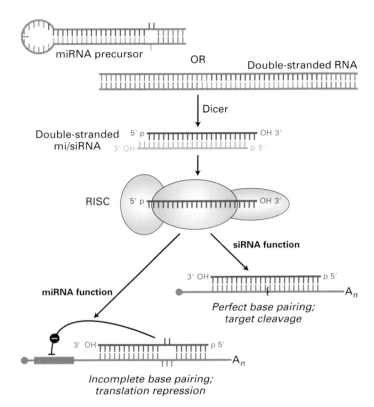

▲ **FIGURE 12-27 Model for miRNA translational repression and RNA interference mediated by the RNA-induced interfering complex (RISC).** Processing of both miRNA precursors into miRNAs and long double-stranded RNAs into short interfering RNAs (siRNAs) requires the Dicer ribonuclease. In both cases, cleavage by Dicer yields a double-stranded RNA intermediate containing 21–23 nucleotides per strand with two-nucleotide 3' single-stranded tails. One strand of this intermediate assembles with multiple proteins to form an RNA-induced silencing complex (RISC). In siRNA function, the target mRNA hybridizes perfectly with the RISC RNA, leading to cleavage of the target. In miRNA function, the RISC RNA forms a hybrid with the target mRNA that contains some base-pair mismatches; in this case translation of the target mRNA is blocked. This is the situation for the *lin-4* and *let-7* miRNAs in *C. elegans.* [Adapted from G. Hutvagner and P. D. Zamore, 2002, *Science* **297**:2056; see also V. Ambrose, 2001, *Cell* **107**:823.]

sequence differs at a few bases from the *let-7* sequence. This finding has led to the hypothesis that RISC complexes have two distinct functions: (1) an siRNA function (i.e., RNA interference) leading to cleavage of precisely complementary mRNAs and (2) an miRNA function leading to translational repression of mRNAs with a few base-pair mismatches (see Figure 12-27).

Mutations in the fragile-X gene *(FMR1),* which encodes a protein containing a KH RNA-binding motif, are associated with the most common form of heritable mental retardation. Recent results indicate that the FMR1 protein is a subunit of human RISC complexes.

Consequently, the mental retardation associated with FMR1 mutations may result from a defect in translational repression of specific target mRNAs during development of the central nervous system. ∎

RNA interference is believed to be an ancient cellular defense against certain viruses and mobile genetic elements in both plants and animals. Plants with mutations in the genes encoding Dicer and RISC proteins exhibit increased sensitivity to infection by RNA viruses and increased movement of **transposons** within their genomes. The double-stranded RNA intermediates generated during replication of RNA viruses are thought to be recognized by the Dicer ribonuclease, inducing an RNAi response that ultimately degrades viral mRNAs. During transposition, transposons are inserted into cellular genes in a random orientation, and their transcription from different promoters produces complementary RNAs that can hybridize with each other, initiating the RNAi system that then interferes with the expression of transposon proteins required for additional transpositions.

In plants and *C. elegans* the RNAi response can be induced in all cells of the organism by introduction of double-stranded RNA into just a few cells. Such organism-wide induction requires a protein produced in plants and in *C. elegans* that is homologous to the RNA replicases of RNA viruses. This finding suggests that double-stranded siRNAs are replicated and then transferred to other cells in these organisms. In plants, transfer of siRNAs might occur through **plasmodesmata**, the cytoplasmic connections between plant cells that traverse the cell walls between them (see Figure 6-34). Organism-wide induction of RNA interference does not occur in *Drosophila* or mammals, presumably because their genomes do not encode RNA replicase homologs. In plants, double-stranded RNA also induces the DNA methylation of genes with the same sequence by an unknown mechanism. Such gene methylation in response to RNAi inhibits transcription of the gene, probably through the binding of histone deacetylases. This type of RNAi-induced gene methylation does not occur in animals.

Cytoplasmic Polyadenylation Promotes Translation of Some mRNAs

In addition to repression of translation by micro RNAs, *protein-mediated* translational control helps regulate expression of some genes. Regulatory sequences, or elements, in mRNAs that interact with specific proteins to control translation generally are present in the untranslated region (UTR) at the 3' or 5' end of an mRNA. Here we discuss a type of protein-mediated translational control involving 3' regulatory elements. A different mechanism involving RNA-binding proteins that interact with 5' regulatory elements is discussed later.

Sequence-specific translation-control proteins may bind cooperatively to neighboring sites in 3' UTRs and function in a combinatorial manner, similar to the cooperative binding of transcription factors to regulatory sites in an enhancer

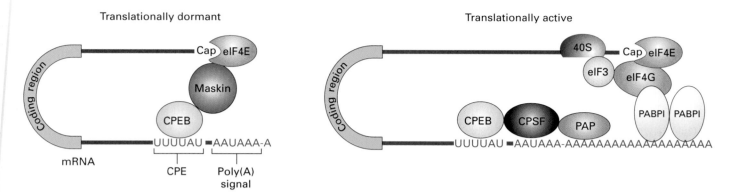

▲ **FIGURE 12-28 Model for control of cytoplasmic polyadenylation and translation initiation.** *Left:* In immature oocytes, mRNAs containing the U-rich cytoplasmic polyadenylation element (CPE) have short poly(A) tails. CPE-binding protein (CPEB) mediates repression of translation through the interactions depicted, which prevent assembly of an initiation complex at the 5' end of the mRNA. *Right:* Hormone stimulation of oocytes activates a protein kinase that phosphorylates CPEB, causing it to release Maskin. The cleavage/polyadenylation specificity factor (CPSF) then binds to the poly(A) site, interacting with both bound CPEB and the cytoplasmic form of poly(A) polymerase (PAP). After the poly(A) tail is lengthened, multiple copies of the cytoplasmic poly(A)-binding protein I (PABPI) can bind to it and interact with eIF4G, which functions with other initiation factors to bind the 40S ribosome subunit and initiate translation. [Adapted from R. Mendez and J. D. Richter, 2001, *Nature Rev. Mol. Cell Biol.* **2**:521.]

region of a gene. In most cases studied, translation is repressed by protein binding to 3' regulatory elements. The mechanism of such repression is best understood for mRNAs that must undergo *cytoplasmic polyadenylation* before they can be translated.

Cytoplasmic polyadenylation is a critical aspect of gene expression in the early embryo. The egg cells (oocytes) of multicellular organisms contain many mRNAs, encoding numerous different proteins, that are not translated until after the egg is fertilized by a sperm cell. Some of these "stored" mRNAs have a short poly(A) tail, consisting of only ≈20–40 A residues, to which just a few molecules of cytoplasmic poly(A)-binding protein (PABPI) can bind. As discussed in Chapter 4, multiple PABPI molecules bound to the long poly(A) tail of an mRNA interact with the eIF4G initiation factor, thereby stabilizing the interaction of the mRNA 5' cap with eIF4E, which is required for translation initiation (see Figure 4-31). Because this stabilization cannot occur with mRNAs that have short poly(A) tails, such mRNAs stored in oocytes are not translated efficiently. At the appropriate time during oocyte maturation or after fertilization of an egg cell, usually in response to an external signal, approximately 150 A residues are added to the short poly(A) tails on these mRNAs in the cytoplasm, stimulating their translation.

Recent studies with mRNAs stored in *Xenopus* oocytes have helped elucidate the mechanism of this type of translational control. Experiments in which short-tailed mRNAs are injected into oocytes have shown that two sequences in their 3' UTR are required for their polyadenylation in the cytoplasm: the AAUAAA poly(A) signal that is also required for the nuclear polyadenylation of pre-mRNAs, and one or more copies of an upstream U-rich *cytoplasmic polyadenyla-*

tion element (CPE). This regulatory element is bound by a highly conserved *CPE-binding protein (CPEB)* that contains an RRM domain and a zinc-finger domain.

According to the current model, in the absence of a stimulatory signal, CPEB bound to the U-rich CPE interacts with the protein Maskin, which in turn binds to eIF4E associated with the mRNA 5' cap (Figure 12-28, *left*). As a result, eIF4E cannot interact with other initiation factors and the 40S ribosomal subunit, so translation initiation is blocked. Signal-induced phosphorylation of CPEB at a specific serine causes the displacement of Maskin, allowing cytoplasmic forms of the cleavage and polyadenylation specificity factor (CPSF) and poly(A) polymerase to bind to the mRNA. Once the poly(A) polymerase catalyzes the addition of A residues, PABPI can bind to the lengthened poly(A) tail, leading to the stabilized interaction of all the participants needed to initiate translation (see Figure 12-28, *right;* see also Figure 4-25). In the case of *Xenopus* oocyte maturation, the protein kinase that phosphorylates CPEB is activated in response to the hormone progesterone. Thus timing of the translation of stored mRNAs encoding proteins needed for oocyte maturation is regulated by this external signal.

Considerable evidence indicates that a similar mechanism of translational control plays a role in learning and memory. In the central nervous system, the axons from a thousand or so neurons can make connections (synapses) with the dendrites of a single postsynaptic neuron (see Figure 7-48). When one of these axons is stimulated, the postsynaptic neuron "remembers" which synapse was stimulated. The next time that synapse is stimulated, the strength of the response triggered in the postsynaptic cell differs from the first time. This change in response has been shown to result largely from the translational activation of mRNAs stored in the region of the synapse, leading to the local synthesis of new proteins that increase the size and alter the neurophysiological characteristics of the synapse. The finding that CPEB is present in neuronal dendrites has led to the proposal that cytoplasmic polyadenylation stimulates translation of specific mRNAs in dendrites, much as it does in oocytes. In this case, presumably, synaptic activity (rather than a hormone) is the signal that induces phosphorylation of CPEB and subsequent activation of translation.

mRNAs Are Degraded by Several Mechanisms in the Cytoplasm

The concentration of an mRNA is a function of both its rate of synthesis and its rate of degradation. For this reason, if two genes are transcribed at the same rate, the steady-state concentration of the corresponding mRNA that is more stable will be higher than the concentration of the other. The stability of an mRNA also determines how rapidly synthesis of the encoded protein can be shut down. For a stable mRNA, synthesis of the encoded protein persists long after transcription of the gene is repressed. Most bacterial mRNAs are unstable, decaying exponentially with a typical half-life of a few minutes. For this reason, a bacterial cell can rapidly adjust the synthesis of proteins to accommodate changes in the cellular environment. Most cells in multicellular organisms, on the other hand, exist in a fairly constant environment and carry out a specific set of functions over periods of days to months or even the lifetime of the organism (nerve cells, for example). Accordingly, most mRNAs of higher eukaryotes have half-lives of many hours.

However, some proteins in eukaryotic cells are required only for short periods of time and must be expressed in bursts. For example, certain signaling molecules called cytokines, which are involved in the immune response of mammals, are synthesized and secreted in short bursts. Similarly, many of the transcription factors that regulate the onset of the S phase of the cell cycle, such as c-Fos and c-Jun, are synthesized for brief periods only (Chapter 21). Expression of such proteins occurs in short bursts because transcription of their genes can be rapidly turned on and off and their mRNAs have unusually short half-lives, on the order of 30 minutes or less.

Cytoplasmic mRNAs are degraded by one of the pathways shown in Figure 12-29. For most mRNAs, the length of the poly(A) tail gradually decreases with time through the action of a deadenylating nuclease. When it is shortened sufficiently, PABPI molecules can no longer bind and stabilize interaction of the 5′ cap and initiation factors (see Figure 4-31). The exposed cap then is removed by a decapping enzyme, and the unprotected mRNA is degraded by a 5′ → 3′ exonuclease. Removal of the poly(A) tail also makes mRNAs susceptible to degradation by cytoplasmic exosomes containing 3′ → 5′ exonucleases. The 5′ → 3′ exonucleases predominate in yeast, and the 3′ → 5′ exosome apparently predominates in mammalian cells.

For mRNAs degraded in these deadenylation-dependent pathways, the rate at which they are deadenylated controls the rate at which they are degraded. The rate of deadenylation varies inversely with the frequency of translation initiation for an mRNA: the higher the frequency of initiation, the slower the rate of deadenylation. This relation probably is due to the reciprocal interactions between initiation factors and PABPI that stabilize the binding of PABPI to the poly(A) tail, thereby protecting it from the deadenylation exonuclease.

Many short-lived mRNAs in mammalian cells contain multiple, sometimes overlapping, copies of the sequence AUUUA in their 3′ untranslated region. Specific RNA-binding

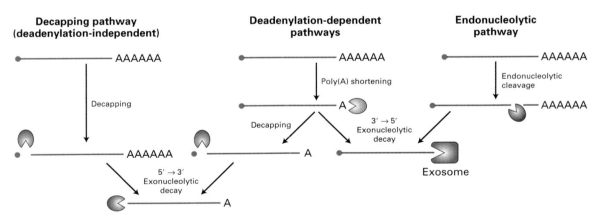

▲ **FIGURE 12-29 Pathways for degradation of eukaryotic mRNAs.** In the deadenylation-dependent (*middle*) pathways, the poly(A) tail is progressively shortened by a deadenylase (orange) until it reaches a length of 20 or fewer A residues at which the interaction with PABPI is destabilized, leading to weakened interactions between the 5′ cap and translation initiation factors. The deadenylated mRNA then may either (1) be decapped and degraded by a 5′ → 3′ exonuclease or (2) be degraded by a 3′ → 5′ exonuclease in cytoplasmic exosomes. Some mRNAs (*right*) are cleaved internally by an endonuclease, and the fragments degraded by an exosome. Other mRNAs (*left*) are decapped before they are deadenylated, and then degraded by a 5′ → 3′ exonuclease. [Adapted from M. Tucker and R. Parker, 2000, *Ann. Rev. Biochem.* **69**:571.]

proteins have been found to bind to these 3′ AU-rich sequences. Recent experiments suggest that the bound proteins interact with a deadenylating enzyme and with the exosome, thereby promoting the rapid deadenylation and subsequent 3′ → 5′ degradation of these mRNAs. In this mechanism, the rate of mRNA degradation is uncoupled from the frequency of translation. Thus mRNAs containing the AUUUA sequence can be translated at high frequency, yet also degraded rapidly, allowing the encoded proteins to be expressed in short bursts.

As shown in Figure 12-29, some mRNAs are degraded in pathways that do not involve significant deadenylation. In one of these, mRNAs are decapped before the poly(A) tail is shortened extensively. It appears that certain mRNA sequences make the cap sensitive to the decapping enzyme, but the precise mechanism is unclear. In the other alternative pathway, mRNAs first are cleaved internally by endonucleases. The RNA-induced silencing complex (RISC) discussed earlier is an example of such an endonuclease (see Figure 12-27). The fragments generated by internal cleavage then are degraded by exonucleases.

An Iron-Sensitive RNA-Binding Protein Regulates mRNA Translation and Degradation

Control of intracellular iron concentrations by the *iron-response element–binding protein (IRE-BP)* is an elegant example of a single protein that regulates the translation of one mRNA and the degradation of another. When intracellular iron stores are low, this dual control system operates to increase the level of free iron ions available for iron-requiring enzymes; when iron is in excess, the system operates to prevent accumulation of toxic levels of free ions. It is one of the simplest and best-understood examples of protein-mediated translational control.

One component in this system is the regulation of production of *ferritin*, an intracellular iron-binding protein. The 5′ untranslated region of ferritin mRNA contains *iron-response elements (IREs)* that have a stem-loop structure. IRE-BP recognizes five specific bases in the IRE loop and the duplex nature of the stem. At low iron concentrations, IRE-BP is in an active conformation that binds to the IREs (Figure 12-30a). The bound IRE-BP blocks the 40S ribosomal subunit from scanning for the AUG start codon (see Figure 4-25), thereby inhibiting translation initiation. The resulting decrease in ferritin means less iron is complexed with the ferritin and is therefore available to iron-requiring enzymes. At high iron concentrations, IRE-BP is in an inactive conformation that does not bind to the 5′ IREs, so translation initiation can proceed. The newly synthesized ferritin then binds free iron ions, preventing their accumulation to harmful levels.

The other part of this regulatory system controls the import of iron into cells. In vertebrates, ingested iron is carried through the circulation bound to a protein called *transferrin*. After binding to the transferrin receptor (TfR) in the plasma membrane, the transferrin-iron complex is brought into cells

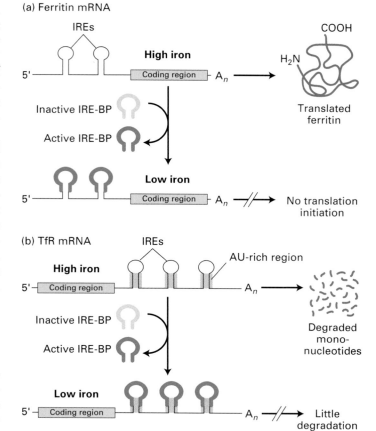

▲ **FIGURE 12-30 Iron-dependent regulation of mRNA translation and degradation.** The iron response element–binding protein (IRE-BP) controls translation of ferritin mRNA (a) and degradation of transferrin-receptor (TfR) mRNA (b). At low intracellular iron concentrations IRE-BP binds to iron-response elements (IREs) in the 5′ or 3′ untranslated region of these mRNAs. At high iron concentrations, IRE-BP undergoes a conformational change and cannot bind either mRNA. The dual control by IRE-BP precisely regulates the level of free iron ions within cells. See the text for discussion.

by receptor-mediated endocytosis (Chapter 17). The 3′-untranslated region of TfR mRNA contains IREs whose stems have AU-rich destabilizing sequences (Figure 12-30b). At high iron concentrations, when the IRE-BP is in the inactive, nonbinding conformation, these AU-rich sequences are thought to promote degradation of TfR mRNA by the same mechanism that leads to rapid degradation of other short-lived mRNAs, as described previously. The resulting decrease in production of the transferrin receptor quickly reduces iron import, thus protecting the cell. At low iron concentrations, however, IRE-BP can bind to the 3′ IREs in TfR mRNA. The bound IRE-BP is thought to block recognition of the destabilizing AU-rich sequences by the proteins that would otherwise rapidly degrade the mRNAs. As a result, production of the transferrin receptor increases and more iron is brought into the cell.

Other regulated RNA-binding proteins may also function to control mRNA translation or degradation, much like the dual-acting IRE-BP. For example, a heme-sensitive RNA-binding protein controls translation of the mRNA encoding aminolevulinate (ALA) synthase, a key enzyme in the synthesis of heme. And in vitro studies have shown that the mRNA encoding the milk protein casein is stabilized by the hormone prolactin and rapidly degraded in its absence.

Nonsense-Mediated Decay and Other mRNA Surveillance Mechanisms Prevent Translation of Improperly Processed mRNAs

Translation of an improperly processed mRNA could lead to production of an abnormal protein that interferes with functioning of the normal protein encoded by the mRNA. (This effect is equivalent to that resulting from dominant-negative mutations, discussed in Chapter 9, although the cause is different.) Several mechanisms collectively termed **mRNA surveillance** help cells avoid the translation of improperly processed mRNA molecules. We have previously mentioned two such surveillance mechanisms: the recognition of improperly processed pre-mRNAs in the nucleus and their degradation by the exosome, and the general restriction against nuclear export of incompletely spliced pre-mRNAs that remain associated with a spliceosome.

Another mechanism called *nonsense-mediated decay* causes degradation of mRNAs in which one or more exons have been skipped during splicing. Except for pre-mRNAs that normally undergo alternative splicing, such exon skipping often will alter the open reading frame of the mRNA 3' to the improper exon junction, resulting in introduction of an incorrect stop codon. For nearly all properly spliced mRNAs, the stop codon is in the last exon. Nonsense-mediated decay results in the rapid degradation of mRNAs with stop codons that occur before the last splice junction in the mRNA.

A search for possible molecular signals that might indicate the positions of splice junctions in a processed mRNA led to the discovery of exon-junction complexes. As noted already, these complexes stimulate export of mRNPs from the nucleus. Analysis of yeast mutants suggests that some of the proteins in exon-junction complexes function in nonsense-mediated decay. One proposal based on these and other findings is that exon-junction complexes interact with a deadenylase that rapidly removes the poly(A) tail from an associated mRNA, leading to its rapid decapping and degradation by a 5 → 3 exonuclease (see Figure 12-29). In the case of properly spliced mRNAs, the exon-junction complexes are thought to be dislodged from the mRNA by passage of the first "pioneer" ribosome to translate the mRNA, thereby protecting the mRNA from degradation. For mRNAs with a stop codon before the final exon junction, however, one or more exon-junction complexes remain associated with the mRNA, resulting in nonsense-mediated decay.

Nonsense-mediated decay occurs in the cytoplasm of yeast cells. Remarkably, in mammalian cells, there is evidence that the pioneer ribosome translates the mRNA while its 5' end is associated with the nuclear cap-binding complex and its poly(A) tail is associated with nuclear PABPII. This finding and other results raise the possibility that in cells of higher organisms, the first round of translation may occur in the nucleus as part of the nonsense-mediated decay mechanism of mRNA surveillance.

Localization of mRNAs Permits Production of Proteins at Specific Regions Within the Cytoplasm

Many cellular processes depend on localization of particular proteins to specific structures or regions of the cell. In later chapters we examine how some proteins are transported *after* their synthesis to their proper cellular location. Alternatively, protein localization might be achieved by localization of mRNAs to specific regions of the cell cytoplasm in which their encoded proteins function. In most cases examined thus far, such mRNA localization is specified by sequences in the 3' untranslated region of the mRNA.

A well-documented example of mRNA localization occurs in mammalian myoblasts (muscle precursor cells) as they differentiate into myotubes, the fused, multinucleated cells that make up muscle fibers. Myoblasts are motile cells that extend cytoplasmic regions, called *lamellipodia*, from the leading edge in the direction of movement. Extension of lamellipodia during cell movement requires polymerization of β-actin (Chapter 19). Sensibly, β-actin mRNA is concentrated in the leading edges of myoblasts, the region of the cell cytoplasm where the encoded protein is needed for motility. When myoblasts fuse into syncytial myotubes, β-actin expression is repressed and the muscle-specific α-actin is induced. In contrast to β-actin mRNA, α-actin mRNA is restricted to the perinuclear regions of myotubes. When cultured myoblasts in the process of differentiating are stained with fluorescent probes specific for α- or β-actin mRNA, both mRNAs are localized to their respective cellular regions.

To test the ability of actin mRNA sequences to direct the cytoplasmic localization of an mRNA, fragments of α- and β-actin cDNAs were inserted into separate plasmid vectors that express β-galactosidase from a strong viral promoter. The resulting plasmids then were transfected into cultured cells, which were assayed for β-galactosidase activity. These experiments showed that inclusion of the 3' untranslated end of α- or β-actin cDNAs directs localization of the expressed β-galactosidase, whereas the 5' untranslated and coding regions do not (Figure 12-31).

Treatment of cultured myoblasts with cytochalasin D, which disrupts actin microfilaments, leads to rapid delocalization of actin mRNAs, indicating that cytoskeletal actin microfilaments participate in the localization process. Disruption of other cytoskeletal components, however, does not alter the localization of actin mRNAs. Other types of evidence also implicate the actin cytoskeleton in mRNA localization. Presumably certain RNA-binding proteins interact

TRANSFECTED ACTIN SEQUENCES

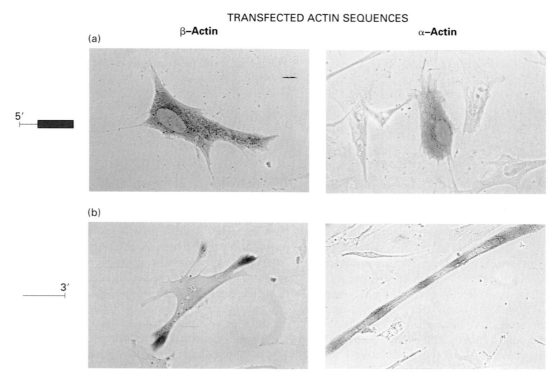

▲ **EXPERIMENTAL FIGURE 12-31 Insertion of various sections of the actin genes in the 3′ untranslated position of an expression vector showed that the 3′ UTRs of α- and β-actin mRNAs direct localization of the reporter mRNA.** Plasmid expression vectors were constructed encoding β-galactosidase mRNAs containing as their 3′ untranslated sequence either the 5′ untranslated sequence and coding region of the α- or β-actin mRNAs (*top panels*), or the 3′ untranslated sequence of these mRNAs (*bottom panels*), as indicated at the left. The recombinant plasmids were transfected separately into differentiating myoblasts. After a period of expression, the cells were fixed and then assayed for β-galactosidase activity by incubating them with X-gal, a substrate that is hydrolyzed by β-galactosidase to yield a blue product. (a) Transfected cells that expressed engineered β-galactosidase mRNAs whose 3′ untranslated region corresponded to the 5′ untranslated sequence and coding region of α- or β-actin mRNA. These actin sequences did not cause localization of the β-galactosidase mRNA, as evidenced by the diffuse blue staining. (b) Transfected cells that expressed engineered β-galactosidase mRNAs whose 3′ untranslated region corresponded to the 3′ untranslated sequences of α- or β-actin mRNA. These sequences led to localization of β-galactosidase mRNA to lamellipodia in myoblasts (β-actin) or perinuclear regions in myotubes (α-actin), as evidenced by the deep staining in these regions. [Micrographs from E. H. Kislaukis et al., 1993, *J. Cell. Biol.* **123**:165.]

with specific sequences in the 3′ untranslated regions of an mRNA and with specific components of microfilaments, including motor proteins that move cargo along the length of microfilaments. The best-understood example of this mechanism of mRNA localization occurs during cell division in *S. cerevisiae*, as we describe in Chapter 22.

KEY CONCEPTS OF SECTION 12.4

Cytoplasmic Mechanisms of Post-transcriptional Control

■ Translation can be repressed by micro RNAs (miRNAs), which form imperfect hybrids with sequences in the 3′ untranslated region (UTR) of specific target mRNAs.

■ The related phenomenon of RNA interference, which probably evolved as an early defense system against viruses and transposons, leads to degradation of mRNAs that form perfect hybrids with short interfering RNAs (siRNAs).

■ Both miRNAs and siRNAs contain 21–23 nucleotides, are generated from longer precursor molecules, and are assembled into a multiprotein RNA-induced silencing complex (RISC) that either represses translation of target mRNAs or cleaves them (see Figure 12-27).

■ Cytoplasmic polyadenylation is required for translation of mRNAs with a short poly(A) tail. Binding of a specific protein to regulatory elements in their 3′ UTRs represses translation of these mRNAs. Phosphorylation of this RNA-binding protein, induced by an external signal, leads to lengthening of the 3′ poly(A) tail and translation (see Figure 12-28).

■ Most mRNAs are degraded as the result of the gradual shortening of their poly(A) tail (deadenylation) followed by exosome-mediated $3' \rightarrow 5'$ digestion or removal of the $5'$ cap and digestion by a $5' \rightarrow 3'$ exonuclease (see Figure 12-29).

■ Eukaryotic mRNAs encoding proteins that are expressed in short bursts generally have repeated copies of an AU-rich sequence in their $3'$ UTR. Specific proteins that bind to these elements also interact with the deadenylating enzyme and cytoplasmic exosomes, promoting rapid RNA degradation.

■ Binding of various proteins to regulatory elements in the $3'$ or $5'$ UTRs of mRNAs regulates the translation or degradation of many mRNAs in the cytoplasm.

■ Translation of ferritin mRNA and degradation of transferrin receptor (TfR) mRNA are both regulated by the same iron-sensitive RNA-binding protein. At low iron concentrations, this protein is in a conformation that binds to specific elements in the mRNAs, inhibiting ferritin mRNA translation or degradation of TfR mRNA (see Figure 12-30). This dual control precisely regulates the iron level within cells.

■ Nonsense-mediated decay and other mRNA surveillance mechanisms prevent the translation of improperly processed mRNAs encoding abnormal proteins that might interfere with functioning of the corresponding normal proteins.

■ Some mRNAs are directed to specific subcellular locations by sequences usually found in the $3'$ UTR, leading to localization of the encoded proteins.

12.5 Processing of rRNA and tRNA

Approximately 80 percent of the total RNA in rapidly growing mammalian cells (e.g., cultured HeLa cells) is rRNA, and 15 percent is tRNA; protein-coding mRNA thus constitutes only a small portion of the total RNA. The primary transcripts produced from most rRNA genes and from tRNA genes, like pre-mRNAs, are extensively processed to yield the mature, functional forms of these RNAs.

Pre-rRNA Genes Are Similar in All Eukaryotes and Function as Nucleolar Organizers

The 28S and 5.8S rRNAs associated with the large (60S) ribosomal subunit and the 18S rRNA associated with the small (40S) ribosomal subunit in higher eukaryotes (and the smaller, functionally equivalent rRNAs in all other eukaryotes) are encoded by a single type of **pre-rRNA** transcription unit. Transcription by RNA polymerase I yields a 45S primary transcript (pre-rRNA), which is processed into the mature 28S, 18S, and 5.8S rRNAs found in cytoplasmic ribosomes. Sequencing of the DNA encoding pre-rRNA from many species showed that this DNA shares several properties in all eukaryotes. First, the pre-rRNA genes are arranged in long tandem arrays separated by nontranscribed spacer regions ranging in length from ≈2 kb in frogs to ≈30 kb in hu-

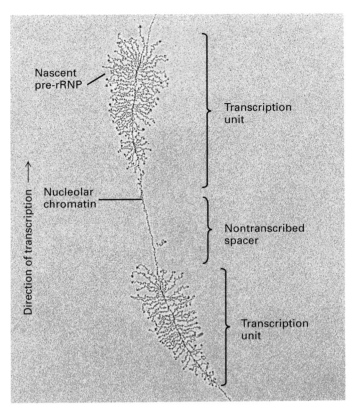

▲ **FIGURE 12-32 Electron micrograph of pre-rRNA transcription units from nucleolus of a frog oocyte.** Each "feather" represents a pre-rRNA molecule associated with protein in a pre-ribonucleoprotein particle (pre-RNP) emerging from a transcription unit. Pre-rRNA transcription units are arranged in tandem, separated by nontranscribed spacer regions of nucleolar chromatin. [Courtesy of Y. Osheim and O. J. Miller, Jr.]

mans (Figure 12-32). Second, the genomic regions corresponding to the three mature rRNAs are always arranged in the same $5' \rightarrow 3'$ order: 18S, 5.8S, and 28S. Third, in all eukaryotic cells (and even in bacteria), the pre-rRNA gene codes for, and the corresponding primary transcript contains, regions that are removed during processing and rapidly degraded. The general structure of pre-rRNAs is diagrammed in Figure 12-33.

Both the synthesis and processing of pre-rRNA occurs in the **nucleolus.** When pre-rRNA genes initially were identified in the nucleolus by in situ hybridization, it was not known whether any other DNA was required to form the nucleolus. Subsequent experiments with transgenic *Drosophila* strains demonstrated that a single complete pre-rRNA transcription unit induces formation of a small nucleolus. Thus a single pre-rRNA gene is sufficient to be a *nucleolar organizer*, and all the other components of the ribosome diffuse to the newly formed pre-rRNA. The structure of the nucleolus observed by light and electron microscopy results from the processing of pre-RNA and the assembly of ribosomal subunits.

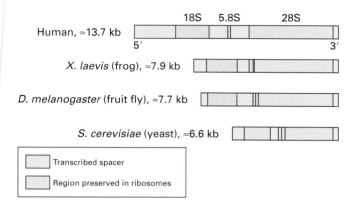

▲ **FIGURE 12-33 General structure of eukaryotic pre-rRNA transcription units.** The three coding regions (blue) encode the 18S, 5.8S, and 28S rRNAs found in ribosomes of higher eukaryotes or their equivalents in other species. The order of these coding regions in the genome is always 5′ → 3′. Variations in the lengths of the transcribed spacer regions (tan) account for the major difference in the lengths of pre-rRNA transcription units in different organisms.

Small Nucleolar RNAs Assist in Processing Pre-rRNAs and Assembling Ribosome Subunits

Following their synthesis in the nucleolus, nascent pre-rRNA transcripts are immediately bound by proteins, forming *pre-ribosomal ribonucleoprotein particles (pre-rRNPs)*. The largest of these (80S) contains an intact 45S pre-rRNA molecule, which is cut in a series of cleavage and exonucleolytic steps that ultimately yield the mature rRNAs found in ribosomes (Figure 12-34). During processing, pre-rRNA also is extensively modified, mostly by methylation of the 2′-hydroxyl group of specific riboses and conversion of specific uridine residues to pseudouridine. Some of the proteins in the pre-rRNPs found in nucleoli remain associated with the mature ribosomal subunits, whereas others are restricted to the nucleolus and assist in assembly of the subunits.

The positions of cleavage sites in pre-rRNA and the specific sites of 2′-O-methylation and pseudouridine formation are determined by approximately 150 different small nucleolus-restricted RNA species, called *small nucleolar RNAs (snoRNAs)*, which hybridize transiently to pre-rRNA molecules. Like the snRNAs that function in pre-mRNA processing, snoRNAs associate with proteins, forming ribonucleoprotein particles called *snoRNPs*. One large class of snoRNPs helps position a methyltransferase enzyme near methylation sites in the pre-mRNA. Another positions the enzyme that converts uridine to pseudouridine. Others function in the cleavage reactions that remove transcribed spacer regions. Once cleaved from pre-rRNAs, these sequences are degraded by the same exosome-associated 3′ → 5′ nuclear exonucleases that degrade introns spliced from pre-mRNAs.

Some snoRNAs are expressed from their own promoters by RNA polymerase II or III. Remarkably, however, the large majority of snoRNAs are spliced-out introns of genes encoding functional mRNAs encoding proteins involved in ribosome synthesis or translation. Some snoRNAs are introns spliced from apparently nonfunctional mRNAs. The genes encoding these mRNAs seem to exist only to express snoRNAs from excised introns.

Unlike pre-rRNA genes, 5S rRNA genes are transcribed by RNA polymerase III in the nucleoplasm outside the nucleolus. Without further processing, 5S RNA diffuses to the nucleolus, where it assembles with the 28S and 5.8S rRNAs and proteins into large ribosomal subunits (see Figure 12-34 and Figure 4-28). When assembly of ribosomal subunits in the nucleolus is complete, they are transported through nuclear pore complexes to the cytoplasm, where they appear first as free subunits. The ribosomal subunits are the largest cellular structures known to be transported through nuclear pore complexes.

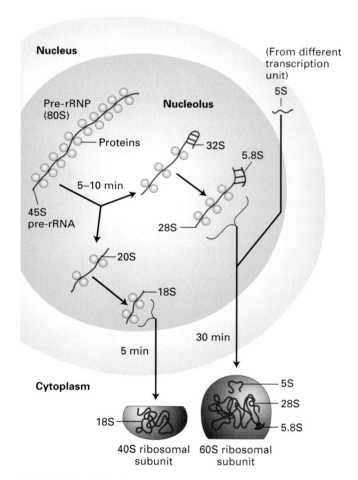

▲ **FIGURE 12-34 Processing of pre-rRNA and assembly of ribosomes in higher eukaryotes.** Ribosomal and nucleolar proteins associate with 45S pre-RNA as it is synthesized, forming an 80S pre-rRNP. Sites of cleavage and chemical modifications are determined by small nucleolar RNAs (not shown). Note that synthesis of 5S rRNA occurs outside the nucleolus.

Self-Splicing Group I Introns Were the First Examples of Catalytic RNA

The DNA in the protozoan *Tetrahymena thermophila* contains an intervening intron in the region that encodes the large pre-rRNA molecule. Careful searches failed to uncover even one pre-rRNA gene without the extra sequence, indicating that splicing is required to produce mature rRNA in these organisms. In vitro studies showing that the pre-rRNA was spliced at the correct sites in the absence of any protein provided the first indication that RNA can function as a catalyst, like enzymes.

A whole raft of self-splicing sequences subsequently were found in pre-rRNAs from other single-celled organisms, in mitochondrial and chloroplast pre-rRNAs, in several pre-mRNAs from certain *E. coli* bacteriophages, and in some bacterial tRNA primary transcripts. The self-splicing sequences in all these precursors, referred to as *group I introns*, use guanosine as a cofactor and can fold by internal base pairing to juxtapose closely the two exons that must be joined. As discussed earlier, certain mitochondrial and chloroplast pre-mRNAs and tRNAs contain a second type of self-splicing intron, designated *group II*.

The splicing mechanisms used by group I introns, group II introns, and spliceosomes are generally similar, involving two transesterification reactions, which require no input of energy (Figure 12-35). Structural studies of the group I intron from *Tetrahymena* pre-rRNA combined with mutational and biochemical experiments have revealed that the RNA folds into a precise three-dimensional structure that, like protein enzymes, contains deep grooves for binding substrates and solvent-inaccessible regions that function in catalysis. The group I intron functions like a metalloenzyme to precisely orient the atoms that participate in the two transesterification reactions adjacent to catalytic Mg^{2+} ions. Considerable evidence now indicates that splicing by group II introns and by snRNAs in the spliceosome also involves bound catalytic Mg^{2+} ions. In both the groups I and II self-splicing introns and probably in the spliceosome, RNA functions as a **ribozyme**, an RNA sequence with catalytic ability.

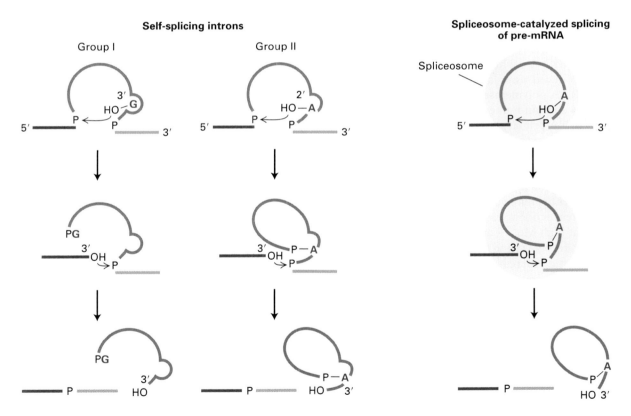

▲ **FIGURE 12-35 Splicing mechanisms in group I and group II self-splicing introns and spliceosome-catalyzed splicing of pre-mRNA.** The intron is shown in gray, the exons to be joined in red. In group I introns, a guanosine cofactor (G) that is not part of the RNA chain associates with the active site. The 3′-hydroxyl group of this guanosine participates in a transesterification reaction with the phosphate at the 5′ end of the intron; this reaction is analogous to that involving the 2′-hydroxyl groups of the branch-site A in group II introns and pre-mRNA introns spliced in spliceosomes (see Figure 12-7). The subsequent transesterification that links the 5′ and 3′ exons is similar in all three splicing mechanisms. Note that spliced-out group I introns are linear structures, unlike the branched intron products in the other two cases. [Adapted from P. A. Sharp, 1987, *Science* **235**:769.]

▶ **FIGURE 12-36 Changes that occur during processing of tyrosine pre-tRNA.** A 14-nucleotide intron (blue) in the anticodon loop is removed by splicing. A 16-nucleotide sequence (green) at the 5′ end is cleaved by RNase P. U residues at the 3′ end are replaced by the CCA sequence (red) found in all mature tRNAs. Numerous bases in the stem-loops are converted to characteristic modified bases (yellow). Not all pre-tRNAs contain introns that are spliced out during processing, but they all undergo the other types of changes shown here. D = dihydrouridine; Ψ = pseudouridine.

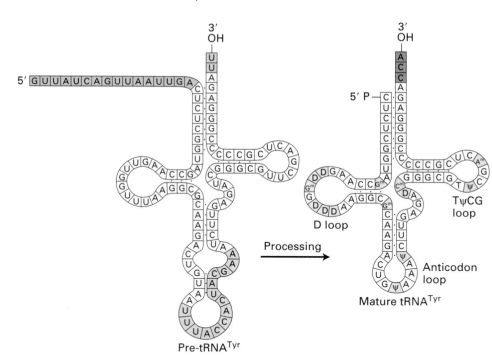

Pre-tRNAs Undergo Cleavage, Base Modification, and Sometimes Protein-Catalyzed Splicing

Mature cytosolic tRNAs, which average 75–80 nucleotides in length, are produced from larger precursors (pre-tRNAs) synthesized by RNA polymerase III in the nucleoplasm. Mature tRNAs also contain numerous modified bases that are not present in tRNA primary transcripts. Cleavage and base modification occur during processing of all pre-tRNAs; some pre-tRNAs also are spliced during processing.

A 5′ sequence of variable length that is absent from mature tRNAs is present in all pre-tRNAs (Figure 12-36). These extra 5′ nucleotides are removed by *ribonuclease P (RNase P),* a ribonucleoprotein endonuclease. Studies with *E. coli* RNase P indicate that at high Mg^{2+} concentrations, the RNA component alone can recognize and cleave *E. coli* pre-tRNAs. The RNase P polypeptide increases the rate of cleavage by the RNA, allowing it to proceed at physiological Mg^{2+} concentrations. A comparable RNase P functions in eukaryotes.

About 10 percent of the bases in pre-tRNAs are modified enzymatically during processing. Three classes of base modifications occur (see Figure 12-36): replacement of U residues at the 3′ end of pre-tRNA with a CCA sequence, which is found at the 3′ end of all tRNAs and is required for their charging by aminoacyl-tRNA synthetases during protein synthesis; addition of methyl and isopentenyl groups to the heterocyclic ring of purine bases and methylation of the 2′-OH group in the ribose of any residue; and conversion of specific uridines to dihydrouridine, pseudouridine, or ribothymidine residues.

As shown in Figure 12-36, the pre-tRNA expressed from the yeast tyrosine tRNA (tRNA^{Tyr}) gene contains a

14-base intron that is not present in mature tRNA^{Tyr}. Some other eukaryotic tRNA genes and some archaeal tRNA genes also contain introns. The introns in nuclear pre-tRNAs are shorter than those in pre-mRNAs and lack the consensus splice-site sequences found in pre-mRNAs (see Figure 12-6). Pre-tRNA introns also are clearly distinct from the much longer self-splicing group I and group II introns found in chloroplast and mitochondrial pre-rRNAs. The mechanism of pre-tRNA splicing differs in three fundamental ways from the mechanisms utilized by self-splicing introns and spliceosomes (see Figure 12-35). First, splicing of pre-tRNAs is catalyzed by proteins, not by RNAs. Second, a pre-tRNA intron is excised in one step that entails simultaneous cleavage at both ends of the intron. Finally, hydrolysis of GTP and ATP is required to join the two tRNA halves generated by cleavage on either side of the intron.

After pre-tRNAs are processed in the nucleoplasm, the mature tRNAs are transported to the cytoplasm through nuclear pore complexes by exportin-t, as discussed previously. In the cytoplasm, tRNAs are passed between aminoacyl-tRNA synthetases, elongation factors, and ribosomes during protein synthesis (Chapter 4). Thus tRNAs generally are associated with proteins and spend little time free in the cell, as is also the case for mRNAs and rRNAs.

KEY CONCEPTS OF SECTION 12.5

Processing of rRNA and tRNA

■ A large precursor pre-rRNA (45S in humans) synthesized by RNA polymerase I undergoes cleavage, exonucleolytic digestion, and base modifications to yield mature 28S, 18S, and 5.8S rRNAs, which associate with ribosomal proteins into ribosomal subunits (see Figure 12-34).

■ Synthesis and processing of pre-rRNA occur in the nucleolus. The 5S rRNA component of the large ribosomal subunit is synthesized in the nucleoplasm by RNA polymerase III and is not processed.

■ Group I and group II self-splicing introns and snRNAs in spliceosomes all function as ribozymes, or catalytically active RNA sequences, that carry out splicing by analogous transesterification reactions requiring bound Mg^{+2} ions (see Figure 12-35).

■ Pre-tRNAs synthesized by RNA polymerase III in the nucleoplasm are processed by removal of the 5'-end sequence, addition of CCA to the 3' end, and modification of multiple internal bases (see Figure 12-36).

■ Some pre-tRNAs contain a short intron that is removed by a protein-catalyzed mechanism distinct from the splicing of pre-mRNA and self-splicing introns.

■ All species of RNA molecules are associated with proteins in various types of ribonucleoprotein particles both in the nucleus and after export to the cytoplasm.

PERSPECTIVES FOR THE FUTURE

In this and the previous chapter, we have seen that in eukaryotic cells, mRNAs are synthesized and processed in the nucleus, transported through nuclear pore complexes to the cytoplasm, and then, in some cases, transported to specific areas of the cytoplasm before being translated by ribosomes. Each of these fundamental processes is carried out by complex macromolecular machines composed of scores of proteins and in many cases RNAs, as well. The complexity of these macromolecular machines ensures accuracy in finding promoters and splice sites in the long length of DNA and RNA sequences and provides various avenues for regulating synthesis of a polypeptide chain. Much remains to be learned about the structure, operation, and regulation of such complex machines as spliceosomes and the cleavage/polyadenylation apparatus.

Recent examples of the regulation of pre-mRNA splicing raise the question of how extracellular signals might control such events, especially in the nervous system of vertebrates. A case in point is the remarkable situation in the chick inner ear where multiple isoforms of the Ca^{2+}-activated K^+ channel called *Slo* are produced by alternative RNA splicing. Cell-cell interactions appear to inform cells of their position in the cochlea, leading to alternative splicing of Slo pre-mRNA. The challenging task facing researchers is to discover how such cell-cell interactions regulate the activity of RNA-processing factors.

The mechanism of mRNP transport through nuclear pore complexes poses many intriguing questions. Future research will likely reveal additional activities of hnRNP and nuclear mRNP proteins and clarify their mechanisms of action. For instance, there is a small gene family encoding proteins homologous to the large subunit of the mRNA-exporter. What are the functions of these related proteins? Do they participate in the transport of overlapping sets of mRNPs? Some hnRNP proteins contain nuclear-retention signals that prevent nuclear export when fused to hnRNP proteins with nuclear-export signals (NESs). How are these hnRNP proteins selectively removed from processed mRNAs in the nucleus, allowing the mRNAs to be transported to the cytoplasm?

The localization of certain mRNAs to specific subcellular locations is fundamental to the development of multicellular organisms. As discussed in Chapter 22, during development an individual cell frequently divides into daughter cells that function differently from each other. In the language of developmental biology, the two daughter cells are said to have different developmental fates. In many cases, this difference in developmental fate results from the localization of an mRNA to one region of the cell before mitosis so that after cell division, it is present in one daughter cell and not the other. Much exciting work remains to be done to fully understand the molecular mechanisms controlling mRNA localization that are critical for the normal development of multicellular organisms.

Some of the most exciting and unanticipated discoveries in molecular cell biology in recent years concern the existence and function of miRNAs and the process of RNA interference. RNA interference (RNAi) provides molecular cell biologists with a powerful method for studying gene function. The discovery of more than a hundred miRNAs in humans and other organisms suggests that multiple significant examples of translational control by this mechanism await to be characterized. Recent studies in plants link similar short nuclear RNAs to the control of DNA methylation and the formation of heterochromatin. Will similar processes control gene expression through the assembly of heterochromatin in humans and other animals? What other regulatory processes might be directed by micro RNAs? Since control by these mechanisms depends on base pairing between miRNAs and target mRNAs or genes, genomic and bioinformatic methods will probably suggest genes that may be controlled by these mechanisms. What other processes in addition to translation control, mRNA degradation, and heterochromatin assembly might be controlled by miRNAs?

These are just a few of the fascinating questions concerning RNA processing, post-transcriptional control, and nuclear transport that will challenge molecular cell biologists in the coming decades. The astounding discoveries of entirely unanticipated mechanisms of gene control by miRNAs remind us that there will likely be many more surprises in the future.

KEY TERMS

5′ cap *493*

alternative splicing *505*

cargo complex *510*

cleavage/polyadenylation
  complex *497*

cross-exon recognition
  complex *501*

exosome *504*

exportin 1 *512*

FG-nucleoporins *510*

group I introns *502*

group II introns *502*

importins *510*

iron-response
  element–binding
  protein (IRE-BP) *522*

karyopherins *513*

micro RNAs
  (miRNAs) *518*

mRNA-exporter *514*

mRNA surveillance *523*

nuclear pore complex
  (NPC) *509*

poly(A) tail *496*

pre-mRNA *494*

pre-rRNA *525*

Ran protein *510*

ribozyme *527*

RNA editing *508*

RNA recognition
  motif (RRM) *495*

RNA splicing *497*

RNA-induced silencing
  complex (RISC) *518*

small nuclear
  RNAs (snRNAs) *499*

spliceosome *499*

SR proteins *501*

REVIEW THE CONCEPTS

1. Describe three types of post-transcriptional regulation of protein-coding genes.

2. What is the evidence that transcription termination by RNA polymerase II is coupled to polyadenylation?

3. It has been suggested that manipulation of HIV anti-termination might provide for effective therapies in combatting AIDS. What effect would a mutation in the TAR sequence that abolishes Tat binding have on HIV transcription after HIV infection and why? A mutation in Cdk9 that abolishes activity?

4. Some heat shock genes encode proteins that act rapidly to protect other proteins from harsh conditions. Describe the mechanism that has evolved to regulate the expression of such genes.

5. You are investigating the transcriptional regulation of genes from a recently discovered eukaryotic organism. You find that this organism has three RNA polymerase–like enzymes and that genes encoding proteins are transcribed using an RNA polymerase II–like enzyme. When examining the full-length mRNAs produced, you find that there is no 5′ structure (i.e., cap). Are you surprised? Why/why not?

6. Why do researchers believe that the fragile-X disorder involves an aberration in RNA binding?

7. What is the difference between hnRNAs, snRNAs, miRNAs, siRNAs, and snoRNAs?

8. Describe how the discovery that introns are removed during splicing was made. How are the locations of exon-intron junctions predicted?

9. What are the mechanistic similarities between Group II intron self-splicing and spliceosomal splicing? What is the evidence that there may be an evolutionary relationship between the two?

10. Where do researchers believe most transcription and RNA-processing events occur? What is the evidence to support this?

11. You obtain the sequence of a gene containing 10 exons, 9 introns, and a 3′ UTR containing a polyadenylation consensus sequence. The fifth intron also contains a polyadenylation site. To test whether both polyadenylation sites are used, you isolate mRNA and find a longer transcript from muscle tissue and a shorter mRNA transcript from all other tissues. Speculate about the mechanism involved in the production of these two transcripts.

12. What is the evidence that an mRNA exporter directs mRNPs through nuclear pores? What is the evidence that most vertebrates also utilize such an exporter?

13. Why is localization of Ran GAP in the nucleus and Ran nucleotide exchange factor (RCC1) in the cytoplasm necessary for unidirectional transport of cargo proteins containing an NES?

14. Speculate about why plants deficient in Dicer activity show increased sensitivity to infection by RNA viruses.

15. What is the evidence that some mRNAs are directed to accumulate in specific subcellular locations?

ANALYZE THE DATA

A specific protein kinase (PK) gene is speculated to be differentially spliced in muscle tissue. This gene comprises three exons and two intron sequences and in fibroblast cells encodes a 38.5-kD protein. Investigators have transfected various portions of a genomic or cDNA copy of the PK gene into both muscle and fibroblast cells (see the constructs in part (a) of the figure). The expression system utilizes a promoter active in both cell types and a C-terminal fusion to a small epitope tag called V5, which contributes ~5.5 kD to the fusion protein. Part (b) of the figure shows the results of an immunoprecipitation experiment designed to analyze the expression products of the transfected cells. A negative control (Neg) of untransfected muscle cells is included. Molecular weight markers in kD are indicated on the left. Immunoprecipitated proteins as shown in part (b) were then placed in protein kinase assays with two different substrates, A and B, to discern activity of the expressed proteins.

(a)

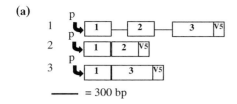

= 300 bp

(b)

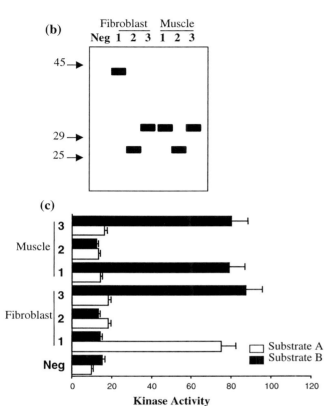

(c)

a. What can be concluded about differential splicing of the PK gene in fibroblast versus muscle cells? Are there other experiments that could confirm these results?

b. What sequence or sequences contribute to regulation of alternative splicing?

c. How does the presence or absence of exon 3 alter the catalytic activity of the encoded PK protein? Which data support your conclusions?

REFERENCES

Processing of Eukaryotic Pre-mRNA

Bentley, D. 2002. The mRNA assembly line: transcription and processing machines in the same factory. *Curr. Opin. Cell Biol.* 14:336–342.

Bonen, L., and J. Vogel. 2001. The ins and outs of group II introns. *Trends Genet.* 17:322–331.

Butler, J. S. 2002. The yin and yang of the exosome. *Trends Cell Biol.* 12:90–96.

Dreyfuss, G., V. N. Kim, and N. Kataoka. 2002. Messenger-RNA-binding proteins and the messages they carry. *Nature Rev. Mol. Cell Biol.* 3:195–205.

Moore, M. J. 2002. Nuclear RNA turnover. *Cell* 108:431–434.

Shatkin, A. J., and J. L. Manley. 2000. The ends of the affair: capping and polyadenylation. *Nature Struct. Biol.* 7:838–842.

Villa, T., J. A. Pleiss, and C. Guthrie. 2002. Spliceosomal snRNAs: Mg(2+)-dependent chemistry at the catalytic core? *Cell* 109:149–152.

Regulation of Pre-mRNA Processing

Black, D. L. 2003. Mechanisms of alternative pre-mRNA splicing. *Ann. Rev. Biochem.* 72:291–336.

Blanc, V., and N. O. Davidson. 2003. C-to-U RNA editing: mechanisms leading to genetic diversity. *J. Biol. Chem.* 278:1395–1398.

Faustino, N. A., and T. A. Cooper. 2003. Pre-mRNA splicing and human disease. *Genes Devel.* 17:419–437.

Maas, S., A. Rich, and K. Nishikura. 2003. A-to-I RNA editing: recent news and residual mysteries. *J. Biol. Chem.* 278:1391–1394.

Maniatis, T., and B. Tasic. 2002. Alternative pre-mRNA splicing and proteome expansion in metazoans. *Nature* 418: 236–243.

Macromolecular Transport Across the Nuclear Envelope

Chook, Y. M., and G. Blobel. 2001. Karyopherins and nuclear import. *Curr. Opin. Struct. Biol.* 11:703–715.

Conti, E., and E. Izaurralde. 2001. Nucleocytoplasmic transport enters the atomic age. *Curr. Opin. Cell Biol.* 13:310–319.

Johnson, A. W., E. Lund, and J. Dahlberg. 2002. Nuclear export of ribosomal subunits. *Trends Biochem Sci.* 27:580–585.

Reed, R., and E. Hurt. 2002. A conserved mRNA export machinery coupled to pre-mRNA splicing. *Cell* 108:523–531.

Ribbeck, K., and D. Gorlich. 2001. Kinetic analysis of translocation through nuclear pore complexes. *Embo. J.* 20:1320–1330.

Rout, M. P., and J. D. Aitchison. 2001. The nuclear pore complex as a transport machine. *J. Biol. Chem.* 276:16593–16596.

Cytoplasmic Mechanisms of Post-transcriptional Control

Ambros, V. 2001. MicroRNAs: tiny regulators with great potential. *Cell* 107:823–826.

Cerutti, H. 2003. RNA interference: traveling in the cell and gaining functions? *Trends Genet.* 19:39–46.

Hannon, G. J. 2002. RNA interference. *Nature* 418:244–251.

Kloc, M., N. R. Zearfoss, and L. D. Etkin. 2002. Mechanisms of subcellular mRNA localization. *Cell* 108:533–544.

Maquat, L. E., and G. G. Carmichael. 2001. Quality control of mRNA function. *Cell* 104:173–176.

Mendez, R., and J. D. Richter. 2001. Translational control by CPEB: a means to the end. *Nat. Rev. Mol. Cell Biol.* 2:521–529.

Van Hoof, A., and R. Parker. 2002. Messenger RNA degradation: beginning at the end. *Curr. Biol.* 12:R285–R287.

Wilkinson, M. F., and A. B. Shyu. 2002. RNA surveillance by nuclear scanning? *Nature Cell Biol.* 4:E144–E147.

Processing of rRNA and tRNA

Fatica, A., and D. Tollervey. 2002. Making ribosomes. *Curr. Opin. Cell Biol.* 14:313–318.

Hopper, A. K., and E. M. Phizicky. 2003. tRNA transfers to the limelight. *Genes Devel.* 17:162–180.

13

SIGNALING AT
THE CELL SURFACE

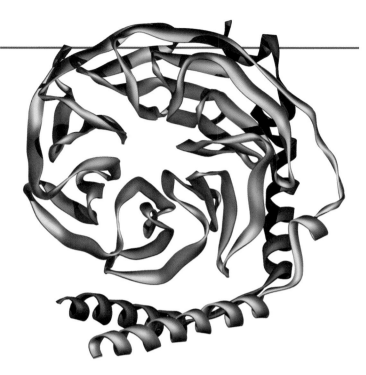

Three-dimensional structure of the G protein β (blue) and γ (purple) complex as obtained by x-ray crystallography.

No cell lives in isolation. In eukaryotic microorganisms such as yeast, slime molds, and protozoans, secreted molecules called **pheromones** coordinate the aggregation of free-living cells for sexual mating or differentiation under certain environmental conditions. Yeast mating-type factors are a well-understood example of pheromone-mediated cell-to-cell signaling (Chapter 22). More important in plants and animals are extracellular **signaling molecules** that function *within* an organism to control metabolic processes within cells, the growth and differentiation of tissues, the synthesis and secretion of proteins, and the composition of intracellular and extracellular fluids. Adjacent cells often communicate by direct cell-cell contact. For example, gap junctions in the plasma membranes of adjacent cells permit them to exchange small molecules and to coordinate metabolic responses. Other junctions between adjacent cells determine the shape and rigidity of many tissues; other interactions adhere cells to the extracellular matrix. Such cell-cell and cell-matrix interactions, which are covered in Chapter 6, may also initiate intracellular signaling via pathways similar to those discussed in this and subsequent chapters.

Extracellular signaling molecules are synthesized and released by *signaling cells* and produce a specific response only in *target cells* that have **receptors** for the signaling molecules. In multicellular organisms, an enormous variety of chemicals, including small molecules (e.g., amino acid or lipid derivatives, acetylcholine), peptides, and proteins, are used in this type of cell-to-cell communication. Some signaling molecules, especially hydrophobic molecules such as steroids, retinoids, and thyroxine, spontaneously diffuse through the plasma membrane and bind to intracellular receptors. Signaling from such intracellular receptors is discussed in Chapter 11.

In this and the next two chapters, we focus on signaling from a diverse group of receptor proteins located in the plasma membrane (Figure 13-1). The signaling molecule acts as a **ligand,** which binds to a structurally complementary site on the extracellular or membrane-spanning domains of the receptor. Binding of a ligand to its receptor causes a conformational change in the cytosolic domain or domains of the receptor that ultimately induces specific cellular responses. The overall process of converting signals into cellular responses, as well as the individual steps in this

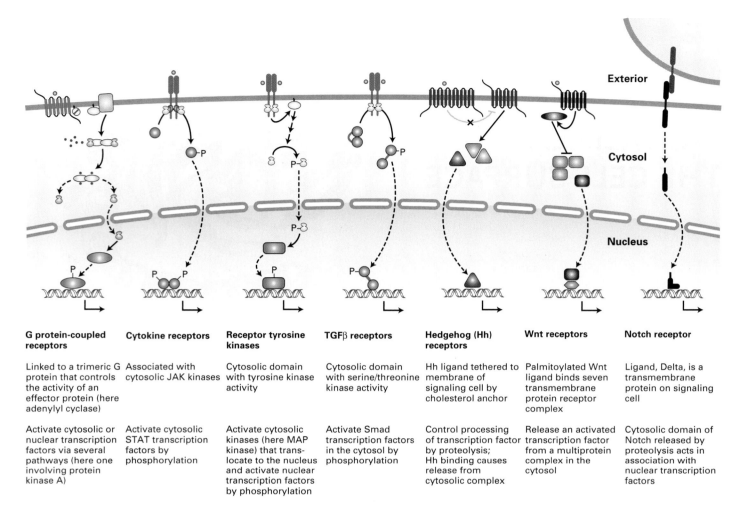

G protein-coupled receptors	Cytokine receptors	Receptor tyrosine kinases	TGFβ receptors	Hedgehog (Hh) receptors	Wnt receptors	Notch receptor
Linked to a trimeric G protein that controls the activity of an effector protein (here adenylyl cyclase)	Associated with cytosolic JAK kinases	Cytosolic domain with tyrosine kinase activity	Cytosolic domain with serine/threonine kinase activity	Hh ligand tethered to membrane of signaling cell by cholesterol anchor	Palmitoylated Wnt ligand binds seven transmembrane protein receptor complex	Ligand, Delta, is a transmembrane protein on signaling cell
Activate cytosolic or nuclear transcription factors via several pathways (here one involving protein kinase A)	Activate cytosolic STAT transcription factors by phosphorylation	Activate cytosolic kinases (here MAP kinase) that translocate to the nucleus and activate nuclear transcription factors by phosphorylation	Activate Smad transcription factors in the cytosol by phosphorylation	Control processing of transcription factor by proteolysis; Hh binding causes release from cytosolic complex	Release an activated transcription factor from a multiprotein complex in the cytosol	Cytosolic domain of Notch released by proteolysis acts in association with nuclear transcription factors

▲ **FIGURE 13-1 Overview of seven major classes of cell-surface receptors discussed in this book.** In many signaling pathways, ligand binding to a receptor leads to activation of transcription factors in the cytosol, permitting them to translocate into the nucleus and stimulate (or occasionally repress) transcription of their target genes. Alternatively, receptor stimulation may lead to activation of cytosolic protein kinases that then translocate into the nucleus and regulate the activity of nuclear transcription factors. Some activated receptors, particularly certain G protein–coupled receptors, also can induce changes in the activity of preexisting proteins. [After A. H. Brivanlou and J. Darnell, 2002, *Science* **295**:813.]

process, is termed **signal transduction.** As we will see, signal-transduction pathways may involve relatively few or many components.

We begin this chapter with two sections that describe general principles and techniques that are relevant to most signaling systems. In the remainder of the chapter, we concentrate on the huge class of cell-surface receptors that activate **trimeric G proteins.** Receptors of this type, commonly called **G protein–coupled receptors (GPCRs),** are found in all eukaryotic cells from yeast to man. The human genome, for instance, encodes several thousand G protein–coupled receptors. These include receptors in the visual, olfactory (smell), and gustatory (taste) systems, many neurotransmitter receptors, and most of the receptors for hormones that control carbohydrate, amino acid, and fat metabolism.

13.1 Signaling Molecules and Cell-Surface Receptors

Communication by extracellular signals usually involves the following steps: (1) synthesis and (2) release of the signaling molecule by the signaling cell; (3) transport of the signal to the target cell; (4) binding of the signal by a specific receptor protein leading to its activation; (5) initiation of one or more intracellular signal-transduction pathways by the activated receptor; (6) specific changes in cellular function, metabolism, or development; and (7) removal of the signal, which often terminates the cellular response (see Figure 13-1). The vast majority of receptors are activated by binding of secreted or membrane-bound molecules (e.g., hormones, growth factors, neurotransmitters, and pheromones).

Some receptors, however, are activated by changes in the concentration of a metabolite (e.g., oxygen or nutrients) or by physical stimuli (e.g., light, touch, heat). In *E. coli*, for instance, receptors in the cell-surface membrane trigger signaling pathways that help the cell respond to changes in the external level of phosphate and other nutrients (see Figure 4-18).

Signaling Molecules in Animals Operate over Various Distances

In animals, signaling by soluble extracellular molecules can be classified into three types—endocrine, paracrine, or autocrine—based on the distance over which the signal acts. In addition, certain membrane-bound proteins act as signals.

In **endocrine** signaling, the signaling molecules, called **hormones,** act on target cells distant from their site of synthesis by cells of the various endocrine organs. In animals, an endocrine hormone usually is carried by the blood or by other extracellular fluids from its site of release to its target.

In **paracrine** signaling, the signaling molecules released by a cell affect target cells only in close proximity. The conduction by a neurotransmitter of a signal from one nerve cell to another or from a nerve cell to a muscle cell (inducing or inhibiting muscle contraction) occurs via paracrine signaling (Chapter 7). Many **growth factors** regulating development in multicellular organisms also act at short range. Some of these molecules bind tightly to the extracellular matrix, unable to signal, but subsequently can be released in an active form. Many developmentally important signals diffuse away from the signaling cell, forming a concentration gradient and inducing various cellular responses depending on their concentration at a particular target cell (Chapter 15).

In **autocrine** signaling, cells respond to substances that they themselves release. Some growth factors act in this fashion, and cultured cells often secrete growth factors that stimulate their own growth and proliferation. This type of signaling is particularly common in tumor cells, many of which overproduce and release growth factors that stimulate inappropriate, unregulated proliferation of themselves as well as adjacent nontumor cells; this process may lead to formation of a tumor mass.

Signaling molecules that are integral membrane proteins located on the cell surface also play an important role in development. In some cases, such membrane-bound signals on one cell bind receptors on the surface of an adjacent target cell to trigger its differentiation. In other cases, proteolytic cleavage of a membrane-bound signaling protein releases the exoplasmic region, which functions as a soluble signaling protein.

Some signaling molecules can act both short range and long range. **Epinephrine,** for example, functions as a neurotransmitter (paracrine signaling) and as a systemic hormone (endocrine signaling). Another example is epidermal growth factor (EGF), which is synthesized as an integral plasma-membrane protein. Membrane-bound EGF can bind to and signal an adjacent cell by direct contact. Cleavage by an extracellular protease releases a soluble form of EGF, which can signal in either an autocrine or a paracrine manner.

Receptors Activate a Limited Number of Signaling Pathways

The number of receptors and signaling pathways that we discuss throughout this book initially may seem overwhelming. Moreover, the terminology for designating pathways can be confusing. Pathways commonly are named based on the general class of receptor involved (e.g., GPCRs, receptor tyrosine kinases), the type of ligand (e.g., TGFβ, Wnt, Hedgehog), or a key intracellular signal transduction component (e.g., NF-κB). In some cases, the same pathway may be referred to by different names. Fortunately, as researchers have discovered the molecular details of more and more receptors and pathways, some principles and mechanisms are beginning to emerge. These shared features can help us make sense of the wealth of new information concerning cell-to-cell signaling.

First, external signals induce two major types of cellular responses: (1) changes in the activity or function of specific pre-existing proteins and (2) changes in the amounts of specific proteins produced by a cell, most commonly as the result of modification of transcription factors leading to activation or repression of gene transcription. In general, the first type of response occurs more rapidly than the second type. Signaling from G protein–coupled receptors, described in later sections, often results in changes in the activity of preexisting proteins, although activation of these receptors on some cells also can induce changes in gene expression.

The other classes of receptors depicted in Figure 13-1 operate primarily to modulate gene expression. In some cases, the activated receptor directly activates a **transcription factor** in the cytosol (e.g., TGFβ and cytokine receptor pathways) or assembles an intracellular signaling complex that activates a cytosolic transcription factor (e.g., Wnt pathways). In yet other pathways, specific proteolytic cleavage of an activated cell-surface receptor or cytosolic protein releases a transcription factor (e.g., Hedgehog, Notch, and NF-κB pathways). Transcription factors activated in the cytosol by these pathways move into the nucleus, where they stimulate (or occasionally inhibit) transcription of specific target genes. Signaling from receptor tyrosine kinases leads to activation of several cytosolic protein kinases that translocate into the nucleus and regulate the activity of nuclear transcription factors. We consider these signaling pathways, which regulate transcription of many genes essential for cell division and for many cell differentiation processes, in the following two chapters.

Second, some classes of receptors can initiate signaling via more than one intracellular signal-transduction pathway, leading to different cellular responses. This complication is typical of G protein–coupled receptors, receptor tyrosine kinases, and cytokine receptors.

Third, despite the huge number of different kinds of ligands and their specific receptors, a relatively small number of signal-transduction mechanisms and highly conserved intracellular proteins play a major role in intracellular signaling pathways. Our knowledge of these common themes has advanced greatly in recent years. For instance, we can trace the entire signaling pathway from binding of ligand to receptors in several classes to the final cellular response.

Before delving into the particulars of individual signaling pathways, we discuss the basic properties of cell-surface receptors, as well as methods for identifying and studying them, in the remainder of this section; important general features of intracellular signal transduction are presented in Section 13.2.

Receptor Proteins Exhibit Ligand-Binding and Effector Specificity

The response of a cell or tissue to specific external signals is dictated by the particular receptors it possesses, by the signal-transduction pathways they activate, and by the intracellular processes ultimately affected. Each receptor generally binds only a single signaling molecule or a group of very closely related molecules (Figure 13-2). In contrast, many signaling molecules bind to multiple types of receptors, each of which can activate different intracellular signaling pathways and thus induce different cellular responses. For instance, different types of acetylcholine receptors are found on the surface of striated muscle cells, heart muscle cells, and pancreatic acinar cells. Release of acetylcholine from a neuron adjacent to a striated muscle cell triggers contraction by activating a ligand-gated ion channel, whereas release adjacent to a heart muscle slows the rate of contraction via activation of a G protein–coupled receptor. Release adjacent to a pancreatic acinar cell triggers exocytosis of secretory granules that contain digestive enzymes. Similarly, epinephrine binds to several different G protein–coupled receptors, each of which induces a distinct cellular response. Thus each receptor protein is characterized by *binding specificity* for a particular ligand, and the resulting receptor-ligand complex exhibits *effector specificity* (i.e., mediates a specific cellular response).

On the other hand, different receptors of the same class that bind different ligands often induce the same cellular responses in a cell. In liver cells, for instance, the hormones epinephrine, glucagon, and ACTH bind to different members of the G protein–coupled receptor family, but all these

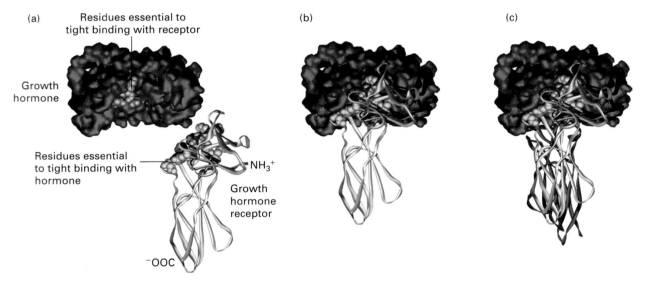

▲ EXPERIMENTAL FIGURE 13-2 Mutational studies have identified the patches of amino acids in growth hormone and its receptor that determine their highly specific mutual interaction. The outer surface of the plasma membrane is toward the bottom of the figure, and each receptor is anchored to the membrane by a hydrophobic membrane-spanning alpha helix that is not shown. As determined from the three-dimensional structure of the growth hormone–growth hormone receptor complex, 28 amino acids in the hormone are at the binding interface with one receptor. Each of these amino acids was mutated, one at a time, to alanine, and the effect on receptor binding was determined. (a) From this study it was found that only eight amino acids on growth hormone (pink) contribute 85 percent of the binding energy; these amino acids are distant in the primary sequence but adjacent in the folded protein. Similar studies showed that two tryptophan residues (blue) in the receptor contribute most of the energy for binding growth hormone, although other amino acids at the interface with the hormone (yellow) are also important. (b) Binding of growth hormone to one receptor molecule is followed by (c) binding of a second receptor to the opposing side of the hormone; this involves the same set of yellow and blue amino acids on the receptor but different residues on the hormone. As we see in the following chapter, such hormone-induced receptor dimerization is a common mechanism for receptor activation. [After B. Cunningham and J. Wells, 1993, *J. Mol. Biol.* **234**:554, and T. Clackson and J. Wells, 1995, *Science* **267**:383.]

receptors activate the same signal-transduction pathway, one that promotes synthesis of **cyclic AMP (cAMP)**. This small signaling molecule in turn regulates various metabolic functions, including glycogen breakdown. As a result, all three hormones have the same effect on liver-cell metabolism.

Maximal Cellular Response to a Signaling Molecule May Not Require Activation of All Receptors

As we've seen, activation of a cell-surface receptor and subsequent signal transduction are triggered by binding of a signaling molecule (ligand) to the receptor. This binding depends on weak, noncovalent forces (i.e., ionic, van der Waals, and hydrophobic interactions) and **molecular complementarity** between the interacting surfaces of a receptor and ligand (Chapter 2). The specificity of a receptor refers to its ability to distinguish closely related substances. The insulin receptor, for example, binds insulin and a related hormone called insulinlike growth factor 1, but no other peptide hormones.

Ligand binding usually can be viewed as a simple reversible reaction,

$$R + L \underset{k_{\text{off}}}{\overset{k_{\text{on}}}{\rightleftharpoons}} RL$$

which can be described by the equation

$$K_{\text{d}} = \frac{[R][L]}{[RL]} \tag{13-1}$$

where $[R]$ and $[L]$ are the concentrations of free receptor and ligand, respectively, at equilibrium, and $[RL]$ is the concentration of the receptor-ligand complex. K_{d}, the **dissociation constant** of the receptor-ligand complex, measures the *affinity* of the receptor for the ligand. This equilibrium binding equation can be rewritten as

$$\frac{[RL]}{R_{\text{T}}} = \frac{1}{1 + \dfrac{K_{\text{d}}}{[L]}} \tag{13-2}$$

where $R_{\text{T}} = [R] + [RL]$, the total concentration of free and bound receptors; therefore, $[RL]/R_{\text{T}}$ is the fraction of receptors that have a bound ligand. The lower the K_{d} value, the higher the affinity of a receptor for its ligand. The K_{d} value is equivalent to the concentration of ligand at which half the receptors contain bound ligand. If $[L] = K_{\text{d}}$, then from Equation 13-2 we can see that $[RL] = 0.5\, R_{\text{T}}$. Equation 13-2 has the same general form as the Michaelis-Menten equation, which describes simple one-substrate enzymatic reactions (Chapter 3). The K_{d} for a binding reaction is equivalent to the Michaelis constant K_{m}, which reflects the affinity of an enzyme for its substrate.

For a simple binding reaction, $K_{\text{d}} = k_{\text{off}}/k_{\text{on}}$, where k_{off} is the rate constant for dissociation of a ligand from its receptor, and k_{on} is the rate constant for formation of a receptor-

ligand complex from free ligand and receptor. The lower k_{off} is relative to k_{on}, the more stable the RL complex, and thus the lower the value of K_{d}. Like all equilibrium constants, however, the value of K_{d} does not depend on the *absolute* values of k_{off} and k_{on}, only on their ratio. For this reason, binding of ligand by two different receptors can have the same K_{d} values but very different rate constants.

In general, the K_{d} value of a cell-surface receptor for a circulating hormone is greater than the normal (unstimulated) blood level of that hormone. Under this circumstance, changes in hormone concentration are reflected in proportional changes in the fraction of receptors occupied. Suppose, for instance, that the normal concentration of a hormone in the blood is 10^{-9} M and that the K_{d} for its receptor is 10^{-7} M; by substituting these values into Equation 13-2, we can calculate the fraction of receptors with bound hormone, $[RL]/R_{\text{T}}$, at equilibrium as 0.0099. Thus about 1 percent of the total receptors will be filled with hormone. If the hormone concentration rises tenfold to 10^{-8} M, the concentration of receptor-hormone complex will rise proportionately, so that about 10 percent of the total receptors would have bound hormone. If the extent of the induced cellular response parallels the amount of RL, as is often the case, then the cellular responses also will increase tenfold.

In many cases, however, the maximal cellular response to a particular ligand is induced when less than 100 percent of its receptors are bound to the ligand. This phenomenon can be revealed by determining the extent of the response and of receptor-ligand binding at different concentrations of ligand (Figure 13-3). For example, a typical erythroid progenitor cell

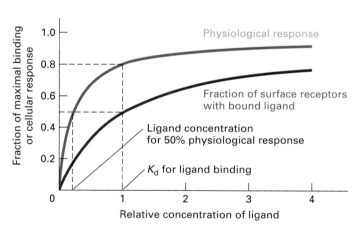

▲ **EXPERIMENTAL FIGURE 13-3 The maximal physiological response to many external signals occurs when only a fraction of the receptor molecules are occupied by ligand.** In this situation, plots of the extent of ligand binding and of physiological response at different ligand concentrations differ. In the example shown here, 50 percent of the maximal physiological response is induced at a ligand concentration at which only 18 percent of the receptors are occupied. Likewise, 80 percent of the maximal response is induced when the ligand concentration equals the K_{d} value, at which 50 percent of the receptors are occupied.

has ≈1000 surface receptors for erythropoietin, which induces progenitor cells to proliferate and differentiate into red blood cells. Because only 100 of these receptors need to bind erythropoietin to induce division of a target cell, the ligand concentration needed to induce 50 percent of the maximal cellular response is proportionally lower than the K_d value for binding. In such cases, a plot of the percentage of maximal binding versus ligand concentration differs from a plot of the percentage of maximal cellular response versus ligand concentration.

Sensitivity of a Cell to External Signals Is Determined by the Number of Surface Receptors

Because the cellular response to a particular signaling molecule depends on the number of receptor-ligand complexes, the fewer receptors present on the surface of a cell, the less *sensitive* the cell is to that ligand. As a consequence, a higher ligand concentration is necessary to induce the usual physiological response than would be the case if more receptors were present.

To illustrate this important point, let's extend our example of a typical erythroid progenitor cell. The K_d for binding of erythropoietin (Epo) to its receptor is about 10^{-10} M. As we noted above, only 10 percent of the ≈1000 cell-surface erythropoietin receptors on the surface of a cell must be bound to ligand to induce the maximal cellular response. We can determine the ligand concentration, [L], needed to induce the maximal response by rewriting Equation 13-2 as follows:

$$[L] = \frac{K_d}{\dfrac{R_T}{[RL]} - 1} \tag{13-3}$$

If $R_T = 1000$ (the total number of Epo receptors per cell), $K_d = 10^{-10}$ M, and [RL] = 100 (the number of Epo-occupied receptors needed to induce the maximal response), then an Epo concentration of 1.1×10^{-11} M will elicit the maximal response. If R_T is reduced to 200/cell, then a ninefold higher Epo concentration (10^{-10} M) is required to occupy 100 receptors and induce the maximal response. If R_T is further reduced to 120/cell, an Epo concentration of 5×10^{-10} M, a 50-fold increase, is necessary to generate the same cellular response.

Regulation of the number of receptors for a given signaling molecule expressed by a cell and thus its sensitivity to that signal plays a key role in directing physiological and developmental events. Alternatively, endocytosis of receptors on the cell surface can sufficiently reduce the number present to terminate the usual cellular response at the prevailing signal concentration.

Binding Assays Are Used to Detect Receptors and Determine Their K_d Values

Cell-surface receptors are difficult to identify and purify, mainly because they are present in such minute amounts. The receptor for a particular signaling molecule commonly con-

stitutes only ≈10^{-6} of the total protein in the cell, or ≈10^{-4} of the plasma-membrane protein. Purification is also difficult because these integral membrane proteins first must be solubilized with a nonionic detergent so they can be separated from other proteins (see Figure 5-40).

Usually, receptors are detected and measured by their ability to bind radioactive ligands to cells or to cell fragments. The results of such a *binding assay* are illustrated and explained in Figure 13-4. Both the number of ligand-binding sites per cell and the K_d value are easily determined from the specific binding curve (Figure 13-4, curve B), which is described by Equation 13-2. Since each receptor generally binds just one ligand molecule, the number of ligand-binding sites equals the number of active receptors per cell. Straight binding assays like the one in Figure 13-4 are feasible with receptors that have a strong affinity for their ligands, such as the erythropoietin receptor ($K_d = 1 \times 10^{-10}$ M) and the insulin receptor on liver cells ($K_d = 1.4 \times 10^{-8}$ M).

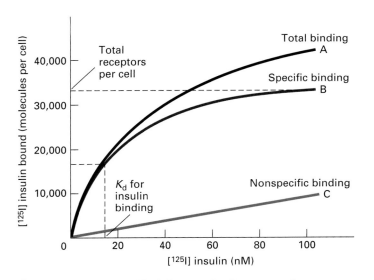

▲ **EXPERIMENTAL FIGURE 13-4 Binding assays for cell-surface receptors can determine the K_d for high-affinity ligands and the number of receptors per cell.** Shown here are data for insulin-specific receptors on the surface of liver cells. A suspension of cells is incubated for 1 hour at 4° C with increasing concentrations of ^{125}I-labeled insulin; the low temperature is used to prevent endocytosis of the cell-surface receptors. The cells are separated from unbound insulin, usually by centrifugation, and the amount of radioactivity bound to them is measured. The total binding curve A represents insulin specifically bound to high-affinity receptors as well as insulin nonspecifically bound with low affinity to other molecules on the cell surface. The contribution of nonspecific binding to total binding is determined by repeating the binding assay in the presence of a 100-fold excess of unlabeled insulin, which saturates all the specific high-affinity sites. In this case, all the labeled insulin binds to nonspecific sites, yielding curve C. The specific binding curve B is calculated as the difference between curves A and C. From curve B, the K_d for insulin binding (≈1.4×10^{-8} M, or 14 nM) and the number of receptor molecules per cell (≈33,000) can be determined. [Adapted from A. Ciechanover et al., 1983, *Cell* **32**:267.]

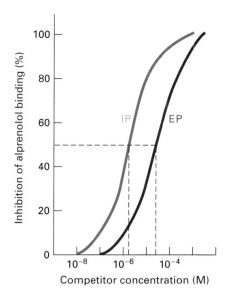

▲ **EXPERIMENTAL FIGURE 13-5 Binding of low-affinity ligands to cell-surface receptors can be detected in competition assays.** In this example, the synthetic ligand alprenolol, which binds with high affinity to the epinephrine receptor on liver cells ($K_d \cong 3 \times 10^{-9}$ M), is used to detect the binding of two low-affinity ligands, the natural hormone epinephrine (EP) and a synthetic ligand called isoproterenol (IP). Assays are performed as described in Figure 13-4 but with a constant amount of [^{3}H]alprenolol to which increasing amounts of unlabeled epinephrine or isoproterenol are added. At each competitor concentration, the amount of bound labeled alprenolol is determined. In a plot of the inhibition of [^{3}H]alprenolol binding versus epinephrine or isoproterenol concentration, such as shown here, the concentration of the competitor that inhibits alprenolol binding by 50 percent approximates the K_d value for competitor binding. Note that the concentrations of competitors are plotted on a logarithmic scale. The K_d for binding of epinephrine to its receptor on liver cells is only ≈5×10^{-5} M and would not be measurable by a direct binding assay with [^{3}H]epinephrine. The K_d for binding of isoproterenol, which induces the normal cellular response, is more than tenfold lower.

Many ligands, however, bind to their receptors with much lower affinity. If the K_d for binding is greater than ≈1×10^{-7} M, any ligand bound to receptors is likely to dissociate in the few seconds it takes to separate the cells (e.g., by centrifugation) from free (unbound) ligand and measure the amount of bound ligand. One way to detect weak binding of a ligand to its receptor is in a *competition assay* with another ligand that binds to the same receptor with high affinity (low K_d value). In this type of assay, increasing amounts of an unlabeled, low-affinity ligand (the competitor) are added to a cell sample with a constant amount of the radiolabeled, high-affinity ligand (Figure 13-5). Binding of unlabeled competitor blocks binding of the radioactive ligand to the receptor; the concentration of competitor required to inhibit binding of half the radioactive ligand approximates the K_d value for binding of the competitor to the receptor.

 Synthetic analogs of natural hormones are widely used in research on cell-surface receptors and as drugs. These analogs fall into two classes: **agonists,** which mimic the function of a natural hormone by binding to its receptor and inducing the normal response, and **antagonists,** which bind to the receptor but induce no response. By occupying ligand-binding sites on a receptor, an antagonist can block binding of the natural hormone (or agonist) and thus reduce the usual physiological activity of the hormone.

For instance, addition of two methyl groups to epinephrine generates isoproterenol, an agonist that binds to epinephrine receptors on bronchial smooth muscle cells about tenfold more strongly than does epinephrine (see Figure 13-5). Because ligand binding to these receptors promotes relaxation of bronchial smooth muscle and thus opening of the air passages in the lungs, isoproterenol is used in treating bronchial asthma, chronic bronchitis, and emphysema. Activation of epinephrine receptors on cardiac muscle cells increases the contraction rate. The antagonist alprenolol and related compounds, referred to as *beta-blockers,* have a very high affinity for these epinephrine receptors. Such antagonists are used to slow heart contractions in the treatment of cardiac arrhythmias and angina. ∎

Receptors Can Be Purified by Affinity Techniques or Expressed from Cloned Genes

Cell-surface receptors often can be identified and followed through isolation procedures by *affinity labeling*. In this technique, cells are mixed with an excess of a radiolabeled ligand for the receptor of interest. After unbound ligand is washed away, the cells are treated with a chemical agent that covalently cross-links bound labeled ligand molecules and receptors on the cell surface. Once a radiolabeled ligand is covalently cross-linked to its receptor, it remains bound even in the presence of detergents and other denaturing agents that are used to solubilize receptor proteins from the cell membrane. The labeled ligand provides a means for detecting the receptor during purification procedures.

Another technique often used in purifying cell-surface receptors that retain their ligand-binding ability when solubilized by detergents is similar to *affinity chromatography* using antibodies (see Figure 3-34). To purify a receptor by this technique, a ligand for the receptor of interest, rather than an antibody, is chemically linked to the beads used to form a column. A crude, detergent-solubilized preparation of membrane proteins is passed through the column; only the receptor binds, and other proteins are washed away. Passage of an excess of the soluble ligand through the column causes the bound receptor to be displaced from the beads and eluted from the column. In some cases, a receptor can be purified as much as 100,000-fold in a single affinity chromatographic step.

Once a receptor is purified, its properties can be studied and its gene cloned. A *functional expression assay* of the cloned cDNA in a mammalian cell that normally lacks the encoded receptor can provide definitive proof that the proper protein indeed has been obtained (Figure 13-6). Such expression assays also permit investigators to study the effects of mutating specific amino acids on ligand binding or on "downstream" signal transduction, thereby pinpointing the receptor amino acids responsible for interacting with the ligand or with critical signal-transduction proteins.

The cell-surface receptors for many signaling molecules are present in such small amounts that they cannot be purified by affinity chromatography and other conventional biochemical techniques. These low-abundance receptor proteins can now be identified and cloned by various recombinant DNA techniques, eliminating the need to isolate and purify them from cell extracts. In one technique, cloned cDNAs prepared from the entire mRNA extracted from cells that produce the receptor are inserted into expression vectors by techniques described in Chapter 9. The recombinant vectors then are transfected into cells that normally do not synthesize the receptor of interest, as in Figure 13-6. Only the very few transfected cells that contain the cDNA encoding the desired receptor synthesize it; other transfected cells produce irrelevant proteins. The rare cells expressing the desired receptor can be detected and purified by various techniques such as fluorescence-activated cell sorting using a fluorescent-labeled ligand for the receptor of interest (see Figure 5-34). Once a cDNA clone encoding the receptor is identified, the sequence of the cDNA can be determined and that of the receptor protein deduced from the cDNA sequence.

Genomics studies coupled with functional expression assays are now being used to identify genes for previously unknown receptors. In this approach, stored DNA sequences are analyzed for similarities with sequences known to encode receptor proteins (Chapter 9). Any putative receptor genes that are identified in such a search then can be tested for their ability to bind a signaling molecule or induce a response in cultured cells by a functional expression assay.

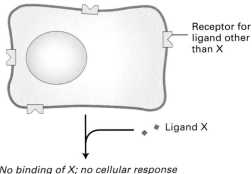

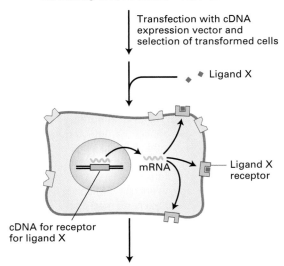

▲ **EXPERIMENTAL FIGURE 13-6 Functional expression assay can identify a cDNA encoding a cell-surface receptor.** Target cells lacking receptors for a particular ligand (X) are stably transfected with a cDNA expression vector encoding the receptor. The design of the expression vector permits selection of transformed cells from those that do not incorporate the vector into their genome (see Figure 9-29b). Providing that these cells already express all the relevant signal-transduction proteins, the transfected cells exhibit the normal cellular response to X if the cDNA in fact encodes the functional receptor.

KEY CONCEPTS OF SECTION 13.1

Signaling Molecules and Cell-Surface Receptors

■ Extracellular signaling molecules regulate interactions between unicellular organisms and are critical regulators of physiology and development in multicellular organisms.

■ Binding of extracellular signaling molecules to cell-surface receptors triggers intracellular signal-transduction pathways that ultimately modulate cellular metabolism, function, or gene expression (Figure 13-1).

■ External signals include membrane-anchored and secreted proteins and peptides, small lipophilic molecules (e.g., steroid hormones, thyroxine), small hydrophilic molecules derived from amino acids (e.g., epinephrine), gases (e.g., nitric oxide), and physical stimuli (e.g., light).

- Signals from one cell can act on nearby cells (paracrine), on distant cells (endocrine), or on the signaling cell itself (autocrine).
- Receptors bind ligands with considerable specificity, which is determined by noncovalent interactions between a ligand and specific amino acids in the receptor protein (see Figure 13-2).
- The maximal response of a cell to a particular ligand generally occurs at ligand concentrations at which most of its receptors are still not occupied (see Figure 13-3).
- The concentration of ligand at which half its receptors are occupied, the K_d, can be determined experimentally and is a measure of the affinity of the receptor for the ligand (see Figure 13-4).
- Because the amount of a particular receptor expressed is generally quite low (ranging from ≈2000 to 20,000 molecules per cell), biochemical purification may not be feasible. Genes encoding low-abundance receptors for specific ligands often can be isolated from cDNA libraries transfected into cultured cells.
- Functional expression assays can determine if a cDNA encodes a particular receptor and are useful in studying the effects on receptor function of specific mutations in its sequence (see Figure 13-6).

13.2 Intracellular Signal Transduction

The various intracellular pathways that transduce signals downstream from activated cell-surface receptors differ in their complexity and in the way they transduce signals. We describe the components and operation of many individual pathways later in this chapter and in other chapters. Some general principles of signal transduction, applicable to different pathways, are covered in this section.

Second Messengers Carry Signals from Many Receptors

The binding of ligands ("first messengers") to many cell-surface receptors leads to a short-lived increase (or decrease) in the concentration of certain low-molecular-weight intracellular signaling molecules termed **second messengers.** These molecules include 3′,5′-cyclic AMP (cAMP), 3′,5′-cyclic GMP (cGMP), 1,2-diacylglycerol (DAG), and inositol 1,4,5-trisphosphate (IP_3), whose structures are shown in Figure 13-7. Other important second messengers are Ca^{2+} and various inositol phospholipids, also called *phosphoinositides,* which are embedded in cellular membranes.

The elevated intracellular concentration of one or more second messengers following binding of an external signaling molecule triggers a rapid alteration in the activity of one or more enzymes or nonenzymatic proteins. In muscle, a signal-induced rise in cytosolic Ca^{2+} triggers contraction (see Figure 19-28); a similar increase in Ca^{2+} induces exocytosis of secretory vesicles in endocrine cells and of neurotransmitter-containing vesicles in nerve cells (see Figure 7-43). Similarly, a rise in cAMP induces various changes in cell metabolism that differ in different types of human cells. The

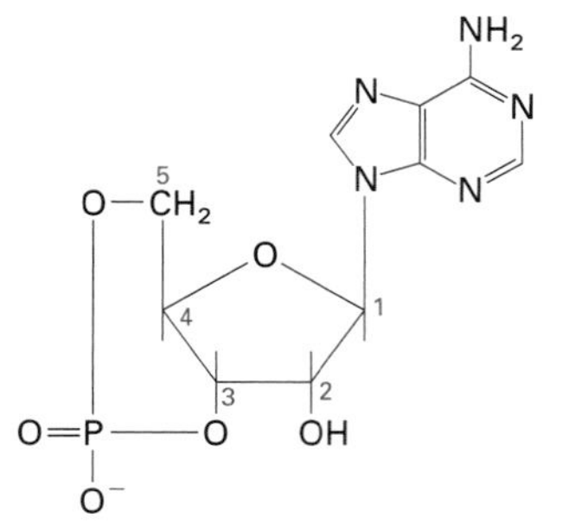

3′,5′-Cyclic AMP (cAMP)

Activates protein kinase A (PKA)

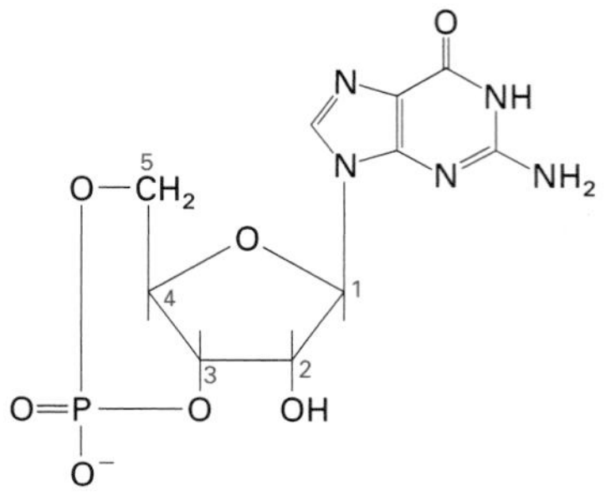

3′,5′-Cyclic GMP (cGMP)

Activates protein kinase G (PKG) and opens cation channels in rod cells

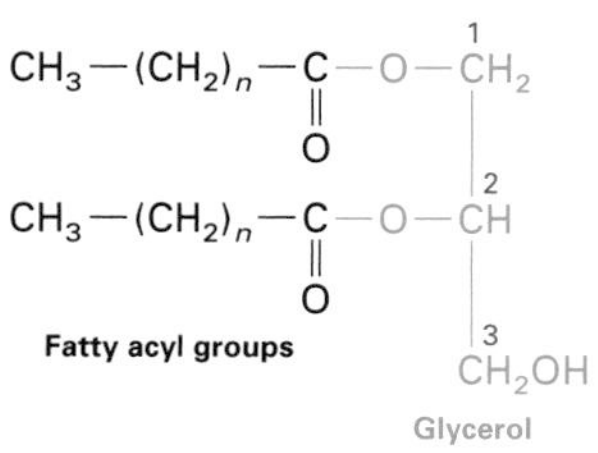

1,2-Diacylglycerol (DAG)

Activates protein kinase C (PKC)

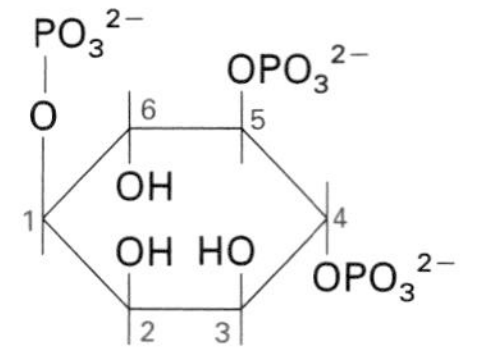

Inositol 1,4,5-trisphosphate (IP_3)

Opens Ca^{2+} channels in the endoplasmic reticulum

▲ **FIGURE 13-7 Four common intracellular second messengers.** The major direct effect or effects of each compound are indicated below its structural formula. Calcium ion (Ca^{2+}) and several membrane-bound phosphoinositides also act as second messengers.

MEDIA CONNECTIONS
Focus Animation: Second Messengers in Signaling Pathways

mode of action of cAMP and other second messengers is discussed in later sections.

Many Conserved Intracellular Proteins Function in Signal Transduction

In addition to cell-surface receptors and second messengers, two groups of evolutionary conserved proteins function in signal-transduction pathways stimulated by extracellular signals. Here we briefly consider these intracellular signaling proteins; their role in specific pathways is described elsewhere.

GTPase Switch Proteins We introduced the large group of intracellular switch proteins that form the **GTPase superfamily** in Chapter 3. These guanine nucleotide–binding proteins are turned "on" when bound to GTP and turned "off" when bound to GDP (see Figure 3-29). Signal-induced conversion of the inactive to active state is mediated by a *guanine nucleotide–exchange factor (GEF)*, which causes release of GDP from the switch protein. Subsequent binding

of GTP, favored by its high intracellular concentration, induces a conformational change in two segments of the protein, termed switch I and switch II, allowing the protein to bind to and activate other downstream signaling proteins (Figure 13-8). The intrinsic GTPase activity of the switch proteins then hydrolyzes the bound GTP to GDP and P_i, thus changing the conformation of switch I and switch II from the active form back to the inactive form. The rate of GTP hydrolysis frequently is enhanced by a *GTPase-accelerating protein (GAP)*, whose activity also may be controlled by extracellular signals. The rate of GTP hydrolysis regulates the length of time the switch protein remains in the active conformation and able to signal downstream.

There are two classes of GTPase switch proteins: *trimeric* (large) G proteins, which as noted already directly bind to and are activated by certain receptors, and *monomeric* (small) G proteins such as Ras and various Ras-like proteins. Ras is linked indirectly to receptors via adapter proteins and GEF proteins discussed in the next chapter. All G proteins contain regions like switch I and switch II that modulate the

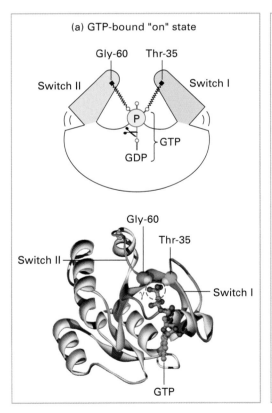
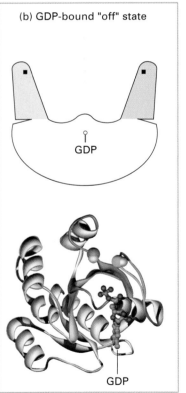

▶ **FIGURE 13-8 Switching mechanism for monomeric and trimeric G proteins.** The ability of a G protein to interact with other proteins and thus transduce a signal differs in the GTP-bound "on" state and GDP-bound "off" state. (a) In the active "on" state, two domains, termed switch I (green) and switch II (blue), are bound to the terminal γ phosphate of GTP through interactions with the backbone amide groups of a conserved threonine and glycine residue. (b) Release of the γ phosphate by GTPase-catalyzed hydrolysis causes switch I and switch II to relax into a different conformation, the inactive "off" state. Shown here as ribbon models are both conformations of Ras, a monomeric G protein. A similar spring-loaded mechanism switches the α subunit in trimeric G proteins between the active and inactive conformations. [Adapted from I. Vetter and A. Wittinghofer, 2001, *Science* **294**:1299.]

activity of specific effector proteins by direct protein-protein interactions when the G protein is bound to GTP. Despite these similarities, these two classes of GTP-binding proteins are regulated in very different ways.

Protein Kinases and Phosphatases Activation of all cell-surface receptors leads directly or indirectly to changes in protein phosphorylation through the activation of protein **kinases** or protein **phosphatases**. Animal cells contain two types of protein kinases: those that add phosphate to the hydroxyl group on tyrosine residues and those that add phosphate to the hydroxyl group on serine or threonine (or both) residues. Phosphatases, which remove phosphate groups, can act in concert with kinases to switch the function of various proteins on or off (see Figure 3-30). At last count the human genome encodes 500 protein kinases and 100 different phosphatases. In some signaling pathways, the receptor itself possesses intrinsic kinase or phosphatase activity; in other pathways, the receptor interacts with cytosolic or membrane-associated kinases.

In general, each protein kinase phosphorylates specific residues in a set of target proteins whose patterns of expression generally differ in different cell types. Many proteins are substrates for multiple kinases, and each phosphorylation event, on a different amino acid, modifies the activity of a particular target protein in different ways, some activating its function, others inhibiting it. The catalytic activity of a protein kinase itself commonly is modulated by phosphorylation by other kinases, by direct binding to other proteins, or by changes in the levels of various second messengers. The activity of all protein kinases is opposed by the activity of protein phosphatases, some of which are themselves regulated by extracellular signals. Thus the activity of a protein in a cell can be a complex function of the activities of the usually multiple kinases and phosphatases that act on it. Several examples of this phenomenon that occur in regulation of the cell cycle are described in Chapter 21.

Some Receptors and Signal-Transduction Proteins Are Localized

Although the epinephrine receptors expressed by adipose (fat-storage) cells appear to be uniformly distributed on the surface of these spherical cells, such a uniform distribution probably is rare. More common is the clustering of receptors and other membrane-associated signaling proteins to a particular region of the cell surface. In this section, we show how multiple protein-protein and protein-lipid interactions can cluster signaling proteins in the plasma membrane and discuss some advantages conferred by such clustering. Other instances of localization of signaling proteins are described elsewhere.

Clustering of Membrane Proteins Mediated by Adapter Domains Perhaps the best example of clustering of receptors and other membrane proteins is the chemical synapse. Recall that synaptic junctions are highly specialized structures at which chemical signals (neurotransmitters) are released from a presynaptic cell and bind receptors on an adjacent postsynaptic cell (see Figure 7-31). Clustering of neurotransmitter receptors in the region of the postsynaptic plasma membrane adjacent to the presynaptic cell promotes rapid and efficient signal transmission. Other proteins in the membrane of the postsynaptic cell interact with proteins in the extracellular matrix in order to "lock" the cell into the synapse.

Proteins containing *PDZ domains* play a fundamental role in organizing the plasma membrane of the postsynaptic cell. The PDZ domain was identified as a common element in several cytosolic proteins that bind to integral plasma-membrane proteins. It is a relatively small domain, containing about 90 amino acid residues, that binds to three-residue sequences at the C-terminus of target proteins (Figure 13-9a). Some PDZ domains bind to the sequence Ser/Thr-X-Φ, where X denotes any amino acid and Φ denotes a hydrophobic amino acid; others bind to the sequence Φ-X-Φ.

Most cell-surface receptors and transporters contain multiple subunits, each of which can bind to a PDZ domain. Likewise, many cytosolic proteins contain multiple PDZ domains as well as other types of domains that participate in protein-protein interactions, and thus can bind to multiple membrane proteins at the same time. These interactions permit the clustering of different membrane proteins into large complexes (Figure 13-9b). Other protein-protein interactions enable these complexes to bind to actin filaments that line the underside of the plasma membrane. Since a single actin filament can bind many clusters of the type depicted in Figure 13-9b, even larger numbers of plasma-membrane proteins can be clustered together specifically. This is one of the mechanisms by which many receptors, binding the same or different ligands, are localized to a specific region of the membrane in postsynaptic cells and other cells as well.

Protein Clustering in Lipid Rafts In Chapter 5, we saw that certain lipids in the plasma membrane, particularly cholesterol and sphingolipids, are organized into aggregates, called **lipid rafts**, that also contain specific proteins (see Figure 5-10). In mammalian cells, lipid rafts termed *caveolae* are of particular interest because they have been found to contain several different receptors and other signal-transducing proteins. These rafts are marked by the presence of *caveolin*, a family of ≈25-kDa proteins. Caveolin proteins have a central hydrophobic segment that is thought to span the membrane twice, and both the N- and C-termini face the cytosol. Large oligomers of caveolin form a proteinaceous coat that is visible on the cytosolic surface of caveolae in the electron microscope. Precisely how certain signaling proteins are anchored in caveolae is unclear. Nonetheless, the proximity of signaling proteins to one another within caveolae may facilitate their interaction, thereby promoting certain signaling pathways that otherwise would operate inefficiently.

(a)

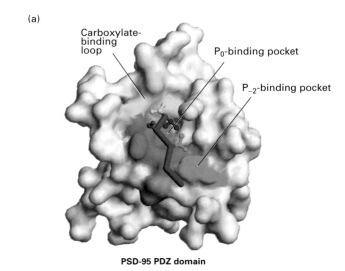

PSD-95 PDZ domain

(b)

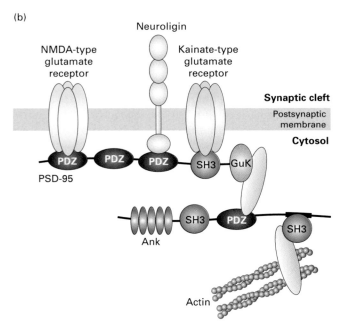

▲ **FIGURE 13-9 Clustering of membrane proteins mediated by cytosolic adapter proteins containing multiple protein-binding domains.** The PDZ domain, which binds to certain C-terminal sequences, and the SH3 domain, which binds to proline-rich sequences, are two of several conserved domains that participate in protein-protein interactions. (a) Three-dimensional surface structure of a PDZ domain showing the backbone of the bound target peptide in red. Regions in the PDZ domain that bind the COO^- group and side chain of the C-terminal residue are colored yellow and blue, respectively. The binding pocket for the residue two distant from the C-terminus (P_{-2}) is green. (b) Schematic diagram of protein-protein interactions that cluster several different membrane proteins in a postsynaptic segment of a nerve cell and anchor the resulting complex to cytoskeletal actin filaments. Within the adapter protein PSD-95, two of the three PDZ domains shown and one SH3 domain bind three different membrane proteins into one complex. The guanylate kinase (GuK) domain of the PSD-95 protein links the complex, via several intervening adapter proteins (including one also containing PDZ and SH3 domains), to fibrous actin underlying the plasma membrane. Neuroligin is an adhesive protein that interacts with components of the extracellular matrix. Ank = ankyrin repeats. Other multibinding adapter proteins localize and cluster different receptors in the synaptic region of the plasma membrane. [Part (a) adapted from B. Harris and W. A. Lim, 2001, *J. Cell Sci.* **114**:3219; part (b) adapted from C. Garner, J. Nash, and R. Huganir, 2000, *Trends Cell Biol.* **10**:274.]

Appropriate Cellular Responses Depend on Interaction and Regulation of Signaling Pathways

In this chapter and the next, we focus primarily on simple signal-transduction pathways triggered by ligand binding to a single type of receptor. Activation of a single type of receptor, however, often leads to production of multiple second messengers, which have different effects. Moreover, the same cellular response (e.g., glycogen breakdown) may be induced by activation of multiple signaling pathways. Such interaction of different signaling pathways permits the fine-tuning of cellular activities required to carry out complex developmental and physiological processes.

The ability of cells to respond appropriately to extracellular signals also depends on regulation of signaling pathways themselves. For example, once the concentration of an external signal decreases, signaling via some intracellular pathways is terminated by degradation of a second messenger; in other pathways, signaling is terminated by deactivation of a signal-transduction protein. Another important mechanism for as-

suring appropriate cellular responses is *desensitization* of receptors at high signal concentrations or after prolonged exposure to a signal. The sensitivity of a cell to a particular signaling molecule can be down-regulated by **endocytosis** of its receptors, thus decreasing the number on the cell surface, or by modifying their activity so that the receptors either cannot bind ligand or form a receptor-ligand complex that does not induce the normal cellular response. Such modulation of receptor activity often results from phosphorylation of the receptor, binding of other proteins to it, or both. We examine the details of various mechanisms for regulating signaling pathways in our discussion of individual pathways.

KEY CONCEPTS OF SECTION 13.2
Intracellular Signal Transduction

■ The level of second messengers, such as Ca^{2+}, cAMP, and IP_3, increases or occasionally decreases in response to binding of ligand to cell-surface receptors (see Figure 13-7). These nonprotein intracellular signaling molecules, in turn, regulate the activities of enzymes and nonenzymatic proteins.

- Conserved proteins that act in many signal-transduction pathways include monomeric and trimeric G proteins (see Figure 13-8) and protein kinases and phosphatases.

- Cytosolic proteins that contain multiple PDZ or other protein-binding domains cluster receptors and other proteins within the plasma membrane, as occurs in postsynaptic cells (see Figure 13-9).

- Many receptors and signal-transduction proteins cluster in caveolin-containing lipid rafts. Such clustering may facilitate interaction between signaling proteins, thus enhancing signal transduction.

- Rapid termination of signaling once a particular ligand is withdrawn and receptor desensitization at high ligand concentrations or after prolonged exposure help cells respond appropriately under different circumstances.

13.3 G Protein–Coupled Receptors That Activate or Inhibit Adenylyl Cyclase

We now turn our attention to the very large group of cell-surface receptors that are coupled to signal-transducing trimeric G proteins. All G protein–coupled receptors (GPCRs) contain seven membrane-spanning regions with their N-terminal segment on the exoplasmic face and their C-terminal segment on the cytosolic face of the plasma membrane (Figure 13-10). The GPCR family includes receptors for numerous hormones and neurotransmitters, light-activated receptors (rhodopsins) in the eye, and literally thousands of odorant receptors in the mammalian nose.

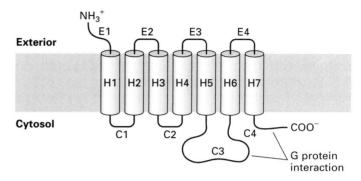

▲ **FIGURE 13-10 Schematic diagram of the general structure of G protein–coupled receptors.** All receptors of this type have the same orientation in the membrane and contain seven transmembrane α-helical regions (H1–H7), four extracellular segments (E1–E4), and four cytosolic segments (C1–C4). The carboxyl-terminal segment (C4), the C3 loop, and, in some receptors, also the C2 loop are involved in interactions with a coupled trimeric G protein.

TABLE 13-1	**Major Classes of Mammalian Trimeric G Proteins and Their Effectors**[*]		
G_α Class	Associated Effector	2nd Messenger	Receptor Examples
$G_{s\alpha}$	Adenylyl cyclase	cAMP (increased)	β-Adrenergic (epinephrine) receptor; receptors for glucagon, serotonin, vasopressin
$G_{i\alpha}$	Adenylyl cyclase K$^+$ channel ($G_{\beta\gamma}$ activates effector)	cAMP (decreased) Change in membrane potential	α$_1$-Adrenergic receptor Muscarinic acetylcholine receptor
$G_{olf\alpha}$	Adenylyl cyclase	cAMP (increased)	Odorant receptors in nose
$G_{q\alpha}$	Phospholipase C	IP$_3$, DAG (increased)	α$_2$-Adrenergic receptor
$G_{o\alpha}$	Phospholipase C	IP$_3$, DAG (increased)	Acetylcholine receptor in endothelial cells
$G_{t\alpha}$	cGMP phosphodiesterase	cGMP (decreased)	Rhodopsin (light receptor) in rod cells

*A given G_α subclass may be associated with more than one effector protein. To date, only one major $G_{s\alpha}$ has been identified, but multiple $G_{q\alpha}$ and $G_{i\alpha}$ proteins have been described. Effector proteins commonly are regulated by G_α but in some cases by $G_{\beta\gamma}$ or the combined action of G_α and $G_{\beta\gamma}$. IP$_3$ = inositol 1,4,5-trisphosphate; DAG = 1,2-diacylglycerol.

SOURCES: See L. Birnbaumer, 1992, *Cell* **71**:1069; Z. Farfel et al., 1999, *New Eng. J. Med.* **340**:1012; and K. Pierce et al., 2002, *Nature Rev. Mol. Cell Biol.* **3**:639.

The signal-transducing G proteins contain three subunits designated α, β, and γ. During intracellular signaling the β and γ subunits remain bound together and are usually referred to as the $G_{\beta\gamma}$ subunit. The G_α subunit is a GTPase switch protein that alternates between an active (on) state with bound GTP and an inactive (off) state with bound GDP (see Figure 13-8). Stimulation of a coupled receptor causes activation of the G protein, which in turn modulates the activity of an associated *effector protein*. Although the effector protein most commonly is activated by G_α·GTP, in some cases it is inhibited. Moreover, depending on the cell and ligand, the $G_{\beta\gamma}$ subunit, rather than G_α·GTP, may transduce the signal to the effector protein. In addition, the activity of several different effector proteins is controlled by different GPCR-ligand complexes. All effector proteins, however, are either membrane-bound ion channels or enzymes that catalyze formation of second messengers (e.g., cAMP, DAG, and IP_3). These variations on the theme of GPCR signaling arise because multiple G proteins are encoded in eukaryotic genomes. The human genome, for example, encodes 27 different G_α, 5 G_β, and 13 G_γ subunits. So far as is known, the different $G_{\beta\gamma}$ subunits function similarly. Table 13-1 summarizes the functions of the major classes of G proteins with different G_α subunits.

In this section, we first discuss how GPCR signals are transduced to an effector protein, a process that is similar for all receptors of this type. Then we focus on pathways in which cAMP is the second messenger, using the epinephrine-stimulated degradation of glycogen as an example.

The G_α Subunit of G Proteins Cycles Between Active and Inactive Forms

Figure 13-11 illustrates how G protein–coupled receptors transduce signals from extracellular hormones to associated effector proteins. Both the G_α and G_γ subunits are linked to the membrane by covalently attached lipids. In the resting state, when no ligand is bound to the receptor, the G_α subunit is bound to GDP and complexed with $G_{\beta\gamma}$. Binding of the normal hormonal ligand (e.g., epinephrine) or an ago-

▶ **FIGURE 13-11 Operational model for ligand-induced activation of effector proteins associated with G protein–coupled receptors.** The G_α and $G_{\beta\gamma}$ subunits of trimeric G proteins are tethered to the membrane by covalently attached lipid molecules (wiggly black lines). Following ligand binding, dissociation of the G protein, and exchange of GDP with GTP (steps **1**–**3**), the free G_α·GTP binds to and activates an effector protein (step **4**). Hydrolysis of GTP terminates signaling and leads to reassembly of the trimeric form, returning the system to the resting state (step **5**). Binding of another ligand molecule causes repetition of the cycle. In some pathways, the effector protein is activated by the free $G_{\beta\gamma}$ subunit.

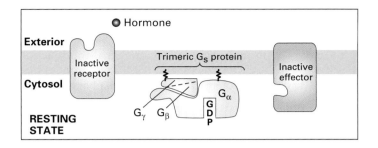

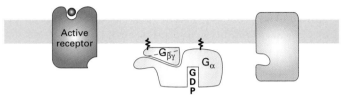

1 Binding of hormone induces a conformational change in receptor

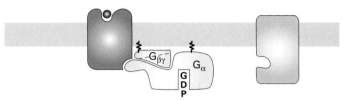

2 Activated receptor binds to G_α subunit

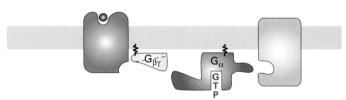

3 Binding induces conformational change in G_α; bound GDP dissociates and is replaced by GTP; G_α dissociates from $G_{\beta\gamma}$

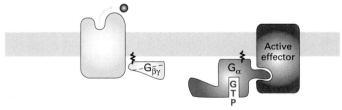

4 Hormone dissociates from receptor; G_α binds to effector, activating it

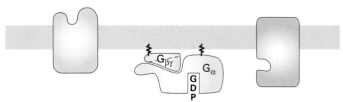

5 Hydrolysis of GTP to GDP causes G_α to dissociate from effector and reassociate with $G_{\beta\gamma}$

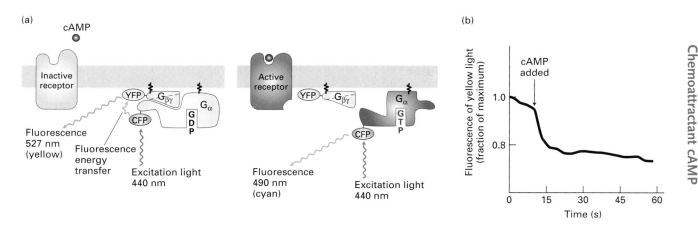

▲ EXPERIMENTAL FIGURE 13-12 Receptor-mediated activation of coupled G proteins occurs within a few seconds of ligand binding in living cells. The amoeba *Dictyostelium discoideum* was transfected with genes encoding two fusion proteins: a G_α fused to cyan fluorescent protein (CFP), a mutant form of green fluorescent protein (GFP), and a G_β fused to another GFP variant, yellow fluorescent protein (YFP). CFP normally fluoresces 490-nm light; YFP, 527-nm light. (a) When CFP and YFP are nearby, as in the resting $G_\alpha \cdot G_{\beta\gamma}$ complex, fluorescence energy transfer can occur between CFP and YFP *(left)*. As a result, irradiation of resting cells with 440-nm light (which directly excites CFP but not YFP) causes emission of 527-nm (yellow) light, char- acteristic of YFP. However, if ligand binding leads to dissociation of the G_α and $G_{\beta\gamma}$ subunits, then fluorescence energy transfer cannot occur. In this case, irradiation of cells at 440 nm causes emission of 490-nm light (cyan) characteristic of CFP *(right)*. (b) Plot of the emission of yellow light (527 nm) from a single transfected amoeba cell before and after addition of cyclic AMP (arrows), the extracellular ligand for the GPCR in these cells. The drop in fluorescence, which results from the dissociation of the G_α-CFP fusion protein from the $G_{\beta\gamma}$-YFP fusion protein, occurs within seconds of cAMP addition. [Adapted from C. Janetopoulos et al., 2001, *Science* **291**:2408.]

nist (e.g., isoproterenol) to the receptor changes its confor- mation, causing it to bind to the G_α subunit in such a way that GDP is displaced from G_α and GTP becomes bound. Thus the activated ligand-bound receptor functions as a GEF for the G_α subunit (see Figure 3-29).

Once the exchange of nucleotides has occurred, the $G_\alpha \cdot$GTP complex dissociates from the $G_{\beta\gamma}$ subunit, but both remain anchored in the membrane. In most cases, $G_\alpha \cdot$GTP then interacts with and activates an associated effector pro- tein, as depicted in Figure 13-11. This activation is short- lived, however, because GTP bound to G_α is hydrolyzed to GDP in seconds, catalyzed by a GTPase enzyme that is an in- trinsic part of the G_α subunit. The resulting $G_\alpha \cdot$GDP quickly reassociates with $G_{\beta\gamma}$, thus terminating effector activation. In many cases, a protein termed *RGS* (*regulator of G protein signaling*) accelerates GTP hydrolysis by the G_α subunit, re- ducing the time during which the effector remains activated.

Early evidence supporting the model shown in Figure 13-11 came from studies with compounds that can bind to G_α subunits as well as GTP does, but cannot be hydrolyzed by the intrinsic GTPase. In these compounds the P–O–P phosphodiester linkage connecting the β and γ phosphates of GTP is replaced by a nonhydrolyzable P–CH₂–P or P–NH–P linkage. Addition of such a GTP analog to a plasma- membrane preparation in the presence of the natural ligand or an agonist for a particular receptor results in a much longer-lived activation of the associated effector protein than

occurs with GTP. That is because once the GDP bound to G_α is displaced by the nonhydrolyzable GTP analog, it remains permanently bound to G_α. Because this complex is as func- tional as the normal $G_\alpha \cdot$GTP complex in activating the ef- fector protein, the effector remains permanently active.

The GPCR-mediated dissociation of trimeric G proteins recently has been detected in living cells. These studies have exploited the phenomenon of *fluorescence energy transfer,* which can change the wavelength of emitted fluorescence when two fluorescent proteins interact. Figure 13-12 shows how this experimental approach has demonstrated the dissociation of the $G_\alpha \cdot G_{\beta\gamma}$ complex within a few seconds of ligand addition, providing further evidence for the model of G protein cycling. This general experimental protocol can be used to follow the formation and dissociation of other protein-protein complexes in living cells.

Epinephrine Binds to Several Different G Protein–Coupled Receptors

Epinephrine is particularly important in mediating the body's response to stress, such as fright or heavy exercise, when all tissues have an increased need to catabolize glucose and fatty acids to produce ATP. These principal metabolic fuels can be supplied to the blood in seconds by the rapid breakdown of glycogen to glucose in the liver *(glycogenolysis)* and of tri- acylglycerols to fatty acids in adipose cells *(lipolysis).*

In mammals, the liberation of glucose and fatty acids can be triggered by binding of epinephrine (or norepinephrine) to *β-adrenergic receptors* on the surface of hepatic (liver) and adipose cells. Epinephrine bound to β-adrenergic receptors on heart muscle cells increases the contraction rate, which increases the blood supply to the tissues. In contrast, epinephrine stimulation of β-adrenergic receptors on smooth muscle cells of the intestine causes them to relax. Another type of epinephrine receptor, the *α₂-adrenergic receptor,* is found on smooth muscle cells lining the blood vessels in the intestinal tract, skin, and kidneys. Binding of epinephrine to these receptors causes the arteries to constrict, cutting off circulation to these peripheral organs. These diverse effects of epinephrine are directed to a common end: supplying energy for the rapid movement of major locomotor muscles in response to bodily stress.

Although all epinephrine receptors are G protein–coupled receptors, the different types are coupled to different G proteins. Thus in addition to their physiological importance, these receptors are of interest because they trigger different intracellular signal-transduction pathways. Both subtypes of β-adrenergic receptors, termed $β_1$ and $β_2$, are coupled to a *stimulatory G protein (G_s)* that activates the membrane-bound enzyme **adenylyl cyclase** (see Table 13-1). Once activated, adenylyl cyclase catalyzes synthesis of the second messenger cAMP. That binding of epinephrine to β-adrenergic receptors induces a rise in cAMP has been demonstrated in functional expression assays like that depicted in Figure 13-6. When cloned cDNA encoding the β-adrenergic receptor is transfected into receptor-negative cells, the transfected cells accumulate cAMP in response to epinephrine stimulation. Similar experiments in which mutant receptors are expressed have helped to define the functions of specific amino acids in binding hormones and activating different G proteins.

The two subtypes of α-adrenergic receptors, $α_1$ and $α_2$, are coupled to different G proteins. The $α_1$-adrenergic receptor is coupled to a G_i protein that inhibits adenylyl cyclase, the same effector enzyme associated with β-adrenergic receptors. In contrast, the G_q protein coupled to the $α_2$-adrenergic receptor activates a different effector enzyme that generates different second messengers (see Section 13.5).

 Some bacterial toxins contain a subunit that penetrates the plasma membrane of cells and catalyzes a chemical modification of $G_{sα}$·GTP that prevents hydrolysis of bound GTP to GDP. As a result, $G_{sα}$ remains in the active state, continuously activating adenylyl cyclase in the absence of hormonal stimulation. Cholera toxin produced by the bacterium *Vibrio cholera* and enterotoxins produced by certain strains of *E. coli* act in this way on intestinal epithelial cells. The resulting excessive rise in intracellular cAMP leads to the loss of electrolytes and water into the intestinal lumen, producing the watery diarrhea characteristic of infection by these bacteria.

Bordetella pertussis, a bacterium that commonly infects the respiratory tract, is the cause of whooping cough. Pertussis toxin catalyzes a modification of $G_{iα}$ that prevents release of bound GDP, thus locking $G_{iα}$ in the inactive state. This inactivation of G_i leads to an increase in cAMP in epithelial cells of the airways, promoting loss of fluids and electrolytes and mucus secretion. ∎

Critical Functional Domains in Receptors and Coupled G Proteins Have Been Identified

As noted already, all G protein–coupled receptors contain seven transmembrane α helices and presumably have a similar three-dimensional structure. Studies with chimeric adrenergic receptors, like those outlined in Figure 13-13, suggest that the long C3 loop between α helices 5 and 6 is important for interactions between a receptor and its coupled G protein. Presumably, ligand binding causes these helices to move relative to each other. As a result, the conformation of the C3 loop connecting these two helices changes in a way that allows the loop to bind and

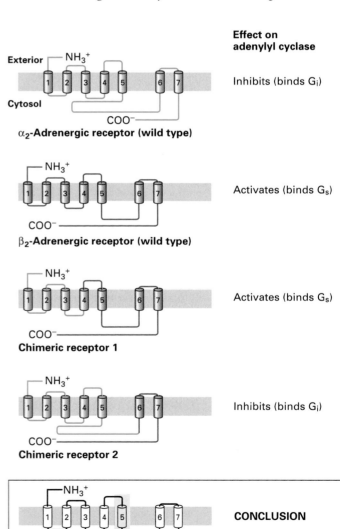

Effect on adenylyl cyclase

Inhibits (binds G_i)

$α_2$-Adrenergic receptor (wild type)

Activates (binds G_s)

$β_2$-Adrenergic receptor (wild type)

Activates (binds G_s)

Chimeric receptor 1

Inhibits (binds G_i)

Chimeric receptor 2

CONCLUSION

Region determining specificity of G protein binding (compare chimeras 1 and 2)

activate the transducing G_α subunit. Specific regions within the C3 loop are thought to assume a unique three-dimensional structure in all receptors that bind the same G protein (e.g., G_s or G_i). Other evidence indicates that the C2 loop, joining helices 3 and 4, also contributes to the interaction of some receptors with a G protein and that residues in at least four transmembrane helices participate in ligand binding.

X-ray crystallographic analysis has pinpointed the regions in $G_{s\alpha} \cdot$GTP that interact with adenylyl cyclase. This enzyme is a multipass transmembrane protein with two large cytosolic segments containing the catalytic domains (Figure 13-14a). Because such transmembrane proteins are notoriously difficult to crystallize, scientists prepared two protein fragments encompassing the catalytic domains of adenylyl cyclase and allowed them to associate in the presence of $G_{s\alpha} \cdot$GTP and forskolin, which stabilizes the catalytic adenylyl cyclase fragments in their active conformations. The complex that formed was catalytically active and showed pharmacological and biochemical properties similar to those of intact full-length adenylyl cyclase. In this complex, two regions of $G_{s\alpha} \cdot$GTP, the switch II helix and the $\alpha3$-$\beta5$ loop, contact the adenylyl cyclase fragments (Figure 13-14b). Recall that switch II is one of the segments of a G protein whose conformation is different in the GTP-bound and GDP-bound states (see Figure 13-8). The GTP-induced conformation of $G_{s\alpha}$ that favors its dissociation from $G_{\beta\gamma}$ is precisely the conformation essential for binding of $G_{s\alpha}$ to adenylyl cyclase. Other studies indicate that $G_{i\alpha}$ binds to a different region of adenylyl cyclase, accounting for its different effect on the effector.

To understand how binding of $G_{s\alpha} \cdot$GTP promotes adenylyl cyclase activity, scientists will first have to solve the structure of the adenylyl cyclase catalytic domains in their unactivated conformations (i.e., in the absence of bound $G_{s\alpha} \cdot$GTP). One hypothesis is that binding of the switch II helix to a cleft in one catalytic domain of adenylyl cyclase leads to rotation of the other catalytic domain. This rotation is proposed to lead to a stabilization of the transition state, thereby stimulating catalytic activity.

◀ **EXPERIMENTAL FIGURE 13-13 Studies with chimeric adrenergic receptors identify the long C3 loop as critical to interaction with G proteins.** *Xenopus* oocytes were microinjected with mRNA encoding a wild-type α_2-adrenergic, β_2-adrenergic, or chimeric α-β receptor. Although *Xenopus* oocytes do not normally express adrenergic receptors, they do express G proteins that can couple to the foreign receptors expressed on the surface of microinjected oocytes. The adenylyl cyclase activity of the injected cells in the presence of epinephrine agonists was determined and indicated whether the adrenergic receptor bound to the stimulatory (G_s) or inhibitory (G_i) type of oocyte G protein. Comparison of chimeric receptor 1, which interacts with G_s, and chimeric receptor 2, which interacts with G_i, shows that the G protein specificity is determined primarily by the source of the cytosol-facing C3 loop (yellow) between α helices 5 and 6. [See B. Kobilka et al., 1988, *Science* **240**:1310.]

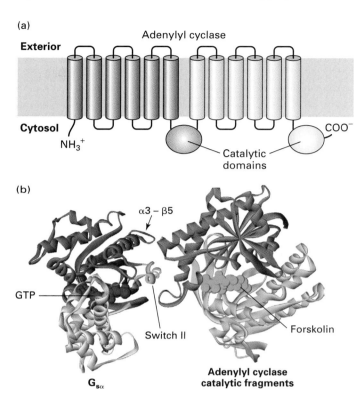

▲ **FIGURE 13-14 Structure of mammalian adenylyl cyclases and their interaction with $G_{s\alpha} \cdot$GTP.** (a) Schematic diagram of mammalian adenylyl cyclases. The membrane-bound enzyme contains two similar catalytic domains on the cytosolic face of the membrane and two integral membrane domains, each of which is thought to contain six transmembrane α helices. (b) Three-dimensional structure of $G_{s\alpha} \cdot$GTP complexed with two fragments encompassing the catalytic domain of adenylyl cyclase determined by x-ray crystallography. The $\alpha3$-$\beta5$ loop and the helix in the switch II region (blue) of $G_{s\alpha} \cdot$GTP interact simultaneously with a specific region of adenylyl cyclase. The darker-colored portion of $G_{s\alpha}$ is the GTPase domain, which is similar in structure to Ras (see Figure 13-8); the lighter portion is a helical domain. The two adenylyl cyclase fragments are shown in orange and yellow. Forskolin (green) locks the cyclase fragments in their active conformations. [Part (a) see W.-J. Tang and A. G. Gilman, 1992, *Cell* **70**:869; part (b) adapted from J. J. G. Tesmer et al., 1997, *Science* **278**:1907.]

Adenylyl Cyclase Is Stimulated and Inhibited by Different Receptor-Ligand Complexes

The versatile trimeric G proteins enable different receptor-hormone complexes to modulate the activity of the same effector protein. In the liver, for instance, glucagon and epinephrine bind to different receptors, but both receptors interact with and activate the same G_s, which activates adenylyl cyclase, thereby triggering the same metabolic responses. Activation of adenylyl cyclase, and thus the cAMP level, is proportional to the total concentration of $G_{s\alpha} \cdot$GTP resulting from binding of both hormones to their respective receptors.

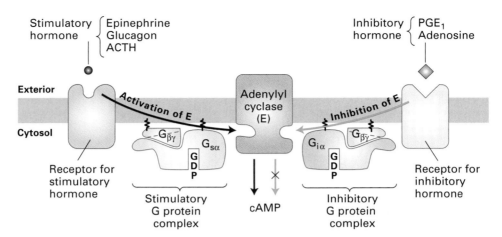

▲ **FIGURE 13-15 Hormone-induced activation and inhibition of adenylyl cyclase in adipose cells.** Ligand binding to G_s-coupled receptors causes activation of adenylyl cyclase, whereas ligand binding to G_i-coupled receptors causes inhibition of the enzyme. The $G_{\beta\gamma}$ subunit in both stimulatory and inhibitory G proteins is identical; the G_α subunits and their corresponding receptors differ. Ligand-stimulated formation of active G_α·GTP complexes occurs by the same mechanism in both G_s and G_i proteins (see Figure 13-11). However, $G_{s\alpha}$·GTP and $G_{i\alpha}$·GTP interact differently with adenylyl cyclase, so that one stimulates and the other inhibits its catalytic activity. [See A. G. Gilman, 1984, *Cell* **36**:577.]

Positive and negative regulation of adenylyl cyclase activity occurs in some cell types, providing fine-tuned control of the cAMP level. For example, stimulation of adipose cells by epinephrine, glucagon, or ACTH activates adenylyl cyclase, whereas prostaglandin PGE1 or adenosine inhibits the enzyme (Figure 13-15). The receptors for PGE1 and adenosine interact with inhibitory G_i, which contains the same β and γ subunits as stimulatory G_s but a different α subunit ($G_{i\alpha}$). In response to binding of an inhibitory ligand to its receptor, the associated G_i protein releases its bound GDP and binds GTP; the active $G_{i\alpha}$·GTP complex then dissociates from $G_{\beta\gamma}$ and inhibits (rather than stimulates) adenylyl cyclase.

cAMP-Activated Protein Kinase A Mediates Various Responses in Different Cells

In multicellular animals virtually all the diverse effects of cAMP are mediated through **protein kinase A (PKA)**, also called *cAMP-dependent protein kinase*. As discussed in Chapter 3, inactive PKA is a tetramer consisting of two regulatory (R) subunits and two catalytic (C) subunits. Each R subunit has two distinct cAMP-binding sites; binding of cAMP to both sites in an R subunit leads to release of the associated C subunit, unmasking its catalytic site and activating its kinase activity (see Figure 3-27a). Binding of cAMP by an R subunit occurs in a cooperative fashion; that is, binding of the first cAMP molecule lowers the K_d for binding of the second. Thus small changes in the level of cytosolic cAMP can cause proportionately large changes in the amount of dissociated C subunits and, hence, in kinase activity. Rapid activation of an enzyme by hormone-triggered dissociation of an inhibitor is a common feature of various signaling pathways.

Most mammalian cells express receptors coupled to G_s protein. Stimulation of these receptors by various hormones leads to activation of PKA, but the resulting cellular response depends on the particular PKA isoform and on the PKA substrates expressed by the cell. For instance, the effects of epinephrine on glycogen metabolism, which are mediated via cAMP and PKA, are confined mainly to liver and muscle cells, which express enzymes for making and degrading glycogen. In adipose cells, epinephrine-induced activation of PKA promotes phosphorylation and activation of the phospholipase that catalyzes hydrolysis of stored triglycerides to yield free fatty acids and glycerol. These fatty acids are released into the blood and taken up as an energy source by cells in other tissues such as the kidney, heart, and muscles. Likewise, stimulation of G protein–coupled receptors on ovarian cells by certain pituitary hormones leads to activation of PKA, which in turn promotes synthesis of two steroid hormones, estrogen and progesterone, crucial to the development of female sex characteristics.

Although PKA acts on different substrates in different types of cells, it always phosphorylates a serine or threonine residue that occurs within the same sequence motif: X-Arg-(Arg/Lys)-X-(Ser/Thr)-Φ, where X denotes any amino acid and Φ denotes a hydrophobic amino acid. Other serine/threonine kinases phosphorylate target residues within different sequence motifs.

Glycogen Metabolism Is Regulated by Hormone-Induced Activation of Protein Kinase A

The first cAMP-mediated cellular response to be discovered—the release of glucose from **glycogen**—occurs in muscle and liver cells stimulated by epinephrine or other hormones whose receptors are coupled to G_s protein. This response exemplifies how activation of PKA can coordinate the activity of a group of intracellular enzymes toward a common purpose.

Glycogen, a large glucose polymer, is the major storage form of glucose in animals. Like all biopolymers, glycogen is synthesized by one set of enzymes and degraded by another (Figure 13-16). Three enzymes convert glucose into uridine diphosphoglucose (UDP-glucose), the primary intermediate in glycogen synthesis. The glucose residue of UDP-glucose is transferred by *glycogen synthase* to the free hydroxyl group on carbon 4 of a glucose residue at the end of a growing glycogen chain. Degradation of glycogen involves the stepwise removal of glucose residues from the same end by a phosphorolysis reaction, catalyzed by *glycogen phosphorylase,* yielding glucose 1-phosphate.

In both muscle and liver cells, glucose 1-phosphate produced from glycogen is converted to glucose 6-phosphate. In muscle cells, this metabolite enters the glycolytic pathway and is metabolized to generate ATP for use in powering muscle contraction (Chapter 8). Unlike muscle cells, liver cells contain a phosphatase that hydrolyzes glucose 6-phosphate to glucose, which is exported from these cells in part by a glucose transporter (GLUT2) in the plasma membrane (Chapter 7). Thus glycogen stores in the liver are primarily broken down to glucose, which is immediately released into the blood and transported to other tissues, particularly the muscles and brain.

The epinephrine-stimulated increase in cAMP and subsequent activation of PKA enhance the conversion of glycogen to glucose 1-phosphate in two ways: by *inhibiting* glycogen synthesis and by *stimulating* glycogen degradation (Figure 13-17a). PKA phosphorylates and thus inactivates glycogen synthase, the enzyme that synthesizes glycogen. PKA promotes glycogen degradation indirectly by phosphorylating and thus activating an intermediate kinase, glycogen phosphorylase kinase (GPK), that in turn phosphorylates and activates glycogen phosphorylase, the enzyme that degrades glycogen. The entire process is reversed when epinephrine is removed and the level of cAMP drops, inactivating PKA. This reversal is mediated by *phosphoprotein phosphatase,* which removes the phosphate residues

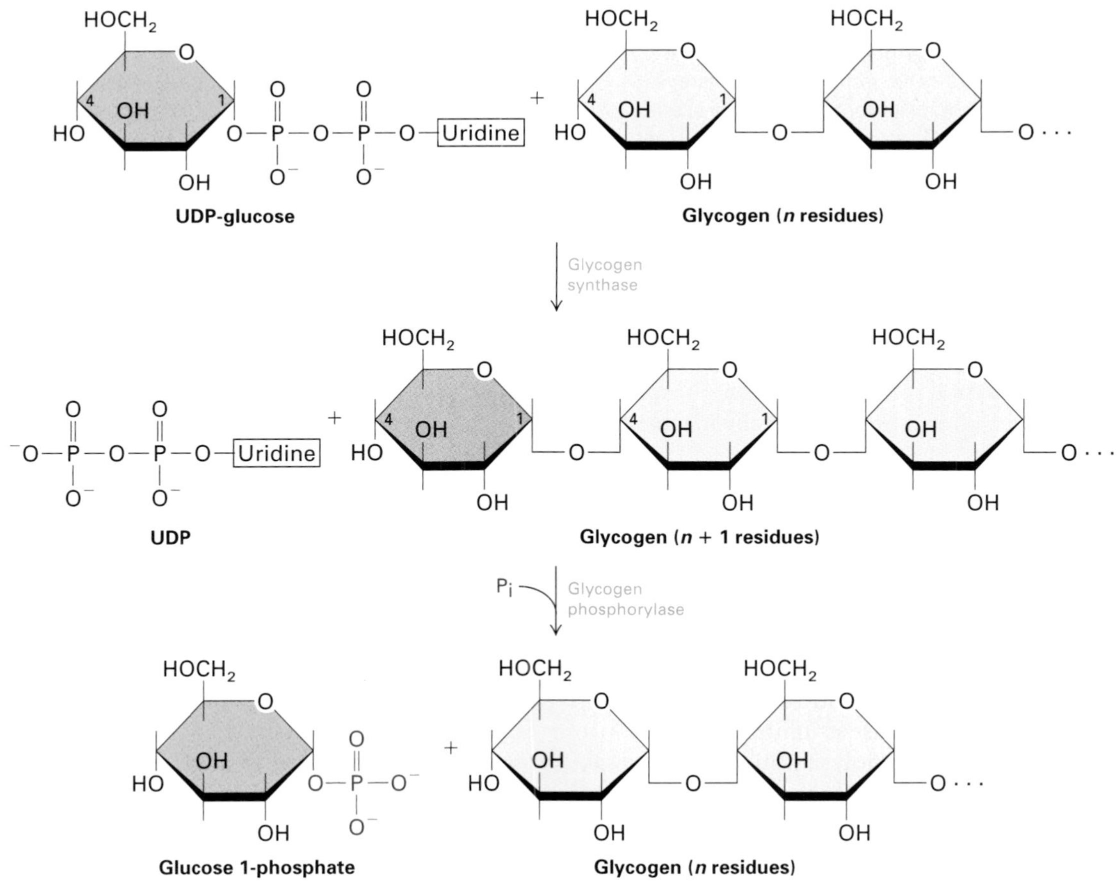

▲ **FIGURE 13-16 Synthesis and degradation of glycogen.** Incorporation of glucose from UDP-glucose into glycogen is catalyzed by glycogen synthase. Removal of glucose units from glycogen is catalyzed by glycogen phosphorylase. Because two different enzymes catalyze the formation and degradation of glycogen, the two reactions can be independently regulated.

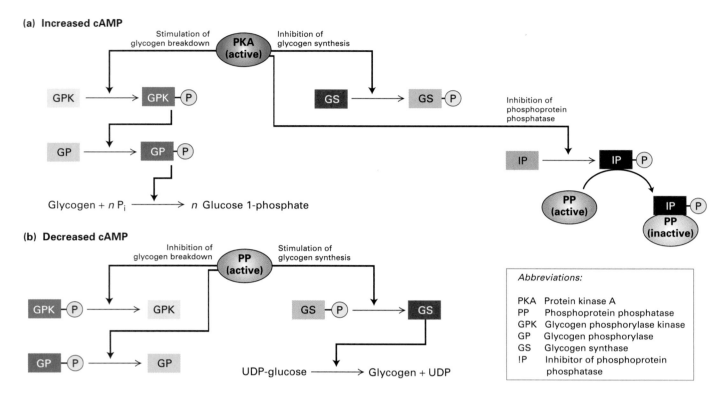

▲ FIGURE 13-17 Regulation of glycogen metabolism by cAMP in liver and muscle cells. Active enzymes are highlighted in darker shades; inactive forms, in lighter shades. (a) An increase in cytosolic cAMP activates PKA, which inhibits glycogen synthesis directly and promotes glycogen degradation via a protein kinase cascade. At high cAMP, PKA also phosphorylates an inhibitor of phosphoprotein phosphatase (PP). Binding of the phosphorylated inhibitor to PP prevents this phosphatase from dephosphorylating the activated enzymes in the kinase cascade or the inactive glycogen synthase. (b) A decrease in cAMP inactivates PKA, leading to release of the active form of phosphoprotein phosphatase. The action of this enzyme promotes glycogen synthesis and inhibits glycogen degradation.

from the inactive form of glycogen synthase, thereby activating it, and from the active forms of glycogen phosphorylase kinase and glycogen phosphorylase, thereby inactivating them (Figure 13-17b).

Phosphoprotein phosphatase itself is regulated by PKA. Activated PKA phosphorylates an inhibitor of phosphoprotein phosphatase; the phosphorylated inhibitor then binds to phosphoprotein phosphatase, inhibiting its activity (see Figure 13-17a). At low cAMP levels, when PKA is inactive, the inhibitor is not phosphorylated and phosphoprotein phosphatase is active. As a result, the synthesis of glycogen by glycogen synthase is enhanced and the degradation of glycogen by glycogen phosphorylase is inhibited.

Epinephrine-induced glycogenolysis thus exhibits dual regulation: activation of the enzymes catalyzing glycogen degradation and inhibition of enzymes promoting glycogen synthesis. Such coordinate regulation of stimulatory and inhibitory pathways provides an efficient mechanism for achieving a particular cellular response and is a common phenomenon in regulatory biology.

Signal Amplification Commonly Occurs Downstream from Cell-Surface Receptors

The cellular responses induced by G protein–coupled receptors that activate adenylyl cyclase may require tens of thousands or even millions of cAMP molecules per cell. Thus the hormone signal must be amplified in order to generate sufficient second messenger from the few thousand receptors for a particular hormone present on a cell. *Signal amplification* is possible because both receptors and G proteins can diffuse rapidly in the plasma membrane. A single receptor-hormone complex causes conversion of up to 100 inactive $G_{s\alpha} \cdot GTP$ molecules to the active form. Each active $G_{s\alpha} \cdot GTP$, in turn, probably activates a single adenylyl cyclase molecule, which then catalyzes synthesis of many cAMP molecules during the time $G_{s\alpha} \cdot GTP$ is bound to it. Although the exact extent of this amplification is difficult to measure, binding of a single hormone molecule to one receptor molecule can result in the synthesis of at least several hundred cAMP molecules before the receptor-hormone complex dissociates and activation of adenylyl cyclase ceases. Similar amplification probably occurs in signaling from receptors coupled to other G proteins and some other types of receptors whose activation induces synthesis of second messengers.

A second level of amplification is illustrated by the cAMP-mediated stimulation of glycogenolysis. As we just discussed, cAMP promotes glycogen degradation via a three-stage *cascade,* that is, a series of reactions in which the enzyme catalyzing one step is activated (or inhibited) by the product of a previous step (see Figure 13-17a). The amplification that occurs in such a cascade depends on the number of steps in it.

Both levels of amplification are depicted in Figure 13-18. For example, blood levels of epinephrine as low as 10^{-10} M can stimulate liver glycogenolysis and release of glucose. An epinephrine stimulus of this magnitude generates an intracellular cAMP concentration of 10^{-6} M, an amplification of 10^4. Because three more catalytic steps precede the release of glucose, another 10^4 amplification can occur. In striated muscle, the concentrations of the three successive enzymes in the glycogenolytic cascade—protein kinase A, glycogen phosphorylase kinase, and glycogen phosphorylase—are in a 1:10:240 ratio, which dramatically illustrates the amplification of the effects of epinephrine and cAMP.

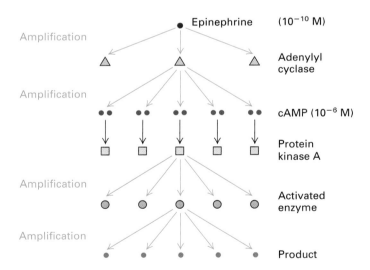

▲ FIGURE 13-18 Amplification of an external signal downstream from a cell-surface receptor. In this example, binding of a single epinephrine molecule to one G_s protein–coupled receptor molecule induces synthesis of a large number of cAMP molecules, the first level of amplification. Four molecules of cAMP activate two molecules of protein kinase A (PKA), but each activated PKA phosphorylates and activates multiple product molecules. This second level of amplification may involve several sequential reactions in which the product of one reaction activates the enzyme catalyzing the next reaction. The more steps in such a cascade, the greater the signal amplification possible.

Although such a cascade may seem overcomplicated, it not only greatly amplifies an external signal but also allows an entire group of enzyme-catalyzed reactions to be coordinately regulated by a single type of signaling molecule. In addition, the multiple steps between stimulus and final response offer possibilities for regulation by other signaling pathways, thereby fine-tuning the cellular response. We will encounter other examples of cascades in signaling pathways discussed in the next chapter.

Several Mechanisms Regulate Signaling from G Protein–Coupled Receptors

Several factors contribute to termination of the response to hormones recognized by β-adrenergic receptors and other receptors coupled to G_s. First, the affinity of the receptor for hormone decreases when the GDP bound to $G_{s\alpha}$ is replaced with a GTP following hormone binding. This increase in the K_d of the receptor-hormone complex enhances dissociation of the hormone from the receptor. Second, the GTP bound to $G_{s\alpha}$ is quickly hydrolyzed, reversing the activation of adenylyl cyclase and production of cAMP (see Figure 13-11). Third, *cAMP phosphodiesterase* acts to hydrolyze cAMP to 5′-AMP, terminating the cellular response. Thus the continuous presence of hormone at a high enough concentration is required for continuous activation of adenylyl cyclase and maintenance of an elevated cAMP level. Once the hormone concentration falls sufficiently, the cellular response quickly terminates.

When a G_s protein–coupled receptor is exposed to hormonal stimulation for several hours, several serine and threonine residues in the cytosolic domain of the receptor become phosphorylated by protein kinase A (PKA). The phosphorylated receptor can bind its ligand, but ligand binding leads to reduced activation of adenylyl cyclase; thus the receptor is desensitized. This is an example of *feedback suppression,* in which the end product of a pathway (here activated PKA) blocks an early step in the pathway (here, receptor activation). Because the activity of PKA is enhanced by the high cAMP level induced by any hormone that activates G_s, prolonged exposure to one such hormone, say, epinephrine, causes desensitization not only of β-adrenergic receptors but also of G_s protein–coupled receptors that bind different ligands (e.g., glucagon). This cross-regulation is called *heterologous desensitization.*

Additional residues in the cytosolic domain of the β-adrenergic receptor are phosphorylated by a receptor-specific enzyme called *β-adrenergic receptor kinase (BARK),* but only when epinephrine or an agonist is bound to the receptor. Because BARK phosphorylates only activated β-adrenergic receptors, this process is called *homologous desensitization.* Prolonged treatment of cells with epinephrine results in extensive phosphorylation and hence desensitization of the β-adrenergic receptor by both PKA and BARK.

Phosphorylated (desensitized) receptors are constantly being resensitized owing to dephosphorylation by constitutive phosphatases. Thus the number of phosphates per receptor molecule reflects how much ligand has been bound in the recent past (e.g., 1–10 minutes). This means that if a cell is constantly being exposed to a certain concentration of a hormone, that hormone concentration will eventually cease to stimulate the receptor. If the hormone concentration is now increased to a new value, the receptor will activate downstream signaling pathways but to a lesser extent than would occur if the cell were switched from a medium without hormone to one with this hormone level. If the hormone is then completely removed, the receptor becomes completely dephosphorylated and "reset" to its maximum sensitivity, in which case it can respond to very low levels of hormone. Thus a feedback loop involving receptor phosphorylation and dephosphorylation modulates the activity of β-adrenergic and related G_s protein–coupled receptors, permitting a cell to adjust receptor sensitivity to the hormone level at which it is being stimulated.

Another key participant in regulation of β-adrenergic receptors is *β-arrestin*. This cytosolic protein binds to receptors extensively phosphorylated by BARK and completely inhibits their interaction with and ability to activate G_s. An additional function of β-arrestin in regulating cell-surface receptors initially was suggested by the observation that loss of cell surface β-adrenergic receptors in response to ligand binding is stimulated by overexpression of BARK and β-arrestin. Subsequent studies revealed that β-arrestin binds not only to phosphorylated receptors but also to clathrin and an associated protein termed AP2, two essential components of the coated vesicles that are involved in one type of endocytosis. These interactions promote the formation of coated pits and endocytosis of the associated receptors, thereby decreasing the number of receptors exposed on the cell surface (Figure 13-19). Eventually the internalized receptors become dephosphorylated in endosomes, β-arrestin dissociates, and the resensitized receptors recycle to the cell surface, similar to recycling of the LDL receptor (Chapter 17). Regulation of other G protein–coupled receptors also is thought to involve endocytosis of ligand-occupied receptors and their sequestration inside the cell.

As we discuss later, β-arrestin also functions as an adapter protein in transducing signals from G_s protein–coupled receptors to the nucleus. The multiple functions of β-arrestin illustrate the importance of adapter proteins in both regulating signaling and transducing signals from cell-surface receptors.

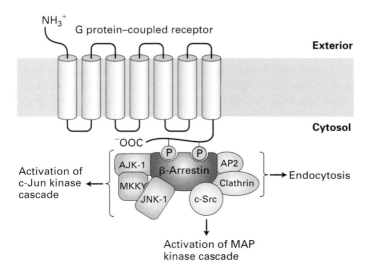

▲ **FIGURE 13-19 Role of β-arrestin in GPCR desensitization and signal transduction.** β-Arrestin binds to phosphorylated serine and tyrosine residues in the C-terminal segment of G protein–coupled receptors (GPCRs). Clathrin and AP2, two other proteins bound by β-arrestin, promote endocytosis of the receptor. β-Arrestin also functions in transducing signals from activated receptors by binding to and activating several cytosolic protein kinases. c-Src activates the MAP kinase pathway, leading to phosphorylation of key transcription factors (Chapter 14). Interaction of β-arrestin with three other proteins, including JNK-1 (a Jun N-terminal kinase), results in phosphorylation and activation of the c-Jun transcription factor. [Adapted from W. Miller and R. J. Lefkowitz, 2001, *Curr. Opin. Cell Biol.* **13**:139, and K. Pierce et al., 2002, *Nature Rev. Mol. Cell Biol.* **3**:639.]

Anchoring Proteins Localize Effects of cAMP to Specific Subcellular Regions

In many cell types, a rise in the cAMP level may produce a response that is required in one part of the cell but is unwanted, perhaps deleterious, in another part. A family of anchoring proteins localizes PKA isoforms to specific subcellular locations, thereby restricting cAMP-dependent responses to these locations. These proteins, referred to as *A kinase–associated proteins (AKAPs)*, have a two-domain structure with one domain conferring a specific subcellular location and another that binds to the regulatory subunit of protein kinase A.

One anchoring protein (AKAP15) is tethered to the cytosolic face of the plasma membrane near a particular type of gated Ca^{2+} channel in certain heart muscle cells. In the heart, activation of β-adrenergic receptors by epinephrine (as part of the fight-or-flight response) leads to PKA-catalyzed phosphorylation of these Ca^{2+} channels, causing them to open; the resulting influx of Ca^{2+} increases the rate of heart muscle contraction. The interaction of AKAP15 with PKA localizes PKA next to these channels, thereby reducing the time that otherwise would be required for diffusion of PKA catalytic subunits from their sites of generation to their Ca^{2+}-channel substrate.

Another A kinase–associated protein (mAKAP) in heart muscle anchors both PKA and cAMP phosphodiesterase (PDE) to the outer nuclear membrane. Because of the close proximity of PDE to PKA, a negative feedback loop provides

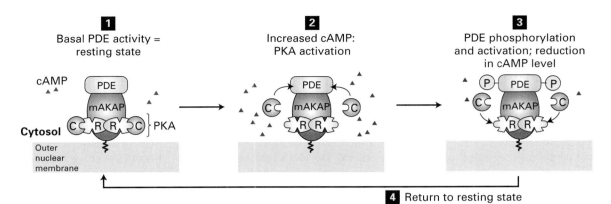

▲ FIGURE 13-20 Localization of protein kinase A (PKA) to the nuclear membrane in heart muscle. This A kinase–associated protein mAKAP anchors both PKA and cAMP phosphodiesterase (PDE) to the nuclear membrane, maintaining them in a negative feedback loop that provides close local control of the cAMP level. Step **1**: The basal level of PDE activity in the absence of hormone (resting state) keeps cAMP levels below those necessary for PKA activation. Step **2**: Activation of β-adrenergic receptors causes an increase in cAMP level in excess of that which can be degraded by PDE. The resulting binding of cAMP to the regulatory (R) subunits of PKA releases the active catalytic (C) subunits. Step **3**: Subsequent phosphorylation of PDE by PKA stimulates its catalytic activity, thereby driving cAMP levels back to basal and causing reformation of the inactive PKA. Subsequent dephosphorylation of PDE (step **4**) returns the complex to the resting state. [Adapted from K. L. Dodge et al., 2001, *EMBO J.* **20**:1921.]

close local control of the cAMP level and hence PKA activity (Figure 13-20). The localization of PKA near the nuclear membrane also facilitates entry of the catalytic subunits into the nucleus, where they phosphorylate and activate certain transcription factors (Section 13.6).

KEY CONCEPTS OF SECTION 13.3

G Protein–Coupled Receptors That Activate or Inhibit Adenylyl Cyclase

■ Trimeric G proteins transduce signals from coupled cell-surface receptors to associated effector proteins, which are either enzymes that form second messengers or cation channel proteins (see Table 13-1).

■ Signals most commonly are transduced by G_α, a GTPase switch protein that alternates between an active ("on") state with bound GTP and inactive ("off") state with GDP. The β and γ subunits, which remain bound together, occasionally transduce signals.

■ Hormone-occupied receptors act as GEFs for G_α proteins, catalyzing dissociation of GDP and enabling GTP to bind. The resulting change in conformation of switch regions in G_α causes it to dissociate from the $G_{\beta\gamma}$ subunit and interact with an effector protein (see Figure 13-11).

■ $G_{s\alpha}$, which is activated by multiple types of GPCRs, binds to and activates adenylyl cyclase, enhancing the synthesis of 3′,5′-cyclic AMP (cAMP).

■ cAMP-dependent activation of protein kinase A (PKA) mediates the diverse effects of cAMP in different cells. The substrates for PKA and thus the cellular response to hormone-induced activation of PKA vary among cell types.

■ In liver and muscle cells, activation of PKA induced by epinephrine and other hormones exerts a dual effect, inhibiting glycogen synthesis and stimulating glycogen breakdown via a kinase cascade (see Figure 13-17).

■ Signaling pathways involving second messengers and kinase cascades amplify an external signal tremendously (see Figure 13-18).

■ BARK phosphorylates ligand-bound β-adrenergic receptors, leading to the binding of β-arrestin and endocytosis of the receptors. The consequent reduction in cell-surface-receptor numbers renders the cell less sensitive to additional hormone.

■ Localization of PKA to specific regions of the cell by anchoring proteins restricts the effects of cAMP to particular subcellular locations.

13.4 G Protein–Coupled Receptors That Regulate Ion Channels

As we learned in Chapter 7, many neurotransmitter receptors are ligand-gated ion channels. These include some types of glutamate and serotonin receptors, as well as the nicotinic acetylcholine receptor found at nerve-muscle synapses. Many neurotransmitter receptors, however, are G protein–coupled receptors. The effector protein for some of these is a Na$^+$ or K$^+$ channel; neurotransmitter binding to these receptors causes the associated ion channel to open or close, leading to changes in the membrane potential. Other neurotransmitter receptors, as well as odorant receptors in the nose and photoreceptors in the eye, are G protein–coupled receptors that indirectly modulate the activity of ion channels via the action of second messengers. In this section, we consider two G protein–coupled receptors that illustrate the direct and indirect mechanisms for regulating ion channels: the muscarinic acetylcholine and G$_t$-coupled receptors.

Cardiac Muscarinic Acetylcholine Receptors Activate a G Protein That Opens K$^+$ Channels

Binding of acetylcholine to nicotinic acetylcholine receptors in striated muscle cells generates an action potential that triggers muscle contraction (see Figure 7-45). In contrast, the *muscarinic acetylcholine receptors* in cardiac muscle are inhibitory. Binding of acetylcholine to these receptors slows the

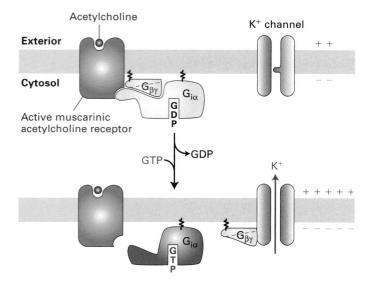

▲ **FIGURE 13-21 Operational model of muscarinic acetylcholine receptor in the heart muscle plasma membrane.** These receptors are linked via a trimeric G protein to K$^+$ channels. Binding of acetylcholine triggers activation of the G$_{i\alpha}$ subunit and its dissociation from the G$_{\beta\gamma}$ subunit in the usual way (see Figure 13-11). In this case, the released G$_{\beta\gamma}$ subunit (rather than G$_{i\alpha}$·GTP) binds to and opens the associated effector, a K$^+$ channel. The increase in K$^+$ permeability hyperpolarizes the membrane, which reduces the frequency of heart muscle contraction. Though not shown here, activation is terminated when the GTP bound to G$_{i\alpha}$ is hydrolyzed to GDP and G$_{i\alpha}$·GDP recombines with G$_{\beta\gamma}$. [See K. Ho et al., 1993, *Nature* **362**:31, and Y. Kubo et al., 1993, *Nature* **362**:127.]

rate of heart muscle contraction by causing a long-lived (several seconds) hyperpolarization of the muscle cell membrane. This can be studied experimentally by direct addition of acetylcholine to heart muscle in culture.

Activation of the muscarinic acetylcholine receptor, which is coupled to a G$_i$ protein, leads to opening of associated K$^+$ channels; the subsequent efflux of K$^+$ ions causes hyperpolarization of the plasma membrane. As depicted in Figure 13-21, the signal from activated receptors is transduced to the effector protein by the released G$_{\beta\gamma}$ subunit rather than by G$_\alpha$·GTP. That G$_{\beta\gamma}$ directly activates the K$^+$ channel was demonstrated by patch-clamping experiments, which can measure ion flow through a single ion channel in a small patch of membrane (see Figure 7-17). When purified G$_{\beta\gamma}$ protein was added to the cytosolic face of a patch of heart muscle plasma membrane, K$^+$ channels opened immediately, even in the absence of acetylcholine or other neurotransmitters.

G$_t$-Coupled Receptors Are Activated by Light

The human retina contains two types of photoreceptors, *rods* and *cones,* that are the primary recipients of visual stimulation. Cones are involved in color vision, while rods are stimulated by weak light like moonlight over a range of wavelengths. The photoreceptors synapse on layer upon layer of interneurons that are innervated by different combinations of photoreceptor cells. All these signals are processed and interpreted by the part of the brain called the *visual cortex.*

Rhodopsin, a G protein–coupled receptor that is activated by light, is localized to the thousand or so flattened membrane disks that make up the outer segment of rod cells (Figure 13-22). The trimeric G protein coupled to rhodopsin, often called *transducin (G$_t$),* is found only in rod cells. A human rod cell contains about 4×10^7 molecules of rhodopsin, which consists of the seven-spanning protein *opsin* to which is covalently bound the light-absorbing pigment 11-*cis*-retinal. Upon absorption of a photon, the retinal moiety of rhodopsin is very rapidly converted to the all-*trans* isomer, causing a conformational change in the opsin portion that activates it (Figure 13-23). This is equivalent to the conformational change that occurs upon ligand binding by other G protein–coupled receptors. The resulting form in which opsin is covalently bound to all-*trans*-retinal is called *meta-rhodopsin II,* or *activated opsin.* Analogous to other G protein–coupled receptors, this light-activated form of rhodopsin interacts with and activates an associated G protein (i.e., G$_t$). Activated opsin is unstable and spontaneously dissociates into its component parts, releasing opsin and all-*trans*-retinal, thereby terminating visual signaling. In the dark, free all-*trans*-retinal is converted back to 11-*cis*-retinal, which can then rebind to opsin, re-forming rhodopsin.

In the dark, the membrane potential of a rod cell is about −30 mV, considerably less than the resting potential (−60 to

(a) (b)

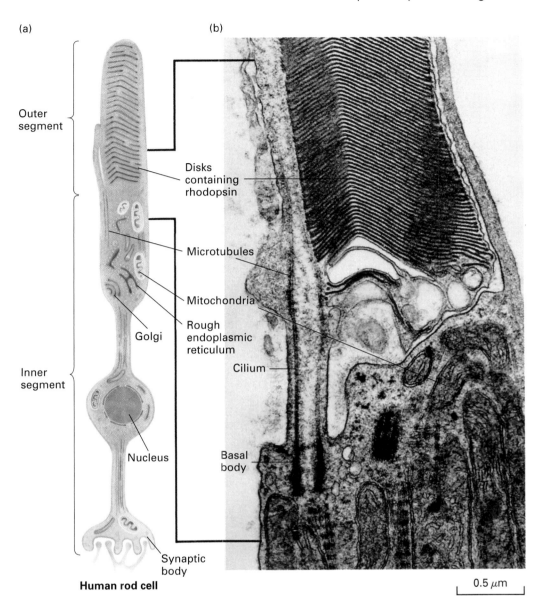

Outer
segment

Disks
containing
rhodopsin

Microtubules

Mitochondria

Golgi

Rough
endoplasmic
reticulum

Cilium

Inner
segment

Nucleus

Basal
body

Synaptic
body

Human rod cell

0.5 μm

▲ **FIGURE 13-22 Human rod cell.** (a) Schematic diagram of an entire rod cell. At the synaptic body, the rod cell forms a synapse with one or more bipolar interneurons. Rhodopsin, a light-sensitive G protein–coupled receptor, is located in the flattened membrane disks of the outer segment. (b) Electron micrograph of the region of the rod cell indicated by the bracket in (a). This region includes the junction of the inner and outer segments. [Part (b) from R. G. Kessel and R. H. Kardon, 1979, *Tissues and Organs: A Text-Atlas of Scanning Electron Microscopy,* W. H. Freeman and Company, p. 91.]

−90 mV) typical of neurons and other electrically active cells. As a consequence of this depolarization, rod cells in the dark are constantly secreting neurotransmitters, and the bipolar interneurons with which they synapse are continually being stimulated. The depolarized state of the plasma membrane of resting rod cells is due to the presence of a large number of open *nonselective* ion channels that admit Na^+ and Ca^{2+}, as well as K^+. Absorption of light by rhodopsin leads to closing of these channels, causing the membrane potential to become *more* negative.

The more photons absorbed by rhodopsin, the more channels are closed, the fewer Na^+ ions cross the membrane from the outside, the more negative the membrane potential becomes, and the less neurotransmitter is released. This change is transmitted to the brain where it is perceived as light. Remarkably, a single photon absorbed by a resting rod cell produces a measurable response, a decrease in the membrane potential of about 1 mV, which in amphibians lasts a second or two. Humans are able to detect a flash of as few as five photons.

11-*cis*-Retinal moiety

H₃C CH₃   CH₃   cis

CH₃   H₃C

                              C=N⁺—(CH₂)₄— Opsin
                              H   H

Lysine side chain

Rhodopsin

Light-induced
isomerization
(<10⁻² s)

H⁺

all-*trans*-Retinal moiety

trans

H₃C CH₃   CH₃   CH₃

                    C=N—(CH₂)₄— Opsin*
                    H

CH₃

***Meta*-rhodopsin II
(activated opsin)**

Activation of Rhodopsin Induces Closing of cGMP-Gated Cation Channels

The key transducing molecule linking activated opsin to the closing of cation channels in the rod-cell plasma membrane is the second messenger **cyclic GMP (cGMP)**. Rod outer segments contain an unusually high concentration (≈ 0.07 mM) of cGMP, which is continuously formed from GTP in a reaction catalyzed by guanylyl cyclase that appears to be unaffected by light. However, light absorption by rhodopsin induces activation of a *cGMP phosphodiesterase*, which hydrolyzes cGMP to 5'-GMP. As a result, the cGMP concentration decreases upon illumination. The high level of cGMP

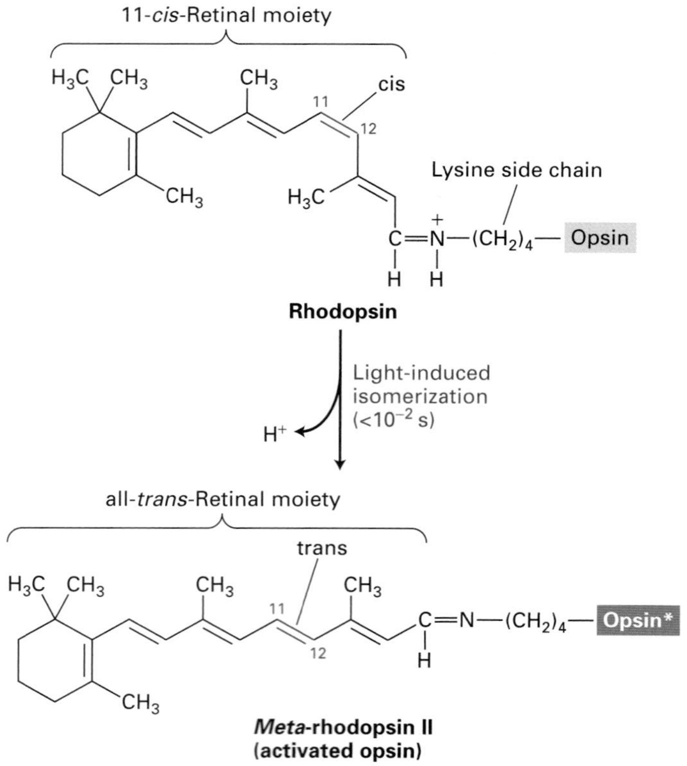

◀ **FIGURE 13-23 The light-triggered step in vision.** The light-absorbing pigment 11-*cis*-retinal is covalently bound to the amino group of a lysine residue in opsin, the protein portion of rhodopsin. Absorption of light causes rapid photoisomerization of the *cis*-retinal to the all-*trans* isomer, forming the unstable intermediate *meta*-rhodopsin II, or activated opsin, which activates G_t proteins. Within seconds all-*trans*-retinal dissociates from opsin and is converted by an enzyme back to the *cis* isomer, which then rebinds to another opsin molecule. [See J. Nathans, 1992, *Biochemistry* **31**:4923.]

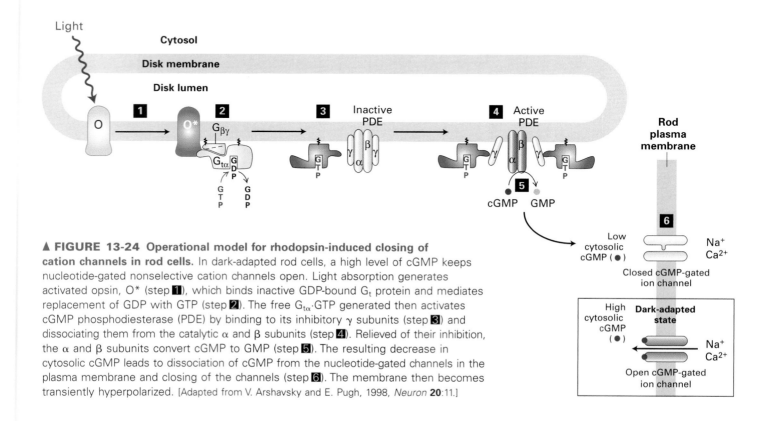

▲ **FIGURE 13-24 Operational model for rhodopsin-induced closing of cation channels in rod cells.** In dark-adapted rod cells, a high level of cGMP keeps nucleotide-gated nonselective cation channels open. Light absorption generates activated opsin, O* (step **1**), which binds inactive GDP-bound G_t protein and mediates replacement of GDP with GTP (step **2**). The free $G_{t\alpha}$·GTP generated then activates cGMP phosphodiesterase (PDE) by binding to its inhibitory γ subunits (step **3**) and dissociating them from the catalytic α and β subunits (step **4**). Relieved of their inhibition, the α and β subunits convert cGMP to GMP (step **5**). The resulting decrease in cytosolic cGMP leads to dissociation of cGMP from the nucleotide-gated channels in the plasma membrane and closing of the channels (step **6**). The membrane then becomes transiently hyperpolarized. [Adapted from V. Arshavsky and E. Pugh, 1998, *Neuron* **20**:11.]

present in the dark acts to keep *cGMP-gated cation channels* open; the light-induced decrease in cGMP leads to channel closing, membrane hyperpolarization, and reduced neurotransmitter release.

As depicted in Figure 13-24, cGMP phosphodiesterase is the effector protein for G_t. The free $G_{t\alpha} \cdot GTP$ complex that is generated after light absorption by rhodopsin binds to the two inhibitory γ subunits of cGMP phosphodiesterase, releasing the active catalytic α and β subunits, which then convert cGMP to GMP. This is another example of how signal-induced removal of an inhibitor can quickly activate an enzyme, a common mechanism in signaling pathways. A single molecule of activated opsin in the disk membrane can activate 500 $G_{t\alpha}$ molecules, each of which in turn activates cGMP phosphodiesterase; this is the primary stage of signal amplification in the visual system. Even though activation of cGMP phosphodiesterase leads to a decrease in a second messenger, cGMP, this activation occurs by the same general mechanism described earlier except that absorption of light by rhodopsin rather than ligand binding is the activating signal (see Figure 13-11).

Conversion of active $G_{t\alpha} \cdot GTP$ back to inactive $G_{t\alpha} \cdot GDP$ is accelerated by a GTPase-activating protein (GAP) specific for $G_{t\alpha} \cdot GTP$. In mammals $G_{t\alpha}$ normally remains in the active GTP-bound state for only a fraction of a second. Thus cGMP phosphodiesterase rapidly becomes inactivated, and the cGMP level gradually rises to its original level when the light stimulus is removed. This allows rapid responses of the eye toward moving or changing objects.

Recent x-ray crystallographic studies reveal how the subunits of G_t protein interact with each other and with light-activated rhodopsin and provide clues about how binding of GTP leads to dissociation of G_α from $G_{\beta\gamma}$ (Figure 13-25). Two surfaces of $G_{t\alpha}$ interact with G_β: an N-terminal region near the membrane surface and the two adjacent switch I and switch II regions, which are found in all G_α proteins. Although G_β and G_γ also contact each other, G_γ does not contact $G_{t\alpha}$.

Studies with adrenergic receptors discussed earlier indicate that ligand binding to a G protein–coupled receptor causes the transmembrane helices in the receptor to slide relative to one another, resulting in conformational changes in the cytosolic loops that create a binding site for the coupled trimeric G protein. The crystallographic structures in Figure 13-25 suggest that the nucleotide-binding domain of $G_{t\alpha}$, together with the lipid anchors at the C-terminus of G_γ and the N-terminus of $G_{t\alpha}$, form a surface that binds to light-activated rhodopsin (O* in Figure 13-24), promoting the release of GDP from $G_{t\alpha}$ and the subsequent binding of GTP. The subsequent conformational changes in $G_{t\alpha}$, particularly those within switches I and II, disrupt the molecular interactions between $G_{t\alpha}$ and $G_{\beta\gamma}$, leading to their dissociation. The structural studies with rhodopsin and G_t are consistent with data concerning other G protein–coupled receptors and are thought to be generally applicable to all receptors of this type.

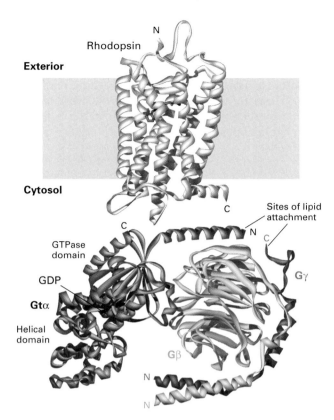

▲ **FIGURE 13-25 Structural models of rhodopsin and its associated G_t protein.** The structures of rhodopsin and the $G_{t\alpha}$ and $G_{\beta\gamma}$ subunits were obtained by x-ray crystallography. The C-terminal segment of rhodopsin is not shown in this model. The orientation of $G_{t\alpha}$ with respect to rhodopsin and the membrane is hypothetical; it is based on the charge and hydrophobicity of the protein surfaces and the known rhodopsin-binding sites on $G_{t\alpha}$. As in other trimeric G proteins, the $G_{t\alpha}$ and G_γ subunits contain covalently attached lipids that are thought to be inserted into the membrane. In the GDP-bound form shown here (GDP, red), the α subunit (gray) and the β subunit (light blue) interact with each other, as do the β and γ (purple) subunits, but the small γ subunit, which contains just two α helices, does not contact the α subunit. Several segments of the α subunit are thought to interact with an activated receptor, causing a conformational change that promotes release of GDP and binding of GTP. Binding of GTP, in turn, induces large conformational changes in the switch regions of $G_{t\alpha}$, leading to dissociation of $G_{t\alpha}$ from $G_{\beta\gamma}$. The structure of a $G_{s\alpha}$ subunit in the GTP-bound form, which interacts with an effector protein, is shown in Figure 13-14b. [Adapted from H. Hamm, 2001, *Proc. Nat'l. Acad. Sci. USA* **98**:4819, and D. G. Lambright et al., 1996, *Nature* **379**:311.]

Direct support for the role of cGMP in rod-cell activity has been obtained in patch-clamping studies using isolated patches of rod outer-segment plasma membrane, which contains abundant cGMP-gated cation channels. When cGMP is added to the cytosolic surface of these patches, there is a rapid increase in the number of open ion channels. The effect occurs in the absence of protein kinases or phosphatases, and cGMP acts directly on the channels to keep them open, indicating that these are nucleotide-gated channels. Like the voltage-gated K^+ channels discussed in Chapter 7, the cGMP-gated channel protein contains four subunits, each of which is able to bind a cGMP molecule (see Figure 7-36a). Three or four cGMP molecules must bind per channel in order to open it; this allosteric interaction makes channel opening very sensitive to small changes in cGMP levels.

Rod Cells Adapt to Varying Levels of Ambient Light

Cone cells are insensitive to low levels of illumination, and the activity of rod cells is inhibited at high light levels. Thus when we move from daylight into a dimly lighted room, we are initially blinded. As the rod cells slowly become sensitive to the dim light, we gradually are able to see and distinguish objects. This process of *visual adaptation* permits a rod cell to perceive contrast over a 100,000-fold range of ambient light levels; as a result, differences in light levels, rather than the absolute amount of absorbed light, are used to form visual images.

One process contributing to visual adaptation involves phosphorylation of activated opsin (O*) by *rhodopsin kinase* (Figure 13-26). This rod-cell enzyme is analogous to β-adrenergic receptor kinase (BARK) discussed previously. Each opsin molecule has three principal serine phosphorylation sites; the more sites that are phosphorylated, the less able O* is to activate G_t and thus induce closing of cGMP-gated cation channels. Indeed, rod cells from mice with mutant rhodopsins bearing zero or only one of these serine residues show a much slower than normal rate of deacti-

vation in bright light. Because the extent of opsin phosphorylation is proportional to the amount of time each opsin molecule spends in the light-activated form, it is a measure of the background (ambient) level of light. Under high-light conditions, phosphorylated opsin is abundant and activation of G_t is reduced; thus, a greater increase in light level will be necessary to generate a visual signal. When the level of ambient light is reduced, the opsins become dephosphorylated and the ability to activate G_t increases; in this case, fewer additional photons will be necessary to generate a visual signal.

At high ambient light (such as noontime outdoors), the level of opsin phosphorylation is such that the protein β-arrestin binds to the C-terminal segment of opsin. The bound β-arrestin prevents interaction of G_t with O*, totally blocking formation of the active $G_{t\alpha} \cdot GTP$ complex and causing a shutdown of all rod-cell activity. The mechanism by which rod-cell activity is controlled by rhodopsin kinase and arrestin is similar to adaptation (or desensitization) of other G protein–coupled receptors to high ligand levels.

A second mechanism of visual adaptation appears unique to rod cells. In dark-adapted cells virtually all the $G_{t\alpha}$ and $G_{\beta\gamma}$ subunits are in the outer segments. But exposure for 10 minutes to moderate daytime intensities of light causes over 80 percent of the $G_{t\alpha}$ and $G_{\beta\gamma}$ subunits to move out of the outer segments into other cellular compartments (Figure 13-27). The mechanism by which these proteins move is not yet known, but as a result of this adaptation G_t proteins are physically unable to bind activated opsin. As occurs in other signaling pathways, multiple mechanisms are thus used to inactivate signaling during visual adaptation, presumably to allow strict control of activation of the signaling pathway over broad ranges of illumination.

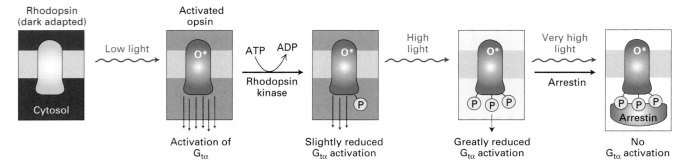

▲ **FIGURE 13-26 Role of opsin phosphorylation in adaptation of rod cells to changes in ambient light levels.** Light-activated opsin (O*), but not dark-adapted rhodopsin, is a substrate for rhodopsin kinase. The extent of opsin phosphorylation is directly proportional to the amount of time each opsin molecule spends in the light-activated form and thus to the average ambient light level over the previous few minutes. The ability of O* to activate $G_{t\alpha}$ is inversely proportional to the number of phosphorylated residues. Thus the higher the ambient light level, the greater the extent of opsin phosphorylation and the larger the increase in light level needed to activate the same number of $G_{t\alpha}$ (transducin) molecules. At very high light levels, arrestin binds to the completely phosphorylated opsin, forming a complex that cannot activate transducin at all. [See L. Lagnado and D. Baylor, 1992, *Neuron* **8**:995, and A. Mendez et al., 2000, *Neuron* **28**:153.]

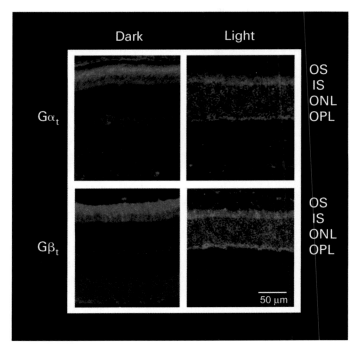

▲ EXPERIMENTAL FIGURE 13-27 Movement of G$_t$ from outer segments of rod cells contributes to visual adaptation. As shown by immunofluorescence staining of retinas of dark-adapted rats, both the α and β subunits of transducin (Gα$_t$ and Gβ$_t$) are localized to the outer segments (OS) of rod cells, where they can be activated by rhodopsin photoreceptors in the membrane disks (see Figure 13-22). After several minutes of bright light most of the transducin α and β subunits have moved to the inner segment (IS) of the rod cells, where they cannot interact with active opsin; this contributes to desensitization of rod cells at high light intensities. [From M. Sokolov et al., 2002, *Neuron* **33**:95. Courtesy of Vadim Arshavsky, Harvard Medical School.]

KEY CONCEPTS OF SECTION 13.4

G Protein–Coupled Receptors That Regulate Ion Channels

■ The cardiac muscarinic acetylcholine receptor is a GPCR whose effector protein is a K$^+$ channel. Receptor activation causes release of the G$_{βγ}$ subunit, which opens K$^+$ channels (see Figure 13-21). The resulting hyperpolarization of the cell membrane slows the rate of heart muscle contraction.

■ Rhodopsin, the photosensitive GPCR in rod cells, comprises the opsin protein linked to 11-*cis*-retinal. The light-induced isomerization of the 11-*cis*-retinal moiety produces activated opsin, which then activates the coupled trimeric G protein transducin (G$_t$) by catalyzing exchange of free GTP for bound GDP on the G$_{tα}$ subunit.

■ The effector protein activated by G$_{tα}$·GTP is cGMP phosphodiesterase. Reduction in the cGMP level by this enzyme leads to closing of cGMP-gated Na$^+$/Ca^{2+} channels, hyperpolarization of the membrane, and decreased release of neurotransmitter (see Figure 13-24).

■ As with other G$_α$ proteins, binding of GTP to G$_{tα}$ causes conformational changes in the protein that disrupt its molecular interactions with G$_{βγ}$ and enable G$_{tα}$·GTP to bind to its downstream effector (see Figure 13-25).

■ Phosphorylation of light-activated opsin by rhodopsin kinase and subsequent binding of arrestin to phosphorylated opsin inhibit its ability to activate transducin (see Figure 13-26). This general mechanism of adaptation, or desensitization, is utilized by other GPCRs at high ligand levels.

13.5 G Protein–Coupled Receptors That Activate Phospholipase C

In this section, we discuss GPCR-triggered signal-transduction pathways involving several other second messengers and the mechanisms by which they regulate various cellular activities. A number of these second messengers are derived from *phosphatidylinositol (PI)*. The inositol group in this phospholipid, which extends into the cytosol adjacent to the membrane, can be reversibly phosphorylated at several positions by the combined actions of various kinases and phosphatases. These reactions yield several different membrane-bound **phosphoinositides**, two of which are depicted in Figure 13-28.

The levels of many phosphoinositides in cells are dynamically regulated by extracellular signals, especially those that bind to receptor tyrosine kinases or cytokine receptors, which we cover in the next chapter. The phosphoinositide PIP$_2$ (PI 4,5-bisphosphate) binds many cytosolic proteins to the plasma membrane. Some of these proteins are required for forming and remodeling the actin cytoskeleton (Chapter 19); others are required for binding of proteins important for endocytosis and vesicle fusions (Chapter 17).

PIP$_2$ is also cleaved by the plasma-membrane–associated enzyme **phospholipase C (PLC)** to generate two important second messengers: **1,2-diacylglycerol (DAG)**, a lipophilic molecule that remains associated with the membrane, and **inositol 1,4,5-trisphosphate (IP$_3$)**, which diffuses in the cytosol (see Figure 13-28). We refer to downstream events involving these two second messengers collectively as the *IP$_3$/DAG pathway*. Hormone binding to receptors coupled to either a G$_o$ or a G$_q$ protein (see Table 13-1) induces activation of the β isoform of phospholipase C (PLCβ) by the general mechanism outlined in Figure 13-11.

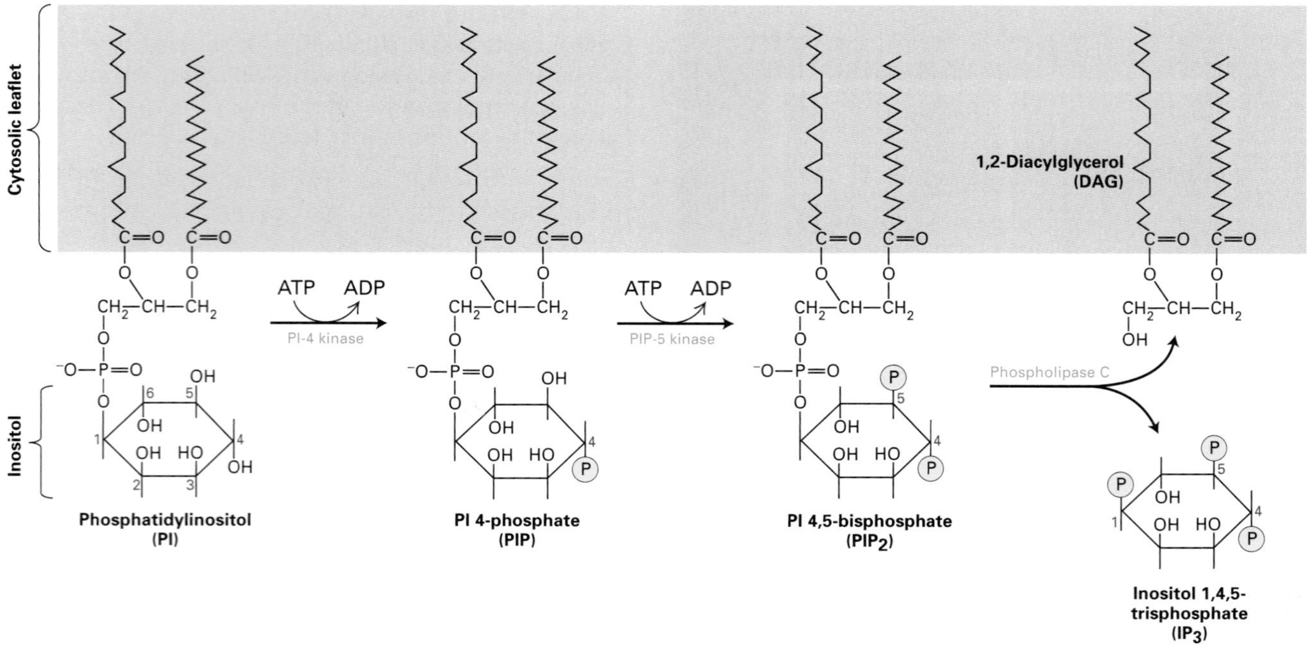

▲ FIGURE 13-28 Synthesis of DAG and IP₃ from membrane-bound phosphatidylinositol (PI). Each membrane-bound PI kinase places a phosphate (yellow circles) on a specific hydroxyl group on the inositol ring, producing the phosphoinositides PIP and PIP₂. Cleavage of PIP₂ by phospholipase C (PLC) yields the two important second messengers DAG and IP₃. [See A. Toker and L. C. Cantley, 1997, *Nature* **387**:673, and C. L. Carpenter and L. C. Cantley, 1996, *Curr. Opin. Cell Biol.* **8**:153.]

Inositol 1,4,5-Trisphosphate (IP₃) Triggers Release of Ca²⁺ from the Endoplasmic Reticulum

Most intracellular Ca²⁺ ions are sequestered in the mitochondria and in the lumen of the endoplasmic reticulum (ER) and other vesicles. Cells employ various mechanisms for regulating the concentration of Ca²⁺ ions in the cytosol, which usually is kept below 0.2 μM. For instance, Ca²⁺ ATPases pump cytosolic Ca²⁺ ions across the plasma membrane to the cell exterior or into the lumens of intracellular Ca²⁺-storing compartments (see Figure 7-7). As we discuss below, a small rise in cytosolic Ca²⁺ induces a variety of cellular responses, and thus the cytosolic concentration of Ca²⁺ is carefully controlled.

Binding of many hormones to their cell-surface receptors on liver, fat, and other cells induces an elevation in cytosolic Ca²⁺ even when Ca²⁺ ions are absent from the surrounding extracellular fluid. In this situation, Ca²⁺ is released into the cytosol from the ER lumen through operation of the *IP₃-gated Ca²⁺ channel* in the ER membrane. This large protein is composed of four identical subunits, each containing an IP₃-binding site in the N-terminal cytosolic domain. IP₃ binding induces opening of the channel, allowing Ca²⁺ ions to exit from the ER into the cytosol (Figure 13-29). When various phosphorylated inositols found in cells are added to preparations of ER vesicles, only IP₃ causes release of Ca²⁺ ions from the vesicles. This simple experiment demonstrates the specificity of the IP₃ effect.

The IP₃-mediated rise in the cytosolic Ca²⁺ level is only transient because Ca²⁺ ATPases located in the plasma membrane and ER membrane actively pump Ca²⁺ from the cytosol to the cell exterior and ER lumen, respectively. Furthermore, within a second of its generation, one specific phosphate on IP₃ is hydrolyzed, yielding inositol 1,4-bisphosphate, which does not stimulate Ca²⁺ release from the ER.

Without some means for replenishing depleted stores of intracellular Ca²⁺, a cell would soon be unable to increase the cytosolic Ca²⁺ level in response to hormone-induced IP₃. Patch-clamping studies have revealed that a plasma-membrane Ca²⁺ channel, called the TRP channel or the store-operated channel, opens in response to depletion of ER Ca²⁺ stores (see Figure 13-29). In a way that is not understood, depletion of Ca²⁺ in the ER lumen leads to a conformational change in the IP₃-gated Ca²⁺ channel that allows it to bind to the TRP Ca²⁺ channel in the plasma membrane, causing the latter to open. Indeed, expression in cells of a specific fragment of the ER membrane IP₃-gated Ca²⁺ channel prevents opening of the TRP channel upon depletion of ER Ca²⁺ stores, implicating an interaction between the two Ca²⁺ channels in opening the TRP channel.

Opening of IP₃-gated Ca²⁺ channels is potentiated by cytosolic Ca²⁺ ions, which increase the affinity of these channel receptors for IP₃, resulting in greater release of stored Ca²⁺. Higher concentrations of cytosolic Ca²⁺, however, inhibit IP₃-induced release of Ca²⁺ from intracellular stores by decreasing the affinity of the receptor for IP₃. This complex regulation of IP₃-gated Ca²⁺ channels in ER membranes by cytosolic Ca²⁺ can lead to rapid oscillations in the cytosolic Ca²⁺ level when the IP₃ pathway in cells is stimu-

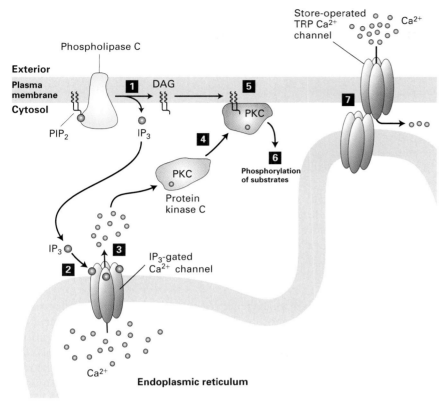

◀ **FIGURE 13-29 IP$_3$/DAG pathway and the elevation of cytosolic Ca^{2+}.** This pathway can be triggered by ligand binding to certain G protein–coupled receptors and several other receptor types, leading to activation of phospholipase C. Cleavage of PIP$_2$ by phospholipase C yields IP$_3$ and DAG (step **1**). After diffusing through the cytosol, IP$_3$ interacts with and opens Ca^{2+} channels in the membrane of the endoplasmic reticulum (step **2**), causing release of stored Ca^{2+} ions into the cytosol (step **3**). One of various cellular responses induced by a rise in cytosolic Ca^{2+} is recruitment of protein kinase C (PKC) to the plasma membrane (step **4**), where it is activated by DAG (step **5**). The activated kinase can phosphorylate various cellular enzymes and receptors, thereby altering their activity (step **6**). As endoplasmic reticulum Ca^{2+} stores are depleted, the IP$_3$-gated Ca^{2+} channels bind to and open store-operated TRP Ca^{2+} channels in the plasma membrane, allowing influx of extracellular Ca^{2+} (step **7**). [Adapted from J. W. Putney, 1999, *Proc. Nat'l. Acad. Sci. USA* **96**:14669.]

lated. For example, stimulation of hormone-secreting cells in the pituitary by luteinizing hormone–releasing hormone (LHRH) causes rapid, repeated spikes in the cytosolic Ca^{2+} level; each spike is associated with a burst in secretion of luteinizing hormone (LH). The purpose of the fluctuations of Ca^{2+}, rather than a sustained rise in cytosolic Ca^{2+}, is not understood. One possibility is that a sustained rise in Ca^{2+} may be toxic to cells.

Diacylglycerol (DAG) Activates Protein Kinase C, Which Regulates Many Other Proteins

After its formation by hydrolysis of PIP$_2$ or other phosphoinositides, DAG remains associated with the plasma membrane. The principal function of DAG is to activate a family of protein kinases collectively termed **protein kinase C (PKC)**. In the absence of hormone stimulation, protein kinase C is present as a soluble cytosolic protein that is catalytically inactive. A rise in the cytosolic Ca^{2+} level causes protein kinase C to bind to the cytosolic leaflet of the plasma membrane, where the membrane-associated DAG can activate it. Thus activation of protein kinase C depends on an increase of both Ca^{2+} ions and DAG, suggesting an interaction between the two branches of the IP$_3$/DAG pathway (see Figure 13-29).

The activation of protein kinase C in different cells results in a varied array of cellular responses, indicating that it plays a key role in many aspects of cellular growth and metabolism. In liver cells, for instance, protein kinase C helps regulate glycogen metabolism by phosphorylating and thus inhibiting glycogen synthase. Protein kinase C also phosphorylates various transcription factors; depending on the cell type; these induce synthesis of mRNAs that trigger cell proliferation.

Ca^{2+}/Calmodulin Complex Mediates Many Cellular Responses to External Signals

Ligand binding to several types of receptors, in addition to G protein–coupled receptors, can activate a phospholipase C isoform, leading to an IP$_3$-mediated increase in the cytosolic level of free Ca^{2+}. Such localized increases in cytosolic Ca^{2+} in specific cell types are critical to its function as a second messenger. For example, acetylcholine stimulation of G protein–coupled receptors in secretory cells of the pancreas and parotid gland induces an IP$_3$-mediated rise in Ca^{2+} that triggers the fusion of secretory vesicles with the plasma membrane and release of their contents into the extracellular space. In blood platelets, the rise in Ca^{2+} induced by thrombin stimulation triggers a conformational change in these cell fragments leading to their aggregation, an important step in plugging holes in blood vessels. Secretion of insulin from pancreatic β cells also is triggered by Ca^{2+}, although the increase in Ca^{2+} occurs by a different mechanism (see Figure 15-7).

A small cytosolic protein called **calmodulin**, which is ubiquitous in eukaryotic cells, functions as a multipurpose switch protein that mediates many cellular effects of Ca^{2+}

ions. Binding of Ca^{2+} to four sites on calmodulin yields a complex that interacts with and modulates the activity of many enzymes and other proteins (see Figure 3-28). Because Ca^{2+} binds to calmodulin in a cooperative fashion, a small change in the level of cytosolic Ca^{2+} leads to a large change in the level of active calmodulin. One well-studied enzyme activated by the Ca^{2+}/calmodulin complex is myosin light-chain kinase, which regulates the activity of myosin in muscle cells (Chapter 19). Another is cAMP phosphodiesterase, the enzyme that degrades cAMP to 5'-AMP and terminates its effects. This reaction thus links Ca^{2+} and cAMP, one of many examples in which two second messengers interact to fine-tune certain aspects of cell regulation.

In certain cells, the rise in cytosolic Ca^{2+} following receptor signaling via PLC-generated IP_3 leads to activation of specific transcription factors. In some cases, Ca^{2+}/calmodulin activates protein kinases that, in turn, phosphorylate transcription factors, thereby modifying their activity and regulating gene expression. In other cases, Ca^{2+}/calmodulin activates a phosphatase that removes phosphate groups from a transcription factor. An important example of this mechanism involves T cells of the immune system in which Ca^{2+} ions enhance the activity of an essential transcription factor, NFAT (nuclear factor of activated T cells). In unstimulated cells, phosphorylated NFAT is located in the cytosol. Following receptor stimulation and elevation of cytosolic Ca^{2+}, the Ca^{2+}/calmodulin complex binds to and activates calcineurin, a protein-serine phosphatase. Activated calcineurin then dephosphorylates key phosphate residues on cytosolic NFAT, exposing a nuclear localization sequence that allows NFAT to move into the nucleus and stimulate expression of genes essential for activation of T cells.

The Ca^{2+}/calmodulin complex also plays a key role in controlling the diameter of blood vessels and thus their ability to deliver oxygen to tissues. This pathway involves a novel signaling molecule and provides another example of cGMP functioning as a second messenger.

Signal-Induced Relaxation of Vascular Smooth Muscle Is Mediated by cGMP-Activated Protein Kinase G

 Nitroglycerin has been used for over a century as a treatment for the intense chest pain of angina. It was known to slowly decompose in the body to *nitric oxide (NO)*, which causes relaxation of the smooth muscle cells surrounding the blood vessels that "feed" the heart muscle itself, thereby increasing the diameter of the blood vessels and increasing the flow of oxygen-bearing blood to the heart muscle. One of the most intriguing discoveries in modern medicine is that NO, a toxic gas found in car exhaust, is in fact a natural signaling molecule. ∎

Definitive evidence for the role of NO in inducing relaxation of smooth muscle came from a set of experiments in which acetylcholine was added to experimental preparations of the smooth muscle cells that surround blood vessels. Direct application of acetylcholine to these cells caused them to contract, the expected effect of acetylcholine on these muscle cells. But addition of acetylcholine to the lumen of small isolated blood vessels caused the underlying smooth muscles to relax, not contract. Subsequent studies showed that in response to acetylcholine the endothelial cells that line the lumen of blood vessels were releasing some substance that in turn triggered muscle cell relaxation. That substance turned out to be NO.

We now know that endothelial cells contain a G_o protein–coupled receptor that binds acetylcholine and activates phospholipase C, leading to an increase in the level of

▶ **FIGURE 13-30 Regulation of contractility of arterial smooth muscle by nitric oxide (NO) and cGMP.** Nitric oxide is synthesized in endothelial cells in response to acetylcholine and the subsequent elevation in cytosolic Ca^{2+}. NO diffuses locally through tissues and activates an intracellular NO receptor with guanylyl cyclase activity in nearby smooth muscle cells. The resulting rise in cGMP leads to activation of protein kinase G (PKG), relaxation of the muscle, and thus vasodilation. The cell-surface receptor for atrial natriuretic factor (ANF) also has intrinsic guanylyl cyclase activity (not shown); stimulation of this receptor on smooth muscle cells also leads to increased cGMP and subsequent muscle relaxation. PP_i = pyrophosphate. [See C. S. Lowenstein et al., 1994, *Ann. Intern. Med.* **120**:227.]

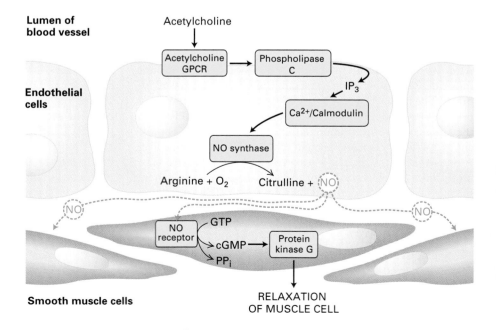

cytosolic Ca^{2+}. After Ca^{2+} binds to calmodulin, the resulting complex stimulates the activity of NO synthase, an enzyme that catalyzes formation of NO from O_2 and the amino acid arginine. Because NO has a short half-life (2–30 seconds), it can diffuse only locally in tissues from its site of synthesis. In particular NO diffuses from the endothelial cell into neighboring smooth muscle cells, where it triggers muscle relaxation (Figure 13-30).

The effect of NO on smooth muscle is mediated by the second messenger cGMP, which can be formed by an intracellular NO receptor expressed by smooth muscle cells. Binding of NO to the heme group in this receptor leads to a conformational change that increases its intrinsic guanylyl cyclase activity, leading to a rise in the cGMP level. Most of the effects of cGMP are mediated by a cGMP-dependent protein kinase, also known as *protein kinase G (PKG)*. In vascular smooth muscle, protein kinase G activates a signaling pathway that results in inhibition of the actin-myosin complex, relaxation of the cell, and dilation of the blood vessel. In this case, cGMP acts indirectly via protein kinase G, whereas in rod cells cGMP acts directly by binding to and thus opening cation channels in the plasma membrane.

Relaxation of vascular smooth muscle also is triggered by binding of atrial natriuretic factor (ANF) and some other peptide hormones to their receptors on smooth muscle cells. The cytosolic domain of these cell-surface receptors, like the intracellular NO receptor, possesses intrinsic guanylyl cyclase activity. When an increased blood volume stretches cardiac muscle cells in the heart atrium, they release ANF. Circulating ANF binds to ANF receptors in smooth muscle cells surrounding blood vessels, inducing activation of guanylyl cyclase activity and formation of cGMP. Subsequent activation of protein kinase G causes dilation of the vessel by the mechanism described above. This vasodilation reduces blood pressure and counters the stimulus that provoked the initial release of ANF.

KEY CONCEPTS OF SECTION 13.5

G Protein–Coupled Receptors That Activate Phospholipase C

■ Simulation of some GPCRs and other cell-surface receptors leads to activation of phospholipase C, which generates two second messengers: diffusible IP_3 and membrane-bound DAG (see Figure 13-28).

■ IP_3 triggers opening of IP_3-gated Ca^{2+} channels in the endoplasmic reticulum and elevation of cytosolic free Ca^{2+}. In response to elevated cytosolic Ca^{2+}, protein kinase C is recruited to the plasma membrane, where it is activated by DAG (see Figure 13-29).

■ The Ca^{2+}/calmodulin complex regulates the activity of many different proteins, including cAMP phosphodiesterase, nitric oxide synthase, and protein kinases or phosphatases that control the activity of various transcription factors.

■ Stimulation of the acetylcholine GPCR on endothelial cells induces an increase in cytosolic Ca^{2+} and subsequent synthesis of NO. After diffusing into surrounding smooth muscle cells, NO activates intracellular guanylate cyclase to synthesize cGMP (see Figure 13-30).

■ Synthesis of cGMP in vascular smooth muscle cells leads to activation of protein kinase G, which triggers a pathway leading to muscle relaxation and vasodilation.

■ cGMP is also produced in vascular smooth muscle cells by stimulation of cell-surface receptors that have intrinsic guanylate cyclase activity. These include receptors for atrial natriuretic factor (ANF).

13.6 Activation of Gene Transcription by G Protein–Coupled Receptors

As mentioned early in this chapter, intracellular signal-transduction pathways can have short-term and long-term effects on the cell. Short-term effects (seconds to minutes) result from modulation of the activity of preexisting enzymes or other proteins, leading to changes in cell metabolism or function. Most of the pathways activated by G protein–coupled receptors fall into this category. However, GPCR signaling pathways also can have long-term effects (hours to days) owing to activation or repression of gene transcription, leading in some cases to cell proliferation or to differentiation into a different type of cell. Earlier we discussed how a signal-induced rise in cytosolic Ca^{2+} can lead to activation of transcription factors. Here we consider other mechanisms by which some G protein–coupled receptors regulate gene expression.

Membrane-Localized Tubby Transcription Factor Is Released by Activation of Phospholipase C

 The *tubby* gene, which is expressed primarily in certain areas of the brain involved in control of eating behavior, first attracted attention because of its involvement in obesity. Mice bearing mutations in the *tubby* gene develop adult-onset obesity, and certain aspects of their metabolism resemble that of obese humans.

Sequencing of the cloned *tubby* gene suggested that its encoded protein contains both a DNA-binding domain and a transcription-activation domain (Chapter 11). However, the Tubby protein was found to be localized near the plasma membrane, making it an unlikely candidate as a transcription factor. Subsequent studies revealed that Tubby binds tightly to PIP_2, anchoring the protein to the plasma membrane (Figure 13-31). Hormone binding to G_o- or G_q-coupled receptors, which activate phospholipase C, leads to hydrolysis of PIP_2 and release of Tubby into the cytosol. Tubby then enters the nucleus and activates transcription of a still unknown gene or genes. Identification of these genes should provide clues about how their encoded proteins relate to obesity. ■

Phospholipase C

PIP₂

DAG   **Exterior**

IP₃

Cytosol

Tubby

Transcriptional
activation
domain

DNA binding
domain

2

Nucleus

3

Transcription

◄ **FIGURE 13-31 Activation of the Tubby transcription factor following ligand binding to receptors coupled to G_o or G_q.** In resting cells, Tubby is bound tightly to PIP₂ in the plasma membrane. Receptor stimulation (not shown) leads to activation of phospholipase C, hydrolysis of PIP₂, and release of Tubby into the cytosol (**1**). Directed by two functional nuclear localization sequences (NLS) in its N-terminal domain, Tubby translocates into the nucleus (**2**) and activates transcription of target genes (**3**). It is not known whether IP₃ remains bound to Tubby. [Adapted from S. Santagata et al., 2001, *Science* **292**:2041.]

▶ **FIGURE 13-32 Activation of gene expression following ligand binding to G_s protein–coupled receptors.** Receptor stimulation (**1**) leads to activation of PKA (**2**). Catalytic subunits of PKA translocate to the nucleus (**3**) and there phosphorylate and activate the transcription factor CREB (**4**). Phosphorylated CREB associates with the co-activator CBP/P300 (**5**) to stimulate various target genes controlled by the CRE regulatory element. See the text for details. [See K. A. Lee and N. Masson, 1993, *Biochim. Biophys. Acta* **1174**:221, and D. Parker et al., 1996, *Mol. Cell Biol.* **16**(2):694.]

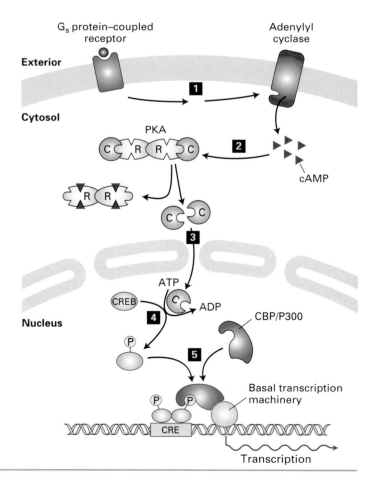

CREB Links cAMP Signals to Transcription

In mammalian cells, an elevation in the cytosolic cAMP level stimulates the expression of many genes. For instance, increased cAMP in certain endocrine cells induces production of somatostatin, a peptide that inhibits release of various hormones; in liver cells, cAMP induces synthesis of several enzymes involved in converting three-carbon compounds to glucose.

All genes regulated by cAMP contain a cis-acting DNA sequence, the *cAMP-response element (CRE)*, that binds the phosphorylated form of a transcription factor called *CRE-binding (CREB) protein*, which is found only in the nucleus. As discussed previously, binding of neurotransmitters and hormones to G_s protein–coupled receptors activates adenylyl cyclase, leading to an increase in cAMP and subsequent release of the active catalytic subunit of PKA. Some of the catalytic subunits then translocate to the nucleus and phosphorylate serine-133 on CREB protein.

Phosphorylated CREB protein binds to CRE-containing target genes and also interacts with a *co-activator* termed *CBP/300*, which links CREB to the basal transcriptional machinery, thereby permitting CREB to stimulate transcription (Figure 13-32). Earlier studies suggested that phosphorylation induced a conformational change in CREB protein, but more recent work indicates that CBP/P300 binds specifically to phosphoserine-133 in activated CREB. As discussed in Chapter 11, other signal-regulated transcription factors rely on CBP/P300 to exert their activating effect. Thus this co-activator plays an important role in integrating signals from multiple signaling pathways that regulate gene transcription.

GPCR-Bound Arrestin Activates Several Kinase Cascades That Control Gene Expression

We saw earlier that binding of β-arrestin to phosphorylated serines in the cytosolic domain of G protein–coupled receptors both blocks activation of G_α and mediates endocytosis of the GPCR-arrestin complex. Perhaps surprisingly, the GPCR-arrestin complex also acts as a scaffold for binding and activating several cytosolic kinases (see Figure 13-19). These include c-Src, which activates the MAP kinase pathway and other pathways leading to transcription of genes needed for cell division. A complex of three arrestin-bound proteins, including a Jun N-terminal kinase (JNK-1), initiates a kinase cascade that ultimately activates the c-Jun transcription factor. Activated c-Jun promotes expression of certain growth-promoting enzymes and other proteins that help cells respond to some stresses.

 Binding of epinephrine to the β-adrenergic receptors in heart muscle stimulates glycogenolysis and enhances the rate of muscle contraction. Prolonged treatment with epinephrine, however, induces proliferation of these cardiac muscle cells. In extreme cases, such *cardiac hypertrophy* causes failure of the heart muscle, a major cause of heart disease. This epinephrine-induced cell proliferation results

in part from activation of the MAP kinase cascade. As just described, the GPCR-arrestin complex can trigger this cascade.

Another, perhaps more important, way that activation of β-adrenergic receptors promotes cardiac hypertrophy involves another type of receptor. The G_s protein activated by β-adrenergic receptors can somehow lead to activation of a specific extracellular metal-containing protease that, in turn, cleaves the transmembrane precursor of epidermal growth factor (EGF). The soluble EGF released into the extracellular space binds to and activates EGF receptors on the same cell in an autocrine fashion. As we learn in the next chapter, the EGF receptor belongs to the receptor tyrosine kinase (RTK) class of receptors, which commonly trigger the MAP kinase cascade leading to cell proliferation. Similar cross-talk between two types of receptors occurs in many other signaling systems. Just as no cell lives in isolation, no receptor and no signal-transduction pathway function by themselves. ∎

KEY CONCEPTS OF SECTION 13.6

Activation of Gene Transcription by G Protein–Coupled Receptors

■ Activation of phospholipase C by receptors coupled to G_o or G_q proteins releases the Tubby transcription factor, which is bound to PIP_2 embedded in the plasma membrane of resting cells (see Figure 13-31).

■ Signal-induced activation of protein kinase A (PKA) often leads to phosphorylation of CREB protein, which together with the CBP/300 co-activator stimulates transcription of many target genes (see Figure 13-32).

■ The GPCR-arrestin complex activates several cytosolic kinases, initiating cascades that lead to transcriptional activation of many genes controlling cell growth.

PERSPECTIVES FOR THE FUTURE

Very soon we will know the identity of all the pieces in many signal-transduction pathways, but putting the puzzle together to predict cellular responses remains elusive. For instance, we can enumerate the G proteins, kinases, phosphatases, arrestins, and other proteins that participate in signaling from β-adrenergic receptors in liver cells, but we are still far from being able to predict, quantitatively, how liver cells react over time to a given dose of adrenaline. In part this is because complex feedback (and in some cases feed-forward) loops regulate the activity of multiple enzymes and other components in the pathway. Although biochemical and cell biological experiments tell us how these interactions occur, we cannot describe quantitatively the rates or extent of these reactions in living cells.

The emerging field of biological systems analysis attempts to develop an integrated view of a cell's response to external signals. Mathematical equations are formulated that incorporate rate constants for enzyme catalysis, formation of protein-protein complexes, and concentrations and diffusion rates of all the various signal-transduction proteins. These models incorporate information about changes in the subcellular localization of proteins with time (e.g., movement of transcription factors into the nucleus or endocytosis of surface receptors) and the effect on the activity of any given protein (e.g., glycogen phosphorylase) of the local Ca^{2+} concentration and the presence of multiple kinases and phosphatases. By comparing the results of such calculations with actual experimental results (say, by increasing or decreasing selectively the concentration of one component of the pathway) we can determine, in principle, whether we have accounted for all of the components of the pathway. Such mathematical modeling will also help the pharmaceutical industry develop new drugs that might activate or inhibit specific pathways. Modeling can enable one to extrapolate the results of experiments on drugs on proteins in test tubes or on cultured cells in order to predict their efficacy and side effects in living organisms.

In this chapter we focused primarily on signal-transduction pathways activated by individual G protein–coupled receptors. However, even these relatively simple pathways presage the more complex situation within living cells. As we've seen, activation of a single type of receptor often leads to production of multiple second messengers or activation of several types of downstream transducing proteins. Moreover, the same cellular response (e.g., glycogen breakdown) is affected by multiple signaling pathways activated by multiple types of receptors. Interaction of different signaling pathways permits the fine-tuning of cellular activities required to carry out complex developmental and physiological processes, and the ability of cells to respond appropriately to extracellular signals also depends on regulation of signaling pathways themselves.

KEY TERMS

adenylyl cyclase 548	paracrine signaling 535
adrenergic receptors 548	PDZ domains 543
agonist 539	phospholipase C (β isoform) 561
autocrine signaling 535	
calmodulin 563	protein kinase A 550
cAMP 537	protein kinase C 563
competition assay 539	rhodopsin 556
desensitization 544	second messengers 541
endocrine signaling 535	signal amplification 552
functional expression assay 540	signal transduction 534
	stimulatory G protein 548
G protein–coupled receptors 534	transducin 556
	trimeric G proteins 534
IP$_3$/DAG pathway 561	visual adaptation 560
muscarinic acetylcholine receptors 556	

REVIEW THE CONCEPTS

1. Signaling by soluble extracellular molecules can be classified into three types: endocrine, paracrine, and autocrine. Describe how these three methods of cellular signaling differ. Growth hormone is secreted from the pituitary, which is located at the base of the brain, and acts through growth hormone receptors located on the liver. Is this an example of endocrine, paracrine, or autocrine signaling? Why?

2. A ligand binds two different receptors with a K_d value of 10^{-7} M for receptor 1 and a K_d value of 10^{-9} M for receptor 2. For which receptor does the ligand show the greater affinity? Calculate the fraction of receptors that have a bound ligand ([RL]/R_T) for ligand with receptor 1 and for ligand with receptor 2, if the concentration of free ligand is 10^{-8} M.

3. A study of the properties of cell-surface receptors can be greatly enhanced by isolation or cloning of the cell-surface receptor. Describe how a cell-surface receptor can be isolated by affinity chromatography. How can you clone a cell surface receptor using a functional-expression assay?

4. Signal-transducing trimeric G proteins consist of three subunits designated α, β, γ. The G_α subunit is a GTPase switch protein that cycles between active and inactive states depending upon whether it is bound to GTP or to GDP. Review the steps for ligand-induced activation of effector proteins mediated by the trimeric-G-protein complex. Suppose that you have isolated a mutant G_α subunit that has an increased GTPase activity. What effect would this mutation have on the G protein and the effector protein?

5. Membrane proteins are often found clustered. Describe how protein clustering can be mediated by adapter proteins or by specialized lipid rafts termed caveolae. What advantage might there be to having a cluster of membrane proteins involved in a signaling pathway rather than spread out in the membrane?

6. Epinephrine binds to both β-adrenergic and α-adrenergic receptors. Describe the opposite actions on the effector protein, adenylyl cyclase, elicited by the binding of epinephrine to these two types of receptors. Describe the effect of adding an agonist or antagonist to a β-adrenergic receptor on the activity of adenylyl cyclase.

7. In liver and muscle cells, epinephrine stimulates the release of glucose from glycogen by inhibiting glycogen synthesis and stimulating glycogen breakdown. Outline the molecular events that occur after epinephrine binds to its receptor and the resultant increase in the concentration of intracellular cAMP. How are the cAMP levels returned to normal? Describe the events that occur after cAMP levels decline.

8. Continuous exposure of a G_s protein–coupled receptor to its ligand leads to a phenomenon known as desensitization. Describe several molecular mechanisms for receptor desensitization. How can a receptor be reset to its original sensitized state? What effect would a mutant receptor lacking serine or threonine phosphorylation sites have on a cell?

9. A number of different molecules act as second messengers. Activation of rhodopsin by light induces the closing of gated cation channels with cyclic GMP as a second messenger. Describe the effect of light on rhodopsin. On what effector protein does the trimeric G protein act? What type of trimeric G protein is involved in this event?

10. Visual adaptation and receptor desensitization involve similar phosphorylation mechanisms. Describe how the β-adrenergic receptor kinase (BARK) and rhodopsin kinase play important roles in these processes. What role does dephosphorylation play in these reactions?

11. Inositol 1,4,5-trisphosphate (IP_3) and diacylglycerol (DAG) are second messenger molecules derived from the cleavage of the phosphoinositide PIP_2 (phosphatidylinositol 4,5-bisphosphate) by activated phospholipase C. Describe the role of IP_3 in the release of Ca^{2+} from the endoplasmic reticulum. How do cells replenish the endoplasmic reticulum stores of Ca^{2+}? What is the principal function of DAG?

12. In 1992, the journal *Science* named nitric oxide the Molecule of the Year. Describe how this important second messenger is synthesized. How does nitric oxide cause relaxation of smooth muscle cells?

13. Ligand binding to G protein–coupled receptors can result in activation of gene transcription. Describe how the second messengers PIP_2 and cAMP can activate transcription of genes.

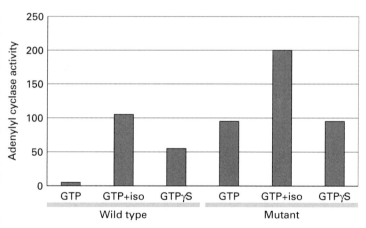

ANALYZE THE DATA

Mutations in the trimeric G proteins can cause many diseases in humans. Patients with acromegaly often have pituitary tumors that oversecrete the pituitary hormone called growth hormone (GH). A subset of these growth hormone (GH)–secreting pituitary tumors result from mutations in G proteins. GH-releasing hormone (GHRH) stimulates GH release from the pituitary by binding to GHRH receptors and stimulating adenylyl cyclase. Cloning and sequencing of the wild-type and mutant $G_{s\alpha}$ gene from normal individuals and patients with the pituitary tumors revealed a missense mutation in the $G_{s\alpha}$ gene sequence.

a. To investigate the effect of the mutation on $G_{s\alpha}$ activity, wild-type and mutant $G_{s\alpha}$ cDNA were transfected into cells that lack the $G_{s\alpha}$ gene. These cells express a $β_2$-adrenergic receptor, which can be activated by isoproterenol, a $β_2$-adrenergic receptor agonist. Membranes were isolated from transfected cells and assayed for adenylyl cyclase activity in the presence of GTP or the hydrolysis-resistant GTP analog, GTP-γS. From the figure above, what do you conclude about the effect of the mutation on $G_{s\alpha}$ activity in the presence of GTP alone compared with GTP-γS alone or GTP plus isoproterenol (iso)?

b. In the transfected cells described in part a, what would you predict would be the cAMP levels in cells transfected with the wild-type $G_{s\alpha}$ and the mutant $G_{s\alpha}$? What effect might this have on the cells?

c. To further characterize the molecular defect caused by this mutation, the intrinsic GTPase activity present in both wild-type and mutant $G_{s\alpha}$ was assayed. Assays for GTPase activity showed that the mutation reduced the $K_{cat\text{-}GTP}$ (catalysis rate constant for GTP hydrolysis) from a wild-type value of $4.1\ min^{-1}$ to the mutant value of $0.1\ min^{-1}$. What do you conclude about the effect of the mutation on the GTPase activity present in the mutant $G_{s\alpha}$ subunit? How do these GTPase results explain the adenylyl cyclase results shown in part a?

REFERENCES

Signaling Molecules and Cell-Surface Receptors

Coughlin, S. R. 2000. Thrombin signaling and protease-activated receptors. *Nature* **407**:258–264.

Farfel, Z., H. Bourne, and T. Iiri. 1999. The expanding spectrum of G protein diseases. *New Eng. J. Med.* **340**:1012–1020.

Simonsen, H., and H. F. Lodish. 1994. Cloning by function: expression cloning in mammalian cells. *Trends Pharmacol. Sci.* **15**:437–441.

Stadel, J. M., S. Wilson, and D. J. Bergsma. 1997. Orphan G protein–coupled receptors: a neglected opportunity for pioneer drug discovery. *Trends Pharmacol. Sci.* **18**:430–437.

Wilson, J. D., D. W. Foster, H. M. Kronenberg, and R. H. Williams. 1998. *Williams Textbook of Endocrinology,* 9th ed. W. B. Saunders.

Intracellular Signal Transduction

Anderson, R. G., and K. Jacobson. 2002. A role for lipid shells in targeting proteins to caveolae, rafts, and other lipid domains. *Science* **296**:1821–1825.

Bourne, H. R. 1997. How receptors talk to trimeric G proteins. *Curr. Opin. Cell Biol.* **9**:134–142.

Galbiati, F., B. Razani, and M. P. Lisanti. 2001. Emerging themes in lipid rafts and caveolae. *Cell* 106:403–411.

Garner, C. C., J. Nash, and R. L. Huganir. 2000. PDZ domains in synapse assembly and signaling. *Trends Cell Biol.* 10:274–280.

Harris, B. Z., and W. A. Lim. 2001. Mechanism and role of PDZ domains in signaling complex assembly. *J. Cell Sci.* 114:3219–3231.

Rebecchi, M. J., and S. Scarlata. 1998. Pleckstrin homology domains: a common fold with diverse functions. *Ann. Rev. Biophys. Biomol. Struc.* 27:503–528.

Van Deurs, B., K. Roepstorff, A. Hommelgaard, and K. Sandvig. 2003. Caveolae: anchored multifunctional platforms in the lipid ocean. *Trends Cell Biol.* 13:92–100.

Vetter, I. R., and A. Wittinghofer. 2001. The guanine nucleotide-binding switch in three dimensions. *Science* 294:1299–1304.

G Protein–Coupled Receptors That Activate or Inhibit Adenylyl Cyclase

Browner, M., and R. Fletterick. 1992. Phosphorylase: a biological transducer. *Trends Biochem. Sci.* 17:66–71.

Diviani, D., and J. D. Scott. 2001. AKAP signaling complexes at the cytoskeleton. *J. Cell Sci.* 114:1431–1437.

Ferguson, S. S., and M. G. Caron. 1998. G protein–coupled receptor adaptation mechanisms. *Semin. Cell Develop. Biol.* 9:119–127.

Hurley, J. H. 1999. Structure, mechanism, and regulation of mammalian adenylyl cyclase. *J. Biol. Chem.* 274:7599–7602.

Johnson, L. N. 1992. Glycogen phosphorylase: control by phosphorylation and allosteric effectors. *FASEB J.* 6:2274–2282.

Luttrell, L. M., and R. J. Lefkowitz. 2002. The role of beta-arrestins in the termination and transduction of G-protein-coupled receptor signals. *J. Cell Sci.* 115:455–465.

Michel, J. J., and J. D. Scott. 2002. AKAP mediated signal transduction. *Ann. Rev. Pharmacol. Toxicol.* 42:235–257.

Perry, S. J., and R. J. Lefkowitz. 2002. Arresting developments in heptahelical receptor signaling and regulation. *Trends Cell Biol.* 12:130–138.

Pierce, K. L., R. T. Premont, and R. J. Lefkowitz. 2002. Seven-transmembrane receptors. *Nature Rev. Mol. Cell Biol.* 3:639–652.

Sprang, S. R. 1997. G protein mechanisms: insights from structural analysis. *Ann. Rev. Biochem.* 66:639–678.

Tesmer, J. J., R. K. Sunahara, A. G. Gilman, and S. R. Sprang. 1997. Crystal structure of the catalytic domains of adenylyl cyclase in a complex with $G_s\alpha\cdot GTP\gamma S$. *Science* 278:1907–1916.

G Protein–Coupled Receptors That Regulate Ion Channels

Borhan, B., et al. 2000. Movement of retinal along the visual transduction path. *Science* 288:2209–2212.

Bourne, H. R., and E. C. Meng. 2000. Structure. Rhodopsin sees the light. *Science* 289:733–734.

Hamm, H. E. 2001. How activated receptors couple to G proteins. *Proc. Nat'l. Acad. Sci. USA* 98:4819–4821.

Hurley, J. H., and J. A. Grobler. 1997. Protein kinase C and phospholipase C: Bilayer interactions and regulation. *Curr. Opin. Struc. Biol.* 7:557–565.

Mendez, A., et al. 2000. Rapid and reproducible deactivation of rhodopsin requires multiple phosphorylation sites. *Neuron* 28:153–164.

Nathans, J. 1999. The evolution and physiology of human color vision: insights from molecular genetic studies of visual pigments. *Neuron* 24:299–312.

Palczewski, K., et al. 2000. Crystal structure of rhodopsin: a G protein-coupled receptor. *Science* 289:739–745.

Sokolov, M., et al. 2002. Massive light-driven translocation of transducin between the two compartments of rod cells: a novel mechanism of light adaptation. *Neuron* 33:95–106.

G Protein–Coupled Receptors That Activate Phospholipase C

Berridge, M. J. 1997. Elementary and global aspects of calcium signaling. *J. Exp. Biol.* 200:315–319.

Chin, D., and A. R. Means. 2000. Calmodulin: a prototypical calcium sensor. *Trends Cell Biol.* 10:322–328.

Czech, M. P. 2000. PIP2 and PIP3: complex roles at the cell surface. *Cell* 100:603–606.

Delmas, P., and D. Brown. 2002. Junctional signaling microdomains: bridging the gap between neuronal cell surface and Ca^{2+} stores. *Neuron* 36:787–790.

Hobbs, A. J. 1997. Soluble guanylate cyclase: the forgotten sibling. *Trends Pharmacol. Sci.* 18:484–491.

Jaken, S. 1996. Protein kinase C isozymes and substrates. *Curr. Opin. Cell Biol.* 8:168–173.

Putney, J. W. 2001. Cell biology: channeling calcium. *Nature* 410:648–649.

Singer, W. D., H. A. Brown, and P. C. Sternweis. 1997. Regulation of eukaryotic phosphatidylinositol-specific phospholipase C and phospholipase D. *Ann. Rev. Biochem.* 66:475–509.

Toker, A., and L. C. Cantley. 1997. Signaling through the lipid products of phosphoinositide-3-OH kinase. *Nature* 387:673–676.

Wedel, B. J., and D. L. Garbers. 1997. New insights on the functions of the guanylyl cyclase receptors. *FEBS Lett.* 410:29–33.

Activation of Gene Transcription by G Protein–Coupled Receptors

Cantley, L. C. 2001. Transcription. Translocating tubby. *Science* 292:2019–2021.

Luttrell, L. M., Y. Daaka, and R. J. Lefkowitz. 1999. Regulation of tyrosine kinase cascades by G-protein-coupled receptors. *Curr Opin. Cell Biol.* 11:177–183.

Mayr, B., and M. Montminy. 2001. Transcriptional regulation by the phosphorylation-dependent factor CREB. *Nature Rev. Mol. Cell Biol.* 2:599–609.

Pierce, K. L., L. M. Luttrell, and R. J. Lefkowitz. 2001. New mechanisms in heptahelical receptor signaling to mitogen activated protein kinase cascades. *Oncogene* 20:1532–1539.

Santagata, S., et al. 2001. G-protein signaling through tubby proteins. *Science* 292:2041–2050.

Shaywitz, A., and M. Greenberg. 1999. CREB: a stimulus-induced transcription factor activated by a diverse array of extracellular signals. *Ann. Rev. Biochem.* 68:821–861.

Vo, N., and R. Goodman. 2001. CREB-binding protein and p300 in transcriptional regulation. *J. Biol. Chem.* 276:13505–13508.

14

SIGNALING PATHWAYS THAT CONTROL GENE ACTIVITY

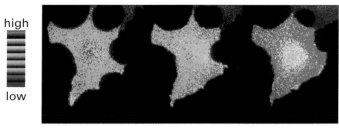

Fluorescence resonance energy transfer (FRET) detects time and location of activation of Ras protein in live cells triggered by epidermal growth factor. [Michiyuki Matsuda, Research Institute for Microbial Diseases, Osaka University.]

The development of all organisms requires execution of a complex program whereby specific genes are activated and repressed in specific sets of cells and in a precise time sequence. Many developmental changes in gene expression are generated by extracellular **signaling molecules** that act on cell-surface **receptors**. Most of these signals are soluble, secreted factors that act in a paracrine fashion on receiving (target) cells near the releasing cell. However, some signaling proteins are themselves attached to the cell surface, where they interact with cell-surface receptors on adjacent cells to alter the receiving cell's pattern of gene expression.

Even mature cells that are part of a differentiated tissue constantly change their patterns of gene expression. In large measure this occurs because of many different cell-surface receptors that continually receive information from extracellular signals and transduce this information into activation of specific transcription factors that stimulate or repress expression of specific target genes. Many such signaling pathways lead to alterations in the cell's metabolic activities. Liver, for example, responds to fluctuations in the levels of many hormones (e.g., insulin, glucagon, and epinephrine) by altering expression of many genes encoding enzymes of glucose and fat metabolism. Other signaling pathways influence the levels of proteins that affect the ability of cells to progress through the cell cycle and divide.

A typical mammalian cell often expresses cell-surface receptors for more than 100 different types of extracellular signaling molecules that function primarily to regulate the activity of **transcription factors** (see Figure 13-1). The signal-induced activation of transcription factors occurs by several mechanisms. In the last chapter, for instance, we saw that stimulation of some G protein–coupled receptors leads to a rise in cAMP and the cAMP-dependent activation of protein kinase A. After translocating to the nucleus, protein kinase A phosphorylates and thereby activates the CREB transcription factor.

In this chapter, we focus on five other classes of cell-surface receptors that illustrate additional signal-induced mechanisms of activating transcription factors. Stimulation of *transforming growth factor β (TGFβ) receptors* and *cytokine receptors* leads directly to activation of cytosolic transcription factors as the result of phosphorylation by a kinase that is part of the receptor or associated with it. The activated transcription factors then translocate into the nucleus and act on specific target genes. In the case of *receptor*

TABLE 14-1	Overview of Major Receptor Classes and Signaling Pathways
Receptor Class/Pathway[*]	**Distinguishing Characteristics**
RECEPTORS LINKED TO TRIMERIC G PROTEINS	
G protein–coupled receptors (13)	*Ligands:* Epinephrine, glucagon, serotonin, vasopressin, ACTH, adenosine, and many others (mammals); odorant molecules, light; mating factors (yeast) *Receptors:* Seven transmembrane α helices; cytosolic domain associated with a membrane-tethered trimeric G protein *Signal transduction:* (1) Second-messenger pathways involving cAMP or IP$_3$/DAG; (2) linked ion channels; (3) MAP kinase pathway
RECEPTORS WITH INTRINSIC OR ASSOCIATED ENZYMATIC ACTIVITY	
TGFβ receptors (14, 15)	*Ligands:* Transforming growth factor β superfamily (TGFβ, BMPs), activin, inhibins (mammals); Dpp (*Drosophila*) *Receptors:* Intrinsic protein serine/threonine kinase activity in cytosolic domain (type I and II) *Signal transduction:* Direct activation of cytosolic Smad transcription factors
Cytokine receptors (14, 15)	*Ligands:* Interferons, erythropoietin, growth hormone, some interleukins (IL-2, IL-4), other cytokines *Receptors:* Single transmembrane α helix; conserved multi-β strand fold in extracellular domain; JAK kinase associated with intracellular domain *Signal transduction:* (1) Direct activation of cytosolic STAT transcription factors; (2) PI-3 kinase pathway; (3) IP$_3$/DAG pathway; (4) Ras-MAP kinase pathway
Receptor tyrosine kinases (14)	*Ligands:* Insulin, epidermal growth factor (EGF), fibroblast growth factor (FGF), neurotrophins, other growth factors *Receptor:* Single transmembrane α helix; intrinsic protein tyrosine kinase activity in cytosolic domain *Signal transduction:* (1) Ras–MAP kinase pathway; (2) IP$_3$/DAG pathway; (3) PI-3 kinase pathway
Receptor guanylyl cyclases (13)	*Ligands:* Atrial natriuretic factor and related peptide hormones *Receptor:* Single transmembrane α helix; intrinsic guanylate cyclase activity in cytosolic domain *Signal transduction:* Generation of cGMP
Receptor phosphotyrosine phosphatases	*Ligands:* Pleiotrophins, other protein hormones *Receptors:* Intrinsic phosphotyrosine phosphatase activity in cytosolic domain inhibited by ligand binding *Signal transduction:* Hydrolysis of activating phosphotyrosine residue on cytosolic protein tyrosine kinases
T-cell receptors	*Ligands:* Small peptides associated with major histocompatability (MHC) proteins in the plasma membrane of macrophages and other antigen-presenting cells *Receptors:* Single transmembrane α helix; several protein kinases associated with cytosolic domain; found only on T lymphocytes *Signal transduction:* (1) Activation of cytosolic protein tyrosine kinases; (2) PI-3 kinase pathway; (3) IP$_3$/DAG pathway; (4) Ras–MAP kinase pathway

TABLE 14-1	Overview of Major Receptor Classes and Signaling Pathways

Receptor Class/Pathway*	Distinguishing Characteristics
RECEPTORS THAT ARE ION CHANNELS	
Ligand-gated ion channels (7, 13)	*Ligands:* Neurotransmitters (e.g., acetylcholine, glutamate), cGMP, physical stimuli (e.g., touch, stretching), IP$_3$ (receptor in ER membrane) *Receptors:* Four or five subunits with a homologous segment in each subunit lining the ion channel *Signal transduction:* (1) Localized change in membrane potential due to ion influx, (2) elevation of cytosolic Ca^{2+}
PATHWAYS INVOLVING PROTEOLYSIS	
Wnt pathway (15)	*Ligands:* Secreted Wnt (mammals); Wg (*Drosophila*) *Receptors:* Frizzled (Fz) with seven transmembrane α helices; associated membrane-bound LDL receptor–related protein (Lrp) required for receptor activity *Signal transduction:* Assembly of multiprotein complex at membrane that inhibits the proteasome-mediated proteolysis of cytosolic β-catenin transcription factor, resulting in its accumulation
Hedgehog (Hh) pathway (15)	*Ligands:* Cell-tethered Hedgehog *Receptors:* Binding of Hh to Patched (Ptc), which has 12 transmembrane α helices; activation of signaling from Smoothened (Smo), with 7 transmembrane α helices *Signal transduction:* Proteolytic release of a transcriptional activator from multiprotein complex in the cytosol
Notch/Delta pathway (14, 15)	*Ligands:* Membrane-bound Delta or Serrate protein *Receptors:* Extracellular subunit of Notch receptor noncovalently associated with transmembrane-cytosolic subunit *Signal transduction:* Intramembrane proteolytic cleavage of receptor transmembrane domain with release of cytosolic segment that functions as co-activator for nuclear trascription factors
NF-κB pathways (14, 15)	*Ligands:* Tumor necrosis factor α (TNF-α), interleukin 1 (mammals); Spätzle (*Drosophila*) *Receptors:* Various in mammals; Toll and Toll-like receptors in *Drosophila* *Signal transduction:* Phosphorylation-dependent degradation of inhibitor protein with release of active NF-κB transcription factor (Dorsal in *Drosophila*) in the cytosol
INTRACELLULAR RECEPTORS PATHWAYS	
Nitric oxide pathway (13)	*Ligands:* Nitric oxide (NO) *Receptor:* Cytosolic guanylyl cyclase *Signal transduction:* Generation of cGMP
Nuclear receptor pathways (11)	*Ligands:* Lipophilic molecules including steroid hormones, thyroxine, retinoids, and fatty acids in mammals and ecdysone in *Drosophila* *Receptors:* Highly conserved DNA-binding domain, somewhat conserved hormone-binding domain, and a variable domain; located within nucleus or cytosol *Signal transduction:* Activation of receptor's transcription factor activity by ligand binding

*Unless indicated otherwise, receptors are located in the plasma membrane. Numbers in parentheses indicate chapters in which a receptor/pathway is discussed in depth.

SOURCES: J. Gerhart, 1999, *Teratology* **60**:226, and A. Brivanlou and J. E. Darnell, 2002, *Science* **295**:813.

tyrosine kinases, binding of a **ligand** to its receptor sets into motion a cascade of intracellular events leading to activation of a cytosolic kinase that moves into the nucleus and activates one or more transcription factors by phosphorylation. Signaling from *tumor necrosis factor α (TNF-α) receptors* generates an active NF-κB transcription factor by proteolytic cleavage of a cytosolic inhibitor protein, and proteolytic cleavage of a *Notch receptor* releases the receptor's cytosolic domain that then functions as a co-activator for transcription factors in the nucleus. Proteolysis also plays a role in the signaling pathways triggered by binding of protein ligands called Wnt and Hedgehog (Hh) to their receptors. We cover these two pathways, which play a major role during **development** and **differentiation,** in Chapter 15.

For simplicity, we often describe the various receptor classes independently, concentrating on the major pathway of **signal transduction** initiated by each class of receptor. However, as shown in Table 14-1, several classes of receptors can transduce signals by more than one pathway. Moreover, many genes are regulated by multiple transcription factors, each of which can be activated by one or more extracellular signals. Especially during early development, such cross talk between signaling pathways and the resultant sequential alterations in the pattern of gene expression eventually can become so extensive that the cell assumes a different developmental fate.

Researchers have employed a variety of experimental approaches and systems to identify and study the function of extracellular signaling molecules, receptors, and intracellular signal-transduction proteins. For instance, the secreted signaling protein Hedgehog (Hh) and its receptor were first identified in *Drosophila* mutants with developmental defects. Subsequently the human and mouse homologs of these proteins were cloned and shown to participate in a number of important signaling events during differentiation. Some signal-transduction proteins were first identified when gain-of-function mutations in the genes encoding them or overexpression of the normal protein caused abnormal cell proliferation leading to malignancy. A mutant Ras protein exhibiting unregulated (i.e., **constitutive**) activity was identified in this way; wild-type Ras later was found to be a key player in many signaling pathways. Numerous extracellular signaling molecules initially were purified from cell extracts based on their ability to stimulate growth and proliferation of specific cell types. These few examples illustrate the importance of studying signaling pathways both genetically—in flies, mice, worms, yeasts, and other organisms—and biochemically.

14.1 TGFβ Receptors and the Direct Activation of Smads

A number of related extracellular signaling molecules that play widespread roles in regulating development in both invertebrates and vertebrates constitute the *transforming growth factor β (TGFβ) superfamily.* One member of this superfamily, *bone morphogenetic protein (BMP),* initially was identified by its ability to induce bone formation in cultured cells. Now called *BMP7,* it is used clinically to strengthen bone after severe fractures. Of the numerous BMP proteins subsequently recognized, many help induce key steps in development, including formation of mesoderm and the earliest blood-forming cells.

Another member of the TGFβ superfamily, now called *TGFβ-1,* was identified on the basis of its ability to induce a transformed phenotype of certain cells in culture. However, the three human TGFβ isoforms that are known all have potent antiproliferative effects on many types of mammalian cells. Loss of TGFβ receptors or certain intracellular signal-transduction proteins in the TGFβ pathway, thereby releasing cells from this growth inhibition, frequently occurs in human tumors. TGFβ proteins also promote expression of cell-adhesion molecules and extracellular-matrix molecules. TGFβ signals certain types of cells to synthesize and secrete growth factors that can, on balance, overcome the normal TFGβ-induced growth inhibition; this explains why TGFβ was originally detected as a growth factor. A *Drosophila* homolog of TGFβ, called *Dpp protein,* controls dorsoventral patterning in fly embryos, as we detail in Chapter 15. Other mammalian members of the TGFβ superfamily, the activins and inhibins, affect early development of the genital tract.

Despite the complexity of cellular effects induced by various members of the TGFβ superfamily, the signaling pathway is basically a simple one. Once activated, receptors for these ligands directly phosphorylate and activate a particular type of transcription factor. The response of a given cell to this activated transcription factor depends on the constellation of other transcription factors it already contains.

TGFβ Is Formed by Cleavage of a Secreted Inactive Precursor

In humans TGFβ consists of three protein isoforms, TGFβ-1, TGFβ-2, and TGFβ-3, each encoded by a unique gene and expressed in both a tissue-specific and developmentally regulated fashion. Each TGFβ isoform is synthesized as part of a larger precursor that contains a pro-domain. This domain is cleaved from but remains noncovalently associated with the mature domain after the protein is secreted. Most secreted TGFβ is stored in the extracellular matrix as a latent, inactive complex containing the cleaved TGFβ precursor and a covalently bound TGFβ-binding protein called *Latent TGFβ Binding Protein,* or *LTBP.* Binding of LTBP by the matrix protein thrombospondin or by certain cell-surface **integrins** triggers a conformational change in LTBP that causes release of the mature, active dimeric TGFβ. Alternatively, digestion of the binding proteins by matrix metalloproteases can result in activation of TGFβ (Figure 14-1a).

The monomeric form of TGFβ growth factors contains 110–140 amino acids and has a compact structure with four antiparallel β strands and three conserved intramolecular

(a) Formation of mature, dimeric TGFβ

Secreted TGFβ precursor

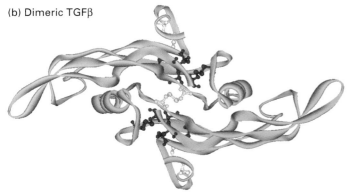

disulfide linkages (Figure 14-1b). These form a structure, called a *cystine knot,* that is relatively resistant to denaturation. An additional N-terminal cysteine in each monomer links TGFβ monomers into functional homodimers and heterodimers. Much of the sequence variation among different TGFβ proteins is observed in the N-terminal regions, the loops joining the β strands, and the α helices. Different heterodimeric combinations may increase the functional diversity of these proteins beyond that generated by differences in the primary sequence of the monomer.

◀ **FIGURE 14-1 Formation and structure of TGFβ superfamily of signaling molecules.** (a) TGFβ precursors are cleaved soon after being secreted. The pro-domain and mature domain are stored in the extracellular matrix in a complex that also contains latent TGFβ-binding protein (LTBP). The mature domain contains six conserved cysteine residues (yellow circles), which form three intrachain disulfide bonds and also a single disulfide bond connecting two monomers. Following proteolysis or a conformational change in LTBP, the active homo- or heterodimeric protein is released. (b) In this ribbon diagram of mature TGFβ dimer, the two subunits are shown in green and blue. Disulfide-linked cysteine residues are shown in ball-and-stick form. The three intrachain disulfide linkages (red) in each monomer form a cystine-knot domain, which is resistant to degradation. [Part (a) see J. Massagué and Y.-G. Chen, 2000, *Genes and Devel.* **14**:627; part (b) from S. Daopin et al., 1992, *Science* **257**:369.]

TGFβ Signaling Receptors Have Serine/Threonine Kinase Activity

To identify the cell-surface TGFβ receptors, investigators first reacted the purified growth factor with the radioisotope iodine-125 (^{125}I) under conditions such that the radioisotope covalently binds to exposed tyrosine residues. The ^{125}I-labeled TGFβ protein was incubated with cultured cells, and the incubation mixture then was treated with a chemical agent that covalently cross-linked the labeled TGFβ to its receptors on the cell surface. Purification of the labeled receptors revealed three different polypeptides with apparent molecular weights of 55, 85, and 280 kDa, referred to as *types RI, RII,* and *RIII* TGFβ receptors, respectively.

The most abundant TGFβ receptor, RIII, is a cell-surface **proteoglycan,** also called *β-glycan,* which binds and concentrates TGFβ near the cell surface. The type I and type II receptors are dimeric transmembrane proteins with serine/threonine kinases as part of their cytosolic domains. RII is a constitutively active kinase that phosphorylates itself in the absence of TGFβ. Binding of TGFβ induces the formation of complexes containing two copies each of RI and RII. An RII subunit then phosphorylates serine and threonine residues in a highly conserved sequence of the RI subunit adjacent to the cytosolic face of the plasma membrane, thereby activating the RI kinase activity.

Activated Type I TGFβ Receptors Phosphorylate Smad Transcription Factors

Researchers identified the transcription factors downstream from TGFβ receptors in *Drosophila* from genetic studies similar to those used to dissect receptor tyrosine kinase pathways (see Section 14.3). These transcription factors in *Drosophila* and the related vertebrate proteins are now called **Smads.** Three types of Smad proteins function in the TGFβ signaling pathway: receptor-regulated Smads (R-Smads), co-Smads, and inhibitory or antagonistic Smads (I-Smads).

As depicted in Figure 14-2, R-Smads contain two domains, MH1 and MH2, separated by a flexible linker region. The N-terminal MH1 domain contains the specific DNA-binding segment and also a sequence called the *nuclear-localization signal (NLS)* that is required for protein transport into the nucleus (Chapter 12). When R-Smads are in their inactive, nonphosphorylated state, the NLS is masked and the MH1 and MH2 domains associate in such a way that they cannot bind to DNA or to a co-Smad. Phosphorylation of three serine residues near the C-terminus of an R-Smad (Smad2 or Smad3) by activated type I TGFβ re-

ceptors separates the domains, permitting binding of importin β to the NLS. Simultaneously a complex containing two molecules of Smad3 (or Smad2) and one molecule of a co-Smad (Smad4) forms in the cytosol. This complex is stabilized by binding of two phosphorylated serines in each Smad3 to phosphoserine-binding sites in both the Smad3 and the Smad4 MH2 domains. The bound importin β then mediates translocation of the heteromeric R-Smad/co-Smad complexes into the nucleus. After importin β dissociates inside the nucleus, the Smad2/Smad4 or Smad3/Smad4 complexes cooperate with other transcription factors to activate transcription of specific target genes.

Within the nucleus R-Smads are continuously being dephosphorylated, which results in the dissociation of the R-Smad/co-Smad complex and export of these Smads from the nucleus. Because of this continuous nucleocytoplasmic shuttling of the Smads, the concentration of active Smads within the nucleus closely reflects the levels of activated TGFβ receptors on the cell surface.

Virtually all mammalian cells secrete at least one TGFβ isoform, and most have TGFβ receptors on their surface. However, because different types of cells contain different sets of transcription factors with which the activated Smads can bind, the cellular responses induced by TGFβ vary among cell types. In epithelial cells and fibroblasts, for example, TGFβ induces expression not only of extracellular-

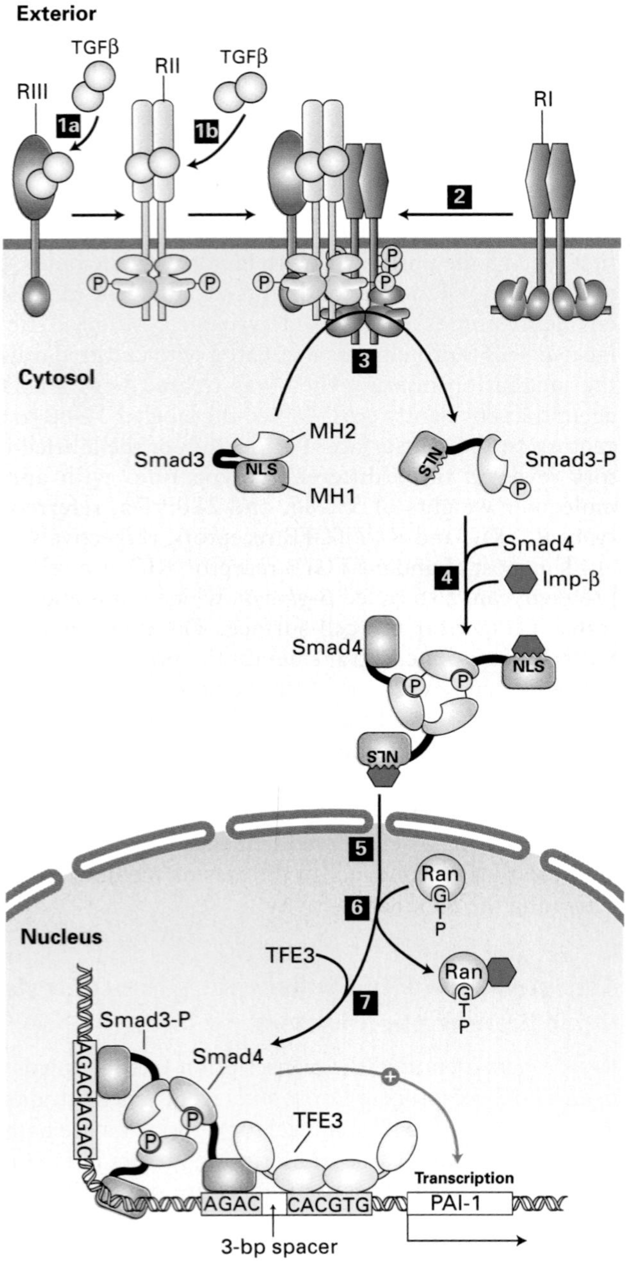

◀ **FIGURE 14-2 TGFβ-Smad signaling pathway.** Step **1a**: In some cells, TGFβ binds to the type III TGFβ receptor (RIII), which presents it to the type II receptor (RII). Step **1b**: In other cells, TGFβ binds directly to RII, a constitutively phosphorylated and active kinase. Step **2**: Ligand-bound RII recruits and phosphorylates the juxtamembrane segment of the type I receptor (RI), which does not directly bind TGFβ. This releases the inhibition of RI kinase activity that otherwise is imposed by the segment of RI between the membrane and kinase domain. Step **3**: Activated RI then phosphorylates Smad3 (shown here) or another R-Smad, causing a conformational change that unmasks its nuclear-localization signal (NLS). Step **4**: Two phosphorylated molecules of Smad3 interact with a co-Smad (Smad4), which is not phosphorylated, and with importin β (Imp-β), forming a large cytosolic complex. Steps **5** and **6**: After the entire complex translocates into the nucleus, Ran·GTP causes dissociation of Imp-β as discussed in Chapter 12. Step **7**: A nuclear transcription factor (e.g., TFE3) then associates with the Smad3/Smad4 complex, forming an activation complex that cooperatively binds in a precise geometry to regulatory sequences of a target gene. Shown at the bottom is the activation complex for the gene encoding plasminogen activator inhibitor (PAI-1). See the text for additional details. [See Z. Xiao et al., 2000, *J. Biol. Chem.* **275**:23425; J. Massagué and D. Wotton, 2000, *EMBO J.* **19**:1745; X. Hua et al., 1999, *Proc. Nat'l. Acad. Sci. USA* **96**:13130; and A. Moustakas and C.-H. Heldin, 2002, *Genes Devel.* **16**:1867.]

matrix proteins (e.g., collagens) but also of proteins that inhibit serum proteases, which otherwise would degrade the matrix. The latter category includes plasminogen activator inhibitor 1 (PAI-1). Transcription of the PAI-1 gene requires formation of a complex of the transcription factor TFE3 with the Smad3/Smad4 complex and binding of all these proteins to specific sequences within the regulatory region of the PAI-1 gene (see Figure 14-2, *bottom*). By partnering with other transcription factors, Smad2/Smad4 and Smad3/Smad4 complexes induce expression of proteins such as p15, which arrests the cell cycle at the G_1 stage and thus blocks cell proliferation (Chapter 21). These Smad complexes also repress transcription of the *myc* gene, thereby reducing expression of many growth-promoting genes whose transcription normally is activated by Myc.

The various growth factors in the TGFβ superfamily bind to their own receptors and activate different sets of Smad proteins, resulting in different cellular responses. The specificity exhibited by these related receptors is a common phenomenon in intercellular signaling, and the TGFβ signaling pathway provides an excellent example of one strategy for achieving such response specificity. As just discussed, for instance, binding of any one TGFβ isoform to its specific receptors leads to phosphorylation of Smad2 or Smad3, formation of Smad2/Smad4 or Smad3/Smad4 complexes, and eventually transcriptional activation of specific target genes (e.g., the PAI-1 gene). On the other hand, BMP proteins, which also belong to the TGFβ superfamily, bind to and activate a different set of receptors, leading to phosphorylation of Smad1, its dimerization with Smad4, and activation of specific transcriptional responses by Smad1/Smad4. These responses are distinct from those induced by Smad2/Smad4 or Smad3/Smad4.

Oncoproteins and I-Smads Regulate Smad Signaling via Negative Feedback Loops

Smad signaling is regulated by additional intracellular proteins, including two cytosolic proteins called *SnoN* and *Ski* (*Ski* stands for "Sloan-Kettering Cancer Institute"). These proteins were originally identified as **oncoproteins** because they cause abnormal cell proliferation when overexpressed in cultured fibroblasts. How they accomplish this was not understood until years later when SnoN and Ski were found to bind to the Smad2/Smad4 or Smad3/Smad4 complexes formed after TGFβ stimulation. SnoN and Ski do not affect the ability of the Smad complexes to bind to DNA control regions. Rather, they block transcription activation by the bound Smad complexes, thereby rendering cells resistant to the growth-inhibitory actions normally induced by TGFβ (Figure 14-3). Interestingly, stimulation by TGFβ causes the rapid degradation of Ski and SnoN, but after a few hours, expression of both Ski and SnoN becomes strongly induced. The increased levels of these proteins are thought to dampen long-term signaling effects due to continued exposure to TGFβ.

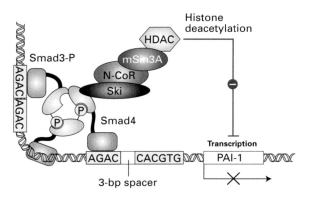

▲ **FIGURE 14-3 Schematic model of Ski-mediated down-regulation of the response to TGFβ stimulation.** Ski binds to Smad4 in Smad3/Smad4 or Smad2/Smad4 (not shown) signaling complexes and may partially disrupt interactions between the Smad proteins. Ski also recruits a protein termed *N-CoR* that binds directly to mSin3A, which in turn interacts with histone deacetylase (HDAC), an enzyme that promotes histone deacetylation (Chapter 11). As a result, transcription activation induced by TGFβ and mediated by Smad complexes is shut down. [See S. Stroschein et al., 1999, *Science* **286**:771; X. Liu et al., 2001, *Cytokine and Growth Factor Rev.* **12**:1; and J.-W. Wu et al., 2002, *Cell* **111**:357.]

Among the proteins induced after TGFβ stimulation are the I-Smads, especially Smad7. Smad7 blocks the ability of activated type I receptors to phosphorylate R-Smad proteins. In this way Smad7, like Ski and SnoN, participates in a negative feedback loop; its induction serves to inhibit intracellular signaling by long-term exposure to the stimulating hormone. In later sections we see how signaling by other cell-surface receptors is also controlled by negative feedback loops.

Loss of TGFβ Signaling Contributes to Abnormal Cell Proliferation and Malignancy

 Many human tumors contain inactivating mutations in either TGFβ receptors or Smad proteins, and thus are resistant to growth inhibition by TGFβ (see Figure 23-20). Most human pancreatic cancers, for instance, contain a deletion in the gene encoding Smad4 and thus cannot induce p15 and other cell-cycle inhibitors in response to TGFβ. This mutation-defined gene originally was called *DPC* (*d*eleted in *p*ancreatic *c*ancer). Retinoblastoma, colon and gastric cancer, hepatoma, and some T- and B-cell malignancies are also unresponsive to TGFβ growth inhibition. This loss of responsiveness correlates with loss of type I or type II TGFβ receptors; responsiveness to TGFβ can be restored by recombinant expression of the "missing" protein. Mutations in Smad2 also commonly occur in several types of human tumors. Not only is TGFβ signaling essential for controlling cell proliferation, as these examples show, but it also causes some cells to differentiate along specific pathways, as discussed in Chapter 15. ∎

KEY CONCEPTS OF SECTION 14.1

TGFβ Receptors and the Direct Activation of Smads

■ TGFβ is produced as an inactive precursor that is stored in the extracellular matrix. Several mechanisms can release the active, mature dimeric growth factor (see Figure 14-1).

■ Stimulation by TGFβ leads to activation of the intrinsic serine/threonine kinase activity in the cytosolic domain of the type I (RI) receptor, which then phosphorylates an R-Smad, exposing a nuclear-localization signal.

■ After phosphorylated R-Smad binds a co-Smad, the resulting complex translocates into the nucleus, where it interacts with various transcription factors to induce expression of target genes (see Figure 14-2).

■ Oncoproteins (e.g., Ski and SnoN) and I-Smads (e.g., Smad7) act as negative regulators of TGFβ signaling.

■ TGFβ signaling generally inhibits cell proliferation. Loss of various components of the signaling pathway contributes to abnormal cell proliferation and malignancy.

14.2 Cytokine Receptors and the JAK-STAT Pathway

We turn now to a second important class of cell-surface receptors, the **cytokine receptors,** whose cytosolic domains are closely associated with a member of a family of cytosolic protein tyrosine kinases, the *JAK kinases.* A third class of receptors, the **receptor tyrosine kinases (RTKs),** contain intrinsic protein tyrosine kinase activity in their cytosolic domains. The mechanisms by which cytokine receptors and receptor tyrosine kinases become activated by ligands are very similar, and there is considerable overlap in the intracellular signal-transduction pathways triggered by activation of receptors in both classes. In this section, we first describe some similarities in signaling from these two receptor classes. We then discuss the JAK-STAT pathway, which is initiated mainly by activation of cytokine receptors.

Cytokine Receptors and Receptor Tyrosine Kinases Share Many Signaling Features

Ligand binding to both cytokine receptors and RTKs triggers formation of functional dimeric receptors. In some cases, the ligand induces association of two monomeric

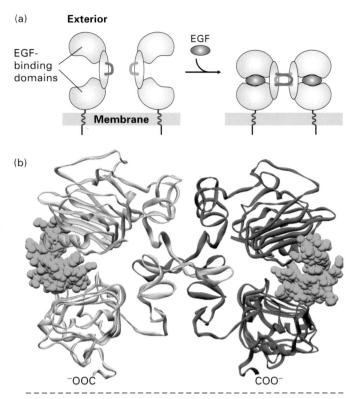

▲ **FIGURE 14-4 Dimerization of the receptor for epidermal growth factor (EGF), a receptor tyrosine kinase.** (a) Schematic depiction of the extracellular and transmembrane domains of the EGF receptor. Binding of one EGF molecule to a monomeric receptor causes an alteration in the structure of a loop located between the two EGF-binding domains. Dimerization of two identical ligand-bound receptor monomers in the plane of the membrane occurs primarily through interactions between the two "activated" loop segments. (b) Structure of the dimeric EGF receptor's extracellular domain bound to transforming growth factor α (TGFα), a homolog of EGF. The EGF receptor extracellular domains are shown in white *(left)* and blue *(right)*. The two smaller TGFα molecules are colored green. Note the interaction between the "activated" loop segments in the two receptors. [Part (a) adapted from J. Schlessinger, 2002, *Cell* **110**:669; part (b) from T. Garrett et al., 2002, *Cell* **110**:763.]

receptor subunits diffusing in the plane of the plasma membrane (Figure 14-4). In others, the receptor is a dimer in the absence of ligand, and ligand binding alters the conformation of the extracellular domains of the two subunits. In either case, formation of a functional dimeric receptor causes one of the poorly active cytosolic kinases to phosphorylate a particular tyrosine residue in the *activation lip* of the second kinase. This phosphorylation activates kinase activity and leads to phosphorylation of the second kinase in the dimer, as well as several tyrosine

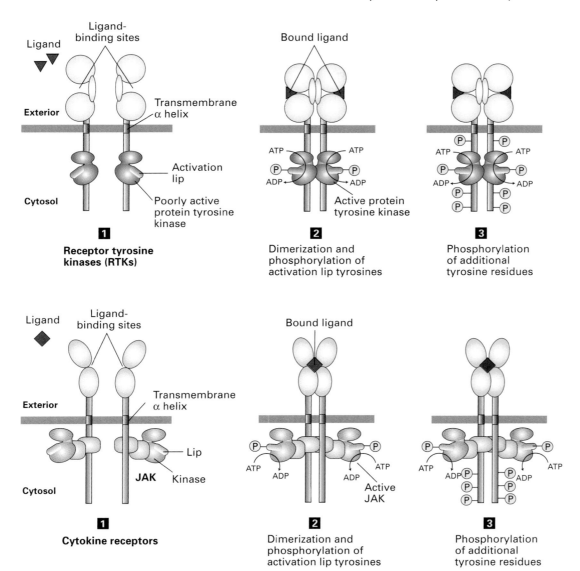

▲ **FIGURE 14-5 General structure and ligand-induced activation of receptor tyrosine kinases (RTKs) and cytokine receptors.** The cytosolic domain of RTKs contains a protein tyrosine kinase catalytic site, whereas the cytosolic domain of cytokine receptors associates with a separate JAK kinase (step **1**). In both types of receptor, ligand binding causes a conformational change that promotes formation of a functional dimeric receptor, bringing together two intrinsic or associated kinases, which then phosphorylate each other on a tyrosine residue in the activation lip (step **2**). Phosphorylation causes the lip to move out of the kinase catalytic site, thus allowing ATP or a protein substrate to bind. The activated kinase then phosphorylates other tyrosine residues in the receptor's cytosolic domain (step **3**). The resulting phosphotyrosines function as docking sites for various signal-transduction proteins (see Figure 14-6).

residues in the cytosolic domain of the receptor (Figure 14-5). As we see later, phosphorylation of residues in the activation loop is a general mechanism by which many kinases are activated.

Certain phosphotyrosine residues formed in activated cytokine receptors and RTKs serve as binding, or "docking," sites for *SH2 domains* or *PTB domains*, which are present in a large array of intracellular signal-transduction proteins.

Once they are bound to an activated receptor, some signal-transduction proteins are phosphorylated by the receptor's intrinsic or associated kinase to achieve their active form. Binding of other signal-transduction proteins, present in the cytosol in unstimulated cells, positions them near their substrates localized in the plasma membrane. Both mechanisms can trigger downstream signaling. Several cytokine receptors (e.g., the IL-4 receptor) and RTKs (e.g., the insulin receptor)

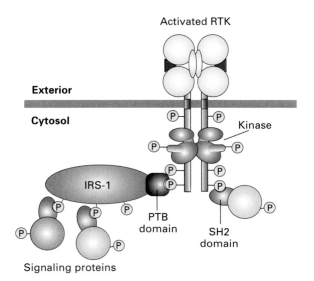

Activated RTK

Exterior

Cytosol

Kinase

IRS-1

PTB domain

SH2 domain

Signaling proteins

▲ **FIGURE 14-6 Recruitment of signal-transduction proteins to the cell membrane by binding to phosphotyrosine residues in activated receptors.** Cytosolic proteins with SH2 (purple) or PTB (maroon) domains can bind to specific phosphotyrosine residues in activated RTKs (shown here) or cytokine receptors. In some cases, these signal-transduction proteins then are phosphorylated by the receptor's intrinsic or associated protein tyrosine kinase, enhancing their activity. Certain RTKs and cytokine receptors utilize multidocking proteins such as IRS-1 to increase the number of signaling proteins that are recruited and activated. Subsequent phosphorylation of the IRS-1 by receptor kinase activity creates additional docking sites for SH2-containing signaling proteins.

▶ **FIGURE 14-7 Role of erythropoietin in formation of red blood cells (erythrocytes).** Erythroid progenitor cells, called *colony-forming units erythroid (CFU-E)*, are derived from hematopoietic stem cells, which also give rise to progenitors of other blood cell types. In the absence of erythropoietin (Epo), CFU-E cells undergo apoptosis. Binding of erythropoietin to its receptors on a CFU-E induces transcription of several genes whose encoded proteins prevent programmed cell death (apoptosis), allowing the cell to survive and undergo a program of three to five terminal cell divisions. Epo stimulation also induces expression of erythrocyte-specific proteins such as the globins, which form hemoglobin, and the membrane proteins glycophorin and anion-exchange protein. The Epo receptor and other membrane proteins are lost from these cells as they undergo differentiation. If CFU-E cells are cultured with erythropoietin in a semisolid medium (e.g., containing methylcellulose), daughter cells cannot move away, and thus each CFU-E produces a colony of 30–100 erythroid cells, hence its name. [See M. Socolovsky et al., 2001, *Blood* **98**:3261.]

bind IRS1 or other multidocking proteins via a PTB domain in the docking protein (Figure 14-6). The activated receptor then phosphorylates the bound docking protein, forming many phosphotyrosines that in turn serve as docking sites for SH2-containing signaling proteins. Some of these proteins in turn may also be phosphorylated by the activated receptor.

Cytokines Influence Development of Many Cell Types

The **cytokines** form a family of relatively small, secreted proteins (generally containing about 160 amino acids) that control many aspects of growth and differentiation of specific types of cells. During pregnancy prolactin, for example, induces epithelial cells lining the immature ductules of the mammary gland to differentiate into the acinar cells that produce milk proteins and secrete them into the ducts. Another cytokine, interleukin 2 (IL-2), is essential for proliferation and functioning of the T cells of the immune system; its close relative IL-4 is essential for formation of functional antibody-producing B cells. Some cytokines, such as interferon α, are produced and secreted by many types of cells fol-

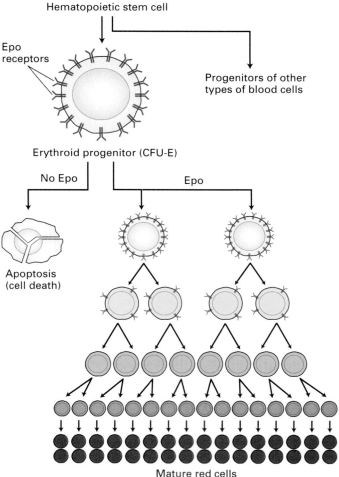

Hematopoietic stem cell

Epo receptors

Progenitors of other types of blood cells

Erythroid progenitor (CFU-E)

No Epo

Epo

Apoptosis (cell death)

Mature red cells

lowing virus infection. The secreted interferons act on nearby cells to induce enzymes that render these cells more resistant to virus infection.

 Many cytokines induce formation of important types of blood cells. For instance, granulocyte colony stimulating factor (G-CSF) induces a particular type of progenitor cell in the bone marrow to divide several times and then differentiate into granulocytes, the type of white blood cell that inactivates bacteria and other pathogens. Because many cancer therapies reduce granulocyte formation by the body, G-CSF often is administered to patients to stimulate proliferation and differentiation of granulocyte progenitor cells, thus restoring the normal level of granulocytes in the blood. Thrombopoietin, a "cousin" of G-CSF, similarly acts on megakaryocyte progenitors to divide and differentiate into megakaryocytes. These then fragment into the cell pieces called *platelets*, which are critical for blood clotting. ∎

Another related cytokine, **erythropoietin (Epo)**, triggers production of red blood cells by inducing the proliferation and differentiation of erythroid progenitor cells in the bone marrow (Figure 14-7). Erythropoietin is synthesized by kidney cells that monitor the concentration of oxygen in the blood. A drop in blood oxygen signifies a lower than optimal level of erythrocytes (red blood cells), whose major function is to transport oxygen complexed to hemoglobin. By means of the oxygen-sensitive transcription factor HIF-1α, the kidney cells respond to low oxygen by synthesizing more erythropoietin and secreting it into the blood (see Figure 15-9). As the level of erythropoietin rises, more and more erythroid progenitors are saved from death, allowing each to produce ≈50 or so red blood cells in a period of only two days. In this way, the body can respond to the loss of blood by accelerating the production of red blood cells.

All Cytokines and Their Receptors Have Similar Structures and Activate Similar Signaling Pathways

Strikingly, all cytokines have a similar tertiary structure, consisting of four long conserved α helices folded together in a specific orientation. Similarly, the structures of all cytokine receptors are quite similar, with their extracellular domains constructed of two subdomains, each of which contains seven conserved β strands folded together in a characteristic fashion. The interaction of erythropoietin with the dimeric erythropoietin receptor (EpoR), depicted in Figure 14-8, exemplifies the binding of a cytokine to its receptor. The structural homology among cytokines is evidence that they all evolved from a common ancestral protein. Likewise, the various receptors undoubtedly evolved from a single common ancestor.

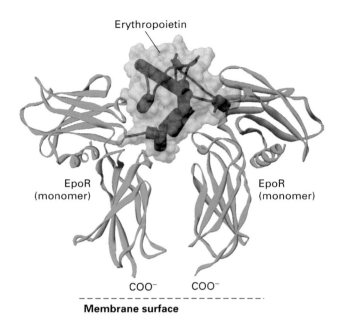

▲ **FIGURE 14-8 Structure of erythropoietin bound to the extracellular domains of a dimeric erythropoietin receptor (EpoR).** Erythropoietin contains four conserved long α helices that are folded in a particular arrangement. The extracellular domain of an EpoR monomer is constructed of two subdomains, each of which contains seven conserved β strands folded in a characteristic fashion. Side chains of residues on two of the helices in erythropoietin contact loops on one EpoR monomer, while residues on the two other Epo helices bind to the same loop segments in a second receptor monomer, thereby stabilizing the dimeric receptor. The structures of other cytokines and their receptors are similar to erythropoietin and EpoR. [Adapted from R. S. Syed et al., 1998, *Nature* **395**:511.]

Whether or not a cell responds to a particular cytokine depends simply on whether or not it expresses the corresponding (cognate) receptor. Although all cytokine receptors activate similar intracellular signaling pathways, the response of any particular cell to a cytokine signal depends on the cell's constellation of transcription factors, chromatin structures, and other proteins relating to the developmental history of the cell. If receptors for prolactin or thrombopoietin, for example, are expressed experimentally in an erythroid progenitor cell, the cell will respond to these cytokines by dividing and differentiating into red blood cells, not into mammary cells or megakaryocytes.

Figure 14-9 summarizes the intracellular signaling pathways activated when the EpoR binds erythropoietin. Stimulation of other cytokine receptors by their specific ligands activates similar pathways. All these pathways eventually lead to activation of transcription factors, causing an increase or decrease in expression of particular target genes. Here we focus on the *JAK-STAT pathway;* the other pathways are discussed in later sections.

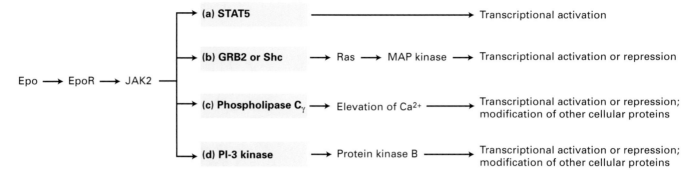

▲ **FIGURE 14-9 Overview of signal-transduction pathways triggered by ligand binding to the erythropoietin receptor (EpoR), a typical cytokine receptor.** Four major pathways can transduce a signal from the activated, phosphorylated EpoR-JAK complex (see Figure 14-5, *bottom*). Each pathway ultimately regulates transcription of different sets of genes. (a) In the most direct pathway, the transcription factor STAT5 is phosphorylated and activated directly in the cytosol. (b) Binding of linker proteins (GRB2 or Shc) to an activated EpoR leads to activation of the Ras–MAP kinase pathway. (c, d) Two phosphoinositide pathways are triggered by recruitment of phospholipase C_γ and PI-3 kinase to the membrane following activation of EpoR. Elevated levels of Ca^{2+} and activated protein kinase B also modulate the activity of cytosolic proteins that are not involved in control of transcription.

Somatic Cell Genetics Revealed JAKs and STATs as Essential Signal-Transduction Proteins

Soon after the discovery and cloning of cytokines, most of their receptors were isolated by expression cloning or other strategies. Elucidation of the essential components of their intracellular signaling pathways, however, awaited develop-ment of new types of genetic approaches using cultured mammalian cells. In these studies, a bacterial **reporter gene** encoding guanine phosphoribosyl transferase (GPRT) was linked to an upstream interferon-responsive promoter. The resulting construct was introduced into cultured mammalian cells that were genetically deficient in the human homolog HGPRT. GPRT or HGPRT is necessary for incorporation of purines

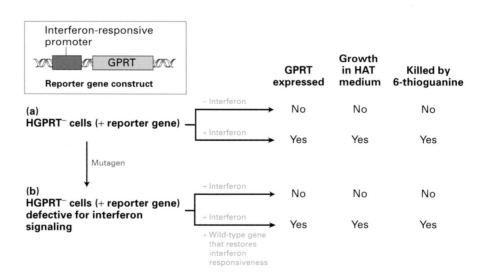

▲ **EXPERIMENTAL FIGURE 14-10 Mutagenized cells carrying an interferon-responsive reporter gene were used to identify JAKs and STATs as essential signal-transduction proteins.** A reporter gene was constructed consisting of an interferon-responsive promoter upstream of the bacterial gene encoding GPRT, a key enzyme in the purine salvage pathway (see Figure 6-39). (a) Introduction of this construct into mammalian cells lacking the mammalian homolog HGPRT yielded reporter cells that grew in HAT medium and were killed by 6-thioguanine in the presence but not the absence of interferon. (b) Following treatment of reporter cells with a mutagen, cells with defects in the signaling pathway initiated by interferon do not induce GPRT in response to interferon and thus cannot incorporate the toxic purine 6-thioguanine. Restoration of interferon responsiveness by functional complementation with wild-type DNA clones identified genes encoding JAKs and STATs. See the text for details. [See R. McKendry et al., 1991, *Proc. Nat'l. Acad. Sci. USA* **88**:11455; D. Watling et al., 1993, *Nature* **366**:166; and G. Stark and A. Gudkov, 1999, *Human Mol. Genet.* **8**:1925.]

in the culture medium into ribonucleotides and then into DNA or RNA. As shown in Figure 14-10a, HGPRT-negative cells carrying the reporter gene responded to interferon treatment by expressing GPRT and thus acquiring the ability to grow in HAT medium. This medium does not allow growth of cells lacking GPRT or HGPRT, since synthesis of purines by the cells is blocked by aminopterin (the A in HAT), and thus DNA synthesis is dependent on incorporation of purines from the culture medium (see Figure 6-39). Simultaneously the cells acquired sensitivity to killing by the purine analog 6-thioguanine, which is converted into the corresponding ribonucleotide by GPRT; incorporation of this purine into DNA in place of guanosine eventually causes cell death.

The reporter cells were then heavily treated with mutagens in an attempt to inactivate both alleles of the genes encoding critical signal-transduction proteins in the interferon signaling pathway. Researchers looked for mutant cells that expressed the interferon receptor (as evidenced by the cell's ability to bind radioactive interferon) but did not express GPRT in response to interferon and thus survived killing by 6-thioguanine when cells were cultured in the presence of interferon (Figure 14-10b). After many such interferon-nonresponding mutant cell lines were obtained, they were used to screen a genomic or cDNA library for the wild-type genes that complemented the mutated genes in nonresponding cells, a technique called **functional complementation** (see Figure 9-20). In this case, mutant cells expressing the corresponding recombinant wild-type gene grew on HAT medium and were sensitive to killing by 6-thioguanine in the presence of interferon. That is, they acted like wild-type cells.

Cloning of the genes identified by this procedure led to recognition of two key signal-transduction proteins: a JAK tyrosine kinase and a STAT transcription factor. Subsequent work showed that one (sometimes two) of the four human JAK proteins and at least one of several STAT proteins are involved in signaling downstream from all cytokine receptors. To understand how JAK and STAT proteins function, we examine one of the best-understood cytokine signaling pathways, that downstream of the erythropoietin receptor.

Receptor-Associated JAK Kinases Activate STAT Transcription Factors Bound to a Cytokine Receptor

The JAK2 kinase is tightly bound to the cytosolic domain of the erythropoietin receptor (EpoR). Like the three other members of the JAK family of kinases, JAK2 contains an N-terminal receptor-binding domain, a C-terminal kinase domain that is normally poorly active catalytically, and a middle domain of unknown function. JAK2, erythropoietin, and the EpoR are all required for formation of adult-type erythrocytes, which normally begins at day 12 of embryonic development in mice. As Figure 14-11 shows, embryonic mice lacking functional genes encoding either the EpoR or JAK2 cannot form adult-type erythrocytes and eventually die owing to the inability to transport oxygen to the fetal organs.

As already noted, erythropoietin binds simultaneously to the extracellular domains of two EpoR monomers on the cell surface (see Figure 14-8). As a result, the associated JAKs are brought close enough together that one can phosphorylate the other on a critical tyrosine in the activation lip. As with other kinases, phosphorylation of the activation lip leads to a conformational change that reduces the K_m for ATP or the substrate to be phosphorylated, thus increasing the kinase activity. One piece of evidence for this activation mechanism comes from study of a mutant JAK2 in which the critical tyrosine is mutated to phenylalanine. The mutant JAK2 binds normally to the EpoR but cannot be phosphorylated.

EpoR

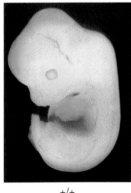 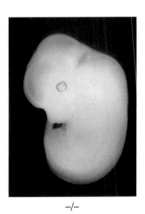

+/+   −/−

JAK2

 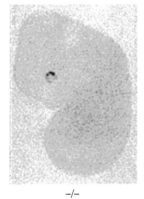

+/+   −/−

▲ **EXPERIMENTAL FIGURE 14-11 Studies with mutant mice reveal that both the erythropoietin receptor (EpoR) and JAK2 are essential for development of erythrocytes.** Mice in which both alleles of the EpoR or JAK2 gene are "knocked out" develop normally until embryonic day 13, at which time they begin to die of anemia due to the lack of erythrocyte-mediated transport of oxygen to the fetal organs. The red organ in the wild-type embryos (+/+) is the fetal liver, the major site of erythrocyte production at this developmental stage. The absence of color in the mutant embryos (−/−) indicates the absence of erythrocytes containing hemoglobin. Otherwise the mutant embryos appear normal, indicating that the main function of the EpoR and JAK2 in early mouse development is to support production of erythrocytes. [EpoR images from H. Wu et al., 1995, *Cell* **83**:59; JAK2 images from H. Neubauer et al., 1998, *Cell* **93**:307.]

Expression of this mutant JAK2 in erythroid cells in greater than normal amounts totally blocks EpoR signaling, as the mutant JAK2 blocks the function of the wild-type protein. This type of mutation, referred to as a *dominant negative,* causes loss of function even in cells that carry copies of the wild-type gene (Chapter 9).

Once the JAK kinases become activated, they phosphorylate several tyrosine residues on the cytosolic domain of the receptor. Certain of these phosphotyrosine residues then serve as binding sites for a group of transcription factors collectively termed *STATs.* All STAT proteins contain an N-terminal SH2 domain that binds to a phosphotyrosine in the receptor's cytosolic domain, a central DNA-binding domain, and a C-terminal domain with a critical tyrosine residue. Once a STAT is bound to the receptor, the C-terminal tyro-

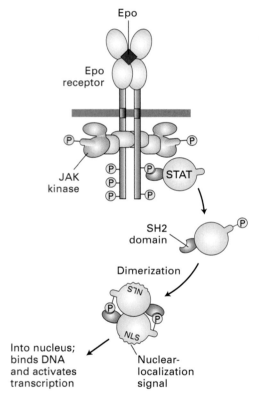

▲ FIGURE 14-12 JAK-STAT signaling pathway. Following ligand binding to a cytokine receptor and activation of an associated JAK kinase, JAK phosphorylates several tyrosine residues on the receptor's cytosolic domain (see Figure 14-5, *bottom*). After an inactive monomeric STAT transcription factor binds to a phosphotyrosine in the receptor, it is phosphorylated by active JAK. Phosphorylated STATs spontaneously dissociate from the receptor and spontaneously dimerize. Because the STAT homodimer has two phosphotyrosine–SH2 domain interactions, whereas the receptor-STAT complex is stabilized by only one such interaction, phosphorylated STATs tend not to rebind to the receptor. The STAT dimer, which has two exposed nuclear-localization signals (NLS), moves into the nucleus, where it can bind to promoter sequences and activate transcription of target genes.

sine is phosphorylated by an associated JAK kinase (Figure 14-12). This arrangement ensures that in a particular cell only those STAT proteins with an SH2 domain that can bind to a particular receptor protein will be activated. A phosphorylated STAT dissociates spontaneously from the receptor, and two phosphorylated STAT proteins form a dimer in which the SH2 domain on each binds to the phosphotyrosine in the other. Because dimerization exposes the nuclear-localization signal (NLS), STAT dimers move into the nucleus, where they bind to specific **enhancer** sequences controlling target genes.

Different STATs activate different genes in different cells. In erythroid progenitors, for instance, stimulation by erythropoietin leads to activation of STAT5. The major protein induced by active STAT5 is Bcl-x_L, which prevents the programmed cell death, or **apoptosis,** of these progenitors, allowing them to proliferate and differentiate into erythroid cells (see Figure 14-7). Indeed, mice lacking STAT5 are highly anemic because many of the erythroid progenitors undergo apoptosis even in the presence of high erythropoietin levels. Such mutant mice produce *some* erythrocytes and thus survive, because the erythropoietin receptor is linked to other anti-apoptotic pathways that do not involve STAT proteins (see Figure 14-9).

SH2 and PTB Domains Bind to Specific Sequences Surrounding Phosphotyrosine Residues

As noted earlier, many intracellular signal-transduction proteins contain an SH2 or PTB domain by which they bind to an activated receptor or other component of a signaling pathway containing a phosphotyrosine residue (see Figure 14-6). The SH2 domain derived its full name, the *Src homology 2 domain,* from its homology with a region in the prototypical cytosolic tyrosine kinase encoded by the *src* gene. The three-dimensional structures of SH2 domains in different proteins are very similar, but each binds to a distinct sequence of amino acids surrounding a phosphotyrosine residue. The unique amino acid sequence of each SH2 domain determines the specific phosphotyrosine residues it binds.

The SH2 domain of the Src tyrosine kinase, for example, binds strongly to any peptide containing a critical four-residue core sequence: phosphotyrosine–glutamic acid–glutamic acid–isoleucine (Figure 14-13). These four amino acids make intimate contact with the peptide-binding site in the Src SH2 domain. Binding resembles the insertion of a two-pronged "plug"—the phosphotyrosine and isoleucine side chains of the peptide—into a two-pronged "socket" in the SH2 domain. The two glutamic acids fit snugly onto the surface of the SH2 domain between the phosphotyrosine socket and the hydrophobic socket that accepts the isoleucine residue.

Variations in the hydrophobic socket in the SH2 domains of different STATs and other signal-transduction proteins allow them to bind to phosphotyrosines adjacent to different sequences, accounting for differences in their binding partners.

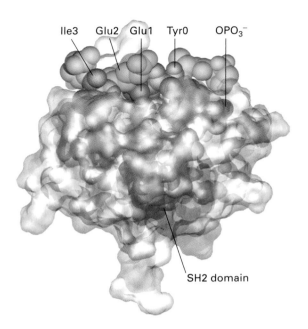

Ile3 Glu2 Glu1 Tyr0 OPO$_3^-$

SH2 domain

▲ **FIGURE 14-13 Surface model of the SH2 domain from Src kinase bound to a phosphotyrosine-containing peptide.** The peptide bound by this SH2 domain (gray) is shown in spacefill. The phosphotyrosine (Tyr0 and OPO$_3^-$, orange) and isoleucine (Ile3, orange) residues fit into a two-pronged socket on the surface of the SH2 domain; the two glutamate residues (Glu1, dark blue; Glu2, light blue) are bound to sites on the surface of the SH2 domain between the two sockets. Nonbinding residues on the target peptide are colored green. [See G. Waksman et al., 1993, *Cell* **72**:779.]

The binding specificity of SH2 domains is largely determined by residues C-terminal to the phosphotyrosine in a target peptide. In contrast, the binding specificity of PTB domains is determined by specific residues five to eight residues N-terminal to a phosphotyrosine residue. Sometimes a PTB domain binds to a target peptide even if the tyrosine is not phosphorylated.

Signaling from Cytokine Receptors Is Modulated by Negative Signals

Signal-induced transcription of target genes for too long a period can be as dangerous for the cell as too little induction. Thus cells must be able to turn off a signaling pathway quickly unless the extracellular signal remains continuously present. In various progenitor cells, two classes of proteins serve to dampen signaling from cytokine receptors, one over the short term (minutes) and the other over longer periods of time.

Short-Term Regulation by SHP1 Phosphatase Mutant mice lacking *SHP1 phosphatase* die because of excess production of erythrocytes and several other types of blood cells. Analysis of these mutant mice offered the first suggestion that SHP1, a phosphotyrosine **phosphatase**, negatively regulates signaling from several types of cytokine receptors in several types of progenitor cells.

How SHP1 dampens cytokine signaling is depicted in Figure 14-14a. In addition to a phosphatase catalytic domain, SHP1 has two SH2 domains. When cells are not stimulated

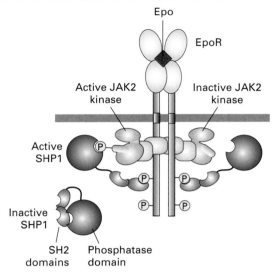

(a) JAK2 deactivation induced by SHP1 phosphatase

Epo

EpoR

Active JAK2 kinase

Inactive JAK2 kinase

Active SHP1

Inactive SHP1

SH2 domains

Phosphatase domain

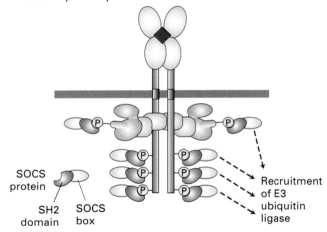

(b) Signal blocking and protein degradation induced by SOCS proteins

SOCS protein

SH2 domain

SOCS box

Recruitment of E3 ubiquitin ligase

▲ **FIGURE 14-14 Two mechanisms for terminating signal transduction from the erythropoietin receptor (EpoR).** (a) SHP1, a protein tyrosine phosphatase, is present in an inactive form in unstimulated cells. Binding of an SH2 domain in SHP1 to a particular phosphotyrosine in the activated receptor unmasks its phosphatase catalytic site and positions it near the phosphorylated tyrosine in the lip region of JAK2. Removal of the phosphate from this tyrosine inactivates the JAK kinase. (b) SOCS proteins, whose expression is induced in erythropoietin-stimulated erythroid cells, inhibit or permanently terminate signaling over longer time periods. Binding of SOCS to phosphotyrosine residues on the EpoR or JAK2 blocks binding of other signaling proteins *(left)*. The SOCS box can also target proteins such as JAK2 for degradation by the ubiquitin-proteasome pathway *(right)*. Similar mechanisms regulate signaling from other cytokine receptors. [Part (a) adapted from S. Constantinescu et al., 1999, *Trends Endocrin. Metabol.* **10**:18; part (b) adapted from B. T. Kile and W. S. Alexander, 2001, *Cell. Mol. Life Sci.* **58**:1.]

by a cytokine (are in the resting state), one of the SH2 domains physically binds to and inactivates the catalytic site in SHP1. In the stimulated state, however, this blocking SH2 domain binds to a specific phosphotyrosine residue in the activated receptor. The conformational change that accompanies this binding unmasks the SHP1 catalytic site and also brings it adjacent to the phosphotyrosine residue in the activation lip of the JAK associated with the receptor. By removing this phosphate, SHP1 inactivates the JAK, so that it can no longer phosphorylate the receptor or other substrates (e.g., STATs) unless additional cytokine molecules bind to cell-surface receptors, initiating a new round of signaling.

Long-Term Regulation by SOCS Proteins Among the genes whose transcription is induced by STAT proteins are those encoding a class of small proteins, termed *SOCS proteins,* that terminate signaling from cytokine receptors. These negative regulators, also known as *CIS proteins,* act in two ways (Figure 14-14b). First, the SH2 domain in several SOCS proteins binds to phosphotyrosines on an activated receptor, preventing binding of other SH2-containing signaling proteins (e.g., STATs) and thus inhibiting receptor signaling. One SOCS protein, SOCS-1, also binds to the critical phosphotyrosine in the activation lip of activated JAK2 kinase, thereby inhibiting its catalytic activity. Second, all SOCS proteins contain a domain, called the *SOCS box,* that recruits components of E3 ubiquitin ligases (see Figure 3-13). As a result of binding SOCS-1, for instance, JAK2 becomes polyubiquitinated and then degraded in **proteasomes,** thus permanently turning off all JAK2-mediated signaling pathways. The observation that proteasome inhibitors prolong JAK2 signal transduction supports this mechanism.

Studies with cultured mammalian cells have shown that the receptor for growth hormone, which belongs to the cytokine receptor superfamily, is down-regulated by another SOCS protein, SOCS-2. Strikingly, mice deficient in this SOCS protein grow significantly larger than their wild-type counterparts and have long bone lengths and proportionate enlargement of most organs. Thus SOCS proteins play an essential negative role in regulating intracellular signaling from the receptors for erythropoietin, growth hormone, and other cytokines.

Mutant Erythropoietin Receptor That Cannot Be Down-Regulated Leads to Increased Hematocrit

In normal adult men and women, the percentage of erythrocytes in the blood (the *hematocrit*) is maintained very close to 45–47 percent. A drop in the hematocrit results in increased production of erythropoietin by the kidney. The elevated erythropoietin level causes more erythroid progenitors to undergo terminal proliferation and differentiation into mature erythrocytes, soon restoring the hematocrit to its normal level. In endurance sports, such as cross-country skiing, where oxygen transport to the muscles may become limiting,

an excess of red blood cells may confer a competitive advantage. For this reason, use of supplemental erythropoietin to increase the hematocrit above the normal level is banned in many athletic competitions, and athletes are regularly tested for the presence of commercial recombinant erythropoietin in their blood and urine.

 Supplemental erythropoietin not only confers a possible competitive advantage but also can be dangerous. Too many red cells can cause the blood to become sluggish and clot in small blood vessels, especially in the brain. Several athletes who doped themselves with erythropoietin have died of a stroke while exercising.

Discovery of a mutant, unregulated erythropoietin receptor (EpoR) explained a suspicious situation in which a winner of three gold medals in Olympic cross-country skiing was found to have a hematocrit above 60 percent. Testing for erythropoietin in his blood and urine, however, revealed lower-than-normal amounts. Subsequent DNA analysis showed that the athlete was heterozygous for a mutation in the gene encoding the erythropoietin receptor. The mutant allele encoded a truncated receptor missing several of the tyrosines that normally become phosphorylated after stimulation by erythropoietin. As a consequence, the mutant receptor was able to activate STAT5 and other signaling proteins normally, but was unable to bind the negatively acting SHP1 phosphatase, which usually terminates signaling (see Figure 14-14a). Thus the very low level of erythropoietin produced by this athlete induced prolonged intracellular signaling in his erythroid progenitor cells, resulting in production of higher-than-normal numbers of erythrocytes. This example vividly illustrates the fine level of control over signaling from the erythropoietin receptor in the human body. ∎

KEY CONCEPTS OF SECTION 14.2

Cytokine Receptors and the JAK-STAT Pathway

■ Two receptor classes, cytokine receptors and receptor tyrosine kinases, transduce signals via their associated or intrinsic protein tyrosine kinases. Ligand binding triggers formation of functional dimeric receptors and phosphorylation of the activation lip in the kinases, enhancing their catalytic activity (see Figure 14-5).

■ All cytokines are constructed of four α helices that are folded in a characteristic arrangement.

■ Erythropoietin, a cytokine secreted by kidney cells, prevents apoptosis and promotes proliferation and differentiation of erythroid progenitor cells in the bone marrow. An excess of erythropoietin or mutations in its receptor that prevent down-regulation result in production of elevated numbers of red blood cells.

■ All cytokine receptors are closely associated with a JAK protein tyrosine kinase, which can activate several downstream signaling pathways leading to changes in transcription of target genes or in the activity of proteins that do not regulate transcription (see Figure 14-9).

■ The JAK-STAT pathway operates downstream of all cytokine receptors. STAT monomers bound to receptors are phosphorylated by receptor-associated JAKs, then dimerize and move to the nucleus, where they activate transcription (see Figure 14-12).

■ Short peptide sequences containing phosphotyrosine residues are bound by SH2 and PTB domains, which are found in many signal-transducing proteins. Such protein-protein interactions are important in many signaling pathways.

■ Signaling from cytokine receptors is terminated by the phosphotyrosine phosphatase SHP1 and several SOCS proteins (see Figure 14-14).

14.3 Receptor Tyrosine Kinases and Activation of Ras

We return now to the receptor tyrosine kinases (RTKs), which have intrinsic protein tyrosine kinase activity in their cytosolic domains. The ligands for RTKs are soluble or membrane-bound peptide or protein hormones including nerve growth factor (NGF), platelet-derived growth factor (PDGF), fibroblast growth factor (FGF), epidermal growth factor (EGF), and insulin. Ligand-induced activation of an RTK stimulates its tyrosine kinase activity, which subsequently stimulates the *Ras–MAP kinase pathway* and several other signal-transduction pathways. RTK signaling pathways have a wide spectrum of functions including regulation of cell proliferation and differentiation, promotion of cell survival, and modulation of cellular metabolism.

 Some RTKs have been identified in studies on human cancers associated with mutant forms of growth-factor receptors, which send a proliferative signal to cells even in the absence of growth factor. For example, a constitutively active mutant form of Her2, a receptor for EGF-like proteins, enables uncontrolled proliferation of cancer cells even in the absence of EGF, which is required for proliferation of normal cells (see Figure 23-14). Alternatively, overproduction of the wild-type receptor for EGF in certain human breast cancers results in proliferation at low EGF levels that do not stimulate normal cells; monoclonal antibodies targeted to the EGF receptor have proved therapeutically useful in these patients. Other RTKs have been uncovered during analysis of developmental mutations that lead to blocks in differentiation of certain cell types in *C. elegans, Drosophila,* and the mouse. ∎

Here we discuss how ligand binding leads to activation of RTKs and how activated receptors transmit a signal to the

Ras protein, the GTPase switch protein that functions in transducing signals from many different RTKs. The transduction of signals downstream from Ras to a common cascade of serine/threonine kinases, leading ultimately to activation of MAP kinase and certain transcription factors, is covered in the following section.

Ligand Binding Leads to Transphosphorylation of Receptor Tyrosine Kinases

All RTKs constitute an extracellular domain containing a ligand-binding site, a single hydrophobic transmembrane α helix, and a cytosolic domain that includes a region with protein tyrosine kinase activity. Most RTKs are monomeric, and ligand binding to the extracellular domain induces formation of receptor dimers, as depicted in Figure 14-4 for the EGF receptor. Some monomeric ligands, including FGF, bind tightly to heparan sulfate, a negatively charged polysaccharide component of the extracellular matrix (Chapter 6); this association enhances ligand binding to the monomeric receptor and formation of a dimeric receptor-ligand complex (Figure 14-15). The ligands for some RTKs are dimeric; their binding brings two receptor monomers together directly. Yet other RTKs, such as the insulin receptor, form disulfide-linked dimers in the absence of hormone; binding of ligand to this type of RTK alters its conformation in such a way that the receptor becomes activated.

Regardless of the mechanism by which ligand binds and locks an RTK into a functional dimeric state, the next step is universal. In the resting, unstimulated state, the intrinsic kinase activity of an RTK is very low. In the dimeric receptor, however, the kinase in one subunit can phosphorylate one or more tyrosine residues in the activation lip near the catalytic site in the other subunit. This leads to a conformational change that facilitates binding of ATP in some receptors (e.g., insulin receptor) and binding of protein substrates in other receptors (e.g., FGF receptor). The resulting enhanced kinase activity then phosphorylates other sites in the cytosolic domain of the receptor. This ligand-induced activation of RTK kinase activity is analogous to the activation of the JAK kinases associated with cytokine receptors (see Figure 14-5). The difference resides in the location of the kinase catalytic site, which is within the cytosolic domain of RTKs, but within a separate JAK kinase in the case of cytokine receptors.

As in signaling by cytokine receptors, phosphotyrosine residues in activated RTKs serve as docking sites for proteins involved in downstream signal transduction. Many phosphotyrosine residues in activated RTKs interact with *adapter proteins,* small proteins that contain SH2, PTB, or SH3 domains but have no intrinsic enzymatic or signaling activities (see Figure 14-6). These proteins couple activated RTKs to other components of signal-transduction pathways such as the one involving Ras activation.

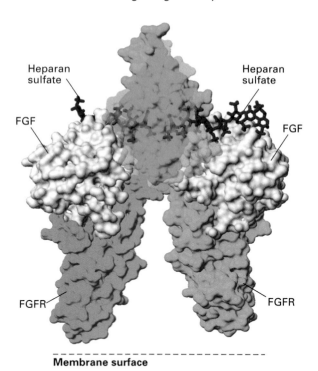

Membrane surface

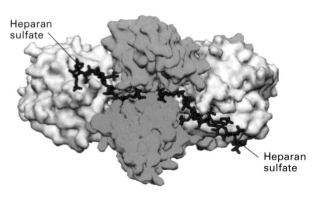

▲ **FIGURE 14-15 Structure of the dimerized ligand-bound receptor for fibroblast growth factor (FGF), which is stabilized by heparan sulfate.** Shown here are side and top views of the complex comprising the extracellular domains of two FGF receptor (FGFR) monomers (green and blue), two bound FGF molecules (white), and two short heparan sulfate chains (purple), which bind tightly to FGF. In the side view, the upper domain of one receptor (blue) is situated behind that of the other (green). In the top view, the heparan sulfate chains thread between and make numerous contacts with the upper domains of both receptor monomers. These interactions promote binding of the ligand to the receptor and receptor dimerization. [Adapted from J. Schlessinger et al., 2000, *Mol. Cell* **6**:743.]

Ras, a GTPase Switch Protein, Cycles Between Active and Inactive States

Ras is a monomeric GTP-binding switch protein that, like the G_α subunits in trimeric G proteins, alternates between an active *on* state with a bound GTP and an inactive *off* state

with a bound GDP. As discussed in Chapter 13, trimeric G proteins are directly linked to cell-surface receptors and transduce signals, via the G_α subunit, to various effectors such as adenylyl cyclase. In contrast, Ras is not directly linked to cell-surface receptors.

Ras activation is accelerated by a *guanine nucleotide–exchange factor (GEF)*, which binds to the Ras·GDP complex, causing dissociation of the bound GDP (see Figure 3-29). Because GTP is present in cells at a higher concentration than GDP, GTP binds spontaneously to "empty" Ras molecules, with release of GEF and formation of the active Ras·GTP. Subsequent hydrolysis of the bound GTP to GDP deactivates Ras. Unlike the deactivation of G_α·GTP, deactivation of Ras·GTP requires the assistance of another protein, a *GTPase-activating protein (GAP)* that binds to Ras·GTP and accelerates its intrinsic GTPase activity by more than a hundredfold. Thus the average lifetime of a GTP bound to Ras is about 1 minute, which is much longer than the average lifetime of G_α·GTP. In cells, GAP binds to specific phosphotyrosines in activated RTKs, bringing it close enough to membrane-bound Ras·GTP to exert its accelerating effect on GTP hydrolysis. The actual hydrolysis of GTP is catalyzed by amino acids from both Ras and GAP. In particular, insertion of an arginine side chain on GAP into the Ras active site stabilizes an intermediate in the hydrolysis reaction.

The differences in the cycling mechanisms of Ras and G_α are reflected in their structures. Ras ($\approx$170 amino acids) is smaller than G_α proteins ($\approx$300 amino acids), but its three-dimensional structure is similar to that of the GTPase domain of G_α (see Figure 13-8). Recent structural and biochemical studies show that G_α also contains another domain that apparently functions like GAP to increase the rate of GTP hydrolysis by G_α. In addition, the direct interaction between an activated receptor and inactive G protein promotes release of GDP and binding of GTP, so that a separate nucleotide exchange factor is not required.

Both the trimeric G proteins and Ras are members of a family of intracellular GTP-binding switch proteins collectively referred to as the **GTPase superfamily**, which we introduced in Chapter 3. The many similarities between the structure and function of Ras and G_α and the identification of both proteins in all eukaryotic cells indicate that a single type of signal-transducing GTPase originated very early in evolution. In fact, their structures are similar to those of the GTP-binding factors involved in protein synthesis, which are found in all prokaryotic and eukaryotic cells. The gene encoding this ancestral protein subsequently duplicated and evolved to the extent that the human genome encodes a superfamily of such GTPases, comprising perhaps a hundred different intracellular switch proteins. These related proteins control many aspects of cellular growth and metabolism.

 Mammalian Ras proteins have been studied in great detail because mutant Ras proteins are associated with many types of human cancer. These mutant proteins, which bind but cannot hydrolyze GTP, are permanently in the "on" state and contribute to neoplastic **transformation** (Chapter 23). Determination of the three-dimensional structure of the Ras-GAP complex explained the puzzling observation that most oncogenic, constitutively active Ras proteins (Ras^D) contain a mutation at position 12. Replacement of the normal glycine-12 with any other amino acid (except proline) blocks the functional binding of GAP, and in essence "locks" Ras in the active GTP-bound state. ∎

An Adapter Protein and Guanine Nucleotide–Exchange Factor Link Most Activated Receptor Tyrosine Kinases to Ras

The first indication that Ras functions downstream from RTKs in a common signaling pathway came from experiments in which cultured fibroblast cells were induced to proliferate by treatment with a mixture of PDGF and EGF. Microinjection of anti-Ras antibodies into these cells blocked cell proliferation. Conversely, injection of Ras^D, a constitutively active mutant Ras protein that hydrolyzes GTP very inefficiently and thus persists in the active state, caused the cells to proliferate in the absence of the growth factors. These findings are consistent with studies showing that addition of FGF to fibroblasts leads to a rapid increase in the proportion of Ras present in the GTP-bound active form.

How does binding of a growth factor (e.g., EGF) to an RTK (e.g., the EGF receptor) lead to activation of Ras? Two cytosolic proteins—GRB2 and Sos—provide the key links (Figure 14-16). An SH2 domain in GRB2 binds to a specific phosphotyrosine residue in the activated receptor. GRB2 also contains two *SH3 domains,* which bind to and activate Sos. GRB2 thus functions as an adapter protein for the EGF receptor. Sos is a guanine nucleotide–exchange protein (GEF), which catalyzes conversion of inactive GDP-bound Ras to the active GTP-bound form. Genetic analyses of mutants in the worm *C. elegans* and in the fly *Drosophila* blocked at particular stages of differentiation were critical in elucidating the roles of these two proteins in linking an RTK to Ras activation. To illustrate the power of this experimental approach, we consider development of a particular type of cell in the compound eye of *Drosophila.*

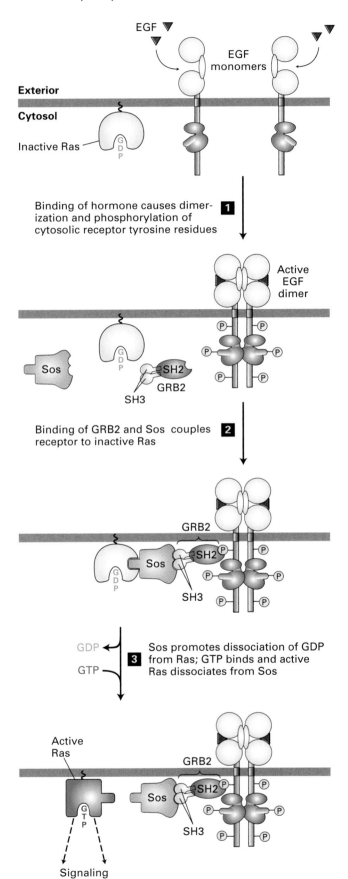

▶ **FIGURE 14-16 Activation of Ras following ligand binding to receptor tyrosine kinases (RTKs).** The receptors for epidermal growth factor (EGF) and many other growth factors are RTKs. The cytosolic adapter protein GRB2 binds to a specific phosphotyrosine on an activated, ligand-bound receptor and to the cytosolic Sos protein, bringing it near its substrate, the inactive Ras·GDP. The guanine nucleotide–exchange factor (GEF) activity of Sos then promotes formation of active Ras·GTP. Note that Ras is tethered to the membrane by a hydrophobic farnesyl anchor (see Figure 5-15). [See J. Schlessinger, 2000, *Cell* **103**:211, and M. A. Simon, 2000, *Cell* **103**:13.]

Genetic Studies in *Drosophila* Identify Key Signal-Transducing Proteins Downstream from Receptor Tyrosine Kinases

The compound eye of the fly is composed of some 800 individual eyes called *ommatidia* (Figure 14-17a). Each ommatidium consists of 22 cells, eight of which are photosensitive neurons called *retinula*, or R cells, designated R1–R8 (Figure 14-17b). An RTK called *Sevenless (Sev)* specifically regulates development of the R7 cell and is not essential for any other known function. In flies with a mutant *sevenless (sev)* gene, the R7 cell in each ommatidium does not form (Figure 14-17c). Since the R7 photoreceptor is necessary for flies to see in ultraviolet light, mutants that lack functional R7 cells but are otherwise normal are easily isolated.

During development of each ommatidium, a protein called *Boss (Bride of Sevenless)* is expressed on the surface of the R8 cell. This membrane-tethered protein is the ligand for the Sev RTK on the surface of the neighboring R7 precursor cell, signaling it to develop into a photosensitive neuron (Figure 14-18a). In mutant flies that do not express a functional Boss protein or Sev RTK, interaction between the Boss and Sev proteins cannot occur, and no R7 cells develop (Figure 14-18b).

To identify intracellular signal-transducing proteins in the Sev RTK pathway, investigators produced mutant flies expressing a temperature-sensitive Sev protein. When these flies were maintained at a permissive temperature, all their ommatidia contained R7 cells; when they were maintained at a nonpermissive temperature, no R7 cells developed. At a par-

ticular intermediate temperature, however, just enough of the Sev RTK was functional to mediate normal R7 development. The investigators reasoned that at this intermediate temperature, the signaling pathway would become defective (and thus no R7 cells would develop) if the level of *another* protein involved in the pathway was reduced, thus reducing the activity of the overall pathway below the level required to form an R7 cell. A recessive mutation affecting such a protein would have this effect because, in diploid organisms like *Drosophila*, a heterozygote containing one wild-type and one mutant allele of a gene will produce half the normal amount of the gene product; hence, even if such a recessive mutation is in an essential gene, the organism will be viable. However, a fly carrying a temperature-sensitive mutation in the *sev* gene and a second mutation affecting another protein in the signaling pathway would be expected to lack R7 cells at the intermediate temperature.

By use of this screen, researchers identified the genes encoding three important proteins in the Sev pathway (see Figure 14-16): an SH2-containing adapter protein exhibiting 64 percent identity to human GRB2; a guanine nucleotide–exchange factor called *Sos (Son of Sevenless)* exhibiting 45 percent identity with its mouse counterpart; and a Ras protein exhibiting 80 percent identity with its mammalian counterparts. These three proteins later were found to function in other signaling pathways initiated by ligand binding to different RTK receptors and used at different times and places in the developing fly.

In subsequent studies, researchers introduced a mutant *ras*[D] gene into fly embryos carrying the *sevenless* mutation.

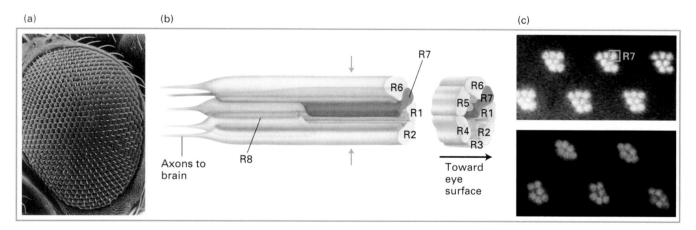

(a)　　　(b)　　　(c)

▲ **FIGURE 14-17 The compound eye of *Drosophila melanogaster.*** (a) Scanning electron micrograph showing individual ommatidia that compose the fruit fly eye. (b) Longitudinal and cutaway views of a single ommatidium. Each of these tubular structures contains eight photoreceptors, designated R1–R8, which are long, cylindrically shaped light-sensitive cells. R1–R6 (yellow) extend throughout the depth of the retina, whereas R7 (brown) is located toward the surface of the eye, and R8 (blue) toward the backside, where the axons exit. (c) Comparison of eyes from wild-type and *sevenless*

mutant flies viewed by a special technique that can distinguish the photoreceptors in an ommatidium. The plane of sectioning is indicated by the blue arrows in (b), and the R8 cell is out of the plane of these images. The seven photoreceptors in this plane are easily seen in the wild-type ommatidia *(top)*, whereas only six are visible in the mutant ommatidia *(bottom)*. Flies with the *sevenless* mutation lack the R7 cell in their eyes. [Part (a) from E. Hafen and K. Basler, 1991, *Development* **1** (suppl.):123; part (b) adapted from R. Reinke and S. L. Zipursky, 1988, *Cell* **55**:321; part (c) courtesy of U. Banerjee.]

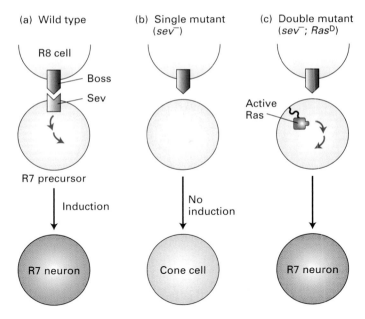

(a) Wild type

R8 cell

Boss
Sev

R7 precursor

Induction

R7 neuron

(b) Single mutant
(*sev⁻*)

No
induction

Cone cell

(c) Double mutant
(*sev⁻; Ras*ᴰ)

Active
Ras

Induction

R7 neuron

▲ EXPERIMENTAL FIGURE 14-18 Genetic studies reveal
that activation of Ras induces development of R7
photoreceptors in the *Drosophila* eye. (a) During larval
development of wild-type flies, the R8 cell in each developing
ommatidium expresses a cell-surface protein, called *Boss*, that
binds to the Sev RTK on the surface of its neighboring R7
precursor cell. This interaction induces changes in gene
expression that result in differentiation of the precursor cell into a
functional R7 neuron. (b) In fly embryos with a mutation in the
sevenless (sev) gene, R7 precursor cells cannot bind Boss and
therefore do not differentiate normally into R7 cells. Rather the
precursor cell enters an alternative developmental pathway and
eventually becomes a cone cell. (c) Double-mutant larvae *(sev⁻;
Ras*ᴰ*)* express a constitutively active Ras (Ras*ᴰ*) in the R7
precursor cell, which induces differentiation of R7 precursor cells
in the absence of the Boss-mediated signal. This finding shows
that activated Ras is sufficient to mediate induction of an R7 cell.
[See M. A. Simon et al., 1991, *Cell* **67**:701, and M. E. Fortini et al., 1992,
Nature **355**:559.]

As noted earlier, the *ras*ᴰ gene encodes a constitutive Ras
protein that is present in the active GTP-bound form even
in the absence of a hormone signal. Although no functional
Sev RTK was expressed in these double-mutants (*sev⁻; ras*ᴰ),
R7 cells formed normally, indicating that activation of Ras is
sufficient for induction of R7-cell development (Figure
14-18c). This finding, which is consistent with the results
with cultured fibroblasts described earlier, supports the conclusion that activation of Ras is a principal step in intracellular signaling by most if not all RTKs.

Binding of Sos Protein to Inactive Ras Causes a Conformational Change That Activates Ras

The adapter protein GRB2 contains two SH3 domains,
which bind to Sos, a guanine nucleotide–exchange factor, in

addition to an SH2 domain, which binds to phosphotyrosine
residues in RTKs. Like phosphotyrosine-binding SH2 and
PTB domains, SH3 domains are present in a large number
of proteins involved in intracellular signaling. Although the
three-dimensional structures of various SH3 domains are
similar, their specific amino acid sequences differ. The SH3
domains in GRB2 selectively bind to proline-rich sequences
in Sos; different SH3 domains in other proteins bind to
proline-rich sequences distinct from those in Sos.

Proline residues play two roles in the interaction between
an SH3 domain in an adapter protein (e.g., GRB2) and a proline-rich sequence in another protein (e.g., Sos). First, the proline-rich sequence assumes an extended conformation that
permits extensive contacts with the SH3 domain, thereby facilitating interaction. Second, a subset of these prolines fit into
binding "pockets" on the surface of the SH3 domain (Figure
14-19). Several nonproline residues also interact with the SH3
domain and are responsible for determining the binding specificity. Hence the binding of proteins to SH3 and to SH2 domains follows a similar strategy: certain residues provide the
overall structural motif necessary for binding, and neighboring residues confer specificity to the binding.

Following activation of an RTK (e.g., Sevenless or the
EGF receptor), a complex containing the activated receptor,
GRB2, and Sos is formed on the cytosolic face of the plasma
membrane (see Figure 14-16). Formation of this complex
depends on the ability of GRB2 to bind simultaneously to
the receptor and to Sos. Thus receptor activation leads to

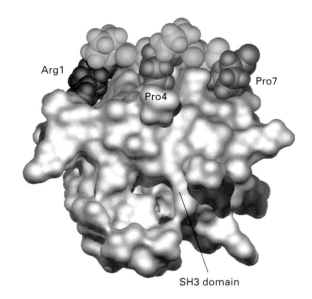

Arg1
Pro4
Pro7

SH3 domain

▲ FIGURE 14-19 Surface model of an SH3 domain bound
to a short, proline-rich target peptide. The target peptide is
shown as a space-filling model. In this target peptide, two
prolines (Pro4 and Pro7, dark blue) fit into binding pockets on the
surface of the SH3 domain. Interactions involving an arginine
(Arg1, red), two other prolines (light blue), and other residues in
the target peptide (green) determine the specificity of binding.
[After H. Yu et al., 1994, *Cell* **76**:933.]

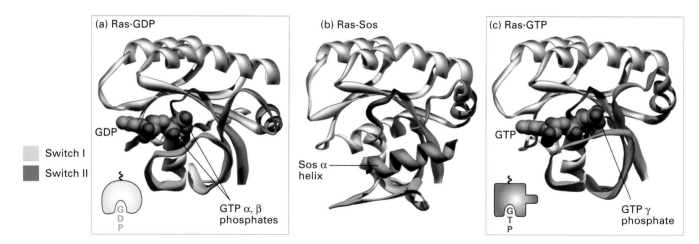

▲ **FIGURE 14-20 Structures of Ras bound to GDP, Sos protein, and GTP.** (a) In Ras·GDP, the Switch I and Switch II segments do not directly interact with GDP (see Figure 13-8). (b) One α helix (orange) in Sos binds to both switch regions of Ras·GDP, leading to a massive conformational change in Ras. In effect, Sos pries Ras open by displacing the Switch I region, thereby allowing GDP to diffuse out. (c) GTP is thought to bind to Ras-Sos first through its base; subsequent binding of the GTP phosphates completes the interaction. The resulting conformational change in Switch I and Switch II segments of Ras, allowing both to bind to the GTP γ phosphate (see Figure 13-8), displaces Sos and promotes interaction of Ras·GTP with its effectors (discussed later). [Adapted from P. A. Boriack-Sjodin and J. Kuriyan, 1998, *Nature* **394**:341.]

relocalization of Sos from the cytosol to the membrane, bringing Sos near to its substrate, namely, membrane-bound Ras·GDP. Biochemical and genetic studies indicate that the C-terminus of Sos inhibits its nucleotide-exchange activity and that GRB2 binding relieves this inhibition.

Binding of Sos to Ras·GDP leads to conformational changes in the Switch I and Switch II segments of Ras, thereby opening the binding pocket for GDP so it can diffuse out (Figure 14-20). Because GTP is present in cells at a concentration some 10 times higher than GDP, GTP binding occurs preferentially, leading to activation of Ras. The activation of Ras and G_α thus occurs by similar mechanisms: a conformational change induced by binding of a protein—Sos and an activated G protein–coupled receptor, respectively—that opens the protein structure so bound GDP is released to be replaced by GTP. Binding of GTP to Ras, in turn, induces a specific conformation of Switch I and Switch II that allows Ras·GTP to activate downstream effector molecules, as we discuss in the next section.

KEY CONCEPTS OF SECTION 14.3

Receptor Tyrosine Kinases and Activation of Ras

■ Receptor tyrosine kinases (RTKs), which bind to peptide and protein hormones, may exist as preformed dimers or dimerize during binding to ligands.

■ Ligand binding leads to activation of the intrinsic protein tyrosine kinase activity of the receptor and phosphorylation of tyrosine residues in its cytosolic domain (see Figure 14-5, *top*). The activated receptor also can phosphorylate other protein substrates.

■ Ras is an intracellular GTPase switch protein that acts downstream from most RTKs. Like G_α, Ras cycles between an inactive GDP-bound form and an active GTP-bound form. Ras cycling requires the assistance of two proteins, a guanine nucleotide–exchange factor (GEF) and a GTPase-activating protein (GAP).

■ RTKs are linked indirectly to Ras via two proteins: GRB2, an adapter protein, and Sos, which has GEF activity (see Figure 14-16).

■ The SH2 domain in GRB2 binds to a phosphotyrosine in activated RTKs, while its two SH3 domains bind Sos, thereby bringing Sos close to membrane-bound Ras·GDP and activating its nucleotide exchange activity.

■ Binding of Sos to inactive Ras causes a large conformational change that permits release of GDP and binding of GTP, forming active Ras (see Figure 14-20). GAP, which accelerates GTP hydrolysis, is localized near Ras·GTP by binding to activated RTKs.

■ Normally, Ras activation and the subsequent cellular response require ligand binding to an RTK or a cytokine receptor. In cells that contain a constitutively active Ras, the cellular response occurs in the absence of ligand binding.

14.4 MAP Kinase Pathways

In mammalian cells all receptor tyrosine kinases (RTKs), as well as most cytokine receptors, appear to utilize a highly conserved signal-transduction pathway in which the signal induced by ligand binding is carried via GRB2 and Sos to Ras, leading to its activation (see Figure 14-16). Activated Ras promotes formation at the membrane of signaling complexes containing three sequentially acting protein kinases that are associated with a scaffold protein. This *kinase cascade* culminates in activation of **MAP kinase**, a serine/threonine kinase also known as *ERK*. After translocating into the nucleus, MAP kinase can phosphorylate many different proteins, including transcription factors that regulate expression of important cell-cycle and differentiation-specific proteins. Activation of MAP kinase in two different cells can lead to similar or different cellular responses, as can its activation in the same cell following stimulation by different hormones.

In this section, we first examine the components of the kinase cascade downstream from Ras in RTK-Ras signaling pathways in mammalian cells. Then we discuss the linkage of other signaling pathways to similar kinase cascades, and we examine recent studies indicating that both yeasts and cells of higher eukaryotes contain multiple MAP kinase pathways.

Signals Pass from Activated Ras to a Cascade of Protein Kinases

A remarkable convergence of biochemical and genetic studies in yeast, *C. elegans*, *Drosophila*, and mammals has revealed a highly conserved cascade of protein kinases that operates in sequential fashion downstream from activated Ras (Figure 14-21). Active Ras·GTP binds to the N-terminal regulatory domain of *Raf*, a serine/threonine kinase, thereby activating it (step 2). Hydrolysis of Ras·GTP to Ras·GDP releases active Raf (step 3), which phosphorylates and thereby activates *MEK* (step 4). Active MEK then phosphorylates and activates MAP kinase, another serine/threonine kinase (step 5). (A dual-specificity protein kinase, MEK phosphorylates its target proteins on both tyrosine and serine or threonine residues.) MAP kinase phosphorylates many different proteins, including nuclear transcription factors, that mediate cellular responses (step 6).

Several types of experiments have demonstrated that Raf, MEK, and MAP kinase lie downstream from Ras and have revealed the sequential order of these proteins in the pathway. For example, mutant Raf proteins missing the N-terminal regulatory domain are constitutively active and induce quiescent cultured cells to proliferate in the absence of stimulation by growth factors. These mutant Raf proteins were initially identified in tumor cells; like the constitutively active Ras^D protein, such mutant Raf proteins are said to be encoded by **oncogenes** (Chapter 23). Conversely, cultured mammalian cells that express a mutant, nonfunctional Raf protein cannot be stimulated to proliferate uncontrollably by a constitutively active Ras^D protein. This finding established

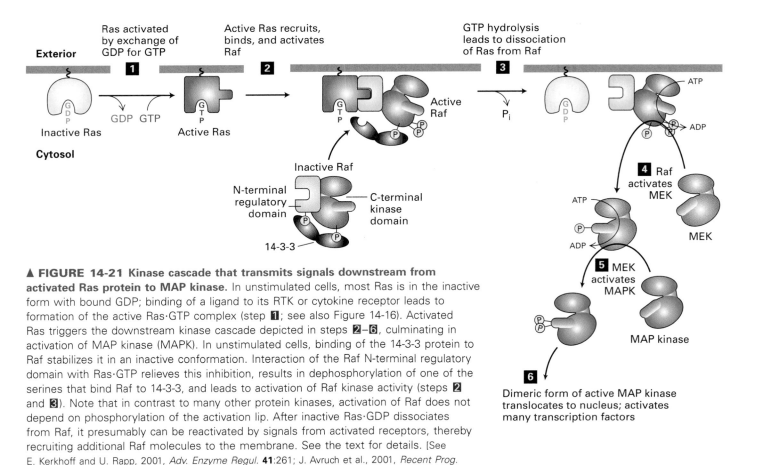

▲ **FIGURE 14-21 Kinase cascade that transmits signals downstream from activated Ras protein to MAP kinase.** In unstimulated cells, most Ras is in the inactive form with bound GDP; binding of a ligand to its RTK or cytokine receptor leads to formation of the active Ras·GTP complex (step 1; see also Figure 14-16). Activated Ras triggers the downstream kinase cascade depicted in steps 2–6, culminating in activation of MAP kinase (MAPK). In unstimulated cells, binding of the 14-3-3 protein to Raf stabilizes it in an inactive conformation. Interaction of the Raf N-terminal regulatory domain with Ras·GTP relieves this inhibition, results in dephosphorylation of one of the serines that bind Raf to 14-3-3, and leads to activation of Raf kinase activity (steps 2 and 3). Note that in contrast to many other protein kinases, activation of Raf does not depend on phosphorylation of the activation lip. After inactive Ras·GDP dissociates from Raf, it presumably can be reactivated by signals from activated receptors, thereby recruiting additional Raf molecules to the membrane. See the text for details. [See E. Kerkhoff and U. Rapp, 2001, *Adv. Enzyme Regul.* **41**:261; J. Avruch et al., 2001, *Recent Prog. Hormone Res.* **56**:127; and M. Yip-Schneider et al., 2000, *Biochem. J.* **351**:151.]

a link between the Raf and Ras proteins. In vitro binding studies further showed that the purified Ras·GTP complex binds directly to the N-terminal regulatory domain of Raf and activates its catalytic activity. An interaction between the mammalian Ras and Raf proteins also was demonstrated in the *yeast two-hybrid system*, a genetic system in yeast used to select cDNAs encoding proteins that bind to target, or "bait," proteins (see Figure 11-39).

That MAP kinase is activated in response to Ras activation was demonstrated in quiescent cultured cells expressing a constitutively active Ras^D protein. In these cells activated MAP kinase is generated in the absence of stimulation by growth-promoting hormones. More importantly, R7 photoreceptors develop normally in the developing eye of *Drosophila* mutants that lack a functional Ras or Raf protein but express a constitutively active MAP kinase. This finding indicates that activation of MAP kinase is sufficient to transmit a proliferation or differentiation signal normally initiated by ligand binding to a receptor tyrosine kinase such as Sevenless (see Figure 14-18). Biochemical studies showed, however, that Raf cannot directly phosphorylate MAP kinase or otherwise activate its activity.

The final link in the kinase cascade activated by Ras·GTP emerged from studies in which scientists fractionated extracts of cultured cells searching for a kinase activity that could phosphorylate MAP kinase and that was present only in cells stimulated with growth factors, not quiescent cells. This work led to identification of MEK, a kinase that specifically phosphorylates one threonine and one tyrosine residue on MAP kinase, thereby activating its catalytic activity. (The acronym MEK comes from *MAP* and *ERK kinase*.) Later studies showed that MEK binds to the C-terminal catalytic domain of Raf and is phosphorylated by the Raf serine/threonine kinase; this phosphorylation activates the catalytic activity of MEK. Hence, activation of Ras induces a kinase cascade that includes Raf, MEK, and MAP kinase: activated RTK → Ras → Raf → MEK → MAP kinase.

Activation of Raf Kinase The mechanism for activating Raf differs from that used to activate many other protein kinases including MEK and MAP kinase. In a resting cell prior to hormonal stimulation, Raf is present in the cytosol in a conformation in which the N-terminal regulatory domain is bound to the kinase domain, thereby inhibiting its activity. This inactive conformation is stabilized by a dimer of the *14-3-3 protein*, which binds phosphoserine residues in a number of important signaling proteins. Each 14-3-3 monomer binds to a phosphoserine residue in Raf, one to phosphoserine-259 in the N-terminal domain and the other to phosphoserine-621 (see Figure 14-21). These interactions are thought to be essential for Raf to achieve a conformational state such that it can bind to activated Ras.

The binding of Ras·GTP, which is anchored to the membrane, to the N-terminal domain of Raf relieves the inhibition of Raf's kinase activity and also induces a conformational change in Raf that disrupts its association with

14-3-3. Raf phosphoserine-259 then is dephosphorylated (by an unknown phosphatase) and other serine or threonine residues on Raf become phosphorylated by yet other kinases. These reactions incrementally increase the Raf kinase activity by mechanisms that are not fully understood.

Activation of MAP Kinase Biochemical and x-ray crystallographic studies have provided a detailed picture of how phosphorylation activates MAP kinase. As in JAK kinases and the cytosolic domain of receptor tyrosine kinases, the catalytic site in the inactive, unphosphorylated form of MAP kinase is blocked by a stretch of amino acids, the activation lip (Figure 14-22a). Binding of MEK to MAP kinase destabilizes the lip structure, resulting in exposure of tyrosine-185, which is buried in the inactive conformation. Following phosphorylation of this critical tyrosine, MEK phosphorylates the neighboring threonine-183 (Figure 14-22b).

Both the phosphorylated tyrosine and the phosphorylated threonine residues in MAP kinase interact with additional amino acids, thereby conferring an altered conformation to the lip region, which in turn permits binding of ATP to the catalytic site. The phosphotyrosine residue (pY185) also plays a key role in binding specific substrate proteins to the surface of MAP kinase. Phosphorylation promotes not only the catalytic activity of MAP kinase but also

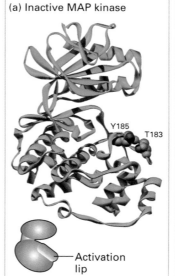
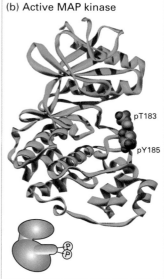

(a) Inactive MAP kinase

(b) Active MAP kinase

▲ **EXPERIMENTAL FIGURE 14-22** Molecular structures of MAP kinase in its inactive, unphosphorylated form (a) and active, phosphorylated form (b). Phosphorylation of MAP kinase by MEK at tyrosine-185 (Y185) and threonine-183 (T183) leads to a marked conformational change in the activation lip. This change promotes dimerization of MAP kinase and binding of its substrates, ATP and certain proteins. A similar phosphorylation-dependent mechanism activates JAK kinases, the intrinsic kinase activity of RTKs, and MEK. [After B. J. Canagarajah et al., 1997, *Cell* **90**:859.]

its dimerization. The dimeric form of MAP kinase (but not the monomeric form) can be translocated to the nucleus, where it regulates the activity of many nuclear transcription factors.

MAP Kinase Regulates the Activity of Many Transcription Factors Controlling Early-Response Genes

Addition of a growth factor (e.g., EGF or PDGF) to quiescent cultured mammalian cells in G_0 causes a rapid increase in the expression of as many as 100 different genes. These are called *early-response genes* because they are induced well before cells enter the S phase and replicate their DNA (see Figure 21-29). One important early-response gene encodes the transcription factor c-Fos. Together with other transcription factors, such as c-Jun, c-Fos induces expression of many genes encoding proteins necessary for cells to progress through the cell cycle. Most RTKs that bind growth factors utilize the MAP kinase pathway to activate genes encoding proteins like c-Fos that propel the cell through the cell cycle.

The enhancer that regulates the c-*fos* gene contains a *serum-response element (SRE)*, so named because it is activated by many growth factors in serum. This complex enhancer contains DNA sequences that bind multiple transcription factors. Some of these are activated by MAP kinase, others by different protein kinases that function in other signaling pathways (e.g., protein kinase A in cAMP pathways and protein kinase C in phosphoinositide pathways).

As depicted in Figure 14-23, activated (phosphorylated) dimeric MAP kinase induces transcription of the c-*fos* gene by modifying two transcription factors, *ternary complex factor (TCF)* and *serum response factor (SRF)*. In the cytosol, MAP kinase phosphorylates and activates another kinase, p90^{RSK}, which translocates to the nucleus, where it phosphorylates a specific serine in SRF. After also translocating to the nucleus, MAP kinase directly phosphorylates specific serines in TCF. Association of phosphorylated TCF with two molecules of phosphorylated SRF forms an active trimeric factor that binds strongly to the SRE DNA segment. As evidence for this model, abundant expression in cultured mammalian cells of a mutant dominant negative TCF that lacks the serine residues phosphorylated by MAP kinase blocks the ability of MAP kinase to activate gene expression driven by the SRE enhancer. Moreover, biochemical studies showed directly that phosphorylation of SRF by active p90^{RSK} increases the rate and affinity of its binding to SRE sequences in DNA, accounting for the increase in the frequency of transcription initiation. Thus both transcription factors are required for maximal growth factor–induced stimulation of gene expression via the MAP kinase pathway, although only TCF is directly activated by MAP kinase.

Phosphorylation of transcription factors by MAP kinase can produce multiple effects on gene expression. For instance, two related *Drosophila* transcription factors, Pointed

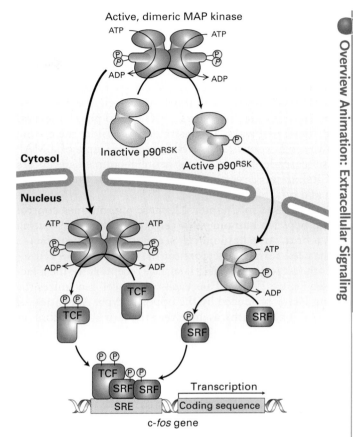

▲ **FIGURE 14-23 Induction of gene transcription by activated MAP kinase.** In the cytosol, MAP kinase phosphorylates and activates the kinase p90^{RSK}, which then moves into the nucleus and phosphorylates the SRF transcription factor. After translocating into the nucleus, MAP kinase directly phosphorylates the transcription factor TCF. Together, these phosphorylation events stimulate transcription of genes (e.g., c-*fos*) that contain an SRE sequence in their promoter. See the text for details. [See R. Marais et al., 1993, *Cell* **73**:381, and V. M. Rivera et al., 1993, *Mol. Cell Biol.* **13**:6260.]

and Yan, which are directly phosphorylated by MAP kinase, are crucial effectors of RTK signaling in the eye and other tissues. Phosphorylation enhances the activity of Pointed, a transcriptional activator. In contrast, unphosphorylated Yan is a transcriptional repressor that accumulates in the nucleus and inhibits development of R7 cells in the eye. Following signal-induced phosphorylation, Yan accumulates in the cytosol and does not have access to the genes it controls, thereby relieving their repression. Mutant forms of Yan that cannot be phosphorylated by MAP kinase are constitutive repressors of R7 development. This example suggests that a complex interplay among multiple transcription factors, regulated by signal-activated kinases, is critical to cellular development.

G Protein–Coupled Receptors Transmit Signals to MAP Kinase in Yeast Mating Pathways

Although many MAP kinase pathways are initiated by RTKs or cytokine receptors, signaling from other receptors can activate MAP kinase in different cell types of higher eukaryotes. Moreover, yeasts and other single-celled eukaryotes, which lack cytokine receptors or RTKs, do possess several MAP kinase pathways. To illustrate, we consider the mating pathway in *S. cerevisiae*, a well-studied example of a MAP kinase cascade linked to G protein–coupled receptors (GPCRs), in this case for two secreted peptide pheromones, the **a** and α factors.

As discussed in Chapter 22, these pheromones control mating between haploid yeast cells of the opposite mating type, **a** or α. An **a** haploid cell secretes the **a** mating factor and has cell-surface receptors for the α factor; an α cell secretes the α factor and has cell-surface receptors for the **a** factor (see Figure 22-13). Thus each type of cell recognizes the mating factor produced by the opposite type. Activation of the MAP kinase pathway by either the **a** or α receptors induces transcription of genes that inhibit progression of the cell cycle and others that enable cells of opposite mating type to fuse together and ultimately form a diploid cell.

Ligand binding to either of the two yeast pheromone receptors triggers the exchange of GTP for GDP on the single G_α subunit and dissociation of $G_\alpha \cdot$GTP from the $G_{\beta\gamma}$ complex. This activation process is identical to that for the GPCRs discussed in the previous chapter (see Figure 13-11). In most, but not all, mammalian GPCR-initiated pathways, the active G_α transduces the signal. In contrast, mutant studies have shown that the dissociated $G_{\beta\gamma}$ complex mediates all the physiological responses induced by activation of the yeast pheromone receptors. For instance, in yeast cells that lack G_α, the $G_{\beta\gamma}$ subunit is always free. Such cells can mate in the absence of mating factors; that is, the mating response is constitutively on. However, in cells defective for the G_β or G_γ subunit, the mating pathway cannot be induced at all. If dissociated G_α were the transducer, the pathway would be expected to be constitutively active in these mutant cells.

In yeast mating pathways, $G_{\beta\gamma}$ functions by triggering a kinase cascade that is analogous to the one downstream from

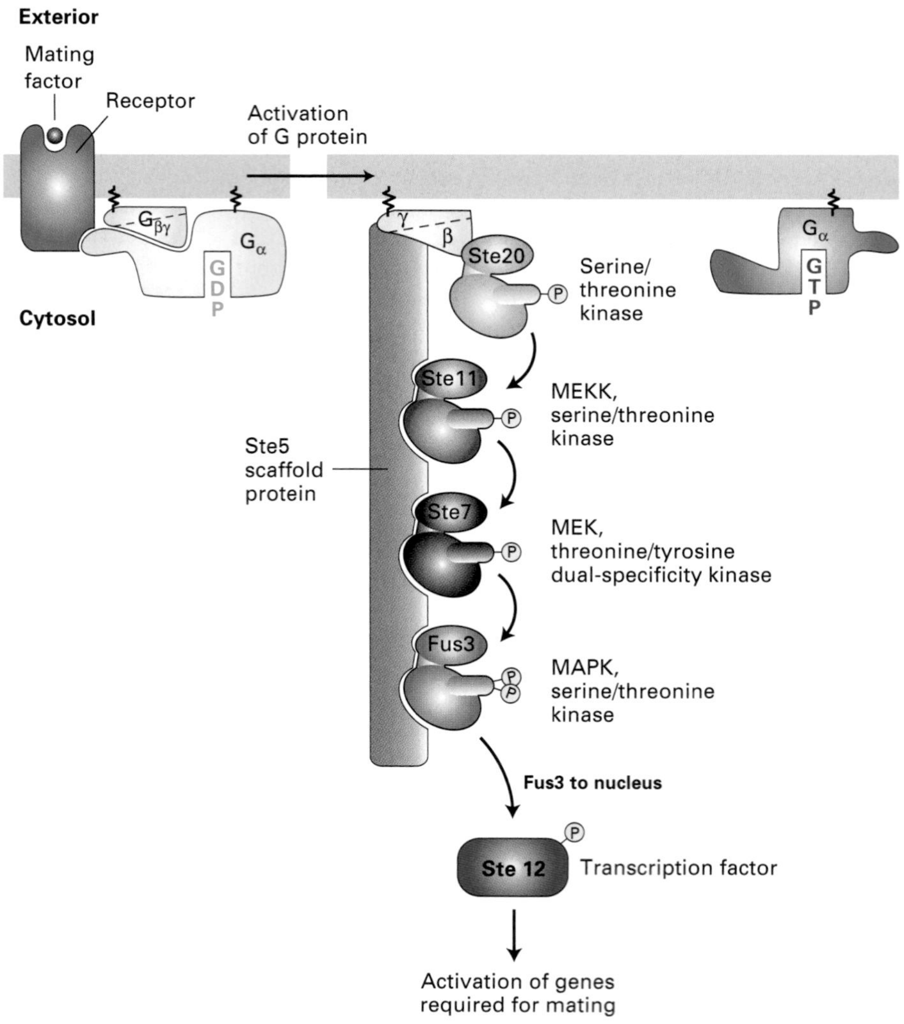

► FIGURE 14-24 **Kinase cascade that transmits signals downstream from mating factor receptors in *S. cerevisiae*.** The receptors for yeast **a** and α mating factors are coupled to the same trimeric G protein. Ligand binding leads to activation and dissociation of the G protein (see Figure 13-10). In the yeast mating pathway, the dissociated $G_{\beta\gamma}$ activates a protein kinase cascade analogous to the cascade downstream of Ras that leads to activation of MAP kinase (see Figure 14-21). The final component, Fus3, is functionally equivalent to MAP kinase (MAPK) in higher eukaryotes. Association of several kinases with the Ste5 scaffold contributes to specificity of the signaling pathway by preventing phosphorylation of other substrates. [See A. Whitmarsh and R. Davis, 1998, *Trends Biochem. Sci.* **23**:481, and H. Dohlman and J. Thorner, 2001, *Ann. Rev. Biochem.* **70**:703.]

Ras. The components of this cascade were uncovered mainly through analyses of mutants that possess functional a and α receptors and G proteins but are sterile (*Ste*), or defective in mating responses. The physical interactions between the components were assessed through immunoprecipitation experiments with extracts of yeast cells and other types of studies. Based on these studies, scientists have proposed the kinase cascade depicted in Figure 14-24. G$_{\beta\gamma}$, which is tethered to the membrane via the γ subunit, binds to and activates Ste20, a protein kinase that in turn phosphorylates and activates Ste11, a serine/threonine kinase analogous to Raf and other mammalian MEKK proteins. Activated Ste11 then phosphorylates Ste7, a dual-specificity MEK that then phosphorylates and activates Fus3, a serine/threonine kinase equivalent to MAP kinase. After translocation to the nucleus, Fus3 promotes expression of target genes by phosphorylating and thus activating nuclear transcription factors (e.g., Ste12) that control expression of proteins involved in mating-specific cellular responses. The other component of the yeast mating cascade, Ste5, interacts with G$_{\beta\gamma}$ as well as Ste11, Ste7, and Fus3. Ste5 has no obvious catalytic function and acts as a scaffold for assembling other components in the cascade.

Scaffold Proteins Isolate Multiple MAP Kinase Pathways in Eukaryotic Cells

In addition to the MAP kinases discussed above, both yeasts and higher eukaryotic cells contain other members of the MAP kinase superfamily. These include mammalian *Jun N-terminal kinases (JNKs)* and *p38 kinases,* which become activated by various types of stresses, and six yeast kinases described below. Collectively referred to as *MAP kinases,* all these proteins are serine/threonine kinases that are activated in the cytosol in response to specific extracellular signals and then translocate to the nucleus. Activation of all known MAP kinases requires phosphorylation of both a tyrosine and a

threonine residue in the lip region (see Figure 14-22). Similarly, all eukaryotic cells contain several members of the dual-specificity MEK kinase superfamily that phosphorylate different members of the MAP kinase superfamily. Thus in all eukaryotic cells, binding of a wide variety of extracellular signaling molecules triggers highly conserved kinase cascades culminating in activation of a particular MAP kinase. The different MAP kinases mediate specific cellular responses, including morphogenesis, cell death, and stress responses.

Current genetic and biochemical studies in the mouse and *Drosophila* are aimed at determining which MAP kinases are required for mediating the response to which signals in higher eukaryotes. This has already been accomplished in large part for the simpler organism *S. cerevisiae*. Each of the six MAP kinases encoded in the *S. cerevisiae* genome has been assigned by genetic analyses to specific signaling pathways triggered by various extracellular signals, such as pheromones, starvation, high osmolarity, hypotonic shock, and carbon/nitrogen deprivation. Each of these MAP kinases mediates very specific cellular responses (Figure 14-25).

In both yeasts and higher eukaryotic cells, different MAP kinase cascades share some common components. For instance, Ste11 functions in the yeast signaling pathways that regulate mating, filamentous growth, and osmoregulation. Nevertheless, each pathway activates its own MAP kinase: Fus3 in the mating pathway, Kss1 in the filamentation pathway, and Hog1 in the osmoregulation pathway. Similarly, in mammalian cells, common upstream signal-transducing proteins participate in activating multiple JNK kinases.

Once the sharing of components among different MAP kinase pathways was recognized, researchers wondered how the specificity of the cellular responses to particular signals could be achieved. Studies with yeast provided the initial evidence that pathway-specific *scaffold proteins* enable the signal-transducing kinases in a particular pathway to interact with one another but not with kinases in other pathways.

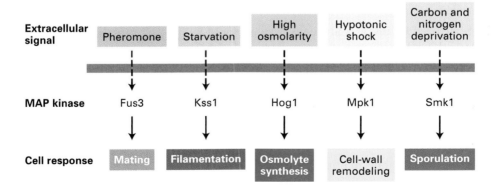

▲ **FIGURE 14-25 Overview of five MAP kinase pathways in *S. cerevisiae*.** Each pathway is triggered by a specific extracellular signal and leads to activation of a single different MAP kinase, which mediates characteristic cellular responses. Formation of pathway-specific complexes of MAP kinases and scaffold proteins prevents "cross talk" between pathways containing a common component such as the MEKK Ste11, which occurs in the mating, filamentation, and osmoregulatory pathways (see Figure 14-24). [Adapted from H. D. Madhani and G. R. Fink, 1998, *Trends Genet.* **14**(4):152.]

For example, the scaffold protein Ste5 stabilizes a large complex that includes Ste11 and other kinases in the mating pathway (see Figure 14-24). Different Ste11-binding scaffold proteins, however, stabilize signaling complexes containing the components of the filamentation and osmoregulation pathways. In each pathway in which Ste11 participates, it is constrained within a large complex that forms in response to a specific extracellular signal, and signaling downstream from Ste11 is restricted to the complex in which it is localized. As a result, exposure of yeast cells to mating factors induces activation of a single MAP kinase, Fus3, whereas exposure to a high osmolarity or starvation induces activation of different MAP kinases (see Figure 14-25).

Scaffolds for MAP kinase pathways are well documented in yeast, fly, and worm cells, but their presence in mammalian cells has been difficult to demonstrate. Perhaps the best documented scaffold protein is *Ksr* (*kinase suppressor of Ras*), which binds both MEK and MAP kinase. Loss of the *Drosophila* Ksr homolog blocks signaling by a constitutively active Ras protein, suggesting a positive role for Ksr in Ras–MAP kinase signaling in fly cells. Although knockout mice that lack Ksr are grossly normal, activation of MAP kinase by growth factors or cytokines is lower than normal in several types of cells in these animals. This finding suggests that Ksr functions as a scaffold that enhances but is not essential for Ras–MAP kinase signaling in mammalian cells. Other proteins also have been found to bind to specific mammalian MAP kinases. Thus the signal specificity of different MAP kinases in animal cells may arise from their association with various scaffold-like proteins, but much additional research is needed to test this possibility.

KEY CONCEPTS OF SECTION 14.4

MAP Kinase Pathways

■ Activated Ras triggers a kinase cascade in which Raf, MEK, and MAP kinase are sequentially phosphorylated and thus activated. Activated MAP kinase dimerizes and translocates to the nucleus (see Figure 14-21).

■ Phosphorylation of one or more residues in a conserved lip region activates MAP kinases and many other protein kinases involved in signal-transduction pathways.

■ Activation of MAP kinase following stimulation of a growth factor receptor leads to phosphorylation and activation of two transcription factors, TCF and SRF. These associate into a trimeric complex that promotes transcription of various early-response genes (see Figure 14-23).

■ Yeast and higher eukaryotes contain multiple MAP kinase pathways that are triggered by activation of various receptor classes including G protein–coupled receptors.

■ Different extracellular signals induce activation of different MAP kinases, which regulate diverse cellular processes (see Figure 14-25).

■ The upstream components of MAP kinase cascades assemble into large pathway-specific complexes stabilized by scaffold proteins (see Figure 14-24). This assures that activation of one pathway by a particular extracellular signal does not lead to activation of other pathways containing shared components.

14.5 Phosphoinositides as Signal Transducers

In previous sections, we have seen how signal transduction from cytokine receptors and receptor tyrosine kinases (RTKs) begins with formation of multiprotein complexes associated with the plasma membrane. Here we discuss how these receptors initiate signaling pathways that involve membrane-bound phosphorylated inositol lipids, collectively referred to as **phosphoinositides**. We begin with the branch of the phosphoinositide pathway that also is mediated by G protein–coupled receptors and then consider another branch that is not shared with these receptors.

Phospholipase C$_\gamma$ Is Activated by Some RTKs and Cytokine Receptors

As discussed in Chapter 13, hormonal stimulation of some G protein–coupled receptors leads to activation of the β isoform of phospholipase C (PLC$_\beta$). This membrane-associated enzyme then cleaves phosphatidylinositol 4,5-bisphosphate (PIP$_2$) to generate two important **second messengers**, 1,2-diacylglycerol (DAG) and inositol 1,4,5-trisphosphate (IP$_3$). Signaling via the *IP$_3$/DAG pathway* leads to an increase in cytosolic Ca^{2+} and to activation of **protein kinase C** (see Figure 13-29).

Many RTKs and cytokine receptors also can initiate the IP$_3$/DAG pathway by activating another isoform of phospholipase C, the γ isoform (PLC$_\gamma$). The SH2 domains of PLC$_\gamma$ bind to specific phosphotyrosines of the activated receptors, thus positioning the enzyme close to its membrane-bound substrate PIP$_2$ (see Figure 13-28). In addition, the receptor kinase activity phosphorylates tyrosine residues on the bound PLC$_\gamma$, enhancing its hydrolase activity. Thus activated RTKs and cytokine receptors promote PLC$_\gamma$ activity in two ways: by localizing the enzyme to the membrane and by phosphorylating it.

Recruitment of PI-3 Kinase to Hormone-Stimulated Receptors Leads to Activation of Protein Kinase B

In addition to initiating the IP$_3$/DAG pathway, some activated RTKs and cytokine receptors can initiate another phosphoinositide pathway, the *PI-3 kinase pathway,* by re-

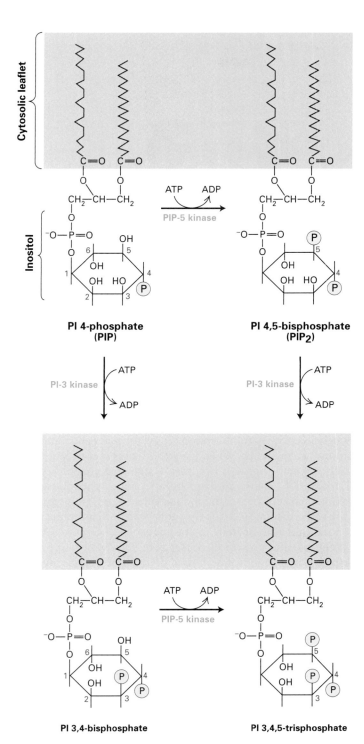

▲ FIGURE 14-26 Generation of phosphatidylinositol 3-phosphates. The enzyme phosphatidylinositol-3 kinase (PI-3 kinase) is recruited to the membrane by many activated receptor tyrosine kinases (RTKs) and cytokine receptors. The 3-phosphate added by this enzyme is a binding site for various signal-transduction proteins. [See L. Rameh and L. C. Cantley, 1999, *J. Biol. Chem.* **274**:8347.]

cruiting the enzyme phosphatidylinositol-3 kinase to the membrane. PI-3 kinase was first identified as a kinase that copurifies with several viral oncoproteins such as the "middle T" protein encoded by polyoma virus. When inactive, dominant negative, versions of PI-3 kinase are expressed in virus-transformed cells, they inhibit the uncontrolled cell proliferation characteristic of virus-transformed cells. This finding suggested that the normal kinase is important in certain signaling pathways essential for cell proliferation or for the prevention of apoptosis. Subsequent work showed that PI-3 kinases participate in many signaling pathways related to cell growth and apoptosis. Of the nine PI-3 kinase homologs encoded by the human genome, the best characterized contains a p110 subunit with catalytic activity and a p85 subunit with an SH2 domain.

The SH2 domain in PI-3 kinase binds to phosphotyrosine residues in the cytosolic domain of many activated RTKs and cytokine receptors. The recruitment of PI-3 kinase to the plasma membrane by activated receptors positions its catalytic domain near its phosphoinositide substrates on the cytosolic face of the plasma membrane, leading to formation of PI 3,4-bisphosphate or PI 3,4,5-trisphosphate (Figure 14-26). By acting as docking sites for various signal-transducing proteins, these membrane-bound PI 3-phosphates in turn transduce signals downstream in several important pathways.

A primary binding target of PI 3-phosphates is **protein kinase B (PKB)**, a serine/threonine kinase. Besides its kinase domain, PKB contains a PH domain that tightly binds the 3-phosphate in both PI 3,4-bisphosphate and PI 3,4,5-trisphosphate. In unstimulated, resting cells, the level of both these compounds is low, and protein kinase B is present in the cytosol in an inactive form. Following hormone stimulation and the resulting rise in PI 3-phosphates, protein kinase B binds to them and is localized at the cell surface membrane.

Binding of protein kinase B to PI 3-phosphates not only recruits the enzyme to the plasma membrane but also releases inhibition of the catalytic site by the PH domain in the cytosol. Maximal activation of protein kinase B, however, depends on recruitment of another kinase, PDK1, to the plasma membrane via binding of its PH domain to PI 3-phosphates. Both membrane-associated protein kinase B and PDK1 can diffuse in the plane of the membrane, bringing them close enough so that PDK1 can phosphorylate protein kinase B (Figure 14-27). PDK1 phosphorylates one serine residue in the activation lip of protein kinase B, providing yet another example of kinase activation by phosphorylation in this segment. Phosphorylation of a second serine, not in the lip segment, is necessary for maximal protein kinase B activity. Thus, as with Raf, an inhibitory domain and phosphorylation by other kinases regulate the activity of protein kinase B. Once fully activated, protein kinase B can dissociate from the plasma membrane and phosphorylate its many target proteins.

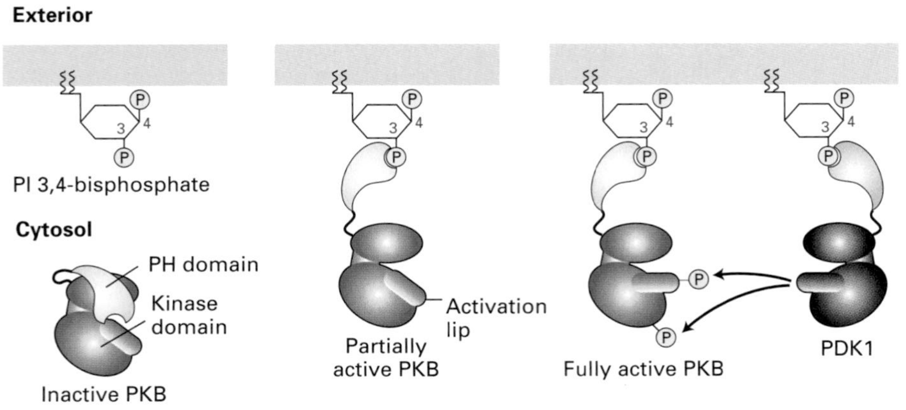

▲ **FIGURE 14-27 Recruitment and activation of protein kinase B (PKB) in PI-3 kinase pathways.** In unstimulated cells, PKB is in the cytosol with its PH domain bound to the catalytic domain, inhibiting its activity. Hormone stimulation leads to activation of PI-3 kinase and subsequent formation of phosphatidylinositol (PI) 3-phosphates (see Figure 14-26). The 3-phosphate groups serve as docking sites on the plasma membrane for the PH domain of PKB and another kinase, PDK1. Full activation of PKB requires phosphorylation both in the activation lip and at the C-terminus by PDK1. [Adapted from A. Toker and A. Newton, 2000, *Cell* **103**:185, and M. Scheid et al., 2002, *Mol. Cell Biol.* **22**:6247.]

The Insulin Receptor Acts Through the PI-3 Kinase Pathway to Lower Blood Glucose

The insulin receptor is a dimeric receptor tyrosine kinase that can initiate the Ras–MAP kinase pathway, leading to changes in gene expression. Insulin stimulation also can initiate the PI-3 kinase pathway just described, leading to activation of protein kinase B. In insulin-stimulated liver, muscle, and fat cells, activated protein kinase B acts in several ways to lower blood glucose and promote glycogen synthesis.

The principal mechanism by which insulin causes a reduction of the blood glucose level is by increasing import of glucose by fat and muscle cells. This effect is mediated by protein kinase B, which through mechanisms that are not fully understood causes movement of the GLUT4 glucose transporter from intracellular membranes to the cell surface (Chapter 15). The resulting increased influx of glucose into these cells lowers blood glucose levels.

In both liver and muscle, insulin stimulation also leads to activation of glycogen synthase (GS), which synthesizes glycogen from UDP-glucose (see Figure 13-16). This represents another mechanism for reducing glucose concentration in the circulation. In resting cells (i.e., in the absence of insulin), glycogen synthase kinase 3 (GSK3) is active and phosphorylates glycogen synthase, thereby blocking its activity. Activated protein kinase B phosphorylates and thereby inactivates GSK3. As a result, GSK3-mediated inhibition of glycogen synthase is relieved, promoting glycogen synthesis.

Activated Protein Kinase B Promotes Cell Survival by Several Pathways

In many cells activated protein kinase B directly phosphorylates pro-apoptotic proteins such as Bad, thereby preventing activation of an apoptotic pathway leading to cell death (Chapter 22). Activated protein kinase B also promotes survival of many cultured cells by phosphorylating the transcription factor Forkhead-1 on as many as three serine or threonine residues. In the absence of growth factors, Forkhead-1 is unphosphorylated and localizes to the nucleus, where it activates transcription of several genes encoding pro-apoptotic proteins. When growth factors are added to the cells, protein kinase B becomes active and phosphorylates Forkhead-1. This allows the cytosolic phosphoserine-binding protein 14-3-3 to bind Forkhead-1 and thus sequester it in the cytosol. (14-3-3 is the same protein that retains phosphorylated Raf protein in the cytosol; see Figure 14-21.) Withdrawal of growth factor leads to inactivation of protein kinase B and dephosphorylation of Forkhead-1, thus favoring apoptosis. A Forkhead-1 mutant in which the three serine target residues for protein kinase B are mutated is "constitutively active" and initiates apoptosis even in the presence of activated protein kinase B. This finding demonstrates the importance of Forkhead-1 in controlling apoptosis of cultured cells.

PTEN Phosphatase Terminates Signaling via the PI-3 Kinase Pathway

Like virtually all intracellular signaling events, phosphorylation by PI-3 kinase is reversible. The relevant phosphatase, termed *PTEN phosphatase,* has an unusually broad specificity. Although PTEN can remove phosphate groups attached to serine, threonine, and tyrosine residues in proteins, its ability to remove the 3-phosphate from PI 3,4,5-trisphosphate is thought to be its major function in cells. Overexpression of PTEN in cultured mammalian cells promotes apoptosis by reducing the level of PI 3,4,5-trisphosphate and hence the activation and anti-apoptotic effect of protein kinase B.

The gene encoding PTEN is deleted in multiple types of advanced human cancers, and its loss is thought to lead to uncontrolled growth. Indeed, cells lacking PTEN have elevated levels of PI 3,4,5-trisphosphate and PKB activity. Since protein kinase B exerts an anti-apoptotic effect, loss of PTEN indirectly reduces the programmed cell death that is the normal fate of abnormally controlled cells. In certain cells, such as neuronal stem cells, absence of PTEN not only prevents apoptosis but also leads to stimulation of cell-cycle progression and an enhanced rate of proliferation. Thus knockout mice that cannot express PTEN have big brains with excess numbers of neurons, attesting to PTEN's importance in control of normal development. ∎

The Receptor for a Particular Growth Factor Often Is Linked to Multiple Signaling Pathways

Interaction of different signaling pathways permits the fine-tuning of cellular activities required to carry out complex developmental and physiological processes. As we have noted previously, both RTKs and cytokine receptors can initiate signaling via the Ras–MAP kinase pathway, DAG/IP_3 pathway, and PI-3 kinase pathway (see Table 14-1). In addition, cytokine receptors can act through their associated JAK kinases to directly activate STAT transcription factors.

Activation of multiple signal-transduction pathways by many receptors allows different sets of genes to be independently controlled by the same or different receptors. Occasionally these pathways can induce opposite effects. For example, genetic manipulation of the Ras–MAP kinase and PI-3 kinase pathways during muscle differentiation indicates that these pathways have opposite phenotypic effects: activation of the Ras–MAP kinase pathway inhibits myocyte differentiation into myotubes, whereas activation of the PI-3 kinase pathway promotes it.

The initiation of tissue-specific signaling pathways by stimulation of the *same* receptor in different cells is exemplified by the EGF receptor. Genetic studies analogous to those described earlier for development of R7 cells in *Drosophila* demonstrated the central importance of EGF-stimulated signaling via the Ras–MAP pathway in development of the vulva in *C. elegans*. Other genetic studies, however, showed that stimulation of the EGF receptor triggers a Ras-independent pathway in some tissues. For example, one of the many functions of EGF in *C. elegans* is to control contractility of smooth muscle, which in turn regulates the extrusion of oocytes from one compartment of the hermaphrodite gonad to another, where they are fertilized. Coupling of the EGF receptor to Ras is not required for the EGF-induced contractions of the gonad. Analysis of several different types of mutations led researchers to conclude that in *C. elegans* smooth muscle, the EGF receptor is linked to the IP_3/DAG pathway. Ligand binding to the receptor leads to activation of PLC_γ activity, an increase in IP_3, and release of intracellular Ca^{2+} stores. The increased cytosolic Ca^{2+} level then promotes muscle contraction.

In Chapter 15 we will encounter several other examples of how stimulation of the same receptor in different cell types activates different signaling pathways that produce very diverse effects on the metabolism and fate of the cell.

KEY CONCEPTS OF SECTION 14.5

Phosphoinositides as Signal Transducers

■ Many RTKs and cytokine receptors can initiate the IP_3/DAG signaling pathway by activating phospholipase Cγ (PLCγ), a different PLC isoform than the one activated by G protein–coupled receptors.

■ Activated RTKs and cytokine receptors can initiate another phosphoinositide pathway by binding PI-3 kinases, thereby allowing the catalytic subunit access to its membrane-bound phosphoinositide (PI) substrates, which are phosphorylated at the 3 position (see Figure 14-26).

■ The PH domain in various proteins binds to PI 3-phosphates, forming signaling complexes associated with the plasma membrane.

■ Protein kinase B (PKB) becomes partially activated by binding to PI 3-phosphates. Its full activation requires phosphorylation by another kinase (PDK1), which also is recruited to the membrane by binding to PI 3-phosphates (see Figure 14-27).

■ Activated protein kinase B promotes survival of many cells by directly inactivating several pro-apoptotic proteins and down-regulating expression of others.

■ Signaling via the PI-3 kinase pathway is terminated by the PTEN phosphatase, which hydrolyzes the 3-phosphate in PI 3-phosphates. Loss of PTEN, a common occurrence in human tumors, promotes cell survival and proliferation.

■ A single RTK or cytokine receptor often initiates different signaling pathways in multiple cell types. Different pathways may be essential in certain cell signaling events but not in others.

14.6 Pathways That Involve Signal-Induced Protein Cleavage

Up to now we have discussed reversible signaling pathways, where inactivation is as important as the initial activation. In contrast are essentially irreversible pathways in which a component is proteolytically cleaved. Here we consider two such pathways: the *NF-κB pathway,* which enables cells to respond immediately and vigorously to a number of stress-inducing conditions, and the *Notch/Delta pathway,* which determines the fates of many types of cells during development. Proteolytic activation of the cell-surface receptor Notch is facilitated by presenilin 1, a membrane protein that also has been implicated in the pathology of Alzheimer's disease.

Signal-Induced Degradation of a Cytosolic Inhibitor Protein Activates the NF-κB Transcription Factor

The examples in previous sections have demonstrated the importance of signal-induced phosphorylation in modulating the activity of many transcription factors. Another mechanism for regulating transcription factor activity in response to extracellular signals was revealed in studies with both mammalian cells and *Drosophila*. This mechanism, which involves phosphorylation and subsequent ubiquitin-mediated degradation of an inhibitor protein, is exemplified by the NF-κB transcription factor.

Originally discovered on the basis of its transcriptional activation of the gene encoding the κ light-chain of antibodies (immunoglobulins) in B cells, NF-κB is now thought to be the master transcriptional regulator of the immune system in mammals. Although flies do not make antibodies, NF-κB homologs in *Drosophila* mediate the immune response to bacterial and viral infection by inducing synthesis of a large number of antimicrobial peptides that are secreted from cells. This indicates that the NF-κB regulatory system is more than half a billion years old. NF-κB is rapidly activated in mammalian immune-system cells in response to infection, inflammation, and a number of other stressful situations, such as ionizing radiation. It also is activated by so-called inflammatory cytokines such as *tumor necrosis factor α (TNF-α)* and interleukin 1 (IL-1), which are released by nearby cells in response to infection.

Biochemical studies in mammalian cells and genetic studies in flies have provided important insights into the operation of the NF-κB pathway (Figure 14-28). The two subunits of heterodimeric NF-κB (p65 and p50) share a region of homology at their N-termini that is required for their dimerization and binding to DNA. In resting cells, NF-κB is sequestered in an inactive state in the cytosol by direct binding to an inhibitor called I-κB. A single molecule of I-κB binds to the N-terminal domains of each subunit in the p50/p65 heterodimer, thereby masking the nuclear-localization signals. A protein kinase complex termed *I-κB kinase* is the point of convergence of all of the extracellular signals that activate NF-κB. Within minutes of stimulation, I-κB kinase becomes activated and phosphorylates two N-terminal serine residues on I-κB. An E3 ubiquitin ligase then binds to these phosphoserines and polyubiquitinates I-κB, triggering its immediate degradation by a proteasome (see Figure 3-13). In cells expressing mutant forms of I-κB in which these two serines have been changed to alanine, and thus cannot be phosphorylated, NF-κB is permanently repressed, demonstrating that phosphorylation of I-κB is essential for pathway activation.

The degradation of I-κB exposes the nuclear-localization signals on NF-κB, which then translocates into the nucleus and activates transcription of a multitude of target genes. Despite its activation by proteolysis, NF-κB signaling eventually is turned off by a negative feedback loop, since one of the genes whose transcription is immediately induced by NF-κB encodes I-κB. The resulting increased levels of the

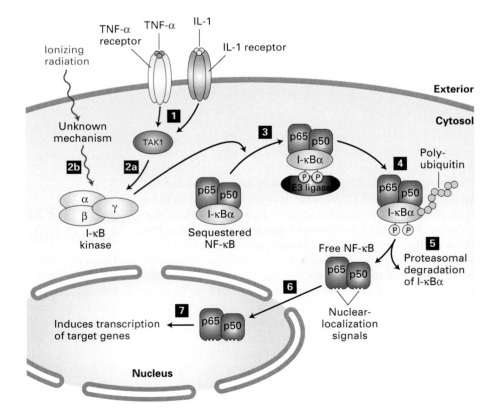

▶ **FIGURE 14-28 NF-κB signaling pathway.** In resting cells, the dimeric transcription factor NF-κB, composed of p50 and p65, is sequestered in the cytosol, bound to the inhibitor I-κB. Stimulation by TNF-α or IL-1 induces activation of TAK1 kinase (step **1**), leading to activation of the trimeric I-κB kinase (step **2a**). Ionizing radiation and other stresses can directly activate I-κB kinase by an unknown mechanism (step **2b**). Following phosphorylation of I-κB by I-κB kinase and binding of E3 ubiquitin ligase (step **3**), polyubiquitination of I-κB (step **4**) targets it for degradation by proteasomes (step **5**). The removal of I-κB unmasks the nuclear-localization signals (NLS) in both subunits of NF-κB, allowing their translocation to the nucleus (step **6**). Here NF-κB activates transcription of numerous target genes (step **7**), including the gene encoding the α subunit of I-κB, which acts to terminate signaling. [See M. Karin and Y. Ben-Neriah, 2000, *Ann. Rev. Immunol.* **18**:621, and R. Khush, F. Leulier, and B. Lemaitre, 2001, *Trends Immunol.* **22**:260.]

I-κB protein bind active NF-κB in the nucleus and return it to the cytosol.

NF-κB stimulates transcription of more than 150 genes, including those encoding cytokines and chemokines that attract other immune-system cells and fibroblasts to sites of infection. It also promotes expression of receptor proteins that enable neutrophils (a type of white blood cell) to migrate from the blood into the underlying tissue (see Figure 6-30). In addition, NF-κB stimulates expression of iNOS, the inducible isoform of the enzyme that produces nitric oxide, which is toxic to bacterial cells, and of several anti-apoptotic proteins, which prevent cell death. Thus this single transcription factor coordinates and activates the body's defense either directly by responding to pathogens and stress or indirectly by responding to signaling molecules released from other infected or wounded tissues and cells.

Besides its roles in inflammation and immunity, NF-κB plays a key role during mammalian development. For instance, mouse embryos that cannot express one of the I-κB kinase subunits die at mid-gestation of liver degeneration caused by excessive apoptosis of cells that would normally survive; thus NF-κB is essential for normal development of this tissue. As we will see in Chapter 21, phosphorylation-dependent degradation of a cyclin kinase–dependent inhibitor plays a central role in regulating progression through the cell cycle in *S. cerevisiae*. It seems likely that phosphorylation-dependent protein degradation may emerge as a common regulatory mechanism in many different cellular processes.

Regulated Intramembrane Proteolysis Catalyzed by Presenilin 1 Activates Notch Receptor

Both Notch and its ligand Delta are transmembrane proteins with numerous EGF-like repeats in their extracellular domains. They participate in a highly conserved and important type of cell differentiation in both invertebrates and vertebrates, called **lateral inhibition,** in which adjacent and developmentally equivalent cells assume completely different fates. This process, discussed in detail in Chapter 15, is particularly important in preventing too many nerve precursor cells forming from an undifferentiated layer of epithelial cells.

Notch protein is synthesized as a monomeric membrane protein in the endoplasmic reticulum, where it binds *presenilin 1*, a multispanning membrane protein; the complex travels first to the Golgi and then on to the plasma membrane. In the Golgi, Notch undergoes a proteolytic cleavage that generates an extracellular subunit and a transmembrane-cytosolic subunit; the two subunits remain noncovalently associated with each other in the absence of interaction with Delta residing on another cell. Binding of Notch to Delta triggers two proteolytic cleavages in the responding cell (Figure 14-29). The second cleavage, within the hydrophobic membrane-spanning region of Notch, is catalyzed by presenilin 1 and releases the Notch cytosolic segment, which immediately translocates to the nucleus. Such signal-induced *regulated intramembrane proteolysis (RIP)* also occurs in the response of cells to high cholesterol (Chapter 18) and to the presence of unfolded proteins in the endoplasmic reticulum (Chapter 16).

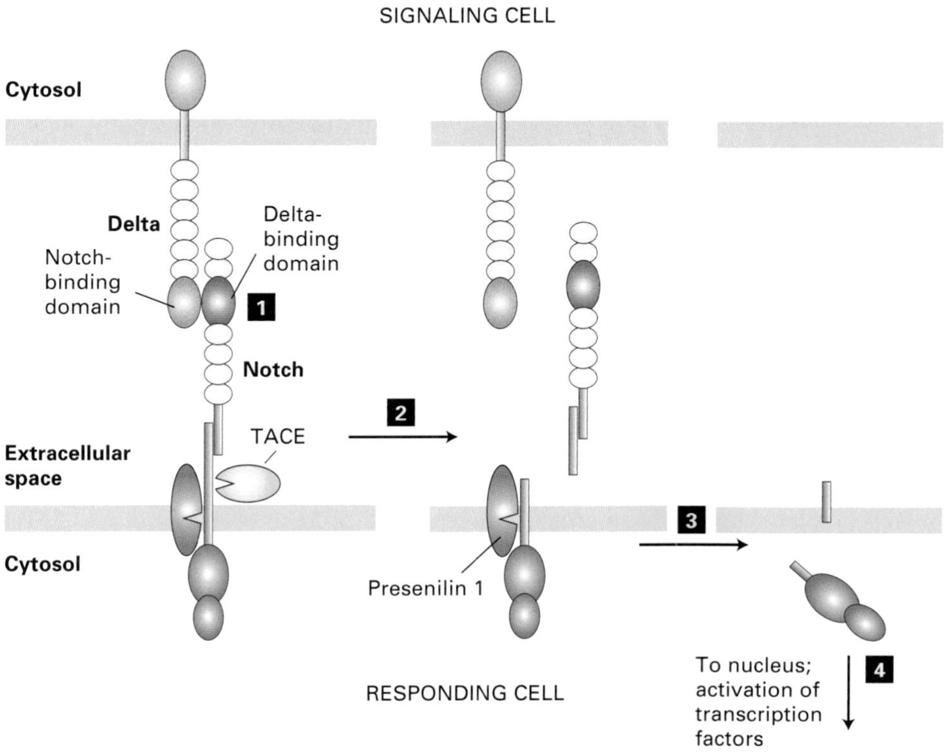

◀ **FIGURE 14-29 Notch/Delta signaling pathway.** The extracellular subunit of Notch on the responding cell is noncovalently associated with its transmembrane-cytosolic subunit. Binding of Notch to its ligand Delta on an adjacent signaling cell (step **1**) first triggers cleavage of Notch by the membrane-bound metalloprotease TACE (*tumor necrosis factor alpha converting enzyme*), releasing the extracellular segment (step **2**). Presenilin 1, an integral membrane protein, then catalyzes an intramembrane cleavage that releases the cytosolic segment of Notch (step **3**). Following translocation to the nucleus, this Notch segment interacts with several transcription factors that act to affect expression of genes that in turn influence the determination of cell fate during development (step **4**). [See M. S. Brown et al., 2000, *Cell* **100**:391, and Y.-M. Chan and Y. Jan, 1999, *Neuron* **23**:201.]

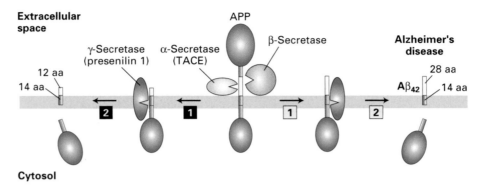

▲ FIGURE 14-30 Proteolytic cleavage of APP, a neuronal plasma membrane protein. (*Left*) Sequential proteolytic cleavage by α-secretase (step **1**) and γ-secretase (step **2**) produces an innocuous membrane-embedded peptide of 26 amino acids. γ-Secretase is a complex of several proteins, but the proteolytic site that catalyzes intramembrane cleavage probably resides within presenilin 1. (*Right*) Cleavage in the extracellular domain by β-secretase (step **1**) followed by cleavage within the membrane by γ-secretase generates the 42-residue Aβ₄₂ peptide that has been implicated in formation of amyloid plaques in Alzheimer's disease. In both pathways the cytosolic segment of APP is released into the cytosol, but its function is not known. [See W. Esler and M. Wolfe, 2001, *Science* **293**:1449, and C. Haass and H. Steiner, 2002, *Trends Cell Biol.* **12**:556.]

In *Drosophila* the released intracellular segment of Notch forms a complex with a DNA-binding protein called Suppressor of Hairless, or Su(H), and stimulates transcription of many genes whose net effect is to influence the determination of cell fate during development. One of the proteins increased in this manner is Notch itself, and Delta production is correspondingly reduced (see Figure 15-38). As we see in Chapter 15, reciprocal regulation of the receptor and ligand in this fashion is an essential feature of the interaction between initially equivalent cells that causes them to assume different cell fates.

 Presenilin 1 (PS1) was first identified as the product of a gene that commonly is mutated in patients with an early-onset autosomal dominant form of Alzheimer's disease. A major pathologic change associated with Alzheimer's disease is accumulation in the brain of *amyloid plaques* containing aggregates of a small peptide containing 42 residues termed Aβ₄₂. This peptide is derived by proteolytic cleavage of APP (amyloid precursor protein), a cell-surface protein of unknown function expressed by neurons. APP actually undergoes cleavage by two pathways (Figure 14-30). In each pathway the initial cleavage occurs within the extracellular domain, catalyzed by α- or β-secretase; γ-secretase then catalyzes a second cleavage at the same intramembrane site in both pathways. The pathway initiated by α-secretase, which involves the same membrane-bound metalloprotease TACE that cleaves Notch, generates a 26-residue peptide that apparently does no harm. The pathway initiated by β-secretase generates the pathologic Aβ₄₂. The missense mutations in presenilin 1 involved in Alzheimer's disease enhance the formation of the Aβ₄₂ peptide, leading to plaque formation and eventually to the death of neurons.

Evidence supporting the involvement of presenilin 1 in Notch signaling (see Figure 14-29) came from genetic studies in the roundworm *C. elegans*. Mutations in the worm homolog of presenilin 1 caused developmental defects similar to those caused by Notch mutations. Later work showed that mammalian Notch does not undergo signal-induced intramembrane proteolysis in mouse neuronal cells genetically missing presenilin 1. But whether presenilin 1 is the actual γ-secretase protease or an essential cofactor of the "real" protease is not yet certain, since presenilin 1 is part of a large complex containing several other integral membrane proteins. Within its membrane-spanning segments, presenilin 1 has two aspartate residues in a configuration that resembles that of the two aspartates in the active site of water-soluble "aspartyl proteases," and mutation of either of these aspartate residues in presenilin 1 abolishes its ability to stimulate cleavage of Notch. Similarly, a battery of chemical protease inhibitors blocks cleavage of Notch and γ-secretase cleavage of APP with the same potency, suggesting that the same protease is involved. Current data are thus consistent with the notion that presenilin 1 is the protease that cleaves both Notch and APP within their transmembrane segments. However, cleavage of both Notch and APP occurs at or near the plasma membrane, whereas the majority of presenilin is found in the endoplasmic reticulum. This finding suggests that presenilin may act in conjunction with other proteins in the unusual intramembrane proteolysis of Notch and APP. ∎

KEY CONCEPTS OF SECTION 14.6

Pathways That Involve Signal-Induced Protein Cleavage

■ The NF-κB transcription factor regulates many genes that permit cells to respond to infection and inflammation.

■ In unstimulated cells, NF-κB is localized to the cytosol, bound to an inhibitor protein, I-κB. In response to extracellular signals, phosphorylation-dependent ubiquitination and degradation of I-κB in proteasomes releases active NF-κB, which translocates to the nucleus (see Figure 14-28).

■ Upon binding to its ligand Delta on the surface of an adjacent cell, the Notch receptor protein undergoes two proteolytic cleavages. The released Notch cytosolic segment then translocates into the nucleus and modulates gene transcription (see Figure 14-29).

■ Presenilin 1, which catalyzes the regulated intramembrane cleavage of Notch, also participates in the cleavage of amyloid precursor protein (APP) into a peptide that forms plaques characteristic of Alzheimer's disease.

14.7 Down-Modulation of Receptor Signaling

We have already seen several ways that signal-transduction pathways can be regulated. The levels of hormones produced and released from signaling cells are adjusted constantly to meet the needs of the organism. For example, kidney cells make and secrete more erythropoietin when the oxygen level is low and more red blood cells are needed. Intracellular proteins such as Ski and SOCS are induced following stimulation by TGFβ or cytokines, and then negatively regulate their respective signal-transduction pathways. Phosphorylation of receptors and downstream signaling proteins are reversed by the carefully controlled action of phosphatases. Here we discuss two other mechanisms by which signaling pathways are down-regulated: removal of receptors from the cell surface by endocytosis, and secretion of proteins that bind and sequester hormones, thus preventing their interaction with cell-surface receptors.

Endocytosis of Cell-Surface Receptors Desensitizes Cells to Many Hormones

In previous sections we discussed several signal-transduction pathways activated immediately after stimulation of cytokine receptors and receptor tyrosine kinases (RTKs). If the level of hormone in the environment remains high for several hours, cells usually undergo *desensitization,* such that they no longer respond to that concentration of hormone. This prevents inappropriate prolonged receptor activity, but under these conditions cells usually will respond if the hormone level is increased further. Ligand-dependent **receptor-mediated endocytosis,** which reduces the number of available cell-surface receptors, is a principal way that cells are desensitized to many peptides and other hormones.

In the absence of EGF ligand, for instance, the EGF receptor is internalized at a relatively slow rate by bulk membrane flow. Besides activating the receptor's protein tyrosine kinase, binding to EGF induces a conformational change in the cytosolic tail of the receptor. This exposes a sorting motif that facilitates receptor recruitment into clathrin-coated pits and subsequent internalization. After internalization, some cell-surface receptors (e.g., the LDL receptor) are efficiently recycled to the surface (see Figure 17-28). In contrast, internalized receptors for many peptide hormones, together with their bound hormone ligands, commonly are transported to lysosomes wherein they are degraded, rather than being recycled to the cell surface.

For example, each time an EGF receptor is internalized with bound EGF, it has about a 50 percent chance of being degraded. Exposure of a fibroblast cell to high levels of EGF for 1 hour induces several rounds of endocytosis, resulting in degradation of most receptor molecules. If the concentration of extracellular EGF is then reduced, the number of EGF receptors on the cell surface recovers by synthesis of new receptor molecules, a slow process that may take more than a day. In this way a cell can become desensitized to a continual high level of hormone and, after hormone removal, reestablish its initial level of cell-surface receptors, thereby becoming sensitive again to a low level of hormone.

Experiments with mutant cell lines demonstrate that internalization of RTKs plays an important role in regulating cellular responses to EGF and other growth factors. For instance, a mutation in the EGF receptor that prevents it from being incorporated into coated pits, and thus makes it resistant to ligand-induced endocytosis, substantially increases the sensitivity of cells to EGF as a mitogenic signal. Such mutant cells are prone to EGF-induced cell transformation. Interestingly, internalized receptors can continue to signal from intracellular compartments prior to their degradation.

 In most cases, peptide hormones that are internalized bound to their receptors are degraded intracellularly. If the initial extracellular hormone level is relatively low, this process may reduce the hormone level sufficiently to terminate cell signaling after a few hours or so. For instance, IL-2, a cytokine that stimulates growth of immune T cells, normally is depleted from the extracellular environment by this mechanism, leading to cessation of signaling. Mutant forms of IL-2 have been obtained that bind to the IL-2 receptor normally at pH 7.5, that of the extracellular medium, but poorly at pH 6, that of the initial endocytic vesicle, or endosome. These mutant IL-2 proteins dissociate from the receptor in the endosome and are "recycled"; that is, they are secreted back into the extracellular medium rather than accompanying the receptor to the lysosome for degradation. Because the lifetime of these mutant IL-2 proteins is longer than normal, they are more potent than their normal counterparts and may be useful therapeutically for stimulating production of T cells. ■

Secreted Decoy Receptors Bind Hormone and Prevent Receptor Activation

Another way of reducing the activity of cell-surface receptors is secretion of a protein that contains a hormone-binding segment but no signal-transducing activity. As might be expected, hormone binding to such proteins, called *decoy receptors,* reduces the amount of hormone available to bind to receptors capable of signaling. This type of regulation is important in controlling bone resorption, a complex physiological process that integrates several molecular mechanisms.

Net bone growth in mammals subsides just after puberty, but a finely balanced, highly dynamic process of disassembly (resorption) and reassembly (bone formation), called *remodeling,* goes on throughout adulthood. Remodeling permits the repair of damaged bones and can release calcium, phosphate, and other ions from mineralized bone into the blood for use elsewhere in the body.

Osteoclasts, the bone-dissolving cells, are a type of macrophage that contain highly dynamic integrin-containing adhesive structures, called *podosomes,* in the plasma membrane (see Figure 6-27). The $\alpha v \beta 3$ integrin in podosomes is crucial to the initial binding of osteoclasts to the surface of bone, since antibodies that bind to and block the activity of this integrin block bone resorption. Following their initial adhesion to bone, osteoclasts form specialized, very tight seals between themselves and bone, creating an enclosed extracellular space (Figure 14-31). An adhered osteoclast then secretes into this space a corrosive mixture of HCl and proteases that dissolves the inorganic components of the bone and digests its protein components. The mechanism of HCl generation and secretion is reminiscent of that used by the stomach to generate digestive juice (see Figure 7-28). As in gastric HCl secretion, carbonic anhydrase and an anion antiport protein are used to generate H^+ ions within osteoclasts. However, osteoclasts employ a V-type proton pump to export H^+ ions into the bone-facing space rather than the P-class ATP-powered H^+/K^+ pump used by gastric epithelial cells (see Figure 7-6).

Bone resorption by osteoclasts is carefully regulated by cell-cell interactions with neighboring *osteoblasts.* These bone-forming cells secrete type I collagen, the major organic component of bones. Osteoblasts express a trimeric cell-surface signaling protein termed *RANKL* that is a member of the TNF-α superfamily of trimeric signaling proteins. RANKL is the ligand for *RANK,* a cell-surface receptor expressed by osteoclasts. Interaction of RANK with RANKL initiates multiple intracellular signaling pathways in osteoclasts, including the NF-κB pathway that also is initiated by stimulation of TNF-α receptors (see Figure 14-28). Collectively, these signals induce the differentiation of osteoclasts and changes in their shape that promote tight binding to bone and thus bone resorption.

Osteoblasts also produce and secrete a soluble decoy receptor protein called *osteoprotegerin (OPG),* named for its ability to "protect bone." Secreted OPG binds to RANKL on

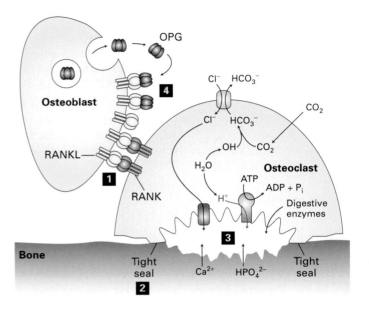

▲ **FIGURE 14-31 Bone resorption and its regulation.** Osteoclasts initially bind to bone via integrin-mediated podosomes. The subsequent activation of an osteoclast by interaction with neighboring osteoblasts via the trimeric membrane proteins RANKL and RANK **1** induces cytoskeletal reorganization, leading to formation of a specialized tight seal with bone **2**. The activated osteoclast secretes into the extracellular space generated by this seal a corrosive mixture of HCl and proteases that resorbs the bone **3**. Osteoblasts can suppress bone resorption by secreting osteoprotegerin (OPG). Binding of this decoy receptor to RANKL **4** blocks RANKL binding to RANK on osteoclasts and thus their activation. See the text for discussion. [Adapted from N. Takahashi et al., 1999, *Biochem. Biophys. Res. Comm.* **256**:449.]

the surface of osteoblasts, thereby preventing the RANKL-RANK interaction and inhibiting osteoclast activation and bone resorption (see Figure 14-31). Mice deleted for the OPG gene have weak, porous bones characteristic of excessive resorption. This finding supports the essential function of OPG in reducing bone resorption.

The rare hereditary disease *osteopetrosis,* marked by increased bone density, is due to abnormally low resorption. Far more common is *osteoporosis,* which is most prevalent among postmenopausal women. This metabolic disorder results from disproportionate bone resorption, leading to porous, less dense bones that are readily broken or fractured.

Many steroid hormones (e.g., estrogen, glucocorticoids), vitamin D, polypeptide hormones, and drugs influence bone metabolism by directly interacting with osteoblasts and altering the RANKL/RANK signaling system. Estrogen, for example, normally induces secretion of OPG and thus inhibits bone resorption. When estrogen is low, as it is in many postmenopausal women, resorption increases and the bones weaken. It may be possible to develop new treatments for osteoporosis based on altering the signaling system that controls bone resorption. ∎

 Because they bind their ligands so tightly, soluble extracellular domains of cell-surface hormone receptors are finding increasing use as therapeutics. Many cell-surface receptors are oriented in the plasma membrane such that the C-terminal signal-transducing domain extends into the cytosol and the N-terminal ligand-binding domain extends into the extracellular space. With recombinant DNA techniques a stop codon can be placed in the cDNA encoding such a receptor so that translation in an appropriate expression system generates a truncated protein corresponding to the receptor's extracellular domain, which will be secreted and can function as a decoy receptor. For example, local increases in TNF-α are frequent in *rheumatoid arthritis,* an inflammatory joint disease. Injection of the recombinant-produced extracellular domain of the TNF-α receptor, which "soaks up" some of the excess TNF-α and reduces inflammation, is now one of the major therapies for severe cases of this disease. ∎

KEY CONCEPTS OF SECTION 14.7

Down-Modulation of Receptor Signaling

∎ Endocytosis of receptor-hormone complexes and their degradation in lysosomes is a principal way of reducing the number of receptor tyrosine kinases and cytokine receptors on the cell surface, thus decreasing the sensitivity of cells to many peptide hormones.

∎ Bone resorption is triggered by binding of RANKL on osteoblasts to its receptor, RANK, on osteoclasts. RANKL/RANK signaling promotes tight adhesion of osteoclasts to bone and secretion of a bone-dissolving mixture of HCl and proteases by osteoclasts (see Figure 14-31).

∎ Osteoprotegerin, a decoy receptor for RANKL, inhibits osteoclast activation and bone resorption.

∎ The extracellular domains of many cell-surface receptors can be produced by recombinant DNA techniques and have potential as therapeutic decoy receptors. Already in use is such a decoy receptor for TNF-α, which binds excess TNF-α associated with rheumatoid arthritis and other inflammatory diseases.

PERSPECTIVES FOR THE FUTURE

The confluence of genetics, biochemistry, and structural biology has given us an increasingly detailed view of how signals are transmitted from the cell surface and transduced into changes in cellular behavior. The multitude of different extracellular signals, receptors for them, and intracellular signal-transduction pathways fall into a relatively small number of classes, and one major goal is to understand how similar signaling pathways often regulate very different cellular processes. For instance, STAT5 activates very different sets of genes in erythroid precursor cells, following stimulation of the erythropoietin receptor, than in mammary epithelial cells, following stimulation of the prolactin receptor. Presumably STAT5 binds to different groups of transcription factors in these and other cell types, but the nature of these proteins and how they collaborate to induce cell-specific patterns of gene expression remain to be uncovered.

Conversely, activation of the same signal-transduction component in the same cell through different receptors often elicits different cellular responses. One commonly held view is that the duration of activation of the MAP kinase and other signaling pathways affects the pattern of gene expression. But how this specificity is determined remains an outstanding question in signal transduction. Genetic and molecular studies in flies, worms, and mice will contribute to our understanding of the interplay between different pathway components and the underlying regulatory principles controlling specificity in multicellular organisms.

Researchers have determined the three-dimensional structures of various signaling proteins during the past several years, permitting more detailed analysis of several signal-transduction pathways. The molecular structures of different kinases, for example, exhibit striking similarities and important variations that impart to them novel regulatory features. The activity of several kinases, such as Raf and protein kinase B (PKB), is controlled by inhibitory domains as well as by multiple phosphorylations catalyzed by several other kinases. But our understanding of how the activity of these and other kinases is precisely regulated to meet the cell's needs will require additional structural and cell biological studies.

Abnormalities in signal transduction underlie many different diseases, including the majority of cancers and many inflammatory conditions. Detailed knowledge of the signaling pathways involved and the structure of their constituent proteins will continue to provide important molecular clues for the design of specific therapies. Despite the close structural relationship between different signaling molecules (e.g., kinases), recent studies suggest that inhibitors selective for specific subclasses can be designed. In many tumors of epithelial origin, the EGF receptor exhibits constitutive (signal-independent) protein tyrosine kinase activity, and a specific inhibitor of this kinase (Iressa™) has proved useful in the treatment of several such cancers. Similarly, monoclonal antibodies or decoy receptors that prevent pro-inflammatory cytokines like IL-1 and TNF-α from binding to their cognate receptors are now being used in treatment of several inflammatory diseases such as arthritis.

Drugs that target other signal-transducing proteins may be useful in controlling their abnormal activities. One example is Ras, which is anchored to cell membranes by farnesyl groups that are linked to Ras by farnesyl transferase. Inhibitors of this enzyme are being tested as therapeutic agents in cancers caused by expression of constitutively active Ras proteins. Detailed structural studies of the interaction between signal-transducing proteins offer exciting possibilities for designing new types of highly specific drugs. For instance, knowledge of the interface between the Sos and Ras proteins or between Ras and Raf could provide the basis of a drug that blocks activation of MAP kinase. As more signaling pathways become understood at a molecular level, additional targets for drug development will undoubtedly emerge.

KEY TERMS

activation lip *578*

constitutive *574*

cytokines *580*

decoy receptors *606*

erythropoietin *581*

JAK-STAT pathway *581*

kinase cascade *593*

MAP kinase *593*

NF-κB pathway *601*

Notch/Delta pathway *601*

nuclear-localization
   signal *576*

phosphoinositides *598*

PI-3 kinase pathway *598*

presenilin 1 *603*

protein kinase B *599*

PTEN phosphatase *600*

Ras protein *587*

receptor tyrosine kinases *578*

regulated intramembrane
   proteolysis *603*

scaffold proteins *597*

SH2 domains *579*

Smads *575*

TGFβ superfamily *574*

REVIEW THE CONCEPTS

1. Binding of TGFβ to its receptors can elicit a variety of responses in different cell types. For example, TGFβ induces plasminogen activator inhibitor in epithelial cells and specific immunoglobulins in B cells. In both cell types, Smad3 is activated. Given the conservation of the signaling pathway, what accounts for the diversity of the response to TGFβ in various cell types?

2. How is the signal generated by binding of TGFβ to cell-surface receptors transmitted to the nucleus where changes in target gene expression occur?

3. Name three features common to the activation of cytokine receptors and receptor tyrosine kinases. Name one difference with respect to the enzymatic activity of these receptors.

4. The intracellular events that proceed when erythropoietin binds to its cell-surface receptor are well-characterized examples of cell-signaling pathways that activate gene expression. What molecule translocates from the cytosol to the nucleus after (a) JAK2 activates STAT5 and (b) GRB2 binds to the Epo receptor?

5. Once an activated signaling pathway has elicited the proper changes in target gene expression, the pathway must be inactivated. Otherwise, pathologic consequences may result, as exemplified by persistent growth factor pathway signaling in many cancers. Many signaling pathways possess intrinsic negative feedback by which a downstream event in a pathway turns off an upstream event. Describe the negative feedback that down-regulates signals induced by (a) TGFβ and (b) erythropoietin.

6. GRB2 is an essential component of the epidermal growth factor (EGF) signaling pathway even though GRB2 lacks intrinsic enzymatic activity. What is the function of GRB2? What role do the SH2 and SH3 domains play in the function of GRB2? Many other signaling proteins possess SH2 domains. What determines the specificity of SH2 interactions with other molecules?

7. A mutation in the Ras protein renders Ras constitutively active (Ras^D). What is *constitutive activation*? How is constitutively active Ras cancer-promoting? What type of mutation might render the following proteins constitutively active: (a) Smad3; (b) MAP kinase; and (c) NF-κB?

8. The enzyme Ste11 participates in several distinct MAP kinase signaling pathways in the budding yeast *S. cerevisiae*. What is the substrate for Ste11 in the mating factor signaling pathway? When a yeast cell is stimulated by mating factor, what prevents the induction of filamentation since Ste11 also participates in the MAP kinase signaling pathway induced by starvation?

9. Describe the events required for full activation of protein kinase B. Name two effects of insulin mediated by protein kinase B in muscle cells.

10. Describe the function of the PTEN phosphatase in the PI-3 kinase signaling pathway. Why is a loss-of-function mutation in PTEN cancer-promoting? Predict the effect of constitutively active PTEN on cell growth and survival.

11. Why is the signaling pathway that activates NF-κB considered to be relatively irreversible compared to cytokine or receptor tyrosine kinase signaling pathways? Nonetheless, NF-κB signaling must be downregulated eventually. How is the NF-κB signaling pathway turned off?

12. What biochemical reaction is catalyzed by the enzyme presenilin 1? What is the role of presenilin 1 in transducing the signal induced by the binding of Delta to its receptor? How are mutations in presenilin 1 thought to contribute to Alzheimer's disease?

13. What is a decoy receptor? How does the decoy receptor osteoprotegerin block bone resorption? Some tumors are characterized by excess production of platelet-derived growth factor (PDGF). How might a recombinant decoy receptor function as a chemotherapeutic agent for such a cancer?

ANALYZE THE DATA

G. Johnson and colleagues have analyzed the kinase cascade in which MEKK2, a MAP kinase kinase kinase, participates in mammalian cells. By a yeast two-hybrid screen (see Chapter 11), MEKK2 was found to bind MEK5, a MAP kinase kinase. To elucidate the signaling pathway transduced by MEKK2 in vivo, the following studies were performed in human embryonic kidney (HEK293) cells in culture.

a. HEK293 cells were transfected with a plasmid encoding recombinant, tagged MEKK2, along with a plasmid encoding MEK5 or a control vector that did not encode a protein (mock). Recombinant MEK5 was precipitated from the cell extract by absorption to a specific antibody. The immunoprecipitated material was then resolved by polyacrylamide gel electrophoresis, transferred to a membrane, and examined by Western blotting with an antibody that recognized tagged MEKK2. The results are shown in the figure below, part a. What information about the MEKK2 kinase cascade do we learn from this experiment? Do the data in part a of the figure prove that MEKK2 activates MEK5, or vice versa?

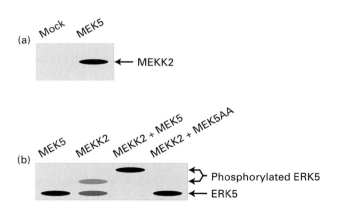

b. ERK5 is a MAP kinase previously shown to be activated when phosphorylated by MEK5. When ERK5 is phosphorylated by MEK5, its migration on a polyacrylamide gel is retarded. In the experiment shown in part b of the figure, HEK293 cells were transfected with a plasmid encoding ERK5 along with plasmids encoding MEK5, MEKK2, MEKK2 and MEK5, or MEKK2 and MEK5AA. MEK5AA is a mutant, inactive version of MEK5 that functions as a dominant-negative. Expression of MEK5AA in HEK293 cells prevents signaling through active, endogenous MEK5. Lysates of transfected cells were analyzed by Western blotting with an antibody against recombinant ERK5. From the data in part b of the figure, what can we conclude about the role of MEKK2 in the activation of ERK5? How do the data obtained when cells are cotransfected with ERK5, MEKK2, and MEK5AA help to elucidate the order of participants in this kinase cascade?

REFERENCES

TGFβ Receptors and the Direct Activation of Smads

Blobe, G., W. Schiemann, and H. Lodish. 2000. Role of transforming growth factor beta in human disease. *N. Eng. J. Med.* **342**:1350–1358.

Liu, X., Y. Sun, R. Weinberg, and H. Lodish. 2001. Ski/Sno and TGF-beta signaling. *Cytokine Growth Factor Rev.* **12**:1–8.

Massagué, J. 2000. How cells read TGF-beta signals. *Nature Rev. Mol. Cell Biol.* **1**:169–178.

Massagué, J., S. Blain, and R. Lo. 2000. TGFbeta signaling in growth control, cancer, and heritable disorders. *Cell* **103**:295–309.

Moustakas, A., S. Souchelnytskyi, and C. Heldin. 2001. Smad regulation in TGF-beta signal transduction. *J. Cell Sci.* **114**:4359–4369.

Shi, Y. 2001. Structural insights on Smad function in TGFbeta signaling. *Bioessays* **23**:223–232.

ten Dijke, P., K. Miyazono, and C. Heldin. 2000. Signaling inputs converge on nuclear effectors in TGF-beta signaling. *Trends Biochem. Sci.* **25**:64–70.

Wotton, D., and J. Massagué. 2001. Smad transcriptional corepressors in TGF beta family signaling. *Curr. Top. Microbiol. Immunol.* **254**:45–164.

Cytokine Receptors and the JAK-STAT Pathway

Alexander, W. 2002. Suppressors of cytokine signalling (SOCS) in the immune system. *Nature Rev. Immunol.* **2**:410–416.

Chatterjee-Kishore, M., F. van den Akker, and G. Stark. 2000. Association of STATs with relatives and friends. *Trends Cell Biol.* **10**:106–111.

Constantinescu, S. N., S. Ghaffari, and H. F. Lodish. 1999. The erythropoietin receptor: structure, activation, and intracellular signal transduction. *Trends Endocrin. Metab.* **10**:8–23.

Kile, B. T., et al. 2002. The SOCS box: a tale of destruction and degradation. *Trends Biochem. Sci.* **27**:235–41.

Leonard, W. 2001. Role of Jak kinases and STATs in cytokine signal transduction. *Int. J. Hematol.* **73**:271–277.

Levy, D. E., and J. E. Darnell, Jr. 2002. Stats: transcriptional control and biological impact. *Nature Rev. Mol. Cell Biol.* **3**:651–62.

Ozaki, K., and W. Leonard. 2002. Cytokine and cytokine receptor pleiotropy and redundancy. *J. Biol. Chem.* **277**:29355–29358.

Ward, A., I. Touw, and A. Yoshimura. 2000. The Jak-Stat pathway in normal and perturbed hematopoiesis. *Blood* **95**:19–29.

Yasukawa, H., A. Sasaki, and A. Yoshimura. 2000. Negative regulation of cytokine signaling pathways. *Ann. Rev. Immunol.* **18**:43–164.

Receptor Tyrosine Kinases and Activation of Ras

Gschwind, A., et al. 2001. Cell communication networks: epidermal growth factor receptor transactivation as the paradigm for interreceptor signal transmission. *Oncogene* **20**:1594–1600.

Harari, D., and Y. Yarden. 2000. Molecular mechanisms underlying ErbB2/HER2 action in breast cancer. *Oncogene* **19**:6102–6114.

Hubbard, S. 1999. Structural analysis of receptor tyrosine kinases. *Prog. Biophys. Mol. Biol.* **71**:343–358.

Huse, M., and J. Kuriyan. 2002. The conformational plasticity of protein kinases. *Cell* **109**:275–282.

Schlessinger, J. 2002. Ligand-induced, receptor-mediated dimerization and activation of EGF receptor. *Cell* **110**:669–672.

Shawver, L., D. Slamon, and A. Ullrich. 2002. Smart drugs: tyrosine kinase inhibitors in cancer therapy. *Cancer Cell* **1**:117–123.

Simon, M. 2000. Receptor tyrosine kinases: specific outcomes from general signals. *Cell* **103**:13–15.

Vetter, I., and A. Wittinghofer. 2001. The guanine nucleotide-binding switch in three dimensions. *Science* **294**:1299–1304.

MAP Kinase Pathways

Avruch, J., et al. 2001. Ras activation of the Raf kinase: tyrosine kinase recruitment of the MAP kinase cascade. *Recent Prog. Horm. Res.* **56**:127–155.

Chang, L., and M. Karin. 2001. Mammalian MAP kinase signalling cascades. *Nature* **410**:37–40.

Cobb, M., and E. Goldsmith. 2000. Dimerization in MAP-kinase signaling. *Trends Biochem. Sci.* **25**:7–9.

Garrington, T., and G. Johnson. 1999. Organization and regulation of mitogen-activated protein kinase signaling pathways. *Curr. Opin. Cell Biol.* **11**:211–218.

Kerkhoff, E., and U. Rapp. 2001. The Ras-Raf relationship: an unfinished puzzle. *Adv. Enz. Regul.* **41**:261–267.

Martin-Blanco, E. 2000. p38 MAPK signalling cascades: ancient roles and new functions. *Bioessays* **22**:637–645.

Tzivion, G., and J. Avruch. 2002. 14-3-3 proteins: active cofactors in cellular regulation by serine/threonine phosphorylation. *J. Biol. Chem.* **277**:3061–3064.

Widmann, C., S. Gibson, M. Jarpe, and G. Johnson. 1999. Mitogen-activated protein kinase: conservation of a three-kinase module from yeast to human. *Physiol. Rev.* **79**:143–180.

Phosphoinositides as Signal Transducers

Belham, C., S. Wu, and J. Avruch. 1999. Intracellular signalling: PDK1—a kinase at the hub of things. *Curr. Biol.* **9**:R93–R96.

Cantley, L. 2002. The phosphoinositide 3-kinase pathway. *Science* **296**:1655–1657.

Cantley, L., and B. Neel. 1999. New insights into tumor suppression: PTEN suppresses tumor formation by restraining the phosphoinositide 3-kinase/AKT pathway. *Proc. Nat'l. Acad. Sci. USA* **96**:4240–4245.

Chung, C., S. Funamoto, and R. Firtel. 2001. Signaling pathways controlling cell polarity and chemotaxis. *Trends Biochem. Sci.* **26**:557–566.

Toker, A., and A. Newton. 2000. Cellular signaling: pivoting around PDK-1. *Cell* **103**:185–188.

Wishart, M., and J. Dixon. 2002. PTEN and myotubularin phosphatases: from 3-phosphoinositide dephosphorylation to disease. *Trends Cell. Biol.* **12**:579–584.

Pathways That Involve Signal-Induced Protein Cleavage

Brown, M., J. Ye, R. Rawson, and J. Goldstein. 2000. Regulated intramembrane proteolysis: a control mechanism conserved from bacteria to humans. *Cell* **100**:391–398.

Esler, W., and M. Wolfe. 2001. A portrait of Alzheimer secretases—new features and familiar faces. *Science* **293**:1449–1454.

Ghosh, S., and M. Karin. 2002. Missing pieces in the NF-kappaB puzzle. *Cell* **109** (suppl.): S81–S96.

Haass, C., and H. Steiner. 2002. Alzheimer disease g-secretase: a complex story of GxGD-type presenilin proteases. *Trends Cell Biol.* **12**:556–561.

Karin, M., and Y. Ben-Neriah. 2000. Phosphorylation meets ubiquitination: the control of NF-[kappa]B activity. *Ann. Rev. Immunol.* **18**:621–663.

Khush, R., and B. Lemaitre. 2000. Genes that fight infection: what the *Drosophila* genome says about animal immunity. *Trends Genet.* **16**:442–449.

Silverman, N., and T. Maniatis. 2001. NF-kappaB signaling pathways in mammalian and insect innate immunity. *Genes Devel.* **15**:2321–2342.

Sisodia, S., W. Annaert, S. Kim, and B. De Strooper. 2001. Gamma-secretase: never more enigmatic. *Trends Neurosci.* **24**:S2–S6.

Weihofen, A., and B. Martoglio. 2003. Intramembrane-cleaving proteases. *Trends Cell Biol.* **13**:71–78.

Down-Modulation of Receptor Signaling

Lauffenburger, D., E. Fallon, and J. Haugh. 1998. Scratching the (cell) surface: cytokine engineering for improved ligand/receptor trafficking dynamics. *Chem. Biol.* **5**:R257–R263.

Theill, L., W. Boyle, and J. Penninger. 2002. RANK-L and RANK: T cells, bone loss, and mammalian evolution. *Ann. Rev. Immunol.* **20**:795–823.

Waterman, H., and Y. Yarden. 2001. Molecular mechanisms underlying endocytosis and sorting of ErbB receptor tyrosine kinases. *FEBS Lett.* **490**:142–152.

15

INTEGRATION OF SIGNALS AND GENE CONTROLS

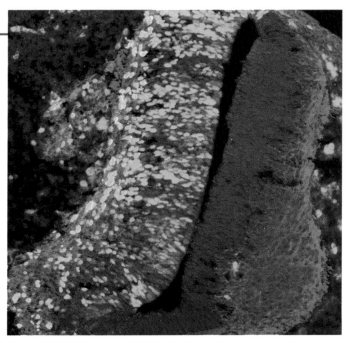

Dividing cells (blue) in the developing spinal cord will differentiate into neurons (red). Cells that were engineered to make a differentiation-inhibiting signal (green) cause persistent cell division and reduce the number of differentiated neurons at the left. [Sean G. Megason and Andrew P. McMahon. Adapted from Sean G. Megason and Andrew P. McMahon, 2002, *Development* **129**:2087–2098.]

Inherent in the full genome contained in most cells is the potential to form vastly diverse cell types, which perform an enormous variety of tasks. Each individual cell, however, employs only part of an organism's complete genetic repertoire. An array of external hormonal, metabolic, developmental, and environmental *signals* influence which genes a cell uses at any given time in its life span. Infections also can trigger many responses. A cell's response to an external signal largely depends on its properties including (1) the inventory, locations, and associations of its proteins and other molecules; (2) its shape and attachments to other cells; and (3) its chromatin structure, which facilitates or blocks access to particular genes. We can think of these properties as a cell's "memory" determined by its history and response to previous signals. Thus, for instance, a cell can respond to a signal only if it possesses a **receptor** for that signal. In addition, a cell typically receives more than one signal at a time: for example, a combination of transforming growth factor β (TGFβ) and fibroblast growth factor (FGF), a hormone signal that is interpreted in light of the ambient temperature, or an electrical pulse that is modulated by local ionic conditions. The response to each signal, or condition, is often influenced by another one. This integration of signals can prevent inappropriate responses and permit more nuanced responses to multiple signals.

To understand a cell's response to one or more signals and the effect of its memory on this response, it is useful to monitor changes in the expression of all genes and changes in the locations of organelles, proteins, or other molecules. Signal-induced changes in the intracellular ionic environment, membrane potential, and cell shape also may be relevant to a cell's response. Two serious limitations have hampered efforts to obtain such a comprehensive view of the nature of cell responses to external signals. First, usually only one or a few aspects of a cell's response to a signal is easily monitored; second, determining responses in living cells in "real time" poses many technical difficulties. Technological advances are beginning to solve these problems, although neither has been completely overcome.

The major question posed in this chapter is how a cell integrates multiple signals and responds in the context of

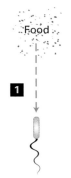 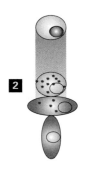 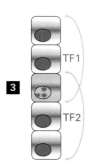 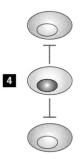 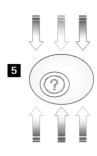

1 Cells adjust to their particular environmental inputs (e.g., oxygen, sugar, and temperature)

2 Graded signals create different cell types

3 Combined actions of transcription factors create different cell types

4 Lateral inhibition signals prevent duplication of unique cell types

5 Integration of signals allows cells to adjust to their neighbors and to change with time

▲ **FIGURE 15-1 Signaling systems and cell responses.** Cells are vibrantly alert detectors, sensing and interpreting information constantly to adjust to the environment (**1**) and coordinate activities with surrounding cells. A cell can respond to signals by changing the genes that it transcribes, altering the cell surface, modifying proteins and enzyme activities, moving materials between compartments, revamping its cytoskeleton, migrating, or dying. A modest number of signaling systems constitutes a core toolbox. Each system is used repeatedly in different organisms, in different tissues, and at different times. Signals are crucial in building multicellular organisms, where different cell types are created by controlled signal transmission and reception. Cells can become different, depending on the amount of a signal (**2**), with a larger amount giving rise to one cell fate and a smaller amount to another. New boundaries form between cells of different types, creating tissues and demarcations within tissues. Different cell types are created by combinations of transcription factors (**3**). Inhibitory signals emitted by cells undergoing a differentiation step can prevent nearby cells from making the same decision (**4**), thus preventing duplication of structures. Cells generally integrate many signals in deciding how to proceed (**5**).

its memory, especially in the course of **development** and cell **differentiation** (Figure 15-1). We begin by looking at various techniques that are beginning to provide a global view of signal-induced responses. In particular, we describe how the determination of whole-genome transcription patterns is a source of new insights into responses to signals. We then consider cell responses to certain environmental perturbations in Section 15.2. The next section introduces the concept of graded regulators that cause different cell responses, depending on their concentration. This type of system allows cells at different distances from the source of a regulatory molecule to become different types of cells. We examine how such regulation creates boundaries within an epithelium in early *Drosophila* development, with cells on one side of the border taking one path of differentiation and those on the other side taking another. The creation of other boundaries by graded transcriptional activators and graded extracellular signals is discussed in Sections 15.4 and 15.5, respectively. As borders form, cells reinforce their decisions by signaling across the borders so that compatible adjacent structures form. As illustrated by the examples in Section 15.6, such signaling can either promote or inhibit particular developmental changes in adjacent cells. In the final section, we take a closer look at how signals are integrated and controlled in different cells.

Although the number of signaling pathways encountered in this chapter and others may seem overwhelming, there are actually a relatively small number of distinct pathways for transducing external signals. Primary among them are the in-

tracellular signal-transduction pathways activated by the various receptor classes listed in Table 14-1. In addition, cell–cell and cell–matrix adhesions mediated by cadherins and integrins can initiate intracellular signaling pathways (Chapter 6). The informational complexity needed to create many cell types and cell properties comes from combining signals. Elucidation of the underlying principles and mechanisms relevant to all signaling pathways forms the foundation for understanding how cells integrate signals to achieve a particular identity or other response.

15.1 Experimental Approaches for Building a Comprehensive View of Signal-Induced Responses

Several technical advances are helping investigators to discern the totality of the cellular response to signals. Perhaps the most significant advance is the sequencing of whole genomes from various organisms and subsequent analysis to identify individual genes and analyze their functions. The data amassed in these genome projects have led to the development of techniques for monitoring the effects of a signal on the expression of the entire gene set. Using gene-inactivation methods discussed in Chapter 9, researchers can mutate specific genes encoding various components of signaling pathways. The phenotypic effects of such mutations often provide clues about the functions of pathway components and the order in which they function. In vitro studies on signaling in a variety of differentiated cell types and even complex tissues are now possible because of recent improvements in cell- and tissue-culture methods. Certain signal-induced responses can be monitored in living cells with the use of various fluorescent agents and the observation of cells in a fluorescence microscope. For instance, this technique can reveal changes in the amounts and localization of specific proteins, as well as fluctuations in H^+ or Ca^{2+} concentrations in the cytosol (see Figures 5-46 and 5-47). Development of additional fluorescent indicator dyes will allow monitoring of other molecules in living cells.

Genomic Analyses Show Evolutionary Conservation and Proliferation of Genes Encoding Signals and Regulators

In Chapter 9, we considered the difficulty and ambiguity in identifying genes within genomic sequences, especially in higher organisms. Despite the limitations, genomic analyses have been sources of exciting and sometimes surprising insights or have confirmed earlier conclusions based on the results of other types of studies.

First, the total number of protein-coding genes does not correlate in any simple way with standard conceptions about animal complexity (see Figure 9-34). Humans, for instance, have only about 1.75 times as many genes as the roundworm *Caenorhabditis elegans*. Likewise, *C. elegans* has about 1.4 times as many genes as the fruit fly *Drosophila*, which exhibits a much more complex body plan and more complex behavior.

Second, genomic comparisons support the conclusion based on two decades of developmental genetics research that many regulatory genes whose encoded proteins control tissue differentiation, organogenesis, and the body plan have been conserved for hundreds of millions of years. For example, the *Pax6* gene is employed in eye development in enormously diverse organisms, such as clams, flies, and humans, and the *tinman* gene is necessary for heart development in flies and humans. As discussed in Section 15.4, the Hox gene cluster controls head-to-tail organization of the body in almost all animals examined to date. Because of the conservation of genes and proteins, the results of experiments on one organism are useful guides for research on other organisms. Indeed much of human biology and medicine has been and continues to be built on knowledge gained from a broad spectrum of experimental systems.

Third, despite the considerable commonality of genes and proteins among different animals, genomic analyses suggest that about 30 percent of the genes of each animal organism are unique to that animal. The invertebrates *Drosophila* and *C. elegans* have in common certain genes that are not recognizable in any of the other genomes analyzed to date. Flies and worms are believed to have a common ancestor that arose from an even more ancient ancestor in common with vertebrates. If this view is correct, any genes present in flies

and humans could be expected to be present in worms; likewise, any genes common to worms and humans would presumably be present in flies. Recent work has revealed that all three species have about 1500 genes in common, as expected (Figure 15-2). Contrary to expectations, however, about 1250 genes common to humans and flies are not found in worms, and about 500 genes common to humans and worms are not found in flies. Thus organism-specific gene loss occurred in the evolution of *C. elegans* and *Drosophila* subsequent to the time when the invertebrate and vertebrate lineages diverged.

Fourth, as noted in preceding chapters, duplication of certain protein-coding genes and subsequent divergence in the course of evolution have given rise to gene families. The members of a **gene family** and corresponding **protein family** have close but nonidentical sequences. Genomic analysis and findings from other studies show that the number of members in a particular protein family varies in different species. For instance, the *transforming growth factor β family* of secreted signaling proteins has 28 members in humans but only 6 in flies and 4 in worms. The semaphorins, which are signals for neural development, form a 22-member family in humans; flies have 6 members and worms have 2. Such proliferation of genes could give rise to signaling proteins that can move different distances through tissue or differ in other properties. Alternatively, the members of a gene family may be differently regulated, thus allowing rather similar proteins to be produced at different times and places. Both types of variation exist, and both allow a moderate number of types of signals to serve a multitude of purposes.

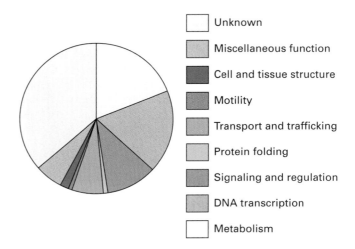

▲ **FIGURE 15-2 Evolutionary conservation of core processes in human, fruit fly (*Drosophila*), and roundworm (*C. elegans*) genomes.** On the basis of fairly stringent criteria for protein similarity, humans, flies, and worms have in common about 1500 genes distributed among the functional classes shown in this pie chart. About 28 percent of this common set of genes encode proteins that function in signaling or gene control. The molecular functions of about one-third of the genes and proteins common to these species are not yet known. [Adapted from J. C. Venter et al., 2001, *Science* **291**:1304.]

In Situ Hybridization Can Detect Transcription Changes in Intact Tissues and Permeabilized Embryos

A common effect of external signals is to alter the pattern of gene expression by a cell. Signal-induced changes in the expression of particular genes is usually monitored by measuring the corresponding mRNAs or proteins in the presence and absence of a signal. The total cellular mRNA can be extracted, separated by gel electrophoresis, and subjected to **Northern blotting,** which detects individual mRNAs by hybridization to labeled complementary DNA probes (see Figure 9-26). Likewise, cellular proteins can be extracted, separated electrophoretically, and subjected to **Western blotting,** a procedure in which individual proteins separated on the blot are detected with specific antibodies (see Figure 3-35). These blotting methods are generally not sensitive enough to determine changes within a single cell. The **polymerase chain reaction** (PCR), however, can amplify a specific mRNA from a single cell so that it is detectable (see Figure 9-24).

Both Northern blotting and PCR amplification require extracting the mRNA from a cell or mixture of cells, which means that the cells are removed from their normal location within an organism or tissue. As a result, the location of a cell and its relation to its neighbors is lost. To retain such positional information, a whole or sectioned tissue or even

a whole permeabilized embryo may be subjected to **in situ hybridization** to detect the mRNA encoded by a particular gene (Figure 15-3). This technique allows gene transcription to be monitored in both time and space. **Immunohistochemistry,** the related technique of staining tissue with fluorescence-labeled antibodies against a particular protein, provides similar information for proteins, an important advantage for obtaining ideas about protein function from its subcellular location (see Figures 5-33 and 5-45).

DNA Microarray Analysis Can Assess Expression of Multiple Genes Simultaneously

A major limitation of in situ hybridization and blotting techniques is that the mRNA or protein product of only a few genes can be examined at a time. Thus monitoring the activity of many genes by these methods requires multiple assays. In contrast, researchers can monitor the expression of thousands of genes at one time with **DNA microarrays** (see Figure 9-35). In this technique, cDNAs labeled with a fluorescent dye are made from the total mRNA extracted from the cells under study. The labeled cDNAs are then hybridized to a microscope slide dotted with spots of DNA. Each DNA spot contains a unique sequence from a particular gene, and tens of thousands of genes can be represented on a standard slide. The fluorescence of spots that

(a)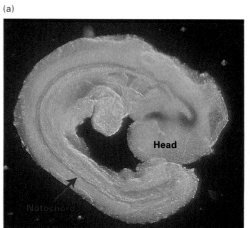

(b)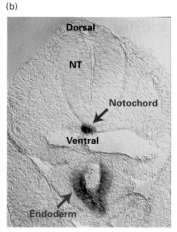

(c)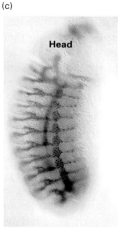

▲ **EXPERIMENTAL FIGURE 15-3 In situ hybridization can detect activity of specific genes in whole and sectioned embryos.** The specimen is permeabilized by treatment with detergent and a protease to expose the mRNA to the probe. A DNA or RNA probe, specific for the mRNA of interest, is made with nucleotide analogs containing chemical groups that can be recognized by antibodies. After the permeabilized specimen has been incubated with the probe under conditions that promote hybridization, the excess probe is removed with a series of washes. The specimen is then incubated in a solution containing an antibody that binds to the probe. This antibody is covalently joined to a reporter enzyme (e.g., horseradish peroxidase or alkaline phosphatase) that produces a colored reaction product. After excess antibody has been removed, substrate for the

reporter enzyme is added. A colored precipitate forms where the probe has hybridized to the mRNA being detected. (a) A whole mouse embryo at about 10 days of development probed for *Sonic hedgehog* mRNA. The stain marks the notochord (red arrow), a rod of mesoderm running along the future spinal cord. (b) A section of a mouse embryo similar to that in part (a). The dorsal/ventral axis of the neural tube (NT) can be seen, with the *Sonic hedgehog*–expressing notochord (red arrow) below it and the endoderm (blue arrow) still farther ventral. (c) A whole *Drosophila* embryo probed for an mRNA produced during trachea development. The repeating pattern of body segments is visible. Anterior (head) is up; ventral is to the left. [Courtesy of L. Milenkovic and M. P. Scott.]

hybridize to a cDNA species is measured with an instrument that scans the slide. Fluorescing spots thus represent active genes, which have been transcribed into their mRNAs (see Figure 1-23).

Microarray experiments are commonly used to compare the mRNAs produced by two different populations of cells: for example, two distinct cell types, the same cell type before and after some treatment, or mutant and normal cells. An example of a microarray-based discovery comes from the results of studies of cultured fibroblast cells, which have long been known to initiate cell division when serum containing growth factors is added to the medium. Microarray analysis of gene expression at different times after treatment of fibroblasts with serum showed that transcription of about 500 of the 8613 genes examined changed substantially over time (see Figure 9-36). Transcriptional changes were detected within 15 minutes, with genes encoding proteins that control progression through the cell cycle becoming active first. Later, genes encoding proteins with roles in wound healing, such as clotting factors and attractants for immune-system cells, became active. The production of these proteins suggests that proliferating fibroblasts are stimulated by serum to participate in wound healing, something that had not been known. In retrospect, it makes sense, because the time during which fibroblasts are exposed to serum in an intact organism is when there is a wound. The results show the usefulness of microarrays in revealing unexpected responses by cells.

The developmental time course of gene transcription has been assessed with DNA microarrays for the nematode *C. elegans* and the fly *Drosophila*. In recent experiments, microarrays representing about 94 percent of the *C. elegans* genes were used to monitor transcription at different stages of development and in both sexes. The results showed that expression of about 58 percent of the monitored genes changes more than twofold during development, and another 12 percent are transcribed in sex-specific patterns. Findings from a similar study assessing about one-third of all *Drosophila* genes showed that transcription of more than 90 percent of them changes by twofold or more during development and that most genes are used repeatedly during development (Figure 15-5). These results clearly show that development is marked by extensive changes in transcription, with few genes exhibiting a monotonous pattern of unchanging transcription.

 In the future, microarray analysis will be a powerful diagnostic tool in medicine. For instance, particular sets of mRNAs have been found to distinguish tumors with a poor prognosis from those with a good prognosis (Chapter 23). Previously indistinguishable disease variations are now detectable. Analysis of tumor biopsies for these distinguishing mRNAs will help physicians to select the most appropriate treatment. As more patterns of gene expression characteristic of various diseased tissues are recognized, the diagnostic use of DNA microarrays will be extended to other conditions. ∎

Protein Microarrays Are Promising Tools for Monitoring Cell Responses That Include Changes in Protein-Binding Patterns

A cell's response to signals can include not only changes in gene expression, but also alterations in the modifications of proteins and the associations between proteins. As discussed in other chapters, the activities of many proteins depend on their association with other proteins or with small intracellular signaling molecules (e.g., cAMP or phosphoinositides). Two common examples are the activation of adenylyl cyclase by interaction with $G_{s\alpha} \cdot GTP$ (see Figure 13-11) and the activation of protein kinase A by binding of cAMP (see Figure 3-27). The activity of some transcriptional regulators (e.g., CREB) also depend on their associating with another protein (see Figure 13-32). The results of systematic studies are beginning to reveal protein–protein associations that are critical for cell functioning and how these associations change in response to signals. For example, scientists have produced large quantities of 5800 yeast proteins (≈80 percent of the total proteins) by cloning them in high-level expression vectors in yeast and purifying the individual proteins. In a technique analogous to DNA microarrays, small samples of the purified yeast proteins can be spotted on microscope slides to produce a **protein microarray**, also called a proteome chip.

To test the efficacy of assaying protein–protein associations on such arrays, researchers exposed the yeast protein microarray to biotin-labeled **calmodulin**, a calcium-binding protein. After excess calmodulin was removed from the microarray, binding of calmodulin to proteins in the array was detected with a fluorescent reagent specific for biotin (Figure 15-4). This experiment succeeded in detecting six proteins already known to bind calmodulin. Six other known calmodulin-binding proteins were not detected, two because they were not included in the array and four that may have been underproduced. In principle, others could be missed because proteins associate only as part of a complex of more than two proteins or because the protein fastened to the chip is in the wrong conformation for binding. Despite these possible problems, 33 other calmodulin-binding yeast proteins not previously recognized also were detected. The gene sequences corresponding to the 39 calmodulin-binding proteins detected indicate that 14 of these proteins have a common motif that may form the binding surface. The results of such experiments show that protein arrays will be a useful, if not completely comprehensive and accurate, tool for monitoring associations of proteins as indicators of cell responses.

Systematic Gene Inactivation by RNA Interference

Changes in transcription at various developmental stages provide one criterion for identifying genes that play a critical role in cell regulation and differentiation. A more important

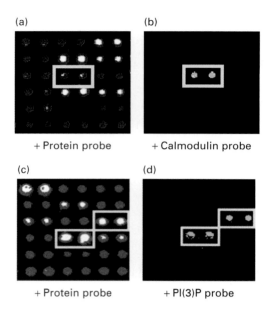

+ Protein probe   + Calmodulin probe

+ Protein probe   + PI(3)P probe

▲ **EXPERIMENTAL FIGURE 15-4** Protein microarray analysis can reveal protein–protein and protein–lipid interactions. High-level expression cloning was used to produce 5800 yeast proteins, which were purified and then spotted in duplicate on microscope slides. Some 13,000 protein samples can be spotted in half the area of a standard microscope slide. A probe was prepared by covalently attaching biotin to calmodulin or to phosphatidylinositol triphosphate (IP_3). The yeast proteome chips were incubated with the biotinylated probe, and then excess probe was washed off. Calmodulin or IP_3 bound to proteins in the microarray was detected with fluorescence-labeled streptavidin, a bacterial protein that strongly and specifically binds biotin. (a) A part of a proteome chip probed to reveal the location and approximate amounts of all proteins, which are spotted in duplicate columns. (b) The same preparation probed with biotinylated calmodulin. The two green signals correspond to a calmodulin-binding protein spotted in duplicate. (c) A different field of protein spots. (d) IP_3 binds to two of the proteins in (c), each in duplicate. [Courtesy of Paul Bertone, Yale University.]

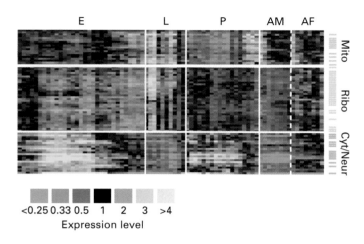

<0.25 0.33 0.5   1   2   3   >4
Expression level

▲ **EXPERIMENTAL FIGURE 15-5** DNA microarray cluster analysis detects global changes in transcription in *Drosophila* development. The life cycle was divided into about 70 periods from early embryos through aged adults. The mRNA was extracted from animals at each period, converted into fluorescence-labeled cDNA, and hybridized to microarrays representing about 5000 genes. Computer analysis of the original microarray data grouped genes showing similar changes in expression relative to a standard reference sample. (See Figure 9-35 for general protocol of DNA microarray experiments.) The three panels shown represent only a small fraction of the genes in the microarray. Each very narrow row in a panel represents a different gene. General periods are labeled E (embryogenesis), L (larval), P (pupal), and A (adult, male or female). Each vertical line represents one of the 70 developmental periods. Yellow indicates an increase in transcription compared with the reference sample; blue, a decrease in expression; and black, no significant change in expression. Genes are grouped, or "clustered," because they have similar rises and falls in RNA abundance during development. Clusters often reveal genes involved in similar processes. The three clusters here are mitochondrial protein genes (*top*), ribosomal protein genes (*middle*), and genes involved in the cytoskeleton and neural development (*bottom*). The clustering of a gene of unknown function with some of known function provides a hypothesis about the function of the former. [Adapted from M. Arbeitman et al., 2002, *Science* **297**:2270.]

criterion, however, is the **phenotype** of cells or animals lacking the gene. Knowledge of the functions of genes and relations among them during development can be obtained by systematically eliminating the function of each gene that is normally expressed in a tissue, one at a time, and observing what goes wrong. This procedure is now theoretically possible in *C. elegans*, which is particularly susceptible to **RNA interference** (RNAi). In this technique, delivery of double-stranded RNA into worms by injection, feeding, or soaking leads to the destruction of the corresponding endogenous mRNA (see Figure 9-43). The phenotype of worms that develop from RNAi-treated embryos is often similar to the one that would result from inactivation of the corresponding gene itself. In initial studies with *C. elegans*, RNA interference with 16,700 genes (about 86 percent of the genome) yielded 1722 visibly abnormal phenotypes. The genes whose

functional inactivation cause particular abnormal phenotypes can be grouped into sets; each member of a set presumably controls the same signals or events. The regulatory relations among the genes in the set—for example, the genes that control muscle development—can then be worked out.

KEY CONCEPTS OF SECTION 15.1

Experimental Approaches for Building a Comprehensive View of Signal-Induced Responses

■ About one-third of the genes in humans, flies (*Drosophila*), and roundworms (*C. elegans*) are unique to each animal.

■ The presence and distribution of specific mRNAs and proteins can be detected in living cells by in situ hybridization and immunohistochemical staining.

■ DNA microarray analysis allows the full complement of genes to be examined for transcriptional changes that occur in response to environmental changes or extracellular signals and in development (see Figure 15-5).

■ Protein microarrays are proving useful in detecting and monitoring changes in protein–protein associations (see Figure 15-4).

■ The function of a gene found to be activated under certain conditions can be tested by inactivating it and observing the resulting phenotype.

■ Signals, altered environmental conditions, or infection generally evoke not a single response by cells but multiple changes in the pattern of gene transcription, protein modifications, and associations between proteins. By monitoring the totality of these individual responses, researchers are developing comprehensive views of how and why cells respond.

15.2 Responses of Cells to Environmental Influences

Much of this chapter deals with signaling pathways that control how cells change in the course of development. The mature cells of some tissues (e.g., blood and skin) have a relatively short life span compared with that of other types of cells and are constantly being replaced by the differentiation and proliferation of stem cells (Chapter 22). In a sense, such tissues never stop developing. The mature cells of other tissues, such as the brain, are long lived; after such tissues reach maturation, there is little additional differentiation. Mature cells in all tissues, however, are changing constantly in response to metabolic or behavioral demands as well as to injury or infection. In this section, we consider several ways that cells respond to variations in the demand for two environmental inputs—glucose and oxygen—or in their levels.

Integration of Multiple Second Messengers Regulates Glycogenolysis

One way for cells to respond appropriately to current physiological conditions is to sense and integrate more than one signal. A good example comes from glycogenolysis, the hydrolysis of glycogen to yield glucose 1-phosphate. In Chapter 13, we saw that a rise in cAMP induced by epinephrine stimulation of β-adrenergic receptors promotes glycogen breakdown in muscle and liver cells (see Figure 13-17). In both muscle and liver cells, other second messengers also produce the same cellular response.

In muscle cells, stimulation by nerve impulses causes the release of Ca^{2+} ions from the sarcoplasmic reticulum and an increase in the cytosolic Ca^{2+} concentration, which triggers muscle contraction. The rise in cytosolic Ca^{2+} also activates *glycogen phosphorylase kinase (GPK)*, thereby stimulating the degradation of glycogen to glucose 1-phosphate, which fuels prolonged contraction. Recall that phosphorylation by cAMP-dependent **protein kinase A** also activates glycogen phosphorylase kinase. Thus this key regulatory enzyme in glycogenolysis is subject to both neural and hormonal regulation in muscle (Figure 15-6a).

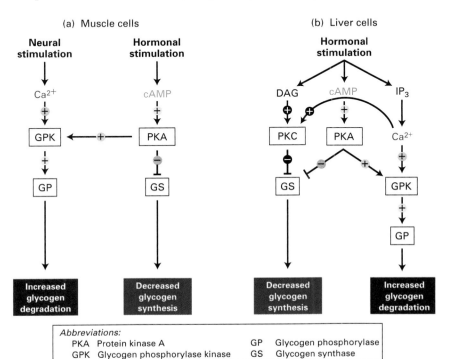

◄ **FIGURE 15-6 Integrated regulation of glycogenolysis mediated by several second messengers.** (a) Neuronal stimulation of striated muscle cells or epinephrine binding to β-adrenergic receptors on their surfaces leads to increased cytosolic concentrations of the second messengers Ca^{2+} or cAMP, respectively. The key regulatory enzyme, glycogen phosphorylase kinase (GPK), is activated by Ca^{2+} ions and by a cAMP-dependent protein kinase A (PKA). (b) In liver cells, β-adrenergic stimulation leads to increased cytosolic concentrations of cAMP and two other second messengers, diacylglycerol (DAG) and inositol 1,4,5-trisphosphate (IP_3). Enzymes are marked by white boxes. (+) = activation of enzyme activity, (−) = inhibition.

In liver cells, hormone-induced activation of phospholipase C also regulates glycogen breakdown and synthesis by the two branches of the inositol-lipid signaling pathway. Phospholipase C generates two second messengers, diacylglycerol (DAG) and inositol 1,4,5-trisphosphate (IP$_3$) (see Figure 13-28). DAG activates **protein kinase C**, which phosphorylates glycogen synthase, yielding the phosphorylated inactive form and thus inhibiting glycogen synthesis. IP$_3$ induces an increase in cytosolic Ca^{2+}, which activates glycogen phosphorylase kinase as in muscle cells, leading to glycogen degradation. In this case, multiple intracellular signal-transduction pathways are activated by the same signal (Figure 15-6b).

The dual regulation of glycogen phosphorylase kinase results from its multimeric subunit structure $(\alpha\beta\gamma\delta)_4$. The γ subunit is the catalytic protein; the regulatory α and β subunits, which are similar in structure, are phosphorylated by protein kinase A; and the δ subunit is the Ca^{2+}-binding switch protein calmodulin. Glycogen phosphorylase kinase is maximally active when Ca^{2+} ions are bound to the calmodulin subunit and at least the α subunit is phosphorylated. In fact, the binding of Ca^{2+} to the calmodulin subunit may be essential to the enzymatic activity of glycogen phosphorylase kinase. Phosphorylation of the α and β subunits increases the affinity of the calmodulin subunit for Ca^{2+}, enabling Ca^{2+} ions to bind to the enzyme at the submicromolar Ca^{2+} concentrations found in cells not stimulated by nerves. Thus increases in the cytosolic concentration of Ca^{2+} or of cAMP or of both induce incremental increases in the activity of glycogen phosphorylase kinase. As a result of the elevated level of cytosolic Ca^{2+} after neuron stimulation of muscle cells, glycogen phosphorylase kinase will be active even if it is unphosphorylated; thus glycogen can be hydrolyzed to fuel continued muscle contraction in the absence of hormone stimulation.

Insulin and Glucagon Work Together to Maintain a Stable Blood Glucose Level

In the regulation of glycogenolysis, neural and hormonal signals regulate the same key multimeric enzyme. In contrast, the maintenance of normal blood glucose concentrations depends on the balance between two hormones that elicit different cell responses. During periods of stress, the epinephrine-induced increase in glycogenolysis in liver cells leads to a rise in blood glucose. During normal daily living, however, the blood glucose level is under the dynamic control of **insulin** and **glucagon**.

Insulin and glucagon are peptide hormones produced by cells within the islets of Langerhans, cell clusters scattered throughout the pancreas. Insulin, which contains two polypeptide chains linked by disulfide bonds, is synthesized by the β cells in the islets; glucagon, a monomeric peptide, is produced by the α cells in the islets. Insulin *reduces* the level of blood glucose, whereas glucagon *increases* blood glucose. The availability of glucose for cellular metabolism is regulated during periods of abundance (following a meal) or

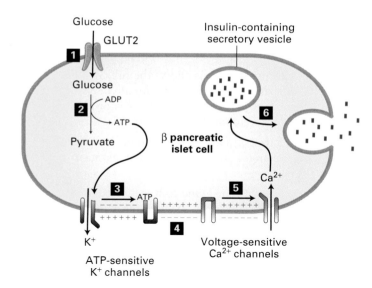

▲ **FIGURE 15-7 Secretion of insulin from pancreatic β cells in response to a rise in blood glucose.** The entry of glucose into β cells is mediated by the GLUT2 glucose transporter (**1**). Because the K_m for glucose of GLUT2 is ≈20 mM, a rise in extracellular glucose from 5 mM, characteristic of the fasting state, causes a proportionate increase in the rate of glucose entry (see Figure 7-3). The conversion of glucose into pyruvate is thus accelerated, resulting in an increase in the concentration of ATP in the cytosol (**2**). The binding of ATP to ATP-sensitive K$^+$ channels closes these channels (**3**), thus reducing the efflux of K$^+$ ions from the cell. The resulting small depolarization of the plasma membrane (**4**) triggers the opening of voltage-sensitive Ca^{2+} channels (**5**). The influx of Ca^{2+} ions raises the cytosolic Ca^{2+} concentration, triggering the fusion of insulin-containing secretory vesicles with the plasma membrane and the secretion of insulin (**6**). Steps **5** and **6** are similar to those that take place at nerve terminals, where a membrane depolarization induced by the arrival of an action potential causes the opening of voltage-sensitive Ca^{2+} channels and exocytosis of vesicles containing neurotransmitters (see Figure 7-43). [Adapted from J.-Q. Henquin, 2000, *Diabetes* **49**:1751.]

scarcity (following fasting) by the adjustment of insulin and glucagon concentrations in the blood.

After a meal, when blood glucose rises above its normal level of 5 mM, the pancreatic β cells respond to the rise in glucose or amino acids by releasing insulin into the blood (Figure 15-7). The released insulin circulates in the blood and binds to insulin receptors on muscle cells and adipocytes (fat-storing cells). The insulin receptor, a **receptor tyrosine kinase (RTK)**, can transduce signals through a phosphoinositide pathway leading to the activation of **protein kinase B** (see Figure 14-27). By an unknown mechanism, protein kinase B triggers the fusion of intracellular vesicles containing GLUT4 glucose transporters with the plasma membrane (Figure 15-8). The resulting increased number of GLUT4 on the cell surface increases glucose influx, thus lowering blood glucose. When insulin is removed, cell-surface GLUT4 is internalized by endocytosis, lower-

(a) Resting cell   (b) 2.5 min   (c) 5 min   (d) 10 min

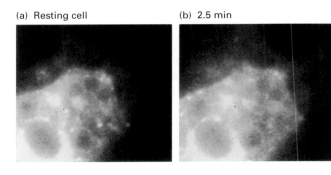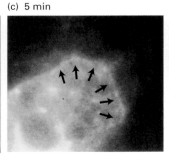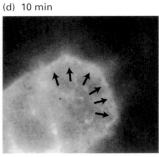

5 μm

▲ **EXPERIMENTAL FIGURE 15-8 Insulin stimulation of fat cells induces translocation of GLUT4 from intracellular vesicles to the plasma membrane.** In this experiment, fat cells were engineered to express a chimeric protein whose N-terminal end corresponded to the GLUT4 sequence, followed by the entirety of the GFP sequence. When a cell is exposed to light of the exciting wavelength, GFP fluoresces yellow-green, indicating the position of GLUT4 within the cell. In resting cells (a), most GLUT4 is in internal membranes that are not connected to the plasma membrane. Successive images of the same cell after treatment with insulin for 2.5, 5, and 10 minutes show that, with time, increasing numbers of these GLUT4-containing membranes fuse with the plasma membrane, thereby moving GLUT4 to the cell surface (arrows) and enabling it to transport glucose from the blood into the cell. Muscle cells also contain insulin-responsive GLUT4 transporters. [Courtesy of J. Bogan.]

ing the level of cell-surface GLUT4 and thus glucose import. Insulin stimulation of muscle cells also promotes the uptake of glucose and its conversion into glycogen, and it reduces the degradation of glucose to pyruvate. Insulin also acts on hepatocytes to inhibit glucose synthesis from smaller molecules, such as lactate and acetate, and to enhance glycogen synthesis from glucose. The net effect of all these actions is to lower blood glucose back to the fasting concentration of about 5 mM.

If the blood glucose level falls below about 5 mM, pancreatic α cells start secreting glucagon. The glucagon receptor, found primarily on liver cells, is coupled to G_s protein, like the epinephrine receptor (Chapter 13). Glucagon stimulation of liver cells activates adenylyl cyclase, leading to the cAMP-mediated cascade that inhibits glycogen synthesis and promotes glycogenolysis, yielding glucose 1-phosphate (see Figure 15-6b). Liver cells can convert glucose 1-phosphate into glucose, which is released into the blood, thus raising blood glucose back toward its normal fasting level.

MEDICINE *Diabetes mellitus* results from a deficiency in the amount of insulin released from the pancreas in response to glucose (type I) or from a decrease in the ability of muscle and fat cells to respond to insulin (type II). In both types, the regulation of blood glucose is impaired, leading to persistent hyperglycemia and numerous other possible complications in untreated patients. Type I diabetes is caused by an autoimmune process that destroys the insulin-producing β cells in the pancreas. Also called insulin-dependent diabetes, this form of the disease is generally responsive to insulin therapy. Most Americans with diabetes mellitus have type II, but the underlying cause of this form of the disease is not well understood. ∎

Oxygen Deprivation Induces a Program of Cellular Responses

In glycogenolysis, the activity of preexisting proteins was regulated by the integration of multiple signals. Organisms that require oxygen respond to oxygen deprivation, a single stimulus, in multiple ways, some occurring rapidly and others taking longer to develop. In addition, over evolutionary time, animals that live at high altitude (e.g., llamas, guanacos, alpacas) became adapted to low oxygen. This adaptation entailed single amino acid changes in the β-globin chain that increased the oxygen affinity of hemoglobin in these animals compared with that of hemoglobin in other animals.

Among the rapid responses to low oxygen (*hypoxia*) is dilation of blood vessels, permitting increased blood flow. This response is regulated by nitric oxide, cyclic GMP, and protein kinase G (see Figure 13-30). A rapid shift in metabolism, called the Pasteur effect, also occurs when cells are deprived of adequate oxygen. First observed in yeast cells, this response accelerates the anaerobic metabolism of glucose when aerobic metabolism and oxidative phosphorylation slows owing to low oxygen. Burning more carbohydrates compensates for the reduced ATP yield from anaerobic metabolism. Phosphofructokinase 1, the third enzyme in glycolysis, is inhibited by ATP and stimulated by AMP; so, when the cell is short on energy, glycolysis increases (see Figure 8-12). The adjustment is rapid, inasmuch as it does not require the synthesis of new molecules.

Slow adaptive responses to low oxygen at the level of the whole organism include increasing the production of erythrocytes, which is stimulated by **erythropoietin** produced in the kidney (see Figure 14-7). Transcription of the erythropoietin gene is regulated primarily by *hypoxia-induced factor 1 (HIF-1)*, a transcriptional activator. The amount of HIF-1 increases drastically as the partial pressure of oxygen decreases from 35 mm Hg to zero, a range typical of normal fluctuations. The nature of the oxygen sensor that causes the increased expression of HIF-1 is not yet known, but it probably requires a protein that has a heme-containing oxygen-binding site somewhat like that in hemoglobin. In addition to regulating the erythropoietin gene, HIF-1 coordinately activates the transcription of several other genes whose encoded proteins help cells respond to hypoxia (Figure 15-9). One of these proteins, vascular endothelial growth factor (VEGF), is secreted by cells lacking oxygen and promotes local *angiogenesis,* the branching growth of blood vessels. Expression of VEGF requires not only HIF-1 but also a Smad transcription factor, which is activated by a TGFβ signal. The ability of HIF-1 to control different genes in different cell types presumably results from this type of combinatorial action.

The results of recent studies revealed that the degradation of HIF-1 is controlled by an oxygen-responsive prolyl hydroxylase. HIF-1 is a dimer composed of two subunits, α and β. The β subunit is abundant in the cytosol under high or low oxygen conditions but, when oxygen is plentiful, the α subunit (HIFα) is ubiquitinated and degraded in proteasomes (Chapter 3). Ubiquitination is promoted by the von Hippel-Lindau protein (pVHL), which binds to a conserved "degradation domain" of HIF-1α. The binding of pVHL in turn is facilitated by hydroxylation of a proline in the pVHL-binding site on HIF-1α. The prolyl hydroxylase catalyzing this reaction requires iron and is most active at high oxygen, leading to degradation of the α subunit and no transcriptional activation by HIF-1. At low oxygen, when hydroxylation does not occur, active dimeric HIF-1 forms and is translocated to the nucleus. The *hypoxia-response pathway* mediated by HIF-1 and its regulation by pVHL have been conserved for more than half a billion years, given that it is the same in mammals, worms, and insects.

 Hypoxia affects the growth of blood vessels, particularly the small capillaries, whose exact pattern, unlike that of major blood vessels like the aorta, is not genetically determined. Angiogenesis, the branching growth of the vasculature, is stimulated by hypoxia, thus ensuring that all cells are in adequate proximity to the oxygenated blood supply. Growing tumors stimulate angiogenesis to ensure their own blood supply (Chapter 23). Understanding the signals that control angiogenesis could potentially lead to the development of therapeutic agents that stimulate angiogenesis in a transplanted or diseased organ that is receiving insufficient blood or that inhibit angiogenesis in developing tumors, thereby suffocating them. ∎

KEY CONCEPTS OF SECTION 15.2

Responses of Cells to Environmental Influences

■ Glycogen breakdown and synthesis is regulated by multiple second messengers induced by neural or hormonal stimulation (see Figure 15-6).

■ A rise in blood glucose stimulates the release of insulin from pancreatic β cells (see Figure 15-7). Subsequent binding of insulin to its receptor on muscle cells and adipocytes leads to the activation of protein kinase B, which promotes glucose uptake and glycogen synthesis, resulting in a decrease in blood glucose.

■ The binding of glucagon to its G protein–coupled receptor on liver cells promotes glycogenolysis and an increase

▶ **FIGURE 15-9 Model of hypoxia response pathway mediated by hypoxia-induced factor 1 (HIF-1).** The local oxygen concentration is sensed by an unknown mechanism probably requiring a heme-associated protein (S-heme). The oxygen-deprived form of the sensor activates intermediates (X), which in turn stimulate increased production of both subunits of hypoxia-induced factor 1 (HIF-1). HIF-1 activates transcription of genes whose encoded proteins mediate short-term and long-term responses to an oxygen deficit. HIF-1 activates different target genes in different cells, indicating that it probably acts in combination with other gene-regulating proteins. VEGF = vascular endothelial growth factor; i-NOS = inducible nitric oxide synthase; HO-1 heme oxygenase-1; EPO = erythropoietin. [Adapted from H. Zhu and H. F. Bunn, 2001, *Science* **292**:449; see also W. G. Kaelin, Jr., 2002, *Genes & Dev.* **16**:1441.]

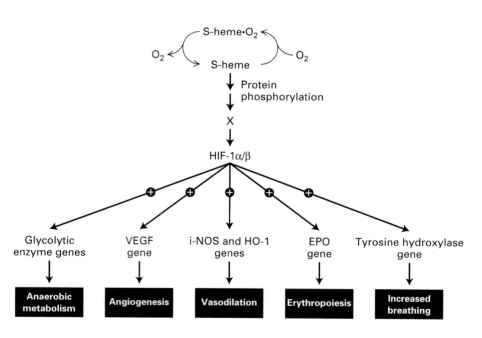

in blood glucose by the cAMP-triggered kinase cascade (similar to epinephrine stimulation under stress conditions).

■ Two oxygen-sensing mechanisms help cells respond to oxygen deprivation: one mechanism activates transcription of hypoxia-induced factor 1 (HIF-1), and the other promotes stabilization of HIF-1 by inhibiting its proteasomal degradation.

■ When cellular oxygen is low, HIF-1 acts with tissue-specific transcription factors to activate the expression of various target genes in different tissues (see Figure 15-9). One protein induced by this pathway is vascular endothelial growth factor, which stimulates local angiogenesis to increase the blood supply and thereby oxygen to oxygen-deprived cells.

15.3 Control of Cell Fates by Graded Amounts of Regulators

In a developing tissue, each cell must learn how to contribute to the overall organization of the tissue. Frequently, cells in a particular position within the developing embryo must divide, move, change shape, or make specialized products, whereas other nearby cells do not. In modern developmental biology, the term **induction** refers to events where one cell population influences the fate of neighboring cells. Figure 15-10 schematically depicts how a series of inductive signals can create several cell types, starting from a population of initially equivalent cells. Induction may create tissue types at specific sites (e.g., formation of a lens near the site at which the retina will grow) or cause changes in the shape of cells at a specific location. For example, changes in the shape of cells in the center but not at the periphery of the neural plate give rise to the neural tube from which our central nervous system develops. Cell orientation is critical, too. If

an epithelium produces appendages, such as feathers or bristles, all of them may need to point in the same direction. All these properties of cells are coordinated by integrating signals in the developing organism, and each cell interprets the signals that it receives in light of its previous experience and state of differentiation.

Some extracellular inductive signals move through tissue and hence can act at a distance from the signaling cell; some signals are tethered to the surface of the signaling cell and thus can influence only the immediate neighboring cells. Still other signals are highly localized by their tight binding to components of the extracellular matrix. The transmission rate of a signal depends on the chemical properties of the signal, the properties of the tissue through which it passes, and the ability of cells along the way to take up or inactivate the signal. The distance that an inductive signal can move influences the size and shape of an organ. For instance, the farther a signal that induces neuron formation can move, the more neurons will form.

In this section and those that follow, we will see how quantitative differences in external signals and transcription factors can determine cell fates and properties. We begin by distinguishing two basic mechanisms of inductive signaling and then, by way of example, examine in some detail early stages of *Drosophila* development. To learn how signals work during cell interactions in development, transgenic animals are used to observe the effects of increasing or decreasing gene function in specific cells. For example, if a cell can send a signal even if a certain gene function is removed, the gene is not required for sending the signal. Removing the same gene function from a cell that normally receives the signal may reveal a requirement of the gene for signal reception or transduction. In this way, even when a novel protein is being studied, it is possible to deduce its place in a signaling pathway. These gene manipulation methods are especially advanced in *Drosophila* (Figure 15-11) but are increasingly being adapted for other experimental organisms.

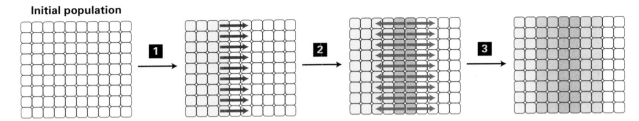

Initial population

▲ **FIGURE 15-10 Simplified model of sequential induction of cell types in an epithelium.** Step **1**: Starting from a population of equivalent cells (white), an initial event (e.g., cell movements or a polarized signal) creates a second population of cells (tan) that secretes a signal (red arrows). This signal reaches only some of the cells in the adjacent field of cells. Step **2**: The

cells capable of receiving and interpreting the red signal now form a new cell type (pink) that secretes a different signal (blue arrows) that moves away from the red cells in both directions. Step **3**: The blue signal induces still more cell types (purple and blue). Note that the effect of the blue signal differs, depending on whether it acts on white cells or on tan cells.

(a) Ectopic gene expression with spatial/temporal control

A fly gene of interest is linked to a regulated promoter in order to express the gene in a time or place that is not normal. If the overexpression is lethal, no transgenic flies can be obtained (see part [b] for a solution to this problem). A variation is to use a heat-inducible promoter from a "heat shock" gene; a pulse of heat (37° C, 30 min) will cause expression in all cells.

(b) Spatially and temporally regulated ectopic gene expression using *GAL4*

One transgenic fly codes for the yeast transcription factor Gal4 under the control of a tissue-specific promoter. Another carries a transgene that can respond to Gal4 because it contains a UAS sequence to which Gal4 binds. After crossing the two flies, any gene that had been attached to the UAS sequence will fall under the control of Gal4 and be expressed in that specific tissue. This allows each transgenic fly line to survive, even if the activation of the gene will prove lethal in the progeny of the cross.

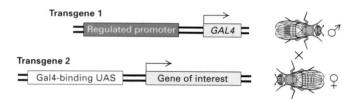

(c) Creating mosaic tissues with clones of cells that lack a gene function

Yeast FLP recombinase, acting upon FRT sequences inserted near the centromere, can be used to create clones of homozygous mutant cells in flies that are heterozygous for a recessive mutation. Recombinase is produced at a specified time using a heat-inducible promoter. Just enough recombinase is induced to cause an occasional recombination event. The effects of lost gene function are then assessed.

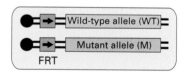

Normal cell division creates more heterozygous cells:

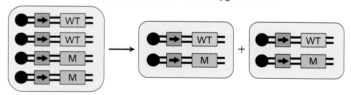

Division after recombination creates a clone of mutant cells and a clone of wild-type cells:

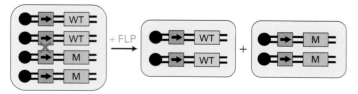

(d) Creating marked mutants to assess the effects of lost gene function in single cells

The goal is to mark mutant cells created by FLP recombination as in (c). Flies are made that carry a gene encoding the yeast protein Gal80 under the control of a *tubulin* promoter that is active in all cells. This transgene is located on the chromosome that carries the wild-type allele of the gene of interest. All cells are engineered to make the yeast transcription factor Gal4 using a constitutive promoter (not shown). After recombination, Gal80 is present in the wild-type cell, where it blocks the activity of Gal4. In the mutant cell, no Gal80 is made and Gal4 acts on a UAS sequence to activate production of a fluorescent protein (GFP). In this way the mutant cells are marked with fluorescence at the same time that they lose their wild-type allele.

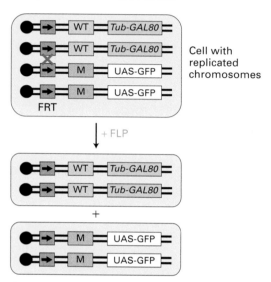

Cell with replicated chromosomes

Daughter cells initiate two clones

(e) Inducibly activating a gene function in clones of cells

The goal is to activate the gene interest in a clone of cells during development. All cells initially express *lacZ*, which is under the control of a constitutive promoter, and therefore stain blue. When a modest pulse of FLP recombinase is produced using a heat-inducible promoter active in all cells, enough recombinase is made in a few cells to act on FRT sites and cause a deletion of the DNA between the FRTs. This removes the *lacZ* gene at the same time that it joins a promoter to the gene of interest. The resulting clone of cells that expresses the gene of interest can be identified because the cells do not express *lacZ*. The effect of the gene of interest is then assessed in the clone.

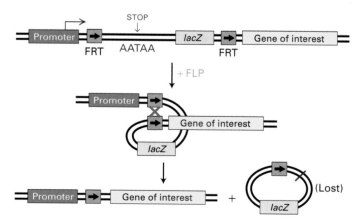

◀ **FIGURE 15-11 Gene manipulation in analysis of signaling systems.** (a) In the simplest case, a gene is activated with the use of a promoter that is specific to certain cells or, as with "heat shock" promoters, that is inducible in all cells. This procedure has the limitation that the transgene may be detrimental and transgenic animals cannot be isolated. (b) An improvement is to make two lines of flies transgenic, one carrying the gene of interest under the control of an upstream activating sequence (UAS, X-ref) from yeast. The UAS is activated when the yeast transcription factor GAL4 is present, as it is when the UAS-bearing flies are crossed with flies having GAL4 expressed in certain cells. (c) The opposite goal is to remove a gene's function selectively from certain cells. The yeast recombinase FLP acts on FRT sequences that have been inserted near the base of a chromosome. Starting from a fly that is heterozygous for a mutation of interest, mutant/wild-type, one can obtain clones of cells that are homozygous for either the mutant or the wild-type allele. (d) It is difficult to recognize and analyze small clones of cells obtained as in part (c) if the cells are not marked. In this refinement, the recombination removes GAL80, a protein that inhibits GAL4 function, at the same time as the mutant allele is made homozygous. The unleashed GAL4 then activates a UAS that drives the production of a fluorescent protein. A mutant cell, like the neuron shown, can be analyzed to see the effect of the mutation on, for example, wiring the brain. (e) To activate a gene in small, randomly generated clones of cells, FLP is again used but this time to remove an intervening transcriptional termination sequence that prevents the gene of interest from being active. At the same time, a *lacZ* or other marker gene is removed; so the clone of cells with the gene turned on is identifiable by the lack of marker gene expression.

Inductive Signaling Operates by Gradient and Relay Mechanisms

In some cases, the induction of cell fates includes a binary choice: in the presence of a signal, the cell is directed down one developmental pathway; in the absence of the signal, the cell assumes a different developmental fate or fails to develop at all. Such signals often work in a *relay mode*. That is, an initial signal induces a cascade of induction in which cells close to the signal source are induced to assume specific fates; they, in turn, produce other signals to organize their neighbors (Figure 15-12a). Alternatively, a signal may induce different cell fates, depending on its concentration. In this *gradient mode*, the fate of a receiving cell is determined by the amount of the signal that reaches it, which is related to its distance from the signal source (Figure 15-12b). Any substance that can induce different responses depending on its concentration is often referred to as a **morphogen.**

The concentration at which a signal induces a specific cellular response is called a *threshold*. A graded signal, or morphogen, exhibits several thresholds, each one corresponding to a specific response in receiving cells. For instance, a low concentration of an inductive signal causes a cell to assume fate A, but a higher signal concentration causes the cell to assume fate B. In the gradient mode of

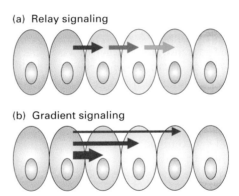

(a) Relay signaling

(b) Gradient signaling

▲ **FIGURE 15-12 Two modes of inductive signaling.** In the relay mode (a), a short-range signal (red arrow) stimulates the receiving cell to send another signal (purple), and so on for one or more rounds. In the gradient mode (b), a signal produced in localized source cells (red arrows) reaches nearby cells in larger amounts than the amounts reaching distant cells. If the receiving cells respond differently to different concentrations of the signal (indicated by width of the arrows), then a single signal may create multiple cell types.

signaling, the signal is newly created, and so it has not built up to equal levels everywhere. Alternatively, the signal could be produced at one end of a field of cells and destroyed or inactivated at the other (the "source and sink" idea), so a graded distribution is maintained.

Mesoderm Cell Fates in *Xenopus* Blastula Studies with activin, a TGFβ-type signaling protein that determines cell fate in early *Xenopus* embryos, have been sources of insight into how cells determine the concentration of a graded inductive signal. Activin helps organize the **mesoderm** along the dorsal/ventral (back/front) axis of an animal. The **endoderm** and **ectoderm** form first after fertilization of a *Xenopus* oocyte; the mesoderm forms slightly later. These three distinct cell populations (germ layers) make up the *blastula*, a hollow ball of cells.

Specific genes are used as indicators of the tissue-creating effects of signals such as activin. For instance, a low concentration of activin induces expression of the *Xenopus brachyury (Xbra)* gene throughout the early mesoderm. Xbra is a transcription factor necessary for mesoderm development. Higher concentrations of activin induce expression of the *Xenopus goosecoid (Xgsc)* gene. Xgsc protein is able to transform ventral into dorsal mesoderm; so the local induction of *Xgsc* by activin causes the formation of dorsal, rather than ventral, mesodermal cells near the activin source. Using ^{35}S-labeled activin, scientists demonstrated that *Xenopus* blastula cells each express some 5000 type II TGFβ-like receptors that bind activin. Findings from additional experiments showed that maximal Xbra expression was achieved when about 100 receptors were occupied. At a concentration of activin at which 300 receptors were occupied, cells began expressing higher levels of Xgsc. Similar results were obtained with blastula cells experimentally manipulated to express sevenfold higher levels of the activin type II receptor. These findings indicate that blastula cells measure the absolute number of ligand-bound receptors rather than the ratio of bound to unbound receptors, and confirm the importance of signal concentration.

Vulva Development in *C. elegans* An example of cell-fate determination by a combination of graded and relayed signals is the development of the vulva of the nematode worm *C. elegans*. This structure develops from a group of epidermal vulval precursor cells (VPCs) whose fates are controlled by an inductive signal from a nearby cell called the anchor cell. All the VPCs have the potential to become any of three different cell types: 1° and 2°, which refer to different vulval cell types, and 3°, which is a nonvulval type. A set of cells, such as the VPCs, is called an *equivalence group* if each cell in the set has equal capacity to form more than one cell type. The inductive signal secreted by the anchor cell is LIN-3, which is similar to vertebrate epidermal growth factor (EGF). Like the EGF receptor, the receptor for LIN-3 is a receptor tyrosine kinase, called LET-23, that acts through a Ras–MAP kinase pathway (see Figure 14-21).

The results of early studies suggested that LIN-3 was a graded signal inducing the 1° fate in the nearest VPC (normally P6.p) and the 2° fate in P5.p and P7.p, which are located slightly farther away from the anchor cell (Figure 15-13a). If determination of the 2° fate depended solely on a graded LIN-3 signal, then mutant P5.p and P7.p cells lacking the receptor for LIN-3 would be expected to assume the nonvulval 3° fate. Surprisingly, when this experiment was done, the mutant cells took on their normal 2° fate (Figure 15-13b). The most likely explanation of these results is that, when the P6.p cell takes on the 1° fate, it sends out a different signal that normally works with moderate levels of LIN-3 to ensure the production of 2° cells. This second, relayed signal appears to be a ligand for LIN-12, a Notch-type receptor. Stimulation of LIN-12 on the P5.p and P7.p cells induces expression of a phosphatase that inactivates MAP kinase and affects other regulators as well, thus preventing the 1° fate choice.

In glycogenolysis, signal integration is at the level of a two-subunit protein, with each signal acting on one of the subunits. In vulval cell-fate determination, the activity of a single kinase, MAP kinase, is controlled by two pathways: signaling from an EGF-type receptor activates MAP kinase; signaling from a Notch-type receptor deactivates it. The convergence of these two pathways on MAP kinase elegantly allows the formation of multiple adjacent cell types.

Morphogens Control Cell Fates in Early *Drosophila* Development

Drosophila has been particularly useful in studying morphogens for three reasons. First, graded regulatory molecules are used extensively in the development of the early *Drosophila* embryo and in growth of the legs and wings. Second, a fertilized egg develops into an adult fly in only about 10 days. Third, powerful genetic screens have identified many developmental mutants with dramatic abnormal phenotypes. Some of these defects have been found to arise from mutations in genes encoding morphogens; others arise in genes encoding signal-transduction proteins.

To understand how morphogens determine cell fates in the early fly embryo, we first need to set the scene. Oogenesis begins with a stem cell that divides asymmetrically to generate a single **germ cell,** which divides four times to generate 16 cells. One of these cells will complete meiosis (see Figure 9-3), becoming an **oocyte;** the other 15 cells become *nurse cells,* which synthesize proteins and mRNAs that are transported through cytoplasmic bridges to the oocyte (Figure 15-14). These molecules are necessary for maturation of the oocyte and for the early stages of **embryogenesis.** At least one-third of the genome is represented in the mRNA contributed by the mother to the oocyte, a substantial dowry. Each group of 16 cells is surrounded by a single layer of somatic cells called the *follicle,* which deposits the eggshell. The mature oocyte, or egg, is released into the oviduct, where it is fertilized; the fertilized egg, or **zygote,** is then laid.

The first 13 nuclear divisions of the *Drosophila* zygote are synchronous and rapid, each division occurring about every 10 minutes. This DNA replication is the most rapid

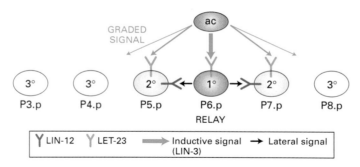

▲ **FIGURE 15-13 Gradient and relay signaling in *C. elegans* vulva development.** The anchor cell (ac) sends a signal to the vulval precursor cells (P3.p–P8.p), all of which have equivalent potential to form any of three cell types—1°, 2°, 3°. The EGF-related signal, LIN-3, is received by the LET-23 receptor. Cells receiving the highest amount of signal form 1°, cells receiving a moderate amount form 2°, and cells receiving little or none form 3°. The three cell fates are distinguished by following their subsequent patterns of cell division. The effects of the graded LIN-3 signal are further controlled by a signal relay. After the anchor cell sends LIN-3 to the nearest cell, normally P6.p, the P6.p cell then sends a different signal to its neighbors, which turns out to be a ligand for LIN-12. The demonstration of the relay effect came from genetically removing LET-23 receptor from just the P5.p and P7.p cells, which prevents them from responding to the LIN-3 signal. If direct LIN-3 graded signal accounted for all the cell types, P5.p and P7.p should take on the 3° fate. However, they still manage to form 2° cells; so another system must operate. This other system is the signal for LIN-12 from the P6.p cell. [Adapted from J. S. Simske and S. K. Kim, 1995, *Nature* **375**:142.]

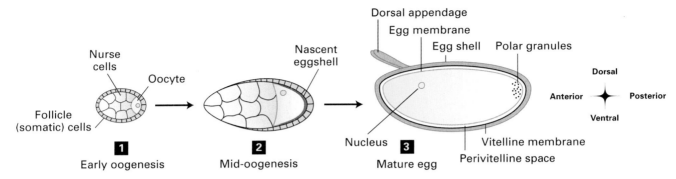

▲ **FIGURE 15-14 Development of a *Drosophila* oocyte into a mature egg.** A single germ cell gives rise to fifteen nurse cells (green) and a single oocyte (yellow) (**1**). The early oocyte is about the same size as the neighboring nurse cells; the follicle, a layer of somatic cells, surrounds the oocyte and nurse cells The nurse cells begin to synthesize mRNAs and proteins necessary for oocyte maturation, and the follicle cells begin to form the egg shell. Midway through oogenesis (**2**), the oocyte has increased in size considerably. The mature egg (**3**) is surrounded by the completed eggshell (gray). The nurse cells have been discarded, but mRNAs synthesized and translocated to the oocyte by the nurse cells function in the early embryo. Polar granules located in the posterior region of the egg cytoplasm mark the region in which germ-line cells will arise. The asymmetry of the mature egg (e.g., the off-center position of the nucleus) sets the stage for the initial cell-fate determination in the embryo. After its release into the oviduct, fertilization of the egg triggers embryogenesis. [Adapted from A. J. F. Griffiths et al., 1993, *An Introduction to Genetic Analysis*, 5th ed., W. H. Freeman and Company, p. 643.]

known for a eukaryote, with the entire 160 Mb of chromosomal DNA copied in a cell-cycle S phase that lasts only 3 minutes. Because these nuclear divisions are not accompanied by cell divisions, they generate a multinucleated egg cell, a **syncytium,** with a common cytoplasm and plasma membrane (Figure 15-15a). As the nuclei divide, they begin to migrate outward toward the plasma membrane. From about 2 to 3 hours after fertilization, the nuclei reach the surface, forming the syncytial blastoderm; during the next hour or so, cell membranes form around the nuclei, generating the cellular blastoderm, or blastula (Figure 15-15b). All future tissues are derived from the 6000 or so epithelial cells on the surface of the blastula. Soon some of these cells move inside, a process termed gastrulation, and eventually develop into the internal tissues.

The syncytial fly embryo is about 100 cells long, head to tail, and about 60 cells around. Within 1 day of fertilization, the zygote develops into a larva, a segmented form that lacks wings and legs. Development continues through three larval stages (~4 days) and the ~5-day pupal stage during which metamorphosis takes place and adult structures are created (Figure 15-16). At the end of pupation, about 10 days after fertilization, the pupal case splits and an adult fly emerges.

The initially equivalent cells of the syncytial embryo rapidly begin to assume different fates, leading to a well-ordered pattern of distinct cell identities. These early patterning events set the stage for the later development and proper placement of different tissues (e.g., muscle, nerve, epidermis) and body parts, as well as the shapes of the appendages and the organization of cell types within them. Because the early embryo is initially symmetric side to side, the creation of differences among cells is a two-axis prob-

lem: *dorsal/ventral* (back/front) and *anterior/posterior* (head/tail). Different sets of genes act on each axis; so every cell learns its initial fate by responding to input from both dorsoventral-acting and anterioposterior-acting regulators in a kind of two-dimensional grid. As we will see, both regulatory systems begin with information and molecules contributed to the oocyte as a dowry from the mother. When the mature egg is laid, it is already asymmetric along both axes (see Figure 15-14).

Because the early fly embryo is a syncytium, regulatory molecules can move in the common cytoplasm without having to cross plasma membranes. Some molecules form gradients, which are used in the earliest stages of cell-fate determination in *Drosophila* before subdivision of the syncytium into individual cells. Thus transcription factors, as well as secreted molecules, can function as morphogens in the syncytial fly embryo. Syncytia are less common in the early development of other animals and in later stages of fly development; in these stages, patterning events are controlled largely by interactions between cells mediated by extracellular signals, which may act in a graded or relay mode.

To decipher the molecular basis of cell-fate determination and global patterning, investigators have (1) carried out massive genetic screens to identify all the genes having roles in the organizing process, (2) cloned mutation-defined genes; (3) determined the spatial and temporal patterns of mRNA production for each gene and the distribution of the encoded proteins in the embryo; and (4) assessed the effects of mutations on cell differentiation, tissue patterning, and the expression of other regulatory genes. The principles of cell-fate determination and tissue patterning learned from *Drosophila* have proved to have broad applicability to animal development.

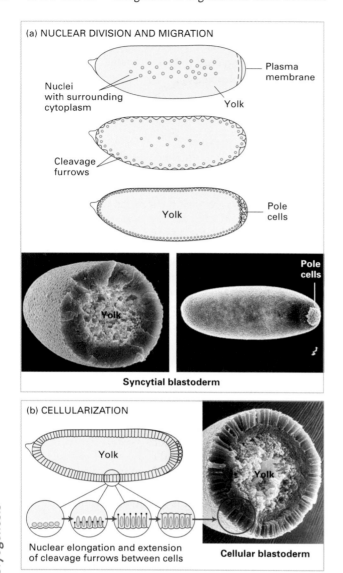

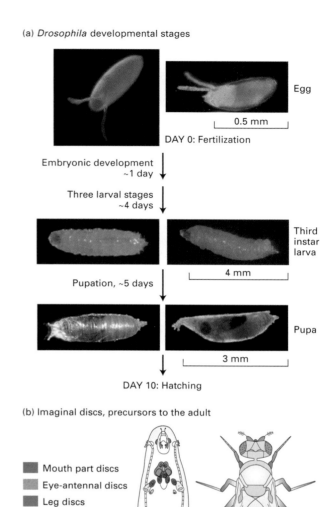

▲ **FIGURE 15-15 Formation of the cellular blastoderm during early *Drosophila* embryogenesis.** Stages from syncytium (a) to cellular blastoderm (b) are illustrated in diagrams and electron micrographs. Nuclear division is not accompanied by cell division until about 6000 nuclei have formed and migrated outward to the plasma membrane. Before cellularization, the embryo displays surface bulges overlying individual nuclei, which remain within a common cytoplasm. No membranes other than that surrounding the entire embryo are present. After cellularization, cell membranes are evident around individual nuclei. Note the segregation of the nuclei of so-called pole cells, which give rise to germ-line cells, at the posterior end of the syncytial blastoderm. [See R. R. Turner and A. P. Mahowald, 1976, *Devel. Biol.* **50**:95; photographs courtesy of A. P. Mahowald; diagrams after P. A. Lawrence, *The Making of a Fly*, 1992, Blackwell Scientific, Oxford.]

▲ **FIGURE 15-16 Major stages in the development of *Drosophila*.** (a) The fertilized egg develops into a blastoderm and undergoes cellularization in a few hours. The larva, a segmented form, appears in about 1 day and passes through three stages (instars) over a 4-day period, developing into a prepupa. Pupation takes ~4–5 days, ending with the emergence of the adult fly from the pupal case. (b) Groups of ectodermal cells called imaginal discs are set aside at specific sites in the larval body cavity. During pupation, these give rise to the various body parts indicated. Other precursor cells give rise to adult muscle, the nervous system, and other internal structures. [Part (a) from M. W. Strickberger, 1985, *Genetics*, 3d ed., Macmillan, p. 38; reprinted with permission of Macmillan Publishing Company. Part (b) Adapted from same source and J. W. Fristrom et al., 1969, in E. W. Hanly, ed., *Park City Symposium on Problems in Biology*, University of Utah Press, p. 381.]

Reciprocal Signaling Between the Oocyte and Follicle Cells Establishes Initial Dorsoventral Patterning in *Drosophila*

Initial dorsal/ventral patterning in *Drosophila* is controlled by the events of oogenesis. Indeed, the shape of the mature oocyte is an accurate predictor of the dorsal/ventral orientation of the embryo. The process begins when the nucleus of the early oocyte moves slightly, perhaps randomly, toward what will become the anterior and dorsal side of the mature egg (Figure 15-17a). That loss of symmetry triggers the polarization of signals that coordinate the dorsal/ventral axes of the oocyte, embryo, and surrounding eggshell. Such coordination is necessary so that the structures on the eggshell become aligned properly with structures of the growing embryo. For example, breathing tubes in the eggshell must connect with appropriate regions of the embryo.

About midway through *Drosophila* oogenesis, the production of Gurken, a signal similar to epidermal growth factor, begins. Because of the off-center location of the nucleus,

Gurken is produced on the dorsal side of the oocyte (Figure 15-17b). The receptor for Gurken, a receptor tyrosine kinase like the EGF receptor, is present on the surfaces of all the follicle cells that abut the oocyte. The dorsal Gurken signal activates its receptors only in dorsal follicle cells, leading to changes in their appearance and to repression of the *pipe* gene within them. Because of this dorsal repression, Pipe protein is produced only in ventral follicle cells. Pipe is an enzyme that catalyzes sulfation of glycosaminoglycans (GAGs), the polysaccharide chains that are added to proteins to form proteoglycans (Chapter 6).

Pipe protein promotes ventral cell fates, probably by activating a still unknown signal that triggers a series of proteolytic cleavages in the perivitelline space on the ventral side of the by now mature egg. The ensuing chain of events has some similarity to the blood-clotting cascade, each protein cleaving and thereby activating the next one in the series. The outcome of the cleavages is the production of a ligand called Spätzle only on the ventral side (Figure 15-17c). By this time, the egg has been fertilized and early nuclear division is

(a) Dorsoventral differentiation

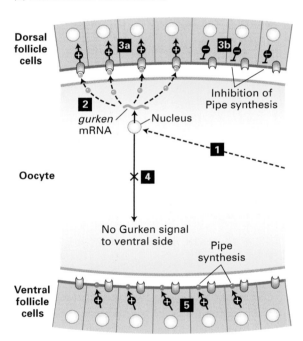

(b) Activation of dorsoventral protease cascade

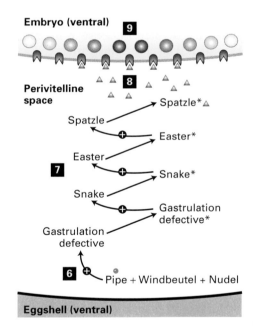

▲ **FIGURE 15-17 Dorsoventral axis determination in *Drosophila*.** This process relies on two signal systems, one in follicle cells and the other in the oocyte or embryo or both, plus a proteolytic cascade within the perivitelline space. (a, b) Movement of the oocyte nucleus creates an initial asymmetry (**1**). The dorsal location of the oocyte nucleus ultimately results in the production of Pipe protein only in ventral follicle cells (**2**–**5**). (c) Subsequent events in the perivitelline space along the ventral surface generate a gradient of active Spätzle (**6**–**9**). By this time, the egg has been fertilized and the embryo is a syncytium with many nuclei; the region around only one nucleus is shown. Activation of Toll on the embryo's surface by Spätzle causes Dorsal protein to enter the nucleus (**10** and **11**) where it activates transcription of specific target genes, depending on its concentration. (d) The concentration of Spätzle—hence Toll activation and nuclear localization of Dorsal—is greatest along the ventral midline, conferring ventral fates (e.g., muscle) on cells in this region. More laterally, less Dorsal enters the nuclei, and in consequence the cell fates are different (e.g., neural). Dorsal cell fates arise where no Dorsal enters the nucleus. Mutants lacking Toll receptor, or Dorsal, form only dorsal cell types. [Adapted from Gilbert and Hashimoto, 1999, *Trends Cell Biol.* **9**:102.]

underway. Spätzle binds to its transmembrane receptor, called Toll, on the ventral embryo surface. Thus the signaling has come full circle: Gurken ligand produced dorsally in oocyte → activation of its EGF-like receptor on dorsal follicle cells, and then back through the protease cascade on the ventral side → Spätzle ligand → activation of its receptor, Toll, on the ventral side of the embryo. The net effect is to coordinate the eggshell structures produced by the follicle cells with the embryo structures produced inside.

Within the embryo, association of the cytosolic domain of activated Toll with two proteins (Tube and Pelle) leads to phosphorylation of Cactus protein. In the absence of Toll signaling, Cactus binds to a transcription factor called Dorsal and traps it, but phosphorylated Cactus is rapidly degraded by the proteasome. The newly freed Dorsal is able to enter the nuclei of the embryo's cells and activate the transcription of different target genes, depending on its concentration (see Figure 15-17c). Spätzle and Dorsal thus function as graded regulators, inducing ventral fates where their concentration is highest and other fates laterally as their concentration diminishes (Figure 15-17d). Dorsal function reaches its peak after cellularization has taken place.

The central features of the *Toll-Dorsal pathway* in flies, which are analogous to those of the mammalian *NF-κB pathway* discussed in Chapter 14, exist in mammals and probably in all animals. Dorsal is similar to the NF-κB transcription factor; Cactus, to its inhibitor, I-κB; and the Toll receptor, to the receptor for interleukin 1, which acts through Tube and Pelle equivalents to cause the phosphorylation of I-κB and the release of NF-κB (see Figure 14-28). NF-κB is a critical regulator of genes required for immune responses in mammals and insects and appears to function in mammalian development as well. It nicely exemplifies the utilization of one signal-transduction pathway to accomplish multiple tasks, such as patterning in development plus the immune response to infection or injury. This phenomenon appears to be fairly common and partly explains the small number of signaling pathways that have evolved over biological time despite the increasing complexity of organisms.

Nuclear Dorsal and Decapentaplegic, a Secreted Signal, Specify Ventral and Dorsal Cell Fates

The remarkable series of steps depicted in Figure 15-17 results in a gradient in the nuclear localization of the transcription factor Dorsal. The concentration of nuclear Dorsal decreases gradually from highest in cells at the ventral midline to lower values in lateral cells and eventually to none in dorsal cells. Mutants lacking *dorsal* function cannot make cells with ventral character; so the entire embryo develops dorsal structures. (Note that fly genes are named according to their mutant phenotypes; thus the *dorsal* gene controls ventral fates.) Once inside the nucleus, Dorsal controls the transcription of specific target genes by binding to distinct high- and low-affinity regulatory sites and by interacting in a combinatorial fashion with other transcription factors. Dorsal represses the transcription

of *decapentaplegic (dpp), tolloid, short gastrulation,* and *zerknüllt* and activates the transcription of *twist, snail, singleminded,* and *rhomboid*. Each of these genes contains a unique combination of cis-acting regulatory sequences to which Dorsal and other transcription factors bind.

Figure 15-18 illustrates how nuclear Dorsal specifies different target-gene expression patterns, depending on its con-

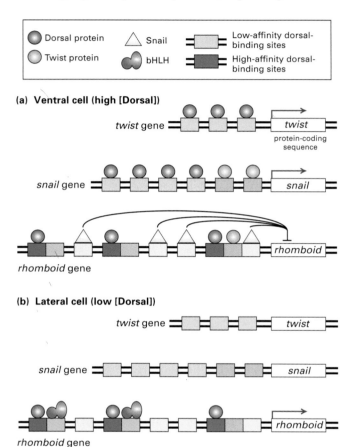

▲ **FIGURE 15-18 Activation of different target genes by Dorsal subsequent to Toll signaling.** Activation of Toll by Spätzle leads to a graded nuclear localization of Dorsal protein (see Figure 15-17d). The resulting Dorsal concentration gradient (ventrally high and dorsally low) can lead to different patterns of gene expression. Shown here are three target genes that have either high-affinity (dark blue) or low-affinity (light blue) Dorsal-binding sites. (a) In ventral regions where the concentration of Dorsal (purple) is high, it can bind to low-affinity sites in *twist* and *snail,* activating the transcription of these genes. Twist protein (orange) also activates the transcription of *snail,* which encodes a repressor (yellow) that prevents the transcription of *rhomboid* in this region. (b) In lateral regions, the Dorsal concentration is not high enough for the binding of Dorsal to the low-affinity sites regulating *twist* and *snail.* The binding of Dorsal to *rhomboid* is facilitated by the presence of high-affinity sites and the synergistic binding of bHLH heterodimeric activators (green) to neighboring sites. The sharp boundary in the expression of Rhomboid causes formation of distinct cell types in ventral vs. lateral regions. [See A. M. Huang et al., 1997, *Genes & Dev.* **11**:1963.]

centration. For instance, the *twist* gene, which contains three low-affinity Dorsal-binding sites, is expressed most ventrally where the Dorsal concentration is highest. As the Dorsal concentration decreases, it falls beneath the threshold necessary to activate the transcription of *twist*. The *rhomboid* gene, which is expressed only in lateral regions, is controlled through a complex cis-acting regulatory region that contains three high-affinity Dorsal-binding sites. Two of these sites are adjacent to regulatory sequences that bind proteins containing the **basic helix-loop-helix** (bHLH) motif, which is present in numerous transcription factors. As Twist contains a bHLH motif, it appears that it acts cooperatively with Dorsal to induce transcription of the *rhomboid* gene in lateral cells. The *rhomboid* control region also contains four binding sites for Snail, a transcriptional repressor. The production of Snail is induced only at high concentrations of Dorsal because the *snail* gene contains only low-affinity Dorsal-binding sites within its control region. Because Snail is localized ventrally, its repressor activity defines a sharp ventral-lateral boundary in the transcription of *rhomboid*. This example nicely illustrates how two transcriptional regulators can collaborate to create a sharp boundary between cell types, something to be discussed further in Sections 15.4 and 15.5.

The dorsal/ventral patterning produced by Dorsal is extended by Decapentaplegic (Dpp). This secreted signaling protein belongs to the TGF β family, which is found in all animals (Chapter 14). Because transcription of the *dpp* gene is repressed by Dorsal, Dpp is produced only in the dorsal-most cells of the early fly embryo, which lack Dorsal in their nuclei. A combination of genetic and molecular genetic evidence suggests that Dpp acts as a morphogen to induce the establishment of different ectoderm cell types in the dorsal region of the embryo. For instance, complete removal of Dpp function leads to a loss of all dorsal structures and their conversion into more-ventral ones. Embryos carrying only one wild-type *dpp* allele show an increase in the number of cells assuming a ventral fate, whereas embryos with three copies of *dpp* form more dorsal cells.

Thus two graded secreted signals, Spätzle and Dpp, play critical roles in determining the dorsal/ventral axis in *Drosophila* and in inducing further patterning within the dorsal and ventral regions. Spätzle, acting through nuclear-localized Dorsal, a transcription factor, induces ventral fates and controls the production of Dpp, which induces dorsal fates. Unlike Dorsal, which functions only in early development, the Dpp signal is used repeatedly in later development, participating in many processes such as appendage development, gut formation, and eye development. The Spätzle signal also has other functions, which are discussed later.

The frog TGFβ family members called BMP2 and BMP4 have inductive effects similar to those of Dpp protein and indeed are the vertebrate proteins most closely related in sequence to Dpp. Most or all components of the TGFβ signaling pathway, including Smad transcription factors, appear to be present and participating in development in all animals

(see Figure 14-2). As discussed in Section 15.5, the vertebrate proteins also control patterning along the dorsal/ventral axis, although the axis is flipped in vertebrates compared with invertebrates. Loss of TGFβ signaling, owing to mutations in TGFβ receptors or Smad proteins, contributes to the onset of cancer (Chapter 23).

Transcriptional Control by Maternally Derived Bicoid Protein Specifies the Embryo's Anterior

We turn now to determination of the anterior/posterior axis in the early fly embryo while it is still a syncytium. As in determination of the dorsal/ventral axis, specification of anterior/posterior cell fate begins during oogenesis. The initial asymmetry also involves so-called *maternal mRNAs*, which are produced by nurse cells and transported into the oocyte. In this case they become localized in discrete spatial domains (see Figure 15-14). For example, *bicoid* mRNA is trapped at the most anterior region, or anterior pole, of the early fly embryo (Figure 15-19). The anterior localization of

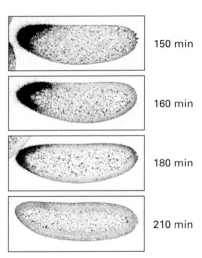

150 min

160 min

180 min

210 min

▲ **EXPERIMENTAL FIGURE 15-19 Maternally derived** *bicoid* **mRNA is localized to the anterior region of early** *Drosophila* **embryos.** All embryos shown are positioned with anterior to the left and dorsal at the top. In this experiment, in situ hybridization with a radioactively labeled RNA probe specific for *bicoid* mRNA was performed on whole-embryo sections 2.5–3.5 hours after fertilization. This time period covers the transition from the syncytial blastoderm to the beginning of gastrulation. After excess probe was removed, probe hybridized to maternal *bicoid* mRNA (dark silver grains) was detected by autoradiography. Bicoid protein is a transcription factor that acts alone and with other regulators to control the expression of certain genes in the embryo's anterior region. [From P. W. Ingham, 1988, *Nature* **335**:25; photographs courtesy of P. W. Ingham.]

bicoid mRNA depends on its 3′-untranslated end and three maternally derived proteins. Embryos produced by female flies that are homozygous for *bicoid* mutations lack anterior body parts, attesting to the importance of Bicoid protein in specifying anterior cell fates.

Bicoid protein, a homeodomain-type transcription factor, activates expression of certain anterior-specific genes discussed later. In the syncytial fly embryo, Bicoid protein spreads through the common cytoplasm away from the an-

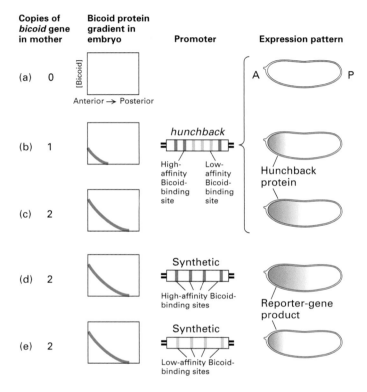

terior end where it is produced from the localized mRNA. As a result, a Bicoid protein gradient is established along the anteroposterior axis of the syncytial embryo. Evidence that the Bicoid protein gradient determines anterior structures was obtained through injection of synthetic *bicoid* mRNA at different locations in the embryo. This treatment led to the formation of anterior structures at the site of injection, with progressively more posterior structures forming at increasing distances from the injection site. Another test was to make flies that produced extra anterior Bicoid protein; in these flies, the anterior structures expanded to occupy a greater proportion of the embryo.

Bicoid protein promotes transcription of the *hunchback* (*hb*) gene from the embryo's genome. Transcription of *hunchback* is greatest in the anterior of the embryo where the Bicoid concentration is highest. Mutations in *hunchback* and several other genes in the embryo's genome lead to large gaps in the anteroposterior pattern of the early embryo; hence these genes are collectively called **gap genes**. Several types of evidence indicate that Bicoid protein directly regulates transcription of *hunchback*. For example, increasing the number of copies of the *bicoid* gene expands the Bicoid and Hunchback (Hb) protein gradients posteriorly in parallel (Figure 15-20a–c). Analysis of the *hunchback* gene revealed that it contains three low-affinity and three high-affinity binding sites for Bicoid protein. The results of studies with synthetic genes containing either all high-affinity or all low-affinity Bicoid-binding sites demonstrated that the affinity of the site determines the threshold concentration of Bicoid at which gene transcription is activated (Figure 15-20d, e). In addition, the number of Bicoid-binding sites occupied at a given concentration has been shown to determine the amplitude, or level, of the transcription response.

Findings from studies of Bicoid's ability to regulate transcription of the *hunchback* gene show that variations in the levels of transcription factors, as well as in the number or affinity of specific regulatory sequences controlling different target genes, or both, contribute to generating diverse patterns of gene expression in development. These findings thus parallel those on the Dorsal transcription factor discussed previously. Similar mechanisms are employed in other developing organisms.

▲ **EXPERIMENTAL FIGURE 15-20 Maternally derived Bicoid controls anterior/posterior expression of the embryonic *hunchback* (*hb*) gene.** (a–c) Increasing the number of *bicoid* genes in mother flies changed the Bicoid gradient in the early embryo, leading to a corresponding change in the gradient of Hunchback protein produced from the *hunchback* gene in the embryo's genome. The *hunchback* promoter contains three high-affinity and three low-affinity Bicoid-binding sites. Transgenic flies carrying a reporter gene linked to a synthetic promoter containing either four high-affinity sites (d) or four low-affinity sites (e) were prepared. In response to the same Bicoid protein gradient in the embryo, expression of the reporter gene controlled by a promoter carrying high-affinity Bicoid-binding sites extended more posteriorly than did transcription of a reporter gene carrying low-affinity sites. This result indicates that the threshold concentration of Bicoid that activates *hunchback* transcription depends on the affinity of the Bicoid-binding site. Bicoid regulates other target genes in a similar fashion. [Adapted from D. St. Johnston and C. Nüsslein-Volhard, 1992, *Cell* **68**:201.]

Maternally Derived Translation Inhibitors Reinforce Bicoid-Mediated Anterioposterior Patterning

Cell types at the posterior end of the fly embryo are controlled by a different mechanism—one in which control is at the translational level rather than the transcriptional level. As just discussed, transcription of the embryo's *hunchback* gene, which promotes anterior cell fates, produces an anteriorly located band of *hunchback* mRNA and Hunchback protein because of the anterior → posterior gradient of maternally derived Bicoid protein. In addition, however, *hunch-*

back mRNA synthesized by nurse cells also is present in the early embryo. Even though this maternal *hunchback* mRNA is uniformly distributed throughout the embryo, its translation is prevented in the posterior region by another maternally derived protein called Nanos, which is localized to the posterior end of the embryo. Nanos protein not only blocks translation of maternal *hunchback* mRNA in the posterior region. The set of genes required for Nanos protein localization is also required for germ-line cells to form at the posterior end of the embryo. The evolutionary conservation of this Nanos function in flies and in worms may indicate an ancient system for forming germ-line cells. Related proteins exist in vertebrates, but their functions are not yet known.

Figure 15-21 illustrates how translational regulation by Nanos helps to establish the anterior → posterior Hunchback gradient needed for normal development. Translational repression of *hunchback* mRNA by Nanos depends on specific sequences in the 3′-untranslated region of the mRNA, called Nanos-response elements (NREs). Along with two other RNA-binding proteins, Nanos binds to the NRE in *hunchback* mRNA. Although the precise mechanism by which repression is achieved is not known, repression inversely correlates with the length of the poly(A) tail in *hunchback* mRNA, which is determined by the balance between the opposite processes of polyadenylation and deadenylation. In wild-type embryos, the length of the poly(A) tail increases immediately before translation of *hunchback* mRNA. The results of genetic and molecular studies suggest that Nanos promotes deadenylation of *hunchback* mRNA and thereby decreases its translation. In the absence of Nanos, an accumulation of maternal Hb protein in the posterior region leads to the failure of the posterior structures to form normally, and the embryo dies. Conversely, if Nanos is produced in the anterior, thereby inhibiting the production of Hb from both maternal and embryonic *hunchback* mRNA, anterior body parts fail to form, again a lethal consequence.

Nanos protein localization in the posterior embryo is intimately coupled to the regulation of translation of *nanos* mRNA. The *nanos* mRNA that is not located at the posterior is not translated due to a protein called Smaug that binds the 3′ UTR of *nanos* mRNA. Localization of *nanos* mRNA at the posterior depends on other proteins as well. One of these is Oskar, whose maternally provided mRNA is transported to the posterior by kinesin, a motor protein that moves along microtubules (Chapter 20). Therefore the kinesin controls, after several intervening steps, the localized activity of a transcription factor (Hunchback).

Translational control due to the action of an inhibitor, mRNA localization, or both, may be widely used strategies for regulating development. For instance, specific mRNAs are localized during the development of muscle cells (see Figure 12-31) and during cell division in the budding yeast *Saccharomyces cerevisiae* (see Figure 22-22). Similar mechanisms operate during the development of *C. elegans*. Even more intriguing is the discovery that Bicoid protein binds

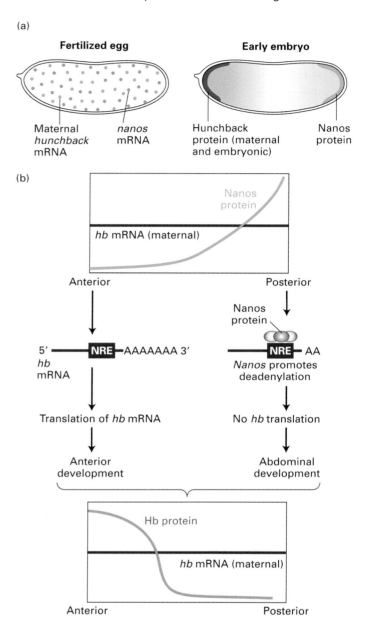

▲ **FIGURE 15-21 Role of Nanos protein in excluding maternally derived Hunchback (Hb) protein from the posterior region of *Drosophila* embryos.** (a) Both *nanos* (red) and *hunchback (hb)* (blue) mRNAs derived from the mother are distributed uniformly in the fertilized egg and early embryo. Nanos protein, which is produced only in the posterior region, subsequently inhibits translation of maternal *hb* (blue) mRNA posteriorly. (b) Diffusion of Nanos protein from its site of synthesis in the posterior region establishes a posterior → anterior Nanos gradient. A complex of Nanos and two other proteins inhibits translation of maternal *hb* mRNA. As a consequence, maternally derived Hb protein is expressed in a graded fashion that parallels and reinforces the Hb protein gradient resulting from Bicoid-controlled transcription of the embryo's *hb* gene (see Figure 15-20). [See C. Wreden et al., 1997, *Development* **124**:3015.]

not only to DNA to promote transcription of the embryonic *hunchback* gene but also to *caudal* mRNA (which encodes another *Drosophila* protein having a role in early-patterning events) and regulates its translation.

Toll-like Signaling Activates an Ancient Defense System in Plants and Animals

Before continuing our account of signaling and gene control in development, we digress briefly to consider the connection between innate immunity and the Toll-Dorsal signaling pathway discussed previously (see Figure 15-17c). Recall that this pathway has many parallels with the NF-κB pathway in mammals, which is intimately involved in immune responses. Discovery of these parallels was the first hint that Toll signaling might function in nondevelopmental contexts. More recently, researchers have found that the Toll receptor, its ligand Spätzle, and other pathway components are required for the expression of an antifungal peptide (drosomycin) in fly larvae and adults. Stimulation of a different Toll-like receptor triggers the production of an antibacterial peptide (diptericin) in flies. Discovery of a mammalian Toll-like receptor that controls the production of anti-inflammatory cytokines stimulated exploration of the whole set of similarities.

Toll signaling now appears to be one of the most evolutionarily conserved processes known. All the components between Toll and the activation of Dorsal have been largely conserved. There are 8 *Drosophila* proteins related to Toll plus Toll itself and 10 Toll-related human proteins that control the production of a wide variety of antimicrobial peptides in flies and cytokines in mammals. These molecules provide a rapid, nonspecific defense against infection by a wide array of pathogens. The adaptive immune response mounted by vertebrates, involving antibodies and T cells, is directed against specific pathogens but is slower to develop than the innate, nonspecific response.

Most remarkably, parts of the Toll pathway and its function in immunity are readily recognizable in plants. For instance, *Arabidopsis* has about 100 proteins containing domains similar to the cytosolic domains of Toll that transduce the intracellular signal. At least some of these proteins are required for resistance to tobacco mosaic virus, and an *Arabidopsis* protein similar in sequence to I-κB is required for resistance to downy mildew fungus. However, the Toll-like signaling in *Arabidopsis* appears to act through transcription factors that are unrelated to Dorsal or NF-κB.

Because the Toll-based innate immunity system appears to be present in both plants and animals, it may be more than a billion years old. In the course of their long evolution, at least some of the genes having roles in the basic survival function of immunity have been adapted to serve as developmental regulators—a nice example of biological parsimony in the use of genetic resources.

KEY CONCEPTS OF SECTION 15.3

Control of Cell Fates by Graded Amounts of Regulators

■ The influence of one cell population on the developmental fate of another one nearby is called induction. Both diffusible signaling molecules and direct cell–cell contacts mediate induction.

■ Morphogens are signals that act in a graded fashion: a cell that receives more of a signal takes on one fate, a cell that receives less takes on a different fate, and so forth. Other signals act in a relay fashion: a signal induces one cell to produce a different signal that instructs cells farther away from the original signal source (see Figure 15-11).

■ In *Drosophila* syncytial embryos, morphogens are used at several stages to induce different cell types along both axes. Asymmetries created in the fly egg during oogenesis trigger events that determine both the dorsoventral and the anterioposterior axes in fly embryos.

■ Two signal systems—one acting during oogenesis and the other acting in the early fly embryo—plus a proteolytic cascade within the perivitelline space lead to the graded nuclear localization of Dorsal, a transcription factor: ventral nuclei receive the most Dorsal; lateral nuclei, less; and dorsal nuclei, none (see Figure 15-17). These early events control development along the dorsoventral axis.

■ In ventral cells, Dorsal turns off genes (e.g., *decapentaplegic*, or *dpp*) needed to make dorsal structures. Dpp, a TGFβ-type signal produced in dorsal cells, functions in graded fashion to specify dorsal cell types. Dpp homologs in vertebrates (BMP proteins) also act in dorsoventral patterning.

■ Early anterioposterior patterning in *Drosophila* produces an anterior → posterior gradient of Hunchback (Hb), a transcription factor that promotes anterior cell fates. Transcription of the embryonic *hb* gene is activated by maternally derived Bicoid protein, which is localized to the anterior (see Figure 15-20).

■ Nanos protein inhibits the translation of maternal *hb* mRNA in the posterior of the fly embryo (see Figure 15-21). Synthesis of Nanos from maternal *nanos* mRNA is restricted to the posterior by translational control linked to motor protein–mediated transport of other regulators to the posterior pole.

15.4 Boundary Creation by Different Combinations of Transcription Factors

In Section 15.3, we saw that maternally derived Bicoid plays a key role in initiating the transcription of gap genes (e.g., *hunchback*) in the anterior region of the early fly embryo; other maternal factors prevent the translation of *hunchback* mRNA in the posterior. Further specification of cell fates in *Drosophila* is controlled by *transcription cascades* in which one transcription factor activates a gene encoding another transcription factor, which in turn acts to promote the expression of a third transcription factor. Such a transcription cascade can generate a population of cells that may all look alike but differ at the transcriptional level.

Transcription cascades have both a temporal and a spatial dimension. At each step in a cascade, for instance, RNA polymerase and ribosomes can take more than an hour to produce a protein, depending on the length of the corresponding gene. Spatial factors come into play when cells at different positions within an embryo synthesize different transcription factors. In this section, we continue the story of early patterning in *Drosophila,* which illustrates both the spatial and the temporal aspects of transcription cascades. The principles learned from the study of *Drosophila* development broadly apply to the creation of form and pattern in all organisms including plants, as discussed at the end of this section. Errors in the genes that control organization, boundary formation, and cell type are associated with many human diseases.

Drosophila Gap Genes Are Transcribed in Broad Bands of Cells and Regulate One Another

The rough outline of cell fates that is laid down in the syncytial fly embryo is refined into a system for precisely controlling the fates of individual cells. The discovery of the relevant regulators came from a genetic screen for mu-

tants with altered embryo body segments. The embryonic body segments go on to grow into the familiar striped pattern seen on any passing hornet. In addition to *hunchback,* four other gap genes—*Krüppel, knirps, giant,* and *tailless*—are transcribed in specific spatial domains, beginning about 2 hours after fertilization and just before cellularization of the embryo is complete (Figure 15-22a).

All the gap-gene proteins are transcription factors. Because these proteins are distributed in broad overlapping peaks (Figure 15-22b), each cell along the anterior/posterior axis contains a particular combination of gap-gene proteins that activates or represses specific genes within that cell. Indeed, something like a battle ensues, because some gap proteins repress the transcription of genes encoding other gap proteins. Although they have no known extracellular ligands, some gap proteins resemble **nuclear receptors,** which are intracellular proteins that bind lipophilic ligands (e.g., steroid hormones) capable of crossing the plasma membrane. Most ligand–nuclear receptor complexes function as transcription factors (see Figure 11-44). The sequence similarity between gap proteins and nuclear receptors suggests that gap genes may have evolved from genes whose transcription was controlled by signals that could cross membranes, such as the steroid hormones. The use of such signal-controlled genes, rather than transcription cascades, could explain how early cell-fate specification operates in animals that do not have a syncytial stage.

(a) Gap-gene proteins

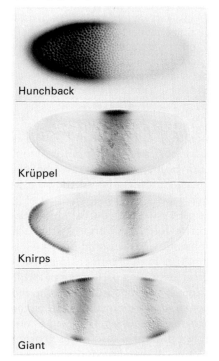

Hunchback

Krüppel

Knirps

Giant

(b) Hunchback and Krüppel

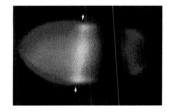

(c) Even-skipped and fushi tarazu

▲ **EXPERIMENTAL FIGURE 15-22 Gap genes and pair-rule genes are expressed in characteristic spatial patterns in early *Drosophila* embryos.** Fixed, permeabilized embryos were stained with fluorescence-labeled antibodies specific for a particular protein. All embryos shown are positioned with anterior to the left and dorsal at the top. (a) These syncytial embryos were stained individually for the indicated gap-gene-encoded proteins. Transcription of the *Krüppel, knirps,* and *giant* gap genes is regulated by Hunchback, Bicoid, and Caudal. (b) This syncytial embryo was doubly stained to visualize Hunchback protein (red) and Krüppel protein (green). The yellow band identifies the region in which expression of these two gap proteins overlaps. (c) Double staining of an embryo at the beginning of gastrulation reveals even-skipped protein (yellow) and Fushi tarazu protein (orange). These two pair-rule gene products are expressed in alternating stripes. See text for discussion. [Part (a) adapted from G. Struhl et al., 1992, *Cell* **69**:237. Parts (b) and (c) courtesy of M. Levine.]

The fates of cells distributed along the anterior/posterior axis are specified early in fly development. At the same time, cells are responding to the dorsal/ventral control system. Each cell is thus uniquely specified along both axes, in a two-dimensional grid. If each of the five gap genes were expressed in its own section of the embryo, at just one concentration, only five cell types could be formed. The actual situation permits far greater diversity among the cells. As noted earlier, the amount of each gap protein varies from low to high to low along the anterior/posterior axis, and the expression domains of different gap genes overlap. This complexity creates combinations of transcription factors that lead to the creation of many more than five cell types. Remarkably, the next step in the *Drosophila* development process generates a repeating pattern of cell types from the rather chaotic nonrepeating pattern of gap-gene expression domains.

Combinations of Gap Proteins Direct Transcription of Pair-Rule Genes in Stripes

Our vertebrae and the body segments of an insect are both examples of a commonly employed tactic in animal structure and music: repeats with variations. The first sign of such repeats in fly embryos is a pattern of repeating stripes of transcription of eight genes collectively called **pair-rule genes.** The body of a larval fly consists of 14 segments, and each pair-rule gene is transcribed in half of its primordia, or seven stripes, separated by "interstripes" where that pair-rule gene is not transcribed (Figure 15-22c). Mutant embryos that lack the function of a pair-rule gene have their body segments fused together in pair-wise fashion—hence the name of this class of genes. The expression stripes for each pair-rule gene partly overlap with those of other pair-rule genes; so each gene must be responding in a unique way to gap-gene and other earlier regulators.

The transcription of three "primary" pair-rule genes is controlled by transcription factors encoded by gap and maternal genes. Because gap and maternal genes are expressed in broad, nonrepeating bands, the question arises: How can such a nonrepeating pattern of gene activities confer a repeating pattern such as the striped expression of pair-rule genes? To answer this question, we consider the transcription of the *even-skipped (eve)* gene in stripe 2, which is controlled by the maternally derived Bicoid protein and the gap proteins Hunchback, Krüppel, and Giant. All four of these transcription factors bind to a clustered set of regulatory sites, or **enhancer,** located upstream of the *eve* promoter (Figure 15-23a). Hunchback and Bicoid activate the transcription of *eve* in a broad spatial domain, whereas Krüppel and Giant repress *eve* transcription, thus creating sharp posterior and anterior boundaries. The combined effects of these proteins, each of which has a unique concentration gradient along the anteroposterior axis, initially demarcates the boundaries of stripe 2 expression (Figure 15-23b).

The initial pattern of pair-rule stripes, which is not very sharp or precise, is sharpened by autoregulation. The Eve

(a) *eve* gene transcription regulation

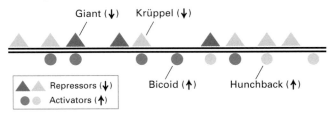

(b) *eve* stripe 2 regulation

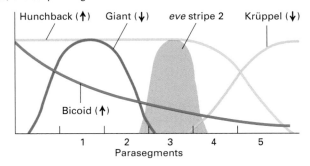

▲ **FIGURE 15-23 Expression of the Even-skipped (Eve) stripe 2 in the *Drosophila* embryo.** (a) Diagram of an 815-bp enhancer controlling transcription of the pair-rule gene *eve*. This regulatory region contains binding sites for Bicoid and Hunchback proteins, which activate the transcription of *eve,* and for Giant and Krüppel proteins, which repress its transcription. Enhancer is shown with all binding sites occupied, but in an embryo occupation of sites will vary with position along the anterior/posterior axis. (b) Concentration gradients and of the four proteins that regulate *eve* stripe 2. The coordinated effect of the two repressors (↓) and two activators (↑) determine the precise boundaries of the second anterior *eve* stripe. Only in the orange region is the combination of regulators correct for the *eve* gene to be transcribed in response to the stripe 2 control element. Further anterior Giant turns it off; further posterior Krüppel turns it off and the level of Bicoid activator is too low. Expression of other stripes is regulated independently by other combinations of transcription factors that bind to enhancers not depicted in part (a). [See S. Small et al., 1991, *Genes & Devel.* **8**:827.]

protein, for instance, binds to its own gene and increases transcription in the stripes, a positive autoregulatory loop. This enhancement does not occur at the edges of stripes where the Eve protein concentration is low; so the boundary between stripe and interstripe is fine-tuned.

Each primary pair-rule gene is regulated by multiple enhancers that are organized in modules. Each stripe is formed in response to a different combination of transcriptional regulators acting on a specific module, so the nonrepeating distributions of regulators can create repeating patterns of pair-rule gene repression and activation. If even one enhancer is bound by an activating combination of transcriptional regulators, the presence of other enhancers in an inactive, "off" state (not bound to a regulator) will not prevent transcription. For instance, in Eve stripe 2, the right combination and amounts of Hunchback and Bicoid create an "on" state that activates transcription even though other enhancers are present in the inactive state. In each stripe, at least one enhancer is bound by an activating combination of regulators. Note that this system of gene control is flexible and could be used to produce nonrepeating patterns of transcription if that were useful to an animal.

Similar responses to gap and maternal proteins govern the striped patterns of transcription of the two other primary pair-rule genes, *runt* and *hairy*. Because the enhancers of *runt* and *hairy* respond to different combinations of regulators, the *eve*, *runt*, and *hairy* expression stripes partly overlap one another, with each stripe for any one gene offset from a stripe

for another gene. Subsequently, other (secondary) pair-rule genes, including *fushi tarazu (ftz)* and *paired*, become active in response to the Eve, Runt, and Hairy proteins, which are transcription factors, as well as to maternal and gap proteins. The outcome is a complex pattern of overlapping stripes.

In early embryos each segment primordium is about four cells wide along the anterioposterior axis, which corresponds to the approximate width of pair-rule expression stripes. With pair-rule genes active in alternating four-on four-off patterns, the repeat unit is about eight cells. Each cell expresses a combination of transcription factors that can potentially distinguish it from any of the other seven cells in the repeat unit. Complete segmentation of the embryo into repeat units in which each cell exhibits a unique transcriptional pattern depends on activation of a third set of genes, the **segment-polarity genes**. These genes, which include *engrailed (en)* and *wingless (wg)*, also are expressed in stripes, but the stripes are narrower and appear once in each segment primordium. Because some segment-polarity genes encode components of cell–cell signaling systems, they are discussed in Section 15.5.

Figure 15-24a schematically depicts the distribution of some key regulators in the *Drosophila* embryo during the first few hours after fertilization. When all the segmentation genes have been turned on, the resulting single-cell accuracy of cell-fate specification is impressive, as the example in Figure 15-24b illustrates.

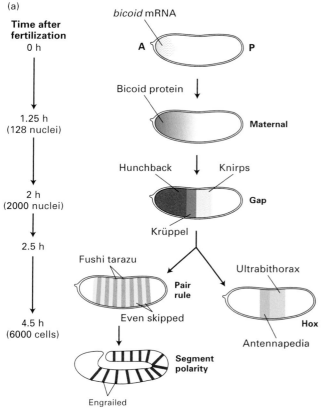

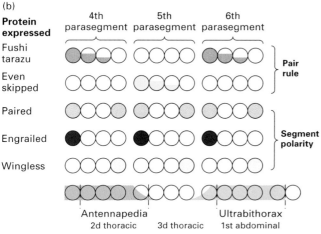

◀ **FIGURE 15-24 Summary of sequential, spatially localized expression of selected genes in early development of the *Drosophila* embryo.** (a) Maternal *bicoid* mRNA is localized at the anterior pole of the egg, but Bicoid protein, which is synthesized soon after fertilization, diffuses to form a gradient in the fly syncytium. In most cases, an mRNA and its corresponding protein are present in the same regions of the embryo. Specific combinations of Bicoid and various gap-gene products, including Hunchback, Krüppel, and Knirps, control transcription of the pair-rule genes such as *fushi tarazu (ftz)* and *even-skipped (eve)*. Gap gene products are shown in discrete bands but actually they overlap. The pair-rule proteins demarcate 14 stripes corresponding to the parasegments (an offset form of the segment primordia). The segment-polarity gene *engrailed (en)* is expressed at the anterior end of each parasegment; it and other segment-polarity genes participate in patterning of each parasegment. Cellularization occurs after 2.5 hours, and gastrulation occurs at about 4.5 hours. By this time, each parasegment consists of four belts of cells. (b) Within a parasegment, each belt of cells (represented by a circle) is characterized by expression of a unique set of proteins encoded by pair-rule and segment-polarity genes. Shown here are the locations of three pair-rule proteins and two segment-polarity proteins in three parasegments (4–6). These expression patterns act as positional values that uniquely characterize each cell belt in a parasegment and determine where Hox genes such as *Antennapedia (Antp)* and *Ultrabithorax (Ubx)* are transcribed. Hox genes give the repeating body segments their distinct shapes and appendages (indicated at the bottom). Hox gene expression is regulated by gap, pair-rule, and segment polarity genes.

MEDIA CONNECTIONS

Video: Expression of Segmentation Genes in a *Drosophila* Embryo

Maternal and Zygotic Segmentation Proteins Regulate Expression of Homeotic (Hox) Genes

As we have just seen, the spatially controlled transcription of pair-rule genes sets up repeating units within the initial sheet of cells composing the early *Drosophila* embryo. Later, these repeating units must diversify: only some of them will produce appendages, and they will specialize internally as well. This early diversification of body segments depends on the **Hox genes,** which are important switches that control cell identities and indeed the identities of whole parts of an animal.

Mutations in Hox genes often cause **homeosis**—that is, the formation of a body part having the characteristics normally found in another part at a different site. For example, flies develop legs on their heads instead of antennae. Loss of function of a particular Hox gene in a location where it is normally active leads to homeosis if a different Hox gene becomes derepressed there; the result is the formation of cells and structures characteristic of the derepressed gene. A Hox gene that is abnormally expressed where it is normally inactive can take over and impose its own favorite developmental pathway on its new location (Figure 15-25).

Hox genes encode highly related transcription factors containing the **homeodomain** motif. Classical genetic studies in *Drosophila* led to the discovery of the first Hox genes (e.g., *Antennapedia* and *Ultrabithorax*). Corresponding genes with similar functions (orthologs) have since been identified in most animal species. Each Hox gene is transcribed in a particular region along the anterioposterior axis in a remarkable arrangement where the order of genes along the chromosomes is colinear with the order in which they are expressed along the anterior/posterior axis. At one end of the complex are head genes, and genes expressed with progressively more posterior boundaries are in order, wih "tail" genes last. Hox-gene expression domains can overlap (see Figure 15-24). In *Drosophila*, the spatial pattern of Hox-gene transcription is regulated by maternal, gap, and pair-rule transcription factors. The protein encoded by a particular Hox gene controls the organization of cells within the region in which that Hox gene is expressed. For example, a Hox protein can direct or prevent the local production of a secreted signaling protein, cell-surface receptor, or transcription factor that is needed to build an appendage on a particular body segment.

Drosophila Hox proteins control the transcription of target genes whose encoded proteins determine the diverse morphologies of body segments. Vertebrate Hox proteins similarly control the different morphologies of vertebrae, of repeated segments of the hindbrain, and of the digits of the limbs. The association of Hox proteins with their binding sites on DNA is assisted by cofactors that bind to both Hox proteins and DNA, adding specificity and affinity to these interactions.

When Hox genes are turned on, their transcription must continue to maintain cell properties in specific locations. As in the *even-skipped* gene, the regulatory regions of some Hox genes contain binding sites for their encoded proteins. Thus Hox proteins can help to maintain their own expression through an autoregulatory loop.

Another mechanism for maintaining normal patterns of Hox-gene expression requires proteins that modulate chromatin structure. These proteins are encoded by two classes of genes referred to as the Trithorax group and Polycomb group. The pattern of Hox-gene expression is initially normal in Polycomb-group mutants, but eventually Hox-gene transcription is derepressed in places where the genes should be inactive. The result is multiple homeotic transformations. This observation indicates that the normal function of Polycomb proteins is to keep Hox genes in a transcriptionally inactive state. The results of immunohistological and biochemical studies have shown that Polycomb proteins bind to multiple chromosomal locations and form large complexes containing different proteins of the Polycomb group. The current view is that the transient repression of genes set up by patterning proteins earlier in development is "locked in" by Polycomb proteins. This stable Polycomb-dependent repression may result from the ability of these proteins to assemble inactive chromatin structures (Chapter 11). Polycomb complexes contain many proteins, including histone

Normal

***Ubx* mutant**

▲ **EXPERIMENTAL FIGURE 15-25** Misexpression of the ***Ultrabithorax (Ubx)* gene leads to development of a second pair of wings in *Drosophila*.** Like other Hox genes, *Ubx* controls the organization of cells within the region in which it is expressed (see Figure 15-24b). Mutations in Hox genes often lead to the formation of a body part where it does not normally exist. In this case the loss of *Ubx* function from the third thoracic segment allows wings to form where normally there are only balancer organs called halteres. [From E. B. Lewis, 1978, *Nature* **276**:565; photographs courtesy of E. B. Lewis. Reprinted by permission from *Nature,* copyright 1978, Macmillan Journals Limited.]

deacetylases, and appear to inactivate transcription by modifying histones to promote gene silencing.

Whereas Polycomb proteins repress the expression of certain Hox genes, proteins encoded by the Trithorax group of genes are necessary for maintaining the expression of Hox genes. Like Polycomb proteins, Trithorax proteins bind to multiple chromosomal sites and form large multiprotein complexes, some with a mass of $\sim 2 \times 10^6$ Da, about half the size of a ribosome. Some Trithorax-group proteins are homologous to the yeast Swi/Snf proteins, which are crucial for transcriptional activation of many yeast genes. Trithorax proteins stimulate gene expression by selectively remodeling the chromatin structure of certain loci to a transcriptionally active form (see Figure 11-37). The core of each complex is an ATPase, often of the Brm class of proteins. There is evidence that many or most genes require such complexes for transcription to take place.

Flower Development Also Requires Spatially Regulated Production of Transcription Factors

PLANTS The basic mechanisms controlling development in plants are much like those in *Drosophila:* differential production of transcription factors, controlled in space and time, specifies cell identities. Our understanding of cell-identity control in plants benefited greatly from the choice of *Arabidopsis thaliana* as a model organism. This plant has many of the same advantages as flies and worms for use as a model system: it is easy to grow, mutants can be obtained, and transgenic plants can be made. We will focus on certain transcription-control mechanisms regulating the formation of cell identity in flowers. These mechanisms are strikingly similar to those controlling cell-type and anteroposterior regional specification in yeast and animals.

Floral Organs A flower comprises four different organs called sepals, petals, stamens, and carpels, which are arranged in concentric circles called whorls. Whorl 1 is the outermost; whorl 4, the innermost. *Arabidopsis* has a complete set of floral organs, including four sepals in whorl 1, four petals in whorl 2, six stamens in whorl 3, and two carpels containing ovaries in whorl 4 (Figure 15-26a). These organs grow from a collection of undifferentiated, morphologically indistinguishable cells called the floral **meristem.** As cells within the center of the floral meristem divide, four concentric rings of primordia form sequentially. The outer-ring primordium, which gives rise to the sepals, forms first, followed by the primordium giving rise to the petals, then the stamen and carpel primordia.

Floral Organ–Identity Genes Genetic studies have shown that normal flower development requires three classes of *floral organ–identity genes*, designated A, B, and C genes. Mutations in these genes produce phenotypes equivalent to those associated with homeotic mutations in flies and mammals; that is, one part of the body is replaced by another. In plants lacking all A, B, and C function, the floral organs develop as leaves (Figure 15-26b).

Figure 15-27 summarizes the loss-of-function mutations that led to the identification of the A, B, and C gene classes. On the basis of these homeotic phenotypes, scientists proposed a model to explain how three classes of genes control floral-organ identity. According to this ABC model for specifying floral organs, class A genes specify sepal identity in whorl 1 and do not require either class B or class C genes to do so. Similarly, class C genes specify carpel identity in whorl 4 and, again, do so independently of class A and B genes. In contrast with these structures, which are specified by only a single class of genes, the petals in whorl 2 are specified by class A and B genes, and the stamens in whorl 3 are specified by class B and C genes. To account for the observed effects of removing A genes or C genes, the model also postulates that A genes repress C genes in whorls 1 and 2 and, conversely, C genes repress A genes in whorls 3 and 4.

To determine if the actual expression patterns of class A, B, and C genes are consistent with this model, researchers

(a)

(b)

◄ EXPERIMENTAL FIGURE 15-26 Mutations in floral organ–identity genes produce homeotic phenotypes. (a) Flowers of wild-type *Arabidopsis thaliana* have four sepals in whorl 1, four petals in whorl 2, six stamens in whorl 3, and two carpels in whorl 4. (b) In *Arabidopsis* with mutations in all three classes of floral organ–identity genes, the four floral organs are transformed into leaf-like structures. [From D. Weigel and E. M. Meyerowitz, 1994, *Cell* **78**:203; courtesy of E. M. Meyerowitz.]

(a) Wild-type floral organs

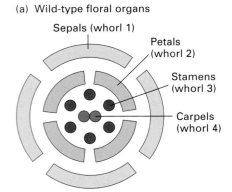

(b) Loss-of-function homeotic mutations

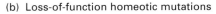
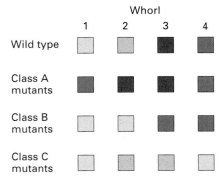

▲ **EXPERIMENTAL FIGURE 15-27 Phenotypic analysis identified three classes of genes that control specification of floral organs in *Arabidopsis*.** (a) Diagram of the arrangement of wild-type floral organs, which are found in concentric whorls. (b) Effect of loss-of-function mutations leading to transformations of one organ into another. Class A mutations affect organ identity in whorls 1 and 2: sepals (green) become carpels (blue) and petals (orange) become stamens (red). Class B mutations cause transformation of whorls 2 and 3: petals become sepals and stamens become carpels. In class C mutations, whorls 3 and 4 are transformed: stamens become petals and carpels become sepals. [See D. Wiegel and E. M. Meyerowitz, 1994, *Cell* **78**:203.]

cloned these genes and assessed the expression patterns of their mRNAs in the four whorls in wild-type *Arabidopsis* plants and in loss-of-function mutants (Figure 15-28a, b). Consistent with the ABC model, A genes are expressed in whorls 1 and 2, B genes in whorls 2 and 3, and C genes in whorls 3 and 4. Furthermore, in class A mutants, class C genes are also expressed in organ primordia of whorls 1 and 2; similarly, in class C mutants, class A genes are also expressed in whorls 3 and 4. These findings are consistent with the homeotic transformations observed in these mutants.

To test whether these patterns of expression are functionally important, scientists produced transgenic *Arabidopsis* plants in which floral organ–identity genes were expressed in inappropriate whorls. For instance, the introduction of a transgene carrying class B genes linked to an A-

class promoter leads to the ubiquitous expression of class B genes in all whorls (Figure 15-28c). In such transgenics, whorl 1, now under the control of class A and B genes, develops into petals instead of sepals; likewise, whorl 4, under the control of both class B and class C genes, gives rise to stamens instead of carpels. These results support the functional importance of the ABC model for specifying floral identity.

Sequencing of floral organ–identity genes has revealed that many encode proteins belonging to the MADS family of transcription factors, which form homo- and heterodimers. Thus floral-organ identity may be specified by a combinatorial mechanism in which differences in the activities of different homo- and heterodimeric forms of various A, B, and C proteins regulate the expression of subordinate downstream genes necessary for the formation of the differ-

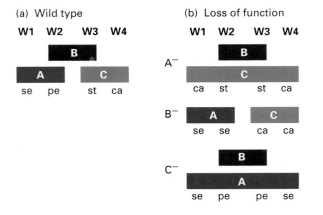

▲ **EXPERIMENTAL FIGURE 15-28 Expression patterns of class A, B, and C genes support the ABC model of floral organ specification.** Depicted here are the observed expression patterns of the floral organ–identity genes in wild-type, mutant, and transgenic *Arabidopsis*. Colored bars represent the A, B, and C mRNAs in each whorl (W1, W2, W3, W4). The observed floral organ in each whorl is indicated as follows: sepal = se; petals = pe; stamens = st; and carpels = ca. See text for discussion. [See D. Wiegel and E. M. Meyerowitz, 1994, *Cell* **78**:203, and B. A. Krizek and E. M. Meyerowitz, 1996, *Development* **122**:11.]

ent cell types in each organ. Other MADS transcription factors function in cell-type specification in yeast and muscle (Chapter 22). ∎

Boundary Creation by Different Combinations of Transcription Factors

■ Gradients of transcription factors, produced from maternal mRNAs in the early *Drosophila* embryo, control the patterned expression of embryonic genes, leading to segmentation of the embryo along the anteroposterior axis.

■ Target genes whose regulatory regions contain multiple enhancers are expressed preferentially in specific regions of the embryo, depending on the amounts and combinations of the transcription factors that control them (see Figure 15-23).

■ Early patterning events, utilizing maternal, gap, pair-rule, and segment-polarity genes, generate a unique pattern of transcription factors in different regions along the anteroposterior axis of *Drosophila* embryos (see Figure 15-24). These transcription factors are expressed transiently and play an essential role in establishing the domains in which different Hox genes are expressed.

■ Hox genes, which encode transcription factors, control the unique morphologic characteristics of different regions along the anteroposterior axis in most or all animals.

■ Misexpression of Hox genes causes homeotic transformations—the development of body parts in abnormal positions.

■ Hox expression patterns are sometimes maintained through positive autoregulatory loops, and through modulation of chromatin by proteins encoded by Polycomb-group and Trithorax-group genes.

■ Three classes of genes (A, B, and C) participate in specifying the identity of the four organs constituting a flower (see Figure 15-27). The patterns of expression of these genes, many of which encode transcription factors, are consistent with the ABC genetic model.

15.5 Boundary Creation by Extracellular Signals

As the syncytial fly embryo becomes cellular and undergoes gastrulation, the movement of proteins and mRNAs through the common cytoplasm of a syncytium is over. Further cell-fate specification is controlled primarily by cells communicating with one another through secreted extracellular signals. In this section, we examine how three signaling pathways, activated by Hedgehog (Hh), Wingless (Wg, a member of the Wnt family), and TGFβ, create boundaries between cell types during *Drosophila* development. The Wingless and Hedgehog proteins are encoded by segment-polarity genes, so named because they affect the orientation of surface features of the cuticle, such as bristles. The events discussed here are representative of what happens in virtually all tissues and all animals to specify cell types and create boundaries between different types.

Two Secreted Signals, Wingless and Hedgehog, Create Additional Boundaries Within Segments of Cellular Fly Embryos

As we saw in Section 15.4, the 14 segment primordia in the early *Drosophila* embryo are defined by various pair-rule proteins, with each protein located in seven stripes that alternate with stripes of cells that do not make the protein. The segment-polarity gene *engrailed*, which encodes a transcription factor, is expressed in the most anterior cell in each primordium, forming 14 Engrailed stripes. Transcription of *engrailed* is activated and repressed by various pair-rule proteins. In each eight-cell repeat unit established by the pair-rule proteins, *engrailed* is transcribed in cells 1 and 5. Recall that the pair-rule proteins produced in cells 1 through 4 differ from those produced in cells 5 through 8 (see Figure 15-24b). Although the transcriptional regulation of *engrailed* is the same in cell 1 in all the eight-cell repeats, it cannot be the same in cells 1 and 5 of a repeat. Thus two different combinations of pair-rule proteins must activate transcription of *engrailed*; so a seemingly simple repeating pattern masks a striking difference in regulation.

Another segment-polarity gene called *wingless* becomes active at about the same time as *engrailed*. It also is expressed in single-cell-wide stripes, adjacent to the Engrailed stripes and just one cell farther anterior (see Figure 15-24b). Wingless is a secreted signaling protein, a member of the Wnt protein family found in most or all animals. With the production of Wingless, the cells of the fly embryo stop ignoring one another and begin communicating through signals. In adjacent Engrailed-producing cells, the Wingless signal maintains the expression of another segment-polarity gene called *hedgehog (hh)*, which also encodes an external signal. Expression of *hedgehog* is initially activated by Engrailed, a transcription factor that has both activating and repressing abilities. Engrailed activates *hedgehog* directly and represses a gene encoding a repressor of *hedgehog*, thereby indirectly promoting *hedgehog* expression. In the fly embryo, the Wingless and Hedgehog signals, produced in adjacent stripes of cells, form a positive feedback loop, with each maintaining expression of the other across the boundary (Figure 15-29).

The Wingless and Hedgehog signals control which cell types form in which positions, creating additional boundaries beyond those established by pair-rule proteins. Even

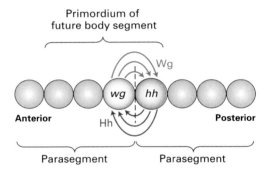

▲ **FIGURE 15-29 Role of Hedgehog (Hh) and Wingless (Wg) in boundary creation between parasegments in *Drosophila* embryo.** Hedgehog is necessary to maintain *wingless* transcription, and, conversely, Wingless is required to maintain *hedgehog*. These two secreted signals play a key role in patterning the epidermis. Both signaling proteins act on cells in addition to those indicated by the arrows. They act through the signal-transduction pathways shown in Figures 15-31 and 15-32. [See M. Hammerschmidt et al., 1997, *Trends Genet.* **13**:14.]

before Wingless- or Hedgehog-induced morphological features are evident, the prospective cell fates can be detected by the production of specific transcription factors. Both Hedgehog and Wingless can act as morphogens, with different concentrations inducing different fates in receiving cells (see Figure 15-11b). Cells that receive a large amount of Wingless turn on certain genes and form certain structures; cells that receive a smaller amount turn on different genes and thus form different structures. The same idea applies to the effects of different amounts of Hedgehog on receiving cells. As Wingless and Hedgehog are secreted from their source cells, they theoretically could move and signal in both directions. Recent work, however, shows that a signal can act mostly in one direction, anterior in the case of Wingless. This directional preference results from active destruction of much of the Wingless protein that moves posteriorly.

Having seen when the *Drosophila* Hedgehog and Wnt-type signals first begin to act in fly development, we take a closer look at the operation of these pathways. Both pathways participate in the development of many different tissues in *Drosophila* and most other animals.

Hedgehog Signaling, Which Requires Two Transmembrane Proteins, Relieves Repression of Target Genes

The Hedgehog signal is secreted from cells as a 45-kDa precursor protein. Cleavage of this secreted precursor produces a 20-kDa N-terminal fragment, which is associated with the plasma membrane and contains the inductive activity, and a 25-kDa C-terminal fragment. A series of elegant experiments demonstrated how the N-terminal Hedgehog fragment, which does not contain any hydrophobic sequences, acquires an affinity for the membrane. As depicted in Figure 15-30,

this process includes adding cholesterol to a glycine residue, splitting the molecule into two fragments, and leaving the N-terminal signaling fragment with an attached hydrophobic cholesterol moiety. The C-terminal domain of the precursor, which catalyzes this reaction, is found in other proteins and may promote the linkage of these proteins to membranes by the same autoproteolytic mechanism. A second modification to Hedgehog, the addition of a palmitoyl group to the N-terminus, makes the protein even more hydrophobic. Together, the two modifications may tether Hedgehog to cells, thereby affecting its range of action in tissue. Spatial restriction plays a crucial role in constraining the effects of powerful inductive signals.

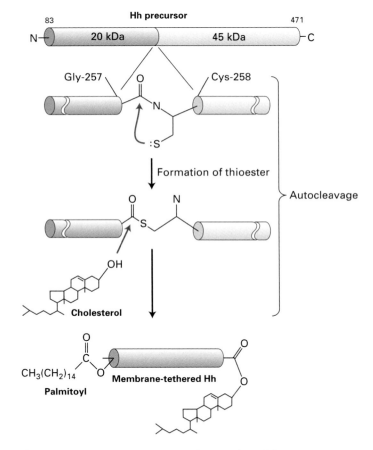

▲ **FIGURE 15-30 Processing of Hedgehog (Hh) precursor protein.** Removal of the N-terminal signal peptide from the initial translation product yields the 45-kDa Hh precursor consisting of residues 83–471 in the original protein. Nucleophilic attack by the thiol side chain of cysteine 258 (Cys-258) on the carbonyl carbon of glycine 257 (Gly-257) forms a thioester intermediate. The C-terminal domain then catalyzes the formation of an ester bond between the β-3 hydroxyl group of cholesterol and glycine 257, cleaving the precursor into two fragments. The N-terminal signaling fragment (tan) retains the cholesterol moiety and is modified by the addition of a palmitoyl group to the N-terminus. These two hydrophobic anchors tether the signaling fragment to the membrane. [Adapted from J. A. Porter et al., 1996, *Science* **274**:255.]

Findings from genetic studies in *Drosophila* indicate that two membrane proteins, Smoothened (Smo) and Patched (Ptc), are required to receive and transduce a Hedgehog signal to the cell interior. Smoothened has 7 membrane-spanning α helices, similarly to G protein–coupled receptors (Chapter 13). Patched is predicted to contain 12 transmembrane α helices and is most similar to Niemann-Pick C1 protein (NPC1). These proteins may act as pumps or transporters. As discussed in Chapter 18, NPC1 protein is necessary for normal intracellular movement of sterols through vesicle-trafficking pathways. In humans, mutations in the NPC1 gene cause a rare, autosomal recessive disorder marked by defects in the lysosomal handling of cholesterol.

Drosophila embryos with loss-of-function mutations in the *smoothened* or *hedgehog* genes have very similar phenotypes. Moreover, both genes are required to activate transcription of the same target genes (e.g., *wingless*) during embryonic development. Loss-of-function mutations in *patched* produce a quite different phenotype, one similar to the effect of flooding the embryo with Hedgehog. Thus Patched appears to antagonize the actions of Hedgehog and vice versa. These findings and analyses of double mutants suggest that, in the absence of Hedgehog, Patched represses target genes by inhibiting a signaling pathway needed for gene activation. The additional observation that Smoothened is required for the transcription of target genes in mutants lacking *patched* function places Smoothened downstream in the pathway. The binding of Hedgehog evidently prevents

Patched from blocking Smoothened action, thus activating the transcription of target genes.

The results of recent studies have shown that, in the absence of Hedgehog, Patched is enriched in the plasma membrane, but Smoothened is in internal vesicle membranes. When cells receive a Hedgehog signal, both Patched and Hedgehog move from the cell surface into internal vesicles, whereas Smoothened moves from internal vesicles to the surface. The similarity of Patched to Niemann-Pick C1 protein, the covalent joining of cholesterol to Hedgehog, and the ability of cholesterol analogs such as cyclopamine to block reception of a Hedgehog signal all suggest a possible link between sterol metabolism and Hedgehog signaling. Indeed, one interesting idea is that developmental regulation by the Hedgehog system evolved from earlier cell components needed to control vesicle composition and movement.

Figure 15-31 depicts a current model of the *Hedgehog pathway*. Although the signal-transduction mechanisms are only partly understood, the pathway includes a cytoplasmic complex of proteins consisting of Fused (Fu), a serine-threonine kinase; Costal-2 (Cos-2), a microtubule-associated kinesin-like protein; and Cubitis interruptus (Ci), a transcription factor. In the absence of Hedgehog, when Patched inhibits Smoothened, these three proteins form a complex that binds to microtubules in the cytoplasm. Proteolytic cleavage of Ci in this complex generates a Ci fragment that translocates to the nucleus and represses target-gene expression. In the presence of Hedgehog, which relieves the

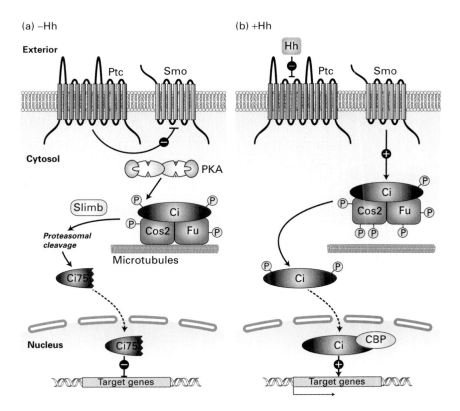

◀ **FIGURE 15-31 Operational model of the Hedgehog (Hh) signaling pathway.** (a) In the absence of Hh, Patched (Ptc) protein inhibits Smoothened (Smo) protein by an unknown mechanism. In the absence of Smo signaling, a complex containing the Fused (Fu), Costal-2 (Cos2), and Cubitis interuptus (Ci) proteins binds to microtubules. Ci is cleaved in a process requiring the ubiquitin/proteasome-related F-box protein Slimb, generating the fragment Ci75, which functions as a transcriptional repressor. (b) In the presence of Hh, inhibition of Smo by Ptc is relieved. Signaling from Smo causes hyperphosphorylation of Fu and Cos2, and disassociation of the Fu/Cos2/Ci complex from microtubules. This leads to the stabilization of a full-length, alternately modified Ci, which functions as a transcriptional activator in conjunction with CREB binding protein (CBP). The exact membrane compartments in which Ptc and Smo respond to Hh and function are unknown; Hh signal causes Ptc to move from the surface to internal compartments while Smo does the opposite. [After K. Nybakken and N. Perrimon, 2002, *Curr. Opin. Genet. Devel.* **12**:503.]

inhibition of Smoothened, the complex of Fu, Cos-2, and Ci is not associated with microtubules, cleavage of Ci is blocked, and an alternatively modified form of Ci is generated. After translocating to the nucleus, this Ci form binds to the transcriptional coactivator CREB-binding protein (CBP), promoting the expression of target genes. In addition to these components, protein kinase A participates in controlling Hedgehog-responsive target genes, which become inappropriately active when protein kinase A is inactivated. Phosphorylation of Ci by protein kinase A appears to stimulate the proteolytic cleavage of Ci.

 Hedgehog signaling, which is conserved throughout the animal kingdom, functions in the formation of many tissues and organs. Mutations in components of the Hedgehog signaling pathway have been implicated in birth defects such as cyclopia, a single eye resulting from union of the right and left brain primordia, and in multiple forms of human cancer. ∎

Wnt Signals Trigger Disassembly of an Intracellular Complex, Releasing a Transcription Factor

As noted previously, the *Drosophila* segment-polarity gene *wingless* encodes a protein that belongs to the Wnt family of secreted signals. Inactivation of *wingless* causes segment-polarity defects very similar to those caused by the loss of *hedgehog* function. This observation is logical because Hedgehog and Wingless form a positive feedback loop, with each protein maintaining production of the other (see Fig-

ure 15-29). The first vertebrate Wnt gene to be discovered was a mouse gene called *Wnt-1* (formerly *int-1*). Activation of *int-1* by insertion of a mouse mammary tumor virus (MMTV) provirus leads to mammary cancer. Hence *Wnt-1* is a **proto-oncogene**, a normal cellular gene whose inappropriate expression promotes the onset of cancer (Chapter 23). The word Wnt is an amalgamation of *wingless*, the corresponding fly gene, with *int* for MMTV integration.

Genetic studies in *Drosophila* and *C. elegans*, studies of mouse proto-oncogenes and tumor-suppressor genes, and studies of cell junction components have all contributed to identifying many components of the Wnt signal-transduction pathway. Like Hedgehog proteins, Wnt proteins are modified by the addition of a hydrophobic palmitate group near their N termini, which may tether them to the plasma membrane of secreting cells and limit their range of action. Wnt proteins act through two cell-surface receptor proteins: Frizzled (Fz), which contains seven transmembrane α helices and directly binds Wnt; and Lrp, which appears to associate with Frizzled in a Wnt signal–dependent manner, at least in frog embryos. Mutations in the genes encoding Wingless, Frizzled, or Lrp (called Arrow in *Drosophila*) all have similar effects on the development of embryos. Frizzled protein and the Smoothened protein in Hedgehog signaling have sequence similarities, and both bear some resemblance to the G protein–coupled receptors discussed in Chapter 13. To date, however, evidence for G protein involvement downstream of Smoothened or Frizzled remains indirect and not compelling.

A current model of the *Wnt pathway* is shown in Figure 15-32. The central player in intracellular Wnt signal transduction is called β-catenin in vertebrates and Armadillo in

▶ **FIGURE 15-32 Operational model of the Wnt signaling pathway.** (a) In the absence of Wnt, the kinase GSK3 constitutively phosphorylates β-catenin. Phosphorylated β-catenin is degraded and hence does not accumulate in cells. Axin is a scaffolding protein that forms a complex with GSK3, β-catenin, and APC, which facilitates phosphorylation of β-catenin by GSK3 by an estimated factor of >20,000. The TCF transcription factor in the nucleus acts as a repressor of target genes unless altered by Wnt signal transduction. (b) Binding of Wnt to its receptor Frizzled (Fz) recruits Dishevelled (Dsh) to the membrane. Activation of Dsh by Fz inhibits GSK3, permitting unphosphorylated β-catenin to accumulate in the cytosol. After translocation to the nucleus, β-catenin may act with TCF to activate target genes or, alternatively cause the export of TCF from the nucleus and perhaps its activation in cytosol. [After R. T. Moon et al., 2002, *Science* **296**:644; see also The Wnt Gene Homepage, www.stanford.edu/~rnusse/wntwindow.html.]

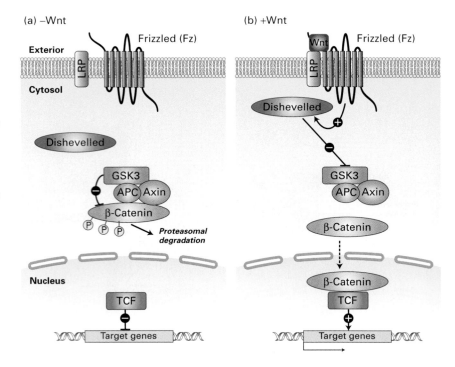

Drosophila. This remarkable protein functions both as a transcriptional activator and as a membrane–cytoskeleton linker protein (see Figure 6-7). In the absence of a Wnt signal, β-catenin is phosphorylated by a complex containing GSK3, a protein kinase; the adenomatosis polyposis coli (APC) protein, an important human tumor suppressor; and Axin, a scaffolding protein. Phosphorylated β-catenin is ubiquitinated and then degraded in proteasomes. In the presence of Wnt, β-catenin is stabilized and translocates to the nucleus. There, it is believed to associate with the TCF transcription factor to activate expression of particular target genes (e.g., *wg, cyclin D1, myc,* and metalloprotease genes), depending on cell type. Recent evidence suggests that β-catenin acts by a different mechanism in which it controls the export of TCF from the nucleus and perhaps its activation in the cytosol.

Findings from genetic studies have shown that Wnt-induced stabilization of β-catenin depends on Dishevelled (Dsh) protein. In the presence of Wnt, Dsh and the Lrp membrane protein appear to interact with components of the phosphorylation complex, thereby inhibiting the phosphorylation and subsequent degradation of β-catenin (see Figure 15-32b). The importance of β-catenin stability and location means that Wnt signals affect a critical balance between the three pools of β-catenin in the cytoskeleton, cytosol, and nucleus.

Wnt signals help control numerous critical developmental events, such as gastrulation, brain development, limb patterning, and organogenesis. The regulated movement of Wnts through tissue is critical to establishing properly placed boundaries between different cell types. As is discussed in Chapter 23, disturbances in signal transduction through the Wnt pathway and many other developmentally important signaling pathways are associated with various human cancers.

Gradients of Hedgehog and Transforming Growth Factor β Specify Cell Types in the Neural Tube

As we have seen in *Drosophila*, many developmental signals act in a graded fashion, inducing different cell fates depending on their concentration. The same phenomenon exists in vertebrates, for example, in the development of the mammalian central nervous system from the neural tube, which forms early in embryogenesis. The neural tube is a simple rolled-up sheet of cells, initially one cell thick. Cells in the ventral part will form motor neurons; lateral cells will form a variety of interneurons. The different cell types can be distinguished prior to morphological differentiation by the proteins that they produce.

Graded concentrations of Sonic hedgehog (Shh), a vertebrate equivalent of *Drosophila* Hedgehog, determine the fates of at least four cell types in the chick ventral neural tube. These cells are found at different positions along the dorsoventral axis in the following order from ventral to dorsal: floor-plate cells, motor neurons, V2 interneurons, and V1 interneurons. During development, Shh is initially expressed at high levels in the notochord, a mesoderm structure in direct contact with the ventralmost region of the neural

tube (Figure 15-33). On induction, floor-plate cells also produce Shh, forming a Shh-signaling center in the ventral-most region of the neural tube. Antibodies to Shh protein block the formation of the different ventral neural-tube cells in the chick, and these cell types fail to form in mice homozygous for mutations in the *Sonic hedgehog (Shh)* gene.

To determine whether Shh-triggered induction of ventral neural-tube cells is through a graded or a relay mechanism, scientists added different concentrations of Shh to chick neural-tube explants. In the absence of Shh, no ventral cells formed. In the presence of very high concentrations of Shh, floor-plate cells formed; whereas, at a slightly lower concentration, motor neurons formed. When the level of Shh was decreased another twofold, only V2 neurons formed. And, finally, only V1 neurons developed when the Shh concentration was decreased another twofold. These data strongly suggest that in the developing neural tube different cell types are formed in response to a ventral → dorsal gradient of Shh. The accumulating evidence for gradients does not rule out additional relay signals that may yet be discovered.

Cell fates in the dorsal region of the neural tube are determined by BMP proteins (e.g., BMP4 and BMP7), which belong to the TGFβ family. Recall that Dpp protein, a *Drosophila* TGFβ signal, is critical in determining dorsal cell fates in early fly embryos. Indeed, TGFβ signaling appears to be an evolutionarily ancient regulator of dorsoventral patterning. In vertebrate embryos, BMP proteins secreted from ectoderm cells overlying the dorsal side of the neural tube promote the formation of dorsal cells such as sensory neurons (see Figure 15-33). Thus cells in the neural tube sense multiple signals that originate at opposite positions on the dorsoventral axis, and measure the signals from both origins to decide on a course of differentiation.

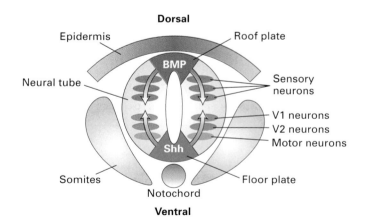

▲ **FIGURE 15-33 Graded induction of different cell types in the neural tube by Sonic hedgehog (Shh) and BMP signaling.** Shh produced in the notochord induces floor-plate development. The floor plate, in turn, produces Shh, which forms a ventral → dorsal gradient that induces additional cell fates. In the dorsal region, BMP proteins secreted from the overlying ectoderm cells act in a similar fashion to create dorsal cell fates. [See T. M. Jessell, 2000, *Nature Rev. Genet.* **1**:20.]

Cell-Surface Proteoglycans Influence Signaling by Some Pathways

How do signals move through or around cells embedded in tissues? The full answer is not known, but the distance that a signal can move has important implications for the size and shape of organs. A signal that causes neurons to form, for example, will create more neurons if its range of movement increases. The binding of signaling proteins to cell-surface **proteoglycans** not only affects the range of signal action but also facilitates signaling in some cases. A proteoglycan consists of a core protein to which is bound **glycosaminoglycan** chains such as heparin sulfate and chondroitin sulfate (see Figure 6-22). Proteoglycans are important components of the extracellular matrix. Some are embedded in the plasma membrane by a hydrophobic transmembrane domain or tethered to the membrane by a lipid anchor.

Evidence for the participation of proteoglycans in signaling comes from Drosophila *sugarless (sgl)* mutants, which lack a key enzyme needed to synthesize heparin (and chondroitin) sulfate. These mutants exhibit the phenotypes associated with defects in Wingless signaling and have greatly depressed levels of extracellular Wingless protein, a Wnt signal. Mutations in *dally* and *dally-like,* both of which encode core proteins of cell-surface proteoglycans, also are associated with defective Wingless signaling.

The Wnt pathway is not the only signaling pathway affected in *sugarless* and other *Drosophila* mutants with defective proteoglycan synthesis. For instance, such mutants have phenotypes (e.g., absence of a heart or trachea) that are associated with loss-of-function of Heartless and Breathless, which are receptor tyrosine kinases that bind FGF-like signaling proteins. These mutants also appear to have defective TGFβ signaling in metamorphosis, though not in embryos, suggesting specific actions of the proteoglycans. In Chapter 14, we saw that the type III TGFβ receptor is a cell-surface proteoglycan. Although not absolutely required for TGFβ signaling, the type III receptor binds and concentrates TGFβ near the surface of a cell in which it is produced, thereby facilitating signaling from the type I and type II receptors (see Figure 14-2).

KEY CONCEPTS OF SECTION 15.5

Boundary Creation by Extracellular Signals

■ Three *Drosophila* segment-polarity genes, *engrailed, wingless,* and *hedgehog,* are expressed in one-cell-wide stripes in each of the 14 body-segment primordia.

■ Engrailed, a transcription factor, activates transcription of *Hedgehog,* which encodes a secreted signaling protein.

■ Wingless, a secreted signaling protein in the Wnt family, is produced in stripes adjacent to the Engrailed/Hedgehog stripes. Wingless and Hedgehog maintain expression of each other's genes in a positive feedback loop (see Figure 15-29).

■ Both Hedgehog and Wingless/Wnt contain lipid anchors that can tether them to cell membranes, thereby reducing their signaling range.

■ The Hedgehog signal acts through two cell-surface proteins, Smoothened and Patched, and an intracellular complex containing the Cubitis interruptus (Ci) transcription factor (see Figure 15-31). An activating form of Ci is generated in the presence of Hedgehog; a repressing form is generated in the absence of Hedgehog. Both Patched and Smoothened change their subcellular location in response to Hedgehog binding to Patched.

■ Wingless and other Wnt signals act through two cell-surface proteins, the receptor Frizzled and coreceptor Lrp, and an intracellular complex containing β-catenin (see Figure 15-32). Wnt signaling promotes the stability and nuclear localization of β-catenin, which either directly or indirectly promotes activation of the TCF transcription factor.

■ Gradients of two external signals, a ventralizing Sonic hedgehog signal and a dorsalizing TGFβ signal, induce different cell types in the vertebrate neural tube (see Figure 15-33).

■ Cell-surface proteoglycans bind some extracellular signaling proteins, restricting their range of action and presenting them to nearby receptor proteins.

■ Cancer and birth defects occur when Hh, TGFβ, or Wnt signaling systems do not work properly.

15.6 Reciprocal Induction and Lateral Inhibition

In the development of an organism, cells must "talk" with one another to ensure a proper division of labor. The outcome of these cell–cell conversations can be an agreement about which cell should follow what differentiation pathway. For instance, negotiation between two initially equivalent cells can send each down distinct developmental paths. One cell preventing the other from following a particular path is called *lateral inhibition,* a process that prevents the duplication of structures at the expense of something not forming. Alternatively, two cells with distinct fates can send and receive signals between themselves, inducing further differentiation. Such *reciprocal induction* is common in the formation of internal organs. In this section, we consider two signaling systems that mediate such dialogues between cells.

Cell-Surface Ephrin Ligands and Receptors Mediate Reciprocal Induction During Angiogenesis

Perhaps the simplest type of reciprocal induction is between cells that interact through two cell-surface proteins, each of

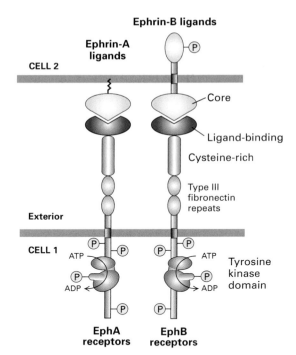

Ephrin-B ligands

Ephrin-A ligands

CELL 2

Core

Ligand-binding

Cysteine-rich

Type III fibronectin repeats

Exterior

CELL 1

ATP   ATP

ADP ←   → ADP

Tyrosine kinase domain

EphA receptors   **EphB receptors**

▲ **FIGURE 15-34 General structure of Eph receptors and their ligands.** The cytosolic domain of Eph receptors has tyrosine kinase activity. Within the Eph receptor family, the receptors exhibit some 30–70 percent homology in their extracellular domains and 65–90 percent homology in their kinase domains. Their ligands, the ephrins, either are linked to the membrane through a hydrophobic GPI anchor (class A) or are single-pass transmembrane proteins (class B). The core domains of various ephrin ligands show 30–70 percent homology. Ephrin-B ligands and their receptors can mediate reciprocal signaling. [Adapted from V. Dodelet and E. Pasquale, 2000, *Oncogene* **19**:5614; see J. G. Flanagan and P. Vanderhaegen, 1998, *Ann. Rev. Neurosci.* **21**:309.]

which can act as a receptor and ligand. To illustrate this phenomenon, we consider the role of the ephrins, a family of cell-surface ligands, and the Eph receptors in the development of mammalian blood vessels.

The Eph receptors, a novel type of receptor tyrosine kinase, have two classes of ligands (Figure 15-34). Ephrin-A ligands are tethered to the plasma membrane by a glycosylphosphatidylinositol (GPI) anchor. These ephrin ligands play a crucial role in forming connections between neurons in the developing nervous system. Ephrin-B ligands are single-pass transmembrane proteins. The results of biochemical experiments showed that ephrin-B ligands stimulate tyrosine phosphorylation of EphB receptors and of their own cytosolic domain. These observations led to the intriguing notion that ephrin-B ligand/EphB receptor complexes promote bidirectional reciprocal interactions. Strong support for this hypothesis has come from the study of blood vessel formation.

Blood vessels, arteries and veins, form a complex network of branched structures in the adult. An early network

of vessels is remodeled during angiogenesis as larger branches assemble from smaller ones and vessels become surrounded by support cells. Knockout mice lacking ephrin-B2 exhibit striking defects in angiogenesis. This finding led scientists to explore the pattern of expression of ephrin-B2 and its receptor, EphB4, in the developing embryo (Figure 15-35). In normal embryos, ephrin-B2 is expressed only in arteries; EphB4, only on veins. Although ephrin-B2 is expressed only on arterial capillaries, venous capillaries also fail to undergo angiogenesis in *ephrin-b2* knockouts. These data suggest that interaction between an arterial cell producing ephrin-B2 and a venous cell producing EphB4 causes the induction of *both* cells (see Figure 15-35c). In other words, ephrin-B2 and EphB4 each functions as both a ligand and a receptor to control the development of both veins and arteries.

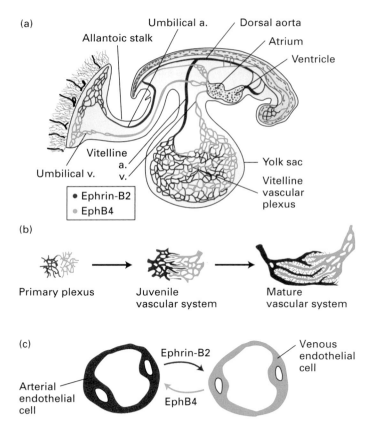

▲ **FIGURE 15-35 Reciprocal induction mediated by ephrin-B2 and its receptor, EphB4, in angiogenesis in the yolk sac.** (a) Ephrin-B2 (red) is expressed on arteries and EphB4 (blue) on veins in the early mouse embryo. (b) The early vascular network is remodeled during angiogenesis. In *ephrin-B2* knockout mice, angiogenesis is blocked at the primary plexus stage. The absence of ephrin-B2 thus interrupts the development of both arteries and veins. (c) Formation of intercalating arteries and veins results from interactions between developing arterial and venous endothelial cells mediated by ephrin-B2 (arterial) and EphB4 (venous). These reciprocal interactions induce the development of both cell types. [Adapted from H. U. Wang et al., 1998, *Cell* **93**:741.]

The Conserved Notch Signaling Pathway Mediates Lateral Inhibition

Now we shift our attention to lateral inhibition, which causes adjacent developmentally equivalent or near-equivalent cells to assume different fates. Genetic analyses in *Drosophila* and *C. elegans* revealed the role of the highly conserved *Notch/Delta pathway* in lateral inhibition. The *Drosophila* proteins Notch and Delta are the prototype receptor and ligand, respectively, in this signaling pathway. Both proteins are large transmembrane proteins whose extracellular domains contain multiple EGF-like repeats and binding sites for the other protein. Although Delta is cleaved to make an apparently soluble version of its extracellular domain, findings from studies with genetically mosaic *Drosophila* have shown that the Delta signal reaches only adjacent cells.

Interaction between Delta and Notch triggers the proteolytic cleavage of Notch, releasing its cytosolic segment, which translocates to the nucleus and regulates the transcription of specific target genes (see Figure 14-29). In particular, Notch signaling activates the transcription of *Notch* itself and represses the transcription of *Delta*, thereby intensifying the difference between the interacting cells (Figure 15-36a). Notch-mediated signaling can give rise to a sharp boundary between two cell populations or can single out one cell from a cluster of cells (Figure 15-36b). Notch signaling controls cell fates in most tissues and has consequences for differentiation, proliferation, the creation of cell asymmetry, and apoptosis. In the immune system, for instance, Notch signaling helps prevent the formation of T cells that attack an individual's own proteins. Here, we describe two examples of Notch signaling in cell-fate determination.

Determination of AC and VU Cell Fates in *C. elegans* Two equivalent cells, designated Z1.ppp and Z4.aaa, in roundworms can give rise to an anchor (AC) cell or a ventral uterine (VU) precursor cell. The results of laser ablation studies showed that, if either the Z1.ppp or Z4.aaa cell is removed, the remaining cell always becomes AC. In worms lacking functional LIN-12, the *C. elegans* homolog of Notch, both cells become AC. Conversely, constitutive activation of LIN-12 in both Z1.ppp and Z4.aaa results in both cells becoming VU. Thus LIN-12 activity levels specify AC and VU cell fates.

Both Z1.ppp and Z4.aaa produce the receptor, LIN-12, and its ligand, the Delta homolog LAG-2, at similar levels (Figure 15-37). As development proceeds, one cell begins to express more receptor through random fluctuations in pro-

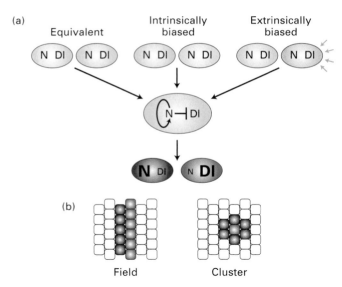

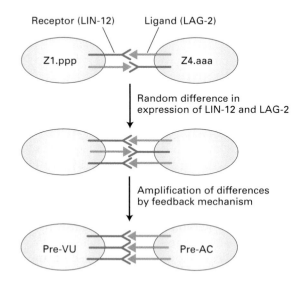

▲ **FIGURE 15-36 Amplification of an initial bias to create different cell types by Notch-mediated lateral inhibition.** (a) A difference between two initially equivalent cells may arise randomly (*left*). Alternatively, interacting cells may have an intrinsic bias (*center*) or an extrinsic bias (*right*). For instance, cells that have received different proteins in an asymmetric cell division will be intrinsically biased; those that have received different signals (orange) will be extrinsically biased. Regardless of how the small initial bias arises, Notch becomes predominant in one of the two cells, promoting its own expression and repressing production of its ligand Delta in that cell. In the other cell, Delta predominates. The outcome is reinforcement of the small initial difference. (b) Notch-mediated lateral inhibition may create a sharp boundary in an initial field of cells, such as along the edge of the developing *Drosophila* wing, or distinguish a central cell from a surrounding cluster of cells, as in neural precursor establishment. [Adapted from S. Artavanis-Tsakonas et al., 1999, *Science* **284**:770.]

▲ **FIGURE 15-37 Determination of different cell fates by lateral inhibition in *C. elegans* development.** LIN-12, a Notch homolog, and LAG-2, a Delta homolog, regulate interactions between two equivalent cells, designated Z1.ppp and Z4.aaa. Either cell can assume a ventral uterine (VU) or anchor (AC) fate. See text for discussion. [Adapted from I. Greenwald, 1998, *Genes & Dev.* **12**:1751.]

tein levels or differences in the ambient level of signaling through the pathway. The cell receiving a slightly higher signal begins to increase its expression of the receptor and decrease its expression of the ligand. In the neighboring cell, now exposed to a reduced level of ligand, expression of the receptor falls and that of the ligand increases. In this way, the initial asymmetry resulting from a random event is amplified, finally leading to the commitment of one cell as a pre-VU cell and its partner as a pre-AC cell. When formed, the AC cell begins sending out a LIN-3 signal that functions in vulva development. Notch-mediated lateral inhibition also operates in that process when the P6.p cell inhibits the neighboring P5.p and P7.p cells (see Figure 15-12b).

Neuronal Development in *Drosophila* and Vertebrates

Loss-of-function mutations in the *Notch* or *Delta* genes produce a wide spectrum of phenotypes in *Drosophila*. One consequence of such mutations in either gene is an increase in the number of neuroblasts in the central nervous system. In *Drosophila* embryogenesis, a sheet of ectoderm cells becomes divided into two populations of cells: those that move inside the embryo eventually develop into neuroblasts; those that remain external form the epidermis and cuticle. As some of the cells enlarge and then loosen from the ectodermal sheet to become neuroblasts, they signal to surrounding cells to prevent their neighbors from becoming neuroblasts—a case of lateral inhibition. Notch signaling is used for this in-

hibition; in embryos lacking the Notch receptor or its ligand, all the ectoderm precursor cells become neural.

The role of Notch signaling in specifying neural cell fates has been studied extensively in the developing *Drosophila* peripheral nervous system. In flies, various sensory organs arise from proneural cell clusters, which produce bHLH transcription factors, such as Achaete and Scute, that promote neural cell fates. In normal development, one cell within a proneural cluster is somehow anointed to become a *sensory organ precursor (SOP)*. In the other cells of a cluster, Notch signaling leads to the repression of proneural genes, and so the neural fate is inhibited; these nonselected cells give rise to epidermis (Figure 15-38). Temperature-sensitive mutations that cause functional loss of either Notch or Delta lead to the development of additional SOPs from a proneural cluster. In contrast, in developing flies that produce a constitutively active form of Notch (i.e., active in the absence of a ligand), all the cells in a proneural cluster develop into epidermal cells.

To assess the role of the Notch pathway during primary neurogenesis in *Xenopus*, scientists injected mRNA encoding different forms of Notch and Delta into embryos. Injection of mRNA encoding the constitutively active cytosolic segment of Notch inhibited the formation of neurons. In contrast, injection of mRNA encoding an altered form of Delta that prevents Notch activation led to the formation of too many neurons. These findings indicate that in vertebrates, as in *Drosophila*, Notch signaling controls neural precursor cell fates.

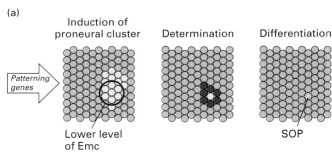

▲ **FIGURE 15-38 Role of Notch-mediated lateral inhibition in formation of sensory organ precursors (SOPs) in *Drosophila*.** (a) Extracellular signaling molecules and transcription factors, encoded by early-patterning genes, control the precise spatiotemporal pattern of proneural bHLH proteins such as Achaete and Scute (yellow). Most cells within the field express Emc (orange), a related protein that antagonizes Achaete and Scute. A small group of cells, a proneural cluster, produce proneural bHLH proteins. The region of a proneural cluster from which an SOP will form expresses lower levels of Emc, giving these cells a bias toward SOP formation. Interactions between these cells, leading to accumulation of E(spl) repressor proteins in neighboring cells (blue), then restrict SOP formation to a single cell (green). (b) Initially, *achaete (ac)* and other proneural genes are transcribed in all the cells within a proneural cluster, as are

Notch and *Delta*. Achaete and other proneural bHLH proteins promote expression of *Delta*. When one cell at random begins to produce slightly more Achaete (*left*), its production of Delta increases, leading to stronger Notch signaling in all its neighboring cells (*right*). In the receiving cells, the Notch signaling pathway activates a transcription factor designated Su(H), which in turn stimulates expression of *E(spl)* genes. The E(spl) proteins specifically repress transcription of *ac* and other proneural genes. The resulting decrease in Achaete leads to a decrease in Delta, thus amplifying the initial random difference among the cells. As a consequence of these interactions and others, one cell of a proneural cluster is selected as a SOP; all the others lose their neural potential and develop into epidermal cells.

15.7 Integrating and Controlling Signals

Cells change their properties rapidly in response to signals, both during and after development. The segregation and progressive restriction of cell potential during development are changes that take place as an organism grows and generates vast numbers of new cells that must be organized into new tissues and shapes. The cells in some adult tissues (e.g., blood, gut epithelia, and skin) also continue to proliferate and differentiate. These cells build on a substantial preexisting framework and have less "original" construction of tissues to do. Both dividing and nondividing cells in adult tissues remain highly responsive to hormones and other signaling molecules and to environmental changes.

Discussion in this chapter and in Chapters 13 and 14 calls attention to the enormous scientific effort that has been made in identifying the components of signaling pathways, how they transduce signals, and the resulting cellular responses. Many current projects are aimed at learning how multiple signaling pathways are mustered to control normal tissue growth and function in embryonic and adult tissues. In many circumstances, the appropriate response depends on the ability of receiving cells to integrate multiple signals and to control the availability of active signals. The examples of signal integration and modulation described in this section illustrate some general mechanisms used in a wide variety of contexts.

Competence Depends on Properties of Cells That Enable Them to Respond to Inductive Signals

Early embryologists noted that cells differ in their ability to respond to various inductive signals. The ability to respond to a particular signaling molecule, referred to as *competence,* depends on several properties of the receiving cell: the presence of receptors specific for the signal, the ability of these receptors to activate specific intracellular pathways, the presence of transcription factors that stimulate the expression of the genes required to implement the appropriate response, and a chromatin structure that makes these genes accessible for transcription.

In some cases, the reception of one signal may make cells competent to receive another. After a part of the liver is damaged or removed surgically, increased amounts of two signals, tumor necrosis factor (TNF) and interleukin-6 (IL-6), are produced as part of the response to liver damage. These signals cause hepatocytes to enter a "primed" state in which the cells increase their production of certain transcription factors (e.g., NF-κB, Stat3, AP1, and CEBP) but do not divide. Primed cells are competent to respond to a combination

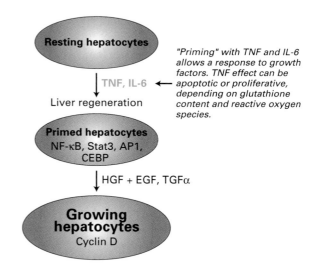

▲ **FIGURE 15-39 Priming of resting hepatocytes for later responses to signaling molecules that induce growth.** Injury to the liver or removal of part of the liver leads to priming of cells in response to interleukin-6 (IL-6) and tumor necrosis factor (TNF) signals. Primed cells increase their production of the indicated transcription factors but do not divide. Subsequent increase in the blood level of hepatocyte growth factor (HGF) induces primed cells to produce cyclin D, which is required for cell division (Chapter 21). HGF acts in conjunction with epidermal growth factor (EGF) and transforming growth factor α (TGFα).

of three signals that together induce the synthesis of cyclin D and mitosis (Figure 15-39).

Liver regeneration is an important survival tactic after a part of the liver is damaged or poisoned, but unregulated liver growth would lead to an unduly large organ or possibly cancer. In the primed state, hepatocytes can measure their own physiological state, their location within a tissue, their proximity to other cells, the need for healing, and the spatial organization of cells within a structure. On the basis of this assessment, the cells can respond most appropriately to subsequent signals.

Some Signals Can Induce Diverse Cellular Responses

Several classes of cell-surface receptors discussed in Chapters 13 and 14 are linked to more than one intracellular signal-transduction pathway (see Table 14-1). Multiple intracellular signaling possibilities are most evident with G protein–coupled receptors, cytokine receptors, and receptor tyrosine kinases. This phenomenon raises a general question: What governs how a cell responds to a signal that can be transduced by multiple pathways? Conversely, if the signaling pathway is the same in many cell types, why does one cell respond by dividing, another by differentiating, and still another by dying? For instance, signaling through the RTK-Ras–MAP kinase pathway (see Figure 14-16) is used repeatedly in the course of development, yet the outcome in regard to cell-fate specification varies in different tissues. If there is no specificity beyond the ligand and receptor, an activated Ras might substitute for any signal. In fact, activated Ras can do so in many cell types. In one DNA microarray study of fibroblasts, for instance, the same set of genes was transcriptionally induced by platelet-derived growth factor (PDGF) and by fibroblast growth factor (FGF), suggesting that exposure to either signaling molecule had similar effects. The PDGF receptor and the FGF receptor are both receptor tyrosine kinases, and the binding of ligand to either receptor can activate Ras.

Several mechanisms for producing diverse cellular responses to a particular signaling molecule seem possible in principle: (1) the strength or duration of the signal governs the nature of the response; (2) the pathway downstream of the receptor is not really the same in different cell types, for example because different complements of transcription factors are present in the receiving cells; and (3) converging inputs from other pathways modify the response to the signal.

Differences in Signal Strength or Duration Evidence supporting the use of the first mechanism comes from studies with PC12 cells, a cultured cell line capable of differentiating into adipocytes or neurons. Nerve growth factor (NGF) promotes the formation of neurons, whereas epidermal growth factor (EGF) promotes the formation of adipocytes.

Strengthening the EGF signal by prolonging exposure to it causes neuronal differentiation. Although both NGF and EGF are RTK ligands, NGF is a much stronger activator of the Ras–MAP kinase transduction pathway than is EGF. The EGF receptor can apparently activate this pathway only after prolonged stimulation.

Differences in Downstream Pathways Signaling through cell type–specific pathways downstream of an RTK has been demonstrated in *C. elegans*. In worms, EGF signals induce at least five distinct responses, each one in a different type of cell. Four of the five responses are mediated by the common Ras–MAP kinase pathway; the fifth, hermaphrodite ovulation, employs a different downstream pathway in which the second messenger inositol trisphosphate is generated. Binding of IP_3 to its receptor (IP3R) in the endoplasmic reticulum membrane leads to the release of stored Ca^{2+} from the ER (see Figure 13-29). The rise in cytosolic Ca^{2+} then triggers ovulation. This alternative pathway was discovered with a genetic screen that implicated IP3R, a Ca^{2+} channel, in EGF signaling—a good example of how a mutation in an unexpected gene can lead to a discovery.

Integration of Signals The third way that the same signaling ligand/receptor pair can produce diverse effects on cells is to integrate more than one signal, as occurs in *Drosophila* muscle development. Figure 15-40 depicts the convergence of signal inputs that leads to the formation of a single muscle precursor cell, which is defined by its ability to transcribe the *even-skipped* gene. Early in muscle development, the *Drosophila* Wnt signal Wingless (Wg) and the TGFβ signal Decapentaplegic (Dpp) prime a cell to make it competent to receive a subsequent signal that is transduced through the MAP kinase pathway downstream of Ras. The Wingless signal is produced in circumferential belts, and the Dpp signal is produced in two longitudinal bands at right angles to the Wnt belts.

One group of cells on each side of each body segment receives both the Wingless and the Dpp signals and thus become competent to respond to an unidentified RTK signal that activates Ras. In these cells, signal integration takes place during transcription of the *eve* gene. The transcription of *eve* is activated when a short 312-bp *eve* enhancer (*not* the same one described in Figure 15-32) is bound by two muscle-specific transcription factors and by three signal-induced transcription factors: TCF factor by Wingless, Mad by Dpp, and Pnt by an RTK acting through Ras. Thus tissue-specific and signal-responsive information is integrated through the action of five regulators on one short piece of DNA in specifying a cell type. As the result of lateral inhibition, eventually only one *eve*-expressing cell in each initial group of competent cells is left (see Figure 15-40). That single cell will develop into a particular muscle fiber by recruiting other cells and fusing with them.

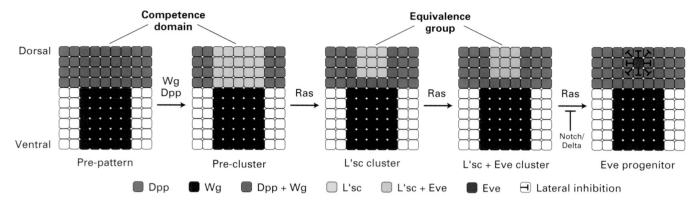

▲ FIGURE 15-40 Sequential action of critical signals in *Drosophila* muscle development. Signal transduction through the RTK pathway is governed by Wnt and TGFβ signals. Wingless (Wg) is produced in a stripe of cells running in a belt around part of each body segment of the embryo (purple). Decapentaplegic (Dpp) is produced in a band of dorsal cells running from head to tail on each side of the embryo (blue), a band that is created by the dorsal/ventral signaling system described in Section 15.3. A patch of cells in each body segment will receive both signals; only these cells (green) are competent to respond to the (unidentified) RTK signal that activates intracellular signaling from Ras. All the cells in the patch activate a gene called *L'sc*, though further signaling restricts *L'sc* and then *eve* transcription to a more restricted set of cells called the pre-cluster (orange). Within the pre-cluster, a central cell begins to use Notch signaling to surrounding cells to repress *L'sc* and *eve* transcription there until only one cell is left making *eve* products (red). That single cell will develop into a particular muscle by recruiting other cells and fusing with them; two such cells are created in each body segment by this elaborate process. Both require RTK-mediated signaling; one cell uses the *Drosophila* EGF receptor (DER) and the EGF-type receptor called Heartless (Htl), and the second cell uses only Htl. [See Halfon et al., 2000, *Cell* **103**:63–74.]

Limb Development Depends on Integration of Multiple Extracellular Signal Gradients

Vertebrate limbs grow from small "buds" composed of an inner mass of mesoderm cells surrounded by a sheath of ectoderm. Secreted signals from both cell layers coordinate limb development and instruct cells about their proper fates within limbs. The first signal, fibroblast growth factor 10 (FGF10) is secreted from the lateral trunk mesoderm and initiates outgrowth of a limb from specific regions of the embryo's flank. Implantation of a bead soaked in FGF10 into places in the flank where a limb does not normally form causes an extra limb to grow; so FGF has remarkable inductive capabilities.

There are three dimensions to a limb: anterior/posterior (thumb to little finger), dorsal/ventral (palm versus back of hand) and proximal/distal (shoulder to fingers). An embryonic cell that knows its position along each of these dimensions is well along toward knowing what to do. A different signaling system operates in each of the three dimensions; so,

▶ FIGURE 15-41 Integration of three signals in vertebrate limb development along proximal/distal and anterior/posterior axes. Each limb bud grows out of the flank of the embryo. (a) A fibroblast growth factor (FGF) signal, probably FGF10, comes from the mesoderm in specific regions of the embryo's flank, one region for each limb. FGF10 acts on a local region of surface ectoderm called the apical ectodermal ridge (AER) because it will form a prominent ridge. (b) The ectoderm that receives a FGF10 signal is induced to produce FGF8, another secreted signal. At the posterior end of the limb bud, FGF8 induces transcription of the *Sonic hedgehog (Shh)* gene. (c) Shh signaling induces transcription of the gene encoding FGF4 in the AER. FGF8 and FGF4 promote continued proliferation of the mesoderm cells, causing outgrowth of the limb bud. Shh also stimulates this outgrowth and confers posterior characteristics on the posterior part of the limb. Development along the dorsal/ventral axis depends on a Wnt signal that is not shown here.

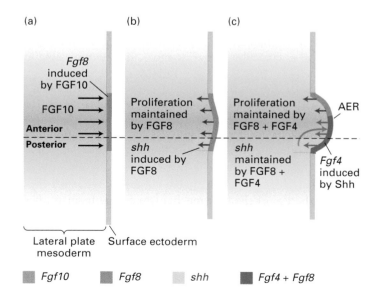

by reading the three signals, a cell appears to learn where it is within the limb bud and to act accordingly.

In response to FGF10, a local region of surface ectoderm becomes a signaling center, the *apical ectodermal ridge (AER)*, at the tip of emerging limb bud (Figure 15-41). This region secretes FGF8 and later FGF4, both of which drive persistent division of mesoderm cells and therefore continued limb outgrowth. The FGF8 signal also induces the production of Sonic hedgehog in the posterior limb bud. The FGF signals tell cells their distance from the distal limb bud, and Shh tells cells that they are posterior. If Shh is added to the anterior part of the bud, the limb that eventually forms will have two posterior patterns of bones and no anterior. Along the dorsal/ventral axis, a Wnt signal instructs cells to form ventral cell types. The Wnt, FGF4, and FGF8 signals promote the transcription of *Shh*, and Shh signaling promotes the transcription of the *Fgf4* and *Fgf8* genes. Thus the signals are mutually reinforcing in cells that are close enough; cells too far from one of the reinforcing signals will cease making their own signal. In this way, the strength and movement of signals is tied to the eventual size and shape of the limb.

A cell in the midst of the limb bud, wondering what to do, is assailed by this brew of signals. By integrating the information from all of them, each cell begins to learn how to proceed. The main developmental task in the formation of limbs and organs is to organize a few cell types (e.g., mesenchyme, vascular, epithelial) into complex multicellular structures. The signals discussed herein provide the initial guidelines for this building process, instructing cells about their location relative to the coordinates of the limb and causing specific transcription factors to become active in proper parts of the limb.

Signals Are Buffered by Intracellular and Extracellular Antagonists

In real life, organisms experience wide variations in their environments and must adapt or die. Most multicellular organisms have highly variable numbers of cells, experience a wide range of temperatures, and must endure periods of nutritional deprivation, environmental toxicity, injury, and competition with their own and other species. Signal systems with rigid requirements for temperature and physiological conditions would be ill adapted to the real world. Scientists are beginning to recognize ways in which signals are regulated so that the outcome in regard to cell fates is correct.

Inducible Antagonists One way that cells can modulate signal activity is by producing *inducible antagonists*. The idea is that a signal induces the transcription of genes in the receiving cells. Among the induced genes is one encoding an antagonist that reduces the effect of the signal. If, by chance, too low an amount of signal gets through, the amount of antagonist also is diminished and the net target-gene induction is preserved. Similarly, too much signal, or signal transduction, will be corrected by greater production of antagonist.

Examples of both intracellular and extracellular (secreted) inducible antagonists are known. Such inducible antagonists have been found for the Hedgehog, Wnt, TGFβ, RTK, and cytokine receptor signaling pathways.

In Chapter 14, we describe how inducible intracellular antagonists modulate signaling from TGFβ receptors (see Figure 14-3) and from cytokine receptors (see Figure 14-14). The Hedgehog pathway, discussed in Section 15.5, also is regulated by an intracellular antagonist. Recall that, in the absence of a Hedgehog signal, Patched inhibits Smoothened and prevents intracellular signaling (see Figure 15-31). One of the target genes transcriptionally activated by Hedgehog signaling is *patched*; the resulting increased level of Patched in the plasma membrane reduces downstream signaling and turns off the transcription of Hedgehog target genes.

In other cases, a signal is controlled by an inducible secreted antagonist that binds the signal's receptor without activating it. For instance, the secreted protein encoded by the *Drosophila argos* gene competes with signals that activate the EGF receptor (EGFR), a receptor tyrosine kinase. Transcription of *argos* is stimulated by EGFR ligands, making Argos a type of inducible antagonist. When bound, Argos blocks dimerization of the EGF receptor, which is necessary for subsequent signal transduction.

The Argos buffering system is used to create different cell types during dorsal/ventral patterning in early fly embryos. Recall that Gurken is a ligand for the EGF receptor on dorsal follicle cells (see Figure 15-17). A consequence of Gurken binding to the EGF receptor is transcriptional activation of the *spitz* and *vein* genes, both of which encode ligands that also bind to the EGF receptor and activate it. In this way, the effect of Gurken is amplified. In dorsal midline cells where the Gurken concentration is highest and EGFR signaling is strongest, the transcription of *argos* is also induced. Argos protein *reduces* signal transduction from EGF receptors in the midline; only in the two flanking regions does EGFR activity continue at high levels. This system creates two distinct populations of cells: those in which the EGF receptor is active flanking a central population of cells in which the EGF receptor is turned off. The highest level of the Gurken signal leads to the lowest level of signal transduction from the EGF receptor.

Antagonists Not Induced by the Signal Signaling pathways are also controlled by secreted antagonists that are not induced by the signal itself. One effect of such an antagonist can be to sharpen or move a boundary between cell types. A signal coming from source cells is progressively less potent with distance; at some point, it falls below a threshold amount and is without effect. If a secreted antagonist comes from the opposite direction, it will block the action of the signal even in cells receiving above-threshold amounts.

We see an example of this effect in the formation of neural cells in vertebrate development. Normally, secreted TGFβ proteins prevent the formation of neural cells in a part of early frog embryos called the animal cap. In *Xenopus* embryos, the

▶ **FIGURE 15-42 Modulation of BMP4 signaling in *Xenopus* by chordin and Xolloid.** (a) Chordin binds BMP4, a TGFβ-class secreted protein signal, and prevents it from binding to its receptor. (b) Xolloid specifically cleaves chordin in the chordin–BMP4 complex, releasing BMP4 in a form that can bind to its receptor and trigger signaling. Similar regulation of Dpp (related to BMP) signaling in *Drosophila* is by Sog (related to chordin) and Tolloid (related to Xolloid). [See S. Piccolo et al., 1997, *Cell* **91**:407.]

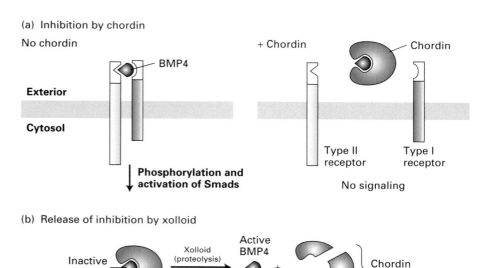

animal cap produces BMP4, a member of the TGFβ family of signals, and will therefore not produce neural tissue. The effect of signals and other regulators on neural induction can be tested by exposing parts of the animal cap in *Xenopus* embryos to individual proteins and seeing whether neural cells form. This type of experiment revealed the ability of *chordin protein* to antagonize BMP4 and induce neural cell identity, an indication of the presence of dorsal mesoderm. The addition of chordin to *Xenopus* animal caps induces the formation of neural cells; this neural induction by chordin is reversed by the addition of an excess of BMP4 protein. On this basis, neural cell fate is looked on as the default state. Only when BMP signaling is successful can other cell types form. Together, these data led to a simple model in which chordin prevents BMP from binding to its receptor. In principle, inhibition could occur by the direct binding of chordin to BMP receptors or to BMP molecules themselves. The results of a series of biochemical studies demonstrated that chordin binds BMP2 and BMP4 homodimers or BMP4/BMP7 heterodimers with high affinity ($K_D = 3 \times 10^{-10}$ M) and prevents them from binding to their receptors (Figure 15-42). Chordin-mediated inhibition of BMP signaling is relieved by Xolloid protein, a protease that specifically cleaves chordin in chordin–BMP complexes, releasing active BMP.

Wnt signaling also is modulated by secreted antagonists, including a special group of secreted proteins, called Frizbees, that are related to the Frizzled receptor in this pathway. Wnt signals bind to a cysteine-rich domain in the extracellular domain of Frizzled receptors, thus activating downstream signaling (see Figure 15-32). Frizbee proteins have the extracellular domain of Frizzled but lack its transmembrane and cytosolic domains. By tightly binding Wnt signals, Frizbees soak up the signal, so less is available to activate Frizzled receptors.

We conclude by mentioning Cerberus, a champion secreted antagonist discovered in frog embryos. This protein is called Cerberus, after the mythological guardian dog with three heads, because it has binding sites for three different types of powerful signals—Wnt, Nodal, and BMP. The binding of these signals by Cerberus prevents activation of their respective receptors. By inactivating Wnt, Nodal, and BMP signals having roles in the development of the trunk and tail of the body, Cerberus promotes head development.

KEY CONCEPTS OF SECTION 15.7

Integrating and Controlling Signals

■ In liver regeneration, early signals change cells into a primed state in which they are competent to respond to subsequent signals that cause growth and mitosis.

■ Most components of the RTK-Ras–MAP kinase signaling pathway are used repeatedly in the course of development and are evolutionarily conserved in a broad spectrum of animals.

■ In embryonic muscle development in *Drosophila*, three types of signals converge to promote differentiation (see Figure 15-40). These signals induce the production of three transcription factors that combine with two tissue-specific transcription factors to activate transcription of the *even-skipped* gene, which marks a muscle precursor cell.

■ At least one signaling protein acts along each axis of the developing limb bud. Cells respond to the combination of signals, and the signals reinforce the production of other signals to coordinate growth and patterning in three dimensions (see Figure 15-41).

■ Inducible antagonists include intracellular and secreted proteins whose production is induced by a particular signal. The antagonist protein then feeds back, opposing the ongoing action of the signal. Inducible antagonists provide buffering in most signaling pathways to compensate for excess or inadequate signals.

■ Signals can also be controlled by antagonists that are not induced by the signal itself. By binding to signals, these antagonists prevent the signals from binding to their specific receptors and activating them (see Figure 15-42).

PERSPECTIVES FOR THE FUTURE

The remarkable interplay of signals and gene controls during development is a process of pattern formation. Simple masses of cells become startlingly beautiful, functional structures such as eyes, lungs, hearts, and wings. For this transformation to happen, cells must respond to a rich mix of short-range and long-range signals and integrate the information to make the right decisions. Only a few signaling systems are used repeatedly to organize cells into tissues. For the most part, the signaling systems are common to all animals, thus vastly simplifying the problem of learning the underlying molecular mechanisms. This manifestation of our common evolutionary origins reduces the problem of understanding biological regulation—but the remaining challenges are still enormous.

To more fully understand how cells acquire their fates and participate in morphogenesis will require a comprehensive view of their molecular responses in vivo. The advent of DNA microarray technology for simultaneously monitoring the transcription of thousands of genes—indeed all of the transcription units in a genome—is a major step forward in "reading the minds" of cells. A signal can now be assessed for its ability to redirect a cell's activities by measuring all the signal-induced changes in gene expression. Exciting as these new insights are, their limitations also are apparent. For instance, transcripts are currently measured in extracts that represent an average of many cells isolated without great temporal precision—from embryos that change in seconds or minutes. Therefore we can look forward to continuing advances as methods improve for observing responses in living cells. Moreover, DNA microarray analysis measures only transcription, not RNA splicing or translation into protein; nor are protein localization and modification and protein–protein associations assessed by this technology. These other levels of gene control are still measured mostly on a gene-by-gene basis. Advances in protein microarray technology that enable monitoring of all proteins at once, as well as protein–protein interactions, will contribute tremendously to our understanding.

The rapid fluxes in cell responses that occur during signaling events, the integration of information from multiple signals, and the dynamic feedback control loops are all difficult to comprehend without precise measurements of signal concentrations, reaction rates, and equilibrium states. More precise quantitation will be necessary to understand signaling circuitry properly. Crude statements of genes being on or off or of proteins being phosphorylated or not will have to be replaced by more precise information, because in fact the changes entail gradations over orders of magnitude, not simply all-or-none phenomena.

The elaborate switching systems in cells, usually many of them working at the same time, have a degree of complexity that requires computer modeling. This field, sometimes referred to as systems biology, aims to produce models that successfully describe the normal responses of cells to stimuli and that predict how cells will respond to additional or different perturbations. An engineering approach of this sort holds the exciting promise that cells could be directed to perform tasks, such as improved immunity or healing, that are useful. The field of molecular cell biology increasingly is becoming a computational science, but with the special fascination that the signals are building life.

KEY TERMS

angiogenesis 620	maternal mRNA 629
chordin protein 652	morphogen 623
development 612	pair-rule gene 634
equivalence group 624	protein microarray
floral organ–identity	(proteome chip) 615
gene 637	reciprocal induction 644
gap gene 630	segment-polarity gene 635
Hedgehog pathway 641	sensory organ
homeosis 636	precursor (SOP) 647
Hox gene 636	syncytium 625
inducible antagonist 651	Toll-Dorsal pathway 628
induction 621	transcription cascade 632
in situ hybridization 614	Wnt pathway 642
lateral inhibition 644	

REVIEW THE CONCEPTS

1. Describe three different pathways in which EGF-type receptors participate.

2. What enzyme catalyzes the degradation of glycogen to glucose 1-phosphate? Describe how the subunit structure of this enzyme allows for integration of multiple signals.

3. People with Type 1 diabetes do not produce insulin. Describe what happens to GLUT4 glucose transporters in Type 1 diabetics just after a meal. Most patients with Type 2 diabetes produce normal amounts of insulin and have insulin receptors present on insulin-responsive cells. Yet these patients are defective in glucose uptake, and hence have abnormally high blood glucose levels. Speculate on the nature of the defect in these patients.

4. Compare and contrast the action of morphogens involved in dorsal-ventral specification in *Xenopus laevis* and *Drosophila melanogaster*.

5. Using in situ hybridization with a dorsal-specific probe, where in the syncytial *Drosophila* embryo would one expect to find *dorsal* expressed? Using immunohistochemistry with an anti-dorsal antibody, where would you expect to find Dorsal protein expressed?

6. A microarray analysis of wildtype vs. *dorsal* mutant embryos could be expected to yield information on all genes regulated by the Dorsal protein. Why? Other than new genes regulated by Dorsal, one would expect to see changes in regulation of previously identified genes. Which genes would be increased or decreased in expression in *dorsal* mutants?

7. Deleting the 3′ UTR of the *bicoid* gene would yield what phenotype in a mutant fly? Why?

8. How does the motor protein kinesin ensure that proper post-translational controls take place in the anterior end of the *Drosophila* embryo? If kinesin functioned only during embryo development, what would the phenotype of a *kinesin* mutant embryo be?

9. What is the evidence that the Toll-based innate immunity system may be more than a billion years old?

10. How can the group of five gap genes specify more than five types of cells in *Drosophila* embryos?

11. What is homeosis? Give an example of a floral homeotic mutation and describe the phenotype of the mutant and the normal function of the wild-type gene product.

12. Compare and contrast the receptor systems responsible for recognition of the Hedgehog and Wnt secreted ligands.

13. What is the evidence that a gradient of Sonic hedgehog leads to development of different cell types within the avian neural tube?

14. The finding that *ephrin b2* knockouts contain striking defects in both arterial and venous capillaries supports the idea that ephrin B2 and its receptor, EphB4, are involved in a reciprocal induction. How?

15. Give a specific example of a signal-buffering inducible antagonist, and describe how its actions lead to signal buffering.

ANALYZE THE DATA

The entire genome sequence of rice (*Oryza sativa*) has been determined, allowing investigators to search the rice genome for gene sequences homologous to *Drosophila* genes involved in developmental processes. A putative *bicoid*

homolog was found in the rice genome that encodes a homeodomain-type transcription factor. This rice gene was used as a probe against rice embryos in an in situ hybridization experiment, shown below. Plant embryos are contained within seeds. The rice embryos used in these experiments were in the globular stage of development. At this stage, the apical and basal ends of the future plant are being specified.

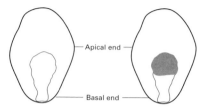

a. What is the expression pattern of the putative rice *bicoid* gene?

b. Speculate as to the function of the putative rice *bicoid* gene.

c. From the above results, what phenotype do you think might develop if one microinjected mRNA molecules transcribed from the putative rice *bicoid* gene into the basal end of the embryo?

d. From the above results, what would the expected expression pattern be in rice if one could identify a putative *nanos* homolog from rice?

REFERENCES

Experimental Approaches for Building a Comprehensive View of Signal-Induced Responses

Arbeitman, M. N., et al. 2002. Gene expression during the life cycle of *Drosophila melanogaster*. *Science* **297**:2270–2275.

Cutler, P. 2003. Protein arrays: the current state-of-the-art. *Proteomics* **3**:3–18.

Kamath, R. S., et al. 2003. Systematic functional analysis of the *Caenorhabditis elegans* genome using RNAi. *Nature* **421**:231–237.

Kim, S. K., et al. 2001. A gene expression map for *Caenorhabditis elegans*. *Science* **293**:2087–2092.

Stathopoulos, A., et al. 2002. Whole-genome analysis of dorsal-ventral patterning in the *Drosophila* embryo. *Cell* **111**:687–701.

Tomancak, P., et al. 2002. Systematic determination of patterns of gene expression during *Drosophila* embryogenesis. *Genome Biol.* **3**:RESEARCH0088–8.

Responses of Cells to Environmental Influences

Downward, J. 1998. Mechanisms and consequences of activation of protein kinase B/Akt. *Curr. Opin. Cell Biol.* **10**:262–267.

Kaelin, W. G., Jr. 2002. How oxygen makes its presence felt. *Genes & Dev.* **16**:1441–1445.

Control of Cell Fates by Graded Amounts of Regulators

Chang, A. J., and D. Morisato. 2002. Regulation of Easter activity is required for shaping the Dorsal gradient in the *Drosophila* embryo. *Development* **129**:5635–5645.

Freeman, M., and J. B. Gurdon. 2002. Regulatory principles of developmental signaling. *Ann. Rev. Cell Dev. Biol.* **18**:515–539.

Johnstone, O., and P. Lasko. 2001. Translational regulation and RNA localization in *Drosophila* oocytes and embryos. *Ann. Rev. Genet.* **35**:365–406.

Stathopoulos, A., and M. Levine. 2002. Dorsal gradient networks in the *Drosophila* embryo. *Devel. Biol.* **246**:57–67.

Takeda, K., T. Kaisho, and S. Akira. 2003. Toll-like receptors. *Ann. Rev. Immunol.* **21**:335–376.

Wang, M., and P. W. Sternberg. 2001. Pattern formation during *C. elegans* vulval induction. *Curr. Topics Devel. Bio.* **51**:189–220.

Boundary Creation by Different Combinations of Transcription Factors

Akam, M. 1987. The molecular basis for metameric pattern in the *Drosophila* embryo. *Development* **101**:1–22.

Andrioli, L. P., et al. 2002. Anterior repression of a *Drosophila* stripe enhancer requires three position-specific mechanisms. *Development* **129**:4931–4940.

Fujioka, M., et al. 1999. Analysis of an even-skipped rescue transgene reveals both composite and discrete neuronal and early blastoderm enhancers, and multi-stripe positioning by gap gene repressor gradients. *Development* **126**:2527–2538.

Houchmandzadeh, B., E. Wieschaus, and S. Leibler. 2002. Establishment of developmental precision and proportions in the early *Drosophila* embryo. *Nature* **415**:798–802.

Boundary Creation by Extracellular Signals

Jessell, T. M. 2000. Neuronal specification in the spinal cord: inductive signals and transcriptional codes. *Nature Rev. Genet.* **1**:20–29.

Moon, R. T., B. Bowerman, M. Boutros, and N. Perrimon. 2002. The promise and perils of Wnt signaling through beta-catenin. *Science* **296**:1644–1646. Review.

Nybakken, K., and N. Perrimon. 2002. Hedgehog signal transduction: recent findings. *Curr. Opin. Genet. Devel.* **12**:503–511.

Pandur, P., D. Maurus, and M. Kuhl. 2002. Increasingly complex: new players enter the Wnt signaling network. *Bioessays* **24**:881–884.

Pires-daSilva, A., and R. J. Sommer. 2003. The evolution of signalling pathways in animal development. *Nature Rev. Genet.* **4**:39–49.

Sanson, B. 2001. Generating patterns from fields of cells: examples from *Drosophila* segmentation. *EMBO Rep.* **2**:1083–1088.

Wilkie, G. S., and I. Davis. 2001. *Drosophila wingless* and pairrule transcripts localize apically by dynein-mediated transport of RNA particles. *Cell* **105**:209–219.

Reciprocal Induction and Lateral Inhibition

Artavanis-Tsakonas, S., M. D. Rand, and R. J. Lake. 1999. Notch signaling: cell fate control and signal integration in development. *Science* **284**:770–776.

Cooke, J. E., and C. B. Moens. 2002. Boundary formation in the hindbrain: Eph only it were simple. *Trends Neurosci.* **25**:260–267.

Grandbarbe, L., et al. 2003. Delta-Notch signaling controls the generation of neurons/glia from neural stem cells in a stepwise process. *Development* **130**:1391–1402.

Ju, B. G., et al. 2000. Fringe forms a complex with Notch. *Nature* **405**:191–195.

Miao, H., et al. 2001. Activation of EphA receptor tyrosine kinase inhibits the Ras/MAPK pathway. *Nature Cell Biol.* **3**:527–530.

Santiago, A., and C. A. Erickson. 2002. Ephrin-B ligands play a dual role in the control of neural crest cell migration. *Development* **129**:3621–3632.

Shaye, D. D., and I. Greenwald. 2002. Endocytosis-mediated downregulation of LIN-12/Notch upon Ras activation in *Caenorhabditis elegans*. *Nature* **420**:686–690.

Integrating and Controlling Signals

Brzozowski, A. M., et al. 1997. Molecular basis of agonism and antagonism in the oestrogen receptor. *Nature* **389**:753–758.

Fambrough, D., K. McClure, A. Kazlauskas, and E. S. Lander. 1999. Diverse signaling pathways activated by growth factor receptors induce broadly overlapping, rather than independent, sets of genes. *Cell* **97**:727–741.

Gerlitz, O., and K. Basler. 2002. Wingful, an extracellular feedback inhibitor of Wingless. *Genes & Dev.* **16**:1055–1059.

Grimm, O. H., and J. B. Gurdon. 2002. Nuclear exclusion of Smad2 is a mechanism leading to loss of competence. *Nature Cell Biol.* **4**:519–522.

Halfon, M. S., et al. 2000. Ras pathway specificity is determined by the integration of multiple signal-activated and tissue-restricted transcription factors. *Cell* **103**:63–74.

Martin, G. 2001. Making a vertebrate limb: new players enter from the wings. *Bioessays* **23**:865–868.

Rubin, C., et al. 2003. Sprouty fine-tunes EGF signaling through interlinked positive and negative feedback loops. *Curr. Biol.* **13**:297–307.

Tickle, C., and A. Munsterberg. 2001. Vertebrate limb development: the early stages in chick and mouse. *Curr. Opin. Genet. Devel.* **11**:476–481.

16

MOVING PROTEINS INTO MEMBRANES AND ORGANELLES

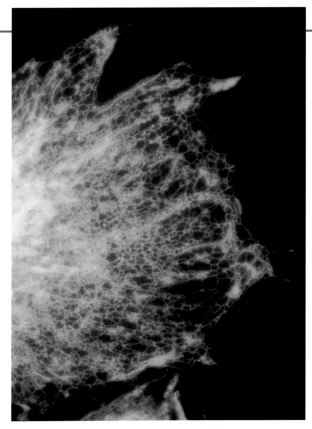

A live bovine endothelial cell stained to reveal different intracellular compartments. The lacelike membranes of the endoplasmic reticulum were stained with a green fluorescent dye, and the wormlike mitochondria were stained with an orange fluorescent dye. [Molecular Probes, Inc.]

A typical mammalian cell contains up to 10,000 different kinds of proteins; a yeast cell, about 5000. The vast majority of these proteins are synthesized by cytosolic ribosomes, and many remain within the cytosol. However, as many as half the different kinds of proteins produced in a typical cell are delivered to a particular cell membrane, an aqueous compartment other than the cytosol, or to the cell surface for secretion. For example, many hormone receptor proteins and transporter proteins must be delivered to the plasma membrane, some water-soluble enzymes such as RNA and DNA polymerases must be targeted to the nucleus, and components of the extracellular matrix as well as polypeptide signaling molecules must be directed to the cell surface for secretion from the cell. These and all the other proteins produced by a cell must reach their correct locations for the cell to function properly.

The delivery of newly synthesized proteins to their proper cellular destinations, usually referred to as *protein targeting* or *protein sorting*, encompasses two very different kinds of processes. The first general process involves targeting of a protein to the membrane of an intracellular organelle and can occur either during or soon after synthesis of the protein by translation at the ribosome. For membrane proteins, targeting leads to insertion of the protein into the lipid bilayer of the membrane, whereas for water-soluble proteins, targeting leads to translocation of the entire protein across the membrane into the aqueous interior of the organelle.

OUTLINE

Proteins are sorted to the endoplasmic reticulum (ER), mitochondria, chloroplasts, peroxisomes, and the nucleus by this general process (Figure 16-1).

A second general sorting process applies to proteins that initially are targeted to the ER membrane, thereby entering

the **secretory pathway.** These proteins include not only soluble and membrane proteins that reside in the ER itself but also proteins that are secreted from the cell, enzymes and other resident proteins in the lumen of the Golgi complex and lysosomes, and integral proteins in the membranes of

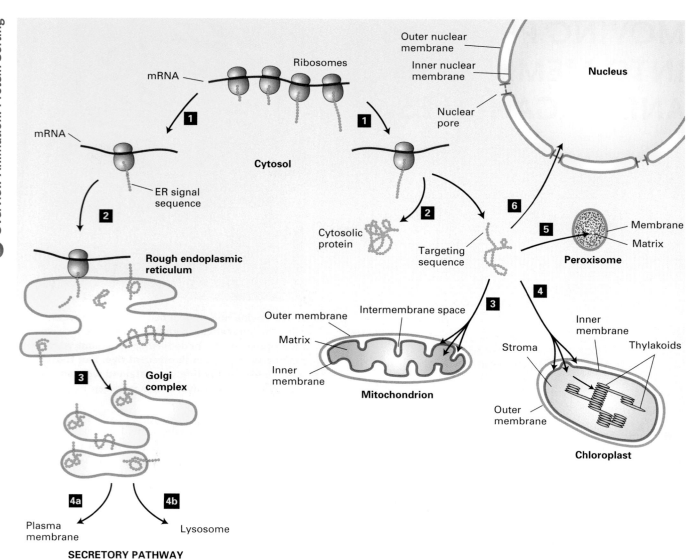

▲ **FIGURE 16-1 Overview of major protein-sorting pathways in eukaryotic cells.** All nuclear-encoded mRNAs are translated on cytosolic ribosomes. *Left (secretory pathway)*: Ribosomes synthesizing nascent proteins in the secretory pathway are directed to the rough endoplasmic reticulum (ER) by an ER signal sequence (pink; steps **1**, **2**). After translation is completed on the ER, these proteins can move via transport vesicles to the Golgi complex (step **3**). Further sorting delivers proteins either to the plasma membrane or to lysosomes (steps **4a**, **4b**). *Right (nonsecretory pathways)*: Synthesis of proteins lacking an ER signal sequence is completed on free ribosomes

(step **1**). Those proteins that contain no targeting sequence are released into the cytosol and remain there (step **2**). Proteins with an organelle-specific targeting sequence (pink) first are released into the cytosol (step **2**) but then are imported into mitochondria, chloroplasts, peroxisomes, or the nucleus (steps **3**–**6**). Mitochondrial and chloroplast proteins typically pass through the outer and inner membranes to enter the matrix or stromal space, respectively. Other proteins are sorted to other subcompartments of these organelles by additional sorting steps. Nuclear proteins enter through visible nuclear pores by processes discussed in Chapter 12.

these organelles and the plasma membrane. Targeting to the ER generally involves *nascent* proteins still in the process of being synthesized. Proteins whose final destination is the Golgi, lysosome, or cell surface are transported along the secretory pathway by small vesicles that bud from the membrane of one organelle and then fuse with the membrane of the next organelle in the pathway (see Figure 16-1, *left*). We discuss vesicle-based protein sorting in the next chapter because mechanistically it differs significantly from protein targeting to the membranes of intracellular organelles.

In this chapter, we examine how proteins are targeted to the membrane of intracellular organelles and subsequently inserted into the organelle membrane or moved into the interior. Two features of this protein-sorting process initially were quite baffling: how a given protein could be targeted to only one specific membrane, and how relatively large protein molecules could be translocated across a membrane without disrupting the bilayer as a barrier to ions and small molecules. Using a combination of biochemical purification methods and genetic screens for identifying mutants unable to execute particular translocation steps, cell biologists have identified many of the cellular components required for translocation across each of the different intracellular membranes. In addition, many of the major translocation processes in the cell have been reconstituted using in vitro systems, which can be freely manipulated experimentally.

These studies have shown that despite some variations, the same basic mechanisms govern protein sorting to all the various intracellular organelles. We now know, for instance, that the information to target a protein to a particular organelle destination is encoded within the amino acid sequence of the protein itself, usually within sequences of 20–50 amino acids, known generically as **signal sequences,** or *uptake-targeting sequences* (see Figure 16-1). Each organelle carries a set of receptor proteins that bind only to specific kinds of signal sequences, thus assuring that the information encoded in a signal sequence governs the specificity of targeting. Once a protein containing a signal sequence has interacted with the corresponding receptor, the protein chain is transferred to some kind of *translocation channel* that allows the protein to pass through the membrane bilayer. The unidirectional transfer of a protein into an organelle, without sliding back out into the cytoplasm, is usually achieved by coupling translocation to an energetically favorable process such as hydrolysis of ATP. Some proteins are subsequently sorted further to reach a subcompartment within the target organelle; such sorting depends on yet other signal sequences and other receptor proteins. Finally, signal sequences often are removed from the mature protein by specific proteases once translocation across the membrane is completed.

For each of the protein-targeting events discussed in this chapter, we will seek to answer four fundamental questions:

1. What is the nature of the *signal sequence*, and what distinguishes it from other types of signal sequences?

2. What is the *receptor* for the signal sequence?

3. What is the structure of the *translocation channel* that allows transfer of proteins across the membrane bilayer? In particular, is the channel so narrow that proteins can pass through only in an unfolded state, or will it accommodate folded protein domains?

4. What is the source of *energy* that drives unidirectional transfer across the membrane?

In the first part of the chapter, we cover targeting of proteins to the ER, including the post-translational modifications that occur to proteins as they enter the secretory pathway. We then look at several mechanisms for exporting proteins from bacteria, some of which are similar to protein sorting in eukaryotic cells. The last two sections describe targeting of proteins to mitochondria, chloroplasts, and peroxisomes. We cover the transport of proteins in and out of the nucleus through nuclear pores in Chapter 12 because nuclear transport is intimately related to post-transcriptional events and involves several nucleus-specific variations on the mechanisms discussed in this chapter.

16.1 Translocation of Secretory Proteins Across the ER Membrane

All eukaryotic cells use essentially the same secretory pathway for synthesizing and sorting secreted proteins and soluble luminal proteins in the ER, Golgi, and lysosomes (see Figure 16-1, *left*). For simplicity, we refer to these proteins collectively as *secretory proteins*. Although all cells secrete a variety of proteins (e.g., extracellular matrix proteins), certain types of cells are specialized for secretion of large

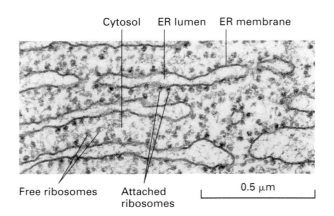

▲ **FIGURE 16-2 Electron micrograph of ribosomes attached to the rough ER in a pancreatic acinar cell.** Most of the proteins synthesized by this type of cell are to be secreted and are formed on membrane-attached ribosomes. A few membrane-unattached (free) ribosomes are evident; presumably, these are synthesizing cytosolic or other nonsecretory proteins. [Courtesy of G. Palade.]

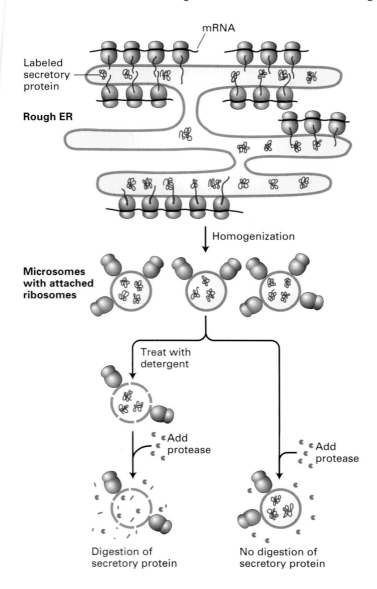

◄ **EXPERIMENTAL FIGURE 16-3 Labeling experiments demonstrate that secretory proteins are localized to the ER lumen shortly after synthesis.** Cells are incubated for a brief time with radiolabeled amino acids, so that only newly synthesized proteins become labeled. The cells then are homogenized, fracturing the plasma membrane and shearing the rough ER into small vesicles called microsomes. Because they have bound ribosomes, *microsomes* have a much greater buoyant density than other membranous organelles and can be separated from them by a combination of differential and sucrose density-gradient centrifugation (Chapter 5). The purified microsomes are treated with a protease in the presence or absence of a detergent. The labeled secretory proteins associated with the microsomes are digested by added proteases only if the permeability barrier of the microsomal membrane is first destroyed by treatment with detergent. This finding indicates that the newly made proteins are inside the microsomes, equivalent to the lumen of the rough ER.

which microsomes isolated from pulse-labeled cells are treated with a protease, demonstrate that although secretory proteins are synthesized on ribosomes bound to the cytosolic face of the ER membrane, they become localized in the lumen of ER vesicles during their synthesis.

A Hydrophobic N-Terminal Signal Sequence Targets Nascent Secretory Proteins to the ER

After synthesis of a secretory protein begins on free ribosomes in the cytosol, a 16- to 30-residue ER signal sequence in the nascent protein directs the ribosome to the ER membrane and initiates translocation of the growing polypeptide across the ER membrane (see Figure 16-1, *left*). An ER signal sequence typically is located at the N-terminus of the protein, the first part of the protein to be synthesized. The signal sequences of different secretory proteins contain one or more positively charged amino acids adjacent to a continuous stretch of 6–12 hydrophobic residues (the core), but otherwise they have little in common. For most secretory proteins, the signal sequence is cleaved from the protein while it is still growing on the ribosome; thus, signal sequences are usually not present in the "mature" proteins found in cells.

The hydrophobic core of ER signal sequences is essential for their function. For instance, the specific deletion of several of the hydrophobic amino acids from a signal sequence, or the introduction of charged amino acids into the hydrophobic core by mutation, can abolish the ability of the N-terminus of a protein to function as a signal sequence. As a consequence, the modified protein remains in the cytosol, unable to cross the ER membrane into the lumen. Using recombinant DNA techniques, researchers have produced cytosolic proteins with added N-terminal amino acid sequences. Provided the added sequence is sufficiently long and hydrophobic, such a modified cytosolic protein is translocated to the ER lumen. Thus the hydrophobic residues

amounts of specific proteins. Pancreatic acinar cells, for instance, synthesize large quantities of several digestive enzymes that are secreted into ductules that lead to the intestine. Because such secretory cells contain the organelles of the secretory pathway (e.g., ER and Golgi) in great abundance, they have been widely used in studying this pathway.

Early pulse-labeling experiments with pancreatic acinar cells showed that radioactively labeled amino acids are incorporated primarily into newly synthesized secretory proteins. The ribosomes synthesizing these proteins are actually bound to the surface of the ER. As a consequence, the portion of the ER that receives proteins entering the secretory pathway is known as the *rough ER* because these membranes are densely studded with ribosomes (Figure 16-2). When cells are homogenized, the rough ER breaks up into small closed vesicles, termed *rough microsomes*, with the same orientation (ribosomes on the outside) as that found in the intact cell. The experiments depicted in Figure 16-3, in

(a) Cell-free protein synthesis; no microsomes present

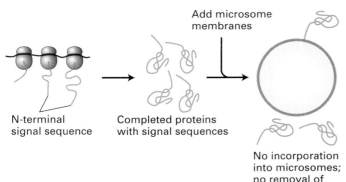

N-terminal
signal sequence

Completed proteins
with signal sequences

Add microsome
membranes

No incorporation
into microsomes;
no removal of
signal sequence

(b) Cell-free protein synthesis; microsomes present

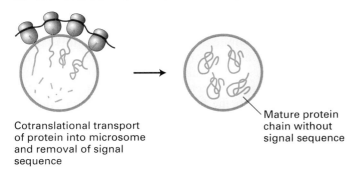

Cotranslational transport
of protein into microsome
and removal of signal
sequence

Mature protein
chain without
signal sequence

◀ **EXPERIMENTAL FIGURE 16-4 Cell-free experiments demonstrate that translocation of secretory proteins into microsomes is coupled to translation.** Treatment of microsomes with EDTA, which chelates Mg^{+2} ions, strips them of associated ribosomes, allowing isolation of ribosome-free microsomes, which are equivalent to ER membranes (see Figure 16-3). Synthesis is carried out in a cell-free system containing functional ribosomes, tRNAs, ATP, GTP, and cytosolic enzymes to which mRNA encoding a secretory protein is added. The secretory protein is synthesized in the absence of microsomes (a), but is translocated across the vesicle membrane and loses its signal sequence only if microsomes are present during protein synthesis (b).

Cotranslational Translocation Is Initiated by Two GTP-Hydrolyzing Proteins

Since secretory proteins are synthesized in association with the ER membrane but not with any other cellular membrane, a signal-sequence recognition mechanism must target them there. The two key components in this targeting are the **signal-recognition particle (SRP)** and its receptor located in the ER membrane. The SRP is a cytosolic ribonucleoprotein particle that transiently binds simultaneously to the ER signal sequence in a nascent protein, to the large ribosomal unit, and to the SRP receptor.

Six discrete polypeptides and a 300-nucleotide RNA compose the SRP (Figure 16-5a). One of the SRP proteins (P54) can be chemically cross-linked to ER signal sequences, evidence that this particular protein is the subunit that binds to the signal sequence in a nascent secretory protein. A region of P54 containing many amino acid residues with hydrophobic side chains is homologous to a bacterial protein known as Ffh, which performs an analogous function to P54 in the translocation of proteins across the inner membrane of bacterial cells. The structure of Ffh contains a cleft whose inner surface is lined by hydrophobic side chains (Figure 16-5b). The hydrophobic region of P54 is thought to contain an analogous cleft that interacts with the hydrophobic N-termini of nascent secretory proteins and selectively targets them to the ER membrane. Two of the SRP proteins, P9 and P14, interact with the ribosome, while P68 and P72 are required for protein translocation.

In the cell-free translation system described previously, the presence of SRP slows elongation of a secretory protein when microsomes are absent, thereby inhibiting synthesis of the complete protein (see Figure 16-4). This finding suggests that interaction of the SRP with both the nascent chain of a secretory protein and with the free ribosome prevents the nascent chain from becoming too long for translocation into the ER. Only after the SRP/nascent chain/ribosome complex has bound to the SRP receptor in the ER membrane does SRP release the nascent chain, allowing elongation at the normal rate.

Figure 16-6 summarizes our current understanding of secretory protein synthesis and the role of the SRP and its

in the core of ER signal sequences form a binding site that is critical for the interaction of signal sequences with receptor proteins on the ER membrane.

Biochemical studies utilizing a cell-free protein-synthesizing system, mRNA encoding a secretory protein, and microsomes stripped of their own bound ribosomes have clarified the function and fate of ER signal sequences. Initial experiments with this system demonstrated that a typical secretory protein is incorporated into microsomes and has its signal sequence removed only if the microsomes are present during protein synthesis (Figure 16-4). Subsequent experiments were designed to determine the precise stage of protein synthesis at which microsomes must be present in order for translocation to occur. In these experiments, a drug that prevents initiation of translation was added to protein-synthesizing reactions at different times after protein synthesis had begun, and then stripped microsomes were added to the reaction mixtures. These experiments showed that microsomes must be added before the first 70 or so amino acids are linked together in order for the completed secretory protein to be localized in the microsomal lumen. At this point, the first 40 amino acids or so protrude from the ribosome, including the signal sequence that later will be cleaved off, and the next 30 or so amino acids are still buried within a channel in the ribosome. Thus the transport of most secretory proteins into the ER lumen occurs while the nascent protein is still bound to the ribosome and being elongated, a process referred to as **cotranslational translocation.**

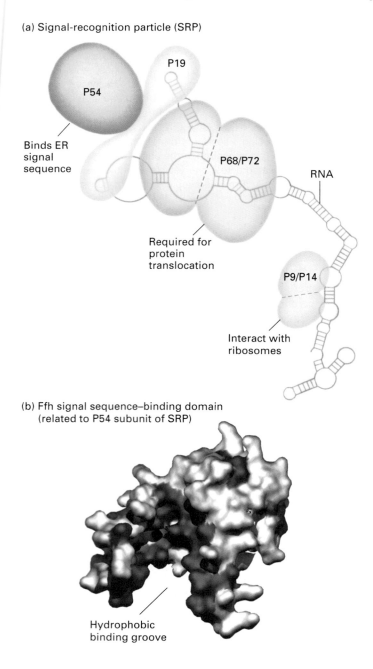

(a) Signal-recognition particle (SRP)

P19

P54

Binds ER
signal
sequence

P68/P72

RNA

Required for
protein
translocation

P9/P14

Interact with
ribosomes

(b) Ffh signal sequence–binding domain
(related to P54 subunit of SRP)

Hydrophobic
binding groove

◄ **FIGURE 16-5 Structure of the signal-recognition particle (SRP).** (a) The SRP comprises one 300-nucleotide RNA and six proteins designated P9, P14, P19, P54, P68, and P72. (The numeral indicates the molecular weight $\times 10^3$.) All proteins except P54 bind directly to the RNA. (b) The bacterial Ffh protein is homologous to the portion of P54 that binds ER signal sequences. This surface model shows the binding domain in Ffh, which contains a large cleft lined with hydrophobic amino acids (purple) whose side chains interact with signal sequences. [Part (a) see K. Strub et al., 1991, *Mol. Cell Biol.* **11**:3949; and S. High and B. Dobberstein, 1991, *J. Cell Biol.* **113**:229. Part (b) adapted from R. J. Keenan et al., 1998, *Cell* **94**:181.]

nascent secretory protein with the ER membrane but also act together to permit elongation and synthesis of complete proteins only when ER membranes are present.

Ultimately, the SRP and SRP receptor function to bring ribosomes that are synthesizing secretory proteins to the ER membrane. The coupling of GTP hydrolysis to this targeting process is thought to contribute to the fidelity by which signal sequences are recognized. Probably the energy from GTP hydrolysis is used to release proteins lacking proper signal sequences from the SRP and SRP receptor complex, thereby preventing their mistargeting to the ER membrane. (A similar coupling of GTP hydrolysis with binding of translation elongation factors to ribosomes increases the fidelity of translation by ejecting aminoacyl-tRNA molecules that cannot form correct base pairs with the codons in mRNA.) Interaction of the SRP/nascent chain/ribosome complex with the SRP receptor is promoted when GTP is bound by both the P54 subunit of SRP and the α subunit of the SRP receptor (see Figure 16-6). Subsequent transfer of the nascent chain and ribosome to a site on the ER membrane where translocation can take place allows hydrolysis of the bound GTP. After dissociating, SRP and its receptor release the bound GDP and recycle to the cytosol ready to initiate another round of interaction between ribosomes synthesizing nascent secretory proteins with the ER membrane.

Passage of Growing Polypeptides Through the Translocon Is Driven by Energy Released During Translation

Once the SRP and its receptor have targeted a ribosome synthesizing a secretory protein to the ER membrane, the ribosome and nascent chain are rapidly transferred to the **translocon**, a protein-lined channel within the membrane. As translation continues, the elongating chain passes directly from the large ribosomal subunit into the central pore of the translocon. The 60S ribosomal subunit is aligned with the pore of the translocon in such a way that the growing chain is never exposed to the cytoplasm and does not fold until it reaches the ER lumen (see Figure 16-6).

The translocon was first identified by mutations in the yeast gene encoding Sec61α, which caused a block in the

receptor in this process. The SRP receptor is an integral membrane protein made up of two subunits: an α subunit and a smaller β subunit. Treatment of microsomes with very small amounts of protease cleaves the α subunit very near its site of attachment to the membrane, releasing a soluble form of the SRP receptor. Protease-treated microsomes are unable to bind the SRP/nascent chain/ribosome complex or support cotranslational translocation. The soluble SRP receptor fragment, however, retains its ability to interact with the SRP/nascent chain/ribosome complex, causing release of SRP and allowing chain elongation to proceed. Thus the SRP and SRP receptor not only help mediate interaction of a

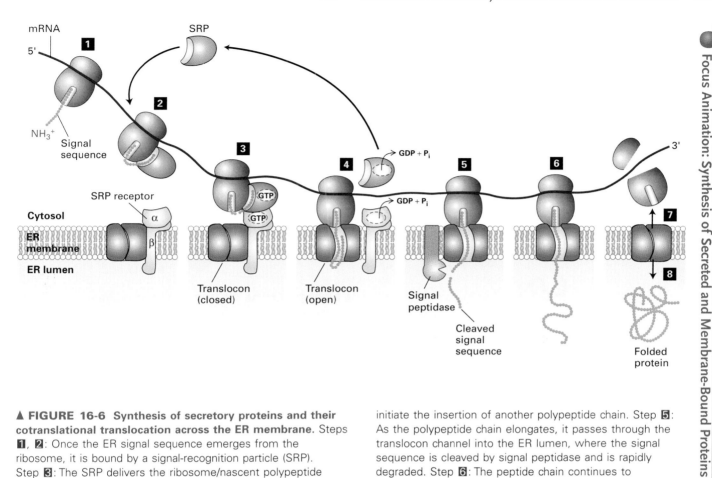

▲ **FIGURE 16-6 Synthesis of secretory proteins and their cotranslational translocation across the ER membrane.** Steps **1**, **2**: Once the ER signal sequence emerges from the ribosome, it is bound by a signal-recognition particle (SRP). Step **3**: The SRP delivers the ribosome/nascent polypeptide complex to the SRP receptor in the ER membrane. This interaction is strengthened by binding of GTP to both the SRP and its receptor. Step **4**: Transfer of the ribosome/nascent polypeptide to the translocon leads to opening of this translocation channel and insertion of the signal sequence and adjacent segment of the growing polypeptide into the central pore. Both the SRP and SRP receptor, once dissociated from the translocon, hydrolyze their bound GTP and then are ready to initiate the insertion of another polypeptide chain. Step **5**: As the polypeptide chain elongates, it passes through the translocon channel into the ER lumen, where the signal sequence is cleaved by signal peptidase and is rapidly degraded. Step **6**: The peptide chain continues to elongate as the mRNA is translated toward the 3' end. Because the ribosome is attached to the translocon, the growing chain is extruded through the translocon into the ER lumen. Steps **7**, **8**: Once translation is complete, the ribosome is released, the remainder of the protein is drawn into the ER lumen, the translocon closes, and the protein assumes its native folded conformation.

translocation of secretory proteins into the lumen of the ER. Subsequently, three proteins called the *Sec61 complex* were found to form the mammalian translocon: Sec61α, an integral membrane protein with 10 membrane-spanning α helices, and two smaller proteins, termed Sec61β and Sec61γ. Chemical cross-linking experiments demonstrated that the translocating polypeptide chain comes into contact with the Sec61α protein in both yeast and mammalian cells, confirming its identity as a translocon component (Figure 16-7).

When microsomes in the cell-free translocation system were replaced with reconstituted phospholipid vesicles containing only the SRP receptor and Sec61 complex, nascent secretory protein was translocated from its SRP/ribosome complex into the vesicles. This finding indicates that the SRP receptor and the Sec61 complex are the only ER-membrane proteins absolutely required for translocation. Thus the energy derived from chain elongation at the ribosome appears to be sufficient to push the polypeptide chain across the membrane in one direction.

Multiple copies of the Sec61 complex assemble to form a translocon channel, which can be visualized by electron microscopy. Images of the channel have been generated by computer averaging of electron micrographs of purified Sec61 channels bound to ribosomes (Figure 16-8). These show the channel as a cylinder, 5–6 nm high and 8.5 nm in

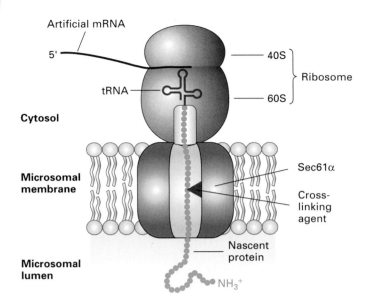

▲ EXPERIMENTAL FIGURE 16-7 Cross-linking experiments show that Sec61α is a translocon component that contacts nascent secretory proteins as they pass into the ER lumen. An mRNA encoding the N-terminal 70 amino acids of the secreted protein prolactin was translated in a cell-free system containing microsomes (see Figure 16-4b). The mRNA lacked a chain-termination codon and contained one lysine codon, near the middle of the sequence. The reactions contained a chemically modified lysyl-tRNA in which a light-activated cross-linking reagent was attached to the lysine side chain. Although the entire mRNA was translated, the completed polypeptide could not be released from the ribosome and thus became "stuck" crossing the ER membrane. The reaction mixtures then were exposed to an intense light, causing the nascent chain to become covalently bound to whatever proteins were near it in the translocon. When the experiment was performed using microsomes from mammalian cells, the nascent chain became covalently linked to Sec61α. Different versions of the prolactin mRNA were used to place the modified lysine residue at different distances from the ribosome; cross-linking to Sec61α was observed only when the modified lysine was positioned within the translocation channel. [Adapted from T. A. Rapoport, 1992, *Science* **258**:931, and D. Görlich and T. A. Rapoport, 1993, *Cell* **75**:615.]

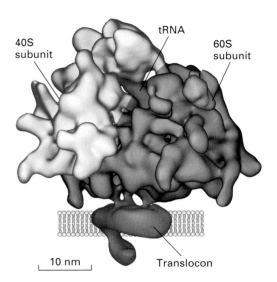

▲ EXPERIMENTAL FIGURE 16-8 Electron microscopy reconstruction reveals that a translocon associates closely with a ribosome. Purified Sec61 complexes were solubilized by treatment of ER membranes with detergents. When ribosomes were added, translocons (blue) reassembled in artificial phospholipid bilayers. The resulting particles were frozen, and electron micrographs of a large number of particles were generated, stored in a computer, and then averaged to produce a single image. A representation of the approximate size and position of the ER lipid bilayer has been added. Note that although the ribosome is firmly attached to the translocon, there is a gap between the two structures. The fingerlike appendage below the translocon channel is thought to be formed from a protein complex that associates with the translocon. [Courtesy Dr. Christopher Akey and Jean-Francois Menetret, Boston University School of Medicine.]

sequence and approximately 30 adjacent amino acids, can insert into the translocon pore (see Figure 16-6).

The mechanism by which the translocon channel opens and closes is controversial at this time. Some evidence suggests that a protein within the ER lumen blocks the translocon pore when a ribosome is not bound to the cytosolic side of the translocon. Other observations, however, indicate that Sec61 complexes may normally reside in the ER membrane in an unassembled state and that the gating process involves the assembly of a translocon channel at the site where the ribosome and nascent chain are brought to the membrane by the SRP and SRP receptor.

As the growing polypeptide chain enters the lumen of the ER, the signal sequence is cleaved by *signal peptidase*, which is a transmembrane ER protein associated with the translocon (see Figure 16-6). This protease recognizes a sequence on the C-terminal side of the hydrophobic core of the signal peptide and cleaves the chain specifically at this sequence once it has emerged into the luminal space of the ER. After the signal sequence has been cleaved, the growing polypeptide moves through the translocon into the ER lumen. The translocon remains open until translation is completed and the entire polypeptide chain has moved into the ER lumen.

diameter, with a central pore, roughly 2 nm in diameter, perpendicular to the plane of the membrane.

If the translocon were always open in the ER membrane, especially in the absence of attached ribosomes and a translocating polypeptide, small molecules such as ATP and amino acids would be able to diffuse freely through the central pore. To maintain the permeability barrier of the ER membrane, the translocon is regulated so that it is open only when a ribosome–nascent chain complex is bound. Thus the translocon is a gated channel analogous to the gated ion channels described in Chapter 7. When the translocon first opens, a loop of the nascent chain, containing the signal

ATP Hydrolysis Powers Post-translational Translocation of Some Secretory Proteins in Yeast

In most eukaryotes, secretory proteins enter the ER by co-translational translocation, using energy derived from translation to pass through the membrane, as we've just described. In yeast, however, some secretory proteins enter the ER lumen after translation has been completed. In such *post-translational translocation*, the translocating protein passes through the same Sec61 translocon that is used in co-translational translocation. However, the SRP and SRP receptor are not involved in post-translational translocation, and in such cases a direct interaction between the translocon and the signal sequence of the completed protein appears to be sufficient for targeting to the ER membrane. In addition, the driving force for unidirectional translocation across the ER membrane is provided by an additional protein complex known as the Sec63 *complex* and a member of the Hsc70 family of molecular **chaperones** known as *BiP*. The tetrameric Sec63 complex is embedded in the ER membrane in the vicinity of the translocon, while BiP is localized to the ER lumen. Like other members of the Hsc70 family, BiP has a peptide-binding domain and an ATPase domain. These chaperones bind and stabilize unfolded or partially folded proteins (see Figure 3-11).

The current model for post-translational translocation of a protein into the ER is outlined in Figure 16-9. Once the N-terminal segment of the protein enters the ER lumen, signal peptidase cleaves the signal sequence just as in cotranslational translocation (step 1). Interaction of BiP·ATP with the luminal portion of the Sec63 complex causes hydrolysis of the bound ATP, producing a conformational change in BiP that promotes its binding to an exposed polypeptide chain (step 2). Since the Sec63 complex is located near the translocon, BiP is thus activated at sites where nascent polypeptides can enter the ER. Certain experiments suggest that in the absence of binding to BiP, an unfolded polypeptide slides back and forth within the translocon channel. Such random sliding motions rarely result in the entire polypeptide's crossing the ER membrane. Binding of a molecule of BiP·ADP to the luminal portion of the polypeptide prevents backsliding of the polypeptide out of the ER. As further inward random sliding exposes more of the polypeptide on the luminal side of the ER membrane, successive binding of

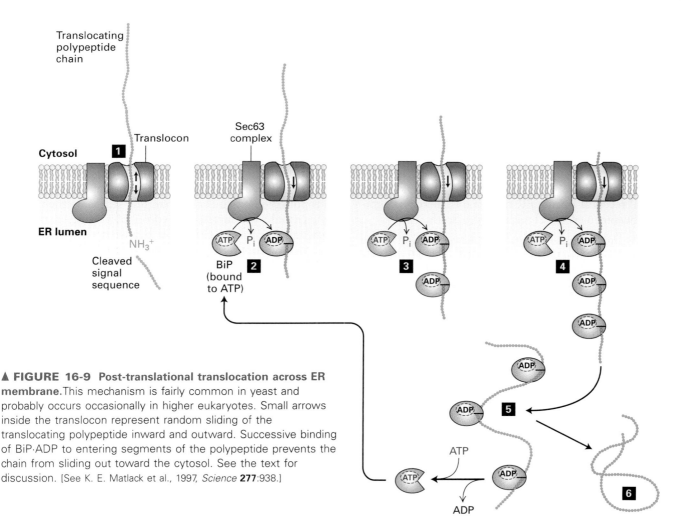

▲ **FIGURE 16-9 Post-translational translocation across ER membrane.** This mechanism is fairly common in yeast and probably occurs occasionally in higher eukaryotes. Small arrows inside the translocon represent random sliding of the translocating polypeptide inward and outward. Successive binding of BiP·ADP to entering segments of the polypeptide prevents the chain from sliding out toward the cytosol. See the text for discussion. [See K. E. Matlack et al., 1997, *Science* **277**:938.]

BiP·ADP molecules to the polypeptide chain acts as a ratchet, ultimately drawing the entire polypeptide into the ER within a few seconds (steps 3 and 4). On a slower time scale, the BiP molecules spontaneously exchange their bound ADP for ATP, leading to release of the polypeptide, which can then fold into its native conformation (steps 5 and 6). The recycled BiP·ATP then is ready for another interaction with Sec63.

The overall reaction carried out by BiP is an important example of how the chemical energy released by the hydrolysis of ATP can power the mechanical movement of a protein across a membrane. Some bacterial cells also use an ATP-driven process for translocating completed proteins across the plasma membrane. However, the mechanism of post-translational translocation in bacteria differs somewhat from that in yeast, as we describe in Section 16.4.

KEY CONCEPTS OF SECTION 16.1

Translocation of Secretory Proteins Across the ER Membrane

■ Synthesis of secreted proteins; enzymes destined for the ER, Golgi complex, or lysosome; and integral plasma-membrane proteins begins on cytosolic ribosomes, which become attached to the membrane of the ER, forming the rough ER (see Figure 16-1, *left*).

■ The ER signal sequence on a nascent secretory protein consists of a segment of hydrophobic amino acids, generally located at the N-terminus.

■ In cotranslational translocation, the signal-recognition particle (SRP) first recognizes and binds the ER signal sequence on a nascent secretory protein and in turn is bound by an SRP receptor on the ER membrane, thereby targeting the ribosome/nascent chain complex to the ER.

■ The SRP and SRP receptor then mediate insertion of the nascent secretory protein into the translocon. Hydrolysis of GTP by the SRP and its receptor drive this docking process (see Figure 16-6). As the ribosome attached to the translocon continues translation, the unfolded protein chain is extruded into the ER lumen. No additional energy is required for translocation.

■ In post-translational translocation, a completed secretory protein is targeted to the ER membrane by interaction of the signal sequence with the translocon. The polypeptide chain is then pulled into the ER by a ratcheting mechanism that requires ATP hydrolysis by the chaperone BiP, which stabilizes the entering polypeptide (see Figure 16-9).

■ In both cotranslational and post-translational translocation, a signal peptidase in the ER membrane cleaves the ER signal sequence from a secretory protein soon after the N-terminus enters the lumen.

16.2 Insertion of Proteins into the ER Membrane

In previous chapters we have encountered many of the vast array of integral (transmembrane) proteins that are present in the plasma membrane and other cellular membranes. Each such protein has a unique orientation with respect to the membrane's phospholipid bilayer. Integral proteins located in ER, Golgi, and lysosomal membranes and in the plasma membrane, which are synthesized on the rough ER, remain embedded in the membrane as they move to their final destinations along the same pathway followed by soluble secretory proteins (see Figure 16-1). During this transport, the orientation of a membrane protein is preserved; that is, the same segments of the protein always face the cytosol, while other segments always face in the opposite direction. Thus the final orientation of these membrane proteins is established during their biosynthesis on the ER membrane. In this section, we first see how integral proteins can interact with membranes. Then we examine how several types of sequences, collectively known as **topogenic sequences**, direct the insertion and orientation of various classes of integral proteins into the membrane. These processes build on and adapt the basic mechanism used to translocate soluble secretory proteins across the ER membrane.

Several Topological Classes of Integral Membrane Proteins Are Synthesized on the ER

The *topology* of a membrane protein refers to the number of times that its polypeptide chain spans the membrane and the orientation of these membrane-spanning segments within the membrane. The key elements of a protein that determine its topology are membrane-spanning segments themselves, which usually contain 20–25 hydrophobic amino acids. Each such segment forms an α helix that spans the membrane, with the hydrophobic amino acid residues anchored to the hydrophobic interior of the phospholipid bilayer.

Scientists have found it useful to categorize integral membrane proteins into the four topological classes illustrated in Figure 16-10. Topological classes I, II, and III comprise *single-pass* proteins, which have only one membrane-spanning α-helical segment. Type I proteins have a cleaved N-terminal signal sequence and are anchored in the membrane with their hydrophilic N-terminal region on the luminal face (also known as the **exoplasmic face**) and their hydrophilic C-terminal region on the cytosolic face. Type II proteins do not contain a cleavable signal sequence and are oriented with their hydrophilic N-terminal region on the cytosolic face and their hydrophilic C-terminal region on the exoplasmic face (i.e., opposite to type I proteins). Type III proteins have the same orientation as type I proteins, but do not contain a cleavable signal sequence. These different topologies reflect distinct mechanisms used by the cell to

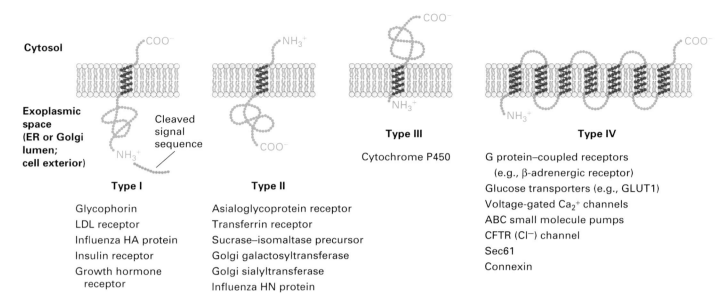

Cytosol

Exoplasmic space (ER or Golgi lumen; cell exterior)

Type I

Glycophorin
LDL receptor
Influenza HA protein
Insulin receptor
Growth hormone receptor

Type II

Asialoglycoprotein receptor
Transferrin receptor
Sucrase–isomaltase precursor
Golgi galactosyltransferase
Golgi sialyltransferase
Influenza HN protein

Type III

Cytochrome P450

Type IV

G protein–coupled receptors (e.g., β-adrenergic receptor)
Glucose transporters (e.g., GLUT1)
Voltage-gated Ca_2^+ channels
ABC small molecule pumps
CFTR (Cl^-) channel
Sec61
Connexin

▲ **FIGURE 16-10 Major topological classes of integral membrane proteins synthesized on the rough ER.** The hydrophobic segments of the protein chain form α helices embedded in the membrane bilayer; the regions outside the membrane are hydrophilic and fold into various conformations. All type IV proteins have multiple transmembrane α helices. The type IV topology depicted here corresponds to that of G protein–coupled receptors: seven α helices, the N-terminus on the exoplasmic side of the membrane, and the C-terminus on the cytosolic side. Other type IV proteins may have a different number of helices and various orientations of the N-terminus and C-terminus. [See E. Hartmann et al., 1989, *Proc. Nat'l. Acad. Sci. USA* **86**:5786, and C. A. Brown and S. D. Black, 1989, *J. Biol. Chem.* **264**:4442.]

establish the membrane orientation of transmembrane segments, as discussed in the next section.

The proteins forming topological class IV contain multiple membrane-spanning segments (see Figure 16-10). Many of the membrane transport proteins discussed in Chapter 7 and the numerous G protein–coupled receptors covered in Chapter 13 belong to this class, sometimes called *multipass* proteins. A final type of membrane protein lacks a hydrophobic membrane-spanning segment altogether; instead, these proteins are linked to an amphipathic phospholipid anchor that is embedded in the membrane.

Internal Stop-Transfer and Signal-Anchor Sequences Determine Topology of Single-Pass Proteins

We begin our discussion with the membrane insertion of integral proteins that contain a single, hydrophobic membrane-spanning segment. Two sequences are involved in targeting and orienting type I proteins in the ER membrane, whereas type II and type III proteins contain a single, internal topogenic sequence.

Type I Proteins All type I transmembrane proteins possess an N-terminal signal sequence that targets them to the ER and an internal hydrophobic sequence that becomes the membrane-spanning α helix. The N-terminal signal sequence on a nascent type I protein, like that of a secretory protein,

initiates cotranslational translocation of the protein through the combined action of the SRP and SRP receptor. Once the N-terminus of the growing polypeptide enters the lumen of the ER, the signal sequence is cleaved, and the growing chain continues to be extruded across the ER membrane. However, unlike the case with secretory proteins, a sequence of about 22 hydrophobic amino acids in the middle of a type I protein stops transfer of the nascent chain through the translocon (Figure 16-11). This internal sequence, because of its hydrophobicity, can move laterally between the protein subunits that form the wall of the translocon and become anchored in the phospholipid bilayer of the membrane, where it remains. Because of its dual function, this sequence is called a *stop-transfer anchor sequence*.

Once translocation is interrupted, translation continues at the ribosome, which is still anchored to the now unoccupied and closed translocon. As the C-terminus of the protein chain is synthesized, it loops out on the cytosolic side of the membrane. When translation is completed, the ribosome is released from the translocon and the C-terminus of the newly synthesized type I protein remains in the cytosol.

Support for this model, depicted in Figure 16-11, has come from studies in which cDNAs encoding various mutant receptors for human growth hormone (HGH) are expressed in cultured mammalian cells. The wild-type HGH receptor, a typical type I protein, is transported normally to the plasma membrane. However, a mutant receptor that has charged residues inserted into the α-helical membrane-spanning segment, or that is missing most of this segment, is translocated

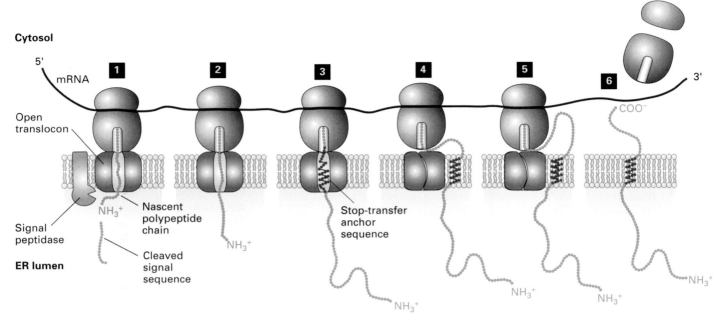

▲ **FIGURE 16-11 Synthesis and insertion into the ER membrane of type I single-pass proteins.** Step **1**: After the ribosome/nascent chain complex becomes associated with a translocon in the ER membrane, the N-terminal signal sequence is cleaved. This process occurs by the same mechanism as the one for soluble secretory proteins (see Figure 16-6). Steps **2**, **3**: The chain is elongated until the hydrophobic stop-transfer anchor sequence is synthesized and enters the translocon, where it prevents the nascent chain from extruding farther into the ER lumen. Step **4**: The stop-transfer anchor sequence moves laterally between the translocon subunits and becomes anchored in the phospholipid bilayer. At this time, the translocon probably closes. Step **5**: As synthesis continues, the elongating chain may loop out into the cytosol through the small space between the ribosome and translocon (see Figure 16-8). Step **6**: When synthesis is complete, the ribosomal subunits are released into the cytosol, leaving the protein free to diffuse in the membrane. [See H. Do et al., 1996, *Cell* **85**:369, and W. Mothes et al., 1997, *Cell* **89**:523.]

entirely into the ER lumen and is eventually secreted from the cell. These findings establish that the hydrophobic membrane-spanning α helix of the HGH receptor and of other type I proteins functions both as a stop-transfer sequence and a membrane anchor that prevents the C-terminus of the protein from crossing the ER membrane.

Type II and Type III Proteins Unlike type I proteins, type II and type III proteins lack a cleavable N-terminal ER signal sequence. Instead, both possess a single internal hydrophobic *signal-anchor sequence* that functions as both an ER signal sequence and membrane-anchor sequence. Recall that type II and type III proteins have opposite orientations in the membrane (see Figure 16-10); this difference depends on the orientation that their respective signal-anchor sequences assume within the translocon.

The internal signal-anchor sequence in type II proteins directs insertion of the nascent chain into the ER membrane so that the N-terminus of the chain faces the cytosol (Figure 16-12). The internal signal-anchor sequence is *not* cleaved and remains in the translocon while the C-terminal region of the growing chain is extruded into the ER lumen by co-translational translocation. During synthesis, the signal-anchor sequence moves laterally between the protein sub-

units forming the translocon wall into the phospholipid bilayer, where it functions as a membrane anchor. Thus this function is similar to the anchoring function of the stop-transfer anchor sequence in type I proteins.

In the case of type III proteins, the signal-anchor sequence, which is located near the N-terminus, inserts the nascent chain into the ER membrane with its N-terminus facing the lumen, just the opposite of type II proteins. The signal-anchor sequence of type III proteins also prevents further extrusion of the nascent chain into the ER lumen, functioning as a stop-transfer sequence. Continued elongation of the chain C-terminal to the signal-anchor/stop-transfer sequence proceeds as it does for type I proteins, with the hydrophobic sequence moving laterally between the translocon subunits to anchor the polypeptide in the ER membrane (see Figure 16-11).

One of the features of signal-anchor sequences that appears to determine their insertion orientation is a high density of positively charged amino acids adjacent to one end of the hydrophobic segment. For reasons that are not well understood these positively charged residues tend to remain on the cytosolic side of the membrane, thereby dictating the orientation of the signal-anchor sequence within the translocon. Thus type II proteins tend to have positively charged

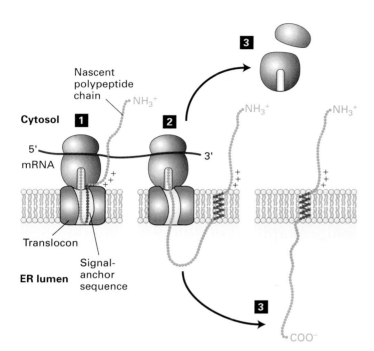

▲ **FIGURE 16-12 Synthesis and insertion into the ER membrane of type II single-pass proteins.** Step **1**: After the internal signal-anchor sequence is synthesized on a cytosolic ribosome, it is bound by an SRP (not shown), which directs the ribosome/nascent chain complex to the ER membrane. This is similar to targeting of soluble secretory proteins except that the hydrophobic signal sequence is not located at the N-terminus and is not subsequently cleaved. The nascent chain becomes oriented in the translocon with its N-terminal portion toward the cytosol. This orientation is believed to be mediated by the positively charged residues shown N-terminal to the signal-anchor sequence. Step **2**: As the chain is elongated and extruded into the lumen, the internal signal-anchor moves laterally out of the translocon and anchors the chain in the phospholipid bilayer. Step **3**: Once protein synthesis is completed, the C-terminus of the polypeptide is released into the lumen, and the ribosomal subunits are released into the cytosol. [See M. Spiess and H. F. Lodish, 1986, *Cell* **44**:177, and H. Do et al., 1996, *Cell* **85**:369.]

residues on the N-terminal side of their signal-anchor sequence, whereas type III proteins tend to have positively charged residues on the C-terminal side of their signal-anchor sequence.

A striking experimental demonstration of the importance of the flanking charge in determining membrane orientation is provided by neuraminidase, a type II protein in the surface coat of influenza virus. Three arginine residues are located just N-terminal to the internal signal-anchor sequence in neuraminidase. Mutation of these three positively charged residues to negatively charged glutamate residues causes neuraminidase to acquire the reverse orientation. Similar experiments have shown that other proteins, with either type II or type III orientation, can be made to "flip" their orientation in the ER membrane by mutating charged residues that flank the internal signal-anchor segment.

Multipass Proteins Have Multiple Internal Topogenic Sequences

Figure 16-13 summarizes the arrangements of topogenic sequences in single-pass and multipass transmembrane proteins. In multipass (type IV) proteins, each of the membrane-spanning α helices acts as a topogenic sequence in the ways that we have already discussed. Multipass proteins fall into one of two types depending on whether the N-terminus extends into the cytosol or the exoplasmic space (i.e., the ER lumen, cell exterior). This N-terminal topology usually is determined by the hydrophobic segment closest to the N-terminus and the charge of the sequences flanking it. If a type IV protein has an *even* number of transmembrane α helices, both its N-terminus and C-terminus will be oriented toward the same side of the membrane (see Figure 16-13d). Conversely, if a type IV protein has an *odd* number of α helices, its two ends will have opposite orientations (see Figure 16-13e).

Proteins with N-Terminus in Cytosol (Type IV-A) Among the multipass proteins whose N-terminus extends into the cytosol are the various glucose transporters (GLUTs) and most ion-channel proteins discussed in Chapter 7. In these proteins, the hydrophobic segment closest to the N-terminus initiates insertion of the nascent chain into the ER membrane with the N-terminus oriented toward the cytosol; thus this α-helical segment functions like the internal signal-anchor sequence of a type II protein (see Figure 16-12). As the nascent chain following the first α helix elongates, it moves through the translocon until the second hydrophobic α helix is formed. This helix prevents further extrusion of the nascent chain through the translocon; thus its function is similar to that of the stop-transfer anchor sequence in a type I protein (see Figure 16-11).

After synthesis of the first two transmembrane α helices, both ends of the nascent chain face the cytosol and the loop between them extends into the ER lumen. The C-terminus of the nascent chain then continues to grow into the cytosol, as it does in synthesis of type I and type III proteins. According to this mechanism, the third α helix acts as another type II signal-anchor sequence, and the fourth as another stop-transfer anchor sequence (see Figure 16-13d). Apparently, once the first topogenic sequence of a multipass polypeptide initiates association with the translocon, the ribosome remains attached to the translocon, and topogenic sequences that subsequently emerge from the ribosome are threaded into the translocon without the need for the SRP and the SRP receptor.

Experiments that use recombinant DNA techniques to exchange hydrophobic α helices have provided insight into the functioning of the topogenic sequences in type IV-A multipass proteins. These experiments indicate that the order of the hydrophobic α helices relative to each other in the growing chain largely determines whether a given helix functions as a signal-anchor sequence or stop-transfer anchor sequence.

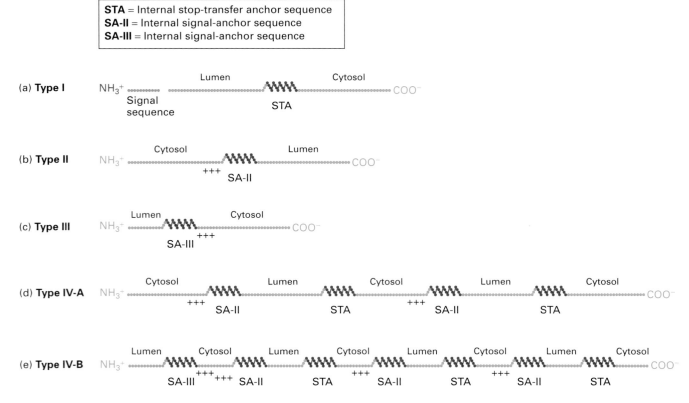

STA = Internal stop-transfer anchor sequence
SA-II = Internal signal-anchor sequence
SA-III = Internal signal-anchor sequence

▲ **FIGURE 16-13 Arrangement of topogenic sequences in single-pass and multipass membrane proteins inserted into the ER membrane.** Topogenic sequences are shown in red; soluble, hydrophilic portions, in blue. The internal topogenic sequences form transmembrane α helices that anchor the proteins or segments of proteins in the membrane. (a) Type I proteins contain a cleaved signal sequence and a single internal stop-transfer anchor (STA). (b, c) Type II and type III proteins contain a single internal signal-anchor (SA) sequence. The difference in the orientation of these proteins depends largely on whether there is a high density of positively charged amino acids (+ + +) on the N-terminal side of the SA sequence (type II) or on the C-terminal side of the SA sequence (type III). (d, e) Nearly all multipass proteins lack a cleavable signal sequence, as depicted in the examples shown here. Type IV-A proteins, whose N-terminus faces the cytosol, contain alternating SA-II sequences and STA sequences. Type IV-B proteins, whose N-terminus faces the lumen, begin with a SA-III sequence followed by alternating SA-II and STA sequences. Proteins of each type with different numbers of α helices (odd or even) are known.

Other than its hydrophobicity, the specific amino acid sequence of a particular helix has little bearing on its function. Thus the first N-terminal α helix and the subsequent odd-numbered ones function as signal-anchor sequences, whereas the intervening even-numbered helices function as stop-transfer anchor sequences.

Proteins with N-Terminus in the Exoplasmic Space (Type IV-B) The large family of G protein–coupled receptors, all of which contain seven transmembrane α helices, constitute the most numerous type IV-B proteins, whose N-terminus extends into the exoplasmic space. In these proteins, the hydrophobic α helix closest to the N-terminus often is followed by a cluster of positively charged amino acids, similar to a type III signal-anchor sequence. As a result, the first α helix inserts the nascent chain into the translocon with the N-terminus extending into the lumen (see Figure 16-13e). As the chain is elongated, it is inserted into the ER membrane by alternating type II signal-anchor sequences and stop-transfer sequences, as just described for type IV-A proteins.

A Phospholipid Anchor Tethers Some Cell-Surface Proteins to the Membrane

Some cell-surface proteins are anchored to the phospholipid bilayer not by a sequence of hydrophobic amino acids but by a covalently attached amphipathic molecule, *glycosylphosphatidylinositol (GPI)* (Figure 16-14a). These proteins are synthesized and initially anchored to the ER membrane exactly like type I transmembrane proteins, with a cleaved N-terminal signal sequence and internal stop-transfer anchor sequence directing the process (see Figure 16-11). However, a short sequence of amino acids in the luminal domain, adjacent to the membrane-spanning domain, is recognized by a transamidase located within the ER membrane. This enzyme simultaneously cleaves off the original

(a)

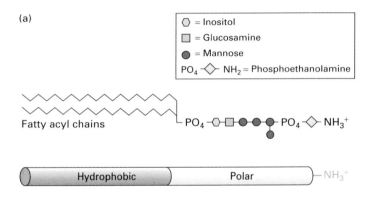

(b)

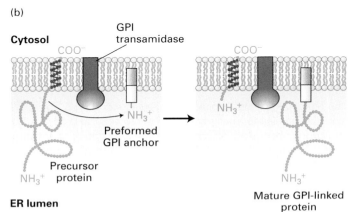

▲ **FIGURE 16-14 GPI-anchored proteins.** (a) Structure of a glycosylphosphatidylinositol (GPI) from yeast. The hydrophobic portion of the molecule is composed of fatty acyl chains, whereas the polar (hydrophilic) portion of the molecule is composed of carbohydrate residues and phosphate groups. In other organisms, both the length of the acyl chains and the carbohydrate moieties may vary somewhat from the structure shown. (b) Formation of GPI-anchored proteins in the ER membrane. The protein is synthesized and initially inserted into the ER membrane as shown in Figure 16-11. A specific transamidase simultaneously cleaves the precursor protein within the exoplasmic-facing domain, near the stop-transfer anchor sequence (red), and transfers the carboxyl group of the new C-terminus to the terminal amino group of a preformed GPI anchor. [See C. Abeijon and C. B. Hirschberg, 1992, *Trends Biochem. Sci.* **17**:32, and K. Kodukula et al., 1992, *Proc. Nat'l. Acad. Sci. USA* **89**:4982.]

stop-transfer anchor sequence and transfers the remainder of the protein to a preformed GPI anchor in the membrane (Figure 16-14b).

Why change one type of membrane anchor for another? Attachment of the GPI anchor, which results in removal of the cytosol-facing hydrophilic domain from the protein, can have several consequences. Proteins with GPI anchors, for example, can diffuse in the plane of the phospholipid bilayer membrane. In contrast, many proteins anchored by membrane-spanning α helices are immobilized in the membrane because their cytosol-facing segments interact with

the cytoskeleton. In addition, the GPI anchor targets the attached protein to the apical domain of the plasma membrane in certain polarized epithelial cells, as we discuss in Chapter 17.

Some membrane proteins are tethered to the cytosolic face of the plasma membrane by other types of lipid anchors (see Figure 5-15). An important example is Ras protein, which plays a key role in intracellular signaling pathways discussed in Chapter 14.

The Topology of a Membrane Protein Often Can Be Deduced from Its Sequence

As we have seen, various topogenic sequences in integral membrane proteins synthesized on the ER govern interaction of the nascent chain with the translocon. When scientists begin to study a protein of unknown function, the identification of topogenic sequences within the corresponding gene sequence can provide important clues about the protein's topological class and function. Suppose, for example, that the gene for a protein known to be required for a cell-to-cell signaling pathway contains nucleotide sequences that encode an apparent N-terminal signal sequence and an internal hydrophobic sequence. These findings would suggest that the protein is a type I integral membrane protein and therefore may be a cell-surface receptor for an extracellular ligand.

Identification of topogenic sequences requires a way to scan sequence databases for segments that are sufficiently hydrophobic to be either a signal sequence or a transmembrane anchor sequence. Topogenic sequences can often be identified with the aid of computer programs that generate a *hydropathy profile* for the protein of interest. The first step is to assign a value known as the *hydropathic index* to each amino acid in the protein. By convention, hydrophobic amino acids are assigned positive values, and hydrophilic amino acids negative values. Although different scales for the hydropathic index exist, all assign the most positive values to amino acids with side chains made up of mostly hydrocarbon residues (e.g., phenylalanine and methionine) and the most negative values to charged amino acids (e.g., arginine and aspartate). The second step is to identify longer segments of sufficient overall hydrophobicity to be N-terminal signal sequences or internal stop-transfer sequences and signal-anchor sequences. To accomplish this, the total hydropathic index for each successive sliding "window" of 20 consecutive amino acids is calculated along the entire length of the protein. Plots of these calculated values against position in the amino acid sequence yield a hydropathy profile.

Figure 16-15 shows the hydropathy profiles for three different membrane proteins. The prominent peaks in such plots identify probable topogenic sequences, as well as their position and approximate length. For example, the hydropathy profile of the human growth hormone receptor reveals the presence of both a hydrophobic signal sequence at the extreme N-terminus of the protein and an internal hydrophobic stop-transfer sequence (see Figure 16-15a). On the basis

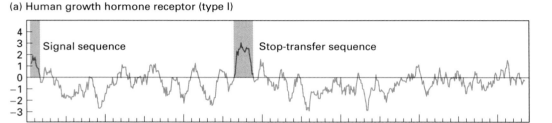

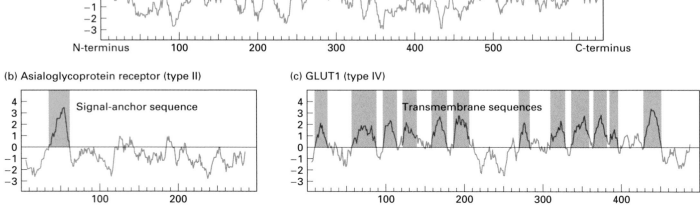

▲ **EXPERIMENTAL FIGURE 16-15 Hydropathy profiles can identify likely topogenic sequences in integral membrane proteins.** Hydropathy profiles are generated by plotting the total hydrophobicity of each segment of 20 contiguous amino acids along the length of a protein. Positive values indicate relatively hydrophobic portions; negative values, relatively polar portions of the protein. Probable topogenic sequences are marked. The complex profiles for multipass (type IV) proteins, such as GLUT1 in part (c), often must be supplemented with other analyses to determine the topology of these proteins. See the text for discussion.

of this profile, we can deduce, correctly, that the human growth hormone receptor is a type I integral membrane protein. The hydropathy profile of the asialoglycoprotein receptor reveals a prominent internal hydrophobic signal-anchor sequence but gives no indication of a hydrophobic N-terminal signal sequence (see Figure 16-15b). Thus we can predict that the asialoglycoprotein receptor is a type II or type III membrane protein. The distribution of charged residues on either side of the signal-anchor sequence often can distinguish between these possibilities, as positively charged amino acids flanking a membrane-spanning segment usually are oriented toward the cytosolic face of the membrane. For instance, in the case of the asialoglycoprotein receptor, a type II protein, the residues on the N-terminal side of the hydrophobic peak carry a net positive charge.

The hydropathy profile of the GLUT1 glucose transporter, a multipass membrane protein, reveals the presence of many segments that are sufficiently hydrophobic to be membrane-spanning helices (Figure 16-15c). The complexity of this profile illustrates the difficulty both in unambiguously identifying all the membrane-spanning segments in a multipass protein and in predicting the topology of individual signal-anchor and stop-transfer sequences. More sophisticated computer algorithms have been developed that take into account the presence of positively charged amino acids adjacent to hydrophobic segments, as well as the length of and spacing between segments. Using all this information, the best algorithms can predict the complex topology of multipass proteins with an accuracy greater than 75 percent.

Finally, sequence homology to a known protein may permit accurate prediction of the topology of a newly discovered multipass protein. For example, the genomes of multicellular organisms encode a very large number of multipass proteins with seven transmembrane α helices. The similarities between the sequences of these proteins strongly suggest that all have the same topology as the well-studied G protein–coupled receptors, which have the N-terminus oriented to the exoplasmic side and the C-terminus oriented to the cytosolic side of the membrane.

KEY CONCEPTS OF SECTION 16.2

Insertion of Proteins into the ER Membrane

■ Integral membrane proteins synthesized on the rough ER fall into four topological classes (see Figure 16-10).

■ Topogenic sequences—N-terminal signal sequences, internal stop-transfer anchor sequences, and internal signal-anchor sequences—direct the insertion and orientation of nascent proteins within the ER membrane. This orientation is retained during transport of the completed membrane protein to its final destination.

■ Single-pass membrane proteins contain one or two topogenic sequences. In multipass membrane proteins, each α-helical segment can function as an internal topogenic sequence, depending on its location in the polypeptide chain and the presence of adjacent positively charged residues (see Figure 16-13).

- Some cell-surface proteins are initially synthesized as type I proteins on the ER and then are cleaved with their luminal domain transferred to a GPI anchor (see Figure 16-14).

- The topology of membrane proteins can often be correctly predicted by computer programs that identify hydrophobic topogenic segments within the amino acid sequence and generate hydropathy profiles (see Figure 16-15).

16.3 Protein Modifications, Folding, and Quality Control in the ER

Membrane and soluble secretory proteins synthesized on the rough ER undergo four principal modifications before they reach their final destinations: (1) addition and processing of carbohydrates *(glycosylation)* in the ER and Golgi, (2) formation of disulfide bonds in the ER, (3) proper folding of polypeptide chains and assembly of multisubunit proteins in the ER, and (4) specific proteolytic cleavages in the ER, Golgi, and secretory vesicles.

One or more carbohydrate chains are added to the vast majority of proteins that are synthesized on the rough ER; indeed, glycosylation is the principal chemical modification to most of these proteins. Carbohydrate chains in glycoproteins may be attached to the hydroxyl group in serine and threonine residues or to the amide nitrogen of asparagine. These are referred to as *O-linked* and *N-linked oligosaccharides*, respectively. O-linked oligosaccharides, such as those found in collagen and glycophorin, often contain only one to four sugar residues. The more common N-linked oligosaccharides are larger and more complex, containing several branches in mammalian cells. In this section we focus on N-linked oligosaccharides, whose initial synthesis occurs in the ER. After the initial glycosylation of a protein in the ER, the oligosaccharide chain is modified in the ER and commonly in the Golgi, as well.

Disulfide bond formation, protein folding, and assembly of multimeric proteins, which take place exclusively in the rough ER, also are discussed in this section. Only properly folded and assembled proteins are transported from the rough ER to the Golgi complex and ultimately to the cell surface or other final destination. Unfolded, misfolded, or partly folded and assembled proteins are selectively retained in the rough ER. We consider several features of such "quality control" in the latter part of this section.

As discussed previously, N-terminal ER signal sequences are cleaved from secretory proteins and type I membrane proteins in the ER. Some proteins also undergo other specific proteolytic cleavages in the Golgi complex or forming secretory vesicles. We cover these cleavages, as well as carbohydrate modifications that occur primarily or exclusively in the Golgi complex, in the next chapter.

A Preformed *N*-Linked Oligosaccharide Is Added to Many Proteins in the Rough ER

Biosynthesis of all *N*-linked oligosaccharides begins in the rough ER with addition of a preformed oligosaccharide precursor containing 14 residues (Figure 16-16). The structure of this precursor is the same in plants, animals, and single-celled eukaryotes—a branched oligosaccharide, containing three glucose (Glc), nine mannose (Man), and two *N*-acetylglucosamine (GlcNAc) molecules, which can be written as $Glc_3Man_9(GlcNAc)_2$. This branched carbohydrate structure is modified in the ER and Golgi compartments, but 5 of the 14 residues are conserved in the structures of all *N*-linked oligosaccharides on secretory and membrane proteins.

The precursor oligosaccharide is linked by a pyrophosphoryl residue to *dolichol*, a long-chain polyisoprenoid lipid that is firmly embedded in the ER membrane and acts as a carrier for the oligosaccharide. The dolichol pyrophosphoryl oligosaccharide is formed on the ER membrane in a complex set of reactions catalyzed by enzymes attached to the cytosolic or luminal faces of the rough ER membrane

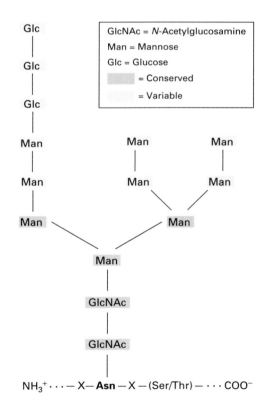

▲ FIGURE 16-16 Common 14-residue precursor of *N*-linked oligosaccharides that is added to nascent proteins in the rough ER. Subsequent removal and in some cases addition of specific sugar residues occur in the ER and Golgi complex. The core region, composed of five residues highlighted in purple, is retained in all *N*-linked oligosaccharides. The precursor can be linked only to asparagine (Asn) residues that are separated by one amino acid (X) from a serine (Ser) or threonine (Thr) on the carboxyl side.

(Figure 16-17). The final dolichol pyrophosphoryl oligosaccharide is oriented so that the oligosaccharide portion faces the ER lumen.

The entire 14-residue precursor is transferred from the dolichol carrier to an asparagine residue on a nascent polypeptide as it emerges into the ER lumen (Figure 16-18, step 1). Only asparagine residues in the tripeptide sequences Asn-X-Ser and Asn-X-Thr (where X is any amino acid except proline) are substrates for *oligosaccharyl transferase*, the enzyme that catalyzes this reaction. Two of the three subunits of this enzyme are ER membrane proteins whose cytosol-facing domains bind to the ribosome, localizing a third subunit of the transferase, the catalytic subunit, near the growing polypeptide chain in the ER lumen. Not all Asn-X-Ser/Thr sequences become glycosylated; for instance, rapid folding of a segment of a protein containing an Asn-X-Ser/Thr sequence may prevent transfer of the oligosaccharide precursor to it.

Immediately after the entire precursor, $Glc_3Man_9(GlcNAc)_2$, is transferred to a nascent polypeptide, three different enzymes remove all three glucose residues and one particular mannose residue (Figure 16-18, steps 2–4). The three glucose residues, which are the last residues added during synthesis of the precursor on the dolichol carrier, appear to act as a signal that the oligosaccharide is complete and ready to be transferred to a protein.

Oligosaccharide Side Chains May Promote Folding and Stability of Glycoproteins

The oligosaccharides attached to glycoproteins serve various functions. For example, some proteins require *N*-linked oligosaccharides in order to fold properly in the ER. This function has been demonstrated in studies with the antibiotic tunicamycin, which blocks the first step in formation of the dolichol-linked precursor of *N*-linked oligosaccharides (see Figure 16-17). In the presence of tunicamycin, for instance, the hemagglutinin precursor polypeptide (HA_0) is synthesized, but it cannot fold properly and form a normal trimer; in this case, the protein remains, misfolded, in the rough ER. Moreover, mutation in the HA sequence of just one asparagine that normally is glycosylated to a glutamine residue, thereby preventing addition of an *N*-linked oligosaccharide to that site, causes the protein to accumulate in the ER in an unfolded state.

In addition to promoting proper folding, *N*-linked oligosaccharides also confer stability on many secreted gly-

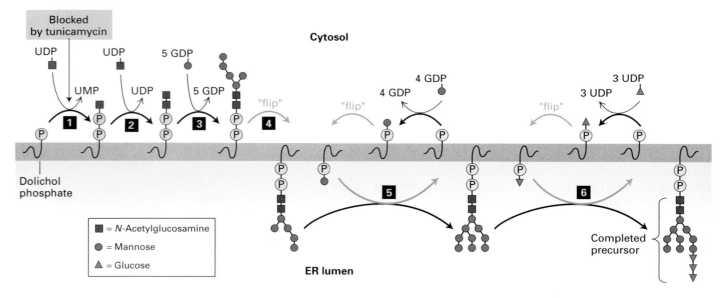

▲ **FIGURE 16-17 Biosynthesis of the dolichol pyrophosphoryl oligosaccharide precursor of *N*-linked oligosaccharides.** Dolichol phosphate is a strongly hydrophobic lipid, containing 75–95 carbon atoms, that is embedded in the ER membrane. Two *N*-acetylglucosamine (GlcNAc) and five mannose residues are added one at a time to a dolichol phosphate on the cytosolic face of the ER membrane (steps **1**–**3**). The nucleotide-sugar donors in these and later reactions are synthesized in the cytosol. Note that the first sugar residue is attached to dolichol by a high-energy pyrophosphate linkage. Tunicamycin, which blocks the first enzyme in this pathway, inhibits the synthesis of

all *N*-linked oligosaccharides in cells. After the seven-residue dolichol pyrophosphoryl intermediate is flipped to the luminal face (step **4**), the remaining four mannose and all three glucose residues are added one at a time (steps **5**, **6**). In the later reactions, the sugar to be added is first transferred from a nucleotide-sugar to a carrier dolichol phosphate on the cytosolic face of the ER; the carrier is then flipped to the luminal face, where the sugar is transferred to the growing oligosaccharide, after which the "empty" carrier is flipped back to the cytosolic face. [After C. Abeijon and C. B. Hirschberg, 1992, *Trends Biochem. Sci.* **17**:32.]

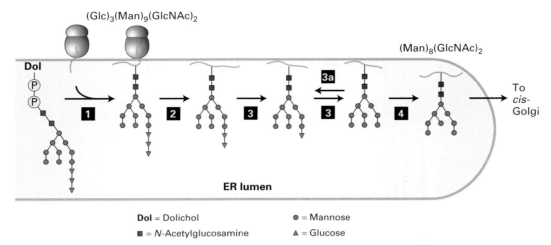

Dol = Dolichol

■ = N-Acetylglucosamine

● = Mannose

▲ = Glucose

▲ **FIGURE 16-18 Addition and initial processing of N-linked oligosaccharides in the rough ER of vertebrate cells.** The $Glc_3Man_9(GlcNAc)_2$ precursor is transferred from the dolichol carrier to a susceptible asparagine residue on a nascent protein as soon as the asparagine crosses to the luminal side of the ER (step **1**). In three separate reactions, first one glucose residue (step **2**), then two glucose residues (step **3**), and finally one mannose residue (step **4**) are removed. Re-addition of one glucose residue (step **3a**) plays a role in the correct folding of many proteins in the ER, as discussed later. [See R. Kornfeld and S. Kornfeld, 1985, *Ann. Rev. Biochem.* **45**:631, and M. Sousa and A. J. Parodi, 1995, *EMBO J.* **14**:4196.]

coproteins. Many secretory proteins fold properly and are transported to their final destination even if the addition of all N-linked oligosaccharides is blocked, for example, by tunicamycin. However, such nonglycosylated proteins have been shown to be less stable than their glycosylated forms. For instance, glycosylated fibronectin, a normal component of the extracellular matrix, is degraded much more slowly by tissue proteases than is nonglycosylated fibronectin.

Oligosaccharides on certain cell-surface glycoproteins also play a role in cell-cell adhesion. For example, the plasma membrane of white blood cells (leukocytes) contains cell-adhesion molecules (CAMs) that are extensively glycosylated. The oligosaccharides in these molecules interact with a sugar-binding domain in certain CAMs found on endothelial cells lining blood vessels. This interaction tethers the leukocytes to the endothelium and assists in their movement into tissues during an inflammatory response to infection (see Figure 6-30). Other cell-surface glycoproteins possess oligosaccharide side chains that can induce an immune response. A common example is the A, B, O blood-group antigens, which are O-linked oligosaccharides attached to glycoproteins and glycolipids on the surface of erythrocytes and other cell types (see Figure 5-16).

Disulfide Bonds Are Formed and Rearranged by Proteins in the ER Lumen

In Chapter 3 we learned that both intramolecular and intermolecular **disulfide bonds** (–S–S–) help stabilize the tertiary and quaternary structure of many proteins. These covalent bonds form by the oxidative linkage of **sulfhydryl groups** (**–SH**), also known as *thiol* groups, on two cysteine residues in the same or different polypeptide chains. This reaction can proceed spontaneously only when a suitable oxidant is present. In eukaryotic cells, disulfide bonds are formed only in the lumen of the rough ER; in bacterial cells, disulfide bonds are formed in the periplasmic space between the inner and outer membranes. Thus disulfide bonds are found only in secretory proteins and in the exoplasmic domains of membrane proteins. Cytosolic proteins and organelle proteins synthesized on free ribosomes lack disulfide bonds and depend on other interactions to stabilize their structures.

The efficient formation of disulfide bonds in the lumen of the ER depends on the enzyme *protein disulfide isomerase* (PDI), which is present in all eukaryotic cells. This enzyme is especially abundant in the ER of secretory cells in such organs as the liver and pancreas, where large quantities of proteins that contain disulfide bonds are produced. As shown in Figure 16-19a, the disulfide bond in the active site of PDI can be readily transferred to a protein by two sequential thiol-disulfide transfer reactions. The reduced PDI generated by this reaction is returned to an oxidized form by the action of an ER-resident protein, called *Ero1*, which carries a disulfide bond that can be transferred to PDI. It is not yet understood how Ero1 itself becomes oxidized. Figure 16-20 depicts the organization of the pathway for protein disulfide-bond formation in the ER lumen and the analogous pathway in bacteria.

In proteins that contain more than one disulfide bond, the proper pairing of cysteine residues is essential for normal structure and activity. Disulfide bonds commonly are

(a) Formation of a disulfide bond

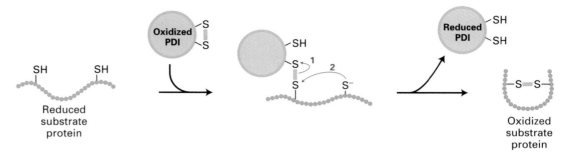

(b) Rearrangement of disulfide bonds

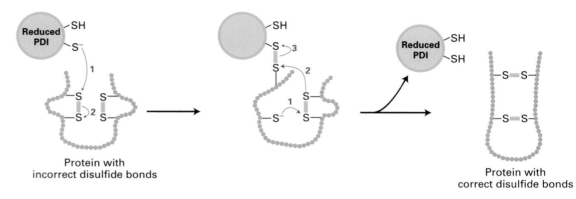

▲ **FIGURE 16-19 Formation and rearrangement of disulfide bonds by protein disulfide isomerase (PDI).** PDI contains an active site with two closely spaced cysteine residues that are easily interconverted between the reduced dithiol form and the oxidized disulfide form. Numbered red arrows indicate the sequence of electron transfers. Yellow bars represent disulfide bonds. (a) In the formation of disulfide bonds, the ionized (–S⁻) form of a cysteine thiol in the substrate protein reacts with the disulfide (S—S) bond in oxidized PDI to form a disulfide-bonded PDI–substrate protein intermediate. A second ionized thiol in the substrate then reacts with this intermediate, forming a disulfide bond within the substrate protein and releasing reduced PDI. (b) Reduced PDI can catalyze rearrangement of improperly formed disulfide bonds by similar thiol-disulfide transfer reactions. In this case, reduced PDI both initiates and is regenerated in the reaction pathway. These reactions are repeated until the most stable conformation of the protein is achieved. [See M. M. Lyles and H. F. Gilbert, 1991, *Biochemistry* **30**:619.]

formed between cysteines that occur sequentially in the amino acid sequence while a polypeptide is still growing on the ribosome. Such sequential formation, however, sometimes yields disulfide bonds between the wrong cysteines. For example, proinsulin has three disulfide bonds that link cysteines 1 and 4, 2 and 6, and 3 and 5. In this case, disulfide bonds initially formed sequentially (e.g., between cysteines 1 and 2) have to be rearranged for the protein to achieve its proper folded conformation. In cells, the rearrangement of disulfide bonds also is accelerated by PDI, which acts on a broad range of protein substrates, allowing them to reach their thermodynamically most stable conformations (Figure 16-19b). Disulfide bonds generally form in a specific order, first stabilizing small domains of a polypeptide, then stabilizing the interactions of more distant segments; this phenomenon is illustrated by the folding of influenza HA protein discussed in the next section.

Most proteins used for therapeutic purposes in humans or animals are secreted glycoproteins stabilized by disulfide bonds. When researchers first tried to synthesize such proteins using plasmid expression vectors in bacterial cells, the results were disappointing. In most cases, the proteins were not secreted (even when a bacterial signal sequence replaced the normal one); instead, they accumulated in the cytosol and often in a denatured state, in part owing to the lack of disulfide bonds. After it became clear that disulfide-bond formation occurs spontaneously only in the ER lumen, biotechnologists eventually developed expression vectors that can be used in animal cells (Chapter 9). Nowadays, such vectors and cultured animal cells are preferred for large-scale production of therapeutic proteins such as tissue plasminogen activator (an anticlotting agent) and erythropoietin, a hormone that stimulates production of red blood cells. ∎

(a) Eukaryotes

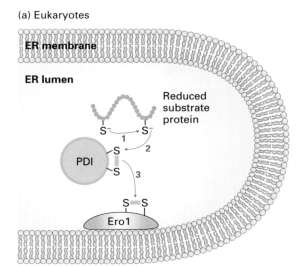

(b) Bacteria

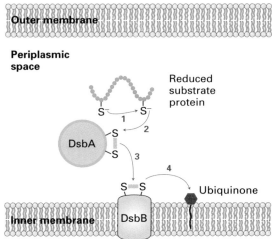

▲ **FIGURE 16-20 Pathways for the formation of disulfide bonds in eukaryotes and bacteria.** Numbered red arrows indicate the sequence of electron flow. A disulfide bond (yellow bar) is formed by loss of a pair of electrons from cysteine thiol (–SH) groups. (a) In the ER lumen of eukaryotic cells, electrons from an ionized thiol in a newly synthesized substrate protein are transferred to a disulfide bond in the active site of PDI (see Figure 16-19). PDI, in turn, transfers electrons to a disulfide bond

in the luminal protein Ero1, thereby regenerating the oxidized form of PDI. The mechanism of reoxidation of Ero1 is not known. (b) In the periplasmic space of bacterial cells, the soluble protein DsbA functions like eukaryotic PDI. Reduced DsbA is reoxidized by DsbB, an inner-membrane protein. Finally, reduced DsbB is reoxidized by transferring electrons to oxidized ubiquinone, a lipid cofactor of the electron-transport chain in the inner membrane (see Chapter 8). [See A. R. Frand et al., 2000, *Trends Cell Biol.* **10**:203.]

Chaperones and Other ER Proteins Facilitate Folding and Assembly of Proteins

Although many reduced, denatured proteins can spontaneously refold into their native state in vitro, such refolding usually requires hours to reach completion. Yet new soluble and membrane proteins produced in the ER generally fold into their proper conformation within minutes after their synthesis. The rapid folding of these newly synthesized proteins in cells depends on the sequential action of several proteins present within the ER lumen.

We have already seen how the chaperone BiP can drive post-translational translocation in yeast by binding fully synthesized polypeptides as they enter the ER (see Figure 16-9). BiP can also bind transiently to nascent chains as they enter the ER during cotranslational translocation. Bound BiP is thought to prevent segments of a nascent chain from misfolding or forming aggregates, thereby promoting folding of the entire polypeptide into the proper conformation. Protein disulfide isomerase (PDI) also contributes to proper folding, which is stabilized by disulfide bonds in many proteins.

As illustrated in Figure 16-21, two other ER proteins, the homologous **lectins** (carbohydrate-binding proteins) *calnexin* and *calreticulin*, bind selectively to certain N-linked oligosaccharides on growing nascent chains. The ligand for these two lectins, which contains a single glucose residue, is generated by a specific glucosyltransferase in the ER lumen (see Figure 16-18, step 3a). This enzyme acts only on polypeptide chains

that are unfolded or misfolded. Binding of calnexin and calreticulin to unfolded nascent chains prevents aggregation of adjacent segments of a protein as it is being made on the ER. Thus calnexin and calreticulin, like BiP, help prevent premature, incorrect folding of segments of a newly made protein.

Other important protein-folding catalysts in the ER lumen are *peptidyl-prolyl isomerases*, a family of enzymes that accelerate the rotation about peptidyl-prolyl bonds in unfolded segments of a polypeptide:

Such isomerizations sometimes are the rate-limiting step in the folding of protein domains. Many peptidyl-prolyl isomerases can catalyze the rotation of exposed peptidyl-prolyl bonds indiscriminately in numerous proteins, but some have very specific protein substrates.

Many important secretory and membrane proteins synthesized on the ER are built of two or more polypeptide subunits. In all cases, the assembly of subunits constituting these

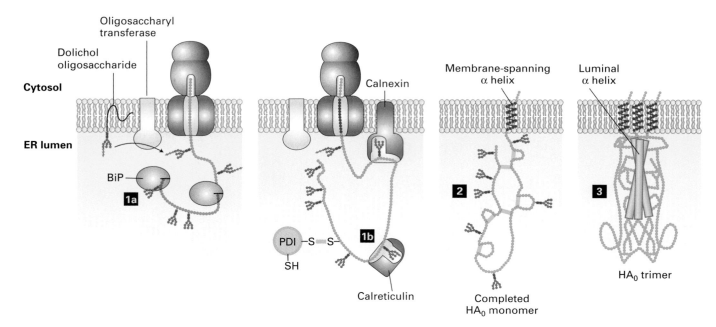

▲ **FIGURE 16-21 Folding and assembly of hemagglutinin (HA₀) trimer in the ER.** Transient binding of the chaperone BiP (step **1a**) to the nascent chain and of two lectins, calnexin and calreticulin, to certain oligosaccharide chains (step **1b**) promotes proper folding of adjacent segments. A total of seven N-linked oligosaccharide chains are added to the luminal portion of the nascent chain during cotranslational translocation, and PDI catalyzes the formation of six disulfide bonds per monomer. Completed HA₀ monomers are anchored in the membrane by a single membrane-spanning α helix with their N-terminus in the lumen (step **2**). Interaction of three HA₀ chains with one another, initially via their transmembrane α helices, apparently triggers formation of a long stem containing one α helix from the luminal part of each HA₀ polypeptide. Finally, interactions between the three globular heads occur, generating a stable HA₀ trimer (step **3**). [See U. Tatu et al., 1995, *EMBO J.* **14**:1340, and D. Hebert et al., 1997, *J. Cell Biol.* **139**:613.]

multisubunit (**multimeric**) proteins occurs in the ER. An important class of multimeric secreted proteins is the immunoglobulins, which contain two heavy (H) and two light (L) chains, all linked by intrachain disulfide bonds. Hemagglutinin (HA) is another multimeric protein that provides a good illustration of folding and subunit assembly (see Figure 16-21). This trimeric protein forms the spikes that protrude from the surface of an influenza virus particle. HA is formed within the ER of an infected host cell from three copies of a precursor protein termed HA₀, which has a single membrane-spanning α helix. In the Golgi complex, each of the three HA₀ proteins is cleaved to form two polypeptides, HA₁ and HA₂; thus each HA molecule that eventually resides on the viral surface contains three copies of HA₁ and three of HA₂ (see Figure 3-7). The trimer is stabilized by interactions between the large exoplasmic domains of the constituent polypeptides, which extend into the ER lumen; after HA is transported to the cell surface, these domains extend into the extracellular space. Interactions between the smaller cytosolic and membrane-spanning portions of the HA subunits also help stabilize the trimeric protein.

The time course of HA₀ folding and assembly in vivo can be determined by pulse-labeling experiments. In a typical experiment, virus-infected cells are pulse-labeled with a radioactive amino acid; at various times during the subsequent chase period, membranes are solubilized by detergent and exposed to monoclonal antibodies specific for either HA₀ monomer or trimer. Immediately after the pulse, the monomer-specific antibody is able to immunoprecipitate all the radioactive HA₀ protein. During the chase period, increasing proportions of the total radioactive HA₀ protein instead react with the trimer-specific monoclonal antibody. Such experiments have shown that each newly made HA₀ polypeptide requires approximately 10 minutes to fold and be incorporated into a trimer in living cells.

Improperly Folded Proteins in the ER Induce Expression of Protein-Folding Catalysts

Wild-type proteins that are synthesized on the rough ER cannot exit this compartment until they achieve their completely folded conformation. Likewise, almost any mutation that prevents proper folding of a protein in the ER also blocks movement of the polypeptide from the ER lumen or membrane to the Golgi complex. The mechanisms for retaining unfolded or incompletely folded proteins within the ER probably increase the overall efficiency of folding by keeping intermediate forms in proximity to folding catalysts, which are most abundant in the ER. Improperly folded proteins retained within the ER generally are found permanently bound

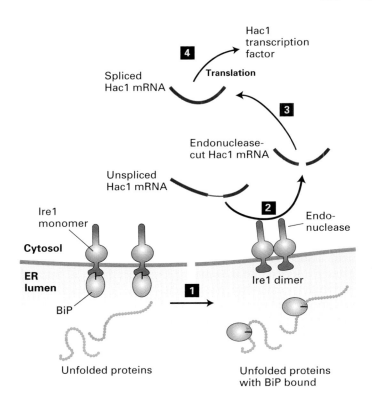

◄ **FIGURE 16-22 The unfolded-protein response.** Ire1, a transmembrane protein in the ER membrane, has a binding site for BiP on its luminal domain; the cytosolic domain contains a specific RNA endonuclease. Step **1**: Accumulating unfolded proteins in the ER lumen bind BiP molecules, releasing them from monomeric Ire1. Dimerization of Ire1 then activates its endonuclease activity. Steps **2**, **3**: The unspliced mRNA precursor encoding the transcription factor Hac1 is cleaved by dimeric Ire1, and the two exons are joined to form functional Hac1 mRNA. Current evidence indicates that this processing occurs in the cytosol, although pre-mRNA processing generally occurs in the nucleus. Step **4**: Hac1 is translated into Hac1 protein, which then moves back into the nucleus and activates transcription of genes encoding several protein-folding catalysts. [See U. Ruegsegger et al., 2001, *Cell* **107**:103; A. Bertolotti et al., 2000, *Nature Cell Biol.* **2**:326; and C. Sidrauski and P. Walter, 1997, *Cell* **90**:1031.]

to the ER chaperones BiP and calnexin. Thus these luminal folding catalysts perform two related functions: assisting in the folding of normal proteins by preventing their aggregation and binding to irreversibly misfolded proteins.

Both mammalian cells and yeasts respond to the presence of unfolded proteins in the rough ER by increasing transcription of several genes encoding ER chaperones and other folding catalysts. A key participant in this *unfolded-protein response* is Ire1, an ER membrane protein that exists as both a monomer and a dimer. The dimeric form, but not the monomeric form, promotes formation of Hac1, a transcription factor in yeast that activates expression of the genes induced in the unfolded-protein response. As depicted in Figure 16-22, binding of BiP to the luminal domain of monomeric Ire1 prevents formation of the Ire1 dimer. Thus the quantity of free BiP in the ER lumen probably determines the relative proportion of monomeric and dimeric Ire1. Accumulation of unfolded proteins within the ER lumen sequesters BiP molecules, making them unavailable for binding to Ire1. As a result the level of dimeric Ire1 increases, leading to an increase in the level of Hac1 and production of proteins that assist in protein folding.

Mammalian cells contain an additional regulatory pathway that operates in response to unfolded proteins in the ER. In this pathway, accumulation of unfolded proteins in the ER triggers proteolysis of ATF6, a transmembrane protein in the ER membrane. The cytosolic domain of ATF6 released by proteolysis then moves to the nucleus, where it stimulates transcription of the genes encoding ER chaperones. Activa-

tion of a transcription factor by such *regulated intramembrane proteolysis* also occurs in the Notch signaling pathway (see Figure 14-29). We will encounter another example of this phenomenon, the cholesterol-responsive transcription factor SREBP, in Chapter 18.

The hereditary form of emphysema illustrates the detrimental effects that can result from misfolding of proteins in the ER. This disease is caused by a point mutation in α_1-antitrypsin, which normally is secreted by hepatocytes and macrophages. The wild-type protein binds to and inhibits trypsin and also the blood protease elastase. In the absence of α_1-antitrypsin, elastase degrades the fine tissue in the lung that participates in the absorption of oxygen, eventually producing the symptoms of emphysema. Although the mutant α_1-antitrypsin is synthesized in the rough ER, it does not fold properly, forming an almost crystalline aggregate that is not exported from the ER. In hepatocytes, the secretion of other proteins also becomes impaired, as the rough ER is filled with aggregated α_1-antitrypsin. ∎

Unassembled or Misfolded Proteins in the ER Are Often Transported to the Cytosol for Degradation

Misfolded secretory and membrane proteins, as well as the unassembled subunits of multimeric proteins, often are degraded within an hour or two after their synthesis in the rough ER. For many years, researchers thought that proteolytic enzymes within the ER catalyzed degradation of misfolded or unassembled polypeptides, but such proteases were never found. More recent studies have shown that misfolded membrane and secretory proteins are transported from the ER lumen, "backwards" through the translocon, into the cytosol, where they are degraded by the ubiquitin-mediated proteolytic pathway (see Figure 3-13).

Ubiquitinylating enzymes localized to the cytosolic face of the ER add ubiquitin to misfolded ER proteins as they exit the ER. This reaction, which is coupled to hydrolysis of ATP, may provide some of the energy required to drag these proteins back to the cytosol. The resulting polyubiquitinylated polypeptides are quickly degraded in proteasomes. Exactly how misfolded soluble and membrane proteins in the ER are recognized and targeted to the translocon for export to the cytosol is not known.

KEY CONCEPTS OF SECTION 16.3

Protein Modifications, Folding, and Quality Control in the ER

■ All *N*-linked oligosaccharides, which are bound to asparagine residues, contain a core of three mannose and two *N*-acetylglucosamine residues and usually have several branches (see Figure 16-16).

■ *O*-linked oligosaccharides, which are bound to serine or threonine residues, are generally short, often containing only one to four sugar residues.

■ Formation of all *N*-linked oligosaccharides begins with assembly of a ubiquitous 14-residue high-mannose precursor on dolichol, a lipid in the membrane of the rough ER (see Figure 16-17). After this preformed oligosaccharide is transferred to specific asparagine residues of nascent polypeptide chains in the ER lumen, three glucose residues and one mannose residue are removed (see Figure 16-18).

■ Oligosaccharide side chains may assist in the proper folding of glycoproteins, help protect the mature proteins from proteolysis, participate in cell-cell adhesion, and function as antigens.

■ Disulfide bonds are added to many secretory proteins and the exoplasmic domain of membrane proteins in the ER. Protein disulfide isomerase (PDI), present in the ER lumen, catalyzes both the formation and the rearrangement of disulfide bonds (see Figure 16-19).

■ The chaperone BiP, the lectins calnexin and calreticulin, and peptidyl-prolyl isomerases work together to assure proper folding of newly made secretory and membrane proteins in the ER. The subunits of multimeric proteins also assemble in the ER.

■ Only properly folded proteins and assembled subunits are transported from the rough ER to the Golgi complex in vesicles.

■ The accumulation of abnormally folded proteins and unassembled subunits in the ER can induce increased expression of ER protein-folding catalysts via the unfolded-protein response (see Figure 16-22).

■ Unassembled or misfolded proteins in the ER often are transported back through the translocon to the cytosol, where they are degraded in the ubiquitin/proteasome pathway.

16.4 Export of Bacterial Proteins

The cell wall surrounding gram-negative bacteria comprises an inner membrane, which is the main permeability barrier for the cytoplasm, a periplasmic space containing various proteins and a layer of peptidoglycan, and an outer membrane, which is permeable to small molecules but not proteins. The peptidoglycan layer gives the cell wall its strength, while the periplasmic proteins function in sensing and importing extracellular molecules and in assembling and maintaining the structural integrity of the cell wall. These proteins (like all bacterial proteins) are synthesized on cytosolic ribosomes and then are translocated in the unfolded state across the inner membrane (also called the cytoplasmic membrane). Most proteins translocated across the inner membrane remain associated with the bacterial cell, either as a membrane protein inserted into the outer or inner membrane or trapped within the periplasmic space. Some bacterial species also possess specialized translocation systems that enable proteins to move through both membranes of the cell wall into the extracellular space. In this section, we describe both types of protein secretion in gram-negative bacteria.

Cytosolic SecA ATPase Pushes Bacterial Polypeptides Through Translocons into the Periplasmic Space

The mechanism for translocating bacterial proteins across the inner membrane shares several key features with the translocation of proteins into the ER of eukaryotic cells. First, translocated proteins usually contain an N-terminal hydrophobic signal sequence, which is cleaved by a signal peptidase. Second, bacterial proteins pass through the inner membrane in a channel, or translocon, composed of proteins that are structurally similar to the eukaryotic Sec61 complex. Third, bacterial cells express two proteins, Ffh and its receptor (FtsY), that are homologs of the SRP and SRP receptor, respectively. In bacteria, however, these latter proteins appear to function mainly in the insertion of hydrophobic membrane proteins into the inner membrane. Indeed, all bacterial proteins that are translocated across the inner membrane do so only after their synthesis in the cytosol is completed but before they are folded into their final conformation.

The post-translational translocation of bacterial proteins across the inner membrane cannot involve a ratchet mechanism similar to that mediated by BiP in the ER lumen (see Figure 16-9) because the ATP required for such a mechanism would be lost by diffusion through the outer membrane. Rather, the driving force for translocation of bacterial proteins is generated by *SecA*, which binds to the cytosolic side of the translocon and hydrolyzes cytosolic ATP. In the model depicted in Figure 16-23, SecA binds to the unfolded translocating polypeptide, and then a conformational change in SecA, driven by the energy released

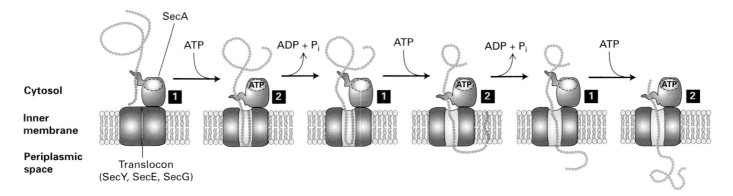

Cytosol

Inner membrane

Periplasmic space

Translocon (SecY, SecE, SecG)

▲ **FIGURE 16-23 Post-translational translocation across inner membrane in gram-negative bacteria.** The bacterial inner membrane contains a translocon channel composed of three subunits that are homologous to the components of the eukaryotic Sec61 complex. Translocation of polypeptides from the cytosol to the periplasmic space is powered by SecA, a cytosolic ATPase that binds to the translocon and to the translocating polypeptide. In the model shown here, binding and hydrolysis of ATP cause conformational changes in SecA that push the bound polypeptide segment through the channel (steps **1**, **2**). Repetition of this cycle results in movement of the polypeptide through the channel in one direction. Current evidence indicates that the N-terminal signal sequence moves from the channel into the bilayer but at some point is cleaved by a signal peptidase, so that the mature polypeptide enters the periplasmic space. [See A. Economou and W. Wickner, 1994, *Cell* **78**:835, and J. Eichler and W. Wickner, 1998, *J. Bacteriol.* **180**:5776.]

from ATP hydrolysis, acts to push the bound polypeptide segment through the translocon pore toward the periplasmic side of the membrane. Repetition of this cycle eventually pushes the entire polypeptide chain through the translocon into the periplasmic space, where disulfide bonds are formed and the polypeptide folds into its proper conformation.

Several Mechanisms Translocate Bacterial Proteins into the Extracellular Space

Quite different mechanisms from the one shown in Figure 16-23 are used to translocate bacterial proteins from the cytosol across both the inner and outer bacterial membranes to the extracellular space. These secretion mechanisms are particularly important for pathogenic bacteria, which commonly use secreted extracellular proteins to colonize specific tissues within the host and to evade host defense mechanisms. Well-known examples of extracellular proteins that promote the growth and dissemination of pathogenic bacteria include protein toxins (e.g., cholera toxin and tetanus toxin) and pili, which are proteinaceous fibers that project from the outer membrane and assist enteric bacteria in adhering to the epithelium of the gut.

The numerous specialized bacterial secretion systems that have been identified can be classified into four general types based on their mechanism of operation. Both the type I and the type II secretion systems involve two steps. First, substrate proteins are translocated across the inner membrane into the periplasmic space, where they fold and often acquire disulfide bonds. Second, the folded proteins are translocated from the periplasmic space across the outer

membrane by complexes of periplasmic proteins that span the inner and outer membranes. The energy for this translocation comes from hydrolysis of ATP in the cytosol, but the mechanisms that couple ATP hydrolysis and translocation across the outer membrane are not well understood.

Translocation by the type III and type IV secretion systems, on the other hand, entails a single step. These systems consist of large protein complexes that span both membranes, allowing proteins to be translocated directly from the cytosol to the extracellular environment. The type III system is adapted not only for secreting proteins but also for injecting them into target cells, a very useful property for pathogenic bacteria.

Pathogenic Bacteria Can Inject Proteins into Animal Cells via Type III Secretion Apparatus

 Yersinia pestis is the bacterial species responsible for the bubonic plague, one of the deadliest diseases in human history. One reason *Yersinia* is such a virulent pathogen lies in its ability to disable host macrophage cells that might otherwise engulf and destroy the invading bacterial cells. The incapacitating effect of *Yersinia* is mediated primarily by a small set of proteins that the bacterial cells inject into macrophage cells.

Various pathogenic bacteria inject proteins into host cells via a complicated syringe-like machine composed of more than 20 different proteins. This type III secretion apparatus, shown in Figure 16-24, has ringlike components embedded in both the inner and outer membranes of the bacterial cell wall and a hollow needlelike structure (pilus) that projects

(a)

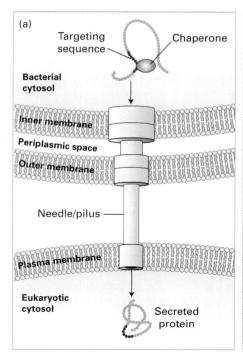

(b)

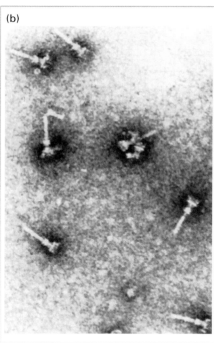

◄ **FIGURE 16-24 Type III secretion apparatus for injecting bacterial proteins into eukaryotic cells.** (a) Schematic diagram of the type III secretion apparatus, which is similar in size and morphology to the bacterial flagellum. Bacterial proteins with targeting sequences (red) that allow interaction with specialized chaperones (orange) enter the cytosol-facing portion of the type III secretion apparatus, travel down the hollow core of the pilus in an ATP-dependent process, and ultimately are delivered to the cytoplasm of the target eukaryotic cell. (b) Electron micrograph of isolated type III secretion apparatuses. Long needlelike pili can be seen extending from the widened basal portions, which are embedded in the outer and inner membranes. See the text for discussion. [Part (a) adapted from D. G. Thanassi and S. J. Hultgren, 2000, *Curr. Opin. Cell Biol.* **12**:420. Part (b) from T. Kubori et al., 1998, *Science* **280**:602.]

out from the outer membrane. Proteins at the outer (distal) end of the pilus can penetrate the plasma membrane of certain mammalian cells, thus creating a conduit that spans three membranes, providing a connection between the bacterial cytoplasm and that of the target host cell.

A key insight into how the type III secretion apparatus may operate came from the observation that many of the components of the secretion apparatus are homologous to proteins in the base of the bacterial flagellum. The flagellar base is located in the inner membrane and functions as a motor to drive rotation of the attached flagellum, which extends outward from the cell surface. The flagellum, which is a hollow helical tube made up entirely of a repeating polymer of the protein flagellin, grows by addition of new flagellin subunits at the distal tip. Several proteins in the flagellar base are thought to use the energy from ATP hydrolysis to push new flagellin subunits through the central channel of the flagellum toward the distal end. Quite possibly, the type III secretion apparatus uses a similar ATP-driven mechanism to push proteins through the central channel of the pilus into target cells.

Recent experiments have identified the signal sequences that target bacterial proteins for transport through the type III secretion apparatus. For instance, recombinant DNA methods were used to express in *Yersinia* cells chimeric proteins containing adenylate cyclase attached to different portions of YopE, which normally is secreted via the type III apparatus. An amphipathic sequence at the N-terminus of YopE was capable of directing adenylate cyclase, which normally resides in the cytosol, to the type III secretion apparatus for injection into mammalian cells. Other experiments showed that YopE proteins in which this amphipathic tar-

geting sequence is mutated still are secreted normally. This surprising finding eventually lead to discovery of a second, independent targeting signal that allows YopE to bind to a small chaperone molecule. The chaperone-YopE complex can be successfully secreted even in the absence of an N-terminal amphipathic sequence. Each of the proteins secreted by the type III apparatus is thought to interact with one of a set of small chaperone proteins (see Figure 16-24). These chaperones may act to keep secreted proteins in a partially unfolded state as they pass through the central channel of the type III apparatus. Once the transported proteins have been released into the target cell, folding can be completed by cytosolic target-cell chaperones. ∎

KEY CONCEPTS OF SECTION 16.4

Export of Bacterial Proteins

■ Gram-negative bacteria translocate completed proteins across the inner membrane through a translocon related to the ER translocon of eukaryotic cells.

■ The driving force for post-translational translocation across the inner membrane of bacteria comes from SecA protein, which uses energy derived from hydrolysis of cytosolic ATP to push polypeptides through the translocon channel (see Figure 16-23).

■ Four bacterial secretion systems exist for translocating proteins from the cytosol across both the inner and outer membranes. Hydrolysis of cytosolic ATP drives secretion in all the systems.

■ The type III secretion apparatus, used by pathogenic bacteria to inject proteins into eukaryotic cells, consists of a basal portion that spans both membranes and an extracellular needlelike structure that can penetrate the plasma membrane of a target cell (see Figure 16-24).

16.5 Sorting of Proteins to Mitochondria and Chloroplasts

In the remainder of this chapter, we examine how proteins synthesized on cytosolic ribosomes are sorted to mitochondria, chloroplasts, and peroxisomes (see Figure 16-1). Both mitochondria and chloroplasts are surrounded by a double membrane and have internal subcompartments, whereas peroxisomes are bounded by a single membrane and have a single luminal compartment known as the *matrix*. Because of these and other differences, we consider peroxisomes separately in the last section.

Besides being bounded by two membranes, both mitochondria and chloroplasts also contain similar types of electron-transport proteins and use an F-class ATPase to synthesize ATP (see Figure 8-2). Remarkably, gram-negative bacteria also exhibit these characteristics. Also like bacterial cells, mitochondria and chloroplasts contain their own DNA, which encodes organelle rRNAs, tRNAs, and some proteins (Chapter 10). Moreover, growth and division of mitochondria and chloroplasts are not coupled to nuclear division. Rather, these organelles grow by the incorporation of cellular proteins and lipids, and new organelles form by division of preexisting organelles—both processes occurring continuously during the interphase period of the cell cycle. The numerous similarities of free-living bacterial cells with mitochondria and chloroplasts have led scientists to hypothesize that these organelles arose by the incorporation of bacteria into ancestral eukaryotic cells, forming endosymbiotic organelles. Striking evidence for this ancient evolutionary relationship can be found in the many proteins of similar sequences shared by mitochondria, chloroplasts, and bacteria, including some of the proteins involved in membrane translocation described in this section.

Proteins encoded by mitochondrial DNA or chloroplast DNA are synthesized on ribosomes within the organelles and directed to the correct subcompartment immediately after synthesis. The majority of proteins located in mitochondria and chloroplasts, however, are encoded by genes in the nucleus and are imported into the organelles after their synthesis in the cytosol. Apparently over eons of evolution much of the genetic information from the ancestral bacterial DNA in these endosymbiotic organelles moved, by an unknown mechanism, to the nucleus. Precursor proteins synthesized in the cytosol that are destined for the matrix of mitochondria or the equivalent space, the stroma, of chloroplasts usually contain specific N-terminal uptake-targeting sequences that specify binding to receptor proteins on the organelle surface. Generally, this sequence is cleaved once it reaches the matrix or stroma. Clearly, these uptake-targeting sequences are similar in their location and general function to the signal sequences that direct nascent proteins to the ER lumen. Although the three types of signals share some common sequence features, their specific sequences differ considerably, as summarized in Table 16-1.

TABLE 16-1	Uptake-Targeting Sequences That Direct Proteins from the Cytosol to Organelles*		
Target Organelle	Location of Sequence Within Protein	Removal of Sequence	Nature of Sequence
Endoplasmic reticulum (lumen)	N-terminus	Yes	Core of 6–12 hydrophobic amino acids, often preceded by one or more basic amino acids (Arg, Lys)
Mitochondrion (matrix)	N-terminus	Yes	Amphipathic helix, 20–50 residues in length, with Arg and Lys residues on one side and hydrophobic residues on the other
Chloroplast (stroma)	N-terminus	Yes	No common motifs; generally rich in Ser, Thr, and small hydrophobic residues and poor in Glu and Asp
Peroxisome (matrix)	C-terminus (most proteins); N-terminus (few proteins)	No	PTS1 signal (Ser-Lys-Leu) at extreme C-terminus; PTS2 signal at N-terminus

*Different or additional sequences target proteins to organelle membranes and subcompartments.
See Chapter 12 for targeting sequences required for uptake of proteins into the nucleus.

In both mitochondria and chloroplasts, protein import requires energy and occurs at points where the outer and inner organelle membranes are in close contact. Because mitochondria and chloroplasts contain multiple membranes and membrane-limited spaces, sorting of many proteins to their correct location often requires the sequential action of two targeting sequences and two membrane-bound translocation systems: one to direct the protein into the organelle, and the other to direct it into the correct organellar compartment or membrane. As we will see, the mechanisms for sorting various proteins to mitochondria and chloroplasts are related to some of the mechanisms discussed previously.

Amphipathic N-Terminal Signal Sequences Direct Proteins to the Mitochondrial Matrix

All proteins that travel from the cytosol to the same mitochondrial destination have targeting signals that share common motifs, although the signal sequences are generally not identical. Thus the receptors that recognize such signals are able to bind to a number of different but related sequences. The most extensively studied sequences for localizing proteins to mitochondria are the *matrix-targeting* sequences. These sequences, located at the N-terminus, are usually 20–50 amino acids in length. They are rich in hydrophobic amino acids, positively charged basic amino acids (arginine and lysine), and hydroxylated ones (serine and threonine), but tend to lack negatively charged acidic residues (aspartate and glutamate).

Mitochondrial matrix-targeting sequences are thought to assume an α-helical conformation in which positively charged amino acids predominate on one side of the helix and hydrophobic amino acids predominate on the other side; thus these sequences are amphipathic. Mutations that disrupt this amphipathic character usually disrupt targeting to the matrix, although many other amino acid substitutions do not. These findings indicate that the amphipathicity of matrix-targeting sequences is critical to their function.

The cell-free assay outlined in Figure 16-25 has been widely used in studies on the import of mitochondrial precursor proteins. In this system, respiring (energized) mitochondria extracted from cells can incorporate mitochondrial precursor proteins carrying appropriate uptake-targeting sequences that have been separately synthesized in the absence of mitochondria. Successful incorporation of the precursor into the organelle can be assayed either by resistance to digestion by an exogenously added protease or, in most cases, by cleavage of the N-terminal targeting sequences by specific proteases. The uptake of completely synthesized mitochondrial precursor proteins by the organelle in this system contrasts with the cell-free cotranslational translocation of secretory proteins, which generally occurs only when microsomal membranes are present during synthesis (see Figure 16-4).

Mitochondrial Protein Import Requires Outer-Membrane Receptors and Translocons in Both Membranes

Figure 16-26 presents an overview of protein import from the cytosol into the mitochondrial matrix, the route into the mitochondrion followed by most imported proteins. We will discuss in detail each step in protein transport into the matrix

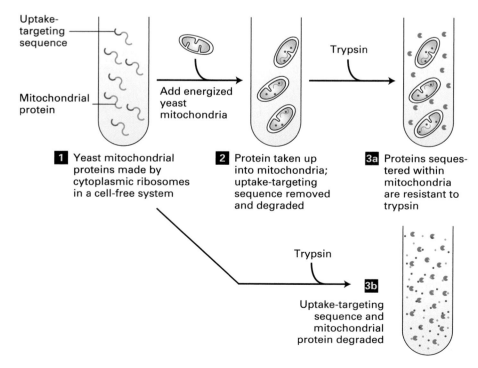

1. Yeast mitochondrial proteins made by cytoplasmic ribosomes in a cell-free system

2. Protein taken up into mitochondria; uptake-targeting sequence removed and degraded

3a. Proteins sequestered within mitochondria are resistant to trypsin

3b. Uptake-targeting sequence and mitochondrial protein degraded

◀ **EXPERIMENTAL FIGURE 16-25 The post-translational uptake of precursor proteins into mitochondria can be assayed in a cell-free system.** The imported protein must contain an appropriate uptake-targeting sequence. Uptake also requires ATP and a cytosolic extract containing chaperone proteins that maintain the precursor proteins in an unfolded conformation. Protein uptake occurs only with energized (respiring) mitochondria, which have a proton electrochemical gradient (proton-motive force) across the inner membrane. This assay has been used to study targeting sequences and other features of the translocation process.

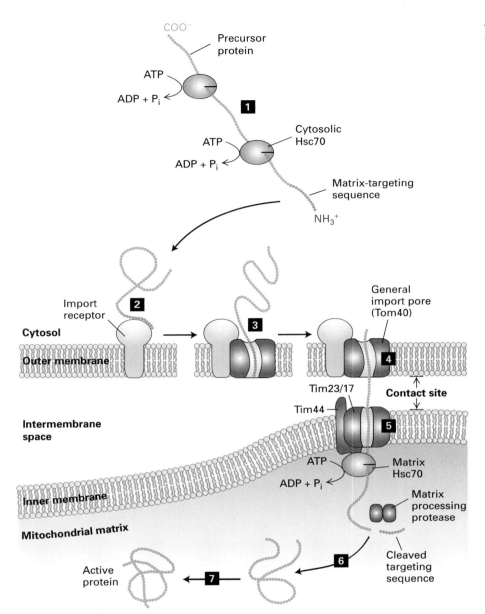

◀ **FIGURE 16-26 Protein import into the mitochondrial matrix.** Precursor proteins synthesized on cytosolic ribosomes are maintained in an unfolded or partially folded state by bound chaperones, such as Hsc70 (step **1**). After a precursor protein binds to an import receptor near a site of contact with the inner membrane (step **2**), it is transferred into the general import pore (step **3**). The translocating protein then moves through this channel and an adjacent channel in the inner membrane (steps **4**, **5**). Note that translocation occurs at rare "contact sites" at which the inner and outer membranes appear to touch. Binding of the translocating protein by the matrix chaperone Hsc70 and subsequent ATP hydrolysis by Hsc70 helps drive import into the matrix. Once the uptake-targeting sequence is removed by a matrix protease and Hsc70 is released from the newly imported protein (step **6**), it folds into the mature, active conformation within the matrix (step **7**). Folding of some proteins depends on matrix chaperonins. See the text for discussion. [See G. Schatz, 1996, *J. Biol. Chem.* **271**:31763, and N. Pfanner et al., 1997, *Ann. Rev. Cell Devel. Biol.* **13**:25.]

and then consider how some proteins subsequently are targeted to other compartments of the mitochondrion.

After synthesis in the cytosol, the soluble precursors of mitochondrial proteins (including hydrophobic integral membrane proteins) interact directly with the mitochondrial membrane. In general, only unfolded proteins can be imported into the mitochondrion. Chaperone proteins such as cytosolic Hsc70 keep nascent and newly made proteins in an unfolded state, so that they can be taken up by mitochondria. Import of an unfolded mitochondrial precursor is initiated by the binding of a mitochondrial targeting sequence to an *import receptor* in the outer mitochondrial membrane. These receptors were first identified by experiments in which antibodies to specific proteins of the outer mitochondrial membrane were shown to inhibit protein import into

isolated mitochondria. Subsequent genetic experiments, in which the genes for specific mitochondrial outer-membrane proteins were mutated, showed that specific receptor proteins were responsible for the import of different classes of mitochondrial proteins. For example, N-terminal matrix-targeting sequences are recognized by Tom20 and Tom22. (Proteins in the outer mitochondrial membrane involved in targeting and import are designated *Tom* proteins for *t*ranslocon of the *o*uter *m*embrane.)

The import receptors subsequently transfer the precursor proteins to an import channel in the outer membrane. This channel, composed mainly of the Tom40 protein, is known as the *general import pore* because all known mitochondrial precursor proteins gain access to the interior compartments of the mitochondrion through this channel. When purified and incorporated into liposomes, Tom40 forms a transmembrane channel with a pore wide enough to accommodate an unfolded polypeptide chain. The general import pore forms a largely passive channel through the outer mitochondrial membrane, and the driving force for unidirectional transport into mitochondria comes from within the mitochondrion. In the case of precursors destined for the mitochondrial matrix, transfer through the outer membrane occurs simultaneously with transfer through an inner-membrane channel composed of the Tim23 and Tim17 proteins. (*Tim* stands for *translocon of the inner membrane.*) Translocation into the matrix thus occurs at "contact sites" where the outer and inner membranes are in close proximity.

Soon after the N-terminal matrix-targeting sequence of a protein enters the mitochondrial matrix, it is removed by a protease that resides within the matrix. The emerging protein also is bound by matrix Hsc70, a chaperone that is localized to the translocation channels in the inner mitochondrial membrane by interacting with Tim44. This interaction stimulates ATP hydrolysis by matrix Hsc70, and together these two proteins are thought to power translocation of proteins into the matrix.

Some imported proteins can fold into their final, active conformation without further assistance. Final folding of many matrix proteins, however, requires a *chaperonin*. As discussed in Chapter 3, chaperonin proteins actively facilitate protein folding in a process that depends on ATP. For instance, yeast mutants defective in Hsc60, a chaperonin in the mitochondrial matrix, can import matrix proteins and cleave their uptake-targeting sequence normally, but the imported polypeptides fail to fold and assemble into the native tertiary and quaternary structures.

Studies with Chimeric Proteins Demonstrate Important Features of Mitochondrial Import

Dramatic evidence for the ability of mitochondrial matrix-targeting sequences to direct import was obtained with chimeric proteins produced by recombinant DNA techniques. For example, the matrix-targeting sequence of alcohol dehydrogenase can be fused to the N-terminus of dihydrofolate reductase (DHFR), which normally resides in the cytosol. In the presence of chaperones, which prevent the C-terminal DHFR segment from folding in the cytosol, cell-free translocation assays show that the chimeric protein is transported into the matrix (Figure 16-27a). The inhibitor methotrexate, which binds tightly to the active site of DHFR and greatly stabilizes its folded conformation, renders the chimeric protein resistant to unfolding by cytosolic chaperones. When translocation assays are performed in the presence of methotrexate, the chimeric protein does not completely enter the matrix. This finding demonstrates that a precursor must be unfolded in order to traverse the import pores in the mitochondrial membranes.

Additional studies revealed that if a sufficiently long spacer sequence separates the N-terminal matrix-targeting sequence and DHFR portion of the chimeric protein, then a stable translocation intermediate forms in the presence of methotrexate (Figure 16-27b). In order for such a stable translocation intermediate to form, the spacer sequence must be long enough to span both membranes; a spacer of 50 amino acids extended to its maximum possible length is adequate to do so. If the chimera contains a shorter spacer—say, 35 amino acids—no stable translocation intermediate is obtained because the spacer cannot span both membranes. These observations provide further evidence that translocated proteins can span both inner and outer mitochondrial membranes and traverse these membranes in an unfolded state.

Microscopic studies of stable translocation intermediates show that they accumulate at sites where the inner and outer mitochondrial membranes are close together, evidence that precursor proteins enter only at such sites (Figure 16-27c). The distance from the cytosolic face of the outer membrane to the matrix face of the inner membrane at these *contact sites* is consistent with the length of an unfolded spacer sequence required for formation of a stable translocation intermediate. Moreover, stable translocation intermediates can be chemically cross-linked to the protein subunits that comprise the translocation channels of both the outer and inner membranes. This finding demonstrates that imported proteins can simultaneously engage channels in both the outer and inner mitochondrial membrane, as depicted in Figure 16-26. Since roughly 1000 stuck translocation intermediates can be observed in a typical yeast mitochondrion, it is thought that mitochondria have approximately 1000 general import pores for the uptake of mitochondrial proteins.

Three Energy Inputs Are Needed to Import Proteins into Mitochondria

As noted previously and indicated in Figure 16-26, ATP hydrolysis by Hsc70 chaperone proteins in both the cytosol and the mitochondrial matrix is required for import of mitochondrial proteins. Cytosolic Hsc70 expends energy to maintain bound precursor proteins in an unfolded state that is competent for translocation into the matrix. The importance of ATP to this function was demonstrated in studies in which a mitochondrial precursor protein was purified and then denatured (unfolded) by urea. When tested in the cell-free mitochondrial translocation system, the denatured protein was incorporated into the matrix in the absence of ATP. In contrast, import of the native, undenatured precursor required ATP for the normal unfolding function of cytosolic chaperones.

The sequential binding and ATP-driven release of multiple matrix Hsc70 molecules to a translocating protein may simply trap the unfolded protein in the matrix. Alternatively, the matrix Hsc70, anchored to the membrane by the Tim44 protein, may act as a molecular motor to pull the protein into the matrix (see Figure 16-26). In this case, the functions of matrix Hsc70 and Tim44 would be analogous to the chaperone BiP and Sec63 complex, respectively, in post-translational translocation into the ER lumen (see Figure 16-9).

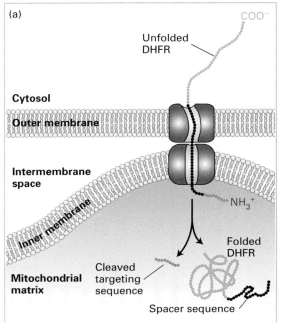

(a)

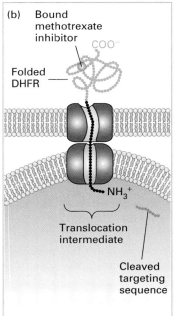

(b)

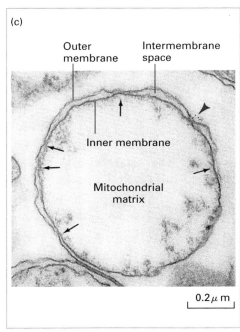

(c)

▲ **EXPERIMENTAL FIGURE 16-27 Experiments with chimeric proteins show that a matrix-targeting sequence alone directs proteins to the mitochondrial matrix and that only unfolded proteins are translocated across both membranes.** The chimeric protein in these experiments contained a matrix-targeting signal at its N-terminus (red), followed by a spacer sequence of no particular function (black), and then by dihydrofolate reductase (DHFR), an enzyme normally present only in the cytosol. (a) When the DHFR segment is unfolded, the chimeric protein moves across both membranes to the matrix of energized mitochondria and the matrix-targeting signal then is removed. (b) When the C-terminus of the chimeric protein is locked in the folded state by binding of methotrexate, translocation is blocked. If the spacer sequence is long enough to extend across the transport channels, a stable translocation intermediate, with the targeting sequence cleaved off, is generated in the presence of methotrexate, as shown here. (c) The C-terminus of the translocation intermediate in (b) can be detected by incubating the mitochondria with antibodies that bind to the DHFR segment, followed by gold particles coated with bacterial protein A, which binds nonspecifically to antibody molecules (see Figure 5-51). An electron micrograph of a sectioned sample reveals gold particles (red arrowhead) bound to the translocation intermediate at a contact site between the inner and outer membranes. Other contact site (black arrows) also are evident. [Parts (a) and (b) adapted from J. Rassow et al., 1990, *FEBS Letters* **275**:190. Part (c) from M. Schweiger et al., 1987, *J. Cell Biol.* **105**:235, courtesy of W. Neupert.]

The third energy input required for mitochondrial protein import is a H^+ electrochemical gradient, or **proton-motive force,** across the inner membrane (Chapter 8). In general, only mitochondria that are actively undergoing respiration, and therefore have generated a proton-motive force across the inner membrane, are able to translocate precursor proteins from the cytosol into the mitochondrial matrix. Treatment of mitochondria with inhibitors or uncouplers of oxidative phosphorylation, such as cyanide or dinitrophenol, dissipates this proton-motive force. Although precursor proteins still can bind tightly to receptors on such poisoned mitochondria, the proteins cannot be imported, either in intact cells or in cell-free systems, even in the presence of ATP and chaperone proteins. Scientists do not fully understand how the proton-motive force is used to facilitate entry of a precursor protein into the matrix. Once a protein is partially inserted into the inner membrane, it is subjected to a transmembrane potential of 200 mV (matrix space negative), which is equivalent to an electric gradient of about 400,000 V/cm. One hypothesis is that the positive charges in the amphipathic matrix-targeting sequence could simply be "electrophoresed," or pulled, into the matrix space by the inside-negative membrane electric potential.

Multiple Signals and Pathways Target Proteins to Submitochondrial Compartments

Unlike targeting to the matrix, targeting of proteins to the intermembrane space, inner membrane, and outer membrane of mitochondria generally requires more than one targeting sequence and occurs via one of several pathways. Figure 16-28 summarizes the organization of targeting sequences in proteins sorted to different mitochondrial locations.

Inner-Membrane Proteins Three separate pathways are known to target proteins to the inner mitochondrial membrane. One pathway makes use of the same machinery that is

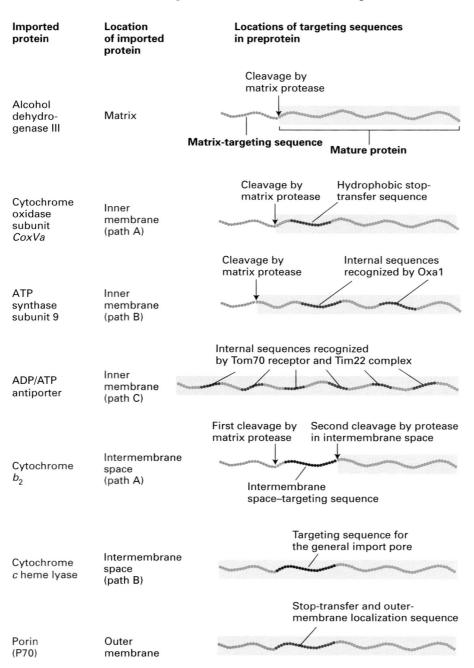

Imported protein	Location of imported protein	Locations of targeting sequences in preprotein
Alcohol dehydrogenase III	Matrix	
Cytochrome oxidase subunit CoxVa	Inner membrane (path A)	
ATP synthase subunit 9	Inner membrane (path B)	
ADP/ATP antiporter	Inner membrane (path C)	
Cytochrome b_2	Intermembrane space (path A)	
Cytochrome c heme lyase	Intermembrane space (path B)	
Porin (P70)	Outer membrane	

◄ **FIGURE 16-28 Arrangement of targeting sequences in imported mitochondrial proteins.** Most mitochondrial proteins have an N-terminal matrix-targeting sequence (pink) that is similar but not identical in different proteins. Proteins destined for the inner membrane, the intermembrane space, or the outer membrane have one or more additional targeting sequences that function to direct the proteins to these locations by several different pathways. The designated pathways in parentheses correspond to those illustrated in Figures 16-29 and 16-30. [See W. Neupert, 1997, *Ann. Rev. Biochem.* **66**:863.]

used for targeting of matrix proteins (Figure 16-29, path A). A cytochrome oxidase subunit called CoxVa is a typical protein transported by this pathway. The precursor form of CoxVa, which contains an N-terminal matrix-targeting sequence recognized by the Tom20/22 import receptor, is transferred through the general import pore of the outer membrane and the inner-membrane Tim23/17 translocation complex. In addition to the matrix-targeting sequence, which is cleaved during import, CoxVa contains a hydrophobic stop-transfer sequence. As the protein passes through the Tim23/17 channel, the stop-transfer sequence blocks translocation of the C-terminus across the inner membrane. The

membrane-anchored intermediate is then transferred laterally into the bilayer of the inner membrane much as type I integral membrane proteins are incorporated into the ER membrane (see Figure 16-11).

A second pathway to the inner membrane is followed by proteins (e.g., ATP synthase subunit 9) whose precursors contain both a matrix-targeting sequence and internal hydrophobic domains recognized by an inner-membrane protein termed *Oxa1*. This pathway is thought to involve translocation of at least a portion of the precursor into the matrix via the Tom20/22 and Tim23/17 channels. After cleavage of the matrix-targeting sequence, the protein is in-

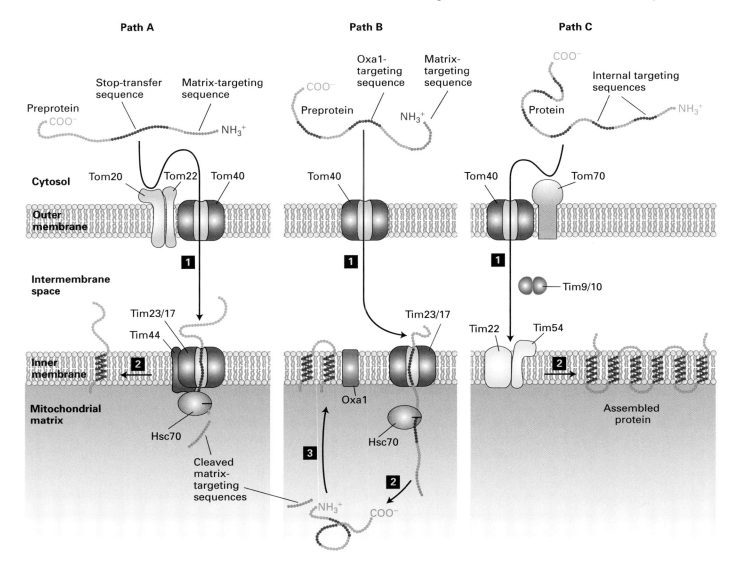

▲ **FIGURE 16-29 Three pathways for transporting proteins from the cytosol to the inner mitochondrial membrane.** Proteins with different targeting sequences are directed to the inner membrane via different pathways. In all three pathways, proteins cross the outer membrane via the Tom40 general import pore. Proteins delivered by pathways A and B contain an N-terminal matrix-targeting sequence that is recognized by the Tom20/22 import receptor in the outer membrane. Although both these pathways use the Tim23/17 inner-membrane channel, they differ in that the entire precursor protein enters the matrix and then is redirected to the inner membrane in pathway B. Matrix Hsc70 plays a role similar its role in the import of soluble matrix proteins (see Figure 16-26). Proteins delivered by pathway C contain internal sequences that are recognized by the Tom70 import receptor. A different inner-membrane translocation channel (Tim22/54) is used in this pathway. Two intermembrane proteins (Tim9 and Tim10) facilitate transfer between the outer and inner channels. See the text for discussion. [See R. E. Dalbey and A. Kuhn, 2000, *Ann. Rev. Cell Devel. Biol.* **16**:51, and N. Pfanner and A. Geissler, 2001, *Nature Rev. Mol. Cell Biol.* **2**:339.]

serted into the inner membrane by a process that requires interaction with Oxa1 and perhaps other inner-membrane proteins (Figure 16-29, path B). Oxa1 is related to a bacterial protein involved in inserting some inner-membrane proteins in bacteria. This relatedness suggests that Oxa1 may have descended from the translocation machinery in the endosymbiotic bacterium that eventually became the mitochondrion. However, the proteins forming the inner-

membrane channels in mitochondria are not related to the SecY protein in bacterial translocons. Oxa1 also participates in the inner-membrane insertion of certain proteins (e.g., subunit II of cytochrome oxidase) that are encoded by mitochondrial DNA and synthesized in the matrix by mitochondrial ribosomes.

The final pathway for insertion in the inner mitochondrial membrane is followed by multipass proteins that contain six membrane-spanning domains, such as the ADP/ATP antiporter. These proteins, which lack the usual N-terminal matrix-targeting sequence, contain multiple internal mitochondrial targeting sequences. After the internal sequences are recognized by Tom70, a second import receptor located in the outer membrane, the imported protein passes through the outer membrane through the general import pore (Figure 16-29, path C). The protein then is transferred to a second translocation complex in the inner membrane composed of the Tim22 and Tim54 proteins. Transfer to the Tim22/54 complex depends on a multimeric complex of two small proteins, Tim9 and Tim10, that reside in the intermembrane space. These may act as chaperones to guide imported proteins from the general import pore to the Tim22/54 complex in the inner membrane. Ultimately the Tim22/54 complex is responsible for incorporating the multiple hydrophobic segments of the imported protein into the inner membrane.

Intermembrane-Space Proteins Two pathways deliver cytosolic proteins to the space between the inner and outer mitochondrial membranes. The major pathway is followed by proteins, such as cytochrome b_2, whose precursors carry two different N-terminal targeting sequences, both of which ultimately are cleaved. The most N-terminal of the two sequences is a matrix-targeting sequence, which is removed by the matrix protease. The second targeting sequence is a hydrophobic segment that blocks complete translocation of the protein across the inner membrane (Figure 16-30, path A). After the resulting membrane-embedded intermediate diffuses laterally away from the Tim23/17 translocation channel, a protease in the membrane cleaves the protein near the hydrophobic transmembrane segment, releasing the mature protein in a soluble form into the intermembrane space. Except for the second proteolytic cleavage, this pathway is similar to that of inner-membrane proteins such as CoxVa (see Figure 16-29, path A).

Cytochrome c heme lyase, the enzyme responsible for the covalent attachment of heme to cytochrome c, illustrates a second pathway for targeting to the intermembrane space. In this pathway, the imported protein is delivered directly to the intermembrane space via the general import pore without involvement of any inner-membrane translocation factors

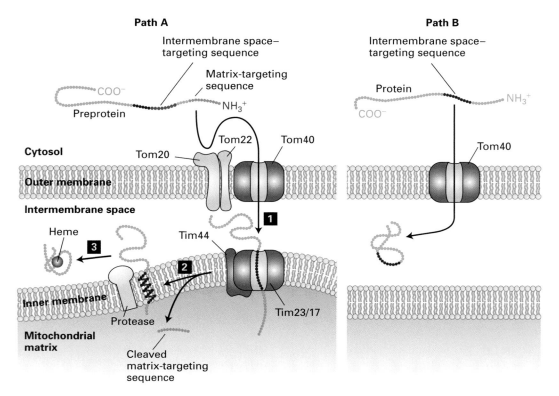

▲ **FIGURE 16-30 Two pathways for transporting proteins from the cytosol to the mitochondrial intermembrane space.** Pathway A, the major one for delivery to the intermembrane space, is similar to pathway A for delivery to the inner membrane (see Figure 16-29). The major difference is that the internal targeting sequence in proteins such as cytochrome b_2 destined for the intermembrane space is recognized by an inner-membrane protease, which cleaves the protein on the intermembrane-space side of the membrane. The released protein then folds and binds to its heme cofactor within the intermembrane space. Pathway B involves direct delivery to the intermembrane space through the Tom40 general import pore in the outer membrane. See the text for further discussion. [See R. E. Dalbey and A. Kuhn, 2000, *Ann. Rev. Cell Devel. Biol.* **16**:51; N. Pfanner and A. Geissler, 2001, *Nature Rev. Mol. Cell Biol.* **2**:339; and K. Diekert et al., 1999, *Proc. Nat'l. Acad. Sci. USA* **96**:11752.]

(Figure 16-30, path B). Since translocation through the Tom40 general import pore does not seem to be coupled to any energetically favorable process such as hydrolysis of ATP or GTP, the mechanism that drives unidirectional translocation through the outer membrane is unclear. One possibility is that cytochrome *c* heme lyase passively diffuses through the outer membrane and then is trapped within the intermembrane space by binding to another protein that is delivered to that location by one of the translocation mechanisms discussed previously.

Outer-Membrane Proteins Experiments with mitochondrial porin (P70) provide clues about how proteins are targeted to the outer mitochondrial membrane. A short matrix-targeting sequence at the N-terminus of P70 is followed by a long stretch of hydrophobic amino acids (see Figure 16-28). If the hydrophobic sequence is experimentally deleted from P70, the protein accumulates in the matrix space with its matrix-targeting sequence still attached. This finding suggests that the long hydrophobic sequence functions as a stop-transfer sequence that both prevents transfer of the protein into the matrix and anchors it as an integral protein in the outer membrane. Normally, neither the matrix-targeting nor stop-transfer sequence is cleaved from the anchored protein. The source of energy to drive outer membrane proteins through the general import pore has not yet been identified.

Targeting of Chloroplast Stromal Proteins Is Similar to Import of Mitochondrial Matrix Proteins

Among the proteins found in the chloroplast stroma are the enzymes of the Calvin cycle, which functions in fixing carbon dioxide into carbohydrates during photosynthesis (Chapter 8). The large (L) subunit of ribulose 1,5-bisphosphate carboxylase (rubisco) is encoded by chloroplast DNA and synthesized on chloroplast ribosomes in the stromal space. The small (S) subunit of rubisco and all the other Calvin cycle enzymes are encoded by nuclear genes and transported to chloroplasts after their synthesis in the cytosol. The precursor forms of these stromal proteins contain an N-terminal *stromal-import* sequence (see Table 16-1).

Experiments with isolated chloroplasts, similar to those with mitochondria illustrated in Figure 16-25, have shown that they can import the S-subunit precursor after its synthesis. After the unfolded precursor enters the stromal space, it binds transiently to a stromal Hsc70 chaperone and the N-terminal sequence is cleaved. In reactions facilitated by Hsc60 chaperonins that reside within the stromal space, eight S subunits combine with the eight L subunits to yield the active rubisco enzyme.

The general process of stromal import appears to be very similar to that for importing proteins into the mitochon-

drial matrix (see Figure 16-26). At least three chloroplast outer-membrane proteins, including a receptor that binds the stromal-import sequence and a translocation channel protein, and five inner-membrane proteins are known to be essential for directing proteins to the stroma. Although these proteins are functionally analogous to the receptor and channel proteins in the mitochondrial membrane, they are not structurally homologous. The lack of homology between these chloroplast and mitochondrial proteins suggests that they may have arisen independently during evolution.

The available evidence suggests that chloroplast stromal proteins, like mitochondrial matrix proteins, are imported in the unfolded state. Import into the stroma depends on ATP hydrolysis catalyzed by a stromal Hsc70 chaperone whose function is similar to Hsc70 in the mitochondrial matrix and BiP in the ER lumen. Unlike mitochondria, chloroplasts cannot generate an electrochemical gradient (proton-motive force) across their inner membrane. Thus protein import into the chloroplast stroma appears to be powered solely by ATP hydrolysis.

Proteins Are Targeted to Thylakoids by Mechanisms Related to Translocation Across the Bacterial Inner Membrane

In addition to the double membrane that surrounds them, chloroplasts contain a series of internal interconnected membranous sacs, the **thylakoids** (see Figure 8-30). Proteins localized to the thylakoid membrane or lumen carry out photosynthesis. Many of these proteins are synthesized in the cytosol as precursors containing multiple targeting sequences. For example, plastocyanin and other proteins destined for the thylakoid lumen require the successive action of two uptake-targeting sequences. The first is an N-terminal stromal-import sequence that directs the protein to the stroma by the same pathway that imports the rubisco S subunit. The second sequence targets the protein from the stroma to the thylakoid lumen. The role of these targeting sequences has been shown in cell-free experiments measuring the uptake into chloroplasts of mutant proteins generated by recombinant DNA techniques. For instance, mutant plastocyanin that lacks the thylakoid-targeting sequence but contains an intact stromal-import sequence accumulates in the stroma and is not transported into the thylakoid lumen.

Four separate pathways for transporting proteins from the stroma into the thylakoid have been identified. All four pathways have been found to be closely related to analogous transport mechanisms in bacteria, illustrating the close evolutionary relationship between the stromal membrane and the bacterial inner membrane. Transport of plastocyanin and related proteins into the thylakoid lumen occurs by an SRP-dependent pathway (Figure 16-31, path A). A second pathway for transporting proteins into the thylakoid lumen

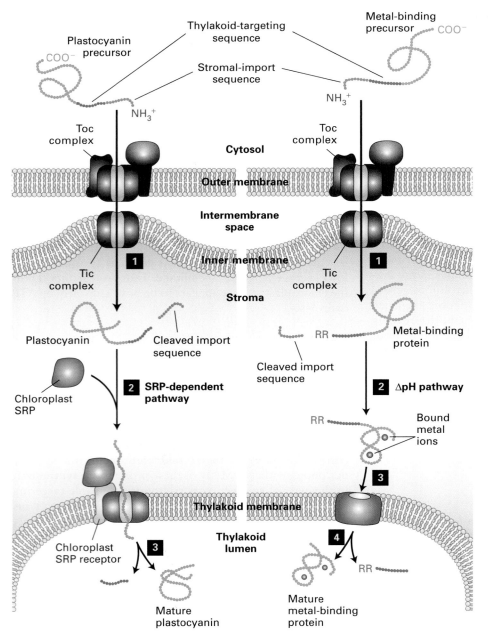

◄ **FIGURE 16-31 Two of the four pathways for transporting proteins from the cytosol to the thylakoid lumen.** In these pathways, unfolded precursors are delivered to the stroma via the same outer-membrane proteins that import stromal-localized proteins. Cleavage of the N-terminal stromal-import sequence by a stromal protease then reveals the thylakoid-targeting sequence. At this point the two pathways diverge. In the SRP-dependent pathway *(left)*, plastocyanin and similar proteins are kept unfolded in the stromal space by a set of chaperones (not shown) and, directed by the thylakoid-targeting sequence, bind to proteins that are closely related to the bacterial SRP, SRP receptor, and SecY translocon, which mediate movement into the lumen. After the thylakoid-targeting sequence is removed in the thylakoid lumen by a separate endoprotease, the protein folds into its mature conformation. In the pH-dependent pathway *(right)*, metal-binding proteins fold in the stroma, and complex redox cofactors are added. Two arginine residues (RR) at the N-terminus of the thylakoid-targeting sequence and a pH gradient across the inner membrane are required for transport of the folded protein into the thylakoid lumen. The translocon in the thylakoid membrane is composed of at least four proteins related to proteins in the bacterial inner membrane. [See R. Dalbey and C. Robinson, 1999, *Trends Biochem. Sci.* **24**:17; R. E. Dalbey and A. Kuhn, 2000, *Ann. Rev. Cell Devel. Biol.* **16**:51; and C. Robinson and A. Bolhuis, 2001, *Nature Rev. Mol. Cell Biol.* **2**:350.]

utilizes a protein related to bacterial SecA and is thought to utilize a mechanism similar to that depicted in Figure 16-23. A third pathway, which targets proteins to the thylakoid membrane, depends on a protein related to the mitochondrial Oxa1 protein and the homologous bacterial protein (see Figure 16-29, path B). Some proteins encoded by chloroplast DNA and synthesized in the stroma or transported into the stroma from the cytosol are inserted into the thylakoid membrane via this pathway.

Finally, thylakoid proteins that bind metal-containing cofactors follow another pathway into the thylakoid lumen (Figure 16-31, ΔpH pathway). The unfolded precursors of these proteins are first targeted to the stroma, where the N-

terminal stromal-import sequence is cleaved off and the protein then folds and binds its cofactor. A set of thylakoid-membrane proteins assists in translocating the folded protein and bound cofactor into the thylakoid lumen, a process powered by the pH gradient normally maintained across the thylakoid membrane. The thylakoid-targeting sequence that directs a protein to this pH-dependent pathway includes two closely spaced arginine residues that are crucial for recognition. Bacterial cells also have a mechanism for translocating folded proteins with a similar arginine-containing sequence across the inner membrane. The molecular mechanism whereby these large folded globular proteins are transported across the thylakoid membrane is currently under intense study.

Sorting of Proteins to Mitochondria and Chloroplasts

■ Most mitochondrial and chloroplast proteins are encoded by nuclear genes, synthesized on cytosolic ribosomes, and imported post-translationally into the organelles.

■ All the information required to target a precursor protein from the cytosol to the mitochondrial matrix or chloroplast stroma is contained within its N-terminal uptake-targeting sequence. After protein import, the uptake-targeting sequence is removed by proteases within the matrix or stroma.

■ Cytosolic chaperones maintain the precursors of mitochondrial and chloroplast proteins in an unfolded state. Only unfolded proteins can be imported into the organelles. Translocation occurs at sites where the outer and inner membranes of the organelles are close together.

■ Proteins destined to the mitochondrial matrix bind to receptors on the outer mitochondrial membrane, and then are transferred to the general import pore (Tom40) in the outer membrane. Translocation occurs concurrently through the outer and inner membranes, driven by the proton-motive force across the inner membrane and ATP hydrolysis by the Hsc70 ATPase in the matrix (see Figure 16-26).

■ Proteins sorted to mitochondrial destinations other than the matrix usually contain two or more targeting sequences, one of which may be an N-terminal matrix-targeting sequence (see Figure 16-28).

■ Some mitochondrial proteins destined for the intermembrane space or inner membrane are first imported into the matrix and then redirected; others never enter the matrix but go directly to their final location.

■ Protein import into the chloroplast stroma occurs through inner-membrane and outer-membrane translocation channels that are analogous in function to mitochondrial channels but composed of proteins unrelated in sequence to the corresponding mitochondrial proteins.

■ Proteins destined for the thylakoid have secondary targeting sequences. After entry of these proteins into the stroma, cleavage of the stromal-targeting sequences reveals the thylakoid-targeting sequences.

■ The three known pathways for moving proteins from the chloroplast stroma to the thylakoid closely resemble translocation across the bacterial inner membrane (see Figure 16-31). One of these systems can translocate folded proteins.

16.6 Sorting of Peroxisomal Proteins

Peroxisomes are small organelles bounded by a single membrane. Unlike mitochondria and chloroplasts, peroxisomes lack DNA and ribosomes. Thus all peroxisomal proteins are encoded by nuclear genes, synthesized on ribosomes free in the cytosol, and then incorporated into preexisting or newly generated peroxisomes. As peroxisomes are enlarged by addition of protein (and lipid), they eventually divide, forming new ones, as is the case with mitochondria and chloroplasts.

The size and enzyme composition of peroxisomes vary considerably in different kinds of cells. However, all peroxisomes contain enzymes that use molecular oxygen to oxidize various substrates, forming hydrogen peroxide (H_2O_2). Catalase, a peroxisome-localized enzyme, efficiently decomposes H_2O_2 into H_2O. Peroxisomes are most abundant in liver cells, where they constitute about 1 to 2 percent of the cell volume.

Cytosolic Receptor Targets Proteins with an SKL Sequence at the C-Terminus into the Peroxisomal Matrix

The import of catalase and other proteins into rat liver peroxisomes can be assayed in a cell-free system similar to that used for monitoring mitochondrial protein import (see Figure 16-25). By testing various mutant catalase proteins in this system, researchers discovered that the sequence Ser-Lys-Leu (SKL in one-letter code) or a related sequence at the C-terminus was necessary for peroxisomal targeting. Further, addition of the SKL sequence to the C-terminus of a normally cytosolic protein leads to uptake of the altered protein by peroxisomes in cultured cells. All but a few of the many different peroxisomal matrix proteins bear a sequence of this type, known as *peroxisomal-targeting sequence 1*, or simply *PTS1*.

The pathway for import of catalase and other PTS1-bearing proteins into the peroxisomal matrix is depicted in Figure 16-32. The PTS1 binds to a soluble receptor protein in the cytosol (Pex5), which in turn binds to a receptor in the peroxisome membrane (Pex14). The soluble and membrane-associated peroxisomal import receptors appear to have a function analogous to that of the SRP and SRP receptor in targeting proteins to the ER lumen. Still bound to Pex5, the imported protein then moves through a multimeric translocation channel, a feature that differs from protein import into the ER lumen. At some stage either during or after entry into the matrix, Pex5 dissociates from the peroxisomal matrix protein and is recycled back to the cytoplasm. In contrast to the N-terminal uptake-targeting sequences on proteins destined for the ER lumen, mitochondrial matrix, and chloroplast stroma, the PTS1 sequence is not cleaved from proteins after their entry into a peroxisome. Protein import into peroxisomes requires ATP hydrolysis, but it is not known how the energy released from ATP is used to power unidirectional translocation across the peroxisomal membrane.

The peroxisome import machinery, unlike most systems that mediate protein import into the ER, mitochondria, and chloroplast, can translocate folded proteins across the

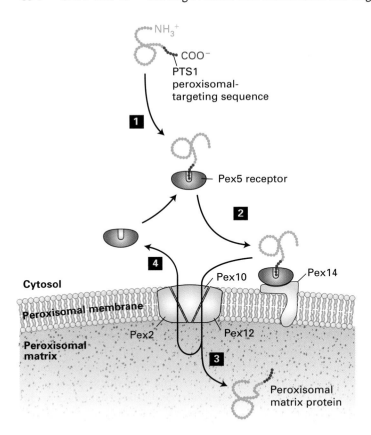

◄ **FIGURE 16-32 Import of peroxisomal matrix proteins directed by PTS1 targeting sequence.** Step **1**: Catalase and most other peroxisomal matrix proteins contain a C-terminal PTS1 uptake-targeting sequence (red) that binds to the cytosolic receptor Pex5. Step **2**: Pex5 with the bound matrix protein interacts with the Pex14 receptor located on the peroxisome membrane. Step **3**: The matrix protein–Pex5 complex is then transferred to a set of membrane proteins (Pex10, Pex12, and Pex2) that are necessary for translocation into the peroxisomal matrix by an unknown mechanism. Step **4**: At some point, either during translocation or in the lumen, Pex5 dissociates from the matrix protein and returns to the cytosol, a process that involves the Pex2/10/12 complex and additional membrane and cytosolic proteins not shown. Note that folded proteins can be imported into peroxisomes and that the targeting sequence is not removed in the matrix. [See P. E. Purdue and P. B. Lazarow, 2001, *Ann. Rev. Cell Devel. Biol.* **17**:701; S. Subramani et al., 2000, *Ann. Rev. Biochem.* **69**:399; and V. Dammai and S. Subramani, 2001, *Cell* **105**:187.]

are eventually degraded. Genetic analyses of cultured cells from different Zellweger patients and of yeast cells carrying similar mutations have identified more than 20 genes that are required for peroxisome biogenesis. ∎

Studies with peroxisome-assembly mutants have shown that different pathways are used for importing peroxisomal matrix proteins versus inserting proteins into the peroxisomal membrane. For example, analysis of cells from some Zellweger patients led to identification of genes encoding the putative translocation channel proteins Pex10, Pex12, and Pex2. Mutant cells defective in any one of these proteins cannot incorporate matrix proteins such as catalase into peroxisomes; nonetheless, the cells contain empty peroxisomes that have a normal complement of peroxisomal membrane proteins (Figure 16-33b). Mutations in any one of three other genes were found to block insertion of peroxisomal membrane proteins as well as import of matrix proteins (Figure 16-33c). These findings demonstrate that one set of proteins translocates soluble proteins into the peroxisomal matrix but a different set is required for insertion of proteins into the peroxisomal membrane. This situation differs markedly from that of the ER, mitochondrion, and chloroplast, for which, as we have seen, membrane proteins and soluble proteins share many of the same components for their insertion into these organelles.

Although most peroxisomes are generated by division of preexisting organelles, these organelles also can arise de novo by the two-stage process depicted in Figure 16-34. In this case, peroxisomal membrane proteins first are targeted to precursor membranes by sequences that differ from both PTS1 and PTS2. Analysis of mutant cells revealed that Pex19 is the receptor protein responsible for targeting of peroxisomal membrane proteins, while Pex3 and Pex16 are necessary for their proper insertion into the membrane. The insertion of peroxisomal membrane proteins generates membranes that have all the components necessary for import of matrix proteins, leading to the formation of mature,

membrane. For example, catalase assumes a folded conformation and binds to heme in the cytoplasm before traversing the peroxisomal membrane. Cell-free studies have shown that the peroxisome import machinery can transport a wide variety of molecules, including very large ones. To explain this unusual ability, scientists have speculated that a translocation channel of variable size may assemble by an unknown mechanism to fit exactly the diameter of the PTS1-bearing substrate molecule and then disassemble once translocation has been completed.

A few peroxisomal matrix proteins such as thiolase are synthesized as precursors with an N-terminal uptake-targeting sequence known as *PTS2*. These proteins bind to a different cytosolic receptor protein, but otherwise import is thought to occur by the same mechanism as for PTS1-containing proteins.

Peroxisomal Membrane and Matrix Proteins Are Incorporated by Different Pathways

Autosomal recessive mutations that cause defective peroxisome assembly occur naturally in the human population. Such defects can lead to severe impairment of many organs and to death. In *Zellweger syndrome* and related disorders, for example, the transport of many or all proteins into the peroxisomal matrix is impaired; newly synthesized peroxisomal enzymes remain in the cytosol and

(a) Wild-type cells

PMP70   Catalase

Peroxisome

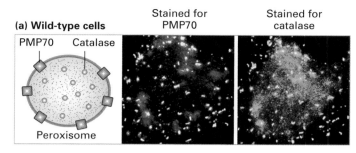

Stained for
PMP70

Stained for
catalase

**(b) Pex1 mutants (deficient
in matrix-protein import)**

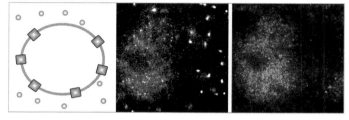

**(c) Pex3 mutants (deficient
in membrane-protein insertion)**

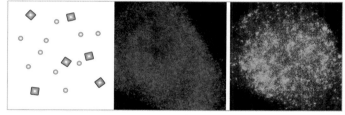

◄**EXPERIMENTAL FIGURE 16-33 Fluorescent-antibody
staining of peroxisomal biogenesis mutants reveals different
pathways for incorporation of membrane and matrix
proteins.** Cells were stained with antibodies to PMP70, a
peroxisomal membrane protein, or with antibodies to catalase, a
peroxisomal matrix protein, then viewed in a fluorescent
microscope. (a) In wild-type cells, both peroxisomal membrane
and matrix proteins are visible as bright foci in numerous
peroxisomal bodies. (b) In cells from a Pex12-deficient patient,
catalase is distributed uniformly throughout the cytosol, whereas
PMP70 is localized normally to peroxisomal bodies. (c) In cells
from a Pex3-deficient patient, peroxisomal membranes cannot
assemble, and as a consequence peroxisomal bodies do not
form. Thus both catalase and PMP70 are mis-localized to the
cytosol. [Courtesy of Stephen Gould, Johns Hopkins University.]

controls the extent of peroxisome division. The small peroxisomes generated by division can be enlarged by incorporation of additional matrix and membrane proteins via
the same pathways described previously.

KEY CONCEPTS OF SECTION 16.6

Sorting of Peroxisomal Proteins

■ All peroxisomal proteins are synthesized on cytosolic
ribosomes and incorporated into the organelle post-
translationally.

■ Most peroxisomal matrix proteins contain a C-terminal
PTS1 targeting sequence; a few have an N-terminal PTS2
targeting sequence. Neither targeting sequence is cleaved
after import.

■ All proteins destined for the peroxisomal matrix bind to
a cytosolic receptor, which differs for PTS1- and PTS2-

functional peroxisomes. Division of mature peroxisomes,
which largely determines the number of peroxisomes
within a cell, depends on still another protein, Pex11. Overexpression of the Pex11 protein causes a large increase in
the number of peroxisomes, suggesting that this protein

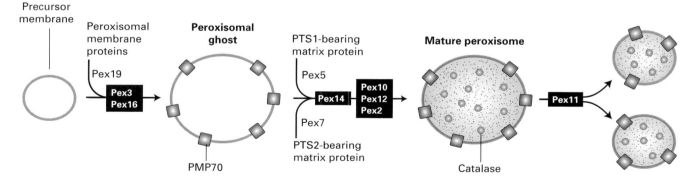

▲ **FIGURE 16-34 Model of peroxisomal biogenesis and
division.** The first stage in the de novo formation of peroxisomes
is the incorporation of peroxisomal membrane proteins into
precursor membranes. Pex19 acts as the receptor for membrane-
targeting sequences. Pex3 and Pex16 are required for the proper
insertion of proteins into the forming peroxisomal membrane.
Insertion of all peroxisomal membrane proteins produces a
peroxisomal ghost, which is capable of importing proteins

targeted to the matrix. The pathways for importing PTS1- and
PTS2-bearing matrix proteins differ only in the identity of the
cytosolic receptor (Pex5 and Pex7, respectively) that binds the
targeting sequence (see Figure 16-32). Complete incorporation of
matrix proteins yields a mature peroxisome. The proliferation of
peroxisomes requires division of mature peroxisomes, a process
that depends on the Pex11 protein.

bearing proteins, and then are directed to common import receptor and translocation machinery on the peroxisomal membrane (see Figure 16-32).

■ Translocation of matrix proteins across the peroxisomal membrane depends on ATP hydrolysis. Many peroxisomal matrix proteins fold in the cytosol and traverse the membrane in a folded conformation.

■ Proteins destined for the peroxisomal membrane contain different targeting sequences than peroxisomal matrix proteins and are imported by a different pathway.

■ Unlike mitochondria and chloroplasts, peroxisomes can arise de novo from precursor membranes, as well as by division of preexisting organelles (see Figure 16-34).

PERSPECTIVES FOR THE FUTURE

As we have seen in this chapter, we now understand many aspects of the basic processes responsible for selectively transporting proteins into the endoplasmic reticulum (ER), mitochondrion, chloroplast, and peroxisome. Biochemical and genetic studies, for instance, have identified cis-acting signal sequences responsible for targeting proteins to the correct organelle membrane and the membrane receptors that recognize these signal sequences. We also have learned much about the underlying mechanisms that translocate proteins across organelle membranes, and have determined whether energy is used to push or pull proteins across the membrane in one direction, the type of channel through which proteins pass, and whether proteins are translocated in a folded or an unfolded state. Nonetheless, many fundamental questions remain unanswered, including how fully folded proteins move across a membrane and how the topology of multipass membrane proteins is determined.

The peroxisomal import machinery provides one example of the translocation of folded proteins. It not only is capable of translocating fully folded proteins with bound cofactors into the peroxisomal matrix but can even direct the import of a large gold particle decorated with a (PTS1) peroxisomal targeting peptide. Some researchers have speculated that the mechanism of peroxisomal import may be related to that of nuclear import, the best-understood example of post-translational translocation of folded proteins (Chapter 12). Both the peroxisomal and nuclear import machinery can transport folded molecules of very divergent sizes, and both appear to involve a component that cycles between the cytosol and the organelle interior—the Pex5 PTS1 receptor in the case of peroxisomal import and the Ran-importin complex in the case of nuclear import. However, there also appear to be crucial differences between the two translocation processes. For example, nuclear pores represent large stable macromolecular assemblies readily observed by electron microscopy, whereas analogous porelike structures have not been observed in the peroxisomal membrane.

Moreover, small molecules can readily pass through nuclear pores, whereas peroxisomal membranes maintain a permanent barrier to the diffusion of small hydrophilic molecules. Taken together, these observations suggest that peroxisomal import may require an entirely new type of translocation mechanism. The evolutionarily conserved mechanisms for translocating folded proteins across the cytoplasmic membrane of bacterial cells and across the thylakoid membrane of chloroplasts also are poorly understood. A better understanding of all of these processes for translocating folded proteins across a membrane will likely hinge on future development of in vitro translocations systems that allow investigators to define the biochemical mechanisms driving translocation and to identify the structures of trapped translocation intermediates.

Compared with our understanding of how soluble proteins are translocated into the ER lumen and mitochondrial matrix, our understanding of how cis-acting sequences specify the topology of multipass membrane proteins is quite elementary. For instance, we do not know how the translocon channel accommodates polypeptides that are oriented differently with respect to the membrane, nor do we understand how local polypeptide sequences interact with the translocon channel both to set the orientation of transmembrane spans and to signal for lateral passage into the membrane bilayer. A better understanding of how the amino acid sequences of membrane proteins can specify membrane topology will be crucial for decoding the vast amount of structural information for membrane proteins contained within databases of genomic sequences.

A more detailed understanding of all translocation processes should continue to emerge from genetic and biochemical studies, both in yeasts and in mammals. These studies will undoubtedly reveal additional key proteins involved in the recognition of targeting sequences and in the translocation of proteins across lipid bilayers. Finally, the now mostly rudimentary structural studies of translocon channels will likely be extended in the future to reveal the structures and conformational states for the channels at resolutions on the atomic scale.

KEY TERMS

chaperones *665*

cotranslational translocation *661*

general import pore *685*

hydropathy profile *671*

N- and O-linked oligosaccharides *673*

post-translational translocation *665*

signal-anchor sequence *668*

signal-recognition particle (SRP) *661*

signal (uptake-targeting) sequences *659*

stop-transfer anchor sequence *667*

topogenic sequences *666*

topology of membrane proteins *666*

translocon *662*

unfolded-protein response *679*

REVIEW THE CONCEPTS

1. Describe the source or sources of energy needed for unidirectional translocation across the membrane in (a) cotranslational translocation into the endoplasmic reticulum (ER); (b) post-translational translocation into the ER; (c) translocation across the bacterial cytoplasmic membrane; and (d) translocation into the mitochondrial matrix.

2. Translocation into most organelles usually requires the activity of one or more cytosolic proteins. Describe the basic function of three different cytosolic factors required for translocation into the ER, mitochondria, and peroxisomes, respectively.

3. Describe the typical principles used to identify topogenic sequences within proteins and how these can be used to develop computer algorithms. How does the identification of topogenic sequences lead to prediction of the membrane arrangement of a multipass protein? What is the importance of the arrangement of positive charges relative to the membrane orientation of a signal anchor sequence?

4. The endoplasmic reticulum (ER) is an important site of "quality control" for newly synthesized proteins. What is meant by quality control in this context? What accessory proteins are typically involved in the processing of newly synthesized proteins within the ER? Cells generally degrade ER-exit-incompetent proteins. Where within the cell does such degradation occur and what is the relationship of Sec61p protein translocon to the degradation process?

5. Temperature-sensitive yeast mutants have been isolated that block each of the enzymatic steps in the synthesis of the dolichol-oligosaccharide precursor for *N*-linked glycosylation (see Figure 16-17). Propose an explanation for why mutations that block synthesis of the intermediate with the structure dolichol-PP-$(GlcNAc)_2Man_5$ completely prevent addition of *N*-linked oligosaccharide chains to secretory proteins, whereas mutations that block conversion of this intermediate into the completed precursor—dolichol-PP-$(GlcNAc)_2Man_9Glc_3$—allow the addition of *N*-linked oligosaccharide chains to secretory glycoproteins.

6. Name four different proteins that facilitate the modification and/or folding of secretory proteins within the lumen of the ER. Indicate which of these proteins covalently modifies substrate proteins and which brings about only conformational changes in substrate proteins.

7. Because you are interested in studying how a particular secretory protein folds within the ER, you wish to determine whether BiP binds to the newly synthesized protein in ER extracts. You find that you can isolate some of the newly synthesized secretory protein bound to BiP when ADP is added to the cell extract but not when ATP is added to the extract. Explain this result based on the mechanism for BiP binding to substrate proteins.

8. Describe how you might use recombinant DNA to engineer a strain of *Yersinia* such that it would be capable of inserting a protein of interest into the cytosol of mammalian macrophage cells.

9. Describe what would happen to the precursor of a mitochondrial matrix protein in the following types of mitochondrial mutants: (a) a mutation in the Tom22 signal receptor; (b) a mutation in the Tom70 signal receptor; (c) a mutation in the matrix Hsc70; and (d) a mutation in the matrix signal peptidase.

10. Describe the similarities and differences between the mechanism of import into the mitochondrial matrix and the chloroplast stroma.

11. Design a set of experiments using chimeric proteins, composed of a mitochondrial precursor protein fused to dihydrofolate reductase (DHFR), that could be used to determine how much of the precursor protein must protrude into the mitochondrial matrix in order for the matrix-targeting sequence to be cleaved by the matrix-processing protease (see Figure 16-27).

12. Protein targeting to both mitochondria and chloroplasts involves the sorting of proteins to multiple sites within the respective organelle. Briefly list these sites. Taking the mitochondrion as an example and the proteins ADP/ATP antiporter and cytochrome b_2 as the specific cases, compare and contrast the extent to which a common mechanism is used for the site-specific targeting of these two proteins.

13. Suppose that you have identified a new mutant cell line that lacks functional peroxisomes. Describe how you could determine experimentally whether the mutant is primarily defective for insertion/assembly of peroxisomal membrane proteins or matrix proteins.

ANALYZE THE DATA

Imagine that you are evaluating the early steps in translocation and processing of the secretory protein prolactin. By using an experimental approach similar to that shown in Figure 16-7, you can use truncated prolactin mRNAs to control the length of nascent prolactin polypeptides that are synthesized. When prolactin mRNA that lacks a chain-termination (stop) codon is translated in vitro, the newly synthesized polypeptide ending with the last codon included on the mRNA will remain attached to the ribosome, thus allowing a polypeptide of defined length to extend from the ribosome. You have generated a set of mRNAs that encode segments of the N-terminus of prolactin of increasing length, and each mRNA can be translated in vitro by a cytosolic translation extract containing ribosomes, tRNAs, aminoacyl-tRNA synthetases, GTP, and translation initiation and elongation factors. When radio-labeled amino acids are included in the translation mixture, only the polypeptide encoded by the added mRNA will be labeled. After completion of translation, each reaction mixture was resolved by SDS polyacrylamide gel electrophoresis, and the labeled polypeptides were identified by autoradiography.

a. The autoradiogram depicted below shows the results of an experiment in which each translation reaction was carried out either in the presence (+) or the absence (−) of microsomal membranes. Based on the gel mobility of peptides synthesized in the presence or absence of microsomes, deduce how long the prolactin nascent chain must be in order for the prolactin signal peptide to enter the ER lumen and to be cleaved by signal peptidase. (Note that microsomes carry significant quantities of SRP weakly bound to the membranes.)

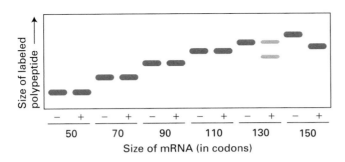

b. Given this length, what can you conclude about the conformational state of the nascent prolactin polypeptide when it is cleaved by signal peptidase? The following lengths will be useful for your calculation: the prolactin signal sequence is cleaved after amino acid 31; the channel within the ribosome occupied by a nascent polypeptide is about 150 Å long; a membrane bilayer is about 50 Å thick; in polypeptides with an α-helical conformation, one residue extends 1.5 Å, whereas in fully extended polypeptides, one residue extends about 3.5 Å.

c. The experiment described in part (a) is carried out in an identical manner except that microsomal membranes are not present during translation but are added after translation is complete. In this case none of the samples shows a difference in mobility in the presence or absence of microsomes. What can you conclude about whether prolactin can be translocated into isolated microsomes post-translationally?

d. In another experiment, each translation reaction was carried out in the presence of microsomes, and then the microsomal membranes and bound ribosomes were separated from free ribosomes and soluble proteins by centrifugation. For each translation reaction, both the total reaction (T) and the membrane fraction (M) were resolved in neighboring gel lanes. Based on the amounts of labeled polypeptide in the membrane fractions in the autoradiogram depicted below, deduce how long the prolactin nascent chain must be in order for ribosomes engaged in

translation to engage the SRP and thereby become bound to microsomal membranes.

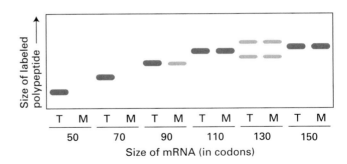

REFERENCES

Translocation of Proteins Across the ER Membrane

Dalbey, R. E., and G. Von Heijne. 1992. Signal peptidases in prokaryotes and eukaryotes: a new protease family. *Trends Biochem. Sci.* **17**:474–478.

Johnson, A. E., and M. A. van Waes. 1999. The translocon: a dynamic gateway at the ER membrane. *Ann. Rev. Cell Devel. Biol.* **15**:799–842.

Keenan, R. J., D. M. Freymann, R. M. Stroud, and P. Walter. 2001. The signal recognition particle. *Ann. Rev. Biochem.* **70**:755–775.

Menetret, J. F., A. Neuhof, D. G. Morgan, K. Plath, M. Radermacher, T. A. Rapoport, and C. W. Akey. 2000. The structure of ribosome-channel complexes engaged in protein translocation. *Mol. Cell* **6**:1219–1232.

Rapoport, T. A., K. E. Matlack, K. Plath, B. Misselwitz, and O. Staeck. 1999. Post-translational protein translocation across the membrane of the endoplasmic reticulum. *Biol. Chem.* **380**: 1143–1150.

Insertion of Proteins into the ER Membrane

Englund, P. T. 1993. The structure and biosynthesis of glycosylphosphatidylinositol protein anchors. *Ann. Rev. Biochem.* **62**:121–138.

Goder, V., and M. Spiess. 2001. Topogenesis of membrane proteins: determinants and dynamics. *FEBS Lett.* **504**:87–93.

Mothes, W., et al. 1997. Molecular mechanism of membrane protein integration into the endoplasmic reticulum. *Cell* **89**:523–533.

von Heijne, G. 1999. Recent advances in the understanding of membrane protein assembly and structure. *Q. Rev. Biophys.* **32**:285–307.

Protein Modifications, Folding, and Quality Control in the ER

Bertolotti, A., Y. Zhang, L. M. Hendershot, H. P. Harding, and D. Ron. 2000. Dynamic interaction of BiP and ER stress transducers in the unfolded-protein response. *Nature Cell Biol.* **2**:326–332.

Ellgaard, L., M. Molinari, and A. Helenius. 1999. Setting the standards: quality control in the secretory pathway. *Science* **286**:1882–1888.

Helenius, A., and M. Aebi. 2001. Intracellular functions of N-linked glycans. *Science* **291**:2364–2369.

Kornfeld, R., and S. Kornfeld. 1985. Assembly of asparagine-linked oligosaccharides. *Ann. Rev. Biochem.* **45**:631–664.

Parodi, A. J. 2000. Protein glucosylation and its role in protein folding. *Ann. Rev. Biochem.* **69**:69–93.

Patil, C., and P. Walter. 2001. Intracellular signaling from the endoplasmic reticulum to the nucleus: the unfolded protein response in yeast and mammals. *Curr. Opin. Cell Biol.* **13**:349–355.

Plemper, R. K., and D. H. Wolf. 1999. Retrograde protein translocation: eradication of secretory proteins in health and disease. *Trends Biochem. Sci.* **24**:266–270.

Sevier, C. S., and C. A. Kaiser. 2002. Formation and transfer of disulphide bonds in living cells. *Nature Rev. Mol. Cell Biol.* **3**:836–847.

Silberstein, S., and R. Gilmore. 1996. Biochemistry, molecular biology, and genetics of the oligosaccharyltransferase. *FASEB J.* **10**:849–858.

Tsai, B., Y. Ye, and T. A. Rapoport. 2002. Retro-translocation of proteins from the endoplasmic reticulum into the cytosol. *Nature Rev. Mol. Cell Biol.* **3**:246–255.

Export of Bacterial Proteins

Danese, P. N., and T. J. Silhavy. 1998. Targeting and assembly of periplasmic and outer-membrane proteins in *Escherichia coli. Ann. Rev. Genet.* **32**:59–94.

Galan, J. E. 2001. Salmonella interactions with host cells: type III secretion at work. *Ann. Rev. Cell Devel. Biol.* **17**:53–86.

Thanassi, D. G., and S. J. Hultgren. 2000. Multiple pathways allow protein secretion across the bacterial outer membrane. *Curr. Opin. Cell Biol.* **12**:420–430.

Wickner, W., and M. R. Leonard. 1996. *Escherichia coli* pre-protein translocase. *J. Biol. Chem.* **271**:29514–29516.

Sorting of Proteins to Mitochondria and Chloroplasts

Chen, K., X. Chen, and D. J. Schnell. 2000. Mechanism of protein import across the chloroplast envelope. *Biochem. Soc. Trans.* **28**:485–491.

Dalbey, R. E., and A. Kuhn. 2000. Evolutionarily related insertion pathways of bacterial, mitochondrial, and thylakoid membrane proteins. *Ann. Rev. Cell Devel. Biol.* **16**:51–87.

Matouschek, A., N. Pfanner, and W. Voos. 2000. Protein unfolding by mitochondria: the Hsp70 import motor. *EMBO Rept.* **1**:404–410.

Neupert, W., and M. Brunner. 2002. The protein import motor of mitochondria. *Nature Rev. Mol. Cell Biol.* **3**:555–565.

Pfanner, N., and A. Geissler. 2001. Versatility of the mitochondrial protein import machinery. *Nature Rev. Mol. Cell Biol.* **2**:339–349.

Robinson, C., and A. Bolhuis. 2001. Protein targeting by the twin-arginine translocation pathway. *Nature Rev. Mol. Cell Biol.* **2**:350–356.

Sorting of Peroxisomal Proteins

Dammai, V., and S. Subramani. 2001. The human peroxisomal targeting signal receptor, Pex5p, is translocated into the peroxisomal matrix and recycled to the cytosol. *Cell* **105**:187–196.

Gould, S. J., and C. S. Collins. 2002. Opinion: peroxisomal-protein import: is it really that complex? *Nature Rev. Mol. Cell Biol.* **3**:382–389.

Gould, S. J., and D. Valle. 2000. Peroxisome biogenesis disorders: genetics and cell biology. *Trends Genet.* **16**: 340–345.

Purdue, P. E., and P. B. Lazarow. 2001. Peroxisome biogenesis. *Ann. Rev. Cell Devel. Biol.* **17**:701–752.

Subramani, S., A. Koller, and W. B. Snyder. 2000. Import of peroxisomal matrix and membrane proteins. *Ann. Rev. Biochem.* **69**:399–418.

17

VESICULAR TRAFFIC, SECRETION, AND ENDOCYTOSIS

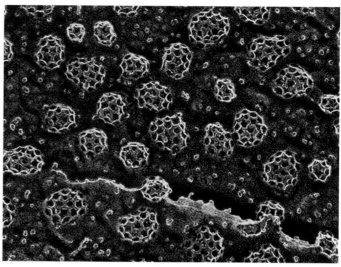

Electron micrograph of clathrin cages, like those that surround clathrin-coated transport vesicles, formed by the in vitro polymerization of clathrin heavy and light chains. [John Heuser, Washington University School of Medicine.]

In the previous chapter we explored how proteins are targeted to and translocated across the membranes of different intracellular organelles. In this chapter we turn our attention to the mechanisms that allow soluble and membrane proteins synthesized on the rough endoplasmic reticulum (ER) to move to their final destinations via the **secretory pathway**. A single unifying principle governs all protein trafficking in the secretory pathway: transport of membrane and soluble proteins from one membrane-bounded compartment to another is mediated by **transport vesicles** that collect *"cargo" proteins* in buds arising from the membrane of one compartment and then deliver these cargo proteins to the next compartment by fusing with the membrane of that compartment. Importantly, as transport vesicles bud from one membrane and fuse with the next, the same face of the membrane remains oriented toward the cytosol. Therefore once a protein has been inserted into the membrane or the lumen of the ER, the protein can be carried along the secretory pathway, moving from one organelle to the next without being translocated across another membrane or altering its orientation within the membrane.

Figure 17-1 outlines the major routes for protein trafficking in the secretory pathway. Once newly synthesized proteins are incorporated into the ER lumen or membrane as discussed in Chapter 16, they can be packaged into *anterograde* (forward-moving) transport vesicles. These vesicles fuse with each other to form a flattened membrane-bounded compartment known as the *cis*-Golgi **cisterna**. Certain proteins, mainly ER-localized proteins, are retrieved from the *cis*-Golgi to the ER via a different set of *retrograde* (backward-moving) transport vesicles. A new *cis*-Golgi cisterna with its cargo of

proteins physically moves from the *cis* position (nearest the ER) to the *trans* position (farthest from the ER), successively becoming first a *medial*-Golgi cisterna and then a *trans*-Golgi cisterna. This process, known as cisternal progression, does not involve the budding off and fusion of anterograde transport vesicles. During cisternal progression, enzymes and other Golgi-resident proteins are constantly being retrieved from later to earlier Golgi cisternae by retrograde transport vesicles, thereby remaining localized to the *cis*-, *medial*-, or *trans*-Golgi cisternae.

Proteins in the secretory pathway that are destined for compartments other than the ER or Golgi eventually reach a complex network of membranes and vesicles termed the ***trans*-Golgi network (TGN)**. From this major branch point in the secretory pathway, a protein can be loaded into one

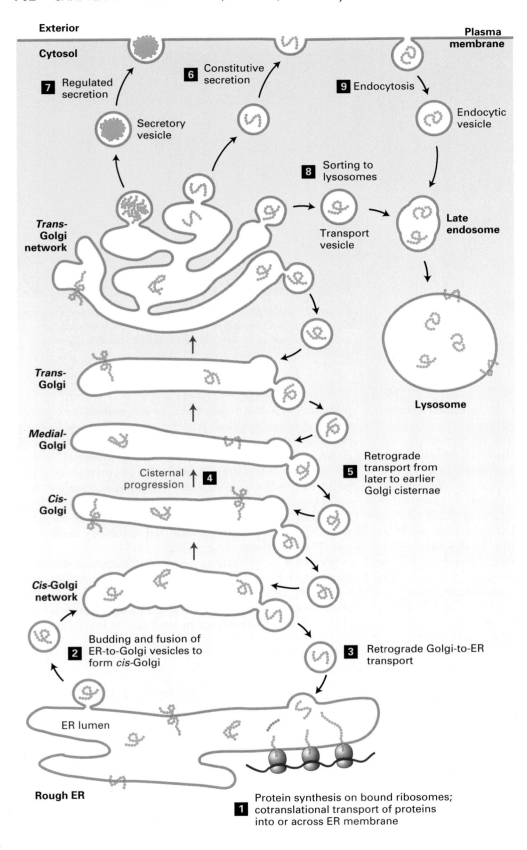

◄ **FIGURE 17-1 Overview of the secretory and endocytic pathways of protein sorting.** *Secretory pathway:* Synthesis of proteins bearing an ER signal sequence is completed on the rough ER (**1**), and the newly made polypeptide chains are inserted into the ER membrane or cross it into the lumen (Chapter 16). Some proteins (e.g., ER enzymes or structural proteins) remain within the ER. The remainder are packaged into transport vesicles (**2**) that bud from the ER and fuse together to form new *cis*-Golgi cisternae. Missorted ER-resident proteins and vesicle membrane proteins that need to be reused are retrieved to the ER by vesicles (**3**) that bud from the *cis*-Golgi and fuse with the ER. Each *cis*-Golgi cisterna, with its protein content, physically moves from the *cis* to the *trans* face of the Golgi complex (**4**) by a nonvesicular process called cisternal progression. Retrograde transport vesicles (**5**) move Golgi-resident proteins to the proper Golgi compartment. In all cells, certain soluble proteins move to the cell surface in transport vesicles (**6**) and are secreted continuously (constitutive secretion). In certain cell types, some soluble proteins are stored in secretory vesicles (**7**) and are released only after the cell receives an appropriate neural or hormonal signal (regulated secretion). Lysosome-destined membrane and soluble proteins, which are transported in vesicles that bud from the *trans*-Golgi (**8**), first move to the late endosome and then to the lysosome. *Endocytic pathway:* Membrane and soluble extracellular proteins taken up in vesicles that bud from the plasma membrane (**9**) also can move to the lysosome via the endosome.

Diagram labels:

Exterior

Cytosol

Plasma membrane

7 Regulated secretion

6 Constitutive secretion

9 Endocytosis

Secretory vesicle

Endocytic vesicle

8 Sorting to lysosomes

Trans-Golgi network

Transport vesicle

Late endosome

Trans-Golgi

Lysosome

Medial-Golgi

Cisternal progression **4**

Retrograde transport from later to earlier Golgi cisternae **5**

Cis-Golgi

Cis-Golgi network

Budding and fusion of ER-to-Golgi vesicles to form *cis*-Golgi **2**

Retrograde Golgi-to-ER transport **3**

ER lumen

Rough ER

Protein synthesis on bound ribosomes; cotranslational transport of proteins into or across ER membrane **1** of at least three different kinds of vesicles. After budding from the *trans*-Golgi network, the first type of vesicle immediately moves to and fuses with the plasma membrane, releasing its contents by **exocytosis**. In all cell types, at least some proteins are loaded into such vesicles and secreted continuously in this manner. Examples of proteins released by such constitutive (or continuous) secretion include collagen by fibroblasts, serum proteins by hepatocytes, and antibodies by activated B lymphocytes. The second type of vesicle to bud from the *trans*-Golgi network, known as **secretory vesicles**, are stored inside the cell until a signal for exocytosis causes release of their contents at the plasma membrane. Among the proteins released by such regulated secretion are peptide hormones (e.g., insulin, glucagon, ACTH) from various endocrine cells, precursors of digestive enzymes from pancreatic acinar cells, milk proteins from the mammary gland, and neurotransmitters from neurons.

The third type of vesicle that buds from the *trans*-Golgi network is directed to the **lysosome**, an organelle responsible for the intracellular degradation of macromolecules, and to lysosome-like storage organelles in certain cells. Secretory proteins destined for lysosomes first are transported by vesicles from the *trans*-Golgi network to a compartment usually called the **late endosome**; proteins then are transferred to the lysosome by a mechanism that is not well understood but may involve direct fusion of the endosome with the lysosomal membrane. Soluble proteins delivered by this pathway include lysosomal digestive enzymes (e.g., proteases, glycosidases, and phosphatases) and membrane proteins (e.g., V-class proton pump) that pump H^+ from the cytosol into the acidic lumen of the endosome and lysosome. As we will see, some of the specific protein-processing and -sorting events that take place within these organelles depend on their low luminal pH.

The endosome also functions in the *endocytic pathway* in which vesicles bud from the plasma membrane bringing membrane proteins and their bound **ligands** into the cell (see Figure 17-1). After being internalized by **endocytosis**, some proteins are transported to lysosomes, while others are recycled back to the cell surface. Endocytosis is a way for cells to take up nutrients that are in macromolecular form—for example, cholesterol in the form of lipoprotein particles and iron complexed with the serum protein transferrin. Endocytosis also can function as a regulatory mechanism to decrease signaling activity by withdrawing receptors for a particular signaling molecule from the cell surface.

17.1 Techniques for Studying the Secretory Pathway

The key to understanding how proteins are transported through the organelles of the secretory pathway has been to develop a basic description of the function of transport vesicles. Many components required for the formation and fu-

sion of transport vesicles have been identified in the past decade by a remarkable convergence of the genetic and biochemical approaches described in this section. All studies of intracellular protein trafficking employ some method for assaying the transport of a given protein from one compartment to another. We begin by describing how intracellular protein transport can be followed in living cells and then consider genetic and in vitro systems that have proved useful in elucidating the secretory pathway.

Transport of a Protein Through the Secretory Pathway Can Be Assayed in Living Cells

The classic studies of G. Palade and his colleagues in the 1960s first established the order in which proteins move from organelle to organelle in the secretory pathway. These early studies also showed that secretory proteins were never released into the cytosol, the first indication that transported proteins are associated with some type of membrane-bounded intermediate. In these experiments, which combined **pulse-chase** labeling (see Figure 3-36) and **autoradiography**, radioactively labeled amino acids were injected into the pancreas of a hamster. At different times after injection, the animal was sacrificed and the pancreatic cells were chemically fixed, sectioned, and subjected to autoradiography to visualize the location of the radiolabeled proteins. Because the radioactive amino acids were administered in a short pulse, only those proteins synthesized immediately after injection were labeled, forming a distinct group, or cohort, of labeled proteins whose transport could be followed. In addition, because pancreatic acinar cells are dedicated secretory cells, almost all of the labeled amino acids in these cells are incorporated into secretory proteins, facilitating the observation of transported proteins.

Although autoradiography is rarely used today to localize proteins within cells, these early experiments illustrate the two basic requirements for any assay of intercompartmental transport. First, it is necessary to label a cohort of proteins in an early compartment so that their subsequent transfer to later compartments can be followed with time. Second, it is necessary to have a way to identify the compartment in which a labeled protein resides. Here we describe two modern experimental procedures for observing the intracellular trafficking of a secretory protein in almost any type of cell.

In both procedures, a gene encoding an abundant membrane glycoprotein (G protein) from vesicular stomatitis virus (VSV) is introduced into cultured mammalian cells either by **transfection** or simply by infecting the cells with the virus. The treated cells, even those that are not specialized for secretion, rapidly synthesize the VSV G protein on the ER like normal cellular secretory proteins. Use of a mutant encoding a temperature-sensitive VSV G protein allows researchers to turn subsequent protein transport on and off. At the restrictive temperature of 40 °C, newly made VSV G protein is misfolded and therefore retained within the ER by quality control mechanisms discussed in Chapter 16, whereas at the permissive temperature of 32 °C, the accumulated protein is correctly folded and is transported through the secretory pathway to the cell surface. This clever use of a temperature-sensitive mutation in effect defines a protein cohort whose subsequent transport can be followed.

In two variations of this basic procedure, transport of VSV G protein is monitored by different techniques. Studies using both of these modern trafficking assays and Palade's early experiments all came to the same conclusion: in mammalian cells vesicle-mediated transport of a protein molecule from its site of synthesis on the rough ER to its arrival at the plasma membrane takes from 30 to 60 minutes.

Microscopy of GFP-Labeled VSV G Protein One approach for observing transport of VSV G protein employs a hybrid gene in which the viral gene is fused to the gene encoding *green fluorescent protein (GFP)*, a naturally fluorescent protein (Chapter 5). The hybrid gene is transfected into cultured cells by techniques described in Chapter 9. When cells expressing the temperature-sensitive form of the hybrid protein (VSVG-GFP) are grown at the restrictive temperature, VSVG-GFP accumulates in the ER, which appears as a lacy network of membranes when cells are observed in a fluorescent microscope. When the cells are subsequently shifted to a permissive temperature, the VSVG-GFP can be seen to move first to the membranes of the Golgi apparatus, which are densely concentrated at the edge of the nucleus, and then to the cell surface (Figure 17-2a). By analyzing the distribution of VSVG-GFP at different times after shifting cells to the permissive temperature, researchers have determined how long VSVG-GFP resides in each organelle of the secretory pathway (Figure 17-2b).

Detection of Compartment-Specific Oligosaccharide Modifications A second way to follow the transport of secretory proteins takes advantage of modifications to their carbohydrate side chains that occur at different stages of the secretory pathway. To understand this approach, recall that many secretory proteins leaving the ER contain one or more copies of the **N-linked oligosaccharide** $Man_8(GlcNAc)_2$, which are synthesized and attached to secretory proteins in the ER (see Figure 16-18). As a protein moves through the Golgi complex, different enzymes localized to the *cis-*, *medial-*, and *trans-*Golgi cisternae catalyze an ordered series of reactions to these core $Man_8(GlcNAc)_2$ chains. For instance, glycosidases that reside specifically in the *cis*-Golgi compartment sequentially trim mannose residues off of the core oligosaccharide to yield a "trimmed" form $Man_5(GlcNAc)_2$ (Figure 17-3, reaction 1). Scientists can use a specialized carbohydrate-cleaving enzyme known as endoglycosidase D to distinguish glycosylated proteins that remain in the ER from those that have entered the *cis*-Golgi: trimmed *cis*-Golgi–specific oligosaccharides are cleaved from proteins by endoglycosidase D, whereas the core (untrimmed) oligosaccharide chains on secretory proteins within the ER are resistant to cleavage by this enzyme. Because a deglycosylated protein produced by endoglycosidase D digestion moves faster on an SDS gel than the corresponding glycosylated protein, they can be readily distinguished.

This type of assay can be used to track movement of VSV G protein in virus-infected cells pulse-labeled with radioactive amino acids. Immediately after labeling, all the extracted labeled VSV G protein is still in the ER and is resistant to digestion by endoglycosidase D, but with time an increasing fraction of the glycoprotein becomes sensitive to digestion

(a)

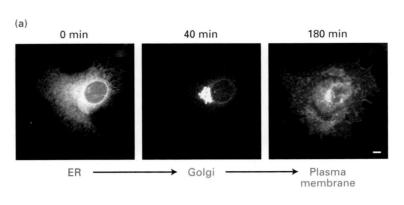

0 min   40 min   180 min

ER ⟶ Golgi ⟶ Plasma membrane

(b)

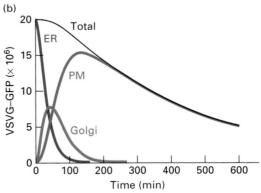

▲ **EXPERIMENTAL FIGURE 17-2 Protein transport through the secretory pathway can be visualized by fluorescence microscopy of cells producing a GFP-tagged membrane protein.** Cultured cells were transfected with a hybrid gene encoding the viral membrane glycoprotein VSV G protein linked to the gene for green fluorescent protein (GFP). A mutant version of the viral gene was used so that newly made hybrid protein (VSVG-GFP) is retained in the ER at 40 °C but is released for transport at 32 °C. (a) Fluorescence micrographs of cells just before and at two times after

they were shifted to the lower temperature. Movement of VSVG-GFP from the ER to the Golgi and finally to the cell surface occurred within 180 minutes. (b) Plot of the levels of VSVG-GFP in the endoplasmic reticulum (ER), Golgi, and plasma membrane (PM) at different times after shift to lower temperature. The kinetics of transport from one organelle to another can be reconstructed from computer analysis of these data. The decrease in total fluorescence that occurs at later times probably results from slow inactivation of GFP fluorescence. [From Jennifer Lippincott-Schwartz and Koret Hirschberg, Metabolism Branch, National Institute of Child Health and Human Development.]

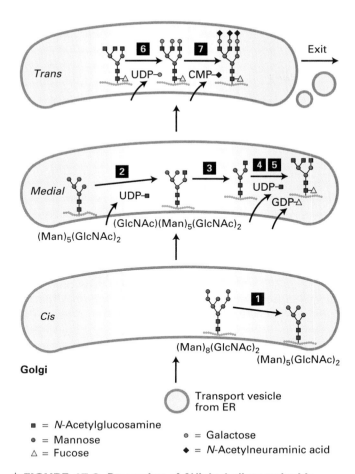

= N-Acetylglucosamine
• = Mannose
△ = Fucose
○ = Galactose
◆ = N-Acetylneuraminic acid

▲ **FIGURE 17-3 Processing of N-linked oligosaccharide chains on glycoproteins within cis-, medial-, and trans-Golgi cisternae in vertebrate cells.** The enzymes catalyzing each step are localized to the indicated compartments. After removal of three mannose residues in the cis-Golgi (step **1**), the protein moves by cisternal progression to the medial-Golgi. Here, three GlcNAc residues are added (steps **2** and **4**), two more mannose residues are removed (step **3**), and a single fucose is added (step **5**). Processing is completed in the trans-Golgi by addition of three galactose residues (step **6**) and finally by linkage of an N-acetylneuraminic acid residue to each of the galactose residues (step **7**). Specific transferase enzymes add sugars to the oligosaccharide, one at a time, from sugar nucleotide precursors imported from the cytosol. This pathway represents the Golgi processing events for a typical mammalian glycoprotein. Variations in the structure of N-linked oligosaccharides can result from differences in processing steps in the Golgi. [See R. Kornfeld and S. Kornfeld, 1985, *Ann. Rev. Biochem.* **45**:631.]

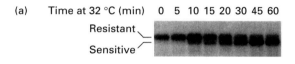

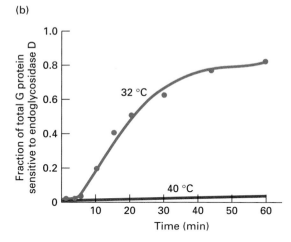

▲ **EXPERIMENTAL FIGURE 17-4 Transport of a membrane glycoprotein from the ER to the Golgi can be assayed based on sensitivity to cleavage by endoglycosidase D.** Cells expressing a temperature-sensitive VSV G protein (VSVG) were labeled with a pulse of radioactive amino acids at the nonpermissive temperature so that labeled protein was retained in the ER. At periodic times after a return to the permissive temperature of 32 °C, VSVG was extracted from cells and digested with endoglycosidase D, which cleaves the oligosaccharide chains from proteins processed in the cis-Golgi but not from proteins in the ER. (a) SDS gel electrophoresis of the digestion mixtures resolves the resistant, uncleaved (slower migrating) and sensitive, cleaved (faster migrating) forms of labeled VSVG. As this electrophoretogram shows, initially all of the VSVG was resistant to digestion, but with time an increasing fraction is sensitive to digestion, reflecting protein transported from the ER to the Golgi and processed there. In control cells kept at 40 °C, only slow-moving, digestion-resistant VSVG was detected after 60 minutes (not shown). (b) Plot of the proportion of VSVG that is sensitive to digestion, derived from electrophoretic data, reveals the time course of ER → Golgi transport. [From C. J. Beckers et al., 1987, *Cell* **50**:523.]

Yeast Mutants Define Major Stages and Many Components in Vesicular Transport

The general organization of the secretory pathway and many of the molecular components required for vesicle trafficking are similar in all eukaryotic cells. Because of this conservation, genetic studies with yeast have been useful in confirming the sequence of steps in the secretory pathway and in identifying many of the proteins that participate in vesicular traffic. Although yeasts secrete few proteins into the growth medium, they continuously secrete a number of enzymes that remain

(Figure 17-4). This conversion of VSV G protein from an endoglycosidase D–resistant form to an endoglycosidase D–sensitive form corresponds to vesicular transport of the protein from the ER to the cis-Golgi. Note that transport of VSV G protein from the ER to the Golgi takes about 30 minutes as measured by either the assay based on oligosaccharide processing or fluorescence microscopy of VSVG-GFP.

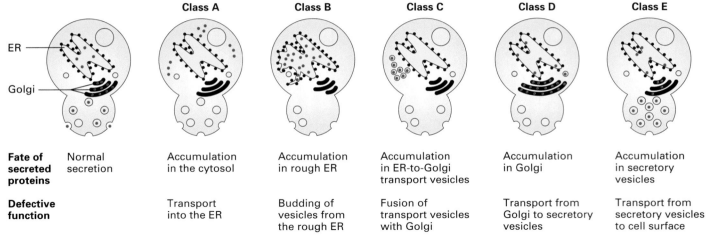

		Class A	Class B	Class C	Class D	Class E
Fate of secreted proteins	Normal secretion	Accumulation in the cytosol	Accumulation in rough ER	Accumulation in ER-to-Golgi transport vesicles	Accumulation in Golgi	Accumulation in secretory vesicles
Defective function		Transport into the ER	Budding of vesicles from the rough ER	Fusion of transport vesicles with Golgi	Transport from Golgi to secretory vesicles	Transport from secretory vesicles to cell surface

▲ **EXPERIMENTAL FIGURE 17-5 Phenotypes of yeast *sec* mutants identified stages in the secretory pathway.** These temperature-sensitive mutants can be grouped into five classes based on the site where newly made secreted proteins (red dots) accumulate when cells are shifted from the permissive temperature to the higher nonpermissive one. Analysis of double mutants permitted the sequential order of the steps to be determined. [See P. Novick et al., 1981, *Cell* **25**:461, and C. A. Kaiser and R. Schekman, 1990, *Cell* **61**:723.]

localized in the narrow space between the plasma membrane and the cell wall. The best-studied of these, invertase, hydrolyzes the disaccharide sucrose to glucose and fructose.

A large number of yeast mutants initially were identified based on their ability to secrete proteins at one temperature and inability to do so at a higher, nonpermissive temperature. When these temperature-sensitive *secretion (sec) mutants* are transferred from the lower to the higher temperature, they accumulate secreted proteins at the point in the pathway blocked by the mutation. Analysis of such mutants identified five classes (A–E) characterized by protein accumulation in the cytosol, rough ER, small vesicles taking proteins from the ER to the Golgi complex, Golgi cisternae, or constitutive secretory vesicles (Figure 17-5). Subsequent characterization of *sec* mutants in the various classes has helped elucidate the fundamental components and molecular mechanisms of vesicle trafficking that we discuss in later sections.

To determine the order of the steps in the pathway, researchers analyzed double *sec* mutants. For instance, when yeast cells contain mutations in both class B and class D functions, proteins accumulate in the rough ER, not in the Golgi cisternae. Since proteins accumulate at the earliest blocked step, this finding shows that class B mutations must act at an earlier point in the secretory pathway than class D mutations do. These studies confirmed that as a secreted protein is synthesized and processed it moves sequentially from the cytosol → rough ER → ER-to-Golgi transport vesicles → Golgi cisternae → secretory vesicles and finally is exocytosed.

Cell-free Transport Assays Allow Dissection of Individual Steps in Vesicular Transport

In vitro assays for intercompartmental transport are powerful complementary approaches to studies with yeast *sec* mu-

tants for identifying and analyzing the cellular components responsible for vesicular trafficking. In one application of this approach, cultured mutant cells lacking one of the enzymes that modify N-linked oligosaccharide chains in the Golgi are infected with vesicular stomatitis virus (VSV). For example, if infected cells lack N-acetylglucosamine transferase I, they produce abundant amounts of VSV G protein but cannot add N-acetylglucosamine residues to the oligosaccharide chains in the *medial*-Golgi as wild-type cells do (Figure 17-6a). When Golgi membranes isolated from such mutant cells are mixed with Golgi membranes from wild-type, uninfected cells, the addition of N-acetylglucosamine to VSV G protein is restored (Figure 17-6b). This modification is the consequence of the retrograde vesicular transport of N-acetylglucosamine transferase I from the wild-type *medial*-Golgi to the *cis*-Golgi compartment from virally infected mutant cells. Successful intercompartmental transport in this cell-free system depends on requirements that are typical of a normal physiological process including a cytosolic extract, a source of chemical energy in the form of ATP and GTP, and incubation at physiological temperatures.

In addition, under appropriate conditions a uniform population of the retrograde transport vesicles that move N-acetylglucosamine transferase I from the *medial*- to *cis*-Golgi can be purified away from the donor wild-type Golgi membranes by centrifugation. By examining the proteins that are enriched in these vesicles, scientists have been able to identify many of the integral membrane proteins and peripheral vesicle coat proteins that are the structural components of this type of vesicle. Moreover, fractionation of the cytosolic extract required for transport in cell-free reaction mixtures has permitted isolation of the various proteins required for formation of transport vesicles and of proteins required for the targeting and fusion of vesicles with appropriate acceptor

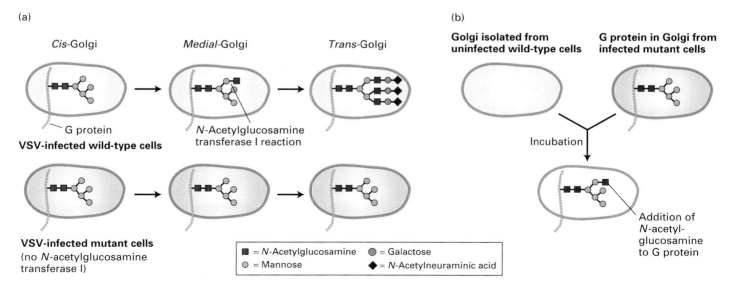

(a)

Cis-Golgi   *Medial*-Golgi   *Trans*-Golgi

G protein
VSV-infected wild-type cells

N-Acetylglucosamine transferase I reaction

VSV-infected mutant cells
(no *N*-acetylglucosamine transferase I)

(b)

Golgi isolated from uninfected wild-type cells   **G protein in Golgi from infected mutant cells**

Incubation

Addition of *N*-acetylglucosamine to G protein

■ = *N*-Acetylglucosamine   ● = Galactose
● = Mannose   ◆ = *N*-Acetylneuraminic acid

▲ **EXPERIMENTAL FIGURE 17-6 Protein transport from one Golgi cisternae to another can be assayed in a cell-free system.** (a) A mutant line of cultured fibroblasts is essential in this type of assay. In this example, the cells lack the enzyme *N*-acetylglucosamine transferase I (step **2** in Figure 17-3). In wild-type cells, this enzyme is localized to the *medial*-Golgi and modifies *N*-linked oligosaccharides by the addition of one *N*-acetylglucosamine. In VSV-infected wild-type cells, the oligosaccharide on the viral G protein is modified to a typical complex oligosaccharide, as shown in the *trans*-Golgi panel. In infected mutant cells, however, the G protein reaches the cell membranes. In vitro assays similar in general design to the one shown in Figure 17-6 have been used to study various transport steps in the secretory pathway.

surface with a simpler high-mannose oligosaccharide containing only two *N*-acetylglucosamine and five mannose residues. (b) When Golgi cisternae isolated from infected mutant cells are incubated with Golgi cisternae from normal, uninfected cells, the VSV G protein produced in vitro contains the additional *N*-acetylglucosamine. This modification is carried out by transferase enzyme that is moved by retrograde transport vesicles from the wild-type *medial*-Golgi cisternae to the mutant *cis*-Golgi cisternae in the reaction mixture. [See W. E. Balch et al., 1984, *Cell* **39**:405 and 525; W. A. Braell et al., 1984, *Cell* **39**:511; and J. E. Rothman and T. Söllner, 1997, *Science* **276**:1212.]

KEY CONCEPTS OF SECTION 17.1

Techniques for Studying the Secretory Pathway

■ All assays for following the trafficking of proteins through the secretory pathway in living cells require a way to label a cohort of secretory proteins and a way to identify the compartments where labeled proteins subsequently are located.

■ Pulse-labeling with radioactive amino acids can specifically label a cohort of newly made proteins in the ER. Alternatively, a temperature-sensitive mutant protein that is retained in the ER at the nonpermissive temperature will be released as a cohort for transport when cells are shifted to the permissive temperature.

■ Transport of a fluorescently labeled protein along the secretory pathway can be observed by microscopy (see Figure 17-2). Transport of a radiolabeled protein commonly is tracked by following compartment-specific covalent modifications to the protein.

■ Many of the components required for intracellular protein trafficking have been identified in yeast by analysis

of temperature-sensitive *sec* mutants defective for the secretion of proteins at the nonpermissive temperature (see Figure 17-5).

■ Cell-free assays for intercompartmental protein transport have allowed the biochemical dissection of individual steps of the secretory pathway. Such in vitro reactions can be used to produce pure transport vesicles and to test the biochemical function of individual transport proteins.

17.2 Molecular Mechanisms of Vesicular Traffic

Small membrane-bounded vesicles that transport proteins from one organelle to another are common elements in the secretory and endocytic pathways (see Figure 17-1). These vesicles bud from the membrane of a particular *"parent" (donor) organelle* and fuse with the membrane of a particular *"target" (destination) organelle*. Although each step in the secretory and endocytic pathways employs a different type of vesicle, studies employing genetic and biochemical techniques described in the previous section have revealed that each of the different vesicular transport steps is simply a variation on a common theme. In this section we explore that common theme, the basic mechanisms underlying vesicle budding and fusion.

(a) Coated vesicle budding

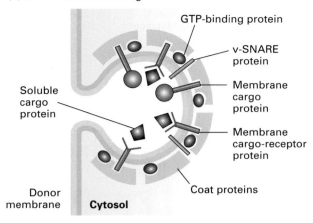

(b) Uncoated vesicle fusion

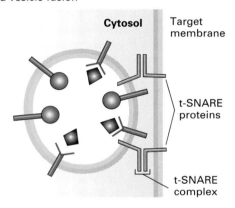

▲ **FIGURE 17-7 Overview of vesicle budding and fusion with a target membrane.** (a) Budding is initiated by recruitment of a small GTP-binding protein to a patch of donor membrane. Complexes of coat proteins in the cytosol then bind to the cytosolic domain of membrane cargo proteins, some of which also act as receptors that bind soluble proteins in the lumen, thereby recruiting luminal cargo proteins into the budding vesicle. (b) After being released and shedding its coat, a vesicle fuses with its target membrane in a process that involves interaction of cognate SNARE proteins.

The budding of vesicles from their parent membrane is driven by the polymerization of soluble protein complexes onto the membrane to form a proteinaceous vesicle coat (Figure 17-7a). Interactions between the cytosolic portions of integral membrane proteins and the vesicle coat gather the appropriate cargo proteins into the forming vesicle. Thus the coat not only adds curvature to the membrane to form a vesicle but also acts as the filter to determine which proteins are admitted into the vesicle.

The integral membrane proteins in a budding vesicle include **v-SNAREs,** which are crucial to eventual fusion of the vesicle with the correct target membrane. Shortly after formation of a vesicle is completed, the coat is shed exposing its v-SNARE proteins. The specific joining of v-SNAREs in

the vesicle membrane with cognate **t-SNAREs** in the target membrane brings the membranes into close apposition, allowing the two bilayers to fuse (Figure 17-7b).

Assembly of a Protein Coat Drives Vesicle Formation and Selection of Cargo Molecules

Three types of coated vesicles have been characterized, each with a different type of protein coat and each formed by reversible polymerization of a distinct set of protein subunits (Table 17-1). Each type of vesicle, named for its primary coat proteins, transports cargo proteins from particular parent organelles to particular destination organelles:

- **COPII** vesicles transport proteins from the rough ER to the Golgi.

- **COPI** vesicles mainly transport proteins in the retrograde direction between Golgi cisternae and from the *cis*-Golgi back to the rough ER.

- **Clathrin** vesicles transport proteins from the plasma membrane (cell surface) and the *trans*-Golgi network to late endosomes.

Researchers have not yet identified the coat proteins surrounding the vesicles that move proteins from the *trans*-Golgi to the plasma membrane during either constitutive or regulated secretion.

The general scheme of vesicle budding shown in Figure 17-7a applies to all three known types of coated vesicles. Experiments with isolated or artificial membranes and purified coat proteins have shown that polymerization of the coat proteins onto the cytosolic face of the parent membrane is necessary to produce the high curvature of the

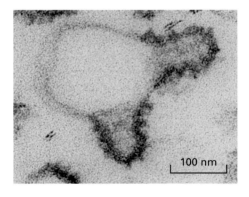

▲ **EXPERIMENTAL FIGURE 17-8 Vesicle buds can be visualized during in vitro budding reactions.** When purified COPII coat components are incubated with isolated ER vesicles or artificial phospholipid vesicles (liposomes), polymerization of the coat proteins on the vesicle surface induces emergence of highly curved buds. In this electron micrograph of an in vitro budding reaction, note the distinct membrane coat, visible as a dark protein layer, present on the vesicle buds. [From K. Matsuoka et al., 1988, *Cell* **93**(2):263.]

TABLE 17-1	Coated Vesicles Involved in Protein Trafficking		
Vesicle Type	**Coat Proteins**	**Associated GTPase**	**Transport Step Mediated**
COPII	Sec23/Sec24 and Sec13/Sec31 complexes, Sec16	Sar1	ER to *cis*-Golgi
COPI	Coatomers containing seven different COP subunits	ARF	*cis*-Golgi to ER Later to earlier Golgi cisternae
Clathrin and adapter proteins[*]	Clathrin + AP1 complexes	ARF	*trans*-Golgi to endosome
	Clathrin + GGA	ARF	*trans*-Golgi to endosome
	Clathrin + AP2 complexes	ARF	Plasma membrane to endosome
	AP3 complexes	ARF	Golgi to lysosome, melanosome, or platelet vesicles

[*]Each type of AP complex consists of four different subunits. It is not known whether the coat of AP3 vesicles contains clathrin.

membrane that is typical of a transport vesicle about 50 nm in diameter. Electron micrographs of in vitro budding reactions often reveal structures that exhibit discrete regions of the parent membrane bearing a dense coat accompanied by the curvature characteristic of a completed vesicle (Figure 17-8). Such structures, usually called *vesicle buds,* appear to be intermediates that are visible after the coat has begun to polymerize but before the completed vesicle pinches off from the parent membrane. The polymerized coat proteins are thought to form some type of curved lattice that drives the formation of a vesicle bud by adhering to the cytosolic face of the membrane.

A Conserved Set of GTPase Switch Proteins Controls Assembly of Different Vesicle Coats

Based on in vitro vesicle-budding reactions with isolated membranes and purified coat proteins, scientists have determined the minimum set of coat components required to form each of the three major types of vesicles. Although most of the coat proteins differ considerably from one type of vesicle to another, the coats of all three vesicles contain a small GTP-binding protein that acts as a regulatory subunit to control coat assembly (see Figure 17-7a). For both COPI and clathrin vesicles, this GTP-binding protein is known as *ARF.* A different but related GTP-binding protein known as *Sar1* is present in the coat of COPII vesicles. Both ARF and Sar1 are monomeric proteins with an overall structure similar to that of Ras, a key intracellular signal-transducing protein (see Figure 14-20). ARF and Sar1 proteins, like Ras, belong to the **GTPase superfamily** of switch proteins that cycle between inactive GDP-bound and active GTP-bound forms (see Figure 3-29).

The cycle of GTP binding and hydrolysis by ARF and Sar1 are thought to control the initiation of coat assembly

as schematically depicted for the assembly of COPII vesicles in Figure 17-9. First, an ER membrane protein known as Sec12 catalyzes release of GDP from cytosolic Sar1 · GDP and binding of GTP. The Sec12 guanine nucleotide–exchange factor apparently receives and integrates multiple, as yet unknown signals, probably including the presence of cargo proteins in the ER membrane that are ready to be transported. Binding of GTP causes a conformational change in Sar1 that exposes its hydrophobic N-terminus, which then becomes embedded in the phospholipid bilayer and tethers Sar1 · GTP to the ER membrane. The membrane-attached Sar1 · GTP drives polymerization of cytosolic complexes of COPII subunits on the membrane, eventually leading to formation of vesicle buds. Once COPII vesicles are released from the donor membrane, the Sar1 GTPase activity hydrolyzes Sar1 · GTP in the vesicle membrane to Sar1 · GDP with the assistance of one of the coat subunits. This hydrolysis triggers disassembly of the COPII coat. Thus Sar1 couples a cycle of GTP binding and hydrolysis to the formation and then dissociation of the COPII coat.

ARF protein undergoes a similar cycle of nucleotide exchange and hydrolysis coupled to the assembly of vesicle coats composed either of COPI or of clathrin and other coat proteins (AP complexes) discussed later. A myristate anchor covalently attached to the N-terminus of ARF protein weakly tethers ARF · GDP to the Golgi membrane. When GTP is exchanged for the bound GDP by a nucleotide-exchange factor attached to the Golgi membrane, the resulting conformational change in ARF allows hydrophobic residues in its N-terminal segment to insert into the membrane bilayer. The resulting tight association of ARF · GTP with the membrane serves as the foundation for further coat assembly.

Drawing on the structural similarities of Sar1 and ARF to other small GTPase switch proteins, researchers have

1 **Sar1 membrane binding, GTP exchange**

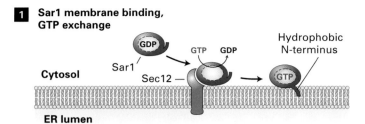

2 **COPII coat assembly**

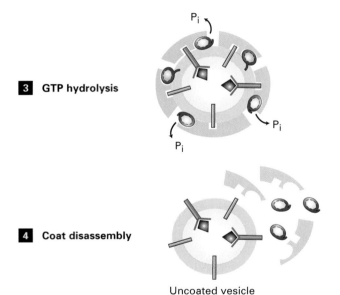

3 **GTP hydrolysis**

4 **Coat disassembly**

Uncoated vesicle

▲ **FIGURE 17-9 Model for the role of Sar1 in the assembly and disassembly of COPII coats.** Step **1**: Interaction of soluble GDP-bound Sar1 with the exchange factor Sec12, an ER integral membrane protein, catalyzes exchange of GTP for GDP on Sar1. In the GTP-bound form of Sar1, its hydrophobic N-terminus extends outward from the protein's surface and anchors Sar1 to the ER membrane. Step **2**: Sar1 attached to the membrane serves as a binding site for the the Sec23/Sec24 coat protein complex. Cargo proteins are recruited to the forming vesicle bud by binding of specific short sequences (sorting signals) in their cytosolic regions to sites on the Sec23/Sec24 complex. The coat is completed by assembly of a second type of coat complex composed of Sec13/and Sec31 (not shown). Step **3**: After the vesicle coat is complete, the Sec23 coat subunit promotes GTP hydrolysis by Sar1. Step **4**: Release of Sar1 · GDP from the vesicle membrane causes disassembly of the coat. [See S. Springer et al., 1999, *Cell* **97**:145.]

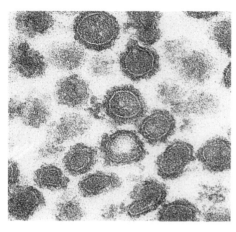

▲ **EXPERIMENTAL FIGURE 17-10 Coated vesicles accumulate during in vitro budding reactions in the presence of a nonhydrolyzable analog of GTP.** When isolated Golgi membranes are incubated with a cytosolic extract containing COPI coat proteins and ATP, vesicles form and bud off from the membranes. Inclusion of a nonhydrolyzable analog of GTP in the budding reaction prevents disassembly of the coat after vesicle release. This micrograph shows COPI vesicles generated in such a reaction and separated from membranes by centrifugation. Coated vesicles prepared in this way can be analyzed to determine their components and properties. [Courtesy of L. Orci.]

constructed genes encoding mutant versions of the two proteins that have predictable effects on vesicular traffic when transfected into cultured cells. For example, in cells expressing mutant versions of Sar1 or ARF that cannot hydrolyze GTP, vesicle coats form and vesicle buds pinch off. However, because the mutant proteins cannot trigger disassembly of the coat, all available coat subunits eventually become permanently assembled into coated vesicles that are unable to fuse with target membranes. Addition of a nonhydrolyzable GTP analog to in vitro vesicle-budding reactions causes a similar blocking of coat disassembly. The vesicles that form in such reactions have coats that never dissociate, allowing their composition and structure to be more readily analyzed. The purified COPI vesicles shown in Figure 17-10 were produced in such a budding reaction.

Targeting Sequences on Cargo Proteins Make Specific Molecular Contacts with Coat Proteins

In order for transport vesicles to move specific proteins from one compartment to the next, vesicle buds must be able to discriminate among potential membrane and soluble cargo proteins, accepting only those cargo proteins that should advance to the next compartment and excluding those that should remain as residents in the donor compartment. In addition to sculpting the curvature of a donor membrane, the vesicle coat also functions in selecting specific proteins as cargo. The primary mechanism by which the vesicle coat selects cargo molecules is by directly binding to specific

TABLE 17-2 Known Sorting Signals That Direct Proteins to Specific Transport Vesicles

Signal Sequence[*]	Proteins with Signal	Signal Receptor	Vesicles That Incorporate Signal-bearing Protein
Lys-Asp-Glu-Leu (KDEL)	ER-resident luminal proteins	KDEL receptor in cis-Golgi membrane	COPI
Lys-Lys-X-X (KKXX)	ER-resident membrane proteins (cytosolic domain)	COPI α and β subunits	COPI
Di-acidic (e.g., Asp-X-Glu)	Cargo membrane proteins in ER (cytosolic domain)	COPII Sec24 subunit	COPII
Mannose 6-phosphate (M6P)	Soluble lysosomal enzymes after processing in cis-Golgi	M6P receptor in trans-Golgi membrane	Clathrin/AP1
	Secreted lysosomal enzymes	M6P receptor in plasma membrane	Clathrin/AP2
Asn-Pro-X-Tyr (NPXY)	LDL receptor in the plasma membrane (cytosolic domain)	AP2 complex	Clathrin/AP2
Tyr-X-X-Φ (YXXΦ)	Membrane proteins in trans-Golgi (cytosolic domain)	AP1 (μ1 subunit)	Clathrin/AP1
	Plasma membrane proteins (cytosolic domain)	AP2 (μ2 subunit)	Clathrin/AP2
Leu-Leu (LL)	Plasma membrane proteins (cytosolic domain)	AP2 complexes	Clathrin/AP2

[*]X = any amino acid; Φ = hydrophobic amino acid. Single-letter amino acid abbreviations are in parentheses.

sequences, or **sorting signals,** in the cytosolic portion of membrane cargo proteins (see Figure 17-7a). The polymerized coat thus acts as an affinity matrix to cluster selected membrane cargo proteins into forming vesicle buds. Soluble proteins within the lumen of parent organelles can in turn be selected by binding to the luminal domains of certain membrane cargo proteins, which act as receptors for luminal cargo proteins. The properties of several known sorting signals in membrane and soluble proteins are summarized in Table 17-2. We describe the role of these signals in more detail in later sections.

Rab GTPases Control Docking of Vesicles on Target Membranes

A second set of small GTP-binding proteins, known as *Rab proteins,* participate in the targeting of vesicles to the appropriate target membrane. Like Sar1 and ARF, Rab proteins belong to the GTPase superfamily of switch proteins. Conversion of cytosolic Rab · GDP to Rab · GTP, catalyzed by a specific guanine nucleotide–exchange factor, induces a conformational change in Rab that enables it to interact with a surface protein on a particular transport vesicle and insert its isoprenoid anchor into the vesicle membrane. Once

Rab · GTP is tethered to the vesicle surface, it is thought to interact with one of a number of different large proteins, known as Rab effectors, attached to the target membrane. Binding of Rab · GTP to a Rab effector docks the vesicle on an appropriate target membrane (Figure 17-11, step 1). After vesicle fusion occurs, the GTP bound to the Rab protein is hydrolyzed to GDP, triggering the release of Rab · GDP, which then can undergo another cycle of GDP-GTP exchange, binding, and hydrolysis.

Several lines of evidence support the involvement of specific Rab proteins in vesicle-fusion events. For instance, the yeast *SEC4* gene encodes a Rab protein, and yeast cells expressing mutant Sec4 proteins accumulate secretory vesicles that are unable to fuse with the plasma membrane (class E mutants in Figure 17-5). In mammalian cells, Rab5 protein is localized to endocytic vesicles, also known as early endosomes. These uncoated vesicles form from clathrin-coated vesicles just after they bud from the plasma membrane during endocytosis (see Figure 17-1, step 9). The fusion of early endosomes with each other in cell-free systems requires the presence of Rab5, and addition of Rab5 and GTP to cell-free extracts accelerates the rate at which these vesicles fuse with each other. A long coiled protein known as EEA1 (*early endosome antigen 1*), which resides on the membrane of the early endosome, functions as the effector for Rab5. In this case, Rab5 · GTP on one endocytic vesicle is thought to specifically bind to EEA1 on the membrane of another endocytic vesicle, setting the stage for fusion of the two vesicles.

Similarly, Rab1 is essential for ER-to-Golgi transport reactions to occur in cell-free extracts. Rab1 · GTP binds to a long coiled-coil protein known as p115, which specifically tethers COPII vesicles carrying Rab1· GTP to the target Golgi membrane. A different type of Rab effector appears to function for each vesicle type and at each step of the secretory pathway. Many questions remain about how Rab proteins are targeted to the correct membrane and how specific complexes form between the different Rab proteins and their corresponding effector proteins.

Paired Sets of SNARE Proteins Mediate Fusion of Vesicles with Target Membranes

As noted previously, shortly after a vesicle buds off from the donor membrane, the vesicle coat disassembles to uncover a vesicle-specific membrane protein, a v-SNARE (see Figure 17-7b). Likewise, each type of target membrane in a cell contains t-SNARE membrane proteins. After Rab-mediated docking of a vesicle on its target (destination) membrane, the interaction of cognate SNAREs brings the two membranes close enough together that they can fuse.

One of the best-understood examples of SNARE-mediated fusion occurs during exocytosis of secreted proteins (Figure 17-11, steps 2 and 3). In this case, the v-SNARE, known as *VAMP* (*v*esicle-*a*ssociated *m*embrane *p*rotein), is incorporated into secretory vesicles as they bud from the *trans*-Golgi network. The t-SNAREs are *syntaxin*, an integral membrane protein in the plasma membrane, and *SNAP-25*, which is attached to the plasma membrane by a hydrophobic lipid anchor in the middle of the protein. The cytosolic region in each of these three SNARE proteins contains a repeating heptad sequence that allows four α helices—one from VAMP, one

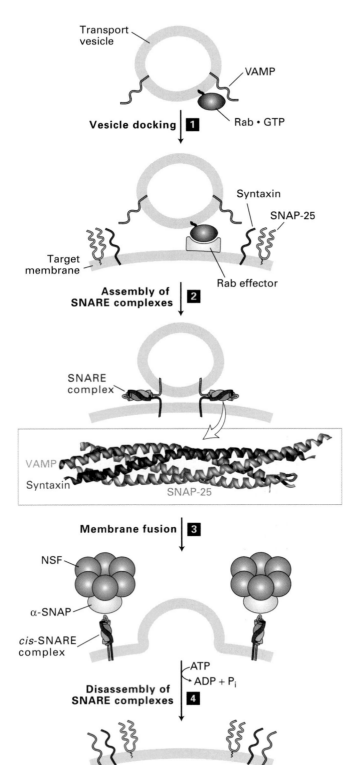

◀ **FIGURE 17-11 Model for docking and fusion of transport vesicles with their target membranes.** The proteins shown in this example participate in fusion of secretory vesicles with the plasma membrane, but similar proteins mediate all vesicle-fusion events. Step **1**: A Rab protein tethered via a lipid anchor to a secretory vesicle binds to an effector protein complex on the plasma membrane, thereby docking the transport vesicle on the appropriate target membrane. Step **2**: A v-SNARE protein (in this case, VAMP) interacts with the cytosolic domains of the cognate t-SNAREs (in this case, syntaxin and SNAP-25). The very stable coiled-coil SNARE complexes that are formed hold the vesicle close to the target membrane. *Inset:* Numerous noncovalent interactions between four long α helices, two from SNAP-25 and one each from syntaxin and VAMP, stabilize the coiled-coil structure. Step **3**: Fusion of the two membranes immediately follows formation of SNARE complexes, but precisely how this occurs is not known. Step **4**: Following membrane fusion, NSF in conjunction with α-SNAP protein binds to the SNARE complexes. The NSF-catalyzed hydrolysis of ATP then drives dissociation of the SNARE complexes, freeing the SNARE proteins for another round of vesicle fusion. [See J. E. Rothman and T. Söllner, 1997, *Science* **276**:1212, and W. Weis and R. Scheller, 1998, *Nature* **395**:328. Inset from Y. A. Chen and R. H. Scheller, 2001, *Nat. Rev. Mol. Cell Biol.* **2**(2):98.]

from syntaxin, and two from SNAP-25—to coil around one another to form a four-helix bundle. The unusual stability of this bundled SNARE complex is conferred by the arrangement of hydrophobic and charged amino residues in the heptad repeats. The hydrophobic amino acids are buried in the central core of the bundle, and amino acids of opposite charge are aligned to form favorable electrostatic interactions between helices. As the four-helix bundles form, the vesicle and target membranes are drawn into close apposition by the embedded transmembrane domains of VAMP and syntaxin.

In vitro experiments have shown that when **liposomes** containing purified VAMP are incubated with other liposomes containing syntaxin and SNAP-25, the two classes of membranes fuse, albeit slowly. This finding is strong evidence that the close apposition of membranes resulting from formation of SNARE complexes is sufficient to bring about membrane fusion. Fusion of a vesicle and target membrane occurs much more rapidly and efficiently in the cell than it does in liposome experiments in which fusion is catalyzed only by SNARE proteins. The likely explanation for this difference is that in the cell the interactions between specific Rab proteins and their effectors promote the formation of specific SNARE bundles by tethering a vesicle to its target membrane.

Yeast cells, like all eukaryotic cells, express more than 20 different related v-SNARE and t-SNARE proteins. Analyses of yeast *sec* mutants defective in each of the SNARE genes have identified the specific membrane-fusion event in which each SNARE protein participates. For all fusion events that have been examined, the SNAREs form four-helix bundled complexes, similar to the VAMP/syntaxin/SNAP-25 complexes that mediate fusion of secretory vesicles with the plasma membrane. However, in other fusion events (e.g., fusion of COPII vesicles with the *cis*-Golgi network), each participating SNARE protein contributes only one α helix to the bundle (unlike SNAP-25, which contributes two helices); in these cases the SNARE complexes comprise one v-SNARE and three t-SNARE molecules.

Using the in vitro liposome fusion assay, researchers have tested the ability of various combinations of individual v-SNARE and t-SNARE proteins to mediate fusion of donor and target membranes. Of the very large number of different combinations tested, only a small number mediated membrane fusion. To a remarkable degree the functional combinations of v-SNAREs and t-SNAREs revealed in these in vitro experiments correspond to the actual SNARE protein interactions that mediate known membrane-fusion events in the yeast cell. Thus the specificity of the interaction between SNARE proteins can account for the specificity of fusion between a particular vesicle and its target membranes.

Dissociation of SNARE Complexes After Membrane Fusion Is Driven by ATP Hydrolysis

After a vesicle and its target membrane have fused, the SNARE complexes must dissociate to make the individual SNARE proteins available for additional fusion events. Be-

cause of the stability of SNARE complexes, which are held together by numerous noncovalent intermolecular interactions, their dissociation depends on additional proteins and the input of energy.

The first clue that dissociation of SNARE complexes required the assistance of other proteins came from in vitro transport reactions depleted of certain cytosolic proteins. The observed accumulation of vesicles in these reactions indicated that vesicles could form but were unable to fuse with a target membrane. Eventually two proteins, designated *NSF* and *α-SNAP*, were found to be required for ongoing vesicle fusion in the in vitro transport reaction. The function of NSF in vivo can be blocked selectively by *N*-ethylmaleimide (NEM), a chemical that reacts with an essential –SH group on NSF (hence the name, *NEM*-sensitive *f*actor).

Among the class C yeast *sec* mutants are strains that lack functional Sec18 or Sec17, the yeast counterparts of mammalian NSF and α-SNAP, respectively. When these class C mutants are placed at the nonpermissive temperature, they accumulate ER-to-Golgi transport vesicles; when the cells are shifted to the lower, permissive temperature, the accumulated vesicles are able to fuse with the *cis*-Golgi.

Subsequent to the initial biochemical and genetic studies identifying NSF and α-SNAP, more sophisticated in vitro transport assays were developed. Using these newer assays, researchers have shown that NSF and α-SNAP proteins are not necessary for actual membrane fusion, but rather are required for regeneration of free SNARE proteins. NSF, a hexamer of identical subunits, associates with a SNARE complex with the aid α-SNAP (soluble *NSF a*ttachment *p*rotein). The bound NSF then hydrolyzes ATP, releasing sufficient energy to dissociate the SNARE complex (Figure 17-11, step ④). Evidently, the defects in vesicle fusion observed in the earlier in vitro fusion assays and in the yeast mutants after a loss of Sec17 or Sec18 were a consequence of free SNARE proteins rapidly becoming sequestered in undissociated SNARE complexes and thus unavailable to mediate membrane fusion.

Conformational Changes in Viral Envelope Proteins Trigger Membrane Fusion

Some animal viruses, including influenza virus, rabies virus, and human immunodeficiency virus (HIV), have an outer phospholipid bilayer membrane, or **envelope,** surrounding the core of the virus particle composed of viral proteins and genetic material. The viral envelope is derived by budding from the host-cell plasma membrane, which contains virus-encoded glycoproteins. Enveloped viruses enter a host cell by endocytosis following binding of one or more viral envelope glycoproteins with a host's cell-surface molecules. Subsequent fusion of the viral envelope with the endosomal membrane releases the viral genome into the cytosol of the host cell, initiating replication of the virus (see Figure 4-41, step ③). The molecular events of this fusion process have been elucidated in considerable detail in the case of influenza virus.

(a) pH ~7.0 (b) pH 5.0–5.5

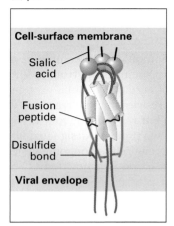
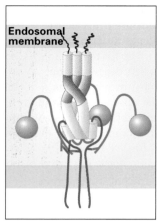

▲ **FIGURE 17-12 Schematic models of the structure of influenza hemagglutinin (HA) at pH 7 and 5.** Three HA₁ and three HA₂ subunits compose a hemagglutinin molecule, which protrudes from the viral envelope like a spike. (a) At pH ≈7, part of each HA₁ subunit forms a globular domain (green) at the tip of the native spike. These domains bind to sialic acid residues on the host-cell plasma membrane, initiating viral entry. Each HA₁ subunit is linked to one HA₂ subunit by a disulfide bond at the base of the molecule near the viral envelope. Each HA₂ subunit contains a fusion peptide (red) at its N-terminus (only two are visible), followed by a short α helix (orange cylinder), a nonhelical loop (brown), and a longer α helix (light purple). The longer α helices from the three HA₂ subunits form a three-stranded coiled-coil structure (see Figure 3-7). In this conformation, the fusion peptides are buried within the molecule. (b) At the acidic pH within a late endosome, the binding of the fusion peptide to other segments of HA₂ is disrupted, inducing major structural rearrangements in the protein. First, the three HA₁ globular domains separate from each other but remain tethered to the HA₂ subunits by the disulfide bonds at the base of the molecule. Second, the loop segment of each HA₂ rearranges into an α helix (brown) and combines with the short and long α-helical segments to form a continuous 88-residue α helix. The three long α helices thus form a 13.5-nm-long three-stranded coiled coil that protrudes outward from the viral envelope. In this conformation, the fusion peptides are at the tip of the coiled coil and can insert into the endosomal membrane. [Adapted from C. M. Carr et al., 1997, *Proc. Nat'l. Acad. Sci.* **94**:14306; courtesy of Peter Kim.]

The predominant glycoprotein of the influenza virus is *hemagglutinin (HA)*, which forms the larger spikes on the surface of the virus. There is considerable evidence that following endocytosis of an influenza virion, the low pH within the enclosing late endosome triggers fusion of its membrane with the viral envelope. For instance, viral infection is inhibited by the addition of lipid-soluble bases, such as ammonia or trimethylamine, which raise the normally acidic pH of late endosomes. Also, a conformational change in the HA protein that is critical for infectivity occurs over a very narrow range in pH (5.0–5.5).

Each HA spike on an influenza virion consists of three HA₁ and three HA₂ subunits. At the N-terminus of HA₂ is a strongly hydrophobic 11-residue sequence, called the fusion

peptide. Structural studies have shown that at pH 7.0, the N-terminus of each HA₂ subunit is tucked into a crevice in the spike (Figure 17-12a). This is the normal HA conformation when a viral particle encounters the surface of a host cell. At the acidic pH characteristic of late endosomes, HA undergoes several conformational changes that cause a major rearrangement of the subunits. As a result, the three HA₂ subunits twist together into a three-stranded coiled-coil rod that protrudes more than 13 nm outward from the viral envelope with the fusion peptides at the tip of the rod (Figure 17-12b). In this conformation, the highly hydrophobic fusion peptides are exposed and can insert into the lipid bilayer of the endosomal membrane, triggering fusion of the viral envelope and the membrane. Thus at pH 7 HA can be said to be trapped in a metastable, "spring-loaded" state, which is converted to the lower-energy fusogenic state by shifting the pH to 5–5.5.

Multiple low pH–activated HA spikes are essential for membrane fusion to occur. Figure 17-13 suggests one way by which the protein scaffold formed by many HA spikes, possibly with the assistance of other cellular proteins, could link together the

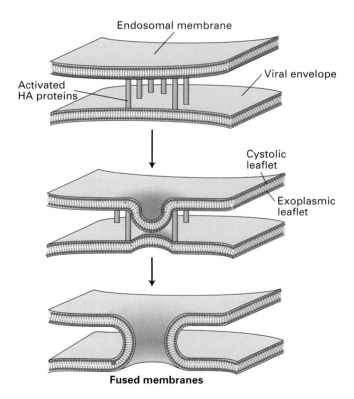

▲ **FIGURE 17-13 Model for membrane fusion directed by hemagglutinin (HA).** A number of low pH–activated HA spikes, possibly in concert with host-cell membrane proteins, form a scaffold that connects a small region of the viral envelope and the endosomal membrane. By unknown mechanisms, the exoplasmic leaflets of the two membranes fuse and then the cytosolic leaflets fuse, forming a pore that widens until the two membranes are completely joined. A similar interaction between membrane bilayers may be brought about during SNARE-mediated vesicle fusion. [Adapted from J. R. Monck and J. M. Fernandez, 1992, *J. Cell Biol.* **119**:1395.]

viral envelope and endosomal membrane and induce their fusion. This figure also illustrates how cellular membranes brought into close apposition by SNARE complexes might fuse. Note that each HA molecule participates in only one fusion event, whereas the cellular fusion proteins, such as SNAREs, are recycled and catalyze multiple cycles of membrane fusion.

KEY CONCEPTS OF SECTION 17.2

Molecular Mechanisms of Vesicular Traffic

■ The three well-characterized transport vesicles—COPI, COPII, and clathrin vesicles—are distinguished by the proteins that form their coats and the transport routes they mediate (see Table 17-1).

■ All types of coated vesicles are formed by polymerization of cytosolic coat proteins onto a donor (parent) membrane to form vesicle buds that eventually pinch off from the membrane to release a complete vesicle. Shortly after vesicle release, the coat is shed exposing proteins required for fusion with the target membrane (see Figure 17-7).

■ Small GTP-binding proteins (ARF or Sar1) belonging to the GTPase superfamily control polymerization of coat proteins, the initial step in vesicle budding (see Figure 17-9). After vesicles are released from the donor membrane, hydrolysis of GTP bound to ARF or Sar1 triggers disassembly of the vesicle coats.

■ Specific sorting signals in membrane and luminal proteins of donor organelles interact with coat proteins during vesicle budding, thereby recruiting cargo proteins to vesicles (see Table 17-2).

■ A second set of GTP-binding proteins, the Rab proteins, regulate docking of vesicles with the correct target membrane. Each Rab appears to bind to a specific Rab effector, a typically long coiled-coil protein, associated with the target membrane.

■ Each v-SNARE in a vesicular membrane specifically binds to a complex of cognate t-SNARE proteins in the target membrane, inducing fusion of the two membranes. After fusion is completed, the SNARE complex is disassembled in an ATP-dependent reaction mediated by other cytosolic proteins (see Figure 17-11).

■ After an enveloped animal virus is endocytosed, the viral envelope fuses with the surrounding endosomal membrane. In the case of influenza virus, the acidic pH within late endosomes causes a conformational change in the HA protein in the viral envelope that permits insertion of HA into the endosomal membrane.

17.3 Early Stages of the Secretory Pathway

In this section we take a closer look at vesicular traffic through the ER and Golgi stages of the secretory pathway

and some of the evidence supporting the general mechanisms discussed in the previous section. Recall that *anterograde transport* from the ER to Golgi, the first step in the secretory pathway, is mediated by COPII vesicles, whereas the reverse *retrograde transport* from the *cis*-Golgi to the ER is mediated by COPI vesicles (Figure 17-14). This retrograde

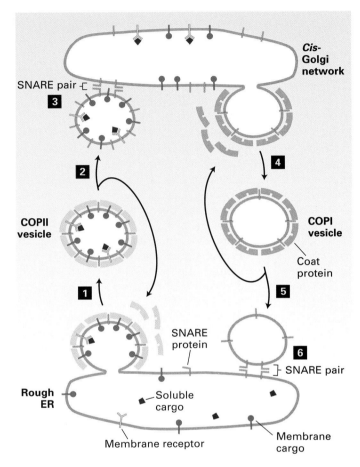

▲ **FIGURE 17-14 Vesicle-mediated protein trafficking between the ER and *cis*-Golgi.** Steps **1**–**3**: Forward (anterograde) transport is mediated by COPII vesicles, which are formed by polymerization of soluble COPII coat protein complexes (blue) on the ER membrane. v-SNAREs (red) and other cargo proteins (green) in the ER membrane are incorporated into the vesicle by interacting with coat proteins. Soluble cargo proteins (purple) are recruited by binding to appropriate receptors in the membrane of budding vesicles. Dissociation of the coat recycles free coat complexes and exposes v-SNARE proteins on the vesicle surface. After the uncoated vesicle becomes tethered to the *cis*-Golgi membrane in a Rab-mediated process, pairing between the exposed v-SNAREs and cognate t-SNAREs in the Golgi membrane allow vesicle fusion, releasing the contents into the *cis*-Golgi compartment (see Figure 17-11). Steps **4**–**6**: Reverse (retrograde) transport, mediated by vesicles coated with COPI proteins (green), recycles the membrane bilayer and certain proteins, such as v-SNAREs and missorted ER-resident proteins (not shown), from the *cis*-Golgi to the ER. All SNARE proteins are shown in red although each v-SNARE and t-SNARE are distinct proteins.

vesicle transport serves to retrieve v-SNARE proteins and the membrane itself back to the ER to provide the necessary material for additional rounds of vesicle budding from the ER. COPI-mediated retrograde transport also retrieves missorted ER-resident proteins from the *cis*-Golgi to correct sorting mistakes. Proteins that have been correctly delivered to the Golgi advance through successive compartments of the Golgi by cisternal progression.

COPII Vesicles Mediate Transport from the ER to the Golgi

COPII vesicles were first recognized when cell-free extracts of yeast rough ER membranes were incubated with cytosol, ATP, and a nonhydrolyzable analog of GTP. The vesicles that formed from the ER membranes had a distinct coat, similar to that on COPI vesicles but composed of different proteins, designated COPII proteins. Yeast cells with mutations in the genes for COPII proteins are class B *sec* mutants and accumulate proteins in the rough ER (see Figure 17-5). Analysis of such mutants has revealed several proteins required for formation of COPII vesicles.

As described previously, formation of COPII vesicles is triggered when Sec12, a guanine nucleotide–exchange factor, catalyzes the exchange of bound GDP for GTP on Sar1. This exchange induces binding of Sar1 to the ER membrane followed by binding of a complex of Sec23 and Sec24 proteins (see Figure 17-9). The resulting ternary complex formed between Sar1· GTP, Sec23, and Sec24 is shown in Figure 17-15. After this complex forms on the ER membrane, a second complex comprising Sec13 and Sec31 proteins then binds to complete the coat structure. A large fibrous protein, called Sec16, which is bound to the cytosolic surface of the ER, interacts with the Sec13/31 and Sec23/24 complexes, and acts to organize the other coat proteins, increasing the efficiency of coat polymerization.

Certain integral ER membrane proteins are specifically recruited into COPII vesicles for transport to the Golgi. The cytosolic segments of many of these proteins contain a *di-acidic sorting signal* (Asp-X-Glu, or DXE in the one-letter code). This sorting signal binds to the Sec24 subunit of the COPII coat and is essential for the selective export of certain membrane proteins from the ER (see Figure 17-15). Biochemical and genetic studies currently are under way to identify additional signals that help direct membrane cargo proteins into COPII vesicles. Other ongoing studies seek to determine how soluble cargo proteins are selectively loaded into COPII vesicles. Although purified COPII vesicles from yeast cells have been found to contain a membrane protein that binds the soluble α mating factor, the receptors for other soluble cargo proteins such as invertase are not yet known.

The experiments described previously in which the transit of VSVG-GFP in cultured mammalian cells is followed by fluorescence microscopy (see Figure 17-2) provided insight into the intermediates in ER-to-Golgi transport. In some cells, small fluorescent vesicles containing VSVG-GFP could be seen to form from the ER, move less than 1 μm, and

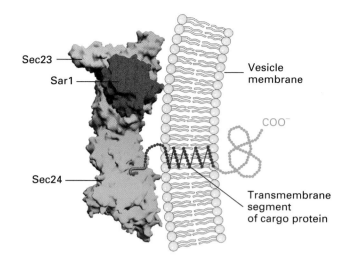

▲ **FIGURE 17-15 Three-dimensional structure of ternary complex comprising the COPII coat proteins Sec23 and Sec24 and Sar1 · GTP.** Early in the formation of the COPII coat, Sec23 (orange)/Sec24 (green) complexes are recruited to the ER membrane by Sar1 (red) in its GTP-bound state. In order to form a stable ternary complex in solution for structural studies, the nonhydrolyzable GTP analog GppNHp was used. A cargo protein in the ER membrane can be recruited to COPII vesicles by interaction of a tripeptide di-acidic signal (purple) in the cargo's cytosolic domain with Sec24. The likely position of the COPII vesicle membrane and the transmembrane segment of the cargo protein are indicated. The N-terminal segment of Sar1 that tethers it to the membrane is not shown. [See X. Bi et al., 2002, *Nature* **419**:271; interaction with peptide courtesy of J. Goldberg.]

then fuse directly with the *cis*-Golgi. In other cells, in which the ER was located several micrometers from the Golgi complex, several ER-derived vesicles were seen to fuse with each other shortly after their formation, forming what is termed the "ER-to-Golgi intermediate compartment." These larger structures then were transported along microtubules to the *cis*-Golgi, much in the way vesicles in nerve cells are transported from the cell body, where they are formed, down the long axon to the axon terminus (Chapter 20). Microtubules function much as "railroad tracks" enabling these large aggregates of transport vesicles to move long distances to their *cis*-Golgi destination. At the time the ER-to-Golgi intermediate compartment is formed, some COPI vesicles bud off from it, recycling some proteins back to the ER.

COPI Vesicles Mediate Retrograde Transport within the Golgi and from the Golgi to the ER

COPI vesicles were first discovered when isolated Golgi fractions were incubated in a solution containing ATP, cytosol, and a nonhydrolyzable analog of GTP (see Figure 17-10). Subsequent analysis of these vesicles showed that the coat is formed from large cytosolic complexes, called *coatomers,* composed of seven polypeptide subunits. Yeast cells containing temperature-sensitive mutations in COPI proteins accumulate proteins in the rough ER at the nonpermissive temperature and thus are categorized as class B *sec* mutants (see Figure 17-5). Although discovery of these mutants initially suggested that COPI vesicles mediate ER-to-Golgi transport, subsequent experiments showed that their main function is retrograde transport, both between Golgi cisternae and from the *cis*-Golgi to the rough ER (see Figure 17-14, *right*). Because COPI mutants cannot recycle key membrane proteins back to the rough ER, the ER gradually becomes depleted of ER proteins such as v-SNAREs necessary for COPII vesicle function. Eventually vesicle formation from the rough ER grinds to a halt; secretory proteins continue to be synthesized but accumulate in the ER, the defining characteristic of class B *sec* mutants.

As discussed in Chapter 16, the ER contains several soluble proteins dedicated to the folding and modification of newly synthesized secretory proteins. These include the chaperone BiP and the enzyme protein disulfide isomerase, which are necessary for the ER to carry out its functions. Although such ER-resident luminal proteins are not specifically selected by COPII vesicles, their sheer abundance causes them to be continuously loaded passively into vesicles destined for the *cis*-Golgi. The transport of these soluble proteins back to the ER, mediated by COPI vesicles, prevents their eventual depletion

Most soluble ER-resident proteins carry a Lys-Asp-Glu-Leu (KDEL in the one-letter code) sequence at their C-terminus (see Table 17-2). Several experiments demonstrated that this *KDEL sorting signal* is both necessary and sufficient for retention in the ER. For instance, when a mutant protein disulfide isomerase lacking these four residues is synthesized in cultured fibroblasts, the protein is secreted. Moreover, if a protein that normally is secreted is altered so that it contains the KDEL signal at its C-terminus, the protein is retained in the ER. The KDEL sorting signal is recognized and bound by the *KDEL receptor*, a transmembrane protein found primarily on small transport vesicles shuttling between the ER and the *cis*-Golgi and on the *cis*-Golgi reticulum. In addition, soluble ER-resident proteins that carry the KDEL signal have oligosaccharide chains with modifications that are catalyzed by enzymes found only in the *cis*-Golgi or *cis*-Golgi reticulum; thus at some time these proteins must have left the ER and been transported at least as far as the *cis*-Golgi network. These findings indicate that the KDEL receptor acts mainly to retrieve soluble proteins containing the KDEL sorting signal that have escaped to the *cis*-Golgi network and return them to the ER (Figure 17-16).

The KDEL receptor and other membrane proteins that are transported back to the ER from the Golgi contain a Lys-Lys-X-X sequence at the very end of their C-terminal segment, which faces the cytosol (see Table 17-2). This *KKXX sorting signal* which binds to a complex of the COPI α and β subunits, is both necessary and sufficient to incorporate membrane proteins into COPI vesicles for retrograde transport to the ER. Temperature-sensitive yeast mutants lacking COPIα or COPIβ not only are unable to bind the KKXX signal but also are unable to retrieve proteins bearing this signal back to the ER, indicating that COPI vesicles mediate retrograde Golgi-to-ER transport.

Clearly, the partitioning of proteins between the ER and Golgi complex is a highly selective and regulated process ultimately controlled by the specificity of cargo loading into both COPII (anterograde) and COPI (retrograde) vesicles. The selective entry of proteins into membrane-bounded transport vesicles, the recycling of membrane phospholipids and proteins, and the recycling of soluble luminal proteins between the two compartments are fundamental features of vesicular protein trafficking that also occur in later stages of the secretory pathway.

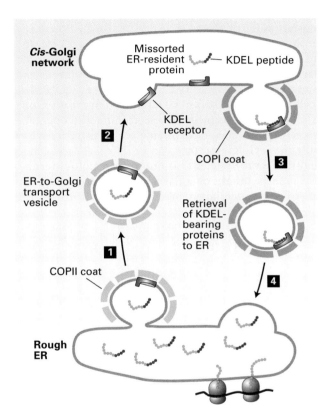

▲ **FIGURE 17-16 Role of the KDEL receptor in retrieval of ER-resident luminal proteins from the Golgi.** ER luminal proteins, especially those present at high levels, can be passively incorporated into COPII vesicles and transported to the Golgi (steps **1** and **2**). Many such proteins bear a C-terminal KDEL (Lys-Asp-Glu-Leu) sequence (red) that allows them to be retrieved. The KDEL receptor, located mainly in the *cis*-Golgi network and in both COPII and COPI vesicles, binds proteins bearing the KDEL sorting signal and returns them to the ER (steps **3** and **4**). This retrieval system prevents depletion of ER luminal proteins such as those needed for proper folding of newly made secretory proteins. The binding affinity of the KDEL receptor is very sensitive to pH. The small difference in the pH of the ER and Golgi favors binding of KDEL-bearing proteins to the receptor in Golgi-derived vesicles and their release in the ER. [Adapted from J. Semenza et al., 1990, *Cell* **61**:1349.]

Anterograde Transport Through the Golgi Occurs by Cisternal Progression

At one time it was thought that small transport vesicles carry secretory proteins from the *cis*- to the *medial*-Golgi and from the *medial*- to the *trans*-Golgi. Indeed, electron microscopy reveals many small vesicles associated with the Golgi complex that move proteins from one Golgi compartment to another (Figure 17-17). However, these vesicles most likely mediate retrograde transport, retrieving ER or Golgi enzymes from a later compartment and transporting them to an earlier compartment in the secretory pathway. In this way enzymes that modify secretory proteins come to be localized in the correct compartment.

The first evidence that the forward transport of cargo proteins from the *cis*- to the *trans*-Golgi occurs by a nonvesicular mechanism, called *cisternal progression*, came from careful microscopic analysis of the synthesis of algal scales. These cell-wall glycoproteins are assembled in the *cis*-Golgi into large complexes visible in the electron microscope. Like other secretory proteins, newly made scales move from the *cis*- to the *trans*-Golgi, but they can be 20 times larger than the usual transport vesicles that bud from Golgi cisternae. Similarly, in the synthesis of collagen by fibroblasts, large aggregates of the procollagen precursor often form in the lumen of the *cis*-Golgi (see Figure 6-20). The procollagen aggregates are too large to be incorporated into small transport vesicles, and investigators could never find such aggregates in transport vesicles. These observations suggested that the forward movement of these and perhaps all secretory proteins from one Golgi compartment to another does *not* occur via small vesicles.

In one test of the cisternal progression model, collagen folding was blocked by an inhibitor of proline hydroxylation, and soon all pre-made, folded procollagen aggregates were secreted from the cell. When the inhibitor was removed, newly made procollagen peptides folded and then formed aggregates in the *cis*-Golgi that subsequently could be seen to move as a "wave" from the *cis*- through the *medial*-Golgi cisternae to the *trans*-Golgi, followed by secretion and incorporation into the extracellular matrix. In these experiments procollagen aggregates were never visible in small transport vesicles. Numerous controversial questions concerning membrane flow within the Golgi stack remain unresolved. Nonetheless, the observed movement of very large macromolecular assemblies through the Golgi stack and the evidence described previously that COPI vesicles mediate retrograde transport have led most researchers in the field to favor the cisternal progression model.

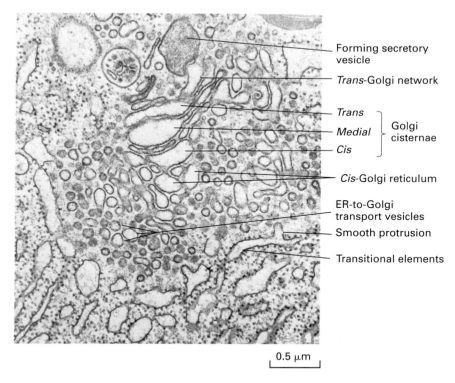

0.5 μm

▲ **EXPERIMENTAL FIGURE 17-17 Electron micrograph of the Golgi complex in an exocrine pancreatic cell reveals both anterograde and retrograde transport vesicles.** A large secretory vesicle can be seen forming from the *trans*-Golgi network. Elements of the rough ER are on the left in this micrograph. Adjacent to the rough ER are transitional elements from which smooth protrusions appear to be budding. These buds form the small vesicles that transport secretory proteins from the rough ER to the Golgi complex. Interspersed among the Golgi cisternae are other small vesicles now known to function in retrograde, not anterograde, transport. [Courtesy G. Palade.]

Vesicle Traffic in the Early Stages of the Secretory Pathway

■ COPII vesicles transport proteins from the rough ER to the *cis*-Golgi; COPI vesicles transport proteins in the reverse direction (see Figure 17-14).

■ COPII coats comprise three components: the small GTP-binding protein Sar1, a Sec23/Sec24 complex, and a Sec13/Sec31 complex.

■ Components of the COPII coat bind to membrane cargo proteins containing a di-acidic or other sorting signal in their cytosolic regions (see Figure 17-15). Soluble cargo proteins probably are targeted to COPII vesicles by binding to a membrane protein receptor.

■ Membrane proteins needed to form COPII vesicles can be retrieved from the *cis*-Golgi by COPI vesicles. One of the sorting signals that directs membrane proteins into COPI vesicles is a KKXX sequence, which binds to subunits of the COPI coat.

■ Many soluble ER-resident proteins contain a KDEL sorting signal. Binding of this retrieval sequence to a specific receptor protein in the *cis*-Golgi membrane recruits missorted ER proteins into retrograde COPI vesicles (see Figure 17-16).

■ COPI vesicles also carry Golgi-resident proteins from later to earlier compartments in the Golgi stack.

■ Soluble and membrane proteins advance through the Golgi complex by cisternal progression, a nonvesicular process of anterograde transport.

17.4 Later Stages of the Secretory Pathway

As cargo proteins move from the *cis* face to the *trans* face of the Golgi complex by cisternal progression, modifications to their oligosaccharide chains are carried out by Golgi-resident enzymes. The retrograde trafficking of COPI vesicles from later to earlier Golgi compartments maintains sufficient levels of these carbohydrate-modifying enzymes in their functional compartments. Eventually, properly processed cargo proteins reach the *trans*-Golgi network, the most-distal Golgi compartment. Here they are sorted into vesicles for delivery to their final destination. In this section we discuss the different kinds of vesicles that bud from the *trans*-Golgi network, the mechanisms that segregate cargo proteins among them, and key processing events that occur late in the secretory pathway. The transport steps mediated by the major types of coated vesicles are summarized in Figure 17-18.

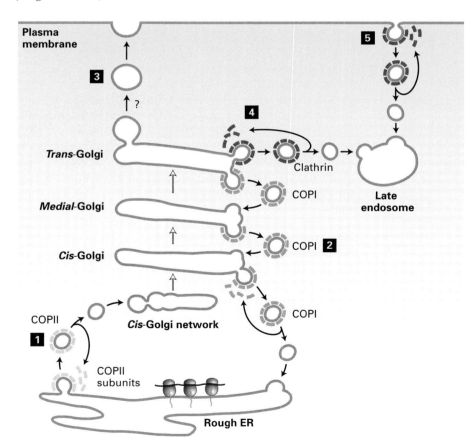

◀ **FIGURE 17-18 Involvement of the three major types of coat proteins in vesicular traffic in the secretory and endocytic pathways.** After formation of vesicles by budding from a donor membrane, the coats depolymerize into their subunits, which are re-used to form additional transport vesicles. COPII vesicles (**1**) mediate anterograde transport from the rough ER to the *cis*-Golgi/*cis*-Golgi network. COPI vesicles (**2**) mediate retrograde transport within the Golgi and from the *cis*-Golgi/*cis*-Golgi network to the rough ER. The coat proteins surrounding secretory vesicles (**3**) are not yet characterized; these vesicles carry secreted proteins and plasma-membrane proteins from the *trans*-Golgi network to the cell surface. Vesicles coated with clathrin (red) bud from the *trans*-Golgi network (**4**) and from the plasma membrane (**5**); after uncoating, these vesicles fuse with late endosomes. The coat on most clathrin vesicles contains additional proteins not indicated here. Note that secretory proteins move from the *cis*- to *trans*-Golgi by cisternal progression, which is not mediated by vesicles. [See H. Pelham, 1997, *Nature* **389**:17, and J. F. Presley et al., 1997, *Nature* **389**:81.]

Vesicles Coated with Clathrin and/or Adapter Proteins Mediate Several Transport Steps

The best-characterized vesicles that bud from the *trans*-Golgi network (TGN) have a two-layered coat: an outer layer composed of the fibrous protein clathrin and an inner layer composed of *adapter protein (AP) complexes.* Purified clathrin molecules, which have a three-limbed shape, are called *triskelions* from the Greek for three-legged (Figure 17-19a). Each limb contains one clathrin heavy chain (180,000 MW) and one clathrin light chain (≈35,000–40,000 MW). Triskelions polymerize to form a polygonal lattice with an intrinsic curvature (Figure 17-19b). When clathrin polymerizes on a donor membrane, it does so in association with AP complexes, which assemble between the clathrin lattice and the membrane. Each AP complex (340,000 MW) contains one copy each of four different adapter subunit proteins. A specific association between the globular domain at the end of each clathrin heavy chain in a triskelion and one subunit of the AP complex both promotes the co-assembly of clathrin triskelions with AP complexes and adds to the stability of the completed vesicle coat (Figure 17-19c).

By binding to the cytosolic face of membrane proteins, adapter proteins determine which cargo proteins are specifically included in (or excluded from) a budding transport vesicle. Each type of AP complex (e.g., AP1, AP2, AP3) and the recently identified GGAs are composed of different, though related, proteins. Vesicles containing each complex have been found to mediate specific transport steps (see Table 17-1). All vesicles whose coats contain one of these complexes utilize ARF to initiate coat assembly onto the donor membrane. As discussed previously, ARF also initiates assembly of COPI coats. The additional features of the membrane or protein factors that determine which type of coat will assemble after ARF attachment are not well understood at this time.

Vesicles that bud from the *trans*-Golgi network en route to the lysosome by way of the late endosome have clathrin coats associated with either AP1 or GGA. Both AP1 and GGA bind to the cytosolic domain of cargo proteins in the donor membrane, but the functional differences between vesicles that contain AP1 or GGA are unclear. Recent studies have shown that membrane proteins containing a Tyr-X-X-Φ sequence, where X is any amino acid and Φ is a bulky hydrophobic amino acid, are recruited into clathrin/AP1 vesicles budding from the *trans*-Golgi network. This *YXXΦ sorting signal* interacts with one of the AP1 subunits in the vesicle coat. As we discuss in the next section, vesicles with clathrin/AP2 coats, which bud from the plasma membrane during endocytosis, also can recognize the YXXΦ sorting signal.

Some vesicles that bud from the *trans*-Golgi network have coats composed of the AP3 complex. These vesicles mediate trafficking to the lysosome, but they appear to bypass the late endosome and fuse directly with the lysosomal membrane. In certain types of cells, such AP3 vesicles mediate protein transport to specialized storage compartments related to the lysosome. For example, AP3 is required for delivery of proteins to melanosomes, which contain the black pigment melanin in skin cells, and to platelet storage vesicles in megakaryocytes, a large cell that fragments into dozens of platelets. Mice with mutations in either of two different subunits of AP3 not only have abnormal skin pigmentation but also exhibit bleeding disorders. The latter occur because tears in blood vessels cannot be repaired without platelets that contain normal storage vesicles.

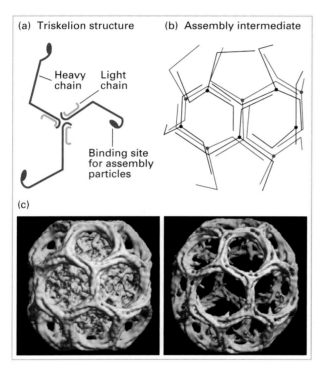

(a) Triskelion structure

Heavy chain Light chain

Binding site for assembly particles

(b) Assembly intermediate

(c)

▲ **FIGURE 17-19 Structure of clathrin coats.** (a) A clathrin molecule, called a triskelion, is composed of three heavy and three light chains. It has an intrinsic curvature due to the bend in the heavy chains. (b) The fibrous clathrin coat around vesicles is constructed of 36 clathrin triskelions. Depicted here is an intermediate in assembly of a clathrin coat, containing 10 of the final 36 triskelions, which illustrates the intrinsic curvature and the packing of clathrin triskelions. (c) Clathrin coats were formed in vitro by mixing purified clathrin heavy and light chains with AP2 complexes in the absence of membranes. Cryoelectron micrographs of more than 1000 assembled particles were analyzed by digital image processing to generate an average structural representation. The left image shows the reconstructed structure of a complete particle with AP2 complexes packed into the interior of the clathrin cage. In the right image, the AP2 complexes have been subtracted to show only the assembled clathrin heavy and light chains. [See B. Pishvaee and G. Payne, 1998, *Cell* **95**:443. Part (c) from Corinne J. Smith, Department of Biological Sciences, University of Warwick.]

Dynamin Is Required for Pinching Off of Clathrin Vesicles

A fundamental step in the formation of a transport vesicle that we have not yet considered is how a vesicle bud is pinched off from the donor membrane. In the case of clathrin/AP-coated vesicles, a cytosolic protein called *dynamin* is essential for release of complete vesicles. At the later stages of bud formation, dynamin polymerizes around the neck portion and then hydrolyzes GTP. The energy derived from GTP hydrolysis is thought to drive "contraction" of dynamin around the vesicle neck until the vesicle pinches off (Figure 17-20). Interestingly, COPI and COPII vesicles appear to pinch off from donor membranes without the aid of a GTPase such as dynamin. At present this fundamental difference in the process of pinching off among the different types of vesicles is not understood.

Incubation of cell extracts with a nonhydrolyzable derivative of GTP provides dramatic evidence for the importance of dynamin in pinching off of clathrin/AP vesicles during endocytosis. Such treatment leads to accumulation of clathrin-coated vesicle buds with excessively long necks that are

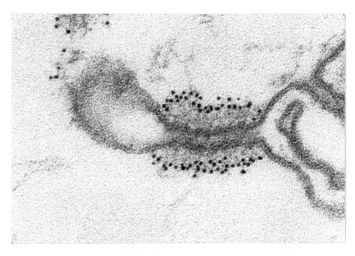

▲ **EXPERIMENTAL FIGURE 17-21 GTP hydrolysis by dynamin is required for pinching off of clathrin-coated vesicles in cell-free extracts.** A preparation of nerve terminals, which undergo extensive endocytosis, was lysed by treatment with distilled water and incubated with GTP-γ-S, a nonhydrolyzable derivative of GTP. After sectioning, the preparation was treated with gold-tagged anti-dynamin antibody and viewed in the electron microscope. This image, which shows a long-necked clathrin/AP-coated bud with polymerized dynamin lining the neck, reveals that buds can form in the absence of GTP hydrolysis, but vesicles cannot pinch off. The extensive polymerization of dynamin that occurs in the presence of with GTP-γ-S probably does not occur during the normal budding process. [From K. Takei et al., 1995, *Nature* **374**:186; courtesy of Pietro De Camilli.]

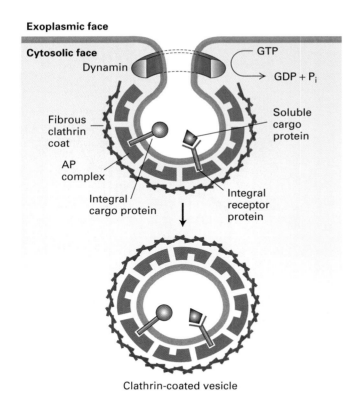

▲ **FIGURE 17-20 Model for dynamin-mediated pinching off of clathrin/AP-coated vesicles.** After a vesicle bud forms, dynamin polymerizes over the neck. By a mechanism that is not well understood, dynamin-catalyzed hydrolysis of GTP leads to release of the vesicle from the donor membrane. Note that membrane proteins in the donor membrane are incorporated into vesicles by interacting with AP complexes in the coat. [Adapted from K. Takei et al., 1995, *Nature* **374**:186.]

surrounded by polymeric dynamin but do not pinch off (Figure 17-21). Likewise, cells expressing mutant forms of dynamin that cannot bind GTP do not form clathrin-coated vesicles, and instead accumulate similar long-necked vesicle buds encased with polymerized dynamin.

As with COPI and COPII vesicles, clathrin/AP vesicles normally lose their coat soon after their formation. Cytosolic Hsc70, a constitutive chaperone protein found in all eukaryotic cells, is thought to use energy derived from the hydrolysis of ATP to drive depolymerization of the clathrin coat into triskelions. Uncoating not only releases triskelions for reuse in the formation of additional vesicles, but also exposes v-SNAREs for use in fusion with target membranes. Conformational changes that occur when ARF switches from the GTP-bound to GDP-bound state are thought to regulate the timing of clathrin coat depolymerization. How the action of Hsc70 might be coupled to ARF switching is not well understood.

Mannose 6-Phosphate Residues Target Soluble Proteins to Lysosomes

Most of the sorting signals that function in vesicular trafficking are short amino acid sequences in the targeted protein. In contrast, the sorting signal that directs soluble

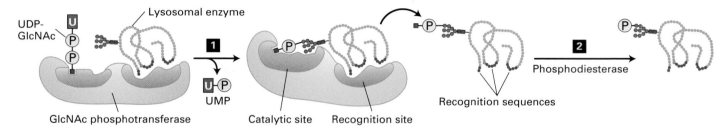

▲ **FIGURE 17-22 Formation of mannose 6-phosphate (M6P) residues that target soluble enzymes to lysosomes.** The M6P residues that direct proteins to lysosomes are generated in the *cis*-Golgi by two Golgi-resident enzymes. Step **1**: An *N*-acetylglucosamine (GlcNAc) phosphotransferase transfers a phosphorylated GlcNAc group to carbon atom 6 of one or more mannose residues. Because only lysosomal enzymes contain sequences (red) that are recognized and bound by this enzyme, phosphorylated GlcNAc groups are added specifically to lysosomal enzymes. Step **2**: After release of a modified protein from the phosphotransferase, a phosphodiesterase removes the GlcNAc group, leaving a phosphorylated mannose residue on the lysosomal enzyme. [See A. B. Cantor et al., 1992, *J. Biol. Chem.* **267**:23349, and S. Kornfeld, 1987, *FASEB J.* **1**:462.]

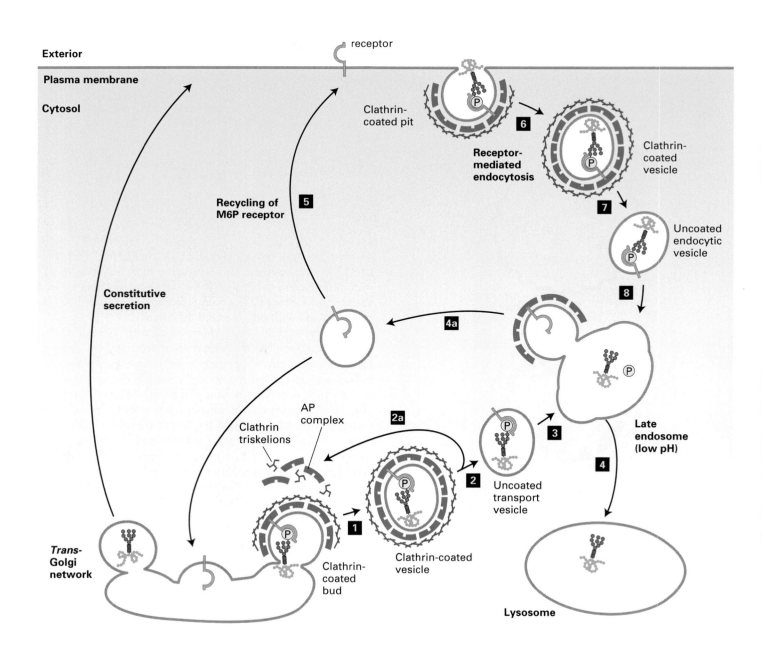

lysosomal enzymes from the *trans*-Golgi network to the late endosome is a carbohydrate residue, *mannose 6-phosphate (M6P)*, which is formed in the *cis*-Golgi. The addition and initial processing of one or more preformed N-linked oligosaccharide precursors in the rough ER is the same for lysosomal enzymes as for membrane and secreted proteins, yielding core $Man_8(GlcNAc)_2$ chains (see Figure 16-18). In the *cis*-Golgi, the N-linked oligosaccharides present on most lysosomal enzymes undergo a two-step reaction sequence that generates M6P residues (Figure 17-22). The addition of M6P residues to the oligosaccharide chains of soluble lysosomal enzymes prevents these proteins from undergoing the further processing reactions characteristic of secreted and membrane proteins (see Figure 17-3).

As shown in Figure 17-23, the segregation of M6P-bearing lysosomal enzymes from secreted and membrane proteins occurs in the *trans*-Golgi network. Here transmembrane *mannose 6-phosphate receptors* bind the M6P residues on lysosome-destined proteins very tightly and specifically. Clathrin/AP1 vesicles containing the M6P receptor and bound lysosomal enzymes then bud from the *trans*-Golgi network, lose their coats, and subsequently fuse with the late endosome by mechanisms described previously. Because M6P receptors can bind M6P at the slightly acidic pH (≈6.5) of the *trans*-Golgi network but not at a pH less than 6, the bound lysosomal enzymes are released within late endosomes, which have an internal pH of 5.0–5.5. Furthermore, a phosphatase within late endosomes usually removes the phosphate from M6P residues on lysosomal enzymes, preventing any rebinding to the M6P receptor that might occur in spite of the low pH in endosomes. Vesicles budding from late endosomes recycle the M6P receptor back to the *trans*-

Golgi network or, on occasion, to the cell surface. Eventually, mature late endosomes fuse with lysosomes, delivering the lysosomal enzymes to their final destination.

The sorting of soluble lysosomal enzymes in the *trans*-Golgi network (see Figure 17-23, steps $\boxed{1}$–$\boxed{4}$) shares many of the features of trafficking between the ER and *cis*-Golgi compartments mediated by COPII and COPI vesicles. First, mannose 6-phosphate acts as a sorting signal by interacting with the luminal domain of a receptor protein in the donor membrane. Second, the membrane-embedded receptors with their bound ligands are incorporated into the appropriate vesicles—in this case, AP1-containing clathrin vesicles—by interacting with the vesicle coat. Third, these transport vesicles fuse only with one specific organelle, here the late endosome, as the result of interactions between specific v-SNAREs and t-SNAREs. And finally, intracellular transport receptors are recycled after dissociating from their bound ligand.

Study of Lysosomal Storage Diseases Revealed Key Components of the Lysosomal Sorting Pathway

MEDICINE A group of genetic disorders, termed *lysosomal storage diseases,* are caused by the absence of one or more lysosomal enzymes. As a result, undigested glycolipids and extracellular components that would normally be degraded by lysosomal enzymes accumulate in lysosomes as large inclusions. *I-cell disease* is a particularly severe type of lysosomal storage disease in which multiple enzymes are missing from the lysosomes. Cells from affected individuals lack the N-acetylglucosamine phosphotransferase that is required for formation of M6P residues on lysosomal enzymes in the *cis*-Golgi (see Figure 17-22). Biochemical comparison of lysosomal enzymes from normal individuals with those from patients with I-cell disease led to the initial discovery of mannose 6-phosphate as the lysosomal sorting signal. Lacking the M6P sorting signal, the lysosomal enzymes in I-cell patients are secreted rather than being sorted to and sequestered in lysosomes.

When fibroblasts from patients with I-cell disease are grown in a medium containing lysosomal enzymes bearing M6P residues, the diseased cells acquire a nearly normal intracellular content of lysosomal enzymes. This finding indicates that the plasma membrane of these cells contain M6P receptors, which can internalize extracellular phosphorylated lysosomal enzymes by receptor-mediated endocytosis. This process, used by many cell-surface receptors to bring bound proteins or particles into the cell, is discussed in detail in the next section. It is now known that even in normal cells, some M6P receptors are transported to the plasma membrane and some phosphorylated lysosomal enzymes are secreted (see Figure 17-23). The secreted enzymes can be retrieved by receptor-mediated endocytosis and directed to lysosomes. This pathway thus scavenges any lysosomal enzymes that escape the usual M6P sorting pathway.

◀ **FIGURE 17-23 Trafficking of soluble lysosomal enzymes from the *trans*-Golgi network and cell surface to lysosomes.** Newly synthesized lysosomal enzymes, produced in the ER, acquire mannose 6-phosphate (M6P) residues in the *cis*-Golgi (see Figure 17-22). For simplicity, only one phosphorylated oligosaccharide chain is depicted, although lysosomal enzymes typically have many such chains. In the *trans*-Golgi network, proteins that bear the M6P sorting signal interact with M6P receptors in the membrane and thereby are directed into clathrin/AP1 vesicles (step ◼**1**). The coat surrounding released vesicles is rapidly depolymerized (step ◼**2**), and the uncoated transport vesicles fuse with late endosomes (step ◼**3**). After the phosphorylated enzymes dissociate from the M6P receptors and are dephosphorylated, late endosomes subsequently fuse with a lysosome (step ◼**4**). Note that coat proteins and M6P receptors are recycled (steps ◼**2a** and ◼**4a**), and some receptors are delivered to the cell surface (step ◼**5**). Phosphorylated lysosomal enzymes occasionally are sorted from the *trans*-Golgi to the cell surface and secreted. These secreted enzymes can be retrieved by receptor-mediated endocytosis (steps ◼**6**–◼**8**), a process that closely parallels trafficking of lysosomal enzymes from the *trans*-Golgi network to lysosomes. [See G. Griffiths et al., 1988, *Cell* **52**:329; S. Kornfeld, 1992, *Ann. Rev. Biochem.* **61**:307; and G. Griffiths and J. Gruenberg, 1991, *Trends Cell Biol.* **1**:5.]

Hepatocytes from patients with I-cell disease contain a normal complement of lysosomal enzymes and no inclusions, even though these cells are defective in mannose phosphorylation. This finding implies that hepatocytes (the most abundant type of liver cell) employ a M6P-independent pathway for sorting lysosomal enzymes. The nature of this pathway, which also may operate in other cells types, is unknown. ∎

Protein Aggregation in the *Trans*-Golgi May Function in Sorting Proteins to Regulated Secretory Vesicles

As noted in the chapter introduction, all eukaryotic cells continuously secrete certain proteins, a process commonly called *constitutive secretion*. Specialized secretory cells also store other proteins in vesicles and secrete them only when triggered by a specific stimulus. One example of such *regulated secretion* occurs in pancreatic β cells, which store newly made insulin in special secretory vesicles and secrete insulin in response to an elevation in blood glucose (see Figure 15-7). These and other secretory cells simultaneously utilize two different types of vesicles to move proteins from the *trans*-Golgi network to the cell surface: regulated transport vesicles, often simply called secretory vesicles, and unregulated transport vesicles, also called constitutive secretory vesicles.

A common mechanism appears to sort regulated proteins as diverse as ACTH (adrenocorticotropic hormone), insulin, and trypsinogen into regulated secretory vesicles. Evidence for a common mechanism comes from experiments in which recombinant DNA techniques are used to induce the synthesis of insulin and trypsinogen in pituitary tumor cells already synthesizing ACTH. In these cells all three proteins segregate into the same regulated secretory vesicles and are secreted together when a hormone binds to a receptor on the pituitary cells and causes a rise in cytosolic Ca^{2+}. Although these three proteins share no identical amino acid sequences that might serve as a sorting sequence, they obviously have some common feature that signals their incorporation into regulated secretory vesicles.

Morphologic evidence suggests that sorting into the regulated pathway is controlled by selective protein aggregation. For instance, immature vesicles in this pathway—those that have just budded from the *trans*-Golgi network—contain diffuse aggregates of secreted protein that are visible in the electron microscope. These aggregates also are found in vesicles that are in the process of budding, indicating that proteins destined for regulated secretory vesicles selectively aggregate together before their incorporation into the vesicles.

Other studies have shown that regulated secretory vesicles from mammalian secretory cells contain three proteins, *chromogranin A*, *chromogranin B*, and *secretogranin II*, that together form aggregates when incubated at the ionic conditions (pH ≈6.5 and 1 mM Ca^{2+}) thought to occur in the *trans*-Golgi network; such aggregates do not form at the neutral pH of the ER. The selective aggregation of regulated secreted proteins together with chromogranin A, chromogranin B, or secretogranin II could be the basis for sorting of these proteins into

regulated secretory vesicles. Secreted proteins that do not associate with these proteins, and thus do not form aggregates, would be sorted into unregulated transport vesicles by default.

Some Proteins Undergo Proteolytic Processing After Leaving the *Trans*-Golgi

For some secretory proteins (e.g., growth hormone) and certain viral membrane proteins (e.g., the VSV glycoprotein), removal of the N-terminal ER signal sequence from the nascent chain is the only known proteolytic cleavage required to convert the polypeptide to the mature, active species (see Figure 16-6). However, some membrane and many soluble secretory proteins initially are synthesized as relatively long-lived, inactive precursors, termed *proproteins,* that require further proteolytic processing to generate the mature, active proteins. Examples of proteins that undergo such processing are soluble lysosomal enzymes, many membrane proteins such as influenza hemagglutinin (HA), and secreted proteins such as serum albumin, insulin, glucagon, and the yeast α mating factor. In general, the proteolytic conversion of a proprotein to the corresponding mature protein occurs after the proprotein has been sorted in the *trans*-Golgi network to appropriate vesicles.

In the case of soluble lysosomal enzymes, the proproteins are called *proenzymes*, which are sorted by the M6P receptor as catalytically inactive enzymes. In the late endosome or lysosome a proenzyme undergoes a proteolytic cleavage that generates a smaller but enzymatically active polypeptide. Delaying the activation of lysosomal proenzymes until they reach the lysosome prevents them from digesting macromolecules in earlier compartments of the secretory pathway.

Normally, mature vesicles carrying secreted proteins to the cell surface are formed by fusion of several immature

▶ **EXPERIMENTAL FIGURE 17-24 Proteolytic cleavage of proinsulin occurs in secretory vesicles after they have budded from the *trans*-Golgi network.** Serial sections of the Golgi region of an insulin-secreting cell were stained with (a) a monoclonal antibody that recognizes proinsulin but not insulin or (b) a different antibody that recognizes insulin but not proinsulin. The antibodies, which were bound to electron-opaque gold particles, appear as dark dots in these electron micrographs (see Figure 5-51). Immature secretory vesicles (closed arrowheads) and vesicles budding from the *trans*-Golgi (arrows) stain with the proinsulin antibody but not with insulin antibody. These vesicles contain diffuse protein aggregates that include proinsulin and other regulated secreted proteins. Mature vesicles (open arrowheads) stain with insulin antibody but not with proinsulin antibody and have a dense core of almost crystalline insulin. Since budding and immature secretory vesicles contain proinsulin (not insulin), the proteolytic conversion of proinsulin to insulin must take place in these vesicles after they bud from the *trans*-Golgi network. The inset in (a) shows a proinsulin-rich secretory vesicle surrounded by a protein coat (dashed line). [From L. Orci et al., 1987, *Cell* **49**:865; courtesy of L. Orci.]

ones containing proprotein. Proteolytic cleavage of proproteins, such as proinsulin, occurs in vesicles after they move away from the *trans*-Golgi network (Figure 17-24). The proproteins of most constitutively secreted proteins (e.g., albumin) are cleaved only once at a site C-terminal to a dibasic

(a) Proinsulin antibody

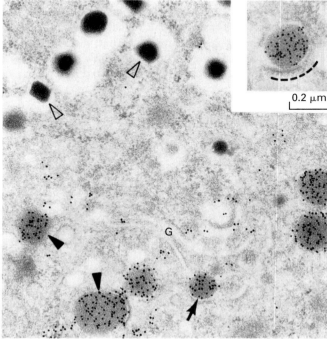

(b) Insulin antibody

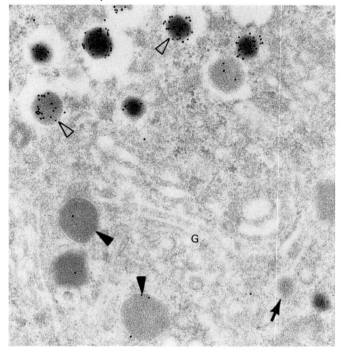

(a) Constitutive secreted proteins

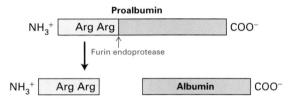

(b) Regulated secreted proteins

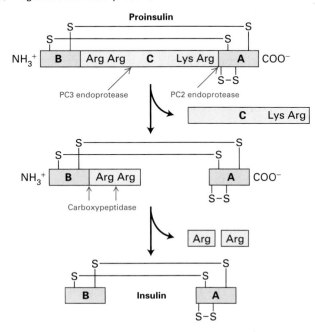

▲ **FIGURE 17-25 Proteolytic processing of proproteins in the constitutive and regulated secretion pathways.** The processing of proalbumin and proinsulin is typical of the constitutive and regulated pathways, respectively. The endoproteases that function in such processing cleave C-terminal to sequences of two consecutive basic amino acids. (a) The endoprotease furin acts on the precursors of constitutive secreted proteins. (b) Two endoproteases, PC2 and PC3, act on the precursors of regulated secreted proteins. The final processing of many such proteins is catalyzed by a carboxypeptidase that sequentially removes two basic amino acid residues at the C-terminus of a polypeptide. [See D. Steiner et al., 1992, *J. Biol. Chem.* **267**:23435.]

recognition sequence such as Arg-Arg or Lys-Arg (Figure 17-25a). Proteolytic processing of proteins whose secretion is regulated generally entails additional cleavages. In the case of proinsulin, multiple cleavages of the single polypeptide chain yields the N-terminal B chain and the C-terminal A chain of mature insulin, which are linked by disulfide bonds, and the central C peptide, which is lost and subsequently degraded (Figure 17-25b).

The breakthrough in identifying the proteases responsible for such processing of secreted proteins came from analysis of yeast with a mutation in the *KEX2* gene. These mutant cells synthesized the precursor of the α mating factor but could not proteolytically process it to the functional form, and thus were unable to mate with cells of the opposite mating type (see Figure 22-13). The wild-type *KEX2* gene encodes an endoprotease that cleaves the α-factor precursor at a site C-terminal to Arg-Arg and Lys-Arg residues. Using the *KEX2* gene as a DNA probe, researchers were able to clone a family of mammalian endoproteases, all of which cleave a protein chain on the C-terminal side of an Arg-Arg or Lys-Arg sequence. One, called *furin*, is found in all mammalian cells; it processes proteins such as albumin that are secreted by the continuous pathway. In contrast, the *PC2* and *PC3* *endoproteases* are found only in cells that exhibit regulated secretion; these enzymes are localized to regulated secretory vesicles and proteolytically cleave the precursors of many hormones at specific sites.

Several Pathways Sort Membrane Proteins to the Apical or Basolateral Region of Polarized Cells

The plasma membrane of polarized epithelial cells is divided into two domains, **apical** and **basolateral**; tight junctions located between the two domains prevent the movement of plasma-membrane proteins between the domains (see Figure 6-5). Several sorting mechanisms direct newly synthesized membrane proteins to either the apical or basolateral domain of epithelial cells, and any one protein may be sorted by more than one mechanism. Although these sorting mechanisms are understood in general terms, the molecular signals underlying the vesicle-mediated transport of membrane proteins in polarized cells are not yet

known. As a result of this sorting and the restriction on protein movement within the plasma membrane due to tight junctions, distinct sets of proteins are found in the apical or basolateral domain. This preferential localization of certain transport proteins is critical to a variety of important physiological functions, such as absorption of nutrients from the intestinal lumen and acidification of the stomach lumen (see Figures 7-27 and 7-28).

Microscopic and cell-fractionation studies indicate that proteins destined for either the apical or the basolateral membranes are initially located together within the membranes of the *trans*-Golgi network. In some cases, proteins destined for the apical membrane are sorted into their own transport vesicles that bud from the *trans*-Golgi network and then move to the apical region, whereas proteins destined for the basolateral membrane are sorted into other vesicles that move to the basolateral region. The different vesicle types can be distinguished by their protein constituents, including distinct Rab and v-SNARE proteins, which apparently target them to the appropriate plasma-membrane domain. In this mechanism, segregation of proteins destined for either the apical or basolateral membranes occurs as cargo proteins are incorporated into particular types of vesicles budding from the *trans*-Golgi network.

Such direct basolateral-apical sorting has been investigated in cultured Madin-Darby canine kidney (MDCK) cells, a line of cultured polarized epithelial cells (see Figure 6-6). In MDCK cells infected with the influenza virus, progeny viruses bud only from the apical membrane, whereas in cells infected with vesicular stomatitis virus (VSV), progeny viruses bud only from the basolateral membrane. This difference occurs because the HA glycoprotein of influenza

▶ **FIGURE 17-26 Sorting of proteins destined for the apical and basolateral plasma membranes of polarized cells.** When cultured MDCK cells are infected simultaneously with VSV and influenza virus, the VSV G glycoprotein (purple) is found only on the basolateral membrane, whereas the influenza HA glycoprotein (green) is found only on the apical membrane. Some cellular proteins (orange circle), especially those with a GPI anchor, are likewise sorted directly to the apical membrane and others to the basolateral membrane (not shown) via specific transport vesicles that bud from the *trans*-Golgi network. In certain polarized cells, some apical and basolateral proteins are transported together to the basolateral surface; the apical proteins (orange oval) then move selectively, by endocytosis and transcytosis, to the apical membrane. [After K. Simons and A. Wandinger-Ness, 1990, *Cell* **62**:207, and K. Mostov et al., 1992, *J. Cell Biol.* **116**:577.]

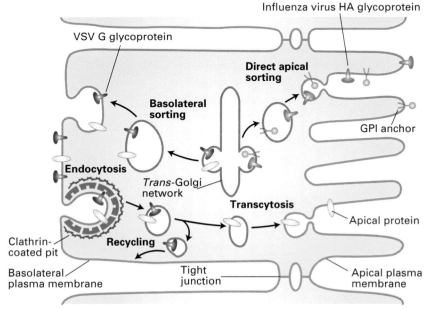

virus is transported from the Golgi complex exclusively to the apical membrane, and the VSV G protein is transported only to the basolateral membrane (Figure 17-26). Furthermore, when the gene encoding HA protein is introduced into uninfected cells by recombinant DNA techniques, all the expressed HA accumulates in the apical membrane, indicating that the sorting signal resides in the HA glycoprotein itself and not in other viral proteins produced during viral infection.

Among the cellular proteins that undergo similar apical-basolateral sorting in the Golgi are those with a *glycosylphosphatidylinositol (GPI) membrane anchor.* In MDCK cells and most other types of epithelial cells, GPI-anchored proteins are targeted to the apical membrane. In membranes GPI-anchored proteins are clustered into **lipid rafts,** which are rich in sphingolipids (see Figure 5-10). This finding suggests that lipid rafts are localized to the apical membrane along with proteins that preferentially partition into them in many cells. However, the GPI anchor is not an apical sorting signal in all polarized cells; in thyroid cells, for example, GPI-anchored proteins are targeted to the basolateral membrane. Other than GPI anchors no unique sequences have been identified that are both necessary and sufficient to target proteins to either the apical or basolateral domain. Instead, each membrane protein may contain multiple sorting signals, any one of which can target it to the appropriate plasma-membrane domain. The identification of such complex signals and of the vesicle coat proteins that recognize them is currently being pursued for a number of different proteins that are sorted to specific plasma-membrane domains of polarized epithelial cells.

Another mechanism for sorting apical and basolateral proteins, also illustrated in Figure 17-26, operates in hepatocytes. The basolateral membranes of hepatocytes face the blood (as in intestinal epithelial cells), and the apical membranes line the small intercellular channels into which bile is secreted. In hepatocytes, newly made apical and basolateral proteins are first transported in vesicles from the *trans*-Golgi network to the basolateral region and incorporated into the plasma membrane by exocytosis (i.e., fusion of the vesicle membrane with the plasma membrane). From there, both basolateral and apical proteins are endocytosed in the same vesicles, but then their paths diverge. The endocytosed basolateral proteins are sorted into transport vesicles that recycle them to the basolateral membrane. In contrast, the apically destined endocytosed proteins are sorted into transport vesicles that move across the cell and fuse with the apical membrane, a process called **transcytosis.** As discussed in the next section, transcytosis also is used to move extracellular materials from one side of an epithelium to another. Even in epithelial cells, such as MDCK cells, in which apical-basolateral protein sorting occurs in the Golgi, transcytosis may provide a "fail-safe" sorting mechanism. That is, an apical protein sorted incorrectly to the basolateral membrane would be subjected to endocytosis and then correctly delivered to the apical membrane.

KEY CONCEPTS OF SECTION 17.4

Protein Sorting and Processing in Later Stages of the Secretory Pathway

■ The *trans*-Golgi network (TGN) is a major branch point in the secretory pathway where soluble secreted proteins, lysosomal proteins, and in some cells membrane proteins destined for the basolateral or apical plasma membrane are segregated into different transport vesicles.

■ Many vesicles that bud from the *trans*-Golgi network as well as endocytic vesicles bear a coat composed of AP (adapter protein) complexes and clathrin (see Figure 17-19).

■ Pinching off of clathrin-coated vesicles requires dynamin, which forms a collar around the neck of the vesicle bud and hydrolyzes GTP (see Figure 17-20).

■ Soluble enzymes destined for lysosomes are modified in the *cis*-Golgi yielding multiple mannose 6-phosphate (M6P) residues on their oligosaccharide chains.

■ M6P receptors in the membrane of the *trans*-Golgi network bind proteins bearing M6P residues and direct their transfer to late endosomes, where receptors and their ligand proteins dissociate. The receptors then are recycled to the Golgi or plasma membrane, and the lysosomal enzymes are delivered to lysosomes (see Figure 17-23).

■ Regulated secreted proteins are concentrated and stored in secretory vesicles to await a neural or hormonal signal for exocytosis. Protein aggregation within the *trans*-Golgi network may play a role in sorting secreted proteins to the regulated pathway.

■ Many proteins transported through the secretory pathway undergo post-Golgi proteolytic cleavages that yield the mature, active proteins. Generally, proteolytic maturation can occur in vesicles carrying proteins from the *trans*-Golgi network to the cell surface, in the late endosome, or in the lysosomal.

■ In polarized epithelial cells, membrane proteins destined for the apical or basolateral domains of the plasma membrane are sorted in the *trans*-Golgi network into different transport vesicles (see Figure 17-26). The GPI anchor is the only apical-basolateral sorting signal identified so far.

■ In hepatocytes and some other polarized cells, all plasma-membrane proteins are directed first to the basolateral membrane. Apically destined proteins then are endocytosed and moved across the cell to the apical membrane (transcytosis).

17.5 Receptor-Mediated Endocytosis and the Sorting of Internalized Proteins

In previous sections we have explored the main pathways whereby secretory and membrane proteins synthesized on the rough ER are delivered to the cell surface or other destinations. Cells also can internalize materials from their surroundings and sort these to particular destinations. A few cell types (e.g., macrophages) can take up whole bacteria and other large particles by **phagocytosis**, a nonselective actin-mediated process in which extensions of the plasma membrane envelop the ingested material, forming large vesicles called phagosomes (see Figure 5-20). In contrast, all eukaryotic cells continually engage in endocytosis, a process in which a small region of the plasma membrane invaginates to form a membrane-limited vesicle about 0.05–0.1 μm in diameter. In one form of endocytosis, called *pinocytosis*, small droplets of extracellular fluid and any material dissolved in it are nonspecifically taken up. Our focus in this section, however, is on **receptor-mediated endocytosis** in which a specific receptor on the cell surface binds tightly to an extracellular macromolecular ligand that it recognizes; the plasma-membrane region containing the receptor-ligand complex then buds inward and pinches off, becoming a transport vesicle.

Among the common macromolecules that vertebrate cells internalize by receptor-mediated endocytosis are cholesterol-containing particles called low-density lipoprotein (LDL); the iron-binding protein transferrin; many protein hormones (e.g., insulin); and certain glycoproteins. Receptor-mediated endocytosis of such ligands generally occurs via clathrin/AP2-coated pits and vesicles in a process similar to the packaging of lysosomal enzymes by mannose 6-phosphate (M6P) in the *trans*-Golgi network (see Figure 17-23). As noted earlier, some M6P receptors are found on the cell surface, and these participate in the receptor-mediated endocytosis of lysosomal enzymes that are secreted. In general, transmembrane receptor proteins that function in the uptake of extracellular ligands are internalized from the cell surface during endocytosis and are then sorted and recycled back to the cell surface, much like the recycling of M6P receptors to the plasma membrane and *trans*-Golgi. The rate at which a ligand is internalized is limited by the amount of its corresponding receptor on the cell surface.

Clathrin/AP2 pits make up about 2 percent of the surface of cells such as hepatocytes and fibroblasts. Many internalized ligands have been observed in these pits and vesicles, which are thought to function as intermediates in the endocytosis of most (though not all) ligands bound to cell-surface receptors (Figure 17-27). Some receptors are clustered over clathrin-coated pits even in the absence of ligand. Other receptors diffuse freely in the plane of the plasma membrane but undergo a conformational change when binding to ligand, so that when the receptor-ligand complex diffuses into a clathrin-coated pit, it is retained there. Two or more types of receptor-bound ligands, such as LDL and transferrin, can be seen in the same coated pit or vesicle.

▶ **EXPERIMENTAL FIGURE 17-27**
The initial stages of receptor-mediated endocytosis of low-density lipoprotein (LDL) particles are revealed by electron microscopy. Cultured human fibroblasts were incubated in a medium containing LDL particles covalently linked to the electron-dense, iron-containing protein ferritin; each small iron particle in ferritin is visible as a small dot under the electron microscope. Cells initially were incubated at 4 °C; at this temperature LDL can bind to its receptor but internalization does not occur. After excess LDL not bound to the cells was washed away, the cells were warmed to 37 °C and then prepared for microscopy at periodic intervals. (a) A coated pit, showing the clathrin coat on the inner (cytosolic) surface of the pit, soon after the temperature was raised. (b) A pit containing LDL apparently closing on itself to form a coated vesicle. (c) A coated vesicle containing ferritin-tagged LDL particles. (d) Ferritin-tagged LDL particles in a smooth-surfaced early endosome 6 minutes after internalization began. [Photographs courtesy of R. Anderson. Reprinted by permission from J. Goldstein et al., *Nature* **279**:679. Copyright 1979, Macmillan Journals Limited. See also M. S. Brown and J. Goldstein, 1986, *Science* **232**:34.]

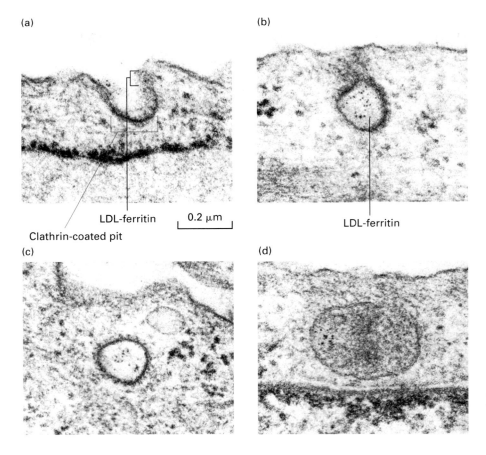

(a)   (b)

LDL-ferritin   0.2 μm

Clathrin-coated pit   LDL-ferritin

(c)   (d)

Receptors for Low-Density Lipoprotein and Other Ligands Contain Sorting Signals That Target Them for Endocytosis

As will be discussed in detail in the next chapter, **low-density lipoprotein (LDL)** is one of several complexes that carry cholesterol through the bloodstream (see Figure 17-28). A LDL particle, a sphere 20–25 nm in diameter, has an outer phospholipid shell containing a single molecule of a large protein known as *apoB-100*; the core of a particle is packed with cholesterol in the form of cholesteryl esters (see Figure 18-12). Most mammalian cells produce cell-surface receptors that specifically bind to apoB-100 and internalize LDL particles by receptor-mediated endocytosis. After endocytosis, the LDL particles are transported to lysosomes via the endocytic pathway and then are degraded by lysosomal hydrolases. LDL receptors, which dissociate from their ligands in the late endosome, recycle to the cell surface.

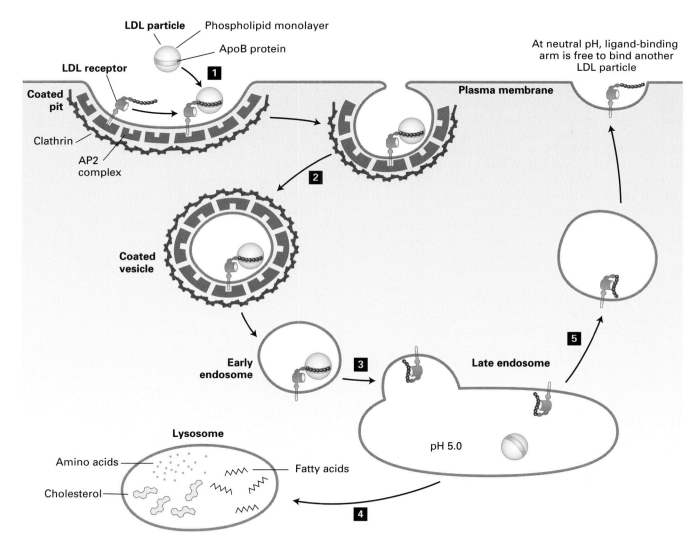

▲ **FIGURE 17-28 Endocytic pathway for internalizing low-density lipoprotein (LDL).** Step **1**: Cell-surface LDL receptors bind to an apoB protein embedded in the phospholipid outer layer of LDL particles. Interaction between the NPXY sorting signal in the cytosolic tail of the LDL receptor and the AP2 complex incorporates the receptor-ligand complex into forming endocytic vesicles. Step **2**: Clathrin-coated pits (or buds) containing receptor-LDL complexes are pinched off by the same dynamin-mediated mechanism used to form clathrin/AP1 vesicles on the *trans*-Golgi network (see Figure 17-20). Step **3**: After the vesicle coat is shed, the uncoated endocytic vesicle (early endosome) fuses with the late endosome. The acidic pH in this compartment causes a conformational change in the LDL receptor that leads to release of the bound LDL particle. Step **4**: The late endosome fuses with the lysosome, and the proteins and lipids of the free LDL particle are broken down to their constituent parts by enzymes in the lysosome. Step **5**: The LDL receptor recycles to the cell surface where at the neutral pH of the exterior medium the receptor undergoes a conformational change so that it can bind another LDL particle. [See M. S. Brown and J. L. Goldstein, 1986, *Science* **232**:34, and G. Rudenko et al., 2002, *Science* **298**:2353.]

Studies of the inherited disorder *familial hypercholes-terolemia* led to discovery of the LDL receptor and the initial understanding of the endocytic pathway. An individual with this disorder produces one of several mutant forms of the LDL receptor, causing impaired endocytosis of LDL and high serum levels of cholesterol (Chapter 18). The major features of the LDL endocytic pathway as currently understood are depicted in Figure 17-28. The LDL receptor is an 839-residue glycoprotein with a single transmembrane segment; it has a short C-terminal cytosolic segment and a long N-terminal exoplasmic segment that contains a β-propeller domain and a ligand-binding domain. Seven cysteine-rich imperfect repeats form the ligand-binding domain, which interacts with the apoB-100 molecule in a LDL particle.

Mutant receptors from some individuals with familial hypercholesterolemia bind LDL normally, but the LDL-receptor complex cannot be internalized by the cell and is distributed evenly over the cell surface rather than being confined to clathrin/AP2-coated pits. In individuals with this type of defect, plasma-membrane receptors for other ligands are internalized normally, but the mutant LDL receptor apparently is not recruited into coated pits. Analysis of this mutant receptor and other mutant LDL receptors generated experimentally and expressed in fibroblasts identified a four-residue motif in the cytosolic segment of the receptor that is crucial for its internalization: Asn-Pro-X-Tyr where X can be any amino acid. This *NPXY sorting signal* binds to the AP2 complex, linking the clathrin/AP2 coat to the cytosolic segment of the LDL receptor in forming coated pits. A mutation in any of the conserved residues of the NPXY signal will abolish the ability of the LDL receptor to be incorporated into coated pits.

A small number of individuals who exhibit the usual symptoms associated with familial hypercholesterolemia produce normal LDL receptors. In these individuals, the gene encoding the AP2 subunit protein that binds the NPXY sorting signal is defective. As a result, LDL receptors are not incorporated into clathrin/AP2 vesicles and endocytosis of LDL particles is compromised. Analysis of patients with this genetic disorder highlights the importance of adapter proteins in protein trafficking mediated by clathrin vesicles. ▮

Mutational studies have shown that other cell-surface receptors can be directed into forming clathrin/AP2 pits by a different sorting signal: Tyr-X-X-Φ, where X can be any amino acid and Φ is a bulky hydrophobic amino acid. This YXXΦ sorting signal in the cytosolic segment of a receptor protein binds to a specific cleft in the μ2 subunit of the AP2 complex. Because the tyrosine and Φ residues mediate this binding, a mutation in either one reduces or abolishes the ability of the receptor to be incorporated into clathrin/AP2-coated pits. Moreover, if influenza HA protein, which is not normally endocytosed, is genetically engineered to contain this four-residue sequence in its cytosolic domain, the mutant HA is internalized. Recall from our earlier discussion that

this same sorting signal recruits membrane proteins into clathrin/AP1 vesicles that bud from the *trans*-Golgi network by binding to the μ1 subunit of AP1 (see Table 17-2). All these observations indicate that YXXΦ is a widely used signal for sorting membrane proteins to clathrin-coated vesicles.

In some cell-surface proteins, however, other sequences (e.g., Leu-Leu) or covalently linked ubiquitin molecules signal endocytosis. Among the proteins associated with clathrin/AP2 vesicles, several contain domains that specifically bind to ubiquitin, and it has been hypothesized that these vesicle-associated proteins mediate the selective incorporation of ubiquitinated membrane proteins into endocytic vesicles. As described later, the ubiquitin tag on endocytosed membrane proteins is also recognized at a later stage in the endocytic pathway and plays a role in delivering these proteins into the interior of the lysosome where they are degraded.

The Acidic pH of Late Endosomes Causes Most Receptor-Ligand Complexes to Dissociate

The overall rate of endocytic internalization of the plasma membrane is quite high; cultured fibroblasts regularly internalize 50 percent of their cell-surface proteins and phospholipids each hour. Most cell-surface receptors that undergo endocytosis will repeatedly deposit their ligands within the

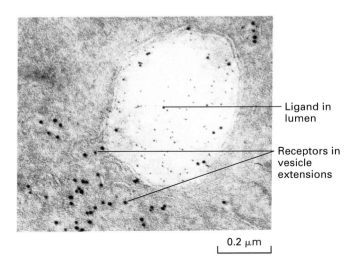

Ligand in lumen

Receptors in vesicle extensions

0.2 μm

▲ **EXPERIMENTAL FIGURE 17-29 Electron microscopy demonstrates that endocytosed receptor-ligand complexes dissociate in late endosomes.** Liver cells were perfused with an asialoglycoprotein ligand and then were fixed and sectioned for electron microscopy. The sections were stained with receptor-specific antibodies, tagged with gold particles 8 nm in diameter, to localize the receptor and with asialoglycoprotein-specific antibody, linked to gold particles 5 nm in diameter, to localize the ligand (see Figure 5-51). As seen in this electron micrograph of a late endosome, the ligand (smaller dark grains) is localized in the vesicle lumen and the asialoglycoprotein receptor (larger dark grains) is localized in the tubular extensions budding off from the vesicle. [Courtesy of H. J. Geuze. Copyright 1983, M.I.T. See H. J. Geuze et al., 1983, *Cell* **32**:277.]

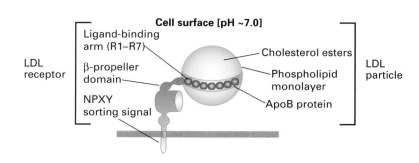

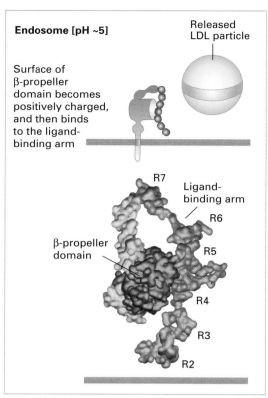

▲ **FIGURE 17-30 Model for pH-dependent binding of LDL particles by the LDL receptor.** Schematic depiction of LDL receptor at neutral pH found at the cell surface (*left*) and at acidic pH found in the interior of the late endosome (*right*). At the cell surface, apoB-100 on the surface of a LDL particle binds tightly to the receptor. Of the seven repeats (R1–R7) in the ligand-binding arm, R4 and R5 appear to be most critical for LDL binding. Within the endosome, histidine residues in the β-propeller domain of the LDL receptor become protonated. The positively charged propeller can bind with high affinity to the ligand-binding arm, which contains negatively charged residues, causing release of the LDL particle. Experimental electron density and C_α trace model of the extracellular region of the LDL receptor at pH 5.3 based on X-ray crystallographic analysis. In this conformation, extensive hydrophobic and ionic interactions occur between the β propeller and the R4 and R5 repeats. Red spheres represent Ca^{2+} ions. [Part (b) from G. Rudenko et al., 2002, *Science* **298**:2353.]

cell and then recycle to the plasma membrane, once again to mediate internalization of ligand molecules. For instance, the LDL receptor makes one round trip into and out of the cell every 10–20 minutes, for a total of several hundred trips in its 20-hour life span.

Internalized receptor-ligand complexes commonly follow the pathway depicted for the M6P receptor in Figure 17-23 and the LDL receptor in Figure 17-28. Endocytosed cell-surface receptors typically dissociate from their ligands within late endosomes, which appear as spherical vesicles with tubular branching membranes located a few micrometers from the cell surface. The original experiments that defined the late endosome sorting vesicle utilized the asialoglycoprotein receptor. This liver-specific protein mediates the binding and internalization of abnormal glycoproteins whose oligosaccharides terminate in galactose rather than the normal sialic acid, hence the name *asialo*glycoprotein. Electron microscopy of liver cells perfused with asialoglycoprotein reveal that 5–10 minutes after internalization, ligand molecules are found in the lumen of late endosomes, while the tubular membrane extensions are rich in receptor and rarely contain ligand (Figure 17-29). These findings indicate that the late endosome is the organelle in which receptors and ligands are uncoupled.

The dissociation of receptor-ligand complexes in late endosomes occurs not only in the endocytic pathway but also in the delivery of soluble lysosomal enzymes via the secretory pathway (see Figure 17-23). As discussed in Chapter 7, the membranes of late endosomes and lysosomes contain V-class

proton pumps that act in concert with Cl^- channels to acidify the vesicle lumen (see Figure 7-10). Most receptors, including the M6P receptor and cell-surface receptors for LDL particles and asialoglycoprotein, bind their ligands tightly at neutral pH but release their ligands if the pH is lowered to 6.0 or below. The late endosome is the first vesicle encountered by receptor-ligand complexes whose luminal pH is sufficiently acidic to promote dissociation of most endocytosed receptors from their tightly bound ligands.

The mechanism by which the LDL receptor releases bound LDL particles is now understood in detail (Figure 17-30). At the endosomal pH of 5.0–5.5, histidine residues in the β-propeller domain of the receptor become protonated, forming a site that can bind with high affinity to the negatively charged repeats in the ligand-binding domain. This intramolecular interaction sequesters the repeats in a conformation that cannot simultaneously bind to apoB-100, thus causing release of the bound LDL particle.

The Endocytic Pathway Delivers Iron to Cells Without Dissociation of Receptor-Transferrin Complex in Endosomes

An exception to the general theme of pH-dependent receptor-ligand dissociation in the late endosome occurs in the endocytic pathway that delivers transferrin-bound iron to cells. A major glycoprotein in the blood, transferrin transports iron to all tissue cells from the liver (the main site of iron storage in the body) and from the intestine (the site of iron absorption). The iron-free form, *apotransferrin*, binds two Fe^{3+} ions very tightly to form *ferrotransferrin*. All mammalian cells contain cell-surface transferrin receptors that avidly bind ferrotransferrin at neutral pH, after which the receptor-bound ferrotransferrin is subjected to endocytosis. Like the components of a LDL particle, the two bound Fe^{3+} atoms remain in the cell, but the apotransferrin part of the ligand does not dissociate from the receptor and is secreted from the cell within minutes after being endocytosed.

Although apotransferrin remains bound to the transferrin receptor at the low pH of late endosomes, changes in pH are critical to functioning of the transferrin endocytic pathway. At a pH below 6.0, the two bound Fe^{3+} atoms dissociate from ferrotransferrin, are reduced to Fe^{2+} by an unknown mechanism, and then are exported into the cytosol by an endosomal transporter specific for divalent metal ions. The receptor-apotransferrin complex remaining after dissociation of the iron atoms is recycled back to the cell surface. Although apotransferrin binds tightly to its receptor at a pH of 5.0 or 6.0, it does not bind at neutral pH. Hence the bound apotransferrin dissociates from the transferrin receptor when the recycling vesicles fuse with the plasma membrane and the receptor-ligand complex encounters the neutral pH of the extracellular interstitial fluid or growth medium. The recycled receptor is then free to bind another molecule of ferrotransferrin, and the released apotransferrin is carried in the bloodstream to the liver or intestine to be reloaded with iron.

Specialized Vesicles Deliver Cell Components to the Lysosome for Degradation

The major function of lysosomes is to degrade extracellular materials taken up by the cell and intracellular components under certain conditions. Materials to be degraded must be delivered to the lumen of the lysosome where the various degradative enzymes reside. As just discussed, endocytosed ligands (e.g., LDL particles) that dissociate from their receptors in the late endosome subsequently enter the lysosomal lumen when the membrane of the late endosome fuses with the membrane of the lysosome (see Figure 17-28). Likewise, phagosomes carrying bacteria or other particulate matter can fuse with lysosomes, releasing their contents into the lumen for degradation. However, the delivery of endocytosed membrane proteins and of cytoplasmic materials to lysosomes for degradation poses special problems and involves two unusual types of vesicles.

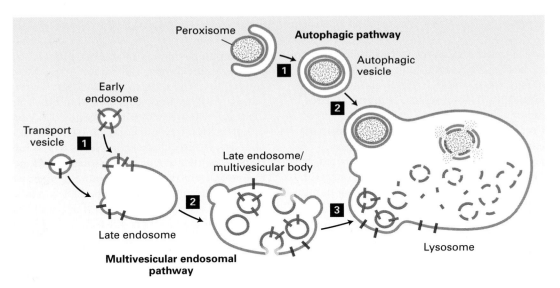

▲ **FIGURE 17-31 Delivery of plasma-membrane proteins and cytoplasmic components to the lysosomal interior for degradation.** *Left:* Early endosomes carrying endocytosed plasma-membrane proteins (blue) and vesicles carrying lysosomal membrane proteins (red) from the *trans*-Golgi network fuse with the late endosome, transferring their membrane proteins to the endosomal membrane (step ■). Proteins to be degraded are incorporated into vesicles that bud *into* the interior of the late endosome, eventually forming a multivesicular endosome containing many such internal vesicles (step ■). Fusion of a multivesicular endosome directly with a lysosome releases the internal vesicles into the lumen of the lysosome where they can be degraded (step ■). Because proton pumps and other lysomal membrane proteins normally are not incorporated into internal endosomal vesicles, they are delivered to the lysosomal membrane and are protected from degradation. *Right:* In the autophagic pathway, a cup-shaped structure forms around portions of the cytosol or an organelle such as a peroxisome, as shown here. Continued addition of membrane eventually leads to the formation of an autophagic vesicle that envelopes its contents by two complete membranes (step ■). Fusion of the outer membrane with the membrane of a lysosome releases a single-layer vesicle and its contents into the lysosome interior (step ■). [See F. Reggiori and D. J. Klionsky, 2002, *Eukaryot. Cell* **1**:11, and D. J. Katzmann et al., 2002, *Nature Rev. Mol. Cell Biol.* **3**:893.]

Multivesicular Endosomes Resident lysosomal proteins, such as V-class proton pumps and other lysosomal membrane proteins, can carry out their functions and remain in the lysosomal membrane where they are protected from degradation by the soluble hydrolytic enzymes in the lumen. Such proteins are delivered to the lysosomal membrane by transport vesicles that bud from the *trans*-Golgi network by the same basic mechanisms described in earlier sections. In contrast, endocytosed membrane proteins to be degraded are transferred in their entirety to the interior of the lysosome by a specialized delivery mechanism. Lysosomal degradation of cell-surface receptors for extracellular signaling molecules is a common mechanism for controlling the sensitivity of cells to such signals (Chapter 13). Receptors that become damaged also are targeted for lysosomal degradation.

Early evidence that membranes can be delivered to the lumen of compartments came from electron micrographs showing membrane vesicles and fragments of membranes within endosomes and lysosomes (see Figure 5-20c). Parallel experiments in yeast revealed that endocytosed receptor proteins targeted to the vacuole (the yeast organelle equivalent to the lysosome) were primarily associated with membrane fragments and small vesicles within the interior of the vacuole rather than with the vacuole surface membrane.

These observations suggest that endocytosed membrane proteins can be incorporated into specialized vesicles that form at the endosomal membrane (Figure 17-31, *left*). Although these vesicles are similar in size and appearance to transport vesicles, they differ topologically. Transport vesicles bud *outward* from the surface of a donor organelle into the cytosol, whereas vesicles within the endosome bud *inward* from the surface into the lumen (away from the cytosol). Mature endosomes containing numerous vesicles in their interior are usually called *multivesicular endosomes* (or bodies). Eventually the surface membrane of a multivesicular endosome fuses with the membrane of a lysosome, thereby delivering its internal vesicles and the membrane proteins they contain into the lysosome interior for degradation. Thus the sorting of proteins in the endosomal membrane determines which ones will remain on the lysosome surface (e.g., pumps and transporters) and which ones will be incorporated into internal vesicles and ultimately degraded in lysosomes.

Autophagic Vesicles The delivery of bulk amounts of cytosol or entire organelles to lysosomes and their subsequent degradation is known as *autophagy* ("eating oneself"). Autophagy is often a regulated process and is typically induced in cells placed under conditions of starvation or other types of stress, allowing the cell to recycle macromolecules for use as nutrients.

The autophagic pathway begins with the formation of a flattened double-membraned cup-shaped structure (Figure 17-31, *right*). This structure can grow by vesicle fusion and eventually seals to form an autophagic vesicle that envelops a region of the cytosol or an entire organelle (e.g., peroxisome, mitochondrion). Unknown at this time is the origin of the membranes that form the initial cup-shaped organelle and the vesicles that are added to it, but the endosome itself is a likely candidate. The outer membrane of an autophagic vesicle can fuse with the lysosome delivering a large vesicle, bounded by a single membrane bilayer, to the interior of the lysosome. The lipases and proteases within the lysosome eventually will degrade this vesicle and its contents into their molecular components.

Retroviruses Bud from the Plasma Membrane by a Process Similar to Formation of Multivesicular Endosomes

The vesicles that bud into the interior of endosomes have a topology similar to that of enveloped virus particles that bud from the plasma membrane of virus-infected cells. Moreover, recent experiments demonstrate that a common set of proteins are required for both types of membrane-budding events. In fact, the two processes so closely parallel one another in mechanistic detail as to suggest that enveloped viruses have evolved mechanisms to recruit the cellular proteins used in inward endosomal budding for their own purposes.

Many of the proteins required for inward budding of the endosomal membrane were first identified by mutations in yeast that blocked delivery of membrane proteins to the interior of the vacuole. More than 10 such "budding" proteins have been identified in yeast, most with significant similarities to mammalian proteins that evidently perform the same function in mammalian cells. The current model of endosomal budding to form multivesicular endosomes in mammalian cells is based primarily on studies in yeast (Figure 17-32). A ubiquitin-tagged peripheral membrane protein of the endosome, known as Hrs, facilitates loading of specific ubiquitinated membrane cargo proteins into vesicle buds directed into the interior of the endosome. The ubiquitinated Hrs protein then recruits a set of three different protein complexes to the membrane. These *ESCRT* (endosomal sorting complexes required for transport) *complexes* include the ubiquitin-binding protein Tsg101. The membrane-associated ESCRT complexes act to complete vesicle budding, leading to release of a vesicle carrying specific membrane cargo into the interior of the endosome. Finally, an ATPase, known as Vps4, uses the energy from ATP hydrolysis to disassemble the ESCRT complexes, releasing them into the cytosol for another round of budding. In the fusion event that pinches off a completed endosomal vesicle, the ESCRT proteins and Vps4 may function like SNAREs and NSF, respectively, in the typical membrane-fusion process discussed previously (see Figure 17-11).

The human immunodeficiency virus (HIV) is an enveloped retrovirus that buds from the plasma membrane of infected cells in a process driven by viral Gag protein, the major structural component of completed virus particles. Gag protein binds to the plasma membrane of an infected cell and ≈4000 Gag molecules polymerize into a spherical

▶ **FIGURE 17-32 Model of the common mechanism for formation of multivesicular endosomes and budding of HIV from the plasma membrane.**
Bottom: In endosomal budding, ubiquitinated Hrs on the endosomal membrane directs loading of specific membrane cargo proteins (blue) into vesicle buds and then recruits cytosolic ESCRT complexes to the membrane (step **1**). Note that both Hrs and the recruited cargo proteins are tagged with ubiquitin. After the set of bound ESCRT complexes mediate membrane fusion and pinching off of the completed vesicle (step **2**), they are disassembled by the ATPase Vps4 and returned to the cytosol (step **3**). *Top:* Budding of HIV particles from HIV-infected cells occurs by a similar mechanism using the virally encoded Gag protein and cellular ESCRT complexes and Vps4 (steps **4**–**6**). Ubiquitinated Gag near a budding particle functions like Hrs. See text for discussion. [Adapted from O. Pornillos et al., 2002, *Trends Cell Biol.* **12**:569.]

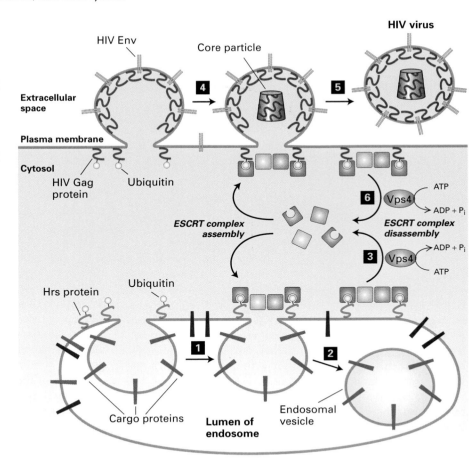

shell, producing a structure that looks like a vesicle bud protruding outward from the plasma membrane. Mutational studies with HIV have revealed that the N-terminal segment of Gag protein is required for association with the plasma membrane, whereas the C-terminal segment is required for pinching off of complete HIV particles. For instance, if the portion of the viral genome encoding the C-terminus of Gag is removed, HIV buds will form in infected cells, but pinching off does not occur and thus no free virus particles are released.

The first indication that HIV budding employs the same molecular machinery as vesicle budding into endosomes came from the observation that Tsg101, a component of the ESCRT complex, binds to the C-terminus of Gag protein. Subsequent findings have clearly established the mechanistic parallels between the two processes (Figure 17-32). For example, Gag is ubiquitinated as part of the process of virus budding, and in cells with mutations in Tsg101 or Vps4, HIV virus buds accumulate but cannot pinch off from the membrane (Figure 17-33). Moreover, when a segment from the cellular Hrs protein is added to a truncated Gag protein, proper budding and release of virus particles is restored. Taken together, these results indicate that Gag protein mimics the function of Hrs, redirecting ESCRT complexes to the

(a)   (b)

▶ **FIGURE 17-33 Electron micrographs of virus budding from wild-type and ESCRT-deficient HIV-infected cells.** (a) In wild-type cells infected with HIV, virus particles bud from the plasma membrane and are rapidly released into the extracellular space. (b) In cells that lack the functional ESCRT protein Tsg101, the viral Gag protein forms dense virus-like structures, but budding of these structures from the plasma membrane cannot be completed and chains of incomplete viral buds still attached to the plasma membrane accumulate. [Wes Sundquist, University of Utah.]

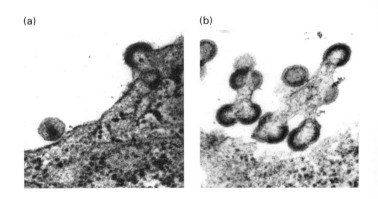

plasma membrane where they can function in the budding of virus particles.

Other enveloped retroviruses such as murine leukemia virus and Rous sarcoma virus also have been shown to require ESCRT complexes for their budding, although each virus appears to have evolved a somewhat different mechanism to recruit ESCRT complexes to the site of virus budding.

Transcytosis Moves Some Endocytosed Ligands Across an Epithelial Cell Layer

As noted previously, transcytosis is used by some cells in the apical-basolateral sorting of certain membrane proteins (see Figure 17-26). This process of transcellular transport, which combines endocytosis and exocytosis, also can be employed to import an extracellular ligand from one side of a cell, transport it across the cytoplasm, and secrete it from the plasma membrane at the opposite side. Transcytosis occurs mainly in sheets of polarized epithelial cells.

Maternal immunoglobulins (antibodies) contained in ingested breast milk are transported across the intestinal epithelial cells of the newborn mouse and human by transcytosis (Figure 17-34). The F_c receptor that mediates this movement binds antibodies at the acidic pH of 6 found in the intestinal lumen but not at the neutral pH of the extracellular fluid on the basal side of the intestinal epithelium. This difference in the pH of the extracellular media on the two sides of intes-

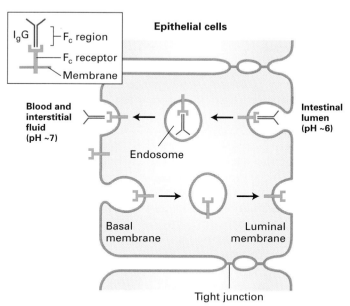

▲ **FIGURE 17-34 Transcytosis of maternal IgG immunoglobulins across the intestinal epithelial cells of newborn mice.** This transcellular movement of a ligand involves both endocytosis and exocytosis. The one-way movement of ligand from the intestinal lumen to the blood depends on the differential affinity of the F_c receptor for antibody at pH 6 (strong binding) and at pH 7 (weak binding). Transcytosis in the opposite direction returns the empty F_c receptor to the luminal membrane. See text for discussion.

tinal epithelial cells allows maternal immunoglobulins to move in one direction—from the lumen to the blood. The same process also moves circulating maternal immunoglobulins across mammalian yolk-sac cells into the fetus.

KEY CONCEPTS OF SECTION 17.5

Receptor-Mediated Endocytosis and the Sorting of Internalized Proteins

■ Some extracellular ligands that bind to specific cell-surface receptors are internalized, along with their receptors, in clathrin-coated vesicles whose coats also contain AP2 complexes.

■ Sorting signals in the cytosolic domain of cell-surface receptors target them into clathrin/AP2-coated pits for internalization. Known signals include the Asn-Pro-X-Tyr, Tyr-X-X-Φ, and Leu-Leu sequences (see Table 17-2).

■ The endocytic pathway delivers some ligands (e.g., LDL particles) to lysosomes where they are degraded. Most receptor-ligand complexes dissociate in the acidic milieu of the late endosome; the receptors are recycled to the plasma membrane, while the ligands are sorted to lysosomes (see Figure 17-28).

■ Iron is imported into cells by an endocytic pathway in which Fe^{3+} ions are released from ferrotransferrin in the late endosome. The receptor-apotransferrin complex is recycled to the cell surface where the complex dissociates, releasing both the receptor and apotransferrin for reuse.

■ Endocytosed membrane proteins destined for degradation in the lysosome are incorporated into vesicles that bud into the interior of the endosome. Multivesicular endosomes, which contain many of these internal vesicles, can fuse with the lysosome to deliver the vesicles to the interior of the lysosome (see Figure 17-31).

■ A portion of the cytoplasm or an entire organelle (e.g., peroxisome) can be enveloped in a flattened membrane and eventually incorporated into a double-membraned autophagic vesicle. Fusion of the outer vesicle membrane with the lysosome delivers the enveloped contents to the interior of the lysosome for degradation.

■ Some of the cellular components (e.g., ESCRT complexes) that mediate inward budding of endosomal membranes are used in the budding and pinching off of enveloped viruses such as HIV from the plasma membrane of virus-infected cells (see Figure 17-32).

17.6 Synaptic Vesicle Function and Formation

In this final section we consider the regulated secretion of **neurotransmitters** that is the basis for signaling by many nerve cells. These small, water-soluble molecules (e.g., acetylcholine, dopamine) are released at chemical **synapses,** specialized sites of contact between a signaling neuron and a receiving cell. Generally signals are transmitted in only one direction: an axon terminal from a *presynaptic cell* releases neurotransmitter molecules that diffuse through a narrow extracellular space (the synaptic cleft) and bind to receptors on a *postsynaptic cell* (see Figure 7-31). The membrane of the postsynaptic cell, which can be another neuron, a muscle cell, or a gland cell, is located within approximately 50 nm of the presynaptic membrane.

Neurotransmitters are stored in specialized regulated secretory vesicles, known as *synaptic vesicles,* which are 40–50 nm in diameter. Exocytosis of these vesicles and release of neurotransmitters is initiated when a stimulatory electrical impulse (**action potential**) travels down the axon of a presynaptic cell to the axon terminal where it triggers opening of voltage-gated Ca^{2+} channels. The subsequent localized rise in the cytosolic Ca^{2+} concentration induces some synaptic vesicles to fuse with the plasma membrane, releasing their contents into the synaptic cleft. We described the major events in signal transmission at chemical synapses and the effects of neurotransmitter binding on postsynaptic cells in Chapter 7. Here we focus on the regulated secretion of neurotransmitters and the formation of synaptic vesicles in the context of the basic principles of vesicular trafficking already outlined in this chapter.

Synaptic Vesicles Loaded with Neurotransmitter Are Localized Near the Plasma Membrane

The exocytosis of neurotransmitters from synaptic vesicles involves targeting and fusion events similar to those that lead to release of secreted proteins in the secretory pathway. However, several unique features permit the very rapid release of neurotransmitters in response to arrival of an action potential at the presynaptic axon terminal. For example, in resting neurons some neurotransmitter-filled synaptic vesicles are "docked" at the plasma membrane; others are in reserve in the *active zone* near the plasma membrane at the synaptic cleft. In addition, the membrane of synaptic vesicles contains a specialized Ca^{2+}-binding protein that senses the rise in cytosolic Ca^{2+} after arrival of an action potential, triggering rapid fusion of docked vesicles with the presynaptic membrane.

A highly organized arrangement of cytoskeletal fibers in the axon terminal helps localize synaptic vesicles in the active zone (Figure 17-35). The vesicles themselves are linked together by *synapsin,* a fibrous phosphoprotein associated with the cytosolic surface of all synaptic-vesicle membranes. Filaments of synapsin also radiate from the plasma membrane and bind to vesicle-associated synapsin. These interactions probably keep synaptic vesicles close to the part of the plasma membrane facing the synapse. Indeed, synapsin knockout mice, although viable, are prone to seizures; during repetitive stimulation of many neurons in such mice, the number of synaptic vesicles that fuse with the plasma mem-

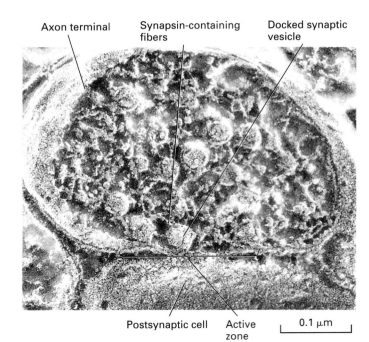

Axon terminal   Synapsin-containing fibers   Docked synaptic vesicle

Postsynaptic cell   Active zone   0.1 μm

▲ **EXPERIMENTAL FIGURE 17-35 Fibrous proteins help localize synaptic vesicles to the active zone of axon terminals.** In this micrograph of an axon terminal obtained by the rapid-freezing deep-etch technique, synapsin fibers can be seen to interconnect the vesicles and to connect some to the active zone of the plasma membrane. Docked vesicles are ready to be exocytosed. Those toward the center of the terminal are in the process of being filled with neurotransmitter. [From D. M. D. Landis et al., 1988, *Neuron* **1**:201.]

brane is greatly reduced. Thus synapsins are thought to recruit synaptic vesicles to the active zone.

Rab3A, a GTP-binding protein located in the membrane of synaptic vesicles, also is required for targeting of neurotransmitter-filled vesicles to the active zone of presynaptic cells facing the synaptic cleft. Rab3A knockout mice, like synapsin-deficient mice, exhibit a reduced number of synaptic vesicles able to fuse with the plasma membrane after repetitive stimulation. The neuron-specific Rab3 is similar in sequence and function to other Rab proteins that participate in docking vesicles on particular target membranes in the secretory pathway.

A Calcium-Binding Protein Regulates Fusion of Synaptic Vesicles with the Plasma Membrane

Fusion of synaptic vesicles with the plasma membrane of axon terminals depends on the same proteins that mediate membrane fusion of other regulated secretory vesicles. The principal v-SNARE in synaptic vesicles (VAMP) tightly binds syntaxin and SNAP-25, the principal t-SNAREs in the plasma membrane of axon terminals, to form four-helix SNARE complexes. After fusion, SNAP proteins and NSF within the axon terminal promote disassociation of VAMP from t-SNAREs, as in the fusion of secretory vesicles depicted previously (see Figure 17-11).

 Strong evidence for the role of VAMP in neurotransmitter exocytosis is provided by the mechanism of action of botulinum toxin, a bacterial protein that can cause the paralysis and death characteristic of *botulism*, a type of food poisoning. The toxin is composed of two polypeptides: One binds to motor neurons that release acetylcholine at synapses with muscle cells, facilitating entry of the other polypeptide, a protease, into the cytosol of the axon terminal. The only protein this protease cleaves is VAMP. After the botulinum protease enters an axon terminal, synaptic vesicles that are not already docked rapidly lose their ability to fuse with the plasma membrane because cleavage of VAMP prevents assembly of SNARE complexes. The resulting block in acetylcholine release at neuromuscular synapses causes paralysis. However, vesicles that are already docked exhibit remarkable resistance to the toxin indicating that SNARE complexes may already be in a partially assembled, protease-resistant state when vesicles are docked on the presynaptic membrane. ∎

The signal that triggers exocytosis of docked synaptic vesicles is a rise in the Ca^{2+} concentration in the cytosol near vesicles from <0.1 μM, characteristic of resting cells, to 1–100 μM following arrival of an action potential in stimulated cells. The speed with which synaptic vesicles fuse with the presynaptic membrane after a rise in cytosolic Ca^{2+} (less than 1 msec) indicates that the fusion machinery is entirely assembled in the resting state and can rapidly undergo a conformational change leading to exocytosis of neurotransmitter. A Ca^{2+}-binding protein called *synaptotagmin*, located in the membrane of synaptic vesicles, is thought to be a key component of the vesicle fusion machinery that triggers exocytosis in response to Ca^{2+} (Figure 17-36).

Several lines of evidence support a role for synaptotagmin as the Ca^{2+} sensor for exocytosis of neurotransmitters. For instance, mutant embryos of *Drosophila* and *C. elegans* that completely lack synaptotagmin fail to hatch and exhibit very reduced, uncoordinated muscle contractions. Larvae with partial loss-of-function mutations of synaptotagmin survive, but their neurons are defective in Ca^{2+}-stimulated vesicle exocytosis. Moreover, in mice, mutations in synaptotagmin that decrease its affinity for Ca^{2+} cause a corresponding

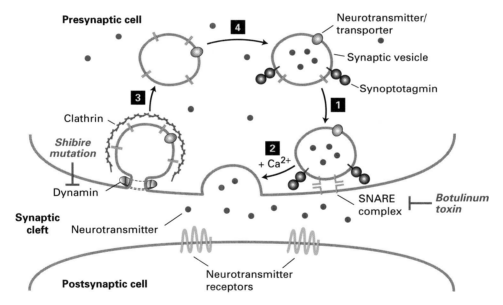

▲ **FIGURE 17-36 Release of neurotransmitters and the recycling of synaptic vesicles.** Step **1**: Synaptic vesicles loaded with neurotransmitter (red circles) move to the active zone and then dock at defined sites on the plasma membrane of a presynaptic cell. Synaptotagmin prevents membrane fusion and release of neurotransmitter. Botulinum toxin prevents exocytosis by proteolytically cleaving VAMP, the v-SNARE on vesicles. Step **2**: In response to a nerve impulse (action potential), voltage-gated Ca^{2+} channels in the plasma membrane open, allowing an influx of Ca^{2+} from the extracellular medium. The resulting Ca^{2+}-induced conformational change in synaptotagmin leads to fusion of docked vesicles with the plasma membrane and release of neurotransmitters into the synaptic cleft. Step **3**: After

clathrin/AP vesicles containing v-SNARE and neurotransmitter transporter proteins bud inward and are pinched off in a dynamin-mediated process, they lose their coat proteins. Dynamin mutations such as *shibire* in *Drosophila* block the re-formation of synaptic vesicles, leading to paralysis. Step **4**: The uncoated vesicles import neurotransmitters from the cytosol, generating fully reconstituted synaptic vesicles and completing the cycle. Most synaptic vesicles are formed by endocytic recycling as depicted here. However, endocytic vesicles containing membrane from the axon terminus can fuse with the endosome; budding from this compartment can then form "new" synaptic vesicles. [See K. Takei et al., 1996, *J. Cell. Biol.* **133**:1237; V. Murthy and C. Stevens, 1998, *Nature* **392**:497; and R. Jahn et al., 2003, *Cell* **112**:519.]

increase in the amount of cytosolic Ca^{2+} needed to trigger rapid exocytosis.

Several hypotheses concerning how synaptotagmin promotes neurotransmitter exocytosis have been proposed, but the precise mechanism of its function is still unresolved. Synaptotagmin is known to bind phospholipids and after undergoing a Ca^{2+}-induced conformational change, it may promote association of the phospholipids in the vesicle and plasma membranes. Synaptotagmin also binds to SNARE proteins and may catalyze a late stage in assembly of SNARE complexes when bound to Ca^{2+}. Finally, synaptotagmin may also act to inhibit inappropriate exocytosis in resting cells. At the low cytosolic Ca^{2+} levels found in resting nerve cells, synaptotagmin apparently binds to a complex of the plasma-membrane proteins neurexin and syntaxin. The presence of synaptotagmin blocks binding of other essential fusion proteins to the neurexin-syntaxin complex, thereby preventing vesicle fusion. When synaptotagmin binds Ca^{2+}, it is displaced from the complex, allowing other proteins to bind and thus initiating membrane docking or fusion. Thus synaptotagmin may operate as a "clamp" to prevent fusion from proceeding in the absence of a Ca^{2+} signal.

Fly Mutants Lacking Dynamin Cannot Recycle Synaptic Vesicles

Synaptic vesicles are formed primarily by endocytic budding from the plasma membrane of axon terminals. Endocytosis usually involves clathrin-coated pits and is quite specific, in that several membrane proteins unique to the synaptic vesicles (e.g., neurotransmitter transporters) are specifically incorporated into the endocytosed vesicles. In this way, synaptic-vesicle membrane proteins can be reused and the recycled vesicles refilled with neurotransmitter (see Figure 17-36).

As in the formation of other clathrin/AP-coated vesicles, pinching off of endocytosed synaptic vesicles requires the GTP-binding protein dynamin (see Figure 17-20). Indeed, analysis of a temperature-sensitive *Drosophila* mutant called *shibire (shi)*, which encodes the fly dynamin protein, provided early evidence for the role of dynamin in endocytosis. At the permissive temperature of 20°C, the mutant flies are normal, but at the nonpermissive temperature of 30°C, they are paralyzed (*shibire*, paralyzed in Japanese) because pinching off of clathrin-coated pits in neurons and other cells is blocked. When viewed in the electron microscope, the *shi* neurons at 30°C show abundant clathrin-coated pits with long necks but few clathrin-coated vesicles. The appearance of nerve terminals in *shi* mutants at the nonpermissive temperature is similar to that of terminals from normal neurons incubated in the presence of a nonhydrolyzable analog of GTP (see Figure 17-21). Because of their inability to pinch off new synaptic vesicles, the neurons in *shi* mutants eventually become depleted of synaptic vesicles when flies are shifted to the nonpermissive temperature, leading to a cessation of synaptic signaling and paralysis.

KEY CONCEPTS OF SECTION 17.6

Synaptic Vesicle Formation and Function

■ Transmission of nerve impulses at chemical synapses depends on the exocytosis of neurotransmitter-filled synaptic vesicles and the regeneration of empty vesicles by endocytosis.

■ Efficient recruitment of vesicles to the presynaptic membrane adjacent to the synaptic cleft requires cytosolic proteins, such as synapsin, and Rab3a, a GTP-binding protein that is tethered to the vesicle membrane.

■ In resting neurons, synaptotagmin in the synaptic-vesicle membrane prevents fusion of docked vesicles with the membrane. The influx of Ca^{2+} following arrival of an action potential at the axon terminus leads to Ca^{2+} binding by synaptotagmin, causing a change in its conformation that permits vesicle fusion to proceed (see Figure 17-36).

■ Synaptic vesicles are rapidly regenerated by endocytic budding of clathrin-coated vesicles from the plasma membrane, a process that requires dynamin. After the clathrin coat is shed, vesicles are refilled with neurotransmitter and move to the active zone for another round of docking and fusion.

PERSPECTIVES FOR THE FUTURE

The biochemical, genetic, and structural information presented in this chapter shows that we now have a basic understanding of how protein traffic flows from one membrane-bounded compartment to another. Our understanding of these processes has come largely from experiments on the function of various types of transport vesicles. These studies have led to the identification of many vesicle components and the discovery of how these components work together to drive vesicle budding, to incorporate the correct set of cargo molecules from the donor organelle, and then to mediate fusion of a completed vesicle with the membrane of a target organelle.

Despite these advances, there remain important stages of the secretory and endocytic pathways about which we know relatively little. For example, we do not yet know what types of proteins form the coats of either the regulated or constitutive secretory vesicles that bud from the *trans*-Golgi network. Indeed, it is not clear whether assembly of a cytosolic coat drives their budding at all. Moreover, the types of signals on cargo proteins that might target them for packaging into secretory vesicles have not yet been defined. Another baffling process is the formation of vesicles that bud away from the cytosol, such as the vesicles that enter multivesicular endosomes. Although some of the proteins that participate in formation of these "internal" endosome vesicles are known, we do not know what determines their shape or what type of process causes them to pinch off from the donor membrane. In the future, it should be possible for these and other poorly understood vesicle-trafficking steps to be dissected through the use of the same powerful combination of biochemical and genetic methods that have delineated the working parts of COPI, COPII, and clathrin/AP vesicles.

Questions still remain about vesicle trafficking between the ER and *cis*-Golgi, between Golgi stacks, and between the *trans*-Golgi and endosome, the best-characterized transport steps. In particular, our understanding of how proteins are actually sorted between these organelles is incomplete largely because of the highly dynamic nature of all the organelles along the secretory pathway. Although we know many of the details of how particular vesicle components function, we cannot account for why their functions are restricted to specific stages in the overall flow of anterograde and retrograde transport steps. For example, we cannot explain why COPII vesicles fuse with one another to form a new *cis*-Golgi stack, whereas COPI vesicles fuse with the membrane of the ER, since both vesicle types appear to contain similar sets of v-SNARE proteins. In the same vein, we do not know what feature of the Golgi membrane actually distinguishes a COPI-coated vesicle bud from a clathrin/AP-coated bud. In both cases binding of ARF protein to the Golgi membrane appears to initiate vesicle budding. The solution to these problems will require a more integrated understanding of the flow of vesicular traffic in the context of the entire secretory pathway. Recent improvements in our ability to image vesicular transport of cargo proteins in live cells gives hope that some of these more subtle aspects of vesicle function may be clarified in the near future.

KEY TERMS

AP (adapter protein)
    complexes *720*
anterograde transport *715*
ARF protein *709*
autophagy *733*
cisternal progression *715*
clathrin *708*
constitutive secretion *724*
COPI *708*
COPII *708*
dynamin *721*
ESCRT complexes *733*
late endosome *702*
mannose 6-phosphate
    (M6P) *723*
multivesicular
    endosomes *733*

Rab proteins *711*
receptor-mediated
    endocytosis *728*
regulated secretion *724*
retrograde transport *715*
sec mutants *706*
secretory pathway *701*
sorting signals *711*
synaptotagmin *737*
transcytosis *727*
trans-Golgi network
    (TGN) *701*
transport vesicles *701*
t-SNAREs *708*
v-SNAREs *708*

REVIEW THE CONCEPTS

1. The studies of Palade and colleagues using pulse-chase labeling with radioactively labeled amino acids and autoradiography to visualize the location of the radiolabeled proteins is a classic case of the experimentalist realizing what works well in his or her biological system. These experiments were done with pancreatic acinar cells. Alternatively, HeLa cells can be used. HeLa cells are a classic human cell line originating from a cervical carcinoma. When applied to HeLa cells, the same experimental protocols are a dismal failure with respect to tracking secretion. What would you expect autoradiography of a HeLa cell to look like after pulse-labeling with radioactive amino acids?

2. *Sec18* is a yeast gene that encodes NSF. It is a class C mutant in the yeast secretory pathway. What is the mechanistic role of NSF in membrane trafficking. As indicated by its class C phenotype, why does an NSF mutation produce accumulation of vesicles at what appears to be only one stage of the secretory pathway?

3. Vesicle budding is associated with coat proteins. What is the role of coat proteins in vesicle budding? How are coat proteins recruited to membranes? What kinds of molecules are likely to be included or excluded from newly formed vesicles? What is the best-known example of a protein likely to be involved in vesicle pinching off?

4. Treatment of cells with the drug brefeldin A (BFA) has the effect of decoating Golgi apparatus membranes, resulting in a cell in which the vast majority of Golgi proteins are found in the ER. What inferences can be made from this observation regarding roles of coat proteins other than promoting vesicle formation? Predict what type of mutation in Arf1 might have the same effect as treating cells with BFA.

5. An antibody to an exposed "hinge" region of βCOPI known as EAGE blocks the function of βCOPI when microinjected into HeLa cells. Predict what the consequences of this functional block might be for anterograde transport from the ER to the plasma membrane. Propose an experiment to test whether the effect of EAGE microinjection is initially on anterograde or retrograde transport.

6. Specificity in fusion between vesicles involves two discrete and sequential processes. Describe the first of the two processes and its regulation by GTPase switch proteins. What effect on the size of early endosomes might result from overexpression of a mutant form of Rab5 that is stuck in the GTP-bound state?

7. Two different protein-mediated membrane fusion processes are described in this chapter, SNARE- and viral HA-mediated fusion. Compare and contrast the two. In each example give particular attention to what the direct effect of polypeptide sequences in membrane fusion is and to what controls the specificity of membrane fusion in each.

8. Sorting signals that cause retrograde transport of a protein in the secretory pathway are sometimes known as retrieval sequences. List the two known examples of retrieval sequences for soluble and membrane proteins of the ER? How does the presence of a retrieval sequence on a soluble ER protein result in its retrieval from the *cis*-Golgi complex? Describe how the concept of a retrieval sequence is essential to the cisternal-progression model.

9. Clathrin adapter protein (AP) complexes bind directly to the cytosolic face of membrane proteins and also interact with clathrin. What are the four known adapter protein complexes? Why may clathrin be considered to be an accessory protein to a core coat composed of adapter proteins?

10. I-cell disease is a classic example of an inherited human defect in protein targeting that affects an entire class of proteins, soluble enzymes of the lysosome. What is the molecular defect in I-cell disease? Why does it affect the targeting of an entire class of proteins? What other types of mutations might produce the same phenotype?

11. The TGN, *trans*-Golgi network, is the site of multiple sorting processes as proteins and lipids exit the Golgi complex. Compare and contrast the sorting of proteins to lysosomes versus the packaging of proteins into regulated secretory granules such as those containing insulin. Compare and contrast the sorting of proteins to the basolateral versus apical cell surfaces in MDCK cells versus hepatocytes.

12. The efficiency of bacterial phagocytosis by macrophages is increased greatly by first binding antibody molecules to the bacterial surface. On the basis of prior descriptions of antibody structure, to what portion of the immunoglobulin molecule do you predict a macrophage receptor for antibody bound to bacteria might be directed? Design an experiment to test this prediction.

13. Describe how pH plays a key role in regulating the interaction between mannose 6-phosphate and the mannose 6-phosphate receptor. Why does elevating endosomal pH lead to the secretion of newly synthesized lysosomal enzymes into the extracellular medium?

14. What mechanistic features are shared by (a) the formation of multivesicular endosomes by budding into the interior of the endosome and (b) the outward budding of HIV virus at the cell surface? You wish to design a peptide inhibitor/competitor of HIV budding and decide to mimic in a synthetic peptide a portion of the HIV Gag protein. Which portion of the HIV Gag protein would be a logical choice? What normal cellular process might this inhibitor block?

15. The exocytosis of neurotransmitter-filled synaptic vesicles is an example of regulated exocytosis. How is the influx of Ca^{2+} following arrival of an action potential at the axon terminus sensed and linked to the exocytosis of synaptic vesicles? Why do normal *Drosophila* neurons incubated in the presence of a nonhydrolyzable analog of GTP have the same appearance as nerve terminals in *shi* mutants?

ANALYZE THE DATA

A variety of protein toxins, such as the bacterial toxin *Pseudomonas* and Shiga toxin and the plant toxin ricin, are heteromeric proteins consisting of A and B subunits. The A subunit is catalytic. For Shiga toxin, the proximal cause of food poisoning due to bacterially contaminated hamburger, the A subunit is an *N*-glycosidase and specifically cleaves 28S ribosomal RNA, thereby intoxicating cells by inhibiting protein synthesis. Amazingly, only one molecule of A subunit when introduced into the cytosol is sufficient to kill a cell. Interestingly the A subunit of Shiga toxin is transferred into the cytosol from the lumen of the ER by the Sec61 protein translocon. The B subunit targets Shiga toxin to the ER by binding to a glycolipid GM3 on the cell surface that acts as the Shiga toxin internalization receptor. Shiga toxin is internalized into endosomes, from endosomes is transferred to the Golgi complex, and from the Golgi complex goes to the ER where the A and B subunits dissociate, permitting the A subunit to translocate into the cytosol.

In a series of experiments designed to characterize the comparative mechanisms of *Pseudomonas* and Shiga toxin transfer from the Golgi complex to the ER, investigators first sequenced the respective targeting subunits. The C-terminal 24 amino acids of the B subunits of *Pseudomonas* toxin and Shiga toxin are shown below:

C-terminal 24 amino acids of *Pseudomonas* toxin B subunit
KEQAISALPD YASQPGKPPR KDEL

C-terminal 24 amino acids of Shiga toxin B subunit
TGMTVTIKTN ACHNGGGFSE VIFR

From inspection of these sequences, what is the probable targeting receptor for transfer of *Pseudomonas* toxins from the Golgi apparatus to the ER?

To test this prediction directly, investigators experimentally characterized the role of COPI coat proteins and KDEL receptors in intoxication. Monkey cells were microinjected with antibodies directed against either COPI coat proteins or the cytosolic domain of KDEL receptors. Cells then were incubated with *Pseudomonas* or Shiga toxin for 4 h. Protein synthesis was determined following a 30-minute pulse labeling with [^{35}S]methionine. Results are shown in the accompanying figure, with controls showing the low level of protein synthesis caused by incubation with either *Pseudomonas* or Shiga toxin without antibody injection.

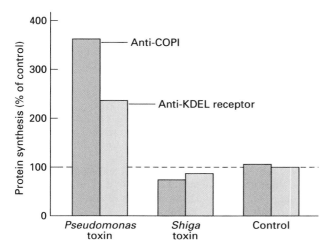

Effect of microinjected antibodies on cell intoxication by *Pseudomonas* and Shiga toxin.

How do these results support your sequence-based predictions and the known role of COPI coat protein in retrograde transport? Can you formulate a hypothesis for how Shiga toxin is transported from the Golgi to the ER?

To explore further whether or not Shiga toxin transfer from the Golgi apparatus to the ER depends on COPI coat proteins, investigators prepared two different fluorescent dye–conjugated Shiga toxin B subunits and then assessed by fluorescence microscopy transport of the B subunits from the Golgi complex to the ER. The first preparation was Cy3-conjugated wild-type B subunit. The second preparation was Cy3-conjugated B subunit in which the C terminus was extended by the four amino acids KDEL (B-KDEL). Cells were microinjected with antibody directed against COPI coat proteins. Following microinjection, cells were incubated with fluorescent B subunit for various periods of time and B subunit distributions scored. The results are shown in the figure below.

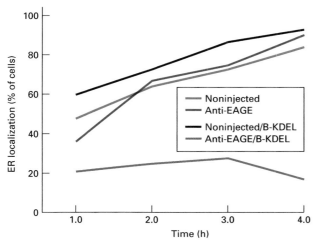

Effect of microinjected antibodies directed against COPI proteins on the transport of wild type Shiga toxin B-subunit or Shiga toxin B-KDEL from Golgi complex to the ER.

What evidence do these results provide for or against transport of wild-type Shiga toxin B subunit from the Golgi complex to the ER in COPI-coated vesicles? What is the importance of the results with B-KDEL in interpreting the overall results of these experiments?

REFERENCES

Techniques for Studying the Secretory Pathway

Beckers, C. J., et al. 1987. Semi-intact cells permeable to macromolecules: use in reconstitution of protein transport from the endoplasmic reticulum to the Golgi complex. *Cell* 50:523–534.

Kaiser, C. A., and R. Schekman. 1990. Distinct sets of SEC genes govern transport vesicle formation and fusion early in the secretory pathway. *Cell* 61:723–733.

Novick, P., et al. 1981. Order of events in the yeast secretory pathway. *Cell* 25:461–469.

Lippincott-Schwartz J., et al. 2001. Studying protein dynamics in living cells. *Nature Rev. Mol. Cell Biol.* 2:444–456.

Orci, L., et al. 1989. Dissection of a single round of vesicular transport: sequential intermediates for intercisternal movement in the Golgi stack. *Cell* 56:357–368.

Palade, G. 1975. Intracellular aspects of the process of protein synthesis. *Science* 189:347–358.

Molecular Mechanisms of Vesicular Traffic

Jahn, R., et al. 2003. Membrane fusion. *Cell* 112:519–533.

Kirchhausen, T. 2000. Three ways to make a vesicle. *Nature Rev. Mol. Cell Biol.* 1:187–198.

McNew, J. A., et al. 2000. Compartmental specificity of cellular membrane fusion encoded in SNARE proteins. *Nature* 407:153–159.

Ostermann, J., et al. 1993. Stepwise assembly of functionally active transport vesicles. *Cell* 75:1015–1025.

Schimmöller, F., I. Simon, and S. Pfeffer. 1998. Rab GTPases, directors of vesicle docking. *J. Biol. Chem.* 273:22161–22164.

Skehel, J. J., and D. C. Wiley. 2000. Receptor binding and membrane fusion in virus entry: the influenza hemagglutinin. *Ann. Rev. Biochem.* 69:531–569.

Söllner, T., et al. SNAP receptors implicated in vesicle targeting and fusion. *Nature* 362:318–324.

Springer, S., et al. 1999. A primer on vesicle budding. *Cell* 97:145–148.

Zerial, M., and H. McBride. 2001. Rab proteins as membrane organizers. *Nature Rev. Mol. Cell Biol.* 2:107–117.

Vesicle Trafficking in the Early Stages of the Secretory Pathway

Barlowe, C. 2002. COPII-dependent transport from the endoplasmic reticulum. *Curr. Opin. Cell Biol.* 14:417–422.

Bi, X., et al. 2002. Structure of the Sec23/24-Sar1 pre-budding complex of the COPII vesicle coat. *Nature* 419:271–277.

Glick, B. S., and V. Malhotra. 1998. The curious status of the Golgi apparatus. *Cell* 95:883–889.

Kuehn, M. J., and R. Schekman. 1997. COPII and secretory cargo capture into transport vesicles. *Curr. Opin. Cell Biol.* 9:477–483.

Letourneur, F., et al. 1994. Coatomer is essential for retrieval of dilysine-tagged proteins to the endoplasmic reticulum. *Cell* **79**:1199–1207.

Pelham, H. R. 1995. Sorting and retrieval between the endoplasmic reticulum and Golgi apparatus. *Curr. Opin. Cell Biol.* **7**:530–535.

Schekman, R., and I. Mellman. 1997. Does COPI go both ways? *Cell* **90**:197–200.

Protein Sorting and Processing in Later Stages of the Secretory Pathway

Bonifacino, J. S., and E. C. Dell'Angelica. 1999. Molecular bases for the recognition of tyrosine-based sorting signals. *J. Cell Biol.* **145**:923–926.

Ghosh, P., et al. 2003. Mannose 6-phosphate receptors: new twists in the tale. *Nature Rev. Mol. Cell Bio.* **4**:202–213.

Hinshaw, J. E. 2000. Dynamin and its role in membrane fission. *Ann. Rev. Cell Dev. Biol.* **16**:483–519.

Kirchhausen, T. 2000. Clathrin. *Ann. Rev. Biochem.* **69**:699–727.

Mostov, K. E., et al. 1995. Regulation of protein traffic in polarized epithelial cells: the polymeric immunoglobulin receptor model. *Cold Spring Harbor Symp. Quant. Biol.* **60**:775–781.

Robinson, M. S., and J. S. Bonifacino. 2001. Adaptor-related proteins. *Curr. Opin. Cell Biol.* **13**:444–453.

Schmid, S. 1997. Clathrin-coated vesicle formation and protein sorting: an integrated process. *Ann. Rev. Biochem.* **66**:511–548.

Simons, K., and E. Ikonen. 1997. Functional rafts in cell membranes. *Nature* **387**:569–572.

Steiner, D. F., et al. 1996. The role of prohormone convertases in insulin biosynthesis: evidence for inherited defects in their action in man and experimental animals. *Diabetes Metab.* **22**:94–104.

Tooze, S. A., et al. 2001. Secretory granule biogenesis: rafting to the SNARE. *Trends Cell Biol.* **11**:116–122.

Receptor-Mediated Endocytosis and the Sorting of Internalized Proteins

Brown, M. S., and J. L. Goldstein. 1986. Receptor-mediated pathway for cholesterol homeostasis. Nobel Prize Lecture. *Science* **232**:34–47.

Katzmann, D. J., et al. 2002. Receptor downregulation and multivesicular-body sorting. *Nature Rev. Mol. Cell Biol.* **3**:893–905.

Khalfan, W. A., and D. J. Klionsky. 2002. Molecular machinery required for autophagy and the cytoplasm to vacuole targeting (Cvt) pathway in *S. cerevisiae. Curr. Opin. Cell Biol.* **14**:468–475.

Lemmon, S. K., and L. M. Traub. 2000. Sorting in the endosomal system in yeast and animal cells. *Curr. Opin. Cell Biol.* **12**:457–466.

Pornillos, O., et al. 2002. Mechanisms of enveloped RNA virus budding. *Trends Cell Biol.* **12**:569–579.

Riezman, H., P. Woodman, G. van Meer, and M. Marsh. 1997. Molecular mechanisms of endocytosis. *Cell* **91**:731–738.

Robinson, M. S., C. Watts, and M. Zerial. 1996. Membrane dynamics in endocytosis. *Cell* **84**:13–21.

Rudenko, G., et al. 2002. Structure of the LDL receptor extracellular domain at endosomal pH. *Science* **298**:2353–2358.

Synaptic Vesicle Function and Formation

Betz, W., and J. Angleson. 1998. The synaptic vesicle cycle. *Ann. Rev. Physiol.* **60**:347–364.

Chapman, E. R. 2002. Synaptotagmin: a Ca^{2+} sensor that triggers exocytosis? *Nature Rev. Mol. Cell Biol.* **3**:498–508.

De Camilli, P., et al. 1995. The function of dynamin in endocytosis. *Curr. Opin. Neurobiol.* **5**:559–565.

Geppert, M., and T. Sudhof. 1998. Rab3 and synaptotagmin: the yin and yang of synaptic membrane fusion. *Ann. Rev. Neurosci.* **21**:75–96.

Neimann, H., J. Blasi, and R. Jahn. 1994. Clostridial neurotoxins: new tools for dissecting exocytosis. *Trends Cell Biol.* **4**:179–185.

18

METABOLISM AND MOVEMENT OF LIPIDS

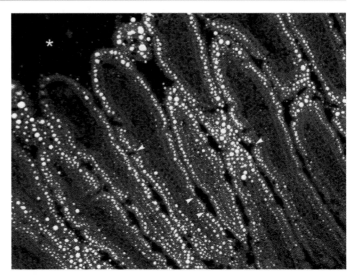

Fluorescence micrograph of hamster intestinal epithelium after cellular uptake into lipid droplets of an orally administered fluorescent analog of cholesterol (fluoresterol, dissolved in corn oil) from the intestinal lumen (upper left, unstained). [C. P. Sparrow et al., 1999, *J. Lipid Res.* **40**:1747–1757.]

In this chapter we consider some of the special challenges that a cell faces in metabolizing and transporting **lipids,** which are poorly soluble in the aqueous interior of cells and in extracellular fluids. Cells use lipids for storing energy, building membranes, signaling within and between cells, sensing the environment, covalently modifying proteins, forming specialized permeability barriers (e.g., in skin), and protecting cells from highly reactive chemicals. **Fatty acids,** which are oxidized in mitochondria to release energy for cellular functions (Chapter 8), are stored and transported primarily in the form of **triglycerides.** Fatty acids are also precursors of **phospholipids,** the structural backbone of cellular membranes (Chapter 5). **Cholesterol,** another important membrane component, is a precursor for steroid hormones and other biologically active lipids that function in cell–cell signaling. Also derived from precursors of cholesterol biosynthesis are the fat-soluble vitamins, which have diverse functions including the detection of light by the retinal form of vitamin A in rhodopsin, the control of calcium metabolism by the active hormone form of vitamin D, protection against oxidative damage to cells by vitamin E, and the cofactor activity of vitamin K in the formation of blood clots.

With the exception of a few specialized cells that store large quantities of lipids, the overwhelming majority of lipids within cells are components of cellular membranes. Therefore we focus our discussion of lipid biosynthesis and movement on the major lipids found in cellular membranes and their precursors (Figure 18-1). In lipid biosynthesis, water-soluble precursors are assembled into membrane-associated intermediates that are then converted into membrane lipid products. The movement of lipids, especially membrane components, between different organelles is critical for maintaining the proper composition and properties of membranes and overall cell structure, but our understanding of such intracellular lipid transport is still rudimentary. In contrast, analysis of the transport of lipids into, out of, and between cells is far more advanced, and we describe in some detail these lipid movements mediated by various cell-surface transport proteins and receptors.

We conclude the chapter by examining the connection between cellular cholesterol metabolism and atherosclerosis, which can lead to cardiovascular disease (e.g., heart attack,

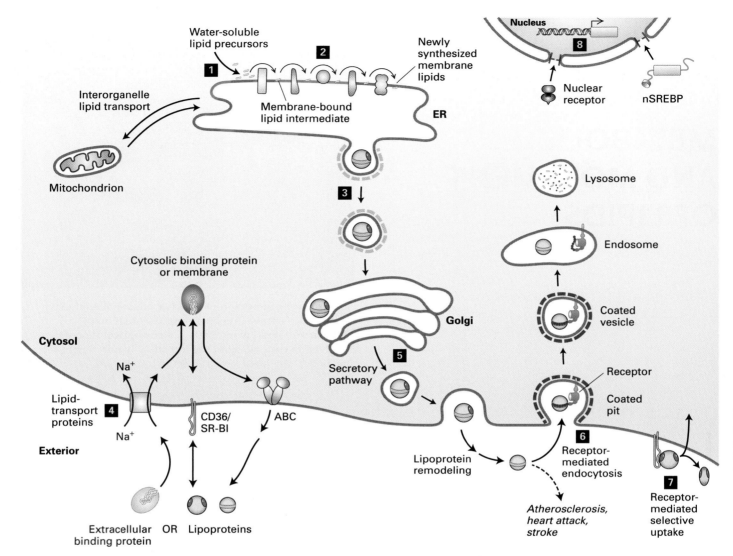

▲ **FIGURE 18-1 Overview of synthesis of major membrane lipids and their movement into and out of cells.** Membrane lipids (e.g., phospholipids, cholesterol) are synthesized through complex multienzyme pathways that begin with sets of water-soluble enzymes and intermediates in the cytosol (**1**) that are then converted by membrane-associated enzymes into water-insoluble products embedded in the membrane (**2**), usually at the interface between the cytosolic leaflet of the endoplasmic reticulum (ER) and the cytosol. Membrane lipids can move from the ER to other organelles (**3**), such as the Golgi apparatus or the mitochondrion, by either vesicle-mediated or other poorly defined mechanisms. Lipids can move into or out of cells by plasma-membrane transport proteins or by lipoproteins. Transport proteins similar to those described in Chapter 7 that move lipids (**4**) include sodium-coupled symporters that mediate import; CD36 and SR-BI superfamily proteins that can mediate unidirectional or bidirectional transport; and ABC superfamily proteins that mediate cellular export, the flipping of lipids from the cytosolic to the exoplasmic leaflet of the membrane, or both. Because lipids are insoluble in water, the transport proteins move lipids from and to carrier proteins, lipoproteins, membranes, or other lipid-binding complexes in the extracellular space and the cytosol. Lipoproteins assembled in the ER carry in their hydrophobic cores large amounts of lipid. They are secreted by the classic Golgi-mediated secretory pathway (**5**), and their lipids are imported either through (**6**) receptor-mediated endocytosis or through (**7**) receptor-mediated selective lipid uptake. Cellular lipid metabolism is regulated (**8**) by nuclear receptor transcription factors that directly bind lipids and by nuclear sterol regulatory element-binding proteins (nSREBPs) that are generated by proteolysis of an integral membrane protein precursor in the Golgi.

stroke), the number one cause of death in Western industrialized societies. We describe current theories about why large arteries can become clogged with cholesterol-containing deposits and how cells recognize the differences between "good" and "bad" cholesterol. As you will see, detailed knowledge of the fundamental cell biology of lipid metabolism has led to the discovery of remarkably effective anti-atherosclerotic drugs.

18.1 Phospholipids and Sphingolipids: Synthesis and Intracellular Movement

A cell cannot divide or enlarge unless it makes sufficient amounts of additional membranes to accommodate the expanded area of its outer surface and internal organelles. Thus the generation of new cell membranes is as fundamentally important to the life of a cell as is protein synthesis or DNA replication. Although the protein components of biomembranes are critical to their biological functions, the basic structural and physical properties of membranes are determined by their lipid components—principally phospholipids, sphingolipids, and sterols such as cholesterol (Table 18-1). Cells must be able to synthesize or import these molecules to form membranes.

A fundamental principle of membrane biosynthesis is that cells synthesize new membranes only by the expansion of existing membranes. Although some early steps in the synthesis of membrane lipids take place in the cytoplasm, the final steps are catalyzed by enzymes bound to preexisting cellular membranes, and the products are incorporated into the membranes as they are generated. Evidence for this phenomenon is seen when cells are briefly exposed to radioactive precursors

(e.g., phosphate or fatty acids): all the phospholipids and sphingolipids incorporating these precursor substances are associated with intracellular membranes; none are found free in the cytosol. After they are formed, membrane lipids must be distributed appropriately both in leaflets of a given membrane and among the independent membranes of different organelles in eukaryotic cells. Here, we focus on the synthesis and distribution of phospholipids and sphingolipids; we cover the synthesis of cholesterol in Section 18.2.

Fatty Acids Are Precursors for Phospholipids and Other Membrane Components

Fatty acids are key components of both phospholipids and sphingolipids; they also anchor some proteins to cellular membranes (see Figure 5-15). Thus the regulation of fatty acid synthesis plays a key role in the regulation of membrane synthesis as a whole. A fatty acid consists of a long hydrocarbon chain with a carboxyl group at one end (Figure 18-2). A *saturated* fatty acid (e.g., palmitate) has only single bonds, and an *unsaturated* fatty acid (e.g., arachidonate) has one or more double bonds in the hydrocarbon chain.

The major fatty acids in phospholipids contain 14, 16, 18, or 20 carbon atoms and include both saturated and unsaturated chains (see Table 18-1). Saturated fatty acids

TABLE 18-1 Synthesis and Transport of Fatty Acids and Major Membrane Lipids in Animal Cells

Lipid Class	Principal Sites of Synthesis	Import or Export Mechanisms	Intercellular Transport
Fatty acids (FAs):* Myristate (C14:0) Palmitate (C16:0) Stearate (C18:0) Oleate (C18:1) Linoleate (C18:2) Arachidonate (C20:4)	Saturated FAs up to 16 carbons long in the cytosol; elongation in the ER and mitochondria; desaturation in the ER	Diffusion and protein-mediated transport of free FAs (FATPs, CD36); secreted in lipoproteins as part of phospholipids, triglycerides and cholesteryl esters	Bound to albumin and other proteins in animal plasma (free FAs); as part of phospholipids, triglycerides, and cholesteryl esters in circulating lipoproteins
Phospholipids (e.g., phosphatidylcholine)†	ER primarily; some in mitochondria	Export by ABC proteins; endocytosis/exocytosis as part of lipoproteins	Packaged into lipoproteins
Plasmalogens	Peroxisomes	—	—
Sphingolipids	ER and Golgi complex	—	Packaged into lipoproteins
Cholesterol	Partly in cytosol and partly in ER	Export by ABC proteins; endocytosis/exocytosis as part of lipoproteins; import by selective lipid uptake from lipoproteins	Packaged into lipoproteins (both unesterified and esterified)

*In Cx:y abbreviation, x is the number of carbons in the chain and y is the number of double bonds. Other abbreviations are: CD36, a multifunctional cell-surface protein; ER, endoplasmic reticulum; FA, fatty acid; FATP, fatty acid transport protein.
†The common diacyl glycerophospholipids also include phosphatidylethanolamine, phosphatidyl serine, and phosphatidylinositol.

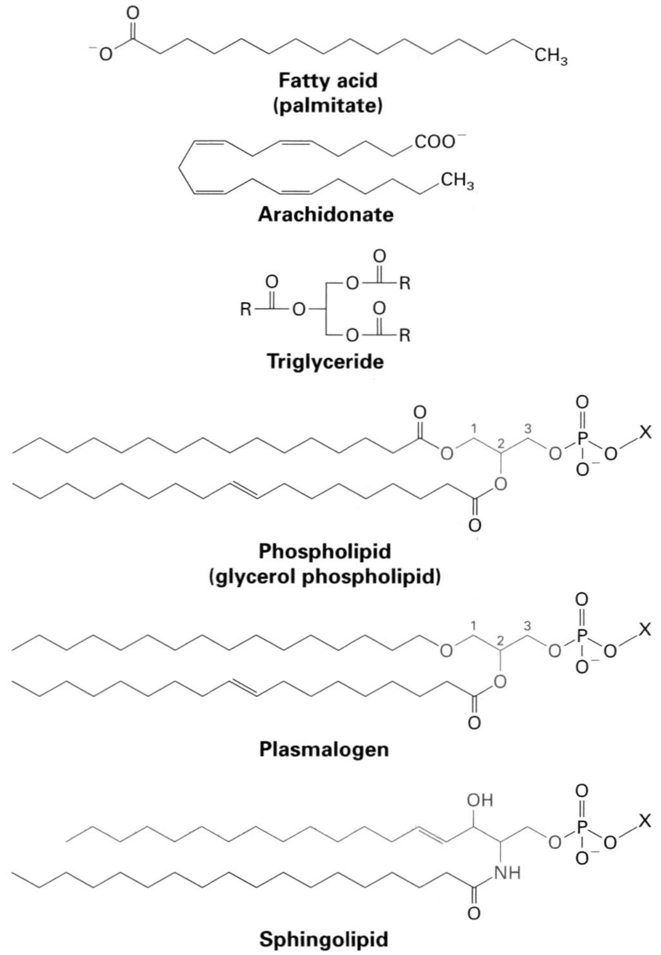

**Fatty acid
(palmitate)**

Arachidonate

Triglyceride

**Phospholipid
(glycerol phospholipid)**

Plasmalogen

Sphingolipid

▲ **FIGURE 18-2 Chemical structures of fatty acids and some of their derivatives.** Palmitate, a saturated fatty acid, contains 16 carbon atoms; arachidonate, a polyunsaturated fatty acid, contains 20 carbons atoms. Both saturated and unsaturated fatty acids are stored as triglycerides in which three fatty acyl chains (R = hydrocarbon portion of fatty acid) are esterified to a glycerol molecule. Fatty acids are also components of phospholipids (glycerol phospholipids, plasmalogens, and sphingolipids), which along with cholesterol are the major lipids present in membranes. The common phospholipids (e.g., phosphatidylcholine) have two acyl chains esterified to glycerol; in plasmalogens, one hydrocarbon chain is attached to glycerol by an ether linkage and the other by an ester linkage. Sphingolipids are built from sphingosine, an amino alcohol that contains a long, unsaturated hydrocarbon chain. Several types of polar X groups are found in all three of these classes of membrane lipids (see Figure 5-5).

Desaturase enzymes, also located in the ER, introduce double bonds at specific positions in some fatty acids. The presence of a double bond creates a kink in the hydrocarbon chain that interrupts intermolecular packing (see Figure 2-18). As a result, membranes or triglyceride droplets whose components are high in unsaturated fatty acids (e.g., liquid corn and olive oils) tend to be more fluid at room temperature than those with a high proportion of saturated fatty acids (e.g., solid animal fats). Because humans cannot synthesize certain essential polyunsaturated fatty acids, such as linoleic acid and linolenic acid, we must obtain them from our diets.

In addition to de novo synthesis from acetyl CoA, fatty acids can be derived from the enzymatic hydrolysis of triglycerides. The primary form in which fatty acids are stored and transported between cells, triglycerides consist of three fatty acyl chains esterified to glycerol; hence they are also called triacylglycerols (see Figure 18-2). Complete hydrolysis of a triglyceride molecule yields three unesterified fatty acid molecules, or free fatty acids (FFAs), and a glycerol molecule.

Unesterified Fatty Acids Move Within Cells Bound to Small Cytosolic Proteins

Unesterified fatty acids within cells are commonly bound by *fatty acid–binding proteins* (FABPs), which belong to a group of small cytosolic proteins that facilitate the intracellular movement of many lipids. These proteins contain a hydrophobic pocket lined by β sheets (Figure 18-3). A long-chain fatty acid can fit into this pocket and interact noncovalently with the surrounding protein.

The expression of cellular FABPs is regulated coordinately with cellular requirements for the uptake and release of fatty acids. Thus FABP levels are high in active muscles

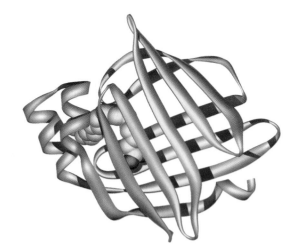

▲ **FIGURE 18-3 Binding of a fatty acid to the hydrophobic pocket of a fatty acid–binding protein (FABP).** The crystal structure of adipocyte FABP (ribbon diagram) reveals that the hydrophobic binding pocket is generated from two β sheets that are nearly at right angles to each other, forming a clam-shell-like structure. A fatty acid (yellow, oxygens red) interacts noncovalently with hydrophobic amino acid residues within this pocket. [See A. Reese-Wagoner et al., 1999, *Biochim. Biophys. Acta* **23**:1441(2–3):106–116.]

containing 14 or 16 carbon atoms are made from acetyl CoA by two enzymes, *acetyl-CoA carboxylase* and *fatty acid synthase*. In animal cells, these enzymes are in the cytosol; in plants, they are found in chloroplasts. Palmitoyl CoA (16 carbons) can be elongated to 18–24 carbons by the sequential addition of two-carbon units in the endoplasmic reticulum (ER) or sometimes in the mitochondrion.

that are using fatty acids for energy and in adipocytes (fat-storing cells) when they are either taking up fatty acids to be stored as triglycerides or releasing fatty acids for use by other cells. The importance of FABPs in fatty acid metabolism is highlighted by the observations that they can compose as much as 5 percent of all cytosolic proteins in the liver and that genetic inactivation of cardiac muscle FABP converts the heart from a muscle that primarily burns fatty acids for energy into one that primarily burns glucose.

Numerous other small water-soluble proteins with lipid-binding hydrophobic pockets are known. Although some evidence suggests that these proteins participate in intracellular lipid transport, their function in this lipid movement remains to be established with certainty.

Incorporation of Fatty Acids into Membrane Lipids Takes Place on Organelle Membranes

Fatty acids are not directly incorporated into phospholipids; rather, they are first converted in eukaryotic cells into CoA esters. The subsequent synthesis of many *diacyl glycerophospholipids* from fatty acyl CoAs, glycerol 3-phosphate, and polar head-group precursors is carried out by enzymes associated with the cytosolic face of the ER membrane, usually the smooth ER, in animal cells (Figure 18-4). Mitochondria synthesize some of their own membrane lipids and import others. In photosynthetic tissues, the chloroplast is the site for the synthesis of all its own lipids. The enzymes that esterify the middle hydroxyl group of glycerol have a preference for adding unsaturated fatty acids.

In addition to diacyl glycerophospholipids, animal cells and some anaerobic microorganisms contain surprisingly large amounts of *plasmalogens*, a different type of glycerol-derived phospholipid. In these molecules, the hydrocarbon chain on carbon 1 of glycerol is attached by an ether linkage, rather than the ester linkage found in diacyl phospholipids (see Figure 18-2). In animal cells, the synthesis of plasmalogens is catalyzed by enzymes bound to the membranes of peroxisomes. Plasmalogens are known to be an important reservoir of arachidonate, a polyunsaturated, long-chain fatty acid that is a precursor for a large group of signaling molecules called eicosanoids (e.g., prostaglandins, thromboxanes, and leukotrienes). The regulated release of arachidonate from membrane glycerophospholipids by the enzyme phospholipase A_2 plays a rate-determining role in many signaling pathways. Also derived from a plasmalogen is platelet-activating factor (PAF), a signaling molecule that plays a key role in the inflammatory response to tissue damage or injury (see Figure 6-32). In addition, plasmalogens may influence the movement of cholesterol within mammalian cells.

Sphingolipids, another major group of membrane lipids, are derivatives of sphingosine, an amino alcohol that contains a long, unsaturated hydrocarbon chain. Sphingosine is

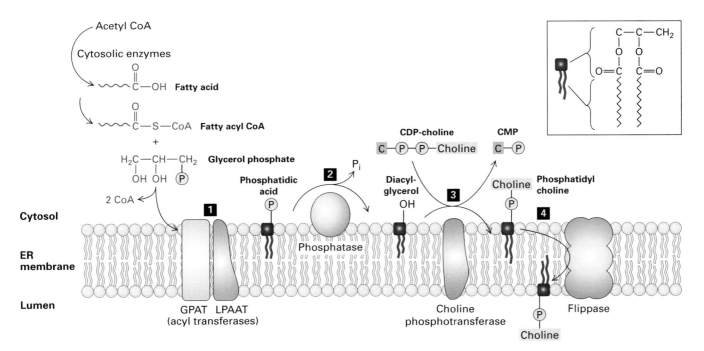

▲ **FIGURE 18-4 Phospholipid synthesis.** Because phospholipids are amphipathic molecules, the last stages of their multistep synthesis take place at the interface between a membrane and the cytosol and are catalyzed by membrane-associated enzymes. Step **1**: Fatty acids from fatty acyl CoA are esterified to the phosphorylated glycerol backbone, forming phosphatidic acid, whose two long hydrocarbon chains anchor the molecule to the membrane. Step **2**: A phosphatase converts phosphatidic acid into diacylglycerol. Step **3**: A polar head group (e.g., phosphorylcholine) is transferred from CDP-choline to the exposed hydroxyl group. Step **4**: Flippase proteins catalyze the movement of phospholipids from the cytosolic leaflet in which they are initially formed to the exoplasmic leaflet.

made in the ER, beginning with the coupling of palmitoyl CoA and serine; the addition of a fatty acyl group to form N-acyl sphingosine (ceramide) also takes place in the ER. The subsequent addition of a polar head group to ceramide in the Golgi yields *sphingomyelin,* whose head group is phosphorylcholine, and various *glycosphingolipids,* in which the head group may be a monosaccharide or a more complex oligosaccharide (see Figure 18-2). Some sphingolipid synthesis can also take place in mitochondria. In addition to serving as the backbone for sphingolipids, ceramide and its metabolic products are important signaling molecules that can influence cell growth, proliferation, endocytosis, resistance to stress, and apoptosis.

After their synthesis is completed in the Golgi, sphingolipids are transported to other cellular compartments through vesicle-mediated mechanisms similar to those discussed in Chapter 17. In contrast, phospholipids, as well as cholesterol, can move between organelles by different mechanisms, described in Section 18.2.

Flippases Move Phospholipids from One Membrane Leaflet to the Opposite Leaflet

Even though phospholipids are initially incorporated into the cytosolic leaflet of the ER membrane, various phospholipids are asymmetrically distributed in the two leaflets of the ER membrane and of other cellular membranes (see Table 5-1). However, phospholipids spontaneously flip-flop from one leaflet to the other only very slowly, although they can rapidly diffuse laterally in the plane of the membrane. For the ER membrane to expand (growth of both leaflets) and have asymmetrically distributed phospholipids, its phospholipid components must be able to rapidly and selectively flip-flop from one membrane leaflet to the other.

The usual asymmetric distribution of phospholipids in membrane leaflets is broken down as cells (e.g., red blood cells) become senescent or undergo apoptosis. For instance, phosphatidylserine and phosphatidylethanolamine are preferentially located in the cytosolic leaflet of cellular membranes. Increased exposure of these anionic phospholipids on the exoplasmic face of the plasma membrane appears to serve as a signal for scavenger cells to remove and destroy old or dying cells. Annexin V, a protein that specifically binds to anionic phospholipids, can be fluorescently labeled and used to detect apoptotic cells in cultured cells and in tissues.

Although the mechanisms employed to generate and maintain membrane phospholipid asymmetry are not well understood, it is clear that **flippases** play a key role. These integral membrane proteins facilitate the movement of phospholipid molecules from one leaflet to the other (see Figure 18-4, step 4). One of the best-studied flippases is the mammalian ABCB4 protein, a member of the **ABC superfamily** of small-molecule pumps. As discussed in Section 18.3, ABCB4 is expressed in certain liver cells (hepatocytes) and moves phosphatidylcholine from the cytosolic to the exoplasmic leaflet of the plasma membrane for subsequent release into the bile in combination with cholesterol and bile acids. Several other ABC superfamily members participate in the cellular export of various lipids (Table 18-2).

ABCB4 was first suspected of having phospholipid flippase activity because mice with homozygous loss-of-function mutations in the *ABCB4* gene exhibited defects in the secretion of phosphatidylcholine into bile. To determine directly if ABCB4 was in fact a flippase, researchers performed experiments on a homogeneous population of purified vesicles with ABCB4 in the membrane and with the cytosolic face directed outward. These vesicles were obtained by introducing cDNA encoding mammalian ABCB4 into a temperature-sensitive

TABLE 18-2	Selected Human ABC Proteins		
Protein	Tissue Expression	Function	Disease Caused by Defective Protein
ABCA1	Ubiquitous	Exports cholesterol and phospholipid for uptake into high-density lipoprotein (HDL)	Tangier's disease
ABCB1 (MDR1)	Adrenal, kidney, brain	Exports lipophilic drugs	
ABCB4 (MDR2)	Liver	Exports phosphatidylcholine into bile	
ABCB11	Liver	Exports bile salts into bile	
CFTR	Exocrine tissue	Transports Cl^- ions	Cystic fibrosis
ABCD1	Ubiquitous in peroxisomal membrane	Influences activity of peroxisomal enzyme that oxidizes very long chain fatty acids	Adrenoleukodystrophy (ADL)
ABCG5/8	Liver, intestine	Exports cholesterol and other sterols	β-Sitosterolemia

yeast *sec* mutant. At the permissive temperature, the ABCB4 protein is expressed by the transfected cells and moves through the secretory pathway to the cell surface (Chapter 17). At the nonpermissive temperature, however, secretory vesicles cannot fuse with the plasma membrane, as they do in wild-type cells; so vesicles containing ABCB4 and other yeast proteins accumulate in the cells. After purifying these secretory vesicles, investigators labeled them in vitro with a fluorescent phosphatidylcholine derivative. The fluorescence-quenching assay outlined in Figure 18-5 was used to demonstrate that the vesicles containing ABCB4 exhibited ATP-dependent some flippase activity, whereas those without ABCB4 did not. The structures and mechanism of action of some flippases are covered in Chapter 7.

Flip-flopping between leaflets, lateral diffusion, and membrane fusion and fission are not the only dynamic processes of phospholipids in membranes. Their fatty acyl chains and, in some cases, their head groups are subject to ongoing covalent remodeling (e.g., hydrolysis of fatty esters by phospholipases and resynthesis by acyl transferases). Another key

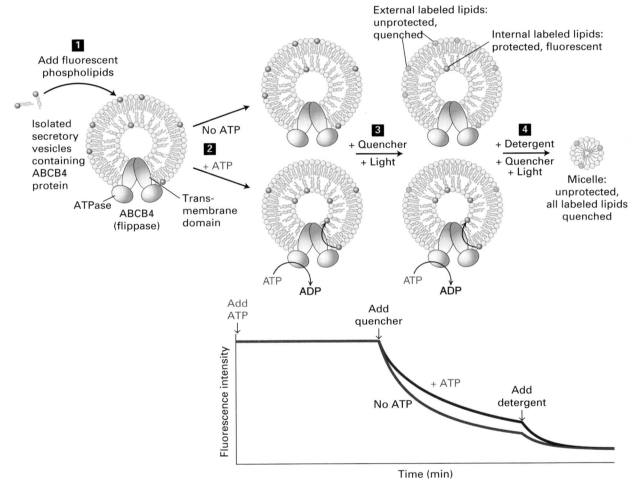

▲ **EXPERIMENTAL FIGURE 18-5 In vitro fluorescence quenching assay can detect phospholipid flippase activity of ABCB4.** A homogeneous population of secretory vesicles containing ABCB4 protein was purified from yeast *sec* mutants transfected with the *ABCB4* gene. Step **1**: Synthetic phospholipids containing a fluorescently modified head group (blue) were incorporated primarily into the outer, cytosolic leaflets of the purified vesicles. Step **2**: If ABCB4 acted as a flippase, then on addition of ATP to the outside of the vesicles a small fraction of the outward-facing labeled phospholipids would be flipped to the inside leaflet. Step **3**: Flipping was detected by adding a membrane-impermeable quenching compound called dithionite to the medium surrounding the vesicles. Dithionite reacts with the fluorescent head group, destroying its ability to fluoresce (gray). In the presence of the quencher, only labeled phospholipid in the protected environment on the inner leaflet will fluoresce. Subsequent to the addition of the quenching agent, the total fluorescence decreases with time until it plateaus at the point at which all external fluorescence is quenched and only the internal phospholipid fluorescence can be detected. The observation of greater fluorescence (less quenching) in the presence of ATP than in its absence indicates that ABCB4 has flipped some of the labeled phospholipid to the inside. Step **4**: Addition of detergent to the vesicles generates micelles and makes all fluorescent lipids accessible to the quenching agent and lowers the fluorescence to baseline values. [Adapted from S. Ruetz and P. Gros, 1994, *Cell* **77**:1071.]

dynamic process is the intracellular movement of phospholipids from one membrane to a different one. Clearly, membranes are dynamic components of the cell that interact with and react to changes in the intracellular and extracellular environments.

KEY CONCEPTS OF SECTION 18.1

Phospholipids and Sphingolipids: Synthesis and Intracellular Movement

■ Saturated and unsaturated fatty acids of various chain lengths are components of phospholipids, sphingolipids, and triglycerides (see Figure 18-2).

■ Fatty acids are synthesized by water-soluble enzymes and modified by elongation and desaturation in the endoplasmic reticulum (ER).

■ The final steps in the synthesis of glycerophospholipids, plasmalogens, and sphingolipids are catalyzed by membrane-

associated enzymes primarily in the ER, but also in the Golgi, mitochondria, and peroxisomes (see Figure 18-4).

■ Each type of lipid is initially incorporated into the pre-existing membranes on which it is made.

■ Most membrane phospholipids are preferentially distributed in either the exoplasmic or the cytosolic leaflet. This asymmetry results in part from the action of flippases such as ABCB4, a phosphatidylcholine flippase contributing to the generation of bile in the liver.

18.2 Cholesterol: A Multifunctional Membrane Lipid

Although phospholipids are critical for the formation of the classic bilayer structure of membranes, eukaryotic cell membranes require other components, including **sterols**. Here, we focus on cholesterol, the principal sterol in animal cells and the most abundant single lipid in the mammalian plasma

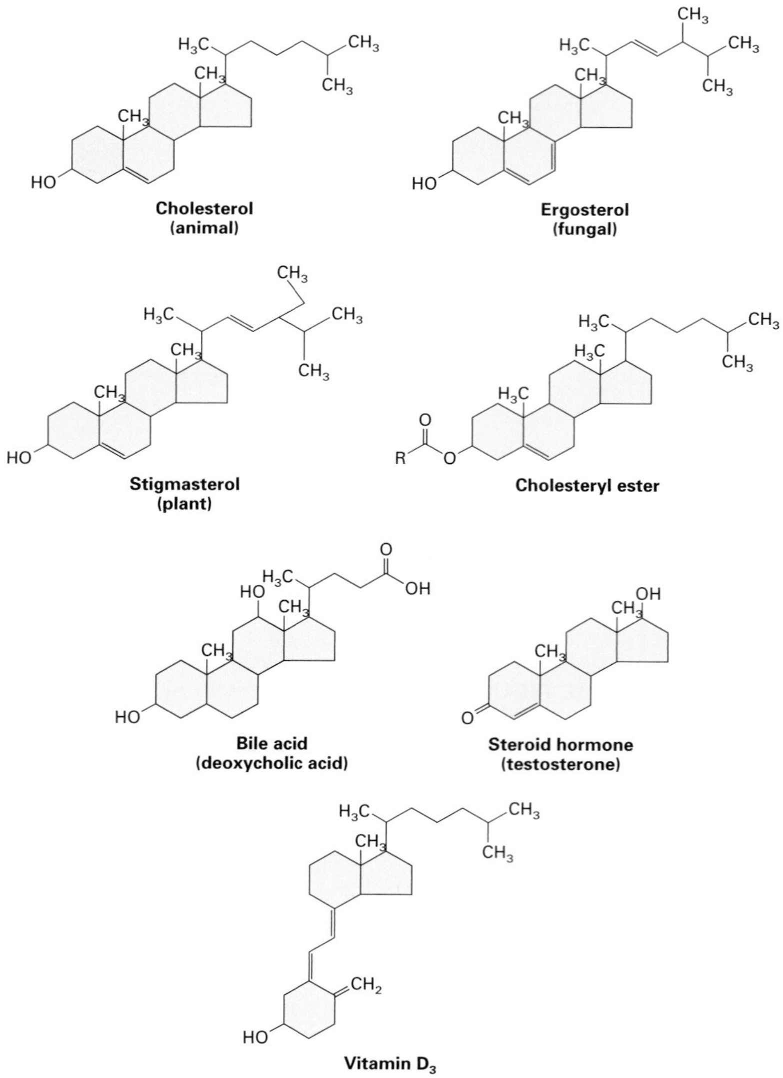

▶ **FIGURE 18-6 Chemical structures of major sterols and cholesterol derivatives.** The major sterols in animals (cholesterol), fungi (ergosterol), and plants (stigmasterol) differ slightly in structure, but all serve as key components of cellular membranes. Cholesterol is stored as cholesteryl esters in which a fatty acyl chain (R = hydrocarbon portion of fatty acid) is esterified to the hydroxyl group. Excess cholesterol is converted by liver cells into bile acids (e.g., deoxycholic acid), which are secreted into the bile. Specialized endocrine cells synthesize steroid hormones (e.g., testosterone) from cholesterol, and photochemical and enzymatic reactions in the skin and kidneys produce vitamin D.

membrane (almost equimolar with all phospholipids). Between 50 and 90 percent of the cholesterol in most mammalian cells is present in the plasma membrane and related endocytic vesicle membranes. Cholesterol is also critical for intercellular signaling and has other functions to be described shortly. The structures of the principal yeast sterol (ergosterol) and plant phytosterols (e.g., stigmasterol) differ slightly from that of cholesterol (Figure 18-6). The small differences in the biosynthetic pathways of fungal and animal sterols and in their structures are the basis of most antifungal drugs currently in use.

Cholesterol Is Synthesized by Enzymes in the Cytosol and ER Membrane

Figure 18-7 summarizes the complex series of reactions that yield cholesterol and several other related biomolecules. The basic features of this pathway are important in the synthesis of other important lipids, and familiarity with these features helps in understanding lipid regulation, discussed later. The first steps of cholesterol synthesis (acetyl CoA → HMG CoA) take place in the cytosol. The conversion of HMG CoA into mevalonate, the key rate-controlling step in cholesterol biosynthesis,

is catalyzed by *HMG-CoA reductase*, an ER integral membrane protein, even though both its substrate and its product are water soluble. The water-soluble catalytic domain of HMG-CoA reductase extends into the cytosol, but its eight transmembrane segments firmly embed the enzyme in the ER membrane and act as a regulatory domain. Five of the transmembrane segments compose the so-called *sterol-sensing domain*. As described later, homologous domains are found in other proteins taking part in cholesterol transport and regulation.

Mevalonate, the 6-carbon product formed by HMG-CoA reductase, is converted in several steps into the 5-carbon isoprenoid compound isopentenyl pyrophosphate (IPP) and its stereoisomer dimethylallyl pyrophosphate (DMPP). These reactions are catalyzed by cytosolic enzymes, as are the subsequent reactions in which six IPP units condense to yield squalene, a branched-chain 30-carbon intermediate. Enzymes bound to the ER membrane catalyze the multiple reactions that convert squalene into cholesterol in mammals or into related sterols in other species.

Because an excessive accumulation of cholesterol can lead to the formation of damaging cholesterol crystals, the production and accumulation of cholesterol is tightly

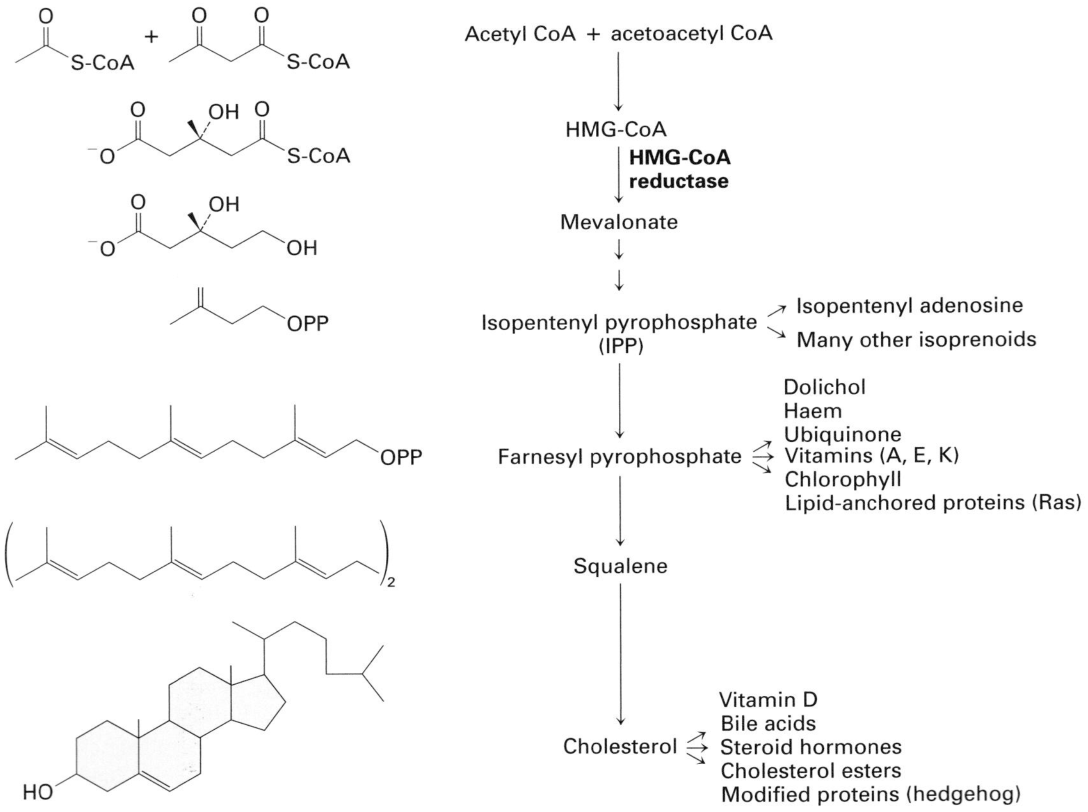

▲ **FIGURE 18-7 Cholesterol biosynthetic pathway.** The regulated rate-controlling step in cholesterol biosynthesis is the conversion of β-hydroxy-β-methylglutaryl CoA (HMG-CoA) into mevalonic acid by HMG-CoA reductase, an ER-membrane protein. Mevalonate is then converted into isopentenyl pyrophosphate (IPP), which has the basic five-carbon isoprenoid structure. IPP can be converted into cholesterol and into many other lipids, often through the polyisoprenoid intermediates shown here. Some of the numerous compounds derived from isoprenoid intermediates and cholesterol itself are indicated. See text for discussion.

controlled. For example, cholesteryl esters (see Figure 18-6) are formed from excess cholesterol and stored as cytosolic lipid droplets. *Acyl:cholesterol acyl transferase (ACAT),* the enzyme that esterifies fatty acyl CoAs to the hydroxyl group of cholesterol, is located in the ER membrane. Substantial amounts of cholesteryl ester droplets are usually found only in cells that produce steroid hormones and in foam cells, which contribute to atherosclerotic disease in artery walls. Intracellular lipid droplets, whether composed of cholesteryl esters or triglycerides, have an outer protein coat that serves as an interface between the aqueous environment of the cytosol and the lipid. The coat proteins on lipid droplets in mammalian cells are called perilipins or perilipin-related proteins, while the oleosins and their related proteins coat the surfaces of lipid droplets called oil bodies in plants.

Many Bioactive Molecules Are Made from Cholesterol and Its Biosynthetic Precursors

In addition to its structural role in membranes, discussed in Chapter 5, cholesterol is the precursor for several important bioactive molecules. They include **bile acids** (see Figure 18-6), which are made in the liver and help emulsify dietary fats for digestion and absorption in the intestines, steroid hormones produced by endocrine cells (e.g., adrenal gland, ovary, testes), and vitamin D produced in the skin and kidneys. Arthropods need cholesterol or other sterols to produce membranes and ecdysteroid hormones, which control development; however, they cannot make the precursor sterols themselves and must obtain these compounds in their diet. Another critical function of cholesterol is its covalent addition to Hedgehog protein, a key signaling molecule in embryonic development (Chapter 15).

Isopentenyl pyrophosphate and other isoprenoid intermediates in the cholesterol pathway also serve as precursors for more than 23,000 biologically active molecules. Some of these molecules are discussed in other chapters: various hemes, including the oxygen-binding component of hemoglobin and electron-carrying components of cytochromes (see Figure 8-15a); ubiquinone, a component of the mitochondrial electron-transport chain (see Figure 8-16); chlorophylls, the light-absorbing pigments in chloroplasts (see Figure 8-31); and dolichol, a polyisoprenoid in the ER membrane that plays a key role in the glycosylation of proteins (see Figure 16-17).

Isoprenoid derivatives are particularly abundant in plants in which they form fragrances and flavors, rubber and latex, hormones and pheromones, various defensive molecules, the active ingredient in marijuana, the cardioprotective natural drug digitalis, anticancer drugs such as taxol, and many others. Given the importance of isoprenoids as biosynthetic precursors, it is not surprising that a second, mevalonate-independent pathway for IPP synthesis evolved in eubacteria (e.g., *E. coli*), green algae, and higher plants. In plants, this pathway is located in organelles called plastids and operates to synthesize carotenoids, phytol (the isoprenoid side chain of chlorophyll), and other isoprenoids. ▌

Cholesterol and Phospholipids Are Transported Between Organelles by Golgi-Independent Mechanisms

As already noted, the final steps in the synthesis of cholesterol and phospholipids take place primarily in the ER, although some of these membrane lipids are produced in mitochondria and peroxisomes (plasmalogens). Thus the plasma membrane and the membranes bounding other organelles (e.g., Golgi, lysosomes) must obtain these lipids by means of one or more intracellular transport processes. For example, in one important pathway, phosphatidylserine made in the ER is transported to the inner mitochondrial membrane where it is decarboxylated to phosphatidylethanolamine, some of which either returns to the ER for conversion into phosphatidylcholine or moves to other organelles.

Membrane lipids do accompany both soluble (luminal) and membrane proteins during vesicular trafficking through the Golgi-mediated secretory pathway (see Figure 17-1). However, several lines of evidence suggest that there is substantial interorganelle movement of cholesterol and phospholipids through other, Golgi-independent mechanisms. For example, chemical inhibitors of the classic secretory pathway and mutations that impede vesicular traffic in this pathway do not prevent cholesterol or phospholipid transport between membranes, although they do disrupt the transport of proteins and Golgi-derived sphingolipids. Furthermore, membrane lipids produced in the ER cannot move to mitochondria by means of classic secretory transport vesicles, inasmuch as no vesicles budding from ER membranes have been found to fuse with mitochondria.

Three mechanisms have been proposed for the transport of cholesterol and phospholipids from their sites of synthesis to other membranes independently of the Golgi-mediated secretory pathway (Figure 18-8). First, some Golgi-independent transport is most likely through membrane-limited vesicles or other protein–lipid complexes. The second mechanism entails direct protein-mediated contact of ER or ER-derived membranes with membranes of other organelles. In the third mechanism, small lipid-transfer proteins facilitate the exchange of phospholipids or cholesterol between different membranes. Although such transfer proteins have been identified in assays in vitro, their role in intracellular movements of most phospholipids is not well defined. For instance, mice with a knockout mutation in the gene encoding the phosphatidylcholine-transfer protein appear to be normal in most respects, indicating that this protein is not essential for cellular phospholipid metabolism.

One well-established component of the intracellular cholesterol-transport system is the *steroidogenic acute regulatory (StAR) protein.* This protein, which is encoded in nuclear DNA, controls the transfer of cholesterol from the cholesterol-rich outer mitochondrial membrane to the cholesterol-poor inner membrane, where it undergoes the first steps in its enzymatic conversion into steroid hormones. StAR-mediated cholesterol transport is a key regulated, rate-controlling step in steroid hormone synthesis. StAR contains

(a)

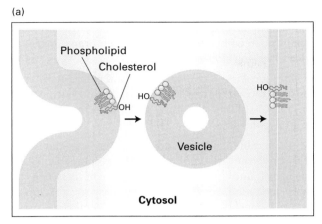

(b)

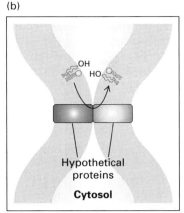

(c)

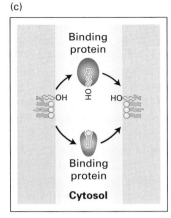

▲ **FIGURE 18-8 Proposed mechanisms of Golgi-independent transport of cholesterol and phospholipids between membranes.** In mechanism (a), vesicles transfer lipids between membranes without passing through the Golgi apparatus. In mechanism (b), lipid transfer is a consequence of direct contact between membranes that is mediated by membrane-embedded proteins. In mechanism (c), transfer is mediated by small, soluble lipid-transfer proteins. Some evidence suggests that this mechanism does not account for a significant part of the Golgi-independent flow of phospholipids between membranes. [Adapted from F. R. Maxfield and D. Wustner, 2002, *J. Clin. Invest.* **110**:891.]

an N-terminal targeting sequence that directs the protein to the mitochondrial outer membrane (Chapter 16) and a C-terminal START (StAR-related transfer) domain that has a cholesterol-binding hydrophobic pocket. Similar START domains are found in several proteins implicated in intracellular cholesterol transport, and these domains have been shown to promote cholesterol transfer in cultured cells. Mutations in the StAR gene can cause congenital adrenal hyperplasia, a lethal disease marked by a drastic reduction in the synthesis of steroid hormones. Some other proteins implicated in lipid transport, including the phosphatidylcholine-transfer protein already mentioned, also contain START domains.

A second well-established contributor to intracellular cholesterol movement is the *Niemann-Pick C1 (NPC1) protein*, an integral membrane protein located in the rapidly moving late endosomal/lysosomal compartment. Some of the multiple membrane-spanning segments of NPC1 form a sterol-sensing domain similar to that in HMG-CoA reductase. Mutations in NPC1 cause defects in intracellular cholesterol and glycosphingolipid transport and consequently in the regulation of cellular cholesterol metabolism. Cells without functional NPC1 or cells treated with a drug that mimics loss of NPC1 function accumulate excess cholesterol in the late endosomal/lysosomal compartment (Figure 18-9). Cholesterol transport in NPC1-deficient cells is restored by overexpression of Rab9, a small GTPase implicated in late endosomal vesicular transport (Chapter 17). This finding suggests that vesicular trafficking plays at least some role in NPC1-dependent cholesterol movement.

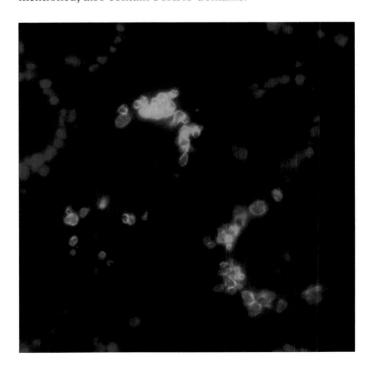

◄ **EXPERIMENTAL FIGURE 18-9 Cells with nonfunctional Niemann-Pick C1 (NPC1) protein accumulate cholesterol in late endosomal/lysosomal vesicles.** In three cells that express a transgene encoding a hybrid NPC1 protein linked to green fluorescent protein (GFP), the hybrid protein is revealed in the late endosomal/lysosomal compartment by its green fluorescence. Shown in this two-color fluorescence micrograph are cells that had been treated with a drug that inhibits NPC1 function. The cells were also stained with a blue-fluorescing cholesterol-binding drug called fillipin. Note the colocalization of cholesterol (blue) with the NPC1-GFP hybrid protein (green) on the surfaces of vesicles adjacent to the nucleus. Cells not expressing the hybrid NPC1 (blue only) are also seen. The accumulation of cholesterol in these vesicles in the absence of functional NPC1 suggests that late endosomal/lysosomal vesicles play a role in intracellular cholesterol trafficking. [From D. C. Ko et al., 2001, *Mol. Biol. Cell* **12**:601; courtesy of D. Ko and M. Scott.]

In humans, defects in NPC1 function cause abnormal lipid storage in intracellular organelles, resulting in neurologic abnormalities, neurodegeneration, and premature death. Indeed, identification of the gene defective in such patients led to discovery of the NPC1 protein. ∎

The lipid compositions of different organelle membranes vary considerably (see Table 5-1). Some of these differences are due to different sites of synthesis. For example, a phospholipid called cardiolipin, which is localized to the mitochondrial membrane, is made only in mitochondria and little is transferred to other organelles. Differential transport of lipids also plays a role in determining the lipid compositions of different cellular membranes. For instance, even though cholesterol is made in the ER, the cholesterol concentration (cholesterol-to-phospholipid molar ratio) is ~1.5–13-fold higher in the plasma membrane than in other organelles (ER, Golgi, mitochondrion, lysosome). Although the mechanisms responsible for establishing and maintaining these differences are not well understood, the distinctive lipid composition of each membrane has a major influence on its physical properties (Chapter 5).

KEY CONCEPTS OF SECTION 18.2

Cholesterol: A Multifunctional Membrane Lipid

■ The initial steps in cholesterol biosynthesis take place in the cytosol, whereas the last steps are catalyzed by enzymes associated with the ER membrane.

■ The rate-controlling step in cholesterol biosynthesis is catalyzed by HMG-CoA reductase, whose transmembrane segments are embedded in the ER membrane and contain a sterol-sensing domain.

■ Cholesterol itself and isoprenoid intermediates in its synthesis are biosynthetic precursors of steroid hormones, bile acids, lipid-soluble vitamins, and numerous other bioactive molecules (see Figure 18-7).

■ Considerable evidence indicates that vesicular trafficking through the Golgi complex is not responsible for much cholesterol and phospholipid movement between membranes. Golgi-independent vesicular transport, direct protein-mediated contacts between different membranes, soluble protein carriers, or all three may account for some interorganelle transport of cholesterol and phospholipids (see Figure 18-8).

■ The StAR protein, which has a hydrophobic cholesterol-binding pocket, plays a key role in moving cholesterol into the mitochondrion for steroid hormone synthesis.

■ The NPC1 protein, a large, multipass transmembrane protein, contains a sterol-sensing domain similar to that in HMG-CoA reductase. NPC1 is required for the normal movement of cholesterol between certain intracellular compartments.

18.3 Lipid Movement into and out of Cells

In multicellular organisms, particularly mammals, lipids are often imported and exported from cells and transported

(a) Transport protein–mediated export and import of lipid

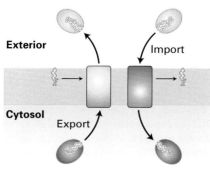

(b) Lipoprotein- and receptor-mediated export and import

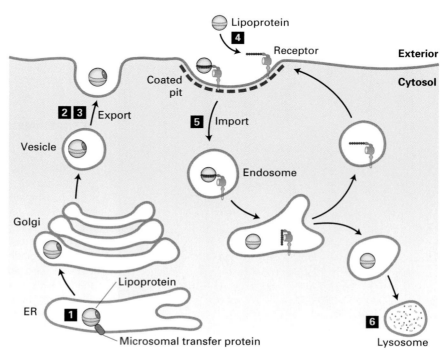

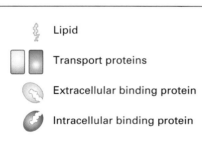

among different tissues by the circulation. Such lipid movements help maintain appropriate intracellular and whole-body lipid levels and have other advantages for an organism. For instance, lipids absorbed from the diet in the intestines or stored in adipose tissue can be distributed to cells throughout the body. In this way, cells can obtain essential dietary lipids (e.g., linoleate) and can avoid wasting energy on the synthesis of lipids (e.g., cholesterol) otherwise available from the diet. The ability of some cells to export lipids permits the excretion of excess lipids from the body or their secretion into certain body fluids (e.g., milk in mammary glands). In addition, the coordination of lipid and energy metabolism throughout the organism depends on intercellular lipid transport.

Not surprisingly, intracellular and intercellular lipid transport and metabolism are coordinately regulated, as discussed in Section 18.4. Here, we review the primary ways in which cells import and export lipids. The two main mechanisms of cellular lipid export and import are summarized in Figure 18-10. In the first mechanism, the import or export of individual lipid molecules is mediated by cell-surface transport proteins. The principles underlying this mechanism are similar to those for small water-soluble molecules such as glucose, discussed in Chapter 7. A noteworthy difference is that the hydrophobic lipids, which are poorly soluble in aqueous solution, often associate with lipid-binding proteins in the extracellular space or the cytosol, rather than remaining free in solution. In the second mechanism of lipid import and export, collections of lipids are packaged with proteins into transport particles called lipoproteins. These lipoproteins are exported from cells through the classic secretory pathway (Chapter 17). Lipids carried by extracellular lipoproteins are taken up by cells through surface receptors.

◀ **FIGURE 18-10 The two major pathways for importing and exporting cellular lipids.** (a) Individual lipid molecules cross the plasma membrane with the assistance of transmembrane transport proteins. Intracellular and extracellular lipid-binding proteins (shown here), lipid micelles, or membranes normally participate as the donors and acceptors of the lipids. In some cases, the concentration gradients of the transported substances are sufficient to drive transport (e.g., import of bile acids in the liver and intestines by Na^+-linked symporters). In others, coupled ATP hydrolysis helps drive lipid transport (e.g., secretion of bile lipids from hepatocytes by ABC proteins). (b) Lipids are also transported as components of lipoproteins. These large assemblies of protein and lipids are put together in the ER with the assistance of microsomal transfer protein (**1**), are exported (**2**) through the secretory pathway (**3**) as water-soluble particles, and then circulate in the blood. After circulating lipoproteins bind to certain cell-surface receptors (**4**), the intact particles can be internalized by endocytosis (**5**) and the lipids are subsequently hydrolyzed in lysosomes (**6**), as depicted here. Other receptors mediate the uptake of individual lipid components from lipoproteins, releasing a lipid-depleted particle into the extracellular space.

Cell-Surface Transporters Aid in Moving Fatty Acids Across the Plasma Membrane

For many years, transmembrane fatty acid transport was thought not to require a protein mediator, because these hydrophobic molecules can diffuse across lipid bilayers. Although some protein-independent fatty acid transport probably does take place, some cells (e.g., intestinal epithelial cells, cardiac muscle cells, and adipocytes) must import or export substantial amounts of fatty acids at rates or with a specificity and regulation or both that are not possible by diffusion alone. Several integral membrane proteins have been shown to participate in fatty acid import in cell culture and whole-animal experiments. These proteins include various *fatty acid transport proteins (FATPs)* and the multifunctional cell-surface protein *CD36*. These transporters mediate the movement of substrates down their concentration gradients. In contrast, transporters that mediate the export of fatty acids have not yet been identified, but at least some transporters may mediate bidirectional transport, depending on the direction of the fatty acid concentration gradient.

Because of their poor water solubility, most fatty acids are bound to a carrier protein in the aqueous cytosol and extracellular space (see Figure 18-10a). The major mammalian extracellular binding protein that donates or accepts fatty acids is serum *albumin*. The most abundant protein in mammalian plasma (fluid part of the blood), albumin has at least one hydrophobic α helix–lined lipid-binding groove or cleft on its surface. In addition to mediating the transport of fatty acids, albumin mediates that of several anionic organic acids (e.g., bilirubin) and other molecules through the bloodstream. The major intracellular carriers of fatty acids are the fatty acid–binding proteins described earlier.

ABC Proteins Mediate Cellular Export of Phospholipids and Cholesterol

The export of phospholipids and cholesterol can be simultaneous owing to the activity of various members of the ABC superfamily (see Table 18-2). The best-understood example of this phenomenon is in the formation of *bile*, an aqueous fluid containing phospholipids, cholesterol, and bile acids, which are derived from cholesterol. After export from liver cells, phospholipids, cholesterol, and bile acids form water-soluble **micelles** in the bile, which is delivered through ducts to the gallbladder, where it is stored and concentrated. In response to a fat-containing meal, bile is released into the small intestine to help emulsify dietary lipids and thus aid in their digestion and absorption into the body. As we shall see later, the alteration of biliary metabolism by drugs can be used to prevent heart attacks.

Figure 18-11 outlines the major transport proteins that mediate the secretion and movement of bile components. Three ABC proteins move phospholipids, cholesterol, and bile acids across the apical surface of liver cells into small ductules (step **1**). One of these proteins, the ABCB4 flippase, flips phosphatidylcholine from the cytosolic leaflet to the

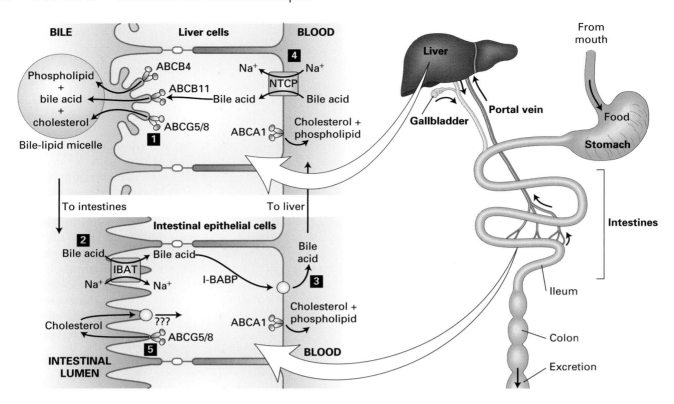

▲ FIGURE 18-11 Major transport proteins in the liver and intestines taking part in the enterohepatic circulation of biliary lipids. The secretion of bile components and recycling of bile acids are mediated by a diverse array of transport proteins in liver cells (hepatocytes) and intestinal epithelial cells. Both of these polarized cell types import lipids across one surface and export them across the opposite surface. Step **1**: Hepatocytes export lipids across their apical membranes into the bile by using three ATP-dependent ABC proteins: ABCB4 (phospholipids), ABCB11 (bile acids), and ABCG5/8 (sterols). Step **2**: Intestinal epithelial cells import bile components and dietary lipids from the intestinal lumen by using the ileal bile acid transporter (IBAT), a Na⁺-linked symporter, and other less well defined transporters located in the apical membrane. Step **3**: Imported bile acids are transported to the basolateral surface bound to intestinal bile acid-binding protein (I-BABP) and are exported into the blood with the aid of unknown transporters. Step **4**: Bile acids returned to the liver in the blood are imported by NTCP, another Na⁺-linked symporter. Step **5**: Absorption of sterols by intestinal cells is reduced by ABCG5/8, which appears to pump plant sterols and cholesterol out of the cells and back into the lumen.

exoplasmic leaflet of the apical membrane in hepatocytes, as described earlier. The precise mechanism by which the excess phospholipid desorbs from the exoplasmic leaflet into the extracellular space is not understood. A related protein, ABCB11, transports bile acids, whereas the ABCG5 and ABCG8 "half" proteins combine into a single ABC protein (ABCG5/8) that exports sterols into the bile.

In the intestine the *ileal bile acid transporter (IBAT)* imports bile acids from the lumen into intestinal epithelial cells (step **2**). IBAT is a Na⁺-linked symporter (see Figure 7-21) that uses the energy released by the movement of Na⁺ down its concentration gradient to power the uptake of about 95 percent of the bile acids. Those bile acids imported on the apical side of intestinal epithelial cells move intracellularly with the aid of *intestinal bile acid–binding protein (I-BABP)* to the basolateral side. There, they are exported into the blood by poorly characterized transport proteins (step **3**) and eventually returned to liver cells by another Na⁺-linked

symporter called NTCP (step **4**). This cycling of bile acids from liver to intestine and back, referred to as the **enterohepatic circulation,** is tightly regulated and plays a major role in lipid homeostasis.

Because the amount of dietary cholesterol is normally low, a substantial fraction of the cholesterol in the intestinal lumen comes from the biliary cholesterol secreted by the liver. ABCG5/8 also is expressed in the apical membrane of intestinal epithelial cells, where it helps control the amounts of cholesterol and plant-derived sterols absorbed apparently by pumping excess or unwanted absorbed sterols out of the epithelial cells back into the lumen (see Figure 18-11, step **5**). Partly as a result of this activity, only about 1 percent of dietary plant sterols, which are not metabolically useful to mammals, enter the bloodstream. Unabsorbed bile acids (normally <5 percent of the luminal bile acids) and unabsorbed cholesterol and plant sterols are eventually excreted in the feces.

Inactivating mutations in the genes encoding ABCG5 or ABCG8 cause *β-sitosterolemia*. Patients with this rare genetic disease absorb abnormally high amounts of both plant and animal sterols and their livers secrete abnormally low amounts into the bile. Indeed, findings from studies with β-sitosterolemia patients first implicated these ABC proteins in cellular sterol export. ∎

Two other cell-surface transport proteins mediate the export of cellular cholesterol, phospholipid, or both: the ABC superfamily member ABCA1 and a homolog of the fatty acid transporter CD36 called SR-BI. These proteins will be described in detail shortly because of their important roles in lipoprotein metabolism.

Lipids Can Be Exported or Imported in Large Well-Defined Lipoprotein Complexes

To facilitate the mass transfer of lipids between cells, animals have evolved an efficient alternative to the molecule-by-molecule import and export of lipids mediated by cell-surface transport proteins, such as those depicted in Figure 18-11. This alternative packages from hundreds to thousands of lipid molecules into water-soluble, macromolecular carriers, called **lipoproteins**, that cells can secrete into the circulation or take up from the circulation as an ensemble.

A lipoprotein particle has a shell composed of proteins (*apolipoproteins*) and a cholesterol-containing phospholipid monolayer (Figure 18-12). The shell is **amphipathic** because its outer surface is hydrophilic, making these particles water soluble, and its inner surface is hydrophobic. Adjacent to the hydrophobic inner surface of the shell is a core of neutral lipids containing mostly cholesteryl esters, triglycerides, or both. Small amounts of other hydrophobic compounds (e.g., vitamin E, carotene) also are carried in the lipoprotein core.

Mammalian lipoproteins fall into four major classes. Three of them—**high-density lipoprotein (HDL), low-density lipoprotein (LDL), and very low density lipoprotein (VLDL)**—are named on the basis of their differing buoyant densities. The lower the protein-to-lipid ratio, the lower the density. The fourth class, the **chylomicrons**, is the least dense and contains the highest proportion of lipids. Each class of lipoproteins has distinctive apolipoprotein and lipid compositions, sizes, and functions (Table 18-3). VLDLs and chylomicrons carry mainly triglycerides in their cores, whereas the cores of LDLs and HDLs consist mostly of cholesteryl esters. Apolipoproteins help organize the structure of a lipoprotein particle and determine its interactions with enzymes, extracellular lipid-transfer proteins, and cell-surface receptors. Each LDL particle contains a single copy of a large (537-kDa) apolipoprotein called apoB-100 embedded in its outer shell (see Figure 18-12). In contrast, several copies of different apolipoproteins are found in each of the other lipoprotein classes.

TABLE 18-3	**Major Classes of Human Plasma Lipoproteins**			
Property	Chylomicron	VLDL	LDL	HDL
Mass, approx. (kDa)	$50–1000 \times 10^3$	$10–80 \times 10^3$	2.3×10^3	$0.175–0.360 \times 10^3$
Diameter (nm)	75–1200	30–80	18–25	5–12
Triglycerides (% of core lipids)	97	75	12	11
Cholesteryl esters (% of core lipids)	3	25	88	89
Protein:lipid mass ratio	1:100	9:100	25:100	90:100
Major apolipoproteins	A, B-48, C, E	B-100, C, E	B-100	A, C
Major physiological function	Transports dietary triglyceride (Tg) from intestines to extrahepatic tissues (e.g., muscle, adipose tissue); Tg-depleted remnants deliver dietary cholesterol and some Tg to the liver	Transports hepatic Tg to extrahepatic tissues; converted into LDL	Transports plasma cholesterol to liver and to extrahepatic tissues	Takes up cholesterol from extrahepatic tissues and delivers it to liver, steroid-producing tissues, and other lipoproteins

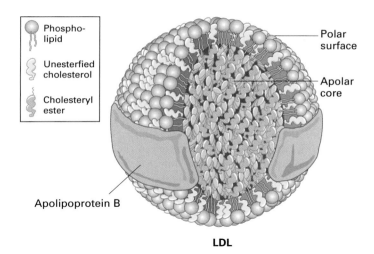

▲ FIGURE 18-12 Model of low-density lipoprotein (LDL). This class and the other classes of lipoproteins have the same general structure: an amphipathic shell, composed of a phospholipid monolayer (not bilayer), cholesterol, and protein, and a hydrophobic core, composed mostly of cholesteryl esters or triglycerides or both but with minor amounts of other neutral lipids (e.g., some vitamins). This model of LDL is based on electron microscopy and other low-resolution biophysical methods. LDL is unique in that it contains only a single molecule of one type of apolipoprotein (apoB), which appears to wrap around the outside of the particle as a band of protein. The other lipoproteins contain multiple apolipoprotein molecules, often of different types. [Adapted from M. Krieger, 1995, in E. Haber, ed., *Molecular Cardiovascular Medicine*, Scientific American Medicine, pp. 31–47.]

Lipoproteins Are Made in the ER, Exported by the Secretory Pathway, and Remodeled in the Circulation

Only two types of lipoproteins, VLDL and chylomicrons, are fully formed within cells by assembly in the ER, a process that requires the activity of *microsomal transfer protein*. The assembled particles move through the secretory pathway to the cell surface and are released by **exocytosis**—VLDL from liver cells and chylomicrons from intestinal epithelial cells (see Figure 18-10b). LDLs, IDLs (intermediate-density lipoproteins), and some HDLs are generated extracellularly in the bloodstream and on the surfaces of cells by the remodeling of secreted VLDLs and chylomicrons. There are four types of modifications:

■ Hydrolysis of triglycerides and phospholipids by lipases and esterification of cholesterol by an acyl transferase

■ Transfer of cholesteryl esters, triglycerides, and phospholipids between lipoproteins by specific lipid-transfer proteins

■ Uptake by some particles of cholesterol and phospholipids exported from cells

■ Association and dissociation of some apolipoproteins from the surfaces of the particles

▶ FIGURE 18-13 Lipoprotein remodeling and interconversions in the circulatory system. Apolipoproteins are indicated by capital letters (e.g., A, B-100) projecting from particles. (a) After having been secreted from the liver **1**, VLDL's triglycerides are hydrolyzed by lipoprotein lipase, an extracellular enzyme attached to the blood-facing surfaces of vessels **2**. The loss of triglycerides and some apolipoproteins from VLDL produces IDL **3**, which is converted into LDL **4**. Both IDL and LDL can be removed from the circulation by endocytosis by means of LDL receptors (LDLR) on liver (hepatic) cells **5** and nonliver (extrahepatic) cells **6**. (b) Dietary lipids are absorbed **1** and packaged into chylomicrons in intestinal epithelial cells, secreted into the lymph, and then enter the bloodstream **2**. In the circulation, they are remodeled similarly to the remodeling of VLDL **3**, forming smaller, cholesterol-enriched chylomicron remnants **4**, which are taken up by hepatocytes through receptor-mediated endocytosis **5**. (c) HDL is thought to be formed after secretion of apolipoprotein A from cells **1** by the formation of preβ-HDL particles, which contain apoA but very little lipid **2**. Those small particles act as acceptors for phospholipid and cholesterol exported from cells (primarily liver and intestine) by the ABCA1 transporter **3**, forming a cholesterol-rich intermediate **4**. Lethicin:cholesterol acyl transferase (LCAT), an enzyme in the plasma, esterifies cholesterol after its incorporation into HDL **5**. Cholesteryl esters in the core of a large HDL particle can be transferred to cells (especially liver **6** and steroidogenic **7** cells) by the receptor SR-BI or to other lipoproteins by cholesteryl ester–transfer protein (CETP) **8** and subsequently to tissues such as the liver **9**. [Adapted from M. S. Brown and J. L. Goldstein, 1984, *Sci. Am.* **251**(5):58, and M. Krieger, 1999, *Ann. Rev. Biochem.* **68**:523.]

For example, VLDL secreted from hepatocytes is converted into IDL and eventually into LDL, which can then deliver its cholesterol to cells through LDL receptors (Figure 18-13a). Similarly, chylomicrons carrying dietary lipids from the intestines are converted by lipase hydrolysis into *chylomicron remnants*, which eventually undergo endocytic uptake by the liver (Figure 18-13b). Small preβ-HDL particles are generated extracellularly from apoA apolipoproteins, secreted mainly by liver and intestinal cells, and from small amounts of cholesterol and phospholipid. They are then further converted into larger, spherical HDL particles, which constitute the bulk of the HDL found in the blood (Figure 18-13c). A major way that preβ-HDL particles grow larger is by accepting phospholipids and cholesterol exported from cells with the aid of yet another ABC protein called *ABCA1* (see Table 18-2). This protein was implicated in the formation of HDL when defects in the *ABCA1* gene were shown to cause Tangier's disease, a very rare genetic disease in which affected persons have almost no HDL in their blood. After being incorporated into HDL, cholesterol is esterified by *lecithin:cholesterol acyl transferase (LCAT)*, an enzyme present in the plasma. Large HDL particles can transfer their cholesteryl esters to other lipoproteins through *cholesteryl ester–transfer protein (CETP)* or to cells (especially the liver and steroidogenic cells) through the receptor SR-BI, discussed later.

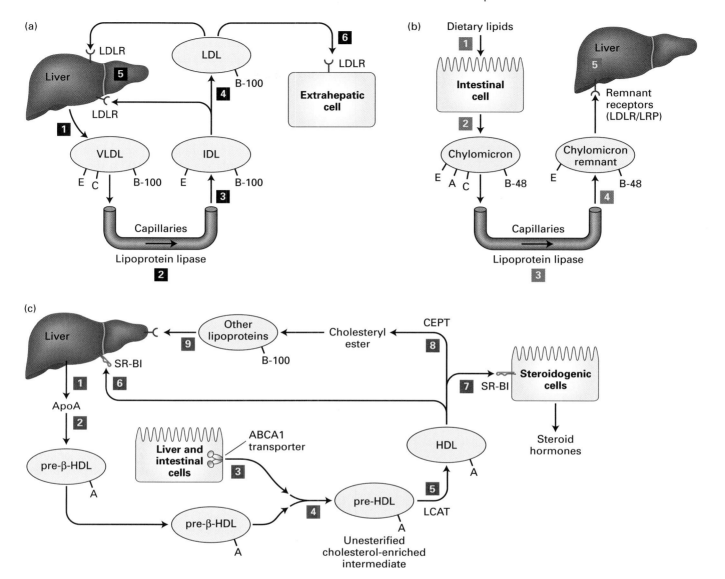

In Section 18.5, we examine in detail the findings that the plasma LDL cholesterol concentration (including both un-esterified and esterified cholesterol) is directly correlated with risk for coronary artery disease, whereas the plasma HDL cholesterol concentration is inversely correlated with risk. For this reason, LDL is popularly called "bad cholesterol" and HDL, "good cholesterol."

Cells Use Several Protein-Mediated Mechanisms to Import Lipoprotein Lipids

For maximum efficiency, the lipids within circulating lipoproteins should be taken up only by those cells that require them for membrane formation (e.g., dividing cells), steroid hormone synthesis (e.g., endocrine cells), energy production (e.g., muscle cells), or storage (adipose cells, endocrine cells). The targeting of lipoprotein lipids to appropriate cells is accomplished in one of two ways: (1) local, partial extracellular hydrolysis of core triglycerides followed by transport protein–mediated uptake of the released fatty acids or (2) the regulated expression of cell-surface lipoprotein receptors that mediate the direct uptake of lipoprotein lipids.

The first targeting mechanism, for example, supplies cells with fatty acids for use as an energy source in muscle and for storage in adipose tissue. The extracellular enzyme lipoprotein lipase is attached by glycosaminoglycan (GAG) chains to the blood-facing surface of endothelial cells in these tissues. Fatty acids released from the hydrolysis of core triglycerides in VLDL and chylomicrons then cross the vessel wall and enter underlying cells through fatty acid transporters such as FATPs and CD36 (see Figure 18-10a). This process results in the delivery of fatty acids to the cells concomitant with remodeling of the lipoprotein particles.

The expression of various lipoprotein receptors by different tissues also ensures that lipids are delivered to cells that need and can use them. In every case of receptor-facilitated delivery, cholesteryl esters and triglycerides in the lipoprotein core must cross two topological barriers to enter the cytoplasmic space: the phospholipid monolayer shell of the lipoprotein particle and the cell's bilayer plasma membrane. At some point in the lipid-delivery process, the esterified transport forms of the core lipids must be hydrolyzed to unesterified forms (cholesterol, fatty acids) to be usable by the importing cells. Cells have evolved two distinct mechanisms for receptor-facilitated uptake of lipids in lipoprotein cores: **receptor-mediated endocytosis** of entire lipoprotein particles and **selective lipid uptake** of certain lipid components of lipoprotein particles.

Analysis of Familial Hypercholesterolemia Revealed the Pathway for Receptor-Mediated Endocytosis of LDL Particles

 There is no better example of the synergistic relation between basic molecular cell biology and medicine than the story of the discovery of the LDL receptor (LDLR) pathway for delivering cholesterol to cells. The series of elegant and Nobel Prize–winning studies leading to this discovery served as sources of insight into the mechanisms underlying LDL metabolism, the functions and properties of several key organelles, cellular systems for coordinately regulating complex metabolic pathways, and new approaches for treating atherosclerosis.

Some of these experiments compared LDL metabolism in normal human cells and in cells from patients with *familial hypercholesterolemia (FH)*, a hereditary disease that is marked by elevated plasma LDL cholesterol and is now known to be caused by mutations in the *LDLR* gene. In patients who have one normal and one defective copy of the *LDLR* gene (heterozygotes), LDL cholesterol is increased about twofold. Those with two defective *LDLR* genes (homozygotes) have LDL cholesterol levels that are from fourfold to sixfold as high as normal. FH heterozygotes commonly develop cardiovascular disease about 10 years earlier than normal people do, and FH homozygotes usually die of heart attacks before reaching their late 20s. ∎

Here, we illustrate how analysis of the cellular defects underlying familial hypercholesterolemia can illuminate normal cellular processes. First let's consider typical cell-culture experiments in which the interactions of LDL with normal and FH homozygous cells were examined as a function of LDL concentration, which defined the high-affinity LDL receptor, and incubation temperature, which established the temperature dependence of LDL uptake. In these experiments, purified LDL was first labeled by the covalent attachment of radioactive ^{125}I to the side chains of tyrosine residues in apoB-100 on the surfaces of the LDL particles. Cultured cells from normal persons and FH patients were incubated for several hours with the labeled LDL. Investigators then de-

termined how much LDL was bound to the surfaces of cells, how much was internalized, and how much of the apoB-100 component of the LDL was degraded by enzymatic hydrolysis to individual amino acids. The degradation of apoB-100 was detected by the release of ^{125}I-tyrosine into the culture medium.

We can see from the results shown in Figure 18-14a that, compared with normal cells, homozygous FH cells are clearly defective in the binding and internalization of added LDL and in the degradation of apoB-100 at the normal physiologic temperature of 37 °C. The homozygous cells exhibit essentially no activity. Heterozygous cells exhibit about half the activity of normal cells. The shape of the binding curve for normal cells is consistent with a receptor that has a high affinity for LDL and is saturable. Note also that the curves for LDL internalization and degradation have the same shape as the binding curve. Moreover, when the experiments were performed with normal cells at 4 °C, LDL binding was observed, but internalization and degradation were inhibited (Figure 18-14b). Low temperature does not normally inhibit the binding of molecules to cell-surface receptors, but it does inhibit processes, such as the internalization and subsequent degradation of molecules, that depend on membrane trafficking (Chapter 17). Thus these results suggest that LDL first binds to cell-surface receptors and is subsequently internalized and degraded. One final feature of these results is worth noting. After cells were incubated for 5 hours at 37 °C, the amounts of internalized LDL and hydrolyzed apoB-100 were substantially greater than those of surface-bound LDL. This result indicates that each receptor molecule bound and mediated the internalization of more than one LDL particle in the incubation period. In other words, the LDL receptor is recycled.

Pulse-chase experiments with normal cells and a fixed concentration of ^{125}I-labeled LDL helped to further define the time course of events in receptor-mediated cellular LDL processing. These experiments clearly demonstrate the order of events: surface binding of LDL → internalization → degradation (Figure 18-15). The results of electron microscopy studies with LDL particles tagged with an electron-dense label revealed that LDL first binds to clathrin-coated endocytic pits that invaginate and bud off to form coated vesicles and then endosomes (see Figure 17-27). Findings from further experiments showed that the LDL receptor recognizes apoB-100 and one or two closely related apolipoproteins; thus binding by this receptor is highly specific for LDL. Binding is also pH dependent: strong binding of LDL occurs at the pH of extracellular fluid (7.4); weak or no binding occurs at the lower pH (≈ 4.5–6) found in some intracellular organelles (e.g., endosomes and lysosomes). Because of this property, the LDL receptor releases bound LDL within intracellular vesicles and can be recycled to the cell surface.

A variety of mutations in the gene encoding the LDL receptor can cause familial hypercholesterolemia. Some mutations prevent the synthesis of the LDLR protein; others prevent proper folding of the receptor protein in the ER,

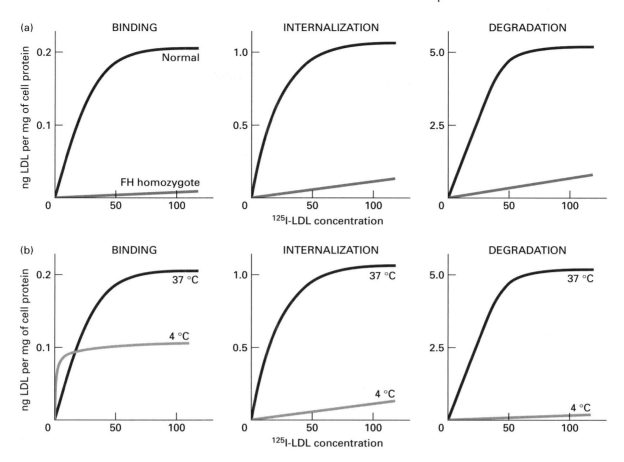

▲ **EXPERIMENTAL FIGURE 18-14 LDL binding, internalization, and degradation are reduced in cells from patients with familial hypercholesterolemia (FH) and are influenced by temperature.** Cultured fibroblasts were incubated for 5 hours in a medium containing LDL whose apolipoprotein B (apoB) component was labeled on tyrosine side chains with ^{125}I. The amounts of LDL bound to the cell surface (binding) and present within cells (internalization), as well as the amount of ^{125}I-tyrosine released into the medium owing to apoB degradation (hydrolysis), were determined as a function of the LDL concentration. (a) The curves shown here represent typical results from experiments at 37 °C with cells from a normal person and from a person homozygous for FH. Note that the y-axis scales are different and that the relative maximal values for binding internalization and hydrolysis are approximately 0.2:1:5. (b) These results for normal cells at 37 °C and 4 °C show that LDL binding is not reduced dramatically at low temperature, whereas subsequent internalization and hydrolysis are blocked. These findings suggest that internalization and hydrolysis entail membrane movements, which are typically inhibited at low temperatures. See text for discussion. [Adapted from M. S. Brown and J. L. Goldstein, 1979, *Proc. Nat'l. Acad. Sci. USA* **76**:3330, and J. L. Goldstein and M. S. Brown, 1989, in C. R. Scriver et al., eds., *The Metabolic Basis of Inherited Disease*, 6th ed, McGraw-Hill, p. 1215.]

leading to its premature degradation (Chapter 16); still other mutations reduce the ability of the LDL receptor to bind LDL tightly. A particularly informative group of mutant receptors are expressed on the cell surface and bind LDL normally but cannot mediate the internalization of bound LDL. Analyses of such defective LDL receptors led to the concept of internalization sequences in cell-surface proteins destined for endocytosis by means of clathrin-coated pits. As discussed in Chapter 17, such sorting signals, located in the cytosolic domains of certain membrane proteins, play a key role in directing these proteins to particular vesicles.

The results of these pioneering studies and other research led to the current model for the receptor-mediated endocy-

tosis of LDL and other receptor–ligand combinations detailed in Chapter 17 (see in particular Figure 17-28) and summarized in Figure 18-10b. After internalized LDL particles reach lysosomes, lysosomal proteases hydrolyze their surface apolipoproteins and lysosomal cholesteryl esterases hydrolyze their core cholesteryl esters. The unesterified cholesterol is then free to leave the lysosome and be used as necessary by the cell in the synthesis of membranes or various cholesterol derivatives. The export of cholesterol from lysosomes depends on the NPC1 protein mentioned previously.

If LDLR-mediated endocytosis were not regulated, cells would continuously take up LDL and accumulate massive amounts of LDL-derived cholesterol because of the recycling

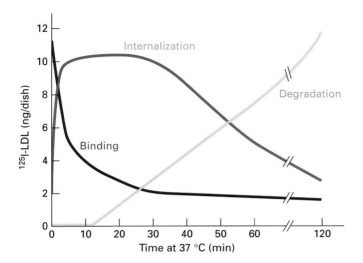

▲ **EXPERIMENTAL FIGURE 18-15 Pulse-chase experiment demonstrates precursor–product relations in cellular uptake of LDL.** Cultured normal human skin fibroblasts were incubated in a medium containing ^{125}I-LDL for 2 hours at 4 °C (the pulse). After excess ^{125}I-LDL not bound to the cells was washed away, the cells were incubated at 37 °C for the indicated amounts of time in the absence of external LDL (the chase). The amounts of surface-bound, internalized, and degraded (hydrolyzed) ^{125}I-LDL were measured as in the experiments presented in Figure 18-14. Binding but not internalization or hydrolysis of LDL apoB-100 occurs during the 4 °C pulse. The data show the very rapid disappearance of bound ^{125}I-LDL from the surface as it is internalized after the cells have been warmed to allow membrane movements. After a lag period of 15–20 minutes, lysosomal degradation of the internalized ^{125}I-LDL commences. [See M. S. Brown and J. L. Goldstein, 1976, *Cell* **9**:663.]

of receptors and the large reservoir of LDL in the bloodstream. However, an elegant regulatory system described in Section 18.4 limits LDL uptake. LDL delivers most of its cholesterol to the liver, which expresses the majority of the body's LDL receptors. The LDL receptor not only interacts with apoB-100 on the surfaces of LDL particles, but also binds to apoE on the surfaces of chylomicron remnants and intermediate-density lipoprotein (IDL) particles. A related receptor called LRP (LDLR-related protein) recognizes apoE but not apoB-100; thus LRP mediates the endocytosis of chylomicron remnants and IDL, but not LDL, by hepatocytes and some other cells (see Figure 18-13b).

Cholesteryl Esters in Lipoproteins Can Be Selectively Taken Up by the Receptor SR-BI

Findings from studies of HDL metabolism led to the discovery of a second, distinct mechanism of receptor-facilitated uptake of lipoprotein lipids. In these studies, experimental animals were injected with purified HDL particles in which the apolipoproteins were labeled with ^{125}I and the core cholesteryl esters were labeled with ^{3}H. The liver and steroid hormone–producing (steroidogenic) tissues in injected animals accumulated substantial amounts of the labeled cholesterol but not the associated labeled apolipoproteins. Conversely, a large amount of the ^{125}I label, but not the ^{3}H label, was ultimately found in the kidneys, where the apolipoproteins are degraded. These findings are inconsistent with receptor-mediated endocytosis of the entire particle. Rather, liver and steroidogenic cells selectively take up cholesteryl esters from the cores of HDL particles without accumulating the components of the outer shells.

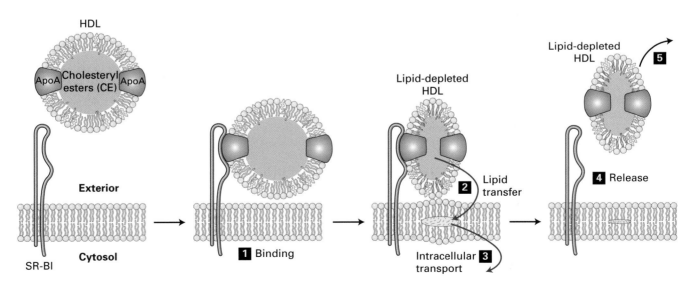

▲ **FIGURE 18-16 Model for the selective uptake of cholesteryl esters from HDL mediated by the receptor SR-BI.** After HDL binding to SR-BI **1**, cholesteryl esters in the core are selectively transferred to a cell's membrane **2** and then into the cytosol **3** by as-yet-unknown mechanisms. The remaining lipid-depleted HDL particle rapidly dissociates from the receptor **4** and eventually returns to the circulation **5**. [See C. Glass et al., 1983, *Proc. Nat'l. Acad. Sci. USA* **80**:5435; Y. Stein et al., 1983, *Biochim. Biophys. Acta* **752**:98; and M. Krieger, 1999, *Ann. Rev. Biochem.* **68**:523.]

Figure 18-16 depicts a model for the selective uptake of cholesteryl esters by a cell-surface receptor called *SR-BI (scavenger receptor, class B, type I)*. SR-BI binds HDL, LDL, and VLDL and can mediate selective uptake from all of these lipoproteins. The detailed mechanism of selective lipid uptake has not yet been elucidated, but it may entail hemifusion of the outer phospholipid monolayer of the lipoprotein and the exoplasmic leaflet of the plasma membrane. The cholesteryl esters initially enter the hydrophobic center of the plasma membrane, are subsequently transferred across the inner leaflet, and are eventually hydrolyzed by cytosolic, not lysosomal, cholesteryl esterases. The lipid-depleted particles remaining after lipid transfer dissociate from SR-BI and return to the circulation; they can then extract more phospholipid and cholesterol from other cells by means of the ABCA1 protein or other cell-surface transport proteins (see Figure 18-13c). Eventually, small lipid-depleted HDL particles circulating in the bloodstream are filtered out by the kidney and bind to a different receptor on renal epithelial cells. After these particles have been internalized by receptor-mediated endocytosis, they are degraded by lysosomes.

The receptor SR-BI differs in two important respects from the LDL receptor. First, SR-BI clusters on microvilli and in cell-surface **lipid rafts** (Chapter 5), not in coated pits as does the LDL receptor. Second, SR-BI mediates the transfer of lipids across the membrane, not endocytosis of entire LDL particles as mediated by the LDL receptor. A multifunctional receptor, SR-BI can mediate the selective uptake from lipoproteins of diverse lipids (e.g., cholesteryl esters, vitamin E); it also functions in the reverse direction to facilitate the export of unesterified cholesterol from cells to bound lipoproteins. SR-BI has a structure similar to that of the fatty acid transporter CD36, and they both belong to the superfamily of *scavenger receptors;* as discussed later, some of these receptors apparently play a role in the onset of atherosclerosis.

KEY CONCEPTS OF SECTION 18.3

Lipid Movement into and out of Cells

- Movement of lipids into and out of cells usually requires either cell-surface transport proteins and water-soluble binding proteins (or micelles), or lipoprotein secretion and lipoprotein receptor–facilitated uptake (see Figure 18-10).

- Fatty acid transporters (FATPs, CD36) in the plasma membrane facilitate the movement of fatty acids between intracellular binding proteins (e.g., FABP) and extracellular carriers (e.g., albumin).

- The ABC superfamily comprises ATP-hydrolyzing small-molecule pumps. Many ABC proteins mediate the export of various lipids from cells (see Table 18-2).

- Several ABC proteins export bile components from hepatocytes: ABCB4 (phospholipids), ABCB11 (bile salts), and ABCG5/8 (cholesterol and plant sterols). Na^+-linked

lipid symporters (e.g., IBAT and NTCP), which mediate the cellular uptake of bile acids, also play a key role in the enterohepatic circulation (see Figure 18-11).

- Lipoproteins are large particles with cores of neutral lipids (cholesteryl esters, triglycerides) and amphipathic shells composed of apolipoproteins, a monolayer of phospholipids, and unesterified cholesterol.

- Each class of lipoprotein has a characteristic protein and lipid composition and functions in the cellular export, extracellular transport through the circulatory system, and receptor-mediated cellular import of lipids (see Table 18-3 and Figure 18-13).

- Familial hypercholesterolemia is caused by mutations in the gene encoding the low-density lipoprotein (LDL) receptor. Persons with this disorder have elevated plasma LDL levels and develop cardiovascular disease at abnormally young ages.

- Receptor-mediated endocytosis of lipoproteins, such as LDL, is one mechanism for delivering cholesterol to cells.

- In a second mechanism for delivering cholesterol to cells, the receptor SR-BI mediates the selective uptake of cholesteryl esters from high-density lipoprotein, LDL, and very low density lipoprotein. The resulting lipid-depleted lipoprotein particle is subsequently released from the cell and can be reused (see Figure 18-16).

18.4 Feedback Regulation of Cellular Lipid Metabolism

As is readily apparent, a cell would soon face a crisis if it did not have enough lipids to make adequate amounts of membranes or had so much cholesterol that large crystals formed and damaged cellular structures. To prevent such disastrous events, cells normally maintain appropriate lipid levels by regulating their supply and utilization of lipids. We have seen how cells acquire lipids by biosynthesis or import and how they export lipids. In this section, we consider the regulation of cellular lipid metabolism, focusing on cholesterol. However, the regulatory pathways that control cellular cholesterol levels also function in controlling fatty acid and phospholipid metabolism. Coordinate regulation of the metabolism of these membrane components is necessary to maintain the proper composition of membranes.

For more than 50 years the cholesterol biosynthetic pathway has been known to be subject to negative feedback regulation by cholesterol. Indeed, it was the first biosynthetic pathway shown to exhibit this type of end-product regulation. As the cellular cholesterol level rises, the need to import cholesterol through the LDL receptor or to synthesize additional cholesterol goes down. As a consequence, transcription of the genes encoding the LDL receptor and cholesterol biosynthetic enzymes decreases. For example, when normal cultured cells are incubated with increasing concentrations of

LDL, the expression and activity of HMG-CoA reductase, the rate-controlling enzyme in cholesterol biosynthesis, is suppressed, whereas the activity of acyl:cholesterol acyl transferase (ACAT), which converts cholesterol into the esterified storage form, is increased. Conversely, when the cellular cholesterol level begins to fall as cells use more cholesterol, expression of the LDL receptor and HMG-CoA reductase increases and the activity of ACAT decreases. Such coordinate regulation is an efficient way for cells to maintain cellular cholesterol homeostasis.

ER-to-Golgi Transport and Proteolytic Activation Control the Activity of SREBP Transcription Factors

Cholesterol-dependent transcriptional regulation often depends on 10-base-pair sterol regulatory elements (SREs), or SRE half-sites, in the promoters of regulated target genes. As you might expect from the discussion of transcriptional control in Chapter 11, the interaction of cholesterol-dependent **SRE-binding proteins (SREBPs)** with these response elements modulates the expression of the target genes. What you

might not expect is that the SREBP-mediated pathway, whereby cholesterol controls the expression of proteins engaged in cholesterol metabolism, begins in the ER and includes at least two other proteins besides SREBP.

When cells have adequate concentrations of cholesterol, SREBP is found in the ER membrane complexed with *SCAP* (SREBP cleavage-activating protein), *insig-1* (or its close homolog insig-2), and perhaps other proteins (Figure 18-17, *left*). SREBP has three distinct domains: an N-terminal cytosolic domain that includes a basic helix-loop-helix (bHLH) DNA-binding motif (see Figure 11-22b) and functions as a transcription factor, a central membrane-anchoring domain containing two transmembrane α helices, and a C-terminal cytosolic regulatory domain. SCAP has eight transmembrane α helices and a large C-terminal cytosolic domain that interacts with the regulatory domain of SREBP. Five of the transmembrane helices in SCAP form a sterol-sensing domain, similar to that in HMG-CoA reductase. The sterol-sensing domain in SCAP binds tightly to insig-1(2), but only at high cellular cholesterol levels. When insig-1(2) is tightly bound to SCAP, it blocks the binding of SCAP to COP II vesicle coat proteins and thus prevents incorporation of the SCAP/SREBP complex into ER-to-Golgi transport vesicles (see Chapter 17).

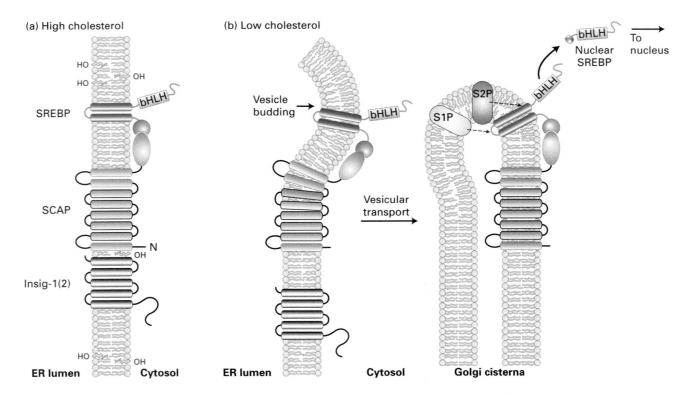

▲ **FIGURE 18-17 Model for cholesterol-sensitive control of SREBP activation mediated by insig-1(2) and SCAP.** The cellular pool of cholesterol is monitored by combined action of insig-1(2) and SCAP, both transmembrane proteins located in the ER membrane. (a) When cholesterol levels are high, insig-1(2) binds to the sterol-sensing domain in SCAP anchoring the SCAP/SREBP complex in the ER membrane. (b) The dissociation of insig-1(2) from SCAP at low cholesterol levels allows the SCAP/SREBP complex to move to the Golgi complex by vesicular transport. (*Right*) The sequential cleavage of SREBP by the site 1 and site 2 proteases (S1P, S2P) associated with the Golgi membrane releases the N-terminal bHLH domain, called nuclear SREBP (nSREBP). After translocating into the nucleus, nSREBP controls the transcription of genes containing sterol regulatory elements (SREs) in their promoters. [Adapted from T. F. Osborne, 2001, *Genes Devel.* **15**:1873; see T. Yang et al., 2002, *Cell* **110**:489.]

When cellular cholesterol levels drop, insig-1(2) no longer binds to SCAP, and the SCAP/SREBP complex can move from the ER to the Golgi apparatus (Figure 18-17, *right*). In the Golgi, SREBP is cleaved sequentially at two sites by two membrane-bound proteases, S1P and S2P. The second intramembrane cleavage at site 2 releases the N-terminal bHLH-containing domain into the cytosol. This fragment, called nSREBP (nuclear SREBP), binds directly to the nuclear import receptor and is rapidly translocated into the nucleus (see Figure 12-21). In the nucleus, nSREBP activates the transcription of genes containing SREs in their promoters, such as those encoding the LDL receptor and HMG-CoA reductase. After cleavage of SREBP in the Golgi, SCAP apparently recycles back to the ER where it can interact with insig-1(2) and another SREBP molecule. High-level transcription of SRE-controlled genes requires the ongoing generation of new nSREBP because it is degraded fairly rapidly by the ubiquitin-mediated proteasomal pathway (Chapter 3). The rapid generation and degradation of nSREBP help cells respond quickly to changes in levels of intracellular cholesterol.

In the *insig-1(2)/SCAP/SREBP pathway* for controlling cellular cholesterol metabolism, the cell exploits intercompartmental movements (ER → Golgi → cytosol → nucleus), regulated by sterol-dependent protein–protein interactions, and post-translational proteolytic cleavage to activate a membrane-bound transcription factor. Cleavage of SREBP in this pathway is one of several known examples of *regulated intramembrane proteolysis (RIP)*. For instance, RIP activates transcription factors in the Notch signaling pathway (Chapter 14) and in the unfolded-protein response (Chapter 16). RIP is also responsible for the generation of the toxic amyloid β peptides that contribute to the onset of Alzheimer's disease (see Figure 14-30).

Multiple SREBPs Regulate Expression of Numerous Lipid-Metabolizing Proteins

Under some circumstances (e.g., during cell growth), cells need an increased supply of all the essential membrane lipids and their fatty acid precursors (coordinate regulation). But

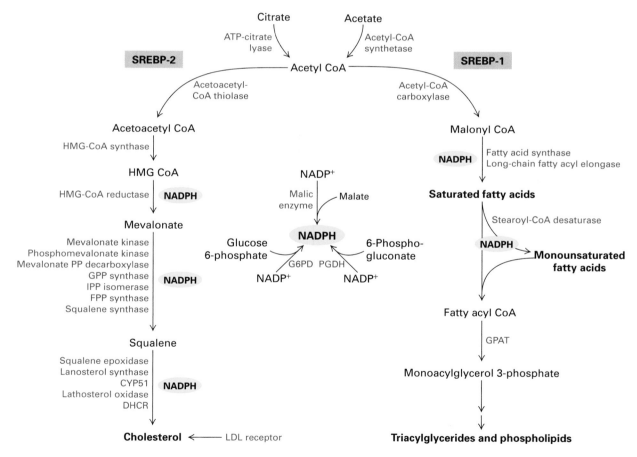

▲ **FIGURE 18-18 Global regulation of cellular lipid metabolism by SREBP.** SREBP controls the transcription of genes indicated here encoding key proteins directly required for the synthesis and import of cholesterol and for the synthesis of fatty acids, phospholipids, and triglycerides. SREBPs also regulate the activity of genes in the production of NADPH, which is an energy source for many of the steps in these biosynthetic pathways. Abbreviations: GPP = geranylgeranyl pyrophosphate; IPP = isopentenyl pyrophosphate; FPP = farnesyl pyrophosphate; CYP51 = lanosterol 14α-demethylase; DHCR = 7-dehydrocholesterol reductase; G6PD = glucose 6-phosphate dehydrogenase; PGDH = 6-phosphogluconate dehydrogenase; GPAT = glycerol 3-phosphate acyl transferase. [Adapted from J. D. Horton, J. L. Goldstein, and M. S. Brown, 2002, *J. Clin. Invest.* **109**:1128.]

cells sometimes need greater amounts of some lipids than others (differential regulation). For example, cells that are producing bile acids or steroid hormones need an increased supply of cholesterol but not of fatty acids or phospholipids. The complex regulation of lipid metabolism characteristic of higher eukaryotes is due largely to a plethora of transcription factors, including multiple SREBPs, that control the expression of proteins taking part in the synthesis, degradation, transport, and storage of lipids.

There are three known isoforms of SREBP in mammals: SREBP-1a and SREBP-1c, which are generated from alternatively spliced RNAs produced from the same gene, and SREBP-2, which is encoded by a different gene. Together these protease-regulated transcription factors control the availability not only of cholesterol but also of fatty acids and their products triglycerides and phospholipids. In mammalian cells, SREBP-1a and SREBP-1c exert a greater influence on fatty acid metabolism than on cholesterol metabolism, whereas the reverse is the case for SREBP-2. As indicated in Figure 18-18, SREBPs can regulate the activities of genes encoding many different proteins. Such proteins include those participating in the cellular uptake of lipids (e.g., the LDL receptor, SR-BI, and lipoprotein lipase) and numerous enzymes in the pathways for synthesizing cholesterol, fatty acids, triglycerides, and phospholipids.

 SREBP-1 may play an important role in the development of *fatty liver,* a major pathologic consequence of alcohol abuse. In fatty liver, abnormally high levels of triglycerides and cholesterol accumulate in the cytosol as lipid droplets, which can contribute to alcoholic hepatitis and cirrhosis. The results of experiments using cultured hepatocytes and mice suggest that the metabolism of alcohol to acetaldehyde by hepatic alcohol dehydrogenase leads to the activation of SREBP-1 and the release of nSREBP-1, which in turn induces the synthesis of excess fatty acids and triglycerides. Consistent with this suggestion is the finding that overexpression of a truncated, constitutively active form of SREBP-1 (i.e., nSREBP-1) in the livers of mice significantly increases both fatty acid and cholesterol synthesis, resulting in a fatty liver. ∎

In contrast with the insig-1(2)/SCAP/SREBP pathway in mammalian cells, the homologous pathway in *Drosophila* does not respond to changes in cellular sterol levels. Instead, the SCAP-dependent proteolytic activation of SREBP is suppressed by high levels of phosphatidylethanolamine, the main phospholipid in fruit flies. This finding, the result of an elegant series of experiments using both enzyme inhibitors and RNA interference (RNAi), indicates that the sterol-sensing domain of *Drosophila* SCAP responds to the cellular level of phosphatidylethanolamine, not cholesterol. Thus so-called sterol-sensing domains might more appropriately be called *lipid-sensing* domains. Whether these domains directly bind to their controlling lipids or mediate interaction with other proteins that directly bind the lipids (i.e., sense the levels of the regulatory lipids) is not yet known.

As mentioned previously, HMG-CoA reductase also contains a sterol-sensing domain. This domain senses high levels of cholesterol, some cholesterol derivatives, and certain nonsteroidal precursors of cholesterol, triggering the rapid, ubiquitin-dependent proteasomal degradation of the enzyme. As a consequence, HMG-CoA reductase activity drops, causing reduced cholesterol synthesis. Like SCAP, HMG-CoA reductase is located in the ER membrane and insig-1(2) binds to its sterol-sensing domain. This binding also is cholesterol dependent and is required for the cholesterol-dependent proteasomal degradation of HMG-CoA reductase. Thus mammalian insig-1(2) and the sterol-sensing domain of SCAP or HMG-CoA reductase apparently combine to form a cholesterol sensor. It seems likely that, in the course of evolution, the sterol-sensing domain and its associated proteins proved effective for recognizing various lipid molecules and were incorporated into a variety of regulatory systems for this purpose.

Members of the Nuclear Receptor Superfamily Contribute to Cellular and Whole-Body Lipid Regulation

In addition to SREBPs, several members of the nuclear receptor superfamily regulate lipid metabolism. **Nuclear receptors** are transcription factors that are generally activated when bound to specific small-molecule ligands (Chapter 11). Certain nuclear receptors influence whole-body lipid metabolism by regulating the absorption of dietary lipids, cellular synthesis of lipids, transport protein–mediated import and export of lipids, levels of lipoproteins and their receptors, and catabolism of lipids (e.g., fatty acid oxidation in the peroxisome) and their secretion from the body.

Some ligands for nuclear receptors are extracellular molecules that diffuse across the plasma membrane (e.g., steroid hormones) or enter cells through transporters (e.g., bile acids, fatty acids). Alternatively, ligands generated within a cell, including oxygen-modified cholesterol (*oxysterols*), bile acids, and certain fatty acids and their derivatives, may bind to nuclear receptors within the same cell. Nuclear receptors sense changes in the levels of all key cellular lipids by binding the lipids themselves or their metabolic products. When activated, these receptors stimulate or suppress gene expression to ensure that the proper physiological levels of lipids are maintained (feedback regulation for cellular homeostasis). The binding of multiple types of lipids to an individual nuclear receptor allows the receptor to coordinately control several metabolic pathways.

For example, hepatocytes express *LXR (liver X receptor),* a nuclear receptor that senses the levels of oxysterols. When cellular cholesterol increases in the liver, oxysterols are generated and activate LXR. Activated LXR stimulates the expression of cholesterol 7α-hydroxylase, the key rate-limiting enzyme in the hepatic conversion of cholesterol into bile acids, a major pathway for disposing of excess cholesterol from the body. LXR also stimulates the expression of the ABC proteins that export cholesterol into the bile (ABCG5/8) or onto lipoproteins in the blood (ABCA1). In addition, LXR promotes lipoprotein production and modification and the expression of SREBP-1c, which then turns on the transcription of genes required for fatty acid synthesis. The resulting increase in fatty acids can contribute to cholesterol esterification and phospholipid synthesis to maintain the proper ratio of cholesterol to phospholipid. Thus sensing of increased cellular cholesterol by LXR results in diverse responses that prevent the accumulation of excess cholesterol.

Another nuclear receptor, called *FXR*, is activated by the binding of bile acids. Expressed in hepatocytes and intestinal epithelial cells, FXR plays a key role in regulating the enterohepatic circulation of bile acids. Bile acid–activated FXR stimulates the expression of intracellular bile acid–binding protein (I-BABP) and of transport proteins (e.g., ABCB11, NTCP) that mediate cellular export and import of bile acids (see Figure 18-11). In contrast, active FXR represses the expression of cholesterol 7α-hydroxylase, thereby decreasing the synthesis of bile acids from cholesterol in the liver—another example of end-product inhibition of a metabolic pathway. Both FXR and LXR function as heterodimers with the nuclear receptor RXR.

In the next section, we will see how an understanding of the SREBP and nuclear receptor regulatory pathways has contributed to effective strategies for reducing the risk of atherosclerosis and cardiovascular disease.

KEY CONCEPTS OF SECTION 18.4

Feedback Regulation of Cellular Lipid Metabolism

■ Two key transcription-control pathways are employed to regulate the expression of enzymes, transporters, receptors, and other proteins taking part in cellular and whole-body lipid metabolism.

■ In the insig-1(2)/SCAP/SREBP pathway, the active nSREBP transcription factor is released from the Golgi membrane by intramembrane proteolysis when cellular cholesterol is low (see Figure 18-17). It then stimulates the expression of genes with sterol regulatory elements (SREs) in their promoters (e.g., the genes for LDLR and HMG-CoA reductase). When cholesterol is high, SREBP is retained in the ER membrane complexed with insig-1(2) and SCAP.

■ SREBP controls the transcription of numerous genes encoding proteins having roles in cellular lipid metabolism (see Figure 18-18). In mammals, SREBP-1a and SREBP-1c have a greater influence on fatty acid metabolism than on cholesterol metabolism; the reverse is the case for SREBP-2.

■ In the nuclear-receptor pathway, transcription factors in the cytosol are activated by intracellular lipids (e.g., high levels of oxysterols or bile acids). The ligand–transcription factor complex then enters the nucleus and regulates the expression of specific target genes that participate in feedback regulation of the synthesis, transport, and catabolism of lipids.

■ Homologous transmembrane sterol-sensing domains are present in several integral membrane proteins participating in lipid metabolism (e.g., SCAP, HMG-CoA reductase, NPC1). These domains appear to help detect and respond to changes in the levels of a variety of lipids, not just sterols. Binding to insig-1(2) is essential for sterol-sensing by at least some of these domains.

18.5 The Cell Biology of Atherosclerosis, Heart Attacks, and Strokes

In this concluding section we examine the relation between lipid metabolism and *atherosclerosis*, the most common cause of heart attacks and strokes. Atherosclerosis accounts for 75 percent of deaths due to cardiovascular disease in the United States. Advances in our understanding of the molecular mechanisms underlying lipid metabolism and its regulation are having an enormous effect on the treatment and prevention of this major health problem.

Frequently called cholesterol-dependent clogging of the arteries, atherosclerosis is characterized by the progressive deposition of lipids, cells, and extracellular matrix material in the inner layer of the wall of an artery. The resulting distortion of the artery's wall can lead, either alone or in combination with a blood clot, to major blockage of blood flow. Thus, to understand the cellular basis of atherosclerosis, we need to first briefly consider the structure of an artery. ■

Specialized epithelial cells called endothelial cells form a thin layer, the **endothelium,** that lines the blood vessel wall immediately adjacent to the lumen through which the blood flows (Figure 18-19a). Beneath the endothelium are several concentric layers of extracellular matrix and cells that make up the artery wall: the *intima,* composed largely of amorphous collagens, proteoglycans, and elastic fibers; the *media,* a well-organized layer of smooth muscle cells whose contraction controls the diameter of the vessel lumen and thus influences blood pressure; and the *adventitia,* a layer of connective tissue and cells that forms the interface between the vessel and the tissue through which it runs.

Under normal circumstances, the plasma (the fluid part of blood) and many types of blood cells flow smoothly and rapidly through the lumen of an artery, a type of movement termed laminar flow. When an infection or traumatic damage occurs within the walls of an artery or in the underlying tissue, a complex series of events permits white blood cells (leukocytes) to initially adhere loosely to the luminal surface of the artery wall and roll, propelled by the laminar flow of the surrounding plasma (see Figure 6-30). Subsequently, adhesion molecules mediate firm attachment of the white cells to the endothelium and their movement across the endothelium into the wall. Within the artery wall, white blood cells called monocytes differentiate into *macrophages,* which fight infection in a number of ways. For instance, macrophages and other leukocytes release proteins and small molecules

(a) Normal artery wall

White blood cells | Rolling white
adhere and migrate | blood cell | Blood flow
into artery wall to | | through lumen
fight infection

Endothelium

Intima

Adventitia | Media
(smooth muscle cells)

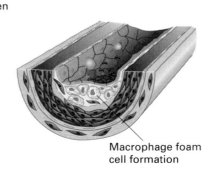

(b) Fatty streak stage

Macrophage foam
cell formation

(c) Atheroslerotic plaque stage

Fibrous cap
formation

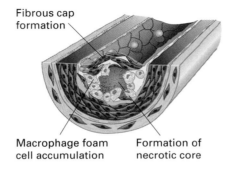

Macrophage foam | Formation of
cell accumulation | necrotic core

(d) Rupture of endothelium and
occlusive blood clot formation

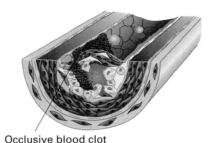

Occlusive blood clot

▲ **FIGURE 18-19 Major stages in the onset and progression of atherosclerosis in the artery wall.** (a) The anatomy of a normal artery wall, which is composed of concentric layers of cells and extracellular matrix, is shown. White blood cells adhere to the endothelium, roll along it, and then migrate into an artery wall to fight infection (see Figure 6-30). (b) When plasma LDL is high or plasma HDL is low, or both, macrophages in the intima can accumulate lipoprotein cholesterol, generating foam cells filled with cholesteryl ester droplets (see Figure 18-20). Accumulation of foam cells produces a fatty streak in the vessel wall that is only visible microscopically. (c) Continued generation of foam cells and migration of smooth muscle cells from the media into the intima is followed by cell death, producing an advanced atherosclerotic plaque. This plaque consists of a necrotic core of lipids (including needlelike cholesterol crystals) and extracellular matrix overlain by a fibrous cap of smooth muscle cells and matrix. (d) As an atherosclerotic plaque grows into the lumen of the artery, it disrupts and reduces the flow of blood. In some cases, the plaque alone can fully occlude the artery. In many cases, the fibrous cap ruptures, inducing formation of a blood clot that can fully occlude the artery. [Adapted from R. Russell, 1999, *N. Engl. J. Med.* **340**(2):115.]

that directly attack bacteria and other pathogens. The cells also secrete proteins that help recruit additional monocytes and other immune cells (e.g., T lymphocytes) to join in the fight. Macrophages also engulf and destroy pathogens, damaged macromolecules, and infected or dead body cells. When the infection has been cured, damaged tissue is repaired and the remaining macrophages and other leukocytes move out of the artery wall and reenter the circulation.

As we will see, atherosclerosis is an "unintended" consequence of this normal physiological *inflammatory response*, which is designed to protect against infection and tissue damage. For this reason and because atherosclerosis most often strikes late in life after the prime reproductive years, there appears to have been little evolutionary selective pressure against the disease. Thus, although atherosclerosis has an enormous negative influence on modern human populations, its high incidence in well-fed, long-lived persons is not surprising.

Arterial Inflammation and Cellular Import of Cholesterol Mark the Early Stages of Atherosclerosis

During an inflammatory response, macrophages in the inflamed artery wall can endocytose substantial amounts of cholesterol from lipoproteins, which accumulate within the artery wall under some circumstances (Figure 18-20a). As macrophages convert the imported cholesterol into the ester

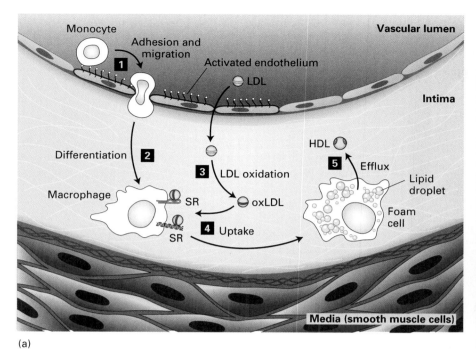

(a)

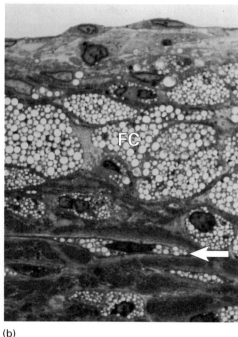

(b)

▲ FIGURE 18-20 Generation of macrophage foam cells in an artery wall. (a) At a site of infection or damage (**1**), monocytes adhere to and migrate across the activated endothelial cell layer into the intima (**2**), where they differentiate into macrophages. When plasma LDL levels are high, the concentration of LDL in the intima is high, and some of the LDL is oxidized to oxLDL or otherwise modified (**3**). Scavenger receptors expressed by macrophages are proposed to bind to and endocytose oxLDL, which is degraded. Its cholesterol accumulates as cholesteryl esters in cytosolic lipid droplets, leading to an accumulation of cholesterol and the formation of foam cells (**4**). Macrophages also express ABCA1 and SR-BI, which can mediate the efflux of excess cellular cholesterol to HDL in the intima (**5**). Thus the amount of cholesterol accumulation is determined by the relative uptake of LDL-derived cholesterol and efflux to HDL. (b) Micrograph of a coronary artery with an atherosclerotic plaque containing many intimal macrophage foam cells (FC) filled with spherical cholesteryl ester lipid droplets (light circles). Some smooth muscle cells also are present and also contain lipid droplets (arrow). [Part (a) adapted from C. K. Glass and J. L. Witztum, 2001, *Cell* **104**:503. Part (b) from H. C. Stary, 2003, *Atlas of Atherosclerosis Progression and Regression*, 2d ed., Parthenon Publishing, p. 61.]

form, they become filled with cholesteryl ester lipid droplets. The resulting lipid-filled macrophages are called *foam cells* because the lipid droplets have a foamy appearance (Figure 18-20b). As macrophage foam cells accumulate in an artery wall, they initially form an early *fatty streak*, the first unique step in atherosclerosis (Figure 18-19b).

The next stage in atherosclerosis is marked by the continued accumulation of macrophage foam cells, proliferation of smooth muscle cells, and migration of these cells from the media into the intima. The smooth muscle cells secrete additional extracellular matrix, and some internalize sufficient amounts of lipoprotein cholesterol to also become foam cells. The initial macroscopically invisible early fatty streak grows bigger as the disease progresses, forming an early **atherosclerotic plaque,** or atheromatous plaque. Cells within the center of the plaque die, producing a necrotic core containing large amounts of cholesteryl esters and unesterified cholesterol (Figure 18-19c). Cholesterol crystals, readily detected microscopically, commonly form within a more advanced plaque, which is eventually covered by a fibrous cap composed of smooth muscle cells and collagen.

Atherosclerotic Plaques Can Impede Blood Flow, Leading to Heart Attacks and Strokes

As an atherosclerotic plaque expands, it projects farther and farther into the lumen of the vessel, narrowing the lumen and distorting the normal shape of the endothelium lining the vessel. Because blood flow through the affected artery is reduced and disturbed, the rate of delivery of nutrient-rich, oxygenated blood to tissues fed by the artery decreases, a condition known as *ischemia*. If sufficiently severe, such partial starvation of the heart can cause pain (angina).

If the endothelial lining covering a plaque ruptures, a large platelet and fibrin blood clot (thrombus) can form very rapidly and block or occlude the artery (Figure 18-19d and Figure 18-21). Tissue downstream of an occlusion soon becomes depleted of oxygen (ischemic hypoxia) and energy sources (e.g., fatty acids in the adult heart, glucose in the brain). The extent of damage, including tissue death, caused by a severe occlusion depends on the length of time that the artery is occluded and the size of the affected area. Severe occlusion of a coronary (heart) artery can cause a

(a)

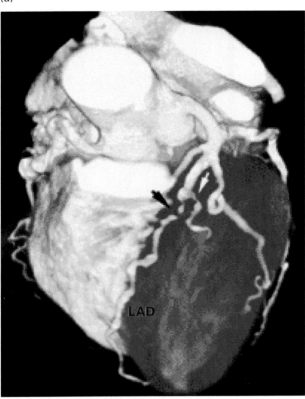

▲ **EXPERIMENTAL FIGURE 18-21 Atherosclerosis narrows and blocks blood flow through coronary arteries.** X-ray multi-slice computed tomographic image of a human heart reveals a major occlusion (black arrow) of the left anterior descending artery (LAD, arrow) and a narrowing of a nearby vessel (white arrow) as a block in the stream of blood (seen in the arteries as a white tube). [From K. Nieman et al., 2001, *Lancet* **357**:599.]

heart attack; occlusion of an artery feeding the brain can cause a stroke.

Atherosclerosis can begin at or even before puberty but usually takes decades to develop into overt disease. In some cases, the growth of new blood vessels permits sufficient blood flow to tissue downstream of a plaque so that major tissue damage does not occur. Balloon stretching, removal of plaques, insertion of metal scaffolds (stents), and grafting of a bypass vessel are among the surgical treatments for advanced blockage of coronary arteries.

LDLR-Independent Uptake of LDL (Bad Cholesterol) Leads to Formation of Foam Cells

As noted, the first unique step of atherosclerosis is the accumulation in the artery wall of macrophage foam cells filled with lipid droplets containing cholesteryl esters. The greater the plasma LDL concentration and the greater the concentration of LDL in the artery wall, the more rapidly foam cells

develop and accumulate to form microscopically visible early fatty streaks. Given these facts, you might initially guess that LDLR-mediated endocytosis is responsible for foam cell formation, but there are two powerful reasons why it cannot be true. First, LDLR activity is under cholesterol-dependent, SREBP-controlled feedback regulation, which maintains cellular cholesterol levels within a narrow range. Intracellular cholesterol levels far below those seen in foam cells prevent transcriptional activation of the LDLR gene by the insig-1(2)/SCAP/SREBP pathway (see Figure 18-17). The consequent low level of LDLR expression prevents massive intracellular cholesterol buildup. Second, in familial hypercholesterolemic patients who lack LDLR activity and consequently have high plasma LDL levels, the onset of atherosclerosis is at much younger ages and its progress is dramatically accelerated compared with normal people. Clearly, the formation of foam cells in these patients does not require LDLR activity.

The obvious conclusion based on this evidence is that an LDLR-independent mechanism is responsible for the cellular uptake of LDL cholesterol that leads to the formation of foam cells. One proposed mechanism is shown in Figure 18-20a. In this model, LDL present in the artery wall at the site of infection or cell damage is subject to oxidation by the products of various oxidative reactions or to other modifications generated by inflammatory cells. For example, the oxidation of unsaturated fatty acyl chains in LDL particles generates reactive aldehyde and other species that can covalently modify the protein and phospholipid components of the outer shells of the lipoprotein particles. Endocytosis of such modified LDL particles by receptors on macrophages not subject to cholesterol-dependent feedback suppression would result in foam cell formation.

Support for this model comes from the finding that macrophages express a diverse array of multiligand receptors belonging to the superfamily of scavenger receptors. Some of these receptors can bind tightly to modified LDLs and, in experiments using transgenic mice, have been implicated in the formation of macrophage foam cells. The normal function of some scavenger receptors (e.g., SR-AI/II) is to recognize and promote the endocytic destruction of damaged macromolecules, pathogens, and injured, aged, or apoptotic cells by macrophages. Because these receptors do not normally take part in cellular cholesterol metabolism, their activities are not subject to regulation by cholesterol. Thus they can mediate the massive accumulation of intracellular cholesterol by macrophages, generally limited only by the amount of extracellular modified LDL.

Reverse Cholesterol Transport by HDL (Good Cholesterol) Protects Against Atherosclerosis

Epidemiological evidence indicates that the concentration of HDL cholesterol in the plasma is inversely correlated with the risk for atherosclerosis and cardiovascular disease. Furthermore, transgenic overexpression of apolipoprotein A-I,

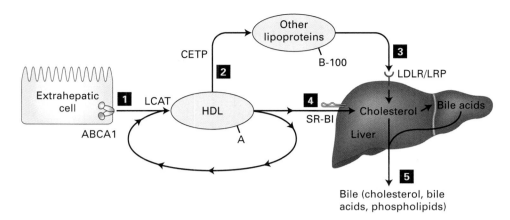

▲ FIGURE 18-22 HDL-mediated reverse cholesterol transport. Cholesterol from peripheral tissues is transferred to circulating HDL by ABCA1 (**1**) and possibly other transporters such as SR-BI and is converted into cholesteryl esters by the plasma enzyme LCAT. Cholesteryl esters in the HDL core can be transferred to other lipoproteins by CETP (**2**) for subsequent endocytosis by lipoprotein receptors expressed primarily by liver cells (**3**). Alternatively, the receptor SR-BI, which is present mostly on liver and steroidogenic tissues, can mediate the selective uptake of cholesteryl esters from HDL (**4**). Cholesterol delivered to the liver and bile acids derived from it are secreted into the bile (**5**). [Adapted from A. Rigotti and M. Krieger, 1999, *N. Engl. J. Med.* **341**:2011; see also M. Krieger, 1999, *Ann. Rev. Biochem.* **68**:523.]

the major apolipoprotein in HDL, suppresses atherosclerosis in animal models of the disease. Several properties of HDL could contribute to its apparent ability to protect against atherosclerosis.

As discussed earlier, HDL can remove cholesterol from cells in extrahepatic tissues, including artery walls, and eventually deliver the cholesterol to the liver either directly by selective lipid uptake mediated by the receptor SR-BI or indirectly by transferring its cholesterol to other lipoproteins that are ligands of hepatic endocytic receptors (see Figure 18-13c). The excess cholesterol can then be secreted into the bile and eventually excreted from the body (see Figure 18-11). Figure 18-22 summarizes this process, called *reverse cholesterol transport,* which lowers both the intracellular cholesterol in macrophages and the total amount of cholesterol carried by the body, thereby directly and indirectly reducing foam cell formation. In a sense, there is a competition between LDL-mediated delivery of cholesterol to cells in the artery wall and HDL-mediated removal of excess cholesterol from those cells. In fact, the ratio of plasma LDL cholesterol to HDL cholesterol is considered a much better indicator of risk for cardiovascular disease than the total plasma cholesterol concentration.

In addition to its role in atheroprotective reverse cholesterol transport, HDL itself and some plasma enzymes associated with HDL can suppress the oxidation of LDL. Decreased LDL oxidation presumably reduces the substrates for scavenger receptors on macrophages, thereby inhibiting their accumulation of LDL cholesterol and thus foam cell formation. HDL also appears to have anti-inflammatory properties, which may contribute to its atheroprotective effect. Finally, the interaction of HDL with the receptor SR-BI can stimulate the activity of endothelial nitric oxide (NO) synthase, leading to increased production of nitric oxide.

This potent atheroprotective signaling molecule can diffuse into nearby vascular smooth muscle and induce its relaxation (see Figure 13-30). Relaxation of the smooth muscle around an artery results in dilation (widening) of the artery lumen and consequently increased blood flow, thereby helping to prevent ischemia and tissue damage.

Two Treatments for Atherosclerosis Are Based on SREBP-Regulated Cellular Cholesterol Metabolism

 The complex pathogenesis of atherosclerosis, which we have only briefly sketched here, presents a daunting challenge to modern molecular medicine. However, as our understanding of the cell biology underlying this complexity has advanced, many opportunities for intervening in the disease process by modulating cellular pathways have arisen. We conclude this chapter by describing two examples of such interventions affecting cellular lipid metabolism.

Because the risk for atherosclerotic disease is directly proportional to the plasma levels of LDL cholesterol and inversely proportional to those of HDL cholesterol, a major public health goal has been to lower LDL and raise HDL cholesterol levels. The most successful drug interventions to date have been aimed at reducing plasma LDL. The steady-state levels of plasma LDL are determined by the relative rates of LDL formation and LDL removal or clearance. LDL receptors, especially those expressed in the liver, play a major role in clearing LDL from the plasma. The liver is key in cholesterol regulation not only because it is the site of about 70 percent of the body's LDL receptors, but also because it is the site where unesterified cholesterol and its bile acid

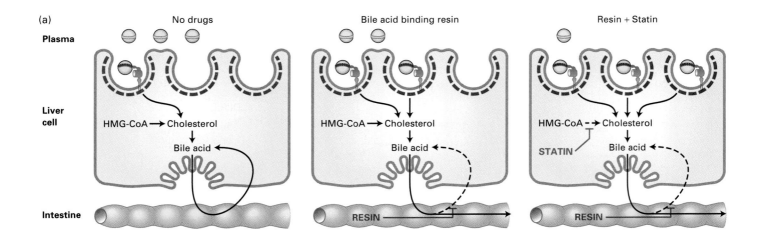

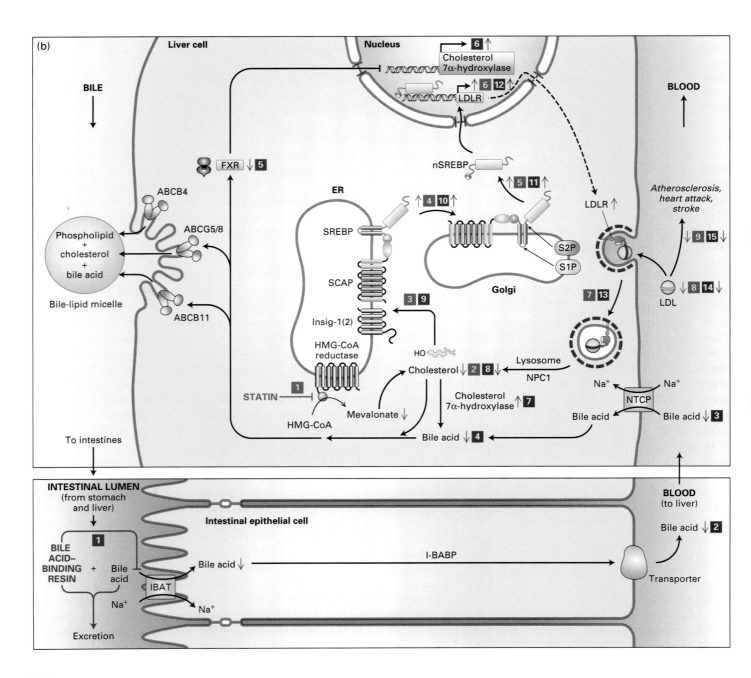

◀ **FIGURE 18-23 Two pharmaceutical approaches for preventing atherosclerosis.** (a) Treatment with statins or bile acid–binding resins lowers hepatic cholesterol levels. This stimulates expression of LDLRs, enhancing LDL removal from the plasma and reducing the risk of atherosclerosis. Combined treatment with both types of drugs often is more effective than using either alone. (b) *Statin therapy:* Step **1**: After oral administration, a statin drug inhibits HMG-CoA reductase in liver (*top*) and other cells, lowering production of mevalonate, a precursor of cholesterol. Step **2**: Cholesterol levels drop, reducing (**3**) the interactions of insig-1(2) with the SCAP/SREBP complex, which can then move (**4**) to the Golgi for processing by the proteases S1P and S2P. As a consequence, there is increased production of the soluble transcription factor nSREBP (**5**), which moves to the nucleus, binds SREs, and increases expression of genes such as the LDL receptor (LDLR) (**6**). nSREBP-mediated increases in LDLR activity result in enhanced LDL endocytosis (**7**), the reduction of plasma LDL

concentrations (**8**), and consequently reduced atherosclerosis, heart attacks, and stroke (**9**).

Bile acid–binding resin therapy: Oral administration of a bile acid–binding resin, or sequestrant (**1**), increases the loss of bile acids from the body by preventing their absorption by intestinal epithelial cells through the IBAT transport protein and reduces bile acids delivered to the blood (**2**) and then to the liver (**3**) by the transporter NTCP. Step **4**: The lower levels of cytoplasmic bile acids reduce the amount of bile acid bound to the nuclear hormone receptor FXR (**5**) and its suppression (**6**) of the expression of cholesterol 7α-hydroxylase. The consequent increased levels of expression and activity of cholesterol 7α-hydroxylase (**7**) reduce the levels of intracellular cholesterol (**8**). As with the statin treatment, the reduced cellular cholesterol levels (**9–15**) increase LDLR activity, lower plasma LDL levels, and protect against atherosclerosis. [Part (a) adapted from M. S. Brown and J. L. Goldstein, 1986, *Science* **232**:34.]

derivatives are secreted into bile, some of which is ultimately excreted from the body (see Figure 18-11).

An increase in LDLR activity, especially in the liver, can lower LDL levels and protect against atherosclerosis. We know the insig-1(2)/SCAP/SREBP pathway normally regulates the expression of the LDL receptor in response to intracellular cholesterol (see Figure 18-17). Thus treatments that lower cellular cholesterol levels in the liver will increase the expression of the LDL receptor and consequently lower plasma LDL levels. Two approaches are widely used to lower steady-state hepatic cholesterol levels: (1) reducing cholesterol synthesis and (2) increasing the conversion of cholesterol into bile acids in the liver (Figure 18-23). Both approaches exploit the regulatory mechanisms for controlling cellular cholesterol already described.

Perhaps the most successful anti-atherosclerosis medications are the *statins*. These drugs bind to HMG-CoA reductase and directly inhibit its activity, thereby lowering cholesterol biosynthesis and the pool of hepatic cholesterol. Activation of SREBP in response to this cholesterol depletion promotes increased synthesis of HMG-CoA reductase and the LDL receptor. Of most importance here is the resulting increased numbers of hepatic LDL receptors, which can mediate increased import of LDL cholesterol from the plasma. Statins also appear to inhibit atherosclerosis by suppressing the inflammation that triggers the process. Although the mechanism of this inhibition is not well understood, it apparently contributes to the atheroprotective effect of statins.

The most common drugs used to lower hepatic cholesterol by increasing the formation of bile acids do so by interrupting the enterohepatic circulation. These *bile acid sequestrants* (e.g., cholestyramine) are insoluble resins that bind tightly to bile acids in the lumen of the intestines, forming complexes that prevent IBAT-mediated absorption by intestinal epithelial cells. The complexes are excreted in the feces. The resulting decrease in the return of bile acids to the liver causes a drop in the hepatic bile acid pool. As

noted previously, when bile acid levels are high, the nuclear receptor FXR represses the expression of cholesterol 7α-hydroxylase activity, the key enzyme in bile acid synthesis. This repression is lifted as the bile acid pool is being depleted by the drug, leading to an increase in the conversion of cholesterol into bile acids. The resulting drop in the cellular cholesterol pool in hepatocytes engages the SREBP pathway, leading to increased hepatic LDLR synthesis and import of LDL cholesterol (see Figure 18-23).

Thus both the statins and the bile acid sequestrants ultimately reduce plasma LDL cholesterol by increasing the SREBP-mediated expression of the LDL receptor in the liver. When a statin and a bile acid sequestrant are administered simultaneously, they act together to substantially lower hepatic cholesterol, stimulate high levels of hepatic LDLR activity, and thus lower plasma LDL concentrations (see Figure 18-23). Unfortunately, statins and bile acid sequestrants are not particularly effective for LDLR-deficient patients who are homozygous for familial hypercholesterolemia, because both therapies depend on stimulating endogenous LDLR activity. Nevertheless, most people at risk for atherosclerosis are not FH homozygotes and the widespread use of these drugs has been shown in many clinical studies to save lives. ∎

KEY CONCEPTS OF SECTION 18.5

The Cell Biology of Atherosclerosis, Heart Attacks, and Strokes

■ Atherosclerosis is the progressive accumulation of cholesterol, inflammatory and other cells, and extracellular matrix in the subendothelial space (intima) of an artery wall, ultimately leading to the formation of a plaque that can occlude the lumen (see Figure 18-19).

■ The partial or complete blockage of coronary arteries by atherosclerotic plaques and associated blood clots can starve heart muscle for oxygen and other nutrients and cause tissue death (heart attack). Similar blockage of arteries supplying the brain causes stroke.

■ Normal inflammatory responses in an artery wall, triggered by infection or injury, may lead to the formation and accumulation of cholesterol-filled macrophage foam cells, the first indication of atherosclerosis.

■ Plasma LDL (bad cholesterol) promotes foam cell formation and thus atherosclerosis by a LDLR-independent mechanism apparently requiring scavenger receptors (see Figure 18-20).

■ Plasma HDL (good cholesterol) reduces the risk for atherosclerosis in part by mediating the reverse transport of cholesterol from peripheral cells to the liver (see Figure 18-22).

■ Two types of drugs are widely used for treating or preventing atherosclerosis: statins, which reduce cholesterol biosynthesis by inhibiting HMG-CoA reductase, and bile acid sequestrants, which prevent the enterohepatic recycling of bile acids from the intestine to the liver. Both treatments lower hepatic cholesterol levels, leading to an SREBP-mediated increase in hepatic expression of LDL receptors, which act to lower plasma LDL levels (see Figure 18-23).

PERSPECTIVES FOR THE FUTURE

Despite considerable progress in our understanding of the cellular metabolism and movement of lipids, the mechanisms for transporting cholesterol and phospholipids between organelle membranes remain poorly characterized. Several recent advances may contribute to progress in this area. First, researchers have partly purified two specialized ER membrane compartments, called MAM (mitochondrion-associated membrane) and PAM (plasma membrane–associated membrane), that may play roles in membrane lipid synthesis and Golgi-independent intracellular lipid transport. Both MAM and PAM have a very high capacity for phospholipid or sterol synthesis or both and directly contact mitochondrial membranes and the plasma membrane, respectively. Second, selection for cells with mutations that interfere with intraorganelle lipid transport opens the door to genetic dissection of this system. The results of such genetic studies in yeast have suggested that ER-to-mitochondrion phospholipid transport is regulated by ubiquitination. Third, the demonstration that some Rab proteins can influence NPC1-dependent cholesterol transport will refocus efforts on vesicular mechanisms for lipid traffic. Finally, cholesterol/sphingolipid-rich lipid rafts and related caveolae are under increased experimental scrutiny that is likely to serve as a source of insight into mechanisms of intracellular lipid transport and control of intracellular signaling pathways.

Another fundamental question concerns the generation, maintenance, and function of the asymmetric distribution of lipids within the leaflets of one membrane and the variation in lipid composition among the membranes of different organelles. What are the mechanisms underlying this complexity and why is such complexity needed? We already know that certain lipids can specifically interact with and influence the activity of some proteins. For example, the large multimeric proteins that participate in oxidative phosphorylation in the inner mitochondrial membrane appear to assemble into supercomplexes whose stability may depend on the physical properties and binding of specialized phospholipids such as cardiolipin.

A detailed molecular understanding of how various transport proteins move lipids from one membrane leaflet to another (flippase activity) and into and out of cells is not yet in hand. Such understanding will undoubtedly require a determination of many high-resolution structures of these molecules, their capture in various stages of the transport process, and careful kinetic and other biophysical analyses of their function, similar to the approaches discussed in Chapter 7 for elucidating the operation of ion channels and ATP-powered pumps.

The fruitful transfer of information between cell biology and medicine can work in both directions. For instance, insights into the insig-1(2)/SCAP/SREBP and nuclear receptor pathways provided the basis for treatment of atherosclerosis with statins and bile acid sequestrants. Continuing efforts are underway to develop even more effective drugs based on these pathways. On the other hand, analysis of familial hypercholesterolemia was a source of major insights into receptor-mediated endocytosis, and a new class of anti-hypercholesterolemia drugs (such as Ezetimibe) that inhibit intestinal cholesterol absorption may serve as a useful tool for identifying the proteins that mediate cholesterol absorption and the mechanisms by which they function.

KEY TERMS

REVIEW THE CONCEPTS

1. Phospholipid biosynthesis at the interface between the endoplasmic reticulum (ER) and the cytosol presents a number of challenges that must be solved by the cell. Explain how each of the following is handled.

a. The substrates for phospholipid biosynthesis are all water soluble, yet the end products are not.

b. The immediate site of incorporation of all newly synthesized phospholipids is the cytosolic leaflet of the ER membrane, yet phospholipids must be incorporated into both leaflets.

c. Many membrane systems in the cell, for example, the plasma membrane, are unable to synthesize their own phospholipids, yet these membranes must also expand if the cell is to grow and divide.

2. What are the common fatty acid chains in glycerophospholipids, and why do these fatty acid chains differ in their number of carbon atoms by multiples of 2?

3. Plants, fungi, and animals all contain sterols in their membranes. Rather than being absorbed from the intestine, most plant sterols are excreted by humans. Why is this element of the human diet not efficiently absorbed?

4. The biosynthesis of cholesterol is a highly regulated process. What is the key regulated enzyme in cholesterol biosynthesis? This enzyme is subject to feedback inhibition. What is feedback inhibition? How does this enzyme sense cholesterol levels in a cell? Cite one other example of a protein whose activity is sensitive to lipid levels.

5. One source of cholesterol for humans is meat. Vegetarians eliminate this cholesterol contribution. Yet vegetarians can still develop atherosclerosis. How can this be?

6. It is evident that one function of cholesterol is structural because it is the most common single lipid molecule in the plasma membrane of mammals. Yet cholesterol may also have other functions. What aspects of cholesterol and its metabolism lead to the conclusion that cholesterol is a multifunctional membrane lipid?

7. Phospholipids and cholesterol must be transported from their site of synthesis to various membrane systems within cells. One way of doing this is through vesicular transport, as is the case for many proteins in the secretory pathway. However, most phospholipid and cholesterol membrane-to-membrane transport in cells is not by Golgi-mediated vesicular transport. What is the evidence for this statement? What appear to be the major mechanisms for phospholipid and cholesterol transport?

8. The intercellular movement of phospholipids and cholesterol is mediated by various proteins. What are the two major mechanisms by which lipids are exported from cells?

What are the three major methods by which lipids are imported into cells?

9. Enterohepatic circulation involves the cycling of bile acids from liver to intestine and back. Explain the role of bile acid cycling in the maintenance of lipid homeostasis.

10. What are the four major classes of lipoproteins in the mammalian circulatory system? Which of these carry mainly cholesterol and which carry mainly triglycerides? Of the cholesterol carriers, a high level of some is considered to be "good" while a high level of some is considered to be "bad." Which are the "good" and "bad" carriers, and why are they considered to be "good" or "bad"?

11. Much health concern is focused on the negative effects of high (or low) levels of lipids in the blood circulatory system. Yet circulating lipids are actually essential to the health and well-being of humans. What are the beneficial effects of circulating lipids, including cholesterol and triglycerides?

12. Atherosclerosis may be thought of as a disease of the organism rather than of the single cell. Do you agree or disagree and why? Explain how hepatic transplantation could, in fact, be one of the strategies for the treatment of familial hypercholesterolemia. How does the fact that atherosclerosis typically manifests itself late in life affect the possibility of whether or not evolution has selected for or against the disease?

13. Increasing concentrations of LDL in the extracellular fluid suppresses HMG-CoA reductase activity and the transcription of genes encoding LDL receptor and HMG-CoA reductase. Conversely, when extracellular LDL levels are low, the transcription of these genes is induced. The responsible transcription factors are SRE-binding proteins (SREBPs). When cells have an adequate level of cholesterol, SREBP is an ER transmembrane protein. To what proteins is SREBP complexed? How does this complex sense the level of sterol? Where within the cell is SREBP released as a soluble protein into the cytosol? What is the mechanism by which SREBP is released? In relation to this process, what is the meaning of RIP?

14. A number of different members of the nuclear-receptor superfamily regulate lipid metabolism. Specific small molecules are typical activators. Such molecules can be either synthesized inside the target cell, diffuse into the cell, or be transported into the cell. One example is FXR, which is activated by binding of bile acids. FXR is expressed in hepatocytes and intestinal epithelial cells. Is the source of bile acid that activates FXR cell synthesized, cell imported, or a combination of the two? Rather than giving specific names of proteins whose synthesis is regulated by FXR, predict classes of proteins whose synthesis should be regulated by a bile acid–activated nuclear receptor.

15. Atherosclerosis may be thought of as an unintended consequence of normal physiological responses designed to protect the body against infection and tissue damage. Explain how inflammation, cholesterol, and foam cells are involved in atherosclerosis.

16. Two different classes of drugs that are in current use to reduce cholesterol levels in the blood have been discussed in detail in this chapter. What are the names of these two drug classes? How does each class act to lower blood cholesterol levels?

ANALYZE THE DATA

You are reviewing the results of a series of recent experiments to determine the regulation of the proteolysis of sterol regulatory element–binding protein (SREBP) by Site-1/2 proteases. In the first experiment, you take an indirect approach to ask if Site-1/2 proteases are components of the Golgi apparatus. You treat CHO cells with the drug brefeldin A (BFA). Brefeldin A induces the redistribution of Golgi proteins to the ER. In the experiment, CHO cells are incubated for 16 hours in the absence of added sterols, then incubated for an additional 5 hours in the absence or presence of added sterols in the absence or presence of BFA (see the first figure). At the end of the 5-hour incubation, the CHO cells are homogenized and separated into a cytoplasm (50 μg protein per gel lane) and nuclear fraction (100 μg protein per gel lane). Proteins are then separated on SDS-polyacrylamide gels and probed for the presence of various molecular forms of SREBP (I = intact, M = intermediate cleavage form, N = nuclear form).

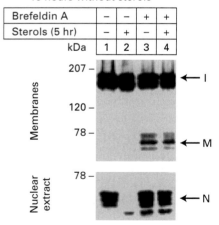

From the observed results, what is the size of M SREBP and N SREBP compared to the size of I SREBP? How can there be an apparent product in this experiment without an apparent loss in intact SREBP? What is the normal effect of added sterols on the production of nuclear SREBP? What is the normal half-life of nuclear SREBP? Why is the instability of nuclear SREBP important to the function of the sterol regulatory pathway? What is the effect of BFA on the level of N SREBP and M SREBP? How does this effect indicate a role of Golgi apparatus in the formation of N SREBP?

REFERENCES

Phospholipids and Sphingolipids: Synthesis and Intracellular Movement

Daleke, D. L. 2003. Regulation of transbilayer plasma membrane phospholipid asymmetry. *J. Lipid Res.* **44**:233–242.

Daleke, D. L., and J. V. Lyles. 2000. Identification and purification of aminophospholipid flippases. *Biochim. Biophys. Acta* **1486**:108–127.

Dowhan, W. 1997. Molecular basis for membrane phospholipid diversity: why are there so many lipids? *Ann. Rev. Biochem.* **66**:199–232.

Henneberry, A. L., M. M. Wright, and C. R. McMaster. 2002. The major sites of cellular phospholipid synthesis and molecular determinants of fatty acid and lipid head group specificity. *Mol. Biol. Cell* **13**:3148–3161.

Kent, C. 1995. Eukaryotic phospholipid biosynthesis. *Ann. Rev. Biochem.* **64**:315–343.

Ostrander, D. B., et al. 2001. Lack of mitochondrial anionic phospholipids causes an inhibition of translation of protein components of the electron transport chain: a yeast genetic model system for the study of anionic phospholipid function in mitochondria. *J. Biol. Chem.* **276**:25262–25272.

Ruetz, S., and P. Gros. 1994. Phosphatidylcholine translocase: a physiological role for the mdr2 gene. *Cell* **77**:1071–1081.

Smit, J. J. M., et al. 1993. Homozygous disruption of the murine mdr2 P-glycoprotein gene leads to a complete absence of phospholipid from bile and to liver disease. *Cell* **75**:451–462.

Storch, J., and A. E. A. Thumser. 2000. The fatty acid transport function of fatty acid-binding proteins. *Biochim. Biophys. Acta* **1486**:28–44.

Thompson, J., A. Reese-Wagoner, and L. Banaszak. 1999. Liver fatty acid binding protein: species variation and the accommodation of different ligands. *Biochim. Biophys. Acta* **1441**(2–3):117–130.

Cholesterol: A Multifunctional Membrane Lipid

Bloch, K. 1965. The biological synthesis of cholesterol. *Science* **150**(692):19–28.

Edwards, P. A., and J. Ericsson. 1999. Sterols and isoprenoids: signaling molecules derived from the cholesterol biosynthetic pathway. *Ann. Rev. Biochem.* **68**:157–185.

Ioannou, Y. A. 2001. Multidrug permeases and subcellular cholesterol transport. *Nature Rev. Mol. Cell Biol.* **2**:657–668.

Maxfield, F. R., and D. Wustner. 2002. Intracellular cholesterol transport. *J. Clin. Invest.* **110**:891–898.

Munn, N. J., et al. 2003. Deficiency in ethanolamine plasmalogen leads to altered cholesterol transport. *J. Lipid Res.* **44**:182–192.

Sacchettini, J. C., and C. D. Poulter. 1997. Creating isoprenoid diversity. *Science* **277**(5333):1788–1789.

Strauss, J. F., P. Liu, L. K. Christenson, and H. Watari. 2002. Sterols and intracellular vesicular trafficking: lessons from the study of NPC1. *Steroids* **67**:947–951.

Vance, D. E., and J. E. Vance. 1996. *Biochemistry of Lipids, Lipoproteins, and Membranes*, 3rd ed. Elsevier.

Vance, D. E., and H. Van den Bosch. 2000. Cholesterol in the year 2000. *Biochim. Biophys. Acta* **1529**(1–3):1–8.

Voelker, D. R. 2003. New perspectives on the regulation of inter-membrane glycerophospholipid traffic. *J. Lipid Res.* **44**:441–449.

Lipid Movement into and out of Cells

Abumrad, N. A., et al. 1993. Cloning of a rat adipocyte membrane protein implicated in binding or transport of long-chain fatty acids that is induced during preadipocyte differentiation: homology with human CD36. *J. Biol. Chem.* **268**:17665–17668.

Acton, S., et al. 1996. Identification of scavenger receptor SR-BI as a high density lipoprotein receptor. *Science* **271**(5248): 518–520.

Borst, P., and R. O. Elferink. 2002. Mammalian ABC transporters in health and disease. *Ann. Rev. Biochem.* **71**:537–592.

Brown, M. S., and J. L. Goldstein. 1986. A receptor-mediated pathway for cholesterol homeostasis. *Science* **232**(4746):34–47.

Glass, C., R. C. Pittman, D. B. Weinstein, and D. Steinberg. 1983. Dissociation of tissue uptake of cholesterol ester from that of apoprotein A-I of rat plasma high density lipoprotein: selective delivery of cholesterol ester to liver, adrenal, and gonad. *Proc. Nat'l. Acad. Sci. USA* **80**:5435–5439.

Goldstein, J. L., and M. S. Brown. 1974. Binding and degradation of low density lipoproteins by cultured human fibroblasts: comparison of cells from a normal subject and from a patient with homozygous familial hypercholesterolemia. *J. Biol. Chem.* **249**: 5153–5162.

Hajri, T., and N. A. Abumrad. 2002. Fatty acid transport across membranes: relevance to nutrition and metabolic pathology. *Ann. Rev. Nutr.* **22**:383–415.

Havel, R. J., and J. P. Kane. 2001. Introduction: Structure and Metabolism of Plasma Lipoproteins. In C. R. Scriver et al., eds., *The Metabolic and Molecular Bases of Inherited Disease.* McGraw-Hill, 8th ed., pp. 2705–2716.

Meier, P. J., and Stieger, B. 2002. Bile salt transporters. *Ann. Rev. Physiol.* **64**:635–661.

Oram, J. F. 2002. ATP-binding cassette transporter A1 and cholesterol trafficking. *Curr. Opin. Lipidol.* **13**:373–381.

Schaffer, J. E., and H. F. Lodish. 1994. Expression cloning and characterization of a novel adipocyte long chain fatty acid transport protein. *Cell* **79**:427–436.

Sharp, D., et al. 1993. Cloning and gene defects in microsomal triglyceride transfer protein associated with a beta lipoproteinaemia. *Nature* **365**(6441):65–69.

Shelness, G. S., and J. A. Sellers. 2001. Very-low-density lipoprotein assembly and secretion. *Curr. Opin. Lipidol.* **12**(2):151–157.

Stahl, A., R. E. Gimeno, L. A. Tartaglia, and H. F. Lodish. 2001. Fatty acid transport proteins: a current view of a growing family. *Trends Endocrinol. Metab.* **12**(6):266–273.

Stein, Y., et al. 1983. Metabolism of HDL-cholesteryl ester in the rat, studied with a nonhydrolyzable analog, cholesteryl linoleyl ether. *Biochim. Biophys. Acta* **752**:98–105.

Wong, H., and M. C. Schotz. 2002. The lipase gene family. *J. Lipid Res.* **43**:993–999.

Feedback Regulation of Cellular Lipid Metabolism

Chawla, A., J. J. Repa, R. M. Evans, and D. J. Mangelsdorf. 2001. Nuclear receptors and lipid physiology: opening the X-files. *Science* **294**(5548):1866–7180.

Dobrosotskaya, Y., et al. 2002. Regulation of SREBP processing and membrane lipid production by phospholipids in *Drosophila*. *Science* **296**(5569):879–883.

Francis, G. A., E. Fayard, F. Picard, and J. Auwerx. 2003. Nuclear receptors and the control of metabolism. *Ann. Rev. Physiol.* **65**:261–311.

Horton, J. D., J. L. Goldstein, and M. S. Brown. SREBPs: activators of the complete program of cholesterol and fatty acid synthesis in the liver. *J. Clin. Invest.* **109**:1125–1131.

Moore, K. J., M. L. Fitzgerald, and M. W. Freeman. 2001. Peroxisome proliferator-activated receptors in macrophage biology: friend or foe? *Curr. Opin. Lipidol.* **12**:519–527.

Repa, J. J., and D. J. Mangelsdorf. 2002. The liver X receptor gene team: potential new players in atherosclerosis. *Nature Med.* **8**:1243–1248.

Sever, N., et al. 2003. Accelerated degradation of HMG CoA reductase mediated by binding of insig-1 to its sterol-sensing domain. *Mol. Cell* **11**:25–33.

Trauner, M., and J. L. Boyer. 2003. Bile salt transporters: molecular characterization, function, and regulation. *Physiol. Rev.* **83**: 633–671.

Yang, T., et al. 2002. Crucial step in cholesterol homeostasis: sterols promote binding of SCAP to INSIG-1, a membrane protein that facilitates retention of SREBPs in ER. *Cell* **110**:489–500.

The Cell Biology of Atherosclerosis, Heart Attacks, and Strokes

Brown, M. S., and J. L. Goldstein. 1983. Lipoprotein metabolism in the macrophage: implications for cholesterol deposition in atherosclerosis. *Ann. Rev. Biochem.* **52**:223–261.

Endo, A., Y. Tsujita, M. Kuroda, and K. Tanzawa. 1977. Inhibition of cholesterol synthesis in vitro and in vivo by ML-236A and ML-236B, competitive inhibitors of 3-hydroxy-3-methylglutaryl-coenzyme A reductase. *Eur. J. Biochem.* **77**:31–36.

Febbraio, M., D. P. Hajjar, and R. L. Silverstein. 2001. CD36: a class B scavenger receptor involved in angiogenesis, athero sclerosis, inflammation, and lipid metabolism. *J. Clin. Invest.* **108**:785–791.

Glomset, J. A. 1968. The plasma lecithin:cholesterol acyltransferase reaction. *J. Lipid Res.* **9**:155–167.

Gotto, A. M., Jr. 2003. Treating hypercholesterolemia: looking forward. *Clin. Cardiol.* **26**(suppl. 1):I21–I28.

Krieger, M., and J. Herz. 1994. Structures and functions of multiligand lipoprotein receptors: macrophage scavenger receptors and LDL receptor-related protein (LRP). *Ann. Rev. Biochem.* **63**:601–637.

Ross, R. 1999. Atherosclerosis: an inflammatory disease. *N. Engl. J. Med.* **340**(2):115–126.

Van Heek, M., et al. 1997. In vivo metabolism-based discovery of a potent cholesterol absorption inhibitor, SCH58235, in the rat and rhesus monkey through the identification of the active metabolites of SCH48461. *J. Pharmacol. Exp. Ther.* **283**:157–163.

Willner, E. L., et al. 2003. Deficiency of acyl CoA:cholesterol acyltransferase 2 prevents atherosclerosis in apolipoprotein E-deficient mice. *Proc. Nat'l. Acad. Sci. USA* **100**(3):1262–1267.

19

MICROFILAMENTS AND INTERMEDIATE FILAMENTS

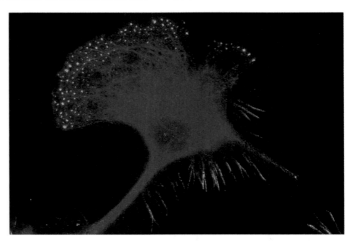

The macrophage cytoskeleton. Prominent structures include a network of intermediate filaments (red) and the punctate distributions of cell adhesions (yellow) containing both actin and vimentin. [Courtesy of J. Evans.]

The ability of cells to migrate is one of the crowning achievements of evolution. Primitive cells probably lacked such self-generated movement (motility), depending on currents in the primordial milieu to move them about. In multicellular organisms, however, the migration of single cells and groups of cells from one part of an embryo to another is critical to the development of the organism. In adult animals, single cells search out foreign invaders as part of a host's defenses against infection; on the other hand, uncontrolled cell migration is an ominous sign of a cancerous cell. Some bacterial cells can move by the beating of flagella powered by a rotary motor in the cell membrane (see Figure 3-22b). Motile eukaryotic cells, however, use different mechanisms to generate movement.

Even stationary cells, which predominate in the body, may exhibit dramatic changes in their morphology—the contraction of muscle cells, the elongation of nerve axons, the formation of cell-surface protrusions, the constriction of a dividing cell in mitosis. Even more subtle than these movements are those that take place within cells—the active separation of chromosomes, the streaming of cytosol, the transport of membrane vesicles. These internal movements are essential elements in the growth and differentiation of cells, carefully controlled by the cell to take place at specified times and in particular locations.

The **cytoskeleton,** a cytoplasmic system of fibers, is critical to cell motility. In Chapter 5, we introduced the three types of cytoskeletal fibers—**microfilaments, intermediate filaments,** and **microtubules**—and considered their roles in supporting cell membranes and organizing the cell contents (see Figure 5-29). All these fibers are polymers built from small protein subunits held together by noncovalent bonds. Instead of being a disordered array, the cytoskeleton is organized into

discrete structures—primarily bundles, geodesic-dome-like networks, and gel-like lattices. In this chapter, we extend our earlier consideration of actin microfilaments and intermediate filaments (Figure 19-1). Both of these cytoskeletal components are usually attached to plasma membrane proteins and form a skeleton that helps support the plasma membrane. However, actin filaments participate in several types of cell movements, whereas intermediate filaments are not directly engaged in cell movements.

All cell movements are a manifestation of mechanical work; they require a fuel (ATP) and proteins that convert the energy stored in ATP into motion. Cells have evolved two basic mechanisms for generating movement. One mechanism entails the assembly and disassembly of microfilaments and microtubules; it is responsible for many changes in cell shape. The other mechanism requires a special class of enzymes called **motor proteins,** first described in Chapter 3. These proteins use energy from ATP to walk or slide along a microfilament or a microtubule and ferry organelles and vesicles with them. A few movements require both the action of motor proteins and cytoskeleton rearrangements. In this chapter, we also cover **myosin,** the motor protein that interacts with actin, building on our earlier description of the

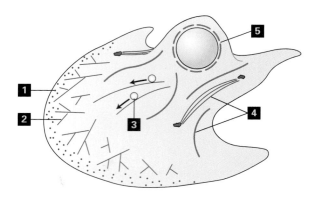

CYTOSKELETAL COMPONENT	CELL FUNCTION
1 Actin dynamics	Membrane extension
2 Filament networks: bundles	Cell structure
3 Myosin motors	Contractility and vesicle transport
4 Actin bundles and intermediate filaments	Cell adhesion
5 Lamin network	Nuclear structure

▲ **FIGURE 19-1 Overview of the actin and intermediate filament cytoskeletons and their functions.** The actin cytoskeletal machinery (red) is responsible for maintaining cell shape and generating force for movements. Polymerization and depolymerization of actin filaments (**1**) drives the membrane forward, whereas actin cross-linking proteins organize bundles and networks of filaments (**2**) that support overall cell shape. Movements within the cell and contractions at the cell membrane (**3**) are produced by myosin motor proteins. The actin (red) and intermediate filament (purple) cytoskeletons integrate a cell and its contents with other cells in tissues (**4**) through attachments to cell adhesions. Another type of intermediate filament, the nuclear lamins (**5**) are responsible for maintaining the structure of the nucleus.

Band of actin and myosin

Actin network

◄ **FIGURE 19-2 Actin cytoskeleton in a moving cell.** Fish keratinocytes are among the fastest crawling cells. Two actin-containing structures work together to generate the force for movement. A network of actin filaments in the front of the cell pushes the membrane forward. Meanwhile, the cell body is pulled by a band of myosin and actin (bracketed). This arrangement of actin and myosin is typical in a moving cell. [From T. M. Svitkina et al., 1997, *J. Cell Biol.* **139**:397.]

structure of myosin II and its role in muscle contraction. Myosin II belongs to a large family of proteins found in both animals and plants. The functions of three myosins (I, II, and V) are well established, but the activities of the others are still largely unknown. A discussion of microtubules, the third type of cytoskeletal fiber, and their motor proteins is deferred until Chapter 20.

19.1 Actin Structures

As we saw in Chapter 5, the *actin cytoskeleton* is organized into various large structures that extend throughout the cell. Because it is so big, the actin cytoskeleton can easily change cell morphology just by assembling or disassembling itself. In preceding chapters, we have seen examples of large protein complexes in which the number and positions of the subunits are fixed. For example, all ribosomes have the same number of protein and RNA components, and their three-dimensional geometry is invariant. However, the actin cytoskeleton is different—the lengths of filaments vary greatly, the filaments are cross-linked into imperfect bundles and networks, and the ratio of cytoskeletal proteins is not rigidly maintained. This organizational flexibility of the actin cytoskeleton permits a cell to assume many shapes and to vary them easily. In moving cells, the cytoskeleton must assemble rapidly and does not always have a chance to form well-organized, highly ordered structures.

In this section, we consider the properties of monomeric and polymeric actin, as well as the various proteins that assemble actin filaments into large structures. With this basic understanding of the actin cytoskeleton established, we examine in Section 19.2 how a cell can tailor this framework to carry out various tasks requiring motion of the entire cell or subcellular parts.

Actin Is Ancient, Abundant, and Highly Conserved

Actin is the most abundant intracellular protein in most eukaryotic cells. In muscle cells, for example, actin comprises 10 percent by weight of the total cell protein; even in nonmuscle cells, actin makes up 1–5 percent of the cellular protein. The cytosolic concentration of actin in nonmuscle cells ranges from 0.1 to 0.5 mM; in special structures such as microvilli, however, the local actin concentration can be 5 mM.

To grasp how much actin cells contain, consider a typical liver cell, which has 2×10^4 insulin receptor molecules but approximately 5×10^8, or half a billion, actin molecules. The high concentration of actin compared with other cell proteins is a common feature of all cytoskeletal proteins. Because they form structures that cover large parts of the cell interior, these proteins are among the most abundant proteins in a cell.

A moderate-sized protein with a molecular weight of 42,000, actin is encoded by a large, highly conserved gene family. Actin arose from a bacterial ancestor and then evolved further as eukaryotic cells became specialized. Some single-celled organisms such as rod-shaped bacteria, yeasts, and amebas have one or two actin genes, whereas many multicellular organisms contain multiple actin genes. For instance, humans have six actin genes, which encode **isoforms** of the protein, and some plants have more than 60 actin genes, although most are **pseudogenes.** In vertebrates, the four α-actin isoforms present in various muscle cells and the β-actin and γ-actin isoforms present in nonmuscle cells differ at only four or five positions. Although these differences among isoforms seem minor, the isoforms have different functions: α-actin is associated with contractile structures; γ-actin accounts for filaments in stress fibers; and β-actin is at the front, or *leading edge*, of moving cells where actin filaments polymerize (Figure 19-2). Sequencing of actins from different sources has revealed that they are among the most conserved proteins in a cell, comparable with histones, the structural proteins of chromatin (Chapter 10). The sequences of actins from amebas and from animals are identical at 80 percent of the positions.

G-Actin Monomers Assemble into Long, Helical F-Actin Polymers

Actin exists as a globular monomer called *G-actin* and as a filamentous polymer called *F-actin*, which is a linear chain of G-actin subunits. (The microfilaments visualized in a cell by electron microscopy are F-actin filaments plus any bound proteins.) Each actin molecule contains a Mg^{2+} ion complexed with either ATP or ADP. Thus there are four states of actin: ATP–G-actin, ADP–G-actin, ATP–F-actin, and ADP–F-actin. Two of these forms, ATP–G-actin and ADP–F-actin, predominate in a cell. The importance of the interconversion between the ATP and the ADP forms of actin in the assembly of the cytoskeleton is discussed later.

Although G-actin appears globular in the electron microscope, x-ray crystallographic analysis reveals that it is separated into two lobes by a deep cleft (Figure 19-3a). The lobes and the cleft compose the *ATPase fold*, the site where ATP

(a)

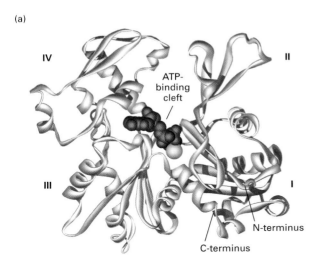

(b)

(c)   (−) end

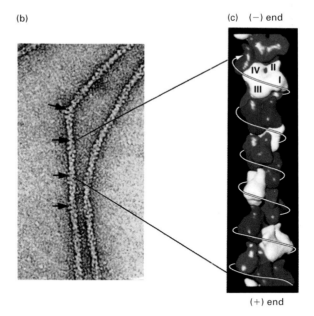

(+) end

▲ **FIGURE 19-3 Structures of monomeric G-actin and F-actin filament.** (a) Model of a β-actin monomer from a nonmuscle cell shows it to be a platelike molecule (measuring 5.5 × 5.5 × 3.5 nm) divided by a central cleft into two approximately equally sized lobes and four subdomains, numbered I–IV. ATP (red) binds at the bottom of the cleft and contacts both lobes (the yellow ball represents Mg^{2+}). The N- and C-termini lie in subdomain I. (b) In the electron microscope, negatively stained actin filaments appear as long, flexible, and twisted strands of beaded subunits. Because of the twist, the filament appears alternately thinner (7 nm diameter) and thicker (9 nm diameter) (arrows). (c) In one model of the arrangement of subunits in an actin filament, the subunits lie in a tight helix along the filament, as indicated by the arrow. One repeating unit consists of 28 subunits (13 turns of the helix), covering a distance of 72 nm. Only 14 subunits are shown in the figure. The ATP-binding cleft is oriented in the same direction (*top*) in all actin subunits in the filament. The end of a filament with an exposed binding cleft is designated the (−)end; the opposite end is the (+) end. [Part (a) adapted from C. E. Schutt et al., 1993, *Nature* **365**:810; courtesy of M. Rozycki. Part (b) courtesy of R. Craig. Part (c) see M. F. Schmid et al., 1994, *J. Cell Biol.* **124**:341; courtesy of M. Schmid.]

and Mg^{2+} are bound. In actin, the floor of the cleft acts as a hinge that allows the lobes to flex relative to each other. When ATP or ADP is bound to G-actin, the nucleotide affects the conformation of the molecule. In fact, without a bound nucleotide, G-actin denatures very quickly.

The addition of ions—Mg^{2+}, K^+, or Na^+—to a solution of G-actin will induce the polymerization of G-actin into F-actin filaments. The process is also reversible: F-actin depolymerizes into G-actin when the ionic strength of the solution is lowered. The F-actin filaments that form in vitro are indistinguishable from microfilaments isolated from cells, indicating that other factors such as accessory proteins are not required for polymerization in vivo. The assembly of G-actin into F-actin is accompanied by the hydrolysis of ATP to ADP and P_i; however, as discussed later, ATP hydrolysis affects the kinetics of polymerization but is not necessary for polymerization to take place.

When negatively stained by uranyl acetate for electron microscopy, F-actin appears as twisted strings of beads whose diameter varies between 7 and 9 nm (Figure 19-3b). From the results of x-ray diffraction studies of actin filaments and the actin monomer structure shown in Figure 19-3a, scientists have produced a model of an actin filament in which the subunits are organized as a tightly wound helix (Figure 19-3c). In this arrangement, each subunit is surrounded by four other subunits, one above, one below, and two to one side. Each subunit corresponds to a bead seen in electron micrographs of actin filaments.

The ability of G-actin to polymerize into F-actin and of F-actin to depolymerize into G-actin is an important property of actin. In this chapter, we will see how the reversible assembly of actin lies at the core of many cell movements.

F-Actin Has Structural and Functional Polarity

All subunits in an actin filament point toward the same end of the filament. Consequently, a filament exhibits **polarity;** that is, one end differs from the other. By convention, the end at which the ATP-binding cleft of the terminal actin subunit is exposed to the surrounding solution is designated the (−) end. At the opposite end, the (+) end, the cleft contacts the neighboring actin subunit and is not exposed (see Figure 19-3c).

Without the atomic resolution afforded by x-ray crystallography, the cleft in an actin subunit and therefore the polarity of a filament are not detectable. However, the polarity of actin filaments can be demonstrated by electron microscopy in "decoration" experiments, which exploit the ability of myosin to bind specifically to actin filaments. In this type of experiment, an excess of myosin S1, the globular head domain of myosin, is mixed with actin filaments and binding is permitted to take place. Myosin attaches to the sides of a filament with a slight tilt. When all the actin subunits are bound by myosin, the filament appears coated ("decorated") with arrowheads that all point toward one end of the filament (Figure 19-4). Because

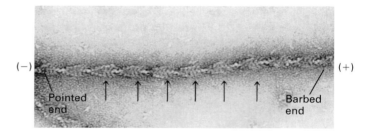

▲ **EXPERIMENTAL FIGURE 19-4 Decoration demonstrates the polarity of an actin filament.** Myosin S1 head domains bind to actin subunits in a particular orientation. When bound to all the subunits in a filament, S1 appears to spiral around the filament. This coating of myosin heads produces a series of arrowhead-like decorations, most easily seen at the wide views of the filament. The polarity in decoration defines a pointed (−) end and a barbed (+) end; the former corresponds to the top of the model in Figure 19-3c. [Courtesy of R. Craig.]

myosin binds to actin filaments and not to microtubules or intermediate filaments, arrowhead decoration is one criterion by which actin filaments are identified among the other cytoskeletal fibers in electron micrographs of thin-sectioned cells.

CH-Domain and Other Proteins Organize Microfilaments into Bundles and Networks

Actin filaments can form a tangled network of filaments in vitro. However, *actin cross-linking proteins* are required to assemble filaments into the stable networks and bundles that provide a supportive framework for the plasma membrane (Figure 19-5). Some actin cross-linking proteins (e.g., fimbrin and fascin) are monomeric proteins that contain two actin-binding domains in a single polypeptide chain. But many cross-linking proteins, particularly those that form networks of filaments, consist of two or more polypeptide chains, each of which contains a single actin-binding domain. Multiple actin-binding sites in these proteins are generated by their assembly into dimers or other oligomers. Actin bundles and networks are often stabilized by several different actin cross-linking proteins.

Many actin cross-linking proteins belong to the *calponin homology–domain superfamily* (Table 19-1). Each of these CH-domain proteins has a pair of actin-binding domains whose sequence is homologous to that of calponin, a muscle protein. The actin-binding domains are separated by repeats of helical coiled-coil or β-sheet immunoglobulin motifs. The organization of the actin-binding sites in these proteins determines whether they organize filaments into bundles or networks. When the binding sites are arranged in tandem, as in fimbrin and α-actinin, the bound actin filaments are packed tightly and align into bundles (see Figure 19-5a) in cell adhesions and extensions. However, if binding sites are spaced apart at the ends of flexible arms, as in filamin, spectrin, and dystrophin, then cross-links can form

TABLE 19-1	Selected Actin-Binding Proteins			
Protein	**MW**	**Domain Organization***		**Location**
CH-DOMAIN SUPERFAMILY				
Fimbrin	68,000			Microvilli, stereocilia, adhesion plaques, yeast actin cables
α-Actinin	102,000			Filopodia, lamellipodia, stress fibers, adhesion plaques
Spectrin	α: 280,000 β: 246,000–275,000			Cortical networks
Dystrophin	427,000			Muscle cortical networks
Filamin	280,000			Filopodia, pseudopodia, stress fibers
OTHERS				
Fascin	55,000			Filopodia, lamellipodia, stress fibers, microvilli, acrosomal process
Villin	92,000			Microvilli in intestinal and kidney brush border

*Blue = actin-binding domains; red = calmodulin-like Ca^{2+}-binding domains; purple = α-helical repeats; green = β-sheet repeats; orange = other domains.

between orthogonally arranged and loosely packed filaments (see Figure 19-5b). The large networks thus formed fill the cytoplasm and give it a gel-like character. Because these proteins also bind membrane proteins, the networks are generally found in the cortical region adjacent to the plasma membrane. In proteins that form networks of filaments, repeats of different protein motifs determine the length of the arms and thus the spacing and orientation between filaments.

Although CH-domain proteins form the majority of actin cross-linking proteins, other proteins that bind to different sites on actin play equally important roles in organizing actin filaments. One such protein, fascin, is found in many actin bundles including stress fibers, cell-surface microvilli, and the sensory bristles that cover the body of the fruit fly *Drosophila*. The important structural role of fascin is illustrated by the effects of mutations in the *singed* gene, which encodes fascin in *Drosophila*. In *singed* mutants, sensory bristles are bent and deformed, evidence that fascin is responsible for maintaining the rigidity of actin bundles in the core of each bristle.

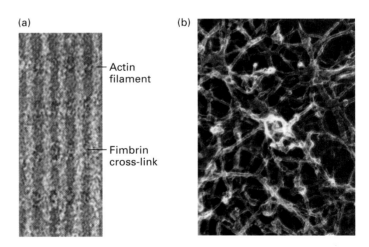

▲ **FIGURE 19-5 Actin cross-linking proteins bridging pairs of actin filaments.** (a) When cross-linked by fimbrin (red), a short protein, actin filaments pack side by side to form a bundle. (b) Long cross-linking proteins such as filamin are flexible and can thus cross-link actin filaments into a network. [Part (a) courtesy of D. Hanein. Part (b) courtesy of J. Hartwig.]

KEY CONCEPTS OF SECTION 19.1

Actin Structures

■ A major component of the cytoskeleton, actin is the most abundant intracellular protein in eukaryotic cells and is highly conserved.

■ The various actin isoforms exhibit minor sequence differences but generally perform different functions.

■ F-actin is a helical filamentous polymer of globular G-actin subunits all oriented in the same direction (see Figure 19-3).

■ Actin filaments are polarized with one end, the (−) end, containing an exposed ATP-binding site.

■ Actin filaments are organized into bundles and networks by a variety of bivalent cross-linking proteins (see Figure 19-5).

■ Many actin cross-linking proteins belong to the CH-domain superfamily (see Table 19-1). They include the bundle-forming proteins fimbrin and α-actinin and the larger network-forming proteins spectrin, dystrophin, and filamin.

19.2 The Dynamics of Actin Assembly

As mentioned previously, the actin cytoskeleton is not a static, unchanging structure consisting of bundles and networks of filaments. Rather, the microfilaments in a cell are constantly shrinking or growing in length, and bundles and meshworks

of microfilaments are continually forming and dissolving. These changes in the organization of actin filaments generate forces that cause equally large changes in the shape of a cell. In this section, we consider the mechanism of actin polymerization and the regulation of this process, which is largely responsible for the dynamic nature of the cytoskeleton.

Actin Polymerization in Vitro Proceeds in Three Steps

The in vitro polymerization of G-actin to form F-actin filaments can be monitored by viscometry, sedimentation, fluorescence spectroscopy, and fluorescence microscopy. When actin filaments become long enough to become entangled, the viscosity of the solution increases, which is measured as a decrease in its flow rate in a viscometer. The basis of the sedimentation assay is the ability of ultracentrifugation (100,000g for 30 minutes) to pellet F-actin but not G-actin. The third assay makes use of G-actin covalently labeled with a fluorescent dye; the fluorescence spectrum of the modified G-actin monomer changes when it is polymerized into F-actin. Finally, growth of the labeled filaments can be imaged with a fluorescence microscope. These assays are useful in kinetic studies of actin polymerization and during purification of actin-binding proteins, which cross-link or depolymerize actin filaments.

The in vitro polymerization of G-actin proceeds in three sequential phases (Figure 19-6a). The first *nucleation phase* is marked by a lag period in which G-actin aggregates into short, unstable oligomers. When the oligomer reaches a cer-

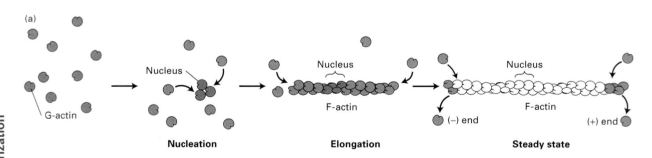

(a)

G-actin — Nucleus — Nucleus — F-actin — Nucleus — F-actin — (−) end — (+) end

Nucleation   **Elongation**   **Steady state**

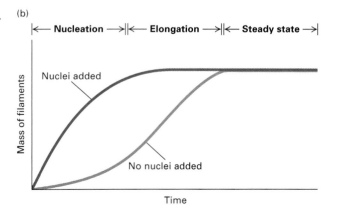

(b)

|← Nucleation →|← Elongation →|← Steady state →|

Mass of filaments

Nuclei added

No nuclei added

Time

▲ **EXPERIMENTAL FIGURE 19-6 Polymerization of G-actin in vitro occurs in three phases.** (a) In the initial nucleation phase, ATP–G-actin monomers (pink) slowly form stable complexes of actin (purple). These nuclei are rapidly elongated in the second phase by the addition of subunits to both ends of the filament. In the third phase, the ends of actin filaments are in a steady state with monomeric ATP–G-actin. After their incorporation into a filament, subunits slowly hydrolyze ATP and become stable ADP–F-actin (white). Note that the ATP-binding clefts of all the subunits are oriented in the same direction in F-actin. (b) Time course of the in vitro polymerization reaction (pink curve) reveals the initial lag period. If some actin filament fragments are added at the start of the reaction to act as nuclei, elongation proceeds immediately without any lag period (purple curve).

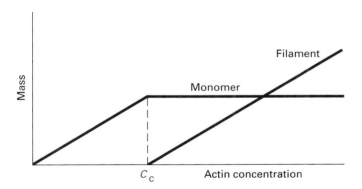

▲ **EXPERIMENTAL FIGURE 19-7 Concentration of G-actin determines filament formation.** The critical concentration (C_c) is the concentration of G-actin monomers in equilibrium with actin filaments. At monomer concentrations below the C_c, no polymerization takes place. At monomer concentrations above the C_c, filaments assemble until the monomer concentration reaches C_c.

tain length (three or four subunits), it can act as a stable seed, or nucleus, which in the second *elongation phase* rapidly increases in length by the addition of actin monomers to both of its ends. As F-actin filaments grow, the concentration of G-actin monomers decreases until equilibrium is reached between filaments and monomers. In this third *steady-state phase*, G-actin monomers exchange with subunits at the filament ends, but there is no net change in the total mass of filaments. The kinetic curves presented in Figure 19-6b show that the lag period can be eliminated by the addition of a small number of F-actin nuclei to the solution of G-actin.

When the steady-state phase has been reached, the concentration of the pool of unassembled subunits is called the

critical concentration, C_c. This parameter is the dissociation constant, the ratio of the "on" and "off" rate constants, and it measures the concentration of G-actin where the addition of subunits is balanced by the dissociation of subunits; that is, the on rate equals the off rate. Under typical in vitro conditions, the C_c of G-actin is 0.1 μM. Above this value, a solution of G-actin will polymerize; below this value, a solution of F-actin will depolymerize (Figure 19-7).

After ATP–G-actin monomers are incorporated into a filament, the bound ATP is slowly hydrolyzed to ADP. As a result of this hydrolysis, most of the filament consists of ADP–F-actin, but some ATP–F-actin is found at one end (see next subsection). However, ATP hydrolysis is not essential for polymerization to take place, as evidenced by the ability of G-actin containing ADP or a nonhydrolyzable ATP analog to polymerize into filaments.

Actin Filaments Grow Faster at (+) End Than at (−) End

We saw earlier that myosin decoration experiments reveal an inherent structural polarity of F-actin (see Figure 19-4). This polarity is also manifested by the different rates at which ATP-G-actin adds to the two ends. One end of the filament, the (+) end, elongates 5–10 times as fast as does the opposite, or (−), end. The unequal growth rates can be demonstrated by a simple experiment in which myosin-decorated actin filaments nucleate the polymerization of G-actin. Electron microscopy of the elongated filaments reveals bare sections at both ends, corresponding to the added undecorated G-actin. The newly polymerized (undecorated) actin is 5–10 times as long at the (+) end as at the (−) end of the filaments (Figure 19-8a).

(a)

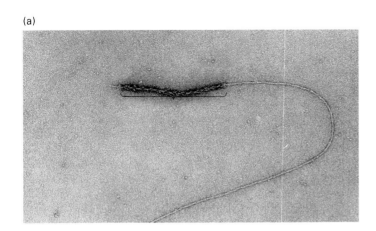

(b)

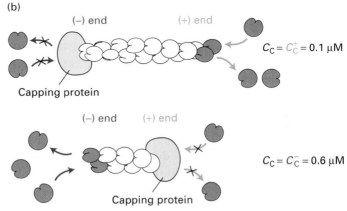

▲ **EXPERIMENTAL FIGURE 19-8 Myosin decoration and capping proteins demonstrate unequal growth rates at the two ends of an actin filament.** (a) When short myosin-decorated filaments are the nuclei for actin polymerization, the resulting elongated filaments have a much longer undecorated (+) end than (−) end. This result indicates that G-actin monomers are added much faster at the (+) end than at the (−) end.

(b) Blocking the (+) or (−) ends of a filament with actin-capping proteins permits growth only at the opposite end. In polymerization assays with capped filaments, the critical concentration (C_c) is determined by the unblocked growing end. Such assays show that the C_c at the (+) end is much lower than the C_c at the (−) end. [Part (a) courtesy of T. Pollard.]

$C_C^- >$ G-actin concentration $> C_C^+$

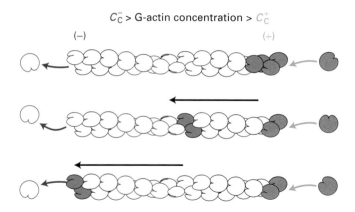

▲ **FIGURE 19-9 Treadmilling of actin filaments.** At G-actin concentrations intermediate between the C_c values for the (−) and (+) ends, actin subunits can flow through the filaments by attaching preferentially to the (+) end and dissociating preferentially from the (−) end of the filament. This treadmilling phenomenon occurs in some moving cells. The oldest subunits in a treadmilling filament lie at the (−) end.

The difference in elongation rates at the opposite ends of an actin filament is caused by a difference in C_c values at the two ends. This difference can be measured by blocking one or the other end with proteins that "cap" the ends of actin filaments. If the (+) end of an actin filament is capped, it can elongate only from its (−) end; conversely, elongation takes place only at the (+) end when the (−) end of a filament is blocked (Figure 19-8b). Polymerization assays of such capped filaments have shown that the C_c is about six times lower for polymerization at the (+) end than for addition at the (−) end.

As a result of the difference in the C_c values for the (+) and (−) ends of a filament, we can make the following predictions: at ATP–G-actin concentrations below C_c^+, there is no filament growth; at G-actin concentrations between C_c^+ and C_c^-, growth is only at the (+) end; and, at G-actin concentrations above C_c^-, there is growth at both ends, although it is faster at the (+) end than at the (−) end. When the steady-state phase is reached at G-actin concentrations intermediate between the C_c values for the (+) and the (−) ends, subunits continue to be added at the (+) end and lost from the (−) end (Figure 19-9). In this situation, the length of the filament remains constant, with the newly added subunits traveling through the filament, as if on a treadmill, until they reach the (−) end, where they dissociate. Turnover of actin filaments at the leading edge of some migrating cells probably occurs by a *treadmilling* type of mechanism, with subunits added to filaments near the leading edge of the cell and lost from the other end toward the rear.

Toxins Perturb the Pool of Actin Monomers

Several toxins shift the usual equilibrium between actin monomers and filaments. Two unrelated toxins, cytochalasin D and latrunculin promote the dissociation of filaments,

though by different mechanisms. Cytochalasin D, a fungal alkaloid, depolymerizes actin filaments by binding to the (+) end of F-actin, where it blocks further addition of subunits. Latrunculin, a toxin secreted by sponges, binds G-actin and inhibits it from adding to a filament end. Exposure to either toxin thus increases the monomer pool. When cytochalasin is added to live cells, the actin cytoskeleton disappears and cell movements such as locomotion and cytokinesis are inhibited. These observations were among the first that implicated actin filaments in cell motility.

In contrast, the monomer–polymer equilibrium is shifted in the direction of filaments by jasplakolinode, another sponge toxin, and by phalloidin, which is isolated from *Amanita phalloides* (the "angel of death" mushroom). Phalloidin poisons a cell by binding at the interface between subunits in F-actin, thereby locking adjacent subunits together and preventing actin filaments from depolymerizing. Even when actin is diluted below its critical concentration, phalloidin-stabilized filaments will not depolymerize. Fluorescence-labeled phalloidin, which binds only to F-actin, is commonly used to stain actin filaments for light microscopy.

Actin Polymerization Is Regulated by Proteins That Bind G-Actin

In the artificial world of a test tube, experimenters can start the polymerization process by adding salts to G-actin or can depolymerize F-actin by simply diluting the filaments. Cells, however, must maintain a nearly constant cytosolic ionic concentration and thus employ a different mechanism for controlling actin polymerization. The cellular regulatory mechanism includes several actin-binding proteins that either promote or inhibit actin polymerization. Here, we consider two such proteins that have been isolated and characterized.

Inhibition of Actin Assembly by Thymosin β_4 Calculations based on the C_c of G-actin (0.1 μM), a typical cytosolic total actin concentration (0.5 mM), and the ionic conditions of the cell indicate that nearly all cellular actin should exist as filaments; there should be very little G-actin. Actual measurements, however, show that as much as 40 percent of actin in an animal cell is unpolymerized. What keeps the cellular concentration of G-actin above its C_c? The most likely explanation is that cytosolic proteins sequester actin, holding it in a form that is unable to polymerize.

Because of its abundance in the cytosol and ability to bind ATP–G-actin (but not F-actin), *thymosin β_4* is considered to be the main actin-sequestering protein in cells. A small protein (5000 MW), thymosin binds ATP–G-actin in a 1:1 complex. The binding of thymosin β_4 blocks the ATP-binding site in G-actin, thereby preventing its polymerization. In platelets, the concentration of thymosin β_4 is 0.55 mM, approximately twice the concentration of unpolymerized actin (0.22 mM). At these concentrations, approximately 70 percent of the monomeric actin in a platelet should be sequestered by thymosin β_4.

Thymosin β_4 (Tβ_4) functions like a buffer for monomeric actin, as represented in the following reaction:

$$F\text{-actin} \rightleftharpoons G\text{-actin} \rightleftharpoons T\beta_4 \, \Delta \, G\text{-actin}/T\beta_4$$

In a simple equilibrium, an increase in the cytosolic concentration of thymosin β_4 would increase the concentration of sequestered actin subunits and correspondingly decrease F-actin, because actin filaments are in equilibrium with actin monomers. This effect of thymosin β_4 on the cellular F-actin level has been experimentally demonstrated in live cells.

Promotion of Actin Assembly by Profilin Another cytosolic protein, *profilin* (15,000 MW), also binds ATP-actin monomers in a stable 1:1 complex. At most, profilin can buffer 20 percent of the unpolymerized actin in cells, a level too low for it to act as an effective sequestering protein. Rather than sequestering actin monomers, the main function of profilin probably is to promote the assembly of actin filaments in cells. It appears to do so by several mechanisms.

First, profilin promotes the assembly of actin filaments by acting as a nucleotide-exchange factor. Profilin is the only actin-binding protein that allows the exchange of ATP for ADP. When G-actin is complexed with other proteins, ATP or ADP is trapped in the ATP-binding cleft of actin. However, because profilin binds to G-actin at a site opposite the ATP-binding cleft, it can recharge ADP-actin monomers released from a filament, thereby replenishing the pool of ATP-actin (Figure 19-10).

Second, as a complex with G-actin, profilin is postulated to assist in the addition of monomers to the (+) end of an actin filament. This hypothesis is consistent with the three-dimensional structure of the profilin–actin complex in which profilin is bound to the part of an actin monomer opposite the ATP-binding end, thereby leaving it free to associate with the (+) end of a filament (see Figure 19-3). After the complex binds transiently to the filament, the profilin dissociates from actin.

Finally, profilin also interacts with membrane components taking part in cell–cell signaling, suggesting that it may be particularly important in controlling actin assembly at the plasma membrane. For example, profilin binds to the membrane phospholipid phosphoinositol 4,5-bisphosphate (PIP$_2$); this interaction prevents the binding of profilin to G-actin. (As discussed in Chapter 13, PIP$_2$ is hydrolyzed in response to certain extracellular signals.) In addition, profilin binds to proline-rich sequences that are commonly found in membrane-associated signaling proteins such as Vasp and Mena. This interaction, which does not inhibit the binding of profilin to G-actin, localizes profilin–actin complexes to the membrane.

Filament-Binding Severing Proteins Create New Actin Ends

A second group of proteins, which bind to actin filaments, control the length of actin filaments by breaking them into shorter fragments and generating new filament ends for polymerization (Table 19-2). A valuable clue that led to the discovery of these severing proteins came from studies of amebas. Viscosity measurements and light-microscope observations demonstrated that during ameboid movement the cytosol flows forward in the center of the cell and then turns into a gel when it reaches the front end of the cell. As discussed later, this "sol to gel" transformation depends on the assembly of new actin filaments in the front part of a moving ameba and the disassembly of old actin filaments in the rear part. Because the actin concentration in a cell favors the formation of filaments, the breakdown of existing actin filaments and filament networks requires the assistance of severing proteins such as *gelsolin* and *cofilin*.

Severing proteins are thought to break an actin filament by stabilizing a change in the conformation of the subunit to which it binds; the resulting strain on the intersubunit bonds leads to its breakage. In support of this hypothesis are electron micrographs showing that an actin filament with bound cofilin is severely twisted. After a severing protein breaks a filament at one site, it remains bound at the (+) end of one of the resulting fragments, where it prevents the addition or exchange of actin subunits, an activity called *capping*. The (−) ends of fragments remain uncapped and are rapidly shortened. Thus severing promotes turnover of actin

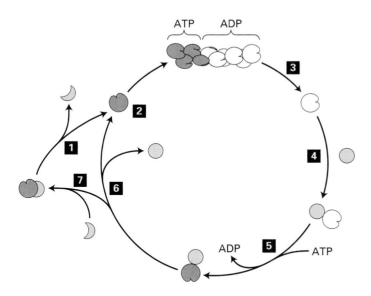

▲ **FIGURE 19-10 Model of the complementary roles of profilin and thymosin β_4 in regulating polymerization of G-actin.** Actin subunits (pink) complexed with thymosin β4 (purple) dissociate (**1**) and add to the end of a filament (**2**). In the filament, ATP is hydrolyzed to ADP, the ADP-associated subunit eventually dissociates from the opposite end of the filament (**3**), the ADP-G-actin forms a complex with profilin (green) (**4**), and ATP exchanges with ADP to form ATP-G-actin (**5**). Profilin delivers actin monomers to the (+) end of actin filaments (**6**) or thymosin β4 sequesters the ATP-G-actin into a polymerization ready pool of subunits (**7**).

TABLE 19-2	Some Cytosolic Proteins That Control Actin Polymerization	
Protein	MW	Activity
Cofilin	15,000	Dissociation from (−) end
Severin	40,000	Severing, capping [(+) end]
Gelsolin	87,000	Severing, capping [(+) end]
CapZ capping protein	36,000 (α) 32,000 (β)	Capping [(+) end]
Tropomodulin	40,000	Capping [(−) end]
Arp2/3 complex	200,000	Capping [(−) end], side binding and nucleation

filaments by creating new (−) ends and causes disintegration of an actin network, although many filaments remain cross-linked. The turnover of actin filaments promoted by severing proteins is necessary not only for cell locomotion but also for cytokinesis.

The capping and severing proteins are regulated by several signaling pathways. For example, both cofilin and gelsolin bind PIP$_2$ in a way that inhibits their binding to actin filaments and thus their severing activity. Hydrolysis of PIP$_2$ by phospholipase C releases these proteins and induces rapid severing of filaments. The reversible phosphorylation and dephosphorylation of cofilin also regulates its activity, and the severing activity of gelsolin is activated by an increase in cytosolic Ca^{2+} to about 10^{-6} M. The counteracting influence of different signaling molecules, Ca^{2+}, and PIP$_2$ permits the reciprocal regulation of these proteins. In Section 19.4, we consider how extracellular signals coordinate the activities of different actin-binding proteins, including severing proteins, in cell migration.

Actin-Capping Proteins Stabilize F-Actin

Another group of proteins can cap the ends of actin filaments but, unlike severing proteins, cannot break filaments to create new ends. One such protein, *CapZ*, binds the (+) ends of actin filaments independently of Ca^{2+} and prevents the addition or loss of actin subunits from the (+) end. Capping by this protein is inhibited by PIP$_2$, suggesting that its activity is regulated by the same signaling pathways that control cofilin and profilin. *Tropomodulin,* which is unrelated to CapZ in sequence, caps the (−) ends of actin filaments. Its capping activity is enhanced in the presence of tropomyosin, which suggests that the two proteins function as a complex to stabilize a filament. An actin filament that is capped at both ends is effectively stabilized, undergoing neither addition nor loss of subunits. Such capped actin filaments are needed in places where the organization of the cytoskeleton is unchanging, as in a muscle sarcomere (Figure 19-11) or at the erythrocyte membrane.

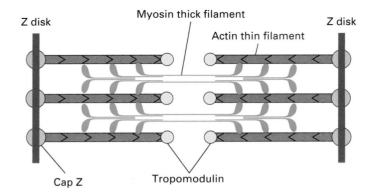

▲ **FIGURE 19-11 Diagram of sarcomere in skeletal muscle showing location of actin-capping proteins.** CapZ (green) caps the (+) ends of actin thin filaments, which are located at the Z disk separating adjacent sarcomeres. Tropomodulin (yellow) caps the (−) ends of thin filaments, located toward the center of a sarcomere. The presence of these two proteins at opposite ends of a thin filament prevents actin subunits from dissociating during muscle contraction.

Arp2/3 Assembles Branched Filaments

A family of *actin-related proteins* (Arps), exhibiting 50 percent sequence similarity with actin, has been identified in many eukaryotic organisms. One group of Arps, a complex of seven proteins called Arp2/3, stimulates actin assembly in vivo. (Another Arp group that is associated with microtubules and a microtubule motor protein is discussed in the next chapter.) Isolated from cell extracts on the basis of its ability to bind profilin, the Arp2/3 complex binds at 70° to the side of an actin filament to nucleate a daughter filament. The combination of mother and daughter filaments creates a branched network in which Arp2/3 is located at the branch points (Figure 19-12). As a result, the newly created ends of filaments elongate and create the force to push the membrane forward.

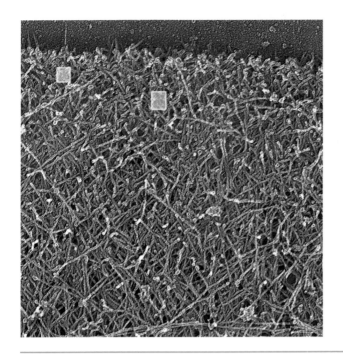

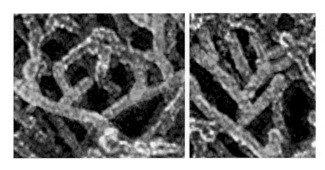

M E D I A   C O N N E C T I O N S

Video: Direct Observation of Actin Filament
Branching Mediated by Arp2/3 Complex

◀ **FIGURE 19-12 Branched actin filaments with Arp2/3 at the branch points.** An extensive network of actin filaments fills the cytoplasm at the leading edge of a keratinocyte. Within selected areas of the network, highly branched filaments (green) are seen. At each branch point lies the Arp2/3 complex. [From T. M. Svitkina and G. G. Borisy, 1999, *J. Cell Biol.* **145**:1009; courtesy of T. M. Svitkina and G. G. Borisy.]

Branching is stimulated by the WASp family of proteins under the control of the Rho GTPases. Actin cross-linking proteins such as filamin stabilize the branched network, whereas actin-severing proteins such as cofilin disassemble the branched structures.

Intracellular Movements and Changes in Cell Shape Are Driven by Actin Polymerization

By manipulating actin polymerization and depolymerization, the cell can create forces that produce several types of movement. As noted previously and described in detail later, actin polymerization at the leading edge of a moving cell is critical to cell migration. Here, we consider other examples of cell movement that most likely result from actin polymerization—one concerning infection and the other blood clotting.

Most infections are spread by bacteria or viruses that are liberated when an infected cell lyses. However, some bacteria and viruses escape from a cell on the end of a polymerizing actin filament. Examples include *Listeria monocytogenes*, a bacterium that can be transmitted from a pregnant woman to the fetus, and vaccinia, a virus related to the smallpox virus. When such organisms infect mammalian cells, they move through the cytosol at rates approaching 11 μm/min. Fluorescence microscopy revealed that a meshwork of short actin filaments follows a moving bacterium or virus like the plume of a rocket exhaust (Figure 19-13). These observations suggested that actin generates the force necessary for movement.

The first hints about how actin mediates bacterial movement were provided by a microinjection experiment in which fluorescence-labeled G-actin was injected into *Listeria*-infected cells. In the microscope, the labeled monomers could be seen incorporating into the tail-like meshwork at the end

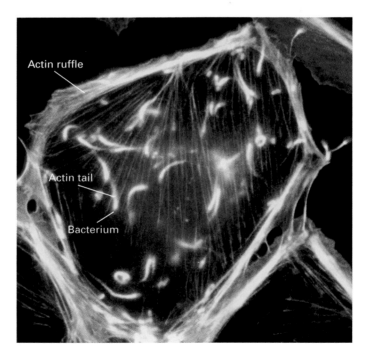

Actin ruffle

Actin tail

Bacterium

▲ **EXPERIMENTAL FIGURE 19-13 Fluorescence microscopy implicates actin in movement of *Listeria* in infected fibroblasts.** Bacteria (red) are stained with an antibody specific for a bacterial membrane protein that binds cellular profilin and is essential for infectivity and motility. Behind each bacterium is a "tail" of actin (green) stained with fluorescent phalloidin. Numerous bacterial cells move independently within the cytosol of an infected mammalian cell. Infection is transmitted to other cells when a spike of cell membrane, generated by a bacterium, protrudes into a neighboring cell and is engulfed by a phagocytotic event. [Courtesy of J. Theriot and T. Mitchison.]

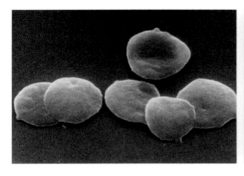
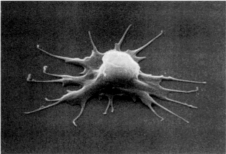
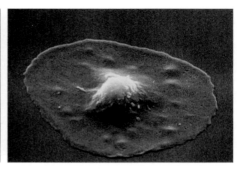

▲ **EXPERIMENTAL FIGURE 19-14 Platelets change shape during blood clotting.** Resting cells have a discoid shape (*left*). When exposed to clotting agents, the cells settle on the substratum, extend numerous filopodia (*center*), and then spread out (*right*). The changes in morphology result from complex rearrangements of the actin cytoskeleton, which is cross-linked to the plasma membrane. [Courtesy of J. White.]

nearest the bacterium, with a simultaneous loss of actin throughout the tail. This result showed that actin polymerizes into filaments at the base of the bacterium and suggested that, as the tail-like meshwork assembles, it pushes the bacterium ahead. Findings from studies with mutant bacteria indicate that the interaction of cellular Arp 2/3 with a bacterial membrane protein promotes actin polymerization at the end of the tail nearest the bacterium. Recent studies have detected rocket tails trailing common cytoplasmic vesicles such as endosomes. Such observations suggest that actin polymerization may underlie the movement of endosomes in the cytoplasm.

During blood clotting, complicated rearrangements of the cytoskeleton in activated platelets dramatically change the cell shape and promote clot formation (Figure 19-14). The cytoskeleton of an unactivated platelet consists of a rim of microtubules (the *marginal band*), a membrane skeleton, and a cytosolic actin network. The membrane skeleton in

(a)

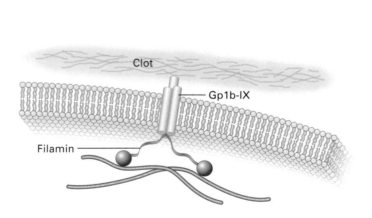

(b)

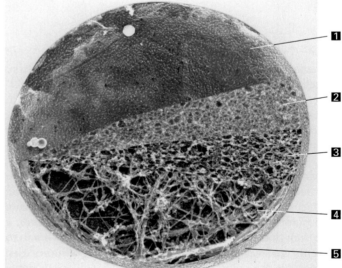

▲ **FIGURE 19-15 Cross-linkage of actin filament networks to the platelet plasma membrane.** In platelets, a three-dimensional network of actin filaments is attached to the integral membrane glycoprotein complex Gp1b-IX by filamin. Gp1b-IX also binds to proteins in a blood clot outside the platelet. Platelets also possess a two-dimensional cortical network of actin and spectrin similar to that underlying the erythrocyte membrane (see Figure 5-31). (b) This composite picture of the actin cytoskeleton in a resting platelet shows the different arrangements of microfilaments. Beneath the plasma membrane (**1**) lies a two-dimensional network of filaments (**2**) cross-linked by spectrin. Filamin organizes the filaments into a three-dimensional gel (**3**), forming the cortex of the cell. A lattice of filament bundles (**4**) forms adhesions to the underlying substratum. The disk shape of the cell is maintained by a ring of microtubules (**5**) at the cell periphery. [Part (b) courtesy of John Hartwig.]

platelets is somewhat similar to the cortical cytoskeleton in erythrocytes (see Figure 5-31). A critical difference between erythrocytes and platelet cytoskeletons is the presence in the platelet of the second network of actin filaments, which are organized by filamin cross-links into a three-dimensional gel (Figure 19-15). The gel fills the cytosol of a platelet and is anchored by filamin to a glycoprotein complex (Gp1b-IX) in the platelet membrane. Gp1b-IX not only binds filamin but also is the membrane receptor for two blood-clotting proteins. Through Gp1b-IX and an integrin receptor, forces generated during rearrangements of the actin cytoskeleton in platelets can be transmitted to a developing clot. Several examples of similar connections between the cytoskeleton and components of the extracellular matrix are described in Chapter 6.

KEY CONCEPTS OF SECTION 19.2

The Dynamics of Actin Assembly

■ Within cells, the actin cytoskeleton is dynamic, with filaments able to grow and shrink rapidly.

■ Polymerization of G-actin in vitro is marked by a lag period during which nucleation occurs. Eventually, a polymerization reaction reaches a steady state in which the rates of addition and loss of subunits are equal (see Figure 19-10).

■ The concentration of actin monomers in equilibrium with actin filaments is the critical concentration (C_c). At a G-actin concentration above C_c, there is net growth of filaments; at concentrations below C_c, there is net depolymerization of filaments.

■ Actin filaments grow considerably faster at their (+) end than at their (−) end, and the C_c for monomer addition to the (+) end is lower than that for addition at the (−) end.

■ The assembly, length, and stability of actin filaments are controlled by specialized actin-binding proteins that can sever filaments or cap the ends or both. These proteins are in turn regulated by various mechanisms.

■ The complementary actions of thymosin β_4 and profilin are critical to regulating the actin cytoskeleton near the cell membrane (see Figure 19-10).

■ The regulated polymerization of actin can generate forces that move certain bacteria and viruses or cause changes in cell shape.

19.3 Myosin-Powered Cell Movements

We now examine the function of different myosin motor proteins in nonmuscle cells and muscle. As discussed in Chapter 3, interactions between myosin II and actin filaments are responsible for muscle contraction. At first, scientists thought that most cell movements were caused by a contrac-

tile mechanism similar to the sliding of actin and myosin filaments in muscle cells. This idea was based on several properties of at least some nonmuscle cells: the ability of cytosolic extracts to undergo contractile-like movements, the presence of actin and myosin II, and the existence of structures similar to muscle sarcomeres. However, the results of later biochemical studies led to the extraction of "unusual" forms of myosin that differed from myosin II in structure, location, and enzymatic properties.

As biologists investigated various types of cell movements, it became clear that myosin II mediates only a few types, such as cytokinesis and muscle contraction. Other types of cell movements, including vesicle transport, membrane extension, and the movement of chromosomes, require either other myosin isoforms, other motor proteins such as kinesin or dynein, or actin polymerization. In this section, we first consider the properties of various myosins and some of their functions in nonmuscle cells. *Contraction*, the special form of movement resulting from the interaction of actin and myosin II, is most highly evolved in skeletal muscle cells. However, somewhat similar contractile events entailing less organized systems are found in nonmuscle cells. After reviewing the highly ordered structure of actin and myosin filaments in the sarcomere of skeletal muscle, we describe the primary mechanisms for regulating contraction.

Myosins Are a Large Superfamily of Mechanochemical Motor Proteins

Eight members of the myosin gene family have been identified by genomic analysis (Chapter 9). Three family members—*myosin I, myosin II,* and *myosin V*—are present in nearly all eukaryotic cells and are the best understood. Although the specific activities of these myosins differ, they all function as motor proteins. As already noted, myosin II powers muscle contraction, as well as cytokinesis. Myosins I and V take part in cytoskeleton–membrane interactions, such as the transport of membrane vesicles.

Researchers are currently uncovering the activities of the remaining myosins. Genetic analysis has revealed that myosins VI, VII, and XV have functions associated with hearing and hair cell stereocilia structure. Plants do not have the same myosins as animal cells. Three myosins (VII, XI, and XIII) are exclusively expressed in plants. Myosin XI, which may be the fastest myosin of all, is implicated in the cytoplasmic streaming seen in green algae and higher plants (Table 19-3).

All myosins consist of one or two heavy chains and several light chains, which generally have a regulatory function. A characteristic head, neck, and tail domain organization is found in all myosin heavy chains. Myosin II and myosin V are dimers in which α-helical sequences in the tail of each heavy chain associate to form a rodlike coiled-coil structure. In contrast some myosins, including myosin I, are monomers because their heavy chains lack this α-helical sequence. All myosin *head domains* have ATPase activity and

TABLE 19-3 **Myosins**

Type	Heavy Chain (MW)	Structure	Step Size (nm)	Activity
I	110,000–150,000		10–14	Membrane binding, endocytic vesicles
II	220,000		5–10	Filament sliding
V	170,000–220,000		36	Vesicle transport
VI	140,000		30	Endocytosis
XI	170,000–260,000		35	Cytoplasmic streaming

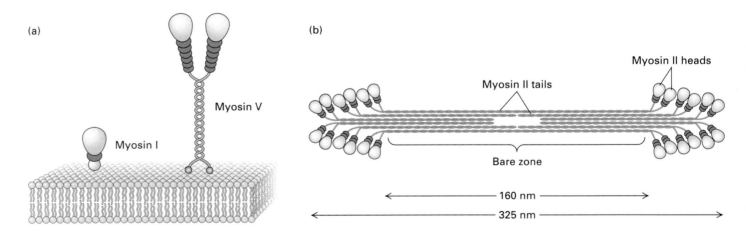

(a)

(b)

▲ **FIGURE 19-16 Functions of myosin tail domains.**
(a) Myosin I and myosin V are localized to cellular membranes by undetermined sites in their tail domains. As a result, these myosins are associated with intracellular membrane vesicles or the cytoplasmic face of the plasma membrane. (b) In contrast, the coiled-coil tail domains of myosin II molecules pack side by side, forming a thick filament from which the heads project. In a skeletal muscle, the thick filament is bipolar. Heads are at the ends of the thick filament and are separated by a bare zone, which consists of the side-by-side tails.

in conjunction with the *neck domain* couple ATP hydrolysis to movement of a myosin molecule along an actin filament via a common mechanism involving cyclical binding and hydrolysis of ATP and attachment/detachment of myosin and actin (see Figures 3-24 and 3-25).

The role of a particular myosin in vivo is related to its *tail domain*. For example, the tail domains of myosins I, V, VI, and XI bind the plasma membrane or the membranes of intracellular organelles; as a result, these molecules have membrane-related activities (Figure 19-16a). In contrast, the coiled-coil tail domains of myosin II dimers associate to form bipolar thick filaments in which the heads are located at both ends of the filament and are separated by a central bare zone devoid of heads (Figure 19-16b). The close packing of myosin molecules into thick filaments, which are a critical part of the contractile apparatus in skeletal muscle, allows many myosin head domains to interact simultaneously with actin filaments.

The number and type of light chains bound in the neck region vary among the different myosins (see Table 19-3). The light chains of myosin I and myosin V are **calmodulin**, a Ca^{2+}-binding regulatory subunit in many intracellular enzymes (see Figure 3-28). Myosin II contains two different light chains called essential and regulatory light chains (see Figure 3-24); both are Ca^{2+}-binding proteins but differ from calmodulin in their Ca^{2+}-binding properties. All myosins are regulated in some way by Ca^{2+}; however, because of the differences in their light chains, the different myosins exhibit different responses to Ca^{2+} signals in the cell.

Myosin Heads Walk Along Actin Filaments in Discrete Steps

Unraveling the mechanism of myosin-powered movement was greatly aided by development of in vitro motility assays. In one such assay, the *sliding-filament assay*, the movement of fluorescence-labeled actin filaments along a bed of myosin molecules is observed in a fluorescence microscope. Because the myosin molecules are tethered to a coverslip, they cannot move; thus any force generated by interaction of myosin heads with actin filaments forces the filaments to move relative to the myosin (Figure 19-17a). If ATP is present, added actin filaments can be seen to glide along the surface of the coverslip; if ATP is absent, no filament movement is observed. This movement is caused by a myosin head (bound to the coverslip) "walking" toward the (+) end of a filament; thus filaments move with the (−) end in the lead. [The one exception is myosin VI, which moves in the opposite direction, toward the (−) end; so the (+) end of a moving filament is in the lead.] The rate at which myosin moves an actin filament can be determined from video camera recordings of sliding-filament assays (Figure 19-17b). The velocity of filament movement can vary widely, depending on the myosin tested and the assay conditions (e.g., ionic strength, ATP and Ca^{2+} concentrations, temperature).

MEDIA CONNECTIONS
Technique Annimation: In Vitro Motility Assay

(a)   Myosin   Actin

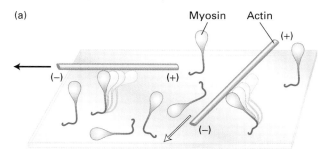

(b)

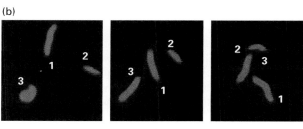

▲ **EXPERIMENTAL FIGURE 19-17 Sliding-filament assay is used to detect myosin-powered movement.**
(a) After myosin molecules are adsorbed onto the surface of a glass coverslip, excess myosin is removed; the coverslip then is placed myosin-side down on a glass slide to form a chamber through which solutions can flow. A solution of actin filaments, made visible by staining with rhodamine-labeled phalloidin, is allowed to flow into the chamber. (The coverslip in the diagram is shown inverted from its orientation on the flow chamber to make it easier to see the positions of the molecules.) In the presence of ATP, the myosin heads walk toward the (+) end of filaments by the mechanism illustrated in Figure 3-25. Because myosin tails are immobilized, walking of the heads causes sliding of the filaments. Movement of individual filaments can be observed in a fluorescence light microscope. (b) These photographs show the positions of three actin filaments (numbered 1, 2, 3) at 30-second intervals recorded by video microscopy. The rate of filament movement can be determined from such recordings. [Part (b) courtesy of M. Footer and S. Kron.]

The most critical feature of myosin is its ability to generate a force that powers movements. Researchers have used a device called an *optical trap* to measure the forces generated by single myosin molecules (Figure 19-18). The results of optical-trap studies show that myosin II moves in discrete steps, approximately 5–10 nm long, and generates 3–5 piconewtons (pN) of force, approximately the same force as that exerted by gravity on a single bacterium. This force is sufficient to cause myosin thick filaments to slide past actin thin filaments during muscle contraction or to transport a membrane-bounded vesicle through the cytoplasm. With a step size of 5 nm, myosin would bind to every actin subunit on one strand of the filament. Some evidence suggests that ATP hydrolysis and myosin walking are closely coupled, with myosin taking a discrete step for every ATP molecule hydrolyzed.

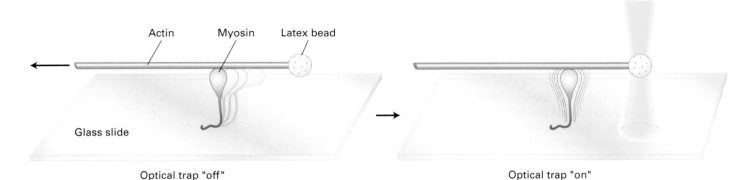

Glass slide

Optical trap "off"

Optical trap "on"

▲ **EXPERIMENTAL FIGURE 19-18 Optical trap determines force generated by a single myosin molecule.** In an optical trap, the beam of an infrared laser focused by a light microscope on a latex bead (or any other object that does not absorb infrared light) captures and holds the bead in the center of the beam. The strength of the force holding the bead is adjusted by increasing or decreasing the intensity of the laser beam. In this experiment, a bead is attached to the end of an actin filament. With the optical trap turned off, the actin filament and its attached bead move in response to the force generated by myosin adsorbed on the coverslip, as in Figure 19-17. When the optical trap is turned on, it captures the filament, through the bead, and holds the filament to the surface of a myosin-coated coverslip. The force exerted by a single myosin molecule on an actin filament is measured from the force needed to hold the bead in the optical trap. A computer-controlled electronic feedback system keeps the bead centered in the trap, and myosin-generated movement of the bead is counteracted by the opposing force of the trap. The distance traveled by the actin filament is measured from the displacement of the bead in the trap.

The neck domains in different myosins vary in length and number of associated light chains. Generally, the longer the neck domain of a myosin, the greater its *step size* (i.e., the distance traveled along an actin filament in one step). Because the neck region is the lever arm of myosin, a longer neck would lead to a longer distance traveled by the arm. For instance, myosin II, with a short neck, has from 5- to 10-nm steps, whereas myosin V, with a long neck, has much longer 36-nm steps. The correlation between step size and neck length has been further supported by experiments in which the neck domain is lengthened by recombinant methods. However, the correlation between neck length and step size is not absolute, as evidenced by myosin VI, which moves in 30-nm steps, although it has a neck domain shorter than that of myosin II.

Myosin-Bound Vesicles Are Carried Along Actin Filaments

Among the many movements exhibited by cells, vesicle translocation has been one of the most fascinating to cell biologists. In early studies of the cytoplasm, researchers found that certain particles, now known to be membrane-bounded vesicles, moved in straight lines within the cytosol, sometimes stopping and then resuming movement, at times after changing direction. This type of behavior could not be caused by diffusion, because the movement was clearly not random. Therefore, researchers reasoned, there must be tracks, most likely actin filaments or microtubules, along which the particles travel, as well as some type of motor to power the movement.

In the sliding-filament assay, walking of the myosin head along an actin filament causes the filament to move because the myosin tail is immobilized. In cells, however, the situation is often reversed: when part of an extensive network, actin filaments are largely immobile, whereas myosin is free to move. In this case, if the tail of a myosin molecule binds to the membrane of a vesicle and the head walks along a filament, the vesicle will be carried along as "cargo." Here, we present evidence that some myosins, including myosins I, V, and VI, do just that. We also consider the related process of cytoplasmic streaming, which is most likely powered by myosin XI.

Vesicle Trafficking (Myosins I, V, and VI) Findings from studies with amebas provided the initial clues that myosin I participates in vesicle transport. Indeed the first myosin I molecule to be identified and characterized was from these organisms; subsequently, the cDNA sequences of three myosin I genes were identified in *Acanthameba*, a common soil ameba. Using antibodies specific for each myosin I isoform, researchers found that the isoforms are localized to different membrane structures in the cell. For example, myosin IA is associated with small cytoplasmic vesicles. Myosin IC, in contrast, is found at the plasma membrane and at the contractile vacuole, a vesicle that regulates the osmolarity of the cytosol by fusing with the plasma membrane. The introduction of antibodies against myosin IC into a living ameba prevents transport of the vacuole to the membrane; as a result, the vacuole expands uncontrollably, eventually bursting the cell. In addition, myosin I in animal cells serves as a membrane–microfilament linkage in microvilli, another example of a membrane-associated function.

Several types of evidence suggest that myosin V also participates in the intracellular transport of membrane-bounded vesicles. For example, mutations in the myosin V gene in yeast disrupt protein secretion and lead to an accumulation of vesicles in the cytoplasm. Vertebrate brain tissue is rich in myosin V, which is concentrated on Golgi stacks. This association with membranes is consistent with the effects of myosin V mutations in mice. Such mutations are associated with defects in synaptic transmission and eventually cause death from seizures. Myosin VI also is implicated in membrane trafficking of vesicles.

Unlike myosin in a thick filament where multiple heads interact with the same actin filament, cytoplasmic myosins work alone in carrying their membrane cargos. How do these myosins move without dissociating from the filament? The answer lies in the *duty ratio*, the fraction of time spent attached to the filament during the ATPase cycle. Myosins with a high duty ratio, such as myosins V and VI, are bound to actin filaments for most of the ATP cycle. Consequently, these myosins process or move along a filament for considerable distances with little danger of falling off.

Cytoplasmic Streaming (Myosin XI) In large, cylindrical green algae such as *Nitella* and *Chara*, cytosol flows rapidly, at a rate approaching 4.5 mm/min, in an endless loop around the inner circumference of the cell (Figure 19-19). This cytoplasmic streaming is a principal mechanism for distributing cellular metabolites, especially in large cells such as plant cells and amebas. This type of movement probably represents an exaggerated version of the smaller-scale movements exhibited during the transport of membrane vesicles.

Close inspection of objects caught in the flowing cytosol, such as the endoplasmic reticulum (ER) and other membrane-bounded vesicles, show that the velocity of streaming increases from the cell center (zero velocity) to the cell periphery. This gradient in the rate of flow is most easily explained if the motor generating the flow lies at the membrane. In electron micrographs, bundles of actin filaments can be seen aligned along the length of the cell, lying across chloroplasts embedded at the membrane. Attached to the actin bundles are vesicles of the ER network. The bulk cytosol is propelled by myosin attached to parts of the ER lying along the stationary actin filaments. The flow

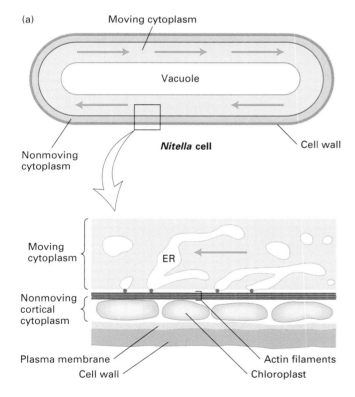

(a)

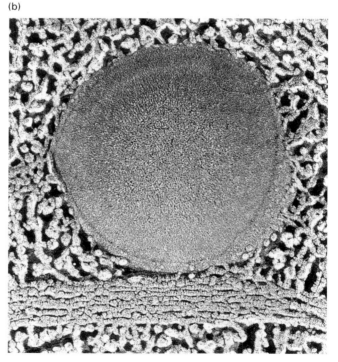

(b)

▲ **FIGURE 19-19 Cytoplasmic streaming in cylindrical giant algae.** (a) The center of a *Nitella* cell is filled with a single large water-filled vacuole, which is surrounded by a layer of moving cytoplasm (indicated by blue arrows). A nonmoving layer of cortical cytoplasm filled with chloroplasts lies just under the plasma membrane (enlarged lower figure). On the inner side of this layer are bundles of stationary actin filaments (red), all oriented with the same polarity. A motor protein (blue dots), most likely myosin XI, carries parts of the endoplasmic reticulum (ER) along the actin filaments. The movement of the ER network propels the entire viscous cytoplasm, including organelles that are enmeshed in the ER network. (b) An electron micrograph of the cortical cytoplasm shows a large vesicle connected to an underlying bundle of actin filaments. This vesicle, which is part of the ER network, contacts the stationary actin filaments and moves along them by a myosin motor protein. [Part (b) from B. Kachar.]

rate of the cytosol in *Nitella* is at least 15 times as fast as the movement produced by any other myosin. This evidence and other evidence suggest that cytoplasmic streaming is powered by myosin XI, one of the fastest moving myosins.

Actin and Myosin II Form Contractile Bundles in Nonmuscle Cells

Nonmuscle cells contain prominent **contractile bundles** composed of actin and myosin II filaments. The contractile bundles of nonmuscle cells, which may be transitory or permanent differ in several ways from the noncontractile bundles of actin described earlier in this chapter (see Figure 19-5). Interspersed among the actin filaments of a contractile bundle is myosin II, which is responsible for their contractility. When isolated from cells, these bundles contract on the addition of ATP. Contractile bundles are always located adjacent to the plasma membrane as a sheet or belt, whereas noncontractile actin bundles form the core of membrane projections (e.g., microvilli and filopodia).

In epithelial cells, contractile bundles are most commonly found as a *circumferential belt,* which encircles the inner surface of the cell at the level of the adherens junction (see Figure 6-5). *Stress fibers,* which are seen along the ventral surfaces of cells cultured on artificial (glass or plastic) surfaces or in extracellular matrices, are a second type of contractile bundle. The ends of stress fibers terminate at **integrin**-containing focal adhesions, special structures that attach a cell to the underlying substratum (see Figure 6-26). Circumferential belts and stress fibers contain several proteins found in smooth muscle, and both exhibit some orga-

nizational features resembling muscle sarcomeres. Thus, both these structures appear to function in cell adhesion and cell movement.

A third type of contractile bundle, referred to as a *contractile ring,* is a transient structure that assembles at the equator of a dividing cell, encircling the cell midway between the poles of the spindle. As division of the cytoplasm (**cytokinesis**) proceeds, the diameter of the contractile ring decreases; so the cell is pinched into two parts by a deepening cleavage furrow. Dividing cells stained with antibodies against myosin I and myosin II show that myosin II is localized to the contractile ring, whereas myosin I is at the cell poles (Figure 19-20). This localization indicates that myosin II but not myosin I takes part in cytokinesis.

The results of experiments in which active myosin II is eliminated from the cell demonstrate that cytokinesis is indeed dependent on myosin II (Figure 19-21). In one type of experiment, anti-myosin II antibodies are microinjected into one cell of a sea urchin embryo at the two-cell stage. In other experiments, expression of myosin II is inhibited by deletion of the myosin gene or by antisense inhibition of myosin mRNA expression. In all cases, a cell lacking myosin II replicates to form a multinucleated syncytium because cytokinesis, but not chromosome separation, is inhibited. Without myosin II, cells fail to assemble a contractile ring, although other events in the cell cycle proceed normally.

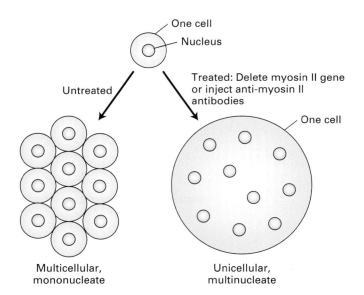

▲ **EXPERIMENTAL FIGURE 19-21 Inhibition of myosin II demonstrates that it is required for cytokinesis.** The activity of myosin II can be inhibited either by deleting its gene or by microinjecting anti-myosin II antibodies into a cell. A cell that lacks myosin II is able to replicate its DNA and nucleus but fails to divide; as a result, the cell forms a large, multinucleate syncytium over a period of time. In comparison, an untreated cell during the same period continues to divide, forming a multicellular ball of cells in which each cell contains a single nucleus.

▲ **EXPERIMENTAL FIGURE 19-20 Fluorescent antibodies reveal the localization of myosin I and myosin II during cytokinesis.** Fluorescence micrograph of a *Dictyostelium* ameba during cytokinesis reveals that myosin II (red) is concentrated in the cleavage furrow, whereas myosin I (green) is localized at the poles of the cell. The cell was stained with antibodies specific for myosin I and myosin II, with each antibody preparation linked to a different fluorescent dye. [Courtesy of Y. Fukui.]

Organized Thick and Thin Filaments in Skeletal Muscle Slide Past One Another During Contraction

Muscle cells have evolved to carry out one highly specialized function—contraction. Muscle contractions must occur quickly and repetitively, and they must occur through long distances and with enough force to move large loads. A typical skeletal muscle cell, called a myofiber, is cylindrical, large (1–40 mm in length and 10–50 μm in width), and multinucleated (containing as many as 100 nuclei). The cytoplasm is packed with a regular repeating array of filament bundles organized into a specialized structure called a **sarcomere**. A chain of sarcomeres, each about 2 μm long in resting muscle, constitutes a *myofibril*. The sarcomere is both the structural and the functional unit of skeletal muscle. During contraction, the sarcomeres are shortened to about 70 percent of their uncontracted, resting length. Electron microscopy and biochemical analysis have shown that each sarcomere contains two types of filaments: *thick filaments*, composed of myosin II, and *thin filaments*, containing actin (Figure 19-22).

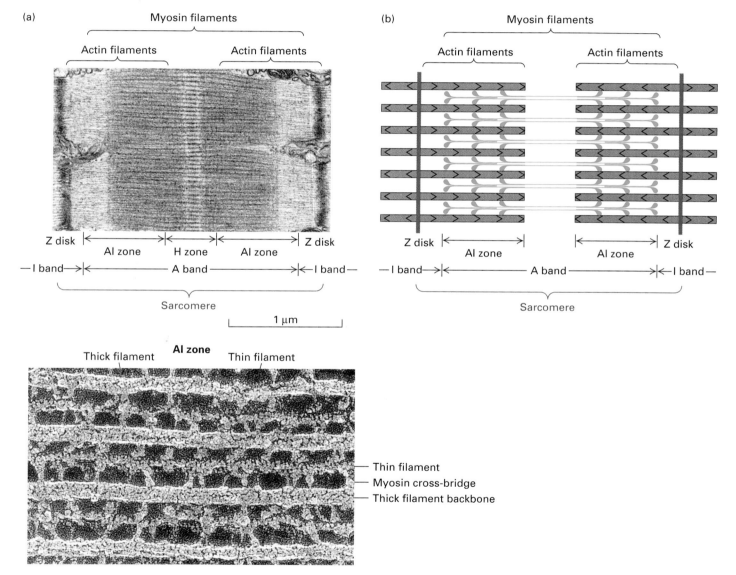

▲ **FIGURE 19-22 Structure of the sarcomere.** (a) Electron micrograph of mouse striated muscle in longitudinal section, showing one sarcomere. On either side of the Z disks are the lightly stained I bands, composed entirely of actin filaments. These thin filaments extend from both sides of the Z disk to interdigitate with the dark-stained myosin thick filaments in the A band. The region containing both thick and thin filaments (the AI zone) is darker than the area containing only myosin thick filaments (the H zone). (b) Diagram of a sarcomere. The (+) ends of actin filaments are attached to the Z disks. (c) Electron micrograph showing actin-myosin cross-bridges in the AI zone of a striated flight muscle of an insect. This image shows a nearly crystalline array of thick myosin and thin actin filaments. The muscle was in the rigor state at preparation. Note that the myosin heads protruding from the thick filaments connect with the actin filaments at regular intervals. [Part (a) courtesy of S. P. Dadoune. Part (c) courtesy of M. Reedy.]

Relaxed

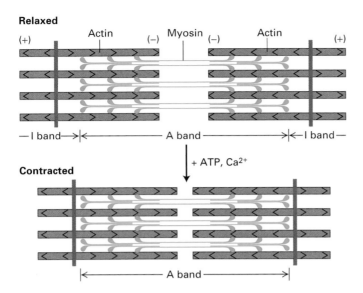

—I band—>|<————————— A band ————————>|<—I band—

+ ATP, Ca²⁺

Contracted

|<———————— A band ————————>|

▲ **FIGURE 19-23 The sliding-filament model of contraction in striated muscle.** The arrangement of thick myosin and thin actin filaments in the relaxed state is shown in the upper diagram. In the presence of ATP and Ca^{2+}, the myosin heads extending from the thick filaments walk toward the (+) ends of the thin filaments. Because the thin filaments are anchored at the Z disks (purple), movement of myosin pulls the actin filaments toward the center of the sarcomere, shortening its length in the contracted state as shown in the lower diagram.

To understand how a muscle contracts, consider the interactions between one myosin head (among the hundreds in a thick filament) and a thin (actin) filament as diagrammed in Figure 3-25. During these cyclical interactions, also called the *cross-bridge cycle*, the hydrolysis of ATP is coupled to the movement of a myosin head toward the Z disk, which corresponds to the (+) end of the thin filament. Because the thick filament is bipolar, the action of the myosin heads at opposite ends of the thick filament draws the thin filaments toward the center of the thick filament and therefore toward the center of the sarcomere (Figure 19-23). This movement shortens the sarcomere until the ends of the thick filaments abut the Z disk or the (−) ends of the thin filaments overlap at the center of the A band. Contraction of an intact muscle results from the activity of hundreds of myosin heads on a single thick filament, amplified by the hundreds of thick and thin filaments in a sarcomere and thousands of sarcomeres in a muscle fiber.

Contraction of Skeletal Muscle Is Regulated by Ca²⁺ and Actin-Binding Proteins

Like many cellular processes, skeletal muscle contraction is initiated by an increase in the cytosolic Ca^{2+} concentration. As described in Chapter 7, the Ca^{2+} concentration of the cytosol is normally kept low, below 0.1 μM. In nonmuscle cells, Ca^{2+} ATPases in the plasma membrane maintain this low concentration. In contrast, in skeletal muscle cells, a

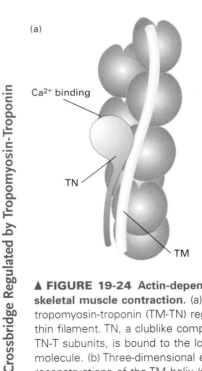

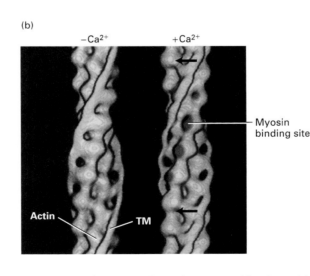

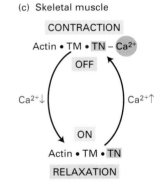

▲ **FIGURE 19-24 Actin-dependent regulation of skeletal muscle contraction.** (a) Model of the tropomyosin-troponin (TM-TN) regulatory complex on a thin filament. TN, a clublike complex of TN-C, TN-I, and TN-T subunits, is bound to the long α-helical TM molecule. (b) Three-dimensional electron-microscopic reconstructions of the TM helix (yellow) on a thin filament from scallop muscle. TM in its "off" state (*left*)

shifts to its new position (arrow) in the "on" state (*right*) when the Ca^{2+} concentration increases. This movement exposes myosin-binding sites (red) on actin. TN is not shown in this representation. (c) Regulation of skeletal muscle contraction by Ca^{2+} binding to TN. Note that the TM-TN complex remains bound to the thin filament whether muscle is relaxed or contracted. [Part (b) adapted from W. Lehman, R. Craig, and P. Vibert, 1993, *Nature* **123**:313; courtesy of P. Vibert.]

low cytosolic Ca^{2+} level is maintained primarily by a unique Ca^{2+} ATPase that continually pumps Ca^{2+} ions from the cytosol into the **sarcoplasmic reticulum (SR)**, a specialized endoplasmic reticulum in the muscle-cell cytosol (see Figure 7-7). This activity establishes a reservoir of Ca^{2+} in the SR.

The arrival of a nerve impulse at a neuromuscular junction leads to the opening of voltage-gated Ca^{2+} channels in the SR membrane (see Figure 7-45). The ensuing release of Ca^{2+} from the SR raises the cytosolic Ca^{2+} concentration surrounding myofibrils sufficiently to trigger contraction. In skeletal muscle, the cytosolic Ca^{2+} concentration influences the interaction of four accessory proteins with actin thin filaments. The position of these proteins on the thin filaments in turn controls myosin–actin interactions.

Tropomyosin (TM) is a ropelike molecule, about 40 nm in length; TM molecules are strung together head to tail, forming a continuous chain along each actin thin filament (Figure 19-24a). Associated with tropomyosin is *troponin* (TN), a complex of the three subunits, TN-T, TN-I, and TN-C. Troponin-C is the calcium-binding subunit of troponin. Similar in sequence to calmodulin and the myosin light chains, TN-C controls the position of TM on the surface of an actin filament through the TN-I and TN-T subunits.

Scientists currently think that, under the control of Ca^{2+} and TN, TM can occupy two positions on a thin filament—an "off" state and an "on" state. In the absence of Ca^{2+} (the off state), myosin can bind to a thin filament, but the TM-TN complex prevents myosin from sliding along the thin filament. Binding of Ca^{2+} ions to TN-C triggers a slight movement of TM that exposes the myosin-binding sites on actin (Figure 19-24b). At Ca^{2+} concentrations $> 10^{-6}$ M, the inhibition exerted by the TM-TN complex is relieved, and con-

traction occurs. The Ca^{2+}-dependent cycling between on and off states in skeletal muscle is summarized in Figure 19-24c.

Myosin-Dependent Mechanisms Regulate Contraction in Smooth Muscle and Nonmuscle Cells

A smooth muscle cell contains large, loosely aligned contractile bundles that resemble the contractile bundles in epithelial cells and contractile ring during cytokinesis. Although specialized for generating force to restrict blood vessels, propel food down the gut, and restrict airway passages, the contractile apparatus of smooth muscle and its regulation constitute a valuable model for understanding how myosin activity is regulated in a nonmuscle cell. As we have just seen, skeletal muscle contraction is regulated by cycling of actin between on and off states. In contrast, smooth muscle contraction is regulated by cycling of myosin II between on and off states. Contraction of smooth muscle and nonmuscle cells is regulated by intracellular Ca^{2+} levels in response to many extracellular signaling molecules.

Calcium-Dependent Activation of Myosin II Contraction of vertebrate smooth muscle is regulated primarily by a complex pathway in which the *myosin regulatory light chain (LC)* undergoes phosphorylation and dephosphorylation. When the regulatory light chain is unphosphorylated, myosin II is inactive. The smooth muscle contracts when the regulatory LC is phosphorylated by the enzyme *myosin LC kinase* (Figure 19-25a). Because this enzyme is activated by Ca^{2+}, the cytosolic Ca^{2+} level indirectly regulates the extent of LC phosphorylation and hence contraction. The Ca^{2+}-dependent regulation of myosin LC kinase activity is mediated through

(a) Phosphorylation of light chains

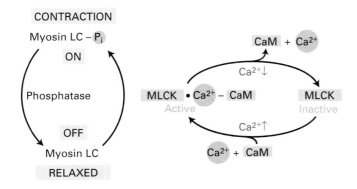

(b) Regulation of myosin LC phosphatase

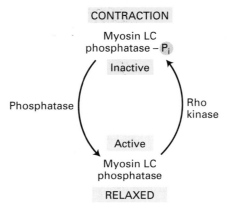

▲ **FIGURE 19-25 Myosin-dependent mechanisms for regulating smooth muscle contraction.** (a) In vertebrate smooth muscle, phosphorylation of the myosin regulatory light chains by Ca^{2+}-dependent myosin LC kinase activates contraction. At Ca^{2+} concentrations $<10^{-6}$M, the myosin LC kinase is inactive, and a myosin LC phosphatase, which is not

dependent on Ca^{2+} for activity, dephosphorylates the myosin LC, causing muscle relaxation. (b) Activation of Rho kinase leads to phosphorylation and inactivation of the myosin LC phosphatase. Various extracellular signaling molecules induce activation of Rho kinase in both smooth muscle and nonmuscle cells.

calmodulin. Calcium first binds to calmodulin, and the Ca^{2+}/calmodulin complex then binds to myosin LC kinase and activates it. Because this mode of regulation relies on the diffusion of Ca^{2+} and the action of protein kinases, contraction is much slower in smooth muscle than in skeletal muscle.

The role of activated myosin LC kinase can be demonstrated by microinjecting a kinase inhibitor into smooth muscle cells. Even though the inhibitor does not block the rise in the cytosolic Ca^{2+} level that follows the arrival of a nerve impulse, injected cells cannot contract. The effect of the inhibitor can be overcome by microinjecting a proteolytic fragment of myosin LC kinase that is active even in the absence of Ca^{2+}-calmodulin (this treatment also does not affect Ca^{2+} levels).

Signal-Induced Activation of Myosin II by Rho Kinase Unlike skeletal muscle, which is stimulated to contract solely by nerve impulses, smooth muscle cells and nonmuscle cells are regulated by many types of external signals in addition to nervous stimuli. For example, norepinephrine, angiotensin, endothelin, histamine, and other signaling molecules can modulate or induce the contraction of smooth muscle or elicit changes in the shape and adhesion of nonmuscle cells by triggering various signal-transduction pathways. Some of these pathways lead to an increase in the cytosolic Ca^{2+} level; as previously described, this increase can stimulate myosin activity by activating myosin LC kinase (see Figure 19-25a).

Other signaling pathways activate *Rho kinase*, which can stimulate myosin activity in two ways. First, Rho kinase can phosphorylate myosin LC phosphatase (see Figure 19-25b), thereby inhibiting its activity. With the phosphatase inactivated, the level of myosin LC phosphorylation and thus myosin activity increase. In addition, Rho kinase directly activates myosin by phosphorylating the regulatory light chain. Note that Ca^{2+} plays no role in the regulation of myosin activity by Rho kinase.

KEY CONCEPTS OF SECTION 19.3

Myosin-Powered Cell Movements

■ All myosin isoforms can interact with actin filaments through their head domains, but their cellular roles differ, depending on their tail domains (see Table 19-3).

■ Movement of actin filaments by myosin can be directly monitored in the sliding-filament assay (see Figure 19-17).

■ Myosins I, V, and VI power intracellular translocation of some membrane-limited vesicles along actin filaments. A similar process is responsible for cytoplasmic streaming, which is probably mediated by myosin XI, one of the fastest moving myosins (see Figure 19-19).

■ In nonmuscle cells, actin filaments and myosin II form contractile bundles that have a primitive sarcomere-like organization. Common examples are the circumferential belt present in epithelial cells and the stress fibers in cells cultured on plastic or glass surfaces; in the latter case, they may be an artifact. Both structures function in cell adhesion.

■ The contractile ring, a transient bundle of actin and myosin II, forms in a dividing cell and pinches the cell into two halves in cytokinesis.

■ In skeletal muscle cells, actin thin filaments and myosin thick filaments are organized into highly ordered structures, called sarcomeres (see Figure 19-22). The (+) end of the thin filaments is attached to the Z disk, the demarcation between adjacent sarcomeres.

■ During skeletal muscle contraction, myosin heads at each end of a thick filament walk along thin filaments toward the Z disks bounding a sarcomere. The force generated by myosin movement pulls the thin filaments toward the center of the sarcomere, shortening its length (see Figure 19-23).

■ The rapid rise in cytosolic Ca^{2+} induced by nerve stimulation of a skeletal muscle changes the interaction between actin filaments and tropomyosin, exposing the myosin-binding sites and thus permitting contraction to occur (see Figure 19-24).

■ Contraction of smooth muscle and nonmuscle cells is triggered by phosphorylation of the myosin regulatory light chains either by myosin LC kinase, in response to a rise in cytosolic Ca^{2+}, or by Rho kinase, in response to external signals (see Figure 19-25).

19.4 Cell Locomotion

We have now examined the different mechanisms used by cells to create movement—from the assembly of actin filaments and the formation of actin-filament bundles and networks to the contraction of bundles of actin and myosin and the sliding of single myosin molecules along an actin filament. These mechanisms are thought to constitute the major processes whereby cells generate the forces needed to migrate. *Cell locomotion* results from the coordination of motions generated by different parts of a cell. These motions are complex, but their major features can be revealed by fluorescent antibody-labeling techniques combined with fluorescence microscopy.

A property exhibited by all moving cells is polarity; that is, certain structures always form at the front of the cell, whereas others are found at the rear. Cell migration is initiated by the formation of a large, broad membrane protrusion at the leading edge of a cell. Video microscopy reveals that a major feature of this movement is the polymerization of actin at the membrane. In addition, actin filaments at the leading edge are rapidly cross-linked into bundles and networks in a protruding region, called a *lamellipodium* in vertebrate cells. In some cases, slender, fingerlike membrane projections, called *filopodia*, also are extended from the leading edge. These structures then form stable contacts with the underlying surface and prevent the membrane from retracting. In this section, we take a closer look at how cells employ the various force-generating processes to move across a

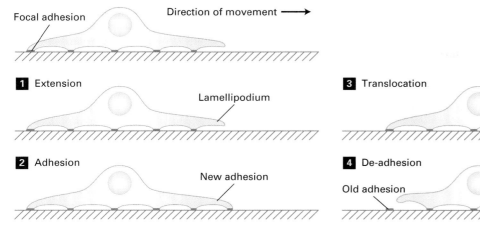

Focal adhesion   Direction of movement ⟶

1 Extension

Lamellipodium

2 Adhesion

New adhesion

3 Translocation

Cell body movement ⟶

4 De-adhesion

Old adhesion

▲ **FIGURE 19-26 Steps in keratinocyte movement.** In a fast-moving cell such as a fish epidermal cell, movement begins with the extension of one or more lamellipodia from the leading edge of the cell (**1**); some lamellipodia adhere to the substratum by focal adhesions (**2**). Then the bulk of the cytoplasm in the cell body flows forward (**3**). The trailing edge of the cell remains attached to the substratum until the tail eventually detaches and retracts into the cell body (**4**). See text for more discussion.

Overview Animation: Cell Motility
Video: Mechanics of Fish Keratinocyte
Migration

MEDIA CONNECTIONS

surface. We also consider the role of signaling pathways in coordinating and integrating the actions of the cytoskeleton, a major focus of current research.

Cell Movement Coordinates Force Generation with Cell Adhesion

A moving keratinocyte (skin cell) and a moving fibroblast (connective tissue cell) display the same sequence of changes in cell morphology—initial extension of a membrane protrusion, attachment to the substratum, forward flow of cytosol, and retraction of the rear of the cell (Figure 19-26).

Membrane Extension The network of actin filaments at the leading edge (see Figure 19-12) is a type of a cellular engine that pushes the membrane forward by an actin polymerization–based mechanism (Figure 19-27a). The key

▶ **FIGURE 19-27 Forces produced by assembly of the actin network.** (a) As shown in this diagram, actin filaments are assembled into a branched network in which the ends of filaments approach the plasma membrane at an acute angle. ATP–G-actin (red) adds to the filament end and pushes the membrane forward (**1**). The Arp2/3 complex (blue) binds to sides of filaments (**2**) and forms a branch at a 70° angle from the filament. With time, filaments ends are capped by capping protein (yellow) (**3**); the ATP–G-actin subunits convert into ADP–G-actin subunits (white) (**4**) and dissociate from the filament through the action of the severing proteins cofilin and gelsolin (gray) (**5**). The released ADP–G-actin subunits form complexes with profilin (green) (**6**) to regenerate ATP–G-actin subunits. (b) The network of actin filaments supports the elongation of filaments and the generation of pushing forces. An actin filament is stiff but can bend from thermal fluctuations. In the elastic Brownian ratchet model, bending of filaments at the leading edge (**1**), where the (+) ends contact the membrane, creates space at the membrane for subunits to bind to the ends of filaments (**2**). The elastic recoil force of the filaments then pushes the membrane forward. [Part (a) adapted from T. M. Svitkina and G. G. Borisy, 1999, *J. Cell Biol.* **145**:1009.]

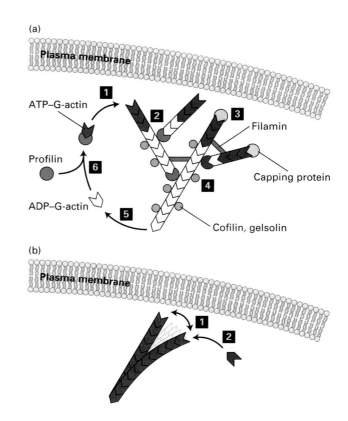

(a)

Plasma membrane

ATP–G-actin

1

2

3

Filamin

Profilin

6

4

Capping protein

ADP–G-actin

5

Cofilin, gelsolin

(b)

Plasma membrane

1

2

step in generating the force is (step 1) the addition of actin subunits at the ends of filaments close to the membrane. New filament ends are created (step 2) by branches formed by the Arp 2/3 complex. The branched network of filaments are stabilized by cross-linking proteins such as filamin. As the filaments grow, the ATP-actin subunits are converted into ADP-actin subunts. Consequently, (step 3) capping protein caps the (+) ends of filaments, and (step 4) cofilin and gelsolin fragment actin filaments and (step 5) cause actin subunits to dissociate. Profilin converts the ADP-actin monomers into a polymerization-competent ATP-actin monomer ready to participate in the next cycle.

A mechanism to explain what propels the membrane forward, called the *elastic Brownian ratchet model,* is based on the elastic mechanical property of an actin filament (Figure 19-27b). Electron micrographs show that the ends of actin filaments abut against the membrane, leaving no space for an actin subunit to bind. However, thermal energy causes a filament to bend, creating room for subunit addition. Because actin filaments have the same stiffness as that of a plastic rod, the energy stored in bending straightens the filament. The concerted action of numerous filaments undergoing similar movements and their cross-linkage into a mechanically strong network generate sufficient force (several piconewtons) to push the membrane forward.

Cell–Substrate Adhesions When the membrane has been extended and the cytoskeleton has been assembled, the membrane becomes firmly attached to the substratum. Time-lapse microscopy shows that actin bundles in the leading edge be-

come anchored to the attachment site, which quickly develops into a focal adhesion. The attachment serves two purposes: it prevents the leading lamella from retracting and it attaches the cell to the substratum, allowing the cell to push forward.

Cell Body Translocation After the forward attachments have been made, the bulk contents of the cell body are translocated forward (see Figure 19-26). How this translocation is accomplished is unknown; one speculation is that the nucleus and the other organelles are embedded in the cytoskeleton and that myosin-dependent cortical contraction moves the cytoplasm forward. The involvement of myosin-dependent cortical contraction in cell migration is supported by the localization of myosin II. Associated with the movement is a transverse band of myosin II and actin filaments at the boundary between the lamellipodia and the cell body (Figure 19-28).

Breaking Cell Attachments Finally, in the last step of movement (de-adhesion), the focal adhesions at the rear of the cell are broken and the freed tail is brought forward. In the light microscope, the tail is seen to "snap" loose from its connections—perhaps by the contraction of stress fibers in the tail or by elastic tension—but it leaves a little bit of its membrane behind, still firmly attached to the substratum.

The ability of a cell to move corresponds to a balance between the mechanical forces generated by the cytoskeleton and the resisting forces generated by cell adhesions. Cells cannot move if they are either too strongly attached or not

(a)

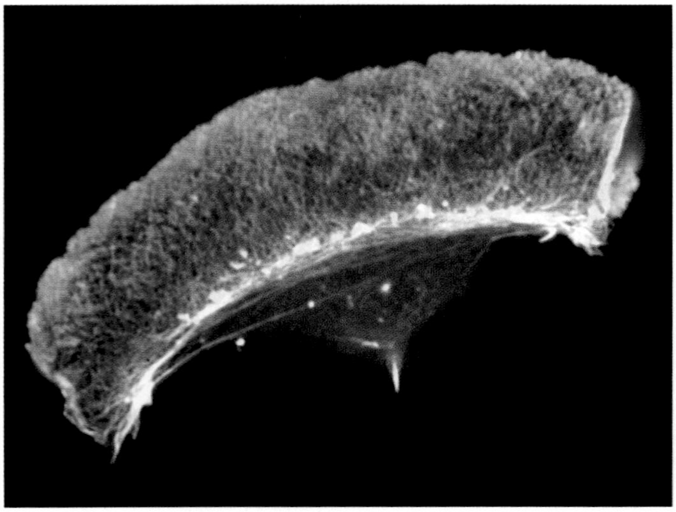

(b)

▲ **EXPERIMENTAL FIGURE 19-28 Contractile forces are generated by a moving cell.** (a) A fluorescence micrograph of a keratinocyte shows that the network of actin filaments (blue) is located at the front of the cell, whereas myosin II (red) is at the rear of the cell. However, both are located in a band (white) that traverses the cell just anterior to the nucleus. Contraction of this band is postulated to pull the cell body forward. (b) A moving cell exerts traction forces on the substratum. A keratinocyte plated on a thin silicon membrane exerts lateral forces from contraction of the cell body and causes the membrane to buckle. [Part (a) from T. M. Svitkina and G. G. Borisy, 1999, *J. Cell Biol.* **145**:1009; courtesy of T. M. Svitkina. Part (b) from K. Burton et al., 1999, *Mol. Biol. Cell* **10**:3745; courtesy of D. L. Taylor.]

attached to a surface. This relation can be demonstrated by measuring the rate of movement in cells that express varying levels of integrins, the cell-adhesion molecules that mediate most cell–matrix interactions (Chapter 6). Such measurements show that the fastest migration occurs at an intermediate level of adhesion, with the rate of movement falling off at high and low levels of adhesion. Cell locomotion thus results from traction forces exerted by the cell on the underlying substratum. The traction forces can be detected by the effects of cells on extremely thin sheets of silicon (Figure 19-28b). As a cell moves forward, contractile forces exerted at the front and the back of the cell cause the membrane to buckle. On a stiffer membrane that resists deformation, the buckling forces will be transformed into the forward movement of the cell.

Ameboid Movement Entails Reversible Gel–Sol Transitions of Actin Networks

Amebas are large, highly motile protozoans whose forward movement exhibits the same basic steps as those characterizing the movement of keratinocytes. Ameboid movement is initiated when the plasma membrane balloons forward to form a *pseudopodium,* or "false foot," which is similar to a lamellipodium in a vertebrate cell. As the pseudopodium attaches to the substratum, it fills with cytosol that is flowing forward through the cell. In the last step in movement, the rear of the ameba is pulled forward, breaking its attachments to the substratum.

Movement of an ameba is accompanied by changes in the viscosity of its cytosol, which cycles between sol and gel states. The central region of cytoplasm, the *endoplasm,* is a fluid, which flows rapidly toward the front of the cell, filling the pseudopodium. Here, the endoplasm is converted into the *ectoplasm,* a gel that forms the cortex, just beneath the plasma membrane. As the cell crawls forward, the ectoplasmic gel at the tail end of the cell is converted back into endoplasmic sol, only to be converted once again into ectoplasm when it again reaches the front of the cell. This cycling between sol and gel states continues only when the cell migrates.

The transformation between sol and gel states results from the disassembly and reassembly of actin microfilament networks in the cytosol. Several actin-binding proteins probably control this process and hence the viscosity of the cytosol. Profilin at the front of the cell promotes actin polymerization, and α-actinin and filamin form gel-like actin networks in the more viscous ectoplasm, as discussed earlier. Conversely, proteins such as cofilin sever actin filaments to form the more fluid endoplasm.

External Signals and Various Signaling Pathways Coordinate Events That Lead to Cell Migration

A striking feature of a moving cell is its polarity: a cell has a front and a back. When a cell makes a turn, a new leading lamellipodium or pseudopodium forms in the new direction.

If these extensions form in all directions, as in myosin I ameba mutants, then the cell is unable to pick a new direction of movement. To sustain movement in a particular direction, a cell requires signals to coordinate events at the front of the cell with events at the back and, indeed, signals to tell the cell where its front is. In this section, we present several examples of how external signals activate cell migration and control the direction of movement.

Activation of Filopodia, Membrane Ruffles, and Stress Fibers by Growth Factors   Certain growth factors in a fresh wound stimulate a quiescent cultured fibroblast to grow and divide by forming filopodia and lamellipodia at its leading edge and later to assemble stress fibers and focal adhesions. Similar signal-induced events are thought to take place in the wound-healing response of fibroblasts in vivo, the development of cells in embryos, and the metastasis of cancer cells. These events require the polymerization of actin filaments, the activation of myosin molecules, and the assembly of actin bundles and networks. The cytoskeletal rearrangements that are a part of the wound-healing response of fibroblasts include intracellular signaling pathways directed by Rac, Rho, and Cdc42, all Ras-like molecules belonging to the **GTPase superfamily** of switch proteins. These pathways are activated by the binding of growth factors to receptor tyrosine kinases, a class of cell-surface receptors described in Chapter 13.

The roles of Ras-related proteins were revealed by simple microinjection experiments. When Rac was microinjected into a fibroblast, the membrane immediately started to form upward projections called *ruffles;* focal adhesions and stress fibers formed 5–10 minutes later. Injection of an inactive form of Rac inhibited all reorganization of the actin cytoskeleton when growth factors were added to the cell. When Rho, rather than Rac, was injected, it mimicked the mitogenic effects of lysophosphatidic acid (LPA), a chemokine in serum and a potent stimulator of platelet aggregation. Both Rho and LPA induced the assembly of stress fibers and focal adhesions within 2 minutes but did not induce membrane ruffling.

These findings lead to a model in which extracellular factors trigger Ras-linked signal-transduction pathways that activate actin polymerization at the leading-edge membrane as an early event and the formation of focal adhesions as a later event (Figure 19-29). If this model is correct, then the inhibition of stress-fiber assembly should not affect membrane ruffling. To test the model, Rac and ADP-ribosylase, an enzyme that inactivates Rho by covalently attaching ADP to it, were co-injected into a fibroblast. As predicted, membrane ruffles were formed, but the assembly of stress fibers was blocked. These observations suggest that Rho-dependent events such as stress-fiber formation are "downstream" of control by Rac. The results of later experiments in which Cdc42 was microinjected into fibroblasts showed that this protein controlled an earlier step, the formation of filopodia. Thus the sequence of events in wound healing begins with the participation of filopodia and lamellipodia during the

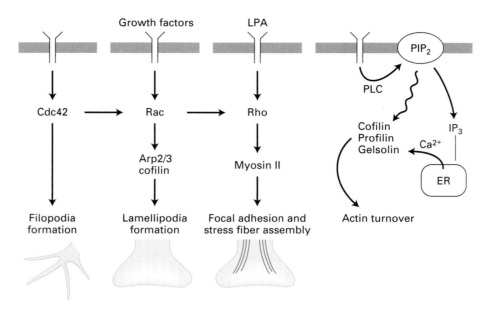

▲ **FIGURE 19-29 Role of signal-transduction pathways in cell locomotion and the organization of the cytoskeleton.** Extracellular signals are transmitted across the plasma membrane by receptors specific for different factors. One set of growth factors induces actin polymerization at the leading edge through a Rac- and Cdc42-dependent pathway (*left*); another set of factors acts downstream through a Rho-dependent pathway to induce the assembly of focal adhesions and cortical contraction (*center*). Adhesion of a cell to the extracellular matrix triggers a parallel signaling pathway that induces the activation of profilin, cofilin, and gelsolin (*right*). Triggering of this pathway activates phospholipase C (PLC), which hydrolyzes PIP₂ in the membrane. The subsequent increase in cytosolic Ca²⁺ stimulates actin turnover.

migration of cells into the wound and the formation of focal adhesions and stress fibers to close the wound.

An important aspect of locomotion is how movement is co-ordinated in response to different stimuli. For example, the assembly of the branched actin network at the membrane is enhanced by the action of several signaling pathways and their adapter proteins. The branching activity of the Arp2/3 complex is activated by an adapter protein, WASp, under the control of the Cdc42 GTPase. In addition, as discussed previously, the hydrolysis of PIP$_2$ by phospholipase C releases profilin, cofilin, and gelsolin from the membrane. In another pathway, inositol 1,4,5-trisphosphate (IP$_3$), a by-product of PIP$_2$ hydrolysis, stimulates the release of Ca^{2+} ions from the endoplasmic reticulum into the cytosol; this increase in Ca^{2+} ions activates myosin II and the severing activity of gelsolin. These parallel pathways thus stimulate both actin severing and filament growth, thereby increasing actin turnover (Figure 19-29, *right*).

Steering of Migrating Cells by Chemotactic Molecules

Under certain conditions, extracellular chemical cues guide the locomotion of a cell in a particular direction. In some cases, the movement is guided by insoluble molecules in the underlying substratum. In other cases, the cell senses soluble molecules and follows them, along a concentration gradient, to their source. The latter response is called **chemotaxis.** One of the best-studied examples of chemotaxis is the migration of *Dictyostelium* amebas along an increasing concentration of cAMP. Following cAMP to its source, the amebas aggregate into a slug and then differentiate into a fruiting body. Many other cells also display chemotactic

movements. For example, leukocytes are guided by a tripeptide secreted by many bacterial cells. In the development of skeletal muscle, a secreted protein signal called scatter factor guides the migration of myoblasts to the proper locations in limb buds (Chapter 22).

Despite the variety of different chemotactic molecules—sugars, peptides, cell metabolites, cell-wall or membrane lipids—they all work through a common and familiar mechanism: binding to cell-surface receptors, activation of intracellular signaling pathways, and re-modeling of the cytoskeleton through the activation or inhibition of various actin-binding proteins. The central question is, How do cell-surface receptors detect as small as a 2 percent difference in the concentration of chemotactic molecules across the length of the cell? To direct cell migration, an external chemoattractant gradient must somehow induce internal gradients that lead to polarization of the actin cytoskeleton.

Coincident Gradients of Chemoattractants, Activated G Proteins, and Ca²⁺ Micrographs of cAMP receptors tagged with green fluorescent protein (GFP) show that the receptors are distributed uniformly along the length of an ameba cell (Figure 19-30). Therefore an internal gradient must be established by another component of the signalling pathway. Because cAMP receptors signal through trimeric G proteins, a subunit of the trimeric G protein and other downstream signaling proteins were tagged with GFP. Fluorescence micrographs show that the concentration of trimeric G proteins is higher in the direction of the chemoattractant. Trimeric G proteins coupled to cAMP receptors can activate pathways

(a) G_β subunit

(b) cAMP receptor

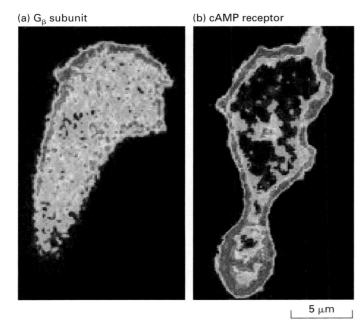

5 μm

◀ **EXPERIMENTAL FIGURE 19-30**
Chemoattractant is used to demonstrate signal-induced gradient of G_β subunit. False color images of amebas expressing (a) a GFP-tagged G_β subunit and (b) a GFP-tagged cAMP receptor, which served as a control. When an external source of chemoattractant was placed near the cells (at the top of the photographs), the cells turned toward the source. (a) The G_β subunit became concentrated at the leading edge of the cell, closest to the chemoattractant, and depleted from the tail. (b) In contrast, the cAMP receptor retained a uniform distribution in the cell membrane. [W. F. Loomis and R. H. Insall, 1999, *Nature* **401**:440–441; courtesy of Peter Devreotes Laboratory.]

leading to the activation of Arp 2/3 through the mediator protein WASp or through other pathways that increase cytosolic Ca^{2+} (see Figure 13-29).

Findings from studies with fluorescent dyes that act as internal Ca^{2+} sensors indicate that a cytosolic gradient of Ca^{2+} also is established in migrating cells, with the lowest concentration at the front of the cell and the highest concentration at the rear. Moreover, if a pipette containing a chemoattractant is placed to the side of a migrating leukocyte, the overall concentration of cytosolic Ca^{2+} first increases and then the Ca^{2+} gradient reorients, with the lowest concentration on the side of the cell closest to the pipette, causing the cell to turn toward the chemotactic source. After the chemoattractant is removed, the cell continues to move in the direction of its newly established Ca^{2+} gradient (see Figure 5-47).

We have seen that many actin-binding proteins, including myosins I and II, gelsolin, α-actinin, and fimbrin, are regulated by Ca^{2+}. Hence the cytosolic Ca^{2+} gradient may regulate the sol-to-gel transitions that take place in cell movement. The low Ca^{2+} concentration at the front of the cell would favor the formation of actin networks by activating myosin I, inactivating actin-severing proteins, and reversing the inhibition of Ca^{2+}-regulated actin cross-linking proteins. The high Ca^{2+} concentration at the rear of the cell would cause actin networks to disassemble and a sol to form by activating gelsolin or would cause cortical actin networks to contract by activating myosin II. Thus an internal gradient of Ca^{2+} would contribute to the turnover of actin filaments in migrating cells.

KEY CONCEPTS OF SECTION 19.4

Cell Locomotion

■ Migrating cells undergo a series of characteristic events: extension of a lamellipodium or pseudopodium, adhesion

of the extended leading edge to the substratum, forward flow (streaming of the cytosol), and retraction of the cell body (see Figure 19-26).

■ Cell locomotion is probably through a common mechanism including actin polymerization and branching–generated movement at the leading edge, assembly of adhesion structures, and cortical contraction mediated by myosin II (see Figure 19-28).

■ External signals (e.g., growth factors and chemoattractants) induce the assembly and organization of the cytoskeleton and the establishment of an internal gradient of trimeric G proteins and calcium (see Figure 19-29). The resulting polarization of the cell leads to locomotion.

19.5 Intermediate Filaments

In the remainder of this chapter, we consider the properties of intermediate filaments (IFs) and the cytoskeletal structures that they form in cells. Intermediate filaments are found in nearly all animals but not in in plants and fungi. The association of intermediate filaments with the nuclear and plasma membranes suggests that their principal function is structural (Figure 19-31). In epithelium, for instance, intermediate filaments provide mechanical support for the plasma membrane where it comes into contact with other cells or with the extracellular matrix. In epidermal cells (outer layer of skin) and the axons of neurons (Figure 19-32), intermediate filaments are at least 10 times as abundant as microfilaments or microtubules, the other components of the cytoskeleton.

Much of the following discussion about intermediate filaments will seem familiar because their cellular organization is similar to that of the actin microfilaments discussed in preceding sections. These two types of cytoskeletal fibers are

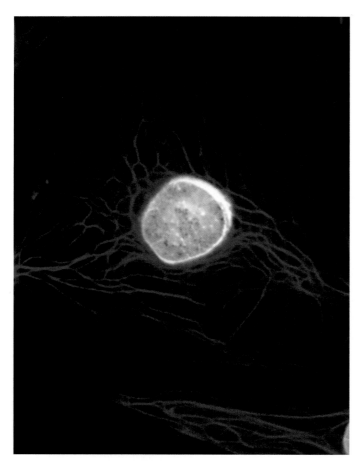

▲ **EXPERIMENTAL FIGURE 19-31 Staining with fluorochrome-tagged antibodies reveals cellular distribution of keratin and lamin intermediate filaments.** In this fluorescent micrograph of a PtK2 cell doubly stained with anti-keratin and anti-lamin antibodies, a meshwork of lamin intermediate filaments (blue) can be seen underlying the nuclear membrane. The cytoplasmic keratin cytoskeleton (red) extends from the nuclear membrane to the plasma membrane. [Courtesy of R. D. Goldman.]

also similar in that they are usually associated with cell membranes. Unlike microfilaments and microtubules, however, intermediate filaments do not contribute to cell motility. There are no known examples of IF-dependent cell movements or of motor proteins that move along intermediate filaments.

Intermediate Filaments Differ in Stability, Size, and Structure from Other Cytoskeletal Fibers

Several physical and biochemical properties distinguish intermediate filaments from microfilaments and microtubules. To begin with, intermediate filaments are extremely stable. Even after extraction with solutions containing detergents and high concentrations of salts, most intermediate filaments in a cell remain intact, whereas microfilaments and microtubules depolymerize into their soluble subunits. In fact, most IF purification methods employ these treatments to free intermediate filaments from other proteins. Intermediate filaments also differ in size from the other two cytoskeletal fibers. Indeed, their name derives from their 10-nm diameter—smaller than microtubules (24 nm) but larger than microfilaments (7 nm) (see Figure 5-29). Moreover, in contrast with the globular actin and tubulin subunits, which polymerize into microfilaments and hollow microtubules, respectively, IF subunits are α-helical rods that assemble into ropelike filaments. Finally, IF subunits do not bind nucleotides, and their assembly into intermediate filaments does not involve the hydrolysis of ATP or GTP, as does the polymerization of G-actin and tubulin. However, many of the details concerning the assembly of intermediate filaments in cells remain speculative.

IF Proteins Are Classified According to Their Distributions in Specific Tissues

In higher vertebrates, the subunits composing intermediate filaments constitute a superfamily of highly α helical proteins that are found in the cytoplasm of different tissues and at the nuclear membrane. The superfamily is divided into four groups on the basis of similarities in sequence and their patterns of expression in cells (Table 19-4). Unlike the actin and tubulin isoforms, the various classes of IF proteins are widely divergent in sequence and vary greatly in molecular weight. We introduce the four groups here and consider their functions in various cells in more detail later.

The most ubiquitous group of IFs are the **lamins**. In contrast with the cytosolic location of the other four classes of IF proteins, lamins are found exclusively in the nucleus. Of the three nuclear lamins, two are alternatively spliced products encoded by a common gene, whereas the third is encoded by a separate gene. A single lamin gene is found in the

▶ **EXPERIMENTAL FIGURE 19-32 Deep-etching reveals microtubules and intermediate filaments in a neuronal axon.** Neurofilaments and microtubules in a quick-frozen frog axon are visualized by the deep-etching technique. Several 24-nm-diameter microtubules run longitudinally; thinner, 10-nm-diameter intermediate filaments also run longitudinally. Occasional connections link the two types of cytoskeletal fibers. [From N. Hirokawa, 1982, *J. Cell Biol.* **94**:129; courtesy of N. Hirokawa.]

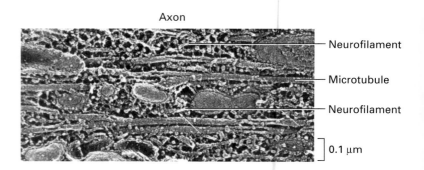

Axon

— Neurofilament

— Microtubule

— Neurofilament

0.1 μm

TABLE 19-4	Primary Intermediate Filaments in Mammals		
IF Protein	**MW (10^{-3})***	**Filament Form**	**Tissue Distribution**
NUCLEAR LAMINS			
Lamin A	70	Homopolymer	Nucleus
Lamin B	67	Homopolymer	Nucleus
Lamin C	67	Homopolymer	Nucleus
KERATINS[†]			
Acidic keratins	40–57	Heteropolymers	Epithelia
Basic keratins	53–67	Heteropolymers	Epithelia
TYPE III INTERMEDIATE FILAMENTS			
Vimentin	57	Homo- and heteropolymers	Mesenchyme (fibroblasts)
Desmin	53	Homo- and heteropolymers	Muscle
Glial fibrillary acidic protein	50	Homo- and heteropolymers	Glial cells, astrocytes
Peripherin	57	Homo- and heteropolymers	Peripheral and central neurons
NEUROFILAMENTS			
NF-L	62	Homopolymers	Mature neurons
NF-M	102	Heteropolymers	Mature neurons
NF-H	110	Heteropolymers	Mature neurons
Internexin	66	—	Developing CNS

*Intermediate filaments show species-dependent variations in molecular weight (MW).
[†]More than 15 isoforms of both acidic and basic keratins are known.

Drosophila genome; none are in the yeast genome. Because the lamin, but not the cytosolic, groups of IFs are expressed in *Drosophila*, lamins are probably the evolutionary precursor of the IF superfamily.

Epithelial cells express acidic and basic **keratins.** They associate in a 1:1 ratio to form heterodimers, which assemble into heteropolymeric keratin filaments; neither type alone can assemble into a keratin filament. The keratins are the most diverse classes of IF proteins, with a large number of keratin isoforms being expressed. These isoforms can be divided into two groups: about 10 keratins are specific for "hard" epithelial tissues, which give rise to nails, hair, and wool; and about 20, called *cytokeratins,* are more generally found in the epithelia that line internal body cavities. Each type of epithelium always expresses a characteristic combination of acidic and basic keratins.

Four proteins are classified as **type III** IF proteins. Unlike the keratins, the type III proteins can form both homo- and heteropolymeric IF filaments. The most widely distributed of all IF proteins is *vimentin,* which is typically expressed in leukocytes, blood vessel endothelial cells, some epithelial cells, and mesenchymal cells such as fibroblasts (see the illustration at the beginning of this chapter). Vimentin filaments help support cellular membranes. Vimentin networks may also help keep the nucleus and other organelles in a defined place within the cell. Vimentin is frequently associated with microtubules and, as noted earlier, the network of vimentin filaments parallels the microtubule network (see Figure 1-15). The other type III IF proteins have a much more limited distribution. *Desmin* filaments in muscle cells are responsible for stabilizing sarcomeres in contracting muscle. *Glial fibrillary acidic protein* forms filaments in the glial cells that surround neurons and in astrocytes. *Peripherin* is found in neurons of the peripheral nervous system, but little is known about it.

The core of neuronal axons is filled with **neurofilaments (NFs),** each a heteropolymer composed of three polypeptides—NF-L, NF-M, and NF-H—which differ greatly in molecular weight (see Figure 19-32 and Table 19-4). Neurofilaments are responsible for the radial growth of an axon and thus determine axonal diameter, which is directly related to the speed at which it conducts impulses. The influence of the number of neurofilaments on impulse conduction is highlighted by a mutation in quails named *quiver,* which blocks the assembly of neurofilaments. As a result, the velocity of nerve conduction is severely reduced. Also present in axons are microtubules, which direct axonal elongation.

Because of their characteristic distributions, IF proteins are useful in the diagnosis and treatment of certain tumors. In a tumor, cells lose their normal appearance, and thus their origin cannot be identified by their morphology. However, tumor cells retain many of the differentiated properties of the cells from which they are derived, including the expression of particular IF proteins. With the use of fluorescence-tagged antibodies specific for those IF proteins, diagnosticians can often determine whether a tumor originated in epithelial, mesenchymal, or neuronal tissue.

For example, the most common malignant tumors of the breast and gastrointestinal tract contain keratins and lack vimentin; thus they are derived from epithelial cells (which contain keratins but not vimentin) rather than from the underlying stromal mesenchymal cells (which contain vimentin but not keratins). Because epithelial cancers and mesenchymal cancers are sensitive to different treatments, identifying the IF proteins in a tumor cell helps a physician select the most effective treatment for destroying the tumor. ∎

All IF Proteins Have a Conserved Core Domain and Are Organized Similarly into Filaments

Besides having in common an ability to form filaments 10 nm in diameter, all IF subunit proteins have a common domain structure: a central α-helical core flanked by globular N- and C-terminal domains. The core helical domain, which is conserved among all IF proteins, consists of four long α helices separated by three nonhelical, "spacer" regions. The α-helical segments pair to form a coiled-coil dimer.

In electron micrographs, an IF-protein dimer appears as a rodlike molecule with globular domains at the ends; two dimers associate laterally into a tetramer (Figure 19-33a, b). The results of labeling experiments with antibodies to the N- or C-terminal domain indicate that the polypeptide chains are parallel in a dimer, whereas the dimers in a tetramer have an antiparallel orientation. The next steps in assembly are not well understood but seem to include the end-to-end association of tetramers to form long protofilaments, which aggregate laterally into a loose bundle of protofibrils. Compaction of a protofibril yields a mature 10-nm-diameter filament with the N- and C-terminal globular domains of the tetramers forming beaded clusters along the surface (Figure 19-33c). Interestingly, because the tetramer is symmetric, an intermediate filament may not have a polarity, as does an actin filament or a microtubule. This idea is supported by findings from experiments showing that vimentin subunits can incorporate along the length, as well as the ends, of a filament.

Although the α-helical core is common to all IF proteins, the N- and C-terminal domains of different types of IF proteins vary greatly in molecular weight and sequence. Partly because of this lack of sequence conservation, scientists initially speculated that the N- and C-terminal domains do not have roles in IF assembly. The results of several subsequent experiments, however, proved this hypothesis to be partly incorrect. For instance, if the N-terminal domain of an IF

▶ **EXPERIMENTAL FIGURE 19-33 Electron microscopy visualizes intermediate structures in the assembly of intermediate filaments.** Shown here are electron micrographs and drawings of IF protein dimers and tetramers and of mature intermediate filaments from *Ascaris,* an intestinal parasitic worm. (a) IF proteins form parallel dimers with a highly conserved coiled-coil core domain and globular tails and heads, which are variable in length and sequence. (b) A tetramer is formed by antiparallel, staggered side-by-side aggregation of two identical dimers. (c) Tetramers aggregate end-to-end and laterally into a protofibril. In a mature filament, consisting of four protofibrils, the globular domains form beaded clusters on the surface. [Adapted from N. Geisler et al., 1998, *J. Mol. Biol.* **282**:601; courtesy of Ueli Aebi.]

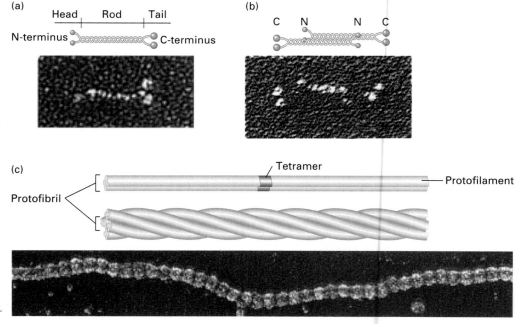

protein is shortened, either by proteolysis or by deletion mutagenesis, the truncated protein cannot assemble into filaments. (Keratins are an exception; they form filaments even if both terminal domains are absent.) The prevailing view now is that the N-terminal domain plays an important role in the assembly of most intermediate filaments. Even though the C-terminal domain is dispensable for IF assembly, it seems to affect the organization of IF cytoskeletons in a cell. Thus these domains may control lateral interactions within an intermediate filament, as well as interactions between intermediate filaments and other cellular components.

Identity of IF Subunits Whether IF monomers, dimers, or tetramers constitute the immediate subunit for assembly of filaments, analogous to G-actin monomers in the assembly of microfilaments, is still unresolved. The main supporting evidence for the involvement of the IF tetramer comes from cell fractionation experiments showing that, although most vimentin in cultured fibroblasts is polymerized into filaments, 1–5 percent of the protein exists as a soluble pool of tetramers. The presence of a tetramer pool suggests that vimentin monomers are rapidly converted into dimers, which rapidly form tetramers.

Homo- and Heteropolymeric Filaments Some IF proteins form homopolymeric filaments; others form only heteropolymeric filaments with other proteins in their class; and some can form both homo- and heteropolymeric filaments. Some IF proteins, but not the keratins, can form heteropolymers with IF proteins in another class. NF-L self-associates

to form a homopolymer, but NF-H and NF-M commonly co-assemble with the NF-L backbone, and so most neurofilaments contain all three proteins. Spacer sequences in the coiled-coil regions of IF dimers or sequences in the diverse N- or C-terminal domains or both are most likely responsible for determining whether particular IF proteins assemble into heteropolymers or homopolymers. In fact, mutations in these regions generate mutated IF polypeptides that can form hetero-oligomers with normal IF proteins. These hybrid molecules often "poison" IF polymerization by blocking assembly at an intermediate stage. The ability of mutated IF proteins to block IF assembly has proved extremely useful in studies of the function of intermediate filaments in a cell. At the end of the chapter, we look at how such mutations in keratins have revealed the role of keratin filaments in the epidermis.

Intermediate Filaments Are Dynamic

Although intermediate filaments are clearly more stable than microtubules and microfilaments, IF proteins have been shown to exchange with the existing IF cytoskeleton. In one experiment, a biotin-labeled type I keratin was injected into fibroblasts; within 2 hours after injection, the labeled protein had been incorporated into the already existing keratin cytoskeleton (Figure 19-34). The results of this experiment and others demonstrate that IF subunits in a soluble pool are able to add themselves to preexisting filaments and that subunits are able to dissociate from intact filaments.

The relative stability of intermediate filaments presents special problems in mitotic cells, which must reorganize all

(a) 20 minutes after injection

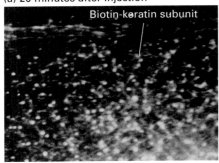 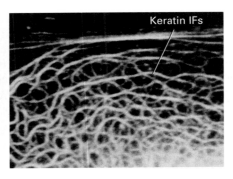

(b) 4 hours after injection

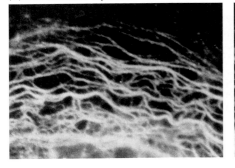 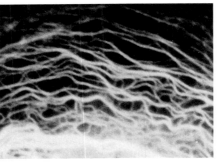

◄ **EXPERIMENTAL FIGURE 19-34
Chemical labeling and fluorescent staining reveal the incorporation of type I keratin into existing IF cytoskeleton.** Monomeric type I keratin was purified, chemically labeled with biotin, and microinjected into living fibroblast cells. The cells were then fixed at different times after injection and stained with a fluorescent antibody to biotin and with antibodies to keratin. (a) At 20 minutes after injection, the injected biotin-labeled keratin is concentrated in small foci scattered through the cytoplasm (*left*) and has not been integrated into the endogenous keratin cytoskeleton (*right*). (b) By 4 hours, the biotin-labeled subunits (*left*) and the keratin filaments (*right*) display identical patterns, indicating that the microinjected protein has become incorporated into the existing cytoskeleton. [From R. K. Miller, K. Vistrom, and R. D. Goldman, 1991, *J. Cell Biol.* **113**:843; courtesy of R. D. Goldman.]

three cytoskeletal networks in the course of the cell cycle. In particular, breakdown of the nuclear envelope early in mitosis depends on the disassembly of the lamin filaments that form a meshwork supporting the membrane. As discussed in Chapter 21, the phosphorylation of nuclear lamins by Cdc2, a **cyclin-dependent kinase** that becomes active early in mitosis (prophase), induces the disassembly of intact filaments and prevents their reassembly. Later in mitosis (telophase), removal of these phosphates by specific phosphatases promotes lamin reassembly, which is critical to reformation of a nuclear envelope around the daughter chromosomes. The opposing actions of kinases and phosphatases thus provide a rapid mechanism for controlling the assembly state of lamin intermediate filaments. Other intermediate filaments undergo similar disassembly and reassembly in the cell cycle.

Various Proteins Cross-Link Intermediate Filaments to One Another and to Other Cell Structures

Intermediate filament–associated proteins (IFAPs) cross-link intermediate filaments with one another, forming a bundle or a network, and with other cell structures, including the plasma membrane. Only a few IFAPs have been identified to date, but many more will undoubtedly be discovered as researchers focus attention on the proteins that control IF organization and assembly. Unlike actin-binding proteins or microtubule-associated proteins, none of the known IFAPs sever or cap intermediate filaments, sequester IF proteins in a soluble pool, or act as a motor protein. Rather, IFAPs appear to play a role in organizing the IF cytoskeleton, integrating the IF cytoskeleton with both the microfilament and the microtubule cytoskeletons, and attaching the IF cytoskeleton to the nuclear membrane and plasma membrane, especially at cell junctions.

A physical linkage between intermediate filaments and microtubules can be detected with certain drugs. Treatment of cells with high concentrations of colchicine causes the complete dissolution of microtubules after a period of several hours. Although vimentin filaments in colchicine-treated cells remain intact, they clump into disorganized bundles near the nucleus. This finding demonstrates that the organization of vimentin filaments is dependent on intact microtubules and suggests the presence of proteins linking the two types of filaments. In other studies, IFs have been shown to be cross-linked to actin filaments.

One family of IFAPs, the **plakins**, is responsible for linking IFs with both microtubules and microfilaments. One plakin family member is *plectin*, a 500,000-MW protein that has been shown to cross-link intermediate filaments with microtubules and actin filaments in vitro. Plectin also interacts with other cytoskeletal proteins, including spectrin, microtubule-associated proteins, and lamin B. Immunoelectron microscopy reveals gold-labeled antibodies to plectin decorating short, thin connections between microtubules and vimentin,

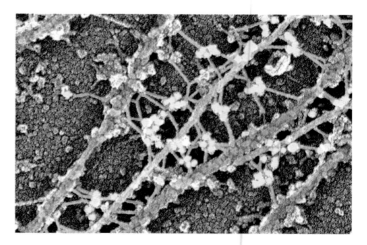

▲ **EXPERIMENTAL FIGURE 19-35 Gold-labeled antibody allows visualization of plectin cross-links between intermediate filaments and microtubules.** In this immunoelectron micrograph of a fibroblast cell, microtubules are highlighted in red; intermediate filaments, in blue; and the short connecting fibers between them, in green. Staining with gold-labeled antibodies to plectin (yellow) reveals that these fibers contain plectin. [From T. M. Svitkina, A. B. Verkhovsky, and G. G. Borisy, 1996, *J. Cell Biol.* **135**:991; courtesy of T. M. Svitkina.]

indicating the presence of plectin in these cross-links (Figure 19-35). The N-terminus of plectin and other plakins contains a calponin-homology (CH) domain similar to that in fimbrin and other actin cross-linking proteins. This finding suggests that some plakins form cross-links between actin microfilaments and intermediate filaments.

Cross-links between microtubules and neurofilaments are seen in micrographs of nerve-cell axons (see Figure 19-32). Although the identity of these connections in axons is unknown, they may be IFAPs whose function is to cross-link neurofilaments and microtubules into a stable cytoskeleton. Alternatively, these connections to microtubules may be the long arms of NF-H, which is known to bind microtubules.

IF Networks Form Various Supportive Structures and Are Connected to Cellular Membranes

A network of intermediate filaments is often found as a laminating layer adjacent to a cellular membrane, where it provides mechanical support. The best example is the **nuclear lamina** along the inner surface of the nuclear membrane (see Figure 21-16). This supporting network is composed of lamin A and lamin C filaments cross-linked into an orthogonal lattice, which is attached by lamin B to the inner nuclear membrane through interactions with a lamin B receptor, an IFAP, in the membrane. Like the membrane skeleton of the plasma membrane, the lamin nuclear skeleton not only supports the inner nuclear membrane but also provides sites where nuclear pores and interphase chromosomes attach. Thus, the nuclear lamins organize the nuclear contents from the outside in.

In addition to forming the nuclear lamina, intermediate filaments are typically organized in the cytosol as an extended system that stretches from the nuclear envelope to the plasma membrane (see Figure 19-31). Some intermediate filaments run parallel to the cell surface, whereas others traverse the cytosol; together they form an internal framework that helps support the shape and resilience of the cell. The results of in vitro binding experiments suggest that, at the plasma membrane, vimentin filaments bind two proteins: ankyrin, the actin-binding protein associated with the Na^+/K^+ ATPase in nonerythroid cells, and plectin, which also binds to $\alpha6\beta4$ integrin in certain cell junctions (Chapter 6). Through these two IFAPs, the vimentin cytoskeleton is attached to the plasma membrane, providing a flexible structural support.

In muscle, a lattice composed of a band of desmin filaments surrounds the sarcomere (Figure 19-36). The desmin filaments encircle the Z disk and are cross-linked to the plasma membrane by several IFAPs, including paranemin and ankyrin. Longitudinal desmin filaments cross to neighboring Z disks within the myofibril, and connections between desmin filaments around Z disks in adjacent myofibrils serve to cross-link myofibrils into bundles within a muscle cell. The lattice is also attached to the sarcomere through interactions with myosin thick filaments. Because the desmin filaments lie outside the sarcomere, they do not actively participate in generating contractile forces. Rather, desmin plays an essential structural role in maintaining muscle integrity. In transgenic mice lacking desmin, for example, this supporting architecture is disrupted and muscles are misaligned.

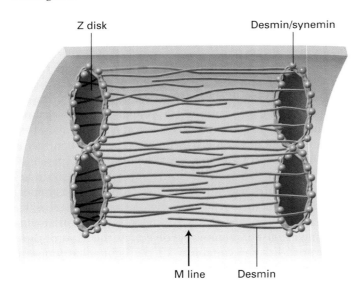

▲ **FIGURE 19-36 Diagram of desmin filaments in muscle.** These type III intermediate filaments encircle the Z disk and make additional connections to neighboring Z disks. The alignment of desmin filaments with the muscle sarcomere is held in place at the Z disk by a collar of desmin/synemin heteropolymers.

In Chapter 6, we describe the linkage between keratin filaments in epithelial cells and two types of anchoring junctions: *desmosomes,* which mediate cell–cell adhesion, and *hemidesmosomes,* which are responsible for attaching cells to the underlying extracellular matrix. In the electron microscope, both junctions appear as darkly staining proteinaceous plaques that are bound to the cytosolic face of the plasma membrane and attached to bundles of keratin filaments (see Figure 6-8). The keratin filaments in one cell are thus indirectly connected to those in a neighboring cell by desmosomes or to the extracellular matrix by hemidesmosomes. As a result of these connections, shearing forces are distributed from one region of a cell layer to the entire sheet of epithelial cells, providing strength and rigidity to the entire epithelium. Without the supporting network of intermediate filaments, an epithelium remains intact, but the cells are easily damaged by abrasive forces. Like actin microfilaments, which are attached to a third type of cell junction in epithelial cells, intermediate filaments form a flexible but resilient framework that gives structural support to an epithelium.

Disruption of Keratin Networks Causes Blistering

The epidermis is a tough outer layer of tissue, which acts as a water-tight barrier to prevent desiccation and serves as a protection against abrasion. In epidermal cells, bundles of keratin filaments are cross-linked by *filaggrin,* an IFAP, and are anchored at their ends to desmosomes. As epidermal cells differentiate, the cells condense and die, but the keratin filaments remain intact, forming the structural core of the dead, keratinized layer of skin. The structural integrity of keratin is essential in order for this layer to withstand abrasion.

 In humans and mice, the K4 and K14 keratin isoforms form heterodimers that assemble into protofilaments. A mutant K14 with deletions in either the N- or the C-terminal domain can form heterodimers in vitro but does not assemble into protofilaments. The expression of such mutant keratin proteins in cells causes IF networks to break down into aggregates. Transgenic mice that express a mutant K14 protein in the basal stem cells of the epidermis display gross skin abnormalities, primarily blistering of the epidermis, that resemble the human skin disease *epidermolysis bullosa simplex* (EBS). Histological examination of the blistered area reveals a high incidence of dead basal cells. Death of these cells appears to be caused by mechanical trauma from rubbing of the skin during movement of the limbs. Without their normal bundles of keratin filaments, the mutant basal cells become fragile and easily damaged, causing the overlying epidermal layers to delaminate and blister (Figure 19-37). Like the role of desmin filaments in supporting muscle tissue, the general role of keratin filaments appears to be to maintain the structural integrity of epithelial tissues by mechanically reinforcing the connections between cells. ∎

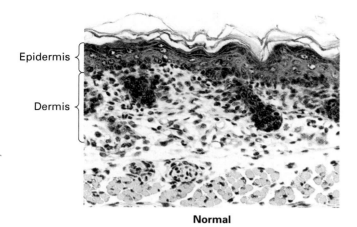

Normal

Mutated

▲ **EXPERIMENTAL FIGURE 19-37 Transgenic mice carrying a mutant keratin gene exhibit blistering similar to that in the human disease epidermolysis bullosa simplex.** Histological sections through the skin of a normal mouse and a transgenic mouse carrying a mutant K14 keratin gene are shown. In the normal mouse, the skin consists of a hard outer epidermal layer covering and in contact with the soft inner dermal layer. In the skin from the transgenic mouse, the two layers are separated (arrow) due to weakening of the cells at the base of the epidermis. [From P. Coulombe et al., 1991, *Cell* **66**:1301; courtesy of E. Fuchs.]

KEY CONCEPTS OF SECTION 19.5

Intermediate Filaments

■ Intermediate filaments are present only in cells that display a multicellular organization. An essential role of intermediate filaments is to distribute tensile forces across cells in a tissue.

■ Unlike microtubules and microfilaments, intermediate filaments are assembled from a large number of different IF proteins. These proteins are divided into four major types based on their sequences and tissue distribution. The lamins are expressed in all cells, whereas the other types are expressed in specific tissues (see Table 19-4).

■ The assembly of intermediate filaments probably proceeds through several intermediate structures, which associate by lateral and end-to-end interactions (see Figure 19-33).

■ Although intermediate filaments are much more stable than microfilaments and microtubules, they readily exchange subunits from a soluble pool.

■ The phosphorylation of intermediate filaments early in mitosis leads to their disassembly; they reassemble late in mitosis after dephosphorylation of the subunits.

■ The organization of intermediate filaments into networks and bundles, mediated by various IFAPs, provides structural stability to cells. IFAPs also cross-link intermediate filaments to the plasma and nuclear membranes, microtubules, and microfilaments.

■ Major degenerative diseases of skin, muscle, and neurons are caused by disruption of the IF cytoskeleton or its connections to other cell structures.

PERSPECTIVES FOR THE FUTURE

With the list of proteins encoded by the genome growing ever larger, a major challenge is to understand how they contribute to intracellular motility and cell movements. Historically, knowledge of motility has been built on a largely biochemical and structural foundation. The major biochemical components of the actin cytoskeleton have been identified, and their roles in defining the structure of the cytoskeleton and regulation are being discovered. However, motility is a dynamic process in which function is also built on the foundation of mechanics.

Dynamics implies a change in structure with time and includes movements of proteins and protein structures. From Newton's laws of physics, movements are linked with forces, but we are only now becoming capable of studying forces at the molecular and cellular levels. Much has changed from the pioneering studies of muscle physiologists who measured muscle fiber contractile forces with tension gauges and of experimental cell physiologists who estimated the contractile force of the contractile ring from the bending of glass needles. These pioneering studies lead directly to modern studies that incorporate elegant physical techniques such as optical traps capable of measuring piconewton forces or that deduce stiffness from images of microtubules and microfilaments buckling against an unmovable object. At the heart of these discoveries lies the capability of the light microscope to image single molecules and to watch the binding of individual ligands and enzyme action on individual nucleotides.

But to understand fully how a cell moves, we must be able to measure forces in the cell. In vivo methods that measure force directly are not yet invented. What is needed? A mechanical sensor, perhaps engineered from a fluorescent protein, that can change fluorescence when a force is applied. Such a technique would be the analog of experiments in which fluorescently labeled actin was microinjected into a live cell and scientists watched for the first time the actin cytoskeleton treadmill at the leading edge of the cell.

A second challenge is to understand how a cell moves in its natural environment—between layers of cells in a three-dimensional extracellular matrix. The story of cell motility is one shaped by studies of cells moving on a flat surface. The flat world imposes a geometry in which interactions with the substratum is through adhesions on the ventral surface of the cell while the free, noninteracting surface defines the dorsal surface. Consider, instead, movement through a gel of matrix. When surrounded by matrix, cells lose their stereotypic flat shape and take on tubular spindle shape. With adhesions on all sides, cells corkscrew through the gel of matrix molecules. The adhesions are smaller than those formed with a flat surface, which itself makes studying adhesions more difficult. Furthermore, the techniques for estimating the forces that a cell must exert to crawl on a flat surface cannot be applied to a three-dimensional situation. Squeezing through a gel adds several dimensions of complexity to our understanding how cell motility and adhesion are coupled.

KEY TERMS

actin-related proteins (Arps) *788*
calponin homology–domain superfamily *782*
cell locomotion *800*
chemotaxis *804*
contractile bundles *796*
critical concentration *785*
cytoskeleton *779*
F-actin *781*
G-actin *781*
intermediate filaments *779*
keratins *807*
lamellipodium *800*
lamins *806*

microfilaments *779*
motor protein *779*
myosin head domain *791*
myosin tail domain *793*
myosin LC kinase *799*
plakins *810*
profilin *787*
sarcomere *797*
sliding-filament assay *793*
thick filaments *797*
thin filaments *797*
thymosin β4 *786*
treadmilling *786*

REVIEW THE CONCEPTS

1. Actin filaments have a defined polarity. What is filament polarity? How is it generated at the subunit level? How is filament polarity detectable?

2. In cells, actin filaments form bundles and/or networks. How do cells form such structures, and what specifically determines whether actin filaments will form a bundle or a network?

3. Much of our understanding of actin assembly in the cell is derived from experiments using purified actin in vitro. What techniques may be used to study actin assembly in vitro? Explain how each of these techniques works.

4. The predominant forms of actin inside a cell are ATP-G-actin and ADP-F-actin. Explain how the interconversion of the nucleotide state is coupled to the assembly and disassembly of actin subunits. What would be the consequence for actin filament assembly/disassembly if a mutation prevented actin's ability to bind ATP? What would be the consequence if a mutation prevented actin's ability to hydrolyze ATP?

5. Actin filaments at the leading edge of a crawling cell are believed to undergo treadmilling. What is treadmilling, and what accounts for this assembly behavior?

6. Although purified actin can reversibly assemble in vitro, various actin-binding proteins regulate the assembly of actin filaments in the cell. Predict the effect on a cell's actin cytoskeleton if function-blocking antibodies against each of the following were independently microinjected into cells: profilin, thymosin β4, gelsolin, tropomodulin, and the Arp2/3 complex.

7. There are at least 17 different types of myosin. What properties do all types share, and what makes them different?

8. The ability of myosin to walk along an actin filament may be observed with the aid of an appropriately equipped microscope. Describe how such assays are typically performed. Why is ATP required in these assays? How may such assays be used to determine the direction of myosin movement or the force produced by myosin?

9. Contractile bundles occur in nonmuscle cells, although the structures are less organized than the sarcomeres of muscle cells. What is the purpose of nonmuscle contractile bundles?

10. Contraction of both skeletal and smooth muscle is triggered by an increase in cytosolic Ca^{2+}. Compare the mechanisms by which each type of muscle converts a rise in Ca^{2+} into contraction.

11. Several types of cells utilize the actin cytoskeleton to power locomotion across surfaces. What types of cells have been utilized as models for the study of locomotion? What sequential morphological changes do each of these model cells exhibit as they move across a surface? How are actin filaments involved in each of these morphological changes?

12. To move in a specific direction, migrating cells must utilize extracellular cues to establish which portion of the cell will act as the front and which will act as the back. Describe how G proteins and Ca^{2+} gradients appear to be involved in the signaling pathways used by migrating cells to determine direction of movement.

13. Unlike actin filaments, intermediate filaments do not exhibit polarity. Explain how the structure of intermediate filament subunits and the relationship between assembled subunits in an intermediate filament produce a filament lacking polarity.

14. Compared to actin filaments, intermediate filaments are relatively stable. However, cells can induce intermediate filament disassembly when needed. How does this disassembly occur, and why is it necessary?

15. Animal cells contain proteins that could be considered to serve as intermediate filament-associated proteins. Several such intermediate filament-associated proteins have now been identified. What functions do these proteins carry out in cells? To what other cellular structures do intermediate filament-associated proteins bind?

ANALYZE THE DATA

Understanding of actin filaments has been greatly facilitated by the ability of scientists to purify actin and actin-binding proteins and the ability to assemble actin filaments in vitro. Following are various experimental approaches designed to characterize actin assembly and the effects of actin-binding proteins on actin assembly.

a. The graph in part (a) of the figure depicts the actin polymerization rate at the plus (+) and minus (−) ends of rabbit actin as a function of actin concentration. Assume that you could add actin filaments of a predefined length to rabbit actin maintained at the concentrations labeled A, B, and C in the figure. Diagram the appearance of the filaments after a 10-minute incubation at each of the indicated actin concentrations, if the original filaments are depicted as follows:

Original filament: +_____−

Make sure to mark the location of the original (+) and (−) ends of the filament on your diagrams.

b. A novel actin-binding protein (X) is overexpressed in certain highly malignant cancers. You wish to determine if protein X caps actin filaments at the (+) or (−) end. You incubate an excess of protein X with various concentrations of G-actin under conditions that induce polymerization. Control samples are incubated in the absence of protein X. The results are shown in part (b) of the figure. How can you conclude from these data that protein X binds to the (+) end of actin filaments? Design an experiment, using myosin S1 fragments and electron microscopy, to corroborate the conclusion that protein X binds to the (+) end. What results would you expect if this conclusion is correct?

c. An in vitro system was developed to study actin assembly and disassembly in nonmuscle cells. In this study, tissue culture cells were incubated for several hours with [^{35}S]methionine so that all the actin monomers in each filament were labeled. Actin filaments were then collected by differential centrifugation and put into a buffer containing one of three

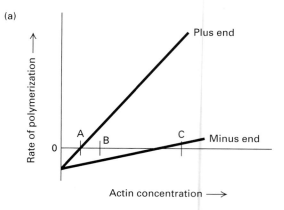

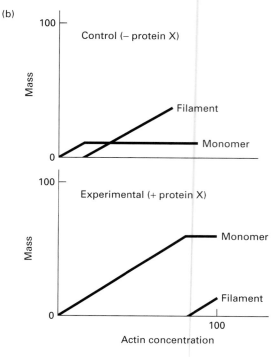

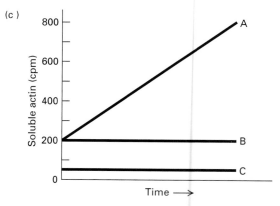

different cytosolic extracts (A, B, or C). The amounts of soluble actin in each sample were monitored over time (see part (c) of the figure). What do these data indicate about the effects of A, B, and C on the assembly and disassembly of actin filaments?

REFERENCES

General References

Bray, D. 2001. *Cell Movements*. Garland.

Howard, J. 2001. *The Mechanics of Motor Proteins and the Cytoskeleton*. Sinauer.

Kreis, T., and R. Vale. 1999. *Guidebook to the Cytoskeletal and Motor Proteins*. Oxford University Press.

Web Sites

The following site contains links to laboratories that study the cytoskeleton

http:/vlib.org/Science/Cell_Biology/labs.shtml

Actin-centric sites containing all things actin

http://www.bms.ed.ac.uk/research/others/smaciver/Cyto-Topics/Topics.htm

http://expmed.bwh.harvard.edu/main/resources.html#cytoskeleton

The myosin home page

http://www.mrc-lmb.cam.ac.uk/myosin/myosin.html

The cytokinetic mafia home page—all things cytokinetic

http://www.unc.edu/depts/salmlab/mafia/mafia.html

Actin Structures

Bretscher, A. 1999. Regulation of cortical structure by the ezrin-radixin-moesin protein family. *Curr. Opin. Cell Biol.* **11**:109–116.

Furukawa, R., and M. Fechheimer. 1997. The structure, function, and assembly of actin filament bundles. *Int'l. Rev. Cytol.* **175**:29–90.

McGough, A. 1998. F-actin-binding proteins. *Curr. Opin. Struc. Biol.* **8**:166–176.

Sheterline, P., J. Clayton, and J. C. Sparrow. 1995. Protein profile. *Actin* **2**:1–103.

Stossel, T. P., et al. 2001. Filamins as integrators of cell mechanics and signalling. *Nature Rev. Cell Biol.* **2**(2):138–145.

Stradal, T., W. Kranewitter, S. J. Winder, and M. Gimona. 1998. CH domains revisited. *FEBS Lett.* **431**:134–137.

The Dynamics of Actin Assembly

Cameron, L. A., P. A. Giardini, F. S. Soo, and J. A. Theriot. 2000. Secrets of actin-based motility revealed by a bacterial pathogen. *Nature Rev. Mol. Cell Biol.* **1**:110–119.

Cooper, J. A., and D. A. Schafer. 2000. Control of actin assembly and disassembly at filament ends. *Curr. Opin. Cell Biol.* **12**:97–103.

Higgs, H. N., and T. D. Pollard. 2000. Regulation of actin filament network formation through ARP2/3 complex: activation by a diverse array of proteins. *Ann. Rev. Biochem.* **70**:619–676.

Kwiatkowski, D. J. 1999. Functions of gelsolin: motility, signaling, apoptosis, cancer. *Curr. Opin. Cell Biol.* **11**:103–108.

Theriot, J. A. 1997. Accelerating on a treadmill: ADF/cofilin promotes rapid actin filament turnover in the dynamic cytoskeleton. *J. Cell Biol.* **136**:1165–1168.

Pollard, T. D., L. Blanchoin, and R. D. Mullins. 2000. Molecular mechanisms controlling actin filament dynamics in nonmuscle cells. *Ann. Rev. Biophys. Biomol. Struc.* **29**:545–576.

Small, J. V., T. Strada, E. Vignal, and K. Rottner. 2002. The lamellipodium: where motility begins. *Trends Cell Biol.* **12**:112–120.

Myosin-Powered Cell Movements

Berg, J. S., B. C. Powell, and R. E. Cheney. 2001. A millennial myosin census. *Mol. Biol. Cell* **12**:780–794.

Bresnick, A. R. 1999. Molecular mechanisms of nonmuscle myosin: II. Regulation. *Curr. Opin. Cell Biol.* **11**:26–33.

Buss, F., J. Luzio, and J. Kendrick-Jones. 2002. Myosin VI, an actin motor for membrane traffic and cell migration. *Traffic* **3**:851–858.

Field, C., R. Li, and K. Oegema. 1999. Cytokinesis in eukaryotes: a mechanistic comparison. *Curr. Opin. Cell Biol.* **11**:68–80.

Gregorio, C. C., H. Granzier, H. Sorimachi, and S. Labeit. 1999. Muscle assembly: a titanic achievement? *Curr. Opin. Cell Biol.* **11**:18–25.

Mermall, V., P. L. Post, and M. S. Mooseker. 1998. Unconventional myosin in cell movement, membrane traffic, and signal transduction. *Science* **279**:527–533.

Rayment, I. 1996. The structural basis of the myosin ATPase activity. *J. Biol. Chem.* **271**:15850–15853.

Simmon, R. 1996. Molecular motors: single-molecule mechanics. *Curr. Biol.* **6**:392–394.

Squire, J. M., and E. P. Morris. 1998. A new look at thin filament regulation in vertebrate skeletal muscle. *FASEB J.* **12**:761–771.

Vale, R. D., and R. A. Milligan. 2000. The way things move: looking under the hood of molecular motor proteins. *Science* **288**:88–95.

Cell Locomotion

Borisy, G. G., and T. M. Svitkina. 2000. Actin machinery: pushing the envelope. *Curr. Opin. Cell Biol.* **12**:104–112.

Chung, C. Y., S. Funamoto, and R. A. Firtel. 2001. Signaling pathways controlling cell polarity and chemotaxis. *Trends Biochem. Sci.* **26**:557–566.

Condeelis, J. 2001. How is actin polymerization nucleated in vivo? *Trends Cell Biol.* **11**:288–93.

Hall, A. 1998. Rho GTPases and the actin cytoskeleton. *Science* **279**:509–514.

Pettit, E. J., and F. S. Fay. 1998. Cytosolic free calcium and the cytoskeleton in the control of leukocyte chemotaxis. *Physiol. Rev.* **78**:949–967.

Pollard, T. D., and G. G. Borisy. 2003. Cellular motility driven by assembly and disassembly of actin filaments. *Cell* **112**:453–465.

Welch, M. D., A. Mallavarapu, J. Rosenblatt, and T. J. Mitchison. 1997. Actin dynamics in vivo. *Curr. Opin. Cell Biol.* **9**:54–61.

Intermediate Filaments

Coulombre, P. A., and B. M. Omary. 2002. "Hard" and "Soft" principles defining the structure, function, and regulation of keratin intermediate filaments. *Curr. Opin. Cell Biol.* **14**:110–122.

Fuchs, E., and D. W. Cleveland. 1998. A structural scaffolding of intermediate filaments in health and disease. *Science* **279**:514–519.

Leung, C. L., K. J. Green, and R. K. Liem. 2002. Plakins: a family of versatile cytolinker proteins. *Trends Cell Biol.* **12**:37–45.

Stuurman, N., S. Heins, and U. Aebi. 1998 Nuclear lamins: their structure, assembly, and interactions. *J. Struc. Biol.* **122**:42–66.

20

MICROTUBULES

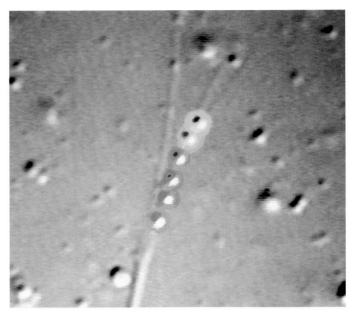

A 3-second time lapse movie captures the kinesin-powered movement of a vesicle along a microtubule. [From N. Pollack et al., 1999, *J. Cell Biol.* **147**:493–506; courtesy of R. D. Vale.]

In Chapter 19, we looked at microfilaments and intermediate filaments—two of the three types of cytoskeletal fibers—and their associated proteins. This chapter focuses on the third cytoskeletal system—**microtubules.** Like microfilaments, microtubules take part in certain cell movements, including the beating of cilia and flagella and the transport of vesicles in the cytoplasm. These movements result from the polymerization and depolymerization of microtubules or the actions of microtubule **motor proteins.** Both processes are required for some other cell movements, such as the alignment and separation of chromosomes in meiosis and mitosis (see Figure 9-3). Microtubules also direct the migration of nerve-cell axons by guiding the extension of the neuronal growth cone.

In addition to contributing to cell motility, microtubules play a major role in organizing the cell through a special structure called the **microtubule-organizing center,** or **MTOC.** Located near the nucleus, the MTOC directs the assembly and orientation of microtubules, the direction of vesicle trafficking, and the orientation of organelles. Because organelles and vesicles are transported along microtubules, the MTOC becomes responsible for establishing the polarity of the cell and the direction of cytoplasmic processes in both interphase and mitotic cells (Figure 20-1).

In this chapter, we build on the general principles learned in Chapter 19 about the structure and function of the microfilament cytoskeleton and show how many of the same concepts also apply to microtubules. We begin the chapter by examining the structure and assembly of microtubules and then consider how microtubule assembly and microtubule motor proteins can power cell movements. The discussion of microtubules concludes with a detailed examination of the translocation of chromosomes in mitosis. Although we consider microtubules, microfilaments, and intermediate filaments individually, the three cytoskeletal systems do not act completely independently of one another. An important example of their interdependence can be found in cell division when interaction between actin microfilaments and microtubules determines the plane of cleavage.

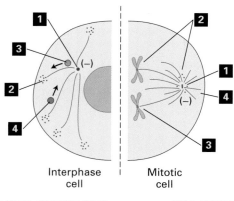

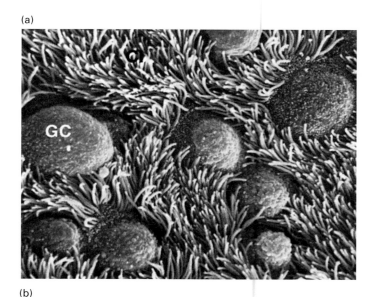

Interphase cell | Mitotic cell

CYTOSKELETAL COMPONENT	CELL FUNCTION
1 MTOC, spindle pole	Organizing cell polarity
2 Microtubule dynamics	Chromosome movements MT assembly
3 Kinesin motors	(+) end–directed vesicle and chromosome transport
4 Dynein motors	(–) end–directed vesicle transport spindle assembly

▲ **FIGURE 20-1 Microtubules (blue) organized around the MTOC and spindle poles (1)** establish an internal polarity to movements and structures in the interphase cell (*left*) and the mitotic cell (*right*). Assembly and disassembly (**2**) cause microtubules to probe the cell cytoplasm and are harnessed at mitosis to move chromosomes. Long-distance movement of vesicles (**3** and **4**) are powered by kinesin and dynein motors. Both motors are critical in the assembly of the spindle and the separation of chromosomes in mitosis.

20.1 Microtubule Organization and Dynamics

A microtubule is a polymer of globular **tubulin** subunits, which are arranged in a cylindrical tube measuring 25 nm in diameter—more than twice the width of an intermediate filament and three times the width of a microfilament (see Figure 5-29). Varying in length from a fraction of a micrometer to hundreds of micrometers, microtubules are much stiffer than either microfilaments or intermediate filaments because of their tubelike construction. A consequence of this tubular design is the ability of microtubules to generate pushing forces without buckling, a property that is critical to the movement of chromosomes and the mitotic spindle in mitosis.

Cells contain two populations of microtubules: stable, long-lived microtubules and unstable, short-lived microtubules. Stable microtubules are generally found in nonreplicating cells. They include a central bundle of microtubules in **cilia** and **flagella,** extensions of the plasma membrane that beat rhythmically to propel materials across epithelial surfaces, to enable sperm to swim, or to push an egg through the oviduct (Figure 20-2a). A marginal band of stable microtubules present in some erythrocytes and platelets enables these cells to pass through small blood vessels. Another example exists in nerve cells (neurons), which must maintain long processes

(a)

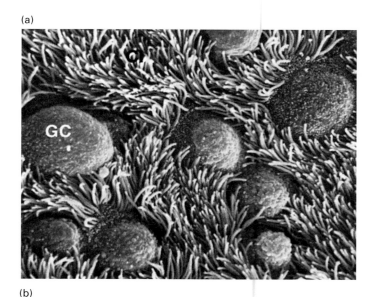

(b)

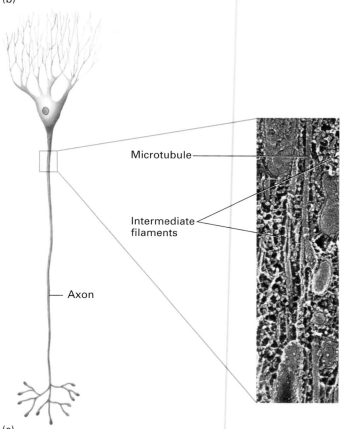

Microtubule

Intermediate filaments

Axon

(c)

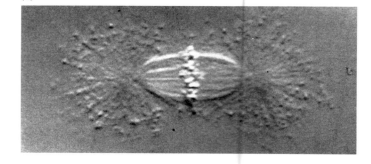

◀ **EXPERIMENTAL FIGURE 20-2** **Various microscopic techniques are used to visualize stable and transient microtubule structures.** (a) Surface of the ciliated epithelium lining a rabbit oviduct viewed in the scanning electron microscope. Microtubule-containing cilia cover ciliated cells, and actin-containing microvilli populate the surface of secretory cells. Beating cilia propel an egg down the oviduct. (b) Microtubules and intermediate filaments in a quick-frozen frog axon visualized by the deep-etching technique (*right*). Several 24-nm-diameter microtubules and thinner, 10-nm-diameter intermediate filaments can be seen. Both types of fibers are oriented longitudinally; they are cross-linked by various proteins. (c) Isolated mitotic apparatus visualized by differential interference contrast (DIC) microscopy. The spindle and asters, which are critical in pulling the chromosomes to the poles, are composed of transient microtubules that assemble early in mitosis and disassemble at its completion. [Part (a) from R. G. Kessels and R. H. Kardon, 1975, *Tissues and Organs*, W. H. Freeman and Company. Part (b) from N. Hirokawa, 1982, *J. Cell Biol.* **94**:129; courtesy of N. Hirokawa. Part (c) from E. D. Salmon and R. R. Segall, 1980, *J. Cell Biol.* **86**:355.]

called **axons** (see Figure 7-29). An internal core of stable microtubules in axons not only supports their structure but also provides tracks along which vesicles move through the axonal cytoplasm (Figure 20-2b). The disassembly of such stable structures would have catastrophic consequences—sperm would be unable to swim, a red blood cell would lose its springlike pliability, and axons would retract.

In contrast with these permanent, stable structures, unstable microtubules are found in cells that need to assemble and disassemble microtubule-based structures quickly. For example, in mitosis, the cytosolic microtubule network characteristic of interphase cells disassembles, and the tubulin from it is used to form the spindle-shaped apparatus that partitions chromosomes equally to the daughter cells (Figure 20-2c). When mitosis is complete, the spindle disassembles and the interphase microtubule network re-forms.

Before proceeding to a discussion of microtubule-based movements, we examine the assembly, disassembly, and polarity of microtubules, as well as a group of proteins that are integrally associated with microtubules. An important property of a microtubule is oscillation between growing and shortening phases. This complex dynamic behavior permits a cell to quickly assemble or disassemble microtubule structures.

Heterodimeric Tubulin Subunits Compose the Wall of a Microtubule

The building block of a microtubule is the tubulin subunit, a heterodimer of α- and β-tubulin. Both of these 55,000-MW monomers are found in all eukaryotes, and their sequences are highly conserved. Although a third tubulin, γ-tubulin, is not part of the tubulin subunit, it probably nucleates the polymerization of subunits to form αβ–microtubules. Encoded by separate genes, the three tubulins exhibit homology with a 40,000-MW bacterial GTPase, called FtsZ (see Figure 5-30b). Like tubulin, this bacterial protein

has the ability to polymerize and participates in cell division. Perhaps the protein carrying out these ancestral functions in bacteria was modified in the course of evolution to fulfill the diverse roles of microtubules in eukaryotes.

Each tubulin subunit binds two molecules of GTP. One GTP-binding site, located in α-tubulin, binds GTP irreversibly and does not hydrolyze it. The second site, located on β-tubulin, binds GTP reversibly and hydrolyzes it to GDP. Thus, tubulin is a GTPase like bacterial FtsZ protein. In the atomic structure of the tubulin subunit, the GTP bound to α-tubulin is trapped at the interface between the α- and β-tubulin monomers and is thus nonexchangeable. The second GTP lies at the surface of the β-tubulin monomer; this GTP is freely exchangeable with GDP (Figure 20-3a). As discussed later, the

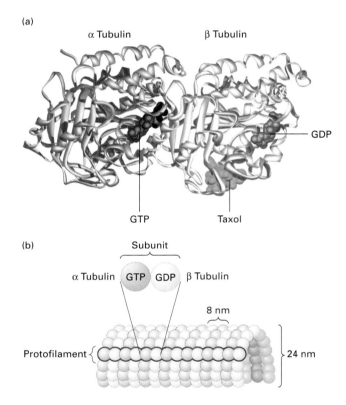

▲ **FIGURE 20-3 Structure of tubulin monomers and their organization in microtubules.** (a) Ribbon diagram of the dimeric tubulin subunit. The GTP (red) bound to the α-tubulin monomer is nonexchangeable, whereas the GDP (blue) bound to the β-tubulin monomer is exchangeable with GTP. The anticancer drug taxol (green) was used in structural studies to stabilize the dimer structure. (b) The organization of tubulin subunits in a microtubule. The subunits are aligned end to end into protofilaments, which pack side by side to form the wall of the microtubule. In this model, the protofilaments are slightly staggered so that α-tubulin in one protofilament is in contact with α-tubulin in the neighboring protofilaments. The microtubule displays a structural polarity in that subunits are added preferentially at the end, designated the (+) end, at which β-tubulin monomers are exposed. [Part (a) modified from E. Nogales et al., 1998, *Nature* **391**:199; courtesy of E. Nogales. Part (b) adapted from Y. H. Song and E. Mandelkow, 1993, *Proc. Nat'l. Acad. Sci. USA* **90**:1671.]

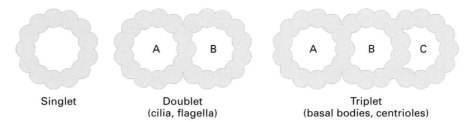

Singlet   Doublet   Triplet
                 (cilia, flagella) (basal bodies, centrioles)

▲ **FIGURE 20-4 Arrangement of protofilaments in singlet, doublet, and triplet microtubules.** In cross section, a typical microtubule, a singlet, is a simple tube built from 13 protofilaments. In a doublet microtubule, an additional set of

10 protofilaments forms a second tubule (B) by fusing to the wall of a singlet (A) microtubule. Attachment of another 10 protofilaments to the B tubule of a doublet microtubule creates a C tubule and a triplet structure.

guanine bound to β-tubulin modulates the addition of tubulin subunits at the ends of a microtubule.

In a microtubule, lateral and longitudinal interactions between the tubulin subunits are responsible for maintaining the tubular form. Longitudinal contacts between the ends of adjacent subunits link the subunits head to tail into a linear *protofilament*. Within each protofilament, the dimeric subunits repeat every 8 nm. Through lateral interactions, protofilaments associate side by side into a sheet or cylinder—a microtubule. In most microtubules, the heterodimers in adjacent protofilaments are staggered only slightly, forming tilted rows of α- and β-tubulin monomers in the microtubule wall. The head-to-tail arrangement of the α- and β-tubulin dimers in a protofilament confers an overall polarity on a microtubule. Because all protofilaments in a microtubule have the same orientation, one end of a microtubule is ringed by α-tubulin, whereas the opposite end is ringed by β-tubulin (Figure 20-3b). As in actin microfilaments, the two ends of a microtubule, designated the (+) and (−) ends, differ in their rates of assembly and critical concentrations (C_c). The (+) end corresponds to the β-tubulin end of a microtubule.

Virtually every microtubule in a cell is a simple tube, a *singlet* microtubule, built from 13 protofilaments. In rare cases, singlet microtubules contain more or fewer protofilaments; for example, certain microtubules in the neurons of nematode worms contain 11 or 15 protofilaments. In addition to the simple singlet structure, *doublet* or *triplet* microtubules are found in specialized structures such as cilia and flagella (doublet microtubules) and centrioles and basal bodies (triplet microtubules). Each doublet or triplet contains one complete 13-protofilament microtubule (A tubule) and one or two additional tubules (B and C) consisting of 10 protofilaments (Figure 20-4).

Microtubule Assembly and Disassembly Take Place Preferentially at the (+) End

Microtubules assemble by the polymerization of dimeric αβ-tubulin. Assembly and stability of microtubules are temperature dependent. For instance, if microtubules are cooled

to 4 °C, they depolymerize into αβ-tubulin dimers (Figure 20-5). When warmed to 37 °C in the presence of GTP, the tubulin dimers polymerize into microtubules. Cycles of heating and cooling are key steps in purifying microtubules and their associated proteins from cell extracts.

Tubulin polymerization has several properties in common with the polymerization of actin to form microfilaments. First, at αβ-tubulin concentrations above the *critical concentration* (C_c), the dimers polymerize into microtubules, whereas at concentrations below the C_c, microtubules depolymerize, similar to the behavior of G-actin and F-actin (see Figure 19-7). Second, the nucleotide, either GTP or GDP, bound to the β-tubulin causes the critical concentration (C_c) for assembly at the (+) and (−) ends of a microtubule to differ. By analogy with F-actin assembly, the preferred assembly end is designated the (+) end. Third,

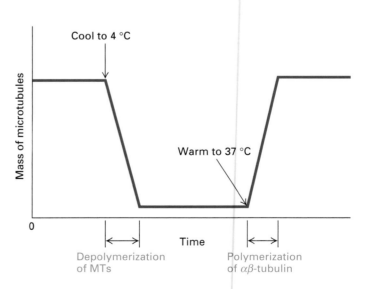

▲ **EXPERIMENTAL FIGURE 20-5 Temperature affects whether microtubules (MTs) assemble or disassemble.** At low temperatures, microtubules depolymerize, releasing αβ-tubulin, which repolymerizes at higher temperatures in the presence of GTP.

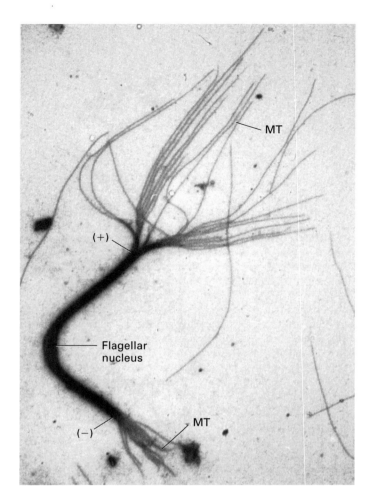

▲ **EXPERIMENTAL FIGURE 20-6** Addition of microtubule fragments demonstrates polarity of tubulin polymerization. Fragments of flagellar microtubules act as nuclei for the in vitro addition of αβ–tubulin. The nucleating flagellar fragment can be distinguished in the electron microscope from the newly formed microtubules (MT) seen radiating from the ends of the flagellar fragment. The greater length of the microtubules at one end indicates that tubulin subunits are added preferentially to this end. [Courtesy of G. Borisy.]

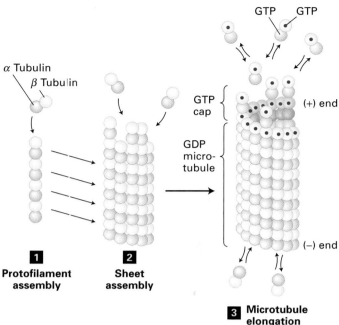

▲ **FIGURE 20-7 Stages in assembly of microtubules.** Free αβ–tubulin dimers associate longitudinally to form short protofilaments (**1**). These protofilaments are probably unstable and quickly associate laterally into more stable curved sheets (**2**). Eventually, a sheet wraps around into a microtubule with 13 protofilaments. The microtubule then grows by the addition of subunits to the ends of protofilaments composing the microtubule wall (**3**). The free tubulin dimers have GTP (red dot) bound to the exchangeable nucleotide-binding site on the β-tubulin monomer. After incorporation of a dimeric subunit into a microtubule, the GTP on the β-tubulin (but not on the α-tubulin) is hydrolyzed to GDP. If the rate of polymerization is faster than the rate of GTP hydrolysis, then a cap of GTP-bound subunits is generated at the (+) end, although the bulk of β-tubulin in a microtubule will contain GDP. The rate of polymerization is twice as fast at the (+) end as at the (−) end.

at αβ–tubulin concentrations higher than the C_c for polymerization, dimers add preferentially to the (+) end. Fourth, when the αβ–tubulin concentration is higher than the C_c at the (+) end but lower than the C_c at the (−) end, microtubules can *treadmill* by adding subunits to one end and dissociating subunits from the opposite end (see Figure 19-9). Because the intracellular concentration of assembly-competent tubulin (10–20 μM) is much higher than the critical concentration (C_c) for assembly (0.03 μM), polymerization is highly favored in a cell. Finally, the initial rate of tubulin polymerization is accelerated in the presence of nuclei—that is, microtubule-based structures or fragments (Figure 20-6).

Microtubule assembly comprises three steps: (1) protofilaments assemble from αβ–tubulin subunits, (2) protofilaments associate to form the wall of the microtubule, and (3) the addition of more subunits to the ends of the protofilaments elongates the microtubule (Figure 20-7). In the electron microscope, the ends of growing microtubules frequently appear uneven because some protofilaments elongate faster than other protofilaments. The appearance of microtubules undergoing shortening is quite different, suggesting that the mechanism of disassembly differs from that of assembly (Figure 20-8). Under shortening conditions, the microtubule ends are splayed, as if the lateral interactions between protofilaments have been broken. When frayed apart and freed from lateral stabilizing interactions, the protofilaments may depolymerize by endwise dissociation of tubulin subunits. The splayed appearance of a shortening microtubule provided clues about the potential instability of a microtubule.

(a) Assembly (elongation)

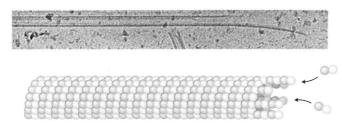

(b) Disassembly (shrinkage)

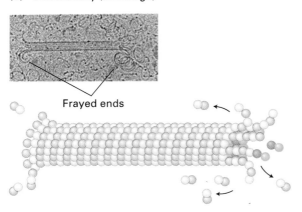

Frayed ends

▲ **EXPERIMENTAL FIGURE 20-8 Cryoelectron microscopy allows observation of disassembled microtubules.** Microtubules undergoing assembly (a) or disassembly (b) can be quickly frozen in liquid ethane and examined in the frozen state in a cryoelectron microscope. In assembly conditions, microtubule ends are relatively smooth; occasionally a short protofilament is seen to extend from one end. In disassembly conditions, the protofilaments splay at the microtubule ends, giving the ends a frayed appearance. Splaying of protofilaments probably promotes the loss of tubulin subunits from their ends, leading to shrinkage of the microtubule. [Micrographs courtesy of E. Mandelkow and E. M. Mandelkow.]

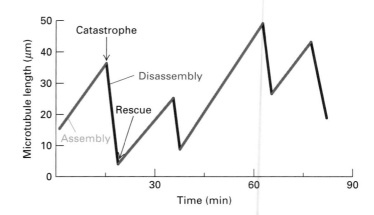

▲ **EXPERIMENTAL FIGURE 20-9 Rate of microtubule growth in vitro is much slower than shrinkage.** Individual microtubules can be observed in the light microscope, and their lengths can be plotted at different times during assembly and disassembly. Assembly and disassembly each proceed at uniform rates, but there is a large difference between the rate of assembly and that of disassembly, as seen in the different slopes of the lines. Shortening of a microtubule is much more rapid (7 μm/min) than growth (1 μm/min). Notice the abrupt transitions to the shrinkage stage (catastrophe) and to the elongation stage (rescue). [Adapted from P. M. Bayley, K. K. Sharma, and S. R. Martin, 1994, in *Microtubules*, Wiley-Liss, p. 118.]

Dynamic Instability Is an Intrinsic Property of Microtubules

Under appropriate in vitro conditions, some individual microtubules oscillate between growth and shortening phases (Figure 20-9). In all cases, the rate of microtubule growth is much slower than the rate of shortening. When first discovered, this behavior of microtubules, termed *dynamic instability*, was surprising to researchers because they expected that under any condition all the microtubules in a solution or the same cytosol would behave identically.

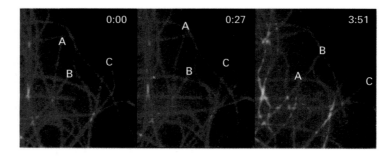

▲ **EXPERIMENTAL FIGURE 20-10 Fluorescence microscopy reveals in vivo growth and shrinkage of individual microtubules.** Fluorescently-labeled tubulin was microinjected into cultured human fibroblasts. The cells were chilled to depolymerize preexisting microtubules into tubulin dimers and were then incubated at 37 °C to allow repolymerization, thus incorporating the fluorescent tubulin into all the cell's microtubules. A region of the cell periphery was viewed in the fluorescence microscope at 0 second, 27 seconds later, and 3 minutes 51 seconds later (left to right panels). In this period, several microtubules elongate and shorten. The letters mark the position of ends of three microtubules. [From P. J. Sammak and G. Borisy, 1988, *Nature* **332**:724.]

The results of subsequent in vivo studies showed that individual cytosolic and mitotic microtubules display dynamic instability. In one set of experiments, fluorescent αβ–tubulin subunits were microinjected into live cultured cells. The cells were chilled to depolymerize preexisting microtubules into tubulin dimers and then incubated at 37 °C to allow repolymerization, thus incorporating the fluorescent tubulin into all the cellular microtubules. Video recordings of a small region in labeled cells showed that some microtubules became longer, others became shorter, and some appeared alternately to grow and to shrink over a period of several minutes (Figure 20-10). Because most microtubules in a cell associate by their (−) ends with MTOCs, their instability is largely limited to the (+) ends.

Two conditions influence the stability of microtubules. First, the oscillations between growth and shrinkage in vitro occur at tubulin concentrations near the C_c. As already stated, at tubulin concentrations above the C_c, the entire population of microtubules grows and, at concentrations below the C_c, all microtubules shrink. At concentrations near the C_c, however, some microtubules grow, whereas others shrink. The second condition affecting microtubule stability is whether GTP or GDP occupies the exchangeable nucleotide-binding site on β-tubulin at the (+) end of a microtubule (Figure 20-11). Because dissociation ("off" rate) of a GDP-tubulin dimer is four orders of magnitude as fast as that of a GTP-tubulin dimer, a microtubule is destabilized and depolymerizes rapidly if the (+) end becomes capped with subunits containing GDP–β-tubulin rather than GTP–β-tubulin. This situation can arise when a microtubule shrinks rapidly, exposing GDP–β-tubulin in the walls of the microtubule, or when a microtubule grows so slowly that the hydrolysis of β-tubulin–bound GTP converts it into GDP before additional subunits can be added to the (+) end of the microtubule. Before a shortening microtubule vanishes entirely, it can be "rescued" and start to grow if tubulin subunits with bound GTP add to the (+) end before the bound GTP hydrolyzes. Thus the parameters that determine the stability of a microtubule are the growth rate, the shrinkage rate, the catastrophe frequency, and the rescue frequency.

Numerous Proteins Regulate Microtubule Dynamics and Cross-Linkage to Other Structures

A large number of proteins influence the assembly and stability of microtubules and their association with other cell structures (Table 20-1). These proteins are collectively called **microtubule-associated proteins (MAPs)** because most copurify with microtubules isolated from cells. The results of immunofluorescence localization studies also have shown a parallel distribution of MAPs and microtubules in cells—strong evidence for their interaction in vivo.

MAPs are classified into two groups on the basis of their function. One group stabilizes microtubules. The stucture of a stabilizing MAP consists of two domains—a basic microtubule-binding domain and an acidic projection domain. In the electron microscope, the projection domain appears as a filamentous arm that extends from the wall of the microtubule. This arm can bind to membranes, intermediate

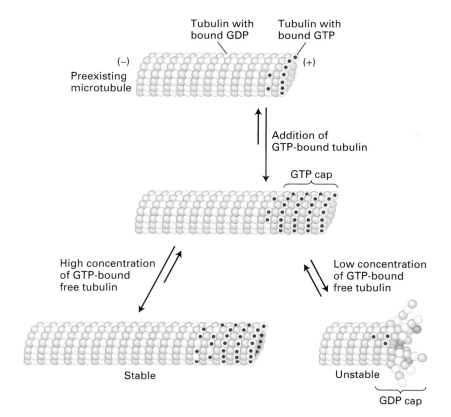

▲ FIGURE 20-11 Dynamic instability model of microtubule growth and shrinkage. GTP-bound αβ–tubulin subunits (red) add preferentially to the (+) end of a preexisting microtubule. After incorporation of a subunit, the GTP (red dot) bound to the β-tubulin monomer is hydrolyzed to GDP. Only microtubules whose (+) ends are associated with GTP-tubulin (those with a GTP cap) are stable and can serve as primers for the polymerization of additional tubulin. Microtubules with GDP-tubulin (blue) at the (+) end (those with a GDP cap) are rapidly depolymerized and may disappear within 1 minute. At high concentrations of unpolymerized GTP-tubulin, the rate of addition of tubulin is faster than the rate of hydrolysis of the GTP bound in the microtubule or the rate of dissociation of GTP-tubulin from microtubule ends; thus the microtubule grows. At low concentrations of unpolymerized GTP-tubulin, the rate of addition of tubulin is decreased; consequently, the rate of GTP hydrolysis exceeds the rate of addition of tubulin subunits and a GDP cap forms. Because the GDP cap is unstable, the microtubule end peels apart to release tubulin subunits. [See T. Mitchison and M. Kirschner, 1984, *Nature* **312**:237; M. Kirschner and T. Mitchison, 1986, *Cell* **45**:329; and R. A. Walker et al., 1988, *J. Cell Biol.* **107**:1437.]

Figure labels: Tubulin with bound GDP; Tubulin with bound GTP; (−) Preexisting microtubule; (+); Addition of GTP-bound tubulin; GTP cap; High concentration of GTP-bound free tubulin; Low concentration of GTP-bound free tubulin; Stable; Unstable; GDP cap

TABLE 20-1 Proteins That Modulate Microtubule (MT) Dynamics

Protein	MW	Location	Function
MICROTUBULE-STABILIZING PROTEINS			
MAP1	250,000–300,000 (heavy chain)	Dendrites and axons; non-neuronal cells	Assembles and stabilizes MTs
MAP2	42,000 and 200,000	Dendrites	Assembles and cross-links MTs to one another and to intermediate filaments
MAP4	210,000	Most cell types	Stabilizes MTs
Tau	55,000–62,000	Dendrites and axons	Assembles, stabilizes, and cross-links MTs
CLIP170	170,000	Most cell types	Cross-links MTs to endosomes and chromosomes
MICROTUBULE-DESTABILIZING PROTEINS			
Katanin	84,000	Most cell types	Microtubule severing
Op18 (stathmin)	18,000	Most cell types	Binds tubulin dimers

filaments, or other microtubules, and its length controls how far apart microtubules are spaced (Figure 20-12).

The microtubule-binding domain contains several repeats of a conserved, positively charged four-residue amino acid sequence that binds the negatively charged C-terminal part of tubulin. This binding is postulated to neutralize the charge repulsion between tubulin subunits within a microtubule, thereby stabilizing the polymer. MAP1A and MAP1B are large, filamentous molecules found in axons and dendrites of neurons as well as in non-neuronal cells. Each of these MAPs is derived from a single precursor polypeptide, which is proteolytically processed in a cell to generate one light chain and one heavy chain.

Other stabilizing MAPs include MAP2, MAP4, Tau, and CLIP170. MAP4, the most widespread of all the MAPs, is found in neuronal and non-neuronal cells. In mitosis, MAP4 regulates microtubule stability, and CLIP170 cross-links microtubules to chromosomes. MAP2 is found only in dendrites, where it forms fibrous cross-bridges between microtubules and links microtubules to intermediate filaments. Tau, which is much smaller than most other MAPs, is present in both axons and dendrites. This protein exists in several iso-

(a)

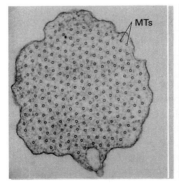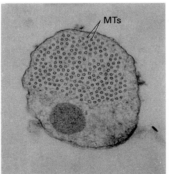

(b)

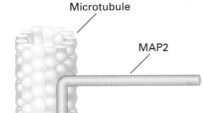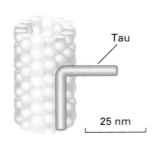

◀ **EXPERIMENTAL FIGURE 20-12 Spacing of microtubules depends on length of projection domain in bound microtubule-associated proteins.** Insect cells transfected with DNA expressing either long-armed MAP2 protein or short-armed Tau protein grow long axonlike processes. (a) Electron micrographs of cross sections through the processes induced by the expression of MAP2 (*left*) or Tau (*right*) in transfected cells. Note that the spacing between microtubules (MTs) in MAP2-containing cells is larger than in Tau-containing cells. Both cell types contain approximately the same number of microtubules, but the effect of MAP2 is to enlarge the caliber of the axonlike process. (b) Diagrams of association between microtubules and MAPs. Note the difference in the lengths of the projection arms in MAP2 and Tau. [Part (a) from J. Chen et al., 1992, *Nature* **360**:674.]

forms derived from alternative splicing of a *tau* mRNA. The ability of Tau to cross-link microtubules into thick bundles may contribute to the stability of axonal microtubules. Findings from gene transfection experiments implicate Tau in axonal elongation. Deletion of the genes encoding Tau and MAPIB leads to more severe phenotypes of axonal dysgenesis and lethality. Furthermore, aberrant polymerization of Tau into filaments is linked to neurodegenerative diseases such as human dementia in Alzheimer's patients.

When stabilizing MAPs coat the outer wall of a microtubule, tubulin subunits are unable to dissociate from the ends of that microtubule. Although bound MAPs generally dampen the rate of microtubule disassembly, the assembly of microtubules is affected to varying degrees: some MAPs, such as Tau and MAP4, stabilize microtubules, whereas other MAPs do not. Because of the effect of assembly MAPs on microtubule dynamics, modulating the binding of MAPs can control the length of microtubules. In most cases, this control is accomplished by the reversible phosphorylation of the MAP projection domain. Phosphorylated MAPs are unable to bind to microtubules; thus they promote microtubule disassembly. **MAP kinase,** a key enzyme for phosphorylating MAPs, is a participant in many signal-transduction pathways (Chapter 14), indicating that MAPs are targets of many extracellular signals. MAPs, especially MAP4, are also phosphorylated by a **cyclin-dependent kinase (CDK)** that plays a major role in controlling the activities of various proteins in the course of the cell cycle (Chapter 21).

A second group of MAPs directly destabilizes microtubules in many cell types. One of this group, called katanin, severs intact cytosolic microtubules by an ATP-dependent process. Internal bonds between tubulin subunits in the microtubule wall are broken, causing microtubules to fragment. This activity may release microtubules at the MTOC. Another protein, called Op18 or stathmin, increases the frequency of rapid disassembly of microtubules in the mitotic spindle. This protein may act by binding tubulin dimers, thereby reducing the pool of dimers available for polymerization. Phosphorylation inactivates Op18 and inhibits its destabilizing effect.

Colchicine and Other Drugs Disrupt Microtubule Dynamics

Some of the earliest studies of microtubules employed several drugs that inhibit mitosis, a cell process that depends on microtubule assembly and disassembly. Two such drugs isolated from plants, *colchicine* and *taxol,* have proved to be very powerful tools for probing microtubule function, partly because they bind only to αβ-tubulin or microtubules and not to other proteins and because their concentrations in cells can be easily controlled.

Colchicine and a synthetic relative, colcemid, have long been used as mitotic inhibitors. In cells exposed to high concentrations of colcemid, cytosolic microtubules depolymerize, leaving an MTOC. However, when plant or animal cells are exposed to low concentrations of colcemid, the microtubules remain and the cells become "blocked" at **meta-**phase, the mitotic stage at which the duplicated chromosomes are fully condensed (see Figure 9-3). When the treated cells are washed with a colcemid-free solution, colcemid diffuses from the cell and mitosis resumes normally. Thus experimenters commonly use colcemid to accumulate metaphase cells for cytogenetic studies; removal of the colcemid leaves a population of cells whose cell cycle is in synchrony. Such synchronous populations are advantageous for studies of the cell cycle (Chapter 21).

The interface between α-tubulin and β-tubulin monomers in dimeric tubulin contains a high-affinity but reversible binding site for colchicine. Colchicine-bearing tubulin dimers, at concentrations much less than the concentration of free tubulin subunits, can add to the end of a growing microtubule. However, the presence of one or two colchicine-bearing tubulins at the end of a microtubule prevents the subsequent addition or loss of other tubulin subunits. Thus colchicine "poisons" the end of a microtubule and alters the steady-state balance between assembly and disassembly. As a result of this disruption of microtubule dynamics, mitosis is inhibited in cells treated with low concentrations of colchicine.

Other drugs bind to different sites on tubulin dimers or to microtubules and therefore affect microtubule stability through different mechanisms. For example, at low concentrations, taxol binds to microtubules and stabilizes them by inhibiting their shortening.

 Drugs that disturb the assembly and disassembly of microtubules have been widely used to treat various diseases. Indeed, more than 2500 years ago, the ancient Egyptians treated heart problems with colchicine. Nowadays, this drug is used primarily in the treatment of gout and certain other diseases affecting the joints and skin. Other inhibitors of microtubule dynamics, including taxol, are effective anticancer agents and are used in the treatment of ovarian cancer. ∎

MTOCs Orient Most Microtubules and Determine Cell Polarity

In an interphase fibroblast cell, cytosolic microtubules are arranged in a distinctive hub-and-spoke array that lies at the center of a cell (Figure 20-13a). The microtubule spokes radiate from a central site occupied by the **centrosome,** which is the primary microtubule-organizing center in many interphase cells. We will use the term *MTOC* to refer to any of the structures used by cells to nucleate and organize microtubules. In animal cells, the MTOC is usually a centrosome, a collection of microtubule-associated proteins that sometimes but not always contains a pair of **centrioles** (Figure 20-13b). The centrioles, each a pinwheel array of triplet microtubules, lie in the center of the MTOC but do not make direct contact with the (−) ends of the cytosolic microtubules. Centrioles are not present in the MTOCs of plants and fungi; moreover, some epithelial cells and newly fertilized eggs from animals also lack centrioles. Thus, it is the associated

(a)

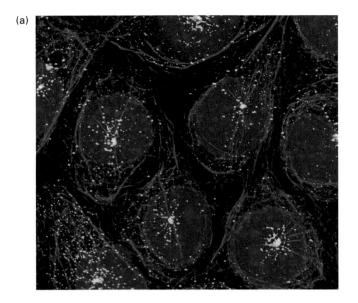

(b)

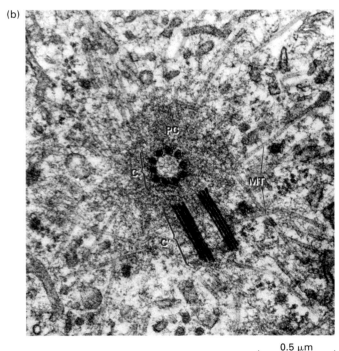

0.5 μm

▲ **EXPERIMENTAL FIGURE 20-13 The centrosome, which functions as a microtubule-organizing center, contains a pair of orthogonal centrioles in most animal cells.** (a) Micrograph showing several cells each with an MTOC identified by fluorescently labeled antibodies against PCM1, a centrosomal protein. (b) Electron micrograph of the MTOC in an animal cell. The pair of centrioles (red), C and C′, in the center are oriented at right angles; thus one centriole is seen in cross section, and the other longitudinally. Surrounding the centrioles is a cloud of material, the pericentriolar (PC) matrix, which contains γ-tubulin and pericentrin. Embedded within the MTOC, but not contacting the centrioles, are the (−) ends of microtubules (MT; yellow). [Part (a) from A. Kubo and S. Tsukita, 2003, *J. Cell Sci.* **116**:919. Part (b) from B. R. Brinkley, 1987, in *Encyclopedia of Neuroscience*, vol. 2, Birkhauser Press, p. 665; courtesy of B. R. Brinkley.]

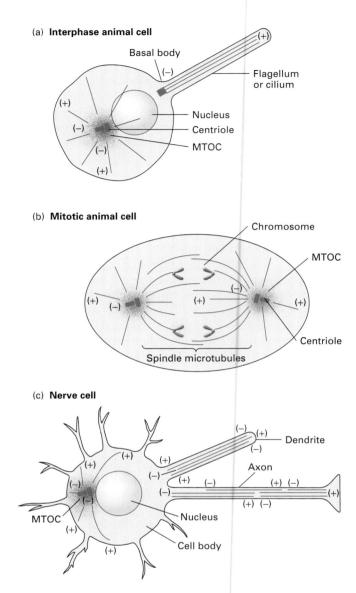

▲ **FIGURE 20-14 Orientation of cellular microtubules.** (a) In interphase animal cells, the (−) ends of most microtubules are proximal to the MTOC. Similarly, the microtubules in flagella and cilia have their (−) ends continuous with the basal body, which acts as the MTOC for these structures. (b) As cells enter mitosis, the microtubule network rearranges, forming a mitotic spindle. The (−) ends of all spindle microtubules point toward one of the two MTOCs, or poles, as they are called in mitotic cells. (c) In nerve cells, the (−) ends of all axonal microtubules are oriented toward the base of the axon, but dendritic microtubules have mixed polarities.

proteins in an MTOC that have the capacity to organize cytosolic microtubules.

Because microtubules assemble from the MTOC, microtubule polarity becomes fixed in a characteristic orientation (Figure 20-14). In most interphase animal cells, for instance, the (−) ends of microtubules are closest to the MTOC. In mitosis, the centrosome duplicates and migrates to new positions flanking the nucleus. There the centrosome becomes the organizing center for microtubules forming the mitotic apparatus, which will separate the chromosomes into the daughter cells in mitosis. The microtubules in the axon of a nerve cell, which help stabilize the long process, are all oriented in the same direction.

In contrast with the single perinuclear MTOC present in most interphase animal cells, plant cells, polarized epithelial cells, and embryonic cells contain hundreds of MTOCs, which are distributed throughout the cell, often near the cell cortex. In plant cells and polarized epithelial cells, a cortical array of microtubules aligns with the cell axis. In both cell types, the polarity of the cell is linked to the orientation of the microtubules.

The γ-Tubulin Ring Complex Nucleates Polymerization of Tubulin Subunits

The MTOC organizes cytosolic microtubules by first nucleating microtubule assembly and then anchoring and releasing microtubules. Despite its amorphous appearance, the pericentriolar material of an MTOC is an ordered lattice that contains many proteins that are necessary for initiating the assembly of microtubules (see Figure 20-13). One of these proteins, γ-tubulin, was first identified in genetic studies designed to discover proteins that interact with β-tubulin. The results of subsequent studies demonstrated that γ-tubulin and the lattice protein pericentrin are part of the pericentriolar material of centrosomes; these proteins have also been detected in MTOCs that lack a centriole. The finding that the introduction of antibodies against γ-tubulin into cells blocks microtubule assembly implicates γ-tubulin as a necessary factor in nucleating the polymerization of tubulin subunits.

Approximately 80 percent of the γ-tubulin in cells is part of a 25S complex, which has been isolated from extracts of frog oocytes and fly embryos. Named the *γ-tubulin ring complex (γ-TuRC)* for its ringlike appearance in the electron microscope, the complex comprises eight polypeptides and measures 25 nm in diameter. Findings from in vitro experiments show that the γ-TuRC can directly nucleate microtubule assembly at subcritical tubulin concentrations—that is, at concentrations below which polymerization would not take place in the absence of the γ-TuRC. To investigate how γ-TuRC associates with microtubules, scientists performed immunolabeling experiments with the use of gold-conjugated antibodies specific for γ-TuRC components, either γ-tubulin or XGRIP. The results of these studies reveal that complexes are localized to one end of a microtubule and are not present along the sides (Figure 20-15a). This location is consistent with a role for

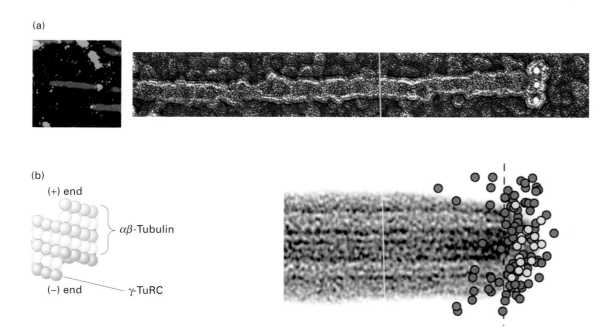

▲ **EXPERIMENTAL FIGURE 20-15 The γ-tubulin ring complex (γ-TuRC) is localized to one end of the microtubule.** (a) A fluorescence micrograph (*left*) and an electron micrograph (*right*) of microtubules stained with gold-labeled antibodies to γ-tubulin or XGRIP, a microtubule-binding protein. Both proteins are components of the γ-TuRC. The labeled proteins are localized to one end of the microtubules. (b) A model of the γ-TuRC. This complex is thought to nucleate microtubule assembly by presenting a row of γ-tubulin subunits, which can directly bind αβ-tubulin subunits. This model is supported by plotting the positions of gold-labeled antibodies to either γ-tubulin (red) or XGRIP109 (yellow) from several experiments onto a microtubule end. [Parts (a) and (b, right) from T. J. Keating and G. G. Borisy, 2000, *Nature Cell Biol.* **2**:352; courtesy of T. J. Keating and G. G. Borisy. Part (b, left) modified from C. Wiese and Y. Zheng, 1999, *Curr. Opin. Struc. Biol.* **9**:250.]

γ-TuRC in nucleating microtubule assembly. A model of γ-TuRC based on electron microscopy shows γ-tubulin in contact with the (−) end of a microtubule (Figure 20-15b).

Cytoplasmic Organelles and Vesicles Are Organized by Microtubules

Fluorescence microscopy reveals that membrane-limited organelles such as the endoplasmic reticulum (ER), Golgi, endosomes, and mitochondria are associated with microtubules. For instance, in cultured fibroblasts stained with anti-tubulin antibodies and $DiOC_6$, a fluorescent dye specific for the ER, the anastomosing ER network in the cytosol is seen to colocalize with microtubules (Figure 20-16). If cells are treated with a microtubule-depolymerizing drug, the ER

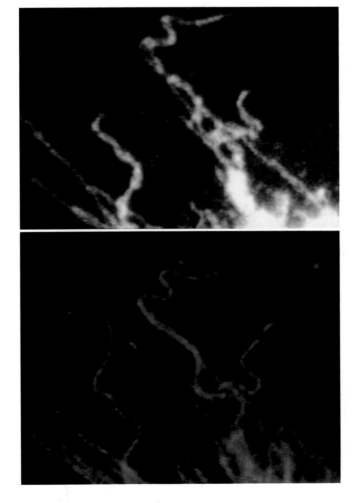

▲ **EXPERIMENTAL FIGURE 20-16 Fluorescence microscopy reveals colocalization of endoplasmic reticulum membranes and cytosolic microtubules.** $DiOC_6$, an ER-binding fluorescent dye (green), and fluorescently-labeled anti-tubulin antibodies (red) were used to stain a cultured frog fibroblast. The alignment of the ER network and microtubules in many but not all regions of the cytoplasm is evident because the cell has sparse microtubules. [Courtesy of M. Terasaki.]

loses its networklike organization. After the drug has been washed from the cells, tubular fingers of ER grow as new microtubules assemble. In cell-free systems, the ER can be reconstituted with microtubules and an ER-rich cell extract. Even under this cell-free regime, ER membranes elongate along microtubules. This close association between the ER and intact microtubules suggests that proteins bind ER membranes to microtubules.

The role of microtubules in organizing the Golgi complex also has been studied extensively. In interphase fibroblasts, the Golgi complex is concentrated near the MTOC. In mitosis (or after the depolymerization of microtubules by colcemid), the Golgi complex breaks into small vesicles that are dispersed throughout the cytosol. When the cytosolic microtubules re-form during interphase (or after removal of the colcemid), the Golgi vesicles move along these microtubule tracks toward the MTOC, where they reaggregate to form large membrane complexes.

These observations were among the first to suggest that microtubules play a role in the intracellular transport of membrane-limited organelles and vesicles. Other examples of such transport and the motor proteins that power them are described in Section 20.2.

KEY CONCEPTS OF SECTION 20.1

Microtubule Organization and Dynamics

■ Tubulins belong to an ancient family of GTPases that polymerize to form microtubules, hollow cylindrical structures 25 nm in diameter.

■ Microtubules, like actin microfilaments, exhibit both structural and functional polarity.

■ Dimeric αβ–tubulin subunits interact end-to-end to form protofilaments, which associate laterally into microtubules (see Figure 20-7).

■ Microtubules exhibit structural polarity. Subunits are added and lost preferentially at one end, the (+) end.

■ Assembly and disassembly of microtubules depends on the critical concentration, C_c, of αβ–tubulin subunits. Above the C_c, microtubules assemble; below the C_c, microtubules disassemble.

■ Microtubules exhibit two dynamic phenomena that are pronounced at tubulin concentrations near the C_c: (1) treadmilling, the addition of subunits at one end and their loss at the other end, and (2) dynamic instability, the oscillation between lengthening and shortening (see Figure 20-9).

■ The balance between growth and shrinkage of unstable microtubules depends on whether the exchangeable GTP bound to β-tubulin is present on the (+) end or whether it has been hydrolyzed to GDP (see Figure 20-11).

■ Microtubule-associated proteins (MAPs) organize microtubules and affect their stability. Some MAPs prevent or promote cytosolic microtubule depolymerization; other MAPs organize microtubules into bundles or cross-link them to membranes and intermediate filaments or both (see Table 20-1).

■ Various drugs, including colchicine and taxol, disrupt microtubule dynamics and have an antimitotic effect. Some of these drugs are useful in the treatment of certain cancers.

■ Cell polarity including the organization of cell organelles, direction of membrane trafficking, and orientation of microtubules is determined by microtubule-organizing centers (MTOCs). Most interphase animal cells contain a single, perinuclear MTOC from which cytosolic microtubules radiate (see Figure 20-13).

■ Because microtubule assembly is nucleated from MTOCs, the (−) end of most microtubules is adjacent to the MTOC and the (+) end is distal (see Figure 20-14).

■ A γ-tubulin–containing complex is a major component of the pericentriolar material and is able to nucleate the polymerization of tubulin subunits to form microtubules in vitro.

20.2 Kinesin- and Dynein-Powered Movements

Within cells, proteins, organelles, and other membrane-limited vesicles, organelles, and proteins are frequently transported distances of many micrometers along well-defined routes in the cytosol and delivered to particular addresses. Diffusion alone cannot account for the rate, directionality, and destinations of such transport processes. Findings from early experiments with fish-scale pigment cells and nerve cells first demonstrated that microtubules function as tracks in the intracellular transport of various types of "cargo." Eventually, two families of motor proteins—**kinesins** and **dyneins**—were found to mediate transport along microtubules.

A second type of movement that depends on microtubule motor proteins is the beating of cilia and flagella. Huge numbers of cilia (more than $10^7/mm^2$) cover the surfaces of mammalian respiratory passages where their beating dislodges and expels particulate matter that collects in the mucus secretions of these tissues. In the oviduct, cilia help transport eggs down the fallopian tube. In contrast, sperm cells and many unicel-

lular organisms have a single flagellum, which propels the cells forward at velocities approaching 1 mm/s.

In this section, we first consider the transport of materials in axons. Studies of such *axonal transport*, a process first discovered more than 50 years ago, have contributed greatly to our understanding of microtubule-associated intracellular transport. We then consider the structure and function of the microtubule motor proteins. A description of the unique microtubule-based structures and motor proteins responsible for the movement of cilia and flagella concludes this section.

Axonal Transport Along Microtubules Is in Both Directions

A neuron must constantly supply new materials—proteins and membranes—to an axon terminal to replenish those lost in the exocytosis of neurotransmitters at the junction (synapse) with another cell. Because proteins and membranes are synthesized only in the cell body, these materials must be transported down the axon, which can be as much as a meter in length, to the synaptic region. This movement of materials is accomplished on microtubules, which are all oriented with their (+) ends toward the terminal (see Figure 20-14c).

The results of classic pulse-chase experiments in which radioactive precursors are microinjected into the dorsal-root ganglia near the spinal cord and then tracked along their nerve axons showed that axonal transport is in both directions. *Anterograde* transport proceeds from the cell body to

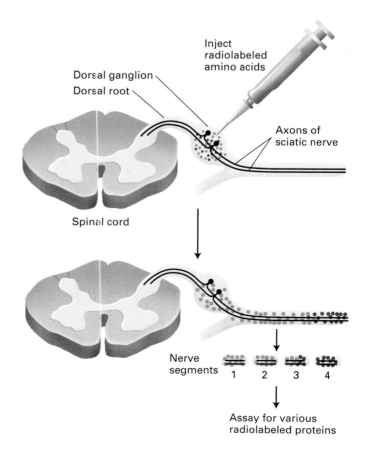

▶ **EXPERIMENTAL FIGURE 20-17 The rate of axonal transport in vivo can be determined by radiolabeling and gel electrophoresis.** The cell bodies of neurons in the sciatic nerve are located in dorsal-root ganglia. Radioactive amino acids injected into these ganglia in experimental animals are incorporated into newly synthesized proteins, which are then transported down the axon to the synapse. Animals are sacrificed at various times after injection and the dissected sciatic nerve is cut into small segments for analysis with the use of gel electrophoresis. The red, blue, and purple dots represent groups of proteins that are transported down the axon at different rates, red most rapidly, purple least rapidly.

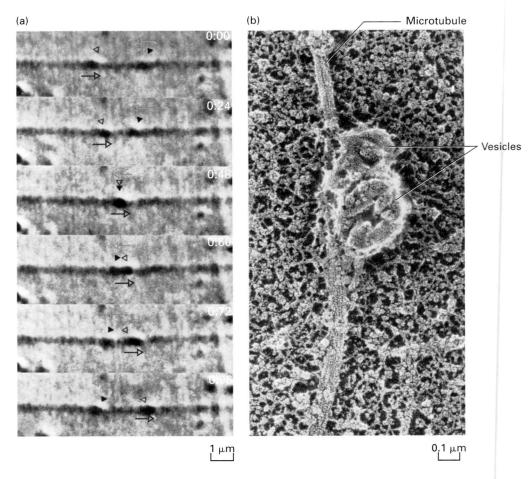

(a)   0:00   0:24   0:48   0:60   0:72   0:96

1 μm

(b)   Microtubule   Vesicles

0.1 μm

▲ **EXPERIMENTAL FIGURE 20-18 DIC microscopy demonstrates microtubule-based vesicle transport in vitro.** (a) The cytoplasm was squeezed from a squid giant axon with a roller onto a glass coverslip. After buffer containing ATP was added to the preparation, it was viewed in a differential interference contrast microscope, and the images were recorded on videotape. In the sequential images shown, the two organelles indicated by open and solid triangles move in opposite directions (indicated by colored arrows) along the same filament, pass each other, and continue in their original directions. Elapsed time in seconds appears at the upper-right corner of each video frame. (b) A region of cytoplasm similar to that shown in part (a) was freeze dried, rotary shadowed with platinum, and viewed in the electron microscope. Two large structures attached to one microtubule are visible; these structures presumably are small vesicles that were moving along the microtubule when the preparation was frozen. [See B. J. Schnapp et al., 1985, *Cell* **40**:455; courtesy of B. J. Schnapp, R. D. Vale, M. P. Sheetz, and T. S. Reese.]

the synaptic terminals and is associated with axonal growth and the delivery of synaptic vesicles. In the opposite, *retrograde,* direction, "old" membranes from the synaptic terminals move along the axon rapidly toward the cell body where they will be degraded in lysosomes. Findings from such experiments also revealed that different materials move at different speeds (Figure 20-17). The fastest-moving material, consisting of membrane-limited vesicles, has a velocity of about 250 mm/day, or about 3 μm/s. The slowest-moving material, comprising tubulin subunits and neurofilaments, moves only a fraction of a millimeter per day. Organelles such as mitochondria move down the axon at an intermediate rate.

Axonal transport can be directly observed by video microscopy of cytoplasm extruded from a squid giant axon. The movement of vesicles along microtubules in this cell-free system requires ATP, its rate is similar to that of fast axonal transport in intact cells, and it can proceed in both the anterograde and the retrograde directions (Figure 20-18a). Electron microscopy of the same region of the axon cytoplasm reveals vesicles attached to individual microtubules (Figure 20-18b). These pioneering in vitro experiments established definitely that organelles move along individual microtubules and that their movement requires ATP. As discussed shortly, these two observations led to the identification of microtubule motor proteins, which generate the movements.

(a)

Gap

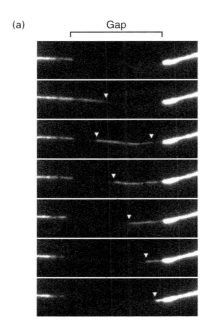

(b)

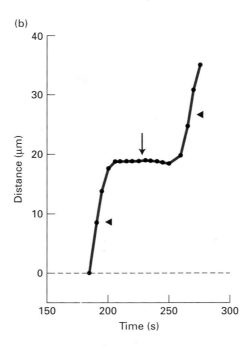

▲ **EXPERIMENTAL FIGURE 20-19** Transport of GFP-tagged neurofilaments down axons exhibits periodic pauses. (a) A segment of an axon is imaged after GFP-labeled neurofilament protein, NF-M, is expressed in a cultured neuronal cell. Bundles of labeled neurofilaments are separated by gaps within an axon. A GFP-labeled neurofilament (arrowhead) is seen to traverse a 15-μm gap between two labeled bundles. In this time series, each frame is taken at 5-second intervals. (b) A plot of the time-dependent distance traveled shows pauses (arrow) in neurofilament transport. Although the peak velocity (arrowheads) is similar to fast axonal transport, the average velocity is much lower. [From L. Wang et al., 2000, *Nature Cell Biol.* **2**:137; courtesy of A. Brown.]

Findings from recent experiments in which neurofilaments tagged with green fluorescent protein (GFP) were injected into cultured cells suggest that neurofilaments pause frequently as they move down an axon (Figure 20-19). Although the peak velocity of neurofilaments is similar to that of fast-moving vesicles, their numerous pauses lower the average rate of transport. These findings suggest that there is no fundamental difference between fast and slow axonal transport, although why neurofilament transport stops periodically is unknown.

Kinesin I Powers Anterograde Transport of Vesicles in Axons

The first microtubule motor protein was identified by using a simple system consisting of microtubules assembled in vitro from purified tubulin subunits and stabilized by the drug taxol. When synaptic vesicles and ATP were added to these microtubules, the vesicles neither bound to the microtubules nor moved along them. However, the addition of a cytosolic extract of squid giant axon (free of tubulin) caused the vesicles to bind to the microtubules and to move along them, indicating that a soluble protein in the axonal cytosol is required for translocation.

When researchers incubated vesicles, axonal cytosol, and microtubules in the presence of AMPPNP, a nonhydrolyzable analog of ATP, the vesicles bound tightly to the microtubules but did not move. However, the vesicles did move when ATP was added. These results suggested that a motor protein in the cytosol binds to microtubules in the presence of ATP or AMPPNP, but movement requires ATP hydrolysis. To purify the soluble motor protein, scientists incubated a mixture of microtubules, cell or tissue extract, and AMPPNP, with the rationale that AMPPNP would promote tight binding between the microtubules and motor proteins in the extract. After incubation, the microtubules with any bound proteins were collected by centrifugation. Treatment of the microtubule-rich material in the pellet with ATP released one predominant protein back into solution; this protein is now known as kinesin I.

Kinesin I isolated from squid giant axons is a dimer of two heavy chains, each complexed to a light chain, with a total molecular weight of 380,000. The molecule comprises a pair of large globular *head domains* connected by a long *central stalk* to a pair of small globular *tail domains*, which contain the light chains (Figure 20-20). Each domain carries out a particular function: the head domain, which binds microtubules and ATP, is responsible for the motor activity of kinesin, and the tail domain is responsible for binding to the membrane of vesicles, most likely through the kinesin light chain.

Kinesin-dependent movement of vesicles can be tracked by in vitro motility assays similar to those used to study myosin-dependent movements. In one type of assay, a vesicle

▶ **FIGURE 20-20 Structure of kinesin.** (a) Schematic model of kinesin showing the arrangement of the two heavy chains (each with a molecular weight of 110,000–135,000) and the two light chains (60,000–70,000 MW). (b) Three-dimensional structure of the kinesin dimer based on x-ray crystallography. Each head is attached to an α-helical neck region, which forms a coiled-coil dimer. Microtubules bind to the helix indicated; this interaction is regulated by the nucleotide (orange) bound at the opposite side of the domain. The distance between microtubule binding sites is 5.5 nm. [Part (b) courtesy of E. Mandelkow and E. M. Mandelkow, adapted from M. Thormahlen et al., 1998, *J. Struc. Biol.* **122**:30.]

(a)

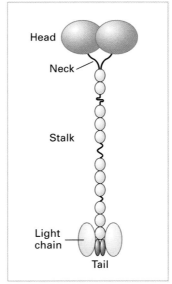

(b)

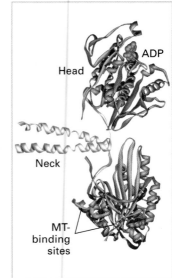

or a plastic bead coated with kinesin is added to a glass slide along with a preparation of microtubules. In the presence of ATP, the beads can be observed microscopically to move along a microtubule in one direction. By determining the polarity of the microtubules, researchers found that the beads coated with kinesin I always moved from the (−) to the (+) end of a microtubule (Figure 20-21). Thus kinesin I is a (+) end–directed microtubule motor protein. Because this direction corresponds to anterograde transport, kinesin I is implicated as a motor protein that mediates anterograde axonal transport.

Most Kinesins Are Processive (+) End–Directed Motor Proteins

To date, approximately 10 different kinesin subfamilies have been identified. All contain a globular head (motor) domain, but they differ in their tail domains and several other properties. In most kinesins, the motor domain is at the N-terminus (N-type) of the heavy chain but, in others, the motor domain is centrally located (M-type) or at the C-terminus (C-type). Both N- and M-type kinesins are *(+) end–directed motors*, whereas C-type kinesins are *(−) end–directed motors*. Although most kinesins have two heavy chains (e.g., kinesin I), others have a single heavy chain (e.g.,

KIF1) or four heavy chains (e.g., BimC). Tetrameric BimC has an unusual bipolar arrangement in which pairs of motor domains lie at opposite ends of a central rod segment.

Kinesins can be divided into two broad functional groups— *cytosolic* and *mitotic* kinesins—on the basis of the nature of the cargo that they transport (Table 20-2). The functional differences between kinesins are related to their unique tail domains, which determine their cargoes. Cytosolic kinesins take part in vesicle and organelle transport; they include the classic axonal kinesin I, which has been shown to transport lysosomes and other organelles. Some cytosolic kinesins, however, transport one specific cargo. For example, KIF1B and its close relative KIF1A transport mitochondria and synaptic vesicles, respectively, to nerve terminals. Other cytosolic kinesins mediate the transport of secretory vesicles to the plasma membrane and the radial movement of ER

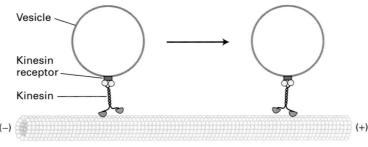

▲ **FIGURE 20-21 Model of kinesin-catalyzed vesicle transport.** Kinesin molecules, attached to unidentified receptors on the vesicle surface, transport the vesicles from the (−) end to the (+) end of a stationary microtubule. ATP is required for movement. [Adapted from R. D. Vale et al., 1985, *Cell* **40**:559; and T. Schroer et al., 1988, *J. Cell Biol.* **107**:1785.]

TABLE 20-2	Functional Classes of Microtubule Motor Proteins		
Class	**Common Members**	**Cargo**	**Direction of Movement***
Cytosolic motors	Kinesins (I, KIFIA, KIFIB)	Cytosolic vesicles/organelles	(+)
	Cytosolic dynein	Cytosolic vesicles/organelles	(−)
	Kinesin II	Cytosolic vesicles/organelles	(+)
Mitotic motors	Kinesin BimC (bipolar)	Spindle and astral MTs	(+)
	Chromokinesins	Chromosomes (arms)	(+)
	MCAK	Kinetochores	(+)
	CENP-E	Kinetochores	(+)
	Kinesin Ncd	Spindle and astral MTs	(−)
	Cytosolic dynein	Kinetochores, centrosomes, cell cortex near spindle poles	(−)
Axonemal motors	Outer-arm and inner-arm dyneins†	Doublet microtubules in cilia and flagella	(−)

*Movement of motor protein toward the (+) end or (−) end of microtubules.
† Outer-arm dyneins have three heavy chains, and inner-arm dyneins have two heavy chains.

membranes and pigment granules. Mitotic kinesins, in contrast, participate in spindle assembly and chromosome segregation in cell division. This group comprises numerous proteins, including the kinetochore-associated protein CENP-E, the bipolar BimC, and a (−) end–directed motor protein called Ncd. The functions of mitotic kinesins are described in more detail in Section 20.3.

A sequence called the *tetratrico peptide sequence* has been recently identified in the light chains of kinesin I and may interact with receptor proteins in the membrane of various cargoes. Such interactions would tether the cargo organelle or vesicle to kinesin. For instance, the tetratrico peptide sequence has been found to bind to several proteins, including the amyloid precursor protein. Other kinesins may have different interaction sequences that bind to other receptors on membranes.

Two fundamental properties of the kinesin motor—its step size and force—have been determined in optical-trap experiments similar to those performed on myosin (see Figure 19-18). Findings from these studies show that a dimeric kinesin molecule (e.g., kinesin I) exerts a force of 6 piconewtons, which is sufficient to pull a bound vesicle through the viscous cytoplasm. The kinesin step size of 8 nm matches the distance between successive α- or β-tubulin monomers in a protofilament, suggesting that kinesin binds to only one or the other monomer. Electron microscopy reconstructions show that kinesin binds primarily to β-tubulin. In other experiments, researchers have established that a double-headed kinesin molecule moves along a single protofilament, with one head always bound to the microtubule. As a result, a kinesin molecule can move along a mi-

crotubule for a long distance without detaching from it, a property referred to as *processivity*. Because of their high processivity, dimeric kinesins are very efficient in transporting cargo from one part of a cell to another.

In Chapter 3, we saw that the neck region of myosin, which acts as a rigid lever arm, is critical in coupling ATP hydrolysis to the movement of myosin along an actin microfilament. In contrast with myosin, kinesin has a flexible neck domain, which links the head domain to the central stalk domain (see Figure 20-20). Current models propose that ATP hydrolysis by kinesin causes movement of the flexible neck, which then positions the head domain into the next step along a microtubule protofilament. According to this model, the direction of kinesin movement depends on neck function, not on the motor domain. This function of the neck is supported by findings from recent domain-replacement experiments. For example, replacing the motor domain in (−) end–directed Ncd with the motor domain from (+) end–directed kinesin I yielded a (−) end–directed chimeric protein. Likewise, swapping the kinesin and Vcd motor domains into kinesin I produced a (+) end–directed protein. These results show that the direction of movement is not an intrinsic property of the motor domain. However, mutations in the neck region of Ncd converted it from a (−) into a (+) end–directed motor protein.

Cytosolic Dyneins Are (−) End–Directed Motor Proteins That Bind Cargo Through Dynactin

The second family of microtubule motor proteins, the dyneins, is responsible for retrograde axonal transport,

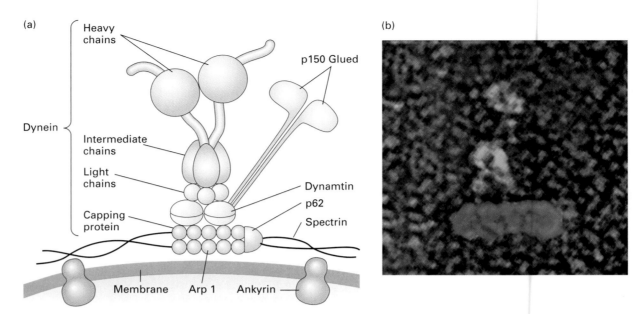

▲ FIGURE 20-22 Cytosolic dynein and the dynactin heterocomplex. (a) Diagram of dynein (green) bound to the dynactin complex (orange) through interactions between the dynein light chains and the dynamtin subunits of dynactin. The Arp1 subunits of dynactin form a minifilament that associates with spectrin underlying the cell membrane. The Glued subunits bind microtubules and vesicles. (b) Electron micrograph of a metal replica of the dynactin complex isolated from brain cells. The Arp1 minifilament (purple) and the dynamtin/Glued side arm (blue) are visible. [Part (a) adapted from N. Hirokawa, 1998, *Science* **279**:518. Part (b) from D. M. Eckley et al., 1999, *J. Cell Biol.* **147**:307.]

transit of Golgi vesicles to the centrosome, and some other (−) end–directed movements. Dyneins are exceptionally large, multimeric proteins, with molecular weights exceeding 1×10^6. They are composed of two or three heavy chains complexed with a poorly determined number of intermediate and light chains. As summarized in Table 20-2, the dyneins are divided into two functional classes. Here we consider *cytosolic* dynein, which has a role in the movement of vesicles and chromosomes. *Axonemal* dyneins, responsible for the beating of cilia and flagella, are considered later.

Like kinesin I, cytosolic dynein is a two-headed molecule, with two identical or nearly identical heavy chains forming the head domains. However, unlike kinesin, dynein cannot mediate cargo transport by itself. Rather, dynein-related transport requires *dynactin,* a large protein complex that links vesicles and chromosomes to the dynein light chains (Figure 20-22). The results of in vitro binding experiments show that dynactin also binds to microtubules, thereby enhancing the processivity of dynein-dependent movement. Dynactin consists of at least eight subunits, including a protein called Glued, which binds microtubules; Arp1, an actin-related protein that binds spectrin; and dynamtin, which interacts with the light chains of dynein. The microtubule-binding site in Glued contains a 57-residue motif that is also present in CLIP170, a microtubule-associated protein that cross-links microtubules and endocytic vesicles (see Table 20-1). One model proposes that dynein generates the force

for vesicle movement but remains tethered to a microtubule through dynactin.

As discussed later, several lines of evidence suggest that the dynein-dynactin complex and another complex, the nuclear/mitotic apparatus (NuMA) protein, mediate the association of microtubules with centrosomes in mitosis. The results of in vitro studies show that truncated NuMA protein binds microtubules if the C-terminal region is retained. As in MAPs, the C-terminal region of NuMA protein is highly acidic, and ionic interactions may mediate its binding to microtubules.

Multiple Motor Proteins Sometimes Move the Same Cargo

Figure 20-23 summarizes the role of kinesins and cytosolic dyneins in intracellular transport along microtubules. Because the orientation of microtubules is fixed by the MTOC, the direction of transport—toward or away from the cell periphery—depends on the motor protein. Some cargoes, such as pigment granules, can alternate their direction of movement along a single microtubule. In this case, both anterograde and retrograde microtubule motor proteins must associate with the same cargo. Recent biochemical experiments have identified dynactin in a complex with kinesin. A model proposes that dynactin is part of the membrane receptor and serves as a common adapter for binding kinesin and cytoplasmic dynein. Thus the direction of movement can be switched by swapping one motor protein for the other.

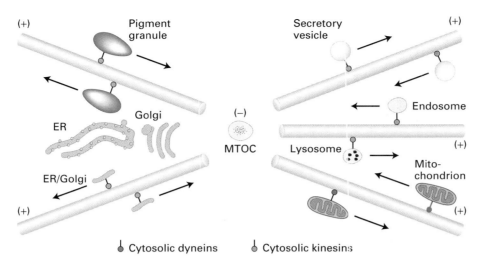

MEDIA CONNECTIONS

Video: Cytoplasmic Dynein Dynamics in Living
Dictyostelium Cells

▲ **FIGURE 20-23 General model of kinesin- and dynein-mediated transport in a typical cell.** The array of microtubules, with their (+) ends pointing toward the cell periphery, radiates from an MTOC in the Golgi region. Kinesin-dependent anterograde transport (red) conveys mitochondria, lysosomes, and an assortment of vesicles to the endoplasmic reticulum (ER) or cell periphery. Cytosolic dynein–dependent retrograde transport (green) conveys mitochondria, elements of the ER, and late endosomes to the cell center. [Adapted from N. Hirokawa, 1998, *Science* **279**:518.]

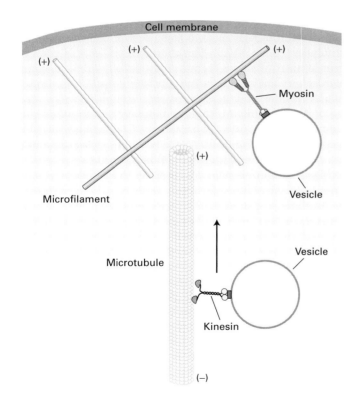

▲ **FIGURE 20-24 Cooperation of myosin and kinesin at the cell cortex.** Microtubules approach the actin-rich cell membrane. Consequently, some cargoes are transported to the cell periphery by kinesin motor proteins on microtubules but complete the journey on microfilaments under the power of myosin motor proteins.

In some cases, a vesicle must traverse microtubule-poor but microfilament-rich regions in the cell. For example, during endocytosis, vesicles from the actin-rich plasma membrane are carried inward, whereas during secretion, vesicles derived from the endoplasmic reticulum and Golgi are moved outward. The results of several complementary experiments imply that microtubule and microfilament motor proteins bind to the same vesicles and cooperate in their transport. One piece of evidence was obtained from microscopy of vesicle movements in extruded cytoplasm from a squid giant axon. As observed many times before, vesicles traveled along microtubule tracks; surprisingly, movement continued at the periphery of the extruded cytoplasm through a region containing microfilaments but no microtubules. Subsequent experiments demonstrated that a given vesicle could move on a microtubule *or* a microfilament. Thus at least two motor proteins, myosin and either kinesin or cytosolic dynein, must be bound to the same vesicle (Figure 20-24). The discovery that a given vesicle can travel along both cytoskeletal systems suggests that, in a neuron, synaptic vesicles are transported at a fast rate by kinesin in the microtubule-rich axon and then travel through the actin-rich cortex at the nerve terminal on a myosin motor.

Eukaryotic Cilia and Flagella Contain a Core of Doublet Microtubules Studded with Axonemal Dyneins

Cilia and flagella are flexible membrane extensions that project from certain cells. They range in length from a few

(a)

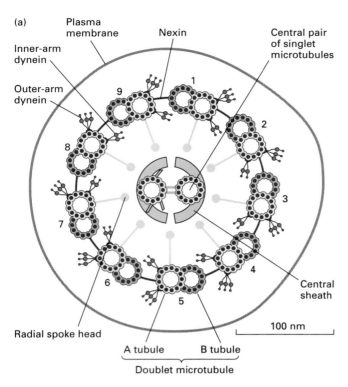

(b)

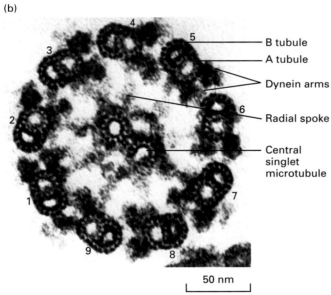

▲ **FIGURE 20-25 Structure of an axoneme.**
(a) Cross-sectional diagram of a typical flagellum showing its major structures. The dynein arms and radial spokes with attached heads surround a central pair of singlet microtubules. (b) Micrograph of a transverse section through an isolated demembranated cilium. [See U. W. Goodenough and J. E. Heuser, 1985, *J. Cell Biol.* **100**:2008. Part (b) courtesy of L. Tilney.]

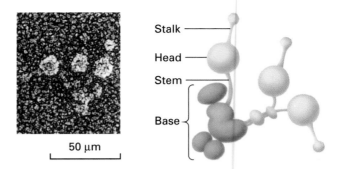

▲ **EXPERIMENTAL FIGURE 20-26 Freeze-etching reveals structure of axonemal dynein.** Electron micrograph of freeze-etched outer-arm dynein from *Tetrahymena* cilia and an artist's interpretation of the structure. The base contains several intermediate and light chains. Attached to the common base are three heavy chains each composed of a long stem, large globular head domain and small globular domain, and short stalk connecting the globular domains. Microtubules bind to the tip of the stalk. All axonemal dyneins are thought to have the general structure shown here, although some outer-arm dyneins contain two heavy chains, and inner-arm dyneins contain one or two heavy chains. [Electron micrograph from U. W. Goodenough and J. E. Heuser, 1984, *J. Mol. Biol.* **18**:1083.]

doublet microtubules surrounding a central pair of singlet microtubules (Figure 20-25). This characteristic "9 + 2" arrangement of microtubules is seen in cross section with the electron microscope. Each doublet microtubule consists of A and B tubules. The (+) end of axonemal microtubules is at the distal end of the axoneme. At its point of attachment to the cell, the axoneme connects with the **basal body**. Containing nine triplet microtubules (see Figure 20-4), the basal body plays an important role in initiating the growth of the axoneme.

The axoneme is held together by three sets of protein cross-links (see Figure 20-25a). The central pair of singlet microtubules is connected by periodic bridges, like rungs on a ladder, and is surrounded by a fibrous structure termed the *inner sheath*. A second set of linkers, composed of the protein *nexin*, joins adjacent outer doublet microtubules. *Radial spokes*, which radiate from the central singlets to each A tubule of the outer doublets, are proposed to regulate dynein.

Permanently attached periodically along the length of the A tubule of each doublet microtubule are *inner-arm* and *outer-arm* dyneins (see Figure 20-25a). These axonemal dyneins are complex multimers of heavy chains, intermediate chains, and light chains. When isolated axonemal dyneins are slightly denatured and spread out on an electron microscope grid, they are seen as a bouquet of two or three "blossoms" connected to a common base (Figure 20-26). Each blossom consists of a large globular head domain attached to a small globular domain through a short stalk; a stem connects one or more blossoms to a common base. The base is thought to attach a dynein to the A tubule, whereas the globular domains project outward toward the B tubule of the neighboring doublet.

micrometers to more than 2 mm for some insect sperm flagella. Virtually all eukaryotic cilia and flagella possess a central bundle of microtubules, called the **axoneme**, which consists of nine

A single dynein heavy chain, which forms each stalk, head, and stem is enormous, approximately 4500 amino acids in length with a molecular weight exceeding 540,000. At least eight or nine different heavy chains have been identified, each capable of hydrolyzing ATP. On the basis of sequence comparisons with the ATP-binding sites in other proteins, the ATP-binding site of axonemal dynein is predicted to lie in the globular head domain of the heavy chain, with the microtubule-binding site being at the tip of the stalk. Inner-arm dyneins are either one- or two-headed structures, containing one or two heavy chains. Outer-arm dyneins contain two heavy chains (e.g., in a sea urchin sperm flagellum) or three heavy chains (e.g., in *Tetrahymena* cilia and *Chlamydomonas* flagella).

The intermediate and light chains in axonemal dynein are thought to form the base region. These chains help mediate the attachment of the dynein arm to the A tubule and may also participate in regulating dynein activity. The base proteins of axonemal dyneins are thus analogous to those composing the dynactin complexes associated with cytosolic dynein.

Ciliary and Flagellar Beating Are Produced by Controlled Sliding of Outer Doublet Microtubules

Ciliary and flagellar beating is characterized by a series of bends, originating at the base of the structure and propagated toward the tip (Figure 20-27). The bends push against the surrounding fluid, propelling the cell forward or moving the fluid across a fixed epithelium. A bend results from the sliding of adjacent doublet microtubules past one another. Because active sliding occurs all along the axoneme, bends can be propagated without damping. Findings from microscopic studies with isolated axonemes from which the cross-linkage proteins (e.g., nexin) are removed have shown that doublet microtubules slide past one another in the presence of ATP but no bending occurs (Figure 20-28a). Thus the

ATP-dependent movement of doublet microtubules must be restricted by cross-linking proteins in order for sliding to be converted into the bending of an axoneme.

On the basis of the polarity and direction of sliding of the doublet microtubules and the properties of axonemal dyneins, the small head domains of the dynein arms on the A tubule of one doublet are thought to "walk" along the adjacent doublet's B tubule toward its base, the (−) end (Figure 20-28b). The force producing active sliding requires ATP and probably entails a conformational change in the head and stem that translocates the stalk. Successive binding and hydrolysis of ATP causes the dynein stalks to successively release from and attach to the adjacent doublet. Although this general model is most likely correct, many important details such as the mechanism of force transduction by dynein are still unknown.

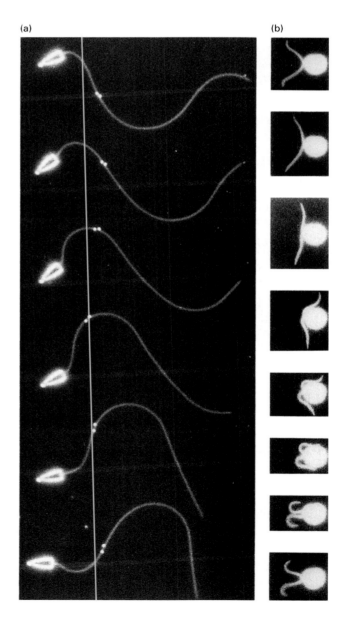

(a)   (b)

▶ **EXPERIMENTAL FIGURE 20-27 Video microscopy shows flagellar movements that propel sperm and *Chlamydomonas* forward.** In both cases, the cells are moving to the left. (a) In the typical sperm flagellum, successive waves of bending originate at the base and are propagated out toward the tip; these waves push against the water and propel the cell forward. Captured in this multiple-exposure sequence, a bend at the base of the sperm in the first (*top*) frame has moved distally halfway along the flagellum by the last frame. A pair of gold beads on the flagellum are seen to slide apart as the bend moves through their region. (b) Beating of the two flagella on *Chlamydomonas* occurs in two stages, called the *effective stroke* (top three frames) and the *recovery stroke* (remaining frames). The effective stroke pulls the organism through the water. During the recovery stroke, a different wave of bending moves outward from the bases of the flagella, pushing the flagella along the surface of the cell until they reach the position to initiate another effective stroke. Beating commonly occurs 5–10 times per second. [Part (a) from C. Brokaw, 1991, *J. Cell Biol.* **114**(6): cover photograph; courtesy of C. Brokaw. Part (b) courtesy of S. Goldstein.]

(a)

Dynein arms

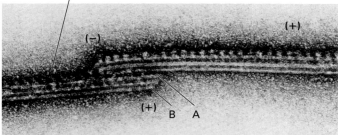

(−)

(+)

(+) B A

(b)

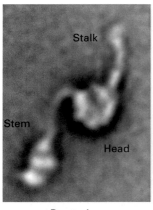

Stalk

Stem

Head

Prestroke

Poststroke

▲ **EXPERIMENTAL FIGURE 20-28 In vitro dynein-mediated sliding of doublet microtubules requires ATP.**
(a) Electron micrograph of two doublet microtubules in a protease-treated axoneme incubated with ATP. In the absence of cross-linking proteins, which are removed by the protease, doublet microtubules slide excessively. The dynein arms can be seen projecting from A tubules and interacting with B tubule of the top microtubule. (b) Single-headed dynein molecules in their prestroke and poststroke states. Thousands of images of purified inner-arm dynein were recorded in an electron microscope and then averaged. A comparison of dynein containing ADP and vanadate, a state mimicking the ADP-Pi state, with dynein absent of any bound nucleotide, suggests that the difference in structure may be related to the conformational changes taking place in the ATP cycle. A model of the force-generation mechanism suggests that the head changes orientation relative to the stem, causing a movement of the microtubule-binding stalk. [Part (a) courtesy of P. Satir. Part (b) from S. A. Burgess et al., 2003, *Nature* **421**:715; courtesy of S. A. Burgess.]

KEY CONCEPTS OF SECTION 20.2

Kinesin- and Dynein-Powered Movements

■ Two families of motor proteins, kinesin and dynein, transport membrane-limited vesicles, proteins, and organelles along microtubules (see Table 20-2).

■ Nearly all kinesins move cargo toward the (+) end of microtubules (anterograde transport), whereas dyneins transport cargo toward the (−) end (retrograde transport).

■ Most kinesins are dimers with a head domain that binds microtubules and ATP and a tail domain that binds vesicles or other cargo (see Figure 20-20). The flexible neck region determines the direction of kinesin movement, and the tail domain determines cargo specificity.

■ Cytosolic dyneins are linked to their cargoes (vesicles and chromosomes) by dynactin, a large multiprotein complex (see Figure 20-22). Dynactin also binds to microtubules, thereby increasing the processivity of dynein-mediated transport.

■ In microtubule-poor regions of the cell, vesicles are probably transported along microfilaments powered by a myosin motor.

■ Flagellar beating propels cells forward, and ciliary beating sweeps materials across tissues.

■ The axoneme in both flagella and cilia contains nine outer doublet microtubules arranged in a circle around two central singlet microtubules (see Figure 20-25).

■ Axonemal dyneins, which are larger and more complex than cytosolic dyneins, are permanently attached to doublet microtubules in axonemes. The dynein arms with their small globular heads project toward the adjacent doublet.

■ Walking of dynein arms extending from one doublet toward the (−) end of a neighboring doublet generates a sliding force in the axoneme (see Figure 20-27). This linear force is converted into a bend by regions that resist sliding.

20.3 Microtubule Dynamics and Motor Proteins in Mitosis

Mitosis is the process that partitions newly replicated chromosomes equally into separate parts of a cell. The last step in the cell cycle, mitosis takes about 1 hour in an actively dividing animal cell (see Figure 1-17). In that period, the cell builds and then disassembles a specialized microtubule structure, the **mitotic apparatus.** Larger than the nucleus, the mitotic apparatus is designed to attach and capture chromosomes, align the chromosomes, and then separate them so that the genetic material is evenly partitioned to each daughter cell. Fifteen hours later, the whole process is repeated by the two daughter cells.

Figure 20-29 depicts the characteristic series of events that can be observed by light microscopy in mitosis in a eukaryotic cell. Although the events unfold continuously, they are conventionally divided into four substages: **prophase, metaphase, anaphase,** and **telophase.** The beginning of mitosis is signaled by the appearance of condensing chromosomes, first visible as thin threads inside the nucleus. By late prophase, each chromosome appears as two identical filaments, the **chromatids** (often called *sister chromatids*), held together at a constricted region, the **centromere.** Each chromatid contains one of the two new daughter DNA molecules produced in the preceding S phase of the cell cycle; thus each cell that enters mitosis has four copies of each chromosomal DNA, designated 4*n*.

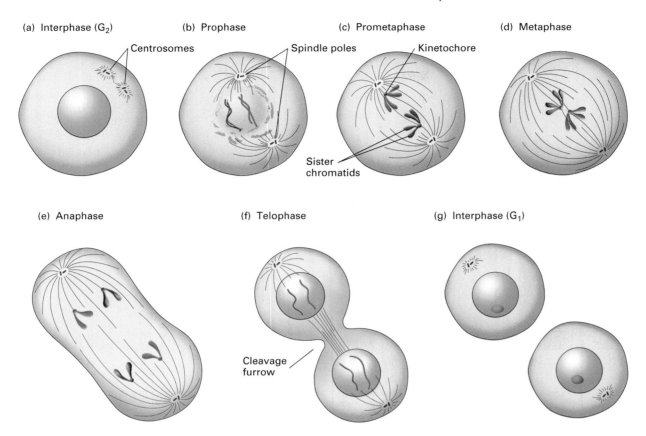

▲ **FIGURE 20-29 The stages of mitosis and cytokinesis in an animal cell.** For simplicity, only two sets of duplicated chromosomes, distinguished by color, are depicted. (a) *Interphase (G₂): DNA and centrosome replication.* After DNA replication during the S phase, the chromosomes, each containing a sister chromatid, are decondensed and not visible as distinct structures. By G₂ the centrioles have replicated to form daughter centrosomes. (b) *Prophase: centrosome migration.* The centrosomes, each with a daughter centriole, begin moving toward opposite poles of the cell. The chromosomes begin to condense, appearing as long threads. (c) *Prometaphase: spindle formation.* The nuclear envelope fragments into small vesicles and spindle microtubules enter the nuclear region. Chromosome condensation is completed; each visible chromosome is composed of two chromatids held together at their centromeres. Kinetochores at centromeres attach chromosomes to spindle microtubules. (d) *Metaphase: chromosome alignment.* The chromosomes move toward the equator of the cell, where they become aligned in the equatorial plane. (e) *Anaphase: chromosome separation.* The two sister chromatids separate into independent chromosomes. Each chromosome, attached to a kinetochore microtubule, moves toward one pole. Simultaneously, the poles move apart. (f) *Telophase and cytokinesis.* Nuclear membranes re-form around the daughter nuclei; the chromosomes decondense and become less distinct. The spindle disappears as the microtubules depolymerize, and cell cleavage proceeds. (g) Interphase (G₁): Following cleavage, the daughter cells enter G₁ of interphase.

In Chapter 21, we consider in detail how progression through the cell cycle, and hence cell replication, is regulated. In this section, we focus on the mechanics of mitosis in a "typical" animal cell. Mistakes in mitosis can lead to missing or extra chromosomes, causing abnormal patterns of development when they occur during embryogenesis and pathologies when they occur after birth. To ensure that mitosis proceeds without errors in the trillions of cell divisions that take place in the life span of an organism, a highly redundant mechanism has evolved in which each crucial step is carried out concurrently by microtubule motor proteins and microtubule assembly dynamics.

The Mitotic Apparatus Is a Microtubule Machine for Separating Chromosomes

The structure of the mitotic apparatus changes constantly during the course of mitosis (Figure 20-30). For one brief moment at metaphase, however, the chromosomes are aligned at the equator of the cell. We begin our discussion by examining the structure of the mitotic apparatus at metaphase and then describe how it captures and organizes chromosomes during prophase, how it separates chromosomes during anaphase, and how it determines where cells divide during telophase.

(a)

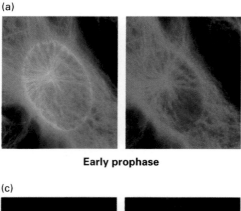

Early prophase

(b)

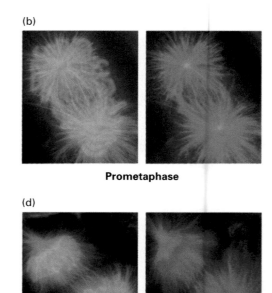

Prometaphase

(c)

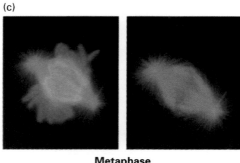

Metaphase

(d)

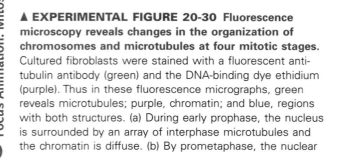

Anaphase

▲ **EXPERIMENTAL FIGURE 20-30 Fluorescence microscopy reveals changes in the organization of chromosomes and microtubules at four mitotic stages.** Cultured fibroblasts were stained with a fluorescent anti-tubulin antibody (green) and the DNA-binding dye ethidium (purple). Thus in these fluorescence micrographs, green reveals microtubules; purple, chromatin; and blue, regions with both structures. (a) During early prophase, the nucleus is surrounded by an array of interphase microtubules and the chromatin is diffuse. (b) By prometaphase, the nuclear membrane has broken down and the replicated centrosomes (centrioles) have migrated to the poles from which microtubules radiate. (c) At metaphase, the fully condensed chromosomes have aligned midway between the poles to form the metaphase plate. Dense bundles of microtubules connect the chromosomes to the poles. (d) In late anaphase, the chromosomes are pulled to the poles along the radiating microtubules. [From J. C. Waters, R. W. Cole, and C. L. Rieder, 1993, *J. Cell Biol.* **122**:361; courtesy of C. L. Rieder.]

At metaphase, the mitotic apparatus is organized into two parts: a central **mitotic spindle** and a pair of **asters** (Figure 20-31a; see also Figure 20-2c). The spindle is a bilaterally symmetric bundle of microtubules and associated proteins with the overall shape of a football; it is divided into opposing halves at the equator of the cell by the metaphase chromosomes. An aster is a radial array of microtubules at each pole of the spindle.

In each half of the spindle, a single centrosome at the pole organizes three distinct sets of microtubules whose (−) ends all point toward the centrosome (Figure 20-31b). One set, the *astral microtubules,* forms the aster; they radiate outward from the centrosome toward the cortex of the cell, where they help position the mitotic apparatus and later help to determine the cleavage plane in cytokinesis. The other two sets of microtubules compose the spindle. The *kinetochore microtubules* attach to chromosomes at specialized attachment sites on the chromosomes called **kinetochores.** *Polar microtubules* do not interact with chromosomes but instead overlap with polar microtubules from the opposite pole. Two types of interactions hold the spindle halves together to form the bilaterally symmetric mitotic apparatus: (1) lateral interactions between the overlapping (+) ends of the polar microtubules and (2) end-on interactions between the kinetochore microtubules and the kinetochores of the sister chromatids. The large protein complexes, called *cohesins,* that link sister chromatids together are discussed in Chapter 21.

The mitotic apparatus is basic to mitosis in all organisms, but its appearance and components can vary widely. In the budding yeast *Saccharomyces cerevisiae,* for instance, the mitotic apparatus consists of just a spindle, which itself is constructed from a minimal number of kinetochore and polar microtubules. These microtubules are organized by *spindle pole bodies,* trilaminated structures located in the nuclear membrane, which do not break down during mitosis. Furthermore, because a yeast cell is small, it does not require well-developed asters to assist in mitosis. Although the spindle pole body and centrosome differ structurally, they have proteins such as γ-tubulin in common that act to organize the mitotic spindle. Like yeast cells, most plant cells do not contain visible centrosomes. We consider the unique features of the mitotic apparatus in plant cells at the end of this section.

(a)

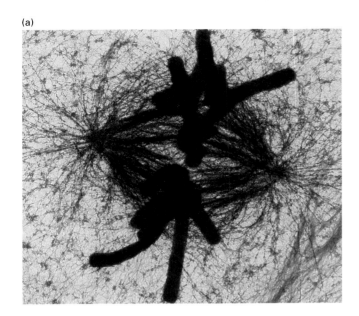

(b)

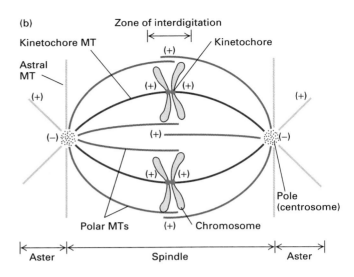

▲ **EXPERIMENTAL FIGURE 20-31 High-voltage electron microscopy visualizes components of the mitotic apparatus in a metaphase mammalian cell.** (a) Microtubules were stained with biotin-tagged anti-tubulin antibodies to increase their size in this electron micrograph. The large cylindrical objects are chromosomes. (b) Schematic diagram corresponding to the metaphase cell in (a). Three sets of microtubules (MTs) make up the mitotic apparatus. All the microtubules have their (−) ends at the poles (centrosomes). Astral microtubules project toward the cortex and are linked to it. Kinetochore microtubules are connected to chromosomes (blue). Polar microtubules project toward the cell center with their distal (+) ends overlapping. [Part (a) courtesy of J. R. McIntosh.]

The Kinetochore Is a Centromere-Based Protein Complex That Captures and Helps Transport Chromosomes

The sister chromatids of a metaphase chromosome are transported to each pole bound to kinetochore microtubules. In regard to their attachment to microtubules and movement, chromosomes differ substantially from the vesicle and organelle cargoes transported along cytosolic microtubules. The linkage of metaphase chromosomes to the (+) ends of kinetochore microtubules is mediated by a large protein complex, the kinetochore, which has several functions: to trap and attach microtubule ends to the chromosomes, to generate force to move chromosomes along microtubules, and to regulate chromosome separation and translocation to the poles. In an animal cell, the kinetochore forms at the centromere and is organized into an inner and outer layer embedded within a fibrous corona (Figure 20-32).

In all eukaryotes, three components participate in attaching chromosomes to microtubules: the centromere, kinetochore and spindle proteins, and the cell-cycle machinery. The location of the centromere and hence that of the kinetochore is directly controlled by a specific sequence of chromosomal DNA termed *centromeric DNA* (Chapter 10). Although the sequences of centromeric DNA and of DNA-binding proteins in the kinetochore are not well conserved through evolution, the cell-cycle proteins and many of the proteins that link the kinetochore to the spindle are homologous in humans and yeast. Microtubule-binding proteins (e.g., CLIP170, CENP-E) and microtubule motor proteins (e.g., the mitotic kinesin MCAK and cytosolic dynein) cooperate in attaching the kinetochore to a microtubule end while tubulin subunits are added or released. The presence of these motor proteins indicates that kinetochores play a role in transporting chromosomes to opposite ends of the cell in mitosis.

Duplicated Centrosomes Align and Begin Separating in Prophase

Because each half of the metaphase mitotic apparatus emanates from a polar centrosome, its assembly depends on duplication of the centrosome and movement of the daughter centrosomes to opposite halves of the cell. This process, known as the *centriole cycle* (or *centrosome cycle*) marks the first steps in mitosis, beginning during G_1 when the centrioles and other centrosome components are duplicated (Figure 20-33). By G_2, the two "daughter" centrioles have reached full length, but the duplicated centrioles are still present within a single centrosome. Early in mitosis, the two pairs of centrioles separate and migrate to opposite sides of the nucleus, establishing the bipolarity of the dividing cell. In some respects, then, mitosis can be understood as the migration of duplicated centrosomes, which along their journey pick up chromosomes, pause in metaphase, and during

(a)

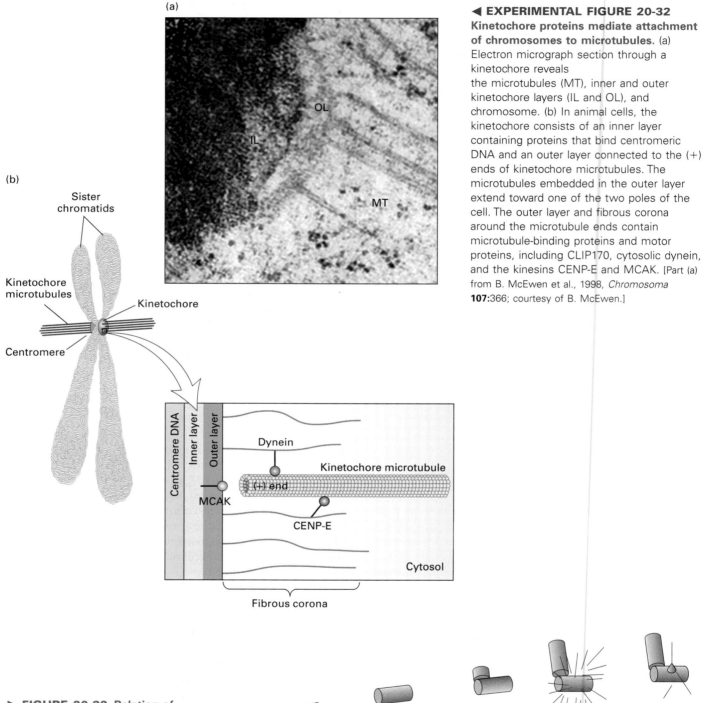

◄ **EXPERIMENTAL FIGURE 20-32**
**Kinetochore proteins mediate attachment
of chromosomes to microtubules.** (a)
Electron micrograph section through a
kinetochore reveals
the microtubules (MT), inner and outer
kinetochore layers (IL and OL), and
chromosome. (b) In animal cells, the
kinetochore consists of an inner layer
containing proteins that bind centromeric
DNA and an outer layer connected to the (+)
ends of kinetochore microtubules. The
microtubules embedded in the outer layer
extend toward one of the two poles of the
cell. The outer layer and fibrous corona
around the microtubule ends contain
microtubule-binding proteins and motor
proteins, including CLIP170, cytosolic dynein,
and the kinesins CENP-E and MCAK. [Part (a)
from B. McEwen et al., 1998, *Chromosoma*
107:366; courtesy of B. McEwen.]

▶ **FIGURE 20-33 Relation of
centrosome duplication to the cell
cycle.** After the pair of parent centrioles
(red) separates slightly, a daughter
centriole (blue) buds from each and
elongates. By G_2, growth of the daughter
centrioles is complete, but the two pairs
remain within a single centrosomal
complex. Early in mitosis, the centrosome
splits, and each centriole pair migrates to
opposite ends of the cell.

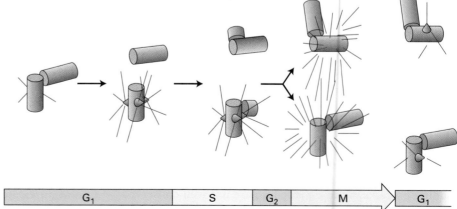

anaphase continue their movement to new locations in the daughter cells, where they release the chromosomes and organize the cytosolic microtubules.

Microtubule dynamics change drastically at the onset of mitosis, when long interphase microtubules disappear and are replaced by astral and spindle microtubules. These mitotic microtubules, which are nucleated from the newly duplicated centrosomes, are more numerous, shorter, and less stable than interphase microtubules. The average lifetime of a microtubule decreases from 10 minutes in interphase cells to 60–90 seconds in the mitotic apparatus. This increase in dynamic instability enables microtubules to assemble and disassemble quickly in mitosis.

The results of genetic and cell biological studies, primarily in yeast and flies, have implicated several kinesins in organ-

izing polar microtubules into a bipolar array, thereby orienting assembly of the spindle and spindle asters. For instance, antibodies against either a (+) or a (−) end–directed kinesin inhibit the formation of a bipolar spindle when they are microinjected into a cell before but not after prophase. A (−) end–directed kinesin protein, such as Kin-C, is thought to help align the oppositely oriented polar microtubules extending from each centrosome. Then a (+) end–directed kinesin, most likely the bipolar BimC, cross-links antiparallel microtubules and pushes them apart. In addition, findings from localization experiments with anti-dynein antibodies have demonstrated the presence of cytosolic dynein in the centrosomes and cortex of dividing animal cells. The results of other studies with yeast mutants lacking cytosolic dynein suggest that dynein at the cortex simultaneously helps tether the astral microtubules and orient the poles of the spindle. Thus the alignment and initial separation of centrosomes at prophase depend on the growth of polar and astral microtubules and on the action of several motor proteins, as depicted in Figure 20-34.

Although centrosomes facilitate the formation of the spindle poles, recent findings show that polar microtubules can be assembled and organized into antiparallel bundles in the absence of centrosomes. For instance, the Ran GTPase that functions in nuclear import and export (Chapter 12) appears to promote the polymerization of tubulin subunits. Ran acts with other proteins, possibly motor proteins, to stabilize microtubules by increasing their frequency of rescue.

Formation of the Metaphase Mitotic Spindle Requires Motor Proteins and Dynamic Microtubules

When the duplicated centrosomes have become aligned, formation of the spindle proceeds, driven by simultaneous events at centrosomes and chromosomes. As just discussed, the centrosome facilitates spindle formation by nucleating the assembly of the spindle microtubules. In addition, the (−) ends of microtubules are gathered and stabilized at the pole by dynein-dynactin working with the nuclear/mitotic apparatus protein. The role of dynein in spindle pole formation has been demonstrated by reconstitution studies in which bipolar spindles form in *Xenopus* egg extracts in the presence of centrosomes, microtubules, and sperm nuclei. The addition of antibodies against cytosolic dynein to this in vitro system releases and splays the spindle microtubules but leaves the centrosomal astral microtubules in position (Figure 20-35).

The dynamic instability of spindle microtubules at the other end, the (+) end, is critical to their capture of chromosomes during late prophase as the nuclear membrane begins to break down. By quickly lengthening and shortening at its (+) end, a dynamic microtubule probes into the chromosome-rich environment of the cell. Sometimes the (+) end of a microtubule directly contacts a kinetochore, scoring a "bull's-eye." More commonly, a kinetochore contacts the side of a microtubule and then slides along the microtubule to the (+) end in a process that includes cytosolic dynein and mitotic kinesins

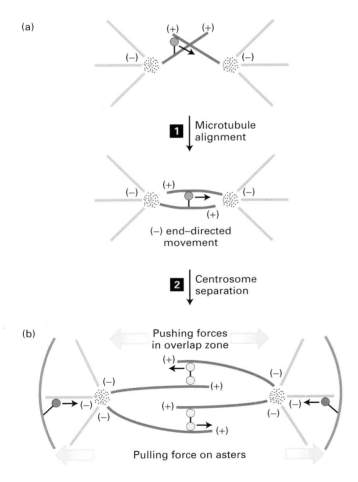

(a)

Microtubule
alignment

(−) end–directed
movement

Centrosome
separation

(b)

Pushing forces
in overlap zone

Pulling force on asters

▲ **FIGURE 20-34 Model for participation of microtubule motor proteins in centrosome movements at prophase.** (a) At prophase, polar microtubules growing randomly from opposite poles are aligned with the aid of (−) end–directed motors (orange). (b) After alignment, (+) end–directed mitotic kinesins (yellow), including the bipolar kinesin BimC, generate pushing forces that separate the poles. In addition, a (−) end–directed force exerted by cytosolic dynein (green) located at the cortex may pull asters toward the poles. Similar forces act later at anaphase.

Control　　　　Addition of dynein antibody

▲ **EXPERIMENTAL FIGURE 20-35 Cytosolic dynein participates in the formation and stabilization of mitotic spindle poles.** In vitro reconstituted spindles were stained with fluorescently-labeled polyclonal antibodies to tubulin (green) and dynein (red) and examined with a fluorescence microsope. In the control spindle (*left*), cytosolic dynein is present at each tapered end, as well as at the equator. Addition of a dynein monoclonal antibody after the formation of the spindle (*right*) disrupts dynein localization and causes the poles to splay. [From R. Heald et al., 1997, *J. Cell Biol.* **138**:615; courtesy of R. Heald.]

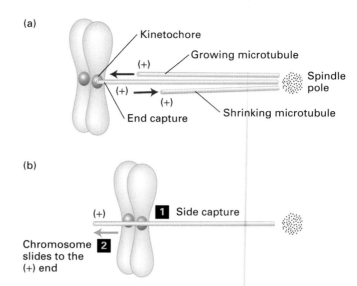

▲ **FIGURE 20-36 Capture of chromosomes by microtubules in prometaphase.** (a) In late prophase, spindle microtubules probe randomly for chromosomes by alternately growing and shrinking at their distal (+) ends. (b) Some chromosomes first encounter the side (**1**), not the end, of a microtubule, interacting with the microtubule through proteins at the kinetochore. Kinetochore-associated (+) end–directed motor proteins (e.g., MCAK) then move the chromosome to the (+) end (**2**), thereby stabilizing the microtubule.

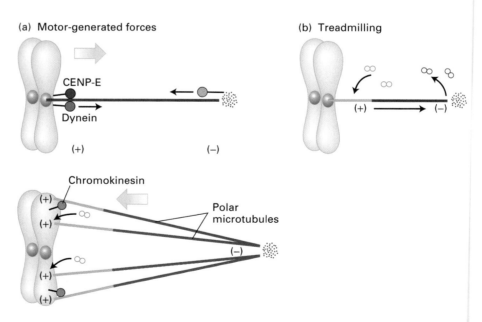

▲ **FIGURE 20-37 Model of the forces stabilizing metaphase chromosomes at the equatorial plate.** (a) Cytoplasmic dynein, a (−) end–directed motor (light green) at the kinetochore and a (+) end–directed motor (pink) at the spindle pole pull chromosomes toward the pole. Chromokinesin, a nonkinetochore (+) end–directed motor on the chromatid arms, exerts an opposite force on polar microtubules, pushing chromosomes away from the pole. CENP-E, a kinesin that does not mediate movement, keeps the kinetochore tethered to the kinetochore microtubule. (b) Treadmilling of tubulin subunits briefly stabilizes the lengths of spindle microtubules at metaphase by balancing assembly at the kinetochore with disassembly at the poles.

(a)　　　　(b)

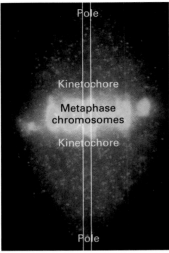

Pole-to-pole image slice

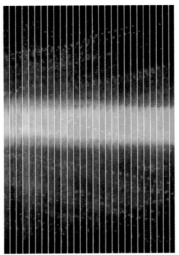

Sequential slices (time) ⟶

◀ **EXPERIMENTAL FIGURE 20-38**
Poleward flux of tubulin subunits during metaphase is visualized by fluorescence speckle microscopy.
(a) When a substoichiometric amount of fluorescent tubulin subunits is assembled into a microtubule, the microtubule appears speckled. (b) In a live cell at metaphase, the speckled microtubules do not have the fibrous appearance observed in conventional fluorescence microscopy. However, the dynamics of subunit translocation is revealed by taking a slice through the same position from each frame of a time-lapse series (boxed area). If the labeled tubulin subunits move, then the slope of the line formed by a speckle in consecutive images corresponds to the translocation rate. Here, the speckles on both sides of the spindle travel toward the poles at a rate of 0.75 μm/s. [Part (b) from T. J. Mitchison and E. D. Salmon, 2001, *Nature Cell Biol.* **3**:E17; courtesy of A. Desai.]

on the kinetochore (Figure 20-36). Whether a chromosome attaches to the (+) end of a spindle microtubule by a direct hit or by the side capture/sliding process, the kinetochore "caps" the (+) end of the microtubule. Eventually, the kinetochore of each sister chromatid in a chromosome is captured by microtubules arising from the nearest spindle poles. Each chromosome arm becomes attached to additional microtubules as mitosis progresses toward metaphase.

During late prophase (prometaphase), the newly condensed chromosomes attached to the (+) ends of kinetochore microtubules move to the equator of the spindle. Along the way, the chromosomes exhibit *saltatory behavior,* oscillating between movements toward and then away from the pole or equator. These oscillations result from alternating depolymerization and polymerization at the (+) ends of kinetochore microtubules. In addition, motor proteins associated with both ends of kinetochore microtubules and with the distal ends of polar microtubules on chromosome arms generate opposing forces that are thought to position captured chromosomes equally between the two spindle poles (Figure 20-37a).

Although the lengths of kinetochore and polar microtubules eventually become stable, there continues to be a flow, or treadmilling, of subunits through the microtubules toward the poles. At metaphase, the loss of tubulin subunits at the (−) ends of spindle microtubules is balanced by the addition of subunits at the (+) ends (Figure 20-37b). The flow of tubulin subunits from kinetochores to the poles can be visualized after a very small "pulse" of fluorescently labeled tubulin subunits has been microinjected into a cell (Figure 20-38). Microtubules appear speckled because very few of the subunits are fluores-

cent. By comparing the positions of each speckle, one can see whether the tubulin subunits are moving in a specific direction. The images show that the mitotic spindle at metaphase is a finely balanced, yet dynamic, structure that holds chromosomes at the equatorial plate. By mechanisms discussed in Chapter 21, the cell cycle is held in check until all chromosomes have been captured and aligned. A single unattached kinetochore is sufficient to prevent entry into anaphase.

Anaphase Chromosomes Separate and the Spindle Elongates

The same forces that form the spindle during prophase and metaphase also direct the separation of chromosomes toward opposite poles at anaphase. Anaphase is divided into two distinct stages, *anaphase A* and *anaphase B* (early and late anaphase). Anaphase A is characterized by the shortening of kinetochore microtubules at their (+) ends, which pulls the chromosomes toward the poles. In anaphase B, the two poles move farther apart, bringing the attached chromosomes with them into what will become the two daughter cells.

Microtubule Shortening in Anaphase A The results of in vitro studies have indicated that the depolymerization of microtubules in *Xenopus* eggs is sufficient to move chromosomes toward the poles. In one such study, purified microtubules were mixed with purified anaphase chromosomes; as expected, the kinetochores bound preferentially to the (+) ends of the microtubules. To induce depolymerization of the

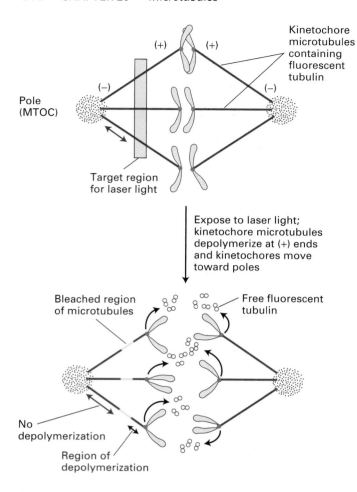

▲ EXPERIMENTAL FIGURE 20-39 Shortening at the (+) end of kinetochore microtubules moves chromosomes poleward in anaphase A. Fibroblasts are injected with fluorescent tubulin and then allowed to enter metaphase so that all the microtubules are fluorescent. Only the kinetochore microtubules are shown. In early anaphase, a band of microtubules (yellow box) is subjected to a laser light, which bleaches the fluorescence but leaves the microtubules continuous and functional across the bleached region. The bleached segment of each microtubule thus provides a marker for the position of that part of the microtubule. In anaphase, the distance between the bleached zone and the adjacent pole (marked by red double-headed arrows) does not change, whereas the distance to the adjacent chromatid (marked by black double-headed arrows) becomes shorter. This finding indicates that during anaphase microtubules disassemble at the (+) end just behind the kinetochores, not at the poles. [Adapted from G. J. Gorbsky et al., 1987, *J. Cell Biol.* **104**:9, and 1988, *J. Cell Biol.* **106**:1185.]

microtubules, the reaction mixture was diluted, thus lowering the concentration of free tubulin dimers. Video microscopy analysis then showed that the chromosomes moved toward the (−) end, at a rate similar to that of chromosome movement during anaphase in intact cells. Because no ATP (or any other energy source) was present in these experiments, chromosome movement toward the (−) end must have been

powered, in some way, by microtubule disassembly and must not have been powered by microtubule motor proteins.

The in vivo fluorescence-tagging experiment depicted in Figure 20-39 provides evidence that shortening of kinetochore microtubules at their (+) ends moves chromosomes toward the poles in mammalian cells. A kinetochore-associated kinesin, MCAK, promotes disassembly at the (+) end while CENP-E, also at kinetochores, binds to the progressively shortening end. According to this model, microtubule disassembly drives poleward movement of chromosomes. Although kinetochore kinesins promote this disassembly or keep chromosomes attached to the shortening (+) end of kinetochore microtubules, they do not actually move chromosomes along the microtubules.

Spindle Elongation in Anaphase B Findings from several types of experiments have implicated three processes in the separation of the poles in anaphase B: a pushing force gen-

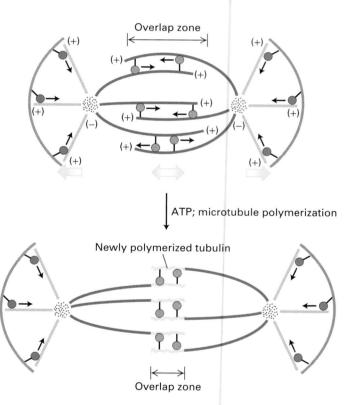

▲ FIGURE 20-40 Model of spindle elongation and movement of poles during anaphase B. One or more (+) end–directed spindle kinesins (orange) bind to antiparallel polar microtubules in the overlap region and then "walk" along a microtubule in the other half-spindle toward its (+) end. In cells that assemble an aster, cytoplasmic dynein, a (−) end–directed motor protein (green) anchored in the cortex of the plasma membrane, walks along astral microtubules, pulling the poles outward. Tubulin subunits are simultaneously added to the plus ends of all polar microtubules, thereby lengthening the spindle. [Adapted from H. Masuda and W. Z. Cande, 1987, *Cell* **49**:193.]

erated by kinesin-mediated sliding of polar microtubules past one another, a pulling force generated by cortex-associated cytosolic dynein, and lengthening of polar microtubules at their (+) ends. The coordinated effect of these processes is depicted in Figure 20-40.

When metaphase cells are depleted of ATP by mild detergent treatment, poleward chromosome movement (anaphase A) proceeds but the spindle poles do not separate (anaphase B). This finding indicates that microtubule motor proteins take part in separating the spindle poles, as they do in centrosome movement in prometaphase (see Figure 20-34). Experiments with artificial spindles assembled from frog egg extracts were sources of further insight into anaphase B movements. In the presence of calcium, the spindle is activated and elongates, simulating anaphase B, and the zone of overlap between the two halves of the spindle decreases in length. The observation that adjacent antiparallel microtubules in this system migrate in the direction of their pole-facing (−) ends suggests that a (+) end–directed kinesin separates spindle poles in anaphase B. In one model, the bipolar kinesin BimC attached to a microtubule in the overlap region walks toward the (+) end of a neighboring but antiparallel microtubule, thus pushing the adjacent microtubule in the direction of its (−) end (see Figure 20-40). Involvement of a kinesin motor is supported by experiments in which antibodies raised against a conserved region of the kinesin superfamily inhibit ATP-induced elongation of diatom spindles in vitro.

In the presence of αβ-tubulin, reactivated frog spindles and isolated diatom spindles add tubulin subunits to the (+) end of polar microtubules, thus lengthening them. The third process taking place in anaphase B can be demonstrated by cutting the spindle in half with a microneedle at anaphase; the resulting half-spindles move quickly to the poles, at a rate faster than usual during anaphase. This observation suggests that cytosolic dynein, a (−) end–directed motor protein associated with the cortex, pulls on astral microtubules, thereby moving the poles farther apart (see Figure 20-40).

▶ **EXPERIMENTAL FIGURE 20-41 Micromanipulation experiments can determine whether the spindle or the asters control location of the cleavage plane during cytokinesis.** A small glass ball is pressed against a fertilized egg until membranes from opposite sides of the cell touch and fuse, thus changing the spherical egg into a doughnut shape. In the first cell division, a normal spindle develops and the doughnut-shaped cell divides to produce a single C-shaped cell with two normal nuclei. This result is expected whether the asters or the spindle determines the cleavage plane. In the second cell division, the two nuclei each produce a normal spindle. If the spindle determined the cleavage plane, then cleavage of the C-shaped cell would yield three cells, one of them with two nuclei (*lower left*). If the asters determined the cleavage plane, then a third cleavage plane would form between asters from two different spindles (*lower right*). Cell division produced four cells, indicating an extra furrow formed between a pair of asters.

Microtubules and Microfilaments Work Cooperatively During Cytokinesis

After the chromosomes have migrated to opposite ends of the cell during anaphase, two closely related events proceed (see Figure 20-29): the nuclear envelope re-forms around each complete set of chromosomes, marking the end of mitosis (telophase), and the cytoplasm divides, a process termed

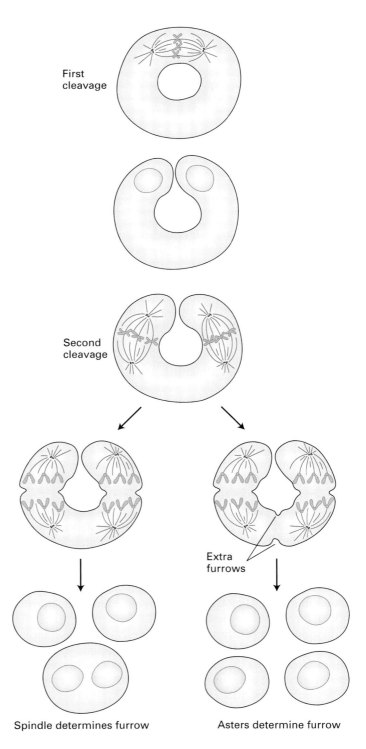

First cleavage

Second cleavage

Extra furrows

Spindle determines furrow   Asters determine furrow

▶ **FIGURE 20-42 Regulation of myosin light chain by mitosis-promoting factor.** The mitosis-promoting factor (MPF) is an activated complex of a cyclin-dependent kinase (CDK1) and a mitotic cyclin protein. Phosphorylation of inhibitory sites on the myosin light chain by the kinase activity of MPF early in mitosis prevents active myosin heavy chains from interacting with actin filaments and sliding along them, a process required for cytokinesis. When MPF activity falls at anaphase, a constitutive phosphatase dephosphorylates the inhibitory sites, permitting cytokinesis to proceed. Another enzyme, myosin light-chain kinase, phosphorylates a different residue, activating myosin to interact with actin filaments. [See L. L. Satterwhite et al., 1992, *J. Cell Biol.* **118**:595; adapted from A. Murray and T. Hunt, 1993, *The Cell Cycle: An Introduction*, W. H. Freeman and Company.]

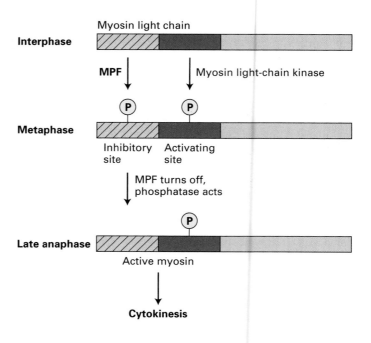

cytokinesis. With completion of cytokinesis, two daughter cells are formed. In Chapter 19, we saw how the contractile ring of actin and myosin II constricts the cell during cytokinesis but does not constrict the mechanism that determines the plane of cleavage through the cell. It is clear that the contractile ring and hence the cleavage furrow always develop where the chromosomes line up during metaphase, but which component of the mitotic apparatus dictates where the contractile ring will assemble is not clear.

Findings from micromanipulation experiments have shown that different cells rely on either the spindle or the asters to determine the cleavage plane (Figure 20-41). For instance, the presence of two asters determines where cleavage occurs in fertilized sand dollar eggs, whereas the spindle determines the cleavage plane in animal cells.

One hypothesis is that astral or spindle microtubules send a signal to the region of the cortex midway between asters. This signal promotes the interaction of actin microfilaments and myosin II, resulting in the formation of the contractile ring followed by the development of the cleavage furrow. The signal is unidentified as yet, but an alluring candidate is CDK1. Not only does activation of this cyclin-dependent kinase trigger entry into mitosis, but the myosin light chain is one of its known substrates. Phosphorylation of the regulatory light chains in myosin II by CDK1 early in mitosis inhibits interaction of myosin with actin filaments, thus preventing premature formation of the contractile ring (Figure 20-42). The decrease in CDK1 activity in late anaphase relieves this inhibition, leading to assembly of the contractile ring and its contraction to form the cleavage furrow (see Figure 19-20). In Chapter 21, we examine how the activity of CDK1 and other cyclin-dependent kinases waxes and wanes in replicating cells.

Plant Cells Reorganize Their Microtubules and Build a New Cell Wall in Mitosis

 As noted previously, interphase plant cells lack a single perinuclear microtubule-organizing center that organizes microtubules into the radiating interphase array typical of animal cells. Instead, numerous MTOCs line the cortex of plant cells and nucleate the assembly of transverse bands of microtubules below the cell wall (Figure 20-43, *left*). These cortical microtubules, which are cross-linked by plant-specific MAPs, aid in laying down extracellular cellulose microfibrils, the main component of the rigid cell wall (see Figure 6-33).

Although mitotic events in plant cells are generally similar to those in animal cells, formation of the spindle and cytokinesis have unique features in plants (see Figure 20-43). Plant cells bundle their cortical microtubules and reorganize them into a spindle at prophase without the aid of centrosomes. At metaphase, the mitotic apparatus appears much the same in plant and animal cells. Golgi-derived vesicles, which appear at metaphase, are transported into the mitotic apparatus along microtubules that radiate from each end of the spindle. At telophase, these vesicles line up near the center of the dividing cell and then fuse to form the *phragmoplast*, a membrane structure that replaces the animal-cell contractile ring. The membranes of the vesicles forming the phragmoplast become the plasma membranes of the daughter cells. The contents of these vesicles, such as polysaccharide precursors of cellulose and pectin, form the early cell plate, which develops into the new cell wall between the daughter cells.

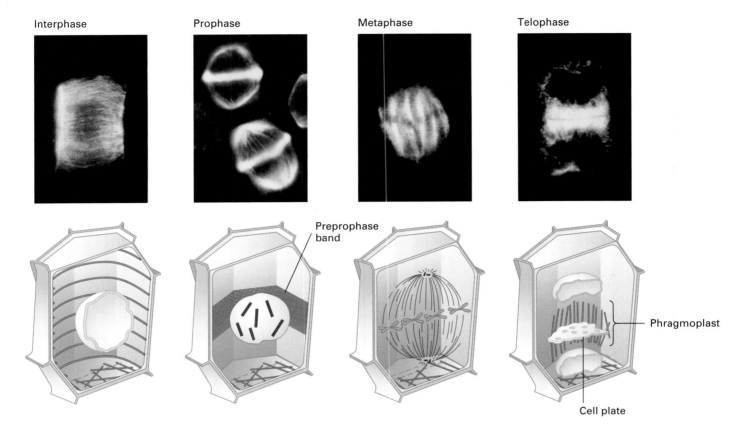

Interphase   Prophase   Metaphase   Telophase

Preprophase band

Phragmoplast

Cell plate

▲ **FIGURE 20-43 Mitosis in a higher plant cell.**
Immunofluorescence micrographs (*top*) and corresponding diagrams (*bottom*) showing arrangement of microtubules in interphase and mitotic plant cells. A cortical array of microtubules girdles a cell during interphase. Webs of microtubules cap the growing ends of plant cells and remain intact during cell division. As a cell enters prophase, the microtubules are bundled around the nucleus and reorganized into a spindle that appears similar to that in metaphase animal cells. By late telophase, the nuclear membrane has re-formed around the daughter nuclei, and the Golgi-derived phragmoplast has assembled at the equatorial plate. Additional small vesicles derived from the Golgi complex accumulate at the equatorial plate and fuse with the phragmoplast to form the new cell plate. [Adapted from R. H. Goddard et al., 1994, *Plant Physiol.* **104**:1; micrographs courtesy of Susan M. Wick.]

KEY CONCEPTS OF SECTION 20.3

Microtubule Dynamics and Motor Proteins in Mitosis

■ In mitosis, the replicated chromosomes are separated and evenly partitioned to two daughter chromosomes. This part of the cell cycle is commonly divided into four sub-stages: prophase, metaphase, anaphase, and telophase (see Figure 20-29).

■ The mitotic apparatus of animal cells comprises the astral microtubules forming the asters, the polar and kine-tochore microtubules forming the football-shaped spindle, the spindle poles derived from the duplicated centrosomes, and chromosomes attached to the kinetochore micro-tubules (see Figure 20-31).

■ A simple set of microtubule motor proteins—BimC, CENP-E, and cytosolic dynein—are conserved in all spin-dles. Additional motor proteins are present in more com-plex organisms.

■ Motor protein–mediated interactions between opposing polar microtubules and between asters and the cell cortex align and separate the centrosomes at prophase, establish-ing the bipolar orientation of the spindle (see Figure 20-34).

■ Kinetochore microtubules capture chromosomes during prophase (see Figure 20-32). Opposing forces generated by motor proteins at both ends of spindle microtubules cen-ter the chromosomes at metaphase (see Figure 20-37).

■ The combined action of microtubule dynamics (poly-merization and depolymerization) and microtubule motors are responsible for the transport of chromosomes to the opposite poles at anaphase (see Figures 20-39 and 20-40).

■ Cleavage during cytokinesis generates two daughter cells. The signal for cleavage is unknown, but it probably relieves the inhibition on interaction between actin fila-ments and myosin II that prevents premature formation of the contractile ring.

PERSPECTIVES FOR THE FUTURE

Dynein is an enigmatic molecular motor for several reasons. First, the head domain is based on distinctive repeats of a β-propeller motif. The β-propeller–based design suggests that the catalytic mechanism may differ from those of kinesin and myosin. Equally distinctive is the microtubule-binding stalk that extends from the catalytic head domain. Electron micrographs suggest that conformation changes in the head are amplified into large motions of the stalk. Clues to the mechanism will emerge from an atomic-level structure of the entire dynein molecule. As in previous studies of myosin, scientists will take a two-pronged approach employing microscopy and crystallography to create a composite model of dynein structure. Three-dimensional reconstructions from cryoelectron micrographs will reveal the overall organization of the stalk, head, and stem domains under conditions that simulate different stages in force generation. High-resolution information about the catalytic core and coupling to the stalk will be obtained by x-ray crystallography. Structures determined by this approach will be fitted into the electron densities from cryoelectron microscopy.

Dynein seems to be a complementary partner of kinesin and myosin in trafficking vesicles and chromosomes. Although an emerging theme is that dynactin couples a membrane cargo to dynein or kinesin, how microtubule-based movements are coordinated with myosin-based movements remains an unresolved question. Several lines of investigation suggest that dynein could play a central role in coordination. Because the (+) ends of actin filaments and microtubules tend to point toward the cell periphery, dynein is the only major class of motors that moves toward the cell interior. The location of cytoplasmic dynein in the actin-rich cortex is similar to that of myosin and distinct from that of kinesin. Whether dynactin also couples myosin to membrane trafficking is unknown.

The most impressive microtubule machine is the mitotic apparatus. In the course of the cell cycle, it assembles and disassembles and faithfully carries out its core function—the separation of chromosomes. The mitotic apparatus does not exist as a program hardwired in the genome. Instead, the instructions for structure are a consequence of chemistry, the assembly encoded in the reversible interactions between proteins and the timing controlled by signaling circuits. Particularly interesting questions concern the tension-sensing mechanisms that center chromosomes at the metaphase plate and translocate chromosomes at anaphase, the positioning mechanisms of the spindle pole body and the contractile ring, and the force generation mechanism at cytokinesis. A combination of biophysical force-measuring methods and molecular biological mutagenesis methods will be employed to extract the answers to these questions.

KEY TERMS

anaphase 838	kinetochores 840
asters 840	metaphase 838
axonal transport 829	microtubule-associated proteins (MAPs) 823
axoneme 836	microtubule-organizing center (MTOC) 817
basal body 836	
centriole cycle 841	microtubules 817
centromere 838	mitosis 838
centrosome 825	mitotic spindle 840
cytokinesis 848	prophase 838
dynamic instability 822	telophase 838
dyneins 829	tubulin 818
kinesins 829	

REVIEW THE CONCEPTS

1. Microtubules are polar filaments; that is, one end is different from the other. What is the basis for this polarity, how is polarity related to microtubule organization within the cell, and how is polarity related to the intracellular movements powered by microtubule-dependent motors?

2. Microtubules both in vitro and in vivo undergo dynamic instability, and this type of assembly is thought to be intrinsic to the microtubule. What is the current model to account for dynamic instability?

3. In cells, microtubule assembly depends on other proteins as well as tubulin concentration and temperature. What types of proteins influence microtubule assembly in vivo, and how does each type affect assembly?

4. Microtubules within a cell appear to be arranged in specific arrays. What cellular structure is responsible for determining the arrangement of microtubules within a cell? How many of these structures are found in a typical cell? Describe how such structures serve to nucleate microtubule assembly.

5. Many drugs that inhibit mitosis bind specifically to tubulin, microtubules, or both. What diseases are such drugs used to treat? Functionally speaking, these drugs can be divided into two groups based on their effect on microtubule assembly. What are the two mechanisms by which such drugs alter microtubule assembly?

6. Kinesin I was the first member of the kinesin motor family to be identified and therefore is perhaps the best-characterized family member. Describe how kinesin I was first isolated and identified. What fundamental property of microtubule-dependent motors was taken advantage of in this procedure?

7. Certain cellular components appear to move bidirectionally on microtubules. Describe how this is possible given that microtubule orientation is fixed by the MTOC. What types of regulatory processes are thought to control which direction a given cellular cargo will move? Is it possible for certain cellular components to move on both microtubules and actin filaments? Explain your answer.

8. The motile properties of kinesin motor proteins involve both the motor domain and the neck domain. Describe the role of each domain in kinesin movement, direction of movement, or both. What is the experimental evidence for the importance of the neck in kinesin motility?

9. Cell swimming depends on appendages containing microtubules. What is the underlying structure of these appendages, and how do these structures generate the force required to produce swimming?

10. The mitotic spindle is often described as a microtubule-based cellular machine. The microtubules that constitute the mitotic spindle can be classified into three distinct types. What are the three types of spindle microtubules, and what is the function of each?

11. Mitotic spindle function relies heavily on microtubule motors. For each of the following motor proteins, predict the effect on spindle formation, function, or both of adding a drug that specifically inhibits only that motor: Kin-C, BimC, cytosolic dynein, and CENP-E.

12. The poleward movement of kinetochores, and hence chromatids, during anaphase A requires that kinetochores maintain a hold on the shortening microtubules. How does a kinetochore hold onto shortening microtubules?

13. Anaphase B involves the separation of spindle poles. What forces have been proposed to drive this separation? What underlying molecular mechanisms are thought to provide these forces?

14. Cytokinesis, the process of cytoplasmic division, occurs shortly after the separated sister chromatids have neared the opposite spindle poles. How is the plane of cytokinesis determined? What are the respective roles of microtubules and actin filaments in cytokinesis?

ANALYZE THE DATA

Motor proteins are generally considered to move cellular components along microtubules. However, some kinesin family motors may have other functions as described below.

a. In one set of experiments, microtubules were observed in a microscope chamber in which the solution could be rapidly exchanged. Part (a) of the accompanying figure depicts the effect on microtubule length as a function of time when the tubulin and buffer in the chamber were replaced first by an identical tubulin and buffer solution and then by

buffer alone. Part (b) depicts the results when the tubulin and buffer solution in the chamber was replaced with a solution containing tubulin and the M-type kinesin motor protein Kin I. What conclusions can you draw from these experiments?

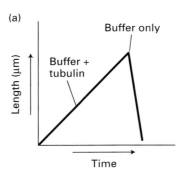

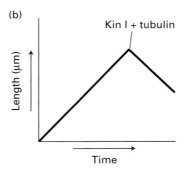

b. In another assay, taxol-stabilized microtubules were incubated alone (Ctrl), with kinesin heavy chain (KHC) and ATP, with Kin I and ATP, or with Kin I and AMPPNP, and then centrifuged to pellet polymerized tubulin. Supernatant (s) and pellet (p) fractions were then separated by SDS-PAGE, and the gel was stained to reveal the position of tubulin, as shown in part (c). What conclusions can you draw from these experiments? What is the significance of the nucleotide present for Kin I activity? Would you expect ATP to be present in the experiments depicted in part (b) of the figure? Why or why not?

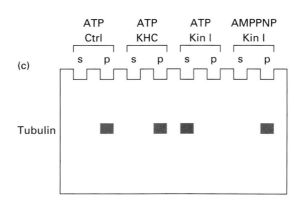

c. Similar results to those shown in part (c) were also obtained when GMPCPP, a nonhydrolyzable GTP analog, was used to assemble stabilized microtubules. What does this result suggest about the action of Kin I?

d. There is additional evidence that Kin I can produce similar results even at low ratios relative to tubulin. Given this fact and what you know about microtubule assembly, where might you expect to find the motor on microtubules?

REFERENCES

Kinesin home page
http://www.proweb.org/kinesin/
Kreis, T., and R. Vale. 1999. *Guidebook to the Cytoskeletal and Motor Proteins*, 2d ed. Oxford University Press.

Microtubule Organization and Dynamics

Bornens, M. 2002. Centrosome composition and microtubule anchoring mechanisms. *Curr. Opin. Cell Biol.* **14**:25–34.

Desai, A., and T. J. Mitchison. 1997. Microtubule polymerization dynamics. *Ann. Rev. Cell Dev. Biol.* **13**:83–117.

Dutcher, S. K. 2001. The tubulin fraternity: alpha to eta. *Curr. Opin. Cell Biol.* **13**:49–54.

Mathur, J., and M. Hulskamp. 2002. Microtubules and microfilaments in cell morphogenesis in higher plants. *Curr. Biol.* **12**:R669–R676.

Nogales, E. 2001. Structural insights into microtubule function. *Ann. Rev. Biomol. Struc.* **30**:397–420.

Vaughn, K. C., and J. D. I. Harper. 1998. Microtubule-organizing centers and nucleating sites in land plants. *Int'l. Rev. Cytol.* **181**:75–149.

Zimmerman, W., C. A. Sparks, and S. J. Doxsey. 1999. Amorphous no longer: the centrosome comes into focus. *Curr. Opin. Cell Biol.* **11**:122–128.

Kinesin- and Dynein-Powered Movements

Allan, V. J., H. M. Thompson, and M. A. McNiven. 2002. Motoring around the Golgi. *Nature Cell Biol.* **4**:E236–E242.

Dell, R. R. 2003. Dynactin polices two-way organelle traffic. *J. Cell Biol.* **160**:291–293.

Dietz, S., W. R. Schief, and J. Howard. 2002. Molecular motors: single-molecule recordings made easy. *Curr. Biol.* **12**:R203–R205.

Dujardin, D. L., and R. B. Vallee. 2002. Dynein at the cortex. *Curr. Opin. Cell Biol.* **14**:44–49.

Dutcher, S. K. 1995. Flagellar assembly in two hundred and fifty easy-to-follow steps. *Trends Genet.* **11**:398–404.

Goldstein, L. S. 2001. Kinesin molecular motors: transport pathways, receptors, and human disease. *Proc. Nat'l. Acad. Sci. USA* **98**:6999–7003.

Higuchi, H., and S. A. Endow. 2002. Directionality and processivity of molecular motors. *Curr. Opin. Cell Biol.* **14**:50–57.

Hirokawa, N. 1998. Kinesin and dynein superfamily proteins and the mechanism of organelle transport. *Science* **279**:519–526.

Holleran, E. A., S. Karki, and E. L. F. Holzbaur. 1998. The role of the dynactin complex in intracellular motility. *Int'l. Rev. Cytol.* **182**:69–109.

Karcher, R. L., S. W. Deacon, and V. I. Gelfand. 2002. Motor-cargo interactions: the key to transport specificity. *Trends Cell Biol.* **12**:21–27.

Kull, F. J., and S. A. Endow. 2002. Kinesin: switch I and II and the motor mechanism. *J. Cell Sci.* **115**:15–23.

Mandelkow, E., and K. A. Johnson. 1998. The structural and mechanochemical cycle of kinesin. *Trends Biochem. Sci.* **23**:429–433.

Moore, J. D., and S. A. Endow. 1996. Kinesin proteins: a phylum of motors for microtubule-based motility. *Bioessays* **18**:207–218.

Sablin, E. P., and R. J. Fletterick. 2001. Nucleotide switches in molecular motors: structural analysis of kinesins and myosins. *Curr. Opin. Struc. Biol.* **11**:716–724.

Shaw, J. V., and D. W. Cleveland. 2002. Slow axonal transport: fast motors in the slow lane. *Curr. Opin. Cell Biol.* **14**:58–62.

Thaler, C. D., and L. T. Haimo. 1996. Microtubules and microtubule motors: mechanisms of regulation. *Int'l. Rev. Cytol.* **164**:269–327.

Vale, R. D., and R. J. Fletterick. 1997. The design plan of kinesin motors. *Ann. Rev. Cell Dev. Biol.* **8**:4–9.

Microtubule Dynamics and Motor Proteins in Mitosis

Compton, D. A. 1998. Focusing on spindle poles. *J. Cell Sci.* **111**:1477–1481.

Cyr, R. J. 2001. Plant cells: mitosis, cytokinesis, and cell plate formation. *Encyclopedia of Life Sciences*.

Doxsey, S. 2001. Re-evaluating centrosome function. *Nature Rev. Mol. Cell. Biol.* **2**:688–698.

Inoue, S. 1996. Mitotic organization and force generation by assembly/disassembly of microtubules. *Cell Struc. Funct.* **21**:375–379.

Kapoor, T. M., and D. A. Compton. 2002. Searching for the middle ground: mechanisms of chromosome alignment during mitosis. *J. Cell Biol.* **157**:551–556.

Karsenti, E., and I. Vernos. 2001. The mitotic spindle: a self-made machine. *Science* **294**:543–547.

Kitagawa, K., and P. Hieter. 2001. Evolutionary conservation between budding yeast and human kinetochores. *Nature Rev. Mol. Cell. Biol.* **2**:678–687.

Mitchison, T. J., and E. D. Salmon. 2001. Mitosis: a history of division. *Nature Cell Biol.* **3**:E17–E21.

Nicklas, R. B. 1997. How cells get the right chromosomes. *Science* **275**:632–637.

Nigg, E. A. 2001. Mitosis. *Nature Encyclopedia of Life Sciences*.

Palazzo, R. E., and T. N. Davis. 2001. Centrosomes and spindle pole bodies. *Methods in Cell Biology*, vol. 67. Academic Press.

Rieder, C. L., and E. D. Salmon. 1998. The vertebrate cell kinetochore and its roles during mitosis. *Trends Cell Biol.* **8**:310–318.

Schimoda, S. L., and F. Solomon. 2002. Integrating functions at the kinetochore. *Cell* **109**:9–12.

Wittmann, T., A. Hyman, and A. Desai. 2001. The spindle: a dynamic assembly of mirotubules and motors. *Nature Cell Biol.* **3**:E28–E34.

Zhou, J., J. Yao, and H. C. Joshi. 2002. Attachment and tension in the spindle assembly checkpoint. *J. Cell Sci.* **115**:3547–3555.

21

REGULATING THE EUKARYOTIC CELL CYCLE

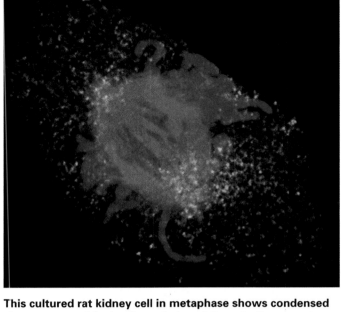

This cultured rat kidney cell in metaphase shows condensed chromosomes (blue), microtubules of the spindle apparatus (red), and the inner nuclear envelope protein POM121 (green). The POM121 staining demonstrates that the inner nuclear envelope proteins retract into the ER during mitosis.

[Brian Burke and Jan Ellenberger, 2002, *Nature Rev. Mol. Cell Biol.* **3**:487.]

The **cell cycle** entails an ordered series of macromolecular events that lead to **cell division** and the production of two daughter cells each containing chromosomes identical to those of the parental cell. Duplication of the parental chromosomes occurs during the S phase of the cycle, and one of the resulting daughter chromosomes is distributed to each daughter cell during **mitosis** (see Figure 9-3). Precise temporal control of the events of the cell cycle ensures that the replication of chromosomes and their **segregation** to daughter cells occur in the proper order and with extraordinarily high fidelity. Regulation of the cell cycle is critical for the normal development of multicellular organisms, and loss of control ultimately leads to cancer, an all-too-familiar disease that kills one in every six people in the developed world (Chapter 23).

In the late 1980s, it became clear that the molecular processes regulating the two key events in the cell cycle—chromosome replication and segregation—are fundamentally similar in all eukaryotic cells. Because of this similarity, research with diverse organisms, each with its own particular experimental advantages, has contributed to a growing understanding of how these events are coordinated and controlled. Biochemical and genetic techniques, as well as recombinant DNA technology, have been employed in studying various aspects of the eukaryotic cell cycle. These studies have revealed that cell replication is primarily controlled by regulating the timing of nuclear DNA replication

and mitosis. The master controllers of these events are a small number of *heterodimeric protein kinases* that contain a regulatory subunit (cyclin) and catalytic subunit (cyclin-dependent kinase). These kinases regulate the activities of multiple proteins involved in DNA replication and mitosis by phosphorylating them at specific regulatory sites, activating some and inhibiting others to coordinate their activities. In this chapter we focus on how the cell cycle is regulated and the experimental systems that have led to our current understanding of these crucial regulatory mechanisms.

21.1 Overview of the Cell Cycle and Its Control

We begin our discussion by reviewing the stages of the eukaryotic cell cycle, presenting a summary of the current model of how the cycle is regulated, and briefly describing key experimental systems that have provided revealing information about cell-cycle regulation.

The Cell Cycle Is an Ordered Series of Events Leading to Cell Replication

As illustrated in Figure 21-1, the cell cycle is divided into four major phases. In cycling (replicating) somatic cells, cells synthesize RNAs and proteins during the **G_1 phase**, preparing for DNA synthesis and chromosome replication during the **S (synthesis) phase**. After progressing through the **G_2 phase**, cells begin the complicated process of mitosis, also called the **M (mitotic) phase**, which is divided into several stages (see Figure 20-29).

In discussing mitosis, we commonly use the term chromosome for the *replicated* structures that condense and become visible in the light microscope during the **prophase** period of mitosis. Thus each chromosome is composed of the two daughter DNA molecules resulting from DNA replication plus the histones and other chromosomal proteins associated with them (see Figure 10-27). The identical daughter DNA molecules and associated chromosomal proteins that form one chromosome are referred to as sister **chromatids**. Sister chromatids are attached to each other by protein cross-links along their lengths. In vertebrates, these become confined to a single region of association called the **centromere** as chromosome condensation progresses.

During **interphase**, the portion of the cell cycle between the end of one M phase and the beginning of the next, the outer nuclear membrane is continuous with the endoplasmic reticulum (see Figure 5-19). With the onset of mitosis in prophase, the nuclear envelope retracts into the endoplasmic reticulum in most cells from higher eukaryotes, and Golgi membranes break down into vesicles. As described in Chapter 20, cellular microtubules disassemble and reassemble into the **mitotic apparatus** consisting of a football-shaped bundle of microtubules (the spindle) with a star-shaped cluster of microtubules radiating from each end, or spindle pole. During the **metaphase** period of mitosis, a multiprotein complex, the **kinetochore**, assembles at each centromere. The kinetochores of sister chromatids then associate with microtubules coming from opposite spindle poles (see Figure 20-31). During the **anaphase** period of mitosis, sister chromatids separate. They initially are pulled by motor proteins along the spindle microtubules toward the opposite poles and then are further separated as the mitotic spindle elongates (see Figure 20-40).

Once chromosome separation is complete, the mitotic spindle disassembles and chromosomes decondense during **telophase**. The nuclear envelope re-forms around the segregated

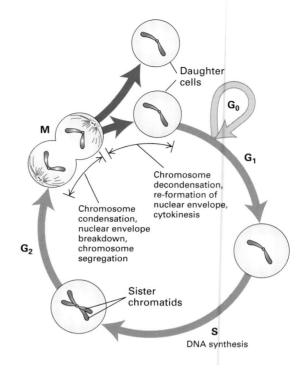

▲ **FIGURE 21-1 Summary of major events in the eukaryotic cell cycle and the fate of a single parental chromosome.** In proliferating cells, G_1 is the period between "birth" of a cell following mitosis and the initiation of DNA synthesis, which marks the beginning of the S phase. At the end of the S phase, a replicated chromosome consists of two daughter DNA molecules and associated chromosomal proteins. Each of the individual daughter DNA molecules and their associated chromosomal proteins (not shown) is called a sister chromatid. The end of G_2 is marked by the onset of mitosis, during which the mitotic spindle (red lines) forms and pulls apart sister chromatids, followed by division of the cytoplasm (cytokinesis) to yield two daughter cells. The G_1, S, and G_2 phases are collectively referred to as interphase, the period between one mitosis and the next. Most nonproliferating cells in vertebrates leave the cell cycle in G_1, entering the G_0 state.

chromosomes as they decondense. The physical division of the cytoplasm, called **cytokinesis,** then yields two daughter cells as the Golgi complex re-forms in each daughter cell. Following mitosis, cycling cells enter the G_1 phase, embarking on another turn of the cycle. In yeasts and other fungi, the nuclear envelope does not break down during mitosis. In these organisms, the mitotic spindle forms within the nuclear envelope, which then pinches off, forming two nuclei at the time of cytokinesis.

In vertebrates and diploid yeasts, cells in G_1 have a **diploid** number of chromosomes ($2n$), one inherited from each parent. In haploid yeasts, cells in G_1 have one of each chromosome ($1n$), the **haploid** number. Rapidly replicating human cells progress through the full cell cycle in about 24 hours: mitosis takes ≈30 minutes; G_1, 9 hours; the S phase, 10 hours; and G_2, 4.5 hours. In contrast, the full cycle takes only ≈90 minutes in rapidly growing yeast cells.

In multicellular organisms, most differentiated cells "exit" the cell cycle and survive for days, weeks, or in some cases (e.g., nerve cells and cells of the eye lens) even the lifetime of the organism without dividing again. Such *postmitotic* cells generally exit the cell cycle in G_1, entering a phase called G_0 (see Figure 21-1). Some G_0 cells can return to the cell cycle and resume replicating; this reentry is regulated, thereby providing control of cell proliferation.

Regulated Protein Phosphorylation and Degradation Control Passage Through the Cell Cycle

The concentrations of the **cyclins,** the regulatory subunits of the heterodimeric protein kinases that control cell-cycle events, increase and decrease as cells progress through the cell cycle. The catalytic subunits of these kinases, called **cyclin-dependent kinases (CDKs),** have no kinase activity unless they are associated with a cyclin. Each CDK can associate with different cyclins, and the associated cyclin determines which proteins are phosphorylated by a particular cyclin-CDK complex.

Figure 21-2 outlines the role of the three major classes of cyclin-CDK complexes that control passage through the cell cycle: the G_1, S-phase, and mitotic cyclin-CDK complexes. When cells are stimulated to replicate, G_1 cyclin-CDK complexes are expressed first. These prepare the cell for the S phase by activating transcription factors that promote

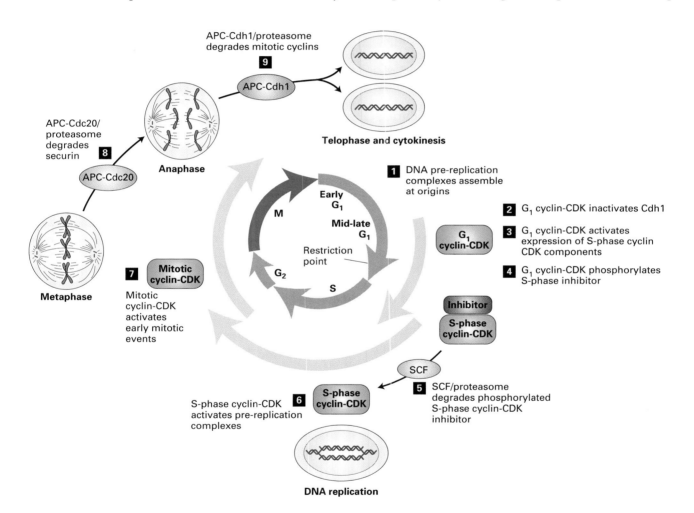

▲ **FIGURE 21-2 Overview of current model for regulation of the eukaryotic cell cycle.** Passage through the cycle is controlled by G_1, S-phase, and mitotic cyclin-dependent kinase complexes (green). These are composed of a regulatory cyclin subunit and a catalytic cyclin-dependent kinase (CDK) subunit. Two ubiquitin ligase complexes (orange), SCF and APC, polyubiquitinate specific substrates including S-phase inhibitors (step 5), securin (step 8), and mitotic cyclins (step 9), marking these substrates for degradation by proteasomes. Proteolysis of the S-phase inhibitor activates S-phase cyclin-CDK complexes, leading to chromosome replication. Proteolysis of securin results in degradation of protein complexes that connect sister chromatids at metaphase, thereby initiating anaphase, the mitotic period in which sister chromatids are separated and moved to the opposite spindle poles. Reduction in the activity of mitotic cyclin-CDK complexes caused by proteolysis of mitotic cyclins permits late mitotic events and cytokinesis to occur. These proteolytic cleavages drive the cycle in one direction because of the irreversibility of protein degradation. See text for further discussion.

transcription of genes encoding enzymes required for DNA synthesis and the genes encoding S-phase cyclins and CDKs. The activity of S-phase cyclin-CDK complexes is initially held in check by inhibitors. Late in G_1, the G_1 cyclin-CDK complexes induce degradation of the S-phase inhibitors by phosphorylating them and consequently stimulating their polyubiquitination by the multiprotein SCF ubiquitin ligase (step 5). Subsequent degradation of the polyubiquitinated S-phase inhibitor by proteasomes releases active S-phase cyclin-CDK complexes.

Once activated, the S-phase cyclin-CDK complexes phosphorylate regulatory sites in the proteins that form DNA pre-replication complexes, which are assembled on *replication origins* during G_1 (Chapter 4). Phosphorylation of these proteins not only activates initiation of DNA replication but also prevents reassembly of new pre-replication complexes. Because of this inhibition, each chromosome is replicated just once during passage through the cell cycle, ensuring that the proper chromosome number is maintained in the daughter cells.

Mitotic cyclin-CDK complexes are synthesized during the S phase and G_2, but their activities are held in check by phosphorylation at inhibitory sites until DNA synthesis is completed. Once activated by dephosphorylation of the inhibitory sites, mitotic cyclin-CDK complexes phosphorylate multiple proteins that promote chromosome condensation, retraction of the nuclear envelope, assembly of the mitotic spindle apparatus, and alignment of condensed chromosomes at the metaphase plate. During mitosis, the anaphase-promoting complex (APC), a multisubunit ubiquitin ligase, polyubiquitinates key regulatory proteins marking them for proteasomal degradation. One important substrate of the APC is securin, a protein that inhibits degradation of the cross-linking proteins between sister chromatids. The polyubiquitination of securin by the APC is inhibited until the kinetochores assembled at the centromeres of all chromosomes have become attached to spindle microtubules, causing chromosomes to align at the metaphase plate. Once all the chromosomes are aligned, the APC polyubiquitinates securin, leading to its proteasomal degradation and the subsequent degradation of the cross-linking proteins connecting sister chromatids (see Figure 21-2, step 8). This sequence of events initiates anaphase by freeing sister chromatids to segregate to opposite spindle poles.

Late in anaphase, the APC also directs polyubiquitination and subsequent proteasomal degradation of the mitotic cyclins. Polyubiquitination of the mitotic cyclins by APC is inhibited until the segregating chromosomes have reached the proper location in the dividing cell (see Figure 21-2, step 9). Degradation of the mitotic cyclins leads to inactivation of the protein kinase activity of the mitotic CDKs. The resulting decrease in mitotic CDK activity permits constitutively active protein phosphatases to remove the phosphates that were added to specific proteins by the mitotic cyclin-CDK complexes. As a result, the now separated chromosomes decondense, the nuclear envelope re-forms around

daughter-cell nuclei, and the Golgi apparatus reassembles during telophase; finally, the cytoplasm divides at cytokinesis, yielding the two daughter cells.

During early G_1 of the next cell cycle, phosphatases dephosphorylate the proteins that form pre-replication complexes. These proteins had been phosphorylated by S-phase cyclin-CDK complexes during the previous S phase, and their phosphorylation was maintained during mitosis by mitotic cyclin-CDK complexes. As a result of their dephosphorylation in G_1, new pre-replication complexes are able to reassemble at replication origins in preparation for the next S phase (see Figure 21-2, step 1). Phosphorylation of Cdh1 by G_1 cyclin-CDK complexes in late G_1 inactivates it, allowing accumulation of S-phase and mitotic cyclins during the ensuing cycle.

Passage through three critical cell-cycle transitions—$G_1 \rightarrow$ S phase, metaphase $\rightarrow$ anaphase, and anaphase $\rightarrow$ telophase and cytokinesis—is irreversible because these transitions are triggered by the regulated degradation of proteins, an irreversible process. As a consequence, cells are forced to traverse the cell cycle in one direction only.

In higher organisms, control of the cell cycle is achieved primarily by regulating the synthesis and activity of G_1 cyclin-CDK complexes. Extracellular **growth factors** function as **mitogens** by inducing synthesis of G_1 cyclin-CDK complexes. The activity of these and other cyclin-CDK complexes is regulated by phosphorylation at specific inhibitory and activating sites in the catalytic subunit. Once mitogens have acted for a sufficient period, the cell cycle continues through mitosis even when they are removed. The point in late G_1 where passage through the cell cycle becomes independent of mitogens is called the restriction point (see Figure 21-2).

Diverse Experimental Systems Have Been Used to Identify and Isolate Cell-Cycle Control Proteins

The first evidence that diffusible factors regulate the cell cycle came from cell-fusion experiments with cultured mammalian cells. When interphase cells in the G_1, S, or G_2 phase of the cell cycle were fused to cells in mitosis, their nuclear envelopes retracted and their chromosomes condensed (Figure 21-3). This finding indicates that some diffusible component or components in the cytoplasm of the mitotic cells forced interphase nuclei to undergo many of the processes associated with early mitosis. We now know that these factors are the mitotic cyclin-CDK complexes.

Similarly, when cells in G_1 were fused to cells in S phase and the fused cells exposed to radiolabeled thymidine, the label was incorporated into the DNA of the G_1 nucleus as well as the S-phase nucleus, indicating that DNA synthesis began in the G_1 nucleus shortly after fusion. However, when cells in G_2 were fused to S-phase cells, no incorporation of labeled thymidine occurred in the G_2 nuclei. Thus diffusible factors in an S-phase cell can enter the nucleus of a G_1 cell and stimulate DNA synthesis, but these factors cannot in-

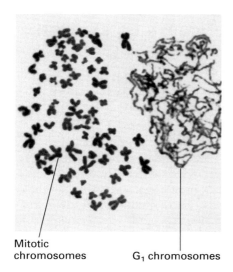

▲ **EXPERIMENTAL FIGURE 21-3 A diffusable factor in mitotic cells can induce mitosis in an interphase cell.** In unfused interphase cells, the nuclear envelope is intact and the chromosomes are not condensed, so individual chromosomes cannot be distinguished (see Figures 1-2b and 5-25). In mitotic cells, the nuclear envelope is absent and the individual replicated chromosomes are highly condensed. This micrograph shows a hybrid cell resulting from fusion of a mitotic cell (left side) with an interphase cell in G_1 (right side). A factor from the mitotic cell cytoplasm has caused the nuclear envelope of the G_1 cell to retract into the endoplasmic reticulum, so that it is not visible. The factor has also caused the G_1 cell chromosomes to partially condense. The mitotic chromosomes can be distinguished because the two sister chromatids are joined at the centromere. [From R. T. Johnson and P. N. Rao, 1970, *Biol. Rev.* **46**:97.]

However, complementation of the recessive mutation by the wild-type allele carried by one of the plasmid clones in the library allows a transformed mutant cell to grow into a colony; the plasmids bearing the wild-type allele can then be recovered from those cells. Because many of the proteins that regulate the cell cycle are highly conserved, human cDNAs cloned into yeast expression vectors often can complement yeast cell-cycle mutants, leading to the rapid isolation of human genes encoding cell-cycle control proteins.

Biochemical studies require the preparation of cell extracts from many cells. For biochemical studies of the cell cycle, the eggs and early embryos of amphibians and marine invertebrates are particularly suitable. In these organisms, multiple synchronous cell cycles follow fertilization of a large egg. By isolating large numbers of eggs from females and

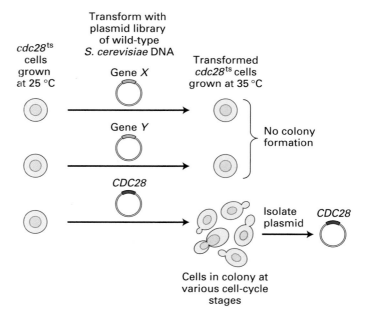

▲ **EXPERIMENTAL FIGURE 21-4 Wild-type cell-division cycle (CDC) genes can be isolated from a *S. cerevisiae* genomic library by functional complementation of *cdc* mutants.** Mutant cells with a temperature-sensitive mutation in a *CDC* gene are transformed with a genomic library prepared from wild-type cells and plated on nutrient agar at the nonpermissive temperature (35 °C). Each transformed cell takes up a single plasmid containing one genomic DNA fragment. Most such fragments include genes (e.g., *X* and *Y*) that do not encode the defective Cdc protein; transformed cells that take up such fragments do not form colonies at the nonpermissive temperature. The rare cell that takes up a plasmid containing the wild-type version of the mutant gene (in this case *CDC28*) is complemented, allowing the cell to replicate and form a colony at the nonpermissive temperature. Plasmid DNA isolated from this colony carries the wild-type *CDC* gene corresponding to the gene that is defective in the mutant cells. The same procedure is used to isolate wild-type *CDC* genes in *S. pombe*. See Figures 9-19 and 9-20 for more detailed illustrations of the construction and screening of a yeast genomic library.

duce DNA synthesis in a G_2 nucleus. We now know that these factors are S-phase cyclin-CDK complexes, which can activate the pre-replication complexes assembled on DNA replication origins in early G_1 nuclei. Although these cell-fusion experiments demonstrated that diffusible factors control entry into the S and M phases of the cell cycle, genetic and biochemical experiments were needed to identify these factors.

The budding yeast *Saccharomyces cerevisiae* and the distantly related fission yeast *Schizosaccharomyces pombe* have been especially useful for isolation of mutants that are blocked at specific steps in the cell cycle or that exhibit altered regulation of the cycle. In both of these yeasts, **temperature-sensitive mutants** with defects in specific proteins required to progress through the cell cycle are readily recognized microscopically and therefore easily isolated (see Figure 9-6). Such cells are called *cdc* (*c*ell-*d*ivision *c*ycle) mutants. The wild-type alleles of recessive temperature-sensitive *cdc* mutant alleles can be isolated readily by transforming haploid mutant cells with a plasmid library prepared from wild-type cells and then plating the transformed cells at the nonpermissive temperature (Figure 21-4). When plated out, the haploid mutant cells cannot form colonies at the nonpermissive temperature.

TABLE 21-1 Selected Cyclins and Cyclin-Dependent Kinases (CDKs) [*]

Organism/Protein	Name
S. POMBE	
CDK (one only)	Cdc2
Mitotic cyclin (one only)	Cdc13
S. CEREVISIAE	
CDK (one only)	Cdc28
Mid G_1 cyclin	Cln3
Late G_1 cyclins	Cln1, Cln2
Early S-phase cyclins	Clb5, Clb6
Late S-phase and early mitotic cyclins	Clb3, Clb4
Late mitotic cyclins	Clb1, Clb2
VERTEBRATES	
Mid G_1 CDKs	CDK4, CDK6
Late G_1 and S-phase CDK	CDK2
Mitotic CDK	CDK1
Mid G_1 cyclins	D-type cyclins
Late G_1 and S-phase cyclin	Cyclin E
S-phase and mitotic cyclin	Cyclin A
Mitotic cyclin	Cyclin B

[*] Those cyclins and CDKs discussed in this chapter are listed and classified by the period in the cell cycle in which they function. A heterodimer composed of a mitotic cyclin and CDK is commonly referred to as a mitosis-promoting factor (MPF).

fertilizing them simultaneously by addition of sperm (or treating them in ways that mimic fertilization), researchers can obtain extracts from cells at specific points in the cell cycle for analysis of proteins and enzymatic activities.

In the following sections we describe critical experiments that led to the current model of eukaryotic cell-cycle regulation summarized in Figure 21-2 and present further details of the various regulatory events. As we will see, results obtained with different experimental systems and approaches have provided insights about each of the key transition points in the cell cycle. For historical reasons, the names of various cyclins and cyclin-dependent kinases from yeasts and vertebrates differ. Table 21-1 lists the names of those that we discuss in this chapter and indicates when in the cell cycle they are active.

Overview of the Cell Cycle and Its Control

■ The eukaryotic cell cycle is divided into four phases: M (mitosis), G_1 (the period between mitosis and the initiation of nuclear DNA replication), S (the period of nuclear DNA replication), and G_2 (the period between the completion of nuclear DNA replication and mitosis) (see Figure 21-1).

■ Cyclin-CDK complexes, composed of a regulatory cyclin subunit and a catalytic cyclin-dependent kinase subunit, regulate progress of a cell through the cell cycle (see Figure 21-2). Large multisubunit ubiquitin ligases also polyubiquitinate key cell-cycle regulators, marking them for degradation by proteasomes.

■ Diffusible mitotic cyclin-CDK complexes cause chromosome condensation and disassembly of the nuclear envelope in G_1 and G_2 cells when they are fused to mitotic cells. Similarly, S-phase cyclin-CDK complexes stimulate DNA replication in the nuclei of G_1 cells when they are fused to S-phase cells.

■ The isolation of yeast cell-division cycle *(cdc)* mutants led to the identification of genes that regulate the cell cycle (see Figure 21-4).

■ Amphibian and invertebrate eggs and early embryos from synchronously fertilized eggs provide sources of extracts for biochemical studies of cell-cycle events.

21.2 Biochemical Studies with Oocytes, Eggs, and Early Embryos

A breakthrough in identification of the factor that induces mitosis came from studies of oocyte maturation in the frog *Xenopus laevis*. To understand these experiments, we must first lay out the events of oocyte maturation, which can be duplicated in vitro. As **oocytes** develop in the frog ovary, they replicate their DNA and become arrested in G_2 for 8 months during which time they grow in size to a diameter of 1 mm, stockpiling all the materials needed for the multiple cell divisions required to generate a swimming, feeding tadpole. When stimulated by a male, an adult female's ovarian cells secrete the steroid hormone progesterone, which induces the G_2-arrested oocytes to enter meiosis I and progress through meiosis to the second meiotic metaphase (Figure 21-5). At this stage the cells are called *eggs*. When fertilized by sperm, the egg nucleus is released from its metaphase II arrest and completes meiosis. The resulting haploid egg pronucleus then fuses with the haploid sperm pronucleus, producing a diploid **zygote** nucleus. DNA replication follows and the first mitotic division of early embryogenesis begins. The resulting embryonic cells then proceed through 11 more rapid, synchronous cell cycles generating a hollow sphere, the blastula. Cell division then

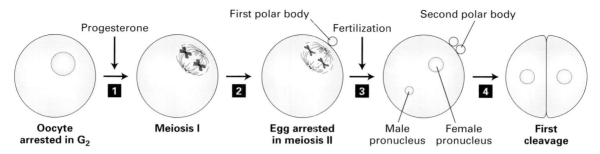

▲ EXPERIMENTAL FIGURE 21-5 Progesterone stimulates meiotic maturation of *Xenopus* oocytes in vitro. Step **1**: Treatment of G_2-arrested *Xenopus* oocytes surgically removed from the ovary of an adult female with progesterone causes the oocytes to enter meiosis I. Two pairs of synapsed homologous chromosomes (blue) connected to mitotic spindle microtubules (red) are shown schematically to represent cells in metaphase of meiosis I. Step **2**: Segregation of homologous chromosomes and a highly asymmetrical cell division expels half the chromosomes into a small cell called the first polar body. The oocyte immediately commences meiosis II and arrests in metaphase to yield an egg. Two chromosomes connected to spindle microtubules are shown schematically to represent egg cells arrested in metaphase of meiosis II. Step **3**: Fertilization by sperm releases eggs from their metaphase arrest, allowing them to proceed through anaphase of meiosis II and undergo a second highly asymmetrical cell division that eliminates one chromatid of each chromosome in a second polar body. Step **4**: The resulting haploid female pronucleus fuses with the haploid sperm pronucleus to produce a diploid zygote, which undergoes DNA replication and the first mitosis of 12 synchronous early embryonic cleavages.

slows, and subsequent divisions are nonsynchronous with cells at different positions in the blastula dividing at different times.

Maturation-Promoting Factor (MPF) Stimulates Meiotic Maturation of Oocytes and Mitosis in Somatic Cells

When G_2-arrested *Xenopus* oocytes are removed from the ovary of an adult female frog and treated with progesterone, they undergo *meiotic maturation,* the process of oocyte maturation from a G_2-arrested oocyte to the egg arrested in metaphase of meiosis II (see Figure 21-5). Microinjection of cytoplasm from eggs arrested in metaphase of meiosis II into G_2-arrested oocytes stimulates the oocytes to mature into eggs in the absence of progesterone (Figure 21-6). This system not only led to the initial identification of a factor in egg cytoplasm that stimulates maturation of oocytes in vitro but also provided an assay for this factor, called *maturation-promoting factor (MPF)*. As we will see shortly, MPF turned out to be the key factor that regulates the initiation of mitosis in all eukaryotic cells.

Using the microinjection system to assay MPF activity at different times during oocyte maturation in vitro, researchers found that untreated G_2-arrested oocytes have low levels of MPF activity; treatment with progesterone induces MPF activity as the cells enter meiosis I (Figure 21-7). MPF activity falls as the cells enter the interphase between meiosis I and II, but then rises again as the cells enter meiosis II and remains high in the egg cells arrested in metaphase II. Following

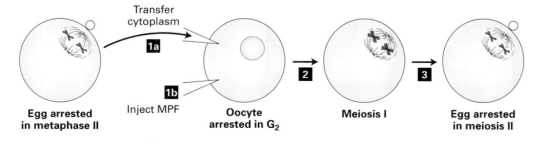

▲ EXPERIMENTAL FIGURE 21-6 A diffusible factor in arrested *Xenopus* eggs promotes meiotic maturation. When ≈5 percent of the cytoplasm from an unfertilized *Xenopus* egg arrested in metaphase of meiosis II is microinjected into a G_2-arrested oocyte (step **1a**), the oocyte enters meiosis I (step **2**) and proceeds to metaphase of meiosis II (step **3**), generating a mature egg in the absence of progesterone. This process can be repeated multiple times without further addition of progesterone, showing that egg cytoplasm contains an oocyte maturation-promoting factor (MPF). Microinjection of G_2-arrested oocytes provided the first assay for MPF activity (step **1b**) at different stages of the cell cycle and in different organisms. [See Y. Masui and C. L. Markert, 1971, *J. Exp. Zool.* **177**:129.]

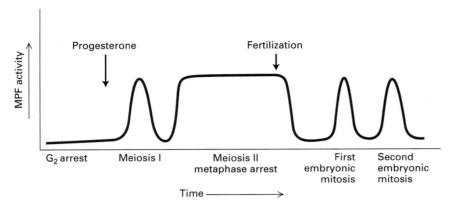

◄ **EXPERIMENTAL FIGURE 21-7** MPF activity in *Xenopus* oocytes, eggs, and early embryos peaks as cells enter meiosis and mitosis. Diagrams of the cell structures corresponding to each stage are shown in Figure 21-5. MPF activity was determined by the microinjection assay shown in Figure 21-6 and quantitated by making dilutions of cell extracts. See text for discussion. [See J. Gerhart et al., 1984, *J. Cell Biol.* **98**:1247; adapted from A. Murray and M. W. Kirschner, 1989, *Nature* **339**:275.]

fertilization, MPF activity falls again until the zygote (fertilized egg) enters the first mitosis of embryonic development. Throughout the following 11 synchronous cycles of mitosis in the early embryo, MPF activity is low in the interphase periods between mitoses and then rises as the cells enter mitosis.

Although initially discovered in frogs, MPF activity has been found in mitotic cells from all species assayed. For example, cultured mammalian cells can be arrested in mitosis by treatment with compounds (e.g., colchicine) that inhibit assembly of microtubules. When cytoplasm from such mitotically arrested mammalian cells was injected into G$_2$-arrested *Xenopus* oocytes, the oocytes matured into eggs; that is, the mammalian somatic mitotic cells contained a cytosolic factor that exhibited frog MPF activity. This finding suggested that MPF controls the entry of mammalian somatic cells into mitosis as well as the entry of frog oocytes into meiosis. When cytoplasm from mitotically arrested mammalian somatic cells was injected into interphase cells, the interphase cells entered mitosis; that is, their nuclear membranes disassembled and their chromosomes condensed. Thus MPF is the diffusible factor, first revealed in cell-fusion experiments (see Figure 21-3), that promotes entry of cells into mitosis. Conveniently, the acronym MPF also can stand for **mitosis-promoting factor,** a name that denotes the more general activity of this factor.

Because the oocyte injection assay initially used to measure MPF activity is cumbersome, several years passed before MPF was purified by column chromatography and characterized. MPF is in fact one of the heterodimeric complexes composed of a cyclin and CDK now known to regulate the cell cycle. Each MPF subunit initially was recognized through different experimental approaches. First we discuss how the regulatory cyclin subunit was identified and then describe how yeast genetic experiments led to discovery of the catalytic CDK subunit.

Mitotic Cyclin Was First Identified in Early Sea Urchin Embryos

Experiments with inhibitors showed that new protein synthesis is required for the increase in MPF during the mitotic phase of each cell cycle in early frog embryos. As in early

(a)

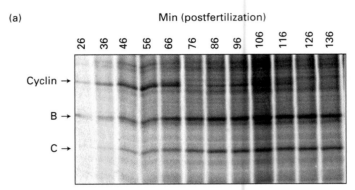

(b)

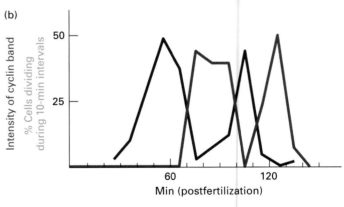

▲ **EXPERIMENTAL FIGURE 21-8** Experimental detection of cyclical synthesis and destruction of mitotic cyclin in sea urchin embryos. A suspension of sea urchin eggs was synchronously fertilized by the addition of sea urchin sperm, and ^{35}S-methionine was added. At 10-minute intervals beginning 16 minutes after fertilization, samples were taken for protein analysis on an SDS polyacrylamide gel and for detection of cell cleavage by microscopy. (a) Autoradiograms of the SDS gels at each sampling time. Most proteins, such as X and Y, continuously increased in intensity. In contrast, cyclin suddenly decreased in intensity at 76 minutes after fertilization and then began increasing again at 86 minutes. The cyclin band peaked again at 106 min and decreased again at 126 min. (b) Plot of the intensity of the cyclin band (red line) and the fraction of cells that had undergone cleavage during the previous 10-minute interval (cyan line). Note that the amount of cyclin fell precipitously just before cell cleavage. [From T. Evans et al., 1983, *Cell* **33**:389; courtesy of R. Timothy Hunt, Imperial Cancer Research Fund.]

frog embryos, the initial cell cycles in the early sea urchin embryo occur synchronously, with all the embryonic cells entering mitosis simultaneously. Biochemical studies with sea urchin eggs and embryos led to identification of the cyclin component of MPF. These studies with synchronously fertilized sea urchin eggs revealed that the concentration of one protein peaked early in mitosis, fell abruptly just before cell cleavage, and then accumulated during the following interphase to peak early in the next mitosis and fall abruptly just before the second cleavage (Figure 21-8). Careful analysis showed that this protein, named *cyclin B*, is synthesized continuously during the embryonic cell cycles but is abruptly destroyed following anaphase. Since its concentration peaks in mitosis, cyclin B functions as a *mitotic cyclin*.

In subsequent experiments, a cDNA clone encoding sea urchin cyclin B was used as a probe to isolate a homologous cyclin B cDNA from *Xenopus laevis*. Pure *Xenopus* cyclin B, obtained by expression of the cDNA in *E. coli*, was used to produce antibody specific for cyclin B. Using this antibody, a polypeptide was detected in Western blots that co-purified with MPF activity from *Xenopus* eggs, demonstrating that one subunit of MPF is indeed cyclin B.

Cyclin B Levels and Kinase Activity of Mitosis-Promoting Factor (MPF) Change Together in Cycling *Xenopus* Egg Extracts

Some unusual aspects of the synchronous cell cycles in early *Xenopus* embryos provided a way to study the role of mitotic cyclin in controlling MPF activity. First, following fertilization of *Xenopus* eggs, the G_1 and G_2 periods are minimized during the initial 12 synchronous cell cycles. That is, once mitosis is complete, the early embryonic cells proceed immediately into the next S phase, and once DNA replication is complete, the cells progress almost immediately into the next mitosis. Second, the oscillation in MPF activity that occurs as early frog embryos enter and exit mitosis is observed in the cytoplasm of a fertilized frog egg even when the nucleus is removed and no transcription can occur. This finding indicates that all the cellular components required for progress through the truncated cell cycles in early *Xenopus* embryos are stored in the unfertilized egg. In somatic cells generated later in development and in yeasts considered in later sections, specific mRNAs must be produced at particular points in the cell cycle for progress through the cycle to proceed. But in early *Xenopus* embryos, all the mRNAs necessary for the early cell divisions are present in the unfertilized egg.

Extracts prepared from unfertilized *Xenopus* eggs thus contain all the materials required for multiple cell cycles, including the enzymes and precursors needed for DNA replication, the histones and other chromatin proteins involved in assembling the replicated DNA into chromosomes, and the proteins and phospholipids required in formation of the nuclear envelope. These egg extracts also synthesize proteins encoded by mRNAs in the extract, including cyclin B.

When nuclei prepared from *Xenopus* sperm are added to such a *Xenopus* egg extract, the highly condensed sperm chromatin decondenses, and the sperm DNA replicates one time. The replicated sperm chromosomes then condense and the nuclear envelope disassembles, just as it does in intact cells entering mitosis. About 10 minutes after the nuclear envelope disassembles, all the cyclin B in the extract suddenly is degraded, as it is in intact cells following anaphase. Following cyclin B degradation, the sperm chromosomes decondense and a nuclear envelope re-forms around them, as in an intact cell at the end of mitosis. After about 20 minutes, the cycle begins again. DNA within the nuclei formed after the first mitotic period (now 2*n*) replicates, forming 4*n* nuclei. Cyclin B, synthesized from the cyclin B mRNA present in the extract, accumulates. As cyclin B approaches peak levels, the chromosomes condense once again, the nuclear envelopes break down, and about 10 minutes later cyclin B is once again suddenly destroyed. These remarkable *Xenopus* egg extracts can mediate several of these cycles, which mimic the rapid synchronous cycles of an early frog embryo.

Studies with this egg extract experimental system were aided by development of a new assay for MPF activity. Using MPF purified with the help of the oocyte injection assay (see Figure 21-6), researchers had found that MPF phosphorylates histone H1. This H1 kinase activity provided a much simpler and more easily quantitated assay for MPF activity than the oocyte injection assay. Armed with a convenient assay, researchers tracked the MPF activity and concentration of cyclin B in cycling *Xenopus* egg extracts. These studies showed that MPF activity rises and falls in synchrony with the concentration of cyclin B (Figure 21-9a). The early events of mitosis—chromosome condensation and nuclear envelope disassembly—occurred when MPF activity reached its highest levels in parallel with the rise in cyclin B concentration. Addition of cycloheximide, an inhibitor of protein synthesis, prevented cyclin B synthesis and also prevented the rise in MPF activity, chromosome condensation, and nuclear envelope disassembly.

To test the functions of cyclin B in these cell-cycle events, all mRNAs in the egg extract were degraded by digestion with a low concentration of RNase, which then was inactivated by addition of a specific inhibitor. This treatment destroys mRNAs without affecting the tRNAs and rRNAs required for protein synthesis, since their degradation requires much higher concentrations of RNase. When sperm nuclei were added to the RNase-treated extracts, the 1*n* nuclei replicated their DNA, but the increase in MPF activity and the early mitotic events (chromosome condensation and nuclear envelope disassembly), which the untreated extract supports, did not occur (Figure 21-9b). Addition of *cyclin B* mRNA, produced in vitro from cloned *cyclin B* cDNA, to the RNase-treated egg extract and sperm nuclei restored the parallel oscillations in the cyclin B concentration and MPF activity and the characteristic early and late mitotic events as observed with the untreated egg extract (Figure 21-9c). Since cyclin B is the only protein synthesized under these conditions, these results demonstrate that it is the

(a) Untreated extract

— = MPF activity
— = Cyclin B concentration

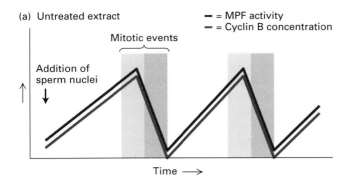

(b) RNase-treated extract

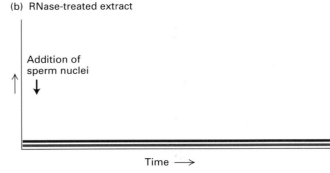

(c) RNase-treated extract + wild-type *cyclin B* mRNA

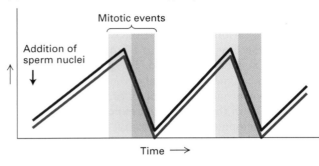

(d) RNase-treated extract + nondegradable *cyclin B* mRNA

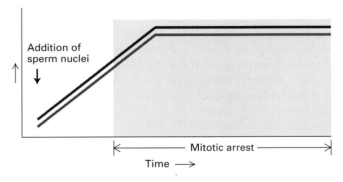

▲ **EXPERIMENTAL FIGURE 21-9 Cycling of MPF activity and mitotic events in *Xenopus* egg extracts depend on synthesis and degradation of cyclin B.** In all cases, MPF activity and cyclin B concentration were determined at various times after addition of sperm nuclei to a *Xenopus* egg extract treated as indicated in each panel. Microscopic observations determined the occurrence of early mitotic events (blue shading), including chromosome condensation and nuclear envelope disassembly, and of late events (orange shading), including chromosome decondensation and nuclear envelope reassembly. See text for discussion. [See A. W. Murray et al., 1989, *Nature* **339**:275; adapted from A. Murray and T. Hunt, 1993, *The Cell Cycle: An Introduction*, W. H. Freeman and Company.]

crucial protein whose synthesis is required to regulate MPF activity and the cycles of chromosome condensation and nuclear envelope breakdown mediated by cycling *Xenopus* egg extracts.

In these experiments, chromosome decondensation and nuclear envelope assembly (late mitotic events) coincided with decreases in the cyclin B level and MPF activity. To determine whether degradation of cyclin B is required for exit from mitosis, researchers added a mutant mRNA encoding a nondegradable cyclin B to a mixture of RNase-treated *Xenopus* egg extract and sperm nuclei. As shown in Figure 21-9d, MPF activity increased in parallel with the level of the mutant cyclin B, triggering condensation of the sperm chromatin and nuclear envelope disassembly (early mitotic events). However, the mutant cyclin B produced in this reaction never was degraded. As a consequence, MPF activity remained elevated, and the late mitotic events of chromosome decondensation and nuclear envelope formation were both blocked. This experiment demonstrates that the fall in MPF activity and exit from mitosis depend on degradation of cyclin B.

The results of the two experiments with RNase-treated extracts show that entry into mitosis requires the accumulation of cyclin B, the *Xenopus* mitotic cyclin, to high levels, and that exit from mitosis requires the degradation of this mitotic cyclin. Since MPF kinase activity varied in parallel with the concentration of the mitotic cyclin, the results implied that high MPF kinase activity results in entry into mitosis and that a fall in MPF kinase activity is required to exit mitosis.

Anaphase-Promoting Complex (APC) Controls Degradation of Mitotic Cyclins and Exit from Mitosis

Further studies revealed that vertebrate cells contain three proteins that can function like cyclin B to stimulate *Xenopus* oocyte maturation: two closely related cyclin Bs and cyclin A. Collectively called *B-type cyclins*, these proteins exhibit regions of high sequence homology. (B-type cyclins are distinguished from G_1 cyclins described in Section 21.5.) In intact cells, degradation of all the B-type cyclins begins after the onset of anaphase, the period of mitosis when sister chromatids are separated and pulled toward opposite spindle poles.

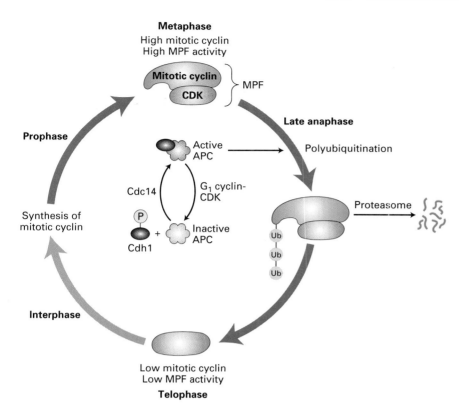

Metaphase
High mitotic cyclin
High MPF activity

Prophase

Synthesis of mitotic cyclin

Interphase

Late anaphase

Polyubiquitination

Proteasome

Active APC

Cdc14

G_1 cyclin-CDK

Inactive APC

Cdh1

Low mitotic cyclin
Low MPF activity

Telophase

◄ **FIGURE 21-10 Regulation of mitotic cyclin levels in cycling *Xenopus* early embryonic cells.** In late anaphase, the anaphase-promoting complex (APC) polyubiquitinates mitotic cyclins. As the cyclins are degraded by proteasomes, MPF kinase activity declines precipitously, triggering the onset of telophase. APC activity is directed toward mitotic cyclins by a specificity factor, called Cdh1, that is phosphorylated and thereby inactivated by G_1 cyclin-CDK complexes. A specific phosphatase called Cdc14 removes the regulatory phosphate from the specificity factor late in anaphase. Once the specificity factor is inhibited in G_1, the concentration of mitotic cyclin increases, eventually reaching a high enough level to stimulate entry into the subsequent mitosis.

Biochemical studies with *Xenopus* egg extracts showed that at the time of their degradation, wild-type B-type cyclins are modified by covalent addition of multiple **ubiquitin** molecules. This process of polyubiquitination marks proteins for rapid degradation in eukaryotic cells by **proteasomes,** multiprotein cylindrical structures containing numerous proteases. Addition of a ubiquitin chain to a B-type cyclin or other target protein requires three types of enzymes (see Figure 3-13). Ubiquitin is first activated at its carboxyl-terminus by formation of a thioester bond with a cysteine residue of *ubiquitin-activating enzyme,* E1. Ubiquitin is subsequently transferred from E1 to a cysteine of one of a class of related enzymes called *ubiquitin-conjugating enzymes,* E2. The specific E2 determines, along with a third protein, *ubiquitin ligase* (E3), the substrate protein to which a ubiquitin chain will be covalently linked. Many ubiquitin ligases are multisubunit proteins.

Sequencing of cDNAs encoding several B-type cyclins from various eukaryotes showed that all contain a homologous nine-residue sequence near the N-terminus called the *destruction box.* Deletion of this destruction box, as in the mutant mRNA used in the experiment depicted in Figure 21-8, prevents the polyubiquitination of B-type cyclins and thus makes them nondegradable. The ubiquitin ligase that recognizes the mitotic cyclin destruction box is a multisubunit protein called the *anaphase-promoting complex (APC),* which we introduced earlier in the chapter (see Figure 21-2, steps ⑧ and ⑨).

Figure 21-10 depicts the current model that best explains the changes in mitotic cyclin levels seen in cycling *Xenopus* early embryonic cells. Cyclin B, the primary mi-

totic cyclin in multicellular animals (metazoans), is synthesized throughout the cell cycle from a stable mRNA. The observed fall in its concentration in late anaphase results from its APC-stimulated degradation at this point in the cell cycle. As we discuss in Section 21.5, genetic studies with yeast led to identification of an *APC specificity factor,* called *Cdh1,* that binds to APC and directs it to polyubiquitinate mitotic cyclins. This specificity factor is active only in late anaphase when the segregating chromosomes have moved far enough apart in the dividing cell to assure that both daughter cells will contain one complete set of chromosomes. Phosphorylation of Cdh1 by other cyclin-CDK complexes during G_1 inhibits its association with the APC and thus degradation of mitotic cyclin (see Figure 21-2, step ②). This inhibition permits the gradual rise in mitotic cyclin levels observed throughout interphase of the next cell cycle.

KEY CONCEPTS OF SECTION 21.2

Biochemical Studies with Oocytes, Eggs, and Early Embryos

■ MPF is a protein kinase that requires a mitotic cyclin for activity. The protein kinase activity of MPF stimulates the onset of mitosis by phosphorylating multiple specific protein substrates, most of which remain to be identified.

■ In the synchronously dividing cells of early *Xenopus* and sea urchin embryos, the concentration of mitotic cyclins (e.g., cyclin B) and MPF activity increase as cells enter mitosis and then fall as cells exit mitosis (see Figures 21-7 and 21-8).

■ The rise and fall in MPF activity during the cell cycle result from concomitant synthesis and degradation of mitotic cyclin (see Figure 21-9).

■ The multisubunit anaphase-promoting complex (APC) is a ubiquitin ligase that recognizes a conserved destruction box sequence in mitotic cyclins and promotes their polyubiquitination, marking the proteins for rapid degradation by proteasomes. The resulting decrease in MPF activity leads to completion of mitosis.

■ The ubiquitin ligase activity of APC is controlled so that mitotic cyclins are polyubiquitinated only during late anaphase (see Figure 21-10). Deactivation of APC in G_1 permits accumulation of mitotic cyclins during the next cell cycle. This results in the cyclical increases and decreases in MPF activity that cause the entry into and exit from mitosis.

21.3 Genetic Studies with *S. pombe*

The studies with *Xenopus* egg extracts described in the previous section showed that continuous synthesis of a mitotic cyclin followed by its periodic degradation at late anaphase is required for the rapid cycles of mitosis observed in early *Xenopus* embryos. Identification of the catalytic protein kinase subunit of MPF and further insight into its regulation came from genetic analysis of the cell cycle in the fission yeast *S. pombe*. This yeast grows as a rod-shaped cell that increases in length as it grows and then divides in the middle during mitosis to produce two daughter cells of equal size (Figure 21-11).

In wild-type *S. pombe*, entry into mitosis is carefully regulated in order to properly coordinate cell division with cell growth. Temperature-sensitive mutants of *S. pombe* with conditional defects in the ability to progress through the cell cycle are easily recognized because they cause characteristic changes in cell length at the nonpermissive temperature. The many such mutants that have been isolated fall into two groups. In the first group are *cdc* mutants, which fail to progress through one of the phases of the cell cycle at the nonpermissive temperature; they form extremely long cells because they continue to grow in length, but fail to divide. In contrast, *wee* mutants form smaller-than-normal cells because they are defective in the proteins that normally prevent cells from dividing when they are too small.

In *S. pombe* wild-type genes are indicated in italics with a superscript plus sign (e.g., $cdc2^+$); genes with a recessive

(a)

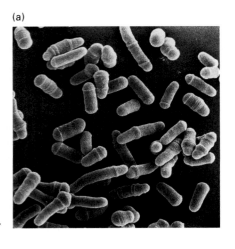

(b)

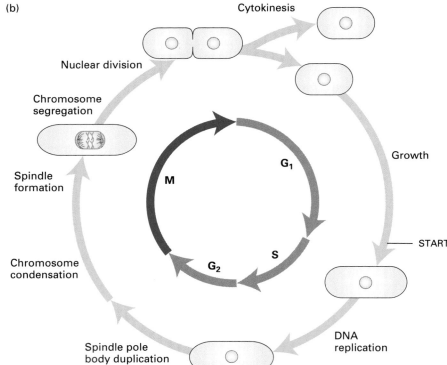

▲ **FIGURE 21-11 The fission yeast *S. pombe*.** (a) Scanning electron micrograph of *S. pombe* cells at various stages of the cell cycle. Long cells are about to enter mitosis; short cells have just passed through cytokinesis. (b) Main events in the *S. pombe* cell cycle. Note that the nuclear envelope does not disassemble during mitosis in *S. pombe* and other yeasts. [Part (a) courtesy of N. Hajibagheri.]

mutation, in italics with a superscript minus sign (e.g., $cdc2^-$). The protein encoded by a particular gene is designated by the gene symbol in Roman type with an initial capital letter (e.g., Cdc2).

A Highly Conserved MPF-like Complex Controls Entry into Mitosis in *S. pombe*

Temperature-sensitive mutations in *cdc2*, one of several different *cdc* genes in *S. pombe*, produce opposite phenotypes depending on whether the mutation is recessive or dominant (Figure 21-12). Recessive mutations ($cdc2^-$) give rise to abnormally long cells, whereas dominant mutations ($cdc2^D$) give rise to abnormally small cells, the wee phenotype. As discussed in Chapter 9, recessive mutations generally cause a *loss* of the wild-type protein function; in diploid cells, both alleles must be mutant in order for the mutant phenotype to be observed. In contrast, dominant mutations generally result in a *gain* in protein function, either because of overproduction or lack of regulation; in this case, the presence of only one mutant allele confers the mutant phenotype in diploid cells. The finding that a loss of Cdc2 activity ($cdc2^-$ mutants) prevents entry into mitosis and a gain of Cdc2 activity ($cdc2^D$ mutants) brings on mitosis earlier than normal identified Cdc2 as a key regulator of entry into mitosis in *S. pombe*.

The wild-type $cdc2^+$ gene contained in a *S. pombe* plasmid library was identified and isolated by its ability to complement $cdc2^-$ mutants (see Figure 21-4). Sequencing showed that $cdc2^+$ encodes a 34-kDa protein with homology to eukaryotic protein kinases. In subsequent studies, researchers identified cDNA clones from other organisms that could complement *S. pombe* $cdc2^-$ mutants. Remarkably, they isolated a human cDNA encoding a protein identical to *S. pombe* Cdc2 in 63 percent of its residues.

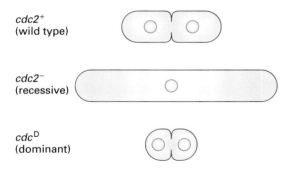

▲ **EXPERIMENTAL FIGURE 21-12 Recessive and dominant *S. pombe* cdc2 mutants have opposite phenotypes.** Wild-type cell ($cdc2^+$) is schematically depicted just before cytokinesis with two normal-size daughter cells. A recessive $cdc2^-$ mutant cannot enter mitosis at the nonpermissive temperature and appears as an elongated cell with a single nucleus, which contains duplicated chromosomes. A dominant $cdc2^D$ mutant enters mitosis prematurely before reaching normal size in G_2; thus, the two daughter cells resulting from cytokinesis are smaller than normal, that is, they have the wee phenotype.

Isolation and sequencing of another *S. pombe cdc* gene ($cdc13^+$), which also is required for entry into mitosis, revealed that it encodes a protein with homology to sea urchin and *Xenopus* cyclin B. Further studies showed that a heterodimer of Cdc13 and Cdc2 forms the *S. pombe* MPF; like *Xenopus* MPF, this heterodimer has protein kinase activity that will phosphorylate histone H1. Moreover, the H1 protein kinase activity rises as *S. pombe* cells enter mitosis and falls as they exit mitosis in parallel with the rise and fall in the Cdc13 level. These findings, which are completely analogous to the results obtained with *Xenopus* egg extracts (see Figure 21-9a), identified Cdc13 as the mitotic cyclin in *S. pombe*. Further studies showed that the isolated Cdc2 protein and its homologs in other eukaryotes have little protein kinase activity until they are bound by a cyclin. Hence, this family of protein kinases became known as cyclin-dependent kinases, or CDKs.

Researchers soon found that antibodies raised against a highly conserved region of Cdc2 recognize a polypeptide that co-purifies with MPF purified from *Xenopus* eggs. Thus *Xenopus* MPF is also composed of a mitotic cyclin (cyclin B) and a CDK (called CDK1). This convergence of findings from biochemical studies in an invertebrate (sea urchin) and a vertebrate (*Xenopus*) and from genetic studies in a yeast indicated that entry into mitosis is controlled by analogous mitotic cyclin-CDK complexes in all eukaryotes (see Figure 21-2, step [7]). Moreover, most of the participating proteins have been found to be highly conserved during evolution.

Phosphorylation of the CDK Subunit Regulates the Kinase Activity of MPF

Analysis of additional *S. pombe cdc* mutants revealed that proteins encoded by other genes regulate the protein kinase activity of the mitotic cyclin-CDK complex (MPF) in fission yeast. For example, temperature-sensitive $cdc25^-$ mutants are delayed in entering mitosis at the nonpermissive temperature, producing elongated cells. On the other hand, overexpression of Cdc25 from a plasmid present in multiple copies per cell decreases the length of G_2 causing premature entry into mitosis and small (wee) cells (Figure 21-13a). Conversely, loss-of-function mutations in the $wee1^+$ gene causes premature entry into mitosis resulting in small cells, whereas overproduction of Wee1 protein increases the length of G_2 resulting in elongated cells. A logical interpretation of these findings is that Cdc25 protein stimulates the kinase activity of *S. pombe* MPF, whereas Wee1 protein inhibits MPF activity (Figure 21-13b).

In subsequent studies, the wild-type $cdc25^+$ and $wee1^+$ genes were isolated, sequenced, and used to produce the encoded proteins with suitable expression vectors. The deduced sequences of Cdc25 and Wee1 and biochemical studies of the proteins demonstrated that they regulate the kinase activity of *S. pombe* MPF by phosphorylating and dephosphorylating specific regulatory sites in Cdc2, the CDK subunit of MPF.

(a)

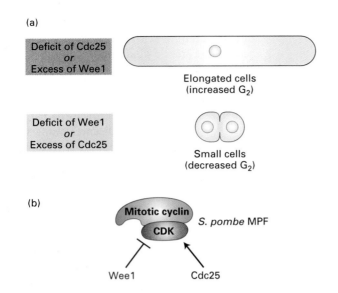

both residues are phosphorylated, MPF is inactive. Finally, the *Cdc25 phosphatase* removes the phosphate from Y15, yielding highly active MPF. Site-specific mutagenesis that changed the Y15 in *S. pombe* CDK to a phenylalanine, which cannot be phosphorylated, produced mutants with the wee phenotype, similar to that of *wee1⁻* mutants. Both mutations prevent the inhibitory phosphorylation at Y15, resulting in the inability to properly regulate MPF activity, and, consequently, premature entry into mitosis.

Conformational Changes Induced by Cyclin Binding and Phosphorylation Increase MPF Activity

Unlike both fission and budding yeasts, each of which produce just one CDK, vertebrates produce several CDKs (see Table 21-1). The three-dimensional structure of one human cyclin-dependent kinase (CDK2) has been determined and provides insight into how cyclin binding and phosphorylation of CDKs regulate their protein kinase activity. Although the three-dimensional structures of the *S. pombe* CDK and most other cyclin-dependent kinases have not been determined, their extensive sequence homology with human CDK2 suggests that all these CDKs have a similar structure and are regulated by a similar mechanism.

Unphosphorylated, inactive CDK2 contains a flexible region, called the T-loop, that blocks access of protein substrates to the active site where ATP is bound (Figure 21-15a). Steric blocking by the T-loop largely explains why free CDK2, unbound to cyclin, has no protein kinase activity. Unphosphorylated CDK2 bound to one of its cyclin partners, cyclin A, has minimal but detectable protein kinase activity in vitro, although it may be essentially inactive in vivo. Extensive interactions between cyclin A and the T-loop cause a dramatic shift in the position of the T-loop, thereby exposing the CDK2 active site (Figure 21-15b). Binding of cyclin A also shifts the position of the α1 helix in CDK2, modifying its substrate-binding surface. Phosphorylation of the activating threonine, located

▲ **EXPERIMENTAL FIGURE 21-13 Cdc25 and Wee1 have opposing effects on *S. pombe* MPF activity.** (a) Cells that lack Cdc25 or Wee1 activity, as a result of recessive temperature-sensitive mutations in the corresponding genes, have the opposite phenotype. Likewise, cells with multiple copies of plasmids containing wild-type *cdc25⁺* or *wee1⁺*, and which thus produce an excess of the encoded proteins, have opposite phenotypes. (b) These phenotypes imply that the mitotic cyclin-CDK complex is activated (→) by Cdc25 and inhibited (—|) by Wee1. See text for further discussion.

Figure 21-14 illustrates the functions of four proteins that regulate the protein kinase activity of the *S. pombe* CDK. First is Cdc13, the mitotic cyclin of *S. pombe* (equivalent to cyclin B in metazoans), which associates with the CDK to form MPF with extremely low activity. Second is the **Wee1 protein-tyrosine kinase,** which phosphorylates an inhibitory tyrosine residue (Y15) in the CDK subunit. Third is another kinase, designated CDK-activating kinase (CAK), which phosphorylates an activating threonine residue (T161). When

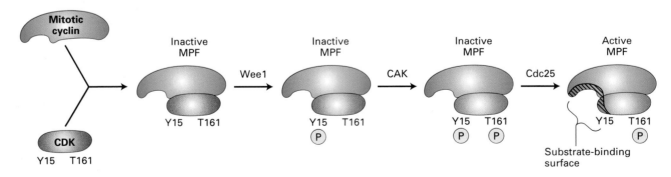

▲ **FIGURE 21-14 Regulation of the kinase activity of *S. pombe* mitosis-promoting factor (MPF).** Interaction of mitotic cyclin (Cdc13) with cyclin-dependent kinase (Cdc2) forms MPF. The CDK subunit can be phosphorylated at two regulatory sites: by Wee1 at tyrosine-15 (Y15) and by CDK-activating kinase (CAK) at threonine-161 (T161). Removal of the phosphate on Y15

by Cdc25 phosphatase yields active MPF in which the CDK subunit is monophosphorylated at T161. The mitotic cyclin subunit contributes to the specificity of substrate binding by MPF, probably by forming part of the substrate-binding surface (cross-hatch), which also includes the inhibitory Y15 residue.

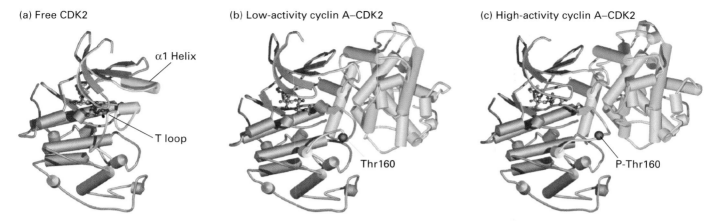

(a) Free CDK2

α1 Helix

T loop

(b) Low-activity cyclin A–CDK2

Thr160

(c) High-activity cyclin A–CDK2

P-Thr160

▲ **FIGURE 21-15 Structural models of human CDK2, which is homologous to the *S. pombe* cyclin-dependent kinase (CDK).** (a) Free, inactive CDK2 unbound to cyclin A. In free CDK2, the T-loop blocks access of protein substrates to the γ-phosphate of the bound ATP, shown as a ball-and-stick model. The conformations of the regions highlighted in yellow are altered when CDK is bound to cyclin A. (b) Unphosphorylated, low-activity cyclin A–CDK2 complex. Conformational changes induced by binding of a domain of cyclin A (green) cause the T-loop to pull away from the active site of CDK2, so that substrate proteins can bind. The α1 helix in CDK2, which interacts extensively with cyclin A, moves several angstroms into the catalytic cleft, repositioning key catalytic side chains required for the phosphorotransfer reaction to substrate specificity. The red ball marks the position equivalent to threonine 161 in *S. pombe* Cdc2. (c) Phosphorylated, high-activity cyclin A–CDK2 complex. The conformational changes induced by phosphorylation of the activating threonine (red ball) alter the shape of the substrate-binding surface, greatly increasing the affinity for protein substrates. [Courtesy of P. D. Jeffrey. See A. A. Russo et al., 1996, *Nature Struct. Biol.* **3**:696.]

in the T-loop, causes additional conformational changes in the cyclin A–CDK2 complex that greatly increase its affinity for protein substrates (Figure 21-15c). As a result, the kinase activity of the phosphorylated complex is a hundredfold greater than that of the unphosphorylated complex.

The inhibitory tyrosine residue (Y15) in the *S. pombe* CDK is in the region of the protein that binds the ATP phosphates. Vertebrate CDK2 proteins contain a second inhibitory residue, threonine-14 (T14), that is located in the same region of the protein. Phosphorylation of Y15 and T14 in these proteins prevents binding of ATP because of electrostatic repulsion between the phosphates linked to the protein and the phosphates of ATP. Thus these phosphorylations inhibit protein kinase activity even when the CDK protein is bound by a cyclin and the activating residue is phosphorylated.

Other Mechanisms Also Control Entry into Mitosis by Regulating MPF Activity

So far we have discussed two mechanisms for controlling entry into mitosis: (a) regulation of the concentration of mitotic cyclins as outlined in Figure 21-10 and (b) regulation of the kinase activity of MPF as outlined in Figure 21-14. Further studies of *S. pombe* mutants with altered cell cycles have revealed additional genes whose encoded proteins directly or indirectly influence MPF activity. At present it is clear that MPF activity in *S. pombe* is regulated in a complex fashion in order to control precisely the timing of mitosis and therefore the size of daughter cells.

Enzymes with activities equivalent to *S. pombe* Wee1 and Cdc25 have been found in cycling *Xenopus* egg extracts. The *Xenopus* Wee1 tyrosine kinase activity is high and Cdc25 phosphatase activity is low during interphase. As a result, MPF assembled from *Xenopus* CDK1 and newly synthesized mitotic cyclin is inactive. As the extract initiates the events of mitosis, Wee1 activity diminishes and Cdc25 activity increases so that MPF is converted into its active form. Although cyclin B is the only protein whose synthesis is required for the cycling of early *Xenopus* embryos, the activities of other proteins, including *Xenopus* Wee1 and Cdc25, must be properly regulated for cycling to occur. In its active form, Cdc25 is phosphorylated. Its activity is also controlled by additional protein kinases and phosphatases.

MPF activity also can be regulated by controlling transcription of the genes encoding the proteins that regulate MPF activity. For example, after the initial rapid synchronous cell divisions of the early *Drosophila* embryo, all the mRNAs are degraded, and the cells become arrested in G_2. This arrest occurs because the *Drosophila* homolog of Cdc25, called String, is unstable. The resulting decrease in String phosphatase activity maintains MPF in its inhibited state, preventing entry into mitosis. The subsequent regulated entry into mitosis by specific groups of cells is then triggered by the regulated transcription of the *string* gene.

KEY CONCEPTS OF SECTION 21.3

Genetic Studies with *S. pombe*

■ In the fission yeast *S. pombe*, the $cdc2^+$ gene encodes a cyclin-dependent protein kinase (CDK) that associates with a mitotic cyclin encoded by the $cdc13^+$ gene. The resulting mitotic cyclin-CDK heterodimer is equivalent to *Xenopus* MPF. Mutants that lack either the mitotic cyclin or the CDK fail to enter mitosis and form elongated cells.

■ The protein kinase activity of the mitotic cyclin-CDK complex (MPF) depends on the phosphorylation state of two residues in the catalytic CDK subunit (see Figure 21-14). The activity is greatest when threonine-161 is phosphorylated and is inhibited by Wee1-catalyzed phosphorylation of tyrosine-15, which interferes with correct binding of ATP. This inhibitory phosphate is removed by the Cdc25 protein phosphatase.

■ A decrease in Wee1 activity and increase in Cdc25 activity, resulting in activation of the mitotic cyclin-CDK complex, results in the onset of mitosis.

■ The human cyclin A–CDK2 complex is similar to MPF from *Xenopus* and *S. pombe*. Structural studies with the human proteins reveal that cyclin binding to CDK2 and phosphorylation of the activating threonine (equivalent to threonine-161 in the *S. pombe* CDK) cause conformational changes that expose the active site and modify the substrate-binding surface so that it has high activity and affinity for protein substrates (see Figure 21-15).

21.4 Molecular Mechanisms for Regulating Mitotic Events

In the previous sections, we have seen that a regulated increase in MPF activity induces entry into mitosis. Presumably, the entry into mitosis is a consequence of the phosphorylation of specific proteins by the protein kinase activity of MPF. Although many of the critical substrates of MPF remain to be identified, we now know of examples that show how regulation by MPF phosphorylation mediates many of the early events of mitosis leading to metaphase: chromosome condensation, formation of the mitotic spindle, and disassembly of the nuclear envelope (see Figure 20-29).

Recall that a decrease in mitotic cyclins and the associated inactivation of MPF coincides with the later stages of mitosis (see Figure 21-9a). Just before this, in early anaphase, sister chromatids separate and move to opposite spindle poles. During telophase, microtubule dynamics return to interphase conditions, the chromosomes decondense, the nuclear envelope re-forms, the Golgi complex is remodeled, and cytokinesis occurs. Some of these processes are triggered by dephosphorylation; others, by protein degradation.

In this section, we discuss the molecular mechanisms and specific proteins associated with some of the events that characterize early and late mitosis. These mechanisms illustrate how cyclin-CDK complexes together with ubiquitin ligases control passage through the mitotic phase of the cell cycle.

Phosphorylation of Nuclear Lamins and Other Proteins Promotes Early Mitotic Events

The nuclear envelope is a double-membrane extension of the rough endoplasmic reticulum containing many nuclear pore

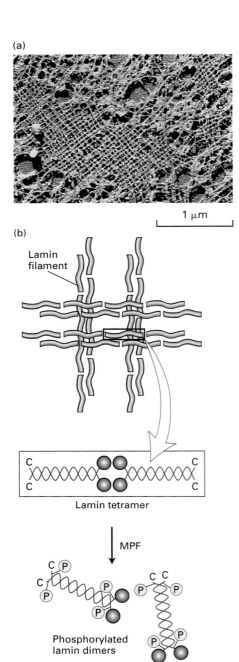

▲ **FIGURE 21-16 The nuclear lamina and its depolymerization.** (a) Electron micrograph of the nuclear lamina from a *Xenopus* oocyte. Note the regular meshlike network of lamin intermediate filaments. This structure lies adjacent to the inner nuclear membrane (see Figure 19-31). (b) Schematic diagrams of the structure of the nuclear lamina. Two orthogonal sets of 10-nm-diameter filaments built of lamins A, B, and C form the nuclear lamina (*top*). Individual lamin filaments are formed by end-to-end polymerization of lamin tetramers, which consist of two lamin dimers (*middle*). The red circles represent the globular N-terminal domains. Phosphorylation of specific serine residues near the ends of the coiled-coil rodlike central section of lamin dimers causes the tetramers to depolymerize (*bottom*). As a result, the nuclear lamina disintegrates. [Part (a) from U. Aebi et al., 1986, *Nature* **323**:560; courtesy of U. Aebi. Part (b) adapted from A. Murray and T. Hunt, 1993, *The Cell Cycle: An Introduction*, W. H. Freeman and Company.]

complexes (see Figure 5-19). The lipid bilayer of the inner nuclear membrane is supported by the **nuclear lamina,** a meshwork of lamin filaments located adjacent to the inside face of the nuclear envelope (Figure 21-16a). The three nuclear **lamins** (A, B, and C) present in vertebrate cells belong to the class of cytoskeletal proteins, the intermediate filaments, that are critical in supporting cellular membranes (Chapter 19).

Lamins A and C, which are encoded by the same transcription unit and produced by alternative splicing of a single pre-mRNA, are identical except for a 133-residue region at the C-terminus of lamin A, which is absent in lamin C. Lamin B, encoded by a different transcription unit, is modified post-transcriptionally by the addition of a hydrophobic isoprenyl group near its carboxyl-terminus. This fatty acid becomes embedded in the inner nuclear membrane, thereby anchoring the nuclear lamina to the membrane (see Figure 5-15). All three nuclear lamins form dimers containing a rodlike α-helical coiled-coil central section and globular head and tail domains; polymerization of these dimers through head-to-head and tail-to-tail associations generates the intermediate filaments that compose the nuclear lamina (see Figure 19-33).

Early in mitosis, MPF phosphorylates specific serine residues in all three nuclear lamins, causing depolymerization of the lamin intermediate filaments (Figure 21-16b). The phosphorylated lamin A and C dimers are released into solution, whereas the phosphorylated lamin B dimers remain associated with the nuclear membrane via their isoprenyl anchor. Depolymerization of the nuclear lamins leads to disintegration of the nuclear lamina meshwork and contributes to disassembly of the nuclear envelope. The experiment summarized in Figure 21-17 shows that disassembly of the nuclear envelope, which normally occurs early in mitosis, depends on phosphorylation of lamin A.

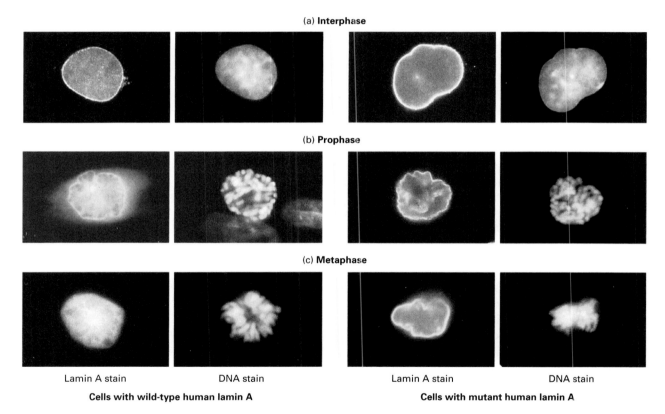

(a) **Interphase**

(b) **Prophase**

(c) **Metaphase**

| Lamin A stain | DNA stain | | Lamin A stain | DNA stain |

Cells with wild-type human lamin A   **Cells with mutant human lamin A**

▲ **EXPERIMENTAL FIGURE 21-17 Transfection experiments demonstrate that phosphorylation of human lamin A is required for lamin depolymerization.** Site-directed mutagenesis was used to prepare a mutant human *lamin A* gene encoding a protein in which alanines replace the serines that normally are phosphorylated in wild-type lamin A (see Figure 21-16b). As a result, the mutant lamin A cannot be phosphorylated. Expression vectors carrying the wild-type or mutant human gene were separately transfected into cultured hamster cells. Because the transfected *lamin* genes are expressed at much higher levels than the endogenous hamster *lamin* gene, most of the lamin A produced in transfected cells is human lamin A. Transfected cells at various stages in the cell cycle then were stained with a fluorescent-labeled monoclonal antibody specific for human lamin A and with a fluorescent dye that binds to DNA. The bright band of fluorescence around the perimeter of the nucleus in interphase cells stained for human lamin A represents polymerized (unphosphorylated) lamin A (a). In cells expressing the wild-type human lamin A, the diffuse lamin staining throughout the cytoplasm in prophase and metaphase (b and c) and the absence of the bright peripheral band in metaphase (c) indicate depolymerization of lamin A. In contrast, no lamin depolymerization occurred in cells expressing the mutant lamin A. DNA staining showed that the chromosomes were fully condensed by metaphase in cells expressing either wild-type or mutant lamin A. [From R. Heald and F. McKeon, 1990, *Cell* **61**:579.]

In addition, MPF-catalyzed phosphorylation of specific nucleoporins (see Chapter 12) causes nuclear pore complexes to dissociate into subcomplexes during prophase. Similarly, phosphorylation of integral membrane proteins of the inner nuclear membrane is thought to decrease their affinity for chromatin and contribute to disassembly of the nuclear envelope. The weakening of the associations between the inner nuclear membrane and the nuclear lamina and chromatin may allow sheets of inner nuclear membrane to retract into the endoplasmic reticulum, which is continuous with the outer nuclear membrane.

Several lines of evidence indicate that MPF-catalyzed phosphorylation also plays a role in chromosome condensation and formation of the mitotic spindle apparatus. For instance, genetic experiments in the budding yeast *S. cerevisiae* identified a family of SMC (*s*tructural *m*aintenance of *c*hromosomes) proteins that are required for normal chromosome segregation. These large proteins ($\approx$1200 amino acids) contain characteristic ATPase domains at their C-terminus and long regions predicted to participate in coiled-coil structures.

Immunoprecipitation studies with antibodies specific for *Xenopus* SMC proteins revealed that in cycling egg extracts some SMC proteins are part of a multiprotein complex called *condensin*, which becomes phosphorylated as cells enter mitosis. When the anti-SMC antibodies were used to deplete condensin from an egg extract, the extract lost its ability to condense added sperm chromatin. Other in vitro experiments showed that phosphorylated purified condensin binds to DNA and winds it into supercoils (see Figure 4-7), whereas unphosphorylated condensin does not. These results have lead to the model that individual condensin complexes are activated by phosphorylation catalyzed by MPF or another protein kinase regulated by MPF. Once activated, condensin complexes bind to DNA at intervals along the chromosome scaffold. Self-association of the bound complexes via their coiled-coil domains and supercoiling of the DNA segments between them is proposed to cause chromosome condensation.

Phosphorylation of microtubule-associated proteins by MPF probably is required for the dramatic changes in microtubule dynamics that result in the formation of the mitotic spindle and asters (Chapter 20). In addition, phosphorylation of proteins associated with the endoplasmic reticulum (ER) and Golgi complex, by MPF or other protein kinases activated by MPF-catalyzed phosphorylation, is thought to alter the trafficking of vesicles between the ER and Golgi to favor trafficking in the direction of the ER during prophase. As a result, vesicular traffic from the ER through the Golgi to the cell surface (Chapter 17), seen in interphase cells, does not occur during mitosis.

Unlinking of Sister Chromatids Initiates Anaphase

We saw earlier that in late anaphase, polyubiquitination of mitotic cyclin by the anaphase-promoting complex (APC) leads to the proteasomal destruction of this cyclin (see Figure 21-10). Additional experiments with *Xenopus* egg extracts provided evidence that degradation of cyclin B, the *Xenopus* mitotic cyclin, and the resulting decrease in MPF activity are required for chromosome decondensation but not for chromosome segregation (Figure 21-18a, b).

▶ **EXPERIMENTAL FIGURE 21-18 Onset of anaphase depends on polyubiquitination of proteins other than cyclin B in cycling *Xenopus* egg extracts.** The reaction mixtures contained an untreated or RNase-treated *Xenopus* egg extract and isolated *Xenopus* sperm nuclei, plus other components indicated below. Chromosomes were visualized with a fluorescent DNA-binding dye. Fluorescent rhodamine-labeled tubulin in the reactions was incorporated into microtubules, permitting observation of the mitotic spindle apparatus. (a, b) After the egg extract was treated with RNase to destroy endogenous mRNAs, an RNase inhibitor was added. Then mRNA encoding either wild-type cyclin B or a mutant nondegradable cyclin B was added. The time at which the condensed chromosomes and assembled spindle apparatus became visible after addition of sperm nuclei is designated 0 minutes. In the presence of wild-type cyclin B (a), condensed chromosomes attached to the spindle microtubules and segregated toward the poles of the spindle. By 40 minutes, the spindle had depolymerized (thus is not visible), and the chromosomes had decondensed (diffuse DNA staining) as cyclin B was degraded. In the presence of nondegradable cyclin B (b), chromosomes segregated to the spindle poles by 15 minutes, as in (a), but the spindle microtubules did not depolymerize and the chromosomes did not decondense even after 80 minutes. These observations indicate that degradation of cyclin B is not required for chromosome segregation during anaphase, although it is required for depolymerization of spindle microtubules and chromosome decondensation during telophase. (c) Various concentrations of a cyclin B peptide containing the destruction box were added to extracts that had not been treated with RNase; the samples were stained for DNA at 15 or 35 minutes after formation of the spindle apparatus. The two lowest peptide concentrations delayed chromosome segregation, and the higher concentrations completely inhibited chromosome segregation. In this experiment, the added cyclin B peptide is thought to competitively inhibit APC-mediated polyubiquitination of cyclin B as well as another target protein whose degradation is required for chromosome segregation. [From S. L. Holloway et al., 1993, *Cell* **73**:1393; courtesy of A. W. Murray.]

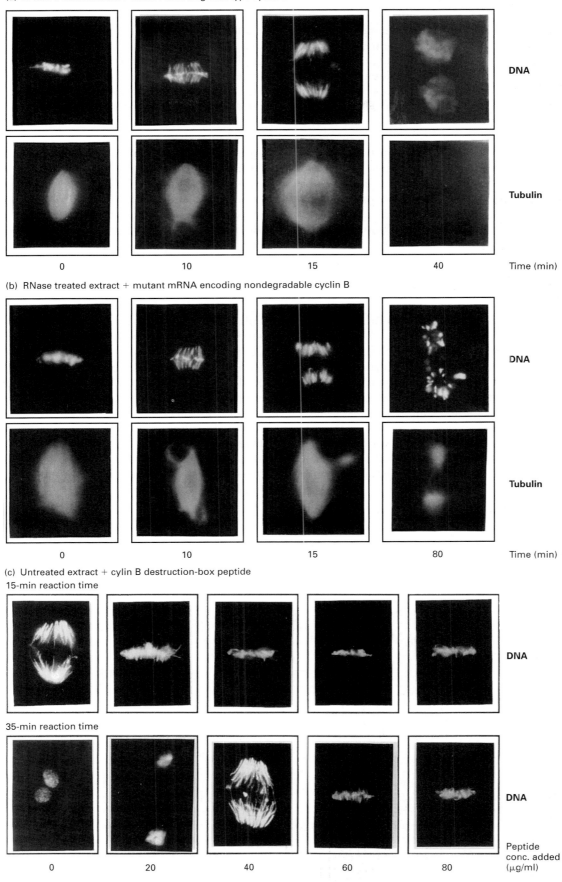

(a) RNase treated extract + mRNA encoding wild-type cyclin B

DNA

Tubulin

0   10   15   40   Time (min)

(b) RNase treated extract + mutant mRNA encoding nondegradable cyclin B

DNA

Tubulin

0   10   15   80   Time (min)

(c) Untreated extract + cylin B destruction-box peptide
15-min reaction time

DNA

35-min reaction time

DNA

0   20   40   60   80   Peptide
                                                conc. added
                                                (μg/ml)

To determine if ubiquitin-dependent degradation of another protein is required for chromosome segregation, researchers prepared a peptide containing the cyclin destruction-box sequence and the site of polyubiquitination. When this peptide was added to a reaction mixture containing untreated egg extract and sperm nuclei, decondensation of the chromosomes and, more interestingly, movement of chromosomes toward the spindle poles were greatly delayed at peptide concentrations of 20–40 μg/ml and blocked altogether at higher concentrations (Figure 21-18c). The added excess destruction-box peptide is thought to act as a substrate for the APC-directed polyubiquitination system, competing with the normal endogenous target proteins and thereby delaying or preventing their degradation by proteasomes. Competition for cyclin B accounts for the observed inhibition of chromosome decondensation. The observation that chromosome segregation also was inhibited in this experiment but not in the experiment with mutant nondegradable cyclin B (see Figure 21-18b) indicated that segregation depends on proteasomal degradation of a different target protein.

As mentioned earlier, each sister chromatid of a metaphase chromosome is attached to microtubules via its kinetochore, a complex of proteins assembled at the centromere. The opposite ends of these kinetochore microtubules associate with one of the spindle poles (see Figure 20-31). At metaphase, the spindle is in a state of tension with forces pulling the two kinetochores toward the opposite spindle poles balanced by forces pushing the spindle poles apart. Sister chromatids do not separate because they are held together at their centromeres and multiple positions along the chromosome arms by multiprotein complexes called *cohesins*. Among the proteins composing the cohesin complexes are members of the SMC protein family discussed in the previous section. When *Xenopus* egg extracts were depleted of cohesin by treatment with antibodies specific for the cohesin SMC proteins, the depleted extracts were able to replicate the DNA in added sperm nuclei, but the resulting sister chromatids did not associate properly with each other. These findings demonstrate that cohesin is necessary for cross-linking sister chromatids.

Recent genetic studies in the budding yeast *S. cerevisiae* have led to the model depicted in Figure 21-19 for how the APC regulates sister chromatid separation to initiate anaphase. Cohesin SMC proteins bind to each sister chromatid; other subunits of cohesin, including *Scc1*, then link the SMC proteins, firmly associating the two chromatids. The cross-linking activity of cohesin depends on *securin*, which is found in all eukaryotes. Prior to anaphase, securin binds to and inhibits separase, a ubiquitous protease related to the caspase proteases that regulate programmed cell death (Chapter 22). Once all chromosome kinetochores have attached to spindle microtubules, the APC is directed by a specificity factor called *Cdc20* to polyubiquitinate securin, leading to the onset of anaphase. (This specificity factor is distinct from Cdh1, which directs the APC to polyubiquitinate B-type cyclins.) Polyubiquitinated securin is rapidly degraded by proteasomes, thereby releasing separase. Free from its inhibitor, separase cleaves Scc1, breaking the protein cross-link between sister chromatids. Once this link is bro-

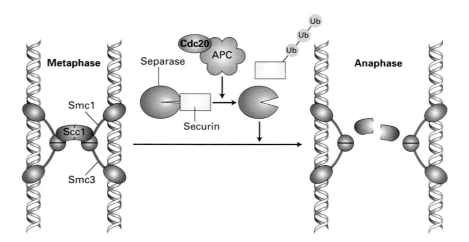

▲ **FIGURE 21-19 Model for control of entry into anaphase by APC-regulated degradation of the cohesin link between sister chromatids.** (*Left*) The multiprotein cohesin complex contains SMC1 and SMC3 (purple), dimeric proteins that bind DNA of each sister chromatid through globular domains at one end. Scc1 (orange) and two other cohesin subunits (not shown) bind to the SMC proteins associated with each chromatid, thus cross-linking the chromatids. (*Center*) Once all chromosome kinetochores have bound to spindle microtubules, the APC specificity factor Cdc20 targets APC to polyubiquitinate securin which is then degraded by the proteasome (not shown). (*Right*) The released separase then cleaves Scc1, severing the cross-link between sister chromatids. See text for discussion. [Adapted from F. Uhlmann, 2001, *Curr. Opin. Cell Biol.* **13**:754; and A. Tomans, *Nature Milestones*, http://www.nature.com/celldivision/milestones/full/milestone23.html.]

ken, the poleward force exerted on kinetochores can move sister chromatids toward the opposite spindle poles.

Because Cdc20—the specificity factor that directs APC to securin—is activated before Cdh1—the specificity factor that directs APC to mitotic cyclins—MPF activity does not decrease until after the chromosomes have segregated (see Figure 21-2, steps 8 and 9). As a result of this temporal order in the activation of the two APC specificity factors (Cdc20 and Cdh1), the chromosomes remain in the condensed state and reassembly of the nuclear envelope does not occur until chromosomes are moved to the proper position. We consider how the timing of Cdh1 activation is regulated in a later section.

Reassembly of the Nuclear Envelope and Cytokinesis Depend on Unopposed Constitutive Phosphatase Activity

Earlier we discussed how MPF-mediated phosphorylation of nuclear lamins, nucleoporins, and proteins in the inner nuclear membrane contributes to the dissociation of nuclear pore complexes and retraction of the nuclear membrane into the reticular ER. Once MPF is inactivated in late anaphase by the degradation of mitotic cyclins, the unopposed action of phosphatases reverses the action of MPF. The dephosphorylated inner nuclear membrane proteins are thought to bind to chromatin once again. As a result, multiple projections of regions of the ER membrane containing these proteins are thought to associate with the surface of the decondensing chromosomes and then fuse with each other to form a continuous double membrane around each chromosome (Figure 21-20). Dephosphorylation of nuclear pore subcomplexes is thought to allow them to reassemble nuclear pore complexes traversing the inner and outer membranes soon after fusion of the ER projections (see Figure 12-20).

The fusion of ER projections depicted in Figure 21-20 occurs by a mechanism similar to that described for the fusion of vesicles and target membranes in the secretory pathway (see Figure 17-11). Proteins with activities similar to those of NSF and α-SNAP in the secretory pathway have been shown to function in the fusion of experimentally produced nuclear envelope vesicles in vitro and are thought to mediate the fusion of ER projections around chromosomes during telophase in the intact cell. These same proteins also function in the fusion events that reassemble the Golgi apparatus. The membrane-associated **SNARE** proteins that direct the fusion of nuclear envelope extensions from the ER have not been identified, but syntaxin 5 has been shown to function as both the V-SNARE and T-SNARE during reassembly of the Golgi. During fusion of nuclear envelope vesicles in vitro, the same Ran GTPase that functions in transport through nuclear pore complexes (Chapter 12) functions similarly to Rab GTPases in vesicle fusion in the secretory pathway. Because the Ran-specific guanosine nucleotide–exchange factor (Ran-GEF) is associated with chromatin, a high local concentration of Ran · GTP is produced around the chromosomes, directing membrane fusions at the chromosome surface.

The reassembly of nuclear envelopes containing nuclear pore complexes around each chromosome forms individual mininuclei called *karyomeres* (see Figure 21-20). Subsequent fusion of the karyomeres associated with each spindle pole generates the two daughter-cell nuclei, each containing a full set of chromosomes. Dephosphorylated lamins A and C appear to be imported through the reassembled nuclear pore complexes during this period and reassemble into a new nuclear lamina. Reassembly of the nuclear lamina in the daughter nuclei probably is initiated on lamin B molecules, which remain associated with the ER membrane via their isoprenyl anchors throughout mitosis and become localized to the inner membrane of the reassembled nuclear envelopes of karyomeres.

During cytokinesis, the final step in cell division, the actin and myosin filaments composing the contractile ring slide past each other to form a cleavage furrow of steadily decreasing

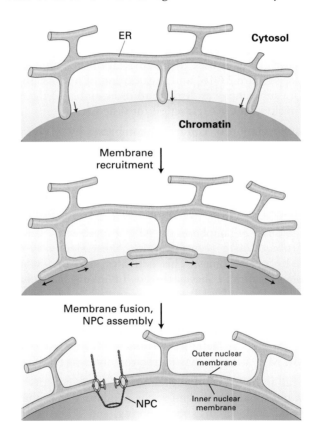

▲ **FIGURE 21-20 Model for reassembly of the nuclear envelope during telophase.** Extensions of the endoplasmic reticulum (ER) associate with each decondensing chromosome and then fuse with each other, forming a double membrane around the chromosome. Nuclear pore subcomplexes reassemble into nuclear pores, forming individual mininuclei called karyomeres. The enclosed chromosome further decondenses, and subsequent fusion of the nuclear envelopes of all the karyomeres at each spindle pole forms a single nucleus containing a full set of chromosomes. Reassembly of the nuclear lamina is not shown. [Adapted from B. Burke and J. Ellenberg, 2002, *Nature Rev. Mol. Cell Biol.* **3**:487.]

diameter (see Figure 19-20). As MPF activity rises early in mitosis, it phosphorylates the regulatory myosin light chain, thereby inhibiting the ability of myosin to associate with actin filaments. The inactivation of MPF at the end of anaphase permits protein phosphatases to dephosphorylate myosin light chain (see Figure 20-42). As a result, the contractile machinery is activated, the cleavage furrow can form, and cytokinesis proceeds. This regulatory mechanism assures that cytokinesis does not occur until the daughter chromosomes have segregated sufficiently toward the opposite poles to assure that each daughter cell receives the proper number of chromosomes.

KEY CONCEPTS OF SECTION 21.4

Molecular Mechanisms for Regulating Mitotic Events

■ Early in mitosis, MPF-catalyzed phosphorylation of lamins A, B, and C and of nucleoporins and inner nuclear envelope proteins causes depolymerization of lamin filaments (see Figure 21-16) and dissociation of nuclear pores into pore subcomplexes, leading to disassembly of the nuclear envelope and its retraction into the ER.

■ Phosphorylation of condensin complexes by MPF or a kinase regulated by MPF promotes chromosome condensation early in mitosis.

■ Sister chromatids formed by DNA replication in the S phase are linked at the centromere by cohesin complexes that contain DNA-binding SMC proteins and other proteins.

■ At the onset of anaphase, the APC is directed by Cdc20 to polyubiquitinate securin, which subsequently is degraded by proteasomes. This activates separase, which cleaves a subunit of cohesin, thereby unlinking sister chromatids (see Figure 21-19).

■ After sister chromatids have moved to the spindle poles, the APC is directed by Cdh1 to polyubiquitinate mitotic cyclins, leading to their destruction and causing the decrease in MPF activity that marks the onset of telophase.

■ The fall in MPF activity in telophase allows constitutive protein phosphatases to remove the regulatory phosphates from condensin, lamins, nucleoporins, and other nuclear membrane proteins, permitting the decondensation of chromosomes and the reassembly of the nuclear membrane, nuclear lamina, and nuclear pore complexes.

■ The association of Ran-GEF with chromatin results in a high local concentration of Ran · GTP near the decondensing chromosomes, promoting the fusion of nuclear envelope extensions from the ER around each chromosome. This forms karyomeres that then fuse to form daughter cell nuclei (see Figure 21-20).

■ The fall in MPF activity also removes its inhibition of myosin light chain, allowing the cleavage furrow to form and cytokinesis to proceed.

21.5 Genetic Studies with S. cerevisiae

In most vertebrate cells the key decision determining whether or not a cell will divide is the decision to enter the S phase. In most cases, once a vertebrate cell has become committed to entering the S phase, it does so a few hours later and progresses through the remainder of the cell cycle until it completes mitosis. S. cerevisiae cells regulate their proliferation similarly, and much of our current understanding of the molecular mechanisms controlling entry into the S phase and the control of DNA replication comes from genetic studies of S. cerevisiae.

S. cerevisiae cells replicate by budding (Figure 21-21). Both mother and daughter cells remain in the G_1 period of the cell cycle while growing, although it takes the initially larger mother cells a shorter time to reach a size compatible with cell division. When S. cerevisiae cells in G_1 have grown sufficiently, they begin a program of gene expression that leads to entry into the S phase. If G_1 cells are shifted from a rich medium to a medium low in nutrients before they reach a critical size, they remain in G_1 and grow slowly until they are large enough to enter the S phase. However, once G_1 cells reach the critical size, they become committed to completing the cell cycle, entering the S phase and proceeding through G_2 and mitosis, even if they are shifted to a medium low in nutrients. The point in late G_1 of growing S. cerevisiae cells when they become irrevocably committed to entering the S phase and traversing the cell cycle is called **START**. As we shall see in Section 21.6, a comparable phenomenon occurs in replicating mammalian cells.

A Cyclin-Dependent Kinase (CDK) Is Critical for S-Phase Entry in S. cerevisiae

All S. cerevisiae cells carrying a mutation in a particular cdc gene arrest with the same size bud at the nonpermissive temperature (see Figure 9-6b). Each type of mutant has a terminal phenotype with a particular bud size: no bud (cdc28), intermediate-sized buds, or large buds (cdc7). Note that in S. cerevisiae wild-type genes are indicated in italic capital letters (e.g., CDC28) and recessive mutant genes in italic lowercase letters (e.g., cdc28); the corresponding wild-type

▶ **FIGURE 21-21 The budding yeast S. cerevisiae.**
(a) Scanning electron micrograph of S. cerevisiae cells at various stages of the cell cycle. The larger the bud, which emerges at the end of the G_1 phase, the further along in the cycle the cell is. (b) Main events in S. cerevisiae cell cycle. Daughter cells are born smaller than mother cells and must grow to a greater extent in G_1 before they are large enough to enter the S phase. As in S. pombe, the nuclear envelope does not break down during mitosis. Unlike S. pombe chromosomes, the small S. cerevisiae chromosomes do not condense sufficiently to be visible by light microscopy. [Part (a) courtesy of E. Schachtbach and I. Herskowitz.]

protein is written in Roman letters with an initial capital (e.g., Cdc28), similar to *S. pombe* proteins.

The phenotypic behavior of temperature-sensitive *cdc28* mutants indicates that Cdc28 function is critical for entry into the S phase. When these mutants are shifted to the non-permissive temperature, they behave like wild-type cells suddenly deprived of nutrients. That is, *cdc28* mutant cells that have grown large enough to pass START at the time of the temperature shift continue through the cell cycle normally and undergo mitosis, whereas those that are too small to have passed START when shifted to the nonpermissive temperature do not enter the S phase even though nutrients are plentiful. Even though *cdc28* cells blocked in G_1 continue to grow in size at the nonpermissive temperature, they cannot pass START and enter the S phase. Thus they appear as large cells with no bud.

The wild-type *CDC28* gene was isolated by its ability to complement mutant *cdc28* cells at the nonpermissive temperature (see Figure 21-4). Sequencing of *CDC28* showed that the encoded protein is homologous to known protein kinases, and when Cdc28 protein was expressed in *E. coli*, it exhibited protein kinase activity. Actually, Cdc28 from *S. cerevisiae* was the first cell-cycle protein shown to be a protein kinase. Subsequently, the wild-type *S. pombe* *cdc2*$^+$ gene was found to be highly homologous to the *S. cerevisiae* *CDC28* gene, and the two encoded proteins—Cdc2 and Cdc28—are functionally analogous. Each type of yeast contains a *single* cyclin-dependent protein kinase (CDK), which can substitute for each other: Cdc2 in *S. pombe* and Cdc28 in *S. cerevisiae* (see Table 21-1).

The difference in the mutant phenotypes of *cdc2*$^-$ *S. pombe* cells and *cdc28 S. cerevisiae* cells can be explained in terms of the physiology of the two yeasts. In *S. pombe* cells growing in rich media, cell-cycle control is exerted primarily at the $G_2 \rightarrow M$ transition (i.e., entry to mitosis). In many *cdc2*$^-$ mutants, including those isolated first, enough Cdc2 activity is maintained at the nonpermissive temperature to permit cells to enter the S phase, but not enough to permit entry into mitosis. Such mutant cells are observed to be elongated cells arrested in G_2. At the nonpermissive temperature, cultures of completely defective *cdc2*$^-$ mutants include some cells arrested in G_1 and some arrested in G_2, depending on their location in the cell cycle at the time of the temperature shift. Conversely, cell-cycle regulation in *S. cerevisiae* is exerted primarily at the $G_1 \rightarrow S$ transition (i.e., entry to the S phase). Therefore, partially defective *cdc28* cells are arrested in G_1, but completely defective *cdc28* cells are arrested in either G_1 or G_2. These observations demonstrate that both the *S. pombe* and the *S. cerevisiae* CDKs are required for entry into both the S phase and mitosis.

Three G_1 Cyclins Associate with *S. cerevisiae* CDK to form S Phase–Promoting Factors

By the late 1980s, it was clear that mitosis-promoting factor (MPF) is composed of two subunits: a CDK and a mitotic

(a)

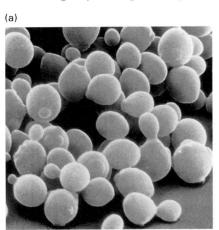

(b)

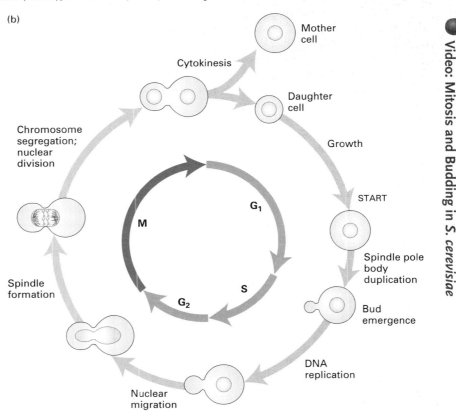

B-type cyclin required to activate the catalytic subunit. By analogy, it seemed likely that *S. cerevisiae* contains an **S phase–promoting factor (SPF)** that phosphorylates and regulates proteins required for DNA synthesis. Similar to MPF, SPF was proposed to be a heterodimer composed of the *S. cerevisiae* CDK and a cyclin, in this case one that acts in G_1 (see Figure 21-2, steps 2–4).

To identify this putative G_1 cyclin, researchers looked for genes that, when expressed at high concentration, could suppress certain temperature-sensitive mutations in the *S. cerevisiae* CDK. The rationale of this approach is illustrated in Figure 21-22. Researchers isolated two such genes designated *CLN1* and *CLN2*. Using a different approach, researchers identified a dominant mutation in a third gene called *CLN3*. Sequencing of the three *CLN* genes showed that they encoded related proteins each of which includes an ≈100-residue region exhibiting significant homology with B-type cyclins from sea urchin, *Xenopus*, human, and *S. pombe*. This region encodes the cyclin domain that interacts with CDKs and is included in the domain of human cyclin A shown in Figure 21-15b,c. The finding that the three Cln proteins contain this region of homology with mitotic cyclins suggested that they were the sought-after *S. cerevisiae* G_1 cyclins. (Note that the homologous CDK-binding domain found in various cyclins differs from the destruction box mentioned earlier, which is found only in B-type cyclins.)

Gene knockout experiments showed that *S. cerevisiae* cells can grow in rich medium if they carry any one of the three G_1 cyclin genes. As the data presented in Figure 21-23 indicate, overproduction of one G_1 cyclin decreases the fraction of cells in G_1, demonstrating that high levels of the G_1 cyclin-CDK complex drive cells through START prematurely. Moreover, in the absence of any of the G_1 cyclins, cells become arrested in G_1, indicating that a G_1 cyclin–CDK heterodimer, or SPF, is required for *S. cerevisiae* cells to enter the S phase. These findings are reminiscent of the results for the *S. pombe* mitotic cyclin (Cdc13) with regard to passage through G_2 and entry into mitosis. Overproduction of the mitotic cyclin caused a shortened G_2 and premature entry into mitosis, whereas inhibition of the mitotic cyclin by mutation resulted in a lengthened G_2 (see Figure 21-12). Thus these results confirmed that the *S. cerevisiae* Cln proteins are G_1 *cyclins* that regulate passage through the G_1 phase of the cell cycle.

In wild-type yeast cells, *CLN3* mRNA is produced at a nearly constant level throughout the cell cycle, but its translation is regulated in response to nutrient levels. The *CLN3* mRNA contains a short upstream open-reading frame that inhibits initiation of translation of this mRNA. This inhibition is diminished when nutrients and hence translation initiation factors are in abundance. Since Cln3 is a highly unstable protein, its concentration fluctuates with the translation rate of *CLN3* mRNA. Consequently, the amount and activity of Cln3-CDK complexes, which depends on the concentration of Cln3 protein, is largely regulated by the nutrient level.

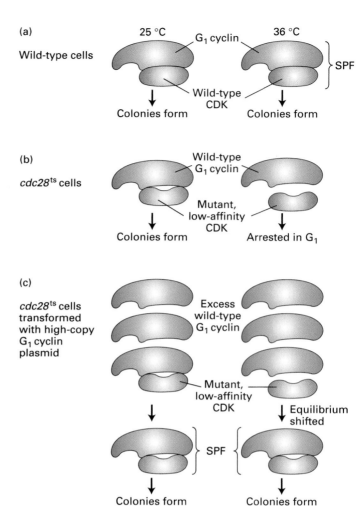

▲ **EXPERIMENTAL FIGURE 21-22 Genes encoding two *S. cerevisiae* G$_1$ cyclins were identified by their ability to supress a temperature-sensitive mutant CDK.** This genetic screen is based on differences in the interactions between G$_1$ cyclins and wild-type and temperature-sensitive (ts) *S. cerevisiae* CDKs. (a) Wild-type cells produce a normal CDK that associates with G$_1$ cyclins, forming the active S phase–promoting factor (SPF), at both the permissive and nonpermissive temperature (i.e., 25 ° and 36 °C). (b) Some *cdc28*^ts mutants express a mutant CDK with low affinity for G$_1$ cyclin at 36 °C. These mutants produce enough G$_1$ cyclin-CDK (SPF) to support growth and colony development at 25 °C, but not at 36 °C. (c) When *cdc28*^ts cells were transformed with a *S. cerevisiae* genomic library cloned in a high-copy plasmid, three types of colonies formed at 36 °C: one contained a plasmid carrying the wild-type *CDC28* gene; the other two contained plasmids carrying either the *CLN1* or *CLN2* gene. In transformed cells carrying the *CLN1* or *CLN2* gene, the concentration of the encoded G$_1$ cyclin is high enough to offset the low affinity of the mutant CDK for a G$_1$ cyclin at 36 °C, so that enough SPF forms to support entry into the S phase and subsequent mitosis. Untransformed *cdc28*^ts cells and cells transformed with plasmids carrying other genes are arrested in G$_1$ and do not form colonies. [See J. A. Hadwiger et al., 1989, *Proc. Nat'l. Acad. Sci. USA* **86**:6255.]

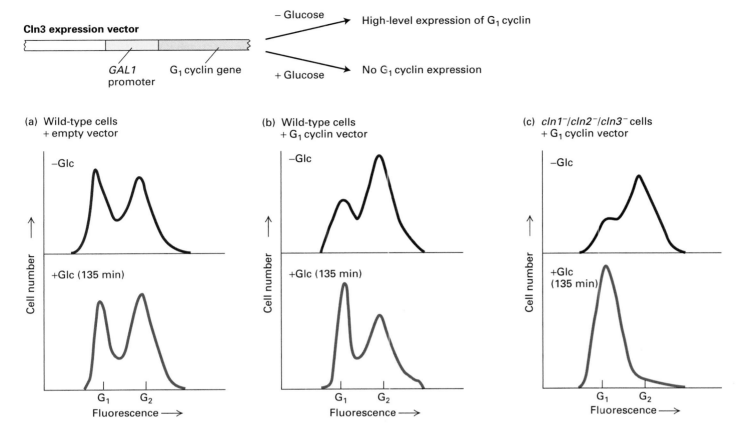

▲ **EXPERIMENTAL FIGURE 21-23 Overexpression of G₁ cyclin prematurely drives *S. cerevisiae* cells into the S phase.** The yeast expression vector used in these experiments (*top*) carried one of the three *S. cerevisiae* G₁ cyclin genes linked to the strong *GAL1* promoter, which is turned off when glucose is present in the medium. To determine the proportion of cells in G₁ and G₂, cells were exposed to a fluorescent dye that binds to DNA and then were passed through a fluorescence-activated cell sorter (see Figure 5-34). Since the DNA content of G₂ cells is twice that of G₁ cells, this procedure can distinguish cells in the two cell-cycle phases. (a) Wild-type cells transformed with an empty expression vector displayed the normal distribution of cells in G₁ and G₂ in the absence of glucose (Glc) and after addition of glucose. (b) In the absence of glucose, wild-type cells transformed with the G₁ cyclin expression vector displayed a higher-than-normal percentage of cells in the S phase and G₂ because overexpression of the G₁ cyclin decreased the G₁ period (top curve). When expression of the G₁ cyclin from the vector was shut off by addition of glucose, the cell distribution returned to normal (bottom curve). (c) Cells with mutations in all three G₁ cyclin genes and transformed with the G₁ cyclin expression vector also showed a high percentage of cells in S and G₂ in the absence of glucose (top curve). Moreover, when expression of G₁ cyclin from the vector was shut off by addition of glucose, the cells completed the cell cycle and arrested in G₁ (bottom curve), indicating that a G₁ cyclin is required for entry into the S phase. [Adapted from H. E. Richardson et al., 1989, *Cell* **59**:1127.]

Once sufficient Cln3 is synthesized from its mRNA, the Cln3-CDK complex phosphorylates and activates two related transcription factors, SBF and MBF. These induce transcription of the *CLN1* and *CLN2* genes whose encoded proteins accelerate entry into the S phase. Thus regulation of *CLN3* mRNA translation in response to the concentration of nutrients in the medium is thought to be primarily responsible for controlling the length of G₁ in *S. cerevisiae*. SBF and MBF also stimulate transcription of several other genes required for DNA replication, including genes encoding DNA polymerase subunits, RPA subunits (the eukaryotic ssDNA-binding protein), DNA ligase, and enzymes required for deoxyribonucleotide synthesis.

One of the important substrates of the late G₁ cyclin–CDK complexes (Cln1-CDK and Cln2-CDK in *S. cerevisiae*) is Cdh1. Recall that this specificity factor directs the APC to polyubiquitinate B-type cyclins during late anaphase of the previous mitosis, marking these cyclins for proteolysis by proteasomes (see Figure 21-10). The MBF transcription factor activated by the Cln3-CDK complex also stimulates transcription of *CLB5* and *CLB6*, which encode cyclins of the B-type, hence their name. Because the complexes formed between these B-type cyclins and the *S. cerevisiae* CDK are required for initiation of DNA synthesis, the Clb5 and Clb6 proteins are called *S-phase cyclins*. Inactivation of the APC earlier in G₁ allows the S-phase cyclin-CDK complexes to accumulate in late G_1. The specificity factor Cdh1 is phosphorylated and inactivated by both late G_1 and B-type cyclin-CDK complexes, and thus remains inhibited throughout S, G_2, and M phase until late anaphase.

Degradation of the S-Phase Inhibitor Triggers DNA Replication

As the S-phase cyclin-CDK heterodimers accumulate in late G_1, they are immediately inactivated by binding of an inhibitor, called *Sic1*, that is expressed late in mitosis and in early G_1. Because Sic1 specifically inhibits B-type cyclin-CDK complexes, but has no effect on the G_1 cyclin-CDK complexes, it functions as an *S-phase inhibitor*.

Entry into the S phase is defined by the initiation of DNA replication. In *S. cerevisiae* cells this occurs when the Sic1 inhibitor is precipitously degraded following its polyubiquitination by a distinct ubiquitin ligase called *SCF* (Figure 21-24; see also Figure 21-2, step ⑤). Once Sic1 is degraded, the S-phase cyclin-CDK complexes induce DNA replication by phosphorylating multiple proteins bound to replication origins. This mechanism for activating the S-phase cyclin-CDK complexes—that is, inhibiting them as the cyclins are synthesized and then precipitously degrading the inhibitor—permits the sudden activation of large numbers of complexes, as opposed to the gradual increase in kinase activity that would result if no inhibitor were present during synthesis of the S-phase cyclins.

We can now see that regulated proteasomal degradation directed by two ubiquitin ligase complexes, SCF and APC, controls three major transitions in the cell cycle: onset of the S phase through degradation of Sic1, the beginning of anaphase through degradation of securin, and exit from mitosis through degradation of B-type cyclins. The APC is directed to polyubiquitinate securin, which functions as an anaphase inhibitor, by the Cdc20 specificity factor (see Figure 21-19). Another specificity factor, Cdh1, targets APC to B-type cyclins (see Figure 21-10). The SCF is directed to

polyubiquitinate Sic1 by a different mechanism, namely, phosphorylation of Sic1 by a G_1 cyclin-CDK (see Figure 21-24). This difference in strategy probably occurs because the APC has several substrates, including securin and B-type cyclins, which must be degraded at different times in the cycle. In contrast, entry into the S phase requires the degradation of only a single protein, the Sic1 inhibitor. An obvious advantage of proteolysis for controlling passage through these critical points in the cell cycle is that protein degradation is an irreversible process, ensuring that cells proceed irreversibly in one direction through the cycle.

Multiple Cyclins Direct the Kinase Activity of *S. cerevisiae* CDK During Different Cell-Cycle Phases

As budding yeast cells progress through the S phase, they begin transcribing genes encoding two additional B-type cyclins, Clb3 and Clb4. These form heterodimeric cyclin-CDK complexes that, together with complexes including Clb5 and Clb6, activate DNA replication origins throughout the remainder of the S phase. The Clb3-CDK and Clb4-CDK complexes also initiate formation of the mitotic spindle at the beginning of mitosis. When *S. cerevisiae* cells complete chromosome replication and enter G_2, they begin expressing two more B-type cyclins, Clb1 and Clb2. These function as mitotic cyclins, associating with the CDK to form complexes that are required for mediating the events of mitosis.

Each group of cyclins thus directs the *S. cerevisiae* CDK to specific functions associated with various cell-cycle phases, as outlined in Figure 21-25. Cln3-CDK induces expression of Cln1, Cln2, and other proteins in mid-late G_1 by phosphorylating and activating the SBF and MBF transcription factors. Cln1-CDK and Cln2-CDK inhibit the APC, allowing B-type cyclins to accumulate; these G_1 cyclin-CDKs also activate degradation of the S-phase inhibitor Sic1. The S-phase CDK complexes containing Clb5, Clb6, Clb3, and Clb4 then trigger DNA synthesis. Clb3 and Clb4 also function as mitotic

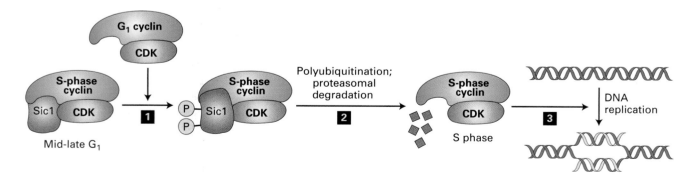

▲ **FIGURE 21-24 Control of the G_1 → S phase transition in *S. cerevisiae* by regulated proteolysis of the S-phase inhibitor Sic1.** The S-phase cyclin-CDK complexes (Clb5-CDK and Clb6-CDK) begin to accumulate in G_1, but are inhibited by Sic1. This inhibition prevents initiation of DNA replication until the cell is fully prepared. G_1 cyclin-CDK complexes assembled in late G_1 (Cln1-CDK and Cln2-CDK) phosphorylate Sic1 (step ❶), marking it for polyubiquitination by the SCF ubiquitin ligase, and subsequent proteasomal degradation (step ❷). The active S-phase cyclin-CDK complexes then trigger initiation of DNA synthesis (step ❸) by phosphorylating substrates that remain to be identified. [Adapted from R. W. King et al., 1996, *Science* **274**:1652.]

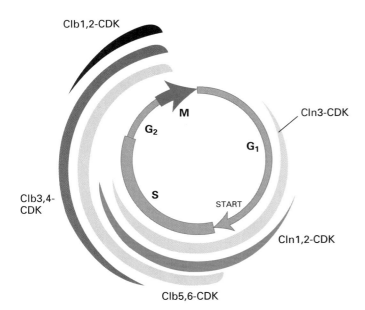

▲ **FIGURE 21-25 Activity of *S. cerevisiae* cyclin-CDK complexes through the course of the cell cycle.** The width of the colored bands is approximately proportional to the demonstrated or proposed protein kinase activity of the indicated cyclin-CDK complexes. *S. cerevisiae* produces a single cyclin-dependent kinase (CDK) whose activity is controlled by the various cyclins, which are expressed during different portions of the cell cycle.

cyclins in that the complexes they form with CDK trigger formation of mitotic spindles. The remaining two *S. cerevisiae* cyclins, Clb1 and Clb2, whose concentrations peak midway through mitosis, function exclusively as mitotic cyclins, forming complexes with CDK that trigger chromosome segregation and nuclear division (see Table 21-1).

Cdc14 Phosphatase Promotes Exit from Mitosis

Genetic studies with *S. cerevisiae* have provided insight into how B-type cyclin-CDK complexes are inactivated in late anaphase, permitting the various events constituting telophase and then cytokinesis to occur. These complexes are the *S. cerevisiae* equivalent of the mitosis-promoting factor (MPF) first identified in *Xenopus* oocytes and early embryos. As mentioned previously, the specificity factor Cdh1, which targets the APC to polyubiquitinate B-type cyclins, is phosphorylated by both late G_1 and B-type cyclin-CDK complexes, thereby inhibiting Cdh1 activity during late G_1, S, G_2, and mitosis before late anaphase.

When daughter chromosomes have segregated properly in late anaphase, the *Cdc14 phosphatase* is activated and dephosphorylates Cdh1, allowing it to bind to the APC. This interaction quickly leads to APC-mediated polyubiquitination and proteasomal degradation of B-type cyclins, and hence MPF inactivation (see Figure 21-10). Since MPF is still active when Cdc14 is first activated in late anaphase, it po-

tentially could compete with Cdc14 by re-phosphorylating Cdh1. However, Cdc14 also induces expression of Sic1 by removing an inhibitory phosphate on a transcription factor that activates transcription of the *SIC1* gene. Sic1 can bind to and inhibit the activity of all B-type cyclin-CDK complexes. Thus, starting late in mitosis, the inhibition of MPF by Sic1 allows the Cdc14 phosphatase to get the upper hand; the B-type cyclin APC specificity factor Cdh1 is dephosphorylated and directs the precipitous degradation of all the *S. cerevisiae* B-type cyclins. In Section 21.7, we will see how the activity of Cdc14 itself is controlled to assure that a cell exits mitosis only when its chromosomes have segregated properly.

As discussed already, Sic1 also inhibits the S-phase cyclin-CDKs as they are formed in mid-G_1 (see Figure 21-24). This inhibitor of B-type cyclins thus serves a dual function in the cell cycle, contributing to the exit from mitosis and delaying entry into the S phase until the cell is ready.

Replication at Each Origin Is Initiated Only Once During the Cell Cycle

As discussed in Chapter 4, eukaryotic chromosomes are replicated from multiple origins. Some of these initiate DNA replication early in the S phase, some later, and still others toward the end. However, no eukaryotic origin initiates more than once per S phase. Moreover, the S phase continues until replication from all origins along the length of each chromosome results in replication of the chromosomal DNA in its entirety. These two factors ensure that the correct gene copy number is maintained as cells proliferate.

Yeast replication origins contain an 11-bp conserved core sequence to which is bound a hexameric protein, the *origin-recognition complex (ORC)*, required for initiation of DNA synthesis. DNase I footprinting analysis (Figure 11-15) and immunoprecipitation of chromatin proteins cross-linked to specific DNA sequences (Figure 11-40) during the various phases of the cell cycle indicate that the ORC remains associated with origins during all phases of the cycle. Several replication initiation factors required for the initiation of DNA synthesis were initially identified in genetic studies in *S. cerevisiae*. These include Cdc6, Cdt1, Mcm10, and the MCM hexamer, a complex of six additional, closely related Mcm proteins; these proteins associate with origins during G_1, but not during G_2 or M. During G_1 the various initiation factors assemble with the ORC into a *pre-replication complex* at each origin. The MCM hexamer is thought to act analogously to SV40 T-antigen hexamers, which function as a helicase to unwind the parental DNA strands at replication forks (see Figure 4-36). Cdc6, Cdt1, and Mcm10 are required to load opposing MCM hexamers on the origin.

The restriction of origin "firing" to once and only once per cell cycle in *S. cerevisiae* is enforced by the alternating cycle of B-type cyclin-CDK activities throughout the cell cycle: low in telophase through G_1 and high in S, G_2, and M through anaphase (see Figure 21-25). As we just discussed, S-phase cyclin-CDK complexes become active at the beginning of the S phase when their specific inhibitor, Sic1, is degraded. In the current model for *S. cerevisiae* replication, pre-replication complexes assemble early in G$_1$ when B-type cyclin activity is low (Figure 21-26, step ☐). Initiation of DNA replication requires an active S-phase cyclin-CDK complex and a second heterodimeric protein kinase, Dbf4-Cdc7, which is expressed in G$_1$ (step ☐). By analogy with cyclin-dependent kinases (CDKs), which must be bound by a partner cyclin to activate their protein kinase activity, the *D*bf4-*d*ependent *k*inase Cdc7 is often called *DDK*. Although the complete set of proteins that must be phosphorylated to activate initiation of DNA synthesis has not yet been determined, there is evidence that phosphorylation of at least one subunit of the hexameric *MCM helicase* and of Cdc6 is required. Another consequence of S-phase cyclin-CDK activation is binding of the initiation factor Cdc45 to the pre-replication complex. Cdc45 is required for the subsequent binding of RPA, the heterotrimeric protein that binds single-stranded DNA generated when the MCM helicase unwinds the parental DNA duplex.

By stabilizing the unwound DNA, RPA promotes binding of the complex between primase and DNA polymerase α (Pol α) that initiates the synthesis of daughter strands (see Figure 21-26, step ☐). Like the MCM helicase, Cdc45 remains associated with the replication forks as they extend in both directions from the origin. Presumably, it functions in the further cycles of RPA and primase–Pol α binding required to prime synthesis of the lagging daughter strand. Subsequent binding of DNA polymerase δ and its accessory cofactors Rfc and PCNA is thought to occur as it does in

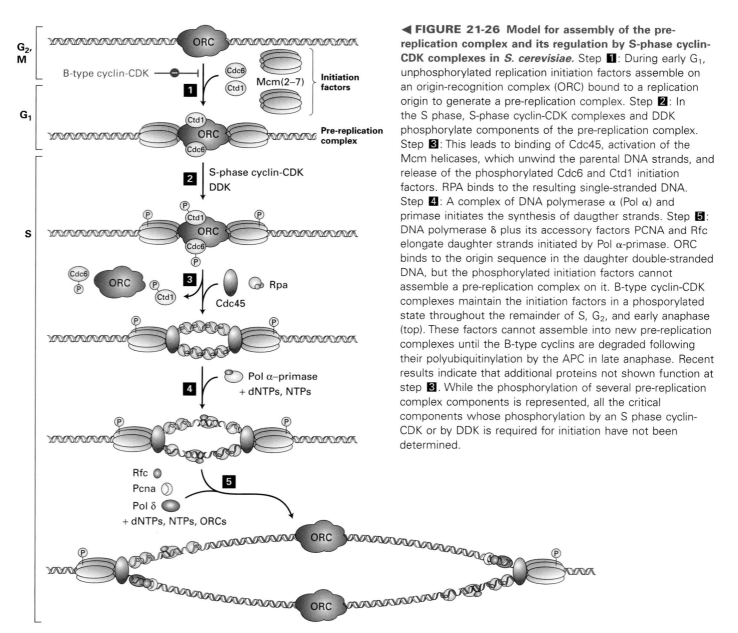

◀ **FIGURE 21-26 Model for assembly of the pre-replication complex and its regulation by S-phase cyclin-CDK complexes in *S. cerevisiae*.** Step ☐: During early G$_1$, unphosphorylated replication initiation factors assemble on an origin-recognition complex (ORC) bound to a replication origin to generate a pre-replication complex. Step ☐: In the S phase, S-phase cyclin-CDK complexes and DDK phosphorylate components of the pre-replication complex. Step ☐: This leads to binding of Cdc45, activation of the Mcm helicases, which unwind the parental DNA strands, and release of the phosphorylated Cdc6 and Ctd1 initiation factors. RPA binds to the resulting single-stranded DNA. Step ☐: A complex of DNA polymerase α (Pol α) and primase initiates the synthesis of daugther strands. Step ☐: DNA polymerase δ plus its accessory factors PCNA and Rfc elongate daughter strands initiated by Pol α-primase. ORC binds to the origin sequence in the daughter double-stranded DNA, but the phosphorylated initiation factors cannot assemble a pre-replication complex on it. B-type cyclin-CDK complexes maintain the initiation factors in a phosporylated state throughout the remainder of S, G$_2$, and early anaphase (top). These factors cannot assemble into new pre-replication complexes until the B-type cyclins are degraded following their polyubiquitinylation by the APC in late anaphase. Recent results indicate that additional proteins not shown function at step ☐. While the phosphorylation of several pre-replication complex components is represented, all the critical components whose phosphorylation by an S phase cyclin-CDK or by DDK is required for initiation have not been determined.

the synthesis of SV40 DNA (see Figure 4-34); these proteins are required for continued synthesis of the leading strand and for synthesis of most of the lagging strand. An additional DNA polymerase, Pol ε, is also required for chromosomal DNA synthesis, but its function is not yet understood.

As the replication forks progress away from each origin, presumably phosphorylated forms of Cdc6, Cdt1, and Mcm10 are displaced from the chromatin. However, ORC complexes immediately bind to the origin sequence in the replicated daughter duplex DNAs and remain bound throughout the cell cycle (see Figure 21-26, step ④). Origins can fire only once during the S phase because the phosphorylated initiation factors cannot reassemble into a pre-replication complex. Consequently, phosphorylation of components of the pre-replication complex by S-phase cyclin-CDK complexes and the DDK complex simultaneously activates initiation of DNA replication at an origin and inhibits re-initiation of replication at that origin. As we have noted, B-type cyclin-CDK complexes remain active throughout the S phase, G₂, and early anaphase, maintaining the phosphorylated state of the replication initiation factors that prevents the assembly of new pre-replication complexes (step ⑤).

Only when the APC triggers degradation of all B-type cyclins in late anaphase and telophase does the then unopposed action of phosphatases remove the phosphates on the initiation factors (Cdc6, Cdt1, and Mcm10), allowing the reassembly of pre-replication complexes during G₁. As discussed previously, the inhibition of APC activity throughout G₁ sets the stage for accumulation of the S-phase cyclins needed for onset of the S phase. This regulatory mechanism has two consequences: (1) pre-replication complexes are assembled only during G₁, when the activity of B-type cyclin-CDK complexes is low, and (2) each origin initiates replication one time only during the S phase, when S phase cyclin-CDK complex activity is high. As a result, chromosomal DNA is replicated only one time each cell cycle.

KEY CONCEPTS OF SECTION 21.5

Genetic Studies with *S. cerevisiae*

■ *S. cerevisiae* expresses a single cyclin-dependent protein kinase (CDK), encoded by *CDC28*, which interacts with several different cyclins during different phases of the cell cycle (see Figure 21-25).

■ Three G₁ cyclins are active in G₁: Cln1, Cln2, and Cln3. The concentration of *CLN3* mRNA does not vary significantly through the cell cycle, but its translation is regulated by the availability of nutrients.

■ Once active Cln3-CDK complexes accumulate in mid-late G₁, they phosphorylate and activate two transcription factors that stimulate expression of Cln1 and Cln2, of enzymes and other proteins required for DNA replication, and of the S-phase B-type cyclins Clb5 and Clb6.

■ The late G₁ cyclin-CDK complexes (Cln1-CDK and Cln2-CDK) phosphorylate and inhibit Cdh1, the specificity factor that directs the anaphase-promoting complex (APC) to B-type cyclins, thus permitting accumulation of S-phase B-type cyclins.

■ S-phase cyclin-CDK complexes initially are inhibited by Sic1. Polyubiquitination of Sic1 by the SCF ubiquitin ligase marks Sic1 for proteasomal degradation, releasing activated S-phase cyclin-CDK complexes that trigger onset of the S phase (see Figure 21-24).

■ B-type cyclins Clb3 and Clb4, expressed later in the S phase, form heterodimers with the CDK that also promote DNA replication and initiate spindle formation early in mitosis.

■ B-type cyclins Clb1 and Clb2, expressed in G₂, form heterodimers with the CDK that stimulate mitotic events.

■ In late anaphase, the specificity factor Cdh1 is activated by dephosphorylation and then directs APC to polyubiquitinate all of the B-type cyclins (Clbs). Their subsequent proteasomal degradation inactivates MPF activity, permitting exit from mitosis (see Figure 21-10).

■ DNA replication is initiated from pre-replication complexes assembled at origins during early G₁. S-phase cyclin-CDK complexes simultaneously trigger initiation from pre-replication complexes and inhibit assembly of new pre-replication complexes by phosphorylating components of the pre-replication complex (see Figure 21-26).

■ Initiation of DNA replication occurs at each origin, but only once, until a cell proceeds through anaphase when activation of APC leads to the degradation of B-type cyclins. The block on reinitiation of DNA replication until replicated chromosomes have segregated assures that daughter cells contain the proper number of chromosomes per cell.

21.6 Cell-Cycle Control in Mammalian Cells

In multicellular organisms, precise control of the cell cycle during development and growth is critical for determining the size and shape of each tissue. Cell replication is controlled by a complex network of signaling pathways that integrate extracellular signals about the identity and numbers of neighboring cells and intracellular cues about cell size and developmental program (Chapter 15). Most differentiated cells withdraw from the cell cycle during G₁, entering the G₀ state (see Figure 21-1). Some differentiated cells (e.g., fibroblasts and lymphocytes) can be stimulated to reenter the cycle and replicate. Many postmitotic differentiated cells, however, never reenter the cell cycle to replicate again. As we discuss in this section, the cell-cycle regulatory mechanisms uncovered in yeasts and *Xenopus* eggs and early embryos also operate in the somatic cells of higher eukaryotes including humans and other mammals.

Mammalian Restriction Point Is Analogous to START in Yeast Cells

Most studies of mammalian cell-cycle control have been done with cultured cells that require certain polypeptide growth factors (mitogens) to stimulate cell proliferation.

Binding of these growth factors to specific receptor proteins that span the plasma membrane initiates a cascade of signal transduction that ultimately influences transcription and cell-cycle control (Chapters 13–15).

Mammalian cells cultured in the absence of growth factors are arrested with a diploid complement of chromosomes

(a)

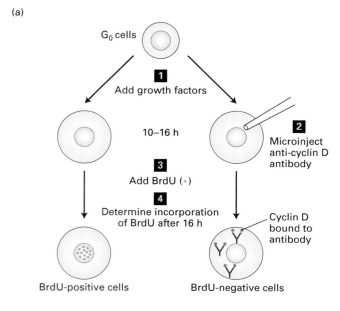

(c)

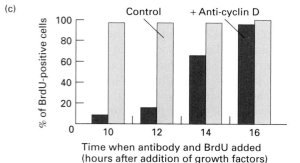

(b)

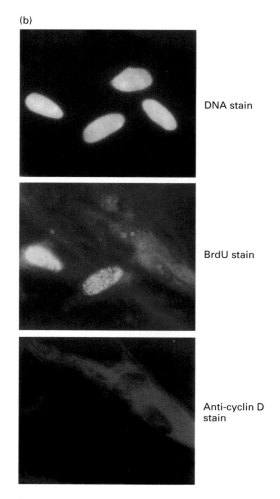

▲ **EXPERIMENTAL FIGURE 21-27 Microinjection experiments with anti-cyclin D antibody demonstrate that cyclin D is required for passage through the restriction point.** The G$_0$-arrested mammalian cells used in these experiments pass the restriction point 14–16 hours after addition of growth factors and enter the S phase 6–8 hours later. (a) Outline of experimental protocol. At various times 10–16 hours after addition of growth factors (**1**), some cells were microinjected with rabbit antibodies against cyclin D (**2**). Bromodeoxyuridine (BrdU), a thymidine analog, was then added to the medium (**3**), and the uninjected control cells (*left*) and microinjected experimental cells (*right*) were incubated for an additional 16 hours. Each sample then was analyzed to determine the percentage of cells that had incorporated BrdU (**4**), indicating that they had entered the S phase. (b) Analysis of control cells and experimental cells 8 hours after addition of growth factors. The three micrographs show the same field of cells stained 16 hours after addition of BrdU to the medium. Cells were stained with different fluorescent agents to visualize DNA (*top*), BrdU (*middle*), and anti-cyclin D antibody (*bottom*). Note that the two cells in this field injected with anti-cyclin D antibody (the red cells in the bottom micrograph) did not incorporate BrdU into nuclear DNA, as indicated by their lack of staining in the middle micrograph. (c) Percentage of control cells (blue bars) and experimental cells (red bars) that incorporated BrdU. Most cells injected with anti-cyclin D antibodies 10 or 12 hours after addition of growth factors failed to enter the S phase, indicated by the low level of BrdU incorporation. In contrast, anti-cyclin D antibodies had little effect on entry into the S phase and DNA synthesis when injected at 14 or 16 hours, that is, after cells had passed the restriction point. These results indicate that cyclin D is required to pass the restriction point, but once cells have passed the restriction point, they do not require cyclin D to enter the S phase 6–8 hours later. [Parts (b) and (c) adapted from V. Baldin et al., 1993, *Genes & Devel.* **7**:812.]

in the G_0 period of the cell cycle. If growth factors are added to the culture medium, these **quiescent cells** pass through the **restriction point** 14–16 hours later, enter the S phase 6–8 hours after that, and traverse the remainder of the cell cycle (see Figure 21-2). Like START in yeast cells, the restriction point is the point in the cell cycle at which mammalian cells become committed to entering the S phase and completing the cell cycle. If mammalian cells are moved from a medium containing growth factors to one lacking growth factors before they have passed the restriction point, the cells do not enter the S phase. But once cells have passed the restriction point, they are committed to entering the S phase and progressing through the entire cell cycle, which takes about 24 hours for most cultured mammalian cells.

Multiple CDKs and Cyclins Regulate Passage of Mammalian Cells Through the Cell Cycle

Unlike *S. pombe* and *S. cerevisiae,* which each produce a single cyclin-dependent kinase (CDK) to regulate the cell cycle, mammalian cells use a small family of related CDKs to regulate progression through the cell cycle. Four CDKs are expressed at significant levels in most mammalian cells and play a role in regulating the cell cycle. Named CDK1, 2, 4, and 6, these proteins were identified by the ability of their cDNA clones to complement certain *cdc* yeast mutants or by their homology to other CDKs.

Like *S. cerevisiae,* mammalian cells express multiple cyclins. Cyclin A and cyclin B, which function in the S phase, G_2, and early mitosis, initially were detected as proteins whose concentration oscillates in experiments with synchronously cycling early sea urchin and clam embryos (see Figure 21-8). Homologous cyclin A and cyclin B proteins have been found in all multicellular animals examined. cDNAs encoding three related human D-type cyclins and cyclin E were isolated based on their ability to complement *S. cerevisiae* cells mutant in all three *CLN* genes encoding G_1 cyclins. The relative amounts of the three D-type cyclins expressed in various cell types (e.g., fibroblasts, hematopoietic cells) differ. Here we refer to them collectively as cyclin D. Cyclin D and E are the mammalian G_1 cyclins. Experiments in which cultured mammalian cells were microinjected with anti-cyclin D antibody at various times after addition of growth factors demonstrated that cyclin D is essential for passage through the restriction point (Figure 21-27).

Figure 21-28 presents a current model for the periods of the cell cycle in which different cyclin-CDK complexes act in G_0-arrested mammalian cells stimulated to divide by the addition of growth factors. In the absence of growth factors, cultured G_0 cells express neither cyclins nor CDKs; the absence of these critical proteins explains why G_0 cells do not progress through the cell cycle and replicate.

Table 21-1, presented early in this chapter, summarizes the various cyclins and CDKs that we have mentioned and the portions of the cell cycle in which they are active. The cyclins fall into two major groups, G_1 cyclins and B-type cy-

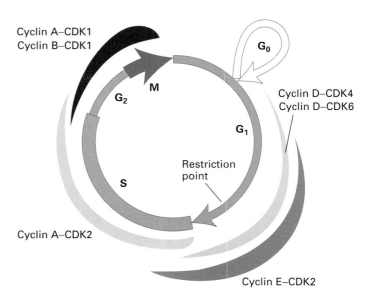

▲ **FIGURE 21-28 Activity of mammalian cyclin-CDK complexes through the course of the cell cycle in cultured G_0 cells induced to divide by treatment with growth factors.** The width of the colored bands is approximately proportional to the protein kinase activity of the indicated complexes. Cyclin D refers to all three D-type cyclins.

clins, which function in S, G_2, and M. Although it is not possible to draw a simple one-to-one correspondence between the functions of the several cyclins and CDKs in *S. pombe, S. cerevisiae,* and vertebrates, the various cyclin-CDK complexes they form can be broadly considered in terms of their functions in mid G_1, late G_1, S, and M phases. All B-type cyclins contain a conserved destruction box sequence that is recognized by the APC ubiquitin ligase, whereas G_1 cyclins lack this sequence. Thus the APC regulates the activity only of cyclin-CDK complexes that include B-type cyclins.

Regulated Expression of Two Classes of Genes Returns G_0 Mammalian Cells to the Cell Cycle

Addition of growth factors to G_0-arrested mammalian cells induces transcription of multiple genes, most of which fall into one of two classes—*early-response* or *delayed-response* genes—depending on how soon their encoded mRNAs appear (Figure 21-29a). Transcription of early-response genes is induced within a few minutes after addition of growth factors by signal-transduction cascades that activate preexisting transcription factors in the cytosol or nucleus. Induction of early-response genes is not blocked by inhibitors of protein synthesis (Figure 21-29b) because the required transcription factors are present in G_0 cells and are activated by phosphorylation or removal of an inhibitor in response to stimulation of cells by growth factors (Chapter 14). Many of the early-response genes encode transcription factors, such as c-Fos and c-Jun, that stimulate transcription of the delayed-

(a)

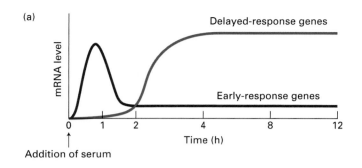

Addition of serum

(b)

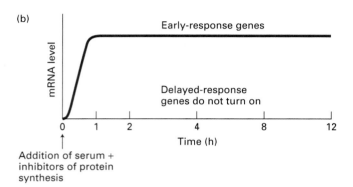

Addition of serum +
inhibitors of protein
synthesis

▲ **FIGURE 21-29 General time course of expression of
early- and delayed-response genes in G$_0$-arrested mammalian
cells after addition of serum.** (a) In the absence of inhibitors of
protein synthesis, expression of early-response genes peaks
about 1 hour after addition of serum, which contains several
mitogens, and then falls as expression of late-response genes
begins. (b) Inhibitors of protein synthesis prevent the drop in
expression of early-response genes and completely block
expression of late-response genes. See text for discussion.
[Adapted from A. Murray and T. Hunt, 1993, *The Cell Cycle: An
Introduction*, W. H. Freeman and Company.]

response genes. Mutant, unregulated forms of both c-Fos
and c-Jun are expressed by oncogenic retroviruses (Chapter
23); the discovery that the viral forms of these proteins
(v-Fos and v-Jun) can transform normal cells into cancer cells
led to identification of the regulated cellular forms of these
transcription factors.

After peaking at about 30 minutes following addition of
growth factors, the concentrations of the early-response
mRNAs fall to a lower level that is maintained as long as
growth factors are present in the medium. Most of the immediate early mRNAs are unstable; consequently, their concentrations fall as their rate of synthesis decreases. This drop
in transcription is blocked by inhibitors of protein synthesis
(see Figure 21-29b), indicating that it depends on production
of one or more of the early-response proteins.

Because expression of delayed-response genes depends on
proteins encoded by early-response genes, delayed-response
genes are not transcribed when mitogens are added to G$_0$-
arrested cells in the presence of an inhibitor of protein synthesis. Some delayed-response genes encode additional
transcription factors (see below); others encode the D-type

cyclins, cyclin E, CDK2, CDK4, and CDK6. The D-type cyclins, CDK4, and CDK6, are expressed first, followed
by cyclin E and CDK2 (see Figure 21-28). If growth factors
are withdrawn before passage through the restriction point,
expression of these G$_1$ cyclins and CDKs ceases. Since these
proteins and the mRNAs encoding them are unstable, their
concentrations fall precipitously. As a consequence, the cells
do not pass the restriction point and do not replicate.

In addition to transcriptional control of the gene encoding cyclin D, the concentration of this mid-G$_1$ cyclin also is
regulated by controlling *translation* of cyclin D mRNA. In
this regard, cyclin D is similar to *S. cerevisiae* Cln3. Addition
of growth factors to cultured mammalian cells triggers signal
transduction via the PI-3 kinase pathway discussed in Chapter 14, leading to activation of the translation-initiation factor eIF4 (Chapter 4). As a result, translation of cyclin D
mRNA and other mRNAs is stimulated. Agents that inhibit
eIF4 activation, such as TGF-β, inhibit translation of cyclin
D mRNA and thus inhibit cell proliferation.

Passage Through the Restriction Point Depends on Phosphorylation of the Tumor-Suppressor Rb Protein

Some members of a small family of related transcription factors, referred to collectively as *E2F factors*, are encoded by
delayed-response genes. These transcription factors activate
genes encoding many of the proteins involved in DNA and deoxyribonucleotide synthesis. They also stimulate transcription
of genes encoding the late-G$_1$ cyclin (cyclin E), the S-phase cyclin (cyclin A), and the S-phase CDK (CDK2). Thus the E2Fs
function in late G$_1$ similarly to the *S. cerevisiae* transcription
factors SBF and MBF. In addition, E2Fs autostimulate transcription of their own genes. E2Fs function as transcriptional
repressors when bound to *Rb protein*, which in turn binds histone deacetylase complexes. As discussed in Chapter 11, transcription of a gene is highest when the associated histones are
highly acetylated; histone deacetylation causes chromatin to
assume a more condensed, transcriptionally inactive form.

 Rb protein was initially identified as the product of
the prototype **tumor-suppressor gene,** *RB*. The
products of tumor-suppressor genes function in
various ways to inhibit progression through the cell cycle
(Chapter 23). Loss-of-function mutations in *RB* are associated with the disease *hereditary retinoblastoma*. A child with
this disease inherits one normal *RB*$^+$ allele from one parent
and one mutant *RB*$^-$ allele from the other. If the *RB*$^+$ allele
in any of the trillions of cells that make up the human body
becomes mutated to a *RB*$^-$ allele, then no functional protein is expressed and the cell or one of its descendants is
likely to become cancerous. For reasons that are not understood, this generally happens in a retinal cell leading to the
retinal tumors that characterize this disease. Also, in most
human cancer cells Rb function is inactivated, either by mutations in both alleles of *RB*, or by abnormal regulation of
Rb phosphorylation. ∎

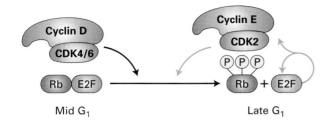

▲ FIGURE 21-30 Regulation of Rb and E2F activities in mid-late G₁. Stimulation of G₀ cells with mitogens induces expression of CDK4, CDK6, D-type cyclins, and the E2F transcription factors, all encoded by delayed-response genes. Rb protein initially inhibits E2F activity. When signaling from mitogens is sustained, the resulting cyclin D–CDK4/6 complexes begin phosphorylating Rb, releasing some E2F, which stimulates transcription of the genes encoding cyclin E, CDK2, and E2F (autostimulation). The cyclin E–CDK4 complexes further phosphorylate Rb, resulting in positive feedback loops (blue arrows) that lead to a rapid rise in the expression and activity of both E2F and cyclin E–CDK2 as the cell approaches the G₁ → S transition.

Rb protein is one of the most significant substrates of mammalian G₁ cyclin-CDK complexes. Phosphorylation of Rb protein at multiple sites prevents its association with E2Fs, thereby permitting E2Fs to activate transcription of genes required for entry into S phase. As shown in Figure 21-30, phosphorylation of Rb protein is initiated by cyclin D–CDK4 and cyclin D–CDK6 in mid G₁. Once cyclin E and CDK2 are induced by phosphorylation of some Rb, the resulting cyclin E–CDK2 further phosphorylates Rb in late G₁. When cyclin E–CDK2 accumulates to a critical threshold level, further phosphorylation of Rb by cyclin E–CDK2 continues even when cyclin D–CDK4/6 activity is removed. This is one of the principle biochemical events responsible for passage through the restriction point. At this point, further phosphorylation of Rb by cyclin E–CDK2 occurs even when mitogens are withdrawn and cyclin D and CDK4/6 levels fall. Since E2F stimulates its own expression and that of cyclin E and CDK2, positive cross-regulation of E2F and cyclin E–CDK2 produces a rapid rise of both activities in late G₁.

As they accumulate, S-phase cyclin-CDK and mitotic cyclin-CDK complexes maintain Rb protein in the phosphorylated state throughout the S, G₂, and early M phases. After cells complete anaphase and enter early G₁ or G₀, the fall in cyclin-CDK levels leads to dephosphorylation of Rb by unopposed phosphatases. As a consequence, hypophosphorylated Rb is available to inhibit E2F activity during early G₁ of the next cycle and in G₀-arrested cells.

Cyclin A Is Required for DNA Synthesis and CDK1 for Entry into Mitosis

High levels of E2Fs activate transcription of the *cyclin A* gene as mammalian cells approach the G₁ → S transition. (Despite its name, cyclin A is a B-type cyclin, not a G₁ cyclin;

see Table 21-1.) Disruption of cyclin A function inhibits DNA synthesis in mammalian cells, suggesting that cyclin A–CDK2 complexes may function like *S. cerevisiae* S-phase cyclin-CDK complexes to trigger initiation of DNA synthesis. There is also evidence that cyclin E–CDK2 may contribute to activation of pre-replication complexes.

Three related CDK inhibitory proteins, or CIPs (p27^{KIP1}, p57^{KIP2}, and p21^{CIP}), appear to share the function of the *S. cerevisiae* S-phase inhibitor Sic1 (see Figure 21-24). Phosphorylation of p27^{KIP1} by cyclin E–CDK2 targets it for polyubiquitination by the mammalian SCF complex (see Figure 21-2, step ⑤). The SCF subunit that targets p27^{KIP1} is synthesized as cells approach the G₁ → S transition. The mechanism for degrading p21^{CIP} and p57^{KIP2} is less well understood. The activity of mammalian cyclin-CDK2 complexes is also regulated by phosphorylation and dephosphorylation mechanisms similarly to those controlling the *S. pombe* mitosis-promoting factor, MPF (see Figure 21-14). The *Cdc25A phosphatase*, which removes the inhibitory phosphate from CDK2, is a mammalian equivalent of *S. pombe* Cdc25 except that it functions at the G₁ → S transition rather than the G₂ → M transition. The mammalian phosphatase normally is activated late in G₁, but is degraded in the response of mammalian cells to DNA damage to prevent the cells from entering S phase (see Section 21.7).

Once cyclin A–CDK2 is activated by Cdc25A and the S-phase inhibitors have been degraded, DNA replication is initiated at pre-replication complexes. The general mechanism is thought to parallel that in *S. cerevisiae* (see Figure 21-26), although small differences are found in vertebrates. As in yeast, phosphorylation of certain initiation factors by cyclin A–CDK2 most likely promotes initiation of DNA replication and prevents reassembly of pre-replication complexes until the cell passes through mitosis, thereby assuring that replication from each origin occurs only once during each cell cycle. In metazoans, a second small protein, geminin, contributes to the inhibition of re-initiation at origins until cells complete a full cell cycle.

The principle mammalian CDK in G₂ and mitosis is CDK1 (see Figure 21-28). This CDK, which is highly homologous with *S. pombe* Cdc2, associates with cyclins A and B. The mRNAs encoding either of these mammalian cyclins can promote meiotic maturation when injected into *Xenopus* oocytes arrested in G₂ (see Figure 21-6), demonstrating that they function as mitotic cyclins. Thus mammalian cyclin A–CDK1 and cyclin B–CDK1 are functionally equivalent to the *S. pombe* MPF (mitotic cyclin-CDK). The kinase activity of these mammalian complexes also appears to be regulated by proteins analogous to those that control the activity of the *S. pombe* MPF (see Figure 21-14). The inhibitory phosphate on CDK1 is removed by *Cdc25C phosphatase*, which is analogous to *S. pombe* Cdc25 phosphatase.

In cycling mammalian cells, cyclin B is first synthesized late in the S phase and increases in concentration as cells proceed through G₂, peaking during metaphase and dropping after late anaphase. This parallels the time course of cyclin B expression in *Xenopus* cycling egg extracts (see Figure 21-9).

In human cells, cyclin B first accumulates in the cytosol and then enters the nucleus just before the nuclear envelope breaks down early in mitosis. Thus MPF activity is controlled not only by phosphorylation and dephosphorylation but also by regulation of the nuclear transport of cyclin B. In fact, cyclin B shuttles between the nucleus and cytosol, and the change in its localization during the cell cycle results from a change in the relative rates of import and export. As in *Xenopus* eggs and *S. cerevisiae,* cyclins A and B are polyubiquitinated by the anaphase-promoting complex (APC) during late anaphase and then are degraded by proteasomes (see Figure 21-2, step [9]).

Two Types of Cyclin-CDK Inhibitors Contribute to Cell-Cycle Control in Mammals

As noted above, three related CIPs—$p21^{CIP}$, $p27^{KIP2}$, and $p57^{KIP2}$—inhibit cyclin A-CDK2 activity and must be degraded before DNA replication can begin. These same CDK inhibitory proteins also can bind to and inhibit the other mammalian cyclin-CDK complexes involved in cell-cycle control. As we discuss later, $p21^{CIP}$ plays a role in the response of mammalian cells to DNA damage. Experiments with knockout mice lacking $p27^{KIP2}$ have shown that this CIP is particularly important in controlling generalized cell proliferation soon after birth. Although $p27^{KIP2}$ knockouts are larger than normal, most develop normally otherwise. In contrast, $p57^{KIP2}$ knockouts exhibit defects in cell differentiation and most die shortly after birth due to defective development of various organs.

A second class of cyclin-CDK inhibitors called *INK4s* (*in*hibitors of *k*inase *4*) includes several small, closely related proteins that interact only with CDK4 and CDK6 and thus function specifically in controlling the mid-G_1 phase. Binding of INK4s to CDK4/6 blocks their interaction with cyclin D and hence their protein kinase activity. The resulting decreased phosphorylation of Rb protein prevents transcriptional activation by E2Fs and entry into the S phase. One INK4 called p16 is a tumor suppressor, like Rb protein discussed earlier. The presence of two mutant p16 alleles in a large fraction of human cancers is evidence for the important role of p16 in controlling the cell cycle (Chapter 23).

■ Mammalian cells use several CDKs and cyclins to regulate passage through the cell cycle. Cyclin D-CDK4/6 function in mid to late G_1; cyclin E-CDK2 in late G_1 and early S; cyclin A-CDK2 in S; and cyclin A/B-CDK1 in G_2 and M through anaphase (see Figure 21-28).

■ Unphosphorylated Rb protein binds to E2Fs, converting them into transcriptional repressors. Phosphorylation of Rb by cyclin D-CDK4/6 in mid G_1 liberates E2Fs to activate transcription of genes encoding cyclin E, CDK2, and other proteins required for the S phase. E2Fs also autostimulate transcription of their own genes.

■ Cyclin E-CDK2 further phosphorylates Rb, further activating E2Fs. Once a critical level of cyclin E-CDK2 has been expressed, a positive feedback loop with E2F results in a rapid rise of both activities that drives passage through the restriction print (see Figure 21-30).

■ The activity of cyclin A-CDK2, induced by high E2F activity, initially is held in check by CIPs, which function like an S-phase inhibitor, and by the presence of an inhibitory phosphate on CDK2. Proteasomal degradation of the inhibitors and activation of the Cdc25A phosphatase, as cells approach the $G_1 \rightarrow$ S transition, generate active cyclin A-CDK2. This complex activates pre-replication complexes to initiate DNA synthesis by a mechanism similar to that in *S. cerevisiae* (see Figure 21-26).

■ Cyclin A/B-CDK1 induce the events of mitosis through early anaphase. Cyclins A and B are polyubiquitinated by the anaphase-promoting complex (APC) during late anaphase and then are degraded by proteasomes.

■ The activity of mammalian mitotic cyclin-CDK complexes also are regulated by phosphorylation and dephosphorylation similar to the mechanism in *S. pombe*, with the Cdc25C phosphatase removing inhibitory phosphates (see Figure 21-14).

■ The activities of mammalian cyclin-CDK complexes also are regulated by CDK inhibitors (CIPs), which bind to and inhibit each of the mammalian cyclin-CDK complexes, and INK4 proteins, which block passage through G_1 by specifically inhibiting CDK4 and CDK6.

KEY CONCEPTS OF SECTION 21.6

Cell-Cycle Control in Mammalian Cells

■ Various polypeptide growth factors called mitogens stimulate cultured mammalian cells to proliferate by inducing expression of early-response genes. Many of these encode transcription factors that stimulate expression of delayed-response genes encoding the G_1 CDKs, G_1 cyclins, and E2F transcription factors.

■ Once cells pass the restriction point, they can enter the S phase and complete S, G_2, and mitosis in the absence of growth factors.

21.7 Checkpoints in Cell-Cycle Regulation

Catastrophic genetic damage can occur if cells progress to the next phase of the cell cycle before the previous phase is properly completed. For example, when S-phase cells are induced to enter mitosis by fusion to a cell in mitosis, the MPF present in the mitotic cell forces the chromosomes of the S-phase cell to condense. However, since the replicating chromosomes are fragmented by the condensation process, such premature entry into mitosis is disastrous for a cell.

Another example concerns attachment of kineto-chores to microtubules of the mitotic spindle during metaphase. If anaphase is initiated before both kine-tochores of a replicated chromosome become attached to micro-tubules from opposite spindle poles, daughter cells are produced that have a missing or extra chromosome (Figure 21-31). When this process, called *nondisjunction*, occurs during the meiotic di-vision that generates a human egg, Down syndrome can occur from trisomy of chromosome 21, resulting in developmental ab-normalities and mental retardation. ∎

To minimize the occurrence of such mistakes in cell-cycle events, a cell's progress through the cycle is monitored at sev-eral **checkpoints** (Figure 21-32). Control mechanisms that operate at these checkpoints ensure that chromosomes are intact and that each stage of the cell cycle is completed before the following stage is initiated. Our understanding of these control mechanisms at the molecular level has advanced con-siderably in recent years.

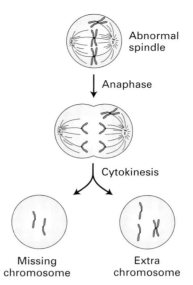

◀ **FIGURE 21-31 Nondisjunction.** This abnormality occurs when chromosomes segregate in anaphase before the kinetochore of each sister chromatid has attached to microtubules (red lines) from the opposite spindle poles. As a result, one daughter cell contains two copies of one chromosome, while the other daughter cell lacks that chromosome. [Adapted from A. Murray and T. Hunt, 1993, *The Cell Cycle: An Introduction*, W. H. Freeman and Company.]

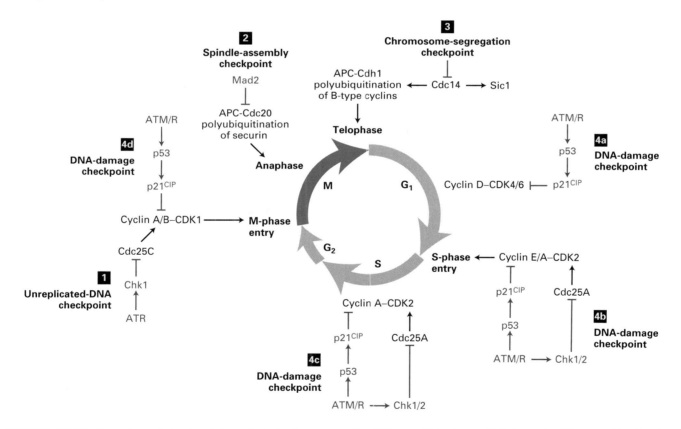

▲ **FIGURE 21-32 Overview of checkpoint controls in the cell cycle.** The unreplicated-DNA checkpoint (**1**) prevents activation of cyclin A-CDK1 and cyclin B-CDK1 (i.e., mitosis-promoting factor, MPF) by activation of an ATR-Chk1 protein kinase cascade that phosphorylates and inactivates Cdc25C, thereby inhibiting entry into mitosis. In the spindle-assembly checkpoint (**2**), Mad2 and other proteins inhibit activation of the APC specificity factor (Cdc20) required for polyubiquitination of securin, thereby preventing entry into anaphase. The chromosome-segregation checkpoint (**3**) prevents release of the Cdc14 phosphatase from nucleoli, thereby blocking activation of

the APC specificity factor (Cdh1) required for polyubiquitination of B-type cyclins as well as induction of Sic1. As a result, the decrease in MPF activity required for the events of telophase does not occur. In the initial phase of the DNA-damage checkpoint (**4**), the ATM or ATR protein kinase (ATM/R) is activated. The active kinases then trigger two pathways: the Chk-Cdc25A pathway (**4b** and **4c**), blocking entry into or through the S phase, and the p53-p21^CIP pathway, leading to arrest in G_1, S, and G_2 (**4a**–**4b**). See text for further discussion. Red symbols indicate pathways that inhibit progression through the cell cycle.

The Presence of Unreplicated DNA Prevents Entry into Mitosis

Cells that fail to replicate all their chromosomes do not enter mitosis. Operation of the *unreplicated-DNA checkpoint* control involves the recognition of unreplicated DNA and inhibition of MPF activation (see Figure 21-32, [1]). Recent genetic studies in *S. pombe* and biochemical studies with *Xenopus* egg extracts suggest that the ATR and Chk1 protein kinases, which also function in the DNA-damage checkpoint, inhibit entry into mitosis by cells that have not completed DNA synthesis.

The association of ATR with replication forks is thought to activate its protein kinase activity, leading to the phosphorylation and activation of the Chk1 kinase. Active Chk1 then phosphorylates and inactivates the Cdc25 phosphatase (Cdc25C in vertebrates), which normally removes the inhibitory phosphate from CDKs that function during mitosis. As a result, the cyclin A/B-CDK1 complexes remain inhibited and cannot phosphorylate targets required to initiate mitosis. ATR continues to initiate this protein kinase cascade until all replication forks complete DNA replication and disassemble.

Improper Assembly of the Mitotic Spindle Prevents the Initiation of Anaphase

The *spindle-assembly checkpoint* prevents entry into anaphase when just a single kinetochore of one chromatid fails to associate properly with spindle microtubules. Clues about how this checkpoint operates has come from isolation of yeast mutants in the presence of benomyl, a microtubule-depolymerizing drug. Low concentrations of benomyl increase the time required for yeast cells to assemble the mitotic spindle and attach kinetochores to microtubules. Wild-type cells exposed to benomyl do not begin anaphase until these processes are completed and then proceed on through mitosis, producing normal daughter cells. In contrast, mutants defective in the spindle-assembly checkpoint proceed through anaphase before assembly of the spindle and attachment of kinetochores is complete; consequently, they mis-segregate their chromosomes, producing abnormal daughter cells that die.

Analysis of these mutants identified a protein called Mad2 and other proteins that regulate Cdc20, the specificity factor required to target the APC to securin (see Figure 21-32, [2]). Recall that APC-mediated polyubiquitination of securin and its subsequent degradation is required for entry into anaphase (see Figure 21-19). Mad2 has been shown to associate with kinetochores that are unattached to microtubules. Experiments with Mad2 fused to green fluorescent protein (GFP) indicate that kinetochore-bound Mad2 rapidly exchanges with a soluble form of Mad2. Current models propose that when Mad2 associates with a kinetochore complex that is not bound by a microtubule, it is converted to a short-lived activated form that can interact with and inhibit Cdc20. Microtubule attachment prevents this activation of Mad2. Consequently, once all kinetochore complexes bind a microtubule, generation of the activated form of Mad2 ceases, the inhibition of Cdc20 is relieved, and Cdc20 is free to direct the APC to polyubiquitinate securin, thereby initiating the onset of anaphase.

Proper Segregation of Daughter Chromosomes Is Monitored by the Mitotic Exit Network

Once chromosomes have segregated properly, telophase commences. The various events of telophase and subsequent cytokinesis, collectively referred to as the exit from mitosis, require inactivation of MPF. As discussed earlier, dephosphorylation of the APC specificity factor Cdh1 by the Cdc14 phosphatase leads to degradation of mitotic cyclins and loss of MPF activity late in anaphase (see Figure 21-10). During interphase and early mitosis, Cdc14 is sequestered in the nucleolus and inactivated. The *chromosome-segregation checkpoint*, which monitors the location of the segregating daughter chromosomes at the end of anaphase, determines whether active Cdc14 is available to promote exit from mitosis (see Figure 21-32, [3]).

Operation of this checkpoint in *S. cerevisiae* depends on a set of proteins referred to as the *mitotic exit network*. A key component is a small (monomeric) GTPase, called Tem1. This member of the **GTPase superfamily** of switch proteins controls the activity of a protein kinase cascade similarly to the way Ras controls MAP kinase pathways (Chapter 14). During anaphase, Tem1 becomes associated with the spindle pole body (SPB) closest to the daughter cell bud. (The SPB, from which spindle microtubules originate, is equivalent to the centrosome in higher eukaryotes.) At the SPB, Tem1 is maintained in the inactive GDP-bound state by a specific GAP (GTPase-accelerating protein). The GEF (guanosine nucleotide–exchange factor) that activates Tem1 is localized to the cortex of the bud and is absent from the mother cell. When spindle microtubule elongation at the end of anaphase has correctly positioned segregating daughter chromosomes into the bud, Tem1 comes into contact with its GEF and is converted into the active GTP-bound state. The terminal kinase in the cascade triggered by Tem1 · GTP then phosphorylates the nucleolar anchor that binds and inhibits Cdc14, releasing it into the cytoplasm and nucleoplasm in both the bud and mother cell (Figure 21-33, [1]). Once active Cdc14 is available, a cell can proceed through telophase and cytokinesis. If daughter chromosomes fail to segregate into the bud, Tem1 remains in its inactive state, Cdc14 is not released from the nucleolus, and mitotic exit is blocked (Figure 21-33, [2]).

In the fission yeast *S. pombe*, formation of the septum that divides daughter cells is regulated by proteins homologous to those that constitute the mitotic exit network in *S. cerevisiae*. Genes encoding similar proteins also have been found in higher organisms where the homologs probably

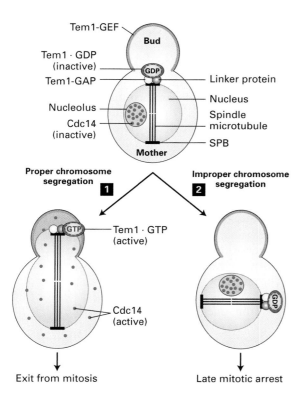

▲ FIGURE 21-33 Operation of the chromosome-segregation checkpoint. In *S. cerevisiae*, Cdc14 phosphatase activity is required for the exit from mitosis. (*Top*) During interphase and early mitosis, Cdc14 is sequestered and inactivated in the nucleolus. Inactive Tem1 · GDP (purple) associates with the spindle pole body (SPB) nearest to the bud early in anaphase with the aid of a linker protein (green) and is maintained in the inactive state by a specific GAP (GTPase-accelerating protein). If chromosome segregation occurs properly (**1**), extension of the spindle microtubules inserts the daughter SPB into the bud, causing Tem1 to come in contact with a specific GEF (guanine nucleotide–exchange factor) localized to the cortex of the bud (brown). This GEF converts inactive Tim1 · GDP to active Tem1 · GTP, which triggers a protein kinase cascade leading to release of active Cdc14 and exit from mitosis. If the spindle apparatus fails to place the daughter SPB in the bud (**2**), Tem1 remains in the inactive GDP-bound state and Cdc14 remains associated with nucleoli. Arrest in late mitosis results. [Adapted from G. Pereira and E. Schiebel, 2001, *Curr. Opin. Cell Biol.* **13**:762.]

function in an analogous checkpoint that leads to arrest in late mitosis when daughter chromosomes do not segregate properly.

Cell-Cycle Arrest of Cells with Damaged DNA Depends on Tumor Suppressors

The *DNA-damage checkpoint* blocks progression through the cell cycle until the damage is repaired. Damage to DNA can result from chemical agents and from irradiation with ultraviolet (UV) light or γ-rays.

Arrest in G_1 and S prevents copying of damaged bases, which would fix mutations in the genome. Replication of damaged DNA also promotes chromosomal rearrangements that can contribute to the onset of cancer. Arrest in G_2 allows DNA double-stranded breaks to be repaired before mitosis. If a double-stranded break is not repaired, the broken distal portion of the damaged chromosome is not properly segregated because it is not physically linked to a centromere, which is pulled toward a spindle pole during anaphase.

As we discuss in detail in Chapter 23, inactivation of tumor-suppressor genes contributes to the development of cancer. The proteins encoded by several tumor-suppressor genes, including *ATM* and *Chk2*, normally function in the DNA-damage checkpoint. Patients with mutations in both copies of *ATM* or *Chk2* develop cancers far more frequently than normal. Both of these genes encode protein kinases.

DNA damage due to UV light somehow activates the ATM kinase, which phosphorylates Chk2, thereby activating its kinase activity. Activated Chk2 then phosphorylates the Cdc25A phosphatase, marking it for polyubiquitination by an as-yet undetermined ubiquitin ligase and subsequent proteasomal degradation. Recall that removal of the inhibitory phosphate from mammalian CDK2 by Cdc25A is required for onset of and passage through the S phase mediated by cyclin E-CDK2 and cyclin A-CDK2. Degradation of Cdc25A resulting from activation of the ATM-Chk2 pathway in G_1 or S-phase cells thus leads to G_1 or S arrest (see Figure 21-32, 4b and 4c). A similar pathway consisting of the protein kinases ATR and Chk1 leads to phosphorylation and polyubiquitination of Cdc25A in response to γ-radiation. As discussed earlier for the unreplicated-DNA checkpoint, Chk1 also inactivates Cdc25C, preventing the activation of CDK1 and entry into mitosis.

Another tumor-suppressor protein, **p53**, contributes to arrest of cells with damaged DNA. Cells with functional p53 arrest in G_1 and G_2 when exposed to γ-irradiation, whereas cells lacking functional p53 do not arrest in G_1. Although the p53 protein is a transcription factor, under normal conditions it is extremely unstable and generally does not accumulate to high enough levels to stimulate transcription. The instability of p53 results from its polyubiquitination by a ubiquitin ligase called Mdm2 and subsequent proteasomal degradation. The rapid degradation of p53 is inhibited by ATM and probably ATR, which phosphorylate p53 at a site that interferes with binding by Mdm2. This and other modifications of p53 in response to DNA damage greatly increase its ability to activate transcription of specific genes that help the cell cope with DNA damage. One of these genes encodes p21^{CIP}, a generalized CIP that binds and inhibits all mammalian cyclin-CDK complexes. As a result, cells are arrested in G_1 and G_2 until the DNA damage is repaired and p53 and subsequently p21^{CIP} levels fall (see Figure 21-32, 4a–4d).

Under some circumstances, such as when DNA damage is extensive, p53 also activates expression of genes that lead to apoptosis, the process of programmed cell death that normally occurs in specific cells during the development of multicellular animals. In vertebrates, the p53 response evolved to induce apoptosis in the face of extensive DNA damage, presumably to prevent the accumulation of multiple mutations that might convert a normal cell into a cancer cell. The dual role of p53 in both cell-cycle arrest and the induction of apoptosis may account for the observation that nearly all cancer cells have mutations in both alleles of the *p53* gene or in the pathways that stabilize p53 in response to DNA damage (Chapter 23). The consequences of mutations in *p53*, *ATM*, and *Chk2* provide dramatic examples of the significance of cell-cycle checkpoints to the health of a multicellular organism. ▮

KEY CONCEPTS OF SECTION 21.7

Checkpoints in Cell-Cycle Regulation

■ Checkpoint controls function to ensure that chromosomes are intact and that critical stages of the cell cycle are completed before the following stage is initiated.

■ The unreplicated-DNA checkpoint operates during S and G_2 to prevent the activation of MPF before DNA synthesis is complete by inhibiting the activation of CDK1 by Cdc25C (see Figure 21-32, ①).

■ The spindle-assembly checkpoint, which prevents premature initiation of anaphase, utilizes Mad2 and other proteins to regulate the APC specificity factor Cdc20 that targets securin for polyubiquitination (see Figures 21-32, ②, and 21-19).

■ The chromosome-segregation checkpoint prevents telophase and cytokinesis until daughter chromosomes have been properly segregated, so that the daughter cell has a full set of chromosomes (see Figure 21-32, ③).

■ In the chromosome-segregation checkpoint, the small GTPase Tem1 controls the availability of Cdc14 phosphatase, which in turn activates the APC specificity factor Cdh1 that targets B-type cyclins for degradation, causing inactivation of MPF (see Figure 21-10).

■ The DNA-damage checkpoint arrests the cell cycle in response to DNA damage until the damage is repaired. Three types of tumor-suppressor proteins (ATM/ATR, Chk1/2, and p53) are critical to this checkpoint.

■ Activation of the ATM or ATR protein kinases in response to DNA damage due to UV light or γ-irradiation leads to arrest in G_1 and the S phase via a pathway that leads to loss of Cdc25A phosphatase activity. A second pathway from activated ATM/R stabilizes p53, which stimulates expression of p21^{CIP}. Subsequent inhibition of multiple CDK-cyclin complexes by p21^{CIP} causes prolonged arrest in G_1 and G_2 (see Figure 21-32, ④a–④d).

■ In response to extensive DNA damage, p53 also activates genes that induce apoptosis.

21.8 Meiosis: A Special Type of Cell Division

In nearly all eukaryotes, **meiosis** generates haploid germ cells (eggs and sperm), which can then fuse to generate a diploid zygote (Figure 21-34). During meiosis, a single round of DNA replication is followed by two cycles of cell division, termed *meiosis I* and *meiosis II*. **Crossing over** of chromatids, visible in the first meiotic metaphase, produces **recombination** between parental chromosomes. This increases the genetic diversity among the individuals of a species. During meiosis I, both chromatids of each homologous chromosome segregate together to opposite spindle poles, so that each of the resulting daughter cells contains one homologous chromosome consisting of two chromatids. During meiosis II, which resembles mitosis, the chromatids of one chromosome segregate to opposite spindle poles, generating haploid germ cells. Meiosis generates four haploid germ cells from one diploid premeiotic cell.

Repression of G_1 Cyclins and Meiosis-Specific Ime2 Prevents DNA Replication in Meiosis II

In *S. cerevisiae* and *S. pombe*, depletion of nitrogen and carbon sources induces diploid cells to undergo meiosis, yielding haploid spores (see Figure 1-5). This process is analogous to

▶ **FIGURE 21-34 Meiosis.** Premeiotic cells have two copies of each chromosome (2*n*), one derived from the paternal parent and one from the maternal parent. For simplicity, the paternal and maternal homologs of only one chromosome are diagrammed. Step ❶: All chromosomes are replicated during the S phase before the first meiotic division, giving a 4*n* chromosomal complement. Cohesin complexes (not shown) link the sister chromatids composing each replicated chromosome along their full lengths. Step ❷: As chromosomes condense during the first meiotic prophase, replicated homologs become paired as the result of at least one crossover event between a paternal and a maternal chromatid. This pairing of replicated homologous chromosomes is called synapsis. At metaphase, shown here, both chromatids of one chromosome associate with microtubules emanating from one spindle pole, but each member of a homologous chromosome pair associates with microtubules emanating from opposite poles. Step ❸: During anaphase of meiosis I, the homologous chromosomes, each consisting of two chromatids, are pulled to opposite spindle poles. Step ❹: Cytokinesis yields the two daughter cells (now 2*n*), which enter meiosis II without undergoing DNA replication. At metaphase of meiosis II, shown here, the chromatids composing each replicated chromosome associate with spindle microtubules from opposite spindle poles, as they do in mitosis. Steps ❺ and ❻: Segregation of chromatids to opposite spindle poles during the second meiotic anaphase followed by cytokinesis generates haploid germ cells (1*n*) containing one copy of each chromosome (referred to as chromatids earlier).

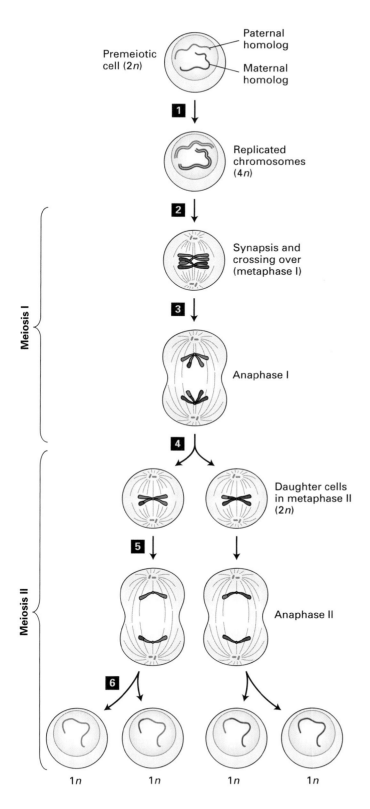

Premeiotic cell (2n)

Paternal homolog

Maternal homolog

1

Replicated chromosomes (4n)

2

Synapsis and crossing over (metaphase I)

3

Anaphase I

Meiosis I

4

Daughter cells in metaphase II (2n)

5

Anaphase II

Meiosis II

6

1n   1n   1n   1n

the formation of germ cells in higher eukaryotes. Multiple yeast mutants that cannot form spores have been isolated, and the wild-type proteins encoded by these genes have been analyzed. These studies have identified specialized cell-cycle proteins that are required for meiosis.

Under starvation conditions, expression of G_1 cyclins (Cln1/2/3) in *S. cerevisiae* is repressed, blocking the normal progression of G_1 in cells as they complete mitosis. Instead, a set of early meiotic proteins are induced. Among these is *Ime2*, a protein kinase that performs the essential G_1 cyclin-CDK function of phosphorylating the S-phase inhibitor Sic1, leading to release of active S-phase cyclin-CDK complexes and the onset of DNA replication in meiosis I (see Figure 21-34, step ⬜). The absence of Ime2 expression during meiosis II and the continued repression of Cln expression account for the failure to replicate DNA during the second meiotic division (steps ⑤ and ⑥).

Crossing Over and Meiosis-Specific Rec8 Are Necessary for Specialized Chromosome Segregation in Meiosis I

Recall that in mitosis, sister chromatids replicated during the S phase are initially linked by cohesin complexes at multiple positions along their full length (see Figure 21-19, *left*). As chromosomes condense, cohesin complexes become restricted to the region of the centromere, and at metaphase, the sister chromatids composing each (replicated) chromosome associate with microtubules emanating from opposite spindle poles. Although motor proteins pull sister chromatids toward opposite spindle poles, their movement initially is resisted by the cohesin complexes linking them at the centromere. The subsequent separase-catalyzed cleavage of the Scc1 cohesin subunit permits movement of the chromatids toward the spindle poles to begin, heralding the onset of anaphase (Figure 21-35a; see also Figure 21-19, *right*).

In metaphase of meiosis I, both sister chromatids in one (replicated) chromosome associate with microtubules emanating from the *same* spindle pole, rather than from opposite poles as they do in mitosis. Two physical links between homologous chromosomes are thought to resist the pulling force of the spindle until anaphase: (a) crossing over between chromatids, one from each pair of homologous chromosomes, and (b) cohesin cross-links between chromatids distal to the crossover point.

Evidence for the role of crossing over in meiosis in *S. cerevisiae* comes from the observation that when recombination is blocked by mutations in proteins essential for the process, chromosomes segregate randomly during meiosis I; that is, homologous chromosomes do not necessarily segregate to opposite spindle poles. Such segregation to opposite spindle poles normally occurs because both chromatids of homologous chromosome pairs associate with spindle fibers emanating from opposite spindle

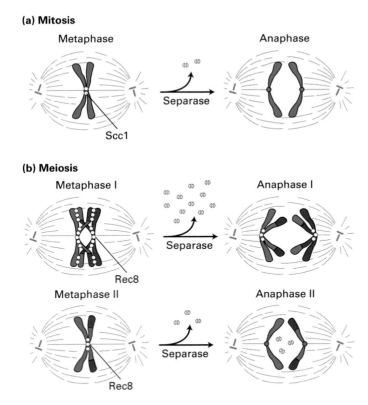

(a) Mitosis

Metaphase   Anaphase

Separase

Scc1

(b) Meiosis

Metaphase I   Anaphase I

Separase

Rec8

Metaphase II   Anaphase II

Separase

Rec8

▲ **FIGURE 21-35 Cohesin function during mitosis and meiosis.** (a) During mitosis, sister chromatids generated by DNA replication in the S phase are initially associated by cohesin complexes along the full length of the chromatids. During chromosome condensation, cohesin complexes (yellow) become restricted to the region of the centromere at metaphase, as depicted here. Once separase cleaves the Scc1 cohesin subunit, sister chromatids can separate, marking the onset of anaphase (see Figure 21-19). (b) In metaphase of meiosis I, crossing over between maternal and paternal chromatids produces synapsis of homologous parental chromosomes. The chromatids of each replicated chromosome are cross-linked by cohesin complexes along their full length. Rec8, a meiosis-specific homolog of Scc1, is cleaved in chromosome arms but not in the centromere, allowing homologous chromosome pairs to segregate to daughter cells. Centromeric Rec8 is cleaved during meiosis II, allowing individual chromatids to segregate to daughter cells. [Modified from F. Uhlmann, 2001, *Curr. Opin. Cell Biol.* **13**:754.]

poles (see Figure 21-34, step ③). This in turn requires that homologous chromosomes pair during meiosis I, a process called *synapsis* that can be visualized microscopically in eukaryotes with large chromosomes. Consequently, the finding that mutations that block recombination also block proper segregation in meiosis I implies that recombination is required for synapsis in *S. cerevisiae*. In higher eukaryotes, processes in addition to recombination and

chromatid linking through cohesin complexes contribute to synapsis in meiosis I.

During meiosis I, the cohesin cross-links between chromosome arms are cleaved by separase, allowing the homologous chromosomes to separate, but cohesin complexes at the centromere remain linked (Figure 21-35b, *top*). The maintenance of centromeric cohesion during meiosis I is necessary for the proper segregation of chromatids during meiosis II. Recent studies with a *S. pombe* mutant have shown that a specialized cohesin subunit, *Rec8*, maintains centromeric cohesion between sister chromatids during meiosis I. Expressed only during meiosis, Rec8 is homologous to Scc1, the cohesin subunit that forms the actual bridge between sister chromatids in mitosis (see Figure 21-19). Immunolocalization experiments in *S. pombe* have revealed that during early anaphase of meiosis I, Rec8 is lost from chromosome arms but is retained at centromeres. However, during early anaphase of meiosis II, centromeric Rec8 is degraded by separase, so the chromatids can segregate, as they do in mitosis (Figure 21-35b, *bottom*). A specific protein expressed during meiosis I, but not during meiosis II, protects centromeric Rec8 from separase cleavage in meiosis I.

S. cerevisiae Rec8 has been shown to localize and function similarly to *S. pombe* Rec8. Homologs of Rec8 also have been identified in higher organisms, and **RNA interference (RNAi)** experiments in *C. elegans* (Chapter 9) indicate that the Rec8 homolog in that organism has a similar function. Recent micromanipulation experiments during grasshopper spermatogenesis support the hypothesis that kinetochore-bound proteins protect centromeric Rec8 from cleavage during meiosis I but not during meiosis II and also direct kinetochore attachment to microtubules emanating from the correct spindle pole (Figure 21-36). Thus crossing over, Rec8, and special kinetochore-associated proteins appear to function in meiosis in all eukaryotes.

Recent **DNA microarray** analyses in *S. cerevisiae* have revealed other proteins that are required for meiosis. As discussed in other chapters, researchers can monitor transcription of thousands of genes, indeed entire genomes, with DNA microarrays (see Figure 9-35). One of the multiple genes found to be expressed in *S. cerevisiae* cells during meiosis but not during mitosis is *MAM1*, which encodes a protein called *monopolin*. Subsequent gene-specific mutagenesis studies revealed that deletion of *MAM1* causes sister chromatids in metaphase of meiosis I to associate with the first meiotic spindle as though they were mitotic chromatids or chromatids in meiosis II. That is, kinetochores of the sister chromatids composing a single (replicated) chromosome attached to microtubules emanating from opposite spindle poles rather than from the same spindle pole. This result indicates that monopolin is required for formation of a specialized kinetochore in meiosis I responsible for the unique co-orientation of sister chromatids of synapsed homologous chromosomes in the first meiotic division.

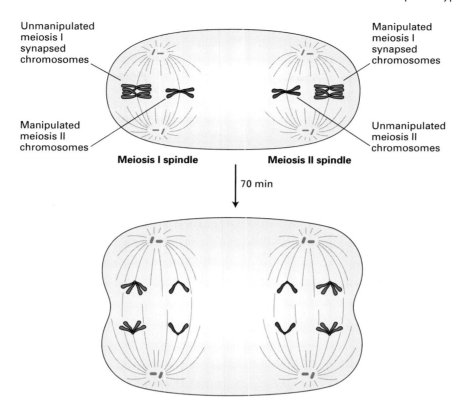

▲ **EXPERIMENTAL FIGURE 21-36 Anaphase movements and cohesion of meiotic chromosomes are determined by proteins associated with the chromosomes.** Grasshopper spermatocytes in meiosis I and II were fused so that both types of spindles with their associated chromosomes were present in a single fused cell. Then a micromanipulation needle was used to move some meiosis I chromosomes (blue) and meiosis II chromosomes (red) from one spindle to another; other chromosomes were left attached to their normal spindles. After 70 minutes, both spindles with their attached chromosomes had completed anaphase movements. The synapsed meiosis I chromosomes (blue) separated normally (i.e., one homologous pair toward one spindle pole and the other homologous pair toward the other) whether attached to the meiosis I spindle (*left*) or meiosis II spindle (*right*). Similarly, the meiosis II chromosomes (red) separated normally (i.e., one chromatid toward one spindle pole and the other chromatid toward the opposite pole) independent of which spindle they were attached to. These results indicate that the association of kinetochores with spindle microtubules and the stability of cohesins linking chromosomes are determined by factors that are associated with the chromosomes and not by the spindles or soluble components of cells in meiosis I and II. [See L. V. Paliulis and R. B. Nicklas, 2000, *J. Cell Biol.* **150**:1223.]

KEY CONCEPTS OF SECTION 21.8

Meiosis: A Special Type of Cell Division

■ Meiosis involves one cycle of chromosome replication followed by two cycles of cell division to produce haploid germ cells from a diploid premeiotic cell (see Figure 21-34).

■ During meiosis I, replicated homologous chromosomes pair along their lengths in a process called synapsis. At least one recombination between chromatids of homologous chromosomes almost invariably occurs.

■ Most of the cell-cycle proteins that function in mitotically dividing cells also function in cells undergoing meiosis, but some proteins are unique to meiosis.

■ In *S. cerevisiae*, expression of G_1 cyclins (Clns) is repressed throughout meiosis. Meiosis-specific Ime2 performs the function of G_1 cyclin-CDK complexes in promoting initiation of DNA replication during meiosis I. DNA replication does not occur during meiosis II because neither Ime2 nor G_1 cyclins are expressed.

■ In *S. cerevisiae*, recombination (crossing over) between chromatids of homologous parental chromatids and cohesin cross-links between chromatids distal to the crossover are responsible for synapsis of homologous chromosomes during prophase and metaphase of meiosis I. A specialized cohesin subunit, Rec8, replaces the Scc1 cohesin subunit during meiosis.

■ During early anaphase of meiosis I, Rec8 in the chromosome arms is cleaved, but a meiosis-specific protein associated with the kinetochore protects Rec8 in the region of the centromere from cleavage. As a result, the chromatids of homologous chromosomes remain associated during segregation in meiosis I. Cleavage of centromeric Rec8 during anaphase of meiosis II allows individual chromatids to segregate into germ cells (see Figure 21-35b).

■ Monopolin, another meiosis-specific protein, is required for both chromatids of homologous chromosomes to associate with microtubules emanating from the same spindle poles during meiosis I.

PERSPECTIVES FOR THE FUTURE

The remarkable pace of cell-cycle research over the last 25 years has led to the model of eukaryotic cell-cycle control outlined in Figure 21-2. A beautiful logic underlies these molecular controls. Each regulatory event has two important functions: to activate a step of the cell cycle and to prepare the cell for the next event of the cycle. This strategy ensures that the phases of the cycle occur in the proper order.

Although the general logic of cell-cycle regulation now seems well established, many critical details remain to be discovered. For instance, although researchers have identified some components of the pre-replication complex that must be phosphorylated by S-phase CDK-cyclin complexes to initiate DNA replication, other components remain to be determined. Also still largely unknown are the substrates phosphorylated by mitotic cyclin-CDK complexes, leading to chromosome condensation and the remarkable reorganization of microtubules that results in assembly of the beautiful mitotic spindle. Much remains to be learned about how the activities of Wee1 kinase and Cdc25 phosphatase are controlled; these proteins in turn regulate the kinase activity of the CDK subunit in most cyclin-CDK complexes.

Much has been discovered recently about operation of the cell-cycle checkpoints, but the mechanisms that activate ATM and ATR in the DNA-damage checkpoint are poorly understood. Likewise, much remains to be learned about the control and mechanism of Mad2 in the spindle-assembly checkpoint and of Cdc14 in the chromosome-segregation checkpoint in higher cells. As we learn in the next chapter, asymmetric cell division plays a critical role in the normal development of multicellular organisms. Many questions remain about how the plane of cytokinesis and the localization of daughter chromosomes are determined in cells that divide asymmetrically. Similarly, the mechanisms that underlie the unique segregation of chromatids during meiosis I have not been elucidated yet.

 Understanding these detailed aspects of cell-cycle control will have significant consequences, particularly for the treatment of cancers. Radiation and many forms of chemotherapy cause DNA damage and cell-cycle arrest in the target cells, leading to their apoptosis. But this induction of apoptosis depends on p53 function. For this reason, human cancers associated with mutations of both *p53* alleles, which is fairly common, are particularly resistant to these therapies. If more were understood about cell-cycle con-

trols and checkpoints, new strategies for treating p53-minus cancers might be possible. For instance, some chemotherapeutic agents inhibit microtubule function, interfering with mitosis. Resistant cells selected during the course of treatment may be defective in the spindle-assembly checkpoint as the result of mutations in the genes encoding the proteins involved. Perhaps the loss of this checkpoint could be turned to therapeutic advantage. Only better understanding of the molecular processes involved will tell. ■

KEY TERMS

anaphase-promoting complex (APC) 863
APC specificity factors 863
Cdc25 phosphatase 866
checkpoints 887
CIPs 885
cohesins 872
condensin 870
crossing over 890
cyclin-dependent kinases (CDKs) 855
destruction box 863
DNA-damage checkpoint 889
E2F proteins 834
G₁ cyclins 876

interphase 854
meiosis 890
mitogens 856
mitosis-promoting factor (MPF) 860
mitotic cyclin 861
p53 protein 889
quiescent cells 883
Rb protein 884
restriction point 883
S-phase inhibitor 868
S phase–phase promoting factor (SPF) 876
securin 872
synapsis 892
Wee1 protein-tyrosine kinase 866

REVIEW THE CONCEPTS

1. What strategy ensures that passage through the cell cycle is unidirectional and irreversible? What is the molecular machinery that underlies this strategy?

2. When fused with an S-phase cell, cells in which of the following phases of the cell cycle will initiate DNA replication prematurely—G1? G2? M? Predict the effect of fusing a cell in G1 and a cell in G2 with respect to the timing of S phase in each cell.

3. In 2001, the Nobel Prize in Physiology or Medicine was awarded to three cell cycle scientists. Sir Paul Nurse was recognized for his studies with the fission yeast *S. pombe*, in particular for the discovery and characterization of the *wee1* gene. What is the wee phenotype? What did the characterization of the *wee1* gene tell us about cell cycle control?

4. Tim Hunt shared the 2001 Nobel Prize for his work in the discovery and characterization of cyclin proteins in eggs and embryos. What experimental evidence indicates that cyclin B is required for a cell to enter mitosis? What evidence indicates that cyclin B must be destroyed for a cell to exit mitosis?

5. Leeland Hartwell, the third recipient of the 2001 Nobel Prize, was acknowledged for his characterization of cell cycle checkpoints in the budding yeast *S. cerevisiae*. What is a cell cycle checkpoint? Where do checkpoints occur in the cell cycle? How do cell cycle checkpoints help to preserve the fidelity of the genome?

6. In *Xenopus*, one of the substrates of MPF is the Cdc25 phosphatase. When phosphorylated by MPF, Cdc25 is activated. What is the substrate of Cdc25? How does this information explain the autocatalytic nature of MPF as described in Figure 21-6?

7. Explain how CDK activity is modulated by the following proteins: (a) cyclin, (b) CAK, (c) Wee1, (d) p21.

8. Three known substrates of MPF or kinases regulated by MPF are the nuclear lamins, subunits of condensin, and myosin light chain. Describe how the phosphorylation of each of these proteins affects its function and progression through mitosis.

9. Describe the series of events by which the APC promotes the separation of sister chromatids at anaphase.

10. A common feature of cell cycle regulation is that the events of one phase ensure progression into a subsequent phase. In *S. cerevisiae*, Cdc28-Clns catalyze progression through G1, and Cdc28-Clbs catalyze progression through S/G2/M. Name three ways in which the activity of Cdc28-Clns promotes the activation of Cdc28-Clbs.

11. For S phase to be completed in a timely manner, DNA replication initiates from multiple origins in eukaryotes. In *S. cerevisiae*, what role do S-phase CDK-cyclin complexes play to ensure that the entire genome is replicated once and only once per cell cycle?

12. What is the functional definition of the restriction point? Cancer cells typically lose restriction point controls. Explain how the following mutations, which are found in some cancer cells, lead to a bypass of restriction point controls: (a) overexpression of cyclin D, (b) loss of Rb function, (c) infection by retroviruses encoding v-Fos and v-Jun.

13. Individuals with the hereditary disorder ataxia telangiectasia suffer from neurodegeneration, immunodeficiency, and increased incidence of cancer. The genetic basis for ataxia telangiectasia is a loss-of-function mutation in the ATM gene (ATM = ataxia telangiectasia-mutated). Name two substrates of ATM. How does the phosphorylation of these substrates lead to inactivation of CDKs to enforce cell cycle arrest at a checkpoint?

14. Meiosis and mitosis are overall analogous processes involving many of the same proteins. However, some proteins function uniquely in each of these cell division events. Explain the meiosis-specific function of the following: (a) Ime2, (b) Rec8, (c) monopolin.

ANALYZE THE DATA

As genes encoding cyclin-dependent kinases were cloned from fission yeast, budding yeast, and animal cells, investigators working in plant systems designed experiments to determine whether plants also possessed CDKs that functioned as the catalysts for cell cycle progression. To isolate a cDNA encoding a CDK in maize, investigators aligned the predicted amino acid sequences of human *cdc2*, *S. pombe cdc2*, and *S. cerevisiae cdc28* genes (see the figure below). Degenerate

```
S. pombe cdc2       ----MENYQKVEKIGE GTYGVVYK ARHKLSG---RIVAMKKIRLEDESEGVPSTAIREIS
human cdc2          ----MEDYTKIEKIGE GTYGVVYK GRHKTTG---QVVAMKKIRLESEEEGVPSTAIREIS
S. cerevisiae cdc28 MSGELANYKRLEKVGE GTYGVVYK ALDLRPGQGQRVVALKKIRLESEDEGVPSTAIREIS
                       :  :*  ::  :**:* ******** .  .   .*   ::**:*****.*.**********

S. pombe cdc2       LLKEVNDENNRSNCVRLLDILHAES-KLYLVFEFLDMDLKKYMDRISETGATSLDPRLVQ
human cdc2          LLKELR----HPNIVSLQDVLMQDS-RLYLIFEFLSMDLKKYLDSIPPG--QYMDSSLVK
S. cerevisiae cdc28 LLKELK----DDNIVRLYDIVHSDAHKLYLFEFLDLDLKRYMEGIPKD--QPLGADIVK
                    ****:.    *  * *  *::   :: :**:****.:***:*:: *.     :.. :*:

S. pombe cdc2       KFTYQLVNGVNFCHSRRII HRDLKPQN LLIDKEGNLKLADFGLARSFGVPLRNYTHEIVT
human cdc2          SYLYQILQGIVFCHSRRVL HRDLKPQN LLIDDKGTIKLADFGLARAFGIPIRVYTHEVVT
S. cerevisiae cdc28 KFMMQLCKGIAYCHSHRIL HRDLKPQN LLINKDGNLKLGDFGLARAFGVPLRAYTHEIVT
                    .:  *  : :*: :***.*:* ********** *:. :* .*****:: ** .:* ****:**

S. pombe cdc2       LWYRAPEVLLGSRHYSTGVDIWSVGCIFAEMIRRSPLFPGDSEIDEIFKIFQVLGTPNEE
human cdc2          LWYRSPEVLLGSARYSTPVDIWSIGTIFAELATKKPLFHGDSEIDQLFRIFRALGTPNNE
S. cerevisiae cdc28 LWYRAPEVLLGGKQYSTGVDTWSIGCIFAEMCNRKPIFSGDSEIDQIFKIFRVLGTPNEA
                    ****:******. :*** ** **:* **** .  :* ** ****::*::**::*.******:

S. pombe cdc2       VWPGVTLLQDYKSTFPRWKRMDLHKVVPNGEEDAIELLSAMLVYDPAHRISAKRALQQNY
human cdc2          VWPEVESLQDYKNTFPKWKPGSLASHVKNLDENGLDLLSKMLIYDPAKRISGKMALNHPY
S. cerevisiae cdc28 IWPDIVYLPDFKPSFPQWRRKDLSQVVPSLDPRGIDLLDKLLAYDPINRISARRAAIHPY
                    :**  :  * *:* :**:*:*:  .  * . *  .::**. :* *** :***.:  *  : *

S. pombe cdc2       LRDFH------
human cdc2          FNDLDNQIKKM
S. cerevisiae cdc28 FQES-------
```

oligonucleotide PCR primers corresponding to the boxed amino acids were generated and a DNA fragment of the expected size was synthesized by PCR using these primers and maize cDNA library as a template. Why were the boxed regions selected as the basis for primer design?

The PCR product was then used as a probe to screen the maize cDNA library, and a full-length cDNA clone was isolated. The clone was sequenced, and the predicted amino acid sequence was 64% identical to human *cdc2* and 63% identical to *S. pombe cdc2* and to *S. cerevisiae cdc28*. What does the level of sequence similarity suggest about the evolution of *cdc2* genes?

A complementation experiment was performed with *S. cerevisiae* cells possessing a temperature-sensitive *cdc28* mutation. Wild-type cells, *cdc28*[ts] cells, and *cdc28*[ts] cells transformed with the maize *cdc2* cDNA under the influence of a strong promoter were grown at the permissive (25 °C) or restrictive (37 °C) temperatures. Cell proliferation was monitored by the growth of colonies on the culture plates (shown in the figure below).

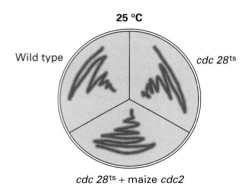

25 °C

Wild type   *cdc 28*[ts]

cdc 28[ts] + maize *cdc2*

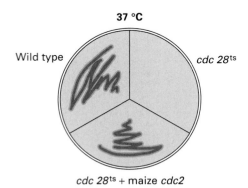

37 °C

Wild type   *cdc 28*[ts]

cdc 28[ts] + maize *cdc2*

Why do the *cdc28*[ts] cells form colonies at 25 °C but not 37 °C? What is the significance of colony formation of the *cdc28*[ts] + maize *cdc2* cells at 37 °C? What does this experiment tell us about the functional homology of cyclin-dependent kinase genes among eukaryotic species?

REFERENCES

Overview of the Cell Cycle and Its Control

Nasmyth, K. 2001. A prize for proliferation. *Cell* **107**: 689–701.

Biochemical Studies with Oocytes, Eggs, and Early Embryos

Doree, M., and T. Hunt. 2002. From Cdc2 to Cdk1: when did the cell cycle kinase join its cyclin partner? *J. Cell Sci.* **115**: 2461–2464.

Fang, G., H. Yu, and M. W. Kirschner. 1999. Control of mitotic transitions by the anaphase-promoting complex. *Philos. Trans. R. Soc. London Ser. B* **354**:1583–1590.

R. Jessberger. 2002. The many functions of SMC proteins in chromosome dynamics. *Nature Rev. Mol. Cell Biol.* **3**:67–78.

Y. Masui. 2001. From oocyte maturation to the in vitro cell cycle: the history of discoveries of Maturation-Promoting Factor (MPF) and Cytostatic Factor (CSF). *Differentiation* **69**:1–17.

Genetic Studies with *S. pombe*

Nurse, P. 2002. Cyclin dependent kinases and cell cycle control (Nobel Lecture). *Chembiochem.* **3**:596–603.

Molecular Mechanisms for Regulating Mitotic Events

Burke, B., and J. Ellenberg. 2002. Remodeling the walls of the nucleus. *Nature Rev. Mol. Cell Biol.* **3**:487–497.

Nelson, W. J. 2000. W(h)ither the Golgi during mitosis? *J. Cell Biol.* **149**:243–248.

Nigg, E. A. 2001. Mitotic kinases as regulators of cell division and its checkpoints. *Nature. Rev. Mol. Cell Biol.* **2**:21–32.

Genetic Studies with *S. cerevisiae*

Bell, S. P., and A. Dutta. 2002. DNA replication in eukaryotic cells. *Ann. Rev. Biochem.* **71**:333–374.

Deshaies, R. J. 1999. SCF and Cullin/Ring H2-based ubiquitin ligases. *Ann. Rev. Cell Devel. Biol.* **15**:435–467.

Diffley, J. F., and K. Labib. 2002. The chromosome replication cycle. *J. Cell Sci.* **115**:869–872.

Hartwell, L. H. 2002. Yeast and cancer (Nobel Lecture). *Biosci. Rep.* **22**:373–394.

Kelly, T. J., and G. W. Brown. 2000. Regulation of chromosome replication. *Ann. Rev. Biochem.* **69**:829–880.

Kitagawa, K., and P. Hieter. 2001. Evolutionary conservation between budding yeast and human kinetochores. *Nature Rev. Mol. Cell Biol.* **2**:678–687.

Lei, M., and B. K. Tye. 2001. Initiating DNA synthesis: from recruiting to activating the MCM complex. *J. Cell Sci.* **114**:1447–1454.

Nasmyth, K. 2001. Disseminating the genome: joining, resolving, and separating sister chromatids during mitosis and meiosis. *Ann. Rev. Genet.* **35**:673–745.

Nasmyth, K. 2002. Segregating sister genomes: the molecular biology of chromosome separation. *Science* **297**:559–565.

Uhlmann, F. 2001. Chromosome cohesion and segregation in mitosis and meiosis. *Curr. Opin. Cell Biol.* **13**:754–761.

Cell-Cycle Control in Mammalian Cells

Ekholm, S. V., and S. I. Reed. 2000. Regulation of G(1) cyclin-dependent kinases in the mammalian cell cycle. *Curr. Opin. Cell Biol.* **12**:676–684.

Harper, J. W., J. L. Burton, and M. J. Solomon. 2002. The anaphase-promoting complex: it's not just for mitosis any more. *Genes Devel.* **16**:2179–2206.

Sears, R. C., and J. R. Nevins. 2002. Signaling networks that link cell proliferation and cell fate. *J. Biol. Chem.* **277**:11617–11620.

Sherr, C. J. 2001. The INK4a/ARF network in tumour suppression. *Nature Rev. Mol. Cell Biol.* **2**:731–737.

Checkpoints in Cell-Cycle Regulation

Bardin, A. J., and A. Amon. 2001. Men and sin: what's the difference? *Nature Rev. Mol. Cell Biol.* **2**:815–826.

Osborn, A. J., S. J. Elledge, and L. Zou. 2002. Checking on the fork: the DNA-replication stress-response pathway. *Trends Cell Biol.* **12**:509–516.

Pereira, G., and E. Schiebel. 2001. The role of the yeast spindle pole body and the mammalian centrosome in regulating late mitotic events. *Curr. Opin. Cell Biol.* **13**:762–769.

Shah, J. V., and D. W. Cleveland. 2000. Waiting for anaphase: Mad2 and the spindle assembly checkpoint. *Cell* **103**:997–1000.

Meiosis: A Special Type of Cell Division

Lee, B., and A. Amon. 2001. Meiosis: how to create a specialized cell cycle. *Curr. Opin. Cell Biol.* **13**:770–777.

Petronczki, M., M. F. Siomos, and K. Nasmyth. 2003. Un menage a quatre: the molecular biology of chromosome segregation in meiosis. *Cell* **112**:423–440.

22

CELL BIRTH, LINEAGE, AND DEATH

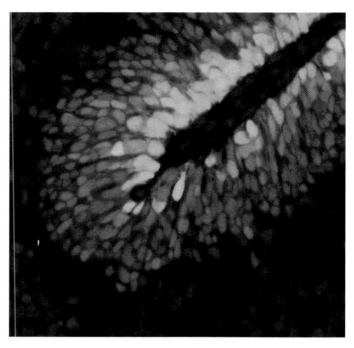

Cells being born in the developing cerebellum. All nuclei are labeled in red; the green cells are dividing and migrating into internal layers of the neural tissue. [Courtesy of Tal Raveh, Matthew Scott, and Jane Johnson.]

During the evolution of multicellular organisms, new mechanisms arose to diversify cell types, to coordinate their production, to regulate their size and number, to organize them into functioning tissues, and to eliminate extraneous or aged cells. Signaling between cells became even more important than it is for single-celled organisms. The mode of reproduction also changed, with some cells becoming specialized as **germ cells** (e.g., eggs, sperm), which give rise to new organisms, as distinct from all other body cells, called **somatic cells.** Under normal conditions somatic cells will never be part of a new individual.

The formation of working tissues and organs during **development** of multicellular organisms depends in part on specific patterns of mitotic cell division. A series of such cell divisions akin to a family tree is called a *cell lineage*, which traces the progressive determination of cells, restricting their developmental potential and their *differentiation* into specialized cell types. Cell lineages are controlled by intrinsic (internal) factors—cells acting according to their history and internal regulators—as well as by extrinsic (external) factors such as cell-cell signals and environmental inputs (Figure 22-1). A cell lineage begins with **stem cells,** unspecialized cells that can potentially reproduce themselves and generate more-specialized cells indefinitely. Their name comes from the image of a plant stem, which grows upward, continuing to form more stem,

while sending off leaves and branches to the side. A cell lineage ultimately culminates in formation of terminally differentiated cells such as skin cells, neurons, or muscle cells. Terminal differentiation generally is irreversible, and the resulting highly specialized cells often cannot divide; they survive, carry out their functions for varying lengths of time, and then die.

Many cell lineages contain intermediate cells, referred to as *precursor cells* or *progenitor cells,* whose potential to form different kinds of differentiated cells is more limited than that of the stem cells from which they arise. (Although some researchers distinguish between precursor and progenitor cells, we will use these terms interchangeably.) Once a new precursor cell type is created, it often produces transcription factors characteristic of its fate. These transcription factors

▶ **FIGURE 22-1 Overview of the birth, lineage, and death of cells.** Following growth, cells are "born" as the result of symmetric or asymmetric cell division. (a) The two daughter cells resulting from symmetric division are essentially identical to each other and to the parental cell. Such daughter cells subsequently can have different fates if they are exposed to different signals. The two daughter cells resulting from asymmetric division differ from birth and consequently have different fates. Asymmetric division commonly is preceded by the localization of regulatory molecules (green) in one part of the parent cell. (b) A series of symmetric and/or asymmetric cell divisions, called a cell lineage, gives birth to each of the specialized cell types found in a multicellular organism. The pattern of cell lineage can be under tight genetic control. Programmed cell death occurs during normal development (e.g., in the webbing that initially develops when fingers grow) and also in response to infection or poison. A series of specific programmed events, called apoptosis, is activated is these situations.

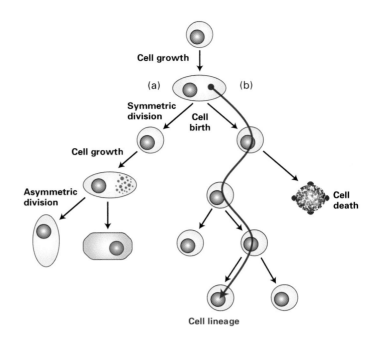

coordinately activate, or repress, batteries of genes that direct the differentiation process. For instance, a few key regulatory transcription factors create the different mating types of budding yeast and coordinate the numerous genes necessary to turn a precursor cell into a muscle cell, two examples that we discuss in this chapter.

Typically we think of cell fates in terms of the differentiated cell types that are formed. A quite different cell fate, **programmed cell death,** also is absolutely crucial in the formation and maintenance of many tissues. A precise genetic regulatory system, with checks and balances, controls cell death just as other genetic programs control cell differentiation. In this chapter, then, we consider the life cycle of cells—their birth, their patterns of division (lineage), and their death. These aspects of cell biology converge with developmental biology and are among the most important processes regulated by various signaling pathways discussed in earlier chapters.

22.1 The Birth of Cells

Many descriptions of cell division imply that the parental cell gives rise to two daughter cells that look and behave exactly like the parental cell, that is, cell division is *symmetric,* and the progeny do not change their properties. But if this were always the case, none of the hundreds of differentiated cell types would ever be formed. Differences among cells can arise when two initially identical daughter cells diverge by receiving distinct developmental or environmental signals. Alternatively, the two daughter cells may differ from "birth," with each inheriting different parts of the parental cell (see Figure 22-1). Daughter cells

produced by such **asymmetric cell division** may differ in size, shape, and/or composition, or their genes may be in different states of activity or potential activity. The differences in these internal signals confer different fates on the two cells.

Here we discuss some general features of how different cell types are generated, culminating with the best-understood complex cell lineage, that of the nematode *Caenorhabditis elegans.* In later sections, we focus on examples of the molecular mechanisms that determine particular cell types in yeast, *Drosophila,* and mammals.

Stem Cells Give Rise to Stem Cells and to Differentiating Cells

Stem cells, which give rise to the specialized cells composing the tissues of the body, exhibit several patterns of cell division. A stem cell may divide symmetrically to yield two daughter stem cells identical to itself (Figure 22-2a). Alternatively, a stem cell may divide asymmetrically to generate a copy of itself and a derivative stem cell that has more-restricted capabilities, such as dividing for a limited period of time or giving rise to fewer types of progeny compared with the parental stem cell (Figure 22-2b). A *pluripotent* (or multipotent) stem cell has the capability of generating a number of different cell types, but not all. For instance, a pluripotent blood stem cell will form more of itself plus multiple types of blood cells, but never a skin cell. In contrast, a *unipotent* stem cell divides to form a copy of itself plus a cell that can form only one cell type. In many cases, asymmetric division of a stem cell generates a progenitor cell, which embarks on a path of differentiation, or even a terminally differentiating cell (Figure 22-2c, d).

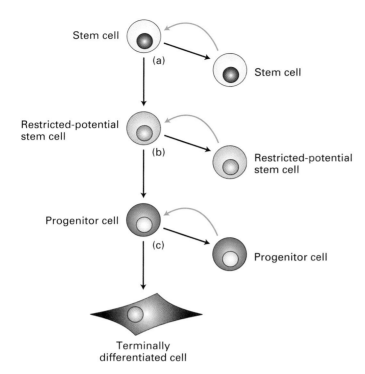

Stem cell

(a)

Stem cell

Restricted-potential stem cell

(b)

Restricted-potential stem cell

Progenitor cell

(c)

Progenitor cell

Terminally differentiated cell

▲ **FIGURE 22-2 Patterns of stem-cell division.** (a) Division of a stem cell produces two cells, one of which is a stem cell like the mother cell. In this way the population of stem cells is maintained. (b) The other daughter cell—a stem cell of more-restricted potential—starts on a pathway toward producing more differentiated cells. When it divides, one of the daughters will be the same sort of restricted-potential stem cell as the mother and the other will be a progenitor cell for a certain type of differentiated cell. Progenitor cells can divide to reproduce themselves and, in response to appropriate signals, can differentiate into a terminally differentiated, nondividing cell.

The two critical properties of stem cells that together distinguish them from all other cells are the ability to reproduce themselves indefinitely, often called *self-renewal,* and the ability to divide asymmetrically to form one daughter stem cell identical to itself and one daughter cell that is different and usually of more restricted potential. In this way, mitotic division of stem cells preserves a population of undifferentiated cells while steadily producing a stream of differentiating cells. Although some types of precursor cells can divide symmetrically to form more of themselves, they do so only for limited periods of time. Moreover, in contrast to stem cells, if a precursor cell divides asymmetrically, it generates two distinct daughter cells, neither of which is identical to the parental precursor cell.

The fertilized egg, or **zygote,** is the ultimate totipotent cell because it has the capability to generate all the cell types of the body. Although not technically a stem cell because it is not self-renewing, the zygote does give rise to

cells with stem-cell properties. For example, the early mouse embryo passes through an eight-cell stage in which each cell can give rise to any cell type of the embryo. If the eight cells are experimentally separated and individually implanted into a suitable foster mother, each can form a whole mouse with no parts missing. This experiment shows that the eight cells are all able to form every tissue; that is, they are *totipotent.* Thus the subdivision of body parts and tissue fates among the early embryonic cells has not irreversibly occurred at the eight-cell stage. At the 16-cell stage, this is no longer true; some of the cells are committed to particular differentiation paths.

Quite different specialized cell types can arise from a common precursor cell. A hematopoietic stem cell can give rise to all the multifarious types of blood cell. However, a whole series of cell divisions is not required; a single division of a pluripotent precursor cell can yield distinct progeny. For instance, lineage studies in which cells are marked by stable infection with a detectable retrovirus have shown that neurons and glial cells can arise from a single division of a particular precursor cell. These cell types are quite different: neurons propagating and transmitting electrical signals and glial cells providing electrical insulation and support. The precursor that generates neurons and glial cells is not a stem cell, since it is incapable of self-renewal; presumably the neuron–glial cell precursor arises from a stem cell further back in the lineage. Recent observations have raised the interesting possibility that stem cells for one tissue may be induced under certain conditions to act as stem cells for a rather different tissue. As we discuss below, postnatal animals contain stem cells for many tissues including the blood, intestine, skin, ovaries and testes, muscle, and liver. Even some parts of the adult brain, where little cell division normally occurs, has a population of stem cells. In muscle and liver, stem cells are most important in healing, as relatively little cell division occurs in the adult tissues otherwise.

Cultured Embryonic Stem Cells Can Differentiate into Various Cell Types

Embryonic stem (ES) cells can be isolated from early mammalian embryos and grown in culture (Figure 22-3a). Cultured ES cells can differentiate into a wide range of cell types, either in vitro or after reinsertion into a host embryo. When grown in suspension culture, human ES cells first differentiate into multicellular aggregates, called embryoid bodies, that resemble early embryos in the variety of tissues they form. When these are subsequently transferred to a solid medium, they grow into confluent cell sheets containing a variety of differentiated cell types including neural cells and pigmented and nonpigmented epithelial cells (Figure 22-3b). Under other conditions, ES cells have been induced to differentiate into precursors for various types of blood cells.

▶ **EXPERIMENTAL FIGURE 22-3**
Embryonic stem (ES) cells can be maintained in culture and form differentiated cell types. (a) Human blastocysts are grown from cleavage-stage embryos produced by in vitro fertilization. The inner cell mass is separated from the surrounding extra-embryonic tissues and plated onto a layer of fibroblast cells that help to nourish the embryonic cells. Individual cells are replated and form colonies of ES cells, which can be maintained for many generations and can be stored frozen. (b) In suspension culture, human ES cells differentiate into multicellular aggregates (embryoid bodies) (*top*). After embryoid bodies are transferred to a gelatinized solid medium, they differentiate further into confluent cell sheets containing a variety of differentiated cell types including neural cells (*middle*), and pigmented and nonpigmented epithelial cells (*bottom*). [Parts (a) and (b) adapted from J. S. Odorico et al., 2001, *Stem Cells* **19**:193–204.]

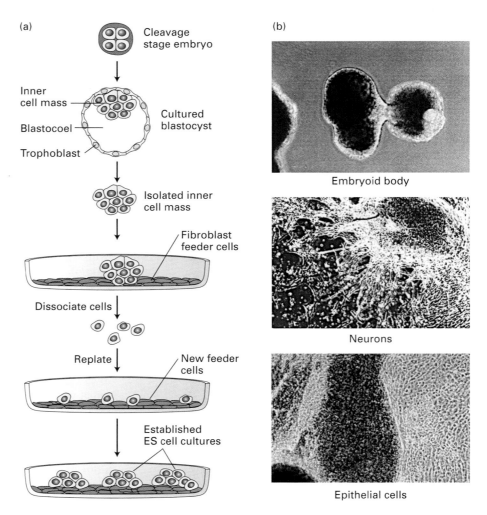

(a)

Cleavage stage embryo

Inner cell mass

Blastocoel

Trophoblast

Cultured blastocyst

Isolated inner cell mass

Fibroblast feeder cells

Dissociate cells

Replate

New feeder cells

Established ES cell cultures

(b)

Embryoid body

Neurons

Epithelial cells

MEDICINE The possibility of using stem cells therapeutically to restore or replace damaged tissue is fueling much research on how to recognize and culture these remarkable cells from embryos and from various tissues in postnatal (adult) animals. For example, if neurons that produce the neurotransmitter dopamine could be generated from stem cells grown in culture, it might be possible to treat people with Parkinson's disease who have lost such neurons. For such an approach to succeed, a way must be found to direct a population of embryonic or other stem cells to form the right type of dopamine-producing neurons, and rejection by the immune system must be prevented. In one ongoing study in which embryonic neurons were transplanted into more than 300 Parkinson's patients, some of the inserted cells have survived for more than 12 years and have provided significant clinical improvement. However, the fetal tissue used in this study is scarce and its use is controversial. Stem cells grown from very early embryos are another option for treating Parkinson's disease and perhaps other neurodegenerative conditions such as Alzheimer's disease. Similar possibilities exist for generating blood, pancreas, and other cell types. Many important questions must be answered before

the feasibility of using stem cells for such purposes can be assessed adequately. ■

Apart from their possible benefit in treating disease, ES cells have already proven invaluable for producing mouse mutants useful in studying a wide range of diseases, developmental mechanisms, behavior, and physiology. By techniques described in Chapter 9, it is possible to eliminate or modify the function of a specific gene in ES cells (see Figure 9-38). Then the mutated ES cells can be employed to produce mice with a **gene knockout** (see Figure 9-39). Analysis of the effects caused by deleting or modifying a gene in this way often provides clues about the normal function of the gene and its encoded protein.

Tissues Are Maintained by Associated Populations of Stem Cells

Many differentiated cell types are sloughed from the body or have life spans that are shorter than that of the organism. Disease and trauma also can lead to loss of differentiated cells. Since these cells generally do not divide, they must be replenished from nearby stem-cell populations.

Our skin, for instance, is a multilayered **epithelium** (the epidermis) underlain by a layer of stem cells that give rise both to more of themselves and to *keratinocytes*, the major cell type in skin. Keratinocytes then move toward the outer surface, becoming increasingly flattened and filled with keratin intermediate filaments. It normally takes about 15–30 days for a newly "born" keratinocyte in the lowest layer to differentiate and move to the topmost layer. The "cells" forming the topmost layer are actually dead and are continually shed from the surface.

In contrast to epidermis, the epithelium lining the small intestine is a single cell thick (Figure 6-4). This thin layer keeps toxins and pathogens from entering our bodies and also transports nutrients essential for survival from the intestinal lumen into the body (Chapter 7). The cells of the intestinal epithelium

continuously regenerate from a stem-cell population located deep in the intestinal wall in pits called *crypts* (Figure 22-4a). The stem cells produce precursor cells that proliferate and differentiate as they ascend the sides of crypts to form the surface layer of the finger-like gut projections called *villi*, across which intestinal absorption occurs. Pulse-chase labeling experiments have shown that the time from cell birth in the crypts to the loss of cells at the tip of the villi is only about 2 to 3 days (Figure 22-4b). Thus enormous numbers of cells must be produced continually to keep the epithelium intact. The production of new cells is precisely controlled: too little division would eliminate villi and lead to breakdown of the intestinal surface; too much division would create an excessively large epithelium and also might be a step toward cancer.

Specific signals are required for creating and maintaining stem-cell populations. In both the skin and the intestinal epithelium, stem-cell growth is regulated in part by β-*catenin*, a protein that helps link certain cell-cell junctions to the cytoskeleton (see Figure 6-7) and also functions as a signal transducer in the Wnt pathway (see Figure 15-32). Activation of β-catenin moves cells from epidermis to hair cell fates. In contrast, removal of β-catenin specifically from the skin of engineered mice eliminates hair cell fates. Skin stem cells then form only epidermis, not hair cells. β-catenin thus acts as a switch between alternative cell fates. Overproduction of active β-catenin leads to excess proliferation of the intestinal epithelium. Blocking the function of β-catenin by interfering with the TCF transcription factor that it activates abolishes the stem cells in the intestine. It seems likely that Wnt signaling, or at least components of the intracellular pathway, is critical for forming, maintaining, or activating stem cells in a variety of tissues.

(a)

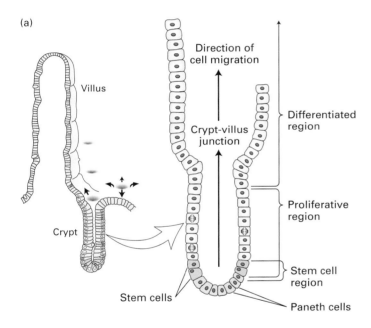

(b)

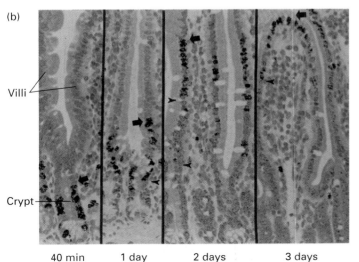

◄ **EXPERIMENTAL FIGURE 22-4 Regeneration of the intestinal epithelium from stem cells can be demonstrated in pulse-chase experiments.** (a) Schematic drawing of the lining of the small intestine, which contains numerous villi formed from a column of cells. These epithelial cells are born near the base of pits (crypts) located between the villi. Located at the very bottom of crypts are Paneth cells, a type of support cell; just above these are four to six stem cells, which divide about once a day, forming precursor cells that also actively divide. As the differentiated cells enter the epithelium of a villus, they stop dividing and begin taking up nutrients from the gut. (b) Results from a pulse-chase experiment in which radioactively labeled thymidine (the pulse) was added to a tissue culture of the intestinal epithelium. Dividing cells incorporated the labeled thymidine into their newly synthesized DNA. The labeled thymidine was washed away and replaced with nonlabeled thymidine (the chase) after a brief period; cells that divided after the chase did not become labeled. These micrographs show that 40 minutes after labeling, all the label is in cells near the base of the crypt. At later times, the labeled cells are seen progressively farther away from their point of birth in the crypt. Cells at the top are shed. This process ensures contant replenishment of the gut epithelium with new cells. [Part (a) adapted from C. S. Potten, 1998, *Philos. Trans. R. Soc. London, Ser. B* **353**:821. Part (b) courtesy of C. S. Potten, from P. Kaur and C. S. Potten, 1986, *Cell Tiss. Kinet.* **19**:601.]

Skin also contains dendritic epidermal T cells, an immune-system cell that produces a certain form of the T-cell receptor (see Table 14-1). When dendritic epidermal T cells are genetically modified so they do not produce T-cell receptors, wound healing is slow and less complete than in normal skin. Normal healing is restored by addition of keratinocyte growth factor. The current hypothesis is that when dendritic epidermal T cells recognize antigens on cells in damaged tissue, they respond by producing stimulating proteins, such as keratinocyte growth factor, that promote keratinocyte growth and wound healing. Many other signals also control the growth of skin cells, including Wnt/β-catenin, Hedgehog, calcium, transforming growth factor α (TGFα), and TGFβ. Discovering how all these signals work together to control growth and stimulate healing is a substantial challenge that will advance our understanding of diseases such as psoriasis and skin cancer and perhaps pave the way for effective treatments. ∎

Another continuously replenished tissue is the blood, whose stem cells are located in the bone marrow in adult animals. The various types of blood cells all derive from a single type of *pluripotent hematopoietic stem cell*, which gives rise to the more-restricted myeloid and lymphoid stem cells (Figure 22-5). The frequency of hematopoietic stem cells is about 1 cell per 10^4 bone marrow cells, even lower than the frequency of intes-

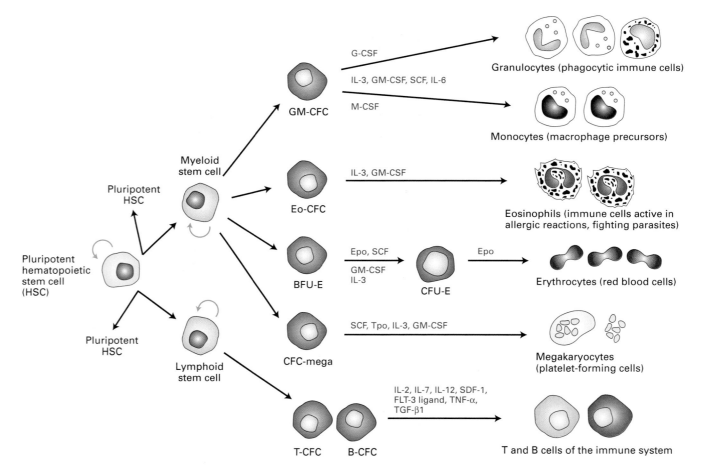

▲ **FIGURE 22-5 Formation of differentiated blood cells from hematopoietic stem cells in the bone marrow.** Pluripotent stem cells may divide symmetrically to self-renew (curved arrow) or divide asymmetrically to form a myeloid or lymphoid stem cell (light green) and a daughter cell that is pluripotent like the parental cell. Although these stem cells are capable of self-renewal, they are committed to one of the two major hematopoietic lineages. Depending on the types and amounts of cytokines present, the myeloid and lymphoid stem cells generate different types of precursor cells (dark green), which are incapable of self-renewal. Precursor cells are detected by their ability to form colonies containing the differentiated cell types shown at right, measured as "colony-forming cells (CFCs)." The colonies are detected on the spleen of animals that have had their own cells eliminated and the precursor cells introduced. Further cytokine-induced proliferation, commitment, and differentiation of the precursor cells give rise to the various types of blood cells. Some of the cytokines that support this process are indicated (red labels). GM = granulocyte-macrophage; Eo = eosinophil; E = erythrocyte; mega = megakaryocyte; T = T-cell; B = B-cell; CFU = colony-forming unit; CSF = colony-stimulating factor; IL = interleukin; SCF = stem cell factor; Epo = erythropoietin; Tpo = thrombopoietin; TNF = tumor necrosis factor; TGF = transforming growth factor; SDF = stromal cell–derived factor; FLT-3 ligand = ligand for fms-like tyrosine kinase receptor 3. [Adapted from M. Socolovsky et al., 1998, *Proc. Nat'l. Acad. Sci. USA* **95**:6573.]

tinal stem cells in crypts. Numerous extracellular growth factors called **cytokines** regulate proliferation and differentiation of the precursor cells for various blood-cell lineages. For example, **erythropoietin** can activate several different intracellular signal-transduction pathways, leading to changes in gene expression that promote formation of erythrocytes (see Figure 14-7).

The hematopoietic lineage originally was worked out by injecting the various types of precursor cells into mice whose precursor cells had been wiped out by irradiation. By observing which blood cells were restored in these transplant experiments, researchers could infer which precursors or terminally differentiated cells (e.g., erythrocytes, monocytes) arise from a particular type of precursor. The first step in these experiments was separation of the different types of hematopoietic precursors. This is possible because each type produces unique combinations of cell-surface proteins that can serve as type-specific markers. If bone marrow extracts are treated with fluorochrome-labeled antibodies for these markers, cells with different surface markers can be separated in a fluorescence-activated cell sorter (see Figure 5-34).

 To date, bone marrow transplants represent the most successful and widespread use of stem cells in medicine. The stem cells in the transplanted marrow can generate new, functional blood cells in patients with certain hereditary blood diseases and in cancer patients who have received irradiation and/or chemotherapy, both of which destroy the bone marrow cells as well as cancer cells. Recent work is directed at exploring whether embryonic stem cells can be induced to differentiate into cells types that would be useful therapeutically. For example, mouse stem cells treated with inhibitors of phosphatidylinositol-3 kinase, a regulator in one of the phosphoinositide signaling pathways (Chapter 14), turn into cells that resemble pancreatic β cells in their production of insulin, their sensitivity to glucose levels, and their aggregation into structures reminiscent of pancreas structures. Implantation of these cells into diabetic mice restored their growth, weight, glucose levels, and survival rates to normal. ∎

 Stem cells in plants are located in **meristems**, populations of undifferentiated cells found at the tips of growing shoots. Shoot apical meristems (SAMs) produce leaves and shoots, and of course more stem cells that constitute the nearly immortal meristems. Meristems can persist for thousands of years in long-lived species such as redwood trees and bristlecone pines. As a plant grows, the cells "left behind" the meristems are encased in rigid cell walls and can no longer grow. SAMs can split to form branches, each branch with its own SAM, or be converted into floral meristems (Figure 22-6). Floral meristems give rise to the four floral organs—sepals, stamens, carpels, and petals—that form flowers. Unlike SAMs, floral meristems are gradually depleted as they give rise to the floral organs.

Numerous genes have been found to regulate the formation, maintenance, and properties of meristems. Many of these genes encode transcription factors that direct progeny of stem cells down different paths of differentiation. For instance, a hierarchy of regulators, particularly transcription factors, controls the separation of differentiating cells from SAMs as leaves form; similarly, three types of regulators control formation of the floral organs from floral meristems (Chapter 15). In both cases, a cascade of gene interactions occurs, with earlier transcription factors causing production of later ones. At the same time, cells are dividing and the differentiating ones are spreading away from their original birth sites. ∎

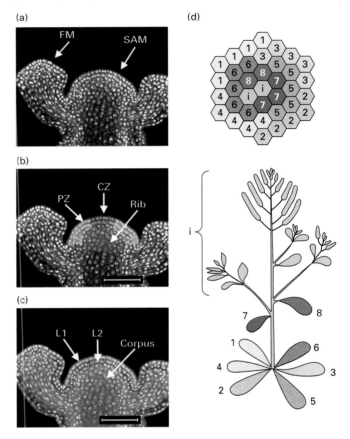

▲ **FIGURE 22-6 Cell fates in meristems of *Arabidopsis*.** In these longitudinal sections through a shoot apical meristem, cell nuclei are revealed by staining with propidium iodide, which binds to DNA. (a) The shoot apical meristem (SAM) produces shoots, leaves, and more meristem. Flower production occurs when the meristem switches from leaf/shoot production to flower production, concomitant with an increase in the number of meristem cells to form floral meristems (FMs), as shown here. (b) Cells in a SAM exhibit different fates and behaviors. Cells divide rapidly in the peripheral zone (PZ) to produce leaves and in the rib zone (Rib) to produce central shoot structures. Cells in the central zone (CZ) divide more slowly, producing an ongoing source of meristem and contributing cells to the PZ and Rib. (c) The layers of the meristem, colored here, are each derived (cloned) from the same precursor cell. The fates of cells in different positions in the L2 layer are shown color-coded in part (d). Scale bars, 50 μm. [Parts (a)–(c) from E. Meyerowitz, 1997, *Cell* **88**:299; micrographs courtesy of Elliot Meyerowitz. Part (d) after C. Wolpert et al., 2002, *Principles of Development*, 2d ed. (Oxford: Oxford University Press).]

Cell Fates Are Progressively Restricted During Development

The eight cells resulting from the first three divisions of a mammalian zygote (fertilized egg) all look the same. As demonstrated experimentally in sheep, each of the cells has the potential to give rise to a complete animal. Additional divisions produce a mass, composed of ≈64 cells, that separates into two cell types: trophectoderm, which will form extra-embryonic tissues like the placenta, and the *inner cell mass*, which gives rise to the embryo proper. The inner cell mass eventually forms three germ layers, each with distinct fates. One layer, the **ectoderm**, will make neural and epidermal cells; another, the **mesoderm**, will make muscle and connective tissue; the third layer, the **endoderm**, will make gut epithelia (Figure 22-7). This conclusion is based on experiments with chimeric animals composed of chicken and quail cells. Embryos composed of cells from both bird species develop fairly normally, yet the cells derived from each donor are distinguishable under the microscope. Thus the contributions of the different donor cells to the final bird can be ascertained. If cells from one germ layer are transplanted into one of the other layers, they do not give rise to cells appropriate to their new location.

Once the three germ layers are established, they subsequently divide into cell populations with different fates. For instance, the ectoderm becomes divided into those cells that are precursors to the skin epithelium and those that are precursors to the nervous system. There appears to be a progressive restriction in the range of cell types that can be formed from stem cells and precursor cells as development proceeds. An early embryonic stem cell, as we've seen, can form every type of cell, an ectodermal cell has a choice between neural and epidermal fates, while a keratinocyte precursor can form skin but not neurons.

Another restriction that occurs early in animal development is the setting aside of cells that will form the **germ line**, that is, the stem cells and precursor cells that eventually will give rise to eggs in a female and sperm in a male. Only the genome of the germ line will ever be passed on to progeny. The setting aside of germ-line cells early in development has been hypothesized to protect chromosomes from damage by reducing the number of rounds of replication they undergo or by allowing special protection of the cells that are critical to heredity. Whatever the reason, the early segregation of the germ line is widespread (though not universal) among animals. In contrast, plants do nothing of the sort; most meristems can give rise to germ-line cells.

One consequence of the early segregation of germ-line cells is that the loss or rearrangement of genes in somatic cells would not affect the inherited genome of a future zygote. Although segments of the genome are rearranged and lost during development of lymphocytes from hematopoietic precursors, most somatic cells seem to have an intact genome, equivalent to that in the germ line. Evidence that at least some somatic cells have a complete and functional genome comes from the successful production of cloned animals by nuclear-transfer cloning. In this procedure, the nucleus of an adult (somatic) cell is introduced into an egg that lacks its nucleus; the manipulated egg, which contains the diploid number of chromosomes and is equivalent to a zygote, then is implanted into a foster mother. The only source of genetic information to guide development of the embryo is the nuclear genome of the donor somatic cell. The frequent failure of such cloning experiments, however, raises

(a)

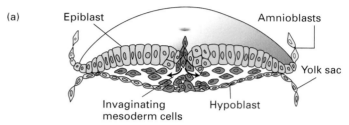

(b)

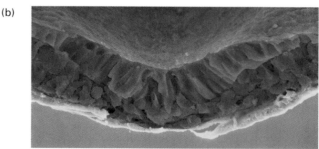

(c)

ECTODERM	**MESODERM**	**ENDODERM**
Central nervous system	Skull	Stomach
Retina and lens	Head, skeletal muscle	Colon
Cranial and sensory	Skeleton	Liver
Ganglia and nerves	Dermis of skin	Pancreas
Pigment cells	Connective tissue	Urinary bladder
Head connective tissue	Urogenital system	Epithelial parts of
Epidermis	Heart	trachea
Hair	Blood, lymph cells	lungs
Mammary glands	Spleen	pharynx
		thyroid
		intestine

▲ **FIGURE 22-7 Fates of the germ layers in animals.** During the early period of mammalian development, cells migrate to create the three germ layers: ectoderm, mesoderm, and endoderm. This process is called gastrulation. Cells in the different layers have largely distinct fates and therefore represent distinct cell lineages. The detailed cell lineage, however, is not exactly the same in different individuals. (a) Shown here is a sketch of a human embryo about 16 days after fertilization of an egg. The first cells to move from the epiblast into the interior form endoderm, and they are followed by invaginating cells that become mesoderm. The remaining epiblast cells become ectoderm. Hypoblast = covering layer. Amnioblasts = cells that will line amniotic cavity. Yolk sac = reminder of our egg-bound ancestors. (b) Scanning electron micrograph of a cross section of a similar-stage embryo. (c) Some of the tissue derivatives of the three germ layers are listed. [Part (a) after T. W. Sadler, 2000, *Langman's Medical Embryology*, 8th ed. (Baltimore: Lippincott Williams and Wilkins), pp. 64–65, Figure 4.3). Part (b) courtesy of Kathy Sulik, University of North Carolina–Chapel Hill.]

questions about how many adult somatic cells do in fact have a complete functional genome. Even the successes, like the famous cloned sheep "Dolly," appear to have some medical problems. Even if differentiated cells have a physically complete genome, clearly only portions of it are transcriptionally active (Chapter 10). Whether the genome of a differentiated cell can revert to having the full developmental potential characteristic of an embryonic cell is a matter of considerable debate. A cell could, for example, have an intact genome, but be unable to properly reactivate it due to inherited chromatin states.

These observations raise two important questions: How are cell fates progressively restricted during development? Are these restrictions irreversible? In addressing these questions, it is important to remember that what a cell in its normal in vivo location will do may differ from what a cell is capable of doing if it is manipulated experimentally. Thus the observed limits to what a cell can do may result from natural regulatory mechanisms or may reflect a failure to find conditions that reveal the cell's full potential.

Although our focus in this chapter is on how cells become different, their ability to remain the same also is critical to the functioning of tissues and the whole organism. Nondividing differentiated cells with particular characteristics must retain them, sometimes for many decades. Stem cells that divide regularly, such as a skin stem cell, must produce one daughter cell with the properties of the parental cell, retaining its characteristic composition, shape, behavior, and responses to specific external signals. Meanwhile, the other daughter cell with its own distinct inheritance, as the result of asymmetric cell division, embarks on a particular differentiation pathway, which may be determined both by the signals the cell receives and by intrinsic bias in the cell's potential, such as the previous activation of certain genes.

The Complete Cell Lineage of *C. elegans* Is Known

In the development of some organisms, cell lineages are under tight genetic control and thus are identical in all individuals of a species. In other organisms the exact number and arrangement of cells vary substantially among different individuals. The best-documented example of a reproducible pattern of cell divisions comes from the nematode *C. elegans*. Scientists have traced the lineage of all the somatic cells in *C. elegans* from the fertilized egg to the mature worm by following the development of live worms using Nomarski interference microscopy (Figure 22-8).

About 10 rounds of cell division, or fewer, create the adult worm, which is about 1 mm long and 70 μm in diameter. The adult worm has 959 somatic cell nuclei (hermaphrodite form) or 1031 (male). The number of somatic cells is somewhat fewer than the number of nuclei because some cells contain multiple nuclei (i.e., they are syncytia). Remarkably, the pattern of cell divisions starting from a *C. elegans* fertilized egg is nearly always the same. As we discuss later in the chapter, many cells that are generated during development undergo programmed cell death and are missing in the adult worm. The consistency of the *C. elegans* cell lineage does not result entirely from each newly born cell inheriting specific information about its destiny. That is, at their birth cells are not necessarily "hard wired" by their own internal inherited instructions to follow a particular path of differentiation. In some cases, various signals direct initially identical cells to different fates, and the outcomes of these signals are consistent from one animal to the next.

The first few cell divisions in *C. elegans* produce six different *founder cells*, each with a separate fate as shown in Figure 22-9a, b. The initial division is asymmetric, giving rise to P1 and the AB founder cell. Further divisions in the P lineage form the other five founder cells. Some of the signals controlling division and fate asymmetry are known. For example, Wnt signals from the P2 precursor control the asymmetric division of the EMS cell into E and MS founder cells. Wnt signaling (see Figure 15-32) is also used in other asymmetric divisions in worms.

The bilateral symmetry of the worm implies duplication of lineages on the two sides; curiously, though, functionally equivalent cells on each side can arise from a pattern of division that is different on the two sides. Some of the embryonic cells function as stem cells, dividing repeatedly to form more of themselves or another type of stem cell, while

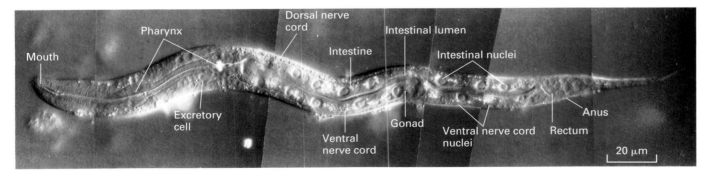

▲ **FIGURE 22-8 Newly hatched larva of *C. elegans*.** Many of the 959 somatic cell nuclei in this hermaphrodite form are visualized in this micrograph obtained by Nomarski interference microscopy. [From J. E. Sulston and H. R. Horvitz, 1977, *Devel. Biol.* **56**:110.]

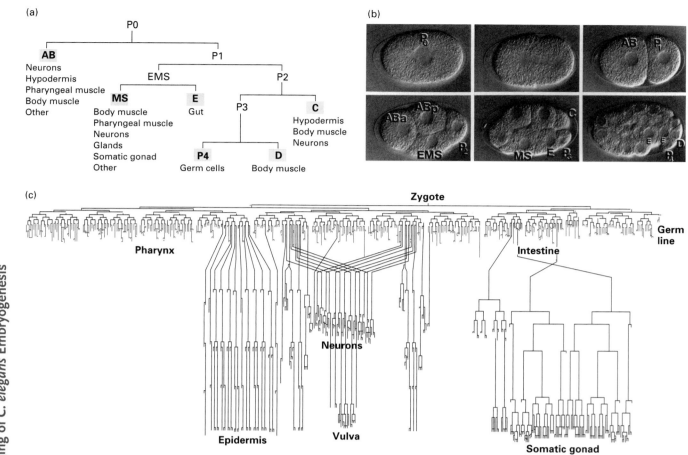

▲ **FIGURE 22-9** *C. elegans* **lineage.** (a) Pattern of the first few divisions starting with P0 (the zygote) and leading to formation of the six founder cells (yellow highlights). The first division, into P1 and AB, is asymmetric. Further divisions in the P lineage generates the other founder cells. Note that more than one lineage can lead to the same tissue type (e.g., muscle or neurons). The EMS cell is so named because it is the precursor to most of the endoderm and mesoderm. The lineage beginning with the P4 cell gives rise to all of the germ-line cells, which are set aside very early, as in most animals. All the other lineages give rise to somatic cells. (b) Light micrographs of the first few divisions of the embryo that generate the founder cells with cells labeled as in part (a). The texture of the cells shows the presence of organelles and is revealed by differential interference contrast microscopy, sometimes called Nomarski microscopy. (c) Full lineage of the entire body of the worm, showing some of the tissues formed. Note that any particular cell undergoes relatively few divisions, typically fewer than 15. [Part (b) from Einhard Schierenberg Zoologisches Institut, Universität Köln.]

spinning off differentiating cells that give rise to a particular tissue. In one pattern of division, one of the daughter cells from a division initiates a repeat of the lineage pattern of the previous few divisions. The complete lineage of *C. elegans* is shown in Figure 22-9c. This organism has been a powerful model system for genetic studies to identify the regulators that control cell lineages in time and space.

Heterochronic Mutants Provide Clues About Control of Cell Lineage

Intriguing evidence for the genetic control of cell lineage has come from isolation and analysis of *heterochronic mutants*. In these mutants, a developmental event typical of one stage

of development occurs too early (precocious development) or too late (retarded development). An example of the former is premature occurrence of a cell division that yields a cell that differentiates and a cell that dies; as a result, the lineage that should have followed from the dead cell never happens. In the latter case, the delayed occurrence of a lineage causes juvenile structures to be produced, incorrectly, in more mature animals. In both cases, the character of a parental cell is, in essence, changed to the character of a cell at a different stage of development.

One example of precocious development in *C. elegans* comes from loss-of-function mutations in the *lin-14* gene, which cause premature formation of the PDNB neuroblast (Figure 22-10a). The *lin-14* gene and several others found to

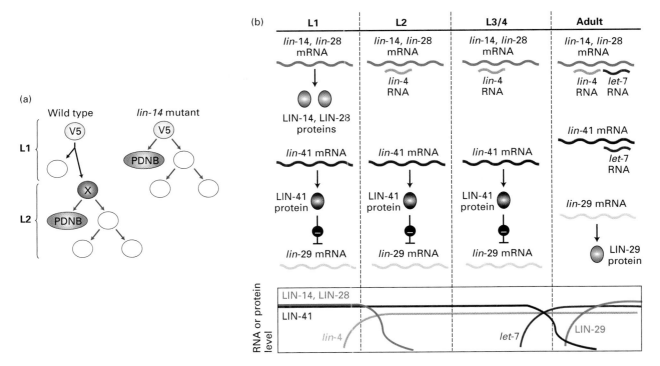

▲ **FIGURE 22-10 Timing of specific types of cell division during development of *C. elegans*.** (a) The pattern of cell division for the V5 cell of *C. elegans* is shown for normal worms and for a heterochronic mutant called *lin-14*. In the *lin-14* mutant, the pattern of cell division (red arrows) that normally occurs only in the second larval stage (L2) occurs in the first larval stage (L1), causing the PDNB neuroblast to be generated prematurely. In the mutant, the V5 cell behaves during L1 like cell "X" (purple) normally does in L2. The inference is that the LIN-14 protein prevents L2-type cell divisions, although precisely how it does so is unknown. (b) Two small regulatory RNAs, *lin-4* and *let-7*, serve as coordinating timers of gene expression. Binding of the *lin-4* RNA to the 3′ untranslated regions (UTRs) of *lin-14* and *lin-28* mRNAs prevents translation of these mRNAs into protein. This occurs following the first larval (L1) stage, permitting development to proceed to the later larval stages. Starting in the fourth larval stage (L4), production of *let-7* RNA begins. It hybridizes to *lin-14*, *lin-28*, and *lin-41* mRNAs, preventing their translation. LIN-41 protein is an inhibitor of translation of the *lin-29* mRNA, so the appearance of *let-7* RNA allows production of LIN-29 protein, which is needed for generation of adult cell lineages. LIN-4 may also bind to *lin-41* RNA at later stages. Only the 3′ UTRs of the mRNAs are depicted. [Adapted from B. J. Reinhart et al., 2000, *Nature* **403**:901.]

be defective in heterochronic worm mutants encode RNA-binding or DNA-binding proteins, which presumably coordinate expression of other genes. However, two other genes (*lin-4* and *let-7*) involved in regulating the timing of cell divisions were initially extremely puzzling, as they appeared to encode small RNAs that do not encode any protein. To discover the products of these genes, scientists first determined which pieces of genomic DNA could restore gene function, and therefore proper cell lineage, to mutants defective in each gene. They then did the same thing with genomic DNA from the corresponding genomic regions of different species of worm. Comparison of the "rescuing" fragments from the different species revealed that they shared common short sequences with little protein-coding potential.

The short RNA molecules encoded by *lin-4* and *let-7* were subsequently shown to inhibit translation of the mRNAs encoded by *lin-14* and other heterochronic genes (Figure 22-10b). These small RNAs, or **micro RNAs (miRNAs)**, are complementary to sequences in the 3′ untranslated parts of target mRNAs and are believed to control translation of the mRNAs by hy-bridizing them. Temporal changes in the production of these and other miRNAs during the life cycle of *C. elegans* serve as a regulatory clock for cell lineage. Molecules related to *let-7* RNA have been identified in many other animals including vertebrates and insects; since their production is temporally regulated in these animals as well, they may serve a similar function in their development as in *C. elegans*. How production of these regulatory miRNAs is temporally controlled is not yet known, but they have turned out to play many roles in regulating gene expression (Chapter 12).

KEY CONCEPTS OF SECTION 22.1

The Birth of Cells

■ In asymmetric cell division, two different types of daughter cells are formed from one mother cell. In contrast, both daughter cells formed in symmetric divisions are identical but may have different fates if they are exposed to different external signals (see Figure 22-1).

■ Pluripotent stem cells can produce more than one type of descendant cell, including in some cases a stem cell with a more-restricted potential to produce differentiated cell types (see Figure 22-2).

■ Cultured embryonic stem cells (ES cells) are capable of giving rise to many kinds of differentiated cell types. They are useful in production of genetically altered mice and offer potential for therapeutic uses.

■ Populations of stem cells associated with most tissues (e.g., skin, intestinal epithelium, blood) regenerate differentiated tissue cells that are damaged or sloughed or become aged (see Figures 22-4 and 22-5).

■ Stem cells are prevented from differentiating by specific controls. A high level of β-catenin, a component of the Wnt signaling pathway, has been implicated in preserving stem cells in the skin and intestine by directing cells toward division rather than differentiation states.

■ Plant stem cells persist for the life of the plant in the meristem. Meristem cells can give rise to a broad spectrum of cell types and structures.

■ During development, precursor cells generally lose potential; that is, they become progressively restricted in the number of different cell types they can form.

■ Early in animal development, the three germ layers—ectoderm, mesoderm, and endoderm—form. Each gives rise to specific tissues and organs (see Figure 22-7).

■ Germ-line cells give rise to eggs or sperm. By definition, all other cells are somatic cells.

■ Embryonic development of *C. elegans* begins with asymmetric division of the fertilized egg (zygote). The lineage of all the cells in adult worms is known and is highly reproducible (see Figure 22-9).

■ Short regulatory RNAs control the timing of developmental cell divisions by preventing translation of mRNAs whose encoded proteins control cell lineages (see Figure 22-10).

22.2 Cell-Type Specification in Yeast

In the previous section, we saw that stem cells and precursor/progenitor cells produce progeny that embark upon specific differentiation paths. The elegant regulatory mechanisms whereby cells become different is referred to as *cell-type specification*. Specification usually involves a combination of external signals with internal signal-transduction mechanisms like those described in Chapters 14 and 15. The transition from an undifferentiated cell to a differentiating one often involves the production of one or a small number of **transcription factors**. The newly produced transcription factors are powerful switches that trigger the activation (and sometimes repression) of large batteries of subservient genes. Thus an initially modest change can cause massive changes in gene expression that confer a new character on the cell.

Our first example of cell-type specification comes from the budding yeast, *S. cerevisiae*. We introduced this useful unicellular eukaryote way back in Chapter 1 and have encountered it in several other chapters. *S. cerevisiae* forms three cell types: haploid **a** and α cells and diploid **a**/α cells. Each type has its own distinctive set of active genes; many other genes are active in all three cell types. In a pattern common to many organisms and tissues, cell-type specification in yeast is controlled by a small number of transcription factors that coordinate the activities of many other genes. Similar regulatory features are found in the responses of higher eukaryotic cells to environmental signals and in the specification and patterning of cells and tissues during development.

Recent **DNA microarray** studies of transcription patterns of ~6000 genes in *S. cerevisiae* have provided a genome-wide picture of the fluctuations of gene expression in the different cell types and different stages of the yeast life cycle (see Figure 9-35 for an explanation of the DNA microarray technique). A powerful advantage of the microarray approach is that it systematically identifies a large fraction of the relevant genes controlling various processes, allowing scientists to concentrate on the most important players that regulate differences in cell types. Among other things, the yeast studies identified 32 genes that are transcribed at more than twofold higher levels in α cells than in **a** cells. Another 50 genes are transcribed at more than twofold higher levels in **a** cells than in α cells. The products of these 82 genes, which initially are activated by cell-type specification transcription regulators, convey many of the critical differences between the two cell types. The results clearly demonstrate that changes in expression of only a small fraction of the genome can significantly alter the behavior and properties of cells. Transcription of a much larger number of genes, about 25 percent of the total assayed, differed substantially in diploid cells compared with haploid cells. **a** and α cells are very similar, and haploid and diploid cells are quite different, so the array results make sense.

Transcription Factors Encoded at the *MAT* Locus Act in Concert with MCM1 to Specify Cell Type

Each of the three *S. cerevisiae* cell types expresses a unique set of regulatory genes that is responsible for all the differences among the three cell types. All haploid cells express certain haploid-specific genes; in addition, **a** cells express **a**-specific genes, and α cells express α-specific genes. In **a**/α diploid cells, diploid-specific genes are expressed, whereas haploid-specific, **a**-specific, and α-specific genes are not. As illustrated in Figure 22-11, three cell type–specific transcription factors (α1, α2, and a1) encoded at the *MAT locus* in combination with a general transcription factor called *MCM1*, which is expressed in all three cell types, mediate cell type–specific gene expression in *S. cerevisiae*. Thus the actions of just three transcription factors can set the yeast cell on a specific differentiation pathway culminating in a particular cell type. From the DNA microarray experiments we know one effect of these key players: the activation or repression of many dozens of genes that control cell characteristics.

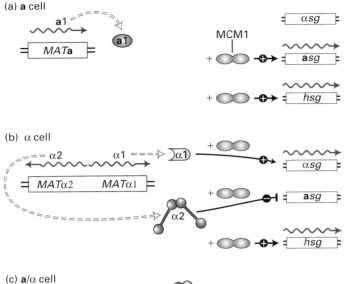

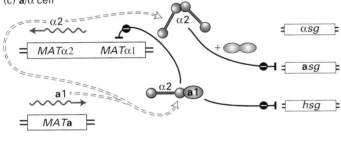

▲ FIGURE 22-11 Transcriptional control of cell type–specific genes in *S. cerevisiae*. The coding sequences carried at the *MAT* locus differ in haploid α and **a** cells and in diploid cells. Three type-specific transcription factors (α1, α2, and **a**1) encoded at the *MAT* locus act with MCM1, a constitutive transcription factor produced by all three cell types, to produce a distinctive pattern of gene expression in each of the three cell types. **a***sg* = **a**-specific genes/mRNAs; α*sg* = α-specific genes/mRNAs; *hsg* = haploid-specific genes/mRNAs.

MCM1 was the first member of the *MADS family* of transcription factors to be discovered. (MADS is an acronym for the initial four factors identified in this family.) The DNA-binding proteins composing this family dimerize and contain a similar N-terminal MADS domain. In Section 22.3 we will encounter other MADS transcription factors that participate in development of skeletal muscle. MADS transcription factors also specify cell types in floral organs (see Figure 15-28). MCM1 exhibits different activity in haploid **a** and α cells due to its association with α1 or α2 protein in α cells. Acting alone, MCM1 activates transcription of **a**-specific genes in **a** cells and of haploid-specific genes in both α and **a** cells (see Figure 22-11a, b). As a result of this combinatorial action, MCM1 promotes transcription of α-specific genes and represses transcription of **a**-specific genes in α cells. Now let's take a closer look at how MCM1 and the *MAT*-encoded proteins exert their effects.

MCM1 and α1-MCM1 Complexes Activate Gene Transcription

In **a** cells, homodimeric MCM1 binds to the so-called P box sequence in the upstream regulatory sequences (URSs) of **a**-specific genes, stimulating their transcription (Figure 22-12a). Transcription of α-specific genes is controlled by two adjacent sequences—the P box and the Q box—located in the URSs associated with these genes. Although MCM1 alone binds to the P box in **a**-specific URSs, it does not bind to the P box in α-specific URSs. Thus **a** cells do not transcribe α-specific genes.

In α cells, which produce the α1 transcription factor encoded by *MAT*α, the simultaneous binding of MCM1 and α1 to PQ sites occurs with high affinity (Figure 22-12b). This binding turns on transcription of α-specific genes. Therefore **a**-specific transcription is a simple matter of a single transcription factor binding to its target genes, while α-specific

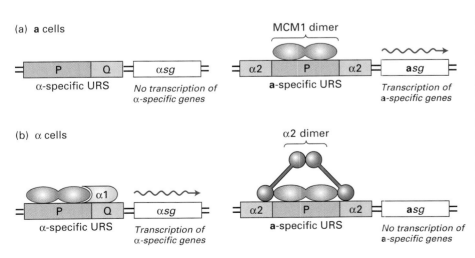

◀ FIGURE 22-12 Activity of MCM1 in a and α yeast cells. MCM1 binds as a dimer to the P site in α-specific and **a**-specific upstream regulatory sequences (URSs), which control transcription of α-specific genes and **a**-specific genes, respectively. (a) In **a** cells, MCM1 stimulates transcription of **a**-specific genes. MCM1 does not bind efficiently to the P site in α-specific URSs in the absence of α1 protein. (b) In α cells, the activity of MCM1 is modified by its association with α1 or α2. The α1-MCM1 complex stimulates transcription of α-specific genes, whereas the α2-MCM1 complex blocks transcription of **a**-specific genes. The α2-MCM1 complex also is produced in diploid cells, where it has the same blocking effect on transcription of **a**-specific genes (see Figure 22-11c).

transcription requires a combination of two factors—neither of which can activate target genes alone.

α2-MCM1 and α2-a1 Complexes Repress Transcription

Highly specific binding occurs as a consequence of the interaction of α2 with other transcription factors at different sites in DNA. Flanking the P box in each a-specific URS are two α2-binding sites. Both MCM1 and α2 can bind independently to an a-specific URS with relatively low affinity. However, in α cells highly cooperative, simultaneous binding of both α2 and MCM1 proteins to these sites occurs with high affinity. This high-affinity binding represses transcription of a-specific genes, ensuring that they are not expressed in α cells and diploid cells (see Figure 22-12b, *right*). MCM1 promotes binding of α2 to an a-specific URS by orienting the two DNA-binding domains of the α2 dimer to the α2-binding sequences in this URS. Since a dimeric α2 molecule binds to both sites in an α-specific URS, each DNA site is referred to as a half-site. The relative positions of both half-sites and their orientation are highly conserved among different a-specific URSs.

Combinations of transcription factors create additional specificity in gene regulation. The presence of numerous α2-binding sites in the genome and the "relaxed" specificity of α2 protein may expand the range of genes that it can regulate. For instance, in a/α diploid cells, α2 forms a heterodimer with a1 that represses both haploid-specific genes and the gene encoding α1 (see Figure 22-11c). The example of α2 suggests that relaxed specificity may be a general strategy for increasing the regulatory range of a single transcription factor.

Pheromones Induce Mating of α and a Cells to Generate a Third Cell Type

An important feature of the yeast life cycle is the ability of haploid a and α cells to mate, that is, attach and fuse giving rise to a diploid a/α cell (see Figure 1-5). Each haploid cell type secretes a different *mating factor*, a small polypeptide **pheromone**, and expresses a cell-surface G protein–coupled receptor that recognizes the pheromone secreted by cells of the other type. Thus a and α cells both secrete and respond to pheromones (Figure 22-13). Binding of the mating factors to their receptors induces expression of a set of genes encoding proteins that direct arrest of the cell cycle in G_1 and promote attachment/fusion of haploid cells to form diploid cells. In the presence of sufficient nutrients, the diploid cells will continue to grow. Starvation, however, induces diploid cells to progress through meiosis, each yielding four haploid spores. If the environmental conditions become conducive to vegetative growth, the spores will germinate and undergo mitotic division.

Studies with yeast mutants have provided insights into how the a and α pheromones induce mating. For instance, haploid yeast cells carrying mutations in the *sterile 12 (STE12)* locus cannot respond to pheromones and do not

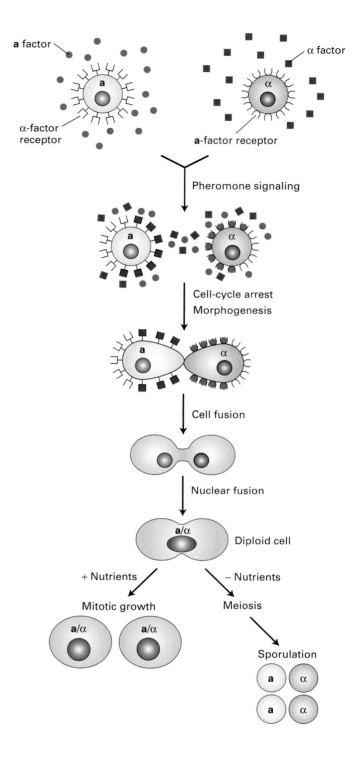

▲ **FIGURE 22-13 Pheromone-induced mating of haploid yeast cells.** The α cells produce α mating factor and **a** receptor; the **a** cells produce **a** factor and α receptor. Binding of the mating factors to their cognate receptors on cells of the opposite type leads to gene activation, resulting in mating and production of diploid cells. In the presence of sufficient nutrients, these cells will grow as diploids. Without sufficient nutrients, cells will undergo meiosis and form four haploid spores.

mate. The *STE12* gene encodes a transcription factor that binds to a DNA sequence referred to as the pheromone-responsive element, which is present in many different **a**- and α-specific URSs. Binding of mating factors to cell-surface receptors induces a cascade of signaling events, resulting in phosphorylation of various proteins including the Ste12 protein (see Figure 14-24). This rapid phosphorylation is correlated with an increase in the ability of Ste12 to stimulate transcription. It is not yet known, however, whether Ste12 must be phosphorylated to stimulate transcription in response to pheromone.

Interaction of Ste12 protein with DNA has been studied most extensively at the URS controlling transcription of *STE2*, an **a**-specific gene encoding the receptor for the α pheromone. Pheromone-induced production of the α receptor encoded by *STE2* increases the efficiency of the mating process. Adjacent to the **a**-specific URS in the *STE2* gene is a pheromone-responsive element that binds Ste12. When **a** cells are treated with α pheromone, transcription of the *STE2* gene increases in a process that requires Ste12 protein. Ste12 protein has been found to bind most efficiently to the pheromone-responsive element in the *STE2* URS when MCM1 is simultaneously bound to the adjacent P site. We saw previously that MCM1 can act as an activator or a repressor at different URSs depending on whether it complexes with α1 or α2. In this case, the function of MCM1 as an activator is stimulated by the binding of yet another transcription factor, Ste12, whose activity is modified by extracellular signals.

KEY CONCEPTS OF SECTION 22.2

Cell-Type Specification in Yeast

■ Specification of each of the three yeast cell types—the **a** and α haploid cells and the diploid **a**/α cells—is determined by a unique set of transcription factors acting in different combinations at specific regulatory sites in the yeast genome (see Figure 22-11).

■ Some transcription factors can act as repressors or activators depending on the specific regulatory sites they bind and the presence or absence of other transcription factors bound to neighboring sites.

■ Binding of mating-type pheromones by haploid yeast cells activates expression of genes encoding proteins that mediate mating, thereby generating the third yeast cell type (see Figure 22-13).

22.3 Specification and Differentiation of Muscle

As indicated by global expression patterns, yeast cells of different mating types are still rather similar. Developmental biologists do not yet know the complete set of molecules that distinguishes any one cell type (e.g., muscle) from all the other cell types in a multicellular organism. The extensive cell specification and differentiation that occur during development of animals and plants depend on both quantitative and qualitative differences in gene expression, controlled at the level of transcription, as well as on cell structures and protein activity states.

An impressive array of molecular strategies, some analogous to those found in yeast cell-type specification, have evolved to carry out the complex developmental pathways that characterize multicellular organisms. Muscle cells have been the focus of many such studies because their development can be studied in cultured cells as well as in intact animals. Early advances in understanding the formation of muscle cells (myogenesis) came from discovery of regulatory genes that could convert cultured cells into muscle cells. Then mouse mutations affecting those genes were created and studied to learn the functions of the proteins encoded by these genes, following which scientists have investigated how the muscle regulatory genes control other genes.

Twist protein is a transcription factor necessary to create muscle cells in flies and other animals. Twist turns on production of other transcription factors that in turn activate genes encoding myosin, actin, and other muscle-specific proteins. DNA microarray analysis has been applied to understand the function of *twist* and other regulatory genes in muscle development. For example, the expression pattern of normal *Drosophila* embryos recently has been compared with that of mutant embryos in which the *twist* gene is defective. To assess how many genes are needed to specify muscle, researchers determined the expression of about 4000 fly genes (about 30 percent of the total) in normal fly embryos and in *twist* mutants. Of the genes included in the microarray, about 130 (3.3 percent), including many known muscle differentiation genes, were transcribed at lower levels (or not at all) in the *twist* mutants. These results suggest that transcriptional changes in at least several hundred genes are associated with differentiation of a highly specialized cell type such as muscle.

Other recent microarray studies have looked for genes whose transcription differs in various subtypes of muscle in mice. These have identified 49 genes out of 3000 genes examined that are transcribed at substantially different levels in red (endurance) muscle and white (fast response) muscle. Clues to the molecular basis of the functional differences between red and white muscle are likely to come from studying those 49 genes and their products.

Here we examine the role of certain transcription factors in creating skeletal muscle in vertebrates. These muscle regulators illustrate how coordinated transcription of sets of target genes can produce differentiated cell types and how a cascade of transcriptional events and signals is necessary to coordinate cell behaviors and functions.

Embryonic Somites Give Rise to Myoblasts, the Precursors of Skeletal Muscle Cells

Vertebrate skeletal myogenesis proceeds through three stages: **determination** of the precursor muscle cells, called *myoblasts;* proliferation and in some cases migration of myoblasts; and their terminal differentiation into mature muscle (Figure 22-14). In the first stage, myoblasts arise from blocks of mesoderm cells, called *somites,* that are located next to the neural tube in the embryo. Specific signals from surrounding tissue play an important role in determining where myoblasts will form in the developing somite. At the molecular level, the decision of a mesoderm cell to adopt a muscle cell fate reflects the activation of genes encoding particular transcription factors.

As myoblasts proliferate and migrate, say, to a developing limb bud, they become aligned, stop dividing, and fuse to form a **syncytium** (a cell containing many nuclei but sharing a common cytoplasm). We refer to this multinucleate cell as a *myotube.* Concomitant with cell fusion is a dramatic rise in the expression of genes necessary for further muscle development and function.

The specific extracellular signals that induce determination of each group of myoblasts are expressed only transiently. These signals trigger production of intracellular factors that maintain the myogenic program after the inducing signals are gone. We discuss the identification and functions of these myogenic proteins, and their interactions, in the next several sections.

Myogenic Genes Were First Identified in Studies with Cultured Fibroblasts

Myogenic genes are a fine example of how transcription factors control the progressive differentiation that occurs in a cell lineage. In vitro studies with the fibroblast cell line designated C3H 10T½ have played a central role in dissecting the transcription control mechanisms regulating skeletal myogenesis. When these cells are incubated in the presence of 5-azacytidine, a cytidine derivative that cannot be methylated and therefore alters transcription, they differentiate into myotubes. Upon entry into cells, 5-azacytidine is converted to 5-azadeoxycytidine triphosphate and then is incorporated into DNA in place of deoxycytidine. Because methylated deoxycytidine residues commonly are present in transcriptionally inactive DNA regions, replacement of cytidine residues with a derivative that cannot be methylated may permit activation of genes previously repressed by methylation.

The high frequency at which azacytidine-treated C3H 10T½ cells are converted into myotubes suggested to early workers that reactivation of one or a small number of closely linked genes is sufficient to drive a myogenic program. To test

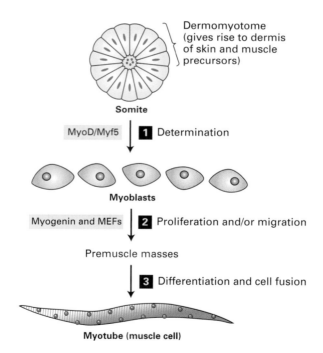

▲ FIGURE 22-14 Three stages in development of vertebrate skeletal muscle. Somites are epithelial spheres of embryonic mesoderm cells, some of which (the myotome) become determined as myoblasts after receiving signals from other tissues (**1**). After the myoblasts proliferate and migrate to the limb buds and elsewhere (**2**), they undergo terminal differentiation into multinucleate skeletal muscle cells, called myotubes (**3**). Key transcription factors that help drive the myogenic program are highlighted in yellow. See also Figure 22-17.

▶ EXPERIMENTAL FIGURE 22-15 Myogenic genes isolated from azacytidine-treated cells can drive myogenesis when transfected into other cells. (a) When C3H 10T½ cells (a fibroblast cell line) are treated with azacytidine, they develop into myotubes at high frequency. To isolate the genes responsible for converting azacytidine-treated cells into myotubes, all the mRNAs from treated cells first were isolated from cell extracts on an oligo-dT column. Because of their poly(A) tails, mRNAs are selectively retained on this column. Steps **1** and **2**: The isolated mRNAs were converted to radiolabeled cDNAs. Step **3**: When the cDNAs were mixed with mRNAs from *untreated* C3H 10T½ cells, only cDNAs derived from mRNAs (light red) produced by both azacytidine-treated cells and untreated cells hybridized. The resulting double-stranded DNA was separated from the unhybridized cDNAs (dark blue) produced only by azacytidine-treated cells. Step **4**: The cDNAs specific for azacytidine-treated cells then were used as probes to screen a cDNA library from azacytidine-treated cells (Chapter 9). At least some of the clones identified with these probes correspond to genes required for myogenesis. (b) Each of the cDNA clones identified in part (a) was incorporated into a plasmid carrying a strong promoter. Steps **1** and **2**: C3H 10T½ cells were cotransfected with each recombinant plasmid plus a second plasmid carrying a gene conferring resistance to an antibiotic called G418; only cells that have incorporated the plasmids will grow on a medium containing G418. One of the selected clones, designated *myoD,* was shown to drive conversion of C3H 10T½ cells into muscle cells, identified by their binding of labeled antibodies against myosin, a muscle-specific protein (step **3**). [See R. L. Davis et al., 1987, *Cell* **51**:987.]

(a) Screen for myogenic genes

Total mRNA from azacytidine-treated cells

1 | Incubate with reverse transcriptase and [^{32}P]dNTPs

2 | Remove mRNAs

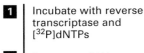

^{32}P-labeled cDNAs

Discard

3 | Hybridize with excess of mRNAs from untreated C3H 10T1/2 cells ()

^{32}P-labeled cDNAs specific for azacytidine-treated cells

4 | Screen cDNA library from treated cells

Isolate cDNAs characteristic of azacytidine-treated cells

(b) Assay for myogenic activity of *myoD* cDNA

C3H 10T1/2 cell

1 | Transfect with a plasmid carrying *myoD* cDNA and a plasmid conferring resistance to G418

2 | Select on G418-containing medium for cells that took up both plasmids

3 | Stain with labeled antimyosin antibody (𝗬)

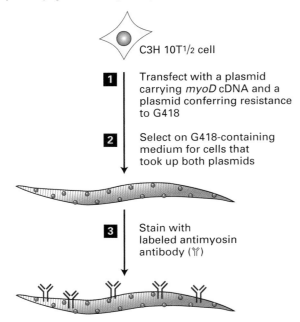

---

this hypothesis, researchers isolated DNA from C3H 10T1/2 cells grown in the presence of 5-azacytidine and transfected it into untreated cells. The observation that 1 in 10^4 cells transfected with this DNA was converted into a myotube is consistent with the hypothesis that one or a small set of closely linked genes is responsible for converting fibroblasts into myotubes.

Subsequent studies led to the isolation and characterization of four different but related genes that can convert C3H 10T1/2 cells into muscle. Figure 22-15 outlines the experimental protocol for identifying and assaying one of these genes, called the *myogenic determination (myoD)* gene. C3H 10T1/2 cells transfected with *myoD* cDNA and those treated with 5-azacytidine both formed myotubes. The *myoD* cDNA also was able to convert a number of other cultured cell lines into muscle. Based on these findings, the *myoD* gene was proposed to play a key role in muscle development. A similar approach identified three other genes—*myogenin, myf5,* and *mrf4*—that also function in muscle development.

Muscle-Regulatory Factors (MRFs) and Myocyte-Enhancing Factors (MEFs) Act in Concert to Confer Myogenic Specificity

The four myogenic proteins—MyoD, Myf5, myogenin, and MRF4—are all members of the **basic helix-loop-helix** (bHLH) family of DNA-binding transcription factors (see Figure 11-22b). Near the center of these proteins is a DNA-binding basic (B) region adjacent to the HLH domain, which mediates dimer formation. Flanking this central DNA-binding/dimerization region are two activation domains. We refer to the four myogenic bHLH proteins collectively as *muscle regulatory factors,* or *MRFs* (Figure 22-16a).

(a) Structure of muscle-regulatory factors (MRFs)

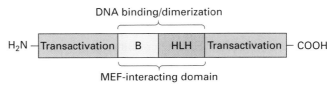

(b) Structure of myocyte-enhancing factors (MEFs)

▲ **FIGURE 22-16 General structures of two classes of transcription factors that participate in myogenesis.** MRFs (muscle regulatory factors) are bHLH (basic helix-loop-helix) proteins produced only in developing muscle. MEFs (myocyte-enhancing factors), which are produced in several tissues in addition to developing muscle, belong to the MADS family. The myogenic activity of MRFs is enhanced by their interaction with MEFs.

bHLH proteins form homo- and heterodimers that bind to a 6-bp DNA site with the consensus sequence CANNTG (N = any nucleotide). Referred to as the E box, this sequence is present in many different locations within the genome (on a purely random basis the E box will be found every 256 nucleotides). Thus some mechanism(s) must ensure that MRFs specifically regulate muscle-specific genes and not other genes containing E boxes in their transcription control regions. One clue to how this myogenic specificity is achieved was the finding that the DNA-binding affinity of MyoD is tenfold greater when it binds as a heterodimer complexed with E2A, another bHLH protein, than when it binds as a homodimer. Moreover, in azacytidine-treated C3H 10T½ cells, MyoD is found as a heterodimer complexed with E2A, and both proteins are required for myogenesis in these cells. The DNA-binding domains of E2A and MyoD have similar but not identical amino acid sequences, and both proteins recognize E box sequences. The other MRFs also form heterodimers with E2A that have properties similar to MyoD-E2A complexes. This heterodimerization restricts activity of the myogenic transcription factors to genes with closely linked E boxes.

Since E2A is expressed in many tissues, the requirement for E2A is not sufficient to confer myogenic specificity. Subsequent studies suggested that specific amino acids in the bHLH domain of all the MRFs confer myogenic specificity by allowing MRF-E2A complexes to bind specifically to another family of DNA-binding proteins called *myocyte enhancing factors*, or *MEFs*. MEFs were considered excellent candidates for interaction with MRFs for two reasons. First, many muscle-specific genes contain recognition sites for both MEFs and MRFs. Second, although MEFs cannot induce myogenic conversion of azacytidine-treated C3H 10T½ cells by themselves, they enhance the ability of MRFs to do so. This enhancement requires physical interaction between a MEF and MRF-E2A heterodimer. MEFs belong to the MADS family of transcription factors and to contain a MEF domain, adjacent to the MADS domain, that mediates interaction with myogenin (Figure 22-16b). The synergistic action of the MEF homodimer and MRF-E2A heterodimer is thought to drive high-level expression of muscle-specific genes.

Knockout mice and *Drosophila* mutants have been used to explore the roles of MRF and MEF proteins in conferring myogenic specificity in intact animals, extending the work in cell culture. These experiments demonstrated the importance of three of the MRF proteins and of MEF proteins for distinct steps in muscle development (see Figure 22-14). The function of the fourth myogenic protein, Mrf4, is not entirely clear; it may be expressed later and help maintain differentiated muscle cells and by combinatorial control to ensure that only muscle-specific genes are activated.

Terminal Differentiation of Myoblasts Is Under Positive and Negative Control

Powerful developmental regulators like the MRFs cannot be allowed to run rampant. In fact, their actions are circum-scribed at several levels. First, production of the muscle regulators is activated only in mesoderm cells, at the right place and time in the embryo, in response to spatial regulators of the sorts that are described in Chapter 15. Other proteins mediate additional mechanisms for assuring tight control over myogenesis: chromatin-remodeling proteins are needed to make target genes accessible to MRFs; inhibitory proteins can restrict when MRFs act; and antagonistic relations between cell-cycle regulators and differentiation factors like MRFs ensure that differentiating cells will not divide, and vice versa. All these factors control when and where muscles form.

Activating Chromatin-Remodeling Proteins MRF proteins control batteries of muscle-specific genes, but can do so only if chromatin factors allow access. Remodeling of chromatin, which usually is necessary for gene activation, is carried out by large protein complexes (e.g., Swi/Snf complex) that have ATPase and perhaps helicase activity. These complexes are thought to recruit histone acetylases that modify chromatin to make genes accessible to transcription factors (Chapter 11). The hypothesis that remodeling complexes help myogenic factors was tested using dominant-negative versions of the ATPase proteins that form the cores of these complexes. (Recall from Chapter 9 that a dominant-negative mutation produces a mutant phenotype even when a normal allele of the gene also is present.) When genes carrying these dominant-negative mutations were transfected into C3H 10T½ cells, the subsequent introduction of myogenic genes no longer converted the cells into myotubes. In addition, a muscle-specific gene that is normally activated did not exhibit its usual pattern of chromatin changes in the doubly transfected C3H 10T½ cells. These results indicate that transcription activation by myogenic proteins depends on a suitable chromatin structure in the regions of muscle-specific genes.

Inhibitory Proteins Screens for genes related to *myoD* led to identification of a related protein that retains the HLH dimerization region but lacks the basic DNA-binding region and hence is unable to bind to E box sequences in DNA. By binding to MyoD or E2A, this protein inhibits formation of MyoD-E2A heterodimers and hence their high-affinity binding to DNA. Accordingly, this protein is referred to as Id, for inhibitor of DNA binding. Id prevents cells that produce MyoD and E2A from activating transcription of the muscle-specific gene encoding creatine kinase. As a result, the cells remain in a proliferative growth state. When these cells are induced to differentiate into muscle (for instance, by the removal of serum, which contains the growth factors required for proliferative growth), the Id concentration falls. MyoD-E2A dimers now can form and bind to the regulatory regions of target genes, driving differentiation of C3H 10T½ cells into myoblast-like cells.

Recent work shows that histone acetylases and deacetylases are also crucial for regulating muscle-specific genes. As explained in Chapter 11, acetylation of histones in chromatin is necessary to activate many genes; in contrast, histone deacetylases cause transcriptional repression (see Figure 11-32). MEF2 recruits histone acetylases such as p300/CBP, through another protein that serves as a mediator, thus activating transcription of target genes. Chromatin immunoprecipitation experiments with antibodies against acetylated histone H4 show that the acetylated histone level associated with MEF2-regulated genes is higher in differentiated myotubes than in myoblasts (see Figure 11-31). The role of histone deacetylases in muscle development was revealed in experiments in which scientists first introduced extra *myoD* genes into cultured C3H 10T½ cells to raise the level of MyoD. This resulted in increased activation of target genes and more rapid differentiation of the cells into myotubes. However, when genes encoding histone deacetylases also were introduced into the C3H 10T½ cells, the muscle-inducing effect of MyoD was blocked and the cells did not differentiate into myotubes.

The explanation for how histone deacetylases inhibit MyoD-induced muscle differentiation came from the surprising finding that MEF2 can bind, through its MADS domain, to a histone deacetylase. This interaction, which can prevent MEF2 function and muscle differentiation, is normally blocked during differentiation because the histone deacetylase is phosphorylated by a calcium/calmodulin-dependent protein kinase; the phosphorylated deacetylase then is moved from the nucleus to the cytoplasm. Taken together, these results indicate that activation of muscle genes by MyoD and MEF2 is in competition with inactivation of muscle genes by repressive chromatin structures and that nuclear versus cytoplasmic localization of chromatin factors is a key regulatory step.

Cell-Cycle Proteins The onset of terminal differentiation in many cell types is associated with arrest of the cell cycle, most commonly in G_1, suggesting that the transition from the determined to differentiated state may be influenced by cell-cycle proteins including **cyclins** and **cyclin-dependent kinases** (Chapter 21). For instance, certain inhibitors of cyclin-dependent kinases can induce muscle differentiation in cell culture, and the amounts of these inhibitors are markedly higher in differentiating muscle cells than in nondifferentiating ones in vivo. Conversely, differentiation of cultured myoblasts can be inhibited by transfecting the cells with DNA encoding cyclin D1 under the control of a constitutively active promoter. Expression of cyclin D1, which normally occurs only during G_1, is induced by mitogenic factors in many cell types and drives the cell cycle (see Figure 21-28). The ability of cyclin D1 to prevent myoblast differentiation in vitro may mimic aspects of the in vivo signals that antagonize the differentiation pathway. The antagonism between negative and positive regulators of G_1 progression is likely to play an important role in controlling myogenesis in vivo.

Cell-Cell Signals Are Crucial for Muscle Cell-Fate Determination and Myoblast Migration

As noted already, after myoblasts arise from somites, they must not only proliferate but also move to their proper locations and form the correct attachments as they differentiate into muscle cells (Figure 22-17). Myogenic gene expression often follows elaborate events that tell certain somite cells to delaminate from the somite epithelium and which way to move. A transcription factor, Pax3, is produced in the subset of somite cells that will form muscle. Pax3 appears to be at the top of the regulatory hierarchy controlling muscle formation in the body wall and limbs. Myoblasts that will migrate, but not cells that remain behind, also produce a transcription factor called Lbx1. If Pax3 is not functional, Lbx1 transcripts are not seen and myoblasts do not migrate. Both Pax3 and Lbx1 can affect expression of *myoD*.

The departure of myoblasts from somites depends upon reception of a secreted protein signal appropriately called *scatter factor*, or *hepatocyte growth factor (SF/HGF)*. This signal is produced by embryonic connective tissue cells (mesenchyme) in the limb buds to which myoblasts migrate. Production of SF/HGF is previously induced by other secreted signals such as fibroblast growth factor and Sonic hedgehog, which are critical to limb development (Chapter 15). The cell-surface receptor for SF/HGF, which is expressed by myoblasts, belongs to the receptor tyrosine kinase (RTK) class of receptors (Chapter 14). Cells migrate from the somites at the regions along the head-to-tail body axis where limbs will form, and not

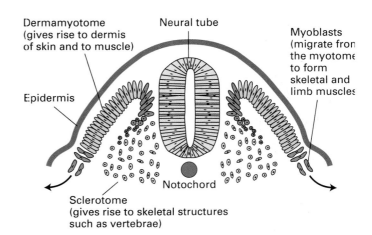

▲ **FIGURE 22-17 Embryonic determination and migration of myoblasts in mammals.** After formation of the neural tube, each somite forms sclerotome, which develops into skeletal structures, and dermomyotome. The dermomyotome gives rise to the dermis of the skin and to the muscles. Lateral myoblasts migrate to the limb bud; medial myoblasts develop into the trunk muscles. The remainder of a dermomyotome gives rise to the connective tissue of the skin. [Adapted from M. Buckingham, 1992, *Trends Genet.* **8**:144.]

elsewhere, due to the presence of SF/HGF at limb locations and not elsewhere. If the SF/HGF signal or its receptor is not functional, somite cells will produce Lbx1 but not go on to migrate; thus no muscles will form in the limbs. Expression of the *myogenin* gene, which is necessary for myotube formation, and of *mrf4*, which is necessary for muscle fiber differentiation, does not begin until migrating myoblasts approach their limb-bud destination; these steps in terminal muscle differentiation are presumably not compatible with migration.

We have touched on just a few of the many external signals and transcription factors that participate in development of a properly patterned muscle. The function of all these regulatory molecules must be coordinated both in space and in time during myogenesis.

bHLH Regulatory Proteins Function in Creation of Other Tissues

Four bHLH transcription factors that are remarkably similar to the myogenic bHLH proteins control neurogenesis in *Drosophila*. Similar proteins appear to function in neurogenesis in vertebrates and perhaps in the determination and differentiation of hematopoietic cells.

The neurogenic *Drosophila* bHLH proteins are encoded by an ≈100-kb stretch of genomic DNA, termed the *achaete-scute complex* (AS-C), containing four genes designated *achaete (ac)*, *scute (sc)*, *lethal of scute (l'sc)*, and *asense (a)*. Analysis of the effects of loss-of-function mutations indicate that the Achaete (Ac) and Scute (Sc) proteins participate in determination of neuronal stem cells, called *neuroblasts*, while the Asense (As) protein is required for differentiation of the progeny of these cells into neurons. These functions are analogous to the roles of MyoD and Myf5 in muscle determination and of myogenin in differentiation. Two other *Drosophila* pro-

teins, designated Da and Emc, are analogous in structure and function to vertebrate E2A and Id, respectively. For example, heterodimeric complexes of Da with Ac or Sc bind to DNA better than the homodimeric forms of Ac and Sc. Emc, like Id, lacks a DNA-binding basic domain; it binds to Ac and Sc proteins, thus inhibiting their association with Da and binding to DNA. The similar functions of these myogenic and neurogenic proteins are depicted in Figure 22-18.

A family of bHLH proteins related to the *Drosophila* Achaete and Scute proteins has been identified in vertebrates. One of these, called neurogenin, which has been identified in the rat, mouse, and frog, controls the formation of neuroblasts. In situ hybridization experiments showed that neurogenin is produced at an early stage in the developing nervous system and induces production of NeuroD, another bHLH protein that acts later (Figure 22-19a). Injection of large amounts of *neurogenin* mRNA into *Xenopus* embryos further demonstrated the ability of neurogenin to induce neurogenesis (Figure 22-19b). These studies suggest that the function of neurogenin is analogous to that of the Achaete and Scute in *Drosophila*; likewise, NeuroD and Asense may have analogous functions in vertebrates and *Drosophila*, respectively.

In addition to neurons, the nervous system contains a large number of glial cells, which also arise from the neuroectoderm. Glial cells support and insulate neurons; they also provide guidance and contact surfaces for migrating neurons during development and send a signal to neurons that promotes formation of synapses. Neurogenins control the fates of precursor cells that are capable of making either neurons or glial cells by promoting neural development and repressing glial development. Neurogenins are therefore switches that control the decision between two alternative cell fates, just as the yeast mating-type regulators select among three cell types.

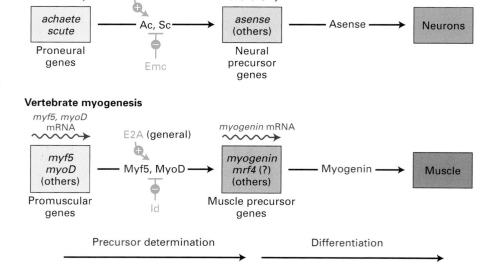

▶ **FIGURE 22-18 Comparison of genes that regulate *Drosophila* neurogenesis and mammalian myogenesis.** bHLH transcription factors have analogous functions in determination of precursor cells (i.e., neuroblasts and myoblasts) and their subsequent differentiation into mature neurons and muscle cells. In both cases, the proteins encoded by the earliest-acting genes (*left*) are under both positive and negative control by other related proteins (blue type). [Adapted from Y. N. Jan and L. Y. Jan, 1993, *Cell* **75**:827.]

(a)

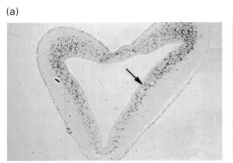 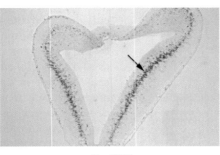

neurogenin mRNA   *neuroD* mRNA

(b)

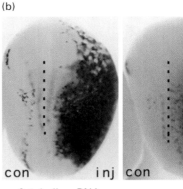 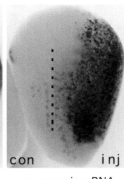

con   inj   con   inj

β-*tubulin* mRNA   *neurogenin* mRNA

▲ **EXPERIMENTAL FIGURE 22-19 In situ hybridization and injection experiments demonstrate that neurogenin acts before NeuroD in vertebrate neurogenesis.** (a) Sections of rat neural tube were treated with a probe specific for *neurogenin* mRNA (*left*) or *neuroD* mRNA (*right*). The open space in the center is the ventricle, and the cells lining this cavity are in the ventricular layer. All the neural cells are born in the ventricular layer and then migrate outward. As illustrated in these micrographs, *neurogenin* mRNA is produced in proliferating neuroblasts in the ventricular layer (*arrow*), whereas *neuroD* mRNA is present in migrating neuroblasts that have left the ventricular zone (*other arrow*). (b) One of the two cells in early *Xenopus* embryos was injected with *neurogenin* mRNA (inj) and then stained with a probe specific for neuron-specific mRNAs encoding β-tubulin (*left*) or NeuroD (*right*). The region of the embryo derived from the uninjected cell served as a control (con). The *neurogenin* mRNA induced a massive increase in the number of neuroblasts expressing *neuroD* mRNA and neurons expressing β-*tubulin* mRNA in the region of the neural tube derived from the injected cell. [From Q. Ma et al., 1996, *Cell* **87**:43; courtesy of D. J. Anderson.]

KEY CONCEPTS OF SECTION 22.3

Specification and Differentiation of Muscle

■ Development of skeletal muscle begins with the signal-induced determination of certain mesoderm cells in somites as myoblasts. Following their proliferation and migration, myoblasts stop dividing and differentiate into multinucleate muscle cells (myotubes) that express muscle-specific proteins (see Figure 22-14).

■ Four myogenic bHLH transcription factors—MyoD, myogenin, Myf5, and MRF4, called muscle-regulatory factors (MRFs)—associate with E2A and MEFs to form large transcriptional complexes that drive myogenesis and expression of muscle-specific genes.

■ Dimerization of bHLH transcription factors with different partners modulates the specificity or affinity of their binding to specific DNA regulatory sites, and also may prevent their binding entirely.

■ The myogenic program driven by MRFs depends on the Swi/Snf chromatin-remodeling complex, which makes target genes accessible.

■ The myogenic program is inhibited by binding of Id protein to MyoD, thereby blocking binding of MyoD to DNA, and by histone deacetylases, which repress activation of target genes by MRFs.

■ Migration of myoblasts to the limb buds is induced by scatter factor/hepatocyte growth factor (SF/HGF), a protein signal secreted by mesenchymal cells (see Figure 22-17). Myoblasts must express both the Pax3 and Lbx1 transcription factors to migrate.

■ Terminal differentiation of myoblasts and induction of muscle-specific proteins do not occur until myoblasts stop dividing and begin migrating.

■ Neurogenesis in *Drosophila* depends upon a set of four neurogenic bHLH proteins that are conceptually and structurally similar to the vertebrate myogenic proteins (see Figure 22-18).

■ A related vertebrate protein, neurogenin, is required for formation of neural precursors and also controls their division into neurons or glial cells.

22.4 Regulation of Asymmetric Cell Division

During **embryogenesis,** the earliest stage in animal development, asymmetric cell division often creates the initial diversity that ultimately culminates in formation of specific differentiated cell types. Even in bacteria, cell division may yield unequal daughter cells, for example, one that remains attached to a stalk and one that develops flagella used for swimming. Essential to asymmetric cell division is *polarization* of the parental cell and then differential incorporation

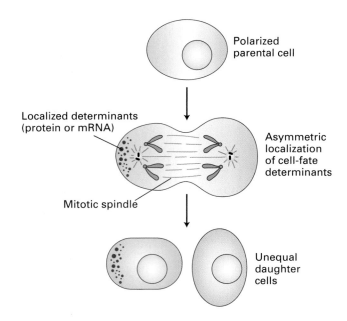

▲ FIGURE 22-20 General features of asymmetric cell division. Various mechanisms can lead to asymmetric distribution of cytoplasmic components, such as particular proteins or mRNAs (red dots) to form a polarized parental cell. Division of a polarized cell will be asymmetric if the mitotic spindle is oriented so that the localized cytoplasmic components are distributed unequally to the two daughter cells, as shown here. However, if the spindle is positioned differently relative to the localized cytoplasmic components, division of a polarized cell may produce equivalent daughter cells.

of parts of the parental cell into the two daughters (Figure 22-20). A variety of molecular mechanisms are employed to create and propagate the initial asymmetry that polarizes the parental cell. In addition to being different, the daughter cells must often be placed in a specific orientation with respect to surrounding structures.

We begin with an especially well-understood example of asymmetric cell division, the budding of yeast cells, and move on to recently discovered protein complexes important for asymmetric cell divisions in multicellular organisms. We see in this example an elegant system that links asymmetric division to the process of controlling cell type.

Yeast Mating-Type Switching Depends upon Asymmetric Cell Division

S. cerevisiae cells use a remarkable mechanism to control the differentiation of the cells as the cell lineage progresses. Whether a haploid yeast cell exhibits the α or **a** mating type is determined by which genes are present at the *MAT* locus (see Figure 22-11). As described in Chapter 11, the *MAT* locus in the *S. cerevisiae* genome is flanked by two "silent," transcriptionally inactive loci containing the alternative α or **a** sequences (see Figure 11-28). A specific DNA rearrangement brings the genes that encode the α-specific or **a**-specific

transcription factors from these silent loci to the active *MAT* locus where they can be transcribed.

Interestingly, some haploid yeast cells can switch repeatedly between the α and **a** types. *Mating-type switching* occurs when the α allele occupying the *MAT* locus is replaced by the **a** allele, or vice versa. The first step in this process is catalyzed by HO endonuclease, which is expressed in mother cells but not in daughter cells. Thus mating-type switching occurs only in mother cells (Figure 22-21). Transcription of the *HO* gene is dependent on the *Swi/Snf chromatin-remodeling complex* (see Figure 11-37), the same complex that we encountered earlier in our discussion of myogenesis. Daughter yeast cells arising by budding from mother cells contain a protein called Ash1 that prevents recruitment of the Swi/Snf complex to the *HO* gene, thereby preventing its transcription. The absence of Ash1 from mother cells allows them to transcribe the *HO* gene.

Recent experiments have revealed how the asymmetry in the distribution of Ash1 between mother and daughter cells is established. *ASH1* mRNA accumulates in the growing bud that will form a daughter cell due to the action of a myosin motor protein (Chapter 19). This motor protein, called Myo4p, moves the *ASH1* mRNA, as a ribonucleoprotein complex, along actin filaments in one direction only, toward the bud (Figure 22-22). Two connector proteins tether *ASH1* mRNA to the motor protein. By the time the bud separates from the mother cell, the mother cell is largely depleted of *ASH1* mRNA and thus can switch mating type in the fol-

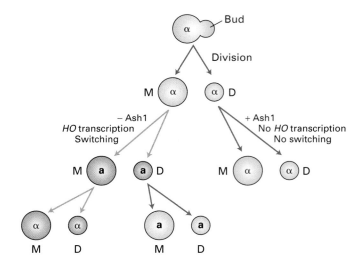

▲ FIGURE 22-21 Specificity of mating-type switching in haploid yeast cells. Division by budding forms a larger mother cell (M) and smaller daughter cell (D), both of which have the same mating type as the original cell (α in this example). The mother can switch mating type during G₁ of the next cell cycle and then divide again, producing two cells of the opposite **a** type. Switching depends on transcription of the *HO* gene, which occurs only in the absence of Ash1 protein. The smaller daughter cells, which produce Ash1 protein, cannot switch; after growing in size through interphase, they divide to form a mother cell and daughter cell. Orange cells and arrows indicate switch events.

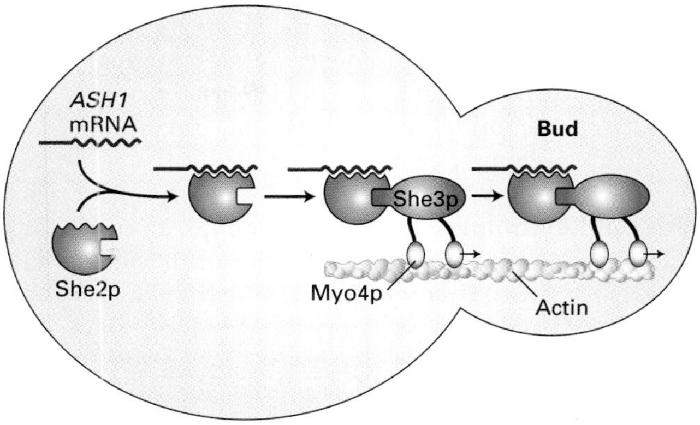

▶ **FIGURE 22-22 Model for restriction of mating-type switching to mother cells in *S. cerevisiae*.** Ash1 protein prevents a cell from transcribing the *HO* gene whose encoded protein initiates the DNA rearrangement that results in mating-type switching from **a** to α or α to **a**. Switching occurs only in the mother cell, after it separates from a newly budded daughter cell, because Ash1 protein is present in the daughter cell but not in the mother cell. The molecular basis for this differential localization of Ash1 is the one-way transport of *ASH1* mRNA into the bud. A linking protein, She2p, binds to specific 3′ untranslated sequences in the *ASH1* mRNA and also binds to She3p protein. This protein in turn binds to a myosin motor, Myo4p, which moves along actin filaments into the bud. [See S. Koon and B. J. Schnapp, 2001, *Curr. Biology* **11**:R166–R168.]

Video: ASH1 mRNA Localization

lowing G₁ before additional *ASH1* mRNA is produced and before DNA replication in the S phase.

Budding yeasts use a relatively simple mechanism to create molecular differences between the two cells formed by division. In higher organisms, polarization of the parental cell involves many more participants, and in addition, as in yeast, the mitotic spindle must be oriented in such a way that each daughter cell receives its own set of cytoplasmic components. To illustrate these complexities, we focus on asymmetric division of neuroblasts in *Drosophila*. Genetic studies in *C. elegans* and *Drosophila* have revealed the key participants, a first step in understanding at the molecular level how asymmetric cell division is regulated in multicellular organisms.

Critical Asymmetry-Regulatory Proteins Are Localized at Opposite Ends of Dividing Neuroblasts in *Drosophila*

Fly neuroblasts, which are stem cells, arise from a sheet of ectoderm cells that is one cell thick. As in vertebrates, the *Drosophila* ectoderm forms both epidermis and the nervous system, and many ectoderm cells have the potential to assume either a neural or epidermal fate. Under the control of genes that become active only in certain cells, some of the cells increase in size and begin to loosen from the ectodermal layer. At this point, the delaminating cells use Notch signal-

ing to mediate **lateral inhibition** of their neighbors, causing them to retain the epidermal fate (see Figures 14-29 and 15-36). The delaminating cells move inside and become spherical neuroblasts, while the prospective epidermal cells remain behind and close up to form a tight sheet.

Once formed, the neuroblasts undergo asymmetric divisions, at each division recreating themselves and producing a *ganglion mother cell (GMC)* at the basal side of the neuroblast (Figure 22-23). A single neuroblast will produce several GMCs; each GMC in turn forms two neurons. Depending on where they form in the embryo and consequent regulatory events, neuroblasts may form more or fewer GMCs.

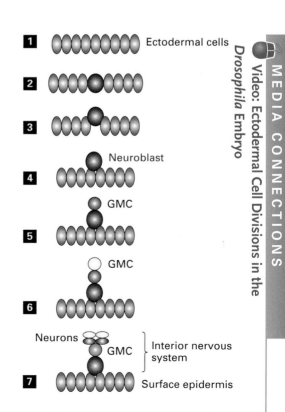

▶ **FIGURE 22-23 Asymmetric cell division during *Drosophila* neurogenesis.** The ectodermal sheet (**1**) of the early embryo gives rise to both epidermal cells and neural cells. Neuroblasts, the stem cells for the nervous system, are formed when ectoderm cells enlarge, separate from the ectodermal epithelium, and move into the interior of the embryo (**2**–**4**). Each neuroblast that arises divides asymmetrically to recreate itself and produce a ganglion mother cell (GMC) (**5**). Subsequent divisions of a neuroblast produce more GMCs, creating a stack of these precursor cells (**6**). Each GMC divides once to give rise to two neurons (**7**). Neuroblasts and their GMC descendants can have different fates depending on their location.

Video: Ectodermal Cell Divisions in the *Drosophila* Embryo

Neuroblasts and GMCs in different locations exhibit different patterns of gene expression, an indicator of their fates. Analysis of fly mutants led to the discovery of key proteins that participate in creating specific neuroblasts, inhibiting epidermal cells from becoming neuroblasts, and directing neuroblasts to divide asymmetrically.

Apical Baz/Par6/aPKC Complex   A ternary complex of proteins, referred to as the *apical complex,* is located at the end of a fly neuroblast that will become the "new" neuroblast

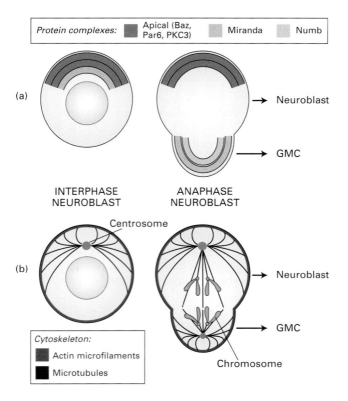

▲ **FIGURE 22-24 Localized proteins that control asymmetric cell division in the *Drosophila* neuroblast.** (a) The apical complex (blue), which consists of three proteins (Baz, Par6, and aPKC), is localized in ectoderm cells and in delaminating neuroblasts. As a neuroblast begins dividing, Miranda protein (orange) accumulates at the apical side and then moves to the basal side, where it will be incorporated into the ganglion mother cell (GMC). The second basal component, the Numb protein (green), is detected only basally. Mutations in genes that encode polarized proteins disrupt asymmetric cell division. (b) Motor protein–mediated transport along cytoskeletal filaments is thought to localize the apical and basal complexes. Actin microfilaments (red) lie just under the cell surface at all times. Microtubules (blue) radiate from the centrosome during interphase and then assemble into the mitotic spindle, attached to the duplicated centrosomes, during cell division. Note that in the neuroblast-GMC division the centrosome is skewed to the neuroblast end of the cell, and the chromosomes (light gray) are diagrammed lined up on the spindle. [Adapted from C. Q. Doe and B. Bowerman, 2001, *Curr. Opin. Cell Biol.* **13**:68.]

in the process of division. This complex is already localized at one end of all ectoderm cells. It stays at the apical end as a prospective neuroblast delaminates from the ectodermal sheet (Figure 22-24a, *left*). Three proteins compose the apical complex: Baz, Par6, and aPKC, the last being an isoform of protein kinase C. The Baz/Par6/aPKC complex persists at the apical end from late interphase (G_2 stage) until late anaphase in mitosis and then disperses or is destroyed (see Figure 20-29 for mitotic stages). After each neuroblast division, the complex re-forms at the apical end of the progeny neuroblast in the next interphase. It remains unclear how polarity information is preserved through telophase, since no known proteins are localized during that time. As noted previously, the very first *C. elegans* cell division after fertilization is asymmetric, forming an AB cell and P1 cell, which have quite different fates (see Figure 22-9). Remarkably, a protein complex like the fly Baz/Par6/aPKC complex controls the asymmetry of the P1-AB division; similar protein complexes exist in mammals as well.

Two mechanisms control the apical localization of the *Drosophila* Baz/Par6/aPKC complex: the first operates in the ectoderm prior to neuroblast delamination; the second is active during the repeating divisions of delaminated neuroblasts. The first mechanism involves at least three proteins, Scribble (Scrib), Discs-large (Dlg), and Lethal giant larvae (Lgl). These proteins are located in the cortical region of ectoderm cells (i.e., the region just below the plasma membrane), but are not polarized along the apical-basal axis of these cells. Thus, though necessary for the apical localization of the Baz/Par6/aPKC complex, the Scrib/Dlg/Lgl proteins probably are not sufficient. Once a neuroblast has separated from the ectoderm, the G_α subunit of the heterotrimeric G protein and two other proteins called Inscuteable and Partner of Inscuteable join the Baz/Par6/aPKC complex, forming a six-protein complex. These two additional proteins probably stabilize the Baz/Par6/aPKC complex. Reassembly of the entire complex at the apical end following each neuroblast division requires actin microfilaments (Figure 22-24b). Generally, asymmetric protein segregation requires actin but not tubulin, suggesting that motor protein–mediated transport of these proteins along microfilaments plays a role in their apical localization.

Basal Proteins   In a dividing neuroblast two basal proteins are found at the end that will form a ganglion mother cell (GMC) at division (see Figure 22-24a, *right*). Localization of these proteins, including Miranda and Numb, at the basal end of neuroblasts requires the apical Baz/Par6/aPKC complex. The apical proteins set up polarity during interphase, and the polarity is read and interpreted by the machinery that transports Numb and Miranda at mitosis. The Numb protein is located throughout the cell during interphase; at prophase it becomes basally located. In contrast, Miranda initially assembles at the apical end; it then moves to the basal end at prophase. The apical Baz/Par6/aPKC complex forms even when either of the basal proteins is missing due to mutations. Mutations in any of the components of the apical complex, however, prevent the Miranda and Numb proteins from moving to the basal region. Instead they accumulate all around the cortical region or in clumps throughout the cell. The simplest idea, that the basal complexes form wherever the apical complex is absent, is probably wrong, since parts of the cell contain neither of these complexes.

Basally located molecules include Prospero protein, a transcription factor that contains a DNA-binding homeodomain; Miranda protein, which has a coiled-coil structure; Staufen, an RNA-binding protein; and *prospero* mRNA. During each round of neuroblast division, new Prospero protein is made during interphase and incorporated into the basal region, which is distributed to the GMC at the time of cleavage. In the absence of Prospero, GMCs do not transcribe the appropriate genes and do not develop into normal neurons. The localization of a transcription factor in the cytoplasm provides a link between the asymmetry of the cytoplasm and subsequent cell-fate determination, potentially explaining how a transcription factor is preferentially distributed to one of two daughter cells.

Orientation of the Mitotic Spindle Is Linked to Cytoplasmic Cell-Asymmetry Factors

For localized protein complexes to be differentially incorporated into two daughter cells requires that the plane of cell division be appropriately oriented. In dividing fly neuroblasts, the **mitotic spindle** first aligns perpendicular to the apical-basal axis and then turns 90 degrees to align with it at the same time that the basal complexes become localized to the basal side (Figure 22-25). The apical complex, which is already in place before spindle rotation, controls the final orientation of the spindle. This is supported by the finding that mutations in any of the components of the apical complex eliminate the coordination of the spindle with apical-basal polarity, causing the spindle orientation to become random.

Spindle orientation is regulated by actin and myosin-related proteins. Mammalian Par6, a component of the apical complex, can bind two small GTPase proteins, Cdc42 and Rac1, that control the arrangement of actin microfilaments in the cytoskeleton (see Figure 19-29). The protein Lethal giant larvae (Lgl), a component of the Scrib/Dlg/Lgl asymmetry pathway, helps to localize Miranda basally in fly neuroblasts.

Lgl binds to myosin II, which functions in cytokinesis (see Figure 19-20). Lgl itself is uniformly localized around the cortex. Lgl is phosphorylated by the apical complex and may be inactivated on the apical side to allow basal Miranda accumulation. Two yeast proteins related to Lgl also bind to myosin II; they have been implicated in exocytosis and secretion, specifically in the docking of post-Golgi vesicles with the plasma membrane (Chapter 17). Mutations in the two yeast proteins suppress mutations in the *myosin II* gene. This observation is interpreted to mean that the function of myosin II in spindle orientation is opposite to that of the Lgl-like proteins: reduction of one protein's function is ameliorated by reduction of the other, restoring a semblance of the normal balance. In this case, therefore, myosin and Lgl probably act in opposite directions: myosin II moving Miranda or other materials to control spindle orientation and Lgl restraining it.

Further evidence for the importance of myosin in spindle orientation comes the finding that another myosin relative,

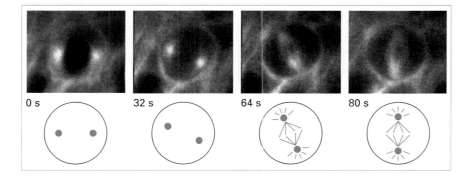

▲ **EXPERIMENTAL FIGURE 22-25 Time-lapse fluorescence imaging reveals rotation of the mitotic spindle in asymmetrically dividing neuroblasts.** Early *Drosophila* embryos were injected with a hybrid gene composed of the gene encoding green fluorescent protein (GFP) fused to the gene encoding Tau, a protein that binds to microtubules. At the top are time-lapse images of a single dividing neuroblast in a live embryo. The basal side is at the top, and the apical side at the bottom. At time 0, equivalent to prophase, the two centrosomes are visible on opposite sides of the cell. These function as the spindle poles; as mitosis proceeds the microtubules forming the mitotic spindle are assembled from the poles (see Figure 20-29). In successive images (at 32, 64, and 80 seconds), the bipolar spindle can be seem to form and rotate 90 degrees to align with the apical-basal axis, as schematically depicted at the bottom. [From J. A. Kaltschmidt et al., 2000, *Nature Cell Biol.* **2**:7–12; courtesy of J. Kaltschmidt and A. H. Brand, Wellcome/CRC Institute, Cambridge University.]

MEDIA CONNECTIONS

Video: Mitotic Spindle Assymetry in *Drosophila* Neuroblast Cell Division myosin VI, binds directly to Miranda. In addition, mutations in myosin VI in flies prevent basal targeting of Miranda and simultaneously block proper spindle orientation. These various findings give us the beginning of a picture of how localized protein complexes are coordinated with cell division, so that each daughter cell receives the appropriate amounts of the various complexes.

At least some of the asymmetric cell division regulators discovered in flies and worms are present in vertebrates and have similar roles there. For example, mammalian Numb is required to maintain the neural stem-cell population. Evolutionary conservation of proteins and mechanisms facilitates rapid progress in research.

KEY CONCEPTS OF SECTION 22.4

Regulation of Asymmetric Cell Division

■ Asymmetric cell division requires polarization of the dividing cell, which usually entails localization of some cytoplasmic components, and then the unequal distribution of these components to the daughter cells (see Figure 22-20).

■ In the asymmetric division of budding yeasts, a myosin-dependent transport system carries *ASH1* mRNA into the bud (see Figure 22-22).

■ Ash1 protein is produced in the daughter cell soon after division and prevents expression of HO endonuclease, which is necessary for mating-type switching. Thus a daughter cell cannot switch mating type, whereas the mother cell from which it arises can (see Figure 22-21).

■ Asymmetric cell division in *Drosophila* neuroblasts depends on localization of the apical protein complex (Baz/Par6/PKC3) and two sets of basal proteins. The basal proteins are incorporated into the ganglion mother cell (GMC) and contain proteins that determine cell fate (see Figure 22-24).

■ The apical Baz/Par6/aPKC complex also controls asymmetric cell division in *C. elegans* and perhaps mammals.

■ Asymmetry factors exert their influence at least in part by controlling the orientation of the mitotic spindle, so that asymmetrically localized proteins and structures are differentially incorporated into the two daughter cells. Myosin proteins bind to proteins that control asymmetry factors of cells to control spindle orientation.

22.5 Cell Death and Its Regulation

Programmed cell death is a cell fate, an odd sort of cell fate but nonetheless one that is essential. Cell death keeps our hands from being webbed, our embryonic tails from persisting, our immune system from responding to our own proteins, and our brain from being filled with useless electrical connections. In fact, the majority of cells generated during brain development die during development.

Cellular interactions regulate cell death in two fundamentally different ways. First, most cells, if not all, in multicellular organisms require signals to stay alive. In the absence of such survival signals, frequently referred to as **trophic factors**, cells activate a "suicide" program. Second, in some developmental contexts, including the immune system, specific signals induce a "murder" program that kills cells. Whether cells commit suicide for lack of survival signals or are murdered by killing signals from other cells, recent studies suggest that death is mediated by a common molecular pathway. In this section, we first distinguish programmed cell death from death due to tissue injury, then consider the role of trophic factors in neuronal development, and finally describe the evolutionarily conserved effector pathway that leads to cell suicide or murder.

Programmed Cell Death Occurs Through Apoptosis

The demise of cells by programmed cell death is marked by a well-defined sequence of morphological changes, collectively referred to as **apoptosis,** a Greek word that means "dropping off" or "falling off," as leaves from a tree. Dying cells shrink and condense and then fragment, releasing small membrane-bound apoptotic bodies, which generally are engulfed by other cells (Figure 22-26; see also Figure 1-19). The nuclei condense and the DNA is fragmented. Importantly, the intracellular constituents are not released into the extracellular milieu where they might have deleterious effects on neighboring cells. The highly stereotyped changes accompanying apoptosis suggested to early workers that this type of cell death was under the control of a strict program. This program is critical during both embryonic and adult life to maintain normal cell number and composition.

The genes involved in controlling cell death encode proteins with three distinct functions:

■ "Killer" proteins are required for a cell to begin the apoptotic process.

■ "Destruction" proteins do things like digest DNA in a dying cell.

■ "Engulfment" proteins are required for phagocytosis of the dying cell by another cell.

At first glance, engulfment seems to be simply an after-death cleanup process, but some evidence suggests that it is part of the final death decision. For example, mutations in killer genes always prevent cells from initiating apoptosis, whereas mutations that block engulfment sometimes allow cells to survive that would normally die. That is, cells with engulfment-gene mutations can initiate apoptosis but then sometimes recover.

In contrast to apoptosis, cells that die in response to tissue damage exhibit very different morphological changes, referred to as **necrosis.** Typically, cells that undergo this process swell and burst, releasing their intracellular contents, which can damage surrounding cells and frequently cause inflammation.

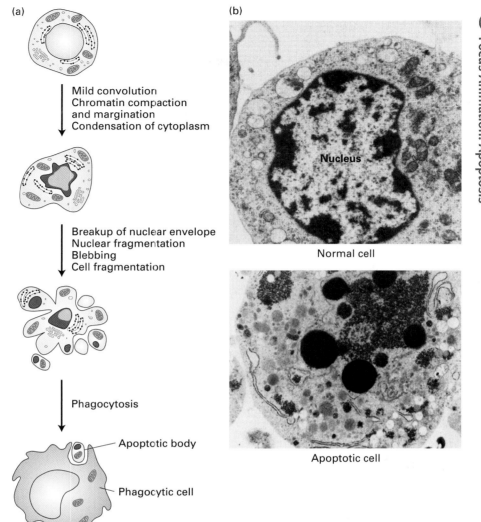

► **FIGURE 22-26 Ultrastructural features of cell death by apoptosis.** (a) Schematic drawings illustrating the progression of morphologic changes observed in apoptotic cells. Early in apoptosis, dense chromosome condensation occurs along the nuclear periphery. The cell body also shrinks, although most organelles remain intact. Later both the nucleus and cytoplasm fragment, forming apoptotic bodies, which are phagocytosed by surrounding cells. (b) Photomicrographs comparing a normal cell (*top*) and apoptotic cell (*bottom*). Clearly visible in the latter are dense spheres of compacted chromatin as the nucleus begins to fragment. [Part (a) adapted from J. Kuby, 1997, *Immunology,* 3d ed., W. H. Freeman & Co., p. 53. Part (b) from M. J. Arends and A. H. Wyllie, 1991, *Int'l. Rev. Exp. Pathol.* **32**:223.]

Mild convolution
Chromatin compaction
and margination
Condensation of cytoplasm

Breakup of nuclear envelope
Nuclear fragmentation
Blebbing
Cell fragmentation

Phagocytosis

Apoptotic body

Phagocytic cell

Nucleus

Normal cell

Apoptotic cell

MEDIA CONNECTIONS
Focus Animation: Apoptosis

Neurotrophins Promote Survival of Neurons

The earliest studies demonstrating the importance of trophic factors in cellular development came from analyses of the developing nervous system. When neurons grow to make connections to other neurons or to muscles, sometimes over considerable distances, more cells grow than will eventually survive. Those that make connections prevail and survive; those that fail to connect die.

In the early 1900s the number of neurons innervating the periphery was shown to depend upon the size of the tissue to which they would connect, the so-called "target field." For instance, removal of limb buds from the developing chick embryo leads to a reduction in the number of sensory neurons and motoneurons innervating the bud (Figure 22-27). Conversely, grafting additional limb tissue to a limb bud leads to an increase in the number of neurons in corresponding regions of the spinal cord and sensory ganglia. Indeed, in-cremental increases in the target-field size are accompanied by commensurate incremental increases in the number of neurons innervating the target field. This relation was found to result from the selective survival of neurons rather than changes in their differentiation or proliferation. The observation that many sensory and motor neurons die after reaching their peripheral target field suggested that these neurons compete for survival factors produced by the target tissue.

Subsequent to these early observations, scientists discovered that transplantation of a mouse sarcoma tumor into a chick led to a marked increase in the numbers of certain types of neurons. This finding implicated the tumor as a rich source of the presumed trophic factor. To isolate and purify this factor, known simply as nerve growth factor (NGF), scientists used an in vitro assay in which outgrowth of neurites from sensory ganglia (nerves) was measured. Neurites are extensions of the cell cytoplasm that can grow to become the long wires of the nervous system, the **axons** and **dendrites**

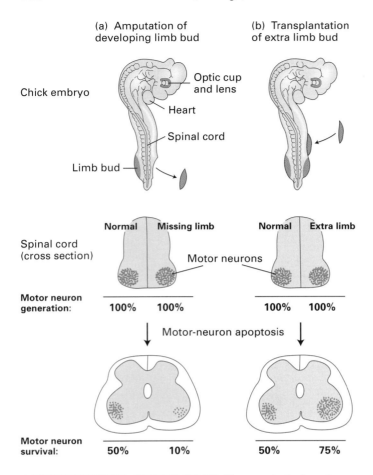

▲ EXPERIMENTAL FIGURE 22-27 The survival of motor neurons depends on the size of the muscle target field they innervate. (a) Removal of a limb bud from one side of a chick embryo at about 2.5 days results in a marked decrease in the number of motor neurons on the affected side. In an amputated embryo, normal numbers of motor neurons are generated on both sides (*middle*). Later in development, many fewer motor neurons remain on the side of the spinal cord with the missing limb than on the normal side (*bottom*). Note that only about 50 percent of the motor neurons that originally are generated normally survive. (b) Transplantation of an extra limb bud into an early chick embryo produces the opposite effect, more motor neurons on the side with additional target tissue than on the normal side. [Adapted from D. Purves, 1988, *Body and Brain: A Trophic Theory of Neural Connections* (Cambridge, MA: Harvard University Press), and E. R. Kandel, J. H. Schwartz, and T. M. Jessell, 2000, *Principles of Neural Science*, 4th ed. (New York: McGraw-Hill), p. 1054, Figure 53-11.]

(see Figure 7-29). The later discovery that the submaxillary gland in the mouse also produces large quantities of NGF enabled biochemists to purify it to homogeneity and to sequence it. A homodimer of two 118-residue polypeptides, NGF belongs to a family of structurally and functionally related trophic factors collectively referred to as *neurotrophins*. Brain-derived neurotrophic factor (BDNF) and neurotrophin-3 (NT-3) also are members of this protein family.

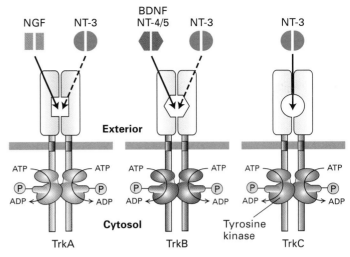

▲ FIGURE 22-28 Specificity of neurotrophins for Trks, a family of receptor tyrosine kinases. Each neurotrophin binds with high affinity to one Trk receptor indicated by the solid arrow from the ligand to the receptor. NT-3 also can bind with lower affinity to both TrkA and TrkB as indicated by the dashed arrow. In addition, neurotrophins bind to a distinct receptor called p75^NTR either alone or in combination with Trks. NGF = nerve growth factor; BDNF = brain-derived neurotrophic factor; NT-3 = neurotrophin-3. [Adapted from W. D. Snider, 1994, *Cell* **77**:627.]

Neurotrophin Receptors Neurotrophins bind to and activate a family of receptor tyrosine kinases called *Trks* (pronounced "tracks"). (The general structure of receptor tyrosine kinases and the intracellular signaling pathways they activate are covered in Chapter 14.) As shown in Figure 22-28, NGF binds to TrkA; BDNF, to TrkB; and NT-3, to TrkC. Binding of these factors to their receptors provides a survival signal for different classes of neurons. A second type of receptor called p75^NTR (NTR = neurotrophin receptor) also binds to neurotrophins, but with lower affinity. However, p75^NTR forms heteromultimeric complexes with the different Trk receptors; this association increases the affinity of Trks for their ligands. Some studies indicate that the binding of NGF to p75^NTR in the absence of TrkA may promote cell death rather than prevent it.

Knockouts of Neurotrophins and Their Receptors To critically address the role of the neurotrophins in development, scientists produced mice with knockout mutations in each of the neurotrophins and their receptors. These studies revealed that different neurotrophins and their corresponding receptors are required for the survival of different classes of sensory neurons, which carry signals from peripheral sensory systems to the brain (Figure 22-29). For instance, pain-sensitive (nociceptive) neurons, which express TrkA, are selectively lost from the dorsal root ganglion of knockout mice lacking NGF or TrkA, whereas TrkB- and TrkC-expressing neurons are unaffected in such knockouts. In contrast, TrkC-expressing proprioceptive neurons, which detect the position of the limbs, are missing from the dorsal root ganglion in *TrkC* and *NT-3* mutants.

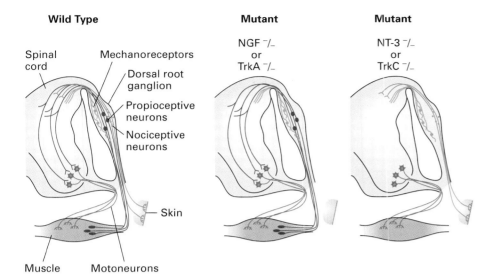

◄ EXPERIMENTAL FIGURE 22-29 Different classes of sensory neurons are lost in knockout mice lacking different trophic factors or their receptors. In animals lacking nerve growth factor (NGF) or its receptor TrkA, small nociceptive (pain-sensing) neurons (blue) that innervate the skin are missing. These neurons express TrkA receptor and innervate NGF-producing targets. In animals lacking either neurotrophin-3 (NT-3) or its receptor TrkC, large propioceptive neurons (red) innervating muscle spindles are missing. Muscle produces NT-3 and the propioceptive neurons express TrkC. Mechanoreceptors (brown), another class of sensory neurons in the dorsal root ganglion, are unaffected in these mutants. [Adapted from W. D. Snider, 1994, *Cell* **77**:627.]

A Cascade of Caspase Proteins Functions in One Apoptotic Pathway

Key insights into the molecular mechanisms regulating cell death came from genetic studies in *C. elegans*. As we noted in Section 22.1, scientists have traced the lineage of all the somatic cells in *C. elegans* from the fertilized egg to the mature worm (see Figure 22-9). Of the 947 nongonadal cells generated during development of the adult hermaphrodite form, 131 cells undergo programmed cell death.

Specific mutations have identified a variety of genes whose encoded proteins play an essential role in controlling programmed cell death during *C. elegans* development. For instance, programmed cell death does not occur in worms carrying loss-of-function mutations in the *ced-3* gene or the *ced-4* gene; as a result, the 131 "doomed" cells survive (Figure 22-30). CED-4 is a protease-activating factor that causes autocleavage of the CED-3 precursor protein, creating an active CED-3 protease that initiates cell death (Figure 22-31). In contrast, in *ced-9* mutants, all cells die during embryonic life, so the adult form never develops. These genetic studies indicate that the CED-3 and CED-4 proteins are required for cell death, that CED-9 suppresses apoptosis, and that the apoptotic pathway can be activated in all cells. Moreover, the finding that cell death does not occur in *ced-9/ced-3* double mutants suggests that CED-9 acts "upstream" of CED-3 to suppress the apoptotic pathway.

The confluence of genetic studies in worms and studies on human cancer cells first suggested that an evolutionarily conserved pathway mediates apoptosis. The first apoptotic gene to be cloned, *bcl-2*, was isolated from human follicular lymphomas. A mutant form of this gene, created in lymphoma cells by a chromosomal rearrangement, was shown to act as an **oncogene** that promoted cell survival rather than cell death (Chapter 23). The human Bcl-2 protein and worm CED-9 protein are homologous, and a *bcl-2* transgene can

(a)
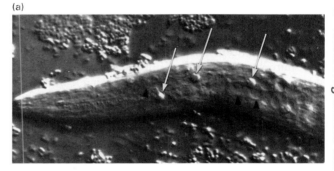

(b)
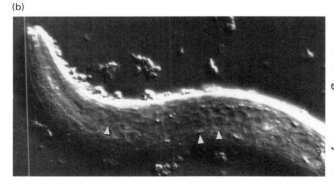

▲ EXPERIMENTAL FIGURE 22-30 Mutations in the *ced-3* gene block programmed cell death in *C. elegans*. (a) Newly hatched larva carrying a mutation in the *ced-1* gene. Because mutations in this gene prevent engulfment of dead cells, highly refractile dead cells accumulate (arrows), facilitating their visualization. (b) Newly hatched larva with mutations in both the *ced-1* and *ced-3* genes. The absence of refractile dead cells in these double mutants indicates that no cell deaths occurred. Thus CED-3 protein is required for programmed cell death. [From H. M. Ellis and H. R. Horvitz, 1986, *Cell* **91**:818; courtesy of Hilary Ellis.]

MEDIA CONNECTIONS

Video: Programmed Cell Death in *C. elegans* Embryonic Development

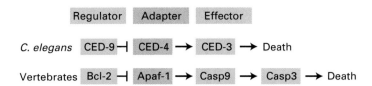

▲ **FIGURE 22-31 Overview of the evolutionarily conserved apoptotis pathway in *C. elegans* and vertebrates.** Three general types of proteins are critical in this conserved pathway. Regulators either promote or suppress apoptosis; two orthologous regulators shown here, CED-9 and Bcl-2, suppress apoptosis in the presence of trophic factors. Adapters interact with both regulators and effectors; in the absence of trophic factors, they promote activation of effectors. Proteases called caspases serve as effector proteins; their activation leads to degradation of various intracellular substrates and eventually cell death. [Adapted from D. L. Vaux and S. J. Korsemeyer, 1999, *Cell* **96**:245.]

block the extensive cell death found in *ced-9* mutant worms. Thus both proteins act as regulators that suppress the apoptotic pathway see (Figure 22-31). In addition, both proteins contain a single transmembrane domain and are localized to the outer mitochondrial, nuclear, and endoplasmic reticulum membranes, where they serve as sensors that control the apoptotic pathway in response to external stimuli. As we discuss below, other regulators promote apoptosis.

The *effector proteins* in the apoptotic pathway are enzymes called **caspases** in vertebrates, so named because they contain a key **c**ysteine residue in the catalytic site and selectively cleave proteins at sites just C-terminal to **as**partate residues. The principal effector protease in *C. elegans* is the CED-3 caspase. Humans have 15 different caspases, all of which are initially made as procaspases that must be cleaved to become active. Such proteolytic processing of proproteins is used repeatedly in blood clotting, generation of digestive enzymes, and generation of hormones (Chapter 17). Initiator caspases (e.g., caspase-9) are activated by autoproteolysis induced by other types of proteins (CED-4 in worms, for example), which help the initiators to aggregate. Activated initiator caspases cleave effector caspases (e.g., caspase-3) and thus quickly amplify the total caspase activity level in the dying cell. The various effector caspases recognize and cleave short amino acid sequences in many different target proteins. They differ in their preferred target sequences. Their specific intracellular targets include proteins of the nuclear lamina and cytoskeleton whose cleavage leads to the demise of a cell. Seven lines of knockout mice have been made, each lacking a particular caspase function. Some homozygotes die as embryos; others are viable but with subtle defects. Strikingly, some exhibit an excessive number of neurons (i.e., neuronal hyperplasia), reaffirming the importance of cell death in development of the nervous system; other mutants have cells that are resistant to apoptosis.

While caspases are critical in many apoptosis events, caspase-independent apoptosis also occurs. Apoptosis-inducing factor (AIF; a flavoprotein) and endonuclease G are two proteins that kill cells on their release from mitochondria.

Cell-culture studies have yielded important insights about a group of *adapter proteins* that coordinate the action of regulators and effectors to control apoptosis (see Figure 22-31). For instance, expression of *C. elegans* CED-4 in a human kidney cell line induces rapid apoptosis. This can be blocked by co-expression of the negative regulator CED-9 (or mammalian Bcl-2), indicating that CED-9 opposes CED-4 action. In addition, CED-9 has been shown to directly bind to CED-4 and move it from the cytosol to intracellular membranes. Thus the pro-apoptotic function of CED-4 is directly suppressed by the anti-apoptotic function of CED-9. CED-4 also binds directly to the CED-3 caspase (and related mammalian caspases) and promotes activation of its protease activity, as noted previously. Biochemical studies have shown that CED-4 can simultaneously bind to both CED-9 and CED-3. These results fit with the genetics, which shows that the absence of CED-9 has no effect if CED-3 is also missing (*ced-3/ced-9* double mutants have no cell death, like *ced-3* mutants).

Pro-Apoptotic Regulators Permit Caspase Activation in the Absence of Trophic Factors

Having introduced the major participants in the apoptotic pathway, we now take a closer look at how the effector caspases are activated in mammalian cells. Although the normal function of CED-9 or Bcl-2 is to suppress the cell-death pathway, other intracellular regulatory proteins promote apoptosis. The first pro-apoptotic regulator to be identified, named Bax, was found associated with Bcl-2 in extracts of cells expressing high levels of Bcl-2. Sequence analysis demonstrated that Bax is related in sequence to CED-9 and Bcl-2, but overproduction of Bax induces cell death rather than protecting cells from apoptosis, as CED-9 and Bcl-2 do. Thus this family of regulatory proteins comprises both *anti-apoptotic* members (e.g., CED-9, Bcl-2) and *pro-apoptotic* members (e.g., Bax). All members of this family, which we refer to as the **Bcl-2 family**, are single-pass transmembrane proteins and can participate in oligomeric interactions. In mammals, six Bcl-2 family members prevent apoptosis and nine promote it. Thus the fate of a given cell—survival or death—may reflect the particular spectrum of Bcl-2 family members made by that cell and the intracellular signaling pathways regulating them.

Some Bcl-2 family members preserve or disrupt the integrity of mitochondria, thereby controlling release of mitochondrial proteins such as cytochrome *c*. In normal healthy cells, cytochrome *c* is localized between the inner and outer mitochondrial membrane, but in cells undergoing apoptosis, cytochrome *c* is released into the cytosol. This release can be blocked by overproduction of Bcl-2; conversely, overproduction of Bax promotes release of cytochrome *c* into the cytosol and apoptosis. Moreover, injection of cytochrome *c* into the cytosol of cells induces apoptosis. A variety of death-inducing stimuli cause Bax monomers to move from the cytosol to the outer mitochondrial membrane where they oligomerize. Bax homodimers, but not Bcl-2 homodimers or Bcl-2/Bax heterodimers, permit influx of ions through the mitochondrial membrane. It remains unclear how this ion influx triggers the release of cytochrome *c*. Once cytochrome *c* is released into the cytosol, it binds to the adapter protein Apaf-1 (the mammalian homolog of CED-4) and promotes activation of a caspase cascade leading to cell death (Figure 22-32a). Bax can cause cell death by a second pathway that triggers mitochondrial dysfunction independently of any caspase action, at least in part by stimulating mitochondrial depolarization.

Gene knockout experiments have dramatically confirmed the importance of both pro-apoptotic and anti-apoptotic Bcl-2 family members in neuronal development. Mice lacking the *bcl-xl* gene, which encodes an anti-apoptotic protein, have massive defects in nervous system development with widespread cell death in the spinal cord, dorsal root ganglion, and brain of developing embryos. In contrast, *bax* knockouts exhibit a marked increase in neurons in some regions of the nervous system. These knockouts also demonstrate that the effects of Bcl-2 family members generally are tissue specific.

Some Trophic Factors Induce Inactivation of a Pro-Apoptotic Regulator

We saw earlier that neurotrophins such as nerve growth factor (NGF) protect neurons from cell death. The intracellular signaling pathways linking such survival factors to inactivation of the cell-death machinery are quite elaborate. The finding that trophic factors appear to work largely independent of protein synthesis suggested that these external signals lead to changes in the activities of preexisting proteins rather than to activation of gene expression. Scientists demonstrated that in the absence of trophic factors, the nonphosphorylated form of Bad is associated with Bcl-2/Bcl-xl at the mitochondrial membrane (see Figure 22-32a). Binding of Bad inhibits the anti-apoptotic function of Bcl-2/Bcl-xl, thereby promoting

(a) **Absence of trophic factor: Caspase activation**

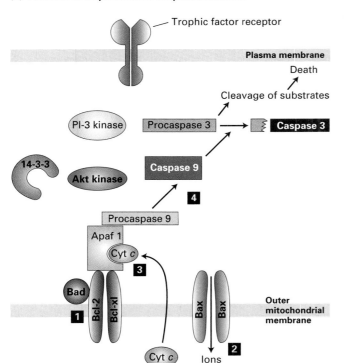

(b) **Presence of trophic factor: Inhibition of caspase activation**

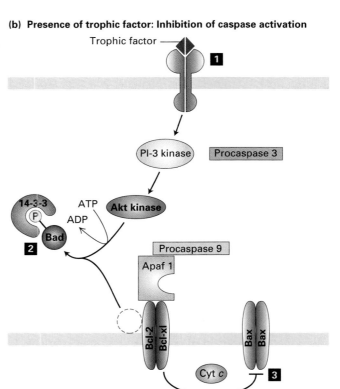

▲ **FIGURE 22-32 Proposed intracellular pathways leading to cell death by apoptosis or to trophic factor–mediated cell survival in mammalian cells.** (a) In the absence of a trophic factor, Bad, a soluble pro-apoptotic protein, binds to the anti-apoptotic proteins Bcl-2 and Bcl-xl, which are inserted into the mitochondrial membrane (**1**). Bad binding prevents the anti-apoptotic proteins from interacting with Bax, a membrane-bound pro-apoptotic protein. As a consequence, Bax forms homo-oligomeric channels in the membrane that mediate ion flux (**2**). Through an as-yet-unknown mechanism, this leads to the release of cytochrome *c* into the cytosol, where it binds to the adapter protein Apaf-1 (**3**), promoting a caspase cascade that leads to cell death (**4**). (b) In some cells, binding of a trophic factor (e.g., NGF) stimulates PI-3 kinase activity, leading to activation of the downstream kinase Akt, which phosphorylates Bad (**1**). Phosphorylated Bad then forms a complex with the 14-3-3 protein (**2**). With Bad sequestered in the cytosol, the anti-apoptotic Bcl-2/Bcl-xl proteins can inhibit the activity of Bax (**3**), thereby preventing the release of cytochrome *c* and activation of the caspase cascade. [Adapted from B. Pettman and C. E. Henderson, 1998, *Neuron* **20**:633.]

cell death. Phosphorylated Bad, however, cannot bind to Bcl-2/Bcl-xl and is found in the cytosol complexed to the phosphoserine-binding protein 14-3-3. Hence, signaling pathways leading to Bad phosphorylation would be particularly attractive candidates for transmitting survival signals.

A number of trophic factors including NGF have been shown to trigger the PI-3 kinase signaling pathway, leading to activation of a downstream kinase called PKB (see Figure 14-27). Activated PKB phosphorylates Bad at sites known to inhibit its pro-apoptotic activity. Moreover, a constitutively active form of PKB can rescue cultured neurotrophin-deprived neurons, which otherwise would undergo apoptosis and die. These findings support the mechanism for the survival action of trophic factors depicted in Figure 22-32b. In other cell types, different trophic factors may promote cell survival through post-translational modification of other components of the cell-death machinery.

Tumor Necrosis Factor and Related Death Signals Promote Cell Murder by Activating Caspases

Although cell death can arise as a default in the absence of survival factors, apoptosis can also be stimulated by positively acting "death" signals. For instance, *tumor necrosis factor (TNF)*, which is released by macrophages, triggers the cell death and tissue destruction seen in certain chronic inflammatory diseases. Another important death-inducing signal, the *Fas ligand*, is a cell-surface protein expressed by activated natural killer cells and cytotoxic T lymphocytes. This signal can trigger death of virus-infected cells, some tumor cells, and foreign graft cells.

Both TNF and Fas ligand act through cell-surface "death" receptors that have a single transmembrane domain and are activated when ligand binding brings three receptor molecules into close proximity. The trimeric receptor complex attracts a protein called FADD (Fas-associated death domain), which serves as an adapter to recruit and in some way activate caspase-8, an initiator caspase, in cells receiving a death signal. The death domain found in FADD is a sequence that is present in a number of proteins involved in apoptosis. Once activated, caspase-8 activates other caspases and the amplification cascade begins. To test the ability of the Fas receptor to induce cell death, researchers incubated cells with antibodies against the receptor. These antibodies, which bind and cross-link Fas receptors, were found to stimulate cell death, indicating that activation of the Fas receptor is sufficient to trigger apoptosis.

KEY CONCEPTS OF SECTION 22.5

Cell Death and Its Regulation

■ All cells require trophic factors to prevent apoptosis and thus survive. In the absence of these factors, cells commit suicide.

■ Genetic studies in *C. elegans* defined an evolutionarily conserved apoptotic pathway with three major components: regulatory proteins, adapter proteins, and effector proteases called caspases in vertebrates (see Figure 22-31).

■ Once activated, apoptotic proteases cleave specific intracellular substrates leading to the demise of a cell. Adapter proteins (e.g., Apaf-1), which bind both regulatory proteins and caspases, are required for caspase activation.

■ Pro-apoptotic regulator proteins (e.g., Bax, Bad) promote caspase activation, and anti-apoptotic regulators (e.g., Bcl-2) suppress activation. Direct interactions between pro-apoptotic and anti-apoptotic proteins lead to cell death in the absence of trophic factors. Binding of extracellular trophic factors can trigger changes in these interactions, resulting in cell survival (see Figure 22-32).

■ The Bcl-2 family contains both pro-apoptotic and anti-apoptotic proteins; all are single-pass transmembrane proteins and engage in protein-protein interactions. Bcl-2 molecules can control the release of cytochrome *c* from mitochondria, triggering cell death.

■ Binding of extracellular death signals, such as tumor necrosis factor and Fas ligand, to their receptors activates an associated protein (FADD) that in turn triggers the caspase cascade leading to cell murder.

PERSPECTIVES FOR THE FUTURE

Cell birth, lineage, and death lie at the heart of the growth of an organism and are also central to disease processes, most notably cancer. Few transformations seem more remarkable than the blooming of cell types during development. A lineage beginning with a fertilized egg, a "plain vanilla" 200-μm sphere, produces neurons a yard long, pulsating multinucleate muscle cells, exquisitely light-sensitive retina cells, ravenous macrophages that recognize and engulf germs, and all the hundreds of other cell types. Regulators of cell lineage produce this rich variety by controlling two critical decisions: (1) when and where to activate the cell division cycle (Chapter 21) and (2) whether the two daughter cells will be the same or different. A cell may be just like its parent, or it may embark on a new path.

Cell birth is normally carefully restricted to specific locales and times, such as the basal layer of the skin or the root meristem. Liver regenerates when there is injury, but liver cancer is prevented by restricting unnecessary growth at other times. Cell lineage is patterned by the asymmetric distribution of key regulators to the daughter cells of a division. Some of these regulators are intrinsic to the parent cell, becoming asymmetrically distributed during polarization of the cell; other regulators are external signals that differentially reach the daughter cells. Asymmetry of cells becomes asymmetry of tissues and whole organisms. Our left and right hands differ only as a result of cell asymmetry.

Some cells persist for the life of the organism, but others such as blood and intestinal cells turn over rapidly. Many cells live for awhile and are then programmed to die and be replaced by others arising from a stem-cell population. Programmed cell death is also the basis for the meticulous elimination of potentially harmful cells, such as autoreactive immune cells, which attack the body's own cells, or neurons that have failed to properly connect. Cell-death programs have also evolved as a defense against infection, and virus-infected cells are selectively murdered in response to death signals. Viruses, in turn, devote much of their effort to evading host defenses. For example, p53, a transcription factor that senses cell stresses and damage and activates transcription of pro-apoptotic members of the *bcl-2* gene family, is inhibited by the adenovirus E1B protein. It has been estimated that about a third of the adenovirus genome is directed at evading host defenses. Cell death is relevant to toxic chemicals as well as viral infections; malformations due to poisons often originate from excess apoptosis.

Failures of programmed cell death can lead to uncontrolled cancerous growth (Chapter 23). The proteins that prevent the death of cancer cells therefore become possible targets for drugs. A tumor may contain a mixture of cells, some capable of seeding new tumors or continued uncontrolled growth, and some capable only of growing in place or for a limited time. In this sense the tumor has its own stem cells, and they must be found and studied, so they become vulnerable to our medicine. One option is to manipulate the cell-death pathway to our own advantage, to send the signals that will make cancer cells destroy themselves.

Much attention is now being given to the regulation of stem cells in an effort to understand how dividing populations of cells are created and maintained. This has clear implications for repair of tissue, for example, to help damaged eyes, torn cartilage, degenerating brain tissue, or failing organs. One interesting possibility is that some populations of stem cells with the potential to generate or regenerate tissue are normally eliminated by cell death during later development. If so, finding ways to selectively block the death of the cells could make regeneration more likely. Could the elimination of such cells during mammalian development be the difference between an amphibian capable of limb regeneration and a mammal that is not?

KEY TERMS

apical complex *922*

apoptosis *924*

asymmetric cell division *900*

β-catenin *903*

Bcl-2 family *928*

caspases *928*

cell lineage *899*

"death" signals *930*

determination *914*

differentiation *899*

ectoderm *906*

embryonic stem (ES) cells *901*

endoderm *906*

ganglion mother cell (GMC) *921*

germ line *906*

heterochronic mutants *908*

MAT locus *910*

mating factor *912*

mating-type switching *920*

meristems *905*

mesoderm *906*

micro RNAs (miRNAs) *909*

muscle regulatory factors (MRFs) *915*

neurotrophins *926*

pluripotent *900*

precursor/progenitor cells *899*

somatic cells *899*

stem cells *899*

trophic factors *924*

REVIEW THE CONCEPTS

1. What two properties define a stem cell? Distinguish between a totipotential stem cell, a pluripotent stem cell, and a progenitor cell.

2. Where are stem cells located in plants? Where are stem cells located in adult animals? How does the concept of stem cell differ between animal and plant systems?

3. In 1997, Dolly the sheep was cloned by a technique called somatic cell nuclear transfer. A nucleus from an adult mammary cell was transferred into an egg from which the nucleus had been removed. The egg was allowed to divide several times in culture, then the embryo was transferred to a surrogate mother who gave birth to Dolly. Dolly died in 2003 after mating and giving birth herself to viable offspring. What does the creation of Dolly tell us about the potential of nuclear material derived from a fully differentiated adult cell? Does the creation of Dolly tell us anything about the potential of an intact, fully differentiated adult cell? Name three types of information that function to preserve cell type. Which of these types of information was shown to be reversible by the Dolly experiment?

4. The roundworm *C. elegans* has proven to be a valuable model organism for studies of cell birth, cell lineage, and cell death. What properties of *C. elegans* render it so well suited for these studies? Why is so much information from *C. elegans* experiments of use to investigators interested in mammalian development?

5. In the budding yeast *S. cerevisiae*, what is the role of the MCM1 protein in the following?

a. transcription of **a**-specific genes in **a** cells

b. blocking transcription of α-specific genes in **a** cells

c. transcription of α-specific genes in α cells

d. blocking transcription of **a**-specific genes in α cells

6. In *S. cerevisiae*, what ensures that **a** and α cells mate with one another rather than with cells of the same mating type (i.e., **a** with **a** or α with α)?

7. Exposure of C3H 10T1/2 cells to 5-azacytidine, a nucleotide analog, is a model system for muscle differentiation. How was 5-azacytidine treatment used to isolate the genes involved in muscle differentiation?

8. Through the experiments on C3H 10T1/2 cells treated with 5-azacytidine, MyoD was identified as a key transcription factor in regulating the differentiation of muscle. To what general class of DNA-binding proteins does MyoD belong? How do the interactions of MyoD with the following proteins affect its function? (a) E2A, (b) MEFs, (c) Id

9. The mechanisms that regulate muscle differentiation in mammals and neural differentiation in *Drosophila* (and probably mammals as well) bear remarkable similarities. What proteins function analogous to MyoD, myogenin, Id, and E2A in neural cell differentiation in *Drosophila*? Based on these analogies, predict the effect of microinjection of MyoD mRNA on the development of *Xenopus* embryos.

10. Predict the effect of the following mutations on the ability of mother and daughter cells of *S. cerevisae* to undergo mating type switching following cell division:

a. loss-of-function mutation in the HO endonuclease

b. gain-of-function mutation that renders HO endonuclease gene constitutively expressed independent of SWI/SNF

c. gain-of-function mutation in SWI/SNF that renders it insensitive to Ash1

11. Asymmetric cell division often relies on cytoskeletal elements to generate or maintain asymmetric distribution of cellular factors. In *S. cerevisiae*, what factor is localized to the bud by myosin motors? In *Drosophila* neuroblasts, what factors are localized apically by microtubules?

12. How do studies of brain development in knockout mice support the statement that apoptosis is a default pathway in neuronal cells?

13. What morphologic features distinguish programmed cell death and necrotic cell death? TNF and Fas ligand bind cell surface receptors to trigger cell death. Although the death signal is generated external to the cell, why do we consider the death induced by these molecules to be apoptotic rather than necrotic?

14. Predict the effects of the following mutations on the ability of a cell to undergo apoptosis:

a. mutation in Bad such that it cannot be phosphorylated by Akt

b. overexpression of Bcl-2

c. mutation in Bax such that it cannot form homodimers

One common characteristic of cancer cells is a loss of function in the apoptotic pathway. Which of the mutations listed might you expect to find in some cancer cells?

ANALYZE THE DATA

To better understand the potential of adult stem cells to differentiate into various cell types, the following studies were performed.

Bone marrow was harvested from adult male mice and then transplanted into isogeneic female recipient mice that had been treated with a level of irradiation sufficient to destroy their own hematopoetic stem cells (see the figure). Although the dose of irradiation given was lethal to mice that did not receive a transplant, the majority of the recipient mice receiving the transplant survived. After 4 weeks, peripheral blood from the recipient mice was analyzed. The composition of the blood was normal with respect to all blood cell types. Every blood cell examined was determined to be positive for the presence of the Y chromosome.

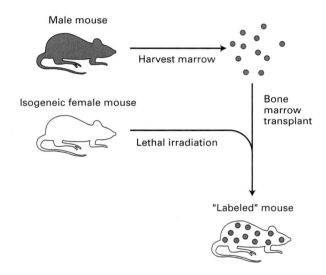

a. What was the purpose of using male mice as the bone marrow donors and female mice as the recipients? Would the converse experiment (female donors, male recipients) have worked?

b. What was the purpose of irradiating the recipient mice prior to transplantation? What outcome might you expect if bone marrow cells were transplanted to nonirradiated mice?

c. What method could have been used to purify hematopoetic stem cells from bone marrow before transplanting these cells into the recipient mouse? How might a purification step have affected the outcome of the experiment?

After 8 weeks, some of the recipient mice were sacrificed and histochemical analysis was performed on tissue sections of various organs. Occasional Y chromosome–positive cells were found in the liver, skeletal muscle, and brain. Furthermore, the Y chromosome–positive cells were positive for markers of liver hepatocytes, skeletal muscle cells, and neurons (albumin, dystrophin, and NeuN, respectively).

d. What does the appearance of Y chromosome–positive cells in multiple organs and the expression of organ-specific markers indicate about the localization of cells derived from bone marrow? About their differentiation potential? About their function?

REFERENCES

The Birth of Cells

Aurelio, O., T. Boulin, and O. Hobert. 2003. Identification of spatial and temporal cues that regulate postembryonic expression of axon maintenance factors in the *C. elegans* ventral nerve cord. *Development* **130**:599–610.

Bach, S. P., A. G. Renehan, and C. S. Potten. 2000. Stem cells: the intestinal stem cell as a paradigm. *Carcinogenesis* **21**:469–476.

Clark, S. E. 2001. Cell signalling at the shoot meristem. *Nature Rev. Mol. Cell Biol.* **2**:276–284.

Edenfeld, G., J. Pielage, and C. Klambt. 2002. Cell lineage specification in the nervous system. *Curr. Opin. Genet. Devel.* **12**:473–477.

Gerlach, M., et al. 2002. Current state of stem cell research for the treatment of Parkinson's disease. *J. Neurol.* **249**(Suppl. 3): III33–III35.

Hori, Y., et al. 2002. Growth inhibitors promote differentiation of insulin-producing tissue from embryonic stem cells. *Proc. Nat'l. Acad. Sci. USA* **99**:16105–16110.

Huelsken, J., et al. 2001. β-Catenin controls hair follicle morphogenesis and stem cell differentiation in the skin. *Cell* **105**:533–545.

Ivanova, N. B., et al. 2002. A stem cell molecular signature. *Science* **298**:601–604.

Lee, R. C., and V. Ambros. 2001. An extensive class of small RNAs in *Caenorhabditis elegans. Science* **294**:862–864.

Lohmann, J. U., et al. 2001. A molecular link between stem cell regulation and floral patterning in *Arabidopsis. Cell* **105**:793–803.

Marshman, E., C. Booth, and C. S. Potten. 2002. The intestinal epithelial stem cell. *Bioessays* **24**:91–98.

Mills, J. C., and J. I. Gordon. 2001. The intestinal stem cell niche: there grows the neighborhood. *Proc. Nat'l. Acad. Sci. USA* **98**:12334–12336.

Niemann, C., and F. M. Watt. 2002. Designer skin: lineage commitment in postnatal epidermis. *Trends Cell Biol.* **12**:185–192.

Okabe, M., et al. 2001. Translational repression determines a neuronal potential in *Drosophila* asymmetric cell division. *Nature* **411**:94–98.

Olsen, P. H., and V. Ambros. 1999. The *lin-4* regulatory RNA controls developmental timing in *Caenorhabditis elegans* by blocking LIN-14 protein synthesis after the initiation of translation. *Devel. Biol.* **216**:671–680.

Orkin, S. H. 2000. Diversification of haematopoietic stem cells to specific lineages. *Nature Rev. Genet.* **1**:57–64.

Pasquinelli, A. E., and G. Ruvkun. 2002. Control of developmental timing by micro RNAs and their targets. *Ann. Rev. Cell Devel. Biol.* **18**:495–513.

Phillips, R. L., et al. 2000. The genetic program of hematopoietic stem cells. *Science* **288**:1635–1640.

Reinhart, B. J., et al. 2000. The 21-nucleotide *let-7* RNA regulates developmental timing in *Caenorhabditis elegans. Nature* **403**:901–906.

Smith, A. G. 2001. Embryo-derived stem cells: of mice and men. *Ann. Rev. Cell Devel. Biol.* **17**:435–462.

Thummel, C. S. 2001. Molecular mechanisms of developmental timing in *C. elegans* and *Drosophila. Devel. Cell* **1**:453–465.

Verfaillie, C. M. 2002. Adult stem cells: assessing the case for pluripotency. *Trends Cell Biol.* **12**:502–508.

Zaret, K. S. 2001. Hepatocyte differentiation: from the endoderm and beyond. *Curr. Opin. Genet. Devel.* **11**:568–574.

Cell-Type Specification in Yeast

Bagnat, M., and K. Simons. 2002. Cell surface polarization during yeast mating. *Proc. Nat'l. Acad. Sci. USA* **99**:14183–14188.

Dittmar, G. A., C. R. Wilkinson, P. T. Jedrzejewski, and D. Finley. 2002. Role of a ubiquitin-like modification in polarized morphogenesis. *Science* **295**:2442–2446.

Dohlman, H. G., and J. W. Thorner. 2001. Regulation of G protein-initiated signal transduction in yeast: paradigms and principles. *Ann. Rev. Biochem.* **70**:703–754.

Hall, I. M., et al. 2002. Establishment and maintenance of a heterochromatin domain. *Science* **297**:2232–2237.

Lau, A., H. Blitzblau, and S. P. Bell. 2002. Cell-cycle control of the establishment of mating-type silencing in *S. cerevisiae. Genes Devel.* **16**:2935–2945.

Miller, M. G., and A. D. Johnson. 2002. White-opaque switching in *Candida albicans* is controlled by mating-type locus homeodomain proteins and allows efficient mating. *Cell* **110**:293–302.

Takizawa, P. A., and R. D. Vale. 2000. The myosin motor, Myo4p, binds Ash1 mRNA via the adapter protein, She3p. *Proc. Nat'l. Acad. Sci. USA* **97**:5273–5278.

Specification and Differentiation of Muscle

Bailey, P., T. Holowacz, and A. B. Lassar. 2001. The origin of skeletal muscle stem cells in the embryo and the adult. *Curr. Opin. Cell Biol.* **13**:679–689.

Buckingham, M. 2001. Skeletal muscle formation in vertebrates. *Curr. Opin. Genet. Devel.* **11**:440–448.

Gustafsson, M. K., et al. 2002. Myf5 is a direct target of long-range Shh signaling and Gli regulation for muscle specification. *Genes Devel.* **16**:114–126.

McKinsey, T. A., C. L. Zhang, and E. N. Olson. 2002. Signaling chromatin to make muscle. *Curr. Opin. Cell Biol.* **14**:763–772.

Perry, R. L., M. H. Parker, and M. A. Rudnicki. 2001. Activated MEK1 binds the nuclear MyoD transcriptional complex to repress transactivation. *Mol. Cell* **8**:291–301.

Puri, P. L., et al. 2001. Class I histone deacetylases sequentially interact with MyoD and pRb during skeletal myogenesis. *Mol. Cell* **8**:885–897.

Wang, D. Z., et al. 2001. The Mef2c gene is a direct transcriptional target of myogenic bHLH and MEF2 proteins during skeletal muscle development. *Development* **128**:4623–4633.

Yan, Z., et al. 2003. Highly coordinated gene regulation in mouse skeletal muscle regeneration. *J. Biol. Chem.* **278**:8826–8836.

Regulation of Asymmetric Cell Division

Adler, P. N., and J. Taylor. 2001. Asymmetric cell division: plane but not simple. *Curr. Biol.* **11**:R233–236.

Ben-Yehuda, S., and R. Losick. 2002. Asymmetric cell division in *B. subtilis* involves a spiral-like intermediate of the cytokinetic protein FtsZ. *Cell* **109**:257–266.

Betschinger, J., K. Mechtler, and J. A. Knoblich. 2003. The Par complex directs asymmetric cell division by phosphorylating the cytoskeletal protein Lgl. *Nature* **422**:326–330.

Bilder, D., M. Li, and N. Perrimon. 2000. Cooperative regulation of cell polarity and growth by *Drosophila* tumor suppressors. *Science* **289**:113–116.

Cayouette, M., and M. Raff. 2002. Asymmetric segregation of Numb: a mechanism for neural specification from *Drosophila* to mammals. *Nature Neurosci.* **5**:1265–1269.

Chia, W., and X. Yang. 2002. Asymmetric division of *Drosophila* neural progenitors. *Curr. Opin. Genet. Devel.* **12**:459–464.

Deng, W., and H. Lin. 2001. Asymmetric germ cell division and oocyte determination during *Drosophila* oogenesis. *Int'l. Rev. Cytol.* **203**:93–138.

Doe, C. Q., and B. Bowerman. 2001. Asymmetric cell division: fly neuroblast meets worm zygote. *Curr. Opin. Cell Biol.* **13**:68–75.

Helariutta, Y., et al. 2000. The SHORT-ROOT gene controls radial patterning of the *Arabidopsis* root through radial signaling. *Cell* **101**:555–567.

Horvitz, H. R., and I. Herskowitz. 1992. Mechanisms of asymmetric cell division: two B's or not two B's, that is the question. *Cell* **68**:237–255.

Knoblich, J. A. 2001. Asymmetric cell division during animal development. *Nature Rev. Mol. Cell Biol.* **2**:11–20.

Lin, D., et al. 2000. A mammalian PAR-3-PAR-6 complex implicated in Cdc42/Rac1 and aPKC signalling and cell polarity. *Nature Cell Biol.* **2**:540–547.

Mahonen, A. P., et al. 2000. A novel two-component hybrid molecule regulates vascular morphogenesis of the *Arabidopsis* root. *Genes Devel.* **14**:2938–2943.

Ohno, S. 2001. Intercellular junctions and cellular polarity: the PAR-aPKC complex, a conserved core cassette playing fundamental roles in cell polarity. *Curr. Opin. Cell Biol.* **13**:641–648.

Petritsch, C., et al. 2003. The *Drosophila* myosin VI Jaguar is required for basal protein targeting and correct spindle orientation in mitotic neuroblasts. *Devel. Cell* **4**:273–281.

Plant, P. J., et al. 2003. A polarity complex of mPar-6 and atypical PKC binds, phosphorylates and regulates mammalian Lgl. *Nature Cell Biol.* **5**:301–308.

Roegiers, F., S. Younger-Shepherd, L. Y. Jan, and Y. N. Jan. 2001. Two types of asymmetric divisions in the *Drosophila* sensory organ precursor cell lineage. *Nature Cell Biol.* **3**:58–67.

Sawa, H., H. Kouike, and H. Okano. 2000. Components of the SWI/SNF complex are required for asymmetric cell division in *C. elegans. Mol. Cell* **6**:617–624.

Shapiro, L., H. H. McAdams, and R. Losick. 2002. Generating and exploiting polarity in bacteria. *Science* **298**:1942–1946.

Cell Death and Its Regulation

Aderem, A. 2002. How to eat something bigger than your head. *Cell* **110**:5–8.

Baehrecke, E. H. 2002. How death shapes life during development. *Nature Rev. Mol. Cell Biol.* **3**:779–787.

Benedict, C. A., P. S. Norris, and C. F. Ware. 2002. To kill or be killed: viral evasion of apoptosis. *Nature Immunol.* **3**:1013–1018.

Bergmann, A. 2002. Survival signaling goes BAD. *Devel. Cell* **3**:607–608.

Cheng, E. H., et al. 2001. BCL-2, BCL-X(L) sequester BH3 domain-only molecules preventing BAX- and BAK-mediated mitochondrial apoptosis. *Mol. Cell* **8**:705–711.

Cory, S., and J. M. Adams. 2002. The Bcl2 family: regulators of the cellular life-or-death switch. *Nature Rev. Cancer* **2**:647–656.

Jacks, T., and R. A. Weinberg. 2002. Taking the study of cancer cell survival to a new dimension. *Cell* **111**:923–925.

Marsden, V. S., and A. Strasser. 2003. Control of apoptosis in the immune system: Bcl-2, BH3-only proteins and more. *Ann. Rev. Immunol.* **21**:71–105.

Martin, S. J. 2002. Destabilizing influences in apoptosis: sowing the seeds of IAP destruction. *Cell* **109**:793–796.

Penninger, J. M., and Kroemer, G. 2003. Mitochondria, AIF, and caspases—rivaling for cell death execution. *Nature Cell Biol.* **5**:97–99.

Ranger, A. M., B. A. Malynn, and S. J. Korsmeyer. 2001. Mouse models of cell death. *Nature Genet.* **28**:113–118.

Vaux, D. L., and S. J. Korsmeyer. 1999. Cell death in development. *Cell* **96**:245–254.

Zuzarte-Luis, V., and J. M. Hurle. 2002. Programmed cell death in the developing limb. *Int'l. J. Devel. Biol.* **46**:871–876.

23

CANCER

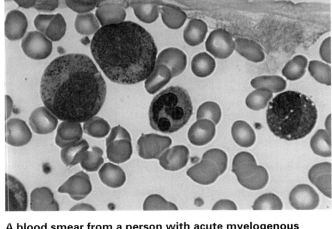

A blood smear from a person with acute myelogenous leukemia. The gigantic cells with irregularly shaped purple nuclei are leukemia cells. The small reddish-gray circular cells are normal red blood cells. [Margaret Cubberly/Phototake.]

Cancer causes about one-fifth of the deaths in the United States each year. Worldwide, between 100 and 350 of each 100,000 people die of cancer each year. Cancer is due to failures of the mechanisms that usually control the growth and proliferation of cells. During normal development and throughout adult life, intricate genetic control systems regulate the balance between cell birth and death in response to growth signals, growth-inhibiting signals, and death signals. Cell birth and death rates determine adult body size, and the rate of growth in reaching that size. In some adult tissues, cell proliferation occurs continuously as a constant tissue-renewal strategy. Intestinal epithelial cells, for instance, live for just a few days before they die and are replaced; certain white blood cells are replaced as rapidly, and skin cells commonly survive for only 2–4 weeks before being shed. The cells in many adult tissues, however, normally do not proliferate except during healing processes. Such stable cells (e.g., hepatocytes, heart muscle cells, neurons) can remain functional for long periods or even the entire lifetime of an organism.

The losses of cellular regulation that give rise to most or all cases of cancer are due to genetic damage (Figure 23-1). Mutations in two broad classes of genes have been implicated in the onset of cancer: **proto-oncogenes** and **tumor-suppressor genes**. Proto-oncogenes are activated to become oncogenes by mutations that cause the gene to be excessively active in growth promotion. Either increased gene expression or production of a hyperactive product will do it. Tumor-suppressor genes normally restrain growth, so damage to them allows inappropriate growth. Many of the genes in both classes encode proteins that help regulate cell birth (i.e., entry into and progression through the cell cycle) or cell death by **apoptosis**; others encode proteins that participate in repairing damaged DNA. Cancer commonly results from mutations that arise during a lifetime's exposure to **carcino-**

gens, which include certain chemicals and ultraviolet radiation. Cancer-causing mutations occur mostly in somatic cells, not in the germ-line cells, and somatic cell mutations are not passed on to the next generation. In contrast, certain inherited mutations, which are carried in the germ line, increase the probability that cancer will occur at some time. In a destructive partnership, somatic mutations can combine with inherited mutations to cause cancer.

Thus the cancer-forming process, called oncogenesis or tumorigenesis, is an interplay between genetics and the environment. Most cancers arise after genes are altered by carcinogens or by errors in the copying and repair of genes. Even if the genetic damage occurs only in one somatic cell, division of this cell will transmit the damage to the daughter cells, giving rise to a **clone** of altered cells. Rarely, however, does mutation in a single gene lead to the onset of cancer. More typically, a series of mutations in multiple genes creates a progressively more rapidly proliferating cell type that escapes normal growth restraints, creating an opportunity for additional mutations. Eventually the clone of cells grows into a **tumor**. In some cases cells from the primary tumor migrate to new sites (*metastasis*), forming secondary tumors that often have the greatest health impact.

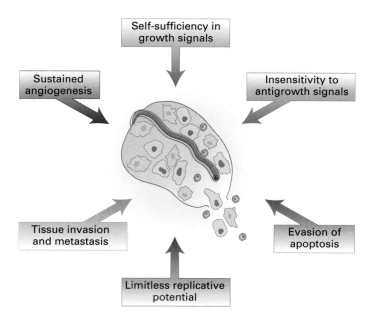

▲ **FIGURE 23-1 Overview of changes in cells that cause cancer.** During carcinogenesis, six fundamental cellular properties are altered, as shown here, to give rise to the complete, most destructive cancer phenotype. Less dangerous tumors arise when only some of these changes occur. In this chapter we examine the genetic changes that result in these altered cellular properties. [Adapted from D. Hanahan and R. A. Weinberg, 2000, *Cell* **100**:57.]

Metastasis is a complex process with many steps. Invasion of new tissues is nonrandom, depending on the nature of both the metastasizing cell and the invaded tissue. Metastasis is facilitated if the tumor cells produce growth and angiogenesis factors (blood vessel growth inducers). Motile, invasive, aggregating, deformable cells are most dangerous. Tissues under attack are most vulnerable if they produce growth factors and readily grow new vasculature. They are more resistant if they produce anti-proliferative factors, inhibitors of proteolytic enzymes, and anti-angiogenesis factors.

Research on the genetic foundations of a particular type of cancer often begins by identifying one or more genes that are mutationally altered in tumor cells. Subsequently it is important to learn whether an altered gene is a contributing cause for the tumor, or an irrelevant side event. Such investigations usually employ multiple approaches: epidemiological comparisons of the frequency with which the genetic change is associated with a type of tumor, tests of the growth properties of cells in culture that have the particular mutation, and the testing of mouse models of the disease to see if the mutation can be causally implicated. A more sophisticated analysis is possible when the altered gene is known to encode a component of a particular molecular pathway (e.g., an intracellular signaling pathway). In this case it is possible to alter other components of the same pathway and see whether the same type of cancer arises.

Because the multiple mutations that lead to formation of a tumor may require many years to accumulate, most cancers develop later in life. The occurrence of cancer after the age of

reproduction may be one reason that evolutionary restraints have not done more to suppress cancer. The requirement for multiple mutations also lowers the frequency of cancer compared with what it would be if tumorigenesis were triggered by a single mutation. However, huge numbers of cells are, in essence, mutagenized and tested for altered growth during our lifetimes, a sort of evolutionary selection for cells that proliferate. Fortunately the tumor itself is not inherited.

23.1 Tumor Cells and the Onset of Cancer

Before examining in detail the genetic basis of cancer, we consider the properties of tumor cells that distinguish them from normal cells and the general process of oncogenesis. The genetic changes that underlie oncogenesis alter several fundamental properties of cells, allowing cells to evade normal growth controls and ultimately conferring the full cancer phenotype (see Figure 23-1). Cancer cells acquire a drive to proliferate that does not require an external inducing signal. They fail to sense signals that restrict cell division and continue to live when they should die. They often change their attachment to surrounding cells or the extracellular matrix, breaking loose to divide more rapidly. A cancer cell may, up to a point, resemble a particular type of normal, rapidly dividing cell, but the cancer cell and its progeny will exhibit inappropriate immortality. To grow to more than a small size, tumors must obtain a blood supply, and they often do so by signaling to induce the growth of blood vessels into the tumor. As cancer progresses, tumors become an abnormal organ, increasingly well adapted to growth and invasion of surrounding tissues.

Metastatic Tumor Cells Are Invasive and Can Spread

Tumors arise with great frequency, especially in older individuals, but most pose little risk to their host because they are localized and of small size. We call such tumors **benign**; an example is warts, a benign skin tumor. The cells composing benign tumors closely resemble, and may function like, normal cells. The cell-adhesion molecules that hold tissues together keep benign tumor cells, like normal cells, localized to the tissues where they originate. A fibrous capsule usually delineates the extent of a benign tumor and makes it an easy target for a surgeon. Benign tumors become serious medical problems only if their sheer bulk interferes with normal functions or if they secrete excess amounts of biologically active substances like hormones. Acromegaly, the overgrowth of head, hands, and feet, for example, can occur when a benign pituitary tumor causes overproduction of growth hormone.

In contrast, cells composing a **malignant** tumor, or **cancer**, usually grow and divide more rapidly than normal, fail to die at the normal rate (e.g., chronic lymphocytic leukemia, a tumor of white blood cells), or invade nearby tissue without a significant change in their proliferation rate (e.g., less

(a)

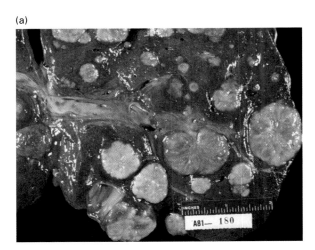

(b)

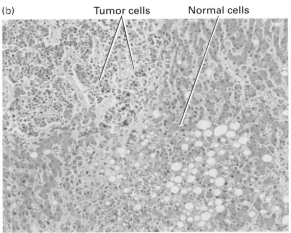

▲ **FIGURE 23-2 Gross and microscopic views of a tumor invading normal liver tissue.** (a) The gross morphology of a human liver in which a metastatic lung tumor is growing. The white protrusions on the surface of the liver are the tumor masses. (b) A light micrograph of a section of the tumor in (a) showing areas of small, dark-staining tumor cells invading a region of larger, light-staining, normal liver cells. [Courtesy of J. Braun.]

harmful tumors of glial cells). Some malignant tumors, such as those in the ovary or breast, remain localized and encapsulated, at least for a time. When these tumors progress, the cells invade surrounding tissues, get into the body's circulatory system, and establish secondary areas of proliferation, a process called **metastasis**. Most malignant cells eventually acquire the ability to metastasize. Thus the major characteristics that differentiate metastatic (or malignant) tumors from benign ones are their invasiveness and spread.

Cancer cells can often be distinguished from normal cells by microscopic examination. They are usually less well differentiated than normal cells or benign tumor cells. In a specific tissue, malignant cells usually exhibit the characteristics of rapidly growing cells, that is, a high nucleus-to-cytoplasm ratio, prominent nucleoli, and relatively little specialized structure. The presence of invading cells in an otherwise normal tissue section is used to diagnose a malignancy (Figure 23-2).

Normal cells are restricted to their place in an organ or tissue by cell-cell adhesion and by physical barriers such as the **basal lamina,** which underlies layers of epithelial cells and also surrounds the endothelial cells of blood vessels (Chapter 6). Cancer cells have a complex relation to the extracellular matrix and basal lamina. The cells must degrade the basal lamina to penetrate it and metastasize, but in some cases cells may migrate along the lamina. Many tumor cells secrete a protein (plasminogen activator) that converts the serum protein plasminogen to the active protease plasmin. Increased plasmin activity promotes metastasis by digesting the basal lamina, thus allowing its penetration by tumor cells. As the basal lamina disintegrates, some tumor cells will enter the blood, but fewer than 1 in 10,000 cells that escape the primary tumor survive to colonize another tissue and form a secondary, metastatic tumor. In addition to escaping the original tumor and entering the blood, cells that will seed new tumors must then adhere to an endothelial cell lining a capillary and migrate across or through it into the underlying tissue. The multiple crossings

of tissue layers that underlie malignancy often involve new or variant surface proteins made by malignant cells.

In addition to important changes in cell-surface proteins, drastic changes occur in the cytoskeleton during tumor-cell formation and metastasis. These alterations can result from changes in the expression of genes encoding Rho and other small GTPases that regulate the actin cytoskeleton (Chapter 19). For instance, tumor cells have been found to over-express the *RhoC* gene, and this increased activity stimulates metastasis.

Cancers Usually Originate in Proliferating Cells

In order for most oncogenic mutations to induce cancer, they must occur in dividing cells so that the mutation is passed on to many progeny cells. When such mutations occur in nondividing cells (e.g., neurons and muscle cells), they generally do not induce cancer, which is why tumors of muscle and nerve cells are rare in adults. Nonetheless, cancer can occur in tissues composed mainly of nondividing differentiated cells such as erythrocytes and most white blood cells, absorptive cells that line the small intestine, and keratinized cells that form the skin. The cells that initiate the tumors are not the differentiated cells, but rather their precursor cells. Fully differentiated cells usually do not divide. As they die or wear out, they are continually replaced by proliferation and differentiation of **stem cells,** and these cells are capable of transforming into tumor cells.

In Chapter 22, we learned that stem cells both perpetuate themselves and give rise to differentiating cells that can regenerate a particular tissue for the life of an organism (see Figure 22-2). For instance, many differentiated blood cells have short life spans and are continually replenished from hematopoietic (blood-forming) stem cells in the bone marrow (see Figure 22-5). Populations of stem cells in the intestine, liver, skin, bone, and other tissues likewise give rise to all or many of the cell types in these tissues, replacing aged and dead cells, by pathways analogous to hematopoiesis in bone marrow. Similarly within a tumor there may be only certain cells with the ability to divide uncontrollably and generate new tumors; such cells are tumor stem cells.

Because stem cells can divide continually over the life of an organism, oncogenic mutations in their DNA can accumulate, eventually transforming them into cancer cells. Cells that have acquired these mutations have an abnormal proliferative capacity and generally cannot undergo normal processes of differentiation. Many oncogenic mutations, such as ones that prevent apoptosis or generate an inappropriate growth-promoting signal, also can occur in more differentiated, but still replicating, progenitor cells. Such mutations in hematopoietic progenitor cells can lead to various types of leukemia.

Normal animal cells are often classified according to their embryonic tissue of origin, and the naming of tumors has followed suit. Malignant tumors are classified as *carcinomas* if they derive from endoderm (gut epithelium) or ectoderm (skin and neural epithelia) and *sarcomas* if they derive from mesoderm (muscle, blood, and connective tissue precursors). The *leukemias*, a class of sarcomas, grow as individual cells in the blood, whereas most other tumors are solid masses. (The name *leukemia* is derived from the Latin for "white blood": the massive proliferation of leukemic cells can cause a patient's blood to appear milky.)

Tumor Growth Requires Formation of New Blood Vessels

Tumors, whether primary or secondary, require recruitment of new blood vessels in order to grow to a large mass. In the absence of a blood supply, a tumor can grow into a mass of about 10^6 cells, roughly a sphere 2 mm in diameter. At this point, division of cells on the outside of the tumor mass is balanced by death of those in the center due to an inadequate supply of nutrients. Such tumors, unless they secrete hormones, cause few problems. However, most tumors induce the formation of new blood vessels that invade the tumor and nourish it, a process called *angiogenesis*. This complex process requires several discrete steps: degradation of the basal lamina that surrounds a nearby capillary, migration of endothelial cells lining the capillary into the tumor, division of these endothelial cells, and formation of a new basement membrane around the newly elongated capillary.

Many tumors produce growth factors that stimulate angiogenesis; other tumors somehow induce surrounding normal cells to synthesize and secrete such factors. Basic fibroblast growth factor (bFGF), transforming growth factor α (TGFα), and vascular endothelial growth factor (VEGF), which are secreted by many tumors, all have angiogenic properties. New blood vessels nourish the growing tumor, allowing it to increase in size and thus increase the probability that additional harmful mutations will occur. The presence of an adjacent blood vessel also facilitates the process of metastasis.

 Several natural proteins that inhibit angiogenesis (e.g., angiogenin and endostatin) or antagonists of the VEGF receptor have excited much interest as potential therapeutic agents. Although new blood vessels are constantly forming during embryonic development, few form normally in adults except after injury. Thus a specific inhibitor of angiogenesis not only might be effective against many kinds of tumors but also might have few adverse side effects. ∎

(a)

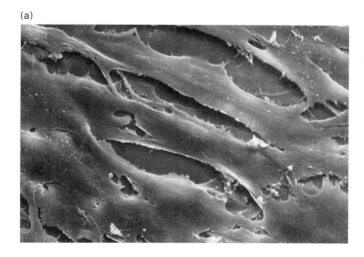

(b)

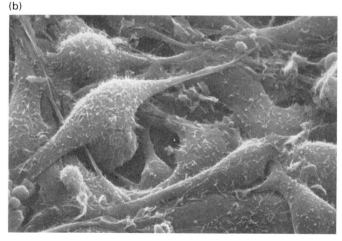

▲ **EXPERIMENTAL FIGURE 23-3 Scanning electron micrographs reveal the organizational and morphological differences between normal and transformed 3T3 cells.** (a) Normal 3T3 cells are elongated and are aligned and closely packed in an orderly fashion. (b) 3T3 cells transformed by an oncogene encoded by Rous sarcoma virus are rounded and covered with small hairlike processes and bulbous projections.

The transformed cells that grow have lost the side-by-side organization of the normal cells and grow one atop the other. These transformed cells have many of the same properties as malignant cells. Similar changes are seen in cells transfected with DNA from human cancers containing the *ras*^D oncogene. [Courtesy of L.-B. Chen.]

Cultured Cells Can Be Transformed into Tumor Cells

The morphology and growth properties of tumor cells clearly differ from those of their normal counterparts; some of these differences are also evident when cells are cultured. That mutations cause these differences was conclusively established by transfection experiments with a line of cultured mouse fibroblasts called 3T3 cells. These cells normally grow only when attached to the plastic surface of a culture dish and are maintained at a low cell density. Because 3T3 cells stop growing when they contact other cells, they eventually form a monolayer of well-ordered cells that have stopped proliferating and are in the quiescent G_0 phase of the cell cycle (Figure 23-3a).

When DNA from human bladder cancer cells is transfected into cultured 3T3 cells, about one cell in a million incorporates a particular segment of the exogenous DNA that causes a distinctive phenotypic change. The progeny of the affected cell are more rounded and less adherent to one another and to the dish than are the normal surrounding cells, forming a three-dimensional cluster of cells (a focus) that can be recognized under the microscope (Figure 23-3b). Such cells, which continue to grow when the normal cells have become quiescent, have undergone oncogenic **transformation**. The transformed cells have properties similar to those of malignant tumor cells, including changes in cell morphology, ability to grow unattached to an extracellular matrix, reduced requirement for growth factors, secretion of plasminogen activator, and loss of actin microfilaments.

Figure 23-4 outlines the procedure for transforming 3T3 cells with DNA from a human bladder cancer and cloning the specific DNA segment that causes transformation. It was remarkable to find a small piece of DNA with this capability; had more than one piece been needed, the experiment would

▶ **EXPERIMENTAL FIGURE 23-4 Transformation of mouse cells with DNA from a human cancer cell permits identification and molecular cloning of the *ras*^D oncogene.** Addition of DNA from a human bladder cancer to a culture of mouse 3T3 cells causes about one cell in a million to divide abnormally and form a focus, or clone, of transformed cells. To clone the oncogene responsible for transformation, advantage is taken of the fact that most human genes have nearby repetitive DNA sequences called *Alu* sequences. DNA from the initial focus of transformed mouse cells is isolated, and the oncogene is separated from adventitious human DNA by secondary transfer to mouse cells. The total DNA from a secondary transfected mouse cell is then cloned into bacteriophage λ; only the phage that receives human DNA hybridizes with an *Alu* probe. The hybridizing phage should contain part of or all the transforming oncogene. This expected result can be proved by showing either that the phage DNA can transform cells (if the oncogene has been completely cloned) or that the cloned piece of DNA is always present in cells transformed by DNA transfer from the original donor cell.

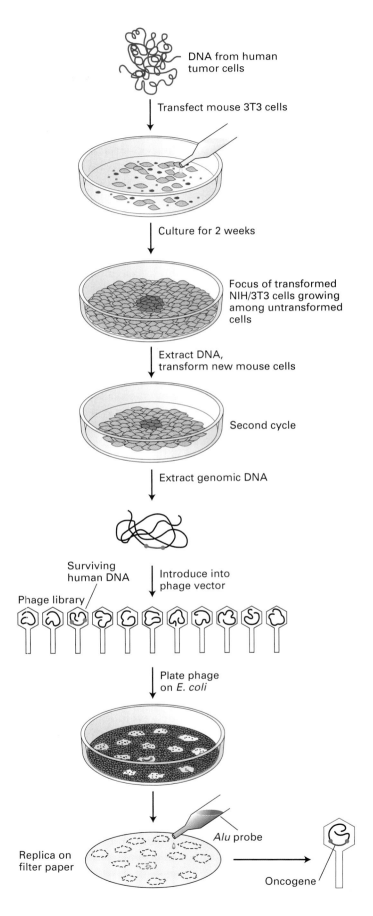

have failed. Subsequent studies showed that the cloned segment included a mutant version of the cellular *ras* gene, designated *ras*^D. Normal **Ras protein**, which participates in many intracellular signal-transduction pathways activated by growth factors, cycles between an inactive, "off" state with bound GDP and an active, "on" state with bound GTP. The mutated Ras^D protein hydrolyzes bound GTP very slowly and therefore accumulates in the active state, sending a growth-promoting signal to the nucleus even in the absence of the hormones normally required to activate its signaling function.

The production and **constitutive** activation of Ras^D protein are not sufficient to cause transformation of normal cells in a primary (fresh) culture of human, rat, or mouse fibroblasts. Unlike cells in a primary culture, however, cultured 3T3 cells have undergone a loss-of-function mutation in the *p16* gene, which encodes a cyclin-kinase inhibitor that restricts progression through the cell cycle. Such cells can grow for an unlimited time in culture if periodically diluted and supplied with nutrients, which normal cells cannot (see Figure 6-37b). These immortal 3T3 cells are transformed into full-blown tumor cells only when they produce a constitutively active Ras protein. For this reason, transfection with the *ras*^D gene can transform 3T3 cells, but not normal cultured primary fibroblast cells, into tumor cells.

A mutant *ras* gene is found in most human colon, bladder, and other cancers, but not in normal human DNA; thus it must arise as the result of a somatic mutation in one of the tumor progenitor cells. Any gene, such as *ras*^D, that encodes a protein capable of transforming cells in culture or inducing cancer in animals is referred to as an **oncogene**. The normal cellular gene from which it arises is called a proto-oncogene. The oncogenes carried by viruses that cause tumors in animals are often derived from proto-oncogenes that were hijacked from the host genome and altered to be oncogenic. When this was first discovered, it was startling to find that these dangerous viruses were turning the animal's own genes against them.

A Multi-hit Model of Cancer Induction Is Supported by Several Lines of Evidence

As noted earlier and illustrated by the oncogenic transformation of 3T3 cells, multiple mutations usually are required to convert a normal body cell into a malignant one. According to this "multi-hit" model, evolutionary (or "survival of the fittest") cancers arise by a process of clonal selection not unlike the selection of individual animals in a large population. A mutation in one cell would give it a slight growth advantage. One of the progeny cells would then undergo a second mutation that would allow its descendants to grow more uncontrollably and form a small benign tumor; a third mutation in a cell within this tumor would allow it to outgrow the others and overcome constraints imposed by the tumor microenvironment, and its

progeny would form a mass of cells, each of which would have these three mutations. An additional mutation in one of these cells would allow its progeny to escape into the blood and establish daughter colonies at other sites, the hallmark of metastatic cancer. This model makes two easily testable predictions.

First, all the cells in a given tumor should contain at least some genetic alterations in common. Systematic analysis of cells from individual human tumors supports the prediction that all the cells are derived from a single progenitor. Recall that during the fetal life of a human female each cell inactivates one of the two X chromosomes. A woman is a genetic mosaic; half the cells have one X inactivated, and the remainder have the other X inactivated. If a tumor did not arise from a single progenitor, it would be composed of a mix of cells with one or the other X inactivated. In fact, the cells from a woman's tumor have the same inactive X chromosome. Different tumors can be composed of cells with either the maternal or the paternal X inactive. Second, cancer incidence should increase with age because it can take decades for the required multiple mutations to occur. Assuming that the rate of mutation is roughly constant during a lifetime, then the incidence of most types of cancer would be independent of age if only one mutation were required to convert a normal cell into a malignant one. As the data in Figure 23-5 show, the incidence of many types of human cancer does indeed increase drastically with age.

More direct evidence that multiple mutations are required for tumor induction comes from transgenic mice

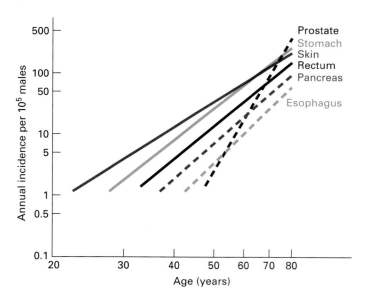

▲ **EXPERIMENTAL FIGURE 23-5 The incidence of human cancers increases as a function of age.** The marked increase in the incidence with age is consistent with the multi-hit model of cancer induction. Note that the logarithm of annual incidence is plotted versus the logarithm of age. [From B. Vogelstein and K. Kinzler, 1993, *Trends Genet.* **9**:101.]

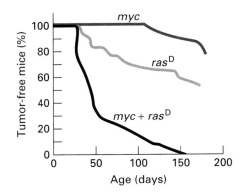

▲ **EXPERIMENTAL FIGURE 23-6 The kinetics of tumor appearance in female mice carrying either one or two oncogenic transgenes shows the cooperative nature of multiple mutations in cancer induction.** The transgenes were driven by the mouse mammary tumor virus (MMTV) breast-specific promoter. The hormonal stimulation associated with pregnancy activated overexpression of the transgenes. The graph shows the time course of tumorigenesis in mice carrying either *myc* or *ras*^D transgenes as well as in the progeny of a cross of *myc* carriers with *ras*^D carriers that contain both transgenes. The results clearly demonstrate the cooperative effects of multiple mutations in cancer induction. [See E. Sinn et al., 1987, *Cell* **49**:465.]

carrying both the mutant *ras*^D oncogene and the c-*myc* proto-oncogene controlled by a mammary cell–specific promoter/enhancer from a retrovirus. When linked to this promoter, the normal c-*myc* gene is overexpressed in breast tissue because the promoter is induced by endogenous hormone levels and tissue-specific regulators. This heightened transcription of c-*myc* mimics oncogenic mutations that turn up c-*myc* transcription, converting the proto-oncogene into an oncogene. By itself, the c-*myc* transgene causes tumors only after 100 days, and then in only a few mice; clearly only a minute fraction of the mammary cells that overproduce the Myc protein become malignant. Similarly, production of the mutant Ras^D protein alone causes tumors earlier but still slowly and with about 50 percent efficiency over 150 days. When the c-*myc* and *ras*^D transgenics are crossed, however, such that all mammary cells produce both Myc and Ras^D, tumors arise much more rapidly and all animals succumb to cancer (Figure 23-6). Such experiments emphasize the synergistic effects of multiple oncogenes. They also suggest that the long latency of tumor formation, even in the double-transgenic mice, is due to the need to acquire additional somatic mutations.

Similar cooperative effects between oncogenes can be seen in cultured cells. Transfection of normal fibroblasts with either c-*myc* or activated *ras*^D is not sufficient for oncogenic transformation, whereas when transfected together, the two genes cooperate to transform the cells. Deregulated levels of c-*myc* alone induce proliferation but also sensitize fibroblasts to apoptosis, and overexpression of activated *ras*^D alone in-

duces senescence. When the two oncogenes are expressed in the same cell, these negative cellular responses are neutralized and the cells undergo transformation.

Successive Oncogenic Mutations Can Be Traced in Colon Cancers

Studies on colon cancer provide the most compelling evidence to date for the multi-hit model of cancer induction. Surgeons can obtain fairly pure samples of many human cancers, but generally the exact stage of tumor progression cannot be identified and analyzed. An exception is colon cancer, which evolves through distinct, well-characterized morphological stages. These intermediate stages—polyps, benign adenomas, and carcinomas—can be isolated by a surgeon, allowing mutations that occur in each of the morphological stages to be identified. Numerous studies show that colon cancer arises from a series of mutations that commonly occur in a well-defined order, providing strong support for the multi-hit model.

Invariably the first step in colon carcinogenesis involves loss of a functional *APC* gene, resulting in formation of polyps (precancerous growths) on the inside of the colon wall. Not every colon cancer, however, acquires all the later mutations or acquires them in the order depicted in Figure 23-7. Thus different combinations of mutations may result in the same phenotype. Most of the cells in a polyp contain the same one or two mutations in the *APC* gene that result in its loss or inactivation; thus they are clones of the cell in which the original mutation occurred. *APC* is a tumor-suppressor gene, and both alleles of the *APC* gene must carry an inactivating mutation for polyps to form because cells with one wild-type *APC* gene express enough APC protein to function normally. Like most tumor-suppressor genes, *APC* encodes a protein that inhibits the progression of certain types of cells through the cell cycle. The APC protein does so by preventing the Wnt signal-transduction pathway from activating expression of proto-oncogenes including the c-*myc* gene. The absence of functional APC protein thus leads to inappropriate production of Myc, a transcription factor that induces expression of many genes required for the transition from the G_1 to the S phase of the cell cycle. Cells homozygous for *APC* mutations proliferate at a rate higher than normal and form polyps.

If one of the cells in a polyp undergoes another mutation, this time an activating mutation in the *ras* gene, its progeny divide in an even more uncontrolled fashion, forming a larger adenoma (see Figure 23-7). Mutational loss of a particular chromosomal region (the relevant gene is not yet known), followed by inactivation of the *p53* gene, results in the gradual loss of normal regulation and the consequent formation of a malignant carcinoma. About half of all human tumors carry mutations in *p53*, which encodes a transcriptional regulator.

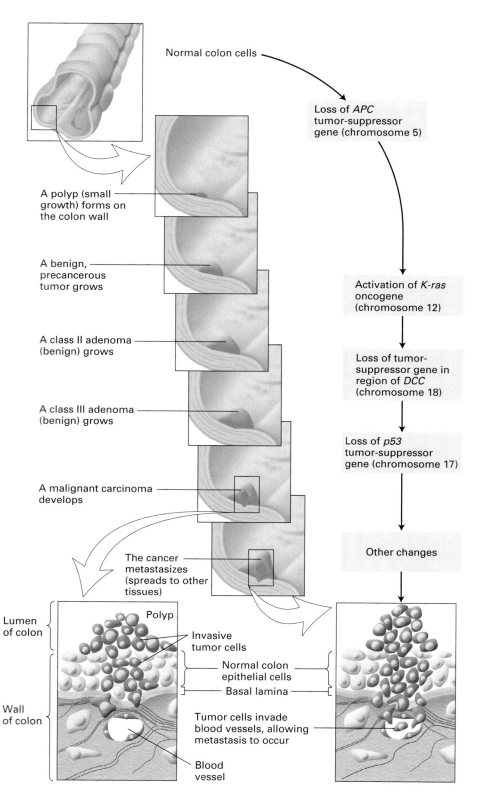

Normal colon cells

A polyp (small growth) forms on the colon wall

A benign, precancerous tumor grows

A class II adenoma (benign) grows

A class III adenoma (benign) grows

A malignant carcinoma develops

The cancer metastasizes (spreads to other tissues)

Loss of *APC* tumor-suppressor gene (chromosome 5)

Activation of *K-ras* oncogene (chromosome 12)

Loss of tumor-suppressor gene in region of *DCC* (chromosome 18)

Loss of *p53* tumor-suppressor gene (chromosome 17)

Other changes

Lumen of colon

Polyp

Invasive tumor cells

Normal colon epithelial cells

Basal lamina

Wall of colon

Tumor cells invade blood vessels, allowing metastasis to occur

Blood vessel

▲ **FIGURE 23-7 The development and metastasis of human colorectal cancer and its genetic basis.** A mutation in the APC tumor-suppressor gene in a single epithelial cell causes the cell to divide, although surrounding cells do not, forming a mass of localized benign tumor cells, or polyp. Subsequent mutations leading to expression of a constitutively active Ras protein and loss of two tumor-suppressor genes—an unidentified gene in the vicinity of *DCC* and *p53*—generate a malignant cell carrying all four mutations. This cell continues to divide, and the progeny invade the basal lamina that surrounds the tissue. Some tumor cells spread into blood vessels that will distribute them to other sites in the body. Additional mutations permit the tumor cells to exit from the blood vessels and proliferate at distant sites; a patient with such a tumor is said to have cancer. [Adapted from B. Vogelstein and K. Kinzler, 1993, *Trends Genet.* **9**:101.]

DNA from different human colon carcinomas generally contains mutations in all these genes—loss-of-function mutations in the tumor suppressors *APC* and *p53*, the as yet mysterious gene, and an activating (gain-of-function) mutation in the dominant oncogene *K-ras*—establishing that multiple mutations in the same cell are needed for the cancer to form. Some of these mutations appear to confer growth advantages at an early stage of tumor development, whereas other mutations promote the later stages, including invasion and metastasis, which are required for the malignant phenotype. The number of mutations needed for colon cancer progression may at first seem surprising, seemingly an effective barrier to tumorigenesis. Our genomes, however, are under constant assault. Recent estimates indicate that sporadically arising polyps have about 11,000 genetic alterations in each cell, though very likely only a few of these are relevant to oncogenesis.

Colon carcinoma provides an excellent example of the multi-hit mode of cancer. The degree to which this model applies to cancer is only now being learned, but it is clear that multiple types of cancer involve multiple mutations. The advent of **DNA microarray** technology is allowing more detailed examination of tumor properties by monitoring the spectrum of mRNA molecules from tens of thousands of genes, and recent data have provided some challenges to the multi-hit model. Not surprisingly, primary tumors can often be distinguishable from metastatic tumors by the pattern of gene expression. More interestingly, a subset of solid primary tumors has been found to have characteristics more typical of metastatic tumors, suggesting that it may be possible to identify primary tumors that have a greater probability of becoming metastatic. This also raises the possibility that, at least for some types of cancer, the initiating events of the primary tumor may set a course toward metastasis. This hypothesis can be distinguished from the emergence of a rare subset of cells within the primary tumor acquiring a necessary series of further mutations.

KEY CONCEPTS OF SECTION 23.1

Tumor Cells and the Onset of Cancer

■ Cancer is a fundamental aberration in cellular behavior, touching on many aspects of molecular cell biology. Most cell types of the body can give rise to malignant tumor (cancer) cells.

■ Cancer cells usually arise from stem cells and other proliferating cells and bear more resemblance to these cells than to more mature differentiated cell types.

■ Cancer cells can multiply in the absence of at least some of the growth-promoting factors required for proliferation of normal cells and are resistant to signals that normally program cell death (apoptosis).

■ Certain cultured cells transfected with tumor-cell DNA undergo transformation (see Figure 23-4). Such transformed cells share certain properties with tumor cells.

■ Cancer cells sometimes invade surrounding tissues, often breaking through the basal laminae that define the boundaries of tissues and spreading through the body to establish secondary areas of growth, a process called metastasis. Metastatic tumors often secrete proteases, which degrade the surrounding extracellular matrix.

■ Both primary and secondary tumors require angiogenesis, the formation of new blood vessels, in order to grow to a large mass.

■ The multi-hit model, which proposes that multiple mutations are needed to cause cancer, is consistent with the genetic homogeneity of cells from a given tumor, the observed increase in the incidence of human cancers with advancing age, and the cooperative effect of oncogenic transgenes on tumor formation in mice.

■ Most oncogenic mutations occur in somatic cells and are not carried in the germ-line DNA.

■ Colon cancer develops through distinct morphological stages that commonly are associated with mutations in specific tumor-suppressor genes and proto-oncogenes (see Figure 23-7).

23.2 The Genetic Basis of Cancer

As we have seen, mutations in two broad classes of genes—proto-oncogenes (e.g., *ras*) and tumor-suppressor genes (e.g., *APC*)—play key roles in cancer induction. These genes encode many kinds of proteins that help control cell growth and proliferation (Figure 23-8). Virtually all human tumors have inactivating mutations in genes that normally act at various cell-cycle **checkpoints** to stop a cell's progress through the cell cycle if a previous step has occurred incorrectly or if DNA has been damaged. For example, most cancers have inactivating mutations in the genes coding for one or more proteins that normally restrict progression through the G_1 stage of the cell cycle. Likewise, a constitutively active Ras or other activated signal-transduction protein is found in several kinds of human tumor that have different origins. Thus malignancy and the intricate processes for controlling the cell cycle discussed in Chapter 21 are two faces of the same coin. In the series of events leading to growth of a tumor, oncogenes combine with tumor-suppressor mutations to give rise to the full spectrum of tumor cell properties described in the previous section (see Figure 23-7).

In this section, we consider the general types of mutations that are oncogenic and see how certain viruses can cause cancer. We also explain why some inherited mutations increase the risk for particular cancers and consider the relation between cancer and developmentally important genes. We conclude this section with a brief discussion of how genomics methods are being used to characterize and classify tumors.

▶ **FIGURE 23-8 Seven types of proteins that participate in controlling cell growth and proliferation.** Cancer can result from expression of mutant forms of these proteins. Mutations changing the structure or expression of proteins that normally promote cell growth generally give rise to dominantly active oncogenes. Many, but not all, extracellular signaling molecules (I), signal receptors (II), signal-transduction proteins (III), and transcription factors (IV) are in this category. Cell-cycle control proteins (VI) that function to restrain cell proliferation and DNA-repair proteins (VII) are encoded by tumor-suppressor genes. Mutations in these genes act recessively, greatly increasing the probability that the mutant cells will become tumor cells or that mutations will occur in other classes. Apoptotic proteins (V) include tumor suppressors that promote apoptosis and oncoproteins that promote cell survival. Virus-encoded proteins that activate signal receptors (Ia) also can induce cancer.

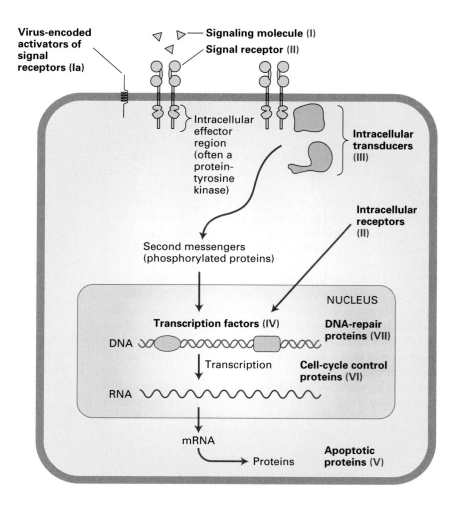

Gain-of-Function Mutations Convert Proto-oncogenes into Oncogenes

Recall that an oncogene is any gene that encodes a protein able to transform cells in culture or to induce cancer in animals. Of the many known oncogenes, all but a few are derived from normal cellular genes (i.e., proto-oncogenes) whose products promote cell proliferation. For example, the *ras* gene discussed previously is a proto-oncogene that encodes an intracellular signal-transduction protein; the mutant *ras*^{D} gene derived from *ras* is an oncogene, whose encoded protein provides an excessive or uncontrolled growth-promoting signal. Other proto-oncogenes encode growth-promoting signal molecules and their receptors, anti-apoptotic (cell-survival) proteins, and some transcription factors.

Conversion, or activation, of a proto-oncogene into an oncogene generally involves a *gain-of-function* mutation. At least four mechanisms can produce oncogenes from the corresponding proto-oncogenes:

- *Point mutation* (i.e., change in a single base pair) in a proto-oncogene that results in a constitutively active protein product

- *Chromosomal translocation* that fuses two genes together to produce a hybrid gene encoding a chimeric protein whose activity, unlike that of the parent proteins, often is constitutive

- *Chromosomal translocation* that brings a growth-regulatory gene under the control of a different promoter that causes inappropriate expression of the gene

- *Amplification* (i.e., abnormal DNA replication) of a DNA segment including a proto-oncogene, so that numerous copies exist, leading to overproduction of the encoded protein

An **oncogene** formed by either of the first two mechanisms encodes an "oncoprotein" that differs from the normal protein encoded by the corresponding proto-oncogene. In contrast, the other two mechanisms generate oncogenes whose protein products are identical with the normal proteins; their oncogenic effect is due to production at higher-than-normal levels or in cells where they normally are not produced.

The localized amplification of DNA to produce as many as 100 copies of a given region (usually a region spanning hundreds of kilobases) is a common genetic change seen in tumors. This anomaly may take either of two forms: the duplicated DNA may be tandemly organized at a single site on a chromosome, or it may exist as small, independent

(a)

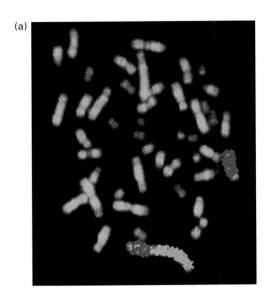

(b)

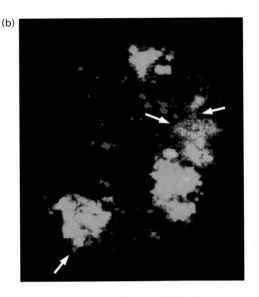

▲ **EXPERIMENTAL FIGURE 23-9 DNA amplifications in stained chromosomes take two forms, visible under the light microscope.** (a) Homogeneously staining regions (HSRs) in a human chromosome from a neuroblastoma cell. The chromosomes are uniformly stained with a blue dye so that all can be seen. Specific DNA sequences were detected using fluorescent in situ hybridization (FISH) in which fluorescently labeled DNA clones are hybridized to denatured DNA in the chromosomes. The chromosome 4 pair is marked (red) by in situ hybridization with a large DNA cosmid clone containing the N-*myc* oncogene. On one of the chromosome 4's an HSR is visible (green) after staining for a sequence enriched in the HSR. (b) Optical sections through nuclei from a human neuroblastoma cell that contain double minute chromosomes. The normal chromosomes are the green and blue structures; the double minute chromosomes are the many small red dots. Arrows indicate double minutes associated with the surface or interior of the normal chromosomes. [Parts (a) and (b) from I. Solovei et al., 2000, *Genes Chromosomes Cancer* **29**:297–308, Figures 4 and 17.]

mini-chromosome-like structures. The former case leads to a homogeneously staining region (HSR) that is visible in the light microscope at the site of the amplification; the latter case causes extra "minute" chromosomes, separate from the normal chromosomes that pepper a stained chromosomal preparation (Figure 23-9).

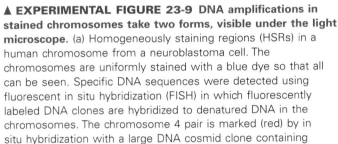 Gene amplification may involve a small number of genes, such as the N-*myc* gene and its neighbor *DDX1* that are amplified in neuroblastoma, or a chromosome region containing many genes. It can be difficult to determine which genes are amplified, a first step in determining which gene caused the tumor. DNA microarrays offer a powerful approach for finding amplified regions of chromosomes. Rather than look at gene expression, the application of microarrays we described earlier, these experiments involve looking for abnormally abundant DNA sequences. Genomic DNA from cancer cells is used to probe arrays containing fragments of genomic DNA and spots with amplified DNA give stronger signals than control spots. Among the amplified genes, the strongest candidates for the relevant ones can be identified by also measuring gene expression. A breast carcinoma cell line, with four known amplified chromosome regions, was screened for amplified genes, and the expression levels of those genes were also studied on microarrays. Fifty genes were found to be amplified, but only five were also highly expressed. These five are new candidates as oncogenes. ∎

However they arise, the gain-of-function mutations that convert proto-oncogenes to oncogenes are genetically **dominant**; that is, mutation in only one of the two alleles is sufficient for induction of cancer.

Cancer-Causing Viruses Contain Oncogenes or Activate Cellular Proto-oncogenes

Pioneering studies by Peyton Rous beginning in 1911 led to the initial recognition that a virus could cause cancer when injected into a suitable host animal. Many years later molecular biologists showed that Rous sarcoma virus (RSV) is a **retrovirus** whose RNA genome is reverse-transcribed into DNA, which is incorporated into the host-cell genome (see Figure 4-43). In addition to the "normal" genes present in all retroviruses, oncogenic transforming viruses like RSV contain the v-*src* gene. Subsequent studies with mutant forms of RSV demonstrated that only the v-*src* gene, not the other viral genes, was required for cancer induction.

In the late 1970s, scientists were surprised to find that normal cells from chickens and other species contain a gene that is closely related to the RSV v-*src* gene. This normal cellular gene, a proto-oncogene, commonly is distinguished from the viral gene by the prefix "c" (c-*src*). RSV and other oncogene-carrying viruses are thought to have arisen by incorporating, or transducing, a normal cellular proto-oncogene into their genome. Subsequent mutation in the transduced gene then converted it into a dominantly acting oncogene, which can induce cell transformation in the presence of the normal c-*src* proto-oncogene. Such viruses are called *transducing retroviruses* because their genomes contain an oncogene derived from a transduced cellular proto-oncogene.

Because its genome carries the potent v-*src* oncogene, the transducing RSV induces tumors within days. In contrast, most oncogenic retroviruses induce cancer only after a period of months or years. The genomes of these *slow-acting retroviruses* differ from those of transducing viruses in one crucial respect: they lack an oncogene. All slow-acting, or "long latency," retroviruses appear to cause cancer by integrating into the host-cell DNA near a cellular proto-oncogene and activating its expression. The long terminal repeat (LTR) sequences in integrated retroviral DNA can act as an enhancer or promoter of a nearby cellular gene, thereby stimulating its transcription. For example, in the cells from tumors caused by avian leukosis virus (ALV), the retroviral DNA is inserted near the c-*myc* gene. These cells overproduce c-Myc protein; as noted earlier, overproduction of c-Myc causes abnormally rapid proliferation of cells. Slow-acting viruses act slowly for two reasons: integration near a cellular proto-oncogene (e.g., c-*myc*) is a random, rare event, and additional mutations have to occur before a full-fledged tumor becomes evident.

In natural bird and mouse populations, slow-acting retroviruses are much more common than oncogene-containing retroviruses such as Rous sarcoma virus. Thus, insertional proto-oncogene activation is probably the major mechanism by which retroviruses cause cancer. Although few human tumors have been associated with any retrovirus, the huge investment in studying retroviruses as a model for human cancer paid off both in the discovery of cellular oncogenes and in the sophisticated understanding of retroviruses, which later accelerated progress on the HIV virus that causes AIDS.

A few DNA viruses also are oncogenic. Unlike most DNA viruses that infect animal cells, oncogenic DNA viruses integrate their viral DNA into the host-cell genome. The viral DNA contains one or more oncogenes, which permanently transform infected cells. For example, many warts and other benign tumors of epithelial cells are caused by the DNA-containing papillomaviruses. Unlike retroviral oncogenes, which are derived from normal cellular genes and have no function for the virus except to allow their proliferation in tumors, the known oncogenes of DNA viruses are integral parts of the viral genome and are required for viral replication. As discussed later, the oncoproteins expressed from integrated viral DNA in infected cells act in various ways to stimulate cell growth and proliferation.

Loss-of-Function Mutations in Tumor-Suppressor Genes Are Oncogenic

Tumor-suppressor genes generally encode proteins that in one way or another inhibit cell proliferation. *Loss-of-function* mutations in one or more of these "brakes" contribute to the development of many cancers. Five broad classes of proteins are generally recognized as being encoded by tumor-suppressor genes:

- Intracellular proteins that regulate or inhibit progression through a specific stage of the cell cycle (e.g., p16 and Rb)
- Receptors or signal transducers for secreted hormones or developmental signals that inhibit cell proliferation (e.g., TGFβ, the hedgehog receptor patched)
- Checkpoint-control proteins that arrest the cell cycle if DNA is damaged or chromosomes are abnormal (e.g., p53)
- Proteins that promote apoptosis
- Enzymes that participate in DNA repair

Although DNA-repair enzymes do not directly inhibit cell proliferation, cells that have lost the ability to repair errors, gaps, or broken ends in DNA accumulate mutations in many genes, including those that are critical in controlling cell growth and proliferation. Thus loss-of-function mutations in the genes encoding DNA-repair enzymes prevent cells from correcting mutations that inactivate tumor-suppressor genes or activate oncogenes.

Since generally one copy of a tumor-suppressor gene suffices to control cell proliferation, both alleles of a tumor-suppressor gene must be lost or inactivated in order to promote tumor development. Thus oncogenic loss-of-function mutations in tumor-suppressor genes are genetically **recessive**. In many cancers, tumor-suppressor genes have deletions or point mutations that prevent production of any protein or lead to production of a nonfunctional protein. Another mechanism for inactivating tumor-suppressor genes is methylation of cytosine residues in the promoter or other control elements. Such methylation is commonly found in nontranscribed regions of DNA.

Inherited Mutations in Tumor-Suppressor Genes Increase Cancer Risk

Individuals with inherited mutations in tumor-suppressor genes have a hereditary predisposition for certain cancers. Such individuals generally inherit a germ-line mutation in one allele of the gene; somatic mutation of the second allele facilitates tumor progression. A classic case is retinoblastoma, which is caused by loss of function of *RB*, the first tumor-suppressor gene to be identified. As we discuss later, the protein encoded by *RB* helps regulate progress through the cell cycle.

Hereditary versus Sporadic Retinoblastoma Children with hereditary retinoblastoma inherit a single defective copy of the *RB* gene, sometimes seen as a small deletion on one of the copies of chromosome 13. The children develop retinal tumors early in life and generally in both eyes. One essential event in tumor development is the deletion or

(a) Hereditary retinoblastoma

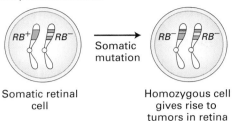

(b) Sporadic retinoblastoma

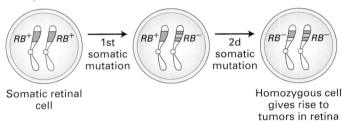

▲ **FIGURE 23-10 Role of spontaneous somatic mutation in retinoblastoma.** This disease is marked by retinal tumors that arise from cells carrying two mutant *RB*⁻ alleles. (a) In hereditary (familial) retinoblastoma, a child inherits a normal *RB*⁺ allele from one parent and a mutant *RB*⁻ allele from the other parent. A single mutation in a heterozygous somatic retinal cell that inactivates the normal allele will produce a cell homozygous for two mutant alleles. (b) In sporadic retinoblastoma, a child inherits two normal *RB*⁺ alleles. Two separate somatic mutations in a particular retinal cell or its progeny are required to produce a homozygous *RB*⁻/*RB*⁻ cell.

mutation of the normal *RB* gene on the other chromosome, giving rise to a cell that produces no functional Rb protein (Figure 23-10). Individuals with sporadic retinoblastoma, in contrast, inherit two normal *RB* alleles, each of which has undergone a loss-of-function somatic mutation in a single retinal cell. Because losing two copies of the *RB* gene is far less likely than losing one, sporadic retinoblastoma is rare, develops late in life, and usually affects only one eye.

If retinal tumors are removed before they become malignant, children with hereditary retinoblastoma often survive until adulthood and produce children. Because their germ cells contain one normal and one mutant *RB* allele, these individuals will, on average, pass on the mutant allele to half their children and the normal allele to the other half. Children who inherit the normal allele are normal if their other parent has two normal *RB* alleles. However, those who inherit the mutant allele have the same enhanced predisposition to develop retinal tumors as their affected parent, even though they inherit a normal *RB* allele from their other, normal parent. Thus the tendency to develop retinoblastoma is inherited as a dominant trait. As discussed below, many human tumors (not just retinal tumors) contain mutant *RB* alleles; most of these arise as the result of somatic mutations.

Inherited Forms of Colon and Breast Cancer Similar hereditary predisposition for other cancers has been associated with inherited mutations in other tumor-suppressor genes. For example, individuals who inherit a germ-line mutation in one *APC* allele develop thousands of precancerous intestinal polyps (see Figure 23-7). Since there is a high probability that one or more of these polyps will progress to malignancy, such individuals have a greatly increased risk for developing colon cancer before the age of 50. Likewise, women who inherit one mutant allele of *BRCA1*, another tumor-suppressor gene, have a 60 percent probability of developing breast cancer by age 50, whereas those who inherit two normal *BRCA1* alleles have a 2 percent probability of doing so. In women with hereditary breast cancer, loss of the second *BRCA1* allele, together with other mutations, is required for a normal breast duct cell to become malignant. However, *BRCA1* generally is not mutated in sporadic, noninherited breast cancer.

Loss of Heterozygosity Clearly, then, we can inherit a propensity to cancer by receiving a damaged allele of a tumor-suppressor gene from one of our parents; that is, we are heterozygous for the mutation. That in itself will not cause cancer, since the remaining normal allele prevents aberrant growth; the cancer is recessive. Subsequent loss or inactivation of the normal allele in a somatic cell, referred to as *loss of heterozygosity (LOH)*, is a prerequisite for cancer to develop. One common mechanism for LOH involves mis-segregation during mitosis of the chromosomes bearing the affected tumor-suppressor gene (Figure 23-11a). This process, also referred to as *nondisjunction,* is caused by failure of the spindle-assembly checkpoint, which normally prevents a metaphase cell with an abnormal mitotic spindle from completing mitosis (see Figure 21-32, ②). Another possible mechanism for LOH is mitotic recombination between a chromatid bearing the wild-type allele and a homologous chromatid bearing a mutant allele. As illustrated in Figure 23-11b, subsequent chromosome segregation can generate a daughter cell that is homozygous for the mutant tumor-suppressor allele. A third mechanism is the deletion or mutation of the normal copy of the tumor-suppressor gene; such a deletion can encompass a large chromosomal region and need not be a precise deletion of just the tumor-suppressor gene.

Hereditary cancers constitute about 10 percent of human cancers. It is important to remember, however, that the inherited, germ-line mutation alone is not sufficient to cause tumor development. In all cases, not only must the inherited normal tumor-suppressor allele be lost or inactivated, but mutations affecting other genes also are necessary for cancer to develop. Thus a person with a recessive tumor-suppressor-gene mutation can be exceptionally susceptible to environmental mutagens such as radiation.

(a) Mis-segregation

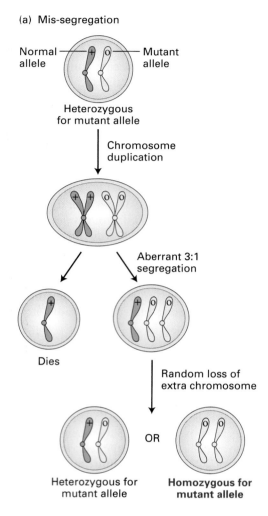

(b) Mitotic recombination

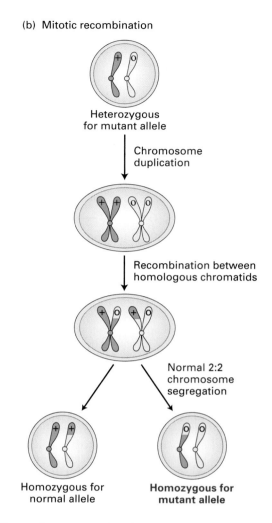

▲ **FIGURE 23-11 Two mechanisms for loss of heterozygosity (LOH) of tumor-suppressor genes.** A cell containing one normal and one mutant allele of a tumor-suppressor gene is generally phenotypically normal. (a) If formation of the mitotic spindle is defective, then the duplicated chromosomes bearing the normal and mutant alleles may segregate in an aberrant 3:1 ratio. A daughter cell that receives three chromosomes of a type will generally lose one, restoring the normal 2n chromosome number. Sometimes the resultant cell will contain one normal and one mutant allele, but sometimes it will be homozygous for the mutant allele. Note that such aneuploidy (abnormal chromosome constitution) is generally damaging or lethal to cells that have to develop into the many complex structures of an organism, but can often be tolerated in clones of cells that have limited fates and duties. (b) Mitotic recombination between a chromosome with a wild-type and a mutant allele, followed by chromosome segregation, can produce a cell that contains two copies of the mutant allele.

Aberrations in Signaling Pathways That Control Development Are Associated with Many Cancers

During normal development secreted signals such as Wnt, TGFβ, and Hedgehog (Hh) are frequently used to direct cells to particular developmental fates, which may include the property of rapid mitosis. The effects of such signals must be regulated so that growth is limited to the right time and place. Among the mechanisms available for reining in the effects of powerful developmental signals are inducible intracellular antagonists, receptor blockers, and competing signals (Chapter 15). Mutations that prevent such restraining mechanisms from operating are likely to be oncogenic, causing inappropriate or cancerous growth.

Hh signaling, which is used repeatedly during development to control cell fates, is a good example of a signaling pathway implicated in cancer induction. In the skin and cerebellum one of the human Hh proteins, Sonic hedgehog, stimulates cell division by binding to and inactivating a membrane protein called Patched1 (Ptc1) (see Figure 15-31). Loss-of-function mutations in *ptc1* permit cell proliferation in the absence of an Hh signal; thus *ptc1* is a tumor-suppressor gene. Not surprisingly, mutations in *ptc1* have been found in tumors of the skin and cerebellum in mice and

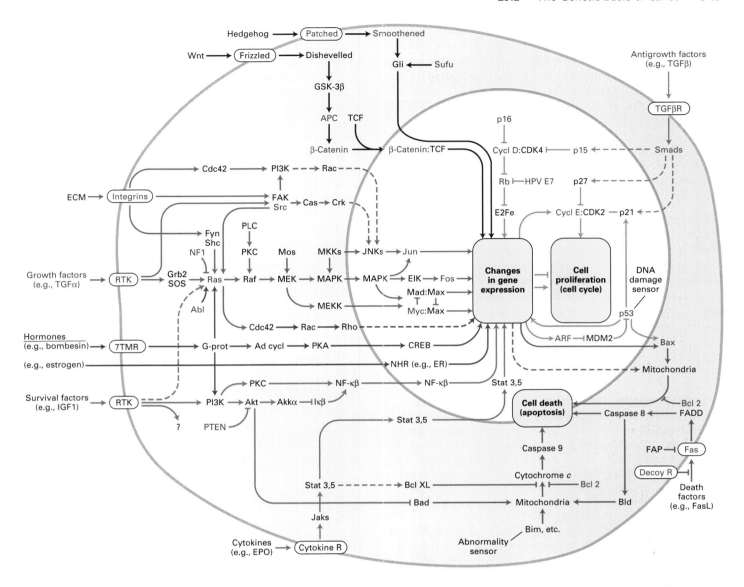

▲ FIGURE 23-12 Cell circuitry that is affected by cancer-causing mutations. Growth control and the cell cycle, the heart of cancer, are influenced by many types of signals, and the external inputs become integrated as the cell makes its decision about whether to divide or to continue dividing.

Developmental pathways give cells their identity, and along with that identity often comes a commitment to proliferate or not. Genes known to be mutated in cancer cells are highlighted in red. Less firmly established pathways are shown with dashed lines. [From D. Hanahan and R. A. Weinberg, 2000, *Cell* **100**:57.]

humans. Mutations in other genes in the Hh signaling pathway are also associated with cancer. Some such mutations create oncogenes that turn on Hh target genes inappropriately; others are recessive mutations that affect negative regulators like Ptc1. As is the case for a number of other tumor-suppressor genes, complete loss of Ptc1 function would lead to early fetal death, since it is needed for development, so it is only the tumor cells that are homozygous *ptc1/ptc1*.

Many of the signaling pathways described in other chapters play roles in controlling embryonic development and cell proliferation in adult tissues. In recent years mutations af-

fecting components of most of these signaling pathways have been linked to cancer (Figure 23-12). Indeed, once one gene in a developmental pathway has been linked to a type of human cancer, knowledge of the pathway gleaned from model organisms like worms, flies, or mice allows focused investigations of the possible involvement of additional pathway genes in other cases of the cancer. For example, *APC*, the critical first gene mutated on the path to colon carcinoma, is now known to be part of the Wnt signaling pathway, which led to the discovery of the involvement of *β-catenin* mutations in colon cancer. Mutations in tumor-suppressor developmental genes promote tumor formation in tissues where the affected gene normally helps restrain growth, and not in cells where the primary role of the developmental regulator is to control cell fate but not growth. Mutations in developmental proto-oncogenes may induce tumor formation in tissues where an affected gene normally promotes growth or in another tissue where the gene has become aberrantly active.

DNA Microarray Analysis of Expression Patterns Can Reveal Subtle Differences Between Tumor Cells

Traditionally the properties of tumor and normal cells have been assessed by staining and microscopy. The prognosis for many tumors could be determined, within certain limits, from their histology. However the appearance of cells alone has limited information content, and better ways to discern the properties of cells are desirable both to understand tumorigenesis and to arrive at meaningful and accurate decisions about prognosis and therapy.

As we've seen, genetic studies can identify the single initiating mutation or series of mutations that cause transforma-

tion of normal cells into tumor cells, as in the case of colon cancer. After these initial events, however, the cells of a tumor undergo a cascade of changes reflecting the interplay between the initiating events and signals from outside. As a result, tumor cells can become quite different, even if they arise from the same initiating mutation or mutations. Although these differences may not be recognized from the appearance of cells, they can be detected from the cells' patterns of gene expression. DNA microarray analysis can determine the expression of thousands of genes simultaneously, permitting complex phenotypes to be defined at the molecular genetic level. (See Figures 9-35 and 9-36 for an explanation of this technique.)

Microarray analysis recently has been applied to diffuse large B cell lymphoma, a disease marked by the presence of abnormally large B lymphocytes throughout lymph nodes. Affected patients have highly variable outcomes, so the disease has long been suspected to be,

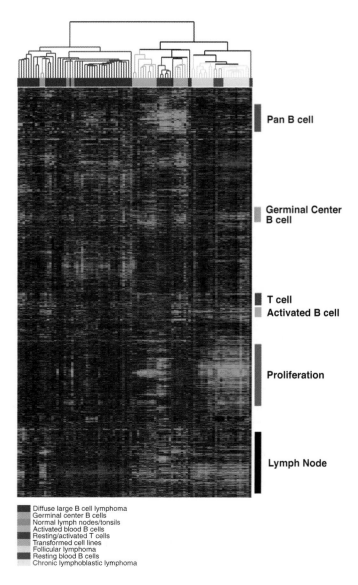

▶ **EXPERIMENTAL FIGURE 23-13 Differences in gene expression patterns determined by DNA microarray analysis can distinguish between otherwise phenotypically similar lymphomas.** Samples of mRNA were extracted from normal lymphocytes at different stages of differentiation and from malignant lymphocytes obtained from patients with three types of lymphoma. DNA microarray analysis of the extracted RNA determined the transcription of about 18,000 genes by each of the 96 experimental samples of normal and malignant lymphocytes relative to a control, reference sample. (See Figures 9-35 and 9-36 for description of microarray analysis.) The cluster diagram shown here includes the data from a selected set of genes whose expression differs the most in the various lymphocyte samples. An intense red color indicates that the experimental cells transcribe the gene represented by a particular DNA spot at a much higher level than the reference cells; intense green indicates the opposite. Black indicates similar transcript levels in the samples compared; gray indicates missing or excluded data. Each vertical column contains the data for a particular lymphocyte sample (see sample key). Each horizontal row contains data for a single gene. The genes were grouped according to their similar patterns of hybridization. For example, genes indicated by the green bar on the right are active in proliferating cells, such as transformed cultured cells (pink bar at top) or lymphoma cells (purple bar at top). The different cell samples (along the top of the diagram) also were grouped according to their similar expression patterns. The resulting dendrogram (tree diagram) shows that the samples from patients with diffuse large B cell lymphoma (purple samples) fall into two groups. One group is similar to relatively undifferentiated B lymphocytes in germinal centers (orange samples); the other is similar to more differentiated B cells (light-purple samples). [From A. A. Alizadeh et al., 2000, *Nature* **403**:505.]

in fact, multiple diseases. Microarray analysis of lymphomas from different patients revealed two groups distinguished by their patterns of gene expression (Figure 23-13). No morphological or visible criteria were found that could distinguish the two types of tumors. Patients with one tumor type defined by the microarray data survived much longer than those with the other type. Lymphomas whose gene expression is similar to that of B lymphocytes in the earliest stages of differentiation have a better prognosis; lymphomas whose gene expression is closer to that of more differentiated B lymphocytes have a worse prognosis. Similar analyses of the gene expression patterns, or "signatures," of other tumors are likely to improve classification and diagnosis, allowing informed decisions about treatments, and also provide insights into the properties of tumor cells. ∎

KEY CONCEPTS OF SECTION 23.2

The Genetic Basis of Cancer

■ Dominant gain-of-function mutations in proto-oncogenes and recessive loss-of-function mutations in tumor-suppressor genes are oncogenic.

■ Among the proteins encoded by proto-oncogenes are growth-promoting signaling proteins and their receptors, signal-transduction proteins, transcription factors, and apoptotic proteins (see Figure 23-8).

■ An activating mutation of one of the two alleles of a proto-oncogene converts it to an oncogene. This can occur by point mutation, gene amplification, and gene translocation.

■ The first human oncogene to be identified encodes a constitutively active form of Ras, a signal-transduction protein. This oncogene was isolated from a human bladder carcinoma (see Figure 23-4).

■ Slow-acting retroviruses can cause cancer by integrating near a proto-oncogene in such a way that transcription of the cellular gene is activated continuously and inappropriately.

■ Tumor-suppressor genes encode proteins that directly or indirectly slow progression through the cell cycle, checkpoint-control proteins that arrest the cell cycle, components of growth-inhibiting signaling pathways, pro-apoptotic proteins, and DNA-repair enzymes.

■ The first tumor-suppressor gene to be recognized, *RB*, is mutated in retinoblastoma and some other tumors.

■ Inheritance of a single mutant allele of *RB* greatly increases the probability that a specific kind of cancer will develop, as is the case for many other tumor-suppressor genes (e.g., *APC* and *BRCA1*).

■ In individuals born heterozygous for a tumor-suppressor gene, a somatic cell can undergo loss of heterozygosity (LOH) by mitotic recombination, chromosome missegregation, mutation, or deletion (see Figure 23-11).

■ Many genes that regulate normal developmental processes encode proteins that function in various signaling pathways (see Figure 23-12). Their normal roles in regulating where and when growth occurs are reflected in the character of the tumors that arise when the genes are mutated.

■ DNA microarray analysis can identify differences in gene expression between types of tumor cells that are indistinguishable by traditional criteria. Some tumor cells appear to be related to specific types of normal cells at certain stages of development based on similarities in their expression patterns.

23.3 Oncogenic Mutations in Growth-Promoting Proteins

Genes encoding each class of cell regulatory protein depicted in Figure 23-8 have been identified as proto-oncogenes or tumor-suppressor genes. In this section we examine in more detail how mutations that result in the unregulated, constitutive activity of certain proteins or in their overproduction promote cell proliferation and transformation, thereby contributing to carcinogenesis. In each case we see how a rare cell that has undergone a very particular sort of mutation becomes abundant owing to its uncontrolled proliferation.

Oncogenic Receptors Can Promote Proliferation in the Absence of External Growth Factors

Although oncogenes theoretically could arise from mutations in genes encoding growth-promoting signaling molecules, this rarely occurs. In fact, only one such naturally occurring oncogene, *sis*, has been discovered. The *sis* oncogene, which encodes a type of platelet-derived growth factor (PDGF), can aberrantly autostimulate proliferation of cells that normally express the PDGF receptor.

In contrast, oncogenes encoding cell-surface receptors that transduce growth-promoting signals have been associated with several types of cancer. The receptors for many such growth factors have intrinsic protein-tyrosine kinase activity in their cytosolic domains, an activity that is quiescent until activated. Ligand binding to the external domains of these **receptor tyrosine kinases (RTKs)** leads to their dimerization and activation of their kinase activity, initiating an intracellular signaling pathway that ultimately promotes proliferation.

In some cases, a point mutation changes a normal RTK into one that dimerizes and is constitutively active in the absence of ligand. For instance, a single point mutation converts the normal Her2 receptor into the Neu oncoprotein, which is an initiator of certain mouse cancers (Figure 23-14, *left*). Similarly, human tumors called multiple endocrine neoplasia type 2 produce a constitutively active dimeric Glia-derived neurotrophic factor (GDNF) receptor that results

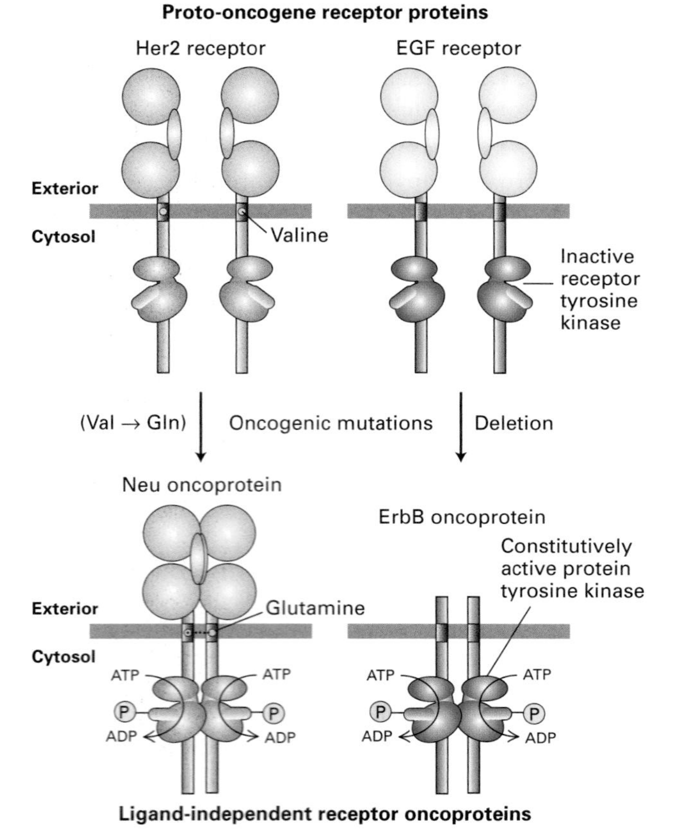

Proto-oncogene receptor proteins

Her2 receptor   EGF receptor

Exterior

Cytosol   Valine

Inactive
receptor
tyrosine
kinase

(Val → Gln)   Oncogenic mutations   Deletion

Neu oncoprotein

ErbB oncoprotein

Constitutively
active protein
tyrosine kinase

Exterior   Glutamine

Cytosol

ATP   ATP   ATP   ATP
P   P   P   P
ADP   ADP   ADP   ADP

Ligand-independent receptor oncoproteins

▲ **FIGURE 23-14 Effects of oncogenic mutations in proto-oncogenes that encode cell-surface receptors.** *Left:* A mutation that alters a single amino acid (valine to glutamine) in the transmembrane region of the Her2 receptor causes dimerization of the receptor, even in the absence of the normal EGF-related ligand, making the oncoprotein Neu a constitutively active kinase. *Right:* A deletion that causes loss of the extracellular ligand-binding domain in the EGF receptor leads, for unknown reasons, to constitutive activation of the kinase activity of the resulting oncoprotein ErbB.

from a point mutation in the extracellular domain. In other cases, deletion of much of the extracellular ligand-binding domain produces a constitutively active oncogenic receptor. For example, deletion of the extracellular domain of the normal EGF receptor converts it to the dimeric ErbB oncoprotein (Figure 23-14, *right*).

Mutations leading to overproduction of a normal RTK also can be oncogenic. For instance, many human breast cancers overproduce a normal Her2 receptor. As a result, the cells are stimulated to proliferate in the presence of very low concentrations of EGF and related hormones, concentrations too low to stimulate proliferation of normal cells.

 A monoclonal antibody specific for Her2 has been a strikingly successful new treatment for the subset of breast cancers that overproduce Her2. Her2 antibody injected into the blood recognizes Her2 and causes it to be internalized, selectively killing the cancer cells without any apparent effect on normal breast (and other) cells that produce moderate amounts of Her2. ∎

Another mechanism for generating an oncogenic receptor is illustrated by the human *trk* oncogene, which was isolated from a colon carcinoma. This oncogene encodes a chimeric protein as the result of a chromosomal translocation that replaced the sequences encoding most of the extracellular domain of the normal Trk receptor with the sequences encoding the N-terminal amino acids of nonmuscle tropomyosin (Figure 23-15). The translocated tropomyosin segment can mediate dimerization of the chimeric Trk receptor by forming a coiled-coil structure, leading to activation of the kinase domains in the absence of ligand. The normal Trk protein is a cell-surface receptor tyrosine kinase that binds a nerve growth factor (Chapter 22). In contrast, the constitutively active Trk oncoprotein is localized in the cytosol, since the N-terminal signal sequence directing it to the membrane has been deleted.

▶ **FIGURE 23-15 Domain structures of normal tropomyosin, the normal Trk receptor, and chimeric Trk oncoprotein.** A chromosomal translocation results in replacement of most of the extracellular domain of the normal human Trk protein, a receptor tyrosine kinase, with the N-terminal domain of nonmuscle tropomyosin. Dimerized by the tropomyosin segment, the Trk oncoprotein kinase is constitutively active. Unlike the normal Trk, which is localized to the plasma membrane, the Trk oncoprotein is found in the cytosol. [See F. Coulier et al., 1989, *Mol. Cell Biol.* **9**:15.]

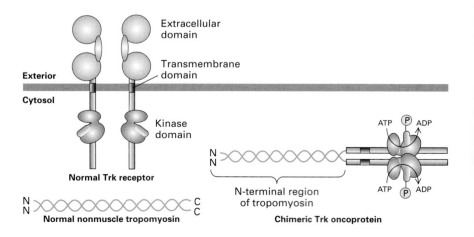

Extracellular
domain

Transmembrane
domain

Exterior

Cytosol

Kinase
domain

Normal Trk receptor

Normal nonmuscle tropomyosin

N-terminal region
of tropomyosin

Chimeric Trk oncoprotein

ATP P ADP

ATP P ADP

Viral Activators of Growth-Factor Receptors Act as Oncoproteins

Viruses use their own tricks to cause cancer, presumably to increase the production of virus from the infected cancer cells. For example, a retrovirus called spleen focus-forming virus (SFFV) induces erythroleukemia (a tumor of erythroid progenitors) in adult mice by manipulating a normal developmental signal. The proliferation, survival, and differentiation of erythroid progenitors into mature red cells absolutely require erythropoietin (Epo) and the corresponding Epo receptor (see Figure 14-7). A mutant SFFV envelope glycoprotein, termed gp55, is responsible for the oncogenic effect of the virus. Although gp55 cannot function as a normal retrovirus envelope protein in virus budding and infection, it has acquired the remarkable ability to bind to and activate Epo receptors in the same cell (Figure 23-16). By inappropriately and continuously stimulating the proliferation of erythroid progenitors, gp55 induces formation of excessive numbers of erythrocytes. Malignant clones of erythroid progenitors emerge several weeks after SFFV infection as a result of further mutations in these aberrantly proliferating cells.

Another example of this phenomenon is provided by human papillomavirus (HPV), a DNA virus that causes genital warts. A papillomavirus protein designated E5, which contains only 44 amino acids, spans the plasma membrane and forms a dimer or trimer. Each E5 polypeptide can form a stable complex with one endogenous receptor for PDGF, thereby aggregating two or more PDGF receptors within the plane of the plasma membrane. This mimics hormone-mediated receptor dimerization, causing sustained receptor activation and eventually cell transformation.

Many Oncogenes Encode Constitutively Active Signal-Transduction Proteins

A large number of oncogenes are derived from proto-oncogenes whose encoded proteins aid in transducing signals from an activated receptor to a cellular target. We describe several examples of such oncogenes; each is expressed in many types of tumor cells.

Ras Pathway Components Among the best-studied oncogenes in this category are the ras^D genes, which were the first nonviral oncogenes to be recognized. A point mutation that substitutes any amino acid for glycine at position 12 in the Ras sequence can convert the normal protein into a constitutively active oncoprotein. This simple mutation reduces the protein's GTPase activity, thus maintaining Ras in the active GTP-bound state. Constitutively active Ras oncoproteins are produced by many types of human tumors, including bladder, colon, mammary, skin, and lung carcinomas, neuroblastomas, and leukemias.

As we saw in Chapter 14, Ras is a key component in transducing signals from activated receptors to a cascade of protein kinases. In the first part of this pathway, a signal from an activated RTK is carried via two adapter proteins to Ras, converting it to the active GTP-bound form (see Figure 14-16). In the second part of the pathway, activated Ras transmits the signal via two intermediate protein kinases to MAP kinase. The activated MAP kinase then phosphorylates a number of transcription factors that induce synthesis of important cell-cycle and differentiation-specific proteins (see Figure 14-21). Activating Ras mutations short-circuit the first part of this pathway, making upstream activation triggered by ligand binding to the receptor unnecessary.

Oncogenes encoding other altered components of the RTK-Ras–MAP kinase pathway also have been identified. One example, found in certain transforming mouse retroviruses, encodes a constitutively activated Raf serine/threonine kinase, which is in the pathway between Ras and MAP kinase. Another is the crk (pronounced "crack") oncogene found in avian sarcoma virus, which causes certain tumors when overexpressed. The Crk protein, which contains one SH2 and two SH3 domains, is similar to the GRB2 adapter protein that functions between an RTK and Ras (see Figure 14-16). The SH2 and SH3 domains in GRB2 and other adapter proteins mediate formation of specific protein aggregates that normally serve as signaling units for cellular events. Overproduction of Crk leads to formation of protein aggregates that inappropriately transduce signals, thus

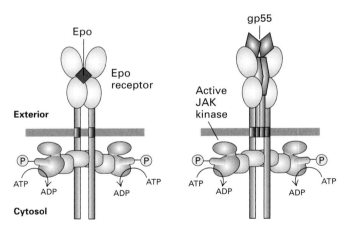

▲ **FIGURE 23-16 Activation of the erythropoietin (Epo) receptor by the natural ligand, Epo, or a viral oncoprotein.** Binding of Epo dimerizes the receptor and induces formation of erythrocytes from erythroid progenitor cells. Normally cancers occur when progenitor cells infected by the spleen focus-forming virus produce the Epo receptor and viral gp55, both localized to the plasma membrane. The transmembrane domains of dimeric gp55 specifically bind the Epo receptor, dimerizing and activating the receptor in the absence of Epo. [See S. N. Constantinescu et al., 1999, *EMBO J.* **18**:3334.]

promoting the growth and metastatic abilities characteristic of cancer cells.

Constitutive Ras activation can also arise from a recessive loss-of-function mutation in a GTPase-accelerating protein (GAP). The normal GAP function is to accelerate hydrolysis of GTP and the conversion of active GTP-bound Ras to inactive GDP-bound Ras (see Figure 3-29). The loss of GAP leads to sustained Ras activation of downstream signal-transduction proteins. For example, neurofibromatosis, a benign tumor of the sheath cells that surround nerves, is caused by loss of both alleles of *NF1*, which encodes a GAP-type protein. Individuals with neurofibromatosis have inherited a single mutant *NF1* allele; subsequent somatic mutation in the other allele leads to formation of neurofibromas. Thus *NF1*, like *RB*, is a tumor-suppressor gene, and neurofibromatosis, like hereditary retinoblastoma, is inherited as an autosomal dominant trait.

Src Protein Kinase Several oncogenes, some initially identified in human tumors, others in transforming retroviruses, encode cytosolic protein kinases that normally transduce signals in a variety of intracellular signaling pathways. Indeed the first oncogene to be discovered, v-*src* from Rous sarcoma retrovirus, encodes a constitutively active protein-tyrosine kinase. At least eight mammalian proto-oncogenes encode a family of nonreceptor tyrosine kinases related to the v-Src protein. In addition to a catalytic domain, these kinases contain SH2 and SH3 protein-protein interaction domains. The kinase activity of cellular Src and related proteins normally is inactivated by phosphorylation of the tyrosine residue at position 527, which is six residues from the C-terminus (Figure 23-17a, b). Hydrolysis of phosphotyrosine 527 by a specific phosphatase enzyme normally activates c-Src. Tyrosine 527 is often missing or altered in Src oncoproteins that have constitutive kinase activity; that is, they do not require activation by a phosphatase. In Rous sarcoma virus, for instance, the *src* gene has suffered a deletion that eliminates the C-terminal 18 amino acids of c-Src; as a consequence the v-Src kinase is constitutively active (Figure 23-17b). Phosphorylation of target proteins by aberrant Src oncoproteins contributes to abnormal proliferation of many types of cells.

Abl Protein Kinase Another oncogene encoding a cytosolic nonreceptor protein kinase is generated by a chromosomal translocation that fuses a part of the c-*abl* gene, which encodes a tyrosine kinase, with part of the *bcr* gene, whose function is unknown. The normal c-Abl protein promotes branching of filamentous actin and extension of cell processes, so it may function primarily to control the cytoskeleton and cell shape. The chimeric oncoproteins encoded by the *bcr-abl* oncogene form a tetramer that exhibits unregulated and continuous Abl kinase activity. (This is similar to dimerization and activation of the chimeric Trk oncoprotein shown in Figure 23-15.) Bcr-Abl can phosphorylate and thereby activate many

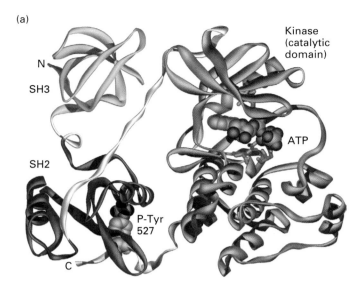

(a)

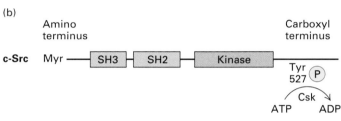

(b)

▲ **FIGURE 23-17 Structure of Src tyrosine kinases and activation by an oncogenic mutation.** (a) Three-dimensional structure of Hck, one of several Src kinases in mammals. Binding of phosphotyrosine 527 to the SH2 domain induces conformational strains in the SH3 and kinase domains, distorting the kinase active site so it is catalytically inactive. The kinase activity of cellular Src proteins is normally activated by removing the phosphate on tyrosine 527. (b) Domain structure of c-Src and v-Src. Phosphorylation of tyrosine 527 by Csk, another cellular tyrosine kinase, inactivates the Src kinase activity. The transforming v-Src oncoprotein encoded by Rous sarcoma virus is missing the C-terminal 18 amino acids including tyrosine 527 and thus is constitutively active. [Part (a) from F. Sicheri et al., 1997, *Nature* **385**:602. See also T. Pawson, 1997, *Nature* **385**:582, and W. Xu et al., 1997, *Nature* **385**:595.]

intracellular signal-transduction proteins; at least some of these proteins are not normal substrates of Abl. For instance, Bcr-Abl can activate JAK2 kinase and STAT5 transcription factor, which normally are activated by binding of growth factors (e.g., erythropoietin) to cell-surface receptors (see Figure 14-12).

The chromosomal translocation that forms *bcr-abl* generates the diagnostic *Philadelphia chromosome*, discovered in 1960 (see Figure 10-29). The identity of the genes involved was discovered by molecular cloning of the relevant translocation "joint," allowing biochemical study of the Bcr-Abl oncoprotein. If this translocation occurs in a hematopoietic cell in the bone marrow, the activity of the chimeric *bcr-abl* oncogene results in the initial phase of human chronic myelogenous leukemia (CML), characterized by an expansion in the number of white blood cells. A second mutation in a cell carrying *bcr-abl* (e.g., in *p53*) leads to acute leukemia, which often kills the patient. The CML chromosome translocation was only the first of a long series of distinctive, or "signature," chromosome translocations linked to particular forms of leukemia. Each one presents an opportunity for greater understanding of the disease and for new therapies. In the case of CML, that second step to successful therapy has already been taken.

 After a painstaking search, an inhibitor of Abl kinase named STI-571 (Gleevec) was identified as a possible treatment for CML in the early 1990s. STI-571 is highly lethal to CML cells while sparing normal cells. After clinical trials showing STI-571 is remarkably effective in treating CML despite some side effects, it was approved by the FDA in 2001, the first cancer drug targeted to a signal-transduction protein unique to tumor cells. STI-571 inhibits several other tyrosine kinases that are implicated in different cancers and has been successful in trials for treating these diseases as well. There are 96 tyrosine kinases encoded in the human genome, so drugs related to Gleevec may be useful in controlling the activities of all these proteins. ▮

The Gleevec story illustrates how genetics—the discovery of the Philadelphia chromosome and the critical oncogene it creates—together with biochemistry—discovery of the molecular action of the Abl protein—can lead to a powerful new therapy. In general, each difference between cancer cells and normal cells provides a new opportunity to identify a specific drug that kills just the cancer cells or at least stops their uncontrolled growth.

Inappropriate Production of Nuclear Transcription Factors Can Induce Transformation

By one mechanism or another, the proteins encoded by all proto-oncogenes and oncogenes eventually cause changes in gene expression. This is reflected in the differences in the proportions of different mRNAs in growing cells and quiescent cells, as well as similar differences between tumor cells and their normal counterparts. As discussed in the last section, we can now measure such differences in the expression of thousands of genes with DNA microarrays (see Figure 23-13).

Since the most direct effect on gene expression is exerted by transcription factors, it is not surprising that many oncogenes encode transcription factors. Two examples are *jun* and *fos*, which initially were identified in transforming

retroviruses and later found to be overexpressed in some human tumors. The c-*jun* and c-*fos* proto-oncogenes encode proteins that sometimes associate to form a heterodimeric transcription factor, called AP1, that binds to a sequence found in promoters and enhancers of many genes (see Figure 11-24). Both Fos and Jun also can act independently as transcription factors. They function as oncoproteins by activating transcription of key genes that encode growth-promoting proteins or by inhibiting transcription of growth-repressing genes.

Many nuclear proto-oncogene proteins are induced when normal cells are stimulated to grow, indicating their direct role in growth control. For example, PDGF treatment of quiescent 3T3 cells induces an ≈50-fold increase in the production of c-Fos and c-Myc, the normal products of the *fos* and *myc* proto-oncogenes. Initially there is a transient rise of c-Fos and later a more prolonged rise of c-Myc (Figure 23-18). The levels of both proteins decline within a few hours, a regulatory effect that may, in normal cells, help to avoid cancer. As discussed in Chapter 21, c-Fos and c-Myc stimulate transcription of genes encoding proteins that promote progression through the G_1 phase of the cell cycle and the G_1 to S transition. In tumors, the oncogenic forms of these or other transcription factors are frequently expressed at high and unregulated levels.

In normal cells, c-Fos and c-Myc mRNAs and the proteins they encode are intrinsically unstable, leading to their rapid loss after the genes are induced. Some of the changes

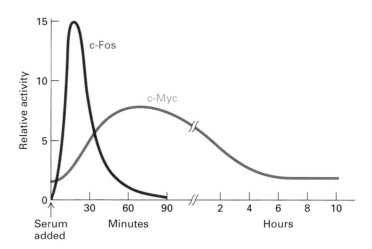

▲ **EXPERIMENTAL FIGURE 23-18 Addition of serum to quiescent 3T3 cells yields a marked increase in the activity of two proto-oncogene products, c-Fos and c-Myc.** Serum contains factors like platelet-derived growth factor (PDGF) that stimulate the growth of quiescent cells. One of the earliest effects of growth factors is to induce expression of c-*fos* and c-*myc*, whose encoded proteins are transcription factors. [See M. E. Greenberg and E. B. Ziff, 1984, *Nature* **311**:433.]

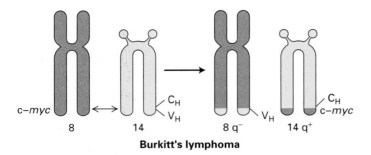

▲ FIGURE 23-19 Chromosomal translocation in Burkitt's lymphoma. As a result of a translocation between chromosomes 8 and 14, the c-*myc* gene is placed adjacent to the gene for part of the antibody heavy chain (C$_H$), leading to overproduction of the Myc transcription factor in lymphocytes and hence their growth into a lymphoma.

that turn c-*fos* from a normal gene to an oncogene involve genetic deletions of sequences that make the Fos mRNA and protein short-lived. Conversion of the c-*myc* proto-oncogene into an oncogene can occur by several different mechanisms. In cells of the human tumor known as *Burkitt's lymphoma*, the c-*myc* gene is translocated to a site near the heavy-chain antibody genes, which are normally active in antibody-producing white blood cells (Figure 23-19). The c-*myc* translocation is a rare aberration of the normal DNA rearrangements that occur during maturation of antibody-producing cells. The translocated *myc* gene, now regulated by the antibody gene enhancer, is continually expressed, causing the cell to become cancerous. Localized amplification of a segment of DNA containing the *myc* gene, which occurs in several human tumors, also causes inappropriately high production of the otherwise normal Myc protein.

KEY CONCEPTS OF SECTION 23.3

Oncogenic Mutations in Growth-Promoting Proteins

■ Mutations or chromosomal translocations that permit RTKs for growth factors to dimerize in the absence of their normal ligands lead to constitutive receptor activity (see Figures 23-14 and 23-15). Such activation ultimately induces changes in gene expression that can transform cells. Overproduction of growth factor receptors can have the same effect and lead to abnormal cell proliferation.

■ Certain virus-encoded proteins can bind to and activate host-cell receptors for growth factors, thereby stimulating cell proliferation in the absence of normal signals.

■ Most tumor cells produce constitutively active forms of one or more intracellular signal-transduction proteins, causing growth-promoting signaling in the absence of normal growth factors.

■ A single point mutation in Ras, a key transducing protein in many signaling pathways, reduces its GTPase activity, thereby maintaining it in an activated state.

■ The activity of Src, a cytosolic signal-transducing protein-tyrosine kinase, normally is regulated by reversible phosphorylation and dephosphorylation of a tyrosine residue near the C-terminus (see Figure 23-17). The unregulated activity of Src oncoproteins that lack this tyrosine promotes abnormal proliferation of many cells.

■ The Philadelphia chromosome results from a chromosomal translocation that produces the chimeric *bcr-abl* oncogene. The unregulated Abl kinase activity of the Bcr-Abl oncoprotein is responsible for its oncogenic effect. An Abl kinase inhibitor (Gleevec) is effective in treating chronic myelogenous leukemia (CML) and may work against other cancers.

■ Inappropriate production of nuclear transcription factors such as Fos, Jun, and Myc can induce transformation. In Burkitt's lymphoma cells, c-*myc* is translocated close to an antibody gene, leading to overproduction of c-Myc (see Figure 23-19).

23.4 Mutations Causing Loss of Growth-Inhibiting and Cell-Cycle Controls

Normal growth and development depends on a finely tuned, highly regulated balance between growth-promoting and growth-inhibiting pathways. Mutations that disrupt this balance can lead to cancer. Most of the mutations discussed in the previous section cause inappropriate activity of growth-promoting pathways. Just as critical are mutations that decrease the activity of growth-inhibiting pathways when they are needed.

For example, transforming growth factor β (TGFβ), despite its name, inhibits proliferation of many cell types, including most epithelial and immune system cells. Binding of TGFβ to its receptor induces activation of cytosolic Smad transcription factors (see Figure 14-2). After translocating to the nucleus, Smads can promote expression of the gene encoding p15, which causes cells to arrest in G$_1$. TGFβ signaling also promotes expression of genes encoding extracellular matrix proteins and plasminogen activator inhibitor 1 (PAI-1), which reduces the plasmin-catalyzed degradation of the matrix. Loss-of-function mutations in either TGFβ receptors or in Smads thus promote cell proliferation and probably contribute to the invasiveness and metastasis of tumor cells (Figure 23-20). Such mutations have in fact been found in a variety of human cancers, as we describe in Chapter 14.

The complex mechanisms for regulating the eukaryotic cell cycle are prime targets for oncogenic mutations. Both positive- and negative-acting proteins precisely control the entry of cells into and their progression through the cell cycle, which consists of four main phases: G_1, S, G_2, and mitosis (see Figure 21-2). This regulatory system assures the proper coordination of cellular growth during G_1 and G_2, DNA synthesis during the S phase, and chromosome segregation and cell division during mitosis. In addition, cells that have sustained damage to their DNA normally are arrested before their DNA is replicated. This arrest allows time for the DNA damage to be repaired; alternatively, the arrested cells are directed to commit suicide via programmed cell death. The whole cell-cycle control system functions to prevent cells from becoming cancerous. As might be expected, mutations in this system often lead to abnormal development or contribute to cancer.

Mutations That Promote Unregulated Passage from G_1 to S Phase Are Oncogenic

Once a cell progresses past a certain point in late G_1, called the **restriction point,** it becomes irreversibly committed to entering the S phase and replicating its DNA (see Figure 21-28). D-type **cyclins, cyclin-dependent kinases (CDKs),** and the Rb protein are all elements of the control system that regulate passage through the restriction point.

The expression of D-type cyclin genes is induced by many extracellular growth factors, or **mitogens.** These cyclins assemble with their partners CDK4 and CDK6 to generate catalytically active cyclin-CDK complexes, whose kinase activity promotes progression past the restriction point. Mitogen withdrawal prior to passage through the restriction point leads to accumulation of p16. Like p15 mentioned above, p16 binds specifically to CDK4 and CDK6, thereby inhibiting their kinase activity and causing G_1 arrest. Under normal circumstances, phosphorylation of Rb protein is initiated midway through G_1 by active cyclin D-CDK4 and cyclin D-CDK6 complexes. Rb phosphorylation is completed by other cyclin-CDK complexes in late G_1, allowing activation of E2F transcription factors, which stimulate transcription of genes encoding proteins required for DNA synthesis. The complete phosphorylation of Rb irreversibly commits the cell to DNA synthesis. Most tumors contain an oncogenic mutation that causes overproduction or loss of one of the components of this pathway such that the cells are propelled into the S phase in the absence of the proper extracellular growth signals (Figure 23-21).

Elevated levels of cyclin D1, for example, are found in many human cancers. In certain tumors of antibody-producing B lymphocytes, for instance, the *cyclin D1* gene is translocated such that its transcription is under control of an antibody-gene enhancer, causing elevated cyclin D1 production throughout the cell cycle, irrespective of extracellular signals. (This phenomenon is analogous to the c-*myc* translocation in Burkitt's lymphoma cells discussed earlier.) That cyclin D1 can function as an oncoprotein was shown by studies with transgenic mice in which the *cyclin D1* gene was

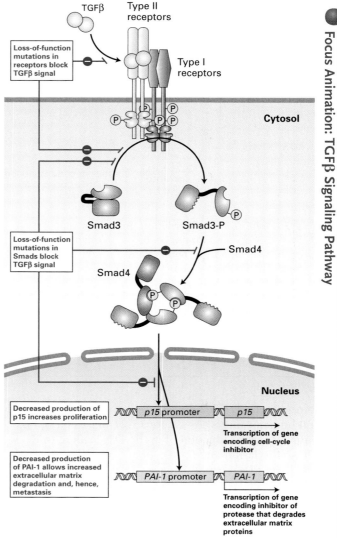

▲ **FIGURE 23-20 Effect of loss of TGFβ signaling.** Binding of TGFβ, an antigrowth factor, causes activation of Smad transcription factors. In the absence of TGFβ signaling due to either a receptor mutation or a SMAD mutation, cell proliferation and invasion of the surrounding extracellular matrix (ECM) increase. [See X. Hua et al., 1998, *Genes & Develop.* **12**:3084.]

placed under control of an enhancer specific for mammary ductal cells. Initially the ductal cells underwent hyperproliferation, and eventually breast tumors developed in these transgenic mice. Amplification of the *cyclin D1* gene and concomitant overproduction of the cyclin D1 protein is common in human breast cancer.

The proteins that function as cyclin-CDK inhibitors play an important role in regulating the cell cycle (Chapter 21). In particular, loss-of-function mutations that prevent p16 from

Focus Animation: TGFβ Signaling Pathway

MEDIA CONNECTIONS

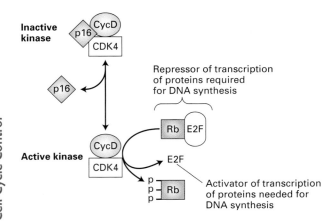

▲ **FIGURE 23-21 Restriction point control.** Unphosphorylated Rb protein binds transcription factors collectively called E2F and thereby prevents E2F-mediated transcriptional activation of many genes whose products are required for DNA synthesis (e.g., DNA polymerase). The kinase activity of cyclin D-CDK4 phosphorylates Rb, thereby activating E2F; this kinase activity is inhibited by p16. Overproduction of cyclin D, a positive regulator, or loss of the negative regulators p16 and Rb, commonly occurs in human cancers.

passage through the restriction point is all that is necessary to subvert normal growth control and set the stage for cancer.

Loss-of-Function Mutations Affecting Chromatin-Remodeling Proteins Contribute to Tumors

Mutations can undermine growth control by inactivating tumor-suppressor genes, but these genes can also be silenced by repressive chromatin structures. In recent years the importance of chromatin-remodeling machines, such as the Swi/Snf complex, in transcriptional control has become increasingly clear. These large and diverse multiprotein complexes have at their core an ATP-dependent helicase and often control acetylation of histones. By causing changes in the positions or structures of nucleosomes, Swi/Snf complexes make genes accessible or inaccessible to DNA-binding proteins that control transcription (Chapter 11). If a gene is normally activated or repressed by Swi/Snf-mediated chromatin changes, mutations in the genes encoding the Swi or Snf proteins will cause changes in expression of the target gene.

Our knowledge of the target genes regulated by Swi/Snf and other such complexes is incomplete, but the targets evidently include some growth-regulating genes. For example, studies with transgenic mice suggest that Swi/Snf plays a role in repressing the *E2F* genes, thereby inhibiting progression through the cell cycle. The relationship between the genes that encode Swi/Snf proteins and the *E2F* gene was discovered in genetic experiments with flies. Transgenic flies were constructed to overexpress *E2F*, which resulted in mild growth defects. A search for mutations that increase the effect of the *E2F* overexpression in these flies identified three components of the Swi/Snf complex. That loss of function of these genes increases the proliferative effects of E2F indicates that Swi/Snf normally counteracts the function of the E2F transcription factor. Thus loss of Swi/Snf function, just like loss of Rb function, can lead to overgrowth and perhaps cancer. Indeed, in mice, Rb protein recruits Swi/Snf proteins to repress transcription of the *E2F* gene.

With chromatin-remodeling complexes involved in so many cases of transcriptional control, it is expected that Swi/Snf and similar complexes will be linked to many cancers. In humans, for example, mutations in *Brg1*, which encodes the Swi/Snf catalytic subunit, have been found in prostate, lung, and breast tumors. Components of the Swi/Snf complex also have been found to associate with BRCA1, a nuclear protein that helps suppress human breast cancer. BRCA-1 is involved in the repair of double-strand DNA breaks (discussed in the final section of this chapter) and in transcriptional control, so the Swi/Snf complex may assist BRCA-1 in these functions.

inhibiting cyclin D-CDK4/6 kinase activity are common in several human cancers. As Figure 23-21 makes clear, loss of p16 mimics overproduction of cyclin D1, leading to Rb hyperphosphorylation and release of active E2F transcription factor. Thus p16 normally acts as a tumor suppressor. Although the *p16* tumor-suppressor gene is deleted in some human cancers, in others the *p16* sequence is normal. In these latter cancers (e.g., lung cancer), the *p16* gene is inactivated by hypermethylation of its promoter region, which prevents transcription. What promotes this change in the methylation of *p16* is not known, but it prevents production of this important cell-cycle control protein.

We've seen already that inactivating mutations in both *RB* alleles lead to childhood retinoblastoma, a relatively rare type of cancer. However, loss of *RB* gene function also is found in more common cancers that arise later in life (e.g., carcinomas of lung, breast, and bladder). These tissues, unlike retinal tissue, most likely produce other proteins whose function is redundant with that of Rb, and thus loss of Rb is not so critical. Several proteins are known that are related in structure and probably function to Rb. In addition to inactivating mutations, Rb function can be eliminated by the binding of an inhibitory protein, designated E7, that is encoded by human papillomavirus (HPV), another nasty viral trick to create virus-producing tissue.

Tumors with inactivating mutations in Rb generally produce normal levels of cyclin D1 and functional p16 protein. In contrast, tumor cells that overproduce cyclin D1 or have lost p16 function generally retain wild-type Rb. Thus loss of only one component of this regulatory system for controlling

Loss of p53 Abolishes the DNA-Damage Checkpoint

A critical feature of cell-cycle control is the G$_1$ checkpoint, which prevents cells with damaged DNA from entering the

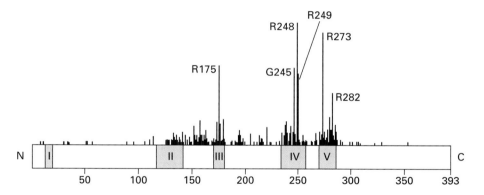

▲ **EXPERIMENTAL FIGURE 23-22 Mutations in human tumors that inactivate the function of p53 protein are highly concentrated in a few residues.** Colored boxes represent sequences in the *p53* gene that are highly conserved in evolution. Vertical lines represent the frequency at which point mutations are found at each residue in various human tumors.

These mutations are clustered in conserved regions II–V. The locations in the protein sequence of the most frequently occurring point mutations are labeled. In these labels, R = arginine; G = glycine. [Adapted from C. C. Harris, 1993, *Science* **262**:1980, and L. Ko and C. Prives, 1996, *Genes & Develop.* **10**:1054.]

S phase (see Figure 21-32, step 4a). The **p53 protein** is a sensor essential for the checkpoint control that arrests cells with damaged DNA in G_1. Although p53 has several functions, its ability to activate transcription of certain genes is most relevant to its tumor-suppressing function. Virtually all p53 mutations abolish its ability to bind to specific DNA sequences and activate gene expression. Mutations in the *p53* tumor-suppressor gene occur in more than 50 percent of human cancers (Figure 23-22).

Cells with functional p53 become arrested in G_1 when exposed to DNA-damaging irradiation, whereas cells lacking functional p53 do not. Unlike other cell-cycle proteins, p53 is present at very low levels in normal cells because it is extremely unstable and rapidly degraded. Mice lacking p53 are viable and healthy, except for a predisposition to develop multiple types of tumors. Expression of the *p53* gene is heightened only in stressful situations, such as ultraviolet or γ irradiation, heat, and low oxygen. DNA damage by γ irradiation or by other stresses somehow leads to the activation of ATM, a serine kinase that phosphorylates and thereby stabilizes p53, leading to a marked increase in its concentration (Figure 23-23). The stabilized p53 activates expression of the gene encoding $p21^{CIP}$, which binds to and inhibits mammalian G_1 cyclin-CDK complexes. As a result, cells with damaged DNA are arrested in G_1, allowing time for DNA repair by mechanisms discussed later. If repair is successful, the levels of p53 and $p21^{CIP}$ will fall, and the cells then can progress into the S phase.

When the p53 G_1 checkpoint control does not operate properly, damaged DNA can replicate, perpetuating mutations and DNA rearrangements that are passed on to daughter cells, contributing to the likelihood of transformation into metastatic cells. In addition, $p21^{CIP}$ and two other proteins induced by p53 inhibit the cyclin B-CDK1 complex required for entry into mitosis, thus causing cells to arrest in G_2 (see Figure 21-32, step 4d). p53 also represses expression

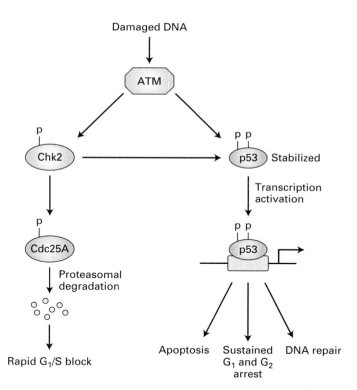

▲ **FIGURE 23-23 G_1 arrest in response to DNA damage.** The kinase activity of ATM is activated in response to DNA damage due to various stresses (e.g., UV irradiation, heat). Activated ATM then triggers two pathways leading to arrest in G_1. Phosphorylation of p53 stabilizes it, permitting p53-activated expression of genes encoding proteins that (**1**) cause arrest in G_1 and in some cases G_2, (**2**) promote apoptosis, or (**3**) participate in DNA repair. In the other pathway phosphorylated Chk2 in turn phosphorylates Cdc25A, thereby marking it for degradation and blocking its role in CDK2 activation. See the text for a discussion.

of the genes encoding cyclin B and topoisomerase II, which also are required for the G$_2$ → mitosis transition. Thus if DNA is damaged following its replication, p53-induced G$_2$ arrest will prevent its transmission to daughter cells.

The active form of p53 is a tetramer of four identical subunits. A missense point mutation in one of the two *p53* alleles in a cell can abrogate almost all p53 activity because virtually all the oligomers will contain at least one defective subunit, and such oligomers cannot function as a transcription factor. Oncogenic *p53* mutations thus act as **dominant negatives,** with mutations in a single allele causing a loss of function. As we learned in Chapter 9, dominant-negative mutations can occur in proteins whose active forms are multimeric or whose function depends on interactions with other proteins. In contrast, loss-of-function mutations in other tumor-suppressor genes (e.g., *RB*) are recessive because the encoded proteins function as monomers and mutation of a single allele has little functional consequence.

Under stressful conditions, the ATM kinase also phosphorylates and thus activates Chk2, a protein kinase that phosphorylates the protein phosphatase Cdc25A, marking it for ubiquitin-mediated destruction. This phosphatase removes the inhibitory phosphate on CDK2, a prerequisite for cells to enter the S phase. Decreased levels of Cdc25A thus block progression into and through the S phase (see Figures 23-23 and 21-32, step 4b). Loss-of-function mutations in the *ATM* or *Chk2* genes have much the same effect as *p53* mutations.

The activity of p53 normally is kept low by a protein called *Mdm2*. When Mdm2 is bound to p53, it inhibits the transcription-activating ability of p53 and catalyzes the addition of ubiquitin molecules, thus targeting p53 for proteasomal degradation. Phosphorylation of p53 by ATM displaces bound Mdm2 from p53, thereby stabilizing it. Because the *Mdm2* gene is itself transcriptionally activated by p53, Mdm2 functions in an autoregulatory feedback loop with p53, perhaps normally preventing excess p53 function. The *Mdm2* gene is amplified in many sarcomas and other human tumors that contain a normal *p53* gene. Even though functional p53 is produced by such tumor cells, the elevated Mdm2 levels reduce the p53 concentration enough to abolish the p53-induced G$_1$ arrest in response to irradiation.

The activity of p53 also is inhibited by a human papillomavirus (HPV) protein called E6. Thus HPV encodes three proteins that contribute to its ability to induce stable transformation and mitosis in a variety of cultured cells. Two of these—E6 and E7—bind to and inhibit the p53 and Rb tumor suppressors, respectively. Acting together, E6 and E7 are sufficient to induce transformation in the absence of mutations in cell regulatory proteins. The HPV E5 protein, which causes sustained activation of the PDGF receptor, enhances proliferation of the transformed cells.

The activity of p53 is not limited to inducing cell-cycle arrest. In addition, this multipurpose tumor suppressor stimulates production of pro-apoptotic proteins and DNA-repair enzymes (see Figure 23-23).

Apoptotic Genes Can Function as Proto-Oncogenes or Tumor-Suppressor Genes

During normal development many cells are designated for **programmed cell death,** also known as apoptosis (see Chapter 22). Many abnormalities, including errors in mitosis, DNA damage, and an abnormal excess of cells not needed for development of a working organ, also can trigger apoptosis. In some cases, cell death appears to be the default situation, with signals required to ensure cell survival. Cells can receive instructions to live and instructions to die, and a complex regulatory system integrates the various kinds of information.

If cells do not die when they should and instead keep proliferating, a tumor may form. For example, chronic lymphoblastic leukemia (CLL) occurs because cells survive when they should be dying. The cells accumulate slowly, and most are not actively dividing, but they do not die. CLL cells have chromosomal translocations that activate a gene called *bcl-2*, which we now know to be a critical blocker of apoptosis (see Figure 22-31). The resultant inappropriate overproduction of Bcl-2 protein prevents normal apoptosis and allows survival of these tumor cells. CLL tumors are therefore attributable to a failure of cell death. Another dozen or so proto-oncogenes that are normally involved in negatively regulating apoptosis have been mutated to become oncogenes. Overproduction of their encoded proteins prevents apoptosis even when it is needed to stop cancer cells from growing.

Conversely, genes whose protein products stimulate apoptosis behave as tumor suppressors. An example is the *PTEN* gene discussed in Chapter 14. The phosphatase encoded by this gene dephosphorylates phosphatidylinositol 3,4,5-trisphosphate, a second messenger that functions in activation of protein kinase B (see Figure 14-27). Cells lacking PTEN phosphatase have elevated levels of phosphatidylinositol 3,4,5-trisphosphate and active protein kinase B, which promotes cell survival and prevents apoptosis by several pathways. Thus PTEN acts as a pro-apoptotic tumor suppressor by decreasing the anti-apoptotic effect of protein kinase B.

The most common pro-apoptotic tumor-suppressor gene implicated in human cancers is *p53*. Among the genes activated by p53 are several encoding pro-apoptotic proteins such as Bax (see Figure 22-32). When most cells suffer extensive DNA damage, the p53-induced expression of pro-apoptotic proteins leads to their quick demise (see Figure 23-23). While this may seem like a drastic response to DNA damage, it prevents proliferation of cells that are likely to accumulate multiple mutations.

 When p53 function is lost, apoptosis cannot be induced and the accumulation of mutations required for cancer to develop becomes more likely. Tumors marked by loss of *p53* or another gene needed for apoptosis are difficult to treat with chemical or radiation therapy, since the resulting DNA damage is not translated into programmed cell death. ∎

Failure of Cell-Cycle Checkpoints Can Also Lead to Aneuploidy in Tumor Cells

It has long been known that chromosomal abnormalities abound in tumor cells. We have already encountered several examples of oncogenes that are formed by translocation, amplification, or both (e.g., c-*myc, bcr-abl, bcl-2,* and *cyclin D1*). Another chromosomal abnormality characteristic of nearly all tumor cells is **aneuploidy,** the presence of an aberrant number of chromosomes—generally too many.

Cells with abnormal numbers of chromosomes form when certain cell-cycle checkpoints are nonfunctional. As discussed in Chapter 21, the unreplicated-DNA checkpoint normally prevents entry into mitosis unless all chromosomes have completely replicated their DNA; the spindle-assembly checkpoint prevents entry into anaphase unless all the replicated chromosomes attach properly to the metaphase mitotic apparatus; and the chromosome-segregation checkpoint prevents exit from mitosis and cytokinesis if the chromosomes segregate improperly (see Figure 21-32, steps 1–3). As advances are made in identifying the proteins that detect these abnormalities and mediate cell-cycle arrest, the molecular basis for the functional defects leading to aneuploidy in tumor cells will become clearer.

KEY CONCEPTS OF SECTION 23.4

Mutations Causing Loss of Growth-Inhibiting and Cell-Cycle Controls

■ Loss of signaling by TGFβ, a negative growth regulator, promotes cell proliferation and development of malignancy (see Figure 23-20).

■ Overexpression of the proto-oncogene encoding cyclin D1 or loss of the tumor-suppressor genes encoding p16 and Rb can cause inappropriate, unregulated passage through the restriction point in late G_1. Such abnormalities are common in human tumors.

■ p53 is a multipurpose tumor suppressor that promotes arrest in G_1 and G_2, apoptosis, and DNA repair in response to damaged DNA (see Figure 23-23). Loss-of-function mutations in the *p53* gene occur in more than 50 percent of human cancers.

■ Overproduction of Mdm2, a protein that normally inhibits the activity of p53, occurs in several cancers (e.g., sarcomas) that express normal p53 protein.

■ Human papillomavirus (HPV) encodes three oncogenic proteins: E6 (inhibits p53), E7 (inhibits Rb), and E5 (activates PDGF receptor).

■ Mutations affecting the Swi/Snf chromatin-remodeling complex, which participates in transcriptional control, are associated with a variety of tumors. In some cases, interaction of the Swi/Snf complex with a nuclear tumor-suppressor protein may have a repressing effect on gene expression.

■ Overproduction of anti-apoptotic proteins (e.g., Bcl-2) can lead to inappropriate cell survival and is associated with chronic lymphoblastic leukemia (CLL) and other cancers. Loss of proteins that promote apoptosis (e.g., p53 transcription factor and PTEN phosphatase) have a similar oncogenic effect.

■ Most human tumor cells are aneuploid, containing an abnormal number of chromosomes (usually too many). Failure of cell-cycle checkpoints that normally detect unreplicated DNA, improper spindle assembly, or mis-segregation of chromosomes permits aneuploid cells to arise.

23.5 The Role of Carcinogens and DNA Repair in Cancer

In this final section, we examine how alterations in the genome arise that may lead to cancer and how cells attempt to correct them. DNA damage is unavoidable and arises by spontaneous cleavage of chemical bonds in DNA, by reaction with genotoxic chemicals in the environment or with certain chemical by-products of normal cellular metabolism, and from environmental agents such as ultraviolet and ionizing radiation. Changes in the DNA sequence can also result from copying errors introduced by DNA polymerases during replication and by mistakes made when DNA polymerase attempts to read from a damaged template. If DNA sequence changes, whatever their cause or nature, are left uncorrected, both proliferating and quiescent somatic cells might accumulate so many mutations that they could no longer function properly. In addition, the DNA in germ cells might incur too many mutations for viable offspring to be formed. Thus the prevention of DNA sequence errors in all types of cells is important for survival, and several cellular mechanisms for repairing damaged DNA and correcting sequence errors have evolved.

As our previous discussion has shown, alterations in DNA that lead to decreased production of functional tumor-suppressor proteins or increased, unregulated production or activation of oncoproteins are the underlying cause of most cancers. These oncogenic mutations in key growth and cell-cycle regulatory genes include insertions, deletions, and point mutations, as well as chromosomal amplifications and translocations. Most cancer cells lack one or more DNA-repair systems, which may explain the large number of mutations that they accumulate. Moreover, some repair mechanisms themselves introduce errors in the nucleotide sequence; such error-prone repair also contributes to oncogenesis. The inability of tumor cells to maintain genomic integrity leads to formation of a heterogeneous population of malignant cells. For this reason, chemotherapy directed toward a single gene or even a group of genes is likely to be ineffective in wiping out all malignant cells. This problem adds to the interest in therapies that interfere with the blood supply to tumors or in other ways act upon multiple types of tumor cells.

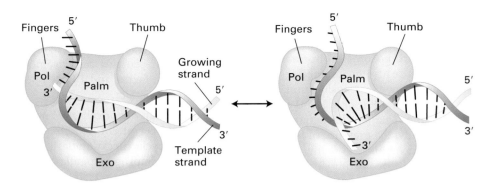

▲ **FIGURE 23-24 Schematic model of the proofreading function of DNA polymerases.** All DNA polymerases have a similar three-dimensional structure, which resembles a half-opened right hand. The "fingers" bind the single-stranded segment of the template strand, and the polymerase catalytic activity (Pol) lies in the junction between the fingers and palm. As long as the correct nucleotides are added to the 3′ end of the growing strand, it remains in the polymerase site. Incorporation of an incorrect base at the 3′ end causes melting of the newly formed end of the duplex. As a result, the polymerase pauses, and the 3′ end of the growing strand is transferred to the 3′ → 5′ exonuclease site (Exo) about 3 nm away, where the mispaired base and probably other bases are removed. Subsequently, the 3′ end flips back into the polymerase site and elongation resumes. [Adapted from C. M. Joyce and T. T. Steitz, 1995, *J. Bacteriol.* **177**:6321, and S. Bell and T. Baker, 1998, *Cell* **92**:295.]

DNA Polymerases Introduce Copying Errors and Also Correct Them

The first line of defense in preventing mutations is DNA polymerase itself. Occasionally, when replicative DNA polymerases progress along the template DNA, an incorrect nucleotide is added to the growing 3′ end of the daughter strand (see Figure 4-34). *E. coli* DNA polymerases, for instance, introduce about 1 incorrect nucleotide per 10^4 polymerized nucleotides. Yet the measured mutation rate in bacterial cells is much lower: about 1 mistake in 10^9 nucleotides incorporated into a growing strand. This remarkable accuracy is largely due to *proofreading* by *E. coli* DNA polymerases.

Proofreading depends on the 3′→5′ exonuclease activity of some DNA polymerases. When an incorrect base is incorporated during DNA synthesis, the polymerase pauses, then transfers the 3′ end of the growing chain to the exonuclease site, where the incorrect mispaired base is removed (Figure 23-24). Then the 3′ end is transferred back to the polymerase site, where this region is copied correctly. All three *E. coli* DNA polymerases have proofreading activity, as do the two DNA polymerases, δ and ε, used for DNA replication in animal cells. It seems likely that proofreading is indispensable for all cells to avoid excessive mutations.

▶ **FIGURE 23-25 Formation of a spontaneous point mutation by deamination of 5-methyl cytosine (C) to form thymine (T).** If the resulting T · G base pair is not restored to the normal C · G base pair by base excision-repair mechanisms (**1**), it will lead to a permanent change in sequence following DNA replication (i.e., a mutation) (**2**). After one round of replication, one daughter DNA molecule will have the mutant T·A base pair and the other will have the wild-type C·G base pair.

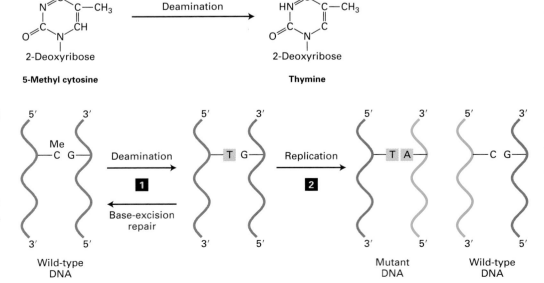

Chemical Damage to DNA Can Lead to Mutations

DNA is continually subjected to a barrage of damaging chemical reactions; estimates of the number of DNA damage events in a single human cell range from 10^4 to 10^6 per day! Even if DNA were not exposed to damaging chemicals, DNA is inherently unstable. For example, the bond connecting a purine base to deoxyribose is prone to hydrolysis, leaving a sugar without an attached base. Thus coding information is lost, and this can lead to a mutation during DNA replication. Normal cellular reactions, including the movement of electrons along the electron-transport chain in mitochondria and lipid oxidation in peroxisomes, produce several chemicals that react with and damage DNA, including hydroxyl radicals and superoxide (O_2^-). These too can cause mutations including those that lead to cancers.

Many spontaneous mutations are **point mutations,** which involve a change in a single base pair in the DNA sequence. One of the most frequent point mutations comes from *deamination* of a cytosine (C) base, which converts it into a uracil (U) base. In addition, the common modified base 5-methyl cytosine forms thymine when it is deaminated. If these alterations are not corrected before the DNA is replicated, the cell will use the strand containing U or T as template to form a U·A or T·A base pair, thus creating a permanent change to the DNA sequence (Figure 23-25).

Some Carcinogens Have Been Linked to Specific Cancers

Environmental chemicals were originally associated with cancer through experimental studies in animals. The classic experiment is to repeatedly paint a test substance on the back of a mouse and look for development of both local and systemic tumors in the animal. Likewise, the ability of ionizing radiation to cause human cancer, especially leukemia, was dramatically shown by the increased rates of leukemia among survivors of the atomic bombs dropped in World War II, and more recently by the increase in melanoma (skin cancer) in individuals exposed to too much sunlight (UV radiation).

The ability of chemical and physical carcinogens to induce cancer can be accounted for by the DNA damage that they cause and by the errors introduced into DNA during the cells' efforts to repair this damage. Thus all carcinogens also are **mutagens.** The strongest evidence that carcinogens act as mutagens comes from the observation that cellular DNA altered by exposure of cells to carcinogens can change cultured cells, such as 3T3 cells, into fast-growing cancer-like cells (see Figure 23-4). The mutagenic effect of carcinogens is roughly proportional to their ability to transform cells and induce cancer in animals.

Although substances identified as chemical carcinogens have a broad range of structures with no obvious unifying features, they can be classified into two general categories. *Direct-acting carcinogens,* of which there are only a few, are mainly reactive electrophiles (compounds that seek out and react with negatively charged centers in other compounds).

By chemically reacting with nitrogen and oxygen atoms in DNA, these compounds can modify bases in DNA so as to distort the normal pattern of base pairing. If these modified nucleotides are not repaired, they allow an incorrect nucleotide to be incorporated during replication. This type of carcinogen includes ethylmethane sulfonate (EMS), dimethyl sulfate (DMS), and nitrogen mustards.

In contrast, *indirect-acting carcinogens* generally are unreactive, water-insoluble compounds that can act as potent cancer inducers only after the introduction of electrophilic centers. In animals, *cytochrome P-450 enzymes* localized to the endoplasmic reticulum of liver cells normally function to add electrophilic centers to nonpolar foreign chemicals, such as certain insecticides and therapeutic drugs, in order to solubilize them so that they can be excreted from the body. But P-450 enzymes also can turn otherwise harmless chemicals into carcinogens.

Although chemical carcinogens are believed to be risk factors for many human cancers, a direct linkage to specific cancers has been established only in a few cases, the most important being lung cancer. Epidemiological studies first indicated that cigarette smoking was the major cause of lung cancer, but why this was so was unclear until the discovery that about 60 percent of human lung cancers contain inactivating mutations in the *p53* gene. The chemical *benzo(a)pyrene* (Figure 23-26), found in cigarette smoke, undergoes metabolic activation in the liver to a potent mutagen that mainly causes conversion of guanine (G) to thymine (T) bases, a transversion mutation. When applied to cultured bronchial epithelial cells,

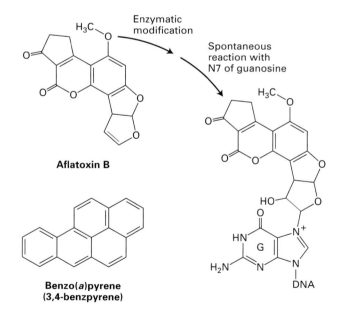

Aflatoxin B

Benzo(*a*)pyrene (3,4-benzpyrene)

▲ **FIGURE 23-26 Two chemical carcinogens that cause mutations in *p53*.** Like all indirect-acting carcinogens, benzo(*a*)pyrene and aflatoxin must undergo enzyme-catalyzed modification before they can react with DNA. In aflatoxin the colored double bond reacts with an oxygen atom, enabling it to react chemically with the N-7 atom of a guanosine in DNA, forming a large bulky molecule. Both compounds mutate the *p53* tumor-suppressor gene and are known risk factors for human cancer.

activated benzo(a)pyrene induces inactivating mutations at codons 175, 248, and 273 of the *p53* gene. These same positions are major mutational hot spots in human lung cancer (see Figure 23-22). Thus, there is a strong correlation between a defined chemical carcinogen in cigarette smoke and human cancer; it is likely that other chemicals in cigarette smoke induce mutations in other genes, as well. Similarly, asbestos exposure is clearly linked to mesothelioma, a type of epithelial cancer.

Lung cancer is not the only major human cancer for which a clear-cut risk factor has been identified. *Aflatoxin*, a fungal metabolite found in moldy grains, induces liver cancer (see Figure 23-26). After chemical modification by liver enzymes, aflatoxin becomes linked to G residues in DNA and induces G-to-T transversions. Aflatoxin also causes a mutation in the *p53* gene. Exposure to other chemicals has been correlated with minor cancers. Hard evidence concerning dietary and environmental risk factors that would help us avoid other common cancers (e.g., breast, colon, and prostate cancer, leukemias) is generally lacking.

Loss of High-Fidelity DNA Excision-Repair Systems Can Lead to Cancer

In addition to proofreading, cells have other repair systems for preventing mutations due to copying errors and exposure to mutagens. Several DNA **excision-repair systems** that normally operate with a high degree of accuracy have been well studied. These systems were first elucidated through a combination of genetic and biochemical studies in *E. coli*. Homologs of the key bacterial proteins exist in eukaryotes from yeast to humans, indicating that these error-free mechanisms arose early in evolution to protect DNA integrity. Each of these systems functions in a similar manner—a segment of the damaged DNA strand is excised, and the gap is filled by DNA polymerase and ligase using the complementary DNA strand as template.

Loss of these systems correlates with increased risk for cancer. For example, humans who inherit mutations in genes that encode a crucial mismatch-repair or excision-repair protein have an enormously increased probability of developing certain cancers (Table 23-1). Without proper DNA repair, people with xeroderma pigmentosum or hereditary nonpolyposis colorectal cancer have a propensity to accumulate mutations in many other genes, including those that are critical in controlling cell growth and proliferation.

We will now turn to a closer look at some of the mechanisms of DNA repair, ranging from repair of single base mutations to repair of DNA broken across both strands. Some of these effect their repairs with great accuracy; others are less precise.

TABLE 23-1 Some Human Hereditary Diseases and Cancers Associated with DNA-Repair Defects

Disease	DNA-Repair System Affected	Sensitivity	Cancer Susceptibility	Symptoms
PREVENTION OF POINT MUTATIONS, INSERTIONS, AND DELETIONS				
Hereditary nonpolyposis colorectal cancer	DNA mismatch repair	UV irradiation, chemical mutagens	Colon, ovary	Early development of tumors
Xeroderma pigmentosum	Nucleotide excision repair	UV irradiation, point mutations	Skin carcinomas, melanomas	Skin and eye photosensitivity, keratoses
REPAIR OF DOUBLE-STRAND BREAKS				
Bloom's syndrome	Repair of double-strand breaks by homologous recombination	Mild alkylating agents	Carcinomas, leukemias, lymphomas	Photosensitivity, facial telangiectases, chromosome alterations
Fanconi anemia	Repair of double-strand breaks by homologous recombination	DNA cross-linking agents, reactive oxidant chemicals	Acute myeloid leukemia, squamous-cell carcinomas	Developmental abnormalities including infertility and deformities of the skeleton; anemia
Hereditary breast cancer, BRCA-1 and BRCA-2 deficiency	Repair of double-strand breaks by homologous recombination		Breast and ovarian cancer	Breast and ovarian cancer

SOURCES: Modified from A. Kornberg and T. Baker, 1992, *DNA Replication*, 2d ed., W. H. Freeman and Company, p. 788; J. Hoeijmakers, 2001, *Nature* **411**:366; and L. Thompson and D. Schild, 2002, *Mutation Res.* **509**:49.

Base Excision Is Used to Repair Damaged Bases and Single-Base Mispairs

In humans, the most common type of point mutation is a C to T, which is caused by deamination of 5-methyl C to T (see Figure 23-25). The conceptual problem with *base excision repair* is determining which is the normal and which is the mutant DNA strand, and repairing the latter so that it is properly base-paired with the normal strand. But since a G · T mismatch is almost invariably caused by chemical conversion of C to U or 5-methyl C to T, the repair system "knows" to remove the T and replace it with a C. The G · T mismatch is recognized by a DNA glycosylase that flips the thymine base out of the helix and then hydrolyzes the bond that connects it to the sugar-phosphate DNA backbone. Following this initial incision, the segment of the damaged strand containing the baseless deoxyribose is excised by an AP endonuclease that cuts the DNA strand near the abasic site. The resultant single-stranded gap in the damaged strand is filled in by a DNA polymerase and sealed by DNA ligase, restoring the original G · C base pair.

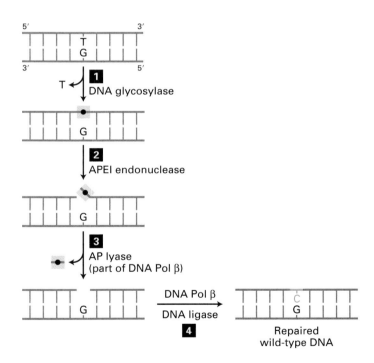

▲ **FIGURE 23-27 Base excision repair of a G·T mismatch**
A DNA glycosylase specific for G · T mismatches, usually formed by deamination of 5-methyl C residues (see Figure 23-25), flips the thymine base out of the helix and then cuts it away from the sugar-phosphate DNA backbone (step **1**), leaving just the deoxyribose (black dot). An endonuclease specific for the resultant baseless site then cuts the DNA backbone (step **2**), and the deoxyribose phosphate is removed by an endonuclease associated with DNA polymerase β (step **3**). The gap is then filled in by DNA Pol β and sealed by DNA ligase (step **4**), restoring the original G · C base pair. [After O. Schärer, 2003, *Angewandte Chemie*, in press.]

Human cells contain a battery of glycosylases, each of which is specific for a different set of chemically modified DNA bases. For example, one removes 8-oxyguanine, an oxidized form of guanine, allowing its replacement by an undamaged G, and others remove bases modified by alkylating agents. By a process similar to that shown in Figure 23-27, the modified base is cleaved (step 1); the damaged strand is then repaired using the "core" enzymes depicted in steps 2 through 4. A similar mechanism repairs lesions resulting from *depurination*, the loss of a guanine or adenine base from DNA resulting from hydrolysis of the glycosylic bond between deoxyribose and the base. Depurination occurs spontaneously and is fairly common in mammals. The resulting abasic sites, if left unrepaired, generate mutations during DNA replication because they cannot specify the appropriate paired base.

Loss of Mismatch Excision Repair Leads to Colon and Other Cancers

Another process, also conserved from bacteria to man, principally eliminates base-pair mismatches, deletions, and insertions that are accidentally introduced by polymerases during replication. As with base excision repair of a T in a T · G mismatch, the conceptual problem with *mismatch excision repair* is determining which is the normal and which is the mutant DNA strand, and repairing the latter. How this happens in human cells is not known with certainty. It is thought that the proteins that bind to the mismatched segment of DNA distinguish the template and daughter strands; then the mispaired segment of the daughter strand—the one with the replication error—is excised and repaired to become an exact complement of the template strand (Figure 23-28).

Hereditary nonpolyposis colorectal cancer, arising from a common inherited predisposition to cancer, results from an inherited loss-of-function mutation in one allele of either the *MLH1* or the *MSH2* gene; the MSH2 and MLH1 proteins are essential for DNA mismatch repair (see Figure 23-28). Cells with at least one functional copy of each of these genes exhibit normal mismatch repair. However, tumor cells frequently arise from those cells that have experienced a somatic mutation in the second allele and thus have lost the mismatch repair system. Somatic inactivating mutations in these genes are also common in noninherited forms of colon cancer.

One gene frequently mutated in colon cancers because of the absence of mismatch repair encodes the type II receptor for TGFβ (see Figure 23-20 and Chapter 14). The gene encoding this receptor contains a sequence of 10 adenines in a row. Because of "slippage" of DNA polymerase during replication, this sequence often undergoes mutation to a sequence containing 9 or 11 adenines. If the mutation is not fixed by the mismatch repair system, the resultant frameshift in the protein-coding sequence abolishes production of the normal receptor protein. As noted earlier, such inactivating mutations make cells resistant to growth inhibition by

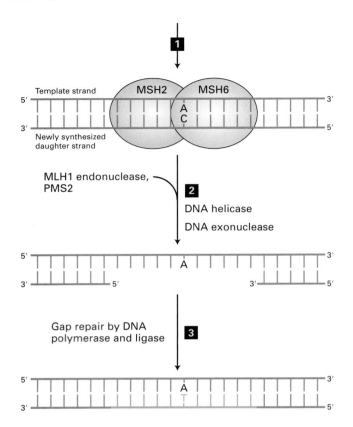

▲ FIGURE 23-28 Mismatch excision repair of newly replicated DNA in human cells. A complex of the MSH2 and MSH6 proteins binds to a mispaired segment of DNA in such a way as to distinguish between the template and newly synthesized daughter strands (step **1**). This triggers binding of the MLH1 endonuclease, as well as other proteins such as PMS2, which has been implicated in oncogenesis through mismatch-repair mutations, although its specific function is unclear. A DNA helicase unwinds the helix and the daughter strand is cut; an exonuclease then removes several nucleotides, including the mismatched base (step **2**). Finally, as with base excision repair, the gap is then filled in by a DNA polymerase (Pol δ, in this case) and sealed by DNA ligase (step **3**).

TGFβ, thereby contributing to the unregulated growth characteristic of these tumors. This finding attests to the importance of mismatch repair in correcting genetic damage that might otherwise lead to uncontrolled cell proliferation.

Nucleotide Excision Repair Was Elucidated Through Study of Xeroderma Pigmentosum, a Hereditary Predisposition to Skin Cancers

Cells use *nucleotide excision repair* to fix DNA regions containing chemically modified bases, often called chemical adducts, that distort the normal shape of DNA locally. A key to this type of repair is the ability of certain proteins to slide along the surface of a double-stranded DNA molecule look-

ing for bulges or other irregularities in the shape of the double helix. For example, this mechanism repairs *thymine-thymine dimers*, a common type of damage caused by UV light (Figure 23-29); these dimers interfere with both replication and transcription of DNA. Nucleotide excision repair also can correct DNA regions containing bases altered by covalent attachment of carcinogens such as benzo(*a*)pyrene and aflatoxin (see Figure 23-26), both of which cause G-to-T transversions.

Figure 23-30 illustrates how the nucleotide excision-repair system repairs damaged DNA. Some 30 proteins are involved in this repair process, the first of which were identified through a study of the defects in DNA repair in cultured cells from individuals with xeroderma pigmentosum, a hereditary disease associated with a predisposition to cancer. Individuals with this disease frequently develop the skin cancers called melanomas and squamous cell carcinomas if their skin is exposed to the UV rays in sunlight. Cells of affected patients lack a functional nucleotide excision-repair system system. Mutations in any of at least seven different genes, called *XP-A* through *XP-G*, lead to inactivation of this repair system and cause xeroderma pigmentosum; all produce the same phenotype and have the same consequences. The

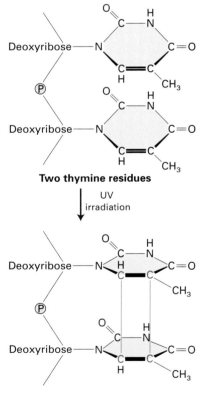

Two thymine residues

UV irradiation

Thymine-thymine dimer residue

▲ FIGURE 23-29 Formation of thymine-thymine dimers. The most common type of DNA damage caused by UV irradiation, thymine-thymine dimers can be repaired by an excision-repair mechanism.

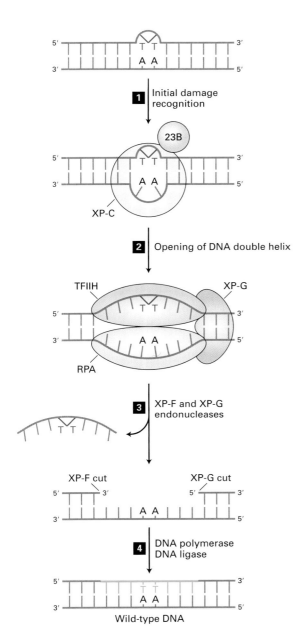

▲ **FIGURE 23-30 Nucleotide excision repair in human cells.**
A DNA lesion that causes distortion of the double helix, such as
a thymine dimer, is initially recognized by a complex of the XP-C
(xeroderma pigmentosum C protein) and 23B proteins (step **1**).
This complex then recruits transcription factor TFIIH, whose
helicase subunits, powered by ATP hydrolysis, partially unwind
the double helix. XP-G and RPA proteins then bind to the
complex and further unwind and stabilize the helix until a bubble
of ≈25 bases is formed (step **2**). Then XP-G (now acting as an
endonuclease) and XP-F, a second endonuclease, cut the
damaged strand at points 24–32 bases apart on each side of
the lesion (step **3**). This releases the DNA fragment with the
damaged bases, which is degraded to mononucleotides. Finally
the gap is filled by DNA polymerase exactly as in DNA replication
(Chapter 4), and the remaining nick is sealed by DNA ligase
(step **4**). [Adapted from J. Hoeijmakers, 2001, *Nature* **411**:366, and
O. Schärer, 2003, *Angewandte Chemie*, in press.]

roles of most of these XP proteins in nucleotide excision
repair are now well understood (see Figure 23-30).

Remarkably, five polypeptide subunits of TFIIH, a general transcription factor, are required for nucleotide excision
repair in eukaryotic cells, including two with homology to
helicases, as shown in Figure 23-30. In transcription, the
helicase activity of TFIIH unwinds the DNA helix at the start
site, allowing RNA polymerase II to begin (see Figure 11-27).
It appears that nature has used a similar protein assembly in
two different cellular processes that require helicase activity.

The use of shared subunits in transcription and DNA repair may help explain the observation that DNA damage in
higher eukaryotes is repaired at a much faster rate in regions
of the genome being actively transcribed than in nontranscribed regions—so-called transcription-coupled repair. Since
only a small fraction of the genome is transcribed in any one
cell in higher eukaryotes, transcription-coupled repair efficiently directs repair efforts to the most critical regions. In
this system, if an RNA polymerase becomes stalled at a lesion on DNA (e.g., a thymine-thymine dimer), a small protein, CSB, is recruited to the RNA polymerase; this triggers
opening of the DNA helix at that point, recruitment of
TFIIH, and the reactions of steps **2** through **4** depicted in
Figure 23-30.

Two Systems Repair Double-Strand Breaks in DNA

Ionizing radiation and some anticancer drugs (e.g.,
bleomycin) cause double-strand breaks in DNA. A cell that
has suffered a particular double-strand break usually contains other breaks. These are particularly severe lesions because incorrect rejoining of double strands of DNA can lead
to gross chromosomal rearrangements and translocations
such as those that produce a hybrid gene or bring a growth-
regulatory gene under the control of a different promoter.
The B and T cells of the immune system are particularly susceptible to DNA rearrangements caused by double-strand
breaks created during rearrangement of their immunoglobulin or T cell receptor genes, explaining the frequent involvement of these gene loci in leukemias and lymphomas.

Two systems have evolved to repair double-strand
breaks—*homologous recombination* and *DNA end-joining*.
The former is used during and after DNA replication, when
the sister chromatid is available for use as a template to repair the damaged DNA strand; homologous recombination
is error-free. The alternative mechanism, DNA end-joining,
is error-prone, since several nucleotides are invariably lost
at the point of repair. This and other error-prone repair systems are thought to mediate much of if not all the carcino-
genic effects of chemicals and radiation.

Error-Free Repair by Homologous Recombination Yeasts
can repair double-strand breaks induced by γ-irradiation.
Isolation and analysis of radiation-sensitive (*RAD*) mutants
that are deficient in this homologous recombination repair

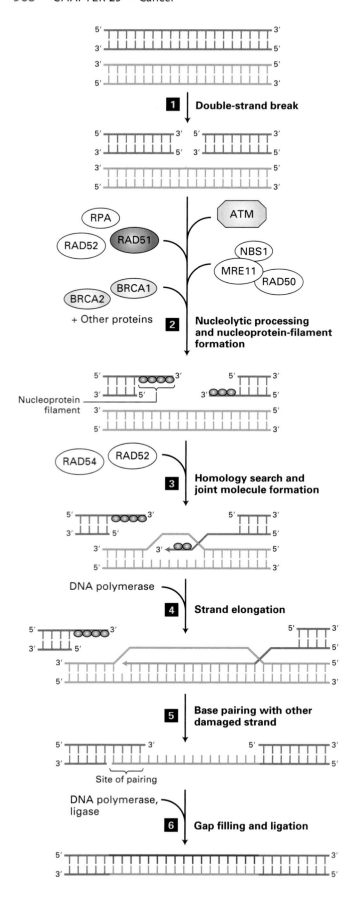

◀ **FIGURE 23-31 Repair of double-strand breaks by homologous recombination.** During S phase cells copy each chromosome to create two identical sister chromatids that later segregate into daughter cells. The black and red DNAs represent the homologous sequences on these sister chromatids. Step **1**: A double-strand DNA break forms in the chromatids. Step **2**: The double-strand break activates the ATM kinase (see Figure 23-23); this leads to activation of a set of exonucleases that remove nucleotides at the break first from the 3′ and then from the 5′ ends of both broken strands, ultimately creating single-stranded 3′ ends. In a process that is dependent on the BRCA1 and BRCA2 proteins, as well as others, the Rad51 protein (green ovals) polymerizes on single-stranded DNA with a free 3′ end to form a nucleoprotein filament. Step **3**: Aided by yet other proteins, one Rad51 nucleoprotein filament searches for the homologous duplex DNA sequence on the sister chromatid, then invades the duplex to form a joint molecule in which the single-stranded 3′ end is base-paired to the complementary strand on the homologous DNA strand. Step **4**: The replicative DNA polymerases elongate this 3′ end of the damaged DNA (green strand), templated by the complementary sequences in the undamaged homologous DNA segment. Step **5**: Next this repaired 3′ end of the damaged DNA pairs with the single-stranded 3′ end of the other damaged strand. Step **6**: Any remaining gaps are filled in by DNA polymerase and ligase (light green), regenerating a wild-type double helix in which an entire segment (dark and light green) has been regenerated from the homologous segment of the sister chromatid. [Adapted from D. van Gant et al., 2001, *Nature Rev. Genet.* **2**:196.]

system facilitated study of the process. Virtually all the yeast Rad proteins have homologs in the human genome, and the human and yeast proteins function in an essentially identical fashion (Figure 23-31). At one time homologous recombination was thought to be a minor repair process in human cells. This changed when it was realized that several human cancers are potentiated by inherited mutations in genes essential for homologous recombination repair (see Table 23-1). For example, the vast majority of women with inherited susceptibility to breast cancer have a mutation in one allele of either the BCRA-1 or the BCRA-2 genes that encode proteins participating in this repair process. Loss or inactivation of the second allele inhibits the homologous recombination repair pathway and thus tends to induce cancer in mammary or ovarian epithelial cells.

Repair of a double-strand break by homologous recombination involves reactions between three DNA molecules—the two DNA ends and the intact DNA strands from the sister chromatid (see Figure 23-31). In this process single-stranded DNAs with 3′ ends are formed from the ends of the broken DNAs and then coated with the Rad51 protein. One Rad51 nucleoprotein filament searches for the homologous duplex DNA sequence in the sister chromatid. This 3′ end is then elongated (green in Figure 23-31) by DNA polymerase, templated by the complementary strand on the homologous DNA. When sufficiently long, this single strand base-pairs with the single-stranded 3' end of the other broken DNA, and DNA polymerase and DNA ligase fill in the gaps. This process regenerates a wild-type double helix with the correct sequence, and in general no mutations are induced during repair by homologous recombination.

Error-Prone Repair by End-Joining In multicellular organisms, the predominant mechanism for repairing double-strand breaks involves rejoining the nonhomologous ends of two DNA molecules. Even if the joined DNA fragments come from the same chromosome, the repair process results in loss of several base pairs at the joining point (Figure 23-32). Formation of such a possibly mutagenic deletion is one example of how repair of DNA damage can introduce mutations.

Since movement of DNA within the protein-dense nucleus is fairly minimal, the correct ends are generally rejoined

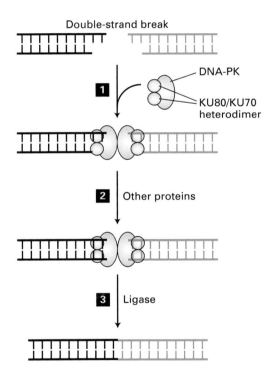

▲ FIGURE 23-32 Repair of double-strand breaks by end-joining. In general, nucleotide sequences are butted together that were not apposed in the unbroken DNA. These DNA ends are usually from the same chromosome locus, and when linked together, several base pairs are lost. Occasionally, ends from different chromosomes are accidentally joined together. A complex of two proteins, Ku and DNA-dependent protein kinase, binds to the ends of a double-strand break (**1**). After formation of a synapse, the ends are further processed by nucleases, resulting in removal of a few bases (**2**), and the two double-stranded molecules are ligated together (**3**). As a result, the double-strand break is repaired, but several base pairs at the site of the break are removed. [Adapted from G. Chu, 1997, *J. Biol. Chem.* **272**:24097; M. Lieber et al., 1997, *Curr. Opin. Genet. Devel.* **7**:99; and D. van Gant et al., 2001, *Nature Rev. Genet.* **2**:196.]

together, albeit with loss of base pairs. However, occasionally broken ends from different chromosomes are joined together, leading to translocation of pieces of DNA from one chromosome to another. As we have seen, such translocations may generate chimeric oncogenes or place a proto-oncogene next to, and thus under the inappropriate control of, a promoter from another gene. The devastating effects of double-strand breaks make these the "most unkindest cuts of all," to borrow a phrase from Shakespeare's *Julius Caesar*.

Telomerase Expression Contributes to Immortalization of Cancer Cells

Telomeres, the physical ends of linear chromosomes, consist of tandem arrays of a short DNA sequence, TTAGGG in vertebrates. Telomeres provide the solution to the end-replication problem—the inability of DNA polymerases to completely replicate the end of a double-stranded DNA molecule. *Telomerase,* a reverse transcriptase that contains an RNA template, adds TTAGGG repeats to chromosome ends to lengthen or maintain the 5- to 20-kb regions of repeats that decorate the ends of human chromosomes (see Figure 10-34). Germ-line cells and rapidly dividing somatic cells (e.g., stem cells) produce telomerase, but most human somatic cells lack telomerase. As a result, their telomeres shorten with each cell cycle. Complete loss of telomeres leads to end-to-end chromosome fusions and cell death. Extensive shortening of telomeres is detected as a kind of DNA damage, with consequent stabilization and activation of p53 protein, leading to p53-triggered apoptosis.

Most tumor cells, despite their rapid proliferation rate, overcome this fate by expressing telomerase. Many researchers believe that telomerase expression is essential for a tumor cell to become immortal, and specific inhibitors of telomerase have been suggested as cancer therapeutic agents. Introduction of telomerase-producing transgenes into cultured human cells that otherwise lack the enzyme can extend their lifespan by more than 20 doublings while maintaining telomere length. The initial finding that mice homozygous for a deletion of the RNA subunit of telomerase are viable and fertile was surprising. However, after four to six generations defects began to appear in the telomerase-null mice as their very long telomeres (40–60 kb) became significantly shorter. The defects included depletion of tissues that require high rates of cell division, like skin and intestine, and infertility.

When treated with carcinogens, telomerase-null mice develop tumors less readily than normal mice do. For example, papillomas induced by a combination of chemical carcinogens occur 20 times less frequently in mice lacking a functional telomerase than in normal mice. Mice with an *APC* mutation normally develop colon tumors, and these too are reduced if the mice lack telomerase. Some other tumors are less affected by loss of telomerase. These studies demonstrate the relevance of telomerase for unbridled cell division and make the enzyme a possible target for chemotherapy.

KEY CONCEPTS OF SECTION 23.5

The Role of Carcinogens and DNA Repair in Cancer

■ Changes in the DNA sequence result from copying errors and the effects of various physical and chemical agents, or carcinogens. All carcinogens are mutagens; that is, they alter one or more nucleotides in DNA.

■ Many copying errors that occur during DNA replication are corrected by the proofreading function of DNA polymerases that can recognize incorrect (mispaired) bases at the 3′ end of the growing strand and then remove them by an inherent 3′→5′ exonuclease activity (see Figure 23-24).

■ Indirect carcinogens, the most common type, must be activated before they can damage DNA. In animals, metabolic activation occurs via the cytochrome P-450 system, a pathway generally used by cells to rid themselves of noxious foreign chemicals.

■ Benzo(a)pyrene, a component of cigarette smoke, causes inactivating mutations in the *p53* gene at the same positions that are mutated in many human lung tumors.

■ Eukaryotic cells have three excision-repair systems for correcting mispaired bases and for removing UV-induced thymine-thymine dimers or large chemical adducts from DNA. Base excision repair, mismatch repair, and nucleotide excision repair operate with high accuracy and generally do not introduce errors.

■ Inherited defects in the nucleotide excision-repair pathway, as in individuals with xeroderma pigmentosum, predispose them to skin cancer. Inherited colon cancer frequently is associated with mutant forms of proteins essential for the mismatch repair pathway.

■ Error-free repair of double-strand breaks in DNA is accomplished by homologous recombination using the undamaged sister chromatid as template.

■ Defects in repair by homologous recombination are associated with inheritance of one mutant allele of the BRCA-1 or BRCA-2 gene and result in predisposition to breast cancer.

■ Repair of double-strand breaks by the end-joining pathway can link segments of DNA from different chromosomes, possibly forming an oncogenic translocation. The repair mechanism also produces a small deletion, even when segments from the same chromosome are joined.

■ Inherited defects in other cellular DNA-repair processes found in certain human diseases are associated with an increased susceptibility for certain cancers (see Table 23-1).

■ Cancer cells, like germ cells and stem cells but unlike most differentiated cells, produce telomerase, which prevents shortening of chromosomes during DNA replication and may contribute to their immortalization. The absence of telomerase is associated with resistance to generation of certain tumors.

PERSPECTIVES FOR THE FUTURE

The recognition that cancer is fundamentally a genetic disease has opened enormous new opportunities for preventing and treating the disease. Carcinogens can now be assessed for their effects on known steps in cell-cycle control. Genetic defects in the checkpoint controls for detecting damaged DNA and in the systems for repairing it can be readily recognized and used to explore the mechanisms of cancer. The multiple changes that must occur for a cell to grow into a dangerous tumor present multiple opportunities for intervention. Identifying mutated genes associated with cancer points directly to proteins at which drugs can be targeted.

Diagnostic medicine is being transformed by our newfound ability to monitor large numbers of cell characteristics. The traditional methods of assessing possible tumor cells, mainly microscopy of stained cells, will be augmented or replaced by techniques for measuring the expression of tens of thousands of genes, focusing particularly on genes whose activities are identified as powerful indicators of the cell's growth properties and the patient's prognosis. Currently, DNA microarray analysis permits measurement of gene transcription. In the future, techniques for systematically measuring protein production, modification, and localization, all important measures of cell states, will give us even more refined portraits of cells. Tumors now viewed as identical or very similar will instead be recognized as distinctly different and given appropriately different treatments. Earlier detection of tumors, based on better monitoring of cell properties, should allow more successful treatment. A focus on that particularly destructive process, metastasis, should be successful in identifying more of the mechanisms used by cells to migrate, attach, and invade. Manipulation of angiogenesis continues to look hopeful as a means of suffocating tumors.

The molecular cell biology of cancer provides avenues for new therapies, but prevention remains crucial and preferable to therapy. Avoidance of obvious carcinogens, in particular cigarette smoke, can significantly reduce the incidence of lung cancer and perhaps other kinds as well. Beyond minimizing exposure to carcinogens such as smoke or sunlight, certain specific approaches are now feasible. New knowledge of the involvement of human papillomavirus 16 in most cases of cervical cancer holds promise for developing a cancer vaccine that will prevent viral action. Antibodies against cell surface markers that distinguish cancer cells are a source of great hope, especially after successes with the clinical use of monoclonal antibodies against human EGF receptor 2 (Her2), a protein involved in some cases of human breast cancer. Further steps must involve medicine and science. Understanding the cell biology of cancer is a critical first step toward prevention and cure, but the next steps are hard. The success with Gleevec (STI-571) against leukemia is exceptional; many cancers remain difficult to treat and cause enormous suffering. Since *cancer* is a term for a group of highly diverse diseases, interventions that are successful for one type may not be useful for others. Despite these daunting realities, we are beginning to reap the benefits of decades of research exploring the molecular biology of the cell. We hope that many of the readers of this book will help to overcome the obstacles that remain.

KEY TERMS

aflatoxin 964

aneuploidy 961

benign 936

Burkitt's lymphoma 956

carcinogens 935

carcinomas 938

deamination 963

DNA end-joining 967

excision-repair systems 964

leukemias 938

loss of heterozygosity (LOH) 947

malignant 936

metastasis 937

mutagens 963

nondisjunction 947

oncogene 940

p53 protein 959

Philadelphia chromosome 954

proto-oncogene 935

sarcomas 938

slow-acting retroviruses 946

thymine-thymine dimers 966

transducing retroviruses 946

transformation 939

tumor-suppressor gene 935

REVIEW THE CONCEPTS

1. What characteristics distinguish benign from malignant tumors? With respect to gene mutations, what distinguishes benign colon polyps from malignant colon carcinoma?

2. Ninety percent of cancer deaths are caused by metastatic rather than primary tumors. Define *metastasis*. Explain the rationale for the following new cancer treatments: (a) batimastat, an inhibitor of matrix metalloproteinases and of the plasminogen activator receptor, (b) antibodies that block the function of integrins, integral membrane proteins that mediate attachment of cells to the basal laminae and extracellular matrices of various tissues, and (c) bisphosphonate, which inhibits the function of bone-digesting osteoclasts.

3. Because of oxygen and nutrient requirements, cells in a tissue must reside within 100 μm of a blood vessel. Based on this information, explain why many malignant tumors often possess gain-of-function mutations in one of the following genes: *bFGF, TGFα,* and *VEGF.*

4. What hypothesis explains the observations that incidence of human cancers increases exponentially with age? Give an example of data that confirm the hypothesis.

5. Distinguish between proto-oncogenes and tumor-suppressor genes. To become cancer promoting, do proto-oncogenes and tumor-suppressor genes undergo gain-of-function or loss-of-function mutations? Classify the following genes as proto-oncogenes or tumor-suppressor genes: *p53, ras, Bcl-2, telomerase, jun,* and *p16.*

6. Hereditary retinoblastoma generally affects children in both eyes, while spontaneous retinoblastoma usually occurs during adulthood only in one eye. Explain the genetic basis for the epidemiologic distinction between these two forms of retinoblastoma. Explain the apparent paradox: loss-of-function mutations in tumor-suppressor genes act recessively, yet hereditary retinoblastoma is inherited as an autosomal dominant.

7. Explain the concept of loss of heterozygosity (LOH). Why do most cancer cells exhibit LOH of one or more genes? How does failure of the spindle assembly checkpoint lead to loss of heterozygosity?

8. Many malignant tumors are characterized by the activation of one or more growth factor receptors. What is the catalytic activity associated with transmembrane growth factor receptors such as the EGF receptor? Describe how the following events lead to activation of the relevant growth factor receptor: (a) expression of the viral protein gp55, (b) translocation that replaces the extracellular domain of the Trk receptor with the N-terminal region of tropomyosin, (c) point mutation that converts a valine to glutamine within the transmembrane region of the Her2 receptor.

9. Describe the common signal-transduction event that is perturbed by cancer-promoting mutations in the genes encoding Ras and NF-1. Why are mutations in Ras more commonly found in cancers than mutations in NF-1?

10. What is the structural distinction between the proteins encoded by c-*src* and v-*src*? How does this difference render v-*src* oncogenic?

11. Describe the mutational event that produces the *myc* oncogene in Burkitt's lymphoma. Why does the particular mechanism for generating oncogenic *myc* result in a lymphoma rather than another type of cancer? Describe another mechanism for generating oncogenic *myc.*

12. Pancreatic cancers often possess loss-of-function mutations in the gene that encodes the SMAD4 protein. How does this mutation promote the loss of growth inhibition and highly metastatic phenotype of pancreatic tumors?

13. Loss of p53 function occurs in the majority of human tumors. Name two ways in which loss of p53 function contributes to a malignant phenotype. Explain the mechanism by which the following agents cause loss of p53 function: (a) human papillomavirus and (b) benzo(*a*)pyrene.

14. DNA-repair systems are responsible for maintaining genomic fidelity in normal cells despite the high frequency with which mutational events occur. What type of DNA mutation is generated by (a) UV irradiation and (b) ionizing radiation? Describe the system responsible for repairing each of these types of mutations in mammalian cells. Postulate why a loss of function in one or more DNA-repair systems typifies many cancers.

15. Which human cell types possess telomerase activity? What characteristic of cancer is promoted by expression of telomerase? What concerns does this pose for medical therapies involving stem cells?

ANALYZE THE DATA

In a recent study by Elizabeth Snyderwine and colleagues, cDNA microarray profiling was used to compare the expression profiles for 6900 genes in normal and malignant breast tissues from rats. RNA was extracted from the following tissues:

a. Breast tissue from virgin rats

b. Breast tissue from pregnant rats

c. Breast tissue from lactating rats

d. Breast carcinoma induced by the meat-derived carcinogen PhIP

e. Breast carcinoma induced by the experimental carcinogen DMBA

After the microarray slides were hybridized with labeled cDNAs derived from these 5 populations of RNAs, the data were analyzed by several different comparisons.

In comparison 1, tissues a, b, and c were grouped together as "normal" tissue samples, and d and e were grouped as "carcinoma" samples. Genes that were either induced (more than twofold increase in expression) or repressed (more than twofold decrease in expression) in both carcinoma samples relative to all three normal samples were determined. A partial listing of differentially expressed genes is shown in the accompanying table.

Induced Genes

Cell-Growth and Cell-Cycle-related Genes

  Platelet-derived growth factor A chain (PDGF-A)

  Cyclin-dependent kinase 4 (Cdk4)

  Cyclin D

Signal-Transduction and Transcription-Related Genes

  STAT5a

Repressed Genes

Extracellular Matrix Genes

  Alpha 1 type V collagen

  Fibronectin 1

  Desmin

1. What was the purpose of analyzing breast tissue from pregnant and lactating rats?

2. What characteristic of cancer is promoted by the overexpression of PDGF-A, Cdk4, and cyclin D?

3. What characteristic of cancer might be promoted by the repression of extracellular matrix genes?

4. STAT5a is a transcription factor that regulates the expression of cyclin D, Bcl-X$_L$, and other genes. Why is it possible for these carcinomas that no mutation occurs in the cyclin D gene despite its overexpression? Why are mutations in transcription-factor genes like STAT5a commonly found in cancer cells?

In comparison 2, expression profiles of carcinomas induced by PhIP and DMBA were compared with each other. In this analysis, some distinctions were found, but the number of differentially expressed genes was far less than the number of genes identified when comparing the grouped "normal" samples (a, b, and c) to the grouped "carcinoma" samples (d and e).

5. On the basis of this information, what can be generalized about the molecular profile of breast carcinomas?

6. Would you anticipate greater or fewer differences in gene expression if two distinct types of cancer (e.g., breast carcinoma vs. B-cell lymphoma) induced by the same carcinogen were compared by microarray analysis?

REFERENCES

Tumor Cells and the Onset of Cancer

Fidler, I. J. 2002. The pathogenesis of cancer metastasis: the "seed and soil" hypothesis revisited. *Nature Rev. Cancer* 3:1–6.

Folkman, J. 2002. Role of angiogenesis in tumor growth and metastasis. *Semin. Oncol.* 29:15–18.

Hanahan, D., and R. A. Weinberg. 2000. The hallmarks of cancer. *Cell* 100:57–70.

Jain, M., et al. 2002. Sustained loss of a neoplastic phenotype by brief inactivation of MYC. *Science* 297:102–104.

Kinzler, K. W., and B. Vogelstein. 1996. Lessons from hereditary colo-rectal cancer. *Cell* 87:159–170.

Klein, G. 1998. Foulds' dangerous idea revisited: the multistep development of tumors 40 years later. *Adv. Cancer Res.* 72:1–23.

Rafii, S., et al. 2002. Vascular and haematopoietic stem cells: novel targets for anti-angiogenesis therapy? *Nature Rev. Cancer* 2:826–835.

Ramaswamy, S., et al. 2003. A molecular signature of metastasis in primary solid tumors. *Nature Genet.* 33:49–54.

Trusolino, L., and P. M. Comoglio. 2002. Scatter-factor and semaphorin receptors: cell signalling for invasive growth. *Nature Rev. Cancer* 2:289–300.

Yancopoulos, G., M. Klagsburn, and J. Folkman. 1998. Vasculogenesis, angiogenesis, and growth factors: ephrins enter the fray at the border. *Cell* 93:661–664.

The Genetic Basis of Cancer

Bienz, M., and H. Clevers. 2000. Linking colorectal cancer to Wnt signaling. *Cell* 103:311–320.

Clark, J., et al. 2002. Identification of amplified and expressed genes in breast cancer by comparative hybridization onto microarrays of randomly selected cDNA clones. *Genes Chrom. Cancer* **34**:104–114.

Classon, M., and E. Harlow. 2002. The retinoblastoma tumour suppressor in development and cancer. *Nature Rev. Cancer* **2**:910–917.

Hesketh, R., ed. 1997. *The Oncogene and Tumor Suppressor Gene Facts Book*, 2d ed. Academic Press.

Hunter, T. 1997. Oncoprotein networks. *Cell* **88**:333–346.

Moon, R. T., et al. 2002. The promise and perils of Wnt signaling through beta-catenin. *Science* **296**:1644–1646.

Nevins, J. R. 2001. The Rb/E2F pathway and cancer. *Hum. Mol. Genet.* **10**:699–703.

Polakis, P. 2000. Wnt signaling and cancer. *Genes Devel.* **14**:1837–1851.

Sasaki, T., et al. 2000. Colorectal carcinomas in mice lacking the catalytic subunit of PI(3)Kgamma. *Nature* **406**:897–902.

Sherr, C. J., and F. McCormick. 2002. The RB and p53 pathways in cancer. *Cancer Cell* **2**:103–112.

Taipale, J., and P. A. Beachy. 2001. The Hedgehog and Wnt signalling pathways in cancer. *Nature* **411**:349–354.

van 't Veer, L. J., et al. 2002. Gene expression profiling predicts clinical outcome of breast cancer. *Nature* **415**:530–536.

West, M., et al. 2001. Predicting the clinical status of human breast cancer by using gene expression profiles. *Proc. Nat'l. Acad. Sci. USA* **98**:11462–11467.

White, R. 1998. Tumor suppressor pathways. *Cell* **92**:591–592.

Oncogenic Mutations in Growth-Promoting Proteins

Capdeville, R., et al. 2002. Glivec (STI571, imatinib), a rationally developed, targeted anticancer drug. *Nature Rev. Drug Discov.* **1**:493–502.

Downward, J. 2003. Targeting RAS signalling pathways in cancer therapy. *Nature Rev. Cancer* **3**:11–22.

Rowley, J. D. 2001. Chromosome translocations: dangerous liaisons revisited. *Nature Rev. Cancer* **1**:245–250.

Sahai, E., and C. J. Marshall. 2002. RHO-GTPases and cancer. *Nature Rev. Cancer* **2**:133–142.

Shaulian, E., and M. Karin. 2002. AP-1 as a regulator of cell life and death. *Nature Cell Biol.* **4**:E131–E136.

Shawver, L. K., D. Slamon, and A. Ullrich. 2002. Smart drugs: tyrosine kinase inhibitors in cancer therapy. *Cancer Cell* **1**:117–123.

Mutations Causing Loss of Growth-Inhibiting and Cell-Cycle Controls

Chau, B. N., and J. Y. Wang. 2003. Coordinated regulation of life and death by RB. *Nature Rev. Cancer* **3**:130–138.

Malumbres, M., and M. Barbacid. 2001. To cycle or not to cycle: a critical decision in cancer. *Nature Rev. Cancer* **1**:222–231.

Mathon, N. F., and A. C. Lloyd. 2001. Cell senescence and cancer. *Nature Rev. Cancer* **1**:203–213.

Paulovich, A. G., D. P. Toczyski, and L. H. Hartwell. 1997. When checkpoints fail. *Cell* **88**:315–321.

Planas-Silva, M. D., and R. A. Weinberg. 1997. The restriction point and control of cell proliferation. *Curr. Opin. Cell Biol.* **9**:768–772.

Sherr, C. J. 2000. Cell cycle control and cancer. *Harvey Lect.* **96**:73–92.

Westphal, C. H. 1997. Cell-cycle signaling: Atm displays its many talents. *Curr. Biol.* **7**:R789–R792.

Zhang, L., et al. 2000. Role of BAX in the apoptotic response to anticancer agents. *Science* **290**:989–992.

The Role of Carcinogens and DNA Repair in Cancer

Batty, D., and R. Wood. 2000. Damage recognition in nucleotide excision repair of DNA. *Gene* **241**:193–204.

D'Andrea, A., and M. Grompe. 2003. The Fanconi anemia/BRCA pathway. *Nature Rev. Cancer* **3**:23–34.

Fishel, R. 1998. Mismatch repair, molecular switches, and signal transduction. *Genes Devel.* **12**:2096–2101.

Flores-Rozas, H., and R. Kolodner. 2000. Links between replication, recombination and genome instability in eukaryotes. *Trends Biochem. Sci.* **25**:196–200.

Friedberg, E. 2003. DNA damage and repair. *Nature* **421**:436–440.

Friedberg, E., G. Walker, and W. Siede. 1995. *DNA Repair and Mutagenesis*. ASM Press.

Hoeijmakers, J. 2001. Genome maintenance mechanisms for preventing cancer. *Nature* **411**:366–374.

Lindahl, T. 2001. Keynote: past, present, and future aspects of base excision repair. *Prog. Nucl. Acid Res. Mol. Biol.* **68**:xvii–xxx.

Muller, A., and R. Fishel. 2002. Mismatch repair and the hereditary non-polyposis colorectal cancer syndrome (HNPCC). *Cancer Invest.* **20**:102–109.

Schärer, O. 2003. Chemistry and biology of DNA repair. *Angewandte Chemie*, in press.

Schärer, O., and J. Jiricny. 2001. Recent progress in the biology, chemistry and structural biology of DNA glycosylases. *Bioessays* **23**:270–281.

Somasundaram, K. 2002. Breast cancer gene 1 (BRCA1): role in cell cycle regulation and DNA repair—perhaps through transcription. *J. Cell Biochem.* **88**:1084–1091.

Thompson, L., and D. Schild. 2002. Recombinational DNA repair and human disease. *Mut. Res.* **509**:49–78.

van Gant, D., J. Hoeijmakers, and R. Kanaar. 2001. Chromosomal stability and the DNA double-stranded break connection. *Nature Rev. Genet.* **2**:196–205.

GLOSSARY

Boldfaced terms within a definition are also defined in this glossary. Figures and tables that illustrate defined terms are noted in parentheses.

ABC superfamily   A large group of integral membrane proteins that often function as ATP-powered **membrane transport proteins** to move diverse molecules (e.g., phospholipids, cholesterol, sugars, ions, peptides) across cellular membranes. (Figure 7-11 and Table 18-2)

acetyl CoA   Small, water-soluble metabolite comprising an acetyl group linked to coenzyme A (CoA); formed during oxidation of pyruvate, fatty acids, and amino acids. Its acetyl group is transferred to citrate in the **citric acid cycle**. (Figure 8-8).

acetylcholine (ACh)   Neurotransmitter that functions at vertebrate neuromuscular junctions and at various neuron-neuron synapses in the brain and peripheral nervous system.

acid   Any compound that can donate a proton (H^+). The carboxyl and phosphate groups are the primary acidic groups in biological macromolecules.

actin   Abundant structural protein in eukaryotic cells that interacts with many other proteins. The monomeric globular form (G-actin) polymerizes to form actin filaments (F-actin). In muscle cells, F-actin interacts with **myosin** during contraction. See also **microfilament**. (Figure 19-3)

action potential   Rapid, transient, all-or-none electrical activity propagated in the plasma membrane of excitable cells (e.g., neurons and muscle cells) as the result of the selective opening and closing of voltage-gated Na^+ and K^+ channels. (Figures 7-30 and 7-35)

activation energy   The input of energy required to (overcome the barrier to) initiate a chemical reaction. By reducing the activation energy, an **enzyme** increases the rate of a reaction. (Figure 3-16)

activator   **Transcription factor** that stimulates **transcription.**

active site   Specific region on the surface of an enzyme that binds a **substrate** molecule and promotes a chemical change in the bound substrate. (Figure 3-17)

active transport   Movement of an ion or small molecule across a membrane against its concentration gradient or electrochemical gradient driven by the coupled hydrolysis of ATP. (Figure 7-2)

adenylyl cyclase   Membrane-bound enzyme that catalyzes formation of **cyclic AMP (cAMP)** from ATP; also called *adenylate cyclase*. Binding of certain ligands to their cell-surface receptors leads to activation of adenylyl cyclase and a rise in intracellular cAMP. (Figure 13-14)

adhesion receptor   Protein in the plasma membrane of animal cells that binds components of the extracellular **matrix,** thereby mediating cell-matrix adhesion. The integrins are the major adhesion receptors. (Figure 6-1)

aerobic   Referring to a cell, organism, or metabolic process that utilizes gaseous oxygen (O_2) or that can grow in the presence of O_2.

aerobic oxidation   Oxygen-requiring metabolism of sugars and fatty acids to CO_2 and H_2O coupled to the synthesis of ATP.

agonist   A molecule, often synthetic, that mimics the biological function of a natural molecule. Commonly used in reference to an analog of a hormone that binds to the hormone's receptor and induces the normal response.

allele   One of two or more alternative forms of a gene. Diploid cells contain two alleles of each gene, located at the corresponding site (locus) on **homologous chromosomes.**

allosteric   Referring to proteins and cellular processes that are regulated by **allostery.**

allostery   Change in the tertiary and/or quaternary structure of a protein induced by binding of a small molecule to a specific regulatory site, causing a change in the protein's activity. Many multisubunit proteins exhibit allosteric regulation.

alpha (α) helix   Common protein **secondary structure** in which the linear sequence of amino acids is folded into a right-handed spiral stabilized by hydrogen bonds between carboxyl and amide groups in the backbone. (Figure 3-3)

amino acid   An organic compound containing at least one amino group and one carboxyl group. In the amino acids that are the **monomers** for building proteins, an amino group and carboxyl group are linked to a central carbon atom, the α carbon, to which a variable side chain is attached. (Figures 2-12 and 2-13)

aminoacyl-tRNA   Activated form of an amino acid, used in protein synthesis, consisting of an amino acid linked via a high-energy ester bond to the 3′-hydroxyl group of a **tRNA** molecule. (Figure 4-21)

amphipathic   Referring to a molecule or structure that has both a **hydrophobic** and a **hydrophilic** part.

anaerobic   Referring to a cell, organism, or metabolic process that functions in the absence of gaseous oxygen (O_2).

anaphase   Mitotic stage during which the sister chromatids (or duplicated homologues in meiosis I) separate and move apart (segregate) toward the spindle poles. (Figure 20-29)

anchoring junctions   Specialized regions on the cell surface, containing **cell-adhesion molecules** or **adhesion receptors,** that interact with other cells or the extracellular matrix and with cytoskeletal fibers. The major types in epithelial cells are *adherens junctions* and *desmosomes,* which mediate cell-cell adhesion, and *hemidesmosomes,* which mediate cell-matrix adhesion. (Figures 6-7 and 6-8)

aneuploidy   Any deviation from the normal **diploid** number of chromosomes in which extra copies of one or more chromosomes are present or one of the normal copies is missing; a common chromosomal aberration in cancer cells.

antagonist A molecule, often synthetic, that blocks the biological function of a natural molecule. Commonly used in reference to an analog of a hormone that binds to the hormone's receptor but induces no response; by preventing binding of the normal hormone, the antagonist reduces its usual physiological effect.

antibody A protein (immunoglobulin) that interacts with a particular site (**epitope**) on an antigen and facilitates clearance of that antigen by various mechanisms. (Figure 3-15)

anticodon Sequence of three nucleotides in a tRNA that is complementary to a **codon** in an mRNA. During protein synthesis, base pairing between a codon and anticodon aligns the tRNA carrying the corresponding amino acid for addition to the growing polypeptide chain.

antigen Any material (usually foreign) that elicits production of and is specifically bound by an **antibody**.

antiport A type of **cotransport** in which a membrane protein (antiporter) transports two different molecules or ions across a cell membrane in opposite directions. See also **symport**.

apical Referring to the tip (apex) of a cell, an organ, or other body structure. In the case of epithelial cells, the apical surface is exposed to the exterior of the body or to an internal open space (e.g., intestinal lumen, duct). (Figure 6-4)

apoptosis Regulated process leading to cell death mediated by a **caspase** cascade and marked by a series of well-defined morphological changes; also called *programmed cell death*. (Figures 22-26 and 22-32a)

archaea Class of **prokaryotes** that constitutes one of the three distinct evolutionary lineages of modern-day organisms; also called *archaebacteria* and *archaeans*. In some respects, archaeans are more similar to eukaryotes than to the so-called true bacteria (eubacteria). (Figure 1-3)

association constant (K_a) See **equilibrium constant**.

aster Structure composed of microtubules (astral fibers) that radiate outward from a **centrosome** during mitosis. (Figure 20-31)

asymmetric carbon atom A carbon atom bonded to four different atoms or chemical groups; also called *chiral carbon atom*. The bonds can be arranged in two different ways, producing **stereoisomers** that are mirror images of each other. (Figure 2-12)

asymmetric cell division Any cell division in which the two daughter cells receive the same genes but otherwise inherit different components (e.g., mRNAs, proteins) from the parental cell. (Figure 22-20)

atherosclerotic plaque A well-demarcated deposition in an artery wall consisting in its early stages of foam cell macrophages and extracellular matrix and in its later stages of a necrotic core of lipids, calcium salts, and extracellular matrix overlain by a fibrous cap of smooth muscle cells and matrix. In advanced atherosclerosis, a plaque may fully block an artery or induce formation of an occlusive blood clot. (Figure 18-19)

ATP (adenosine 5′-triphosphate) A nucleotide that is the most important molecule for capturing and transferring **free energy** in cells. Hydrolysis of each of the two **phosphoanhydride bonds** in ATP releases a large amount of free energy that is used to drive energy-requiring cellular processes. (Figures 1-14 and 2-24)

ATP synthase Multimeric protein complex bound to inner mitochondrial membranes, thylakoid membranes of chloroplasts, and the bacterial plasma membrane that catalyzes synthesis of ATP during oxidative phosphorylation and photosynthesis; also called *F_0F_1 complex*. (Figure 8-24)

ATPase One of a large group of enzymes that catalyze hydrolysis of **ATP** to yield ADP and inorganic phosphate with release of free energy. See also **Na$^+$/K$^+$ ATPase** and **pumps**.

autocrine Referring to signaling mechanism in which a cell binds and responds to a signaling molecule (e.g., growth factor) that it produces itself.

autoradiography Technique for visualizing radioactive molecules in a sample (e.g., a tissue section or electrophoretic gel) by exposing a photographic film or emulsion to the sample. The exposed film is called an autoradiogram or autoradiograph.

autosome Any chromosome other than a sex chromosome.

axon Long process extending from the cell body of a neuron that is capable of conducting an electric impulse (**action potential**) generated at the junction with the cell body toward its distal, branching end (the axon terminus). (Figure 7-29)

axoneme Bundle of **microtubules** and associated proteins present in **cilia** and **flagella** and responsible for their movement. (Figure 20-25)

bacteriophage (phage) Any virus that infects bacterial cells. Some bacteriophages are widely used as **vectors** in **DNA cloning**.

basal body Structure at the base of **cilia** and **flagella** from which microtubules forming the axoneme radiate; structurally similar to a **centriole**.

basal lamina (pl. basal laminae) A thin sheetlike network of extracellular-matrix components that underlies most animal epithelia and other organized groups of cells (e.g., muscle), separating them from connective tissue or other cells. (Figure 6-13)

base Any compound, usually containing nitrogen, that can accept a proton (H$^+$). Commonly used to denote the **purines** and **pyrimidines** in DNA and RNA.

base pair Association of two complementary **nucleotides** in a DNA or RNA molecule stabilized by hydrogen bonding between their base components. Adenine pairs with thymine or uracil (A · T, A · U) and guanine pairs with cytosine (G · C). (Figure 4-3b)

basic helix-loop-helix (bHLH) Structural **motif** of the DNA-binding domain of a class of dimeric eukaryotic **transcription factors**. (Figure 11-22b)

basolateral Referring to the base and side of a cell, an organ, or other body structure. In the case of epithelial cells, the basolateral surface abuts adjacent cells and the underlying **basal lamina**. (Figure 6-5)

Bcl-2 family A family of proteins usually localized to the mitochondrion that either promote or inhibit **apoptosis**.

benign Referring to a tumor containing cells that closely resemble normal cells. Benign tumors stay in the tissue where they originate. See also **malignant**.

beta (β) sheet A planar **secondary structure** of proteins that is created by hydrogen bonding between the backbone atoms in two different polypeptide chains or segments of a single folded chain. (Figure 3-4)

bilayer See **phospholipid bilayer.**

bile acids Derivatives of **cholesterol,** produced in the liver and secreted into the bile, that help emulsify dietary fats for digestion and absorption in the intestines; are recycled via the **enterohepatic circulation.** (Figure 18-6)

biomembrane Permeability barrier, surrounding cells or organelles, that consists of a **phospholipid bilayer** and associated membrane proteins, as well as cholesterol and glycolipids in many cases. (Figure 5-2)

buffer A solution of the acid (HA) and base (A$^-$) form of a compound that undergoes little change in pH when small quantities of strong acid or base are added.

cadherins A family of dimeric **cell-adhesion molecules** that aggregate in adherens junctions and desmosomes and mediate Ca^{2+}-dependent cell-cell homophilic interactions. (Figure 6-2)

calmodulin A small cytosolic regulatory protein that binds four Ca^{2+} ions. The Ca^{2+}/calmodulin complex binds to many proteins, thereby activating or inhibiting them. (Figure 3-28)

calorie A unit of heat (thermal energy). One calorie is the amount of heat needed to raise the temperature of 1 gram of water by 1 °C. The *kilocalorie* (kcal) commonly is used to indicate the energy content of foods and the **free energy** of a system.

Calvin cycle The major metabolic pathway that fixes CO$_2$ into carbohydrates during photosynthesis; also called *carbon fixation.* It is indirectly dependent on light but can occur both in the dark and light. (Figure 8-42)

cAMP-dependent protein kinase See **protein kinase A.**

cancer General term denoting any of various malignant tumors, whose cells grow and divide more rapidly than normal, invade surrounding tissue, and generally spread (metastasize) to other sites.

capsid The outer proteinaceous coat of a **virus,** formed by multiple copies of one or more protein subunits and enclosing the viral nucleic acid.

carbohydrate General term for certain polyhydroxyaldehydes, polyhydroxyketones, or compounds derived from these usually having the formula (CH$_2$O)$_n$. Primary type of compound used for storing and supplying energy in animal cells. (Figure 2-16)

carbon fixation See **Calvin cycle.**

carcinogen Any chemical or physical agent that can cause cancer when cells or organisms are exposed to it.

caspase One of several vertebrate protein-degrading enzymes (proteases) that function in programmed cell death. (Figure 22-32)

catabolism Cellular degradation of complex molecules to simpler ones usually accompanied by the release of energy. Anabolism is the reverse process in which energy is used to synthesize complex molecules from simpler ones.

catalyst A substance that increases the rate of a chemical reaction without undergoing a permanent change in its structure. **Enzymes** are protein catalysts that catalyze most biochemical reactions, although some RNAs have catalytic activity.

CDK See **cyclin-dependent kinase.**

cDNA (complementary DNA) DNA molecule copied from an mRNA molecule by **reverse transcriptase** and therefore lacking the introns present in the DNA of the **genome.** Sequencing of a cDNA permits the amino acid sequence of the encoded protein to be deduced; expression of cloned cDNAs can be used to produce large quantities of their encoded proteins in vitro.

cell-adhesion molecules (CAMs) Proteins in the plasma membrane of cells that bind similar proteins on other cells, thereby mediating cell-cell adhesion. The four major classes of CAMs are the cadherins, selectins, immunoglobulin (Ig) superfamily members, and integrins, which also function in cell-matrix interactions. (Figure 6-2)

cell cycle Ordered sequence of events in which a eukaryotic cell duplicates its chromosomes and divides into two; normally consists of four phases: G$_1$ before DNA synthesis occurs; S when DNA replication occurs; G$_2$ after DNA synthesis; and M when **cell division** occurs, yielding two daughter cells. Under certain conditions, cells exit the cell cycle during G$_1$ and remain in the G$_0$ state as nondividing cells. (Figure 21-1)

cell division Separation of a cell into two daughter cells. In higher eukaryotes, it involves division of the nucleus (**mitosis**) and of the cytoplasm (**cytokinesis**); mitosis often is used to refer to both nuclear and cytoplasmic division.

cell junctions Specialized regions on the cell surface through which cells are joined to each other or to the extracellular matrix. (Figure 6-5)

cell line A population of cultured cells, of plant or animal origin, that has undergone a genetic change allowing the cells to grow indefinitely. Cell lines can result from chemical or viral **transformation** and are said to be immortal.

cell strain A population of cultured cells, of plant or animal origin, that has a finite life span and eventually dies, commonly after 25–50 generations.

cell wall A specialized, rigid extracellular matrix that lies next to the plasma membrane, protecting a cell and maintaining its shape. It is prominent in most fungi, plants, and prokaryotes, but is not present in most multicellular animals. (Figure 6-33)

cellulose A structural polysaccharide made of glucose units linked together by β(1 → 4) **glycosidic bonds.** It forms long microfibrils, which are the major component of plant cell walls. (Figure 6-33)

centriole Either of two cylindrical structures within the **centrosome** of animal cells and containing nine sets of triplet microtubules; structurally similar to a **basal body.** (Figure 20-13b)

centromere DNA sequence required for proper **segregation** of **chromosomes** during **mitosis** and **meiosis;** the region of mitotic chromosomes where the **kinetochore** forms and that appears constricted. (Figures 10-27, 10-28, and 10-32b)

centrosome (cell center) Structure located near the nucleus of animal and plant cells that is the primary microtubule-organizing center (MTOC); in animals it contains a pair of **centrioles** embedded in a protein matrix. It divides during mitosis, forming the spindle poles.

chaperone Collective term for two types of proteins that prevent misfolding of a target protein or actively facilitate its proper folding. (Figure 3-11)

checkpoint Any of several points in the eukaryotic **cell cycle** at which progression of a cell to the next stage can be halted until conditions are suitable. (Figure 21-32)

chemical equilibrium The state of a chemical reaction in which the concentration of all products and reactants is constant because the rates of the forward and reverse reactions are equal.

chemiosmosis Process whereby an electrochemical proton gradient (pH plus electric potential) across a membrane is used to drive an energy-requiring process such as ATP synthesis or transport of molecules across a membrane against their concentration gradient; also called *chemiosmotic coupling*. (Figure 8-1)

chemotaxis Movement of a cell or organism toward or away from certain chemicals.

chimera (1) An animal or tissue composed of elements derived from genetically distinct individuals; a hybrid. (2) A protein molecule containing segments derived from different proteins.

chlorophylls A group of light-absorbing porphyrin pigments that are critical in **photosynthesis.** (Figure 8-31)

chloroplast A specialized organelle in plant cells that is surrounded by a double membrane and contains internal chlorophyll-containing membranes (**thylakoids**) where the light-absorbing reactions of **photosynthesis** occur. (Figure 8-30)

cholesterol A lipid containing the four-ring steroid structure with a hydroxyl group on one ring; a component of many eukaryotic membranes and the precursor of steroid hormones, bile acids, and vitamin D. (Figures 18-6 and 18-7)

chromatid One copy of a replicated chromosome, formed during the S phase of the cell cycle, that is joined at the **centromere** to the other copy; also called *sister chromatid*. During **mitosis,** the two chromatids separate, each becoming a chromosome of one of the two daughter cells. (Figure 10-27)

chromatin Complex of DNA, histones, and nonhistone proteins from which eukaryotic chromosomes are formed. Condensation of chromatin during mitosis yields the visible **metaphase** chromosomes. (Figures 10-19 and 10-24)

chromatography, liquid Group of biochemical techniques for separating mixtures of molecules based on their mass (gel filtration chromatography), charge (ion-exchange chromatography), or ability to bind specifically to other molecules (affinity chromatography). These techniques commonly are used to separate and purify proteins. (Figure 3-34)

chromosome In eukaryotes, the structural unit of the genetic material consisting of a single, linear double-stranded DNA molecule and associated proteins. In most prokaryotes, a single, circular double-stranded DNA molecule constitutes the bulk of the genetic material. See also **chromatin** and **karyotype.**

chylomicron A class of **lipoproteins** that is secreted by intestinal epithelial cells and transports dietary triglycerides (fat) and

cholesterol from the intestines to other tissues. (Table 18-3 and Figure 18-13b)

cilium (pl. **cilia**) Short, membrane-enclosed structure extending from the surface of eukaryotic cells and containing a core bundle of **microtubules.** Cilia usually occur in groups and beat rhythmically to move a cell (e.g., single-celled organism) or to move small particles or fluid along the surface (e.g., trachea cells). See also **axoneme** and **flagellum.**

cisterna (pl. **cisternae**) Flattened membrane-bounded compartment, as found in the Golgi complex and endoplasmic reticulum.

citric acid cycle A set of nine coupled reactions occurring in the matrix of the **mitochondrion** in which acetyl groups derived from food molecules are oxidized, generating CO_2 and reduced intermediates used to produce ATP; also called *Krebs cycle* and *tricarboxylic acid (TCA) cycle*. (Figure 8-9)

clathrin A fibrous protein that with the aid of assembly proteins polymerizes into a lattice-like network at specific regions on the cytosolic side of a membrane, thereby forming a clathrin-coated pit that buds off to form a vesicle. (Figures 17-18 and 17-19)

cleavage/polyadenylation complex Large, multiprotein complex that catalyzes the cleavage of **pre-mRNA** at a 3′ poly(A) site and the initial addition of adenylate (A) residues to form the poly(A) tail. (Figure 12-4)

clone (1) A population of genetically identical cells or organisms descended from a common ancestor. (2) Multiple identical copies of a gene or DNA fragment generated and maintained via **DNA cloning.**

cloning vector See **vector.**

codon Sequence of three nucleotides in DNA or mRNA that specifies a particular amino acid during protein synthesis; also called *triplet*. Of the 64 possible codons, three are stop codons, which do not specify amino acids and cause termination of synthesis. (Table 4-1)

coenzyme A (CoA) See **acetyl CoA.**

coiled-coil A protein structural **motif** marked by amphipathic α-helical regions that can self-associate to form stable, rodlike oligomeric proteins; commonly found in fibrous proteins and certain transcription factors. (Figure 3-6c)

collagen A triple-helical glycoprotein rich in glycine and proline that is a major component of the **extracellular matrix** and connective tissues. The numerous subtypes differ in their tissue distribution and the extracellular components and cell-surface proteins with which they associate. (Table 6-1)

complementary (1) Referring to two nucleic acid sequences or strands that can form perfect **base pairs** with each other. (2) Describing regions on two interacting molecules (e.g., an enzyme and its substrate) that fit together in a lock-and-key fashion.

complementary DNA (cDNA) See **cDNA.**

complementation, functional Procedure for screening a DNA library to identify the wild-type gene that restores the function of a defective gene in a particular mutant. (Figure 9-20)

complementation, genetic Restoration of a wild-type function in diploid heterozygous cells generated from haploid cells, each of which carries a mutation in a different gene whose encoded protein is required for the same biochemical pathway. If two mutants with the same mutant phenotype (e.g., inability to grow on galactose or to divide) can complement each other, then their mutations are in different genes. (Figure 9-7)

conformation   The precise shape of a protein or other macromolecule in three dimensions resulting from the spatial location of the atoms in the molecule. A small change in the conformations of some proteins affects their activity considerably.

connexin   One of a family of transmembrane proteins that form **gap junctions** in vertebrates.

constitutive   Referring to the continuous production or activity of a cellular molecule or the continuous operation of a cellular process (e.g., constitutive secretion) that is not regulated by internal or external signals.

contractile bundle   Bundles of **actin** and **myosin** in nonmuscle cells that function in cell adhesion (e.g., stress fibers) or cell movement (contractile ring in dividing cells).

cooperativity   Property exhibited by some proteins with multiple ligand-binding sites whereby binding of one ligand molecule increases (positive cooperativity) or decreases (negative cooperativity) the binding affinity for successive ligand molecules.

COP proteins   Two groups of proteins that form the coats around transport vesicles in the **secretory pathway.** COPII-coated vesicles move proteins from the endoplasmic reticulum to the Golgi; COPI-coated vesicles move proteins from the Golgi to the endoplasmic reticulum and from later to earlier Golgi cisternae. (Figure 17-18)

cotranslational translocation   Simultaneous transport of a secretory protein into the endoplasmic reticulum as the nascent protein is still bound to the ribosome and being elongated. (Figure 16-6)

cotransport   Protein-mediated transport of an ion or small molecule across a membrane against a concentration gradient driven by coupling to movement of a second molecule down its concentration gradient in the same (**symport**) or opposite (**antiport**) direction. (Table 7-1 and Figure 7-2)

covalent bond   Stable chemical force that holds the atoms in molecules together by sharing of one or more pairs of electrons. Such a bond has a strength of 50-200 kcal/mol, much greater than that of **noncovalent interactions.**

cross-exon recognition complex   Large assembly including RNA-binding SR proteins and other components that helps delineate exons in the **pre-mRNAs** of higher eukaryotes and assure correct **RNA splicing.** (Figure 12-11)

crossing over   Exchange of genetic material between maternal and paternal **chromatids** during **meiosis** to produce recombined chromosomes. See also **recombination.** (Figure 9-45)

cyclic AMP (cAMP)   A **second messenger,** produced in response to hormonal stimulation of certain G protein-coupled receptors, that activates protein kinase A. (Figure 13-7)

cyclic GMP (cGMP)   A **second messenger** that opens cation channels in rod cells and activates protein kinase G in vascular smooth muscle and other cells. (Figure 13-7)

cyclin   Any of several related proteins whose concentrations rise and fall during the course of the eukaryotic cell cycle. Cyclins form complexes with **cyclin-dependent kinases,** thereby activating and determining the substrate specificity of these enzymes.

cyclin-dependent kinase (CDK)   A protein kinase that is catalytically active only when bound to a cyclin. Various cyclin-CDK complexes trigger progression through different stages of the eukaryotic cell cycle by phosphorylating specific target proteins. (Figure 21-28)

cytochromes   A group of colored, heme-containing proteins that function as **electron carriers** during cellular respiration and photosynthesis. (Figure 8-15a)

cytokine   Any of numerous small, secreted proteins (e.g., erythropoietin, G-CSF, interferons, interleukins) that bind to cell-surface receptors on certain cells to trigger their differentiation or proliferation.

cytokine receptor   Member of major class of cell-surface signaling receptors including those for erythropoietin, growth hormone, interleukins, and interferons. Ligand binding leads to receptor dimerization and activation of cytosolic JAK kinases associated with the intracellular domain of the receptor, thereby initiating intracellular signaling pathways. (Figure 14-12)

cytokinesis The division of the cytoplasm following mitosis to generate two daughter cells, each with a nucleus and cytoplasmic organelles.

cytoplasm   Viscous contents of a cell that are contained within the plasma membrane but, in eukaryotic cells, outside the nucleus.

cytoskeleton   Network of fibrous elements, consisting primarily of **microtubules,** actin **microfilaments,** and **intermediate filaments,** found in the cytoplasm of eukaryotic cells. The cytoskeleton provides structural support for the cell and permits directed movement of organelles, chromosomes, and the cell itself. (Figure 5-29)

cytosol   Unstructured aqueous phase of the cytoplasm excluding organelles, membranes, and insoluble cytoskeletal components.

cytosolic face   The face of a cell membrane directed toward the cytosol. (Figure 5-4)

DAG   See **diacylglycerol.**

dalton   Unit of molecular mass approximately equal to the mass of a hydrogen atom (1.66×10^{-24} g).

denaturation   Drastic alteration in the **conformation** of a protein or nucleic acid due to disruption of various noncovalent bonds caused by heating or exposure to certain chemicals; usually results in loss of biological function.

dendrite   Process extending from the cell body of a neuron that is relatively short and typically branched and receives signals from **axons** of other neurons. (Figure 7-29)

deoxyribonucleic acid   See **DNA.**

depolarization   Change in the cytosolic-face negative electric potential that normally exists across the plasma membrane of a cell at rest, resulting in a less negative **membrane potential.**

desmosomes   See **anchoring junctions.**

determination   In embryogenesis, a change in a cell that commits the cell to a particular developmental pathway (cell fate).

development   Overall process involving growth and **differentiation** by which a fertilized egg gives rise to an adult plant or animal, including the formation of individual cell types, tissues, and organs.

diacylglycerol (DAG)   Membrane-bound **second messenger** produced by cleavage of **phosphoinositides** in response to stimulation of certain cell-surface receptors. (Figures 13-7 and 13-28)

differentiation   Process usually involving changes in gene expression by which a precursor cell becomes a distinct specialized cell type.

diploid   Referring to an organism or cell having two full sets of **homologous chromosomes** and hence two copies (**alleles**) of each gene or genetic locus. Somatic cells contain the diploid number of chromosomes ($2n$) characteristic of a species. See also **haploid.**

disaccharide   A small carbohydrate (sugar) composed of two monosaccharides covalently joined by a **glycosidic bond.** Common examples are lactose (milk sugar) and sucrose, a major photosynthetic product in higher plants. (Figure 2-17)

dissociation constant (K_d)   See **equilibrium constant.**

disulfide bond (—S—S—)   A common covalent linkage between the sulfur atoms on two cysteine residues in different proteins or in different parts of the same protein; found commonly in extracellular proteins or protein domains and very rarely in cytosolic proteins.

DNA (deoxyribonucleic acid)   Long linear polymer, composed of four kinds of deoxyribose **nucleotides**, that is the carrier of genetic information. In its native state, DNA is a double helix of two antiparallel strands held together by hydrogen bonds between complementary purine and pyramidine bases. (Figure 4-3)

DNA cloning   Recombinant DNA technique in which specific **cDNA**s or fragments of **genomic DNA** are inserted into a cloning **vector,** which then is incorporated into cultured host cells (e.g., *E. coli* cells) and maintained during growth of the host cells; also called *gene cloning.* (Figures 9-13 and 9-15)

DNA library   Collection of cloned DNA molecules consisting of fragments of the entire genome (*genomic library*) or of DNA copies of all the mRNAs produced by a cell type (*cDNA library*) inserted into a suitable cloning **vector.**

DNA microarray   An ordered set of thousands of different nucleotide sequences arrayed on a microscope slide or other solid surface. A microarray corresponding to different genes can be used to determine the pattern of gene expression in a particular cell population by **hybridization** with fluorescently labeled cDNA prepared from the total mRNA isolated from the cells. (Figures 9-35 and 9-36)

DNA polymerase   An enzyme that copies one strand of DNA (the template strand) to make the complementary strand, forming a new double-stranded DNA molecule. All DNA polymerases add deoxyribonucleotides one at a time in the $5' \rightarrow 3'$ direction to the $3'$ end of a short preexisting primer strand of DNA or RNA.

DNase hypersensitive site   Region in chromatin that is unusually sensitive to digestion by DNase I when isolated nuclei are treated with the enzyme. Usually indicates a region where transcription factors are bound.

domain   Region of a protein with a distinct tertiary structure (e.g., globular or rodlike) and characteristic activity; homologous domains may occur in different proteins.

dominant   In genetics, referring to that allele of a gene expressed in the **phenotype** of a heterozygote; the nonexpressed allele is **recessive.** Also referring to the phenotype associated with a dominant allele. Mutations that produce dominant alleles generally result in a gain of function. (Figure 9-2)

dominant negative   In genetics, an allele that acts in a **dominant** manner but produces an effect similar to a loss of function; generally is an allele that encodes a mutant protein that blocks function of the normal protein by binding either to it or to a protein **upstream** or **downstream** of it in a pathway.

dorsal   Relating to the back of an animal or the upper surface of a structure (e.g., leaf, wing).

double helix, DNA   The most common three-dimensional structure for cellular DNA in which the two polynucleotide strands are antiparallel and wound around each other with complementary bases hydrogen-bonded. (Figure 4-3)

downstream   (1) For a gene, the direction RNA polymerase moves during transcription, which is toward the end of the template DNA strand with a $5'$-hydroxyl group. By convention, the $+1$ position of a gene is the first transcribed nucleotide; nucleotides downstream from the $+1$ position are designated $+2$, $+3$, etc. (2) Events that occur later in a cascade of steps (e.g., signaling pathway). See also **upstream.**

duplex DNA   See **double helix, DNA.**

dyneins   A class of **motor proteins** that use the energy released by ATP hydrolysis to move toward the ($-$) end of **microtubules.** Dyneins can transport vesicles and organelles, are responsible for the movement of cilia and flagella, and play a role in chromosome movement during mitosis. (Figure 20-22 and Table 20-2)

ectoderm   Outermost of the three primary cell layers of the animal embryo; gives rise to epidermal tissues, the nervous system, and external sense organs. See also **endoderm** and **mesoderm.** (Figure 22-7)

electrochemical gradient   The driving force that determines the energetically favorable direction of transport of an ion (or charged molecule) across a membrane. It represents the combined influence of the ion's concentration gradient across the membrane and the membrane potential.

electron carrier   Any molecule or atom that accepts electrons from donor molecules and transfers them to acceptor molecules in coupled **oxidation** and **reduction** reactions. Both small metal-containing groups or ions (e.g., heme, copper, iron-sulfur clusters) associated with membrane-bound proteins and diffusible molecules (e.g., NAD^+, FAD, and ubiquinone (coenzyme Q) function as electron carriers.

electron transport   Flow of electrons via a series of electron carriers from reduced electron donors (e.g., NADH) to O_2 in the inner mitochondrial membrane, or from H_2O to $NADP^+$ in the thylakoid membrane of plant chloroplasts. (Figure 8-13)

electron-transport chain   See **respiratory chain.**

electrophoresis   Any of several techniques for separating macromolecules based on their migration in a gel or other medium subjected to a strong electric field. (Figure 3-32)

electrophoretic mobility shift assay (EMSA)   A semiquantitative method for determining proteins that bind to specific DNA sequences based on the decreased migration in gel electrophoresis of a labeled DNA fragment when bound to protein. (Figure 11-14)

elongation factor   One of a group of nonribosomal proteins required for continued **translation** of mRNA (protein synthesis) following initiation. (Figure 4-26)

embryogenesis   Early development of an individual from a fertilized egg (zygote). In animals, following cleavage of the zygote, the major axes are established during the *blastula* stage; in the subsequent *gastrula* stage, the early embryo invaginates and acquires three cell layers.

embryonic stem (ES) cells   A line of cultured cells derived from very early embryos that can differentiate into a wide range of cell types either in vitro or after reinsertion into a host embryo. (Figure 22-3)

endocrine   Referring to signaling mechanism in which target cells bind and respond to a **hormone** released into the blood by distant specialized secretory cells usually present in a gland (e.g., pituitary or thyroid gland).

endocytosis   See **receptor-mediated endocytosis.**

endoderm   Innermost of the three primary cell layers of the animal embryo; gives rise to the gut and most of the respiratory tract. See **ectoderm** and **mesoderm.** (Figure 22-7)

endoplasmic reticulum (ER)   Network of interconnected membranous structures within the cytoplasm of eukaryotic cells contiguous with the outer nuclear envelope. The rough ER, which is associated with **ribosomes,** functions in the synthesis and processing of secreted and membrane proteins; the smooth ER, which lacks ribosomes, functions in lipid synthesis. (Figure 5-19)

endosome, late   Membrane-bounded compartment that participates in sorting of lysosomal enzymes and in recycling of receptors endocytosed from the plasma membrane. The acidic internal pH of late endosomes causes bound ligands to dissociate from their membrane-bound receptor proteins. (Figures 17-1 and 17-28)

endothelium   Layer of highly flattened cells that forms the inner lining of all blood vessels and regulates exchange of materials between the bloodstream and surrounding tissues; it usually is underlain by a **basal lamina.**

enhancer   A regulatory sequence in eukaryotic DNA that may be located at a great distance from the gene it controls or even within the coding sequence. Binding of specific proteins to an enhancer modulates the rate of transcription of the associated gene. (Figure 11-12)

enhancesome   Large nucleoprotein complex that assembles from **transcription factors** (activators and repressors) as they bind cooperatively to their multiple binding sites in an **enhancer** with the assistance of DNA-bending proteins. (Figure 11-26)

enterohepatic circulation   The cycling of **bile acids** from the liver to intestine (via bile ducts) and back to the liver (via blood). This circulation is tightly regulated and plays a major role in lipid homeostasis. (Figure 18-11)

enthalpy (H)   Heat; in a chemical reaction, the enthalpy of the reactants or products is equal to their total bond energies.

entropy (S)   A measure of the degree of disorder or randomness in a system; the higher the entropy, the greater the disorder.

envelope   See **nuclear envelope** or **viral envelope.**

enzyme   A protein that catalyzes a particular chemical reaction involving a specific **substrate** or small number of related substrates.

epidermal growth factors (EGF)   A family of secreted signaling proteins that is used in the development of most tissues in most or all animals. The name reflects the mode of discovery; the functions are far more diverse. Receptors are tyrosine kinases. Mutations in EGF signal transduction components are implicated in human cancer, including brain cancer.

epinephrine   A catecholamine secreted by the adrenal gland and some neurons in response to stress; also called *adrenaline*. It functions as both a hormone and neurotransmitter, mediating "fight or flight" responses including increased blood glucose levels and heart rate. (Figure 7-42)

epithelium (pl. epithelia)   Sheetlike covering, composed of one or more layers of tightly adhered cells, on external and internal body surfaces. (Figure 6-4)

epitope   The part of an antigen molecule that binds to an **antibody;** also called *antigenic determinant*.

equilibrium constant (K)   Ratio of forward and reverse rate constants for a reaction. For a binding reaction, A + B $\rightleftharpoons$ AB, it equals the association constant, K_a; the higher the K_a, the tighter the binding between A and B. The reciprocal of the K_a is the dissociation constant, K_d; the higher the K_d, the weaker the binding between A and B.

erythropoietin (Epo)   A cytokine, secreted by kidney cells in response to low blood oxygen levels, that triggers production of red blood cells by inducing the proliferation and differentiation of erythroid progenitor cells in the bone marrow. (Figures 14-7 and 22-5)

eubacteria   Class of **prokaryotes** that constitutes one of the three distinct evolutionary lineages of modern-day organisms; also called the *true bacteria* or simply *bacteria*. Phylogenetically distinct from **archaea** and **eukaryotes.** (Figure 1-3)

euchromatin   Less condensed portions of **chromatin,** including most transcribed regions, present in interphase chromosomes. See also **heterochromatin.**

eukaryotes   Class of organisms, composed of one or more cells containing a membrane-enclosed nucleus and organelles, that constitutes one of the three distinct evolutionary lineages of modern-day organisms; also called *eukarya*. Includes all organisms except viruses and **prokaryotes.** (Figure 1-3)

excision-repair system   One of several mechanisms for repairing DNA damaged due to spontaneous depurination or deamination or exposure to **carcinogens**. These repair systems normally operate with a high degree of fidelity and their loss is associated with increased risk for certain cancers. (Table 23-1)

exocytosis Release of intracellular molecules (e.g., hormones, matrix proteins) contained within a membrane-bounded vesicle by fusion of the vesicle with the plasma membrane of a cell. This is the process whereby most molecules are secreted from eukaryotic cells.

exon Segment of a eukaryotic gene (or of its **primary transcript**) that reaches the cytoplasm as part of a mature mRNA, rRNA, or tRNA molecule. See also **intron**.

exon shuffling Evolutionary process for creating new genes (i.e., new combinations of exons) from preexisting ones by recombination between interspersed repeat sequences within introns of two separate genes or by transposition of mobile DNA elements. (Figures 10-17 and 10-18)

exoplasmic face The face of a cell membrane directed away from the cytoplasm. The exoplasmic face of the plasma membrane faces the cell exterior, whereas the exoplasmic face of organelles bounded by a single membrane (e.g., endoplasmic reticulum, lysosome, transport vesicle) faces their lumen. (Figure 5-4)

exosomes Large exonuclease-containing complexes that degrade spliced out introns and improperly processed pre-mRNAs in the nucleus and mRNAs with shortened poly(A) tails in the cytoplasm.

exportin Protein that binds a "cargo" protein in the nucleus and with the aid of Ran (a member of the **GTPase superfamily**) transports the cargo through a nuclear pore complex to the cytoplasm. See also **importin**. (See Figure 12-23)

expression See **gene expression**.

expression vector A modified **plasmid** or virus that carries a gene or cDNA into a suitable host cell and there directs synthesis of the encoded protein. Expression vectors can be used to screen DNA libraries for a gene of interest or to produce large amounts of a protein from its cloned gene (Figures 9-28 and 9-29)

extracellular matrix A usually insoluble network consisting of polysaccharides, fibrous proteins, and adhesive proteins that are secreted by animal cells. It provides structural support in tissues and can affect the development and biochemical functions of cells.

extrinsic protein See **peripheral membrane protein**.

F_0F_1 complex See **ATP synthase**.

facilitated diffusion Protein-aided transport of an ion or small molecule across a cell membrane down its concentration gradient at a rate greater than that obtained by **passive diffusion**; also called *facilitated transport*. The glucose transporters (GLUTs) are well-studied examples of proteins that mediate facilitated diffusion. (Figure 7-3)

FAD (flavin adenine dinucleotide) A diffusible electron carrier that accepts two electrons from a donor molecule and two H^+ from the solution. In mitochondria, the reduced form, $FADH_2$, transfers electrons to carriers that function in **oxidative phosphorylation**. (Figure 2-26b)

fatty acid Any long hydrocarbon chain that has a carboxyl group at one end; a major source of energy during metabolism and a precursor for synthesis of phospholipids, triglycerides, and cholesteryl esters. (Figure 2-18 and Table 18-1)

fibroblast A common type of connective tissue cell that secretes **collagen** and other components of the **extracellular matrix**. It migrates and proliferates during wound healing and in tissue culture.

fibroblast growth factors (FGF) A family of secreted signaling proteins that are used in the development of most tissues in most or all animals. The name reflects the mode of discovery; the functions are far more diverse. FGFs are often angiogenic or mitogenic. Receptors are tyrosine kinases, and some receptor mutations are associated with inherited human birth defects.

fibronectin See **multiadhesive matrix proteins**.

FISH See **fluorescence in situ hybridization**.

flagellum (pl. **flagella**) Long locomotory structure, extending from the surface of a eukaryotic cell, whose whiplike bending propels the cell forward or backward. Usually there is only one flagellum per cell (as in sperm cells). Bacterial flagella are smaller and much simpler structures. See also **axoneme** and **cilium**.

flippase Protein that facilitates the movement of membrane lipids from one leaflet to the other leaflet of a phospholipid bilayer. (Figures 7-12 and 18-4)

fluorescence in situ hybridization (FISH) Any of several related techniques for detecting specific DNA or RNA sequences in cells and tissues by treating samples with fluorescent **probes** that hybridize to the sequence of interest and observing the samples by fluorescence microscopy.

fluorescent staining General technique for visualizing cellular components by treating cells or tissues with a **fluorochrome**-labeled agent (e.g., antibody) that binds specifically to a component of interest and observing the samples by fluorescence microscopy.

fluorochrome Molecule that absorbs light at one wavelength and emits light at another (longer) wavelength; also called *fluorescent dye*. Fluorescein, which emits green light, and rhodamine, which emits red light, are two commonly used fluorochromes.

footprinting Technique for identifying protein-binding regions of DNA or RNA based on the ability of a protein bound to a region of DNA or RNA to protect that region from digestion. (Figure 11-13a)

free energy (G) A measure of the potential energy of a system, which is a function of the **enthalpy** (H) and **entropy** (S).

free-energy change (ΔG) The difference in the total free energy of the product molecules and of the starting molecules (reactants) in a chemical reaction. A large negative value of ΔG indicates that a reaction has a strong tendency to occur; that is, at chemical equilibrium the concentration of products will be much greater than the concentration of reactants.

G_0, G_1, G_2 phase See **cell cycle**.

G protein, monomeric (small). See **GTPase superfamily**.

G protein, trimeric (large)   Any of numerous heterotrimeric GTP-binding proteins that function in intracellular signaling pathways; usually activated by ligand binding to a coupled seven-spanning receptor on the cell surface. See also **GTPase superfamily.** (Table 13-1)

G protein–coupled receptor (GPCR)   Member of a large class of cell-surface signaling receptors, including those for epinephrine, glucagon, ACTH, serotonin, and mating factors (yeast). All GPCRs contain seven transmembrane α helices. Ligand binding leads to activation of a coupled trimeric G protein, thereby initiating intracellular signaling pathways. (Figure 13-11)

GAG   See **glycosaminoglycan.**

gamete   Specialized haploid cell (in animals either a sperm or an egg) produced by **meiosis** of germ cells; in sexual reproduction, union of a sperm and an egg initiates the development of a new individual.

gap genes   In *Drosophila,* group of genes (e.g., *hunchback, Krüppel, knirps*) that are activated in the early embryo by transcription factors produced from maternal mRNAs in the zygote; all encode transcription factors that function in early patterning along the anterioposterior axis. (Figure 15-22a)

gap junction   Protein-lined channel connecting the cytoplasms of adjacent animal cells that allows passage of ions and small molecules between the cells. (Figure 6-32)

gene   Physical and functional unit of heredity, which carries information from one generation to the next. In molecular terms, it is the entire DNA sequence—including **exons, introns,** and noncoding transcription-control regions—necessary for production of a functional protein or RNA. See also **transcription unit.**

gene cloning   See **DNA cloning.**

gene control   All of the mechanisms involved in regulating **gene expression.** Most common is regulation of transcription, although mechanisms influencing the processing, stabilization, and **translation** of mRNAs help control expression of some genes.

gene expression   Overall process by which the information encoded in a gene is converted into an observable **phenotype** (most commonly production of a protein).

gene family   Set of genes that arose by duplication of a common ancestral gene and subsequent divergence due to small changes in the nucleotide sequence. (Figure 9-32)

general transcription factor (GTF)   See **transcription factor.**

genetic code   The set of rules whereby nucleotide triplets (**codons**) in DNA or RNA specify amino acids in proteins. (Table 4-1)

genetic markers   Alleles associated with an easily detectable **phenotype** that are used experimentally to identify or select for a linked gene, a chromosome, a cell, or an individual. Noncoding DNA sequences that vary among individuals (*DNA polymorphisms*) can serve as genetic markers.

genome   Total genetic information carried by a cell or organism.

genomic DNA   All the DNA sequences composing the **genome** of a cell or organism. See also **cDNA.**

genomics   Comparative analyses of the complete genomic sequences from different organisms; used to assess evolutionary relations among species and to predict the number and general types of proteins produced by an organism.

genotype   Entire genetic constitution of an individual cell or organism, usually with emphasis on the particular alleles at one or more specific loci.

germ cell   Any precursor cell that can give rise to gametes. See also **somatic cell.**

germ line   Lineage of germ cells, which give rise to **gametes** and thus participate in formation of the next generation of organisms; also, the genetic material transmitted from one generation to the next through the gametes.

glucagon   A peptide hormone produced in the α cells of pancreatic islets that triggers the conversion of glycogen to glucose by the liver; acts with **insulin** to control blood glucose levels.

glucose   Six-carbon monosaccharide (sugar) that is the primary metabolic fuel in most cells. The large glucose polymers, glycogen and starch, are used to store energy in animal cells and plant cells, respectively.

glycogen   A very long, branched polysaccharide, composed exclusively of glucose units, that is the primary storage carbohydrate in animals. It is found primarily in liver and muscle cells.

glycolipid   Any lipid to which a short carbohydrate chain is covalently linked; commonly found in the plasma membrane.

glycolysis   Metabolic pathway in which sugars are degraded anaerobically to lactate or pyruvate in the cytosol with the production of ATP. (Figure 8-4)

glycoprotein   Any protein to which one or more oligosaccharide chains are covalently linked. Most secreted proteins and many membrane proteins are glycoproteins.

glycosaminoglycan (GAG)   A long, linear, highly charged polymer of a repeating disaccharide in which one member of the pair usually is a sugar acid (uronic acid) and the other is an amino sugar. Usually many residues are sulfated. GAGs are major components of the extracellular matrix, usually as components of **proteoglycans.** (Figures 6-17 and 6-19)

glycosidic bond   The covalent linkage between two monosaccharide residues formed by a dehydration reaction in which one carbon, usually carbon-1, of one sugar reacts with a hydroxyl group on a second sugar with the loss of a water molecule. (Figure 2-11)

Golgi complex   Stacks of flattened, interconnected membrane-bounded compartments (cisternae) in eukaryotic cells that function in processing and sorting of proteins and lipids destined for other cellular compartments or for secretion; also called *Golgi apparatus.* (Figure 5-23)

growing fork   See **replication fork.**

growth factor   An extracellular polypeptide molecule that binds to a cell-surface receptor, triggering an intracellular signaling pathway generally leading to cell proliferation.,

GTP (guanosine 5′-triphosphate)   A nucleotide that is a precursor in RNA synthesis and also plays a special role in protein synthesis, signal-transduction pathways, and microtubule assembly.

GTPase superfamily   Group of intracellular switch proteins that cycle between an inactive state with bound GDP and an active state with bound GTP. These proteins regulate various cellular processes and include the G_α subunit of trimeric (large) **G proteins** and monomeric (small) G proteins such as **Ras,** Rab, Ran, and Rac as well as certain **elongation factors** used in protein synthesis. (Figure 3-29)

haploid   Referring to an organism or cell having only one member of each pair of **homologous chromosomes** and hence only one copy (**allele**) of each gene or genetic locus. Gametes and bacterial cells are haploid. See also **diploid.**

Hedgehog (Hh)   A family of secreted signaling proteins that are used in the development of most tissues in most or all animals. Hh proteins are initially made as longer precursors. The C-terminal part of the precursor has autoproteolytic activity, cleaving the molecule to release the N-terminal signaling moiety. In this process a cholesterol molecule is covalently joined to the carboxy terminus of the signaling peptide. Mutations in Hh signal transduction components are implicated in human cancer and birth defects. The receptor is the Patched transmembrane protein.

helicase   Any enzyme that moves along a DNA duplex using the energy released by ATP hydrolysis to separate (unwind) the two strands. Required for DNA replication.

helix-loop-helix   A conserved structural **motif** found in many monomeric Ca^{2+}-binding proteins and dimeric eukaryotic transcription factors. (Figures 3-6a and 11-22b)

hemidesmosome   See **anchoring junctions.**

heterochromatin   Regions of **chromatin** that remain highly condensed and transcriptionally inactive during interphase.

heterozygous   Referring to a diploid cell or organism having two different **alleles** of a particular gene.

hexose   A six-carbon **monosaccharide.**

high-density lipoprotein (HDL)   A class of **lipoproteins** whose major protein component is apolipoprotein A and that is generated extracellularly and transports cholesterol from tissues to the liver, steroid-producing tissues, and other lipoproteins. The plasma HDL level is inversely related to the risk for coronary artery disease; hence it is popularly called "good cholesterol." (Table 18-3 and Figure 18-13c)

high-energy bond   Covalent bond that releases a large amount of energy when hydrolyzed under the usual intracellular conditions. Examples include the phosphoanhydride bonds in ATP, thioester bond in acetyl CoA, and various phosphate ester bonds.

histone   One of several small, highly conserved basic proteins, found in the **chromatin** of all eukaryotic cells, that associate with DNA in the **nucleosome.** (Figures 10-20 and 10-21)

homeodomain   Conserved DNA-binding **motif** found in many developmentally important transcription factors.

homeosis   Transformation of one body part into another arising from mutation in or misexpression of certain developmentally critical genes.

homologous chromosome   One of the two copies of each morphologic type of chromosome present in a **diploid** cell; also called *homolog.* Each homolog is derived from a different parent.

homologs   Proteins or genes that exhibit **homology.**

homology   Similarity in characteristics (e.g., protein and nucleic acid sequences or the structure of an organ) that reflects a common evolutionary origin. Molecules or sequences that exhibit homology are referred to as *homologs.* In contrast, analogy is a similarity in structure or function that does not reflect a common evolutionary origin.

homozygous   Referring to a diploid cell or organism having two identical **alleles** of a particular gene.

hormone   General term for any extracellular substance that induces specific responses in target cells; specifically, those signaling molecules that circulate in the blood and mediate **endocrine** signaling.

Hox genes   Group of developmentally important genes that encode homeodomain-containing transcription factors and help determine the body plan in animals. Mutations in Hox genes often cause **homeosis.**

hyaluronan   A large, highly hydrated glycosaminoglycan (GAG) that is a major component of the extracellular matrix; also called *hyaluronic acid* and *hyaluronate.* It imparts stiffness and resilience as well as a lubricating quality to many types of connective tissue. (Figure 6-17)

hybridization   Association of two **complementary** nucleic acid strands to form double-stranded molecules, which can contain two DNA strands, two RNA strands, or one DNA and one RNA strand. Used experimentally in various ways to detect specific DNA or RNA sequences.

hybridoma   A clone of hybrid cells that are immortal and produce **monoclonal antibody;** formed by fusion of normal antibody-producing B lymphocytes with myeloma cells. (Figure 6-38)

hydrogen bond   A **noncovalent interaction** between an atom (commonly oxygen or nitrogen) carrying a partial negative charge and a hydrogen atom carrying a partial positive charge. Particularly important in stabilizing the three-dimensional structure of proteins and in formation of **base pairs** between nucleic acid strands. (Figure 2-6)

hydrophilic   Interacting effectively with water. See also **polar.**

hydrophobic   Not interacting effectively with water; in general, poorly soluble or insoluble in water. See also **nonpolar.**

hydrophobic effect   The tendency of nonpolar molecules or parts of molecules to associate with each other in aqueous solution; particularly important in stabilizing the phospholipid bilayer. It is commonly called a *hydrophobic interaction* or *bond.* (Figure 2-9)

hypertonic   Referring to an external solution whose solute concentration is high enough to cause water to move out of cells due to **osmosis.**

hypotonic   Referring to an external solution whose solute concentration is low enough to cause water to move into cells due to **osmosis.**

IgCAMs   A family of **cell-adhesion molecules** that contain multiple immunoglobulin (Ig) domains and mediate Ca^{2+}-independent cell-cell interactions. IgCAMs are produced in a variety of tissues and are components of **tight junctions.** (Figure 6-2)

immunohistochemistry   Staining of cells and tissues with a **fluorochrome**-labeled antibody specific for a particular protein; common technique for localizing specific proteins in fixed samples. (Figure 5-45)

importin   Protein that binds a "cargo" protein in the cytoplasm and transports the cargo through a nuclear pore complex into the nucleus. Return of importin to the cytoplasm requires the aid of Ran (a member of the **GTPase superfamily**) See also **exportin.** (Figure 12-21)

in situ hybridization   Any technique for detecting specific DNA or RNA sequences in cells and tissues by treating samples with single-stranded RNA or DNA **probes** that hybridize to the sequence of interest. (Figure 15-3)

in vitro   Denoting a reaction or process taking place in an isolated cell-free extract; sometimes used to distinguish cells growing in culture from those in an organism.

in vivo   In an intact cell or organism.

induction   (1) In embryogenesis, a change in the developmental fate of one cell or tissue caused by direct interaction with another cell or tissue or with an extracellular signaling molecule. (2) In metabolism, an increase in the synthesis of an enzyme or series of enzymes mediated by a specific molecule (inducer).

initiation factor   One of a group of nonribosomal proteins that promote the proper association of ribosomes and mRNA and are required for initiation of **translation** (protein synthesis). (Figure 4-25)

inositol 1,4,5-trisphosphate (IP$_3$)   Intracellular **second messenger** produced by cleavage of **phosphoinositides** in response to stimulation of certain cell-surface receptors. IP$_3$ triggers release of Ca^{2+} stored in the endoplasmic reticulum. (Figures 13-7 and 13-29)

insulin   A protein hormone produced in the β cells of the pancreatic islets that stimulates uptake of glucose into muscle and fat cells; acts with **glucagon** to help regulate blood glucose levels. Insulin also functions as a growth factor for many cells.

integral membrane protein   Any protein containing one or more hydrophobic segments embedded within the core of the **phospholipid bilayer** and can be removed from the membrane only by extraction with detergent; also called *intrinsic membrane protein.* Four major classes, differing in the number and orientation of membrane-spanning segments, are synthesized on the endoplasmic reticulum. (Figure 16-10)

integrins   A large family of heterodimeric transmembrane proteins that promote adhesion of cells to the extracellular matrix or to the surface of other cells. (Table 6-2)

interferons (IFNs)   Small group of cytokines that bind to cell-surface receptors on target cells inducing changes in gene expression that lead to an antiviral state or other cellular responses important in functions of the immune system.

intermediate filament   Cytoskeletal fiber (10 nm in diameter) formed by polymerization of several types of subunit proteins including keratins, lamins, and vimentin. Intermediate filaments constitute the major structural proteins of skin and hair; form the scaffold that holds Z disks and myofibrils in place in muscle; and provide support for cellular membranes. (Figure 19-33)

interphase   Long period of the cell cycle, including the G$_1$, S, and G$_2$ phases, between one M (mitotic) phase and the next. (Figure 21-1)

intrinsic protein   See **integral membrane protein.**

intron   Part of a **primary transcript** (or the DNA encoding it) that is removed by splicing during RNA processing and is not included in the mature, functional mRNA, rRNA, or tRNA.

ionic interaction   A **noncovalent interaction** between a positively charged ion (cation) and negatively charged ion (anion); commonly called *ionic bond.*

ionophore   Any small lipid-soluble molecule that selectively binds a specific ion and carries it across otherwise impermeable membranes.

IP$_3$ See **inositol 1,4,5-trisphosphate.**

isoelectric focusing   Technique for separating molecules by gel **electrophoresis** in a pH gradient subjected to an electric field. A protein migrates to the pH at which its overall net charge is zero.

isoelectric point (pI)   The pH of a solution at which a dissolved protein or other potentially charged molecule has a net charge of zero and therefore does not move in an electric field.

isoform   One of several forms of the same protein whose amino acid sequences differ slightly and whose general activities are similar. Isoforms may be encoded by different genes or by a single gene whose primary transcript undergoes alternative splicing.

isotonic   Referring to a solution whose solute concentration is such that it causes no net movement of water in or out of cells.

karyopherin   One of a family of nuclear transport proteins that functions as an **importin, exportin,** or occasionally both. Each karyopharin binds to a specific signal sequence in cargo proteins.

karyotype   Number, sizes, and shapes of the entire set of **metaphase** chromosomes of a eukaryotic cell. (Chapter 10 opening figure)

keratins   A group of intermediate filament proteins found in epithelial cells that assemble into heteropolymeric filaments.

kinase   An enzyme that transfers the terminal (γ) phosphate group from ATP to a substrate. Protein kinases, which phosphorylate specific serine, threonine, or tyrosine residues in target proteins, play a critical role in regulating the activity of many cellular proteins. See also **phosphatases.** (Figure 3-30)

kinesins   A class of **motor proteins** that use energy released by ATP hydrolysis to move toward the (+) end of a **microtubule.** Kinesins can transport vesicles and organelles and play a role in chromosome movement during mitosis. (Figure 20-20 and Table 20-2).

kinetochore   A multilayer protein structure located at or near the **centromere** of each mitotic chromosome from which microtubules extend toward the spindle poles of the cell; plays an active role in movement of chromosomes toward the poles during anaphase. (Figure 20-32)

K_m A parameter that describes the affinity of an enzyme for its substrate and equals the substrate concentration that yields the half-maximal reaction rate; also called the *Michaelis constant.* A similar parameter describes the affinity of a transport protein for the transported molecule or the affinity of a receptor for its ligand. (Figure 3-19)

knockout, gene Selective inactivation of a specific gene by replacing it with a nonfunctional (disrupted) allele in an otherwise normal organism.

Krebs cycle See **citric acid cycle.**

label A chemical group or radioactive atom incorporated into a molecule in order to spatially locate the molecule or follow it through a reaction or purification scheme. As a verb, to add such a group or atom to a cell or molecule.

lagging strand One of the two daughter DNA strands formed at a **replication fork** as short, discontinuous segments (Okazaki fragments), which are later joined. Although overall lagging-strand synthesis occurs in the $3' \rightarrow 5'$ direction, each segment is synthesized in the $5' \rightarrow 3'$ direction. See also **leading strand.** (Figure 4-33)

laminin See **multiadhesive matrix protein.**

lamins A group of intermediate filament proteins that form a fibrous network, the **nuclear lamina,** on the inner surface of the nuclear envelope.

lateral inhibition Important developmental process in which Notch or other type of signaling causes adjacent developmentally equivalent or near-equivalent cells to assume different fates. (Figure 15-36)

leading strand One of the two daughter DNA strands formed at a **replication fork** by continuous synthesis in the $5' \rightarrow 3'$ direction. The direction of leading-strand synthesis is the same as movement of the replication fork. See also **lagging strand.** (Figure 4-33)

lectin Any protein that binds tightly to specific sugars. Lectins can be used in affinity chromatography to purify glycoproteins or as reagents to detect them in situ.

leucine zipper A type of coiled coil of two α helixes that forms specific homo- or heterodimers. Common motif in many eukaryotic transcription factors.

library See **DNA library.**

ligand Any molecule, other than an enzyme **substrate,** that binds tightly and specifically to a macromolecule, usually a protein, forming a macromolecule-ligand complex.

ligase, DNA An enzyme that links together the 3' end of one DNA fragment with the 5' end of another, forming a continuous strand.

linkage In genetics, the tendency of two different loci on the same chromosome to be inherited together. The closer two loci are, the lower the frequency of **recombination** between them and the greater their linkage.

lipid Any organic molecule that is poorly soluble or virtually insoluble in water but is soluble in nonpolar organic solvents. Major classes include **fatty acids, phospholipids, steroids, polyisoprenoids,** and **triglycerides.**

lipophilic See **hydrophobic.**

lipid rafts Microdomains in the plasma membrane that are enriched in cholesterol, sphingomyelin, and certain proteins. (Figure 5-10)

lipid-anchored membrane protein Any protein that is tethered to a cellular membrane by one or more covalently attached lipid groups, which are embedded in the phospholipid bilayer. (Figure 5-15)

lipoproteins Large, water-soluble protein and lipid complexes that function in mass transfer of lipids throughout the body. The outer shell is a phospholipid monolayer containing cholesterol, and one or more protein molecules called apolipoproteins (e.g., apoA, apoB, etc.); the core contains triglycerides, cholesteryl esters, and small amounts of other hydrophobic substances, such as lipid-soluble vitamins. (Figure 18-12 and Table 18-3)

liposome Artificial spherical **phospholipid bilayer** structure with an aqueous interior that forms in vitro from phospholipids and may contain membrane proteins. (Figure 7-5)

locus In genetics, the specific site of a gene on a chromosome. All the **alleles** of a particular gene occupy the same locus.

long terminal repeat (LTR) Direct repeat sequence, containing up to 600 base pairs, that flanks the coding region of integrated retroviral DNA and viral **retrotransposons.**

low-density lipoprotein (LDL) A class of **lipoproteins** containing apolipoprotein B-100 that is a primary transporter of cholesterol in the form of cholesteryl esters. It is generated extracellularly from **very low density lipoprotein (VLDL)** and transports cholesterol between tissues, especially to the liver. Plasma LDL cholesterol levels are directly related to the risk for coronary artery disease; hence it is popularly called "bad cholesterol." (Table 18-3 and Figure 18-13a)

lumen The space within a tubular structure (e.g., a vessel or the intestine) or the interior volume of a membrane-bounded compartment within a cell.

lymphocytes Two classes of white blood cells that can recognize foreign molecules (**antigens**) and mediate immune responses. B lymphocytes are responsible for production of antibodies; T lymphocytes are responsible for destroying virus- and bacteria-infected cells, foreign cells, and cancer cells.

lysis Destruction of a cell by rupture of the plasma membrane and release of the contents.

lysogeny Phenomenon in which the DNA of a bacterial virus (bacteriophage) is incorporated into the host-cell genome and replicates along with the bacterial DNA but is not expressed. Subsequent activation of the virus leads to replication and expression of the viral DNA and formation of new viral particles, eventually causing lysis of the cell.

lysosome Small organelle having an internal pH of 4–5 and containing hydrolytic enzymes.

M (mitotic) phase See **cell cycle.**

macromolecule Any large, usually polymeric molecule (e.g., a protein, nucleic acid, polysaccharide) with a molecular mass greater than a few thousand daltons.

malignant Referring to a tumor or tumor cells that can invade surrounding normal tissue and/or undergo **metastasis.** See also **benign.**

MAP kinase Protein kinase that is activated in response to cell stimulation by many different growth factors and that mediates cellular responses by phosphorylating specific transcription factors and other target proteins.

mediator A very large multiprotein complex that forms a molecular bridge between transcriptional activators bound to an **enhancer** and to RNA polymerase II bound at a promoter; functions as a co-activator in stimulating **transcription.** (Figures 11-35 and 11-36)

meiosis In eukaryotes, a special type of cell division that occurs during maturation of germ cells; comprises two successive nuclear and cellular divisions with only one round of DNA replication resulting in production of four genetically nonequivalent haploid cells (**gametes**) from an initial diploid cell. (Figure 9-3)

membrane See **biomembrane.**

membrane potential Voltage difference across a membrane due to the slight excess of positive ions (cations) on one side and negative ions (anions) on the other.

membrane transport protein Collective term for any integral membrane protein that mediates movement of one or more specific ions or small molecules across a cellular membrane regardless of the transport mechanism.

meristem Organized group of undifferentiated, dividing cells that are maintained at the tips of growing shoots and roots in plants. All the adult structures arise from meristems.

mesenchyme Immature embryonic connective tissue derived from either the **mesoderm** or **ectoderm** in animals.

mesoderm The middle of the three primary cell layers of the animal embryo, lying between the **ectoderm** and **endoderm;** gives rise to the notochord, connective tissue, muscle, blood, and other tissues. (Figure 22-7)

messenger RNA See **mRNA.**

metabolism The sum of the chemical processes that occur in an organism.

metaphase Mitotic stage at which chromosomes are fully condensed and attached to the mitotic spindle but have not yet started to segregate toward the opposite spindle poles. (Figure 20-29)

metastasis Spread of cancer cells from their site of origin and establishment of areas of secondary growth.

micelle A water-soluble spherical aggregate of phospholipids or other amphipathic molecules that form spontaneously in aqueous solution. (Figure 2-20)

Michaelis constant See K_m.

micro RNA (miRNA) Small RNA, containing 21–23 nucleotides, that associates with multiple proteins in a **RNA-induced silencing complex (RISC)** and represses translation of a specific target mRNA by hybridizing to its 3′ untranslated region. (Figure 12-27)

microfilament Cytoskeletal fiber ($\approx$7 nm in diameter) that is formed by polymerization of monomeric globular (G) **actin;** also called *actin filament.* Microfilaments play an important role in muscle contraction, cytokinesis, cell movement, and other cellular functions and structures. (Figure 19-3)

microtubule Cytoskeletal fiber ($\approx$24 nm in diameter) that is formed by polymerization of α,β-**tubulin** monomers and exhibits structural and functional polarity. Microtubules are important components of cilia, flagella, the mitotic spindle, and other cellular structures. (Figure 20-1)

microtubule-associated protein (MAP) Any protein that binds to microtubules in a constant ratio and determines the unique properties of different types of microtubules. (Table 20-1)

microtubule-organizing center See **MTOC.**

microvillus (pl. microvilli) Small, membrane-covered projection on the surface of an animal cell containing a core of actin filaments. Numerous microvilli are present on the absorptive surface of intestinal epithelial cells, increasing the surface area for transport of nutrients. (Figure 5-28)

mitochondrion (pl. mitochondria) Large organelle that is surrounded by two phospholipid bilayer membranes, contains DNA, and carries out **oxidative phosphorylation,** thereby producing most of the ATP in eukaryotic cells. (Figures 5-26 and 8-6)

mitogen Any extracellular molecule, such as a growth factor, that promotes cell proliferation.

mitosis In eukaryotic cells, the process whereby the nucleus is divided to produce two genetically equivalent daughter nuclei with the diploid number of chromosomes. See also **cytokinesis** and **meiosis.** (Figure 20-29)

mitosis-promoting factor. See **MPF.**

mitotic apparatus A specialized temporary structure, present in eukaryotic cells during mitosis, that captures the chromosomes and then pushes and pulls them to opposite sides of the dividing cell. It consists of a central bilaterally symmetric bundle of microtubules with the overall shape of an American football (the mitotic spindle) and two tufts of microtubules (the asters), one at each pole of the spindle. (Figure 20-31)

mitotic spindle See **mitotic apparatus.**

mobile DNA element Any DNA sequence that is not present in the same chromosomal location in all individuals of a species and can move to a new position by **transposition;** also called *transposable element* and *interspersed repeat.* (Table 10-1)

molecular complementarity Lock-and-key kind of fit between the shapes, charges, hydrophobicity, and/or other physical properties of two molecules or portions thereof that allow formation of multiple **noncovalent interactions** between them at close range. (Figure 2-10)

monoclonal antibody Antibody produced by the progeny of a single B cell and thus a homogeneous protein that recognizes a single antigen. It can be produced experimentally by use of a **hybridoma.** (Figure 6-38)

monomer Any small molecule that can be linked chemically with others of the same type to form a **polymer.** Examples include amino acids, nucleotides, and monosaccharides.

monomeric For proteins, consisting of a single polypeptide chain.

monosaccharide Any simple sugar with the formula $(CH_2O)_n$ where $n = 3–7$.

morphogen A molecule that specifies cell identity during development, often as a function of its concentration.

motif In proteins, a short, conserved structure that often can be recognized in the primary amino acid sequence. A motif usually is indicative of a particular three-dimensional architecture and usually is associated with a specific functional property. For example, a **leucine zipper** motif indicates the presence of a **coiled-coil** domain with the ability to bind DNA.

motor protein Any member of a special class of mechanochemical enzymes that use energy from ATP hydrolysis to generate either linear or rotary motion. See also **dyneins, kinesins,** and **myosins.** (Table 3-2; Figures 3-22 and 3-23)

MPF (mitosis-promoting factor) A heterodimeric protein, composed of a mitotic **cyclin** and **cyclin-dependent kinase (CDK),** that triggers entrance of a cell into mitosis by phosphorylating multiple specific proteins, resulting in chromosome condensation, assembly of the mitotic apparatus, and nuclear envelope breakdown (except in fungi).

mRNA (messenger RNA) Any RNA that specifies the order of amino acids in a protein (i.e., the primary structure). It is produced by **transcription** of DNA by RNA polymerase. In eukaryotes, the initial RNA product (primary transcript) undergoes processing to yield functional mRNA. See also **translation.** (Figure 4-14)

mRNA-exporter A heterodimeric protein that binds to mRNA-containing ribonucleoprotein particles (mRNPs) and directs their export from the nucleus to cytoplasm by interacting transiently with **nucleoporins** in the nuclear pore complex. (Figure 12-24)

mRNA surveillance Collective term for several cellular mechanisms that help prevent translation of improperly processed mRNA molecules.

MTOC (microtubule-organizing center) General term for any structure (e.g., the **centrosome**) that organizes microtubules in nonmitotic (interphase) cells. (Figures 5-33 and 20-14)

multiadhesive matrix proteins Group of long flexible proteins that bind to other components of the extracellular matrix (collagen, polysaccharides) and to cell-surface receptors, thereby cross-linking matrix components to the cell membrane. Examples include laminin, a major component of the basal lamina, and fibronectin, present in many tissues. (Figures 6-16 and 6-23)

multimeric For proteins, containing several polypeptide chains (or subunits).

mutagen A chemical or physical agent that induces mutations.

mutation In genetics, a permanent, heritable change in the nucleotide sequence of a chromosome, usually in a single gene; commonly causes an alteration in the function of the gene product.

myelin sheath Stacked specialized cell membrane that forms an insulating layer around vertebrate **axons** and increases the speed of impulse conduction. (Figures 7-39 and 7-40)

myosins A class of **motor proteins** that have actin-stimulated ATPase activity. Myosins move along actin **microfilaments** during muscle contraction and cytokinesis and also mediate vesicle translocation. The tail region determines the function of a particular myosin. (Figure 3-24 and Table 19-3)

NAD⁺ (nicotinamide adenine dinucleotide) A widely used diffusible electron carrier that accepts two electrons from a donor molecule and one H^+ from the solution. In mitochondria, the reduced form, NADH, transfers electrons to carriers that function in **oxidative phosphorylation.** (Figure 2-26a)

NADP⁺ (nicotinamide adenine dinucleotide phosphate) Phosphorylated form of **NAD⁺;** used extensively as an electron carrier in biosynthetic pathways and during photosynthesis.

Na⁺/K⁺ ATPase A P-class ion **pump** present in the plasma membrane of all animal cells that couples hydrolysis of one ATP molecule to export of three Na^+ ions and import of two K^+ ions. This pump is largely responsible for maintaining the normal intracellular concentrations of Na^+ (low) and K^+ (high) in animal cells. (Figure 7-9)

necrosis Cell death resulting from tissue damage or other pathology; usually marked by swelling and bursting of cells with release of their contents. Contrast with **apoptosis.**

neurofilaments (NFs) A class of intermediate filaments, found only in neurons, that determine the diameter of axons. (Figure 19-32)

neuron (nerve cell) Any of the impulse-conducting cells of the nervous system. A typical neuron contains a cell body; multiple short, branched processes (**dendrites**); and one long process (**axon**). (Figure 7-29)

neurotransmitter Extracellular signaling molecule that is released by the presynaptic neuron at a chemical **synapse** and relays the signal to the postsynaptic cell. The response elicited by a neurotransmitter, either excitatory or inhibitory, is determined by its receptor on the postsynaptic cell. (Figure 7-42)

N-linked oligosaccharides Branched oligosaccharide chains that are attached to the side-chain amino group in asparagine residues in glycoproteins. Synthesis of all N-linked oligosaccharides begins with linkage of a preformed 14-residue precursor to a nascent protein in the endoplasmic reticulum. See also **O-linked oligosaccharides.** (Figures 16-16 and 16-18)

noncovalent interaction Any relatively weak chemical interaction that does not involve an intimate sharing of electrons. Multiple noncovalent interactions often stabilize the conformation of macromolecules and mediate highly specific binding between proteins (e.g., an antigen and antibody)

nonpolar   Referring to a molecule or structure that lacks any net electric charge or asymmetric distribution of positive and negative charges. Nonpolar molecules generally are insoluble in water.

Northern blotting   Technique for detecting specific RNAs separated by electrophoresis by hybridization to a labeled DNA **probe**. See also **Southern blotting**. (Figure 9-27)

nuclear envelope   Double-membrane structure surrounding the nucleus; the outer membrane is continuous with the endoplasmic reticulum and the two membranes are perforated by **nuclear pore complexes**. (Figure 5-19)

nuclear lamina   Fibrous network on the inner surface of the nuclear envelope composed of lamin intermediate filaments. (Figure 21-16)

nuclear pore complex (NPC)   Large, multiprotein structure, composed largely of nucleoporins, that extends across the nuclear envelope. Ions and small molecules freely diffuse through NPCs; large proteins and ribonucleoprotein particles are selectively transported through NPCs with the aid of soluble proteins that bind macromolecules and also interact with certain nucleoporins. (Figure 12-18)

nuclear receptor   Member of a class of intracellular receptors that bind lipid-soluble molecules (e.g., steroid hormones, bile acids); also called *steroid receptor superfamily*. Following ligand binding, the steroid hormone-receptor complex translocates to the nucleus and functions as a transcription activator. Other nuclear receptors bind to their response elements in DNA in the nucleus in the absence of ligand-repressing transcription and are converted to **activators** when they bind ligand. (Figure 11-44)

nucleic acid   A polymer of **nucleotides** linked by **phosphodiester bonds**. DNA and RNA are the primary nucleic acids in cells.

nucleocapsid   A viral **capsid** plus the enclosed nucleic acid.

nucleolus   Large structure in the nucleus of eukaryotic cells where rRNA synthesis and processing occurs and ribosome subunits are assembled. (Figure 5-25)

nucleoporins   Large group of proteins that make up the nuclear pore complex. One class (FG-nucleoporins) participates in nuclear import and export.

nucleoside   A small molecule composed of a **purine** or **pyrimidine** base linked to a pentose (either ribose or deoxyribose). (Table 2-2)

nucleosome   Structural unit of **chromatin** consisting of a disk-shaped core of **histone** proteins around which a 147-bp segment of DNA is wrapped. (Figures 10-20 and 10-21)

nucleotide   A **nucleoside** with one or more phosphate groups linked via an ester bond to the sugar moiety, generally to the 5′ carbon atom. DNA and RNA are polymers of nucleotides containing deoxyribose and ribose, respectively. (Figure 2-14 and Table 2-2)

nucleus   Large membrane-bounded organelle in eukaryotic cells that contains DNA organized into chromosomes; synthesis and processing of RNA and ribosome assembly occur in the nucleus.

Okazaki fragments   Short (<1000 bases), single-stranded DNA fragments that are formed during synthesis of the **lagging strand** in DNA replication and are rapidly joined by DNA ligase to form a continuous DNA strand. (Figure 4-33)

O-linked oligosaccharides   Oligosaccharide chains that are attached to the side-chain hydroxyl group in serine or threonine residues in glycoproteins. See also **N-linked oligosaccharides**.

oncogene   A gene whose product is involved either in transforming cells in culture or in inducing cancer in animals. Most oncogenes are mutant forms of normal genes (**proto-oncogenes**) that encode proteins involved in the control of cell growth or division.

oncoprotein   A protein encoded by an oncogene that causes abnormal cell proliferation; may be a mutant unregulated form of a normal protein, or a normal protein that is produced in excess or in the wrong time or place in an organism.

oocyte   Immature egg cell (ovum).

open reading frame (ORF)   Region of sequenced DNA that is not interrupted by stop codons in one of the triplet reading frames. An ORF that begins with a start codon and extends for at least 100 codons has a high probability of encoding a protein.

operator   Short DNA sequence in a bacterial or bacteriophage genome that binds a repressor protein and controls transcription of an adjacent gene. (Figure 4-16)

operon   In bacterial DNA, a cluster of contiguous genes transcribed from one **promoter** that gives rise to an mRNA containing coding sequences for multiple proteins. (Figure 4-12a)

ORF   See **open reading frame**.

organelle   Any membrane-limited subcellular structure found in eukaryotic cells. (Figure 5-19)

osmosis   Net movement of water across a semipermeable membrane from a solution of lesser to one of greater solute concentration. The membrane must be permeable to water but not to solute molecules.

osmotic pressure   Hydrostatic pressure that must be applied to the more concentrated solution to stop the net flow of water across a semipermeable membrane separating solutions of different concentrations. (Figure 7-24)

oxidation   Loss of electrons from an atom or molecule as occurs when a hydrogen atom is removed from a molecule or oxygen is added. The opposite of **reduction**.

oxidation potential   The voltage change when an atom or molecule loses an electron, a measure of the tendency of a molecule to loose an electron.

oxidative phosphorylation   The phosphorylation of ADP to form ATP driven by the transfer of electrons to oxygen (O_2) in bacteria and mitochondria. This process involves generation of a **proton-motive force** during electron transport and its subsequent use to power ATP synthesis. (Figure 8-7)

P element   A **DNA transposon** in *Drosophila* that is used experimentally in production of transgenic flies.

p53 protein   The product of a tumor-suppressor gene that plays a critical role in the arrest of cells with damaged DNA, thereby preventing perpetuation of mutations or other chromosomal ab-normalities in daughter cells. Inactivating mutations in the *p53* gene are found in many human cancers. (Figures 21-23 and 23-23)

pair-rule genes   In *Drosophila*, group of genes (e.g., *even-skipped, fushi tarazu,* and *paired*) that are expressed in alternat-ing stripes along the anterioposterior axis in early *Drosophila* embryos. All encode transcription factors and function, along with gap genes and segment-polarity genes, in determining the body segments in flies. (Figure 15-24)

paracrine   Referring to signaling mechanism in which a target cell responds to a signaling molecule (e.g., growth factor) that is produced by a nearby cell(s) and reaches the target by diffusion. Paracrine signaling is important in development, with the re-sponse of target cells to a particular signal often varying with dis-tance from the signal source.

passive (simple) diffusion   Net movement of a molecule across a membrane down its concentration gradient at a rate propor-tional to the gradient and the permeability of the membrane.

patch clamping   Technique for determining ion flow through a single ion channel or across the membrane of an entire cell by use of a micropipette whose tip is applied to a small patch of the cell membrane. (Figure 7-17)

PCR (polymerase chain reaction)   Technique for amplifying a specific DNA segment in a complex mixture by multiple cycles of DNA synthesis from short oligonucleotide primers followed by brief heat treatment to separate the complementary strands. (Fig-ure 9-24)

pentose   A five-carbon **monosaccharide.** The pentoses ribose and deoxyribose are present in RNA and DNA, respectively.

peptide   A small polymer usually containing fewer than 30 amino acids connected by peptide bonds.

peptide bond   The covalent amide linkage between amino acids formed by a dehydration reaction between the amino group of one amino acid and the carboxyl group of another with release of a water molecule. (Figure 2-11)

peripheral membrane protein   Any protein that associates with the cytosolic or exoplasmic face of a membrane but does not enter the hydrophobic core of the phospholipid bilayer; also called ex-trinsic protein. See also **integral membrane protein.** (Figure 5-11)

peroxisome   Small organelle that contains enzymes for degrad-ing fatty acids and amino acids by reactions that generate hydro-gen peroxide, which is converted to water and oxygen by catalase.

pH   A measure of the acidity or alkalinity of a solution defined as the negative logarithm of the hydrogen ion concentration in moles per liter: $pH = -\log[H^+]$. Neutrality is equivalent to a pH of 7; values below this are acidic and those above are alkaline.

phage   See **bacteriophage.**

phagocytosis   Process by which relatively large particles (e.g., bacterial cells) are specifically internalized by certain eukaryotic cells; distinct from **receptor-mediated endocytosis.** (Figure 5-20)

phenotype   The detectable physical and physiological character-istics of a cell or organism determined by its **genotype;** also, the specific trait associated with a particular **allele.**

pheromone   A signaling molecule released by an individual that can alter the behavior or gene expression of other individuals of the same species. The yeast α and **a** mating-type factors are well-studied examples.

phosphatase   An enzyme that removes a phosphate group from a substrate by hydrolysis. Phosphoprotein phosphatases act with protein kinases to control the activity of many cellular proteins. (Figure 3-30)

phosphoanhydride bond   A type of **high-energy bond** formed be-tween two phosphate groups, such as the γ and β phosphates and the β and α phosphates in ATP. (Figure 2-24)

phosphodiester bond   Chemical linkage between adjacent nu-cleotides in DNA and RNA; actually consists of two phosphoester bonds, one on the 5′ side of the phosphate and another on the 3′ side. (Figure 4-2)

phosphoglycerides   Amphipathic derivatives of glycerol 3-phosphate that are the most abundant lipids in biomembranes. Generally consist of two hydrophobic fatty acyl chains esterified to the glycerol hydroxyl groups and a polar head group attached to the phosphate. (Figure 5-5a)

phosphoinositides   A group of membrane-bound lipids contain-ing phosphorylated inositol derivatives; some function as **second messengers** in several signal-transduction pathways. (Figures 14-26 and 14-27)

phospholipase   One of several enzymes that cleave various bonds in the hydrophilic end of **phospholipids.** Phospholipase C acts on a phosphatidylinositol derivative to generate two second messengers, DAG and IP₃. (Figures 5-9 and 13-28)

phospholipid bilayer   A two-layer structure, which is the foun-dation for all biomembranes, in which the polar head groups of phospholipids are exposed to the aqueous media on either side, while the nonpolar hydrocarbon chains of the fatty acids are in the center. (Figure 2-11)

phospholipids   The major class of lipids present in biomem-branes, including **phosphoglycerides** and sphingomyelin. (Figure 5-5)

photoelectron transport   Light-driven electron transport involv-ing reaction center chlorophylls that is the primary event in pho-tosynthesis. This process generates the initial charge separation across the **thylakoid** membrane that drives subsequent events in photosynthesis.

photorespiration   A reaction pathway that competes with CO_2 fixation (**Calvin cycle**) by consuming ATP and generating CO_2, thus reducing the efficiency of photosynthesis. (Figure 8-43)

photosynthesis   Complex series of reactions occurring in some bacteria and in plant **chloroplasts** in which light energy is used to generate carbohydrates from CO_2, usually with the consumption of H_2O and evolution of O_2.

photosystems   Multiprotein complexes present in all photosyn-thetic organisms that absorb light energy and convert it into chemical energy. A photosystem consists of light-harvesting com-plexes containing **chlorophylls** and a reaction center containing two chlorophyll molecules where **photoelectron transport** occurs.

pI   See **isoelectric point.**

plaque assay Technique for determining the number of infectious viral particles in a sample by culturing a diluted sample on a layer of susceptible host cells and then counting the clear areas of lysed cells (plaques) that develop. (Figure 4-39)

plasma membrane The membrane surrounding a cell that separates the cell from its external environment, consisting of a **phospholipid bilayer** and associated membrane lipids and proteins.

plasmid Small, circular extrachromosomal DNA molecule capable of autonomous replication in a cell. Commonly used as a **vector** in **DNA cloning.**

plasmodesmata (sing. **plasmodesma**) Tubelike cell junctions that interconnect the cytoplasms of adjacent plant cells and are functionally analogous to **gap junctions** in animal cells. (Figure 6-34)

point mutation Change of a single nucleotide in DNA, especially in a region coding for protein; can result in formation of a codon specifying a different amino acid or a stop codon. An addition or deletion of a single nucleotide will cause a shift in the **reading frame.**

polar Referring to a molecule or structure with a net electric charge or asymmetric distribution of positive and negative charges. Polar molecules are usually soluble in water.

polarity In cell biology, presence of functional and/or structural differences in distinct regions of a cell or cellular component.

polarized cell Any cell marked by stable functional and structural asymmetries. In epithelial cells, for example, the apical and basolateral regions of the plasma membrane have different lipid compositions, contain different sets of proteins, and assume different shapes.

polymer Any large molecule composed of multiple identical or similar units (**monomers**) linked by covalent bonds.

polymerase chain reaction See **PCR.**

polypeptide Linear polymer of amino acids connected by peptide bonds. A protein is a large polypeptide folded into a distinct three-dimensional shape.

polyribosome A complex containing several ribosomes, all translating a single messenger RNA; also called *polysome.* (Figure 4-31)

polysaccharide Linear or branched polymer of monosaccharides, linked by glycosidic bonds, usually containing more than 15 residues. Examples include glycogen, cellulose, and glycosaminoglycans (GAGs).

polytene chromosome Enlarged chromosome composed of many parallel copies of itself formed by multiple cycles of DNA replication without chromosomal separation. Polytene chromosomes, which have highly reproducible banding patterns, are found in the salivary glands and some other tissues of *Drosophila* and other dipteran insects. (Figures 10-30 and 10-31)

porins Transmembrane proteins present in outer mitochondrial and chloroplast membranes and in the outer membrane of gram-negative bacteria through which small water-soluble molecules can cross the membrane. The structure of a porin monomer consists of beta (β) strands arranged to form a barrel-shaped cylinder with a central pore. (Figure 5-14)

preinitiation complex Complex of RNA polymersase II, general transcription factors, and promoter DNA. (Figure 11-27)

pre-mRNA Precursor messenger RNA; the **primary transcript** and intermediates in RNA processing. (Figure 12-2)

pre-rRNA Large precursor ribosomal RNA that is synthesized in the nucleolus of eukaryotic cells and processed to yield three of the four RNAs present in ribosomes. (Figure 12-34)

primary structure In proteins, the linear arrangement (sequence) of amino acids and the location of covalent (mostly disulfide) bonds within a polypeptide chain.

primary transcript In eukaryotes, the initial RNA product, containing **introns** and **exons,** produced by transcription of DNA. Many primary transcripts must undergo RNA processing to form the physiologically active RNA species.

primase A specialized RNA polymerase that synthesizes short stretches of RNA used as primers for DNA synthesis.

primer A short nucleic acid sequence containing a free 3′-hydroxyl group that forms **base pairs** with a complementary template strand and functions as the starting point for addition of nucleotides to copy the template strand.

probe Defined RNA or DNA fragment, radioactively or chemically labeled, that is used to detect specific nucleic acid sequences by **hybridization.**

programmed cell death See **apoptosis.**

prokaryotes Class of organisms, including the **eubacteria** and **archaea,** that lack a true membrane-limited nucleus and other organelles. See also **eukaryotes.**

promoter DNA sequence that determines the site of **transcription** initiation for a RNA polymerase.

promoter-proximal element Any regulatory sequence in eukaryotic DNA that is located within ≈200 base pairs of the transcription start site. Transcription of many genes is controlled by multiple promoter-proximal elements. (Figure 11-12)

prophase Earliest stage in mitosis, during which the chromosomes condense, the centrioles begin moving toward the spindle poles, and the nuclear envelope breaks down (except in fungi). (Figure 20-29)

proteasome Large multifunctional protease complex in the cytosol that degrades intracellular proteins marked for destruction by attachment of multiple **ubiquitin** molecules. (Figure 3-13)

protein A linear polymer usually containing 50 or more **amino acids** linked together by peptide bonds and folded into a characteristic three-dimensional shape (conformation). Proteins form the key structural elements in cells and participate in nearly all cellular activities.

protein family Set of homologous proteins encoded by a **gene family.**

protein kinase A (PKA) Cytosolic protein kinase that is activated by **cyclic AMP (cAMP)** and functions to phosphorylate and thus regulate the activity of numerous cellular proteins; also called *cAMP-dependent protein kinase.* PKA generally is activated in response to a rise in cAMP level resulting from stimulation of certain **G protein-coupled receptors.** (Figure 13-32)

protein kinase B (PKB; also called Akt) Cytosolic protein kinase that is recruited to the plasma membrane by signal-induced phosphoinositides and subsequently activated. (Figure 14-27)

protein kinase C (PKC) Cytosolic enzyme that is recruited to the plasma membrane in response to signal-induced rise in cytosolic Ca^{2+} level and is activated by the membrane-bound second messenger diacylglycerol (DAG). (Figure 13-29)

protein microarray A microscope slide or other solid surface to which small samples of purified proteins are affixed; also called *proteome chip*. Protein microarrays are being used to assess protein-protein associations in an approach analogous to the use of **DNA microarrays** to monitor gene expression. (Figure 15-4)

proteoglycans A group of glycoproteins that contain a core protein to which is attached one or more **glycosaminoglycans**. They are found in nearly all animal extracellular matrices, and some are integral membrane proteins. (Figure 6-22)

proteome The entire complement of proteins produced by a cell.

proteomics Comprehensive analysis of the identity, interactions, and locations of proteins within a cell.

proton-motive force The energy equivalent of the proton (H^+) concentration gradient and electric potential gradient across a membrane; used to drive ATP synthesis by **ATP synthase**, transport of molecules against their concentration gradient, and movement of bacterial flagella. (Figure 8-1)

proto-oncogene A normal cellular gene that encodes a protein usually involved in regulation of cell growth or differentiation and that can be mutated into a cancer-promoting oncogene, either by changing the protein-coding segment or by altering its expression.

pseudogenes DNA sequences that are similar to functional genes but do not express a functional product; probably arose by sequence drift of duplicated genes.

pulse-chase A type of experiment in which a radioactive small molecule is added to a cell for a brief period (the pulse) and then is replaced with an excess of the unlabeled form of the same small molecule (the chase). Used to detect changes in the cellular location of a molecule or its metabolic fate over time. (Figure 3-36)

pump Any transmembrane protein that has ATPase activity and couples hydrolysis of ATP to the active transport of an ion or small molecule across a biomembrane against its electrochemical gradient. (Figure 7-6)

purines A class of nitrogenous compounds containing two fused heterocyclic rings. Two purines, adenine (A) and guanine (G), are the base components of nucleotides found in DNA and RNA. See also **base pair.** (Figure 2-15)

pyrimidines A class of nitrogenous compounds containing one heterocyclic ring. Two pyrimidines, cytosine (C) and thymine (T), are the base components of nucleotides found in DNA; in RNA, uracil (U) replaces thymine. See also **base pair.** (Figure 2-15)

quaternary structure The number and relative positions of the polypeptide chains in multisubunit proteins.

quiescent Referring to a cell that has exited the **cell cycle** and is in the G_0 state.

radioisotope Unstable form of an atom that emits radiation as it decays. Several radioisotopes are commonly used experimentally as **labels** in biological molecules. (Table 3-3)

Ras protein A monomeric GTP-binding protein that is tethered to the plasma membrane by a lipid anchor and functions in intracellular signaling pathways. Activation of Ras is induced by ligand binding to **receptor tyrosine kinases** and some other cell-surface receptors. See also **GTPase superfamily.** (Figures 14-16 and 14-20)

reading frame The sequence of nucleotide triplets (**codons**) that runs from a specific **translation** start codon in an mRNA to a stop codon. Some mRNAs can be translated into different polypeptides by reading in two different reading frames. (Figure 4-20)

receptor Any protein that specifically binds another molecule to mediate cell-cell signaling, adhesion, endocytosis, or other cellular process. Most commonly denotes a protein located in the plasma membrane or cytoplasm that is activated by binding a specific extracellular **signaling molecule** (ligand), thereby initiating a cellular response. See also **adhesion receptor** and **receptor-mediated endocytosis.** (Table 14-1)

receptor-mediated endocytosis Uptake of extracellular materials bound to specific cell-surface receptors by invagination of the plasma membrane to form a small membrane-bounded vesicle (early endosome). (Figure 17-28)

receptor tyrosine kinase (RTK) Member of a large class of cell-surface signaling receptors including those for insulin and many growth factors. Ligand binding leads to receptor dimerization and activation of tyrosine-specific protein kinase activity in the receptor's cytosolic domain, thereby initiating intracellular signaling pathways. (Figure 14-16)

recessive In genetics, referring to that allele of a gene that is not expressed in the **phenotype** when the **dominant** allele is present. Also refers to the phenotype of an individual (homozygote) carrying two recessive alleles. Mutations that produce recessive alleles generally result in a loss of function. (Figure 9-2)

recombinant DNA Any DNA molecule formed in vitro by joining DNA fragments from different sources. Commonly produced by cutting DNA molecules with **restriction enzymes** and then joining the resulting fragments from different sources with DNA ligase.

recombination Any process in which chromosomes or DNA molecules are cleaved and the fragments are rejoined to give new combinations. It occurs naturally in cells as the result of the exchange (**crossing over**) of DNA sequences on maternal and paternal chromatids during **meiosis**; also is carried out in vitro with purified DNA and enzymes.

reduction Gain of electrons by an atom or molecule as occurs when a hydrogen atom is added to a molecule or oxygen is removed. The opposite of **oxidation.**

reduction potential (*E*) The voltage change when an atom or molecule gains an electron.

release factor One of two types of nonribosomal proteins that recognize stop codons in mRNA and promote release of the completed polypeptide chain, thereby terminating **translation** (protein synthesis). (Figure 4-29)

replication fork   Site in double-stranded DNA at which the two strands are separated and templated by these strands and addition of deoxyribonucleotides to each newly formed chain occurs; also called *growing fork*. (Figure 4-33)

replication origin   Unique DNA segments present in an organism's genome at which DNA replication begins. Eukaryotic chromosomes contain multiple origins, whereas bacterial chromosomes and plasmids usually contain just one.

reporter gene   A gene encoding a protein that is easily assayed (e.g., β-galactosidase, luciferase). Reporter genes are used in various types of experiments to indicate the activation of a promoter to which it is linked.

repressor   Transcription factor that inhibits transcription.

resolution   The minimum distance between two objects that can be distinguished by an optical apparatus; also called *resolving power*.

respiration, cellular   General term for any cellular process involving the uptake of oxygen (O_2) coupled to production of carbon dioxide (CO_2).

respiratory chain   Set of four large multiprotein complexes in the inner mitochondrial membrane plus diffusible cytochrome *c* in the intermembranous space and coenzyme Q through which electrons flow from reduced electron donors (e.g., NADH) to O_2. Each member of the chain contains one or more bound **electron carriers** that sequentially accept and donate electrons. (Figure 8-17 and Table 8-2)

respiratory control   Dependence of mitochondrial oxidation of NADH and $FADH_2$ on the supply of ADP and P_i for ATP synthesis. It occurs because of the obligatory coupling of electron transport along the respiratory chain to transport of protons across the inner mitochondrial membrane.

resting K^+ channels   Nongated K^+ ion channels in the plasma membrane that, in conjunction with the high cytosolic K^+ concentration produced by the **Na^+/K^+ ATPase,** are primarily responsible for generating the inside-negative resting membrane potential in animal cells.

restriction enzyme (endonuclease)   Any enzyme that recognizes and cleaves a specific short sequence, the restriction site, in double-stranded DNA molecules. These enzymes are widespread in bacteria and function to degrade foreign DNA taken up by a bacterial cell. They are used extensively in vitro to produce **recombinant DNA.** (Table 9-1 and Figure 9-10)

restriction fragment   A defined DNA fragment resulting from cleavage with a particular **restriction enzyme.** These fragments are used in the production of recombinant DNA molecules and DNA cloning.

restriction fragment length polymorphisms   See **RFLPs.**

restriction point   The point in late G_1 of the cell cycle at which mammalian cells become committed to entering the S phase and completing the cycle even in the absence of growth factors; functionally equivalent to START in yeast. (Figure 21-28)

retrotransposon   Type of eukaryotic **mobile DNA element** whose movement in the genome is mediated by an RNA intermediate and involves a reverse transcription step. See also **transposon.** (Figure 10-8)

retrovirus   A type of eukaryotic virus containing an RNA genome that replicates in cells by first making a DNA copy of the RNA. This proviral DNA is inserted into cellular chromosomal DNA and gives rise to further genomic RNA as well as the mRNAs for viral proteins. (Figure 4-43)

reverse transcriptase   Enzyme found in retroviruses that catalyzes a complex reaction in which a double-stranded DNA is synthesized from a single-stranded RNA template. (Figure 10-13)

RFLPs (restriction fragment length polymorphisms)   Differences among individuals in the sequence of genomic DNA that create or destroy sites recognized by particular **restriction enzymes.** RFLPs, detected as differences in the lengths of fragments produced when genomic DNA is digested with a **restriction enzyme,** are one of several types of sequence differences between individuals that can serve as molecular genetic markers in human **linkage** studies. (Figure 9-46)

ribosomal RNA   See **rRNA.**

ribosome   A large complex comprising several different **rRNA** molecules and more than 50 proteins, organized into a large subunit and small subunit; the site of **translation** (protein synthesis). (Figures 4-27 and 4-28)

ribozyme   An RNA molecule with catalytic activity.

RNA (ribonucleic acid)   Linear, single-stranded polymer, composed of ribose **nucleotides.** Three types of cellular RNA—mRNA, rRNA, and tRNA—play different roles in protein synthesis. (Figure 4-19)

RNA editing   Unusual type of RNA processing in which the sequence of a pre-mRNA is altered.

RNA-induced silencing complex (RISC)   Large multiprotein complex associated with a short single-stranded RNA that mediates degradation or translational repression of a complementary or near-complementary mRNA. (Figure 12-27)

RNA interference (RNAi)   Functional inactivation of a specific gene by experimental introduction of a corresponding double-stranded RNA, which induces degradation of the complementary single-stranded mRNA encoded by the gene but not that of mRNAs with a different sequence. (Figures 9-43 and 12-27)

RNA polymerase   An enzyme that copies one strand of DNA (the template strand) to make the **complementary** RNA strand using as substrates ribonucleoside triphosphates. (Figure 4-10)

RNA processing   Various modifications that are made within the nucleus to many but not all **primary transcripts** to yield functional RNA molecules in eukaryotes. Pre-mRNAs undergo three major processing events: capping at the 5′ end, cleavage and polyadenylation at the 3′ end, and RNA splicing (Figure 12-2).

RNA splicing   A process that results in removal of **introns** and joining of **exons** in pre-mRNAs. See also **spliceosome.** (Figure 12-7)

rRNA (ribosomal RNA)   Any one of several large RNA molecules that are structural and functional components of **ribosomes.** Often designated by their sedimentation coefficient: 28S, 18S, 5.8S, and 5S rRNA in higher eukaryotes. (Figure 4-24)

S (synthesis) phase   See **cell cycle.**

sarcomere   Repeating structural unit of striated (skeletal) muscle composed of organized, overlapping thin (**actin**) filaments and thick (**myosin**) filaments and extending from one Z disk to an adjacent one; shortens during contraction. (Figures 19-22 and 19-23)

sarcoplasmic reticulum   Network of membranes in cytoplasm of a muscle cell that sequesters Ca^{2+} ions. Stimulation of a muscle cell induces release of Ca^{2+} ions into the cytosol, triggering coordinated contraction along the length of the cell.

second messenger   A small intracellular molecule whose concentration increases (or decreases) in response to binding of an extracellular signal and that functions in **signal transduction.** Examples include cAMP, cGMP, Ca^{2+}, DAG, IP_3, and several phosphoinositides.

secondary structure   In proteins, local folding of a polypeptide chain into regular structures including the **α helix, β sheet,** and U-shaped turns and loops.

secretory pathway   Cellular pathway for synthesizing and sorting soluble and membrane proteins localized to the endoplasmic reticulum, Golgi, and lysosomes; plasma membrane proteins; and proteins eventually secreted from the cell. (Figures 16-1 and 17-1)

secretory vesicle   Small membrane-bound vesicle derived from the *trans*-Golgi network containing molecules destined to be released from the cell.

segment-polarity genes   In *Drosophila*, group of genes (e.g., *engrailed, hedgehog, wingless*) that function in final patterning events within body segments in early embryogenesis. The pattern of expression of these and other early-patterning genes confers unique cell fates along the anterioposterior axis. (Figure 15-24)

segregation   The process that distributes an equal complement of chromosomes to daughter cells during mitosis and meiosis.

selectins   A family of **cell-adhesion molecules** that mediate Ca^{2+}-dependent heterophilic interactions with specific oligosaccharide moieties of glycoproteins and glycolipids on the surface of adjacent cells. Selectins function in interactions between leukocytes and endothelial cells and also for binding of extracellular glycoproteins to cells.

selective lipid uptake   Receptor-dependent import into cells of specific components (e.g., cholesteryl esters) from certain lipoproteins; distinguished from **receptor-mediated endocytosis** of entire lipoprotein particles. (Figure 18-16)

shuttle vector   Plasmid vector capable of propagation in two different hosts (e.g., *E. coli* and yeast, or *E. coli* and mammalian cells). Shuttle vectors are useful in screening a yeast DNA library by functional complementation and in expression systems based on transient transfection of animal cells. (Figures 9-19 and 9-29)

signal-recognition particle (SRP)   A cytosolic ribonucleoprotein particle that binds to the ER signal sequence in a nascent secretory protein and delivers the nascent chain/ribosome complex to the ER membrane where synthesis of the protein and translocation into the ER are completed. (Figure 16-5)

signal sequence   A relatively short amino acid sequence that directs a protein to a specific location within the cell; also called *signal peptide* and *uptake-targeting sequence.* (Table 16-1)

signal transduction   Conversion of a signal from one physical or chemical form into another. In cell biology commonly refers to the sequential process initiated by binding of an extracellular signal to a receptor and culminating in one or more specific cellular responses.

signaling molecule   General term for any extracellular or intracellular molecule involved in mediating the response of a cell to its external environment or to other cells.

silencer sequence   A sequence in eukaryotic DNA that promotes formation of condensed chromatin structures in a localized region, thereby blocking access of proteins required for transcription of genes within several hundred base pairs of the silencer sequence; also called simply *silencer.*

simple-sequence DNA   Short, tandemly repeated sequences that are found in **centromeres** and **telomeres** as well as at other chromosomal locations and are not transcribed.

Smads   Class of cytosolic transcription factors that are activated by phosphorylation following binding of hormones in the transforming growth factor β (TGFβ) superfamily to their cell-surface receptor serine kinases. (Figure 14-2)

small nuclear RNA   See **snRNA.**

SNAREs   Cytosolic and integral membrane proteins that promote fusion of vesicles with target membranes. Interaction of v-SNAREs on a vesicle with cognate t-SNAREs on a target membrane forms very stable complexes, bringing the vesicle and target membranes into close apposition. (Figure 17-11)

snRNA (small nuclear RNA)   One of several small, stable RNAs localized to the nucleus. Five snRNAs are components of the **spliceosome** and function in splicing of pre-mRNA. (Figures 12-8 and 12-9)

somatic cell   Any plant or animal cell other than a **germ cell** or germ-cell precursor.

sorting signal   A relatively short amino acid sequence that directs a protein to particular transport vesicles as they bud from a donor membrane in the secretory or endocytic pathway. (Table 17-2)

Southern blotting   Technique for detecting specific DNA sequences separated by electrophoresis by hybridization to a labeled nucleic acid **probe.** (Figure 9-26)

SPF (S phase–promoting factor)   A heterodimeric protein, composed of a G_1 cyclin and **cyclin-dependent kinase (CDK),** that triggers entrance of a eukaryotic cell into the S phase of the cell cycle by phosphorylating specific proteins, thereby inducing expression of proteins required for DNA replication and passage through the S phase.

sphingolipids   Major group of membrane lipids that are derived from sphingosine; contain two long hydrocarbon chains and either a phosphorylated head group (sphingomyelin) or carbohydrate head group (cerebrosides, gangliosides). (Figure 5-5)

spliceosome   Large ribonucleoprotein complex that assembles on a pre-mRNA and carries out **RNA splicing.** (Figure 12-9)

SREBP (SRE-binding protein)   Transcription factor that is activated by intracellular movement and proteolysis in response to low cellular cholesterol levels and stimulates expression of genes encoding proteins required for the synthesis and import of cholesterol and synthesis of fatty acids, phospholipids, and triglycerides. (Figures 18-17 and 18-18)

starch   A very long, branched polysaccharide, composed exclusively of glucose units, that is the primary storage carbohydrate in plant cells.

START   A point in the G_1 stage of the yeast cell cycle that controls entry of cells into the S phase; functionally equivalent to the restriction point in mammalian cells. Passage of a cell through **START** commits a cell to proceed through the remainder of the cell cycle. (Figure 21-25)

STATs   Class of transcription factors that are activated in the cytosol following ligand binding to **cytokine receptors.** (Figure 14-12)

steady state   In cellular metabolic pathways, the condition when the rate of formation and rate of consumption of a substance are equal, so that its concentration remains constant. (Figure 2-21)

stem cell   A self-renewing cell that can divide symmetrically to give rise to two daughter cells with a developmental potential identical to the parental stem cell or asymmetrically to generate one daughter stem cell and one with a more restricted developmental potential than the parental cell. (Figure 22-2)

stereoisomers   Two compounds with identical molecular formulas whose atoms are linked in the same order but in different spatial arrangements. In optical isomers, designated D and L, the atoms bonded to an **asymmetric carbon atom** are arranged in a mirror-image fashion. Geometric isomers include the *cis* and *trans* forms of molecules containing a double bond.

steroids   A group of four-ring hydrocarbons including **cholesterol** and related compounds. Many important hormones (e.g., estrogen and progesterone) are steroids.

sterols   Steroids containing one hydroxyl group that are components of eukaryotic membranes. (Figure 18-6)

substrate   Molecule that undergoes a change in a reaction catalyzed by an enzyme.

substrate-level phosphorylation   Formation of ATP from ADP and P_i catalyzed by cytosolic enzymes in reactions that do not depend on a proton-motive force.

sulfhydryl group (—SH)   A substituent group present in the amino acid cysteine and other molecules consisting of a hydrogen atom covalently bonded to a sulfur atom; also called a *thiol group.*

suppressor mutation   A mutation that reverses the phenotypic effect of a second mutation. Suppressor mutations are frequently used to identify genes encoding interacting proteins. (Figure 9-9a)

symport   A type of **cotransport** in which a membrane protein (symporter) transports two different molecules or ions across a cell membrane in the same direction. See also **antiport.**

synapse   Specialized region between an axon terminal of a neuron and an adjacent neuron or other excitable cell (e.g., muscle cell) across which impulses are transmitted. At a chemical synapse, the impulse is conducted by a **neurotransmitter;** at an electric synapse, impulse transmission occurs via **gap junctions** connecting the pre- and postsynaptic cells. (Figure 7-31)

syncytium   A multinucleated mass of cytoplasm enclosed by a single plasma membrane.

syndecans   A class of **proteoglycans** that are integral membrane proteins. They function in cell-matrix adhesion, interact with the cytoskeleton, and may bind external signals, thereby participating in cell-cell signaling.

synthetic lethal mutation   A mutation that increases the phenotypic effect of another mutation in the same or a related gene. (Figure 9-9b, c)

TATA box   A conserved sequence in the **promoter** of many eukaryotic protein-coding genes where the transcription-initiation complex assembles. (Figure 11-9)

telomere   Region at each end of a eukaryotic chromosome containing multiple tandem repeats of a short telomeric (TEL) sequence. Telomeres are required for proper chromosome **segregation** and are replicated by a special process, thereby counteracting the tendency of a chromosome to be shortened during each round of DNA replication. (Figure 10-34)

telophase   Final mitotic stage during which the nuclear envelope re-forms around the two sets of separated chromosomes, the chromosomes decondense, and division of the cytoplasm (cytokinesis) is completed. (Figure 20-29)

temperature-sensitive (ts) mutation   A mutation that produces a wild-type phenotype at one temperature (the permissive temperature) but a mutant phenotype at another temperature (the nonpermissive temperature). This type of mutation is especially useful in identification of genes essential for life.

termination factor   See **release factor.**

tertiary structure   In proteins, overall three-dimensional form of a polypeptide chain, which is stabilized by multiple noncovalent interactions between side chains.

thylakoids   Flattened membranous sacs in a chloroplast that are arranged in stacks and contain the photosynthetic pigments. (Figure 8-30)

tight junction   A type of cell-cell junction between adjacent epithelial cells that prevents diffusion of macromolecules and many small molecules and ions in the spaces between cells and diffusion of membrane components between the apical and basolateral regions. (Figure 6-9)

topogenic sequences   Segments within a protein whose sequence, number, and arrangement direct the insertion and orientation of various classes of transmembrane proteins into the endoplasmic reticulum membrane. (Figure 16-13)

transcript   See **primary transcript.**

transcription   Process in which one strand of a DNA molecule is used as a template for synthesis of a **complementary** RNA by RNA polymerase. (Figures 4-9 and 4-10)

transcription-control region   Collective term for all the DNA regulatory sequences that regulate transcription of a particular gene.

transcription factor (TF)   General term for any protein, other than RNA polymerase, required to initiate or regulate transcription in eukaryotic cells. General factors, required for transcription of all genes, participate in formation of the transcription-preinitiation complex near the start site. Specific factors stimulate (activators) or inhibit (repressors) transcription of particular genes by binding to their regulatory sequences.

transcription unit A region in DNA, bounded by an initiation (start) site and termination site, that is transcribed into a single **primary transcript.**

transcytosis Mechanism for transporting certain substances across an epithelial sheet that combines **receptor-mediated endocytosis** and **exocytosis.** (Figure 17-34)

transfection Experimental introduction of foreign DNA into cells in culture, usually followed by expression of genes in the introduced DNA. (Figure 9-29)

transfer RNA See **tRNA.**

transformation (1) Permanent, heritable alteration in a cell resulting from the uptake and incorporation of a foreign DNA into the host-cell genome; also called *stable transfection.* (2) Conversion of a "normal" mammalian cell into a cell with cancer-like properties usually induced by treatment with a virus or other cancer-causing agent.

transforming growth factor beta (TGFβ) A family of secreted signaling proteins that are used in the development of most tissues in most or all animals. The name reflects the mode of discovery; the functions are far more diverse. TGFβ more often inhibits growth than stimulates it. Mutations in TGFβ signal transduction components are implicated in human cancer, including breast cancer.

transgene A cloned gene that is introduced and stably incorporated into a plant or animal and is passed on to successive generations.

transgenic Referring to any plant or animal carrying a **transgene.**

trans-**Golgi network (TGN)** Complex network of membranes and vesicles that serves as a major branch point in the **secretory pathway.** Vesicles budding from this most-distal Golgi compartment carry membrane and soluble proteins to the cell surface or to lysosomes. (Figure 17-1)

translation The **ribosome**-mediated production of a polypeptide whose amino acid sequence is specified by the nucleotide sequence in an mRNA. (Figure 4-19)

translocon Multiprotein complex in the membrane of the rough endoplasmic reticulum through which a nascent secretory protein enters the ER lumen as it is being synthesized. (Figure 16-6)

transport vesicle A small membrane-bounded compartment that carries soluble and membrane "cargo" proteins in the forward or reverse direction in the **secretory pathway.** Vesicles form by budding off from the donor organelle and release their contents by fusion with the target membrane.

transposition Movement of a **mobile DNA element** within the genome; occurs by a cut-and-paste mechanism or copy-and-paste mechanism depending on the type of element. (Figure 10-8)

transposon, DNA A **mobile DNA element** present in prokaryotes and eukaryotes that moves in the genome by a mechanism involving DNA synthesis and transposition. See also **retrotransposon.** (Figure 10-8)

triacylglycerol See **triglyceride.**

tricarboxylic acid cycle See **citric acid cycle.**

triglyceride Major form in which fatty acids are stored and transported in animals; consists of three fatty acyl chains esterified to a glycerol molecule. (Figure 18-2)

tRNA (transfer RNA) A group of small RNA molecules that function as amino acid donors during protein synthesis. Each tRNA becomes covalently linked to a particular amino acid, forming an **aminoacyl-tRNA.** (Figures 4-21 and 4-22)

tubulin A family of globular cytoskeletal proteins that polymerize to form **microtubules.** (Figure 20-3)

tumor A mass of cells, generally derived from a single cell, that arises due to loss of the normal regulators of cell growth; may be **benign** or **malignant.**

tumor-suppressor gene Any gene whose encoded protein directly or indirectly inhibits progression through the cell cycle and in which a loss-of-function mutation is oncogenic. Inheritance of a single mutant allele of many tumor-suppressor genes (e.g., *RB, APC,* and *BRCA1*) greatly increases the risk for developing certain types of cancer.

ubiquitin A small, highly conserved protein that becomes covalently linked to lysine residues of other intracellular proteins. Tagging a protein with a covalently added ubiquitin molecule can cause the protein to be degraded by the **proteasome,** can alter the function of the protein, or can cause a membrane protein to be sorted to the lysosome. (Figure 3-13)

uncoupler An agent that dissipates the **proton-motive force** across the inner mitochondrial membrane or thylakoid membrane of chloroplasts, thereby inhibiting ATP synthesis.

uniport Protein-mediated **facilitated diffusion** of a small molecule across a membrane down its concentration gradient. The glucose transporters (GLUTs) are well-studied examples of uniport proteins. (Figure 7-4)

upstream (1) For a gene, the direction opposite to that in which RNA polymerase moves during transcription. By convention, the +1 position in a gene is the first transcribed base; nucleotides upstream from the +1 position are designated −1, −2, etc. (2) Events that occur earlier in a cascade of steps (e.g., signaling pathway). See also **downstream.**

upstream activating sequence (UAS) Any protein-binding regulatory sequence in the DNA of yeast and other simple eukaryotes that is necessary for maximal gene expression; equivalent to an enhancer or promoter-proximal element in higher eukaryotes. (Figure 11-12)

van der Waals interaction A weak **noncovalent interaction** due to small, transient asymmetric electron distributions around atoms (dipoles).

vector In cell biology, an autonomously replicating genetic element used to carry a cDNA or genomic DNA fragment into a host cell for the purpose of gene cloning. Commonly used vectors are bacterial plasmids and modified bacteriophage genomes. See also **expression vector** and **shuttle vector.** (Figures 9-12 and 9-15)

ventral   Relating to the front of an animal or lower surface of a structure (e.g., wing or leaf).

very low density lipoprotein (VLDL)   A class of **lipoproteins** that is secreted by liver cells and transports **triglycerides** from the liver to extrahepatic tissues and serves as a precursor for the synthesis of **low-density lipoproteins (LDLs).** (Table 18-3 and Figure 18-13a)

viral envelope   A phospholipid bilayer forming the outer covering of some viruses (e.g., influenza and rabies viruses) derived by budding from a host-cell membrane and containing virus-encoded glycoproteins. (Figure 4-41)

virion   An individual viral particle.

virus   A small intracellular parasite consisting of nucleic acid (RNA or DNA) enclosed in a protein coat that can replicate only in a susceptible host cell; widely used in cell biology research. (Figures 1-25 and 4-37)

V_{max}   Parameter that describes the maximal velocity of an enzyme-catalyzed reaction or other process such as protein-mediated transport of molecules across a membrane. (Figures 3-19 and 7-3)

Western blotting   Technique for detecting specific proteins separated by electrophoresis by use of labeled antibodies. (Figure 3-35)

wild type   Normal, nonmutant form of a gene, protein, cell, or organism.

Wnt   A family of secreted signaling proteins that are used in the development of most tissues in most or all animals. The name "Wnt" was derived from the first two such molecules found, Int-1 (for integration of a mammary tumor virus) and Wingless (a *Drosophila* gene). Wnt proteins are palmitoylated near the N terminus. Mutations in Wnt signal transduction components are implicated in human cancer. Receptors are Frizzled-class proteins with 7 transmembrane segments.

x-ray crystallography   Most commonly used technique for determining the three-dimensional structure of macromolecules (particularly proteins and nucleic acids) by passing x-rays through a crystal of the purified molecules and analyzing the diffraction pattern of discrete spots that results. (Figure 3-38)

yeast two-hybrid system   Widely used experimental technique for identifying genes whose encoded protein interacts with a specific protein of interest. (Figure 11-39)

zinc finger   Several related DNA-binding **motifs** composed of secondary structures folded around a zinc ion; present in numerous eukaryotic **transcription factors.** (Figures 3-6b and 11-21)

zygote   A fertilized egg; diploid cell resulting from fusion of a male and female **gamete.**

INDEX

Note: Page numbers followed by f indicate figures; those followed by t indicate tables.

Overview Animation: Extracellular Signaling (Figure 13-11)

Video: Chemotaxis of a Single *Dictyostelium* Cell to the Chemoattractant cAMP (Figure 13-12)

Overview Animation: Extracellular Signaling (Figure 13-26)

Focus Animation: Second Messengers in Signaling Pathways (Figure 13-29)

Overview Animation: Extracellular Signaling (Figure 13-32)

Chapter 14   Signaling Pathways that Control Gene Activity

Focus Animation: TGFβ Signaling Pathway (Figure 14-2)

Overview Animation: Extracellular Signaling (Figure 14-23)

Video: Protein Dynamics in Response to cAMP Stimulation of a *Dictyostelium* Cell (Figure 14-27)

Video: Localized Activation of PI-3 Kinase in a Chemotactic *Dictyostelium* Cell (Section 14.5)

Video: Chemotaxis of a Single *Dictyostelium* Cell to the Chemoattractant cAMP (Section 14.5)

Chapter 15   Integration of Signals and Gene Controls

Technique Animation: Reporter Constructs (Figure 15-8)

Video: Time-Lapse Imaging of *Drosophila* Embryogenesis (Figure 15-15)

Overview Animation: Gene Control in Embryonic Development (Figure 15-19)

Video: Establishing Eve Expression in *Drosophila* Embryogenesis (Section 15.4)

Video: Expression of Segmentation Genes in a *Drosophila* Embryo (Figure 15-22)

Video: Expression of Segmentation Genes in a *Drosophila* Embryo (Figure 15-24)

Chapter 16   Moving Proteins into Membranes and Organelles

Video: Three-Dimensional Model of a Protein Translocation Channel (Section 16.1)

Overview Animation: Protein Sorting (Figure 16-1)

Focus Animation: Synthesis of Secreted and Membrane-Bound Proteins (Figure 16-6)

Chapter 17   Vesicular Trafficking, Secretion, and Endocytosis

Overview Animation: Protein Secretion (Figure 17-1)

Video: Transport of VSVG-GFP Through the Secretory Pathway (Figure 17-2)

Video: Three-Dimensional Model of a Golgi Complex (Figure 17.4)

Video: KDEL Receptor Trafficking (Figure 17-16)

Video: Birth of a Clathrin Coat (Figure 17-19)

Video: Segregation of Apical and Basolateral Cargo in the Golgi of Live Cells (Figure 17-26)

Chapter 19   Microfilaments and Intermediate Filaments

Video: Actin Filaments in a Lamellipodium of a Fish Keratinocyte (Figure 19-2)

Focus Animation: Actin Polymerization (Figure 19-6)

Video: Direct Observation of Actin Filament Branching Mediated by Arp2/3 Complex (Figure 19-12)

Video: Animated Model for Myosin-Based Motility (Section 19.3)

Technique Animation: In Vitro Motility Assay (Figure 19-17)

Video: Three-Dimensional Animation of an Actin-Myosin Crossbridge Regulated by Tropomyosin-Troponin (Figure 19-24)

Overview Animation: Cell Motility (Figure 19-26)

Video: Mechanics of Fish Keratinocyte Migration (Figure 19-26)